The Mathematica GuideBook

for Numerics

Michael Trott

The Mathematica GuideBook
for Numerics

With 1364 Illustrations

 **INCLUDES DVD**

 Springer

Michael Trott
Wolfram Research
Champaign, Illinois

Mathematica is a registered trademark of Wolfram Research, Inc.

Library of Congress Control Number: 2005928494

ISBN-10: 0-387-95011-7   Printed on acid-free paper.
ISBN-13: 978-0387-95011-2
e-ISBN 0-387-28814-7

Printed in the United States of America.   (HAM)

9 8 7 6 5 4 3 2 1

springeronline.com

Preface

Bei mathematischen Operationen kann sogar eine gänzliche Entlastung des Kopfes eintreten, indem man einmal ausgeführte Zähloperationen mit Zeichen symbolisiert und, statt die Hirnfunktion auf Wiederholung schon ausgeführter Operationen zu verschwenden, sie für wichtigere Fälle aufspart.

When doing mathematics, instead of burdening the brain with the repetitive job of redoing numerical operations which have already been done before, it's possible to save that brainpower for more important situations by using symbols, instead, to represent those numerical calculations.

— Ernst Mach (1883) [45]

Computer Mathematics and Mathematica

Computers were initially developed to expedite numerical calculations. A newer, and in the long run, very fruitful field is the manipulation of symbolic expressions. When these symbolic expressions represent mathematical entities, this field is generally called computer algebra [8]. Computer algebra begins with relatively elementary operations, such as addition and multiplication of symbolic expressions, and includes such things as factorization of integers and polynomials, exact linear algebra, solution of systems of equations, and logical operations. It also includes analysis operations, such as definite and indefinite integration, the solution of linear and nonlinear ordinary and partial differential equations, series expansions, and residue calculations. Today, with computer algebra systems, it is possible to calculate in minutes or hours the results that would (and did) take years to accomplish by paper and pencil. One classic example is the calculation of the orbit of the moon, which took the French astronomer Delaunay 20 years [12], [13], [14], [15], [11], [26], [27], [53], [16], [17], [25]. (The *Mathematica GuideBooks* cover the two other historic examples of calculations that, at the end of the 19th century, took researchers many years of hand calculations [1], [4], [38] and literally thousands of pages of paper.)

Along with the ability to do symbolic calculations, four other ingredients of modern general-purpose computer algebra systems prove to be of critical importance for solving scientific problems:

- a powerful high-level programming language to formulate complicated problems
- programmable two- and three-dimensional graphics
- robust, adaptive numerical methods, including arbitrary precision and interval arithmetic
- the ability to numerically evaluate and symbolically deal with the classical orthogonal polynomials and special functions of mathematical physics.

The most widely used, complete, and advanced general-purpose computer algebra system is *Mathematica*. *Mathematica* provides a variety of capabilities such as graphics, numerics, symbolics, standardized interfaces to other programs, a complete electronic document-creation environment (including a full-fledged mathematical typesetting system), and a variety of import and export capabilities. Most of these ingredients are necessary to coherently and exhaustively solve problems and model processes occurring in the natural sciences [41], [58], [21], [39] and other fields using constructive mathematics, as well as to properly represent the results. Consequently, *Mathematica*'s main areas of application are presently in the natural sciences, engineering, pure and applied mathematics, economics, finance, computer graphics, and computer science.

Mathematica is an ideal environment for doing general scientific and engineering calculations, for investigating and solving many different mathematically expressable problems, for visualizing them, and for writing notes, reports, and papers about them. Thus, *Mathematica* is an integrated computing environment, meaning it is what is also called a "problem-solving environment" [40], [23], [6], [48], [43], [50], [52].

Scope and Goals

The *Mathematica GuideBooks* are four independent books whose main focus is to show how to solve scientific problems with *Mathematica*. Each book addresses one of the four ingredients to solve nontrivial and real-life mathematically formulated problems: programming, graphics, numerics, and symbolics. The Programming and the Graphics volume were published in autumn 2004.

The four *Mathematica GuideBooks* discuss programming, two-dimensional, and three-dimensional graphics, numerics, and symbolics (including special functions). While the four books build on each other, each one is self-contained. Each book discusses the definition, use, and unique features of the corresponding *Mathematica* functions, gives small and large application examples with detailed references, and includes an extensive set of relevant exercises and solutions.

The *GuideBooks* have three primary goals:

- to give the reader a solid working knowledge of *Mathematica*
- to give the reader a detailed knowledge of key aspects of *Mathematica* needed to create the "best", fastest, shortest, and most elegant solutions to problems from the natural sciences
- to convince the reader that working with *Mathematica* can be a quite fruitful, enlightening, and joyful way of cooperation between a computer and a human.

Realizing these goals is achieved by understanding the unifying design and philosophy behind the *Mathematica* system through discussing and solving numerous example-type problems. While a variety of mathematics and physics problems are discussed, the *GuideBooks* are not mathematics or physics books (from the point of view of content and rigor; no proofs are typically involved), but rather the author builds on *Mathematica*'s mathematical and scientific knowledge to explore, solve, and visualize a variety of applied problems.

The focus on solving problems implies a focus on the computational engine of *Mathematica*, the kernel—rather than on the user interface of *Mathematica*, the front end. (Nevertheless, for a nicer presentation inside the electronic version, various front end features are used, but are not discussed in depth.)

The *Mathematica GuideBooks* go far beyond the scope of a pure introduction into *Mathematica*. The books also present instructive implementations, explanations, and examples that are, for the most part, original. The books also discuss some "classical" *Mathematica* implementations, explanations, and examples, partially available only in the original literature referenced or from newsgroups threads.

In addition to introducing *Mathematica*, the *GuideBooks* serve as a guide for generating fairly complicated graphics and for solving more advanced problems using graphical, numerical, and symbolical techniques in cooperative ways. The emphasis is on the *Mathematica* part of the solution, but the author employs examples that are not uninteresting from a content point of view. After studying the *GuideBooks*, the reader will be able to solve new and old scientific, engineering, and recreational mathematics problems faster and more completely with the help of *Mathematica*—at least, this is the author's goal. The author also hopes that the reader will enjoy using *Mathematica* for visualization of the results as much as the author does, as well as just studying *Mathematica* as a language on its own.

In the same way that computer algebra systems are not "proof machines" [46], [9], [37], [10], [54], [55], [56] such as might be used to establish the four-color theorem ([2], [22]), the Kepler [28], [19], [29], [30], [31], [32], [33], [34], [35], [36] or the Robbins ([44], [20]) conjectures, proving theorems is not the central theme of the *GuideBooks*. However, powerful and general proof machines [9], [42], [49], [24], [3], founded on *Mathematica'* s general programming paradigms and its mathematical capabilities, have been built (one such system is *Theorema* [7]). And, in the *GuideBooks*, we occasionally prove one theorem or another theorem.

In general, the author's aim is to present a realistic portrait of *Mathematica*: its use, its usefulness, and its strengths, including some current weak points and sometimes unexpected, but often nevertheless quite "thought through", behavior. *Mathematica* is not a universal tool to solve arbitrary problems which can be formulated mathematically—only a fraction of all mathematical problems can even be formulated in such a way to be efficiently expressed today in a way understandable to a computer. Rather, it is often necessary to do a certain amount of programming and occasionally give *Mathematica* some "help" instead of simply calling a single function like `Solve` to solve a system of equations. Because this will almost always be the case for "real-life" problems, we do not restrict ourselves only to "textbook" examples, where all goes smoothly without unexpected problems and obstacles. The reader will see that by employing *Mathematica's* programming, numeric, symbolic, and graphic power, *Mathematica* can offer more effective, complete, straightforward, reusable, and less likely erroneous solution methods for calculations than paper and pencil, or numerical programming languages.

Although the *Guidebooks* are large books, it is nevertheless impossible to discuss all of the 2,000+ built-in *Mathematica* commands. So, some simple as well as some more complicated commands have been omitted. For a full overview about *Mathematica's* capabilities, it is necessary to study *The Mathematica Book* [60] in detail. The commands discussed in the *Guidebooks* are those that an scientist or research engineer needs for solving *typical* problems, if such a thing exists [18]. These subjects include a quite detailed discussion of the structure of *Mathematica* expressions, *Mathematica* input and output (important for the human–*Mathematica* interaction), graphics, numerical calculations, and calculations from classical analysis. Also, emphasis is given to the powerful algebraic manipulation functions. Interestingly, they frequently allow one to solve analysis problems in an algorithmic way [5]. These functions are typically not so well known because they are not taught in classical engineering or physics-mathematics courses, but with the advance of computers doing symbolic mathematics, their importance increases [47].

A thorough knowledge of:

- machine and high-precision numbers, packed arrays, and intervals
- machine, high-precision, and interval arithmetic
- process of compilation, its advantages and limits
- main operations of numerical analysis, such as equation solving, minimization, summation
- numerical solution of ordinary and partial differential equations

is needed for virtually any nontrivial numeric calculation and frequently also in symbolic computations. *The Mathematica GuideBook for Numerics* discusses these subjects.

The current version of the *Mathematica GuideBooks* is tailored for *Mathematica* Version 5.1.

Content Overview

The Mathematica GuideBook for Numerics has two chapters. Each chapter is subdivided into sections (which occasionally have subsections), exercises, solutions to the exercises, and references.

This volume deals with *Mathematica*'s numerical mathematics capabilities, the indispensable tools for dealing with virtually any "real life" problem. Fast machine, exact integer, and rational and verified high-precision arithmetic is applied to a large number of examples in the main text and in the solutions to the exercises.

Chapter 1 deals with numerical calculations, which are important for virtually all *Mathematica* users. This volume starts with calculations involving real and complex numbers with an "arbitrary" number of digits. (Well, not really an "arbitrary" number of digits, but on present-day computers, many calculations involving a few million digits are easily feasible.). Then follows a discussion of significance arithmetic, which automatically keeps track of the digits which are correct in calculations with high-precision numbers. Also discussed is the use of interval arithmetic. (Despite being slow, exact and/or inexact, interval arithmetic allows one to carry out validated numerical calculations.) The next important subject is the (pseudo)compilation of *Mathematica* code. Because *Mathematica* is an interpreted language that allows for "unforeseeable" actions and arbitrary side effects at runtime, it generally cannot be compiled. Strictly numerical calculations can, of course, be compiled.

Then, the main numerical functions are discussed: interpolation, Fourier transforms, numerical summation, and integration, solution of equations (root finding), minimization of functions, and the solution of ordinary and partial differential equations. To illustrate *Mathematica*'s differential equation solving capabilities, many ODEs and PDEs are discussed. Many medium-sized examples are given for the various numerical procedures. In addition, *Mathematica* is used to monitor and visualize various numerical algorithms.

The main part of Chapter 1 culminates with two larger applications, the construction of Riemann surfaces of algebraic functions and the visualization of electric and magnetic field lines of some more complicated two- and three-dimensional charge and current distributions. A large, diverse set of exercises and detailed solutions ends the first chapter.

Chapter 2 deals with exact integer calculations and integer-valued functions while concentrating on topics that are important in classical analysis. Number theory functions and modular polynomial functions, currently little used in most natural science applications, are intentionally given less detailed treatment. While some of the functions of this chapter have analytic continuations to complex arguments and could so be considered as belonging to Chapter 3 of the Symbolics volume, emphasis is given to their combinatorial use in this chapter.

This volume explains and demonstrates the use of the numerical functions of *Mathematica*. It only rarely discusses the underlying numerical algorithms themselves. But occasionally *Mathematica* is used to monitor how the algorithms work and progress.

The Books and the Accompanying DVDs

Each of the *GuideBooks* comes with a multiplatform DVD. Each DVD contains the fourteen main notebooks, the hyperlinked table of contents and index, a navigation palette, and some utility notebooks and files. All notebooks are tailored for *Mathematica* 5.1. Each of the main notebooks corresponds to a chapter from the printed books. The notebooks have the look and feel of a printed book, containing structured units, typeset formulas, *Mathemat-*

ica code, and complete solutions to all exercises. The DVDs contain the fully evaluated notebooks corresponding to the chapters of the corresponding printed book (meaning these notebooks have text, inputs, outputs and graphics). The DVDs also include the unevaluated versions of the notebooks of the other three *GuideBooks* (meaning they contain all text and *Mathematica* code, but no outputs and graphics).

Although the *Mathematica GuideBooks* are printed, *Mathematica* is "a system for doing mathematics by computer" [59]. This was the lovely tagline of earlier versions of *Mathematica,* but because of its growing breadth (like data import, export and handling, operating system-independent file system operations, electronic publishing capabilities, web connectivity), nowadays *Mathematica* is called a "system for technical computing". The original tagline (that is more than ever valid today!) emphasized two points: doing mathematics and doing it on a computer. The approach and content of the *GuideBooks* are fully in the spirit of the original tagline: They are centered around *doing* mathematics. The second point of the tagline expresses that an electronic version of the *GuideBooks* is the more natural medium for *Mathematica*-related material. Long outputs returned by *Mathematica,* sequences of animations, thousands of web-retrievable references, a 10,000-entry hyperlinked index (that points more precisely than a printed index does) are space-consuming, and therefore not well suited for the printed book. As an interactive program, *Mathematica* is best learned, used, challenged, and enjoyed while sitting in front of a powerful computer (or by having a remote kernel connection to a powerful computer).

In addition to simply showing the printed book's text, the notebooks allow the reader to:

- experiment with, reuse, adapt, and extend functions and code
- investigate parameter dependencies
- annotate text, code, and formulas
- view graphics in color
- run animations.

The Accompanying Web Site

Why does a printed book need a home page? There are (in addition to being just trendy) two reasons for a printed book to have its fingerprints on the web. The first is for (*Mathematica*) users who have not seen the book so far. Having an outline and content sample on the web is easily accomplished, and shows the look and feel of the notebooks (including some animations). This is something that a printed book actually cannot do. The second reason is for readers of the book: *Mathematica* is a large modern software system. As such, it ages quickly in the sense that in the timescale of $10^{1 \cdot smallInteger}$ months, a new version will likely be available. The overwhelmingly large majority of *Mathematica* functions and programs will run unchanged in a new version. But occasionally, changes and adaptations might be needed. To accommodate this, the web site of this book—http://www.MathematicaGuideBooks.org—contains a list of changes relevant to the *GuideBooks*. In addition, like any larger software project, unavoidably, the *GuideBooks* will contain suboptimal implementations, mistakes, omissions, imperfections, and errors. As they come to his attention, the author will list them at the book's web site. Updates to references, corrections [51], hundreds of pages of additional exercises and solutions, improved code segments, and other relevant information will be on the web site as well. Also, information about OS-dependent and *Mathematica* version-related changes of the given *Mathematica* code will be available there.

Evolution of the Mathematica GuideBooks

A few words about the history and the original purpose of the *GuideBooks*: They started from lecture notes of an *Introductory Course in Mathematica 2* and an advanced course on the *Efficient Use of the Mathematica Programming System*, given in 1991/1992 at the Technical University of Ilmenau, Germany. Since then, after each release of a new version of *Mathematica*, the material has been updated to incorporate additional functionality. This electronic/printed publication contains text, unique graphics, editable formulas, runable, and modifiable programs, all made possible by the electronic publishing capabilities of *Mathematica*. However, because the structure, functions and examples of the original lecture notes have been kept, an abbreviated form of the *GuideBooks* is still suitable for courses.

Since 1992 the manuscript has grown in size from 1,600 pages to more than three times its original length, finally "weighing in" at nearly 5,000 printed book pages with more than:

- 18 gigabytes of accompanying *Mathematica* notebooks
- 22,000 *Mathematica* inputs with more than 13,000 code comments
- 11,000 references
- 4,000 graphics
- 1,000 fully solved exercises
- 150 animations.

This first edition of this book is the result of more than eleven years of writing and daily work with *Mathematica*. In these years, *Mathematica* gained hundreds of functions with increased functionality and power. A modern year-2005 computer equipped with *Mathematica* represents a computational power available only a few years ago to a select number of people [57] and allows one to carry out recreational or new computations and visualizations—unlimited in nature, scope, and complexity— quickly and easily. Over the years, the author has learned a lot of *Mathematica* and its current and potential applications, and has had a lot of fun, enlightening moments and satisfaction applying *Mathematica* to a variety of research and recreational areas, especially graphics. The author hopes the reader will have a similar experience.

Disclaimer

In addition to the usual disclaimer that neither the author nor the publisher guarantees the correctness of any formula, fitness, or reliability of any of the code pieces given in this book, another remark should be made. No guarantee is given that running the *Mathematica* code shown in the *GuideBooks* will give identical results to the printed ones. On the contrary, taking into account that *Mathematica* is a large and complicated software system which evolves with each released version, running the code with another version of *Mathematica* (or sometimes even on another operating system) will very likely result in different outputs for some inputs. And, as a consequence, if different outputs are generated early in a longer calculation, some functions might hang or return useless results.

The interpretations of *Mathematica* commands, their descriptions, and uses belong solely to the author. They are not claimed, supported, validated, or enforced by Wolfram Research. The reader will find that the author's view on *Mathematica* deviates sometimes considerably from those found in other books. The author's view is more on the formal than on the pragmatic side. The author does not hold the opinion that any *Mathematica* input has to have an immediate semantic meaning. *Mathematica* is an extremely rich system, especially from the language point of view. It is instructive, interesting, and fun to study the behavior of built-in *Mathematica* functions when called with a variety of arguments (like unevaluated, hold, including undercover zeros, etc.). It is the author's strong belief that doing this and being able to explain the observed behavior will be, in the long term, very fruitful for the reader because it develops the ability to recognize the uniformity of the principles underlying *Mathematica* and to make constructive, imaginative, and effective use of this uniformity. Also, some exercises ask the reader to investigate certain "unusual" inputs.

From time to time, the author makes use of undocumented features and/or functions from the `Developer` and `Experimental` contexts (in later versions of *Mathematica* these functions could exist in the `System` context or could have different names). However, some such functions might no longer be supported or even exist in later versions of *Mathematica*.

Acknowledgements

Over the decade, the *GuideBooks* were in development, many people have seen parts of them and suggested useful changes, additions, and edits. I would like to thank Horst Finsterbusch, Gottfried Teichmann, Klaus Voss, Udo Krause, Jerry Keiper, David Withoff, and Yu He for their critical examination of early versions of the manuscript and their useful suggestions, and Sabine Trott for the first proofreading of the German manuscript. I also want to thank the participants of the original lectures for many useful discussions. My thanks go to the reviewers of this book: John Novak, Alec Schramm, Paul Abbott, Jim Feagin, Richard Palmer, Ward Hanson, Stan Wagon, and Markus van Almsick, for their suggestions and ideas for improvement. I thank Richard Crandall, Allan Hayes, Andrzej Kozlowski, Hartmut Wolf, Stephan Leibbrandt, George Kambouroglou, Domenico Minunni, Eric Weisstein, Andy Shiekh, Arthur G. Hubbard, Jay Warrendorff, Allan Cortzen, Ed Pegg, and Udo Krause for comments on the prepublication version of the *GuideBooks*. I thank Bobby R. Treat, Arthur G. Hubbard, Murray Eisenberg, Marvin Schaefer, Marek Duszynski, Daniel Lichtblau, Devendra Kapadia, Adam Strzebonski, Anton Antonov, and Brett Champion for useful comments on the *Mathematica* Version 5.1 tailored version of the *GuideBooks*.

My thanks are due to Gerhard Gobsch of the Institute for Physics of the Technical University in Ilmenau for the opportunity to develop and give these original lectures at the Institute, and to Stephen Wolfram who encouraged and supported me on this project.

Concerning the process of making the *Mathematica GuideBooks* from a set of lecture notes, I thank Glenn Scholebo for transforming notebooks to $T_{E}X$ files, and Joe Kaiping for $T_{E}X$ work related to the printed book. I thank John Novak and Jan Progen for putting all the material into good English style and grammar, John Bonadies for the chapter-opener graphics of the book, and Jean Buck for library work. I especially thank John Novak for the creation of *Mathematica* 3 notebooks from the $T_{E}X$ files, and Andre Kuzniarek for his work on the stylesheet to give the notebooks a pleasing appearance. My thanks go to Andy Hunt who created a specialized stylesheet for the actual book printout and printed and formatted the $4 \times 1000+$ pages of the *Mathematica GuideBooks*. I thank Andy Hunt for making a first version of the homepage of the *GuideBooks* and Amy Young for creating the current version of the homepage of the *GuideBooks*. I thank Sophie Young for a final check of the English. My largest thanks go to Amy Young, who encouraged me to update the whole book over the years and who had a close look at all of my English writing and often improved it considerably. Despite reviews by many individuals any remaining mistakes or omissions, in the *Mathematica* code, in the mathematics, in the description of the *Mathematica* functions, in the English, or in the references, etc. are, of course, solely mine.

Let me take the opportunity to thank members of the Research and Development team of Wolfram Research whom I have met throughout the years, especially Victor Adamchik, Anton Antonov, Alexei Bocharov, Arnoud Buzing, Brett Champion, Matthew Cook, Todd Gayley, Darren Glosemeyer, Roger Germundsson, Unal Goktas, Yifan Hu, Devendra Kapadia, Zbigniew Leyk, David Librik, Daniel Lichtblau, Jerry Keiper, Robert Knapp, Roman Mäder, Oleg Marichev, John Novak, Peter Overmann, Oleksandr Pavlyk, Ulises Cervantes–Pimentel, Mark Sofroniou, Adam Strzebonski, Oyvind Tafjord, Robby Villegas, Tom Wickham–Jones, David Withoff, and Stephen Wolfram for numerous discussions about design principles, various small details, underlying algorithms, efficient implementation of various procedures, and tricks concerning *Mathematica*. The appearance of the notebooks profited from discussions with John Fultz, Paul Hinton, John Novak, Lou D'Andria, Theodore Gray, Andre Kuzniarek, Jason Harris, Andy Hunt, Christopher Carlson, Robert Raguet–Schofield, George Beck, Kai Xin, Chris Hill, and Neil Soiffer about front end, button, and typesetting issues.

I'm grateful to Jeremy Hilton from the Corporation for National Research Initiatives for allowing the use of the text of Shakespeare's *Hamlet* (to be used in Chapter 1 of *The Mathematica GuideBook to Numerics*).

It was an interesting and unique experience to work over the last 12 years with five editors: Allan Wylde, Paul Wellin, Maria Taylor, Wayne Yuhasz, and Ann Kostant, with whom the *GuideBooks* were finally published. Many book-related discussions that ultimately improved the *GuideBooks*, have been carried out with Jan Benes from TELOS and associates, Steven Pisano, Jenny Wolkowicki, Henry Krell, Fred Bartlett, Vaishali Damle, Ken Quinn, Jerry Lyons, and Rüdiger Gebauer from Springer New York.

The author hopes the *Mathematica GuideBooks* help the reader to discover, investigate, urbanize, and enjoy the computational paradise offered by *Mathematica*.

Wolfram Research, Inc.   Michael Trott
April 2005

References

1   A. Amthor. *Z. Math. Phys.* 25, 153 (1880).

2   K. Appel, W. Haken. *J. Math.* 21, 429 (1977).

3   A. Bauer, E. Clarke, X. Zhao. *J. Automat. Reasoning* 21, 295 (1998).

4   A. H. Bell. *Am. Math. Monthly* 2, 140 (1895).

5   M. Berz. *Adv. Imaging Electron Phys.* 108, 1 (2000).

6   R. F. Boisvert. *arXiv:cs.MS*/0004004 (2000).

7   B. Buchberger. *Theorema Project* (1997).   ftp://ftp.risc.uni-linz.ac.at/pub/techreports/1997/97-34/ed-media.nb

8   B. Buchberger. *SIGSAM Bull.* 36, 3 (2002).

9   S.-C. Chou, X.-S. Gao, J.-Z. Zhang. *Machine Proofs in Geometry*, World Scientific, Singapore, 1994.

10   A. M. Cohen. *Nieuw Archief Wiskunde* 14, 45 (1996).

11   A. Cook. *The Motion of the Moon*, Adam-Hilger, Bristol, 1988.

12   C. Delaunay. *Théorie du Mouvement de la Lune*, Gauthier-Villars, Paris, 1860.

13   C. Delaunay. *Mem. de l' Acad. des Sc. Paris* 28 (1860).

14   C. Delaunay. *Mem. de l' Acad. des Sc. Paris* 29 (1867).

15   A. Deprit, J. Henrard, A. Rom. *Astron. J.* 75, 747 (1970).

16   A. Deprit. *Science* 168, 1569 (1970).

17   A. Deprit, J. Henrard, A. Rom. *Astron. J.* 76, 273 (1971).

18   P. J. Dolan, Jr., D. S. Melichian. *Am. J. Phys.* 66, 11 (1998).

19   S. P. Ferguson, T. C. Hales. *arXiv:math.MG/* 9811072 (1998).

20   B. Fitelson. *Mathematica Educ. Res.* 7, n1, 17 (1998).

21   A. C. Fowler. *Mathematical Models in the Applied Sciences,* Cambridge University Press, Cambridge, 1997.

22   H. Fritsch, G. Fritsch. *The Four-Color Theorem*, Springer-Verlag, New York, 1998.

23   E. Gallopoulos, E. Houstis, J. R. Rice (eds.). *Future Research Directions in Problem Solving Environments for Computational Science: Report of a Workshop on Research Directions in Integrating Numerical Analysis, Symbolic Computing, Computational Geometry, and Artificial Intelligence for Computational Science*, 1991. http://www.cs.purdue.edu/research/cse/publications/tr/92/92-032.ps.gz

24   V. Gerdt, S. A. Gogilidze in V. G. Ganzha, E. W. Mayr, E. V. Vorozhtsov (eds.). *Computer Algebra in Scientific Computing*, Springer-Verlag, Berlin, 1999.

25   M. C. Gutzwiller, D. S. Schmidt. *Astronomical Papers: The Motion of the Moon as Computed by the Method of Hill, Brown, and Eckert*, U.S. Government Printing Office, Washington, 1986.

26   M. C. Gutzwiller. *Rev. Mod. Phys.* 70, 589 (1998).

27   Y. Hagihara. *Celestial Mechanics* vII/1, MIT Press, Cambridge, 1972.

28   T. C. Hales. *arXiv:math.MG/* 9811071 (1998).

29   T. C. Hales. *arXiv:math.MG/* 9811073 (1998).

30   T. C. Hales. *arXiv:math.MG/* 9811074 (1998).

31   T. C. Hales. *arXiv:math.MG/* 9811075 (1998).

32   T. C. Hales. *arXiv:math.MG/* 9811076 (1998).

33   T. C. Hales. *arXiv:math.MG/* 9811077 (1998).

34   T. C. Hales. *arXiv:math.MG/* 9811078 (1998).

35   T. C. Hales. *arXiv:math.MG/*0205208 (2002).

36   T. C. Hales in L. Tatsien (ed.). *Proceedings of the International Congress of Mathematicians* v. 3, Higher Education Press, Beijing, 2002.

37   J. Harrison. *Theorem Proving with the Real Numbers*, Springer-Verlag, London, 1998.

38   J. Hermes. *Nachrichten Königl. Gesell. Wiss. Göttingen* 170 (1894).

39   E. N. Houstis, J. R. Rice, E. Gallopoulos, R. Bramley (eds.). *Enabling Technologies for Computational Science*, Kluwer, Boston, 2000.

40   E. N. Houstis, J. R. Rice. *Math. Comput. Simul.* 54, 243 (2000).

41   M. S. Klamkin (eds.). *Mathematical Modelling*, SIAM, Philadelphia, 1996.

42   H. Koch, A. Schenkel, P. Wittwer. *SIAM Rev.* 38, 565 (1996).

43   Y. N. Lakshman, B. Char, J. Johnson in O. Gloor (ed.). *ISSAC 1998*, ACM Press, New York, 1998.

44   W. McCune. *Robbins Algebras Are Boolean*, 1997.   http://www.mcs.anl.gov/home/mccune/ar/robbins/

45   E. Mach (R. Wahsner, H.-H. von Borszeskowski eds.). *Die Mechanik in ihrer Entwicklung*, Akademie-Verlag, Berlin, 1988.

46   D. A. MacKenzie. *Mechanizing Proof: Computing, Risk, and Trust*, MIT Press, Cambridge, 2001.

47   B. M. McCoy. *arXiv:cond-mat/*0012193 (2000).

48   K. J. M. Moriarty, G. Murdeshwar, S. Sanielevici. *Comput. Phys. Commun.* 77, 325 (1993).

49   I. Nemes, M. Petkovšek, H. S. Wilf, D. Zeilberger. *Am. Math. Monthly* 104, 505 (1997).

50   W. H. Press, S. A. Teukolsky. *Comput. Phys.* 11, 417 (1997).

51   D. Rawlings. *Am. Math. Monthly* 108, 713 (2001).

52   *Problem Solving Environments Home Page.*   http://www.cs.purdue.edu/research/cse/pses

53   D. S. Schmidt in H. S. Dumas, K. R. Meyer, D. S. Schmidt (eds.). *Hamiltonian Dynamical Systems*, Springer-Verlag, New York, 1995.

54   S. Seiden. *SIGACT News* 32, 111 (2001).

55   S. Seiden. *Theor. Comput. Sc.* 282, 381 (2002).

56   C. Simpson. *arXiv:math.HO/*0311260 (2003).

57   A. M. Stoneham. *Phil. Trans. R. Soc. Lond.* A 360, 1107 (2002).

58   M. Tegmark. *Ann. Phys.* 270, 1 (1999).

59   S. Wolfram. *Mathematica: A System for Doing Mathematics by Computer*, Addison-Wesley, Redwood City, 1992.

60   S. Wolfram. *The Mathematica Book*, Wolfram Media, Champaign, 2003.

Contents

CHAPTER *2*

Computations with Exact Numbers

Introduction and Orientation to
The Mathematica GuideBooks

0.1 Overview

0.1.1 Content Summaries

The *Mathematica GuideBooks* are published as four independent books: *The Mathematica GuideBook to Programming, The Mathematica GuideBook to Graphics, The Mathematica GuideBook to Numerics*, and *The Mathematica GuideBook to Symbolics*.

■ The Programming volume deals with the structure of *Mathematica* expressions and with *Mathematica* as a programming language. This volume includes the discussion of the hierarchical construction of all *Mathematica* objects out of symbolic expressions (all of the form *head* [*argument*]), the ultimate building blocks of expressions (numbers, symbols, and strings), the definition of functions, the application of rules, the recognition of patterns and their efficient application, the order of evaluation, program flows and program structure, the manipulation of lists (the universal container for *Mathematica* expressions of all kinds), as well as a number of topics specific to the *Mathematica* programming language. Various programming styles, especially *Mathematica'* s powerful functional programming constructs, are covered in detail.

■ The Graphics volume deals with *Mathematica*'s two-dimensional (2D) and three-dimensional (3D) graphics. The chapters of this volume give a detailed treatment on how to create images from graphics primitives, such as points, lines, and polygons. This volume also covers graphically displaying functions given either analytically or in discrete form. A number of images from the *Mathematica* Graphics Gallery are also reconstructed. Also discussed is the generation of pleasing scientific visualizations of functions, formulas, and algorithms. A variety of such examples are given.

■ The Numerics volume deals with *Mathematica*'s numerical mathematics capabilities—the indispensable sledgehammer tools for dealing with virtually any "real life" problem. The arithmetic types (fast machine, exact integer and rational, verified high-precision, and interval arithmetic) are carefully analyzed. Fundamental numerical operations, such as compilation of programs, numerical Fourier transforms, minimization, numerical solution of equations, and ordinary/partial differential equations are analyzed in detail and are applied to a large number of examples in the main text and in the solutions to the exercises.

■ The Symbolics volume deals with *Mathematica*'s symbolic mathematical capabilities—the real heart of *Mathematica* and the ingredient of the *Mathematica* software system that makes it so unique and powerful. Structural and mathematical operations on systems of polynomials are fundamental to many symbolic calculations and are covered in detail. The solution of equations and differential equations, as well as the classical calculus operations, are exhaustively treated. In addition, this volume discusses and employs the classical orthogonal polynomials and special functions of mathematical physics. To demonstrate the symbolic mathematics power, a variety of problems from mathematics and physics are discussed.

The four *GuideBooks* contain about 25,000 *Mathematica* inputs, representing more than 75,000 lines of commented *Mathematica* code. (For the reader already familiar with *Mathematica*, here is a more precise measure: The LeafCount of all inputs would be about 900,000 when collected in a list.) The *GuideBooks* also have more than 4,000 graphics, 150 animations, 11,000 references, and 1,000 exercises. More than 10,000 hyperlinked index entries and hundreds of hyperlinks from the overview sections connect all parts in a convenient way. The evaluated notebooks of all four volumes have a cumulative file size of about 20 GB. Although these numbers may sound large, the *Mathematica GuideBooks* actually cover only a portion of *Mathematica*'s functionality and features and give only a glimpse into the possibilities *Mathematica* offers to generate graphics, solve problems, model systems, and discover new identities, relations, and algorithms. The *Mathematica* code is explained in detail throughout all chapters. More than 13,000 comments are scattered throughout all inputs and code fragments.

0.1.2 Relation of the Four Volumes

The four volumes of the *GuideBooks* are basically independent, in the sense that readers familiar with *Mathematica* programming can read any of the other three volumes. But a solid working knowledge of the main topics discussed in *The Mathematica GuideBook to Programming*—symbolic expressions, pure functions, rules and replacements, and list manipulations—is required for the Graphics, Numerics, and Symbolics volumes. Compared to these three volumes, the Programming volume might appear to be a bit "dry". But similar to learning a foreign language, before being rewarded with the beauty of novels or a poem, one has to sweat and study. The whole suite of graphical capabilities and all of the mathematical knowledge in *Mathematica* are accessed and applied through lists, patterns, rules, and pure functions, the material discussed in the Programming volume.

Naturally, graphics are the center of attention of the *The Mathematica GuideBook to Graphics*. While in the Programming volume some plotting and graphics for visualization are used, graphics are not crucial for the Programming volume. The reader can safely skip the corresponding inputs to follow the main programming threads. The Numerics and Symbolics volumes, on the other hand, make heavy use of the graphics knowledge acquired in the Graphics volume. Hence, the prerequisites for the Numerics and Symbolics volumes are a good knowledge of *Mathematica*'s programming language and of its graphics system.

The Programming volume contains only a few percent of all graphics, the Graphics volume contains about two-thirds, and the Numerics and Symbolics volume, about one-third of the overall 4,000+ graphics. The Programming and Graphics volumes use some mathematical commands, but they restrict the use to a relatively small number (especially Expand, Factor, Integrate, Solve). And the use of the function N for numerical ization is unavoidable for virtually any "real life" application of *Mathematica*. The last functions allow us to treat some mathematically not uninteresting examples in the Programming and Graphics volumes. In addition to putting these functions to work for nontrivial problems, a detailed discussion of the mathematics functions of *Mathematica* takes place exclusively in the Numerics and Symbolics volumes.

The Programming and Graphics volumes contain a moderate amount of mathematics in the examples and exercises, and focus on programming and graphics issues. The Numerics and Symbolics volumes contain a substantially larger amount of mathematics.

Although printed as four books, the fourteen individual chapters (six in the Programming volume, three in the Graphics volume, two in the Numerics volume, and three in the Symbolics volume) of the *Mathematica Guide-Books* form one organic whole, and the author recommends a strictly sequential reading, starting from Chapter 1 of the Programming volume and ending with Chapter 3 of the Symbolics volume for gaining the maximum benefit. The electronic component of each book contains the text and inputs from all the four *GuideBooks*, together with a comprehensive hyperlinked index. The four volumes refer frequently to one another.

0.1.3 Chapter Structure

A rough outline of the content of a chapter is the following:

■ The main body discusses the *Mathematica* functions belonging to the chapter subject, as well their options and attributes. Generically, the author has attempted to introduce the functions in a "natural order". But surely, one cannot be axiomatic with respect to the order. (Such an order of the functions is not unique, and the author intentionally has "spread out" the introduction of various *Mathematica* functions across the four volumes.) With the introduction of a function, some small examples of how to use the functions and comparisons of this function with related ones are given. These examples typically (with the exception of some visualizations in the Programming volume) incorporate functions already discussed. The last section of a chapter often gives a larger example that makes heavy use of the functions discussed in the chapter.

■ A programmatically constructed overview of each chapter functions follows. The functions listed in this section are hyperlinked to their attributes and options, as well as to the corresponding reference guide entries of *The Mathematica Book*.

■ A set of exercises and potential solutions follow. Because learning *Mathematica* through examples is very efficient, the proposed solutions are quite detailed and form up to 50% of the material of a chapter.

■ References end the chapter.

Note that the first few chapters of the Programming volume deviate slightly from this structure. Chapter 1 of the Programming volume gives a general overview of the kind of problems dealt with in the four *GuideBooks*. The second, third, and fourth chapters of the Programming volume introduce the basics of programming in *Mathematica*. Starting with Chapters 5 of the Programming volume and throughout the Graphics, Numerics, and Symbolics volumes, the above-described structure applies.

In the 14 chapters of the *GuideBooks* the author has chosen a "we" style for the discussions of how to proceed in constructing programs and carrying out calculations to include the reader intimately.

0.1.4 Code Presentation Style

The typical style of a unit of the main part of a chapter is: Define a new function, discuss its arguments, options, and attributes, and then give examples of its usage. The examples are virtually always *Mathematica* inputs and outputs. The majority of inputs is in `InputForm` are the notebooks. On occasion `StandardForm` is also used. Although `StandardForm` mimics classical mathematics notation and makes short inputs more readable, for "program-like" inputs, `InputForm` is typically more readable and easier and more natural to align. For the outputs, `StandardForm` is used by default and occasionally the author has resorted to `InputForm` or `FullForm` to expose digits of numbers and to `TraditionalForm` for some formulas. Outputs are mostly not programs, but nearly always "results" (often mathematical expressions, formulas, identities, or lists of numbers rather than program constructs). The world of *Mathematica* users is divided into three groups, and each of them has a nearly religious opinion on how to format *Mathematica* code [1], [2]. The author follows the `InputForm` cult(ure) and hopes that the *Mathematica* users who do everything in either StandardForm or Tradition‑ alForm will bear with him. If the reader really wants to see all code in either StandardForm or Tradition‑ alForm, this can easily be done with the **Convert To** item from the **Cell** menu. (Note that the relation between InputForm and StandardForm is not symmetric. The InputForm cells of this book have been line-broken and aligned by hand. Transforming them into StandardForm or TraditionalForm cells works well because one typically does not line-break manually and align *Mathematica* code in these cell types. But converting StandardForm or TraditionalForm cells into InputForm cells results in much less pleasing results.)

In the inputs, special typeset symbols for *Mathematica* functions are typically avoided because they are not monospaced. But the author does occasionally compromise and use Greek, script, Gothic, and doublestruck characters.

In a book about a programming language, two other issues come always up: indentation and placement of the code.

■ The code of the *GuideBooks* is largely consistently formatted and indented. There are no strict guidelines or even rules on how to format and indent *Mathematica* code. The author hopes the reader will find the book's formatting style readable. It is a compromise between readability (mental parsabililty) and space conservation, so that the printed version of the *Mathematica GuideBook* matches closely the electronic version.

■ Because of the large number of examples, a rather imposing amount of *Mathematica* code is presented. Should this code be present only on the disk, or also in the printed book? If it is in the printed book, should it be at the position where the code is used or at the end of the book in an appendix? Many authors of *Mathematica* articles and books have strong opinions on this subject. Because the main emphasis of the *Mathematica GuideBooks* is on *solving* problems *with Mathematica* and not on the actual problems, the *GuideBooks* give all of the code at the point where it is needed in the printed book, rather than "hiding" it in packages and appendices. In addition to being more straightforward to read and conveniently allowing us to refer to elements of the code pieces, this placement makes the correspondence between the printed book and the notebooks close to 1:1, and so working back and forth between the printed book and the notebooks is as straightforward as possible.

0.2 Requirements

0.2.1 Hardware and Software

Throughout the *GuideBooks*, it is assumed that the reader has access to a computer running a current version of *Mathematica* (version 5.0/5.1 or newer). For readers without access to a licensed copy of *Mathematica*, it is possible to view all of the material on the disk using a trial version of *Mathematica*. (A trial version is download‑ able from http://www.wolfram.com/products/mathematica/trial.cgi.)

The files of the *GuideBooks* are relatively large, altogether more than 20 GB. This is also the amount of hard disk space needed to store uncompressed versions of the notebooks. To view the notebooks comfortably, the reader's computer needs 128 MB RAM; to evaluate the evaluation units of the notebooks 1 GB RAM or more is recommended.

In the *GuideBooks*, a large number of animations are generated. Although they need more memory than single pictures, they are easy to create, to animate, and to store on typical year-2005 hardware, and they provide a lot of joy.

0.2.2 Reader Prerequisites

Although prior *Mathematica* knowledge is not needed to read *The Mathematica GuideBook to Programming*, it is assumed that the reader is familiar with basic actions in the *Mathematica* front end, including entering Greek characters using the keyboard, copying and pasting cells, and so on. Freely available tutorials on these (and other) subjects can be found at http://library.wolfram.com.

For a complete understanding of most of the *GuideBooks* examples, it is desirable to have a background in mathematics, science, or engineering at about the bachelor's level or above. Familiarity with mechanics and electrodynamics is assumed. Some examples and exercises are more specialized, for instance, from quantum mechanics, finite element analysis, statistical mechanics, solid state physics, number theory, and other areas. But the *GuideBooks* avoid very advanced (but tempting) topics such as renormalization groups [6], parquet approximations [27], and modular moonshines [14]. (Although *Mathematica* can deal with such topics, they do not fit the character of the *Mathematica GuideBooks* but rather the one of a *Mathematica Topographical Atlas* [a monumental work to be carried out by the *Mathematica*–Bourbakians of the 21st century]).

Each scientific application discussed has a set of references. The references should easily give the reader both an overview of the subject and pointers to further references.

0.3 What the GuideBooks Are and What They Are Not

0.3.1 Doing Computer Mathematics

As discussed in the Preface, the main goal of the *GuideBooks* is to demonstrate, showcase, teach, and exemplify scientific problem solving with *Mathematica*. An important step in achieving this goal is the discussion of *Mathematica* functions that allow readers to become fluent in programming when creating complicated graphics or solving scientific problems. This again means that the reader must become familiar with the most important programming, graphics, numerics, and symbolics functions, their arguments, options, attributes, and a few of their time and space complexities. And the reader must know which functions to use in each situation.

The *GuideBooks* treat only aspects of *Mathematica* that are ultimately related to "doing mathematics". This means that the *GuideBooks* focus on the functionalities of the kernel rather than on those of the front end. The knowledge required to use the front end to work with the notebooks can easily be gained by reading the corresponding chapters of the online documentation of *Mathematica*. Some of the subjects that are treated either lightly or not at all in the *GuideBooks* include the basic use of *Mathematica* (starting the program, features, and special properties of the notebook front end [16]), typesetting, the preparation of packages, external file operations, the communication of *Mathematica* with other programs via *MathLink*, special formatting and string manipulations, computer- and operating system-specific operations, audio generation, and commands available in various packages. "Packages" includes both, those distributed with *Mathematica* as well as those available from the *Mathematica* Information Center (http://library.wolfram.com/infocenter) and commercial sources, such as MathTensor for doing general relativity calculations (http://smc.vnet.net/MathTensor.html) or FeynCalc for doing high-energy physics calculations (http://www.feyncalc.org). This means, in particular, that probability and statistical calculations are barely touched on because most of the relevant commands are contained in the packages. The *GuideBooks* make little or no mention of the machine-dependent possibilities offered by the various *Mathematica* implementations. For this information, see the *Mathematica* documentation.

Mathematical and physical remarks introduce certain subjects and formulas to make the associated *Mathematica* implementations easier to understand. These remarks are not meant to provide a deep understanding of the (sometimes complicated) physical model or underlying mathematics; some of these remarks intentionally oversimplify matters.

The reader should examine all *Mathematica* inputs and outputs carefully. Sometimes, the inputs and outputs illustrate little-known or seldom-used aspects of *Mathematica* commands. Moreover, for the efficient use of *Mathematica*, it is very important to understand the possibilities and limits of the built-in commands. Many commands in *Mathematica* allow different numbers of arguments. When a given command is called with fewer than the maximum number of arguments, an internal (or user-defined) default value is used for the missing arguments. For most of the commands, the maximum number of arguments and default values are discussed.

When solving problems, the *GuideBooks* generically use a "straightforward" approach. This means they are not using particularly clever tricks to solve problems, but rather direct, possibly computationally more expensive, approaches. (From time to time, the *GuideBooks* even make use of a "brute force" approach.) The motivation is that when solving new "real life" problems a reader encounters in daily work, the "right mathematical trick" is seldom at hand. Nevertheless, the reader can more often than not rely on *Mathematica* being powerful enough to often succeed in using a straightforward approach. But attention is paid to *Mathematica*-specific issues to find time- and memory-efficient implementations—something that should be taken into account for any larger program.

As already mentioned, all larger pieces of code in this book have comments explaining the individual steps carried out in the calculations. Many smaller pieces of code have comments when needed to expedite the understanding of how they work. This enables the reader to easily change and adapt the code pieces. Sometimes, when the translation from traditional mathematics into *Mathematica* is trivial, or when the author wants to emphasize certain aspects of the code, we let the code "speak for itself". While paying attention to efficiency, the *GuideBooks* only occasionally go into the computational complexity ([8], [40], and [7]) of the given implementations. The implementation of very large, complicated suites of algorithms is not the purpose of the *GuideBooks*. The *Mathematica* packages included with *Mathematica* and the ones at *MathSource* (http://library.wolfram.com/ database/MathSource) offer a rich variety of self-study material on building large programs. Most general guidelines for writing code for scientific calculations (like descriptive variable names and modularity of code; see, e.g., [19] for a review) apply also to *Mathematica* programs.

The programs given in a chapter typically make use of *Mathematica* functions discussed in earlier chapters. Using commands from later chapters would sometimes allow for more efficient techniques. Also, these programs emphasize the use of commands from the current chapter. So, for example, instead of list operation, from a complexity point of view, hashing techniques or tailored data structures might be preferable. All subsections and sections are "self-contained" (meaning that no other code than the one presented is needed to evaluate the subsections and sections). The price for this "self-containedness" is that from time to time some code has to be repeated (such as manipulating polygons or forming random permutations of lists) instead of delegating such programming constructs to a package. Because this repetition could be construed as boring, the author typically uses a slightly different implementation to achieve the same goal.

0.3.2 Programming Paradigms

In the *GuideBooks,* the author wants to show the reader that *Mathematica* supports various programming paradigms and also show that, depending on the problem under consideration and the goal (e.g., solution of a problem, test of an algorithm, development of a program), each style has its advantages and disadvantages. (For a general discussion concerning programming styles, see [3], [41], [23], [32], [15], and [9].) *Mathematica* supports a functional programming style. Thus, in addition to classical procedural programs (which are often less efficient and less elegant), programs using the functional style are also presented. In the first volume of the *Mathematica GuideBooks*, the programming style is usually dictated by the types of commands that have been discussed up to that point. A certain portion of the programs involve recursive, rule-based programming. The choice of programming style is, of course, partially (ultimately) a matter of personal preference. The *GuideBooks'* main aim is to explain the operation, limits, and efficient application of the various *Mathematica* commands. For certain commands, this dictates a certain style of programming. However, the various programming styles, with their advantages and disadvantages, are not the main concern of the *GuideBooks*. In working with *Mathematica*, the reader is likely to use different programming styles depending if one wants a quick one-time calculation or a routine that will be used repeatedly. So, for a given implementation, the program structure may not always be the most elegant, fastest, or "prettiest".

The *GuideBooks* are not a substitute for the study of *The Mathematica Book* [45] http://documents. wolfram.com/mathematica). It is impossible to acquire a deeper (full) understanding of *Mathematica* without a *thorough* study of this book (reading it twice from the first to the last page is highly recommended). It *defines* the language and the spirit of *Mathematica*. The reader will probably from time to time need to refer to parts of it, because not all commands are discussed in the *GuideBooks*. However, the story of what can be done with *Mathematica* does not end with the examples shown in *The Mathematica Book*. The *Mathematica GuideBooks* go beyond *The Mathematica Book*. They present larger programs for solving various problems and creating complicated graphics. In addition, the *GuideBooks* discuss a number of commands that are not or are only fleetingly mentioned in the manual (e.g., some specialized methods of mathematical functions and functions from the `Developer`` and `Experimental`` contexts), but which the author deems important. In the notebooks, the author gives special emphasis to discussions, remarks, and applications relating to several commands that are typical for *Mathematica* but not for most other programming languages, e.g., `Map`, `MapAt`, `MapIndexed`, `Distribute`, `Apply`, `Replace`, `ReplaceAll`, `Inner`, `Outer`, `Fold`, `Nest`, `Nest`` `List`, `FixedPoint`, `FixedPointList`, and `Function`. These commands allow to write exceptionally elegant, fast, and powerful programs. All of these commands are discussed in *The Mathematica Book* and others that deal with programming in *Mathematica* (e.g., [33], [34], and [42]). However, the author's experience suggests that a deeper understanding of these commands and their optimal applications comes only after working with *Mathematica* in the solution of more complicated problems.

Both the printed book and the electronic component contain material that is meant to teach in detail how to use *Mathematica* to solve problems, rather than to present the underlying details of the various scientific examples. It cannot be overemphasized that to master the use of *Mathematica,* its programming paradigms and individual functions, the reader must experiment; this is especially important, insightful, easily verifiable, and satisfying with graphics, which involve manipulating expressions, making small changes, and finding different approaches. Because the results can easily be visually checked, generating and modifying graphics is an ideal method to learn programming in *Mathematica*.

0.4 Exercises and Solutions

0.4.1 Exercises

Each chapter includes a set of exercises and a detailed solution proposal for each exercise. When possible, all of the purely *Mathematica*-programming related exercises (these are most of the exercises of the Programming volume) should be solved by every reader. The exercises coming from mathematics, physics, and engineering should be solved according to the reader's interest. The most important *Mathematica* functions needed to solve a given problem are generally those of the associated chapter.

For a rough orientation about the content of an exercise, the subject is included in its title. The relative degree of difficulty is indicated by level superscript of the exercise number ([L1] indicates easy, [L2] indicates medium, and [L3] indicates difficult). The author's aim was to present understandable interesting examples that illustrate the *Mathematica* material discussed in the corresponding chapter. Some exercises were inspired by recent research problems; the references given allow the interested reader to dig deeper into the subject.

The exercises are intentionally not hyperlinked to the corresponding solution. The independent solving of the exercises is an important part of learning *Mathematica*.

0.4.2 Solutions

The *GuideBooks* contain solutions to each of the more than 1,000 exercises. Many of the techniques used in the solutions are not just one-line calls to built-in functions. It might well be that with further enhancements, a future version of *Mathematica* might be able to solve the problem more directly. (But due to different forms of some results returned by *Mathematica*, some problems might also become more challenging.) The author encourages the reader to try to find shorter, more clever, faster (in terms of runtime as well complexity), more general, and more elegant solutions. *Doing* various calculations is the most effective way to learn *Mathematica*. A proper *Mathematica* implementation of a function that solves a given problem often contains many different elements. The function(s) should have sensibly named and sensibly behaving options; for various (machine numeric, high-precision numeric, symbolic) inputs different steps might be required; shielding against inappropriate input might be needed; different parameter values might require different solution strategies and algorithms, helpful error and warning messages should be available. The returned data structure should be intuitive and easy to reuse; to achieve a good computational complexity, nontrivial data structures might be needed, etc. Most of the solutions do not deal with all of these issues, but only with selected ones and thereby leave plenty of room for more detailed treatments; as far as limit, boundary, and degenerate cases are concerned, they represent an outline of how to tackle the problem. Although the solutions do their job in general, they often allow considerable refinement and extension by the reader.

The reader should consider the given solution to a given exercise as a proposal; quite different approaches are often possible and sometimes even more efficient. The routines presented in the solutions are not the most general possible, because to make them foolproof for every possible input (sensible and nonsensical, evaluated and unevaluated, numerical and symbolical), the books would have had to go considerably beyond the mathematical and physical framework of the *GuideBooks*. In addition, few warnings are implemented for improper or improperly used arguments. The graphics provided in the solutions are mostly subject to a long list of refinements. Although the solutions do work, they are often sketchy and can be considerably refined and extended by the reader. This also means that the provided solutions to the exercises programs are not always very suitable for solving larger classes of problems. To increase their applicability would require considerably more code. Thus, it is not guaranteed that given routines will work correctly on related problems. To guarantee this generality and scalability, one would have to protect the variables better, implement formulas for more general or specialized cases, write functions to accept different numbers of variables, add type-checking and error-checking functions, and include corresponding error messages and warnings.

To simplify working through the solutions, the various steps of the solution are commented and are not always packed in a `Module` or `Block`. In general, only functions that are used later are packed. For longer calculations, such as those in some of the exercises, this was not feasible and intended. The arguments of the functions are not always checked for their appropriateness as is desirable for robust code. But, this makes it easier for the user to test and modify the code.

0.5 The Books Versus the Electronic Components

0.5.1 Working with the Notebooks

Each volume of the *GuideBooks* comes with a multiplatform DVD, containing fourteen main notebooks tailored for *Mathematica* 4 and compatible with *Mathematica* 5. Each notebook corresponds to a chapter from the printed books. (To avoid large file sizes of the notebooks, all animations are located in the Animations directory and not directly in the chapter notebooks.) The chapters (and so the corresponding notebooks) contain a detailed description and explanation of the *Mathematica* commands needed and used in applications of *Mathematica* to the sciences. Discussions on *Mathematica* functions are supplemented by a variety of mathematics, physics, and graphics examples. The notebooks also contain complete solutions to all exercises. Forming an electronic book, the notebooks also contain all text, as well as fully typeset formulas, and reader-editable and reader-changeable input. (Readers can copy, paste, and use the inputs in their notebooks.) In addition to the chapter notebooks, the DVD also includes a navigation palette and fully hyperlinked table of contents and index notebooks. The *Mathematica* notebooks corresponding to the printed book are fully evaluated. The evaluated chapter notebooks also come with hyperlinked overviews; these overviews are not in the printed book.

When reading the printed books, it might seem that some parts are longer than needed. The reader should keep in mind that the primary tool for working with the *Mathematica* kernel are *Mathematica* notebooks and that on a computer screen and there "length does not matter much". The *GuideBooks* are basically a printout of the notebooks, which makes going back and forth between the printed books and the notebooks very easy. The *GuideBooks* give large examples to encourage the reader to investigate various *Mathematica* functions and to become familiar with *Mathematica* as a system for doing mathematics, as well as a programming language. Investigating *Mathematica* in the accompanying notebooks is the best way to learn its details.

To start viewing the notebooks, open the table of contents notebook TableOfContents.nb. *Mathematica* notebooks can contain hyperlinks, and all entries of the table of contents are hyperlinked. Navigating through one of the chapters is convenient when done using the navigator palette GuideBooksNavigator.nb.

When opening a notebook, the front end minimizes the amount of memory needed to display the notebook by loading it incrementally. Depending on the reader's hardware, this might result in a slow scrolling speed. Clicking the "Load notebook cache" button of the GuideBooksNavigator palette speeds this up by loading the complete notebook into the front end.

For the vast majority of sections, subsections, and solutions of the exercises, the reader can just select such a structural unit and evaluate it (at once) on a year-2005 computer ($\geq$512 MB RAM) typically in a matter of minutes. Some sections and solutions containing many graphics may need hours of computation time. Also, more than 50 pieces of code run hours, even days. The inputs that are very memory intensive or produce large outputs and graphics are in inactive cells which can be activated by clicking the adjacent button. Because of potentially overlapping variable names between various sections and subsections, the author advises the reader not to evaluate an entire chapter at once.

Each smallest self-contained structural unit (a subsection, a section without subsections, or an exercise) should be evaluated within one *Mathematica* session starting with a freshly started kernel. At the end of each unit is an input cell. After evaluating all input cells of a unit in consecutive order, the input of this cell generates a short summary about the entire *Mathematica* session. It lists the number of evaluated inputs, the kernel CPU time, the wall clock time, and the maximal memory used to evaluate the inputs (excluding the resources needed to evaluate the `Program` cells). These numbers serve as a guide for the reader about the to-be-expected running times and memory needs. These numbers can deviate from run to run. The wall clock time can be substantially larger than the CPU time due to other processes running on the same computer and due to time needed to render graphics. The data shown in the evaluated notebooks came from a 2.5 GHz Linux computer. The CPU times are generically proportional to the computer clock speed, but can deviate within a small factor from operating system to operating system. In rare, randomly occurring cases slower computers can achieve smaller CPU and wall clock times than faster computers, due to internal time-constrained simplification processes in various symbolic mathematics functions (such as `Integrate`, `Sum`, `DSolve`, …).

The Overview Section of the chapters is set up for a front end and kernel running on the same computer and having access to the same file system. When using a remote kernel, the directory specification for the package `Overview.m` must be changed accordingly.

References can be conveniently extracted from the main text by selecting the cell(s) that refer to them (or parts of a cell) and then clicking the "Extract References" button. A new notebook with the extracted references will then appear.

The notebooks contain color graphics. (To rerender the pictures with a greater color depth or at a larger size, choose **Rerender Graphics** from the **Cell** menu.) With some of the colors used, black-and-white printouts occasionally give low-contrast results. For better black-and-white printouts of these graphics, the author recommends setting the `ColorOutput` option of the relevant graphics function to `GrayLevel`. The notebooks with animations (in the printed book, animations are typically printed as an array of about 10 to 20 individual graphics) typically contain between 60 and 120 frames. Rerunning the corresponding code with a large number of frames will allow the reader to generate smoother and longer-running animations.

Because many cell styles used in the notebooks are unique to the *GuideBooks*, when copying expressions and cells from the *GuideBooks* notebooks to other notebooks, one should first attach the style sheet notebook GuideBooksStylesheet.nb to the destination notebook, or define the needed styles in the style sheet of the destination notebook.

0.5.2 Reproducibility of the Results

The 14 chapter notebooks contained in the electronic version of the *GuideBooks* were run mostly with *Mathematica* 5.1 on a 2 GHz Intel Linux computer with 2 GB RAM. They need more than 100 hours of evaluation time. (This does not include the evaluation of the currently unevaluatable parts of code after the **Make Input** buttons.) For most subsections and sections, 512 MB RAM are recommended for a fast and smooth evaluation "at once" (meaning the reader can select the section or subsection, and evaluate all inputs without running out of memory or clearing variables) and the rendering of the generated graphic in the front end. Some subsections and sections need more memory when run. To reduce these memory requirements, the author recommends restarting the *Mathematica* kernel inside these subsections and sections, evaluating the necessary definitions, and then continuing. This will allow the reader to evaluate all inputs.

In general, regardless of the computer, with the same version of *Mathematica*, the reader should get the same results as shown in the notebooks. (The author has tested the code on Sun and Intel-based Linux computers, but this does not mean that some code might not run as displayed (because of different configurations, stack size settings, etc., but the disclaimer from the Preface applies everywhere). If an input does not work on a particular machine, please inform the author. Some deviations from the results given may appear because of the following:

- Inputs involving the function `Random[...]` in some form. (Often `SeedRandom` to allow for some kind of reproducibility and randomness at the same time is employed.)
- *Mathematica* commands operating on the file system of the computer, or make use of the type of computer (such inputs need to be edited using the appropriate directory specifications).

- Calculations showing some of the differences of floating-point numbers and the machine-dependent representation of these on various computers.
- Pictures using various fonts and sizes because of their availability (or lack thereof) and shape on different computers.
- Calculations involving `Timing` because of different clock speeds, architectures, operating systems, and libraries.
- Formats of results depending on the actual window width and default font size. (Often, the corresponding inputs will contain `Short`.)

Using anything other than *Mathematica* Version 5.1 might also result in different outputs. Examples of results that change form, but are all mathematically correct and equivalent, are the parameter variables used in underdetermined systems of linear equations, the form of the results of an integral, and the internal form of functions like `InterpolatingFunction` and `CompiledFunction`. Some inputs might no longer evaluate the same way because functions from a package were used and these functions are potentially built-in functions in a later *Mathematica* version. *Mathematica* is a very large and complicated program that is constantly updated and improved. Some of these changes might be design changes, superseded functionality, or potentially regressions, and as a result, some of the inputs might not work at all or give unexpected results in future versions of *Mathematica*.

0.5.3 Earlier Versions of the Notebooks

The first printing of the Programming volume and the Graphics volumes of the *Mathematica GuideBooks* were published in October 2004. The electronic components of these two books contained the corresponding evaluated chapter notebooks as well as unevaluated versions of preversions of the notebooks belonging to the Numerics and Symbolics volumes. Similarly, the electronic components of the Numerics and Symbolics volume contain the corresponding evaluated chapter notebooks and unevaluated copies of the notebooks of the Programming and Graphics volumes. This allows the reader to follow cross-references and look up relevant concepts discussed in the other volumes. The author has tried to keep the notebooks of the *GuideBooks* as up-to-date as possible. (Meaning with respect to the efficient and appropriate use of the latest version of *Mathematica*, with respect to maintaining a list of references that contains new publications, and examples, and with respect to incorporating corrections to known problems, errors, and mistakes). As a result, the notebooks of all four volumes that come with later printings of the Programming and Graphics volumes, as well with the Numerics and Symbolics volumes will be different and supersede the earlier notebooks originally distributed with the

Programming and Graphics volumes. The notebooks that come with the Numerics and Symbolics volumes are genuine *Mathematica* Version 5.1 notebooks. Because most advances in *Mathematica* Version 5 and 5.1 compared with *Mathematica* Version 4 occurred in functions carrying out numerical and symbolical calculations, the notebooks associated with Numerics and Symbolics volumes contain a substantial amount of changes and additions compared with their originally distributed version.

0.6 Style and Design Elements

0.6.1 Text and Code Formatting

The *GuideBooks* are divided into chapters. Each chapter consists of several sections, which frequently are further subdivided into subsections. General remarks about a chapter or a section are presented in the sections and subsections numbered 0. (These remarks usually discuss the structure of the following section and give teasers about the usefulness of the functions to be discussed.) Also, sometimes these sections serve to refresh the discussion of some functions already introduced earlier.

Following the style of *The Mathematica Book* [45], the *GuideBooks* use the following fonts: For the main text, Times; for *Mathematica* inputs and built-in *Mathematica* commands, Courier plain (like `Plot`); and for user-supplied arguments, Times italic (like *userArgument$_1$*). Built-in *Mathematica* functions are introduced in the following style:

`MathematicaFunctionToBeIntroduced[`*typeIndicatingUserSuppliedArgument(s)*`]`
　　is a description of the built-in command `MathematicaFunctionToBeIntroduced` upon its first
　　appearance. A definition of the command, along with its parameters is given. Here, *typeIndicatingUserSupplied-*
　　Argument(s) is one (or more) user-supplied expression(s) and may be written in an abbreviated form or in a
　　different way for emphasis.

The actual *Mathematica* inputs and outputs appear in the following manner (as mentioned above, virtually all inputs are given in `InputForm`).

```
(* A comment. It will be/is ignored as Mathematica input:
   Return only one of the solutions *)
Last[Solve[{x^2 - y == 1, x - y^2 == 1}, {x, y}]]
```

When referring in text to variables of *Mathematica* inputs and outputs, the following convention is used: Fixed, nonpattern variables (including local variables) are printed in Courier plain (the equations solved above contained the variables `x` and `y`). User supplied arguments to built-in or defined functions with pattern variables are printed in Times italic. The next input defines a function generating a pair of polynomial equations in x and y.

```
equationPair[x_, y_] := {x^2 - y == 1, x - y^2 == 1}
```

x and y are pattern variables (usimng the same letters, but a different font from the actual code fragments `x_` and `y_`) that can stand for any argument. Here we call the function `equationPair` with the two arguments `u + v` and `w - z`.

```
equationPair[u + v, w - z]
```

Occasionally, explanation about a mathematics or physics topic is given before the corresponding *Mathematica* implementation is discussed. These sections are marked as follows:

Mathematical Remark: Special Topic in Mathematics or Physics

A *short* summary or review of mathematical or physical ideas necessary for the following example(s).

From time to time, *Mathematica* is used to analyze expressions, algorithms, etc. In some cases, results in the form of English sentences are produced programmatically. To differentiate such automatically generated text from the main text, in most instances such text is prefaced by "∘" (structurally the corresponding cells are of type `"PrintText"` versus `"Text"` for author-written cells).

Code pieces that either run for quite long, or need a lot of memory, or are tangent to the current discussion are displayed in the following manner.

```
        mathematicaCodeWhichEitherRunsVeryLongOrThatIsVeryMemoryIntensive⌐
        OrThatProducesAVeryLargeGraphicOrThatIsASideTrackToTheSubjectUnder⌐
        Discussion
        (* with some comments on how the code works *)
```

To run a code piece like this, click the **Make Input** button above it. This will generate the corresponding input cell that can be evaluated if the reader's computer has the necessary resources.

The reader is encouraged to add new inputs and annotations to the electronic notebooks. There are two styles for reader-added material: `"ReaderInput"` (a *Mathematica* input style and simultaneously the default style for a new cell) and `"ReaderAnnotation"` (a text-style cell type). They are primarily intended to be used in the `Reading` environment. These two styles are indented more than the default input and text cells, have a green left bar and a dingbat. To access the `"ReaderInput"` and `"ReaderAnnotation"` styles, press the system-dependent modifier key (such as Control or Command) and 9 and 7, respectively.

0.6.2 References

Because the *GuideBooks* are concerned with the solution of mathematical and physical problems using *Mathematica* and are not mathematics or physics monographs, the author did not attempt to give complete references for each of the applications discussed [38], [20]. The references cited in the text pertain mainly to the applications under discussion. Most of the citations are from the more recent literature; references to older publications can be found in the cited ones. Frequently URLs for downloading relevant or interesting information are given. (The URL addresses worked at the time of printing and, hopefully, will be still active when the reader tries them.) References for *Mathematica*, for algorithms used in computer algebra, and for applications of computer algebra are collected in the Appendix A.

The references are listed at the end of each chapter in alphabetical order. In the notebooks, the references are hyperlinked to all their occurrences in the main text. Multiple references for a subject are not cited in numerical order, but rather in the order of their importance, relevance, and suggested reading order for the implementation given.

In a few cases (e.g., pure functions in Chapter 3, some matrix operations in Chapter 6), references to the mathematical background for some built-in commands are given—mainly for commands in which the mathematics required extends beyond the familiarity commonly exhibited by non-mathematicians. The *GuideBooks* do not discuss the algorithms underlying such complicated functions, but sometimes use *Mathematica* to "monitor" the algorithms.

References of the form *abbreviationOfAScientificField/yearMonthPreprintNumber* (such as quant-ph/0012147) refer to the arXiv preprint server [43], [22], [30] at http://arXiv.org. When a paper appeared as a preprint and (later) in a journal, typically only the more accessible preprint reference is given. For the convenience of the reader, at the end of these references, there is a **Get Preprint** button. Click the button to display a palette notebook with hyperlinks to the corresponding preprint at the main preprint server and its mirror sites. (Some of the older journal articles can be downloaded free of charge from some of the digital mathematics library servers, such as http://gdz.sub.uni-goettingen.de, http://www.emis.de, http://www.numdam.org, and http://dieper.aib.uni-linz.ac.at.)

As much as available, recent journal articles are hyperlinked through their digital object identifiers (http://www.doi.org).

0.6.3 Variable Scoping, Input Numbering, and Warning Messages

Some of the *Mathematica* inputs intentionally cause error messages, infinite loops, and so on, to illustrate the operation of a *Mathematica* command. These messages also arise in the user's practical use of *Mathematica*. So, instead of presenting polished and perfected code, the author prefers to illustrate the potential problems and limitations associated with the use of *Mathematica* applied to "real life" problems. The one exception are the spelling warning messages `General::spell` and `General::spell1` that would appear relatively frequently because "similar" names are used eventually. For easier and less defocused reading, these messages are turned off in the initialization cells. (When working with the notebooks, this means that the pop-up window asking the user "Do you want to automatically evaluate all the initialization cells in the notebook?" should be evaluated should always be answered with a "yes".) For the vast majority of graphics presented, the picture is the focus, not the returned *Mathematica* expression representing the picture. That is why the `Graphics` and `Graphics3D` output is suppressed in most situations.

To improve the code's readability, no attempt has been made to protect all variables that are used in the various examples. This protection could be done with `Clear`, `Remove`, `Block`, `Module`, `With`, and others. Not protecting the variables allows the reader to modify, in a somewhat easier manner, the values and definitions of variables, and to see the effects of these changes. On the other hand, there may be some interference between variable names and values used in the notebooks and those that might be introduced when experimenting with the code. When readers examine some of the code on a computer, reevaluate sections, and sometimes perform subsidiary calculations, they may introduce variables that might interfere with ones from the *GuideBooks*. To partially avoid this problem, and for the reader's convenience, sometimes `Clear[`*sequenceOfVariables*`]` and `Remove[`*sequenceOfVariables*`]` are sprinkled throughout the notebooks. This makes experimenting with these functions easier.

The numbering of the *Mathematica* inputs and outputs typically does not contain all consecutive integers. Some pieces of *Mathematica* code consist of multiple inputs per cell; so, therefore, the line numbering is incremented by more than just 1. As mentioned, *Mathematica* should be restarted at every section, or subsection or solution of an exercise, to make sure that no variables with values get reused. The author also explicitly asks the reader to restart *Mathematica* at some special positions inside sections. This removes previously introduced variables, eliminates all existing contexts, and returns *Mathematica* to the typical initial configuration to ensure reproduction of the results and to avoid using too much memory inside one session.

0.6.4 Graphics

In *Mathematica* 5.1, displayed graphics are side effects, not outputs. The actual output of an input producing a graphic is a single cell with the text `-Graphics-` or `-Graphics3D-` or `-GraphicsArray-` and so on. To save paper, these output cells have been deleted in the printed version of the *GuideBooks*.

Most graphics use an appropriate number of plot points and polygons to show the relevant features and details. Changing the number of plot points and polygons to a higher value to obtain higher resolution graphics can be done by changing the corresponding inputs.

The graphics of the printed book and the graphics in the notebooks are largely identical. Some printed book graphics use a different color scheme and different point sizes and line and edge thicknesses to enhance contrast and visibility. In addition, the font size has been reduced for the printed book in tick and axes labels.

The graphics shown in the notebooks are PostScript graphics. This means they can be resized and rerendered without loss of quality. To reduce file sizes, the reader can convert them to bitmap graphics using the Cell$\longrightarrow$ Convert To$\longrightarrow$Bitmap menu. The resulting bitmap graphics can no longer be resized or rerendered in the original resolution.

To reduce file sizes of the main content notebooks, the animations of the *GuideBooks* are not part of the chapter notebooks. They are contained in a separate directory.

0.6.5 Notations and Symbols

The symbols used in typeset mathematical formulas are not uniform and unique throughout the *GuideBooks*. Various mathematical and physical quantities (such as normals, rotation matrices, and field strengths) are used repeatedly in this book. Frequently the same notation is used for them, but depending on the context, also different ones are used, e.g. sometimes bold is used for a vector (such as $\mathbf{r}$) and sometimes an arrow (such as $\vec{r}$). Matrices appear in bold or as doublestruck letters. Depending on the context and emphasis placed, different notations are used in display equations and in the *Mathematica* input form. For instance, for a time-dependent scalar quantity of one variable $\psi(t; x)$, we might use one of many patterns, such as $\psi[t][x]$ (for emphasizing a parametric t-dependence) or $\psi[t, x]$ (to treat t and x on an equal footing) or $\psi[t, \{x\}]$ (to emphasize the one-dimensionality of the space variable x).

Mathematical formulas use standard notation. To avoid confusion with *Mathematica* notations, the use of square brackets is minimized throughout. Following the conventions of mathematics notation, square brackets are used for three cases: a) Functionals, such as $\mathcal{F}_t[f(t)](\omega)$ for the Fourier transform of a function $f(t)$. b) Power series coefficients, $[x^k](f(x))$ denotes the coefficient of x^k of the power series expansion of $f(x)$ around $x = 0$. c) Closed intervals, like $[a, b]$ (open intervals are denoted by (a, b)). Grouping is exclusively done using parentheses. Upper-case double-struck letters denote domains of numbers, $\mathbb{Z}$ for integers, $\mathbb{N}$ for nonnegative integers, $\mathbb{Q}$ for rational numbers, $\mathbb{R}$ for reals, and $\mathbb{C}$ for complex numbers. Points in $\mathbb{R}^n$ (or $\mathbb{C}^n$) with explicitly given coordinates are indicated using curly braces $\{c_1, ..., c_n\}$. The symbols $\wedge$ and $\vee$ for And and Or are used in logical formulas.

For variable names in formula- and identity-like *Mathematica* code, the symbol (or small variations of it) traditionally used in mathematics or physics is used. In program-like *Mathematica* code, the author uses very descriptive, sometimes abbreviated, but sometimes also slightly longish, variable names, such as `buildBril\`louinZone` and `FibonacciChainMap`.

0.6.6 Units

In the examples involving concepts drawn from physics, the author tried to enhance the readability of the code (and execution speed) by not choosing systems of units involving numerical or unit-dependent quantities. (For more on the choice and treatment of units, see [39], [4], [5], [10], [13], [11], [12], [36], [35], [31], [37], [44], [21], [25], [18], [26], [24].) Although *Mathematica* can carry units along with the symbols representing the physical quantities in a calculation, this requires more programming and frequently diverts from the essence of the problem. Choosing a system of units that allows the equations to be written without (unneeded in computations) units often gives considerable insight into the importance of the various parts of the equations because the magnitudes of the explicitly appearing coefficients are more easily compared.

0.6.7 Cover Graphics

The cover graphics of the *GuideBooks* stem from the *Mathematica GuideBooks* themselves. The construction ideas and their implementation are discussed in detail in the corresponding *GuideBook*.

■ The cover graphic of the Programming volume shows 42 tori, 12 of which are in the dodecahedron's face planes and 30 which are in the planes perpendicular to the dodecahedron's edges. Subsections 1.2.4 of Chapter 1 discusses the implementation.

■ The cover graphic of the Graphics volume first subdivides the faces of a dodecahedron into small triangles and then rotates randomly selected triangles around the dodecahedron's edges. The proposed solution of Exercise 1b of Chapter 2 discusses the implementation.

■ The cover graphic of the Numerics volume visualizes the electric field lines of a symmetric arrangement of positive and negative charges. Subsection 1.11.1 discusses the implementation.

■ The cover graphic of the Symbolics volume visualizes the derivative of the Weierstrass $\wp'$ function over the Riemann sphere. The "threefold blossoms" arise from the poles at the centers of the periodic array of period parallelograms. Exercise 3j of Chapter 2 discusses the implementation.

■ The four spine graphics show the inverse elliptic nome function q^{-1}, a function defined in the unit disk with a boundary of analyticity mapped to a triangle, a square, a pentagon, and a hexagon. Exercise 16 of Chapter 2 of the Graphics volume discusses the implementation.

0.7 Production History

The original set of notebooks was developed in the 1991–1992 academic year on an Apple Macintosh IIfx with 20 MB RAM using *Mathematica* Version 2.1. Over the years, the notebooks were updated to *Mathematica* Version 2.2, then to Version 3, and finally for Version 4 for the first printed edition of the Programming and Graphics volumes of the *Mathematica GuideBooks* (published autumn 2004). For the Numerics and Symbolics volumes, the *GuideBooks* notebooks were updated to *Mathematica* Version 5 in the second half of 2004. Historically, the first step in creating the book was the translation of a set of Macintosh notebooks used for lecturing and written in German into English by Larry Shumaker. This was done primarily by a translation program and afterward by manually polishing the English version. Then the notebooks were transformed into T_EX files using the program nb2tex on a NeXT computer. The resulting files were manually edited, equations prepared in the original German notebooks were formatted with T_EX, and macros were added corresponding to the design of the book. (The translation to T_EX was necessary because *Mathematica* Version 2.2 did not allow for book-quality printouts.) They were updated and refined for nearly three years, and then *Mathematica* 3 notebooks were generated from the T_EX files using a preliminary version of the program `tex2nb`. Historically and technically, this was an important step because it transformed all of the material of the *GuideBooks* into *Mathematica* expressions and allowed for automated changes and updates in the various editing stages. (Using the *Mathematica* kernel allowed one to process and modify the notebook files of these books in a uniform and time-efficient manner.) Then, the notebooks were expanded in size and scope and updated to *Mathematica* 4. In the second half of the year 2003, and first half of the year 2004, the *Mathematica* programs of the notebooks were revised to be compatible with *Mathematica* 5. In October 2004, the Programming and the Graphics volumes were published. In the last quarter of 2004, all four volumes of the *GuideBooks* were updated to be tailored for *Mathematica* 5.1 A special set of styles was created to generate the actual PostScript as printouts from the notebooks. All inputs were evaluated with this style sheet, and the generated PostScript was directly used for the book production. Using a little *Mathematica* program, the index was generated from the notebooks (which are *Mathematica* expressions), containing all index entries as cell tags.

0.8 Four General Suggestions

A reader new to *Mathematica* should take into account these four suggestions.

■ There is usually more than one way to solve a given problem using *Mathematica*. If one approach does not work or returns the wrong answer or gives an error message, make every effort to understand what is happening. Even if the reader has succeeded with an alternative approach, it is important to try to understand why other attempts failed.

■ Mathematical formulas, algorithms, and so on, should be implemented as directly as possible, even if the resulting construction is somewhat "unusual" compared to that in other programming languages. In particular, the reader should not simply translate C, Pascal, Fortran, or other programs line-by-line into *Mathematica*, although this is indeed possible. Instead, the reader should instead reformulate the problem in a clear mathematical way. For example, Do, While, and For loops are frequently unnecessary, convergence (for instance, of sums) can be checked by *Mathematica*, and If tests can often be replaced by a corresponding pattern. The reader should start with an exact mathematical description of the problem [28], [29]. For example, it does not suffice to know which transformation formulas have to be used on certain functions; one also needs to know how to apply them. "The power of mathematics is in its precision. The precision of mathematics must be used precisely." [17]

■ If the exercises, examples, and calculation of the *GuideBooks* or the listing of calculation proposals from Exercise 1 of Chapter 1 of the Programming volume are not challenging enough or do not cover the reader's interests, consider the following idea, which provides a source for all kinds of interesting and difficult problems: The reader should select a built-in command and try to reconstruct it using other built-in commands and make it behave as close to the original as possible in its operation, speed, and domain of applicability, or even to surpass it. (Replicating the following functions is a serious challenge: N, Factor, FactorInteger, Integrate, NIntegrate, Solve, DSolve, NDSolve, Series, Sum, Limit, Root, Prime, or PrimeQ.)

■ If the reader tries to solve a smaller or larger problem in *Mathematica* and does not succeed, keep this problem on a "to do" list and periodically review this list and try again. Whenever the reader has a clear strategy to solve a problem, this strategy can be implemented in *Mathematica*. The implementation of the algorithm might require some programming skills, and by reading through this book, the reader will become able to code more sophisticated procedures and more efficient implementations. After the reader has acquired a certain amount of *Mathematica* programming familiarity, implementing virtually all "procedures" which the reader can (algorithmically) carry out with paper and pencil will become straightforward.

References

1 P. Abbott. *The Mathematica Journal* 4, 415 (2000).

2 P. Abbott. *The Mathematica Journal* 9, 31 (2003).

3 H. Abelson, G. Sussman. *Structure and Interpretation of Computer Programs*, MIT Press, Cambridge, MA, 1985.

4 G. I. Barenblatt. *Similarity, Self-Similarity, and Intermediate Asymptotics*, Consultants Bureau, New York, 1979.

5 F. A. Bender. *An Introduction to Mathematical Modeling*, Wiley, New York, 1978.

6 G. Benfatto, G. Gallavotti. *Renormalization Group*, Princeton University Press, Princeton, 1995.

7 L. Blum, F. Cucker, M. Shub, S. Smale. *Complexity and Real Computation*, Springer, New York, 1998.

8 P. Bürgisser, M. Clausen, M. A. Shokrollahi. *Algebraic Complexity Theory*, Springer, Berlin, 1997.

9 L. Cardelli, P. Wegner. *Comput. Surveys* 17, 471 (1985).

10 J. F. Carinena, M. Santander in P. W. Hawkes (ed.). *Advances in Electronics and Electron Physics* 72, Academic Press, New York, 1988.

11 E. A. Desloge. *Am. J. Phys.* 52, 312 (1984).

12 C. L. Dym, E. S. Ivey. *Principles of Mathematical Modelling*, Academic Press, New York, 1980.

13 A. C. Fowler. *Mathematical Models in the Applied Sciences*, Cambridge University Press, Cambridge, 1997.

14 T. Gannon. *arXiv:math.QA*/9906167 (1999).

15 R. J. Gaylord, S. N. Kamin, P. R. Wellin. *An Introduction to Programming with Mathematica*, TELOS/Springer-Verlag, Santa Clara, 1993.

16 J. Glynn, T. Gray. *The Beginner's Guide to Mathematica Version 3*, Cambridge University Press, Cambridge, 1997.

17 D. Greenspan in R. E. Mickens (ed.). *Mathematics and Science*, World Scientific, Singapore, 1990.

18 G. W. Hart. *Multidimensional Analysis*, Springer-Verlag, New York, 1995.

19 A. K. Hartman, H. Rieger. *arXiv:cond-mat*/0111531 (2001).

20 M. Hazewinkel. *arXiv:cs.IR*/0410055 (2004).

21 E. Isaacson, M. Isaacson. *Dimensional Methods in Engineering and Physics*, Edward Arnold, London, 1975.

22 A. Jackson. *Notices Am. Math. Soc.* 49, 23 (2002).

23 R. D. Jenks, B. M. Trager in J. von zur Gathen, M. Giesbrecht (eds.). *Symbolic and Algebraic Computation*, ACM Press, New York, 1994.

24 C. G. Jesudason. *arXiv:physics*/0403033 (2004).

25 C. Kauffmann in A. van der Burgh (ed.). *Topics in Engineering Mathematics*, Kluwer, Dordrecht, 1993.

26 R. Khanin in B. Mourrain (ed.). *ISSAC 2001*, ACM, Baltimore, 2001.

27 P. Kleinert, H. Schlegel. *Physica* A 218, 507 (1995).

28 D. E. Knuth. *Am. Math. Monthly* 81, 323 (1974).

29 D. E. Knuth. *Am. Math. Monthly* 92, 170 (1985).

30 G. Kuperberg. *arXiv:math.HO*/0210144 (2002).

31 J. D. Logan. *Applied Mathematics*, Wiley, New York, 1987.

32 K. C. Louden. *Programming Languages: Principles and Practice*, PWS-Kent, Boston, 1993.

33 R. Maeder. *Programming in Mathematica*, Addison-Wesley, Reading, 1997.

34 R. Maeder. *The Mathematica Programmer*, Academic Press, New York, 1993.

35   B. S. Massey. *Measures in Science and Engineering*, Wiley, New York, 1986.

36   G. Messina, S. Santangelo, A. Paoletti, A. Tucciarone. *Nuov. Cim.* D 17, 523 (1995).

37   J. Molenaar in A. van der Burgh, J. Simonis (eds.). *Topics in Engineering Mathematics*, Kluwer, Dordrecht, 1992.

38   E. Pascal. *Repertorium der höheren Mathematik* Theil 1/1 [page V, paragraph 3], Teubner, Leipzig, 1900.

39   S. H. Romer. *Am. J. Phys.* 67, 13 (1999).

40   R. Sedgewick, P. Flajolet. *Analysis of Algorithms*, Addison-Wesley, Reading, 1996.

41   R. Sethi. *Programming Languages: Concepts and Constructions*, Addison-Wesley, New York, 1989.

42   D. B. Wagner. *Power Programming with Mathematica: The Kernel*, McGraw-Hill, New York, 1996.

43   S. Warner. *arXiv:cs.DL*/0101027 (2001).

44   H. Whitney. *Am. Math. Monthly* 75, 115, 227 (1968).

45   S. Wolfram. *The Mathematica Book*, Wolfram Media, Champaign, 2003.

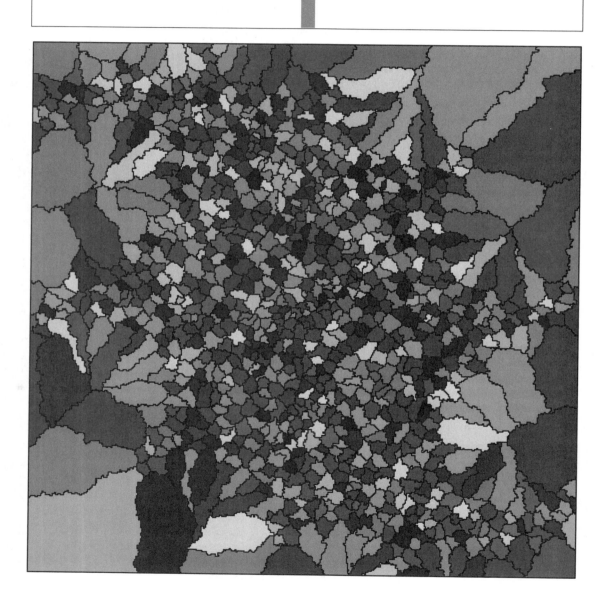

Numerical Computations

1.0 Remarks

This chapter deals with numerical methods in *Mathematica*. *The Mathematica Book* [1919] does not discuss all options and option settings of the numerical functions in great depth, so as an important additional source about many details of selected numerical functions we strongly recommend the Advanced Documentation located in the help browser.

For numerical analysis itself, see the special textbooks, such as [1812], or [481]. (For some general issues on how to do numerical analysis in nonprocedural languages, see [1730].) In this chapter, we discuss numerics within *Mathematica*. Concerning the use of computer algebra for deriving stable numerical methods, see [659] and [244]. For the application of numerical methods to carry out mathematical proofs, see [797], [588].

We discussed numerical methods stemming from problems in linear algebra in Chapter 6 of the Programming volume [1793], so there is no need to repeat the discussion here extensively.

Before proceeding to a discussion of numbers with an arbitrary number of digits, we make a few comments about computing with machine numbers. In the nine chapters of the Programming volume [1793] and Graphics volume [1794], we made repeated use of N[*expr*]. N[*expr*] calculates a numerical value of *expr* by using machine arithmetic. If we add some numbers that differ greatly in size, it is possible to get a very poor result. Here is an example.

```
In[1]:= -10.0^22 + 2.0 - 1.0 + 10.0^22
Out[1]= 0.
```

The result is 0.0, even though the "expected" result is 1.0. We can get various "incorrect" results if we change the order of the summands. Here all possibilities are investigated.

```
In[2]:= Apply[Plus, Permutations[{-10.0^22, 10.0^22, 2.0, -1.0}], {1}]
Out[2]= {1., 1., -1., 0., 2., 0., 1., 1., -1., 0.,
        2., 0., -1., 0., -1., 0., 0., 0., 2., 0., 2., 0., 0., 0.}
```

The summation process loses the "remaining" −1.0 and 2.0 because the number of digits used in machine arithmetic calculations is less than 22.

```
In[3]:= $MachinePrecision
Out[3]= 15.9546
```

The machine numbers are summed in the order given. Here, one summation is done step by step.

```
In[4]:= Fold[Plus, 0, #]& /@ Permutations[{-10.0^22, 10.0^22, 2.0, -1.0}]
Out[4]= {1., 1., -1., 0., 2., 0., 1., 1., -1., 0.,
        2., 0., -1., 0., -1., 0., 0., 0., 2., 0., 2., 0., 0., 0.}
```

From the last experiment, we have to conclude that for machine numbers (for a better overall efficiency of *Mathematica*) the `Orderless` attribute of `Plus` does not go into effect before the summation is carried out.

For exact numbers or high-precision numbers, the order of the summands does, of course, not matter. (In the following, we will use the words "high-precision number" and "bignumber", as well as "high-precision arithmetic" and "bignum arithmetic" interchangeably.)

```
In[5]:= Apply[Plus, Permutations[{-10^22, 10^22, 2, -1}], {1}]
Out[5]= {1, 1, 1, 1, 1, 1, 1, 1, 1, 1, 1, 1, 1, 1, 1, 1, 1, 1, 1, 1, 1, 1, 1, 1}
```

With this amount of digits, it is not always possible to correctly keep track of the last digit. The automatic precision control of *Mathematica* shows that not a single digit is known and estimates the order of magnitude of the result.

```
In[6]:= Apply[Plus, Permutations[N[{-10^22, 10^22, 2, -1},
                        $MachinePrecision + 1]], {1}]
```
$$Out[6]= \{0. \times 10^5, 0. \times 10^5, 0. \times 10^5, 0. \times 10^5, 0. \times 10^5, 0. \times 10^5, 0. \times 10^5, 0. \times 10^5,$$
$$0. \times 10^5, 0. \times 10^5, 0. \times 10^5, 0. \times 10^5, 0. \times 10^5, 0. \times 10^5, 0. \times 10^5, 0. \times 10^5,$$
$$0. \times 10^5, 0. \times 10^5, 0. \times 10^5, 0. \times 10^5, 0. \times 10^5, 0. \times 10^5, 0. \times 10^5, 0. \times 10^5\}$$

Using 30 digits allows obtaining the correct result for all permutations.

```
In[7]:= Apply[Plus, Permutations[N[{-10^22, 10^22, 2, -1}, 30]], {1}]
Out[7]= {1.0000000, 1.0000000, 1.0000000, 1.0000000, 1.0000000, 1.0000000,
         1.0000000, 1.0000000, 1.0000000, 1.0000000, 1.0000000, 1.0000000,
         1.0000000, 1.0000000, 1.0000000, 1.0000000, 1.0000000, 1.0000000,
         1.0000000, 1.0000000, 1.0000000, 1.0000000, 1.0000000, 1.0000000}
```

Note that inserting parentheses does not help because the brackets only influence the parsing process, but not the evaluation process. It is possible to force *Mathematica* to give the correct result by explicitly enforcing a certain evaluation order.

```
In[8]:= Identity[-10.0^22 + 10.0^22] + 2.0 - 1.0
Out[8]= 1.
```

The function `Total` (to be discussed below) cannot give a unique result here either.

```
In[9]:= Map[Total[#] &, Permutations[{-10.0^22, 10.0^22, 2.0, -1.0}], {1}]
Out[9]= {1., 1., -1., 0., 2., 0., 1., 1., -1., 0.,
         2., 0., -1., 0., -1., 0., 0., 0., 2., 0., 2., 0., 0., 0.}
```

For more on this and related issues, see [887], [563], [1016], [1540], [945], and [832].

The capability of *Mathematica* to calculate numerical quantities with an arbitrary number of digits is an important ingredient in "experimental mathematics". Given a conjectured identity, the possibility to check that it is correct to 10, to 100, to 1000 or even more digits at "randomly" selected points does not constitute a proof, but is often very strong evidence that often opens the possibility for a symbolic proof. The probability to accidentally get hundreds of digits that are identical in two expressions, or to get zero to hundreds of digits in an expression of order 1, is far smaller than is the probability of an error in a human-generated proof [458], [479]. (This is a kind of "experimental mathematics" [257], [115], [256], [254], [116], [278], [117], [1178].)

As a first example of the usefulness of being able to carry out calculations with many hundred of digits, let us just "check" an identity numerically. Let us deal with the following statement: The function

$$w(\tau) = \frac{1}{2} \frac{\partial \log(J'(\tau)^6 \, J(\tau)^{-4} \, (J(\tau) - 1)^{-3})}{\partial \tau}$$

is a solution of the Chazy equation [794], [201], [407]

$$3\,w'(\tau)^2 - 2\,w(\tau)\,w''(\tau) + w^{(3)}(\tau) = 0.$$

Here, $J(\tau)$ is Klein's modular function, in *Mathematica* `KleinInvariantJ`. While there is a closed form for the derivative of Klein's modular function through elliptic integrals $K(z)$, $E(z)$ and the modular function $\lambda(z)$

$$J'(\tau) = -\frac{8\,i\,K(\lambda(\tau))^2\,(\lambda(\tau) - 2)\,(\lambda(\tau) + 1)\,(2\,\lambda(\tau) - 1)\,((\lambda(\tau) - 1)\,\lambda(\tau) + 1)^2}{27\,(K(1 - \lambda(\tau))\,(E(\lambda(\tau)) - K(\lambda(\tau))) + E(1 - \lambda(\tau))\,K(\lambda(\tau)))\,(\lambda(\tau) - 1)^2\,\lambda(\tau)^2},$$

we do not carry the differentiations symbolically because the resulting expressions are very large.

This is the conjectured solution of the Chazy equation.

```
In[10]:= w[τ_] = 1/2 D[Log[J'[τ]^6/(J[τ]^4(J[τ] - 1)^3)], τ]
```

$$\text{Out[10]=}\quad \frac{(-1 + J[\tau])^3\,J[\tau]^4\,\left(-\frac{4\,J'[\tau]^7}{(-1+J[\tau])^3\,J[\tau]^5} - \frac{3\,J'[\tau]^7}{(-1+J[\tau])^4\,J[\tau]^4} + \frac{6\,J'[\tau]^5\,J''[\tau]}{(-1+J[\tau])^3\,J[\tau]^4}\right)}{2\,J'[\tau]^6}$$

We substitute the solution in the differential equation, and write the result over a common denominator.

```
In[11]:= ode = w'''[τ] - 2 w[τ] w''[τ] + 3 w'[τ]^2;
```

```
In[12]:= (odeNum = Numerator[Together[ode]]) // Simplify
```

$$\begin{aligned}
\text{Out[12]=}\quad &(32 - 80\,J[\tau] + 120\,J[\tau]^2 - 80\,J[\tau]^3 + 35\,J[\tau]^4)\,J'[\tau]^8 - 108\,(-1 + J[\tau])^4\,J[\tau]^4\,J''[\tau]^4 + \\
&6\,(-1 + J[\tau])^2\,J[\tau]^2\,(24 - 40\,J[\tau] + 35\,J[\tau]^2)\,J'[\tau]^5\,J^{(3)}[\tau] + \\
&144\,(-1 + J[\tau])^4\,J[\tau]^4\,J'[\tau]\,J''[\tau]^2\,J^{(3)}[\tau] + (-1 + J[\tau])^2\,J[\tau]^2\,J'[\tau]^4 \\
&(-9\,(24 - 40\,J[\tau] + 35\,J[\tau]^2)\,J''[\tau]^2 + 10\,J[\tau]\,(4 - 11\,J[\tau] + 7\,J[\tau]^2)\,J^{(4)}[\tau]) - \\
&12\,(-1 + J[\tau])^3\,J[\tau]^3\,J'[\tau]^2\,\left(-5\,(-4 + 7\,J[\tau])\,J''[\tau]^3 - \right. \\
&\left. 6\,(-1 + J[\tau])\,J[\tau]\,J^{(3)}[\tau]^2 + 10\,(-1 + J[\tau])\,J[\tau]\,J''[\tau]\,J^{(4)}[\tau]\right) + \\
&12\,(-1 + J[\tau])^3\,J[\tau]^3\,J'[\tau]^3\,(-5\,(-4 + 7\,J[\tau])\,J''[\tau]\,J^{(3)}[\tau] + (-1 + J[\tau])\,J[\tau]\,J^{(5)}[\tau])
\end{aligned}$$

Now, we define a function `d` that calculates an approximation of the derivative of a function f at $x0$ using symmetric differences.

```
In[13]:= d[f_, x0_, ε_] = (f[x0 + ε] - f[x0 - ε])/(2ε);
```

Because the above polynomial `odeNum` contains up to fifth derivatives, we calculate the derivatives of $J(z)$ using the function `d`. `d` implements recursively a simple finite difference approximation.

```
In[14]:= j[0][x_, ε_] = KleinInvariantJ[x];
        Do[j[i][x_, ε_] = Together[d[j[i - 1][#, ε]&, x, ε]], {i, 1, 5}]
```

Here are the definitions of `j`.

```
In[16]:= ??j
```

```
        Global`j
```

$j[0][x\_, \varepsilon\_] = \text{KleinInvariantJ}[x]$

$j[1][x\_, \varepsilon\_] = \frac{-\text{KleinInvariantJ}[x-\varepsilon]+\text{KleinInvariantJ}[x+\varepsilon]}{2\varepsilon}$

$j[2][x\_, \varepsilon\_] = \frac{-2\,\text{KleinInvariantJ}[x]+\text{KleinInvariantJ}[x-2\varepsilon]+\text{KleinInvariantJ}[x+2\varepsilon]}{4\varepsilon^2}$

$j[3][x\_, \varepsilon\_] =$
$\frac{-\text{KleinInvariantJ}[x-3\varepsilon]+3\,\text{KleinInvariantJ}[x-\varepsilon]-3\,\text{KleinInvariantJ}[x+\varepsilon]+\text{KleinInvariantJ}[x+3\varepsilon]}{8\varepsilon^3}$

$j[4][x\_, \varepsilon\_] = \frac{1}{16\varepsilon^4}$
$\quad (6\,\text{KleinInvariantJ}[x] + \text{KleinInvariantJ}[x-4\varepsilon] - 4\,\text{KleinInvariantJ}[x-2\varepsilon] -$
$\quad 4\,\text{KleinInvariantJ}[x+2\varepsilon] + \text{KleinInvariantJ}[x+4\varepsilon])$

$j[5][x\_, \varepsilon\_] = \frac{1}{32\varepsilon^5} \, (-\text{KleinInvariantJ}[x-5\varepsilon] + 5\,\text{KleinInvariantJ}[x-3\varepsilon] -$
$\quad 10\,\text{KleinInvariantJ}[x-\varepsilon] + 10\,\text{KleinInvariantJ}[x+\varepsilon] -$
$\quad 5\,\text{KleinInvariantJ}[x+3\varepsilon] + \text{KleinInvariantJ}[x+5\varepsilon])$

If we now use high-precision values for *x0* and small values of *εs*, we will get reliable values for the derivatives. We display the first 50 digits of the derivatives.

```
In[17]:= z0 = 1/7 + I/3;
        Table[N[j[3][N[z0, 400], 10^-k], 50], {k, 10, 50, 10}]
```

Out[18]= {1.9491815278678998385779209611995390716036646976072× 10^8 +
    6.6931642042995765002200844562245560872132319110987× 10^8 i,
    1.9491815278678997056851738096547832251773036170940× 10^8 +
    6.6931642042995765636469838413684337616098626718096× 10^8 i,
    1.9491815278678997056851738096547832238483761455785× 10^8 +
    6.6931642042995765636469838413684337622441316656611× 10^8 i,
    1.9491815278678997056851738096547832238483761455785× 10^8 +
    6.6931642042995765636469838413684337622441316656611× 10^8 i,
    1.9491815278678997056851738096547832238483761455785× 10^8 +
    6.6931642042995765636469838413684337622441316656611× 10^8 i}

```
In[19]:= Last[%] - Drop[%, -1]
```

Out[19]= {−1.3289274715154475584775528855205520287× 10^{-8} +
    6.3426899385143877675030899754562× 10^{-9} i,
    −1.3289274715154× 10^{-28} + 6.342689938514× 10^{-29} i,
    0.× 10^{-42} + 0.× 10^{-42} i, 0.× 10^{-42} + 0.× 10^{-42} i}

We now convert the sum `odeNum` into a list and substitute numerical values for the derivatives at the random point $x0 = 1/7 + i/3$.

```
In[20]:= z0 = 1/7 + I/3;
        (odeNumTerms = (List @@ odeNum) /.
            (* use numerical values for derivatives *)
            Table[Derivative[k][J][τ] -> j[k][N[z0, 400], 10^-40],
                  {k, 0, 5}]  // N) // Short[#, 6]&
```

Out[21]//Short=
    {−4.2528× 10^{43} − 2.28422× 10^{44} i,
    −9.69544× 10^{47} + 2.59522× 10^{48} i, 1.55149× 10^{52} − 1.23336× 10^{52} i,
    −6.24876× 10^{55} + 8.17256× 10^{54} i, 1.20518× 10^{59} + 5.26045× 10^{58} i,
    ≪46≫, −1.39674× 10^{47} + 4.12372× 10^{48} i, 1.60648× 10^{52} − 2.47643× 10^{52} i,
    −8.43616× 10^{55} + 4.11394× 10^{55} i, 1.1148× 10^{59} + 9.82261× 10^{57} i}

```
In[22]:= {Min[#], Max[#]}& @ Abs[odeNumTerms]
```

Out[22]= {2.16278× 10^{44}, 2.34245× 10^{60}}

The magnitudes of the single terms for the last list are between 44 and 60. Adding the terms gives a result of order 44. This means that all higher terms are cancelled to the precision 16.

```
In[23]:= Plus @@ %
Out[23]= 1.11504 × 10^44 + 1.25442 × 10^43 i
```

Using better approximations for the derivatives, the sum can be reduced to about 10^{-40}.

```
In[24]:= z0 = 1/7 + I/3;
         odeNum /. Table[Derivative[k][J][τ] -> j[k][N[z0, 500], 10^-50],
                         {k, 0, 5}]  // N
Out[25]= -3.69059 × 10^-41 + 6.1683 × 10^-42 i
```

Using still better approximations for the derivatives, the sum can be further reduced to about 10^{-240}. This means about 300 digits cancelled each other. The probability of this happening "accidentally" is astronomically small.

```
In[26]:= z0 = 1/7 + I/3;
         odeNum /.  Table[Derivative[k][J][τ] -> j[k][N[z0, 1200], 10^-150],
                          {k, 0, 5}]  // N
Out[27]= -3.69059 × 10^-241 + 6.1683 × 10^-242 i
```

As a security check, let us try two other values *x0*.

```
In[28]:= Function[z0, {{Min[Abs[#]], Max[Abs[#]]} // N, (Plus @@ #) // N}&[
           (List @@ odeNum) /.
                   Table[Derivative[k][J][τ] -> j[k][N[z0, 1200], 10^-150],
                         {k, 0, 5}]]] /@ {1/5 + I 2/9, 13/11 + I 53/71}
Out[28]= {{{3.30221 × 10^44, 1.19972 × 10^60}, 2.98747 × 10^-241 + 2.94343 × 10^-241 i},
          {{3.06251 × 10^11, 7.63516 × 10^13}, -3.34019 × 10^-288 - 1.22776 × 10^-287 i}}
```

Again, a few hundred digits cancel. This gives strong evidence that the function $w(\tau)$ from above *is* a solution of the Chazy equation $w^{(3)}(\tau) + 3\, w'(\tau)^2 - 2\, w(\tau)\, w''(\tau) = 0$. This kind of high-precision checking is not a mathematical proof, but the probability that for random numbers one achieves hundreds of coinciding digits is overwhelming. (For some examples where this kind of reasoning can go wrong, see [249].)

Using the well-known differential equation for Klein's modular function $J(z)$, it is also possible to symbolically prove that $w(\tau) = 1/2\ \partial \log(J'(\tau)^6\, J(\tau)^{-4}\, (J(\tau) - 1)^{-3})/\partial \tau$ is a solution of the Chazy equation $w^{(3)}(\tau) + 3\, w'(\tau)^2 - 2\, w(\tau)\, w''(\tau) = 0$. To do this, we show that odeNum lies in the ideal generated by the differential equation for Klein's modular function $J(z)$ and its derivatives. The following code contains the corresponding *Mathematica* inputs (we discuss the function GroebnerBasis, which is used in the following inputs in Chapter 1 of the Symbolics volume [1795]).

```
In[29]:= Clear[w];
         (* the differential equation for KleinInvariantJ *)
         odeJ = -32 w'[z]^4 + 41 w[z] w'[z]^4 - 36 w[z]^2 w'[z]^4 +
                108 w[z]^2 w''[z]^2 - 216 w[z]^3 w''[z]^2 +
                108 w[z]^4 w''[z]^2 - 72 w[z]^2 w'[z] w'''[z] +
                144 w[z]^3 w'[z] w'''[z] - 72 w[z]^4 w'[z] w'''[z];

In[32]:= (* the conjectured solution of the differential equation
             w'''[z] - 2 w[z]w''[z] + 3w'[z]^2 == 0 *)
         w[z_] = 1/2 D[Log[J'[z]^6/(J[z]^4(J[z] - 1)^3)], z];

In[34]:= (* the differential equation *)
         ode = w'''[z] - 2 w[z] w''[z] + 3 w'[z]^2;

In[36]:= Clear[w];
         (* the numerator of the resulting identity *)
         odeNum = Numerator[Together[ode]] /. J -> w;
```

```
In[39]:= (* a basis for the ideal generated by
             odeJ, D[odeJ, z], and D[odeJ, z, z] *)
         gb = GroebnerBasis[{odeJ, D[odeJ, z], D[odeJ, z, z]},
                               (* derivatives are the variables *)
                               Table[Derivative[k][w][z], {k, 0, 5}],
                               MonomialOrder -> DegreeReverseLexicographic];

In[41]:= (* odeNum lies in the above ideal *)
         PolynomialReduce[odeNum, gb, Table[Derivative[k][w][z], {k, 0, 5}],
                            MonomialOrder -> DegreeReverseLexicographic][[2]]

Out[42]= 0
```

The 0 in the second element of the last list returned by `PolynomialReduce` shows that `odeNum` follows from `odeJ`.

We give another introductory example for the use of high-precision arithmetic. Here is a trajectory (starting at the origin) of the famous Rössler system,

$$x'(t) = -y(t) - z(t)$$
$$y'(t) = x(t) + a\,y(t)$$
$$z'(t) = b - c\,z(t) + x(t)\,z(t)$$

a coupled system of three nonlinear ordinary differential equations. (We use the parameter values $a = 1/2$, $b = 2$, and $c = 4$.)

```
In[43]:= Module[{a = 1/2, b = 2, c = 4, T = 500},
           ndsolRössler = NDSolve[{x'[t] == -y[t] - z[t], y'[t] == x[t] + a y[t],
                             z'[t] == b + x[t] z[t] - c z[t],
                             (* start at the origin *)
                             x[0] == 0, y[0] == 0, z[0] == 0},
                             {x, y, z}, {t, 0, T}, MaxSteps -> 10^5];
           (* show trajectory *)
           ParametricPlot3D[Evaluate[{x[t], y[t], z[t]} /. ndsolRössler[[1]]], {t, 0, T},
                             PlotPoints -> 10000, PlotRange -> All]];
```

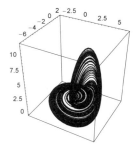

While the built-in function `NDSolve` can easily solve these differential equations, we will quickly investigate the value of the largest step size h that can be used in a simple forward Euler discretization so that the trajectories do not escape to infinity [1131]:

$$x_{n+1}^{(h)} = x_n + h\,(-y_n^{(h)} - z_n^{(h)})$$
$$y_{n+1}^{(h)} = y_n + h\,(x_n^{(h)} + a\,y_n^{(h)})$$
$$z_{n+1}^{(h)} = z_n + h\,(b - c\,z_n^{(h)} + x_n^{(h)}\,z_n^{(h)})$$

```
In[44]:= hpEuler[n_, h_, xyz0_:{0, 0, 0}] :=
        Module[{a = 1/2, b = 2, c = 4, x = xyz0[[1]], y = xyz0[[2]], z = xyz0[[3]]},
            Table[{x, y, z} = {x + h (-y - z), y + h (x + a y),
                    z + h (b + x z - c z)}, {n}]]
```

The next series of five graphics shows the solution obtained from using the Euler discretization for step sizes $k \, 10^{-2}$ for $k = 1, ..., 6$. We follow the solution over a time interval of length 100. We use a step size of precision 1000. This guarantees that we have at $t = 100$ still about 800 valid digits left.

```
In[45]:= Show[GraphicsArray[
        Table[Graphics3D[Line[N[hpEuler[Round[10^4/k], N[k 10^-2, 10^3]]]],
                    PlotRange -> All, PlotLabel -> k], {k, 6}]]]
```

Starting with an upper bound for the maximal step size (meaning a step size so that the trajectory surely diverges—we use the value 1 here), it is straightforward to implement a simple bisection algorithm findMaxi‹ malStepSize that brackets the maximal step size in an interval of length ε. To get a sufficiently small interval for the step size, we use high-precision arithmetic with initial 4000 digits to evolve the $\{x_n^{(h)}, y_n^{(h)}, z_n^{(h)}\}$. We consider a trajectory that escapes to infinity if its coordinates becomes larger than 10^6.

```
In[46]:= (* message in case all precision is lost *)
        findMaximalStepSize::eps =
        "Excessive precision loss in iterations at step size `1`.";

        findMaximalStepSize[hp_, ε_, n_:4000, prec_:4000, max_:10^6, hMax0_:1] :=
        Module[{hMin = 0, hMax = hMax0, hNew, res, escape = 10^6, msg},
            (* turn off message *)
            msg = General::ovfl; Off[General::ovfl];
            (* iterate forward at interval midpoint *)
            Do[hNew = (hMin + hMax)/2; res = hp[n, N[hNew, prec]][[-1]];
                (* enough precision left? *)
                If[Precision[DeleteCases[res, Overflow[], Infinity]] < 20,
                    Message[findMaximalStepSize::eps, hNew]];
                (* new interval for the maximal step size *)
                If[MemberQ[res, Overflow[], Infinity] || Max[Abs[res]] > escape,
                    hMax = hNew, hMin = hNew]; print[{k, N[hMax - hMin]}],
                {k, Ceiling[Log[1/2, ε]] + 1}];
            (* turn on message *)
            If[Head[msg] === String, On[General::ovfl]];
            {hMin, hMax}]
```

For a given bracketing interval ($hMin$, $hMax$) for the maximal step size, the function makeTrajectory‹ Graphics visualizes various properties of the corresponding forward iterated solutions. The first row of graphics shows the trajectories for step sizes $hMin/10$ (surely nondiverging), $hMin$ (on the edge of diverging), and $hMax$ (barely diverging). The second row shows the precision of the $\{x_n^{(h)}, y_n^{(h)}, z_n^{(h)}\}$ as a function of n (making sure that in all iteration steps we still have digits left), the components $x_n^{(h)}$, $y_n^{(h)}$, and $z_n^{(h)}$ as a function of n and the last graphic shows $\{|x_n^{(h_{max})} x_n^{(h_{min})}|, |y_n^{(h_{max})} - y_n^{(h_{min})}|, |z_n^{(h_{max})} - z_n^{(h_{min})}|\}$ as a function of n.

```
In[49]:= makeTrajectoryGraphics[hp_, {hMin_, hMax_}, o_, prec_:4000] :=
        Module[{cOpts, dataMinP, dataMin, dataMaxP, dataMax, dataSmall, max, pos, pr},
```

```
cOpts[x_] := Sequence[PlotRange -> All, BoxRatios -> {1, 1, 1},
                      Axes -> True, PlotRange -> All,
                      PlotLabel -> Subscript["h", x]];
(* iterate forward propagation *)
{dataMinP, dataMaxP, dataSmall} = hp[o, N[Rationalize[#, 0], prec]]& /@
                                   {hMin, hMax, (hMin + hMax)/10};
(* extract nondiverging part of diverging trajectory *)
max = 2 Max[Abs[dataMinP]];
pos = Position[dataMaxP, _?(Max[Abs[#]] > max&), {1}, 1][[1, 1]];
dataMax = Take[dataMaxP, pos]; dataMin = Take[dataMinP, pos];
(* make two graphics arrays *)
GraphicsArray /@
Block[{$DisplayFunction = Identity,
       pr = FullOptions[Graphics3D[Line[dataMinP], PlotRange -> All],
                        PlotRange],
       xTicks = Table[k 1000, {k, 0, Round[Log[10, o + 1]]}],
       rgb = {RGBColor[1, 0, 0], RGBColor[0, 1, 0], RGBColor[0, 0, 1]}},
       Print[xTicks];
       (* show trajectories for three step sizes *)
      {Graphics3D[Line[#1], PlotRange -> 3/2 pr, cOpts[#2]]& @@@
         {{dataSmall, "small"}, {dataMinP, "min"}, {dataMax, "max"}},
         (* precision of iterates *)
         {Graphics[{PointSize[0.003],
           Transpose[{rgb, MapIndexed[Point[{#2[[1]] - 1,
                           Precision[#1]}]&, #]& /@
                                      {dataSmall, dataMinP, dataMax}}]}],
                 PlotRange -> {All, {0, 1.1 prec}}, Frame -> True,
                 PlotLabel -> "Precision"],
         (* x, y, and z-component of iterates *)
         Graphics[{PointSize[0.003],
           Transpose[{rgb, MapIndexed[Point[{#2[[1]] - 1, #1}]&, #]& /@
                                           Transpose[dataMax]}]}],
                 Frame -> True, PlotRange -> {{0, o}, Automatic},
                 FrameTicks -> {xTicks, Automatic, None, None},
                 PlotLabel -> (Subscript[#, n]& /@ {x, y, z})],
         (* difference in trajectories does bracketing step sizes *)
         Graphics[{PointSize[0.003],
           Transpose[{rgb, MapIndexed[If[# != 0, Point[{#2[[1]] - 1,
                           Log[10, Abs[#1]]}], {}]&, #]& /@
                      Transpose[N[dataMax - Take[dataMin, pos]]]}]}],
                 Frame -> True, PlotRange -> {{0, o}, Automatic},
                 FrameTicks -> {xTicks, Automatic, None, None},
                 PlotLabel -> (Subscript[#, n]& /@ {δx, δy, δz})]}}]]
```

Here are the just-described graphics for the Euler iterations for a starting interval of length 10^{-25}. We obtain `General::ovfl` messages from the diverging trajectories. The maximal step length is $h_{\max} = 0.06832597\ldots$.

```
In[50]:= findMaximalStepSize[hpEuler, 10^-25]
```

$$\text{Out[50]}= \left\{ \frac{1321616582571470688732441}{1934281311383406679529816}, \frac{2643233165142941377464883}{3868562622776813359059763\underline{2}} \right\}$$

```
In[51]:= Show /@ makeTrajectoryGraphics[hpEuler,
           (* findMaximalStepSize[hpEuler, 10^-25] = *)
           {1321616582571470688732441/1934281311383406679529816,
            2643233165142941377464883/3868562622776813359059763\underline{2}}, 4000]
           General::ovfl : Overflow occurred in computation. More…
```

{0, 1000, 2000, 3000, 4000}

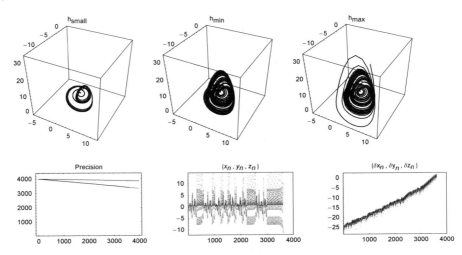

The range of possible step sizes can be increased by using different discretization schemes. Here is a rather sophisticated-looking nonstandard discretization scheme [1131], [1270].

```
In[52]:= hpMickens[n_, h_, xyz0_:{0, 0, 0}] :=
    Module[{a = 1/2, b = 2, c = 4, φ1, φ2, φ3, xc,
         x = xyz0[[1]], y = xyz0[[2]], z = xyz0[[3]]},
      φ1 = h; φ2 = (1 - Exp[-h a])/a;
      xc = (-c + Sqrt[c^2 - 4 a b])/2; φ3 = (1 - Exp[-h xc])/xc;
      Table[x = x - φ1 (y + z);
            {y, z} = {(1 + a φ2) y + φ2 x, b φ3 + (1 + φ3 (x - c)) z};
            {x, y, z}, {n}]]
```

And here are the resulting trajectories at the edge of stability—they look quite different from the ones of the Euler discretization.

```
In[53]:= findMaximalStepSize[hpMickens, 10^-25]
```
$$\text{Out[53]= } \left\{ \frac{1683055133138314330 6135827}{3868562622766813359 0597632}, \frac{4207637832845785826 533957}{9671406556917033397 649408} \right\}$$

```
In[54]:= Show /@ makeTrajectoryGraphics[hpMickens,
         (* findMaximalStepSize[hpMickens, 10^-25] = *)
         {1683055133138314330 6135827/3868562622766813359 0597632,
          4207637832845785826 533957/9671406556917033397 649408}, 4000]
```

    General::ovfl : Overflow occurred in computation. More…

    {0, 1000, 2000, 3000, 4000}

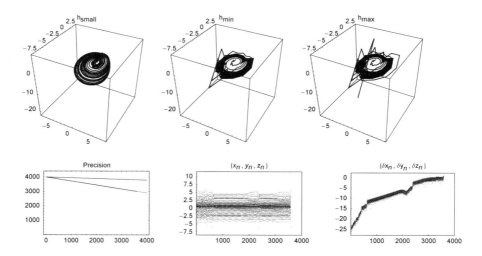

One could now go on and investigate the dependence of the maximal step size on the initial conditions. The following input calculates the maximal step size for a dense grid over the $x(0), y(0)$-plane (we keep $z(0) = 0$) and will run a few hours. A structure mildly resembling the projected attractor arises. (The corresponding graphic for the nonstandard discretization `hpMickens` looks qualitatively similar.)

```
(* calculate maximal step sizes *)
With[{L = 10, pp = 120},
dataEuler = Table[{x0, y0, Plus @@ findMaximalStepSize[
                hpEuler[#1, #2, {x0, y0, 0}]&, 10^-4, 1000, 600]/2},
            {x0, -L, L, 2L/pp}, {y0, -L, L, 2L/pp}]];

(* visualize maximal step sizes *)
Show[GraphicsArray[
  Block[{$DisplayFunction = Identity},
    {ListDensityPlot[Transpose[Map[Last, dataEuler, {2}]],
                Mesh -> False, PlotRange -> All, FrameTicks -> None],
      ListPlot3D[Transpose[Map[Last, dataEuler, {2}]], Mesh -> False,
            PlotRange -> All, Axes -> False]}]]]
```

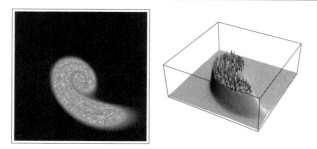

So far, we have discussed the usefulness of high-precision arithmetic for numerical calculations. High-precision arithmetic allows one to obtain reliable results with many digits that sometimes are impossible to obtain using machine arithmetic. On the other hand, for many calculations, machine precision is sufficient and speed is

essential. *Mathematica* frequently compiles numerical calculations invisible to the user to accelerate calculations and also allows for explicit compilation. Because of the importance of compilation, let us give one typical example in this introductory section. We will encounter many more invocations of the `Compile` function throughout this chapter.

We consider a simple 1D molecular dynamics simulation [661], [1897], [1548], [1505], [766]. Between two walls, located at $z = 0$ and $z = L$ we have n particles with potentially different masses. If two (or more) particles meet, and have a center of mass energy less than V, they undergo an elastic collision. If they have a center of mass energy greater than V, they just move through each other (this is a very crude model of a realistic collision process). The two walls are held at temperatures T_0 and T_L and if particles collide with the walls they thermalize, meaning, after the wall collision, their velocities are Maxwellian with temperature T_0 or T_L. In the following, we will assume the data for each particle in the form {*particleNumber*, *mass*, *startTime*, *startPoint*, *velocity*} and the data for all particles is a list of such 5-element lists. The function `collisionStep` evolves a system of particles until the next collision and returns the particle data with the new *startTimes*, *startPoints*, and *velocities*. We account for the possibility that more than one collision could happen at one time.

```mathematica
In[55]:= collisionStep[particleData_?(MatrixQ[#, NumberQ]&),
                L_, V_, Ts_?(VectorQ[#, NumberQ]&)] :=
    Module[{d, n, n, tOld, cts, δxδvs, leftt, rightt, ts, tC, cList, c,
            m1, v1, m2, v2, Tc, V1, V2, I, T0 = Ts[[1]], TL = Ts[[2]]},
        d = particleData; n = d; I = $MaxMachineNumber/10;
        n = Length[d]; tOld = d[[1, 3]];
        (* collision of all adjacent pairs *)
        δxδvs = {#[[2, 4]] - #[[1, 4]], #[[1, 5]] - #[[2, 5]]}& /@
                                        Partition[d, 2, 1];
        cts = If[#[[2]] == 0., I, #[[1]]/#[[2]]]& /@ δxδvs;
        {leftt, rightt} = {0 - d[[1, 4]]/d[[1, 5]], (L - d[[n, 4]])/d[[n, 5]]};
        ts = Join[{leftt}, cts, {rightt}];
        (* time of next collision *)
        tC = Max[ts];
        Do[If[0. < ts[[j]] < tC, tC = ts[[j]]], {j, Length[ts]}];
        (* list of colliding pairs; multiple collision can (depending
           on time granularity) happen at the same time *)
        (* which particles collide? *)
        cList = {-1}; Do[If[ts[[j]] == tC, AppendTo[cList, j - 1]], {j, n + 1}];
        (* new collision times and new positions *)
        Do[n[[j, 3]] = tOld + tC; n[[j, 4]] = n[[j, 4]] + tC n[[j, 5]], {j, n}];
        (* carry out the collision(s) *)
        Do[c = cList[[j]];
        Which[(* left wall collision *) c == 0, n[[1, 4]] = 0.;
              (* thermalize on left wall *)
              n[[1, 5]] = Sqrt[2 T0/n[[1, 2]]] Abs[InverseErf[2 Random[] - 1]],
              (* two particle collision *)
              0 < c < n,
              {m1, v1} = n[[c, {2, 5}]]; {m2, v2} = n[[c + 1, {2, 5}]];
              If[(* high-energy collision *)
                  Tc = 1/2 m1 m2/(m1 + m2) (v1 - v2)^2; Tc > V,
                  {n[[c]], n[[c + 1]]} = {n[[c + 1]], n[[c]]};
                  n[[c + 1, 4]] = n[[c, 4]],
                  (* else -- a real collision *)
                  (* velocities after collision *)
                  {V1, V2} = {v1, v2} + 2 {m2, -m1}(v2 - v1)/(m1 + m2);
                  {n[[c, 5]], n[[c + 1, 5]]} = {V1, V2};
                  n[[c + 1, 4]] = n[[c, 4]]],
```

```
              (* right wall collision *) c == n, n[[n, 4]] = L;
              (* thermalize on right wall *)
              n[[n, 5]] = -Sqrt[2 TL/n[[n, 2]]] Abs[InverseErf[2 Random[] - 1]]],
         {j, 2, Length[cList]}];
    (* do nearby in time collisions *)
    Do[If[n[[j, 4]] == n[[j + 1, 4]] && n[[j, 5]] > n[[j + 1, 5]],
         n[[j + 1, 4]] = n[[j, 4]];
         {n[[j]], n[[j + 1]]} = {n[[j + 1]], n[[j]]}],
      {j, n - 1}];
    If[n[[1, 4]] == 0. && n[[1, 5]] < 0., n[[1, 5]] = -n[[1, 5]]];
    If[n[[n, 4]] == L  && n[[n, 5]] > 0., n[[n, 5]] = -n[[n, 5]]];
    (* return new state *) n]
```

To visualize the trajectories of the particles, we implement a function collisionHistoryGraphics that shows the space-time curves of the particles with time running upwards. We color the particles according to their mass.

```
In[56]:= collisionHistoryGraphics[collisions_, L_, opts___] :=
    Module[{segments = Transpose /@ Partition[Sort /@ collisions, 2, 1]},
    Graphics[{(* the two walls *)
              {GrayLevel[0.5], Thickness[0.02],
              Line[{{0, #1}, {0, #2}}], Line[{{L, #1}, {L, #2}}]}& @@
                ({Min[#], Max[#]}&[#[[1, 3]]& /@ collisions]),
              (* the particle trajectories *)
              {Map[{Hue[#[[1, 2]]], Line[{{#[[1, 4]], #[[1, 3]]},
                                          {#[[2, 4]], #[[2, 3]]}}]}&,
                segments, {2}]}}, opts, FrameTicks -> None, Frame -> True,
              PlotRange -> {{-L/10, 1.1 L}, All}, AspectRatio -> 1]]
```

Here is an example. We start with 30 particles with random masses, starting points, and starting velocities in a well of length 100. The leftmost graphic uses $V = \infty$, meaning all collisions are elastic. As a result, only after the first particle hits a wall (with temperature 10; the Boltzmann constant is set to 1), we see an interesting structure of the trajectories evolving, propagating from the right to the left through successive collisions. The middle and right graphics use $V = 2$ and $V = 1$. Because now some collisions do not happen, the disturbance can propagate more quickly to the left through a particle not colliding with each particle on the way.

```
In[57]:= Module[{n = 30, Ts = {10, 10}, L = 100, cs = 1000,
            pd, data, particleData},
    SeedRandom[33];
    (* initial particle positions and velocities *)
    particleData = Sort[Table[{k, 0.8 Random[], 0, L Random[],
                        1/100 (2 Random[Integer] - 1)}, {k, n}],
                    (* sort left to right *) #1[[4]] < #2[[4]]&];
    (* show the three graphics *)
    Show[GraphicsArray[
       Function[V, (* initialize random number generation *) SeedRandom[111];
         (* carry out cs collisions *)
         data = NestList[collisionStep[#, L, V, Ts]&, particleData, cs];
         collisionHistoryGraphics[data, L]] /@ {Infinity, 2, 1}]]]
```

Next, we will use two kinds of particles (meaning two masses) and walls of different temperatures. We expect to see the Ludwig–Soret effect [761] in the resulting spatial distribution of the particles, meaning the lighter particles should accumulate near the cold wall.

We now carry out 1000 collisions for 2×50 particles with masses $m = 1$ and $m = 0.12$, wall temperatures $T_0 = 0$ and $T_{200} = 10$. `particleData` is the random initial configuration of the particles.

```
In[58]:= particleData =
    With[{n = 50, Ta = 5.5, L = 200, μ1 = 1., μ2 = 0.12},
    Sort[Join[(* initial conditions for heavy particles *)
            Table[{k, μ1, 0., L Random[],
                    Sqrt[2 Ta/μ1] InverseErf[2 Random[] - 1]}, {k, n}],
                (* initial conditions for light particles *)
            Table[{n + k, μ2, 0., L Random[],
                    Sqrt[2 Ta/μ2] InverseErf[2 Random[] - 1]}, {k, n}]],
        (* sort left to right *) #1[[4]] < #2[[4]]&]];
```

We carry out 100 collisions using the initial configuration `particleData` and measure how long this takes.

```
In[59]:= L = 200; n = 50; SeedRandom[111];
    (dataLS1 = NestList[collisionStep[#, L, 2, {1, 10}]&,
                    particleData, 100]); // Timing
Out[60]= {0.45 Second, Null}
```

Invoking now the *Mathematica* compiler explicitly, we can considerably speed-up this calculation. To compile *Mathematica* code explicitly, we invoke the function `Compile`. We want to compile the definition of the function `collisionStep` and to access the body of the function `collisionStep`; we use its downvalues. For compilation, the type of the input arguments must be known. The above pattern for the particle data `particleData_?(MatrixQ[#, NumberQ]&)` translates (for our purposes here) to `{particleData, _Real, 2}` and the pattern `Ts_?(VectorQ[#, NumberQ]&)` for the wall temperatures translates to `{Ts, _Real, 1}`. The inverse error function needed to generate the Maxwellian velocity distribution is not compilable; so, we tell the `Compile` function that the inverse error function returns a single real number here.

```
In[61]:= collisionStepC =
    Function[body, Compile @@
        Join[Hold[{{particleData, _Real, 2}, L, V, {Ts, _Real, 1}}], body,
            (* evaluate outside the compiler *) Hold[{{InverseErf[_], _Real}}]],
            (* extracted and held original definition *)
            {HoldAll}] @ MapAt[Hold, DownValues[collisionStep], {1, 2}][[1, 2]];
```

Now, we run the compiled function `collisionStepC` and measure again how long it takes.

```
In[62]:= SeedRandom[111];
    (dataLS1C = NestList[collisionStepC[#, L, 2, {1, 10}]&,
                    particleData, 100]); // Timing
```

Out[63]= {0.07 Second, Null}

The result of the compiled run agrees with the result of the uncompiled run (we initialized the random number generator identical in the two runs).

In[64]:= **dataLS1 == dataLS1C**

Out[64]= True

Using the faster function `collisionStepC`, we now carry out 100000 collision steps. (Carrying out all collisions within a compiled function and keeping a list of next collision times would eliminate some of the repeated calculation of the next collision times and would further speed-up the calculation.) Instead of graphing the individual trajectories, we display only the centers of the light and heavy particles. The function `centersOf Gravity` calculates the center of masses for the $2n$ particles.

In[65]:= **centersOfGravity[pd_, n_] := Plus @@@ Map[#[[4]]&, {Take[#, {1, n}],**
       Take[#, {n + 1, 2n}]}&[(* heavy particles first *) **Sort[pd]], {2}]/n**

We start again with the initial particle configuration `particleData`. We clearly see how the light particles accumulate near the cold wall. (The center of gravity of the heavy particles does not change much because the light particles accumulate at a relatively small distance from the cold wall.)

In[66]:= **SeedRandom[111];**
       (* list of center of gravity data *)
       cogDataLS2 = Block[{pd = particleData, cs = 100000},
       Table[pd = collisionStepC[pd, L, 2, {1, 10}];
                        {{pd[[1, 3]], #[[1]]}, {pd[[1, 3]], #[[2]]}}&[
                        centersOfGravity[pd, n]], {cs}]];

In[69]:= **Show[ListPlot[#1 /@ cogDataLS2, PlotStyle -> {#2},**
                DisplayFunction -> Identity]& @@@
                (* light particles in red; heavy particles in blue *)
                {{First, RGBColor[0, 0, 1]}, {Last, RGBColor[1, 0, 0]}},
       DisplayFunction -> $DisplayFunction, AspectRatio -> 1/4,
       PlotRange -> All, Axes -> False, Frame -> True]

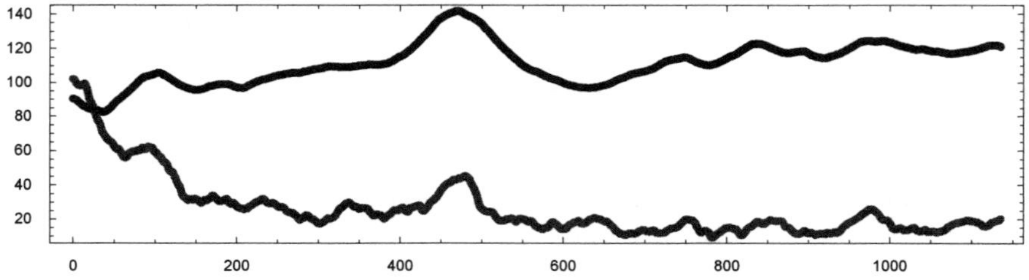

Many related simulations could be carried out, such as modeling the gas temperature for finite reservoirs [1115].

1.1 Approximate Numbers

1.1.0 Remarks

This subsection will cover the basics of the usage and underlying principles of machine and high-precision numbers. While calculations with machine numbers have been used already extensively in the chapters of the Programming [1793] and Graphics [1794] volumes, high-precision arithmetic was used only rarely. So in the next subsection we will discuss the notations of precision and accuracy of a real and a complex number, the principles of automatic precision control and its consequences, the phenomena of precision gain and loss in calculations, and the possibilities to raise or lower the precision or accuracy of a number. Later we will investigate the internal structure of high-precision numbers and how high-precision numbers are treated special in functions like `Union` and which system options are of relevance here. In discussing all these subjects, we will provide a large number of examples to demonstrate all statements. The second subsection will deal with intervals. Intervals provide a method to carry out validated numerical calculations. The third subsection will discuss continued fractions and related expansions in detail. Continued fractions are relevant for converting approximate numbers to nearby rational numbers. The last subsection of this section deals with a more advanced topic for machine number calculations—packed arrays. While packed arrays are in many instances automatically generated by *Mathematica* without the users explicit influences, for more complicated calculations it is sometimes necessary for the user to pack and unpack lists of machine numbers under his guidance to achieve optimal performance.

1.1.1 Numbers with an Arbitrary Number of Digits

For most numerical calculations, around 14 to 19 digits of (hardware-dependent) machine precision is adequate. Machine arithmetic is used when evaluating expressions of the form `N[`*expr*`]`. However, sometimes we would like to work with more digits to ensure that a certain number of interesting digits is correct. (For several examples, it is necessary to use more than around 20 digits in applications; see [1904], [1667], [1349], [453], [503], and [413].) In such situations, *Mathematica*'s high-precision numbers, together with their automatic precision control, come in very handy. We will use *Mathematica*'s high-precision capabilities many times in this and the following chapters.

We have already introduced `N[`*expression*`,` *digits*`]` in Chapter 2 of the Programming volume [1793]. It computes the expression *expression* to a precision of *digits* digits.

Mathematica's high-precision arithmetic is based on significance arithmetic (see [1266] and [1267]). This means that the result of a high-precision calculation is a number (composed of digits) including the knowledge of which of the digits are correct. In `OutputForm` and `StandardForm` only the certified digits are printed. `InputForm` and `FullForm` will display all digits together with the information of how many of them are correct.

In a nutshell, here is how *Mathematica* can be used to carry out high-precision arithmetic. In principle, there are three possibilities:

■ Floating-point arithmetic with a fixed number of digits. This can be forced by choosing `$MinPrecision` = `$MaxPrecision` = *number* (which is better used inside a `Block` by using localized versions of `$MinPreci`↘

`sion` and `$MaxPrecision`).

- Floating-point arithmetic with a variable number of digits. This is the method implemented in the current *Mathematica* kernel. Roughly speaking, the main idea is to simulate interval arithmetic by constantly maintaining a "few more digits" than needed (hereafter called guard digits), and to use them to analyze the error.

- Interval arithmetic. This stricter method is implemented in *Mathematica* for real numbers (see the subsection below).

For more detailed information on *Mathematica*'s implementation of bignum arithmetic and interval arithmetic (see below), as well as a discussion of the general problem of calculating with approximate numbers and intervals, see [967], [830], [964], [887], [697], [1667], [1293], [1294], [36], [959], [1850], [968], and [969].

In the last section, we used high-precision arithmetic to "check" an identity numerically. Because we obtained zero to hundreds of digits we did not have to worry about the correctness of the digits. (The probability that the identity would be wrong and that accumulated errors in the carried out arithmetic make a wrong identity correct is vanishing small.) In the next two introductory examples, we are not primarily dealing with many digits, but rather how much we can trust the digits calculated. Here is a first simple example using high-precision arithmetic. This is an iterated modified logistic map [735], [539].

```
In[1]:= modifiedLogisticMap[x_, λ_, ε_, n_] = (λ + (-1)^n ε) x (1 - x);

      modifiedLogisticMapData[x0_, λ_, ε_, l_, 10_] :=
        Drop[FoldList[modifiedLogisticMap[#, λ, ε, #2]&, x0, Range[l]], 10]
```

Making a plot using machine arithmetic yields a "typical" looking picture for a logistic equation [949].

```
In[3]:= Show[Graphics[{PointSize[0.003], Table[Point[{λ, #}]& /@
          modifiedLogisticMapData[1/210, λ, 0.2, 300, 100],
          {λ, 2.6, 3.8, 0.01}}]], PlotRange -> All, Frame -> True]
```

Using bignum arithmetic yields a qualitatively similar looking picture.

```
In[4]:= Show[Graphics[{PointSize[0.003],
        Table[Point[{λ, #}]& /@
          modifiedLogisticMapData[N[1/210, 200], λ, 2/10, 300, 100],
          {λ, 26/10, 38/10, 1/100}}]], PlotRange -> All, Frame -> True]
```

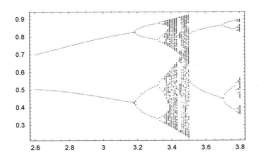

But the actual orbit for a fixed λ shows that after around 50 iterations, the machine result is wrong.

```
In[5]:= ListPlot[
    (* difference between machine-precision and high-precision version *)
    modifiedLogisticMapData[1/210, 3.49, 0.2, 300, 100] -
    modifiedLogisticMapData[N[1/210, 200], 349/100, 2/10, 300, 100],
       PlotJoined -> True, PlotRange -> All, Frame -> True, Axes -> False]
```

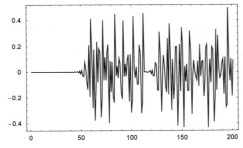

As long as the calculated high-precision numbers have explicitly displayed digits, we can trust them to be correct. If we would carry out the last iteration a bit longer, no correct digit remains. The numbers of the form 0×10^n indicate that the result is somewhere in the interval $[-10^n, +10^n]$. (Of course, by starting with a higher initial precision we can get verified correct results for any number of iterations.)

```
In[6]:= Take[modifiedLogisticMapData[N[1/210, 200], 359/100, 2/10, 400, 100],
       {-45, - 30}]
           General::unfl : Underflow occurred in computation. More…
```

Out[6]= $\{0.4906, 0.8472, 0.491, 0.847, 0.49, 0.85, 0.5, 0.8,$
$0. \times 10^{-1}, 0., 0., 0. \times 10^{1}, 0. \times 10^{4}, 0. \times 10^{8}, 0. \times 10^{17}, 0. \times 10^{36}\}$

For an interesting two-dimensional (2D) version of a logistic map, see [349]; for the logistic map in general, see [788].

Here is another small example. It shows a too pessimistic result. The following input implements the iterative calculation of numerical values of the Weierstrass function $\wp(z; g_2, g_3)$ [431] based on argument reduction and series expansion.

$$\wp(z; g_2, g_3) = \lim_{n \to \infty} p^{(n)}(z; g_2, g_3)$$

$$p^{(n)}(z; g_2, g_3) = \tilde{p}(p^{(n-1)}(z; g_2, g_3); g_2, g_3)$$

$$\tilde{p}(z; g_2, g_3) = \frac{(6z^2 - \frac{g_2}{2})^2}{4(4z^3 - g_2 z - g_3)} - 2z$$

$$p^{(0)}(z; g_2, g_3) = \left(\frac{2^n}{z}\right)^2 + \frac{g_2}{20}\left(\frac{z}{2^n}\right)^2 + \frac{g_3}{28}\left(\frac{z}{2^n}\right)^4$$

The function `weierstrass℘List` returns a list of approximations.

```
In[7]:= weierstrass℘List[z_, {g2_, g3_}, o_] :=
First /@ NestList[Function[{w, n},
   {Nest[-2# + (6#^2 - g2/2)^2/(4 (4#^3 - g2 # - g3))&,
      1/#^2 + g2/20 #^2 + g3/28 #^4&[z/2^n], n], n + 1}] @@ #&, {z, 1}, o]
```

Using the function with machine numbers gives a converging (looking) result.

```
In[8]:= weierstrass℘List[N[2], {2, 3}, 40]
```
```
Out[8]= {2., 6.82755, 6.44608, 6.44555, 6.44555, 6.44555, 6.44555, 6.44555, 6.44555,
   6.44555, 6.44555, 6.44555, 6.44555, 6.44555, 6.44555, 6.44555, 6.44555,
   6.44555, 6.44555, 6.44555, 6.44555, 6.44555, 6.44555, 6.44555, 6.44555,
   6.44555, 6.44555, 6.44555, 6.44555, 6.44555, 6.44555, 6.44555, 6.44555,
   6.44555, 6.44555, 6.44555, 6.44555, 6.44555, 6.44555, 6.44555, 6.44555}
```

The result actually agrees with the value calculated by *Mathematica*.

```
In[9]:= WeierstrassP[2., {2, 3}]
```
```
Out[9]= 6.44555 + 0. i
```

Using high-precision arithmetic and its automatic precision control in the iterations shows that 1) massive calculations occurred in the process of calculations, and that 2) we cannot trust the above result completely blindly and a more detailed analysis of the stability of the iterations is needed. (In the last inputs, all occurrences of `Slot[1]` were treated independently, although they all represent the same number; we will come back to this example later in this subsection.)

```
In[10]:= weierstrass℘List[N[2, 20], {2, 3}, 20]
```

    Power::infy : Infinite expression $\frac{1}{0. \times 10^3}$ encountered. More…

    ∞::indet : Indeterminate expression $0. \times 10^4$ ComplexInfinity encountered. More…

    Power::infy : Infinite expression $\frac{1}{0. \times 10^5}$ encountered. More…

    ∞::indet : Indeterminate expression $0. \times 10^6$ ComplexInfinity encountered. More…

    Power::infy : Infinite expression $\frac{1}{0. \times 10^6}$ encountered. More…

    General::stop :
     Further output of Power::infy will be suppressed during this calculation. More…

    ∞::indet : Indeterminate expression $0. \times 10^9$ ComplexInfinity encountered. More…

    General::stop :
     Further output of ∞::indet will be suppressed during this calculation. More…

```
Out[10]= {2.0000000000000000000, 6.82755307730362983, 6.446077715813308,
   6.4455510992952, 6.44554936048, 6.4455493540, 6.44554935, 6.445549, 6.4455,
   6.45, 6., 0., Indeterminate, Indeterminate, Indeterminate, Indeterminate,
   Indeterminate, Indeterminate, Indeterminate, Indeterminate, Indeterminate}
```

Mathematica has a high-precision arithmetic for calculating with numbers with an arbitrary number of digits in the interval $MaxNumber $< |x| <$ $MaxNumber (provided the final answers and all intermediate results stay in this range). The current value of $MaxNumber is the following.

In[11]:= **$MaxNumber**

Out[11]= $1.920224672692357 \times 10^{646456887}$

The user can establish an upper bound on the number of digits (the precision) carried by setting the value of $MaxPrecision.

$MaxPrecision

    gives the maximum number of digits *digits* used in numerical calculations with N[*expression*, *digits*].

By default, the value of $MaxPrecision is infinity. (Practically this means that as much precision, as is possible within the memory and operating system constraints as possible will be used. In some rare cases, one might want to use a finite upper limit to avoid long-running calculations.)

In[12]:= **defaultMaxPrecisionValue = $MaxPrecision**

Out[12]= ∞

If we attempt to evaluate an expression with more digits, such as N[2, 2 $MaxPrecision], we get an error message.

In[13]:= **Block[{$MaxPrecision = 100}, N[2, 2 $MaxPrecision]]**

        N::preclg : Requested precision 200 is larger than $MaxPrecision.
            Using current $MaxPrecision of 100.` instead. $MaxPrecision =
            Infinity specifies that any precision should be allowed. More…

Out[13]= 2.0000000000000000000000000000000000000000000000000000000000000000000000 \
       00000000000000000000000000

In such cases, the computation continues with $MaxPrecision digits. The analogous lower bound is $Min\. Precision.

$MinPrecision

    gives the minimum number of digits *digits* used in numerical calculations with N[*expression*, *digits*].

Here is its current (user-changeable) value.

In[14]:= **$MinPrecision**

Out[14]= $-\infty$

This is in comparison to the $MachinePrecision value.

In[15]:= **$MachinePrecision**

Out[15]= 15.9546

To uniquely differentiate between a high-precision calculation that uses $MachinePrecision precision, the built-in command MachinePrecision exists.

MachinePrecision

    represents a symbol to indicate machine precision.

While `MachinePrecision` does not evaluate to a number directly, it evaluates to a number after applying `N`.

In[16]:= **{MachinePrecision, N[MachinePrecision]}**

Out[16]= {MachinePrecision, 15.9546}

Sometimes, we would like to carry out parts of a calculation with more than `$MachinePrecision`. To avoid using `N[..., digits]` for each intermediate expression, we can use a construction of the form

```
Block[{$MinPrecision = digits},
      Module[{ allOtherVariables}, theCalculation]]
```

Sometimes, one also wants to do calculations with more digits than provided by the hardware, but with a fixed precision instead of using the automatic precision control of *Mathematica* (for instance to avoid excessive loss of precision). This can be achieved via the following construction.

```
Block[{$MinPrecision = digits, $MaxPrecision = digits},
      Module[{ allOtherVariables}, theCalculation]]
```

But the reader should be careful when using such constructions. In such situations, the reader has to do all error analysis alone because such constructions disable the automatic precision control.

Occasionally one does not have to know any digits of the numerical value of an expression, but knowledge that the value is less than a certain given size is needed. This can be realized with `N` too.

`N[expression, {precision, accuracy}]`

 tries to numericalize expression to precision *precision* or accuracy *accuracy*.

Here is a small quantity (of approximate size $4.3 \, 10^{-1656521}$).

In[17]:= **smallQuantity = Exp[-Exp[Exp[E]]];**

The next input establishes that `smallQuantity` has an absolute value less than $10^{-1000000}$.

In[18]:= **N[smallQuantity, {Infinity, 10^6}]**

Out[18]= $0. \times 10^{-1000001}$

Because the absolute value is larger than $10^{-10000000}$, for the next input four digits are calculated.

In[19]:= **N[smallQuantity, {4, 10^7}]**

Out[19]= $4.289 \times 10^{-1656521}$

There are relatively few instances where the use of fixed precision arithmetic is preferable over the use of the default significance arithmetic. Basically, significance arithmetic can never be wrong (as long as one uses enough guaranteed digits, say at least 10). But because the method assumes that all errors are completely independent, it might overestimate the precision loss in calculations. Fixed precision arithmetic on the other side becomes error-free in the limit of totally correlated errors in the limit of many operations. Here is an example of the use of fixed precision arithmetic. We carry out the recursion $z_n = \pm z_{n-1} \pm z_{n-2}$ [1854]. The coefficients ± 1 of z_{n-1} are chosen randomly in each recursion step.

In[20]:= **randomFibonacciFP[n_, prec_] :=**
```
    Block[(* fix precision inside Block *)
          {$MinPrecision = prec, $MaxPrecision = prec,
           (* the random ±1 *) r := 2 Random[Integer] - 1},
          Last /@ NestList[{#[[2]], r #[[1]] + r #[[2]]}&, N[{1, 1}, prec], n]]
```

Here is the result of 1000 steps of the recursion. (We use `SeedRandom` to generate reproducible results.)

```
In[21]:= SeedRandom[11111];
       randomFibonacciFP[1000, $MachinePrecision + 1][[-1]]
Out[22]= -2.5055139869821551×10^55
```

Using significance arithmetic and starting with 155 digits gives only a single guaranteed digit after carrying out 1000 iterations. But all of the guard digits (see below) are correct anyway.

```
In[23]:= randomFibonacciSA[n_] :=
       Block[{r := 2 Random[Integer] - 1},
             Last /@ NestList[{#[[2]], r #[[1]] + r #[[2]]}&, N[{1, 1}, 155], n]]

In[24]:= SeedRandom[11111];
       {#, InputForm[#]}& @ randomFibonacciSA[1000][[-1]]
Out[25]= {-3.×10^55, -2.505513986982155`1.3427663032154002*^55}
```

Here is a more extreme example. Using pseudorandom numbers, we carry out one million recursions for this sequence. The result obtained in exact integer arithmetic is identical to the result using only $MachinePreci√sion + 1 digits. In addition, we carry out the calculation using machine precision. Because we add numbers of similar magnitude at each step of the recursion, even the machine arithmetic result is correct.

```
In[26]:= With[{prec = $MachinePrecision + 1},
       Block[{n = 10^6, $MinPrecision = prec, $MaxPrecision = prec, r1, r2},
             Nest[((* two pseudorandom numbers *)
                   r1 = 2 Abs[Round[Sin[#[[4]]]]] - 1;
                   r2 = 2 Abs[Round[Cos[#[[4]]]]] - 1;
                   (* machine value, fixed precision value, exact value,
                     and counter *)
                   {{#[[1, 2]], r1 #[[1, 1]] + r2 #[[1, 2]]},
                    {#[[2, 2]], r1 #[[2, 1]] + r2 #[[2, 2]]},
                    {#[[3, 2]], r1 #[[3, 1]] + r2 #[[3, 2]]}, #[[4]] + 1})&,
                   {N[{1, 1}], N[{1, 1}, prec], {1, 1}, 0}, n]]] // N
Out[26]= {{-1.916001878177566×10^36474, 5.052944278059832×10^36473},
         {-1.916001878177571×10^36474, 5.052944278059843×10^36473},
         {-1.916001878177571×10^36474, 5.052944278059843×10^36473}, 1.×10^6}
```

This random Fibonacci recursion has the interesting feature that the quantity $|z_n|^{1/n}$ approaches, for almost all realizations of the random ± 1 choices, a constant limit. The next graphic shows three examples of carrying out 10000 recursion steps.

```
In[27]:= SeedRandom[11111];
       Show[Table[ListPlot[MapIndexed[Abs[#1]^(1/#2[[1]])& ,
                                      randomFibonacciFP[10^5, 20]],
                  (* set options properly *)
                  Frame -> True, PlotStyle -> {PointSize[0.003], Hue[k/3]},
                  DisplayFunction -> Identity, PlotRange -> {1.10, 1.20}], {k, 0, 2}],
             DisplayFunction -> $DisplayFunction]
```

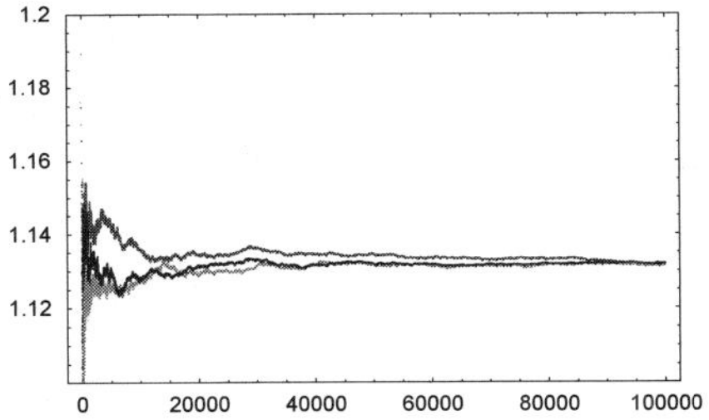

(For related random recurrences, see [1052], [176].)

In regard to machine precision, note the following observation: N[*expression*, *digits*] with machine precision numbers present in *expression* typically results in a machine precision result.

In[29]:= **machineNumber = 1.2345678987654321;**

    {N[machineNumber, 10], N[machineNumber, 20]}
Out[30]= {1.23457, 1.23457}

N[*expression*, *digits*] with *digits* < *machinePrecision* is still using high-precision in all calculations. But with only a few digits in use, the calculations become less reliable.

In[31]:= **exactNumber = 12345678987654321/10^12;**

    {N[exactNumber, 2], N[exactNumber, 4], N[exactNumber, 6]}
Out[32]= $\{1.2 \times 10^4, 1.235 \times 10^4, 12345.7\}$

Typically, for machine-precision numbers, only about six digits are displayed, so the numericalized version of the following fractions displays in the output as 1.

In[33]:= **N[111111111111/111111111112, (* not needed *) MachinePrecision]**
Out[33]= 1.

In[34]:= **1 - %**
Out[34]= 9.00002×10^{-12}

InputForm and FullForm show all digits.

In[35]:= **FullForm[%%]**
Out[35]//FullForm=
    0.999999999991`

In[36]:= **InputForm[%%%]**
Out[36]//InputForm=
    0.999999999991

We make one more remark concerning expressions of the form N[*something*]. After calculating *something* as far as possible using standard evaluation, N is invoked (N does not have any Hold-like attribute). For some choices of *something* (e.g., Solve, Sum, Integrate), not only are the corresponding variables converted to numbers, but N also activates a corresponding numerical command (e.g., NSolve, NSum, NIntegrate).

Here is an example of an integral that cannot be computed symbolically in closed form (currently).

In[37]:= **Integrate[Sin[x^3]/(x^3 + x^2 + 1), {x, 0, Pi}]**
Out[37]= $\int_0^\pi \frac{\text{Sin}[x^3]}{1 + x^2 + x^3}\, dx$

N[%] does not produce the essentially worthless result
Integrate[Sin[x^3]/(x^3 + x^2 + 1.), {x, 0, 3.14159}]
but instead uses NIntegrate (see below) to numerically compute the following.

In[38]:= **N[%]**
Out[38]= 0.179654

But, for such expressions, N[*symbolicNumericalExpression*, *prec*] typically does not give a result with precision *prec*. Typically, the expression will be evaluated with working precision *prec* and this typically results in a precision less than *prec*. Here is an example for such a situation.

In[39]:= **N[Sum[k^-k, {k, Infinity}], 100]**

N::meprec : Internal precision limit

$MaxExtraPrecision = 50.` reached while evaluating $\sum_{k=1}^{\infty} k^{-k}$. More…

Out[39]= 1.2912859970626635404072825905956005414986193684

In[40]:= **Precision[%]**
Out[40]= 46.1588

As we will see later, the second argument of N influenced the WorkingPrecision option setting, not the setting of the PrecisionGoal option.

In[41]:= **NSum[k^-k, {k, Infinity}, WorkingPrecision -> 100]**
Out[41]= 1.2912859970626635404072825905956005414986193684

In[42]:= **Precision[%]**
Out[42]= 46.1588

Now, let us discuss what exactly characterizes a high-precision number. With high-precision arithmetic, it is important to carefully distinguish between accuracy and precision. In principle, this distinction is also important in calculations using machine precision, but unfortunately in this case, we do not have full control over these values.

The accuracy of a number is (roughly) defined to be the number of correct digits to the right of the decimal point.

The precision of a number is (roughly) defined to be the total number of correct significant digits.

(The difference $p - a$ is sometimes called scale or order of magnitude of a number.)

Here are the most important cases. For brevity, let p stand for precision and a for accuracy.

- $0 < a < p$. These numbers have the form: $d_1 d_2 \ldots d_s.d_{s+1} d_{s+2} \ldots d_p$. Here, $a + s = p$.

- $a \geq p > 0$. These numbers have the form: $0.000000 \ldots 0000000 d_1 d_2 \ldots d_p$.

- $a < 0$ and $p > 0$. These numbers have the form: $d_1 d_2 \ldots d_p 000000 \ldots 000000.0$.

- $p \leq 0$. These numbers have no guaranteed digits at all. (For this case, a can be positive or negative.) But nevertheless, a indicates the order of magnitude of the number. Numbers with a negative precision are automatically converted to numbers with the corresponding (negative or positive) accuracy.

In *Mathematica*, the concepts accuracy and precision are realized by the two functions Precision and Accuracy.

Precision[*number*]

   gives the precision of the number *number* (the total approximative number of significant digits in *number*).

Accuracy[*number*]

   gives the accuracy of the number *number* (the approximative number of significant digits to the right of the decimal point in *number*).

In general, the accuracy and precision of a number are not integers; so the above rough definitions apply after appropriate rounding. (There are two reasons for this: 1) a number is internally represented through bits and one bit is a fraction of a decimal digit and 2) when doing an arithmetic operation the error estimation might involve a fraction of a bit.)

The following definitions of precision and accuracy are mathematically stricter. Suppose the representation of a number x has an error of size ϵ. Then, the accuracy of $x \pm \epsilon/2$ is defined to be $-\log_{10}|\epsilon|$, and its precision is defined by $-\log_{10}|\epsilon/x|$ (be aware that for $x = 0$, this definition has a singularity).

Reverting these definitions, we can say that a number z with accuracy a and precision p will lie with certainty in the interval $(x - 10^{-a}/2, x + 10^{-a}/2) = (x - x\,10^{-p}/2, x + x\,10^{-p}/2)$ (a distinction between open and closed intervals is not useful for approximative numbers).

The next examples involve Precision and Accuracy. We make use of a function that produces a two-element list containing the precision and the accuracy for each input number.

In[43]:= $\mathcal{PA}$[x_] := {Precision[x], Accuracy[x]}

Here is a number with machine precision, that corresponds to the second case above, and so has an accuracy greater than its precision.

In[44]:= $\mathcal{PA}$[0.000123]

Out[44]= {MachinePrecision, 19.8647}

Here is a longer number (it is not a high-precision number, despite the many zeros that are input to the right of the decimal point).

In[45]:= With[{x = 0.0000000000000000000000000000000000000000000123},
            {$\mathcal{PA}$[x], MachineNumberQ[x]}]
Out[45]= {{MachinePrecision, 58.8647}, True}

Here are a total of approximately $MachinePrecision digits after the first nonzero digit is recognized. So the second of the next numbers is a high-precision number.

In[46]:= {$\mathcal{PA}$[0.00000123], $\mathcal{PA}$[0.000001230000000000000000000000000000000000]}
Out[46]= {{MachinePrecision, 21.8647}, {35.0899, 41.}}

Explicit zeros at the end of a number (after the decimal point) are kept. Here is a machine number.

In[47]:= $\mathcal{PA}$[123.45]

Out[47]= {MachinePrecision, 13.8631}

Here is a comparison with a high-precision number.

In[48]:= $\mathcal{PA}$[123.45600000000000000000000000000000000000000000000]
Out[48]= {49.0915, 47.}

In the next example, we get a negative accuracy because 45 is larger than is $MachinePrecision and shows that we have uncertain digits to the left of the decimal point.

In[49]:= $\mathcal{PA}$[3.56 10^45]

Out[49]= {MachinePrecision, -29.5969}

According to the above definitions, the following relation holds between the precision and the accuracy of a real number x: precision(x) = accuracy(x) + $\log_{10}(|x|)$ (the quantity $\log_{10}(|x|)$ is sometimes called the scale of the number x; available as the experimental function Experimental`$NumberScale[$x$]). Here are some examples demonstrating this identity.

In[50]:= Function[x, {Precision[x], Accuracy[x] + Log[10, Abs[x]]}] /@
              (* numbers of different accuracy *)
              {N[1, 100], N[10^100, 30], N[2^-300, 40], N[10^20, 3000]}

Out[50]= {{100., 100.}, {30., 30.}, {40., 40.}, {3000., 3000.}}

Here are two examples for numbers with zero precision. The accuracy of such numbers can be positive or negative.

In[51]:= With[{x = (* a high precision zero with positive accuracy
                    arising from subtraction *)
              N[10^10, 50] - N[10^10, 50]},
         {x, $\mathcal{PA}$[x]}]

Out[51]= {0. × 10^{-40}, {0., 39.699}}

In[52]:= With[{x = (* a high precision zero with negative accuracy
                    arising from subtraction *)
              N[10^100, 50] - N[10^100, 50]},
         {x, $\mathcal{PA}$[x]}]

Out[52]= {0. × 10^{50}, {0., -50.301}}

The precision and accuracy can also be defined for complex numbers. Here is a complex number that is a machine number.

In[53]:= $\mathcal{PA}$[2.1 + 7.8 I]

Out[53]= {MachinePrecision, 15.0473}

A complex number is considered to be of high precision if the real and imaginary parts are of high precision.

In[54]:= $\mathcal{PA}$[2.1 + 7.80000000000000000000000022222222222222222 I]

Out[54]= {MachinePrecision, 15.6324}

In[55]:= $\mathcal{PA}$[2.1 + 7.8`4 I]

Out[55]= {MachinePrecision, 3.10791}

In[56]:= $\mathcal{PA}$[2.100000000000000000000000000000444444444444444444 +
          7.800000000000000000000000022222222222222222222 I]

Out[56]= {47.7568, 46.8495}

For complex numbers, the values of Precision and Accuracy are determined in the following way: If $z = (x \pm \epsilon_x/2) + i(y \pm \epsilon_y/2)$, the uncertainty of z is determined by the size of the real and imaginary parts as well as by their precision. Here are some typical situations for Accuracy and Precision of some complex numbers composed from real and imaginary numbers of different values for accuracy and precision. (We give a detailed form in a moment.)

In the next input, a dominates the precision and the accuracy of the sum.

In[57]:= $\mathcal{PA}$ /@ {a = N[3 10^35, 50], b = N[I 3 10^-35, 20], a + b}

Out[57]= {{50., 14.5229}, {20., 54.5229}, {50., 14.5229}}

In the next input, a again determines the precision and the accuracy of the result.

In[58]:= `PA /@ {a = N[3 10^35, 20], b = N[I 3 10^-35, 50], a + b}`

Out[58]= `{{20., -15.4771}, {50., 84.5229}, {20., -15.4771}}`

Now, both numbers a and b influence the precision of the result.

In[59]:= `PA /@ {a = N[3 10^35, 120], b = N[I 3 10^-35, 30], a + b}`

Out[59]= `{{120., 84.5229}, {30., 64.5229}, {100., 64.5229}}`

The following two pictures show the precision and the accuracy of $a + ib$ as a function of the precision of a and b. We use again the values $a = 3 \times 10^{35}$ and $b = 3 \times 10^{-35}$.

In[60]:=
```
With[{pMax = 120},
    Show[GraphicsArray[
    ListPlot3D[(* the real and imaginary parts of different sizes
                 and precisions *)
            Table[#[N[3 10^35, precA] + I N[3 10^-35, precB]],
                  {precB, 1, pMax}, {precA, 1, pMax}],
            PlotRange -> All, MeshRange -> {{1, pMax}, {1, pMax}},
            Mesh -> False, DisplayFunction -> Identity,
            AxesLabel -> {"Precision[a]", "Precision[b]", None},
            PlotLabel -> StyleForm[ToString[#] <> "[a + I b]",
                            FontFamily -> "Courier"]]& /@
    (* accuracy and precision *) {Precision, Accuracy}]]]
```

In the next two pictures, we fix the precision of the summands and then plot the precision and the accuracy of the sum as a function of the size of the real and imaginary parts. We vary the size of the summands on an exponential scale. Obviously, the precision is constant in this case and the accuracy is symmetric in the two summands.

In[61]:=
```
With[{ε = 100, pp = 61},
    Show[GraphicsArray[
    ListPlot3D[(* the real and imaginary parts of different sizes
                 and precisions *)
            Table[#[N[10^k, 30] + I N[10^l, 30]],
                  {l, -ε, ε, 2ε/pp}, {k, -ε, ε, 2ε/pp}],
            DisplayFunction -> Identity, PlotRange -> All,
            AxesLabel -> {"10^k", "10^l", None},
            MeshRange -> {{-100, 100}, {-100, 100}},
            Mesh -> False, DisplayFunction -> Identity,
            AxesLabel -> {"10^k", "10^l", None}, Mesh -> False,
            PlotLabel -> StyleForm[ToString[#] <> "[a + I b]",
                FontFamily -> "Courier"]]& /@
            (* accuracy and precision *) {Precision, Accuracy}]]]
```

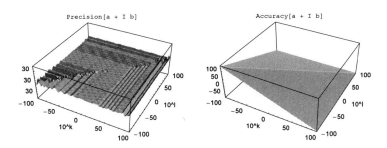

The real and the imaginary parts of a complex number of a given precision might have different precisions than does the number itself. This happens when the magnitudes of the real and the imaginary parts are (greatly) different.

```
In[62]:= With[{p = Precision}, {p[#], p[Re[#]], p[Im[#]]}&[
             (* the complex number *) N[1, 20] + N[10^10, 40] I]]
Out[62]= {30., 20., 40.}
```

Here is the "same" number with the real and imaginary parts having the same precision.

```
In[63]:= With[{p = Precision}, {p[#], p[Re[#]], p[Im[#]]}&[
             N[1, 40] + N[10^10, 40] I]]
Out[63]= {40., 40., 40.}
```

The two last numbers are indeed considered as identical by *Mathematica*.

```
In[64]:= N[1, 20] + N[10^10, 40] I === N[1, 40] + N[10^10, 40] I
Out[64]= True
```

The following plot shows the precision of the real and imaginary parts for different ratios of the real and imaginary parts. (We will discuss the function SetPrecision in a moment.)

```
In[65]:= With[{p = Precision},
             Show[Graphics[Line /@ Transpose[
                 Table[({i, #}& /@ {p[Re[#]], p[Im[#]]}&)[
                     (* make an approximate number with precision 6 *)
                     SetPrecision[SetPrecision[1 + I 10^i, 6], 6]],
                     {i, -10, 10, 1/20}]]],
             Frame -> True]]
```

The accuracy of a complex number $z = x + i\,y$, x and y being real, is defined in the following way: accuracy$(z) = -1/2 \log_{10} (10^{-2\,\text{accuracy}(x)} + 10^{-2\,\text{accuracy}(y)})$. Here is a randomly chosen example demonstrating this formula.

```
In[66]:= With[{x = N[Pi^300, 120], y = N[E^-234, 400]},
           {Accuracy[x + I y],
            N[-Log[10, 10^(-2 Accuracy[x]) + 10^(-2 Accuracy[y])]/2]}]
Out[66]= {-29.145, -29.145}
```

The precision of a complex number $z = x + i\,y$, x and y being real, is defined in the following way: precision(z) = accuracy(z)+$\log_{10}(|z|)$. This is the exact equivalent of the corresponding formula for real numbers. Here is a randomly chosen example demonstrating this formula.

```
In[67]:= With[{x = N[(E/Pi)^333, 300], y = N[(Pi/E)^555, 600]},
           {Precision[x + I y],
            Accuracy[x + I y] + N[Log[10, Abs[x + I y]]]}]
Out[67]= {355.816, 355.816}
```

Now, we continue with infinite precision and accuracy values.

> For exact numbers and mathematical constants, the result of either `Precision` or `Accuracy` is `Infinity`.

Here are some examples.

```
In[68]:= {𝒫𝒜[3], 𝒫𝒜[3/7], 𝒫𝒜[3 + 4/5 I], 𝒫𝒜[E]}
Out[68]= {{∞, ∞}, {∞, ∞}, {∞, ∞}, {∞, ∞}}
```

For any exact quantity with the head `Symbol` (or `String`), we get the following accuracy and precision.

```
In[69]:= {𝒫𝒜[Sin], 𝒫𝒜[veryInexactNumber], 𝒫𝒜["aStringHasNoPrecision"]}
Out[69]= {{∞, ∞}, {∞, ∞}, {∞, ∞}}
```

Also, `Infinity` and `Indeterminate` are exact quantities.

```
In[70]:= {𝒫𝒜[Infinity], 𝒫𝒜[Indeterminate]}
Out[70]= {{∞, ∞}, {∞, ∞}}
```

Looking at the `Precision` and `Accuracy` of `0.0` and related quantities shows the following values. (We mentioned the singularity in the definition.)

```
In[71]:= {𝒫𝒜[0], 𝒫𝒜[0.0], 𝒫𝒜[0``50]}
Out[71]= {{∞, ∞}, {MachinePrecision, 307.653}, {0., 50.}}
```

Note that `N[0]` gives `0.0`, but that `N[0, `*precisionOtherThanMachinePrecisionToken*`]` stays as an exact 0. The numericalization with N happens to allow packing of lists of numbers (see below) and the nonnumericalization with N[..., *prec*] happens because it is not possible to give any nontrivial digit for 0.

```
In[72]:= {N[0, MachinePrecision], N[0, $MachinePrecision]}
Out[72]= {0., 0}
```

In carrying out a numerical calculation, we may lose precision. Here is an example in which we iterate $f(x) = (x^2 - 1)/(x - 1) - 1$ ($= x$ up to numerical precision loss) [1953].

```
In[73]:= Block[{i = -1},
    Nest[(* print out current precision of the result *) i = i + 1;
    (CellPrint[Cell["∘ Iteration: " <> ToString[i] <> "  Precision: " <>
                ToString[Precision[#]] <> "  Accuracy: " <>
                ToString[Accuracy[#]], "PrintText"]];
     (#^2 - 1)/(# - 1) - 1)&, N[2, 200], 12]]
```

○ Iteration: 0 Precision: 200. Accuracy: 199.699

○ Iteration: 0 Precision: 199.155 Accuracy: 198.854

○ Iteration: 0 Precision: 198.31 Accuracy: 198.009

○ Iteration: 0 Precision: 197.465 Accuracy: 197.164

○ Iteration: 0 Precision: 196.62 Accuracy: 196.319

○ Iteration: 0 Precision: 195.775 Accuracy: 195.473

○ Iteration: 0 Precision: 194.929 Accuracy: 194.628

○ Iteration: 0 Precision: 194.084 Accuracy: 193.783

○ Iteration: 0 Precision: 193.239 Accuracy: 192.938

○ Iteration: 0 Precision: 192.394 Accuracy: 192.093

○ Iteration: 0 Precision: 191.549 Accuracy: 191.248

○ Iteration: 0 Precision: 190.704 Accuracy: 190.403

Out[73]= 2.0000000000000000000000000000000000000000000000000000000000000000000000000000000000 ⁝
00000000000000000000000000000000000000000000000000000000000000000000000000000000000 ⁝
0000000000000000000000000000000000000000

In[74]:= $\mathcal{PA}$[%]

Out[74]= {189.859, 189.558}

If we carry this calculation still further, we end up with less than one correct digit.

In[75]:= **Nest[(#^2 - 1)/(# - 1) - 1&, N[2, 200], 236]**

Out[75]= 2.

In[76]:= $\mathcal{PA}$[%]

Out[76]= {0.556863, 0.255833}

Performing still one more iteration yields 0.

In[77]:= **Nest[(#^2 - 1)/(# - 1) - 1&, N[2, 200], 237]**

Out[77]= 0.

In[78]:= $\mathcal{PA}$[%]

Out[78]= {0., -0.589265}

The appearance of 0 as the precision and accuracy here indicates a potential problem for the reliability of the calculation carried out. And performing one more iteration step results in the value `Indeterminate`. This result arises from the uncertainty of the last zero that is occurs in a denominator.

In[79]:= **Nest[(#^2 - 1)/(# - 1) - 1&, N[2, 200], 238]**

Power::infy : Infinite expression $\frac{1}{0.}$ encountered. More…

∞::indet : Indeterminate expression $0. \times 10^1$ ComplexInfinity encountered. More…

Out[79]= Indeterminate

Of course, the opposite can also happen, and one can gain precision in a calculation. To gain precision the function value should not change much when the argument changes. Here is a simple example.

In[80]:= **var1 = N[999999/100000, 20];**
Precision[var1]

In[81]:= 20.

In[82]:= `1 - Exp[-var1^2] // Precision`

Out[82]= 61.1283

Here is a still simpler example, involving just the addition of an exact number.

In[83]:= `{Precision[N[10^-30, 100]], Precision[1 + N[10^-30, 100]]}`

Out[83]= {100., 130.}

Now let us deal with calculating out mathematical functions of high-precision numbers. When carrying out an arithmetic operation, *Mathematica* keeps track of the precision of the result. So, how exactly does this happen? Let f be the (smooth) univariate function under consideration and δx be the (nonnegative) uncertainty of the argument x. This means the argument is known as $x \pm \delta x$. Then, the precision of the input is given as $-\log_{10} \frac{\delta x}{x}$. The uncertainty of the output δf can then be approximated as $\delta f \approx \left| \frac{\partial f(x)}{\partial x} \delta x \right|$. From this follows that the precision of the output $-\log_{10} \frac{\delta f}{f(x)}$ can be approximated as $-\log_{10} \frac{\delta f}{f(x)} \approx -\log_{10}(\left| \frac{\partial f(x)}{\partial x} \frac{x}{f(x)} \right| \frac{\delta x}{x})$. The condition number $\left| \frac{\partial f(x)}{\partial x} \frac{x}{f(x)} \right|$ is what determines the precision of the result. This quantity can be less or greater than unity, and as a result, the precision of the output can be smaller or larger than the precision of the input x.

Here is a plot of the precision of $\sin(x)$ as a function of x and the precision of x. The right graphic visualizes the above precision formula $f(x) \approx -\log_{10}(\left| \frac{\partial f(x)}{\partial x} \frac{x}{f(x)} \right| \frac{\delta x}{x})$, and we obtain the same picture.

In[84]:=
```
Show[GraphicsArray[
    Block[{$DisplayFunction = Identity},
    {(* left: measure precision of the expression *)
     Plot3D[Precision[Sin[SetPrecision[x, p]]], {x, 1/2, 3},
            {p, 1, 20}, Compiled -> False, PlotPoints -> 41],
     (* right: model precision of the expression *)
     Plot3D[-Log[10, Abs[Cos[x] x/Sin[x] 10^-p]], {x, 1/2, 3}, {p, 1, 20},
            Compiled -> False, PlotPoints -> 41]}]]]
```

The small difference between the precision of `Precision[Sin[SetPrecision[x, p]]]` and `-Log[10, Abs[Cos[x] x/Sin[x] 10^-p]]` is practically zero. It is caused by the internal numerical methods used to calculate the sine function. Here we show the difference between the actual precision and the calculated precision for a fixed p and varying x. The agreement is excellent.

In[85]:=
```
With[{p = 2 $MachinePrecision},
    Plot[Precision[Sin[SetPrecision[x, p]]] +
        Log[10, Abs[Cos[x] x/Sin[x] 10^-p]], {x, -10, 10},
        Compiled -> False, PlotPoints -> 41, PlotRange -> All,
        Frame -> True, Axes -> False]]
```

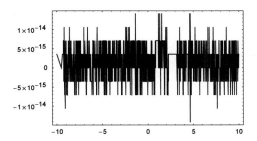

Observe the small peak around $x = \pi/2$ in the above 3D graphic. Here, $\frac{\partial f(x)}{\partial x}$ vanishes for $f(x) = \sin(x)$, and the precision has a singularity. The next picture shows in detail the region $x \approx \pi/2$ for a fixed precision.

```
In[86]:= With[{δ = 10^-1, prec = 50},
    ListPlot[Table[{x, Precision[Sin[SetPrecision[Pi/2 + x, prec]]]},
            {x, -δ, δ, 2δ/1001}],
        PlotRange -> All, PlotJoined -> True,
        Frame -> True, Axes -> False]]
```

The formula for determining the precision of the output is based on the first-order Taylor series of $f(x)$. This means that this precision model will be accurate for $\frac{\delta x}{x} \ll 1$. For arguments of $f(x)$ with a small precision, *Mathematica*'s high-precision arithmetic might give either too pessimistic or even wrong results. In the following picture, the two blue lines bound $\sin(one)$, where *one* is 1 with precision p, $one = 1 \pm 10^{-p}$ as a function of the precision p. The red lines represent the exact bound for $\sin(one)$. They are obtained using interval arithmetic; see below. For precision greater than two, there is virtually no difference between the two curves anymore.

```
In[87]:= Show[GraphicsArray[
    {#, (* magnification for larger values of p *)
    MapAt[(PlotRange -> {{2, 3}, {0.836, 0.846}})&, #, {-1}]}&[
    With[{(* value and function *) one = 1, f = Sin},
     Graphics[{{PointSize[0.008], Hue[0],
            (* exact interval arithmetic in red *)
            Table[Point[{p, #}]& /@ f[Interval[one + 1/2 10^-p {-1, 1}]]][[1]],
                {p, 10^-2, 3, 10^-2}]},
            {PointSize[0.004], Hue[0.7],
            (* high-precision arithmetic in blue *)
            Table[{Point[{p, f[one] - #}], Point[{p, f[one] + #}]}&[
                (* estimated uncertainty of f[one] *)
                f[one] 1/2 10^-Precision[f[SetPrecision[one, p]]]],
                {p, 10^-2, 3, 10^-2}]}} // N,
        Frame -> True, PlotRange -> All]]]]]
```

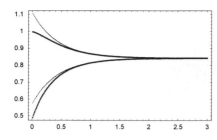
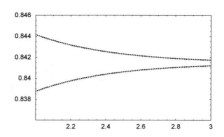

Now, let us come back to the Weierstrass $\wp$ calculation from above. The iterated expressions contain many occurrences of z. Here is the result after the first iteration.

```
In[88]:= weierstrasspList[z, {2, 3}, 1][[-1]]
```

$$
Out[88]= \;-2\left(\frac{4}{z^2}+\frac{z^2}{40}+\frac{3z^4}{448}\right)+\frac{\left(-1+6\left(\frac{4}{z^2}+\frac{z^2}{40}+\frac{3z^4}{448}\right)^2\right)^2}{4\left(-3-2\left(\frac{4}{z^2}+\frac{z^2}{40}+\frac{3z^4}{448}\right)+4\left(\frac{4}{z^2}+\frac{z^2}{40}+\frac{3z^4}{448}\right)^3\right)}
$$

Substituting an approximative value for z and evaluating the above iteration effectively treats every z as a different number. As a result, we see the loss of nine digits in the following input.

```
In[89]:= {#, Precision[#]}& @ weierstrasspList[N[2, 30], {2, 3}, 5][[-1]]
Out[89]= {6.4455493539655083655 9, 20.9498}
```

A more careful analysis that identifies all of the z occurring in the expression yields a much smaller loss of precision (the symbolic expression D[$\mathcal{P}$, z] in the following input in quite large, its size is about 5 MB).

```
In[90]:= 𝒫 = Last[weierstrasspList[z, {2, 3}, 5]];
        D[𝒫, z] z/𝒫 δz/z /. {z -> 2, δz -> 10^-30} // N[#, 22]&
Out[91]= 5.039811020046271555398× 10^-30
```

The precision model described, generalizes in an obvious way to multivariate functions. Let, for instance, f be a bivariate function of x and y and δx, δy be the uncertainties of the arguments. The uncertainty of the output δf can then be approximated as $\delta f \approx \left|\frac{\partial f(x,y)}{\partial x}\right|\delta x+\left|\frac{\partial f(x,y)}{\partial y}\right|\delta y$. From this follows that the precision of the output $-\log_{10}\frac{\delta f}{f(x,y)}$ can be approximated as $-\log_{10}\frac{\delta f}{f(x,y)}\approx-\log_{10}(\left|\frac{\partial f(x,y)}{\partial x}\right|\frac{x}{f(x,y)}\left|\frac{\delta x}{x}\right|+\left|\frac{\partial f(x,y)}{\partial y}\right|\frac{y}{f(x,y)}\left|\frac{\delta y}{y}\right|)$. Here are two examples of a built-in bivariate function—arctan and log.

```
In[92]:= modelPrecision[f_, {x_, y_}, {x0_, y0_}] :=
        (* form symbolic derivatives *)
        (Abs[D[f[x, y], x] x/f[x, y] 10^-px] +
         Abs[D[f[x, y], y] y/f[x, y] 10^-py]) /.
        (* substitute actual values *)
        {x -> x0, px -> Precision[x0], y -> y0, py -> Precision[y0]} //
                                        N // -Log[10, #]&
```

```
In[93]:= Function[{x0, y0},
        {Precision[ArcTan[x0, y0]], modelPrecision[ArcTan, {x, y}, {x0, y0}]}][
                        SetPrecision[10^-50, 20], SetPrecision[2, 30]]
Out[93]= {70.4971, 70.4971}
```

Expressing arctan through logarithm and square root gives a lower precision because of the repeated (independent) appearance of x and y. (We use the function TrigToExp, to be discussed in Chapter 1 of the Symbolics volume [1795], to rewrite ArcTan in logarithms.)

```
In[94]:= f = Function[{x, y}, Evaluate[TrigToExp[ArcTan[x, y]]]]
```

Out[94]= $\text{Function}\left[\{x, y\}, -i \text{ Log}\left[\frac{x + i y}{\sqrt{x^2 + y^2}}\right]\right]$

In[95]:= `Function[{x0, y0},`
`    {Precision[f[x0, y0]], modelPrecision[f, {x, y}, {x0, y0}]}][`
`                 SetPrecision[10^-50, 20], SetPrecision[2, 30]]`

Out[95]= `{29.8951, 70.4971}`

Here is the corresponding univariate version, which for $x \neq 0$, allows expressing the bivariate arctan.

In[96]:= `With[{f = Function[{x, y}, ArcTan[y/x]]},`
`    Function[{x0, y0},`
`    {Precision[f[x0, y0]], modelPrecision[f, {x, y}, {x0, y0}]}][`
`                 SetPrecision[10^-50, 20], SetPrecision[2, 30]]]`

Out[96]= `{70.4971, 70.4971}`

As the second bivariate function, we will choose the two-argument form of `Log`. Again, the modeled and the actual precision agree.

In[97]:= `Function[{x0, y0},`
`    {Precision[Log[x0, y0]], modelPrecision[Log, {x, y}, {x0, y0}]}][`
`                 SetPrecision[10^50, 10], SetPrecision[2 10^60, 12]]`

Out[97]= `{12.0576, 12.0576}`

The precision of multivariate functions with more than two arguments arise as obvious generalizations of the described model.

The precision of (nested) expression is the minimal precision of all of its subparts. Here are two examples. In the second example, the two high-precision zeros result in precision `0.`.

In[98]:= `{Precision[F[1.`20, 1.`30, 1.`40]], Precision[F[0``50, 0``20, 1]]}`

Out[98]= `{20., 0.}`

The accuracy of an expression is the minimal accuracy of all of its subparts.

In[99]:= `{Accuracy[F @@ Table[N[10^-k, 1], {k, 20, 50, 10}, {1, 20, 50, 10}]],`
`    Accuracy[N[10^-20, 20]]}`

Out[99]= `{40., 40.}`

Let us make the following definition for `f`. We give `f` the `NumericFunction` attribute.

In[100]:= `Remove[f]`

`    SetAttributes[f, NumericFunction]`

`    f[x_?InexactNumberQ] := Abs[x]`

Adding a real number to `f[I]` results in *one* real number for the whole expression (be aware that `f` has never been explicitly called with an inexact argument; but internally it is called with an approximative argument).

In[103]:= `2. + f[I]`

Out[103]= `3.`

The numericalization happens because `f[I]` is considered as a numeric expression.

In[104]:= `f[I] // NumericQ`

Out[104]= `True`

In the next input, the precision of the inexact number `N[2, 22]` is used for the numericalization of `f[I]`.

In[105]:= **N[2, 22] + f[I]**

Out[105]= 3.000000000000000000000

Now, using the functions `Precision` and `Accuracy` we can quantify a previous remark concerning precision loss in calculations.

> Because a loss of precision can happen in numerical calculations (starting with an inexact number), *Mathematica* varies the `Precision` of the resulting numbers depending on the numerical operation being performed. Precisions produced in longer calculations by *Mathematica* are usually too pessimistic (which is better than overly optimistic ones or no estimates at all).

In most calculations, one typically loses precision. Here, this is shown for nesting the tangent function.

In[106]:= **ListPlot[Precision /@ NestList[Tan, N[10 + 1I, 20], 100]]**

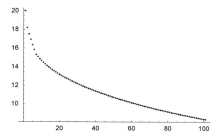

In the following repeated application of the cosine function and the square root function, one gains precision. (The behavior depends on the starting values.)

In[107]:= **Show[GraphicsArray[**
    Block[{$DisplayFunction = Identity},
     {(* iterated cosinus function *)
      ListPlot[Precision /@ NestList[Cos, N[Pi/2 +10^-5, 20], 100]],
      (* iterated square root function *)
      ListPlot[Precision /@ NestList[Power[#, 1/2]&, N[10 + 1 I, 20],
                    100]]}]]]

We now give some additional examples in which precision and accuracy are varied (and not just reduced) during a numerical calculation. (We use the following values $a = 10^{-10}$ with precision 10, $b = 10^{10}$ with precision 10, *one* $= 1$ with precision 10, *mOne* $= 1$ as a machine number, and *zero* $= 0$ to accuracy 10.) The second element of the returned list is again the list with the precision and accuracy of the result. (We encourage the reader to have a careful look at each row of the following table.)

In[108]:= **Block[{zero, one, mOne, a, b},**
        (* the values for zero, one, mOne, a, b *)

```
zero = SetAccuracy[0, 10];
one = SetPrecision[1, 10];
mOne = 1.0;
a = SetPrecision[10^-10, 10];
b = SetPrecision[10^+10, 10];
{#[[1]], #[[2]], {Precision[#[[2]]], Accuracy[#[[2]]]}}& /@
Transpose[{#, ToExpression[#]}&[
(* all expressions to be investigated *)
{"a + b", "a - b", "a/b", "b/a", "a^b", "b^a",
 "1 + a", "1 + b", "1/a", "1/b", "1^a", "1^b",
 "mOne + a", "mOne + b", "mOne/a", "mOne/b", "mOne^a", "mOne^b",
 "one + a", "one + b", "one/a", "one/b", "one^a", "one^b",
 "zero + a", "zero + b", "zero/a", "zero/b", "zero^a", "zero^b",
 "Sin[a]", "Sin[b]"}]]] //
                    (* make a table *) TableForm[#, TableDepth -> 1]&
        General::unfl : Underflow occurred in computation. More…
```

Out[108]//TableForm=

$\{a + b, 1.0000000000 \times 10^{10}, \{10., 0.\}\}$

$\{a - b, -1.0000000000 \times 10^{10}, \{10., 0.\}\}$

$\{a/b, 1.000000000 \times 10^{-20}, \{9.69897, 29.699\}\}$

$\{b/a, 1.000000000 \times 10^{20}, \{9.69897, -10.301\}\}$

$\{a\^b, \text{Underflow}[], \{0., \infty\}\}$

$\{b\^a, 1.000000002302585096, \{18.6193, 18.6193\}\}$

$\{1 + a, 1.00000000010000000000, \{20., 20.\}\}$

$\{1 + b, 1.000000000 \times 10^{10}, \{10., 0.\}\}$

$\{1/a, 1.0000000000 \times 10^{10}, \{10., 0.\}\}$

$\{1/b, 1.0000000000 \times 10^{-10}, \{10., 20.\}\}$

$\{1\^a, 1, \{\infty, \infty\}\}$

$\{1\^b, 1, \{\infty, \infty\}\}$

$\{\text{mOne} + a, 1., \{\text{MachinePrecision}, 15.9546\}\}$

$\{\text{mOne} + b, 1. \times 10^{10}, \{\text{MachinePrecision}, 5.95459\}\}$

$\{\text{mOne}/a, 1. \times 10^{10}, \{\text{MachinePrecision}, 5.95459\}\}$

$\{\text{mOne}/b, 1. \times 10^{-10}, \{\text{MachinePrecision}, 25.9546\}\}$

$\{\text{mOne}\^a, 1., \{\text{MachinePrecision}, 15.9546\}\}$

$\{\text{mOne}\^b, 1., \{\text{MachinePrecision}, 15.9546\}\}$

$\{\text{one} + a, 1.000000000, \{10., 10.\}\}$

$\{\text{one} + b, 1.000000000 \times 10^{10}, \{10., -4.34292 \times 10^{-11}\}\}$

$\{\text{one}/a, 1.000000000 \times 10^{10}, \{9.69897, -0.30103\}\}$

$\{\text{one}/b, 1.000000000 \times 10^{-10}, \{9.69897, 19.699\}\}$

$\{\text{one}\^a, 1.00000000000000000000, \{20., 20.\}\}$

$\{\text{one}\^b, 0., \{4.27789 \times 10^{-15}, -0.434294\}\}$

$\{\text{zero} + a, 0. \times 10^{-10}, \{0., 10.\}\}$

$\{\text{zero} + b, 1.000000000 \times 10^{10}, \{10., -4.34292 \times 10^{-11}\}\}$

$\{\text{zero}/a, 0., \{0., 0.\}\}$

$\{\text{zero}/b, 0. \times 10^{-20}, \{0., 20.\}\}$

$\{\text{zero}\^a, 0. \times 10^{-1}, \{0., 1. \times 10^{-9}\}\}$

$\{\text{zero}\^b, 0. \times 10^{-20201778}, \{0., 2.02018 \times 10^{7}\}\}$

$\{\text{Sin}[a], 1.0000000000 \times 10^{-10}, \{10., 20.\}\}$

$\{\text{Sin}[b], 0., \{0., 0.\}\}$

It is a general rule that the high-precision arithmetic is used whenever 1) the numbers that appear in the expression to be calculated are already high-precision numbers, 2) if the use of high-precision numbers is explicitly forced via N[*expr*, *prec*] or 3) via the WorkingPrecision option of the N-functions such as FindRoot, NSolve, ... So, for instance, in N[Sin[2.0], 200], the 2.0 is only a machine number for which digits beyond $MachinePrecision are not known, and so it is impossible to generate a meaningful 200-digit result for N[Sin[2.0], 200]. Consequently, every time we want a high-precision result, we must take care that the starting expression has sufficient precision (typically, we will only convert exact symbolic expressions, such as integers, algebraic numbers, or transcendental constants into high-precision numbers). One exception to this rule is the use of high-precision arithmetic in expressions whose magnitude is becoming bigger than $MaxMachineNumber. In this case, *Mathematica* automatically switches to high-precision numbers. Here, this is visualized. We display the precision of 10^{10^i} as a function of *i*.

```
In[109]:= ListPlot[Function[i, {i, Precision[10.^(10^i)]}] /@
                    Table[i, {i, 0, 8.5, 0.01}],
            Frame -> True, PlotStyle -> {PointSize[0.005]},
            PlotRange -> All, Axes -> False,
            FrameLabel -> (StyleForm[#, FontFamily -> "Courier"]& /@
                    {"Log[10, Log[10, x]]", "Precision[x]"}),
            (* vertical line *)
            GridLines -> {{Log[10, Log[10, $MaxMachineNumber]] // N}, None}]
```

We clearly see the decrease in Precision[x] around Log[10, Log[10, $MaxMachineNumber]].

```
In[110]:= Log[10, Log[10, $MaxMachineNumber]]
Out[110]= 2.48891
```

Frequently, we would like to change the accuracy and/or the precision of a number (or a larger expression) in the course of a calculation. This can be accomplished as follows. (We used the function SetPrecision already repeatedly.)

SetAccuracy[*expression*, *accuracy*]

　replaces all numbers and constants appearing in *expression* by their numerical values with accuracy *accuracy*. If this requires an increase in accuracy, zeros will be added in base 2 after using any digits in the internal representation, if present. The resulting number is a high-precision number, even if the number of digits requested is smaller than the machine precision. If *accuracy* is Infinity, all approximate numbers in *expression* will be converted to exact numbers.

SetPrecision[*expression*, *precision*]

　replaces all numbers and constants in *expression* by their numerical values with precision *precision*. If the precision is increased, zeros will be introduced in base 2 after using any digits in the internal representation, if present. The resulting number is a high-precision number, even if the required number of digits is smaller than the machine precision. If *precision* is Infinity, all approximate numbers in *expression* will be converted to exact numbers.

Because the functions SetPrecision and SetAccuracy operate on the numbers inside an expression and the resulting approximate numbers might cause further evaluations, the result returned might have a lower precision/accuracy than specified in the second argument of SetPrecision/SetAccuracy. The difference between N[*expr*, *prec*] and SetPrecision[*expr*, *prec*] is best demonstrated in case *expr* is numerically zero. N[*expr*, *prec*] will give a message because no reliable digit can be calculated. SetPrecision[*expr*, *prec*] just replaces the numbers inside with approximate ones, and then the expression evaluates.

```
In[111]:= zero = Sin[Pi/16] - Sqrt[2 - Sqrt[2 + Sqrt[2]]]/2;

        N[zero, 2 MachinePrecision]
```
$$N::meprec : \text{Internal precision limit } \$MaxExtraPrecision = 50.\text{`}$$
$$\text{reached while evaluating } -\frac{1}{2}\sqrt{2 - \sqrt{2 + \text{Power}[\ll 2\gg]}} + \sin\left[\frac{\pi}{16}\right]. \text{ More...}$$

$$Out[112]= \; 0. \times 10^{-82}$$

```
In[113]:= {SetPrecision[zero, 2 MachinePrecision],
        SetAccuracy[ zero, 2 MachinePrecision]}
```
$$Out[113]= \; \{0. \times 10^{-32}, \; 0. \times 10^{-32}\}$$

By avoiding the evaluation, we see the numericalized numbers and constants.

```
In[114]:= Hold[Evaluate[zero]] // SetPrecision[#, Round[MachinePrecision/2]]&
```
$$Out[114]= \; \text{Hold}\left[-0.50000000 \sqrt{2.0000000 - 1.0000000 \sqrt{2.0000000} + 1.41421356} + \right.$$
$$\left. \sin[0.062500000 \; 3.1415927]\right]$$

It follows from the definitions of precision and accuracy that the accuracy of a sum and the precision of a product depend only on the accuracy of the summands or the precision of the factors, but not from the precision of the summands or the accuracy of the factors. Here this is shown. (We use SetAccuracy and SetPreci⌐ sion to generate the numbers that are added and multiplied.)

```
In[115]:= With[{x = SetAccuracy[12478923 10^100, 12],
            y = SetAccuracy[  848964 10^-30, 63]},
            {Accuracy[x + y], Precision[x + y], Accuracy[x y], Precision[x y]}]
Out[115]= {12., 119.096, -44.0962, 38.9289}

In[116]:= (* same accuracy as last input; accuracy of sum is the same *)
        With[{x = SetAccuracy[12478923 10^140, 12],
            y = SetAccuracy[  848964 10^-60, 63]},
            {Accuracy[x + y], Precision[x + y], Accuracy[x y], Precision[x y]}]
Out[117]= {12., 159.096, -84.0962, 8.92889}

In[118]:= With[{x = SetPrecision[12478923 10^100, 12],
            y = SetPrecision[  848964 10^-30, 63]},
            {Accuracy[x + y], Precision[x + y], Accuracy[x y], Precision[x y]}]
Out[118]= {-95.0962, 12., -71.0251, 12.}

In[119]:= (* same precision as last input; precision of product is the same *)
        With[{x = SetPrecision[12478923 10^150, 12],
            y = SetPrecision[  848964 10^-80, 63]},
            {Accuracy[x + y], Precision[x + y], Accuracy[x y], Precision[x y]}]
Out[120]= {-145.096, 12., -71.0251, 12.}
```

To check if a number is a machine number, we make use of MachineNumberQ. (There is no built-in function HighPrecisionNumberQ, but it is straightforward to implement such a function.)

MachineNumberQ[*number*]
> is True if *number* is a number with machine precision, and is otherwise False.

We draw a picture, similar to the above one, to show switching to high-precision numbers by testing if a number is a machine number. Again, we see the switch at Log[10, Log[10, $MaxMachineNumber]]≈2.49.

```
In[121]:= ListPlot[Function[i, {i, If[MachineNumberQ[10.^(10^i)], 1, -1]}] /@
                Table[i, {i, 0, 8.5, 0.01}],
            Frame -> True, PlotStyle -> {PointSize[0.01]}, PlotRange -> All,
            Axes -> False, FrameLabel -> {"Log[10, Log[10, x]]", None}]
```

We now start with an integer and manipulate its accuracy and precision. The function testNumber gives various information about the number *number*.

```
In[122]:= (* show the input form of the number, its precision, accuracy and
            if it is a machine number *)
        testNumber[number_] := {FullForm[number], Precision[number],
                            Accuracy[number], MachineNumberQ[number]}
```

```
In[124]:= testNumber[SetPrecision[5, 3]]
Out[124]= {5.`2.9999999999999973, 3., 2.30103, False}
```

```
In[125]:= testNumber[SetPrecision[5, 30]]
Out[125]= {5.`30., 30., 29.301, False}
```

```
In[126]:= testNumber[SetAccuracy[5, 3]]
Out[126]= {5.`3.6989700043360183, 3.69897, 3., False}
```

```
In[127]:= testNumber[SetAccuracy[5, 30]]
Out[127]= {5.`30.698970004336022, 30.699, 30., False}
```

We repeat the last four tests with real numbers.

```
In[128]:= testNumber[SetPrecision[5.0, 3]]
Out[128]= {5.`2.9999999999999973, 3., 2.30103, False}
```

```
In[129]:= testNumber[SetPrecision[5.0, 30]]
Out[129]= {5.`30., 30., 29.301, False}
```

```
In[130]:= testNumber[SetAccuracy[5.0, 3]]
Out[130]= {5.`3.6989700043360183, 3.69897, 3., False}
```

```
In[131]:= testNumber[SetAccuracy[5.0, 30]]
Out[131]= {5.`30.698970004336022, 30.699, 30., False}
```

Here is what happens when we begin with a high-precision number.

```
In[132]:= testNumber[SetPrecision[5, 100]]
Out[132]= {5.`100., 100., 99.301, False}
```

```
In[133]:= testNumber[SetAccuracy[5, 100]]
```

Out[133]= {5.`100.69897000433602, 100.699, 100., False}

In[134]:= **testNumber[SetPrecision[N[5 10^34, 100], 10]]**

Out[134]= {5.`10.*^34, 10., -24.699, False}

In[135]:= **testNumber[SetAccuracy[N[5 10^34, 100], 10]]**

Out[135]= {5.`44.698970004336026*^34, 44.699, 10., False}

We can also make a high-precision number with $MachinePrecision.

In[136]:= **MachineNumberQ[SetPrecision[1, $MachinePrecision]]**

Out[136]= False

But this does give a machine number.

In[137]:= **MachineNumberQ[SetPrecision[1, MachinePrecision]]**

Out[137]= True

Using SetPrecision, we can avoid the difficulty in the above iteration (but this does not guarantee that the result is correct).

In[138]:= (* now "regain" lost precision in each step *)
    Nest[SetPrecision[(#^2 - 1)/(# - 1) - 1, 200]&, N[2, 200], 237]

Out[139]= 2.000000000000000000000000000000000000000000000000000000000000000000000000000000 ⸗
    000000000000000000000000000000000000000000000000000000000000000000000000000000 ⸗
    00000000000000000000000000000000000000000000000

In[140]:= *PR*[%]

Out[140]= {200., 199.699}

Note that when doing a calculation with higher precision, it is better (for readability and speed) to use a construction like the following. (Now SetPrecision is automatically invoked when needed.)

In[141]:= **Block[{$MinPrecision = 200}, Nest[(#^2 - 1)/(# - 1) - 1&, N[2, 200], 237]]**

Out[141]= 2.000000000000000000000000000000000000000000000000000000000000000000000000000000 ⸗
    000000000000000000000000000000000000000000000000000000000000000000000000000000 ⸗
    00000000000000000000000000000000000000000000000

In[142]:= **Precision[%]**

Out[142]= 200.

We reiterate that in such cases, one has to be aware that this causes the loss of any control over the precision of the result. Depending on its accuracy, a real 0 (head Real) will be output differently.

In[143]:= **$MinPrecision = -100;**
    Table[SetAccuracy[0, -i], {i, 10, -10, -1}]

Out[144]= {$0. \times 10^{10}$, $0. \times 10^{9}$, $0. \times 10^{8}$, $0. \times 10^{7}$, $0. \times 10^{6}$, $0. \times 10^{5}$,
    $0. \times 10^{4}$, $0. \times 10^{3}$, $0. \times 10^{2}$, $0. \times 10^{1}$, 0., $0. \times 10^{-1}$, $0. \times 10^{-2}$, $0. \times 10^{-3}$,
    $0. \times 10^{-4}$, $0. \times 10^{-5}$, $0. \times 10^{-6}$, $0. \times 10^{-7}$, $0. \times 10^{-8}$, $0. \times 10^{-9}$, $0. \times 10^{-10}$}

Further, these zeros also explain the above iteration problem with the value 0 for the Precision. And this converts a real 0.0 to an exact one.

In[145]:= **SetAccuracy[0.0, Infinity]**

Out[145]= 0

A nonzero number will always be represented in essentially the same way, assuming that it has some meaningful digits. For negative accuracy, we obtain an approximate zero with zero precision. Next, we use the numbers 10^{-10}, 1, and 10^{10} and numericalize these numbers to various accuracies.

```
In[146]:= Table[SetAccuracy[10^-10, -i], {i, 10, -10, -1}]
```
$$Out[146]= \{0.\times 10^{10}, 0.\times 10^{9}, 0.\times 10^{8}, 0.\times 10^{7}, 0.\times 10^{6}, 0.\times 10^{5},$$
$$0.\times 10^{4}, 0.\times 10^{3}, 0.\times 10^{2}, 0.\times 10^{1}, 0., 0.\times 10^{-1}, 0.\times 10^{-2}, 0.\times 10^{-3},$$
$$0.\times 10^{-4}, 0.\times 10^{-5}, 0.\times 10^{-6}, 0.\times 10^{-7}, 0.\times 10^{-8}, 0.\times 10^{-9}, 0.\times 10^{-10}\}$$

```
In[147]:= Table[SetAccuracy[1, -i], {i, 10, -10, -1}]
```
$$Out[147]= \{0.\times 10^{10}, 0.\times 10^{9}, 0.\times 10^{8}, 0.\times 10^{7}, 0.\times 10^{6}, 0.\times 10^{5},$$
$$0.\times 10^{4}, 0.\times 10^{3}, 0.\times 10^{2}, 0.\times 10^{1}, 0., 1., 1.0, 1.00, 1.000,$$
$$1.0000, 1.00000, 1.000000, 1.0000000, 1.00000000, 1.0000000000\}$$

```
In[148]:= Table[SetAccuracy[10^+10, -i], {i, 10, -10, -1}]
```
$$Out[148]= \{0.\times 10^{10}, 1.\times 10^{10}, 1.0\times 10^{10}, 1.00\times 10^{10}, 1.000\times 10^{10}, 1.0000\times 10^{10},$$
$$1.00000\times 10^{10}, 1.000000\times 10^{10}, 1.0000000\times 10^{10}, 1.00000000\times 10^{10},$$
$$1.0000000000\times 10^{10}, 1.00000000000\times 10^{10}, 1.000000000000\times 10^{10},$$
$$1.0000000000000\times 10^{10}, 1.00000000000000\times 10^{10}, 1.000000000000000\times 10^{10},$$
$$1.0000000000000000\times 10^{10}, 1.00000000000000000\times 10^{10}, 1.000000000000000000\times 10^{10},$$
$$1.0000000000000000000\times 10^{10}, 1.00000000000000000000\times 10^{10}\}$$

```
In[149]:= {Precision /@ %, Accuracy /@ %}
```
$$Out[149]= \{\{0., 1., 2., 3., 4., 5., 6., 7., 8., 9., 10., 11., 12., 13.,$$
$$14., 15., 16., 17., 18., 19., 20.\}, \{-10., -9., -8., -7., -6., -5.,$$
$$-4., -3., -2., -1., 0., 1., 2., 3., 4., 5., 6., 7., 8., 9., 10.\}\}$$

In the next input, we try to generate numbers with negative precision. Such numbers are converted to zero precision numbers with an accuracy corresponding to the intended negative precision.

```
In[150]:= SetPrecision[#, -3]& /@ {10^-10, Pi, Pi, 10^10 Pi}
```
$$Out[150]= \{0.\times 10^{-7}, 0., 0., 0.\times 10^{13}\}$$

```
In[151]:= Precision /@ %
```
$$Out[151]= \{0., 0., 0., 0.\}$$

```
In[152]:= Accuracy /@ %%
```
$$Out[152]= \{7., -0.49715, -0.49715, -13.4971\}$$

The corresponding accuracy results from the above-mentioned identity *precision − accuracy = scale*.

```
In[153]:= Experimental`NumberScale /@ %%%
```
$$Out[153]= \{-7., 0.49715, 0.49715, 13.4971\}$$

Here is an example involving the above-mentioned added binary zeros. In binary, 0.2 is equal to $0.\overline{0011}$. (The command BaseForm[*number, base*] represents the number *number* in the number system with base *base*.)

```
In[154]:= BaseForm[0.20000000000000000000, 2]
```
Out[154]//BaseForm=
```
0.001100110011001100110011001100110011001100110011001100110011001100110011001100112
```

Thus, we have the following string of digits. (The nonzero digits start at $MachinePrecision + 1).

```
In[155]:= SetPrecision[0.2, 50]
```
$$Out[155]= 0.20000000000000001110223024625156540423631668090820$$

The number produced by *Mathematica* is not 0.200000000000000000000000000000000000000000000000000.... In base 2, this is the difference. We use RealDigits and BaseForm to display the bits.

```
In[156]:= {BaseForm[1/5 - %, 2], RealDigits[1/5 - %, 2]}
```

Out[156]= {-1.1001100110011001100110011001100110011001100110011001100110011001100110011001100110 ·
                01100110011001100110011001100110011001101₂ × 2⁻⁵⁷,
            {{1, 1, 0, 0, 1, 1, 0, 0, 1, 1, 0, 0, 1, 1, 0, 0, 1, 1, 0, 0, 1, 1, 0, 0, 1, 1, 0, 0, 1,
              1, 0, 0, 1, 1, 0, 0, 1, 1, 0, 0, 1, 1, 0, 0, 1, 1, 0, 0, 1, 1, 0, 0, 1, 1, 0, 0, 1, 1,
              0, 0, 1, 1, 0, 0, 1, 1, 0, 0, 1, 1, 0, 0, 1, 1, 0, 0, 1, 1, 0, 0, 1, 1, 0, 0, 1, 1,
              0, 0, 1, 1, 0, 0, 1, 1, 0, 0, 1, 1, 0, 0, 1, 1, 0, 0, 1, 1, 0, 0, 1, 1, 0, 1}, -56}}

Here, we make use of SetPrecision with an exact number as its argument.

In[157]:= **SetPrecision[1/5, 50] // FullForm**

Out[157]//FullForm=
        0.2`50.

The added binary zeros become visible only if all other bits of the number are already displayed. Here, this is visualized (for $i \geq 6$, the result does not change anymore). (Here, we use the lower precision 6 to avoid lengthy outputs.) With increasing precision, we see the validated digits, the guard digits, and the base 2 zeros.

In[158]:= **Table[SetPrecision[N[1/3, 6], 10 i], {i, 1, 7}] // TableForm**

Out[158]//TableForm=
        0.3333333333
        0.33333333333333333332
        0.333333333333333333315263297125
        0.3333333333333333333152632971252415927665
        0.33333333333333333331526329712524159276654245331883
        0.333333333333333333315263297125241592766542453318834304809570
        0.3333333333333333333152632971252415927665424533188343048095703125000000

The internal representation of N[1/3, 20] has slightly more than 20 correct digits.

In[159]:= **N[SetPrecision[N[1/3, 20], Infinity] - 1/3, 20]**

Out[159]= 5.6468863150286689271 × 10⁻²²

We now use SetPrecision to elaborate on a remark from above concerning the precision of functions of high-precision numbers. If we mix high-precision numbers with different precisions in a calculation (assuming no machine numbers are involved and no function that raises the precision is present), the number with the smallest precision will typically dominate the precision of the result.

In[160]:= **(SetPrecision[1, 20] + SetPrecision[2, 40])^3 // Precision**

Out[160]= 20.

In[161]:= **(SetPrecision[1, 2] + SetPrecision[2, 4])^3 // Precision**

Out[161]= 1.9914

When one of the numbers is a machine number (that is smaller than $MaxMachineNumber), the result is also a machine number, independently of whether the precision of the high-precision number is greater or less than $MachinePrecision.

In[162]:= **(1.0 + SetPrecision[2, 40])^3 // Precision**

Out[162]= MachinePrecision

In[163]:= **(1.0 + SetPrecision[2, 04])^3 // Precision**

Out[163]= MachinePrecision

If we use Infinity for the second variable of SetAccuracy or SetPrecision, the first argument will be converted to an exact expression. In view of the way in which approximate numbers are input, this does not always yield the "expected" result, because of the internal representation in base 2. (We search here for one of these numbers.)

In[164]:= (* find rational number such that numericalization and
              making and exact number are not idempotent *)
          For[i = 1, 1/i - SetPrecision[N[1/i], Infinity] === 0, i = i + 1, Null];
          {1/i, SetPrecision[N[1/i], Infinity]}

Out[166]= $\left\{\frac{1}{3}, \frac{6004799503160661}{18014398509481984}\right\}$

The two fractions are, however, almost equal.

In[167]:= **N[Subtract @@ %, 50]**

Out[167]= $1.8503717077085942340393861134847005208333333333333\times 10^{-17}$

In[168]:= **RealDigits[%, 2]**

Out[168]= {{1, 0, 1, 0, 1, 0, 1, 0, 1, 0, 1, 0, 1, 0, 1, 0, 1, 0, 1, 0, 1, 0, 1, 0, 1, 0, 1, 0,
           1, 0, 1, 0, 1, 0, 1, 0, 1, 0, 1, 0, 1, 0, 1, 0, 1, 0, 1, 0, 1, 0, 1, 0, 1, 0, 1, 0,
           1, 0, 1, 0, 1, 0, 1, 0, 1, 0, 1, 0, 1, 0, 1, 0, 1, 0, 1, 0, 1, 0, 1, 0, 1, 0, 1, 0,
           1, 0, 1, 0, 1, 0, 1, 0, 1, 0, 1, 0, 1, 0, 1, 0, 1, 0, 1, 0, 1, 0, 1, 0, 1, 0, 1, 0,
           1, 0, 1, 0, 1, 0, 1, 0, 1, 0, 1, 0, 1, 0, 1, 0, 1, 0, 1, 0, 1, 0, 1, 0, 1, 0, 1, 0, 1,
           0, 1, 0, 1, 0, 1, 0, 1, 0, 1, 0, 1, 0, 1, 0, 1, 0, 1, 0, 1, 0, 1, 0, 1, 1}, -55}

Next, we repeat the last search for high-precision numbers. We use twice the precision of machine numbers.

In[169]:= **For[i = 1, 1/i - SetPrecision[N[1/i, 2 $MachinePrecision], Infinity] === 0,**
              i = i + 1, Null];
          {1/i, SetPrecision[N[1/i, 2$MachinePrecision], Infinity]}

Out[170]= $\left\{\frac{1}{3}, \frac{108172851219475575594385340192085}{324518553658426726783156020576256}\right\}$

In[171]:= **N[Subtract @@ %, 50]**

Out[171]= $1.0271626370065257882965215693786279032203208790482\times 10^{-33}$

In[172]:= **RealDigits[%, 2]**

Out[172]= {{1, 0, 1, 0, 1, 0, 1, 0, 1, 0, 1, 0, 1, 0, 1, 0, 1, 0, 1, 0, 1, 0, 1, 0, 1, 0, 1, 0,
           1, 0, 1, 0, 1, 0, 1, 0, 1, 0, 1, 0, 1, 0, 1, 0, 1, 0, 1, 0, 1, 0, 1, 0, 1, 0, 1, 0,
           1, 0, 1, 0, 1, 0, 1, 0, 1, 0, 1, 0, 1, 0, 1, 0, 1, 0, 1, 0, 1, 0, 1, 0, 1, 0, 1, 0,
           1, 0, 1, 0, 1, 0, 1, 0, 1, 0, 1, 0, 1, 0, 1, 0, 1, 0, 1, 0, 1, 0, 1, 0, 1, 0, 1, 0,
           1, 0, 1, 0, 1, 0, 1, 0, 1, 0, 1, 0, 1, 0, 1, 0, 1, 0, 1, 0, 1, 0, 1, 0, 1, 0, 1,
           0, 1, 0, 1, 0, 1, 0, 1, 0, 1, 0, 1, 0, 1, 0, 1, 0, 1, 0, 1, 0, 1, 0, 1, 1}, -109}

We now examine graphically the errors of several fractions generated by SetPrecision[N[i/j], Infinity].

In[173]:= **ListPlot[Transpose[{#,** (* make numerical and after that again exact *)
              10^MachinePrecision N[SetPrecision[N[#], Infinity]& /@ # - #]}],
                  (* setting all options for the plot *)
              PlotStyle -> {PointSize[0.01]}, AxesLabel -> {"n", None},
              PlotLabel -> StyleForm[
                      "10^MachinePrecision N[SetPrecision[N[n], Infinity]",
                      FontFamily ->"Courier", FontWeight -> "Bold",
                      FontSize -> 6], GridLines -> {None, {0}},
                      PlotRange -> All, AxesOrigin -> {0, -1/2},
                      AspectRatio -> 1/4,
                      (* a lot of fractions as ticks *)
                      Ticks -> {{#, StyleForm[#, FontSize -> 5]}& /@ #,
                          Automatic}]&[
                              (* the fractions to be investigated *)
                          Union[Flatten[Table[j/i, {i, 16}, {j, i}]]]]]

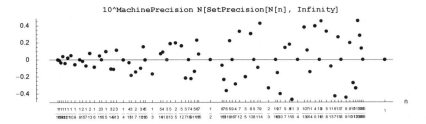

The precision of variables (head `Symbol`) and of 0 (head `Integer`) cannot be altered with `SetPrecision`.

```
In[174]:= {#, Precision[#]}& @ SetPrecision[0, 12]
Out[174]= {0, ∞}
```

```
In[175]:= {#, Accuracy[#]}& @ SetAccuracy[abcxyz, 4]
Out[175]= {abcxyz, ∞}
```

We now present an illustration of the idea of "a few more digits than required". In this example, there are more than thirty 3's in `SetPrecision[SetPrecision[a, 30], 50]`.

```
In[176]:= a = 1/3;

       a30 = SetPrecision[a, 30];
       {N[a, 50], SetPrecision[a30, 50]}
Out[178]= {0.33333333333333333333333333333333333333333333333333,
       0.33333333333333333333333333333333333333333333333311}
```

```
In[179]:= Position[RealDigits[%[[2]]][[1]], _?(# =!= 3&)][[2]]
Out[179]= {49}
```

Analogously, if we reduce the precision, we do not lose so many digits. Indeed, `a10` still has more than ten 3's.

```
In[180]:= {N[a, 50], a10 = SetPrecision[a30, 10]; SetPrecision[a10, 50]}
Out[180]= {0.33333333333333333333333333333333333333333333333333,
       0.33333333333333333333333329126075172154603704471}
```

We can get knowledge of all (base 2) digits (the explicitly shown ones and the hidden ones) of a number in the form of their binary representation by using the undocumented function `$NumberBits`.

`$NumberBits[`*number*`]`

   gives a list of the form {*sign*, *listOfCorrectBits*, *listOfGuardBits*, *base2Exponent*} of the approximate real number *approximateRealNumber*.

We put the context that contains the symbol `$NumberBits` on the context path.

```
In[181]:= ??*`*NumberBits

              NumericalMath`$NumberBits

              Attributes[NumericalMath`$NumberBits] = {Listable, Protected}
```

```
In[182]:= (* in case the function "hides" in some context other than System` *)
       If[FreeQ[$ContextPath, #], AppendTo[$ContextPath, #]]& /@ Context /@
              Union[Join[Names["*`*Number*Bits*"], Names["*`*Bits*Number*"]]]
Out[183]= {{Global`, System`, NumericalMath`}}
```

Let us look at some examples. For a machine number, no precision analysis is done, and so the second list containing the correct digits is an empty list.

In[184]:= **$NumberBits[-12.34]**

Out[184]= {-1, {}, {1, 1, 0, 0, 0, 1, 0, 1, 0, 1, 1, 1, 0, 0, 0, 0, 1, 0, 1, 0, 0, 0, 1, 1, 1, 1, 0, 1, 0, 1, 1, 1, 0, 0, 0, 0, 1, 0, 1, 0, 0, 0, 1, 1, 1, 1, 0, 1, 0, 1, 1, 1, 0}, 4}

Here are two high-precision numbers. Now, the first list is nonempty, and the third list shows the guard digits in binary form, which are hidden in the ordinary output form.

In[185]:= **$NumberBits[SetPrecision[1/3, 4]]**

Out[185]= {1, {1, 0, 1, 0, 1, 0, 1, 0, 1, 0, 1, 0, 1},
{0, 1, 0, 1, 0, 1, 0, 1, 0, 1, 0, 1, 0, 1, 0, 1, 0, 1, 0, 1, 0, 1, 0, 1, 0, 1,
0, 1, 0, 1, 0, 1, 0, 1, 0, 1, 0, 1, 0, 1, 0, 1, 0, 1, 0, 1, 0, 1}, -1}

In[186]:= **$NumberBits[SetPrecision[1/5, 40]]**

Out[186]= {1, {1, 1, 0, 0, 1, 1, 0, 0, 1, 1, 0, 0, 1, 1, 0, 0, 1, 1, 0, 0, 1, 1, 0, 0, 1, 1, 0,
0, 1, 1, 0, 0, 1, 1, 0, 0, 1, 1, 0, 0, 1, 1, 0, 0, 1, 1, 0, 0, 1, 1,
0, 0, 1, 1, 0, 0, 1, 1, 0, 0, 1, 1, 0, 0, 1, 1, 0, 0, 1, 1, 0, 0, 1,
1, 0, 0, 1, 1, 0, 0, 1, 1, 0, 0, 1, 1, 0, 0, 1, 1, 0, 0, 1, 1, 0,
0, 1, 1, 0, 0, 1, 1, 0, 0, 1, 1, 0, 0, 1, 1, 0, 0, 1, 1, 0, 0, 1},
{1, 0, 0, 1, 1, 0, 0, 1, 1, 0, 0, 1, 1, 0, 0, 1, 1, 0, 0, 1, 1, 0, 0, 1, 1,
0, 0, 1, 1, 0, 0, 1, 1, 0, 0, 1, 1, 0, 0, 1, 1, 0, 0, 1, 1, 0, 0, 1, 1}, -2}

The next input shows that for 1 with precision 10 one has to add about 2.3×10^{-10} to change the last significant bit of this number.

In[187]:= **Block[{one = N[1, 10], k = 0},**
     (* add small fraction until last significant bit changes *)
     While[$NumberBits[one][[2]] ===
          $NumberBits[one + k 10^-14][[2]], k++];
     k]

Out[187]= 23284

Let us now look graphically at the number of correct and guard digits stored in the number generated by SetPrecision[1/3, *i*].

In[188]:= **Show[GraphicsArray[**
     MapThread[ListPlot[#2, PlotLabel -> #1, AxesLabel -> {"i", None},
                    DisplayFunction -> Identity, PlotRange -> All,
                    AxesOrigin -> {0, 0}, PlotRange -> All]&,
     (* labels *)
     {{"number of correct bits", "number of guard bits",
       "number of all bits"},
     (* the data to visualize *)
     {Length /@ #[[2]], Length /@ #[[3]],
       Length /@ #[[2]] + Length /@ #[[3]]}&[
       Transpose[Table[$NumberBits[SetPrecision[1/3, i]], {i, 100}]]]}]]]]

Because the result of $NumberBits contains all of the binary information on the number, we can easily reconstruct the corresponding number in the decimal system. Here, this is implemented (the result of toExact⁚ NumberGuaranteed and toExactNumberAll are typically fractions and sometimes integers).

```
In[189]:= toExactNumberGuaranteed[l_List] := (* use only guaranteed bits *)
          First[l] 2^Last[l] l[[2]].Table[2^-i, {i, Length[l[[2]]]}]
```

```
In[190]:= toExactNumberAll[l_List] := (* use all bits *)
          First[l] 2^Last[l] Join[l[[2]], l[[3]]].
                         Table[2^-i, {i, Length[l[[2]]] + Length[l[[3]]]}]
```

```
In[191]:= (* for convenience, accept also a real numbers as argument *)
          toExactNumberGuaranteed[z_Real?(Not[MachineNumberQ[#]]&)] :=
                 toExactNumberGuaranteed[$NumberBits[z]]

          toExactNumberAll[z_Real?(Not[MachineNumberQ[#]]&)] :=
                 toExactNumberAll[$NumberBits[z]]
```

In the following example, we use only the correct bits of SetPrecision[1/3, i]. The resulting number has i 3's.

```
In[194]:= Table[toExactNumberGuaranteed[$NumberBits[SetPrecision[1/3, i]]], {i, 2, 16}]
```

$$Out[194]= \left\{ \frac{85}{256}, \frac{341}{1024}, \frac{5461}{16384}, \frac{87381}{262144}, \frac{349525}{1048576}, \frac{5592405}{16777216}, \frac{89478485}{268435456}, \right.$$
$$\frac{357913941}{1073741824}, \frac{5726623061}{17179869184}, \frac{91625968981}{274877906944}, \frac{366503875925}{1099511627776},$$
$$\frac{5864062014805}{17592186044416}, \frac{93824992236885}{281474976710656}, \frac{375299968947541}{1125899906842624}, \left. \frac{6004799503160661}{18014398509481984} \right\}$$

```
In[195]:= N[%, 60]
```

Out[195]= {0.332031250000000000000000000000000000000000000000000000000000`
  0.333007812500000000000000000000000000000000000000000000000000`
  0.333312988281250000000000000000000000000000000000000000000000`
  0.333332061767578125000000000000000000000000000000000000000000`
  0.333333015441894531250000000000000000000000000000000000000000`
  0.333333313465118408203125000000000000000000000000000000000000`
  0.333333332091569900512695312500000000000000000000000000000000`
  0.333333333302289247512817382812500000000000000000000000000000`
  0.333333333331393077969551086425781250000000000000000000000000`
  0.333333333332120673730969429016113281250000000000000000000000`
  0.333333333333303016843274235725402832031250000000000000000000`
  0.333333333333314385527046397328376770019531250000000000000000`
  0.333333333333333214909544039983302354812622070312500000000000`
  0.333333333333333303727386009995825588703155175578125000000000`
  0.333333333333333331482961625624739099293947219848632812500000`}

If we use all stored bits, we see that, in fact, groups of numbers are equal because the guard digits get added in blocks (this feature was already visible from the above pictures for the i used here).

```
In[196]:= Table[toExactNumberAll[$NumberBits[SetPrecision[1/3, i]]], {i, 2, 16}]
```

$$Out[196]= \left\{ \frac{6148914691236517205}{18446744073709551616}, \frac{6148914691236517205}{18446744073709551616}, \frac{6148914691236517205}{18446744073709551616}, \right.$$
$$\frac{6148914691236517205}{18446744073709551616}, \frac{6148914691236517205}{18446744073709551616}, \frac{6148914691236517205}{18446744073709551616},$$
$$\frac{6148914691236517205}{18446744073709551616}, \frac{6148914691236517205}{18446744073709551616}, \frac{26409387504754779197847983445}{79228162514264337593543950336},$$

$$\frac{26409387504754779197847983445}{79228162514264337593543950336}, \quad \frac{26409387504754779197847983445}{79228162514264337593543950336},$$
$$\frac{26409387504754779197847983445}{79228162514264337593543950336}, \quad \frac{26409387504754779197847983445}{79228162514264337593543950336},$$
$$\left. \frac{26409387504754779197847983445}{79228162514264337593543950336}, \quad \frac{26409387504754779197847983445}{79228162514264337593543950336} \right\}$$

In[197]:= **N[Union[%], 60]**

Out[197]= {0.33333333333333333333152632971252415927665424533188343048095`70,
   0.33333333333333333333333333333333329126075172154603704470780985158}

Because most rational numbers are not exactly representable as a base 2 floating point number, we do not get 1/7 back in the following example [784].

In[198]:= **{toExactNumberGuaranteed[#], toExactNumberAll[#]}&[**
            $NumberBits[N[1/7, 20]]]

Out[198]= $\left\{ \frac{10540996613548315209}{73786976294838206464}, \quad \frac{48611766702991209066196372490252601637}{340282366920938463463374607431768211456} \right\}$

Let us come again back to the iteration NestList[(#^2-1)/(#-1)-1&, N[2, 200], 200] used above. Here are pictures of the number of correct and the number of guard digits.

In[199]:= **manyTwos = NestList[(#^2 - 1)/(# - 1) - 1&, N[2, 200], 200];**

In[200]:= **With[{tr = Transpose[$NumberBits /@ manyTwos]},**
        Show[GraphicsArray[{
          (* number of significant bits *)
          ListPlot[Length /@ tr[[2]], DisplayFunction -> Identity,
                PlotLabel -> "number of significant bits"],
          (* number of guard bits *)
          ListPlot[Length /@ tr[[3]], PlotJoined -> True, PlotRange -> All,
                AxesOrigin -> {0, 0}, PlotLabel -> "number of guard bits",
                DisplayFunction -> Identity]}]]]

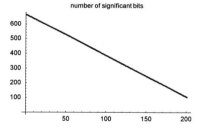
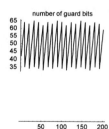

Although the number of correct digits continuously decreases, the number itself (generated using toExactNumberAll) is actually the same every time (but *Mathematica* cannot guarantee that the guard digits are correct).

In[201]:= **toExactNumberAll[$NumberBits[#]]& /@ manyTwos**

Out[201]= {2, 2, 2, 2, 2, 2, 2, 2, 2, 2, 2, 2, 2, 2, 2, 2, 2, 2, 2, 2, 2, 2, 2, 2, 2, 2, 2, 2,
   2, 2, 2, 2, 2, 2, 2, 2, 2, 2, 2, 2, 2, 2, 2, 2, 2, 2, 2, 2, 2, 2, 2, 2, 2, 2, 2, 2,
   2, 2, 2, 2, 2, 2, 2, 2, 2, 2, 2, 2, 2, 2, 2, 2, 2, 2, 2, 2, 2, 2, 2, 2, 2, 2, 2, 2,
   2, 2, 2, 2, 2, 2, 2, 2, 2, 2, 2, 2, 2, 2, 2, 2, 2, 2, 2, 2, 2, 2, 2, 2, 2, 2, 2, 2,
   2, 2, 2, 2, 2, 2, 2, 2, 2, 2, 2, 2, 2, 2, 2, 2, 2, 2, 2, 2, 2, 2, 2, 2, 2, 2, 2,
   2, 2, 2, 2, 2, 2, 2, 2, 2, 2, 2, 2, 2, 2, 2, 2, 2, 2, 2, 2, 2, 2, 2, 2, 2, 2,
   2, 2, 2, 2, 2, 2, 2, 2, 2, 2, 2, 2, 2, 2, 2, 2, 2, 2, 2, 2, 2, 2, 2, 2, 2, 2}

Let us use toExactNumberAll to investigate what happens when one raises the precision of a number. We start with the machine number 1./3.

In[202]:= **ξ = 1./3;**

In[203]:= **FullForm[ξ]**
Out[203]//FullForm=
         0.3333333333333333`

In[204]:= **$NumberBits[ξ]**
Out[204]= {1, {}, {1, 0, 1, 0, 1, 0, 1, 0, 1, 0, 1, 0, 1, 0, 1, 0, 1, 0, 1, 0, 1, 0, 1, 0, 1, 0, 1,
         0, 1, 0, 1, 0, 1, 0, 1, 0, 1, 0, 1, 0, 1, 0, 1, 0, 1, 0, 1, 0, 1, 0, 1}, -1}

Raising the precision of the last number shows added zeros in base 2.

In[205]:= **ξP = SetPrecision[ξ, 70]**
Out[205]= 0.333333333333333331482961625624739099293947219848632812500000000000000000

In[206]:= **$NumberBits[ξP]**
Out[206]= {1, {1, 0, 1, 0, 1, 0, 1, 0, 1, 0, 1, 0, 1, 0, 1, 0, 1, 0, 1, 0, 1, 0, 1, 0, 1, 0,
         1, 0, 1, 0, 1, 0, 1, 0, 1, 0, 1, 0, 1, 0, 1, 0, 1, 0, 1, 0, 1, 0, 1, 0, 1, 0, 1, 0,
         1, 0, 0, 0, 0, 0, 0, 0, 0, 0, 0, 0, 0, 0, 0, 0, 0, 0, 0, 0, 0, 0, 0, 0, 0, 0, 0, 0,
         0, 0, 0, 0, 0, 0, 0, 0, 0, 0, 0, 0, 0, 0, 0, 0, 0, 0, 0, 0, 0, 0, 0, 0, 0, 0, 0, 0,
         0, 0, 0, 0, 0, 0, 0, 0, 0, 0, 0, 0, 0, 0, 0, 0, 0, 0, 0, 0, 0, 0, 0, 0, 0, 0, 0, 0,
         0, 0, 0, 0, 0, 0, 0, 0, 0, 0, 0, 0, 0, 0, 0, 0, 0, 0, 0, 0, 0, 0, 0, 0, 0, 0, 0, 0,
         0, 0, 0, 0, 0, 0, 0, 0, 0, 0, 0, 0, 0, 0, 0, 0, 0, 0, 0, 0, 0, 0, 0, 0, 0, 0, 0, 0,
         0, 0, 0, 0, 0, 0, 0, 0, 0, 0, 0, 0, 0, 0, 0, 0, 0, 0, 0, 0, 0, 0, 0, 0, 0, 0, 0, 0,
         0, 0, 0, 0, 0, 0, 0, 0, 0, 0, 0, 0, 0, 0, 0, 0, 0, 0, 0, 0, 0, 0, 0, 0, 0, 0},
         {0, 0, 0, 0, 0, 0, 0, 0, 0, 0, 0, 0, 0, 0, 0, 0, 0, 0, 0, 0, 0, 0, 0, 0, 0, 0, 0,
         0, 0, 0, 0, 0, 0, 0, 0, 0, 0, 0, 0, 0, 0, 0, 0, 0, 0, 0, 0, 0, 0, 0, 0, 0, 0}, -1}

Here are the added zeros counted.

In[207]:= **{Union[#], Length[#]}& @**
               Drop[$NumberBits[ξP][[2]], Length[$NumberBits[ξ][[3]]]]
Out[207]= {{0}, 180}

The additional digits displayed in ξP arise from the precision raising of the value of ξ.

In[208]:= **toExactNumberAll[$NumberBits[ξ]] // N[#, 70]&**
Out[208]= 0.333333333333333331482961625624739099293947219848632812500000000000000000

Be aware that $NumberBits returns the full internal binary representation of a real floating-point number. RealDigits returns the base 2 representation of the correct digits of a number. This means that $Number⸱ Bits[*number*][[2]] and RealDigits[*number*, 2][[1]] might be slightly different (remember that the precision of a number is an general not in integer in base 10 and/or base 2). Here is such a case.

In[209]:= **With[{pi = N[Pi, $MachinePrecision + 1]},**
          {$NumberBits[pi][[2]], RealDigits[pi, 2][[1]]}]
Out[209]= {{1, 1, 0, 0, 1, 0, 0, 1, 0, 0, 0, 0, 1, 1, 1, 1, 1, 1, 0, 1, 1, 0, 1, 0, 1, 0, 1, 0, 0, 0,
         1, 0, 0, 0, 1, 0, 0, 0, 0, 1, 0, 1, 1, 0, 1, 0, 0, 0, 1, 1, 0, 0, 0, 0, 1, 0},
         {1, 1, 0, 0, 1, 0, 0, 1, 0, 0, 0, 0, 1, 1, 1, 1, 1, 1, 0, 1, 1, 0, 1, 0, 1, 0, 1, 0,
         0, 0, 1, 0, 0, 0, 1, 0, 0, 0, 0, 1, 0, 1, 1, 1, 0, 1, 0, 0, 0, 1, 1, 0, 0, 0, 0, 1, 0}}

The following graphic shows the number of certified bits of SetPrecision[*x, prec*] as a function of *prec*. We use four different exact *x*: the integer 2, and the three irrational numbers $2^{1/2}$, π, and cos(1). The red curve is the nearest integer to *prec* / $\log_{10} 2$.

In[210]:= **Show[GraphicsArray[**
          Plot[(* estimated rounded precision in red *)

```
{Round[prec/Log[10, 2]],
 Length[$NumberBits[SetPrecision[#, prec]][[2]]]},
{prec, $MachinePrecision - 2, $MachinePrecision + 2},
Compiled -> False, Frame -> True, Axes -> False,
PlotLabel -> StyleForm[#, FontFamily -> "Courier"],
PlotPoints -> 200, DisplayFunction -> Identity,
PlotStyle -> {{Hue[0], Thickness[0.015]}, {GrayLevel[0]}}]& /@
            (* four numbers *) {2, Sqrt[2], Pi, Cos[1]}]]
```

We now look at a small application of high-precision arithmetic in the calculation of π. It is well known that $\sum_{k=0}^{\infty}(-1)^k/(2k+1)=\pi/4$, but the convergence is very slow. To check the speed of convergence, we find the sum of the first 10 terms in the series, maintaining 1000-digit precision, and compare the result with the one above to see how many digits are correct. The resulting list gives roughly the number of correct digits of the ith partial sum.

In[211]:= `arcTanSeriesTerms[x_, o_] := Table[(-1)^k/(2k + 1) x^(2k + 1), {k, 0, o}]`

In[212]:= `series1 = Table[N[4 (-1)^k/(2k + 1), 1000], {k, 0, 10}];`

In[213]:= `(* for comparison purposes *)`
`pi = N[Pi, 1000];`

The function `correctPartialSumDigits` gives a list of the number of correct digits of the partial sums.

In[215]:= `correctPartialSumDigits[series_, exactValue_:Pi] :=`
`    Abs[Log[10, (* difference to Pi/4 *) Abs[N[exactValue - #, 4]]]]& /@`
`        (* form partial sums *) Drop[FoldList[Plus, 0, series], 1]`

In[216]:= `correctPartialSumDigits[series1]`

Out[216]= `{0.0663, 0.3234, 0.4880, 0.6084, 0.7031,`
`    0.7811, 0.8473, 0.9048, 0.9556, 1.0011, 1.0423}`

The series expansion of arctan itself is $\sum_{k=0}^{\infty}(-1)^k/(2k+1)\,x^{2k+1}=\arctan(x)$. Using the alternative series

$$\arctan(x) = \frac{1}{x}\sum_{k=0}^{o}\frac{k!^2\,4^k}{(2k+1)!}\left(1+\frac{1}{x^2}\right)^{k+1}$$

we obtain a much faster converging sequence.

In[217]:= `arcTanSeriesTermsFast[x_, o_] :=`
`        Table[(k!)^2 4^k/(2k + 1)!(1/x^2 + 1)^(k + 1), {k, 0, o}]/x`

In[218]:= `series1Fast = arcTanSeriesTermsFast[N[1, 100], 10];`
`    correctPartialSumDigits[series1Fast]`

Out[219]= `{0.0575, 0.1833, 0.7628, 1.1175, 1.4168,`
`    1.6968, 1.9703, 2.2427, 2.5162, 2.7916, 3.0690}`

Most of the classic series representations of π are based on series expansions of arctan around 0. For example, we have:

$$16 \arctan(1/5) - 4 \arctan(1/239) = \pi$$
$$4 \arctan(1/2) + 4 \arctan(1/5) + 4 \arctan(1/8) = \pi.$$

(For many more of this type of formulae, see [69], [1122], [1521], [1475], [354], [355], and [1770]. For the construction of such type identities, see [1641].)

The next series converges considerably faster than does the above one.

$$\pi = 88 \arctan\left(\frac{1}{28}\right) + 8 \arctan\left(\frac{1}{443}\right) - 20 \arctan\left(\frac{1}{1393}\right) - 40 \arctan\left(\frac{1}{11018}\right)$$

```
In[220]:= series2 = N[88 arcTanSeriesTerms[1/28, 10] +
                8 arcTanSeriesTerms[1/443, 10] -
                20 arcTanSeriesTerms[1/1393, 10] -
                40 arcTanSeriesTerms[1/11018, 10], 1000];
```

```
          correctPartialSumDigits[series2]
Out[222]= {2.8744, 5.9907, 9.0312, 12.0346, 15.0161,
           17.9830, 20.9395, 23.8881, 26.8308, 29.7686, 32.7024}
```

Using the alternative expansion for arctan for the last linear combination gives us about 94 correct digits for the first 10 terms.

```
In[223]:= series2Fast = N[88 arcTanSeriesTermsFast[1/28, 10] +
                  8 arcTanSeriesTermsFast[1/443, 10] -
                  20 arcTanSeriesTermsFast[1/1393, 10] -
                  40 arcTanSeriesTermsFast[1/11018, 10], 1000];
```

```
          correctPartialSumDigits[series2Fast]
Out[225]= {13.7288, 21.6365, 29.6238, 37.6410, 45.6741,
           53.7169, 61.7663, 69.8206, 77.8785, 85.9392, 94.0022}
```

In 1914, Ramanujan (see [246], [367] and [250]) gave the following series, which converges very fast.

$$\frac{2^{3/2}}{9801} \sum_{k=0}^{\infty} (4n)! \, \frac{1103 + 26390\,k}{4^{4k}\,k!^4\,99^{4k}} = \frac{1}{\pi}$$

```
In[226]:= series3 = N[2 Sqrt[2]/9801 *
            Table[(4k)!/4^(4k)/k!^4 * (1103 + 26390k)/99^(4k), {k, 0, 10}], 1000];
```

```
          correctPartialSumDigits[1/series3]
Out[228]= {7.1168, 8.1111, 16.1884, 24.2398, 32.2751,
           40.3002, 48.3184, 56.3316, 64.3411, 72.3476, 80.3517}
```

```
In[229]:= correctPartialSumDigits[series2]
Out[229]= {2.8744, 5.9907, 9.0312, 12.0346, 15.0161,
           17.9830, 20.9395, 23.8881, 26.8308, 29.7686, 32.7024}
```

This gives us approximately eight additional correct digits for each term include! A still better series was found by D. V. Chudnovsky and G. V. Chudnovsky [247]:

$$12 \sum_{k=0}^{\infty} (-1)^k \, (6k)! \, \frac{13591409 + 545140134\,k}{k!^3\,(3k)!\,(262537412640768000)^{k+1/2}} = \frac{1}{\pi}$$

Here, each term increases the number of correct digits by approximately 14.

```
In[230]:= series4 = 12 Table[N[(-1)^k (6k)! (13591409 + k 545140134)/(k!^3 (3k)! *
                      (640320^3)^(k + 1/2)), 1000], {k, 0, 10}];

        correctPartialSumDigits[1/series4]
Out[232]= {13.2289, 14.2232, 28.5060, 42.7586, 56.9942,
          71.2191, 85.4369, 99.6496, 113.8584, 128.0641, 142.2675}
```

This last formula is also used inside *Mathematica* to calculate π to arbitrary digits. For more on efficient series expansions of π, see [403], [247], [248], [182], [1831], [245], [1500], and http://www.cecm.sfu.ca/pi/. For an interesting method to compute the digits of π digit by digit, see [1866]. For a billiard-based method to calculate π, see [654]. For the relativistic invariance of π (when defined by the circumference and radius of a disk, see [1348]). If one needs a large number of digits of π, N[Pi, 10^6] calculates in about 12 seconds on a 2 GHz computer.

```
In[233]:= N[Pi, 10^6] // Precision // Timing
Out[233]= {9.46 Second, 1. × 10^6}
```

Now, we have finished discussing high-precision numbers themselves. We continue to discuss some functions that handle high-precision numbers in a special way. We illustrate the use of the option SameTest for Equal, Union, and Sort, which can be used to compare two numbers that differ from each other by less than their respective accuracies. We start by demonstrating the "obvious" behavior. The following two numbers are neither equal (in the *Mathematica* sense of Equal) nor identical (in the *Mathematica* sense of SameQ).

```
In[234]:= Union[N[{2, 2 + 10^-35}, 40], SameTest -> SameQ]
Out[234]= {2.000000000000000000000000000000000000000,
          2.000000000000000000000000000000000010000}
```

```
In[235]:= Union[N[{2, 2 + 10^-35}, 40], SameTest -> Equal]
Out[235]= {2.000000000000000000000000000000000000000,
          2.000000000000000000000000000000000010000}
```

The length two output no longer happens when we use our own (less restrictive) test in the following example.

```
In[236]:= Union[N[{2, 2 + 10^-35}, 40], SameTest -> (Abs[#1 - #2] < 10^-34&)]
Out[236]= {2.000000000000000000000000000000000000000}
```

In the next input, the two numbers are identified by Union with the option setting Equal.

```
In[237]:= Union[N[{2, 2 + (* smaller *) 10^-38}, 40], SameTest -> Equal]
Out[237]= {2.000000000000000000000000000000000000000}
```

Here we compare some more numbers.

```
In[238]:= Union[Table[#[3, SetPrecision[3, i]], {i, 1000}]]& /@ {Equal, SameQ}
Out[238]= {{True}, {False}}
```

```
In[239]:= Union[Table[#[3, SetPrecision[3., i]], {i, 1000}]]& /@ {Equal, SameQ}
Out[239]= {{True}, {False}}
```

```
In[240]:= Union[Table[#[3., SetPrecision[3., i]], {i, 1000}]]& /@ {Equal, SameQ}
Out[240]= {{True}, {True}}
```

As we see, SameQ takes into account the type of a number. An integer is never the same as an approximative number.

The a's and b's in the following table are equal (in the sense of Equal), because they differ from each other by less than the precision of a.

```
In[241]:= Block[{(* make it local here *) $MinPrecision, a, b},
    Table[{$MinPrecision = i - 1,
          a = SetPrecision[2, i]; b = SetPrecision[2 + 10^(-i - 6), i];
          a == b, a - b}, {i, 5, 15}] //
         TableForm[#, TableSpacing -> {0, 3}, TableHeadings -> {{},
          StyleForm[#, FontWeight -> "Bold"]& /@ {"prec", "a == b","a - b"}}]&]
```

Out[241]//TableForm=

prec	a == b	a - b
4	True	$0. \times 10^{-5}$
5	True	$0. \times 10^{-6}$
6	True	$0. \times 10^{-7}$
7	True	$0. \times 10^{-8}$
8	True	$0. \times 10^{-9}$
9	True	$0. \times 10^{-10}$
10	True	$0. \times 10^{-11}$
11	True	$0. \times 10^{-12}$
12	True	$0. \times 10^{-13}$
13	True	$0. \times 10^{-14}$
14	True	$0. \times 10^{-15}$

Comparison of high-precision numbers is set up in such a way that (despite the potential uncertainty contained in high-precision numbers) two high-precision numbers are never equal and not equal (in the sense of Less and Greater) at the same time. Here is an example demonstrating this property.

```
In[242]:= ξ = SetPrecision[10, 1];  η = SetPrecision[11, 1];
    {ξ === η, ξ == η, ξ <= η, ξ < η, ξ >= η, ξ > η, ξ != η, ξ =!= η}
Out[243]= {True, True, True, False, True, False, False, False}
```

The behavior of Equal and SameQ with respect to floating point numbers can be influenced by the two functions Experimental`$EqualTolerance and Experimental`$SameQTolerance. They represent the number of digits such that two numbers are considered to be equal or the same. The current values are as follows.

```
In[244]:= {Experimental`$EqualTolerance, Experimental`$SameQTolerance}
Out[244]= {2.10721, 0.30103}
```

In binary, this means that, by default, Equal ignores the last seven bits and SameQ ignores the last bit.

```
In[245]:= Log[2, 10^%]
Out[245]= {7., 1.}
```

The values of Experimental`$EqualTolerance and Experimental`$SameQTolerance can be changed. In the following setting, all binary digits are ignored. As a result, the number 1. becomes equal to the number 2..

```
In[246]:= Block[{Experimental`$EqualTolerance = Length[$NumberBits[1.][[3]]]}, 1. == 2.]
Out[246]= True
```

Setting Experimental`$EqualTolerance to zero yields strict comparisons.

```
In[247]:= Block[{Experimental`$EqualTolerance = 0}, 1`30 == 1`30 + 10^-29]
Out[247]= False
```

Using a negative value for Experimental`$EqualTolerance allows even to take guard digits into account.

```
In[248]:= Table[{k, Block[{Experimental`$EqualTolerance = -3}, 1`30 == 1`30 + 10^-k]},
            {k, 25, 35}]
```
```
Out[248]= {{25, False}, {26, False}, {27, False}, {28, False}, {29, False},
          {30, True}, {31, True}, {32, True}, {33, True}, {34, True}, {35, True}}
```

Frequently, it is useful to have a way to substitute 0 for numbers whose absolute values are smaller than some given bound. This can be accomplished using Chop.

Chop[*expression, bound*]

> substitutes 0 for all numbers occurring in *expression* whose absolute values are smaller than *bound*. If the second variable is not present, the default value is 10^{-10}.

Here are some examples.

```
In[249]:= Chop[{10.0^-3, 10.0^-21, 10.0^-22, 10.0^-23,
             0.000000000000000000000099999999999999999999,
             0.000000000000000000001, 0.000000000000000000002,
             10^-28, N[1 + 10^-55, 100], -10.0^-34, 1.001 10^-22}, 10^-22]
```
$$Out[249]= \Big\{0.001, \ 1.\times10^{-21}, \ 1.\times10^{-22}, \ 0, \ 9.9999999999999999999\times10^{-22},$$
$$1.\times10^{-22}, \ 2.\times10^{-22}, \ \frac{1}{1000000000000000000000000000000},$$
$$1.00000000000000000000000000000000000000000000000000000010000000000000000000\cdot$$
$$000000000000000000000000, \ 0, \ 1.001\times10^{-22}\Big\}$$

For complex numbers, Chop works with the real and imaginary parts, rather than with the absolute value of the number.

```
In[250]:= Chop[{10^-12, 10^-12 I, 10^-12 + 10^-12 I,
             10^-12 - 10^-12 I, 10^-12 + 10^-11 I,
             10^-11 - 10^-12 I, 1.001 10^-11 - 10^-12 I} // N, 1.2 10^-12]
```
$$Out[250]= \{0, \ 0, \ 0, \ 0, \ 1.\times10^{-11} \ i, \ 1.\times10^{-11}, \ 1.001\times10^{-11}\}$$

When carrying out approximative calculations using machine arithmetic, it might happen that a number smaller than $MinMachineNumber, or a number larger than $MaxMachineNumber, will be encountered. In such situations, *Mathematica* will automatically switch to its high-precision arithmetic. Here is such a case.

```
In[251]:= smallMachineInteger = $MinMachineNumber
```
$$Out[251]= 2.22507\times10^{-308}$$

```
In[252]:= MachineNumberQ[%]
```
```
Out[252]= True
```

```
In[253]:= $MinMachineNumber^2
```
$$Out[253]= 4.95095367581213\times10^{-616}$$

```
In[254]:= {MachineNumberQ[%], Precision[%]}
```
```
Out[254]= {False, 15.6536}
```

Taking the square root of the above $MinMachineNumber^2 does not give the original $MinMachineNumber. $MinMachineNumber^2 is a high-precision number and so will be its square root.

```
In[255]:= Sqrt[%%]
```
$$Out[255]= 2.225073858507201\times10^{-308}$$

Numbers might be so big, that they cannot even be represented as high-precision numbers. All numbers larger than $MaxNumber yield Overflow[].

In[256]:= **$MaxNumber^2**

        General::ovfl : Overflow occurred in computation. More…

Out[256]= Overflow[]

All numbers smaller than $MinNumber are represented as Underflow[].

In[257]:= **1/%**

Out[257]= Underflow[]

Here is an exact input that results in Overflow[].

In[258]:= **10^10^10**

        General::ovfl : Overflow occurred in computation. More…

Out[258]= Overflow[]

Because no (complex) sign information is contained in Overflow[], the following gives the result Indeterminate.

In[259]:= **Exp[-%]**

Out[259]= Indeterminate

The automatic switch to high-precision numbers in case of the occurrence of numbers that are too small or too large to be represented as machine numbers is in many situations very useful. But from time to time, this is not what one wants (because either using the high-precision arithmetic is just too expensive or because for all practical purposes, such a small number can be considered as 0 anyway). If one does not want the switch to happen, one can use the system option "CatchMachineUnderflow" to tell *Mathematica* that any too small quantity that develops from a machine number calculation is to be set to 0.0. Here is a simple example. In the following iteration, in the first few steps, the machine arithmetic is used. Then, the high-precision arithmetic is used, and finally, the numbers become smaller than $MinNumber and Underflow[] is produced.

In[260]:= **Developer`SetSystemOptions["CatchMachineUnderflow" -> True];**
      FixedPointList[#^8&, 0.1, 50]

        General::unfl : Underflow occurred in computation. More…

Out[261]= $\{0.1, 1. \times 10^{-8}, 1. \times 10^{-64}, 1.000000000000055 \times 10^{-512}, 1.00000000000044 \times 10^{-4096},$
    $1.0000000000035 \times 10^{-32768}, 1.000000000028 \times 10^{-262144}, 1.000000000225 \times 10^{-2097152},$
    $1.00000000180 \times 10^{-16777216}, 1.0000000144 \times 10^{-134217728}, Underflow[], Underflow[]\}$

Setting the system option "CatchMachineUnderflow" to False results in a direct switch from the machine numbers to 0.0.

In[262]:= **Developer`SetSystemOptions["CatchMachineUnderflow" -> False];**
      FixedPointList[#^4&, 0.1, 50]

Out[263]= $\{0.1, 0.0001, 1. \times 10^{-16}, 1. \times 10^{-64}, 1. \times 10^{-256}, 0., 0.\}$

High-precision arithmetic is not influenced by the setting the system option "CatchMachineUnderflow".

In[264]:= **FixedPointList[#^4&, SetPrecision[1/10, $MachinePrecision]]**

        General::unfl : Underflow occurred in computation. More…

Out[264]= $\{0.1000000000000000, 0.0001000000000000000,$
    $1.00000000000000 \times 10^{-16}, 1.00000000000000 \times 10^{-64},$
    $1.0000000000000 \times 10^{-256}, 1.000000000000 \times 10^{-1024}, 1.000000000000 \times 10^{-4096},$
    $1.00000000000 \times 10^{-16384}, 1.00000000000 \times 10^{-65536}, 1.0000000000 \times 10^{-262144},$
    $1.000000000 \times 10^{-1048576}, 1.000000000 \times 10^{-4194304}, 1.00000000 \times 10^{-16777216},$
    $1.00000000 \times 10^{-67108864}, 1.0000000 \times 10^{-268435456}, Underflow[], Underflow[]\}$

1.1.2 Interval Arithmetic

Mathematica includes an implementation of interval arithmetic for simple arithmetic operations with real numbers. This is useful when manipulating data subject that are to measurement errors, requiring as good an estimate as possible of the final error (see, e.g., [412], [1059], [1807], [721], [268], and the references mentioned in the last subsection). The basic object in interval arithmetic is `Interval`. For details on interval arithmetic, see [1222], [1656], [1468], [1438], [900], [965], and http://cs.utep.edu/interval-comp/main.html. Interval arithmetic can be considered as a made rigorous version of the significance arithmetic discussed in the last subsection. As such, it comes with a price: It is slower and, especially for lower precision, its results are more "pessimistic" than the significance arithmetic results. Also, interval arithmetic is currently available only for elementary functions and for real arguments. But is powerful enough to provide the rigor of a mathematical proof [1800], [1598], [862], [472], [756].

`Interval[{x_{l1}, x_{u1}}, {x_{l2}, x_{u2}}, ...]`

> represents subintervals (x_{li}, x_{ui}) of the real line. Here, x_{li} and x_{ui} can be either approximate or exact real numbers. `Interval[x]` is equivalent to `Interval[{x − ϵ, x + ϵ}]`, where $ϵ$ is the uncertainty in x.

Here are two examples.

```
In[1]:= Interval[{1, 2}]
Out[1]= Interval[{1, 2}]
```

```
In[2]:= Interval[{Pi // N, 3.15}]
Out[2]= Interval[{3.14159, 3.15}]
```

For `Interval`-objects with inexact limits, the notion of open or closed intervals does not apply. `Interval`-objects with exact limits are often considered to be closed intervals. The order of the variables in the argument of `Interval` is arbitrary. They will be reordered so that the smaller interval endpoint is to the left.

```
In[3]:= Interval[{Pi, E}]
Out[3]= Interval[{e, π}]
```

Intervals can extend to infinity.

```
In[4]:= Interval[{0, Infinity}]
Out[4]= Interval[{0, ∞}]
```

Here, *Mathematica* cannot numerically determine which of the two interval endpoints is larger. *Mathematica* is not being able to order the two endpoints of the interval in increasing order. We will come back to such problems in detail in the last subsection of this section.

```
In[5]:= Interval[{1, 1 + (* a hidden zero *)
              Sin[Pi/16] - 1/2 Sqrt[2 - Sqrt[2 + Sqrt[2]]]}]
```

$$N::meprec : \text{Internal precision limit } \$MaxExtraPrecision = 50.\grave{}$$
$$\text{reached while evaluating } \frac{1}{2}\sqrt{2 - \sqrt{2 + \text{Power}[\ll2\gg]}} - \sin\left[\frac{\pi}{16}\right]. \text{ More…}$$

$$Out[5]= \text{Interval}\left[\left\{1, 1 - \frac{1}{2}\sqrt{2 - \sqrt{2 + \sqrt{2}}} + \sin\left[\frac{\pi}{16}\right]\right\}\right]$$

Intervals can be added and multiplied.

```
In[6]:= Interval[{1, 2}] + Interval[{5, 6}]
```

Out[6]= `Interval[{6, 8}]`

In[7]:= `Interval[{1, 2}] - Interval[{5, 6}]`

Out[7]= `Interval[{-5, -3}]`

In[8]:= `Interval[{1, 2}] Interval[{5, 6}]`

Out[8]= `Interval[{5, 12}]`

In[9]:= `Interval[{1, 2}]/Interval[{5, 6}]`

Out[9]= $\text{Interval}\left[\left\{\frac{1}{6}, \frac{2}{5}\right\}\right]$

The next input subtracts two `Interval`-objects, each containing two intervals. The resulting `Interval`-object contains three disjoint intervals.

In[10]:= `Interval[{-2, -1}, {1, 2}] - Interval[{-2, -1}, {1, 2}]`

Out[10]= `Interval[{-4, -2}, {-1, 1}, {2, 4}]`

Here is a symbolic proof of the last result. It uses the function `Reduce` (to be discussed in the Symbolic volume [1795]).

In[11]:= `Exists[{s1, s2}, Element[{s1, s2, d}, Reals] &&`
`(-2 < s1 < -1 || 1 < s1 < 2) &&`
`(-2 < s2 < -1 || 1 < s2 < 2),`
`d == s1 - s2] // Reduce`

Out[11]= $-4 < d < -2 \ || \ -1 < d < 1 \ || \ 2 < d < 4$

Intervals do not obey the same rules as "ordinary" numbers. For instance, the difference between two identical intervals is not the zero interval.

In[12]:= `Interval[{1, 3}] - Interval[{1, 3}]`

Out[12]= `Interval[{-2, 2}]`

Also, the distributive law does not hold anymore.

In[13]:= `Function[{a, b, c}, {a (b + c), a b + a c}][`
`Interval[{1, 2}], Interval[{-3, 8}], Interval[{5, 6}]]`

Out[13]= `{Interval[{2, 28}], Interval[{-1, 28}]}`

When one divides by an `Interval`-object, the result contains two disjoint intervals if the original interval contained 0.

In[14]:= `1/Interval[{-1, 1}]`

Out[14]= `Interval[{-∞, -1}, {1, ∞}]`

The following input does not make sense in the present implementation because intervals must be subintervals of the real axis.

In[15]:= `Interval[{-I, I}]`

Out[15]= `Interval[{-i, i}]`

We can take unions and intersections of intervals.

`IntervalUnion[`*interval*₁, *interval*₂, ..., *interval*ₙ`]`
   represents the (set theoretic) union of the intervals *interval*ᵢ.

`IntervalIntersection[`*int*₁, *int*₂, ..., *int*ₙ`]`
   represents the (set theoretic) intersection of the intervals *int*ᵢ.

And we can test, whether a number or an interval is contained in a given interval.

> ```
> IntervalMemberQ[interval, numberOrInterval]
> ```
> tests if *numberOrInterval* is contained in *interval*.

We now give a few simple examples of these operations on interval objects. The result of the following union of two intervals is one interval with a single argument.

```
In[16]:= IntervalUnion[Interval[{1, 3}], Interval[{2, 4}]]
Out[16]= Interval[{1, 4}]
```

The following two intervals have a finite overlap.

```
In[17]:= IntervalIntersection[Interval[{1, 3}], Interval[{2, 4}]]
Out[17]= Interval[{2, 3}]
```

Mathematica thinks that the following intervals (with exact limits) do not overlap.

```
In[18]:= IntervalIntersection[Interval[{1, 2}], Interval[{2, 4}]]
Out[18]= Interval[]
```

But if we work with approximate numbers, 2.0 is a common point.

```
In[19]:= IntervalIntersection[Interval[{1, 2.0}], Interval[{2.0, 4}]]
Out[19]= Interval[{2., 2.}]
```

Several intervals can be contained into one `Interval`-object, in particular, if they do not intersect.

```
In[20]:= IntervalUnion[Interval[{1, 2.0}], Interval[{3.0, 4}]]
Out[20]= Interval[{1, 2.}, {3., 4}]
```

E is inside the interval `Interval[2, 3]`.

```
In[21]:= IntervalMemberQ[Interval[{2, 3}], E]
Out[21]= True
```

However, $1 + \$MachineEpsilon$ is not in `Interval[0, 1]`.

```
In[22]:= IntervalMemberQ[Interval[{0, 1}], 1 + $MachineEpsilon]
Out[22]= False
```

But $1 + \$MachineEpsilon$ is contained in `Interval[0, 1.]`.

```
In[23]:= IntervalMemberQ[Interval[{0, 1.}], 1 + $MachineEpsilon]
Out[23]= True
```

Next, we generate the number 10 with precision 3. According to the last subsection it represents a number in the interval $(10(1 - 10^{-3}/2), 10(1 + 10^{-3}/2))$.

```
In[24]:= prec = 3; Ξ = 10;
         ξ = SetPrecision[Ξ, prec];
         (* the interval that represents x *)
         int = Interval[Ξ {1 - 10^-prec/2, 1 + 10^-prec/2}];
```

Because of the above-discussed behavior of comparison functions for high-precision numbers, `IntervalMemberQ` will return `True` for (inexact, low-precision numbers) that are outside of the actual interval.

```
In[28]:= Plot[{If[IntervalMemberQ[int, SetPrecision[ζ, prec]], 1, 0]}, {ζ, 9.95,
        10.05},
            PlotStyle -> {GrayLevel[0]},
            Epilog -> {Hue[0], Thickness[0.02],
                        Line[{{int[[1, 1]], 0}, {int[[1, 2]], 0}}]}]]
```

We can also apply elementary numerical operations to intervals.

```
In[29]:= 2 Interval[{1, 2}]
Out[29]= Interval[{2, 4}]
```

```
In[30]:= Interval[{1, 2}] Interval[{-2, 2}]
Out[30]= Interval[{-4, 4}]
```

```
In[31]:= Sin[Interval[{1, 2}]]
Out[31]= Interval[{Sin[1], 1}]
```

In calculating with interval arithmetic, "every approximate symbol/number is always uncertain". So, the difference of an `Interval`-object x with itself is not zero. When the subtraction is actually carried out, the x already evaluated (see Chapter 4 of the Programming volume [1793]) to an `Interval`-object and *Mathematica* no longer knows that these two `Interval`-objects originated from the same variable. (In affine arithmetic [465], and in generalized interval arithmetic [785], [1804] it is possible to keep track of correlations in the errors of quantities.)

```
In[32]:= x = Interval[{-1, 1}];

        x - x
Out[33]= Interval[{-2, 2}]
```

But as a *Mathematica* expression (meaning structurally), x is of course equal to itself.

```
In[34]:= {x == x, x === x}
Out[34]= {True, True}
```

But with symbolic arguments in `Interval`, we get the behavior discussed in Chapter 2 of the Programming volume [1793] and not a result with head `Interval`.

```
In[35]:= Interval[{i, j}] - Interval[{i, j}]
Out[35]= 0
```

Here is a somewhat more complicated example. Let $f(x) = \sin^2(2x + 1/(x+1))/(x^2 + 1)$. The variable x enters $f(x)$ in three different places.

```
In[36]:= Clear[x];
        f[x_] := Sin[2 x + 1/(x + 1)]^2/(x^2 + 1)
```

In the following plot, the corresponding values of the intervals are shown in red; the horizontal width corresponds to the associated *x*-interval. In areas where the function is monotonically increasing or decreasing, the function often passes through the red rectangle at the corner points.

```
In[38]:= ePlot[function_, curveInterval_,
            rectangleInterval_, intervalLength_, opts___] :=
      Module[{ξ}, Show[Graphics[
      {Function[x, (* make rectangles from intervals *)
       {Hue[0], Rectangle @@ Transpose[{x[[1, 1]], x[[2, 1]]}]}] /@
                 (* the x-intervals and the y-intervals *)
                 ({#, function[#]}& /@ (* the x-intervals *)
       (Table @@ {Interval[{ξ, ξ + intervalLength}],
                   Prepend[rectangleInterval, ξ]}) // N),
       {Thickness[0.01], (* the original curve *)
       Line[{#, function[#]}& /@ Range @@ curveInterval]}} // N],
       opts, Axes -> True, AxesOrigin -> {0.4, 0}, PlotRange -> All]]
```

```
In[39]:= ePlot[f, {0.1, Pi, Pi/400}, {0.2, Pi, 0.2}, 0.16]
```

Here, we should note that the curve does not always pass through the diagonal of the red rectangles. This behavior of `Interval[...]` is caused by the fact that in this example *x* makes several (for the interval arithmetic independent) appearances in $\sin(2x + 1/(x + 1))^2/(x^2 + 1)$ [358]. Using more intervals, we can make sure that the given curve is "bracketed". For comparison, we now plot the maximal deviations of $f(x)$ for the case of "the same" *x* (dashed black line) [1336] and "independent" *x* (blue line), based on a naive maximum error estimate calculated via the corresponding total differential [1495]. (Because of *Mathematica*'s evaluation strategy—as discussed in Chapter 4 of the Programming volume [1793]—it is impossible for *Mathematica* to recognize that the *x* is always the same and to nevertheless get smaller intervals as the result of the calculation.)

```
In[40]:= Show[{ePlot[f, {0, Pi, Pi/40}, {0, Pi, 0.01}, 0.1,
                DisplayFunction -> Identity],
           Plot[Evaluate[{f[x] + #, f[x] - #}&[0.1*
                (* total differential approximation for
                   x == y == z independent *)
                (Abs[D[#, x]] + Abs[D[#, y]] + Abs[D[#, z]])&[
                    Sin[2x + 1/(y + 1)]^2/(z^2 + 1)] //.
                {y -> x, z -> x}]], {x, 0, Pi},
                DisplayFunction -> Identity,
                PlotStyle -> {{RGBColor[0, 0, 1], Thickness[0.005]}}],
           Plot[Evaluate[{f[x] + #, f[x] - #}&[0.1*
                (* total differential approximation for one x only *)
                Abs[D[Sin[2x + 1/(x + 1)]^2/(x^2 + 1), x]]]], {x, 0, Pi},
                DisplayFunction -> Identity,
                PlotStyle -> {{GrayLevel[0], Thickness[0.002],
                        Dashing[{0.01, 0.01}]}}]},
            DisplayFunction -> $DisplayFunction]
```

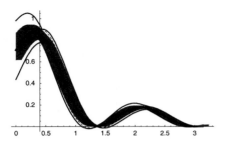

We clearly see the differing sensitivities of the function values to errors, depending on the values of the arguments.

`Interval` is also useful in solving some problems in linear algebra. We calculate the determinant of a 2×2 Hilbert matrix whose matrix elements have "small" errors.

In[41]:= **Table[Interval[1/(i + j + 1) + {-1/1000, 1/1000}], {i, 2}, {j, 2}]**

Out[41]= $\left\{\left\{\text{Interval}\left[\left\{\frac{997}{3000}, \frac{1003}{3000}\right\}\right], \text{Interval}\left[\left\{\frac{249}{1000}, \frac{251}{1000}\right\}\right]\right\},\right.$
$\left.\left\{\text{Interval}\left[\left\{\frac{249}{1000}, \frac{251}{1000}\right\}\right], \text{Interval}\left[\left\{\frac{199}{1000}, \frac{201}{1000}\right\}\right]\right\}\right\}$

In[42]:= **{Det[%], N[Det[%]]}**

Out[42]= $\left\{\text{Interval}\left[\left\{\frac{47}{15000}, \frac{13}{2500}\right\}\right], \text{Interval}[\{0.00313333, 0.0052\}]\right\}$

Here is a comparison with the numericalized version of the exact determinant. (But be aware that using intervals in other *Mathematica* functions just carries out calculations with intervals, it does not carry out calculations that minimize the intervals by reordering the operations occurring in calculating the determinant.)

In[43]:= **N[Det[Array[1/(#1 + #2 + 1)&, {2, 2}]]]**

Out[43]= 0.00416667

In the next example, no comparison is possible because the power of the interval remains partially uncomputed, leaving, for example, expressions of the following form.

In[44]:= **Interval[{3.0, 1.0}]^(1/Pi) < 100**

Out[44]= $\text{Interval}[\{1., 3.\}]^{\frac{1}{\pi}} < 100$

For more complicated functions such as the special functions of mathematical physics, interval arithmetic is also not yet available.

In[45]:= **Gamma[Interval[{0.3, 1.3}]]**

Out[45]= Gamma[Interval[{0.3, 1.3}]]

Intervals formed from numbers with higher accuracy have an interesting feature in the last few digits in order to correctly represent the interval. Here is such an example.

In[46]:= **Interval[{N[2, 50], N[3, 50]}]**

Out[46]= Interval[{2.0000000000000000000000000000000000000000000000000q
3.0000000000000000000000000000000000000000000000000}]

In[47]:= **%^3**

Out[47]= Interval[{7.9999999999999999999999999999999999999999999999999q
27.000000000000000000000000000000000000000000000003}]

For machine numbers, we do not see the interval ends directly in output form.

```
In[48]:= Interval[{N[2], N[3]}]
Out[48]= Interval[{2., 3.}]
```

But when we subtract the endpoints, we see the outward rounding.

```
In[49]:= First[%] - {2, 3}
Out[49]= {-4.44089×10^-16, 4.44089×10^-16}
```

Here is the interval around a machine zero.

```
In[50]:= Interval[{0., 0.}]
Out[50]= Interval[{-2.22507×10^-308, 2.22507×10^-308}]
```

We now use *Mathematica*'s interval arithmetic to follow the inaccuracies in the last subsection `Nest[(#^2 - 1)/(# - 1) - 1&` ... example. We begin with the interval $(2 - 10^{-200}, 2 + 10^{-200})$, and watch how it grows (on a logarithmic scale) as the calculation proceeds with 500 digits.

```
In[51]:= ListPlot[(* use log in base 10 *)
    Log[10, -N[Subtract @@ #[[1]]]]& /@
    NestList[(* the function to be iterated *) (#^2 - 1)/(# - 1) - 1&,
        (* starting interval with enough digits:
            N[{2 - 10^-200, 2 + 10^-200}, 500] *)
        Interval[N[{2 - 10^-200, 2 + 10^-200}, 500]], 237],
    FrameLabel -> {"iteration", "log(interval width)"}, Frame -> True,
    RotateLabel -> True, PlotStyle -> {PointSize[0.003]}]
```

Using `Intervals`, we can also directly compare the growth of the fuzziness of the above iteration for high-precision numbers.

```
In[52]:= (* to watch for all details in the precision *)
    Precision /@ NestList[(#^2 - 1)/(# - 1) - 1&, N[2, 200], 20]
Out[53]= {200., 199.155, 198.31, 197.465, 196.62, 195.775, 194.929,
    194.084, 193.239, 192.394, 191.549, 190.704, 189.859, 189.014,
    188.169, 187.324, 186.478, 185.633, 184.788, 183.943, 183.098}
```

```
In[54]:= (* according to definition of precision *)
    N[-Log[10, Abs[Subtract @@ #[[1]]]]]& /@
    NestList[(#^2 - 1)/(# - 1) - 1&,
            (* starting interval equivalent to N[2, 200] *)
            Interval[{SetPrecision[2 - 5 10^-201, 500],
                    SetPrecision[2 + 5 10^-201, 500]}], 20]
Out[55]= {200., 199.155, 198.31, 197.465, 196.62, 195.775, 194.929,
    194.084, 193.239, 192.394, 191.549, 190.704, 189.859, 189.014,
    188.169, 187.324, 186.478, 185.633, 184.788, 183.943, 183.098}
```

For not too small precisions, the high-precision arithmetic is an excellent approximation to interval arithmetic. We generate a random rational function to demonstrate this.

```
In[56]:= (* make a "random" rational function *)
    randomProduct[n_, c_, x_] :=
        Product[x - SetPrecision[Random[Real, {-c, c}], Infinity], {n}]

    SeedRandom[111];
    fr[x_] = Expand[randomProduct[20, 5, x]]/Expand[randomProduct[20, 5, x]];
```

The function is quite large when written out and contains numbers with nearly 300 digits.

```
In[61]:= {ByteCount[fr[x]], N @ Max[Flatten[{Numerator[#], Denominator[#]}& /@
                            Cases[fr[x], _Rational, Infinity]]]}
Out[61]= {12144, 2.09534×10^299}
```

The following graphic shows the precision of $fr[N[x, 30]]$ as a red line and the corresponding quantity calculated using exact interval arithmetic as a black line. The points indicate the corresponding values for the quality of the high-precision values. As visible, the two lines agree perfectly and the points are all above the two lines—as it should be for significance arithmetic with sufficiently many digits.

```
In[62]:= (* precision from high-precision arithmetic *)
    pSA[f_, x_, prec_] := Precision[f[N[SetPrecision[x, Infinity],
                                    prec]]]

In[64]:= (* precision from interval arithmetic *)
    pI[f_, x_, prec_] := With[{f =
    f[Interval[N[SetPrecision[x, Infinity]*
            {1 - 1/2 10^-prec, 1 + 1/2 10^-prec}, 3 prec]]]},
        (* -Log[10, intervalLength/intervalCenter] *)
        -Log[10, Abs[Subtract @@ f[[1]]]/Abs[Plus @@ f[[1]]/2]]]

In[66]:= With[{prec = 30},
    Plot[{pSA[fr, x, prec], pI[fr, x, prec]}, {x, -6, 6},
        (* -Log[10, relativeError] *)
        Epilog -> {PointSize[0.003], RGBColor[0, 0, 1],
            Table[Point[{x, -Log[10, Abs[(fr[x] -
                SetPrecision[fr[SetPrecision[x, prec]], Infinity])/
                fr[x]]]}], {x, -6, 6, 12/101}] // N},
        PlotRange -> All, PlotPoints -> 100,
        Frame -> True, Axes -> False,
        PlotStyle -> {{Thickness[0.02], Hue[0]},
                        {Thickness[0.002], GrayLevel[0]}}]]
```

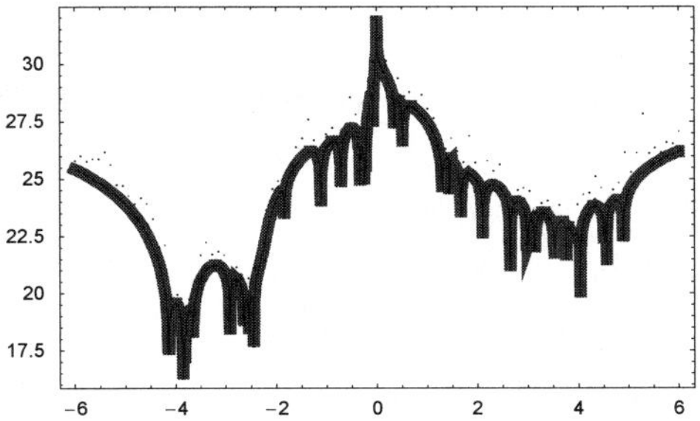

The next input shows the precision of $\cos(k)$ where the integers k have precision 20 versus the corresponding lengths of intervals representing these fuzzy integers. The rightmost plot shows the differences between the two data sets.

```
In[67]:= Show[GraphicsArray[
    Block[{$DisplayFunction = Identity, o = 2^12, p = 20, f = Cos, t1, t2,
        opts = Sequence[PlotRange -> {All, {14, 22}}, Frame -> True,
                        Axes -> False, PlotStyle -> PointSize[0.002]]},
        {(* using high-precision arithmetic *)
        ListPlot[t1 = Table[Precision[f[SetPrecision[k, p]]], {k, o}], opts],
        (* using interval arithmetic *)
        ListPlot[t2 = Table[-Log[10, Abs[Subtract @@
                            f[Interval[N[k {1 - 10^-p/2, 1 + 10^-p/2},
                                5p]]][[1]]/f[k]]], {k, o}], opts],
        (* difference between the two results *)
        ListPlot[Reverse @ Sort[t1 - t2], PlotRange -> All, opts]}]]]
```

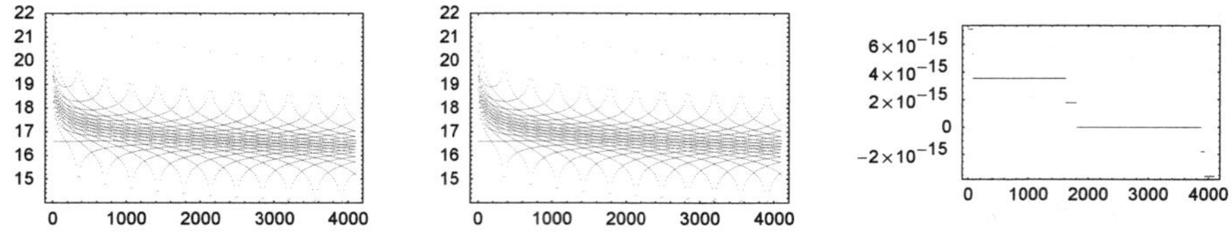

We end this subsection with an application of interval arithmetic.

Computing Relative Global Attractors of Rationals Maps

We consider a rational function $f(x_1, x_2, ..., x_d)$ from $\mathbb{R}^d$ to $\mathbb{R}^d$ and a compact region V of $\mathbb{R}^d$. The global attractor f relative to V is the intersection of all $f(V)$, $f(f(V))$, $f(f(f(V)))$, ... [1067], [474], [1113].

The global attractor f relative to V can be calculated in the following way: We start with a subdivision $\{S_k^{(0)}\}$ ($k = 1, 2, ..., n$) of V. The union of regions $S_1^{(0)}, S_2^{(0)}, ..., S_n^{(0)}$ covers V. We then form the images $T_k^{(0)} = f^{-1}(S_k^{(0)})$ and select the $S_k^{(0)}$ having a nonempty intersection with any of the $T_j^{(0)}$. Then we subdivide these selected $S_k^{(0)}$ to obtain a subdivision $\{S_k^{(1)}\}$. We continue and form the $T_k^{(1)} = f^{-1}(S_k^{(1)})$ and select the $S_k^{(1)}$ having a nonempty intersection with any of the $T_j^{(1)}$. Continuing this way yields a better and better approximation $\{S_k^{(n)}\}$ of the global attractor f relative to V. For a more detailed mathematical description, see [475], [477], and [476].

In the following, we will consider the case $d = 2$. For the $S_k^{(n)}$, we will use rectangles in $\mathbb{R}^2$. The most difficult step in the described algorithm is the calculation of the $T_k^{(n)}$. We will use intervals to describe the $S_k^{(n)}$ and interval arithmetic with machine-real number intervals to calculate the $T_j^{(n)}$. To select the relevant $S_k^{(n)}$ at each step we cover the $T_k^{(n)}$ with the $S_k^{(n)}$.

Here is an example map f.

```
In[68]:= Clear[x, y, ξ, η];
    f1[{x_, y_}] = {-29/39 + 1637/857 x^6 + 59/238 y, x};
```

Direct numerical forward iteration yields the following graphic of the attractor.

```
In[70]:= iterationGraphics[f_, {n_, k_}, opts___] :=
    Graphics[{PointSize[0.003], Table[Point /@ Drop[NestList[f,
                (* k random starting points *)
                Table[Random[Real, {-1, 1}], {2}], n],
            Floor[n/100]], {k}]}, opts, Frame -> True]

In[71]:= SeedRandom[111];
    Show @ iterationGraphics[f1, {10^4, 5}]
```

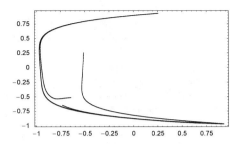

This is the inverse map f^{-1}. (The function f has been selected in such a way that f^{-1} does not contain radicals that are currently nonexhaustively-handled `Interval`-objects.)

```
In[73]:= f1Inv[{x_, y_}] = {ξ, η} /. Solve[{x, y} == f1[{ξ, η}], {ξ, η}][[1]]
```

$$\text{Out[73]= } \left\{ y, \frac{238\,(24853 + 33423\,x - 63843\,y^6)}{1971957} \right\}$$

The next inputs implement the above-described procedure for calculating $\{S_k^{(n)}\}$. For the initial region, we use the square $(-L, L) \times (-L, L)$. We subdivide each square at each step into four new squares of identical size and edge length δ. To find the intersections without testing each $S_k^{(n)}$ with each $T_k^{(n)}$, we discretize the $T_k^{(n)}$ through the $S_k^{(n)}$ and define a function `active` that marks the current $S_k^{(n)}$ inside the $(-L, L) \times (-L, L)$ square.

The three functions `intervalToSquareBounds`, `intervalPairToCoveringSquares`, and `inter-valPairToCoveringSquares` convert covering intervals to pairs of integers representing the position of a subdivision square in the $(-L, L) \times (-L, L)$ square.

```
In[74]:= (* bounds for squares *)
        intervalToSquareBounds[Interval[{a_, b_}], {x1_, x2_}, δ_] :=
        With[{n = Round[(x2 - x1)/δ], L = x2 - x1},
            {Max[Ceiling[(a - x1)/L n], 1], Min[Ceiling[(b - x1)/L n], n]}]
```

```
In[76]:= (* all squares corresponding to a pair rectangular 2D interval *)
        intervalPairToCoveringSquares[{i1:Interval[{_, _}],
                                       i2:Interval[{_, _}]}, {x1_, x2_}, δ_] :=
        Outer[List, Range @@ intervalToSquareBounds[i1, {x1, x2}, δ],
                    Range @@ intervalToSquareBounds[i2, {x1, x2}, δ]] //
                                                            Flatten[#, 1]&
```

```
In[78]:= (* reduce multiintervals to individual intervals *)
        intervalPairToCoveringSquares[{Interval[l1__], Interval[l2__]},
                                       {x1_, x2_}, δ_] :=
        Join @@ (intervalPairToCoveringSquares[#, {x1, x2}, δ]& /@
                Flatten[Outer[List, Interval /@ {l1}, Interval /@ {l2}], 1])
```

`subDivideSquareInterval` subdivides a square into four new ones.

```
In[80]:= (* subdivide rectangular 2D interval into four new ones *)
        subDivideSquareInterval[{Interval[{a_, b_}], Interval[{c_, d_}]}] :=
        Module[{xN = (a + b)/2, yN = (c + d)/2, i = Interval},
            {{i[{a, xN}], i[{c, yN}]}, {i[{xN, b}], i[{c, yN}]},
             {i[{a, xN}], i[{yN, d}]}, {i[{xN, b}], i[{yN, d}]}}]
```

```
In[82]:= (* square bounds corresponding to a pair rectangular 2D interval *)
        intervalPairToSquareRanges[{Interval[l1__], Interval[l2__]},
                                    {x1_, x2_}, δ_] :=
        Map[intervalToSquareBounds[#, {x1, x2}, δ]&,
            Flatten[Outer[List, Interval /@ {l1}, Interval /@ {l2}], 1], {2}]
```

The function `isOverlappingQ` determines if the interval pair $\{i1, i2\}$ overlaps any of the currently marked as active squares. To do this, it converts the square intervals into square numbers and compares these square numbers against the active ones.

```
In[84]:= (* does 2D interval overlap with any active 2D interval? *)
     isOverlappingQ[{{i1_Interval, i2_Interval}, {x1_, x2_}, δ_}, active_] :=
     Module[{squareRanges = intervalPairToSquareRanges[{i1, i2}, {x1, x2}, δ],
             ll, flag = False, xl, xu, yl, yu, xc, yc, cl = 1},
       ll = Length[squareRanges];
       While[flag === False && cl <= ll, (* initial square counting *)
         {{xl, xu}, {yl, yu}} = squareRanges[[cl]]; {xc, yc} = {xl, yl};
         (* loop over subsquares until active one is found *)
         While[yc <= yu && Not[flag], xc = xl;
             While[xc <= xu &&
                     (If[TrueQ[active[[xc, yc]]], flag = True]; Not[flag]),
                     xc++]; yc++]; cl++]; flag]
```

The function `refine` carries out the main work: it subdivides the squares, maps them using the function `fInv`, and selects the new global relative attractor approximations.

```
In[86]:= (* refine subdivision, map squares and select relevant ones *)
     refine[{activeSquares_, {x1_, x2_}, δ_}, fInv_] :=
     Module[{currentSquares = Flatten[subDivideSquareInterval /@
                                          activeSquares, 1],
             currentSquareNumbers, active, numberedMappedSquares},
       currentSquareNumbers = Union[Flatten[intervalPairToCoveringSquares[#,
                               {x1, x2}, δ/2]& /@ currentSquares, 1]];
       (* mark current squares as active *)
       (active[#] = True)& /@ currentSquareNumbers;
       numberedMappedSquares = MapIndexed[{#2[[1]], fInv[#1]}&, currentSquares];
       (* select squares with intersections *)
       currentSquares[[First /@ Select[numberedMappedSquares,
                         isOverlappingQ[{#[[2]], {x1, x2}, δ/2}, active]&]]]]
```

`globalRelativeAttractorList` finally returns a list of the first o approximations of the global relative attractor.

```
In[88]:= globalRelativeAttractorList[fInv_, L_, o_] :=
     Module[{activeSquares, δ = 2},
             activeSquares[0] = {{Interval[{-L, L}], Interval[{-L, L}]}} // N;
             Table[activeSquares[k] = refine[{activeSquares[k - 1],
                             {-L, L}, δ/2^(k - 1)}, fInv], {k, o}]]
```

We calculate the first 12 approximations to the global relative attractor shown above.

```
In[89]:= gral1 = globalRelativeAttractorList[f1Inv, 2., 12];
     Length /@ gral1
Out[90]= {4, 12, 25, 55, 134, 316, 721, 1651, 3746, 8540, 19523, 43434}
```

The function `globalRelativeAttractorListGraphic` visualizes the various approximations given by the list of interval pairs returned by `globalRelativeAttractorList`.

```
In[91]:= intervalPairToPolygon[{Interval[{a_, b_}], Interval[{c_, d_}]}] :=
                     Polygon[{{a, c}, {b, c}, {b, d}, {a, d}}]

In[92]:= globalRelativeAttractorListGraphic[l_, L_] :=
     With[{n = Length[l]},
          Graphics[MapIndexed[{Hue[0.8 (#2[[1]] - 1)/n],
```

```
                                        intervalPairToPolygon /@ #1}&, 1],
         PlotRange -> L{{-1, 1}, {-1, 1}}, AspectRatio -> Automatic]]
```

The so-calculated attractor covers the points from above, that were obtained by forward iteration of single points. Typically, the global relative attractor is larger than the one obtained from forward iteration of points.

ln[93]:= **Show[globalRelativeAttractorListGraphic[gral1, 2]]**

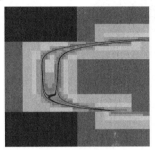

Here are two further polynomial maps f2 and f3. Their inverses are quite complicated rational functions.

ln[94]:= **f2[{x_ , y_}] = {178/339 - 1299/889x^4 + (16/45 + 349/443 x^2)y, x};**
　　　(* calculate inverse map *)
　　　f2Inv[{x_ , y_}] = {ξ, η} /. Solve[{x, y} == f2[{ξ, η}], {ξ, η}][[1]]

Out[96]= $\left\{ y, \dfrac{6645\,(-158242 + 301371\,x + 440361\,y^4)}{100457\,(7088 + 15705\,y^2)} \right\}$

ln[97]:= **f3[{x_ , y_}] = {-152/193 - 337/475 x + 20/113 x^2 + 10/101 x^3 - 198/257 y +**
　　　　　　　　233/273 x y - 1/161 x^2 y - 71/435 x^3 y, x};
　　　(* calculate inverse map *)
　　　f3Inv[{x_ , y_}] = {ξ, η} /. Solve[{x, y} == f3[{ξ, η}], {ξ, η}][[1]]

Out[99]= $\left\{ y, \dfrac{46797387\,(-824018600 - 1046286775\,x - 742312933\,y + 185183500\,y^2 + 103592750\,y^3)}{209257355\,(180270090 - 199703135\,y + 1453335\,y^2 + 38190971\,y^3)} \right\}$

And here are the corresponding graphics for the global attractors.

ln[100]:= **SeedRandom[111];**
　　　Show[GraphicsArray[iterationGraphics[#, {10^4, 5}]& /@ {f2, f3}]]

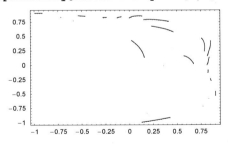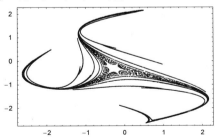

ln[102]:= **gral2 = globalRelativeAttractorList[f2Inv, 2., 9];**
　　　gral3 = globalRelativeAttractorList[f3Inv, 3., 9];
　　　{Length /@ gral2, Length /@ gral3}

Out[104]= {{4, 16, 54, 155, 406, 1111, 3167, 8555, 22725},
　　　　{4, 16, 61, 227, 792, 2733, 9307, 30184, 93193}}

ln[105]:=
　　　Show[GraphicsArray[globalRelativeAttractorListGraphic @@@
　　　　　　　　　　　　{{gral2, 2}, {gral3, 3}}]]

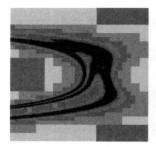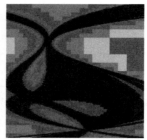

1.1.3 Converting Approximate Numbers to Exact Numbers

Sometimes, we need the "inverse" of N.

> Rationalize[*number*, *maxError*]
>
> converts the floating-point number *number* to an exact rational number, such that the difference between the two numbers is smaller than a function of the exact rational number and *maxError*. If only the first variable is present, the conversion will take place only if a "sufficiently close" fraction exists. The command Rationalize[*number*, 0] always results in a conversion.

The following example shows a number that converts nicely.

In[1]:= **Rationalize[4.68]**

Out[1]= $\dfrac{117}{25}$

This is not the case for the machine-precision approximation of π.

In[2]:= **Rationalize[N[Pi]]**

Out[2]= 3.14159

However, we can force a conversion.

In[3]:= **Rationalize[N[Pi], 0]**

Out[3]= $\dfrac{245850922}{78256779}$

We now look at a series of approximations of π.

In[4]:= **pi = N[Pi, 150];**

(piTab = Table[Rationalize[pi, 10^-i], {i, 50}]) // Short[#, 4]&

Out[5]//Short= $\left\{ \dfrac{22}{7}, \dfrac{22}{7}, \dfrac{201}{64}, \dfrac{333}{106}, \dfrac{355}{113}, \dfrac{355}{113}, \ll 39 \gg, \dfrac{23710859962467670903211 6}{7547401135972883479126 7}, \right.$
$\dfrac{4366816097360900760108 71}{1390000734936493284823 41}, \dfrac{2857198258041217165097 342}{9094744523216248056853 13},$
$\left. \dfrac{17579871157983393066594 923}{55958467874233981625942 19}, \dfrac{26151465932107044561886 949}{83242701443882725796501 58} \right\}$

The following plot shows the error in these approximations of π by fractions.

In[6]:= **ListPlot[(Log[10, Abs[N[#, 300] - pi]])& /@ piTab,**
            PlotRange -> All, AxesLabel -> {"i", "digits"}]

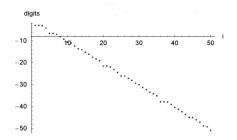

Rationalize also works with complex arguments. Here it recovers the given numericalized fraction.

In[7]:= **Rationalize[N[637/534 + I 991/451]]**

Out[7]= $\dfrac{637}{534} + \dfrac{991\,i}{451}$

There is one other obvious method for changing approximate numbers to exact ones—using SetPrecision[..., Infinity]. But this method typically results in quite large denominators.

In[8]:= **x = 0.75458;**

{Rationalize[x], SetPrecision[x, Infinity]}

Out[9]= $\left\{ \dfrac{37729}{50000},\ \dfrac{3398326206821229}{4503599627370496} \right\}$

The two rationals are equal within machine precision.

In[10]:= **N[Subtract @@ %]**

Out[10]= -2.84928×10^{-17}

xPrec is obtained by the direct transformation from the binary representation of x (all guard bits are taken into account).

In[11]:= **BaseForm[x, 2]**

Out[11]//BaseForm=
0.1100000100101100001_2

Rationalize will typically not convert a randomly chosen, machine-real number to a rational number. If Rationalize succeeds, the resulting denominators will typically have only five or six digits.

In[12]:= **Module[{c}, SeedRandom[111];**

Table[c = 0; (* how many trials? *)

While[c++; Head[Rationalize[r = Random[]]] == Real, Null];

{c, Rationalize[r]}, {3}]]

Out[12]= $\left\{ \left\{ 9849,\ \dfrac{54713}{84946} \right\},\ \left\{ 5185,\ \dfrac{24452}{210421} \right\},\ \left\{ 5731,\ \dfrac{75799}{84439} \right\} \right\}$

Very intimately related (and being the algorithmic underpinning of the function Rationalize) to the conversion of floating point numbers or exact irrational numbers to rational numbers are continued fractions [1050]. Regular continued fractions are of the form

$$a_0 + \cfrac{1}{a_1 + \cfrac{1}{a_2 + \cfrac{1}{a_3 + \cfrac{1}{a_4 + \cdots}}}}.$$

In *Mathematica*, they can be obtained using the function ContinuedFraction.

ContinuedFraction[*realNumber, order*]

    gives *order* terms a_i of the continued fraction approximation of *realNumber*.

Given the first n of the a_k, the function `FromContinuedFractions` generates the corresponding rational number.

FromContinuedFraction[*listOfIntegers*]

    calculates the convergent corresponding to the continued fraction *listOfIntegers*.

The `GoldenRatio` has a particularly simple continued fraction.

In[13]:= **ContinuedFraction[GoldenRatio, 50]**

Out[13]= {1, 1, 1, 1, 1, 1, 1, 1, 1, 1, 1, 1, 1, 1, 1, 1, 1, 1, 1, 1, 1, 1, 1, 1,
    1, 1, 1, 1, 1, 1, 1, 1, 1, 1, 1, 1, 1, 1, 1, 1, 1, 1, 1, 1, 1, 1, 1, 1, 1, 1}

The continued fraction of *e* shows also some regularities.

In[14]:= **ContinuedFraction[E, 50]**

Out[14]= {2, 1, 2, 1, 1, 4, 1, 1, 6, 1, 1, 8, 1, 1, 10, 1, 1, 12, 1, 1, 14, 1, 1, 16, 1, 1,
    18, 1, 1, 20, 1, 1, 22, 1, 1, 24, 1, 1, 26, 1, 1, 28, 1, 1, 30, 1, 1, 32, 1, 1}

Certain rational functions of *e* also have continued fraction representations with simple patterns.

In[15]:= **ContinuedFraction[Tanh[1/2], 50]**

    ContinuedFraction::terms :
      Warning: ContinuedFraction only obtained 38 of 50 requested
        terms using precision 134. Try increasing $MaxExtraPrecision. More…

Out[15]= {0, 2, 6, 10, 14, 18, 22, 26, 30, 34, 38, 42, 46, 50, 54, 58, 62, 66, 70, 74, 78, 82,
    86, 90, 94, 98, 102, 106, 110, 114, 118, 122, 126, 130, 134, 138, 142, 146}

The continued fraction of π does not show any obvious (known) pattern.

In[16]:= **ContinuedFraction[Pi, 50]**

Out[16]= {3, 7, 15, 1, 292, 1, 1, 1, 2, 1, 3, 1, 14, 2, 1, 1, 2, 2, 2, 2, 1, 84, 2, 1, 1,
    15, 3, 13, 1, 4, 2, 6, 6, 99, 1, 2, 2, 6, 3, 5, 1, 1, 6, 8, 1, 7, 1, 2, 3, 7}

Here is another continued fraction with an interesting pattern [606], [1032].

In[17]:= **ContinuedFraction[BesselJ[0, 2]/BesselJ[1, 2], 60]**

Out[17]= {0, 2, 1, 1, 2, 1, 3, 1, 4, 1, 5, 1, 6, 1, 7, 1, 8, 1, 9, 1,
    10, 1, 11, 1, 12, 1, 13, 1, 14, 1, 15, 1, 16, 1, 17, 1, 18, 1, 19, 1,
    20, 1, 21, 1, 22, 1, 23, 1, 24, 1, 25, 1, 26, 1, 27, 1, 28, 1, 29, 1}

Here is an interesting continued fraction of a rational number [420]. It is the order-8 approximation to the Liouville constant $\sum_{k=0}^{\infty} 10^{-k!}$ [315]. For brevity we use 9_k to represents the integer 9...9 consisting of k 9's.

In[18]:= **ContinuedFraction[Sum[10^-k!, {k, 0, 8}]] /. i_Integer :>**
        (If[Union[#] === {9}, Subscript[9, Length[#]], i]&[IntegerDigits[i]])

Out[18]= {0, 4, 1, 3, 5, 9_2, 1, 4, 3, 1, 4, 9_{12}, 1, 3, 1, 3, 4, 1, 9_2, 5, 3, 1, 4, 9_{72}, 1, 3, 1,
    3, 5, 9_2, 1, 4, 3, 1, 4, 9_{12}, 4, 1, 3, 4, 1, 9_2, 5, 3, 1, 4, 9_{480}, 1, 3, 1, 3, 5,
    9_2, 1, 4, 3, 1, 4, 9_{12}, 1, 3, 1, 3, 4, 1, 9_2, 5, 3, 1, 3, 1, 9_{72}, 4, 1, 3, 5, 9_2,
    1, 4, 3, 1, 3, 1, 9_{12}, 4, 1, 3, 4, 1, 9_2, 5, 3, 1, 4, 9_{3600}, 1, 3, 1, 3, 5, 9_2, 1,
    4, 3, 1, 4, 9_{12}, 1, 3, 1, 3, 4, 1, 9_2, 5, 3, 1, 4, 9_{72}, 1, 3, 1, 3, 5, 9_2, 1, 4, 3,
    1, 3, 1, 9_{12}, 4, 1, 3, 4, 1, 9_2, 5, 3, 1, 3, 1, 9_{480}, 4, 1, 3, 5, 9_2, 1, 4, 3, 1,

$4, 9_{12}, 1, 3, 1, 3, 4, 1, 9_2, 5, 3, 1, 3, 1, 9_{72}, 4, 1, 3, 5, 9_2, 1, 4, 3, 1, 3, 1,$
$9_{12}, 4, 1, 3, 4, 1, 9_2, 5, 3, 1, 4, 9_{30240}, 1, 3, 1, 3, 5, 9_2, 1, 4, 3, 1, 4, 9_{12},$
$1, 3, 1, 3, 4, 1, 9_2, 5, 3, 1, 4, 9_{72}, 1, 3, 1, 3, 5, 9_2, 1, 4, 3, 1, 3, 1, 9_{12}, 4,$
$1, 3, 4, 1, 9_2, 5, 3, 1, 4, 9_{480}, 1, 3, 1, 3, 5, 9_2, 1, 4, 3, 1, 4, 9_{12}, 1, 3, 1,$
$3, 4, 1, 9_2, 5, 3, 1, 3, 1, 9_{72}, 4, 1, 3, 5, 9_2, 1, 4, 3, 1, 3, 1, 9_{12}, 4, 1, 3, 4,$
$1, 9_2, 5, 3, 1, 3, 1, 9_{3600}, 4, 1, 3, 5, 9_2, 1, 4, 3, 1, 4, 9_{12}, 1, 3, 1, 3, 4, 1,$
$9_2, 5, 3, 1, 4, 9_{72}, 1, 3, 1, 3, 5, 9_2, 1, 4, 3, 1, 3, 1, 9_{12}, 4, 1, 3, 4, 1, 9_2,$
$5, 3, 1, 3, 1, 9_{480}, 4, 1, 3, 5, 9_2, 1, 4, 3, 1, 4, 9_{12}, 1, 3, 1, 3, 4, 1, 9_2, 5, 3,$
$1, 3, 1, 9_{72}, 4, 1, 3, 5, 9_2, 1, 4, 3, 1, 3, 1, 9_{12}, 4, 1, 3, 4, 1, 9_2, 5, 3, 1, 4\}$

For a nice collection of closed-form continued fractions, see [1870] and [1173]. Rational numbers have finite continued fraction representations. Suppressing the second argument in `ContinuedFraction` returns, for an exact rational first argument, all terms of the continued fraction expansion.

In[19]:= `ContinuedFraction[181916/140563315]`

Out[19]= `{0, 772, 1, 2, 6, 1, 2, 23, 1, 10, 1, 9}`

In[20]:= `FromContinuedFraction[%]`

Out[20]= $\dfrac{181916}{140563315}$

Here is a slightly larger calculation involving the continued fraction of rational numbers. The length of the continued fraction expansion of $(p/q)^k$ (where $\gcd(p, q) = 1$) is conjectured to approach $12/\pi^2 \ln(2) \ln(q)$ [629]. Here we display the length of $(p/11)^k$ for $p = 1, \ldots, 10$ and $k \le 500$ together with the conjectured asymptotic value.

In[21]:=
```
Show[Graphics[
    With[{q = 11, n = 500},
    Table[{Hue[p/q], Line[{{n/2, 12/Pi^2 Log[2] Log[p]},
                           {n,    12/Pi^2 Log[2] Log[p]}}],
        PointSize[0.003],
        Table[Point @ {k, 1/k Length[ContinuedFraction[(p/q)^k]]}, {k, n}]},
        {p, q - 1}] // N]], Frame -> True, PlotRange -> {0, 2.5}] // Timing
```

For floating point numbers, `ContinuedFraction` sometimes returns fewer terms than expected or even no terms at all.

In[22]:= `ContinuedFraction /@ {2., N[2, 22], 2}`

```
ContinuedFraction::start :
  Warning: ContinuedFraction was unable to obtain at least
    one term using precision 15.954589770191005`. More…

ContinuedFraction::start : Warning: ContinuedFraction was
    unable to obtain at least one term using precision 22.`. More…
```

Out[22]= `{ContinuedFraction[2.], ContinuedFraction[2.000000000000000000000], {2}}`

The reason is that small changes in the input number may give different integers of the continued fraction expansion.

```
In[23]:= {ContinuedFraction[2 + 10^-22], ContinuedFraction[2 - 10^-22]}

Out[23]= {{2, 10000000000000000000000}, {1, 1, 99999999999999999999999}}
```

Finite continued fractions are basically rational functions in the a_j. For a given set of numbers a_j, and nonnegative integers $k \le l \le m \le n$, the following identities hold for continued fractions [1299]:

$$\left(\prod_{j=k}^{l-1}(a_j + 1/(a_{j+1} + 1/(\cdots + 1/a_n)))\right)\left(\prod_{j=m+1}^{n}(a_j + 1/(a_{j-1} + 1/(\cdots + 1/a_l)))\right) =$$

$$\left(\prod_{j=m+1}^{n}(a_j + 1/(a_{j-1} + 1/(\cdots + 1/a_k)))\right)\left(\prod_{j=k}^{l-1}(a_j + 1/(a_{j+1} + 1/(\cdots + 1/a_m)))\right).$$

Here this identity is implemented.

```
In[24]:= cfIdentity[aList_, k_, l_, m_, n_] :=
    Module[{K},
      (* define the continued fractions from the list aList *)
      K[j0_, j1_, δj_:1] :=
                   FromContinuedFraction[Table[aList[[j]], {j, j0, j1, δj}]];
      (* the identity *)
      Product[K[j, n], {j, k, l - 1}] Product[K[j, l, -1], {j, m + 1, n}] -
      Product[K[j, k, -1], {j, m + 1, n}] Product[K[j, m], {j, k, l - 1}]]
```

Here is a quick check for all possible 330 realizations of the above identity for eight a_j.

```
In[25]:= With[{o = 8},
        Union[Factor //@ (cfIdentity[Table[a[j], {j, o}], ##]& @@@
        Flatten[Table[{k, l, m, n},
                      {k, o}, {l, k, o}, {m, l, o}, {n, m, o}], 3])]]

Out[25]= {0}
```

How does the length of a continued fraction of a rational number behave as a function of this rational number? Here we investigate all rational numbers with denominators less than or equal to 256. For some theoretical results, see [628].

```
In[26]:= With[{n = 256},
        ListPlot[{#, Length[ContinuedFraction[#]]}& /@
                Union[Flatten[Table[i/j, {j, n}, {i, 0, j}]]],
              PlotRange -> All, PlotJoined -> True, Frame -> True,
              PlotStyle -> {Thickness[0.002]}, Axes -> False]]
```

Square roots of integers have continued fraction representations that are ultimately periodic [1286], [1285]. (For closed forms of nonsimple continued fractions for higher roots of rationals numbers, see [1106].)

```
In[27]:= ContinuedFraction[2 + 3 Sqrt[79], 80]
```

```
Out[27]= {28, 1, 1, 1, 52, 1, 1, 1, 52, 1, 1, 1, 52, 1, 1, 1, 52, 1, 1, 1, 52, 1, 1,
         1, 52, 1, 1, 1, 52, 1, 1, 1, 52, 1, 1, 1, 52, 1, 1, 1, 52, 1, 1, 1, 52,
         1, 1, 1, 52, 1, 1, 1, 52, 1, 1, 1, 52, 1, 1, 1, 52, 1, 1, 1, 52, 1, 1, 1}
```

In this case, the second argument can be suppressed. `ContinuedFraction` then returns the periodic part enclosed in an additional list.

```
In[28]:= ContinuedFraction[2 + 3 Sqrt[79]]
```

```
Out[28]= {28, {1, 1, 1, 52}}
```

```
In[29]:= FromContinuedFraction[%]
```

$$Out[29]= 2 + 3\sqrt{79}$$

The length of the periodic part of the continued fraction expansion of $\sqrt{k}$ varies greatly as a function of k [1907], [1189]. The right graphic shows the length of the periodic parts for all rational numbers with denominators less than or equal to 128.

```
In[30]:= Show[GraphicsArray[
          Block[{$DisplayFunction = Identity, opts = Sequence[Axes -> False,
                  Frame -> True, PlotRange -> All, PlotStyle -> {PointSize[0.003]}]},
           {(* length of period for square root of integers *)
           ListPlot[Table[{k, If[IntegerQ[Sqrt[k]], 0,
                          (* length of periodic part *)
                          Length[Last[ContinuedFraction[Sqrt[k]]]]]},
                      {k, 10000}], opts],
           (* length of period for square root of fractions *)
           ListPlot[{#, Length[ContinuedFraction[Sqrt[#]][[-1]]]}& /@
                 Union[Flatten[Table[i/j, {j, 128}, {i, 0, j}]]], opts]}]]]
```

Continued fraction expansions of rational numbers and square roots can quite frequently explain various properties of number theoretical functions and related objects. For instance, the only slightly different continued fraction expansions of $\sqrt{2}$ and $\sqrt{3}$ can be used to explain the grossly different behaviors of the Cesaro means of the cumulative sums of the average fractional parts ($m = 2$, $m = 3$) [150]

$$f_n(m) = \frac{1}{n+1} \sum_{l=1}^{n} \sum_{k=1}^{l} \left(\text{frac}\left(k \sqrt{m}\right) - \frac{1}{2} \right).$$

Here are the periodic continued fractions of $\sqrt{2}$ and $\sqrt{3}$.

```
In[31]:= ContinuedFraction /@ {Sqrt[2], Sqrt[3]}
```

In[31]:= {{1, {2}}, {1, {1, 2}}}

This implements the function $\mathfrak{f}_n(m)$.

```
In[32]:= f[m_, n_] := MapIndexed[#1/(#2[[1]] + 1)&,
              Rest[FoldList[Plus, 0, Rest[FoldList[Plus, 0,
                  Table[FractionalPart[k m] - 1/2, {k, n}]]]]]]
```

And the following graphic shows $\mathfrak{f}_n(m)$ as a function of n for $m = \sqrt{2}$ and $m = \sqrt{3}$.

```
In[33]:= Show[GraphicsArray[
            ListPlot[f[#, 10^5], Frame -> True, Axes -> False, PlotRange -> All,
                    DisplayFunction -> Identity]& /@ Sqrt[{2., 3.}]]]
```

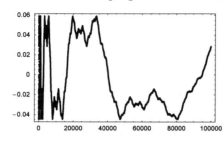 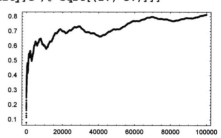

Only solutions of quadratic polynomials have periodic continued fraction expansions. So the following expansion of the real root of $x^3 - 8x - 10 = 0$ fails. (For some discussion of this special continued fraction, see [1705] and [404].)

```
In[34]:= Ξ = (135 - 3 Sqrt[489])^(1/3)/3 + (45 + Sqrt[489])^(1/3)/3^(2/3);
        ContinuedFraction[Ξ]
```

ContinuedFraction::noterms :

$\frac{1}{3}\left(135 - 3\sqrt{489}\right)^{1/3} + \frac{(45 + \sqrt{489})^{1/3}}{3^{2/3}}$ does not have a terminating
or periodic continued fraction expansion; specify
an explicit number of terms to generate. More...

Out[35]= ContinuedFraction$\left[\frac{1}{3}\left(135 - 3\sqrt{489}\right)^{1/3} + \frac{(45 + \sqrt{489})^{1/3}}{3^{2/3}}\right]$

Specifying an explicit number of terms works, of course.

```
In[36]:= ContinuedFraction[Ξ, 150]
```

Out[36]= {3, 3, 7, 4, 2, 30, 1, 8, 3, 1, 1, 1, 9, 2, 2, 1, 3, 22986, 2, 1, 32, 8, 2, 1, 8, 55, 1,
    5, 2, 28, 1, 5, 1, 1501790, 1, 2, 1, 7, 6, 1, 1, 5, 2, 1, 6, 2, 2, 1, 2, 1, 1, 3,
    1, 3, 1, 2, 4, 3, 1, 35657, 1, 17, 2, 15, 1, 1, 2, 1, 1, 5, 3, 2, 1, 1, 7, 2, 1, 7,
    1, 3, 25, 49405, 1, 1, 3, 1, 1, 4, 1, 2, 15, 1, 2, 83, 1, 162, 2, 1, 1, 1, 2, 2,
    1, 53460, 1, 6, 4, 3, 4, 13, 5, 15, 6, 1, 4, 1, 4, 1, 1, 2, 1, 16467250, 1, 3, 1,
    7, 2, 6, 1, 95, 20, 1, 2, 1, 6, 1, 1, 8, 1, 48120, 1, 2, 17, 2, 1, 2, 1, 4, 2, 3}

There is an intimate relation between the function `Rationalize` and continued fractions: Let t be a positive real number. Every "best" approximation to t is a convergent of the regular continued fraction approximation of t.

Here, the first 200 convergents of π are calculated.

```
In[37]:= convergents = With[{cf = ContinuedFraction[Pi, 200]},
                    Table[FromContinuedFraction[Take[cf, i]], {i, 200}]];
```

The absolute value of the difference between the convergent A_n and π is a monotonic function of n.

In[38]:= **N[Pi - Take[convergents, 20], 22] // N**

Out[38]= $\{0.141593, -0.00126449, 0.0000832196, -2.66764 \times 10^{-7}, 5.77891 \times 10^{-10},$
$-3.31628 \times 10^{-10}, 1.22357 \times 10^{-10}, -2.91434 \times 10^{-11}, 8.71547 \times 10^{-12}, -1.61074 \times 10^{-12},$
$4.04067 \times 10^{-13}, -2.21448 \times 10^{-14}, 5.79087 \times 10^{-16}, -1.64084 \times 10^{-16}, 7.81794 \times 10^{-17},$
$-1.93638 \times 10^{-17}, 3.07008 \times 10^{-18}, -5.51366 \times 10^{-19}, 7.62659 \times 10^{-20}, -3.12317 \times 10^{-20}\}$

In[39]:= **$MaxExtraPrecision = 1000;**

In[40]:= **ListPlot[Log[10, Abs[N[(Pi - #)& /@ convergents, 22] // N]]]**

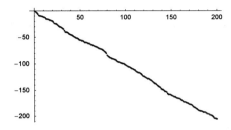

If $A_k = p_k / q_k$ is a convergent of the continued fraction approximation of t, then $|p_k / q_k - t| < 1 / q_k^2$ [1543], [259] (the opposite does not hold). Here, for the above convergents this is visualized.

In[41]:= **ListPlot[N[Denominator[#]^2 (Pi - #)& /@ convergents, 22] // N,**
       PlotRange -> All]

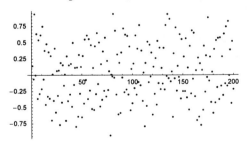

The following graphic shows the difference between π and its nearest rational approximation for denominators less than 10^5 in steps of 10 [800].

In[42]:= **ListPlot[Table[{q,**
    With[{p = IntegerPart[q Pi]},
          (* select nearest rational number for a given denominator *)
          Log[10, N[Min[Abs[{Pi - p/q, Pi - (p + 1)/q}]], 22]]]},
            {q, 1, 10^5, 10}],
        Frame -> True, Axes -> False, PlotRange -> All,
        PlotStyle -> {PointSize[0.003]}]

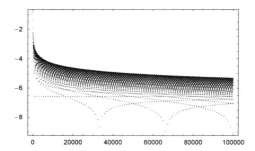

The clearly visible minima in the above picture occur at the denominators of the consecutive continued fraction approximations.

```
In[43]:= Table[FromContinuedFraction[Take[
              ContinuedFraction[Pi, 20], k]], {k, 8}]
```
$$Out[43]= \left\{3, \frac{22}{7}, \frac{333}{106}, \frac{355}{113}, \frac{103993}{33102}, \frac{104348}{33215}, \frac{208341}{66317}, \frac{312689}{99532}\right\}$$

The fact that the continued fraction convergents are the good approximations to a given number is used inside `Rationalize`. Here, the difference $\pi - e$ is rationalized.

```
In[44]:= Table[Rationalize[N[Pi - E, 30], 10^-i], {i, 20}]
```
$$Out[44]= \left\{\frac{1}{2}, \frac{3}{7}, \frac{11}{26}, \frac{58}{137}, \frac{69}{163}, \frac{908}{2145}, \frac{1253}{2960}, \frac{6334}{14963}, \frac{10093}{23843},\right.$$
$$\frac{79491}{187784}, \frac{119863}{283156}, \frac{469359}{1108781}, \frac{589222}{1391937}, \frac{13432243}{31731395}, \frac{22270573}{52610450},$$
$$\left.\frac{23449017}{55394324}, \frac{141283324}{333757881}, \frac{306015665}{722910086}, \frac{2000826331}{4726612721}, \frac{2612857661}{6172432893}\right\}$$

The same fractions appear in the list of all possible continued fraction convergents.

```
In[45]:= Table[FromContinuedFraction[Take[#, i]], {i, 20}]&[
              ContinuedFraction[N[Pi - E, 30]]]
```
$$Out[45]= \left\{0, \frac{1}{2}, \frac{2}{5}, \frac{3}{7}, \frac{11}{26}, \frac{69}{163}, \frac{1253}{2960}, \frac{8840}{20883}, \frac{10093}{23843},\right.$$
$$\frac{109770}{259313}, \frac{119863}{283156}, \frac{469359}{1108781}, \frac{589222}{1391937}, \frac{23449017}{55394324}, \frac{141283324}{333757881},$$
$$\left.\frac{164732341}{389152205}, \frac{306015665}{722910086}, \frac{2306841996}{5449522807}, \frac{2612857661}{6172432893}, \frac{7532557318}{17794388593}\right\}$$

While the convergents of a continued fraction form excellent approximations to a number, without an additional constraint on the size of the denominator of a fraction, the so-called pseudoconvergents are the best. The following graphic shows the logarithm of the denominator of successive approximations of π and $1/\pi$ as a function of α in `Rationalize[ξ, 10^-α]`. The blue horizontal lines indicate the denominators of the continued fraction convergents. Clearly, some of the data points are not lying on the blue lines—these are the pseudoconvergents.

```
In[46]:= Show[GraphicsArray[
          Function[ξ, With[{A = 20},
          ListPlot[(* denominators *)
                  Table[{α, Log[10, Denominator[Rationalize[ξ, 10^-α]]]},
                      {α, 0, A, 1/100}] // N,
                   (* continued fraction convergents *)
                  GridLines -> {None, Log[10,
                    Select[Denominator[Table[FromContinuedFraction[
                     Take[ContinuedFraction[ξ, 20], k]], {k, 20}]],
                        (* with relevant denominators *)
```

```
        (# < Denominator[Rationalize[ξ, 10^-20]])&]] // N},
    Axes -> False, DisplayFunction -> Identity,
    PlotStyle -> {PointSize[0.006]},
    Frame -> True, PlotLabel -> InputForm[ξ]]]] /@
                    (* the two numbers *) {Pi, 1/Pi}]]
```

Here is the concrete example $\alpha = 7.8$. The fraction returned from `Rationalize` agrees with the best fraction found by searching and is not a continued fraction convergent.

```
In[47]:= {(* use fast Rationalize *) Rationalize[1/Pi, 10^-7.8],
        (* exhaustive search over all denominators *)
        Module[{den = 1},
            While[δ = Abs[1/Pi - Round[den/Pi]/den]; δ > 10^-7.8, den++];
            Round[den/Pi]/den],
        (* the continued fraction convergents *)
        Table[FromContinuedFraction[ContinuedFraction[1/Pi, k]], {k, 8}]]
```

$$Out[47]= \left\{ \frac{21011}{66008}, \frac{21011}{66008}, \left\{0, \frac{1}{3}, \frac{7}{22}, \frac{106}{333}, \frac{113}{355}, \frac{33102}{103993}, \frac{33215}{104348}, \frac{66317}{208341} \right\} \right\}$$

The pseudoconvergent and the convergent deviate from each other only in the last digit of their continued fraction approximation. And the pseudoconvergent "interpolates" between two continued fraction convergents.

```
In[48]:= {ContinuedFraction /@ {21011/66008, 33102/103993, 33215/104348},
        (* denominator divisibility *) (103993 - 66008)/355}
Out[48]= {{0, 3, 7, 15, 1, 185}, {0, 3, 7, 15, 1, 292}, {0, 3, 7, 15, 1, 293}}, 107}
```

The continued fraction expansion of a real number from the interval (0, 1) has (for almost all real numbers) a very interesting property: The probability of the occurrence of the integer k is $-\log_2(1 - (k+1)^{-2})$ [983], [873]. In the next input, we use the fractional part of π^n as the "random" numbers for $n = 1, 2, \dots, 100$. The red line is the theoretical distribution.

```
In[49]:= data = Table[Rest[ContinuedFraction[FractionalPart[Pi^n], 100]], {n, 100}];
```

```
In[50]:= frequencies = {#[[1]], Length[#]}& /@ Split[Sort[Flatten[data]]];
```

```
In[51]:= ListPlot[Take[Apply[{#1, #2/9900}&, frequencies, {1}], 100],
            PlotRange -> All, Frame -> True, Axes -> False,
            PlotStyle -> {GrayLevel[0], PointSize[0.01]},
            (* theoretical frequencies *)
            Prolog -> {Hue[0], Thickness[0.002],
                        Line[Table[{k, -Log[2, 1. - 1/(k + 1)^2]}, {k, 120}]]}]
```

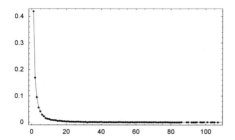

Let us analyze the number of times a certain integer appears in the continued fraction approximation of π. We start by calculating 100000 terms of the continued fraction expansion of π.

```
In[52]:= (cf = ContinuedFraction[Pi, 10^5];) // Timing
Out[52]= {0.97 Second, Null}
```

This shows the probability distribution of the integer i in the continued fraction expansion. The red underlying curve is the theoretical distribution, which holds for almost all real numbers.

```
In[53]:= Needs["Graphics`Graphics`"]

In[54]:= LogLogListPlot[Map[{First[#], Length[#]}&, Split[Sort[cf]]] // N,
    PlotRange -> All, PlotJoined -> True,
      Prolog -> {{Hue[0], Thickness[0.012],
        Line[Table[{Log[10, k], Log[10, -10^5 Log[2, 1 - 1/(k + 1)^2]]} // N,
                 {k, 10^3}] // N]}}]
```

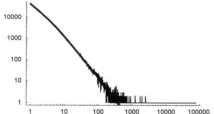

It is conjectured that the same distribution holds for the average over many fractions [96]. Next, we analyze the digits in all proper fractions with denominators less than or equal 1000. This can be done in a fraction of a minute and results in 1784866 continued fraction digits.

```
In[55]:= (* proper fractions with denominator n *)
    properNFractions[den_] := Select[Range[den]/den, Denominator[#] == den&]

In[57]:= (* digit counts of the continued fractions of
      all proper fractions with denominator n *)
    cfDigitData[n_] := {First[#], Length[#]}& /@
              Split[Sort[Flatten[DeleteCases[ContinuedFraction[#], 0]& /@
                                   properNFractions[n]]]]

In[59]:= (* function to add counts *)
    updateCounts[cfDigitData_] := (count[#1] =
        If[Head[count[#1]] === Integer, count[#1], 0] + #2)& @@@ cfDigitData

In[61]:= (* analyze all fractions up to denominator 1000 *)
    Clear[count];
```

```
        {Do[updateCounts @ cfDigitData[n], {n, 2, 1000}] // Timing,
         totalDigits = Plus @@ (Last /@ DownValues[count])}
Out[63]= {{7.37 Second, Null}, 1784866}
```

Here are the counted digit probabilities and the theoretical probabilities shown.

```
In[64]:= Show[Block[{$DisplayFunction = Identity},
           {(* actual digit counts *)
            ListPlot[{#1[[1, 1]], Log[#2/totalDigits]}& @@@ DownValues[count],
                PlotRange -> {{0, 200}, All}],
            (* theoretical distribution *)
            Plot[Log[1/Log[2] Log[1 + 1/(k k +2)]], {k, 1, 200},
                PlotStyle -> {Hue[0]}]}],
         Frame -> True, Axes -> False]
```

Unlike decimal fractions, numbers represented as continued fractions are quite unevenly distributed in the real axis. Here, the cumulative probability distribution of all numbers of the form $a/(b + c/(d + e))$, $1 \le a, b, c, d, e \le 9$ are shown.

```
In[65]:= ListPlot[MapIndexed[{#1, #2[[1]]}]&, Sort[Flatten[
             Table[(* all continued fractions *)
                FromContinuedFraction[{0, a, b, c, d, e}],
                {a, 9}, {b, 9}, {c, 9}, {d, 9}, {e, 9}]]]],
           PlotStyle -> {PointSize[0.003]},
           PlotRange -> All, Axes -> False, Frame -> True]
```

The following constant is related to continued fractions.

```
Khinchin
    represents the Khinchin constant K.
```

```
In[66]:= N[Khinchin, 22]
Out[66]= 2.685452001065306445310
```

The importance of this constant stems from the following interesting fact: For almost all real numbers t with the regular continued fraction expansion

$$t = a_0 + \cfrac{1}{a_1 + \cfrac{1}{a_2 + \cfrac{1}{a_3 + \cfrac{1}{a_4 + \cdots}}}}$$

the following holds [1543], [916] :

$$\lim_{n\to\infty} \sqrt[n]{a_1\,a_2\cdots a_n} = \prod_{k=1}^{\infty}\left(1 + \frac{1}{k(k+2)}\right)^{\log_2 k} = K = 2.6854\ldots$$

Let us pick a number at "random", say, $t = \phi^\pi$. We calculate its first 1000 terms of the continued fraction expansion.

```
In[67]:= cf = Rest @ ContinuedFraction[GoldenRatio^Pi, 1001];
```

The following graphic shows how the expression $\sqrt[k]{a_1\,a_2\cdots a_k}$ behaves [1924], [113].

```
In[68]:= ListPlot[Table[(Times @@ Take[cf, i])^(1/i), {i, 1000}] // N]
```

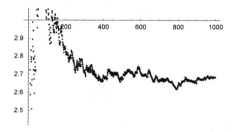

The next graphic shows $\sqrt[n]{a_1\,a_2\cdots a_n}$ for K itself. This time we use a more efficient procedure to calculate the means.

```
In[69]:= cf = Rest @ ContinuedFraction[Khinchin, 4001];
         fl = FoldList[(#1^(#2[[1]] - 1) #2[[2]])^(1/#2[[1]])&, 1,
                    Transpose[{Range[4000], N[cf]}]];
         ListPlot[fl]
```

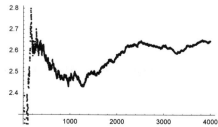

In a similar way, the nth root of the denominator of the nth convergent approaches the limit $\exp(\pi^2/(12\ln(2)))$. (This fact is the Khinchin–Lévy theorem [1543].)

```
In[72]:= ListPlot[MapIndexed[#1^(1./#2[[1]])&, Denominator[
             Table[FromContinuedFraction[Take[cf, i]], {i, 1000}]]]]
```

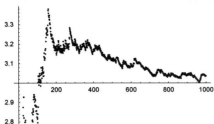

```
In[73]:= Exp[Pi^2/12/Log[2]] // N
Out[73]= 3.27582
```

The expression $n / \sum_{k=1}^{n} a_k^{-1}$ also approaches a universal constant as $n \to \infty$ [983].

```
In[74]:= ListPlot[MapIndexed[#2[[1]]/#1&, Rest[
             FoldList[Plus, 0, 1./Rest[ContinuedFraction[Pi + Log[2] + 3^(1/3),
                 20000]]]]], Frame -> True, Axes -> False]
```

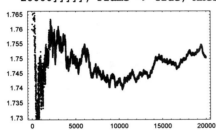

An interesting question about the connection of decimal approximations and continued fraction approximations is the following: Given an n-digit approximation $a_n(x)$ to an irrational number x, how many terms $k_n(x)$ does the continued fraction representation contain on average [1165], [577], [576], [57]?

Let us do a numerical experiment with some "random" irrational numbers.

```
In[75]:= Length[ContinuedFraction[#]]& /@
            N[{Pi, E, Sqrt[2], 3^(1/5), Sin[1/3], Log[7], Tan[1/9],
               Csc[EulerGamma], Log[Log[Log[100]]], ArcSin[1/2]^ArcSin[1/3]}, 1000]
Out[75]= {967, 603, 1306, 963, 1013, 974, 329, 966, 978, 949}
```

The number of continued fraction terms in this list varies greatly. In the next line, we take 90 square roots and calculate the average number of terms.

```
In[76]:= Function[l, (Plus @@ l)/Length[l]][Length[ContinuedFraction[#]]& /@
            N[DeleteCases[Sqrt[Range[100]], _Integer], 1000]] // N
Out[76]= 963.767
```

This number is near to some of the numbers from above. Actually, it turns out that for almost all irrational x in $(0,1)$, the following holds: $\lim_{n \to \infty} k_n(x)/n = 6 \log(2) \log(10) \pi^{-2}$. This means that for a 1000-digit numerical approximation, we expect on average about 970 correct terms in the continued fraction approximation.

```
In[77]:= 1000 N[(6 Log[2] Log[10])/Pi^2]
Out[77]= 970.27
```

Let us look at the continued fraction form of the following real number: $0.\lfloor 1\,\pi\rfloor\lfloor 2\,\pi\rfloor\lfloor 3\,\pi\rfloor\ldots$ (the integer parts of multiples of π) [1022]. We generate a high-precision approximation of this number by using string operations.

In[78]:= `normalNumber = ToExpression["0." <> StringJoin[`
        `Table[ToString[IntegerPart[Pi i]], {i, 300}]]]`

Out[78]= 0.369121518212528313437404347505356596265697275788184879194971001031061091131 1
6119122125128131135138141144147150153157160163166169172175179182185188191194 1
9720120420721021321621922322622923223253238241245248251254257260263267270273276 1
2792822852892922952983013043073113143173203233263293333363393423453483513545435 1
8361364367370373376380383386389392395398402405408411414417420424427430433436 4
3944244646494524554584614644684714744774804834864904934964995025055085125155 18
5215245275305345375405435465495525556595625655685715745785815845875905935966 0
0603606660961261561862262562863163463763464646476506536656696626666696726756786 1
8168468869169469770070370670970971371671972727257287317357387417447477507537576 0
7637667697727757779782785788791794797801804807810813816819823826829832835838 84
1845848851854857860863867870873876879882885889892895898901904907911914917920 9
2392692993393693994

In[79]:= `Precision[normalNumber]`

Out[79]= 865.567

The rationalized version looks quite simple.

In[80]:= `normalNumberR = Rationalize[normalNumber]`

Out[80]= 18437619834715789256198347157892561983471578925619834715788015015015015020015 1
0150150150150150200150150150150150150200150150150150150200150150150150150150 1
5020015015015015015015020015015015015015015020015015015015015020015015015015015015020015 0
1501501501501501502001501501483063881667285712245303154343609439518029922575 1
8816672857127/
4994999999999999999999999999999999999999999999999999999999999999999999999999 1
9999999999999999999999999999999999999999999999999999999999999999999999999999 1
9999999999999999999999999999999999999999999999999999999999999999999999999999 1
9999999999999999999999999999999999999999999999999999999999999999999999999999 1
99999999999999999999999999999999995005000000000000000000000000000000000000000 1
00000000000000000000

`normalNumberR` and `normalNumber` agree to about 900 digits.

In[81]:= `normalNumberR - normalNumber`

Out[81]= $0.\times10^{-867}$

The continued fraction expansion of `interestingNumber` shows many small terms and a few very large ones.

In[82]:= `ContinuedFraction[normalNumber]`

Out[82]= {0, 2, 1, 2, 2, 3, 1, 1, 6, 1, 17, 3, 1, 1483, 3, 13, 3, 1, 1, 1, 2, 4, 3, 5, 1,
12, 1, 6, 1, 8, 1, 216994394417308015029172926, 1, 598, 1, 1, 23, 1, 2, 2,
8, 3, 2, 1, 2, 1, 1, 1, 14, 3, 38, 1, 3, 2, 1, 7, 4, 1, 9, 10, 8, 2, 1, 38, 1,
1, 1, 173, 2, 4, 5, 9, 1, 25, 6, 1, 1, 56, 10, 1, 4, 1, 3, 13, 574, 4, 1, 46, 3,
40080120160200240280360440520600680760840961081201321441561681842002162322482 1
6428030032034036038040042044446849251654056458616644672700728755679, 6,
1, 1, 1, 2, 1, 7, 10, 1, 2, 2, 6, 1, 1, 21, 1, 6, 5, 3, 8, 1, 4, 1, 3, 5, 1, 2, 3,
2, 1, 43, 11, 1, 6, 3, 1, 1, 1, 1, 1, 25, 2, 1, 3, 6, 1, 12, 1, 21, 2, 2, 2, 4, 4,
1, 2, 2, 3, 3, 1, 18, 3, 1, 6, 1, 2, 1, 32, 1, 6, 1, 1, 1, 3, 2, 1, 254, 7, 2, 1,

```
39, 6, 1, 2, 3, 1, 3, 17, 25, 12, 3, 3, 55, 4, 98, 1, 1, 1, 11, 1, 3, 1, 2, 1, 2,
5, 4, 2, 17, 1, 2, 1, 2, 39, 1, 7, 2, 40, 1, 1, 4, 19, 1, 1, 1, 2, 1, 6, 10, 3, 1,
251, 1, 2, 2, 1, 2, 1, 2, 1, 1, 5, 2, 5, 1, 1, 45, 1, 1, 8, 2, 4, 2, 1, 17, 1, 1,
1002003004005006007009011012 9, 1, 1, 1, 3, 131, 1, 3, 1, 4, 1, 195, 2, 2, 1,
100, 4, 1, 1, 5, 2, 2, 3, 2, 1, 8, 22, 5, 8, 6, 1, 2, 2, 1, 1, 1, 5, 1, 3, 3, 7, 3,
1, 3, 10, 1, 4, 2, 86, 49, 1, 3, 1, 1, 4, 7, 4, 1, 6, 1, 23, 8, 5, 2, 1, 1, 1, 65,
1, 1, 2, 1, 11, 2, 1, 19, 1, 2, 1, 1, 5, 1, 1, 2, 10, 4, 7, 2, 2, 4, 2, 3, 6, 1, 3,
2, 2, 1, 5, 2, 12, 1, 3, 6, 1, 3, 1, 3, 2, 1, 1, 3, 2, 1, 1, 2, 1, 1, 8, 1, 11, 1}
```

Using $a + b = a + 1/(0 + 1/b)$ it is possible to rewrite every regular continued fraction in a so-called canonical continued fraction, that, with the possible exception of its first digit, consists only of 0s and 1s [287].

```
In[83]:= CanonicalContinuedFraction[x_] :=
    With[{cf = ContinuedFraction[x]},
        Flatten[{cf[[1]], {Table[{1, 0}, {# - 1}], 1}& /@ Rest[cf]}]]
```

Here is the canonical continued fraction built from the first 10 digits of the regular continued fraction.

```
In[84]:= CanonicalContinuedFraction[N[1/E, 10]]

Out[84]= {0, 1, 0, 1, 1, 1, 0, 1, 1, 1, 1, 0, 1, 0, 1, 0, 1, 1, 1, 1, 0, 1, 0,
    1, 0, 1, 0, 1, 0, 1, 1, 1, 1, 0, 1, 0, 1, 0, 1, 0, 1, 0, 1, 0, 1, 0, 1}
```

We can recover the corresponding convergent using `FromContinuedFraction`.

```
In[85]:= FromContinuedFraction[%] - 1/E // N[#, 30]&

Out[85]= -9.1310001804331709631861910 2107×10⁻¹⁰
```

`ContinuedFraction` does currently not accept complex arguments [1605], [1606], [1607], [1326], [1325].

```
In[86]:= ContinuedFraction[1 + I Sqrt[3]]
```

    ContinuedFraction::noterms :
      $1 + i\sqrt{3}$ does not have a terminating or periodic continued fraction
       expansion; specify an explicit number of terms to generate. More...

```
Out[86]= ContinuedFraction[1 + i √3 ]
```

The digits of the continued fraction expansion $\{a_0, a_1, \ldots\}$ of a real number $x \in (0, 1)$ can be used to define the values of the Minkowski function $¿$ at x [1408], [1501]. (We visualized the Minkowski function in Subsection 1.2.2 of the Graphics volume [1794].)

$$¿(x) = 2 - 2\sum_{n=1}^{\infty} (-1)^n \, 2^{-\sum_{k=1}^{n} a_k}$$

```
In[87]:= ¿[x_] := 2 - 2 Plus @@ MapIndexed[(-1)^(#2[[1]] + 1) 2.^-#1&,
    FoldList[Plus, 0, Rest[ContinuedFraction[x]]]]
```

```
In[88]:= Plot[¿[x], {x, 0, 1}, Frame -> True, PlotRange -> All, PlotPoints -> 100]
```

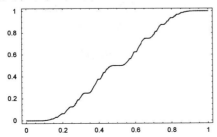

Here is a related function.

```
In[89]:= H[n_Integer, x_Real] := FromDigits[{#, 0}, n]& @
              Flatten[MapIndexed[Table[n - Mod[#2[[1]], n] - 1, {#1}]&,
                  ContinuedFraction[x]]]
```

```
In[90]:= Plot[Evaluate[Table[H[n, x], {n, 2, 12}]], {x, 0, 10},
          PlotPoints -> 200, PlotRange -> {0, 1}, Frame -> True,
          Axes -> False, FrameLabel -> {"x", None},
          PlotStyle -> Table[Hue[0.8 (n - 2)/12], {n, 2, 12}]]
```

Other applications of continued fractions in numerical analysis can be found in [916]. For doing arithmetic with continued fractions, see [770], [1864]; for differentiation see [1445]. For some very interesting continued fractions, see [985], [1825], and [1428]. For more on continued fractions in *Mathematica*, see the package Number Theory`ContinuedFractions` and [915] and [1173].

We could also visualize continued fractions by associating with each continued fraction a curve in the complex plane via

$$x = a_0 + \cfrac{1}{a_1 + \cfrac{1}{a_2 + \cfrac{1}{a_3 + \cfrac{1}{a_4 + \cdots}}}} \Longrightarrow x(\varphi) = a_0 + \cfrac{1}{a_1 e^{1 i \varphi} + \cfrac{1}{a_2 e^{2 i \varphi} + \cfrac{1}{a_3 e^{3 i \varphi} + \cfrac{1}{a_4 e^{4 i \varphi} + \cdots}}}}.$$

Here are the resulting curves shown for $x = e$, $x = \pi$, $x = K$, $x = 2^{1/2}$, and $x = 3^{1/3}$.

```
In[91]:= cfSpirographGraphic[x_, n_, φMax_, pp_, f_:Identity] :=
        Graphics[{PointSize[0.0025],
            Table[Point[{Im[#], Re[#]}]& @ FromContinuedFraction[
                (* parametrized curves *)
                MapIndexed[N[#1 Exp[f[#2[[1]]] I φ]]&,
                    (* cf digits *) ContinuedFraction[x, 10]]],
                {φ, 0, φMax, φMax/pp}]}, FrameTicks -> None,
            PlotRange -> Automatic, Frame -> True, AspectRatio -> 1]
```

```
In[92]:= Show[GraphicsArray[cfSpirographGraphic[#, 10, 2Pi, 10000]& /@
                    (* the five numbers to be used *)
                    {E, Pi, Khinchin, Sqrt[2], 3^(1/3)}]]
```

The next graphic uses $\varphi^{1/2}$ instead of φ. As a result, the curves are no longer periodic.

```
In[93]:= Show[GraphicsArray[cfSpirographGraphic[#, 10, 10 Pi, 20000, Sqrt]& /@
                (* the five numbers to be used *)
                {E, Pi, Pi/E, 3^(1/3), Cos[1]}]]
```

From a broader context, continued fraction expansions and base b expansions are special cases of iterations of bivariate rational functions $r(x, y)$ [1122], [1232], [1627], [1450], [662], [1231]. $r(x, y) = y + 1/x$ results in continued fractions.

```
In[94]:= Fold[Function[{x, y}, y + 1/x], z5, {z4, z3, z2, z1}]
```

$$Out[94]= z1 + \cfrac{1}{z2 + \cfrac{1}{z3 + \cfrac{1}{z4 + \frac{1}{z5}}}}$$

$r(x, y) = y + x/10$ results in decimal expansions.

```
In[95]:= Fold[Function[{x, y}, y + x/10], z5, {z4, z3, z2, z1}] // Expand
```

$$Out[95]= z1 + \frac{z2}{10} + \frac{z3}{100} + \frac{z4}{1000} + \frac{z5}{10000}$$

$r(x, y) = (xy + 1)/(x - y)$ results in continued cotangents. $(xy + 1)/(x - y)$ can be rewritten as $\cot(\text{arccot}(x) - \text{arccot}(y))$ [1123], [1645].

```
In[96]:= Clear[x, y];
        Together //@ TrigToExp[Cot[ArcCot[y] - ArcCot[x]]]
```

$$Out[97]= \frac{1 + x y}{x - y}$$

Here is a "one-line" implementation of the continued cotangent expansion.

```
In[98]:= ContinuedCotangent[x_, n_] :=
        Cot[Plus @@ MapIndexed[(-1)^(#2[[1]] - 1) #&, ArcCot[Last /@
        NestList[Function[{a, b}, {#, Floor[#]}&[(a b + 1)/(a - b)]] @@ #&,
                {x, Floor[x]}, n]]]]
```

Here are the first terms of the continued cotangent expansion of Euler's constant e.

```
In[99]:= ContinuedCotangent[N[E, 200], 8]
```

```
Out[99]= Cot[ArcCot[2] - ArcCot[8] + ArcCot[75] - ArcCot[8949] + ArcCot[119646723] -
         ArcCot[15849841722437093] + ArcCot[70865758016338206583629213377 4995] - ArcCot[
           529026553215766321676623343348414600292754204772300344704877695232] + ArcCot[
           515242553865726035688292036059153719454546961626988807336855829294899338742 ·
           145207767435587561691068553657192491959216959592398323714]]
```

The expansion shows excellent numerical agreement.

In[100]:= **N[% - E, 22]**

Out[100]= $2.67036174016952481 0912 \times 10^{-210}$

We now study the rounding of numbers to integers. This is accomplished with Round, Floor, and Ceiling.

Round[*number*]

    finds the integer that is closest to *number*.

If *number* is halfway between two integers and is a fraction (head Rational or Real), Round gives the closest even integer. This happens for exact numbers, machine numbers, and high-precision numbers.

In[101]:= **Union[(Round /@ N[{3/2, 5/2, -3/2, -5/2}, #])& /@**
           (* exact, machine, and high-precision numbers *)
                  {Infinity, MachinePrecision, $MachinePrecision}]

Out[101]= {{2, 2, -2, -2}}

Note that rounding is not a completely trivial problem, for example, if the numbers to be rounded are percentages, and we want the final set to still add to 100%. For more on this problem, see [122], [123], and [520] and the references therein.

Floor[*number*]

    produces the largest integer that is less than or equal to *number*.

Ceiling[*number*]

    produces the smallest integer that is greater than or equal to *number*.

Here are a few examples in tabular form.

In[102]:= **TableForm[**
    {#, Round[#], Floor[#], Ceiling[#]}& /@
    {-3, -2.8, -2.5, -5/2, -2.1, -2, -0.5, -1/2, 0,
     1/2, 0.5, 11/10, 3/2, 1.5, 1.8, 2.5},
    TableHeadings ->
    {None, StyleForm[#, FontWeight -> "Bold"]& /@
        {"x", "Round[x]", "Floor[x]", "Ceiling[x]"}},
     TableAlignments -> {Center, Center}, TableSpacing -> {1, 1}]

Out[102]//TableForm=

x	**Round[x]**	**Floor[x]**	**Ceiling[x]**
-3	-3	-3	-3
-2.8	-3	-3	-2
-2.5	-2	-3	-2
$-\frac{5}{2}$	-2	-3	-2
-2.1	-2	-3	-2
-2	-2	-2	-2
-0.5	0	-1	0
$-\frac{1}{2}$	0	-1	0
0	0	0	0
$\frac{1}{2}$	0	0	1
0.5	0	0	1
$\frac{11}{10}$	1	1	2

$\frac{3}{2}$	2	1	2
1.5	2	1	2
1.8	2	1	2
2.5	2	2	3

We now look at these three functions graphically.

```
In[103]:= Show[GraphicsArray[
        Plot[{x, #[x]}, {x, -3, 3},
            DisplayFunction -> Identity, PlotLabel ->
            StyleForm[ToString[#] <> "[x]",
                FontFamily -> "Courier", FontSize -> 5],
            TextStyle -> {FontWeight -> "Bold", FontSize -> 5},
            PlotStyle -> {{Thickness[0.002], Dashing[{0.015, 0.015}]},
                        {Thickness[0.03]}}]& /@
        (* the three functions of interest *) {Round, Floor, Ceiling}]]
```

Here are some more interesting pictures based on the `Floor` function. We visualize the nontrivial q-fold iterates of the map $\{m,\ n\} \to \{\lfloor 2\cos(2\pi p/q)\,m \rfloor - n,\ m\}$ [1176], [1960].

```
In[104]:= (* the map *)
        φ[v_][{m_, n_}] := {Floor[2 Cos[2 Pi v] m] - n, m}
        (* the iterated map *)
        Φ[v_Rational][{m0_, n0_}] := Nest[φ[v], {m0, n0}, Denominator[v]]

In[108]:= Show[GraphicsArray[
        Graphics[{Thickness[0.002],
        (* start from all integer pairs within a square *)
        DeleteCases[Table[Line[{{m0, n0}, Φ[#][{m0, n0}]}],
                    {m0, -40, 40}, {n0, -40, 40}],
                Line[{a_, a_}], Infinity]},
            AspectRatio -> Automatic]& /@
        (* three selected p/q *) {1/5, 2/9, 11/14}]]
```

The `Floor` and `Ceiling` functions can be used to construct closed-form expressions for many "unusual" sequences [1361]. $\lceil \sqrt{\lceil 2k \rceil} + 1/2 \rceil - 1$ gives the sequence consisting of one 1, two 2's, three 3's and so on.

```
In[109]:= Table[Ceiling[Sqrt[Ceiling[2k]] + 1/2] - 1, {k, 16}]
```

Out[109]= {1, 2, 2, 3, 3, 3, 4, 4, 4, 4, 5, 5, 5, 5, 5, 6}

Here is another possibility to encode this sequence.

In[110]:= **Table[Floor[(Sqrt[8k - 7] - 1)/2] + 1, {k, 16}]**
Out[110]= {1, 2, 2, 3, 3, 3, 4, 4, 4, 4, 5, 5, 5, 5, 5, 6}

The sequence of all nonsquare numbers can be written as $\lfloor k + (k + k^{1/2})^{1/2} \rfloor$.

In[111]:= **Table[Floor[k + Sqrt[k + Sqrt[k]]], {k, 16}]**
Out[111]= {2, 3, 5, 6, 7, 8, 10, 11, 12, 13, 14, 15, 17, 18, 19, 20}

Be aware that the functions Ceiling, Floor, and Round, when operating on high precision numbers, take the guard digits into account. This means, when carrying out such operations, one might obtain potentially wrong answers. Here is an example where the upwards rounding is not justified by the known-to-be-correct digits.

In[112]:= **Round[N[5/2, 30] + 10^-40]**
Out[112]= 3

The potential failure of significance arithmetic in this example is not unexpected. One of its assumptions—continuous dependence of functions from their arguments—is violated here.

Here is a little calculation that, although using high-precision numbers gives a wrong result. The following is a (slightly unusual) closed form formula for the *n*th prime number [1564].

In[113]:= **myPrime[n_] := Sum[1 - Floor[**
            Sum[1 + Floor[(2 - Sum[Floor[j/i] -
                Floor[(j - 1)/i], {i, 1, j}])/j],
                   {j, 2, k}]/n], {k, 2, Floor[2 n Log[n] + 2]}]

The 17th prime is 57. Using exact rational arithmetic, we get this result. With 30-digit high-precision numbers, we get the result 59.

In[114]:= **{myPrime[17], myPrime[N[17, 30]]}**
Out[114]= {57, 59}

The inner sum for $k = 59$ gives, in exact arithmetic, the result 0; in floating point arithmetic the result 1.

In[115]:= **ιΣ[k_, n_] := Sum[1 + Floor[(2 - Sum[Floor[j/i] -**
                           Floor[(j - 1)/i], {i, 1, j}])/j], {j, 2, k}]/n
In[116]:= **{ιΣ[59, 17], ιΣ[59, N[17, 30]]}**
Out[116]= {1, 1.00000000000000000000000000000}

The actual difference of the high-precision one to 1 is less than 10^{-45}, far less than the precision of the number.

In[117]:= **((* load context of function $NumberBits *)**
         If[FreeQ[$ContextPath, #], AppendTo[$ContextPath, #]]& /@ Context /@
              Union[Join[Names["*\`\*Number\*Bits\*"], Names["\*\`*Bits*Number*"]]];
         (* definition from above *)
         toExactNumberAll[l_List] := (* use all bits *)
         First[l] 2^Last[l] Join[l[[2]], l[[3]]].
                     Table[2^-i, {i, Length[l[[2]]] + Length[l[[3]]]}])
In[118]:= **{Floor[%%], N[1 - toExactNumberAll[$NumberBits[%%[[2]]]]]}**
Out[118]= {{1, 0}, 6.84228×10^-49}

Repeating the calculation with 1000 digits still gives a wrong result (and even using one million digits would still give a wrong result).

```
In[119]:= {Floor[/Σ[59, N[17, 1000]]],
        N[1 - toExactNumberAll[$NumberBits[/Σ[59, N[17, 1000]]]]]}
Out[119]= {0, 3.461103360517829×10^-1012}
```

Let us use `Floor` to construct an interesting function, one that is monotonic, nondecreasing, and noncontinuous at all rational points. It is the so-called Frisch function [571]. We start by implementing the nth-order approximation `FrischF[n, x]` to the Frisch function.

```
In[120]:= FrischF[x_, n_] := Sum[If[Floor[k x] == Floor[(k + 1) x], 0, 1] 2^-k, {k, n}]
```

Here is a graphic of the first 20 approximations. We use all fractions with denominators less than or equal to 50.

```
In[121]:= fractions = Union[Flatten[Table[i/j, {j, 50}, {i, 0, j}]]];
```

```
In[122]:= Show[Graphics[Reverse[Table[{Hue[n/20 0.79],
        Line[{#, FrischF[#, n]}& /@ fractions]}, {n, 20}]]], Frame -> True]
```

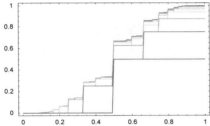

For rational x, it is possible to calculate exact values (equal to the limit $n \to \infty$ of the above approximations of the Frisch function).

```
In[123]:= FrischF[x_Rational] :=
        Module[{q = Denominator[x]},
            (Sum[If[Floor[k x] == Floor[(k + 1) x], 0, 1] 2^-k, {k, q - 2}] +
            2^(1 - q))/(1 - 2^-q)] /; 0 < x < 1
```

```
In[124]:= FrischF[0] := 0;
        FrischF[1] := 1
```

```
In[126]:= ListPlot[{#, FrischF[#]}& /@ fractions,
            PlotJoined -> True, PlotStyle -> {Thickness[0.002], Hue[0]}]
```

By combining the commands `Floor`, `Ceiling`, and `Round` in a linear map, a large variety of functions can be constructed [535].

For complex arguments, the commands Floor, Ceiling, and Round also work. In this case, the real and the imaginary parts are treated independently. Here, this is visualized by lines connecting the complex *z* with *floorOrCeilingOrRound*[z].

```
In[127]:= Show[GraphicsArray[
    Function[{floorOrCeilingOrRound},
    (* the three pictures for Floor, Ceiling, and Round *)
    Graphics[
    Table[Line[{{Re[#], Im[#]}, {Re[#], Im[#]}&[
        floorOrCeilingOrRound[#]]}]]&[
        (* random complex arguments *) Random[Complex, {-6 - 6I, 6 + 6I}]],
        {2500}],
        PlotLabel -> StyleForm[ToString[floorOrCeilingOrRound] <> "[x + I y]",
                            FontFamily -> "Courier"],
        AspectRatio -> Automatic, PlotRange -> {{-6, 6}, {-6, 6}},
        Frame -> True, FrameLabel -> {"x", "y"},
        GridLines -> ({#, #}&[{#, {GrayLevel[0.7]}}& /@ Range[-6, 6]])]] /@
            (* the three integer-valued functions *) {Floor, Ceiling, Round}]]
```

After the discussion of the foundations of Rationalize—continued fractions—we will continue to discuss the function Rationalize itself. Let us now use Rationalize to have a closer look at Plus and Times for machine numbers. Machine arithmetic is (in distinction to exact arithmetic) not associative. In the following, we will look for explicit examples of numbers exhibiting this nonassociativeness in Plus and Times. Rationalize is a convenient way to test weather two numbers agree in all bits.

The following While loop loops until it finds three numbers that exhibit the mentioned nonassociativeness.

```
In[128]:= SeedRandom[1];
    While[a = Random[]; b = Random[]; c = Random[];
        aPLUSb = a + b; bPLUSc = b + c;
        Rationalize[aPLUSb + c, 0] === Rationalize[a + bPLUSc, 0], Null]
```

These are the numbers found.

```
In[130]:= {a, b, c} // InputForm
Out[130]//InputForm=
    {0.12463419022903224, 0.9345372830959425, 0.6002520340808316}
```

In InputForm, one does not see a difference.

```
In[131]:= {a + b + c, aPLUSb + c, a + bPLUSc} // InputForm
Out[131]//InputForm=
    {1.6594235074058066, 1.6594235074058066, 1.6594235074058064}
```

The difference of the two numbers is in their last bit.

```
In[132]:= {#, N[Subtract @@ #]}& @ Rationalize[{aPLUSb + c, a + bPLUSc}, 0]
```
$$Out[132]= \left\{\left\{\frac{243680247}{146846327}, \frac{564919750}{340431329}\right\}, 2.60046 \times 10^{-16}\right\}$$

```
In[133]:= #[[3]]& /@ $NumberBits /@ {aPLUSb + c, a + bPLUSc}
```
```
Out[133]= {{1, 1, 0, 1, 0, 1, 0, 0, 0, 1, 1, 0, 0, 1, 1, 1, 1, 1, 1, 1, 1, 1, 0, 1, 0, 1, 0,
          0, 1, 1, 1, 1, 0, 1, 0, 0, 0, 0, 1, 0, 1, 1, 0, 0, 0, 0, 1, 0, 0, 1, 0, 0, 0},
         {1, 1, 0, 1, 0, 1, 0, 0, 0, 1, 1, 0, 0, 1, 1, 1, 1, 1, 1, 1, 1, 1, 0, 1, 0, 1, 0,
          0, 1, 1, 1, 1, 0, 1, 0, 0, 0, 0, 1, 0, 1, 1, 0, 0, 0, 0, 1, 0, 0, 0, 1, 1, 1}}
```

For randomly chosen triplets of numbers, the associative property will seldom hold. We generate 1000 triples of numbers randomly and count for how many the identity $(a + b) + c \neq a + (b + c)$ holds.

```
In[134]:= counter = 0;
          (* 1000 time pick three random number and test associativity *)
          Do[a = Random[]; b = Random[]; c = Random[];
            aPLUSb = a + b; bPLUSc = b + c;
            If[Rationalize[aPLUSb + c, 0] === Rationalize[a + bPLUSc, 0],
              counter = counter + 1], {1000}];
          counter
Out[137]= 772
```

The following input repeats all of the above steps for multiplication.

```
In[138]:= While[a = Random[]; b = Random[]; c = Random[];
            aTIMESb = a b; bTIMESc = b c;
            Rationalize[aTIMESb c, 0] === Rationalize[a bTIMESc, 0], Null]
```

```
In[139]:= {a, b, c} // InputForm
Out[139]//InputForm=
          {0.5820543182231254, 0.8508073527787985, 0.8097237287551128}
```

```
In[140]:= {aTIMESb c, a bTIMESc} // InputForm
Out[140]//InputForm=
          {0.4009882218986335, 0.40098822189863353}
```

```
In[141]:= {#, N[Subtract @@ #]}& @ Rationalize[{aTIMESb c, a bTIMESc}, 0]
```
$$Out[141]= \left\{\left\{\frac{176375967}{439853236}, \frac{72321951}{180359290}\right\}, -7.56319 \times 10^{-17}\right\}$$

```
In[142]:= #[[3]]& /@ $NumberBits /@ {aTIMESb c, a bTIMESc}
```
```
Out[142]= {{1, 1, 0, 0, 1, 1, 0, 1, 0, 1, 0, 0, 1, 1, 1, 0, 0, 1, 0, 1, 0, 1, 0, 0, 0, 0, 0,
          0, 0, 1, 1, 0, 0, 1, 0, 0, 0, 1, 0, 1, 1, 0, 0, 0, 1, 0, 1, 0, 0, 1, 1, 1, 0},
         {1, 1, 0, 0, 1, 1, 0, 1, 0, 1, 0, 0, 1, 1, 1, 0, 0, 1, 0, 1, 0, 1, 0, 0, 0, 0, 0,
          0, 0, 1, 1, 0, 0, 1, 0, 0, 0, 1, 0, 1, 1, 0, 0, 0, 1, 0, 1, 0, 0, 1, 1, 1, 1}}
```

```
In[143]:= counter = 0;
          Do[a = Random[]; b = Random[]; c = Random[];
            aTIMESb = a b; bTIMESc = b c;
            If[Rationalize[aTIMESb c, 0] === Rationalize[a bTIMESc, 0],
              counter = counter + 1], {1000}];
          counter
Out[145]= 720
```

Another application for `Rationalize` is to have a "close" look at the result of applying simple functions to numbers that are very near to each other with respect to machine arithmetic. The function `maDiff` gives the difference between the machine arithmetic and the high-precision result measured in units of `$MachineEpsi-lon`.

```
In[146]:= $machineEpsilon = Rationalize[$MachineEpsilon, 0];

    (* difference in units of $MachineEpsilon *)
    maDiff[func_, x0_, i_] :=
        (Rationalize[func[N[x0 + i $machineEpsilon]], 0] -
         Rationalize[func[N[x0 + i $machineEpsilon, 50]], 0])/
         $machineEpsilon

In[149]:= diffPlot[f_, opts___] :=
    ListPlot[Table[{i, maDiff[f, 1, i]}, {i, -250, 250}],
            DisplayFunction -> Identity, PlotRange -> All,
            PlotLabel -> StyleForm[ToString[InputForm[f]],
            FontFamily -> "Courier", FontSize -> 6]];

In[150]:= Show[GraphicsArray[diffPlot /@ #]]& /@
        {{Sin, Log}, {Exp, Sqrt}, {#^3&, 1/(1 + #)&}}
```

We conclude this section with an application of Ceiling to Egyptian fractions. Although it has no direct relevance to numerics, it makes heavy use of Ceiling.

Mathematical Application: Egyptian Fractions

The ancient Egyptians were familiar only with "unit fractions", that is, those whose numerator is 1, such as $1/2, 1/3, 1/4, \ldots$. General fractions can always be written as sums of distinct unit fractions. Starting with a given fraction p/q ($p < q$), we first find the largest unit fraction $1/r$ less than or equal to p/q, and write $p/q = 1/r + p'/q'$. We then repeat this decomposition for p'/q', p''/q'', etc., until the remainder is 0.

The largest fraction $1/r$ less than or equal to a given fraction p/q can be found in *Mathematica* using $r =$ Ceiling[q/p].

Here is the decomposition of a fraction into the largest unit fraction plus a remainder.

```
In[151]:= fractionDecomposition[0] = 0;

        fractionDecomposition[pq_Rational?(# < 1&)] :=
                {1/Ceiling[1/pq], pq - 1/Ceiling[1/pq]}
```

Here is an example.

```
In[153]:= fractionDecomposition[31/789]
```

$$\text{Out[153]}= \left\{ \frac{1}{26}, \frac{17}{20514} \right\}$$

Here is a test.

```
In[154]:= Plus @@ %
```

$$\text{Out[154]}= \frac{31}{789}$$

Next, we show the result of repeating this fractional decomposition (until it stops naturally), using fractionDe composition. Observe the use of FixedPointList. At the outset, we do not know how many terms will be needed in the Egyptian decomposition. The use of Drop[..., -2] in the following is necessary to remove the last two zeros.

```
In[155]:= decomposition[pq_Rational?(# < 1&)] :=
        Drop[FixedPointList[If[# =!= 0, fractionDecomposition[#[[2]]], 0]&,
                        {0, pq} ], -2]
```

Now, we have the complete decomposition.

```
In[156]:= {#, (* check *) Plus @@ (First /@ #)}& @ decomposition[31/789]
```

$$\text{Out[156]}= \left\{ \left\{ \left\{0, \frac{31}{789}\right\}, \left\{\frac{1}{26}, \frac{17}{20514}\right\}, \left\{\frac{1}{1207}, \frac{5}{24760398}\right\}, \right. \right.$$
$$\left. \left. \left\{\frac{1}{4952080}, \frac{1}{61307735863920}\right\}, \left\{\frac{1}{61307735863920}, 0\right\} \right\}, \frac{31}{789} \right\}$$

Finally, we have the definition.

```
In[157]:= EgyptianFractions[pq_Rational?(# < 1&)] := Rest[First /@ decomposition[pq]]
```

Here is an example illustrating a weak point (at least for practical applications) of EgyptianFractions.

```
In[158]:= (ef = EgyptianFractions[36/457]);
        Take[ef, 6]
```

$$\text{Out[159]}= \left\{ \frac{1}{13}, \frac{1}{541}, \frac{1}{321409}, \frac{1}{114781617793}, \frac{1}{14821672255960844346913}, \right.$$
$$\left. \frac{1}{251065106814993628596500876449600804290086881} \right\}$$

For brevity, we display the large integers of the denominators of `ef` in abbreviated form. $\mathcal{R}_{k(n)}$ stands for $1/n$ where n is an integer with $k(n)$ decimal digits.

```
In[160]:= abbreviateLargeUnitFractions[expr_] := expr //.
              Rational[1, i_?(# > 10^10&)] :> Subscript[R, Floor[Log[10, i]]]
```

```
In[161]:= ef // abbreviateLargeUnitFractions
```

$$Out[161]= \left\{ \frac{1}{13}, \frac{1}{541}, \frac{1}{321409}, \mathcal{R}_{11}, \mathcal{R}_{22}, \mathcal{R}_{44}, \mathcal{R}_{88}, \mathcal{R}_{177}, \mathcal{R}_{355}, \mathcal{R}_{711}, \mathcal{R}_{1423}, \mathcal{R}_{2846} \right\}$$

This example shows that even for relatively small denominators in the initial fraction, we can get extremely large denominators in the unit fractions.

```
In[162]:= Length[Last[IntegerDigits[Denominator[ef]]]]
```

```
Out[162]= 2847
```

Once again, here is a check.

```
In[163]:= Plus @@ ef
```

$$Out[163]= \frac{36}{457}$$

It is now not too difficult to find automatically such fractions with huge denominators. For a given set of fractions, we look for those whose last summand is smallest. We take this set of fractions to be those whose denominators are smaller than a prescribed maximum value.

```
In[164]:= biggestDenominator[maxDenominator_] :=
          (* take "best" one *)
          #[[1, Position[#[[2]], Min[#[[2]]]][[1, 1]]]]&[
          (* EgyptianFraction-ize all fractions *)
            {#, Last[EgyptianFractions[#]]& /@ #}&[
          (* all possible fractions *)
              Union[Flatten[Table[i/j, {j, maxDenominator}, {i, j - 1}]]]]]
```

Here is an example.

```
In[165]:= biggestDenominator[25]
```

$$Out[165]= \frac{4}{25}$$

```
In[166]:= EgyptianFractions[%]
```

$$Out[166]= \left\{ \frac{1}{7}, \frac{1}{59}, \frac{1}{5163}, \frac{1}{53307975} \right\}$$

Here is an example showing a denominator with 149 digits.

```
In[167]:= EgyptianFractions[8/97]
```

$$Out[167]= \left\{ \frac{1}{13}, \frac{1}{181}, \frac{1}{38041}, \frac{1}{1736503177}, \right.$$

$$\frac{1}{3769304102927363485}, \frac{1}{18943537893793408504192074528154430149},$$

$$\frac{1}{53828644190038021136581728510490708634743974613022697325377813249422 5813153},$$

$$1 /$$

$$5795045870675428017131031918599186082510302919521954235835293576538994186863 \cdot$$

$$\left. 4236036179868905327374937261504366181022837189853958386201142499390978 9665 \right\}$$

For fractions with denominators less than 100, 8/97 is "best"—the smallest fraction contains a 150-digit number in its denominator.

The representation of a fraction as an Egyptian fraction is not unique; here, the so-called splitting method

[1710], [151] is implemented (the implementation is such that the method becomes obvious; from a computational point of view, it can be considerably accelerated using functions like `Split` instead of pattern matching of sequences of undetermined length).

```
In[168]:= EgyptianFractionsViaSplitting[Rational[p_, q_]?Positive] :=
          1/FixedPoint[Flatten[Sort[# //. {a___, b_, b_, c___} :>
                      {a, b, b + 1, b(b + 1), c}]]&, Table[q, {p}]]
```

Here is an example and the corresponding result of `EgyptianFractions`.

```
In[169]:= EgyptianFractionsViaSplitting[5/7]
```

$$
Out[169]= \left\{ \frac{1}{7}, \frac{1}{8}, \frac{1}{9}, \frac{1}{10}, \frac{1}{11}, \frac{1}{56}, \frac{1}{57}, \frac{1}{58}, \frac{1}{59}, \frac{1}{72}, \frac{1}{73}, \frac{1}{74}, \frac{1}{90}, \frac{1}{91}, \frac{1}{110}, \right.
$$
$$
\frac{1}{3192}, \frac{1}{3193}, \frac{1}{3194}, \frac{1}{3306}, \frac{1}{3307}, \frac{1}{3422}, \frac{1}{5256}, \frac{1}{5257}, \frac{1}{5402}, \frac{1}{8190},
$$
$$
\left. \frac{1}{10192056}, \frac{1}{10192057}, \frac{1}{10198442}, \frac{1}{10932942}, \frac{1}{27630792}, \frac{1}{103878015699192} \right\}
$$

```
In[170]:= Plus @@ %
```

$$
Out[170]= \frac{5}{7}
$$

```
In[171]:= EgyptianFractions[5/7]
```

$$
Out[171]= \left\{ \frac{1}{2}, \frac{1}{5}, \frac{1}{70} \right\}
$$

If one allows negative fractions [1014], it is possible to decompose a given fraction into Egyptian fractions with much smaller denominators. The important function in the following method [220] is the use of `Continued`. `Fraction`.

```
In[172]:= EgyptianFractionsViaContinuedFractions[Rational[p_, q_]?Positive] :=
          Module[{makeUnitFraction},
            (* extract one unit fraction *)
            makeUnitFraction[ Rational[1, r_]] := { 1/r};
            makeUnitFraction[-Rational[1, r_]] := {-1/r};
            makeUnitFraction[f_Rational] :=
            {f - #, #}&[Numerator[#]/Denominator[#]&[
            FromContinuedFraction[Take[#, Length[#] - 1]]&[ContinuedFraction[f]]]];
            (* iterate the application of makeUnitFraction *)
               First /@ Rest[NestWhileList[makeUnitFraction[#[[2]]]&,
                              {Null, p/q}, Length[#1] === 2&]]]
```

Here are some of the above fractions treated with `EgyptianFractionsViaContinuedFractions`. Now, we get negative fractions and 1, but have much smaller denominators.

```
In[173]:= {#, Plus @@ #}& @ EgyptianFractionsViaContinuedFractions[5/7]
```

$$
Out[173]= \left\{ \left\{ \frac{1}{21}, -\frac{1}{3}, 1 \right\}, \frac{5}{7} \right\}
$$

```
In[174]:= {#, Plus @@ #}& @ EgyptianFractionsViaContinuedFractions[8/97]
```

$$
Out[174]= \left\{ \left\{ -\frac{1}{1164}, \frac{1}{12} \right\}, \frac{8}{97} \right\}
$$

```
In[175]:= {#, Plus @@ #}& @ EgyptianFractionsViaContinuedFractions[36/457]
```

$$
Out[175]= \left\{ \left\{ -\frac{1}{75405}, -\frac{1}{6270}, \frac{1}{494}, \frac{1}{13} \right\}, \frac{36}{457} \right\}
$$

For more on Egyptian fractions, see [558], [221], [1865], [849], [220], [1714], [1826], [552], and http://www.ics.uci.edu/eppstein/.

1.1.4 When N Does Not Succeed

As already mentioned, N[*expression, precision*] calculates *expression* to precision *precision*. Here is a simple example.

```
In[1]:= N[(Sqrt[2] - 3)/(Pi + E), 200]
Out[1]= -0.27061781655644828026000719249218870714570984222587876667976980643198441253\
        4485271993747823709967966983358930079614284440691659543581721893374452505679\
        3693187332135557895453304827109255227475240809358
```

```
In[2]:= Precision[%]
Out[2]= 200.
```

For the majority of all inputs N will succeed in achieving the required precision. But sometimes it might fail to achieve its goal.

Here is a small number representing the difference between π and a rational approximation of π to 100 digits.

```
In[3]:= diff = SetPrecision[N[Pi, 100], Infinity] - Pi
Out[3]= 137429016170745219382391623152025666371669894951468010510891139570723086718398\
        5741844377679649595395/
        437450144956602384874500445423524273070633886178642487285154121281990599839\
        7518464470263540461076488 - π
```

Trying to calculate the diff to 100 digits fails.

```
In[4]:= N[diff, 200]
        N::meprec :
          Internal precision limit $MaxExtraPrecision = 50.` reached while evaluating
          1374290161707452193823916231520256 ≪33≫ 2308671839857418443776796495953395
          ───────────────────────────────────────────────────────────────────────────
          4374501449566023848745004454235242 ≪32≫ 9905998398751846447026354046107648 − π. More…
Out[4]= 8.28551615444109862276848680624048607110739111267411803983267415393818416257052\
        2271778083204311450552542896084126587051784324557325526734291767777482109× 10⁻¹⁰¹
```

The function in the message issued by the last input is $MaxExtraPrecision.

$MaxExtraPrecision

     specifies the additional precision, to be used in internal calculations.

The current value of $MaxExtraPrecision is 50 (compared to the default value infinity of $MaxPrecision).

```
In[5]:= $MaxExtraPrecision
Out[5]= 50.
```

If we increase this value, we can calculate the diff to 100 digits.

```
In[6]:= $MaxExtraPrecision = 200;
        N[diff, 200]
Out[7]= 8.28551615444109862276848680624048607110739111267411803983267415393818416257052\
        22717780832043114505525428960841265870517843245573255267342917677748210922481\
        050676676602050850770128836090549208447675278 3× 10⁻¹⁰¹
```

The following input defines a function `lowerPrecision` that returns a number with a much lower precision than its input. Printing the precision of the input allows us to monitor the way *Mathematica* internally tries to increase the precision of the input to achieve the required precision.

```
In[8]:= $MaxExtraPrecision = 20000;

    SetAttributes[lowerPrecision, NumericFunction];

    lowerPrecision[x_?InexactNumberQ] :=
        (Print[Precision[x]];
         (* output precision is a slowly increasing function of input precision *)
         SetPrecision[x, Round[Sqrt[Precision[x]]]]);

    (* try to obtain the function value at 1 to 100 digits *)
    N[lowerPrecision[1], 100]
```

```
        100.

        209.934

        429.802

        869.538

        1749.01

        3507.95

        7025.84

        9277.89

        10029.2
```

Out[13]= 1.000000000000000000000000000000000000000000000000000000000000000000000000000000 00000000000000000000000000

For expressions that are not functions, we cannot use the `NumericFunction` attribute. In such cases, we can directly set the `NumericQ` property. A typical situation are symbolic derivatives of functions.

Numerical techniques, including `$MaxExtraPrecision` are not only used in N, but in some other functions, such as `Equal`, `Less`, `Greater`, `Floor`, `Ceiling`, and so on. The next input shows that the numerical comparison functions (like `Less`) also use numerical techniques.

```
In[14]:= $MaxExtraPrecision = 50;
       E < 100286898419598513365/36893488147419103232
Out[15]= False
```

A mathematical equality of not structurally-identical exact quantities can never return `True` based on numerical verification methods. Numericalization can only show that an equality does not hold. With the default setting of `$MaxExtraPrecision`, *Mathematica* cannot prove by validated numericalization that the left-hand side and the right-hand side are numerically different.

```
In[16]:= EA = Rationalize[E, 10^-80];
       E == EA
```

N::meprec : Internal precision limit $MaxExtraPrecision = 50.` reached
while evaluating $-\dfrac{9836510838135482898661487252572 4254586156}{3618650110210117185514126098867 8169767695} + $ e. More...

Out[17]= $\mathbb{e} == \dfrac{9836510838135482898661487252572424254586156}{3618650110210117185514126098867816769695}$

The left-hand side of the following equation is identically zero.

In[18]:= `Sqrt[2] + Sqrt[3] - Sqrt[5 + 2 Sqrt[6]] == 0`

    N::meprec : Internal precision limit $MaxExtraPrecision =
        50.` reached while evaluating $\sqrt{2} + \sqrt{3} - \sqrt{5 + 2\sqrt{6}}$. More...

Out[18]= $\sqrt{2} + \sqrt{3} - \sqrt{5 + 2\sqrt{6}} == 0$

Because the precision of (the numerically calculated) 0 is always zero and not a single significant digit can be found, increasing $MaxExtraPrecision does not help in this example.

In[19]:= `$MaxExtraPrecision = 500;`
       `Sqrt[2] + Sqrt[3] - Sqrt[5 + 2 Sqrt[6]] == 0`

    N::meprec : Internal precision limit $MaxExtraPrecision =
        500.` reached while evaluating $\sqrt{2} + \sqrt{3} - \sqrt{5 + 2\sqrt{6}}$. More...

Out[20]= $\sqrt{2} + \sqrt{3} - \sqrt{5 + 2\sqrt{6}} == 0$

If one writes routines for numerical calculation, it is often useful to make leverage from the built-in precision-raising mechanism of N. The following function `badlyConditioned` mimics a (potentially large) calculation in which a lot of precision is lost. We restrict its input to approximative numbers. (For exact input, many numerical routines will take a much longer time.)

In[21]:= `$MaxExtraPrecision = 50;`
       `badlyConditioned[x_?InexactNumberQ] := SetPrecision[x, Precision[x]/2]`

Now, we try to get a 100-digit result for `badlyConditioned[2]`.

In[23]:= `N[badlyConditioned[2], 100]`

    N::meprec : Internal precision limit $MaxExtraPrecision =
        50.` reached while evaluating badlyConditioned[2]. More...

Out[23]= `2.00000000000000000000000000000000000000000000000000000000000000000000000000000`

It did not work!

In[24]:= `Precision[%]`

Out[24]= `75.`

Why? N numericalizes on numeric quantities properly, but `badlyConditioned[2]` is not a numeric quantity.

In[25]:= `NumericQ[badlyConditioned[2]]`

Out[25]= `False`

This means we must give the function `badlyConditioned` the attribute `NumericFunction`.

In[26]:= `Remove[badlyConditioned];`

       `SetAttributes[badlyConditioned, NumericFunction];`

       `badlyConditioned[x_?InexactNumberQ] := SetPrecision[x, Precision[x]/2]`

Now, `badlyConditioned[2]` is a numeric object.

In[29]:= `NumericQ[badlyConditioned[2]]`

Out[29]= `True`

The built-in precision-raising technology now goes to work.

In[30]:= **N[badlyConditioned[2], 100]**

       N::meprec : Internal precision limit $MaxExtraPrecision =
          50.` reached while evaluating badlyConditioned[2]. More…

Out[30]= 2.000000000000000000000000000000000000000000000000000000000000000000000000000

In[31]:= **Precision[%]**

Out[31]= 75.

We did not succeed in getting the required precision, but this can now be cured by raising the value of $MaxEx ⋮
traPrecision.

In[32]:= **$MaxExtraPrecision = 200;**
    N[badlyConditioned[2], 100]

Out[33]= 2.0000000000000000000000000000000000000000000000000000000000000000000000000000 ⋮
    000000000000000000000000

In[34]:= **Precision[%]**

Out[34]= 100.

Sometimes when dealing a lot with high-precision numbers, it might occur that the same calculation gives different results when carried out at different times. Let us construct such a situation. We start with an expression that is numerically zero.

In[35]:= **zero = Sin[Pi/32] - Sqrt[2 - Sqrt[2 + Sqrt[2 + Sqrt[2]]]]/2;**

Adding a small rational number to `zero` and testing if the resulting expression is identically zero does not succeed with the default value of $MaxExtraPrecision.

In[36]:= **$MaxExtraPrecision = 50;**
    zero + 10^-100 == 0

       N::meprec :
         Internal precision limit $MaxExtraPrecision = 50.` reached while evaluating

$$\frac{1}{1000000000000000000000000000000000 \ll 33\gg 000000000000000000000000000000000} - \frac{1}{2}\sqrt{2 - \sqrt{2 + \text{Power}[\ll 2\gg]}} + \text{Sin}\left[\frac{\pi}{32}\right]. \text{ More…}$$

Out[37]= 1 /
    1000000000000000000000000000000000000000000000000000000000000000000000000000000 ⋮

$$000000000000000000000000000 - \frac{1}{2}\sqrt{2 - \sqrt{2 + \sqrt{2 + \sqrt{2}}}} + \text{Sin}\left[\frac{\pi}{32}\right] == 0$$

Now, we temporarily set $MaxExtraPrecision to a higher value and calculate a 500-digit approximation of zero + 10^-100.

In[38]:= **$MaxExtraPrecision = 2000;**
    N[zero + 10^-100, 500]

Out[39]= 1.00000000000000000000000000000000000000000000000000000000000000000000000000000 ⋮
    00000000000000000000000000000000000000000000000000000000000000000000000000000000 ⋅
    00000000000000000000000000000000000000000000000000000000000000000000000000000000 ⋅
    00000000000000000000000000000000000000000000000000000000000000000000000000000000 ⋅
    00000000000000000000000000000000000000000000000000000000000000000000000000000000 ⋅
    00000000000000000000000000000000000000000000000000000000000000000000000000000000 ⋅
    00000000000000000000000000000000000000000000000× 10^{-100}

In[40]:= **zero + 10^-100 == 0**

Out[40]= False

Now, we are setting the value of $MaxExtraPrecision back to 50.

In[41]:= **$MaxExtraPrecision = 50;**
        zero + 10^-100 == 0

    N::meprec :
      Internal precision limit $MaxExtraPrecision = 50.` reached while evaluating

$$\frac{1}{100000000000000000000000000000000 \ll 33 \gg 00000000000000000000000000000000000000} - $$

$$\frac{1}{2}\sqrt{2 - \sqrt{2 + \text{Power}[\,\ll 2 \gg\,]}} + \text{Sin}\left[\frac{\pi}{32}\right]. \text{ More...}$$

Out[42]= 1 /

$$1000000000000000000000000000000000000000000000000000000000000000000000000000000000000\,\text{\.{}}$$

$$00000000000000000000000000 - \frac{1}{2}\sqrt{2 - \sqrt{2 + \sqrt{2 + \sqrt{2}}}} + \text{Sin}\left[\frac{\pi}{32}\right] == 0$$

Mathematica keeps internal cache tables with certain symbolic and numeric quantities that it has already calculated. Often, this caching of values is quite useful and can reduce timings considerably. In the following two inputs, the real work is done only once.

In[43]:= **N[Pi, 100000]; // Timing**

Out[43]= {0.37 Second, Null}

In[44]:= **N[Pi, 100000]; // Timing**

Out[44]= {0. Second, Null}

When one wants absolutely reproducible timings and values, one can erase all previously calculated values using the function Developer`ClearCache[].

In[45]:= **Developer`ClearCache[]**

Here is an example where the autonumericalization fails. Typically, sums of inexact and exact numerical quantities collapse to an inexact number. In the following, this does not happen, because no reliable numerical value of ln(*zero*) can be determined.

In[46]:= **With[{zero = Sin[Pi/16] - Sqrt[2 - Sqrt[2 + Sqrt[2]]]/2},**
        1``100 + Log[10^10 zero]]

    N::meprec : Internal precision limit $MaxExtraPrecision = 50.` reached while

$$\text{evaluating Log}\left[10000000000 \left(-\frac{1}{2}\sqrt{2 - \text{Power}[\,\ll 2 \gg\,]} + \text{Sin}\left[\frac{\pi}{16}\right]\right)\right]. \text{ More...}$$

    N::meprec : Internal precision limit $MaxExtraPrecision = 50.` reached while

$$\text{evaluating Log}\left[10000000000 \left(-\frac{1}{2}\sqrt{2 - \text{Power}[\,\ll 2 \gg\,]} + \text{Sin}\left[\frac{\pi}{16}\right]\right)\right]. \text{ More...}$$

Out[46]= 1.0000000000000000000000000000000000000000000000000000000000000000000000000000000000000\,\text{\.{}}

$$00000000000000000000000000 + \text{Log}\left[10000000000 \left(-\frac{1}{2}\sqrt{2 - \sqrt{2 + \sqrt{2}}} + \text{Sin}\left[\frac{\pi}{16}\right]\right)\right]$$

Explicit use of N forces numericalization (N does not return nonnumericalized numeric expressions). Because no digits of the argument of the logarithm can be determined, the result is Indeterminate (it could be -∞ if the argument of Log would be zero and a finite number otherwise).

In[47]:= **N[%, 22]**

```
N::meprec :
  Internal precision limit $MaxExtraPrecision = 50.` reached while evaluating
    1.000000000000000000000000000000000000000000000000000000000000000000000
    0000000000000000000000000 + Log[10000000000 (-1/2 √(2 + Times[≪2≫]) + Sin[π/16])].
  More…
```

Out[47]= Indeterminate

Above, we saw occurrences of the N::meprec message when using N. Some symbolic functions (meaning they get exact input and return exact output) use numerical techniques to do their job. (Examples of such functions are Equal, Unequal, Floor, Ceiling, and Round.) These functions too (sometimes unexpectedly) can issue the N::meprec message. Here is a definition for a function $p[n]$ that returns the nth prime number [1522].

In[48]:= `p[n_] := 1 + Sum[Floor[(n/Sum[Floor[Cos[Pi ((j - 1)! + 1)/j]^2],`
`                {j, m}])^(1/n)], {m, 1, 2^n}]`

The calculation of the first five primes succeeds without problems.

In[49]:= `Table[p[n], {n, 1, 5}]`
Out[49]= {2, 3, 5, 7, 11}

Calculating the eighth prime number results in N::meprec messages.

In[50]:= `p[8];`

```
Floor::meprec : Internal precision limit $MaxExtraPrecision = 50.` reached
  while evaluating Floor[Cos[ (89461821307 ≪94≫ 000000000001 π)/80 ]^2]. More…

Floor::meprec : Internal precision limit $MaxExtraPrecision = 50.` reached
  while evaluating Floor[Cos[ (89461821307 ≪94≫ 000000000001 π)/80 ]^2]. More…

Floor::meprec : Internal precision limit $MaxExtraPrecision = 50.` reached
  while evaluating Floor[Cos[ (71569457046 ≪96≫ 000000000001 π)/81 ]^2]. More…

General::stop :
  Further output of Floor::meprec will be suppressed during this calculation. More…
```

Setting to a higher value allows the calculation of $p[8]$.

In[51]:= `$MaxExtraPrecision = 1000;`
`p[8]`
Out[52]= 19

Here is a similar example. The function generalizedCantorProductTerms calculates the first n terms in the Cantor product representation [1089], [1930] of the real number x.

In[53]:= `generalizedCantorProductTerms[n_, k_Integer?Positive,`
`            x_?(Precision[#] == Infinity && 0 <= # < 1&)] :=`
`  #/(# + k)& /@ ((Floor[k #/(1 - #)] + 1)& /@`
`            NestList[(# + # k/(1 + Floor[k #/(1 - #)]))&, x, n])`

Because the difference nth term from 1 decays double exponentially with n, setting $MaxExtraPrecision to Infinity is a good choice for exact x that are not rational numbers. Here are the differences of the first Cantor product terms from 1 for $x = 1/\pi$.

In[54]:= `$MaxExtraPrecision = Infinity;`
`(1 - generalizedCantorProductTerms[16, 2, 1/Pi]) // N`
Out[55]= {0.666667, 0.0444444, 0.000654879, 1.2973×10^{-7}, 3.88894×10^{-15},
        6.98165×10^{-30}, 1.58238×10^{-59}, 7.73652×10^{-119}, 9.60689×10^{-238},

$$3.734973141925541 \times 10^{-475},\ 2.642440594048152 \times 10^{-951},\ 1.534888963582161 \times 10^{-1902},$$
$$1.138305115731868 \times 10^{-3804},\ 4.500558112597279 \times 10^{-7609},\ 3.777238050023831 \times 10^{-15218},$$
$$1.313309993766191 \times 10^{-30436},\ 5.865124822604842 \times 10^{-60873}\}$$

The product equals the original number within 16 digits.

```
In[56]:= Times @@ (1 - %) - 1/Pi
```
```
Out[56]= 5.55112 × 10^-17
```

The following graphic shows the Cantor product terms for rational numbers.

```
In[57]:= With[{pp = 1000},
    Show[Graphics[(* color all nth product term with same color *)
    MapIndexed[{PointSize[0.0025], Hue[#2[[1]]/16], #}&, Transpose[
        Table[Point[{x, #}]& /@ (* display Log[|Log[1 - nthTerm]|] *)
            Log[10., -Log[1 - generalizedCantorProductTerms[12, 1, x]]],
            {x, 1/pp, 1 - 1/pp, 1/pp}]]]],
    PlotRange -> All, Frame -> True]]
```

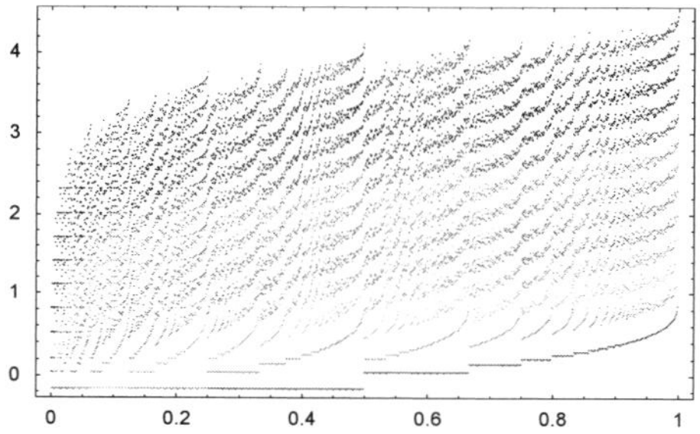

```
In[58]:= (* reset $MaxExtraPrecision value *)
    $MaxExtraPrecision = 1000;
```

At this point, let us revisit the Egyptian fractions from the last subsection. The simple algorithm given above to write a fraction as a sum of positive unit fractions always terminated and required no approximative arithmetic to be carried out. This expansion can straightforwardly be generalized to any real x from the interval $(0, 1]$. Then we have $x = \sum_{k=1}^{\infty} 1/d_k$ where $d_n = \lceil x - \sum_{k=1}^{n-1} 1/d_k \rceil$. This is the so-called Sylvester expansion [650]. The function SylvesterDigits[x, n] returns a list of the first n of the $d_k(x)$.

```
In[60]:= SylvesterDigits[x_?((0 < # < 1 && Head[#] =!= Rational)&),
                    n_Integer?Positive] :=
        First /@ Rest[NestList[
            Function[ξ, {#, ξ[[2]] - 1/#}&[Ceiling[1/ξ[[2]]]]],
                {0, x}, n]];
```

From the above experience with Egyptian fractions, we expect the d_k to be fast growing functions of k. Here are the first seven Sylvester digits of $x = 1/\pi$.

```
In[61]:= SylvesterDigits[1/Pi, 7]
```
```
Out[61]= {4, 15, 609, 845029, 1010073215739, 1300459886313272270974271,
    1939680952094609786557359582286462958434022504402}
```

The summed form of these seven unit fractions agree to nearly 100 digits with $1/\pi$.

```
In[62]:= N[1/Pi - Plus @@ (1/%), 22]
```
```
Out[62]= 9.881496721966501970053 × 10^-98
```

While the last input did not explicitly call for any approximate function evaluation, approximative numerical calculations were used to calculate the value of the Ceiling function. The input Sylvester: Digits[1/Pi, 8] would generate N::meprec messages with the default value of $MaxExtraPreci: sion. Using the value for 10^5 for $MaxExtraPrecision, we can, without problems, extract 16 Sylvester digits for most numbers from the unit interval. Here are six examples.

```
In[63]:= $MaxExtraPrecision = 10^5;
        SeedRandom[123];
        sds = SylvesterDigits[#, 16]& /@
              (* seven "random" numbers *)
              {1/Pi, 1/E, 1/Sqrt[3], 1/3^(1/3), Log[2], Sin[1],
               Random[Real, {0, 1}, 10^6]};
```

The last Sylvester digits of the above expansions have about 50000 digits.

```
In[66]:= Round[Log[Max[#]]]& /@ sds
Out[66]= {57404, 85516, 75100, 36063, 40806, 50538, 66446}
```

On average, the size of the Sylvester digits grows double exponentially, meaning $\ln(\ln(d_k)) \sim k$. This dependence is shown in the left graphic. The Sylvester digits have an interesting statistical property: for almost all x, the following limit holds: $\lim_{n\to\infty} (d_{n+1} / \prod_{k=1}^{n} d_k)^{1/n} = e$ [650], [1929]. The right graphic shows the corresponding quantities for the first few n of the above Sylvester digits.

```
In[67]:= Show[GraphicsArray[
        Block[{prods},
          {Graphics[(* digits growth *)
            Table[{Hue[k/8], Line[MapIndexed[{#2[[1]], #1}&,
                                   Log[10, Log[10, N[sds[[k]]]]]]]},
                  {k, 6}], Frame -> True, PlotRange -> All],
          Graphics[(* average quotients *)
            Table[prods = Rest[FoldList[Times, 1., sds[[j]]]];
                  {Hue[j/8], Line[
                    Table[{k, (sds[[j, k + 1]]/prods[[k]])^(1/k)},
                          {k, 4, Length[sds[[j]]] - 1}]]},
                  {j, Length[sds]}], Frame -> True, PlotRange -> All]}]]]
```

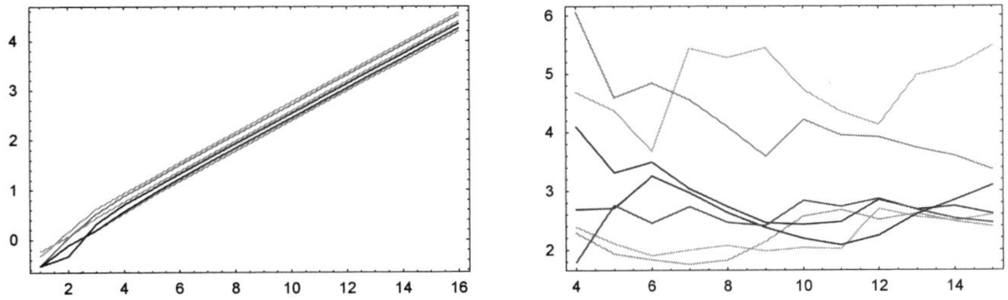

1.1.5 Packed Arrays

As discussed in Chapter 2 of the Programming volume [1793], everything in *Mathematica* is an expression. An expression can contain or be anything: integers, floating-point numbers, strings, other raw types, and symbolic entries. And as such, its size is not limited, but can vary. The following picture shows the memory used by 2^i as a function of i.

```
In[1]:= ListPlot[Table[{i, ByteCount[2^i]}, {i, 1000}],
            PlotRange -> All, Frame -> True, Axes -> False]
```

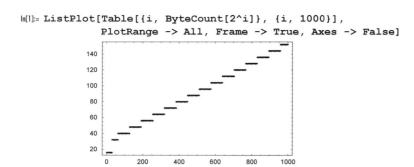

The single number 2 needs 16 Bytes in *Mathematica* on a 32 bit operating system.

```
In[2]:= ByteCount[2]
Out[2]= 16
```

And a list of five 2's needs 120 bytes. This seems expensive in comparison with other computer languages.

```
In[3]:= ByteCount[{2, 2, 2, 2, 2}]
Out[3]= 120
```

For memory use comparable to that of compiled languages, we can "pack" lists of machine integers, machine real numbers, and machine complex numbers, in *Mathematica*. This packing sometimes happens "automatically" , and it can also be done by hand. The functions for operations related to packed arrays are located in the `Developer` context [1009].

```
In[4]:= Names["Developer`*Pack*"]
Out[4]= {Developer`FromPackedArray, Developer`PackedArrayForm,
         Developer`PackedArrayQ, Developer`ToPackedArray}
```

The three functions that create, destruct, and test packed arrays are the following.

> `Developer`ToPackedArray[`*list*`]`
>   creates a packed version of the list *list*.

> `Developer`FromPackedArray[`*list*`]`
>   creates an expression version of the packed array *list*.

> `Developer`PackedArrayQ[`*list*`]`
>   returns `True` if the list *list* is a packed array, `False` otherwise.

Not all lists are packable—only the ones that are tensors. Tensors are rectangular arrays of dimensions $n_1 \times n_2 \times \ldots \times n_k$ (the degenerate case of a list with a single number can also be packed). And all elements of such a tensor must be numbers that are machine numbers or again tensors of appropriate dimensions. Array produces such tensors; this explains the `Array` in the last function names. Only lists (head `List`) are packable. No other head can replace `List`.

Here is a $3 \times 3 \times 3$ array of integers.

```
In[5]:= integerTensor = Table[i j k, {i, 3}, {j, 3}, {k, 3}]

Out[5]= {{{1, 2, 3}, {2, 4, 6}, {3, 6, 9}},
        {{2, 4, 6}, {4, 8, 12}, {6, 12, 18}}, {{3, 6, 9}, {6, 12, 18}, {9, 18, 27}}}
```

By default, a small list generated by Table is not packed.

```
In[6]:= Developer`PackedArrayQ[integerTensor]

Out[6]= False
```

The memory needed for storing integerTensor is slightly less than 1 kB.

```
In[7]:= ByteCount[integerTensor]

Out[7]= 848
```

And the storage requirements of our starting list of five integers shrinks now to less than 100 Bytes, compared with the more than 100 Bytes needed for the unpacked list.

```
In[8]:= ByteCount[Developer`ToPackedArray[{2, 2, 2, 2, 2}]]

Out[8]= 76
```

The packed version displays in the same form as the unpacked one in the output.

```
In[9]:= packedIntegerTensor = Developer`ToPackedArray[integerTensor]

Out[9]= {{{1, 2, 3}, {2, 4, 6}, {3, 6, 9}},
        {{2, 4, 6}, {4, 8, 12}, {6, 12, 18}}, {{3, 6, 9}, {6, 12, 18}, {9, 18, 27}}}

In[10]:= Developer`PackedArrayQ[packedIntegerTensor]

Out[10]= True
```

Now, the memory needed is much smaller—by approximately a factor five.

```
In[11]:= ByteCount[packedIntegerTensor]

Out[11]= 172
```

A packed version of a (nested) list will always behave the same as an ordinary expression version. It will give the same result for all calculations. When it is appropriately used, it will give the result much faster. Also, for comparison purposes with Equal and SameQ, the two lists integerTensor and packedIntegerTen‹ sor are the same. (Of course, internally there is a difference between a packed array and a nonpacked array. But this difference does not influence the result of any mathematical operation or structural operation (such as Part or Length), just the time and memory consumption for this operation.)

```
In[12]:= integerTensor == packedIntegerTensor

Out[12]= True
```

The memory savings are only one part of the advantages of packed arrays. A second one is speed. Here a simple addition is carried out 10^5 times. The packed version is approximately an order of magnitude faster.

```
In[13]:= Module[{sum = integerTensor},
            Do[sum = sum + integerTensor, {i, 10^5}];
            sum] // Timing

Out[13]= {3.75 Second,
        {{{100001, 200002, 300003}, {200002, 400004, 600006}, {300003, 600006, 900009}},
         {{200002, 400004, 600006}, {400004, 800008, 1200012},
          {600006, 1200012, 1800018}}, {{300003, 600006, 900009},
          {600006, 1200012, 1800018}, {900009, 1800018, 2700027}}}}
```

```
In[14]:= Module[{sum = packedIntegerTensor},
          Do[sum = sum + packedIntegerTensor, {i, 10^5}];
          sum] // Timing
Out[14]= {0.48 Second,
         {{{100001, 200002, 300003}, {200002, 400004, 600006}, {300003, 600006, 900009}},
          {{200002, 400004, 600006}, {400004, 800008, 1200012},
           {600006, 1200012, 1800018}}, {{300003, 600006, 900009},
           {600006, 1200012, 1800018}, {900009, 1800018, 2700027}}}}
```

Using one single function to carry out the summation also speeds up things. (We cover the reason of this acceleration in a moment.)

```
In[15]:= Nest[(# + packedIntegerTensor)&, packedIntegerTensor, 10000] // Timing
Out[15]= {0.01 Second,
         {{{10001, 20002, 30003}, {20002, 40004, 60006}, {30003, 60006, 90009}},
          {{20002, 40004, 60006}, {40004, 80008, 120012}, {60006, 120012, 180018}},
          {{30003, 60006, 90009}, {60006, 120012, 180018}, {90009, 180018, 270027}}}}
```

The following input shows iterations of the logistic map at the onset of chaos μ_0. The inner `Nest[$\mu$ # (1 - #) &, #, Is]` is automatically compiled. As a result, the five millions applications of μ # $(1 - \#)$ & are carried out in a few seconds. The graphic shows the trajectories for the orbits starting at $1/\pi$ for the first 10000 iterations in a neighborhood of size 10^{-6} around μ_0.

```
In[16]:= Module[{μ0 = Rationalize[3.569945672, 0], μδ = 10^-6, ppμ = 500,
          Is = 100, ppI = 100, data, align},
          (* iterate 100×100 times for each value of μ;
             keep every 100th value *)
          data = Table[MapIndexed[{μ - μ0, #2[[1]] Is, #1}&,
                              NestList[Nest[μ # (1 - #)&, #, Is]&,
                                       N[1/Pi], ppI]],
                       {μ, μ0 - μδ/2, μ0 + μδ/2, μδ/ppμ}];
          (* align values at the onset of chaos *)
          align[l_] := With[{z = l[[Round[Length[l]/2], 3]]},
                            {#1, #2, #3 - z}& @@@ l];
          (* show orbits in the μ,iterations,value-space *)
          Show[Graphics3D[{PointSize[0.003],
                  (* color points equal for fixed number of iterations *)
                              MapIndexed[{Hue[0.8 #2[[1]]/ppI], Point @ #}&,
                                       align /@ Transpose[data], {2}]}],
                  BoxRatios -> {2, 1, 1}, Axes -> True, ViewPoint -> {0, -3, 1},
                  PlotRange -> 10^-5 {-1, 1}, AxesLabel -> {"μ-μ0", "i", None}]]
```

The following list shows the memory savings of a rank *n* tensor, *n* ranging from 1 to 16.

```
In[17]:= Table[
    With[{list = Developer`FromPackedArray[
        Table[1, Evaluate[Sequence @@ Table[{2}, {n}]]]]},
     {n, ByteCount[list],
        ByteCount[Developer`ToPackedArray[list]]}], {n, 16}]
Out[17]= {{1, 64, 64}, {2, 160, 76}, {3, 352, 96}, {4, 736, 132}, {5, 1504, 200},
    {6, 3040, 332}, {7, 6112, 592}, {8, 12256, 1108}, {9, 24544, 2136},
    {10, 49120, 4188}, {11, 98272, 8288}, {12, 196576, 16484}, {13, 393184, 32872},
    {14, 786400, 65644}, {15, 1572832, 131184}, {16, 3145696, 262260}}
```

Real numbers, as well as integers, can be packed. Here is a 100×100 array of real numbers.

```
In[18]:= packedRealMatrix = Table[Random[], {100}, {100}];
```

This time, `Table` created a packed list automatically. (We will come back to this issue in a moment).

```
In[19]:= Developer`PackedArrayQ[packedRealMatrix]
Out[19]= True
```

Using `Developer`FromPackedArray`, we generate an unpacked version of the matrix.

```
In[20]:= realMatrix = Developer`FromPackedArray[packedRealMatrix];
```

Again, the packed version used less memory than the unpacked version.

```
In[21]:= {ByteCount[realMatrix], ByteCount[packedRealMatrix]}
Out[21]= {202824, 80060}
```

And as long as we stay within machine arithmetic, all numerical calculations will be much faster using packed arrays. For this example, it is faster by more than an order of magnitude.

```
In[22]:= Timing[Plus @@ Flatten[Sum[Sin[x realMatrix], {x, 1., 100.}]]]
Out[22]= {2.16 Second, 50598.9}
```

```
In[23]:= Timing[Plus @@ Flatten[Sum[Sin[x packedRealMatrix], {x, 1., 100.}]]]
Out[23]= {0.18 Second, 50598.9}
```

Complex machine numbers can also be packed.

```
In[24]:= Developer`ToPackedArray[Table[Random[Complex], {2}, {2}]]
Out[24]= {{0.138439 + 0.152107 i, 0.982459 + 0.858234 i},
    {0.921821 + 0.749206 i, 0.503525 + 0.593595 i}}
```

```
In[25]:= Developer`PackedArrayQ[%]
Out[25]= True
```

But only complex numbers with approximate real and imaginary parts can be packed. Complex numbers with exact integer real and imaginary parts cannot be packed.

```
In[26]:= Developer`ToPackedArray[
        Table[Random[] + 2 I, {2}, {2}]] // Developer`PackedArrayQ
Out[26]= False
```

```
In[27]:= Developer`ToPackedArray[
        Table[1 + 2 I, {2}, {2}]] // Developer`PackedArrayQ
Out[27]= False
```

Here is a definition that constructs iteratively the Lenard sequence [1128]. In the limit of infinite iterations, the Lenard sequence is an infinite sequence of nonvanishing numbers a_j, such that $\sum_{j=1}^{\infty} a_j^r = 0$ holds for all positive integers r.

```
In[28]:= (* pack initial sequence approximation *)
         LenardSequence[0] := Developer`ToPackedArray @ {1. + 0. I};

         LenardSequence[n_] := LenardSequence[n] =
           Join[LenardSequence[n - 1],
             Sequence @@ Table[Exp[I Pi/N[n]]/n LenardSequence[n - 1], {n^n}]]
```

Constructing the sequence and checking the defining property can be done quite fast (`LenardSequence[4]` has 71960 elements).

```
In[31]:= Table[Plus @@ (LenardSequence[n]^n), {n, 4}] // Timing
Out[31]= {0.15 Second, {0. + 1.22465×10^-16 i, 0. + 2.44929×10^-16 i,
            -7.55123×10^-18 + 2.75821×10^-16 i, -1.8696×10^-12 + 1.71693×10^-13 i}}
```

Now, let us come back to the question raised above: Why does `Table` sometimes create a packed array, and sometimes not?

Packed arrays are seamlessly integrated into *Mathematica* to make larger numerical calculations that stay inside the arena of machine numbers more efficient. Many *Mathematica* functions, such as `Plus`, `Times`, `Sin`, `Cos`, `Flatten`, `Union`, `Join`, `Partition`, and so on recognize them and have special code to manipulate them quickly. But not all functions (see examples below) can handle packed arrays effectively, and so sometimes packed arrays have to be converted to their expression equivalent. This unpacking takes some time, and so it is not always advantageous to have all in principle packable lists packed from the beginning.

The next input determines the minimal length of a packed list generated by `Table`.

```
In[32]:= Module[{n = 1},
           While[Not[Developer`PackedArrayQ[Table[1, {n}]]],
             n = n + 1]; n]
Out[32]= 250
```

This number can be changed. It is a system option. Here are all system options related to compilation (automatic compilation and explicit compilation through the function `Compile`).

```
In[33]:= Cases[Developer`SystemOptions[], HoldPattern["CompileOptions" -> _]]
Out[33]= {CompileOptions → {ApplyCompileLength → ∞,
            ArrayCompileLength → 250, AutoCompileAllowCoercion → False,
            AutoCompileProtectValues → False, AutomaticCompile → False,
            CompileAllowCoercion → True, CompileConfirmInitializedVariables → True,
            CompiledFunctionArgumentCoercionTolerance → 2.10721,
            CompileEvaluateConstants → True, CompileReportCoercion → False,
            CompileReportExternal → False, CompileReportFailure → False,
            CompileValuesLast → True, FoldCompileLength → 100,
            InternalCompileMessages → False, MapCompileLength → 100,
            NestCompileLength → 100, NumericalAllowExternal → True,
            SystemCompileOptimizations → All, TableCompileLength → 250}}
```

"TableCompileLength"
    is a system option and determines the length at which `Table` generates packed lists, if this is possible.

The current value of the system option "TableCompileLength" is 250.

We can reset this number to a smaller value.

```
In[34]:= Developer`SetSystemOptions["CompileOptions" -> {"TableCompileLength" -> 100}];
```

Now, the switch already happens at length 100.

```
In[35]:= Module[{n = 1},
              While[Not[Developer`PackedArrayQ[Table[1, {n}]]], n = n + 1]; n]
Out[35]= 100
```

Packed lists cannot only be generated by Developer`ToPackedArray and functions such as Table and Range, but also some list manipulating functions sometimes return packed lists. We start with an unpacked list.

```
In[36]:= unpackedList = Developer`FromPackedArray[
                                Table[N[x], {x, 1/10, 1, 9/10/1000}]];
```

```
In[37]:= Developer`PackedArrayQ[unpackedList]
Out[37]= False
```

We map the function #^2& over this list. The result is a packed list.

```
In[38]:= Developer`PackedArrayQ[list1 =  Map[#^2&, unpackedList]]
Out[38]= True
```

In evaluating the last expression, *Mathematica* internally compiled the whole expression Map[#^2&, unpackedList]. The compilation was successful because the pure function #^2& could be compiled. If we define a function outside of Map, the compilation cannot succeed (because only square[x] evaluates nontrivially, but not square itself), and so the list returned by Map is not compiled. (When Map evaluates its first argument square no nontrivial value is found, only when square is called with an argument will it evaluate nontrivially.)

```
In[39]:= square[x_] := x^2
```

```
In[40]:= Developer`PackedArrayQ[list2 = Map[square, unpackedList]]
Out[40]= False
```

The two lists list1 and list2 are the same in an arithmetical sense.

```
In[41]:= list1 == list2
Out[41]= True
```

The switch from returning a packed or an unpacked list happened at the list length 100.

```
In[42]:= Module[{n = 1},
              While[Not[Developer`PackedArrayQ[
                 Map[#^2&, Developer`FromPackedArray[
                    Table[N[x], {x, 1/10, 1, 9/10/n}]]]]], n = n + 1]; n + 1]
Out[42]= 100
```

Again, a system option determines the last value. The functions that compile at a certain list length are Apply, Array, Fold, FoldList, Map, Nest, NestList, and Table. The default values for their compilation are given below.

```
In[43]:= Cases[Developer`SystemOptions[],
              _String?(StringMatchQ[#, "*CompileLength"]&), {-1}]
Out[43]= {ApplyCompileLength, ArrayCompileLength, FoldCompileLength,
         MapCompileLength, NestCompileLength, TableCompileLength}
```

Note that `Range` was not in the above list, because it always returns a packed list (already for list of length 1).

```
In[44]:= Developer`PackedArrayQ @ Range[1]
Out[44]= True
```

There are other functions that return packed arrays—plotting functions, for instance.

```
In[45]:= plot = ParametricPlot[{Cos[s] + Sin[4s]/12, Sin[3s] - Cos[7s]/4},
                {s, 0, 2Pi}, Frame -> True, Axes -> False]
```

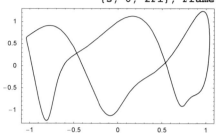

The list of *x*- and *y*-coordinates in the `Line` of plot is a packed array.

```
In[46]:= Developer`PackedArrayQ[plot[[1, 1, 1, 1]]]
Out[46]= True
```

Now, let us briefly discuss what circumstances potentially block packing of (large) lists. Because a packed list can only contain machine numbers, appending a nonmachine number to the following list will make it an unpacked list.

```
In[47]:= testList = Table[N[i], {i, 10^4}];
```

```
In[48]:= Developer`PackedArrayQ[testList]
Out[48]= True
```

```
In[49]:= Developer`PackedArrayQ[Append[testList, 10001.]]
Out[49]= True
```

```
In[50]:= Developer`PackedArrayQ[Append[testList, 10.^500]]
Out[50]= False
```

Packed arrays can only contain raw data of one type. So either adding an integer, or a high-precision real number, or a complex number to `testList` will cause the creation of an unpacked version.

```
In[51]:= Developer`PackedArrayQ[Append[testList, 10001]]
Out[51]= False
```

```
In[52]:= Developer`PackedArrayQ[Append[testList, N[10001, 40]]]
Out[52]= False
```

```
In[53]:= Developer`PackedArrayQ[Append[testList, 10001. + 0.0 I]]
Out[53]= False
```

When it is possible that the resulting list is packable, it will often be packed (also when the operation includes a change of the type of the elements of the list).

```
In[54]:= Developer`PackedArrayQ[testList + 2]
Out[54]= True
```

In[55]:= **Developer`PackedArrayQ[testList + 1. I]**

Out[55]= True

In[56]:= **Developer`PackedArrayQ[Round[testList]]**

Out[56]= True

When doing a (larger) calculation, if possible, one should stay within packed arrays. To monitor if at internal steps unpacking becomes necessary, one can use the system option "UnpackMessage".

In[57]:= **"UnpackMessage" /. Developer`SystemOptions[]**

Out[57]= False

Setting the option to True will generate messages every time a packed list gets unpacked.

In[58]:= **Developer`SetSystemOptions["UnpackMessage" -> True];**

Here is a $3 \times 3 \times 3$ tensor.

In[59]:= **testList = Table[1., {10}, {10}, {10}];**

In[60]:= **Developer`PackedArrayQ[testList]**

Out[60]= True

The following input unpacks the tensor because potentially an element with head Integer gets returned.

In[61]:= **SeedRandom[111]**
Map[If[Random[] < 1/2, 0, 1.]&, testList, {3}];

Developer`FromPackedArray::punpack1 : Unpacking array to level 3. More…

Note that the unpacking already happens in the beginning of the calculation, not at the time the first explicit integer is encountered.

In[63]:= **SeedRandom[111]**
Map[If[Random[] < 10^-1000, 0, 1.]&, testList, {3}];

Developer`FromPackedArray::punpack1 : Unpacking array to level 3. More…

In[65]:= **Union[Head /@ Flatten[%]]**

Out[65]= {Real}

For pattern matching, it is often necessary to unpack packed arrays. Here is a very simple function f1. No unpacking is needed when applying f1.

In[66]:= **argList = N[Range[100]];**

In[67]:= **Developer`PackedArrayQ[argList]**

Out[67]= True

In[68]:= **f1[x_] := x^2**

In[69]:= **Developer`PackedArrayQ[f1[argList]]**

Out[69]= True

The function definition for the function f2 contains a more complicated test. To carry out the test, *Mathematica* unpacks the packed array argList.

In[70]:= **f2[x_?(Plus @@ # < 10000&)] := x^2**

In[71]:= **Developer`PackedArrayQ[f2[argList]]**

Developer`FromPackedArray::punpack1 : Unpacking array to level 1. More…

Out[71]= True

The following definition also unpacks the list `argList`.

```
In[72]:= f3[x_] := x^2 /; (Plus @@ x < 10000)
```

```
In[73]:= Developer`PackedArrayQ[f3[argList]]
```

       Developer`FromPackedArray::punpack1 : Unpacking array to level 1. More…

Out[73]= True

We reset the value of the system option `"UnpackMessage"` to its original value.

```
In[74]:= Developer`SetSystemOptions["UnpackMessage" -> False]
```

Out[74]= UnpackMessage → False

Here is a small example showing the efficiency of packed arrays. We generate sets of size n of random integers between 1 and m. Then, we calculate the number of all possible values of sums, products, differences, and quotients within each set. The function carries out o counts for the functions *fList*.

```
In[75]:= setSize[n_, m_, o_, fList_] :=
    Module[{λ, T0, T, c, set},
      (* count different results for operation o *)
      λ[o_, set_] := Length[Union[Flatten[
            Table[o[set[[i]], set[[j]]], {i, n}, {j, n}]]]];
      (* empty packed arrays *)
      T0 = Developer`ToPackedArray @ Table[0, {m}];
      (* make list of [exactly] n random numbers [in O(n) time] *)
      Table[T = T0; c = 0;
        While[c < n, v = Random[Integer, {1, m}];
                If[T[[v]] === 0, T[[v]] = v; c++]];
        set = Rest[Union[T]];
      (* count different values *)
        λ[#, set]& /@ fList, {o}]]
```

Because forming the quotients needs unpacking the second timing is considerably slower (in addition, calculating exact quotients is slower than multiplication).

```
In[76]:= With[{n = 100, m = 1000, o = 1000},
      (* carry out ten times more counts for sums/products *)
      {(plusTimesCount = setSize[n, m, 10 o, {Plus, Times}]); // Timing,
       (subtractDivideCount = setSize[n, m, o, {Subtract, Divide}]); // Timing}]
```

Out[76]= {{167.05 Second, Null}, {138.28 Second, Null}}

Generically there are more different products than sums and more different quotients than differences [540], [1333], [541].

```
In[77]:= Show[GraphicsArray[
        Graphics[Point /@ #, PlotRange -> All, Frame -> True]& /@
        {plusTimesCount, subtractDivideCount}]]
```

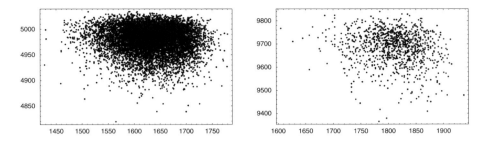

As a computational-intensive application of packed arrays, let us deal with long-range correlations in natural and machine-generated texts.

Long-Range Correlations in Natural Texts

The basic idea of finding correlations in texts is to identify the sequence of letters in a text with a random walk and then to analyze the statistics of the walk.

We proceed as follows: We start by choosing a (largely arbitrary) binary encoding and apply the map $\{1 \rightarrow 1, 0 \rightarrow -1\}$ for the letters and symbols of the text. We replace all letters and symbols by their encoded version and concatenate all strings. As a result, we get a long sequence s_i of -1's and 1's. (Typically, the absolute number of 0's and 1's will not be random, and on average, -1's or 1's will dominate.) Now, we look for correlations in the sequence s_i. The position x_i of our random walker is given by $x_i = \sum_{j=1}^{i} s_j$, and its walked distance between step k and $k + l$ is $d(k, l) = x_{k+l} - x_k$. For a true random walk with no correlation between the steps, we expect the mean square fluctuation of $f(l)^2 = \langle d(k, l)^2 \rangle_k - \langle d(k, l) \rangle_k^2$ to have the form $f(l) \sim \sqrt{l}$. Here, $\langle \ldots \rangle_k$ denotes the averaging over k. For human-written texts, one finds $f(l) \sim l^\alpha$ with $\alpha > 0.5$. This indicates long-range correlations are present in the text. (An ideal random walk would have $\alpha = 0.5$.) The correlations are believed to be universally present, independent of the language of the text, and largely independent of the encoding used [530], [1597], [769], [49], [1290], [1655], [1260]. (For general patterns in large strings, see [1345].)

We will use Shakespeare's *Hamlet* here. An on-line version can be found at http://the-tech.mit.edu/Shakespeare. I am grateful to Jeremy Hilton at the Corporation for National Research Initiatives for allowing accessing and excerpting from the text.

```
In[78]:= (* read in the webpage with the Hamlet text *)
    hamletWebPage =
     Import["http://www-tech.mit.edu/Shakespeare/hamlet/full.html", "Text"];

In[80]:= (* make plain ASCII text *)
    completeHamleteText =
     (* skip introduction *) StringDrop[#, {41, 102}]& @
    StringTake[#,{StringPosition[#, "The Tragedy of Hamlet"][[-1, 1]],
            StringLength[#]}]&[(* remove html-formatting *)
          StringReplace[hamletWebPage,
                ShortestMatch["<" ~~ __ ~~ ">"] -> ""]];
```

Here are the first few lines of the text.

```
    Print @ StringTake[completeHamleteText, 167]

        The Tragedy of Hamlet, Prince of Denmark
```

```
ACT I
SCENE I. Elsinore. A platform before the castle.

FRANCISCO at his post. Enter to him BERNARDO


BERNARDO

Who's there?

The Tragedy of Hamlet, Prince of Denmark\n\nACT I\nSCENE
    I. Elsinore. A platform before the castle.\n\nFRANCISCO at
    his post. Enter to him BERNARDO\n\nBERNARDO\nWho's there?
```

Let us see if the text really contains the famous words "To be, or not to be".

StringPosition[completeHamleteText, "To be, or not to be"]

{{76080, 76098}}

It does. Here is the complete famous monologue.

StringTake[#, {StringPosition[#, "To be, or not to be"][[1, 1]],
        StringPosition[#, "Be all my sins remember'd."][[1, -1]]}]&[
        completeHamleteText]

```
To be, or not to be: that is the question:
Whether 'tis nobler in the mind to suffer
The slings and arrows of outrageous fortune,
Or to take arms against a sea of troubles,
And by opposing end them? To die: to sleep;
No more; and by a sleep to say we end
The heart-ache and the thousand natural shocks
That flesh is heir to, 'tis a consummation
Devoutly to be wish'd. To die, to sleep;
To sleep: perchance to dream: ay, there's the rub;
For in that sleep of death what dreams may come
When we have shuffled off this mortal coil,
Must give us pause: there's the respect
That makes calamity of so long life;
For who would bear the whips and scorns of time,
The oppressor's wrong, the proud man's contumely,
The pangs of despised love, the law's delay,
The insolence of office and the spurns
That patient merit of the unworthy takes,
When he himself might his quietus make
With a bare bodkin? who would fardels bear,
To grunt and sweat under a weary life,
But that the dread of something after death,
The undiscover'd country from whose bourn
No traveller returns, puzzles the will
And makes us rather bear those ills we have
Than fly to others that we know not of?
Thus conscience does make cowards of us all;
And thus the native hue of resolution
Is sicklied o'er with the pale cast of thought,
And enterprises of great pith and moment
With this regard their currents turn awry,
And lose the name of action.--Soft you now!
```

```
The fair Ophelia! Nymph, in thy orisons
Be all my sins remember'd.
```

To make sure that all packed arrays always stay packed, we will monitor their unpacking.

```
In[85]:= SetSystemOptions["UnpackMessage" -> True];
```

This text has about 180000 characters (including white space and punctuation).

```
In[86]:= StringLength[completeHamleteText]
Out[86]= 179371
```

These are the different characters used in our Hamlet text.

```
In[87]:= allCharacters = Characters[completeHamleteText];
        usedChars = Union[allCharacters]
Out[88]= {!, &, -, [, ], ., ,, ;, ?, ', :,
         , , a, A, b, B, c, C, d, D, e, E, f, F, g, G, h, H, i, I, j, J, k, K, l, L, m,
         M, n, N, o, O, p, P, q, Q, r, R, s, S, t, T, u, U, v, V, w, W, x, y, Y, z, Z}
```

Here is the number of their occurrences.

```
In[89]:= (characterDistribution =
        Sort[{#, Count[allCharacters, #]}& /@ usedChars,
            #1[[2]] > #2[[2]]&]) // InputForm
Out[89]//InputForm=
        {{" ", 27591}, {"e", 14422}, {"t", 10942}, {"o", 10394}, {"\n", 8862}, {"a",
        8715}, {"s", 7899}, {"n", 7757}, {"h", 7713}, {"i", 7409}, {"r", 7227},
        {"l", 5338}, {"d", 4869}, {"u", 4085}, {"m", 3629}, {",", 3269}, {"y",
        3092}, {"w", 2671}, {"f", 2491}, {"c", 2386}, {"g", 2061}, {"p", 1714},
        {"b", 1577}, {"A", 1539}, {"T", 1511}, {"I", 1497}, {"E", 1428}, {".",
        1301}, {"v", 1187}, {"L", 1152}, {"k", 1127}, {"O", 1058}, {"'", 962}, {"H",
        928}, {"R", 876}, {"N", 823}, {"S", 769}, {"U", 653}, {"M", 612}, {";",
        582}, {":", 535}, {"D", 508}, {"C", 455}, {"W", 439}, {"G", 433}, {"?",
        428}, {"-", 421}, {"!", 301}, {"P", 293}, {"B", 239}, {"F", 223}, {"x",
        177}, {"K", 131}, {"q", 123}, {"Y", 107}, {"j", 101}, {"Q", 95}, {"Z", 69},
        {"z", 51}, {"]", 38}, {"[", 38}, {"V", 32}, {"J", 9}, {"&", 5}, {"\t", 2}}
```

The distribution of the letters obeys a Zipf law. (See Section 6.6 of the Programming volume [1806★] and [775★]).

```
In[90]:= ListPlot[Log[10, N[(#/Plus @@ #)&[Last /@
            characterDistribution]]], PlotRange -> All, Frame -> True]
```

Just on the side, let us look at the single words, too.

```
In[91]:= allWords = DeleteCases[#, ""]& @
             StringSplit[ToLowerCase[completeHamleteText], (* split at *)
                 {" " | "\n" | "\t" | "," | "." | ":" | "?" | "!" |
                  "(" | ")" | "-" | "[" | "]" | ";" | "|" | "&"}];
```

The text has about 30000 words.

```
In[92]:= Length[allWords]
Out[92]= 32254
```

Hamlet's name appears about 400 times.

```
In[93]:= Count[allWords, "Hamlet" | "hamlet"]
Out[93]= 461
```

By sorting the list of words allWords, equal words become neighbors and we can use Split to count the frequency of the words.

```
In[94]:= wordFrequencies =
             Sort[{#[[1]], Length[#]}& /@ Split[Sort[allWords]], #1[[2]] > #2[[2]]&];
```

Here are the most frequently used words.

```
In[95]:= {StringJoin[#[[1]]], #[[2]]}& /@ Take[wordFrequencies, 12] //
                                TableForm[#, TableAlignments -> {Right}]&
Out[95]//TableForm=
           the    1138
           and     965
            to     754
            of     669
           you     550
             i     542
             a     542
            my     514
        hamlet     461
            in     436
            it     416
          that     391
```

Here is the Zipf plot corresponding to the word frequency.

```
In[96]:= Needs["Graphics`Graphics`"]

          LogLogListPlot[N[(#/Plus @@ #)&[Last /@ wordFrequencies]],
                  Frame -> True, FrameLabel -> {"k", "p(k)"},
                  PlotJoined -> True, PlotRange -> All,
                  PlotStyle -> {Hue[0], Thickness[0.004]}]
```

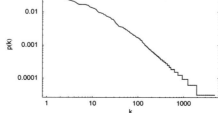

Now, let us come back to the characters. Less than 70 different characters are used in *Hamlet*.

```
In[98]:= characters = First /@ characterDistribution;
```

```
In[99]:= Length[characters]
```
```
Out[99]= 65
```

This means we need a binary number of length 7 ($2^7 = 128$) to decode the 68 characters. The function `makeEn` `coding` generates a (dispatched) list representing the encoding.

```
In[100]:= makeEncoding[chars_] :=
    With[{(* length of the code *)codeLength = Ceiling[Log[2, Length[chars]]]},
    Dispatch[Apply[Rule, Transpose[{usedChars,
      (* unique sequence of -1 and 1 *)
    Take[Flatten[Outer[List, ##]& @@ Table[{-1, 1},
      {codeLength}], codeLength - 1], Length[chars]]}], {1}]]]
```

Here is the actual encoding used in the following to binarize *Hamlet*.

```
In[101]:= (coding = makeEncoding[characters]) // Short[#, 8]&
```
```
Out[101]//Short=
    Dispatch[{! → {-1, -1, -1, -1, -1, -1, -1}, & → {-1, -1, -1, -1, -1, -1, 1},
     - → {-1, -1, -1, -1, -1, 1, -1}, [ → {-1, -1, -1, -1, -1, 1, 1},
     ] → {-1, -1, -1, -1, 1, -1, -1}, . → {-1, -1, -1, -1, 1, -1, 1},
     , → {-1, -1, -1, -1, 1, 1, -1}, ; → {-1, -1, -1, -1, 1, 1, 1},
     ? → {-1, -1, -1, 1, -1, -1, -1}, ' → {-1, -1, -1, 1, -1, -1, 1},
     : → {-1, -1, -1, 1, -1, 1, -1}, ≪43≫, u → {-1, 1, 1, -1, 1, 1, -1},
     U → {-1, 1, 1, -1, 1, 1, 1}, v → {-1, 1, 1, 1, -1, -1, -1},
     V → {-1, 1, 1, 1, -1, -1, 1}, w → {-1, 1, 1, 1, -1, 1, -1},
     W → {-1, 1, 1, 1, -1, 1, 1}, x → {-1, 1, 1, 1, 1, -1, -1},
     y → {-1, 1, 1, 1, 1, -1, 1}, Y → {-1, 1, 1, 1, 1, 1, -1},
     z → {-1, 1, 1, 1, 1, 1, 1}, Z → {1, -1, -1, -1, -1, -1, -1}}, -DispatchTables-]
```

To model a random 1D walk, we replace the characters by their code.

```
In[102]:= randomWalkHamlet = Flatten[allCharacters /. coding];
```

```
In[103]:= ByteCount[randomWalkHamlet]
```
```
Out[103]= 25111968
```

```
In[104]:= Take[randomWalkHamlet, 100]
```
```
Out[104]= {-1, 1, 1, -1, 1, -1, 1, -1, -1, 1, 1, 1, -1, -1, -1, -1, 1, -1, 1, 1,
    -1, -1, -1, -1, 1, 1, -1, 1, -1, 1, 1, -1, 1, -1, 1, -1, 1, 1, -1, -1,
    -1, -1, -1, -1, -1, 1, 1, 1, -1, -1, -1, 1, 1, -1, 1, -1, -1, -1, 1, -1,
    1, 1, -1, -1, -1, 1, -1, 1, -1, -1, -1, 1, 1, 1, 1, -1, 1, -1, -1, -1,
    1, 1, -1, 1, -1, 1, -1, 1, -1, 1, -1, -1, -1, 1, 1, -1, -1, -1, -1, -1}
```

The resulting random walk has more than one million steps.

```
In[105]:= Length[randomWalkHamlet]
```
```
Out[105]= 1255597
```

Now, we pack this long list.

```
In[106]:= packedRandomWalkHamlet = Developer`ToPackedArray[randomWalkHamlet];
```

This makes this list much more memory efficient—we saved about 75% memory.

In[107]:= **ByteCount [packedRandomWalkHamlet]**

Out[107]= 5022444

To save even more memory, we delete the unpacked lists that are no longer needed.

In[108]:= **MemoryInUse []**

Out[108]= 32375928

In[109]:= **Clear [binaryHamlet, randomWalkHamlet]**

In[110]:= **MemoryInUse []**

Out[110]= 32375928

Here, we determine the number of $+1$ and -1 in the resulting list.

In[111]:= **{Count [packedRandomWalkHamlet, +1], Count [packedRandomWalkHamlet, -1]}**

Out[111]= {520509, 735088}

The numbers are quite different, but because we are interested in correlations, this does not matter. Here is a part of the binary sequence representing a random walk.

In[112]:= **ListPlot [FoldList [Plus, 0, Take [packedRandomWalkHamlet, 10^3]],**
PlotStyle -> {{Thickness [0.002], GrayLevel [0]}},
PlotJoined -> True, Frame -> True]

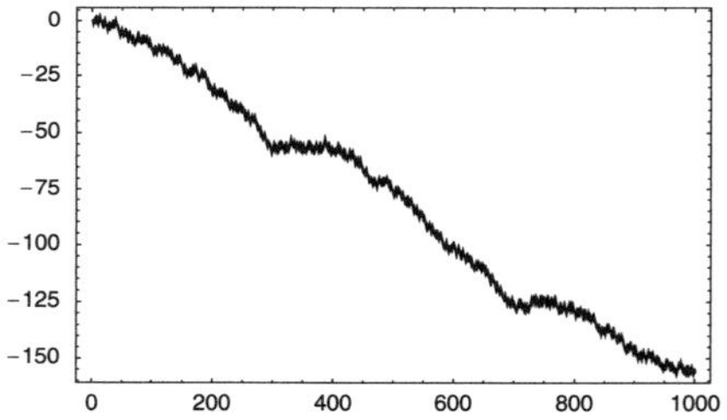

The list `packedRandomWalkHamletCumulative` represents the current position $x_0 = d(0, l)$ of the random walker. Having this list precalculated allows us to calculate $d(k, l)$ in a fast way by using $d(k, l) = d(0, k + l) - d(0, k)$.

In[113]:= **packedRandomWalkHamletCumulative = FoldList [Plus, 0, packedRandomWalkHamlet];**

Given the list of positions of a random walker $sum = x_i$, the function `meanSquareFluctuation` calculates the mean square fluctuations $f(l)^2 = \langle (d(k, l)^2 \rangle_k - \langle d(k, l) \rangle_l^2$.

In[114]:= **meanSquareFluctuation [walkSum_, l_] :=**
```
    Module [{s1 = 0., s2 = 0., d, ls = Length [walkSum] - 1},
            (* walk difference *)
            Do [d = N [walkSum [[k + 1]] - walkSum [[k]]];
            (* sum of squares *)
              s1 = s1 + d^2;
            (* square of sums *)
              s2 = s2 + d, {k, 1, ls}];
            (* mean square fluctuation *)
            s1/ls - (s2/ls)^2;
```

A more functional programming oriented version of the function meanSquareFluctuation is given in the next unevaluated input. It uses the function ListConvolve (to be discussed below in the section of Fourier transforms).

```
meanSquareFluctuationF[walkSum_, l_] :=
With[{ls = Length[walkSum] - 1},
     (Fold[Plus, 0, #^2]/ls - (Fold[Plus, 0, #]/ls)^2)&[
        ListConvolve[MapAt[-1&, MapAt[1&, Array[0.&, l + 1], 1], -1],
                     walkSum]]]
```

Here is an example for window size 10.

```
In[115]:= meanSquareFluctuation[N @ packedRandomWalkHamletCumulative, 10] // Timing
Out[115]= {13.42 Second, 5.45558}
```

Some time was needed to calculate the last result. If we want to calculate, say, 100 points, we would have to wait a few hours. Fortunately, packed arrays work seamlessly together with compiled functions. We already encountered autocompilation in functions like Map above. meanSquareFluctuation is a program, and compiling it now requires an explicit call to Compile (the function Compile will be discussed soon). The next version of meanSquareFluctuation, again called meanSquareFluctuation, uses the *Mathematica* compiler.

```
In[116]:= meanSquareFluctuation =
    Compile[{{walkSum, _Integer, 1}, {l, _Integer, 0}},
            Module[{s1 = 0., s2 = 0., d, ls = Length[walkSum] - 1},
                    (* walk difference *)
                    Do[d = N[walkSum[[k + 1]] - walkSum[[k]]];
                       (* sum of squares *)
                       s1 = s1 + d^2;
                       (* square of sums *)
                       s2 = s2 + d, {k, 1, ls}];
                    (* mean square fluctuation *)
                    s1/ls - (s2/ls)^2]];
```

Because meanSquareFluctuation is a CompiledFunction-object, the calculation of a single value is now quite fast—about 20 times faster.

```
In[117]:= meanSquareFluctuation[packedRandomWalkHamletCumulative, 10] // Timing
Out[117]= {0.57 Second, 5.45558}
```

Sometimes, one considers only nonoverlapping windows. The change in the code from meanSquareFluctuation to meanSquareFluctuationNonOverlapping is minimal.

```
In[118]:= meanSquareFluctuationNonOverlapping =
    Compile[{{walkSum, _Integer, 1}, {l, _Integer, 0}},
            Module[{s1 = 0., s2 = 0., d,
                    ls = Length[walkSum] - 1},
                    (* walk difference *)
                    Do[d = N[walkSum[[k + 1]] - walkSum[[k]]];
                       (* sum of squares *)
                       s1 = s1 + d^2;
                       (* square of sums *)
                       s2 = s2 + d,
                       {k, 1, ls, 1}];
```

```
                                    (* mean square fluctuation *)
                          s1/Ceiling[ls/l] - (s2/Ceiling[ls/l])^2]];
```

Because fewer steps now have to be carried out, it is much faster to calculate the mean square fluctuations.

In[119]:= `meanSquareFluctuationNonOverlapping[`
`                              packedRandomWalkHamletCumulative, 10] // Timing`

Out[119]= {0.07 Second, 5.43121}

Here is a list of window lengths that we will use to calculate the corresponding mean square fluctuations. To cover all length scales uniformly, we use $j \times 10^i$ as our window lengths.

In[120]:= `windowLengthsList[list_] :=`
`    With[{λ = Length[list]},`
`        Select[Sort[Union[Flatten[Table[i 10^j, {i, 10},`
`                                   {j, 0, Floor[Log[10, λ]]}]]]], # < λ&]]`

In[121]:= `windows = windowLengthsList[packedRandomWalkHamletCumulative];`

`    Short[windows, 4]`

Out[122]//Short=
```
{1, 2, 3, 4, 5, 6, 7, 8, 9, 10, 20, 30, 40, 50, 60, 70, 80, 90, 100, 200, 300, 400,
  500, 600, 700, 800, 900, 1000, 2000, 3000, 4000, 5000, 6000, 7000, 8000,
  9000, 10000, 20000, 30000, 40000, 50000, 60000, 70000, 80000, 90000, 100000,
  200000, 300000, 400000, 500000, 600000, 700000, 800000, 900000, 1000000}
```

Now, we calculate the mean square fluctuations for `windows`.

In[123]:= `data = {#, meanSquareFluctuation[`
`            packedRandomWalkHamletCumulative, #]}& /@ windows;`

To display the $f(l) \sim l^\alpha$-dependency, we take logarithms on both sides. This gives $\log_{10} f(l) = \alpha \log_{10} l + const.$

In[124]:= `(* form {Log[10, k], Log[10, Sqrt[x[k]]} *)`
`    toLogData[data_] := Apply[{Log[10, #1], Log[10, Sqrt[#2]]}&, data, {1}];`

Here is the resulting dependency between the logarithm of the square root of the mean square fluctuations and the window length.

In[126]:= `ListPlot[toLogData[data] // N,`
`            PlotRange -> All, PlotJoined -> True, Frame -> True,`
`            Axes -> False, Epilog -> {PointSize[0.01], Hue[0],`
`                              Point /@ N[toLogData[data]]}]`

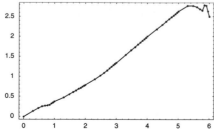

The data seem to behave nearly linearly in the interval $1 \lesssim \log_{10}(l) \lesssim 5$.

In[127]:= `middleData = Select[data, 1 <= Log[10, #[[1]]] <= 5&];`

```
In[128]:= ListPlot[toLogData[middleData] // N,
              PlotRange -> All, PlotJoined -> True, Frame -> True,
              Axes -> False, Epilog -> {PointSize[0.01], Hue[0],
                              Point /@ N[toLogData[middleData]]}]
```

Using the function `Fit` (to be discussed in this chapter), we find the exponent α to be about 0.59. This is clearly distinct from $\alpha = 1/2$ and shows long-range correlation within *Hamlet*. (The exact numerical value depends on the edition, the language, and the encoding used.)

```
In[129]:= Fit[toLogData[middleData], {1, x}, x]
```

```
Out[129]= -0.383861 + 0.593093 x
```

Using nonoverlapping windows results in a much faster calculation that basically gives the same result.

```
In[130]:= data = {#, meanSquareFluctuationNonOverlapping[
              packedRandomWalkHamletCumulative, #]}& /@ windows;
```

```
In[131]:= ListPlot[toLogData[DeleteCases[data, {_, 0.}]] // N,
              PlotRange -> All, PlotJoined -> True, Frame -> True,
              Axes -> False, Epilog -> {PointSize[0.01], Hue[0],
                            Point /@ N[toLogData[DeleteCases[data, {_, 0.}]]]}]
```

```
In[132]:= middleData = Select[data, 1 <= Log[10, #[[1]]] <= 5&];
```

```
In[133]:= Fit[toLogData[middleData], {1, x}, x]
```

```
Out[133]= -0.382217 + 0.591715 x
```

Now let us chop the *Hamlet* text into 1000 pieces of equal size and repeat the calculation above. The function `randomPermutation` calculates a random permutation of a given list. (Be aware that we can only exchange an element with a still unused element, else we will get a nonuniform distribution that, for larger *n*, favors no shuffling at all [717*].) The function `reorderText[`*text*, *n*`]` splits the text into *n* pieces, randomly reorders these pieces, and then glues the reordered pieces together again.

```
In[134]:= SeedRandom[111]
```

```
In[135]:= randomPermutation[l_List] :=
        Module[{lTemp = l, ll = Length[l], tmp1, tmp2, j},
```

```
                Do[tmp1 = lTemp[[i]];
                   j = Random[Integer, {i, 11}];
                   tmp2 = lTemp[[j]];
                   lTemp[[i]] = tmp2;
                   lTemp[[j]] = tmp1, {i, Length[l]}];
                lTemp]
```

In[136]:=
```
reorderText[text_, n_] :=
    Module[{l = StringLength[text], averagePieceLength, pieces},
           (* make pieces *)
           averagePieceLength = Floor[l/n];
           pieces = Append[Table[StringTake[text,
               {k averagePieceLength + 1,
               (k + 1) averagePieceLength}], {k, 0, n - 1}],
                          StringTake[text, {n averagePieceLength + 1, l}]];
           (* join permuted pieces *)
           StringJoin[randomPermutation[pieces]]]
```

In[137]:=
```
(* reorder the text *)
reorderdRandomWalkHamlet =
    Flatten[Characters[reorderText[completeHamleteText, 1000]] /. coding];

(* repeat steps from above *)
packedReorderedRandomWalkHamlet =
        Developer`ToPackedArray[reorderdRandomWalkHamlet];

packedReorderedRandomWalkHamletCumulative =
        FoldList[Plus, 0, packedReorderedRandomWalkHamlet];

data = {#, meanSquareFluctuation[
        packedReorderedRandomWalkHamletCumulative, #]}& /@ windows;
```

In[146]:=
```
(* middle data *)
middleData = Select[data, 0 <= Log[10, #[[1]]] <= 5&];
```

In[148]:=
```
(* show all data and middle data *)
Show[GraphicsArray[
Block[{$DisplayFunction = Identity, opts = Sequence[PlotRange -> All,
        PlotJoined -> True, Frame -> True, Axes -> False]},
 {(* all data *)
  ListPlot[toLogData[data] // N, Evaluate[opts], Epilog ->
          {PointSize[0.01], Hue[0], Point /@ N[toLogData[data]]}],
  (* middle data *)
  ListPlot[toLogData[middleData] // N, Evaluate[opts], Epilog ->
          {PointSize[0.01], Hue[0], Point /@ N[toLogData[middleData]]}]}]]]
```

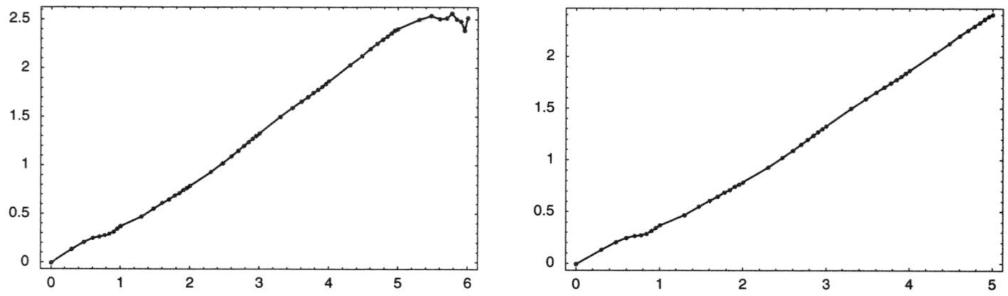

As expected, this time we see less long-range correlations.

In[150]:= **Fit[toLogData[middleData], {1, x}, x]**

Out[150]= $-0.1433 + 0.49952 x$

For hidden combinatorial messages in *Hamlet*, see [608], [609]. For analyzing the role of the words in *Hamlet*, see [1289]; for letter correlations, see [534].

Not only human texts (meaning texts written in natural languages) but also computer code shows long range-correlations. This is not completely unexpected—the syntax forces certain correlations, like closed parentheses [93].

Let us analyze the long-range correlations in a *Mathematica* notebook, say, the notebook containing Chapter 1 of the Programming volume [1793].

In[151]:= **chapterP1 = Get[ToFileName[ReplacePart[**
"FileName" /. NotebookInformation[EvaluationNotebook[]],
"1_Programming_1.nb", 2]]];

We eliminate the GraphicsData from the notebook for two reasons. First, this makes the data smaller and faster to analyze, and second, they contain PostScript, which is not what we want to analyze at the moment. We will discuss PostScript at the end of this subsection.

In[152]:= **chapterP1 = chapterP1 //. _GraphicsData -> {};**

In[153]:= **ByteCount[chapterP1]**

Out[153]= 2590560

The notebook contains mainly boxes with list and string arguments.

In[154]:= **(* count strings in the chapter *)**
Count[chapterP1[[1]], _String, {-1}]

Out[155]= 41940

In[156]:= **(* count boxes in the chapter *)**
Sort[Select[{#, Count[Level[chapterP1[[1]], {-1},
Heads -> True], #]}& /@ (ToExpression /@ Names["*Box*"]),
#[[2]] =!= 0&], #1[[2]] > #2[[2]]&]

Out[157]= {{ButtonBox, 4025}, {RowBox, 3110}, {CounterBox, 2856}, {StyleBox, 2639},
{BoxData, 701}, {FormBox, 628}, {SuperscriptBox, 400}, {SubscriptBox, 275},
{FractionBox, 120}, {SqrtBox, 33}, {SubsuperscriptBox, 31},
{UnderoverscriptBox, 28}, {OverscriptBox, 20}, {RadicalBox, 17},
{TagBox, 12}, {GridBox, 10}, {UnderscriptBox, 5}, {InterpretationBox, 1}}

We transform the *Mathematica* expression into a long string.

In[158]:= **chapterP1 = ToString[chapterP1, FormatType -> InputForm];**

In[159]:= **StringLength[chapterP1]**

Out[159]= 1058517

These are the first few lines of the string corresponding to the notebook.

In[160]:= **StringTake[chapterP1, 500]**

Out[160]= Notebook[{Cell["P   R   O   G   R   A
     M   M   I   N   G", "VolumeLabel"], Cell[TextData[
     {StyleBox["CHAPTER ", "CHP"], "1"}], "ChapterNumber"], Cell[
     TextData[{"Introduction to ", StyleBox["*Mathematica*", Rule[
     FontSlant, "Italic"]], " "}], "Chapter", CellTags -> {"Index[01,

```
{Mathematica, introduction into ~}]", "Index[01, {Mathematica, overview
of ~}]", "T[C[1]]"}], Cell[CellGroupData[{Cell["1.0 Remarks", "Section",
CellTags -> {"T[S[1.0]]", "Index[01, {general, overview}]"}], Cell[Tex
```

These are the different characters used in the text. This time, we have many non-ASCII characters.

```
In[161]:= allCharacters = Characters[chapterP1];
          usedChars = Union[allCharacters]
Out[162]= {!, @, #, %, ^, &, *, (, ), _, -, +, =, ~, `,
           {, [, }, ], |, \, <, >, ., ,, ;, ", ?, ', /, :,        ,
           ,  ,  ,  ,
             , ■, 0, 1, 2, 3, 4, 5, 6, 7, 8, 9, a, á, ä, A, Á, Å, b, B, c, ć, č, ç, C, Č, d,
           D, e, é, è, E, É, f, F, g, G, h, H, i, í, ì, I, j, J, k, K, l, L, m, M, n, ñ,
           N, o, ó, ô, ö, O, Ø, p, P, q, Q, r, R, s, š, ß, S, Š, t, T, u, ú, ü, U, Ü, v,
           V, w, W, x, X, y, Y, z, Z, 𝑎, ℬ, 𝑐, 𝐶, 𝑑, 𝑒, ℰ, 𝑓, ℱ, 𝑔, ℎ, ℋ, ℓ, 𝑚, 𝑛, 𝑝, 𝒫, 𝑄,
           𝑟, ℛ, 𝑢, 𝑣, 𝑥, 𝒴, B, ℒ, 𝕒, 𝔸, 𝔹, 𝕔, ℂ, 𝕕, 𝕖, 𝕗, 𝕚, 𝕝, 𝕃, 𝕞, ℕ, 𝕠, ℙ, 𝕣, ℝ, 𝕥,
           𝕋, 𝕪, 𝕫, ℤ, α, β, B, γ, Γ, δ, Δ, ϵ, ε, ζ, η, θ, ϑ, ι, κ, ϰ, λ, Λ, μ, ν, ξ, ο,
           π, ϖ, Π, ρ, ϱ, σ, ς, τ, υ, ϕ, φ, χ, ψ, Φ, ω, Ω, ς, ϝ, ϙ, ϶, $, ⋯, −, °, ″, …,
           ▪, -, ⅈ, ∞, —, , ′, ®, ✸, ℘, ”, ’, 〈, ∣, ⌈, 〚, ⌊, “, ‘, 〉, ∤, ⌉, 〛, ⌋, ⋱, , -,
           ⟶, ↑, ⅆ, ∂, ×, ×, ∏, ∫, ∑, ==, ≥, ≳, ≤, ≠, ∼, ≃, ≈, ∈, →, ⧴, ∧, ∨}
```

To model a random 1D walk, we again replace the characters by their character encoding in base 2 and then change all 0's to -1's. (The results are largely independent of the applied encoding.)

```
In[163]:= randomWalkChapterP1 =
          Flatten[(2 IntegerDigits[ToCharacterCode[#], 2] - 1)& /@ allCharacters];

In[164]:= Length[randomWalkChapterP1]
Out[164]= 7131038

In[165]:= ByteCount[randomWalkChapterP1]
Out[165]= 142620792
```

Now, we pack this long list.

```
In[166]:= packedRandomWalkChapterP1 = Developer`ToPackedArray[randomWalkChapterP1];
```

Again, this makes this list much more memory efficient.

```
In[167]:= ByteCount[packedRandomWalkChapterP1]
Out[167]= 28524208
```

To save memory, we again delete unpacked lists that are no longer needed.

```
In[168]:= Clear[randomWalkChapterP1]
```

Here we determine the number of $+1$ and -1 in the resulting list.

```
In[169]:= {Count[packedRandomWalkChapterP1, +1], Count[packedRandomWalkChapterP1, -1]}
Out[169]= {3778501, 3352537}
```

The first 10000 steps of the corresponding random walk are visualized in the next picture.

```
In[170]:= ListPlot[FoldList[Plus, 0, Take[packedRandomWalkChapterP1, 10^4]],
                   PlotStyle -> {{Thickness[0.002], GrayLevel[0]}},
                   PlotJoined -> True, Frame -> True]
```

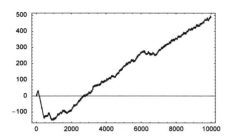

Out[170]= - Graphics -

In[171]:= **packedRandomWalkChapterP1Cumulative =**
FoldList[Plus, 0, packedRandomWalkChapterP1];

This time, the walk has more than five million steps.

In[172]:= **Length[packedRandomWalkChapterP1Cumulative]**
Out[172]= 7131039

We proceed as above and calculate the mean square fluctuations.

In[173]:= **windows = windowLengthsList[packedRandomWalkChapterP1Cumulative];**

In[174]:= **data = {#, meanSquareFluctuation[**
packedRandomWalkChapterP1Cumulative, #]}& /@ windows;

In[175]:= **ListPlot[toLogData[data] // N,**
PlotRange -> All, PlotJoined -> True, Frame -> True, Axes -> False,
Epilog -> {PointSize[0.01], Hue[0], Point /@ N[toLogData[data]]}]

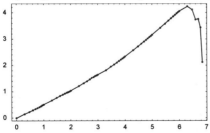

Out[179]= - Graphics -

This time, a fit yields $\alpha \approx 0.7$, showing pronounced long-range correlations.

In[176]:= **Fit[toLogData[Select[data, 1 <= Log[10, #[[1]]] <= 6&]], {1, x}, x]**
Out[180]= -0.436678 + 0.721491 x

The reader interested in this kind of investigation can go further and analyze long-range correlations in some *Mathematica* graphics. The following input produces a long string, being the PostScript code of the picture.

```
SeedRandom[111];

picture = Graphics[
ContourPlot[Evaluate[Sum[Random[Real, {-1, 1}] Sin[i x] Sin[j y],
                        {i, 8}, {j, 8}]],
            {x, 0, 2Pi}, {y, 0, 2Pi}, ColorFunction -> (Hue[Random[]]&),
            ContourStyle -> {{Thickness[0.002], GrayLevel[0]}},
            PlotPoints -> 50, Contours -> 20]]

postScriptString = DisplayString[picture];
```

For long-range correlations in genomes, see [1265], [996]. For missing words in texts, see [1497].

1.2 Fitting and Interpolating Functions

Frequently, we are given a possibly large set of numerical data (e.g., measurements) that we wish to process analytically or graphically. It often turns out to be convenient to fit this data with a function. One way to accomplish this approximation is to use the Fit command.

Fit[{{x_1, y_1, ..., *functionValue*$_1$}, {x_2, y_2, ..., *functionValue*$_2$}, ...,
   {x_n, y_n, ..., *functionValue*$_n$}}, {*function*$_1$, *function*$_2$, ..., *function*$_m$},
    {*var*$_1$, *var*$_2$, ..., *var*$_o$}]

computes a fit to the data
   {{x_1, y_1, ..., *functionValue*$_1$}, {x_2, y_2, ..., *functionValue*$_2$}, ...,
   {x_n, y_n, ..., *functionValue*$_n$}},
by a linear combination of the functions {*function*$_1$, *function*$_2$, ..., *function*$_m$} that depend on the variables
{*var*$_1$, *var*$_2$, ..., *var*$_o$}. The function f is chosen to minimize the sum of the squares of the differences
between $f(x_i, y_i, ...)$ and the given values *functionValue*$_i$.

To illustrate, we first create a set of "measurement values" that originate from a simulated experiment involving dropping an object from a height of 5 m with an initial velocity 0 m/sec. The list contains values of the form {*time*, *position*}.

```
In[1]:= measurementValues =
    Table[{t, -9.81 t^2/2 + 5 +
            (* modeling some measurement error *) 1/100 Random[]},
        {t, 0.1, 0.9, 0.01}];
```

```
    Short[measurementValues, 4]
```
Out[2]//Short= {{0.1, 4.95658}, {0.11, 4.94372}, {0.12, 4.93079},
    {0.13, 4.92696}, {0.14, 4.90902}, ≪71≫, {0.86, 1.38055},
    {0.87, 1.28958}, {0.88, 1.20905}, {0.89, 1.11635}, {0.9, 1.03452}}

Fit produces the following values for the gravitational acceleration and position.

```
In[3]:= Fit[measurementValues, {t^2, 1}, {t}]
```
Out[3]= $5.00513 - 4.90528 t^2$

Of course, we could have found this result "ourselves", but it would have taken longer. We would form the sum of the squares of the deviations in terms of the unknown position and acceleration, and then compute the derivatives with respect to these two variables to get a homogeneous linear system of two equations with two unknowns.

```
In[4]:= errorSum = Sum[(measurementValues[[i, 2]] - (g/2 t^2 + h) /.
                        {t -> measurementValues[[i, 1]]})^2,
                    {i, Length[measurementValues]}] // Simplify
```
$$Out[4]= 1109.5 + 3.03503\,g^2 - 568.726\,h + 81.\,h^2 + g\,(-63.9659 + 24.678\,h)$$

```
In[5]:= Solve[{D[errorSum, g] == 0, D[errorSum, h] == 0}, {g, h}]
```
$$Out[5]= \{\{g \to -9.81056,\ h \to 5.00513\}\}$$

```
In[6]:= (g/2 t^2 + h) /. %
```
$$Out[6]= \{5.00513 - 4.90528\,t^2\}$$

Another way to solve this problem would be the use of `PseudoInverse`, which is a generalization of `Inverse` for nonsquare matrices; for details, see [332] and [696].

```
In[7]:= PseudoInverse[{1, First[#]^2/2}& /@ measurementValues].
                        (Last /@ measurementValues)
```
$$Out[7]= \{5.00513, -9.81056\}$$

(The last example had errors in the dependent quantity, but not in the independent quantity t. If also the independent quantity has errors, the so-called total least-squares method has to be employed, instead of the least-squares method; see [468], [1350], [1292], [1829], [1951], and [215] for details. For fits of data with errors, see [755].)

Let us now deal with a more interesting example now: What is the complexity of calculating the sin function to precision 2^k? The function `getTime` yields a reliable timing for the function f called with a random argument from the domain *domain* of precision *prec*.

```
In[8]:= getTime[f_, prec__, domain__] :=
    Module[{n = 1, time},
            While[(* until we get reliable timings *)
                    xList = Table[Random[domain, prec], {n}];
                    (time = Timing[f[xList]][[1, 1]]) < 1/10,
                    n = 2 n];
                time/n]
```

Here are the timings.

```
In[9]:= data = Table[{2^k, getTime[Sin, 2^k, Real, {0, 100}]}, {k, 5, 15}]
```
$$Out[9]= \{\{32, 0.0000317383\}, \{64, 0.0000415039\}, \{128, 0.0000585938\},$$
$$\{256, 0.000136719\}, \{512, 0.000429687\}, \{1024, 0.0015625\},$$
$$\{2048, 0.0059375\}, \{4096, 0.0225\}, \{8192, 0.1\}, \{16384, 0.32\}, \{32768, 0.91\}\}$$

We conjecture a dependence of the form $t \sim \log(prec)$ [952], [283], [1615], [901] and fit the data. (Be aware that this is not equivalent to carrying out a least-squares fit for the corresponding exponentiated form [1566]. For the complexity of general real functions, see [282].)

```
In[10]:= logData = {Log[2, #[[1]]], Log[2, #[[2]]]}& /@ data;
```

```
In[11]:= timeFit[prec_] = Fit[logData, {1, prec}, prec]
```
$$Out[11]= -24.6622 + 1.60374\,prec$$

Here is a graphic showing the measured timings and the fit.

```
In[12]:= ListPlot[logData,
           PlotRange -> All, Frame -> True, PlotJoined -> True, Axes -> False,
           (* the measured timings *)
           Prolog -> {{GrayLevel[0], PointSize[0.02], Point /@ logData},
             {Hue[0.8], Thickness[0.01],
              Line[Table[{prec, timeFit[prec]},
                    {prec, logData[[1, 1]], logData[[-1, 1]],
                     (logData[[-1, 1]] - logData[[1, 1]])/100}]]}}]
```

Here is another example where fitting a function of the data instead of the original data is easy. We want to approximate the number of zeros $\mathcal{O}_b(n!)$ of $n!$ in base b as a function of n [1177].

```
In[13]:= nonzeroDigitCount[n_, b_] :=
         With[{digits = IntegerDigits[n!, b]},
              Length[digits] - Count[digits, 0]]
```

Graphing $\ln(\mathcal{O}_b(n!))$ versus $\ln(n)$ suggests $\mathcal{O}_b(n!) \sim n^\alpha$ over a wide range of n.

```
In[14]:= {data2, data10} = Table[{Log @ n, Log @ nonzeroDigitCount[n, #]},
                                  {n, 2^6, 2^12}]& /@ {2, 10};
```

```
In[15]:= Show[GraphicsArray[ListPlot[#, DisplayFunction -> Identity]& /@
                            N[{data2, data10}]]]
```

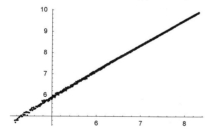 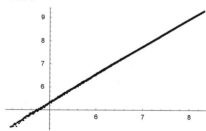

Fitting the data reveals a small quadratic component indicating that the relation is more complicated than $\mathcal{O}_b(n!) \sim n^\alpha$.

```
In[16]:= Fit[#, {1, x, x^2}, x]& /@ {data2, data10}
Out[16]= {-0.975503 + 1.4722 x - 0.0200989 x^2, -1.42731 + 1.43786 x - 0.0180319 x^2}
```

Using larger n for approximating $\mathcal{O}_2(n!)$ allows approximating $\mathcal{O}_2(10^6)$ remarkably well.

```
In[17]:= With[{n = 10^6},
         {Exp[Fit[Drop[data2, 2167], {1, x, x^2}, x] /. x -> Log[n]],
          nonzeroDigitCount[n, 2] // N}]
Out[17]= {8.74295×10^6, 8.7442×10^6}
```

Here is an example for fitting 2D data (be aware of the two functions y and y^2 in the list of functions; these functions do not appear in the model of the data).

```
In[18]:= Fit[Flatten[Table[{x, y, x + 4 Sin[x] - 7/4 Log[y]},
                {x, 1, 12}, {y, 1, 23}], 1],
        {x, y, y^2, Sin[x], Log[y]}, {x, y}] // Chop
Out[18]= 1. x - 1.75 Log[y] + 4. Sin[x]
```

As a more challenging example for Fit, let us consider the following problem. The functions

$$\Psi_\varphi^{(k)}(x, y) = \sin(\cos(\varphi)\, k\, x) \sin(\sin(\varphi)\, k\, y)$$

are for all values of the real parameter φ eigenfunctions of the 2D Laplace operator with eigenvalue k^2 (Helmholtz equation). In addition, they fulfill the boundary condition $\Psi_\varphi^{(k)}(0, y) = \Psi_\varphi^{(k)}(x, 0) = 0$. (Such a system could be realized as a thin metallic plate with clamped boundaries [1706], [1074].)

We will find an approximate solution of the Dirichlet problem of the Laplace operator for the following domain.

```
In[19]:= Show[Graphics[{Disk[{1/2, 0}, 1], Disk[{-1/2, 0}, 1],
                GrayLevel[1], Disk[{ 1/2, 0}, 1/4], Disk[{-1/2, 0}, 1/4]}],
        AspectRatio -> Automatic, Frame -> True]
```

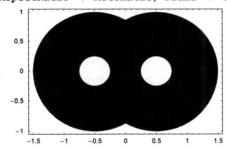

Because of the symmetry of the domain, we restrict ourselves to the first quadrant. For the antisymmetric solutions we make an ansatz of the form [811], [414], [415] (a superposition of plane waves in various directions)

$$\Psi^{(k)}(k; x, y) = \sum_{j=0}^{n} c_j(k)\, \Psi_{j/n\,2\pi}^{(k)}(x, y).$$

We will determine the $c_j(k)$ so that $\Psi^{(k)}(k; x, y)$ fulfills the Dirichlet boundary conditions on the two circle parts as well as possible. (For a generic, symmetric, polygonal 2D region, the system $\Psi_{j/n\,2\pi}(x, y)$ does not contain enough functions to exactly fulfill the boundary conditions; also not in the limit $n \to \infty$ [754], [47], [1839], [1900], [195], [1874], [1823].) Here are parametrizations for the two boundary curves.

```
In[20]:= outerCircle[t_] := {1/2 + Cos[t], Sin[t]} // N
        innerCircle[t_] := {1/2 + 1/4 Cos[t], 1/4 Sin[t]} // N
```

These are the basis functions $\Psi_{j/n}(x, y)$.

```
In[22]:= Φ[n_, k_][j_, {x_, y_}] := Sin[Cos[j/n 2Pi] k x] Sin[Sin[j/n 2Pi] k y]
```

We try to make the function $\Psi^{(k)}(k; x, y)$ vanish on 600 points on each of the two circles. To avoid the trivial solution $c_j(k) = 0$ for all j, we use one "random" interior point. At this point, we force $\Psi^{(k)}(k; x, y)$ to be nonvanishing.

```
In[23]:= n = 600;
     data = Join[Table[Append[outerCircle[t], 0.], {t, 0, 2/3Pi, 2./3Pi/n}],
              Table[Append[innerCircle[t], 0.], {t, 0, Pi, Pi/n}],
              {{1., 1/2., 1.}}];
```

Depending on k, the boundary conditions can be fulfilled with varying quality. For k being an eigenvalue, the boundary conditions can be fulfilled quite accurately. In the following example, we will use 300 basis functions and $k = 33.63$. This means we have about four times as many equations as unknowns $c_j(k)$.

```
In[25]:= SinSinFit[{x_, y_}] =
     With[{n = 300, k = 33.63, normS = 0.0131717},
          1/Sqrt[normS] Fit[data, Table[Ψ[n, k][l, {x, y}], {l, 0, n}] // N,
              {x, y}]];
```

To check the quality of the solution, we use about 10000 points on the boundaries. The sum of squared values at the boundaries is quite small.

```
In[26]:= checkPoints = Join[Table[outerCircle[t], {t, 0, 2/3Pi, 2./3Pi/8/n}],
                        Table[innerCircle[t], {t, 0, Pi, Pi/8/n}]] // N;
     Plus @@ ((SinSinFit /@ checkPoints)^2)
Out[27]= 67.4884
```

The following graphic shows how the $\Psi^{(k)}(k; x, y) = 0$ contour lines approach the two circles.

```
In[28]:= cp = ContourPlot[Evaluate @ SinSinFit[{x, y}],
              {x, -0.1, 1.6}, {y, -0.1, 1.1}, PlotPoints -> 250,
              Prolog -> {GrayLevel[0.7], Thickness[0.02],
               (* the two circle pieces *)
               Line[Table[outerCircle[t], {t, 0, 2/3 Pi, 2/3Pi/100}]],
               Line[Table[innerCircle[t], {t, 0, Pi, Pi/100}]]},
              PlotRange -> All, AspectRatio -> Automatic,
              Contours -> {0}, ContourShading -> False,
              ContourStyle -> {Hue[0]}]
```

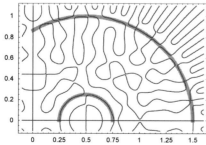

The next graphic shows a contour plot of the approximated eigenfunction.

```
In[29]:= cp1 = ContourPlot[Evaluate @ SinSinFit[{x, y}],
              {x, -0, 1.5}, {y, -0, 1}, PlotPoints -> 50,
              AspectRatio -> Automatic, ContourLines -> False,
              DisplayFunction -> Identity, PlotRange -> {-2.5, 2.5},
              Contours -> Table[c, {c, -2.5, 2.5, 5/50}],
              ColorFunction -> (Hue[0.8 #]&), Epilog -> {GrayLevel[1],
               (* overlay polygons over "not needed" parts *)
               Polygon[Join[Table[outerCircle[t], {t, 0, 2/3 Pi, 2/3Pi/100}],
                        {{0, 1.}, {1.5, 1}, {1.5, 0}}]],
               Polygon[Table[innerCircle[t], {t, 0, Pi + 0.1, Pi/100}]],
```

```
                 Polygon[{{0, 0}, {1.51, 0}, {1.51, 0}, {0, 0}}],
                 Polygon[{{0, 0}, {0, 1.51}, {0, 1.51}, {0, 0}}]}];

In[30]:= Show[Graphics[cp1] /. p_Polygon :> {p, Map[{-1, 1}#&, p, {-2}],
                 Map[{1, -1}#&, p, {-2}], Map[{-1, -1}#&, p, {-2}]}],
           PlotRange -> {{-3/2, 3/2}, {-1, 1}},
           Frame -> False, DisplayFunction -> $DisplayFunction]
```

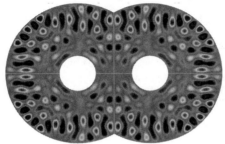

For appropriate numerical methods to solve the Helmholtz equation in general 2D domains, see, for instance, [105]. A 3D plot of the calculated function shows nicely how well the Dirichlet boundary conditions are fulfilled.

```
In[31]:= (* 3D data *)
       𝒯 = With[{ℛ = Function[φ, If[φ <= 2Pi/3, 1, -1/(2 Cos[φ])]]},
         Table[#[r, φ], {φ, 0., 1.Pi, Pi/300.},
               {r, 1./4, 1. ℛ[φ], (ℛ[φ] - 1/4)/160.}]&[
               Compile[{r, φ}, Evaluate[N[{1/2 + r Cos[φ], r Sin[φ],
                       SinSinFit[{1/2 + r Cos[φ], r Sin[φ]}]}]]]]];

In[33]:= (* form polygons *)
       polys = Table[Polygon[{𝒯[[i, j]], 𝒯[[i + 1, j]],
                 𝒯[[i + 1, j + 1]], 𝒯[[i, j + 1]]}],
                 {i, #1 - 1}, {j, #2 - 1}]& @@ Dimensions[𝒯];

In[35]:= (* color polygons *)
       colorPoly[p:Polygon[l_]] :=
       Module[{zMax = Pi, z = Plus @@ (Last /@ l)/4},
              {SurfaceColor[#, #, 2.6]&[
                 If[z < 0, Hue[1 + 0.5 z/zMax], Hue[0.5 z/zMax]]], p}]

In[37]:= Show[Graphics3D[{EdgeForm[], colorPoly /@ Flatten[polys]}] /.
               (* make polygons in remaining three quadrants *)
               p_Polygon :> {p, Map[{-1, 1, 1}#&, p, {-2}],
                 Map[{1, -1, 1}#&, p, {-2}], Map[{-1, -1, 1}#&, p, {-2}]},
           BoxRatios -> {3/2, 1, 0.7}, Axes -> False]
```

The related problem of finding the best fit of given data using functions in which the parameters enter nonlinearly (not just as coefficients in a linear combination of basis functions) can be solved using the function `FindFit`.

`FindFit[`*data, model, parameters, variables*`]`

    finds values of the parameters *parameters* that fit the data *data* optimal (in a least squares sense) the model *model.*

The "Find" in the naming of the function `FindFit` indicates that the function attempts to find a good fit, but cannot guarantee to find a good one (or even the best possible one). For the general case, finding optimal parameters is a nonlinear global optimization problem that can contain many local minima. The function `FindFit` has many options.

```
In[38]:= Options[FindFit]
```

```
Out[38]= {AccuracyGoal → Automatic, Compiled → Automatic,
    EvaluationMonitor → None, Gradient → Automatic, MaxIterations → 100,
    Method → Automatic, NormFunction → Norm, PrecisionGoal → Automatic,
    StepMonitor → None, WorkingPrecision → Automatic}
```

Most of these options are similar to the functions `FindRoot` and `FindMinimum`. We will discuss these two functions and their options in more detail later. Here we give only a simple example of the use of the function `FindFit`.

We create test data using the function $\cos(a\,x + b\,x^2 + c)$.

```
In[39]:= data = Table[{x, Cos[1.2 x + 0.00023 x^2 + 0.345]} // N,
              {x, 0, 2Pi, 2Pi/60}];
```

The following plot presents a comparison of the data with a pure $\cos(1.2\,x)$ function.

```
In[40]:= Show[(* the data *)
       {ListPlot[data, DisplayFunction -> Identity],
           (* the Cos[1.2 x] picture *)
        Plot[Cos[1.2 x], {x, 0, 2Pi}, DisplayFunction -> Identity]},
          DisplayFunction -> $DisplayFunction]
```

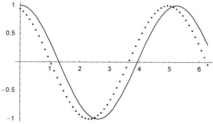

We are now ready to compute the fit. Without any prior information on the size of the parameters, `FindFit` does not succeed in finding a good fit.

```
In[41]:= FindFit[data, Cos[α x + β x^2 + γ], {α, β, γ}, x]
```

           FindFit::cvmit : Failed to converge to the
                requested accuracy or precision within 100 iterations. More…

```
Out[41]= {α → -1.6191, β → 1.4527, γ → 1.22264}
```

In using FindFit, we can (and if possible should) provide starting values for the parameters to be determined.

In[42]:= **FindFit[data, Cos[α x + β x^2 + γ], {{α, 1}, {β, 0}, {γ, 1/2}}, x]**

Out[42]= $\{\alpha \rightarrow 1.2,\ \beta \rightarrow 0.00023,\ \gamma \rightarrow 0.345\}$

The resulting fitting function approximates the data extremely well.

In the next input, we compute fits to the function $\cos(\alpha + x + \gamma x^2)$. In the first pair of plots, we use no starting values for the fit and in the second pair of plots, we use starting values that are within 10% of the exact values. The left graphic show the squared parameter differences and the right graphics show the squared value differences. The graphics show that it is typically advantageous to give starting values in case good starting values are known. Because the fit procedure does not succeed to produce good fits for all parameter values, we obtain some messages.

In[43]:= **Needs["Graphics`Graphics3D`"]**

In[44]:= **Module[{pp = 20, δs, datar, ff, Δ, ε = 0.1},**
    Function[αγ, δs =
  Table[(* "exact" data *)
        datar = Table[{x, Cos[αr + x + γr x^2]} // N,
                    {x, 0, 2Pi, 2Pi/60}];
        (* fitted data parameters *)
        ff = FindFit[datar, Cos[α + x + γ x^2], αγ, x,
             MaxIterations -> 200];
        (* square of difference of original and fitted data *)
        Δ = Sum[Abs[Cos[α + x + γ x^2] - Cos[αr + x + γr x^2]]^2,
                {x, 0, 2Pi, 2Pi/60}] //. ff;
        (* {αr, γr, parameterDifferenceSumOfSquares,
                sumOfSquaresOfValues *)
        {αr, γr, (αr - α)^2 + (γr - γ)^2 //. ff, Δ},
        {αr, 0, 2, 2/pp}, {γr, 0, 2, 2/pp}];
  (* show squared differences over the α,γ-plane *)
  Show[GraphicsArray[Function[g,
  ListSurfacePlot3D[Map[Delete[#, g]&, δs, {2}],
                BoxRatios -> {1, 1, 0.6}, Axes -> True,
                DisplayFunction -> Identity, PlotRange -> All]] /@
                {4, 3}]], {HoldAll}] @@@
    (* no starting values and starting values
        within ε of the exact values *)
    {Hold[{α, γ}],
    Hold[{{α, Random[Real, {1 - ε, 1 + ε} αr]},
        {γ, Random[Real, {1 - ε, 1 + ε} γr]}}]}]]

    FindFit::cvmit : Failed to converge to the
      requested accuracy or precision within 200 iterations. More…

    FindFit::cvmit : Failed to converge to the
      requested accuracy or precision within 200 iterations. More…

    FindFit::cvmit : Failed to converge to the
      requested accuracy or precision within 200 iterations. More…

    General::stop :
     Further output of FindFit::cvmit will be suppressed during this calculation. More…

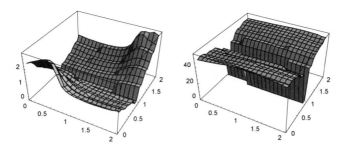

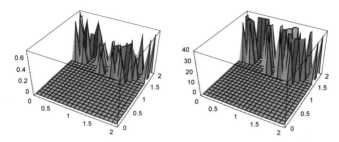

```
FindFit::fmgz :
  Encountered a gradient which is effectively zero. The result returned
    may not be a minimum; it may be a maximum or a saddle point.
```

Here is a small application example for `FindFit`. The *Mathematica* directory contains a few thousand files (notebooks, data files, binary files, and so on).

```
In[45]:= allFileNames = FileNames["*", $InstallationDirectory, Infinity];
         Length[allFileNames]
Out[46]= 2370
```

We show the distribution of the file sizes in a log-log-plot in the left graphic and the distribution of the first digit (Benford's law) in the right graphic. We see that the file sizes span about eight orders of magnitude. Over some intermediate file size interval [1 kB, 1 MB], the log-log data are approximately linear.

```
In[47]:= Needs["Graphics`Graphics`"]

In[48]:= allFileSizes =
           Sort[DeleteCases[FileByteCount /@ Take[allFileNames, All], 0]];

In[49]:= Show[GraphicsArray[
         {(* file size distribution in a log-log plot *)
         LogLogListPlot[allFileSizes, DisplayFunction -> Identity],
         (* histogram of first digit of file sizes *)
         Graphics[Rectangle[{#1 - 1/3, 0}, {#1 + 1/3, #2}]]& @@@
         ({First[#], Length[#]}& /@ Split[Sort[RealDigits[#][[1, 1]]& /@
                                                   allFileSizes]]),
                 Frame -> True, PlotRange -> All,
                 FrameTicks -> {Range[9], Automatic}]}]]
```

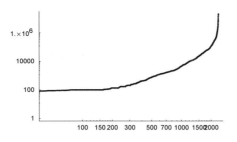 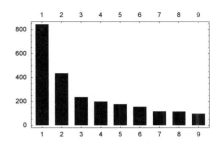

On average, the distribution of the file sizes of a larger set of files can be assumed to be lognormal distributed [1281]. We obtain the corresponding cumulative distribution function from the package `Statistics`Continuous`Distributions``.

```
In[50]:= Needs["Statistics`ContinuousDistributions`"]
```

```
In[51]:= cdf[x_, {μ_, σ_}] = CDF[LogNormalDistribution[μ, σ], x]
```

$$Out[51]= \frac{1}{2} \left(1 + \text{Erf}\left[\frac{-\mu + \text{Log}[x]}{\sqrt{2}\,\sigma} \right] \right)$$

`cdfData` is a list of the cumulative distribution of the observed file sizes.

```
In[52]:= (* form pairs {fileSize, probability FILESIZE < fileSize} *)
        cdfData = With[{λ = Length[allFileSizes]},
                    MapIndexed[{#1, #2[[1]]/λ}&, Take[N[allFileSizes], λ]]];
```

We now use `FindFit` to find optimal values of the two parameters μ and σ.

```
In[54]:= FindFit[cdfData, cdf[x, {μ, σ}],
                (* use good starting values for the fit *)
                {{σ, StandardDeviation[Log[First /@ cdfData]]},
                 {μ, Mean[Log[First /@ cdfData]]}}, x]
```

$$Out[54]= \{\sigma \to 2.84937, \mu \to 8.96429\}$$

```
In[55]:= cdfOpt[x_] = cdf[x, {μ, σ}] /. %;
```

We show the observed data and the best-fit lognormal distribution. We use again a log-log-plot. The overall fit is good and the fit for large file sizes is excellent.

```
In[56]:= Show[
        Block[{$DisplayFunction = Identity},
          {LogLogListPlot[{#1, cdfOpt[#1]}& @@@ cdfData, PlotJoined -> True,
                        PlotStyle -> {Hue[0], Thickness[0.006]}],
           LogLogListPlot[cdfData, PlotStyle -> {GrayLevel[0],
                                                PointSize[0.004]}]}]]
```

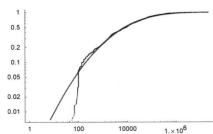

For the distribution of print job arrival times, see [791].

We now introduce the next important function for interpolating and fitting data.

```
InterpolatingPolynomial[{{x₁, {f₁, f₁′, f₁″, …, f₁⁽ⁿ¹⁾}},
    {x₂, {f₂, f₂′, f₂″, …, f₂⁽ⁿ²⁾}}, …,
    {xₘ, {fₘ, fₘ′, fₘ″, …, fₘ⁽ⁿᵐ⁾}}}, var]
```

constructs an interpolating polynomial in the variable *var* with the property that at each point x_i, it matches the function value f_i, the value of the first derivative f_i', etc., up to the n_ith derivative $f_i^{(n_i)}$, for $i = 1, …, m$. The interpolating polynomial is written in a form that is very convenient for further numerical calculations. The degree of the polynomial is $k - 1$, where $k = n_1 + n_2 + \cdots + n_m + m$ is the total number of individual data.

Here is an example. Suppose at x_1 we are given the function value, at x_2 the function value and derivative, at x_3 the function value, first, and second derivative, etc. Thus, at the ith point x_i, we are given the function value, first derivative, second derivative, third derivative,…, $i - 1$th derivative.

```
In[57]:= data = Table[{ξ, Table[i, {i, ξ}]}, {ξ, 6}] // N;
```

```
    MatrixForm[data]
```
Out[58]//MatrixForm=
$$\begin{pmatrix} 1. & \{1.\} \\ 2. & \{1., 2.\} \\ 3. & \{1., 2., 3.\} \\ 4. & \{1., 2., 3., 4.\} \\ 5. & \{1., 2., 3., 4., 5.\} \\ 6. & \{1., 2., 3., 4., 5., 6.\} \end{pmatrix}$$

Here is the associated interpolating polynomial.

```
In[59]:= intPol[v_] = InterpolatingPolynomial[data, v]
```
Out[59]= 1. + (2. + (-2. + (3. +
    (-3.75 + (1.75 + (-0.0833333 + (-1.59722 + (3.50463 + (-2.28241 + (1.49074 +
        (-1.09144 + (1.05083 + (-1.36353 + (0.804599 +
            (-0.486061 + (0.294132 + (-0.171083 +
                (0.0828598 - 0.00560712 (-6. + v)) (-6. +
                    v)) (-6. + v)) (-6. + v)) (-6. + v)) (-5. +
                    v)) (-5. + v)) (-5. + v)) (-5. + v))
            (-5. + v)) (-4. + v)) (-4. + v)) (-4. + v)) (-4. + v))
        (-3. + v)) (-3. + v)) (-3. + v)) (-2. + v)) (-2. + v) (-1. + v)

If we multiply this polynomial out, it looks a little friendlier (but it actually is not, as we shall see).

```
In[60]:= intPolEx[v_] = Expand[%]
```
Out[60]= $-1.75027 \times 10^{10} + 9.87748 \times 10^{10}\, v - 2.57994 \times 10^{11}\, v^2 + 4.15967 \times 10^{11}\, v^3 -$
$4.6536 \times 10^{11}\, v^4 + 3.84654 \times 10^{11}\, v^5 - 2.44068 \times 10^{11}\, v^6 + 1.21858 \times 10^{11}\, v^7 -$
$4.86613 \times 10^{10}\, v^8 + 1.57042 \times 10^{10}\, v^9 - 4.12007 \times 10^9\, v^{10} + 8.8048 \times 10^8\, v^{11} -$
$1.53017 \times 10^8\, v^{12} + 2.1506 \times 10^7\, v^{13} - 2.41988 \times 10^6\, v^{14} + 214526.\, v^{15} -$
$14620.2\, v^{16} + 737.307\, v^{17} - 25.8373\, v^{18} + 0.559465\, v^{19} - 0.00560712\, v^{20}$

We now check whether this polynomial actually interpolates, as it should. We consider only the first four interpolating points.

```
In[61]:= Do[intPolDer[i][v_] = D[intPol[v], {v, i}], {i, 0, 3}];
```

```
      Table[{ξ, Table[intPolDer[i][ξ], {i, 0, ξ - 1}]},
          {ξ, 4}] // (* for better readability *) MatrixForm
```
Out[62]//MatrixForm=
$$\begin{pmatrix} 1 & \{1.\} \\ 2 & \{1., 2.\} \\ 3 & \{1., 2., 3.\} \\ 4 & \{1., 2., 3., 4.\} \end{pmatrix}$$

If we perform the same check using the multiplied-out form of the polynomial, the importance of the phrase "very convenient for further numerical calculations" becomes clearer. The large coefficients in front of the high powers of v cause major problems with the numerical precision of the result.

In[63]:= `Do[intPolExder[i][v_] = D[intPolEx[v], {v, i}], {i, 0, 3}];`

```
      Table[{ξ, Table[intPolExder[i][ξ], {i, 0, ξ - 1}]}, {ξ, 4}] // MatrixForm
```
Out[64]//MatrixForm=
$$\begin{pmatrix} 1 & \{1.\} \\ 2 & \{0.999694, 1.99137\} \\ 3 & \{1.00864, 1.95276, 2.80698\} \\ 4 & \{1.23963, 1.44191, 6.12875, 7.36182\} \end{pmatrix}$$

`InterpolatingPolynomial` also works with exact (and with symbolic) data. Here is what happens if we do not use N.

In[65]:= **data // Rationalize**

Out[65]= {{1, {1}}, {2, {1, 2}}, {3, {1, 2, 3}},
        {4, {1, 2, 3, 4}}, {5, {1, 2, 3, 4, 5}}, {6, {1, 2, 3, 4, 5, 6}}}

In[66]:= **InterpolatingPolynomial[%, v] // Short[#, 6]&**

Out[66]//Short=
$$1 + \left(2 + \left(-2 + \left(3 + \left(-\frac{15}{4} + \left(\frac{7}{4} + \left(-\frac{1}{12} + \left(-\frac{115}{72} + \left(\frac{757}{216} + \left(-\frac{493}{216} + \left(\frac{161}{108} + \left(-\frac{943}{864} + \right.\right.\right.\right.\right.\right.\right.\right.\right.\right.\right.\right.$$
$$\left(\frac{10895}{10368} + \left(-\frac{169645}{124416} + \left(\frac{100105}{124416} + \left(-\frac{302369}{622080} + \left(\frac{914869}{3110400} + \left(-\frac{5321363}{31104000} + \right.\right.\right.\right.\right.\right.$$
$$\left.\left(\frac{12886363}{155520000} - \frac{209284549\,(-6+v)}{37324800000}\right)(-6+v)\right)(-6+v)\right)(-6+v)\right)(-6+v)\right)$$
$$\left.\left.(-5+v)\right)(-5+v)\right)(-5+v)\right)(-5+v)\right)(-5+v)\right)(-4+v)\right)(-4+v)\right)(-4+v)\right)$$
$$\left.(-4+v)\right)(-3+v)\right)(-3+v)\right)(-3+v)\right)(-2+v)\right)(-2+v)\right)(-1+v)$$

The independent (as well as the dependent) variable data can also contain symbolic (nonnumeric) entries. Here are two simple examples of this case.

In[67]:= **Clear[x, f, z, z0];**
        InterpolatingPolynomial[Table[{x[i], f[i]}, {i, 4}], z]

Out[68]= $f[1] + (z - x[1])\left(\frac{-f[1] + f[2]}{-x[1] + x[2]} + (z - x[2])\right.$
$$\left(\frac{-\frac{-f[1]+f[2]}{-x[1]+x[2]} + \frac{-f[2]+f[3]}{-x[2]+x[3]}}{-x[1] + x[3]} + (z - x[3])\left(\frac{-\frac{-\frac{-f[1]+f[2]}{-x[1]+x[2]} + \frac{-f[2]+f[3]}{-x[2]+x[3]}}{-x[1]+x[3]} + \frac{-\frac{-f[2]+f[3]}{-x[2]+x[3]} + \frac{-f[3]+f[4]}{-x[3]+x[4]}}{-x[2]+x[4]}}{-x[1] + x[4]}\right)\right)\right)$$

In[69]:= **InterpolatingPolynomial[{{z0, Table[f[i], {i, 6}]}}, z]**

Out[69]= $f[1] + (z - z0)$
$$\left(f[2] + (z - z0)\left(\frac{f[3]}{2} + (z - z0)\left(\frac{f[4]}{6} + (z - z0)\left(\frac{f[5]}{24} + \frac{1}{120}(z - z0)\,f[6]\right)\right)\right)\right)$$

The Neville algorithm [1460], [1748] shows nicely how the Lagrange interpolating polynomial of n points can be hierarchically constructed from the first $n-1$ and last $n-1$ points.

```
In[70]:= (* single point interpolation *)
        ipNeville[{j_}, {ξ_, φ_}][x_] := φ[j]

        (* recursive multiple point interpolation *)
        ipNeville[js_List, {ξ_, φ_}][x_] :=
        Module[{j1 = js[[+1]], J1 = Drop[js, -1],
               jm = js[[-1]], Jm = Drop[js, +1]},
             (* joint interpolating functions for
                beginning and ending sequence of points *)
             ((x - ξ[j1]) ipNeville[Jm, {ξ, φ}][x] -
              (x - ξ[jm]) ipNeville[J1, {ξ, φ}][x])/(ξ[jm] - ξ[j1])]
```

Here are four points interpolated.

```
In[74]:= ip4[x_] = ipNeville[{1, 2, 3, 4}, {ξ, φ}][x]
```

$$
Out[74]= \frac{1}{-\xi[1]+\xi[4]}
$$

$$
\left(-\frac{(x-\xi[4])\left(-\frac{(x-\xi[3])(-(x-\xi[2])\phi[1]+(x-\xi[1])\phi[2])}{-\xi[1]+\xi[2]}+\frac{(x-\xi[1])(-(x-\xi[3])\phi[2]+(x-\xi[2])\phi[3])}{-\xi[2]+\xi[3]}\right)}{-\xi[1]+\xi[3]} + \right.
$$

$$
\left. \frac{(x-\xi[1])\left(-\frac{(x-\xi[4])(-(x-\xi[3])\phi[2]+(x-\xi[2])\phi[3])}{-\xi[2]+\xi[3]}+\frac{(x-\xi[2])(-(x-\xi[4])\phi[3]+(x-\xi[3])\phi[4])}{-\xi[3]+\xi[4]}\right)}{-\xi[2]+\xi[4]} \right)
$$

```
In[75]:= (* quick check of the last interpolating polynomial *)
        Table[ip4[ξ[j]], {j, 4}] // Simplify
Out[76]= {φ[1], φ[2], φ[3], φ[4]}
```

Let us shortly discuss the "quality" of the interpolation carried out by `InterpolatingPolynomial`. We start by interpolating the function $f(x) = \cos(\pi x/2)^2$ in the interval $\{-1, 1\}$ on n ($= 50$ in the next plot) equidistant points.

```
In[77]:= n = 50;
        f[x_] := Cos[Pi x/2]^2;
        data = Table[{x, f[x]}, {x, -1, 1, 2/n}];
In[80]:= ipo[x_] = InterpolatingPolynomial[data // N, x];
```

The interpolating polynomial looks like the original function everywhere in the interval $\{-1, 1\}$.

```
In[81]:= Show[GraphicsArray[
        Plot[#, {x, -1, 1}, Frame -> True, Axes -> False,
            DisplayFunction -> Identity]& /@ {f[x], ipo[x]}]]
```

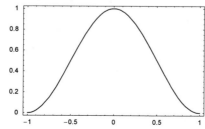 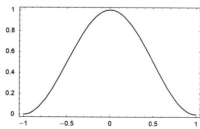

To quantify the quality of the approximation, we show the logarithm of the absolute value of the difference as a function of *n*. For all $-1 \le x \le 1$, the error is a decreasing function of *n*.

```
In[82]:= Module[{ε = 10^-10, data, ipo},
    ListPlot3D[Table[data = Table[N[{x, f[x]}, 100], {x, -1, 1, 2/n}];
                     ipo[x_] = InterpolatingPolynomial[data, x];
                     Table[Log[10, Abs[ipo[x] - f[x]]],
                            {x, -1 - ε, 1 - ε, (2 - 2ε)/31}], {n, 60}],
              PlotRange -> All]]
```

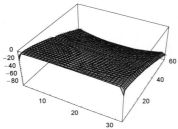

In the complex *x*-plane, for fixed *n*, the quality of approximation is best in the interval $\{-1, 1\}$.

```
In[83]:= Module[{n = 50, data, ipo},
    data = Table[N[{x, f[x]}, 100], {x, -1, 1, 2/n}];
    ipo[x_] = InterpolatingPolynomial[data, x];
                   (* high-precision definition for δ *)
    δ[z_?NumberQ] := With[{ζ = SetPrecision[z, 100]}, ipo[ζ] - f[ζ]]]
```

```
In[84]:= Plot3D[Log[10, Abs[δ[x + I y]]], {x, -4, 4}, {y, -3, 3},
           PlotPoints -> 30, Compiled -> False]
```

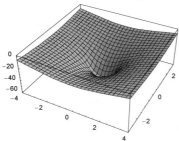

Now, let us consider the function $f(x) = 2 \arctan(8\,x)$. Like $\cos(\pi x / 2)^2$, it is a smooth function on the interval $\{-1, 1\}$. But the function $2 \arctan(8\,x)$ has a singularity near the interval $\{-1, 1\}$. As a result, of this singularity, we encounter the so-called Runge phenomena—near ± 1 the convergence of the interpolating polynomial to the original function breaks down. (For a detailed quantitative discussion about the relation between the nonconvergence of the interpolating polynomial and the location of the singularity, see [622], [919]; for general orthogonality properties of Lagrange interpolants, see [1965].)

```
In[85]:= n = 40;
    f[x_] := 2 ArcTan[8 x]/Pi;
    data = Table[{x, f[x]}, {x, -1, 1, 2/n}];
```

```
In[88]:= ipo[x_] = InterpolatingPolynomial[data // N, x];
```

```
In[89]:= Show[GraphicsArray[
          Plot[#, {x, -1, 1}, Frame -> True, Axes -> False,
              DisplayFunction -> Identity]& /@ {f[x], ipo[x]}]]
```

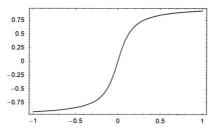 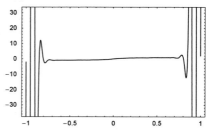

As a function of *n*, the Runge phenomenon is clearly visible.

```
In[90]:= Module[{ε = 10^-10, data, ipo},
        ListPlot3D[Table[
        data = Table[N[{x, f[x]}, 100], {x, -1, 1, 2/n}];
        ipo[x_] = InterpolatingPolynomial[data, x];
        Table[Log[10, Abs[ipo[x] - f[x]]],
                {x, -1 - ε, 1 - ε, (2 - 2ε)/31}], {n, 60}],
            PlotRange -> All, AxesLabel -> {"x", "n", None}]]
```

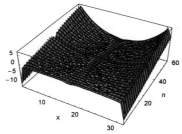

One can avoid the Runge phenomena by using nonequidistant *x*-values. It can be shown that *x*-values with an asymptotic density ($n \to \infty$) of the form $(1 - x^2)^{-1/2}/\pi$ are especially suited [218], [1332], [1149], [1673]. The points $x_j^{(n)} = -\cos(j/n\pi)$ exhibit this density and show a clustering at the endpoints of the interval $\{-1, 1\}$. (For every continuous function *f* on $[-1, 1]$ there exists a sequence of points $x_j^{(n)}$, such that as $n \to \infty$ the sequence of interpolating polynomials converges uniformly to *f*, but this sequence is not universal [544], [1454], [824], [1519], [1236], [988].) The following graphics shows the $x_j^{(n)}$ for $1 \le n \le 120$. The clustering of the zeros near ± 1 is clearly visible.

```
In[91]:= ChebyshevX[j_, n_] := -Cos[j/n Pi]
```

```
In[92]:= Show[Graphics[{PointSize[0.003], Table[Point[{n, ChebyshevX[j, n]}],
                    {n, 120}, {j, 0, n}]}], Frame -> True];
```

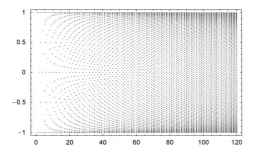

An interpolating polynomial formed with the $x_j^{(n)}$ shows a much better convergence.

```
In[93]:= n = 40;
      data = Table[{ChebyshevX[j, n], f[ChebyshevX[j, n]]}, {j, 0, n}];

In[95]:= ipo[x_] = InterpolatingPolynomial[data // N, x];

In[96]:= Plot[ipo[x], {x, -1, 1}, PlotRange -> {-1, 1},
         Frame -> True, Axes -> False]
```

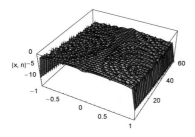

As a function of *n* now, we again have a decreasing error inside the whole interval {−1, 1}.

```
In[97]:= Module[{ε = 10^-10, data, ipo},
      ListPlot3D[Table[
      data = Table[N[{ChebyshevX[j, n], f[ChebyshevX[j, n]]}, 100], {j, 0, n}];
      ipo[x_] = InterpolatingPolynomial[data, x];
      Table[Log[10, Abs[ipo[x] - f[x]]],
              {x, -1 - ε, 1 - ε, (2 - 2ε)/31}], {n, 60}],
           PlotRange -> All, AxesLabel -> {"x", "n"},
           MeshRange -> {{-1, 1}, {1, 60}}]]
```

The next input shows the interpolating polynomials of increasing order for the nonanalytic function $f(x) = |x|$ in $[-1, 1]$. For increasing interpolation order, the approximation property breaks down for smaller and smaller x. In the limit, only the points $x = 0$ and $x = \pm 1$ are pointwise approached.

```
In[98]:= Module[{o = 40, ip},
         Show[Graphics[{Thickness[0.002],
             Table[(* interpolate Abs at n + 1 points *)
                 ip[x_] = InterpolatingPolynomial[
                       Table[{x, Abs[x]}, {x, -1, 1, 2/n}], x];
                 {Hue[0.78 (n - 1)/o], (* interpolating polynomial *)
                  Line[Table[{x, ip[x]}, {x, -1, 1, 1/(5n)}]]},
                 {n, o, 1, -1}]}],
           PlotRange -> {-1/2, 3/2}, Frame -> True]]
```

Let us give another application of `InterpolatingPolynomial`: The Newton–Cotes formulas for integrating functions given at equidistant abscissas can easily be derived from the interpolating (Lagrange) polynomials. Let a function $f(x)$ be defined in the interval $[0, 1]$. Let f_k be the value of the function to be integrated at $x_k = (k - 1)/n$, denoted by $f_k = f(x_k)$. Then, the Newton–Cotes weights are obtained by integrating the corresponding interpolating polynomial. Here is a straightforward implementation. (Using the binomial coefficients the interpolating polynomial $p_n(x)$ with $p_n(k) = \delta_{n,k}$, $k = 0, \ldots, n$ can be written in closed form as $p_n(k) = \binom{x}{k}\binom{n - x}{n - k}$ [682], [548].)

```
In[99]:= NewtonCotesCoefficient1[n_, f_] :=
         Integrate[InterpolatingPolynomial[
               Array[f, n], x], {x, 1, n}]/(n - 1) // Factor
```

The next output contains a list of the first 10 Newton–Cotes formulas.

```
In[100]:= Table[Timing[NewtonCotesCoefficient1[n, f]], {n, 2, 10}] /.
                              f[i_] -> Subscript[f, i]
```

$$Out[100]= \left\{\left\{0.06\,\text{Second}, \frac{1}{2}\,(f_1 + f_2)\right\},\right.$$

$$\left\{0.08\,\text{Second}, \frac{1}{6}\,(f_1 + 4f_2 + f_3)\right\}, \left\{0.15\,\text{Second}, \frac{1}{8}\,(f_1 + 3f_2 + 3f_3 + f_4)\right\},$$

$$\left\{0.19\,\text{Second}, \frac{1}{90}\,(7f_1 + 32f_2 + 12f_3 + 32f_4 + 7f_5)\right\},$$

$$\left\{0.31\,\text{Second}, \frac{1}{288}\,(19f_1 + 75f_2 + 50f_3 + 50f_4 + 75f_5 + 19f_6)\right\},$$

$$\left\{0.47\,\text{Second}, \frac{1}{840}\,(41f_1 + 216f_2 + 27f_3 + 272f_4 + 27f_5 + 216f_6 + 41f_7)\right\}, \left\{0.66\,\text{Second},\right.$$

$$\frac{751f_1 + 3577f_2 + 1323f_3 + 2989f_4 + 2989f_5 + 1323f_6 + 3577f_7 + 751f_8}{17280}\right\}, \left\{1.04\,\text{Second},\right.$$

$$\frac{989\,f_1 + 5888\,f_2 - 928\,f_3 + 10496\,f_4 - 4540\,f_5 + 10496\,f_6 - 928\,f_7 + 5888\,f_8 + 989\,f_9}{28350}\Big\},$$

$$\Big\{1.25\ \text{Second},\ \frac{1}{89600}\,(2857\,f_1 + 15741\,f_2 + 1080\,f_3 + 19344\,f_4 +$$
$$5778\,f_5 + 5778\,f_6 + 19344\,f_7 + 1080\,f_8 + 15741\,f_9 + 2857\,f_{10})\Big\}\Big\}$$

The timings in the last output indicate a very fast growth. This timing growth is caused by the fast size growth of the interpolating polynomial with symbolic f_i.

```
In[101]:= (* use three measures for the size *)
        Table[{ByteCount[#], ByteCount[Expand[#]], Depth[#]}&[
            InterpolatingPolynomial[Array[f, n], x]], {n, 12}]
Out[102]= {{40, 40, 2}, {312, 376, 6}, {656, 1136, 8}, {1584, 2312, 12}, {3456, 3904, 16},
        {7216, 5832, 20}, {14752, 8128, 24}, {29840, 10856, 28}, {60032, 13904, 32},
        {120432, 17384, 36}, {241248, 21136, 40}, {482896, 25368, 44}}
```

Using an interpolation technique, we can derive the Newton–Cotes formulas faster. For random numeric realizations of the f_k, we calculate the interpolating polynomials. For n "different" realizations, we have then enough data to interpolate the multivariate linear function $\sum_{k=0}^{n} c_k\,f_k$ (the linearity follows from the linearity of integration). Unfortunately, `InterpolatingPolynomial` works only for univariate polynomials; so, we have to carry out the multivariate interpolation by ourselves. But because of the linearity in each f_k, the interpolation reduces to solving a linear system of equations.

```
In[103]:= NewtonCotesCoefficient2[n_, f_] :=
        Module[{x, var, eqs, polyData, ipo, int, nc},
        (* list of the f[k] *)
        vars = Array[c, n];
        (* linear independent equations;
            (with a very small probability, we will have linear dependent ones) *)
        While[(* did we encounter linear dependent equations? *)
         Check[
         eqs = Table[
           (* random realizations of the f[k] *)
           polyData = Table[Random[Integer, {-10^5, 10^5}], {n}];
           (* the interpolated polynomial; one variable is x *)
           ipo = InterpolatingPolynomial[polyData, x];
           (* the integral *)
           int = Integrate[Expand[ipo], {x, 1, n}]/(n - 1);
           vars.polyData == int, {n}];
           (* solve equations *)
           sol = Solve[eqs, vars][[1]], True, Solve::"svars"],
           Null];
        (* return result in factored form *)
        Factor[Array[f, n].(vars /. sol)]]
```

Using `NewtonCotesCoefficient2`, we can calculate higher-order Newton–Cotes formulas much faster.

```
In[104]:= SeedRandom[999];

        Table[Timing[NewtonCotesCoefficient2[n, f]], {n, 2, 10}] /.
                                f[i_] -> Subscript[f, i]
```
$$Out[105]= \Big\{\Big\{0.01\ \text{Second},\ \frac{1}{2}\,(f_1 + f_2)\Big\},$$
$$\Big\{0.04\ \text{Second},\ \frac{1}{6}\,(f_1 + 4\,f_2 + f_3)\Big\},\ \Big\{0.06\ \text{Second},\ \frac{1}{8}\,(f_1 + 3\,f_2 + 3\,f_3 + f_4)\Big\},$$
$$\Big\{0.09\ \text{Second},\ \frac{1}{90}\,(7\,f_1 + 32\,f_2 + 12\,f_3 + 32\,f_4 + 7\,f_5)\Big\},$$

$$\left\{0.12 \text{ Second}, \frac{1}{288} (19 f_1 + 75 f_2 + 50 f_3 + 50 f_4 + 75 f_5 + 19 f_6)\right\},$$

$$\left\{0.18 \text{ Second}, \frac{1}{840} (41 f_1 + 216 f_2 + 27 f_3 + 272 f_4 + 27 f_5 + 216 f_6 + 41 f_7)\right\}, \left\{0.24 \text{ Second}, \right.$$

$$\left. \frac{751 f_1 + 3577 f_2 + 1323 f_3 + 2989 f_4 + 2989 f_5 + 1323 f_6 + 3577 f_7 + 751 f_8}{17280}\right\}, \left\{0.3 \text{ Second}, \right.$$

$$\left. \frac{989 f_1 + 5888 f_2 - 928 f_3 + 10496 f_4 - 4540 f_5 + 10496 f_6 - 928 f_7 + 5888 f_8 + 989 f_9}{28350}\right\},$$

$$\left\{0.37 \text{ Second}, \frac{1}{89600} (2857 f_1 + 15741 f_2 + 1080 f_3 + 19344 f_4 + \right.$$

$$\left. 5778 f_5 + 5778 f_6 + 19344 f_7 + 1080 f_8 + 15741 f_9 + 2857 f_{10})\right\}\right\}$$

Starting with $n = 11$, something unfortunate happens—some of the coefficients become negative.

`In[106]:=` **NewtonCotesCoefficient2[11, f]**

`Out[106]=` $\frac{1}{598752} (16067 f[1] + 106300 f[2] - 48525 f[3] + 272400 f[4] - 260550 f[5] +$
$427368 f[6] - 260550 f[7] + 272400 f[8] - 48525 f[9] + 106300 f[10] + 16067 f[11])$

Here is a graphic of the Newton–Cotes weights for $2 \le n \le 16$. They grow in size without bounds and show sign variations—this makes them numerically unstable [557], [459].

`In[107]:=` **Show[Graphics[**
    Table[With[{nc = NewtonCotesCoefficient2[n, f]},
        {Hue[n/19], Line[MapIndexed[{(#2[[1]] - 1)/(n - 1), #1}&,
                      Table[D[nc, f[k]], {k, n}]]]}], {n, 2, 16}]],
    PlotRange -> All, Frame -> True]

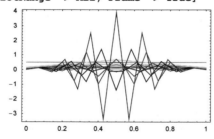

For the use of the Lagrange interpolating polynomial to carry out differentiation efficiently, see [436], [1886].

`InterpolatingPolynomial` allows interpolating data with polynomials. A more versatile, and probably the most important *Mathematica* function for fitting and interpolation is `InterpolatingFunction`. (While `Fit` is an important function input by users, the `InterpolatingFunction`-objects are frequently output by *Mathematica*.)

`InterpolatingFunction[`*interval, details*`]`

    represents the approximation of a function. Here, *interval* describes the interval on which the interpolating function is to be defined, whereas *details* contains the explicit values of the data.

`InterpolatingFunction` is a basic object for many larger numerical calculations in *Mathematica*. It works for arbitrarily many arguments and can be used with complex function values. Relative to the arguments, it behaves as a built-in function and can (to the extent that it makes sense mathematically) work in and with other built-in functions.

To solve an interpolation problem for given data, an `InterpolatingFunction`-object is produced as follows.

```
Interpolation[{{x₁, {f₁, f₁', f₁'', ..., f₁⁽ⁿ¹⁾}},
    {x₂, {f₂, f₂', f₂'', ..., f₂⁽ⁿ²⁾}}, ...,
    {xₘ, {fₘ, fₘ', fₘ'', ..., fₘ⁽ⁿᵐ⁾}}}]
```

produces a function that interpolates the function value f_i and successive derivatives up to the n_ith derivative $f_i^{(n_i)}$ at the point x_i, for $i = 1, ..., m$ (the derivatives are optional and need not to be given). In the multidimensional case, we can specify the function values and the first partial derivatives with respect to each of the variables. In this case, the input has the form

```
Interpolation[{{{x₁, y₁, ...}, {f₁, {dfx₁, dfy₁, ...}}},
    {{x₂, y₂, ...}, {f₂, {dfx₂, dfy₂, ...}}}, ...,
    {{xₙ, yₙ, ...}, {fₙ, {dfxₙ, dfyₙ, ...}}}}]
```

`Interpolation` works with both exact and approximate numbers as variables. The data points in the 2D case $\{x_i, y_j, ...\}$ must lie on a tensor product lattice.

We first investigate a relatively simple example (for later use, we name the result).

In[108]:= `ipo = Interpolation[Table[N[{x, x^2/2}], {x, 0, 5, 1/2}]]`

Out[108]= `InterpolatingFunction[{{0., 5.}}, <>]`

Here is the generated object in detail. We do not need to be concerned with this form at the moment. As with `Graphics`, `Graphics3D`, etc., not all internal available data are represented in the `OutputForm`.

In[109]:= `InputForm[ipo]`

Out[109]//InputForm=
```
InterpolatingFunction[{{0., 5.}}, {2, 0, True, Real, {3}, {0}},
    {{0., 0.5, 1., 1.5, 2., 2.5, 3., 3.5, 4., 4.5, 5.}}, {{0, 1, 2, 3, 4, 5, 6,
7, 8, 9, 10, 11},
    {0., 0.125, 0.5, 1.125, 2., 3.125, 4.5, 6.125, 8., 10.125, 12.5}},
{Automatic}]
```

We can work with the generated `InterpolatingFunction`-object just as any other function. For example, we can prescribe arguments.

In[110]:= `{ipo[0], ipo[1], ipo[3.567885]}`

Out[110]= `{0., 0.5, 6.3649}`

However, these must lie in the interpolating interval (the function is called `Interpolation`, not `Extrapolation`). A warning message is generated, and extrapolation is used.

In[111]:= `{ipo[-0.5], ipo[5.1]}`

```
InterpolatingFunction::dmval :
    Input value {-0.5} lies outside the range of data in
        the interpolating function. Extrapolation will be used. More...
InterpolatingFunction::dmval :
    Input value {5.1} lies outside the range of data in the
        interpolating function. Extrapolation will be used. More...
```

Out[111]= `{0.125, 13.005}`

If we input a symbolic variable, we do not get an explicit result.

In[112]:= `ipo[xyz]`

Out[112]= `InterpolatingFunction[{{0., 5.}}, <>][xyz]`

An `InterpolatingFunction` function can also be plotted. The second argument of `Plot` in the following `{x, ##}& @@ ipo[[1]]` yields the interpolation interval in the interpolating function.

In[113]:= `Plot[ipo[x], Evaluate[Flatten[{x, ##}]& @@ ipo[[1]]]]`

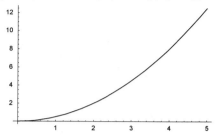

Functions produced by `Interpolation` can also be differentiated just like any other function (we give details in the next chapter). The function itself (as a pure function) can be differentiated by means of `'`. The result is again a pure function. Here are two examples.

In[114]:= `Sin'`
Out[114]= `Cos[#1] &`

In[115]:= `#^2&'`
Out[115]= `2 #1 &`

In[116]:= `ipoPrime = ipo'`
Out[116]= `InterpolatingFunction[{{0., 5.}}, <>]`

The result is again a pure function that works as usual.

In[117]:= `ipoPrime[1]`
Out[117]= `1.`

Here, we could also apply `D`.

In[118]:= `fs[x_] = D[ipo[x], x]`
Out[118]= `InterpolatingFunction[{{0., 5.}}, <>][x]`

As expected, the result of the differentiation is a straight line.

In[119]:= `Plot[{ipoPrime[x], fs[x]}, {x, 1, 4}]`

As with `Fit`, the variables of `Interpolation` can also be exact numbers. (`NumberQ` must yield `True`; so nonnumeric symbolic quantities are now not allowed.)

In[120]:= **Interpolation[{{1, 1}, {2, 4}, {3, 9}, {4, 16}, {5, 25}}]**

Out[120]= InterpolatingFunction[{{1, 5}}, <>]

With exact numbers as input, the resulting InterpolatingFunction also produces exact numbers as the result.

In[121]:= **%[5/4]**

Out[121]= $\dfrac{25}{16}$

While Interpolation forms a piecewise polynomial interpolation, the function PiecewiseExpand does currently not expand InterpolatingFunction-objects.

In[122]:= **PiecewiseExpand[Interpolation[Range[5]]]**

Out[122]= InterpolatingFunction[{{1, 5}}, <>]

The function Interpolation has one option.

In[123]:= **Options[Interpolation]**

Out[123]= {InterpolationOrder → 3, PeriodicInterpolation → False}

InterpolationOrder

    is an option for Interpolation. It specifies the degree of the polynomials that will be used to construct a piecewise approximation.

    Default:

        3

    Admissible values:

        0, 1, 2, …, as far as the data allows

We now compute interpolating values of the function $\sin(2\,x)$ at 11 distinct points using various orders.

In[124]:= **Do[ℓp[i] = Interpolation[Table[N[{x, Sin[2x]}], {x, 0, 5, 1/2}],**
                                InterpolationOrder -> i],
      {i, 0, 11}]

              Interpolation::inhr :
                Requested order is too high; order has been reduced to {10}. More…

For a fixed number of data points, the interpolation order cannot be arbitrarily large. This is why we get an error message for order 11. We now graphically examine the results for various interpolation orders. A clear difference is only visible for the interpolation orders 0, 1, and 2. To better compare the higher orders, look at the error between the interpolating values and $\sin(2\,x)$ as shown in the plots on the right.

In[125]:= **Show[GraphicsArray[#]]& /@ Map[Flatten,**
    Table[{Plot[ℓp[2i + j][x], {x, 0, 5},
             PlotLabel -> "Order = " <> ToString[2i + j],
             DisplayFunction -> Identity],
             Plot[(Sin[2x] - ℓp[2i + j][x]), {x, 0, 5},
             PlotLabel -> "Order = " <> ToString[2i + j],
             PlotRange -> All, DisplayFunction -> Identity]},
        {i, 0, 4}, {j, 0, 1}], 1]

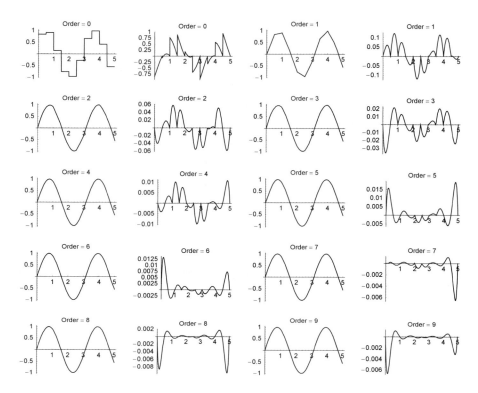

Here, we see clearly that (in the sense of the L^2-norm) the quality of the approximations is improving with increasing order. We can find numerical values for these norms by integration. (We come back to the command NIntegrate later in this chapter. Here, we are computing the sum of the integrals between successive data points.)

```
In[126]:= Table[NIntegrate[(Sin[2x] - ip[i][x])^2,
                Evaluate[Join[{x}, Table[x, {x, 0, 5, 1/2}]]]],
            {i, 0, 9}]
Out[126]= {0.826637, 0.0188685, 0.00490201, 0.0010122, 0.000120672,
        0.000172104, 0.0000614648, 0.0000101144, 0.0000389734, 0.0000124812}
```

Next, we investigate the higher derivatives of *ip*[3] in somewhat more detail. For convenience, we graph all curves on one plot. To distinguish between the various curves, we use the capabilities of the package Graphics`Legend`. The (higher) derivatives of an InterpolatingFunction are typically only piecewise smooth functions and have points of discontinuity.

```
In[127]:= Needs["Graphics`Legend`"]

In[128]:= Plot[Evaluate[{ip[3][x], ip[3]'[x], ip[3]''[x], ip[3]'''[x]}],
            {x, 0, 5}, PlotPoints -> 100,
            (* give each derivative its own style *)
            PlotStyle -> (Append[{Thickness[0.002]}, #]& /@ {{},
                Dashing[{0.01, 0.02}], Dashing[{0.01, 0.01}], Dashing[{0.02,
            0.01}]}),
            PlotRange -> All, FontFamily -> {"Helvetica", FontSize -> 5},
            PlotLegend -> (StyleForm[#, FontFamily -> "Courier", FontSize -> 5]& /@
```

```
                            {"ip[3]", "ip[3]'", "ip[3]''", "ip[3]'''"}),
                  LegendPosition -> {0.55, -0.9}, LegendSize -> {1.1, 0.6}]
```

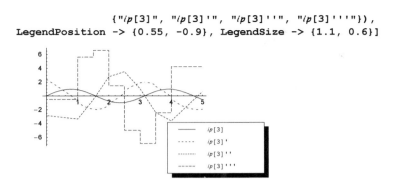

For a higher order `InterpolatingFunction`, the derivatives become smoother, but are still discontinuous. Using a still higher `InterpolationOrder` (order 8 in the right graphics) results in smooth derivatives up to higher order.

```
In[129]:= Show[GraphicsArray[Function[o,
      Plot[Evaluate[{ip[o][x], ip[o]'[x], ip[o]''[x], ip[o]'''[x],
                  ip[o]''''[x], ip[o]'''''[x]}],
          {x, 0, 5}, PlotPoints -> 100,
          (* give each derivative its own style *)
          PlotStyle -> (Append[{Thickness[0.002]}, #]& /@ {{},
            Dashing[{0.01 , 0.02 }], Dashing[{0.01 , 0.01 }],
            Dashing[{0.02 , 0.01 }], Dashing[{0.02 , 0.02 }],
            Dashing[{0.008, 0.008}]}),
          DisplayFunction -> Identity, PlotRange -> All,
          FontFamily -> {FontFamily -> "Helvetica", FontSize -> 5},
          PlotLegend -> (StyleForm[#, FontFamily -> "Courier", FontSize -> 5]& /@
                  Table["ip[" <> ToString[6] <> "]" <> Table["'", {k}],
                    {k, 0, 5}]),
          LegendPosition -> {0.45, -0.8}, LegendSize -> {0.9, 0.9}]] /@ {6, 8}]]
```

When differentiating an `InterpolatingFunction`-object, we have

$$(InterpolatingFunctionObject^{(n)})^{(m)} = InterpolatingFunctionObject^{(n+m)}$$

Here, interpolations of sin at 25 points for various `InterpolationOrder`s are calculated. The picture shows the logarithmic difference between the interpolated data and the sin function. (To avoid messages issued in case the difference becomes 0, we turn off a message.)

```
In[130]:= tab = Table[ipo =
      Interpolation[N @ Table[{x, Sin[x]}, {x, 0, Pi, Pi/25}],
```

```
                    InterpolationOrder -> i];
        {Hue[i/35], Line @ Table[N[{x, Log[10, Abs[ipo[x] - Sin[x]]]}],
                           {x, 0, Pi, Pi/251}]},
                {i, 25}];
```

In[131]:= `Off[Graphics::"gptn"];`
`Show[Graphics[tab], Frame -> True]`

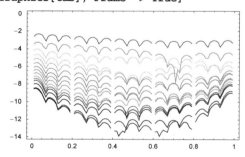

The last picture shows that within machine arithmetic, we cannot always resolve the differences. The next picture shows the difference of the rational function $(x^3 - 2x + 8)/(x^5 + x^3 + 2x + 2)$ and its interpolation on 15 points within rational arithmetic.

In[133]:= `tab = Table[ipo =`
`Interpolation[Table[{x, (x^3 - 2x + 8)/(x^5 + x^3 + 2x + 2)},`
`{x, 0, 1, 1/15}], InterpolationOrder -> i];`
`{Hue[i/20], Line @ Table[N[{x,`
`Log[10, Abs[ipo[x] - (x^3 - 2x + 8)/(x^5 + x^3 + 2x + 2)]]}],`
`{x, 0, 1, 1/100}]},`
`{i, 15}];`

In[134]:= `Show[Graphics[tab], Frame -> True]`

The data to be interpolated can contain symbolic entities. (This is similar to the `InterpolatingPolyno`‑
`mial` case.)

In[135]:= `Clear[a, α, ipo];`
`ipo[a_] = Interpolation[{{1, 1}, {2, 4}, {3, 9}, {4, 16},`
`{5, a},`
`{6, 36}, {7, 49}, {8, 64}, {9, 81}, {10, 100}}]`

Out[136]= `InterpolatingFunction[{{1, 10}}, <>]`

In[137]:= `ipo[α][Pi]`

Out[137]= $16 + (-4 + \pi)$

$$\left(7 + (-3 + \pi) \left(-7 + \frac{1}{2}(-5 + \pi)\left(-7 + \frac{9 - \alpha}{2} + \frac{1}{2}(-9 + \alpha) + \frac{1}{3}(-4 + \alpha)\right) + \frac{1}{2}(-9 + \alpha)\right)\right)$$

Here, the resulting interpolating function as a function of *a* is shown. We plot over the *x*-range {−1, 11}. All data were contained in the range {1, 10}. Outside, the last domain extrapolation is used and a warning message is issued.

```
In[138]:= Plot3D[ipo[a][x], {x, -1, 11}, {a, 0, 50},
              Compiled -> False, PlotPoints -> 50]
```

```
       InterpolatingFunction::dmval :
        Input value {-1.} lies outside the range of data in the
           interpolating function. Extrapolation will be used. More…

       InterpolatingFunction::dmval :
        Input value {-0.755102} lies outside the range of data in
           the interpolating function. Extrapolation will be used. More…

       InterpolatingFunction::dmval :
        Input value {-0.510204} lies outside the range of data in
           the interpolating function. Extrapolation will be used. More…

       General::stop : Further output of
           InterpolatingFunction::dmval will be suppressed during this calculation. More…
```

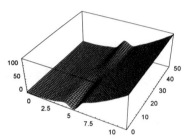

Interpolation has a second option, called `PeriodicInterpolation`. When this option is set to `True`, the resulting `InterpolatingFunction` will represent a periodic function. In this case, the first function value of the data and the last function value of the data must be identical.

The next input constructs a univariate periodic function `ifPeriodic` over the interval [0, 1].

```
In[139]:= ifPeriodic =
     Module[{λ = 12}, SeedRandom[123];
          Interpolation[MapIndexed[{(#2[[1]] - 1)/λ, #1}&, Append[#, First[#]]& @
                            Table[Random[Real, {-1, 1}], {λ}]],
                      PeriodicInterpolation -> True, InterpolationOrder -> 6]]
Out[139]= InterpolatingFunction[{{..., 0., 1., ...}}, <>]
```

Displaying the autocorrelation {`ifPeriodic[s]`, `ifPeriodic[s + Δ]`} nicely shows the periodic nature of the function.

```
In[140]:= Show[Graphics3D[{Thickness[0.002],
          Table[{Hue[Δ], Line[Table[{Δ, ifPeriodic[s], ifPeriodic[s + Δ]},
                         {s, 0, 1, 1/300}]]}, {Δ, 0, 1, 1/48}]},
                  PlotRange -> All, BoxRatios -> {3, 1, 1}]]
```

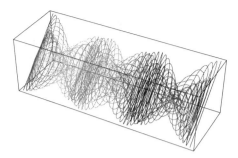

At this point, we will add a word of caution. When repeatedly interpolating in a calculation, it is often advantageous to change the interpolation points. The following input is a crude model of the evolution of a random 2D curve $c(t; s)$ (t being time and s being the arclength) according to [357], [343], [1658], [1524], [1393], [397], [729], [106], [1589]

$$\frac{\partial c(t; s)}{\partial t} = \arctan(\kappa_c(t; s)) \, \mathbf{n}_c(t; s)$$

Here $\kappa_c(t; s)$ is the curvature of $c(t; s)$ and $\mathbf{n}_c(t; s)$ is its normal. We solve this partial differential equation by iteratively carrying out $c(t + \delta t; s) \approx c(t; s) + \varepsilon \arctan(\kappa_c(t; s)) \, \mathbf{n}_c(t; s)$ and interpolate $c(t + \delta t; s)$ with respect to s (for more appropriate numerical methods to solve such-type equations, see [1642], [342]; for various final curve shapes of related models, see [73]). The left graphic uses always the same values of s for interpolation and the right picture chooses the interpolation points slightly randomly. The left picture shows clearly numerical artifacts from the repeated interpolation at the same points.

```
In[141]:= step[{x_, y_}, ε_, pp_, io_] :=
      Module[{d0, d1, d2, rs, newPoints},
        (* curve and derivatives *)
        {d0, d1, d2} = Table[#, {s, -1/2, 1/2, 1/pp}]& /@
                          Table[D[{x[s], y[s]}, {s, j}], {j, 0, 2}];
        (* curvature values *)
        rs = Table[(d1[[k, 2]] d2[[k, 1]] - d1[[k, 1]] d2[[k, 2]])/
                  (d1[[k]].d1[[k]])^(3/2), {k, Length[d0]}];
        (* move in direction of the normal *)
        newPoints = Table[d0[[k]] + ε ArcTan[rs[[k]]] *
                      (#/Sqrt[#.#]&[Reverse[d1[[k]]]{-1, 1}]), {k, Length[d0]}];
        (* reinterpolate *)
        Interpolation[Transpose[{Table[s, {s, -1/2, 1/2, 1./pp}], #}],
                  InterpolationOrder -> io
                  (* , PeriodicInterpolation -> True *)]& /@
                                        Transpose[newPoints]]

In[142]:= xGraphics[n_, ε_, pp_, ipoOrder_, ipoPoints_, opts___] :=
      Module[{(* not included for later use: xy0, *)xy, pp1},
        Show[GraphicsArray[
        Function[io, (* seed random number generation *) SeedRandom[1000];
        (* initial curve; use a random trigonometric sum for x, y *)
        xy0 = Function /@
              Table[Sum[Random[] Cos[k 2Pi # + 2Pi Random[]], {k, 5}], {2}];
        xy = xy0;
        Show[Reverse @ Table[(* iterate *)
        (* interpolation points *)
        pp1 = If[ipoPoints === Equal, pp, Round[Random[Real, {3/4, 5/4}] pp]];
```

```
xy = step[xy, ε, pp1, io];
(* display curve *)
ParametricPlot[Evaluate[#[s]& /@ xy], {s, -1/2, 1/2}, opts,
            PlotPoints -> 100, Axes -> False, AspectRatio -> Automatic,
            DisplayFunction -> Identity, PlotStyle -> Hue[0.8 j/n],
            Frame -> True], {j, n}]] /@ {ipoOrder}]]];
```

```
In[143]:= Show[GraphicsArray[
    Block[{$DisplayFunction = Identity, κg = κGraphics},
        {κg[60, 0.02, 100, 4, Equal], κg[60, 0.02, 100, 4, Random]}]]]
```

For a reliable numerical solution of the last partial differential equation, such a straightforward Euler-type time advancing is, of course, unsuited. Later in this chapter, we will discuss how to solve such parabolic-type differential equations using the function `NDSolve`. The next input solves the PDE numerically.

```
In[144]:= ndsol =
    Module[{d0, d1, d2, λ, normal},
        (* curve, tangent, and second derivative as a function of s *)
        {d0, d1, d2} = Table[D[{xy0[[1]] @ s, xy0[[2]] @ s}, {s, j}], {j, 0, 2}];
        (* curvature as a function of s *)
        κ = (d1[[2]] d2[[1]] - d1[[1]] d2[[2]])/(d1.d1)^(3/2);
        (* normal as a function of s *)
        normal = #/Sqrt[#.#]&[Reverse[d1]{-1, 1}];
        (* solve differential equation numerically *)
        NDSolve[Flatten[{
        {(* the partial differential equation *)
        Thread[D[{cx[s, t], cy[s, t]}, t] == ArcTan[κ] normal],
        (* initial condition *)
        Thread[{cx[s, 0], cy[s, 0]} == d0]},
        (* use periodicity with respect to s *)
        Thread[{cx[-1/2, t], cy[-1/2, t]} == {cx[1/2, t], cy[1/2, t]}]}],
        {cx, cy}, {s, -1/2, 1/2}, {t, -3, 3},
        AccuracyGoal -> 3, PrecisionGoal -> 3,
        (* set options for the numerical solution *)
        Method -> {"MethodOfLines", "SpatialDiscretization" ->
                    {"TensorProductGrid",
                    "DifferenceOrder" -> "Pseudospectral",
                    "MaxPoints" -> {300}, "MinPoints" -> {300}}}]]
```
```
Out[144]= {{cx → InterpolatingFunction[{{..., -0.5, 0.5, ...}, {-3., 3.}}, <>],
          cy → InterpolatingFunction[{{..., -0.5, 0.5, ...}, {-3., 3.}}, <>]}}
```

The left graphics shows the solution for $0 \le t \le 1$ and the right graphic shows the solution over the larger time interval $-2 \le t \le 2$. While the above right graphic is clearly a better solution than the above left graphic, there are visible differences to the solution returned from `NDSolve`.

```
In[145]:= Show[GraphicsArray[Show[Table[
       ParametricPlot[Evaluate[{cx[s, t], cy[s, t]} /. ndsol[[1]]],
                    {s, -1/2, 1/2}, PlotPoints -> 100, Axes -> False,
                    AspectRatio -> Automatic,  PlotStyle -> Hue[0.8 t],
                    DisplayFunction -> Identity],
              {t, #1, #2, (#2 - #1)/60}]]& @@@ {{0, 1}, {-2, 2}}]]
```

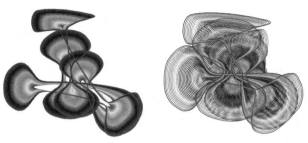

The above interpolation of the function $\sin(2\,x)$ can more conveniently be done with `FunctionInterpola`tion.

`FunctionInterpolation[`*function*`, {`x`, `x_{min}`, `x_{max}`}]`

    returns an `InterpolatingFunction`-object that approximates the function *function*(x) in the interval $\{x_{min},\ x_{max}\}$.

```
In[146]:= fip = FunctionInterpolation[Sin[2 x], {x, 0, 5}]
Out[146]= InterpolatingFunction[{{0., 5.}}, <>]
```

The quality of approximation of `fip` is much better than the quality of the above-calculated `ipoRel[i]`.

```
In[147]:= Plot[Evaluate[fip[x] - Sin[2x]], {x, 0, 5}, PlotRange -> All]
```

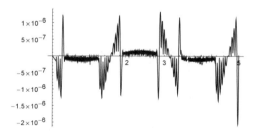

About 150 points were used to approximate the function $\sin(2\,x)$.

```
In[148]:= fip[[3, 1]] // Length
Out[148]= 151
```

The third derivative shows nicely the discrete sample points. In the regions where the function does not change much we see a smaller sampling rate.

```
In[149]:= Plot[Evaluate[fip'''[x]], {x, 0, 2},
          PlotRange -> All, Frame -> True, Axes -> False,
          PlotStyle -> {Thickness[0.006], GrayLevel[0]},
```

```
(* the sampling points *)
Prolog -> {Hue[0], Line[{{#, -8}, {#, 8}}]& /@ fip[[3, 1]]},
AspectRatio -> 1/3]
```

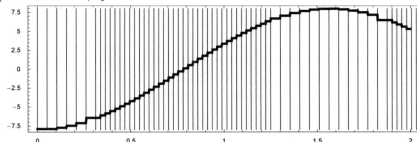

Here is the difference between fip, and the above *ip*[*i*] is shown.

In[150]:= `Plot[Evaluate[Table[ip[i][x] - fip[x], {i, 11}]], {x, 0, 5}, PlotRange -> All,`
        `PlotStyle -> Table[{Thickness[0.002], Hue[i/14]}, {i, 11}]]`

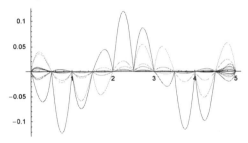

The use of FunctionInterpolation is recommended when the calculation of a function takes a relatively long time and function values are repeatedly used. One such function is FresnelC (see Chapter 3 of the Symbolics volume [1795]). We start with a plot of this function in the interval {0, 10}.

In[151]:= `Plot[FresnelC[x], {x, 0, 10}, PlotRange -> All] // Timing`

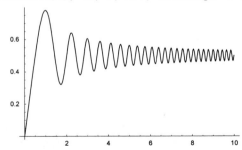

Around 600 points were needed for the generation of this plot.

In[152]:= `%[[2, 1, 1, 1, 1]] // Length`

Out[152]= 586

To achieve a six-digit interpolation, we only need about twice as many points.

In[153]:= **(fipFresnelC = FunctionInterpolation[FresnelC[x], {x, 0, 10},**
                                                MaxRecursion -> 8]) // Timing

Out[153]= {2.95 Second, InterpolatingFunction[{{0., 10.}}, <>]}

In[154]:= **fipFresnelC[[3, 1]] // Length**

Out[154]= 1048

Using the InterpolatingFunction fipFresnelC, plotting can now be done much faster.

In[155]:= **Plot[fipFresnelC[x], {x, 0, 10}, PlotRange -> All] // Timing**

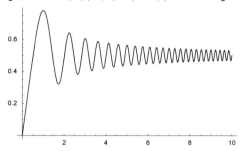

To complete this section, we consider the 2D interpolation problem. We first prescribe function values $f(x, y)$ and derivatives $\partial_x f(x, y), \partial_y f(x, y)$: $f(x, y) = x\,y$, $\partial f(x, y)/\partial x = y^2$, and $\partial f(x, y)/\partial y = x^2$.

In[156]:= **multidimPol = Interpolation[Flatten[**
              Table[{x, y, {x y, {y^2, x^2}}}, {x, -3, 3}, {y, -3, 3}], 1]]

Out[156]= InterpolatingFunction[{{-3, 3}, {-3, 3}}, <>]

The following plots show a clear difference between the interpolating values and $f(x, y) = x\,y$. The plot on the left shows $x\,y$, the one in the middle shows the interpolated data, and the one on the right shows the difference between the two functions.

In[157]:= **Show[GraphicsArray[{**
      (* the x,y-surface *)
      Plot3D[x y, {x, -3, 3}, {y, -3, 3},
            PlotPoints -> 30, DisplayFunction -> Identity],
      (* the surface of the interpolation *)
      Plot3D[Evaluate[multidimPol[x, y]], {x, -3, 3}, {y, -3, 3},
            PlotPoints -> 30, DisplayFunction -> Identity],
      (* the difference *)
      Plot3D[Evaluate[multidimPol[x, y] - x y], {x, -3, 3}, {y, -3, 3},
            PlotPoints -> 30, DisplayFunction -> Identity]}]]

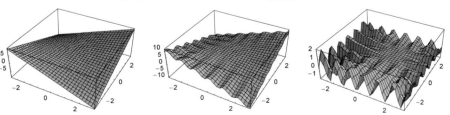

The two functions coincide at the data points.

In[158]:= **Table[multidimPol[x, y] - x y, {x, -3, 3}, {y, -3, 3}]**

```
Out[158]= {{0, 0, 0, 0, 0, 0, 0}, {0, 0, 0, 0, 0, 0, 0},
         {0, 0, 0, 0, 0, 0, 0}, {0, 0, 0, 0, 0, 0, 0},
         {0, 0, 0, 0, 0, 0, 0}, {0, 0, 0, 0, 0, 0, 0}, {0, 0, 0, 0, 0, 0, 0}}
```

Next, we generate a two-dimensional array of random data. The option setting for `InterpolationOrder` determines the overall smoothness of the result. The first row of 3D plots shows the interpolated function and the second row shows the first derivative of the interpolated function with respect to x. Observe the different axes scalings on the derivative plots.

```
In[159]:= With[{n = 8},
         (* random data *)
         SeedRandom[12345];
         data = Flatten[Table[{i, j, Random[Real, {-1, 1}]},
                        {i, 0, n}, {j, 0, n}], 1];
         Function[do, Show[GraphicsArray[
         Table[(* the interpolation *)
              ipo = Interpolation[data, InterpolationOrder -> k];
              (* graph the derivative of the interpolation;
                 use many plot points *)
              Plot3D[Evaluate[D[ipo[x, y], {x, do}]], {x, 0, n}, {y, 0, n},
                     PlotPoints -> 120, Mesh -> False, PlotLabel -> k,
                     PlotRange -> All, DisplayFunction -> Identity],
           {k, 2, 8, 2}]]]] /@ {0, 1}]
```

To end our discussion of interpolation with the function `Interpolation`, we will make a picture of a torus whose cross section changes gradually from a triangle to a square to a pentagon to a hexagon and then back to a triangle. Here is the implementation. It makes heavy use of `Interpolation`.

```
In[160]:= Needs["Graphics`Graphics3D`"]

         Module[{g, dm, h, p, x, y},
         (* parts of 2D n-gons with appropriate point numbers;
            LCM[2, 3, 4, 5, 6] = 60 *)
         g[n_] := g[n] =
         N[Table[{-Sin[Pi(1/2 - 1/n)] Csc[a + Pi(1/2 - 1/n)] Sin[a],
                   Cos[a] Sin[Pi(1/2 - 1/n)] Csc[a + Pi(1/2 - 1/n)]},
              {a, 0, 2Pi/n, 2Pi/n/(60/n - 1)}]];
         (* rotation matrices for making other pieces of the n-gons *)
         R[n_, i_] := R[n, i] = {{Cos[2Pi/n i], -Sin[2Pi/n i]},
                                  {Sin[2Pi/n i], Cos[2Pi/n i]}};
         (* a complete 2D n-gon *)
         h[n_] := h[n] = Join @@ Table[R[n, i].#& /@ g[n], {i, 0, n - 1}];
```

```
p = Table[N[φ], {φ, 0, 2Pi + Pi/2, Pi/2}];
(* interpolate x- and y-coordinates between n and (n+1)-gon
   (n = 2, ..., 5) and 6 and 3 *)
 (* x-values and y-values *)
Function[{xy, p}, MapIndexed[Set[xy[#2[[1]]], #1]&,
Interpolation[#, InterpolationOrder -> 5]& /@ (Transpose[{p, #}]& /@
(Join[#, Take[#, 2]]& /@ Transpose[Table[Transpose[
            N @ h[n]][[p]], {n, 3, 6}]]))]] @@@ {{x, 1}, {y, 2}};
(* wrap interpolated n-gons along torus and display them *)
Show[Graphics3D[{EdgeForm[Thickness[0.002], GrayLevel[
Sin[2ArcTan @@ (Plus @@ (Drop[#, -1]& /@ #[[1]]))]^2]], #}& /@
ListSurfacePlot3D[Table[(* the modified torus *)
{3 Cos[φ] + Cos[φ] x[ϑ][φ], 3 Sin[φ] + Sin[φ] x[ϑ][φ], y[ϑ][φ]} // N,
 {φ, 0, 2Pi, 2Pi/120}, {ϑ, 60}], DisplayFunction -> Identity][[1]]],
 Boxed -> False]]
```

One can use Interpolation to create a lot more nice and interesting graphics. For instance one can construct a contour plot of a random, but nevertheless, smooth function, like with the following.

```
randomContourPlot[randomPoints:pp_, plotPoints:pp1_, contours:cls_] :=
Module[{ipo, li, cfDead, cfUsed},
(* create a smooth (by use of Interpolation) random function *)
ipo = Interpolation[Flatten[Table[{i, j, Random[] (* the randomness *)},
                                  {i, 0, pp}, {j, 0, pp}], 1]];
(* collect needed values for ColorFunction *)
li = {}; cfDead[x_] := (AppendTo[li, x]; GrayLevel[x]);
(* make preliminary ContourPlot *)
ContourPlot[ipo[x, y], {x, 0, pp}, {y, 0, pp},
          Compiled -> False, Contours -> Table[i, {i, 0, 1, 1/cls}],
          PlotPoints -> pp1, ColorFunction -> cfDead];
  (* construct color function that gives alternating black and white *)
cfUsed[x_] = Which @@
  Flatten[MapIndexed[{x <= #1, If[EvenQ[#2[[1]]], 1, 0]}&, li]];
(* show the contour graphics *)
ContourPlot[ipo[x, y], {x, 0, pp}, {y, 0, pp},
          Compiled -> False, ContourLines -> False,
          Contours -> Table[i, {i, 0, 1, 1/cls}], PlotPoints -> pp,
          ColorFunction -> (GrayLevel[cfUsed[#]]&), Frame -> None]];

(* show two examples *)
Show[GraphicsArray[
     {randomContourPlot[20, 100, 20], randomContourPlot[160, 80, 80]}]]
```

While it is possible to manipulate `InterpolatingFunction`-objects directly (for instance extracting domains, and values), the inner details of `InterpolatingFunction`-objects are not documented in the *Mathematica* book. This means that there is no guarantee that future versions of *Mathematica* will have identical structures. A convenient way to extract the details of `InterpolatingFunction`-objects is provided through a set of functions from the `DifferentialEquations`InterpolatingFunctionAnatomy`` package.

The functions provided by this package have self-explanatory names.

```
In[162]:= Needs["DifferentialEquations`InterpolatingFunctionAnatomy`"]
```

```
In[163]:= Rest[Names["InterpolatingFunction*"]]
```

```
Out[163]= {InterpolatingFunctionCoordinates, InterpolatingFunctionDerivativeOrder,
    InterpolatingFunctionDomain, InterpolatingFunctionGrid,
    InterpolatingFunctionInterpolationOrder, InterpolatingFunctionValuesOnGrid}
```

Here is a 2D `InterpolatingFunction`-object that was created by differentiating an interpolating function.

```
In[164]:= if2DExact =
    With[{pp = 3, L = 3}, D[#, x, x, y, y]& @
        Interpolation[Flatten[Table[{x, y, 2. Sin[x y]},
                        {x, 0, L, L/pp}, {y, 0, L, L/pp}],
                    1]][x, y]][[0]]
```

```
Out[164]= InterpolatingFunction[{{0., 3.}, {0., 3.}}, <>]
```

And here are the grid points, grid point values, and derivatives extracted from `if2DExact`.

```
In[165]:= {#, #[if2DExact]}& /@
    ToExpression[Rest[Names["InterpolatingFunction*"]]]
```

```
Out[165]= {{InterpolatingFunctionCoordinates, {{0., 1., 2., 3.}, {0., 1., 2., 3.}}},
    {InterpolatingFunctionDerivativeOrder, Derivative[2, 2]},
    {InterpolatingFunctionDomain, {{0., 3.}, {0., 3.}}},
    {InterpolatingFunctionGrid, {{{0., 0.}, {0., 1.}, {0., 2.}, {0., 3.}},
        {{1., 0.}, {1., 1.}, {1., 2.}, {1., 3.}}, {{2., 0.}, {2., 1.},
        {2., 2.}, {2., 3.}}, {{3., 0.}, {3., 1.}, {3., 2.}, {3., 3.}}}},
    {InterpolatingFunctionInterpolationOrder, {3, 3}},
    {InterpolatingFunctionValuesOnGrid, {{-46.7706, -11.7434, 23.2838, 58.311},
        {-11.7434, -2.05622, 7.63099, 17.3182}, {23.2838, 7.63099, -8.02182, -23.6746},
        {58.311, 17.3182, -23.6746, -64.6674}}}}}
```

The famous cubic spline interpolation of a collection of data points $\{x_i, y_i\}$ can be constructed using the package `NumericalMath`SplineFit``.

```
In[166]:= Needs["NumericalMath`SplineFit`"]
```

```
In[167]:= ?SplineFit
```

```
SplineFit[points, type] generates a SplineFunction
    object of the given type from the points.  Supported
    types are Cubic, CompositeBezier, and Bezier. More...
```

We now look at an example using the three interpolation methods implemented in this package. (For more on spline interpolation, see [1272], [1545], [1015], [1271], [1533], [1695], [493], [94], [231], [416], and [78].) In comparison to the built-in Interpolation, with these functions, we can immediately model multivalued functions. (With the built-in Interpolation, we would have to make a separate Interpolating ⋮ Function-object for the *x*- and *y*-values.) Suppose the data to be interpolated is the following.

```
In[168]:= data = Table[{Sin[x] + 1/4 Sin[8x] + 1/15 Sin[20x],
                Sin[(x + Pi/3)] + 1/4 Cos[8x] + 1/15 Cos[20x]},
               {x, 0.0, 1.0, 0.1}] // N;
```

For later use in a plot, we compute values of this function on a set of points.

```
In[169]:= exactData = Table[{Sin[x] + 1/4 Sin[8x] + 1/15 Sin[20x],
                    Sin[(x + Pi/3)] + 1/4 Cos[8x] + 1/15 Cos[20x]},
                   {x, 0.0, 1.0, 0.01}] // N;
```

Here are the three spline interpolating values and their plots.

```
In[170]:= cubSpline = SplineFit[data, Cubic]
Out[170]= SplineFunction[Cubic, {0., 10.}, <>]
```

```
In[171]:= bezSpline = SplineFit[data, Bezier]
Out[171]= SplineFunction[Bezier, {0., 10.}, <>]
```

```
In[172]:= combezSpline = SplineFit[data, CompositeBezier]
Out[172]= SplineFunction[CompositeBezier, {0., 10.}, <>]
```

As is to be expected, only the cubic spline interpolating values agree exactly with the data at the data points. (For making these pictures, we could have also used the Spline command from the Graphics`Spline` package.)

```
In[173]:= Show[Graphics[{{PointSize[0.025], Point /@ data},
          {GrayLevel[0], Line[exactData]},
          (* the three splines *)
          {Hue[0.0], Line[cubSpline    /@ Range[0, 10, 1/10]]},
          {Hue[0.3], Line[bezSpline    /@ Range[0, 10, 1/10]]},
          {Hue[0.6], Line[combezSpline /@ Range[0, 10, 1/10]]}}],
          PlotRange -> All, AspectRatio -> Automatic]
```

We should mention that the package NumericalMath`SplineFit` also works with exact data as arguments and, in this case, produces exact results.

In[174]:= **SplineFit[{{1, E}, {Sqrt[5] + Pi/2, 3/56}, {45, 7/Pi},**
**   {4, 7/EulerGamma}}, Cubic]**

Out[174]= SplineFunction[Cubic, {0., 3.}, <>]

In[175]:= **%[5/6] // Expand**

Out[175]= $\left\{ -\dfrac{53}{12} + \dfrac{71\sqrt{5}}{72} + \dfrac{71\pi}{144}, \ \dfrac{71}{1344} + \dfrac{8\,\mathrm{e}}{81} + \dfrac{77}{648\,\mathrm{EulerGamma}} - \dfrac{77}{108\,\pi} \right\}$

For an overview of interpolation methods, see [1255]. For analytic interpolations of an infinite number of data, see [489].

1.3 Compiled Programs

To speed up the execution of large, purely numerical calculations, *Mathematica* allows the option of compiling parts of programs or even entire programs. Because all symbols appearing in the program will be explicitly assigned either numbers with machine accuracy, integers, or the logical values True or False, or lists of such objects (and not other symbols, high-precision numbers, or strings), execution times are up to about 2 to 50 times faster than in the general case. (This is because it is not necessary to carry out the entire standard evaluation of expressions and to check for all upvalues, downvalues, and subvalues of all symbols present in these cases.) Here are three important principles concerning the *Mathematica* compiler.

> The use of the *Mathematica* compiler has no influence on the result. This means Compile[{}, *program*][] and *program* give the same result.

This means that (modulo some type coercion—see below) one will always get the same result by using the compiler, as one would get through ordinary evaluation. Especially in cases where the compilation fails, *Mathematica* resorts to the evaluation of the program through the ordinary evaluation process.

> The *Mathematica* compiler does not produce machine code, but instead produces a (reversible) pseudocode for a virtual computer. This pseudocode runs on every implementation of *Mathematica*, independent of the type of computer.

We do not discuss how a *Mathematica* program can be reconstructed from compiled code. Indeed, this is not of great practical interest and, moreover, the compiled object contains a copy of the *Mathematica* source code (see below).

> If compiled code returns lists, they are packed.

This means packed arrays and compiled *Mathematica* code work together in a seamless way.

It is exactly this compilation that is used in Plot, Plot3D, ParametricPlot, ParametricPlot3D, and similar routines (see below) by using the option Compiled. The compilation by itself is called as follows.

Compile[{{*var*$_1$, *type*$_1$, *rank*$_1$}, {*var*$_2$, *type*$_2$, *rank*$_2$}, ..., {*var*$_n$, *type*$_n$, *rank*$_n$}},
  toBeCompiled,
  {{*extraVar*$_1$, *extraType*$_1$, *extraRank*$_1$, *notEvaluated*$_1$},
  {*extraVar*$_2$, *extraType*$_2$, *extraRank*$_2$, *notEvaluated*$_2$}, ...,
  {*extraVar*$_m$, *extraType*$_m$, *extraRank*$_m$, *notEvaluated*$_m$}}]

  compiles *toBeCompiled*, where the input variables *var*$_i$ have type *type*$_i$. Temporarily generated variables *extraVar*$_i$ (or those produced by calls on external functions during the computation) should be of type *extraType*$_i$. The possible types for *type*$_i$ and *extraType*$_i$ are as follows:

  _Integer

  for integers

  _Real

  for real numbers

  _Complex

  for complex numbers

  True | False

  for logical variables

  rank$_k$ specifies the tensor rank of the *var*$_k$.

  If an external function *extraVar*$_i$ does not return a result (e.g., Print[...]), then its type is Null.

  The arguments of *extraVar*$_i$ in the case *notEvaluated*$_i$ are passed to external functions in unevaluated form.

  If the types of variables are not explicitly prescribed, variables will be regarded as real-valued.

If uncompilable expressions are detected in code to be compiled, they are left in their original form. (In general this reduces the efficiency of the compiled code.)

We now give some practical applications of the compiler. Here is a simple function. The default argument type for compiled functions is real (in distinction to the general complex number paradigm of *Mathematica*). So instead of {x, _Real} just x could be used here.

In[1]:= **compTest = Compile[{{x, _Real}, {y, _Integer}}, (x^3 + 7x^4 + 8)^y]**

Out[1]= CompiledFunction[{x, y}, $(x^3 + 7\,x^4 + 8)^y$, -CompiledCode-]

The result of compiling a program is a CompiledFunction object (which in OutputForm appears in a somewhat abbreviated form).

CompiledFunction[{*var*$_1$, *var*$_2$, ..., *var*$_n$}, *compiledCode*, -CompiledCode-, *runTimeErrorHandler*]
  is the result of the compilation
  Compile[{{*var*$_1$, *type*$_1$, *rank*$_1$}, {*var*$_2$, *type*$_2$, *rank*$_2$}, ..., {*var*$_n$, *type*$_n$, *rank*$_n$}},
    compiledCode,
        {{*extraVar*$_1$, *extraType*$_1$, *notEvaluated*$_1$},
    {*extraVar*$_2$, *extraType*$_2$, *notEvaluated*$_2$}, ...,
        {*extraVar*$_m$, *extraType*$_m$, *notEvaluated*$_m$}}].
  The resulting CompiledFunction object is machine independent.

We now look at the above compTest in more detail.

In[2]:= **??compTest**

> Global`compTest

> compTest = CompiledFunction[{x, y}, $(x^3 + 7\,x^4 + 8)^y$, –CompiledCode–]

In[3]:= **InputForm[compTest]**

Out[3]//InputForm=

```
CompiledFunction[{_Real, _Integer}, {{3, 0, 0}, {2, 0, 0}, {3, 0, 3}},
                                    {0, 3, 4, 0, 0},
  {{1, 5}, {4, 3, 1}, {89, 264, 3, 0, 0, 2, 0, 1, 3, 0, 1}, {4, 7, 1},
   {4, 4, 2}, {89, 264, 3, 0, 0, 2, 0, 2, 3, 0, 2}, {14, 1, 1, 3},
   {29, 3, 2, 3}, {4, 8, 1}, {14, 0, 1, 2}, {25, 1, 3, 2, 1},
   {89, 264, 3, 0, 1, 2, 0, 0, 3, 0, 3}, {2}},
  Function[{x, y}, (x^3 + 7*x^4 + 8)^y], Evaluate]
```

Its structure is a bit complicated. The first component is a list of the types of variables that are involved.

In[4]:= **compTest[[1]]**

Out[4]= {_Real, _Integer}

The second and third components describe the registers of type integer, real number, complex number, and logical variable needed.

In[5]:= **compTest[[2]]**

Out[5]= {{3, 0, 0}, {2, 0, 0}, {3, 0, 3}}

The fourth component is a list of the operations to be carried out. If this list contains sublists of integers (and sometimes floating-point numbers and Boolean variables), the compilation succeeded.

In[6]:= **compTest[[4]]**

Out[6]= {{1, 5}, {4, 3, 1}, {89, 264, 3, 0, 0, 2, 0, 1, 3, 0, 1}, {4, 7, 1}, {4, 4, 2},
   {89, 264, 3, 0, 0, 2, 0, 2, 3, 0, 2}, {14, 1, 1, 3}, {29, 3, 2, 3}, {4, 8, 1},
   {14, 0, 1, 2}, {25, 1, 3, 2, 1}, {89, 264, 3, 0, 1, 2, 0, 0, 3, 0, 3}, {2}}

Each operation is coded in the form of a list of one or more numbers or operations.

The fifth component is the function to be compiled itself.

In[7]:= **compTest[[5]]**

Out[7]= Function[{x, y}, $(x^3 + 7\,x^4 + 8)^y$]

Finally, the sixth argument is a function to be applied when the evaluation fails and the pure function version of the original (potentially slightly modified due to variable renaming) Compile body has to be used.

In[8]:= **compTest[[6]]**

Out[8]= Evaluate

Here is a simple compiled function that has real numbers and Boolean variables as parts of its fourth argument.

In[9]:= **Compile[{{x, _Real}}, If[False, 2, If[True, x + Pi, 1]]] // InputForm**

Out[9]//InputForm=

```
CompiledFunction[{_Real}, {{3, 0, 0}, {3, 0, 2}}, {2, 2, 4, 0, 0},
  {{1, 5}, {3, False, 0}, {41, 0, 5}, {4, 2, 0}, {14, 0, 0, 3}, {10, 3, 2},
   {42, 11}, {3, True, 1},
   {41, 1, 5}, {5, 3.141592653589793, 1}, {25, 0, 1, 2}, {10, 2, 1}, {42, 4},
```

```
{4, 1, 1}, {14, 0, 1, 3},
   {10, 3, 1}, {10, 1, 2}, {2}}, Function[{x}, If[False, 2, If[True, x + Pi,
1]]], Evaluate]
```

This part is extracted when the compiled code produces an error, and further evaluation is only possible using the original uncompiled expression.

> In applying compiled functions, they behave as any other functions (head Function or user-defined functions).

Here is how compTest works.

In[10]:= **compTest[3.98, 3]**

Out[10]= 6.10313×10^9

In[11]:= **Head[%]**

Out[11]= Real

If variables of the wrong type are input, CompiledFunction will give an error message, and the computation will continue with the uncompiled version (which is stored in the penultimate component of the Compiled Function code). The first variable in compTest must be a real number. If it is a complex number, we get an error message.

In[12]:= **compTest[1 + 2I, 1]**

```
        CompiledFunction::cfsa :
           Argument 1 + 2 i at position 1 should be a machine-size real number. More…
```
Out[12]= $-52 - 170\,i$

If possible, arguments are silently coerced.

In[13]:= **compTest[2, 1]**

Out[13]= 128.

In the next example, we compile a module containing the localized variable invisibleVar. In addition, the variable visibleVar, which is not locally scoped within the Module, appears.

In[14]:= **subOptiomalCF = Compile[{}, Module[{invisibleVar = 2},**
                                         visibleVar = invisibleVar^3]]

Out[14]= CompiledFunction[{},
         Module[{invisibleVar = 2}, visibleVar = invisibleVar³], -CompiledCode-]

The presence of a variable whose value must be known globally after the compiled function would have been executed can be easily seen in the fourth argument of the compiled function subOptiomalCF. The explicit Function[*potentialVars*, *body*] indicates that using *potentialVars* the code *body* has to be executed outside of the compiler and that *body* will undergo the general *Mathematica* evaluation mechanism described in Chapter 4 of the Programming volume [1793].

In[15]:= **subOptiomalCF[[4]]**

Out[15]= {{1, 5}, {4, 2, 0}, {21, Function[{}, visibleVar = invisibleVar³],
         {invisibleVar, 2, 0, 0, Module}, 2, 0, 1}, {2}}

Here we use the compiled function subOptiomalCF. After its execution, the variable visibleVar has the global value 8.

In[16]:= **{subOptiomalCF[], invisibleVar, visibleVar}**

Out[16]= {8, invisibleVar, 8}

The following symbolic calculation cannot be compiled. We get a corresponding message, and the evaluation is carried out through the ordinary *Mathematica* evaluator.

```
In[17]:= Compile[{}, a + b][]

        CompiledFunction::cfse :
         Compiled expression a should be a machine-size real number. More…

        CompiledFunction::cfex : External evaluation error
            at instruction 2; proceeding with uncompiled evaluation. More…

Out[17]= a + b
```

We now compare speed of execution. Here is an example of evaluating one root of $a x^3 + b x^2 + c x + d = 0$.

```
In[18]:= (* from Solve[x^3 + b x^2 + c x + d == 0, x] *)
        longgg = -b/(3 a) - (2^(1/3) (-b^2 + 3 a c))/
                (3 a (-2 b^3 + 9 a b c - 27 a^2 d +
                  3^(3/2) a (-(b^2 c^2) + 4 a c^3 + 4 b^3 d -
                  18 a b c d + 27 a^2 d^2)^(1/2))^(1/3)) +
                (-2 b^3 + 9 a b c - 27 a^2 d +
                  3^(3/2) a (-(b^2 c^2) + 4 a c^3 + 4 b^3 d -
                  18 a b c d + 27 a^2 d^2)^(1/2))^(1/3)/(3 2^(1/3) a);
```

This is a relatively large expression.

```
In[20]:= LeafCount[longgg]
Out[20]= 179
```

We now carry out the computation of the root as a function of the variables *a*, *b*, *c*, and *d* without compilation.

```
In[21]:= withoutComp[a_, b_, c_, d_] = N[longgg];
```

Here is the computation with compilation. The compilation process is very fast for such short inputs. The last expression can be compiled 1000 times in less than one second.

```
In[22]:= Timing[Do[withComp =
        Compile[{{a, _Complex}, {b, _Complex}, {c, _Complex}, {d, _Complex}},
                Evaluate[longgg]], {1000}]]
Out[22]= {0.83 Second, Null}
```

The size of the compiled code is about 1 kB in the current example.

```
In[23]:= LeafCount[withComp]
Out[23]= 827
```

We now compare the times and results. For a better granularity of the timings, we repeat the calculation 1000 times.

```
In[24]:= Timing[Table[withoutComp[1.3, 8.9, 7.9I, 6.8 + 8.4I], {10000}];]
Out[24]= {1.39 Second, Null}

In[25]:= Timing[Table[withComp[1.3, 8.9, 7.9I, 6.8 + 8.4I], {10000}];]
Out[25]= {0.19 Second, Null}
```

The compiled version was about six times faster. The results that `withComp` and `withoutComp` calculated were the same.

```
In[26]:= (withComp[##] == withoutComp[##])&[1.3, 8.9, 7.9 I, 6.8 + 8.4 I]
Out[26]= True
```

Here is a slightly larger example showing the speed-up. We calculate the Julia set for the map $z \longrightarrow z^2 + c$ for $c = 0.2593 + 0.00046\,i$ [1653]. This is the compiled function. It counts the number of iterations that do not lead to numbers greater than 10^{100}.

```
In[27]:= cf = Compile[{{z, _Complex}},
            Module[{c = 0, s = z},
                While[s = s^2 + 0.25393 + 0.00046I;
                    Abs[s] < 10.^100 && c < 1000, c++]; c]];
```

Carrying out more than 10^6 iterations of the map takes only a few minutes on a year-2005 computer.

```
In[28]:= (data = With[{pp1 = 650, pp2 = 325},
                Table[cf[x + I y],
                    {y, 0, 1.1, 1.1/pp2}, {x, -0.88, 0.88, 1.76/pp1}]]); // Timing
Out[28]= {142.94 Second, Null}
```

```
In[29]:= Plus @@ Flatten[data]
Out[29]= 102972669
```

Here are the just-calculated data displayed.

```
In[30]:= ListDensityPlot[(* add mirrored version *)
            Transpose[Join[Reverse /@ (Reverse @ data), data]],
                ColorFunction -> (GrayLevel[1 - #/1000]&),
                ColorFunctionScaling -> False, Mesh -> False,
                AspectRatio -> 1.1/1.76, PlotRange -> All,
                Frame -> True, FrameTicks -> False]
```

This is a speed-up of about 25, compared with the uncompiled version. Here is the uncompiled version of the function. We extract it from the `CompiledFunction`-object.

```
In[31]:= pf = Extract[cf, -2]
Out[31]= Function[{z}, Module[{c = 0, s = z},
            While[s = s^2 + 0.25393 + 0.00046 i; Abs[s] < 1. × 10^100 && c < 1000, c++]; c]]
```

```
In[32]:= (* use only 1% of the above points *)
        With[{pp1 = 65, pp2 = 32},
            Table[pf[x + I y],
                {y, 0, 1.1, 1.1/pp2}, {x, -0.88, 0.88, 1.76/pp1}]]; // Timing
Out[33]= {12.22 Second, Null}
```

The following input generates a graphic showing the contour lines $\mathrm{Re}(f^{(4)}(z)) = 0$ and $\mathrm{Im}(f^{(4)}(z)) = 0$ where $f(z) = \csc(\csc(\csc(z)))$. We compile the large expression for the fourth derivative (leaf count 1247). For small imaginary z, $f^{(4)}(z)$ has a very large imaginary part ($f(10^{-5}\,i) \approx -1.4\,10^{43469}\,i$). Such large numbers can only be represented as high-precision numbers and we obtain `CompiledFunction::cfn` messages. But for the majority of z-values the compiled version works fine and again, we obtain a plot quite quickly.

```
In[34]:= Module[{pp = 120, X = 3/2 Pi, Y = 5/4 Pi, ε = 10^-5,
           z, f, fC, data, contours},
    (* the function *)
    f[z_] = D[Csc[Csc[Csc[z]]], {z, 4}];
    (* compiled version of the function *)
    fC = Compile[{{z, _Complex}}, Evaluate[f[z]]];
    (* array of function values in first quadrant *)
    data = Table[fC[N[x + I y]], {y, -ε, Y, (Y + ε)/pp},
                                 {x, -ε, X, (X + ε)/pp}];
    (* replace large high-precision numbers
       by large machine numbers of same sign *)
    data = data /. x_?(Not[MachineNumberQ[#]]&) :>
                        Sqrt[$MaxMachineNumber] Sign[x];
    (* Re[f[z]] == 0 and Im[f[z]] == 0 contours *)
    contours = (Cases[Graphics[#], _Line, Infinity]& @
      ListContourPlot[# @ data,  Contours -> {0},
      ContourShading -> False, DisplayFunction -> Identity,
      MeshRange -> {{-ε, X}, {-ε, Y}}])& /@ {Re, Im};
    (* show contours *)
    Show[Graphics[
    Transpose[{{RGBColor[1, 0, 0], RGBColor[0, 0, 1]},
      (* map contour lines to other three quadrants *)
     {#, Map[# {-1, +1}&, #, {-2}], Map[# {+1, -1}&, #, {-2}],
         Map[# {-1, -1}&, #, {-2}]}& /@ contours}]],
         AspectRatio -> 1/2]]
```

          CompiledFunction::cfn : Numerical error encountered
              at instruction 3; proceeding with uncompiled evaluation. More…

          CompiledFunction::cfn : Numerical error encountered
              at instruction 3; proceeding with uncompiled evaluation. More…

> In spite of the possible speed-up using compiled programs, in numerical computations, one should use
> built-in functions whenever possible. Usually, they are faster than user-defined compiled code, and
> moreover, they have much better error and accuracy controls than are likely to be achieved without a lot
> of (user programming) effort. Use Compile for your own programs that do not have built-in equivalents.

Block, Module, and With can be used in code to be compiled.

```
In[35]:= Compile[{x}, Block[{ξ = x^2}, ξ + 1]][2]
```
Out[35]= 5.

```
In[36]:= Compile[{x}, Module[{ξ = x^2}, ξ + 1]][2]
```
Out[36]= 5.

```
In[37]:= Compile[{x}, With[{ξ = x^2}, ξ + 1]][2]
Out[37]= 5.
```

Because the first argument of `Block`, `Module`, and `With` must be a list of symbols, indexed variables must be treated especially inside `Compile`. While a direct input of indexed variables in scoping constructs is not possible, `Compile` could get indexed variables from functions like `Plot`, `NSum`, and `NIntegrate`. In the next input, we supply indexed variables as the first argument to `Compile`.

```
In[38]:= cf = Compile[{x[1], x[1, 1], x[1][1], 1[y]}, x[1] + x[1, 1] + x[1][1] + 1[y]]
Out[38]= CompiledFunction[{x$33336, x$33337, x$33338, Integer$33339},
         x$33336 + x$33337 + x$33338 + Integer$33339, -CompiledCode-]
```

We see that the resulting `CompiledFunction` renamed the indexed variables to *variable$Integer*. As a result, all indexed variables disappeared and this case has been reduced to the standard case of symbols.

Many other functions (as `Map`, `Fold`, `Function`, ...) can be compiled too. If the arguments are lists, matrices, or tensors, the arguments are specified in the following way: {*varName*, *type*, *tensorRank*}. Scalars have tensor rank 0, lists have tensor rank 1, matrices have tensor rank 2, etc.
Here is a toy example—a matrix product.

```
In[39]:= Compile[{{x, _Real, 1}, {A, _Real, 2}}, A.x]
Out[39]= CompiledFunction[{x, A}, A.x, -CompiledCode-]
```

```
In[40]:= InputForm[%]
Out[40]//InputForm=
        CompiledFunction[{{_Real, 1}, {_Real, 2}}, {{3, 1, 0}, {3, 2, 1}, {3, 1, 2}},
        {0, 0, 0, 0, 3},
        {{1, 5}, {86, 1, 0, 2}, {2}}, Function[{x, A}, A . x], Evaluate]
```

```
In[41]:= %%[{1., 2., 3.}, {{1., 4., 5.}, {6., -2., 2.}, {0., 8., 11.}}]
Out[41]= {24., 8., 49.}
```

We use the third argument in `Compile` for a recursive definition that we will compile.

```
In[42]:= fc = Compile[{{x, _Real, 0}}, If[x > 1, x fc[x - 1], x], {{fc, _Real, 0}}];
```

The symbol `fc` in the fourth argument of `fc` indicates that inside the compiled function a call to the function `fc` happens. Because the compiled function `fc` does not know that `fc` (meaning itself) is a compiled function, this call to the external evaluation of `fc` is slower than an evaluation within the compiled function.

```
In[43]:= InputForm[fc]
Out[43]//InputForm=
        CompiledFunction[{_Real}, {{3, 0, 0}, {3, 0, 1}}, {1, 1, 3, 0, 0},
        {{1, 5}, {4, 1, 0}, {14, 0, 0, 1}, {49, 1, 0, 0}, {41, 0, 8}, {4, -1, 0},
        {14, 0, 0, 1},
        {25, 0, 1, 2}, {22, fc, 3, 0, 2, 3, 0, 1}, {29, 0, 1, 2}, {10, 2, 1}, {42,
        2}, {10, 0, 1}, {2}},
        Function[{x}, If[x > 1, x*fc[x - 1], x]], Evaluate]
```

```
In[44]:= fc[10]
Out[44]= 3.6288×10^6
```

Comparing with the uncompiled version shows a clear speed-up. But it cannot compete with the built-in function `Factorial`.

```
In[45]:= fF[x_] := If[x > 1, x fF[x - 1], x]
```

```
In[46]:= {Timing[Do[fF[100],  {1000}]],
      Timing[Do[fc[100],  {1000}]],
      Timing[Do[   100.!, {1000}]]}
```

Out[46]= {{0.52 Second, Null}, {0.28 Second, Null}, {0. Second, Null}}

If a function is called from inside `Compile`, its numerical type must be explicitly given unless it is of type `Real`. In the following example, `int[3]` produces a complex number.

```
In[47]:= int[x_] := x + I x^2;

      Compile[{x}, int[x]][3]
```

       CompiledFunction::cfse :
        Compiled expression 3. + 9. i should be a machine-size real number. More…

       CompiledFunction::cfex : External evaluation error
         at instruction 2; proceeding with uncompiled evaluation. More…

Out[48]= 3 + 9 i

It is the intermediate `int[3]` that causes the problem, rather than the argument. If we declare the argument of the compiled function to be a complex number, then the evaluation proceeds in an error-free manner. The input number is numericalized and so the external function `int` returns an approximative result.

```
In[49]:= Compile[{{x, _Complex}}, int[x]][3]
```

Out[49]= 3. + 9. i

If we explicitly prescribe the type of the third variable of `Compile`, the compiled code runs also error-free for an explicit real argument.

```
In[50]:= Compile[{x}, int[x^2], {{int[_], _Complex}}][3]
```

Out[50]= 9. + 81. i

However, we note that the following example does not produce an exact number.

```
In[51]:= Compile[{{x, _Integer}}, int[x^2], {{int[_], _Complex}}][3]
```

Out[51]= 9. + 81. i

For integers, the following `CompiledFunction` produces an exact integer result.

```
In[52]:= int[x_] := x + x^2;
      Compile[{{x, _Integer}}, int[x^2], {{int[_], _Integer}}][3]
```

Out[53]= 90

However, rational numbers are not available as their own class in `Compile` (nor are they present in the hardware arithmetic), and if they appear they will be immediately converted to reals.

```
In[54]:= int[x_] := x + x^2;
      Compile[{{x, _Integer}}, int[x^2]/int[x^3], {{int[_], _Integer}}][3]
```

Out[55]= 0.119048

In the next input, we define a function `maxValueList` that returns a list of the cumulative maxima of $f(k)$ for $k = 1, \ldots, o_{\max}$. We calculate real numbers, compare their size, and assemble a list L of integers.

```
In[56]:= maxValueList[f_, oMax_Integer] :=

      Compile[{}, Module[{L = {0}, max = 0., ξ},
            Do[ξ = Abs[f[k]];
                (* keep k if f[k] is larger than all earlier f[j] *)
                If[ξ > max, max = Abs[ξ]; AppendTo[L, k]], {k, oMax}];
```

Rest[L]]] (* carry out compiled function;
no arguments needed *) **[]**

Here are the cumulative maxima for the family of functions $k \to f(k) = |\sin(i^{1/j} k)|$ for $1 \le i, j \le 24$. We display
the value $\log(1 - f(k))$ as a function of k.

```
In[57]:= Show[Graphics[Table[{RGBColor[i/24, 0, j/24],
                    Line[{#, N[Log[10, 1 - Abs[Sin[i^(1/j) #]]], 22]}& /@
                        maxValueList[Evaluate[Sin[i^(1/j) #]]&, 10^4]]},
             {i, 24}, {j, 24}]], PlotRange -> All, Frame -> True, Axes -> False]
```

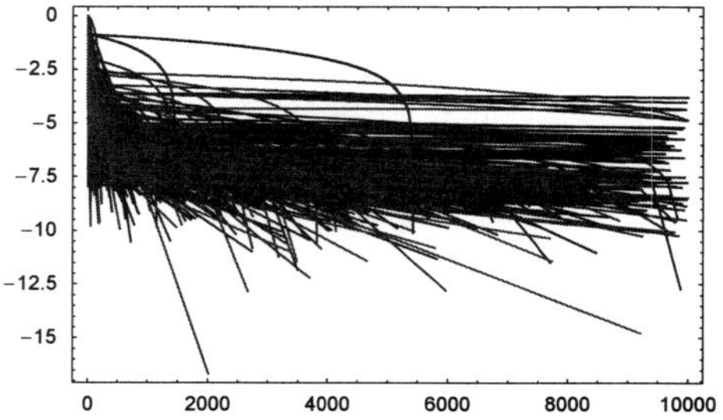

The very small value near $k \approx 2000$ is $\sin(2^{1/5}\, 2017) = -0.999999999999999978\ldots$.

Complex numbers with exact integer real and imaginary parts are also not supported by `Compile`. Either we
have to resort to exact evaluation or we have to numericalize the complex numbers.

```
In[58]:= Compile[{{x, _Integer}}, int[x^2]/int[x^3], {{int[_], _Integer}}][3 + I]
```

            CompiledFunction::cfsa :
               Argument 3 + i at position 1 should be a machine-size integer. More…

Out[58]= $\dfrac{861}{10370} - \dfrac{687\, i}{10370}$

```
In[59]:= Compile[{{x, _Complex}}, int[x^2]/int[x^3], {{int[_], _Complex}}][3 + I]
```

Out[59]= $0.083028 - 0.0662488\, i$

In intermediate steps, integers can arise in a compiled function that uses real numbers. These integers are
automatically converted to real numbers. Here is an example: the compiled function $\mathcal{PRW}$ calculates a list of the
maximal values of the partial sums of $\sum_{k=1}^{o} (-1)^{\lfloor k\,\alpha \rfloor}$ [1365].

```
In[60]:= PRW = Compile[{{o, _Integer}, α},
              Module[{extremeList = {{0., 0.}}, sum = 0., max = 0.},
                     Do[(* next partial sum *)
                         sum = sum + (-1)^Floor[k α];
                         If[(* new maximal value? *)
                             Abs[sum] > max, max = Abs[sum];
                             (* keep maximal values *)
                             AppendTo[extremeList, {k, max}]], {k, o}];
                     Rest[extremeList]]];
```

The following graphic shows the resulting extreme values for various α formed from constants and for
random-value realizations of α. The maximal values are very slowly growing functions.

```
In[61]:= Show[GraphicsArray[Function[xs,
      Graphics[(* color for a guide for the eye *)
      MapIndexed[{Hue[0.8 #2[[1]]/20], #1}&,
      Line[{Log10, #1}, #2]& @@@ PRW[10^6, #]]& /@ xs],
          PlotRange -> All, Frame -> True]] /@
```

```
(* two lists of real numbers *)
{{Pi, 1/Pi, Pi^Pi, E, 1/E, E^E, E/Pi, Pi/E,
  E^Pi, Pi^E, Sqrt[2], Sqrt[3], 3^(1/3), 5^(1/5),
  Log[2], Log[3], Tan[1], Cos[1], Sin[1], 1/17},
  (* seed random number generator *) SeedRandom[9];
  Table[10^Random[Real, {-1, 1}], {20}]}]]
```

We note that, analogous to Function, nothing is computed before the application of Compile, because of the HoldAll attribute.

```
In[62]:= co1 = Compile[{x}, x + 3.0 + 7.0]
Out[62]= CompiledFunction[{x}, x + 3. + 7., -CompiledCode-]
```

```
In[63]:= co2 = Compile[{x}, x + 10.0]
Out[63]= CompiledFunction[{x}, x + 10., -CompiledCode-]
```

```
In[64]:= co1 === co2
Out[64]= False
```

The error message in the following example clearly demonstrates the HoldAll attribute. The first attempt fails in substituting the actual definition of circle[*t*] in c1, and the second example succeeds.

```
In[65]:= circle[t_] = {Cos[t], Sin[t]}
Out[65]= {Cos[t], Sin[t]}
```

```
In[66]:= c1 = Compile[{{t, _Real}}, circle[t]]
Out[66]= CompiledFunction[{t}, circle[t], -CompiledCode-]
```

```
In[67]:= c1[1.0]
```

> CompiledFunction::cfse : Compiled expression
>   {0.540302, 0.841471} should be a machine-size real number. More…
>
> CompiledFunction::cfex : External evaluation error
>   at instruction 2; proceeding with uncompiled evaluation. More…

```
Out[67]= {0.540302, 0.841471}
```

```
In[68]:= c1 = Compile[{{t, _Real}}, Evaluate[circle[t]]]
Out[68]= CompiledFunction[{t}, {Cos[t], Sin[t]}, -CompiledCode-]
```

```
In[69]:= c1[1.0]
Out[69]= {0.540302, 0.841471}
```

The third argument in Compile permits type declarations for expressions that may be external function calls, provided the argument of the externally defined function does not need to be computed inside the compiled function. The function g absolutely requires uncomputed variables to work correctly.

```
In[70]:= g = Function[x, Length[Unevaluated[x]], HoldAll]
```

Out[70]= `Function[x, Length[Unevaluated[x]], HoldAll]`

In[71]:= `g[4 + 4 + 4]`

Out[71]= 3

Arguments of a function called from inside a compiled function are evaluated (note the point following the 0).

In[72]:= `Compile[{x}, g[x + 1 + 4]][8]`

Out[72]= 0.

If we explicitly choose g[_] for the fourth variable (the third one is the tensor rank of the return type), the arguments will be passed on to g unevaluated.

In[73]:= `Compile[{x}, g[x + 1 + 4], {{g[_], _Integer, 0, g[_]}}][8]`

Out[73]= 3

Compile can compile quite complicated programs containing pure functions, functional constructs, and so on. To make sure that the compilation succeeded, it is best to look at the InputForm of a CompiledFunction object and to verify that the fourth argument does not contain any Function. Here we compile a Table construct with a Fold body.

In[74]:= `Compile[{{x, _Real, 1}, {n, _Integer, 0}},`
`        Table[Sin @@ {Fold[ArcTan[#1, #2^2]&, i, x]}, {i, n}]] // InputForm`

Out[74]//InputForm=

```
CompiledFunction[{{_Real, 1}, _Integer}, {{3, 1, 0}, {2, 0, 0}, {3, 1, 1}},
                                          {0, 8, 5, 0, 3},
    {{1, 5}, {9, 0, 1}, {4, 0, 3}, {58, 1, 3, 1}, {4, 0, 4}, {9, 4, 2},
     {79, 2, 1, 17}, {5, 0., 1}, {14, 0, 2, 3}, {25, 3, 1, 3}, {12, 0, 2},
     {57, 2, 5}, {4, 0, 6}, {79, 6, 5, 7}, {64, 2, 0, 6, 0, 1}, {4, 2, 7},
     {89, 264, 3, 0, 1, 2, 0, 7, 3, 0, 4},
     {89, 266, 3, 0, 3, 3, 0, 4, 3, 0, 1}, {10, 1, 3}, {42, -6},
     {88, 1, 3, 0, 3, 3, 0, 2}, {59, 3, 2, 1}, {42, -16}, {2}},
    Function[{x, n}, Table[Sin @@ {Fold[ArcTan[#1, #2^2] & , i, x]}, {i, n}]],
    Evaluate]
```

The function Prime is not compilable currently. In the fourth argument of the CompiledFunction, one sees the call to the evaluator with Function[{x, n}, Prime[n]].

In[75]:= `Compile[{{x, _Real, 1}, {n, _Integer, 0}},`
`        Prime[n] + Plus @@ Sin[x]] // InputForm`

```
CompiledFunction[{{_Real, 1}, _Integer}, {{3, 1, 0}, {2, 0, 0}, {3, 0, 2}},
    {0, 5, 3, 0, 3}, {{1, 5}, {21, Function[{x, n}, Prime[n]], 3, 1, 0, 2, 0, 0, 2, 0, 1},
     {88, 1, 3, 1, 0, 3, 1, 2}, {5, 0., 0}, {57, 2, 3}, {4, 0, 4}, {79, 4, 3, 5},
     {64, 2, 0, 4, 0, 1}, {25, 0, 1, 2}, {10, 2, 0}, {42, -4}, {14, 0, 1, 2},
     {25, 2, 0, 2}, {2}}, Function[{x, n}, Prime[n] + Plus @@ Sin[x]], Evaluate]
```

Although the application of this compiled function works, for clarity, it is recommended to add {Prime[_], _Integer, 0} as a third argument in Compile.

In[76]:= `Compile[{{x, _Real, 1}, {n, _Integer, 0}},`
`        Prime[n] + Plus @@ Sin[x]][{2., 3.}, 4]`

Out[76]= 8.05042

In[77]:= `(* preferable for clarity *)`
`    Compile[{{x, _Real, 1}, {n, _Integer, 0}},`
`        Prime[n] + Plus @@ Sin[x],`
`        {{Prime[_], _Integer, 0}}][{2., 3.}, 4]`

Out[78]= 8.05042

As already mentioned, the packed arrays go hand in hand with compilation. Whenever a `CompiledFunction` returns a tensor, it will be packed. So, in the following, the list of five elements returned from `CompiledFunction` is packed, although an uncompiled version would give an unpacked result.

```
In[79]:= Developer`PackedArrayQ[Compile[{}, Table[1, {5}]][]]
Out[79]= True
```

```
In[80]:= Developer`PackedArrayQ[Function[{}, Table[1, {5}]][]]
Out[80]= False
```

Frequently one has a larger *Mathematica* expression that should be compiled. Instead of being explicitly inlined into the body of `Compile`, it is often convenient to inline it through a `Function[p, Compile[vars, body(p)]][program]` construction (possibly giving the pure function the `HoldAll` attribute). Here is an example of such a construction. We construct random rational function, compile them, and iterate them.

```
In[81]:= randomRationalIterations[seed_, opts___] :=
    Module[{ri, rr, rc, rpol, rquot, nl, tab},
    Block[{z},
      (* random numbers, polynomials, and rational functions *)
      ri[n_] := Random[Integer, n {-1, 1}];
      rr[x_] := Random[Real, x {-1, 1}];
      rc[x_] := rr[x] + I rr[x];
      rpol[o_, n_, x_, z_] := Sum[rc[x] z^ri[n] Conjugate[z]^ri[n], {o}];
      rquot[o_, n_, x_, z_] := rpol[o, n, x, z]/rpol[o, n, x, z];
      (* seed random number generator *)
      SeedRandom[seed];
    (* search for rational function *)
    While[(* compile for each rational function *)
    nl = Compile[{},
    Module[{tab = Table[0. + 0. I, {2 10^5}], z = 1. I, k = 1},
        While[Abs[Log[10, Abs[z]]] < 10. && k < 2 10^5,
            z = (* inline function here *) #;
            tab[[k]] = z; k++];
        If[(* long enough list? *) k > 1.6 10^5, tab, {}]]]&[
          (* the rational function to be inlined *)
          rquot[6, 6, 6, z]][];
        nl === {}, Null];
    (* make graphics *)
    Graphics[{PointSize[0.003], Point[{Re[#], Im[#]}]& /@
            Union[Drop[nl, 10^5]]}, opts,
          AspectRatio -> 1, PlotRange -> All]]]
```

```
In[82]:= (* four example graphics *)
    Show[GraphicsArray[#]]& @ Apply[randomRationalIterations,
          {{2847}, {26909}, {142573},
          {5518644, PlotRange -> {{-11, 13}, {-13, 7}}}}, {1}]
```

We make one additional remark concerning compilable structures: As already mentioned, functional constructions, such as `Outer`, `Fold`, `Map`, `FoldList`, and pure functions can be compiled too. However, with respect
to speed, using compiled code for a functional construction is, of course, not always preferable. The relative
speed-up obtained with `Compile` is typically larger for procedural code than for functional code. To illustrate,
we look at the cross section of a potential field of a point charge in the vicinity of a metal half-plane
($z = 0$, $x > 0$). The potential is given by:

$$\phi(x, y, z) = \frac{1}{R} \arccos\left(-\frac{\cos((\theta - \gamma)/2)}{\cos(\eta/2)}\right) - \frac{1}{R'} \arccos\left(-\frac{\cos((\theta + \gamma)/2)}{\cos(\eta/2)}\right)$$

$$R = \sqrt{r_0^2 + r^2 + z^2 - 2\, r\, r_0 \cos(\theta - \gamma)}$$

$$R' = \sqrt{r_0^2 + r^2 + z^2 - 2\, r\, r_0 \cos(\theta + \gamma)}$$

$$\cosh(\eta) = \frac{r_0^2 + r^2 + z^2}{2\, r\, r_0} \qquad r = \sqrt{x^2 + y^2}.$$

Here, $\{r_0, \gamma, 0\}$ is the position of the charge in cylindrical coordinates (see [1098] and [1117]). (For the potential
of a point charge near a 3D corner, see [1595].) We now give a procedural and a functional implementation.

```
In[84]:= equiPotP[{x_, y_, z_}, {r0_, γ_}] :=
      Module[{r, R, Rs, θ, θ1, e},
        r = Sqrt[x^2 + y^2];
        θ1 = ArcTan[x, y];
        θ = If[θ1 < 0, 2Pi + θ1, θ1];
        R  = Sqrt[r0^2 + r^2 + z^2 - 2r r0 Cos[γ - θ]];
        Rs = Sqrt[r0^2 + r^2 + z^2 - 2r r0 Cos[γ + θ]];
        e = ArcCosh[(r0^2 + r^2 + z^2)/(2 r0 r)];
        1/R  ArcCos[-Cos[(θ - γ)/2]/Cosh[e/2]] -
        1/Rs ArcCos[-Cos[(θ + γ)/2]/Cosh[e/2]]]
```

Here is a more functional implementation.

```
In[85]:= equiPotF[{x_, y_, z_}, {r0_, γ_}] :=
      (* nest pure function to avoid a) naming and
         b) multiple evaluation of the same expression *)
      (1/#1 ArcCos[-Cos[(#4 - γ)/2]/#3] -
       1/#2 ArcCos[-Cos[(#4 + γ)/2]/#3]& @@
      {Sqrt[r0^2 + #1^2 + z^2 - 2#1 r0 Cos[γ - #2]],
       Sqrt[r0^2 + #1^2 + z^2 - 2#1 r0 Cos[γ + #2]],
       Cosh[ArcCosh[(r0^2 + #1^2 + z^2)/(2 r0 #1)]/2],
       #2}&[Sqrt[x^2 + y^2], If[# < 0, 2Pi + #, #]&[ArcTan[x, y]]])
```

Both functions produce the same result.

```
In[86]:= (equiPotP[##] == equiPotF[##])&[{2., 2., 1.}, {1., Pi/4.}]
Out[86]= True
```

The functional version `equiPotF` is faster.

```
In[87]:= {Timing[Do[equiPotP[{2., 2., 1.}, {1., Pi/4.}], {1000}]],
         Timing[Do[equiPotF[{2., 2., 1.}, {1., Pi/4.}], {1000}]]}
Out[87]= {{0.16 Second, Null}, {0.12 Second, Null}}
```

We now look at the potential distribution using `ContourPlot`. We also measure the times needed to generate
the plot.

```
In[88]:= ContourPlot[equiPotP[{x, y, 1.}, {1., Pi/4.}],
             {x, -3/2, 3/2}, {y, -3/2, 3/2},
             Compiled -> False, PlotPoints -> 60,
             ColorFunction -> Hue, FrameTicks -> None,
             Epilog -> {Thickness[0.02], GrayLevel[0],
                        Line[{{0, 0}, {3/2, 0}}]}]] // Timing
```

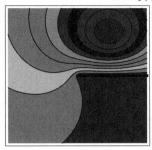

```
In[89]:= ContourPlot[equiPotF[{x, y, 1.}, {1., Pi/4.}],
             {x, -3/2, 3/2}, {y, -3/2, 3/2},
             Compiled -> False, PlotPoints -> 60,
             ColorFunction -> Hue, FrameTicks -> None,
             Epilog -> {Thickness[0.02], GrayLevel[0],
                        Line[{{0, 0}, {3/2, 0}}]}]] // Timing
```

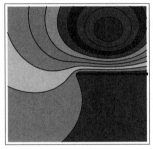

Again, the functional version was faster. Now let us make compiled versions of the procedural and functional programs.

```
In[90]:= equiPotPCF =
    Compile[{{xyz, _Real, 1}, {params, _Real, 1}},
    Module[{x = xyz[[1]], y = xyz[[2]], z = xyz[[3]], r0 = params[[1]],
            γ = params[[2]], r, R, Rs, θ, θ1, e},
    r = Sqrt[x^2 + y^2];
    θ1 = ArcTan[x, y];
    θ = If[θ1 < 0, 2Pi + θ1, θ1];
    R  = Sqrt[r0^2 + r^2 + z^2 - 2r r0 Cos[γ - θ]];
    Rs = Sqrt[r0^2 + r^2 + z^2 - 2r r0 Cos[γ + θ]];
    e = ArcCosh[(r0^2 + r^2 + z^2)/(2 r0 r)];
    1/R  ArcCos[-Cos[(θ - γ)/2]/Cosh[e/2]] -
    1/Rs ArcCos[-Cos[(θ + γ)/2]/Cosh[e/2]]]];

In[91]:= (* was everything compiled? *)
    FreeQ[equiPotPCF[[4]], Function, Infinity]

Out[92]= True
```

```
In[93]:= equiPotFCF =
    Compile[{{xyz, _Real, 1}, {params, _Real, 1}},
    Module[{x = xyz[[1]], y = xyz[[2]], z = xyz[[3]],
            r0 = params[[1]], γ = params[[2]]},
     (1/#1 ArcCos[-Cos[(#4 - γ)/2]/#3] -
      1/#2 ArcCos[-Cos[(#4 + γ)/2]/#3]& @@
    {Sqrt[r0^2 + #1^2 + z^2 - 2#1 r0 Cos[γ - #2]],
     Sqrt[r0^2 + #1^2 + z^2 - 2#1 r0 Cos[γ + #2]],
     Cosh[ArcCosh[(r0^2 + #1^2 + z^2)/(2 r0 #1)]/2],
     #2}&[Sqrt[x^2 + y^2], If[# < 0, 2 Pi + #, #]&[ArcTan[x, y]]])]];

In[94]:= (* was everything compiled? *)
    FreeQ[equiPotFCF[[4]], Function, Infinity]
Out[95]= True
```

The two functions now have about equal absolute speed (and, of course, still give the same results).

```
In[96]:= (equiPotPCF[##] == equiPotFCF[##])&[{2., 2., 1.}, {1., Pi/4.}]
Out[96]= True

In[97]:= {Timing[Do[equiPotPCF[{2., 2., 1.}, {1., Pi/4.}], {10^5}]],
    Timing[Do[equiPotFCF[{2., 2., 1.}, {1., Pi/4.}], {10^5}]]}
Out[97]= {{1.94 Second, Null}, {2.12 Second, Null}}

In[98]:= ContourPlot[equiPotPCF[{x, y, 1.}, {1., Pi/4.}],
                {x, -3/2, 3/2}, {y, -3/2, 3/2},
                Compiled -> False, PlotPoints -> 60,
                DisplayFunction -> Identity] // Timing
Out[98]= {0.09 Second, - ContourGraphics -}

In[99]:= ContourPlot[equiPotFCF[{x, y, 1.}, {1., Pi/4.}],
                {x, -3/2, 3/2}, {y, -3/2, 3/2},
                Compiled -> False, PlotPoints -> 60,
                DisplayFunction -> Identity] // Timing
Out[99]= {0.09 Second, - ContourGraphics -}
```

This shows that the speed-up for procedural programs is in this example (and typically) larger than the speed-up for functional programs.

Random numbers can, of course, also be generated within compiled functions. The sequence of random numbers generated with compiled functions is identical to the ones generated without compilation.

```
In[100]:= SeedRandom[12345];
    {Random[], Random[Integer], Random[Complex]}
Out[101]= {0.214347, 0, 0.539981 + 0.0875722 i}

In[102]:= SeedRandom[12345];
    {Compile[{}, Random[]][], Compile[{}, Random[Integer]][],
     Compile[{}, Random[Complex]][]}
Out[103]= {0.214347, 0, 0.539981 + 0.0875722 i}
```

Here is a compiled function of the iteration $x_{n+1} = x_n \pm \beta x_{n-1}$ (where $\pm \beta$ means $+\beta$ and $-\beta$ with equal probability) [1665]. We continue the iteration until $|x_n| > 10^{10}$ or at most 1000 times. The function randomFibonic $\cdot$ ciCF returns the number of iterations for a given β.

```
In[104]:= randomFibonacciCF = Compile[{β},
            Module[{x = 1., y = 1., Y, c = 0},
                While[(* stop iteration? *)
```

```
                         c < 1000 && Abs[y] < 10.^10,
                         (* update x and y *)
                         {x, y} = {y, y + (2 Random[Integer] - 1) β x}; c++];
                         (* return y and c *) {y, c}]];
```

The following graphics show that, for $\beta \lesssim 0.7$, the x_n begin to grow exponentially. (With minor modifications to the function `randomFibonacciCF`, we could also study the convergence for complex β.)

```
In[105]:= Show[GraphicsArray[
         Block[{(* 250 runs for each β *)
                data = Table[{β, Table[randomFibonacciCF[β]/250, {250}]},
                             {β, 0, 3/2, 1/100}],
                $DisplayFunction = Identity},
          {(* average number of iterations *)
           ListPlot[{#1, Plus @@ Log[Abs[First /@ #2]]}& @@@ data],
           (* average logarithmic value of xEnd *)
           ListPlot[{#1, Plus @@ (Last /@ #2)}& @@@ data]}]]]
```

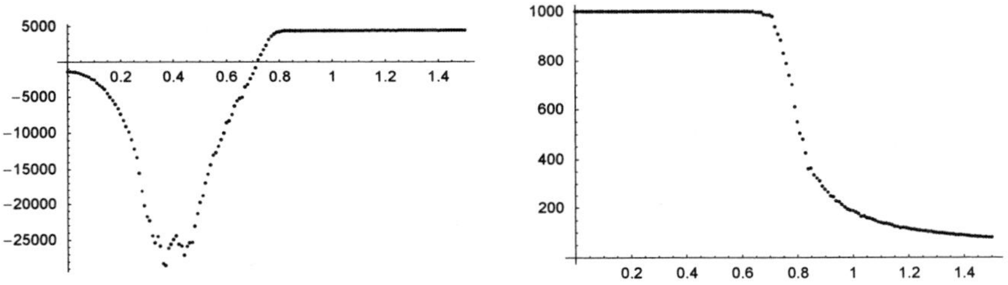

Here is another example. Given a sequence of random independent and identically distributed positive random variables x_k with mean μ and variance σ, the following expression containing the products of the partial sums $s_k = \sum_{j=1}^{k} x_k$ is normally distributed: $2^{-1/2} \ln\!\big((1/(n!\,\mu^n)) \prod_{k=1}^{n} s_k)^{\mu/\sigma\, n^{1/2}}\big)$ [1490]. The function `cumulativeSum`﹅ `ProductsCF` calculates the binned counts (using $2l+1$ bins of width δ) of n draws of a random variable uniformly distributed in $[0, 2]$.

```
In[106]:= cumulativeSumProductsCF =
         Compile[{{n, _Integer}, {l, _Integer}, δ},
          Module[{μ = 1., γ = Sqrt[1./3], j, Π = 1., Σ = 0.,
                  (* count table *) T = Table[0, {2l + 1}], m = 2l + 1},
           Do[(* cumulative sum and product *)
              Σ = Σ + 2. Random[]; Π = Π Σ/(k μ);
              j = Round[1/(Sqrt[2] γ Sqrt[k]) Log[Π]/δ] + 1;
              (* add count *)
              If[1 <= j <= m, T[[j]] = T[[j]] + 1], {k, n}]; T]];
```

Next, we carry out 10^4 random runs, each generating 10^4 partial sums and products. The resulting distribution approximates the Gaussian distribution quite well.

```
In[107]:= SeedRandom[111];
         With[{n = 10^4, o = 10^4, l = 500, δ = 0.05},
          Module[{data = Sum[cumulativeSumProductsCF[n, l, δ], {o}]},
          Show[Block[{$DisplayFunction = Identity},
           {(* theoretical distribution *)
            Plot[1/Sqrt[2 Pi] Exp[-x^2/2], {x, -5, 5}, PlotStyle -> {Hue[0]}],
            (* experimental frequencies *)
            ListPlot[MapIndexed[{(#2[[1]] - l) δ, #1/(n o δ)}&, data]]},
             PlotRange -> {{-5, 5}, All}, Frame -> True, Axes -> False]]]
```

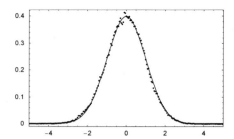

Next, we give two compiled functions that deal with integers. We will model the following two birthday problems: How many people does one need in average to have two with the same birthday and how many people does one need in average such that all possible birthdays are covered? (Here we assume the birthday probability for each day of the year being identical). The following two functions oneBirthday and allBirthdays randomly select birthdays until two birthdays coincide or all possible birthdays are covered, respectively. l is the list of the 365 possible birthdays, c the number of persons, and r is a random birthday (an integer between 1 and 365).

```
In[109]:= oneBirthday = Compile[{{n, _Integer}},
             Module[{l = Table[0, {n}], c = 0, r = 0},
                    (* add new person until one birthday appears twice *)
                    While[r = Random[Integer, {1, n}];
                        l[[r]] == 0, l[[r]] = 1; c++];
                    c + 1]];
```

```
In[110]:= allBirthdays = Compile[{{n, _Integer}},
             Module[{l = Table[0, {n}], c = 0, r = 0,
                    (* number of not assumed birthdays *) v = n},
                    (* add new person until all birthday appear *)
                    While[v != 0,
                        r = Random[Integer, {1, n}];
                        If[l[[r]] == 0, v--];
                        l[[r]] = l[[r]] + 1; c++];
                    c]];
```

Carrying out 10000 runs of the two functions oneBirthday and allBirthdays yields the following distributions birthdayData for the number of people needed.

```
In[111]:= birthdayData =
             With[{n = 365, o = 10^4},
                 ({First[#], Length[#]/o}& /@
                         Split[Sort[Table[#[n], {o}]]])& /@
                         {oneBirthday, allBirthdays}];
```

```
In[112]:= Show[GraphicsArray[
             ListPlot[#, PlotRange -> All, DisplayFunction -> Identity,
                     Frame -> True, Axes -> False]& /@
                     birthdayData]]
```

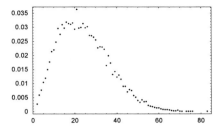

The averages obtained in this simulation agree well with the theoretical expectation values.

In[113]:= **Plus @@@ Apply[Times, birthdayData, {2}] // N**

Out[113]= {24.6131, 2363.37}

In[114]:= **(* theoretical expectation values *)**
        {1 + n Exp[n] ExpIntegralE[1 - n, n],
        n (EulerGamma + PolyGamma[0, 1 + n])} /. n -> 365.

Out[115]= {24.6166, 2364.65}

Inside Compile, iterators have to be explicit; meaning constructions such as Sequence @@ Table[{k[j], kMin, kMax}, {k, n}] will not work. If such "variable" iterators are needed, one can generate them outside of Compile and substitute them. The following input demonstrates this for the alternating harmonic sums $\sum_{j_1, j_2, \ldots, j_n} \varepsilon_{j_1}/1 + \varepsilon_{j_2}/2 + \ldots + \varepsilon_{j_n}/n$, where $\varepsilon_{j_k} = \pm 1$. alternatingHarmonicSums[n][$bins$] calculates a binned probability distribution with *bins* bins.

In[116]:= **alternatingHarmonicSums[n_] := alternatingHarmonicSums[n] =**
      (* apply the pure function Compile[...]& to the iterators *)
      Compile[{{bins, _Integer}},
            Module[{js = 1/Range[#1], **(* bin size *) δ = 8./bins, k, Λ,**
                (* calculate sum and bin it *)
                ξ = Round[If[OddQ[bins], (bins - 1.)/2., bins/2.]]},
                (* counters *) Λ = Table[0, {bins}];
                Do[k = Round[#2.js/δ] + ξ;
                  Λ[[k]] = Λ[[k]] + 1, ##3];
                (* add abscissas *)
                Transpose[{Table[j δ - 4, {j, bins}],
                      (* scale *) Λ/2^n/δ}]]]&[n,
           (* generate iterators outside compile *)
           Table[ε[j], {j, n}], Sequence @@ Table[{ε[j], -1, 1, 2}, {j, n}]]];

Here are the resulting distributions for $n = 18$ and $n = 24$. Interestingly, the distribution has a constant value ($< 1/4$) for sum values from the interval $[-1, 1]$ [1613].

In[117]:= **Show[GraphicsArray[**
      Function[{n, bins}, ListPlot[alternatingHarmonicSums[n][bins],
                       Frame -> True, Axes -> False,
                       DisplayFunction -> Identity,
                       PlotStyle -> PointSize[0.003]]] @@@
           {{18, 1001}, {24, 2001}}]]

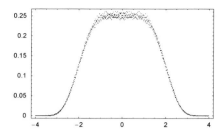
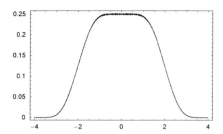

The function `Compile` has two options.

In[118]:= **Options[Compile]**

Out[118]= {CompileOptimizations → Automatic, RuntimeErrorHandler → Evaluate}

The option `CompileOptimizations` determines if the body of the compiled function should be optimized before compilation. The pure function that is the fifth argument of a `CompiledFunction` contains the (potentially optimized) body as a pure function. The next input shows the optimized and the unoptimized (meaning the original) body.

In[119]:= **Compile[{}, Evaluate[Nest[Sqrt[Sqrt[2] + #]&, 0, 6]],**
          CompileOptimizations -> #][[5]]& /@
                (* should body be optimized *) {Automatic, False}

Out[119]= $\{$ Function$\big[\{\}$, Block$\big[\{$\$\$49600$\}$, \$\$49600 = $\sqrt{2}$;

$$\sqrt{\$\$49600 + \sqrt{\$\$49600 + \sqrt{\$\$49600 + \sqrt{\$\$49600 + \sqrt{2^{1/4} + \$\$49600}}}}}\ \big]\big],$$

$$\text{Function}\big[\{\}, \sqrt{\sqrt{2} + \sqrt{\sqrt{2} + \sqrt{\sqrt{2} + \sqrt{\sqrt{2} + \sqrt{2^{1/4} + \sqrt{2}}}}}}\ \big]\}$$

The option `"RuntimeErrorHandler"` is reflected in the sixth argument of a `CompiledFunction` and determines the action to be taken when a run-time error occurs in the evaluation of the compiled code. Here is a sample example.

In[120]:= **Compile[x, x^x, RuntimeErrorHandler :>**
                (("Runtime error at x = " <> ToString[#1])&)][I]
                CompiledFunction::cfsa :
                    Argument i at position 1 should be a machine-size real number. More...

Out[120]= Runtime error at x = I

When doing more advanced operations using the *Mathematica* compiler, some functions from the Developer` SystemOptions[] are useful. The system functions dealing with the direct use of `Compile` are as follows.

In[121]:= **Select[Developer`SystemOptions[],**
                StringMatchQ[#[[1]], "Compile*"]&]

Out[121]= {CompileOptions → {ApplyCompileLength→ ∞,
          ArrayCompileLength→ 250, AutoCompileAllowCoercion→ False,
          AutoCompileProtectValues→ False, AutomaticCompile→ False,
          CompileAllowCoercion→ True, CompileConfirmInitializedVariables→ True,
          CompiledFunctionArgumentCoercionTolerance→ 2.10721,
          CompileEvaluateConstants→ True, CompileReportCoercion→ False,

```
CompileReportExternal→ False, CompileReportFailure→ False,
CompileValuesLast → True, FoldCompileLength→ 100,
InternalCompileMessages→ False, MapCompileLength→ 100,
NestCompileLength→ 100, NumericalAllowExternal→ True,
SystemCompileOptimizations→ All, TableCompileLength→ 250}}
```

We will not discuss them all in detail. Just a few remarks are in order. In the following example, we specify that cf takes real arguments.

In[122]:= **cf = Compile[{{x, _Real}}, Sin[x]]**

Out[122]= CompiledFunction[{x}, Sin[x], –CompiledCode–]

But cf does not complain with the following complex argument.

In[123]:= **cf[2. + 10^-16 I]**

Out[123]= 0.909297

The setting of the system option "CompiledFunctionArgumentCoercionTolerance" measures the number of decimal digits the type rule can be "violated". Now, we demand that the inputs strictly have the correct type.

In[124]:= **Developer`SetSystemOptions["CompileOptions" ->**
         "CompiledFunctionArgumentCoercionTolerance" -> 0];

The above input carried out again leads to a message.

In[125]:= **cf[2. + 10^-16 I]**

CompiledFunction::cfsa : Argument 2. + $\frac{i}{1000000000000000}$
        at position 1 should be a machine-size real number. More…

Out[125]= $0.909297 - 4.16147 \times 10^{-17}$ i

We reset the option to its old value.

In[126]:= **Developer`SetSystemOptions[**
         "CompiledFunctionArgumentCoercionTolerance" -> 2.10721]
        Developer`SetSystemOptions::sysname :
          CompiledFunctionArgumentCoercionTolerance is not a known SystemOption. More…

Out[126]= Developer`SetSystemOptions[CompiledFunctionArgumentCoercionTolerance→ 2.10721]

A programmatic way to check if a compilation was successful is to inspect the content of the fourth argument of the generated CompiledFunction. For interactive use, we can use the "CompileReportExternal" system option.

Here is a try that will certainly fail. Symbolic integration and rule substitutions can currently not be compiled. But no message is issued at the time the CompiledFunction is generated.

In[127]:= **cf = Compile[{{n, _Integer}, {x, _Real}}, Integrate[y^n, y] /. y -> x]**

Out[127]= CompiledFunction[{n, x}, $\int y^n \, dy$ /. y → x, –CompiledCode–]

Calling the compiled function with explicit arguments gives a message and the evaluation proceeds outside of the compiler.

In[128]:= **cf[Pi, 2]**

CompiledFunction::cfsa :
  Argument π at position 1 should be a machine-size integer. More…

Out[128]= $\frac{2^{1+\pi}}{1 + \pi}$

Setting now the system option `"CompileReportExternal" ->` True yields a warning message at the time the `CompiledFunction` is generated.

```
In[129]:= Developer`SetSystemOptions["CompileOptions" ->
                          {"CompileReportExternal" -> True
                          (* (* useful for some other cases *) ,
                          "CompileReportFailure" -> True,
                          "InternalCompileMessages" -> True *)}];
```

```
In[130]:= Compile[{{x, _Real}, {n, _Integer}}, Integrate[y^y, y] /. y -> x];
```

        Compile::extscalar : y cannot be compiled and will be
            evaluated externally. The result is assumed to be of type Real. More…

        Compile::extscalar : y cannot be compiled and will be
            evaluated externally. The result is assumed to be of type Real. More…

        Compile::extscalar : y cannot be compiled and will be
            evaluated externally. The result is assumed to be of type Real. More…

        General::stop : Further output of
            Compile::extscalar will be suppressed during this calculation. More…

We reset the `"CompileReportExternal"` system option value to its original value.

```
In[131]:= Developer`SetSystemOptions["CompileOptions" ->
                          {"CompileReportExternal" -> False}];
```

Now, let us give some more examples for the use of `Compile`. Let us start by having a look at the convergence of the Hansen–Patrick root-finding method [786]. It is built from the following iterations for finding the (simple) root of $f(z)$ (α is a free parameter):

$$z \to z - \frac{(\alpha - 1)\, f(z)}{\alpha\, f'(z) + \sqrt{f'(z)^2 - (\alpha + 1)\, f(z)\, f''(z)}}$$

We take $f(z) = \cos(z) - z$ as our example function. Here is an uncompiled version of a function measuring the number of iterations needed.

```
In[132]:= uc[α_, z0_] := Length @ FixedPointList[
        Function[z, (* use Chop to avoid very small imaginary parts *)
                  Chop[z - ((1 + α) (z - Cos[z]))/
                      (α (1 + Sin[z]) + Sqrt[-(1 + α) (z - Cos[z]) Cos[z] +
                      (1 + Sin[z])^2])]], z0, 100]
```

```
In[133]:= uc[-3. + 1. I, 3.]
```
```
Out[133]= 7
```

And here is the corresponding compiled version. We use the down values of `uc` to avoid retyping the actual function.

```
In[134]:= cf = ReleaseHold[Hold[Compile[{{α, _Complex}, {z0, _Complex}},
                          uc[α, z0]]] /. DownValues[uc]];
```

The compiled version gives the same result as the uncompiled version.

```
In[135]:= cf[-3. + 1. I, 3.]
```
```
Out[135]= 7
```

The speed difference between the two versions is about a factor of 3.

```
In[136]:= {Timing[Do[uc[-1. + I, 1.], {10^4}]],
          Timing[Do[cf[-1. + I, 1.], {10^4}]]}
Out[136]= {{2.03 Second, Null}, {0.58 Second, Null}}
```

Using the compiled version allows us to more quickly make a picture of the iteration speed depending on the complex parameter α.

```
In[137]:= Show[GraphicsArray[
          Block[{$DisplayFunction = Identity, pp = 360, L = 8},
          (* display contour plot and 3D plot *)
          {ContourPlot[cf[αr + I αi, -20], {αr, -L, L}, {αi, -L, L},
                      PlotPoints -> pp, Compiled -> False, PlotRange -> All,
                      ColorFunction -> (Hue[0.1 + 0.2 #]&), Contours -> 20,
                      ContourStyle -> {{Thickness[0.002], GrayLevel[0]}}],
          (* use smaller value for z0 for 3D plot *)
          Plot3D[cf[αr + I αi, -2], {αr, -L, L}, {αi, -L, L},
                 PlotPoints -> pp, Compiled -> False, PlotRange -> All,
                 Mesh -> False]}]]]
```

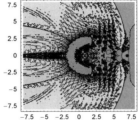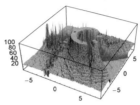

In the next graphics, we fix the value of α and show which of the four roots of $z^4 = 1$ is found. Due to the square root and its branch cut along the negative real axis, we obtain more complicated fractals than from the Newton method.

```
In[138]:= cfzP4 = Compile[{{α, _Complex}, {z0, _Complex}},
          Module[{fp, (* the roots of z^4 == 1 *)
                  roots = Table[Exp[2 Pi I k/4], {k, 0, 3}]},
          (* use z^4 == 1 as the equation *)
          fp = FixedPoint[Function[z, Chop[z - ((1 + α) (z^4 - 1))/
                          (α (4 z^3) + Sqrt[-(1 + α) (z^4 - 1) 12 z^2 +
                          (4 z^3)^2])]], z0, (* limit iterations *) 100];
          (* which root was computed? *)
          Sort[Transpose[{Abs[fp - roots], {1, 2, 3, 4}}]][[1, 2]]]];

In[139]:= Show[GraphicsArray[
          Block[{$DisplayFunction = Identity, L = 8},
          DensityPlot[cfzP4[#, x + I y], {x, -L, L}, {y, -L, L},
                      PlotPoints -> 400, Mesh -> False,
                      FrameTicks -> None, ColorFunction -> (Hue[0.8 #]&)]& /@
          (* values for α *)
          {-2.72894 + 0.828214 I, 0.103488 + 0.091653 I, 1.3565 + 2.08514 I}]]]
```

Let us consider an application of Compile to graphics generation. Given is a list of lines *lines* in the form $\{line_1, line_2, \ldots, line_n\}$, where each *line$_i$* is of the form $\{x_1, y_1\}, \{x_2, y_2\}$. This means *lines* is a tensor of rank 3. Given a point p, the function distanceCF calculates the sum of the reciprocals of the minimal distances of the point to all lines.

```
In[140]:= distanceCF =
    Compile[{{lines, _Real, 3}, {p, _Real, 1}},
    Plus @@ (Function[line,
        Module[{vec, lineDirection, normal, lineLength},
            vec = p - line[[1]];
            If[lineDirection = #/Sqrt[#.#]&[line[[2]] - line[[1]]];
                lineLength = Sqrt[#.#]&[line[[2]] - line[[1]]];
                0 <= vec.lineDirection <= lineLength,
                (* nearest point is in the interior of the line *)
                normal = vec - vec.lineDirection lineDirection;
                1/Sqrt[normal.normal],
                (* nearest point is one of the endpoints *)
                1/Sqrt[Min[{(p - line[[1]]).(p - line[[1]]),
                        (p - line[[2]]).(p - line[[2]])}]]]] /@ lines)];
```

The function was successfully compiled.

```
In[141]:= FreeQ[distanceCF[[4]], _Function, Infinity]
Out[141]= True
```

Here are two simple examples of distanceCF visualized—a straight line and two perpendicular lines.

```
In[142]:= Show[GraphicsArray[
    Block[{$DisplayFunction = Identity, opts = Sequence[
            PlotPoints -> 50, Compiled -> False, ColorFunction -> Hue]},
    {(* one horizontal line segment *)
    ContourPlot[distanceCF[{{{-1/2, 0}, {1/2, 0}}}, {x, y}],
                {x, -1, 1}, {y, -1, 1}, Evaluate[opts]],
    (* two crossing line segment *)
    ContourPlot[distanceCF[{{{-1/2, 0}, {1/2, 0}},
                        {{0, -1/2}, {0, 1/2}}}, {x, y}],
                {x, -1, 1}, {y, -1, 1}, Evaluate[opts], Contours -> 20]}]]]
```

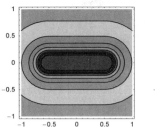
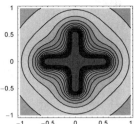

Let us consider a more complicated example, say, the Gosper curve from Chapter 1 of the Graphics volume [1794].

```
In[143]:= line = Partition[N[
        {{0, 0}, {2, 0}, {3, s}, {1, s}, {0, 2 s}, {2, 2 s}, {4, 2 s}, {5, s},
         {7, s}, {8, 2 s}, {9, 3 s}, {7, 3 s}, {6, 2 s}, {5, 3 s}, {6, 4 s},
         {5, 5 s}, {3, 5 s}, {2, 4 s}, {4, 4 s}, {3, 3 s}, {1, 3 s},
         {0, 4 s}, {-2, 4 s}, {-1, 3 s}, {-2, 2 s}, {-3, 3 s}, {-4, 4 s},
         {-3, 5 s}, {-1, 5 s}, {0, 6 s}, {-2, 6 s}, {-3, 7 s}, {-1, 7 s},
         {1, 7 s}, {2, 6 s}, {4, 6 s}, {5, 7 s}, {3, 7 s}, {2, 8 s}, {4, 8 s},
         {6, 8 s}, {7, 7 s}, {6, 6 s}, {7, 5 s}, {8, 4 s}, {9, 5 s}, {8, 6 s},
         {10, 6 s}, {11, 5 s}} /. s -> Sqrt[3]]/2, 2, 1];
```

Because of the more complicated nature of this curve, we use more plot points.

```
In[144]:= Show[GraphicsArray[
        Block[{$DisplayFunction = Identity},
          {(* 3D plot of the distances *)
          Plot3D[distanceCF[line, {x, y}], {x, -3, 6}, {y, -1, 8},
                PlotPoints -> 150, Compiled -> False, Mesh -> False],
          (* contour plot of the distances *)
          (* name it for further use *) gosperCp =
          ContourPlot[distanceCF[line, {x, y}], {x, -3, 6}, {y, -1, 8},
                PlotPoints -> 250, Compiled -> False, Contours -> 20,
                ColorFunction -> (Hue[3 #]&), FrameTicks -> None,
                AspectRatio -> Automatic]}]]]
```

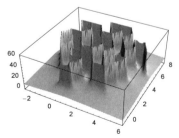

Using a more homogeneous spacing between the contour lines gives an interesting-looking picture.

```
In[145]:= Show[GraphicsArray[
        Block[{$DisplayFunction = Identity},
          (* homogeneous contour lines *)
          cls = (#[[36]]& /@ Partition[Sort[Flatten[gosperCp[[1]]]], 750]);
```

```
(* alternating black-white contour plot color function *)
cf = Function[Evaluate[Which @@ Flatten[
  MapIndexed[Function[{v, p}, {# < v, GrayLevel[Mod[p[[1]], 2]]}],
            Append[cls, True]]]]];
{(* randomly colored contour plot *)
 Show[gosperCp, Contours -> cls,
    ColorFunction -> (Hue[Random[]]&), ContourLines -> False],
 (* alternating black-white contour plot *)
 Show[gosperCp, Contours -> cls,
    ColorFunction -> cf, ContourLines -> False,
    ColorFunctionScaling -> False]}]]]
```

It is also straightforward to calculate the minimal distance of a point to a circle segment. The function dis⸱ tanceSumCF calculates the sum of the roots inverse distances of a point *point* to a set of circles *circlePiece-Data*. The circle segments are specified in the form {*midpoint, angleRange, beginningPoint, endPoint*}.

```
In[146]:= distanceSumCF =
    Compile[{{circlePieceData, _Real, 3}, {point, _Real, 1}},
            Module[{cpd, dir = {0., 0.}, Σ = 0., ω, ρ},
                   Do[cpd = circlePieceData[[k]];
                      dir = point - cpd[[1]];
                      ω = If[dir == {0., 0.}, 0, ArcTan[dir[[1]], dir[[2]]]];
                      ρ = If[cpd[[2, 1]] <= ω <= cpd[[2, 2]],
                             (* shortest distance to inner points *)
                             Abs[Sqrt[dir.dir] - 1.],
                             (* shortest distance to end points *)
                             Sqrt[Min[(point - cpd[[3]]).(point - cpd[[3]]),
                                      (point - cpd[[4]]).(point - cpd[[4]])]]];
                      Σ = Σ + 1/Sqrt[ρ], {k, Length[circlePieceData]}]; Σ]];
```

We use a Truchet pattern from Subsection 1.5.6 from the Graphics volume for the circle segments. Here is a visualization of the resulting sums of distances.

```
In[147]:= Module[{L = 12, contours = 30, pp = 240,
           circleParts, circlePieceData, cp, contourValues, LCP},
    (* quarter circles of a Truchet pattern *)
    SeedRandom[1];
    circleParts =  Table[{{Circle[{i, j} + {-1, +1}, 1, {-1/2, 0} Pi],
                           Circle[{i, j} + {+1, -1}, 1, {1/2, 1} Pi]},
                          {Circle[{i, j} + {+1, -1}, 1, {-1, -1/2} Pi],
                           Circle[{i, j} + {-1, -1}, 1, {0, 1/2} Pi]}}[[
                                     Random[Integer, {1, 2}]]]],
           {i, -L, L, 2}, {j, -L, L, 2}] // Flatten;
    (* {midpoint, angleRange, beginningPoint, endPoint} *)
    circlePieceData = {#1, #3, #1 + #2 {Cos[#3[[1]]], Sin[#3[[1]]]},
```

```
                              #1 + #2 {Cos[#3[[2]]], Sin[#3[[2]]]}}& @@@
                       circleParts // N;
(* contour plot of the sums of inverse roots of distances *)
cp = ContourPlot[distanceSumCF[circlePieceData, {x, y}],
                    {x, -L, L}, {y, -L, L},
                    PlotPoints -> pp, DisplayFunction -> Identity];
(* homogeneously spaced contour values *)
contourValues[cp_, n_] :=
Module[{l = Sort[Flatten[cp[[1]]]], λ = Times @@ Dimensions[cp[[1]]]},
       #[[Round[λ/n]]]& /@ Partition[Sort[Flatten[cp[[1]]]], Round[λ/n]]];
(* list contour plot for 30 homogeneously spaced contour values *)
LCP[cp_] := ListContourPlot[cp[[1]], Contours -> contourValues[cp, 30],
                            ContourLines -> False, FrameTicks -> False,
                            ColorFunction -> (Hue[30 #]&)];
(* contour plot and 3D plot of distance sums *)
Show[GraphicsArray[
  Block[{$DisplayFunction = Identity},
  {LCP[cp], ListPlot3D[Log[cp[[1]]], Mesh -> False,
                       PlotRange -> All, Axes -> False]}]]];
(* two contour plots of averaged distance sums *)
Show[GraphicsArray[
Block[{$DisplayFunction = Identity, d = 3, d = 2},
{LCP[ContourGraphics[
     ListConvolve[Table[((i - (d + 1)))^2  ((j - (d + 1))^2 + 1)^1,
                    {i, 2 d + 1}, {j, 2 d + 1}],
                cp[[1]], {d + 1, d + 1}]]],
 LCP[ContourGraphics[
     ListConvolve[Table[Random[Real, {-1, 1}],
                    {i, 2 d + 1}, {j, 2 d + 1}],
                cp[[1]], {d + 1, d + 1}]]]}]]]]
```

(For the use of different distance functions, see [855].)

As a further application of the use of the *Mathematica* compiler to generate graphics, let us produce an interesting fractal [120], [1603], [138], [1431], [1531], [1092], [554], [1344], [575]: The iteration of $z \to z^z$ leads to periodic attractors. The iteration converges for real z in the interval $e^{-e} < z < e^{1/e}$, for complex values in the region of the complex z-plane, bounded by $z\,e^{-z}$ with $|z| < 1$.

The iteration of the `Power` function can have one fixed point.

```
In[148]:= With[{z = Sqrt[2]}, Take[NestList[z^#&, N[z, 22], 50], -10]]
```

      General::ovfl : Overflow occurred in computation. More…

      General::ovfl : Overflow occurred in computation. More…

      General::ovfl : Overflow occurred in computation. More…

      General::stop :
        Further output of General::ovfl will be suppressed during this calculation. More…

```
Out[148]= {0., 0., 0.×10^1, 0.×10^6, 0.×10^486855, Overflow[],
       0.×10^20201777, Overflow[], 0.×10^20201777, Overflow[]}
```

The iteration can also have period 2.

```
In[149]:= With[{z = 2/100}, Take[NestList[z^#&, N[z, 22], 50], -10]]
Out[149]= {0.88419438478537555550631238, 0.031461560668676259445963366,
       0.88419438399276271889677704, 0.031461560766229738965890551,
       0.88419438365532632868585422, 0.031461560807760852963106470,
       0.88419438351167067177016540, 0.031461560825441754141534250,
       0.88419438345051263316025696, 0.031461560832968984607620882}
```

In the complex plane, we have 2 periods, 3 periods, and so on.

The evaluation of nested powers can take quite a long time because very large and very small numbers can build up.

```
In[150]:= fp[z_, n_, prec_] := NestList[SetPrecision[z^#, prec]&, N[z, prec], n]
```

```
In[151]:= period[z_] :=
     Module[{nl = Take[fp[z, 100, 22], -10], last, ε = 10^-3},
     If[MatchQ[nl[[-1]], Overflow[] | Underflow[] | Indeterminate],
       (* no period found *) 0,
       last = nl[[-1]];
       (* find second occurrence of last element from the end *)
       If[# === {}, 0, #[[1, 1]]]&[
         Position[Rest[Reverse[nl]], _?(Abs[(# - last)/#] < ε&), {1}, 1]]]]
```

```
In[152]:= Table[period[x + I y], {x, -2, 2}, {y, -2, 2}] // Timing
```

      Power::indet : Indeterminate expression 0^0 encountered. More…

      General::ovfl : Overflow occurred in computation. More…

```
Out[152]= {0.3 Second,
       {{3, 3, 8, 3, 3}, {3, 0, 1, 0, 3}, {3, 1, 0, 1, 3}, {0, 1, 1, 1, 0}, {9, 0, 0, 0, 9}}}
```

Using `Compile`, speeds up the calculation considerably.

```
In[153]:= cf = With[{prec = $MachinePrecision - 1},
             Compile[{{z, _Complex}}, (* do the iterated powering *)
                 NestList[Which[# === 0. + 0.I, 0. + 0.I,
```

```
                              Re[#] > 100., 0. + 0.I,
                              True, N[z^#, prec]]&, z, 200]]];
```

```
In[154]:= periodcf[z_] :=
     Module[{nl = Take[cf[z], -10], last, ε = 10^-3},
     If[nl[[-1]] === True, 0, last = nl[[-1]];
        If[# === {}, 0, #[[1, 1]]]&[
          Position[Rest[Reverse[nl]], _?(Abs[(# - last)/#] < ε&), {1}, 1]]]]
```

```
In[155]:= (* turn off various messages *)
     Off[Power::infy]; Off[Infinity::indet]; Off[General::ovfl];
     Off[General::unfl]; Off[CompiledFunction::cfn]
```

```
In[158]:= Table[periodcf[x + I y], {x, -2, 2}, {y, -2, 2}] // Timing
```

```
Out[158]= {0.05 Second,
     {{3, 3, 8, 3, 3}, {3, 0, 1, 0, 3}, {3, 1, 0, 1, 3}, {0, 1, 1, 1, 0}, {9, 0, 0, 0, 9}}}
```

Here, the iterations in a part of the complex plane are shown. The color determines the length of the period.

```
In[159]:= tabcf = With[{pp = 200},
     Table[periodcf[x + I y], {x, -4., 4., 8./pp}, {y, 0., 8., 8./pp}]];
```

```
In[160]:= ListDensityPlot[Join[Reverse[#], #]&[Transpose[tabcf]],
              Mesh -> False, ColorFunction -> (Hue[0.8 #]&),
              MeshRange -> {{-4, 4}, {-8, 8}},
              AspectRatio -> Automatic, PlotRange -> All]
```

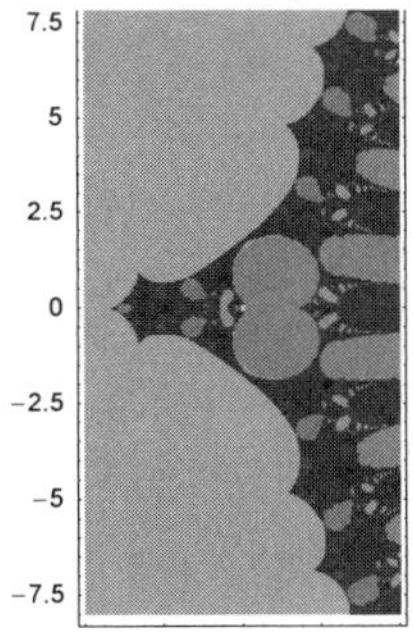

The next graphic shows the lines of vanishing real and imaginary parts of the map $z - z^{z^z}$ over of the complex z-plane. This yields a more detailed picture of some of the structures of the last graphic.

```
In[161]:= With[{o = 2, pp = 201, wp = 40},
     Module[{g, data, sp = SetPrecision},
        (* iterate power function *)
        g[n_, z_] := With[{ζ = sp[z, wp]}, Sign @ (ζ - Nest[sp[ζ^#, wp]&, ζ, n])];
        (* use mirror symmetry with respect real axis *)
        data = Join[Reverse[Rest[#]], #]&[
           Table[g[o, x + I y], {y, 0, 0.8, 0.8/pp}, {x, -1, 4, 5/(2 pp)}]] /.
                                         Indeterminate -> 0;
     (* display vanishing real and imaginary parts *)
     Show[Block[{$DisplayFunction = Identity},
        ListContourPlot[#1[data], MeshRange -> {{-1, 4}, 0.8 {-1, 1}},
                    Contours -> {0}, ContourShading -> False,
                    ContourStyle -> {{Thickness[0.002], #2}}]& @@@
           (* vanishing real part as red lines;
              vanishing imaginary as blue lines *)
```

```
        {{Re, Hue[0]}, {Im, Hue[0.76]}}],
        PlotRange -> All, AspectRatio -> Automatic]]]
```

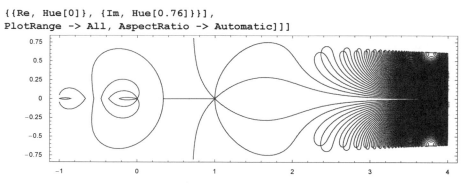

A map closely related to $z \to z^z$ is the Ikeda map $z \to a + b\,z\,\exp(i\,|z|^2)$ [870], [776], [1086], [495]. Similar to the last map, it becomes periodic. The following graphic shows the period in the a, b-plane.

```
In[162]:= Module[{ppx = 600, ppy = 400, pMax = 16, cf},
          (* compiled function for period calculation *)
          cf = Compile[{a, b},
            Module[{z, ζ, c, ε = 10^-10.},
                   (* ignore initial iterations *)
                   z = Nest[a + b # Exp[I Abs[#]^2]&, 0. + 0. I, 1000];
                   (* now watch for period *) ζ = z; c = 1;
                   While[ζ = a + b ζ Exp[I Abs[ζ]^2];
                         Abs[z/ζ - 1.] > ε && c < pMax, c++]; c]];
          (* visualize results *)
          ListDensityPlot[Table[cf[a, b], {b, 1/128, 1/4, 31/128/ppx},
                          {a, 3/2, 6, 9/2/ppy}],
                    AspectRatio -> 1/3, PlotRange -> All,
                    Mesh -> False, ColorFunctionScaling -> False,
                    FrameTicks -> None,
                    (* color according to period *) ColorFunction ->
                    (Which[# == pMax, RGBColor[0, 0, 0],
                           Mod[#, 3] == 2, RGBColor[1, 0, 0],
                           Mod[#, 3] == 1, RGBColor[0, 1, 0],
                           Mod[#, 3] == 0, RGBColor[0, 0, 1]]&)]]
```

A very natural application of compilation is for iterated maps, as in the logistic equation. Here, we will deal with a slightly more complicated example of a piecewise smooth map that does not have periodic windows. This map was introduced in [128].

```
In[163]:= update[ivn_List, R_] :=
          Module[{C = 220 10^-6, Iref = 0.5, Vin = 30, T = 400 10^-6,
```

```
        L = 0.1, in = ivn[[1]], vn = ivn[[2]], Ton},
        Ton = L(Iref - in)/Vin;
    If[Ton < T,
       {Iref + 1/L (Vin - vn + vn Ton/(C R))(T - Ton),
        vn - vn Ton/(C R) + (Iref/C - vn/(C R) +
        vn Ton/(C^2  R^2))(T - Ton)},
       {in + Vin/L T, vn - vn T/(C R)}]];
```

Here is the map.

```
In[164]:= Show[GraphicsArray[
      Plot3D[update[{in, vn}, 300][[#]], {in, 0, 1}, {vn, 0, 100},
              PlotPoints -> 60,
              DisplayFunction -> Identity]& /@ {1, 2}]]
```

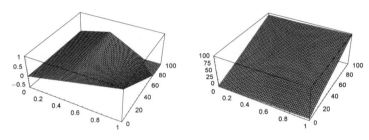

A compiled version is about 30 times faster.

```
In[165]:= updateCompiled = Compile[{{ivn, _Real, 1}, {R, _Real, 0}},
      Module[{C = 220 10^-6, Iref = 0.5, Vin = 30, T = 400 10^-6,
              L = 0.1, in = ivn[[1]], vn = ivn[[2]], Ton},
              Ton = L(Iref - in)/Vin;
       If[Ton < T,
          {Iref + 1/L (Vin - vn + vn Ton/(C R))(T - Ton),
           vn - vn Ton/(C R) + (Iref/C - vn/(C R) +
           vn Ton/(C^2  R^2))(T - Ton)},
           {in + Vin/L T, vn - vn T/(C R)}]]];
```

```
In[166]:= Timing[Nest[update[#, 250]&, {1., 1.}, 1000]]
```

```
Out[166]= {0.14 Second, {0.438727, 61.4073}}
```

```
In[167]:= Timing[Nest[updateCompiled[#, 250]&, {1., 1.}, 1000]]
```

```
Out[167]= {0.01 Second, {0.438954, 61.4064}}
```

This speed increase makes it possible to generate the corresponding bifurcation diagram(s) quickly.

```
In[168]:= Show[Graphics[{PointSize[0.003],
      Table[Point[{R, #[[1]]}]]& /@
      Drop[NestList[updateCompiled[#, R]&,
                  {Random[], Random[]}, 2000], 1000],
                  {R, 200, 500}]}], PlotRange -> All, Frame -> True]
```

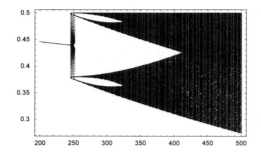

Here is a parametrized period-doubling map in two variables [993], [652], [1512], [1708].

```
In[169]:= lm3Dcf = Compile[{a, α, {n, _Integer}, {p, _Integer}},
            Module[{L = Table[{0., 0.}, {n + p}], xy,
                M = 10.^12, m = 10.^-12, k = 0, x = 0.2, y = 0.3},
                (* iterate maps until values diverge or converge *)
                While[(* two coupled maps *)
                    {x, y} = {a - y (α x + (1. - α) y), x + y^2/100.};
                    xy = Abs[{x, y}]; (k++) < n + p &&
                    Max[xy] < M && Min[xy] > m,
                    L[[k]] = {x, y}]; Union[Take[L, -n]]]];
```

Displaying the two variables as a function of the coupling parameter gives a 3D graphic.

```
In[170]:= periodDoubling3DGraphics[α_, n_:20, p_:500, ppa_:1000] :=
        Graphics3D[{PointSize[0.002], (* coupling parameter α runs upward *)
        Table[{Hue[0.8 a/2], Point[{##, a}]& @@@ lm3Dcf[a, α, n, p]},
                {a, 1/2, 2, 3/2/ppa}]}, Axes -> False,
                BoxRatios -> {1, 1, 2}, ViewPoint -> {1, -2, 0},
                PlotRange -> {{-2, 2}, {-2, 2}, {1/2, 2}}]
```

As the parameter α changes, the shape of the 3D bifurcation diagram changes.

```
In[171]:= Show[GraphicsArray[periodDoubling3DGraphics /@ {0, 1/4, 1/3, 1/2}]]
```

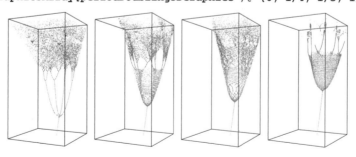

```
Do[Show[periodDoubling3DGraphics[α]], {α, 0, 3/4, 3/4/60}]
```

As a larger application of `Compile`, let us model a nice example of a cellular automaton. It is the model of a sandpile [440], [441], [1254], [1284].

Physical Remark: Sandpile Model

The following sandpile model intends to describe the behavior of a sandpile when some grains of sand are added to it. Either the grains will stick on the pile or cause an avalanche. In the cellular automata, we work on a square grid of dimension *dimx* × *dimy*. The value at each lattice point is a nonnegative integer. When the value is greater than 4 (grains of sand), the sand will flow to its four neighboring sides, one grain flows to each neighbor. Sand that flows out of the grid is lost.

The following routine `update` implements the updating of a given sand pile *matrix*. All lattice points are updated simultaneously.

```
In[172]:= update (* makes one step in the flow of a sandpile *)=
    Compile[{{matrix, _Integer, 2}},
        Module[{dimx = 0, dimy = 0, a, m1, m2, m3, m4},
                {dimx, dimy} = Dimensions[matrix];
                (* the large elements *)
                a = Map[If[# >= 4, 1, 0]&, matrix, {-1}];
                (* the neighbor matrices *)
                m1 = RotateRight[a];
                Do[m1[[1, j]] = 0, {j, dimy}];
                m2 = RotateLeft[a];
                Do[m2[[dimx, j]] = 0, {j, dimy}];
                m3 = Transpose[RotateRight[Transpose[a]]];
                Do[m3[[i, 1]] = 0, {i, dimx}];
                m4 = Transpose[RotateLeft[Transpose[a]]];
                Do[m4[[i, dimy]] = 0, {i, dimx}];
                (* form new matrices *)
                matrix - 4a + m1 + m2 + m3 + m4]];
```

The function `update` was successfully compiled. No `Function` is present in the resulting `CompiledFunction` object.

```
In[173]:= MatchQ[update[[4]], {{__Integer}..}]
Out[173]= False
```

Let us look at an example. The number of grains at the 18 × 18 lattice points is randomly selected. We repeat the application of `update` until a stationary state is reached.

```
In[174]:= SeedRandom[33729];
    sandPileList = FixedPointList[update, initialSandPile =
                    Table[Random[Integer, {0, 12}], {18}, {18}]];
```

In this case, 200 applications of update were needed.

```
In[176]:= Length[sandPileList]
Out[176]= 200
```

The next graphics arrays show the first ten and the last ten sandpile configurations.

```
In[177]:= makePicture[m_, opts___] :=
        ListDensityPlot[m, opts, Mesh -> False, FrameTicks -> None,
                    ColorFunction -> (Hue[0.8 #]&),
                    PlotRange -> All, ColorFunctionScaling -> False]
```

```
In[178]:= Show[GraphicsArray[#]] & /@
           Map[makePicture[#, DisplayFunction -> Identity]&,
               Partition[sandPileList, 10][[{1, 2, -1, -2}]], {2}]
```

The following animation shows all steps.

```
        makePicture /@ sandPileList;
```

The next pictures show a sandpile arising from a 31×31 grid with a large amount of sand grains at the center grid point.

```
In[179]:= fp3131Data = FixedPointList[update, Table[If[i == j == 16, #, 0],
                                       {i, 31}, {j, 31}]]& /@
                                       {1200, 12000, 24000};
In[180]:= (* number if iterations needed *)
          -1 + Length /@ fp3131Data
Out[181]= {752, 8438, 17005}

In[182]:= Show[GraphicsArray[
          ListDensityPlot[#[[-1]], Mesh -> False, FrameTicks -> None,
                          ColorFunction -> (Hue[0.8 #]&),
                          DisplayFunction -> Identity]& /@ fp3131Data]];
```

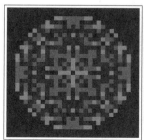  

Here is the time-development of a sandpile starting from a symmetric initial state.

```
In[183]:= Module[{d = 120, o = 12, i, j},
          (* initial state *)
```

```
M = Table[If[i = Abs[Min[Abs[i] - d/3]]; j = Abs[Min[Abs[j] - d/3]];
            i < o || j < o, Min[i, j], 0], {i, -d, d}, {j, -d, d}];
(* show time development *)
Show[GraphicsArray[ListDensityPlot[#, Mesh -> False, FrameTicks -> False,
                                   DisplayFunction -> Identity]& /@
            FoldList[(* evolve *) Nest[update, #1, #2]&, M,
            (* time steps *) {10, 100, 1000, 5000}]]]]
```

The above routine `update` can be used to calculate a particularly interesting sandpile pattern. If this sandpile is added to *any* given other sandpile, it does not change the final equilibrium state of the sandpile [440], [432], [242]. This special sandpile is also called an identity sandpile. To construct this special state, one starts with a matrix that has 1's at all four edges and 4's at the four corners. Then, we evolve this state using the repeated application of update until it reached its final position. This matrix is multiplied by 2, and again, let this state evolve using the repeated application of update until it reaches its final position. The result is again multiplied by 2 and so on until the procedure ends in a fixed point. The procedure `initialIdentitySandPile` generates the matrix with the 1's and 2's located at the edges and corners.

```
In[184]:= initialIdentitySandPile[{dx_, dy_}] :=
        Table[Which[(* the four corners *)
                ix == 1 && iy == 1 || ix == dx && iy == 1 ||
                ix == 1 && iy == dy || ix == dx && iy == dy, 2,
                (* the four edges *)
                ix == 1 || ix == dx || iy == 1 || iy == dy, 1,
                True, 0], {ix, dx}, {iy, dy}]
```

Here, an example of the starting matrix is shown.

```
In[185]:= TableForm[initialIdentitySandPile[{12, 12}], TableSpacing -> {0, 0.4}]
Out[185]//TableForm=
        2 1 1 1 1 1 1 1 1 1 1 2
        1 0 0 0 0 0 0 0 0 0 0 1
        1 0 0 0 0 0 0 0 0 0 0 1
        1 0 0 0 0 0 0 0 0 0 0 1
        1 0 0 0 0 0 0 0 0 0 0 1
        1 0 0 0 0 0 0 0 0 0 0 1
        1 0 0 0 0 0 0 0 0 0 0 1
        1 0 0 0 0 0 0 0 0 0 0 1
        1 0 0 0 0 0 0 0 0 0 0 1
        1 0 0 0 0 0 0 0 0 0 0 1
        1 0 0 0 0 0 0 0 0 0 0 1
        2 1 1 1 1 1 1 1 1 1 1 2
```

The identity state is reached by carrying out the relaxation implemented through `update` and the ongoing doubling of the number of grains until we reach the identity state. The function `identityState` implements this.

```
In[186]:= identityState =
        Compile[{{matrix, _Integer, 2}},
                FixedPoint[2 FixedPoint[update, #]&, matrix],
                {{update[_], _Integer, 2}}];
```

Here, the identity state for an 18×18 grid is calculated.

```
In[187]:= (identityState18 = identityState[initialIdentitySandPile[{18, 18}]]) //
            (TableForm[#, TableSpacing -> 0.4]&)
```
Out[187]//TableForm=
```
0 4 2 6 4 6 0 6 4 4 6 0 6 4 6 2 4 0
4 4 6 6 4 6 4 6 2 2 6 4 6 4 6 6 4 4
2 6 4 4 2 6 6 2 6 6 2 6 6 2 4 4 6 2
6 6 4 4 4 0 4 6 6 6 6 4 0 4 4 4 6 6
4 4 2 4 4 6 6 6 6 6 6 6 6 4 4 2 4 4
6 6 6 0 6 4 4 4 4 4 4 4 6 0 6 6 6
0 4 6 4 6 4 4 4 4 4 4 4 6 4 6 4 0
6 6 2 6 6 4 4 4 4 4 4 4 6 6 2 6 6
4 2 6 6 6 4 4 4 4 4 4 4 6 6 6 2 4
4 2 6 6 6 4 4 4 4 4 4 4 6 6 6 2 4
6 6 2 6 6 4 4 4 4 4 4 4 6 6 2 6 6
0 4 6 4 6 4 4 4 4 4 4 4 6 4 6 4 0
6 6 6 0 6 4 4 4 4 4 4 4 6 0 6 6 6
4 4 2 4 4 6 6 6 6 6 6 6 6 4 4 2 4 4
6 6 4 4 4 0 4 6 6 6 6 4 0 4 4 4 6 6
2 6 4 4 2 6 6 2 6 6 2 6 6 2 4 4 6 2
4 4 6 6 4 6 4 6 2 2 6 4 6 4 6 6 4 4
0 4 2 6 4 6 0 6 4 4 6 0 6 4 6 2 4 0
```

Adding this sandpile to the sandpile from above does not change the final state.

```
In[188]:= FixedPoint[update, initialSandPile + identityState18] == sandPileList[[-1]]
```
Out[188]= True

Now, let us calculate the identity state for a larger, 100×100, grid. After every relaxation step, because of the repeated application of `update`, we display the resulting sandpile. We also count the number of applications of `update` by incrementing the value of the counter `counter`.

```
In[189]:= picturedIdentityStateEvolution =
      Compile[{{matrix, _Integer, 2}},
              Module[{counter = 0},
              FixedPoint[(AppendTo[sandPilePictureList,
                                   makePicture[#, DisplayFunction -> Identity]];
                    (* iterate *)
                    2 FixedPoint[(counter++; update[#])&, #])&, matrix];
                  (* count *) counter],
              {{update[_], _Integer, 2}}];
```

```
In[190]:= sandPilePictureList = {};
      With[{dimx = 100, dimy = 100},
          picturedIdentityStateEvolution[initialIdentitySandPile[{dimx, dimy}]]]
```
Out[191]= 6804

```
In[192]:= Show[GraphicsArray[#]]& /@ Partition[Rest[sandPilePictureList], 4]
```

The last calculation needed 6804 calls to the routine update. We end with the calculation on a 200×200 grid. This can be done in a fraction of an hour on a year-2005 computer.

```
In[193]:= ListDensityPlot[identityState[initialIdentitySandPile[{200, 200}]],
            PlotRange -> All, Mesh -> False, FrameTicks -> None,
            ColorFunction -> (* better contrast for printed version *)
            If[Options[Graphics3D, ColorOutput][[1, 2]] === GrayLevel,
              GrayLevel, (RGBColor[#, 0, 1 - #]&)]]
```

Here is a continuous version of a sandpile model [1374], [1273], [728], [264], [1159], [519]. Again, we start with a square matrix of values, but this time we allow continuous values. If a lattice point has a value greater than 1, a fraction of this value is redistributed to the four neighbors. This process is iterated until all values are less than 1.

```
In[194]:= makeCriticalAndRelax = Compile[{{M, _Real, 2}, α},
      FixedPoint[(* iterate until no high sand pile left *)
      Function[m,
       Module[{M = m, n = Length[m], ξ, αξ, ip, im, jp, jm},
         If[(* redistribution needed? *) Max[m] >= 1.,
          Do[If[m[[i, j]] >= 1.,
              ξ = m[[i, j]]; αξ = α ξ;
              (* subtract distributed sand *) M[[i, j]] = M[[i, j]] - ξ;
              {ip, im, jp, jm} = Mod[{i + 1, i - 1, j + 1, j - 1}, n, 1];
              (* redistribute *)
              M[[ip, j]] = M[[ip, j]] + αξ; M[[im, j]] = M[[im, j]] + αξ;
```

```
              M[[i, jp]] = M[[i, jp]] + αξ; M[[i, jm]] = M[[i, jm]] + αξ],
      {i, n}, {j, n}]]; M]], (* make critical *) M + (1. - Max[M])]];
```

Here are two final states of this sandpile model. The first starts from random initial values, and the second one from pseudorandom initial values.

```
ln[195]:= Module[{n = 400,  α = 0.1},
        Show[GraphicsArray[
        ListDensityPlot[makeCriticalAndRelax[#, α],
                    Mesh -> False, DisplayFunction -> Identity,
                    FrameTicks -> False, PlotRange -> All]& /@
        {(* random initial sand distribution *)
        SeedRandom[222]; Table[Random[Real, {0.4, 0.6}], {n}, {n}],
        (* pseudorandom initial sand distribution *)
        Table[If[# < 1, #, 1/#]&[Abs[Tan[i  j]]], {i, n}, {j, n}]
        (*   (* quarter ring *)
        Table[If[1/4 n < Sqrt[i^2 + j^2] < 3/4 n, 1., 0], {i, n}, {j, n}]
        *)}]]]
```

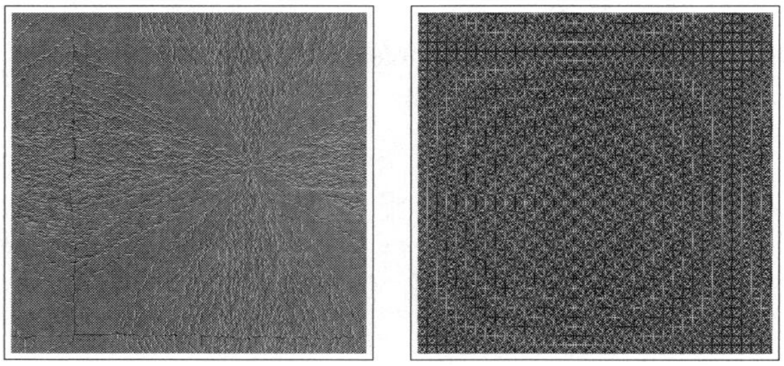

For modeling real sandpiles, see [1483].

The above graphics showed crystal-like structures with a high degree of symmetry. If we not only take nearest neighbors into account in the updating steps, but also allow for more nonlocal interactions, such cellular models can show a very rich, "biology-like" set of "shapes". In the following, we will use a $d \times d$ with $d = 1024$ square grid with periodic boundary conditions. We will update the cells $c_{i,j}^{(n)}$ (the upper index n indicates the time/generation) according to [1397], [1648]

$$c_{i,j}^{(n+1)} = (1 - \varepsilon_1 - \varepsilon_2)\, f\!\left(c_{i,j}^{(n)}\right) + \frac{\varepsilon_2}{d^2} \sum_{i,j=1}^{d} f\!\left(c_{i,j}^{(n)}\right) + \varepsilon_1\!\left(\frac{1}{6}\left(f\!\left(c_{i+1,j}^{(n)}\right) + f\!\left(c_{i-1,j}^{(n)}\right) + f\!\left(c_{i,j+1}^{(n)}\right) + f\!\left(c_{i,j+1}^{(n)}\right)\right) + \right.$$

$$\left. \frac{1}{12}\left(f\!\left(c_{i+1,j+1}^{(n)}\right) + f\!\left(c_{i+1,j-1}^{(n)}\right) + f\!\left(c_{i-1,j+1}^{(n)}\right) + f\!\left(c_{i-1,j-1}^{(n)}\right)\right)\right).$$

(For similar nonlocally interacting cellular automata, see [1327], [229]; for time-delayed versions, see [906]; for the richness compared with PDEs, see [686]; for added noise, see [601]; for coupled maps in general, see [937].) The first term is a local, nonlinear one, the second term is an averaging, nonlocal interaction, and the third term describes a nonlinear nearest neighbor interaction. For the function f, we use $f(x) = 1 - a\,x^2$. To implement the nearest neighbor sums efficiently, we use the function `ListConvolve`.

We show some of the resulting cell values for selected n between 1 and 2500. The initial values $c_{i,j}^{(0)}$ are zero with the exception of a set of 10000 randomly selected points lying approximatively on a piece of a spiral-shaped curve. (Because the following calculation will take a few hours, it is not evaluated by default.) The majority of the code below deals with the generation, grouping, and display of the graphics; the actual model is quite short to implement—the function `stepCF` does all the relevant work, and it does this quite efficiently.)

```
stepCF = Compile[{{L, _Real, 2}, ε1, ε2, a},
Module[{l = 1. - a L^2, n = Length[L]},
(* nonlinear local term *)
(1 - ε1 - ε2) l +
(* nearest neighbor interactions *)
ε1 (ListConvolve[N @ {{0, 1, 0}, {1., 0, 1}, {0, 1, 0}}, l, {2, 2}]/6 +
    ListConvolve[N @ {{1, 0, 1}, {0, 0, 0}, {1, 0, 1}}, l, {2, 2}]/12) +
(* global average *)
ε2/n^2 Sum[l[[i, j]], {i, n}, {j, n}] ]];

(* parameters used; similar parameters give qualitatively similar results *)
n = 1024;
SeedRandom[123456789];
(* the starting spiral *)
L0 = ReplacePart[Table[0., {k, n}, {l, n}], 1.,
        (Round[Table[N[{n, n}/2 + n/2 #/(5 Pi)*
        {Cos[#], Sin[#]}&[5/2 Pi Random[]]], {10000}]] //
                                    Union) /. 0 -> 1];

(* selected frames to be displayed;
   quantitative results are machine dependent due to error amplification *)
frames = {1, 157, 248, 341, 448, 636, 919, 1450, 1564, 1694, 2246, 2495};

(* partition frames for 1×framesPerRow graphics arrays *)
framesPerRow = 3;
framesBag = Partition[frames, framesPerRow];
framesBagCounter = 1; framesCounter = 0; pictureBag = {};
frameRow = framesBag[[framesBagCounter]];

Do[(* update L0 *)
    L0 = stepCF[L0, 0.2, 0.3, 1.8];
    (* should graphics be made? *)
    If[MemberQ[frameRow, k],
        (* show number 341 magnified *)
        If[k == 341,
           ListDensityPlot[Take[L0, {320, 800}, {270, 850}],
                Mesh -> False, Frame -> None, PlotRange -> All,
                AspectRatio -> Automatic]];
        framesCounter = framesCounter + 1;
        AppendTo[pictureBag, (* make graphics *)
           ListDensityPlot[L0, Mesh -> False, Frame -> None,
                            PlotRange -> All, Axes -> False,
                            DisplayFunction -> Identity]];
        If[framesCounter === framesPerRow,
            (* display set of four frames *)
            Show[GraphicsArray[pictureBag], GraphicsSpacing -> 0.01];
            (* prepare for next framesPerRow pictures *)
            pictureBag = {};
            framesBagCounter = framesBagCounter + 1;
            If[framesBagCounter <= Length[framesBag],
                framesCounter = 0;
                frameRow = framesBag[[framesBagCounter]]]]],
    {k, (* number of evolution steps to follow *) 2500}]
```

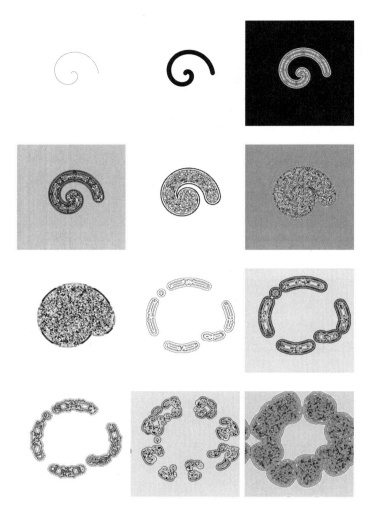

We end the compilation discussion with a quite different application of Compile that also produces some nice graphics. We consider the intensity of rays refracted by a random smooth surface [883]. Given a function *function* over the *x,y*-plane and its normal vector *normal*, the index of refraction *n*, and the height Z of the plane where the refracted rays are measured, makePathTraceCF generates a compiled function that is then used to follow pp^2 rays on a grid of size $pp \times pp$.

```
In[196]:= (* function to generate the compiled function *)
    makePathTraceCF[{function_, normal_, n_, Z_, k_:0}] :=
    Compile[{{pp, _Integer}, {pp, _Integer}},
    Module[{ez = {0., 0., 1.}, M = Table[0. + 0. I, {pp + 1}, {pp + 1}],
           p, q, λ, n, α, β, d, ℓ, xy, i, j, L = 1.5},
      (* loop over the rays *)
    Do[p = function; n = normal; n = n/Sqrt[n.n];
       (* refract ray *) α = ArcCos[n.ez]; β = ArcSin[Sin[α]/n];
       d = n - n.ez ez; d = d/Sqrt[d.d]; ℓ = -Cos[β] n + Sin[β] d;
       (* calculate intersection of ray with plane z == Z *)
       xy = {p[[1]], p[[2]]} + (Z - p[[3]])/ℓ[[3]] {ℓ[[1]], ℓ[[2]]};
       (* count binned ray intersection with cell in plane z == Z *)
       {i, j} = Round[pp (xy + L)/(2L)];
       If[0 <= i <= pp && 0 <= j <= pp,
         q = p - {xy[[1]], xy[[2]], Z}; λ = Sqrt[q.q];
         M[[i + 1, j + 1]] = M[[i + 1, j + 1]] + Exp[1. I k λ]],
      {x, -1, 1, 2/pp}, {y, -1, 1, 2/pp}]; M]]
```

The function makeRandomData generates random smooth functions and their normals, as well as *n* and *Z*.

```
In[198]:= (* parametrization and normal vector of a random smooth function *)
    makeRandomData[factorType_, seed_] :=
    Module[{f, normal, o, dMax, Z, n},
    (* seed random number generator *) SeedRandom[seed];
    o = Random[Integer, {24, 60}]; dMax = Random[Integer, {6, 24}];
    f[x_, y_] = {x, y, Sum[Random[Real, {-1, 1}]*
                      Sin[Random[factorType, dMax {-1, 1}] x +
                          Random[factorType, dMax {-1, 1}] y +
                          2Pi Random[]], {k, o}]};
    (* the normal *) normal[x_, y_] = Cross[D[f[x, y], x], D[f[x, y], y]];
    Z = Random[Real, {-4, 4}]; n = Random[Real, {1, 3}];
```

```
(* list of random function, normal, Z, and n *)
{f[x, y], normal[x, y], n, Z}]
```

Here are four examples. While already *pp* = 4000, and pp=400 give quite good resolutions in about half an hour; higher values of *pp* give, of course, better results. We clearly see cusps and folds as expected from catastrophe theory in 2D [192], [1474], [70], [356], [1763], [1816], [1469].

```
In[200]:= Module[{pp = 400, pp = 4000, rayData},
    Show[GraphicsArray[
    (Function[{type, seed},
      (* make graphics array of two graphics *)
      rayData = makePathTraceCF[makeRandomData[Real, seed]][pp, pp];
      ListDensityPlot[Log[rayData + 1.], DisplayFunction -> Identity,
                PlotRange -> All, FrameTicks -> False, Mesh -> False,
                ColorFunction -> (GrayLevel[1 - #]&)]] @@@ #)]& /@
                (* coefficient types and seeds *)
                {{{Real, 639508823225}, {Real, 827010877375}},
                 {{Integer, 148496114544}, {Integer, 824085936152}}}]]
```

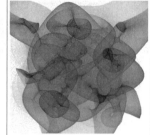

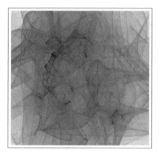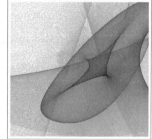

By changing the intensity along each ray as exp($i\,k\,l$) for a fixed k, we obtain pictures like the following. (This time we use a 600×600 discretization and ten billion rays.)

```
Module[{pp = 600, pp = 10000, rayData},
  Show[GraphicsArray[(Function[{type, seed},
    (* make graphics array of two graphics *)
  rayData = makePathTraceCF[Append[makeRandomData[type, seed],
                                    (* random value for k *)
                                    10^Random[Real, {1, 2}]]]][pp, pp];
    ListDensityPlot[Log[Abs[rayData] + 1.], DisplayFunction -> Identity,
              PlotRange -> All, FrameTicks -> False, Mesh -> False,
              ColorFunction -> (GrayLevel[1 - #]&)]] @@@ #]]& /@
              (* coefficient types and seeds *)
          {{{Real, 6216585876}, {Real, 8225265753}},
           {{Real, 6539889434}, {Real, 6087423247}}}]]
```

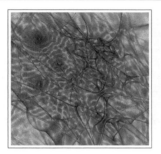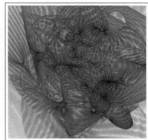

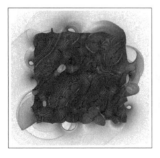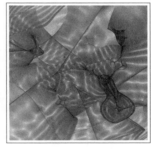

So far, we have been dealing with compiling, meaning the use of fast machine arithmetic (fast compared to high-precision arithmetic) to carry out numerical calculations. An additional possibility to speed up calculations (that is also useful for high-precision and sometimes even symbolic calculations), is optimization, meaning to write an expression in a form that is more efficient for numerical calculations (mainly by avoiding recalculating identical subexpressions). A quick search through the available functions shows that in the `Experimental` context we find a function that will optimize an expression.

In[201]:= **Names["*`*Optimiz*"]**

Out[201]= {CompileOptimizations, OptimizationLevel,
      Experimental`OptimizationSymbol, Experimental`OptimizedExpression,
      Experimental`OptimizeExpression, Reduce`UnivariateOptimize}

The function `Experimental`OptimizeExpression` is the actual optimizer.

In[202]:= **??Experimental`OptimizeExpression**

```
Experimental`OptimizeExpression

Attributes[Experimental`OptimizeExpression] = {Protected}
Options[Experimental`OptimizeExpression] = {ExcludedForms → {},
   OptimizationLevel → 1, Experimental`OptimizationSymbol → $,
   Experimental`StructuralForms → {_List}}
```

Here is an example of a larger expression built from nested square roots and nested fractions.

```
In[203]:= nestedSquareRoots = Nest[Sqrt[(# + Sqrt[#])/(# - Sqrt[#])]&, 2, 5];
         Short[nestedSquareRoots, 4]
Out[204]//Short=
```

Its optimized form has the head `Experimental`OptimizedExpression` and contains a `Block` inside. The named common subexpressions are clearly visible.

```
In[205]:= optimizedNestedSquareRoots =
             Experimental`OptimizeExpression[nestedSquareRoots];
         Short[optimizedNestedSquareRoots, 10]
Out[206]//Short=
         Experimental`OptimizedExpression[
           Block[{$$139778, $$139779, $$139780, $$139781, $$139782, $$139783, $$139784,
             $$139786, $$139785, $$139787, $$139788, $$139789, $$139790, $$139791, $$139793,
             $$139792, $$139794, $$139795, $$139796, $$139797, $$139798, $$139800,
             $$139799, $$139801, $$139802, $$139803, $$139804, $$139805, $$139807}, ≪1≫]]
```

The optimized form is frequently much shorter.

```
In[207]:= {ByteCount[#], LeafCount[#]}& /@
                 {nestedSquareRoots, optimizedNestedSquareRoots}
Out[207]= {{163744, 6823}, {3232, 211}}
```

Evaluating the inner `Block` gives the original expression back.

```
In[208]:= optimizedNestedSquareRoots[[1]] === nestedSquareRoots
Out[208]= True
```

The numerical evaluation of the optimized version is much faster than the one of the original version. (To obtain more reliable timings, we repeat the calculation 100 times and to have a convenient way to obtain a high-precision result with N, we define a function oNSR.)

```
In[209]:= (* make a numeric function to allow automatic precision control *)
         SetAttributes[oNSR, NumericFunction];

         (* make function definition based on optimized version *)
         Function[x, SetDelayed @@ Hold[oNSR[prec_?InexactNumberQ], x],
             {HoldAll}] @@ (optimizedNestedSquareRoots /.
                     Sqrt[2] :> N[Sqrt[2], Precision[prec]]);
In[213]:= (* clear the cache between calculations for correct timings *)
         {Timing[Union @ Table[Developer`ClearCache[];
```

```
                          N[nestedSquareRoots, 22], {100}]],
        Timing[Union @ Table[Developer`ClearCache[];
                          N[oNSR[1], 22], {100}]]]}
Out[214]= {{0.28 Second, {2.262876358768136694642}},
          {0.06 Second, {2.262876358768136694642}}}
```

The function `Compile` knows about the special head `Experimental`OptimizedExpression` and strips it automatically. Here this is demonstrated using `optimizedNestedSquareRoots`.

```
In[215]:= cf = Compile[{}, Evaluate[optimizedNestedSquareRoots]];
        Short[cf, 6]
Out[216]//Short=
        CompiledFunction[{},
          Block[{$$139778, $$139779, $$139780, $$139781, $$139782, $$139783,
            $$139784, $$139786, $$139785, $$139787, $$139788, $$139789,
            $$139790, $$139791, $$139793, $$139792, $$139794, $$139795,
            $$139796, $$139797, $$139798, $$139800, $$139799, $$139801, $$139802,
            $$139803, $$139804, $$139805, $$139807}, ≪1≫], -CompiledCode-]
```

```
In[217]:= cf[]
Out[217]= 2.26288
```

Many numeric functions compile their input for a faster evaluation. The option that controls the compilation is `Compiled`. The possible settings for this option are `True`, `False` and `Automatic`. Here are all the system functions that have the `Compiled` option.

```
In[218]:= Select[DeleteCases[Names["*"], "$Failed" | "$DefaultFont"],
            MemberQ[Options @@ ToExpression[#, StandardForm, Hold],
                Compiled, Infinity]&]
Out[218]= {ContourPlot, DensityPlot, FindFit, FindMaximum, FindMinimum, FindRoot, NDSolve,
          NIntegrate, NProduct, NSum, ParametricPlot, ParametricPlot3D, Play, Plot, Plot3D}
```

But some of these functions allow also for a working precision that will force the use of high-precision arithmetic. To influence the optimization and the compilation independently, one can use a length two list for this options setting:

`Compiled -> {True | False | Automatic, CompileOptimizations -> All | None}`.

1.4 Linear Algebra

The solution of systems of linear equations and the computation of eigenvalues have already been discussed in our treatment of vectors and matrices in Chapter 1 of the Programming volume [1793]; so, we do not pursue these subjects in detail here. We should mention, however, that for matrices with numerical entries, several special *Mathematica* commands are available. Three of the most important of these are the following:

- `LUDecomposition` for finding the LU-decomposition of a matrix
- `SchurDecomposition` for finding the Schur-decomposition of a matrix
- `SingularValues` for finding the singular values of a matrix
- `NullSpace` for finding the null space of a matrix

In `NSolve`, *Mathematica* has access to sparse matrix techniques for the solution of linear systems of equations. Sparse matrices arise frequently in applications (e.g., in the solution of partial differential equations, network analysis, etc.) and can be effectively solved by this method.

In the following, we will mainly discuss a few examples that make use of the linear algebra functions of *Mathematica*.

We start with an example that yields a system of linear equations: Given a finite rectangular array of resistors [483], [903], we connect two neighboring lattice points of this resistor lattice with some external current source and want to determine the distribution of currents in the various resistors (for the infinite case, see [1844], [610], [611], [86], [1961], [1962], [1963], [443], [890], and [1675]). First, let us generate the relevant equations. They are given by Kirchhoff's theorems [994], [1114], [1970]. currentInput and currentOutput determine the two lattice points $\{x1,\ y1\}$ and $\{x2,\ y2\}$ through which the current enters and leaves the network.

```
In[1]:= ElectricCurrentEquationsOnRectangularResistorGrid[I_,
        resistanceFunction: R_,
        currentInput : {x1_Integer, y1_Integer},
        currentOutput: {x2_Integer, y2_Integer},
        gridSize:{n_Integer?(# >= 2&), m_Integer?(# >= 2&)}] :=
    Module[{currents, meshes, knots},
    (* I[{j, k}, {js, ks}] is the current from the mesh
       point {j, k} to the neighboring one {js, ks} *)
    (* all currents in the network *)
    currents = Flatten[Join[
    Table[I[{j, k}, {j + 1, k}], {j, n - 1}, {k, m}],
    Table[I[{j, k}, {j, k + 1}], {j, n}, {k, m - 1}]]];
    (* equations for the voltages in all loops *)
    meshes = Flatten[
      Table[I[{j, k}, {j + 1, k}] R[{j, k}, {j + 1, k}] +
            I[{j + 1, k}, {j + 1, k + 1}] R[{j + 1, k}, {j + 1, k + 1}] -
            I[{j, k + 1}, {j + 1, k + 1}] R[{j, k + 1}, {j + 1, k + 1}] -
            I[{j, k}, {j, k + 1}] R[{j, k}, {j, k + 1}] == 0,
            {j, n - 1}, {k, m - 1}]];
    (* the current source and sink *)
    Ω[{j_, k_}] := Which[{j, k} == currentInput,   1,
                         {j, k} == currentOutput, -1, True, 0];
    (* equations of the currents at all nodes *)
    knots = Union[Flatten[
    (* inner nodes *)
    Table[I[{j - 1, k}, {j, k}] - I[{j, k}, {j + 1, k}] +
        I[{j, k - 1}, {j, k}] - I[{j, k}, {j, k + 1}] == -Ω[{j, k}],
        {j, 2, n - 1}, {k, 2, m - 1}]],
    (* edge nodes *)
    Table[I[{j - 1, 1}, {j, 1}] - I[{j, 1}, {j + 1, 1}] -
        I[{j, 1}, {j, 2}] == -Ω[{j, 1}], {j, 2, n - 1}],
    Table[I[{j - 1, m}, {j, m}] - I[{j, m}, {j + 1, m}] +
        I[{j, m - 1}, {j, m}] == -Ω[{j, m}], {j, 2, n - 1}],
    Table[I[{1, k - 1}, {1, k}] - I[{1, k}, {1, k + 1}] -
        I[{1, k}, {2, k}] == -Ω[{1, k}], {k, 2, m - 1}],
    Table[I[{n, k - 1}, {n, k}] - I[{n, k}, {n, k + 1}] +
        I[{n - 1, k}, {n, k}] == -Ω[{n, k}], {k, 2, m - 1}],
    (* corner nodes *)
    {-I[{1, 1}, {2, 1}] - I[{1, 1}, {1, 2}] ==  -Ω[{1, 1}],
     I[{n - 1, 1}, {n, 1}] - I[{n, 1}, {n, 2}] ==  -Ω[{n, 1}],
     -I[{1, m}, {2, m}] + I[{1, m - 1}, {1, m}] ==  -Ω[{1, m}],
     I[{n - 1, m}, {n, m}] + I[{n, m - 1}, {n, m}] ==  -Ω[{n, m}]}]];
    (* combine all equations *)
    eqns = Join[meshes, knots];
    (* list of equations and currents *)
```

```
{eqns, currents}] /; (* makes input sense? *)
        x1 >= 1 && y1 >= 1 && x1 <= n && y1 <= n &&
        x2 >= 1 && y2 >= 1 && x2 <= n && y2 <= n
```

Here are the equations for a very small lattice. The second element of the list returned by `ElectricCurrent`-`EquationsOnRectangularResistorGrid` is a list of the currents. We use $R = 1$.

```
In[2]:= ElectricCurrentEquationsOnRectangularResistorGrid[
                    I, R&, {1, 1}, {2, 2}, {3, 3}]
```
```
Out[2]= {{-R I[{1, 1}, {1, 2}] + R I[{1, 1}, {2, 1}] - R I[{1, 2}, {2, 2}] + R I[{2, 1}, {2, 2}] ==
     0, -R I[{1, 2}, {1, 3}] + R I[{1, 2}, {2, 2}] -
     R I[{1, 3}, {2, 3}] + R I[{2, 2}, {2, 3}] == 0, -R I[{2, 1}, {2, 2}] +
     R I[{2, 1}, {3, 1}] - R I[{2, 2}, {3, 2}] + R I[{3, 1}, {3, 2}] == 0,
     -R I[{2, 2}, {2, 3}] + R I[{2, 2}, {3, 2}] - R I[{2, 3}, {3, 3}] + R I[{3, 2}, {3, 3}] ==
     0, -I[{1, 1}, {1, 2}] - I[{1, 1}, {2, 1}] == -1,
     I[{1, 1}, {1, 2}] - I[{1, 2}, {1, 3}] - I[{1, 2}, {2, 2}] == 0,
     I[{1, 2}, {1, 3}] - I[{1, 3}, {2, 3}] == 0,
     I[{1, 1}, {2, 1}] - I[{2, 1}, {2, 2}] - I[{2, 1}, {3, 1}] == 0,
     I[{1, 2}, {2, 2}] + I[{2, 1}, {2, 2}] - I[{2, 2}, {2, 3}] - I[{2, 2}, {3, 2}] == 1,
     I[{1, 3}, {2, 3}] + I[{2, 2}, {2, 3}] - I[{2, 3}, {3, 3}] == 0,
     I[{2, 1}, {3, 1}] - I[{3, 1}, {3, 2}] == 0,
     I[{2, 2}, {3, 2}] + I[{3, 1}, {3, 2}] - I[{3, 2}, {3, 3}] == 0,
     I[{2, 3}, {3, 3}] + I[{3, 2}, {3, 3}] == 0},
     {I[{1, 1}, {2, 1}], I[{1, 2}, {2, 2}], I[{1, 3}, {2, 3}], I[{2, 1}, {3, 1}],
     I[{2, 2}, {3, 2}], I[{2, 3}, {3, 3}], I[{1, 1}, {1, 2}], I[{1, 2}, {1, 3}],
     I[{2, 1}, {2, 2}], I[{2, 2}, {2, 3}], I[{3, 1}, {3, 2}], I[{3, 2}, {3, 3}]}}
```

Because we have used all possible equations, we now have an (slightly) overdetermined system of equations. We see this clearly by looking at the following, slightly larger example.

```
In[3]:= Length /@ (eqsAndVars =
    ElectricCurrentEquationsOnRectangularResistorGrid[
                    I, 1&, {1, 1}, {6, 6}, {16, 16}])
```
```
Out[3]= {481, 480}
```

We have just one equation "too many".

```
In[4]:= (n m + (n - 1) (m - 1)) - (n(m - 1) + m (n - 1)) // Simplify
```
```
Out[4]= 1
```

The exact solution required is quite fast. *Mathematica* uses the sparsity of the coefficient matrix in the internal calculations.

```
In[5]:= Short[Timing[solExact = Solve[Drop[eqsAndVars[[1]], -1], eqsAndVars[[2]]]], 5]
```
```
Out[5]//Short= {0.37 Second, {{I[{1, 1}, {2, 1}] → 1/2,
```
$$I[\{1, 2\}, \{2, 2\}] \to \frac{7374980334725646987856177218345974899149 5}{3712922812516487830234724704080673821666 56},$$
$$I[\{1, 3\}, \{2, 3\}] \to \frac{3560326941594501812394908134634555589115 7}{3712922812516487830234724704080673821666 56}, \ll 475 \gg,$$
$$I[\{16, 14\}, \{16, 15\}] \to \frac{1091500878838022613331870517465043725 51}{3712922812516487830234724704080673821666 56},$$
$$I[\{16, 15\}, \{16, 16\}] \to 0\}\}\}$$

A numerical solution, which in our case also makes use of sparse code, is still faster. To prevent the indices from being transformed into numbers with head `Real` by application of `N`, we give i the attribute `NHoldAll`, which we mentioned in Chapter 3 of the Programming volume [1793].

In[6]:= **SetAttributes[I, NHoldAll]**

In[7]:= **Short[Timing[solApprox =**
   NSolve[(* remove one equation; not quite any can be removed *)
     N[Drop[eqsAndVars[[1]], -1]], eqsAndVars[[2]]]], 12]

Out[7]//Short= $\{0.07$ Second, $\{\{I[\{1, 1\}, \{2, 1\}] \rightarrow 0.5, I[\{1, 2\}, \{2, 2\}] \rightarrow 0.19863,$
   $I[\{1, 3\}, \{2, 3\}] \rightarrow 0.0958901, I[\{1, 4\}, \{2, 4\}] \rightarrow 0.0565837,$
   $I[\{1, 5\}, \{2, 5\}] \rightarrow 0.0399245, I[\{1, 6\}, \{2, 6\}] \rightarrow 0.0311365,$
   $I[\{1, 7\}, \{2, 7\}] \rightarrow 0.0238055, I[\{1, 8\}, \{2, 8\}] \rightarrow 0.0172727,$
   $I[\{1, 9\}, \{2, 9\}] \rightarrow 0.0120094, I[\{1, 10\}, \{2, 10\}] \rightarrow 0.00814897,$
   $I[\{1, 11\}, \{2, 11\}] \rightarrow 0.00549207, \ll458\gg, I[\{16, 5\}, \{16, 6\}] \rightarrow 0.00370536,$
   $I[\{16, 6\}, \{16, 7\}] \rightarrow 0.00336343, I[\{16, 7\}, \{16, 8\}] \rightarrow 0.00280383,$
   $I[\{16, 8\}, \{16, 9\}] \rightarrow 0.00215684, I[\{16, 9\}, \{16, 10\}] \rightarrow 0.0015301,$
   $I[\{16, 10\}, \{16, 11\}] \rightarrow 0.00099366, I[\{16, 11\}, \{16, 12\}] \rightarrow 0.00058008,$
   $I[\{16, 12\}, \{16, 13\}] \rightarrow 0.00029299, I[\{16, 13\}, \{16, 14\}] \rightarrow 0.000117589,$
   $I[\{16, 14\}, \{16, 15\}] \rightarrow 0.0000293973, I[\{16, 15\}, \{16, 16\}] \rightarrow 2.234\times10^{-16}\}\}\}$

The results of the numerical and exact solutions are the same.

In[8]:= **Module[{exactCurrents = eqsAndVars[[2]] /. solExact[[1]],**
   approxCurrents = eqsAndVars[[2]] /. solApprox[[1]]},
    {(* maximal absolute difference *)
    Max[Abs[exactCurrents - approxCurrents]],
    (* sum of squares of differences *)
    Norm[exactCurrents - approxCurrents]^2}]

Out[8]= $\{1.11022\times10^{-15}, 1.14408\times10^{-29}\}$

A high-precision solution of the system of equations is, of course, more time-consuming.

In[9]:= **Timing[NSolve[N[Drop[eqsAndVars[[1]], -1], 20], eqsAndVars[[2]]]][[1]]**

Out[9]= 9.11 Second

When a system of equations can be "naturally" ordered in a 2D way, the explicit introduction of "solve variables" is sometimes a waste of time. All one really needs is a coefficient matrix and a right-hand side vector. For the equations under consideration, we determine the coefficient matrix (the following way is an inefficient one; using hashing techniques this can be considerably accelerated).

In[10]:= **cfMat = Function[eqs, Coefficient[eqs[[1]], #]& /@**
      eqsAndVars[[2]]] /@ Rest[eqsAndVars[[1]]];

The coefficient matrix is often a very sparse matrix. The following density plot confirms this property for our resistor network equations. Less than 1% of all entries are nonzero.

In[11]:= **ListDensityPlot[cfMat, Mesh -> False]**

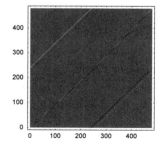

Because the numerical solution of the above system was quite fast, we can treat a bigger example and visualize the current distribution via different line thicknesses (be aware that the directive Thickness[0] results in a visible line). Using this graphic, we could also apply the thickness and the gray level of the line segments as a measure for the current. Using the height of 3D polygons as the measure for the current, we obtain the rightmost of the following pictures. We position the current input and output in one corner and at the center of the square grid.

```
In[12]:= (* the currents are proportional to the thickness *)
    currentGraphics =
    Module[{solApprox, solAbs, eqsAndVars},
     eqsAndVars = ElectricCurrentEquationsOnRectangularResistorGrid[
                    I, 1&, {1, 1}, {16, 16}, {31, 31}];
     (* solve the system of equations for the currents *)
     solApprox = Solve[N[Drop[eqsAndVars[[1]], -1]], eqsAndVars[[2]]];
     freeI = Union[Cases[Last /@ solApprox[[1]], _I, Infinity]];
     solApprox = solApprox /. ((# -> 1)& /@ freeI);
     solAbs = MapAt[Abs, #, 2]& /@ solApprox[[1]];
     (* minimal and maximal current--needed for scaling;
        not a local variable *)
     {minI, maxI} = #[Last /@ solAbs]& /@ {Min, Max};
     (* make line thickness proportional to current *)
     Graphics[{{Thickness[(#[[2]] - minI)/(maxI - minI) 0.025],
               Line[List @@ #[[1]]]}& /@ solAbs},
          PlotRange -> All, AspectRatio -> Automatic]];

    (* the currents are proportional to the grayness and thickness *)
    currentGraphicsGray = Function[max,
    currentGraphics /. {Thickness[i_], l_Line} ->
            {Hue[0.78 40 i], Thickness[0.02 i/maxI], l}][
            Max[First /@ Cases[currentGraphics, _Thickness, {1, Infinity}]]];

    (* the currents are proportional to the height *)
    currentGraphics3D = Show[Graphics3D[{EdgeForm[Thickness[0.001]],
    currentGraphics[[1]] /. {Thickness[z_], Line[{{x1_, y1_}, {x2_, y2_}}]} ->
            Polygon[{{x1, y1, 0}, {x1, y1, z}, {x2, y2, z}, {x2, y2, 0}}]}},
        BoxRatios -> {2, 2, 1}, Boxed -> False,
        PlotRange -> All, DisplayFunction -> Identity];

    (* display all three graphics *)
    Show[GraphicsArray[{currentGraphics, currentGraphicsGray,
    currentGraphics3D}]]
```

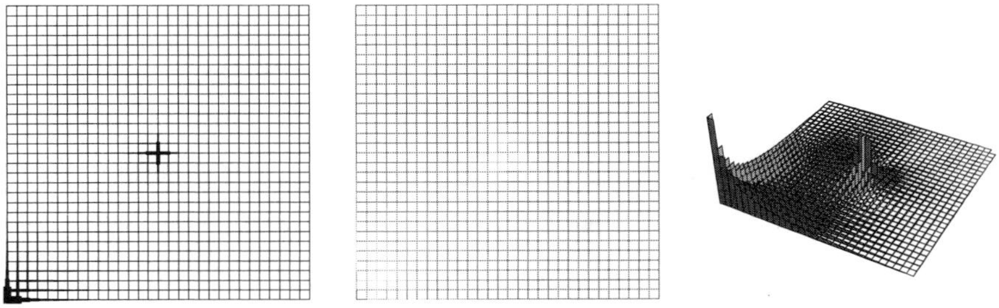

Interestingly, also the archetypical for linear algebra calculations function Eigensystem can be used to carry out resistance calculations for networks. Given a network with resistors of conductivities c_{ij} between the nodes i and j, the resistance between nodes i and j can be expressed as

$$R_{i,j} = {\sum_k}' \frac{1}{\lambda_k} |\psi_k(i) - \psi_k(j)|^2$$

where λ_k are the kth eigenvalue and $\psi_k(i)$ is the ith component of the kth eigenvector of the matrix $\mathbf{C}$ and the prime on the sum indicates the omission zero eigenvalue [1926]. The matrix $\mathbf{C}$ has nondiagonal elements $-c_{ij}$ and its diagonal elements are the negative sum of its nondiagonal row elements. The next input defines the construction of the matrix $\mathbf{C}$ for a square lattice resistor network of dimension $d \times d$.

```
In[23]:= squareResistorNetworkMatrix[d_] :=
    Module[{nodes = Flatten[Table[{i, j}, {i, d}, {j, d}], 1],
            λ = d^2, M},
        (* number nodes nodeNumber with hashed values *)
        Do[nodeNumber[nodes[[i]]] = i, {i, λ}];
        neighborNodeNumbers[{i_, j_}] := nodeNumber /@
          Select[({i, j} + #)& /@ {{1, 0}, {-1, 0}, {0, -1}, {0, 1}},
                 (Min[#] >= 1 && Max[#] <= d)&];
        (* connectivity matrix *)
        M = Table[0, {λ}, {λ}];
        Do[(M[[i, #]] = -1)& /@ neighborNodeNumbers[nodes[[i]]], {i, λ}];
        (* diagonals for current conservation *)
        MapIndexed[(M[[#2[[1]], #2[[1]]]] = -Plus @@ #1)&, M]; M]
```

Here is the resistance value between the nodes (1, 1) and (6, 6).

```
In[24]:= resistorM16 = squareResistorNetworkMatrix[16];
         {evalsM16, evecsM16} = Eigensystem[N[resistorM16]];

In[26]:= {α, β} = {nodeNumber[{1, 1}], nodeNumber[{6, 6}]};
         Sum[1/evalsM16[[j]] Abs[evecsM16[[j, α]] - evecsM16[[j, β]]]^2, {j, 16^2 - 1}]
Out[27]= 1.89137
```

While calculating the resistance value between two individual nodes using the eigenfunction approach is computationally more expensive than solving the linear system as done above, once the eigensystem is calculated, it is easy to calculate many resistance values. So, we calculate all 65536 resistance values for the 16×16 network.

```
In[28]:= (* use a compiled function to calculate all resistance values *)
         allResistanceValuesCF =
         Compile[{{evals, _Real, 1}, {evecs, _Real, 2}},
                 Module[{λ = Length[evals]},
                        Table[Sum[1/evals[[j]] Abs[evecs[[j, α]] - evecs[[j, β]]]^2,
                              {j, λ - 1}],
                        {α, λ}, {β, λ}] // Flatten]];
```

The next graphic shows the resistance values sorted by size on a linear and a logarithmic scale.

```
In[30]:= Module[{allRs = Sort[allResistanceValuesCF[evalsM16, evecsM16]]},
         Show[GraphicsArray[
               ListPlot[#, PlotRange -> All, DisplayFunction -> Identity]& /@
               {allRs, Log[MapIndexed[{#2[[1]], #1}&, DeleteCases[allRs, 0.]]]}]]]
```

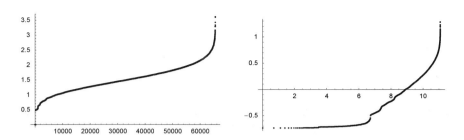

More complicated resistor networks will produce more interesting current distributions; for Sierpinski triangles see, e.g., [888], and [889]; for diode networks, see [838];

As a more challenging example for the linear solver, let us deal with the following example: the triadic Tagaki function $\mathcal{T}(x)$. $\mathcal{T}(x)$ fulfills the functional equation [683], [684], [1750], [929], [408], [930]:

$$\mathcal{T}(x) = \begin{cases} \frac{1}{3}\,\mathcal{T}(3\,x) + x & 0 \le x \le \frac{1}{3} \\ \frac{1}{3}\,\mathcal{T}(3\,x - 1) + \frac{1}{3} & \frac{1}{3} \le x \le \frac{2}{3} \\ \frac{1}{3}\,\mathcal{T}(3\,x - 2) - x + 1 & \frac{2}{3} \le x \le 1 \end{cases}$$

We discretize the functional equation and get a sparse linear system of equations. Solving the resulting system for 10000 points takes just a few seconds.

```
In[31]:= Block[{n = 10^4, 𝒯},
        (* avoid numericalization of indices *)
        SetAttributes[𝒯, NHoldAll];
        (* discretized functional equations *)
        eqs𝒯 = Table[𝒯[x] - Which[x < 1/3, 𝒯[3 x]/3 + x,
                               x < 2/3, 𝒯[3 x - 1]/3 + 1/3,
                               True,    𝒯[3 x - 2]/3 - x + 1],
              {x, 0, 1, 1/n}];
        (* solve sparse equations *)
        Solve[# == 0& /@ N[eqs𝒯],
            Table[𝒯[x], {x, 0, 1, 1/n}]]] // (sol𝒯 = #)&; // Timing
Out[31]= {1.03 Second, Null}
```

Here is the function $\mathcal{T}(x)$ shown. It is a continuous, but nowhere differentiable function.

```
In[32]:= Show[Graphics[Polygon[{#[[1, 1]], #[[2]]}& /@ sol𝒯[[1]]]],
            Frame -> True, AspectRatio -> Automatic]
```

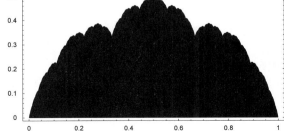

Remark: The triadic Tagaki can also be represented as the following infinite sum [758].

$$t_1(x) = (1/3 - |x - 2/3|)\,\theta(x - 1/3) + (1/3 - |x - 1/3|)\,\theta(2/3 - x);$$
$$t_k(x) = t_{k-1}(3x - \lfloor 3x \rfloor)/3$$
$$\mathcal{T}(x) = \sum_{k=1}^{\infty} t_k(x).$$

One remark is in order concerning the use of linear algebra functions on matrices containing high-precision numbers. In larger linear algebra calculations, massive cancellations would occur frequently when applying the automatic precision control model at each arithmetic operation. But the uncertainties of the numbers are not independent from each other (as the model assumes) and virtually always cancel each other to a very large degree. That is why (and also because of speed reasons) linear algebra functions use automatically fixed precision in internal calculations. The following simple eigenvalue calculation of a 50-digit high-precision matrix shows that the eigenvalues calculated using Eigenvalues always have precision 50, but that already for such small matrices a direct calculation loses about seven digits.

```
In[33]:= Clear[a];
         dim  = 3;
         evs = Table[a[i, j], {i, dim}, {j, dim}] // Eigenvalues;

In[36]:= (* a random number *)
         rand := (2 Random[Integer] - 1) Random[Real, {0, 1}, 50] *
                                 10^Random[Integer, {-10, 10}]

In[38]:= Table[Precision /@ {(* use Eigenvalues *)
                 Eigenvalues[Table[a[i, j] = rand + I rand,
                                 {i, dim}, {j, dim}]],
                 (* direct calculation *) evs},
             {100}] // Transpose // Plus @@@ #/100&

Out[38]= {50., 48.8829}
```

A related remark concerns high-precision zeros of low accuracy. Such quantities can frequently dramatically lower the precision of the result of a linear algebra function. Here is an exact 3×3 matrix with one symbolic element badZero. The eigenvalues of the matrix do not depend on the value of badZero.

```
In[39]:= tM = {{1, 0, 0}, {0, 2, 0}, {badZero, 0, 1}};
         Eigenvalues[tM]

Out[40]= {2, 1, 1}
```

But the result for the eigenvalues of the matrix tM with a low-precision zero instead of the symbol badZero gives quite different eigenvalues.

```
In[41]:= Eigenvalues[tM /. badZero -> SetAccuracy[0, -2]] // InputForm
Out[41]//InputForm=
         {0``0, 0``0, 0``-0.3010299956639812}
```

This result arises from unifying the precision of the whole matrix to its precision.

```
In[42]:= SetPrecision[{{1, 0, 0}, {0, 2, 0}, {0, 0, 1}},
                 Precision[SetAccuracy[0, -2]]] //
                                 Eigenvalues // InputForm
Out[42]//InputForm=
         {0``0, 0``0, 0``-0.3010299956639812}
```

The previous result means that the last digits of the result returned by Eigenvalues are not certified in the sense that typical *Mathematica* high-precision calculations are certified. So for high-precision linear algebra calculations, the reader should use more than the minimal number of digits. But the use of fixed precision arithmetic does not automatically mean that all control over the precision is lost. Various calculations are internally carried out to estimate the precision of the result. The following calculates 100 determinants of 3×3 matrices. In average about seven digits are lost.

```
In[43]:= Sum[Table[rand + I rand, {i, dim}, {j, dim}] //
                    Det // Precision, {100}]/100
```

Out[43]= 43.8763

As an example of the use of high-precision arithmetic in solving linear systems, we will use numerical techniques to derive a modular equation. The modular function `KleinInvariantJ`, denoted by $J(\tau)$ ($\mathrm{Im}(\tau) > 0$), fulfills functional equations of the following form:

$$\sum_{k,l=1}^{n(m)} c_{kl} \, (1728 \, J(\tau))^k \, (1728 \, J(m \, \tau))^l = 0.$$

In the last equation m, $n(m) \in \mathbb{N}$ and all c_{kl} are integers [1100], [411], [438], [752], [826], [1955], [418], [419].

In the following example, we will derive the modular equation for $m = 3$. We start by making the following symmetric polynomial ansatz.

```
In[44]:= Clear[u, v, c];
        poly[u_, v_] = u^3 + v^3 + c[1] u^2 v^2 +
                    c[2](u^2 v + v^2 u) + c[3] (u^2 + v^2) +
                    c[4] u v + c[5] (u + v) + c[6]
```

Out[45]= $u^3 + v^3 + u^2 \, v^2 \, c[1] + (u^2 \, v + u \, v^2) \, c[2] + (u^2 + v^2) \, c[3] + u \, v \, c[4] + (u + v) \, c[5] + c[6]$

To determine the six unknowns `c[1]`, `c[2]`, `c[3]`, `c[4]`, `c[5]`, and `c[6]`, we calculate high-precision values of `poly` for six different choices for τ. (The choice of the values $k/11 + i/13$ is "accidental"; any set of generic values for τ would work).

```
In[46]:= eqs = Table[poly[N[1728 KleinInvariantJ[τ + 1/13 I], 300],
                    N[1728 KleinInvariantJ[2(τ + 1/13 I)], 300]],
                {τ, 1/11, 6/11, 1/11}];
```

Looking at the solutions of these six equations gives a strong hint that all unknowns take integer values.

```
In[47]:= sol = Solve[# == 0& /@ eqs, Table[c[i], {i, 6}]]
```

Out[47]= {{c[1] →
            −1.000000000000000000000000000000000000000000000000000000000000000000000000000000000000000 ·.
            000000000000000000000000000000000000000000000000000000000000000000000000000000000000000000 ·.
            000000000000000000000000000000000000000000000000000000000000000000000000000000000000000000 ·.
            000000000000000000000000000000000000000000000000000000000000000000000000000000000000000000 ·.
            00000000000 + 0. × 10^−300 i, c[2] →
            1488.0000000000000000000000000000000000000000000000000000000000000000000000000000000000000 ·.
            000000000000000000000000000000000000000000000000000000000000000000000000000000000000000000 ·
            000000000000000000000000000000000000000000000000000000000000000000000000000000000000000000 ·
            000000000000000000000000000000000000000000000000000000000000000000000000000000000000000000 ·
            000 + 0. × 10^−297 i, c[3] →
            −162000.00000000000000000000000000000000000000000000000000000000000000000000000000000000000 ·.
            000000000000000000000000000000000000000000000000000000000000000000000000000000000000000000 ·.
            000000000000000000000000000000000000000000000000000000000000000000000000000000000000000000 ·.
            000000000000000000000000000000000000000000000000000000000000000000000000000000000000000000 ·.
            0000000000 + 0. × 10^−295 i, c[4] →
```
```

$$4.0773375000000000000000000000000000000000000000000000000000000000000000 \\ 00000000000000000000000000000000000000000000000000000000000000000000000 \\ 00000000000000000000000000000000000000000000000000000000000000000000000 \\ 00000000000000000000000000000000000000000000000000000000000000000000000 \\ 000 \times 10^7 + 0. \times 10^{-292} \, i, \; c[5] \rightarrow$$

$$8.7480000000000000000000000000000000000000000000000000000000000000000000 \\ 00000000000000000000000000000000000000000000000000000000000000000000000 \\ 00000000000000000000000000000000000000000000000000000000000000000000000 \\ 00000000000000000000000000000000000000000000000000000000000000000000000 \\ 000 \times 10^9 + 0. \times 10^{-290} \, i, \; c[6] \rightarrow$$

$$-1.574640000000000000000000000000000000000000000000000000000000000000000 \\ 00000000000000000000000000000000000000000000000000000000000000000000000 \\ 00000000000000000000000000000000000000000000000000000000000000000000000 \\ 00000000000000000000000000000000000000000000000000000000000000000000000 \\ 0000000000 \times 10^{14} + 0. \times 10^{-286} \, i \} \}$$

Using Round, we extract the corresponding integer values.

```
In[48]:= solInteger = sol /. c_Complex :> Round[c]
Out[48]= {{c[1] → -1, c[2] → 1488, c[3] → -162000,
          c[4] → 40773375, c[5] → 8748000000, c[6] → -157464000000000}}
```

So, we are led to the following modular equation.

```
In[49]:= (modularEq[τ_] = Factor[
            poly[1728 KleinInvariantJ[τ], 1728 KleinInvariantJ[2τ]] /.
              solInteger[[1]]] /. i_Integer*r_Plus -> r) // TraditionalForm
```
Out[49]//TraditionalForm=

$$64 J(\tau)^3 - 110592 J(2\,\tau)^2 J(\tau)^2 + 95232 J(2\,\tau) J(\tau)^2 - 6000 J(\tau)^2 + 95232 J(2\,\tau)^2 J(\tau) + \\ 1510125 J(2\,\tau) J(\tau) + 187500 J(\tau) + 64 J(2\,\tau)^3 - 6000 J(2\,\tau)^2 + 187500 J(2\,\tau) - 1953125$$

A quick numerical check for some random values from the upper half-plane confirms the correctness of the derived equation.

```
In[50]:= Table[modularEq[SetPrecision[
                    Random[Complex, {-1, 1 + 2I}], 200]], {20}] // Abs // Max
Out[50]= 0. × 10⁻¹⁸³
```

For methods to derive higher order modular functions, see [878], and for determining relations between numbers with rational coefficients, see also [254].

Here is a small example that uses again the function Eigenvalues. Given a sequence of ± 1, we form a parametrized tridiagonal matrix with diagonal elements $\pm\varepsilon$ and nondiagonal matrix elements -1. Then, we calculate and visualize the eigenvalues as a function of ε. The left graphic results from the Fibonacci sequence [1162], and the right from a random sequence.

```
In[51]:= Module[{toMatrix, M, n, o = 10},
    (* convert rules to a dense matrix *)
    toMatrix[rules_] :=
    Module[{d = Max[First /@ rules], M}, M = Table[0, {d}, {d}];
      (M[[#1, #2]] = #3)& @@@ (Flatten /@ (List @@@ rules)); M];
    (* show eigenvalues for Fibonacci sequence and random sequence *)
    Show[GraphicsArray[(M[e_] = toMatrix[#];
    Graphics[{(* show eigenvalues as points *) PointSize[0.003],
      Table[Point[{#, e}]& /@ Eigenvalues[N @ M[e]],
          {e, -2., 2., 1/100.}]}, Frame -> True, PlotRange -> All,
          FrameLabel -> {None, "ε"}])& /@
```

```
(* matrices of the two sequences *)
{(n = Length[#];
  Table[{{i, i} -> If[#[[i]] === A, 1, -1] e,
          If[i > 1, {i, i - 1} -> -1, Sequence @@ {}],
          If[i < n, {i, i + 1} -> -1, Sequence @@ {}]},
         {i, n}] // Flatten)&[
             Flatten[Nest[(# /. {A -> {A, B}, B -> A})&, {B}, o]]],
  (Table[{{i, i} -> If[Random[] < 1/2, 1, -1] e,
          If[i > 1, {i, i - 1} -> -1, Sequence @@ {}],
          If[i < n, {i, i + 1} -> -1, Sequence @@ {}]},
         {i, n}] // Flatten)&[
         2 RealDigits[Pi, 2, n][[1]] - 1]}]]]
```

We will discuss one function which we did not cover in Chapter 6 of the Programming volume: NullSpace.

NullSpace[*squareMatrix*]

calculates the null space of *squareMatrix* (the nontrivial, linear independent solutions *x* of *squareMatrix*.*x* = 0).

Here is a square matrix with determinant 0.

```
In[52]:= {{ 1,  3, -5,  0,  2,  3},
        {-3, -6, -5, -1,  0,  0},
        { 0,  5,  1, -2, -6, -1},
        {-1, -3,  0, -5, -6, -5},
        {-4,  1, -6, -3,  3, -1},
        { 4, -1,  6,  3, -3,  1}} // (M = #)& // Det
Out[52]= 0
```

LinearSolve of the matrix *M* returns only a trivial solution with the zero vector for the right-hand side.

```
In[53]:= LinearSolve[M, {0, 0, 0, 0, 0, 0}]
Out[53]= {0, 0, 0, 0, 0, 0}
```

Solve returns a parametrization of all solutions.

```
In[54]:= eqs = Array[α, 6].# == 0& /@ M
Out[54]= {1[1] + 3 1[2] - 5 1[3] + 2 1[5] + 3 1[6] == 0, -3 1[1] - 6 1[2] - 5 1[3] - 1[4] == 0,
         5 1[2] + 1[3] - 2 1[4] - 6 1[5] - 1[6] == 0, -1[1] - 3 1[2] - 5 1[4] - 6 1[5] - 5 1[6] == 0,
         -4 1[1] + 1[2] - 6 1[3] - 3 1[4] + 3 1[5] - 1[6] == 0,
         4 1[1] - 1[2] + 6 1[3] + 3 1[4] - 3 1[5] + 1[6] == 0}
```

```
In[55]:= nsSol = Solve[eqs, Array[α, 6]]
         Solve::svars : Equations may not give solutions for all "solve" variables. More…
```

Out[55]= $\left\{\left\{1[1] \rightarrow -\dfrac{289\ 1[6]}{1412},\ 1[2] \rightarrow -\dfrac{445\ 1[6]}{2824},\right.\right.$
$\left.\left.1[3] \rightarrow \dfrac{1455\ 1[6]}{2824},\ 1[4] \rightarrow -\dfrac{2871\ 1[6]}{2824},\ 1[5] \rightarrow \dfrac{179\ 1[6]}{1412}\right\}\right\}$

In many cases, one needs all linear independent solutions **x** of **A.x** = 0 for a singular matrix **A**. This is what `NullSpace` returns.

In[56]:= **NullSpace[M]**

Out[56]= {{-578, -445, 1455, -2871, 358, 2824}}

(This is a special solution of the result from `Solve`.)

In[57]:= **Append[nsSol[[1]] /. α[6] -> 2824, α[6] -> 2824]**

Out[57]= {1[1] → -578, 1[2] → -445, 1[3] → 1455, 1[4] → -2871, 1[5] → 358, 1[6] → 2824}

For a nonsingular matrix, `NullSpace` returns an empty list. Here is a simple example.

In[58]:= **NullSpace[{{1, 0}, {0, 1}}]**

Out[58]= {}

Let us deal with a slightly more complicated application of the function `NullSpace`. Frequently one encounters the situation where one has a parametrized matrix $\mathbf{M}(\varepsilon)$ and searches for ε, such that $\det(\mathbf{M}(\varepsilon_0)) = 0$ and one is interested in the null space of $\mathbf{M}(\varepsilon_0)$ too. The following example leads to this kind of problem. We will determine the lowest eigenvalue and the corresponding eigenfunction in a waveguide crossing. This means we want to solve $-\Delta\psi_0(x, y) = \varepsilon_0\,\psi_0(x, y)$ in the domain $\{\Omega\,|\,\{x, y\},\ |x| \le 1 \vee |y| \le 1\}$ and Dirichlet boundary conditions [1620]. Here is a sketch of the domain.

In[59]:= **Plot3D[If[Min[Abs[{x, y}]] > 1, 1, 0], {x, -4, 4}, {y, -4, 4},**
          Mesh -> False, PlotPoints -> 120, Axes -> False]

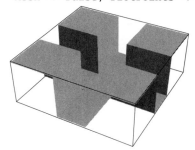

Interestingly, in the intersection area of the two canals, a localized bound state can exist. Inside each of the four arms, we expand the function $\psi_0(x, y)$ into a series $\psi_0(x, y) = \sum_{j=0}^{\infty} X_j(x)\,Y_j(y)$. Because of the symmetry of the problem and the fact that we are only interested in the ground sate, we can restrict ourselves to the right arm along the x-axis. The eigenfunctions in y-direction are $\cos((j + 1/2)\pi y)$ and in x-direction, we obtain exponential functions after separation of variables. Because we are only interested in localized states, we take only exponentially decreasing terms into account:

$$\psi_A(x, y) = \sum_{j=0}^{\infty} \alpha_j \cos((j + 1/2)\,\pi\, y)\,\exp(-\kappa_j\, x)$$

Here $\kappa_j = \left(\pi^2(j + 1/2)^2 - \varepsilon\right)^{1/2}$ and α_j is a function of ε.

In a similar way, we expand the function $\psi_0(x, y)$ in the intersection area. Now, we must also take exponentially increasing terms into account; because of the symmetry of the problem this means terms of the form $\cosh(\kappa_j\,x\,Or\,y)$. (Again, the β_j are functions of ε.)

$$\psi_S(x, y) = \sum_{j=0}^{\infty} \beta_j(\cos((j + 1/2)\,\pi\,y)\cosh(\kappa_j\,x) + \cos((j + 1/2)\,\pi\,x)\cosh(\kappa_j\,y))$$

Here we define the expansions in the right arm and the intersection.

```
In[60]:= ψArm[x_, y_, o_] := Sum[α[j] Cos[(j + 1/2) Pi y] Exp[-κ[j] x],
                                 {j, 0, o}]

       ψSquare[x_, y_, o_] :=
       Sum[β[j] Cos[(j + 1/2) Pi y] Cosh[κ[j] x] +
           β[j] Cos[(j + 1/2) Pi x] Cosh[κ[j] y], {j, 0, o}]

       κ[j_] := Sqrt[(Pi + 2 j Pi)^2/4 - ε]
```

At the border between the arms and the intersection area, we need the function $\psi_0(x, y)$ to be smooth (meaning continuous and differentiable). This means at the line segment $x = 1$, $-1 \le y \le 1$, we the following two identities must hold (mode matching [344], [1928], [1403], [90], [1942], [144], [1885]):

$$\psi_A(1, y) = \psi_S(1, y)$$

$$\frac{\partial \psi_A(x, y)}{\partial x}\Big|_{x=1} = \frac{\partial \psi_S(x, y)}{\partial x}\Big|_{x=1} .$$

Within any numerical approximation, the last two identities can only hold approximately. We could demand the last two identities to fold for a certain number of points $\{1, y_k\}$, $k = 1, \ldots, n$ exactly. Instead, we will expand $\psi_A(1, y)$ and $\psi_S(1, y)$ in a Fourier series and equate the corresponding Fourier coefficients. Taking the orthogonality of the $Y_j(y)$ into account

$$\int_{-1}^{1} \cos((\mu + 1/2)\,\pi\,y)\cos((\nu + 1/2)\,\pi\,y)\,dy = \delta_{\mu\nu}$$

we obtain the following system of equations for the unknown coefficients $\alpha_j(\varepsilon)$ and $\beta_j(\varepsilon)$.

$$+\alpha_i \exp(-\kappa_i) = \beta_i \cosh(\kappa_i) + \sum_{j=0}^{o-1} \beta_j \cos((j + 1/2)\pi)\,g(\kappa_j)$$

$$-\alpha_i \exp(-\kappa_i) = \beta_i \kappa_i \sinh(\kappa_i) - \sum_{j=0}^{o-1} \beta_j(j + 1/2)\pi \sin((j + 1/2)\pi)\,g(\kappa_j)$$

Here $g(\kappa)$ is $g(\kappa) = 4\,(-1)^j\,(2\,j + 1)\,\pi/((2\,\pi\,j + \pi)^2 + 4\,\kappa^2)\cosh(\kappa)$.

```
In[65]:= g[j_, κ_] = Integrate[Cos[(j + 1/2) Pi y] Cosh[κ y], {y, -1, 1}] //
                                 Simplify[#, Element[j, Integers]]&
```

$$Out[65]= \frac{4\,(-1)^j\,(1 + 2\,j)\,\pi\,\text{Cosh}[\kappa]}{(\pi + 2\,j\,\pi)^2 + 4\,\kappa^2}$$

We now order the $2o$ unknowns $\alpha_j(\varepsilon)$ and $\beta_j(\varepsilon)$ linearly according to $z_{k+1} = \alpha_k$, and $z_{o+k+1} = \beta_k$, $k = 0, \ldots, o-1$. The coefficient matrix of the resulting linear system $\mathbf{M}(\varepsilon)$ for the z_j allows for a nontrivial solution when $\det(\mathbf{M}(\varepsilon_0))$. The function M calculates the $2o \times 2o$ matrix $\mathbf{M}(\varepsilon)$.

```
In[66]:= M[o_] := Module[{m = Table[0, {2o}, {2o}]},
          (* continuity *)
          Do[m[[i + 1, i + 1]] = Exp[-κ[i]];
            m[[i + 1, o + i + 1]] = -Cosh[κ[i]];
            Do[m[[i + 1, o + j + 1]] = m[[i + 1, o + j + 1]] -
                                Cos[(j + 1/2) Pi] g[i, κ[j]],
              {j, 0, o - 1}], {i, 0, o - 1}];
          (* derivative continuity *)
          Do[m[[o + i + 1, i + 1]] = -κ[i] Exp[-κ[i]];
            m[[o + i + 1, o + i + 1]] = -κ[i] Sinh[κ[i]];
            Do[m[[o + i + 1, o + j + 1]] = m[[o + i + 1, o + j + 1]] +
                                (j + 1/2) Pi Sin[(j + 1/2) Pi] g[i, κ[j]],
              {j, 0, o - 1}], {i, 0, o - 1}]; m]
```

The function MDet calculates $\det(\mathbf{M}(\varepsilon_0))$.

```
In[67]:= Mo[ε_, o_] := (Mo[ε_, o] = M[o] /. ε -> ε);
         MDet[ε_?NumberQ, o_] := Det @ Mo[ε, o];
         Mo[ε, 8]; Mo[ε, 12]; Mo[ε, 24];
```

The following two graphics show $\det(\mathbf{M}(\varepsilon_0))$ for $o = 8$ and $o = 16$. We conjecture $\varepsilon_0 \approx 1.6$.

```
In[70]:= Show[GraphicsArray[
           Plot[MDet[ε, #], {ε, 0, Pi^2/4}, PlotRange -> All,
             Compiled -> False, DisplayFunction -> Identity]& /@ {8, 12}]]
```

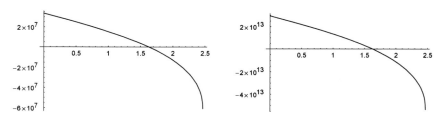

To calculate a more precise value for $\det(\mathbf{M}(\varepsilon_0))$, we implement a small bisection root-finding routine, bisect. (We could, of course, also use the function FindRoot to be discussed below.)

```
In[71]:= bisect[f_, {ε_, εl_, εu_}, tol_] :=
         Module[{εm = (εl + εu)/2, fm, εInt},
                fm = f /. ε -> εm;
                εInt = If[fm > 0, {εm, εu}, {εl, εm}];
                If[-Subtract @@ εInt < tol, Plus @@ εInt/2,
                   bisect[f, {ε, Sequence @@ εInt}, tol]]]
```

Here is the eigenvalue ε_0 for $o = 24$ to 50 digits. (Only about four digits are correct for the eigenvalue itself, but to guarantee the singular nature of $\mathbf{M}(\varepsilon_0)$ we need more digits.)

```
In[72]:= o = 24; Mo[ε, o];
         $RecursionLimit = 500;
         {ε0 = bisect[MDet[ε, o], {ε, N[3/2, 200], 2}, 10^-50],
          MDet[ε0, o] // N}
```

Out[74]= {1.62834079181743089364771908903880158534711513542320812488395403969124717272 9 ⋅
983336045592351418370913395767118016374024533151503892070000034664190025068 8 ⋅
1952285766601562500000000000000000000000000000000000, 2.29714 × 10⁻¹⁷}

The null space of $M(\varepsilon_0)$ are the coefficients α_j and β_j that form the eigenfunctions. We get a 1D null space.

In[75]:= **Length[ns = NullSpace[Mo24 = Mo[N[ε0, 100], o]]]**

Out[75]= 1

The following graphics show the matrix $M(\varepsilon_0)$ and the size of the coefficients α_j and β_j.

In[76]:= **Show[GraphicsArray[**
    Block[{$DisplayFunction = Identity},
      {ListDensityPlot[Log[10, Abs[Mo24] + 1], Mesh -> False],
      ListPlot[Log[10, Abs[ns[[1]]]], PlotRange -> All,
        Frame -> True, AspectRatio -> 1]}]]]

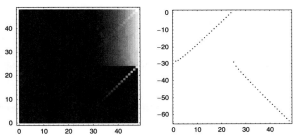

Using the just-calculated null space, we can now construct the lowest eigenfunction.

In[77]:= **ψArmN[x_, y_] = -ψArm[x, y, o - 1] //. ε -> ε0 //.**
    (Rule @@@ Transpose[{Table[α[j], {j, 0, o - 1}],
                Take[ns[[1]], o]}]) // N;

ψSquareN[x_, y_] = -ψSquare[x, y, o - 1] //. ε -> ε0 //.
    (Rule @@@ Transpose[{Table[β[j], {j, 0, o - 1}],
                Take[ns[[1]], {o + 1, 2o}]}]) // N;

In[79]:= **(* unit in one definition for all x, y *)**
    ψAllN[x_, y_] = Which @@
    {Abs[x] >= 1 && Abs[y] <= 1, ψArmN[Abs[x], y],
    Abs[y] >= 1 && Abs[x] <= 1, ψArmN[Abs[y], x],
    Abs[x] < 1 && Abs[y] < 1, ψSquareN[x, y], True, 0};

Here are a 3D plot and a contour plot of the corresponding eigenfunction.

In[81]:= **Show[GraphicsArray[**
    Block[{$DisplayFunction = Identity, pl3d},
    {(* 3D plot *)
    pl3d = Plot3D[(* scale *) 10^29 ψAllN[x, y], {x, -4, 4}, {y, -4, 4},
            PlotRange -> {10^-16, 2}, ClipFill -> None,
            PlotPoints -> 160, Mesh -> False, Axes -> False],
    (* contour plot *)
    ListContourPlot[pl3d[[1]], Contours -> 30, FrameTicks -> None,
            PlotRange -> All, ContourShading -> False,
            ContourStyle -> {Thickness[0.002]}]}]]]

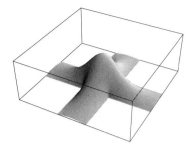 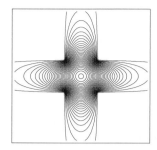

For a similar problem with a repulsive potential, see [1968]; for a smooth potential, see [1661]. For the application of the mode matching technique to the problem of sound transmission through a door in a thick wall, see [1640].

Until now, we have discussed and used matrices that had all their elements explicitly present. In many applications (most notably the numerical solution of elliptic partial differential equations with finite difference or finite element methods), one has to deal with large matrices having only a small number of nonvanishing elements. Just storing or even carrying out calculations with the nonvanishing elements greatly enhances the performance and allows treating matrices of much larger dimensions. The *Mathematica* form of a sparse matrix is `SparseArray`.

`SparseArray[{`*indices*$_1$ `-> ` *value*$_1$`, ` *indices*$_2$ ` -> ` *value*$_2$`, ...}]`

    represent a rectangular tensor with nonvanishing elements with value *value*$_k$ at position *value*$_k$.

Many arithmetic functions (like `Plus`, and `Times`) and linear algebra functions (like `Dot`, `LinearSolve`, and `Eigensystem`) work with sparse arrays in pretty much the same manner as they work with their dense relatives.

Here are two symmetric tridiagonal matrices of size 400×400. `fullTriDiagonalMatrix` is a dense matrix and `sparseTriDiagonalMatrix` is a sparse matrix.

```
In[82]:= fullTriDiagonalMatrix =
    Table[Which[i == j, 1, i == j - 1 || i == j + 1, (-1)^j j, True, 0],
        {i, 400}, {j, 400}];
```

```
    Short[fullTriDiagonalMatrix, 12]
```
Out[84]//Short=
```
{{1, 2, 0, 0, 0, 0, 0, 0, 0, 0, 0, 0, 0, 0, 0, 0, 0, 0, 0, 0, 0, 0, 0, 0, 0, 0, 0, 0, 0, 0, 0,
  0, 0, 0, 0, 0, 0, 0, 0, 0, 0, 0, 0, 0, 0, 0, 0, 0, 0, 0, 0, 0, 0, 0, 0, 0, 0, 0, 0,
  0, 0, 0, 0, 0, 0, 0, 0, 0, 0, 0, 0, 0, 0, 0, 0, 0, 0, 0, 0, 0, 0, 0, 0, 0, 0, 0, 0,
  0, 0, 0, 0, 0, 0, 0, 0, 0, 0, 0, 0, 0, 0, 0, 0, 0, 0, 0, 0, 0, 0, 0, 0, 0, 0, 0, 0,
  0, 0, 0, 0, 0, 0, 0, 0, 0, 0, 0, 0, 0, 0, 0, 0, 0, 0, 0, 0, 0, 0, 0, 0, 0, 0, 0, 0,
  0, 0, 0, 0, 0, 0, 0, 0, 0, 0, 0, 0, 0, 0, 0, 0, 0, 0, 0, 0, 0, 0, 0, 0, 0, 0, 0, 0,
  0, 0, 0, 0, 0, 0, 0, 0, 0, 0, 0, 0, 0, 0, 0, 0, 0, 0, 0, 0, 0, 0, 0, 0, 0, 0, 0, 0,
  0, 0, 0, 0, 0, 0, 0, 0, 0, 0, 0, 0, 0, 0, 0, 0, 0, 0, 0, 0, 0, 0, 0, 0, 0, 0, 0, 0,
  0, 0, 0, 0, 0, 0, 0, 0, 0, 0, 0, 0, 0, 0, 0, 0, 0, 0, 0, 0, 0, 0, 0, 0, 0, 0, 0, 0,
  0, 0, 0, 0, 0, 0, 0, 0, 0, 0, 0, 0, 0, 0, 0, 0, 0, 0, 0, 0, 0, 0, 0, 0, 0, 0, 0, 0,
  0, 0, 0, 0, 0, 0, 0, 0, 0, 0, 0, 0, 0, 0, 0, 0, 0, 0, 0, 0, 0, 0, 0, 0, 0, 0, 0, 0,
  0, 0, 0, 0, 0, 0, 0, 0, 0, 0, 0, 0, 0, 0, 0, 0, 0, 0, 0, 0, 0, 0, 0, 0, 0, 0, 0, 0,
  0, 0, 0, 0, 0, 0, 0, 0, 0, 0, 0, 0, 0, 0, 0, 0, 0, 0, 0, 0, 0, 0, 0, 0, 0, 0, 0, 0,
```

```
        0, 0, 0, 0, 0, 0, 0, 0, 0, 0, 0, 0, 0, 0, 0, 0, 0, 0, 0, 0, 0, 0, 0, 0, 0},
        <<398>>, {0, 0, 0, 0, 0, 0, 0, 0, 0, 0, 0, 0, 0, 0, 0, 0, 0, 0, 0, 0,
        0, 0, 0, 0, 0, 0, 0, 0, 0, <<342>>, 0, 0, 0, 0, 0, 0, 0, 0, 0,
        0, 0, 0, 0, 0, 0, 0, 0, 0, 0, 0, 0, 0, 0, 0, 0, 0, 0, -399, 1}}
```

In[85]:= **sparseTriDiagonalMatrix = SparseArray @ Flatten[**
       Table[{i, j} -> Which[i == j, 1, i == j - 1 || i == j + 1, (-1)^j j],
          {i, 400}, {j, 400}] /. (_ -> Null) -> Sequence[]]

Out[85]= SparseArray[<1198>, {400, 400}]

Like other functions that return potentially large outputs (for instance the function InterpolatingFunc⌐
tion from above), in outputs, sparse arrays are shown in abbreviated form. The first element indicates the
number of nonvanishing elements and the second the tensor dimension.

Mathematically the two matrices are identical.

In[86]:= **sparseTriDiagonalMatrix == fullTriDiagonalMatrix**

Out[86]= True

While Equal checks mathematical equivalence, SameQ checks structural equivalence. Structurally the two
matrices are not identical. One is a dense matrix with head List and the other is a sparse array with head
SparseArray.

In[87]:= **sparseTriDiagonalMatrix === fullTriDiagonalMatrix**

Out[87]= False

Because the sparse matrix does not store the 158802 zeros, it has a much smaller ByteCount.

In[88]:= **ByteCount /@ {fullTriDiagonalMatrix, sparseTriDiagonalMatrix}**

Out[88]= {640060, 11560}

Most numerical algorithms dealing with sparse matrices use floating point numbers. So, we create approximate
versions of the last two matrices. (Be aware that N is only numericalizing the matrix elements of a sparse array,
not the positions of the nonvanishing elements.)

In[89]:= **{fullTriDiagonalMatrixN, sparseTriDiagonalMatrixN} =**
      N @ {fullTriDiagonalMatrix, sparseTriDiagonalMatrix};

Again, the sparse matrix is much smaller.

In[90]:= **ByteCount /@ {fullTriDiagonalMatrixN, sparseTriDiagonalMatrixN}**

Out[90]= {1280060, 16352}

Here are some timings for adding and multiplying the dense and the sparse matrices. The sparse matrix operations are substantially faster.

In[91]:= **Function[{op, o}, {op, Function[m, Timing[Nest[op[#1, m]&, m, o]][[1]]] /@**
              {fullTriDiagonalMatrixN, sparseTriDiagonalMatrixN}}] @@@
              {{Plus, 100}, {Times, 100}, {Dot, 10}}

Out[91]= {{Plus, {0.67 Second, 0.01 Second}},
       {Times, {0.86 Second, 0. Second}}, {Dot, {0.47 Second, 0.01 Second}}}

Given a sparse matrix, one might want to convert it either to a dense matrix or to a list of rules indicating the
nonvanishing elements. The two functions Normal and ArrayRules carry out these conversions.

Normal[*sparseArray*]
    gives the sense version of the array *sparseArray* with head List.

ArrayRules[*sparseArray*]
    gives a list of the positions and values of the elements of the sparse array *sparseArray*.

Here is a small sparse vector.

In[92]:= **saV = SparseArray[{1, 0, 0, 0, 0, 0, 0, 0, 1}]**

Out[92]= SparseArray[<2>, {9}]

And here is the full vector recovered.

In[93]:= **Normal[saV]**

Out[93]= {1, 0, 0, 0, 0, 0, 0, 0, 1}

Now, we convert the sparse vector into a list of rules exhibiting the nonvanishing elements. The last rule { _ } -> 0 indicates the not-explicitly-declared background values.

In[94]:= **ArrayRules[saV]**

Out[94]= {{1} → 1, {9} → 1, {_} → 0}

Similar to rational numbers and complex numbers, that both display as compound objects, sparse arrays are atomic objects. They cannot be structurally taken apart.

In[95]:= **AtomQ[saV]**

Out[95]= True

In[96]:= **{Level[saV, {-Infinity, Infinity}], Depth[saV]}**

Out[96]= {{SparseArray[<2>, {9}]}, 2}

But functions like Map act on the elements of sparse arrays (so, in the next example, after applying $\mathcal{F}$ to each element, the background value is $\mathcal{F}[0]$).

In[97]:= **Map[$\mathcal{F}$, saV]**

Out[97]= SparseArray[<2>, {9}, $\mathcal{F}[0]$]

In[98]:= **Normal[%]**

Out[98]= {$\mathcal{F}[1]$, $\mathcal{F}[0]$, $\mathcal{F}[0]$, $\mathcal{F}[0]$, $\mathcal{F}[0]$, $\mathcal{F}[0]$, $\mathcal{F}[0]$, $\mathcal{F}[0]$, $\mathcal{F}[1]$}

Here is a sparse $10^6 \times 10^6$ matrix with two nonvanishing elements.

In[99]:= **SparseArray[{{1, 1} -> 1, {10^6, 10^6} -> 1}]**

Out[99]= SparseArray[<2>, {1000000, 1000000}]

Because the dense form of the matrix would be enormously huge, *Mathematica* does not even try to create it.

In[100]:= **Normal[%]**

           SparseArray::ntb :
          Cannot convert the sparse array SparseArray[Automatic, ≪3≫] to
            an ordinary array because the 1000000000000 elements
            required exceeds the current size limit. More…

Out[100]= SystemException[SparseArrayNormalLimit,
       Normal[SparseArray[<2>, {1000000, 1000000}]]]

Nontrivial linear algebra operations (meaning other than Plus, Times, and Dot) currently work only with machine precision sparse matrices. No exact or high-precision matrices can be dealt with.

We now give two examples of the use of sparse matrices. We start with a larger version of the above electrical grid example. This time, we use a 120×120 grid with the drain and source at opposite corners. We treat two instances, one with all edges having resistance 1 and one with randomly chosen resistances 1 and 2 [1933], [995], [1722].

```
In[101]:= Module[{RCache, g = 120},
        (* all resistances are 1 *)
        ece1 = ElectricCurrentEquationsOnRectangularResistorGrid[
                      I, 1.&, {1, 1}, {g, g}, {g, g}];
        (* resistances are 1 or 2 with equal probability *)
        RCache[x_, y_] := (RCache[x, y] = RCache[y, x] =
                                      If[Random[] < 0.5 , 1., 2.]);
        ece2 = ElectricCurrentEquationsOnRectangularResistorGrid[
                      I, RCache, {1, 1}, {g, g}, {g, g}]];
```

Because ElectricCurrentEquationsOnRectangularResistorGrid returns equations, and because we want to use sparse arrays, we convert the equations to a sparse coefficient matrix and a (inhomogeneous right-hand side) vector using the function toSparseArrays.

```
In[102]:= toSparseArrays[{eqs_, vars_}] :=
        Module[{eqs1 = Most[eqs], varsToIntegerRules, rhs, L2, L3},
         (* number the I-variables *)
         varsToIntegerRules = Dispatch[MapIndexed[(#1 -> #2[[1]])&, vars]];
         (* form right-hand side *)
         rhs = Last /@ eqs1;
         (* form coefficient matrix *)
         L2 = (First /@ eqs1) /. Plus -> List /.
                                 (c_. i_I) :> (i -> c) /. varsToIntegerRules;
         L3 = MapIndexed[Function[{eq, p}, ({p[[1]], #[[1]]} -> #[[2]])& /@ eq], L2];
         {SparseArray[Flatten[L3]], SparseArray[rhs]}]
```

Here is a quick look at the nonvanishing elements.

```
In[103]:= {tsa1, tsa2} = toSparseArrays /@ {ece1, ece2}

Out[103]= {{SparseArray[<113762>, {28560, 28560}], SparseArray[<1>, {28560}]},
           {SparseArray[<113762>, {28560, 28560}], SparseArray[<1>, {28560}]}}
```

```
In[104]:= Show[Graphics[Point[#1]& @@@ Take[Most[ArrayRules[tsa1[[1]]]], All]],
          PlotRange -> All, Frame -> True, AspectRatio -> Automatic]
```

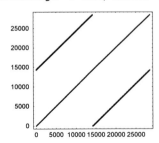

Solving now the two systems of size 28560 × 28560 takes about a second.

```
In[105]:= Timing[{ls1, ls2} = LinearSolve @@@ #;]& @
                       ({tsa1, tsa2} = toSparseArrays /@ {ece1, ece2})

Out[105]= {1.35 Second, Null}
```

The quality of the solution is quite good. Here we calculate the residuals.

```
In[106]:= {Max[tsa1[[1]].ls1 - tsa1[[2]]], Max[tsa2[[1]].ls2 - tsa2[[2]]]}

Out[106]= {2.05847×10^-13, 1.32551×10^-12}
```

This time, we visualize the resulting current distribution using colors. Because the current values extend over several orders of magnitude, we group the currents into approximately equal groups and color the groups consecutively from red to blue.

```
In[107]:= scaledPositions[ls_, ε_:10^-6] :=
      Module[{l1, l2, l3, λ},
              (* sort by current size *)
              l1 = Sort[MapIndexed[{Abs[#1], #2[[1]]}&, ls]];
              (* group *)
              l2 = Split[l1, Abs[#1[[1]]]/#2[[1]] - 1] < ε&];
              λ = Length[l2];
              (* reorder *)
              l3 = MapIndexed[Function[{e, p},
                              {#[[2]], (p[[1]] - 1)/(λ - 1)}& /@ e], l2];
              Last /@ Sort[Flatten[l3, 1]]]
```

The grid with uniform resistance values gives a smooth coloring and the random resistances show a fractal-like current distribution.

```
In[108]:= Show[GraphicsArray[
      Graphics[{Thickness[0.002],
                      Transpose[{Hue[0.78 (1 - #)]& /@ scaledPositions[#2],
                              Line[{##}]& @@@ Last[#1]}]},
              PlotRange -> All, AspectRatio -> Automatic]& @@@
                                              {{ece1, ls1}, {ece2, ls2}}]]
```

 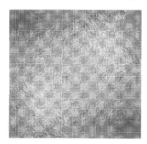

We continue with an example that calculates eigenvalues of a sparse matrix. The celebrated Anderson model represents a disordered system [746]. The function ε represents the site energies and the function V the nearest neighbor interaction energies.

```
In[109]:= AndersonHamiltonianNN[d_, ε_, V_] :=
      Module[{sites, εR, nearestPairs, VR, nonVanishingElements, toSiteNumber,
              renumberedNonVanishingElements},
              (* the sites *)
              sites = Flatten[Table[{i, j}, {i, d}, {j, d}], 1];
              (* site energies *)
              εR = ({#, #} -> ε[#])& /@ sites;
              (* nearest neighbor transfer energies *)
              nearestPairs = Union[Sort /@ Flatten[
                      {{{#1, #2}, {Mod[#2 - 1, d, 1], #1}},
```

```
                      {{#1, #2}, {#1, Mod[#2 - 1, d, 1]}},
                      {{#1, #2}, {Mod[#2 + 1, d, 1], #1}},
                      {{#1, #2}, {#1, Mod[#2 + 1, d, 1]}}}& @@@ sites, 1]];
         VR = Join[#, (Reverse[#1] -> #2)& @@@ #]&[
                      ((# -> V[#])& /@ nearestPairs)];
      (* construct Hamilton matrix *)
      nonVanishingElements = Join[εR, VR];
      (* number the sites *)
      toSiteNumber[{i_Integer, j_Integer}] := (i - 1) d + j;
      (* construct Hamilton matrix *)
      renumberedNonVanishingElements =
                 ((toSiteNumber /@ #1) -> #2)& @@@ nonVanishingElements;
      SparseArray[renumberedNonVanishingElements]]
```

For a 2D 200×200 lattice, we obtain a matrix of size 40000×40000 with 279200 nonvanishing elements.

```
In[110]:= SeedRandom[123];
      AH = AndersonHamiltonianNN[200, Random[Real, {1, 2}]&, Random[Real, {2, 4}]&];
```

Because calculating all eigenvalues of a large matrix is very expensive, the function Eigenvalues (and the sister functions Eigenvectors and Eigensystem) takes a second argument indicating the number of eigenvalues (eigenvectors) to be returned. We explicitly specify the method to be used to calculate the eigenvalues. The criteria "BothEnds" gives the lowest and the highest eigenvalues. We also give a value for the Tolerance option to specify that we are not looking for very very precise values of the eigenvalues.

```
In[112]:= esData = Eigensystem[AH, 6, Method -> {"Arnoldi", BasisSize -> 200,
                 Criteria -> "BothEnds", MaxIterations -> 10^4, Tolerance ->
         10^-4}];
```

We see that the spectrum extends approximately from -17 to $+20$.

```
In[113]:= esData[[1]]
Out[113]= {19.8991, 19.8518, 19.8326, -16.8548, -16.8244, -16.822}
```

Specifying an explicit shift returns the eigenvalues in the neighborhood of a given value. Here we calculate and visualize the three eigenvectors for each of a set of intermediate energies. The eigenvectors show complicated spatial patterns.

```
In[114]:= Function[λ,
         (* calculate eigenvectors near a given spectrum value *)
         esData =
         Eigensystem[AH, 3, Method -> {"Arnoldi", BasisSize -> 200, Shift -> λ,
                           MaxIterations -> 10^4, Tolerance -> 10^-5}];
         (* visualize eigenvectors *)
         Show[GraphicsArray[
          ListDensityPlot[(* form matrix *) Abs[Partition[#, Sqrt[Length[#]]]],
                     Mesh -> False, DisplayFunction -> Identity,
                     ColorFunction -> (Hue[1 - #]&), FrameTicks -> None,
                     PlotRange -> All]& /@ esData[[2]]]]] /@
                 (* selected spectrum values *)
                 {-4.78268, 13.764, 17.7492, 19.249, 19.7145}
```

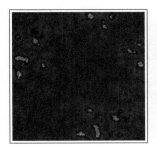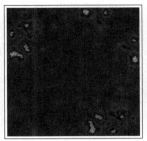

For a detailed discussion of the linear algebra functions, and their options, and sparse arrays, see the Advanced Linear Algebra documentation.

1.5 Fourier Transforms

For many purposes (e.g., analyzing data), Fourier analysis is an extremely valuable tool. Given a periodic function $f(x)$ given on $0 \le x < L$ (with period L), whose values $f_k = f(x_k)$ are given at the n points $x_k = (k-1)L/n$, $k = 1, 2, \ldots, n$ (and nowhere else), we seek to interpolate these function values using trigonometric functions, that is, as a linear combination of the functions $\phi_l(x) = n^{-1/2} e^{-2\pi i (l-1)x/L}$, $j = 1, \ldots, m$. Thus, the expansion has the form $f(x) = \sum_{j=1}^{n} c_j \phi_j(x)$.

Because $\sum_{j=1}^{n} \phi_l(x_j) \phi_k(x_j) = \delta_{lk}$, we get easily $c_k = n^{-1/2} \sum_{j=1}^{n} f_j \exp(2\pi i (j-1)(k-1)/n)$.

These coefficients correspond to the best approximation of $f(x)$ by a function of the form above, relative to the L^2-norm. See [894], [1476], [1868], [1732], [663], [293], [562], [907], [676], [462], [1297], [1101], [823], and [399] for more details as well as applications and [95] for some more theoretical considerations. The discrete Fourier transform is implemented in *Mathematica* as follows.

```
Fourier[{x₁, x₂, ..., xₙ}]
```
> produces the list $\{y_1, y_2, \ldots, y_n\}$ that is the discrete Fourier transform of the list $\{x_1, x_2, \ldots, x_n\}$, defined by $y_j = n^{-1/2} \sum_{k=1}^{n} x_k \exp(2\pi i(k-1)(j-1)/n)$.
> Multidimensional data can be transformed by
> ```
> Fourier[{{x₁₁, x₁₂, ..., x₁ₙ},
>     {x₂₁, x₂₂, ..., x₂ₙ}, ...,
>     {xₘ₁, xₘ₂, ..., xₘₙ}}].
> ```
> If we input exact data, the data will be numericalized to machine precision.

This means that for a vector l of length n, `Fourier[l]` effectively carries out the matrix product **F.**l, where the (j, k) element of the square matrix **F** is $n^{-1/2} \exp(2\pi i(k-1)(j-1)/n)$. (For properties of this matrix, see [504], [1239].)

Be aware that this definition of the Fourier transform is slightly different from what is typically used in engineering. (*Mathematica* uses a different normalization and a different sign of the exponential kernel.) (For a listing of exactly doable discrete Fourier transforms, see [294].)

We start with a simple example. Here is the Fourier spectrum of the function $\cos(500\,t) \sum_{k=5}^{15} \cos(10\,k\,t)/k$ sampled at the 32769 data points $t_k = k\,2\pi/32768$ (in engineering terms, the frequency spectrum of an amplitude-modulated carrier). The calculation of the data points, the Fourier transformation, and the display of the result take less than a second on a 2 GHz computer.

```
In[1]:= ListPlot[Abs[Take[Fourier[
        Table[Evaluate[Cos[500. t] Sum[Cos[k t]/k, {k, 50., 150., 10}]],
            {t, 0., 2.Pi, 2.Pi/32768}]], 1000]],
        PlotRange -> All, PlotJoined -> True,
        Frame -> True, Axes -> False] // Timing
```

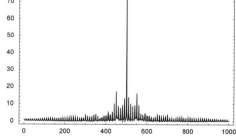

In the next graphics, we vary the frequency according to $\cos((500 + 10 \sum_{k=5}^{15} \cos(10\,k\,t)/k)\,t)$.

```
In[2]:= ListPlot[Abs[Take[Fourier[
        Table[Evaluate[Cos[(500. + 10 Sum[Cos[k t]/k, {k, 50., 150., 10}]) t]],
            {t, 0., 2.Pi, 2.Pi/32768}]], 1000]], PlotRange -> All,
        PlotJoined -> True, Frame -> True, Axes -> False]
```

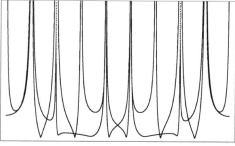

The following graphic shows the Fourier transform of the base 10 digits of $(\sum_{k=0}^{65536} 10^k)^2 = 111\ldots111^2$.

```
In[3]:= ListPlot[Abs[Fourier[IntegerDigits[(* form the large integer *)
            FromDigits[Flatten[Table[1, {2^16}]]]]^2, 10]],
        PlotStyle -> {PointSize[0.003]},
        Axes -> None, Frame -> True, FrameTicks -> None]
```

And the next graphic shows the absolute value of the Fourier transform of $2^{s_2(k)}$ [409], where $s_2(k)$ is the sums of digits of the integer k in base 2.

```
In[4]:= ListPlot[Abs @ Fourier[2^Table[DigitCount[k, 2, 1], {k, 2^16}]],
            PlotStyle -> {PointSize[0.003]}, PlotRange -> {0, 1000},
            Axes -> None, Frame -> True, FrameTicks -> None]
```

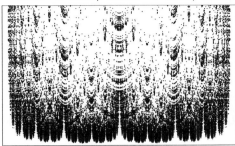

Here is a more complicated-looking Fourier transform. The piecewise character of the function results in a "many individual curves" appearance. (A similar curve was considered in [1739].)

```
In[5]:= ListPlot[Abs[Fourier[Table[Round[x] Sin[x]/x,
                        {x, 1, 66, 66/(2^16 - 1)}]]],
            PlotStyle -> {PointSize[0.002]},
            Frame -> True, PlotRange -> {0, 0.02}, FrameTicks -> None]
```

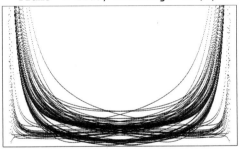

Now let us look at the approximation through trigonometric functions in more detail. Here is a more simple example using the values of $f(x) = x^2$ on the interval [0, 4].

```
In[6]:= δx = 0.1;
        values = Table[N[x^2], {x, 0, 4, δx}]
Out[7]= {0., 0.01, 0.04, 0.09, 0.16, 0.25, 0.36, 0.49, 0.64, 0.81, 1., 1.21, 1.44, 1.69,
        1.96, 2.25, 2.56, 2.89, 3.24, 3.61, 4., 4.41, 4.84, 5.29, 5.76, 6.25, 6.76, 7.29,
        7.84, 8.41, 9., 9.61, 10.24, 10.89, 11.56, 12.25, 12.96, 13.69, 14.44, 15.21, 16.}
```

Here is the approximation in terms of trigonometric functions.

```
In[8]:= n = Length[values];
        L = δx n;
        trigExpBasis = Table[N[1/Sqrt[n] Exp[-2Pi I x/L (l - 1)]], {l, n}];

        quadApprox[x_] = Dot[Fourier[values], trigExpBasis];
```

This approximation gives the input values at the data points. (We use Chop to eliminate small imaginary components which arise in the computation.)

```
In[12]:= Table[Chop[quadApprox[x]], {x, 0, 4, δx}]
Out[12]= {0, 0.01, 0.04, 0.09, 0.16, 0.25, 0.36, 0.49, 0.64, 0.81, 1., 1.21, 1.44, 1.69, 1.96,
         2.25, 2.56, 2.89, 3.24, 3.61, 4., 4.41, 4.84, 5.29, 5.76, 6.25, 6.76, 7.29, 7.84,
         8.41, 9., 9.61, 10.24, 10.89, 11.56, 12.25, 12.96, 13.69, 14.44, 15.21, 16.}
```

Between the data points, the function oscillates strongly.

```
In[13]:= makeFourierGraphicsPair[approxFunction_] :=
         Show[GraphicsArray[
         Block[{$DisplayFunction = Identity},
         {(* real part *)
          Plot[{Re[approxFunction[x]], x^2}, {x, 0, 4}, PlotPoints -> 200,
              PlotRange -> {-5, 20}, Frame -> True, Axes -> False,
              PlotStyle -> {Thickness[0.002]},
              Epilog -> {PointSize[0.015],
                              Point /@ Table[{x, x^2}, {x, 0, 4, 0.1}]}]},
          (* imaginary part *)
          Plot[Im[approxFunction[x]], {x, 0, 4}, PlotPoints -> 200,
              PlotStyle -> {Thickness[0.002]}, PlotRange -> {-10, 10},
              Frame -> True, Axes -> False, Epilog -> {PointSize[0.015],
                              Point /@ Table[{x, 0}, {x, 0, 4, 0.1}]}}]]]]

In[14]:= makeFourierGraphicsPair[quadApprox]
```

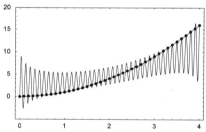 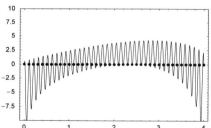

Although the approximation is exact at the given data points, the difference of the two curves between the data points is considerable. The approximation can be greatly improved by observing that we only took into account positive Fourier frequencies in the periodically continued Fourier series expansion of x^2 [1460]. We symmetrize and use positive and negative l by using the identity $\exp(-2\pi i x_j/4.1\,(41 - l)) = \exp(-2\pi i x_j/4.1\,(-l))$ for the data points x_j. This does not change the values at the data point, but greatly improves the approximation between them [1588], [341], [88] (this process is also called unaliasing). Here, the results for the above example are shown.

```
In[15]:= trigBasisExpPos = Table[N[1/Sqrt[n] Exp[-2Pi I x/L l]], {l, 0, 20}];
         trigBasisExpNeg = Table[N[1/Sqrt[n] Exp[-2Pi I x/L l]], {l, -1, -20, -1}];

         quadApprox1[x_] =
         Function[fourier, (* positive l *)
           Take[fourier, 21].trigBasisExpPos + (* negative l *)
           Reverse[Take[fourier, {22, n}]]. trigBasisExpNeg][Fourier[values]];

In[18]:= (* no change at the data points *)
         Table[Chop[quadApprox1[x]], {x, 0, 4, δx}]
Out[19]= {0, 0.01, 0.04, 0.09, 0.16, 0.25, 0.36, 0.49, 0.64, 0.81, 1., 1.21, 1.44, 1.69, 1.96,
         2.25, 2.56, 2.89, 3.24, 3.61, 4., 4.41, 4.84, 5.29, 5.76, 6.25, 6.76, 7.29, 7.84,
         8.41, 9., 9.61, 10.24, 10.89, 11.56, 12.25, 12.96, 13.69, 14.44, 15.21, 16.}
```

We display the two graphics from above, but this time we use the approximation `quadApprox1`. The real part approximates much better now and the imaginary part now vanishes identically—as it should for a good approximation.

In[20]:= **makeFourierGraphicsPair[quadApprox1]**

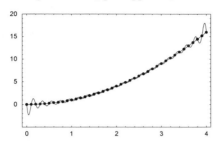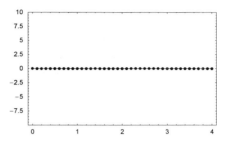

Numerical Fourier transforms are very fast; several (tens of) thousand data points can be handled on virtually any modern computer. Here is the Fourier transform corresponding to the first 100000 digits of π. There does not appear to be any particular pattern; this means there is no hidden periodicity.

In[21]:= **ListPlot[Take[Abs[Fourier[N[First[RealDigits[N[Pi, 2^17]]]]]], {2, 2^16}],**
       PlotRange -> All, PlotStyle -> {PointSize[0.003]}] // Timing

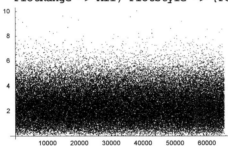

Most of the well-known theorems and properties of the continuous Fourier transform have equivalents for the discrete Fourier transform. Here, we give just one example: The famous Heisenberg uncertainty relation for the discrete Fourier transform reads $n_x\, n_\omega \geq l$, where n_x is the number of nonvanishing elements of a list *list*, n_ω is the number of nonvanishing elements of a `Fourier[list]`, and l is the length of *list* and not all elements of list are identical zero [513], [297], [296], [620], [621], [745], [336]. The corresponding additive theorem is $n_x + n_\omega \geq \sqrt{2}\, l$. Here is an example in which the lower bound is achieved.

In[22]:= (* an eigenfunction of Fourier *)
       Fourier[{1, 0, 0, 1, 0, 0, 1, 0, 0}] // Chop
Out[23]= {1., 0, 0, 1., 0, 0, 1., 0, 0}

(For the discrete equivalent of the Poisson formula, see [1820].)

Looking at the Fourier transform of data is often a very good idea. Regular patterns in data as well as sometimes irregular patterns in data will often exhibit some of their structure in the Fourier transform. The following shows some generalizations of Strang's strange figures [1720], [1528] and their Fourier transform. The function `strangPlot` makes a pair of plots, one for the data and one for the absolute values of the Fourier transform of the data.

```
In[24]:= strangPlot[functionPair_, n_] :=
    Module[{data = N @ Table[functionPair[i], {i, n}]},
           Show[GraphicsArray[
           Block[{$DisplayFunction = Identity},
           {ListPlot[(* the points *) data,
                   PlotStyle -> {PointSize[0.003]},
                   Frame -> True, Axes -> False, PlotRange -> {All, All}],
               ListPlot[(* Fourier transform of the points *) Abs[Fourier[data]],
                   PlotStyle -> {PointSize[0.003]},
                   Frame -> True, Axes -> False, PlotRange -> Automatic]}]],
                   DisplayFunction -> $DisplayFunction]]
```

```
In[25]:= strangPlot[Function[n, {n^0.7, Sin[n^0.92]}], 10^4]
```

```
In[26]:= strangPlot[Function[n, {n^0.51, Sin[n^1.11]}], 10^4]
```

```
In[27]:= strangPlot[Function[n, {n^0.02, Sin[n^1.02]}], 10^4]
```

```
In[28]:= strangPlot[Function[n, {Log[n], Sin[n^(0.9)]}], 10^4]
```

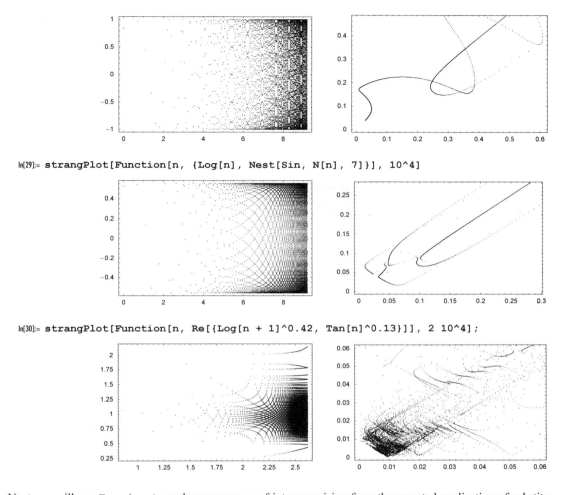

```
In[29]:= strangPlot[Function[n, {Log[n], Nest[Sin, N[n], 7]}], 10^4]
```

```
In[30]:= strangPlot[Function[n, Re[{Log[n + 1]^0.42, Tan[n]^0.13}]], 2 10^4];
```

Next, we will use `Fourier` to analyze sequences of integers arising from the repeated application of substitution rules. The function `substitutionFourierPlot` plots the Fourier transform of the sequence arising from *o* applications of the rules *rules* starting with the sequence {1}.

```
In[31]:= substitutionFourierPlot[rules_, o_, pr_, opts___] :=
    ListPlot[Take[Abs[Fourier[Nest[Flatten[FoldList[(#1 /. #2)&, #,
                                        rules]]&, {1}, o]]],
                {2, -2}], opts, PlotRange -> pr, FrameTicks -> False,
                    Frame -> True, Axes -> False, AspectRatio -> 1,
                    PlotStyle -> {PointSize[0.003]}];
```

Here are some examples. Depending on the rules, the Fourier transforms of the resulting sequences show a wide variety of shapes.

```
In[32]:= Show[GraphicsArray[
    substitutionFourierPlot[##, DisplayFunction -> Identity]& @@@
    {(* "random" Fourier transform *)
     {{1 -> {2, 1}, 2 -> {1}, 3 -> {1, 1, 4, 4, 5}, 4 -> {4, 4},
       5 -> {5, 2, 1}}, 4, {0, 0.2}},
```

```
(* "hierarchical" Fourier transform *)
{{{1 -> {5}, 2 -> {5}, 3 -> {5}, 4 -> {6}, 5 -> {2},
    6 -> {1}}, {1 -> {1}}}, 10, {0, 1}},
(* "uniformly distributed" Fourier transform *)
{{{1 -> {3}, 2 -> {1, 2}, 3 -> {1, 2}, 4 -> {5, 4}, 5 -> {4, 3}}},
 10, {0, 0.1}},
(* "uniformly distributed" Fourier transform *)
{{{1 -> {3, 1}, 2 -> {2}, 3 -> {1}}}, 11, {0, 0.05}},
(* "periodic" Fourier transform *)
{{{1 -> {1}}, {1 -> {4, 4}, 2 -> {2, 1}, 3 -> {3, 4, 4},
    4 -> {4, 4}, 5 -> {1, 5}}, {1 -> {2, 1, 2},
    2 -> {1, 1, 2, 1}}, {1 -> {1, 2}, 2 -> {2}}},
   {1 -> {1, 1}, 2 -> {1, 2, 2, 1, 2}}}, 4, {0, 0.3}}}]]
```

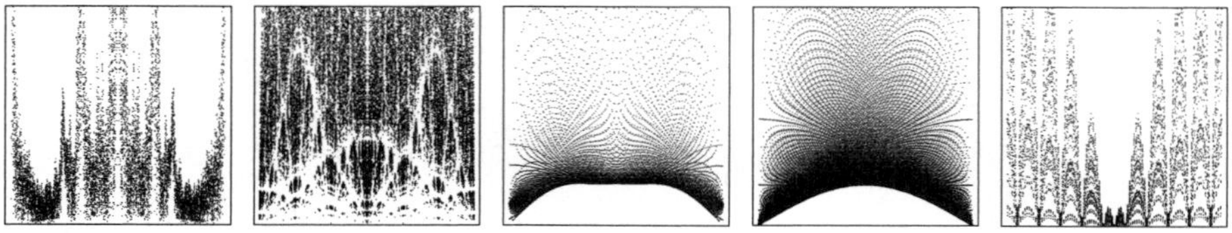

```
In[33]:= Show[(* now show all values *)
        Map[Show[#, PlotRange -> All]&, %, {2}]]
```

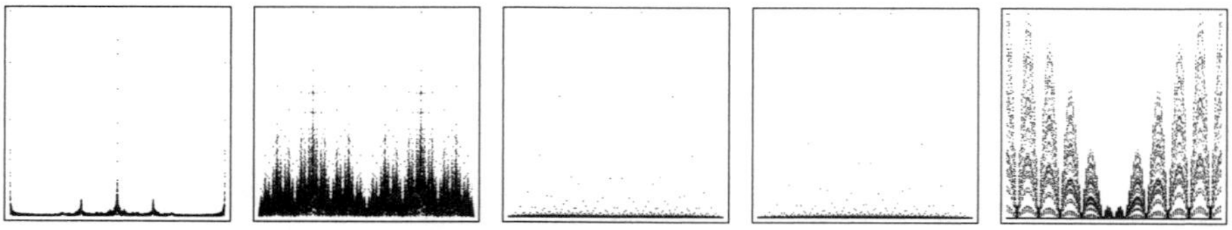

Numerical Fourier transforms can be carried out in lists of any length. (For multidimensional lists, they must be tensors.) The timing for lists of length n and length $n + 1$ may differ considerably. The smallest timings are achieved for n being a power of 2. The next graphic shows the timings for carrying out a Fourier transform of an $n \times n$ matrix with random entries. The left picture shows the timing as a function of n. The right graphic shows the timing as a function of $n^2 / numberOfDivisorsOfn$. The more divisors n, the faster the Fourier transform can be carried out.

```
In[34]:= SeedRandom[111];
        timings = With[{o = 512},
        Module[{mat = Table[Random[], {o}, {o}]},
            (* make table of timing *)
            Table[{n, Timing[Fourier[Take[mat, {1, n}, {1, n}]]][[1, 1]]},
                {n, o}]]];
In[36]:= Show[GraphicsArray[{
        ListPlot[timings, DisplayFunction -> Identity],
        ListPlot[Apply[{#1^2/DivisorSigma[0, #1]/10^4, #2}&, timings, {1}],
                PlotRange -> All, DisplayFunction -> Identity]}]]
```

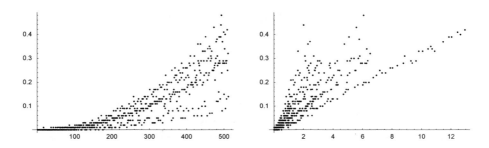

If the number of elements in the input list is highly composite, which means that it factors in many prime factors, the Fourier transform can be done more quickly. Here is a graph of the times needed by `Fourier` to transform `Range[n]` depending on the number of prime factors (we discuss in the next chapter how to determine them) for $2 \le n \le 256$. (Every point corresponds typically to more than one Fourier transformation, and the clustering is from the granularity of the time measurement.)

```
In[37]:= data = Table[{j, (* how compound is j *)
                Plus @@ Last[Transpose[FactorInteger[j]]],
                (* timing for the Fourier analysis;
                   repeat for reliable timings *)
                Timing[Do[Fourier[N[Range[j]]], {10}]][[1, 1]]},
              {j, 2, 2^12}];
```

```
In[38]:= Show[GraphicsArray[ListPlot[#1 /@ data, AxesOrigin -> {0, 0},
            PlotRange -> All, AxesLabel -> {#2, "t/s"},
            DisplayFunction -> Identity]& @@@
              {{Rest, "prime factors"}, {Delete[#, 2]&, "list length"}}]]]
```

Mathematica uses the FFT algorithm for a fast calculation of the numerical Fourier transform. While typically the resulting Fourier transform does not show imprints of the algorithm, we can make them visible. In the following, we first calculate the high-precision Fourier transform of a list of *l* ones. Then, ignoring the first element, we determine the accuracy of the resulting zeros. A Fourier transform of these accuracies shows a clear pattern. The pattern is distinctly different for *l* being a power of two, a highly composite number, or a prime.

```
In[39]:= Show[GraphicsArray[Function[l,
        ListPlot[(* Fourier transform of the accuracies *)
                Rest[Abs[Fourier[Accuracy /@
            (* Fourier transform of the list of high-precision ones *)
          Rest[Fourier[Table[N[1, $MachinePrecision + 1], {l}]]]]]],
                PlotRange -> All, PlotStyle -> {PointSize[0.003]},
                DisplayFunction -> Identity,
                Ticks -> {{2000, 4000}, {1, 2}}]] /@
                    (* the four lengths (of about 4000) --
```

from highly composite to prime numbers *)
{2^12, 2^12 + 1, 2^2 3 5 7 11, Prime[565]}]]

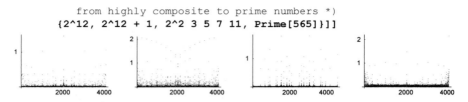

The inverse Fourier transform can be obtained as follows.

InverseFourier[{y_1, y_2, ..., y_n}]

produces the list {x_1, x_2, ..., x_n} that is the discrete inverse Fourier transform of {y_1, y_2, ..., y_n}, defined by $x_j = n^{-1/2} \sum_{k=1}^{n} y_k \exp(-2\pi i(k-1)(j-1)/n)$.

InverseFourier[{{y_{11}, y_{12}, ..., y_{1n}},
        {y_{21}, y_{22}, ..., y_{2n}}, ...,
        {y_{m1}, y_{m2}, ..., y_{mn}}}]

is used for multidimensional data.

Thus (up to numerical inaccuracies),
Fourier[InverseFourier[*data*]] = InverseFourier[Fourier[*data*]] = *data*
holds.

High-precision numbers are Fourier-transformed into high-precision numbers. Here is an example in which the input consists of exact numbers.

In[40]:= **InverseFourier[Table[i, {i, 12}]]**

Out[40]= {22.5167 + 0. i, −1.73205 + 6.4641 i, −1.73205 + 3. i, −1.73205 + 1.73205 i,
        −1.73205 + 1. i, −1.73205 + 0.464102 i, −1.73205 + 0. i, −1.73205 − 0.464102 i,
        −1.73205 − 1. i, −1.73205 − 1.73205 i, −1.73205 − 3. i, −1.73205 − 6.4641 i}

The exact input was automatically converted to machine-precision approximative numbers. Here is the result in which the "same" numbers are given with 40 digits.

In[41]:= **InverseFourier[Table[N[i, 40], {i, 4}]]**

Out[41]= {5.0000000000000000000000000000000000000000,
        −1.0000000000000000000000000000000000000000+
        1.0000000000000000000000000000000000000000i,
        −1.0000000000000000000000000000000000000000,
        −1.0000000000000000000000000000000000000000−
        1.0000000000000000000000000000000000000000i}

In[42]:= **Precision[%]**

Out[42]= 39.2218

We now perform a Fourier analysis of a model signal that consists of six frequencies of different amplitudes.

In[43]:= **fourierSignal =
    Fourier[** (* six sine waves of different frequencies and amplitudes *)
            **signal = Table[N[Sum[1/(i + 2) Sin[20 i 2 Pi n/512], {i, 6}]],
                {n, 512}]];**

The following plot shows the Fourier-transformed signal. The six different frequencies and the relative absolute values of the amplitudes are clearly visible. The right graphics shows the signal as recovered from the Fourier transform.

```
In[44]:= Show[GraphicsArray[
    ListPlot[Take[#, 200], DisplayFunction -> Identity,
        PlotRange -> All, PlotJoined -> True]& /@
    {Abs /@ fourierSignal, InverseFourier[fourierSignal]}]]
```

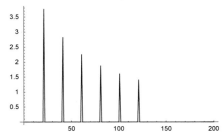 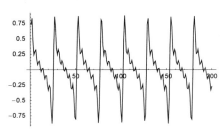

We now look at the Fourier transform of a more complicated object: the outline of the number 4. The following list gives the representation of an idealized 4 (the first number provides the arc length of the boundary, and the following list contains the coordinates of the true points).

```
In[45]:= four = {{ 0, {0, 2}}, { 1, {1, 2}}, { 2, {2, 2}}, { 3, {3, 2}},
    { 4, {3, 1}}, { 5, {3, 0}}, { 6, {4, 0}}, { 7, {4, 1}},
    { 8, {4, 2}}, { 9, {5, 2}}, {10, {6, 2}}, {11, {6, 3}},
    {12, {5, 3}}, {13, {4, 3}}, {14, {4, 4}}, {15, {3, 4}},
    {16, {3, 3}}, {17, {2, 3}}, {18, {1, 3}}, {19, {1, 4}},
    {20, {1, 5}}, {21, {1, 6}}, {22, {1, 7}}, {23, {0, 7}},
    {24, {0, 6}}, {25, {0, 5}}, {26, {0, 4}}, {27, {0, 3}},
    { 0, {0, 2}}} // N;
```

We now examine the 4 graphically.

```
In[46]:= Show[Graphics[{Hue[0], Polygon[#[[2]]& /@ four]}],
    Axes -> True, AspectRatio -> Automatic]
```

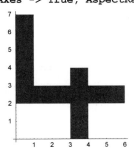

Next, we extract the *x*- and *y*-coordinates, and compute their Fourier transforms.

```
In[47]:= Clear[ξ, η, xf, yf, num, appx, x, appy, y];

    (* x- and y-coordinates of the polygon vertices *)
    ξ = #[[2, 1]]& /@ four; η = #[[2, 2]]& /@ four;

    (* Fourier transforms of the x- and y-coordinates *)
    xf = Fourier[ξ]; yf = Fourier[η];

    num = Length[four];
```

Because we have only real values, we compute a real-valued Fourier transform, in which we replace $e^{2i\pi\cdots}$, $e^{-2i\pi\cdots}$ by $n^{-1/2}\sin(2\pi x/L\,(l-1))$ and $n^{-1/2}\cos(2\pi x/L\,(l-1))$, observing that $y_i = \bar{y}_{n-i+2}$ ($i \geq 2$, x_i real-valued). We will use the inverse transform not only at the data points, (where the values are just the prescribed input), but also in between. These values can be found using the above basis functions that are functions of the continuous variable x. We are led to the following approximations for `appx[i][x]` and `appy[i][y]` involving sin and cos terms.

```
In[53]:= Function[{appxy, xyf},
    (* approximation of various orders *)
    appxy[1][xy_] = N[1/Sqrt[num] Re[xyf[[1]]]];
    Do[appxy[i][xy_] = appxy[i - 1][xy] +
      N[1/Sqrt[num] (2 Re[xyf[[i]]] Cos[2Pi (xy - 1)/num (i - 1)] +
              2 Im[xyf[[i]]] Sin[2Pi (xy - 1)/num (i - 1)])],
     {i, 2, (num + 1)/2}]] @@@ {{appx, xf}, {appy, yf}};
```

The two functions `appx[i]` and `appy[i]` approximate the true values at the data points to order 15 exactly. (Here, we need only about half as large an n as compared with $e^{2i\pi\cdots}$, because for given n, we have two basis functions, sin and cos.)

```
In[54]:= Max[Abs[Table[Chop[#[15][xy]], {xy, 29}] - #[2]]]& @@@
                                        {{appx, ξ}, {appy, η}}
Out[54]= {3.10862×10^-15, 2.66454×10^-15}
```

If we use fewer Fourier components, the 4 is no longer so easy to identify. We also display the approximations to 4 in three dimensions by stacking them on top of each other. We make use of the function `ListSurface`-`Plot3D` from the package `Graphics`Graphics3D``.

```
In[55]:= Needs["Graphics`Graphics3D`"]

In[56]:= Show[GraphicsArray[
    Block[{$DisplayFunction = Identity},
      {(* 2D plot of approximations of the 4 *)
       Graphics[Table[{Hue[(o - 1)/18], Line[Append[#, First[#]]& @
                    Thread[{Table[appx[o][x], {x, 29}],
                            Table[appy[o][y], {y, 29}]}]]}, {o, 15}],
                  AspectRatio -> Automatic],
       (* 3D plot of staggered approximations of the 4 *)
       ListSurfacePlot3D[Table[{appx[order][t], appy[order][t], order},
                 {order, 2, 15}, {t, 29}],
               Boxed -> False, BoxRatios -> {1, 1, 0.8},
               PlotLabel -> StyleForm["Genesis of 4",
                     FontFamily -> "Helvetica",
                     FontWeight -> "Bold", FontSize -> 7]]}]]]
```

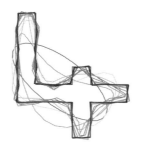

Genesis of 4

In case we are only interested in the Fourier approximations at integer *t* values, there is actually no need to construct functions appx[*order*][*t*] and appy[*order*][*t*], but we can just drop the higher Fourier modes and transform back.

```
In[57]:= fastApp[n_] := Transpose[Re[InverseFourier /@
         (Join[Take[#, n], Table[0, {num - n}]]& /@ {xf, yf})]]

       Show[GraphicsArray[#]]& /@ Partition[
       Table[Graphics[Polygon[fastApp[n]],
               AspectRatio -> Automatic], {n, 2, (num + 1)/2}], 7]
```

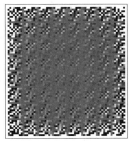

```
Do[Show[Graphics[Polygon[fastApp[n]]],
       AspectRatio -> Automatic], {n, 2, (num + 1)/2}]
```

We can also make related graphics of Fourier transforms from DensityPlot. Here is the 2D Fourier transformation of the above 4.

```
In[59]:= Show[GraphicsArray[ListDensityPlot[#,
         AspectRatio -> 9/8, Mesh -> False, FrameTicks -> None,
         DisplayFunction -> Identity]& /@
      (* the 4 and its Fourier transform *) {#, Im[Fourier[N @ #]]}]]&[
      Function[n, (* discretization of the 4 area *)
      Table[Function[{x, y}, (* the area of 4 *)
      If[(0 < x < 1 && 2 < y < 7) || (0 < x < 6 && 2 < y < 3) ||
         (3 < x < 4 && 0 < y < 4), 0, 1]][x, y],
               {y, 1/(2n) - 1, 7 + 1, 1/n},
               {x, 1/(2n) - 1, 6 + 1, 1/n}]][8]]
```

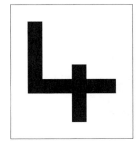

We can also animate how taking into account more and more Fourier modes results finally in the exact 4.

```
With[{steps = 100},
 Module[{tab, ftab, max},
 tab = Function[n,  (* discretization of the 4 area *)
 Table[Function[{x, y},  (* the area of 4 *)
 If[(0 < x < 1 && 2 < y < 7) || (0 < x < 6 && 2 < y < 3) ||
    (3 < x < 4 && 0 < y < 4), 0, 1]][x, y],
               {y, 1/(2n) - 1, 7 + 1, 1/n}, {x, 1/(2n) - 1, 6 + 1,
1/n}]][8];
 (* the Fourier transform *)
 ftab = Fourier[N @ tab];
 (* taking into account only a limited number of modes *)
 max = Plus @@ (Dimensions[tab]^2);
 Do[ListDensityPlot[Re[InverseFourier[
    MapIndexed[If[Plus @@ (#2^2) - i max/steps > 0, 0, #]&,
               ftab, {2}]]], Mesh -> False, FrameTicks -> None],
    {i, steps}]]];
```

 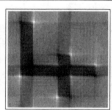  

For more on Fourier transforms of planar curves, see [60], [1247], and [1539]. For DensityPlot-type decompositions of a "G", see [630] and [631]. For a discussion of how to write entire texts by this method, see [706].

A visually interesting, more complicated example is the Gosper curve discussed in Section 1.5.9 of the Graphics volume [1794]. A similar picture by R. W. Gosper can be found on page 24 of [1918]. The following example is an animation of the genesis of the Gosper curve Gosper[3]. For convenience, we give all necessary code. Because the animation contains 184 pictures and needs a lot of memory, we do not carry it out by default.

```
LSystemWithFlAndFr[axiom_, rules_, iter_, δ_] :=
Module[{minus, plus, fastRules, last, dir, c = Cos[δ], s = Sin[δ]},
{minus, plus} = {{{c, s}, {-s, c}}, {{c, -s}, {s, c}}} // N;
fastRules = rules /. {(a_ -> b_) -> (a :> Sequence @@ b)};
last = {0.0, 0.0};   dir = {1.0, 0.0};
Select[Prepend[(Which[# == "l", last = last + dir,
                      # == "r", last = last + dir,
                      # == "+", dir = plus.dir; ,
                      # == "-", dir = minus.dir; ]& /@
    Nest[(# /. fastRules)&, axiom, iter]), {0, 0}], # =!= Null&]]
```

(* the Gosper curve *)
```
Gosper[i_Integer] := LSystemWithFlAndFr[{"l"},
{"l" -> {"l", "+", "r", "+", "+", "r", "-", "l",
         "-", "-", "l", "l", "-", "r", "+"},
 "r" -> {"-", "l", "+", "r", "r", "+", "+", "r",
         "+", "l", "-", "-", "l", "-", "r"}}, i, N[2Pi/6]]
```

(* closing the Gosper curve by a part of a
  circle with appropriate length to use periodicity *)
```
closedGosper =
Module[{l = Gosper[3], mp, d, o, r, τ, α, ll, n = #/Sqrt[#.#]&},
 li = Rest[Gosper[3]]; mp = (First[l] + Last[l])/2;
 d = n[(First[l] - Last[l])]; o = -n[{d[[2]], -d[[1]]}];
 r = Sqrt[#.#]&[(First[l] - Last[l])]/2;
 τ = FindRoot[2ArcTan[r/t] Sqrt[r^2 + t^2] == 24, {t, 10}][[1, 2]];
 α = ArcTan[r/τ];
 (* the circle part *)
 ll = Table[mp - τ o + Sqrt[r^2 + τ^2](Cos[φ] o + Sin[φ] d),
            {φ, α, -α, -2α/24}] // N;
 Join[l, Reverse[ll]]];
```

(* the Fourier transform of the x- and y-coordinates *)
```
{xf, yf} = Fourier[# /@ closedGosper]& /@ {First, Last};
num = Length[xf];
```
(* the approximation *)
```
app[n_] := Transpose[Re[InverseFourier /@
 {Join[Take[xf, n], Table[0, {num - n}]],
  Join[Take[yf, n], Table[0, {num - n}]]}]];
```

(* the pictures as an array of graphics; display only selected ones *)
```
Show[GraphicsArray[#, GraphicsSpacing -> -0.5]]& /@
  Table[Graphics[{Polygon[app[30 i + j + 1]]},
        AspectRatio -> Automatic, PlotRange -> {{-10, 15}, {4, 16}}],
            {i, 0, 4}, {j, 1, 6}]
```

```
(* the animation *)
Do[Show @ Graphics[{Hue[i/(num/2)], Polygon[app[i]]},
        AspectRatio -> Automatic, PlotRange -> {{-10, 15}, {-1, 22}}],
    {i, 2, num/2}]
```

Those who still have memory available can try to make the following 3D graphics, remembering R. W. Gosper's picture of [1918].

```
Needs["Graphics`Graphics3D`"]

Show[Graphics3D[{EdgeForm[],
MapIndexed[{SurfaceColor[Hue[#2[[1]]/90000], Hue[#2[[1]]/90000], 2.5],
#}&,
ListSurfacePlot3D[Table[Append[#, i]& /@ app[i], {i, 3, num/2}],
                DisplayFunction -> Identity][[1]]]}],
    ViewPoint -> {-3, 3, 1.9}, BoxRatios -> {1, 1, 6},
        Boxed -> False, ViewVertical -> {1, 0, 0}]
```

Or the following graphic where we wrap the curve around a cylinder.

```
Module[{ω = 60, λ = Length[app[1]], mp, rAv, ξ, η, points},
    mp = Plus @@ app[1]/λ;
    (* curve to cylinder *)
    wrapAround[k_, xyl_] :=
        (rAv = Plus @@ (Sqrt[#.#]& /@ ((# - mp)& /@ xyl))/λ;
         Table[{ξ, η} = xyl[[j]] - mp; (* keep direction *)
               Append[k #/Sqrt[#.#]&[{ξ, η}], (* radius -> z *)
                     Sqrt[ξ^2 + η^2] - rAv], {j, λ + 1}]);
    (* points of the surface *)
    points = Table[wrapAround[k, Append[#, First[#]]&[app[k]]], {k, 2, ω}];
    (* show the surface *)
    Show[Graphics3D[{EdgeForm[],
    (* polygons of the surface *)
    Table[{SurfaceColor[Hue[j/ω], Hue[j/ω], 2.4],
          Table[Polygon[{points[[j, k]], points[[j, k + 1]],
                     points[[j + 1, k + 1]], points[[j + 1, k]]}],
              {k, λ}]}, {j, ω - 2}]}],
    BoxRatios -> {1, 1, 0.8}, PlotRange -> All,
    Boxed -> False, ViewPoint -> {-3, 3, 1.4}]]
```

Periodic data have peaked Fourier spectra, and random data have typically nonsmooth Fourier spectra showing the amount of noise at various frequencies. Quasiperiodic data (as originating from the Penrose tilings discussed in Chapter 1 of the Graphics volume [1794]) show interesting hierarchical Fourier spectra [695], [1639], [103], [1121]. Here is the Fourier transform of a simple 1D example of a period doubling map [1493], [270].

```
In[60]:= ListPlot[Abs[Fourier[
    Nest[Flatten[# /. {0 -> {0, 1}, 1 -> {0, 0}}]&,
              {0, 1}, 10]]], PlotJoined -> True,
        Frame -> True, PlotRange -> All, Axes -> False]
```

The next graphics show the absolute values and arguments of the Fourier transform of two Cantor-like sets.

```
In[61]:= cantorFTPlots[base_, nons_] :=
    Module[{o = Floor[Log[base, 10^5]], T, F},
      (* form Cantor-like set and remove terms *)
      T = Table[If[MemberQ[RealDigits[k base^-o, base][[1]], Alternatives @@
      nons],
                      0, 1], {k, 0, base^o}];
      (* form Fourier transform *)
      F = Fourier[T];
      Show[GraphicsArray[
       (* show absolute value and argument *)
       ListPlot[# @ F, PlotStyle -> {PointSize[0.003]},
            PlotRange -> All, Frame -> True, Axes -> False,
            DisplayFunction -> Identity]& /@ {Abs, Arg}]]]

In[62]:= cantorFTPlots @@@ {{11, {2, 3, 4, 5, 6, 7, 8, 9, 10}}, {4, {2, 3, 4}}}
```

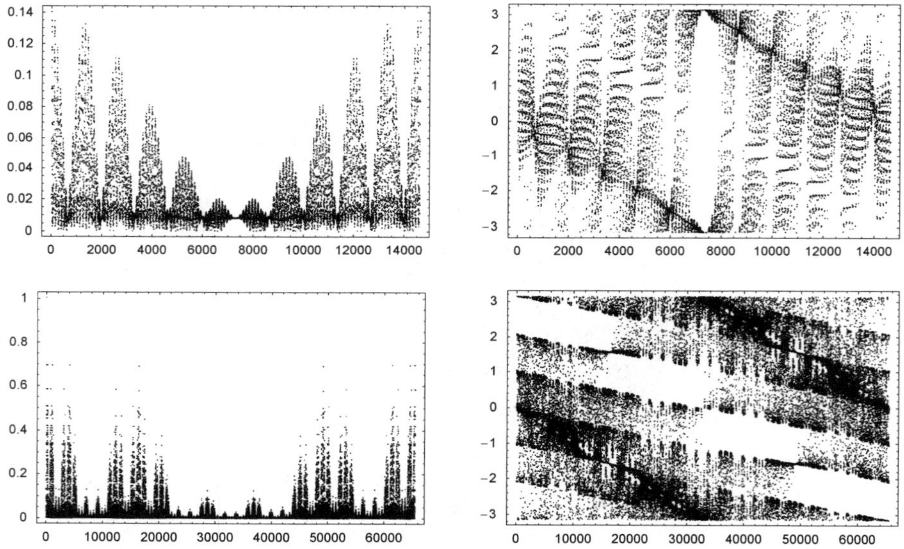

As mentioned, Fourier operates on tensors of any (finite) dimension. Here is one more example of a 2D Fourier transform, for i^j mod 23.

```
In[63]:= With[{n = 128},
      Module[{data = Table[Mod[Power[i, j], 23], {i, n}, {j, n}]},
        Show[GraphicsArray[
          DensityGraphics[#, Mesh -> False, ColorFunction -> Hue,
                      FrameTicks -> None]& /@ {data, Abs[Fourier[data]]}]]]]
```

 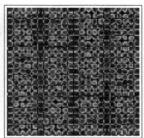

Next, we choose one of the simplest 2D examples, the subdivided L from Subsection 1.5.4 of the Graphics volume [1794]. The left picture shows the vertices of all L's and the right picture shows the corresponding Fourier spectra.

```
In[64]:= Module[{n = 7, Ls, c, points, data, fdata},
    Ls = Nest[Flatten[
    (* make 4 new Ls *)
    Module[{p1, p2, p3, p4, p5, p6, p7, p8, p9,
           p10, p11, p12, p13, p14, p15, p16},
          {p1, p2, p3, p4, p5, p6, p7} = #;
          p8  = (p2 + p3)/2;  p9  = (p1 + p2)/2;  p10 = (p4 + p8)/2;
          p11 = (p2 + p5)/2;  p12 = (p6 + p9)/2;  p13 = (p5 + p8)/2;
          p14 = (p5 + p9)/2;  p15 = (p4 + p5)/2;  p16 = (p5 + p6)/2;
          {{p9, p2, p8, p13, p11, p14, p9},
           {p12, p11, p10, p15, p5, p16, p12},
           {p8, p3, p4, p15, p10, p13, p8},
           {p6, p1, p9, p14, p12, p16, p6}}]& /@ #, 1]&,
      (* two starting L's *)
      {{{0, 2}, {0, 0}, {2, 0}, {2, 1}, {1, 1}, {1, 2}, {0, 2}},
       {{0, 3}, {2, 3}, {2, 1}, {1, 1}, {1, 2}, {0, 2}, {0, 3}}}/2,
       n - 1];
    (* the points appearing as vertices *)
    points = Union[Level[Ls, {-2}]];
    (* a fast way to identify the points *)
    (c[#] = 1)& /@ points;
    (* array of 0's and 1's *)
    data = Table[c[{i/2^n, j/2^n}], {i, 0, 2^n}, {j, 0, 3/2 2^n}] /.
                      (* not a vertex *) _c -> 0;
    fdata = Fourier[data];
    (* the pictures *)
    Show[GraphicsArray[
    ListDensityPlot[Transpose[#],
                ColorFunction -> (Hue[0.7 #]&), Mesh -> False,
                AspectRatio -> Automatic, FrameTicks -> False,
                DisplayFunction -> Identity]& /@
    {(* the vertices of the L's *) data,
     (* the magnitude of the Fourier spectrum *) Abs[fdata],
     (* the argument of the Fourier spectrum *)  Arg[fdata]}]]]
```

 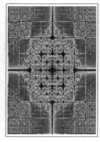 

Typically, only periodic functions give rise to Fourier spectra, which are a sum of Dirac delta functions. An interesting example of a nonperiodic function whose Fourier spectrum is a sum of Dirac delta functions are the points of a lattice visible from the origin [1304], [102]. The Fourier spectrum is periodic, and the next picture shows one period.

```mathematica
In[65]:= Module[{n = 128, ft},
    ft = (* drop boundaries *) Rest /@ Rest[
        Fourier[(* visible lattice points *)
            Table[If[GCD[i, j] === 1, 0, 1], {i, -n, n}, {j, -n, n}]]];
    ListPlot3D[Abs[ft], (* color according to height *)
        Map[Hue, Log[10, 1 + Abs[ft]]/1.5, {2}],
        Mesh -> False, Axes -> False, PlotRange -> All, Boxed -> False]]
```

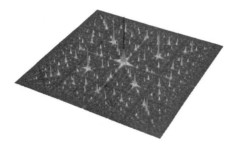

The function `Fourier` (and the function `InverseFourier`) has one option, namely, `FourierParame`‐ ters. The definition of a discrete Fourier transform given above was $y_j = n^{-1/2} \sum_{k=1}^{n} x_k \, \omega_{j,k}^{(n)}$ with $\omega_{j,k}^{(n)} = \exp(2\pi i (k-1)(j-1)/n)$. This can naturally be generalized to

$$y_j = \frac{1}{\sqrt{n^{1-a}}} \sum_{k=1}^{n} x_k \, e^{2\pi i b \frac{(k-1)(j-1)}{n}}$$

with arbitrary parameters a and b.

> `FourierParameters`
>
> > determines the value of the constants *a* and *b* in the Fourier transform.
> >
> > **Default:**
> > > `{0, 1}`
> >
> > **Admissible:**
> > > {*complexNumber, complexNumber*}

Various fields of natural sciences, applied mathematics, and engineering use various standard sets for *a* and *b*. The most frequently used choice for *a* is 0 or 1 and for *b* is ±1. Whereas *a* influences only the normalization, the parameter *b* is more interesting. It leads to the so-called fractional Fourier transform [111] (not to be confounded with the fractional order Fourier transform [1401], [1400], [1077], [1659], [1168], [1084], [1329], [82], [484], [1425], [39], which is a quadratic transformation; a generalization of a Fresnel transform [718], [1407]).

```
In[66]:= FractionalFourier[x_, α_] :=
            Fourier[x, FourierParameters -> {1, -α Length[x]}]
```

```
In[67]:= (* turn off the option which warns about a potential
            inconsistency of the inverse transform *)
         Off[Fourier::"fpopt2"];
```

Here is a data list.

```
In[69]:= data = Table[Sin[k/33 Pi], {k, 256}] // N;
```

Depending on the value of α, the fractional Fourier transform behaves differently. The right graphic is a view from far above on the left graphic.

```
In[70]:= fftData = Table[FractionalFourier[data, α], {α, 0, 3/256, 3/256^2}];
```

```
In[71]:= Show[GraphicsArray[
         ListPlot3D[(* show absolute value and color by phase *)
                Abs[fftData], Map[Hue, (Arg[fftData] + Pi)/(2Pi), {2}],
                Mesh -> False, PlotRange -> All, ViewPoint -> #,
                Axes -> False, DisplayFunction -> Identity]& /@
                {{3, -0.2, 1.5}, {0, 0, 100}}]]
```

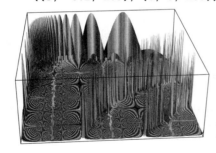

The fractional Fourier transformation can effectively solve various problems, such as high-precision trigonometric interpolation [111], numerical continuous Laplace and Fourier transforms [112], and the fast Fourier transform of nonequispaced data [524]. Here, we will use it in the following example: Given a sampled noisy periodic signal of unknown frequency, we will determine its frequency as precisely as possible [111]. Let ω_0 be the "unknown" frequency.

```
In[72]:= SeedRandom[1439721301];
        ω0 = Random[Real, {20, 50}]
Out[73]= 38.4445
```

data is the list of 4096 sample points of the signal. To emulate noise, we add the term `0.1 Random[Real, {-1, 1}]`.

```
In[74]:= m = 2^12;
        data = Table[Exp[I 2Pi j ω0/m] + 0.1 Random[Real, {-1, 1}], {j, 0, m - 1}];
```

Here is a graphic of the signal.

```
In[76]:= ListPlot[Re[data], PlotJoined -> True]
```

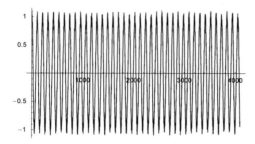

Here is the Fourier transform of the data. In the right graphic, we are zooming into the left plot.

```
In[77]:= ftd = FractionalFourier[data, 1/m];

        With[{lp = ListPlot[Abs[ftd], PlotRange -> All,
                    DisplayFunction -> Identity, PlotJoined -> True,
                    Frame -> True, Axes -> False]},
            Show[GraphicsArray[{lp, Show[lp, (* zooming *)
                                    PlotRange -> {{20, 50}, All}]}]]]
```

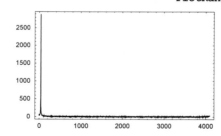
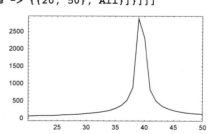

The last plot shows that $\omega \approx 38$.

```
In[79]:= Ω = Position[#, Max[#]]&[Abs[ftd]][[1, 1]] - 1
Out[79]= 38
```

To determine a more precise value of ω, we zoom into the spectrum near Ω. To zoom in, we basically carry out a Fourier transformation of dataΩ, in which dataΩ has the main frequency taken out to better see the difference $\omega - \Omega$.

```
In[80]:= (* take out Fourier modes *)
        dataΩ = MapIndexed[Exp[-2 Pi I Ω (#2[[1]] - 1)/m] #&, data];
        δ = 4/m;
```

```
(* apply fractional Fourier transform *)
ftdΩ = FractionalFourier[dataΩ, δ/m];
ListPlot[Abs[ftdΩ], PlotRange -> All, PlotJoined -> True,
                    Frame -> True, Axes -> False]
```

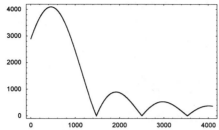

Extracting the maximum of the Fourier amplitudes from the last plot yields a very accurate value for ω.

```
In[86]:= Ω + (δ Position[#, Max[#]]&[Abs[ftdΩ]][[1, 1]]) // N
Out[86]= 38.4453
```

Here is another example of the explicit use of the `FourierParameters` option: Above, we calculated the Fourier coefficients of the function $y = x^2$ and then calculated values between the data points by explicitly summing terms of the form $f_k \cos(\frac{x}{L} k \pi)$. We can use the function `Fourier` to calculate values of the interpolating series between the data points. To do this, we need the so-called unaliased Fourier transform (similar to above). Instead of having sums of the form $\sum_{k=1}^{n} x_k \exp(2 \pi i(k - 1) (j - 1) / n)$ we will deal with sums of the form $\sum_{k=-n/2}^{n/2-1} x_k \exp(2 \pi i(k - 1) (j - 1) / n)$, (here we let n be even, and the case n odd is straightforward too). Using the built-in aliased Fourier transform, we can implement the direct and inverse unaliased Fourier transform `UnaliasedFourier` and `UnaliasedInverseFourier`.

```
In[87]:= UnaliasedFourier[l_] :=
    MapIndexed[(-1)^First[#2] #&,
      Fourier[MapIndexed[(-1)^(First[#2] - 1) #&, l],
                    FourierParameters -> {1, -1}]] /; EvenQ[Length[l]]
```

```
In[88]:= UnaliasedInverseFourier[l_] :=
    MapIndexed[(-1)^First[#2] #&,
      Fourier[MapIndexed[(-1)^(First[#2] - 1) #&, l],
                    FourierParameters -> {-1, 1}]] /; EvenQ[Length[l]]
```

The function `extendList` basically adds zeros to a given list and uses the first element in a symmetrical way [111].

```
In[89]:= extendList[l_, n_?EvenQ] :=
    With[{δ = n/2 - Length[l]/2},
        Flatten[{Table[0, {δ}], l[[1]]/2, Rest[l], l[[1]]/2,
                    Table[0, {δ - 1}]}]]
```

Using the above-defined two functions `UnaliasedFourier` and `UnaliasedInverseFourier`, we can implement the function `highResolutionTrigInterpolation[`*data*`, k]`. Given a list of data points, `highResolutionTrigInterpolation` calculates a list of interpolated values of these data points with k points between each two original data points.

```
In[90]:= highResolutionTrigInterpolation[data_, k_] :=
    With[{m = Length[data]},
        Drop[UnaliasedFourier[ (* extend data *)
            extendList[UnaliasedInverseFourier[data], m (k + 1)]], -k] /;
    EvenQ[m]]
```

In the next graphic, we use the function `highResolutionTrigInterpolation` to redo our example from above and to find points interpolating between the points $\{x_j, x_j^2\}$ with $x_j = \{0, \frac{1}{10}, \frac{2}{10}, ..., \frac{39}{10}\}$. (For the Gibbs phenomena of Fourier interpolation, see [814] and [815]; for determining the period of a signal in case the signal is only known for a small unknown fraction of a period, see [955].)

```
In[91]:= With[{k = 5, v = 39},
    ListPlot[
    MapIndexed[{v/10 (#2[[1]] - 1)/((k + 1) v), #1}&,
                (* trig-interpolated data points *)
                highResolutionTrigInterpolation[
                    Table[x, {x, 0, v/10, 1/10}]^2, k]],
    PlotRange -> All, Frame -> True, Axes -> False,
    PlotStyle -> {GrayLevel[0], PointSize[0.006]},
    Prolog -> {(* the curve y = x^2 *)
                {Hue[0], Line[Table[{x, x^2}, {x, 0, v/10, 1/100}]]},
                (* the original data points *)
                {Hue[0.8], Thickness[0.002],
                 Table[Rectangle[{x, x^2} - {0.02, 0.15},
                             {x, x^2} + {0.02, 0.15}],
                                {x, 0, v/10, 1/10}]}}]]
```

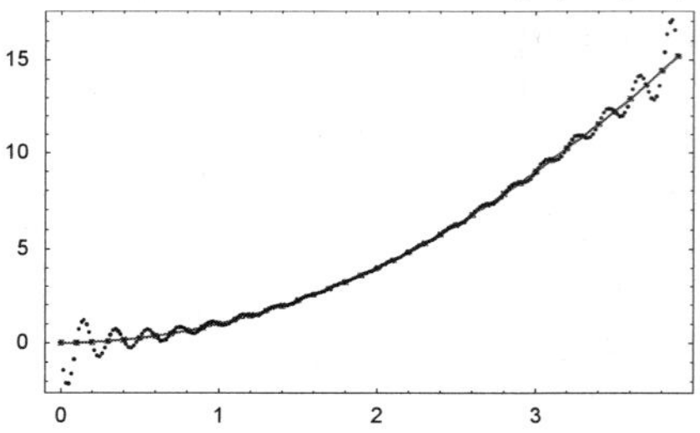

An important application of the fractional Fourier transform is approximating the continuous Fourier transform [111]. Consider a function $f(t)$ that vanishes outside the interval $(-L, L)$. Its continuous Fourier transform is $\mathcal{F}_t[f(t)](\omega) = (2\pi)^{-1/2} \int_{-\infty}^{\infty} e^{i\omega t} f(t)\,dt$. We denote the fractional Fourier transform with parameters a, and b of a sequence $\{u_r\}_{r=1,...,n}$ by $\mathcal{FFT}_{\{a,b\}}[\{u_r\}_{r=1,...,n}]$. Then an approximation of $\mathcal{F}_t[f(t)](\omega)$ in the interval $(-\Omega, \Omega]$ at the points $\omega_s = -\Omega + 2\Omega s/n$ is given by

$$\mathcal{F}_t[f(t)](\omega_s) \approx$$
$$\frac{\Delta t}{\sqrt{2\pi}}\,\exp\!\left(\frac{i\,L\,\Omega}{n^2}\left((n^2-4)+2(2-n)s\right)\right)\times\mathcal{FFT}_{\{1, \frac{2L\Omega}{n\pi}\}}\!\left[\left\{\exp\!\left(\frac{2i\,(2-n)\,L\,\Omega}{n^2}\,r\right)f(t_r)\right\}_{r=1,...,n}\right].$$

Here $t_r = -L + 2Lr/n, r = 1, ..., n$ are the n equidistant discretization points in $(-L, L)$.

```
In[92]:= FourierTransformApproximationList[{f_, t_Symbol}, {L_?Positive, Ω_?Positive},
                                n_Integer?Positive] :=
    Module[{f = Function[t, f], Δt = 2L/n, fList, fFFT, sFactor},
                (* list for the fractional Fourier transform *)
                fList = Table[Exp[2 I (2 - n) L Ω n^-2 r]f[-L + 2L r/n], {r, n}];
                (* carry out fractional Fourier transform *)
                fFFT = Fourier[fList, FourierParameters -> {1, 2 L Ω/(n Pi)}];
                (* s-dependent prefactor *)
                sFactor = Table[Δt/Sqrt[2Pi] Exp[I (n^2 - 4) L Ω n^-2]*
```

```
                              Exp[2 I (2 - n) L Ω n^-2 s], {s, n}];
          (* return approximate Fourier transform *) sFactor fFFT]
```

Here is an example. We approximate the Fourier transform of $f(t) = \theta(1 - t) \theta(t + 1)$. The next input calculates the approximation in the *t*-range $[-10, 10]$ using 512 points and the exact continuous Fourier transform.

```
In[93]:= (* exact continuous Fourier transform *)
        approxFT = FourierTransformApproximationList[
                                {If[Abs[t] < 1, 1, 0], t}, {2, 10}, 512];

        (* exact continuous Fourier transform *)
        exactFT[ω_] = FourierTransform[If[Abs[t] < 1, 1, 0], t, ω] //
                                        ExpToTrig // Simplify
```

$$Out[97]= \quad \frac{\sqrt{\frac{2}{\pi}} \, Sin[\omega]}{\omega}$$

The following graphic shows the real and the imaginary parts of the exact result and the approximation. The agreement is excellent.

```
In[98]:= Show[GraphicsArray[
        Block[{$DisplayFunction = Identity},
        Show[
        {(* exact continuous Fourier transform as a red line *)
        Plot[#[exactFT[ω]], {ω, -10, 10}, PlotStyle -> {Hue[0], Thickness[0.016]}],
        (* approximation as red points *)
        ListPlot[MapIndexed[{-10 + #2[[1]]/512 20, #1}&, #[approxFT]]]},
            PlotRange -> All, Frame -> True, Axes -> False]]& /@ {Re, Im}]]
```

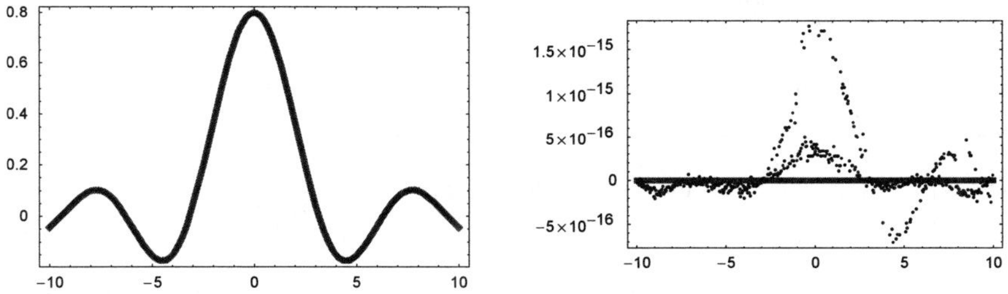

Let us use the function `Fourier` to produce some more pictures. Here is a contour plot of a product of 30 implicitly given circles of random midpoints and radii.

```
In[99]:= SeedRandom[55555];
        circles = Table[Circle[{Random[], Random[]}, Random[]], {30}];

In[101]:= product[x_, y_] =
        Times @@ Apply[(x - #1[[1]])^2 + (y - #1[[2]])^2 - #2^2&,
                        circles, {1}];

In[102]:= cp = ContourPlot[Evaluate[product[x, y]], {x, 0, 1}, {y, 0, 1},
                        Contours -> {0}, PlotPoints -> 401, Frame -> None,
                        Epilog -> {Hue[0], Thickness[0.002], circles}]
```

We extract the data from the last picture (an array of 400×400 real numbers) and display the same picture (now looking a bit grainier) as a density graphics. The matrix data is a 400×400 array of ± 1.

```
In[103]:= ListDensityPlot[data = Sign[cp[[1]]], Mesh -> False, Frame -> None]
```

ftData is the Fourier transform of data.

```
In[104]:= ftData = Fourier[data];
```

The function ftPicture basically displays the first *n* Fourier coefficients of data.

```
In[105]:= ftPicture[n_, f_, opts___] :=
    ListDensityPlot[f[InverseFourier[
                    MapIndexed[If[Max[#2] > n, 0., #1]&, ftData, {2}]]],
                opts, Mesh -> False, Frame -> False]
```

Taking more and more Fourier coefficients and using the real parts shows nicely how the original picture reemerges.

```
In[106]:= Show[GraphicsArray[#]]& /@
    Partition[ftPicture[#, Re, DisplayFunction -> Identity]& /@
                {10, 50, 100, 200, 300, 350, 390, 399, 400}, 3]
```

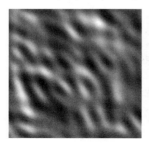  

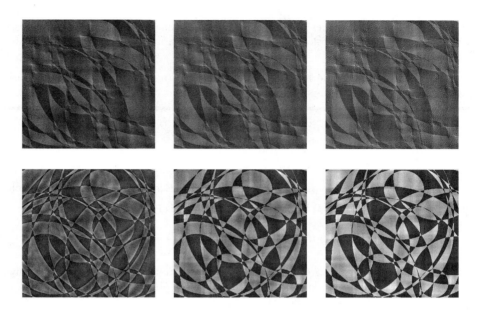

But also, the imaginary part, the absolute value, and even the argument contain visually recognizable information of our circles.

```
In[107]:= Show[GraphicsArray[
    Block[{$DisplayFunction = Identity},
        {ftPicture[300, Im], ftPicture[380, Abs]}]]]
```

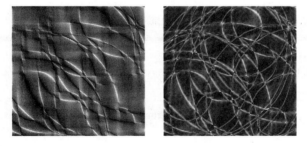

```
In[108]:= Show[GraphicsArray[
    ftPicture[#, Arg, DisplayFunction -> Identity]& /@ {10, 100, 360, 390}]]
```

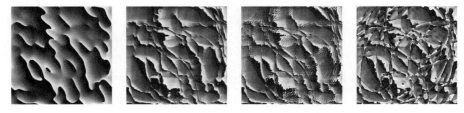

For some application of Fourier transforms in optics, including further interesting pictures, see [1158] and [790]. Sometimes, one is interested in the real symmetric part; for details of the corresponding discrete Fourier cos transform, see [1721].

Two closely related functions to `Fourier` and `InverseFourier` are `ListConvolve` and `ListCorre⌐late` [1917].

`ListConvolve[`*kernel*`,` *data*`]`

    calculates the convolution of the lists *kernel* and *data*.

For a 1D list data of the form $\{x_1, x_2, \ldots, x_n\}$ and a kernel of the form $\{y_1, y_2, \ldots, y_m\}$, the elements z_k of the convolution $\{z_1, z_2, \ldots, z_{n-m+1}\}$ are $z_k = \sum_{j=1}^{m} y_j x_{k-j}$. Here is a simple symbolic example.

```
In[109]:= ListConvolve[Table[k[i], {i, 3}], Table[a[i], {i, 8}]]
```
```
Out[109]= {a[3] k[1] + a[2] k[2] + a[1] k[3], a[4] k[1] + a[3] k[2] + a[2] k[3],
           a[5] k[1] + a[4] k[2] + a[3] k[3], a[6] k[1] + a[5] k[2] + a[4] k[3],
           a[7] k[1] + a[6] k[2] + a[5] k[3], a[8] k[1] + a[7] k[2] + a[6] k[3]}
```

The elements of a convolution c_k of a kernel κ_k and a list a_k arise as coefficients when multiplying two univariate polynomials where κ_k and a_k are the lists of coefficients [1274].

```
In[110]:= ListConvolve[Array[κ, 3], Array[a, 6]]
```
```
Out[110]= {a[3] κ[1] + a[2] κ[2] + a[1] κ[3], a[4] κ[1] + a[3] κ[2] + a[2] κ[3],
           a[5] κ[1] + a[4] κ[2] + a[3] κ[3], a[6] κ[1] + a[5] κ[2] + a[4] κ[3]}
```

```
In[111]:= Table[(Plus @@ Cases[Expand[Sum[κ[k] x^k, {k, 3}] Sum[a[k] x^k, {k, 6}]],
                (* could use CoefficientList here *) _ x^n]) // Factor, {n, 4, 7}]
```
```
Out[111]= {x^4 (a[3] κ[1] + a[2] κ[2] + a[1] κ[3]), x^5 (a[4] κ[1] + a[3] κ[2] + a[2] κ[3]),
           x^6 (a[5] κ[1] + a[4] κ[2] + a[3] κ[3]), x^7 (a[6] κ[1] + a[5] κ[2] + a[4] κ[3])}
```

To align elements of the kernel and the data, we can use the following form of `ListConvolve`.

`ListConvolve[`*kernel*`,` *data*`,` *j*`]`

    calculates the correlation of the lists *kernel* and *data*, where the *j*th element of kernel is aligned with each element of *data*.

Here are, for a simple example, all possible alignments for a kernel with three elements and a data list with four elements.

```
In[112]:= Table[ListConvolve[Table[k[i], {i, 3}], Table[a[i], {i, 4}], j], {j, 4}]
```
```
Out[112]= {{a[1] k[1] + a[4] k[2] + a[3] k[3], a[2] k[1] + a[1] k[2] + a[4] k[3],
            a[3] k[1] + a[2] k[2] + a[1] k[3], a[4] k[1] + a[3] k[2] + a[2] k[3]},
           {a[2] k[1] + a[1] k[2] + a[4] k[3], a[3] k[1] + a[2] k[2] + a[1] k[3],
            a[4] k[1] + a[3] k[2] + a[2] k[3], a[1] k[1] + a[4] k[2] + a[3] k[3]},
           {a[3] k[1] + a[2] k[2] + a[1] k[3], a[4] k[1] + a[3] k[2] + a[2] k[3],
            a[1] k[1] + a[4] k[2] + a[3] k[3], a[2] k[1] + a[1] k[2] + a[4] k[3]},
           {a[4] k[1] + a[3] k[2] + a[2] k[3], a[1] k[1] + a[4] k[2] + a[3] k[3],
            a[2] k[1] + a[1] k[2] + a[4] k[3], a[3] k[1] + a[2] k[2] + a[1] k[3]}}
```

Aligning the first element, we can nicely demonstrate the relationship between `Fourier` and `ListConvolve`. The Fourier transformation is unitary. This means the norm is conserved. Multiplication transforms into convolution. Here, this is demonstrated for some "random" vectors.

```
In[113]:= Module[{vec = N[Range[100], 22], vecF},
            vecF = Fourier[vec];
            (* norm conserving *)
            {vec.vec, vecF.Conjugate[vecF]}]
Out[113]= {338350.0000000000000000, 338350.00000000000000 + 0. × 10⁻¹⁵ i}
```

```
In[114]:= Module[{vec1 = N[Range[100], 22],
            vec2 = N[Range[101, 200], 22], vecF1, vecF2},
            {vecF1, vecF2} = Fourier /@ {vec1, vec2};
            (* multiplication --> convolution *)
            {vec1.vec2, ListConvolve[vecF1, vecF2, 1][[1]]}]
Out[114]= {843350.000000000000000, 843350.000000000000 + 0. × 10⁻¹⁴ i}
```

```
In[115]:= Module[{vec1 = N[Range[100], 22],
            vec2 = N[Range[101, 200], 22], vecF1, vecF2},
            {vecF1, vecF2} = Fourier /@ {vec1, vec2};
            (* convolution ⟹ multiplication *)
            {vecF1.vecF2, ListConvolve[vec1, vec2, 1][[1]]}]
Out[115]= {681650.00000000000000 + 0. × 10⁻¹⁵ i, 681650.000000000000000}
```

Carrying out convolutions is algorithmically a very fast process. Here is a timing comparison between the built-in `ListConvolve` and a compiled version of forming the convolution.

```
In[116]:= kernel = Table[Random[], {100}];
         list = Table[Random[], {1000}];
```

```
In[118]:= Timing[Do[lc = ListConvolve[kernel, list], {100}]]
Out[118]= {0.01 Second, Null}
```

```
In[119]:= listConvolve = (* model ListConvolve through other functions *)
         Compile[{{kernel, _Real, 1}, {list, _Real, 1}},
             Module[{rKernel = Reverse[kernel]},
                 rKernel.#& /@ Partition[list, Length[kernel], 1]]];
```

```
In[120]:= Timing[Do[lcc = listConvolve[kernel, list], {100}]]
Out[120]= {0.68 Second, Null}
```

The two lists calculated were the same.

```
In[121]:= lc == lcc
Out[121]= True
```

Convolutions have important applications in image processing and signal processing. They also allow for a fast multiplication of polynomials. Let `coeffsA` and `coeffsB` be two lists of coefficients of univariate polynomials of degree 5.

```
In[122]:= coeffsA = Array[A, 5, 0];
         coeffsB = Array[B, 5, 0];
```

Forming polynomials, multiplying them, and expanding them leads to the following list of coefficients of the resulting degree 10 polynomial (here, we use the function `CoefficientList`, which gives the coefficient of a univariate polynomial—we will discuss it in Chapter 1 of the Symbolics volume [1795]).

```
In[124]:= (Plus @@ MapIndexed[#1 x^(#2[[1]] - 1) &, coeffsA]) *
         (Plus @@ MapIndexed[#1 x^(#2[[1]] - 1) &, coeffsB]) //
                              Expand // CoefficientList[#, x]&
Out[124]= {A[0] B[0], A[1] B[0] + A[0] B[1], A[2] B[0] + A[1] B[1] + A[0] B[2],
         A[3] B[0] + A[2] B[1] + A[1] B[2] + A[0] B[3],
```

```
            A[4] B[0] + A[3] B[1] + A[2] B[2] + A[1] B[3] + A[0] B[4],
            A[4] B[1] + A[3] B[2] + A[2] B[3] + A[1] B[4],
            A[4] B[2] + A[3] B[3] + A[2] B[4], A[4] B[3] + A[3] B[4], A[4] B[4]}
```

Carrying out a convolution between the (properly aligned) two lists of coefficients `coeffsA` and `coeffsB` gives exactly the same result.

In[125]:= **ListConvolve[coeffsA, coeffsB, {1, -1}, 0]**

Out[125]= {A[0] B[0], A[1] B[0] + A[0] B[1], A[2] B[0] + A[1] B[1] + A[0] B[2],
            A[3] B[0] + A[2] B[1] + A[1] B[2] + A[0] B[3],
            A[4] B[0] + A[3] B[1] + A[2] B[2] + A[1] B[3] + A[0] B[4],
            A[4] B[1] + A[3] B[2] + A[2] B[3] + A[1] B[4],
            A[4] B[2] + A[3] B[3] + A[2] B[4], A[4] B[3] + A[3] B[4], A[4] B[4]}

Similar to `Fourier`, `ListConvolve` takes also higher-dimensional tensors as input. Here are two examples.

In[126]:= **ListConvolve[Table[k[i, j], {i, 3}, {j, 3}], Table[a[i, j], {i, 4}, {j, 4}]]**

Out[126]= {{a[3, 3] k[1, 1] + a[3, 2] k[1, 2] + a[3, 1] k[1, 3] + a[2, 3] k[2, 1] + a[2, 2] k[2, 2] +
            a[2, 1] k[2, 3] + a[1, 3] k[3, 1] + a[1, 2] k[3, 2] + a[1, 1] k[3, 3],
           a[3, 4] k[1, 1] + a[3, 3] k[1, 2] + a[3, 2] k[1, 3] + a[2, 4] k[2, 1] + a[2, 3] k[2, 2] +
            a[2, 2] k[2, 3] + a[1, 4] k[3, 1] + a[1, 3] k[3, 2] + a[1, 2] k[3, 3]},
          {a[4, 3] k[1, 1] + a[4, 2] k[1, 2] + a[4, 1] k[1, 3] + a[3, 3] k[2, 1] +
            a[3, 2] k[2, 2] + a[3, 1] k[2, 3] + a[2, 3] k[3, 1] + a[2, 2] k[3, 2] + a[2, 1] k[3, 3],
           a[4, 4] k[1, 1] + a[4, 3] k[1, 2] + a[4, 2] k[1, 3] + a[3, 4] k[2, 1] + a[3, 3] k[2, 2] +
            a[3, 2] k[2, 3] + a[2, 4] k[3, 1] + a[2, 3] k[3, 2] + a[2, 2] k[3, 3]}}

In[127]:= **Dimensions[%]**

Out[127]= {2, 2}

Here are two first graphics that are based on the function `ListConvolve`. We take a kernel that models a "quantum random walk" [1187] (not unitary) starting with a matrix with one 1 and iterate 16 and 90 times.

In[128]:= **Show[GraphicsArray[**
      Module[{d = 11, n = 240, o = #, m},
            (* the kernel with complex entries *)
            𝓕 = Table[If[i == j == 0, 1, 1/Sqrt[i^2 + j^2]*
                        Exp[I ArcTan[j, i]]], {i, -d, d}, {j, -d, d}] // N;
            (* initial matrix of 0s with central 1 *)
            m = Table[If[k == l == n/2, 1. + 0. I, 0. + 0. I],
                        {k, n - 1}, {l, n - 1}];
            (* iterate list convolution *)
            m = Nest[(* normalize and list convolve *)
                        #/Max[Abs[#]]&[ListConvolve[𝓕, #, {d + 1, d + 1}]]&, m, o];
            (* display contour plot of resulting matrix *)
            Show[(ListContourPlot[Abs[Transpose[m]], PlotRange -> All,
                FrameTicks -> None, Contours -> 60, ContourLines -> False,
                DisplayFunction -> Identity] // Graphics) /.
                    (* random coloring *) **_GrayLevel :> Hue[Random[]]]]& /@**
                        (* number of iterations *) **{16, 90}]]**
```
```

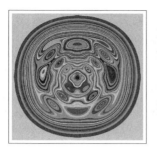 

We continue with a variation on the bifurcation diagrams from above. This time we consider a set of coupled elements $z_{ij}^{(n)}$ (i, j index the elements and n indexes the time step) whose evolution equations contain time delays [85].

$$z_{ij}^{(n)} = (1 - \varepsilon)\, f_\rho\big(z_{ij}^{(n-1)}\big) + \varepsilon \sum_{m=1}^{d} \sum_{k,l} \alpha_{ijkl}^{(m)}\, f_\rho\big(z_{kl}^{(n-m)}\big).$$

Here $f_\rho(z) = \rho\, z\,(1 - z)$. The next graphic shows an example that exhibits a quite complicated bifurcation diagram. We display the normalized average $\sum_{k,l} z_{kl}^{(n-m)}$ as a function of ρ. The inner double sum over k and l we carry out in an efficient way using ListConvolve.

```
In[129]:= Module[{f, m0 = 2000, m = 100, o, k, d, ε, 𝒦, data, xL, xLL,
            tdp = Developer`ToPackedArray},
     (* the map *)
     f[ρ_, x_] := ρ x (1 - x);
     SeedRandom[4293305149477815];
     (* the system parameters *)
     (* system size *)          o = Random[Integer, {3, 10}];
     (* neighborhood size *) k = Random[Integer, {1, (o + 1)/2}];
     (* maximal delay *)       d = Random[Integer, {1, 6}];
     (* coupling value *)      ε = Random[];
     (* couplings to neighbors for various delays *)
     Do[𝒦[j] = MapAt[0.&, Table[Random[], {2 k + 1}, {2 k + 1}],
                 {k + 1, k + 1}];
        𝒦[j] = tdp @ (𝒦[j]/(Plus @@ Flatten[𝒦[j]])), {j, d}];
     data =
     With[{body = Unevaluated[(* iterate map application *)
         xL = (1 - ε) f[ρ, xLL[[1]]] +
         ε/d Sum[ListConvolve[𝒦[j], f[ρ, xLL[[j + 1]]], {2, 2}], {j, d}];
         xLL = RotateRight[xLL]; xLL[[1]] = Mod[xL, 1.]]},
       (* loop over ρ *)
      Table[xLL = tdp /@ Table[Random[], {d + 1}, {o}, {o}];
           (* discarded initial steps *) Do[body, {m0}];
           (* recorded steps *)
           Table[body; (* keep average *)
                 {ρ, Plus @@ Flatten[xL]/o^2}, {j, m}],
           {ρ, 3.2, 4.5, 1.3/1200}]];
     (* show resulting averages *)
     Show[Graphics[{PointSize[0.003], Map[Point, data, {2}]},
         PlotRange -> All, Frame -> True]]]
```

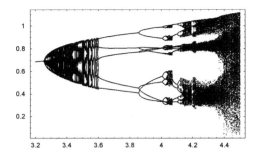

Next, we consider the random sequence $s_k = s_{j(k)} + 1$, $s_0 = 0$, where $s(j)$ is a random integer from the interval $[0, k-1]$ [176], [1623]. We calculate the average over 1000 realizations of sequences of length 1000.

```
In[130]:= SeedRandom[50];
    randomSequenceAverages =
    Module[{o = 10^3, m = 10^3, L},
        Sum[L = Table[0, {k, o}];
            Do[L[[n]] = L[[Random[Integer, {1, n - 1}]]] + 1, {n, 2, o}];
            L, {m}]/m];
```

A graphic of the average sequence terms suggests a logarithmic growth $\overline{s_k} \sim \ln(k)$.

```
In[132]:= ListPlot[randomSequenceAverages]
```

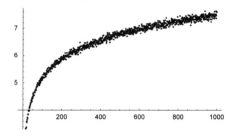

We average the differences $\delta s_k = s_k - \ln(k)$ with convolution kernels of increasing width.

```
In[133]:= (* differences to Log[k] *)
    δrandomSequenceAverages = Rest[randomSequenceAverages] -
                    Log[Range[Length[randomSequenceAverages] - 1]] // N;

In[135]:= Module[{lc, m = Length[δrandomSequenceAverages]/2},
        (* make graphics *)
        Show[Graphics[{PointSize[0.006],
            (* form averages *)
        Table[lc = ListConvolve[Table[1/(2j + 1), {2 j + 1}],
                                δrandomSequenceAverages];
            {Hue[0.78 j/m], MapIndexed[Point[{#2[[1]] + j + 1, #1}]&, lc]},
            {j, 0, m, 5}]}], Frame -> True]]
```

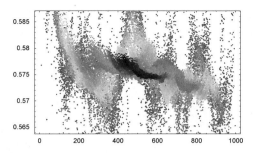

The last graphic suggests that $\overline{s_k} = \ln(k) + \gamma$. Forming the average over all averages series terms yields the value of γ within 0.004%.

```
In[136]:= 1 - EulerGamma /(Total[δrandomSequenceAverages]/
                          (Length[δrandomSequenceAverages] + 1))
Out[136]= -0.0000365355
```

The second function related to `Fourier` and `InverseFourier` is `ListCorrelate`.

> `ListCorrelate[`*kernel*, *data*`]`
>
>   calculates the correlation of the lists *kernel* and *data*.

For a 1D data list of the form $\{x_1, x_2, \ldots, x_n\}$ and a kernel of the form $\{y_1, y_2, \ldots, y_m\}$, the elements z_k of the correlation $\{z_1, z_2, \ldots, z_{n-m+1}\}$ are $z_k = \sum_{j=1}^{m} y_j\, x_{k+j}$.

Here are the two examples from above used in `ListCorrelate` instead of `ListConvolve`. Again, the arguments can be tensors of any rank.

```
In[137]:= ListCorrelate[Table[k[i], {i, 3}], Table[a[i], {i, 8}]]
Out[137]= {a[1] k[1] + a[2] k[2] + a[3] k[3], a[2] k[1] + a[3] k[2] + a[4] k[3],
           a[3] k[1] + a[4] k[2] + a[5] k[3], a[4] k[1] + a[5] k[2] + a[6] k[3],
           a[5] k[1] + a[6] k[2] + a[7] k[3], a[6] k[1] + a[7] k[2] + a[8] k[3]}
```

```
In[138]:= ListCorrelate[Table[k[i, j], {i, 3}, {j, 3}], Table[a[i, j], {i, 4}, {j, 4}]]
Out[138]= {{a[1, 1] k[1, 1] + a[1, 2] k[1, 2] + a[1, 3] k[1, 3] + a[2, 1] k[2, 1] + a[2, 2] k[2, 2] +
            a[2, 3] k[2, 3] + a[3, 1] k[3, 1] + a[3, 2] k[3, 2] + a[3, 3] k[3, 3],
           a[1, 2] k[1, 1] + a[1, 3] k[1, 2] + a[1, 4] k[1, 3] + a[2, 2] k[2, 1] + a[2, 3] k[2, 2] +
            a[2, 4] k[2, 3] + a[3, 2] k[3, 1] + a[3, 3] k[3, 2] + a[3, 4] k[3, 3]},
          {a[2, 1] k[1, 1] + a[2, 2] k[1, 2] + a[2, 3] k[1, 3] + a[3, 1] k[2, 1] +
            a[3, 2] k[2, 2] + a[3, 3] k[2, 3] + a[4, 1] k[3, 1] + a[4, 2] k[3, 2] + a[4, 3] k[3, 3],
           a[2, 2] k[1, 1] + a[2, 3] k[1, 2] + a[2, 4] k[1, 3] + a[3, 2] k[2, 1] + a[3, 3] k[2, 2] +
            a[3, 4] k[2, 3] + a[4, 2] k[3, 1] + a[4, 3] k[3, 2] + a[4, 4] k[3, 3]}}
```

Now, let us use the function `ListConvolve` to manipulate our circle picture from above. Carrying out the convolution of a kernel with `data` means to "mix" neighboring values into a given colored square [317]. By choosing different explicit forms for kernel, we can produce a variety of different pictures. The first two kernels mix in neighbors with a cos-modulation. The rightmost graphic uses a sin-modulation that favors neighboring values over the center value.

```
In[139]:= lcGraphics[f_, {x_, y_}, {α_, pp_}, opts___] :=
            Block[{kernel = Table[f, {x, -α Pi, α Pi, 2 α Pi/pp},
                                    {y, -α Pi, α Pi, 2 α Pi/pp}] // N},
```

```
        ListDensityPlot[ListConvolve[kernel, data], Mesh -> False,
                    Frame -> None, DisplayFunction -> Identity]]
```

In[140]:= (* show resulting graphic for three trigonometric products *)
```
        Show[GraphicsArray[
                {lcGraphics[Cos[x] Cos[y], {x, y}, {  1,  4}],
                 lcGraphics[Cos[x] Cos[y], {x, y}, {  1, 10}],
                 lcGraphics[Sin[x] Sin[y], {x, y}, {3/2, 10}]}]]
```

Here is a more interesting looking kernel. Because of the singularities of the tangent function, some elements are weighted much more. This time we use ListCorrelate.

In[142]:= Show[GraphicsArray[
```
        Apply[Function[{β, α},
        ListDensityPlot[ListCorrelate[
                N[Table[Tan[x] Cot[y], {x, -α Pi, α Pi, 2α Pi/β},
                                        {y, -α Pi, α Pi, 2α Pi/β}]],
                data], Mesh -> False, ColorFunction -> GrayLevel,
                    Frame -> None, DisplayFunction -> Identity]],
            (* three sets of parameters *)
            {{11, 0.49999}, {11, 0.99999}, {30, 0.49999}}, {1}]]]
```

The following three pictures have random kernels. We display them using the color function Hue. ListCorre⁚ late and ListConvolve produce similar pictures.

In[143]:= Show[GraphicsArray[
```
        ListDensityPlot[ListCorrelate[Table[Random[Real, {-1, 1}], {#}, {#}], data],
                    Mesh -> False, Frame -> None, ColorFunction -> (Hue[2#]&),
                    DisplayFunction -> Identity]& /@ {4, 8, 16, 32}]]
```

 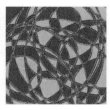  

Here are three individual convolutions with kernels of different size. To obtain matrices of the same size, we pad cyclically. We display them in one picture using the array values inside `RGBColor`. The right graphic uses a kernel that contains a dominating smooth part and a small fluctuating part.

```
In[144]:= Show[GraphicsArray[
          {(* different kernels for red, green blue part *)
          Graphics[Raster[Transpose[
           ListConvolve[Table[Random[Real, {-1, 1}], {#}, {#}],
                        data, 4]& /@ {8, 16, 32}, {3, 1, 2}],
                          ColorFunction -> RGBColor],
               AspectRatio -> Automatic],
          (* kernel with small fluctuating part *)
          With[{pp = 24, α = 3/2},
            kernel = Table[Cos[x y] + Random[Real, {-1, 1}]/3,
                          {x, -α Pi, α Pi, 2 α Pi/pp},
                          {y, -α Pi, α Pi, 2 α Pi/pp}] // N;
            ListDensityPlot[ListCorrelate[kernel, data],
                           Mesh -> False, Frame -> None,
                           DisplayFunction -> identity,
                           ColorFunction -> (Hue[1.6 #]&)]]}]]
```

We end this set of graphic examples with some more variations of kernels.

```
In[145]:= Show[GraphicsArray[
          ListDensityPlot[ListConvolve[#, data],
                         Mesh -> False, Frame -> None, ColorFunction -> Hue,
                         DisplayFunction -> Identity]& /@
          (* three different kernels *)
          N[{Table[-1^(i + j), {i, 6}, {j, 6}],
             Table[1/(i^2 + j^2 + 10), {i, -14, 14}, {j, -14, 14}],
             Table[Sin[i j] i j, {i, -14, 14}, {j, -14, 14}]}]]]
```

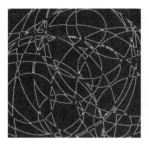 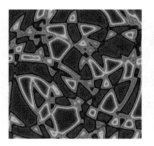 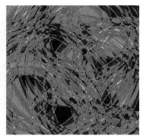

Many more interesting graphics can be generated using `Fourier` and `ListConvolve` [531]. Besides starting from an array of gray levels or colors, a wide range of possibilities opens when we convert generic *Mathematica* graphics into bitmaps and then use the bitmap data within `ListConvolve`. The next picture does this with the expression $e^{i\pi}$. The resulting gray level values are interpreted as heights for a 3D graphic.

```
In[146]:= Module[{n = 25, data, kernel, lc, polys},
     (* rasterize graphics *)
     data = ImportString[ExportString[Graphics[
         {Text[StyleForm[(* the expression Exp[I Pi] *)"\!\(e\^\(i π\)\)",
             FontSize -> 180, FontWeight -> "Plain"],
             {0, 0}]}, AspectRatio -> 1, ColorOutput ->
             GrayLevel, FormatType -> TraditionalForm], "PGM",
                     ImageResolution -> 63], "PGM"][[1, 1]];
     (* the smoothing kernel *)
     kernel = Table[-(Abs[n - i]^2 + Abs[n - j]^2 + 0.5)^(-0.7),
                 {i, 2n - 1}, {j, 2n - 1}];
     If[Length[Dimensions[data]] == 3, data = Map[First, data, {2}]];
     (* the smoothed data *)
     lc = 0.8 #/Max[#]&[(# - Min[#])&[ListConvolve[kernel, N[data]]]];
     (* make 3D polygons *)
     polys = Cases[Graphics3D[ListPlot3D[lc,
                 Mesh -> False, PlotRange -> All,
                 DisplayFunction -> Identity]], _Polygon, Infinity];
     (* display smoothed 3D expression *)
     Show[Graphics3D[{EdgeForm[],
      {SurfaceColor[Hue[#[[3]]], Hue[#[[3]]], 2.6]&[
                     Plus @@ #[[1]]/4], #}& /@ polys}],
         Boxed -> False, BoxRatios -> {1, 1, 0.5},
         ViewPoint -> {-1, 0.2, 3}, PlotRange -> All,
         ViewVertical -> {0, 1, 0}]]
```

Here is a more complex picture using trigonometric identities.

```
Off[N::"meprec"];
SeedRandom[314159265358979323846];

Module[{makeRaster, randomNontrivialIdentity,
        gr, tab, dimx, dimy, pos1, pos2, data},
(* bitmap make from Graphics *)
makeRaster[gr_Graphics, r_Integer] :=
    ImportString[ExportString[gr, "PGM", ImageResolution -> r],
                "PGM"][[1]];
(* random trigonometric identity *)
randomNontrivialIdentity :=
Module[{randomIdentity},
        randomIdentity := Module[{r, e},
                r = {3, 4, 5, 6, 8, 12, 15, 20}[[
                        Random[Integer, {1, 8}]]];
                e = Exp[I Random[Integer, {1, r - 1}]/r Pi];
                e == FunctionExpand[ExpToTrig[e]]];
        While[e = randomIdentity, Null]; e];
(* a collection of identities *)
gr = Show[Graphics[
            Table[{GrayLevel[Random[]], Text[StyleForm[
                (* avoid evaluation *)
                Function[c, ToString[Unevaluated[c], StandardForm],
                        {HoldAll}] @@ randomNontrivialIdentity,
                FontSize -> Random[Integer, {4, 20}],
                FontWeight -> "Bold"], {Random[], Random[]}]}, {120}]],
            DisplayFunction -> Identity];
(* making a bitmap *)
tab = makeRaster[gr, 120];
(* extracting interesting parts *)
{dimx, dimy} = tab[[2, 2]];
cM = tab[[3, 2]];
pos1 = Position[tab[[1]], _?(# =!= cM&), {2}, 1, Heads -> False][[1, 1]];
pos2 = dimy - Position[Reverse[tab[[1]]], _?(# =!= cM&),
                    {2}, 1, Heads -> False][[1, 1]];
data = Map[(* color or gray? *) If[Head[#] === List, First[#], #]&,
            Take[tab[[1]], {pos1, pos2}, {1, pos2 - pos1 + 1}], {2}];
(* display original and Listconvolve'd version *)
Show[GraphicsArray[
  ListDensityPlot[#, DisplayFunction -> Identity, FrameTicks -> None,
            Mesh -> False, AspectRatio -> 1,
            ColorFunction -> (Hue[2#]&)]& /@
        {Reverse /@ data, ListConvolve[Table[1., {3}, {3}], data]}]]]
```

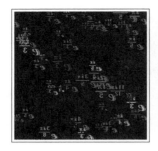 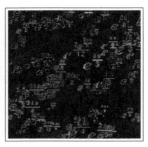

Convolutions are sometimes also useful for manipulating 3D graphics. Here is a random curve in $\mathbb{R}^3$.

```
In[147]:= SeedRandom[987654321];
     With[{n = 8},
         curve[t_] = Table[Sum[Random[Real, {-1, 1}] Cos[i t + 2Pi Random[]],
                         {i, n}], {3}]];
```

We discretize the curve using 250 points.

```
In[149]:= data = Table[Evaluate[curve[t]], {t, 0., 2.Pi, 2.Pi/250}];
```

We smooth (and contract) the curve by averaging over sets of five neighboring points. `ListConvolve` allows us to do this averaging process quickly.

```
In[150]:= uniformKernel = Table[1, {5}]/5;
```

```
In[151]:= mesh = Append[#, First[#]]& /@ NestList[
             Function[data, Transpose[ListConvolve[
              (* the structured kernel *)
             uniformKernel, #, 3]& /@  Transpose[data]]], data, 100];
```

```
In[152]:= Show[Graphics3D[{Thickness[0.002],
                     MapIndexed[{Hue[#2[[1]]/120], Line[#]}&, mesh]}],
         Boxed -> False, PlotRange -> All]
```

Next, we use a convolution kernel with seven random elements. As a result, the curve still shrinks, but as smoothly as before.

```
In[153]:= randomKernel = #/Plus @@ #&[Table[Random[], {7}]];
```

```
In[154]:= mesh = Append[#, First[#]]& /@ NestList[
             Function[data, Transpose[ListConvolve[(* the random kernel *)
                 randomKernel, #, 4]& /@ Transpose[data]]], data, 100];
```

```
In[155]:= Show[GraphicsArray[
     Graphics3D[{Thickness[0.002],
```

```
        MapIndexed[{Hue[#2[[1]]/120], Line[#]}&, #]},
  Boxed -> False, PlotRange -> All]& /@ {mesh, Transpose[mesh]}]]
```

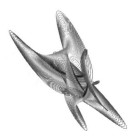

There are, similar to convolutions, many relations between the Fourier transform and the correlation of a list. One is the fact that the autocorrelation list is (modulo a prefactor) just the inverse Fourier transform of the square of the absolute value of the Fourier transform of a list. Here is a random walk. The local stepping direction is within a cone of the direction of the last step.

```
In[156]:= SeedRandom[54321];
      walk1 =
      Module[{δω = 1., dir = {1, 0}},
      NestList[((* the new direction *)
              dir = {{Cos[#], Sin[#]}, {-Sin[#], Cos[#]}}&[
                  Random[Real, {-δω, δω}]].dir;
              (* new step *) # + dir)&,
              {0, 0}, 2^12]];
```

```
In[158]:= Show[Graphics[Line[walk1]], PlotRange -> All, Frame -> True]
```

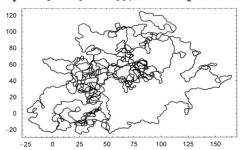

Here is the autocorrelation function of the *x*-values of the walk calculated in the two described ways.

```
In[159]:= Show[
      Block[{xWalk1 = First /@ walk1, $DisplayFunction = Identity},
      GraphicsArray[
        {(* autocorrelation calculated as correlation *)
         ListPlot[ListCorrelate[xWalk1, xWalk1, 1]],
         (* autocorrelation calculated as inverse Fourier transform *)
         ListPlot[Sqrt[Length[xWalk1]]*
                 InverseFourier[Abs[Fourier[xWalk1]]^2]]}]]]
```

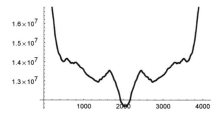 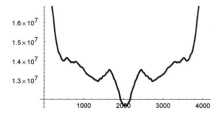

1.6 Numerical Functions and Their Options

The most important numerical functions for numerical summation (NSum), numerical product formation (NProduct), numerical integration (NIntegrate), numerical equation solving (FindRoot), numerical minimization (FindMinimum, NMinimize), and numerical differential equation solving (NDSolve) have a good amount of options to control their behavior in detail.

```
In[1]:= numericalFunctions =
          {NSum, NProduct, NIntegrate, FindRoot, FindMinimum, NMinimize, NDSolve};
```

```
In[2]:= {#, Length[Options[#]]}& /@ numericalFunctions
```
```
Out[2]= {{NSum, 10}, {NProduct, 10}, {NIntegrate, 11},
          {FindRoot, 10}, {FindMinimum, 9}, {NMinimize, 7}, {NDSolve, 14}}
```

There are five options common to all of the numerical functions.

```
In[3]:= Intersection @@ ((First /@ Options[#])& /@ numericalFunctions)
```
```
Out[3]= {AccuracyGoal, EvaluationMonitor, Method, PrecisionGoal, WorkingPrecision}
```

Because these options are quite important for an efficient use of the numerical functions, we will discuss them in this section. We will use examples of various numerical functions to be discussed in greater detail in the following sections.

We start with the WorkingPrecision option.

WorkingPrecision

   is an option to numerical (and other) functions that determines the precision to be used in the numerical computations.

By default, all numerical calculations are carried out in machine precision. This is efficient, and for most cases, also sufficient.

```
In[4]:= {#, WorkingPrecision /. Options[#]}& /@ numericalFunctions
```
```
Out[4]= {{NSum, MachinePrecision},
          {NProduct, MachinePrecision}, {NIntegrate, MachinePrecision},
          {FindRoot, MachinePrecision}, {FindMinimum, MachinePrecision},
          {NMinimize, MachinePrecision}, {NDSolve, MachinePrecision}}
```

Here is an innocent looking function. It has the structure $f(x) = largeNumber\, hiddenZero + smallNumber$.

```
In[5]:= f[x_] := 10^10 (Sqrt[2 - Sqrt[2 + Sqrt[2]]]/2 - Sin[Pi/16]) x + 10^-10
```

A direct call to `NIntegrate` without any custom option setting yields an obviously wrong result for the integral $\int_0^\pi f(x)\,dx$, the returned number is off by many orders of magnitude.

In[6]:= **NIntegrate[f[x], {x, 0, Pi}]**

Out[6]= 1.37×10^{-6}

We have a quick look at a plot (using `Plot`, which uses machine arithmetic too to calculate function values, to visualize the function) of $f(x)$. The hidden zero does not result in an exact zero when calculating a numerical value, but rather in a result of the form *smallInteger* $10^{-\text{\$MachinePrecision}}$. With a machine precision of about 16, the first term is of the order 10^{-6}, as visible in the plot. The contribution of the second term of the order 10^{-10} is so much smaller and not visible to `NIntegrate` within machine arithmetic.

In[7]:= **Plot[f[x], {x, 0, Pi}, PlotRange -> All];**

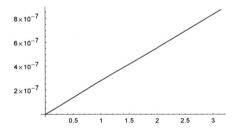

Carrying out the calculation using high-precision arithmetic yields the expected result.

In[8]:= **NIntegrate[f[x], {x, 0, Pi}, WorkingPrecision -> 30]**

Out[8]= $3.14159265358979323846264338328 \times 10^{-10}$

Be aware that with the setting `WorkingPrecision -> MachinePrecision`, machine arithmetic is used, but with the setting `WorkingPrecision -> $MachinePrecision`, high-precision arithmetic is used (`$MachinePrecision` evaluates to a number and any positive number indicates the use of high-precision arithmetic).

In[9]:= **{#, NIntegrate[f[x], {x, 0, Pi}, WorkingPrecision -> #]}& /@**
                                   {MachinePrecision, $MachinePrecision}

Out[9]= $\{\{\text{MachinePrecision}, 1.37 \times 10^{-6}\}, \{15.9546, 3.141592653589793 \times 10^{-10}\}\}$

Here is the result of using small working precision. Even with a working precision setting of 5, we obtain a few correct digits and with a working precision of 10^{-3}, we do not get a correct digit, but still obtain the correct order of magnitude of the result.

In[10]:= **{#, NIntegrate[f[x], {x, 0, Pi}, WorkingPrecision -> #]}& /@ {5, 10^-3}**

         NIntegrate::tmap :
           NIntegrate is unable to achieve the tolerances specified by the PrecisionGoal
             and AccuracyGoal options because the working precision is
             insufficient. Try increasing the setting of the WorkingPrecision option.

Out[10]= $\left\{\{5, 3.1416 \times 10^{-10}\}, \left\{\dfrac{1}{1000}, 0. \times 10^{-10}\right\}\right\}$

In general, calculations carried out with the default value `MachinePrecision` are faster than high-precision calculations, but less reliable.

We continue with the `PrecisionGoal` option.

```
PrecisionGoal
```
    is an option to numerical (and other) functions that determines the precision of the numerical result requested.

In the following input, we ask *Mathematica* to calculate a numerical approximation of the integral $\int_0^\infty x^{-x}\,dx$ to 15 digits. Using machine arithmetic for the function evaluations is not sufficient to obtain 15 digits in the result. The message that is generated indicates this.

```
In[11]:= NIntegrate[x^(-x), {x, 0, Infinity}, PrecisionGoal -> 15]
```
              NIntegrate::tmap :
               NIntegrate is unable to achieve the tolerances specified by the PrecisionGoal
                  and AccuracyGoal options because the working precision is
                  insufficient. Try increasing the setting of the WorkingPrecision option.

```
Out[11]= 1.99546
```

Carrying out the individual calculations using high-precision arithmetic results in the desired 15 digit result.

```
In[12]:= NIntegrate[x^(-x), {x, 0, Infinity}, PrecisionGoal -> 15,
                WorkingPrecision -> 50]
Out[12]= 1.9954559575001380
```

The value of the `PrecisionGoal` option is typically set to be `Automatic`. In most cases, this means about 10 for a working precision `MachinePrecision` or *workingPrecision*/2 if the `WorkingPrecision` option is set differently.

```
In[13]:= {#, PrecisionGoal /. Options[#]}& /@ numericalFunctions
Out[13]= {{NSum, Automatic}, {NProduct, Automatic},
          {NIntegrate, Automatic}, {FindRoot, Automatic},
          {FindMinimum, Automatic}, {NMinimize, Automatic}, {NDSolve, Automatic}}
```

When calculating numerical problems, in most cases, one is looking for a result with a certain number of correct digits. But if the result is zero, it will never be possible to obtain only a single correct digit. In such situations (or in the numerically very similar situation of a very small result), a better measure is the accuracy of the result. The accuracy goal option determines the accuracy of the result of a numerical computation.

```
AccuracyGoal
```
    is an option to numerical (and other) functions that determines the accuracy of the numerical result requested.

In the following attempt to calculate the root near $x = 0$ of $\sin(x) + 3\sin(x^3) - 5\sin^5(x) = 0$ to five correct digits, a message is issued because this is not possible to obtain only a single correct digit for the exact result 0.

```
In[14]:= g[x_] := Sign[Sin[x]] Abs[Sin[x]]^(1/3) +
                Sin[x] + 3 Sin[x^3] - 5 Sin[x]^5

In[15]:= FindRoot[g[x], {x, 1/2, 1},
                PrecisionGoal -> 5, WorkingPrecision -> 40,
                AccuracyGoal -> Infinity]
```
              FindRoot::frdig :
                40.` working digits is insufficient to achieve the absolute tolerance 0. More…

```
Out[15]= {x -> -6.0995088424564721509399304316102294570093×10^-43}
```

The next input gives an approximation to the root that has an accuracy greater than 20.

In[16]:= **FindRoot[g[x], {x, 1/2, 1},**
        PrecisionGoal -> 5, WorkingPrecision -> 40,
        AccuracyGoal -> 20]

Out[16]= $\{x \to 6.6923093744088587177420059490468004453810 \times 10^{-41}\}$

In most cases, one is interested in the value of the independent variables, meaning the value of x in the last example. But sometimes it is more important to solve (minimize) equations to a given accuracy (making a residue smaller than a given value). Frequently this can be done by specifying a two-element list for the Accura‑cyGoal option setting.

In the next input, we search for a root of $f \sum_{k=1}^{o} x^{2k+1}$ that uses a large value of f.

In[17]:= **f[x_?InexactNumberQ] :=**
    With[{o = 5, f = 10^300}, f x^3 (x^o - 1)(x^o + 1)/((x - 1)(x + 1))]

In[18]:= **FindRoot[f[x], {x, 1/2, 3/4},**
        PrecisionGoal -> 10, WorkingPrecision -> 200,
        AccuracyGoal -> {20, 100}, MaxIterations -> 2000]

    N::meprec : Internal precision limit $MaxExtraPrecision = 50.`
        reached while evaluating $\frac{1}{2}\sqrt{2 - \sqrt{2 + \text{Power}[\ll 2\gg]}} - \text{Sin}\left[\frac{\pi}{16}\right]$. More…

    N::meprec : Internal precision limit $MaxExtraPrecision = 50.`
        reached while evaluating $\frac{1}{2}\sqrt{2 - \sqrt{2 + \text{Power}[\ll 2\gg]}} - \text{Sin}\left[\frac{\pi}{16}\right]$. More…

    N::meprec : Internal precision limit $MaxExtraPrecision = 50.`
        reached while evaluating $\frac{1}{2}\sqrt{2 - \sqrt{2 + \text{Power}[\ll 2\gg]}} - \text{Sin}\left[\frac{\pi}{16}\right]$. More…

    General::stop :
     Further output of N::meprec will be suppressed during this calculation. More…

    FindRoot::cvmit : Failed to converge to the
     requested accuracy or precision within 2000 iterations. More…

Out[18]= $\{x \to$
    0.750000000000000000000000000000000000000000000000000000000000000000000000000000000․
    000000000000000000000000000000000000000000000000000000000000000000000000000000000․
    00000000000000000000000000000000000000000000000000000000000000000$\}$

The residue is, as was specified through the AccuracyGoal option, smaller than 10^{-100}.

In[19]:= **f[x /. %] // N**

Out[19]= 9.09983×10^{299}

The value of the AccuracyGoal option is typically set to automatic. In most cases, this means about 10 for a working precision MachinePrecision or *workingPrecision*/2 in case the WorkingPrecision option is set differently. Some functions though, have the accuracy goal set to infinity. This means all possible attempts are made to give correct digits in the result.

In[20]:= **{#, AccuracyGoal /. Options[#]}& /@ numericalFunctions**

Out[20]= {{NSum, ∞}, {NProduct, ∞}, {NIntegrate, ∞}, {FindRoot, Automatic},
    {FindMinimum, Automatic}, {NMinimize, Automatic}, {NDSolve, Automatic}}

In general, when both a precision goal *pg* and an accuracy goal *ag* are specified, then the absolute error of the result x, the quantity $\|x - \hat{x}\|$ (where $\hat{x}$ is the, in general unknown, exact value) is attempted to be made smaller than $\max(10^{-ag}, \|\hat{x}\| 10^{-pg})$ or $10^{-ag} + \|\hat{x}\| 10^{-pg}$. So in the following integration, the setting AccuracyGoal -> 4 dominates over the PrecisionGoal -> 10000 setting.

```
In[21]:= NIntegrate[Evaluate @ Sum[x^k, {k, 100}], {x, 0, 1},
              WorkingPrecision -> 20, PrecisionGoal -> 10000,
              AccuracyGoal -> 4]
```

```
Out[21]= 4.1973
```

By setting the `AccuracyGoal` option to the value `Infinity`, we force to obtain correct digits. For the integral $\int_0^1 (x/1/2)\,dx = 0$, this is not possible. As a result, a message is issued from the next input.

```
In[22]:= NIntegrate[x - 1/2, {x, 0, 1},
              WorkingPrecision -> 20, PrecisionGoal -> 5,
              AccuracyGoal -> Infinity]
```

```
        NIntegrate::ploss :
          Numerical integration stopping due to loss of precision. Achieved
            neither the requested PrecisionGoal nor AccuracyGoal; suspect one of
            the following: highly oscillatory integrand or the true value of the
            integral is 0. If your integrand is oscillatory on a (semi-)infinite
            interval try using the option Method->Oscillatory in NIntegrate. More…
```

```
Out[22]= 0. × 10^{-21}
```

The last of the options that are common to all numerical functions is `EvaluationMonitor`.

`EvaluationMonitor`

   is an option to numerical (and other) functions that determines an expression to be evaluated at each function evaluation.

The exact definition of "step" in the last definition depends on the concrete numerical function. For `NInte⸱grate` for instance, it is a function evaluation of the integrand. Here is a simple integrand $h(x)$, a smooth single bump with five superimposed small bumps in the interval $[0, \pi]$.

```
In[23]:= h[x_?InexactNumberQ] := Sin[x] + 1/3 Cos[4 x]^100;
```

We numerically carry out the integral over the bump and inside the evaluation monitor; we collect the values of x, for which $g(x)$ was numerically evaluated.

```
In[24]:= {integralValue, usedxValues} =
              Reap[NIntegrate[h[x], {x, 0, Pi},
                              WorkingPrecision -> 20, PrecisionGoal -> 10,
                              Method -> GaussKronrod,
                              EvaluationMonitor :> Sow[x]]];

       {integralValue, λ = Length[usedxValues[[1]]]}
```

```
Out[25]= {2.083345654, 1339}
```

A graph of the x-values collected clearly shows that the numerical integration algorithm sampled the integrand more frequently near the small bumps.

```
In[26]:= Show[Block[{$DisplayFunction = Identity},
            {(* plot of the integrand in red *)
             Plot[h[x], {x, 0, Pi}, PlotStyle -> Hue[0]],
             (* cumulative number of samples x-values *)
             ListPlot[MapIndexed[{#1, #2[[1]]/λ}&, Sort[usedxValues[[1]]]]]}]]
```

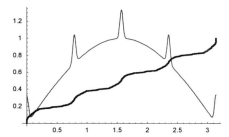

While all numerical functions have the evaluation monitor, some have in addition a step monitor.

`StepMonitor`

   is an option to numerical (and other) functions that determines an expression to be evaluated at each step.

For some of the more complicated numerical algorithms more than one function evaluation might be needed to carry out one "step" of an algorithm. Here are the built-in functions that have the `StepMonitor` option.

```
In[27]:= Function[monitor,
             {monitor, Select[DeleteCases[Names["*"],
                                          "$Failed" | "$DefaultFont"],
             MemberQ[First /@ (Options @@
                               ToExpression[#, StandardForm, Unevaluated]),
                     monitor]&]}] /@
                     (* look for evaluation and step monitor option presence *)
                     {EvaluationMonitor, StepMonitor}
Out[27]= {{EvaluationMonitor, {FindFit, FindMaximum, FindMinimum, FindRoot, NDSolve,
           NIntegrate, NMaximize, NMinimize, NProduct, NSum}}, {StepMonitor,
           {FindFit, FindMaximum, FindMinimum, FindRoot, NDSolve, NMaximize, NMinimize}}}
```

Here is a simple example. We solve the differential equation $x'''(t) = \exp(1/\ln(x(t)))$. The number of function evaluations of the right-hand side of the differential equation is approximately twice the number of steps carried out.

```
In[28]:= (* right-hand side of the differential equation *)
      Λ[x_?NumericQ] := Exp[1/Log[x]];

      (* solve differential equation and reap collected values *)
      {#, Length[Reap[
      NDSolve[{x'''[t] == Λ[x[t]], x[0] == 2, x'[0] == 1, x''[0] == 1},
              x, {t, 0, 1}, # :> Sow[t]]][[2, 1]]]}& /@
                                   {EvaluationMonitor, StepMonitor}
Out[31]= {{EvaluationMonitor, 77}, {StepMonitor, 37}}
```

After this discussion about the common options of the numerical functions, we have a quick look at the attributes of the numerical functions. Five out of the seven numerical functions have the attribute `HoldAll`.

```
In[32]:= Select[numericalFunctions, MemberQ[Attributes[#], HoldAll]&]
Out[32]= {NSum, NProduct, NIntegrate, FindRoot, FindMinimum}
```

The reason for the `HoldAll` attribute of these functions is the presence of local variables. `NSum`, `NProduct`, and `NIntegrate` have genuine dummy variables that do not appear in the result and whose name is irrelevant. `FindRoot`, `FindMinimum` also have local variables (that are present in the inputs and outputs of these functions too). To localize these variables within the numerical functions, the numerical functions must have a

`HoldAll` attribute. Else, the arguments of the function would evaluate and values of these variables outside the function would immediately be taken. Instead, the `HoldAll` attributes allows finding the (dummy) variables, scope them in a `Block`-like manner, then evaluate the body of the numerical function. After the body has been evaluated, the numerical functions typically analyze the body symbolically. As a result of this analysis, the functions carry out useful transformations (for instance using a symmetry such as $f(z) = f(-z)$ for halving the work for a numerical integration), derive derivatives if needed (say for root-finding and minimization problems), and also select an algorithm suitable for the evaluated body at hand. After these transformations, the numerical function carries out the actual numerical algorithm selected to obtain the requested result.

In the next input, the dummy integration variable x has the global value 1. Because `NIntegrate` uses a local scoped copy of the variable x, no problems arise.

```
In[33]:= x = 1;
      NIntegrate[x^2, {x, 0, 1}]
Out[34]= 0.333333
```

A `Block`-like scoping is needed for the scoping. `Block` directly would not work because its local variables must be symbols. So the following would not work.

```
In[35]:= ξ[1] = 0;
      NSum[ξ[1], {ξ[1], 1, 10}]
Out[36]= 55
```

And a `Module`-like scoping is not possible to use to allow bodies of numerical functions to be defined outside the numerical function. (A `Module`-like scoping would generate a local variable $ξ2\$number$ which is not even present in `summand`.)

```
In[37]:= summand = ξ2;

      NSum[summand, {ξ2, 1, 10}]
Out[38]= 55.
```

The evaluation of the body of the numerical functions is sometimes inconvenient. This is especially the case when the body of the numerical function evaluates to a large expression. Here is a typical example. Because of the large expression that results after the evaluation of the body, the following root-finding procedure is relatively slow.

```
In[39]:= FindRoot[Nest[Sin[# + #^2]&, y, 10], {y, 1, 2}] // Timing
Out[39]= {0.01 Second, {y → 1.34163}}
```

In such situations, it is sometimes useful to define the body as a independent function that evaluates only for numerical arguments.

```
In[40]:= f[y_?NumberQ] := Nest[Sin[# + #^2]&, y, 10]

      FindRoot[f[y], {y, 1, 2}] // Timing
Out[41]= {0. Second, {y → -0.000360288}}
```

But such an independent definition also has disadvantages. For instance, it does not allow for the compilation within the numerical function (because the numerical function never has access to the symbolic function). In such cases, one can use the system option `"EvaluateNumericalFunctionArgument"`. By default, this option is set to be `True`, meaning the body will be evaluated.

```
In[42]:= Cases[Developer`SystemOptions[], _["EvaluateNumericalFunctionArgument", _]]
Out[42]= {EvaluateNumericalFunctionArgument → True}
```

Setting this system option to be `False` avoids the evaluation of the body. Now the calculation of the root is much faster.

```
In[43]:= Developer`SetSystemOptions["EvaluateNumericalFunctionArgument" -> False];
```

```
In[44]:= FindRoot[Nest[Sin[# + #^2]&, y, 10], {y, 1, 2}] // Timing
Out[44]= {0. Second, {y → 1.34163}}
```

We restore the evaluation of the arguments of numerical functions.

```
In[45]:= Developer`SetSystemOptions["EvaluateNumericalFunctionArgument" -> True];
```

The examples given so far always used scalar variables. Most of the numerical functions can be used with vector-variables to allow for a short and convenient notation. In the following root finding call, we use the vector-valued variable `xVector` and give a starting value that itself is a vector.

```
In[46]:= FindRoot[xVector.{1, 2, 3} + {1, 1, 1}, {xVector, {0, 0, 0}}]
Out[46]= {xVector → {-0.0714286, -0.142857, -0.214286}}
```

And here is a matrix-valued differential equation in the variable `yMatrix`. The right-hand side of the initial condition is a 3×3 matrix.

```
In[47]:= NDSolve[{yMatrix'[t] == yMatrix[t],
                  yMatrix[0] == {{1, 2, 3}, {4, 5, 6}, {7, 8, 9}}},
                 yMatrix, {t, 0, 1}]
Out[47]= {{yMatrix → InterpolatingFunction[{{0., 1.}}, <>]}}
```

The resulting interpolating function evaluates to a 3×3 matrix for a given argument.

```
In[48]:= yMatrix[1] /. %
Out[48]= {{{2.71828, 5.43656, 8.15485},
          {10.8731, 13.5914, 16.3097}, {19.028, 21.7463, 24.4645}}}
```

Some of the numerical functions allow their body to be a (symbolic, pure, or compiled) function. This means no (dummy) variable has to be present in the body and in the initial/starting conditions. The next input solves the root-finding problem for the function $z \to \sin(z) + \cos(z)$.

```
In[49]:= FindRoot[(Cos[#] + Sin[#])&, {-1}]
Out[49]= {-0.785398}
```

Also, compiled functions can be used in a variable-free body. In the next example, we look for the minimum of the compiled map $\{x, y\} \to (x - 2)^2 + (y - 4)^2$.

```
In[50]:= cf = Compile[{x, y}, (x - 2)^2 + (y - 4)^2];
```

```
         FindMinimum[cf, {1, 2}, {2, 3}]
Out[51]= {1.23487 × 10^{-26}, {2., 4.}}
```

(Be aware that because of the absence of a dummy variable in the last two examples, the form of the output is different from the case where variables are given—no natural choice for the right-hand side of the corresponding `Rule` exists.)

No good excuse is available why `NDSolve` and `NMinimize` do not have the `HoldAll` attribute. Both functions have dummy variables that should be scoped from the surrounding.

Most of the numerical functions have the option `Compiled`.

```
In[52]:= Select[numericalFunctions, MemberQ[Options[#], Compiled -> _]&]
```

In[52]:= {NSum, NProduct, NIntegrate, FindRoot, FindMinimum, NDSolve}

```
Compiled
```
is an option to numerical (and other) functions that determines if the body should be compiled.

Compilation typically does not change the result of a calculation, but allows for a much faster execution. In the following example of a numerical integration, the compiled version is approximately 70 times faster.

```
In[53]:= Timing[NIntegrate[Nest[Sin, x, 1000], {x, 0, Pi},
                          Compiled -> #]]& /@
                          (* compile or not compile? *) {True, False}
Out[53]= {{0.05 Second, 0.165926}, {10.13 Second, 0.165926}}
```

As mentioned in the section about the `Compile` function, a subpart of the compilation process is optimization. Using the suboption `CompileOptimizations` we can optimize and not compile or compile and not optimize (optimizing and not compiling is useful for using a higher working precision). The next examples show that optimizing can have a dramatic influence on the execution speed.

```
In[54]:= (* machine arithmetic; do not compile *)
        Timing[NIntegrate[Nest[Sin[# + x #]&, x, 20], {x, 0, 1},
                          Compiled -> {True, CompileOptimizations -> #}]]& /@
                          (* optimize or not *) {All, None}
Out[55]= {{2.81 Second, 0.867292}, {51.44 Second, 0.867292}}
```

```
In[56]:= (* use high-precision arithmetic; do not compile *)
        Timing[NIntegrate[Nest[Sin[# + x #]&, x, 10], {x, 0, 1},
                          WorkingPrecision -> 20, PrecisionGoal -> 10,
                          Compiled -> {False, CompileOptimizations -> #}]]& /@
                          (* optimize or not *) {All, None}
Out[56]= {{0.08 Second, 0.8319358183}, {5.8 Second, 0.8319358183}}
```

We end this section by remarking that in the `Experimental`` context we find some `"*NumericalFunc·tion*"` named functions. This type of numerical function is a generalization of a compiled function and allows for a convenient form to define a (vector-valued) function and its derivatives with (vector-)arguments.

```
        Names["*`*NumericalFunction*"]
Out[57]= {Experimental`CreateNumericalFunction, Experimental`NumericalFunction,
         NDSolve`NumericalFunctionToEquations, NDSolve`ValidNumericalFunctionQ}
```

1.7 Sums and Products

This section deals with numerical summation and computation of products. We begin with products.

```
NProduct[term_i, iterator, options]
```
computes the numerical product of the factors *term_i*, in which the terms are defined by *iterator*. Here, *iterator* is given in the usual iterator notation.

Numerically, this is the standard factorial function.

In[1]:= `nfac[x_Integer] := NProduct[i, {i, x}];`

    `Table[nfac[i], {i, 20}]`

Out[2]= {1., 2., 6., 24., 120., 720., 5040., 40320., 362880., 3.6288×10^6,
3.99168×10^7, 4.79002×10^8, 6.22702×10^9, 8.71783×10^{10}, 1.30767×10^{12},
2.09228×10^{13}, 3.55687×10^{14}, 6.40237×10^{15}, 1.21645×10^{17}, 2.4329×10^{18}}

If the factors approach 1 asymptotically, we can (sometimes) also evaluate infinite products such as $\prod_{k=1}^{\infty} (1 - k^{-2})$.

In[3]:= `NProduct[1 - 1/k^2, {k, 2, Infinity}]`

Out[3]= $0.5 + 0. \, i$

This is actually the exact result ($= 1/2$) to the specified precision. Here is another example.

In[4]:= `cosProduct[x_] := NProduct[Cos[x/2^k], {k, 1, Infinity}]`

In[5]:= `{cosProduct[100], cosProduct[0.5]}`

Out[5]= {-0.00506366, 0.958851}

Here is a comparison with the exact value of the product, which is $\sin(x)/x$.

In[6]:= `{Sin[100]/100, Sin[1/2]/(1/2)} // N`

Out[6]= {-0.00506366, 0.958851}

Here is another infinite product that we compare with the exact result: $\prod_{k=1}^{\infty} (1 + k^{-3}) = \pi^{-1} \cosh\left(\pi \sqrt{3}\,/2\right)$.

In[7]:= `{NProduct[1 + 1/k^3, {k, 1, Infinity}], 1/Pi Cosh[Pi Sqrt[3]/2] // N}`

Out[7]= {2.42819 + 0. i, 2.42819}

We now give a more difficult example and the corresponding exact result (*n* integer, $n > 2$).

$$\prod_{k=1}^{\infty} \left(1 - \frac{x^n}{k^n}\right) = \frac{1}{x^n} \prod_{k=0}^{n-1} \frac{1}{\Gamma(-e^{2\pi ik/n} x)}$$

In[8]:= `ΠN[x_, n_Integer] := NProduct[1 - x^n/k^n, {k, 1, Infinity}];`

    `ΠEx[x_, n_Integer] := -1/x^n Product[1/Gamma[-Exp[2Pi I k/n] x],`
                                               `{k, 0, n - 1}] // N[#, 22]& // N;`

For various *n* and *x* we will now compare the exact result with the result from `NProduct`.

In[10]:= `With[{n = 02, x = 1}, {ΠN[x, n], ΠEx[x, n]}]`

Out[10]= {0., 0.}

In[11]:= `With[{n = 02, x = 10 I}, {ΠN[x, n], ΠEx[x, n]}]`

Out[11]= {7.00783×10^{11}, $7.00783 \times 10^{11} + 0. \, i$}

In[12]:= `With[{n = 02, x = 50 I}, {ΠN[x, n], ΠEx[x, n]}]`

Out[12]= {5.26826×10^{65}, $5.26827 \times 10^{65} + 0. \, i$}

In[13]:= `With[{n = 20, x = 3}, {ΠN[x, n], ΠEx[x, n]}]`

Out[13]= {0., 0.}

In some cases, *Mathematica* does not find the correct solution. Typically, in such cases, it issues messages. Here is an example.

In[14]:= `With[{n = 3, x = 23.3}, {ΠN[x, n], ΠEx[x, n]}]`

```
NIntegrate::ncvb :
  NIntegrate failed to converge to prescribed accuracy after 7 recursive
    bisections in k near k = 23.258064516129032`. More…
```
Out[14]= $\{-1.07569 \times 10^{15} + 8.03053 \times 10^{14}\, i,\ -2.06363 \times 10^{15} + 0.\, i\}$

(We shall see several more examples of failures below in connection with NSum.) In such cases, it may be possible to work with one of the options NProductFactors or NProductExtraFactors. Here are the options of NProduct.

In[15]:= **Options[NProduct]**

Out[15]= {AccuracyGoal → ∞, Compiled → True, EvaluationMonitor → None, Method → Automatic,
    NProductExtraFactors → 15, NProductFactors → 15, PrecisionGoal → Automatic,
    VerifyConvergence → True, WorkingPrecision → MachinePrecision, WynnDegree → 1}

Because we typically do not need to evaluate large numerical products often, we do not go into a detailed discussion of all of the options. We are already familiar with Compiled as a Plot command. We have discussed the options AccuracyGoal, PrecisionGoal, and WorkingPrecision in the last section. We discuss Method, WynnDegree, and VerifyConvergence in connection with sums. That leaves for now the discussion of NProductFactors and NProductExtraFactors—two options exclusively belonging to the function NProduct.

NProductFactors

    is an option for NProduct, and it specifies the number of factors to be explicitly evaluated and multiplied together.
    Default:
        15

    Admissible:
        arbitraryPositiveInteger

NProductExtraFactors

    is an option for NProduct, and it specifies the number of factors to be evaluated in computing a limiting value.
    Default:
        12

    Admissible:
        arbitraryPositiveInteger

Making use of the option NProductFactors with a higher than the default setting yields the correct result and no message for the example from above.

In[16]:= **With[{n = 3, x = 23.3},**
        NProduct[1 - x^n/k^n, {k, 1, Infinity}, NProductFactors -> 30]]
Out[16]= $-2.06363 \times 10^{15} + 0.\, i$

Numerical summation is essentially analogous to numerical computation of products. Sums arise much more frequently in practice than do products, for example, in the evaluation of infinite series solutions of ordinary and partial differential equations [125]. Before we discuss the *Mathematica* command NSum, let us have a short look at the interesting subject of adding floating point numbers [1252], [832], [945], [482] in more detail (we mentioned this subject already briefly in the beginning of this chapter).

This is a list of 100 random numbers to be summed.

In[17]:= **SeedRandom[888];**
    summands = Table[Random[], {1000}];

s1 is the cumulative sum obtained using straightforward summation.

In[19]:= **s1 = Rest[FoldList[Plus, 0, summands]];**

s2 is the corresponding "exact" result.

In[20]:= **s2 = Rest[FoldList[Plus, 0, SetPrecision[summands, Infinity]]];**

Here the (quite large) difference of the two cumulative sums in units of $MachineEpsilon is shown.

In[21]:= **ListPlot[(s1 - s2)/$MachineEpsilon, PlotRange -> All, Frame -> True]**

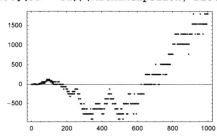

Note that the order of summing numbers is only an issue for machine numbers; using *Mathematica*'s high-precision arithmetic, we always get a unique result. Here, we generate 100 random permutations of high-precision copies of summands.

In[22]:= **(* make a random permutation of the integers 1 to n *)**
    randomPermutation :=
    Compile[{{n, _Integer}},
      Module[{l = Range[n], tmp1, tmp2, j},
        Do[(* randomly swap elements *)
          tmp1 = l[[i]]; j = Random[Integer, {i, n}]; tmp2 = l[[j]];
          l[[i]] = tmp2; l[[j]] = tmp1, {i, Length[l]}]; l]]

In[24]:= **SeedRandom[1];**

    With[{preciseSummands = SetPrecision[summands, 30],
        λ = Length[summands]},
    (* return exact result to detect any difference *)
    SetPrecision[Table[Fold[Plus, 0,
                        preciseSummands[[randomPermutation[λ]]]],
                    {100}], Infinity] // (* remove multiples *) Union]

Out[25]= $\left\{ \dfrac{226518077844760744875\mathbf{1}}{4611686018427387904} \right\}$

If we take high-precision numbers varying in size over many orders of magnitudes, then it is possible that we get slightly different results depending on the order of the summands. In the next example, we use 100 30-digit random numbers with magnitudes varying over 100 orders of magnitude. We obtain about 10 different results for the sums and their relative difference is in the order 10^{-40}.

In[26]:= **SeedRandom[2];**

    {Length[#], N @ Abs[(Max[#] - Min[#])/Plus @@ #]}& @
    (* different sums from 100 random permutations of 100 summands *)
    Module[{preciseSummands =

```
        Table[Random[Real, {-1, 1}, 30] 10^Random[Integer, {-50, 50}],
              {100}]},
    SetPrecision[Table[Fold[Plus, 0,
                       preciseSummands[[randomPermutation[100]]]],
               {100}], Infinity] // Union]
```

Out[27]= $\{9, 3.44602 \times 10^{-41}\}$

We now use our list summands again. Sorting the numbers in increasing order before summing does not reduce the error by much.

In[28]:= `s3 = Table[Fold[Plus, 0, Sort[Take[summands, i]]], {i, Length[summands]}];`

In[29]:= `ListPlot[(s3 - s2)/$MachineEpsilon, PlotRange -> All, Frame -> True]`

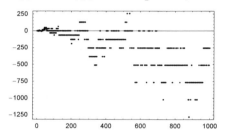

Sorting in decreasing order typically increases the error.

In[30]:= `s4 = Table[Fold[Plus, 0, Reverse[Sort[Take[summands, i]]]],`
          `{i, Length[summands]}];`

In[31]:= `ListPlot[(s4 - s2)/$MachineEpsilon, PlotRange -> All, Frame -> True]`

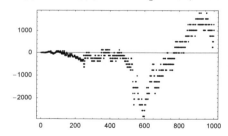

Now, let us sum adjacent pairs of numbers. (In case of a list with an odd number of elements, we keep the first element.)

In[32]:= `pairSum[list_ ? (EvenQ[Length[#]]&)] :=`
          `pairSum[Apply[Plus, Partition[list, 2], {1}]];`

         `pairSum[list_ ? (OddQ[Length[#]]&)] :=`
          `pairSum[Prepend[Apply[Plus, Partition[Rest[list], 2], {1}], First[list]]];`

         `pairSum[{n_}] := n;`

In[36]:= `s5 = Table[pairSum[Take[summands, i]], {i, Length[summands]}];`

Now, the error is considerably smaller.

In[37]:= `ListPlot[(s5 - s2)/$MachineEpsilon, PlotRange -> All, Frame -> True]`

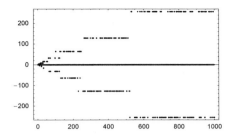

One of the best (and with a complexity linear in the number of terms to be summed) methods is the so-called Kahan summation. In the intermediate stages, it keeps track of the approximate error.

```
In[38]:= kahanSummation[{sum_, error_}, newSummand_] :=
    Function[y, {#, Identity[# - sum] - y}&[sum + y]][newSummand - error]
```

```
In[39]:= s6 = First /@ FoldList[kahanSummation, {summands[[1]], 0}, Rest[summands]];
```

```
In[40]:= ListPlot[(s6 - s2)/$MachineEpsilon, PlotRange -> All, Frame -> True]
```

The error accumulated in straightforward summation is an interesting function and typically is nonmonotonic, even for identical summands. The following function cf collects all instances where the error of summing δ is nonmonotonic.

```
In[41]:= cf = Compile[{δ, {n, _Integer}},
    Module[{sum = 0., Δ = {}, diff, penultimateDiff = 0., lastDiff = 0.},
    Do[(* current sum value *) sum = sum + δ;
        (* difference to exact sum *) diff = sum - k δ;
        (* collect occurrences of relative maxima in the error *)
        If[diff <= lastDiff <= penultimateDiff ||
            diff >= lastDiff >= penultimateDiff, Null,
            Δ = Join[Δ, {k, diff}]];
        penultimateDiff = lastDiff;
        lastDiff = diff, {k, n}];
        (* return error maxima positions and values *)
        Partition[Δ, 2]]];
```

Adding $11/17$ shows that the absolute value of the error is in average increasing, but changes sign. In summing 10^8 terms, we obtain local maxima in the errors and the final error is of size 5×10^{-3}.

```
In[42]:= {Round[#1], #2}& @@@ cf[11/17, 10^8]
```

```
Out[42]= {{7, 0.}, {25, -3.55271×10^-15}, {198, -4.83169×10^-13},
    {396, 1.13687×10^-12}, {1583, -2.251×10^-11}, {6332, -2.97405×10^-10},
    {50643, -3.21306×10^-8}, {101284, 7.62229×10^-8},
    {3241054, 0.00011365}, {3241055, 0.00011365}, {25928425, 0.00499646}}
```

There are two quite different ways one might want to sum a sequence. Either one wants to add one element after another to the running sum or one wants to add all numbers at once. To add a given sequence of length *l*, the function Total is most suited.

Total[*list, options*]

> gives the sum of the elements of *l*.

Here is a simple example for a list of five integers.

In[43]:= **Total[{1, 2, 3, 4, 5}]**

Out[43]= 15

Here Total is applied to symbolic elements.

In[44]:= **Total[Table[NaN[k], {k, 5}]]**

Out[44]= NaN[1] + NaN[2] + NaN[3] + NaN[4] + NaN[5]

Setting the option method to "CompensatedSummation" minimizes the numerical error in the sum. We use this option setting also explicitly for our list summands from above. The result is excellent.

In[45]:= **{ (Total[summands] - Last[s2])/\$MachineEpsilon,**
      (Total[summands, Method -> "CompensatedSummation"] -
      Last[s2])/\$MachineEpsilon}

Out[45]= {1792., 0.}

After these preliminary remarks about summation, let us turn to the function NSum.

NSum[*term_i, iterator, options*]

> produces the numerical sum of the terms *term_i*, where the terms are defined by the iterator. Here, *iterator* is given in the usual iterator notation.

Here are some examples. First, we sum 100 terms of the harmonic series (sequence): $\sum_{n=1}^{100} 1/n$.

In[46]:= **NSum[1/n, {n, 1, 100}]**

Out[46]= 5.18738

Here is an infinite sum whose value is *e*: $\sum_{n=0}^{\infty} 1/n\,! = e$

In[47]:= **NSum[1/n!, {n, 0, Infinity}]**

Out[47]= 2.71828

The following sum (which we already encountered in the first section of this chapter) has value $\pi/4$: $\sum_{n=0}^{\infty} (-1)^n/(2n+1) = \pi/4$

In[48]:= **NSum[(-1)^n/(2n + 1), {n, 0, Infinity}] - Pi/4**

Out[48]= -4.44089×10^{-16}

Divergent series are usually identified (although the actual value of the next sum is Infinity, using only numerical techniques, one might not get an explicit Infinity). The warning messages generated in the next example indicate a potentially wrong answer. (In distinction to Sum to be discussed in Chapter 1 of the Symbolics volume [1795], the numerical function NSum will not carry out any regularization procedure for divergent sums.) Next, we consider $\sum_{n=1}^{\infty} 1/n = \infty$.

In[49]:= **NSum[1/n, {n, 1, Infinity}]**

```
        NIntegrate::slwcon :
         Numerical integration converging too slowly; suspect one of the
            following: singularity, value of the integration being 0, oscillatory
            integrand, or insufficient WorkingPrecision. If your integrand is
            oscillatory try using the option Method->Oscillatory in NIntegrate. More…
        NIntegrate::ncvb :
         NIntegrate failed to converge to prescribed accuracy after 7 recursive
            bisections in n near n = 2.288332793335697`*^56. More…
```
Out[49]= 23953.7

Sometimes *Mathematica* is not able to find the correct values of sums.

```
In[50]:= summand[i_] := Which[i <= 50, 0,
                              i > 50 && i < 100, 1,
                              i >= 100, 0]
```

```
In[51]:= NSum[summand[i], {i, 100}]
```
Out[51]= 0.

NSum has the following options. We will discuss those specific to NSum that are not options of all numerical functions.

```
In[52]:= Options[NSum]
```
Out[52]= {AccuracyGoal → ∞, Compiled → True, EvaluationMonitor → None, Method → Automatic,
          NSumExtraTerms → 12, NSumTerms → 15, PrecisionGoal → Automatic,
          VerifyConvergence → True, WorkingPrecision → MachinePrecision, WynnDegree → 1}

The reason for the above result is a straightforward consequence of the way NSum works: NSum currently does not use any smart algorithm to determine the sums of a summation index-dependent symbolically given summand. It rather carries out the following algorithm: a number of initial terms are explicitly calculated, the tail is estimated, and the convergence is checked (first, if specified). By default, only the first 27 values of a sum will ever be explicitly calculated. Using the options NSumTerms and NSumExtraTerms, we can include more terms.

NSumTerms

    is an option for NSum, and it specifies the number of terms to be evaluated and summed.

    Default:

        15

    Admissible:

        arbitraryPositiveInteger

NSumExtraTerms

    is an option for NSum, and it specifies the number of terms to be explicitly evaluated and used for the determination of the limiting value of the sum.

    Default:

        12

    Admissible:

        arbitraryPositiveInteger

Here, it can be seen that all further terms are also assumed to be exactly 1.

```
In[53]:= NSum[summand[i], {i, 100}, NSumTerms -> 50]
```

Out[53]= `49.5`

Next, we explicitly add all 49 terms that are not equal to 0.

In[54]:= `NSum[summand[i], {i, 100}, NSumTerms -> 100]`

Out[54]= `49`

Most of the remaining options are self-explanatory or were already discussed, but `VerifyConvergence`, `Method`, and `WynnDegree` probably are not.

`VerifyConvergence`

> is an option for `NSum`, and it determines whether convergence of the sum according to the d'Alembert criterion (ratio test) should be checked.
> **Default:**
>> `True`
>
> **Admissible:**
>> `False` (saves computation time)

`Method`

> is an option for `NSum` and `NProduct`, and it determines the method for finding the limit value of the sum or product, respectively. The limit value will be computed using the number of terms specified by the options `NSumExtraTerms` or `NProductExtraFactors`, respectively.
> **Default:**
>> `Automatic`
>
> **Admissible:**
>> `Method -> Integrate`
>>
>> or
>>
>> `Method -> NIntegrate`
>>
>> or
>>
>> `Method -> SequenceLimit`

The `VerifyConvergence` option is set to `True` by default. Because the convergence test can sometimes be quite expensive, we recommend setting it to `False` for sums known to be convergent.

The following extremely slowly divergent sum [23] $\sum_{k=1}^{\infty} \tilde{\ln}(k)$ (where $\tilde{\ln}(k) = k \ln(k) \ln(\ln(k))$ ... and we take logarithms as long as the result is greater than one cannot be recognized as divergent. But the message indicates a potential problem.

In[55]:= `iteratedLog[x_?NumberQ] := If[x > E, x iteratedLog[Log[x]], x]`
      `(* or, more procedural,`
         `Times @@ Drop[NestWhileList[Log, n, # >= 1&], -1] *)`

In[57]:= `NSum[1/iteratedLog[k], {k, Infinity}, VerifyConvergence -> True]`

        `NIntegrate::slwcon :`
          `Numerical integration converging too slowly; suspect one of the`
            `following: singularity, value of the integration being 0, oscillatory`
            `integrand, or insufficient WorkingPrecision. If your integrand is`
            `oscillatory try using the option Method->Oscillatory in NIntegrate. More…`

```
NIntegrate::ncvb :
  NIntegrate failed to converge to prescribed accuracy after 7 recursive
    bisections in k near k = 2.288332793335697`*^56. More…
```
Out[57]= 4.43542

Mathematica's current NSum will not succeed evaluating very slowly (meaning logarithmically slowly) convergent sums [470], [222]; for such sums and other special cases, see the more specialized literature, such as [1184].

In the first case, Method -> Integrate, the sum will be transformed into an integral with correction terms. *Mathematica* attempts to compute the resulting integral analytically. If this fails, it is computed numerically. If it is obvious from the beginning that the integral cannot be evaluated analytically (e.g., if the function to be summed does not have an analytic continuation, as happens when it contains a number-theoretic function), the choice Method -> NIntegrate may be appropriate. The advantage of Method -> Integrate and Method -> NIntegrate is that divergent sums can be identified. The third possible choice is based on the command SequenceLimit, which we will discuss in more detail in a moment. (Method -> Fit is identical with Method -> SequenceLimit.) With the default Method -> Automatic in NSum, *Mathematica* chooses between SequenceLimit and Integrate. The following example illustrates the possibilities:

$$\sum_{i=1}^{\infty} \frac{1}{i^2 + 2} = -\frac{1}{4} + 2^{-3/2} \pi \coth(\sqrt{2}\,\pi) = 0.861028100573727\ldots$$

Note both the results and the computational times. Because of the slow convergence of the sequence of partial sums, the settings SequenceLimit and Fit give a relatively poor result.

In[58]:= **Timing[NSum[1/(i^2 + 2), {i, Infinity}, Method -> #]]& /@**
             {Integrate, NIntegrate, SequenceLimit, Fit} // InputForm
Out[58]//InputForm=
        {{0.480000000000004*Second, 0.8610281004917036}, {0.*Second,
        0.8610281004917042},
          {0.*Second, 0.859776878710365}, {0.009999999999990905*Second,
        0.859776878710365}}

Computing the same sum without explicitly setting the options, we can see which method NSum chooses for this sum: NIntegrate.

In[59]:= **NSum[1/(i^2 + 2), {i, Infinity}] // InputForm // Timing**
Out[59]= {0. Second, 0.8610281004917042}

In the next three inputs, we compare the function evaluations for the three methods for the sum $\sum_{k=0}^{\infty} \cos(k)/k^4$. Using the method Integrate, the function $\cos(k)/k^4$ is symbolically integrated and no function evaluation for explicit values of k other than the first NSumTerms ones are needed.

In[60]:= **NSum[k^-4 Cos[k], {k, Infinity}, Method -> Integrate,**
             EvaluationMonitor :> Sow[k]] // Reap // Short[#, 6]&
Out[60]//Short=
        {0.500822, {{1., 2., 3., 4., 5., 6., 7., 8., 9., 10., 11., 12., 13., 14., 15.}}}

Using the method NIntegrate, the function $\cos(k)/k^4$ is numerically integrated and a few hundred function evaluations for values of k ranging over many orders of magnitude are needed. The message indicates that NIntegrate was not able to achieve the requested precision of the result.

In[61]:= **NSum[k^-4 Cos[k], {k, Infinity}, Method -> NIntegrate,**
             EvaluationMonitor :> Sow[k]] // Reap // Short[#, 6]&

```
NIntegrate::ncvb :
    NIntegrate failed to converge to prescribed accuracy after 7 recursive
        bisections in k near k = 148.55854776681167`. More…
```

Out[61]//Short=

```
{0.500822, {{1., 2., 3., 4., 5., 6., 7., 8., 9., 10., 11., 12.,
    13., 14., 15., 17., 16.008, ≪421≫, 141.236, 156.605, 143.261,
    154.224, 145.764, 151.453, 130.507, 138.848, 123.065, 138.151,
    123.624, 136.811, 124.73, 134.951, 126.332, 132.78, 128.307}}}
```

Using the method Fit (or SequenceLimit) the function $\cos(k)/k^4$ is sampled at NSumExtraTerms values of k. The message indicates that the fit was not able to determine the general form of the sequence (and so the returned result returned by NSum is not fully trustworthy). It is interesting to remark that the additional terms were tested before the main terms (if the sum would diverge, this test would save the evaluation of the first terms).

In[62]:= **NSum[k^-4 Cos[k], {k, Infinity}, Method -> Fit,**
   EvaluationMonitor :> Sow[k]] // Reap // Short[#, 6]&

```
SequenceLimit::seqlim : The general form of the sequence
    could not be determined, and the result may be incorrect. More…
```

Out[62]//Short=

```
{0.500829, {{16., 17., 18., 19., 20., 21., 22., 23., 24., 25., 26.,
    27., 1., 2., 3., 4., 5., 6., 7., 8., 9., 10., 11., 12., 13., 14., 15.}}}
```

The function Prime is defined only for integer arguments. NSum substitutes an approximate value of the iterator variable in the function. So the following input generates messages. But we also get a result.

In[63]:= **NSum[1/(Prime[k]^3 + 2), {k, Infinity}]**

```
Prime::intpp : Positive integer argument expected in Prime[16.]. More…

Prime::intpp : Positive integer argument expected in Prime[16.]. More…

Prime::intpp : Positive integer argument expected in Prime[17.]. More…

General::stop :
    Further output of Prime::intpp will be suppressed during this calculation. More…
```

Out[63]= 0.147061

We insert an additional Round to make sure Prime gets an integer argument. Now, we get messages because of a call of Prime with a very large argument in the tail approximation.

In[64]:= **NSum[1/(Prime[Round[k]]^3 + 2), {k, Infinity}]**

```
NIntegrate::singd :
    NIntegrate's singularity handling has failed at point {k}={2.70134×10^150} for the
        specified precision goal. Try using larger values for any of $MaxExtraPrecision
        or the options WorkingPrecision, or SingularityDepth and MaxRecursion. More…
NIntegrate::inum :
    Integrand $\dfrac{1}{2 + \text{Prime}[\text{Round}[k]]^3}$ is not numerical at {k} = {2.70134×10^150}. More…
```

Out[64]= $\text{NSum}\left[\dfrac{1}{\text{Prime}[\text{Round}[k]]^3 + 2}, \{k, \infty\}\right]$

The option setting Integrate gives the same message. Because the function Prime is not defined for each real argument, it cannot be integrated. So the symbolic integration fails and N[*symbolicUnevaluatedIntegral*] will be evaluated. This results in the numerical integration, and we encounter the same problem as in the last input.

```
In[65]:= (* two failing attempts *)
    {NSum[1/(Prime[Round[k]]^3 + 2), {k, Infinity}, Method ->  Integrate],
     NSum[1/(Prime[Round[k]]^3 + 2), {k, Infinity}, Method -> NIntegrate]}
```

NIntegrate::singd :
   NIntegrate's singularity handling has failed at point {k}={2.70134×10^150} for the
      specified precision goal. Try using larger values for any of $MaxExtraPrecision
      or the options WorkingPrecision, or SingularityDepth and MaxRecursion. More...

NIntegrate::inum :
   Integrand $\dfrac{1}{2 + \text{Prime}[\text{Round}[k]]^3}$ is not numerical at {k} = {2.70134×10^150}. More...

NIntegrate::singd :
   NIntegrate's singularity handling has failed at point {k}={2.70134×10^150} for the
      specified precision goal. Try using larger values for any of $MaxExtraPrecision
      or the options WorkingPrecision, or SingularityDepth and MaxRecursion. More...

NIntegrate::inum :
   Integrand $\dfrac{1}{2 + \text{Prime}[\text{Round}[k]]^3}$ is not numerical at {k} = {2.70134×10^150}. More...
```

$$Out[66]= \left\{ \text{NSum}\left[ \frac{1}{\text{Prime}[\text{Round}[k]]^3 + 2}, \{k, \infty\}, \text{Method} \to \text{Integrate}\right],\right.$$
$$\left.\text{NSum}\left[ \frac{1}{\text{Prime}[\text{Round}[k]]^3 + 2}, \{k, \infty\}, \text{Method} \to \text{NIntegrate}\right]\right\}$$

Using the method option setting Fit does not require values at large iterator variables. Because the function Prime does not evaluate for such large numbers, the sum finally returns a result without an error message.

```
In[67]:= NSum[1/(Prime[Round[k]]^3 + 2), {k, Infinity}, Method -> Fit]
Out[67]= 0.147061
```

Next, we will use NSum to exhibit a series with nonvanishing Taylor coefficients at $x = 0$. This series has the property that the Taylor series does not converge to the original function. The original function is the following sum (which cannot be expressed in closed form though named functions) [1485].

```
In[68]:= s[x_] = Sum[(-1)^k/k! 1/(1 + 2^k x), {k, 0, Infinity}];
```

The *j*th Taylor coefficient at $x = 0$ is $(-1)^k k! (1/e)^{2^k}$.

```
In[69]:= Table[Evaluate //@ D[s[x], {x, j}] /. x -> 0, {j, 0, 5}]
```

$$Out[69]= \left\{ \frac{1}{e}, -\frac{1}{e^2}, \frac{2}{e^4}, -\frac{6}{e^8}, \frac{24}{e^{16}}, -\frac{120}{e^{32}} \right\}$$

For $x = 10^{-3}$, we clearly see that the original function and the summed Taylor series have different values. The two functions sumSeries and sumTaylorSeries sum the two relevant series numerically using high-precision arithmetic and sequence extrapolation.

```
In[70]:= (* original and Taylor series; use appropriate option settings *)
    sumSeries[x_] := NSum[(-1)^k/k! 1/(1 + 2^k x), {k, 0, Infinity},
                          PrecisionGoal -> 20, WorkingPrecision -> 60,
                          VerifyConvergence -> False, Method -> Fit]

    sumTaylorSeries[x_] := NSum[(-1)^k k! (1/E)^(2^k) x^k, {k, 0, Infinity},
                          PrecisionGoal -> 20, WorkingPrecision -> 60,
                          VerifyConvergence -> False, Method -> Fit]
```

```
In[74]:= {sumSeries[10^-3], sumTaylorSeries[10^-3]} // N[#, 20]&
Out[74]= {0.36774412421264200300 + 0. × 10^-21 i, 0.36774414251747071331 + 0. × 10^-21 i}
```

Frequently, especially for complicated sums, the convergence test can take a very long time compared with the actual summation of the terms. So, if we know that a sum converges (or do not care about a finite result for a
```

divergent sum), we can turn off the convergence checking. The following function $u[c]$ defines a function that has the value $c/2$ for $0 \le c \le 1$ [179] (the Borel–Tanner distribution [765]). The function $uD[c]$ implements the derivative of $u[c]$ with respect to c.

```
In[75]:= u[c_] := 1 - 1/c NSum[k^(k - 2)/k! (c Exp[-c])^k, {k, Infinity},
                    Method -> Fit, VerifyConvergence -> False]
```

```
In[76]:= uD[c_] := -NSum[(c Exp[-c])^k k^(k - 2) (1 + k (c - 1))/(c^2 k!),
                    {k, Infinity}, Method -> Fit, VerifyConvergence -> False]
```

Calculating a few ten thousand sums can be done in second with the convergence checking turned off. The following graphic shows the function $u[c]$ and its derivative over the complex z-plane. We clearly see a region where the sum converges and outside of this region the sum diverges.

```
In[77]:= (* suppress messages *) Off[SequenceLimit::seqlim];
         grA = GraphicsArray[
         (* display real part of function and derivative *)
         Plot3D[Re @ #[cx + I cy], {cx, -2, 2}, {cy, -2, 2},
                 PlotRange -> {-2, 2}, ClipFill -> None, PlotPoints -> 180,
                 Mesh -> False, DisplayFunction -> Identity]& /@ {u, uD}];
         On[SequenceLimit::seqlim];
         Show[grA]
```

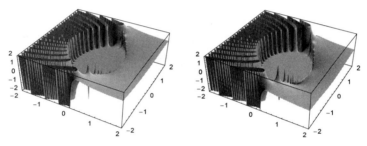

The right plot shows that the derivative of the sum is constant within the convergence region. There the sum has the value $c/2$. As a quick check, we carry out the following call to `Series` that takes 201 terms of the series into account.

```
In[81]:= Series[1 - 1/c Sum[k^(k - 2)/k! (c Exp[-c])^k, {k, 201}], {c, 0, 200}]
```

$$Out[81]= \frac{c}{2} + O[c]^{201}$$

We now discuss the command `SequenceLimit`.

`SequenceLimit[`*sequenceInListForm*`]`

    produces the "limit value" of the sequence *sequenceInListForm*, assuming it is "continued".

This command forms the limiting value of a finite sequence, assuming it is "continued" for an infinite number of terms. (This process only works under appropriate assumptions for the terms in the sum, essentially, that they have the form $\exp(index) \times polynomial(index)$.) We now examine a few examples.

```
In[82]:= SequenceLimit[{1, 2, 3, 4, 5}]
```

```
Out[82]= ComplexInfinity
```

```
In[83]:= SequenceLimit[{1 + 1/2^1, 1 + 1/2^2, 1 + 1/2^3, 1 + 1/2^4, 1 + 1/2^5}]
```

```
Out[83]= 1
```

`SequenceLimit` has one option.

In[84]:= **Options[SequenceLimit]**

Out[84]= {Method → Fit, WynnDegree → 1}

`WynnDegree`

is an option for `SequenceLimit`, as well as `NSum` and `NProduct`, when `Method -> SequenceLimit`. It determines the number of iterations to be used in finding the `SequenceLimit`.

Default:

1

Admissible:

an arbitrary integer *n* with $2n + 2 \le lengthOfTheSequence$, or `Infinity`

We create the following sequence.

In[85]:= **seqValues = Table[1 + 1/i^2, {i, 1, 12}]**

Out[85]= $\{2, \frac{5}{4}, \frac{10}{9}, \frac{17}{16}, \frac{26}{25}, \frac{37}{36}, \frac{50}{49}, \frac{65}{64}, \frac{82}{81}, \frac{101}{100}, \frac{122}{121}, \frac{145}{144}\}$

Now, we compute its "limit value" using various values for the option `WynnDegree`.

In[86]:= **Table[SequenceLimit[seqValues, WynnDegree -> k], {k, 5}] - 1**

Out[86]= $\{0. \times 10^{-4}, 0.0002, 0. \times 10^{-4}, 0. \times 10^{-4}, 0.001\}$

Often, `SequenceLimit` results in remarkable accuracy. Six terms give five correct digits (the exact infinite sum is $\pi/4$, as discussed above).

In[87]:= **{SequenceLimit[FoldList[Plus, 0,**
                        Table[(-1)^k/(2k + 1), {k, 0, 6}]]],
        N[Pi/4]}

Out[87]= {0.7854, 0.785398}

For comparison, the explicit sum of the six terms used is very different from the extrapolated value.

In[88]:= **N @ (Plus @@ Table[(-1)^k/(2k + 1), {k, 0, 6}])**

Out[88]= 0.820935

To see how transformations with `SequenceLimit` work, we call it with symbolic variables.

In[89]:= **SequenceLimit[{a, b, c, d, e, f}, WynnDegree -> 1] // Short**

Out[89]//Short=

$$\frac{a c^5 - 2 b c^5 + \ll 48 \gg + 2 b^2 c d^2 e - a c^2 d^2 e}{(b - 2 c + d) (\ll 38 \gg + 4 b c d e - c^2 d e)}$$

In[90]:= **SequenceLimit[{a, b, c, d, e, f}, WynnDegree -> 2]**

Out[90]= $\frac{c^3 - 2 b c d + a d^2 + b^2 e - a c e}{b^2 - a c - 2 b c + 3 c^2 + 2 a d - 2 b d - 2 c d + d^2 - a e + 2 b e - c e}$

Every application of Wynn's ϵ-algorithm (see exercise) shortens the sequence, which may lead to a sequence that is too short.

In[91]:= **SequenceLimit[{a, b, c, d, e, f}, WynnDegree -> 3]**

SequenceLimit::seqw :
    Sequence of length 6 is too short for use with WynnDegree -> 3. More…

Out[91]= SequenceLimit[{a, b, c, d, e, f}, WynnDegree → 3]

The following defines a function $\zeta N[n, \lambda]$ [50]. For all values of λ, the value of $\zeta N[n, \lambda]$ is the Riemann Zeta function $\zeta(n)$. (But the convergence depends on the value of λ; for small integer n, we have $\lambda_{opt} \approx 0.5$ for optimal convergence.)

```
In[92]:= ζN[n_, λ_, opts___] := 2^(n - 1)/(2^(n - 1) - 1) *
         NSum[1/(1 + λ)^(k + 1) k!/(j! (k - j)!) λ^(k - j) (-1)^j/(1 + j)^n,
              {k, 0, Infinity}, {j, 0, k}, opts]
```

The next input uses NSum to estimate the value of the sum for various values of the option setting of WynnDe﹀gree. Because the extrapolated sequence uses the summand values from the NSumExtraTerms option setting, we increase this value with increasing Wynn degree. And because the sequence limit calculation typically loses many digits, we use a relatively large setting for the WorkingPrecision option. The resulting difference between the NSum result and the exact value clearly shows how the result improves with more terms taken into account. We use $n = 5$ in the next input.

```
In[93]:= (* turn off message *)
         Internal`DeactivateMessages[
         Table[{m, (ζN[5, (* suboptimal value to stress test NSum *) 1/10,
                        WorkingPrecision -> 100, Method -> SequenceLimit,
                        PrecisionGoal -> 10, NSumTerms -> 30,
                        NSumExtraTerms -> Round[10 m], WynnDegree -> m] -
                        (* exact value *) Zeta[5]) // Abs // N}, {m, 1, 5}]]
Out[94]= {{1, 1.26229×10^-12}, {2, 5.53255×10^-14},
          {3, 2.97848×10^-15}, {4, 1.84773×10^-16}, {5, 1.27151×10^-17}}
```

We will give one nontrivial example that uses the WynnDegree option. In addition to extrapolating the values of slowly converging sequences, the function SequenceLimit can sometimes sum formally divergent series numerically (the so-called antilimit). Here is a simple toy example: $\sum_{k=0}^{\infty} (-1)^k 2^k$.

```
In[95]:= (* turn on message *) On[SequenceLimit::seqlim];

         SequenceLimit[FoldList[Plus, 0, Table[(-1)^k 2^k, {k, 0, 100}]]]
                SequenceLimit::seqlim : The general form of the sequence
                    could not be determined, and the result may be incorrect. More…
```

$$Out[96]= \frac{1}{3}$$

This result can be understood as the analytical continuation of $\sum_{k=0}^{\infty} (-1)^k \zeta^k$.

```
In[97]:= Sum[(-1)^k ζ^k, {k, 0, Infinity}] /. ζ -> 2
```

$$Out[97]= \frac{1}{3}$$

Here is a similar example involving logarithm: $\sum_{k=0}^{\infty} (-1)^k \ln(k)$. Again, by analytical continuation of $\sum_{k=0}^{\infty} (-1)^k \ln(k) t^k$ as $t \to 1$, one finds the value $\ln(\pi/2)/2$ for this sum.

```
In[98]:= SymbolicSum`SymbolicSum[(-1)^k Log[k], {k, Infinity},
                              GenerateConditions -> False] // Simplify
```

$$Out[98]= \frac{1}{2} Log\left[\frac{\pi}{2}\right]$$

Using the function SequenceLimit, we can confirm this value to more than 100 decimal digits. The following graphic shows the logarithm of the difference in base 10 as a function of the Wynn degree.

```
In[99]:= logSumSequence = Rest[FoldList[Plus, 0, Table[(-1)^k Log[k] // N[#, 1000]&,
                                                        {k, 120}]]];

In[100]:= ListPlot[#, PlotRange -> All, Frame -> True, Axes -> False]& @
          Table[{wd, If[# == 0, Accuracy[#], Log[10, Abs[#]]]}&[
```

```
SequenceLimit[logSumSequence, WynnDegree -> wd] -
            Log[Pi/2]/2]}, {wd, 1, 20}]
```

```
SequenceLimit::seqlim : The general form of the sequence
    could not be determined, and the result may be incorrect. More…

SequenceLimit::seqlim : The general form of the sequence
    could not be determined, and the result may be incorrect. More…

SequenceLimit::seqlim : The general form of the sequence
    could not be determined, and the result may be incorrect. More…

General::stop : Further output of
    SequenceLimit::seqlim will be suppressed during this calculation. More…
```

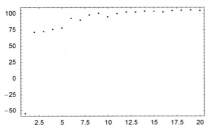

Here is a much more complicated example. The integral $\int_0^\infty \cos(x\,e^x)\,dx$ can be brought into the form $\int_0^\infty g(x)\cos(x)\,dx$ and then repeatedly partially integrated to yield a sum of derivatives of $g(x)$ at $x = 0$. This leads to the diverging (and oscillating) sum $-\sum_{k=0}^\infty (-1)^k (2\,k)^{2\,k-1}$. Here are the first five digits of the resummed result (equal to the convergent original integral with value $0.32336743167777\ldots$).

In[101]:= **SequenceLimit[FoldList[Plus, 0, N[#, 8000]& @**
                        Table[-(-1)^k (2k)^(2k - 1), {k, 800}]],
                WynnDegree -> 70] // InputForm

```
SequenceLimit::seqlim : The general form of the sequence
    could not be determined, and the result may be incorrect. More…
```

Out[101]//InputForm=
    0.3233674332960893587`7.5257498915995305

A detailed discussion of sequence transformations is beyond the scope of this book. For details, see [967], [290], [1888], [1909], [471], [1887], [1889], [288], [289], and [286]. For the choice of the "right" sequence transformation, see [585]. For the optimal precision to be used in sequence calculations, see [908].

Remark: If one has a few terms of a general sequence and one looks for a general form to build the sequence, a good source is *The On-Line Encyclopedia of Integer Sequences* [405], [1670] by N. J. A. Sloane, http://www.research.att.com/~njas/sequences.

The tail estimation of NSum (when using the NIntegrate or Integrate option setting) is often very accurate, much more accurate than it would be to take into account millions of terms. As an example, let us consider the sum [831], [925], [737], [738], [739]

$$\varphi(r) = \sum_{k=0}^\infty \left(\arctan\!\left(\frac{1}{\sqrt{k+1}} \right) - \arctan\!\left(\frac{1}{\sqrt{r^2+k}} \right) \right).$$

In[102]:= **φ[r_, opts___] := NSum[ArcTan[1/Sqrt[1 + k]] - ArcTan[1/Sqrt[r^2 + k]],**
                    {k, 0, Infinity}, opts, VerifyConvergence -> False,
                    Method -> NIntegrate (* or Integrate *)**]**

$\varphi(r)$ fulfills the following functional equation:

$$\phi\left(\sqrt{r^2 + 1}\,\right) - \phi(r) - \arctan\left(\frac{1}{r}\right) = 0.$$

Using NSum with the default parameters (meaning summing only 15 + 12 terms) yields a residual of about 10^{-8}.

```
In[103]:= With[{r = 2}, φ[Sqrt[r^2 + 1]] - φ[r] - ArcTan[1/r]]
Out[103]= -1.55571×10⁻⁸ + 0. i
```

Carrying out the summation by taking into account the first 10^7 terms gives a residual that is about 30000 times larger.

```
In[104]:= φ := Compile[{{n, _Integer}, {r, _Real}},
                    Sum[ArcTan[1/Sqrt[1 + k]] - ArcTan[1/Sqrt[r^2 + k]], {k, 0, n}]]
In[105]:= With[{r = 2, n = 10^7}, φ[n, Sqrt[r^2 + 1]] - φ[n, r] - ArcTan[1/r]]
Out[105]= -0.000316228
```

Using high-precision arithmetic we can, of course, fulfill the functional identity more precisely.

```
In[106]:= With[{r = 2, opts = Sequence[PrecisionGoal -> 15, WorkingPrecision -> 100]},
            φ[Sqrt[r^2 + 1], opts] - φ[r, opts] - ArcTan[1/r]]
Out[106]= 0. ×10⁻¹⁶
```

(The direct summation method could be improved by expanding the summands around $k = \infty$ and summing the resulting terms exactly.)

The function $\varphi(r)$ is a continuous description of the integer square root spiral obtained by placing the integer n in $\mathbb{R}^2$ at distance $\sqrt{n}$ from the origin and distance 1 from the point of the integer $n - 1$.

```
In[107]:= tab = Table[{Sqrt[r], φ[Sqrt[r]]}, {r, 2, 30}];
         points = Apply[{#1 Cos[#2], #1 Sin[#2]}&, tab, {1}];
         (* check *)
         Abs[1 - #]& /@ (Sqrt[#.#]& /@ (Subtract @@@ Partition[points, 2, 1]))
Out[110]= {2.19644×10⁻⁸, 2.04505×10⁻⁸, 3.11142×10⁻⁸, 2.64485×10⁻⁸,
         3.02262×10⁻⁸, 4.35727×10⁻⁸, 4.39616×10⁻⁸, 1.19173×10⁻⁷,
         4.90515×10⁻⁸, 5.3024×10⁻⁸, 5.37976×10⁻⁸, 4.26599×10⁻⁸, 4.43037×10⁻⁸,
         6.00504×10⁻⁸, 4.74497×10⁻⁸, 6.40234×10⁻⁸, 4.83124×10⁻⁸, 5.38236×10⁻⁸,
         7.14465×10⁻⁸, 7.13779×10⁻⁸, 7.49658×10⁻⁸, 1.92675×10⁻⁷, 8.069×10⁻⁸,
         7.74838×10⁻⁸, 6.03818×10⁻⁸, 6.41473×10⁻⁸, 6.28154×10⁻⁸, 8.38182×10⁻⁸}
In[111]:= (* show points and lines from the origin to the points *)
         Show[Graphics[{{GrayLevel[0], Line[points]},
                        MapIndexed[{Hue[(#2[[1]] - 1)/30],
                                    Line[{{0, 0}, #}]}&, points]}],
              Frame -> True, AspectRatio -> Automatic]
```

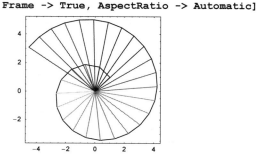

1.8 Integration

Because of the relatively small number of functions that are exactly integrable, NIntegrate is one of the more important *Mathematica* functions for practical applications.

NIntegrate[*integrand*, {*x*, x_0, x_1}, *options*]

computes a numerical approximation to the integral: $\int_{x_0}^{x_1} integrand\, dx$. The options *options* serve to take account of special properties of the integral.

For computation of integrals along straight line segments in the complex plane, one can use the notation {*x*, x_{start}, $x_{between1}$, $x_{between2}$, ..., $x_{between\,n}$, x_{end}}. A similar notation can be used to indicate points where singularities may occur. Multidimensional integrals can be calculated using NIntegrate[*integrand*, {*x*, x_0, x_1}, {*y*, y_0, y_1}, ..., *options*].

We begin with a very simple example.

```
In[1]:= NIntegrate[Exp[x], {x, 0, 2}]
Out[1]= 6.38906
```

Here is the exact result for comparison.

```
In[2]:= Exp[2] - 1 // N
Out[2]= 6.38906
```

Here is a numerical integral over an infinite interval. (The exact result is $\pi^{1/2}\, e^{-1/4} = 1.380388....$)

```
In[3]:= NIntegrate[Exp[-x^2] Cos[x], {x, -Infinity, Infinity}]
Out[3]= 1.38039
```

And here is a piecewise continuous function with 18 points of discontinuity in the integration interval. (The exact result is an algebraic number of quite high degree.)

```
In[4]:= NIntegrate[FractionalPart[x Max[x^3 + 4, (x - 2)/(x + 4)]]^2,
            {x, 1, 2}]
Out[4]= 0.325304
```

And here is a 2D integral. (The exact result can be expressed as a hypergeometric function.)

```
In[5]:= NIntegrate[Sin[x y] Exp[-y^2], {x, 0, Pi}, {y, 0, Infinity}]
Out[5]= 1.29031
```

In the next 3D integral, the limits of the inner integrals depend on the outer integration variables.

```
In[6]:= NIntegrate[Sin[x y z] Exp[-x^2 + y],
            {x, 0, Pi}, {y, 0, Cos[x]}, {z, -Abs[x], x y}]
Out[6]= -0.0284358
```

Even integrals with integrable singularities can easily be computed (using appropriate transformations of variables, not visible to the user).

```
In[7]:= NIntegrate[1/Sqrt[x], {x, 0, 1}]
Out[7]= 2.
```

Next, we present a series of singular integrands of the form x^{-p}. Up to integrands with singularities of the form $x^{-0.8}$, there is no problem. Singular integrals with stronger singularities will produce a message. As always, when a numeric function issues a message, one should be alert that the result might be wrong. In the case $p = 0.9$ in the following, it is still correct (but we should not blindly trust the result), but in the case $p = 0.99$, the last digit displayed is incorrect.

```
In[8]:= {NIntegrate[x^-#, {x, 0, 1}], 1/(1 - #)}& /@ {0.2, 0.4, 0.6, 0.8}

Out[8]= {{1.25, 1.25}, {1.66667, 1.66667}, {2.5, 2.5}, {5., 5.}}
```

```
In[9]:= {NIntegrate[x^-0.9, {x, 0, 1}], 1/(1 - 0.9)}
```
```
          NIntegrate::ncvb :
            NIntegrate failed to converge to prescribed accuracy after 7 recursive
              bisections in x near x = 4.369993747903698`*^-57. More…
```
```
Out[9]= {10., 10.}
```

```
In[10]:= {NIntegrate[x^-0.99, {x, 0, 1}], 1/(1 - 0.99)}
```
```
          NIntegrate::ncvb :
            NIntegrate failed to converge to prescribed accuracy after 7 recursive
              bisections in x near x = 4.369993747903698`*^-57. More…
```
```
Out[10]= {100.006, 100.}
```

If the singularity is not at the endpoint of the interval of integration, it is advisable to choose the form $\{x, x_{start}, x_1, x_2, \ldots, x_n, x_{end}\}$ for the domain of integration.

```
In[11]:= f[x_?NumberQ] := 1/Sqrt[Abs[Sin[x]]]

         NIntegrate[f[x], {x, -8, 7}]
```
```
          NIntegrate::slwcon :
            Numerical integration converging too slowly; suspect one of the
              following: singularity, value of the integration being 0, oscillatory
              integrand, or insufficient WorkingPrecision. If your integrand is
              oscillatory try using the option Method->Oscillatory in NIntegrate. More…
          NIntegrate::ncvb : NIntegrate failed to converge to prescribed
              accuracy after 7 recursive bisections in x near x = 0.02734375`. More…
```
```
Out[12]= 23.6521
```

In this example, the problematic points are at $n\pi$, $n \in \mathbb{Z}$.

```
In[13]:= NIntegrate[f[x], {x, -8, -2Pi, -Pi, 0, Pi, 2Pi, 7}]
Out[13]= 25.4529
```

In the last example, we defined the integrand in such a manner that NIntegrate did not have access to its symbolic form. If NIntegrate has access to the symbolic form of the integrand and can analyze the integrand symbolically, it can detect the singularities and insert the intermediate points automatically (we come back to this feature in a moment).

```
In[14]:= NIntegrate[1/Sqrt[Abs[Sin[x]]], {x, -8, 7}]
Out[14]= 25.4529 - 1.05528 × 10^{-10} i
```

The intermediate points can also lie in the complex plane, and in this case, they are connected with straight line segments. The Laurent expansion of $1/\sin(x)$ about $x = 0$ is $x^{-1} + x/6 + O(x^3)$. Consequently, by the residue theorem, the following contour integral of $1/\sin(x)$ has value $2\pi i$. First, we look at the path of integration and the integrand using DensityPlot around the point 0 in the complex plane.

```
In[15]:= intPath = Table[If[EvenQ[i], 2, 1] Exp[I i/16 2Pi], {i, 0, 16}]
```

Out[15]= $\left\{2,\ e^{\frac{i\pi}{8}},\ 2\,e^{\frac{i\pi}{4}},\ e^{\frac{3\,i\,\pi}{8}},\ 2\,i,\ e^{\frac{5\,i\,\pi}{8}},\ 2\,e^{\frac{3\,i\,\pi}{4}},\ e^{\frac{7\,i\,\pi}{8}},\right.$

$\left.-2,\ e^{-\frac{7\,i\,\pi}{8}},\ 2\,e^{-\frac{3\,i\,\pi}{4}},\ e^{-\frac{5\,i\,\pi}{8}},\ -2\,i,\ e^{-\frac{3\,i\,\pi}{8}},\ 2\,e^{-\frac{i\pi}{4}},\ e^{-\frac{i\pi}{8}},\ 2\right\}$

In[16]:= `DensityPlot[1/Abs[Sin[x + I y]], {x, -2, 2}, {y, -2, 2},`
`                Mesh -> False, PlotPoints -> 26,`
`                (* the integration path *)`
`                Epilog -> {GrayLevel[1], Thickness[0.01],`
`                            Line[N[{Re[#], Im[#]}]& /@ intPath]}]`

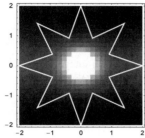

Here are the attributes of `NIntegrate`.

In[17]:= `Attributes[NIntegrate]`

Out[17]= `{HoldAll, Protected}`

Thus, we use `Evaluate` in the second argument. (While `NIntegrate` evaluates the first argument after the localization of the integration variable, the iterator `Join[{x}, intPath]` does not have the correct shape to even determine the dummy integration variable.)

In[18]:= `NIntegrate[1/Sin[x], Evaluate[Join[{x}, intPath]]]`

Out[18]= `0. + 6.28319 i`

In a way analogous to how we "fooled" `Plot` in Section 1.2.1 of the Graphics volume [1794] with poly: flopp, we can construct an integrand, which is 1 at the points evaluated by `NIntegrate`, but 2 everywhere else. Because the points used for a numerical integration are typically not equidistant, we must collect them on the fly.

In[19]:= `(* collect points touched by NIntegrate *)`
`     Module[{f, λNI = {}},`
`             (* each call to fFoo adds the argument x to the list λNI *)`
`             f[x_?NumberQ] := (AppendTo[λNI, x]; 1);`
`             NIntegrate[f[x], {x, 0, 1}];`
`     (* making definition for fFoo with the collected points *)`
`     SetDelayed @@ ReplacePart[`
`         Hold @@ {fFoo[x_?NumberQ], If @@ {{Union[λNI], x}, 1, 2}},`
`                                         MemberQ, {2, 1, 0}]]`

Here are the touched *x*-values for this simple integrand. 11 points were sampled.

In[21]:= `??fFoo`

`            Global`fFoo`

`            fFoo[x_ ?NumberQ] :=`
`             If[MemberQ[{0.00795732, 0.0469101, 0.122917, 0.230765, 0.360185,`
`                 0.5, 0.639815, 0.769235, 0.877083, 0.95309, 0.992043}, x], 1, 2]`

Integrating now the carefully (and artificially) constructed function fFoo, we get the result determined by the exceptional points that were collected in lNI.

```
In[22]:= NIntegrate[fFoo[x], {x, 0, 1}]
Out[22]= 1.
```

But already a tiny change in the limits causes the algorithm to evaluate the integrand at different points and the result of the numerical integration is 2. (as expected).

```
In[23]:= {NIntegrate[fFoo[x], {x, $MachineEpsilon, 1}],
        NIntegrate[fFoo[x], {x, 0, 1 + $MachineEpsilon}]}
Out[23]= {2., 2.}
```

The body of NIntegrate can be a (nested) list. In this case, the numerical integration is carried out for each list element. Here is a small example: The function $f(\mathbf{A})$ of a square matrix $\mathbf{A}$ can be defined as (see [1097], [1523], and [1715])

$$f(\mathbf{A}) = \oint_C f(z)\,(1 - z\,\mathbf{A})^{-1}\,dz.$$

where C is a closed contour encircling all eigenvalues of $\mathbf{A}$, $\mathbf{1}$ is the identity matrix of the same dimensions as $\mathbf{A}$ and $(1 - z\,\mathbf{A})^{-1}$ is the inverse of the matrix $(1 - z\,\mathbf{A})^{-1}$. (This definition becomes obvious in a basis where $\mathbf{A}$ is diagonal.) The function matrixExp is a straightforward implementation of the last formula.

```
In[24]:= matrixExp[mat_] :=
      Module[{R = 2 Max[Abs[Eigenvalues[mat]]],
              m = Inverse[z IdentityMatrix[Length[mat]] - mat]},
              z = R Exp[I φ];
         (* integrate along circle of Radius R *)
         1/(2 Pi I) NIntegrate[Evaluate[m Exp[z]  D[z, φ]], {φ, 0, 2Pi}]]
```

Within the default precision goal of NIntegrate, the function matrixExp yields the same results as the built-in function MatrixExp.

```
In[25]:= matrixExp[{{1, 2}, {3, 4.}}]
Out[25]= {{51.969 - 2.89501×10⁻¹³ i, 74.7366 + 2.65894×10⁻¹³ i},
         {112.105 + 1.06018×10⁻¹³ i, 164.074 - 5.79002×10⁻¹³ i}}
```

```
In[26]:= MatrixExp[{{1, 2}, {3, 4.}}]
Out[26]= {{51.969, 74.7366}, {112.105, 164.074}}
```

After these introductory remarks about the function NIntegrate, we will now discuss the options of NIntegrate.

```
In[27]:= Options[NIntegrate]
Out[27]= {AccuracyGoal → ∞, Compiled → True, EvaluationMonitor → None,
         GaussPoints → Automatic, MaxPoints → Automatic, MaxRecursion → 6,
         Method → Automatic, MinRecursion → 0, PrecisionGoal → Automatic,
         SingularityDepth → 4, WorkingPrecision → MachinePrecision}
```

We are already very familiar with the option Compiled, and so, we do not discuss it further here. We also have already discussed the options WorkingPrecision, AccuracyGoal and PrecisionGoal. NIntegrate is one of the numerical functions where the values for the AccuracyGoal and the PrecisionGoal options can be set to Infinity and the default value for the accuracy goal is actually by default set to be infinity.

The numerical integration algorithm is terminated as soon as its results fulfill the `PrecisionGoal` or `Accuracy`cyGoal goals (the word "goal" is to be taken verbatim). That both options are needed from time to time can be seen from the following: Frequently, one has no idea of how large the integral of a given function should be (not even the order of magnitude). For example, if it is exactly 0, it is impossible to satisfy a requirement of `Preci`sionGoal for any positive number, and in this case, in principle, it is impossible to compute even one correct digit. (One could enforce a nonzero result by adding a nonzero value.)

```
In[28]:= (* exact value of the integral is 0 *)
        NIntegrate[x + x^2, {x, -1, 1/2}, PrecisionGoal -> 1]
                    NIntegrate::ploss :
                     Numerical integration stopping due to loss of precision. Achieved
                        neither the requested PrecisionGoal nor AccuracyGoal; suspect one of
                        the following: highly oscillatory integrand or the true value of the
                        integral is 0. If your integrand is oscillatory on a (semi-)infinite
                        interval try using the option Method->Oscillatory in NIntegrate. More…
Out[29]= 2.42861 × 10^{-17}
```

To get at least an upper bound for the value of the integrals, we use the following input.

```
In[30]:= NIntegrate[x + x^2, {x, -1, 1/2}, AccuracyGoal -> 15]
Out[30]= 2.08167 × 10^{-17}
```

Here the same integral is calculated using a higher `WorkingPrecision`.

```
In[31]:= NIntegrate[x + x^2, {x, -1, 1/2}, PrecisionGoal -> Infinity,
                AccuracyGoal -> 30, WorkingPrecision -> 50]
Out[31]= 0. × 10^{-51}
```

In the next example, it is not easy to even estimate the order of magnitude of the result. To get some idea of the order of magnitude of the result, we require at least five correct digits.

```
In[32]:= NIntegrate[Exp[-x^2] Sin[Pi Log[x]]/Log[Log[x]], {x, 10, 20},
                PrecisionGoal -> 5, AccuracyGoal -> Infinity]
Out[32]= 1.82073 × 10^{-45}
```

As mentioned above, when a finite value is set for both the `AccuracyGoal` *and* the `PrecisionGoal`, the result will be correct to $10^{-accuracyGoal} + |estimatedExactResult|\, 10^{-precisionGoal}$. Here is an exactly doable integral.

```
In[33]:= exactResult = Integrate[x^3 Sin[x], {x, 0, 2}]
Out[33]= 4 Cos[2] + 6 Sin[2]
```

To demonstrate the dependence of the result on the `AccuracyGoal` and the `PrecisionGoal` options, we implement a function `compare` that returns a list of two elements. The first is the difference of the exact value of the integral and the numerically calculated value of the integral $\int_0^2 \alpha\, x^3 \sin(x)\, dx$ using the specified options. The second element of the list is the value of $10^{-accuracyGoal} + |exactResult|\, 10^{-precisionGoal}$. The numerical factor α scales the absolute value of the integral.

```
In[34]:= compare[α_, {ag_, pg_, wp_}] :=
        {(* the actual difference *) α exactResult -
          NIntegrate[α x^3 Sin[x], {x, 0, 2}, WorkingPrecision -> wp,
                PrecisionGoal -> pg, AccuracyGoal -> ag],
           (* the prescribed difference; display only two digits *)
          SetPrecision[10^-ag + Abs[α exactResult] 10^-pg, 2]}
```

In the next input, the `PrecisionGoal` option setting dominates.

```
In[35]:= compare[10^-10, {Infinity, 30, 60}]
```

Out[35]= $\{0. \times 10^{-66}, 4. \times 10^{-40}\}$

In the next input, the AccuracyGoal option setting dominates.

In[36]:= **compare[10^-10, {30, Infinity, 60}]**
Out[36]= $\{0. \times 10^{-66}, 1.0 \times 10^{-30}\}$

Using a working precision setting of 30 and the value $\alpha = 10^{10}$ does not allowed us to achieve the required accuracy goal.

In[37]:= **compare[10^10, {30, Infinity, 30}]**

> NIntegrate::tmap :
>   NIntegrate is unable to achieve the tolerances specified by the PrecisionGoal
>   and AccuracyGoal options because the working precision is
>   insufficient. Try increasing the setting of the WorkingPrecision option.

Out[37]= $\{0. \times 10^{-20}, 1.0 \times 10^{-30}\}$

> If we want a very accurate result for an integral, we need to set WorkingPrecision as well as PrecisionGoal and AccuracyGoal. In this case, all numbers (e.g., limits of integration, parameters in the integrand) must possess the corresponding accuracy.

Using machine arithmetic, we cannot get 20 digits in the following example.

In[38]:= **NIntegrate[Exp[-Sin[x]^2], {x, 0, Pi}, PrecisionGoal -> 20]**

> NIntegrate::tmap :
>   NIntegrate is unable to achieve the tolerances specified by the PrecisionGoal
>   and AccuracyGoal options because the working precision is
>   insufficient. Try increasing the setting of the WorkingPrecision option.

Out[38]= 2.02644

Increasing the WorkingPrecision allows us to get 20 correct digits in the result.

In[39]:= **NIntegrate[Exp[-Sin[x]^2], {x, 0, Pi},**
            PrecisionGoal -> 20, WorkingPrecision -> 30]
Out[39]= 2.0264380669493553051434

Here are three further options of NIntegrate defined.

MinRecursion

is an option for NIntegrate that determines the minimum number of recursive subdivisions of the interval of integration to be used.
Default:

0

Admissible:

arbitrary integer smaller than MaxRecursion

MaxRecursion

is an option for NIntegrate that determines the maximum number of recursive subdivisions of the interval of integration to be used.
Default:

6

Admissible:

arbitraryPositiveInteger

SingularityDepth

is an option for NIntegrate that determines the number of recursive subdivisions of the interval of integration near the endpoints and specified internal points, before a transformation of variables of the endpoints and the specified internal points is to be performed (provided the required accuracy or precision has not yet been achieved).
Default:

4

Admissible:

arbitraryPositiveInteger

Using larger number for the option MaxRecursion allows carrying out the two integrals from the beginning of this section without messages being generated.

```
In[40]:= NIntegrate[x^-0.9, {x, 0, 1}, MaxRecursion -> 20]
Out[40]= 10.
```

```
In[41]:= NIntegrate[x^-0.99, {x, 0, 1}, MaxRecursion -> 100]
Out[41]= 100.
```

```
In[42]:= NIntegrate[x^-0.9, {x, 0, 1}, MaxRecursion -> 6]
          NIntegrate::ncvb :
           NIntegrate failed to converge to prescribed accuracy after 7 recursive
            bisections in x near x = 4.369993747903698`*^-57. More...
Out[42]= 10.
```

In the next input, we vary the option value of SingularityDepth. Switching after two subdivisions allows obtaining a correct result for the integral.

```
In[43]:= NIntegrate[x^-0.9, {x, 0, 1}, MaxRecursion -> 10,
                SingularityDepth -> #]& /@ {2, 20}
             NIntegrate::ncvb : NIntegrate failed to converge to prescribed accuracy
                after 11 recursive bisections in x near x = 0.000244140625`. More...
Out[43]= {10., 7.56061}
```

The next input calculates the much slower converging integral $\int_0^1 x^{9999/10000}\,dx$. We use the singularity depths 4 and 100 and display the sample points. We clearly see how the sampling of points near the origin happens much later in case of a larger value of the `SingularityDepth` option setting. As a result, much more sample points are needed.

```
In[44]:= Module[{data},
    (* integrate function with integrable singularity *)
    data = Reap[Timing @
    NIntegrate[x^-(9999/10000), {x, 0, 1}, Method -> GaussKronrod,
            WorkingPrecision -> 100,
            PrecisionGoal -> 6, AccuracyGoal -> Infinity,
            MaxRecursion -> 100000, SingularityDepth -> #,
            Compiled -> False, EvaluationMonitor :> Sow[x]]]& /@
                                            {4, 100};
    (* sample points in double logarithmic plot *)
    Show[GraphicsArray[
      ListPlot[Log[10, Abs @ Log[10, N @ #[[2, 1]]]],
            PlotRange -> All, Frame -> True, Axes -> False,
            DisplayFunction -> Identity]& /@ data]];
    First /@ data]
```

NIntegrate::slwcon :
  Numerical integration converging too slowly; suspect one of the
    following: singularity, value of the integration being 0, oscillatory
    integrand, or insufficient WorkingPrecision. If your integrand is
    oscillatory try using the option Method->Oscillatory in NIntegrate. More…

NIntegrate::slwcon :
  Numerical integration converging too slowly; suspect one of the
    following: singularity, value of the integration being 0, oscillatory
    integrand, or insufficient WorkingPrecision. If your integrand is
    oscillatory try using the option Method->Oscillatory in NIntegrate. More…

```
Out[44]= {{0.51 Second, 10000.00}, {2.23 Second, 10000.00}}
```

The next two options to discuss are `Method` and `GaussPoints`.

Method

is an option for NIntegrate that determines the integration method.

Default:

Automatic, that is, GaussKronrod for 1D integrals that do not contain piecewise functions, SymbolicPiecewiseSubdivision for integrals that do contain piecewise functions and MultiDimensional (the last for multidimensional integrals only).

Admissible:

GaussKronrod

or

DoubleExponential

or

Trapezoidal

or

MonteCarlo

or

QuasiMonteCarlo

or

Oscillatory

or

SymbolicPiecewiseSubdivision

or

EvenOddSubdivision

or

MultiDimensional for multidimensional integrals

GaussPoints

is an option for NIntegrate that determines the number of points where the integrand is to be evaluated initially (assuming the Method is GaussKronrod [1071], [1109], [665]).

Default:

Automatic

Admissible:

arbitrary positive integer

Consider the following integral whose integrand is nonzero only in a small interval. Without properly set options, we cannot get a reasonable result.

```
In[45]:= (* function that is nonvanishing in the interval [-1, 1] *)
         bump[z_] := Im[z + Sqrt[z - 1] Sqrt[z + 1]]

In[47]:= (* nonvanishing only in the interval [0.005, 0.015] *)
         integrand[x_] := bump[200 (x - 1/100)]

         NIntegrate[integrand[x], {x, 0, 1}]

                 NIntegrate::ploss :
                  Numerical integration stopping due to loss of precision. Achieved
                     neither the requested PrecisionGoal nor AccuracyGoal; suspect one of
                     the following: highly oscillatory integrand or the true value of the
                     integral is 0. If your integrand is oscillatory on a (semi-)infinite
                     interval try using the option Method->Oscillatory in NIntegrate. More…

Out[49]= 0.
```

Without properly chosen options, we may even get a wrong result with no warning, as shown in the following example (NIntegrate assumes sufficient smoothness from the integrand—a feature not satisfied by the current integrand).

```
In[50]:= NIntegrate[integrand[x], {x, 0, 1}, AccuracyGoal -> 30]
Out[50]= 0.
```

Here is a somewhat better result that is correct up to the sixth place after (rounded) the decimal point. (Without explicit specification of the points of discontinuity, integrals with discontinuous integrands are notoriously difficult to calculate numerically.)

```
In[51]:= NIntegrate[integrand[x], {x, 0, 1},
                 MaxRecursion -> 18, MinRecursion -> 5, AccuracyGoal -> 6,
                 PrecisionGoal -> Infinity]
Out[51]= 0.00785417
```

If we knew where the integrand had a discontinuity or a singularity, it would, of course, be best to specify this point explicitly in the second argument of NIntegrate, together with appropriately chosen options).

```
In[52]:= NIntegrate[integrand[x], {x, 0, 0.005, 0.015, 1}]
Out[52]= 0.00785398
```

The method option settings of NIntegrate can be subdivided in four groups. The first group contains the main basic numerical method for doing the numerical integration of 1D integrals: GaussKronrod, Trapezoi: dal, DoubleExponential. The second group contains the Monte Carlo-type methods MonteCarlo and QuasiMonteCarlo for high-dimensional integrals. The third group contains the method MultiDimen: sional for low dimensional integrals and the fourth group contains the symbolic preprocessing options SymbolicPiecewiseSubdivision and EvenOddSubdivision.

We now graphically examine the position of the data points used in the three basic numerical methods for 1D integrals. We will do this for a 1D example first. To this end, we first define an integrand.

```
In[53]:= fu[x_] := Exp[-5 x^2] + 1/2;
```

The exact value of the integral of $\exp(-5 x^2)$ between -3 and 3 is given by $\sqrt{\pi/5}\ \mathrm{erf}(3\sqrt{5})$. (Here erf$(z)$ denotes the error function; see Chapter 3 of the Symbolics volume [1795].)

```
In[54]:= exact = N[Sqrt[Pi/5] Erf[3 Sqrt[5]] + 3, 30]
Out[54]= 3.79266545952120220266711794084
```

We now integrate using the three possible methods. We write out the relative error and the number of data points used, and look at the data points used in the integration graphically. First, we define an auxiliary function `nodesAndResult`. It returns a list of the evaluated *x*-values and the value of the integral. To do this we use the `EvaluationMonitor` option with a `Sow[`*samplePoint*`]` setting.

```
In[55]:= nodesAndResult[method_] := nodesAndResult[method] =
    Module[{res, nodes},
            {res, nodes} = Reap[NIntegrate[fu[x], {x, -3, 3}, Method -> method,
                                           EvaluationMonitor :> Sow[x]]];
            {1 - res/exact, nodes[[1]]}]
```

We use it to compute the above integral with the three sensible methods for this kind of integral.

```
In[56]:= baseMethodOptions = {GaussKronrod, Trapezoidal, DoubleExponential};
```

```
In[57]:= (* do the calculation; remember values in the bag 1 *)
    nodesAndResult /@ baseMethodOptions;
```

Here are the numbers of data points used and the relative deviations from the exact result. All three methods perform quite well for this simple integral.

```
In[59]:= (* table of the results *)
    TableForm[Transpose[
       (* the three methods compared *)
       {Length[nodesAndResult[#][[2]]], N[nodesAndResult[#][[1]], 2]}& /@
                                         baseMethodOptions],
         TableHeadings -> {None, StyleForm[#, FontWeight -> "Bold"]& /@
                                         baseMethodOptions}]
```

Out[60]//TableForm=

GaussKronrod	**Trapezoidal**	**DoubleExponential**
121	33	217
7.17204×10^{-14}	0.	-4.44089×10^{-16}

Here is a graphical representation of the data points used to calculate the integrals.

```
In[61]:= Show[GraphicsArray[
    (* plots for the three method options *)
    Graphics[Line[{{#, 0}, N[{#, fu[#]}]}]& /@
                         nodesAndResult[#][[2]],
            PlotRange -> All, PlotLabel ->
            StyleForm[#, FontFamily -> "Courier", FontSize -> 6]]& /@
                         baseMethodOptions]]
```

Note that the accuracy and efficiency of each of these particular results do not indicate the quality of the three methods for an arbitrary function.

We continue with a slightly different visualization of the points used by `NIntegrate`. For a smooth integrand with two sharp peaks, we again monitor the points used and show them in the order that they are evaluated. The points used in the `GaussKronrod` method show the zooming in of the algorithm into the regions of large changes of the curve. The points used in the `Trapezoidal` method are uniformly distributed and the points used in the `DoubleExponential` method show a clustering at the integration endpoints. All three returned results agree to the requested 20 digits with the exact result.

```
In[62]:= Module[{integrand, (* the three methods *)
             methods = {GaussKronrod, Trapezoidal, DoubleExponential}},
      (* store points used in the list l *)
      integrand[x_] := (* integrand with two sharp peaks *) Sin[4x]^2 +
                 Exp[-2000 (x - 1 Pi/4)^2] + Exp[-2000 (x - 3 Pi/4)^2];
      (* integration results and points used *)
      data = ({#[[1]], #[[2, 1]]}&[Reap[NIntegrate[integrand[x],
                      (* intermediate points at peaks *) {x, 0, Pi/4, 3Pi/4, Pi},
                      PrecisionGoal -> 2, WorkingPrecision -> 30,
                      EvaluationMonitor :> Sow[x], Method -> #]] //
                                    SetPrecision[#, 20]&)& /@ methods;
      (* show points used *)
      Show[GraphicsArray[Function[{λ, μ},
       Graphics[{(* integrand as red curve *)
                {Hue[0], l = {}; Line[Table[{x, integrand[N @ x]},
                                   {x, 0, Pi, Pi/300}]]},
                (* points in the order of their evaluation *)
                {PointSize[0.003], MapIndexed[Point[{#1, #2[[1]]/
                                   Length[λ]}]&, λ]}},
                Frame -> True, FrameTicks -> False, PlotLabel ->
                StyleForm[μ, FontFamily -> "Courier", FontSize -> 6]]] @@@
                          Transpose[{Last /@ data, methods}]]]];
      (* return numerical and exact integration results *)
      {First /@ data, N[Integrate[integrand[x], {x, 0, Pi}], 20]}]
```

GaussKronrod	Trapezoidal	DoubleExponential

```
Out[62]= {{1.6506586179510246871, 1.6500848184520898568, 1.6504692626045726120},
      1.650628727470168395}
```

As mentioned, `NIntegrate` can also integrate multivariate functions. The following inputs calculate the Gauss
linking number [962], [81], [570], [71], [1110], [842], [1462], [339], [1331], [570], [499], [1449] for two
interlinked circles.

```
In[63]:= Clear[x, y, φ, ϑ]
      (* a circle in the x,y-plane *)
      {x[1][φ_], x[2][φ_], x[3][φ_]} = {Cos[φ], Sin[φ], 0};
      (* a circle in the x,z-plane *)
      {y[1][ϑ_], y[2][ϑ_], y[3][ϑ_]} = {Cos[ϑ] + 1, 0, Sin[ϑ]};

      (* the Gauss linking number *)
      Sum[Signature[{a, b, c}] D[x[a][φ], φ] D[y[b][ϑ], ϑ]*
         (x[c][φ] - y[c][ϑ]), {a, 3}, {b, 3}, {c, 3}]/
         Sum[(x[j][φ] - y[j][ϑ])^2, {j, 3}]^(3/2)/(4Pi) // Simplify
```

$$
Out[69]= -\frac{\cos[\vartheta]\,(-1+\cos[\varphi])+\cos[\varphi]}{4\,\pi\,(3-2\cos[\vartheta]\,(-1+\cos[\varphi])-2\cos[\varphi])^{3/2}}
$$

```
In[70]:= NIntegrate[%, {φ, 0, 2Pi}, {ϑ, 0, 2Pi}]
Out[70]= -1.
```

The following small example calculates the area $\mathcal{A}(n)$ of a supersphere $|x|^n + |y|^n + |z|^n = 1$, $n \in \mathbb{R}$, $n > 0$. To avoid cusps in the bivariate function to be integrated, we integrate only over 1/48th of the surface. The inner integration depends on the outer integration variable in a nontrivial manner.

```
In[71]:= superSphereArea[n_, opts___] := 48 NIntegrate[
          (* integrand comes from:
            Module[{r, e, f, g, dA},
            (* radius vector in spherical coordinates *)
              r = 1/(Cos[ϑ]^n + Cos[φ]^n Sin[ϑ]^n + Sin[ϑ]^n Sin[φ]^n)^(1/n)*
                    {Cos[φ] Sin[ϑ], Sin[φ] Sin[ϑ], Cos[ϑ]};
            (* surface area element *)
              e = D[r, φ].D[r, φ]; f = D[r, φ].D[r, ϑ]; g = D[r, ϑ].D[r, ϑ];
              dA = Sqrt[e g - f^2] // Simplify] *)
          Sqrt[(Cos[ϑ]^(2n - 2) Sin[ϑ]^2 + Sin[ϑ]^(2n)*
              (Cos[φ]^(2n - 2) + Sin[φ]^(2n - 2)))/
              (Cos[ϑ]^n + Sin[ϑ]^n (Cos[φ]^n + Sin[φ]^n))^((2(2 + n))/n)],
          (* integrate over 1/48th of the solid angle *)
          {φ, 0, Pi/4}, {ϑ, 0, ArcCos[1/Sqrt[2 + Tan[φ]^2]]}, opts,
          (* option values appropriate for a plot *)
          PrecisionGoal -> 4, AccuracyGoal -> 6, Method -> GaussKronrod]
```

For an octahedron ($n = 1$), a sphere ($n = 2$), and a cube ($n = \infty$), we obtain the well known areas.

```
In[72]:= (superSphereArea[#, WorkingPrecision -> 30,
                          PrecisionGoal -> 20, AccuracyGoal -> 20]& /@
                {1, 2, 10^6}) - {4 Sqrt[3], 4Pi, 24}
Out[72]= {0. × 10^-19, 0. × 10^-22, 0. × 10^-19}
```

Here is a plot for the normalized area ($\mathcal{A}(n)/\mathcal{A}(\infty)$) as a function of n. The black curve is the corresponding normalized volume.

```
In[73]:= Plot[{(* normalized area *) superSphereArea[n]/24,
              (* normalized volume *) Gamma[1 + 1/n]^2 Gamma[1/n]/Gamma[3/n]/3},
              {n, 0, 10}, PlotPoints -> 60, PlotStyle -> {Hue[0], GrayLevel[0]},
              AspectRatio -> 1/2]
```

Applying any of the three 1D methods to compute multidimensional integrals will place the data points on the corresponding tensor-product grid. With `Method -> MultiDimensional`, however, the integral is computed over a nontrivial adaptively refined grid. The exact computation of the integral $\int_0^1 \int_0^1 \cos(xy)\, dy\, dx$ gives $\text{Si}(\pi) = 1.851937\ldots$ ($\text{Si}(z)$ is the sine integral function.)

```
In[74]:= nodesAndResult2D[method_] := nodesAndResult2D[method] =
          Module[{res, nodes},
                  {res, nodes} = Reap[NIntegrate[Cos[x y], {x, 0, Pi}, {y, 0, 1},
                                      Method -> method,
```

```
                                              EvaluationMonitor :> Sow[{x, y}]]];
        {1 - res/SinIntegral[Pi], nodes[[1]]}]
```

We use the (tensor product version of the) three 1D integration methods and the genuine multidimensional method `MultiDimensional`.

```
In[75]:= baseMethodOptions2D = Prepend[baseMethodOptions, MultiDimensional]

Out[75]= {MultiDimensional, GaussKronrod, Trapezoidal, DoubleExponential}

In[76]:= (* do the numerical integration for the four methods *)
        nodesAndResult2D /@ baseMethodOptions2D;

            NIntegrate::slwcon :
             Numerical integration converging too slowly; suspect one of the
               following: singularity, value of the integration being 0, oscillatory
               integrand, or insufficient WorkingPrecision. If your integrand is
               oscillatory try using the option Method->Oscillatory in NIntegrate. More…

            NIntegrate::ncvi :
             NIntegrate failed to converge to prescribed accuracy after 14 iterated
               refinements in x in the interval {{x, 0., 3.14159}, {y, 0., 1.}}. More…
```

Here are the numbers of data points used and the relative deviations from the exact result. We see that the Gauss–Kronrod and the multidimensional method needed just a few hundred points and achieved an excellent result (for this very simple integrand, the Gauss–Kronrod is the clear winner). And the `Trapezoidal` method setting needed more than half a million points and had the largest error.

```
In[78]:= (* table of the results *)
        TableForm[Transpose[
          (* the three methods compared *)
          {Length[nodesAndResult2D[#][[2]]], N[nodesAndResult2D[#][[1]], 2]}& /@
                                                    baseMethodOptions2D],
          TableHeadings -> {None, StyleForm[#, FontWeight -> "Bold"]& /@
                                                    baseMethodOptions2D}]
```

Out[79]//TableForm=

MultiDimensional	**GaussKronrod**	**Trapezoidal**	**DoubleExponential**
629	121	525825	3063
1.34791×10^{-9}	-2.22045×10^{-16}	6.74081×10^{-7}	-2.22045×10^{-16}

Here is a graphical representation of the data points used to calculate the integrals.

```
In[80]:= Show[GraphicsArray[
          (* plots for the three method options *)
          Graphics[Point /@ nodesAndResult2D[#][[2]],
                 PlotRange -> {{0, Pi}, {0, 1}}, PlotLabel ->
                 StyleForm[#, FontFamily -> "Courier", FontSize -> 6]]& /@
                                                    baseMethodOptions2D]]
```

Of course, we could also iterate the double integration and carry out two 1D integrations (this time, we use a higher working precision).

```
In[81]:= nodesAndResult2DTensorProduct[{methodx_, methody_}] :=
        Module[{res, nodes, f},
               (* define function to be integrated to avoid evaluation *)
               f[x_?NumberQ, y_?NumberQ] := Cos[x y];
```

```
{res, nodes} = (* two nested integrations *)
Reap[NIntegrate[NIntegrate[f[x, y], {y, 0, 1}, Method -> methody,
                WorkingPrecision -> 20, PrecisionGoal -> 8,
                EvaluationMonitor :> Sow[{x, y}]],
            {x, 0, Pi}, Method -> methodx,
            WorkingPrecision -> 20, PrecisionGoal -> 6]];
(* return relative error and node counts *)
{N[1 - res/SinIntegral[Pi], 3], Length[nodes[[1]]]}}
```

The results obtained in this manner are qualitatively similar to the last results. (We suppress some messages indicating that the default goals for accuracy and precision were not achieved.)

```
In[82]:= TableForm[{ToString[#1[[1]]] <> "<>" <> ToString[#1[[2]]],
            #2[[1]], #2[[2]]}& @@@
        Internal`DeactivateMessages[{#, nodesAndResult2DTensorProduct[#]}& /@
        (* the nine combined methods *)
        Flatten[Outer[List, baseMethodOptions, baseMethodOptions], 1]],
            TableHeadings -> {None, StyleForm[#, FontWeight -> "Bold"]& /@
                                {"methods", "error", "points"}}]
```

Out[82]//TableForm=

methods	error	points
GaussKronrod<>GaussKronrod	$0. \times 10^{-12}$	169
GaussKronrod<>Trapezoidal	5.39×10^{-7}	6285
GaussKronrod<>DoubleExponential	$0. \times 10^{-12}$	900
Trapezoidal<>GaussKronrod	$0. \times 10^{-6}$	6669
Trapezoidal<>Trapezoidal	$0. \times 10^{-6}$	259441
Trapezoidal<>DoubleExponential	$0. \times 10^{-6}$	32790
DoubleExponential<>GaussKronrod	$0. \times 10^{-10}$	715
DoubleExponential<>Trapezoidal	5.39×10^{-7}	17575
DoubleExponential<>DoubleExponential	$0. \times 10^{-10}$	4190

Here is a more complicated 2D integration problem. We integrate a bivariate function that has 24 bumps of different width and height. We also monitor the sampled points. The left graphic shows the function to be integrated, the middle picture shows the sample points over a density plot of the function and the right graphic shows the sample points over a density plot of the norm of the gradient of the function.

```
In[83]:= Module[{d = 24, f},
    SeedRandom[12];
    (* sum of broad and narrow peaks *)
    f = Sum[Random[] Exp[-Random[Integer, {0, 1200}] *
                ((x - Random[])^2 + (y - Random[])^2)^(4/5)], {d}];
    (* 3D plot of the function *)
    p3D = Plot3D[Evaluate[f], {x, 0, 1}, {y, 0, 1}, PlotRange -> All,
                PlotPoints -> 120, Mesh -> False,
                DisplayFunction -> Identity, Axes -> False];
    (* density plot of function and gradient *)
    {dp, dpG} =
    DensityPlot[#, {x, 0, 1}, {y, 0, 1}, PlotRange -> All,
                PlotPoints -> 240, Mesh -> False,
                DisplayFunction -> Identity, FrameTicks -> None,
                ColorFunction -> (Hue[0.78 (1 - #)]&)]& /@
                                {f, Norm[D[f, #]& /@ {x, y}]};
    (* do numerical integration and collect sample points *)
    data = Reap[NIntegrate[Evaluate[f], {x, 0, 1}, {y, 0, 1},
                Method -> MultiDimensional, MaxRecursion -> 20,
```

```
                  PrecisionGoal -> 2, AccuracyGoal -> Infinity,
                  EvaluationMonitor :> Sow[{x, y}]]];
      (* show 3D plot and density plots with sample points *)
      Show[GraphicsArray[Flatten[{p3D,
            Show[{#, Graphics[{PointSize[0.002], GrayLevel[0],
                           Point /@ data[[2, 1]]}]}]& /@ {dp, dpG}}]]]]
```

 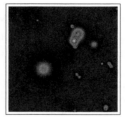 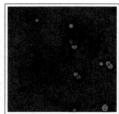

Setting options appropriately, we can use the built-in algorithms to investigate their convergence behaviors. The picture shows the number of correct digits obtained in the numerical integration of $\int_0^1 \sin(x)\, dx$ using the method GaussKronrod as a function of the number of Gauss points.

```
In[84]:= (* shut off messages *)
      Off[NIntegrate::"ncvi"]; Off[NIntegrate::"ncvb"]; Off[Accuracy::"mnprec"];

      ListPlot[
      (* list of {numberOfGaussPoints, accuracyOfIntegral} *)
      Table[{k, Accuracy[(* absolute error *)
      (1 - Cos[1]) - NIntegrate[Sin[x], {x, 0, 1},
               Method -> GaussKronrod, GaussPoints -> k,
               MinRecursion -> 0, MaxRecursion -> 0, SingularityDepth -> 0,
               (* use sufficient precision *)
               WorkingPrecision -> 100]]},
             {k, 20}], Frame -> True, PlotRange -> All, Axes -> False]

      (* turn off messages again *)
      On[NIntegrate::"ncvi"]; On[NIntegrate::"ncvb"]; On[Accuracy::"mnprec"];
```

Here are some rules of thumb for choosing one of the basic methods in NIntegrate (after the settings SymbolicPiecewiseSubdivision and/or EvenOddSubdivision have been applied):

DoubleExponential

  for analytic, or for strongly oscillating functions (good convergence, high accuracy attainable).

Trapezoidal

  for analytic, periodic integrands that are to be integrated over one period.

GaussKronrod

  if accurate results are needed (the initial computation of the Gauss points takes some time, but they are internally stored for further calculation).

Oscillatory

  for integrals of the form $\int_a^\infty \sin(f(x))\,g(x)\,dx$ or $\int_a^\infty \cos(f(x))\,g(x)\,dx$ or $\int_a^\infty J_n(f(x))\,g(x)\,dx$ or $\int_a^\infty Y_n(f(x))\,g(x)\,dx$.

MultiDimensional

  for multidimensional integrals of relatively small dimensions (say, between 2 and 6).

MonteCarlo and QuasiMonteCarlo

  especially useful for multidimensional integrals in many dimensions.

For a more detailed discussion of when to apply each integration method, and of the advantages and disadvantages of the various numerical integration methods for various classes of functions, see [1070], [459], [777], [1676], [892], [1972], [566], [536], and [1672]. For the interesting question "how good, on average is an integration scheme", see [311]. For complexity issues, see [1893], [808], and [1535].

We now give an example of the application of the DoubleExponential method [1742]. The method is based on a change of variables of the form $z \longrightarrow (\tanh(\alpha \sinh(\beta\,((1-z)^{-1} - z^{-1}))) + 1)/2$ (for a discussion why double exponential is generically better than single, or say triple, exponential see [1729], [1728]). This transformation makes all derivatives vanish at $z = 0$ and $z = 1$. Let us use the method to compute the integral $\int_1^9 \exp(i\,x^3)/x\,dx$. We can get a result by setting no options explicitly. But we do get messages that should make us skeptical of the result.

```
In[90]:= NIntegrate[Exp[I x^3]/x, {x, 1, 8}] // Timing
```
>   NIntegrate::ncvb : NIntegrate failed to converge to prescribed
>   accuracy after 7 recursive bisections in x near x = 7.53515625`. More…

```
Out[90]= {0.03 Second, -0.112415 + 0.208887 i}
```

This result of the Method -> DoubleExponential option setting coincides with the one produced above.

```
In[91]:= NIntegrate[Exp[I x^3]/x, {x, 1, 8}, Method -> DoubleExponential] // Timing
Out[91]= {0.03 Second, -0.112415 + 0.208887 i}
```

However, in the second case, we are much more likely to be persuaded that the result is correct than in the case in which messages were generated. (Finally, the fact that two completely different methods gave the same result is convincing.)

The variable change used in the DoubleExponential method transforms the values of the integrand at the integration limits to zero. This means that this method is very well suited for integrands such as the following rapidly increasing function.

```
In[92]:= NIntegrate[x^x^x, {x, 0, 7}]
```
>   NIntegrate::ncvb :
>   NIntegrate failed to converge to prescribed accuracy after 7 recursive
>   bisections in x near x = 6.9999992081554945`. More…

Out[92]= 7.770499322529536× 10⁶⁹⁵⁹⁶⁷

In[93]:= **NIntegrate[x^x^x, {x, 0, 7}, Method -> DoubleExponential]**
Out[93]= 7.770499286761009× 10⁶⁹⁵⁹⁶⁷

Of course, simpler singularities are also suppressed.

In[94]:= **NIntegrate[x^-0.999, {x, 0, 1}]**

> NIntegrate::slwcon :
> Numerical integration converging too slowly; suspect one of the
>   following: singularity, value of the integration being 0, oscillatory
>   integrand, or insufficient WorkingPrecision. If your integrand is
>   oscillatory try using the option Method->Oscillatory in NIntegrate. More…

> NIntegrate::ncvb :
> NIntegrate failed to converge to prescribed accuracy after 7 recursive
>   bisections in x near x = 4.369993747903698`*^-57. More…

Out[94]= 973.841

In[95]:= **NIntegrate[x^-0.999, {x, 0, 1}, Method -> DoubleExponential]**
Out[95]= 1000.

In the next example [909] theDoubleExponential method needs much less function evaluations than the GaussKronrod method.

In[96]:= **g[a_, opts___] := NIntegrate[(Exp[x] + Exp[-x])^a - Exp[a x] - Exp[-a x],**
                                    {x, 0, Infinity}, opts,
                                    WorkingPrecision -> 200, PrecisionGoal -> 20] //
                                                                        N[#, 10]&

In[97]:= **Function[method, {method, #1, Length[#2[[1]]]}]& @@**
             Reap[g[5/3, Method -> method,
                      EvaluationMonitor :> Sow[x]]]] /@
                      {GaussKronrod, DoubleExponential}
Out[97]= {{GaussKronrod, 4.626291112, 931}, {DoubleExponential, 4.626291112, 545}}

For a special, but practically important, class of integrals, one previously unused method option setting is available: Oscillatory. A way (that works frequently) to find out about all possible option settings is to call a function with an unknown option setting.

In[98]:= **NIntegrate[x, {x, 0, 1}, Method -> FastGoodPreciseAndStable]**

> NIntegrate::bdmtd :
> The method specified by the Method option should be one of Automatic,
>   GaussKronrod, DoubleExponential, Trapezoidal, Oscillatory, MonteCarlo,
>   QuasiMonteCarlo, EvenOddSubdivision, or SymbolicPiecewiseSubdivision. Only
>   methods EvenOddSubdivision and SymbolicPiecewiseSubdivision can be specified
>   with submethods in the nested form Method->{method, Method->submethod}. More…

Out[98]= NIntegrate[x, {x, 0, 1}, Method → FastGoodPreciseAndStable]

The option setting Oscillatory is useful for integrands of the form $\int_a^\infty sinOrCosOrABesselJ(f(x)) g(x) dx$ [1592]. Here is a simple example.

In[99]:= **NIntegrate[Sin[x^3 + 1] 1/(x^5 + 4x + 4), {x, 0, Infinity},**
                   Method -> Oscillatory]
Out[99]= 0.157046

The method can—as any other method option setting—be used in conjunction with any other option setting for other options of NIntegrate.

```
In[100]:= NIntegrate[Sin[x^3 + 1] 1/(x^5 + 4x + 4), {x, 0, Infinity},
              Method -> Oscillatory,
              WorkingPrecision -> 30, PrecisionGoal -> 12]
Out[100]= 0.157046263601555
```

Within the `Oscillatory` option setting, no convergence checks are currently done. So the following wrong result is silently returned.

```
In[101]:= NIntegrate[Sin[x] Exp[Sin[x]], {x, 0, Infinity}, Method -> Oscillatory]
Out[101]= 43.3871
```

Be aware that within the `Oscillatory` option setting, the integrand and the integration limits must match the above description exactly.

```
In[102]:= NIntegrate[Sin[x + Log[x]] Exp[-x], {x, 0, Infinity}, Method -> Oscillatory]

          NIntegrate::ierr : The argument in the oscillatory function
                must be of the form a + b x^n, where a and b are constants. More…

Out[102]= NIntegrate[Sin[x + Log[x]] e^-x, {x, 0, ∞}, Method → Oscillatory]
```

```
In[103]:= NIntegrate[Sin[x] Exp[-x], {x, 0, 10^100}, Method -> Oscillatory]

          NIntegrate::osrn :
           For oscillating integrals, the range has to contain one Infinity (
              or -Infinity) and be of non-zero length. The integration
              continues with the Gauss-Kronrod method. More…

          NIntegrate::ploss :
           Numerical integration stopping due to loss of precision. Achieved
              neither the requested PrecisionGoal nor AccuracyGoal; suspect one of
              the following: highly oscillatory integrand or the true value of the
              integral is 0. If your integrand is oscillatory on a (semi-)infinite
              interval try using the option Method->Oscillatory in NIntegrate. More…

Out[103]= 0.
```

Let us monitor `NIntegrate`'s behavior when using the option `Oscillatory`. We will use the following example.

```
In[104]:= NIntegrate[1/x Sin[x], {x, 0, Infinity}, Method -> Oscillatory]
Out[104]= 1.5708
```

The next graphic neatly shows what `NIntegrate` is doing. It starts by carrying out the numerical integration near the origin $0 \le x \le 3\pi$. Then it moves on for larger values of x, roughly $18\pi \le x \le 30\pi$. Inside this region, we see a periodic structure resulting from the integration between $k\pi$ and $(k+1)\pi$. The partial sum of these values is then used to determine the limit of the sequence when $k \to \infty$. Finally, for the remaining interval $(3\pi, 18\pi)$ the integrations are performed between $k\pi$ and $(k+1)\pi$.

```
In[105]:= Module[{f, bag = {}},
          (* each call to f adds the argument x to the list bag *)
          f[x_?InexactNumberQ] := (AppendTo[bag, x]; 1/x);
          NIntegrate[f[x] Sin[x], {x, 0, ∞},
                  Method -> Oscillatory, GaussPoints -> 4];
          (* display content of bag *)
          ListPlot[bag, PlotStyle -> {PointSize[0.003]}]]
```

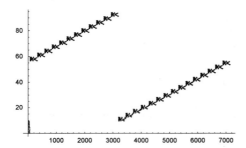

The above graphic indicates that for NIntegrate, with the method setting Oscillatory to calculate a correct result, the function should have no "unexpected" structure for $x \gtrsim 30\,\pi$.

The next two possible option settings for the Method option in NIntegrate are to be discussed: Monte·Carlo and QuasiMonteCarlo. [672]

They are especially useful for (the much harder to carry out [1669]) high-dimensional integrals. When such an option setting is used, all other option settings are ignored (the method is only available for machine arithmetic). Up to 50000 sample points will be initially used and then *Mathematica* tries to achieve around two digits of precision. mcInt in the following input evaluates the integral

$$\int_{-5}^{5}\int_{-5}^{5}\cdots\int_{-5}^{5} e^{-x_1^2-x_2^2-\cdots-x_n^2}\,dx_n \ldots dx_2\,dx_1.$$

```
In[106]:= mcInt[dim_, opts___] :=
    NIntegrate[Evaluate[Exp[-Sum[x[i]^2, {i, dim}]]],
               Evaluate[Sequence @@ Table[{x[i], -5, 5}, {i, dim}]], opts]
```

Here, a simple 1D Gaussian is integrated using the (quasi-) Monte–Carlo method.

```
In[107]:= Timing[mcInt[1, Method -> QuasiMonteCarlo]]
Out[107]= {0.16 Second, 1.77245}
```

Already for two dimensions, we need more than 100000 sample points to get about two correct digits.

```
In[108]:= Timing[mcInt[2, Method -> QuasiMonteCarlo, MaxPoints -> 10^5]]
            NIntegrate::mccnv : The integral failed to converge after 100000 iterations. More…
Out[108]= {0.91 Second, 3.14159}

In[109]:= Timing[mcInt[2, Method -> QuasiMonteCarlo, MaxPoints -> 2 10^5]]
Out[109]= {1.26 Second, 3.14159}
```

Next, we use the MonteCarlo option setting.

```
In[110]:= Timing[mcInt[1, Method -> MonteCarlo]]
Out[110]= {0.13 Second, 1.80247}

In[111]:= Timing[mcInt[2, Method -> MonteCarlo]]
            NIntegrate::mccnv : The integral failed to converge after 50000 iterations. More…
Out[111]= {0.29 Second, 3.04193}

In[112]:= Timing[mcInt[2, Method -> MonteCarlo, MaxPoints -> 2 10^5]]
Out[112]= {0.84 Second, 3.10937}
```

We calculate an integral over the function $\exp(-x^2 - y^2 + \sin(x\,y))$ using the `MonteCarlo` and the `QuasiMonteCarlo` option settings. As a side effect, we monitor the points used to sample the function.

```
In[113]:= {pointsMC, pointsQMC} =
    Reap[NIntegrate[Exp[-x^2 - 2 y^2 + Sin[x y]], {x, -1, 1}, {y, -1, 1},
            Method -> #, Compiled -> False, MaxPoints -> 10^5,
            EvaluationMonitor :> Sow[{x, y}]]][[2, 1]]& /@
                                {MonteCarlo, QuasiMonteCarlo};

    Length /@ {pointaMC, pointsQMC}
Out[114]= {3200, 4369}
```

Here, the points used are shown. The distribution of the sample points of the `QuasiMonteCarlo` option setting is much more homogeneous.

```
In[115]:= Show[GraphicsArray[
    Function[points, Graphics[{PointSize[0.003], Point /@ points},
        Frame -> True, PlotRange -> {{-1, 1}, {-1, 1}},
        AspectRatio -> Automatic]] /@ {pointsMC, pointsQMC}]]
```

Connecting the nearest neighbors shows the structure of the `QuasiMonteCarlo`-method points more clearly. The next input uses fewer points for better graphics.

```
In[116]:= Show[GraphicsArray[Graphics[
    Module[{λ = Select[#, Norm[#] < 1/2&]},
        Table[Line[{λ[[j]], Sort[{Norm[λ[[j]] - #], #}& /@
                            Delete[λ, j]][[1, -1]]}],
            {j, Length[λ]}]], Frame -> True,
        PlotRange -> 1/2 {{-1, 1}, {-1, 1}},
        AspectRatio -> 1]& /@ {pointsMC, pointsQMC}]]
```

Let us calculate how many sample points are on average in a disk of radius r around a given point. These are the points we will use to sample the pair correlation function. We will investigate all points in distance 1/4 around a given point from the set `middlePoints`. (We will use a simple, straightforward method to calculate the nearby points; for faster methods, see [638].)

```
In[117]:= neighborDataLine[mcpoints_] :=
    Module[{δ = 1/2, middlePoints, n, data},
            (* extract points in the center of the integration region *)
            middlePoints = Select[mcpoints, (Max[Abs[#]] < δ)&];
            n = Length[middlePoints];
            (* extract neighbors *)
            data = Function[point, Sort[Select[#.#& /@ ((# - point)& /@
                    DeleteCases[mcpoints, point]), # < 1/4&]]] /@ middlePoints;
            Line[Reverse /@ MapIndexed[{#2[[1]]/n, Sqrt[#1]}&,
                                            Sort[Flatten[data]]]]];
```

Now, we compare the two results. Zooming into the left picture shows that two nearby points seldom occur in the `QuasiMonteCarlo` method [1300], [1783].

```
In[118]:= With[{gr = Graphics[{{Hue[0.0], neighborDataLine[pointsMC]},
                            {Hue[0.7], neighborDataLine[pointsQMC]}},
                Frame -> True, Axes -> False, DisplayFunction -> Identity]},
        (* show with different plot ranges *)
        Show[GraphicsArray[{Show[gr, PlotRange -> All],
                        Show[gr, PlotRange -> {{0, 0.1}, {0, 10}}]}]]]
```

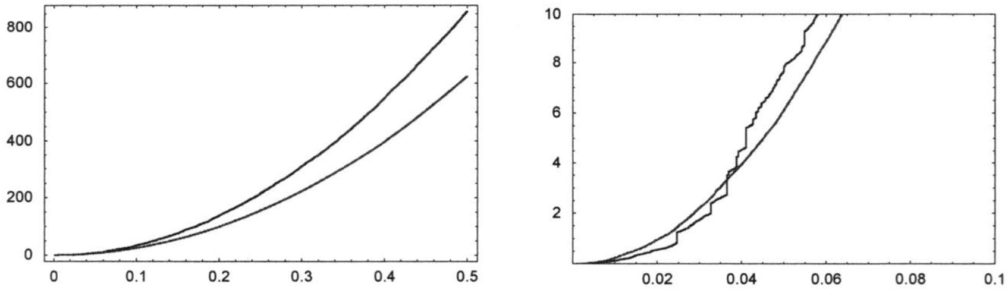

"Quasi-random" numbers (meaning maximal "uniform" distributed numbers) can be produced in various ways [1070], [1238], [1881]. One way is to "mirror" the base b digits of consecutive integers. Here is the result of this operation. The left graphic shows the points for $b = 2$ and the right shows points for $b = 2, 3, \ldots, 60$ in different colors.

```
In[119]:= VanDerCorputSequence[n_, b_] :=
        Table[FromDigits[{Reverse[IntegerDigits[k, b]], 0}, b], {k, n}]


    Show[GraphicsArray[
     Block[{$DisplayFunction = Identity},
      {(* base = 2 *)
       ListPlot[VanDerCorputSequence[2000, 2]],
       (* many bases in different colors *)
       Graphics[{PointSize[0.004], Table[{Hue[k/80],
               MapIndexed[Point[{#2[[1]], #1}]&,
                   VanDerCorputSequence[1000, k]]},
                                   {k, 2, 60}]}, Axes -> True]}]]]
```

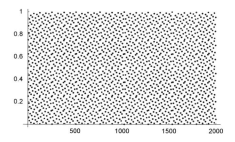 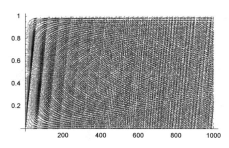

Although the error of a Monte Carlo integration for a sufficiently smooth function is $n^{-1/2}$ (independent of the dimension), where n is the number of points sampled, the error of quasi-Monte-Carlo methods is roughly n^{-1}, which is significantly better (see [1070], [1301], [1784], [1881], [1405], and [1356]). (The probability distribution of Monte-Carlo integrals converges in a more complicated fashion [1136].)

We repeat the side remark from above. Because all integration methods discussed so far require some continuity or analyticity (in the function theoretic sense of the word), if possible we should do the following (for all possible method option settings).

> Always compute integrals with discontinuous integrands piecewise over smooth parts.

We now come to the last group of option settings, the `SymbolicPiecewiseSubdivision` and `EvenOdd` `Subdivision` setting. These option settings determine the methods to be used in the symbolic preprocessing of the integrand before the actual numerical integrations are carried out.

Here is a simple integral over a discontinuous function. Because each basic numerical integration algorithm needs a certain amount of smoothness, the calculation of the integral to the required (default) precision goal fails.

```
In[122]:= fDiscontinuous[x_?NumberQ] := Floor[x]

NIntegrate[fDiscontinuous[x], {x, -E, Pi}]
        NIntegrate::slwcon :
         Numerical integration converging too slowly; suspect one of the
           following: singularity, value of the integration being 0, oscillatory
           integrand, or insufficient WorkingPrecision. If your integrand is
           oscillatory try using the option Method->Oscillatory in NIntegrate. More…
        NIntegrate::ncvb : NIntegrate failed to converge to prescribed
           accuracy after 7 recursive bisections in x near x = -2.00869. More…
Out[123]= -1.68927
```

Also, when we force the use of a basic 1D integration method directly on the integrand, we get the same result.

```
In[124]:= NIntegrate[Floor[x], {x, -E, Pi}, Method -> GaussKronrod]
        NIntegrate::slwcon :
         Numerical integration converging too slowly; suspect one of the
           following: singularity, value of the integration being 0, oscillatory
           integrand, or insufficient WorkingPrecision. If your integrand is
           oscillatory try using the option Method->Oscillatory in NIntegrate. More…
        NIntegrate::ncvb : NIntegrate failed to converge to prescribed
           accuracy after 7 recursive bisections in x near x = -2.00869. More…
Out[124]= -1.68927
```

If NIntegrate has access to the integrand and the integrand contains piecewise continuous functions that the function PiecewiseExpand can handle, then the regions where the function is continuous are automatically determined and the numerical integration is carried out in each of these regions. The Automatic option setting of the Method option allows for analyzing (and potentially manipulate) the integrand and after this analysis (and to possible rewrite of the integrand) the calls to the basic numerical integration routines are carried out.

```
In[125]:= NIntegrate[Floor[x], {x, -E, Pi}]
Out[125]= -1.73007
```

The next two inputs emulate the just-described steps.

```
In[126]:= PiecewiseExpand @ If[-E < x < Pi, Floor[x], 0] // InputForm
Out[126]//InputForm=
    Piecewise[{{-3, Inequality[-E, Less, x, Less, -2]}, {-2, Inequality[-2,
    LessEqual, x, Less, -1]},
        {-1, Inequality[-1, LessEqual, x, Less, 0]}, {1, Inequality[1, LessEqual,
        x, Less, 2]},
        {2, Inequality[2, LessEqual, x, Less, 3]}, {3, Inequality[3, LessEqual, x,
        Less, Pi]}}, 0]
```

```
In[127]:= Plus @@ (NIntegrate[#1, {x, #2[[1]], #2[[-1]]}]& @@@ %[[1]])
Out[127]= -1.73007
```

Because after splitting up the integrand into continuous parts, the actual numerical integration still has to be performed, the SymbolicPiecewiseSubdivision method option setting allows setting a suboption for the method to be used for the actual numerical integration. The next input specifies the use of the Trapezoi. dal method for carrying out the integration over the six regions.

```
In[128]:= NIntegrate[Floor[x], {x, -E, Pi},
            Method -> {SymbolicPiecewiseSubdivision,
                            Method -> Trapezoidal}]
Out[128]= -1.73007
```

The last possible method option setting to be discussed is the EvenOddSubdivision setting. It again is a symbolic preprocessing option and allows using a potential mirror symmetry of the integrand. So, from $\sin(-x) = \sin(x)$ it follows immediately that the following integral is zero.

```
In[129]:= NIntegrate[Sin[x], {x, -1, 1}]
Out[129]= 0.
```

From the fact that the integrand is antisymmetric over the integration region does not automatically follow that the integral is zero. The integrand could exhibit nonintegrable singularities. So for the following example, we obtain a message warning us that this could be the case.

```
In[130]:= NIntegrate[Tan[x], {x, -Pi, Pi}]
            NIntegrate::ploss :
            Numerical integration stopping due to loss of precision. Achieved
                neither the requested PrecisionGoal nor AccuracyGoal; suspect one of
                the following: highly oscillatory integrand or the true value of the
                integral is 0. If your integrand is oscillatory on a (semi-)infinite
                interval try using the option Method->Oscillatory in NIntegrate. More…
            NIntegrate::oidiv : The odd integrand Tan[x] is being considered
                as zero over the specified region, but may actually be divergent.
Out[130]= 0.
```

Here is a 3D integral (the integral of r^2 over the unit sphere) with the exact value $5\pi/4 = 2.513274...$. Using the symmetry properties of the integrand yields a tenfold speed-up of the numerical integration.

```
In[131]:= (NIntegrate[x^2 + y^2 + z^2,
                    {x, -1, 1}, {y, -Sqrt[1 - x^2], Sqrt[1 - x^2]},
                    {z, -Sqrt[1 - x^2 - y^2], Sqrt[1 - x^2 - y^2]},
                    Method -> #] // Timing) & /@
                  {Automatic, EvenOddSubdivision, MultiDimensional}
Out[131]= {{0.2 Second, 2.51327}, {0.21 Second, 2.51327}, {1.93 Second, 2.51327}}
```

We rewrite the last integral as an integral of a discontinuous function over the whole $\mathbb{R}^3$. The symbolically processed integral returns as quickly as in the last example. But now the direct numerical integration needs orders of magnitude longer. (The difference between the returned timing for the direct integration and the actual constraining time arises from the difference between wall-clock time and CPU-time.)

```
In[132]:= (TimeConstrained[
        NIntegrate[If[x^2 + y^2 + z^2 < 1, x^2 + y^2 + z^2, 0],
                    {x, -Infinity, Infinity}, {y, -Infinity, Infinity},
                    {z, -Infinity, Infinity}, Method -> #], 100] // Timing) & /@
                  {Automatic, EvenOddSubdivision, MultiDimensional}
        NIntegrate::slwcon :
        Numerical integration converging too slowly; suspect one of the
            following: singularity, value of the integration being 0, oscillatory
            integrand, or insufficient WorkingPrecision. If your integrand is
            oscillatory try using the option Method->Oscillatory in NIntegrate. More…
Out[132]= {{0.2 Second, 2.51327}, {1.96 Second, 2.51327}, {100.02 Second, $Aborted}}
```

We end the discussion of the symbolic preprocessing options with an example that contains triply nested options. We consider a double integral over the whole $\mathbb{R}^2$. We first split up the integrand in regions where no discontinuities occur. Then we use the symmetry of the resulting pieces and finally we carry out the actual integration using the Gauss–Kronrod method.

```
In[133]:= NIntegrate[If[x^2 + y^2  < 1, x^2 + y^2 , 0],
                    {x, -Infinity, Infinity}, {y, -Infinity, Infinity},
                    WorkingPrecision -> 30, PrecisionGoal -> 20,
                    (* multi-nested option settings *)
                    Method -> {SymbolicPiecewiseSubdivision,
                              Method -> EvenOddSubdivision,
                                    Method -> GaussKronrod}] // Timing
Out[133]= {0.95 Second, 1.57079632679489661923}
```

The result has the required precision.

```
In[134]:= %[[2]] - Pi/2
Out[134]= 0. × 10^{-21}
```

We end this section with an application of NIntegrate: the Picard–Lindelöf iterative method for solving an ordinary differential equation of first order, which is given in explicit form; that is, it has been solved for the first derivative. (Note: For the purpose of calculating numerically a solution of a differential equation, NDSolve is naturally more appropriate.)

Mathematical Remark: Picard–Lindelöf Iteration

Suppose we are given an ordinary differential equation of first order in the form $y'(x) = f(x, y(x))$ with initial condition $y(x_0) = \tilde{y}_0$. This differential equation can be rewritten as the integral equation

$$y(x) = \bar{y}_0 + \int_{x_0}^{x} f(t, y(t)) \, dt.$$

Because of the contractive properties of the associated integral operator for sufficiently smooth functions, applying the iteration (where the superscript here indicates the iteration number)

$$y_{k+1}(x) = \bar{y}_0 + \int_{x_0}^{x} f(t, y_k(t)) \, dt, \quad y_0(t) = \bar{y}_0$$

to this integral equation will converge to $y(x)$ as $i \to \infty$. For the exact conditions of convergence, see, for instance, [299] and [1719].

Here is a possible implementation. The input parameters of `PicardLindeloefIteration` are a function $f(x, y)$ in the form of a pure function $f(\#1, \#2)$, the starting value *y0*, and the integration interval in the form of node points *nodePoints*. We integrate using `NIntegrate` on each of these subintervals. The actual function that we integrate is obtained by interpolation of the values obtained in the previous step.

We begin with the approximate solution $y_0(x) = y_0$. We iterate until nothing changes. To watch the convergence, we use `FixedPointList`.

```
In[135]:= PicardLindeloefIteration[rightHandSide_, y0_,
                        nodePoints_List, intOpts___Rule] :=
    Module[{t}, (* drop last element *) Most @
    FixedPointList[ (* until nothing changes anymore *)
      Interpolation[ (* make an interpolation from the results *)
          Transpose[{nodePoints,
                  FoldList[Plus, (* add up the results *)
                      y0     (* the starting function
                                value to be added *),
              (* map NIntegrate to each interval *)
              Map[Function[interval, NIntegrate[rightHandSide[t, #[t]],
              (* NIntegrate has HoldAll *)
              Evaluate[Prepend[interval, t]], intOpts]],
          (* NIntegrate every interval individually *)
                      Partition[nodePoints, 2, 1]]]}]]&,
              (* start with a pure function *) y0&]]
```

We now solve the differential equation $y'(x) = -2 x y(x)$, $y(0) = 1$ using `PicardLindeloefIteration`. The exact solution is e^{-x^2}.

```
In[136]:= (sol = PicardLindeloefIteration[-2#1 #2&, 1,
                        Range[0.0, 1.0, 0.1]]) // Short[#, 5]&
Out[136]//Short=
      {1 &, InterpolatingFunction[{{0., 1.}}, <>],
       InterpolatingFunction[{{0., 1.}}, <>],
       InterpolatingFunction[{{0., 1.}}, <>], InterpolatingFunction[{{0., 1.}}, <>],
       <<8>>, InterpolatingFunction[{{0., 1.}}, <>],
       InterpolatingFunction[{{0., 1.}}, <>], InterpolatingFunction[{{0., 1.}}, <>],
       InterpolatingFunction[{{0., 1.}}, <>], InterpolatingFunction[{{0., 1.}}, <>]}
```

We obtain convergence after just a few iterations. We now look graphically at the relative and absolute errors for several iterations. For better visibility, we superimpose vertical lines (`GridLines`) at the *x*-values of the data points.

```
In[137]:= opts = Sequence[PlotRange -> All, DisplayFunction -> Identity,
              GridLines -> {Range[0.0, 1.0, 0.1], None},
              AxesOrigin -> {0, 0}];

    Table[Show[GraphicsArray[
      (* the absolute error *)
      {Plot[(sol[[i]][x] - Exp[-x^2]), {x, 0.001, 0.999}, Evaluate[opts],
          PlotLabel -> StyleForm["abs. error ", FontSize -> 6]],
      (* the relative error *)
      Plot[(sol[[i]][x] - Exp[-x^2])/Exp[-x^2], {x, 0.001, 0.999},
          Evaluate[opts],
          PlotLabel -> StyleForm["rel. error ", FontSize -> 6]]},
          PlotLabel -> StyleForm[If[i === 1, "1st",
            ToString[i] <> "th iteration"], FontSize -> 6],
          DisplayFunction -> $DisplayFunction]], {i, 1, 16, 3}]
```

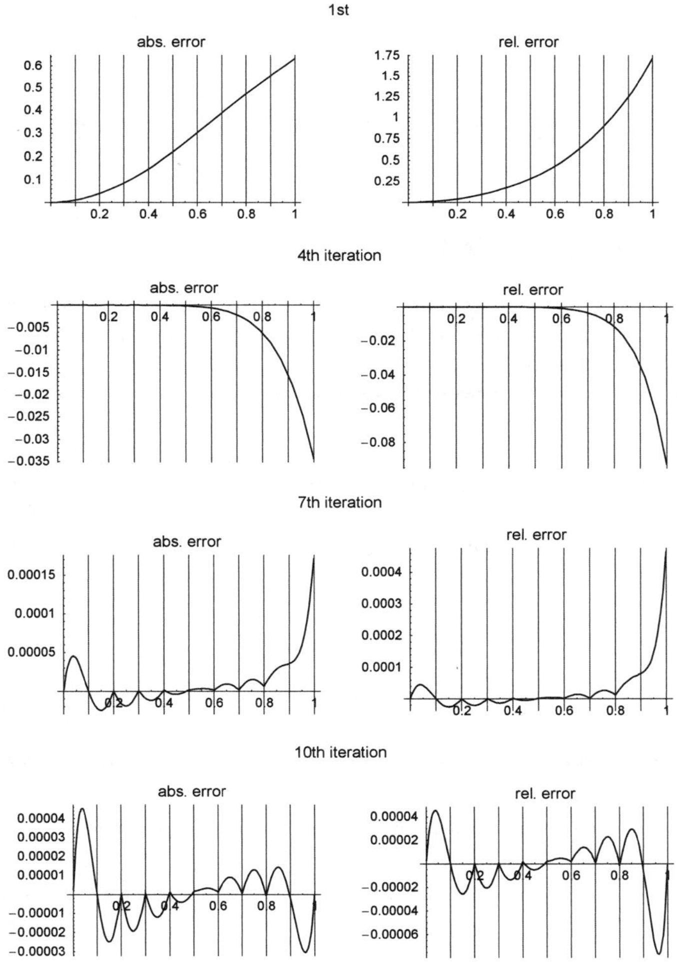

1st

4th iteration

7th iteration

10th iteration

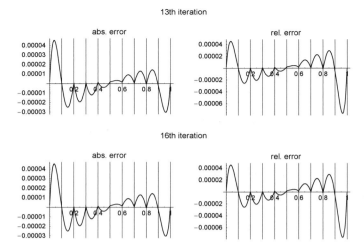

The method converged; the remaining error is caused by the insufficient discretization.

1.9 Solution of Equations

Finding the solutions of nonlinear equations and systems of equations is a very common numerical procedure. *Mathematica* includes three routines for this purpose: NSolve, NRoots, and FindRoot. We start with the two functions that deal mainly for polynomials and polynomial systems.

NSolve[*polynomialEquations*, *variables*]

    produces a numerical solution to the polynomial (or rational) equation(s) *polynomialEquations* in the variables *variables*. The result is returned in the form of a nested list of rules.

NRoots[*polynomialEquation*, *variables*]

    produces a numerical solution to the polynomial equation *polynomialEquation* in the variable *variable*. The result is returned in form of a logical alternative.

NRoots[*polynomialEquation*, *variables*, *numDigits*]

    computes the solution using precision *numDigits* in the calculation.

A rough guideline for the use of the three functions NSolve, NRoots, and FindRoot (to be discussed later in this section) is the following.

NRoots

    for the calculation of all roots of an (explicitly written as such) univariate polynomial.

NSolve

    for the calculation of all roots of univariate rational functions and multivariate polynomial systems with as many variables as equations.

FindRoot

    for the calculation of single roots of transcendental equations and single roots of high–order polynomials.

For polynomials over rationals in one variable, all roots will be found by NSolve. For polynomials in several variables, this is the case too, but it may take a considerable amount of time.

The capability to numerically solve systems of polynomial equations is very important. Polynomial systems appear frequently in applications, and, interestingly, solving transcendental equations can be reduced to solving systems of polynomial equations by using contour integration and Newton relations (for details, see [1058]). Polynomial systems can be solved fully algorithmically. This guarantees that no solution will be missed.

Let us start discussing the two polynomial solving functions of this section by treating univariate polynomials. Here is a simple example with a polynomial of degree five. The fundamental theorem of algebra tells us that we have to expect five solutions.

```
In[1]:= poly = x^5 + 2x^4 + 3x^3 + 4x^2 + 5x + 6;

       solution = NSolve[poly == 0, x]
Out[2]= {{x → -1.4918}, {x → -0.805786 - 1.2229 i}, {x → -0.805786 + 1.2229 i},
        {x → 0.551685 - 1.25335 i}, {x → 0.551685 + 1.25335 i}}
```

We could also set the option WorkingPrecision to obtain a more precise solution.

```
In[3]:= NSolve[poly == 0, x, WorkingPrecision -> 50]
Out[3]= {{x → -1.4917979881399007106088100134486543604404282872373},
        {x → -0.8057864693890312241074714735636824932180297538638-
            1.2229047133744098601545412950565017487407163694045i},
        {x → -0.8057864693890312241074714735636824932180297538638+
            1.2229047133744098601545412950565017487407163694045i},
        {x → 0.5516854634589815794118764802880096734382438974824-
            1.2533488602772061362503184219058377508732302494878i},
        {x → 0.5516854634589815794118764802880096734382438974824+
            1.2533488602772061362503184219058377508732302494878i}}
```

We can also solve the system to very low precision (although for most applications, this is not very useful).

```
In[4]:= NRoots[poly == 0, x, 0.5] // InputForm
Out[4]//InputForm=
       x == -1.4917979881399006103`0.6020599913279624 ||
        x == -0.8057864693890312013`0.4931027971823613 -
       1.2229047133744097575`0.674275445535706*I ||
        x == -0.8057864693890312013`0.4931027971823613 +
       1.2229047133744097575`0.674275445535706*I ||
        x == 0.5516854634589815065`0.35773807577133077 -
       1.253348860277206045`0.7141185054705512*I ||
        x == 0.5516854634589815065`0.35773807577133077 +
       1.253348860277206045`0.7141185054705512*I
```

The differences between NRoots and NSolve lie mainly (in the case of one variable for which NRoots only works) in the form of the result and in the time (because of preprocessing steps done by NSolve but not done by NRoots). NSolve produces a List of rules, and NRoots produces a logical Or. We check the solution by substitution.

```
In[5]:= N[poly /. solution, 50]
Out[5]= {1.77636×10^-15, -2.22045×10^-16 + 0. ×10^-66 i, -2.22045×10^-16 + 0. ×10^-66 i,
        1.77636×10^-15 - 2.66454×10^-15 i, 1.77636×10^-15 + 2.66454×10^-15 i}
```

In[6]:= (* check precision of the solution *)
N[poly /. Map[If[NumberQ[#], # + 10^-50, #]&, solution, {-1}], 50]

Out[7]= {1.77636×10^{-15}, $-2.22045 \times 10^{-16} + 0. \times 10^{-66}$ i, $-2.22045 \times 10^{-16} + 0. \times 10^{-66}$ i,
$1.77636 \times 10^{-15} - 2.66454 \times 10^{-15}$ i, $1.77636 \times 10^{-15} + 2.66454 \times 10^{-15}$ i}

For polynomials of higher order, it might happen that the precision of the calculated zeros is less than *numDigits*. (*numDigits* specifies the working precision not the precision goal). Here is an example of such a situation.

In[8]:= **Precision /@ (List @@ NRoots[Sum[(23 - i)^8 x^i, {i, 0, 20}] == 0, x, 100])**

Out[8]= {97.5757, 97.5757, 97.5682, 97.5682, 97.5388, 97.5388,
97.4184, 97.4184, 97.4373, 97.4373, 97.7593, 97.7593, 98.2086,
98.2086, 98.6344, 98.6344, 99.1158, 99.1158, 99.7227, 99.7227}

NRoots is a purely numerical function. It does not carry out any symbolic transformations. While the roots of the following polynomial are easy to recognize exactly, when carrying out the numerical root finding process we lose precision.

In[9]:= **Precision /@ (List @@ NRoots[Expand[Product[x - i, {i, 1, 30}]] == 0, x, 100])**

Out[9]= {96.7305, 94.365, 92.3536, 90.5948, 89.0337, 87.6358, 86.3776,
85.2423, 84.2173, 83.293, 82.4619, 81.7183, 81.0575, 80.4758,
79.9707, 79.54, 79.1827, 78.8981, 78.6862, 78.5479, 78.4848, 78.4994,
78.5955, 78.7784, 79.0558, 79.4386, 79.9433, 80.5971, 81.45, 82.6261}

The timing difference between NRoots and NSolve results from the fact that NRoots only treats expressions that are explicitly a polynomial in one variable, and NSolve treats expressions that reduce to polynomials in several variables; so in the following case some additional symbolic work needs to be done. The following example shows the difference clearly.

In[10]:= **NRoots[(z^2 - 1)/(z - 1) == 0, z]**

Out[10]= NRoots$\left[\frac{-1 + z^2}{-1 + z} == 0, z \right]$

In[11]:= **NSolve[(z^2 - 1)/(z - 1) == 0, z]**

Out[11]= {{$z \rightarrow -1.$}}

Be aware that polynomial roots are typically very sensitive functions of their coefficients. As such, the form in which a polynomial is written might matter a lot for numerical issues. Let us compare the results of NRoots for the factored and expanded forms of $(1 + x)^{12} = 0$.

In[12]:= **poly = (x - 1)^12;**
NRoots[poly == 0, x]

Out[13]= x == 1. || x == 1. || x == 1. || x == 1. || x == 1. ||
x == 1. || x == 1. || x == 1. || x == 1. || x == 1. || x == 1. || x == 1.

In[14]:= **NRoots[Expand[poly] == 0, x]**

Out[14]= x == 0.957922 || x == 0.966174 || x == 0.966438 || x == 0.966516 - 0.0441191 i ||
x == 0.966516 + 0.0441191 i || x == 1.00509 || x == 1.0051 ||
x == 1.01844 || x == 1.01844 || x == 1.02059 || x == 1.046 || x == 1.06277

The roots of the last results deviate clearly from 1. Even when carrying out the calculation with 50-digit working precision we get only four correct digits.

In[15]:= **NRoots[Expand[poly] == 0, x, 50]**

Out[15]= x == 1.000 || x == 1.000 || x == 1.000 || x == 1.000 || x == 1.000 || x == 1.000 ||
x == 1.000 || x == 1.000 || x == 1.000 || x == 1.000 || x == 1.000 || x == 1.000

The errors, about 5% in the expanded form, arise due the sensitivity of the roots as a function of the coefficients. The next input calculates the roots using numericalization after using exact arithmetic where the coefficient c_d of the polynomial `poly` has been modified to $(1 + 10^{-\$MachinePrecision})c_d$. The absolute value of the largest difference to 1 is displayed. These roots clearly show the dramatic influence of the exact coefficient value.

```
In[16]:= Table[{d, N @ Max[Abs[1 - (x /. N[{ToRules[Roots[
                poly + ε Coefficient[poly, x, d] x^d == 0, x]]} /.
                      ε -> 10^-$MachinePrecision, 30])]]},
            {d, 0, 12}]
Out[16]= {{0, 0.0468221}, {1, 0.0510283}, {2, 0.0308881}, {3, 0.126914},
        {4, 0.00252173}, {5, 0.0295531}, {6, 0.124463}, {7, 0.127821}, {8, 0.029654},
        {9, 0.0350009}, {10, 0.0999796}, {11, 0.0713957}, {12, 0.111595}}
```

Because NRoots is faster than NSolve, we will preferably use NRoots when many roots have to be determined. Here are two examples of the use of NRoots. The first example is the iterated calculation (and the ongoing visualization) of the roots of the polynomial $4z^3 + z^2/2 + z + i/2 = previousRootsOfRightHandSidePolynomial$.

```
In[17]:= Show[Graphics[{PointSize[0.003], (* make points for graphic *)
        Map[Point[{Re[#], Im[#]}]&, (* the iteration of the root calculation *)
        NestList[Flatten[Cases[NRoots[4 C^3 + C^2/2 + C + I/2 == #, C],
            _?NumberQ, {-1}]& /@ #]&,
            {0}, 10 (* this makes in summary 88573 roots *)], {-1}]]}],
        Frame -> True, PlotRange -> All]
```

In the second example, we again iteratively solve a polynomial equation and substitute the value of the last root on the right-hand side of this equation. Again, because the calculation results in many roots, we graphically present the result.

```
In[18]:= iteratedRootPicture[poly_, rhs0_, opts___] :=
    Show[Graphics[{MapIndexed[
    Function[{arg, pos}, {Thickness[0.002],
      (* lines to the next generation of roots *)
      Apply[Function[{z1, z2}, Line[If[(-1)^pos[[1]] === 1,
        {{Re[z1], Im[z1]}, {Re[z1], Im[z2]}, {Re[z2], Im[z2]}},
        {{Re[z1], Im[z1]}, {Re[z2], Im[z1]}, {Re[z2], Im[z2]}}]]],
        arg, {1}]}], Reverse[Rest[NestList[Flatten[Function[rhs,
            (* one-time iteration of root findings *)
                    {rhs, #}& /@ Cases[NRoots[poly[C] == rhs, C],
        _?NumberQ, {-1}]][[Last[#]]]& /@ #, 1]&,
                {{0, rhs0}}, 4]]], {1}]}], opts]

In[19]:= Show[GraphicsArray[
        (* start with a random polynomial *)
        Table[iteratedRootPicture[Function @@
```

```
{z, Sum[Random[Complex, {-1 - I, 1 + I}] z^i,
    {i, 0, 6}]}, 1 + I, DisplayFunction -> Identity], {3}]]]
```

Given a univariate polynomial with one or more parameters, it is interesting to consider the minimal distance between the roots as a function of these parameters [224], [1479]. Because roots of polynomials are piecewise continuous functions of their parameters, but they are not differentiable everywhere. As a result, the root distances will also be not smooth everywhere with respect to the parameters. Here is a random sixth order polynomial in z with a parameter c.

```
In[20]:= randomComplexPoly[z_, c_, n_] :=
            Sum[Random[Integer, {-10, 10}] z^i c^j, {i, n - 1}, {j, n}] + z^n
```

```
In[21]:= SeedRandom[7618666700];
         R[ξ_, c_] = randomComplexPoly[ξ, c, 6]
```

$$Out[22]= \; 9\,c\,\xi + 4\,c^2\,\xi - 7\,c^3\,\xi - 10\,c^4\,\xi + 4\,c^5\,\xi + 3\,c^6\,\xi + 5\,c\,\xi^2 - 4\,c^2\,\xi^2 - 10\,c^3\,\xi^2 + c^4\,\xi^2 +$$
$$10\,c^5\,\xi^2 + 10\,c^6\,\xi^2 + 3\,c\,\xi^3 - 7\,c^2\,\xi^3 - 10\,c^3\,\xi^3 + 4\,c^4\,\xi^3 + 3\,c^5\,\xi^3 + 4\,c\,\xi^4 + 9\,c^2\,\xi^4 -$$
$$7\,c^3\,\xi^4 - 7\,c^4\,\xi^4 - 7\,c^6\,\xi^4 + 2\,c\,\xi^5 + 3\,c^2\,\xi^5 + c^3\,\xi^5 - 2\,c^4\,\xi^5 + 2\,c^5\,\xi^5 + 6\,c^6\,\xi^5 + \xi^6$$

The next two graphics show the minimal root distance as a function of the complex variable c. This function shows an interesting, moon surface-like structure. To generate the graphics, we have to solve 62500 equations of the form *sixthOrderPolynomial* = 0.

```
In[23]:= δMin[c_] := (* calculate all roots and all distances *)
         With[{roots = Last /@ (List @@ NRoots[R[z, c] == 0, z])},
             Min[Abs[Flatten[Table[roots[[i]] - roots[[j]], {i, 6}, {j, i - 1}]]]]]
```

```
In[24]:= Show[GraphicsArray[
         {#, (* make contour plot from 3D graphics data *)
          ListContourPlot[#[[1]], ColorFunction -> (Hue[0.8#]&),
                     MeshRange -> {{-2, 2}, {-2, 2}},
                     Contours -> 20, DisplayFunction -> Identity]}]&[
          (* 3D graphics of the distance *)
          Plot3D[δMin[cx + I cy], {cx, -2, 2}, {cy, -2, 2},
              PlotPoints -> 250, Mesh -> False, Compiled -> False,
              DisplayFunction -> Identity]]]
```

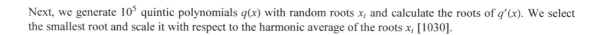

Next, we generate 10^5 quintic polynomials $q(x)$ with random roots x_i and calculate the roots of $q'(x)$. We select the smallest root and scale it with respect to the harmonic average of the roots x_i [1030].

```
In[25]:= With[{n = 5, o = 10^5},
    data = Table[roots = Table[Random[], {n}];
            1/Plus @@ (1/roots)/Min[Last /@ List @@ NRoots[
            D[Times @@ (x - roots), x] == 0, x]], {o}]];
```

Binning the scaled roots gives the following sharply peaked histogram for the distribution.

```
In[26]:= With[{δ = 0.005, o = 10^5},
    ListPlot[{δ First[#], Length[#]/o}& /@ Split[Sort[Round[data/δ]]],
            PlotRange -> All]]
```

We now give a somewhat more complicated example that uses NRoots from solid-state physics: Hofstadter's butterfly.

Physical Remark: Hofstadter's Butterfly

In 1976, D. Hofstadter published a paper entitled *Energy levels and wave functions of Bloch electrons in rational and irrational magnetic fields* [850], discussing the following problem. Given a crystal in a magnetic field, find the quantum-mechanical energy spectrum of the system of electrons, that is, the permissible energy states.

In a periodic potential (without magnetic field), the permissible energy levels are bands. This is a consequence of the Bloch–Floquet theorem [75], which holds for the Schrödinger equation

$$\left(\frac{\mathbf{p}^2}{2\,m} + V(\mathbf{r})\right)\Psi_E(\mathbf{r}) = E\,\Psi_E(\mathbf{r})$$

with periodic potential $V(\mathbf{r}) = V(\mathbf{r} + \mathbf{R})$, where $\mathbf{R}$ is a lattice vector and $\mathbf{p} = -i\,\hbar\,\boldsymbol{\nabla}$ is the momentum operator.

m is the (effective) electron mass, and E is the (continuous or discrete) energy eigenvalue to be determined. In those energy bands of practical interest, we can use a semiclassical approximation for investigating the behavior of lattice electrons. In this case, we have the following equation for the band structure (with $\epsilon(\mathbf{p})$ the dispersion relation that in the general case is a pseudodifferential operator):

$$(\epsilon(\mathbf{p}) + V_{ext}(\mathbf{r}))\,\Phi_E(\mathbf{r}) = E\,\Phi_E(\mathbf{r}).$$

Here, $V_{ext}(\mathbf{r})$ is the potential corresponding to external disturbances (e.g., boundary surfaces, electric fields).

If we apply a magnetic field $\mathbf{B}(\mathbf{r})$, it will be "minimally (invariant) coupled" in the Schrödinger equation (in view of Peierl's theorem). Technically, this means that in the Schrödinger equation, the kinematic impulse $\mathbf{p}$ is replaced by the generalized impulse $\boldsymbol{\pi} = \mathbf{p} - e\,\mathbf{A}(\mathbf{r})$, where $\mathbf{A}(\mathbf{r})$ is the vector potential with curl $\mathbf{A}(\mathbf{r}) = \mathbf{B}(\mathbf{r})$.

This results in a so-called pseudodifferential equation in which the differential operators appear as arguments inside nonpolynomial functions. The simplest approach is to assume the band structure of a square lattice is of tight-binding form (we only consider the behavior in the plane perpendicular to the applied magnetic field):

$$\epsilon(\mathbf{p}) = 2\,\epsilon_0\,(\cos(p_x\,a/\hbar) + \cos(p_y\,a/\hbar)) \quad (a = \text{lattice constant}).$$

and that the magnetic field is homogeneous in the z-direction $\mathbf{B}(\mathbf{r}) = \{0, 0, B\}$ (for hexagonal lattices, see [750], [1373]). Applying the Landau gauge $\mathbf{A}(\mathbf{r}) = \{0, B\,x, 0\}$, we are led to an equation of the form:

$$2\,\epsilon_0\,(\cos(i\,\partial_x\,a) + \cos((i\,\partial_y - e\,B\,x)\,a))\,\Phi_E(x, y) = E\,\Phi_E(x, y).$$

The problem now is to solve this equation. Scaling the equation with $x = m\,a$, $y = n\,a$, $\alpha = eBa^2/h$, $\epsilon = E/\epsilon_0$ and inserting the ansatz $\Phi_E(m\,a, n\,a) = e^{-i\nu n}\,g(m)$, we get the following 1D difference equation for $g(m)$:

$$g(m + 1) + 2\cos(2\pi m\alpha - \nu)\,g(m) = \epsilon\,g(m) - g(m - 1).$$

We write this equation in recursive form using a 2×2 matrix:

$$\begin{pmatrix} g(m+1) \\ g(m) \end{pmatrix} = \begin{pmatrix} \epsilon - 2\cos(2\pi m\alpha - \nu) & -1 \\ 1 & 0 \end{pmatrix} \begin{pmatrix} g(m) \\ g(m-1) \end{pmatrix}.$$

We denote the 2×2 matrix on the right-hand side by $\mathbf{A}(\epsilon, m, \alpha, \nu)$. By known properties of the Harper equation (for details, see the references cited below), we get the following condition for the eigenvalue spectrum: ϵ is a permissible energy eigenvalue, provided $|\mathrm{Tr}\,\mathbf{Q}(\epsilon, \alpha, q)| \leq 4$, where

$$\mathbf{Q}(\epsilon, \alpha, q) = \prod_{k=0}^{q-1} \mathbf{A}\left(\epsilon, k, \alpha, \frac{\pi}{2q}\right).$$

Here, $q = \text{denominator}(\alpha)$ for rational α.

The spectrum is periodic in α with $0 \leq \alpha \leq 1$ as the interval of periodicity and the spectrum is symmetric with respect to $\alpha = 1/2$.

For more details, see the references given below.

Now, we proceed to the implementation. First, we look at the number of (proper) fractions with denominators smaller than n, which are $\leq 1/2$.

```
In[27]:= ListPlot[Table[Length[Union[ (* eliminate multiples *)
              Flatten[ (* all proper fractions *)
                Table[j/i, {i, maxDenominator},
                    {j, i/2}]]]], {maxDenominator, 30}]]
```

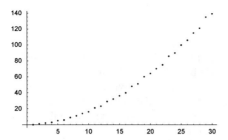

Note that a faster method [1617] for generating the first *n* terms of these series follows (we come back to the command `EulerPhi` in the next chapter).

```
FareySequence[maxDenominator_Integer?Positive, n_] :=
Apply[Rational, Drop[#, 2]& /@ NestList[
Function[{x2, y2, x1, y1}, {x1, y1, # x1 - x2, # y1 - y2}&[
        Floor[(y2 + maxDenominator)/y1]]] @@ ##&,
        {0, 1, 1, maxDenominator}, n], {1}] /;
            n <= Sum[EulerPhi[i], {i, maxDenominator}] - 1

FareySequence[maxDenominator_Integer?Positive] :=
FareySequence[maxDenominator, Sum[EulerPhi[i], {i, maxDenominator}] - 1]
```

Here is a similar recursive method. [197]

```
FareySequenceRecursive[maxDenominator_Integer?Positive] :=
Block[{x, y},
        {x[0], y[0], x[1], y[1]} = {0, 1, 1, n};
        x[k_] := x[k] = Floor[(y[k - 2] + n)/y[k - 1]] x[k - 1] - x[k - 2];
        y[k_] := y[k] = Floor[(y[k - 2] + n)/y[k - 1]] y[k - 1] - y[k - 2];
        Table[x[k]/y[k], {k, Sum[EulerPhi[i], {i, n}]}]]]
```

Next, we give the defining equation for the matrix $\mathbf{A}(\epsilon, m, \alpha, \nu)$.

```
In[28]:= A[ε_, m_, α_, ν_] := {{ε - 2 Cos[2Pi m α - ν], -1}, {1, 0}};
```

We denote the trace of the above product by $\mathbf{Q}(\epsilon, \alpha)$. We compute it in the following steps: 1) Compute a list of the **A**-matrices. 2) Alter the head `List` to `Dot`. 3) Multiply out all products. 4) Form the trace.

To speed up the calculation, at this point, we apply `N`.

```
In[29]:= Q[ε_, α_] := Dot @@ (* the matrix product *)
                Table[N[A[ε, m, α, Pi/(2 Denominator[α])]],
                    {m, 0, Denominator[α] - 1}] // Expand // Tr
```

The result of this computation of **Q** is a polynomial in the scaled energy ϵ depending on the rational scaled magnetic field value α. Here are a few examples.

```
In[30]:= {Q[ε, 1], Q[ε, 1/2], Q[ε, 1/3], Q[ε, 1/4]} // Chop
Out[30]= {ε, -4. + ε², -6. ε + ε³, 4. - 8. ε² + ε⁴}
```

$\alpha = 2/4$ reduces to the case $\alpha = 1/2$. Similar reductions happen for other reducible fractions.

```
In[31]:= Q[ε, 2/4] // Chop
```
Out[31]= $-4. + \epsilon^2$

Now, we proceed to the computation of the energy spectrum for a given α. The key is the use of NRoots. First, we find the energy values for which $\mathbf{Q}(\epsilon, \alpha)$ takes on the value ± 4. We put these solutions in one list. The solutions produced by NRoots have the form Or[Equal[...], ...]. We take just the real part of the solutions and sort them by size. We use the real parts, because even though the exact solutions are real, roundoff errors in the coefficients can cause NRoots to produce approximate solutions with small complex parts. Next, we divide this list into intervals. These are the allowed energy intervals.

```
In[32]:= Spectrum[α_] :=
    Module[{poly},
        (* the polynomial in e *) poly = Q[ε, α];
        Partition[ (* every two zeros make a line *)
            Sort[Re[#]& /@ (* sorting the zeros *)
            (#[[2]]& /@ ((* the zeros of poly == ±4 *)
        Join[If[Head[#] === Equal, {#}, List @@ #]&[NRoots[poly == -4, ε]],
            If[Head[#] === Equal, {#}, List @@ #]&[NRoots[poly == +4, ε]]])], 2]]
```

Here are the permissible energy intervals corresponding to the above **Q**'s.

```
In[33]:= Spectrum[1]
```
Out[33]= {{-4., 4.}}

```
In[34]:= Spectrum[1/2]
```
Out[34]= {{-2.82843, 0.}, {0., 2.82843}}

```
In[35]:= Spectrum[1/3]
```
Out[35]= {{-2.73205, -2.}, {-0.732051, 0.732051}, {2., 2.73205}}

```
In[36]:= Spectrum[1/4]
```
Out[36]= {{-2.82843, -2.61313}, {-1.08239, -2.10734×10^{-8}},
         {2.10734×10^{-8}, 1.08239}, {2.61313, 2.82843}}

We see that for a given scaled magnetic field value α, there are exactly denominator(α) energy intervals.

Finally, we plot the Hofstadter butterfly. It lies in the energy magnetic field plane, called the ϵ, α-plane. In plotting it, we take account of the symmetry about $\alpha = 1/2$, as discussed above. The permissible energy bands are shown as lines corresponding to a given α.

```
In[37]:= SpectrumGraph[α_] := {Line[{{#[[1]], α}, {#[[2]], α}}],
                                Line[{{#[[1]], 1 - α}, {#[[2]], 1 - α}}]}& /@
                                                        Spectrum[α]
```

To get the entire butterfly, we just have to combine several magnetic field values in one plot. We do this by giving the maximum denominator, which occurs as an argument to HofstadterButterfly.

```
In[38]:= HofstadterButterfly[maxDenominator_] :=
    Show[Graphics[{Thickness[0.002],
        (* the allowed energy values *)
        SpectrumGraph /@ (* all fractions of interest *)
        Union[Flatten[Table[j/i, {i, maxDenominator},
                            {j, If[EvenQ[i], i/2, (i - 1)/2]}]]]}],
            (* make plot nice looking *)
            PlotRange -> {{-4.12, 4.12}, {-0.04, 1.04}},
            PlotLabel -> StyleForm["Hofstadter's butterfly",
```

```
              FontWeight -> "Bold"], Axes -> False, Frame -> True,
              FrameLabel -> {"ε", "α"}]
```

Here is the butterfly with maximal denominator of size 15.

```
In[39]:= HofstadterButterfly[15];
```

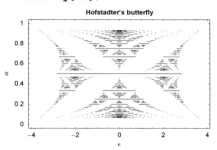

Asymptotically the total width l of the spectrum is conjectured to be $l \underset{q \to \infty}{\sim} q^{-1}$ for $\alpha = p/q$ [1764], [1765], [779], [1875], [1157], [1053]. The next graphic shows the inverse of the total width of the spectrum as a function of the denominator q.

```
In[40]:= ListPlot[{Denominator[#[[1, 1]]], 1/(Plus @@ (Last /@ #))}& /@
          Split[Sort[Cases[%, _Line, Infinity] /. Line[{{a_, b_}, {c_, b_}}] :>
                      {b, Abs[a - c]}, #1[[1]] < #2[[1]]&],
           #1[[1]] === #2[[1]]&]]
```

As an exercise in the use of pure functions, we now present a somewhat more compact code for the Hofstadter butterfly. It is a direct nested combination of the above partial definitions. Unfortunately, it is impossible to completely avoid the use of named variables. The polynomial must be a polynomial in one variable, which we call C (which has attribute `Protected`, see Section 1.8 of the Symbolics volume [1795]). For the sake of efficiency, we alter the order of execution of the code. For a larger number of matrices, `Expand[Dot @@ ...]` is extremely slow. The reason for this is that elements of the matrix containing the variable e are not multiplied out, and at the end, we have to deal with four huge expressions using `Expand`. Here, we instead use `Fold[` `Expand[#1.#2]&, First[#], Rest[#]]&[...]`, only multiply two matrices with each other, and then use `Expand`. This approach results in a much faster code.

Here is a compact one-liner for `HofstadterButterfly`.

```
In[41]:= HofstadterButterfly[maxDenominator_Integer?(# > 1&)] :=
          Show[Graphics[{Thickness[0.002],
            Function[α, {Line[{{#[[1]], α}, {#[[2]], α}}],
                    Line[{{#[[1]], 1 - α}, {#[[2]], 1 - α}}]}]& /@
            Partition[Sort[Re[#]& /@ Select[Level[
```

```
            {NRoots[# == -4, C], NRoots[# == +4, C]}, {-1}], NumberQ]], 2]&[
            (#[[1, 1]] + #[[2, 2]])&[Fold[Expand[#1.#2]&, First[#], Rest[#]]&[
            Function[x, N[{{C - 2 Cos[2Pi x α - Pi/(2 Denominator[α])]], -1},
                        {1, 0}}], {Listable}][Range[0, Denominator[α] - 1]]]]]]] /@
        Union[Flatten[Range[#/2]/#& /@ Range[2, maxDenominator]]]}],
            PlotRange -> {{-4.12, 4.12}, {-0.04, 1.04}},
            PlotLabel -> StyleForm["Hofstadter's butterfly", FontWeight -> "Bold"],
            Axes -> False, Frame -> True, FrameLabel -> {"ε", "α"}]
```

Here is a more detailed version of the Hofstadter butterfly. (For higher values of *maxDenominator*, we might want to change the last code to use high-precision arithmetic to obtain numerically correct roots also for the higher-order polynomials.)

In[42]:= **HofstadterButterfly[32] // Timing**

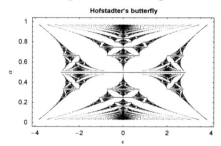

We could now go on and visualize, not the spectrum, but the gaps between the allowed bands. Coloring them "appropriately" (we do not discuss the details of the coloring here, see [1391], [97], [98], and [664] for details) yields another neat graphic. We modify the function Q[ε, α] to incorporate the above Fold-related remark and use high-precision arithmetic to reliably calculate (using NRoots again) the gaps for larger values of the denominators of α.

```
In[43]:= (* expand intermediate results and use high-precision arithmetic *)
        Q[ε_, α_] := (Fold[Expand[Dot[#1, #2]]&, {{1, 0}, {0, 1}},
        Table[N[A[ε, m, α, Pi/(2 Denominator[α])], 40],
                {m, 0, Denominator[α] - 1}]] // Tr) /. _?(# == 0&) -> 0

In[45]:= modInv[{p_, q_}, j_] :=
            Block[{k = Ceiling[-q/2]}, While[Mod[p k, q] =!= j, k++]; k]

In[46]:= coloredGaps[α_] :=
        Module[{p, q, s = Spectrum[α], gaps},
            (* numerator and denominator of α *)
            {p, q} = {Numerator[α], Denominator[α]};
            (* gaps between the spectrum *)
            gaps = Partition[Drop[Drop[Flatten[s], 1], -1], 2];
            (* color gaps; ignore zero gap case *)
            MapIndexed[If[EvenQ[q] && #2[[1]] === q/2, {},
                        {Hue[modInv[{p, q}, #2[[1]]]],
                         Line[{#, α}& /@ #]}]&, gaps]]

In[47]:= Module[{maxDenominator = 50, pqs, data},
            (* all α fractions *)
            pqs = Cases[Union[Flatten[Table[i/j,
                        {j, maxDenominator}, {i, 0, j}]]], _Rational];
            (* the colored gaps *)
            data = coloredGaps /@ pqs;
```

```
(* display colored gaps; use pseudorandom coloring *)
Show[Graphics[{Thickness[0.002], data} /.
                       Hue[k_] :> Hue[Sin[Log[3] k]]],
    Frame -> False, PlotRange -> All]];
```

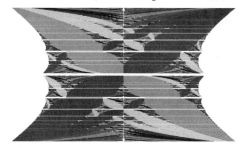

The shown fractal energy spectrum was for an infinitely extended system. By considering a finite system, we can see how the empty regions in the ϵ, α-plane arise. In a finite system, the density of the eigenvalues varies strongly with α and ϵ [1044], [51]. Here is an example shown. The function makeLhsMatrix forms the left-hand side of the above eigenvalue equation of a $n \times n$ lattice with Dirichlet boundary conditions on the four edges.

```
In[48]:= makeLhsMatrix[n_, α_] :=
    Module[{ψ, eqs, ψNumber, makeEntry, M},
      (* left-hand sides of the eigenvalue equation *)
      eqs = DeleteCases[Flatten[Table[ψ[x + 1, y] + ψ[x - 1, y] +
          Exp[-2 Pi I α x] ψ[x, y + 1] + Exp[+2 Pi I α x] ψ[x, y - 1],
              {x, 1, n - 1}, {y, 1, n - 1}] /.
                  (* Dirichlet boundary conditions *) (ψ[x_, y_] :> 0 /;
                      (Min[{x, y}] == 0 || Max[{x, y}] == n))], 0];
      (* convert double indexed ψs to single index *)
      MapIndexed[(ψNumber[#] = #2[[1]])&, Union[Cases[eqs, _ψ, Infinity]]];
      (* convert equation term to matrix entry *)
      makeEntry[j_, c_. Φ_ψ] := (M[[j, ψNumber[Φ]]] = M[[j, ψNumber[Φ]]] + c);
      makeEntry[j_, p_Plus] := makeEntry[j, #]& /@ (List @@ p);
      (* empty matrix *)
      M = Table[0, {(n - 1)^2}, {(n - 1)^2}];
      (* fill elements of matrix *)
      Do[makeEntry[j, eqs[[j]]], {j, (n - 1)^2}]; M]

In[49]:= With[{n = 14 (* n = 25 for high resolution *)},
    Module[{M, α},
    (* matrix to be diagonalized for various Φ *)
    Set @@ {M[α_], N[makeLhsMatrix[n, Φ]]};
    (* show graphic of eigenvalues *)
    Show[Graphics[{Thickness[0.002], PointSize[0.003],
        (* mirror at Φ = 1/2 *)
        {#, Apply[{#1, 1 - #2}&, #, {-2}]}&[Line /@ Transpose[
        Table[Re[{#, Φ}]& /@ Sort[Eigenvalues[M[Φ]]],
            {Φ, 0., 1/2., 1/2./(n - 1)^2}]]]} /.
                (* for better visibility on screen *)
                Line[l_] :> Point /@ l,
        Frame -> True, PlotRange -> All, FrameLabel -> {"ε", "α"}]]]
    (* for print:
       Show[% /. Point[p_] :> p, FrameTicks -> False] *)
```

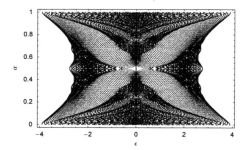

It would be interesting to continue our discussion (covering such topics as "smoothing" behavior between neighboring rational magnetic fields with very different denominators; experimental observations; generalizations; etc.), but this is a book on *Mathematica* and not one on crystals in magnetic fields. For details and more examples of butterflies, see Hofstadter's original paper as well as [681], [977], [134], [1078], [1898], [1503], [679], [1061], [133], [748], [53], [795], [135], [1967], [1446], [674], [158], [1076], [1363], [1080], [976], [1813], [1604], [801], [99], [159], [1849], [1079], and references therein. For a neat experimental version of the Hofstadter butterfly, see [1075]. For the energy spectrum of finite crystals, see [1848].

We consider another system that can be calculated through the roots of univariate polynomials. The following graphic shows the energy spectrum of a Sierpinski triangle in a magnetic field [1163], [1142]. This time, we do not solve algebraic equations explicitly, but use `Eigenvalues` to diagonalize a matrix (we could have done this also in the above Hofstadter butterfly example).

```
In[51]:= With[{o = 4, pp = 1000},
    Module[{d = 3/2 (3^o + 1), SierpinskiTriangle, SierpinskiSegments,
            vertices, vertexNumber, numberedSegments, M, z},
     (* the line segments of a Sierpinski triangle of order o *)
     SierpinskiSegments =
     Sort /@ Flatten[Partition[Append[#, #[[1]]], 2, 1]& /@
      Nest[Flatten[Map[(# /. {a_, b_, c_} ->
       {{2a, a + b, a + c}, {a + b, 2b, b + c}, {a + c, b + c, 2c}}/2)&,
                    #], 1]&, {{{-1, 0}, {1, 0}, {0, Sqrt[3]}}}, o], 1];
     (* the line segments of a Sierpinski triangle of order o *)
     vertices = Union[Level[SierpinskiSegments, {2}]];
     (* number vertices *)
     Do[vertexNumber[vertices[[k]]] = k, {k, d}];
     numberedSegments = Map[vertexNumber, SierpinskiSegments, {2}];
     (* form matrix of connected vertices; z measured flux *)
     M = Table[0, {d}, {d}];
     Do[{i, j} = numberedSegments[[k]];
        M[[i, j]] = z; M[[j, i]] = Conjugate[z], {k, d}];
     (* show spectrum as a function of z *)
     Show[Graphics[{PointSize[0.003],
         Table[Point[{φ, Re[#]}]& /@ Eigenvalues[M /. z -> N[Exp[I φ]]],
             {φ, 0, 2Pi, 2Pi/pp}] // N}],
         PlotRange -> All, Frame -> True]]]
```

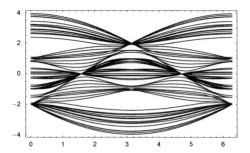

Note that the above-developed technique for visualizing the energy spectrum of an electron in a magnetic field can be easily adapted to use similar problems of the energy spectrum of quasi-periodic systems [1559]. Here, the energy spectrum of the so-called Kohmoto model [157], [1338], [1023] is shown.

```
In[52]:= KohmotoHamiltonianSpectrum[V_: -1, maxDenominator_Integer?Positive] :=
    Show[Graphics[{Thickness[0.002],
    Function[α, (* make lines *)
      Line /@ Map[{#, α}&, Partition[
        (* solve for regions so that the absolute value of the trace
           of the product of all transfer matrices is less than 2 *)
        Sort[Cases[{NRoots[# == +2, C], NRoots[# == -2, C]},
              _?NumberQ, {-1}]], 2]&[
      (* take the trace *) #[[1, 1]] + #[[2, 2]]&[
        (* fast nesting of the matrix multiplication *)
        Fold[Expand[#1.#2]&, First[#], Rest[#]]&[
          (* the individual transfer matrix *)
          Array[{{C - V If[1 - α <= Mod[# α, 1] < 1, 1, 0], -1}, {1, 0}}&,
            Denominator[α], 0]]]], {-1}]] /@
        (* all fractions *) Union[Flatten[Range[# - 1]/#& /@
          Range[maxDenominator]]]}],
      (* set options for a nice graphic *)
      PlotRange -> All, Frame -> True, AspectRatio -> 1.2,
      RotateLabel -> False, FrameLabel -> {"ε", "α"}]

In[53]:= KohmotoHamiltonianSpectrum[17]
```

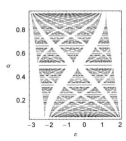

We now end our small physics examples sessions and continue with the solution of polynomial systems. NSolve also solves polynomial equations in several variables. (We use variables of the form x[i], which allows us to construct many equations in a unified way.) Here is a series of equations.

```
In[54]:= eqn[o_] := Table[Sum[(-1)^i x[i]^j, {i, o}] == j^2, {j, o}]
```

Here is an example for $o = 3$.

In[55]:= **eqn[3] // TableForm**

Out[55]//TableForm=

$$-x[1] + x[2] - x[3] == 1$$
$$-x[1]^2 + x[2]^2 - x[3]^2 == 4$$
$$-x[1]^3 + x[2]^3 - x[3]^3 == 9$$

Here are their numerical solutions. Each solution can be obtained from any other by permutations.

In[56]:= **NSolve[eqn[2], Table[x[i], {i, 2}]] // Timing**

Out[56]= $\{0.01 \text{ Second}, \{\{x[1] \to 1.5, x[2] \to 2.5\}\}\}$

In[57]:= **NSolve[eqn[3], Table[x[i], {i, 3}]] // Timing**

Out[57]= $\{0.02 \text{ Second},$
$\quad \{\{x[1] \to 0.533333 - 0.385861\,i, x[2] \to 2.06667, x[3] \to 0.533333 + 0.385861\,i\},$
$\quad \{x[1] \to 0.533333 + 0.385861\,i, x[2] \to 2.06667, x[3] \to 0.533333 - 0.385861\,i\}\}\}$

In[58]:= **NSolve[eqn[4], Table[x[i], {i, 4}]] // Timing**

Out[58]= $\{0.03 \text{ Second},$
$\quad \{\{x[1] \to 0.826923 + 0.793915\,i, x[2] \to 1.86801, x[3] \to 0.826923 - 0.793915\,i,$
$\quad x[4] \to 0.785835\}, \{x[1] \to 0.826923 - 0.793915\,i, x[2] \to 1.86801,$
$\quad x[3] \to 0.826923 + 0.793915\,i, x[4] \to 0.785835\},$
$\quad \{x[1] \to 0.826923 + 0.793915\,i, x[2] \to 0.785835, x[3] \to 0.826923 - 0.793915\,i,$
$\quad x[4] \to 1.86801\}, \{x[1] \to 0.826923 - 0.793915\,i,$
$\quad x[2] \to 0.785835, x[3] \to 0.826923 + 0.793915\,i, x[4] \to 1.86801\}\}\}$

The running time of NSolve typically increases quickly with the number of equations and their (total) degree.

In[59]:= **Table[{o, Timing[NSolve[eqn[o], Table[x[i], {i, o}]]][[1]]},**
$\quad\quad$ **{o, 7}]**

Out[59]= $\{\{1, 0.\text{ Second}\}, \{2, 0.01\text{ Second}\}, \{3, 0.02\text{ Second}\},$
$\quad \{4, 0.03\text{ Second}\}, \{5, 0.12\text{ Second}\}, \{6, 0.53\text{ Second}\}, \{7, 4.87\text{ Second}\}\}$

Remark: In the univariate polynomial case, the number of solutions is easy to determine—it is just the degree of the polynomial. In the multivariate case, the determination of the number of solutions is a nontrivial issue. The Bézout number (the product of the total degree of all equations) is an upper bound for the affine solutions that is seldom reached for "real-life" polynomial systems. For methods to determine the number of roots of polynomial systems, see [1709], [1547], [188], [189], [699], [715], [1691], [671], [859], [1546], [1479], [439], [1847], [986], and [1726]. (The corresponding question for real roots is more complicated; see [1251]). So the following system, which has a Bézout bound of 168, has only four solutions.

In[60]:= **NSolve[{x^6 y^8 - x^8 y^6 == 1, x^5 y^7 - x^7 y^5 == 1}, {x, y},**
$\quad\quad$ **WorkingPrecision -> 60] // Chop[#, 10^-50]&**

Out[60]= $\{\{x \to -1.272019649514068964252422461737491491715608041840096248616641\,i,$
$\quad y \to 0.786151377757423286069558585842958929523122057837723237664900\,i\},$
$\quad \{x \to 1.272019649514068964252422461737491491715608041840096248616641\,i,$
$\quad y \to -0.786151377757423286069558585842958929523122057837723237664900\,i\},$
$\quad \{x \to -0.786151377757423286069558585842958929523122057837723237664900,$
$\quad y \to -1.272019649514068964252422461737491491715608041840096248616641\},$
$\quad \{x \to 0.786151377757423286069558585842958929523122057837723237664900,$
$\quad y \to 1.272019649514068964252422461737491491715608041840096248616641\}\}$

Here is a more complicated polynomial system in three variables. Given the following implicitly defined surface f (from [140]), we will visualize the position of its singular points, where all of its partial derivatives vanish.

```
In[61]:= f = With[{φ = GoldenRatio},
           4 (φ^2 x^2 - y^2) (φ^2 y^2 - z^2) (φ^2 z^2 - x^2) -
           (1 + 2φ) (x^2 + y^2 + z^2 - 1)^2];
```

Here is a sketch of the surface defined implicitly by $f = 0$.

```
In[62]:= Needs["Graphics`ContourPlot3D`"]

         Show[Graphics3D[{EdgeForm[], (* reflect polygons in other octants *)
           {#, Map[{1, 1, -1}#&, #, {-2}]}&[
           {#, Map[{1, -1, 1}#&, #, {-2}]}&[{#, Map[{-1, 1, 1}#&, #, {-2}]}&[
           Select[(* contour plot inside one octant *)
             ContourPlot3D[Evaluate[f], {x, 0, 2}, {y, 0, 2}, {z, 0, 2},
               Contours -> {0}, MaxRecursion -> 1, PlotPoints -> {30, 4},
               DisplayFunction -> Identity][[1]],
                     (Max[#.#& /@ First[#]] < 5)&]]]]}], Boxed -> False]
```

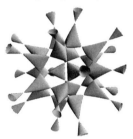

This is the resulting system of equations for the singular points.

```
In[64]:= (eqs = {f, D[f, x], D[f, y], D[f, z]}) // TraditionalForm
Out[64]//TraditionalForm=
```

$$\{4\,(\phi^2\,x^2 - y^2)\,(\phi^2\,y^2 - z^2)\,(\phi^2\,z^2 - x^2) - (1 + 2\,\phi)\,(x^2 + y^2 + z^2 - 1)^2,$$
$$-8\,x\,(\phi^2\,x^2 - y^2)\,(\phi^2\,y^2 - z^2) + 8\,\phi^2\,x\,(\phi^2\,z^2 - x^2)\,(\phi^2\,y^2 - z^2) - 4\,(1 + 2\,\phi)\,x\,(x^2 + y^2 + z^2 - 1),$$
$$-4\,(1 + 2\,\phi)\,y\,(x^2 + y^2 + z^2 - 1) + 8\,\phi^2\,y\,(\phi^2\,x^2 - y^2)\,(\phi^2\,z^2 - x^2) - 8\,y\,(\phi^2\,y^2 - z^2)\,(\phi^2\,z^2 - x^2),$$
$$8\,\phi^2\,(\phi^2\,x^2 - y^2)\,z\,(\phi^2\,y^2 - z^2) - 4\,(1 + 2\,\phi)\,z\,(x^2 + y^2 + z^2 - 1) - 8\,(\phi^2\,x^2 - y^2)\,z\,(\phi^2\,z^2 - x^2)\}$$

In using NSolve with polynomial systems, it is important to use a sufficiently large working precision. Here, 20 digits are enough.

```
In[65]:= nsol = {x, y, z} /. NSolve[eqs, {x, y, z}, WorkingPrecision -> 30];
```

Mathematica's high-precision arithmetic is quite fast and using a working precision of a few hundred digits typically does not constitute a time problem. In general, if multiplicity is of importance, symbolical solution techniques are preferable.

We found 50 solutions, and all of them are real.

```
In[66]:= {Length[nsol], Union[Im /@ Flatten[nsol]]}
Out[66]= {50, {0}}
```

Backsubstituting the solutions into the original equations gives zeros with accuracies of about 27.

```
In[67]:= {Min[#], Max[#]}&[
            Accuracy /@ (eqs /. (Thread[{x, y, z} -> #]& /@ nsol))]
Out[67]= {26.75, 28.1689}
```

Now, we display the position of the singularities. They have icosahedral symmetry. For easier viewing, we do not just display the points, but join nearest neighbors by a line.

```
In[68]:= Show[Graphics3D[{{Thickness[0.005], Hue[Random[]], #}& /@
         (Map[First, Split[(* groups of lines of equal length *)
          Sort[Flatten[
          Table[{Line[{nsol[[k]], First[#]}], Last[#]}& /@
                First[Split[(* select shortest lines *)
                Sort[(* sort with respect to length *)
                     {#, #.#&[nsol[[k]] - #1]}& /@ Delete[nsol, k],
                     ((#.#&[nsol[[k]] - #1[[2]]]) <
                     (#.#&[nsol[[k]] - #2[[2]]]))&], #1[[2]] == #2[[2]]&]],
                {k, Length[nsol]}], 1], (* equal length *)
                      #1[[2]] < #2[[2]]&], #1[[2]] == #2[[2]]&], {2}])}],
                   PlotRange -> All, Boxed -> False]
```

For other low-degree other algebraic surfaces with many singular points, see [1088].

Let us use NSolve to calculate the quadrature abscissas and weights for five points in the interval $[0, 1]$. $\int_0^1 f(x)\, dx \approx \sum_{k=1}^n w_k\, f(x_k)$. By demanding that the quadrature is exact for polynomials up to degree $2\,n-1$ [557], [1112], [1111], we get the following system of nonlinear equations for the w_k and x_k. We consider the case $n = 5$. (The following is not the most efficient method to calculate quadrature weights.)

```
In[69]:= n = 5;
         eqs = Table[Sum[w[i] x[i]^k - Integrate[x^k, {x, 0, 1}],
                     {i, n}], {k, 0, 2n - 1}]
```

$$
\text{Out[70]=}\ \Big\{-5 + w[1] + w[2] + w[3] + w[4] + w[5],
$$
$$
-\frac{5}{2} + w[1]\, x[1] + w[2]\, x[2] + w[3]\, x[3] + w[4]\, x[4] + w[5]\, x[5],
$$
$$
-\frac{5}{3} + w[1]\, x[1]^2 + w[2]\, x[2]^2 + w[3]\, x[3]^2 + w[4]\, x[4]^2 + w[5]\, x[5]^2,
$$
$$
-\frac{5}{4} + w[1]\, x[1]^3 + w[2]\, x[2]^3 + w[3]\, x[3]^3 + w[4]\, x[4]^3 + w[5]\, x[5]^3,
$$
$$
-1 + w[1]\, x[1]^4 + w[2]\, x[2]^4 + w[3]\, x[3]^4 + w[4]\, x[4]^4 + w[5]\, x[5]^4,
$$
$$
-\frac{5}{6} + w[1]\, x[1]^5 + w[2]\, x[2]^5 + w[3]\, x[3]^5 + w[4]\, x[4]^5 + w[5]\, x[5]^5,
$$
$$
-\frac{5}{7} + w[1]\, x[1]^6 + w[2]\, x[2]^6 + w[3]\, x[3]^6 + w[4]\, x[4]^6 + w[5]\, x[5]^6,
$$
$$
-\frac{5}{8} + w[1]\, x[1]^7 + w[2]\, x[2]^7 + w[3]\, x[3]^7 + w[4]\, x[4]^7 + w[5]\, x[5]^7,
$$
$$
-\frac{5}{9} + w[1]\, x[1]^8 + w[2]\, x[2]^8 + w[3]\, x[3]^8 + w[4]\, x[4]^8 + w[5]\, x[5]^8,
$$
$$
-\frac{1}{2} + w[1]\, x[1]^9 + w[2]\, x[2]^9 + w[3]\, x[3]^9 + w[4]\, x[4]^9 + w[5]\, x[5]^9\Big\}
$$

NSolve is able to solve this system of equations.

```
In[71]:= nsol = NSolve[eqs, Flatten[Table[{w[i], x[i]}, {i, n}]]];
```

Because of the permutation invariance of `eqs`, we have 120 solutions.

```
In[72]:= Length[nsol]
Out[72]= 120
```

Here is the first solution.

```
In[73]:= nsol[[1]] // Short[#, 4]&
Out[73]//Short=
```

$\{w[1] \to$
    1.4222222222222222222222222222222222222222222222222222222222222222222222222222222222 ⋮
    2222222222222222222222222, ≪8≫, x[5] →
    0.2307653449471584544818427896498955975163566965472200218988841864702644073 ⋮
    61223544820981663747145345$\}$

We now end the discussion of purely polynomial systems. Above, we stated that `NSolve` will solve polynomial systems. Actually, it solves more. `NSolve` has special algorithms for the numerical solution of polynomial systems and else it will try to call `Solve` to solve a set of equations and then numericalize the symbolic solution. In the next input `NSolve` solves a transcendental equation.

```
In[74]:= NSolve[x Exp[x] == 2, x]
```

InverseFunction::ifun : Inverse functions are
    being used. Values may be lost for multivalued inverses. More…

Solve::ifun : Inverse functions are being used by Solve, so some solutions
    may not be found; use Reduce for complete solution information. More…

```
Out[74]= {{x → 0.852606}}
```

For comparison, here is the exact solution obtained by `Solve` and its numerical value. (The function `Product ⋮ Log` will be discussed in Chapter 3 of the Symbolics volume [1795].)

```
In[75]:= {#, N @ #}&[Solve[x Exp[x] == 2, x]]
```

InverseFunction::ifun : Inverse functions are
    being used. Values may be lost for multivalued inverses. More…

Solve::ifun : Inverse functions are being used by Solve, so some solutions
    may not be found; use Reduce for complete solution information. More…

```
Out[75]= {{{x → ProductLog[2]}}, {{x → 0.852606}}}
```

Next, we examine the function `FindRoot`, the third important command in this subsection. Its main use is for nonpolynomial functions in which one knows an approximate solution of the equation(s). For an overview on numerical root finding, see [37].

FindRoot[*equations*, *variablesAndStartValues*, *options*]

finds the numerical solution of the equations *equations* in the variables *variablesAndStartValues*, starting with the given *startValue*s and using the options *options*.

If it makes sense (from the point of view of efficiency and doability) to compute derivatives of the functions, the input of *variablesAndStartValues* should be in the form $\{\{x, x_{start}\}, \{y, y_{start}\}, \ldots\}$. If not, they should be in the form $\{\{x, x_{start1}, x_{start2}\}, \{y, y_{start1}, y_{start2}\}, \ldots\}$.

To constrain the search to a given search interval, one can use

$\{\{x, x_{start}, x_{min}, x_{max}\}, \{y, y_{start}, y_{min}, y_{max}\}, \ldots\}$

or respectively:

$\{\{x, x_{start1}, x_{start2}, x_{min}, x_{max}\}, \{y, y_{start1}, y_{start2}, y_{min}, y_{max}\}, \ldots\}$.

We make a short general remark about `Find*`-named functions. Here are all functions matching such a name.

```
In[76]:= Names["Find*"]
```

```
Out[76]= {Find, FindFit, FindInstance, FindList,
          FindMaximum, FindMinimum, FindRoot, FindSettings}
```

The numerical functions that start with `Find` will try to "find" values satisfying given criteria. Typically, it is not possible to predict which of the possible values will be found and sometimes the search might not even succeed in finding any value.

If real starting values are specified, `FindRoot` tries to get a real solution; if this does not seem possible, it tries to bracket a complex solution.

Here is an example where the derivatives can and should be computed (internally) by `FindRoot`.

```
In[77]:= FindRoot[Exp[x] == 5 x, {x, 1.0}]
```

```
Out[77]= {x → 0.259171}
```

Here, the computation of the derivative is not immediately possible.

```
In[78]:= piecewiseFunc[x_?(# >= 0&)] := x^2 + 1;
```

```
        piecewiseFunc[x_?(# < 0&)] := x + 1;
```

In view of `PatternTest` for which the symbolic variables fail, `fr` cannot be explicitly differentiated.

```
In[80]:= piecewiseFunc'[x]
```

```
Out[80]= piecewiseFunc′[x]
```

As a consequence, `FindRoot` must use a finite difference approximation of the gradient.

```
In[81]:= FindRoot[piecewiseFunc[x] == 0, {x, 0}]
```

```
Out[81]= {x → -1.}
```

Inputting two start values, and using a secant-type search method, is in most cases a preferable input for such cases.

```
In[82]:= FindRoot[piecewiseFunc[x] == 0, {x, 0.5, 1.5}]
```

```
Out[82]= {x → -1.}
```

We illustrate the situation in which we wish to constrain ourselves to a given search interval. We define a function f that is only defined for $3.1 < x < 4$.

```
In[83]:= (* outside the "allowed" interval *)
        f2[x_?(# < 3.1&)] := (Print[x]; noNumber);
        f2[x_?(# > 4.0&)] := (Print[x]; againNoNumber);
        (* inside the "allowed" interval *)
        f2[x_?(((# >= 3.1) && (# <= 4))&)] :=
          23922.0000 - 31484.7777 x + 16246.7592 x^2 -
          4087.96296 x^3 + 497.685185 x^4 - 23.1481481 x^5;
```

Here is a graph of this function using `Plot`. This leads to multiple printouts of the *x*-values, because the function is sampled at points where this printout happens according to the definition of `f`.

```
In[88]:= Plot[f2[x], {x, 3.00, 4.1},
          Compiled -> False, AxesOrigin -> {3, 0},
          Epilog -> {GrayLevel[1/2], Line[{{3, 1}, {4, 1}}]}]
          3.
```

```
Plot::plnr :
 f2[x] is not a machine-size real number at x = 3.000000045833333`. More…

3.04462

Plot::plnr :
 f2[x] is not a machine-size real number at x = 3.0446236907302073`. More…

3.09329

Plot::plnr :
 f2[x] is not a machine-size real number at x = 3.093289679845311`. More…

General::stop :
 Further output of Plot::plnr will be suppressed during this calculation. More…

3.09928

4.00698

4.00154

4.00031

4.05054

4.09694

4.1
```

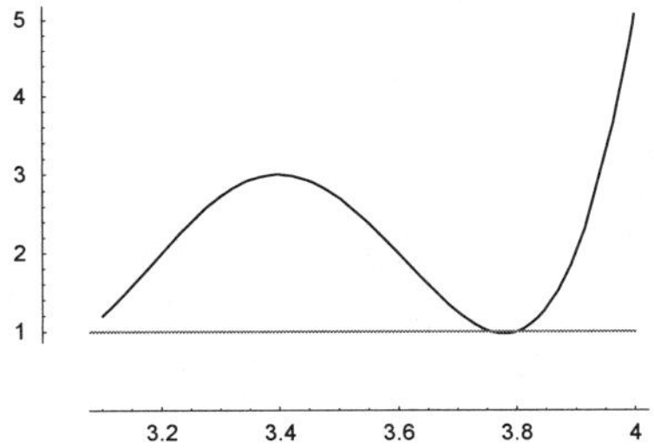

For this function and poor starting values, FindRoot is not able to find a solution, but it does not leave the search interval. With the following starting values, everything still works (but note that FindRoot always finds only one of the possible solutions).

In[89]:= **FindRoot[f2[x] == 1, {x, 3.9, 3.91, 3.1, 4}]**

Out[89]= $\{x \rightarrow 3.80016\}$

In the following example with starting values 3.2 and 4.0, no solution is found, and as soon as we get a value of the independent variable lying outside the search interval, the computation stops, without ever evaluating the function there (the Print[x] in the definition of f does not execute). The two starting values to the left of the fill avoid that the solution to the right of the hill is found.

In[90]:= **FindRoot[f2[x] == 1, {x, 3.2, 3.3, 3.1, 4}]**

FindRoot::reged :
 The point {3.1} is at the edge of the search region {3.1, 4.} in coordinate
  1 and the computed search direction points outside the region. More…

Out[90]= $\{x \rightarrow 3.1\}$

As already mentioned, `FindRoot` sometimes fails with poor starting values and without using its options. Starting with $x = 1$, `FindRoot` finds a solution in the following example.

```
In[91]:= FindRoot[x^3 + Sin[x] + x^x == 1, {x, 1}]
Out[91]= {x → 0.281509}
```

However, starting with $x = 100$, it does not.

```
In[92]:= FindRoot[x^3 + Sin[x] + x^x == 1, {x, 100}]
        FindRoot::cvmit : Failed to converge to the
            requested accuracy or precision within 100 iterations. More…
Out[92]= {x → 81.8491}
```

This illustrates a basic characteristic for `FindRoot`.

> `FindRoot` essentially searches (in a smart way) for the solution along the path with the largest gradient in the direction of the previously computed value, and thus it may not find any solution in the given interval, even if one exists. In such cases, the start values should be changed.

We now look at an example in two variables. For a better visualization of the solution, we plot the contours $x^3 - y^3 = 2$, $x^2 + y = 3$ and a contour plot of the sum of the deviations of the left-hand side of the equation from the right-hand side.

```
In[93]:= (* the bag *) Λ = {};

        (* Λ keeps track of the points used;
           could use EvaluationMonitor here *)
        compl[x_?NumberQ, y_?NumberQ] := (AppendTo[Λ, {x, y}]; x^3 - y^3)
        (* use default option settings *)
        FindRoot[{compl[x, y] == 2, x^2 + y == 3}, {x, 5}, {y, 10}];
        (* common options for c1 and c2 *)
        opts = Sequence[ContourShading -> False, DisplayFunction -> Identity,
                    ContourStyle -> {{GrayLevel[0], Thickness[0.01]}}];
        (* the contour of x^2 + y == 3 *)
        c1 = Graphics[ContourPlot[x^2 + y, {x, 0, 5}, {y, 0, 10},
                            Contours -> {3}, Evaluate @ opts]];
        (* the contour of x^3 - y^3 == 2 *)
        c2 = Graphics[ContourPlot[x^3 - y^3, {x, 0, 5}, {y, 0, 10},
                            Contours -> {2}, Evaluate @ opts]];
        (* ContourPlot of "how well are equations fulfilled" *)
        c3 = Graphics[ContourPlot[Abs[x^3 - y^3 - 2]/30 + Abs[x^2 + y - 3],
                        {x, 0, 5}, {y, 0, 10},
                        Contours -> 20, DisplayFunction -> Identity,
                        ColorFunction -> (GrayLevel[(# + 2)/3]&)]];
        (* showing all together *)
        Show[{c3, c1, c2, Graphics[{Thickness[0.01],
            (* the search path *)
            Thread[{Table[Hue[0.8 i /Length[Λ]], {i, 0, Length[Λ] - 2}],
            (* the search path *) Line /@ Partition[Λ, 2, 1]}]}]},
            Frame -> True, DisplayFunction -> $DisplayFunction]
```

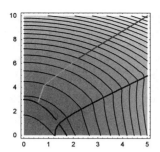

In order to be able to intervene "by hand" in the search for a zero (with the desired accuracy), there are 10 options for FindRoot. We see the general options of a numeric functions AccuracyGoal, Precision-Goal, WorkingPrecision, Compiled, and EvaluationMonitor, the Method option and the options specific to FindRoot: DampingFactor, MaxIterations, and Jacobian.

In[108]:= **Options[FindRoot]**

Out[108]= {AccuracyGoal → Automatic, Compiled → True,
    DampingFactor → 1, EvaluationMonitor → None, Jacobian → Automatic,
    MaxIterations → 100, Method → Automatic, PrecisionGoal → Automatic,
    StepMonitor → None, WorkingPrecision → MachinePrecision}

The options AccuracyGoal and PrecisionGoal refer to the values of the independent variables. But sometimes, one needs the equations to be fulfilled to a certain accuracy. In this case, one can use a two-element list as the AccuracyGoal option setting. The first element will refer to the accuracy of the independent variables and the second argument will refer to the residue of the equations. In the next input, we use an explicit setting of AccuracyGoal and PrecisionGoal.

In[109]:= **f[x_] := Exp[x] + Sin[x] + x^3 - Log[x] - 8;**

sol = FindRoot[f[x] == 0, {x, 4}, AccuracyGoal -> 6, PrecisionGoal -> 6];

{InputForm[sol], f[x] /. sol}
Out[111]= {{x -> 1.4566822931188068}, 6.66134 × 10^{-15}}

By manipulating the option WorkingPrecision (which we have already discussed above in connection with NIntegrate), we can get a 25-digit accuracy.

In[112]:= **sol = FindRoot[f[x] == 0, {x, 4}, PrecisionGoal -> 25,**
                                WorkingPrecision -> 30];
    {InputForm[sol], f[x] /. sol // FullForm}
Out[113]= {{x -> 1.45668229311880621350294798240393396588926093585`30.0000000000000004,
    0``28.777532965763374}

In the following example, we specify the residue of the equation to have accuracy 10.

In[114]:= **{#, 10^50 Sin[x] /. #}& @**
    FindRoot[10^50 Sin[x] == 0, {x, 3}, PrecisionGoal -> 5,
            AccuracyGoal -> {Infinity, 10}, WorkingPrecision -> 100]
Out[114]= {{x ->
        3.14159265358979323846264338327950288419716939937510582097494459230781640625
        8620899862803465000544296}, 1.75336674105 × 10^{-38}}

We make one remark about the solution of equations to very high precision. The WorkingPrecision option setting determines the precision to be used for the function evaluations. Sometimes high precision function evaluation is needed to obtain reliable function values everywhere and sometimes a high precision is (only) needed to obtain a high-precision root. In the latter case, it is sometimes advantageous to raise the actual precision used in the calculations as one approaches the root. While there is no explicit option present to influence if adaptive precision or working precision should be used, the precision technique to use is influenced by the precision of the starting values. In the following input, we look for the largest positive root of $H_{100}(x) = 0$. $H_n(x)$ is the Hermite polynomial of order n (to be discussed in the Symbolics volume [1795]). For $n = 100$, this is a polynomial of degree 100 with coefficients with up to 98 digits. We use exact, high-precision, and machine precision starting values for the Newton root finding procedure. As a side effect, we collect the x-values at which the Hermite polynomial was evaluated.

```
In[115]:= {reapWP, reapWPHP, reapAdaptive} =
      Reap[Timing[FindRoot[HermiteH[200, x] == 0, {x, #},
                  WorkingPrecision -> 500,  PrecisionGoal -> 10,
                  EvaluationMonitor :> Sow[x]]]]& /@
                     (* three starting values of different precision *)
                                           {25, 25`500, 25.};
```

The calculation times show that the adaptive precision raising technique was much faster and obtained a as good result as the calculation that used the 500-digit working precision for all function evaluations. The actual number of function evaluations is nearly identical for the two cases.

```
In[116]:= (* timing, root, number of function evaluations,
           and minimal precision of x-values *)
      {N[#1, 10], Length[#2[[1]]], Min[Precision[#2]]}& @@@
                                {reapWP, reapWPHP, reapAdaptive}
Out[117]= {{{1.47 Second, {x → 19.33924867}}, 88, 500.},
          {{1.47 Second, {x → 19.33924867}}, 88, 500.},
          {{0.38 Second, {x → 19.33924867}}, 91, MachinePrecision}}
```

Generically, we can specify the accuracy goal option through `AccuracyGoal -> {`*rootAG*`,` *functionValueAG*`}`, and the precision goal option through `PrecisionGoal -> `*rootPG* (because the function value should assume the value 0, no precision goal can be meaningfully specified for the function value). In this case, `FindRoot` attempts to fulfill both, the conditions on the root and the condition on the function value, meaning $\|root\text{-}estimatedExactRoot\| < \max(10^{-rootAG}, \|root\| \, 10^{-rootPG})$ and $\|functionValue\| < \|10^{-functionValueAG}\|$ are attempted to be fulfilled.

Now let us look at the `Method` option, an option shared by many numerical functions. The most important methods are the following.

```
Method
     is an option for FindRoot that determines the method used for the root-finding process.
  Default:
       Automatic

  Admissible:
       Newton

       or

       Secant

       or

       Brent
```

The `Newton` method requires the calculation of second derivatives and one starting value (per problem dimension) is sufficient. The `Secant` method needs two starting values. The `Brent` method is used for 1D problems with two initial points bracketing the root.

Here is a very simple example of a 1D problem with a complex root. Both, the `Newton` and the `Secant` method give the same root in approximately the same time, with the `Newton` method being faster. (We repeat the calculation to obtain more reliable timings.)

```
In[118]:= Do[frN = FindRoot[Sin[x] == 4, {x, 2 + I}, Method -> Newton],
         {10^3}] // Timing
Out[118]= {1.19 Second, Null}
```

```
In[119]:= Do[frS = FindRoot[Sin[x] == 4, {x, 2 + I, 3 + I}, Method -> Secant],
         {10^3}] // Timing
Out[119]= {1.57 Second, Null}
```

Both methods obtained reliable solutions.

```
In[120]:= (1 - #[[1, 2]]/Conjugate[ArcSin[4]]) & /@ {frN, frS}
Out[120]= {0. + 3.88578×10^-16 i, 0. + 0. i}
```

The next option to be discussed is `MaxIterations`.

`MaxIterations`

    is an option for `FindRoot`, and it sets the maximum number of steps allowed.

    Default:

        15

    Admissible:

        positiveInteger

We return once more to the above example in which no solution could be found. Using the option `MaxIterations`, we can get a solution even with poor starting values.

```
In[121]:= FindRoot[x^3 + Sin[x] + x^x == 1, {x, 10}, MaxIterations -> 50]
Out[121]= {x → 0.281509}
```

In order to use Newton's method [651], we need the Jacobian matrix. We can either input it explicitly, or leave its calculation to *Mathematica*.

```
Jacobian
```
is an option for `FindRoot`, and it determines the Jacobian matrix needed for Newton's method.

Default:
```
Automatic
```

Admissible:

ownJacobiMatrix

or

`"Symbolic"` (to carry out differentiation symbolically)

or

`"FiniteDifference"` (to carry out differentiation numerically)

Typically, we should input the Jacobian matrix explicitly in the following two instances: 1) When we know it, in a much simpler form than *Mathematica* would compute it, or *Mathematica* cannot find it at all and we know the derivatives needed or sufficiently good approximations of it. 2) When `FindRoot` will be called several times for the same function but with different parameters or different starting values (this can save a significant amount of computational time, because in this case the Jacobian matrix is computed only once and compiled).

Here is an example where the function and its derivative are computed and compiled 10 times.

```
In[122]:= Table[FindRoot[(2 + x^2 - x)/(3x^2 + 5x + 6) == i, {x, 1/2}],
            {i, 0.15, 0.20, 10^-4}]; // Timing
Out[122]= {0.88 Second, Null}
```

Here, the function is compiled only once, and its derivative is computed just once. We see a clear speed advantage.

```
In[123]:= Clear[f, fN, jacN, x];

        (* make function definition *)
        f = Compile[x, (2 + x^2 - x)/(3x^2 + 5x + 6)];
        fN[x_?NumberQ] := f[x];

        (* make Jacobian definition *)
        jacN = Compile[x, Evaluate[{{D[(2 + x^2 - x)/
                                    (3x^2 + 5x + 6), x] // Simplify}}]];

        Table[FindRoot[fN[x] == i, {x, 1/2}, Jacobian :> jacN[x]],
            {i, 0.15, 0.20, 10^-4}]; // Timing
Out[129]= {0.59 Second, Null}
```

Next, we discuss an option that is very important, especially for zeros of multiplicity larger than 1.

```
DampingFactor
```
is an option for `FindRoot`, and it defines the damping factor δ for the iteration. For the Newton method, for example, $x_{n+1} = x_n - \delta \frac{f(x_n)}{f'(x_n)}$.

Default:

    1

Possible:

    arbitraryNumber or *listOfArbitraryNumbers*; in the case of a list, the damping factors are used one after the other for each iteration

The damping factor should always be chosen (if known) to be the multiplicity of the desired zero, because this guarantees good convergence.

Here is a function with a triple zero. (To recognize that it is a triple zero, the reader should think of the Taylor series of $\sin(x) - x$ about $x = 0$.) Using a damping factor equal to 1 (default), we get poor convergence.

In[130]:= `FindRoot[Sin[x] - x == 0, {x, 0.2}, AccuracyGoal -> 12]`

Out[130]= $\{x \to 1.60912 \times 10^{-8}\}$

If we set `DampingFactor -> 3`, we again get good convergence and need fewer iterations.

In[131]:= `FindRoot[Sin[x] - x == 0, {x, 0.2},`
                `AccuracyGoal -> 12, DampingFactor -> 3]`

Out[131]= $\{x \to 8.84853 \times 10^{-13}\}$

We can demonstrate the optimality of the damping factor 3 for the last example by counting the steps necessary to reach the solution as a function of the damping factor setting. Because for values of the damping factor that are substantially different from 3 the convergence of the algorithm will be poor, corresponding messages will be issued. We suppress these messages in the following because we are only interested in the number of function evaluations. We see the optimality of the damping factor 3 very pronounced for the `Newton` method and less pronounced for the `Secant` method.

In[132]:=
```
Show[GraphicsArray[
   Block[{$DisplayFunction = Identity},
   ListPlot[
    Function[dp, Module[{l = {}, f, jac},
    (* define function *)
    f[x_?NumberQ] := (AppendTo[l, x]; Sin[x] - x);
    (* counting the iterations needed *)
    Internal`DeactivateMessages[
     FindRoot[f[x] == 0, {x, ##2},
            AccuracyGoal -> 12, DampingFactor -> dp,
            MaxIterations -> 100, Method -> #1];
            {dp, Length[l]}]]] /@
                    (* the damping factors to be tested *)
                  Table[i, {i, 1/4, 23/4, 1/8}],
            AxesLabel -> {"damping", "iterations"}]& @@@
         (* use Newton and Secant method option setting *)
         {{Newton, 0.2}, {Secant, -0.2, 0.2}}]]]
```

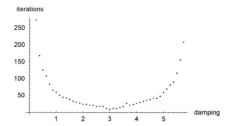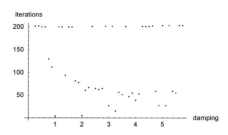

The following example shows the influence of the `DampingFactor` option on the number of steps and the quality of the solution for a function with a zero of very high multiplicity. `poly` is the expanded version of $(x + 1)^{12}$.

```
In[133]:= poly2[x_?NumberQ] := (i = i + 1;
    1 + 12 x + 66 x^2 + 220 x^3 + 495 x^4 + 792 x^5 + 924 x^6 +
    792 x^7 + 495 x^8 + 220 x^9 + 66 x^10 + 12 x^11 + x^12)
```

```
In[134]:= tab = Internal`DeactivateMessages[Table[i = 0;
        {dp, (* the root for a given damping factor *)
        x /. FindRoot[poly2[x], {x, 1, 2},
        DampingFactor -> dp, MaxIterations -> 300,
        WorkingPrecision -> 30], i}, {dp, 1, 22, 1/5}]];
```

We clearly see the optimality of the damping factor setting 12 in the next picture. The right graphic shows number of iterations as a function of the damping factor.

```
In[135]:= Show[GraphicsArray[
    Block[{$DisplayFunction = Identity},
        {ListPlot[{#[[1]], Log[10, Abs[#[[2]] + 1]]}& /@ tab,
            PlotJoined -> True, PlotRange -> All],
        ListPlot[{#[[1]], #[[3]]}& /@ tab,
            PlotJoined -> True, PlotRange -> All]}]]]
```

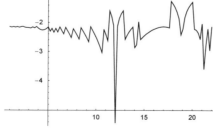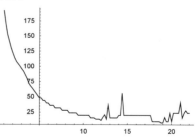

Now let us deal with a zero of fractional order. The root obtained using the default `DampingFactor` setting is quite good. But the number of function evaluations needed to obtain the result is about an order of magnitude more than with a tailored value for the `DampingFactor` option value.

```
In[136]:= FindRoot[x^(1/3) == 0, {x, 4}, AccuracyGoal -> 6, DampingFactor -> 1]
```

$Out[136]= \{x \to 4.52217 \times 10^{-16} - 7.3188 \times 10^{-30} \, i\}$

```
In[137]:= ({#[[1]], Length[#[[2, 1]]]}& @
    Reap[FindRoot[Sign[x] Abs[x]^(1/Pi) == 0, {x, 4},
        AccuracyGoal -> 6, DampingFactor -> #,
        EvaluationMonitor :> Sow[x]])& /@ {1, 1/Pi}
```

Out[137]= {{{x → -2.45707×10⁻¹⁶}, 95}, {{x → -1.76333×10⁻¹⁶}, 9}}

More details about the suboptions of the `Newton` and `Secant` method could be added here. We refer the reader to the Advanced Documentation within the help browser.

While it is often convenient to specify the variables to be solved for explicitly, there is no need to specify them. The equation(s) can be specified as a pure function and the list of variables and starting values can have the variables missing. Of course, in these cases the returned result cannot be of the form of lists of rules, the values instead are returned only. Here are two simple examples.

```
In[138]:= FindRoot[{Sin[#1] - #1, Cos[#2] - #2}&, {3}, {2},
              WorkingPrecision -> 10]
Out[138]= {0.00002550946441, 0.7390851332}
```

```
In[139]:= FindRoot[{Sin[#1] - #1, Cos[#2] - #2}&, {3, 4}, {2, 3},
              WorkingPrecision -> 10]
Out[139]= {0.04156354857, 0.7390851331}
```

We end with two applications of numerical root finding.

We now come back to the Voderberg polygons discussed in Chapter 1 of the Graphics volume [1794]. We were interested in the fact that these polygons could be nested. In this case, we are interested in using them to construct a spiral-shaped, area-filling covering of the plane. We do not present the construction rule, but instead build suitable polygons corresponding to such a spiral. To do this, we have to slightly modify those in Chapter 1 of the Graphics volume and then numerically solve a nonlinear system of equations in two variables. Suppose we are given a line ABCDEF with inversion symmetry of the following form:

By a clockwise rotation of this line by an integer part of 2π around the point F, the point B should be mapped to the point A. In terms of equations, we need the following: Let ϑ be the angle that the line segment $\overline{BC}$ makes with the positive x-axis, let φ be the angle between the line segment $\overline{BA}$ and the x-axis, and let e be the angle of rotation. Without loss of generality, we may choose the coordinates of the point C to be $\{-1.65, 0\}$. We denote the points A, B, … in *Mathematica* using small letters.

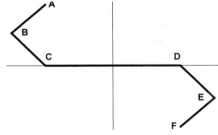

```
In[140]:= Clear[a, b, c, d, e, f, φ, ϑ, ω, ℛ, pa, pb, pc, pd, pe, pf]

        c = {-1.65, 0};
        b = c + {Cos[ϑ], Sin[ϑ]};
        a = b + {Cos[φ], Sin[φ]};
        f = -a;
        ℛ = {{Cos[ω], Sin[ω]}, {-Sin[ω], Cos[ω]}};
```

Then, the following equations must hold.

```
In[146]:= equations = f + ℛ.(b - f) == a // Thread
```

Out[146]= $\{1.65 - \text{Cos}[\vartheta] - \text{Cos}[\varphi] + (-3.3 + 2\,\text{Cos}[\vartheta] + \text{Cos}[\varphi])\,\text{Cos}[\omega] + (2\,\text{Sin}[\vartheta] + \text{Sin}[\varphi])\,\text{Sin}[\omega] ==$
$-1.65 + \text{Cos}[\vartheta] + \text{Cos}[\varphi], \ -\text{Sin}[\vartheta] - \text{Sin}[\varphi] + \text{Cos}[\omega]\,(2\,\text{Sin}[\vartheta] + \text{Sin}[\varphi]) -$
$(-3.3 + 2\,\text{Cos}[\vartheta] + \text{Cos}[\varphi])\,\text{Sin}[\omega] == \text{Sin}[\vartheta] + \text{Sin}[\varphi]\}$

We numerically solve this system of equations using `FindRoot` for one choice of rotation angle.

In[147]:= (* possible rotation angle *) **ω = Pi/15;**

      numSol = FindRoot[Evaluate[equations], {ϑ, 140 Degree}, {φ, 70 Degree},
                    AccuracyGoal -> 10]

Out[148]= $\{\vartheta \to 2.42932, \ \varphi \to 1.00021\}$

This leads to the following explicit points.

In[149]:= **{pa, pb, pc} = {a, b, c} /. numSol;**
     {pd, pe, pf} = -{pc, pb, pa};
     {pes, pds, pcs} = (pf + R.(# - pf))& /@ {pe, pd, pc};

The resulting Voderberg polygon has the following shape.

In[152]:= **VP = Polygon[{pf, pe, pd, pc, pb, pa, pcs, pds, pes}];**

     Show[Graphics[VP], PlotRange -> All, AspectRatio -> Automatic, Axes -> True]

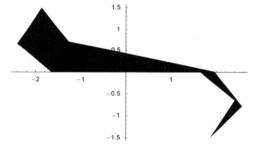

Now, we turn to the spirals. We only graph them; for details on their construction, and why the *i*th Voderberg polygon has to be positioned in this way, see [1858] and [1859]. In order to get various Voderberg polygons in the correct position, we make use of parts of the routine `PolygonPut` (here named `PP`) from Section 1.1.4 of the Graphics volume [1794].

In[154]:= **Module[{PP** (* short for PolygonPut *)**, s, t, f},**
     (* how to fit two polygons together *)
     PP[poly_Polygon, we_List, wo_List] :=
     Module[{p1, p2, p1s, p2s, eqn, trafo, c, s, tx, ty},
          p1 = poly[[1, we[[1]]]]; p2 = poly[[1, we[[2]]]];
          p1s = poly[[1, wo[[1]]]]; p2s = poly[[1, wo[[2]]]];
          eqn = Flatten[Thread /@ ({{c, s}, {-s, c}}.#1 +
               {tx, ty} == #2)& @@@ {{p1, p1s}, {p2, p2s}}];
          trafo = ({{{c, s}, {-s, c}}, {tx, ty}} /.
               Solve[eqn, {c, s, tx, ty}])[[1]];
        Polygon[(trafo[[1]].# + trafo[[2]])& /@ poly[[1]]]];
     s[1, 1] = VP;
     (* the rules how to build the spiral; nontrivial to calculate *)
     Do[s[1, i] = PP[s[1, i - 1], {4, 5}, {7, 6}], {i, 2, 16}];
     Do[s[1, i] = PP[s[1, i - 1], {6, 7}, {9, 8}];
       s[1, i + 1] = PP[s[1, i], {5, 6}, {2, 1}];
       s[1, i + 2] = PP[s[1, i + 1], {3, 4}, {8, 7}], {i, 17, 59, 3}];

```
   Do[s[1, i] = PP[s[1, i - 1], {6, 7}, {9, 8}];
      s[1, i + 1] = PP[s[1, i], {5, 6}, {2, 1}];
      s[1, i + 2] = PP[s[1, i + 1], {3, 4}, {8, 7}];
      s[1, i + 3] = PP[s[1, i + 2], {6, 7}, {9, 8}];
      s[1, i + 4] = PP[s[1, i + 3], {1, 2}, {6, 5}], {i, 59, 120, 5}];
      s[2, 1] = VP;
      s[2, 2] = PP[s[2, 1], {5, 6}, {2, 1}];
   Do[s[2, i] = PP[s[2, i - 1], {4, 5}, {7, 6}], {i, 3, 17}];
   Do[s[2, i] = PP[s[2, i - 1], {6, 7}, {9, 8}];
      s[2, i + 1] = PP[s[2, i], {5, 6}, {2, 1}];
      s[2, i + 2] = PP[s[2, i + 1], {1, 2}, {1, 9}], {i, 18, 60, 3}];
   Do[s[2, i] = PP[s[2, i - 1], {6, 7}, {9, 8}];
      s[2, i + 1] = PP[s[2, i], {5, 6}, {2, 1}];
      s[2, i + 2] = PP[s[2, i + 1], {1, 2}, {1, 9}];
      s[2, i + 3] = PP[s[2, i + 2], {7, 8}, {8, 7}];
      s[2, i + 4] = PP[s[2, i + 3], {1, 2}, {6, 5}], {i, 60, 120, 5}];
   (* the picture *)
   Show[Graphics[{ (* making nonagons alternating black and white *)
      MapIndexed[If[EvenQ[#2[[1]]], {GrayLevel[0], #1}, {GrayLevel[1], #1}]&,
                 Table[s[1, i], {i, 1, 120}]],
      MapIndexed[If[EvenQ[#2[[1]]], {GrayLevel[1], #1}, {GrayLevel[0], #1}]&,
                 Table[s[2, i], {i, 2, 120}]]}],
      AspectRatio -> Automatic, PlotRange -> All]]
```

For similar spirals made from polygonal tiles, see [1717], [709].

We end with a multivariate example: a numerical root search for a polynomial system of 15 (16) unknowns. While we could use NSolve to obtain all solutions for such a system, we will use FindRoot here because we are only interested in one solution and we know good starting values for the numerical root finding procedure. Given a circle Circle[{xM, yM}, r] and three points at {$x1$, $y1$}, {$x2$, $y2$}, and {$x3$, $y3$} on its circumference, the function newThreeCirclesAndTouchingPoints calculates three new circles inside the first that are tangential to the original circle and to each other. (The last equation containing the auxiliary variable ζ ensures that each of the three new circles is distinct from the original one.)

```
In[155]:= newThreeCirclesAndTouchingPoints[{Circle[{xM_, yM_}, r_],
          {Point[{x1_, y1_}], Point[{x2_, y2_}], Point[{x3_, y3_}]}},
          opts___] :=
        {{Circle[{xM1, yM1}, r1], Circle[{xM2, yM2}, r2],
          Circle[{xM3, yM3}, r3]},
         {Point[{xP12, yP12}], Point[{xP13, yP13}],
          Point[{xP23, yP23}]}} /.
        FindRoot[SetPrecision[{
         (* the system of equations for the arrangement of circles *)
           (* outer points lie on the three new circles *)
```

```
(x1 - xM1)^2 + (y1 - yM1)^2 - r1^2 == 0,
(x2 - xM2)^2 + (y2 - yM2)^2 - r2^2 == 0,
(x3 - xM3)^2 + (y3 - yM3)^2 - r3^2 == 0,
(* touching points lie on the three new circles  *)
(xP12 - xM1)^2 + (yP12 - yM1)^2 - r1^2 == 0,
(xP13 - xM3)^2 + (yP13 - yM3)^2 - r3^2 == 0,
(xP23 - xM2)^2 + (yP23 - yM2)^2 - r2^2 == 0,
(* center distances of the three new circles *)
(xM2 - xM1)^2 + (yM2 - yM1)^2 - (r1 + r2)^2 == 0,
(xM3 - xM1)^2 + (yM3 - yM1)^2 - (r1 + r3)^2 == 0,
(xM3 - xM2)^2 + (yM3 - yM2)^2 - (r2 + r3)^2 == 0,
(* three new circles are tangential to outer circle *)
(x1 - xM)(y1 - yM1) - (x1 - xM1)(y1 - yM) == 0,
(x2 - xM)(y2 - yM2) - (x2 - xM2)(y2 - yM) == 0,
(x3 - xM)(y3 - yM3) - (x3 - xM3)(y3 - yM) == 0,
(* three new circles are tangential at touching points *)
(xP12 - xM1)(yP12 - yM2) - (xP12 - xM2)(yP12 - yM1) == 0,
(xP13 - xM1)(yP13 - yM3) - (xP13 - xM3)(yP13 - yM1) == 0,
(xP23 - xM2)(yP23 - yM3) - (xP23 - xM3)(yP23 - yM2) == 0,
(* each of the three new circles is different from original *)
1 - ζ (r - r1) (r - r2) (r - r3) == 0}, 50],
  (* approximate starting values for
     circle centers, radii, and touching points *)
  {xM1, (x1 + xM)/2}, {xM2, (x2 + xM)/2}, {xM3, (x3 + xM)/2},
  {yM1, (y1 + yM)/2}, {yM2, (y2 + yM)/2}, {yM3, (y3 + yM)/2},
  {xP12, (x1 + x2)/2}, {xP13, (x1 + x3)/2}, {xP23, (x2 + x3)/2},
  {yP12, (y1 + y2)/2}, {yP13, (y1 + y3)/2}, {yP23, (y2 + y3)/2},
  {r1, r/2}, {r2, r/2}, {r3, r/2}, {ζ, 1}, opts]
```

Iterating the process of forming three new circles by choosing as the three new points two of the new touching points and the original point on the outer circle yields the function newThreeCirclesAndPoints.

```
In[156]:= (* project a point p onto the circle c *)
         projectPoint[c:Circle[mp_, r_], p_] := mp + r #/Sqrt[#.#]&[p - mp]
```

```
In[158]:= newThreeCirclesAndPoints[data: {c_, {p1_, p2_, p3_}}, opts___] :=
         Module[{c1, c2, c3, p12, p13, p23, p112, p113, p212, p223, p313, p323},
           (* calculate three new circles *)
           {{c1, c2, c3}, {p12, p13, p23}} =
                       newThreeCirclesAndTouchingPoints[data, opts];
           (* project touching points onto the new circles *)
           {p1, p112, p113} = Point[projectPoint[c1, #[[1]]]]& /@ {p1, p12, p13};
           {p2, p212, p223} = Point[projectPoint[c2, #[[1]]]]& /@ {p2, p12, p23};
           {p3, p313, p323} = Point[projectPoint[c3, #[[1]]]]& /@ {p3, p13, p23};
           (* three new circle become outer circles with given points *)
           {{c1, {p1, p112, p113}}, {c2, {p2, p212, p223}}, {c3, {p3, p313, p323}}}]
```

Iterating this process six times yields a nested arrangement of 1093 circles. Here they are calculated and visualized for two sets of initial touching points. (We use the "Newton" method because for this low-order system of polynomial equations the symbolic calculation of the Jacobian is very easy.)

```
In[159]:= nestedCircles = Function[iPs,
           (* iterate process six times *)
             NestList[Flatten[newThreeCirclesAndPoints[#,
               AccuracyGoal -> {120, 8}, PrecisionGoal -> 10,
               (* get precise solutions *)
               Method -> {"Newton", "StepControl" -> None},
```

```
        MaxIterations -> 100, WorkingPrecision -> 50]& /@ #, 1]&,
              {{Circle[{0, 0}, 1], Point /@ iPs}}, 6]] /@
    {(* regular triangle vertices *)
    {{0, 2}, {Sqrt[3], -1}, {-Sqrt[3], -1}}/2,
    (* isosceles triangle vertices *)
    {{1, 0}, {-1, 0}, {0, 1}}}];
```

In[160]:= **Show[GraphicsArray[**
    Graphics[(* color circles of each generation differently *)
                Transpose[{Hue /@ Take[{0, 0.3, 0.7, 0.84, 0.5,
                                        0.17, 0.92, 0.9, 0.2}, Length[#]],
                            DeleteCases[#, _Point, Infinity]}] /.
                Circle -> Disk,
        AspectRatio -> Automatic, PlotRange -> All,
        Frame -> True, FrameTicks -> None]& /@ nestedCircles]]

To complete this section, we mention two very useful packages.

■ The package `NumericalMath`Bessel`` implements a method for numerically computing the zeros of Bessel functions, their derivatives, and their linear combinations.

$$J_\nu(x), \; Y_\nu(x), \; J'_\nu(x), \; Y'_\nu(x),$$
$$J'_\nu(x)\, Y'_\nu(\lambda x) - J'_\nu(x)\, Y'_\nu(\lambda\, x) \text{ and } J'_\nu(x)\, Y_\nu(\lambda\, x) - J_\nu(x)\, Y'_\nu(\lambda\, x)$$

These zeros are frequently needed because they are the eigenvalues of 1D Sturm–Liouville differential operators.

■ The package `NumericalMath`InterpolateRoot``, like `FindRoot`, is designed to find the zeros of functions of one variable. It differs from `FindRoot` in that in searching for a zero, it makes use of all previously computed function values. This method usually needs fewer function calls than does `FindRoot`, but with some loss of stability. It is particularly useful for very high-precision calculations.

1.10 Minimization

The problem of finding a minimum or maximum of a function is closely related to the solution of nonlinear systems of equations. However, if the function to be minimized is not straightforwardly differentiable, the search for a minimum cannot be reduced to the solution of a system of equations.

The `FindMinimum` command is available in *Mathematica* for finding a single local minima.

FindMinimum[*toMinimize*, *variablesAndStartValues*, *options*]

> finds the minimum of the function *toMinimize*, as well as the values of the independent variables for which this minimum is assumed under the options *options*. The search for a local minimum begins at the prescribed start values from *variablesAndStartValues*.
>
> If it makes sense (and is possible) to compute derivatives of the functions, we input *variablesAndStartValues* in the form $\{\{x, x_{start}\}, \{y, y_{start}\}, \ldots\}$. If not, they should be input in the form
>
> $\{\{x, x_{start1}, x_{start2}\}, \{y, y_{start1}, y_{start2}\}, \ldots\}$.
>
> To constrain the search to a given search interval, we write
>
> $\{\{x, x_{start}, x_{min}, x_{max}\}, \{y, y_{start}, y_{min}, y_{max}\}, \ldots\}$ or
>
> $\{\{x, x_{start1}, x_{start2}, x_{min}, x_{max}\}, \{y, y_{start1}, y_{start2}, y_{min}, y_{max}\}, \ldots\}$.

There is also a *Mathematica* command for searching for a local maximum, called FindMaximum. Its syntax, options, and behavior is completely analogous to the one of FindMinimum. In the following, we will only use the function FindMinimum. The "Find" in FindMinimum indicates again that an attempt is made to find one minimum, but not all and not always the same.

The search for a minimum proceeds basically along a path of steepest descent. We begin with a simple example.

```
In[1]:= FindMinimum[x^2, {x, 0, 2}]
```
```
Out[1]= {0., {x → 0.}}
```

The minimum of x^2 is 0, and it occurs at $x = 0$. The function $|z|^2$ of a complex variable z cannot be analytically differentiated. In this case, *Mathematica* resorts to finite difference approximations for the derivatives.

```
In[2]:= FindMinimum[Abs[x + I y]^2, {x, 1}, {y, 1}]
```
```
Out[2]= {6.16298 × 10^{-33}, {x → 5.55112 × 10^{-17}, y → 5.55112 × 10^{-17}}}
```

One can also use two starting values.

```
In[3]:= FindMinimum[Abs[x + I y]^2, {x, 1, 2}, {y, 1, 1/2}]
```
```
Out[3]= {7.7309 × 10^{-37}, {x → 6.21728 × 10^{-19}, y → 6.21727 × 10^{-19}}}
```

Here is a piecewise linear function that is unbounded as $x \to \infty$. It has a minimum at $x = -1$. FindMinimum has no problems finding the unique minimum of this function.

```
In[4]:= fP[x_] := Piecewise[{{-x - 1, x < -1}, {1 + x, -1 < x < 0},
                            {1 - x, x > 0}}];
```
```
      FindMinimum[fP[x], {x, -Pi, E}]
```
```
Out[5]= {0., {x → -1.}}
```

The start value and the corresponding minimal value for FindMinimum must be real.

```
In[6]:= FindMinimum[Im[y^2], {y, I}]
```
```
            FindMinimum::fmns : Starting value {0. + 1. i} contains numbers which are not real.
```
```
Out[6]= FindMinimum[Im[y^2], {y, i}]
```

And *Mathematica* will search for the minimum only for real variables (there are no imaginary minima). (This is different from FindRoot's behavior. FindRoot will also search for complex roots. But it is straightforward to use the real and imaginary part as independent variables.)

The start value for FindMinimum should not be a maximum or a saddle point.

```
In[7]:= FindMinimum[Cos[x], {x, 0}]
```

```
FindMinimum::fmgz :
  Encountered a gradient which is effectively zero. The result returned
    may not be a minimum; it may be a maximum or a saddle point. More…
```

Out[7]= {1., {x → 0.}}

We can explicitly constrain the search domain. In the following example, no *x*-values outside [0,1] will be evaluated.

```
In[8]:= {Hold[#1], (* range of x-values *) {Min[#], Max[#]}&[#2]}& @@
          (hh = Reap[FindMinimum[Sqrt[x], {x, 0.1, 0.2, 0, 1},
                      EvaluationMonitor :> Sow[x]]])
```

```
FindMinimum::regex :
  Reached the point {-0.0618034} which is outside the region {{0.}, {1.}}. More…
FindMinimum::regex :
  Reached the point {-0.0618034} which is outside the region {{0.}, {1.}}. More…
FindMinimum::regex :
  Reached the point {-0.0618034} which is outside the region {{0.}, {1.}}. More…
General::stop : Further output of
    FindMinimum::regex will be suppressed during this calculation. More…
```

Out[8]= {Hold[FindMinimum[$\sqrt{x}$, {x, 0.1, 0.2, 0, 1}, EvaluationMonitor:↦ Sow[x]]],
          {0., 0.1}}

These are the possible options for FindMinimum.

In[9]:= **Options[FindMinimum]**

Out[9]= {AccuracyGoal → Automatic, Compiled → True,
          EvaluationMonitor → None, Gradient → Automatic,
          MaxIterations → 100, Method → Automatic, PrecisionGoal → Automatic,
          StepMonitor → None, WorkingPrecision → MachinePrecision}

Except for Gradient, we are already familiar with all of these. MaxIterations is the maximum number of steps allowed in each dimension (variable). Similar to FindRoot, the options AccuracyGoal and Preci⌇ sionGoal refer to the values of the independent variables, not to the function value at the minimum. To specify the values for the minimum too, a pair of values should be used. The default values for AccuracyGoal and PrecisionGoal are WorkingPrecision/2. FindMinimum ends its search when the goals for AccuracyGoal or PrecisionGoal are achieved. So in the next input, the value of the minimum is found to accuracy 20.

```
In[10]:= FindMinimum[x^4, {x, 100}, WorkingPrecision -> 50,
                AccuracyGoal -> 20, PrecisionGoal -> Infinity]
```

Out[10]= {1.2567279271162179975925176204764865646016551711999×10^{-80},
          {x → 1.0587911840678754238354031258495524525642395019531×10^{-20}}}

Because the true minimum is at *x* = 0, the following precision goal is unfulfillable.

```
In[11]:= FindMinimum[x^4, {x, 100}, WorkingPrecision -> 50,
                AccuracyGoal -> Infinity, PrecisionGoal -> 20]
```

```
FindMinimum::cvmit : Failed to converge to the
    requested accuracy or precision within 100 iterations. More…
```

Out[11]= {3.8725919148493182728180306332863518475702191920488×10^{-113},
          {x → 7.8886090522101180541172865528278622967320643510902×10^{-29}}}

Allowing for more iterations (as the last issued message suggests) does, of course, not cure the problem. This time, FindMinimum recognizes that within the specified working precision it is not possible to obtain 20 correct digits for the position of the minimum. (Raising the value of the WorkingPrecision option will, of course, also not help.)

```
In[12]:= FindMinimum[x^4, {x, 100}, WorkingPrecision -> 100,
            AccuracyGoal -> Infinity, PrecisionGoal -> 20,
            MaxIterations -> Infinity]
```

FindMinimum::fmdig : 100.` working digits is
    insufficient to achieve the requested accuracy or precision. More...

Out[12]= {1.62766807356514044634164592696250600357261023059664190168538350604684934822 3 `
         84365474399245924553933×10^{-398}, {x →
         3.5718355977571093194243235884541904982872187338212458162972682600863519872 `
         08556861485775068313037593×10^{-100}}}

We now come to the discussion of the Method option and its possible settings. The main methods, applicable to most problems are the following.

Method

    is an option for FindMinimum that determines the minimization method.

    Default:

        Automatic, that is, "QuasiNewton" for one starting value per dimension and "Principal `
        Axis" else. For sums of squares, "LevenbergMarquardt" is used.

    Admissible:

        "Gradient" (a special version of the "ConjugateGradient" method)

        or

        "Newton"

        or

        "QuasiNewton"

        or

        "ConjugateGradient"

        or

        "PrincipalAxis"

        or

        "LevenbergMarquardt"

We compare the four applicable methods for a 1D problem with one starting value.

```
In[13]:= baseMethods = {"Gradient", "Newton", "QuasiNewton", "ConjugateGradient"};
```

We try to find the minimum of a single well function $f(x)$. We use 22 different starting values per method. For some starting values, the minimization will not converge picture-perfect and we potentially obtain some warning messages about this fact.

```
In[14]:= f[x_] := ArcTan[x - 5] + ArcTan[-x - 5]

    (* find minimum; monitor function evaluations *)
    data = Table[Reap[FindMinimum[f[x], {x, x0}, Method -> #,
                            EvaluationMonitor :> Sow[x]]],
                {x0, -5, 5, 10/21}]& /@ baseMethods;
```

```
FindMinimum::lstol :
 The line search decreased the step size to within tolerance specified
   by AccuracyGoal and PrecisionGoal but was unable to find a sufficient
   decrease in the function.  You may need more than MachinePrecision
   digits of working precision to meet these tolerances. More…

FindMinimum::lstol :
 The line search decreased the step size to within tolerance specified
   by AccuracyGoal and PrecisionGoal but was unable to find a sufficient
   decrease in the function.  You may need more than MachinePrecision
   digits of working precision to meet these tolerances. More…

FindMinimum::lstol :
 The line search decreased the step size to within tolerance specified
   by AccuracyGoal and PrecisionGoal but was unable to find a sufficient
   decrease in the function.  You may need more than MachinePrecision
   digits of working precision to meet these tolerances. More…

General::stop : Further output of
   FindMinimum::lstol will be suppressed during this calculation. More…
```

The 88 minima found agree with each other to about eight digits.

```
In[17]:= {Min[#], Max[#]}& @ Map[#[[1, 1]]&, data, {2}] // InputForm
Out[17]//InputForm=
       {-2.746801533890032, -2.7468015338900313}
```

Here are the average numbers of function evaluations per minimization. The `"Newton"` method was clearly the fastest.

```
In[18]:= Transpose[{baseMethods,
              Round[Length[Flatten[#]]/22]& /@ Map[#[[2]]&, data, {2}]}]
Out[18]= {{Gradient, 15}, {Newton, 6}, {QuasiNewton, 10}, {ConjugateGradient, 16}}
```

And here are the search paths. Each path is colored according to its starting value. We see how the `"Gradi`\`
`ent"` method stays near to the function and the conjugate gradient method has the largest step variations.

```
In[19]:= pathGraphic[mData_, label_] :=
   Module[{xValues, xyValues, coloredPaths, pl},
     (* make lines of search path *)
     xValues = #[[2, 1]]& /@ mData;
     xyValues = Map[{#, f[#]}&, xValues, {2}];
     coloredPaths = MapIndexed[{Hue[#2[[1]]/24], Line[#1]}&, xyValues];
     (* function whose minimum we look for *)
     pl = Plot[ArcTan[x - 5] + ArcTan[-x - 5], {x, -20, 20},
             DisplayFunction -> Identity];
     (* function and search paths *)
     Graphics[{{Thickness[0.006], GrayLevel[0], pl[[1, 1. 1]]},
             {Thickness[0.002], coloredPaths}},
             Frame -> True, PlotRange -> {{-20, 20}, {-3, 0}},
             PlotLabel -> label]]

In[20]:= Show[GraphicsArray[pathGraphic @@@ Transpose[{data, baseMethods}]]]
```

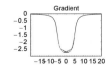 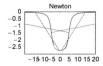 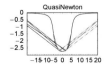 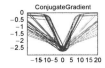

We repeat the last calculations with a function that is not symbolically accessible to `FindMinimum`.

```
In[21]:= (* no access to symbolic form of the function *)
        fN[x_?NumberQ] := ArcTan[x - 5] + ArcTan[-x - 5]

        (* find minimum; monitor function evaluations *)
        dataN = Table[Reap[FindMinimum[fN[x], {x, x0}, Method -> #,
                            EvaluationMonitor :> Sow[x]]],
                    {x0, -5, 5, 10/21}]& /@ baseMethods;
        FindMinimum::lstol :
          The line search decreased the step size to within tolerance specified
            by AccuracyGoal and PrecisionGoal but was unable to find a sufficient
            decrease in the function.  You may need more than MachinePrecision
            digits of working precision to meet these tolerances. More...
```

The results still have the same quality as above, but the number of function evaluations needed is now higher because derivatives must be formed in a numerical way through finite differences.

```
In[25]:= {Min[#], Max[#]}& @ Map[#[[1, 1]]&, dataN, {2}] // InputForm
Out[25]//InputForm=
        {-2.746801533890032, -2.7468015251313593}

In[26]:= Transpose[{baseMethods,
                    Round[Length[Flatten[#]]/22]& /@ Map[#[[2]]&, dataN, {2}]}]
Out[26]= {{Gradient, 37}, {Newton, 51}, {QuasiNewton, 23}, {ConjugateGradient, 37}}
```

Here are some rules of thumb for choosing one of the basic methods in FindMinimum.

"Newton"

    for analytically easily differentiable objective functions and/or when a high precision result is needed.

"QuasiNewton"

    for avoiding symbolic differentiation; when a high precision result is needed.

"Gradient"

    to follow the steepest descent of a function.

"LevenbergMarquardt"

    for sums of squares.

"ConjugateGradient"

    for high-dimensional problems.

"PrincipalAxis"

    for avoiding any (symbolic and numeric) derivative.

For a more detailed discussion of these methods, see [1354], [685], [867].

We now carry out a 2D problem. We search for the minimum of the following function $g(x, y)$ using the four methods from above.

```
In[27]:= g[x_, y_] := x^2 + y^4 + y^6/12

        (* find minimum; monitor function evaluations *)
        data2D = Reap[FindMinimum[g[x, y], {x, 1}, {y, 2}, Method -> #,
                            EvaluationMonitor :> Sow[{x, y}]]]& /@
                                            baseMethods;
```

The minimum found for all four methods is less than 10^{-12}.

```
In[30]:= {Min[#], Max[#]}& @ Map[#[[1, 1]]&, data2D, {1}]
Out[30]= {7.69295×10^-31, 7.42502×10^-13}
```

Here is the number of function evaluations needed for each method.

```
In[31]:= Transpose[{baseMethods,
                Round[Length[Flatten[#]]]& /@ Map[#[[2]]&, data2D, {1}]}]
Out[31]= {{Gradient, 154}, {Newton, 92}, {QuasiNewton, 134}, {ConjugateGradient, 66}}
```

We again visualize the search path. The function pathGraphic2D shows the search path as a white line on top of a contour plot of the function whose minimum we are looking for.

```
In[32]:= pathGraphic2D[mData_, g_, label_, opts___] :=
    Module[{xyValues, cp, cls},
      (* make lines of search path *)
      path = Line[mData[[2, 1]]];
      (* contour plot of the function whose minimum we look for;
         blue for low values, red for high values *)
      cp = ContourPlot[g[x, y], {x, -0.6, 1.2}, {y, -0.7, 2.1},
          ColorFunction -> (Hue[0.7 (1 - #)]&), PlotPoints -> 60,
          ContourStyle -> {Thickness[0.002]}, DisplayFunction -> Identity];
      (* homogeneously spaced contour levels *)
      cls = #[[90]]& /@ Partition[Sort[Flatten[cp[[1]]]], 180];
      (* function and search paths *)
      Show[{Show[cp, Contours -> cls],
          Graphics[{Thickness[0.006], GrayLevel[1], path}]},
          PlotRange -> {{-0.6, 1.2}, {-0.7, 2.1}}, opts, PlotLabel -> label]]
```

Here are the search paths for the four methods.

```
In[33]:= Show[GraphicsArray[pathGraphic2D[#1, g, #2]& @@@
                  Transpose[{data2D, baseMethods}]]]
```

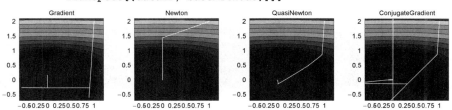

For the simple single-minimum function g from above, we can also search for the minimum iteratively in the horizontal and the vertical direction (not as a sensible minimization technique, but rather as some programming fun with minimization functions). The function nestedFindMinimum finds the minimum using *method2* in *x*-direction and *method1* in *y*-direction.

```
In[34]:= nestedFindMinimum[{method1_, method2_}] :=
    Module[{gy, aux},
      (* minimize with respect to x for fixed y *)
      gy[y_?NumberQ] := (aux1 = FindMinimum[x^2 + y^4 + y^6/12, {x, 1},
                              Method -> method1,
                              Gradient -> "FiniteDifference",
                              EvaluationMonitor :> Sow[{x, y}]])[[1]];
      (* form properly formed 2D result and return search path too *)
      MapAt[Join[aux, #]&,
```

```
         Reap[FindMinimum[gy[y], {y, 2}, Method -> method2,
                           Gradient -> "FiniteDifference"]], {{1, 2}}]]
```

In[35]:= `productMethods = Flatten[Outer[List, baseMethods, baseMethods], 1] ;`

In[36]:= `data2DT = {#, nestedFindMinimum[#]}& /@ productMethods;`

The minimal function values found are quite good.

In[37]:= `{Min[#], Max[#]}& @ Map[#[[1, 1]]&, Last /@ data2DT, {1}]`

Out[37]= $\{7.69744 \times 10^{-31}, 1.18119 \times 10^{-9}\}$

Of course, the number of function evaluation is now much larger.

In[38]:= `Transpose[{productMethods,`
        `Round[Length[Flatten[#]]]& /@ Map[#[[2]]&, data2DT, {1}]}]`

Out[38]= `{{{Gradient, Gradient}, 2162}, {{Gradient, Newton}, 12882},`
       `{{Gradient, QuasiNewton}, 7562}, {{Gradient, ConjugateGradient}, 2522},`
       `{{Newton, Gradient}, 2480}, {{Newton, Newton}, 12854},`
       `{{Newton, QuasiNewton}, 8186}, {{Newton, ConjugateGradient}, 2814},`
       `{{QuasiNewton, Gradient}, 974}, {{QuasiNewton, Newton}, 5798},`
       `{{QuasiNewton, QuasiNewton}, 3404}, {{QuasiNewton, ConjugateGradient}, 1136},`
       `{{ConjugateGradient, Gradient}, 1742}, {{ConjugateGradient, Newton}, 12614},`
       `{{ConjugateGradient, QuasiNewton}, 7382},`
       `{{ConjugateGradient, ConjugateGradient}, 1748}}`

And here are the 16 search paths used in these tensor product-type searches.

In[39]:= `Show[GraphicsArray[#]]& /@`
    `Partition[pathGraphic2D[#2, g, #1, FrameTicks -> None]& @@@`
               `(data2DT /. s_String :> StringTake[s, 1]), 4]`

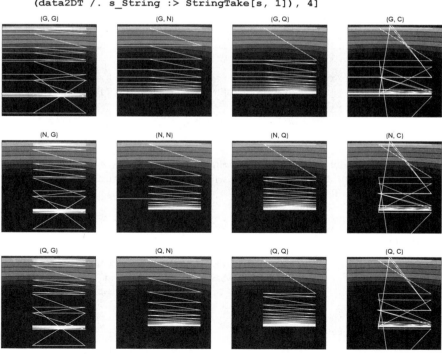

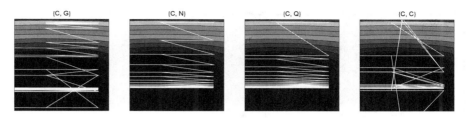

At this point, we should mention that the package `Optimization`UnconstrainedProblems`` has a (option-free) built-in function `FindMinimumPlot` that for 1D and 2D minimization problems visualizes the function and the search path, and returns information about the number of function evaluations and related data. The next inputs show this function at work at the last objective function $g(x, y)$ with the two options `"Leven⋮ bergMarquardt"` and `"PrincipalAxis"`.

In[40]:= `Needs["Optimization`UnconstrainedProblems`"]`

In[41]:= `Module[{fmData},`
`        fmData = Block[{$DisplayFunction = Identity},`
`        (* two calls to FindMinimumPlot for the two options *)`
`        {FindMinimumPlot[g[x, y], {{x, 1.}, {y, 2.}},`
`                        Method -> "LevenbergMarquardt"],`
`         FindMinimumPlot[g[x, y], {{x, 1., 1.001}, {y, 2., 2.001}},`
`                        Method -> "PrincipalAxis"]}];`
`        (* display minimum search path and return evaluation data *)`
`        Show[GraphicsArray[Last /@ fmData]]; Most /@ fmData]`

Out[41]= $\{\{\{1.21965 \times 10^{-31}, \{x \to 0., y \to 1.86878 \times 10^{-8}\}\},$
$\{Steps \to 27, Residual \to 28, Jacobian \to 28\}\},$
$\{\{1.94584 \times 10^{-22}, \{x \to -1.26978 \times 10^{-11}, y \to -2.4031 \times 10^{-6}\}\},$
$\{Steps \to 2, Function \to 96\}\}\}$

Now, let us discuss the gradient option. It is very similar to the `Jacobian` option of `FindRoot`.

Gradient

is an option for FindMinimum, and it defines the gradient to be used in carrying out the method.

Default:

Automatic

Admissible:

ownGradientVector

or

"Symbolic" (to carry out differentiation symbolically)

or

"FiniteDifference" (to carry out differentiation numerically)

When the function to be minimized is a sum of squares $\sum_{k=0}^{n} c_k f_k^2(parameter)$ (where the f_k do not have to all be of the same form and the c_k are numerical prefactors), one other Method option setting is available, the "LevenbergMarquardt" method.

Let us give a simple example that uses this option. Here are 1000 points scattered around a circle.

```
In[42]:= SeedRandom[999] ;

With[{δ = 0.1},
circleData = Table[
  Function[φ, {Cos[φ], Sin[φ]} +
            {Random[Real, {-δ, δ}], Random[Real, {-δ, δ}]}][
                      Random[Real, {0, 2Pi}]], {1000}]];
```

```
In[44]:= Show[Graphics[Point /@ circleData],
        PlotRange -> All, Frame -> True, AspectRatio -> Automatic]
```

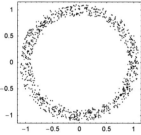

We want to find the circle, such that the sum of the squares of the distances of the points to this circle becomes minimal. The function distance measures the distance between a circle and a point [658], [1480], [1553], [1696], [1833].

```
In[45]:= distance[Circle[{mpx_, mpy_}, r_], point:{x_, y_}] :=
    With[{R = Sqrt[(mpx - x)^2 + (mpy - y)^2]}, Sqrt[(r - R)^2]]
```

These data yield the following function to be minimized.

```
In[46]:= toMinimize = Plus @@ (distance[Circle[{mpx, mpy}, r], #]^2& /@ circleData);
```

Now, we will compare the timings for various option settings of FindMinimum. While all option settings give a good result, the "LevenbergMarquardt" is clearly the fastest.

```
In[47]:= {#, Timing[FindMinimum[toMinimize, {mpx, 1}, {mpy, 1}, {r, 2},
                      Method -> #]]}& /@
            Append[{"Gradient", "Newton", "QuasiNewton", "ConjugateGradient"},
                "LevenbergMarquardt"]

     FindMinimum::lstol :
        The line search decreased the step size to within tolerance specified
          by AccuracyGoal and PrecisionGoal but was unable to find a sufficient
          decrease in the function.  You may need more than MachinePrecision
          digits of working precision to meet these tolerances. More…

Out[47]= {{Gradient, {1.82 Second,
        {3.57338, {mpx → -0.000104214, mpy → 0.000960085, r → 1.00334}}}}, {Newton,
        {4.37 Second, {3.57338, {mpx → -0.000104475, mpy → 0.000960514, r → 1.00334}}}},
      {QuasiNewton, {1.69 Second, {3.57338,
          {mpx → -0.000104475, mpy → 0.000960515, r → 1.00334}}}}, {ConjugateGradient,
        {1.9 Second, {3.57338, {mpx → -0.000104475, mpy → 0.000960515, r → 1.00334}}}},
      {LevenbergMarquardt, {1.44 Second,
        {3.57338, {mpx → -0.000104477, mpy → 0.000960511, r → 1.00334}}}}}
```

Here is the calculated circle together with the original points.

```
In[48]:= Show[Graphics[{{Hue[0], Thickness[0.005],
                      Circle[{mpx, mpy}, r] /. %[[-1, -1, -1, -1]]},
                  {GrayLevel[0], Point /@ circleData}}],
            PlotRange -> All, Frame -> True, AspectRatio -> Automatic]
```

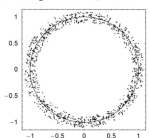

Now, let us tackle a slightly more challenging problem for the Levenberg–Marquardt method. Given the area among all triples of three points of the polyhedron, we will reconstruct the polyhedron. We will use a Platonic solid—the icosahedron. Here are the vertices of an ideal icosahedron.

```
In[49]:= Needs["Graphics`Polyhedra`"]
        vertices = Vertices[Icosahedron];
```

These are all 220 nontrivial triples of three points.

```
In[51]:= (triples =  Select[Union[
            Union /@ Flatten[Table[{i, j, k}, {i, 12}, {j, 12}, {k, 12}], 2]],
                      Length[#] === 3&]) // Length
Out[51]= 220
```

This is a list of the 36 coordinates to be determined.

```
In[52]:= Table[Vertices[i] = {x[i], y[i], z[i]}, {i, 12}];
```

Our input data are the values of the areas of sets of three points. We calculate them using the values of the vertices from the package Graphics`Polyhedra`.

```
In[53]:= areaSquare[{p1_, p2_, p3_}] := #.#& @ Cross[p1 - p2, p3 - p1]/4
```

```
In[54]:= eqs = Apply[
    (areaSquare[{Vertices[#1], Vertices[#2], Vertices[#3]}] -
     areaSquare[{vertices[[#1]], vertices[[#2]], vertices[[#3]]} // N])&,
              triples, {1}];
```

To avoid the continuous degrees of freedom caused by translational invariance, we fix the position of the first vertex.

```
In[55]:= toMinimize = Plus @@ (eqs^2 //. {x[1] -> 0, y[1] -> 0, z[1] -> 0});
```

Here is the last summand of the function to be minimized. It is a complicated function of the coordinates of the vertices.

```
In[56]:= toMinimize[[-1]]
```

$$
\text{Out[56]= } \Big(-0.437694 +
$$
$$
\frac{1}{4} \, ((x[11]\, y[10] - x[12]\, y[10] - x[10]\, y[11] + x[12]\, y[11] + x[10]\, y[12] - x[11]
$$
$$
y[12])^2 + (-x[11]\, z[10] + x[12]\, z[10] + x[10]\, z[11] - x[12]\, z[11] -
$$
$$
x[10]\, z[12] + x[11]\, z[12])^2 + (y[11]\, z[10] - y[12]\, z[10] -
$$
$$
y[10]\, z[11] + y[12]\, z[11] + y[10]\, z[12] - y[11]\, z[12])^2) \Big)^2
$$

vars are the 33 coordinates that remain to be determined.

```
In[57]:= vars = Flatten[Table[Vertices[i], {i, 2, 12}]];
```

Now, we search for the coordinates of the vertices. To find a good global maximum, we use an additional While loop and start the searches at randomly chosen starting values.

```
In[58]:= resData =
    ((* generate reproducible start data *) SeedRandom[3498];
    Module[{fm},
    While[fm = FindMinimum[Evaluate[toMinimize], Evaluate[Sequence @@
                     ({#, Random[Real, {-1, 1}]}& /@ vars)],
                   Method -> #, MaxIterations -> 60];
          Abs[fm[[1]]] > 10^-6, Null]; fm] // Timing)& /@
    {"LevenbergMarquardt","QuasiNewton", "ConjugateGradient"};
        FindMinimum::cvmit : Failed to converge to the
           requested accuracy or precision within 60 iterations. More…
```

```
In[59]:= {#1, #2[[1]]}& @@@ resData
Out[59]= {{1.26 Second, 3.11199×10^-29},
        {1.51 Second, 2.19734×10^-18}, {2.33 Second, 2.70019×10^-9}}
```

Modulo the orientation, we exactly recover an icosahedron for all three methods.

```
In[60]:= (* cut a hole in a polygon *)
    makeHole[Polygon[l_], f_] :=
    Module[{mp = Plus @@ l/Length[l], ℓ}, ℓ = (mp + f*(# - mp))& /@ l;
        {MapThread[Polygon[Join[#1, Reverse[#2]]]&,
           Partition[Append[#, First[#]], 2, 1]& /@ {l, ℓ}],
         Line[Append[#, First[#]]]&[ℓ]}]
```

```
In[62]:= Show[GraphicsArray[
    Graphics3D[makeHole[Polygon[#], 0.8]& /@ (
      Map[{x[#], y[#], z[#]}&, Faces[Icosahedron], {2}] /.
        Join[#[[2, 2]], {x[1] -> 0, y[1] -> 0, z[1] -> 0}])]& /@
        resData]]
```

But even for this function, we can find the minimum using appropriate option settings.

```
In[63]:= FindMinimum[f[x], {x, -1, 2}, MaxIterations -> 300,
                WorkingPrecision -> 100, AccuracyGoal -> 10]
Out[63]= {-2.746801533889991538035331110223294949479690953483660911296997772434239442647
        3824062806862134946132333, {x -> 1.64816×10⁻⁶}}
```

Sometimes FindMinimum might be very slow. Although the strategy of moving roughly in the direction of steepest descent is acceptable, it might be ineffective. We now give a somewhat unpleasant problem (unpleasant for FindMinimum): Let $f(x, y)$ be a spiral-shaped function defined on the x,y-plane that represents a "slide" of finite width with infinitely high walls that spirals down to zero. Suppose $f(x, y)$ is defined at every point in the x,y-plane. Here is a plot of such a function.

```
In[64]:= slideSlip[φ_, s_] = If[φ > 2Pi,
    {(φ + (s - 1) Pi) Cos[φ], (φ + (s - 1) Pi) Sin[φ], φ + Tan[Pi/2 s]^2},
    {(s + 1) φ/2 Cos[φ], (s + 1) φ/2 Sin[φ], φ + Tan[Pi/2 s]^2}];
    (* the Tan[pi/2 s]^2 makes the steep walls *)
```

We can easily check that the x,y-coordinate formula for the last function slideSlip is given by the following function slideSlipCartesian.

```
In[66]:= (* argument continuous in (0, 2Pi) instead of (-Pi, Pi) *)
    myArg[x_, y_] := If[# >= 0, #, 2Pi + #]&[Arg[x + I y]]

In[68]:= slideSlipCartesian[x_?NumberQ, y_?NumberQ] :=
    Module[{φrel = myArg[x, y], r = 1. Sqrt[x^2 + y^2], φn, φ, s},
    (* use polar coordinates *)
    φn = Quotient[r - φrel, 2Pi];
    (* make the spiral-like function *)
    Which[φn >= 0,
        φ = (φn + 1) 2Pi + φrel;
        s = (r - ((φn 2Pi) - Pi + φrel))/Pi,
        φn == -1,
        φ = φrel; s = (r - φ/2)/(φ/2)];
    φ + Tan[s Pi/2.]^2]
```

To understand this function better, we now present it as 3D graphic based on slideSlip and a gray-scale density plot based on slideSlipCartesian.

```
In[69]:= Show[GraphicsArray[
    Block[{$DisplayFunction = Identity},
     {(* use parametrization in polar coordinates *)
      ParametricPlot3D[Evaluate[slideSlip[φ, s]], {φ, 0, 6 Pi}, {s, -0.5, 0.5},
                Compiled -> False, PlotPoints -> {60, 5},
                ViewPoint -> {1.3, -1.6, 0.9}],
      (* use parametrization in Cartesian coordinates *)
      dp = DensityPlot[slideSlipCartesian[x, y], {x, -30, 30}, {y, -30, 30},
                PlotPoints -> 150, Mesh -> False]}]]]
```

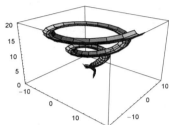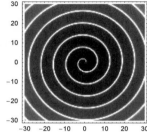

In terms of the parameters φ and s, the minimum is easy to find.

```
In[70]:= FindMinimum[Which[φ > 2Pi, φ + Tan[Pi/2 s]^2,
                 0 <= φ <= 2Pi, φ + Tan[Pi/2 s]^2,
                 φ < 0, 10^10], {φ, 12, 12.001}, {s, -0.001, 0.001}]
Out[70]= {4.48728×10⁻¹⁸, {φ → 4.48055×10⁻¹⁸, s → -5.22148×10⁻¹¹}}
```

Now, the search for a minimum is much more complicated. We have to raise the MaxIterations option value to catch the minimum. We monitor the steps and the function evaluations.

```
In[71]:= slideData = {#, Timing[Reap[
    FindMinimum[slideSlipCartesian[x, y], {x, 9Pi}, {y, 0.02},
                (* use higher value *) MaxIterations -> 1000,
                StepMonitor :> Sow[{x, y}, StepMonitor],
                EvaluationMonitor :> Sow[{x, y}, EvaluationMonitor],
                Method -> #], {StepMonitor, EvaluationMonitor}]]}& /@ baseMethods;
```

```
        FindMinimum::lstol :
         The line search decreased the step size to within tolerance specified
            by AccuracyGoal and PrecisionGoal but was unable to find a sufficient
            decrease in the function.  You may need more than MachinePrecision
            digits of working precision to meet these tolerances. More…

        FindMinimum::cvmit : Failed to converge to the
            requested accuracy or precision within 1000 iterations. More…

        FindMinimum::sdprec :
         Line search unable to find a sufficient decrease in the function
            value with MachinePrecision digit precision. More…
```

```
In[72]:= {#1, #2[[1]], #2[[2, 1]], Length[#2[[2, 2, 1]]]}& @@@ slideData
Out[72]= {{Gradient, 2.81 Second, {0.0000171746, {x → 8.5867×10⁻⁶, y → 1.47459×10⁻¹⁰}}, 1},
    {Newton, 2.38 Second, {6.13677×10⁻⁷, {x → 2.7487×10⁻⁷, y → 1.51083×10⁻¹³}}, 1},
    {QuasiNewton, 1.11 Second, {2.35549×10⁻⁶, {x → 1.17722×10⁻⁶, y → 2.77176×10⁻¹²}},
     1}, {ConjugateGradient, 1.33 Second,
     {0.000506108, {x → 0.000253067, y → 1.28072×10⁻⁷}}, 1}}
```

Note that for functions of several variables, `MaxIterations` does not control the maximum number of function calls, but instead controls the number of "effective calls per search direction". Here are the found minima and the number of steps and function evaluations for each of the methods.

```
In[73]:= {#1, #2[[1]], #2[[2, 1]],
       Length[#2[[2, 2, 1, 1]]], Length[#2[[2, 2, 2, 1]]]}& @@@ slideData
```

Out[73]= {{Gradient, 2.81 Second,
      {0.0000171746, {x → 8.5867×10⁻⁶, y → 1.47459×10⁻¹⁰}}, 363, 16588}, {Newton,
      2.38 Second, {6.13677×10⁻⁷, {x → 2.7487×10⁻⁷, y → 1.51083×10⁻¹³}}, 820, 13645},
      {QuasiNewton, 1.11 Second, {2.35549×10⁻⁶, {x → 1.17722×10⁻⁶, y → 2.77176×10⁻¹²}},
      1000, 5988}, {ConjugateGradient, 1.33 Second,
      {0.000506108, {x → 0.000253067, y → 1.28072×10⁻⁷}}, 501, 7580}}

Here is a plot of the search path. The global spiral-shaped search path is clearly visible on top of the repeated tangential search directions along the valley. The first row of pictures uses the recorded steps (from `StepMoni` `tor`) and the second row uses the recorded function evaluations (from `EvaluationMonitor`).

```
In[74]:= Function[{label, k},
     Show[GraphicsArray[
     Block[{$DisplayFunction = Identity},
          (* show density plot of surface and path *)
          Show[dp, Epilog -> {Hue[0], Line[#1[[2, 2, 2, k, 1]]]},
              PlotLabel -> #2]& @@@
              Transpose[{slideData, baseMethods}]],
          PlotLabel -> label]] @@@
     (* steps and function evaluations *)
     {{StepMonitor, 1}, {EvaluationMonitor, 2}}
```

So far, we have been discussing finding a local minimum. It would be possible to discuss many more details of the various methods, especially their suboptions. But we end here. See the Advanced Documentation for this purpose.

Sometimes one has to find global minima. This is typically a much more complicated problem. The *Mathematica* function that tries to find a global minimum numerically is `NMinimize`. (Similarly to `FindMaximum`, the function `NMaximize` tries to find the global maximum of a function.)

NMinimize[*toMinimize*, *variablesAndPotentialStartValues*, *options*]

    tries to find the global minimum of the function *toMinimize*, as well as the values of the independent variables for which this minimum is assumed under the options *options*. The first argument can be a list of the function to be minimized and the constraints on the independent variables.

In[75]:= **fOsc3D[x_, y_] := -4 Exp[(-x^2 - y^2)] + Sin[6 x] Sin[5 y]**

In[76]:= **Plot3D[fOsc3D[x, y], {x, -3, 3}, {y, 0, 3},**
        PlotPoints -> 120, PlotRange -> All, Mesh -> False]

A direct call with the default option settings and no starting region for the search yields a minimum, but not the global minimum.

In[77]:= **NMinimize[{fOsc3D[x, y], y > 0}, {x, y}]**

Out[77]= {-2.74362, {x → 0.228026, y → 0.799865}}

The currently implemented methods for finding the global minimum are "NelderMead", "Differential Evolution", "SimulatedAnnealing", and "RandomSearch". For the example function at hand, we will compare the four methods. The function fgm calls NMinimize with the method *method* and returns the found minimum as well as all the points used in the search.

In[78]:= **fgm[method_] := Reap[Timing @**
    NMinimize[{fOsc3D[x, y], y > 0}, {{x, -4, 4}, {y, 0, 4}},
        MaxIterations -> 1000,
        PrecisionGoal -> 10, Method -> method,
        StepMonitor :> Sow[{x, y}]]]

Here are the results for the four methods. We see that the "NelderMead" was used by the Automatic option setting. The method "DifferentialEvolution" and the method "RandomSearch" found the global minimum. In general, the method "DifferentialEvolution" will frequently find better minima but will typically be one of the slowest methods.

In[79]:= **data = {#, fgm[#]}& /@**
        {"NelderMead", "DifferentialEvolution",
        "SimulatedAnnealing", "RandomSearch", Automatic};

In[80]:= **{#1, #2[[1]]}& @@@ data**

Out[80]= {{NelderMead, {0.07 Second, {-1.00073, {x → 0.785366, y → 2.82727}}}},
    {DifferentialEvolution, {3.38 Second, {-4.50107, {x → -0.215018, y → 0.240356}}}},
    {SimulatedAnnealing, {0.05 Second, {-1., {x → 13.8754, y → 3.45575}}}},
    {RandomSearch, {10.85 Second, {-4.50107, {x → -0.215018, y → 0.240356}}}},
    {Automatic, {0.07 Second, {-1.00073, {x → 0.785366, y → 2.82727}}}}}

And here are the search paths for the four methods shown.

```
In[81]:= Module[{cp},
    (* contour plot of the objective function *)
    cp = Block[{$DisplayFunction = Identity},
        ContourPlot[fOsc3D[x, y], {x, -4, 4}, {y, 0, 4},
                PlotPoints -> 240, PlotRange -> All,
                ContourLines -> False]];
    Show[GraphicsArray[
            (* show contour plot and search path *)
            Show[{cp, Graphics[{Hue[0], Thickness[0.005], Line[#[[2, 2, 1]]]}]},
                    PlotRange -> {{-4, 4}, {0, 4}}, PlotLabel -> #[[1]],
                    DisplayFunction -> Identity, Frame -> False]& /@
                                                Take[data, 4]]]]
```

We repeat the last calculations and visualizations and use the evaluation monitor instead of the step monitor. The number of function evaluations is for some methods much larger than the number of steps.

```
In[82]:= (DownValues[fgm] = DownValues[fgm] /. StepMonitor -> EvaluationMonitor;
    In[$Line - 3]; Print[{#1, Length[#2[[2, 1]]]}]& @@@ data]; In[$Line - 1])
            {{NelderMead, 112}, {DifferentialEvolution, 10026},
                {SimulatedAnnealing, 51}, {RandomSearch, 5075}, {Automatic, 112}}
```

We will end this section with a computationally more challenging example: an energy minimization problem.

Physical Remark: Minimum Energy Configuration of *n* Electrons in a Disk

In 1985, Berezin [180] considered the following problem: Let n electrons be confined inside the unit disk. What is the configuration of these n electrons such that their electrostatic energy is minimal. Up to $n = 11$, all electrons will be located at the outer boundary of the disk and form there a regular n-gon. But for $n = 12$, the minimum energy configuration is such that one electron is in the center of the disk.

For more than 12 electrons, it is conjectured that the minimal energy configuration is such that the electrons form (nearly) concentric circles with a nonincreasing number of electrons on the inner circles. For details, see [1905], [1258], [560], [1415], and [1360]; for thin parallel magnets, see [1081].

Here, the electrostatic energies for the two configurations for $n = 11$ and $n = 12$ are calculated.

```
In[83]:= position[i_, n_] = {Cos[i/n 2Pi], Sin[i/n 2Pi]};

In[84]:= energyAllOutside[n_] :=
    Sum[1/Sqrt[#.#]&[position[i, n] - position[j, n]], {i, n}, {j, i - 1}];

In[85]:= energyAllButOneOutside[n_] :=
    Sum[1/Sqrt[#.#]&[position[i, n - 1] - position[j, n - 1]],
            {i, n - 1}, {j, i - 1}] +
    Sum[1/Sqrt[#.#]&[position[i, n - 1] - {0, 0}], {i, n - 1}]

In[86]:= N[{energyAllOutside[11], energyAllButOneOutside[11]}]

Out[86]= {48.5757, 48.6245}

In[87]:= N[{energyAllOutside[12], energyAllButOneOutside[12]}]

Out[87]= {59.8074, 59.5757}
```

The configuration with one charge in the center is clearly favored. In the following, we will determine the minimum energy configuration for $n = 21$.

The function `energy` takes as its argument a list of integers, representing the number of electrons in the ith circle. It returns a list with its first argument being the energy and its second argument being the list of undetermined radii and angles. All of the outermost electrons are assumed to have radius 1. The inner electrons are supposed to lie on concentric circles.

```
In[88]:= energy[l_] :=
    Module[{length, coords, sum, last1Q, λ = Length[l]},
    MapIndexed[(length[#2[[1]]] = #1)&, l];
    (* coordinates of the charges *)
    coords = Flatten[MapIndexed[r[#2[[1]]] *
            {Cos[#2[[2]]/length[#2[[1]]] 2π + φ0[#2[[1]]]],
             Sin[#2[[2]]/length[#2[[1]]] 2π + φ0[#2[[1]]]]}&,
                        Range /@ l, {2}], 1];
    (* the electrostatic energy *)
    sum = Sum[1/Sqrt[#.#]&[coords[[i]] - coords[[j]]],
            {i, Length[coords]}, {j, i - 1}];
    (* is there exactly one charge in the middle? *)
    last1Q = Last[l] === 1;
    sum = sum //. {φ0[1] -> 0, r[1] -> 1} //.
                    If[last1Q, {r[Length[l]] -> 0}, {}];
    {sum, Transpose[{Table[ r[i], {i, 2, If[last1Q, λ - 1, λ]}],
                    Table[φ0[i], {i, 2, If[last1Q, λ - 1, λ]}]}]}]
```

Here are two examples.

```
In[89]:= energy[{3, 1}]

Out[89]= {3 + √3 , {}}

In[90]:= energy[{2, 2}]
```

$$\text{Out[90]= } \Big\{ \frac{1}{2} + \frac{1}{\sqrt{(-1 - \text{Cos}[\varphi 0[2]] \, r[2])^2 + r[2]^2 \, \text{Sin}[\varphi 0[2]]^2}} +$$

$$\frac{1}{\sqrt{(1 - \text{Cos}[\varphi 0[2]] \, r[2])^2 + r[2]^2 \, \text{Sin}[\varphi 0[2]]^2}} +$$

$$\frac{1}{\sqrt{(-1 + \text{Cos}[\varphi 0[2]] \, r[2])^2 + r[2]^2 \, \text{Sin}[\varphi 0[2]]^2}} +$$

$$\frac{1}{\sqrt{(1 + \text{Cos}[\varphi 0[2]] \, r[2])^2 + r[2]^2 \, \text{Sin}[\varphi 0[2]]^2}} +$$

$$\frac{1}{\sqrt{4 \, \text{Cos}[\varphi 0[2]]^2 \, r[2]^2 + 4 \, r[2]^2 \, \text{Sin}[\varphi 0[2]]^2}}, \ \{\{r[2], \varphi 0[2]\}\}\Big\}$$

The function `makeInitialConditions` generates initial conditions for the to-be-determined variables of the energy to be minimized. The radii are decreasing, and the initial angles are such that regular *n*-gons with random initial orientation are formed.

```
In[91]:= makeInitialConditions[l_] :=
    Module[{ll = Length[l], rInits},
        (* initial radii *)
        rInits = Apply[{#1 + 1/3(#2 - #1), #1 + 2/3(#2 - #1)}&,
        Partition[Flatten[{0, Sort[Table[Random[], {ll - 1}]], 1}], 2, 1], {1}];
    Join[Flatten /@
        (Flatten[{#, 0, 1}]& /@ Transpose[{First /@ l, rInits}]),
                (* initial angles *)
            {#, Random[] 2Pi, Random[] 2Pi}& /@ (Last /@ l)]]
```

An example for the result of `makeInitialConditions`.

```
In[92]:= makeInitialConditions[{{r[2], φ0[2]}}]
```

$$\text{Out[92]= } \Big\{ \Big\{ r[2], \frac{1}{3}, \frac{2}{3}, 0, 1 \Big\}, \{\varphi 0[2], 1.44259, 2.84835\} \Big\}$$

Now, we must generate all possible nonincreasing sequences of numbers representing the occupation of the rings. The function `concentricCircles[m, n]` returns a list of such sequences of length *n* and sum *m*.

```
In[93]:= concentricCircles[m_, 1] := {{m}}

    concentricCircles[m_, n_] :=
    Module[{body, iterator},
        (* form all appropriate partitions by recursive subtraction *)
        body = Append[Table[i[j], {j, n - 1}], m - Sum[i[j], {j, n - 1}]];
        iterator = Table[{i[j], Which[j === 1, 1,
                                    j === 2, Max[2, i[1]],
                                    True, i[j - 1]],
                    (m - Sum[i[k], {k, 1, j - 1}])/2}, {j, n - 1}];
        Reverse /@ Flatten[Table[Evaluate[body],
                        Evaluate[Sequence @@ iterator]], n - 2]]
```

Here are two examples.

```
In[95]:= concentricCircles[21, 2]
```
```
Out[95]= {{20, 1}, {19, 2}, {18, 3}, {17, 4},
    {16, 5}, {15, 6}, {14, 7}, {13, 8}, {12, 9}, {11, 10}}
```

```
In[96]:= concentricCircles[21, 4] // Short[#, 4]&
```
```
Out[96]//Short=
    {{16, 2, 2, 1}, {15, 3, 2, 1}, {14, 4, 2, 1}, {13, 5, 2, 1},
    {12, 6, 2, 1}, {11, 7, 2, 1}, {10, 8, 2, 1}, {9, 9, 2, 1}, {14, 3, 3, 1},
```

```
{13, 4, 3, 1}, {12, 5, 3, 1}, {11, 6, 3, 1}, {10, 7, 3, 1}, {9, 8, 3, 1},
≪35≫, {8, 7, 3, 3}, {10, 4, 4, 3}, {9, 5, 4, 3}, {8, 6, 4, 3},
{7, 7, 4, 3}, {8, 5, 5, 3}, {7, 6, 5, 3}, {6, 6, 6, 3}, {9, 4, 4, 4},
{8, 5, 4, 4}, {7, 6, 4, 4}, {7, 5, 5, 4}, {6, 6, 5, 4}, {6, 5, 5, 5}}
```

Altogether, for 21 charges, there are 302 possible arrangements in concentric circles, taking only the topology of the circles into account.

```
In[97]:= allPossibilities = Join @@ Table[concentricCircles[21, k], {k, 11}];
```

```
In[98]:= Table[{k, Length @ concentricCircles[21, k]}, {k, 11}]
```

```
Out[98]= {{1, 1}, {2, 10}, {3, 36}, {4, 63}, {5, 71},
         {6, 56}, {7, 35}, {8, 18}, {9, 8}, {10, 3}, {11, 1}}
```

```
In[99]:= Length[allPossibilities]
```

```
Out[99]= 302
```

Now, for a given occupation of circles, let us actually find the minimum energy. Here are two examples.

```
In[100]:= es = energy[{18, 3}];
         FindMinimum[Evaluate[es[[1]]],
                     Evaluate[Sequence @@ makeInitialConditions[es[[2]]]],
                     MaxIterations -> 50]
```

```
Out[101]= {216.185, {r[2] → 0.377905, φ0[2] → 4.7124}}
```

```
In[102]:= es = energy[{16, 3, 2}];
         FindMinimum[Evaluate[es[[1]]],
                     Evaluate[Sequence @@ makeInitialConditions[es[[2]]]],
                     MaxIterations -> 50]
```

```
Out[103]= {219.802, {r[2] → 0.357717, r[3] → 0.646641, φ0[2] → 1.50515, φ0[3] → 4.12321}}
```

Let us automate this process within the function `findMinimumEnergy`. Not every occupation will be such that we can find a minimum of the energy compatible with the constraints. In such cases, we do ten trials and then give up.

```
In[104]:= findMinimumEnergy[l_] :=
         Module[{counter = 0, maxTrials = 10, es = energy[l], res},
         If[(* no free variables *)
            es[[2]] === {}, N[es[[1]], 6],
         res = FindMinimum[Evaluate[es[[1]]],
                Evaluate[Sequence @@ makeInitialConditions[es[[2]]]],
                MaxIterations -> 50];
         While[(* was a minimum found? *)
               Head[res] === FindMinimum &&
               (* are the radii nonincreasing? *)
               Less @@ ((First /@ es[[2]]) /. res[[-1]]) &&
               (* fewer than five trials? *)
               counter < maxTrials,
               counter = counter + 1; res = FindMinimum[Evaluate[es[[1]]],
               Evaluate[Sequence @@ makeInitialConditions[es[[2]]]],
               MaxIterations -> 50]];
         If[Head[res] === FindMinimum, $Failed, N[res[[1]], 6]]]]]
```

Now, we do the actual minimization for all 302 configurations and return the ones with the lowest energy. It is not possible to find a minimum obeying the parameter constraints for every configuration. As a result, we get some `FindMinimum::lsbrak` and `FindMinimum::regex` messages.

```
In[105]:= (* seed the random number generator;
          use reproducible starting values *)
       SeedRandom[12345];

       (* turn off messages from nonconvergence
          and potential division by zero *)
       Internal`DeactivateMessages[
       (Take[#, 12]& @ Sort[DeleteCases[({#, findMinimumEnergy[#]}& /@
                           allPossibilities), {_, $Failed}],
                       #1[[2]] < #2[[2]]&)]] // Timing
Out[108]= {125.18 Second, {{{18, 3}, 216.185}, {{17, 2, 2}, 216.478}, {{18, 2, 1}, 217.102},
          {{17, 3, 1}, 217.262}, {{16, 5}, 217.681}, {{16, 4, 1}, 218.184},
          {{16, 2, 2, 1}, 218.184}, {{20, 1}, 218.690}, {{15, 3, 3}, 219.668},
          {{15, 6}, 219.668}, {{15, 2, 2, 2}, 219.668}, {{16, 3, 2}, 219.802}}}
```

Now, having found that the {18, 3} configuration is the best one, let us try to further minimize its energy by allowing the outer 18 electrons to move azimuthally and the inner 3 to move azimuthally and radially.

```
In[109]:= coords = Join[Table[{Cos[φ1[i]], Sin[φ1[i]]}, {i, 18}],
                   Table[r2[i] {Cos[φ2[i] + φ0[2]],
                           Sin[φ2[i] + φ0[2]]}, {i, 3}]];

In[110]:= sum = Sum[1/Sqrt[#.#]&[coords[[i]] - coords[[j]]],
               {i, Length[coords]}, {j, i - 1}];

In[111]:= inits = {#[[1]], #[[2]] 0.999, #[[2]] 1.001}& /@
       Join[Table[{φ1[i], i/18 2Pi}, {i, 18}],
           Table[{φ2[i], 2.26893 + (i - 1)/3 2Pi}, {i, 3}],
           Table[{r2[i], 0.377905}, {i, 3}], {{φ0[2], 2.26893}}]];

In[112]:= sol1 = FindMinimum[Evaluate[sum], Evaluate[Sequence @@ inits],
                   PrecisionGoal -> 10, AccuracyGoal -> 10]
Out[112]= {216.179, {φ1[1] → 0.344367, φ1[2] → 0.698024, φ1[3] → 1.04854, φ1[4] → 1.3948,
          φ1[5] → 1.73987, φ1[6] → 2.08694, φ1[7] → 2.4386, φ1[8] → 2.79227, φ1[9] → 3.14278,
          φ1[10] → 3.48913, φ1[11] → 3.83419, φ1[12] → 4.18137, φ1[13] → 4.533,
          φ1[14] → 4.88679, φ1[15] → 5.23727, φ1[16] → 5.58365, φ1[17] → 5.92871,
          φ1[18] → 6.27593, φ2[1] → 2.26834, φ2[2] → 4.36249, φ2[3] → 6.45722,
          r2[1] → 0.378957, r2[2] → 0.378865, r2[3] → 0.378858, φ0[2] → 2.39759}}
```

Here, this configuration is shown.

```
In[113]:= Show[Graphics[{
       (* confining circle *)
       {GrayLevel[0.7], Thickness[0.002], Circle[{0, 0}, 1]},
       (* positioned charges *)
       {PointSize[0.02], Point /@
         Join[Table[{Cos[φ1[i]], Sin[φ1[i]]}, {i, 18}],
             Table[r2[i]{Cos[φ2[i]], Sin[φ2[i]]}, {i, 3}]] /. sol1[[2]]}}],
           PlotRange -> All, Frame -> True, FrameTicks -> None,
           AspectRatio -> Automatic]
```

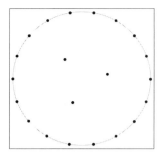

To refine the last result, we switch to high-precision arithmetic and refine the last minimum once more by using higher values for the AccuracyGoal and PrecisionGoal.

```
In[114]:= refinedInitialConditions =
      {#[[1]], SetPrecision[(1 - 10^-8) #[[2]], 22],
           SetPrecision[(1 + 10^-8) #[[2]], 22]}& /@ sol1[[2]];
```

```
In[115]:= sol2 = FindMinimum[Evaluate[sum],
                    Evaluate[Sequence @@ refinedInitialConditions],
                    PrecisionGoal -> 8, AccuracyGoal -> 8,
                    WorkingPrecision -> 22];
```

```
      sol2[[1]]
Out[117]= 216.1791133262891045166
```

We achieved a slightly smaller energy. The positions are nearly unchanged.

```
In[118]:= Max[Abs[(Last /@ sol1[[2]]) - (Last /@ sol2[[2]])]]
Out[118]= 0.0000511984
```

The energy 216.17911 agrees well with the one obtained in the above-cited references. Other numbers of charges can be treated similarly. For corresponding experimental results, see [1570].

In a similar way, we could treat the problem of arranging n points in the unit square in such a way that the minimal distance between them is maximal [442]. For the treatment of point charges on a round conducting disk, see [1399], [185]. For the dependence on the exponent γ in the interaction force r^γ, see [1133].

1.11 Solution of Differential Equations

1.11.1 Ordinary Differential Equations

In the category of numerical methods, surely the most important built-in command is NDSolve for solving ordinary (also nonlinear and stiff) systems of differential equations. For this reason, this subsection is longer than the earlier sections that discussed other numerical functions and has more examples. (For a nice exposition on the use of NDSolve with many examples, see [1624]; for the solution of some more tricky ordinary differential equations, see [1008].) In addition, NDSolve is currently the single most complicated and most user-extensible function of *Mathematica*. It has tens of possible suboptions and option settings. In this subsection, we will discuss the most important of natural science applications options. For an exhaustive discussion see the online Advanced Documentation, which contains hundreds of pages of detailed material about NDSolve.

NDSolve[*odeList*, *functionList*, *independentVariableAndDomain*, *options*]

> computes the numerical solution of the system of ordinary differential equations *odeList* (consisting of the differential equations and the corresponding initial conditions) using the options *options*.

> The desired solution is returned for the functions from *functionList*, whereas *independentVariableAndDomain* defines the independent variable and its domains. The form of *independentVariablesAndDomain* can be used in a form analogous to the second argument of NIntegrate: {*x*, x_0, x_1, ..., x_n}.

What are the classes of ordinary differential equations handled by NDSolve? (We discuss partial differential equations, which are also solved by NDSolve, in the next subsection.)

> In *Mathematica* 5.1, NDSolve solves initial-value problems with regular initial conditions, linear boundary-value problems, and differential-algebraic systems. Sturm–Liouville eigenvalue problems and delay differential equations are currently not solved.

We begin with a very simple example, the differential equation for harmonic motion.

```
In[1]:= harmonicEquation[x0_, v0_] = {x''[t] == -x[t], x[0] == x0, x'[0] == v0}
Out[1]= {x''[t] == -x[t], x[0] == x0, x'[0] == v0}
```

Here is the solution for a particular initial condition.

```
In[2]:= sol = NDSolve[harmonicEquation[1, 0], x[t], {t, 0, 10}]
Out[2]= {{x[t] → InterpolatingFunction[{{0., 10.}}, <>][t]}}
```

The returned outputs of NDSolve are lists of InterpolatingFunction-objects, and we can look at them using Plot. Because the result has the form List[List[Rule[...]]], we have to replace the x[t] in Plot by use of ReplaceAll. Evaluating the returned interpolating function can be done quite quickly.

```
In[3]:= (* do 1000 plots for reproducible timings *)
        Timing[Do[Plot[Evaluate[x[t] /. sol], {t, 0, 10},
                    DisplayFunction -> Identity], {1000}]]

        Plot[Evaluate[x[t] /. sol], {t, 0, 10}]
Out[4]= {1.51 Second, Null}
```

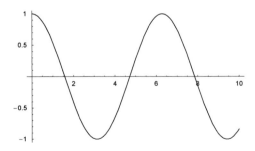

We use Evaluate because Plot has the attribute HoldAll. (Without the Evaluate, the plots would be slower because the replacement would be carried out for each value of t inside Plot.) We could have also used Part to extract the InterpolatingFunction-object from sol.

We can, of course, unite the solving step and the plotting step by inlining the NDSolve call. Here this is done for a harmonic oscillator with a nonlinear damping [1319].

```
In[7]:= Module[{η = 10^-1, T = 80},
    Plot[Evaluate[x[t] /. (* use constant-period damping *)
        NDSolve[{x''[t] == -(x[t] + η Abs[x[t]] x'[t]/Abs[x'[t]]),
            x[0] == 0, x'[0] == 1}, x, {t, 0, T}][[1]]], {t, 0, T},
        Axes -> False, Frame -> True, PlotStyle -> {Hue[0]}]]
```

To plot the numerical solution of a differential equation, we use

Plot[Evaluate[*function*[*param*] /. *solution*], *domain*, *options*]

The analog holds for Plot3D, ParametricPlot, ParametricPlot3D, and other plotting functions or numerical functions such as NIntegrate.

A slight issue can occur if we use ParametricPlot, however. Here is a system of differential equations. The solution returned by NDSolve is again a list of lists. Because the solution is unique, the outer list has just one element. This element contains four Rule-objects, one for each of the variables to be solved for.

```
In[8]:= sol =
    NDSolve[{x'[t] == y[t]^3, y'[t] == x[t]^3,
        z'[t] == -w[t]^2, w'[t] == -z[t]^2,
        x[0] == 1, y[0] == -1, w[0] == 1, z[0] == -1},
        {x[t], y[t], w[t], z[t]}, {t, 0, 1}]
Out[8]= {{x[t] → InterpolatingFunction[{{0., 1.}}, <>][t],
    y[t] → InterpolatingFunction[{{0., 1.}}, <>][t],
    w[t] → InterpolatingFunction[{{0., 1.}}, <>][t],
    z[t] → InterpolatingFunction[{{0., 1.}}, <>][t]}}
```

But a direct adaptation of the input from above in the form ParametricPlot[Evaluate[{{x[t], y[t]}, {z[t], w[t]}} /. sol], {t, 0, 1}] would not produce a plot. After carrying out x[t] /. sol above, we have a list of depth one. But in the parametric plot example, we would have a list of depth five. So, we have to use ... /. sol[[1]] instead of just ... /. sol.

Here is a simple linear boundary-value problem solved.

```
In[9]:= NDSolve[{wb''[ξ] + wb[ξ] == Cos[ξ], wb[0] == -2, wb[2] == 3},
        wb[ξ], {ξ, 0, 2}]
Out[9]= {{wb[ξ] → InterpolatingFunction[{{0., 2.}}, <>][ξ]}}
```

Here is a plot of the solution.

```
In[10]:= Plot[Evaluate[wb[ξ] /. %], {ξ, 0, 2}]
```

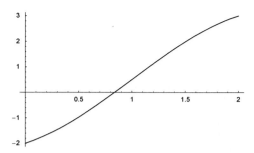

A check shows that the boundary values are fulfilled excellently.

In[11]:= `{{-2}, {3}} - (wb[ξ] /. %% /. {{ξ -> 0}, {ξ -> 2}})`

Out[11]= $\{\{-3.3907 \times 10^{-8}\}, \{0.\}\}$

And here is a simple differential-algebraic equation.

In[12]:=
```
ndsol =
    NDSolve[{x'[t] == 2 x[t] + 3 z[t], y'[t] == -x[t] + y[t] + Sin[y[t]],
             (* no z derivative *) z[t] == -(x[t]^2 + y[t]^2) + z[t]^3,
             (* no initial z-value *) x[0] == 1, y[0] == -1}, {x, y, z},
             {t, -1, 1}]
```

Out[12]=
```
{{x → InterpolatingFunction[{{-1., 1.}}, <>],
  y → InterpolatingFunction[{{-1., 1.}}, <>],
  z → InterpolatingFunction[{{-1., 1.}}, <>]}}
```

Here is a plot it its solution.

In[13]:=
```
Plot[Evaluate[{x[t], y[t], z[t]} /. ndsol[[1]]], {t, -1, 1},
     Frame -> True, Axes -> False, PlotRange -> All,
     PlotStyle -> Table[MapAt[1&, RGBColor[0, 0, 0], k], {k, 3}]]
```

We can also solve more complicated differential equations. Here is a relative of the harmonic differential equation, but this time $\sin(x(t))$ is not replaced by $x(t)$. Now, the initial conditions play a major role in the form of the solution. For small initial velocities, we get qualitatively the same solution as above.

In[14]:=
```
pendulumSol[φ0_, φ0s_, tmax_] :=
    φ[t] /. NDSolve[{φ''[t] == -Sin[φ[t]], φ[0] == φ0, φ'[0] == φ0s},
                    φ[t], {t, 0, tmax}]
```

In[15]:= `Plot[Evaluate[pendulumSol[0, 1, 8]], {t, 0, 8}]`

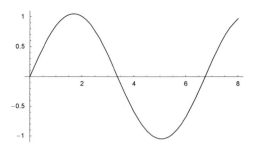

For larger initial conditions, we get an "overshoot" (the reader should think of a pendulum; here we ignore friction).

In[16]:= **Plot[Evaluate[pendulumSol[0, 2.01, 16]], {t, 0, 16}]**

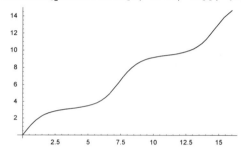

To examine the velocity of a pendulum, we have the following "problem".

In[17]:= **x'[t] /. NDSolve[{x''[t] == -x[t], x[0] == 1, x'[0] == 0}, x[t], {t, 0, 2}]**

Out[17]= {x′[t]}

In[18]:= **FullForm[%]**

Out[18]//FullForm=
      List[Derivative[1][x][t]]

No substitution has taken place, because the rule for x'[t] does not contain x[t]. (We come back to this in Chapter 1 of the Symbolics volume [1795].) In order to generate a suitable rule, the desired functions can be given without arguments to NDSolve. The solutions are given in the form of interpolating functions (which, similarly to pure functions can be applied to arguments). And the derivative of an interpolating function InterpolatingFunction[...]' evaluates again to an InterpolatingFunction-object.

In[19]:= **NDSolve[{x''[t] == -x[t], x[0] == 1, x'[0] == 0}, x[t], {t, 0, 2}]**

Out[19]= {{x[t] → InterpolatingFunction[{{0., 2.}}, <>][t]}}

In[20]:= **NDSolve[{x''[t] == -x[t], x[0] == 1, x'[0] == 0}, x, {t, 0, 2}]**

Out[20]= {{x → InterpolatingFunction[{{0., 2.}}, <>]}}

Now, derivatives can also be replaced. First x is replaced by the interpolating function. Then the derivative is evaluated and finally the resulting interpolating function is applied to the symbolic argument t, which cannot evaluate to an explicit number.

In[21]:= **x'[t] /. %**

Out[21]= {InterpolatingFunction[{{0., 2.}}, <>][t]}

Then they can be plotted.

```
In[22]:= Plot[Evaluate[x'[t] /. NDSolve[{x''[t] == -x[t], x[0] == 1, x'[0] == 0},
                x, {t, 0, 20}]], {t, 0, 20}]
```

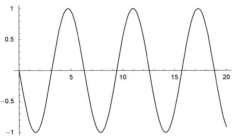

> If the arguments of the desired functions are not given, the functions are output in the form of (pure) interpolating functions, which have the greatest flexibility for later use.

Because the numerical values of the solution can be quickly extracted from the interpolating functions returned by NDSolve, we can easily and quickly carry out many recursive calls to NDSolve. In the following example, we solve the differential equation for the position and the velocity of an anharmonic oscillator with random external forcing $x''(t) + x(t)^3 = 1/2 \sin(\omega(t) t + \varphi(t)) x(t)$ with piecewise constant $\omega(t)$ and $\varphi(t)$. We repeatedly randomly select $\omega(t) = \omega_k$ and $\varphi(t) = \varphi_k$ and evolve the oscillator over one period of the forcing. We see that the amplitude of the oscillator is initially increasing in average [1688] and asymptotically becomes (approximately) in resonance with the average frequency of the external driving force (for periodic square-wave forcing, see [322]).

```
In[23]:= (* extract extrema positions from a curve segment *)
      extrema[Line[l_]] := Point[#[[2]]]& /@
            Select[Partition[l, 3, 1],
                  (Abs[#[[2, 2]]] > Max[Abs[{#[[1, 2]], #[[3, 2]]}]])&]

      Module[{ω0 = 10, x0, v0, φ, ω, δT, T0, ndsol, pl, o = 2, n = 20000},
      Show[
      (* two random realizations of the time evolution *)
       Table[
       (* seed random number generator *) SeedRandom[r];
       (* initial time *) T0 = 0;
       (* start position and velocity *) {x0, v0} = {0., 1.};
       (* show graphics of maximal elongations *)
      Graphics[{PointSize[0.005], Hue[(r - 1)/(o - 1) 0.78],
         (* follow oscillator over n periods of the field oscillations *)
        Table[If[(* exit gracefully in case of a solution blow-up *)
              MatchQ[{x0, v0}, {_Real, _Real}],
              (* random phase and frequency *)
              φ = Random[Real, {0, 2Pi}]; ω = Random[Real, {3/4 ω0, 5/4 ω0}];
              (* one period *) δT = 2Pi/ω;
              (* solve differential equations of motion *)
              ndsol = Check[NDSolve[{v[t] == x'[t],
                            v'[t] + x[t]^3 + 1/2 Sin[ω t + φ] x[t] == 0,
                            x[T0] == x0, v[T0] == v0},
                            {x, v}, {t, T0, T0 + δT}], $Failed];
```

```
(* extract final position and velocity *)
If[ndsol === $Failed, {x0, v0} = {$Failed, $Failed},
   {x0, v0} = {x[T0 + δT], v[T0 + δT]} /. ndsol[[1]];
   (* plot x[t] *)
   pl = Plot[Evaluate[x[t] /. ndsol[[1]]], {t, T0, T0 + δT},
          PlotPoints -> Ceiling[5 + 10 δT],
          DisplayFunction -> Identity];
   (* extract extremas from lines from the plot *)
   T0 = T0 + δT; extrema /@ Cases[pl, _Line, Infinity]]], {n}]} /.
   (* show on logarithmic scale *)
   Point[{t_, x_}] :> Point[{t, Log[Abs[x]]}]],
   {r, o}], Frame -> True, PlotRange -> All]]
```

In the next example, we solve the differential equation

$$\frac{\partial(l(t)^2\,\theta'(t))}{\partial t} + \gamma\,l(t)^2\,\theta'(t) + l(t)\sin(\theta(t)) = 0$$

for the elongation $\theta'(t)$ more than 200 times for $l(t)$ of the form $l(t) = 1 + \varepsilon\sin(\omega\,t + 9/8\,\pi)^7$. This differential equation describes a massless rod pendulum with time-dependent rod length and a mass at the end of the rod—an idealization of the process of squatting when being on a swing [1643]. We recognize the main resonance at twice the frequency of the pendulum with constant rod length and glimpses of the subsidiary resonances at integer fractions of the main resonance.

```
In[27]:= (* load package for function ListSurfacePlot3D *)
       Needs["Graphics`Graphics3D`"]

In[29]:= Module[{γ = 0.16, ε = 0.2, ω = 1, T = 60, ppω = 240, ppT = 480, ℓ, data},
           (* time dependent periodic rod length with main frequency ω *)
           ℓ[t_] := 1 + ε Sin[ω t + 9/8 Pi]^7;
           data = Table[(* solve differential equation for fixed ω *)
               ndsol = NDSolve[{D[ℓ[t]^2 θ'[t], t] + γ ℓ[t]^2 θ'[t] +
                              ℓ[t] Sin[θ[t]] == 0,
                          θ[0] == -1, θ'[0] == 1}, θ, {t, 0, T}];
               (* elongation data for fixed ω *)
               Table[Evaluate[{t, ω, θ[t]} /. ndsol[[1]]], {t, 0, T, T/ppT}],
                   {ω, 0, 3, 2/ppω}];
           (* show elongation over the t,ω-plane *)
           Show[Graphics3D[{EdgeForm[], (* form polygons *)
               Cases[ListSurfacePlot3D[data, DisplayFunction -> Identity],
                          _Polygon, Infinity]}],
               PlotRange -> All, BoxRatios -> {2.8, 1, 0.6}, Axes -> True,
               AxesLabel -> {"t", "ω", None}]]
```

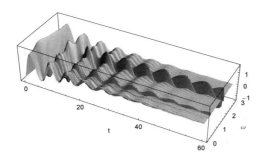

Here is another example: Given a polynomial $p(x)$, the function $-p(x)/p'(x)$ is called the Newtonian vector field of $p(x)$ [1047], [917], [506], [1808], [918], [26], [27]. The name originates from the fact that the Newton method for root findings uses a discrete version of the Newtonian vector field. Starting from a generic point $z_0 = z(0)$ and solving the differential equation $z'(t) = -p(z(t))/p'(z(t))$, one finds a path to a root of $p(x)$. Here, this is demonstrated for a randomly chosen polynomial.

```
In[30]:= SeedRandom[555];
```

```
In[31]:= f[x_] = Sum[Random[Integer, {-5, 5}] x^i, {i, 0, 8}]
```

$$\text{Out[31]= } 1 - 4\,x + 3\,x^2 - 2\,x^3 + 2\,x^4 - 5\,x^5 - x^6 + 3\,x^7 - 2\,x^8$$

```
In[32]:= df[x_] = -f[x]/D[f[x], x]
```

$$\text{Out[32]= } -\frac{1 - 4\,x + 3\,x^2 - 2\,x^3 + 2\,x^4 - 5\,x^5 - x^6 + 3\,x^7 - 2\,x^8}{-4 + 6\,x - 6\,x^2 + 8\,x^3 - 25\,x^4 - 6\,x^5 + 21\,x^6 - 16\,x^7}$$

```
In[33]:= polyRoots = x /. NSolve[f[x] == 0, x]
```

$$\text{Out[33]= } \{-1.29978, -0.44976 - 0.761506\,i,$$
$$-0.44976 + 0.761506\,i, 0.307289, 0.581793 - 0.579026\,i,$$
$$0.581793 + 0.579026\,i, 1.11421 - 1.06489\,i, 1.11421 + 1.06489\,i\}$$

For 250 randomly chosen initial values, we numerically solve the corresponding differential equation.

```
In[34]:= Off[NDSolve::mxst];
      flowLines =
      Table[NDSolve[{z'[t] == df[z[t]],
                  (* use random complex numbers as initial conditions *)
                  z[0] == Random[Complex, {-2 - 2I, 2 + 2I}]},
                  z, {t, 0, 1000}, MaxStepSize -> 1], {250}];
      On[NDSolve::mxst];
```

The following picture shows the resulting flow lines.

```
In[37]:= Show[{Graphics[{PointSize[0.03],
                  (* the roots *)
                  GrayLevel[1/2], Point[{Re[#], Im[#]}]& /@ polyRoots}],
                  (* the flow lines *)
              ParametricPlot[Evaluate[{Re[z[t]], Im[z[t]]} /. #],
                          {t, 0, #[[1, 1, 2, 1, 1, 2]]},
                  PlotStyle -> {Hue[Random[]], Thickness[0.002]},
                  PlotPoints -> 200, DisplayFunction -> Identity]& /@ flowLines},
              DisplayFunction -> $DisplayFunction, PlotRange -> All,
              Frame -> True, Axes -> False, AspectRatio -> Automatic]
```

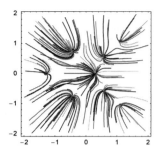

In the last example, there was no (simple) connection between the arc-length of the curves and their parametrizations. For differential equations describing trajectories, it is often convenient to use parametrizations that represent the arc length of the curve. In the problem under consideration, this can be easily achieved by normalizing the right-hand side of the differential equation. As an example, we calculate the Newton flow for the function $x \to \exp((x^3 - 1)^{-1})$. (We turn off some messages resulting from NDSolve using the maximal allowed number of steps; we come back to the corresponding option in a moment.)

```
In[38]:= SeedRandom[123];
     Module[{n = 200, df, newtonFlow, flowLines},
     (* the rhs *)
     df[x_] = Function[f, -f'[x]/f[x]][Exp[1/(#^3 + 1)]&];
     (* the normalized rhs *)
     newtonFlow[x_?InexactNumberQ] := #/Abs[#]&[df[x]];
     (* calculate flow lines *)
     Off[NDSolve::mxst]; Off[NDSolve::ndcf];
     flowLines = Table[NDSolve[{z'[t] == newtonFlow[z[t]],
                 (* use random starting values *)
                 z[0] == Random[Complex, 2 {-1 - I, 1 + I}]},
                 z, {t, -3, 3}], {n}];
     On[NDSolve::mxst]; On[NDSolve::ndcf];
     (* show trajectories and contour plot of function *)
     Show[{(* the contour plot *)
         ContourPlot[Evaluate[Arg[newtonFlow[x + I y]]^2],
             {x, -3, 3}, {y, -3, 3},
             PlotPoints -> 200, ColorFunction -> (Hue[0.8 #]&),
             ContourLines -> False, Contours -> 30,
             DisplayFunction -> Identity],
         (* the trajectories *)
     ParametricPlot[Evaluate[{Re[z[t]], Im[z[t]]} /. #],
             (* plot as far as the solution was carried out *)
             Evaluate[Flatten[{t, #[[1, 1, 2, 1, 1]]}]]],
         PlotStyle -> {GrayLevel[0], Thickness[0.002]},
         DisplayFunction -> Identity]& /@ flowLines},
     DisplayFunction -> $DisplayFunction,
     PlotRange -> {{-3, 3}, {-3, 3}}]]
```

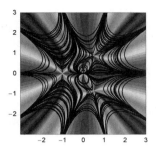

Here is a similar ODE of second order $x''(t) = p(x(t))$ where $p(\xi)$ is a (random) complex-valued polynomial.

```
In[40]:= complexODETrajectory[d_, T_] :=
    Module[{rc, F, ndsol, x, t, p3d, line, λ, ppT = 60},
            rc := Random[Real, {0, 2}] Exp[2 Pi I Random[]];
            (* right-hand side is random complex polynomial *)
            F[x_] := Sum[rc x^k, {k, 0, d}];
            (* solve differential equation *)
            ndsol = NDSolve[{x''[t] == F[x[t]], Sequence @@
                        Table[Derivative[j][x][0] == rc, {j, 0, 1}]},
                    x, {t, 0, T}, MaxSteps -> 10^5];
            (* make plot in complex plane *)
            p3d = ParametricPlot[Evaluate[{Re[#], Im[#]}&[x[t] /. ndsol[[1]]]],
                        {t, 0, T}, PlotPoints -> ppT T,
                        DisplayFunction -> Identity];
            (* color trajectory from red to blue *)
            line = p3d[[1, 1, 1, 1]]; λ = Length[line];
            Graphics[{Thickness[0.002],
                    MapIndexed[{Hue[0.76 #2[[1]]/λ], Line[#1]}&,
                        Partition[line, 2, 1]]},
                PlotRange -> All, AspectRatio -> 1]]
```

This time, the solution curves are nonintersecting and folded (around the complex classical turning points of the corresponding potential).

```
In[41]:= Show[GraphicsArray[(SeedRandom[#2]; complexODETrajectory[#1, 60])& @@@
            (* two example data sets *) {{8, 1}, {12, 28}, {55, 12}}]]
```

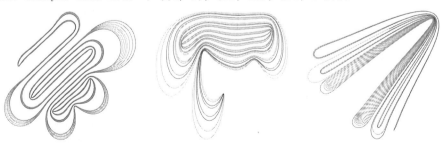

Now let us deal with a more complicated system of two coupled differential equations. The right hand sides are complicated rational functions in the state variables. The following system models the formation of spiral waves in chemical reactions [787].

```
In[42]:= phasePortrait[{l_, ω_, p_, d_, q_, e_, c_}, tMax_, pp_] :=
    Module[{x0, y0, x, y, t, nsol},
      Do[(* solve differential equations *)
        nsol[x0, y0] = NDSolve[Flatten[
        {x'[t] == l x[t] (1 - x[t]^2 - (q + (1 + p) y[t])^2) -
                ω y[t] (1 - q^2 - (d ((1 + p)^2 - c (1 - q)) y[t]
                (1 - e + e y[t]))/(1 + p - e (p + q)) -
                c (x[t]^2 + y[t]^2)),
         y'[t] == l y[t] (1 - x[t]^2 - (q + (1 + p) y[t])^2) +
                ω x[t] (1 - q^2 - (d ((1 + p)^2 - c (1 - q)) y[t]
                (1 - e + e y[t]))/(1 + p - e (p + q)) -
                c (x[t]^2 + y[t]^2)),
         x[0] == x0, y[0] == y0}], {x, y}, {t, 0, tMax}],
                {x0, -2, 2, 4/pp}, {y0, -3, 1, 4/pp}];
      (* generate visualizations *)
      Show[Table[
      ParametricPlot[Evaluate[{x[t], y[t]} /. nsol[x0, y0][[1]]],
                  {t, 0, tMax}, PlotStyle -> {Thickness[0.002]},
                  DisplayFunction -> Identity],
              {x0, -2, 2, 4/pp}, {y0, -3, 1, 4/pp}],
      (* set graphic options *)
      DisplayFunction -> $DisplayFunction, AspectRatio -> Automatic,
      PlotRange -> All, Frame -> True, Axes -> False,
      PlotRange -> {{-3, 1}, {-3, 3}}]]

In[43]:= phasePortrait[{1, 1/2, 0, 1/2, 11/10, 0, 1}, 10, 12]
```

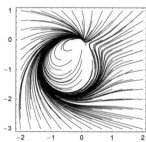

Here is another system of four coupled nonlinear ordinary differential equations with polynomial right-hand sides [1947], [1492]. The solution curve forms a complicated chaotic attractor in 4D. We display two 3D projections of this attractor.

```
In[44]:= Module[{α, β, x0, y0, vx0, vy0, T = 200, ndsol, path4D, colors,
            ℛ, viewGraphics},
      (* parameters *)
      {α, β, x0, y0, vx0, vy0} =
      {770/471, 24/187, -312/583, -18/37, -324/403, 104/785};
        (*   or {197/22, -301/87, 445/11, 9/83, -335/834, -7/38}
          or {41/103, 393/116, -16/75, 297/658, -188/929, -7/31}
          or {5/922, 106/299, 1/50, 171/509, 7/3, -119/993} *)
      (* solve differential equations *)
      ndsol = NDSolve[{x'[t] ==  vx[t], y'[t] == vy[t],
                  vx'[t] == -(α + y[t]^2) x[t] + y[t],
                  vy'[t] == -(β + x[t]^2) y[t] + x[t],
                  x[0] == x0, y[0] == y0, vx[0] == vx0, vy[0] == vy0},
```

```
                                    {x, y, vx, vy}, {t, 0, T}, MaxSteps -> 10^5];
      (* path in 4D phase space *)
      path4D = Table[Evaluate[{x[t], y[t], vx[t], vy[t]} /. ndsol[[1]]],
                         {t, 0, T, 1/20}];
      (* color of line segments *)
      colors = Table[Hue[0.78 x], {x, 0, 1, 1/(Length[path4D] - 2)}];
      (* 4D rotation matrix *)
      ℛ[φ_] := N @ {{Cos[φ], Sin[φ], 0, 0}, {-Sin[φ], Cos[φ], 0, 0},
                      {0, 0, 1, 0}, {0, 0, 0, 1}};
         (* make 3D projection graphics *)
      viewGraphics[φ_] := With[{rm = ℛ[φ]},
      Graphics3D[{Thickness[0.002],
                    Transpose[{colors, Line /@ Partition[Take[rm.#, 3]& /@
                                                            path4D, 2, 1]}]},
                     PlotRange -> All, SphericalRegion -> True,
                    Boxed -> False, BoxRatios -> {1, 1, 1}]];
      (* show three different projections *)
      Show[GraphicsArray[viewGraphics /@ {0, Pi/4, Pi/2}],
          GraphicsSpacing -> -0.1]]
```

NDSolve also solves large systems of differential equations in an acceptable time. Here is a system of 10 second-order differential equations that are coupled nonlinearly with one another (at the end of this subsection we will solve a system of thousands of differential equations). We chose the constants and the exponents randomly.

```
In[45]:= num = 10;
        SeedRandom[888];
        (diffEqn = Table[x[i]''[t] == Sum[Random[Integer, {-2, 2}] *
                             x[j][t]^Random[Integer, {0, 3}], {j, num}],
                      {i, num}]) // Short[#, 16]&
```

Out[47]//Short=

$\{x[1]''[t] == 1 - 2\,x[1][t]^3 + 2\,x[2][t]^2 - x[3][t]^2 + x[4][t]^2 - 2\,x[5][t]^2 - x[10][t],$
$\quad x[2]''[t] == 1 + 2\,x[1][t] + 2\,x[2][t] + 2\,x[5][t] + x[6][t]^2 - 2\,x[7][t] +$
$\quad\quad x[8][t] + x[9][t]^3 - 2\,x[10][t]^2, \; x[3]''[t] == 1 - x[1][t]^3 + 2\,x[2][t] +$
$\quad\quad x[3][t] - x[4][t] - 2\,x[6][t] + 2\,x[7][t]^2 + x[8][t] - 2\,x[9][t]^3,$
$\quad x[4]''[t] == 4 + 2\,x[1][t] + x[5][t]^2 - x[6][t] + x[7][t] + x[8][t]^3,$
$\quad x[5]''[t] == 2\,x[1][t]^3 - x[2][t]^3 - x[3][t] - 2\,x[4][t] -$
$\quad\quad 2\,x[5][t]^3 - x[6][t] - 2\,x[7][t] + 2\,x[8][t]^3 + 2\,x[10][t]^3,$
$\quad x[6]''[t] == -3 + 2\,x[1][t]^2 - 2\,x[5][t]^3 - x[10][t]^3,$
$\quad x[7]''[t] == -1 - x[1][t] + x[2][t]^2 + x[3][t] - 2\,x[4][t]^2 -$
$\quad\quad x[5][t]^3 - x[6][t] - 2\,x[8][t] + x[9][t]^3 - x[10][t]^2,$
$\quad x[8]''[t] == 1 - 2\,x[1][t]^3 - 2\,x[2][t]^2 - 2\,x[5][t]^3 - 2\,x[6][t]^3 + 2\,x[8][t]^2,$
$\quad x[9]''[t] == x[1][t]^2 - x[2][t]^2 + x[3][t]^3 + 2\,x[4][t]^2 -$
$\quad\quad 2\,x[5][t] - x[6][t] - 2\,x[7][t] + 2\,x[8][t] + x[9][t]^2,$
$\quad x[10]''[t] == 2 - 2\,x[3][t] + x[7][t]^2 + x[8][t]^3 - x[10][t]^2\}$

Next, we add initial conditions.

```
In[48]:= eqn = Join[diffEqn, Table[x[i][0] == 1, {i, num}],
                     Table[x[i]'[0] == 1, {i, num}]];
```

Here is the solution of the system of differential equations.

```
In[49]:= Timing[sol = NDSolve[eqn, Table[x[i], {i, num}], {t, 0, 1/2}]]

Out[49]= {0.02 Second, {{x[1] → InterpolatingFunction[{{0., 0.5}}, <>],
          x[2] → InterpolatingFunction[{{0., 0.5}}, <>],
          x[3] → InterpolatingFunction[{{0., 0.5}}, <>],
          x[4] → InterpolatingFunction[{{0., 0.5}}, <>],
          x[5] → InterpolatingFunction[{{0., 0.5}}, <>],
          x[6] → InterpolatingFunction[{{0., 0.5}}, <>],
          x[7] → InterpolatingFunction[{{0., 0.5}}, <>],
          x[8] → InterpolatingFunction[{{0., 0.5}}, <>],
          x[9] → InterpolatingFunction[{{0., 0.5}}, <>],
          x[10] → InterpolatingFunction[{{0., 0.5}}, <>]}}}
```

Here is a plot.

```
In[50]:= Plot[Evaluate[Table[x[i][t] /. sol, {i, num}]], {t, 0, 1/2},
          PlotStyle -> Table[Hue[j/10 0.7], {j, num}], PlotRange -> All]]
```

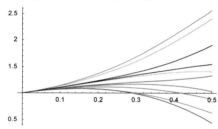

(For ODE systems with rational right-hand sides, see [844].)

NDSolve can also quickly solve complex-valued differential equations. Next, we will use it to find the Schrödinger equation-based allowed energy bands in a complex-valued periodic potential $V(x)$ with period 1. (For the meaning of a complex potential, see [1313].) For a solution $\psi_\varepsilon(x)$ of $-\psi_\varepsilon''(x) + V(x)\psi_\varepsilon(x) = \varepsilon\psi_\varepsilon(x)$ to fulfill $|\psi_\varepsilon(x)| = |\psi_\varepsilon(x+1)|$, the energy ε must fulfill the conditions $\mathrm{Im}(\Delta(\varepsilon)) = 0$ and $-1 \le \Delta(\varepsilon) \le 1$ [1652]. The Floquet discriminant $\Delta(\varepsilon)$ can be calculated through two special initial value problems of the Schrödinger equation.

```
In[51]:= Δ[ε_] := 1/2 (ψ1[1] + ψ2'[1]) /.
         NDSolve[{-ψ1''[x] + V[x] ψ1[x] == ε ψ1[x],
                  -ψ2''[x] + V[x] ψ2[x] == ε ψ2[x],
                  ψ1[0] == 1, ψ1'[0] == 0, ψ2[0] == 0, ψ2'[0] == 1},
                 {ψ1, ψ2}, {x, 0, 1}, PrecisionGoal -> 5, AccuracyGoal -> 5][[1]]
```

The function energyBandGraphic solves 36×72 pairs of the Schrödinger equation for a random trigonometric potential of approximate variation *VMax*. *seed* seeds the random number generator for the potential generation and *d* determines the degree of the rational trigonometric function. Because we want to generate a visualization, we use lower than default values for the precision goal and the accuracy goal.

```
In[52]:= energyBandGraphic[d_, VMax_, seed_] :=
         Module[{ε = 4 VMax, ppε = 36,
```

```
            randomTrigPoly, cp, lineSegments, allowedLineSegments},
 (* random complex 1D potential (rational function in trigs) *)
 randomTrigPoly[d_] := Sum[Random[] Exp[2 Pi I Random[]]*
                            Cos[k 2Pi x + 2Pi Random[]], {k, d}];
 V[x_] = VMax randomTrigPoly[d]/randomTrigPoly[d];
 (* show curves Im[Δ[ε]] == 0 *)
 cp = ContourPlot[Im[Δ[εr + I εi]], {εr, -ℰ, ℰ}, {εi, -ℰ/3, ℰ/3},
                  Contours -> {0}, DisplayFunction -> Identity,
                  ContourShading -> False, PlotPoints -> {2, 1} ppε];
 (* extract linesegments that form the curves *)
 lineSegments = Line /@ Flatten[Partition[First[#], 2, 1]& /@
                        Cases[Graphics[cp], _Line, Infinity], 1];
 (* extract linesegments corresponding to allowed energy bands *)
 allowedLineSegments =  Select[lineSegments,
                        Abs[Re[Δ[{1, I}.(Plus @@ #[[1]]/2)]]] < 1&];
 (* show allowed energy bands in red *)
 Show[Graphics[{{Thickness[0.004], GrayLevel[0],
                Complement[lineSegments, allowedLineSegments]},
            Thickness[0.004], Hue[0], allowedLineSegments}],
        PlotRange -> All, AspectRatio -> Automatic, Frame -> True]]
```

Within a few minutes, we can calculate and visualize the energy spectrum shown as red arcs in the complex plane. The black curves are defined by the condition $Im(\Delta(\varepsilon)) = 0$.

In[53]:= **energyBandGraphic[10, 97, 269889295]**

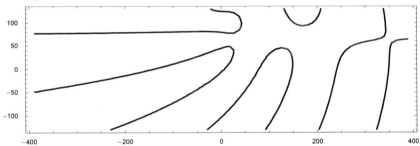

| NDSolve also solves systems of stiff differential equations.

Roughly speaking, systems of stiff differential equations are such that the components of the solution vary on very different scales. Nonstiff systems are frequently solved with the Adams predictor–corrector method, whereas stiff ones are solved with the "BDF" method of the "StiffnessSwitching" method. The following system of differential equations essentially has one cos function and one $\exp(-x^2)$ as solutions. However, because of the small terms y[t] and x[t] y[t], the two components of the solution are not independent.

In[54]:= **Plot[Evaluate[{x[t], y[t]} /.**
```
        NDSolve[{x''[t] == -300 x[t] + 1/(1 + y[t]),
                y'[t] == -t y[t] + x[t] y[t],
                x[0] == 1, x'[0] == 0, y[0] == 1},
                {x[t], y[t]}, {t, 0, 4}]],
        {t, 0, 4}, PlotRange -> All]
```

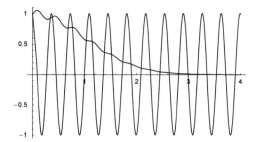

Now, we look at the options of NDSolve.

In[55]:= **Options[NDSolve]**

Out[55]= $\{$AccuracyGoal → Automatic, Compiled → True, DependentVariables → Automatic,

EvaluationMonitor → None, MaxStepFraction → $\frac{1}{10}$, MaxSteps → Automatic,

MaxStepSize → Automatic, Method → Automatic, NormFunction → Automatic,

PrecisionGoal → Automatic, SolveDelayed → False, StartingStepSize → Automatic,

StepMonitor → None, WorkingPrecision → MachinePrecision$\}$

We have already encountered many of these in connection with other numerical routines. MaxSteps, MaxStep Fraction, SolveDelayed, and StartingStepSize are among the new ones. Most of the discussed options, such as Compiled, and WorkingPrecision behave for NDSolve in an analogous way as for other numerical functions. So, we give just one simple example for WorkingPrecision here.

Because of the default option settings in NDSolve, PrecisionGoal = WorkingPrecision /2 and AccuracyGoal = WorkingPrecision /2, increasing the working precision will automatically increase the precision goal and accuracy goal options. Without explicit settings for PrecisionGoal and Accuracy·. Goal, using a higher working precision will typically produce a better solution (for the importance of high-precision solutions of differential equations, see [1609], [1610]). In the next picture, we look at $\log_{10}(w(k\pi))$, where $w(z)$ is the solution obtained with NDSolve of $w''(z) = -w(z)$, $w(0) = 0$, $w'(0) = 1$ as a function of the WorkingPrecision. Every curve is for one k, $k = 1, ..., 8$ with small k being colored red and larger k being colored blue. We see that for WorkingPrecision < $MachinePrecision, all calculations are carried out using machine arithmetic. We also nicely see the decrease of the error with increasing working precision and the general increase of the error with k.

In[56]:= ```
Module[{n = 8, data}, (* the values w[k Pi] *)
    data = Table[{wp, #}& /@
    (Table[y[k Pi], {k, n}] /. (* solve ode with working precision wp *)
        NDSolve[{y''[x] == - y[x], y[0] == 0, y'[0] == 1},
            y, {x, 0, n Pi}, WorkingPrecision -> wp][[1]]),
            {wp, MachinePrecision - 5, $MachinePrecision + 20}];
    (* display logarithmic errors at k Pi *)
    Show[Graphics[{Thickness[0.002],
        MapIndexed[{Hue[#2[[1]]/10], #}&, Line /@ Transpose[Apply[{#1,
            Log[10, Abs[#2]]}&, data, {2}]]]]},
                PlotRange -> All, Frame -> True,
                FrameLabel -> {"WorkingPrecision", "Log[10, w[k Pi]]"}]]
```
```

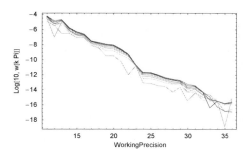

Sometimes it is convenient to set the value of the `AccuracyGoal` option to `Infinity`. This is especially the case when it is not known how small the solution becomes. The following example solves a differential equation with a small parameter that becomes very small in the interval (0, 2) [1385]. With the default option setting `AccuracyGoal -> Automatic`, the solution is not correct for $t > 2$. The right column of graphics shows the solution with the `AccuracyGoal -> Infinity` option setting and the lower row of graphics shows the logarithm of the solutions. As one sees in the lower right graphics, the solution decreases quite quickly and becomes at $z = 1$ very small, about 10^{-200}.

```
In[57]:= Module[{ε = 10^-3, nsol1, nsol2, x, t},
        (* default AccuracyGoal setting *)
        nsol1 = NDSolve[{ε x'[t] == (t - 1) x[t] - x[t]^2, x[0] == 1},
                    x, {t, 3}];
        (* Infinity AccuracyGoal setting *)
        nsol2 = NDSolve[{ε x'[t] == (t - 1) x[t] - x[t]^2, x[0] == 1},
                    x, {t, 3}, AccuracyGoal -> Infinity];
        (* show solution and logarithm of solution *)
        Function[f, Show[GraphicsArray[
        Plot[Evaluate[f @ x[t] /. #], {t, 0, 3}, PlotRange -> All,
            PlotStyle -> {{Thickness[0.005], Hue[0]}},
            DisplayFunction -> Identity]& /@ {nsol1, nsol2}]]] /@
        {Identity, Log[10, Abs[#]]&}]
```

In the following example, we will demonstrate that `PrecisionGoal` and `AccuracyGoal` might not be satisfied globally for a given differential equation. As already mentioned, they are local specifications. The differential equation of interest is a linear third-order differential equation with polynomial coefficients.

```
In[58]:= Clear[x, a, f, x0, x1, sol];

       ode[x_, a_] =
       (1 + 2 a x + 2 x^2) f'''[x] + 4 x (a + x) (a + 2 x) f''[x] +
       2 (3 - 2 a^2 + 4 a^2 x^2 + 8 a x^3 + 4 x^4) f'[x] == 0;
```

One solution of this differential equation is given by the sum of two Gaussians plus a constant. Here this solution is verified.

```
In[60]:= sol[x_, a_] = Exp[-x^2] + Exp[-(x + a)^2] + 1;

       (* check the solution *)
       (ode[x, a] /. f -> Function[x, Evaluate[sol[a, x]]])[[1]] // Simplify
Out[62]= 0
```

`numSol` solved the above differential equation `ode` with initial conditions corresponding to the solution `sol`.

```
In[63]:= numSol[a_, x0_, x1_, opts___] :=
       NDSolve[{ode[x, a], f[x0] == sol[x0, a],
               f'[x0] == D[sol[x, a], x] /. x -> x0,
               f''[x0] == D[sol[x, a], {x, 2}] /. x -> x0},
           f, {x, x0, x1}, opts, MaxSteps -> Infinity]
```

The function `errorPlot` plots the numerical solution of the differential equation and its absolute error.

```
In[64]:= errorPlot[numSol_, a_, X_] :=
       Show[GraphicsArray[
       Block[{$DisplayFunction = Identity},
       {(* the calculated solution *)
       Plot[Evaluate[f[x] /. numSol], {x, 0, X},
           PlotRange -> All, Frame -> True, Axes -> False,
           PlotLabel -> StyleForm["f[x]", FontFamily -> "Courier"]],
       (* difference to the exact solution *)
       Plot[Evaluate[(f[x] /. numSol) - sol[x, a]], {x, 0, X},
           PlotRange -> All, Frame -> True,
           PlotLabel -> "error", Axes -> False]}]]]

In[65]:= errorPlotU[a_, X_, opts___] := errorPlot[numSol[a, 0, X, opts], a, X]
```

If the two Gaussian peaks are located near each other, the result is quite precise.

```
In[66]:= errorPlotU[-1/2, 3]
```

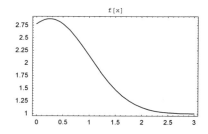 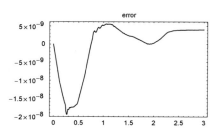

For a larger separation of the two peaks, we obtain a much larger error with the default maximal step size. With a smaller step size, we obtain a much better solution. (Changing the maximal step size is not the best possible approach to obtain a precise solution for this problem; using a higher working precision and a different method are more appropriate.)

```
In[67]:= errorPlotU[-424/100, 8, MaxStepSize -> #]& /@
            (* default and small step size *) {Automatic, 10^-4}
```

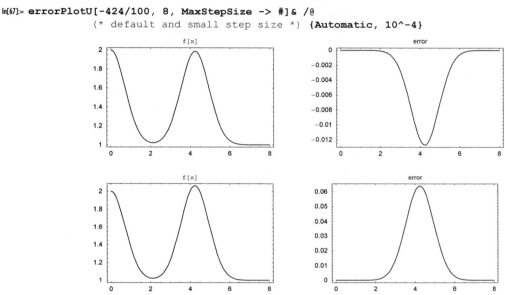

Next, we consider the following nonlinear third-order differential equation [327]

$$z'''(\tau) - 6\,i\,z''(\tau) - 11\,z'(\tau) + 6\,i\,z(\tau) = 2\,z'(\tau)^2 + 2\,z(\tau)\,z''(\tau) - 6\,z(\tau)^2 - 10\,i\,z'(\tau)\,z(\tau).$$

This differential equation has many periodic solutions with period 2π.

```
In[68]:= nsol[{z0_, z0p_, z0pp_}, opts___] :=
    NDSolve[{z'''[τ] - 6 I z''[τ] - 11 z'[τ] + 6 I z[τ] ==
            2(z[τ] z''[τ] + z'[τ]^2) - 10 I z[τ] z'[τ] - 6 z[τ]^2,
            z[0] == z0, z'[0] == z0p, z''[0] == z0pp}, z,
            {τ, 0, 2Pi}, opts, MaxSteps -> 10000, MaxStepSize -> 1/100]
```

For most initial conditions, the default values of `PrecisionGoal` and `AccuracyGoal` give a visually periodic solution.

```
In[69]:= Show[#, DisplayFunction -> $DisplayFunction]& @
    (ParametricPlot[Evaluate[{Re @ z[τ], Im @ z[τ]} /. nsol[#]],
                {τ, 0, 2Pi}, DisplayFunction -> Identity,
                PlotStyle -> {{Thickness[0.002], Hue[Random[]]}},
                PlotRange -> All, Frame -> True, Axes -> False,
                PlotPoints -> 1000]& /@
    {{-I, 1/2, I}, {3 + 3I, 8/3 + 6/5 I, 10/3 + 12/11 I},
     {4 + 2 I, 4 + 4 I, 4/3 + 4/7 I},
     {24/7 + 4/3 I, 4 + 18/5 I, 4 + 16/9 I},
     {7/2 + 14/5 I, 3 + 8/7 I, 4 + 4 I}})
```

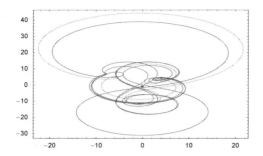

But to obtain a periodic solution for $z(0) = 4 + 4i$, $z'(0) = z''(0) = 4 + 8/3i$, we need to use a higher working precision and have to specify much higher and precision goal (because the solution is nowhere small we do not have to specify a higher accuracy goal).

```
In[70]:= Show[GraphicsArray[
        ParametricPlot[Evaluate[{Re @ z[τ], Im @ z[τ]} /.
                nsol[{4 + 4 I, 4 + 8/3 I, 4 + 8/3 I}, ##]],
                {τ, 0, 2Pi}, DisplayFunction -> Identity,
                PlotRange -> All, Frame -> True, Axes -> False,
                PlotPoints -> 1000]& @@@
    {{}, {PrecisionGoal -> 20, WorkingPrecision -> 40}}]]
```

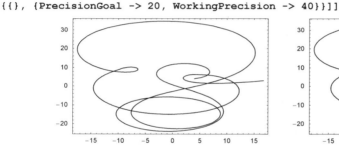 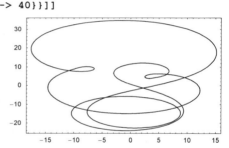

In the next example, we will again use high-precision arithmetic for solving a differential equation. We solve the Buchstab delay-differential equation $zw'(z) + w(z) = w(z-1)$ [161], [381], [1227], [981], [366]. The starting solution is $w(z) = 1/z$ in the interval [1, 2]. The solution of the Buchstab is known to approach a constant as $z \to \infty$. We iterate the numerical solution until we reach identical values at the endpoints of integration. In each step (with the exception of the first), we use the earlier-obtained interpolating functions from NDSolve to calculate $w(z-1)$.

```
In[71]:= BuchstabFunctionList =
    FixedPointList[{(* integration interval start point *) #[[1]] + 1,
            NDSolve[{z w'[z] + w[z] == #[[2]][[1, 1, 2]][z - 1],
                    (* solve differential equation *)
                    w[#[[1]] + 1] == #[[2]][[1, 1, 2]][#[[1]] + 1]}, w,
                    {z, #[[1]] + 1, #[[1]] + 2},
                    WorkingPrecision -> 40, MaxSteps -> 10^4,
                    PrecisionGoal -> 20, AccuracyGoal -> 20]}&,
            (* solution in interval [1, 2] *)
            {1, {{w[z] -> (1/#&)}}},
            (* solve until values at endpoints coincide *)
            SameTest -> ((#1[[2, 1, 1, 2]][#1[[1]] + 1] ==
                         #2[[2, 1, 1, 2]][#2[[1]] + 1])&)];
```

Here is the resulting Buchstab function and the logarithm of its absolute difference to $\exp(-\gamma)$ shown. The right picture shows that we obtain about twenty correct digits for the asymptotic value. This is to be expected, according to the `PrecisionGoal` option setting that was used in `NDSolve`.

```
In[72]:= BuchstabFunction[x_] :=
            BuchstabFunctionList[[IntegerPart[x]]][[2, 1, 1, 2]][x]
```

```
In[73]:= Show[GraphicsArray[
        Block[{$DisplayFunction = Identity},
          {(* the Buchstab function *)
           Plot[BuchstabFunction[x], {x, 1, Length[BuchstabFunctionList] + 1},
              PlotRange -> All],
           (* difference to asymptotic value Exp[-EulerGamma] *)
           ListPlot[Table[Log[10, Abs[BuchstabFunction[x] - Exp[-EulerGamma]]],
                    {x, 1, Length[BuchstabFunctionList], 1/10}],
                    PlotRange -> All]}]]]
```

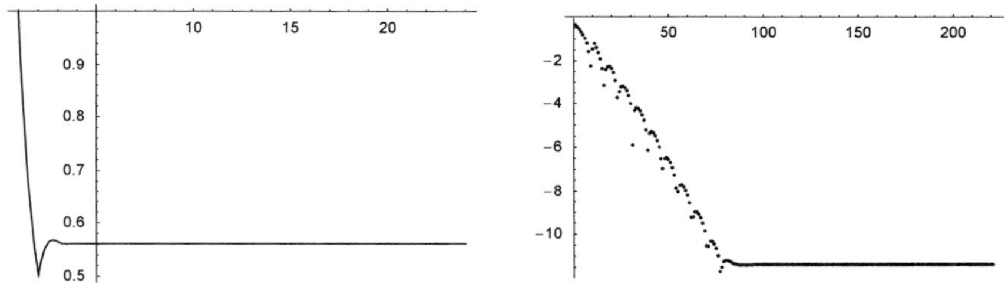

If we have a higher order differential equation and are interested in derivatives of the solution, we should introduce the derivative as a separate function in the list of functions to be computed, instead of differentiating the resulting `InterpolatingFunction`-object. The solution of the differential equation $x''''(t) = x(t)$ with initial conditions $x(0) = 1$, $x'(t) = 0$, $x''(t) = -1$, $x'''(t) = 0$ is $x(t) = \cos(t)$. Here, $x(t)$ is directly computed, and $x'''(t) - \sin(t)$ is plotted.

```
In[74]:= nsol = NDSolve[{D[x[t], {t, 4}] == x[t],
                    x[0] == 1, x'[0] == 0, x''[0] == -1, x'''[0] == 0},
                    x, {t, 0, 4Pi}]
```

```
Out[74]= {{x → InterpolatingFunction[{{0., 12.5664}}, <>]}}
```

```
In[75]:= Plot[Evaluate[(x'''[t] /. nsol) - Sin[t]], {t, 0, 2Pi}]
```

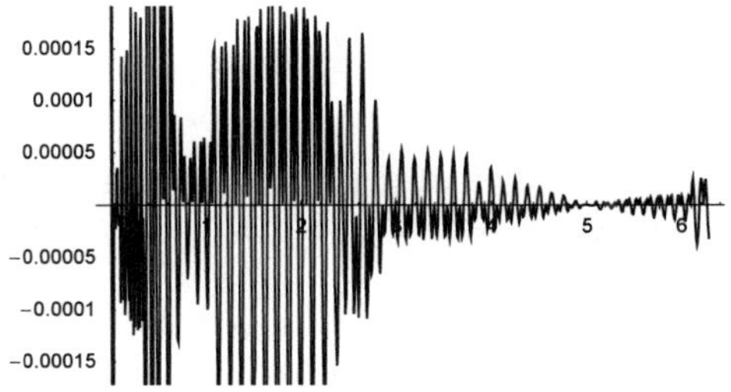

Actually, the error displayed in the last graphic is a combination of three errors. The first is the error in the solution of the differential equation. The second one is introduced in the differentiation. And the third one originates from the interpolation process. As it turns out, the last one is often dominating. In the next picture, we extract the points used by `NDSolve` from `nsol` and display the error at these points only. Now the error is much smaller. (We load the package DifferentialEquations`InterpolatingFunctionAnat ⟍ omy` to conveniently get access to the sample points.

In[76]:= **Needs["DifferentialEquations`InterpolatingFunctionAnatomy`"]**

In[77]:= **ListPlot[({#, Sin[#] - x'''[#]}& /@**
            Flatten[InterpolatingFunctionGrid[nsol[[1, 1, 2]]]]) /.
            nsol[[1]], PlotJoined -> True, PlotRange -> All]

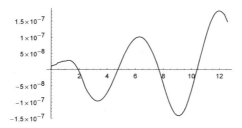

We get a significantly smaller error by direct calculation of $x'''(t)$ (note the function values) instead of differentiating an InterpolatingFunction-object.

In[78]:= **nsolp = NDSolve[{p1[t] == x'[t], p2[t] == p1'[t],**
             p3[t] == p2'[t], p3'[t] == x[t],
             x[0] == 1, p1[0] == 0, p2[0] == -1, p3[0] == 0},
             {x, p1, p2, p3}, {t, 0, 4Pi}]

Out[78]= {{x → InterpolatingFunction[{{0., 12.5664}}, <>],
          p1 → InterpolatingFunction[{{0., 12.5664}}, <>],
          p2 → InterpolatingFunction[{{0., 12.5664}}, <>],
          p3 → InterpolatingFunction[{{0., 12.5664}}, <>]}}

In[79]:= **Plot[Evaluate[(p3[t] /. nsolp) - Sin[t]], {t, 0, 2Pi}, PlotRange -> All]**

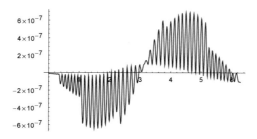

The following data show that nontrivial data for derivative information up to order 11 are stored in the resulting InterpolatingFunction-objects. The error of the derivatives increases quite quickly with the order.

In[80]:=
        Plot[Evaluate[(* show logarithm of absolute error *) **Log @ Abs @**
                  Table[D[nsol[[1, 1, 2]][x] - Cos[x], {x, k}], {k, 12}]],
            {x, 0, 3Pi}, PlotStyle -> Table[Hue[(k - 1)/14], {k, 12}],
            Frame -> True, PlotRange -> {-12, 25}] // Internal`DeactivateMessages

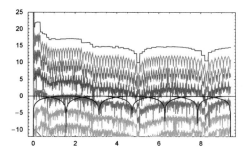

As a side remark, note that NDSolve is currently not able to handle Interval-objects: not in the differential equation, and not in the initial or boundary conditions.

Now, we come to the first of the options unique to NDSolve, namely MaxSteps.

MaxSteps

    is an option for NDSolve, and it determines the maximum number of steps to be taken.

    Default:

        Automatic

    Admissible:

        arbitraryPositiveInteger or Infinity

Next, we model a forced harmonic oscillator with damping. We use a spiky (sometimes negative) time-dependent frequency. To achieve the integration over the time interval (0, 1000), we have to use a higher setting of the option MaxSteps. (For similar increasing amplitude effects due to a time-dependent frequency, see [363].)

```
In[81]:= Module[{nsol, ω, γ = 0.012, T = 120},
    (* the time-dependent frequency *)
    ω2[t_] = 1/2 + Sin[2Pi t/T]^999;
    (* solve the differential equation *)
    nsol = NDSolve[
      {ψ''[t] + γ ψ'[t] + ω2[t] ψ[t] == 1, ψ[0] == 0, ψ'[0] == 1},
      ψ, {t, 0, 1000}, MaxSteps -> 2 10^4];
    (* display the solution *)
    Plot[Evaluate[ψ[t] /. nsol], {t, 0, 500},
        PlotPoints -> 2000, PlotRange -> All,
        Axes -> False, AspectRatio -> 1/4, Frame -> True]]
```

Next, we consider the differential equation for a generalized damped system in a "potential" V. We limit the numerical solution of the differential equation to 20000 steps.

$$\mathbf{r}^{(n)}(t) = -\operatorname{grad} V(\mathbf{r}^{(n)}(t)) - \sum_{k=1}^{n-1} \mathbf{r}^{(k)}(t) \, |\mathbf{r}^{(k)}(t)|$$

```
In[82]:= Show[GraphicsArray[#]]& @
    Block[{V, $DisplayFunction = Identity, cp, p3D, R = 15},
            (* the potential used—cone-shaped with some periodic oscillations *)
            V[{x_, y_}] := Sqrt[x^2 + y^2] - Cos[x] Cos[y];
            (* 3D plot of the potential surface *)
            p3D = Plot3D[V[{x, y}], {x, -6Pi, 6Pi}, {y, -6Pi, 6Pi},
                PlotPoints -> 120, Mesh -> False, Axes -> False];
            (* density plot of the surface *)
            cp = DensityPlot[V[{x, y}], {x, -6Pi, 6Pi}, {y, -6Pi, 6Pi},
                            PlotPoints -> 240, Mesh -> False,
                            FrameTicks -> None];
            (* show potential surface and paths *)
    latten[{p3D, Function[{n, γ},
            (* equations of motions *)
        eqs = D[{x[t], y[t]}, {t, n}] ==
            -{D[V[{x[t], y[t]}], x[t]], D[V[{x[t], y[t]}], y[t]]} -
            γ Sum[# Sqrt[#.#]& @ D[{x[t], y[t]}, {t, j}], {j, 1, n - 1}];
    Show[{cp,
    Table[(* solve differential equations *)
            ndsol = Internal`DeactivateMessages @ NDSolve[Thread /@
            Flatten[{eqs,  {x[0], y[0]} == R {Cos[φ], Sin[φ]},
            Table[{Derivative[j][x][0],
                Derivative[j][y][0]} == {Sin[φ], Cos[φ]},
                {j, n - 1}]}], {x, y}, {t, 0, 100},
                PrecisionGoal -> 8, AccuracyGoal -> 8,
                WorkingPrecision -> MachinePrecision,
                (* number of steps *) MaxSteps -> 2 10^4];
            (* plot trajectories *)
        T = ndsol[[1, 1, 2, 1, 1, 2]];
        ParametricPlot[Evaluate[{x[t], y[t]} /. ndsol],
                    {t, 0, T}, PlotPoints -> Ceiling[20 T] + 10,
                    PlotStyle -> Hue[φ/(2Pi)]],
                    {φ, 0, 2 Pi, 2Pi/17}]},
            PlotRange -> 6 Pi {{-1, 1}, {-1, 1}}]] @@@
            (* ODE degrees and damping constants *)
            {{2, 0.05}, {3, 0.4}, {4, 4}}}]]
```

This option setting `MaxSteps -> Infinity` allows the solution of differential equations and systems of differential equations over long time intervals. For example, a nondefault option setting for `MaxSteps` is needed to understand the phase diagram for the Duffing oscillator [67], [177].

```
In[83]:= ParametricPlot[Evaluate[{x[t], v[t]} /.
            NDSolve[{x'[t] == v[t],
                v'[t] + 1/100 v[t] + 10 x[t] + 6 x[t]^3 ==
                    8 Sin[3/100 t], x[0] == 1/2, v[0] == 0},
                {x, v}, {t, 0, 600}, MaxSteps -> Infinity]], {t, 0, 600},
            Axes -> None, PlotPoints -> 5000, AspectRatio -> 1/3,
            PlotStyle -> {Thickness[0.002]}]
```

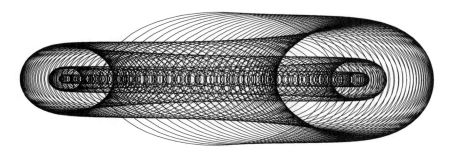

Next, we solve a larger system of equations: the nonlinear coupled equations of the Ablowitz–Ladik chain [1209]. `AblowitzLadikEqs` is a list of the $4\,o + 2$ equations together with (localized) initial conditions.

```
In[84]:= AblowitzLadikEqs[o_, {u_, v_}, ε_, X_] :=
    Join[(* the coupled ODEs *)
        Table[{u[n]'[t] == -I (1 + Abs[u[n][t]]^2)((u[n + 1][t] + u[n - 1][t]) +
                                    ε(v[n + 1][t] + v[n - 1][t])),
               v[n]'[t] == -I (1 + Abs[v[n][t]]^2)((v[n + 1][t] + v[n - 1][t]) +
                                    ε(u[n + 1][t] + u[n - 1][t]))},
            {n, - o, o}] /. (* periodic boundary conditions *)
                        uv_[n_?(Abs[#] > o&)][t] :> uv[-Sign[n](Abs[n] -
    1)][t],
            (* localized initial conditions *)
        Table[Evaluate //@ {u[n][0] == If[Abs[n] < X, Cos[n/X (Pi/2)], 0],
                            v[n][0] == If[Abs[n] < X, Cos[n/X (Pi/2)], 0]},
            {n, - o, o}]]
```

We visualize the absolute value of the $u_n(t)$ and $v_n(t)$. We again use the `Infinity` setting of the `MaxSteps` option.

```
In[85]:= AblowitzLadikEvolutionGraphics[o_, ε_, X_, T_] :=
    Module[{dsol, uData, vData},
        (* solve differential equations *)
        ndsol = NDSolve[AblowitzLadikEqs[o, {u, v}, ε, X],
                    Flatten[Table[{u[n], v[n]}, {n, -o, o}]],
                    {t, 0, T}, MaxSteps -> Infinity];
        (* data of u- and v- values *)
        {uData, vData} = Table[Evaluate[Table[Abs[#[n][t]]^2, {n, -o, o}] /.
                        ndsol[[1]]], {t, 0, T, T/200}]& /@ {u, v};
        (* show absolute value of u- and v- values *)
        Show[GraphicsArray[
        ListContourPlot[#, ColorFunction -> (Hue[0.8 #]&),
                    MeshRange -> {{-o, o}, {0, T}},
                    DisplayFunction -> Identity,
                    ContourLines -> False]& /@ {uData, vData}]]]
```

The next two inputs solve the coupled Ablowitz–Ladik chain equations for 2×61 sites. The equations allow single solitonic solutions and the splitting of one originally localized structure into multiple localized structures is clearly visible.

```
In[86]:= AblowitzLadikEvolutionGraphics[30, -4.8, 3, 30]
```

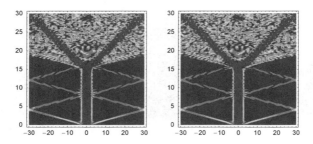

In[87]:= **AblowitzLadikEvolutionGraphics[30, 4.12, 15, 30]**

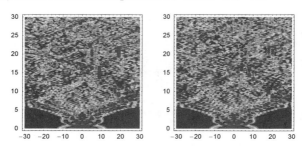

Here is a relatively complicated system of three coupled nonlinear differential equations describing the motion of a charged relativistic particle in an electromagnetic wave [233], [565], [1029], [1756], [1781], [762]. The wave is propagating along the *x*-axis.

In[88]:= **NWEqs[{x_, y_, z_}, t_, {m_, c_, α_, β_, ω_}] =**
```
Module[{(* position and velocity vectors *)
        R = {x[t], y[t], z[t]}, U = {x'[t], y'[t], z'[t]}, k = {1, 0, 0}},
    Apply[Equal, Solve[Thread[
    D[(* relativistic momentum *) m U/(1 - U.U/c^2), t] ==
       (* electric and magnetic field component *)
    α {Cos[k.R - ω t], 0, 0} + β Cross[U, {0, Cos[k.R - ω t], 0}]],
                        {x''[t], y''[t], z''[t]}] // Simplify, {2}]]
```

Out[88]= $\left\{ \left\{ x''[t] \right. \right.$ ==

$-(\text{Cos}[t \omega - x[t]] (c^2 - x'[t]^2 - y'[t]^2 - z'[t]^2) (x'[t]^2 (\alpha + \beta z'[t]) - (\alpha - \beta z'[t])$
$(c^2 + y'[t]^2 + z'[t]^2))) / (c^2 m (c^2 + x'[t]^2 + y'[t]^2 + z'[t]^2)),$

$y''[t] == \dfrac{2 \alpha \text{Cos}[t \omega - x[t]] x'[t] y'[t] (-c^2 + x'[t]^2 + y'[t]^2 + z'[t]^2)}{c^2 m (c^2 + x'[t]^2 + y'[t]^2 + z'[t]^2)},$

$z''[t] == -(\text{Cos}[t \omega - x[t]] x'[t] (-c^2 + x'[t]^2 + y'[t]^2 + z'[t]^2) (c^2 \beta + \beta x'[t]^2 +$
$\left. \left. \beta y'[t]^2 - 2 \alpha z'[t] + \beta z'[t]^2)) / (c^2 m (c^2 + x'[t]^2 + y'[t]^2 + z'[t]^2)) \right\} \right\}$

Next, we numerically solve the last equations for randomly chosen parameters and initial conditions. The left graphic shows the trajectory of the particle, and the right graphic shows the autocorrelation $\{x(t), x(t + 1)\}$. We again have to carry out potentially many integration steps and use a nondefault setting of the MaxSteps option.

In[89]:= **Module[{T = 24, p3D, lines3D, λ3D, p2D, lines2D, λ2D,**
```
                (* extract line segments from an expression *)
                lineSegments = (Line /@ Flatten[Partition[First[#], 2, 1]& /@
                                        Cases[#, _Line, Infinity], 1])&},
    (* seed random number generator *) SeedRandom[563609686962];
    (* solve differential equations *)
```

```
ndsol = NDSolve[Join[NWEqs[{x, y, z}, t, (* random parameters *)
                Append[Table[10^Random[Real, 3 {-1, 1}], {4}],
                        (* use ω = 1 for fixed time scale *) 1]],
                Thread[{x[0], y[0], z[0], x'[0], y'[0], z'[0]} ==
                        Array[Random[Real, {-1, 1}]&, 6]]],
                {x, y, z}, {t, 0, T}, MaxSteps -> Infinity];
(* 3D plot of trajectory {x[t], y[t], z[t]} *)
p3D = ParametricPlot3D[Evaluate[{x[t], y[t], z[t]} /. ndsol], {t, 0, T},
                PlotPoints -> 200 Round[T], PlotRange -> All,
                BoxRatios -> {1, 1, 1}, DisplayFunction -> Identity];
lines3D = lineSegments[p3D]; λ3D = Length[lines3D];
(* 3D plot of autocorrelation {x[t], x[t + 1] *)
p2D = ParametricPlot[Evaluate[{x[t], x[t + 1]} /. ndsol], {t, 0, T - 1},
                PlotPoints -> 200 T, DisplayFunction -> Identity];
lines2D = lineSegments[p2D]; λ2D = Length[lines2D];
(* show the two graphics *)
Show[GraphicsArray[{
 Graphics3D[{Thickness[0.002], (* color trajectory *)
                MapIndexed[{Hue[0.8 #2[[1]]/λ3D], #1}&, lines3D]},
            BoxRatios -> {1, 1, 1}, PlotRange -> All],
 Graphics[{Thickness[0.002], (* color trajectory *)
                MapIndexed[{Hue[0.8 #2[[1]]/λ2D], #1}&, lines2D]},
            PlotRange -> All, Frame -> True, Axes -> False,
            FrameTicks -> None, AspectRatio -> 1]}]]]
```

For the motion of a harmonic oscillator in a wave, see [932]; for the motion in a superposition of plane waves, see [362], [1068]; for the inclusion of spin, see [1549].

In the following input, we will use `NDSolve` to solve a very "bad" differential equation, the Chazy equation. (We encountered it already in the beginning of this chapter.) It has the form $y'''(x) = 2\,y(x)\,y''(x) - 3\,y'(x)^2$ [1072], [310], [1678].

The special property of this differential equation is that its solution contains singular lines (natural boundaries if analyticity). (These boundaries depend on the initial conditions [922].) Across these lines, the solution cannot be (analytically) continued. This time, we do not want to spend a large amount of time approaching the natural boundaries if analyticity; and so, we use a smaller setting of the `MaxSteps` option.

We change the independent variable $x \to r\,e^{i\varphi}$ and integrate the differential equation radially outward starting from the origin for some randomly chosen initial conditions.

```
In[90]:= ndsols = Table[{φ, y /.
        NDSolve[{y'''[r] Exp[-3 I φ] == 2 y[r] y''[r] Exp[-2 I φ] -
                                3 y'[r]^2 Exp[-2 I φ],
                y[0] == 1, y'[0] == 1, y''[0] == 1},
```

```
        {y}, {r, 0, 10}, MaxSteps -> 1000][[1]]},
        {φ, 0, 2Pi, 2Pi/60}];

NDSolve::mxst :
 Maximum number of 1000 steps reached at the point r == 2.1529900913754783`. More…

NDSolve::mxst :
 Maximum number of 1000 steps reached at the point r == 2.1474604769778405`. More…

NDSolve::mxst :
 Maximum number of 1000 steps reached at the point r == 2.169042700206838`. More…

General::stop :
 Further output of NDSolve::mxst will be suppressed during this calculation. More…
```

The messages indicate that in the solution process the natural barrier was encountered. Here, the region in which the numerical solution succeeded is shown. The right graphic shows the scaled real part of the solution.

```
In[91]:= Needs["Graphics`Graphics3D`"]

In[92]:= Show[GraphicsArray[
    Block[{$DisplayFunction = Identity, φ, ipo},
     {(* the maximal radial points reached *)
      Graphics[Line[Apply[N[{#2 Cos[#1], #2 Sin[#1]}]&,
               {#[[1]], #[[2, 1, 1, 2]]}& /@ ndsols, {1}]],
        AspectRatio -> Automatic, Frame -> True],
      (* 3D plot of the scaled real part of the solution *)
      ListSurfacePlot3D[Table[{φ, ipo} = ndsols[[k]];
        Table[{ρ Cos[φ], ρ Sin[φ], (* scale *) ArcTan @ Re @ ipo[ρ]},
               {ρ, 0, ipo[[1, 1, 2]], ipo[[1, 1, 2]]/60}],
          {k, Length[ndsols]}], BoxRatios -> {1, 1, 0.7}]}]]]
```

When the independent variable approaches the natural barrier, the solution starts to oscillate violently. Here this is shown (the bluer the curve, the nearer to the boundary the function is evaluated).

```
In[93]:= points[{φ_, ipo:InterpolatingFunction[{{_, R_}}, __]},
         f_, n_] := Table[{φ, f[ipo[i/n R]]}, {i, 0, n}]

In[94]:= Show[GraphicsArray[Function[reIm,
    Graphics[MapIndexed[{Hue[#2[[1]]/60], Line[#1]}&,
             Transpose[points[#, reIm, 50]& /@ ndsols]] // N,
       Frame -> True, PlotRange -> {-100, 100}, PlotLabel -> reIm]] /@
        (* real and imaginary part *) {Re, Im}]]
```

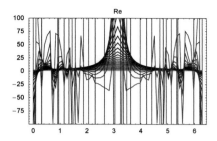
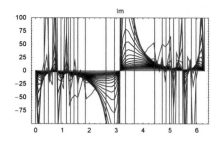

Here is another example where we use a lower than the default value for `MaxSteps`. In the next example, we solve the differential equation $w'(s) = \exp(s^5 + z\,s)$, $w(0) = 0$ with z as a parameter for exactly 100 steps. For larger values of s, the solution again becomes very oscillating which graphically shows as the curls at the tips. To avoid spending larger times calculating the graphically unresolvable inner details of the curls, we use a low value of the `MaxSteps` option setting. Using the first element of the resulting `InterpolatingFunction` gives the information for the s interval that the differential equation was solved in.

```
In[95]:= Show[Graphics[{Thickness[0.002], Table[
         (* make a plot of the solution *)
         ParametricPlot[{Re[#[[1, 1, 2]][s]], Im[#[[1, 1, 2]][s]]},
                       Evaluate[Join[{s}, Flatten[#[[1, 1, 2, 1]]]]],
                       PlotRange -> All, DisplayFunction -> Identity]&[
         (* solve the differential equation;
            instead of s^5 try, say, Tan[α s] as a function of α *)
         NDSolve[{C'[s] == Exp[I(s^5 + z s)], C[0] == 0}, C, {s, -10, 10},
         (* 100 steps and then end *) MaxSteps -> 200]]
                                     [[(* extract the curve *) 1]],
                       {z, -8, 8, 1/2}]}], PlotRange -> All]

         NDSolve::mxst :
          Maximum number of 200 steps reached at the point s == -2.21222. More…

         NDSolve::mxst :
          Maximum number of 200 steps reached at the point s == 2.212221057841809`. More…

         NDSolve::mxst :
          Maximum number of 200 steps reached at the point s == -2.21371. More…

         General::stop :
          Further output of NDSolve::mxst will be suppressed during this calculation. More…
```

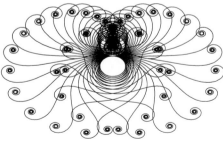

The next option we discuss is `StartingStepSize`.

StartingStepSize

is an option for NDSolve, and it defines the initial step size.

Default:

Automatic

Admissible:

arbitrary positive number

The step size is adaptively selected so that PrecisionGoal and AccuracyGoal are satisfied locally (globally, this might not be the case!). If we choose too large an initial step size (so large that the required precision or accuracy cannot be achieved) using StartingStepSize, this size will be reduced for the next steps, and "skipped" argument values will be evaluated. We now look at such a situation, using a function *func* that not only gives the right-hand side of the differential equation, but also notes the value of the argument. Here, the differential equation is $x''(t) = t/\sin^2(x(t))$.

```
In[96]:= tValuesUsed = (Reap @
         NDSolve[{x''[t] == Sin[t]/x[t]^2, x[1] == 1, x'[1] == 1},
            x, {t, 1, 5}, StartingStepSize -> 0.3,
            EvaluationMonitor :> Sow[t]])[[2, 1]];
```

Here are the data points used in the *i*th step. After the initial large specified step sizes, we see how the step size control of NDSolve requires the use of much smaller step sizes.

```
In[97]:= Show[Graphics[Line[MapIndexed[{#2[[1]], #1}&, tValuesUsed]]],
            Axes -> True, PlotRange -> {{0, 100}, {1, 2}},
            AxesLabel -> {" step ", "t"}]
```

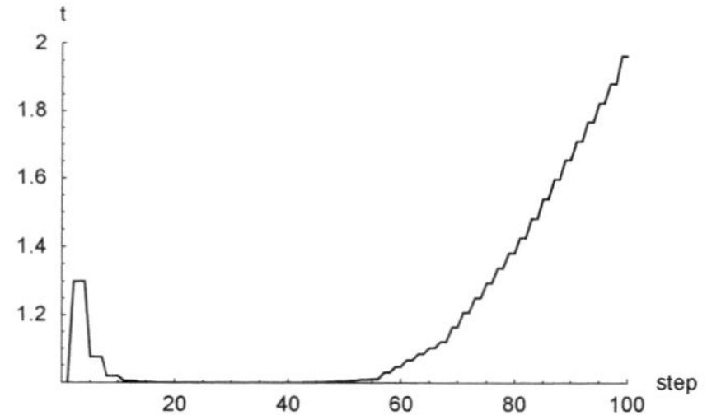

```
In[98]:= Options[NDSolve]
```

Out[98]= {AccuracyGoal → Automatic, Compiled → True, DependentVariables → Automatic,

EvaluationMonitor → None, MaxStepFraction → $\frac{1}{10}$, MaxSteps → Automatic,

MaxStepSize → Automatic, Method → Automatic, NormFunction → Automatic,

PrecisionGoal → Automatic, SolveDelayed → False, StartingStepSize → Automatic,

StepMonitor → None, WorkingPrecision → MachinePrecision}

To keep the adaptive method from taking steps that are too large, we can adjust for the maximum step size.

MaxStepSize

is an option for NDSolve, and it defines the maximum allowable step size.

Default:

    Automatic

possible:

    arbitrary positive number

Sometimes, it is more convenient to specify not an absolute maximum step size, but the maximum step size expressed in the length of the interval where the differential equation is to be solved. A closely related function to MaxStepSize is MaxStepFraction. It allows us to control the maximum relative step size.

MaxStepFraction

is an option for NDSolve, and it defines the maximal relative length (relative to the range of the independent variable) of the allowed step sizes.

Default:

    1

Admissible:

    arbitrary positive number between 0 and 1

The following differential equation has a small (in width) discontinuity on the right side. Without a special choice for MaxStepSize, it remains unnoticed (left graphic). But using a small enough value for MaxStep‹ Size allows us to detect the peak in the right-hand side of the differential equation, and the solution now comes out correctly (right graphic).

```
In[99]:= Show[GraphicsArray[
    Block[{$DisplayFunction = Identity},
    Plot[Evaluate[y[x] /.
        NDSolve[{y'[x] == If[0.5 < x < 0.51, 1000, 1],
            y[0] == 0}, y[x], {x, 0, 1}, MaxStepSize -> #]],
    {x, 0, 1}]& /@
        (* the two step size settings *) {Automatic, 0.0005}]]]
```

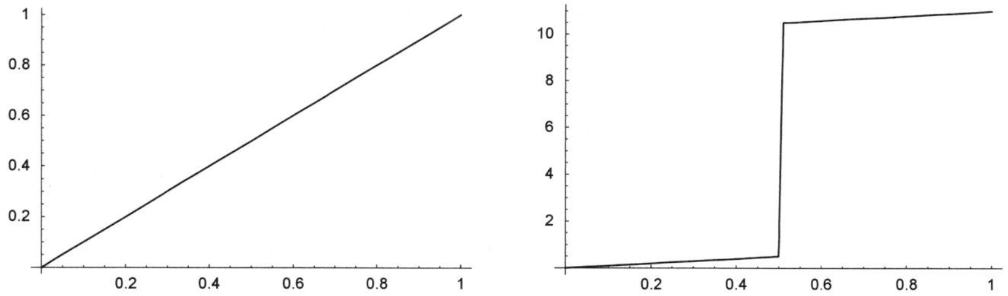

An alternative possibility (and for the simple example from the last input a more appropriate one) would have been the following input:

```
NDSolve[{y'[x] == If[0.5 < x < 0.51, 1000, 1], y[0] == 0},
    y[x], {x, 0, 0.5, 1}]
```

The next input solves a more complicated system of differential equations. We consider 50 coupled pendulums [137], [1138], [456], [1724], [1489]. To obtain the solution over the specified *t*-interval, we have to use a larger value for the MaxSteps option. To obtain an appropriately resolved "focusing" structures, we use a small value for MaxStepSize.

```
In[100]:= Module[{n = 50, T = 120, θ, eqs, nsol},
         (* equations *)
         eqs = Join[(* equations of motion *)
                   Table[θ[k]''[t] + 1/(2n) Sum[Sin[θ[k][t] - θ[j][t]],
                                               {j, n}] == 0, {k, n}],
                   (* initial conditions *)
                   Table[θ[k][0] == 2Pi k/n, {k, n}],
                   Table[θ[k]'[0] == 0, {k, n}]];
         (* solve system of ODEs numerically *)
         nsol = NDSolve[eqs, Table[θ[k], {k, n}], {t, 0, T},
                       (* small maximal step size; many allowed steps *)
                       MaxStepSize -> 0.1, MaxSteps -> 10^5];
         (* visualize solutions *)
         Show[Graphics[{Thickness[0.002],
         MapIndexed[{Hue[#2[[1]]/60], #1}&, Map[Line,
         (* remove jumps from 0 to 2Pi and 2Pi to 0 *)
         Select[Partition[#[[1, 1, 1]], 2, 1],
               Abs[#[[1, 2]] - #[[2, 2]]] < 2&]& /@
               (* plot solution curves *)
               Plot[Evaluate[Table[Mod[θ[k][t], 2Pi], {k, n}] /. nsol],
                   {t, 0, T}, PlotPoints -> 500,
                   DisplayFunction -> Identity][[1]], {-3}]]}],
         Frame -> True, AspectRatio -> 1/3,
         (* show detail *) PlotRange -> {{80, 120}, {Pi, 2Pi}}]]
```

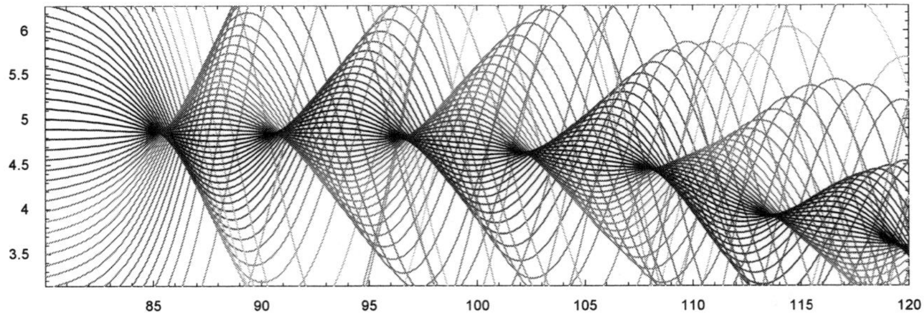

The next option we will discuss is SolveDelayed.

SolveDelayed

is an option for NDSolve, and it determines whether in the preprocessing stage to solve for the highest derivatives.

Default:
    False

Admissible:
    True

By default, *Mathematica* tries to solve for the highest derivatives before starting the actual numerical solution of the differential equation. Sometimes this might not be possible, for instance, in the following example. The `If` in the following cannot be symbolically inverted.

`In[101]:= NDSolve[{If[x[t] > E, 3 x'[t], x'[t]] == x[t], x[0] == 1}, {x}, {t, 0, 2}]`

> Solve::dinv : The expression If[#2 > e, 3 #3, #3] involves unknowns in
> more than one argument, so inverse functions cannot be used. More…

> NDSolve::ntdv : Cannot solve to find an explicit formula for the
> derivatives. Consider using the option setting SolveDelayed->True. More…

`Out[101]= NDSolve[{If[x[t] > e, 3 x'[t], x'[t]] == x[t], x[0] == 1}, {x}, {t, 0, 2}]`

If `x[t]` has a numerical value (this means at the time the numerical solution is in progress), it is of course possible to solve for `x'[t]`. The next two inputs solve the differential equation and plot the resulting solution function.

`In[102]:= NDSolve[{If[x[t] > E, 3 x'[t], x'[t]] == x[t], x[0] == 1}, {x}, {t, 0, 2},`
`          SolveDelayed -> True]`

`Out[102]= {{x → InterpolatingFunction[{{0., 2.}}, <>]}}`

`In[103]:= Plot[Evaluate[x[t] /. %], {t, 0, 2}]`

Here is a slightly more complicated example: $\cos(x''(t)) = x''(t) + x(t)^2$. Without setting the option `SolveDe`-`layed` to `True`, the solution is not accomplished.

`In[104]:= NDSolve[{Cos[x''[t]] == x''[t] + x[t]^2, x[0] == 1, x'[0] == 1}, x, {t, 0, 1}]`

> Solve::tdep : The equations appear to involve the variables
> to be solved for in an essentially non-algebraic way. More…

> NDSolve::ntdv : Cannot solve to find an explicit formula for the
> derivatives. Consider using the option setting SolveDelayed->True. More…

`Out[104]= NDSolve[{Cos[x''[t]] == x[t]² + x''[t], x[0] == 1, x'[0] == 1}, x, {t, 0, 1}]`

With the setting `SolveDelayed -> True`, `NDSolve` has to solve a nonlinear equation at each quadrature step.

`In[105]:= NDSolve[{Cos[x''[t]] == x''[t] + x[t]^2, x[0] == 1, x'[0] == 1},`
`          x, {t, 0, 1}, SolveDelayed -> True]`

`Out[105]= {{x → InterpolatingFunction[{{0., 1.}}, <>]}}`

Conceptually related to the `SolveDelayed` option is the following remark about delaying the evaluation of function evaluations until numerical arguments are supplied.

> Several numerical routines, including `NDSolve`, perform tests and reformulations aimed at optimizing the computation before starting. To avoid this, we can constrain the evaluation to numerical arguments using `?NumberQ`.

Here is an example involving a system of equations arising in electrostatics of semiconductor devices (see, e.g., [1082], [1218], [1219], and [1771]). The Poisson equation for the electrostatic potential $V(x)$, taking into account the charge of free movable electrons, is [1578]

$$V''(x) = \int_0^\infty \frac{\sqrt{\varepsilon}}{1 + e^{\varepsilon - V(x)}} \, d\varepsilon + otherCharges.$$

```
In[106]:= NDSolve[{v''[x] ==
              NIntegrate[Sqrt[ε]/(1 + Exp[ε - v[x]]), {ε, 0, Infinity}],
              v[1] == 1, v'[1] == 1}, v[x], {x, 1, 2}]
```

NIntegrate::inum : Integrand $\frac{\sqrt{\varepsilon}}{1 + e^{\varepsilon - v[x]}}$ is not numerical at $\{\varepsilon\} = \{1.\}$. More…

NIntegrate::inum :

Integrand $\frac{\sqrt{\varepsilon}}{1 + e^{\varepsilon - \text{TemporaryVariable}[1][x]}}$ is not numerical at $\{\varepsilon\} = \{1.\}$. More…

NIntegrate::inum : Integrand $\frac{\sqrt{\varepsilon}}{1 + e^{\varepsilon - \#2}}$ is not numerical at $\{\varepsilon\} = \{1.\}$. More…

General::stop : Further output of
    NIntegrate::inum will be suppressed during this calculation. More…

```
Out[106]= {{v[x] → InterpolatingFunction[{{1., 2.}}, <>][x]}}
```

While we obtained a solution, the messages issued are distracting. If we define the right side of the differential equation only for variables with explicit numerical values, no messages are issued.

```
In[107]:= rightHandSide[v_?NumberQ] :=
              NIntegrate[Sqrt[ε]/(1 + Exp[ε - v]), {ε, 0, Infinity}];

          (NDSolve[{v''[x] == rightHandSide[v[x]]},
              v[1] == 1, v'[1] == 1}, v[x], {x, 1, 2}])
Out[109]= {{v[x] → InterpolatingFunction[{{1., 2.}}, <>][x]}}
```

Here is a problem that is frequently of practical interest. We want to follow the solution of a differential equation only as long as the solution satisfies certain conditions. With the present version of NDSolve, this is not directly possible. In such cases, we can usually proceed along the following lines. Suppose, for demonstration, the differential equation to be solved is $y'(x) = x\,y(x)$, $y(0) = 1$, and that we only want a solution as long as $y(x) < 3$. We can implement this as follows.

```
In[110]:= Clear[f, x];

          f[y_, x_] := If[y < 3, x y]
```

Here is the solution of the differential equation.

```
In[112]:= NDSolve[{y'[x] == f[y[x], x], y[0] == 1}, y[x], {x, 0, 2}]
```

NDSolve::nlnum : The function value {Null} is not a list
    of numbers with dimensions {1} at x = 1.5084242394881038`. More…

```
Out[112]= {{y[x] → InterpolatingFunction[{{0., 1.44854}}, <>][x]}}
```

Indeed, we get error messages because of the Null, which is produced by f[y, x] for $y > 3$, but the result satisfies the desired requirements. If one wants to solve a differential equation only as long as some conditions on the dependent variables are satisfied, the option StoppingTest is a better choice.

StoppingTest

> is an option for NDSolve, that ends the differential equation solving if the option setting no longer evaluates to True.

As an example for the use of the StoppingTest option in NDSolve, let us take the following pursuit problem [187], [142], [647], [1525], [1526]. A point {x[t], y[t]} is moving along the unit circle with unit speed. Another point is approaching the first one in such a manner that its velocity is at every time directed toward the first point. The speed of the second point is 1, too. Here, the corresponding differential equations are solved. We end solving the differential equations when the distance between the two points is less than 10^{-3}.

Let us start the second point particle at {2, 2}.

```
In[113]:= (* point on the circle *)
        {xC[t_], yC[t_]} = {Cos[t], Sin[t]};

        pursuitODEs[κ_:1] := Sequence @@ Thread[
        {sx'[t], sy'[t]} == κ {xC[t] - sx[t], yC[t] - sy[t]}/
                        Sqrt[(xC[t] - sx[t])^2 + (yC[t] - sy[t])^2]]

        pursuitODENSol[{x0_, y0_}, T_, κ_:1] :=
        NDSolve[(* always point toward the point on the circle *)
                    {pursuitODEs[κ],   sx[0] == x0, sy[0] == y0},
                    {sx, sy}, {t, 0, T}, MaxSteps -> 10000, StoppingTest :>
                                ((xC[t] - sx[t])^2 + (yC[t] - sy[t])^2 < 10^-3)]

        nsol = pursuitODENSol[{2, 2}, 100]
Out[118]= {{sx → InterpolatingFunction[{{0., 61.6846}}, <>],
        sy → InterpolatingFunction[{{0., 61.6846}}, <>]}}
```

Here is a picture showing the distance between the two points. The right logarithmic plot shows better when the stopping criteria was achieved.

```
In[119]:= Show[GraphicsArray[
        Block[{$DisplayFunction = Identity},
        {Plot[(* quality of stopping *)
                Evaluate[(xC[t] - sx[t])^2 + (yC[t] - sy[t])^2 /. nsol],
                {t, 0, nsol[[1, 1, 2, 1, 1, 2]]}],
            Plot[Evaluate[Log[10, (xC[t] - sx[t])^2 + (yC[t] - sy[t])^2] /. nsol],
                {t, 0, nsol[[1, 1, 2, 1, 1, 2]]}]}]]]
```

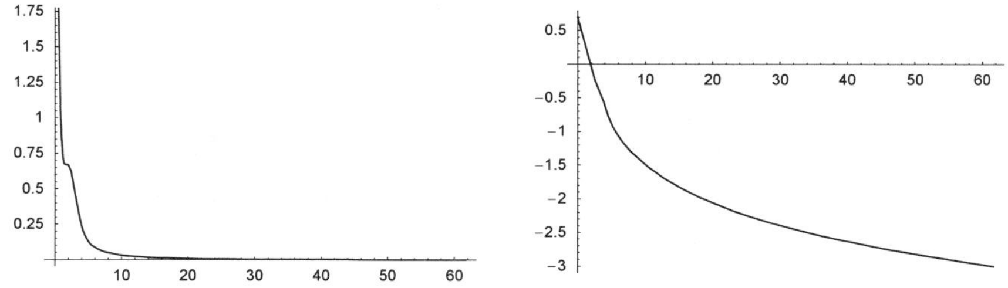

The following graphic shows the second point approaching the first one. The gray lines indicate the tangent vectors of the path of the second point.

```
In[120]:= pursuitGraphics[nsol_, pp_, col_] :=
    Module[{tMax = nsol[[1, 1, 2, 1, 1, 2]], tab1, tab2},
       (* the trajectory *)
       tab1 = Table[Evaluate[{sx[t], sy[t]} /. nsol[[1]]],
                 {t, 0, tMax, tMax/pp}];
       (* the circle *)
       tab2 = Table[{xC[t], yC[t]}, {t, 0, tMax, tMax/pp}];
       Show[Graphics[{(* connect trajectory to circle *)
                      {Thickness[0.002], col,
                       Line /@ Transpose[{tab2, tab1}]},
                      {Thickness[0.003], GrayLevel[0.4], Line[tab2]},
                      {Thickness[0.006], GrayLevel[0], Line @ tab1}}],
            PlotRange -> All, AspectRatio -> Automatic]]
In[121]:= pursuitGraphics[nsol, 800, Hue[0]]
```

We use the just-defined functions `pursuitGraphics` and `pursuitODENSol` to display 16 trajectories simultaneously. The right graphic shows the paths traced out by followers of different speeds.

```
In[122]:= Show[GraphicsArray[
    Block[{$DisplayFunction = Identity, pp = 16},
      {(* different distances of the starting points *)
       Show[Table[pursuitGraphics[pursuitODENSol[
          (12/10 + φ/2) {Cos[φ], Sin[φ]}, 10], 60, Hue[0.8 φ/(2Pi)]],
          {φ, 0, 2Pi, 2Pi/pp}], PlotRange -> All],
       (* different speeds of the followers *)
       Show[Table[pursuitGraphics[pursuitODENSol[
          3 {Cos[φκ], Sin[φκ]}, 10, φκ/Pi], 60, Hue[0.8 φκ/(2Pi)]],
          {φκ, 0, 2Pi (1 - 1/pp), 2Pi/pp}], PlotRange -> All]}]]]
```

As another example of the use of the `StoppingTest` option, we calculate an approximative value of $y'(0)$ such that $y''(x) = x^{-1/2} y(x)^{3/2}$ and $y(0) = 1$, $y(\infty) = 0$. (For details on the Thomas–Fermi equation, see [170] and [589]; for the nonextensive version, see [1233].)

We do this by starting very near to the origin (because at the origin we have a singularity) and solve the differential equation until either the solution becomes negative or starts to increase. maxX returns the furthest point reached on a sensible solution. We use the now familiar construction *function* [*var_* ?NumberQ] := ... to avoid the symbolic evaluation of −maxX[*ys0*] inside FindRoot.

```
In[123]:= maxX[ys0_?NumberQ, ndsolveOpts___] :=
          Module[{x, y},
            With[{wp = Precision[ys0]},
                NDSolve[{y''[x] == y[x]^(3/2)/Sqrt[x],
                        y[10^(-wp)] == 1, y'[10^(-wp)] == ys0},
                        y, {x, 10^(-wp), 10^6}, StoppingTest -> (* stop? *)
                          If[Re[y'[x]] > 0 || Re[y[x]] < 0, True, False],
                        ndsolveOpts, MaxSteps -> 2000, AccuracyGoal -> Infinity,
                        WorkingPrecision -> wp][[1, 1, 2, 1, 1, 2]]]]
```

Now, maximizing the largest reachable *x*-value, we find an approximation for $y'(0)$.

```
In[124]:= Timing[FindMinimum[-maxX[ys0], {ys0, -1, 2},
                          PrecisionGoal -> 10, MaxIterations -> 100]]
Out[124]= {2.94 Second, {-454.246, {ys0 → -1.58807}}}
```

As the last message suggests, we now repeat the last calculation using high-precision arithmetic. Now, we can obtain a 10-digit approximation of $y'(0)$ [1779].

```
In[125]:= Timing[FindMinimum[-maxX[ys0, PrecisionGoal -> 15, MaxSteps -> 5000,
                            WorkingPrecision -> 30],
                    {ys0, -1588/1000, -15881/10000},
                    WorkingPrecision -> 40,
                    PrecisionGoal -> 10, MaxIterations -> 100] // N[#, 10]&]
Out[125]= {132.5 Second, {-8238.653798, {ys0 → -1.588071023}}}
```

(For a similar technique to determine the bound states of the Schrödinger–Newton equation, see [1259].)

Here is a slightly more complicated example using the StoppingTest option. We consider the set of coupled differential equations [581], [166]

$$x_i'(t) = (\beta \cos(t\,\omega + i) - 1)\,x_i(t) - (2\,x_i(t) - x_{i-1}(t) - x_{i+1}(t)).$$

It is known that for $\beta > \beta_n^*$, the solutions $x_i(t)$ will blow up. To find an approximate value of β_n^*, we solve the system until one component, say $|x_1(t)|$, becomes large (say 10^3, meaning we use the StoppingTest option to carry out the stop of the numerical integration). The next inputs solve the system for $2 \le \beta \le 4$ for $n = 30$.

```
In[126]:= Clear[x];
          odes[β_, ω_, n_] :=
          Table[x[i]'[t] == (-1 + β Cos[ω t + i]) x[i][t] -
                          (2 x[i][t] - x[i + 1][t] - x[i - 1][t]),
                {i, n}] /. {x[0][t] -> 0, x[n + 1][t] -> 0};

In[128]:= data = With[{n = 30}, Table[
          nsol = NDSolve[Join[odes[β, 1, n], Table[x[i][0] == 1, {i, n}]],
                          Table[x[i], {i, n}], {t, 0, 100}, MaxSteps -> 10^4,
                          StoppingTest :> (Abs[x[1][t]] > 10^3)];
                      {β, nsol[[1, 1, 2, 1, 1, 2]]}, {β, 2, 4, 0.01}]];
```

The following graphic shows that $\beta_{30}^* \approx 2.52$.

```
In[129]:= ListPlot[data, PlotJoined -> True]
```

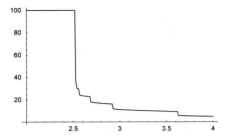

And here is one more application of the `StoppingTest` option. We consider the scattering of particles on the 2D potential $\mathcal{V}(x, y) = x^2 y^2 \exp(-x^2 + y^2)$ [219], [217], [1943].

```
In[130]:= Clear[x, y];
         𝒱[x_, y_] := x^2 y^2 Exp[-(x^2 + y^2)]
```

We consider the particles with initial velocity $v0$ moving towards the center of the potential from a direction φ. The potential $\mathcal{V}(x, y)$ has four peaks at $\{\pm 1, \pm 1\}$. For certain initial velocities and impact angles, a particle can get "caught" and bounce around between the four hills multiple times. We integrate the equations of motion until the particle has definitely left the center part of the potential.

```
In[132]:= trajectory[φ_, R0_, v0_] :=
     Module[{(* maximal time to integrate the ODEs *) T = 20 R0/v0},
        NDSolve[{x''[t] == -D[𝒱[x[t], y[t]], x[t]],
                 y''[t] == -D[𝒱[x[t], y[t]], y[t]],
                 (* initial direction toward origin *)
                 x[0] == R0 Cos[φ], x'[0] == -v0 Cos[φ],
                 y[0] == R0 Sin[φ], y'[0] == -v0 Sin[φ]},
                {x, y}, {t, 0, T}, MaxStepSize -> 0.1, MaxSteps -> 10^5,
                (* stop when the core of the potential is left *)
                StoppingTest :> If[t < R0/v0, False, x[t]^2 + y[t]^2 > R0^2]]]
```

The following two graphics show the scattering on the potential for $v0 = 0.22$. The left graphic shows 120 trajectories. One sees that for $25° \lesssim \varphi \lesssim 45°$ (and for the other seven symmetrically positioned φ-ranges) the particle enters the inside of the four hills and bounces multiple times around. The right graphic shows the integration times T of the differential equation. At the confining angles, the time $T(\varphi)$ is a complicated function of φ, with values up to twice as large as outside.

```
In[133]:= Module[{v0 = 0.22, pt = 120, pT = 100},
     Show[GraphicsArray[
     Block[{$DisplayFunction = Identity, ndsol},
     {Show[(* plot of the pt trajectories and the potential *)
     {(* the potential; hills are white *)
     ContourPlot[Evaluate[𝒱[x, y]], {x, -5, 5}, {y, -5, 5},
             PlotPoints -> 40, ColorFunction -> (GrayLevel[1 - #]&),
             PlotRange -> All, Contours -> 100, ContourLines -> False],
     (* the particle trajectories *)
     Table[ndsol = trajectory[φ, 10, v0];
         ParametricPlot[Evaluate[{x[t], y[t]} /. ndsol],
                 {t, 0, ndsol[[1, 1, 2, 1, 1, 2]]},
                 PlotStyle -> {{Hue[φ/(2Pi)], Thickness[0.002]}},
                 AspectRatio -> Automatic],
         {φ, 0, 2 Pi, 2Pi/pt}]}, PlotRange -> {{-5, 5}, {-5, 5}}],
     (* plot of the integration times *)
     ListPlot[Table[{φ, trajectory[φ, 10, v0][[1, 1, 2, 1, 1, 2]]} // N,
```

```
{φ, 0, Pi/4, Pi/4/pT}], PlotRange -> All,
   AxesOrigin -> {0, 0}]}]]]]
```

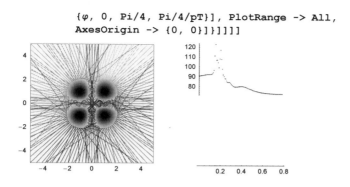

In the above applications, we used the option `StoppingTest` to end the numerical integration of the differential equations. To achieve the exact stopping point was not essential. Sometimes it is important to stop a differential equation as precisely as possible. This can be done with the event detection method option of `NDSolve`.

`EventLocator`

   is a method option for `NDSolve` that determines the action to be taken when certain events happen.

The next input calculates the periods of a generalized harmonic oscillator with a force of the form $x(t)^\alpha + \alpha^{x(t)}$ and initial condition $x(0) = 0$, $x'(0) = 1$. We use the option `StoppingTest` to end the integration of the equation of motion as soon as $x(t)$ becomes negative.

```
In[134]:= (ahoDataSTFR =
   Module[{α, ThApprox, x, t, T = 6},
   Table[(* vary α exponentially *) α = 10^β;
     (* solve differential equation until elongation becomes negative *)
     ndsol = x[t] /. NDSolve[{x''[t] == -Sign[x[t]] Abs[x[t]]^α -
                                   Sign[x[t]] α^Abs[x[t]],
                        x[0] == 0, x'[0] == 1},
                   x, {t, 0, T}, StoppingTest -> x[t] < -1/100,
                   MaxSteps -> 10^5, PrecisionGoal -> 10][[1]];
     (* approximate half period *)
     ThApprox = ndsol[[0, 1, 1, 2]];
     {α, 2 t /. FindRoot[Evaluate[ndsol == 0], {t, 0.9 ThApprox, ThApprox}]}, //
                                    {β, -3, 12/10, 1/100}]]); //
                                                   Timing
Out[134]= {7.78 Second, Null}
```

This family of oscillators has a local minimal period of 2.72... at $\alpha_{min} = 0.072...$ and a local maximal period of 3.14... at $\alpha_{min} = 1.19...$ and

```
In[135]:= ListPlot[ahoDataSTFR // N, AxesLabel -> {"α", "T"}]
```

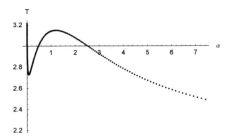

The use of the postprocessing call to FindRoot worked fine in the last example, but is inelegant to program. A much more convenient way to obtain the period (and also a slightly faster and numerically more reliable way) is to let NDSolve stop the integration at the right moment. With the suboption set to "Event" -> *term*, NDSolve will stop so that term will be zero. Using the "EvenLocationMethod" allows specifying how exactly the zero-valuedness of the specified expression should be determined as a function of the independent variable. In the following, we use the "Brent" setting which specifies to use the 1D bracketing Brent root-finding method.

```
In[136]:= (ahoDataEvent =
        Module[{α, ThApprox, x, t, T = 6},
        Table[(* vary α exponentially *) α = 10^β;
          (* solve differential equation until elongation becomes negative *)
          ndsol = x[t] /.
          NDSolve[{x''[t] == -Sign[x[t]] Abs[x[t]]^α - Sign[x[t]] α^Abs[x[t]],
                x[0] == 0, x'[0] == 1}, x, {t, 0, 6},
              MaxSteps -> 10^5, PrecisionGoal -> 10,  Method ->
              {EventLocator, "Event" -> x[t], "EvenLocationMethod" -> "Brent"}];
            {α, 2 ndsol[[1, 0, 1, 1, 2]]}, {β, -3, 12/10, 1/100}]]); // Timing
Out[136]= {6.53 Second, Null}
```

The two sets of periods calculated agree excellently with each other.

```
In[137]:= Max[Abs[Last /@ (ahoDataEvent - ahoDataSTFR)]]
Out[137]= 1.01503 × 10^{-8}
```

NDSolve can also deal with vector- and matrix-valued equations. In the next example, we solve a matrix-valued first-order ODE. Only the initial conditions indicate the matrix-valued nature of the equations because the right-hand side does not evaluate to a matrix for symbolic input. The function oscillatorPhas‹ esDensityGraphics generates a list of graphics or shows single graphic for a $d \times d$ array of coupled oscillator phases $y_{i,j}(t)$ [992].

$$y'_{i,j}(t) = \frac{1}{4} \left(\sin(y_{i,j}(t) - y_{i+1,j}(t) - \alpha) + \sin(y_{i,j}(t) - y_{i,j+1}(t) - \alpha) + \right.$$
$$\left. \sin(y_{i,j}(t) - y_{i-1,j}(t) - \alpha) + \sin(y_{i,j}(t) - y_{i,j-1}(t) - \alpha) \right)$$

So far, we have solved all differential equation by using NDSolve[*equations, variables, timeRange, options*]. The *timeRange* can specify arbitrary times independent from the times at which the initial conditions are given. This means it can be of the form $\{t, T, T\}$ which implies that only a solution at a specific time is returned (and such solutions are typically much smaller than large interpolating functions that approximate the solution over longer time intervals). Another method, especially suited for repeatedly propagating solutions in "time" (forward or backward), is the NDSolve`Iterate function. After setting up a numerical differential equation problem with NDSolve`ProcessEquations, the NDSolve`StateData-object can be propagated using `NDSolve`Iterate`. Compared with a repeated call to `NDSolve`, this method has the advantages of avoiding the (potentially quite expensive) preprocessing step of the problem at hand and allows keeping derivative information in addition to the initial conditions for a restart of the differential equation solver. At the end, we convert a `NDSolve`StateData`-object to a list with a rule using `NDSolve`ProcessSolu`tions`. The next input uses these two functions for the just-described system of coupled oscillators.

```
In[138]:= oscillatorPhasesDensityGraphics[d_, α_, times_, DoTable_,
                                  f_:(Random[Real, {-Pi, Pi}]&)] :=
      Module[{rhsMatrix, rhsMatrixC, state, ndsol},
         (* right hand side of the matrix equation *)
         rhsMatrix[y_?MatrixQ, a_] := rhsMatrixC[y, a];
         rhsMatrixC = Compile[{{y, _Real, 2}, a},
           Module[{d = Length[y], y = 0. y, i, j,
                   nList = {{-1, 0}, {1, 0}, {0, -1}, {0, 1}}},
               Do[y[[i, j]] = Sum[{i, j} = Mod[{i, j} + nList[[k]], d, 1];
                                  Sin[y[[i, j]] - y[[i, j]] - a],
                                  {k, Length[nList]}],
                  {i, d}, {j, d}]; y/Length[nList]]];
         (* set up equations *)
         {state} = NDSolve`ProcessEquations[
                       {Y'[t] == rhsMatrix[Y[t], α],
                        Y[0] == Table[f[i/d, j/d], {i, d}, {j, d}]},
                        Y, t, PrecisionGoal -> 6, AccuracyGoal -> 6];
         (* iterate to the next time *)
         DoTable[NDSolve`Iterate[state, times[[j]]];
            (* visualize solution *)
            ndsol = NDSolve`ProcessSolutions[state, "Forward"];
            ListDensityPlot[Mod[ndsol[[1, 2]], 2Pi, -Pi],
                      PlotRange -> {-Pi, Pi}, Mesh -> False,
                      ColorFunction -> Hue, FrameTicks -> False],
                  {j, Length[times]}]]
```

This example starts with random initial phases. We clearly see the evolution of local spirals.

```
In[139]:= SeedRandom[1];

        Show[GraphicsArray[
          Block[{$DisplayFunction = Identity},
              oscillatorPhasesDensityGraphics[128, # Pi/4, {0, 5, 50},
                                        Table]]]]& /@ {-1, +1}
```

 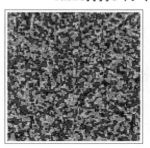 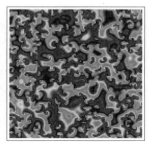

  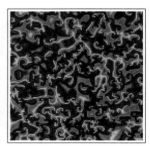

```
SeedRandom[1];
oscillatorPhasesDensityGraphics[128, Pi/4, Range[100], Do]
```

The next two graphics start with a discretized radial symmetric initial condition. Depending on the value of α, we see quite different structures evolve.

```
In[141]:= SeedRandom[1];

Show[GraphicsArray[
Block[{$DisplayFunction = Identity},
oscillatorPhasesDensityGraphics[128, #, {60}, Table,
        (Pi/2 Sign[Sin[100 Sqrt[(#1 - 1/2)^2 + (#2 - 1/2)^2]]])&][[-1]]& /@
                                                {Pi/4, 3Pi/4}]]]
```

Lastly, we will discuss the possible basic integration methods that can be used inside NDSolve.

Here are some rules of thumb for choosing a method option (in case the Automatic option setting results in a suboptimal choice of the option). (These are some of the methods available, for a complete listing, see the Advanced Documentation. Many of these method options have suboptions, like the order for Runge–Kutta methods, we do not go into such details here.)

```
"Adams"

    for large nonstiff systems.
"BDF"

    for stiff system.
"LSODA"

    for obtaining solutions of potentially stiff problems.
"ExplicitRungeKutta"

    for nonstiff systems.
"ImplicitRungeKutta"

    for stiff systems.
"Extrapolation"

    for obtaining very high–precision solutions.
"StiffnessSwitching"

    for obtaining high–precision of potentially stiff problems.
"Projection"

    for problems where maintaining invariants is important.
"SymplecticRungeKutta"

    for problems arising from Hamiltonian systems.
"Chasing"

    for boundary value problems.
"IDA"

    for differential-algebraic systems (used with SolveDelayed -> True).
```

For most purposes, one wants to use the method option setting Automatic. In this mode, *Mathematica* switches between the various methods in such a way that the solution is fast and reliable at the same time. Here is the result of solving $\varepsilon\, y'(t) = t\,(1 - y(t)^2)$ (known for its "jumpy" behavior [1196], [1384], [1383]) for various ε and the various option settings displayed. The set of solutions obtained from the Automatic is a correct one (as are the other solutions from the first row).

```
In[143]:= (* compare the methods *)
    Module[{nsol, y},
    Show[GraphicsArray[#]]& /@
    Partition[(Show[Table[
    (* avoid message generation *) Internal`DeactivateMessages[
    (* solve odes *)
    nsol = NDSolve[{10^-ε y'[t] == t (1 - y[t]^2), y[-1] == -1/2},
                    y, {t, -2, 1}, Method -> #];
    (* generate plots *)
    Plot[Evaluate[y[x] /. nsol],
        {x, -2, nsol[[1, 1, 2, 1, 1, 2]]},
        PlotRange -> All, DisplayFunction -> Identity,
        PlotStyle -> {{Thickness[0.002], Hue[ε/4]}}]],
            {ε, 0, 2, 2/30}] //.
            (* take out too large numbers *)
            r_Real :> If[Abs[r] > 2, 2 Sign[r], r], PlotLabel ->
                    StyleForm[ToString[#], FontFamily -> "Courier"],
                    Frame -> True, Axes -> False,
```

```
PlotRange -> {-2, 2}, AspectRatio -> 1]) & /@
{Automatic, "Adams", "Extrapolation", "LSODA",
  "BDF","ExplicitRungeKutta", "ImplicitRungeKutta",
  "StiffnessSwitching"}, 4]]
```

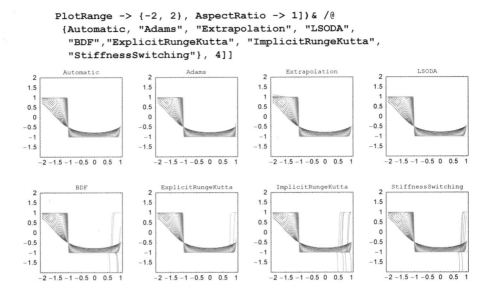

Now, let us have a more detailed look at how the various methods available perform on the following system of first-order differential equations:

$$x_1'(t) = \frac{7}{5} x_2(t)^2 x_4(t)^3 - \frac{3}{5} x_2(t)$$

$$x_2'(t) = -\frac{81}{2} x_1(t) x_2(t)^2 x_4(t) x_3(t)^2 - \frac{1}{2} x_2(t)$$

$$x_3'(t) = 5 x_1(t) x_3(t) x_4(t)^3 + \frac{25}{36} x_3(t)$$

$$x_4'(t) = -\frac{16}{9} x_1(t) + \frac{78}{29} x_2(t) - \frac{38}{83} x_3(t)$$

We start by implementing the right-hand sides of the differential equations.

```
In[145]:= eqs =
    {x[1]'[t] == (-3/5 x[2][t] + 7/5 x[2][t]^2 x[4][t]^3),
     x[2]'[t] == (-1/2 x[2][t] - 81/2 x[1][t] x[2][t]^2 x[3][t]^2 x[4][t]),
     x[3]'[t] == (25/36 x[3][t] + 5 x[1][t] x[3][t] x[4][t]^3),
     x[4]'[t] == (-16/9 x[1][t] + 78/29 x[2][t] - 38/83 x[3][t])};
```

The function `solveAndAnalyze` solves the system using the method option setting *method* and displays an array of three plots: the four solution curves, the residual, and the *t*-values that were used in the integration (we collect these *t*-values using the `EvaluationMonitor` option).

```
In[146]:= solveAndAnalyze[method_, otherNDSolveOptions___] :=
    Module[{tMax = 2, nsol, tValues, plotOpts},
      (* solve ODEs and extract t-values used *)
      {nsol, tValues} =
      Internal`DeactivateMessages[MapAt[First, #, 2]& @
      Reap[NDSolve[Join[eqs, {x[1][0] == -2/17, x[2][0] == 92,
                          x[3][0] == -1/2, x[4][0] == -9/7}],
                 {x[1], x[2], x[3], x[4]}, {t, 0, tMax},
                 otherNDSolveOptions, Method -> method,
```

```
            MaxSteps -> 20000, EvaluationMonitor :> Sow[t]]]];
(* a function for frequent use in the following *)
plotOpts[label_, pr_] =
Sequence[Frame -> True, Axes -> False, PlotLabel -> label,
        PlotStyle -> Table[{Thickness[0.002], Hue[x]},
                        {x, 0, 3/4, 1/4}],
        FrameTicks -> {{0.5, 1., 1.5, 2.}, Automatic, None, None},
        PlotRange -> {{0, tMax}, pr}];
(* maximal time achieved for the numerical solution *)
T = nsol[[1, 1, 2, 1, 1, 2]];
(* graphs of functions, logarithms of errors, and function evaluations *)
GraphicsArray[
 Block[{$DisplayFunction = Identity},
    {Plot[Evaluate[{x[1][t], x[2][t], x[3][t], x[4][t]} /. nsol],
        {t, 0, T}, Evaluate[plotOpts["values", {-10, 10}]]],
     Plot[Evaluate[Log[10, Abs[(Subtract @@@ eqs) /. nsol]]],
        {t, 0, T}, Evaluate[plotOpts["ln(errors)", {-12, 6}]]],
     ListPlot[MapIndexed[{#1, #2[[1]]}&, Sort[Flatten[tValues]]],
            Evaluate[plotOpts["steps", All]]]}]]]
```

Here are the resulting graphics for nine of the possible option settings.

```
In[147]:= Show[solveAndAnalyze[#], PlotLabel -> StyleForm[#, "MR"]]& /@
    (* the method option settings to be compared *)
    {Automatic, "Adams", "Extrapolation", "LSODA", "BDF",
     "ExplicitRungeKutta", "ImplicitRungeKutta", "Extrapolation",
     "StiffnessSwitching"}
```

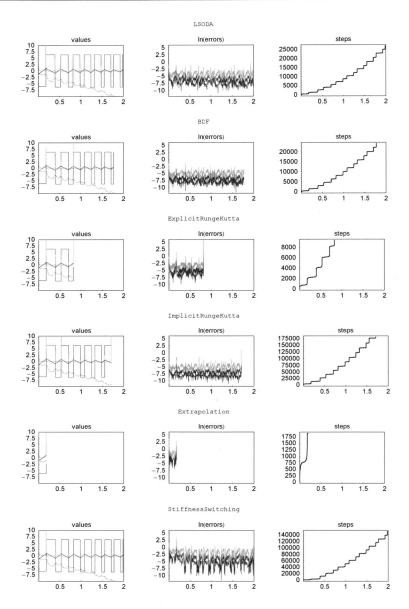

We continue with the integration of a differential equation with a fast varying right-hand side. We approximate a Brownian motion $\mathcal{B}(t)$ [953], [636] using the function BrownianMotion.

```
In[148]:= BrownianMotion[o_Integer?Positive] :=
    Module[{SchauderS, ξ},
        (* Schauder basis functions *)
        SchauderS[k_, n_, t_] := If[(k - 1)/2^n < t < (k + 1)/2^n,
                            2^-((n + 1)/2) (1 - Abs[t - k/2^n]/2^-n), 0];
        SchauderS[1, 0, t_] := t;
        ξ[k_, m_] := Sqrt[2] InverseErf[0, -1 + 2 Random[]];
```

```
(* form efficiently evaluatable pure function version of
   Sum[ξ[k, m] SchauderS[k, m, t], {m, 0, n}, {k, 1, 2^m, 2}] *)
Plus @@@ Function[Evaluate[Table[Interpolation[
Table[{k/2^m, If[EvenQ[k], 0, ξ[k, m] SchauderS[k, m, k/2^m]]},
      {k, 0, 2^m}], InterpolationOrder -> 1][#], {m, 0, o}]]]]
```

Here are three realizations of the coarse-grained Brownian motion.

```
In[149]:= Module[{o = 14, pp = 2^15},
    Show[GraphicsArray[Table[
    Graphics[(* color each consecutive sum differently *)
          MapIndexed[{Hue[#2[[1]]]/18},
                Line[MapIndexed[{(#2[[1]] - 1)/pp, #1}&, #1]]}&,
          Rest[Rest[FoldList[Plus, 0, Transpose[
          Table[(* evaluate each summand individually *)
                Evaluate[(List @@@ BrownianMotion[o])[t]],
                {t, 0, 1, 1/pp}]]]]]], Frame -> True], {3}]]]]
```

Next, we calculate the quantity $\exp\left(-\vartheta^2 \int_0^1 W_1(t)^2 \, dt/2\right)$ where $W_1(t) = \int_0^t \mathcal{B}(\tau) \, d\tau$ for two dozen realizations of the Brownian motions. For the integration of the differential equations, we use the fast "ExplicitRunge‹ Kutta" method.

```
In[150]:= exponentialAverage[ϑ_] =
    Module[{o = 14, μ = 24, W, int, nds},
    SeedRandom[7];
    Sum[(* solve differential equation for W[1] and ℐ
          simultaneously *)
        nds = NDSolve[{W[1]'[t] == BrownianMotion[o][t],
                  W[1][0] == 0,
                  ℐ'[t] == W[1][t]^2, ℐ[0] == 0}, {W[1], ℐ},
        (* need only integral value at t == 1 *) {t, 1, 1},
        (* reach each linear interval *)  MaxStepFraction -> 2^-o,
        MaxSteps -> 10^6, Method -> "ExplicitRungeKutta",
        PrecisionGoal -> 6, AccuracyGoal -> 6];
        Exp[-ϑ^2/2 ℐ[1]] /. nds[[1]], {μ}]/μ;
```

The resulting averaged expectation as a function of the parameter ϑ agrees well with the theoretical value [984], [1920], [387].

```
In[151]:= Plot[Evaluate @ {exponentialAverage[ϑ],
                  (* exact average as a function of ϑ *)
                  Sqrt[2/(Cosh[Sqrt[ϑ/2]]^2 + Cos[Sqrt[ϑ/2]]^2)]},
        {ϑ, 0, 20}, PlotStyle -> {Hue[0], GrayLevel[0]}]
```

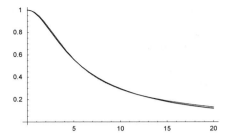

For specialized methods for the solution of stochastic differential equations, see [1277], [1007], [1459].

In general, in case one has a complicated differential equation (or a system of differential equations) and one needs a reliable solution, it is strongly advised to calculate the solution using different Method option settings and compare the results. In the next input, we use three different methods to calculate the solutions of an ingeniously modified Lorenz system [1026]. We use relatively high precision/accuracy goals and use high-precision arithmetic. The three solutions show that none of them should be trusted without further detailed study. (To plot the solutions we do not use Plot or ParametricPlot3D, but rather extract the samples *t*-values from the InterpolatingFunction that is returned by NDSolve. This ensures we do not miss any detail in the visualization of the calculated solutions.) We use four different option settings.

```
In[152]:= Function[method,
    Module[{σ = 10, b = 8/3, r = 28, T0 = 40, T1 = 60,
            ax = 1, ay = 1, az = 80, α = 1, ε = 3/100,
            μ = 4/100, r1 = 1, r0 = 2, ψ0, ψ1, ρS},
    (* perturbation functions *)
    ψ0[s_] := If[s < 1, -100 Abs[Cos[Pi s] + 1], 0];
    ψ1[s_] := s - 1;
    (* distance to critical point *)
    ρS = (x[t] - ax)^2 + (y[t] - ay)^2 + (z[t] - az)^2;
    (* solve differential equation *)
    ndsol = NDSolve[(* modified Lorenz system *)
     {x'[t] == σ (y[t] - x[t]) - α Y[t]^2 (x[t] - ax),
      y'[t] == r x[t] - y[t] - x[t] z[t] - α Y[t]^2 (y[t] - ay),
      z'[t] == -b z[t] + x[t] y[t] - α Y[t]^2 (z[t] - az),
      Y'[t] == (Y[t] ψ1[ρS/r1^2] + ψ0[ρS/r0^2] + μ) Y[t]/ε,
      x[0] == 0, y[0] == 0, z[0] == 1, Y[0] == 1},
     {x, y, z, Y}, {t, T0, T1}, Method -> method,
     MaxSteps -> Infinity, PrecisionGoal -> 12, AccuracyGoal -> 12,
     WorkingPrecision -> 30, MaxStepSize -> 10^-2];
    (* explicitly sampled t-values *)
    tValues = ndsol[[1, 1, 2, 3, 1]]; λt = Length[tValues];
    (* points {x[t], y[t], z[t]} *)
    xyzData = ({x[#], y[#], z[#]}& /@ tValues) /. ndsol[[1]];
    (* scaled Y[t] *)
    YData = ({#, Log[Abs[Y[#]]]}& /@ tValues) /. ndsol[[1]];
     (* show solutions for {x[t], y[t], z[t]} and Y[t] *)
    Show[GraphicsArray[
     {Graphics3D[{Thickness[0.002], MapIndexed[{Hue[0.78 #2[[1]]/λt], #1}&,
                 Line /@ Partition[xyzData, 2, 1]]}, Boxed -> False,
              BoxRatios -> {1, 1, 1}, PlotRange -> All],
      Graphics[Line[YData], PlotRange -> All, Frame -> True]},
     PlotLabel -> StyleForm[Method -> method, "MR"]]]]] /@
```

```
(* the three methods *)
{"BDF", "Adams", "ExplicitRungeKutta", "StiffnessSwitching"}
```

Method → BDF

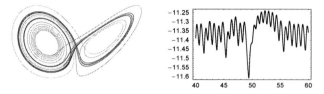

Method → Adams

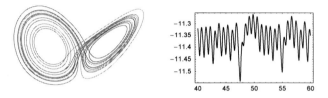

Method → ExplicitRungeKutta

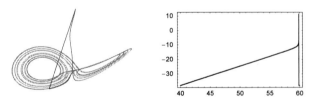

Method → StiffnessSwitching

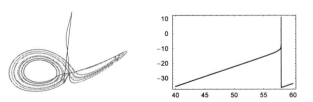

The method options have themselves suboptions. But because the method option settings are strings, we do not expect these options through `Options["`*method*`"]`. But we can get a list of the suboptions programmatically using `Options[NDSolve`*method*`]`. Here are all suboptions for all method option settings.

```
In[153]:= {#1, First /@ #2}& @@@
    Select[{#, Options[#]}& /@ (ToExpression /@
```

```
        DeleteCases[Names["NDSolve`*"],
                    (* PDE method only *) "NDSolve`MethodOfLines"]),
        (Last[#] =!= {})&]
Out[153]= {{NDSolve`Adams, {DenseOutput, MaxDifferenceOrder, VariableStepCoefficients}},
    {NDSolve`BDF,
     {DenseOutput, ImplicitSolver, MaxDifferenceOrder, VariableStepCoefficients}},
    {NDSolve`Chasing, {Method, ExtraPrecision, ChasingType}},
    {NDSolve`Composition, {Coefficients, DifferenceOrder, Method}},
    {NDSolve`DoubleStep, {LocalExtrapolation, Method, StepSizeRatioBounds,
      StepSizeSafetyFactors, StiffnessTest}}, {NDSolve`EventLocator,
     {Direction, Event, EventAction, EventLocationMethod, Method}},
    {NDSolve`ExplicitRungeKutta, {Coefficients, DifferenceOrder,
      EmbeddedDifferenceOrder, StepSizeControlParameters, StepSizeRatioBounds,
      StepSizeSafetyFactors, StepSizeStartingError, StiffnessTest}},
    {NDSolve`Extrapolation, {ExtrapolationSequence, MaxDifferenceOrder,
      Method, MinDifferenceOrder, OrderSafetyFactors, StartingDifferenceOrder,
      StepSizeRatioBounds, StepSizeSafetyFactors, StiffnessTest}},
    {NDSolve`FiniteDifferenceDerivative, {DifferenceOrder, PeriodicInterpolation}},
    {NDSolve`FixedStep, {Method}},
    {NDSolve`GMRES, {Preconditioner, OrthogonalizationType,
      MaxKrylovSubspaceDimension, MaxKrylovRestarts}},
    {NDSolve`IDA, {DenseOutput, MaxDifferenceOrder, ImplicitSolver}},
    {NDSolve`ImplicitRungeKutta,
     {Coefficients, DifferenceOrder, ImplicitSolver, StepSizeControlParameters,
      StepSizeRatioBounds, StepSizeSafetyFactors, StepSizeStartingError}},
    {NDSolve`ImplicitRungeKuttaCoefficients, {Solve}},
    {NDSolve`LinearlyImplicitEuler, {LinearSolveMethod, StabilityCheck}},
    {NDSolve`LinearlyImplicitMidpoint, {LinearSolveMethod, StabilityCheck}},
    {NDSolve`LinearlyImplicitModifiedMidpoint,
     {LinearSolveMethod, StabilityCheck}},
    {NDSolve`LocallyExact, {SimplificationFunction}},
    {NDSolve`LSODA, {DenseOutput, LinearSolveMethod, MaxDifferenceOrder}},
    {NDSolve`Newton, {LinearSolveMethod}}, {NDSolve`OrthogonalProjection,
     {Dimensions, IterationSafetyFactor, MaxIterations, Method}},
    {NDSolve`ProcessEquations, {AccuracyGoal, Compiled, DependentVariables,
      EvaluationMonitor, MaxStepFraction, MaxSteps, MaxStepSize, Method,
      NormFunction, PrecisionGoal, SolveDelayed, StartingStepSize,
      StepMonitor, WorkingPrecision}}, {NDSolve`Projection,
     {Invariants, IterationSafetyFactor, LinearSolveMethod, MaxIterations, Method}},
    {NDSolve`Splitting, {Coefficients, DifferenceOrder, Equations, Method}},
    {NDSolve`StiffnessSwitching, {Method}},
    {NDSolve`StiffnessTest, {MaxStiffnessTestFailures, StiffnessTestSafetyFactor}},
    {NDSolve`SymplecticPartitionedRungeKutta,
     {Coefficients, DifferenceOrder, PositionVariables}}}}
```

The method options "Projection" and "SymplecticRungeKutta" are very useful in many special ODE systems because they try to fulfill the conservation of certain quantities. We give two examples here.

Because of its practical importance, we give one example of the method option setting "SymplecticPartitionedRungeKutta". This method is especially suited for Hamiltonian systems because it preserves flow invariants. Hamiltonian systems appear naturally in classical mechanics. Here we will discuss a little known application from quantum mechanics [1491]. It will result in a time-dependent Hamiltonian problem. We consider the time-independent 1D Schrödinger equation $-\psi''(x) + V(x)\psi(x) = \varepsilon\psi(x)$. Rewriting the equation as

$\psi''(x) = -2(\varepsilon - V(x))\psi(x)$ or $\psi''(x) = B(x)\psi(x)$ results in a form of Newton equations. After a Legendre transformation, we obtain the Hamiltonian $H = \varphi^2/2 + B(x)\psi^2/2$. Here $\varphi(x)$ is the canonically conjugated momentum. The Hamilton equations of motion for this explicitly "time"-dependent Hamiltonian are

$$\frac{\partial\psi(x)}{\partial x} = \frac{\partial H}{\partial\varphi} = \varphi$$

$$\frac{\partial\varphi(x)}{\partial x} = -\frac{\partial H}{\partial\psi} = -B(x)\psi(x)$$

The next input defines the potential, the function $B(x)$, and the Hamiltonian. We use the Pöschl–Teller potential for $V(x)$.

```
In[154]:= (* Pöschl-Teller potential *)
    V[x_, α_] := -1/Cosh[x]^2
    (* "time"-dependence of "force" *)
    B[x_, ε_, α_] := ε - V[x, α]
    (* Hamiltonian *)
    H[ψ_, φ_, x_, ε_, α_] := φ^2/2 + B[x, ε, α] ψ^2/2
```

We will solve for symmetric eigenfunctions (meaning $\psi(0) = 1$, $\psi'(0) = 0$) for this symmetric potential. The exact solution can be expressed in Legendre functions.

```
In[160]:= ψExact[α_, ε_][x_] :=
    With[{P = LegendreP, Q = LegendreQ,
        s1 = Sqrt[1 - 4 α] - 1, s3 = Sqrt[1 - 4 α] - 3},
        (P[s3/2, I Sqrt[ε], 0] Q[s1/2, I Sqrt[ε], Tanh[x]] -
        P[s1/2, I Sqrt[ε], Tanh[x]] Q[s3/2, I Sqrt[ε], 0])/
        (P[s3/2, I Sqrt[ε], 0] Q[s1/2, I Sqrt[ε], 0] -
        P[s1/2, I Sqrt[ε], 0] Q[s3/2, I Sqrt[ε], 0])]
```

We solve the equations using five different methods. `Automatic` is the first setting, and `"SymplecticPartitionedRungeKutta"` is the last.

```
In[161]:= numSolutions = Function[method, Timing @
    NDSolve[{ψ'[x] == φ[x], φ'[x] == -B[x, 2, -1] ψ[x],
        ψ[0] == 1, φ[0] == 0},
        {ψ, φ}, {x, 0, 100}, Method -> method,
        MaxStepSize -> 1, MaxSteps -> Infinity]] /@
    (* five different methods *)
    {Automatic, "StiffnessSwitching",
     "ExplicitRungeKutta", "ImplicitRungeKutta",
     {"SymplecticPartitionedRungeKutta", (* suboptions *)
      "DifferenceOrder" -> 4, "PositionVariables" -> {ψ[x]}}};

    (* return calculation times *)
    First /@ numSolutions
Out[164]= {0.01 Second, 0.03 Second, 0.02 Second, 0.08 Second, 0.11 Second}
```

The function plots the logarithm of the absolute error of the Hamiltonian $\mathcal{H}$ for the calculated solutions.

```
In[165]:= errorHPlot[if_, color_, opts___] :=
    Plot[Evaluate[Log[10, Abs[(* difference of Hamiltonian *)
        H[ψExact[-1, 2][x], D[ψExact[-1, 2][x], x], x, 2, -1] -
        H[ψ[x], φ[x], x, 2, -1] /. if[[1]]]]],
        {x, 1, 10}, opts, PlotStyle -> {color}]
```

In some rare circumstances one might not want to use an adaptive step size, but rather a fixed step size. Using the method `"FixedStep"` this can be realized. We consider the following modified differential equation for a forced anharmonic oscillator (for the linear case, see [42]):

$$x''(t) + \left| \alpha \max_{0 \leq \tau \leq t} (x(\tau)) + \beta \min_{0 \leq \tau \leq t} (x(\tau)) \right| x(t) - \left| \alpha \max_{0 \leq \tau \leq t} (x(\tau)) - \beta \min_{0 \leq \tau \leq t} (x(\tau)) \right| x(t)^3 = \cos(t).$$

The following plot shows the maximal values of t, such that $|x(t)| < 100$ over the α, β-plane. (We use the initial conditions $x(0) = x'(0) = 1$.)

```
In[166]:= Module[{T = 1000, x0 = 1, L = 6, pp = 240, λ = 20,
              fmax, fmin, data, nds, cls, colorC},
         (* functions for maximum and minimum taken so far *)
         fmax[x_?InexactNumberQ] :=  (max = Max[max, x]);
         fmin[x_?InexactNumberQ] :=  (min = Min[min, x]);
         data = Table[(* initialize values of min and max *) max = min = x0;
           nds = NDSolve[{x''[t] + Abs[α fmax[x[t]] + β fmin[x[t]]] x[t] -
                            Abs[α fmax[x[t]] - β fmin[x[t]]] x[t]^3 == Cos[t],
                      x[0] == x0, x'[0] == 1}, x, {t, 0, T},
                      MaxSteps -> 10^2, StoppingTest -> (Abs[x[t]] > 100),
                      AccuracyGoal -> 3, PrecisionGoal -> 3,
                      StartingStepSize -> 1/10,
                      Method -> {(* use fixed step size *) "FixedStep",
                          Method -> {"ExplicitRungeKutta", DifferenceOrder -> 4}}];
                      nds[[1, 1, 2]][[1, 1, 2]],
                {β, -L, L, 2L/pp}, {α, -L, L, 2L/pp}];
         (* homogeneous color scaling *)
         cls = #[[Round[pp^2/λ/2]]]& /@
               Partition[Sort[Flatten[data]], Round[pp^2/λ]];
         (* construct color function *)
         colorC = Compile[{x}, Evaluate[Which @@ Join[Flatten[
                     MapIndexed[{x < #1, (#2[[1]] - 1)/λ}&, cls]], {True,
1}]]]];
         (* make density plot *)
         ListDensityPlot[data, Mesh -> False, PlotRange -> All,
                     ColorFunction -> (Hue[0.78 colorC[#]]&),
                     ColorFunctionScaling -> False,
                     MeshRange -> L {{-1, 1}, {-1, 1}}]]

         NDSolve::mxst :
         Maximum number of 100 steps reached at the point t == 10.100000000000001`. More…
```

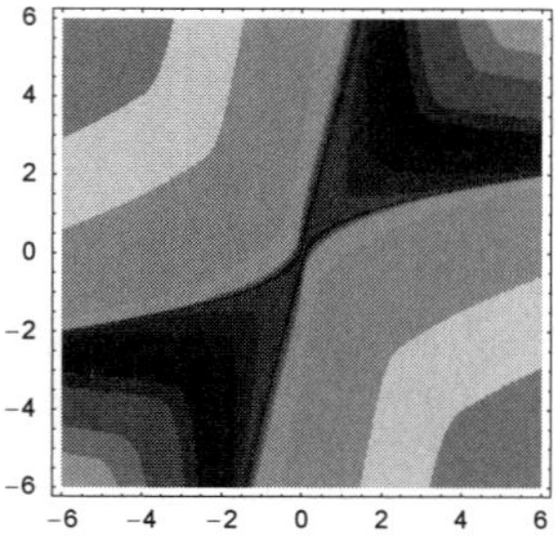

The next graphic shows the errors for the five methods. The `"SymplecticPartitionedRungeKutta"` method (blue curve) performs by far the best.

```
In[167]:= Show[Table[(* color consecutively from red to blue *)
                errorHPlot[numSolutions[[k, 2]], Hue[(k - 1)/7],
                        DisplayFunction -> Identity],
                {k, Length[numSolutions]}],
        Axes -> False, Frame -> True, PlotRange -> {-10, 0},
        DisplayFunction -> $DisplayFunction]
```

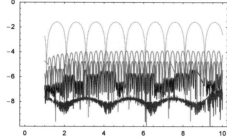

We continue with an example of the `"Projection"` method option setting. We consider the following clover-shaped curve $x^3 (x + 1) + (2x - 3) x y^2 + y^4 = 0$. Using the function `ContourPlot`, we can easily generate a plot of this implicitly defined curve.

```
In[168]:= clover[x_, y_] := x^3 (1 + x) + x (2 x - 3) y^2 + y^4
```

```
In[169]:= ContourPlot[Evaluate[clover[x, y]], {x, -1, 1}, {y, -1, 1},
                Contours -> {0}, PlotPoints -> 60, ContourShading -> False,
                ContourStyle -> {{Thickness[0.016], GrayLevel[0]}}]
```

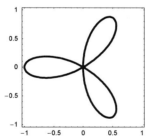

Assuming a parametric representation of the form $\{\xi_x(s), \eta_y(s)\}$ with s the arc length, we obtain the following pair of coupled nonlinear differential equations for the $\xi_x(s)$ and $\eta_y(s)$.

```
In[170]:= cloverEqs = Thread[{ξx'[s], ηy'[s]} == (* direction cosines *)
                (#/Sqrt[#.#]&[{D[clover[ξx, ηy], ηy], -D[clover[ξx, ηy], ξx]} /.
                        {ξx -> ξx[s], ηy -> ηy[s]}])] // FullSimplify
```

$$
\begin{aligned}
Out[170]= \quad & \big\{\xi x'[s] == (2\,\eta y[s]\,(2\,\eta y[s]^2 + \xi x[s]\,(-3 + 2\,\xi x[s]))) \big/ \big(\sqrt{((\eta y[s]^2 + \xi x[s]^2)} \\
& (16\,\eta y[s]^4 + \xi x[s]^2\,(3 + 4\,\xi x[s])^2 + \eta y[s]^2\,(9 + 8\,\xi x[s]\,(-9 + 4\,\xi x[s]))))\big)), \\
& \eta y'[s] == (\eta y[s]^2\,(3 - 4\,\xi x[s]) - \xi x[s]^2\,(3 + 4\,\xi x[s])) \big/ \big(\sqrt{((\eta y[s]^2 + \xi x[s]^2)} \\
& (16\,\eta y[s]^4 + \xi x[s]^2\,(3 + 4\,\xi x[s])^2 + \eta y[s]^2\,(9 + 8\,\xi x[s]\,(-9 + 4\,\xi x[s]))))\big))\big\}
\end{aligned}
$$

Starting at the point $\{-1, 0\}$, we now try to traverse the curve for three times. We use two method option settings. A generic solution method (here we use the generically quite reliable `"StiffnessSwitching"` method) has two drawbacks. The residual error is on average higher and increases with arc length and at the origin, where the curve has selfintersections, the method does not always follow the solution on the smooth continuation. The "Projection" method on the other hand has no problems following the curve and the residuals are orders of magnitudes smaller. By construction, the solution is guaranteed to stay at the solution curve, even after many traversals.

```
In[171]:= Show[GraphicsArray[#]]& /@
    Function[method,
    Block[{$DisplayFunction = Identity, r = 3,
        (* arc length for one complete traversal *) λΣ = 2 EllipticE[-8]},
      (* solve differential equations *)
    nds = NDSolve[Join[cloverEqs,
                {ξx[0] == -1, ηy[0] == 0}], {ξx, ηy}, {s, 0, r λΣ},
                (* use high-precision arithmetic *)
                Method -> method, WorkingPrecision -> 30,
                PrecisionGoal -> 10, AccuracyGoal -> 10];
      (* show resulting curve and errors *)
      {(* resulting curve *)
      ParametricPlot[Evaluate[{ξx[s], ηy[s]} /. nds[[1]]], {s, 0, r λΣ},
                PlotPoints -> 600, PlotRange -> {{-1.1, 0.6}, {-0.9, 0.9}},
                Axes -> False, Frame -> True, AspectRatio -> Automatic],
        (* accuracy of the residual;
          use only grid points to avoid interpolation errors *)
      ListPlot[Transpose[{Flatten[InterpolatingFunctionGrid[nds[[1, 1, 2]]]],
            Accuracy /@ (clover @@@ Transpose[
                InterpolatingFunctionValuesOnGrid[#[[2]]]& /@ nds[[1]]])}],
          PlotRange -> All]}]] /@
        (* method option settings *)
        {"StiffnessSwitching",
        {"Projection", Method -> "StiffnessSwitching",
        (* stay on curve *) "Invariants" -> {clover[ξx[s], ηy[s]]}}}]
```

We give another example of a system of nonlinear differential equations with constraints. Here are two differential equations for $x''(\tau)$ and $y''(\tau)$.

```
In[172]:= gEq[{x_, y_, z_}, τ_] := x''[τ] == (Cos[x[τ]] Sec[z[τ]]^2*
        ((Sin[x[τ]] + Cos[x[τ]]^2 Sec[z[τ]] Tan[z[τ]]) x'[τ]^2 +
         2 Cos[x[τ]] Cos[y[τ]] Sec[z[τ]] Tan[z[τ]] x'[τ] y'[τ] +
         (Sin[y[τ]] + Cos[y[τ]]^2 Sec[z[τ]] Tan[z[τ]]) y'[τ]^2))/
        (1 + (Cos[x[τ]]^2 + Cos[y[τ]]^2) Sec[z[τ]]^2)
```

These differential equations describe curves on the surface $\sin(x(\tau)) + \sin(y(\tau)) + \sin(z(\tau)) = 0$.

```
In[173]:= sEq[{x_, y_, z_}, τ_] := Sin[x[τ]] + Sin[y[τ]] + Sin[z[τ]] == 0
```

The function sinsinsinSurfaceTrio solves the equations gEq supplemented by f applied to sEq. Then the solutions are visualized as colored curves and are shown from three different viewpoints. A second set of graphic shows the distance of the solutions from the origin as a function of τ and show well the curves lie on the surface.

```
In[174]:= sinsinsinSurfaceTrio[T_, o_, f_:Identity, {ndsOptions___},
                            plots_:{True, True}] :=
      Module[{p3D, Tmax, nsol, Tmin, cOpts},
      p3D = Show[Table[
      nsol[φ] = NDSolve[{gEq[{x, y, z}, τ], gEq[{y, x, z}, τ],
                       f @ sEq[{x, y, z}, τ],
              (* starting values in plane z == -x - y *)
                      x[0] == 0, y[0] == 0, z[0] == 0,
                      x'[0] == -Cos[φ]/Sqrt[2] - Sin[φ]/Sqrt[6],
                      y'[0] ==  Cos[φ]/Sqrt[2] - Sin[φ]/Sqrt[6]},
                 {x, y, z}, {τ, 0, T}, ndsOptions, MaxSteps -> 10^5,
                  PrecisionGoal -> 8, AccuracyGoal -> 8];
      (* maximal t-vale reached *)
      Tmax[φ] = 0.9 nsol[φ][[1, 1, 2, 1, 1, 2]];
      (* make 3D plot of trajectories*)
      ParametricPlot3D[Evaluate[{x[τ], y[τ], z[τ],
                          {Thickness[0.001], Hue[0.78 φ/(Pi/2)]}} /.
                           nsol[φ][[1]]],
                    {τ, 0, Tmax[φ]}, PlotPoints -> Round[Tmax[φ] 24],
                    PlotRange -> 2/3 T {{-1, 1}, {-1, 1}, {-1, 1}},
                    DisplayFunction -> Identity, Axes -> False],
                    {φ, 0, 2Pi, 2Pi/o}]];
      (* show 3D plot from three orthogonal directions *)
      If[plots[[1]],
         Show[GraphicsArray[Show[p3D, ViewPoint -> 3 #/Norm[#]]& /@
                            {{1, 1, 1}, {-1, -1, 2},  {-1, +1, 0}}]]];
      Tmin = Min[Table[Tmax[φ], {φ, 0, 2Pi, 2Pi/o}]];
      cOpts = Sequence[PlotPoints -> Round[Tmin 24],
              PlotStyle -> Table[Hue[φ/(2Pi)], {φ, 0, 2Pi, 2Pi/o}],
              Frame -> True, Axes -> False, DisplayFunction -> Identity];
      (* make two 2D plots *)
      If[plots[[2]],
      Show[GraphicsArray[Internal`DeactivateMessages @
          {(* distance from origin *)
          Plot[Evaluate[Table[Norm[{x[τ], y[τ], z[τ]}] /. nsol[φ][[1]],
                          {φ, 0, 2Pi, 2Pi/o}]],
               {τ, 0, Tmin}, Evaluate[cOpts]],
          (* how good say solutions on the surface? *)
          Plot[Evaluate[Table[Log[10, Abs[sEq[{x, y, z}, τ][[1]]]] /.
                  nsol[φ][[1]], {φ, 0, 2Pi, 2Pi/o}]],
               {τ, 0, Tmin}, PlotDivision -> 1,
```

```
            PlotPoints -> 10, PlotRange -> {0, -20},
            Evaluate[cOpts]]}]]]]
```

With machine arithmetic, we can just use sEq as an additional algebraic equation. The solution works well for a while, but fails for some initial conditions for large values of τ.

In[175]:= **sinsinsinSurfaceTrio[16, 60, {Method -> "IDA"}]**

            NDSolve::ndcf : Repeated convergence test
               failure at τ == 15.146154847978867`; unable to continue. More...

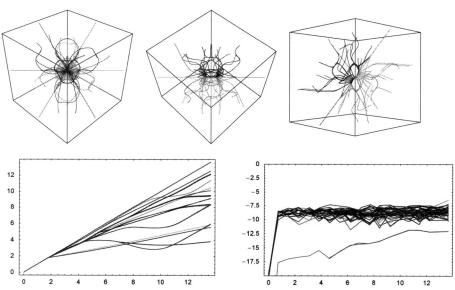

If we differentiate sEq, we can use the "Projection" method setting, which allows using a higher working precision. Now, the solution proceeds flawless for much larger values of τ. The solution curves seem in average not to lie on the surface to the same quality as the last solution. (This deviation is mainly caused by the interpolation from the sampled data. The method setting "StiffnessSwitching" itself, without enforcing explicitly the solution curves to lie on the surface, would in this case similar good results.)

In[176]:= **sinsinsinSurfaceTrio[36, 60, D[#, τ]&,**
            {Method -> {"Projection", Method -> "StiffnessSwitching",
               (* stay on surface *) **"Invariants" -> {sEq[{x, y, z}, τ][[1]]}}}]**

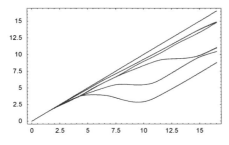 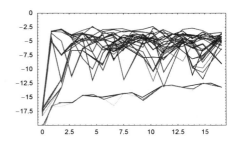

Many more details about `NDSolve`, like step size control, stability of solutions, and so on could now be discussed. But as already mentioned in the beginning of this subsection, `NDSolve` is the single most versatile function of *Mathematica* and the Advanced Documentation is the place to look up much more details about it.

Now, let us consider some more applications of `NDSolve`. A very interesting application of the numerical solution of differential equations is the solution of polynomial equations. Here, we give a simple example for a univariate polynomial; for the general case, see [1147], [1144], [1145], [1296], [1932], [1846], [1146], and [1847]. We want to solve a polynomial $p(x) = 0$ of degree n. The idea behind the homotopy method is to take a simpler, easily solvable polynomial of the $q(x) = 0$, also of degree n. Then, we form the polynomial $r(x(t)) = (1 - t) q(x(t)) + t p(x(t))$. The function $r(x(t))$ interpolates smoothly between the simple and the destination polynomials. Considering x as a function of t, we get a first-order nonlinear differential equation for $x'(t)$ by differentiating $r(x(t))$ with respect to t. Using the known roots of the simple polynomial $q(x)$ as the initial conditions for the differential equation for $x(t)$, we can solve the differential equations for t in the range $[0, 1]$, and $x(1)$ will then be a root of the polynomial $p(x)$; this means $p(x(1)) = 0$.

Here is an example polynomial of degree 25.

```
In[177]:= poly =
    41 + 39 x + 55 x^2 + 68 x^3 - 49 x^4 - 55 x^5 - 69 x^6 + 81 x^7 +
    4 x^8 + 53 x^9 - 4 x^10 - 84 x^11 + 66 x^12 + 50 x^13 + 49 x^14 +
    40 x^15 - 76 x^16 + 33 x^17 + 84 x^18 + 77 x^19 + 49 x^20 +
    67 x^21 + 52 x^22 + 100 x^23 + 48 x^24 + 73 x^25;

    degree = Exponent[poly, x];
```

As the simple system, we just use $x^n = 1$ and form the interpolating polynomial.

```
In[179]:= interpolationPoly = t poly + (1 - t) (x^degree - 1) /. x -> x[t]
```

$$Out[179]= (1 - t) (-1 + x[t]^{25}) +$$
$$t (41 + 39 x[t] + 55 x[t]^2 + 68 x[t]^3 - 49 x[t]^4 - 55 x[t]^5 - 69 x[t]^6 + 81 x[t]^7 +$$
$$4 x[t]^8 + 53 x[t]^9 - 4 x[t]^{10} - 84 x[t]^{11} + 66 x[t]^{12} + 50 x[t]^{13} +$$
$$49 x[t]^{14} + 40 x[t]^{15} - 76 x[t]^{16} + 33 x[t]^{17} + 84 x[t]^{18} + 77 x[t]^{19} +$$
$$49 x[t]^{20} + 67 x[t]^{21} + 52 x[t]^{22} + 100 x[t]^{23} + 48 x[t]^{24} + 73 x[t]^{25})$$

Here is the corresponding differential equation.

```
In[180]:= ode = D[interpolationPoly, t] == 0;
```

The initial values are simply $e^{2\pi i k/n}$.

```
In[181]:= startRoots = Table[Exp[I 2Pi k/degree], {k, 0, degree - 1}];
```

Solving the differential equation does not make any difficulties.

```
In[182]:= nsol = NDSolve[{ode, x[0] == #}, x, {t, 0, 1},
                MaxSteps -> 4000, PrecisionGoal -> 20,
                AccuracyGoal -> 10]& /@ startRoots;
```

Comparing the resulting values for the roots of `poly` with the result of `NSolve` shows a very good agreement.

```
In[183]:= With[{ex = x /. NSolve[poly == 0, x],
           ap = Flatten[x[1] /. nsol]},
      diff = Sum[Min[Abs[ex[[i]] - ap]^2], {i, degree}]]
Out[183]= 5.52879×10^-19
```

The next picture shows the path of the roots as a function of the parameter *t* (running upwards) over the complex *x*-plane.

```
In[184]:= Show[{Graphics3D[(* the starting points *)
          {PointSize[0.01], Point[N[{Re[#], Im[#], 0}]]}& /@ startRoots],
        (* the trajectories *)
        ParametricPlot3D[Evaluate[{Re[x[t]], Im[x[t]], t,
                        {Hue[Random[]], Thickness[0.002]}} /. #[[1]]],
                {t, 0, 1}, DisplayFunction -> Identity,
                        PlotPoints -> 300]& /@ nsol},
        DisplayFunction -> $DisplayFunction,
        Axes -> True, BoxRatios -> Automatic]
```

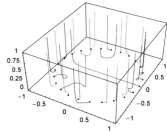

Now, having discussed the most important options of `NDSolve` for solving ordinary differential equations, we will continue this section by presenting a larger example: the ball pendulum. We recommend restarting *Mathematica* here.

Physical Example: Ball Pendulum (Newton's Cradle)

The so-called ball pendulum consists of *n* metal balls (typically, between five and eight) independently suspended by cords of the same length from a common support arranged in a horizontal line. We swing out *m* < *n* of the balls, and release them so that they hit the remaining *n* − *m* balls. This causes *m* of the balls on the opposite side of the original *n* − *m* balls to move. Here is a sketch of the ball pendulum for *m* = 2, *n* = 5.

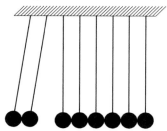

To model by first principles (in the framework of classical mechanics) of this process comes from Newton's law of motion for the n balls

$$m\, x_i''(t) = F_i(x_{i-1}(t),\, x_i(t),\, x_{i+1}(t)) \quad i = 1,\, \dots\, n.$$

$x_i(t)$ is the position of the center of the ith ball as a function of time t. The force between two balls should be 0 if they do not touch. If they touch, the force should have the form

$$F(x_i,\, x_{i-1}) = k\, |x_i - x_{i-1} + 2\, R|^\epsilon.$$

It follows from linear elasticity theory for the motion of two balls ([863], [1099], [375], [1725], [951], [802], [1419]) that $\epsilon = 3/2$ and $k = (R/2)^{1/2}\, 4\, E/(3\,(1 - \sigma^2))$, where R is the radius of the balls, σ is the Poisson number, and E is the modulus of elasticity. Typical values for steel balls are $r \approx 1\,\text{cm}$, $\delta \approx 0.29$, $E = 2.2\,10^{12}\,(\text{g}/(\text{cm}\,s))^2$.

For simplicity, we treat the problem in one dimension, which means a real ball pendulum with long strings. (It is not too hard to make a more realistic model, but the essential physics remains unchanged.) We begin by finding the equations needed to apply `NDSolve`. Because we are interested in the position x and the velocity v, we introduce the differential equation $x'(t) = v(t)$ by which x and v are connected. Let n be the total number of balls.

```
In[185]:= Clear[n, x, R, k, ε, m, e]
         xvEquations[n_] := Array[x[#]'[t] == v[#][t]&, n]
```

Here is an example with five balls.

```
In[187]:= xvEquations[5] // TableForm
Out[187]//TableForm=
         x[1]'[t] == v[1][t]
         x[2]'[t] == v[2][t]
         x[3]'[t] == v[3][t]
         x[4]'[t] == v[4][t]
         x[5]'[t] == v[5][t]
```

Now, we deal with the right-hand side in Newton's second law. `force` gives the formula for the force exerted on the ith ball. The first and last balls must be treated separately because they only have one neighbor. Suppose the balls are numbered from right to left (along the negative x-axis). Here is the situation for six balls:

```
    •   •   •   •   •   •
    6   5   4   3   2   1
```

Note the use of `Evaluate` in the use of `If` in what follows to make sure that the arguments of `If` (which has the attribute `HoldRest`) get evaluated. The result of `force` should explicitly contain `If` with the corresponding variables. Let R be the radius of all balls.

```
In[188]:= force[i_, n_, R_, k_, ε_] := Evaluate //@
         (If[i == 1, (* left-most ball *) 0, 1]*
         If[x[i - 1][t] < x[i][t] + 2R, -k (x[i][t] - x[i - 1][t] + 2R)^ε, 0] +
         If[i == n, (* right-most ball *) 0, 1]*
         If[x[i][t] < x[i + 1][t] + 2R, k (x[i + 1][t] - x[i][t]   + 2R)^ε, 0])
```

Here are a few examples of the resulting forces.

```
In[190]:= Table[force[i, 3, R, k, ε], {i, 3}]
```

Out[190]= $\{$ If $[x[1][t] < 2R + x[2][t], k(2R - x[1][t] + x[2][t])^{\epsilon}, 0]$,
     If $[x[1][t] < 2R + x[2][t], -k(2R - x[1][t] + x[2][t])^{\epsilon}, 0] +$
       If $[x[2][t] < 2R + x[3][t], k(2R - x[2][t] + x[3][t])^{\epsilon}, 0]$,
     If $[x[2][t] < 2R + x[3][t], -k(2R - x[2][t] + x[3][t])^{\epsilon}, 0]\}$

We can now formulate Newton's equations of motion for the ball pendulum.

In[191]:= `newton2[n_, R_, k_, ε_, m_] := Array[v[#]'[t] == force[#, n, R, k, ε]/m&, n]`

Here are the equations for the case of six balls.

In[192]:= `newton2[6, R, k, ε, m] // Short[#, 4]&`
Out[192]//Short=

$$\Big\{ v[1]'[t] == \frac{\text{If}[x[1][t] < 2R + x[2][t], k(2R - x[1][t] + x[2][t])^{\epsilon}, 0]}{m},$$

$$v[2]'[t] == \frac{\text{If}[x[1][t] < 2R + x[2][t], -k(2R - x[1][t] + x[2][t])^{\epsilon}, 0] + \text{If}[\ll 1 \gg]}{m},$$

$$v[3]'[t] == \frac{\text{If}[\ll 1 \gg] + \ll 1 \gg}{m}, \quad v[4]'[t] == \frac{\text{If}[\ll 1 \gg] + \ll 1 \gg}{m},$$

$$v[5]'[t] == \frac{\text{If}[x[4][t] < 2R + x[5][t], -k(\ll 1 \gg)^{\epsilon}, 0] + \text{If}[\ll 1 \gg]}{m},$$

$$v[6]'[t] == \frac{\text{If}[x[5][t] < 2R + x[6][t], -k(2R - x[5][t] + x[6][t])^{\epsilon}, 0]}{m} \Big\}$$

For the first argument of `NDSolve`, we still need to supply the initial conditions.

In[193]:= `initialConditions[x0_List, v0_List] :=`
     `Flatten[`(* make indexed variables *)
         `Array[{x[#][0] == x0[[#]], v[#][0] == v0[[#]]}&, Length[x0]]]`

We can now give the complete set of equations.

In[194]:= `equations[n_, R_, k_, exp_, m_, x0_List, v0_List] :=`
     `Join[xvEquations[n], newton2[n, R, k, exp, m], initialConditions[x0, v0]]`

For six balls, this is already a relatively long list.

In[195]:= `equations[6, R, k, exp, m, {x01, x02, x03, x04, x05, x06},`
                `{v01, v02, v03, v04, v05, v06}] // Short[#, 6]&`
Out[195]//Short=

$\{x[1]'[t] == v[1][t], x[2]'[t] == v[2][t], x[3]'[t] == v[3][t],$
   $x[4]'[t] == v[4][t], x[5]'[t] == v[5][t], x[6]'[t] == v[6][t],$

$$v[1]'[t] == \frac{\text{If}[x[1][t] < 2R + x[2][t], k(2R - x[1][t] + x[2][t])^{\exp}, 0]}{m},$$

   $\ll 10 \gg, v[3][0] == v03, x[4][0] == x04, v[4][0] == v04,$
   $x[5][0] == x05, v[5][0] == v05, x[6][0] == x06, v[6][0] == v06\}$

The numerical solution of this system of equations can be found relatively quickly. In view of the piecewise smooth nature of the force functions and the rapid growth of the force when the deformation becomes large, we have to compute many data points; and so, we choose `MaxSteps` appropriately.

In[196]:= `SolutionBallPendulum[n_, R_, k_, ε_, m_, x0_List, v0_List, time_] :=`
     `NDSolve[equations[n, R, k, ε, m, x0, v0],`
           `Join[Table[x[i], {i, n}], Table[v[i], {i, n}]],`
           `{t, 0, time}, MaxSteps -> 10^6]`

Because the result of `SolutionBallPendulum` is a list of `InterpolatingFunction`-objects whose behavior is not clear from looking at the `FullForm` (or `InputForm`, not to speak about `OutputForm`) of these objects, we now present our results graphically.

First, we prepare a graph of the problem, in which each ball has its color and its velocity arrow.

```
In[197]:= sketch[n_, R_, x0_, v0_] :=
        Show[{Graphics[ (* the balls *)
                     Table[{Hue[(i - 1)/(n - 1) 0.7],
                            Disk[{x0[[i]], 0}, R]}, {i, n}]],
              Graphics[ (* the initial velocity arrows *)
                  Table[Map[({x0[[i]], 0} +
                         (v0[[i]]/Max[Abs[v0]] # R/2))&,
                     Polygon[{{0, -0.8}, {1.6, 0}, {0, 0.8}, {0, 0.3},
                              {-1, 0.3}, {-1, -0.3}, {0, -0.3}}]], {2}], {i, n}]]},
              PlotRange -> All, AspectRatio -> Automatic,
              Axes -> {True, None}, Ticks -> Automatic,
              AxesOrigin -> {Min[x0] - (Max[x0] - Min[x0])/10, -3/2R},
              PlotLabel -> StyleForm["Sketch", FontWeight -> "Bold", FontSize -> 7]]
```

Next, we present the graphs of the corresponding position versus time curves.

```
In[198]:= (* function for adding labels *)
        label[plotLabel_, {xLabel_, yLabel_}] :=
        Sequence[PlotLabel -> StyleForm[TraditionalForm[plotLabel],
            FontWeight -> "Bold", FontSize -> 7],
                AxesLabel -> (StyleForm[TraditionalForm[#],
            FontWeight -> "Plain", FontSize -> 6]& /@ {xLabel, yLabel})]

In[200]:= positionPlot[sol_, n_, time_, opts___] :=
        Plot[Evaluate[Table[x[i][t], {i, n}] /. sol], {t, 0, time},
            opts, PlotStyle -> Table[Hue[i/(n - 1) 0.7], {i, 0, n - 1}],
            PlotRange -> All, Evaluate[label["x(t)", {"t", "x"}]]]
```

Here are the corresponding velocity versus time curves.

```
In[201]:= (* use last definition with x ⟷ v *)
        DownValues[velocityPlot] = DownValues[positionPlot] /.
                {x -> v, "x" -> "v"} /. positionPlot -> velocityPlot;
```

Here are the paths $(x(t), x'(t))$ in phase space.

```
In[203]:= phasePlot[sol_, n_, time_, opts___] :=
        ParametricPlot[Evaluate[(Table[{x[i][t], v[i][t]},
                            {i, n}] /. sol)[[1]]], {t, 0, time},
            opts, PlotStyle -> Table[Hue[i/(n - 1) 0.7], {i, 0, n - 1}],
            PlotRange -> All, Evaluate[label["v(x)", {"x", "v"}]]]
```

Finally, we present plots of the forces acting on the individual balls.

```
In[204]:= forcePlot[sol_, n_, time_, opts___] :=
        Plot[Evaluate[Table[force[i, n, R, k, exp], {i, n}] /. sol],
            {t, 0, time}, opts,
            PlotStyle -> Table[Hue[i/(n - 1) 0.7], {i, 0, n - 1}],
            PlotRange -> All, Evaluate[label["F(t)", {"t", "F"}]]]
```

We now combine all of these plots in a single routine. At the same time, we test whether the arguments make sense, and implement a check on the conservation of energy and momentum. These two tests give us some indication of the quality of the solution of the system of differential equations. We consciously use Block instead of Module, because the subroutines names are not scoped (because we constructed them separately to enhance understanding), and the relevant variables of the special data are also globally visible.

```
In[205]:= BallPendulum[initialPosition_List,
                    initialVelocities_List,
                    ballMass_?(# > 0&),
                    ballRadius_?(# > 0&),
                    springConstant_?(# > 0&),
                    forceExponent_?(# > 0&),
                    timeInterval_?(# > 0&)] :=
     Block[{(* number of balls *) n = Length[initialPosition],
             (* for brevity, use some abbreviations *)
             x0 = initialPosition, v0 = initialVelocities,
             R = ballRadius, m = ballMass, k = springConstant,
             exp = forceExponent, time = timeInterval, sol, vF},
       (* sketch the ball situation at hand *)
       sketch[n, R, x0, v0];
       (* the solution of the system of ODEs *)
       sol1 = SolutionBallPendulum1[n, R, k, exp, m, x0, v0, time];
       sol = SolutionBallPendulum[n, R, k, exp, m, x0, v0, time];
       (* final velocities *)
       vF = Table[v[i][time] /. sol[[1]], {i, n}];
       (* print out energy and momentum conservation *)
       CellPrint[Cell["o Impulse before: " <> ToString[m Plus @@ v0] <> "; " <>
                    "Impulse after: " <> ToString[m Plus @@ vF],
                    "PrintText"]];
       CellPrint[Cell["o Energy before: " <> ToString[m/2 Plus @@ v0^2] <>
                      "; " <> "Energy after: " <> ToString[m/2 Plus @@ vF^2],
                    "PrintText"]];
       (* make the four pictures in a GraphicsArray *)
       Show[GraphicsArray[#]]& /@
           Map[#[sol, n, time, DisplayFunction -> Identity]&,
              {{positionPlot, velocityPlot}, {phasePlot, forcePlot}}, {2}]]
```

We are now ready for some results. We begin with "classical" springs between the balls; that is, the exponent in the force law is chosen to be 1. We deal with four balls being struck by two balls from the right.

```
In[206]:= BallPendulum[{2, 1, -1, -2, -3, -4},
                               (* initial position   *)
                   {-1, -1, 0, 0, 0, 0},
                               (* initial velocities *)
                   1,          (* ball mass          *)
                   1/2,        (* ball radius        *)
                   1000,       (* spring constant    *)
                   1,          (* force exponent     *)
                   2]          (* time               *)
```

Sketch

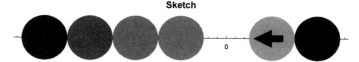

o Impulse before: -2; Impulse after: -2.

o Energy before: 2; Energy after: 2.

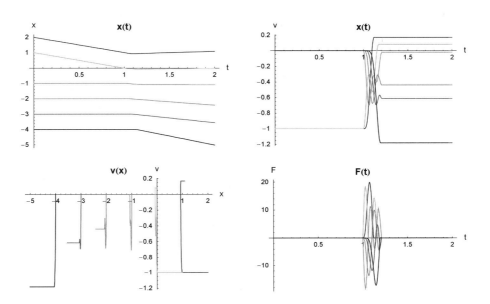

The "expected" did not occur: After the collision, *all* balls are in motion. Now, we repeat the computation with realistic parameters for steel balls.

```
In[207]:= BallPendulum[{2, 1, -1, -2, -3, -4},
                           (* initial position    *)
             {-10, -10, 0, 0, 0, 0},
                           (* initial velocities  *)
             4.1,          (* ball mass           *)
             1/2,          (* ball radius         *)
             1.6 10^12,    (* spring constant     *)
             3/2,          (* force exponent      *)
             0.2]          (* time                *)
```

Sketch

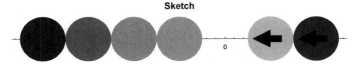

∘ Impulse before: -82.; Impulse after: -82.

∘ Energy before: 820.; Energy after: 820.

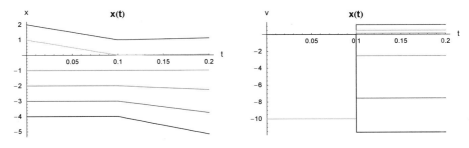

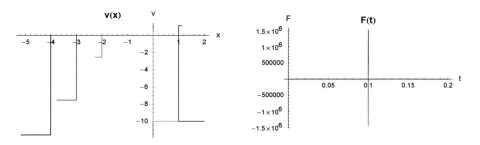

We now examine this practical case in slow motion ($x(t)$ and $v(x)$ are no longer especially interesting to plot). Note the different scale used in the following $F(t)$ plot compared with the earlier one. NDSolve has indeed done a good job in detecting the small time interval of interest, but Plot does not show it. Because of the small time interval, we start the two swinging balls closer to the stationary four (here, we would have to reduce the size of the letters further).

```
In[208]:= BallPendulum[{1.001, 0.001, -1, -2, -3, -4},
                        (* initial position   *)
           {-10, -10, 0, 0, 0, 0},
                        (* initial velocities *)
           4.1,         (* ball mass          *)
           1/2,         (* ball radius        *)
           1.6 10^12,   (* spring constant    *)
           3/2,         (* force exponent     *)
           0.0003]      (* time               *)
```

Sketch

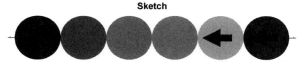

∘ Impulse before: -82.; Impulse after: -82.

∘ Energy before: 820.; Energy after: 820.

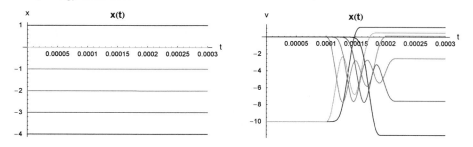

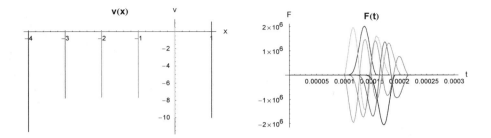

For details, see the references cited below. (The maximum deformation of the balls in this example is approximately 2 μm.)

We can model the expected result using larger force exponents (interestingly, the velocity distribution after the collision depends almost totally on the force exponents and very little on the spring constant and the initial velocities).

```
In[209]:= BallPendulum[{1.1, 0.1, -1, -2, -3, -4},
                          (* initial position    *)
           {-10, -10, 0, 0, 0, 0},
                          (* initial velocities *)
           4.1,           (* ball mass           *)
           1/2,           (* ball radius         *)
           1.6 10^12,     (* spring constant     *)
           4,             (* force exponent      *)
           0.03]          (* time                *)
```

Sketch

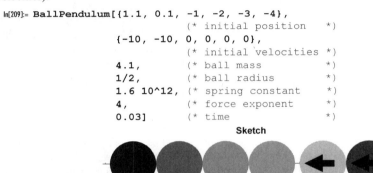

○ Impulse before: -82.; Impulse after: -82.

○ Energy before: 820.; Energy after: 820.

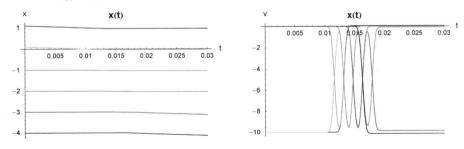

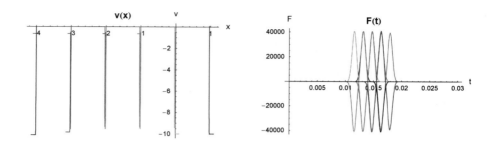

The implementation of `BallPendulum` described here can easily be generalized to the case of balls of different radii and masses. For a detailed discussion of this very interesting physical problem, see [1515], [827], [828], [1055], [1409], [92], [149], [1453], [667], [1725], [318], [370], [1341], [1694], [1323], and [1473], [861] for many balls (and [1102], [291], [292], [44] for infinitely many balls). For solitons in bead chains, see [1637].

For further potential mechanical applications of `NDSolve` to problems of classical theoretical mechanics, for example, the rotating double pendulum, see [961] and [91]. Many other simple but interesting examples can easily be computed with `NDSolve`. This includes a classical double pendulum [1966], a set of coupled oscillators [124], [1090], vortex pairs [17], a pendulum with a spring instead of a string [1359], an excited spherical pendulum [934], a spring attached to a mass and a barrier present [1513], the movement of a particle on absorbing surfaces [1246], and the movement of a space ship [936]. An especially "beautiful" (and interesting) 2D force field for a point mass is presented in [1689]. Additional interesting differential equations can be found in [196], [1690], [1949], [1651], [1215], [1699], [1573], [666], [1424], and [1228].

Also, some generalizations of the two-body Kepler problems circumventing Bertrand's theorem (see [882], [1934], [1195], [1194], [1388], and references therein) or systems of coupled oscillators [280], [979], [1041] or the motion in monopole and dipole fields [1806], [369], [923], [2], [100] are interesting. Aggregates of point masses that can carry out internal motions and move within a given force field, are a further not uninteresting subject [1170].

The three-body (of more general the n-body) problem is also very interesting (especially for larger times or including relativistic velocity-dependent masses [275], retardation [1347], [392], [393], [1799], [1536], [1134], [835], for large n [732], or in a noncommutative space [1550]). As a function of the initial conditions, we get very interesting paths (see [1471], [1760], [678], [1944], [763], [1216], [48], [1939], [76], [1472], and [1568]; for choreographic solutions of the n-body problem, see [1660], [597], [1045], [388], [598], [385], for 1D and $n = \infty$, see [1103]). Here are the relevant equations for the three- (or more) body problem in the plane. For a two-body force of the form $F \sim \mathbf{r} \, |\mathbf{r}|^{\alpha}$, the following animation shows the trajectories for two, three, and four point masses with random initial conditions. The special status of the harmonic oscillator ($\alpha = 0$), the Kepler problem in 3D ($\alpha = -3$), and the Kepler problem in 2D ($\alpha = -2$) [693], [1755] are clearly visible.

```
(* equations of motion of n point masses in d dimension
   interacting via a pairwise distance depending force r^α *)
nBodyProblemEquations[n_, d_, α_, {x_, mL_}] :=
Table[mL[[k]] x[k][j]''[t] == (* force of the n - 1 masses on m[k] *)
      Sum[If[l === k, 0, mL[[k]] mL[[l]] (x[l][j][t] - x[k][j][t])*
            Sqrt[Sum[(x[l][j][t] - x[k][j][t])^2, {j, d}]]^α],
         {l, n}], {k, n}, {j, d}]

(* initial conditions; r is a function *)
rICs[n_, d_, {x_}, r_] :=
Table[(# == r[#])&[Derivative[o][x[k][j]][0]], {k, n}, {j, d}, {o, 0, 1}]

(* the average velocity vector for the n point masses *)
averageVelocities[n_, d_, rics_, mL_] :=
Module[{M = Plus @@ mL,
        R = Dispatch[Flatten[Apply[Rule, rics, {3}]]]},
        1/M Table[Sum[mL[[k]] x[k][j]'[0], {k, n}], {j, d}] /. R]

(* solve equations of motion and plot path of the n point masses
   in the 1,2-coordinate plane *)
solveEqationsAndMakeGraphics[n_, d_, α_, T_, opts___] :=
Module[{mL, rics, ndsol, T, vs},
  (* all masses are equal *) mL = Table[1, {n}];
  (* random initial conditions *)
  rics = rICs[n, d, {x}, Random[Real, {-1, 1}]&];
  (* solve equations of motions *)
  ndsol = NDSolve[Flatten[(* use function from above *)
                  {nBodyProblemEquations[n, d, α, {x, mL}], rics}],
                  Flatten[Table[x[k][j], {k, n}, {j, d}]], {t, 0, T},
                  PrecisionGoal -> 8, AccuracyGoal -> 8, MaxSteps -> 2 10^4];
  (* time achieved for the maximal number of steps *)
  T = ndsol[[1, 1, 2, 1, 1, 2]];
  (* average velocities in coordinate directions *)
  vs = Take[averageVelocities[n, d, rics, mL], 2];
  (* show trajectories in center of mass system *)
  Show[Table[ParametricPlot[
             Evaluate[{x[k][1][t], x[k][2][t]} - vs t /. ndsol[[1]]],
                  {t, 0, T}, PlotPoints -> 400, PlotRange -> All,
                  DisplayFunction -> Identity, (* color differently *)
                  PlotStyle -> {Hue[0.78 (k - 1)/n], Thickness[0.003]}],
        {k, n}],
      opts, Frame -> True, Axes -> False, AspectRatio -> 1,
      FrameTicks -> False, DisplayFunction -> $DisplayFunction]]

Off[NDSolve::mxst];
Do[Show[GraphicsArray[
   Table[SeedRandom[j];
         solveEqationsAndMakeGraphics[j, 2, α, 30,
             DisplayFunction -> Identity], {j, 2, 4}]],
           PlotLabel -> "α=" <> ToString[N[α]]],
     {α, 7/2, -7/2, -1/10}]
```

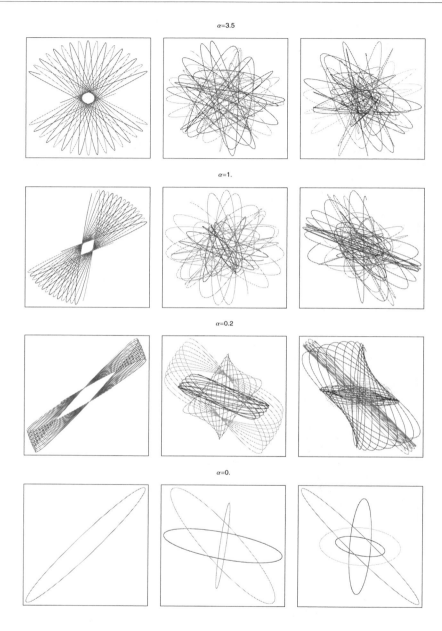

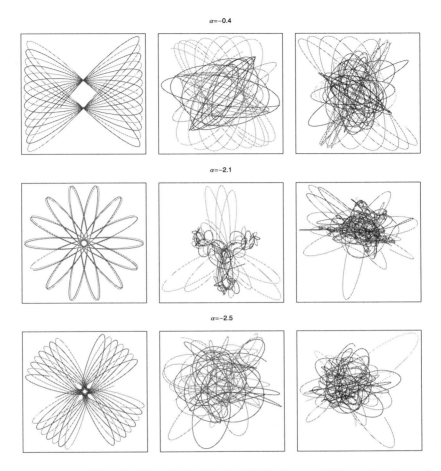

Here is a look at the scattering problem for three bodies for the 3D Kepler case. We use two heavy bodies (masses m_1 and m_3) and a light body (mass m_2) that initially is near mass m_1. We use random values for m_2 and the initial positions and velocities of m_2 and m_3. Depending on the parameters, the lighter body sometimes stays with m_1, the lighter body sometimes switches over to m_3 [646], the lighter body escapes and the two heavy bodies form a bound system, and sometimes all three bodies fly asymptotically away in different directions.

```
threeBodyScattering[{m1_, m2_, m3_}, {R2_, R3_},
                    {V2_, V3_}, {φ2_, φ3_}] :=
Flatten[{nBodyProblemEquations[3, 2, -3, {x, {m1, m2, m3}}],
         {x[1][1][0] == 0, x[1][2][0] == 0,
          x[1][1]'[0] == 0, x[1][2]'[0] == 0,
          x[2][1][0] == 0, x[2][2][0] == R2,
          x[2][1]'[0] == V2 Cos[φ2], x[2][2]'[0] == V2 Sin[φ2],
          x[3][1][0] == R3 Cos[φ3], x[3][2][0] == R3 Sin[φ3],
          x[3][1]'[0] == -V3 Cos[φ3], x[3][2]'[0] == -V3 Sin[φ3]}}]

randomThreeBodyKeplerScattering[seed_, T_] :=
Module[{m1, m2, m3, R2, R3, V2, V3, φ2, φ3, ndsol, tMax},
 SeedRandom[seed];
 (* random parameter values *)
 {m1, m2, m3} = {1, 10^Random[Real, {-2, -1}], 1};
 {R2, R3} = {10^Random[Real, {-1, 1/2}], 10^Random[Real, {1, 2}]};
 {V2, V3} = {10^Random[Real, {-2, 1}], Random[Real, {1/2, 2}]};
 {φ2, φ3} = {Random[Real, {3/4, 5/4} Pi], Random[Real, {-Pi, Pi}]};
 (* solve ODEs *)
 ndsol = NDSolve[threeBodyScattering[{m1, m2, m3}, {R2, R3},
                                     {V2, V3}, {φ2, φ3}],
          {x[1][1], x[1][2], x[2][1], x[2][2], x[3][1], x[3][2]},
          {t, 0, T}, MaxSteps -> 2 10^5];
 (* display trajectories *)
 tMax = ndsol[[1, 1, 2, 1, 1, 2]];
 Show[Graphics[{{(* starting points *) {PointSize[0.01], GrayLevel[0],
                Table[Point[{x[k][1][0], x[k][2][0]}] /. ndsol[[1]],
                      {k, 3}]},
  (* trajectories *) Thickness[0.002],
  Transpose[{{RGBColor[1, 0, 0], RGBColor[0, 1, 0], RGBColor[0, 0, 1]},
   Cases[ParametricPlot[Evaluate[{x[#][1][t] + t, x[#][2][t]} /.
                                                  ndsol[[1]]],
                 {t, 0, tMax},
                 PlotPoints -> 10 Ceiling[tMax],
                 DisplayFunction -> Identity], _Line, Infinity]& /@
                                                {1, 2, 3}}]}}],
     PlotRange -> All, Frame -> True, FrameTicks -> None]]

Show[GraphicsArray[
Block[{$DisplayFunction = Identity},
     randomThreeBodyKeplerScattering[#, 120]& /@ #]]]& /@
  {{(* stay with original heavy body *) 7794057862, 6458773008},
   {(* move over to second heavy body *) 3972351378, 1560261572},
   {(* two heavy bodies form a bound system *) 101076702, 4989711205},
   {(* all three bodies fly away separately *) 1889116017, 6142224887}}
```

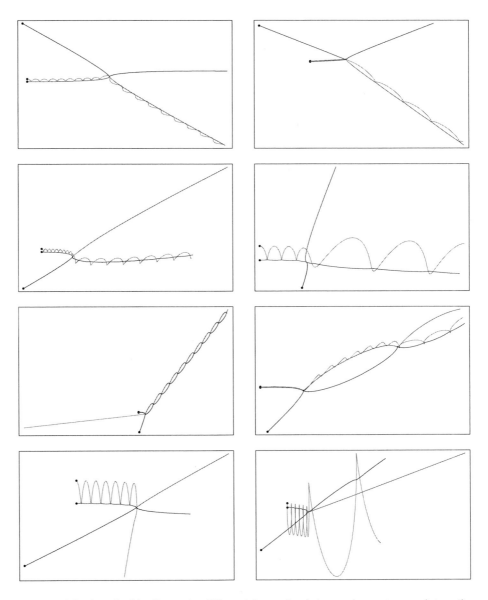

Another, slightly simpler (while described by first order differential equations), interesting system are interacting vortices. A system of n vortices of strength Γ_k that are confined to the unit disk (by placing mirror vortices outside the unit disk) is described by the following equations for the complex variables $z_j(t)$ [1954], [1342], [61], [62], [1357], [1328], [1569].

$$\frac{\partial z_j(t)}{\partial t} = \frac{i}{2\pi}\left(\sum_{\substack{k=1 \\ k\neq j}}^{n} \Gamma_k \frac{z_j(t) - z_k(t)}{|z_j(t) - z_k(t)|^2} - \sum_{k=1}^{n} \Gamma_k \frac{z_j(t) - z_k(t)\,|z_k(t)|^{-2}}{|z_j(t) - z_k(t)\,|z_k(t)|^{-2}|^2} \right)$$

Similarly to the gravitational three-body problem, we get unpredictable motions for $n = 3$ and higher. The function `vortexEquationsNSolution` solves the equation of motion for the time T for n vortices with randomly chosen initial conditions and vortex strengths.

```
In[210]:= vortexEquations[n_] := Table[
    D[z[j][t], t] == I/(2 Pi)(
    Sum[If[k =!= j, Γ[k] (z[j][t] - z[k][t])/
                        Abs[z[j][t] -  z[k][t]]^2, 0], {k, n}] -
    Sum[            Γ[k] (z[j][t] - z[k][t]/Abs[z[k][t]]^2)/
                        Abs[z[j][t] -  z[k][t]/Abs[z[k][t]]^2]^2, {k, n}]),
                        {j, n}]
```

```
In[211]:= vortexEquationsNSolution[n_, T_] := NDSolve[
    Join[vortexEquations[n] /. (* random strengths *)
            Table[Γ[i] -> 1 Random[Real, {-1, 1}], {i, n}],
        (* random initial conditions *)
        Table[z[i][0] == Random[] Exp[2Pi I Random[]], {i, n}]],
        Table[z[i], {i, n}], {t, 0, T},
        MaxStepSize -> 0.1, MaxSteps -> 10^5]
```

Here are six paths for three vortices. We use `SeedRandom` to get reproducible trajectories.

```
In[212]:= showVortexGraphics[nsol_, opts___] :=
    With[{T = nsol[[1, 1, 2, 1, 1, 2]], n = Length[First[nsol]]},
        Show[Function[{k, col},
        (* plot path of the vortices *)
        ParametricPlot[Evaluate[{Re[z[k][t]], Im[z[k][t]]} /. nsol], {t, 0, T},
                AspectRatio -> Automatic, Axes -> False, PlotPoints -> 500,
                (* the unit circle *)
                Prolog -> {Thickness[0.02], GrayLevel[1/2],
                            Circle[{0, 0}, 1]},
                PlotRange -> 1.05 {{-1, 1}, {-1, 1}}, PlotStyle ->
                {{col, Thickness[0.002]}},
                DisplayFunction -> Identity]] @@@
            (* three vortices are colored red, green, blue;
                more are colored in a rainbow fashion *)
            If[n === 3, {{1, RGBColor[1, 0, 0]}, {2, RGBColor[0, 1, 0]},
                    {3, RGBColor[0, 0, 1]}},
                Table[{k, Hue[0.78 (k - 1)/(n - 1)]}, {k, n}]], opts,
            DisplayFunction -> $DisplayFunction]]
```

```
In[213]:= Show[GraphicsArray[ (SeedRandom[#];
    showVortexGraphics[vortexEquationsNSolution[3, 300],
                DisplayFunction -> Identity])& /@ #]]& /@
    (* six seed for the random number generator *)
    {{8216344413, 7007375890, 3890522980},
     {7969730955, 5965017411, 1701994967}}]
```

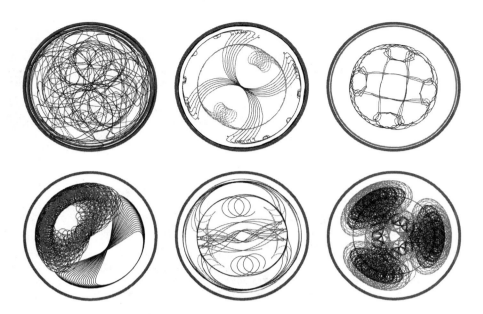

Instead of the above-mentioned full-blown gravitational three-body problem, let us have a brief look at the restricted three-body problem. We assume two masses move according to their mutual gravitational influence along circle and a third mass with coordinates $\{\xi(t), \eta(t)\}$ moves in their resulting force field. Then the equations of motions for the third mass are [1463], [757], [304], [490]

$$\xi''(t) = \xi(t) + 2\eta'(t) - (1-\mu)\frac{(\xi(t)+\mu)}{\rho_1} - \mu\frac{(\xi(t)-(1-\mu))}{\rho_2}$$

$$\eta''(t) = \eta(t) - 2\xi'(t) - (1-\mu)\frac{\eta(t)}{\rho_1} - \mu\frac{\eta(t)}{\rho_2}$$

$$\rho_1 = \left((\xi(t)+\mu)^2 + \eta(t)^2\right)^{3/2}$$

$$\rho_2 = \left((\xi(t)-(1-\mu))^2 + \eta(t)^2\right)^{3/2}.$$

Especially interesting are periodic solutions of the restricted three-body problem. We will search for symmetric periodic solutions [1956], [822], [1334], meaning $\{\xi(-t), \eta(-t)\} = \{\xi(t), -\eta(t)\}$. So, we use $\xi'(0) = \eta(0) = 0$ and $\xi(0)$ and $\eta'(0)$ as free parameters.

It is straightforward to combine various numerical functions in *Mathematica*. For the search for periodic orbits, we will make heavy use of the combination of NDSolve and FindMinimum. We will choose random initial conditions $\xi(0)$ and $\eta'(0)$ and solve the equations of motions until we find ones whose orbit comes close to the initial conditions again. Then we will use FindMinimum to minimize the distance to the starting initial conditions. The function to be minimized results from solving the above set of coupled nonlinear differential equations. Because the solution of each differential equation can be carried out in a fraction of a second, the whole two-parameter minimization will just take a few seconds. The following functions solveODEs, min⁝ Diff, orbitLengthEstimate, findGoodStartξ0H0, and findBestPeriodicξ0H0 implement such a search.

```
In[214]:= (* solve the differential equations for given initial conditions *)
        solveODEs[{ξ0_, H0_}, T_, opts___] :=
        Block[{(* moon-earth value *) μ = 0.01227747},
```

```
      eqs = Module[{ν = 1 - μ, ρ1, ρ2},
        {ρ1, ρ2} = ((ξ[t] + {μ, -ν})^2 + η[t]^2)^(3/2);
        (* use first order system *)
        {ξ'[t] == Ξ[t], η'[t] == H[t],
         Ξ'[t] == ξ[t] + 2 H[t] - ν(ξ[t] + μ)/ρ1 - μ(ξ[t] - ν)/ρ2,
         H'[t] == η[t] - 2 Ξ[t] - ν η[t]/ρ1 - μ η[t]/ρ2}];
      Check[NDSolve[Flatten[{eqs,
          ξ[0] == ξ0, Ξ[0] == 0, η[0] == 0, H[0] == H0}],
          {ξ, η, Ξ, H}, {t, 0, T}, opts, MaxSteps -> 10^4,
          PrecisionGoal -> 10, AccuracyGoal -> 10], $Failed]]
```

```
In[216]:= (* determine minimum distance in initial condition space *)
      minDiff[{ξ0_, H0_}, T_, nsol_] :=
      Module[{δ, pl, δs, fm},
        (* summed squared difference to initial conditions *)
        δ[t_?NumberQ] = (ξ[t] - ξ0)^2 + η[t]^2 +
                        Ξ[t]^2 + (H[t] - H0)^2 /. nsol[[1]];
        (* find global minimum using Plot's adaptive refinement *)
        pl = Plot[Evaluate[δ[t]], {t, 5, T},
              AxesOrigin -> {0, 0}, PlotRange -> All,
              PlotStyle -> {Hue[0]}, DisplayFunction -> Identity];
        points = Cases[pl, _Line, Infinity][[1, 1]];
        δs = points[[Position[points, Min[Last /@ points]][[1, 1]], 1]];
        (* refine minimum if possible *)
        Off[InterpolatingFunction::dmval];
        If[δs < 6., $Failed,
          fm = FindMinimum[δ[t], {t, δs, δs + 10^-3}];
          {fm[[2, 1, 2]], fm[[1]]}]]
```

```
In[218]:= minDiff[{ξ0_, H0_}, T_, $Failed] = $Failed;
```

```
In[219]:= (* estimate length of the orbit *)
      orbitLengthEstimate[nsol_, T_] :=
      (* some length of discretized orbit *)
      Plus @@ (Sqrt[#.#]& /@ (Subtract @@@
          Partition[Table[Evaluate[{ξ[t], η[t]} /. nsol[[1]]],
                    {t, 0, T, 10^-2}], 2, 1]));
```

```
In[221]:= (* use random starting values until one finds an orbit that
    comes near to the starting point again *)
      findGoodStartξ0H0[r_, T_, l_] :=
      Module[{ξ0, H0, nsol, md, res},
      (* the seed *) SeedRandom[r];
      While[(* until we find a good starting point *)
        While[{ξ0, H0} = {Random[Real, {0, +3}], Random[Real, {0, -3}]};
              nsol = solveODEs[{ξ0, H0}, T];
              md = minDiff[{ξ0, H0}, T, nsol];
              md === $Failed || md[[2]] > 1/2,
            Null];
            res = {{ξ0, H0}, md};
            orbitLengthEstimate[nsol, md[[1]]] > l, Null];
          res]
```

```
In[223]:= (* the minimum distance between initial and later *)
      minDistance[{ξ0_?NumberQ, H0_?NumberQ}, T_] :=
          minDiff[{ξ0, H0}, T, solveODEs[{ξ0, H0}, T]][[2]]
```

```
In[225]:= findBestPeriodicξ0H0[{ξ0v_, H0v_}, T_, δ_] :=
      Module[{fm},
```

```
       fm = FindMinimum[minDistance[{ξ0, H0}, T],
                        {ξ0, 0.99 ξ0v, 1.01 ξ0v}, {H0, 0.99 H0v, 1.01 H0v}];
           If[Head[fm] =!= FindMinimum,
               If[fm[[1]] < δ, {ξ0, H0} /. fm[[2]], $Failed], $Failed]]
```

For displaying the orbit $\{\xi(t), \eta(t)\}$ (and $\{\xi'(t), \eta'(t)\}$), we will use the function showPeriodicOrbit.

```
In[226]:= colorLine[gr_Graphics] := Function[l,
              MapIndexed[{Hue[(#2[[1]] - 1)/(Length[l] - 1)],
                        Line[#]}&, Partition[l, 2, 1]]][
                             Cases[gr, _Line, Infinity][[1, 1]]]

In[227]:= showPeriodicOrbit[{ξ0_, H0_}, T_] :=
          Module[{nsol, 𝒯, μ = 0.01227747},
                 (* solve differential equations *)
                 nsol = solveODEs[{ξ0, H0}, T];
                 𝒯 = minDiff[{ξ0, H0}, T, nsol][[1]];
                 (* display orbit *)
               Show[GraphicsArray[
                     Graphics[{Thickness[0.006], colorLine[
                     ParametricPlot[Evaluate[# /. nsol], {t, 0, 𝒯},
                     PlotPoints -> 1000, DisplayFunction -> Identity]]},
                     FrameTicks -> {Automatic, Automatic, None, None},
                     PlotRange -> All, AspectRatio -> Automatic,
                     Frame -> True, Axes -> True]& /@
                        {{ξ[t], η[t]}, {Ξ[t], H[t]}}]]]
```

Assembling now all the above functions, we can implement findPeriodicξ0H0. Evaluated with the random number seed r, findPeriodicξ0H0 returns a periodic orbit of length $\leq l$. The differential equations will be solved up to time T and the squared sum of the differences of the initial conditions and later orbit coordinates will be $\leq \delta$. We use a seeding for the random numbers to easily reproduce the results. Here is find‐Periodicξ0H0 implemented and shown at work.

```
In[228]:= findPeriodicξ0H0[r_, T_, l_, δ_] :=
          Module[{fp},
              While[(* find candidate and improve it *)
                  fp = findBestPeriodicξ0H0[
                             findGoodStartξ0H0[r, T, l][[1]], T, δ];
                  fp === $Failed, Null]; fp]

In[229]:= findPeriodicξ0H0[11111, 20, 20, 10^-8]

Out[229]= {0.824168, -0.144718}

In[230]:= showPeriodicOrbit[%, 20]
```

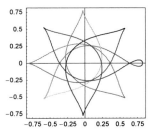 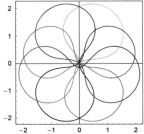

The function doAll finally unites all of the above functions for searching and displaying closed orbits.

```
In[231]:= doAll[r_, T_, l_, δ_] := showPeriodicOrbit[findBestPeriodicξ0H0[
                  findGoodStartξ0H0[r, T, l][[1]], T, δ], T]
```

The periodic orbits can have quite complicated shapes [764], [304], [1585]. Here are some more examples. For a complete classification of the possible orbits, see [764], [304], and [820]; for algorithms to compute periodic orbits, see [1452], [1140].

```
In[232]:= doAll[58440623274390470756, 30, 20, 10^-6]
```

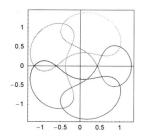 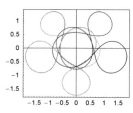

```
In[233]:= doAll[13938310832482415053, 20, 10, 10^-6]
```

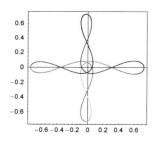 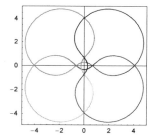

```
In[234]:= doAll[ 7184662868728565889, 30, 50, 10^-6]
```

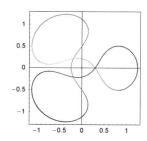 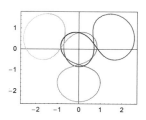

```
In[235]:= doAll[69133196196102120169, 30, 50, 10^-6]
```

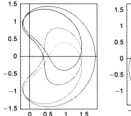
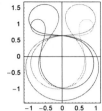

In[236]:= **doAll[25060060673945256124, 30, 50, 10^-6]**

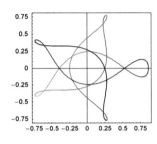
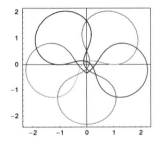

In[237]:= **doAll[36241940556126524981, 30, 20, 10^-6]**

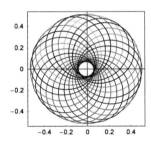
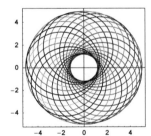

For a general investigation of the possible motions in the restricted three-body problem, see [1324]. For the still simpler, but nevertheless interesting problem of one mass in the field of two fixed masses, see [1837]. For periodic orbits of more than two masses, see [1291].

In a similar manner, one could find periodic orbits of the Lorenz system [1855], [1856]. (For their importance, see [1684], [1801].) Here are initial conditions for the simplest orbits. (When numerically searching for initial conditions that yield periodic orbits of the Lorenz system one can use the fact that for a fixed z_0, the $\{x_0, y_0\}$ coordinates cluster in thin slanted stripes in the x_0, y_0-plane.)

```
In[238]:= periodicLorenzICs = (* all start at z[0] == 27 *)
          {(* {1 left, 1 right} *) { 2.14736763108117, -2.07804821234662, 27},
           (* {1 left, 2 right} *) {-1.10196591102154,  3.42176028989961, 27},
           (* {2 left, 1 right} *) {-2.52882238160245,  1.55505505747398, 27},
           (* {1 left, 3 right} *) { 4.54156878404161,  1.47261307171282, 27},
           (* {2 left, 2 right} *) { 1.43458687357288, -3.00891350123217, 27},
           (* {3 left, 1 right} *) {-3.86105744302905, -0.40050023440817, 27}};
```

The function `periodicLorenzGraphics` visualizes the periodic orbits (either as one graphic or as an animation).

```
In[239]:= periodicLorenzGraphics[makeAnimation_] :=
    Module[{T = 5, pp = 1800, ndsol, lines, colors,
            λ = Length[periodicLorenzICs]},
      (* solve Lorenz system *)
    ndsol[{x0_, y0_, z0_}, T_] :=
    NDSolve[{x'[t] == 10 (-x[t] + y[t]),
             y'[t] == -x[t] z[t] + 28 x[t] - y[t],
             z'[t] ==  x[t] y[t] - 8/3 z[t],
             x[0] == x0, y[0] == y0, z[0] == z0},
            {x, y, z}, {t, 0, T}, MaxSteps -> 10^5,
            PrecisionGoal -> 8, AccuracyGoal -> 8];
      (* solve Lorenz system for initial conditions periodicLorenzICs *)
    lines = (Table[{x[t], y[t], z[t]}, {t, 0, T, T/pp}] /.
                 ndsol[#, T][[1]])& /@ periodicLorenzICs;
      (* colors for orbits *)
    colors = Table[{Hue[0.8 (j - 1)/(λ - 1)], Thickness[0.002],
                   PointSize[0.03]}, {j, λ}];
      (* make animation/one 3D graphic *)
    Do[Show[Graphics3D[
       {If[makeAnimation, (* show endpoints in animation *)
           Flatten /@ {Transpose[{colors, Point[#[[k]]]& /@ lines}]}, {}],
           (* orbits *)
           Flatten /@  Transpose[{colors, Line[Take[#, k]]& /@ lines}]},
                   PlotRange -> {{-17, 17}, {-23, 23}, {8, 43}}]],
       {k, If[makeAnimation, 1, pp], pp, 30}]];
```

Here are the shortest periodic orbits shown.

```
In[240]:= periodicLorenzGraphics[False]
```

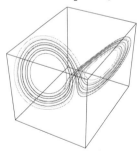

```
      periodicLorenzGraphics[True]
```

As mentioned in the beginning of this section, the function `NDSolve` is probably the most important numerics function in *Mathematica*. After the last example from the oldest and most established parts of the exact natural sciences, namely, classical mechanics, let us continue this subsection with a more modern and speculative subject, Bohm's version of a hidden variable theory underlying quantum mechanics. (The character of the *Mathematica GuideBooks* does not allow an in-depth mathematical, physical, and philosophical discussion of the fascinating subject of local and nonlocal hidden variable theories; for a start, see [308], [156], [164], [228], [852], [492], [1062], [895], [1386], [155], and [213].) In the widely accepted Copenhagen interpretation, the complex-valued wave function $\psi(x, t)$ (for simplicity, we assume one particle in one dimension) determines the state of a quantum system completely and $|\psi(x, t)|^2$ represents the probability density to find a microparticle at position x at time t. $\psi(x, t)$ obeys the Schrödinger equation (in appropriate units)

$$i\,\frac{\partial \psi(x, t)}{\partial t} = -\frac{\partial^2 \psi(x, t)}{\partial x^2} + V(x)\,\psi(x, t).$$

David Bohm [225], [226], [227], [163] added a more detailed description of the position $q(t)$ of the particle [1803]. Starting by rewriting $\psi(x, t)$ in the form $\psi(x, t) = R(x, t) \exp(i\,S(x, t)/\hbar)$, he derives a Hamilton–Jacobi equation for $S(x, t)$ that in addition to the classical terms contains a "quantum-potential" term

$$V_{\mathrm{qu}}(x, t) = -\frac{1}{R(x, t)}\,\frac{\partial^2 R(x, t)}{\partial x^2}$$

[523], [1811], [1355], [834], [43], [1612]. (For the construction of genuine quantum actions, see [1069].) Given the solution $\psi(x, t)$ of the Schrödinger equation, this yields the following equation of motion for the micro-particle:

$$\frac{\partial q(t)}{\partial t} = 2\,\mathrm{Im}\Big(\frac{1}{\psi(q(t), t)}\,\frac{\partial \psi(x, t)}{\partial x}\Big|_{x=q(t)}\Big)$$

Given an initial ensemble of particles $q^{(k)}(0)$, such that their spatial probability agrees with $|\psi(x, t)|^2$, the two probability densities will agree also later.

Here, we will deal with the simplest possible case—a particle in a potential well (for a well with a barrier, see [1718]). The inpenetrable walls are located at $x = 0$ and $x = \pi$. To avoid the numerical solution of the time-dependent Schrödinger equation (this is the subject of the next section), we will assume the initial wave function to be a superposition $\psi(x, 0) = \sum_{k=1}^{n} c_k\,\phi_k(x)$ of the first few eigenstates $\varphi_k(x) = \sqrt{2/\pi}\,\sin(k\,x)$. Then, the wavefunction at time t is given as $\psi(x, t) = \sum_{k=1}^{n} c_k\,\phi_k(x) \exp(-i\,k^2\,t)$.

We start by selecting a random number of terms and random coefficients c_k.

```
In[241]:= Clear[x, t, q, n, m, P, ψ];
         SeedRandom[8549202731]

In[243]:= n = Random[Integer, {4, 9}];

In[244]:= coeffList = Table[Random[Complex, {-1 - I, 1 + I}], {k, n}];
```

This is the resulting time-dependent wave function (we drop the normalization factor $\sqrt{2/\pi}$ because it cancels anyway later).

```
In[245]:= ψ[x_, t_] := Table[Sin[k x] Exp[-I k^2 t], {k, n}].coeffList;
```

We integrate the initial probability density, and, by inverting the cumulative density numerically, we obtain appropriately distributed points.

```
In[246]:= Psol = NDSolve[{P'[x] == Abs[ψ[x, 0]]^2, P[0] == 0},
                  P, {x, 0, Pi}, MaxStepSize -> 0.01];

In[247]:=
      m = 50; PEnd = Psol[[1, 1, 2]][Pi];
      q0s = Table[x /. FindRoot[{Psol[[1, 1, 2]][x] == k/m PEnd},
                        {x, 0, Pi}], {k, 1, m - 1}];
```

This is the quantum potential $V_{\mathrm{qu}}(x, t)$.

```
In[249]:= Vqu[x_, t_] =
    With[{R = Sqrt[ψ[x, t] (ψ[x, t] /. c_Complex :> Conjugate[c])]},
        -D[R, x, x]/R];
```

This is right-hand side of $\partial q(t)/\partial t = 2\,\mathrm{Im}(1/\psi(q(t),\,t)\,\partial\psi(q(t),\,t)/\partial x)$.

```
In[250]:= rhs[x_, t_] = 2 Im[D[ψ[x, t], x]/ψ[x, t]];
```

Now, we calculate the trajectories of the imagined particles (we color the trajectories, alternating black and white).

```
In[251]:= trajectories =
    Table[(* solve differential equation *) nsol =
    NDSolve[{q'[t] == rhs[q[t], t], q[0] == q0s[[k]]},
            q, {t, 0, Pi}, MaxSteps -> 10000];
    {(* color *) GrayLevel[(1 - (-1)^k)/2],
     Cases[(* trajectory as a line *)
            ParametricPlot[Evaluate[{q[t], t} /. nsol], {t, 0, Pi},
                        PlotPoints -> 200, DisplayFunction -> Identity],
            _Line, Infinity]}, {k, m - 1}];
```

For a nicer picture, we show the trajectories together with a contour plot of the "quantum potential" as the background.

```
In[252]:= (* avoid singularities and the wall boundaries;
        avoid spurious imaginary parts *)
    cp1 = With[{ε = 10^-5},
    ContourPlot[Evaluate[Re[N[Vqu[x, t]]]], {x, ε, Pi - ε}, {t, 0, Pi},
                    PlotPoints -> 240, ContourLines -> False,
                    ColorFunction -> (Hue[0.8 #]&), Contours -> 100,
                    DisplayFunction -> Identity]]
```

Now, we have all ingredients together to display the trajectories and the quantum potential. The position runs horizontally and the time vertically.

```
In[254]:= Show[{cp1,
            Graphics[{Thickness[0.002], trajectories}]},
            PlotRange -> All, Frame -> True, FrameTicks -> None,
            Axes -> False, AspectRatio -> 3/2,
            DisplayFunction -> $DisplayFunction]
```

Although the trajectories look complicated [1199], [579], [1910], there is a lot of order in them [711], [228], [1927], [668], [669], [1586], [847], [851], [186], [522]. (As an ensemble, their density just represents $|\psi(x,\,t)|^2$.) Using more complicated (multidimensional) wavefunctions $\psi(\vec{x},\,t)$, the resulting Bohm trajectories exhibit a very interesting behavior [1649], [488], [461], [1822], [1611], [422], [423], [556], [1063] (we encourage the reader to calculate the trajectories in a wavefunction build by superimposing the first hydrogen eigenstates). For the trajectories of photons, see [691], [692]. For trajectories identical to classical mechanics particles, see [1198], [1200]. For fractal trajectories, see [1587].

We will end this subjection by solving large systems of coupled linear differential equations. In the spirit of the last few examples, we will deal with a problem that has a classical and a quantum variant: a continuous time random walk on a graph [971], [1315], [394], [972], [1316], [973], [168], [319]. For the graph, we will take the edges of an icosahedron with o-times recursively subdivided faces. We start with the construction of the graph Laplacian, a matrix **A** of dimension $d \times d$ where d is the number of nodes of the graph [320]. The elements a_{ij} of **A** of are -1 if the nodes i and j are connected by an edge; a_{ii} is the number of neighboring vertices of the vertex i and all other elements are zero.

```
In[255]:= o = 5;
```

```
In[256]:= Needs["Graphics`Polyhedra`"]
```

```
In[257]:= (* subdivide a triangle into four triangles *)
        subdividePolygon[Polygon[{p1_, p2_, p3_}]] :=
        With[{p12 = (p1 + p2)/2, p23 = (p2 + p3)/2, p31 = (p3 + p1)/2},
            Polygon /@ {{p1, p12, p31}, {p12, p2, p23},
                        {p23, p3, p31}, {p12, p23, p31}}]
```

```
In[259]:= (* the triangles of the recursively subdivided icosahedron surface *)
        smallFaces = Map[#/Sqrt[#.#]&,
        Nest[Flatten[subdividePolygon /@ #]&, Polygon /@
            N[Map[Vertices[Icosahedron][[#]]&, Faces[Icosahedron],
            {-1}], 50], o], {-2}];
```

```
In[261]:= (* the edges of the graph *)
        edges = Sort /@ Flatten[N[Partition[
                Append[#[[1]], #[[1, 1]]], 2, 1]& /@ smallFaces], 1];
        (* the vertices of the graph *)
        vertices = Union[Flatten[edges, 1]];
```

After five recursive subdivisions of the icosahedron, we have a graph with 10242 nodes (twelve with five neighbors and all others with six neighbors) and 61440 edges.

```
In[265]:= {d, λ} = Length /@ {vertices, edges}
```

```
Out[265]= {10242, 61440}
```

We represent the graph Laplacian as a sparse array.

```
In[266]:= vertexNumberRules = MapIndexed[(#1 -> #2[[1]])&, vertices];
        bothEdges = Flatten[{#, Reverse[#]}& /@
            Union[Sort /@ (edges //. Dispatch[vertexNumberRules])], 1];
```

```
In[268]:= vertexFunctionality = {#[[1, 1]], Length[#]}& /@
            Split[Sort[bothEdges], #1[[1]] === #2[[1]]&];
```

```
In[269]:= adjacencyMatrix = SparseArray[Join[
        Table[{i, aii} = vertexFunctionality[[k]]; {i, i} -> aii, {k, d}],
        Table[{i, j} = bothEdges[[k]]; {i, j} -> -1, {k, λ}]]]
```

```
Out[269]= SparseArray[<71682>, {10242, 10242}]
```

We have a quick look at the matrix **A**. Because the matrix **A** has all its nonvanishing elements in the neighborhood of the main diagonal, we display just the part along the main diagonal.

```
In[270]:= Show[Graphics[{PointSize[0.002], Point /@
                (* use main diagonal as x-axis;
                    y-axis is in direction subdiagonal *)
                        ({#1 + #2, #1 - #2}& @@@
```

```
                    First /@ Most[ArrayRules[adjacencyMatrix]])}],
        AspectRatio -> 1/6, PlotRange -> All];
```

To later visualize the time-development of the random walk, we will graphically visualize each vertex as set of polygons centered at the vertex. The list `vertexSurroundingPolygons` is a list of these polygons. Its *k*th element contains the polygons surrounding the *k*th vertex.

```
In[271]:= (* list of the form {vertexNumber, polygonsSharingThisVertex} *)
      neighborPolygons = Sort[{#[[1, 1]], #[[2]]& /@ #}& /@
        Split[Sort[Sort[Flatten[Function[p, {#, p}& /@ p[[1, 1]]] /@
            Transpose[{N[smallFaces] //. Dispatch[vertexNumberRules],
                        N[smallFaces]}], 1]]],
          #1[[1]] == #2[[1]]&]];

In[273]:= (* form the Voronoi polygons around vertex n *)
      makeSurroundingPolys[n_, {Polygon[l_], Polygon[lN_]}] :=
      Module[{p1, p2, p3, mp = Mean[lN]},
        p = Position[l, n][[1, 1]];
        {q, r} = Complement[{1, 2, 3}, {p}];
        {p1, p2, p3} = lN[[{p, q, r}]];
        Polygon[{p1, (p1 + p2)/2, mp, (p1 + p3)/2}]]

In[275]:= (* list of polygons around vertex n *)
      vertexSurroundingPolygons = Function[{v, ps},
          makeSurroundingPolys[v, #]& /@ ps] @@@ neighborPolygons;
```

Here is a quick look at the original graph (with polygons filled in) and the polygons centered at the vertices.

```
In[277]:= Show[GraphicsArray[
      Graphics3D[{EdgeForm[], Thickness[0.002], #}, Boxed -> False]& /@
        {{smallFaces /. p:Polygon[l_] :> {p, Line[Append[l, First[l]]]}},
         vertexSurroundingPolygons /.
                    {p:Polygon[l_] :> {p, Line[Take[l, {2, 4}]]}}}}]]
```

Given the ODE solution vector *nds*, the function `coloredSphereGraphic3D` displays (the function *cf* of) the solution at time τ as colored vertices. Blue represent a small value, and red a high value.

```
In[278]:= (* use cut-off logarithmic scale based on 1/d is average *)
      color[x_] := Hue[0.68 (1 - (ArcTan[Log[x d]]/(Pi/2) + 1)/2)]

      coloredSphereGraphic3D[nds_, τ_, cf_, opts___] :=
```

```
Graphics3D[{EdgeForm[], Transpose[{color /@ cf /@ nds[τ],
                            vertexSurroundingPolygons}]},
        opts, Boxed -> False, Lighting -> False,
        PlotRange -> 1.05 {{-1, 1}, {-1, 1}, {-1, 1}}]
```

Now, we have all ingredients together to actually solve the differential equations. For a classical continuous time random walk, the differential equation is (after appropriate rescaling of time) $p'_k(t) = -\sum_{j=0}^{d} a_{jk} p_j(t)$. For the quantum continuous time random walk, the differential equation is (again after appropriate rescaling of time) $\psi'_k(t) = i \sum_{j=0}^{d} a_{jk} \psi_j(t)$. For the classical case, $p_k(t)$ is the probability to find a walker at vertex k and for the quantum case the probability is $|\psi_k(t)|^2$. We will start the walks with the initial condition $p_k(0) = \delta_{k\hat{k}}$ and $\psi_k(0) = \delta_{k\hat{k}}$. For $\hat{k}$, we will choose the uppermost vertex (one of the original vertices of the icosahedron).

```
In[281]:= sPos = Position[vertices, {0., 0., 1.}][[1, 1]];
```

```
In[282]:= Tmax = 20;
        (* use Adams method in the following *)
        SetOptions[NDSolve, Method -> "Adams"];
        Developer`SetSystemOptions["CatchMachineUnderflow" -> False];

        Timing[ndsC = NDSolve[{p'[t] == -adjacencyMatrix.p[t],
                        p[0] == MapAt[1&, Table[0, {d}], sPos]},
                        p, {t, 0, Tmax}]]
Out[286]= {5.29 Second, {{p → InterpolatingFunction[{{0., 20.}}, <>]}}}
```

```
In[287]:= Timing[ndsQ = NDSolve[{ψ'[t] == I adjacencyMatrix.ψ[t],
                        ψ[0] == MapAt[1&, Table[0, {d}], sPos]},
                        ψ, {t, 0, Tmax}]]
Out[287]= {68.8 Second, {{ψ → InterpolatingFunction[{{0., 20.}}, <>]}}}
```

We have the two probability conservation laws $\sum_{j=0}^{d} p_j(t) = 1$ and $\sum_{j=0}^{d} |\psi_j(t)|^2$. They are fulfilled to a good quality. The complex solution is considerably more expensive (longer calculation times and larger memory use).

```
In[288]:= 1 - {Plus @@ ndsC[[1, 1, 2]][10],
        Plus @@ (Abs[ndsQ[[1, 1, 2]][10]]^2)}
Out[288]= {-2.44249×10^{-15}, -3.08606×10^{-6}}
```

```
In[289]:= ByteCount[#]/10.^6 "MB"& /@ {ndsC, ndsQ}
Out[289]= {37.8937 MB, 287.907 MB}
```

And here are the resulting probabilities shown for various times. We see that initially the classical walker spreads faster but soon after the quantum walkers walk much faster [395] (depending on the graph and the initial conditions, quantum walks can also be slower than classical walks).

```
In[290]:= Function[τ, Show[GraphicsArray[
        {coloredSphereGraphic3D[ndsC[[1, 1, 2]], τ, #&],
        coloredSphereGraphic3D[ndsQ[[1, 1, 2]], τ, Abs[#]^2&]}]]] /@
                        (* show state at four times *) {1/2, 6, 10, 15};
```

To follow the time evolution of the probability densities over a larger time, we will solve the differential equations over a small time interval and then visualize the solutions. This avoids the construction of large (memory constrained) interpolating functions.

```
ΔT = 1;
p0 = MapAt[1&, Table[0, {d}], sPos]; ψ0 = p0;
Do[(* propagate solutions *)
    ndsC = NDSolve[{p'[t] == -adjacencyMatrix.p[t],
                    p[0] == p0}, p, {t, ΔT, ΔT}];
    ndsQ = NDSolve[{ψ'[t] == I adjacencyMatrix.ψ[t],
                    ψ[0] == ψ0}, ψ, {t, ΔT, ΔT}];
    (* show solutions *)
    Show[GraphicsArray[
      {coloredSphereGraphic3D[ndsC[[1, 1, 2]], ΔT, #&],
       coloredSphereGraphic3D[ndsQ[[1, 1, 2]], ΔT, Abs[#]^2&]}],
      PlotLabel -> jk];
    (* update initial conditions *)
    p0 = p[ΔT] /. ndsC[[1]]; ψ0 = ψ[ΔT] /. ndsQ[[1]], {jk, 100}]
```

We end now our walk (pun intended) through the syntax, use, and examples of NDSolve. There is a multitude of other interesting applications of NDSolve, like charging the walls of the mazes constructed in Chapter 1 of the Graphics volume [1794] to find the way out of the maze by following the electric field lines, and so on. We leave it to the reader to explore other situations using NDSolve.

1.11.2 Partial Differential Equations

Mathematica currently is able to solve partial differential equations in 1+1 dimensions (using the method of lines [1824]). (This does not include Laplace equations that are rephrased as a Cauchy problem [1064], [789], [337], [165], [1948].)

NDSolve[*pdesAndBoundaryAndInitialConditions*, *function*(s),
            independentVariablesAndDomain, *options*]

computes the numerical solution of the system of partial differential equations *pdesAndBoundaryAndInitialConditions* (consisting of the differential equation and the corresponding initial and boundary conditions) for the dependent function(s) *function(s)* using the options *options*.

The numerical solution of partial differential equations is more expensive than the solution of ordinary differential equations. So the examples in this subsection will run longer than the ones from the last subsection. Most of the options available in NDSolve for ordinary differential equations are also valid for partial differential equations. The explicit values used for the options AccuracyGoal and PrecisionGoal will frequently be lower for PDEs than for ODEs (to allow for a faster execution).

As some first examples, let us deal with two classical linear PDEs: we will model the quantum-mechanical scattering of a Gaussian wave packet on a Gaussian potential bump governed by the 1D time-dependent Schrödinger equation and the propagation of a cos-pulse governed by the 1D wave equation.

We start with the Schrödinger equation example (we set $\hbar = 1$, $m = 1$):

$$i\,\frac{\partial\,\Psi(x,\,t)}{\partial\,t} = -\frac{1}{2}\,\frac{\partial^2\,\Psi(x,\,t)}{\partial x^2} + V(x)\,\Psi(x,\,t)$$
$$\Psi(-x_M,\,t) = \Psi(x_M,\,t) = 0.$$

The function `enforceDirichletBCs` adds a cos-shaped function to ψ to make sure that at $\pm$xM the resulting wave function vanishes identically.

```
In[1]:= (* numerical value for xM *)
    xM = 15;

    enforceDirichletBCs[ψ_, x_, xM_] :=
    ψ - (((ψ /. x -> -xM) - (ψ /. x -> xM))/2*
        (Cos[(x + xM)/(2xM) Pi] + 1) + (ψ /. x -> xM))

    (* check that the initial condition fulfills the boundary condition *)
    {enforceDirichletBCs[Exp[-(x + 3)^2], x, xM] /. x -> -xM,
     enforceDirichletBCs[Exp[-(x + 3)^2], x, xM] /. x -> +xM}
Out[6]= {0, 0}
```

The kinetic and potential energy (in a Gaussian potential well) of the Gaussian wave packet $e^{-(x+3)^2 + 3ix}$ are as follows.

```
In[7]:= With[{ψ = enforceDirichletBCs[Exp[-(x + 3)^2], x, xM] Exp[3 I x]},
    {-1/2 NIntegrate[Evaluate[D[ψ, {x, 2}] Conjugate[ψ]], {x, -xM, xM}],
    VMax NIntegrate[Evaluate[ψ Exp[-x^2] Conjugate[ψ]],
                {x, -xM, xM}]}] // Chop[#, 10^-6]&
Out[7]= {6.26657, 0.00253657 VMax}
```

We will choose the prefactor of the potential to be 6; this means the energy of the incoming wave packet is about the maximum value of the potential. As a result, the kinetic energy is much larger than the potential energy at $t = 0$.

```
In[8]:= schrödingerEq =
    I D[Φ[x, t], {t, 1}] == -1/2 D[Φ[x, t], {x, 2}] + 6 Exp[-x^2] Φ[x, t]
Out[8]= i Φ^(0,1)[x, t] == 6 e^{-x^2} Φ[x, t] - 1/2 Φ^(2,0)[x, t]
```

The initial velocity of our Gaussian wave packet is about 3.75.

```
In[9]:= With[{ψ = enforceDirichletBCs[Exp[-(x + 3)^2], x, 12] Exp[3 I x]},
    NIntegrate[Evaluate[-I D[ψ, x] Conjugate[ψ]],
                {x, -xM, xM}]] // Chop[#, 10^-6]&
Out[9]= 3.75994
```

Now, we solve the Schrödinger equation for a time of 5 (units)—this takes about $1\,s$ on a 2 GHz computer. For partial differential equations, the settings for the `PrecisionGoal` and `AccuracyGoal` options should be much smaller than for ordinary differential equations.

```
In[10]:= (nsol =
    NDSolve[{schrödingerEq,
            Φ[x, 0] == enforceDirichletBCs[
                    Exp[-(x + 3)^2], x, xM] Exp[3 I x],
        Φ[ xM, t] == 0, Φ[-xM, t] == 0},
        Φ[x, t], {x, -xM, xM}, {t, 0, 5},
        AccuracyGoal -> 3, PrecisionGoal -> 3]) // Timing
Out[10]= {1.56 Second,
        {{Φ[x, t] → InterpolatingFunction[{{-15., 15.}, {0., 5.}}, <>][x, t]}}}
```

The resulting `InterpolatingFunction` has a size of about 15 MB.

```
In[11]:= ByteCount[nsol]/10.^6
Out[11]= 15.3259
```

Now, let us have a look at the reflection and transmission of the wave packet. We show first a density plot and then a contour plot. (We also see how the transmitted part of the wave packet is slightly faster than "expected" —meaning the maximum of the transmitted part appears earlier than the linearly extrapolated impinging maximum [884], [1510], [491], [1376], [1377], [1139], [505], [1066], [1441], [463], [1318], [1683], [1378].)

```
In[12]:= Show[GraphicsArray[
         Block[{$DisplayFunction = Identity, dp},
           {(* density plot *)
            dp = DensityPlot[Evaluate[Abs[Ψ[x, t]] /. nsol[[1]]],
                    {x, -xM, xM}, {t, 0, 5},
                    PlotPoints -> 200, Mesh -> False,
                    ColorFunction -> (Hue[0.78 #]&)],
            (* make contour plot from the density plot; reuse data from dp *)
            ListContourPlot[dp[[1]], FrameTicks -> None,
                        ColorFunction -> (Hue[0.78 #]&), Contours -> 30,
                        ContourStyle -> {{Thickness[0.002]}}]}]]]
```

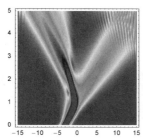 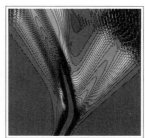

The pictures show that up to about $t \approx 3$ the boundaries do not play an important role for the scattering process. At $t \approx 3$, the transmitted part of the wave packet reaches the right boundary and gets reflected there. Here is an animation of the scattering process until $t = 3$.

```
In[13]:= potentialWell = {GrayLevel[1/2],
         Polygon[N[Table[{x, 6 Exp[-x^2]}, {x, -xM, xM, 2 xM/200}]]]};

In[14]:= Show[GraphicsArray[#]]& /@ Partition[
     Table[Plot[Evaluate[6.3 + 3 Abs[Ψ[x, t]] /. nsol], {x, -xM, xM},
               PlotPoints -> 100, PlotRange -> {0, 12}, Frame -> True,
               Axes -> False, Prolog -> potentialWell,
               PlotStyle -> {{Hue[0], Thickness[0.002]}},
               DisplayFunction -> Identity,
               Epilog -> {GrayLevel[0], (* time *)
                      Text["t = " <> ToString[InputForm[t]], {-9, 3}]},
               FrameTicks -> {Automatic, None, None, None}],
         {t, 0, 3, 3/15}], 4]
```

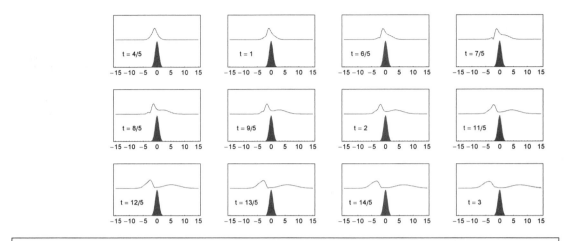

```
Do[Plot[Evaluate[6.3 + 3 Abs[Ψ[x, t]] /. nsol],
        {x, -xM, xM}, PlotPoints -> 100,
        PlotRange -> {0, 12}, Frame -> True, Axes -> False,
        PlotStyle -> {{Hue[0], Thickness[0.002]}},
        Epilog -> {GrayLevel[0], Text["t = " <> ToString[t], {-9, 3}]},
        FrameTicks -> {Automatic, None, None, None}, Prolog ->
potentialWell],
    {t, 0., 3., 0.1}]
```

We can use the function `Fourier` to visualize the momentum distribution of the wave packet [886].

```
In[15]:= momentumDistribution[t_, n_] :=
    With[{f = nsol[[1, 1, 2, 0]]},
        Fourier[Table[f[x, t], {x, -xM, xM, 2xM/n}]]]
```

In the beginning, we have a peaked momentum distribution of the packet moving to the right. Then the scattering process happens and we see a second peak (the reflected packet) evolve. Because of energy conservation, the transmitted packet gets a smaller average momentum.

```
In[16]:= ListPlot3D[Abs[Table[momentumDistribution[t, 100], {t, 0, 3, 3/100}]],
            Axes -> False, Mesh -> False]
```

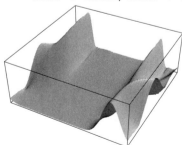

Finally, let us quickly check the correctness of the solution. The normalization integral $\int_{-x_M}^{x_M} |\Psi(x, t)|^2\, dx$ is a conserved quantity. For all t, this quantity is correct to about 1%.

```
In[17]:= Function[t, {t, NIntegrate[Abs[nsol[[1, 1, 2, 0]][x, t]]^2,
                          {x, -xM, xM}, MaxRecursion -> 18,
                          PrecisionGoal -> 4]}] /@ {0, 1, 2, 3}
Out[17]= {{0, 1.2533}, {1, 1.2511}, {2, 1.25108}, {3, 1.25057}}
```

Calculating the local residue $\delta\Delta\,[x,\,t]$ of the differential operator shows that the error is largest at the part of the wave packet that tunneled through the barrier.

```
In[18]:= δΔ[x_, t_] = With[{ψ = nsol[[1, 1, 2, 0]]},
            Abs[(I Derivative[0, 1][ψ][x, t] -
            (-1/2 Derivative[2, 0][ψ][x, t] + 6 Exp[-x^2] ψ[x, t]))]];
```

```
In[19]:= Show[GraphicsArray[
        Plot[{(* magnified error *) 10^2 Abs[δΔ[x, #]],
            (* solution *) Abs[nsol[[1, 1, 2, 0]][x, #]]},
        {x, -xM, xM}, PlotRange -> {0, 1}, Frame -> True,
        DisplayFunction -> Identity, Axes -> False,
        PlotStyle -> {{Thickness[0.002], Hue[0]},
                    {Thickness[0.02], GrayLevel[1/2]}}]& /@
        (* three times *) {1/2, 3/2, 5/2}]]
```

(For the reflection of a Schrödinger wave packet on a wall, see also [55], [126], [1353], [511], [148], [434], and [931]; for the treatment of reflectionless potentials, see [997]; for two-particle scattering, see [1190], [1116]. For absorbing boundary conditions see, [1083], [599], [1443], [742], [910], and [1593]. For a potential back-flow, see [277], [572]. For the relativistic situation, see [281], [1852], [1758], [1065]. For a high-precision, *Mathematica*-based solution, see, [578]. For time-dependent barriers, see [127].)

As a second example, let us choose a hyperbolic equation—the wave equation (we set the phase velocity c to 1);

$$\frac{\partial^2 \Psi(x,\,t)}{\partial t^2} - \frac{\partial^2 \Psi(x,\,t)}{\partial x^2} = 0.$$

```
In[20]:= waveEq = D[Ψ[x, t], {t, 2}] - D[Ψ[x, t], {x, 2}] == 0
Out[20]= Ψ^(0,2)[x, t] - Ψ^(2,0)[x, t] == 0
```

For hyperbolic and parabolic PDEs, the method of lines is used. By explicitly specifying it, it is possible to allow the specification of suboptions specific to the method of lines.

MethodOfLines

    is a method option of NDSolve to indicate the use of the method of lines.

The most important suboptions of settings of the "MethodOfLines" method option setting are (be aware that most of them are strings):

"SpatialDiscretization" -> "TensorProductGrid"

"MinPoints" -> *listOfSiscretizationPointsForEachSpatialVariable*

"MaxPoints" -> *listOfSiscretizationPointsForEachSpatialVariable*

`"DifferenceOrder" -> positiveInteger | "PseudoSpectral"`
Method -> *ODEMethodForTemporalODEs*.

```
In[21]:= Options[NDSolve`MethodOfLines]

Out[21]= {DifferentiateBoundaryConditions-> True, DiscretizedMonitorVariables-> False,
          ExpandEquationsSymbolically-> False, Method -> Automatic,
          SpatialDiscretization-> TensorProductGrid, TemporalVariable -> Automatic}
```

As the initial condition, we choose a cos-shaped initial elongation (and vanishing initial velocity). The following example calculates the amplitude for Dirichlet boundary conditions. We now use explicit (sub)option settings specific for the method of lines. The message generated indicates that we might need more spatial discretization points to obtain a solution that has the intended precision and accuracy.

```
In[22]:= (* solve the 1D wave equation for given boundary conditions *)
         solve1DWaveEquation[{bcs__}, ics_:Derivative[0, 1][Ψ][x, 0] == 0] :=
         Block[{xM = 5, T = 10},
         (* solve the wave equation with specified initial and boundary conditions *)
         NDSolve[{waveEq, Ψ[x, 0] == If[Abs[x] > Pi/2, 0, Cos[x]^2], ics, bcs},
                 Ψ[x, t], {x, -xM, xM}, {t, 0, T},
                 AccuracyGoal -> 3, PrecisionGoal -> 3,
                 Method -> {"MethodOfLines", "SpatialDiscretization" ->
                             {"TensorProductGrid",
                              "MaxPoints" -> {600}, "MinPoints" -> {600}}}]]

In[24]:= Clear[xM];


         nsol = solve1DWaveEquation[{Ψ[ xM, t] == 0, Ψ[-xM, t] == 0}]

         NDSolve::eerri :
            Warning: Estimated initial error on the specified spatial grid in the direction
               of independent variable x exceeds prescribed error tolerance. More…

         NDSolve::eerr :
            Warning: Scaled local spatial error estimate of 191.32863756758078`
               at t = 10.` in the direction of independent variable x is much
               greater than prescribed error tolerance. Grid spacing with 600 points
               may be too large to achieve the desired accuracy or precision.  A
               singularity may have formed or you may want to specify a smaller
               grid spacing using the MaxStepSize or MinPoints options. More…

Out[25]= {{Ψ[x, t] -> InterpolatingFunction[{{-5., 5.}, {0., 10.}}, <>][x, t]}}
```

A picture of the solution shows in the beginning the expected splitting of the initial pulse. When the pulses hit the spatial boundaries, they become reflected. Because of the Dirichlet boundary conditions, we have an additional phase jump.

```
In[26]:= combined3DAndDensityPlot[nsol_, xM_:5, T_:10] :=
         Show[GraphicsArray[
         Block[{dp, $DisplayFunction = Identity},
         {dp = DensityPlot[Evaluate[Ψ[x, t] /. nsol[[1]]], {x, -xM, xM}, {t, 0, T},
                           PlotPoints -> 200, Mesh -> False,
                           ColorFunction -> (Hue[0.78 #]&)],
         ListPlot3D[dp[[1]], Mesh -> False, PlotRange -> All,
                    MeshRange -> {{-xM, xM}, {0, T}}]}]]]

In[27]:= combined3DAndDensityPlot[nsol]
```

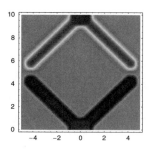 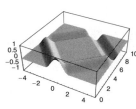

Theoretically, at time $t = 10$, the solution is exactly the negative of the original starting condition. The following picture shows that the differences between the numerical solution and the exact one are in the order of the specified precision goal. The difference is slightly larger than the specified precision goal; this was to be expected due to the issued messages.

```
In[28]:= Block[{xM = nsol[[1, 1, 2, 0, 1, 1, 2]]},
        Plot[Evaluate[(Ψ[x, t] /. nsol[[1]] /. t -> 10) +
                If[Abs[x] > Pi/2, 0, Cos[x]^2]],
            {x, -xM, xM}, PlotRange -> All]]
```

We repeat the above calculation, but this time we are using Neumann boundary conditions.

```
In[29]:= nsol = solve1DWaveEquation[{Derivative[1, 0][Ψ][ xM, t] == 0,
                            Derivative[1, 0][Ψ][-xM, t] == 0}]
```

NDSolve::eerri :
   Warning: Estimated initial error on the specified spatial grid in the direction
       of independent variable x exceeds prescribed error tolerance. More…

NDSolve::eerr :
   Warning: Scaled local spatial error estimate of 123.36262904116403`
       at t = 10.` in the direction of independent variable x is much
       greater than prescribed error tolerance. Grid spacing with 600 points
       may be too large to achieve the desired accuracy or precision. A
       singularity may have formed or you may want to specify a smaller
       grid spacing using the MaxStepSize or MinPoints options. More…

```
Out[29]= {{Ψ[x, t] → InterpolatingFunction[{{-5., 5.}, {0., 10.}}, <>][x, t]}}
```

In case of Neumann boundary conditions, there is no phase jump when the pulse is reflected at the boundaries.

```
In[30]:= combined3DAndDensityPlot[nsol]
```

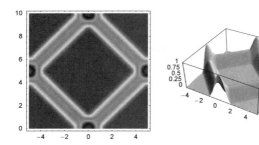

Next, we choose the left boundary of type Neumann and the right one of type Dirichlet.

```
In[31]:= nsol = solve1DWaveEquation[{Ψ[ xM, t] == 0,
                              Derivative[1, 0][Ψ][-xM, t] == 0}]
```

NDSolve::eerri :
    Warning: Estimated initial error on the specified spatial grid in the direction
        of independent variable x exceeds prescribed error tolerance. More…

NDSolve::eerr :
    Warning: Scaled local spatial error estimate of 69.96579259954042` at t = 10.` in
        the direction of independent variable x is much greater than prescribed error
        tolerance. Grid spacing with 600 points may be too large to achieve the desired
        accuracy or precision. A singularity may have formed or you may want to
        specify a smaller grid spacing using the MaxStepSize or MinPoints options. More…

```
Out[31]= {{Ψ[x, t] → InterpolatingFunction[{{-5., 5.}, {0., 10.}}, <>][x, t]}}
```

```
In[32]:= combined3DAndDensityPlot[nsol]
```

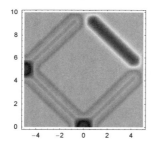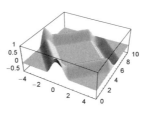

The next example for the 1D wave equation has Robin boundary conditions [753], a nonvanishing initial elongation, and a nonvanishing initial velocity (for a transformation from 1D Neumann and Dirichlet problems to Robin problems, see [238]).

```
In[33]:= nsol = solve1DWaveEquation[{Ψ[ xM, t] + Derivative[1, 0][Ψ][ xM, t] == 0,
                              Ψ[-xM, t] - Derivative[1, 0][Ψ][-xM, t] == 0},
                              Derivative[0, 1][Ψ][x, 0] ==
                              If[Pi/4 < Abs[x] < 3Pi/4, -Sin[2(x - Pi/4)]^2, 0]]
```

NDSolve::eerri :
    Warning: Estimated initial error on the specified spatial grid in the direction
        of independent variable x exceeds prescribed error tolerance. More…

NDSolve::eerr :
Warning: Scaled local spatial error estimate of 240.420502202003` at t = 10.` in
the direction of independent variable x is much greater than prescribed error
tolerance. Grid spacing with 600 points may be too large to achieve the desired
accuracy or precision. A singularity may have formed or you may want to
specify a smaller grid spacing using the MaxStepSize or MinPoints options. More…

Out[33]= {{Ψ[x, t] → InterpolatingFunction[{{-5., 5.}, {0., 10.}}, <>][x, t]}}

In[34]:= **combined3DAndDensityPlot[nsol]**

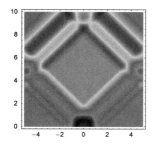 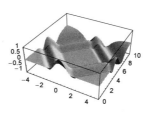

And in the next input we use a boundary condition that is nonlinear in derivatives of Ψ.

In[35]:= **nsol = solve1DWaveEquation[**
   {Derivative[0, 1][Ψ][-xM, t] + Derivative[1, 1][Ψ][-xM, t]^2 == 0,
   Derivative[0, 1][Ψ][+xM, t] + Derivative[1, 1][Ψ][+xM, t]^2 == 0}]

NDSolve::eerri :
Warning: Estimated initial error on the specified spatial grid in the direction
of independent variable x exceeds prescribed error tolerance. More…

NDSolve::eerr :
Warning: Scaled local spatial error estimate of 115.90281025651866`
at t = 10.` in the direction of independent variable x is much
greater than prescribed error tolerance. Grid spacing with 600 points
may be too large to achieve the desired accuracy or precision. A
singularity may have formed or you may want to specify a smaller
grid spacing using the MaxStepSize or MinPoints options. More…

Out[35]= {{Ψ[x, t] → InterpolatingFunction[{{-5., 5.}, {0., 10.}}, <>][x, t]}}

In[36]:= **combined3DAndDensityPlot[nsol]**

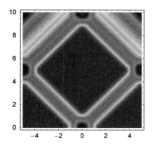 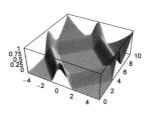

We now consider some explicitly time-dependent boundary conditions. We start with the wave equation in a
1 + 1D rectangular domain. At the left and right edges of the spatial domain, we enforce a boundary condition
that is a time-dependent linear combination of the function value and the first spatial derivative. Because at $t = 0$
the boundary condition and the initial conditions specified are not compatible at $x = \pm x_M$, we obtain a message.

In[37]:= **nsol = solve1DWaveEquation[**
       {Φ[+xM, t] + Derivative[1, 0][Φ][+xM, t] == -Pi/10 Cos[6 t/xM Pi],
         Φ[-xM, t] - Derivative[1, 0][Φ][-xM, t] == -Pi/10 Cos[6 t/xM Pi]}]

      NDSolve::ibcinc : Warning: Boundary and initial conditions are inconsistent. More…

      NDSolve::eerri :
       Warning: Estimated initial error on the specified spatial grid in the direction
        of independent variable x exceeds prescribed error tolerance. More…

      NDSolve::eerr :
       Warning: Scaled local spatial error estimate of 106.65358749137464`
        at t = 10.` in the direction of independent variable x is much
        greater than prescribed error tolerance. Grid spacing with 600 points
        may be too large to achieve the desired accuracy or precision. A
        singularity may have formed or you may want to specify a smaller
        grid spacing using the MaxStepSize or MinPoints options. More…

Out[37]= {{Φ[x, t] → InterpolatingFunction[{{-5., 5.}, {0., 10.}}, <>][x, t]}}

The next problem also contains time-dependent boundary conditions.

In[38]:= **nsol = solve1DWaveEquation[{Φ[xM, t] == Sin[t^2], Φ[-xM, t] == Sin[t^2]}]**

      NDSolve::eerri :
       Warning: Estimated initial error on the specified spatial grid in the direction
        of independent variable x exceeds prescribed error tolerance. More…

      NDSolve::eerr :
       Warning: Scaled local spatial error estimate of 48.16973670418257` at t = 10.` in
        the direction of independent variable x is much greater than prescribed error
        tolerance. Grid spacing with 600 points may be too large to achieve the desired
        accuracy or precision. A singularity may have formed or you may want to
        specify a smaller grid spacing using the MaxStepSize or MinPoints options. More…

Out[38]= {{Φ[x, t] → InterpolatingFunction[{{-5., 5.}, {0., 10.}}, <>][x, t]}}

In[39]:= **combined3DAndDensityPlot[nsol]**

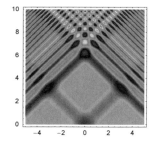 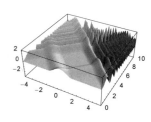

Boundary conditions must be local functions on each boundary segment. There is one exception to this rule—periodic boundary conditions are allowed.

In[40]:= **nsol = solve1DWaveEquation[{Φ[xM, t] == Φ[-xM, t]}]**

      NDSolve::eerri :
       Warning: Estimated initial error on the specified spatial grid in the direction
        of independent variable x exceeds prescribed error tolerance. More…

```
          NDSolve::eerr :
            Warning: Scaled local spatial error estimate of 129.88019080849497`
              at t = 10.` in the direction of independent variable x is much
              greater than prescribed error tolerance. Grid spacing with 600 points
              may be too large to achieve the desired accuracy or precision.  A
              singularity may have formed or you may want to specify a smaller
              grid spacing using the MaxStepSize or MinPoints options. More…
```

Out[40]= {{Φ[x, t] → InterpolatingFunction[{{..., -5., 5., ...}, {0., 10.}}, <>][x, t]}}

In[41]:= **combined3DAndDensityPlot[nsol]**

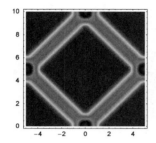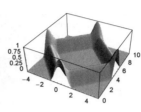

The following antiperiodic boundary conditions do currently not result in a solution.

In[42]:= **solve1DWaveEquation[Φ[xM, t] == -Φ[-xM, t]] // Head**

Out[42]= solve1DWaveEquation

(For chaos-causing boundary condition for the 1D wave equation, see [1687].)

The method used to solve 1+1-dimensional differential equations assumes that the initial conditions are sufficiently smooth. In case of nonsmooth initial conditions, often many small oscillations develop. Here we give such an example for the wave equation. The initial condition for Ψ has discontinuities and its time-derivative has poles [956]. We choose the initial points so, that

$$\Psi(x, 0) = \theta(|x| - 1)/2 \ \wedge \ \frac{\partial \Psi(x, t)}{\partial t}\Big|_{t=0} = \frac{i}{4\pi}\left(\frac{1}{x - 1} - \frac{1}{x + 1}\right).$$

Not unexpectedly, for these nonsmooth initial conditions having poles we obtain warning messages from NDSolve.

In[43]:= **Block[{xM = 3, T = 3},**
      nsol = NDSolve[{waveEq,
              Φ[x, 0] == If[Abs[x] > 1, 0, 1/2],
              Derivative[0, 1][Φ][x, 0] == I/(4Pi)(2/(x^2 - 1)),
              Φ[xM, t] == 0, Φ[-xM, t] == 0},
            Φ[x, t], {x, -xM, xM}, {t, 0, T},
            Method -> {"MethodOfLines", "SpatialDiscretization" ->
                    {"TensorProductGrid",
                     "MaxPoints" -> {600}, "MinPoints" -> {600}}}]];

```
          NDSolve::ibcinc : Warning: Boundary and initial conditions are inconsistent. More…

          NDSolve::eerri :
            Warning: Estimated initial error on the specified spatial grid in the direction
              of independent variable x exceeds prescribed error tolerance. More…
```

```
NDSolve::eerr :
  Warning: Scaled local spatial error estimate of 4363.627985251988` at t = 3.` in
    the direction of independent variable x is much greater than prescribed error
    tolerance. Grid spacing with 600 points may be too large to achieve the desired
    accuracy or precision. A singularity may have formed or you may want to
    specify a smaller grid spacing using the MaxStepSize or MinPoints options. More…
```

The following graphics clearly show how discretization-based oscillations develop for increasing *t*.

```
In[44]:= Block[{xM = 3, T = 3},
    Show[GraphicsArray[
    DensityPlot[#[nsol[[1, 1, 2]]], {x, -xM, xM}, {t, 0, T},
            DisplayFunction -> Identity,
            PlotPoints -> 200, Mesh -> False,
            ColorFunction -> (Hue[0.78 #]&)]& /@ {Re, Im, Abs}]]]
```

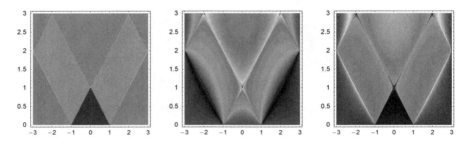

```
In[45]:= Block[{xM = 3, T = 3},
    Show[GraphicsArray[
    Plot3D[Evaluate[#[Ψ[x, t]] /. nsol[[1]]], {x, -xM, xM}, {t, 0, T},
            DisplayFunction -> Identity,
            Axes -> {Automatic, Automatic, None},
            PlotPoints -> 200, Mesh -> False]& /@ {Re, Im, Abs}]]]
```

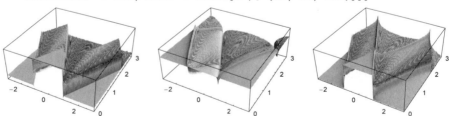

Smoothing these oscillations yields reasonable solutions of the wave equation.

```
In[46]:= Module[{pp = 500, n = 8, ΨData, xM = 3, T = 3},
    ΨData = Table[nsol[[1, 1, 2]], {t, 0, T, T/pp}, {x, -xM, xM, 2xM/pp}];
    Show[Table[
    ListPlot[MapIndexed[{-xM + 2xM (#2[[1]] - 1)/pp, #1}&,
                (* smooth through averaging *)
                Abs[ListConvolve[Table[1, {n}]/n, ΨData[[k]]]]],
            PlotJoined -> True, Frame -> True, Axes -> False,
            PlotStyle -> {Thickness[0.002], Hue[k/500 0.8]},
            DisplayFunction -> Identity], {k, 1, 400, 20}],
            DisplayFunction -> $DisplayFunction]]
```

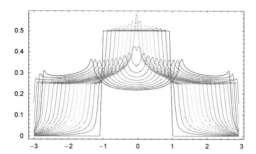

We continue with a more complicated example: the time-dependent Schrödinger equation in an interval whose length is time-dependent. NDSolve in Version 5.1 does not allow for spatial domains whose boundaries are time dependent. If the domain length is given by $x(t) = c(t)\xi$ with $\xi \in [0, 1]$, we can introduce new variables ξ and τ through $x = c(\tau)\xi$, $t = \tau$. In the ξ,τ-coordinates, the 1+1D domain is rectangular and we can use NDSolve to solve the differential equation.

```
In[47]:= movingWallSchrödingerEq = (* in transformed variables ξ and τ *)
        I (D[ψ[ξ, τ], τ] - ξ c'[τ]/c[τ] D[ψ[ξ, τ], ξ]) ==
        -D[ψ[ξ, τ], ξ, ξ]/c[τ]^2
```

$$Out[47]= \ \mathrm{i}\ \left(\psi^{(0,1)}[\xi,\tau] - \frac{\xi\, c'[\tau]\, \psi^{(1,0)}[\xi,\tau]}{c[\tau]}\right) == -\frac{\psi^{(2,0)}[\xi,\tau]}{c[\tau]^2}$$

Here is an example. We use an interval whose length oscillates harmonically with time. We enforce homogeneous Dirichlet boundary conditions on both interval ends.

```
In[48]:= (* time dependence of the well width *)
        c[τ_] := Pi + Pi/2 Cos[2 Pi τ]

        (nsol = (* solve over two full periods *)
        NDSolve[{movingWallSchrödingerEq,
            ψ[ξ, 0] == Sin[Pi ξ], ψ[0, τ] == 0, ψ[1, τ] == 0},
            ψ[ξ, τ], {ξ, 0, 1}, {τ, 0, 3},
            AccuracyGoal -> 4, PrecisionGoal -> 4, MaxSteps -> 10^5,
            Method -> {"MethodOfLines", Method -> "StiffnessSwitching",
                "SpatialDiscretization" ->
                {"TensorProductGrid", "DifferenceOrder" -> "Pseudospectral",
                "MaxPoints" -> {101}, "MinPoints" -> {101}}}]) // Timing
```

$$Out[50]= \ \{27.34\,\mathrm{Second}, \{\{\psi[\xi,\tau] \to \mathrm{InterpolatingFunction}[\{\{0.,1.\},\{0.,3.\}\}, <>][\xi,\tau]\}\}\}$$

The next two graphics show the absolute value and the argument of the resulting wave function. We see a shredding effect at the times where the domain has smallest extensions. To visualize the solution we first generate a contour plot of the solution in the rectangular ξ,τ-domain and then map the domain to the time-dependent interval length.

```
In[51]:= toTimeDependetWidthGraphic[cp_ContourGraphics] :=
            Graphics[cp] /. (* map to x, t-coordinates *)
                        Polygon[l_] :> Polygon[{#1, #2 c[#1]}& @@@
                                            addPoints[l, 0.1]]

In[52]:= (* insert points into edges of a polygon *)
        addPoints[l_, δε_] := Module[{n, λ}, Join @@ (
            Function[s, (* segment s too short? *)
                    If[(λ = Sqrt[#. #]&[Subtract @@ s]) < δε, s,
            n = Floor[λ/δε] + 1; (* form segments *)
```

```
            Table[# + k/n (#2 - #1), {k, 0, n - 1}]& @@ s]] /@
                Partition[Append[1, First[1]], 2, 1])]

In[54]:= Show[toTimeDependetWidthGraphic @
           ContourPlot[Evaluate[#1[ψ[ξ, τ]]^2 /. nsol[[1]]], {τ, 0, 3}, {ξ, 0, 1},
                  PlotPoints -> {480, 160}, ##3, AspectRatio -> 1/3,
                  ColorFunction -> #2, ContourLines -> False,
                  Contours -> 50, DisplayFunction -> Identity],
              DisplayFunction -> $DisplayFunction]& @@@
              (* show absolute value and argument;
                 use different coloring scheme *)
              {{Abs, (Hue[0.78 (# - 1)]&)},
               {If[# == 0., 0, Arg[#]]&, Hue, PlotRange -> All}};
```

Can one use the transformation $x \to \tan(\xi)$, $-\pi/2 < \xi < \pi/2$ to treat the above Schrödinger scattering problem over an infinite domain using $(\psi_{xx} \to -2\cos^3(\xi)\sin(\xi)\psi_\xi + \cos^4(\xi)\psi_{\xi\xi})$? Can this be used to see a wave packet in the potential $V(x) = -\exp(x)$ to move to infinity and back in finite time?

Before leaving the wave equation, we will calculate some solutions for the 2D case.

$$\frac{\partial^2 \Psi(t, x, y)}{\partial t^2} = \frac{\partial^2 \Psi(t, x, y)}{\partial x^2} + \frac{\partial^2 \Psi(t, x, y)}{\partial y^2}$$

This time, we will make an animation. To avoid generating huge `InterpolatingFunction` objects, we now use `NDSolve`ProcessEquations` to set up the problem and `NDSolve`Iterate` to evolve it in time.

For comparison, we will use three different boundary conditions: homogeneous Dirichlet boundary conditions on all four edges, homogeneous Neumann on all four edges, and homogeneous Dirichlet and Neumann on opposite edges.

```
In[55]:= (* Dirichlet case *)
    bcD = {u[t, -1, y] == u[t, +1, y] == 0, u[t, x, -1] == u[t, x, +1] == 0};

    (* Neumann case *)
    bcN = {Derivative[0, 1, 0][u][t, -1, y] ==
            Derivative[0, 1, 0][u][t, +1, y] == 0,
            Derivative[0, 0, 1][u][t, x, -1] ==
            Derivative[0, 0, 1][u][t, x, +1] == 0};

    (* mixed case *)
    bcM = {u[t, -1, y] == u[t, +1, y] == 0,
            Derivative[0, 0, 1][u][t, x, -1] ==
            Derivative[0, 0, 1][u][t, x, +1] == 0};
```

This creates the three `NDSolve`StateData` objects for the three cases. We start with a smooth localized hump in the center of a square with edge lengths 2.

```
In[62]:= makeNDSolveStateDataObject[{bcs__}, {ics0_, ics1_}, ppL_, do_, opts___] :=
    NDSolve`ProcessEquations[{D[u[t, x, y], t, t] ==
                    D[u[t, x, y], x, x] + D[u[t, x, y], y, y],
            bcs, u[0, x, y] == ics0, Derivative[1, 0, 0][u][0, x, y] == ics1},
            (* empty list here! *) {}, t, {x, -1, 1}, {y, -1, 1}, opts,
            MaxSteps -> 10^5, PrecisionGoal -> 2, AccuracyGoal -> 2,
            Method -> {"MethodOfLines", "SpatialDiscretization" ->
                        {"TensorProductGrid", "DifferenceOrder" -> do,
                        "MaxPoints" -> {ppL, ppL}, "MinPoints" -> {ppL, ppL}}}]
```

```
In[63]:= {{stateD}, {stateN}, {stateM}} =
    makeNDSolveStateDataObject[{#}, {If[Sqrt[x^2 + y^2] < 1/4,
                    Cos[Pi/2 Sqrt[x^2 + y^2]/(1/4)]^2, 0], 0},
                    100, 4]& /@ {bcD, bcN, bcM};
```

```
        NDSolve::eerri :
         Warning: Estimated initial error on the specified spatial grid in the direction
            of independent variable x exceeds prescribed error tolerance. More…

        NDSolve::eerri :
         Warning: Estimated initial error on the specified spatial grid in the direction
            of independent variable y exceeds prescribed error tolerance. More…

        NDSolve::eerri :
         Warning: Estimated initial error on the specified spatial grid in the direction
            of independent variable x exceeds prescribed error tolerance. More…

        General::stop :
         Further output of NDSolve::eerri will be suppressed during this calculation. More…
```

The function `equalPMPlot` visualized the solutions.

```
In[64]:= equalPMPlot[f_, L_] :=
    Block[{p3d, pr, $DisplayFunction = Identity},
            p3d = Plot3D[f, {x, -L, L}, {y, -L, L}, Mesh -> False,
                    PlotPoints -> 60, PlotRange -> All];
            pr = MapAt[{-1, 1} Max[Abs[#]]&, FullOptions[p3d, PlotRange], 3];
            Show[p3d, PlotRange -> pr, Axes -> False]]
```

The function `waveEquation2DAnimation` generates *frames* frames for time steps of ΔT.

```
In[65]:= waveEquation2DAnimation[ΔT_, frames_] :=
    Module[{uCurrent},
    (* initial state *)
```

```
uCurrent = NDSolve`ProcessSolutions[#, "Forward"]& /@
                                     {stateD, stateN, stateM};
Show[GraphicsArray[equalPMPlot[#[[1, 2]], 1]& /@ uCurrent]];
(* propagate solution in time and make visualizations *)
Do[(* visualize solution *)
   (* solve wave equation in time steps *)
   NDSolve`Iterate[stateD, j ΔT];
   NDSolve`Iterate[stateN, j ΔT];
   NDSolve`Iterate[stateM, j ΔT];
   (* form current solution *)
   uCurrent = NDSolve`ProcessSolutions[#, "Forward"]& /@
                                       {stateD, stateN, stateM};
   Show[GraphicsArray[equalPMPlot[#[[1, 2]], 1]& /@ uCurrent]],
   {j, frames}]]
```

We nicely see how the superposition of the reflected waves forms interesting symmetric patterns in the pure Dirichlet and pure Neumann case. We also see, that inside the light cone, the elongation does not become zero (a general feature of even-dimensional spaces).

In[66]:= **waveEquation2DAnimation[1/2, 8]**

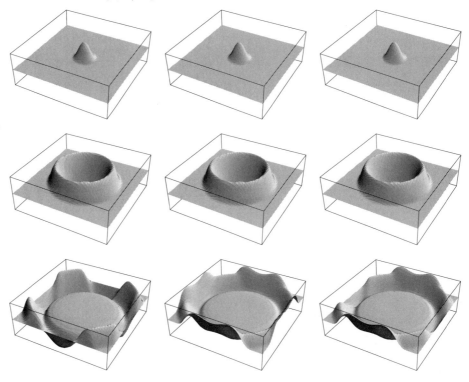

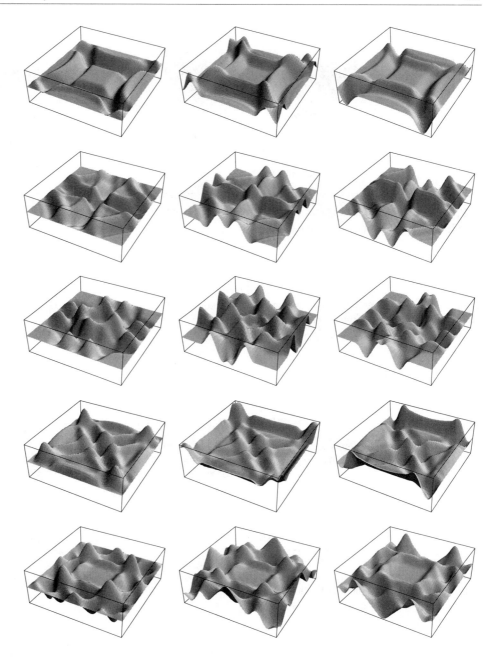

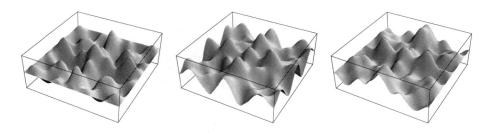

```
waveEquation2DAnimation[0.1, 100]
```

We continue with a linear parabolic differential equation in 2+1D. We will consider the Fokker–Planck equation [1532] for a damped harmonic oscillator [368]:

$$\frac{\partial P(x, v; t)}{\partial t} = \frac{\partial}{\partial v}\left(\gamma v P(x, v; t) + d \frac{\partial P(x, v; t)}{\partial v}\right) - v \frac{\partial P(x, v; t)}{\partial x} + x \frac{\partial P(x, v; t)}{\partial v}$$

Here $P(x, v; t)$ is the probability density that a damped harmonic oscillator under the influence of a stochastic force with zero mean and delta correlation has the position x and velocity v at time t. Here the diffusion coefficient d is related to the damping factor γ through $d = \gamma k_B T / m$.

```
In[67]:= FokkerPlanckEq = With[{p = P[x, v, t]},
                D[p, t] == D[γ v p + d D[p, v], v] - v D[p, x]  + x D[p, v]]
Out[67]= P^(0,0,1)[x, v, t] == γ P[x, v, t] + x P^(0,1,0)[x, v, t] +
         v γ P^(0,1,0)[x, v, t] + d P^(0,2,0)[x, v, t] - v P^(1,0,0)[x, v, t]
```

```
In[68]:= X = 5; V = 5; T = 4Pi;
     Block[{d = 0.2, γ = 0.2, (* initial distribution data *)
            x0 = 2, σx = 1/Sqrt[5], v0 = 0, σv = 1/Sqrt[5]},
     ndsol =
     NDSolve[{FokkerPlanckEq,
             P[x, v, 0] == Exp[-(x - x0)^2/σx^2] Exp[-(v - v0)^2/σv^2],
             P[-X, v, t] == P[+X, v, t] == P[x, -V, t] == P[x, +V, t] == 0},
             {P}, {x, -X, X}, {v, -V, V}, {t, 0, T},
             PrecisionGoal -> 2.5, AccuracyGoal -> 2.5,
             Method -> {"MethodOfLines", "SpatialDiscretization" ->
                        {"TensorProductGrid", "DifferenceOrder" -> 4,
                        "MaxPoints" -> 120 {1, 1}, "MinPoints" -> 120 {1, 1}}}]]
Out[69]= {{P → InterpolatingFunction[{{-5., 5.}, {-5., 5.}, {0., 12.5664}}, <>]}}
```

The interpolating function of a time-dependent PDE solution in two spatial dimensions is a relatively large object due to the large amount of data stored in it.

```
In[70]:= ByteCount[ndsol]/10.^6 "MB"
Out[70]= 248.839 MB
```

For each time t, we calculate the marginal probabilities $P_x(t) = \int_{-\infty}^{\infty} P(x, v; t)\, dv$ and $P_v(t) = \int_{-\infty}^{\infty} P(x, v; t)\, dx$ by numerical integration and visualize the resulting probabilities as a function of t. We clearly recognize the oscillations of a damped harmonic oscillator.

```
In[71]:= (* define marginal probabilities for position and velocity *)
     (#1 := NIntegrate[Evaluate[P[x, v, τ] /. τ -> t /. ndsol[[1]]], #2,
             PrecisionGoal -> 4, AccuracyGoal -> 4,
```

```
                Method -> Trapezoidal])& @@@ (* integration data *)
                {{xP[x_, t_], {v, -V, V}}, {vP[v_, t_], {x, -X, X}}};
In[73]:= Show[GraphicsArray[
    Block[{$DisplayFunction = Identity, opts},
            opts[xv_] = Sequence[PlotRange -> All, PlotPoints -> 60,
                                AxesLabel -> {"t", xv, None}, Mesh -> False];
        {(* plot marginal probability for position *)
        Plot3D[xP[x, t], {t, 0, T}, {x, -X, X}, Evaluate @ opts["x"]],
        (* plot marginal probability for velocity *)
        Plot3D[vP[v, t], {t, 0, T}, {v, -V, V}, Evaluate @ opts["v"]]}]]]
```

We clearly see how the initially localized distribution spreads out more and more due to the diffusive terms. To display the evolution in phase space, we construct contour plots of $P(x, v; t) = c$ in x,v,t-space for various values of c.

```
In[74]:= Needs["Graphics`ContourPlot3D`"]

In[75]:= (* cut a hole in a polygon *)
    makeHole[Polygon[l_], f_] :=
    Module[{mp = Plus @@ l/Length[l], ℓ}, ℓ = (mp + f*(# - mp))& /@ l;
        {MapThread[Polygon[Join[#1, Reverse[#2]]]&,
            Partition[Append[#, First[#]], 2, 1]& /@ {l, ℓ}]}]

In[77]:= Show[Graphics3D[
    MapIndexed[{EdgeForm[], Thickness[0.001],
                SurfaceColor[Hue[#2[[1]]/5], Hue[#2[[1]]/5], 2.4],
        (* cut a hole in the polygons *) makeHole[#, 0.6]& /@ Cases[
        (* contour plots of P[x, v, t] == c *)
        ContourPlot3D[Evaluate[P[x, v, t] /. ndsol[[1]]],
                {t, 0, T}, {x, -X, X}, {v, -X, X},
                PlotPoints -> {24, 3}, MaxRecursion -> 1,
                DisplayFunction -> Identity, Contours -> {#1}],
                        _Polygon, Infinity]}&,
            (* values of c *) {0.02, 0.05, 0.1, 0.2, 0.3}]],
        PlotRange -> All, BoxRatios -> {3, 1, 1}, Axes -> True,
        AxesLabel -> {"t", "x", "v"}]
```

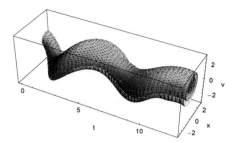

In the absence of a stochastic force, the Fokker–Planck equation reduces to the Liouville equation. Here is the Liouville equation for an anharmonic oscillator with no damping.

```
In[78]:= LiouvilleEq = With[{p = P[x, v, t]},
              D[p, t] == (-2 x + x^3/2) D[p, v] - v D[p, x]]
```

$$Out[78]= \; P^{(0,0,1)}[x, v, t] == \left(-2 x + \frac{x^3}{2}\right) P^{(0,1,0)}[x, v, t] - v\, P^{(1,0,0)}[x, v, t]$$

Because of the absence of the diffusive term, the probability density for the Liouville equation can develop much more easily fine structures. Despite the relatively large number of discretization points and relatively small final time T, we obtain messages that the solution probably does not satisfy the required precision and accuracy goals. To avoid constructing and storing a potentially very large interpolating function, we only calculate the time-independent state of the system at the final time T. From the resulting interpolating function, we make a plot of the probability density over the x,v-plane. Because of the relative small final time T, we still have a very pronounced peak in the probability density, but we clearly see nested figure eight-like structures that are typical for phase space trajectories in double minima potentials.

```
In[79]:= Developer`SetSystemOptions["CatchMachineUnderflow" -> False];
```

```
In[80]:= Block[{d = 0, γ = 0, X = 12, V = 12, T = 6 Pi,
            x0 = 2, σx = 1/4, v0 = 0, σv = 1/3, pp = 200, do = 4},
        ndsol =
        NDSolve[{LiouvilleEq,
                P[x, v, 0] == Exp[-(x - x0)^2/σx^2] Exp[-(v - v0)^2/σv^2],
                P[-X, v, t] == P[+X, v, t] == P[x, -V, t] == P[x, +V, t] == 0},
                {P}, {x, -X, X}, {v, -V, V}, {t, T, T},
                PrecisionGoal -> 3, AccuracyGoal -> 3,
                Method -> {"MethodOfLines", Method -> "StiffnessSwitching",
                        "SpatialDiscretization" ->
                        {"TensorProductGrid", "DifferenceOrder" -> do,
                        "MaxPoints" -> pp {1, 1}, "MinPoints" -> pp {1, 1}}}];
        (* plot resulting distribution *)
        Plot3D[Evaluate[P[x, v, T] /. ndsol[[1]]], {x, -X/2, X/2}, {v, -V/2, V/2},
            PlotPoints -> 360, Mesh -> False, PlotRange -> All]]
```

```
        NDSolve::eerri :
        Warning: Estimated initial error on the specified spatial grid in the direction
            of independent variable x exceeds prescribed error tolerance. More…

        NDSolve::eerri :
        Warning: Estimated initial error on the specified spatial grid in the direction
            of independent variable v exceeds prescribed error tolerance. More…
```

```
NDSolve::eerr :
 Warning: Scaled local spatial error estimate of 26.990171843563985`
   at t = 18.84955592153876` in the direction of independent variable
   x is much greater than prescribed error tolerance. Grid spacing
   with 200 points may be too large to achieve the desired accuracy or
   precision.  A singularity may have formed or you may want to specify a
   smaller grid spacing using the MaxStepSize or MinPoints options. More…
```

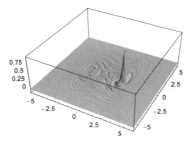

We return to the 1+1D case and continue with another hyperbolic equation, the 1D Klein–Gordon equation [216], [1880], [1738]

$$\frac{\partial^2 \Psi(x, t)}{\partial t^2} - \frac{\partial^2 \Psi(x, t)}{\partial x^2} + m^2 \Psi(x, t) = 0.$$

For positive m^2, this equation has plane wave solution with a group velocity less than the speed of light 1. But for negative m^2, it has tachyonic solutions with group velocities greater than the speed of light. But independently from the sign of m^2, a signal does not propagate with a velocity greater than the speed of light [625], [1541], [1295], [1413]. The following function KleinGordonGraphics solves the Klein–Gordon equation with a localized initial condition and displays the resulting solution as a contour plot. We use two different initial conditions. In the left graphic $\Psi_t(x, t = 0)$ vanishes and in the right $\Psi(x, t = 0)$ vanishes. This time, we use the "Pseudospectral" discretization setting.

```
In[81]:= KleinGordonGraphics[m_, {a0_, a1_, b_}, {L_, T_}] :=
   Module[{startPacket, Ψ, ndsol},
   (* bump-like localized initial conditions *)
   startPacket[{a_, b_}, x_] := If[Abs[x] <= b, a Cos[Pi/2 x/b]^4, 0];
   (* solve PDEs *)
   ndsol = NDSolve[{(* Klein-Gordon equation *)
                   D[Ψ[x, t], t, t] - D[Ψ[x, t], x, x] - m^2 Ψ[x, t] == 0,
                   (* localized initial conditions *)
                   Ψ[x, 0] == startPacket[{a0, b}, x/L],
                   Derivative[0, 1][Ψ][x, 0] == startPacket[{a1, b}, x/L],
                   (* hard-wall boundary conditions *)
                   Ψ[ L, t] == 0, Ψ[-L, t] == 0},
           {Ψ}, {x, -L, L}, {t, 0, T},
           Method -> {"MethodOfLines", "SpatialDiscretization" ->
                   {"TensorProductGrid",
                    "DifferenceOrder" -> "Pseudospectral",
                    "MaxPoints" -> Ceiling[L/0.1] + 1,
                    "MinPoints" -> Ceiling[L/0.1] + 1}}];
   (* visualize solution *)
   ContourPlot[Evaluate[Abs[Ψ[x, t]] /. ndsol[[1]]], {x, -L, L}, {t, 0, T},
       PlotRange -> All, PlotPoints -> 120, Contours -> 20,
```

```
        ContourLines -> False, FrameTicks -> False,
        ColorFunction -> (Hue[0.8 #]&)]]
```

Here are three examples. The first pair shows the bradyonic/ittyonic situation [1295], [625], [24], the second the classical wave equation ($m^2 = 0$), and the third the tachyonic case [625]. It is clearly visible that also in the tachyonic case the signal wave front does not propagate superluminally.

```
In[82]:= Function[m, Show[GraphicsArray[
        Block[{$DisplayFunction = Identity},
        KleinGordonGraphics[m, #, {10, 12}]& /@ {{1, 0, 1/4}, {0, 1, 1/4}}]],
                PlotLabel -> "m^2 == " <> ToString[InputForm[m^2]]]] /@
                                (* three values for m *) {1/3, 0, I Sqrt[2]}
```

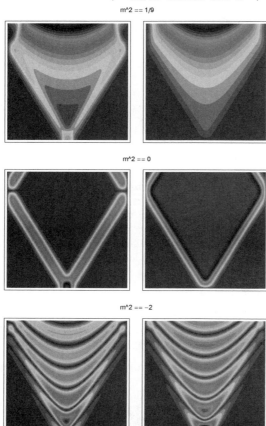

The partial derivatives can also appear in mixed form in the differential equations. Here are two partial differential equations with mixed temporal-spatial derivatives, one linear one and one nonlinear one.

$$\frac{\partial u(x, t)}{\partial t} + \frac{39}{100} \frac{\partial^3 u(x, t)}{\partial x^2 \partial t} - \frac{1}{32} \frac{\partial^3 u(x, t)}{\partial x^3} = 0$$

$$\frac{\partial u(x, t)}{\partial t} + \frac{287}{185} \frac{\partial^3 u(x, t)}{\partial x^2 \partial t} - \frac{728}{333} u(x, t)^2 \frac{\partial u(x, t)}{\partial x} = 0$$

```
In[83]:= Module[{xM = 10, T = 150, pp = 301},
    Show[GraphicsArray[
    Block[{$DisplayFunction = Identity},
     ((* solve PDE *)
     ndsol = NDSolve[{# == 0, (* localized initial bump *)
         u[x, 0] == Exp[-x^2], u[ xM, t] == u[-xM, t]},
         u[x, t], {x, -xM, xM}, {t, 0, T},
         PrecisionGoal -> 4, AccuracyGoal -> 4, Method ->
          {"MethodOfLines", "SpatialDiscretization" ->
           {"TensorProductGrid", "DifferenceOrder" -> Pseudospectral,
            "MaxPoints" -> pp, "MinPoints" -> pp}}];
     (* make contour plots *)
     DensityPlot[Evaluate[Abs[u[x, t]] /. ndsol[[1]]], {x, -xM, xM}, {t, 0, T},
         PlotPoints -> 200, Mesh -> False,  PlotRange -> All,
         ColorFunction -> (Hue[0.78 #]&)])& /@
     (* the two PDEs *)
     {D[u[x, t], t] + 39/100 D[u[x, t], x, x, t] -
             1/32 D[u[x, t], x, x, x],
      D[u[x, t], t]  + 287/185 D[u[x, t], x, x, t] -
             728/333 u[x, t]^2 D[u[x, t], x]}]]]]
```

We continue with some nonlinear partial differential equations (another interesting linear partial differential equation is the telegrapher equation [1884], [619], [970]; see also [460], [1583]). As a further example treated here, we choose the following nonlinear equation:

$$\frac{\partial^2 \Psi(x, t)}{\partial t^2} - \sin\left(x \Psi(x, t) \frac{\partial \Psi(x, t)}{\partial x}\right) = 0.$$

```
In[84]:= nlnEq = D[Ψ[x, t], {t, 2}] - Sin[x Ψ[x, t] D[Ψ[x, t], x]] == 0;
```

We solve this differential equation for an initial condition having two peaks. Again, we use the "Pseudospectral" setting, which is especially suited for periodic boundary conditions and allows obtaining quite precise solutions.

```
In[85]:= xM = 2Pi;
    nsol =
    NDSolve[{nlnEq,
            Ψ[x, 0] == If[Abs[x] < Pi, Sin[x]^2, 0],
            Derivative[0, 1][Ψ][x, 0] == 0,
            Ψ[ xM, t] == 0, Ψ[-xM, t] == 0},
            Ψ, {x, -xM, xM}, {t, 0, 5},
            PrecisionGoal -> 6, AccuracyGoal -> 6,
            Method -> {"MethodOfLines", "SpatialDiscretization" ->
                    {"TensorProductGrid",
```

```
                          "DifferenceOrder" -> "Pseudospectral",
                          "MaxPoints" -> 201, "MinPoints" -> 201}}]
```

Out[86]= {{Ψ → InterpolatingFunction[{{-6.28319, 6.28319}, {0., 5.}}, <>]}}

Mathematica again warned us that the error might be larger than prescribed within the AccuracyGoal and PrecisionGoal setting. Let us visualize the solution.

```
In[87]:= Show[GraphicsArray[
          Block[{dp, $DisplayFunction = Identity},
          {DensityPlot[Evaluate[Ψ[x, t] /. nsol[[1]]],
                    {x, -xM, xM}, {t, 0, 5},
                    PlotPoints -> 200, Mesh -> False,
                    ColorFunction -> (Hue[0.78 #]&)],
          Plot3D[Evaluate[Ψ[x, t] /. nsol[[1]]], {x, -xM, xM}, {t, 0, 5},
                    PlotPoints -> 100, Mesh -> False, PlotRange -> All]}]]]
```

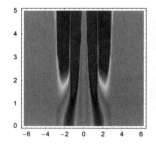 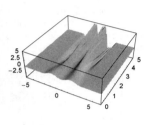

Looking at the residual, we see that this solution becomes locally unreliable for $t \gtrsim 2.5$.

```
In[88]:= Plot3D[Evaluate[Log[1 + Abs[nlnEq[[1]] /. nsol[[1]]]]],
          {x, -xM, xM}, {t, 0, 5},
          PlotPoints -> 100, Mesh -> False, PlotRange -> All]
```

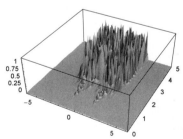

Here are two nonlinear wave equations in 1+1 dimensions. Depending on the form of the nonlinearities quite different behaviors of the solutions emerge.

```
In[89]:= Show[GraphicsArray[
          Function[{pde, xM, T, pp},
          Module[{nsol = (* solve nonlinear wave equation *) NDSolve[{pde,
                    Ψ[x, 0] == If[Abs[x] > Pi, 0, Cos[x/2]^2],
                    Derivative[0, 1][Ψ][x, 0] == 0,
                    (* periodic boundary conditions *) Ψ[-xM, t] == Ψ[+xM, t]},
                 Ψ[x, t], {x, -xM, xM}, {t, 0, T},
                 AccuracyGoal -> 6, PrecisionGoal -> 6, MaxSteps -> 2 10^4,
```

```
         Method -> {"MethodOfLines", "SpatialDiscretization" ->
                 {"TensorProductGrid", DifferenceOrder -> Pseudospectral,
                  "MaxPoints" -> {pp}, "MinPoints" -> {pp}}}]},
      (* make 3D plots *)
         Plot3D[Evaluate[Φ[x, t] /. nsol[[1]]], {x, -xM, xM}, {t, 0, T},
             Mesh -> False, PlotRange -> All, PlotPoints -> 360,
             Axes -> False, DisplayFunction -> Identity]]] @@@
  (* nonlinear wave equations and parameters *)
  {{D[Φ[x, t], {t, 2}] - D[Φ[x, t], {x, 2}] + 45/8 Φ[x, t] +
    307/92 Φ[x, t]^2 - 5/31 Φ[x, t]^2 D[Φ[x, t], {x, 1}]^2 == 0,
    6 Pi, 24, 601},
   {D[Φ[x, t], {t, 2}] - D[Φ[x, t], {x, 2}] + 179/20 Φ[x, t]^5 -
    118/13 D[Φ[x, t], {x, 1}] == 0, 3 Pi, 3, 1801}}]]
```

Very interesting (and complicated) solutions can be obtained from the nonlinear Schrödinger equation [300], [1276], [359].

$$i\,\epsilon\,\frac{\partial\,\Psi(x,\,t)}{\partial\,t} = -\epsilon^2\,\frac{\partial^2\,\Psi(x,\,t)}{\partial x^2} + |\Psi(x,\,t)|^2\,\Psi(x,\,t).$$

```
In[90]:= Clear[x, t, ε, Φ];
      nonlinearSchrödingerEq = I ε D[Φ[x, t], {t, 1}] ==
             -1/2 ε^2 D[Φ[x, t], {x, 2}] - Abs[Φ[x, t]]^2 Φ[x, t];
```

In the following inputs, we start with a single smooth symmetric bump centered at $x = 0$. At $t \approx 0.4$, the solution starts to become quite complex [13], [324].

```
In[92]:= xM = 3; ε = 0.06;
      ψ0[x_] = Exp[-3 x^2];

In[94]:= nsol =
      NDSolve[{nonlinearSchrödingerEq, Φ[x, 0] == ψ0[x],
             Φ[+xM, t] == ψ0[xM], Φ[-xM, t] == ψ0[-xM]},
           Φ[x, t], {x, -xM, xM}, {t, 0, 2},
           AccuracyGoal -> 3, PrecisionGoal -> 2, MaxSteps -> 10^5,
           Method -> {"MethodOfLines", "SpatialDiscretization" ->
                   {"TensorProductGrid",
                    (* "DifferenceOrder" -> "Pseudospectral", *)
                    "MaxPoints" -> 400, "MinPoints" -> 400}}];

In[95]:= Show[GraphicsArray[
      Block[{$DisplayFunction = Identity},
      {Plot3D[Evaluate[Abs[Φ[x, t]] /. nsol[[1]]], {x, -3, 3}, {t, 0, 2},
           PlotPoints -> 251, Mesh -> False, PlotRange -> All],
       ContourPlot[Evaluate[Abs[Φ[x, t]] /. nsol[[1]]],
              {x, -xM, xM}, {t, 0, 2},
              PlotPoints -> 251, ContourLines -> False,
              ColorFunction -> (Hue[0.8 #]&), Contours -> 50}]]]]
```

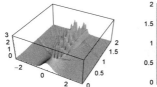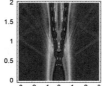

(Nonlinear Klein–Gordon equations also can give very complicated solutions; see [673].)

For partial differential equations, a similar remark for ordinary differential equations has to be made: The values for the `AccuracyGoal` and `PrecisionGoal` options are local goals. This means that (much) larger global errors could arise. Often in such cases, *Mathematica* will issue an `NDSolve::eerrss` message, but not always. So, when solving partial differential equations using `NDSolve` it is always wise to carefully check the solutions; for instance, by using different number of points for the spatial discretization and different values for the `DifferenceOrder` option. (`DifferenceOrder` is an option of `NDSolve` that determines the order of the finite difference approximation to be used for the spatial discretization when using the method of lines. See [1591] and [1601] for a discussion of optimal settings.)

The following two instances of initial value problems of the complex Ginzburg–Landau equation [59], [1832], [1828], [32], [1288], [1494], [1560], [1033] show such a situation [680], [374], [874]. Although the qualitative features of the solutions (the nested triangles [1017], [1018]) are correct, a correct quantitative solution requires many plotpoints and a relatively high difference order. We show the absolute value and the argument of the solutions over the *x,t*-plane.

$$\frac{\partial \psi(x,\,t)}{\partial t} = (1 + i\,\nu)\,\psi(x,\,t) + \frac{\partial^2 \psi(x,\,t)}{\partial x^2} - (1 + i\,\alpha)\,\psi(x,\,t)\,|\psi(x,\,t)|^2 + B$$

```
In[96]:= CGL = D[ψ[x, t], t] == (1 + I ν) ψ[x, t] -
              (1 + I α) Abs[ψ[x, t]]^2 ψ[x, t] + D[ψ[x, t], x, x] + B;

In[97]:= CGLSolutionAndGraphics[{α_, ν_, B_}, ψ0_,
                               {x_, t_}, {xM_, T_, pp_},
                               (* function and corresponding color function
                                  of the solution to be visualized *)
                               fList_] :=
         Module[{nsol, CGL, ψ},
         (* the forced complex Ginzburg-Landau equation *)
         CGL = D[ψ[x, t], t] == (1 + I ν) ψ[x, t] -
               (1 + I α) Abs[ψ[x, t]]^2 ψ[x, t] + D[ψ[x, t], x, x] + B;
         (* solve equation; use periodic boundary conditions *)
         nsol = NDSolve[{CGL, ψ[x, 0] == ψ0, ψ[xM, t] == ψ[-xM, t]},
                        ψ[x, t], {x, -xM, xM}, {t, 0, T},
                 AccuracyGoal -> 2.5, PrecisionGoal -> 2.5,
                 Method -> {"MethodOfLines", "SpatialDiscretization" ->
                            {"TensorProductGrid", "DifferenceOrder" -> 6,
                            "MaxPoints" -> pp, "MinPoints" -> pp}}];
         (* show contour plot(s) of solution *)
         Show[GraphicsArray[
         DensityPlot[Evaluate[#1[ψ[x, t]] /. First[nsol]],
                 {x, -xM, xM}, {t, 2, T}, Mesh -> False,
                 PlotRange -> All, PlotPoints -> 300,
                 DisplayFunction -> Identity, FrameTicks -> None,
                 ColorFunction -> #2]& @@@ fList]]]
```

```
In[98]:= With[{xM = 120},
         CGLSolutionAndGraphics[{-2.27699, 0, 0},
             2/3 Cos[x/xM] + 3/5 Cos[2x/xM] + 1/2 Cos[3x/xM] +
             34/37 Cos[4x/xM] - 91/92 Cos[5x/xM] + 17/29 Cos[6x/xM],
             {x, t}, {xM, 120, 300}, {{Abs, Hue[0.78 #]&}, {Arg, Hue}}]]
```

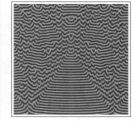

```
In[99]:= With[{xM = 200},
         CGLSolutionAndGraphics[{1, 1.1, 0.073},
             Sum[(-1)^k Cos[k] Cos[k x/xM], {k, 6}], {x, t}, {xM, 250, 200},
             {{Abs[# + 1]&, Hue[0.78 #]&}, {Arg, Hue}}]]
```

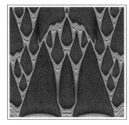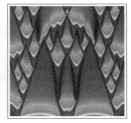

`NDSolve` can also solve systems of $n + 1$-dimensional partial differential equations. Here is a vector-valued version of the nonlinear Schrödinger equation. This time we include a quartic nonlinearity.

$$i\epsilon \frac{\partial \Psi_1(x,\,t)}{\partial t} = -\epsilon^2 \frac{\partial^2 \Psi_1(x,\,t)}{\partial x^2} + |\Psi_2(x,\,t)|^4 \Psi_1(x,\,t)$$

$$i\epsilon \frac{\partial \Psi_2(x,\,t)}{\partial t} = -\epsilon^2 \frac{\partial^2 \Psi_2(x,\,t)}{\partial x^2} + |\Psi_1(x,\,t)|^4 \Psi_2(x,\,t)$$

```
In[100]:= nonlinearSchrödingerSystemEqs =
          {I ε D[Ψ[1][x, t], {t, 1}] == -1/2 ε^2 D[Ψ[1][x, t], {x, 2}] -
                                        Abs[Ψ[2][x, t]]^4 Ψ[1][x, t],
           I ε D[Ψ[2][x, t], {t, 1}] == -1/2 ε^2 D[Ψ[2][x, t], {x, 2}] -
                                        Abs[Ψ[1][x, t]]^4 Ψ[2][x, t]};
```

We use different initial and boundary values for the two components $\Psi_1(x,\,t)$ and $\Psi_2(x,\,t)$.

```
In[101]:= xM = 1; ε = 0.06; Clear[ψ0];
          ψ0[1][x_] = Cos[Pi/2 x]; ψ0[2][x_] = 1;
```

Similarly to the scalar case, we see that, after a time, singularities form and the solution of the system of differential equations exhibits complicated behavior.

```
In[103]:= nsol =
          With[{pp = 81},
          NDSolve[{nonlinearSchrödingerSystemEqs,
                   Ψ[1][x, 0] == ψ0[1][x], Ψ[2][x, 0] == ψ0[2][x],
```

```
          Φ[1][+xM, t] == 0, Φ[1][-xM, t] == 0,
          Φ[2][+xM, t] == 1, Φ[2][-xM, t] == 1},
          {Φ[1], Φ[2]}, {x, -xM, xM}, {t, 0, 3/4},
          AccuracyGoal -> 2.5, PrecisionGoal -> 2.5,
          MaxSteps -> 10^5,
          Method -> {"MethodOfLines", "SpatialDiscretization" ->
                     {"TensorProductGrid",
                      "DifferenceOrder" -> "Pseudospectral",
                      "MaxPoints" -> pp, "MinPoints" -> pp}}]]
```

Out[103]= {{Φ[1] → InterpolatingFunction[{{-1., 1.}, {0., 0.75}}, <>],
           Φ[2] → InterpolatingFunction[{{-1., 1.}, {0., 0.75}}, <>]}}

In[104]:= **Show[GraphicsArray[Table[**
```
     ContourPlot[Evaluate[Abs[Φ[k][x, t]] /. nsol[[1]]],
                 {x, -xM, xM}, {t, 0, 3/4},
                 PlotPoints -> 151, Contours -> 50,
                 ContourLines -> False, DisplayFunction -> Identity,
                 ColorFunction -> (Hue[0.8 #]&)], {k, 2}]]]
```

The next system, we will solve is the Zakharov equations [1537], [805]

$$
i\,\frac{\partial \mathcal{E}(x,\,t)}{\partial t} + \frac{\partial^2 \mathcal{E}(x,\,t)}{\partial x^2} = \mathcal{E}(x,\,t)\,n(x,\,t)
$$

$$
\frac{\partial n(x,\,t)}{\partial t} - \frac{\partial^2 n(x,\,t)}{\partial x^2} = \frac{\partial^2 |\mathcal{E}(x,\,t)|^2}{\partial x^2}.
$$

To avoid differentiating the nondifferentiable absolute value function, we write $\mathcal{E}(x,\,t)$ in the form $\mathcal{E}(x,\,t) = \mathrm{Re}(\mathcal{E}(x,\,t)) + i\,\mathrm{Im}(\mathcal{E}(x,\,t))$ to obtain three coupled nonlinear partial differential equations in three real-valued functions. We use random oscillating initial conditions and periodic boundary conditions in the following numerical solution. In a generically random background, we see solitary structures evolving and interacting.

In[105]:= **Module[{r, xM = 6 Pi, pp = 150, T = 100, T1},**
```
     (* random oscillating function *)
     r[n_] := Sum[Random[Real, {-1, 1}] Cos[k x + 2Pi Random[]],
                  {k, 0, n}];
     (* seed random number generator *) SeedRandom[156];
     (* solve differential equations numerically *)
     nsol = NDSolve[{(* Re[ε[x, t]] *)
                     D[-εi[x, t], t] + D[εr[x, t], x, x] == n[x, t] εr[x, t],
                     (* Im[ε[x, t]] *)
                     D[+εr[x, t], t] + D[εi[x, t], x, x] == n[x, t] εi[x, t],
                     (* n[x, t] *)
                     D[n[x, t], t, t] - D[n[x, t], x, x] ==
```

```
                    D[Ɛr[x, t]^2 + Ɛi[x, t]^2, x, x],
              (* initial values *)
              Ɛr[x, 0] == r[2], Ɛi[x, 0] == 0,
              n[x, 0] == r[2], Derivative[0, 1][n][x, 0] == 0,
              (* periodic boundary conditions *)
              Ɛr[+xM, t] == Ɛr[-xM, t], Ɛi[+xM, t] == Ɛi[-xM, t],
              n[+xM, t] == n[-xM, t]},
              {Ɛr, Ɛi, n}, {x, -xM, xM}, {t, 0, T},
              MaxSteps -> 5 10^4, AccuracyGoal -> 3, PrecisionGoal -> 3,
              Method -> {"MethodOfLines", "SpatialDiscretization" ->
                       {"TensorProductGrid", "DifferenceOrder" -> Pseudospectral,
                        "MaxPoints" -> pp, "MinPoints" -> pp}}];
   (* in case a singularity formed was too small *)
   T1 = nsol[[1, 1, 2, 1, 2, 2]]; If[T1 < 0.99 T, T1 = 0.9 T1];
   (* data of Abs[Ɛ[x, t]]^2 *)
   data = Table[Evaluate[Abs[Ɛr[x, t] + I Ɛi[x, t]] /. nsol[[1]]],
                {x, -xM, xM, 2xM/120}, {t, 0, T1, T1/120}];
   (* visualize data *)
   ListContourPlot[Transpose[Flatten /@ data],
                   PlotRange -> All, ColorFunction -> (Hue[0.8#]&),
                   MeshRange -> {{-xM, xM}, {0, T}}]]
```

The next example is a coupled system of two nonlinear second order partial differential equations, the Prague model [803], [958].

$$\frac{\partial u(x, t)}{\partial t} = \frac{\partial^2 u(x, t)}{\partial x^2} + \frac{1}{\epsilon}\, u(x, t)\, (c - u(x, t) + (1 - c)\, v(x, t)) \left(u(x, t) - \frac{b + v(x, t)}{a}\right)$$

$$\frac{\partial v(x, t)}{\partial t} = \delta\, \frac{\partial^2 v(x, t)}{\partial x^2} + u(x, t) - v(x, t).$$

For appropriate parameter values a, b, c, δ, and ϵ, the solution forms a regular pattern over the x,t-plane. For $u(x, 0)$, we use an exponentially localized function. To avoid the switch to high-precision numbers inside NDSolve we set the system option "CatchMachineUnderflow" to False.

```
In[106]:= Module[{pdeU, pdeV, nsol, xM = 150, T = 150, pp = 450,
            δ = 0.5, ε = 0.01, a = 0.99, b = 0.01, c = 0.2},
       Developer`SetSystemOptions["CatchMachineUnderflow" -> False];
       (* differential equations *)
       pdeU = D[u[x, t], t] == D[u[x, t], {x, 2}] +
              1/ε u[x, t](c - u[x, t] + (1 - c) v[x, t])*
                       (u[x, t] - (v[x, t] + b)/a);
       pdeV = D[v[x, t], t] == δ D[v[x, t], {x, 2}] + u[x, t] - v[x, t];
       (* initial conditions *)
       u0[x_] := Exp[-x^2]; v0[x_] := 0;
```

```
(* solve differential equations numerically *)
nsol = NDSolve[{pdeU, pdeV, u[x, 0] == u0[x], v[x, 0] == v0[x],
        (* periodic boundary conditions *)
        u[+xM, t] == u[-xM, t], v[+xM, t] == v[-xM, t]},
        {u, v}, {x, -xM, xM}, {t, 0, T},
        AccuracyGoal -> 2, PrecisionGoal -> 2,
        Method -> {"MethodOfLines", "SpatialDiscretization" ->
                  {"TensorProductGrid",
                   "DifferenceOrder" -> "Pseudospectral",
                   "MaxPoints" -> pp, "MinPoints" -> pp}}];
Developer`SetSystemOptions["CatchMachineUnderflow" -> True];
(* display density plot of v[x, t] *)
DensityPlot[Evaluate[u[x, t] /. nsol[[1]]],
        {x, -xM, xM}, {t, 0, T}, Mesh -> False,
        PlotPoints -> 200, PlotRange -> All,
        ColorFunction -> (Hue[0.78 #]&)]] // Timing
```

We end with a differential equation that is first order in time and sixth order in x [178], [1937], [1208], [1787], [1788]. (ε is a free parameter.)

$$\frac{\partial \Psi(x, t)}{\partial t} = -\frac{\partial \Psi(x, t)}{\partial x} \Psi(x, t) + \frac{\partial^2}{\partial x^2}\left(\varepsilon \Psi(x, t) - \left(2 \frac{\partial^2 \Psi(x, t)}{\partial x^2} + \frac{\partial^4 \Psi(x, t)}{\partial x^4} + \Psi(x, t)\right)\right) - \sin(x)$$

```
In[107]:= Clear[Ψ, x, xM, t, T, ε];
     eq = D[Ψ[x, t], t] + D[ε Ψ[x, t] -
       (Ψ[x, t] + 2 D[Ψ[x, t], {x, 2}] + D[Ψ[x, t], {x, 4}]), {x, 2}] +
       Ψ[x, t] D[Ψ[x, t], x] + Sin[x] == 0;
```

Starting from quite regular initial conditions, the solutions show a chaotic behavior at larger times.

```
In[109]:= Block[{xM = 4Pi, ε = 2, T = 50, pp = 121, ndsol},
       (* solve PDE *)
       ndsol = NDSolve[Flatten[{eq,
               Ψ[x, 0] == Cos[x] + 1/3 Sin[2x],
               Table[Derivative[k, 0][Ψ][+xM, t] ==
                   Derivative[k, 0][Ψ][-xM, t], {k, 0, 5}]}],
           Ψ[x, t], {x, -xM, xM}, {t, 0, T},
           AccuracyGoal -> 4, PrecisionGoal -> 4,
           Method -> {"MethodOfLines",
                     "SpatialDiscretization" ->
                         {"TensorProductGrid", "DifferenceOrder" ->
Pseudospectral,
                         "MaxPoints" -> pp, "MinPoints" -> pp}}];
       (* visualize solution as 3D plot *)
```

```
Plot3D[Evaluate[Ψ[x, t] /. ndsol[[1]]], {x, -4Pi, 4Pi}, {t, 0, T},
       PlotPoints -> {200, 200}, Mesh -> False, PlotRange -> All]]
```

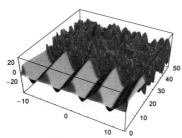

Now, we leave it to the reader to explore more (non)linear partial differential equations, such as simple nonlinear differential equations exhibiting complex behavior [1154], [1768], [348], [1735], [208], [1647], [173], [845], [846], [1264], [1027], [704], [713], [1700], [1118], [1160], [500], [819], [400], [1421], [174], [1212], [203], [373], [371], [1129], [65], [66], [1339], [1253], [1256], [1019], [1150], [1835], [204], [1283], [1205], and [1938]. Interesting 2+1D equations include [1873], [1042] for snow avalanches, [1120] for the growth and shape of bacterial colonies. But already linear equations can show quite intricate behavior, see for instance [624], [1488], and [591]. Even the wave equation can show unexpected behavior, for instance for piecewise periodically changing phase velocities [856], [1234]. For blowups in nonlinear wave equations, see [214]. For the inverse problem, of finding a PDE that models a given complex function, see [107], [1602], [725].

1.12 Two Applications

1.12.0 Remarks

Following the style of the larger chapters of the Chapters 1 and 2 of the Graphics volume [1794], in the third part, we will end the larger chapters (as we will in the Chapters 1 and 3 of the Symbolics volume [1795]) with some more complex examples demonstrating the functions in the chapters at work. At the end of Section 1.9, we already gave a slightly larger example for FindMinimum. The two following subsections will, at their heart, contain NDSolve, NIntegrate, and Compile and NSolve and NDSolve. These functions will carry out all of the hard number-crunching work. In addition, we have to implement a certain amount of code dealing with data manipulation and visualization.

1.12.1 Visualizing Electric and Magnetic Field Lines

In this subsection, we will calculate and visualize electric and magnetic field lines for some simple electrostatic and magnetostatic field configurations.

Field lines are curves that are parallel to the local field direction. Let $\vec{C}(\vec{x})$ be the vector field describing the electric field $\vec{E}(\vec{x})$ or magnetic field $\vec{B}(\vec{x})$ at the point $\vec{x}$, and let $\vec{c}(t)$ be a parametrized field line. The tangent of the field line points in the direction of the field at each point; this means $\vec{c}'(t) \sim \vec{C}(\vec{c}(t))$ [1674], [1013]. The set of coupled nonlinear ordinary differential equations is the defining set of equations for the field lines that we will solve numerically in this subsection.

In the electrostatic case, given the charge density $\rho(\vec{y})$, the electric field $\vec{E}(\vec{x})$ is given as the Poisson integral (ignoring units and powers of 4π; $d\vec{y}$ is the 3D volume element)

$$\vec{E}(\vec{x}) = \int_{\mathbb{R}^3} \frac{(\vec{x} - \vec{y})\,\rho(\vec{y})}{|\vec{x} - \vec{y}|^3}\,d\vec{y}.$$

For a point charge q at the point $\vec{r}$, this simplifies to $\vec{E}(\vec{x}) = q(\vec{x} - \vec{r})/|\vec{x} - \vec{r}|^3$.

Often in this subsection, we will have to normalize vectors. We will repeatedly make use of the two functions norm and n.

```
In[1]:= (* length of a vector *)
       norm[vector_] := Sqrt[vector.vector]
       (* normalizing vectors *)
       n[vector_] := vector/Sqrt[vector.vector]
```

Let us start the field line visualization with a simple example: point charges in two dimensions. (We imagine them embedded in three dimensions and use the r^{-1} potential instead of a pure 2D, this means $\varphi(r) = -\ln(r)$ potential.) Here, we will calculate the field lines by explicitly solving the corresponding differential equations; for 2D configurations, it is often possible to use complex variable techniques to obtain the equipotential and the field lines [1674], [309], [1379], [1380], [205], [1876].

Here is a square array of 5×5 alternatingly charged points.

```
In[5]:= charges = Flatten[Table[{(-1)^(i + j), {i, j}},
                        {i, -2, 2}, {j, -2, 2}], 1];
```

φCoulomb calculates the (normalized—no units) the Coulomb potential of a single charge q at position *pos*. We will use φCoulomb for 2D as well as for 3D charge configurations.

```
In[6]:= φCoulomb[{q_, pos_}, xyz_] := q/norm[xyz - pos]
```

By the superposition principle, the electrostatic potential is simply the sum of the potentials of all charges.

```
In[7]:= φ[{x_, y_}] = Plus @@ (φCoulomb[#, {x, y}]& /@ charges);
```

We generate a contour plot of the charge array.

```
In[8]:= cp = ContourPlot[Evaluate[φ[{x, y}]], {x, -4, 4}, {y, -4, 4},
                  PlotPoints -> 200, Contours -> 30,
                  ContourStyle -> {Thickness[0.002]},
                  DisplayFunction -> Identity];
```

For a nicer contour plot, we implement a function makeContours that will display a given contour graphic with n contour lines in such a way that the n contour lines are spaced as evenly as possible.

```
In[9]:= makeContours[ContourGraphics[data_, ___], n_] :=
       With[{λ = Sort[Flatten[data]]},
            (* generate about n equal length sets *)
            #[[Floor[Length[λ]/n/2]]]& /@
                  Partition[λ, Floor[Length[λ]/n]]]
In[10]:= equiPotentialPicture =
       ListContourPlot[cp[[1]], MeshRange -> {{-4, 4}, {-4, 4}},
          ColorFunction -> (Hue[# 0.75]&),  Contours -> makeContours[cp, 30],
             ContourStyle -> {{Thickness[0.002]}}]
```

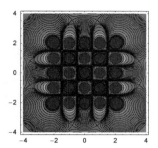

This is the electric field corresponding to the charge array. We obtain it by using the formula $\vec{E}(\vec{x}) = -\vec{\nabla}\varphi(\vec{x})$.

In[11]:= **EField[{x_, y_}] = -{D[φ[{x, y}], x], D[φ[{x, y}], y]};**

For the calculation of the field lines $\vec{c}(t)$, we will not use $\vec{c}'(t) = \vec{E}(\vec{c}(t))$, but rather $\vec{c}'(t) = \vec{E}(\vec{c}(t))/|\vec{E}(\vec{c}(t))|$. This normalization $|\vec{c}'(t)| = 1$ of the field strength automatically gives the arc length of the field lines as the parameter t and gives a more intuitive measure for the t-range required in solving the differential equations. The resulting field lines then have the natural parametrization.

Instead of defining $Ex[\{x\_, y\_\}] = n[EField[\{x, y\}]][[1]]$, $Ey[\{x\_, y\_\}] =...$ for the components of the electric field, we define a more general function `makeFieldComponentDefinitions`. This function will create such definitions for a given list of field components *fieldComponents* and a given vector-valued field *field*. Because we will define a variety of different charge and current configurations in this subsection, the function `makeFieldComponentDefinitions` will in the long run make the inputs shorter and more readable. The optional last argument type *varType* has the following reason. When we do not supply it, the created definitions are of the form $Ex[\{x\_, y\_\}] := fieldComponent$. If we later would use $Ex[\{x, y\}]$ on the right-hand side of a differential equation, *fieldComponent* will be substituted. NDSolve will compile *fieldComponent* and solve the differential equations. When calculating the x-, y-, and z-component of the fields, much of the computational work has to be repeated. To avoid redoing computational work, we will store the fields as vectors and access only their components. In such cases, we do not want $ex[\{x, y\}]$ to evaluate for symbols (head Symbol) x and y. We avoid this by instead creating the definition $Ex[\{x\_Real, y\_Real\}] := fieldComponent$. Using Real for *varType* will generate such definitions.

In[12]:= **makeFieldComponentDefinitions[**
        fieldComponents_, field_, vars_, varType___] :=
    ((* remove existing definitions *)
    Clear /@ fieldComponents;
    (* create new definitions *)
    MapIndexed[(SetDelayed @@
            (Hold[#1[Pattern[#, Blank[varType]]]& /@ vars],
                n[field[vars]][[C]]] /. C -> #2[[1]]))&,
        fieldComponents])

Now, we invoke `makeFieldComponentDefinitions` to make the definitions for the field components.

In[13]:= **makeFieldComponentDefinitions[{Ex, Ey}, EField, {x, y}];**

These are the current definitions for the field components.

In[14]:= **DownValues /@ {Ex, Ey}**
Out[14]= {{HoldPattern[Ex[{x_, y_}]] :→ n[EField[{x, y}]][[1]]},
        {HoldPattern[Ey[{x_, y_}]] :→ n[EField[{x, y}]][[2]]}}

We separate the positive and the negative charge positions.

```
In[15]:= {positiveCharges, negativeCharges} =
            (Last /@ Cases[charges, {_?#, _}]) & /@ {Positive, Negative};
```

The function `calculateFieldLine` calculates the field lines as `InterpolatingFunctions` by solving their differential equation numerically. To be flexible and to be able to use the function `calculateField⌐ Line` for the 2D and 3D case, and for the electric and for the magnetic field cases, we use the arguments *fieldLineVector*, *startPoint*, and *fields*. This allows us to specify the dimensions and the kind of field later.

```
In[16]:= calculateFieldLine[fieldLineVector_, startPoint_,
                            fields_, {t, t0_:0, t1_}, opts___] :=
        With[{flvt = #[t]& /@ fieldLineVector},
        NDSolve[Join[(* differential equation *)
                    Thread[D[flvt, t] == (#[flvt]& /@ fields)],
                    (* initial conditions *)
                    Thread[(flvt /. t -> t0) == startPoint]],
                fieldLineVector, {t, t0, t1}, opts]]
```

Now, let us calculate the first set of field lines. We start near the positive charges and move toward the negative ones. To avoid integrating the differential equations near the singularity of the negative charges (which would need many steps without making progress in the field line), we stop the integration process (using the option `StoppingTest`) when we are too near to a negative charge.

```
In[17]:= startPoints[charges_, n_] :=
        Flatten[Table[charges[[j]] + ε {Cos[φ], Sin[φ]}, {j, Length[charges]},
                    {φ, 0, 2Pi (1 - 1/n), 2Pi/n}], 1]
```

```
In[18]:= sTest[cs_, pn_] := (Sqrt[Min[#.#& /@ ((cs - #)& /@ pn)]] < ε)
```

```
In[19]:= ε = 10^-2; n = 12;

        fieldLines["+" -> "-"] =
        calculateFieldLine[{cx, cy}, #, {Ex, Ey}, {t, 3},
            StoppingTest :> sTest[{cx[t], cy[t]}, negativeCharges]]& /@
        startPoints[positiveCharges, n];
```

Given the field lines in the form of `InterpolatingFunctions`, we use `ParametricPlot` to generate a line representing the field line.

```
In[21]:= fieldLineGraphic[cxcy_] :=
        ParametricPlot[Evaluate[{cx[t], cy[t]} /. cxcy],
                    Evaluate[Flatten[{t, cxcy[[1, 1, 2, 1, 1]]}]],
                    PlotPoints -> 300, DisplayFunction -> Identity]
```

```
In[22]:= fl["+" -> "-"] = fieldLineGraphic /@ fieldLines["+" -> "-"];
```

(As a side remark, we want to mention that the local density of the field lines does not reflect the local value of the field strength [1915], [634], [829].) Now, we do the same calculation, but we start at the negative charges. We could either integrate the equation $\vec{c}'(t) = \vec{E}(\vec{c}(t))\big/\big|\vec{E}(\vec{c}(t))\big|$ to move toward the positive charges or we move backwards with respect to *t*. We choose the latter approach; so, we can reuse the above definition of `calcu⌐ lateFieldLine`.

```
In[23]:= fieldLines["-" -> "+"] =
        calculateFieldLine[{cx, cy}, #, {Ex, Ey},
                        {t, (* go backwards *) -3},
            StoppingTest :> sTest[{cx[t], cy[t]}, positiveCharges]]& /@
                            startPoints[negativeCharges, n];
```

```
In[24]:= fl["-" -> "+"] = fieldLineGraphic /@ fieldLines["-" -> "+"];
```

Now, we have all ingredients together to display the equicontour lines with the field lines (in white).

```
In[25]:= Show[{equiPotentialPicture,
         Graphics[{Thickness[0.002], GrayLevel[1],
                    Cases[{fl["+"] -> "-"], fl["-"] -> "+"]}, _Line, Infinity]}]},
             FrameTicks -> None, PlotRange -> {{-4, 4}, {-4, 4}}]
```

In a similar way, we could treat nonsymmetric arrangements of charges. Here is an example made from 21 point charges. Each magnitude and position of the charges is randomly chosen. For the few charges under consideration here, we just sum their potentials (for more charges one could use faster methods [389], [1387], [580], [1427]; for visualizations of gradient fields in general, see [1561]).

```
In[26]:= SeedRandom[9999];
         n = 21;
         charges = Table[{Random[Real, {-1, 1}], {Random[], Random[]}}, {n}];

         (* split into positive and negative charges *)
         {positiveCharges, negativeCharges} =
           (Last /@ Cases[charges, {_?#, _}])& /@ {Positive, Negative};

         φ[{x_, y_}] = Plus @@ (φCoulomb[#, {x, y}]& /@ charges);

In[34]:= equiPotentialPicture = Function[cp, (* add features *)
                     ListContourPlot[cp[[1]],
                     MeshRange -> {{-1/4, 5/4}, {-1/4, 5/4}},
                     ColorFunction -> (Hue[# 0.75]&),
                     Contours -> makeContours[cp, 30],
                     ContourStyle -> {Thickness[0.002]},
                     DisplayFunction -> Identity]][
             (* the contour plot *)
             ContourPlot[Evaluate[φ[{x, y}]], {x, -1/4, 5/4}, {y, -1/4, 5/4},
                     PlotPoints -> 200, Contours -> 50,
                     DisplayFunction -> Identity]];
```

We just repeat the above calculations.

```
In[35]:= EField[{x_, y_}] = -{D[φ[{x, y}], x], D[φ[{x, y}], y]};

In[36]:= {fieldLines["+"] -> "-"], fieldLines["-"] -> "+"]} =
         Apply[Function[{np, pn, T},
         calculateFieldLine[{cx, cy}, #, {Ex, Ey}, {t, T},
             StoppingTest :> sTest[{cx[t], cy[t]}, np]]& /@
                                     startPoints[pn, n]],
         {{negativeCharges, positiveCharges, +3},
          {positiveCharges, negativeCharges, -3}}, {1}];
```

```
In[37]:= Show[{equiPotentialPicture,
        Graphics[{Thickness[0.002], GrayLevel[1],
                Cases[{fieldLineGraphic /@ fieldLines["-" -> "+"],
                        fieldLineGraphic /@ fieldLines["+" -> "-"]},
                    _Line, Infinity]}]},
            FrameTicks -> None, PlotRange -> {{-1/4, 5/4}, {-1/4, 5/4}},
            DisplayFunction -> $DisplayFunction]
```

Now, let us deal with a slightly more complicated example: a set of charged straight wires in two dimensions. We obtain the potential by integration of the general formula along a parametrized line segment.

```
In[38]:= Simplify[((# /. t -> 1) - (# /. t -> 0))&[
                Integrate[1/Sqrt[(x - (p1x + t (p2x - p1x)))^2 +
                                (y - (p1y + t (p2y - p1y)))^2], t]]]
```

$$Out[38]= \left(-Log\left[\left(2\left(-p1x^2 - p1y^2 - p2x\,x + p1x\,(p2x + x) - p2y\,y + \right.\right.\right.\right.$$
$$\left.\left.p1y\,(p2y + y) + \frac{\sqrt{p1x^2 + p1y^2 - 2\,p1x\,p2x + p2x^2 - 2\,p1y\,p2y + p2y^2}}{\sqrt{p1x^2 + p1y^2 - 2\,p1x\,x + x^2 - 2\,p1y\,y + y^2}}\right)\right)\bigg/$$
$$\left(\sqrt{p1x^2 + p1y^2 - 2\,p1x\,p2x + p2x^2 - 2\,p1y\,p2y + p2y^2}\right)\bigg] +$$
$$Log\left[\left(2\left(p2x^2 - p1y\,p2y + p2y^2 - p2x\,x + p1x\,(-p2x + x) + p1y\,y - p2y\,y + \right.\right.\right.$$
$$\left.\frac{\sqrt{p1x^2 + p1y^2 - 2\,p1x\,p2x + p2x^2 - 2\,p1y\,p2y + p2y^2}}{\sqrt{p2x^2 + p2y^2 - 2\,p2x\,x + x^2 - 2\,p2y\,y + y^2}}\right)\right)\bigg/$$
$$\left(\sqrt{p1x^2 + p1y^2 - 2\,p1x\,p2x + p2x^2 - 2\,p1y\,p2y + p2y^2}\right)\bigg]\right)\bigg/$$
$$\left(\sqrt{p1x^2 + p1y^2 - 2\,p1x\,p2x + p2x^2 - 2\,p1y\,p2y + p2y^2}\right)$$

For a faster numerical evaluation, we implement a compiled form of the last result. The last result is simple enough that we can implement a procedural version of it "by hand".

```
In[39]:= φFiniteStraightWire =
        Compile[{{xy, _Real, 1}, {p1, _Real, 1}, {p2, _Real, 1}},
        Module[{a, b, c, ax, ay, az, bx, by, bz, as, cs, abcs, α1, α2},
        {a, b, c} = {(xy - p1).(xy - p1), -2(p2 - p1).(xy - p1),
                    (p2 - p1).(p2 - p1)};
        {as, cs} = {Sqrt[a], Sqrt[c]}; abcs = Sqrt[a + b + c];
        {α1, α2} = {(2as + b/cs), (2abcs +(b + 2c)/cs)};
        (Log[α2] - Log[α1])/cs]];
```

The compilation was successful.

```
In[40]:= Union[Head /@ Flatten[φFiniteStraightWire[[4]]]]
Out[40]= {Integer}
```

We will take the following set of wires as our example configuration. It is a square surrounded by eight wires of opposite charge.

```
In[41]:= wires = N @
    {Partition[Table[{Cos[φ], Sin[φ]}/2, {φ, 0, 2Pi, 2Pi/4}], 2, 1],
     Partition[Table[{Cos[φ], Sin[φ]}  , {φ, 0, 2Pi, 2Pi/17}], 2]};
```

For the total potential φ, we obtain the following expression. We weight the contribution of all line segments according to their length.

```
In[42]:= {l1, l2} = (Plus @@ (norm /@ Apply[Subtract, #, {1}]))& /@ wires;

    φ[{x_, y_}] := Subtract @@ MapThread[
        (Plus @@ Apply[φFiniteStraightWire[{x, y}, ##]&,
            wires[[#1]], {1}])/#2&, {{1, 2}, {l1, l2}}]
```

Here is a contour plot showing the distribution of the potential.

```
In[45]:= equiPotentialPicture =
    Function[cp, ListContourPlot[cp[[1]],
        MeshRange -> {{-3/2, 3/2}, {-3/2, 3/2}},
        ColorFunction -> (Hue[# 0.75]&),
        Contours -> makeContours[cp, 30],
        ContourStyle -> {{Thickness[0.002]}},
        Epilog -> {GrayLevel[1/2], Thickness[0.01],
            Map[Line, wires, {2}]}]][
    ContourPlot[φ[{x, y}], {x, -3/2, 3/2}, {y, -3/2, 3/2},
        PlotPoints -> 50, Compiled -> False,
        DisplayFunction -> Identity]]
```

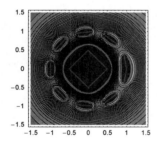

For the electric field, we differentiate the above expression. For speed reasons, again, we use a compiled version.

```
In[46]:= EFieldFiniteStraightWire =
    Compile[{{xy, _Real, 1}, {p1, _Real, 1}, {p2, _Real, 1}},
        Module[{a, b, c, ax, ay, az, bx, by, bz, as, cs, abcs, α1, α2},
        {a, b, c} = {(xy - p1).(xy - p1), -2(p2 - p1).(xy - p1),
                    (p2 - p1).(p2 - p1)};
        {ax, ay, bx, by} = {{2, 0}.(xy - p1), {0, 2}.(xy - p1),
                        -(p2 - p1).{2, 0}, -(p2 - p1).{0, 2}};
        {as, cs, abcs} = {Sqrt[a], Sqrt[c], Sqrt[a + b + c]};
        {α1, α2} = {(2as + b/cs), (2abcs +(b + 2c)/cs)};
        {(-((ax/as + bx/cs)/α1) +
            ((ax + bx)/abcs + bx/cs)/α2)/cs,
         (-((ay/as + by/cs)/α1) + ((ay + by)/abcs + by/cs)/α2)/cs}]];
```

We get the electric field caused by all charged pieces by adding all of their contributions.

```
In[47]:= EField[{x_?NumberQ, y_?NumberQ}] := Subtract @@ MapThread[
         (Plus @@ Apply[EFieldFiniteStraightWire[{x, y}, ##]&,
              wires[[#1]], {1}])/#2&, {{1, 2}, {11, 12}}]

In[48]:= makeFieldComponentDefinitions[{Ex, Ey}, EField, {x, y}, Real];
```

We integrate the field line differential equation starting from *n* points in distance *ε* to both sides of the lines.

```
In[49]:= startPoints[{p1_, p2_}, n_] :=
         With[{normal = n[Reverse[p2 - p1]{1, -1}], ε = 0.02},
              Table[p1 + i/n(p2 - p1) + ε normal, {i, n}]]
```

Here is the resulting picture of the potential and of the field lines. We integrate the field lines until we reach the end of the integration domain or until the field strength becomes quite large (this means until we are very near a charge).

```
In[50]:= Show[{(* potential as background *)
               equiPotentialPicture,
               Graphics[{Thickness[0.002], GrayLevel[1],
                Cases[Function[ε, fieldLineGraphic /@
                (* calculate the field lines *)
             (calculateFieldLine[{cx, cy}, #, {Ex, Ey}, {t, (-1)^ε},
                          MaxSteps -> 1000,
                          PrecisionGoal -> 3, AccuracyGoal -> 3,
                StoppingTest :> (* end integrating near to a charge *)
                (#.#&[EField[{cx[t], cy[t]}]] > 10^6)]& /@
             Flatten[startPoints[#, 13]& /@ wires[[ε]], 1])] /@
                 {1, 2}, _Line, Infinity]}]},
                    PlotRange -> {{-3/2, 3/2}, {-3/2, 3/2}},
                    AspectRatio -> Automatic, FrameTicks -> None,
                    DisplayFunction -> $DisplayFunction]
```

Now, let us calculate the electric field lines of a set of charged spheres in three dimensions. We start by generating a set of spheres of random diameter and at random positions. makeCharges yields a list of *pos* positive and *neg* negative spheres representing the charges. The While loops searches for a sphere configuration in which the individual spheres are nonoverlapping. We assume that the spheres are homogeneously charged (*charge ~ radius³*).

```
In[51]:= makeCharges[{pos_, neg_}, {rMin_, rMax_, minDist_}] :=
         Module[{p, n},
         While[(* random values for charges and positions *)
                {p, n} = Table[
                (* charge *)  {Random[Real, {rMin, rMax}],
                (* position *) Table[Random[Real, {-1, 1}], {3}]},
                    {#}]& /@ {pos, neg};
```

```
                (* do the charged spheres intersect? *)
                Not[And @@ Flatten[
                Table[norm[#[[i, 2]] - #[[j, 2]]] > #[[i, 1]] + #[[j, 1]],
                      {i, pos + neg}, {j, i - 1}]]]&[
                Join[p, n]], Null];
                (* charge ~ radius^3 *)
         Join[{+#[[1]]^3, #[[2]]}& /@ p, {-#[[1]]^3, #[[2]]}& /@ n]]
In[52]:= SeedRandom[1111111111111];
         charges = makeCharges[{6, 5}, {0.04, 0.12, 0.3}];
```

We color the positive and negative spheres differently.

```
In[54]:= {color["⊕"], color["⊖"]} = SurfaceColor[GrayLevel[#]]& /@ {0.3, 0.9};
```

Here is the configuration of charges.

```
In[55]:= chargePic :=
         With[{(* a sphere *) sphere =
               Cases[ParametricPlot3D[
                   {Cos[φ] Sin[θ], Sin[φ] Sin[θ], Cos[θ]},
                   {φ, 0, 2Pi}, {θ, 0, Pi}, PlotPoints -> 20,
                   DisplayFunction -> Identity], _Polygon, Infinity]},
          (* color positive and negative charges *)
          Graphics3D[{EdgeForm[],
           Function[c, {If[Sign[c[[1]]] < 0, color["⊖"], color["⊕"]],
           Map[(c[[2]] + Abs[c[[1]]]^(1/3) #)&, sphere, {-2}]}] /@ charges}]];
In[56]:= pic = Show[chargePic, Axes -> True,
                PlotRange -> 1.2 {{-1, 1}, {-1, 1}, {-1, 1}}]
```

In the following example, we are only interested in the electric potential and the electric field outside of the spheres. In this case, the electric field coincides with the one from a point charge of the same charge at the position of the center of the sphere.

```
In[57]:= φ[{x_, y_, z_}] = Plus @@ (φCoulomb[#, {x, y, z}]& /@ charges);
In[58]:= EField[{x_, y_, z_}] = -{D[#, x], D[#, y], D[#, z]}&[φ[{x, y, z}]];
In[59]:= makeFieldComponentDefinitions[{Ex, Ey, Ez}, EField, {x, y, z}];
```

Similar to the 2D case, we calculate the field lines by solving their characteristic differential equation. We start the integration at a set of roughly uniformly distributed points just outside of the spheres.

```
In[60]:= {positiveCharges, negativeCharges} =
              Cases[charges, {_?#, _}]& /@ {Positive, Negative};
In[61]:= startPoints[charges_, {ppφ_, ppθ_}, γ_] :=
         Flatten[Table[#[[2]] + γ Abs[#[[1]]]^(1/3) *
```

```
        {Cos[φ] Sin[ϑ], Sin[φ] Sin[ϑ], Cos[ϑ]},
                 {φ, 0, 2Pi(1 - 1/ppφ), 2Pi/ppφ},
        {ϑ, Pi/ppϑ, Pi(1 - 1/ppϑ), Pi/ppϑ}]& /@ charges, 2]
```

Now, we calculate the field lines. We end integrating the differential equations either when we are near the surface of another sphere or when the magnitude of the field strength becomes too large. The calculation of the field lines can be accomplished quickly

```
In[62]:= sTest[cs_, pn_] := (Min[(norm[cs - #[[2]]] -
                  Abs[#[[1]]]^(1/3))& /@ pn] < 0 || (#.#&[cs] > 4))
```

```
In[63]:= (fieldLines["⊕" -> "⊖"] =
       calculateFieldLine[{cx, cy, cz}, #, {Ex, Ey, Ez}, {t, 2},
          StoppingTest :> sTest[{cx[t], cy[t], cz[t]}, negativeCharges]]& /@
       startPoints[positiveCharges, {5, 4}, 1.02];) // Timing
Out[63]= {6.31 Second, Null}
```

The function `fieldLineGraphic3D` uses `ParametricPlot3D` to generate a line approximation of the field lines.

```
In[64]:= fieldLineGraphic3D[cxcycz_] :=
       ParametricPlot3D[Evaluate[{cx[t], cy[t], cz[t]} /. cxcycz],
                    Evaluate[Flatten[{t, cxcycz[[1, 1, 2, 1, 1]]}]],
                    PlotPoints -> 300, DisplayFunction -> Identity]
```

In an analogous manner, we calculate the field lines from the negative to the positive charges.

```
In[65]:= fieldLines["⊖" -> "⊕"] =
       calculateFieldLine[{cx, cy, cz}, #, {Ex, Ey, Ez}, {t, -2},
        StoppingTest :> sTest[{cx[t], cy[t], cz[t]}, positiveCharges]]& /@
       startPoints[negativeCharges, {5, 4}, 1.02];
```

Here is the picture of the charges and the field lines between them.

```
In[66]:= Show[{pic,
            Graphics3D[{Hue[0], Thickness[0.002], Cases[
               fieldLineGraphic3D /@ #, _Line, Infinity]}]}& /@
             {fieldLines["⊕" -> "⊖"], fieldLines["⊖" -> "⊕"]}},
        PlotRange -> 3/2 Table[{-1, 1}, {3}]]
```

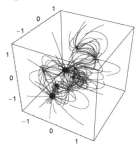

In the 2D case, the field lines gave a good visual representation of the field lines; this is not so much the case for the 3D picture. The main reason is that lines (head `Line`) are purely 1D objects and do not give any "depth feeling" [1969]. Thickening the lines to thin tubes gives a much better visualization. Let us do this. The function `makeTube` generates a tube (in the form of a list of polygons) along the line `Line[points]`, where *rFunction* determines its local radius. *startCrossSection* determines the initial cross section of the tube. Later, we will reuse these functions to graph wires.

```
In[67]:= makeTube[Line[points_], startCrossSection_, rFunction_] :=
     MapThread[Polygon[Join[#1, Reverse[#2]]]&, #1]& /@
     (Map[Partition[#, 2, 1]&, Partition[
     MapIndexed[(* change tube diameter *)
               rFunction[#1, #2, Length[points]]&,
     FoldList[Function[{p, t},
     Module[{o = orthogonalDirections[t]},
           (* move along the line *)
           prolongate[#, t[[2]], t[[2]] - t[[1]], o]& /@ p]],
     startCrossSection,
     Rest[Partition[points, 3, 1]]]], 2, 1], {2}]);
```

The function makeTube works similarly to the one discussed in Subsection 2.3.1 of the Graphics volume [1794]. It prolongates a given cross section *startCrossSection* along a given set of points *points* (the set of points *points* can either form a smooth or a nonsmooth curve). The two functions prolongate and orthogonalDirections do most of the work involved. prolongate moves the cross section of the tube from one end of a line segment to the end of the next line segment, and orthogonalDirections creates a pair of orthogonal directions in the plane between two line segments.

```
In[68]:= prolongate[p_, q_, d_, {dirx_, diry_}] :=
     Module[{s, u, v}, First[p + s d /.
           Solve[Thread[(* line intersects plane *)
                       p + s d == q + u dirx + v diry], {s, u, v}]]];
In[69]:= orthogonalDirections[{p1_, p2_, p3_}] :=
     Module[{d},
            If[Abs[#1.#2] == 1, (* parallel case *)
                d = If[Abs[#1[[3]]] < 1, {-#1[[2]], #1[[1]], 0},
                      {0, #1[[3]], -#1[[2]]}],
                d = (#1 + #2)/2];
            n /@ {d, Cross[#1, d]}]&[n[p3 - p2], n[p1 - p2]];
```

We will change the diameter of the tubes in such a way that at the starting and ending points of the corresponding lines, they thin out. The function ρFunc determines the tube diameter along the line.

```
In[70]:= ρFunc[l_, {pos_}, n_] :=
     Function[mp, mp + Sin[(pos - 1)/(n - 1) Pi]^2 *
                 (# - mp)& /@ l][(* cross-section center *)
                 (Plus @@ Rest[l])/(Length[l] - 1)]
```

The function tubeGraphics3D calculates the tubes.

```
In[71]:= tubeGraphics3D[line_Line, {r_:0.015, rFunction_:(#1&)},
                       color_, startCrossSection_:Automatic] :=
     Module[{dir, dir1, dir2, scs, ppSCS = 12},
     If[startCrossSection === Automatic,
        (* make start cross section *)
        dir = n[line[[1, 2]] - line[[1, 1]]];
        dir1 = n[# - # dir]&[Table[Random[], {3}]];
        dir2 = n[Cross[dir, dir1]];
        scs = Table[N[line[[1, 1]] + r (Cos[φ] dir1 + Sin[φ] dir2)],
                   {φ, 0, 2π, 2π/ppSCS}],
        scs = startCrossSection[[1]]];
     Graphics3D[{EdgeForm[], color, (* make tube *)
     makeTube[Line[Join[{line[[1, -2]]}, line[[1]],
                       {line[[1, 2]]}]], scs, rFunction]}]]
```

Using this function, we implement a function `fieldLine3D` that will convert an `InterpolatingFunc`tion representing the field line in a thin tube.

```
In[72]:= fieldLine3D[fieldLineIF_, color_] :=
    Module[{tMax, n = 60},
      tMax = DeleteCases[fieldLineIF[[1, 1, 2, 1, 1]], 0.][[1]];
      If[Abs[tMax] > 0,
        tubeGraphics3D[Line[Table[Evaluate[{cx[t], cy[t], cz[t]} /.
          fieldLineIF[[1]]], {t, 0, tMax, tMax/n}]],
            {0.016, ρFunc}, color], {}]]

In[73]:= Length[Join[fieldLines["⊕" -> "⊖"], fieldLines["⊖" -> "⊕"]]]

Out[73]= 165
```

Now, we have everything together to generate a "real" 3D picture of the field lines. To enhance the appearance, we color each of the tubes differently. Now, we get a "3D" feeling for the field lines. We have, of course, to pay a price for this improved picture—it requires much more memory than does the line picture from above.

```
In[74]:= fieldLineColor :=
        SurfaceColor[Hue[Random[]], Hue[Random[]], 3 Random[]];

In[75]:= Show[{chargePic, fieldLine3D[#, fieldLineColor]& /@
            Join[fieldLines["⊕" -> "⊖"], fieldLines["⊖" -> "⊕"]]},
          PlotRange -> All, Boxed -> False, ViewPoint -> {1, 1, 1}]
```

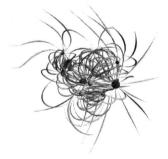

We could now make the last graphic more interesting graphically (but less physically realistic) by adding random translations along each of the field line tubes. To make such a graphic, we have to change the function ρFunc from above. In addition to forming a tube, we shift the centers of the tube cross sections randomly in the plane perpendicular to the field line.

```
In[76]:= (* make a zig-zag filed line tube *)
    ρFuncZigZag[l_, {pos_}, n_] := Function[{mp, shift},
    (mp + shift + Sin[(pos - 1)/(n - 1) Pi]^2 (# - mp))& /@ l][
            (* center and random shift of tube cross section *)
          Module[{R = 0.012, mp, dir1, dir2, φR, rR, shift},
              mp = (Plus @@ Rest[l])/(Length[l] - 1);
              dir1 = n[l[[1]] - mp]; dir2 = n[l[[2]] - mp];
              dir2 = n[dir2 - dir2.dir1 dir1];
              φR = 2Pi Random[]; rR = 2 R Random[];
              shift = Sin[(pos - 1)/(n - 1) Pi] *
                      rR (Cos[φR] dir1 + Sin[φR] dir2);
              Sequence @@ {mp, shift}]]

In[78]:= SeedRandom[123];
    Block[{ρFunc = ρFuncZigZag},
```

```
(* show two realizations of the randomized field lines *)
Show[GraphicsArray[Table[
  Show[{chargePic, fieldLine3D[#, fieldLineColor]& /@
      Take[Join[fieldLines["⊕" -> "⊖"], fieldLines["⊖" -> "⊕"]], All]},
    PlotRange -> All, Boxed -> False, ViewPoint -> {1, 1, 1},
    DisplayFunction -> Identity], {2}]]]]
```

Now, let us calculate the electric field lines of a regular array of charged spheres. We will position the charges at the lattice points of a cubic lattice. To achieve charge neutrality of the whole cluster of $3 \times 3 \times 3$ charges, the central charge will be made twice as large as the other ones.

```
In[80]:= {color["⊕"], color["⊖"]} = SurfaceColor[GrayLevel[#]]& /@ {0.3, 0.9};
```

```
In[81]:= charges = Flatten[Table[{0.001 (-1)^(i + j + k), {i, j, k}},
                         {i, -1, 1}, {j, -1, 1}, {k, -1, 1}], 2];
```

```
In[82]:= charges = charges /. {f_, {0, 0, 0}} :> {2 f, {0, 0, 0}};
```

```
In[83]:= {positiveCharges, negativeCharges} =
           Cases[charges, {_?#, _}]& /@ {Positive, Negative};
```

```
In[84]:= pic = Show[chargePic, Axes -> True,
               PlotRange -> 1.3 {{-1, 1}, {-1, 1}, {-1, 1}}]
```

We just repeat the calculations from above for the current charge distribution.

```
In[85]:= φ[{x_, y_, z_}] = Plus @@ (φCoulomb[#, {x, y, z}]& /@ charges);
```

```
In[86]:= EField[{x_, y_, z_}] = -{D[#, x], D[#, y], D[#, z]}&[φ[{x, y, z}]];
```

```
In[87]:= makeFieldComponentDefinitions[{Ex, Ey, Ez}, EField, {x, y, z}];
```

We will use another distribution for the starting points of the field lines around the charged spheres. The field lines will start at the vertices of a cube around the charges.

```
In[88]:= startPoints[charges_, γ_] :=
    (dirList = {{-1, -1, -1}, {-1, -1, 1}, {-1, 1, -1}, {-1, 1, 1},
               {1, -1, -1}, {1, -1, 1}, {1, 1, -1}, {1, 1, 1}}/Sqrt[3.];
     Flatten[Table[charges[[j, 2]] +
               γ Abs[charges[[j, 1]]]^(1/3) dirList[[k]],
                   {j, Length[charges]}, {k, Length[dirList]}], 1])

In[89]:= {fieldLines["⊕" -> "⊖"], fieldLines["⊖" -> "⊕"]} =
    Apply[Function[{np, pn, T},
      calculateFieldLine[{cx, cy, cz}, #, {Ex, Ey, Ez}, {t, T},
        StoppingTest :> sTest[{cx[t], cy[t], cz[t]}, np]]& /@
                                        startPoints[pn, 1.02]],
      {{negativeCharges, positiveCharges, +2},
       {positiveCharges, negativeCharges, -2}}, {1}];
```

We will also use another coloring for the field lines. Instead of coloring each field line with one color, we will color the field lines locally according to their distance from the origin. (Because this assigns a color to each single polygon, the following picture is more memory-consuming than the above one.)

```
In[90]:= fieldLineColor := {};

    colorFieldLine[gr3d_] := gr3d /. p_Polygon :>
      ({SurfaceColor[Hue[#], Hue[#], 2], p}&[norm[Plus @@ p[[1]]]/4])

In[92]:= Show[{pic, colorFieldLine[fieldLine3D[#, {}]]& /@
        Join[fieldLines["⊕" -> "⊖"], fieldLines["⊖" -> "⊕"]]},
        PlotRange -> All, Boxed -> False, Axes -> False]
```

```
In[93]:= (* to save memory *)
    Clear[fieldLines];
```

Here ends the electric field line case—let us now deal with the more interesting case of magnetic field lines. Compared with the electric case, it is often much more difficult to intuitively "sketch" the behavior of magnetic field lines [1442]. Roughly speaking, the reason for the more predictable behavior of the electric field lines is the existence of the electrostatic potential φ. The set of differential equations for magnetic field lines can be rewritten as a time-dependent Hamiltonian system [1451], [896], [1579], [1780], [351], and as such, the field lines can exhibit chaotic behavior (we will see such examples later).

Given the (time-independent) current density $\vec{J}(\vec{y})$, the magnetic field $\vec{B}(\vec{x})$ is given by the Biot–Savart formula [340] (also applicable to electrostatics [1375]):

$$\vec{B}(\vec{x}) = \int_{\mathbb{R}^3} \frac{\vec{J}(\vec{y}) \times (\vec{x} - \vec{y})}{|\vec{x} - \vec{y}|^3} \, d\vec{y}.$$

For the simplest case of a 1D wire $\vec{\gamma}$ with parametric representation $\vec{\gamma}(t)$, the current is given by $\vec{j}(\vec{x}) = \int_{-\infty}^{\infty} \delta(\vec{x} - \vec{\gamma}(s)) \, \vec{\gamma}'(s) \, ds$ (where s is the curve length parameter) and the above reduces to

$$\vec{B}(\vec{x}) = \int_{-\infty}^{\infty} \frac{\vec{\gamma}'(s) \times (\vec{x} - \vec{\gamma}(s))}{|\vec{x} - \vec{\gamma}(s)|^3} \, ds = \int_{-\infty}^{\infty} \frac{\vec{\gamma}'(t) \times (\vec{x} - \vec{\gamma}(t))}{|\vec{x} - \vec{\gamma}(t)|^3} \, dt.$$

We start with the simplest possible configuration—a straight, infinitely long, and infinitely thin current carrying wire. Here is a parametrization of the wire.

```
In[95]:= infiniteWire[t_] = {0, 0, t};
```

Using the Biot–Savart law, we obtain the well-known magnetic field $\vec{B}(\vec{x}) \sim \{-y, x, 0\}/(x^2 + y^2)$.

```
In[96]:= Integrate[Cross[D[infiniteWire[t], t], {x, y, z} - infiniteWire[t]]/
              norm[{x, y, z} - infiniteWire[t]]^3,
            {t, -Infinity, Infinity}, GenerateConditions -> False]
```

$$\text{Out[96]}= \left\{ -\frac{2\,y}{x^2 + y^2}, \ \frac{2\,x}{x^2 + y^2}, \ 0 \right\}$$

```
In[97]:= BInfiniteWire[{x_, y_}] = 2 {-y, x}/(x^2 + y^2);
```

Using the last result, we define the components of the magnetic field.

```
In[98]:= makeFieldComponentDefinitions[{Bx, By}, BInfiniteWire, {x, y}];
```

In the following example, we will display all current carrying wires as thin tubes with a yellow–orange color.

```
In[99]:= currentColor = SurfaceColor[Hue[0.12], Hue[0.22], 2.12];
```

For 12 randomly selected initial points, we calculate the field lines.

```
In[100]:= tMax = 12;
        fieldLines =
        calculateFieldLine[{cx, cy}, #, {Bx, By}, {t, tMax},
              MaxSteps -> 10^4]& /@ Table[Random[], {12}, {2}];
```

As expected, the field lines are concentric circles around the infinite wire. (For particle trajectories in the magnetic field of a long straight current-carrying wire, see [1340].)

```
In[102]:= Show[{tubeGraphics3D[Line[{{0, 0, -1}, {0, 0, 0}, {0, 0, 1}}],
                            {}, currentColor],
        ParametricPlot3D[Evaluate[{cx[t], cy[t], 0,
              {Thickness[0.002]}} /. #], {t, 0, tMax},
              PlotPoints -> 200, PlotRange -> All,
              DisplayFunction -> Identity]& /@ fieldLines},
            DisplayFunction -> $DisplayFunction]
```

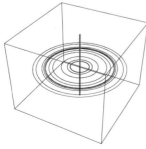

Instead of one infinite straight current, let us now deal with many of them. We choose 15 parallel currents at random positions and of random strengths. This system is translation-invariant along the wires. So, we restrict ourselves to the plane $z = 0$.

```
In[103]:= SeedRandom[83569034823]
        BField[{x_, y_}] = Plus @@
        ((Random[Real, {-1, 1}] BInfiniteWire[{x, y} - #])& /@
          (mps = Table[{Random[Real], Random[Real]}, {15}]));
```

For a faster numerical calculation, we compile the last expression.

```
In[105]:= BFieldC = Compile[{x, y}, Evaluate[BField[{x, y}]]];
```

```
In[106]:= BField[{x_, y_}] := BFieldC[x, y]
```

To get an impression about the magnetic field of the 15 currents, we visualize it with little arrows pointing in the direction of the field.

```
In[107]:= (* a 2D arrow *)
        arrow2D[{p1_, d_}, l_] :=
        Module[{α = 0.14, β = 0.7, γ = 0.35, d1, d2},
              d1 = n[d]; d2 = Reverse[d1]{-1, 1};
           Polygon[{p1 + α l d2, p1 + α l d2 + β l d1,
                    p1 + γ l d2 + β l d1, p1 + l d1,
                    p1 - γ l d2 + β l d1, p1 - α l d2 + β l d1, p1 - α l d2}]]
```

The red points are the positions of the currents (which flow perpendicular to the graphics plane).

```
In[109]:= fieldPic = Module[{point},
        Show[Graphics[{{PointSize[0.02], Hue[0], Point /@ mps},
            {Thickness[0.002],
        Table[arrow2D[{#, BFieldC @@ #}&[
            {Random[Real, {-1, 2}], Random[Real, {-1, 2}]}]], 0.05],
            {2500}]}}],
        DisplayFunction -> $DisplayFunction, PlotRange -> All,
        AspectRatio -> Automatic, Frame -> True, FrameTicks -> None]]
```

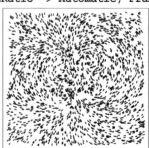

Now, we define the components of the magnetic field and calculate 100 field lines.

```
In[110]:= makeFieldComponentDefinitions[{Bx, By}, BField, {x, y}, Real];
```

```
In[111]:= tMax = 20;
        (fieldLines =
         Table[calculateFieldLine[{cx, cy},
                 {Random[], Random[]}, {Bx, By}, {t, tMax},
                 MaxSteps -> 10^4, Compiled -> False,
                 AccuracyGoal -> 6, PrecisionGoal -> 6], {100}];) // Timing
```

Out[112]= {42.06 Second, Null}

All field lines in the arrangement of parallel currents are closed curves.

```
In[113]:= Show[{fieldPic,
            ParametricPlot[Evaluate[{cx[t], cy[t]} /. fieldLines],
                          {t, 0, tMax}, PlotPoints -> 200,
                          PlotStyle ->
                           Table[{Thickness[0.002], Hue[Random[]]},
                                 {Length[fieldLines]}],
                          Compiled -> False, DisplayFunction -> Identity]},
                   PlotRange -> {{-1, 2}, {-1, 2}}]
```

After an infinite straight line, the next simplest current is described by a circle. Although the integrations involved can be carried out in closed form using elliptic integrals, here we will use purely numerical techniques—namely, NIntegrate to calculate the magnetic field. cfx, cfy, and cfz are the compiled versions of the integrands for the field components.

```
In[114]:= {cfx, cfy, cfz} =
       Compile[{t, x, y, z},
              Module[{c = Cos[t], s = Sin[t]},
                    #/(z^2 + (x - c)^2 + (y - s)^2)^(3/2)]& /@
              {z c, z s, (c (c - x) + s (s - y)^2)}];
```

In the solution of the differential equations of the field lines, we need the components of the field at the same point. We calculate all components at once and store the already-calculated field values via a SetDelayed[; Set [...]] construction.

```
In[115]:= (* avoid automatic switching to high-precision numbers *)
       Developer`SetSystemOptions["CatchMachineUnderflow" -> False];

       (* avoid evaluation of arguments of NIntegrate *)
       Developer`SetSystemOptions["EvaluateNumericalFunctionArgument" -> False]
Out[118]= EvaluateNumericalFunctionArgument→ False

In[119]:=
       Clear[BField];

       BField[{x_, y_, z_}] := BField[{x, y, z}] =
       NIntegrate[{cfx[t, x, y, z], cfy[t, x, y, z], cfz[t, x, y, z]},
                  {t, 0, 2 Pi}, Compiled -> False,
                  AccuracyGoal -> 3, PrecisionGoal -> 6]
```

Proceeding as above, we make the field component assignments and calculate some field lines. Because of the obvious rotational invariance of the magnetic field, we calculate all field lines in the *x,z*-plane.

```
In[121]:= makeFieldComponentDefinitions[{Bx, By, Bz}, BField, {x, y, z}, Real];
```

```
In[122]:= tMax = 6;
        fieldLines = calculateFieldLine[{cx, cy, cz}, #, {Bx, By, Bz}, {t, tMax},
                                MaxSteps -> 10^5]& /@
                     Table[{x, 0., 0.}, {x, 0.3, 0.9, 0.1}];
```

As is well known, the field lines form concentric closed curves around the wire.

```
In[124]:= Show[{tubeGraphics3D[Line[
          Table[{Cos[φ], Sin[φ], 0}, {φ, 0, 2Pi, 2Pi/60}]], {0.02, #1&},
                                currentColor],
          ParametricPlot3D[Evaluate[{cx[t], cy[t], cz[t],
                {Thickness[0.002]}} /. #], {t, 0, tMax},
                PlotPoints -> 200,
                DisplayFunction -> Identity]& /@ fieldLines},
          DisplayFunction -> $DisplayFunction, PlotRange -> All]
```

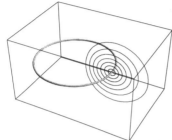

The next example is the combination of an infinite straight current and a circular current in the plane perpendicular to the infinite wire. Instead, using an exact circle, we will approximate the circle with straight pieces. This has two advantages: First, it allows for a faster numerical calculation of the magnetic field; and second, using straight line segments allows us to model arbitrary current configurations. The qualitative behavior of the field is the same for the exact and the approximated circles.

```
In[125]:= lineSegment[t_, {p1_, p2_}] = p1 + t(p2 - p1);
```

```
In[126]:= BFiniteWire[{x_, y_, z_}, {{p1x_, p1y_, p1z_}, {p2x_, p2y_, p2z_}}] =
        With[{l = lineSegment[t, {{p1x, p1y, p1z}, {p2x, p2y, p2z}}]},
        Integrate[Cross[D[l, t], {x, y, z} - l]/
                norm[{x, y, z} - l]^3, {t, 0, 1}, GenerateConditions -> False]];
```

The expression for `BFiniteWire` is relatively large.

```
In[127]:= BFiniteWire[{x, y, z}, {{p1x, p1y, p1z}, {p2x, p2y, p2z}}] // LeafCount
Out[127]= 1519
```

Using the closed-form formula in `BFiniteWire` directly is relatively slow. Calculating 10000 field values takes a few seconds on a 2 GHz computer.

```
In[128]:= Timing[Do[BFiniteWire[Table[Random[], {3}],
                                Table[Random[], {2}, {3}]], {10000}]]
Out[128]= {8.06 Second, Null}
```

To speed up the calculation carried out by `BFiniteWire`, we simplify the expression [736], use the package `NumericalMath`OptimizeExpression`` to rewrite the expression in a more effective way, and compile the resulting expression. Using this package avoids the time-consuming and cumbersome work of generating a procedural version of BFiniteWire "by hand".

```
In[129]:= BFiniteWireC =  (* use flat argument structure *)
            Compile[{x, y, z, p1x, p1y, p1z, p2x, p2y, p2z},
                (* use function from Experimental` context *)
                Evaluate[Experimental`OptimizeExpression[
                  Simplify[BFiniteWire[{x, y, z},
                            {{p1x, p1y, p1z}, {p2x, p2y, p2z}}]],
                    OptimizationLevel -> 1, ExcludedForms -> {}]]];
```

BFiniteWireC is about 20 times faster than BFiniteWire.

```
In[130]:= Timing[Do[BFiniteWireC @@ Table[Random[], {9}], {10000}]]
Out[130]= {0.29 Second, Null}
```

Both functions calculate the same values. We check it for a set of "random" values.

```
In[131]:= {BFiniteWire[{#1, #2, #3}, {{#4, #5, #6}, {#7, #8, #9}}],
          BFiniteWireC[##]}& @@ Table[N[Sin[k]], {k, 9}]
Out[131]= {{-1.82111, 1.9679, -1.82111}, {-1.82111, 1.9679, -1.82111}}
```

This is the wire approximating the circular current.

```
In[132]:= wire = N[Table[{Cos[φ], Sin[φ], 0}, {φ, 0, 2Pi, 2Pi/12}]];
```

To calculate the magnetic field of all wire pieces, we just sum the contributions from the 12 finite pieces and the infinite wire.

```
In[133]:= wireData = Flatten /@ Partition[wire, 2, 1];
```

```
In[134]:= BField[{x_, y_, z_}] := 2 {-y, x, 0.}/(x^2 + y^2) +
            (Plus @@ Apply[BFiniteWireC[x, y, z, ##]&, wireData, {1}]);
```

These are the definitions for the three components of the magnetic field.

```
In[135]:= makeFieldComponentDefinitions[{Bx, By, Bz}, BField, {x, y, z}, Real];
```

```
In[136]:= SeedRandom[123454321];
          startPoints = Table[Random[Real, {-2, 2}], {9}, {3}];
```

```
In[138]:= tMax = 100;
          Timing[fieldLines = calculateFieldLine[{cx, cy, cz}, #, {Bx, By, Bz},
                                        {t, 0, tMax}, MaxSteps -> 10^5,
                              PrecisionGoal -> 12]& /@  startPoints];
```

Because some of the following pictures of magnetic field lines might look unexpected, we should check for the quality of the result of NDSolve. We do this by running the just-calculated field lines backward. We could also change the Method option setting to NDSolve to make sure we carry out two independent calculations.

```
In[140]:= Norm /@ (startPoints - (({cx[0], cy[0], cz[0]} /.
            calculateFieldLine[{cx, cy, cz},
                  {cx[#], cy[#], cz[#]}&[tMax] /. #[[1]],
                  {Bx, By, Bz}, {t, tMax, 0},
                   MaxSteps -> 10^5, StartingStepSize -> 10^-2,
                   PrecisionGoal -> 12][[1]])& /@ fieldLines))
Out[140]= {6.22746×10^-6, 0.0000387776, 0.00011681, 0.000169618,
          0.0000301555, 4.83168×10^-6, 0.0000403706, 1.04161×10^-6, 0.0000122832}
```

As the last result shows, the differential equations were correctly solved within the resolution of the graphic shown.

```
In[141]:= With[{gr1 = tubeGraphics3D[Line[{{0, 0, -1}, {0, 0, 0}, {0, 0, 1}}],
                             {}, currentColor],
          gr2 = tubeGraphics3D[Line[wire], {0.02, #1&}, currentColor]},
     Show[GraphicsArray[#]]& /@ Partition[
     Show[{gr1, gr2, ParametricPlot3D[
       Evaluate[{{cx[t], cy[t], cz[t],
          {Thickness[0.002], Hue[Random[]]}} /. #], {t, 0, tMax},
          PlotPoints -> 10000, DisplayFunction -> Identity]},
               DisplayFunction -> Identity, BoxRatios -> {1, 1, 1}]& /@
                                                  fieldLines, 3]]
```

Now let us look at the field lines of a much more complicated wire—the wire will be shaped like a 2D Peano curve. Here are the points of this wire.

```
In[142]:= wire = Append[#, 0]& /@
     {{0, 0}, {1, 0}, {1, 2}, {2, 2}, {2, 0}, {4, 0}, {4, 1}, {3, 1},
      {3, 2}, {4, 2}, {4, 3}, {1, 3}, {1, 6}, {2, 6}, {2, 4}, {4, 4},
      {4, 5}, {3, 5}, {3, 6}, {4, 6}, {4, 7}, {1, 7}, {1, 11}, {2, 11},
      {2, 10}, {3, 10}, {3, 11}, {4, 11}, {4, 9}, {2, 9}, {2, 8},
      {5, 8}, {5, 11}, {6, 11}, {6, 10}, {7, 10}, {7, 11}, {8, 11},
      {8, 9}, {6, 9}, {6, 8}, {8, 8}, {8, 5}, {7, 5}, {7, 7}, {5, 7},
      {5, 6}, {6, 6}, {6, 5}, {5, 5}, {5, 4}, {8, 4}, {8, 1}, {7, 1},
```

```
           {7, 3}, {5, 3}, {5, 2}, {6, 2}, {6, 1}, {5, 1}, {5, 0}, {9, 0},
           {9, 3}, {10, 3}, {10, 2}, {11, 2}, {11, 3}, {12, 3}, {12, 1},
           {10, 1}, {10, 0}, {13, 0}, {13, 2}, {14, 2}, {14, 0}, {16, 0},
           {16, 1}, {15, 1}, {15, 2}, {16, 2}, {16, 3}, {13, 3}, {13, 6},
           {14, 6}, {14, 4}, {16, 4}, {16, 5}, {15, 5}, {15, 6}, {16, 6},
           {16, 7}, {12, 7}, {12, 4}, {11, 4}, {11, 5}, {10, 5}, {10, 4},
           {9, 4}, {9, 6}, {11, 6}, {11, 7}, {9, 7}, {9, 11}, {10, 11},
           {10, 10}, {11, 10}, {11, 11}, {12, 11}, {12, 9}, {10, 9},
           {10, 8}, {13, 8}, {13, 10}, {14, 10}, {14, 8}, {16, 8}, {16, 9},
           {15, 9}, {15, 10}, {16, 10}, {16, 11}, {13, 11}, {13, 14},
           {14, 14}, {14, 12}, {16, 12}, {16, 13}, {15, 13}, {15, 14},
           {16, 14}, {16, 15}, {12, 15}, {12, 12}, {11, 12}, {11, 13},
           {10, 13}, {10, 12}, {9, 12}, {9, 14}, {11, 14}, {11, 15},
           {8, 15}, {8, 12}, {7, 12}, {7, 13}, {6, 13}, {6, 12},
           {5, 12}, {5, 14}, {7, 14}, {7, 15}, {4, 15}, {4, 12},
           {3, 12}, {3, 13}, {2, 13}, {2, 12}, {1, 12}, {1, 14},
           {3, 14}, {3, 15}, {0, 15}, {0, 0}};
```

In[143]:= **Show[tubeGraphics3D[Line[wire], {0.06, #1&}, currentColor]]**

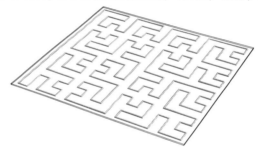

In[144]:= **wireData = Flatten /@ Partition[N[wire], 2, 1];**

We define the magnetic field generated by this wire. We add the contributions from all wire segments using the function BFiniteWireC.

In[145]:= **BField[{x_, y_, z_}] := (Plus @@ Apply[BFiniteWireC[x, y, z, ##]&,
                                        wireData, {1}]);**

In[146]:= **makeFieldComponentDefinitions[{Bx, By, Bz}, BField, {x, y, z}, Real];**

We use four selected start points and calculate the field lines over a long "time".

In[147]:= **startPoints =
          {{15.68110, 6.45551, 0.87960}, {11.85503, 0.72703, 0.87980},
           {10.72057, 12.10141, 0.70727}, {9.94457, 3.24733, 0.58636}};**

In[148]:= **tMax = 100;
          Timing[fieldLines =
                calculateFieldLine[{cx, cy, cz}, #, {Bx, By, Bz},
                   {t, 0, tMax}, MaxSteps -> 10^4,
                      PrecisionGoal -> 6]& /@ startPoints;]**

Out[149]= {141.5 Second, Null}

Despite the complicated shape of the current-carrying wire, the resulting field lines are closed curves, winding just once around.

In[150]:= **With[{gr = tubeGraphics3D[Line[wire], {0.06, #1&}, currentColor]},
          Show[{gr, ParametricPlot3D[
            Evaluate[{cx[t], cy[t], cz[t],
               {Thickness[0.003], Hue[Random[]]}} /. #], {t, 0, tMax},**

```
            PlotPoints -> 10000, DisplayFunction -> Identity]& /@ fieldLines},
                DisplayFunction -> $DisplayFunction,
                PlotRange -> All, Boxed -> False]]
```

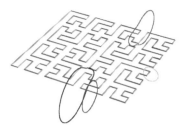

The field lines for every planar set of currents have the topological shape, such as the three field lines in the last picture.

Now, let us calculate and visualize the magnetic field lines in the simplest nonplanar case—a wire made from four finite straight line segments.

```
In[151]:= wire = {{0, -1, 0}, {-1, 0, 0}, {0, 1, 0}, {0, 0, 1}, {0, -1, 0}};
```

```
In[152]:= Show[tubeGraphics3D[Line[wire], {0.02, #1&}, currentColor]]
```

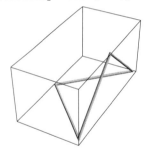

Proceeding as above, we define the field components.

```
In[153]:= wireData = Flatten /@ Partition[N[wire], 2, 1];

        BField[{x_, y_, z_}] :=
            (Plus @@ Apply[BFiniteWireC[x, y, z, ##]&, wireData, {1}]);
```

```
In[155]:= makeFieldComponentDefinitions[
            {Bx, By, Bz}, BField, {x, y, z}, Real];
```

For two initial points, we solve the field line differential equation for an arc length of 200.

```
In[156]:= tMax = 200;
        fieldLines =
                calculateFieldLine[{cx, cy, cz}, #, {Bx, By, Bz},
                {t, 0, tMax}, MaxSteps -> 10^4,
                    PrecisionGoal -> 8, AccuracyGoal -> 8]& /@
                    {{-0.767890, 0.4911667, -1.415637},
                    {-1.402164, 1.9174317, -1.398096}};
```

The field lines show a rather complicated behavior. They are no longer closed lines. (Sometimes one reads the statement that magnetic field lines are always closed because of the Maxwell equation div $\vec{B}(\vec{x}) = 0$. This equation says nothing about a single field line; it only makes a statement about the *density* of field lines in some volume element [1668]. While there exists some closed field lines [1220], "most" field lines are not closed for a current arrangement.)

In[158]:=
```
With[{gr = tubeGraphics3D[Line[wire], {0.06, #1&}, currentColor]},
Show[GraphicsArray[
Show[{gr, (* the field lines *)
ParametricPlot3D[Evaluate[{cx[t], cy[t], cz[t],
                {Thickness[0.002], Hue[0]}} /. #],
                {t, 0, tMax}, PlotPoints -> 2500,
                DisplayFunction -> Identity]},
        DisplayFunction -> Identity, PlotRange -> All,
        Boxed -> True, Axes -> False]& /@ fieldLines]]]
```

If we allow for infinite wires again, already two perpendicular wires generate interesting field lines. Here is the magnetic field generated by a current along the *z*-axis and a second current parallel to the *y*-axis through the point {1, 0, 0}.

In[159]:=
```
BField[{x_, y_, z_}] := {-y, x, 0}/(x^2 + y^2) -
                        {-z, 0, (x - 1)}/((x - 1)^2 + z^2)
```

In[160]:=
```
makeFieldComponentDefinitions[{Bx, By, Bz}, BField, {x, y, z}, Real];
```

The field line calculated and displayed winds around both wires in a "symmetric" way.

In[161]:=
```
tMax = 200;
fieldLine =  calculateFieldLine[{cx, cy, cz},
        {0.027622, 0.551576, 0.378692}, {Bx, By, Bz},
        {t, 0, tMax}, MaxSteps -> 10^4];
```

In[163]:=
```
With[{grs = tubeGraphics3D[Line[#], {0.1}, currentColor]& /@
        {{{0, 0, -10}, {0, 0, 0}, {0, 0, 20}},
         {{1, -10, 0}, {1, 0, 0}, {1, 20, 0}}}},
    Show[{grs, ParametricPlot3D[
        Evaluate[{cx[t], cy[t], cz[t],
        {Thickness[0.002], Hue[0]}} /. fieldLine], {t, 0, tMax},
        PlotPoints -> 2000, DisplayFunction -> Identity]},
            DisplayFunction -> $DisplayFunction,
            PlotRange -> All, Boxed -> True,
            ViewPoint -> {3, -1, 2}, BoxRatios -> Automatic]]
```

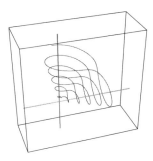

From the last picture, one should not infer that any nonplanar or any nonparallel arrangement of currents causes a complicated field line pattern. Here, we consider a set of wires arranged in the form of the edges of a cube. The current flows from one vertex to the opposite one and back along the diagonal. These are the single straight wires forming the wire-cube.

```
In[164]:= wirePieces =
    {(* "diagonal wire" *)
    {{{1, 1, 1}, {0, 0, 0}}},
    (* from {0, 0, 0} and to {1, 1, 1} *)
    {{{0, 0, 0}, {1, 0, 0}}, {{0, 0, 0}, {0, 1, 0}},
     {{0, 0, 0}, {0, 0, 1}}, {{0, 1, 1}, {1, 1, 1}},
     {{1, 0, 1}, {1, 1, 1}}, {{1, 1, 0}, {1, 1, 1}}},
    (* the six "middle" pieces *)
    {{{1, 0, 0}, {1, 1, 0}}, {{1, 0, 0}, {1, 0, 1}},
     {{0, 1, 0}, {1, 1, 0}}, {{0, 1, 0}, {0, 1, 1}},
     {{0, 0, 1}, {1, 0, 1}}, {{0, 0, 1}, {0, 1, 1}}}} // N;
```

```
In[165]:= gr = Show[tubeGraphics3D[Line[{#[[1]], (#[[1]] + #[[2]])/2, #[[2]]}],
                    {}, currentColor]& /@ Flatten[wirePieces, 1]]
```

For the calculation of the magnetic field, we have to take into account that all of the current is flowing along the diagonal, but that the current along the cube edges is only a fraction of the first.

```
In[166]:= BField[{x_, y_, z_}] :=
    1/1 (Plus @@ Apply[BFiniteWireC[x, y, z, ##]&,
                    Flatten /@ wirePieces[[1]], {1}]) +
    1/3 (Plus @@ Apply[BFiniteWireC[x, y, z, ##]&,
                    Flatten /@ wirePieces[[2]], {1}]) +
    1/6 (Plus @@ Apply[BFiniteWireC[x, y, z, ##]&,
                    Flatten /@ wirePieces[[3]], {1}])
    (* or, shorter, but less readable:
    Sum[2/(k (k + 1)) (Plus @@ Apply[BFiniteWireC[x, y, z, ##]&,
```

```
                    Flatten /@ wirePieces[[k]], {1}]), {k, 3}] *)
```

We generate the definitions for the magnetic field and calculate two field lines for two selected initial conditions.

```
In[168]:= makeFieldComponentDefinitions[
            {Bx, By, Bz}, BField, {x, y, z}, Real];

In[169]:= tMax = 50;
        fieldLines = calculateFieldLine[{cx, cy, cz}, #, {Bx, By, Bz},
                {t, 0, tMax}, MaxSteps -> 10^5, PrecisionGoal -> 6]& /@
                {{0.038, 1.029, 0.4565}, {0.74097, -0.0416, 1.10925}};
```

All field lines are now closed curves.

```
In[171]:= Show[GraphicsArray[
        Show[{gr, ParametricPlot3D[
          Evaluate[{cx[t], cy[t], cz[t],
            {Thickness[0.002], Hue[0]}} /. #], {t, 0, tMax},
            PlotPoints -> 10000, DisplayFunction -> Identity]},
                DisplayFunction -> Identity,
                PlotRange -> All, Boxed -> False]& /@  fieldLines]]
```

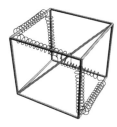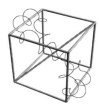

In the last example, we will calculate magnetic field lines of a ringcoil. We use the following parametrization for the coil.

```
In[172]:= spaceCurve[ϑ_] = {3 Cos[ϑ] + Cos[ϑ] Cos[20 ϑ],
                            3 Sin[ϑ] + Sin[ϑ] Cos[20 ϑ], Sin[20 ϑ]};
```

We will start by calculating the field lines for a 200-piece approximation to spaceCurve.

```
In[173]:= wire = Table[spaceCurve[ϑ], {ϑ, 0, 2. Pi, 2. Pi/200}];

In[174]:= gr = Show[tubeGraphics3D[Line[wire], {0.03, #1&}, currentColor]]
```

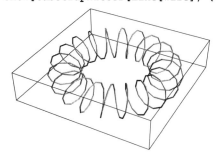

Proceeding as above, we define the field components. This time, the calculation of `BField` is more expensive—because of the 200 finite straight wires involved. We use a `SetDelayed[Set[...]]` construction to avoid the recalculation of the field when calculating the field components.

```
In[175]:= wireData =
          Developer`ToPackedArray[Flatten /@ Partition[N[wire], 2, 1]];

In[176]:= Clear[BField];
          BField[{x_, y_, z_}] := BField[{x, y, z}] =
              (Plus @@ Apply[BFiniteWireC[x, y, z, ##]&, wireData, {1}]);

In[178]:= makeFieldComponentDefinitions[
              {Bx, By, Bz}, BField, {x, y, z}, Real];
```

For two initial conditions, we solve the field line differential equation for an arc length of 100. We choose three initial conditions representing different "types" of field lines.

```
In[179]:= (tMax = 100;
          fieldLines = calculateFieldLine[{cx, cy, cz}, #, {Bx, By, Bz},
                          {t, 0, tMax}, MaxSteps -> 10^5,
                          PrecisionGoal -> 6, AccuracyGoal -> 6]& /@
              {{3.1, 0., 0.}, {3.7, 0., -0.667}, {1.55, 0., 0.}};) // Timing

Out[179]= {77.64 Second, Null}
```

To better see the behavior field lines inside the ringcoil, we color the field line from red to blue.

```
In[180]:= colorLines[l_List] :=
          With[{n = Length[l]}, MapIndexed[{Hue[#2[[1]]/n 0.8], #1}&, l]]
```

Here are the field lines. The first field line starts near the center of the cross section of the coil. After winding around, it leaves the coil and comes back. The second field line starts near the wire forming the coil and winds around the wire. The third line starts in the middle of the ring and surrounds the cross section (the whole cross section acts as a current-carrying wire for it). None of the field lines displayed is closed.

```
In[181]:= Show[GraphicsArray[
          Show[{gr, Graphics3D[{Thickness[0.002],
            colorLines[ParametricPlot3D[
              Evaluate[{cx[t], cy[t], cz[t]} /. #], {t, 0, tMax},
              PlotPoints -> 2500, DisplayFunction -> Identity][[1, 1]]]}]},
                  DisplayFunction -> Identity, PlotRange -> All,
                  Boxed -> False, Axes -> False]& /@ fieldLines],
              GraphicsSpacing -> -0.1]
```

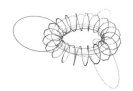  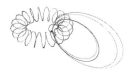

The first of the last pictures might seem unexpected. To show that it is not an artifact based on the discretization of the wire, we will calculate the field for the nondiscretized version of `spaceCurve`. We will use `NInte grate` to calculate the field values. To avoid compiling the integrand at each call to `NIntegrate`, we compile the integrand once in the beginning and then call `NIntegrate` with the option setting `Compiled -> False`. Of course, the calculation of the magnetic field will take longer than in the discretized case.

```
In[182]:= {cfx, cfy, cfz } =
    Compile[{t, x, y, z},
        Module[{c = Cos[t], s = Sin[t],
                c20 = Cos[20t], s20 = Sin[20t]}, #]]& /@
        (Cross[spaceCurve'[𝜃], {x, y, z} - spaceCurve[𝜃]]/
            (#.#)^(3/2)&[{x, y, z} - spaceCurve[𝜃]] /.
    {Cos[𝜃] -> c, Sin[𝜃] -> s, Cos[20𝜃] -> c20, Sin[20𝜃] -> s20});

In[183]:= Clear[BField];

    BField[{x_, y_, z_}] := (BField[{x, y, z}] =
    NIntegrate[{cfx[t, x, y, z], cfy[t, x, y, z], cfz[t, x, y, z]},
            (* use intermediate points *)
            Evaluate[Flatten[{t, Table[𝜃, {𝜃, 0, 2Pi, 2Pi/20}]}]],
            Compiled -> False, Method -> GaussKronrod,
            AccuracyGoal -> 4, PrecisionGoal -> 6])

In[185]:= makeFieldComponentDefinitions[{Bx, By, Bz}, BField, {x, y, z}, Real];
```

We calculate a field line starting exactly at the center of the coil's cross section.

```
In[186]:= (tMax = 60;
    fieldLine = calculateFieldLine[{cx, cy, cz},
            {3., 0., 0.}, {Bx, By, Bz},
            {t, 0, tMax}, MaxSteps -> 10^5,
                PrecisionGoal -> 6, AccuracyGoal -> 6];) // Timing
Out[186]= {117.25 Second, Null}
```

Because of the finite number of revolutions and the finite slope of the revolutions, the magnetic field is not rotational invariant around the *z*-axis. The next three graphics show the radial, the azimuthal, and the *z*-component of the field along the center of the cross section. The azimuthal periodicity is clearly visible.

```
In[187]:= BFieldData = Table[{φ, BField[{3 Cos[φ], 3 Sin[φ], 0}]} // N,
                {φ, 0, Pi/2, Pi/2/200}];

In[188]:= Show[GraphicsArray[
    Apply[Function[{f, l},
    ListPlot[{#[[1]], f}& /@ BFieldData,
            PlotRange -> All, Frame -> True, Axes -> False,
            PlotJoined -> True, PlotLabel -> l,
            DisplayFunction -> Identity]],
        (* the components *)
        {{Unevaluated[#[[2, 1]] Sin[-#[[1]]] +
                    #[[2, 2]] Cos[ #[[1]]]], "Br"},
         {Unevaluated[#[[2, 1]] Cos[ #[[1]]] +
                    #[[2, 2]] Sin[ #[[1]]]], "Bφ"},
         {Unevaluated[#[[2, 3]]], "Bz"}}, {1}]]]
```

The field line that starts at the center of the cross section moves away from the center and actually leaves the coil after a while.

```
In[189]:= ringCoilGraphics = tubeGraphics3D[
      Line[Table[spaceCurve[ϑ], {ϑ, 0, 2. Pi, 2. Pi/600}]],
            {0.03, #1&}, currentColor];
```

```
In[190]:= Show[{ringCoilGraphics, Graphics3D[{Thickness[0.002],
      colorLines[ParametricPlot3D[
        Evaluate[{cx[t], cy[t], cz[t]} /. fieldLine], {t, 0, tMax},
        PlotPoints -> 2500, DisplayFunction -> Identity][[1, 1]]]}]},
            PlotRange -> All, Boxed -> False, Axes -> False]
```

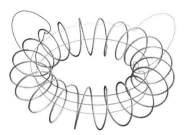

Here we end calculating and visualizing field lines and leave it to the reader to explore further configurations like molecules [782]. Current-carrying conductors of various shapes [1261], current-carrying 3D Peano curves, edges of polyhedra, and so on will yield interesting field lines. Knotted wires [1890], [1891], [1892] are another rich source of unexpected field line configurations. Are there closed field lines around them that are knotted too, [1815]? Then one could go on and investigate the field lines of time-dependent currents [643], [1504], [1838], [153], rotating magnets [1414], (super-)shielded magnetic fields [302] and so on.

1.12.2 Riemann Surfaces of Algebraic Functions

In Subsection 2.3.7 of the Graphics volume [1794], we discussed the visualization of Riemann surfaces of some simple, multivalued functions [1241]. Here, we will deal with a much more general class of functions: all algebraic functions. Given a (irreducible) polynomial p in two variables $p = \sum_{j,k=0}^{n} a_{ij} z^j w^k$, the equation $p = 0$ defines $w(z)$ implicitly as a (generically multivalued) function. This class of functions includes elliptic and hyperelliptic curves. Graphing the real or the imaginary part of $w(z)$ represents a faithful representation of the Riemann surface of $w(z)$ (for nonalgebraic functions, the faithfulness issue is more complicated). (For introductions on Riemann surfaces, see [152], [1237], [1287], [1392], and [1221]; for a more detailed mathematical discussion, see [28], [1520], [262], [582], [836], [1249], [1448], and [1896]. For some issues special to Riemann surfaces of algebraic functions, see [913], [295], [338], [924], [269], and [1507]. For the relevance of Riemann surfaces to physics, see, for instance, [727], [1230], [162], [263], and [1171].)

A numerical algorithm for the construction of the Riemann surface is given in the following and will be used to visualize the Riemann surfaces of some sample functions.

The key point of the functions used in Subsection 2.3.7 of the Graphics volume was the fact that their branch points and branch cuts were easy to determine and to exclude in the graphics to construct smooth surfaces.

For the function $w(z) = \sqrt[3]{z^4 + z^{-4}}$ (or $1 + w^3 z^4 - z^8 = 0$), the determination of the form of the branch cuts becomes a bit more complicated; the discontinuities will no longer be located on a straight line. Here is the principal sheet.

```
In[1]:= Plot3D[Evaluate[Im[(z^4 + z^-4)^(1/3) /. z -> x + I y]],
            {x, -2, 2}, {y, -2, 2}, PlotPoints -> 120, Mesh -> False]
```

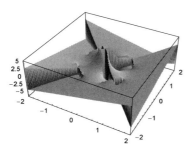

Using a `ContourPlot` shows the branch cuts nicely; they form clusters of contour lines.

```
In[2]:= ContourPlot[Evaluate[Im[(z^4 + z^-4)^(1/3) /. z -> x + I y]],
            {x, -3/2, 3/2}, {y, -3/2, 3/2},
            PlotPoints -> 200, Contours -> 30,
            ColorFunction -> (Hue[Random[]]&),
            ContourStyle -> {{Thickness[0.002], GrayLevel[1]}}]
```

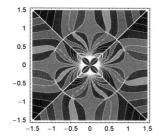

All of the branch cuts are caused by the branch cut of the `Power` function and occur when $z^4 + z^{-4}$ is equal to a negative real number v.

Analyzing the solutions of $z^4 + z^{-4} = v$ in detail shows that they form rays originating from $e^{i(\pi/4 + k\,\pi/2)}$, $k = 0$, 1, 2, 3 and going radially inward and outward, and arcs of the unit circle.

Knowing the location of the branch cuts now allows us to split the complex z-plane into regions such that inside each region we can use `ParametricPlot3D` to get a smooth surface of the imaginary part. We generate all three possible values for the cube root and display all 24 (= 3 (different roots) × 4 (inside the unit circle) + 4 (outside the unit circle)) pieces.

```
In[3]:= With[{ε = 10^-6, ε = Exp[I 2Pi/3]},
    Show[Graphics3D[{EdgeForm[], Thickness[0.002],
        SurfaceColor[Hue[0.26], Hue[0.31], 2.15],
      Cases[Table[Apply[(* inside each sector *)
      ParametricPlot3D[{r Cos[φ], r Sin[φ],
        Im[ε^j ((r E^(I φ))^4 + (r E^(I φ))^-4)^(1/3)]},
            #1, #2, PlotPoints -> {12, 22},
            DisplayFunction -> Identity]&,
    Join @@ (* inside and outside the unit circle *)
      Table[{{r, ρ + ε, ρ + 1 - ε},
            {φ, -Pi/4 + ε + k Pi/2, +Pi/4 - ε + k Pi/2}},
            {ρ, 0, 1}, {k, 0, 3}], {1}], {j, 3}], _Polygon, Infinity]}],
        PlotRange -> {-4, 4}, BoxRatios -> {1, 1, 1.9}, Boxed -> False]]
```

When the polynomial $p = 0$ can be solved for z, it is possible to obtain a parametrization of the Riemann surface. Let us consider the polynomial $w^5 + w^3 = z$. The following parametric plot shows $\mathrm{Im}(w)$ parametrized in the form $\mathrm{Im}(w) = \mathrm{Im}(w(z)) = \mathrm{Im}(w(z(w))) = \mathrm{Im}(w(z(\mathrm{Re}(w) + i\,\mathrm{Im}(w))))$. One clearly sees that the whole z-plane is covered multiple times.

```
In[4]:= ParametricPlot3D[
    Evaluate[{Re[(wr + I wi)^5 + (wr + I wi)^3],
              Im[(wr + I wi)^5 + (wr + I wi)^3], wi,
              {EdgeForm[], SurfaceColor[Hue[wr/6]]}}],
       {wr, -6, 6}, {wi, -6, 6}, PlotPoints -> 100,
       BoxRatios -> {1, 1, 3/2}, PlotRange -> All,
       Axes -> False, Boxed -> False]
```

For more complicated polynomials, which cannot be solved nicely for w or z, it turns out to be very difficult to determine the location of the branch cuts either numerically or in closed form. Here is an example of a more complicated polynomial.

```
In[5]:= ContourPlot[Evaluate[
        Im[(-5 + 4z + 6z^2 + z^3 + 5z^4 + 7z^5 -
           7z^6 + 6z^7 + 4z^8 + 4z^9 - 4z^10)^(1/7) /. z -> x + I y]],
        {x, -3/2, 3/2}, {y, -3/2, 3/2}, PlotPoints -> 200,
        ColorFunction -> (Hue[0.7 #]&), Contours -> 30,
        ContourStyle -> {{Thickness[0.002], GrayLevel[0]}}]
```

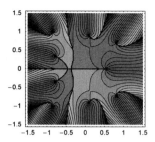

When we cannot solve $p = \sum_{j,k=0}^{n} a_{ij} z^j w^k$ with respect to w [1534], we cannot describe the branch cuts at all. (Using the in Chapter 1 of the Symbolics volume [1795] discussed `Root`-objects, it is theoretically possible to describe the branch cuts, but for more complicated examples it will be prohibitively complicated.) (The function `Root` can always solve bivariate polynomials, but the resulting branch cuts—better lines of discontinuity—are quite complicated and do not solve the problem at hand.)

To avoid discontinuities in the graphics of the Riemann surface, we would have to restrict carefully the regions where to plot the function. On the other hand, branch cuts can be chosen to a large extent arbitrarily, and only the location of the branch points, where two sheets are glued together, represents a property inherent and well defined for an algebraic function. (Branch cuts have to connect branch points. But how exactly a branch cut is chosen in the principal sheet between the branch points does not matter). That is why we will avoid using closed-form radicals (or `Root`s) formulas and branch cuts at all, and instead use `NDSolve` to calculate smooth patches inside regions of the complex z-plane that are free of branch points [77]. This treatment completely *avoids* dealing with (on a Riemann surface anyway) nonexisting branch cuts. The patches will be chosen in such a way that the branch points (which must be treated carefully—here, two or more sheets are glued together) are always at their vertices. This also allows for a good spatial resolution near the branch points, which is important because typically $w(z)$ varies most near a branch point. This will result in a smooth in 3D self-intersecting surface that represents the Riemann surface of the polynomial under consideration.

To construct the Riemann surface, we will proceed with the following steps:

1) We use `NSolve` to calculate all potential branch points p.

2) We divide a circular part of the complex z-plane radially and azimuthally into sectors generated by the outer product of the radial and azimuthal coordinates of the branch points, so that every sector has no branch point inside.

3) We derive a differential equation for $w'(z)$ from p.

4) By starting (quite) near the branch points, we solve the differential equation radially outward for all sheets of every sector.

5) Starting from the radial solutions of the differential equations, we solve the differential equation azimuthally inside the sectors constructed in 3). This results in parametrized (by the radius r and the angle φ) parts of all sheets for each sector.

6) We generate polygons from the parametrizations for all sheets and all sectors and display all things together (and maybe cut holes into the polygons for better visibility of the inner parts).

Now, let us implement this construction.

1) At a branch point, the equation $p = 0$ has a multiple solution for $w(z)$. This means p and $\frac{\partial p}{\partial w}$ must both be zero [269]. We use `NSolve` to calculate a high-precision approximation of potential branch points.

```
In[6]:= branchPoints[poly_, {w_, z_}] :=
         Union[z /. NSolve[{poly == 0, D[poly, w] == 0}, {z, w},
```

```
                       WorkingPrecision -> 200],
            SameTest -> (Abs[#1 - #2] < 10^-6&)]
```

As an example for the polynomial *p*, let us take the following bivariate polynomial. (We will use this polynomial throughout implementing the approach sketched above for exemplifying the various functions defined below.)

```
In[7]:= pExample = 2z^5 - z^3 + z^4 - 2z^2 w - z w^2 + 2z w^4 + w^5;
```

This function has 12 potential branch points. (For a detailed discussion of the properties of branch points of Riemann surfaces of algebraic functions, see [1650], [352], [605], [402], [1650], [1244], [836], [1245], [584], [9], [10], [525], [1869], [338], [607], [1280], and [1754].)

```
In[8]:= Off[NSolve::"zerosol"];
       (potentialBranchPointsExample = branchPoints[pExample, {w, z}]) // N
Out[9]= {-1.28509, -0.864198 - 0.308813 i, -0.864198 + 0.308813 i,
         -0.307643, -0.288979 - 0.139225 i, -0.288979 + 0.139225 i,
         -1.57035×10^-134, 0.206333 - 0.186587 i, 0.206333 + 0.186587 i,
         0.448111 - 0.450203 i, 0.448111 + 0.450203 i, 0.873818}
```

Here, their location in the complex *z*-plane is shown.

```
In[10]:= Show[Graphics[{PointSize[0.02], Point[{Re[#], Im[#]}]& /@
                  potentialBranchPointsExample}],
          PlotRange -> All, AspectRatio -> Automatic,
          Frame -> True, Axes -> False]
```

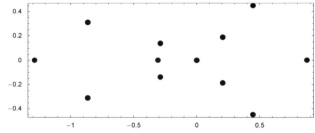

2) Given the branch points, we calculate their distances from the origin and their phase and divide the r,φ-plane into radial and azimuthal sectors that are formed between radially and azimuthally adjacent branch points.

```
In[11]:= sectors[poly_, {w_, z_}] :=
    Module[{ε = 10^-5, (* the branch points *)
            bps = branchPoints[poly, {w, z}], arg, rMax, rList, φList},
            (* a relative of the Arg function;
               defined at the origin *)
            arg[ξ_] := If[ξ == 0., 0``25, Arg[ξ]];
            rMax = (If[#1 == 0., 1, 12/10 #1] &)[
            Max[Abs[rList = Abs /@ bps]]];
            (* r- and φ-values *)
            rList = Union[Prepend[Append[rList, rMax], 0],
                          SameTest -> (Abs[#1 - #2] < ε&)];
            (* sort counter-clockwise *)
            φList = Sort[Union[arg /@ bps, SameTest -> (Abs[#1 - #2] < ε&)],
                    #1 < #2&];
            φList = Append[φList, First[φList] + 2Pi];
            (* the sectors *)
            N[#1, 25]&[Flatten[Table[
             {rList[[i]] + ε, φList[[j]] + ε,
```

```
                rList[[i + 1]] - ε, φList[[j + 1]] - ε},
            {i, Length[rList] - 1}, {j, Length[φList] - 1}], 1]]]
```

Here, the sectors for our example polygon pExample are calculated.

```
In[12]:= Short[s1 = sectors[pExample, {w, z}], 8]
```
Out[12]//Short=

```
    {{0.00001000000000000000000000000000, -2.798383782691386556312530,
      0.2781771470121340848515355, -2.692636823461273542810424},
     {0.00001000000000000000000000000000, -2.692616823461273542810424,
      0.2781771470121340848515355, -0.7877372343419659419495382},
     ≪77≫, {1.285101306170283296935922, 3.141602653589793238462643,
      1.542099567404339956323107, 3.484781524488199920612757}}
```

Let us visualize the sectors together with the branch points by coloring each sector with a randomly chosen color.

```
In[13]:= sectorPolygon[{r1_, φ1_, r2_, φ2_}] :=
      With[{pp = 12}, Polygon[
        Join[Table[r1 {Cos[φ], Sin[φ]}, {φ, φ1, φ2, (φ2 - φ1)/pp}],
            Table[r {Cos[φ2], Sin[φ2]}, {r, r1, r2, (r2 - r1)/pp}],
            Table[r2 {Cos[φ], Sin[φ]}, {φ, φ2, φ1, (φ1 - φ2)/pp}],
            Table[r {Cos[φ1], Sin[φ1]}, {r, r2, r1, (r1 - r2)/pp}]]]]
```

```
In[14]:= Show[Graphics[{{Hue[Random[]], sectorPolygon[#1]}& /@ s1,
                    PointSize[0.02], Point[{Re[#1], Im[#1]}]& /@
                                    potentialBranchPointsExample}],
          PlotRange -> All, Frame -> True, AspectRatio -> Automatic]
```

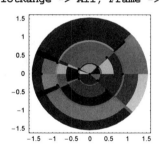

3) Now, let us derive a differential equation that is equivalent to the polynomial under consideration.

The solution of the differential equation we are looking for should reproduce the root of the original polynomial. If we integrate the differential equation and avoid branch points along the integration path, this solution will be a smooth function representing the original root. We will derive a first-order nonlinear differential equation. (We could also generate higher order linear differential equations with polynomial coefficients [see [1402] for details], but for higher order equations, constructing the initial conditions is no longer so easy—for the first-order one just has to solve numerically the polynomial under consideration for a given z with respect to w.)

If we view w as a function of z, we can differentiate the polynomial with respect to z and solve the resulting linear equation in $w'(z)$.

```
In[15]:= ode[poly_, {w_, z_}] :=
        Equal @@ Solve[D[poly /. w -> w[z], z] == 0, w'[z]][[1, 1]]
```

The nonlinear first-order differential equation we get this way is the following for our example polynomial.

```
In[16]:= odepExample = ode[pExample, {w, z}]
```

Out[16]= $w'[z] == \dfrac{-3 z^2 + 4 z^3 + 10 z^4 - 4 z w[z] - w[z]^2 + 2 w[z]^4}{2 z^2 + 2 z w[z] - 8 z w[z]^3 - 5 w[z]^4}$

Later, we need to solve this differential equation in azimuthal and radial directions in a polar coordinate system. Transforming the independent variable from z to φ, we get the differential equation in azimuthal direction.

```
In[17]:= odeφ[ode_, {w_, z_}, {ψ_, φ_}, r_] :=
            ode //. {(w'[z] == rhs_) -> (ψ'[φ] == rhs D[r Exp[I φ], φ]),
                     w[z] -> ψ[φ], z -> r Exp[I φ]}
```

For our example polynomial, we get the following differential equation.

```
In[18]:= odeφExample = odeφ[odepExample, {w, z}, {ψ, φ}, r]
```

Out[18]= $\psi'[\varphi] == \dfrac{i\, e^{i \varphi}\, r\, (-3\, e^{2 i \varphi}\, r^2 + 4\, e^{3 i \varphi}\, r^3 + 10\, e^{4 i \varphi}\, r^4 - 4\, e^{i \varphi}\, r\, \psi[\varphi] - \psi[\varphi]^2 + 2\, \psi[\varphi]^4)}{2\, e^{2 i \varphi}\, r^2 + 2\, e^{i \varphi}\, r\, \psi[\varphi] - 8\, e^{i \varphi}\, r\, \psi[\varphi]^3 - 5\, \psi[\varphi]^4}$

Transforming the independent variable from z to r, we get the differential equation in radial direction d.

```
In[19]:= oder[ode_, {w_, z_}, {ψ_, ρ_}, d_] :=
            ode //. {(w'[z] == rhs_) -> (ψ'[ρ] == rhs D[ρ d, ρ]),
                     w[z] -> ψ[ρ], z -> d ρ}
```

For our example polynomial, we get the following differential equation.

```
In[20]:= oderExample = oder[odepExample, {w, z}, {ψ, ρ}, d]
```

Out[20]= $\psi'[\rho] == \dfrac{d\, (-3\, d^2\, \rho^2 + 4\, d^3\, \rho^3 + 10\, d^4\, \rho^4 - 4\, d\, \rho\, \psi[\rho] - \psi[\rho]^2 + 2\, \psi[\rho]^4)}{2\, d^2\, \rho^2 + 2\, d\, \rho\, \psi[\rho] - 8\, d\, \rho\, \psi[\rho]^3 - 5\, \psi[\rho]^4}$

We will not carry out the points 4), 5), and 6) sequentially, but for each sector at once. The advantage of this approach is that we do not have to store a large number of `InterpolatingFunction`-objects (which can be quite memory intensive).

For the numerical solution of the obtained differential equations, we set some options in `NDSolve`. (The following option settings will be suitable for the examples treated in the following—more complicated polynomials might need larger values of the maximum number of steps as well of the precision/accuracy goals.)

```
In[21]:= SetOptions[NDSolve, PrecisionGoal -> 12, AccuracyGoal -> 12,
                    MaxSteps -> 10000];
```

The function `patch` calculates polygons of all sheets of the polynomial *poly* using *ppφ* azimuthal and *ppr* radial plot points in a given sector. (We give the radial as well as the azimuthal differential equation as arguments to `patch`. Although they could be calculated from *poly*, later we have to call `patch` on every one of the sectors, and we would like to avoid recalculating the differential equation for every sector.)

```
In[22]:= patch[reIm_, {poly_, {w_, z_}},
              {odeφ_, {ψ_, φ_}, r_}, {oder_, {ψ_, ρ_}, dir_},
              {r1_, φ1_, r2_, φ2_}, {ppφ_, ppr_}] :=
        Module[{radialStartingValues, radialIFs, azimuthalIFs, points, ra, φa},
                (* solve poly == 0 to get initial values
                   for the differential equation *)
                radialStartingValues = w /.
                    Solve[poly == 0 /. {z -> r1 Exp[I φ1]}, w];
                (* solve the differential equation radially *)
                radialIFs = NDSolve[{oder /. dir -> Exp[I φ1],
                            ψ[r1] == #}, ψ, {ρ, r1, r2}][[1, 1, 2]]& /@
                                radialStartingValues;
                (* solve the differential equation azimuthally *)
                azimuthalIFs = Table[ra = r1 + i/ppr(r2 - r1);
                NDSolve[{odeφ /. r -> ra, ψ[φ1] == radialIFs[[j]][ra]},
```

```
              ψ, {φ, φ1, φ2}][[1, 1, 2]],
                        {j, Length[radialIFs]}, {i, 0, ppr}];
        (* calculate points for all sheets *)
        points = Table[ra = r1 + i/ppr(r2 - r1);
                       φa = φ1 + k/ppφ(φ2 - φ1);
                       {ra Cos[φa], ra Sin[φa],
                        reIm[azimuthalIFs[[j, i + 1]][φa]]},
                       {j, Length[radialIFs]},
                       {i, 0, ppr}, {k, 0, ppφ}];
        (* make polygons for all sheets *)
        Function[s, Apply[Polygon[Join[#1, Reverse[#2]]]]&,
          Transpose /@ Partition[Partition[#, 2, 1]& /@ s, 2, 1],
                                                  {2}]] /@ points]
```

Here, the six sheets inside three of the sectors of our example polynomial are calculated. Each of the three examples contains a clearly visible branch point.

```
In[23]:= Show[GraphicsArray[
            (Graphics3D[{EdgeForm[Thickness[0.002]],
              patch[Im, {pExample, {w, z}},
              {odeφExample, {ψ, φ}, r}, {oderExample, {ψ, ρ}, d},
              s1[[#1]], {8, 8}]},
              BoxRatios -> {1, 1, 2}, ViewPoint -> {2, 1, 0.6}]&) /@
                        {2, 10, 21}]]
```

For the whole Riemann surface, we want the mesh in the patches as uniform as possible, and at the same time, we want the branch points always near mesh points. The function `patchPlotPoints` calculates the number of radial and azimuthal points inside a given sector, provided the whole number of radial and azimuthal points should be around *ppr* and *ppφ*. (The actual number of plot points may slightly deviate from *ppr* and *ppφ*.)

```
In[24]:= patchPlotPoints[{r1_, φ1_, r2_, φ2_}, rMax_, {ppφ_, ppr_}] :=
    Max[#, 1]& /@ Round[{ppφ, ppr} {φ2 - φ1, r2 - r1}/{2Pi, rMax}]
```

Now, we have all of the pieces together to implement the function `RiemannSurface`. Its first argument *poly* is a bivariate polynomial in *w* and *z*, the second argument is the function to be shown (in most cases Re or Im), and the last argument is the number of radial and azimuthal plot points. The function `RiemannSurface` returns a list of polygons of all sheets.

```
In[25]:= RiemannSurface[poly_, {w_, z_}, reIm_, {ppφ_, ppr_}] :=
    Module[{s, rMax, odePoly, odePolyφ, odePolyr, ψ, φ, r, dir},
                (* subdivide the complex plane *)
                s = sectors[poly, {w, z}];
                rMax = Max[#[[3]]& /@ s];
                (* calculate differential equations *)
                odePoly = ode[poly, {w, z}];
```

```
odePolyφ = odeφ[odePoly, {w, z}, {ψ, φ}, r];
odePolyr = oder[odePoly, {w, z}, {ψ, ρ}, dir];
(* calculating the patches *)
Table[patch[reIm, {poly, {w, z}},
        {odePolyφ, {ψ, φ}, r}, {odePolyr, {ψ, ρ}, dir}, s[[i]],
        patchPlotPoints[s[[i]], rMax, {ppφ, ppr}]],
    {i, Length[s]}]]]
```

Here, one-half of the Riemann surface of pExample is shown. A cross-sectional view along the plane $y = 0$ nicely shows the branching at the four branch points along the real axis.

```
In[26]:= Show[Graphics3D[{EdgeForm[{Thickness[0.002], GrayLevel[0.8]}],
        SurfaceColor[Hue[0.35], Hue[0.35], 2.2],
        RiemannSurface[pExample, {w, z}, Im, {50, 36}]}],
    PlotRange -> {All, {0, 3/2}, All},
    ViewPoint -> {0.9, -3.3, 1.6}, Boxed -> False]
```

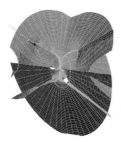

By using only one plot point in the azimuthal direction and replacing polygons with lines, we can make a picture showing cross sections along all directions that contain branch points. One nicely sees the splitting of the branches at the branch points (generically, one expects most branch points to be of the square-root type [1971]). Both the real as well as the imaginary parts are shown.

```
In[27]:= Show[GraphicsArray[(Function[l, Function[reIm,
            Graphics3D[{Thickness[0.002], Hue[0],
                        Map[If[Head[#1] === List,
            MapAt[reIm, #, 3], #]&, l, {-2}]}],
        ViewPoint -> {-2, -1, 0.7}, BoxRatios -> {1, 1, 2},
        PlotRange -> {All, All, {-3, 3}}]] /@
        {Re, Im}])[RiemannSurface[pExample, {w, z}, Identity,
        {1, 30}] /. Polygon[{p1_, p2_, p3_, p4_}] :>
        {Line[{p2, p3}], Line[{p4, p1}]}]]]
```

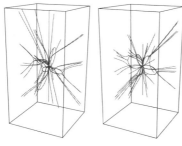

Now, let us use the function `RiemannSurface` to generate a few Riemann surfaces of algebraic functions. To better "see inside" the surface, we will replace all rectangles with "diamonds".

```
In[28]:= makeDiamond[Polygon[l_]] := Polygon[Apply[Plus,
                Partition[Append[l, First[l]], 2, 1], {1}]/2]
```

The first example is a quartic polynomial in *w*. We see four sheets.

```
In[29]:= pExample = -1 + 2 w^2 + 5 w^3 + 3 w^4 z + 3 z^2 + 8 z^5;

    Show[Graphics3D[{EdgeForm[Thickness[0.002]],
          SurfaceColor[Hue[0.12], Hue[0.22], 2.3],
          RiemannSurface[pExample, {w, z}, Im, {60, 30}]} /.
                            p_Polygon :> makeDiamond[p]],
        PlotRange -> {-3, 3}, BoxRatios -> {1, 1, 1.3},
        Boxed -> False]
```

The next example is also a quartic polynomial in *w*.

```
In[31]:= pExample = -1 - 4 w^3 - 3 z - w^4 z^2 - 4 z^3 + 3 w^2 z^5;

    Show[Graphics3D[{EdgeForm[{Hue[0], Thickness[0.002]}],
          SurfaceColor[Hue[0.12], Hue[0.12], 2.2],
          RiemannSurface[pExample, {w, z}, Im, {60, 30}]} /.
                        p_Polygon :> makeDiamond[p]],
        PlotRange -> {-4, 6}, BoxRatios -> {1, 1, 1.6},
        Boxed -> False]
```

Before the next degree seven polynomial, let us display the Riemann surfaces for the simple polynomials $w^2 - w - z = 0$, $w^3 - w - z = 0$, and $w^4 - w - z = 0$. We clearly see two, three, and four sheets in the pictures.

```
In[33]:= Show[GraphicsArray[Table[
    Show[Graphics3D[{EdgeForm[],
```

```
        SurfaceColor[Hue[0.12], Hue[0.8], 2.6],
        RiemannSurface[w^k + w + z, {w, z}, Im, {60, 30}]}] /.
                          p_Polygon :> makeDiamond[p],
      PlotRange -> All, BoxRatios -> {1, 1, 2},
      DisplayFunction -> Identity, ViewPoint -> {2, 2, 1}],
              {k, 2, 4}]]]
```

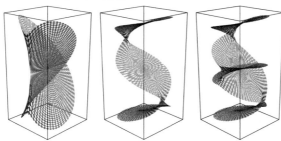

Here is the simplest form of a nontrivial septic: $w^7 - w - z = 0$. The seven sheets are nicely visible when viewed from the side.

```
In[34]:= pExample = w^7 - w - z;

    Show[Graphics3D[{EdgeForm[],
          SurfaceColor[Hue[0.1], Hue[0.1], 2.01],
          RiemannSurface[pExample, {w, z}, Im, {60, 30}]} /.
                          p_Polygon :> makeDiamond[p]],
        PlotRange -> Automatic, BoxRatios -> {1, 1, 2},
        ViewPoint -> {-4, 2, 2}, Boxed -> False]
```

Here is a quadratic with nine branch points (also called a hyperelliptic curve). We color the polygons according to their azimuthal position.

```
In[36]:= pExample = w^2 - z^9 + 1;

    col[Polygon[l_]] :=
    SurfaceColor[Hue[#], Hue[#], 2]&[
          (Pi + ArcTan @@ Take[Plus @@ l/4, 2])/(2 Pi)];

    Show[Graphics3D[{EdgeForm[],
          SurfaceColor[Hue[0.3], Hue[0.25], 0.5],
          RiemannSurface[pExample, {w, z}, Im, {60, 30}]} /.
                          p_Polygon :> {col[p], makeDiamond[p]}],
        PlotRange -> All, BoxRatios -> {1, 1, 0.8}, Boxed -> False]
```

The next picture shows a quintic.

```
In[39]:= pExample = w^5 - 1 + z^2 - z^3 + z^4 - z^5;

Show[Graphics3D[{EdgeForm[Thickness[0.002]],
        SurfaceColor[Hue[0.3], Hue[0.1], 2.4],
        RiemannSurface[pExample, {w, z}, Im, {60, 30}]} /.
                        p_Polygon :> makeDiamond[p]],
    PlotRange -> All, BoxRatios -> {1, 1, 1},
    ViewPoint -> {2, 2, 1}, Boxed -> False]
```

The last picture also shows a quintic, but a much more complicated one. This time, we also have poles.

```
In[41]:= pExample = 4 - 2 w^4 - 4 w^5 + 5 z + 2 w^4 z + 2 w^5 z - z^2 -
            4 w z^2 - w^2 z^2 - z^3 + 4 w^2 z^3 +  4 w^3 z^3 +
            2 z^4 - 4 w z^4 - 4 w^2 z^4 + 4 w^4 z^4 + 3 z^5 +
            2 w^4 z^5 - 5 w^5 z^5;

Show[Graphics3D[{EdgeForm[Thickness[0.002]],
        SurfaceColor[Hue[0.2], Hue[0.2], 2.3],
        RiemannSurface[pExample, {w, z}, Im, {60, 30}]} /.
                        p_Polygon :> makeDiamond[p]],
    PlotRange -> {-3, 3}, BoxRatios -> {1, 1, 1}, Boxed -> False]
```

The interested reader can continue to visualize many more Riemann surfaces. For a larger selection of surfaces (including compactified versions over the Riemann sphere) produced along the procedure outlined above, see [1790]. For Riemann surfaces of nonalgebraic functions, see [1791], [1792], and [1797]. For some pictures of physical models of Riemann surfaces, see [527], [604], and [433].

Exercises

1.^{L1} Logistic Map, Iterations with Noise

a) In connection with the study of bifurcations, it is interesting to investigate the result of iterating the mapping f: $z \to z^2 - c$. We call z^* a period n fixed point ([1867], [1330] and [1302]), provided the nth iteration of f satisfies $f_n(z^*) = z^*$ where $f_n(z)$ is defined recursively by $f_n(z) = f(f_{n-1}(z))$, $f_1(z) = f(z)$.

Determine graphically the dependence of the (complex) period 1, 2, 3, and 4 fixed points on the real value c.

b) Visualize the real period 1, ..., 7 fixed points of the map $x \to x^2 - c$ in the real c,x-plane [653], [1664], [926], [841].

c) Consider the sequence $\{x_0, x_1, x_2, ...\}$ where $x_n = f(x_{n-1})$ and $f(x) = \exp((x - 1/2)^2/(3/10))$.

What happens with the sequences as a function of $x_0 \in \mathbb{R}^+$ when we disturb the iterations and use $x_n = f(x_{n-1} + \xi_n)$ (ξ_n is a uniformly distributed random quantity from $(-1, 1)$) [1774], [1776], [1775]?

2.^{L2} Analytic Functions with Finite Domains of Analyticity, $\cos_q(z)$ and $\sin_q(z)$

a) Visualize the function

$$f(z) = \sum_{k=1}^{\infty} z^{k!}.$$

Visualize how the function behaves when approaching the unit circle.

b) The q-trigonometric functions $\cos_q(z)$ and $\sin_q(z)$ are defined by the following series [1734], [1894], [1895], [1733], [561], [705], [1040], [237], [1172], [1628], [1381], [241], [1562], [1563]:

$$\cos_q(z) = \sum_{k=0}^{\infty} (-1)^k \, \frac{z^{2k}}{[2k]_q!} \, \frac{1}{q^k(q - 1/q)^{2k}}$$

$$\sin_q(z) = \sum_{k=0}^{\infty} (-1)^k \, \frac{z^{2k+1}}{[2k+1]_q!} \, \frac{q^{k+1}}{(q - 1/q)^{2k+1}}.$$

Here $[n]_q!$ is the q factorial defined as

$$[n]_q \mathrel{!=} \prod_{k=1}^{n} \frac{q^k - q^{-k}}{q - q^{-1}}.$$

The q-trigonometric functions $\cos_q(z)$ and $\sin_q(z)$ obey the following identities similar to the identities for $\cos(z)$ and $\sin(z)$:

$$\frac{1}{z}\left(\sin_q(z) - \sin_q(q^{-2}z)\right) = \cos_q(z)$$

$$\frac{1}{z}\left(\cos_q(z) - \cos_q(q^{-2}z)\right) = -q^{-2}\cos_q(z)$$

$$\cos_q(z)\cos_q(qz) + q^{-1}\sin_q(z)\sin_q(q^{-1}z) = 1.$$

Implement the q-trigonometric functions $\cos_q(z)$ and $\sin_q(z)$ numerically for $|z| < \pi$, $|q| < \pi$. Check the above-mentioned identities for some randomly chosen numerical values of z and q to at least 100 digits. For a fixed z, visualize $\cos_q(z)$ as a function of q in the complex q-plane.

3.^{L2} NDSolve Fails, Integro-Differential Equation, Franel Identity, Bloch Particle in an Oscillating Field, Coupled and Accelerated Particles

a) Find an ordinary differential equation of first order that NDSolve is not able to solve.

b) Find a numerical approximation for the solution of the following integro-differential equation in the region $0 \le x \le 3\pi$:

$$y'(x) = \sin(x) + \frac{1}{\left(\arctan\left(1 + \left(\int_0^x y(t)\,dt\right)^2 + \int_0^x y(t)\,dt\right)\right)}.$$

c) Numerically determine the "exact value" of the constant α in the following identity:

$$\frac{1}{ab}\int_0^{ab}\left(\left\{\frac{x}{a}\right\} - \frac{1}{2}\right)\left(\left\{\frac{x}{b}\right\} - \frac{1}{2}\right)dx = \alpha\,\frac{\gcd(a,b)}{\mathrm{lcm}(a,b)}.$$

Here, a and b are positive integers and $\{x\}$ denotes the fractional part.

d) Calculate a reliable phase-space trajectory (in the $x(t), x'(t)$-plane) for $0 \le t \le 200$ for the following differential equation: $x''(t) = -F + \sin(x(t) - \varepsilon\cos(\omega t))$, $x(0) = 0$, $p(0) = 10$ [708]. Use the following parameter values $\varepsilon = 3/2$, $\omega = 5/3$, $F = 13/100$.

e) For a system of n cyclically coupled, accelerated particles, described by the following system for their velocities $v_k(t)$

$$v_k'(t) = \frac{k-1}{n-1} + \varepsilon\left(\sin(v_{k+1}(t) - v_k(t)) + \sin(v_{k-1}(t) - v_k(t))\right), \quad k = 1, \ldots, n,$$

$$v_0(t) = v_n(t), \quad v_{n+1}(t) = v_1(t)$$

calculate the average velocity of the particles $\bar{v}_k$ ($\bar{v}_k \approx (v_k(T) - v_k(0))/T$ for large enough T) as a function of the coupling parameter ε [449], [551].

4.^{L2} Trick of James of Courtright, Hannay Angle, Harmonic Nonlinear Oscillator, Closed Orbits

a) Create a visualization of the Trick of James of Courtright with *Mathematica*. The trick is the following: Take a light object (e.g., a match box), and a heavier object (e.g., a set of keys), and tie them together with a string (of length about 1 m). Now, hang the string over a horizontal pencil so that the heavy object is as high as possible. Do this by holding the light object approximately at the height of the pencil ($\pm \approx 45°$, measured against the horizontal). What happens if we now release the light object? Does the heavy object fall to the floor?

Here is the author carrying out this interesting experiment.

b) A bead is moving frictionless around an ellipse with half axes $\{1, 2\}$. Assume the ellipse is rotated with angular velocity ω around an axis that is perpendicular to the ellipse and goes through the focus. Calculate numerically to 20 digits how much the position of the bead of the rotated ellipse deviates from the position of a bead on a nonrotating ellipse in the limit $\omega \to 0$ ([193], [325], [1229]).

c) All generic solutions $\mathbf{M}(t)$ of the following nonlinear matrix oscillator are periodic (with period $T = 2\pi/\omega$) [328] (harmonic nonlinear oscillators [614], [329], [307]).

$$\mathbf{M}''(t) = 2\,\omega^2\,\mathbf{M}(t) + 3\,i\,\omega\,\mathbf{M}'(t) + c\,\mathbf{M}(t).\mathbf{M}(t).\mathbf{M}(t)$$

Here $\omega \in \mathbb{R}$, $c \in \mathbb{C}$, and $\mathbf{M}(t)$ is a $d \times d$ matrix. Visualize some solutions of this matrix differential equation.

d) The parametrized Hamiltonian $\mathcal{H} = \sigma\,p^{\alpha_1}\,r^{\beta_1} + r^{\beta_2}\,p^{\alpha_2}$ [542], [543] can be used to smoothly interpolate between the Hamiltonians of a harmonic oscillator $p^2 + r^2$ and the Kepler problem $p^2 - 1/r$ (here $p = |\boldsymbol{p}|$ and $r = |\boldsymbol{r}|$ and units were chosen to eliminate all prefactors). For the 2D case, construct a parametrized interpolating Hamiltonian and the corresponding equations of motion and make an animation how a family of orbits changes as one transits from the harmonic oscillator to the Kepler problem.

5.¹² Eigenvalue Problems

Currently, `NDSolve` only solves initial-value problems. It does not work for eigenvalue problems, which arise frequently.

a) Using a shooting method (see, for instance, [1153]) (this means, try to find iteratively the initial conditions for the corresponding initial value problem) obtained by a combination of `NDSolve` and `FindRoot`, find the smallest eigenvalue of

$$-y''(x) + x^4\,y(x) = \lambda\,y(x)$$

(the quantum-mechanical steady state energy of an anharmonic oscillator). (For some more advanced methods to calculate energy values of anharmonic oscillators, see [1861], [1860], and [1185].)

b) Discretizing the eigenvalue problem [1663]

$$-y''(x) + f(x)\,y(x) - \lambda\,y(x) = 0,\ y(x_0) = y(x_{n+1}) = 0$$

using $n + 2$ points $x_0, x_1, \ldots, x_n, x_{n+1}$, we get the following system of equations:

$$-\frac{1}{\Delta^2}(y_{i-1} - 2\,y_i + y_{i+1}) + f(x_i)\,y_i - \lambda\,y_i = 0$$

$$\Delta = \frac{x_{n+1} - x_0}{n+1}$$

$$x_i = x_0 + i\,\Delta, \ i = 1, \ldots, n+1$$

$$y_i = y(x_i), \ y_0 = y_{n+1} = 0.$$

This linear system in the y_i is solvable, provided that the coefficient determinant D_n of the y_i vanishes: $D_n = 0$. The direct computation of this determinant from the coefficient matrix is inefficient for large values of n, because the number of nonvanishing elements in the matrix grows like n, but the total number of elements grows like n^2. The determinant of this tridiagonal symmetric system can be computed easily by recursion [1819], [1207], [1054], [868], [549], [1921], [1851], [1278], [550] (for simplicity, we multiply by Δ^2):

$$D_1 = w_1$$

$$D_2 = w_1\,w_2 - 1$$

$$D_i = w_i\,D_{i-1} - D_{i-2}$$

$$w_i = 2 + (f(x_i) - \lambda)\,\Delta^2.$$

Implement the approximate computation of the eigenvalues of the above differential operator for arbitrary, nonsingular f, and moderate n by determining the zeros of D_n. Use the program to find the eigenvalues for some examples.

c) The eigenvalue problem

$$-y''(x) + V(x)\,y(x) = \lambda\,y(x), \quad y(-\infty) = y(\infty) = 0$$

for $V(x) = x^2 + \sum_{k=0}^{o} c_k\,x^k$ can be solved by solving the finite eigenvalue problem for the matrix $(h_{i,j})_{i,j=-n,\ldots,n}$ [891]:

$$h_{i,j} = \delta_{i,j-2}(1 - \omega_0^2)\frac{\sqrt{j(j-1)}}{2\,\omega_0} + \delta_{i,j+2}(1 - \omega_0^2)\frac{\sqrt{(j+1)(j+2)}}{2\,\omega_0} + \delta_{i,j}\left(\frac{j(\omega_0^2 + 1)}{\omega_0} + \frac{\omega_0^2 + 1}{2\,\omega_0}\right) +$$

$$\sum_{m=0}^{o}\sum_{k=0}^{\lfloor m/2 \rfloor}\sum_{n=0}^{m-2k}\delta_{i,j-2k+m-2n}\frac{c_m\,m!\,\sqrt{j!\,(j-2k+m-2n)!}}{(2\,\omega_0)^{m/2}\,2^k\,k!\,n!\,(j-n)!\,(m-n-2k)!}$$

Here ω_0 is implicitly defined through (optimized harmonic oscillator basis [897])

$$\sum_{\substack{m=2 \\ \Delta m=2}}^{o}\frac{c_m\,m!}{2^{m-1}\,\omega_0^{m/2+1}\,(m/2-1)!} = 1 - \frac{1}{\omega_0^2}.$$

Use these formulas to calculate the two lowest eigenvalues for $V(x) = x^4 - 4\,x^2$ to 10 correct digits.

d) In a $2\,n + 1$-dimensional representation of the Weyl system (Schwinger representation [1862], [648], [1567], [767], [1863], [473]) where differentiation operators are represented as $\mathcal{F}_{2n+1}^{-1}\,X\,\mathcal{F}_{2n+1}$ (X being the multiplication operator and $\mathcal{F}_{2n+1}$ the discrete Fourier transform of order $2\,n+1$), the approximate eigenvalues of the differential operator $\frac{\partial^2}{\partial x^2} + V(x)$ ($V(x)$ being sufficiently smooth) are given by the eigenvalues of the matrix $\mathcal{M} = (m_{k,l})_{k,l=-n,\ldots,n}$ [507], [508], [1625]:

$$m_{k,l} = \begin{cases} \frac{(-1)^{l-k}}{v} \, \pi \cot\left(\frac{(l-k)\pi}{v}\right) \csc\left(\frac{(l-k)\pi}{v}\right) & k \neq l \\ \frac{(v^2-1)}{6v} \, \pi + V\left(k \, \sqrt{\frac{2\pi}{2v+1}}\right) & k = l. \end{cases}$$

For $V(x) = x^2$ and $n = 10, 50$ calculate the first few eigenvalues and compare with the precision of the result with the case from the last subexercise where the differentiation operator was represented using a finite difference approach. Which n is needed to obtain 20 correct digits for the ground-state of the quartic oscillator $V(x) = x^4$?

e) For the time-independent Schrödinger equation $-\psi_j''(\alpha; x) + V(\alpha; x)\psi_j(\alpha; x) = \varepsilon_j(\alpha)\psi_j(\alpha; x)$ with the piecewise constant "Möbius potential"

$$V(\alpha; x) = \alpha \sum_{k=0}^{n} (2\,\mu(k) + 1)\,\theta(x - x_k)\,\theta(x_{k+1} - x),$$

where $\mu(k)$ is the value of the Möbius function $\mu(k)$ (in *Mathematica* MoebiusMu[k]) at the integer k, determine the dependence of the first ten eigenvalues in the range $0 \leq \alpha \leq 2$. $\psi_\alpha(x)$ fulfills Dirichlet boundary conditions $\psi_\alpha(\alpha; 0) = \psi_\alpha(\alpha; n+1) = 0$, and let $n = 50$. Determine for which α^* the lowest eigenvalue ε_1 coincides with α; this means $\varepsilon_1(\alpha^*) = \alpha^*$. Visualize the dependence of the $\psi_j(\alpha; x)$ on α. (For the physics of 1D nonperiodic, piecewise constant potentials, see [583], [1242], [1179], [360], [35], [642], [1051], [839], [1580], [1024], [305], [74], [1370], [898], [1275], [1736], [660], [1372], [480], [364], and [1757].)

f) Investigate the symmetric solutions of $-\psi_\varepsilon''(x) + V(x)\psi_\varepsilon(x) = \varepsilon\psi_\varepsilon(x)$ for the potential

$$V(x) = -\theta(|x|)\,\theta\left(\frac{\pi}{2} - |x|\right) - 4\,\theta\left(|x| - \frac{\pi}{2}\right)\theta(\pi - |x|) - \sum_{k=3}^{\infty} k^2\,\theta(|x| - (k-2)\,\pi)\,\theta((k-1)\,\pi - |x|).$$

Visualize the solutions for some ε. For which ε will the solutions be square integrable?

6.^{L2} Wynn's ϵ-Algorithm, Aitken Algorithm, Numerical Regularization

To illustrate how to find the infinite limiting value of a finite sequence, we consider Wynn's ϵ-algorithm and the Aitken algorithm (these are very near relatives of each other). The idea is to successively reduce the length of the sequence in such a way that the resulting terms approach the infinite limiting value [361], [1888], [1889].

a) We begin with a sequence ϵ_0, whose nth element is $\epsilon_0^{(n)}$. We define $\epsilon_{-1}^{(n)} = 0$ for all n. The (shortened) sequence at the kth step is in Wynn's ϵ-algorithm obtained by (see [1935], [1909], [743], [1646], [471], [1887], [905], [286] and [749] for details):

$$\epsilon_k^{(n)} = \epsilon_{k-2}^{(n+1)} + \frac{1}{\epsilon_{k-1}^{(n+1)} - \epsilon_{k-1}^{(n)}}.$$

(We suppose in the latter formula that this operation is well defined, i.e., there is no division by zero.) The results make sense only for even k. Implement this form of Wynn's ϵ-algorithm, and apply it to some test sequences.

b) Aitken transformation [30], [285], [207], and [1581] can be written in the following form:

$$\epsilon_k^{(1)} = \epsilon_0^{(k)}$$

$$\epsilon_k^{(n)} = \frac{\epsilon_{k-1}^{(n+1)} \, \epsilon_{k-1}^{(n-1)} - \left(\epsilon_{k-1}^{(n)}\right)^2}{\epsilon_{k-1}^{(n+1)} + \epsilon_{k-1}^{(n-1)} - 2\,\epsilon_{k-1}^{(n)}} \qquad n \geq 2.$$

Here again $\epsilon_k^{(n)}$ is the nth element of the kth transformation of the given sequence ϵ_0 with terms $\epsilon_0^{(k)}$. Write a functional program that carries out the Aitken's transformation as long as possible. Again, test the transformation for some examples.

c) Consider the following divergent sum [172]:

$$\lim_{n\to\infty} \sum_{k=1}^{n} (k^2 + k)\ln(k).$$

Use purely numerical techniques to calculate a five-digit approximation of Hadamard-regularized version of this sum (meaning that all divergencies of the form $n^k \ln^l(n)$, $k, l \in \mathbb{N}$ are subtracted).

7.¹² Scherk's Fifth Surface, Clebsch's Surface, Holed Dodecahedron, Smoothed Dodecahedron

a) Plot the surface defined implicitly by $\sin(z) = \sinh(x)\sinh(y)$ (Scherk's fifth surface) over a large domain of z-values. What peculiarities does it exhibit? Do *not* just use

```
ContourPlot3D[Sin[z] - Sinh[x] Sinh[y], {x, -3, 3}, {y, -3, 3}, {z, 0, 4Pi},
    Contours -> {0.0}, MaxRecursion -> 1,
    PlotPoints -> {{6, 5}, {6, 5}, {12, 5}}].
```

b) Make a pretty picture of the surface defined by the implicit equation

$$32 - 216\,x^2 + 648\,x^2\,y - 216\,y^2 - 216\,y^3 - 150\,z + 216\,x^2\,z + 216\,y^2\,z + 231\,z^2 - 113\,z^3 = 0$$

using ContourPlot3D and using the symmetry of the surface. Try to make the picture of this surface without using ContourPlot3D.

c) Given the following definitions (after loading the package Graphics`Polyhedra`):

```
polyWithPlatoSymmetry[plato_, {x_, y_, z_}, n_] :=
Plus @@ (({x, y, z}.#)^n & /@ ((Plus @@ #/Length[#]& /@
(First /@ (Polyhedron[plato][[1]])))))) - 1

dodePoly = Chop[polyWithPlatoSymmetry[Dodecahedron, {x, y, z}, 14]]

cylinderAxisDirections = #/Sqrt[#.#]& /@ Chop[Apply[Plus,
   First /@ Take[Polyhedron[Dodecahedron][[1]], 6], {1}]/5]

cylinderPoly = Times @@ (x^2 + y^2 + z^2 - ({x, y, z}.#)^2 - 0.01 & /@
                                          cylinderAxisDirections);

thePoly[{x_, y_, z_}] = dodePoly cylinderPoly + 3 10^-2;
```

What is the form of the surface implicitly defined by thePoly[{x, y, z}] == 0? Make a picture of the inner part of this surface without using ContourPlot3D.

d) Construct a thickened wireframe version of a dodecahedron made from triangles. The polygonal mesh forming the thickened wireframe should have at least 10^4 triangles. Smooth the thickened wireframe by replacing the coordinates of each point with the average of all neighboring points. Iterate this smoothing procedure 1000 times.

8.^{L1} A Convergent Sequence for π, Contracting Interval Map

a) Investigate the convergence of the sequence $1/\alpha_n$ to π, where

$$y_{n+1} = \frac{1 - \sqrt[4]{1 - y_n^4}}{1 + \sqrt[4]{1 - y_n^4}}$$

$$\alpha_{n+1} = (1 + y^{n+1})^4 \, \alpha_n - 2^{2n+3} \, y_{n+1} \, (1 + y_{n+1} + y_{n+1}^2)$$

$$y_0 = \sqrt{2} - 1$$

$$\alpha_0 = 6 - 4\sqrt{2}.$$

Do the same for the following sequence.

$$y_{n+1} = \frac{25 \, v_n^2}{(v_n^2 + u_n + v_n)^2 \, y_n}$$

$$u_n = \frac{5}{y_n} - 1$$

$$v_n = \sqrt[5]{\frac{u_n}{2} \left((u_n - 1)^2 + 7 + \sqrt{((u_n - 1)^2 + 7)^2 - 4 \, u_n^3} \right)}$$

$$\alpha_{n+1} = y_n^2 \, \alpha_n - \frac{5^n}{2} \left(y_n^2 - 5 - 2 \sqrt{y_n(y_n^2 - 2 \, y_n + 5)} \right)$$

$$y_0 = 5 \left(\sqrt{5} - 2 \right)$$

$$\alpha_0 = \frac{1}{2}.$$

For derivations of these iterations, see [114], [247], [864].

b) Consider the map $x \to \{\alpha x + \beta\}$ with $0 < \alpha, \beta < 1$ and $\{x\}$ denoting the fractional part of x [313]. Starting with the interval $[0, 1]$, visualize the repeated application of the map for various α, β.

9.^{L1} Standard Map, Stochastic Webs, Iterated Cubics, Hénon Map, Triangle Map

a) The so-called standard mapping is an iterative mapping of $(0, 1) \times (0, 1)$ into itself. It is defined by

$$x_{i+1} = \left(x_i + y_i - \frac{K}{2\pi} \sin(2\pi x_i) \right) \bmod 1$$

$$y_{i+1} = \left(y_i - \frac{K}{2\pi} \sin(2\pi x_i) \right) \bmod 1.$$

Depending on the choice of K and the starting point $\{x_0, y_0\}$, we get very different "movements" of the points $\{x_i, y_i\}$. For randomly chosen starting points, examine the movement for $0 \le K \le 3$.

b) A related mapping is

$$x_{i+1} = \sin(\alpha)\, y_i + \cos(\alpha)\, (x_i + K \sin(2\,\pi\, y_i))$$
$$y_{i+1} = \cos(\alpha)\, y_i + \sin(\alpha)\, (x_i + K \sin(2\,\pi\, y_i)).$$

For various choices of α (e.g., rational approximations of π with small denominators) and randomly chosen starting points, we get very interesting patterns, called stochastic webs [1958], [80], [22], [1957], [1952]. Examine some examples. In addition, make an animation that shows how an initially localized distribution diffuses into the observed structures from iterating points [350].

c) Given the following two maps [243]

$$\text{map 1}: \quad \begin{aligned} p_{i+1} &= (p_i + K \sin(\theta_i)) \bmod 2\,\pi \\ \theta_{i+1} &= \theta_i + p_{i+1} \end{aligned}$$

$$\text{map 2}: \quad \begin{aligned} p_{i+1} &= (p_i + K \sin(\theta_i)\, \mathrm{sign}(\cos(\theta_i))) \bmod 2\,\pi \\ \theta_{i+1} &= \theta_i + p_{i+1}. \end{aligned}$$

Visualize how the two mappings transform into each other by varying the parameter t $(0 \le t \le 1)$ in

$$p_{i+1} = (p_i + K\, ((1-t) \sin(\theta_i) + t \sin(\theta_i)\, \mathrm{sign}(\cos(\theta_i)))) \bmod 2\,\pi$$
$$\theta_{i+1} = \theta_i + p_{i+1}.$$

Choose K small, say, $K \approx 1/100$.

d) The map [1629]

$$x_{i+1} = \sin(\omega)\, (p_i - \mu\, \mathrm{sgn}(x_i)) + \cos(\omega)\, x_i$$
$$y_{i+1} = \cos(\omega)\, (p_i - \mu\, \mathrm{sgn}(x_i)) - \sin(\omega)\, x_i$$

exhibits a wide variety of possible phase space patterns. Find at least 10 pairs of values for the parameters $\{\omega, \mu\}$ that show "different" patterns.

e) The forces logistic map [455]

$$x_{i+1} = \alpha\, x_i(1 - x_i) + \varepsilon \sin(2\,\pi\, \vartheta_i)$$
$$\vartheta_{i+1} = (\vartheta_i + \omega) \bmod 1$$

with parameters ω, α, and ε exhibits a variety of possible patterns. Visualize some of the resulting patterns in the ϑ,x-plane.

f) The following implementation of the so-called web map [1498], [1959], [1960]

$$x_{n+1} = y_n$$
$$y_{n+1} = -x_n - \kappa \sin(y_n)$$

```
iList[κ_, {x0_, y0_}, n_] := First /@
     NestList[{#[[2]], -#[[1]] - κ Sin[#[[2]]]}&, {x0, y0}, n]

ListPlot[iList[2.178805237252476,
             {-2.9085181016961186, 0.8317263342785042}, 20000],
        Frame -> True, Axes -> False];
```

shows four nearly parallel "lines" for $5500 \lesssim n \lesssim 8500$. Are these nearly parallel "lines" numerical artifacts or are they real? If they are real, find other κ, $\{x_0, y_0\}$-values so that `ListPlot[iList[κ, {x₀, y₀}, n]]` shows regions of four parallel "lines".

g) The points in $\mathbb{R}^2$ obtained from the iteration of the bivariate polynomial map

$$\{x, y\} \longrightarrow \left\{ \sum_{k,l=0}^{n} c_{k,l}^{(x)} x^k y^l, \ \sum_{k,l=0}^{n} c_{k,l}^{(y)} x^k y^l \right\}$$

lead, for appropriately chosen $c_{k,l}^{(x,y)}$, to strange attractors [398], [1698], [602], [1697], [687]. For $n = 3, 4, 5$ and $-1 \le c_{k,l}^{(x,y)} \le 1$, find such strange attractors by carrying out automatic searches. For randomly chosen $c_{k,l}^{(x,y)}$, how frequently does one encounter a strange attractor?

h) Consider the following input (the map comes from [793]).

```
step[x_, {P_, e_}] :=
Module[{p = If[Random[] < P, 1/2 + e, 1 - (1/2 + e)],
        η = If[Random[] < P, 1, 0]},
       η (1 - p)/(1 - (1 - p) x) + (1 - η) (1 - p)/(1 - p x)]

SeedRandom[82313041463260884204]
NestList[step[#, {0.47629739820406214, 0.40133038775291635}]&,
         0.01011450108382423, 10^4];
```

Is the result mathematically correct? Write a compiled function that correctly carries out this iteration.

i) Given two lists of length l of real numbers $\{u_i\}$, $\{v_i\}$, $i = 1, \ldots, l$, consider the mapping [230], [372]

$$u_i^{(n+1)} = f_r\big(u_i^{(n)}\big) + \frac{\varepsilon}{2} \Big(f_r\big(u_{i-1}^{(n)}\big) - 2 f_r\big(u_i^{(n)}\big) + f_r\big(u_{i+1}^{(n)}\big) \Big) + v_i^{(n)}$$
$$v_i^{(n+1)} = b\big(u_i^{(n+1)} - u_i^{(n)}\big).$$

Here the function f is defined as

$$f_r(x) = \begin{cases} r\,x & \text{if } x \le \frac{1}{2} \\ r(1-x) & \text{if } \frac{1}{2} \le x \le 1 \\ x & \text{if } x > 1 \end{cases}$$

and r, ε, and b are real parameters. Find triples of values $\{r, \varepsilon, b\}$ such that the data $u_i^{(n)}$ show qualitatively different behavior over the discrete i,n-plane. Use periodic boundary conditions $u_i^{(n)} = u_{i+l}^{(n)}$, $v_i^{(n)} = v_{i+l}^{(n)}$ and random initial $u_i^{(0)}$ and $v_i^{(0)}$.

j) The iterations of the Hénon map [821], [1551]

$$\begin{aligned} x_{i+1} &= A + B\, y_i - x_i^2 \\ y_{i+1} &= x_i \end{aligned}$$

exhibit a variety of possible patterns. Show various "types" of patterns that arise from starting with $\{x_0, y_0\}$ located on a circle. Conduct an automated search for "interesting" sets of parameters A and B and initial circles.

k) The map $z \to z^2 - (1 - i\lambda)\bar{z}$, where λ is the largest positive root of $3\lambda^2 (9\lambda^4 + 3\lambda^2 + 11) = 77$, has in the complex plane the line $\{-(3\lambda^2 + 1)/4 + i\lambda/2, (9\lambda^4 + 6\lambda^2 + 5)/16 + i\lambda/2\}$ and its two images under counter-clockwise rotation by $2\pi/3$ and $4\pi/3$ as attractors [38]. The basins of attraction of these three lines are intermingled. Sketch these intermingled basins. Carry out the calculations using machine and high-precision arithmetic.

l) Consider the map [34]

$$x_m(s) = \beta_0 + \sum_{j=1}^{n} \beta_j \tanh\left(s\,\omega_j + \sum_{k=1}^{d} \omega_{jk}\, x_{m-k}(s)\right).$$

For each map, β_0 is a uniformly distributed variable from [0, 1], the β_j are rescaled (to $\sum_{j=1}^{n} \beta_j^2 = n$) uniformly distributed variable from [0, 1], and the ω_j and ω_{jk} are independently and identically distributed Gaussian random variables with mean 0 and variance 1. The d initial values $x_1, x_2, \ldots, x_d$ are assumed to be uniformly distributed variable from $[-1, 1]$.

For $n \approx d \approx 2^6$ and a concrete realization of the random variables, make an animation showing how the set of points $\{x_{m-1}(s), x_m(s)\}$ evolve as a function of the real parameter s.

10.^{12} Movement in a 2D Periodic and Egg Crate Potential, Pearcey Integral, Subdivision of Surface, Construction of Part of Surface, Electric Field Visualizations

a) Suppose an electrically charged particle moves in a periodic potential of the form

$$V(x, y) = V_m \cos(x) + V_m \cos(y).$$

In addition, suppose we impose a static homogeneous electric field in the x-direction, and a static homogeneous magnetic field in the z-direction. Compute some resulting paths of the particle, and plot them.

b) Suppose a particle moves in a periodic potential of the form [670], [1003], [1960]

$$V(x, y) = \frac{1}{2}\,(5 + 3\,(\cos(x) + \cos(y)) + \cos(x)\cos(y)).$$

Show some "qualitatively different" trajectories. How sensitive are these trajectories qualitatively and quantitatively depending on the initial conditions?

c) Make a contour plot of the absolute value and the phase of the following function $P(x, y)$ (Pearcey integral)

$$P(x, y) = \int_{-\infty}^{+\infty} e^{i(u^4 + x\,u^2 + y\,u)}\, du$$

for real values of $\{x, y\}$ around $\{0, 0\}$. $P(x, y)$ can be effectively computed by solving the following two ordinary differential equations:

$$4\,\frac{\partial^2}{\partial x^2}\,P(x, 0) + 2\,i\,x\,\frac{\partial}{\partial x}\,P(x, 0) + i\,P(x, 0) \;= 0$$

$$P(x, 0)|_{x=0} \;=\; \frac{(-1)^{(1/8)}}{2}\,\Gamma(1/4)$$

$$\frac{\partial}{\partial x}\,P(x, 0)\bigg|_{x=0} \;=\; \frac{i\,(-1)^{(3/8)}}{2}\,\Gamma(3/4)$$

$$4 \frac{\partial^3}{\partial y^3} P(x, y) - 2x \frac{\partial^2}{\partial y^2} P(x, y) - i y P(x, y) \quad = 0$$

$$P(x, y)|_{y=0} \quad = P(x, 0)$$

$$\frac{\partial}{\partial y} P(x, y)\bigg|_{y=0} \quad = 0$$

$$\frac{\partial^2}{\partial y^2} P(x, y)\bigg|_{y=0} \quad = i \frac{\partial}{\partial x} P(x, 0).$$

(see [1282], [1422], [1362], [194], [427], [425], [1060], [426], [1704], [933], [192], [428], and [1410] for details).

d) Given the parametrically defined surface (the Klein bottle discussed in Chapter 2 of the Graphics volume [1794])

$$x(s, t) = (2 + \cos(s/2) \sin(t) - \sin(s/2) \sin(2t)) \cos(s)$$
$$y(s, t) = (2 + \cos(s/2) \sin(t) - \sin(s/2) \sin(2t)) \sin(s)$$
$$z(s, t) = \sin(s/2) \sin(t) + \cos(s/2) \sin(2t)$$

divide the s,t-plane in n_s, n_t parts, so that the corresponding pieces of the surface all have the same area.

e) Make a picture of the part of the surface that is visible in the following picture (do not use the `PlotRange -> ...` restriction).

```
r[φ_, z_] = (1 + 2 Sin[z/Pi]^2 Cos[2φ]^6);

ParametricPlot3D[Evaluate[{r[φ, z] Cos[φ], r[φ, z] Sin[φ], z,
{EdgeForm[Thickness[0.002]],
 SurfaceColor[RGBColor[0, 1, 0], RGBColor[1, 1, 0], 2.6]}}],
            {φ, 0, 2Pi}, {z, 0, Pi},
            PlotPoints -> {36, 46}, BoxRatios -> {1, 1, 2},
            PlotRange -> {{-5/4, 5/4}, {-5/4, 5/4}, {0, Pi}},
            Boxed -> False, Axes -> False];
```

f) Given a random trigonometric function (such as

```
Sum[Random[] Cos[i x + 2 Pi Random[]] Cos[j y + 2 Pi Random[]],
 {i, 0, n}, {j, 0, n}])
```

calculate curves that start and end at extremal points and follow the gradient.

g) Make an animation that shows the equipotential lines of two superimposed finite square (hexagonal) grids made from line segments. Let the angle the two grids form be the animation parameter. Try for an efficient implementation.

11.[L2] A Ruler on the Fingers, Trajectories in Highly Oscillating Potential, Branched Flows

a) Describe the following experiment via a numerical solution of Newton's equation of motion. Hold a ruler horizontally on your index fingers (about 20 to 40 cm apart). Now move the fingers horizontally toward each other at a nearly uniform velocity. This causes the ruler to move alternately left and right. In computing the solution of this problem, be especially careful where the direction of movement changes. What would be the result if there were no static friction?

b) Calculate and visualize 100 trajectories (with randomly chosen initial conditions) of a particle moving in the 2D potential $V(x, y) = \sin(\cot(y - x^2) + \tan(x + y))$.

c) Visualize the motion of many particles moving under the simultaneous influence of a smooth, random potential and a monotonic potential.

12.^{L1} Maxwell's Line, Quartic Determinant

a) Plot the isothermals of a van der Waal's gas on a p,V-diagram (compute Maxwell's line). For details on the construction of Maxwell's line, see, for example, [853], [1417], [301], [1073], [1871], [79], and [640]. For a general treatment of van der Waal's gas, see [1814], [928], [816], [817], and [4].

b) Calculate realizations for $a_{i,j}$, $b_{i,j}$, and $c_{i,j}$ such that the following determinantal identity holds [533]:

$$\begin{vmatrix} 0 & a_{1,2}\,x + b_{1,2}\,y + c_{1,2}\,z & a_{1,3}\,x + b_{1,3}\,y + c_{1,3}\,z & a_{1,4}\,x + b_{1,4}\,y + c_{1,4}\,z \\ a_{1,2}\,x + b_{1,2}\,y + c_{1,2}\,z & 0 & a_{2,3}\,x + b_{2,3}\,y + c_{2,3}\,z & a_{2,4}\,x + b_{2,4}\,y + c_{2,4}\,z \\ a_{1,3}\,x + b_{1,3}\,y + c_{1,3}\,z & a_{2,3}\,x + b_{2,3}\,y + c_{2,3}\,z & 0 & a_{3,4}\,x + b_{3,4}\,y + c_{3,4}\,z \\ a_{1,4}\,x + b_{1,4}\,y + c_{1,4}\,z & a_{2,4}\,x + b_{2,4}\,y + c_{2,4}\,z & a_{3,4}\,x + b_{3,4}\,y + c_{3,4}\,z & 0 \end{vmatrix} = $$

$$2\,i\,(x^4 + y^4 + z^4).$$

It is known that some of the $a_{i,j}$, $b_{i,j}$, and $c_{i,j}$ are ± 1.

13.^{L2} Smoothing Functions, Secant Method Iterations, Unit Sphere Inside a Unit Cube

a) Plot the function

$$h_\epsilon(x) = \int_{-\infty}^{\infty} g_\epsilon(y)\, f(x - y)\, dy$$

with

$$g_\epsilon(x) = \begin{cases} \exp(1/(x^2 - \epsilon^2))/\int_{-\epsilon}^{\epsilon} \exp(1/(x^2 - \epsilon^2))\, dx & -\epsilon \le x \le \epsilon \\ 0 & \text{otherwise} \end{cases}$$

$$f(x) = \begin{cases} \sin(x) & 0 \le x \le 2\pi \\ 0 & \text{otherwise} \end{cases}$$

for $\epsilon = 4, 2, 1, 1/2$, $0 \le x \le 2\pi$. Be careful that no error messages are generated during the computation.

b) How does the pattern in the following graphic arise?

```
DensityPlot[x /. FindRoot[Cos[x] - x, {x, x1, x2}, Method -> Secant,
            MaxIterations -> 30], {x1, -10, 10}, {x2, -10, 10},
            Compiled -> False, PlotPoints -> 200,
            Mesh -> False, ColorFunction -> (Hue[0.8 #]&)]
```

c) The part $a_d(\rho)$ of a dD sphere of radius ρ that is inside a dD unit cube centered at the origin (and, without loss of generality, oriented along the coordinate axes) is given by

$$a_d(\rho) = \int_{-1/2}^{1/2} \int_{-1/2}^{1/2} \cdots \int_{-1/2}^{1/2} \delta\left(\rho - \sqrt{x_1^2 + x_2^2 + \ldots + x_d^2}\right) dx_1\, dx_2 \ldots dx_d.$$

Integrating the Dirac delta function gives the following recursion relation for the area [1028]

$$a_d(\rho) = 2\,\rho \int_{\max\left(0,(\rho^2-1/4)^{1/2}\right)}^{\min\left(\rho,(d-1)^{1/2}/2\right)} \frac{a_{d-1}(r)}{\sqrt{\rho^2 - r^2}}\, dr.$$

Starting with $a_1(\rho) = 2\,\theta(\rho)\,\theta(1/2 - \rho)$, calculate numerically and visualize $a_d(\rho)$ for $1 \le d \le 25$.

14.^{L1} Computation of Determinants, Numerical Integration, Binary Trees, Matrix Eigenvalues

a) Implement the computation of determinants using Laplace expansion by minors [1192], and using this implementation, find the determinant of

```
Array[Which[#2 - 2 <= #1 <= #2 + 2, N[#1 + #2], True, 0] &, {5, 5}].
```

b) Decide purely numerically if the value of the following integral is $1/2$ [251], [252], [255], [279].

$$\int_0^\infty \left(\prod_{k=0}^{7} \frac{(2k+1)}{x} \, \sin\!\left(\frac{(2k+1)}{x}\right) \right) dx \overset{?}{=} \frac{1}{2}.$$

c) Confirm to at least 10 digits that the integral $I = \int_0^\infty x \ln(x)\,(1 - y(x)^2)\, dx$ has the value $I = 1/4 + 7/12 \ln(2) - 3 \ln(G)$ where G is the Glaisher constant (in *Mathematica* `Glaisher`) [136], [1931]. Here $y(x)$ is the solution of the nonlinear differential equation (Painlevé 3)

$$y''(x) = y(x)^3 + \frac{y'(x)^2}{y(x)} - \frac{y'(x)}{x} - \frac{1}{y(x)}$$

obeying the boundary conditions $y(x) \underset{x\to 0}{\approx} -x\,(\ln(x/4) + \gamma)$ and $y(x) \underset{x\to\infty}{\approx} 1 - (\pi x)^{-1/2} \exp(-2\,x)$.

d) Generate "random binary trees" (meaning start with a binary tree of fixed depth and then randomly delete branches). Generate the corresponding adjacency matrices (containing a 1 in position (i, j) if the node i of the random binary tree is connected to the node j and 0 else). Analyze the eigenvalue distribution of the adjacency matrices for 100 random binary trees that were derived from full binary trees of depth eight.

e) Carry out some numerical experiments to conjecture an approximate formula describing the value of the kth eigenvalue (sorted by decreasing magnitude) $\lambda_k^{(n)}$ of an $n \times n$ left triangular matrix with elements 1. Consider the case $n \gg 1$.

f) The expectation value of the kth power of the determinant of a $n \times n$ matrix with entries ± 1 is [160]

$$\overline{\det\!\left((\pm 1)_{\substack{i=1,\ldots,n \\ j=1,\ldots,n}}\right)} = i^{k\,n} \left(\left(\det\!\left(\left(\frac{\partial}{\partial z_{i,j}}\right)_{\substack{i=1,\ldots,n \\ j=1,\ldots,n}} \right) \right)^{k} \prod_{i=1}^{n} \prod_{j=1}^{n} \cos(z_{i,j}) \right)\Bigg|_{z_{i,j}=0}.$$

where $(\det | \ldots |)^k$ represents the k-fold application of the differential operator obtained from taking the determinant and the overbar denotes averaging. For $1 \le k,\, n \le 4$, calculate the expectation values and compare them with the average of 10^6 random realizations of the matrices.

15.^{11} Root Pictures, Weierstrass Iterations, Taylor Series Remainders

a) In analogy with the Goffinet picture of Chapter 1 of the Graphics volume [1794], visualize all of the following sums:

$$\sum_{i=1}^{m} j_i x_i \quad i = 1, \ldots n,$$

where the x_i are the zeros of a given polynomial of nth degree and the j_i are independently taken from the possible values $0 \le j_i \le n$.

b) Construct n random polynomials $p_j(z)$ of degree j. Then, starting from a polynomial $q(z)$ of degree n, make the following substitution in $q(z)$: $z^j \to p_j(z)$. Iterate this process, and make a picture of the zeros of these iterated polynomials. (For details on this process with $p_j(z)$ as the orthogonal polynomials, see [875].)

c) The polynomial $x^2 + x + 1 = 0$ has two roots $x_{1,2}$. Adding a term εx^3 to this polynomial results in a third root x_3. For small ε, the new root x_3 will be much larger than the two starting roots $|x_3| \gg |x_{1,2}|$. Visualize how the "third root" depends on the (complex) parameter ε.

d) Implement a one-liner that finds all of the roots of a polynomial $p(x)$ of degree n via the following iteration (see [1087], [210], [974], [1638], [1337], [1346], [1731], [1439], [1936], [730], [941], [1845], [778], [860], [1433], [1435], [1940], [84], [1878], [780], [1777], [854], [1636], [1434], [1436], [1000], [1769], and [209], and for higher-order convergent methods using the same principle, see [869], [1404], [1435], [1440], and [83]).

$$x_j^{(i+1)} = x_j^{(i)} - \frac{p\left(x_j^{(i)}\right)}{\displaystyle\prod_{\substack{k=1 \\ k \ne j}}^{n} \left(x_j^{(i)} - x_k^{(i)}\right)}$$

$j = 1, \ldots, n$ where $x_j^{(i)}$ is the ith iterate of the jth root. The starting roots $x_j^{(0)}$ might be arbitrary complex numbers (for polynomials with only real coefficients having complex roots, some of the starting values must be complex numbers).

In Section 3.7 of the Programming volume [1793], some graphics showing fractals arising from iterating the Newton method were shown [1010], [312]. Can the Weierstrass iteration also produce fractals?

e) Take a polynomial and calculate its roots. Then, take these roots (in all possible combinations) as coefficients for new polynomials. Calculate their roots, and take them as coefficients for new polynomials, and so on. Iterate this process a few times and show graphically all of the roots in the complex plane.

f) Use a "random" parametrized polynomial $p_\tau(z)$ to make an animation of the convergence of the Newton iterations $z \to z - p_\tau(z) / p_\tau'(z)$ (the prime denotes differentiation with respect to z) as a function of the starting value z_0.

g) The Taylor series expansion of a function $f(x)$ around x_0 using the Lagrange form of the remainder is

$$f(x) = \sum_{k=0}^{n} f^{(k)}(x_0) \frac{(x - x_0)^k}{k!} + \frac{(x - x_0)^{n+1}}{(n+1)!} f^{(n+1)}(\tilde{x}).$$

where $x_0 \le \tilde{x} \le x$.

Choose random polynomials of degree d for $f(x)$. Fix x and make histograms that show the distribution of $\bar{x}$ as a function of n.

16.[12] Nodal Line of $\sin(24\,x)\sin(y) + 6/5\sin(x)\sin(24\,y) = 0$

a) The curve defined implicitly by $\sin(24\,x)\sin(y) + 6/5\sin(x)\sin(24\,y) = 0$ is free of self-intersections inside the region $0 < x < \pi$, $0 < y < \pi$. Use this fact to construct a very high-resolution contour plot analogous to

```
ContourPlot[Sin[24 x] Sin[y] + 6/5 Sin[x] Sin[24 y],
            {x, 0, Pi}, {y, 0, Pi}, Contours -> {0},
            PlotPoints -> {bigIntegerSay1000}];
```

without using `ContourPlot` (and without reprogramming `ContourPlot`) by solving differential equations for the nodal curve(s). (The function to be visualized is an eigenfunction of the Helmholtz operator on the square; in comparison to most examples treated in Solution 3 of Chapter 3 of the Graphics volume [1794], this times the weights of the two eigenfunctions (with the same eigenvalue) are slightly different, and, as a result, self-intersections of the nodal curve become atypical in this case; see [437], [1250], and [768].)

a) Consider the nonlinear Bloch equations [1834]

$$S_x'(t) = \kappa\, S_y(t)\, S_z(t)$$
$$S_y'(t) = S_z(t) + \kappa\, S_x(t)\, S_z(t)$$
$$S_z'(t) = -S_y(t).$$

Show how the solution curves $\{S_x(t), S_y(t), S_z(t)\}$ depend on κ. Choose random initial conditions on the unit sphere $S_x(0)^2 + S_y(0)^2 + S_z(0)^2 = 1$.

17.[11] Branch Cuts of an Elliptic Curve, Strange 4D Attractors

a) Make a picture of the branch cuts of the function

$$\sqrt{z \prod_{j=0}^{6}\left(z - \frac{1}{4}\,e^{\frac{2i\pi j}{7}}\right) \prod_{j=0}^{4}\left(z - \frac{1}{2}\,e^{\frac{2i\pi j}{5} + \frac{i\pi}{10}}\right) \prod_{j=0}^{2}\left(z - \frac{3}{4}\,e^{\frac{2i\pi j}{3} + \frac{i\pi}{6}}\right)}.$$

First, solve it by using `ContourPlot`. Second, solve a differential equation for the location of the branch cut and then display the result of solving the differential equation.

b) Find systems of coupled nonlinear ODEs of first order $x_i'(t) = p_i(x_1(t), x_2(t), x_3(t), x_4(t))$, $i = 1, 2, 3, 4$ whose solutions exhibit strange attractors in 4D. Use low-order polynomials for the p_i.

18.[11] Differently Colored Spikes, Billiard with Gravity

a) Color the various spikes in the following picture differently.

```
f[φ_, ϑ_] := Function[r, (Sign[#] Abs[#]^(5/3))& /@
             N[r {Cos[φ] Sin[ϑ], Sin[φ] Sin[ϑ], Cos[ϑ]}]][N @
             Abs[(3465/16(1 - Cos[ϑ]^2)^(3/2)*(221Cos[ϑ]^6 -
             195Cos[ϑ]^4 + 39Cos[ϑ]^2 - 1)) Cos[4φ]]]

ParametricPlot3D[f[φ, ϑ], {φ, 0, 2Pi}, {ϑ, 0, Pi},
                PlotPoints -> {41, 41}, Compiled -> False,
```

```
PlotRange -> All, BoxRatios -> {1, 1, 1},
Boxed -> False, Axes -> False];
```

b) Consider a point particle in 2D under the influence of gravity (acting downward) being repeatedly and ideally reflected from the curve

$$y(x) = -\sum_{k=-\infty}^{\infty} \theta\left(x - k + \frac{1}{2}\right) \theta\left(k + \frac{1}{2} - x\right)(-1)^k \sqrt{\frac{1}{4} - (x - k)^2}.$$

Visualize some qualitatively different trajectories for various initial conditions.

19.[L2] Schwarz–Riemann Minimal Surface, Jorge–Meeks Trinoid, Random Minimal Surfaces

a) Make a picture of the following Schwarz–Riemann minimal surface [1351], [1626], [616], [617], and [1303] (see also Subsection 1.5.2 of the Symbolics volume [1795]):

$$\{x(s, t), y(s, t), z(s, t)\} = \text{Re}\left(\int_0^{s+it} \frac{1}{\sqrt{1 - 14\,\omega^4 + \omega^8}} \{1 - \omega^2, i(1 + \omega^2), 2\,\omega^2\} d\omega\right).$$

The parameters s and t are from the region of the s,t-plane where the following four circles overlap: $(s \pm 2^{-1/2})^2 + (t \pm 2^{-1/2})^2 = 2$. Carry out all calculations numerically. The resulting surface can be smoothly continued by reflecting the surface across the lines that form its boundary. Carry out this continuation.

b) Make a picture of the following minimal surface (Jorge–Meeks trinoid) [132], [1555], [1171] and [1369] (see also Subsection 1.5.2 of the Symbolics volume [1795]):

$$\{x(s, t), y(s, t), z(s, t)\} = \text{Re}\left(\int_0^{s+it} \frac{1}{(\omega^3 - 1)^2} \{1 - \omega^4, i(1 + \omega^4), \omega^2\} d\omega\right).$$

$\omega = s + i\,t, s, t \in \mathbb{R}$.

Carry out all calculations numerically. The Jorge-Meeks trinoid has a threefold rotational symmetry and four-mirror symmetry plane. Make use of this symmetry in the construction. A region in the s,t-plane that generates one part of the surface is in polar coordinates given by $0 \le r < 1$, $0 \le \varphi \le \pi/3$.

c) For the following pairs of functions $f(\xi)$ and $g(\xi)$, calculate numerically the parts of the surfaces (for some suitable r_0) that are defined via

$$\{x(r, \varphi), y(r, \varphi), z(r, \varphi)\} = \text{Re}\left(\int_{r_0}^{r\exp(i\varphi)} \{(1 - g(\xi)^2), i(1 + g(\xi)^2)\, f(\xi), 2\, f(\xi)\, g(\xi)\} d\xi\right).$$

If the functions $f(\xi)$ and $g(\xi)$ have branch cuts, continue the functions to the next Riemann sheet to avoid discontinuities.

This is the list of function pairs:

$$f(\xi) = \xi - \frac{67}{78} \qquad\qquad g(\xi) = \left(\frac{1}{\xi} - 1\right)^2 + \xi^2 + 2\,\xi + \frac{44}{31}$$

$$f(\xi) = \xi + \frac{14}{27} + \frac{1}{\xi^2} \qquad\qquad g(\xi) = \frac{1}{\xi^{5/3}}$$

$$f(\xi) = \xi^3 + 2\xi + \left(\frac{1}{\xi^{4/3}} - 1\right)^2 + \frac{2}{3} \quad g(\xi) = 2\xi + \frac{112}{69}$$

$$f(\xi) = -2\xi^4 \qquad\qquad g(\xi) = \frac{\left(\frac{1}{\xi^{5/3}} - 1\right)^2 (\xi - 2)}{\xi^3} - 2$$

$$f(\xi) = -\frac{1173}{310}\xi^{-3} \qquad\qquad g(\xi) = \xi^3$$

$$f(\xi) = \left(2 + \frac{1}{\xi^6}\right)^2 \qquad\qquad g(\xi) = -\frac{28}{55}\xi(\xi^{5/3} + 1)^2$$

$$f(\xi) = \sqrt[3]{\xi + \frac{28}{9}} + 2 \qquad\qquad g(\xi) = \sinh\left(\frac{1}{\xi^{4/3}}\right)$$

$$f(\xi) = \frac{\xi - \frac{7}{12}}{\xi^3} \qquad\qquad g(\xi) = \frac{61}{56}\xi^{7/3}$$

$$f(\xi) = -\frac{\ln(2)}{\ln\left(\frac{1}{\xi^2 + 1}\right)} \qquad\qquad g(\xi) = -\frac{20}{17\xi^2}$$

$$f(\xi) = \frac{i\pi}{\ln((\xi^{5/3} + 1)^{-2})} \qquad\qquad g(\xi) = \left(2\xi + \frac{4}{5}\right)^{\frac{1}{\xi^2} - 1}$$

$$f(\xi) = \left(\frac{\ln^{\frac{32}{9}}(\xi)}{3000000} + 1\right)^{-2} \qquad\qquad g(\xi) = \frac{\xi^3}{\left(\frac{1}{4\xi^{4/3}} - 1\right)^2}.$$

20.[12] Precision Modeling, `GoldenRatio` Code from the Tour, Resistor Network

a) Model the following curve (as a function[al] of `f [x]`):

```
f[x_] := 2 - x - 5x^2 + 4x^3 + 3x^4 - 2x^5;

SetPrecision[Round, False];

ListPlot[Table[{x, Precision[f[SetPrecision[x, 25]]]}, {x, -5, 5, 10/1001}],
    PlotRange -> {{-5, 5}, {20, 30}},
        Frame -> True, PlotJoined -> True, Axes -> False];
```

b) Why do the following two definitions below from page 18 of *The Mathematica Book* [1919] really work and give as a result the value of `GoldenRatio` to *k* digits? Should not there be a small loss of precision in every step of the `FixedPoint` calculation?

```
g1[k_] := FixedPoint[N[Sqrt[1 + #], k] &, 1]
g2[k_] := 1 + FixedPoint[N[1/(1 + #), k] &, 1]
```

c) Predict the first few digits of the number calculated by the following sequence of inputs.

```
data = Table[If[Not[IntegerQ[n/10]],
 {n, Coefficient[Fit[Table[Plus @@ IntegerDigits[n^k], {k, 100}],
      {1, x}, x], x]}, Sequence @@ {}], {n, 1000}];

fit = Fit[data, {Log[10, n]}, n]
```

d) The resistance $R_{m,n}$ between the lattice point $\{0, 0\}$ and the lattice point $\{m, n\}$ of an infinite square lattice with unit resistors between lattice points obeys the following set of equations for nonnegative n, m [443], [444], [445], [1840]:

$$R_{m+2,m+2} = 4 \, \frac{m+1}{2\,m+3} \, R_{m+1,m+1} - \frac{2\,m+1}{2\,m+3} \, R_{m,m}$$

$$R_{n+2,n+1} = 2 \, R_{n+1,n+1} - R_{n+1,n}$$

$$R_{m+2,0} = -R_{m,0} + 4 \, R_{m+1,0} - 2 \, R_{m+1,1}$$

$$R_{m+1,n} = -R_{m,n} - R_{m+1,n-1} - R_{m+1,n+1} + 4 \, R_{m+1,n} \text{ if } 2 < n < m + 1$$

$$R_{n,m} = R_{m,n}$$

The initial conditions for the recursion are $R_{0,0} = 0$, $R_{1,0} = 1/2$, and $R_{1,1} = 2/\pi$.

Visualize $R_{n,m}$ for $0 \le n$, $m \le 200$. In which direction is the resistance (for a fixed distance) the largest? For large distances, the following asymptotic expansion holds:

$$R_{n,m} \xrightarrow[\sqrt{n^2+m^2} \to \infty]{} \frac{1}{\pi} \left(\log\!\left(\sqrt{m^2 + n^2} \right) + \gamma + \frac{\log(8)}{2} \right).$$

In which direction does this expansion hold best?

21.^{13} Auto-Compiling Functions, Card Game

a) Given some function definitions for a symbol f (such as $f[x\_, y\_] := ...$, $f[x\_, y\_, z\_] := ...$), implement a function to be called on f such that subsequent calls to f with specific numeric arguments generate and use compiled versions of the appropriate definitions. Calls to f with uncompilable arguments should use the original definitions for f.

b) Consider the following card game [175]: Two players each get n cards with unique values between 1 and $2\,n$. In each round, each player selects one card randomly from their pile. The player with the smaller card value wins both cards. The game ends when one player runs out of cards.

If possible, speed up the following implementation of the modeled game (A and B are the two initial lists of card values).

```
cardGameSteps1 = Compile[{{A, _Integer, 1}, {B, _Integer, 1}},
Module[{a = A, b = B, ra, rb, (* round counter *) c = 0},
      While[a != {} && b != {}, c++;
            (* select two random cards *)
            ra = Random[Integer, {1, Length[a]}];
            rb = Random[Integer, {1, Length[b]}];
            (* compare cards and add new card to one player;
              remove second card from other player *)
            If[a[[ra]] > b[[rb]],
               b = Append[b, a[[ra]]]; a[[ra]] = a[[-1]];
                                       a = Drop[a, -1],
               a = Append[a, b[[rb]]]; b[[rb]] = b[[-1]];
                                       b = Drop[b, -1]]];
      (* return number of rounds *) c]];
```

For $n = 10$ carry out 10^6 games and calculate the average length of the game.

22.^{L2} Path of Steepest Descent, Arclength of Fourier Sum, Minimum-Energy Charge Configuration

a) For the spiral minimum search problem from Section 1.9, find the minimum by following the path of steepest descent until one reaches the minimum.

b) Consider the partial sums $\sum_{k=1}^{n} \sin(k\,x)/k$ of the Fourier series of the function $f(x) = \pi/2 - x/2$. As $n \to \infty$, the arclength of the graph of the Fourier series diverges [1723], [1482]. How many terms does one have to take into account so that the arclength of the graph of the Fourier series is equal to or greater than twice the arclength of $f(x)$?

c) Consider $48\,n$ ($n = 1, 2, \ldots$) point charges on a sphere. Enforce the symmetry group of the cube (that has 48 elements) on the charges and find minimum energy configurations for the charge positions for small n. Compare results and timings for various method option settings. Calculate the minimum energy configuration for $n = 36$.

23.^{L2} N[*expr*, *prec*] Questions and Compile Questions

a) Might it happen that N[*expr*, *prec*] (*prec* < ∞) returns a result with infinite precision for a NumericQ *expr*?

b) Predict the result of the following input.

```
Precision[SetAccuracy[10^+30 Pi, 50]] -
Accuracy[SetPrecision[10^+30 Pi, 50]] +
Precision[SetAccuracy[10^-30 Pi, 50]] -
Accuracy[SetPrecision[10^-30 Pi, 50]]
```

c) Predict the result of the following input.

```
Log[10, Abs[N[SetPrecision[SetPrecision[Pi, 50], Infinity]/Pi - 1, 30]]] < -50
```

d) Construct two functions built from elementary functions (like Log, Exp, Sqrt, Power, Sin, Cos, …) $f_1(x)$ and $f_2(x)$, such that the precision $f_1(x)$ is more than ten times the precision of the argument x, such that the precision of $f_2(x)$ is less than one-tenth of the precision of the argument x.

e) For a numerical expression *expr*, N[*expr*, *prec*] (with *prec* > $MachinePrecision) typically gives a result that is correct to precision *prec*. Try to construct an expression *expr* such that:
- returns true for NumericQ[*expr*]
- is built from elementary functions (like Log, Exp, Sqrt, Power, Sin, Cos, …)
- is not identically zero
- does not give any N::meprec messages when N[*expr*, 50] is evaluated
- gives a result for N[*expr*, 50] that is wrong in the first digit already.
(Do not use Unprotect, or set unusual UpValues, ….)

f) Typically, doing a calculation with machine numbers is faster than doing a calculation with high-precision numbers. Find a counter example to this statement.

g) Find a symbolic numeric expression *expr* (meaning NumericQ[*expr*] yields True and Precision[*expr*] gives Infinity), that contains only analytic functions and that, when evaluated, gives N::meprec or Divide::infy messages.

h) Why do the following two inputs give different results?

```
Compile[{x}, 2/Exp[x]][1000]

Compile[{x}, Evaluate[2/Exp[x]]][1000]
```

i) Explain the look of the following three plots.

```
Plot[FractionalPart[Exp[n]],{n, 0, 200},
  Frame -> True, Axes -> False, PlotRange -> All];


f[n_?NumberQ] := FractionalPart[Exp[SetPrecision[n, Infinity]]];

Plot[f[n], {n, 0, 200}, Frame -> True, Axes -> False, PlotRange -> All];


$MaxExtraPrecision = 1000;
Plot[f[n], {n, 0, 2000}, Frame -> True, Axes -> False, PlotRange -> All];
```

j) For most inputs *input*, the compiled version `Compile[{}, input][]` will give the same result as the uncompiled one. Find an example where the compiled version gives a different result.

k) Predict the result of the following input.

```
Round[E - w[1] /. NDSolve[{w'[z] == (x /. FindRoot[
   (y /. FindMinimum[-Cos[y - x], {y, x + Pi/8}][[2]])] == w[z],
   {x, 0, 1}]), w[0] == 1}, w, {z, 0, 1}][[1]]]
```

Avoid the premature evaluation of the arguments of the numerical functions.

l) Find three real numbers *a*, *b*, and *c* such that three expressions $a === b$, $a === c$, and $b === c$ all give `True`, but `Union[{a, b, c}]` returns $\{a, b, c\}$.

m) Predict the result of the following input.

```
Precision[Im[SetPrecision[N[#, 200] & /@
   Unevaluated[10^100 + 10^-10 I], 200]]]
```

n) Find an algebraic expression (containing arithmetic operations and one-digit integers) ξ that is zero such that $N[\xi]$ gives a result value whose magnitude is larger than 1.

o) Predict the result of the following input.

```
Compile[{{Pi, _Real}}, Pi][2]
```

p) Will evaluating the following input give `True`?

```
 N[FindRoot[1]] - (FindRoot[N[1]]) === 0
```

q) Given the following definition for the function f, find an argument *x*, such that evaluating `f[x]` emits a `N::meprec` message when evaluating the right-hand side of the function definition.

```
f[x_Real] := N[x, $MachinePrecision]
```

r) Find an (analytic in the function-theoretic sense) integrand *int(x)*, such that `NIntegrate[int(x), {x, 0, Infinity}]` does not issue any messages and returns a result that is twice the correct value.

s) Devise exact rational numbers x_k, such that the expression
$N[1, 20] + N[x_1, 20] + \cdots + N[x_n, 20] == N[1 + x_1 + \ldots + x_n, 20]$
would evaluate to `False`. (Assume that `$MachinePrecision` is less than 20.)

t) Guess the shape of the graphic produced by the following input. What exactly does the input do?

```
squareRootOf3CF = With[{p = $MachinePrecision - 2},
With[{r = (# (1 + Random[Real, {-1., 1.} 10.^-p])) &},
    Compile[x, FixedPoint[
```

```
        Function[ξ, r[r[r[ξ]/r[2.]] + r[r[3.]/r[2.]/r[ξ]]]], x]]]];

roots = Table[squareRootOf3CF[1.], {10^5}]

Show[Graphics[
Polygon[{{#1 - 1/2, 0}, {#1 + 1/2, 0}, {#1 + 1/2, #2},
        {#1 - 1/2, #2}}]&[First[#], Length[#]]& /@
           Split[Round[(Sqrt[3] - Sort[roots])*
                       10^($MachinePrecision - 3/2)]]],
      Frame -> True];
```

u) Implement an optimized version of the following function f. Visualize `f[1/GoldenRatio, 10^5, 10^5]`, `f[1/Pi, 10^5, 10^5]`, and `f[1/E, 10^5, 10^5]`.

```
f[x_?(0 < # < 1&), p_Integer, n_Integer] :=
MapIndexed[#1^(1/#2[[1]])&, Rest[
FoldList[Times, 1, Rest[First /@ Rest[
NestList[{FromDigits[{Last[#], 0}, 2.],
   Drop[Last[#], Position[Last[#], 1, {1}, 1][[1, 1]]]}&,
   {1, RealDigits[x, 2, p][[1]]}, n]]]]]];
```

v) Find two approximative numbers z_1 and z_2 and a numerical function f, such that $z_1 === z_2$ returns `True`, but $f(z_1) === f(z_2)$ returns `False`. Can one find examples for high-precision numbers z_1 and z_2 and a function f continuous in the neighborhood of z_1, z_2?

w) Find a short (shorter than 20 characters) input that issues a `N::meprec` message.

x) As mentioned in the main text, linear algebra functions operating on high-precision matrices use internally fixed-precision to carry out the calculations to avoid excessive cancellations. As a result of using fixed-precision arithmetic, the resulting digits of all numbers are no longer guaranteed to be correct. Find an example of a 3×3 high-precision matrix, where `Inverse` applied to this matrix results in matrix elements with incorrect digits.

y) Predict the result of the following input.

```
(ArcTan[10^100] - Pi/2)^0``100
```

z) Predict the result of the following input:

```
f[x_] := x/Sqrt[x^2] Exp[-1/Sqrt[x^2]]
f[x /; x == 0] = 0
```

```
FindRoot[f[x/10] == 0, {x, 1}]
```

How does one set the options of `FindRoot` to get the zero $x_0 = 0$ of `f[x/10]` within $|x_0| < 10^{-6}$?

α) Why does the following input give a message?

```
Module[{c = 0.543626895591537208986, g},
       g[x_] := x + c;
       NIntegrate[g[x] - 1/g[x], {x, 0, 1}]]
```

24.^{L2} Series Expansion for Anharmonic Oscillator

a) After making the substitution $y(x) = \exp(-x^3/3)\, u(x)$ [688], [58], [773], [701] for $y(x)$ in

$$-y''(x) + x^4\, y(x) = \lambda\, y(x)$$

we get a new differential equation for $u(x)$. Making for $u(x)$ a power series ansatz of the form $u(x) = \sum_{i=0}^{\infty} a_i x^i$ calculate the first 100 a_i. Determine λ such that the first zero in x of $\exp(-x^3/3) \sum_{i=0}^{100} a_i x^i$ is a double zero.

b) After making the substitution $y(x) = \exp(-x^2/2) u(x)$ for $y(x)$ in

$$-y''(x) + (x^2 + x^4) y(x) = \lambda y(x)$$

we get a new differential equation for $u(x)$ [1809], [1810]. Making for $u(x)$ a power series ansatz of the form $u(x) = \sum_{i=0}^{\infty} a_i(\lambda) x^i$ calculate the first 200 a_i. Determine λ such that the first zero in λ of $a_i(\lambda)$. How many correct digits does one get from a_{200}?

c) Generate n terms of the power series solution $y_n(x) = \sum_{i=0}^{n} a_i(\lambda) x^i$ of the differential equation $-y''(x) + x^4 y(x) = \lambda y(x)$, $y(\pm\infty) = 0$. Determine the upper and lower bounds for the lowest possible λ by finding high-precision approximations for the zeros of $y_n(x^*)$ and $y_n'(x^*)$ for "suitably chosen" x^*. Find an approximation for the lowest possible λ that is correct to 1000 digits [1796].

25.^{11} Gibbs Distributions, Optimal Bin Size, Rounded Sums, Odlyzko-Stanley Sequences

a) Model the approach to equilibrium of an ensemble of pairwise interacting particles. Let a set of n particles, each initially having energy e_0, be given. For all particles, allow only (positive) integer-valued energies. Model a two-body collision by taking two randomly selected particles and randomly redistribute their (common) energy between these two particles [592], [516], [365].

b) In [595], [596], the following statistical mechanics-inspired method for the generation of n normally distributed random numbers was given: Prepare a list of n approximative 1's. Randomly (with uniform probability distribution) select two nonidentical integers i and j. Update the ith and jth elements l_i and l_j according to $l_i' = \mathfrak{r}(l_i + l_j)/\sqrt{2}$, $l_j' = (2\,\mathfrak{r}^2\,l_j - l_i - l_j)/\sqrt{2}$. Here $\mathfrak{r}$ is a random variable with values ± 1. Repeat this updating process about $3/2 \log(n) n$ times. Then, one obtains n normally distributed random numbers with probability distribution $p(x) = (2\pi)^{-1/2} \exp(-x^2/2)$.

Implement a compiled version of this method for generating normally distributed random numbers. Compare the resulting distribution with the ideal one. How long does it take to generate 10^6 numbers in this way? Compare with the direct method that uses the inverse error function `InverseErf`.

c) Generate 10^4 sums of 10^3 uniformly distributed numbers. Use the central limit theorem to approximate the distribution function of the sums. Which bin size results in a minimal mean square difference between an approximative histogram density and the limit distribution?

d) Typically, the sum of rounded summands does not coincide with the rounded sum of summands. Let ξ_k be uniformly distributed random variables from $[-L, L]$ and ζ_k their rounded nearest integers in base b (assume $L \geq 1/b$). Then the probability $p_n^{(b)}$ that the two sums $\sum_{k=1}^{n} \xi_k$ and $\sum_{k=1}^{n} \zeta_k$ rounded to the nearest multiple of b coincide is [1175]

$$p_n^{(b)} = \frac{2}{\pi b} \int_0^{\infty} \left(\frac{\sin(x)}{x}\right)^{n-1} \frac{\sin^2(b\,x)}{x^2} (1 + \delta_{n \bmod 2, 0} \cos(x)) \, dx.$$

Form a random sum with $n = 10^6$ terms that confirms this probability for $b = 10$ within 10^{-5}.

e) Implement a function that returns all elements of the Odlyzko-Stanley sequence $\mathcal{S}_k$ that are less than a given integer n. The Odlyzko-Stanley sequence $\mathcal{S}_k$ is the sequence of integers $\{a_0, a_1, \ldots, a_j, \ldots\}$ with $a_0 = 0$, $a_1 = k$, and a_j defined implicitly through the condition that it is the smallest integer, such that the sequence $\{a_0, a_1, \ldots, a_{j-1}\}$ does not contain any subsequence of three elements that form an arithmetic progression [677]. Visualize the resulting sequences and local averages of the sequences for $k = 1, \ldots, 13$ and $n = 10^6$.

26.^{L1} Nesting Tan, Thompson's Lamp, Digit Jumping

a) Explain the "steps" visible in the following picture.

```
nl = NestList[Tan, N[2, 200], 2000];
precList = Precision /@ nl;
ListPlot[precList];
```

b) Explain some characteristics of the following graphic.

```
fl = Rest[FoldList[Times, 1., Table[Tan[k], {k, 2 10^5}] // N]];
ListPlot[Log @ Abs[fl], PlotRange -> All, PlotStyle -> {PointSize[0.002]}]
```

c) Predict whether in the following attempt to model Thompson's lamp [1762], [167], [190], [775], [744], [529], [457], [1737], [1456], [1644], [430] the lamp be on or off "at the end"? Do not run the code to find out.

```
t = 0; δt = 1/2; lampOnQ = True;
While[t != 1``10000, t = t + δt; δt = δt/2; lampOnQ = Not[lampOnQ]]
lampOnQ
```

What happens when one replaces 1``10000 by the exact integer 1?

d) What does the following code do and what is the expected shape of the resulting plots?

```
cf = Compile[{{n, _Integer}},
Table[Module[{digits, λ = 100, k = 1, sOld = 1, sNew = 1},
            digits = Join[{Random[Integer, {1, 9}]},
                          Table[Random[Integer, {0, 9}], {λ}]];
            While[k++; If[sOld > λ,
                          digits = Join[digits,
                              Table[Random[Integer, {0, 9}], {sOld}]];
                    λ = Length[digits]];
                  sNew = sOld + digits[[sOld]];
                  sOld =!= sNew, sOld = sNew];
      {k, sNew}], {n}]];

data = cf[10^6];

Show[GraphicsArray[
ListPlot[{First[#], Log @ Length[#]}& /@ Split[Sort[# /@ data]],
            PlotRange -> All, DisplayFunction -> Identity]& /@ {First,
Last}]];
```

27.^{L2} Parking Cars, Causal Network, Seceder Model, Run Lengths, Cycles in Random Permutations, Iterated Inner Points, Exchange Shuffling, Frog Model, Second Arcsine Law, Average Brownian Excursion Shape

a) Make a Monte-Carlo simulation of the following problem: Cars of length 1 are parked randomly inside a linear parking lot of length l (consider the cases $l = 100$, $l = 1000$). What is the expected number of cars that fit into the lot? Compare with the theoretical result [1005], [1426], [1466] for large l

$$expectedNumberOfCars = c\,l - (1 - c) + O\!\left(\frac{1}{l}\right)$$

where c is given by

$$c = \int_0^\infty \exp\left(-2 \int_0^t \frac{1 - e^{-u}}{u}\, du\right) dt.$$

Write a version of the simulation that makes use of `Compile`.

b) Find a fast method to calculate the ψ_n^m ($-1000 \le m \le m$, $0 \le n \le 1000$) that obey the following equations [899]:

$$\psi_n^m\, \psi_n^{m+1}\, \psi_n^{m-1}\, \psi_{n+1}^m = 1 \qquad 0 \le n < \infty,\ -\infty < m < \infty$$
$$\psi_0^m = 1 - 2\,\delta_{m,0}$$
$$\psi_1^m = 1.$$

c) The seceder model [509], [1693] (used to model the spontaneous formation of groups) is the following: Starting with a list l of n zeros, one step consists of n iterations of an update step. In the update step, randomly three elements of l are selected. Then the element of the three that has the largest distance from the average is chosen. A random real number (say uniformly distributed in $[-1, 1]$) is added the selected element and the resulting value replaces a randomly chosen element of l. All other elements of l stay unchanged.

Implement the seceder model efficiently. Run and visualize 1000 update steps of a starting list of length 1000. Is a run and a visualization of 10000 update steps of a starting list of length 10000 doable?

d) A run in a list of numbers $\{n_1, n_2, \ldots, n_k\}$ is a sublist of consecutive increasing numbers $\{n_i, n_{i+1}, \ldots, n_{i+l}\}$. For 1000 random permutations, each of length 10000, calculate how many runs of length λ occurred. Use `Compile` for the generation and analysis of the random permutation.

e) Calculate 1000 random permutations of length 1000, and analyze the number of cycles of these permutations. Use `Compile` for the generation and analysis of the random permutation.

f) Given a permutation $\sigma = \{\sigma_1, \sigma_2, \ldots, \sigma_n\}$ of the integers $\{1, 2, \ldots, n\}$, and a permutation $\tau = \{\tau_1, \tau_2, \ldots, \tau_n\}$ of the integers $\{1, 2, \ldots, k\}$ (where $k \le n$), one says the pattern τ occurs in the permutation σ if there exists a sequence of indices $\{j_1, j_2, \ldots, j_k\}$, such that $\sigma_{j_m} \lessgtr \sigma_{j_n}$ whenever $\tau_{j_m} \lessgtr \tau_{j_n}$ for all $1 \le m < n \le k$. For each of the 4! different patterns of $\{1, 2, 3, 4\}$, find how many of the 8! permutations of $\{1, 2, \ldots, 8\}$ do not include the search pattern? Can you calculate the same for the 10! permutations of $\{1, 2, \ldots, 10\}$? Aim for a memory- and time-efficient implementation.

g) The cut sequence [1093], [1094] of a permutation of the first n integers $\{k_1, k_2, \ldots, k_n\}$ is a list of lists of the form $\{\{k_1, k_2, \ldots, k_{i_1}\}, \{k_{i_1+1}, \ldots, k_{i_2}\}, \{k_{i_2+1}, \ldots, k_{i_3}\}, \ldots, \{k_{i_{m+1}}, \ldots, k_n\}\}$ such that $k_{i_l} < k_j$ for all $i_l < j$. (So, for example the cut sequence of $\{1, 2, 6, 7, 4, 5, 3\}$ is $\{\{1\}, \{2\}, \{6, 7, 4, 5, 3\}\}$.) Given a permutation, implement an efficient calculation of its cut sequence. For 10000 random permutations of 1000 integers, calculate the average length of the cut sequence.

h) Generate an $n \times n$ matrix $\mathbf{A}$ (say, $n = 256$) with $p\%$ 0's and $(1 - p)\%$ 1's. Now iterate the following process until all 0's have been transformed to 1's: For each element a_{ij} of the matrix that is 0, check if its two left and right and its two upper and lower neighbors are 0. If they are not, change the element to 1; if they are, do nothing. Visualize how the 1's form in the iterations. Use $p = 90$, $p = 99$, and $p = 99.9$.

i) Model the following (exchange) shuffle process: Given a list of integers $\{1, 2, \ldots, n\}$, exchange the first integer with a randomly selected one from the list (maybe with itself). Then exchange the second one with a randomly selected one, $\ldots$, then exchange the last one with a randomly selected one [1538], [1608]. Model 10^6 such exchange shuffles and analyze the probabilities of the resulting permutations for $n = 4, 5, 6$. Compare with the exact probabilities.

j) Consider the so-called frog model [45], [1467] of statistical mechanics. Assume sleeping frogs are placed on the lattice points of a $n \times n$ square lattice. Then one frog awakes and jumps to one of the four neighboring sites. This wakes up the frog on this site and in the next step each of the two frogs jump to a randomly selected neighboring site. Then this process continues. Make an animation that shows how the awakened frogs spread out and carry out enough steps to estimate the distribution of the long-time limit of the probability to find 1 $0 \le n \le 10$ frogs per site.

k) The so-called second arc-sine law says that in a 1D Brownian motion that starts at time 0 at the origin and stops at time T, the probability distribution $p(\tau)$ for the largest time τ the particle visited the origin is $p(\tau/T) = 2/\pi \arcsin((\tau/T)^{1/2})$ [1950], [424]. Model a 1D Brownian motion with constant step size and calculate the theoretical probability density with the modeled one.

l) An excursion of a 1D Brownian motion $x(t)$ is defined to be a maximal connected part $\{x(t)\}_{T_0 \le t \le T_1}$ of the same sign. The average shape of the (scaled) excursions is $|\bar{x}(\tau)|(T_1 - T_0)^{-1/2} = (8/\pi\tau(1-\tau))^{1/2}$ where $\tau = (t - T_0)/(T_1 - T_0)$ [121], [421], [1743]. Model a 1D Brownian motion by using constant step size (and ignoring very small excursions) and compare the theoretical probability density with the modeled one.

28.[13] Poincaré Sections, Random Stirring, ABC-System, Vortices on a Sphere, Oscillations of a Triangular Spring Network, Lorenz System

a) Make a Poincaré section plot for the following coupled nonlinear system of differential equations:

$$x'(t) = y(t)$$
$$y'(t) = y(t)\, z(t) - x(t)$$
$$z'(t) = 1 - (y(t))^2.$$

(A Poincaré section [927] here means a plot of points in the x,y-plane (this means $z(t) = 0$) formed by the solutions of the differential equations [517].) Use two different approaches to obtain the Poincaré sections.

b) Predict what the following code will do.

```
rotate = Compile[{{center, _Real, 1}, φ, {points, _Real, 2}},
         Module[{r, p, φ, R}, Function[point,
                 p = point - center; r = Sqrt[p.p]; φ = Exp[-r] φ;
                 R = {{Cos[φ], Sin[φ]}, {-Sin[φ], Cos[φ]}};
                 R.p + center] /@ points]];

pp = 10000; R = 3/4; n = 3; m = 3;
centers = Table[N[{Cos[j/n 2Pi], Sin[j/n 2Pi]}], {j, 0, n - 1}];
L = Table[R {Cos[φ], Sin[φ]}, {φ, 0, 2Pi, 2Pi/pp}] // N;

SeedRandom[333];
Do[center = centers[[Random[Integer, {1, n}]]];
   φ = Random[Real, Pi {0, 1}];
   Do[L = rotate[center, k/m φ, L];
     Show[Graphics[Polygon[L],
                   PlotRange -> Max[Abs[L]]{{-1, 1}, {-1, 1}},
                   Frame -> True, FrameTicks -> None,
                   AspectRatio -> Automatic]], {k, 0, m - 1}];
     L = rotate[center, φ, L], {j, 100}];
```

Generalize whatever the code is doing to 3D case.

c) Consider the system of coupled differential equations (Arnold–Beltrami–Childress system)

$$x'(t) = \gamma \cos(y(t)) + \alpha \sin(z\,(t))$$
$$y'(t) = \alpha \cos(x(t)) + \beta \sin(x\,(t))$$
$$z'(t) = \beta \cos(x(t)) + \gamma \sin(y(t)).$$

Make an animation showing how the family of solutions depends on $0 \le \gamma \le 1$ [707], [229].

d) The equations of motion of n point vortices of strength Γ_k on a unit sphere are [1423], [987], [1107], [1576]:

$$\mathbf{x}'_j(t) = \sum_{\substack{k=1 \\ k \ne j}}^{n} \Gamma_k \, \frac{\mathbf{x}_k(t) \times \mathbf{x}_j(t)}{1 - \mathbf{x}_k(t).\mathbf{x}_j(t)}.$$

Calculate and visualize some example trajectories of three and four vortices.

e) Consider a mechanical system built from point masses and springs. The springs are located along the edges of a recursively subdivided regular triangle and the point masses are the corresponding vertices. For a system consisting of 45 point masses (meaning 108 springs), fix the outermost three point masses and visualize the oscillations of the system for some chosen initial conditions.

Linearize the resulting equations (meaning small oscillations) and visualize some of the eigenoscillations [587].

f) Consider the classic Lorenz system [1174], [1279], [89], [1740], [1091]

$$x'(t) = \sigma\,(y(t) - x(t))$$
$$y'(t) = \rho\,x(t) - y(t) - x(t)\,z(t)$$
$$z'(t) = x(t)\,y(t) - \beta\,z(t)$$

with parameters $\sigma = 10$, $\rho = 28$, $\beta = 8/3$ and initial conditions $x(0) = x_0$, $y(0) = y_0$, $z(0) = z_0$. Sketch how the surface $x(t) = 0$ (as a function of the initial conditions) in x_0, y_0, z_0-space evolves as a function of t.

29.^{12} Polynomial Coefficient, Fourier Differentiation

a) Calculate numerically via contour integration the coefficient of x^{3000} in the expansion of the polynomial $p(x) = (x + 1)^{2000}\,(x^2 + x + 1)^{1000}\,(x^4 + x^3 + x^2 + x + 1)^{500}$ to 13 significant digits. Calculate the coefficient in the saddle point approximation [1444], [1382], [1516], [722] of the integral. (This problem comes from [586].)

b) Given the continuous Fourier transform $\hat{f}(k)$ of a function $f(x)$ restricted to the interval $[0, L]$,

$$\hat{f}(k) = \frac{1}{\sqrt{2\pi}} \int_{-\infty}^{\infty} [f(k)]_{[0,L]}\, e^{ikx}\, dx = \frac{1}{\sqrt{2\pi}} \int_{0}^{L} f(k)\, e^{ikx}\, dx$$

($[f]_I$ indicates the function f restricted to I), how can one approximate the νth element $\hat{f}_\nu$, $\nu = 1, \ldots, n$ of the discrete Fourier transform

$$\hat{f}_\nu = \frac{1}{\sqrt{n}} \sum_{\mu=1}^{n} f_\mu \exp(2\pi i\,(\mu - 1)\,(\nu - 1)/n)$$

through $\hat{f}(k)$ in the large n limit (L/n sufficiently small)? Here $f_\mu = f(x_\mu)$ and $x_\mu = \mu/n\,L$. Compare the direct result with the asymptotic result for some examples.

For the continuous Fourier transform the differentiation of the original function means multiplication by $(-ik)$ for the transformed function. How does this translate to the discrete Fourier transform?

30.[L1] Abel Differentials, Implicit Parameter Dependence

a) Calculate a numerical approximation for $s(x)$, $0 \le x \le 2$

$$s(x) = \int_0^x \left(\sum_{i=1}^6 (y_i(x) - 1)/(y_i(x) + 1) \right) dx.$$

Here, $y_i(x)$ are the six independent roots of the polynomial $y(x)^6 + x^3 y(x)^2 + y(x) - 4x^4 - 2x - 3 = 0$ considered as smooth functions of x.

b) Given the ordinary differential equation $w'_\zeta(z) = \cos(w_\zeta(z)) + \sin(\zeta)$, $w_\zeta(0) = \zeta^2$ (prime denotes differentiation with respect to z) with parametric dependence on ζ, calculate $\partial^2 w_\zeta(1)/\partial \zeta^2 |_{\zeta=1}$ to 20 digits.

31.[L2] Modular Function Fourier Series, Differential Equation for $J(z)$, Singular Moduli

a) The Dedekind eta function $\eta(z)$ (in *Mathematica* called `DedekindEta`) has a series representation of the form $\eta(z) = \sum_{k=1}^\infty a_k e^{k i \pi z/12}$ ($a_k = -1$ or 1 or 0). Calculate numerically the first 50 nonvanishing terms of this series. ($\eta(z)$ is defined only in the upper half-plane.)

The Dedekind eta function has the product representation $\eta(z) = e^{\frac{\pi i z}{12}} \prod_{k=1}^\infty (1 - e^{2\pi i k z})$. Use this product representation to calculate the first 50 nonvanishing series terms.

b) The Klein modular function $J(z)$ has a series representation of the form $J(z) = 1/1728 \, (e^{-2i\pi z} + 744 + \sum_{k=1}^\infty a_k e^{2ki\pi z})$ with all a_k being integers. Calculate the first 100 (nonvanishing) terms of this series. ($J(z)$ is defined only in the upper half-plane).

Use two different methods to calculate the terms to make sure that they are correct.

The coefficients a_k of the series can be represented in the following way:

$$a_k = \frac{2\pi}{\sqrt{k}} \sum_{j=1}^\infty \frac{1}{j} A_j(k) I_1\left(4\pi \frac{\sqrt{k}}{j}\right)$$

where $I_1(z)$ is the Bessel function `BesselI[1,z]` and

$$A_j(k) = \sum_{\substack{h=0 \\ \gcd(h,j)=1}}^{j-1} \exp(-2\pi i (hk + H(j,h))/j)$$

$$h H(j,h) = -1 \bmod j.$$

Calculate a_{100} using this formula.

c) The function $w(z)$ (in *Mathematica* `KleinInvariantJ[z]`) fulfills a nonlinear differential equation of the following form ($c_k \in \mathbb{Z}$, $\mathrm{Im}(z) > 0$):

$$c_1 w'(z)^4 + c_2 w'(z)^4 w(z) + c_3 w'(z)^4 w(z)^2 + c_4 w''(z)^2 w(z)^2 + c_5 w''(z)^2 w(z)^3 +$$
$$c_6 w''(z)^2 w(z)^4 + c_7 w'(z) w^{(3)}(z) w(z)^2 + c_8 w'(z) w^{(3)}(z) w(z)^3 + c_9 w'(z) w^{(3)}(z) w(z)^4 = 0.$$

Find the c_k.

d) For certain values $q_{n,j}^*$ of $q = \exp(2 k i \pi z)$, the Klein modular function $J(q)$ (in *Mathematica,* as a function of z, `KleinInvariantJ[z]`) has the property $J(q_{n,j}^*) = J(n q_{n,j}^*)$ where n is a positive integer (because the values are not unique, we use the additional index j). These are the so-called singular moduli and they have the form $q^* = r_1 + i\sqrt{r_2}$ where r_1 and r_2 are rational numbers, $r_2 \geq 0$ [202], [740], [494], [991], [514], [417]. For $n = 1, \ldots, 20$, find such $q^*_{n,j}$.

32.[12] Curve Thickening, Textures, Seed Growth

a) In [14], [1169], [15], the following formula for generating a function $f_\rho[c(t)](z)$ whose absolute value represents a "thickened" version of the 2D curve $c(t) = c_x(t) + i c_y(t)$, $0 \leq t \leq T$ (represented in the complex plane) was given:

$$f_\rho[c(t)](z) = \exp\left(-\frac{z\bar{z}}{\rho^2}\right) \int_0^T |c'(\tau)| \exp\left(-\frac{c(t)\overline{c(t)}}{\rho^2} + \frac{2 z \overline{c(t)}}{\rho^2} + \frac{1}{\rho^2}\int_0^t (c(\tau)\overline{c'(\tau)} - c'(\tau)\overline{c(\tau)})\,d\tau\right) dt.$$

The parameter ρ controls the "thickness" of the curve. Visualize the real part, the imaginary part, the absolute value and the argument of the function $f_1[t\,e^{it}](z)$, $0 \leq t \leq 2\pi$ (a spiral) and $f_1[e^{it}](z)$, $0 \leq t \leq 2\pi$ (a circle).

b) Iteratively carrying out Fourier transforms, list convolution, and other arithmetic operations on lists of numbers can be used to generate textures (for texture mappings, for instance [531]). Find examples of such procedures that generate "nice" textures.

c) Given a square matrix $\mathbf{A}$ with nonnegative integer entries $a_{i,j} = k_m(i, j)$, implement a compiled function `growSeeds` that randomly replaces eventually present zeros in $\mathbf{A}$ with the values of one of the neighboring elements until all zeros are gone. Apply the function `growSeeds` to some matrices of size 100×100 to 400×400 that initially contain a few ten to a few hundred randomly distributed nonzero integers, and to matrices whose initial nonzeros form various regular patterns and visualize the results. Highlight the boundaries between matrix elements with different $k_m(i, j)$.

d) Given a matrix with 0s and 1s, replace each 0 by the number of neighborhood enlargements it minimally takes until this 0 is reached from an existing 1. For various random and structured initial matrices, visualize the resulting matrices.

33.[12] First Digit Frequencies in Mandelbrot Set Calculation

Implement the numerical calculation of the Mandelbrot set defined as all points c of the complex plane such that the iteration of the map $z \to z^2 - c$ stays bounded. Monitor the first digit of (the real and imaginary parts of) z in all iteration steps for all z's. (This should be done as a "side effect" in the calculation, but it is the main point within this exercise). Calculate how many times the digit i ($i = 1, 2, \ldots, 9$) occurred. Compare the result with Benford's rule [181] (see Exercise 1 from Chapter 6 of the Programming volume [1793]).

34.[11] Interesting Jerk Functions

A jerk function [1616], [1154], [1699], [1700], [1155], [266], [1201], [1701], [1702], [537], [1203], [538], [1202], [719], [1925], [720] is a polynomial $p(t, x(t), x'(t), x''(t))$ used on the right-hand side of the differential equation $x'''(t) = p(t, x(t), x'(t), x''(t))$. (Sometimes jerk functions are defined to be explicitly independent of the independent variable, this means of the form $p(x(t), x'(t), x''(t))$.) Find some jerk functions that generate an "interesting" phase portrait (graph of $\{x(t), x'(t)\}$). Find some jerk functions that have "interesting" extended phase portraits $\{x(t), x'(t), x''(t)\}$.

35.[12] Initial Value Problems for the Schrödinger Equation

In scaled coordinates ξ, τ, (ξ being the position, $0 \leq \xi \leq 1$, τ being the time), the Schrödinger equation of a free particle in a potential well can be written in the form

$$i \, \frac{\partial \psi(\xi, \tau)}{\partial \tau} = -\frac{1}{4\pi} \, \frac{\partial^2 \psi(\xi, \tau)}{\partial \xi^2}.$$

Solve numerically this equation and then use the obtained solution to visualize the result for the four initial-boundary value problems:

1) $\psi(0, \tau) = \psi(1, \tau) = 0$ and $\psi(\xi, 0) = \sum_{k=1}^{5} \sin(k\pi\xi)$ and $0 \leq \tau \leq 1$

2) $\psi(0, \tau) = \psi(1, \tau) = 0$ and $\psi(\xi, 0) = \sum_{k=1}^{20} \sin(k\pi\xi)$ and $0 \leq \tau \leq 1$

3) $\psi(0, \tau) = \psi(1, \tau) = 0$ and $\psi(\xi, 0) = \theta(\xi)\,\theta(\frac{1}{2} - \xi)\,e^{5\,i\,\xi} \exp\!\left(\left((\xi - \frac{1}{4})^2 - \frac{1}{16}\right)^{-1}\right)$ and $0 \leq \tau \leq 1$

4) $\psi(0, \tau) = \psi(1, \tau) = 0$ and $\psi(\xi, 0) = 1$ and $0 \leq \tau \leq 1$

5) $\psi(0, \tau) = \psi(1, \tau) = 0$ and $\psi(\xi, 0) = \sum_{k=\overline{k}-\delta k}^{\overline{k}+\delta k} \exp\!\left(-(k - \overline{k})^2 / (4\,\sigma^2)\right) \exp(-i\,\pi\,k^2\,\tau/4)\sin(k\pi\xi)$ and
   $10^{-3} \leq \tau \leq 40 \times 10^{-3}$ (chose $\overline{k} \approx 10^3$, $\delta k \approx 10^2$, and $\sigma \approx 10^0$)

For each of the four problems, compare the solution generated by the separation of variable method with the one generated by `NDSolve`.

36.[13] Initial Value Problems for the Wave Equation

Use the function `NDSolve` to numerically solve the following initial value problem for the wave equation in *dim* dimensions ($dim = 1, 2, 3$)

$$\frac{\partial^2 u(t, \, \boldsymbol{x})}{\partial t^2} = \Delta\, u(t, \, \boldsymbol{x}) = \sum_{k=1}^{dim} \frac{\partial^2 u(t, \, \boldsymbol{x})}{\partial x_k^2}$$

$$\boldsymbol{x} = \{x_1, x_2, \ldots, x_{dim}\}$$

in the domains $0 \leq t \leq 3$ and $0 \leq |\boldsymbol{x}| \leq 3$. Let $u_0(\boldsymbol{x}) = u(0, \, \boldsymbol{x})$ be concentrated in a ball of radius $R = 1$ around the origin, and let $u_0(\boldsymbol{x}) = \cos^2(\pi\,|\boldsymbol{x}|/2)$. Let $\frac{\partial}{\partial t} u(t, \, \boldsymbol{x})\big|_{t=0} = 0$. This choice of the function $u_0(\boldsymbol{x})$ goes smoothly to zero so that singularities will form—contrasted to constant initial conditions inside a ball.

Compare the `NDSolve` solution with the solution obtained by using the following representations of the solutions [1857], [1789], [618], [940], [978], [1366], [139], [1141], [1465]:

One dimension:

$$u(t, \, x) = \frac{1}{2}\left(u_0(x + t) + u_0(x - t)\right)$$

Two dimensions:

$$u(t, x) = \frac{1}{2\pi} \frac{\partial}{\partial t} \int_{B_t(x)} \frac{u_0(\xi)}{\sqrt{t^2 - |x - \xi|^2}} \, d\xi$$

Here, the integration is carried out over the sphere $B_t(x)$ of radius t around x.

Three dimensions:

$$u(t, x) = \frac{1}{4\pi} \frac{\partial}{\partial t} \left(\int_{\partial B_1(x)} t \, u_0(x + t\xi) \, d\sigma_\xi \right)$$

Here, the integration is carried out over the surface of a sphere $\partial B_1(x)$ of radius 1 around x. (For n dimensions, see [569].)

37.[1.1] Kleinian Group, Continued Fractions, Lüroth Expansions, Lehner Fractions, Brjuno Function, Bolyai Expansion

a) The two matrices

$$\begin{pmatrix} \gamma & 0 & 0 \\ 0 & \gamma^4 & 0 \\ 0 & 0 & \gamma^2 \end{pmatrix} \quad \text{and} \quad \begin{pmatrix} a & b & c \\ b & c & a \\ c & a & b \end{pmatrix}$$

where

$$\gamma = e^{\frac{2\pi i}{7}}, \quad a = \frac{\gamma^5 - \gamma^2}{\sqrt{-7}}, \quad b = \frac{\gamma^3 - \gamma^4}{\sqrt{-7}}, \quad c = \frac{\gamma^6 - \gamma}{\sqrt{-7}}$$

are generators of a finite group with the group composition being matrix multiplication (see, for instance, [464]). Calculate numerically the order of the group as well as the complete group multiplication table.

b) Define a function `continuedFraction[`*realNumber, order*`]` that approximates a positive real number *realNumber* (head `Real`) by a continued fraction of depth *order*. Compare with the built-in function `Contin¬uedFraction`.

For the first 20 integers n, determine numerically approximate values for the probability that the integer n appears in the continued fraction of a randomly chosen real number.

Implement a similar function `continuedInverseRoot` that, for a given m, generates the c_k in the expansion [1231]

$$realNumber = c_0 + \cfrac{1}{\sqrt[m]{c_1 + \cfrac{1}{\sqrt[m]{c_2 + \cfrac{1}{\sqrt[m]{c_3 + \cfrac{1}{\sqrt[m]{c_4 + \cdots}}}}}}}}.$$

c) Let $\{a_1(x), a_2(x), \ldots, a_n(x)\}$ be the first n terms of the continued fraction expansion of the real number x. For almost all $x \in (0, 1)$ and $n \to \infty$, the probability that $\ln(2)/n \max(\{a_1(x), a_2(x), \ldots, a_n(x)\}) < X$ is given by $\exp(-1/X)$ [649]. Carry out a numerical simulation that confirms this result.

d) The Lüroth expansion of a real number x ($0 < x < 1$) is of the form [1429], [650], [657], [1627]

$$x = \frac{1}{a_1} + \sum_{k=2}^{\infty} \frac{1}{a_k} \prod_{j=1}^{k-1} \frac{1}{a_j(a_j - 1)} =$$

$$= \frac{1}{a_1} + \frac{1}{a_1(a_1 - 1) a_2} + \frac{1}{a_1(a_1 - 1) a_2(a_2 - 1) a_3} +$$

$$\cdots + \frac{1}{a_1(a_1 - 1) a_2(a_2 - 1) \ldots a_{n-1}(a_{n-1} - 1) a_n} + \ldots.$$

The Lüroth terms a_n can be calculated via:

$$a_n = \alpha(\mathcal{T}^{n-1}(x))$$

$$\alpha(x) = \left\lfloor \frac{1}{x} \right\rfloor + 1$$

$$\mathcal{T}(x) = \left\lfloor \frac{1}{x} \right\rfloor \left(\left\lfloor \frac{1}{x} \right\rfloor + 1 \right) x - \left\lfloor \frac{1}{x} \right\rfloor.$$

How many Lüroth terms are needed to approximate $1/\pi$ to 1000 digits?

e) The Lehner continued fraction expansion of a real number x ($1 < x < 2$) is of the form [1124]

$$x = b_0 + \cfrac{c_1}{b_1 + \cfrac{c_2}{b_2 + \cfrac{c_3}{b_3 + \cdots}}}.$$

The integers $\{b_k, c_{k+1}\}$ can be calculated via

$$\{b_k, c_{k+1}\} = \beta(\mathcal{L}^{k+1}(x))$$

$$\beta(x) = \begin{cases} \{2, -1\} & 1 \le x < \frac{3}{2} \\ \{1, \ 1\} & \frac{3}{2} \le x < 2 \end{cases}$$

$$\mathcal{L}(x) = \begin{cases} \frac{1}{2-x} & 1 \le x < \frac{3}{2} \\ \frac{1}{x-1} & \frac{3}{2} \le x < 2. \end{cases}$$

Investigate the structure of $\mathcal{L}^k(x)$ for various real x ($1 < x < 2$). For high-precision approximations of $\pi/2$ and $e/2$, investigate the probability that $\{b_k, c_{k+1}\} = \{1, 1\}$. For some randomly chosen real numbers, calculate the geometric mean $(\prod_{k=1}^{n} a_k)^{1/n}$ as a function of n. Does it exist?

f) The Brjuno function $\mathcal{B}(x)$ of an (irrational) real number x (if it exists) is defined in the following way [345], [1226], [1307], [1308], [1309], [1224], [1164], [239], [191], [1204], [1225]:

$$\mathcal{B}(x) = \sum_{k=0}^{\infty} \beta_{k-1}(x) \ln\left(\frac{1}{\xi_k} \right)$$

$$\beta_{-1}(x) = 1$$

$$\beta_k(x) = |p_k(x) - x \, q_k(x)|$$

$$\xi_0 = x - \lfloor x \rfloor$$
$$\xi_k = \frac{p_k(x) - x\, q_k(x)}{x\, q_{k-1}(x) - p_{k-1}(x)}.$$

Here $p_k(x)$ and $q_k(x)$ are the numerators and denominators of the convergent fraction expansion of x. The Brjuno function $\mathcal{B}(x)$ obeys the following functional equation for $0 < x < 1$:

$$\mathcal{B}(x) = -\ln(x) + x\, \mathcal{B}\!\left(\frac{1}{x}\right).$$

Make a sketch of the graph of $\mathcal{B}(x)$ and check the functional equation for some random value of x to at least 100 digits.

g) Let $p_n(x)/q_n(x)$ be the sequence of convergents of the continued fraction approximation of x. Define the function sum of errors function $\mathcal{P}(x)$ as [1529]

$$\mathcal{P}(x) = \sum_{k=0}^{\infty} \left(x - \frac{p_n(x)}{q_n(x)}\right)$$

where the sum might terminate for rational x. Visualize $P(x)$ and calculate a 20-digit approximation of the integral $\int_0^\infty P(x)\, dx$.

Carry out a numerical simulation demonstrating to at least three digits the following limit (valid for almost all x): $\lim_{n\to\infty} n^{-1} \sum_{k=1}^{n} \theta_n(x) = 1/\ln(16)$ [258], [260]. Here $\theta_n(x)$ is the scaled error of the rational approximations of x, meaning $\theta_n(x) = q_n(x)^2 |x - p_n(x)/q_n(x)|$.

h) The iteration

$$x_1 = \xi$$
$$x_{n+1} = \left(1 + \frac{1}{x_n}\right)^n$$

increases monotonic for increasing n for exactly one positive $\xi = \tilde{\xi}$ [573]. Find this $\tilde{\xi}$ to 100 digits.

i) The square root-Bolyai expansion of a real number x ($0 < x < 1$) is of the form [1518], [261]

$$x = -1 + \sqrt{a_1 + \sqrt{a_2 + \sqrt{a_3 + \cdots}}}\ .$$

The Bolyai digits $a_n \in \{0, 1, 2\}$ can be calculated via:

$$a_n = \alpha(\mathcal{T}^{n-1}(x))$$
$$\mathcal{T}(x) = (x + 1)^2 - 1 - \alpha(x)$$

$$\alpha(x) = \begin{cases} 0 \text{ if } 0 \le x < \sqrt{2} - 1 \\ 1 \text{ if } \sqrt{2} - 1 \le x < \sqrt{3} - 1 \\ 2 \text{ if } \sqrt{3} - 1 \le x \le 1. \end{cases}$$

How many Bolyai digits are needed to approximate $1/\pi$ and $1/e$ to 1000 digits? Find exact values of the numbers that have the following repeating sequences of Bolyai digits: $\{1, 2\}$, $\{2, 1\}$, and $\{0, 1, 2\}$.

j) For a real number x greater than 1, define the symmetric continued fraction expansion by

$$x = d_0(x) + \cfrac{d_0(x)}{d_1(x) + \cfrac{d_1(x)}{d_2(x) + \cfrac{d_2(x)}{d_3(x) + \cfrac{d_3(x)}{d_4(x) + \cfrac{d_4(x)}{d_5(x) + \cfrac{d_5(x)}{\cdots}}}}}}.$$

Here the $d_k(x)$ are positive integers. Implement a function `symmetricContinuedFraction` that calculates the first o terms of the symmetric continued fraction expansion of a given number x. Calculate the first 100 terms of the symmetric continued fraction expansion of π, e, and $\sqrt{2}$. How precise are these approximations?

Solutions

1. Logistic Map, Iterations with Noise

a) Here is the equation for the fixed points of the iterated polynomial in *var*.

```
In[1]:= polyEq[c_, var_, iter_] := Nest[(#^2 + c)&, var + c, iter] == var
```

Written explicitly, it is already quite large for fourth order.

```
In[2]:= polyEq[x, c, 4]
```

$$\text{Out[2]= } x + \left(x + \left(x + (x + (c+x)^2)^2\right)^2\right)^2 == c$$

```
In[3]:= Expand /@ % // Short[#, 4]&
```

$$\text{Out[3]//Short= } c^{16} + x + 2\,c^8\,x + 4\,c^{12}\,x + 8\,c^{14}\,x + 16\,c^{15}\,x + x^2 + 4\,c^4\,x^2 + 8\,c^6\,x^2 + 16\,c^7\,x^2 + 6\,c^8\,x^2 +$$
$$24\,c^{10}\,x^2 + 48\,c^{11}\,x^2 + 28\,c^{12}\,x^2 + 112\,c^{13}\,x^2 + 120\,c^{14}\,x^2 + 2\,x^3 + \ll 86 \gg + 8008\,c^4\,x^{11} +$$
$$4368\,c^5\,x^{11} + 94\,x^{12} + 608\,c\,x^{12} + 1848\,c^2\,x^{12} + 2912\,c^3\,x^{12} + 1820\,c^4\,x^{12} + 60\,x^{13} +$$
$$336\,c\,x^{13} + 728\,c^2\,x^{13} + 560\,c^3\,x^{13} + 28\,x^{14} + 112\,c\,x^{14} + 120\,c^2\,x^{14} + 8\,x^{15} + 16\,c\,x^{15} + x^{16} == c$$

Because of this size (and because this problem belongs in a chapter on numerical methods), we find the fixed point using NRoots instead of using Solve to get an exact solution. Because NRoots does not always deliver the solutions that "belong together" at the same position, we represent the results as points (in the plot) instead of connecting them by lines.

```
In[4]:= points[cMin_, cMax_, step_, iter_] :=
        Flatten[Table[ (* make points in 3D space *)
                Point[{Re[#], Im[#], c}]& /@
            (* solve polynomial *) Cases[NRoots[polyEq[N[c], z, iter], z],
                                _?NumberQ, {-1}],
                {c, cMin, cMax, (cMax - cMin)/step}], 1]
In[5]:= (* make and display 3D graphics *)
        fixPointMapGraphics3D[cMin_, cMax_, num_, iter_, opts___] :=
        Show[Graphics3D[{PointSize[0.003], points[cMin, cMax, num, iter]}],
            opts, Axes -> True, AxesLabel -> {"Re(z)", "Im(z)", "c"},
            AxesEdge -> {{-1, -1}, {1, -1}, {-1, -1}}, Boxed -> True,
            BoxRatios -> {1, 1, 2}, (* indicate order *)
            PlotLabel ->  ToString[iter] <> ". order "]
```

Here are the images for the first four orders. (Higher orders also look very interesting, but require large computing times.)

```
In[7]:= Show[GraphicsArray[{fixPointMapGraphics3D[-2, 2, 200, #,
                            DisplayFunction -> Identity]& /@ Range[4]}]]
```

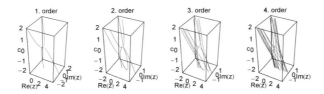

For the next iteration level, we will use another approach. Instead of repeatedly (for each value of *c*) solving the polynomial, we will construct a differential equation for the zeros $z_i(c)$ and only once use NRoots to calculate the initial conditions of this differential equation. Here is the differential equation for the zeros.

```
In[8]:= ode = D[Subtract @@ polyEq[z[c], c, 5], c] == 0
```

$$\text{Out[8]= } -1 + z'[c] + 2\left(z[c] + \left(z[c] + \left(z[c] + (z[c] + (c + z[c])^2)^2\right)^2\right)^2\right)$$
$$\left(z'[c] + 2\left(z[c] + \left(z[c] + (z[c] + (c + z[c])^2)^2\right)^2\right)\right)\left(z'[c] + 2\left(z[c] + (z[c] + (c + z[c])^2)^2\right)\right)$$
$$(z'[c] + 2\,(z[c] + (c + z[c])^2)\,(z'[c] + 2\,(c + z[c])\,(1 + z'[c]))))) == 0$$

And these are the initial conditions for the zeros. (To achieve sufficient precision, we will carry out all calculations with high-precision numbers.)

```
In[9]:= initialConditions = List @@ (Last /@ NRoots[polyEq[z, -4, 5], z, 64]);
```

We now solve the differential equation for the zeros for c in the interval (-4, 4).

```
In[10]:= nsols = (z[c] /. NDSolve[{ode, z[-4] == #}, z[c], {c, -4, 4},
                        (* use high precision *) WorkingPrecision -> 24,
                        PrecisionGoal -> 10, AccuracyGoal -> 10,
                        MaxSteps -> 50000][[1]])& /@ initialConditions;
```

Let us check the quality of the zeros at the end of the integration interval.

```
In[11]:= With[{nroots = List @@ (Last /@ NRoots[polyEq[z, 4, 5], z, 64])},
            Function[x, Min[Abs[x - #& /@ nroots]]] /@
                                    (nsols /. c -> 4)] // Abs // Max
```
$$Out[11]= 2.802914421088858 \times 10^{-8}$$

Using `ParametricPlot3D`, we can now display the zeros $z_i(c)$ as continuous functions of c.

```
In[12]:= pic = Show[Table[ParametricPlot3D[
                    Evaluate[(* parametrized zeros *)
                          {Re[nsols[[i]]], Im[nsols[[i]]], c,
                          {Thickness[0.002], Hue[Random[]]}}],
                          {c, -4, 4}, PlotPoints -> 500,
                        DisplayFunction -> Identity], {i, 32}],
                    DisplayFunction -> $DisplayFunction,
                    Axes -> True, Boxed -> True, BoxRatios -> {1, 1, 2},
                    AxesLabel -> {"Re(z)", "Im(z)", "c"},
                    AxesEdge -> {{-1, -1}, {1, -1}, {-1, -1}}]
```

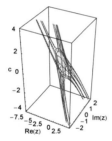

To determine approximative values for the bifurcation points (branch points of the multivalued algebraic function $z(c)$), we search for points such that the distance between two zeros becomes small. `bifurcations` gives a list of lists, each list containing the indices of the two roots that meet each other and information about the corresponding c-value.

```
In[13]:= nsolsData = Table[Table[Evaluate[nsols[[i]]], {c, -4, 4, 8/100}], {i, 32}];
```

```
In[14]:= bifurcations = DeleteCases[Flatten[
                Table[{{i, j}, Position[nsolsData[[i]] - nsolsData[[j]],
                    _?(Abs[#] < 10^-4&), {1}, 1]},
                    {i, 32}, {j, i + 1, 32}], 1], {_, {}}]
```
```
Out[14]= {{{1, 2}, {{51}}}, {{3, 4}, {{49}}}, {{5, 6}, {{44}}}, {{7, 8}, {{58}}}, {{9, 10}, {{42}}},
         {{11, 12}, {{54}}}, {{15, 16}, {{66}}}, {{17, 18}, {{74}}}, {{19, 20}, {{87}}},
         {{21, 22}, {{71}}}, {{23, 24}, {{98}}}, {{25, 26}, {{87}}}, {{31, 32}, {{94}}}}
```

To get precise values for the bifurcation points, we have to look for such values of c, such that `poly` has a double root in z.

```
In[15]:= eqs = {#, D[#, z]}&[Subtract @@ polyEq[z, c, 5]]
```
$$Out[15]= \left\{-c + z + \left(z + \left(z + \left(z + (c + z)^2\right)^2\right)^2\right)^2,\right.$$
$$1 + 2 \left(1 + 2 \left(1 + 2 \left(1 + 2 (1 + 2 (c + z)) (z + (c + z)^2)\right) \left(z + \left(z + (c + z)^2\right)^2\right)\right)\right.$$
$$\left.\left(z + \left(z + (c + z)^2\right)^2\right)\right) \left.\left(z + \left(z + \left(z + (c + z)^2\right)^2\right)^2\right)\right\}$$

Using `FindRoot`, we calculate precise values for the bifurcation points, starting with the values contained in the list `bifurcations`.

```
In[16]:= findC[{{i_, j_}, {{c0_}}}] :=
           {{i, j}, FindRoot[Evaluate[eqs], {z, nsolsData[[i, c0]]},
                            {c, -4 + 8(c0 - 1)/100},
                            AccuracyGoal -> 30, WorkingPrecision -> 22,
                            MaxIterations -> 200]}

In[17]:= cs = Select[Check[findC[#], Sequence @@ {}, FindRoot::cvnwt]& /@
                       bifurcations, (* approximate zero imaginary part *)
                       Abs[Im[#[[2, 2, 2]]]/Re[#[[2, 2, 2]]]] < 10^-20&];

         (* display sorted version *)
         {#1, N[#2]}& @@@ Chop[cs]
Out[20]= {{{1, 2}, {z -> -1.75885, c -> -0.0031681}},
          {{3, 4}, {z -> -0.924367, c -> -0.200433}}, {{5, 6}, {z -> -0.711172, c -> -0.600805}},
          {{7, 8}, {z -> -2.00327, c -> 0.529347}}, {{9, 10}, {z -> -0.733166, c -> -0.732302}},
          {{11, 12}, {z -> -1.22405, c -> 0.23974}}, {{15, 16}, {z -> -1.94394, c -> 1.17006}},
          {{17, 18}, {z -> -2.17027, c -> 1.83064}}, {{19, 20}, {z -> -2.24873, c -> 2.80645}},
          {{21, 22}, {z -> -1.9451, c -> 1.56424}}, {{23, 24}, {z -> -2.49721, c -> 3.73662}},
          {{25, 26}, {z -> -2.07614, c -> 2.86137}}, {{31, 32}, {z -> -2.04158, c -> 3.43857}}}
```

At these values for c, we have a double root. Within the precision of the result, the roots are all real. Here is a quick check for the first value.

```
In[21]:= doubleRoots[c -> c0_] :=
           With[{(* calculate high-precision values for the roots *)
                 nroots = List @@ (Last /@ NRoots[(eqs[[1]] /. c -> c0) == 0, z, 32])},
                 (* find identical roots *)
                 Position[Table[Abs[nroots[[i]] - nroots[[j]]],
                               {i, 32}, {j, i + 1, 32}], _?(# == 0&), {2}]]

In[22]:= Length[doubleRoots[cs[[1, 2, 2]]]]

Out[22]= 17
```

At a generic value for c, all roots are simple.

```
In[23]:= doubleRoots[c -> -1/3]

Out[23]= {}
```

Finally, we have a close look at a bifurcation point by displaying the parts of the above picture near to it.

```
In[24]:= zoomIn[{{i_, j_}, {z -> z0_, c -> c0_}}, opts___] :=
           Show[pic, opts, BoxRatios -> {1, 1, 1}, PlotRange ->
                {{0.9 Re[z0], 1.1 Re[z0]}, {-0.1, 0.1}, {0.9 c0, 1.1 c0}}];

In[25]:= zoomIn[cs[[8]]]
```

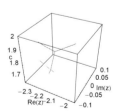

We end with a graphic showing the solutions of `polyEq[x, c, 4]`=0 for complex values of c over the complex x-plane.

```
In[26]:= Show[Graphics3D[{PointSize[0.004],
           Table[{Hue[φ/(2 Pi)], Point[{Re[#], Im[#], ρ}]}& /@
                      (* use values from concentric circles for c *)
                      (x /. NSolve[polyEq[x, N[ρ Exp[I φ]], 4], x])},
```

```
                    {ρ, 1/4, 2, 7/4/30}, {φ, 0, 2Pi, 2Pi/120}}],
                BoxRatios -> {1, 1, 3/2}]
```

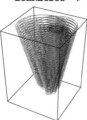

For the mathematical details of these kinds of iterated mappings, see, for example, [110].

b) This defines the iterated logistic map.

```
In[1]:= f[x_, c_] = x^2 - c;
        poly[c_, n_] := Factor[Nest[f[#, c]&, x, n] - x];
```

We start by determining all different period-n fixed points for $c = 2$ (all period-n fixed points are real there).

```
In[3]:= periods[c_, n_] :=
        Module[{roots, δ, next, cycles, rootGroups},
          (* calculate fixed points; use sufficient precision *)
          roots = x /. {ToRules[NRoots[Factor[poly[c, n]] == 0, x, 50]]};
          (* find period n fixed points *)
          (* minimal root separation *)
          δ = Min[Abs[Apply[Subtract, Partition[Sort[roots], 2, 1], {1}]]];
          If[δ < 10^-10, Print["Root separation too small"]];
          (* find cycles by applying f once *)
          Apply[Set[next[#1], #2]&, Transpose[{Range[2^n],
                Flatten[Position[roots - #, _?((# == 0.)&)]& /@
                    (f[#, 2]& /@ roots)]}], {1}];
          cycles = Select[Union[RotateLeft[#, Position[#, Min[#]][[1, 1]] - 1]& /@
            (Rest[NestWhileList[next, #, UnsameQ, All]]& /@ Range[2^n])],
              (Length[#] === n)&];
          (* return groups of period n fixed points *)
          rootGroups = Map[roots[[#]]&, cycles, {-1}]]
```

Here is the number of period-n fixed points for $n \le 7$.

```
In[4]:= NMax = 7;
        Table[Length[periodData[n] = periods[2, n]], {n, NMax}]
Out[5]= {2, 1, 2, 3, 6, 9, 18}
```

Using the calculated period-n fixed points as starting values we numerically solve the differential equation $\partial(poly(x(c), c))/\partial c = 0$ along the trajectories of the fixed points $x^*(c)$.

```
In[6]:= Do[ode = D[Nest[f[#, c]&, x[c], n] - x[c], c];
          rootTrajectories[n] = MapIndexed[
            (* watch progress Print[{n, #2}]; *)
            (* avoid messages from potential singularities *)
            Internal`DeactivateMessages[
              NDSolve[{ode == 0, x[2] == #}, x[c], {c, -2, 4},
                  WorkingPrecision -> 22]]&, periodData[n], {2}],
        {n, NMax}]
```

rootTrajectoryGraphic uses the interpolating function from *nsol* to generate a graph of the period-n fixed points.

```
In[7]:= rootTrajectoryGraphic[nsol_, col_] :=
        Plot[Evaluate[x[c] /. nsol], Evaluate[Flatten[{c, nsol[[1, 1, 2, 0, 1]]}]],
            PlotPoints -> 100, DisplayFunction -> Identity,
            PlotStyle -> {{Thickness[0.002], col}}]
```

Here is a graphic of the resulting period-n fixed points in the c,x-plane [1554].

```
In[8]:= Show[Table[rootTrajectoryGraphic[#, Hue[0.77 (n - 1)/(NMax - 1)]]& /@
                        Flatten[rootTrajectories[n], 1], {n, NMax}],
```

```
        DisplayFunction -> $DisplayFunction, Frame -> True,
        Axes -> False, PlotRange -> {{-2, 4}, {-3, 3}}]
```

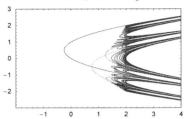

If we are only interested in a visualization, we can use purely graphical techniques instead of solving differential equations.

```
In[9]:= Show[Table[With[{n = k},
            (* substitute value of k before auto-compilation *)
        ContourPlot[Nest[(#^2 - c)&, x, n] - x, {c, -2, 4}, {x, -3, 3},
            (* use many points *) PlotPoints -> 1000, Contours -> {0},
            ContourShading -> False, AspectRatio -> 1/GoldenRatio,
            ContourStyle -> {{Hue[0.77(n - 1)/6], Thickness[0.002]}},
            DisplayFunction -> Identity]], {k, 7, 1, -1}],
            DisplayFunction -> $DisplayFunction]
```

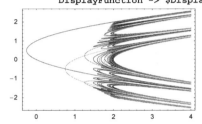

For a symbolic method to determine the extremal *c*-values for *r* periods, see [316].

We end with the equivalent graphic for the logistic map of the form $x_n = \mu \, x_{n-1}(1 - x_{n-1})$.

```
In[10]:= Show[Table[With[{n = k},
            (* substitute value of k before auto-compilation *)
        ContourPlot[Nest[(μ # (1 - #))&, x, n] - x, {μ, 2.8, 4}, {x, 10^-10, 1},
            (* use many points *) PlotPoints -> 1200, Contours -> {0},
            ContourShading -> False, AspectRatio -> 1/GoldenRatio,
            ContourStyle -> {{Hue[0.8 (n - 1)/15], Thickness[0.002]}},
            DisplayFunction -> Identity]], {k, 16, 1, -1}],
            DisplayFunction -> $DisplayFunction, PlotRange -> {{2.8, 4}, {0, 1}}]
```

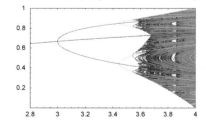

c) This is the map under consideration.

```
In[1]:= f[x_] := Exp[-((x - 1/2)/(3/10))^2]
```

Iterating the map 1000 times shows no regularity. To have numerically reliable results we use high-precision numbers with sufficiently many digits.)

```
In[2]:= nl1 = NestList[f, N[1, 300], 1000];
```

```
In[3]:= Min[Precision /@ nl1]
```

```
Out[3]= 3.78887
```

```
In[4]:= ListPlot[nl1]
```

Iterating the map from another starting point x_0 shows no relation to the first list nl1.

```
In[5]:= nl2 = NestList[f, N[1/3, 300], 1000];
```

```
In[6]:= Show[GraphicsArray[ListPlot[#, Frame -> True, Axes -> False,
                                    DisplayFunction -> Identity]& /@
              (* second list and difference *) {nl2, nl1 - nl2}]]
```

Adding now a random number at each iteration step yields again "random" sequences.

```
In[7]:= SeedRandom[111];
        nl1 = NestList[f[# + Random[Real, {-1, 1}, 300]]&, N[1/5, 300], 500];
```

```
In[9]:= SeedRandom[111];
        nl2 = NestList[f[# + Random[Real, {-1, 1}, 300]]&, N[7/9, 300], 500];
```

But now the sequences converge for increasing n [223].

```
In[11]:= Show[GraphicsArray[
         Block[{$DisplayFunction = Identity,
                opts = Sequence[PlotRange -> All, Frame -> True, Axes -> False]},
              {(* the two individual sequences *)
               ListPlot[nl1, opts], ListPlot[nl2, opts],
               (* the difference of the two individual sequences *)
               Off[Graphics::gptn]; ListPlot[Log[10, Abs[nl1 - nl2]], opts]}]]]
```

```
In[12]:= Show[GraphicsArray[ListPlot[#,
          DisplayFunction -> Identity]& /@ {nl1, nl2}]]
```

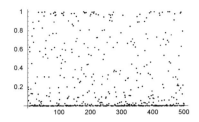 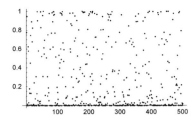

For human balancing applications of added noise, see [323].

2. Analytic Functions with Finite Domains of Analyticity, $\cos_q(z)$ and $\sin_q(z)$

a) The direct numerical computation of the sum $z^{k!}$ using NSum without special option settings fails. This is because of the huge exponents corresponding to $k!$ that occur already for default values of the NSumTerms and NSumExtraTerms option setting: the *NSumTermsDefaultValue* + *NSumExtraTermsDefaultValue*=27.

```
In[1]:= fF[z_?NumericQ, k_?NumericQ] := z^(k!)

       NSum[fF[0.4 Exp[2.87 I], k], {k, 1, Infinity}]
```
  General::unfl : Underflow occurred in computation. More…

  General::unfl : Underflow occurred in computation. More…

  General::unfl : Underflow occurred in computation. More…

  General::stop :
   Further output of General::unfl will be suppressed during this calculation. More…

```
Out[2]= -0.248608 + 0.020519 i
```

Even if we take only a very few terms in the summation, we still have problems.

```
In[3]:= NSum[fF[0.4 Exp[2.87 I], k], {k, 1, Infinity},
            NSumTerms -> 6, NSumExtraTerms -> 6]
```
  NIntegrate::ploss :
   Numerical integration stopping due to loss of precision. Achieved
    neither the requested PrecisionGoal nor AccuracyGoal; suspect one of
    the following: highly oscillatory integrand or the true value of the
    integral is 0. If your integrand is oscillatory on a (semi-)infinite
    interval try using the option Method->Oscillatory in NIntegrate. More…

```
Out[3]= -0.248608 + 0.020519 i
```

We can get an answer with the appropriate choice of a method. For $|z| < 1$, we need not bother with checking convergence, and moreover, we do not really need to include 25 terms as long as do not come too close to the unit circle.

```
In[4]:= f[z_] := NSum[z^(k!), {k, 1, Infinity},
                   (* appropriate options for z^(k!) *)
                   Method -> SequenceLimit, NSumTerms -> 6,
                   VerifyConvergence -> False, NSumExtraTerms -> 5]

In[5]:= f[0.4 Exp[2.87 I]]

Out[5]= -0.248608 + 0.020519 i
```

We define the following obvious special rules for f.

```
In[6]:= f[_?((# == 0.)&)] = 0;
```

In the following plots, the height corresponds to the real part and the absolute value, and the color corresponds to the phase of the function.

```
In[7]:= Show[GraphicsArray[
        Block[{h, $DisplayFunction = Identity},
        ParametricPlot3D[{r Cos[φ], r Sin[φ], #[h = f[r Exp[I φ]]]},
                         {EdgeForm[], Hue[(* modified Arg *)
```

```
          If[(a = If[h == 0, 0, Arg[h]]) > 0, a, 2Pi + a]/(2Pi)]}},
     {r, 0, 0.99}, {φ, 0, 2Pi}, BoxRatios -> {1, 1, 0.6},
     PlotPoints -> {90, 160}, Lighting -> False,
     Boxed -> True, Axes -> True, Compiled -> False,
     ViewPoint -> {-2, -2, 2}, PlotRange -> All]& /@
     (* real part and absolute value *) {Re, Abs}]]]
```

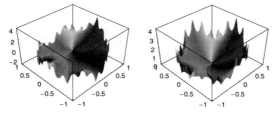

For $|z| < 0.999$, we get a precise enough result for graphics by summing just the first eight terms. To see the region near $|z| = 1$ better, we rescale $r = |z|$ so that we approach logarithmically to $|z| = 1$.

```
In[8]:= f[R_, φ_, o_:8] = Sum[#^(k!), {k, o}]&[N[(1 - 10^(-R)) Exp[I φ]]];
```

Approaching the unit circle results in a rapidly fluctuating behavior of f, here visualized by showing the real part of f by surrounding zero with three different radii.

```
In[9]:= Show[GraphicsArray[{
     Plot[Re[f[#, φ]], {φ, 0, 2Pi}, PlotPoints -> 40, PlotRange -> All,
          DisplayFunction -> Identity, PlotLabel -> "r = " <> ToString[
          N[(1 - 10^(-#))]], AxesLabel -> {"R", "f"}]& /@ {1, 2, 3}}]]
```

To see the behavior over a broader range of r-values, we use a `ContourPlot`. The vertical axis measures the logarithmic distance to the unit circle.

```
In[10]:= Module[{ppφ = 301, ppr = 240, cts = 60, values, z,
     s, i, R, h, conts, p, li, df, m = 5.},
     (* array of function values *)
     values = Join[{Table[0, {ppφ + 1}]},
                    Table[Re[f[R, φ, 9]], {R, m/ppr, m, m/ppr},
                          {φ, 0, N[2Pi], N[2Pi/ppφ]}]];
     (* the contours *) conts = Table[i, {i, -4, 4, 8/cts}];
     (* remember the requested values for graylevels *)
     h[x_?NumberQ] := (AppendTo[li, x]; GrayLevel[1]); li = {};
     (* the following calls two times ListContourPlot
        to generate alternating black and white stripes *)
     Graphics[ListContourPlot[values, ColorFunction -> h,
     Contours -> conts, DisplayFunction -> Identity]];
     (* construct black-white coloring color function *)
     MapIndexed[(df[#1] = GrayLevel[If[EvenQ[#2[[1]]], 0, 1]])&, Sort[li]];
     ListContourPlot[values, ColorFunction -> df, Contours -> conts,
                     AspectRatio -> 1/Pi, FrameTicks -> None,
                     ContourLines -> False]]
```

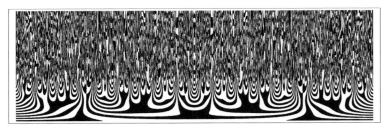

Other functions "similar" to the one discussed are [792] $f(z) = \sum_{k=1}^{\infty} k^{-2} z^{k^k}$ or [1464] ($\sigma_{10}(k)$ is the sum of the digits of k in base 10) $f(z) = \sum_{k=1}^{\infty} \sigma_{10}(k) z^k$ (For the use of such functions for generating interesting space-filling curves, see [1455], [1582], and [1594].)

Rewriting the truncated power series as a rational function in the form of the diagonal Padé approximation allows frequently to obtain accurate numerical values within a disk bounded by the first zero of the denominator [1727]. The following input uses the [200/200] Padé approximation of the series $\sum_{k=1}^{\infty} k^2 z^{k^2}$ to visualize the resulting function. Again, a dense set of singularities forms on the unit circle.

```
In[11]:= Needs["Calculus`Pade`"]

In[12]:= Module[{op = 200, pp = 401, L = 7/4, pNAB, pNABCF, data},
          (* calculate diagonal Padé approximation *)
          pNAB[z_] = Pade[Sum[N[k^2, 300] z^(k^2), {k, 0, 3/2 Sqrt[op]}],
                        {z, 0, op, op}];
          (* compile diagonal Padé approximation *)
          pNABCF = Compile[{{z, _Complex}}, Evaluate[N[pNAB[z]]]];
          (* values of the Padé approximation *)
          data = Table[pNABCF[x + I y], {y, -L, L, 2L/pp}, {x, -L, L, 2L/pp}];
          (* homogeneous contour spacing *)
          contourValues[λ_, n_] := Function[l, #[[Round[Length[#]/n/2]]]& /@
                        Partition[l, Round[Length[l]/n]]][Sort[Flatten[λ]]];
          (* make graphics of argument and real part *)
          Show[GraphicsArray[
           Block[{$DisplayFunction = Identity},
            ListContourPlot[#1[data], Contours -> contourValues[#1[data], #2],
                        ColorFunctionScaling -> #3, ContourLines -> False,
                        Frame -> True, FrameTicks -> False,
                        PlotRange -> #4, ColorFunction -> #5]& @@@
                             (* option details *)
                             {{Arg, 30, False, {-Pi, Pi}, (Hue[0.78#]&)},
                              {Re, 60, True, All, (Hue[Random[]]&)}}]]]]
```

We potentially have boundaries of analyticity not only for power series, but also for continued fractions [1766]. The next graphic shows the real part of $1 + 1/(z + 1/(z^2 + 1/(z^3 + ...)))$ truncated after 100 denominators over the complex z-plane. This continued fraction can be written as the power series $1 + z^2 + z^4 - z^5 + z^6 - 2z^7 + 2z^8 - 3z^9 + 4z^{10} + ...$ for odd o [1031]. (Here we just take enough terms in FromContinuedFraction to obtain a reliable numerical result; for speeding up the convergence of continued fractions, see [1416].)

```
In[13]:= cf1[z_, o_] := FromContinuedFraction[Table[N[z]^k, {k, 0, o}]]

            (* small modification for large k *)
```

```
cf2[z_, o_] := FromContinuedFraction[Table[N[z]^(k + I Cos[k]/k),
                                            {k, o}]]
```

```
In[16]:= With[{o = 100, ε = 10^-6},
         Show[GraphicsArray[
         Block[{$DisplayFunction = Identity},
         (* 3D plots of the two continued fractions *)
         ParametricPlot3D[{r Cos[φ], r Sin[φ], Re @ #[r Exp[I φ], o],
                         {EdgeForm[], SurfaceColor[#, #, 2.4]&[
                            Hue[Arg[#[r Exp[I φ], o]]^2/Pi^2]]}},
                         {r, 10^-6, 1.5}, {φ, -Pi + ε, Pi - ε},
                         PlotRange -> {-5, 5}, PlotPoints -> {180, 240},
                         Boxed -> False, Axes -> False,
                         BoxRatios -> {1, 1, 1}]]& /@
         (* the two continued fractions *) {cf1, cf2}]]]
```

The similar continued fraction $1/(1 + z/(1 + z^2/(1 + z^3/\ldots)))$ [1104], [1105] shows also a boundary of analyticity.

In truncations of the infinite sum or the continued fraction boundaries of analyticity result in clustering of zeros along closed curves. Here are the zeros of the degree 120 (numerator) polynomials from the sum and the continued fraction shown.

```
In[17]:= Show[GraphicsArray[
         (* show zeros in the complex plane *)
         Graphics[Point[{Re[#], Im[#]}]& /@ (z /. NSolve[# == 0, z]),
                 Frame -> True, PlotRange -> All, AspectRatio -> Automatic]& /@
         (* truncated forms; degree 120 *) {Sum[z^k!, {k, 5}], cf1[z, 15]}]]]
```

Infinite products can too exhibit boundaries of analyticity. An example is the following product of Riemann zeta functions: $\prod_{k=1}^{\infty} \zeta(k\,z)$. The boundary of analyticity is the imaginary axis and the functions exists only in the right half plane.

We end with a "random analytic function" of the form $w(z) = \sum_{k=1}^{\infty} k^{1/2}\, c_k\, z^k$ [502], [1679], [1680], [865], [1681]. Here the c_k are Gaussian random variables and we will truncate the sums after 1000 terms. Given the list of the c_k, the compiled function `raf` calculates the value $w(z)$.

```
In[18]:= raf = Compile[{{cs, _Complex, 1}, {z, _Complex}},
           Module[{sum = 0. + 0. I, λ = Length[cs], k = 1,
                 ζ = 1. + 0. I, μ = Max[Abs[cs]] 10.^-10},
                 (* sum until terms summands are small enough *)
                 While[k < λ && Abs[ζ] > μ,
                     ζ = z ζ; sum = sum + Sqrt[k] cs[[k]] ζ; k++];
                 sum]];
```

Here is an array of function values of a "random analytic function" in polar coordinates.

```
In[19]:= SeedRandom[777];
         (* list of random coefficients *)
```

```
cList = Table[Sqrt[2] Exp[I Random[Real, {0, 2Pi}]]*
              InverseErf[Random[Real, {-1, 1}]], {1000}];

Module[{ρ = 10.^-6, (* keep distance to unit circle *) R = 0.99,
        (* enough plot points *) ppr = 300, ppφ = 600},
  data = Table[{r Cos[φ], r Sin[φ], raf[cList, r Exp[I φ]]},
               {r, ρ, R, (R - ρ)/ppr}, {φ, 0., 2.Pi, 2.Pi/ppφ}]];
```

The next graphics show the real and imaginary parts, the absolute value, and the argument. For coloring, we scale the first three quantities. For most realizations of the c_k, the unit circle is again a boundary of analyticity.

```
In[23]:= (* make colored polygons from array data *)
        makeColoredPolygons[reIm_, scaleFunction_, data_] :=
        Table[{{Hue[scaleFunction[Plus @@ reIm[Last /@ #]/4]],
             Polygon[Drop[#, -1]& /@ #]}&[
               {data[[i, j]], data[[i + 1, j]],
                data[[i + 1, j + 1]], data[[i, j + 1]]}]},
             {i, Length[data] - 1}, {j, Length[data[[1]]] - 1}]

In[25]:= Show[GraphicsArray[Graphics[#, AspectRatio -> Automatic]& /@
           (* show Re, Im, Abs, Arg; scale appropriately *)
           {makeColoredPolygons[ Re, 0.78 (ArcTan[#] + Pi/2)/Pi&, data],
            makeColoredPolygons[ Im, 0.78 (ArcTan[#] + Pi/2)/Pi&, data],
            makeColoredPolygons[Abs, 0.78 (ArcTan[#]/(Pi/2))&, data],
            makeColoredPolygons[Arg, (# + Pi)/(2Pi)&, data]},
           GraphicsSpacing -> 0.01]]
```

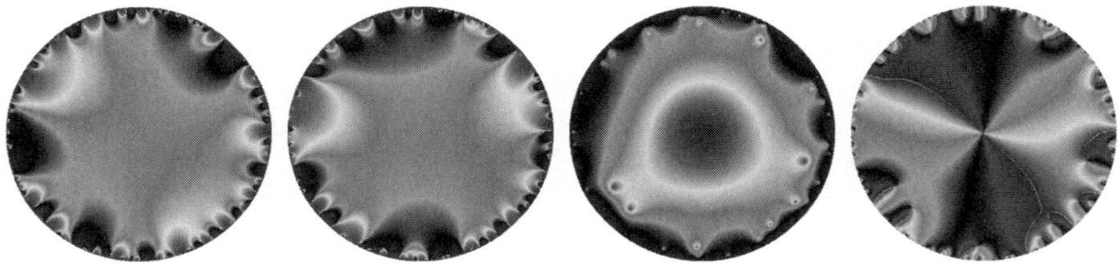

For series that converge to different functions in different regions, see [1484], [265].

b) The implementation of the given series are straightforward. We sum the series as long as the terms are increasing in size and as long as we have not reached the desired precision. (For more advanced methods to calculate related products, see [1682].)

```
In[1]:= qCos[z_, q_, prec_] :=
       Module[{sum = 1, old = 1, new, qfac = 1, k = 0},
       While[k = k + 1;
             (* term of the series *)
             qfac = qfac*(q^(2k - 1) - q^-(2k - 1))/(q - 1/q)*
                    (q^(2k) - q^-(2k))/(q - 1/q);
             new = (-1)^k z^(2k)/qfac q^-k/(q - 1/q)^(2k);
             (* more terms needed? *)
             Abs[new] > Abs[old] || Log[10, Abs[new]] > -2 prec || k < 6,
             sum = sum + new; old = new]; sum + new]

In[2]:= (* get precision from input precision *)
       qCos[z_, q_] := qCos[z, q, Precision[{z, q}]] /;
                       InexactNumberQ[z] || InexactNumberQ[q]

In[4]:= qSin[x_, q_, prec_] :=
       Module[{old = x q/(q - 1/q), new, sum, qfac = 1, k = 0},
       sum = old;
       While[k = k + 1;
             (* term of the series *)
             qfac = qfac*(q^(2k) - q^-(2k))/(q - 1/q)*
                    (q^(2k + 1) - q^-(2k + 1))/(q - 1/q);
             new = (-1)^k x^(2k + 1)/qfac q^(k + 1)/(q - 1/q)^(2k + 1);
             (* more terms needed? *)
```

```
            Abs[new] > Abs[old] || Log[10, Abs[new]] > - 5/4 prec || k < 5,
            sum = sum + new; old = new]; sum + new]
In[5]:= qSin[z_, q_] := qSin[z, q, Precision[{q, z}]] /;
            InexactNumberQ[z] || InexactNumberQ[q]
```

Now, let us check the three mentioned identities for "random" values of z and q.

```
In[6]:= With[{z = N[1 + 2/3 I, 200], q = 13/10},
            (1/z (qSin[z, q] - qSin[z/q^2, q])) - qCos[z, q]]
Out[6]= 0. × 10^{-199} + 0. × 10^{-199} i
```

```
In[7]:= With[{z = N[1/4 + 11/5 I, 200], q = 2 + I},
            (1/z (qCos[z, q] - qCos[z/q^2, q])) - (-1/q^2 qSin[z/q^2, q])]
Out[7]= 0. × 10^{-201} + 0. × 10^{-201} i
```

```
In[8]:= With[{z = N[-17/24 + 1/8 I, 200], q = -17/7 - 2/11 I},
            qCos[z, q] qCos[q z, q] + 1/q qSin[z, q] qSin[z/q, q] - 1]
Out[8]= 0. × 10^{-201} + 0. × 10^{-201} i
```

For a visualization, we will implement a compiled version of qCos. We rewrite the form of the terms in such a manner that the largest occurring exponents are as small as possible to allow for the use of compiled code for a large range of arguments.

```
In[9]:= qCosC =
        Compile[{{z, _Complex}, {q, _Complex}, {prec, _Integer}},
        Module[{sum = 1. + 0.I, old = 1. + 0.I, new = 1. + 0.I, k = 0},
        While[k = k + 1;
            new = -z^2/(-1 - q^2 + q^(2 - 4k) + q^(4k)) old;
            Abs[new] > Abs[old] || Log[10, Abs[new]] > -2 prec || k < 6,
            sum = sum + new; old = new]; sum + new]];
```

qCos and qCosC give the same results.

```
In[10]:= Table[Function[{z, q}, qCos[z, q] - qCosC[z, q, 10 ]][
            Random[Complex, Pi{-1 - I, 1 + I}],
            Random[Complex, Pi{-1 - I, 1 + I}]], {100}] // Abs // Max
Out[10]= 0.0000223504
```

Using qCosC, we can quickly generate some visualizations of the q deformed cosine function. We start with $\cos_q(x)$ over the x,q-plane. Because of the symmetry $\cos_q(x) = \cos_q(-x) = \cos_{-q}(x)$, we display the function only in the first quadrant.

```
In[11]:= Plot3D[qCosC[N[x], N[q], 6], {x, 0.0001, Pi}, {q, 0.001, 2},
            PlotPoints -> 400, Compiled -> False,
            Mesh -> False, PlotRange -> {-1, 1}]
```

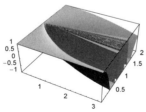

The next picture shows a contour plot of the region with the most structure from the last picture—around the line $q = 1$.

```
In[12]:= ContourPlot[qCosC[N[x], N[q], 6], {x, 0.001, Pi - 0.001}, {q, 0.5, 1.5},
                PlotPoints -> {200, 100},
                ColorFunction -> (Hue[Random[]]&), Contours -> 30,
                ContourLines -> False, AspectRatio -> Automatic]
```

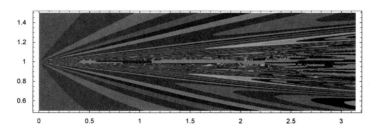

Now, let us show $\cos_q(z)$ over the complex q-plane. We will use $z = 1$, although other values of z will give very similar pictures.

```
In[13]:= Needs["Graphics`Graphics3D`"]
```

```
In[14]:= {ppr, ppφ} = {100, 200};
        {r, R} = {0.01, 2};
        data = Table[{qr Cos[qφ], qr Sin[qφ], qCosC[1, N[qr Exp[I qφ]], 6]},
                    {qr, r, R - r, (R - r)/ppr}, {qφ, 0, 2Pi, 2Pi/ppφ}];
```

Around the unit circle $|q| = 1$, we see a lot of structure in $\mathrm{Re}(\cos_q(1))$ and $\mathrm{Im}(\cos_q(1))$.

```
In[17]:= Show[GraphicsArray[
        Function[reIm, Graphics3D[{EdgeForm[],
        SurfaceColor[Hue[0.8], Hue[0.1], 2.3], Cases[#, _Polygon, Infinity]&[
        ListSurfacePlot3D[Apply[{#1, #2, reIm[#3]}&, data, {2}],
                    DisplayFunction -> Identity]]},
                    PlotRange -> {-3/2, 2}, BoxRatios -> {1, 1, 0.6}]] /@
                    (* real and imaginary part *) {Re, Im}]]
```

A circular contour plot (see Exercise 16 of Chapter 3 of the Graphics volume [1794]) for $\mathrm{Re}(\cos_q(1))$ and $\mathrm{Im}(\cos_q(1))$ also shows a dense set of singularities along the unit circle. (But because of the $[n]_q!$, the function values for $|q| > 1$ are not the analytical continuation of the values for $|q| < 1$.)

```
In[18]:= (* make contour plot of real and imaginary part *)
        lcps = Function[reIm,
            ListContourPlot[Apply[reIm[#3]&, data, {2}],
                    ColorFunction -> (Hue[2#]&), Contours -> 50,
                    MeshRange -> {{0, 2Pi}, {r, R - r}},
                    DisplayFunction -> Identity]] /@ {Re, Im};
In[20]:= (* map a polygon to a disk *)
        toDisk[Polygon[l_], δ_] :=
        Module[{refinedLine, upperHalfDisk},
        refinedLine = Join @@
          (If[(* add new points if needed *)
            (φDist = Sqrt[#.#&[#[[2]] - #[[1]]]]) < δ, #,
            δl = #[[2]] - #[[1]]; φl = Ceiling[φDist/δ];
            Table[#[[1]] + i/φl δl, {i, 0, φl }]& /@
            (* line segments *)
                    (N[Partition[Append[l, First[l]], 2, 1]]));
        Polygon[Apply[#2 {Cos[#1], Sin[#1]}&, refinedLine, {1}]]]
In[22]:= (* show circular contour plots of real and imaginary parts *)
        Show[GraphicsArray[
        Show[DeleteCases[Graphics[#], _Line, Infinity] /.
```

```
(* map to disk *) p_Polygon :> toDisk[p, 0.1],
    AspectRatio -> Automatic, Frame -> False]& /@ lcps]]
```

The following limit holds: $\lim_{q \to 0} \cos_q((q - q^{-1}) z) = \cos(z)$. (For visualizations of q-exponential functions, see [1132].)

3. NDSolve **Fails, Integro-Differential Equation, Franel Identity, Bloch Particle in an Oscillating Field, Coupled and Accelerated Particles**

a) Currently, NDSolve solves equations that can be solved analytically for the highest derivative (not in the preprocessing stage and also not at runtime in case the option SolveDelayed is set to True). We define f[z].

```
In[1]:= f[z_] := z^5 + Sin[5z + 1] + 2
```

Then the following differential equation can be solved (actually it is already in canonical form).

```
In[2]:= NDSolve[{y'[x] == f[y[x]], y[0] == 1}, y, {x, 0, 0.1}]
Out[2]= {{y → InterpolatingFunction[{{0., 0.1}}, <>]}}
```

However, the equation obtained by exchanging $y(x) \leftrightarrow y'(x)$ cannot be solved.

```
In[3]:= NDSolve[{y[x] == f[y'[x]], y[0] == 1}, y, {x, 0, 0.1}]
```

> Solve::tdep : The equations appear to involve the variables
>   to be solved for in an essentially non-algebraic way. More…
>
> NDSolve::ntdv : Cannot solve to find an explicit formula for the
>   derivatives. Consider using the option setting SolveDelayed->True. More…

```
Out[3]= NDSolve[{y[x] == 2 + Sin[1 + 5 y'[x]] + y'[x]^5, y[0] == 1}, y, {x, 0, 0.1}]
```

Also, the SolveDelayed -> True option setting does not help here. NDSolve cannot find appropriate starting values.

```
In[4]:= NDSolve[{y[x] == f[y'[x]], y[0] == 1}, y, {x, 0, 0.1}, SolveDelayed -> True]
Out[4]= {{y → InterpolatingFunction[{{0., 0.1}}, <>]}}
```

For other starting values, $y(0)$ NDSolve does succeed finding a starting value for $y'(0)$.

```
In[5]:= NDSolve[{y[x] == f[y'[x]], y[0] == 2}, y, {x, 0, 0.1}, SolveDelayed -> True]
Out[5]= {{y → InterpolatingFunction[{{0., 0.1}}, <>]}}
```

(For methods to solve such-type equations, see [843].)

Note that when it is possible to solve the differential equation for the highest power, frequently there are several solutions for this highest power and NDSolve will compute solutions corresponding to each of them. This is the case for the following differential equation of second order.

```
In[6]:= NDSolve[{y''[x]^3 == y[x]^3 + y[x]^2 - Sin[5 y[x]],
            y[0] == 1 + I, y'[0] == 1 - I}, y, {x, 0, 1.2}]
Out[6]= {{y → InterpolatingFunction[{{0., 1.2}}, <>]},
    {y → InterpolatingFunction[{{0., 1.2}}, <>]},
    {y → InterpolatingFunction[{{0., 1.2}}, <>]}}
```

Here, the solution consists of three independent InterpolatingFunction-objects. (For multivalued solutions of Newton's equations, see [501].)

```
In[7]:= ParametricPlot[Evaluate[Transpose[{Re[y[x]] /. %, Im[y[x]] /. %}]],
            {x, 0, 1.2}, PlotRange -> All,
            PlotStyle -> {{Thickness[0.01], Hue[0]},
```

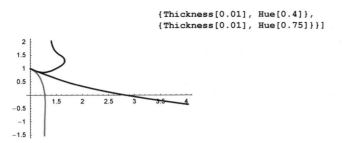

```
                                              {Thickness[0.01], Hue[0.4]},
                                              {Thickness[0.01], Hue[0.75]}}]
```

Using `SolveDelayed -> True` results in no solution.

```
In[8]:= NDSolve[{If[x > -1, y''[x]^3, y''[x]^3] == y[x]^3 + y[x]^2 - Sin[5 y[x]],
            y[0] == 1 + I, y'[0] == 1 - I}, y, {x, 0, 1.2}, SolveDelayed -> True]
        NDSolve::mconly :
            For the method NDSolve`IDA, only machine real code is available. Unable to
                continue with complex values or beyond floating point exceptions. More…

Out[8]= {}
```

Even if the differential equation can be solved for the highest derivative, `NDSolve` might still fail to calculate a solution. This happens, for instance, if the coefficients are singular at the starting point. Here is an example.

```
In[9]:=  NDSolve[{y''[x] ==  y[x]^(3/2)/Sqrt[x],
              y[0] == 1, y'[0] == 1}, y, {x, 0, 1}]

        Power::infy : Infinite expression  1/√0.  encountered. More…

        NDSolve::ndnum :
            Encountered non-numerical value for a derivative at x == 0.`. More…
```

$$Out[9]= \text{NDSolve}\left[\left\{y''[x] = \frac{y[x]^{3/2}}{\sqrt{x}},\ y[0] = 1,\ y'[0] = 1\right\}, y, \{x, 0, 1\}\right]$$

By moving an ε away from the singular point (and using a series solution for the dependent function), `NDSolve` will typically succeed.

```
In[10]:= yTFSeries[x_] :=
            1 + 4/3 x^(3/2) + x^3/3 + 2/27 x^(9/2) + 4/405 x^6  (* + ... *)
In[11]:= With[{ε = 10^-12},
            NDSolve[{y''[x] == If[x < ε, yTFSeries[x], y[x]^(3/2)/Sqrt[x]],
                y[0] == 1, y'[0] == 1}, y, {x, 0, 1}]]
Out[11]= {{y → InterpolatingFunction[{{0., 1.}}, <>]}}
```

Here is another differential equation that the built-in `NDSolve` cannot solve. Effectively this is a delay-differential equation [1657], [1148].

```
In[12]:= NDSolve[{x''[t] == x[x[t]], x[0] == 1, x'[0] == 0}, x, {t, 0, 2}]

          NDSolve::nestdv : The expression x[t] has nested dependent variables. More…

Out[12]= NDSolve[{x''[t] = x[x[t]], x[0] = 1, x'[0] = 0}, x, {t, 0, 2}]
```

Another class of differential equations that cannot be solved accurately by `NDSolve` is discussed in [1477], [1478], [1879], [8], [1457].

b) We solve the problem by iteration, starting with sin(x). We do not use `NIntegrate` to calculate the integral on the right-hand side of the equation, but obtain it by solving the corresponding differential equation (the resulting `Interpolat` `ingFunction`-objects can be used to get many needed values for the integral in a fast way). Here is the iteration implemented with `FixedPointList`.

```
In[1]:= fpl = FixedPointList[Function[y, Module[{int, int1, y1},
            (* the integral *)
            int = NDSolve[{int1'[x] == y[x], int1[0] == 0},
                    int1, {x, 0, 3Pi}][[1, 1, 2]];
            (* solve the differential equation *)
```

```
NDSolve[{y1'[x] == Sin[x] + 1/(ArcTan[int[x]] + int[x]^2 + 1), y1[0] == 1},
         y1, {x, 0, 3Pi}, Method -> {"LSODA", "DenseOutput" -> True},
         PrecisionGoal -> 10][[1, 1, 2]]]],
    Function[x, Sin[x]], (* convergence test *)
    SameTest -> (Max[Abs[Table[#1[x] - #2[x],
                                {x, 0, 3Pi, 3Pi/1000}]]] < 10^-7&)];
```

After only a few iterations, the process has already converged.

In[2]:= **Length[fp1]**

Out[2]= 8

Here, the iterative solutions are visualized.

In[3]:= **Plot[Evaluate[#[x]& /@ fp1], {x, 0, 3Pi}, PlotRange -> All]**

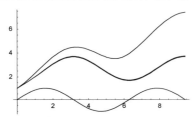

c) The integrand $(\{x/a\} - 1/2)(\{x/b\} - 1/2)$ has many discontinuities. To carry out a reliable numerical calculation, we use the points at which discontinuities appear inside the second argument of NIntegrate (we could use the Symbolic↴ PiecewiseSubdivision, but giving the points explicitly is faster).

In[1]:= (* in case no intermediate points are specified
         $MaxPiecewiseCases = Infinity *)

```
integrate[a_, b_, opts___] :=
NIntegrate[Evaluate[(FractionalPart[x/a] - 1/2) (FractionalPart[x/b] - 1/2)],
   (* intermediate points are not needed for method
      SymbolicPiecewiseSubdivision *)
   Evaluate[Prepend[Union[Flatten[{Table[k/a, {k, 0, a^2 b}],
                          Table[k/b, {k, 0, a b^2}]}]], x]], opts]
```

Some "random" values for *a* and *b* let us conjecture that $1/\alpha = 12$.

In[3]:= **{integrate[4, 5], integrate[16, 17]}**

Out[3]= {0.0833333, 0.0833333}

Carrying out a high-precision calculation for 36 different *a,b*-pairs yields a "numerical confirmation" for the result $1/\alpha = 12$. This is indeed the correct result [1908].

In[4]:= **1/Table[1/(a b) NIntegrate[**(* evaluate the integrand *)
    Evaluate[(FractionalPart[x/a] - 1/2) (FractionalPart[x/b] - 1/2)],
    (* evaluate the integration region *)
    Evaluate[Prepend[Union[Flatten[{Table[k/a, {k, 0, a^2 b}],
                          Table[k/b, {k, 0, a b^2}]}]], x]],
            WorkingPrecision -> 30, PrecisionGoal -> 20]/
            (GCD[a, b]/LCM[a, b]), {a, 8}, {b, a}] - 12 // Abs // Max

Out[4]= 0. × 10^{-29}

d) The function calculateAndGraph implements the solution of the differential equation as well as a plot of the phase-space trajectory. To be able to quantitatively judge the quality of the solution, we also return the values *x*(200) and *v*(200).

In[1]:= **calculateAndGraph[opts___] :=**
    Module[{ε = 3/2, ω = 5/3, F = 13/100},
     (* solve differential equation *)
     nsol = NDSolve[{x'[t] == v[t], v'[t] == -F + Sin[x[t]] - ε Cos[t ω]],
                     x[0] == 0, v[0] == 10}, {x, v}, {t, 200},
                     MaxSteps -> 10^5, opts];

```
(* visualize solution *)
ParametricPlot[Evaluate[{x[t], v[t]} /. nsol], {t, 0, 200},
               PlotRange -> All, PlotPoints -> 2000,
               Frame -> True, Axes -> False];
{(* return position and velocity *)
 {x[200], v[200]} /. nsol[[1]]}]
```

Here is a first try to calculate the phase-space trajectory. We use the default option values of NDSolve.

In[2]:= **calculateAndGraph[] // Timing**

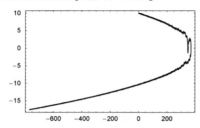

Out[2]= {0.45 Second, {{-768.729, -17.5633}}}

Using a slightly higher precision and accuracy goal results in completely different values.

In[3]:= **calculateAndGraph[PrecisionGoal -> 10, AccuracyGoal -> 10] // Timing**

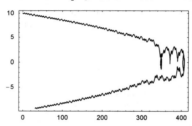

Out[3]= {0.32 Second, {{31.3382, -9.2363}}}

Using high-precision numbers instead of machine numbers gives again a different result. The time to calculate the solution is now much larger.

In[4]:= **calculateAndGraph[WorkingPrecision -> $MachinePrecision + 1,**
               PrecisionGoal -> 10, AccuracyGoal -> 10] // Timing

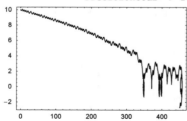

Out[4]= {3.87 Second, {{450.2377988034477, -2.669742982174119}}}

To get a reliable result for the phase-space trajectory, we have to increase the working precision and the precision and accuracy goal until we get a result no longer changing.

In[5]:= **calculateAndGraph[WorkingPrecision -> $MachinePrecision + 10,**
               PrecisionGoal -> 10, AccuracyGoal -> 10] // Timing

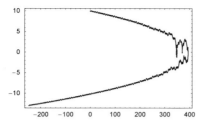

Out[5]= {4.99 Second, {{-247.7816758580770571654114, -13.00235493294069453095164}}}

Doubling the accuracy and precision goals dramatically increases the calculation times, but it makes the result much more reliable.

In[6]:= calculateAndGraph[WorkingPrecision -> 40,
                        PrecisionGoal -> 20, AccuracyGoal -> 20] // Timing

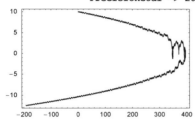

Out[6]= {34.5 Second, {{-192.51528465902120814729325421670 6543285,
          -12.47735610597910250431465119528 49696334}}}

Using the extrapolation method yields the same result for the first few digits, but faster.

In[7]:= calculateAndGraph[Method -> "Extrapolation", WorkingPrecision -> 60,
                        PrecisionGoal -> 30, AccuracyGoal -> 30] // Timing

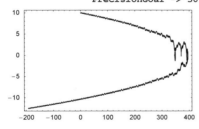

Out[7]= {25.09 Second, {{-192.5422694074260882334381190181 0392564899597836307 49936768,
          -12.47618728938990434798525549183 98958975331153784970 59088035}}}

The last two results agree to about four digits, and this indicates that the solution is now correct. We could go on and carry out more validation by using different Method option settings.

e) The equations, their numerical solution, and the calculation of the average velocity are straightforward to implement. We use $T = 999$ and 1000 values for ε. We start with random initial conditions. To get reproducible results, we seed the random number generator.

```
In[1]:= With[(* parameter values *) {n = 10, pp = 1000, T = 999},
    (* seed random number generation *) SeedRandom[111];
    Module[{f = Sin, vsEnd, vsOld, vsNew, vsStart, data},
    (* the coupling function *)
    eqs[ε_] = Table[v[k]'[t] == (k - 1)/(n - 1) +
                    ε (f[v[Mod[k + 1, n, 1]]][t] - v[k][t]] +
                        f[v[Mod[k - 1, n, 1]]][t] - v[k][t]]), {k, n}];
    (* solve odes and return the v[k][T] *)
    Internal`DeactivateMessages[
```

```
vsEnd[ε_, ics_List] := Table[v[k][T], {k, n}] /.
  NDSolve[Join[eqs[ε], MapIndexed[(v[#2[[1]]][0] == #1)&, ics]],
          Table[v[k], {k, n}], {t, T, T}, MaxSteps -> 10^6][[1]];
(* generate the data; start with random values *)
vsOld = Table[Random[], {n}];
data = Table[vsNew = vsEnd[ε, vsStart = vsOld - Min[vsOld]];
              points = Point[{ε, #}]& /@ ((vsNew - vsStart)/T);
              (* reuse relative phase knowledge *)
              vsOld = vsNew; points, {ε, 0, 9/8, 9/8/pp}]];
(* show result *)
Show[Graphics[{PointSize[0.003], data}],
     PlotRange -> All, Frame -> True]] // Timing
```

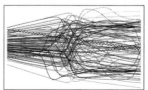

We see an interesting clustering/bifurcation behavior as a function of the coupling between neighboring particles. The next graphic show the $v_k(t)$ for $n = 100$ for various values of ε.

```
In[2]:= Module[{n = 100, T = 600},
        (* differential equations *)
        eqs[ε_] = Table[Derivative[1][v[k]][t] == (k - 1)/(n - 1) +
                        ε (Sin[v[Mod[k + 1, n, 1]][t] - v[k][t]] +
                           Sin[v[Mod[k - 1, n, 1]][t] - v[k][t]]), {k, n}];
        (* show graphics *)
        Show[GraphicsArray[#]]& /@
        Partition[Function[ε, SeedRandom[1];
        (* solve differential equations *)
        ndsol = NDSolve[Join[eqs[ε], MapIndexed[(v[#2[[1]]][0] == #1)&,
                                               Table[Random[], {n}]]],
                        Table[v[k], {k, n}], {t, 0, T}, MaxSteps -> 10^6,
                        PrecisionGoal -> 8, AccuracyGoal -> 8];
        (* the velocity averages *)
        avs = Table[v[k][T]/T, {k, n}] /. ndsol[[1]];
        (* make plots of solution curves *)
        Plot[Evaluate[Table[v[k][t] - avs[[k]] t, {k, n}] /. ndsol[[1]]],
                    {t, T/4, 3T/4}, PlotPoints -> 120, DisplayFunction -> Identity,
                    PlotStyle -> Table[{Thickness[0.003], Hue[0.76 k/n]}, {k, n}],
                    PlotRange -> All, Frame -> True, Axes -> False,
                    FrameTicks -> None]] /@
                                (* values for ε *) {0, 0.05, 0.25, 0.5, 1, 4}, 3]]
```

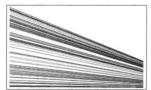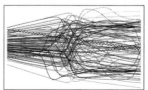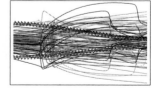

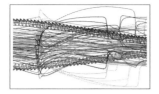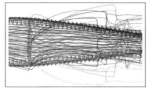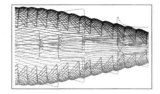

4. Trick of James of Courtright, Hannay Angle, Harmonic Nonlinear Oscillator, Closed Orbits

a) For the purposes of visualizing this trick, we neglect both the finite diameter of the pencil and the friction. (Of course, friction is actually responsible for the fact that the system stops, but even without friction, the heavy object would not fall to the ground.) First, we derive the equations of motion of the system. The Lagrangian [325] is

$$\mathcal{L} = \frac{m}{2}\left(x'(t)^2 + y'(t)^2\right) + \frac{M}{2}\, Y'(t)^2 - m\, g\, y(t) - M\, g\, Y(t).$$

Here, $x(t)$ and $y(t)$ are the Cartesian coordinates of the lighter object, m is its mass, $Y(t)$ is the coordinate of the heavier object (which can only move vertically), M is its mass, and g is the acceleration because of gravity. The system is subject to the side condition $-Y(t) + (x(t)^2 + y(t)^2)^{1/2} = l$ (l is the length of the string). We change to polar coordinates $\{r, \varphi\}$ using $x(t) = r(t)\cos(\varphi(t))$ and $y(t) = r(t)\sin(\varphi(t))$ which leads to the following Lagrangian `lagrangianL`.

```
In[1]:= lagrangianL = Simplify[Expand[
            m/2 (D[r[t] Cos[φ[t]], t]^2 + D[r[t] Sin[φ[t]], t]^2) +
            M/2  D[r[t] - 1, t]^2 - m g r[t] Sin[φ[t]] - M g (r[t] - 1)]]

Out[1]= 1/2 (2 g l M - 2 g r[t] (M + m Sin[φ[t]]) + (m + M) r'[t]² + m r[t]² φ'[t]²)
```

We now get the equations of motion from the Lagrangian via

$$\frac{d}{dt}\frac{\partial\mathcal{L}}{\partial r'(t)} - \frac{\partial\mathcal{L}}{\partial r(t)} = 0, \qquad \frac{d}{dt}\frac{\partial\mathcal{L}}{\partial\varphi'(t)} - \frac{\partial\mathcal{L}}{\partial\varphi(t)} = 0.$$

We can simplify this somewhat by dividing out common factors and introducing the mass relationship $\mu = M/m$.

```
In[2]:= {radialEq, tangentialEq} = (Simplify[
            D[D[lagrangianL, #'[t]], t] - D[lagrangianL, #[t]]] == 0)& /@ {r, φ}

Out[2]= {g (M + m Sin[φ[t]]) - m r[t] φ'[t]² + (m + M) r''[t] == 0,
         m r[t] (g Cos[φ[t]] + 2 r'[t] φ'[t] + r[t] φ''[t]) == 0}
```

We can now create the following program for visualizing the motion. Here, μ is the mass relation, $r0$ is the initial distance of the lighter object from the center of rotation, and $\varphi0$ is the initial angle in degrees.

```
In[3]:= CourtrightPath[μ_?((NumberQ[#] && # > 1)&), r0_?((NumberQ[#] && # > 0)&),
            φ0_?((NumberQ[#] && - 60 < # < 240)&)] :=
        With[{g = 9.81},
        Module[{tMax = 1.12 Sqrt[2 r0/(g (μ + 1)/(μ - 1))], sol, r, φ, opts},
        (* solve the equations of motions *)
        sol = NDSolve[
        {g μ + g Sin[φ[t]] - r[t] φ'[t]^2 + r''[t] + μ r''[t] == 0,
        g Cos[φ[t]] + 2 φ'[t] r'[t] + r[t] φ''[t] == 0,
        r[0] == r0, r'[0] == 0, φ[0] == φ0 2Pi/360, φ'[0] == 0},
        {r, φ}, {t, 0, tMax}, MaxSteps -> 3000];
        (* options for all three graphics *)
        opts[axesLabels_, plotLabel_] :=
        Sequence[AxesLabel -> axesLabels, PlotRange -> All, AxesOrigin -> {0, 0},
                DisplayFunction -> Identity, PlotLabel -> StyleForm[plotLabel,
                                    FontWeight -> "Bold", FontSize -> 6]];
        (* make some pictures *)
        Show[GraphicsArray[{(* (x(t), y(t)) *)
        ParametricPlot[Evaluate[{r[t] Cos[φ[t]], r[t] Sin[φ[t]]} /. sol],
                {t, 0, tMax}, Evaluate[opts[{"x", "y"}, "y(x)"]]],
            (* r(t) *) Plot[Evaluate[r[t] /. sol], {t, 0, tMax},
                    Evaluate[opts[{"t", "r"}, "r(t)"]]],
```

```
(* φ(t) *) Plot[Evaluate[φ[t] /. sol], {t, 0, tMax},
              Evaluate[opts[{"t", "φ"}, "φ(t)"]]]}]]]]
```

Here is a typical example.

In[4]:= CourtrightPath[60, 1.3, 40]

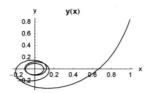 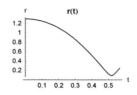 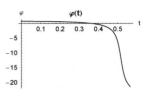

We clearly see the rotation, as well as the small movement of the light object toward the center of rotation, and then away from it. For a detailed discussion, see [1223], [734], and [1631].

b) r is the radius vector of the bead, $\Omega[t]$ is the rotation angle of the ellipse, and $\varphi[t]$ the rotation angle of the bead.

```
In[1]:= r = {{Cos[Ω[t]], Sin[Ω[t]]}, {-Sin[Ω[t]], Cos[Ω[t]]}}.
          {2 Cos[φ[t]] - Sqrt[3], Sin[φ[t]]};
```

The kinetic energy of the bead is given by the following expression.

```
In[2]:= L = D[r, t].D[r, t] // Simplify
```

$$Out[2]= \frac{1}{2} \left((5 - 3 \cos[2\,\varphi[t]]) \, \varphi'[t]^2 + \right.$$
$$\left. 4 \left(-2 + \sqrt{3} \cos[\varphi[t]]\right) \varphi'[t] \, \Omega'[t] + \left(11 - 8\sqrt{3} \cos[\varphi[t]] + 3 \cos[2\,\varphi[t]]\right) \Omega'[t]^2 \right)$$

```
In[3]:= L /. (* limit value *) φ[t] -> 0 /. Ω'[t] -> 0
```

$$Out[3]= \varphi'[t]^2$$

The kinetic energy of the bead is constant. We will call it $\mathcal{E}$. Energy conservation allows us to obtain the equation of motion on the form $\varphi'[t] == \text{rhs}[\varphi[t], \omega, \mathcal{E}]$. The function rhs follows from solving $\mathcal{L} == \mathcal{E}$ with respect to $\varphi'[t]$. Here we assume that the ellipse is rotated with a constant angular velocity $\Omega[t] = \omega$.

```
In[4]:= rhs[φ_, ω_, E_] =
          (4(-2 + Sqrt[3] Cos[φ]) ω - Sqrt[8 E (5 - 3 Cos[2φ]) -
          24(11 - 8 Sqrt[3] Cos[φ] + 3Cos[2φ]) Sin[φ]^2 ω^2])/(6 Cos[2φ] - 10);
```

Using the differential equation $\varphi'[t] == \text{rhs}[\varphi[t], \omega, \mathcal{E}]$, we can calculate the difference in the angle $\delta\varphi$ for the position of the bead on a rotating versus a nonrotating ellipse.

```
In[5]:= δφ[ω_, T_, E_, odeOpts___] :=
          ((* angle of the bead on the rotating ellipse *)
          φ[T] /. (nsol1 = NDSolve[{φ'[t] == rhs[φ[t], ω, E], φ[0] == 0},
                            φ, {t, 0, T}, odeOpts][[1]])) -
          (* angle of the bead on the nonrotating ellipse *)
          (φ[T] /. (nsol2 = NDSolve[{φ'[t] == rhs[φ[t], 0, E], φ[0] == 0},
                            φ, {t, 0, T}, odeOpts][[1]]))
```

Using an angular velocity of $\omega = 1/100$, we follow the angle of the bead over one rotation of the ellipse.

```
In[6]:= δφ[10^-2, 2Pi/10^-2, 2, MaxStepSize -> 1, PrecisionGoal -> 10,
                            AccuracyGoal -> 10, MaxSteps -> 10^6]
```

Out[6]= 5.30213

As a function of time, the angle difference is an oscillating function (as is the angle $\varphi[t]$ of the bead).

```
In[7]:= ListPlot[Table[Evaluate[{t, (φ[t] /. nsol1) - (φ[t] /. nsol2)}],
                            {t, 0, 2Pi/10^-2, 2Pi/10^-2/2000}] // N,
                            PlotStyle -> {PointSize[0.003]}]
```

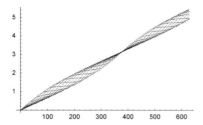

Due to the angle-dependent curvature of the ellipse, a more natural measure than the angle difference is the arclength difference. $\lambda[\varphi]$ expresses the arclength of the ellipse as a function of φ.

```
In[8]:= λ[φ_] = With[{r = r /. _φ -> t /. Ω[t] -> 0},
                Integrate[Sqrt[D[r, t].D[r, t]], {t, 0, φ}]]
Out[8]= EllipticE[φ, -3]
```

$l[\varphi]$ expresses the periodic part of the arclength of the bead position. Its exact value is given by the following expression.

```
In[9]:= l[φ_] =   4 EllipticE[-3] IntegerPart[φ/(2Pi)] + EllipticE[Mod[φ, 2Pi], -3];
```

The difference in arclength is a monotonic function of φ.

```
In[10]:= ListPlot[Table[Evaluate[{t, l[φ[t] /. nsol1] - l[φ[t] /. nsol2]}],
                {t, 0, 2Pi/10^-2, 2Pi/10^-2/100}] // N,
                PlotStyle -> {PointSize[0.003]}, PlotRange -> {0, 9}]
```

δl calculates the difference of the arc lengths.

```
In[11]:= δl[{φ1_, φ2_}] := l[φ2] - l[φ1]
```

To calculate a high-precision value of δl in the limit $\omega \to 0$, we will follow the bead over one revolution (of the bead). The time for one revolution of the bead is $T = \lambda(2\pi)\mathcal{E}^{1/2}$. Multiplying the resulting arclength difference by $2\pi\mathcal{E}^{1/2}/(\omega\lambda(2\pi))$ gives the arclength difference for one revolution of the ellipse. Because we are only interested in the final angle difference, we change the definition for $\delta\varphi$ to only return the value of $\varphi[T]$ from NDSolve. (For the high-precision calculations to be carried out below, this saves a lot of time and memory.)

```
In[12]:= DownValues[δφList]   = DownValues[δφ] /. {t, 0, T} :> {t, T, T} /.
                                a_ - b_ :> {a, b} /. δφ -> δφList
Out[12]= {HoldPattern[δφList[ω_, T_, ε_, odeOpts___]] :>
            {φ[T] /. (nsol1 = NDSolve[{φ'[t] == rhs[φ[t], ω, ε], φ[0] == 0}, φ, {t, T, T}, odeOpts][1]),
             φ[T] /. (nsol2 = NDSolve[{φ'[t] == rhs[φ[t], 0, ε], φ[0] == 0}, φ, {t, T, T}, odeOpts][1])}}
```

In the limit $\omega \to 0$, the energy $\mathcal{E}$ of the bead does not matter; the arclength difference will be independent of $\mathcal{E}$. Using appropriately set options in NDSolve, we now calculate the arc length difference for $\omega = 10^{-k}$ for k = -2, -4, ..., -26.

```
In[13]:= δls = Table[
        Module[{ε = 100, ω = 10^k, T},
                (* period *) T = λ[2Pi]/Sqrt[ε];
                {k, 2Pi/ω/T δl[δφList[ω, λ[2Pi]/Sqrt[ε], ε,
                MaxStepSize -> 10^-3/2, PrecisionGoal -> 40,
                AccuracyGoal -> 25, MaxSteps -> Infinity,
                WorkingPrecision -> 60]]}], {k, -2, -26, -2}]
Out[13]= {{-2, -8.1272593344694401350686030306586145964698832030362154145},
          {-4, -8.1493614209813660252383240532811462466316128453069826},
```

```
{-6, -8.149583062244016123924898638372074968186087525337147},
{-8, -8.1495852787186967542117653425050890081002271285082},
{-10, -8.1495853008834497659408810895352342932584497435},
{-12, -8.1495853011050972966787148850201398991929483 9},
{-14, -8.14958530110731377198615527723880372216947 2},
{-16, -8.1495853011073359367392296873664167418892},
{-18, -8.149585301107336158386760431468313414472},
{-20, -8.1495853011073361606032357389093324 44},
{-22, -8.149585301107336160625400491983742 6},
{-24, -8.1495853011073361606256221395144 9}, {-26, -8.14958530110733616062556243559 90}}
```

The values δls converge nicely. The following graphic shows that δls[[-1, 2]] is correct to about 20 digits.

In[14]:= `ListPlot[Apply[{#1, Log[10, Abs[δls[[-1, 2]] - #2]]}&, Drop[δls, -1], {1}],`
       `Frame -> True, Axes -> False]`

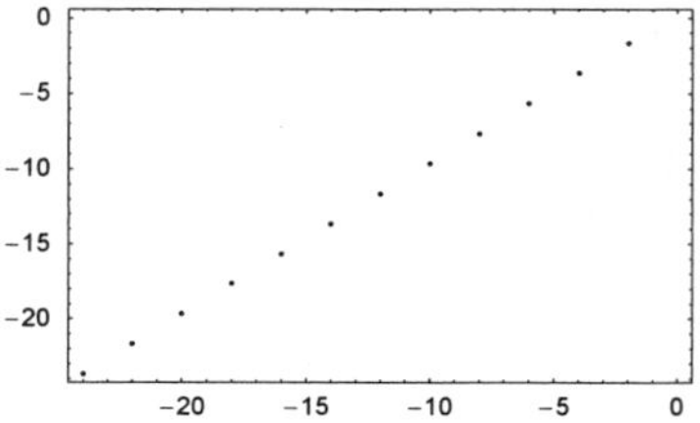

In[15]:= `δls[[-1, 2]]`

Out[15]= `-8.149585301107336160625624355990`

The exact value of the arc length difference can be calculated using the theory of adiabatic invariants [325], and [921] and geometric phases [774] (the Hannay's angle [783], [712], [1206], the classical limit of Berry's phase [193]). It is $-4\pi A/l$ where A is the area and l the circumference of the ellipse. The exact result agrees favorably with the above-calculated numerical one.

In[16]:= `δlExact = With[{A = 1 2 Pi, l = λ[2Pi]}, -4 Pi A/l];`
       `δlExact - δls[[-1, 2]]`

Out[17]= $1.6340159406 \times 10^{-20}$

For the rotating bead under the additional influence of gravity, see [232].

c) For an explicit matrix $\mathbf{M}(t)$, we obtain d^2 coupled nonlinear differential equations for the matrix elements $m_{i,j}(t)$. The following function `nonlinearHarmonicOscillatorGraphic` solves these equations and returns a graphics array showing the parametrized curves $\{\text{Re}(m_{i,j}(t)), \text{Im}(m_{i,j}(t))\}$ and $\{\text{Re}(m_{i,j}(t)), \text{Re}(m'_{i,j}(t))\}$. The solution curve for each element is colored differently. The arguments of `nonlinearHarmonicOscillatorGraphic` are the parameters ω, c, and the initial conditions $\mathbf{M}(0)$ and $\mathbf{M}'(0)$.

In[1]:= `nonlinearHarmonicOscillatorGraphic[ω_, c_, M0_, Mp0_] :=`
```
    Module[{d = Length[M0], M, m, T = 2Pi/ω},
      (* the matrix M[t] *)
      M[t_] = Table[m[i, j][t], {i, d}, {j, d}];
      Show[GraphicsArray[
      (* show real and imaginary parts of solutions and
         real part of function and first derivative *)
      Function[{f1, g1, f2, g2}, Show[MapIndexed[
      ParametricPlot[{f1[g1[#[[2]]]][t]],
                      f2[g2[#[[2]]]][t]]}, {t, 0, T},
          PlotStyle -> {Hue[0.8 #2[[1]]/d^2]}, PlotPoints -> 300,
          PlotRange -> All, DisplayFunction -> Identity]&,
      (* solve differential equations for matrix elements *)
      NDSolve[# == 0& /@ Flatten[
            {M''[t] - 3 I ω M'[t] - 2 ω^2 M[t] - c M[t].M[t].M[t],
             M[0] - M0, M'[0] - Mp0}],
             #[[0]]& /@ Flatten[M[t]], {t, 0, T},
             MaxSteps -> 10^5][[1]]],
```

```
                    PlotRange -> All, Frame -> True, Axes -> False,
                    FrameTicks -> None, AspectRatio -> 1]] @@@
                    (* functions of solution *)
                {{Re, Identity, Im, Identity}, {Re, Identity, Re, Derivative[1]}}]]]
```

We give two explicit examples. The first is a 3×3 matrix with regular initial conditions.

```
In[2]:= Module[{d = 3, c = 1},
        nonlinearHarmonicOscillatorGraphic[11/10, (1 + I)/8,
          Table[c Exp[I c++ 2Pi/d^2], {i, d}, {j, d}],
          Table[0, {d}, {d}]]]
```

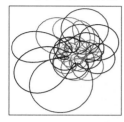 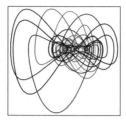

The second is a 4×4 matrix with random initial conditions.

```
In[3]:= SeedRandom[999];
        Module[{d = 4},
          nonlinearHarmonicOscillatorGraphic[1, 2,
                Table[Random[Complex, {-1 - I, 1 + I}], {d}, {d}],
                Table[Random[Complex, {-1 - I, 1 + I}], {d}, {d}]]]
```

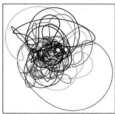 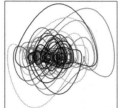

Despite the strong nonlinearity of the original equation and the complicated shape of the solution curves, we clearly see that the solutions are strictly periodic. For other isochronous systems, see [330].

d) We use the simplest possible interpolation between the harmonic oscillator and the Kepler problem, namely a linear one. τ is the interpolation parameter. $\mathcal{H}(0)$ gives the harmonic oscillator and $\mathcal{H}(1)$ gives the Kepler problem. To not just interpolate the potential, we make σ a function of τ, so effectively switching the coordinates and the momenta as τ varies between 0 and 1.

```
In[1]:= α1 = 2 - 2τ; β1 = -τ; α2 = 2τ; β2 = 2 - 2τ; σ = 1 - 2τ;
        H[τ_] = σ p^α1 r^β1 + p^α2 r^β2;

        {H[0], H[1/2], H[1]}
Out[3]= {p² + r², p r, p² - 1/r}
```

Using $\mathcal{H}[\tau]$ in 2D gives formally the following equations of motion.

```
In[4]:= H2D[τ_] = H[τ] /. {r -> Sqrt[qx^2 + qy^2], p -> Sqrt[px^2 + py^2]} /.
        {qx -> qx[t], qy -> qy[t], px -> px[t], py -> py[t]};
In[5]:= eqs[τ_] = {qx'[t] == D[H2D[τ], px[t]], qy'[t] == D[H2D[τ], py[t]],
                px'[t] == -D[H2D[τ], qx[t]], py'[t] == -D[H2D[τ], qy[t]]} //
                                                            Simplify
```

$$Out[5]= \left\{ qx'[t] == px[t] \, (px[t]^2 + py[t]^2)^{-\tau} \, (qx[t]^2 + qy[t]^2)^{-\tau} \right.$$
$$\left(2\,\tau\,(px[t]^2 + py[t]^2)^{-1+2\,\tau}\,(qx[t]^2 + qy[t]^2) + 2\,(-1+\tau)\,(-1+2\,\tau)\,(qx[t]^2 + qy[t]^2)^{\tau/2} \right),$$
$$\left. qy'[t] == py[t]\,(px[t]^2 + py[t]^2)^{-\tau}\,(qx[t]^2 + qy[t]^2)^{-\tau} \right.$$

$$\left(2\,\tau\,(px[t]^2 + py[t]^2)^{-1+2\,\tau}\,(qx[t]^2 + qy[t]^2) + 2\,(-1+\tau)\,(-1+2\,\tau)\,(qx[t]^2 + qy[t]^2)^{\tau/2}\right),$$
$$px'[t] = (px[t]^2 + py[t]^2)^{-\tau}\,qx[t]\,(qx[t]^2 + qy[t]^2)^{-\tau}$$
$$\left(2\,(-1+\tau)\,(px[t]^2 + py[t]^2)^{2\,\tau} + (1-2\,\tau)\,\tau\,(px[t]^2 + py[t]^2)\,(qx[t]^2 + qy[t]^2)^{-1+\frac{\tau}{2}}\right),$$
$$py'[t] = (px[t]^2 + py[t]^2)^{-\tau}\,qy[t]\,(qx[t]^2 + qy[t]^2)^{-\tau}$$
$$\left.\left(2\,(-1+\tau)\,(px[t]^2 + py[t]^2)^{2\,\tau} + (1-2\,\tau)\,\tau\,(px[t]^2 + py[t]^2)\,(qx[t]^2 + qy[t]^2)^{-1+\frac{\tau}{2}}\right)\right\}$$

Now, it is straightforward to implement the animation. The function `animationFrame` solves the equations of motion for a family of orbits and displays the orbits in position and momentum space. All *qx0pp* orbits start on the *x*-axis and their initial momentum is mostly in *y*-direction.

```
In[6]:= animationFrame[τ_, T_, qx0pp_, opts___] :=
    Module[{nsol, T}, Off[NDSolve::ndsz]; Off[NDSolve::mxst];
    (* position and momentum space orbits *)
    GraphicsArray[Show /@ Transpose[
    Table[nsol = (* solve equations of motion numerically *)
          NDSolve[Join[eqs[τ], {qx[0] == (* family parameter *) qx0,
                               qy[0] == 0, px[0] == 0.02, py[0] == 0.2}],
                  {qx, qy, px, py}, {t, 0, T}, MaxSteps -> 10^5];
    (* maximal t *) T = nsol[[1, 1, 2, 1, 1, 2]];
    (* plot solution curve; color according to qx0-value *)
    ParametricPlot[Evaluate[# /. nsol[[1]]], {t, 0, T},
                  PlotPoints -> Round[60 T] + 20,
                  PlotStyle -> {{Hue[0.8 (qx0 - 1/2)], Thickness[0.002]}},
                  DisplayFunction -> Identity, FrameTicks -> False,
                  Frame -> True, PlotRange -> 2 {{-1, 1}, {-1, 1}},
                  AspectRatio -> Automatic, Axes -> False]& /@
                  (* show r and p *) {{qx[t], qy[t]}, {px[t], py[t]}},
                  {qx0, 1/2, 3/2, 1/(qx0pp - 1)}]], GraphicsSpacing -> 0]]
In[7]:= Show[GraphicsArray[#]]& /@
    Partition[animationFrame[#, 12, 20]& /@
             Table[τ, {τ, -0.2, 1.2, 1.4/11}], 2]
```

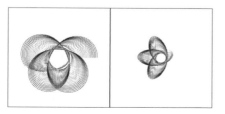

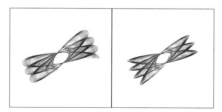
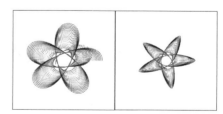

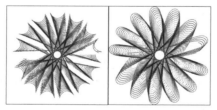
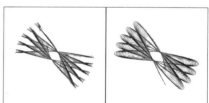

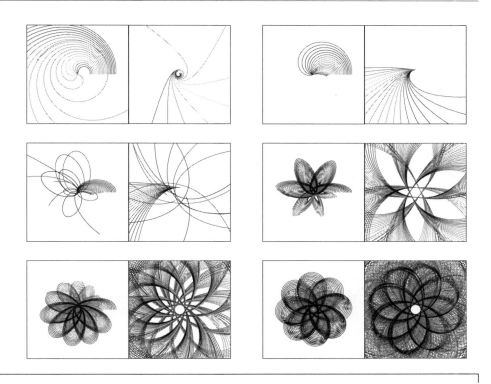

```
Do[Show[animationFrame[τ, 20, 12, PlotLabel -> N[τ]]], {τ, -0.2, 1.2, 1.4/120}]
```

Interestingly, all orbits seem to be closed for $\tau \approx 0.23$ and $\tau \approx 0.364$. And around $\tau \approx 0.8$, a singularity forms.

In[8]:= **Show[animationFrame[0.23, 20, 12]]**

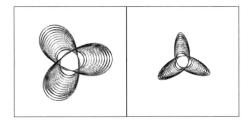

In[9]:= **Show[animationFrame[0.364, 20, 12]]**

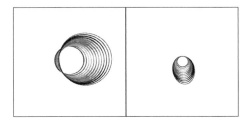

Using a different parametrization, we can generate more closed orbits. For other Hamiltonians with closed orbits, see [881], [448], [882], [1934].

5. Eigenvalue Problems

a) After some thought, and/or a study of the references on Sturm–Liouville eigenvalue problems (see, e.g., [1857], [273], [1217], and [1137]), we come to the following conclusions:

1) The eigenfunction corresponding to the smallest eigenvalue is symmetric about $x = 0$, that is, $y'(0) = 0$.

2) For $|x| \ll 1$, the solution oscillates, and for $|x| \gg 1$, it is exponentially decreasing like $\exp(-x^3/3)$.

3) The smallest eigenvalue is of order 1.

In view of these properties, we first look for a rough approximation for λ_0. (A simple renormalization group approach at $x = \infty$ yields the value $\lambda_0 \approx 0$ [632].)

```
In[1]:= sol[λ_?NumberQ] :=
        (y[3 (* should be some number far enough in the classically forbidden region
                (to use quantum mechanics slang) *)] /.
        NDSolve[{-y''[x] + x^4 y[x] == λ y[x], y[0] == 1, y'[0] == 0},
                y, {x, 0, 3}])[[1]]

In[2]:= Plot[sol[λ], {λ, 0.1, 2}]
```

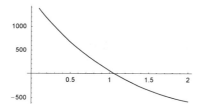

Using `FindRoot`, we can now find a more accurate solution.

```
In[3]:= FindRoot[sol[λ] == 0, {λ, 1.0, 1.1}, AccuracyGoal -> 8]

Out[3]= {λ → 1.06036}
```

Next, we graph the associated eigenfunction.

```
In[4]:= Plot[Evaluate[y[x] /. NDSolve[{-y''[x] + x^4 y[x] == 1.06036 y[x],
                                        y[0] == 1, y'[0] == 0}, y, {x, 0, 3}]],
        {x, 0, 3}]
```

Now let us determine more digits of λ_1. To do this, we solve the differential equation as long as the (numerical) solution is positive and decreasing. `xMax[λ]` gives the value of x where this was no longer the case. Because we want to extend the integration as far as possible to the right and are not interested in the exact stopping point, we do not use any event location method.

```
In[5]:= xMax[λ_?NumberQ] :=
        NDSolve[{- y''[x] + x^4 y[x] == SetPrecision[λ, Infinity] y[x],
                y[0] == 1, y'[0] == 0},
                y, {x, 0, (* this should be enough *) 100},
                (* a rough stopping test *)
```

```
                    StoppingTest -> (y[x] < 0 || y'[x] > 0),
                    WorkingPrecision -> 30][[(* take out the x-value *) 1, 1, 2, 1, 1, 2]]
```

As a function of λ, the function xMax[λ] has a minimum around $\lambda \approx 1.06$.

In[6]:= `Plot[xMax[λ], {λ, 1/2, 2}, Axes -> False, Frame -> True]`

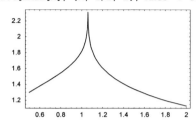

Using the above-calculated value for λ, we search for the maximum of xMax[λ]. As a side effect, we monitor the λ and xMax[λ] values.

In[7]:= `toMax[λ_?NumberQ] := Module[{res}, AppendTo[l, {λ, res = xMax[λ]}]; res]`

`l = {};`

```
FindMaximum[toMax[λ], {λ, 106035/100000, 106037/100000},
          WorkingPrecision -> 100, MaxIterations -> 100,
          PrecisionGoal -> 8, AccuracyGoal -> 8]
```
Out[9]= {4.855650050996382833196182583300929696435180362037892260865898525948038614961596893`
        726409756497344138, {λ →
        1.060362090484177214414798718615078725371038809881708631424439434850275759602159 2`
        6476819540182 0271809}}

Looking at the change in λ during the minimization process shows that about nine digits of the result are reliable.

In[10]:= `ListPlot[Abs[Log[10, (First /@ Drop[l, -1]) - l[[-1, 1]]]]]`

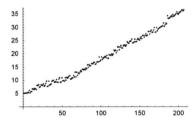

Looking now at xMax[λ] shows that to achieve a better value for λ, we need a higher WorkingPrecision in the numerical differential equation-solving process inside xMax.

In[11]:= `ListPlot[Last /@ l]`

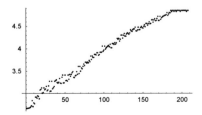

The StoppingTest option is a relatively crude way to determine x_{max}. To extract the largest nonnegative x where $y(x)$ is nonincreasing more precisely we use the InterpolatingFunction itself. This gives three more correct digits at the same working precision and less time. The next input uses a working precision of 30.

```
In[12]:= xMax[λ_?NumberQ] :=
    Module[{nsol, f, g, xu, xl},
     nsol = NDSolve[{y'[x] == ys[x], -ys'[x] + x^4 y[x] ==
                          SetPrecision[λ, Infinity] y[x],
            y[0] == 1, ys[0] == 0},
            {y, ys}, {x, 0, (* this should be enough *) 100},
            MaxStepSize -> 1/10, MaxSteps -> 10^4,
            PrecisionGoal -> 20, AccuracyGoal -> 20,
            StoppingTest -> (y[x] < -10^-3 || ys[x] > 10^-3),
            WorkingPrecision -> 30][[1]];
     (* function that should vanish *)
     f[x_?NumberQ] = y[x] ys[x] /. nsol;
     g[x_?NumberQ] := f[SetPrecision[x, 30]];
     (* starting value for FindRoot *)
     xu = nsol[[1, 2, 1, 1, 2]]; xl = 99/100 xu;
     (* find the root *)
     x /. FindRoot[g[x], {x, xl, xu}, AccuracyGoal -> 60,
                      WorkingPrecision -> 30, MaxIterations -> 50]]

In[13]:= FindMaximum[xMax[λ], {λ, 106035/100000, 106037/100000},
               WorkingPrecision -> 50, MaxIterations -> 100,
               PrecisionGoal -> 12, AccuracyGoal -> 12]

    FindRoot::cvmit : Failed to converge to the
       requested accuracy or precision within 50 iterations. More…

    FindRoot::cvmit : Failed to converge to the
       requested accuracy or precision within 50 iterations. More…

    FindRoot::cvmit : Failed to converge to the
       requested accuracy or precision within 50 iterations. More…

    General::stop : Further output of
       FindRoot::cvmit will be suppressed during this calculation. More…

    FindMaximum::fmdig : 50.` working digits is
       insufficient to achieve the requested accuracy or precision. More…

Out[13]= {4.6452624756366182169581252932337428032673584544699,
      {λ → 1.060362090484182900053397645148262401112455498454б}}
```

The correct result is $1.060362090484182899\ldots$ (see Chapter 2 of the Symbolics volume [1795]). This time, we got about 14 correct digits.

For some high-order perturbation theory techniques to calculate the eigenvalues under consideration, see [593] and Chapter 2 of the Symbolics volume [1795].

b) The implementation of the computation of the coefficient determinant and its solution is straightforward. (Note the application of Expand to prevent the blowup of deter[i].)

```
In[1]:= eigenvalues[{x0_, x1_}, func_Function, number_Integer, points_Integer] :=
    Module[{$recli = $RecursionLimit, x, δ, w, deter, e},
           (* locally change $RecursionLimit *)
           $RecursionLimit = Infinity;
           (* the definitions *)
           δ = (x1 - x0)/(points + 1);
           x[i_] := x[i] = x0 + i δ;
           w[i_] := w[i] = (func[x[i]] - λ) δ^2 + 2;
           (* initial conditions for deter *)
           deter[1] = w[1]; deter[2] = w[1] w[2] - 1;
           (* the recursive definition of the determinant *)
           deter[i_] := deter[i] =
           Expand[w[i] deter[i - 1] - deter[i - 2]];
           (* solve the eigenvalue equation *)
           sol = λ /. (List /@ ((List @@ (Take[NRoots[deter[points] == 0, λ],
                      number])) /. Equal -> Rule));
           (* reset $RecursionLimit before leaving Block *)
           $RecursionLimit = $recli; sol] /; number <= points  && x0 < x1
```

A test using the simplest possible problem $-y''(x) = λ y(x)$, $y(0) = y(1) = 0$ shows the expected results.

```
In[2]:= eigenvalues[{0, 1}, 0&, 6, 50]
Out[2]= {9.86648, 39.4285, 88.5739, 157.116, 244.796, 351.279}
```

This coincides well with the following exact result for the lower eigenvalues.

```
In[3]:= Table[i^2 Pi^2, {i, 6}] // N
Out[3]= {9.8696, 39.4784, 88.8264, 157.914, 246.74, 355.306}
```

For this simple case, it is also possible to calculate the eigenvalues exactly [733].

```
In[4]:= Module[{x0 = 0, x1 = 1, points = 50, δ, mii, mij},
            δ = (x1 - x0)/(points + 1); mii = 2/δ^2; mij = -1/δ^2;
            Table[mii + 2 mij Cos[j Pi/(points + 1)], {j, 6}] // N]
Out[4]= {9.86648, 39.4285, 88.5739, 157.116, 244.796, 351.279}
```

We can also solve more difficult problems. The use of NRoots basically determines how much time will be required. Unfortunately, there is no built-in or package function for computing the smallest and/or largest zeros, which are generally the most interesting ones. Here are the eigenvalues of a confined harmonic oscillator.

```
In[5]:= eigenvalues[{-1, 1}, 100 #^2&, 3, 20]
Out[5]= {9.94642, 29.7682, 49.6376}
```

The results agree with the ones obtained by diagonalizing the tridiagonal matrix.

```
In[6]:= eigenvaluesMat[{x0_, x1_}, func_Function, number_Integer, points_Integer] :=
        Module[{x, δ, n = points, mat },
                (* the definitions *) δ = (x1 - x0)/(n + 1);
                (* the coefficient matrix *)
                mat = Table[(* all possible cases *)
                        Which[i == j, func[x0 + i δ] + 2/δ^2,
                              Abs[i - j] == 1, -1/δ^2,
                              True, 0], {i, n}, {j, n}];
                (* calculate eigenvalues *)
                Take[Sort[Eigenvalues[N[mat]]], number]] /;
                            number <= points  && x0 < x1

In[7]:= eigenvaluesMat[{0, 1}, 0&, 6, 50]
Out[7]= {9.86648, 39.4285, 88.5739, 157.116, 244.796, 351.279}

In[8]:= eigenvaluesMat[{-1, 1}, 100 #^2&, 3, 20]
Out[8]= {9.94642, 29.7682, 49.6376}
```

For a fairly large collection of exactly solvable eigenvalue problems, see, for instance, [615] and [429]; for general numerical solution techniques of Sturm-Liouville problems, see [1487] and [989]. For an especially fast method to calculate such eigenvalue problems, see [1406].

c) We start by defining the potential $V(x) = x^4 - 4 x^2$ through $c_2 = -5$ and $c_4 = 1$.

```
In[1]:= o = 4;
        {c[2], c[4], c[_]} = {-5, 1, 0};
```

Next, we calculate the optimal value of ω_0.

```
In[3]:= eqω0 = 1 - 1/ω0^2 - Sum[c[m] m!/
            (2^(m - 1) ω0^(m/2 + 1) ((m/2) - 1)!), {m, 2, o, 2}];

        {Ω0} = Cases[ω0 /. Solve[eqω0 == 0, ω0] // N[#, 30]&, _Real?(# > 0&)]
Out[4]= {0.673593058218709979627647841749}
```

For the efficient calculation of the matrix elements $h_{i,j}(\omega_0)$, we use nested sums and If-statements (instead of multiplying later not needed expressions by zero).

```
In[5]:= h[{i_, j_}, ω0_] :=
        If[i == j, (1 + ω0^2)/(2ω0) + (1 + ω0^2)/(ω0) j, 0] +
        If[i == j - 2, (1 - ω0^2)/(2ω0) Sqrt[j (j - 1)] , 0] +
        If[i == j + 2, (1 - ω0^2)/(2ω0) Sqrt[(j + 1)(j + 2)] , 0] +
```

```
Sum[Sum[Sum[If[i == j + m - 2k - 2n,
              c[m] m! Sqrt[j! (j + m - 2k - 2n)!]/
              ((2ω0)^(m/2) 2^k k! n! (j - n)! (m - 2k - n)!), 0],
              {n, 0, m - 2k}], {k, 0, Floor[m/2]}], {m, 0, o}]
```

Using matrices of dimensions 50, 100, and 200 shows a good convergence of the lowest eigenvalues.

```
In[6]:= lowestEigenvalues[dim_, k_, ω0_:Ω0] := Sort[
          Eigenvalues[Table[h[{i, j}, ω0], {i, 0, dim}, {j, 0, dim}]]][[Range[k]]]

In[7]:= {1e50, 1e100, 1e200} = lowestEigenvalues[#, 2]& /@ {50, 100, 200}

Out[7]= {{-1.710350379774648944907733963, -1.247921840808687106368979122},
         {-1.710350450132301130196365629, -1.247922492065826733054749341},
         {-1.710350450132639012458042121, -1.247922492066213924852025953}}
```

For $n = 50$, we calculate the lowest eigenvalues as a function of ω_0 and compare with `1e200`. Because of the double well nature of the potential, the above value of $\Omega0$ is not optimal, but too small. The above optimal ω_0 formula is best suited for single minima potentials.

```
In[8]:= data = Table[{ω0, lowestEigenvalues[50, 2, ω0]}, {ω0, 0.1, 6, 0.1}];

In[9]:= Show[GraphicsArray[Table[
          ListPlot[{#[[1]], Log[10, Abs[(1e200 - #[[2]])[[k]]]]}& /@ data,
                   DisplayFunction -> Identity, PlotRange -> All], {k, 2}]]]
        Graphics::gptn : Coordinate Indeterminate
            in {5.6, Indeterminate} is not a floating-point number. More…
```

We end by calculating the two lowest eigenvalues by solving the differential equation directly.

```
In[10]:= V[x_] := x^4 - 4 x^2;

In[11]:= Y[ε_?NumberQ, {y0_, y0p_}] := (* solve until x == 6 where y[x] is small *)
          y[6] /. NDSolve[{-y''[x] + V[x] y[x] == ε y[x], y[0] == y0, y'[0] == y0p},
                           y, {x, 0, 6}, MaxSteps -> 10^4, WorkingPrecision -> 30,
                           PrecisionGoal -> 12, AccuracyGoal -> 12][[1]]

In[12]:= Off[FindRoot::frdig];
         With[{opts = Sequence[WorkingPrecision -> 35, AccuracyGoal -> 25,
                               MaxIterations -> 50]},
             (* root finding for the two initial conditions *)
             {FindRoot[Y[ε, {1, 0}] == 0, {ε, -16/10, -18/10}, opts],
              FindRoot[Y[ε, {0, 1}] == 0, {ε, -12/10, -13/10}, opts]}]

Out[13]= {{ε -> -1.71035045013577753523718235100019304},
          {ε -> -1.24792249206712294290811481479993758}}
```

The so-obtained differential equation-based values agree with the ones from the diagonalization of the matrix $h_{i,j}$.

```
In[14]:= (ε /. %) - 1e200

Out[14]= {-3.138522779140230×10^{-12}, -9.09018056088862×10^{-13}}
```

d) $m[\{k, l\}, V, n]$ gives the matrix elements of the matrix $\mathcal{M}$ for a given potential V.

```
In[1]:= m[{k_, l_}, V_, n_] := With[{v = 2 n + 1, ε = Sqrt[2Pi/(2 n + 1)]},
          If[k =!= l, (-1)^(l - k) Pi/v Cot[(l - k)/v Pi] Csc[(l - k)/v Pi],
                      Pi/(6 v) (v^2 - 1) + V[k ε]]]
```

Calculating the lowest eigenvalues of value $2k+1$ for the harmonic oscillator $V(x) = x^2$ shows that this discretization method is far superior to the finite difference based approach. Because the results are so precise, we use high-precision arithmetic for the computations.

```
In[2]:= n = 10;
        M = Table[m[{k, 1}, #^2&, n], {k, -n, n}, {l, -n, n}];
        Take[MapIndexed[N[((2 #2[[1]] - 1) - #1)]&,
                        Reverse[Eigenvalues[N[M, 100]]]], 10]
Out[4]= {1.27382×10⁻¹³, -7.92294×10⁻¹², 2.53888×10⁻¹⁰, -4.87304×10⁻⁹, 7.37275×10⁻⁸,
         -7.78352×10⁻⁷, 7.37694×10⁻⁶, -0.0000499011, 0.000334711, -0.0015398}
```

```
In[5]:= n = 50;
        M = Table[m[{k, 1}, #^2&, n], {k, -n, n}, {l, -n, n}];
        Take[MapIndexed[N[((2 #2[[1]] - 1) - #1)]&,
                        Reverse[Eigenvalues[N[M, 100]]]], 10]
Out[7]= {7.19766×10⁻⁶⁸, -2.25698×10⁻⁶⁵, 3.55574×10⁻⁶³, -3.66796×10⁻⁶¹, 2.85303×10⁻⁵⁹,
         -1.74232×10⁻⁵⁷, 8.91892×10⁻⁵⁶, -3.83742×10⁻⁵⁴, 1.45407×10⁻⁵², -4.79828×10⁻⁵¹}
```

Ignoring a zero-mode state and degeneracy, this method calculates (by construction) the exact values for $V(x) = 0$.

```
In[8]:= n = 10;
        M = Table[m[{k, 1}, 0&, n], {k, -n, n}, {l, -n, n}];
        evs = Eigenvalues[N[M, 30]];
        (First /@ Partition[Rest[Reverse[evs]], 2])/(2Pi/(2n + 1))
Out[11]= {1.000000000000000000000000000000, 4.000000000000000000000000000000,
          9.000000000000000000000000000000, 16.00000000000000000000000000000,
          25.00000000000000000000000000000, 36.00000000000000000000000000000,
          49.00000000000000000000000000000, 64.00000000000000000000000000000,
          81.00000000000000000000000000000, 100.0000000000000000000000000000}
```

To get 20 correct digits for the quartic oscillator we increase n until the first 20 digits no longer change. This happens near $n \approx 70$.

```
In[12]:= Do[M = Table[m[{k, 1}, #^4&, n], {k, -n, n}, {l, -n, n}];
           Print[{n, Eigenvalues[N[M, 50]] // Last // N[#, 30]&}],
           {n, 10, 80, 10}]
         {10, 1.060350643090168293093957872l9}
         {20, 1.0603620894585466177585384252³}
         {30, 1.0603620904858148884369061273⁹}
         {40, 1.0603620904841852l7714387506⁴²}
         {50, 1.060362090484182903488576066⁹³}
         {60, 1.06036209048418289965539188⁶⁹²}
         {70, 1.060362090484182899647065451l0}
         {80, 1.06036209048418289964704604l30}
```

For a very similar approach, see [326], [333], [306], [334], and [335]. For discretizations through discrete singular convolutions, see [1877], [1946].

e) Because of the piecewise constant nature of the potential, it is straightforward to solve the differential equation symbolically; we will not use `NDSolve` for this purpose. Within each "cell" $x_l < x < x_{l+1}$, the potential is constant and has the value $V_k(\alpha)$. The solution of the Schrödinger equation within each cell can be written (the subscript now labels the cells, not the eigenvalues) [1634], [1430], [980] in the following form:

$$\psi_l(\alpha; x) = s_l \sin(k_l(x - x_l)) + c_k \cos(k_l(x - x_l)) \text{ where } k_l = \sqrt{\varepsilon - V_l(\alpha)} \text{ in case of } \varepsilon - V_l(\alpha) > 0 \text{ or}$$

$$\psi_l(\alpha; x) = i\, s_l \sinh(\kappa_l(x - x_l)) + c_l \cosh(\kappa_l(x - x_l)) \text{ where } \kappa_l = \sqrt{V_l - \varepsilon} \text{ in case of } \varepsilon - V_l(\alpha) < 0.$$

(The special case $\varepsilon - V_l(\alpha) = 0$ can be treated as a limiting case; we will not deal with it here.)

Because of the linear nature of the governing differential equation, the relation between $\begin{pmatrix} c_{l+1} \\ s_{l+1} \end{pmatrix}$ and $\begin{pmatrix} c_l \\ s_l \end{pmatrix}$ is linear:

$$\begin{pmatrix} c_{l+1} \\ s_{l+1} \end{pmatrix} = \begin{pmatrix} a_{11}\, c_l + a_{12}\, s_l \\ a_{21}\, c_l + a_{22}\, s_l \end{pmatrix}$$

Let us calculate the coefficients a_{ij}: At $x = x_{l+1}$, the two functions $\psi_l(\alpha; x)$ and $\psi_{l+1}(\alpha; x)$ must be continuous and differentiable. The values for the a_{ij} can be found from the following solution.

```
In[1]:= ψ[z_, j_] := s[j] Sin[k[j] (z - x[j]) ] + c[j] Cos[k[j] (z - x[j])]
```

```
In[2]:= sol = Solve[{ψ[x[j + 1], j] == ψ[x[j + 1], j + 1],
              D[ψ[z, j], z] == D[ψ[z, j + 1], z]} /. z -> x[j + 1],
              {c[j + 1], s[j + 1]}]
```

```
Out[2]= {{c[1 + j] → c[j] Cos[k[j] (-x[j] + x[1 + j])] + s[j] Sin[k[j] (-x[j] + x[1 + j])],
```
$$s[1 + j] \to \frac{\text{Cos}[k[j]\ (-x[j] + x[1 + j])]\ k[j]\ s[j] - c[j]\ k[j]\ \text{Sin}[k[j]\ (-x[j] + x[1 + j])]}{k[1 + j]} \}\}$$

Now let us turn to the specific potential under consideration. We assume the potential is given in the form $\{\{x_0, x_1, V_0\}, \{x_1, x_2, V_1\}, , ..., \{x_n, x_{n+1}, V_n\}\}$. `potentialGraphics` is a routine for displaying a piecewise constant potential.

```
In[3]:= potentialGraphics[pot_] :=
        Graphics[{Hue[0], Thickness[0.002],
                 (* piecewise horizontal and vertical *)
                 Line[{{pot[[1, 1]], pot[[1, 3]]}, {pot[[1, 2]], pot[[1, 3]]}}],
                 Line[{{pot[[-1, 1]], pot[[-1, 3]]}, {pot[[-1, 2]], pot[[-1, 3]]}}],
                 Map[Line[{{{#[[2, 1]], #[[1, 3]]}, {#[[2, 1]], #[[2, 3]]},
                           {#[[2, 2]], #[[2, 3]]}, {#[[2, 2]], #[[3, 3]]}}}]&,
                 (* potential segments *) Partition[pot, 3, 1], {1}]}]
```

This is the Möbius potential.

```
In[4]:= potential[α_] = Table[{k, k + 1, α (MoebiusMu[k] + 1)}, {k, 0, 50}];
```

```
In[5]:= Show[potentialGraphics[potential[1]],
            Frame -> True, PlotRange -> All, AspectRatio -> 1/3]
```

First, we will determine the eigenvalues and afterward the corresponding eigenfunctions. For an eigenfunction, we have $\psi(0) = 0$. Choosing the function in the leftmost cell to be of the form $\psi(\alpha; x) = 1 \times \sin(k_0(x - x_0)) + 0 \times \cos(k_0(x - x_0))$ ensures this condition. And we must find such an ε, such that $\psi(\alpha; L = n + 1) = 0$ too. The function calculates ψOfL for a given energy ε. Its argument is the potential *pot* and the energy ε. We use such starting conditions in the leftmost cell such that the resulting wavefunction is purely real everywhere.

```
In[6]:= ψOfL[pot_, ε_] :=
        Module[{δList, kList, cs, kj, kj1, cos, sin},
          (* lengths of the cells *)
          δList = -Apply[Subtract, Take[#, 2]& /@ pot, {1}];
          (* wave vector values inside the cells *)
          kList = Sqrt[ε - (Last /@ pot)];
          (* starting value, such that the wave function is real *)
          cs = If[ε < pot[[1, -1]], {0, -1 I}, {0, 1}];
          Do[kj = kList[[j]]; kj1 = kList[[j + 1]]; δj = δList[[j]];
             cos = Cos[kj δj]; sin = Sin[kj δj];
             (* the connection across a cell boundary *)
             cs = {{cos, sin}, kj/kj1 {-sin, cos}}.cs, {j, Length[pot] - 1}];
          (* value of the wave function at the right boundary *)
          cs[[1]] Cos[kList[[-1]] δList[[-1]]] + cs[[2]] Sin[kList[[-1]] δList[[-1]]]]]
```

For speed reasons, we also implement a compiled version ψOfLC of the function ψOfL.

```
In[7]:= ψOfLC =
    Compile[{{pot, _Real, 2}, {ε, _Real, 0}},
    Module[{δList, kList, cs, kj, kj1, δj, cos, sin},
     δList = (#[[2]] - #[[1]])& /@ (Take[#, 2]& /@ pot);
     kList = Sqrt[ε - (Last /@ (pot + (* complex type *) 0.I))]);
     cs = If[ε < pot[[1, -1]], {0., -1. I}, {0., 1. + 0. I}];
     Do[kj = kList[[j]]; kj1 = kList[[j + 1]]; δj = δList[[j]];
       cos = Cos[kj δj]; sin = Sin[kj δj];
       cs = {{cos, sin}, kj/kj1 {-sin, cos}}.cs, {j, Length[pot] - 1}];
     cs[[1]] Cos[kList[[-1]] δList[[-1]]] + cs[[2]] Sin[kList[[-1]] δList[[-1]]]]];
```

ψOfLC compiled successfully.

```
In[8]:= And @@ (NumberQ /@ Flatten[ψOfLC[[4]]])
Out[8]= True
```

For $\alpha = 0$, the eigenfunctions are $\psi(0; x) \propto \sin(k\,\pi/L)$ and the eigenvalues ε_k are $k^2\,\pi^2/L^2$, $k = 1, 2, \ldots$. The red points in the following graphic indicate these energies. The function ψOfLC$[L]$ is zero at these energies—it fulfills the right boundary condition.

```
In[9]:= Plot[Re[ψOfLC[potential[0], ε]], {ε, 0, 3},
    Compiled -> False, PlotPoints -> 500,
    PlotRange -> {-5, 5}, Frame -> True, Axes -> False,
    PlotStyle -> {Thickness[0.002], GrayLevel[0]},
    Prolog -> {PointSize[0.02], Hue[0],
        Table[Point[{k^2 Pi^2/51^2, 0}] // N, {k, 28}]}]
```

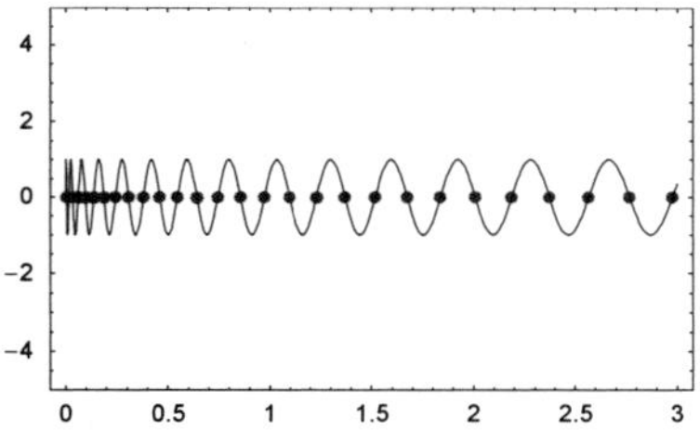

For $\alpha = 1, 2$, the values of ψOfLC$[L]$ show a more complicated dependence on ε.

```
In[10]:= Show[GraphicsArray[
    Plot[Re[ψOfLC[potential[#], ε]], {ε, 0, 3},
        Compiled -> False, PlotPoints -> 500,
        DisplayFunction -> Identity, PlotRange -> {-5, 5},
        Frame -> True, Axes -> False]& /@ {1, 2}]]
```

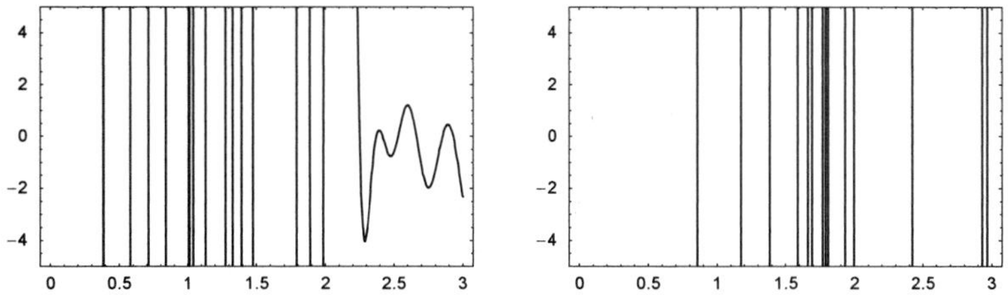

Here, for $\alpha = 2$ the graph of ψOfLC$[L]$ as a function of ε is shown. For $\varepsilon \approx \alpha$, the function ψOfLC$[L]$ does not have real zeros, but just touches the real axis from one side. (This case is shown in the right graphic.)

```
In[11]:= Show[GraphicsArray[
    Block[{$DisplayFunction = Identity},
    {(* case α = 2 *)
    Plot[Re[ψOfLC[potential[2], ε]], {ε, 0, 3},
        Compiled -> False, PlotPoints -> 500,
```

```
          PlotRange -> {-5, 5}, Frame -> True, Axes -> False],
    (* case α ≈ ε *)
    Plot[Re[ψOfLC[potential[1], ε]], {ε, 0.99, 1.01},
        Compiled -> False, PlotPoints -> 1000,
        PlotRange -> {-100, 100}, Frame -> True, Axes -> False]}]]]
```

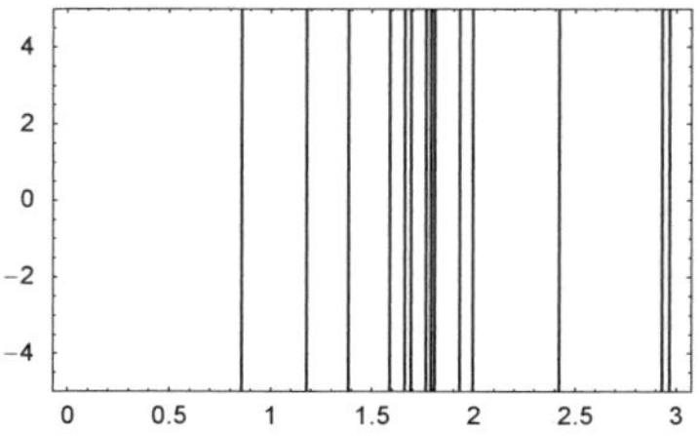

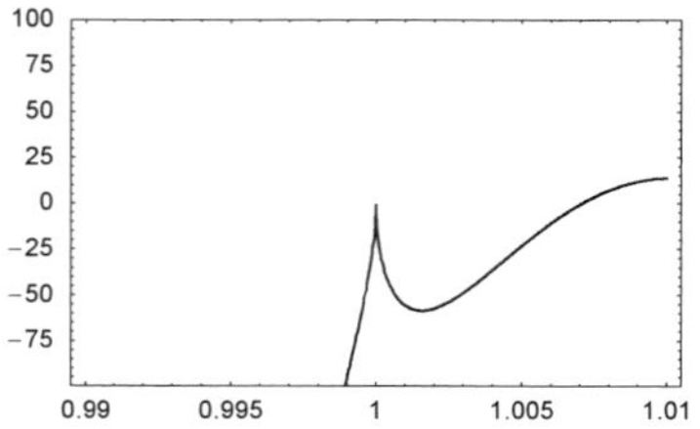

Now, let us determine the dependence of the eigenvalues on α. We get a rough idea about the dependence by graphing a contour plot of the zero values of ψOfLC$[L]$ in the α,ε-plane.

```
In[12]:= With[{ε = 10^-7},
            ContourPlot[Re[ψOfLC[potential[α], ε]], {α, ε, 3/2}, {ε, 0.9 ε, 2},
                Compiled -> False, Contours -> {0}, PlotPoints -> 200,
                ContourShading -> False, AspectRatio -> Automatic]]
```

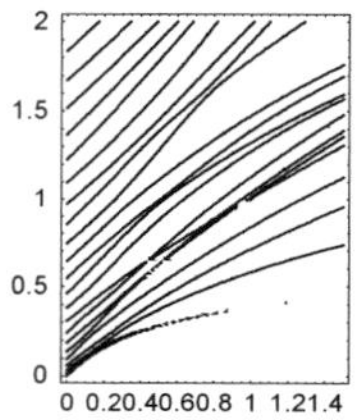

The last picture shows that the energy range $0 \le \varepsilon \le 2$ includes at least the first 10 eigenvalues for the α-range of interest. To get a more detailed picture, we calculate numerically the above-pictured zeros of ψOfLC$[L]$. We start with plots (using the adaptive function Plot) to get rough estimations of the zeros. The function zeroIntervals extracts the intervals within which the line segments of l cross the real axis.

```
In[13]:= zeroIntervals[Line[l_]]  :=
            Map[First, Select[Partition[l, 2, 1], #[[1, 2]] #[[2, 2]] < 0&], {2}]
```

Using the endpoints of these intervals as starting points, we use bisectFindRoot to get better numerical approximations for the zeros. We use the faster version ψOfLC in the next input. The list evTab1 will contain a list of the eigenvalues in the energy range $0 \le \varepsilon \le 2$ for the α-range $0 \le \alpha \le 3/2$.

```
In[14]:= (* robust 1D root finder *)
        bisectFindRoot[f_, {αl_, αu_}, δ_] :=
        Module[{l = αl, u = αu, fl = f[αl], fu = f[αu]},
            While[u - l > δ, m = (l + u)/2; fm = f[m];
                    If[fl fm < 0, u = m, l = m]]; (l + u)/2]

In[16]:= evTab1 =
    Module[{thePot, F, pl, zI},
    Table[thePot = potential[α];
            (* function  whose zeros are needed *)
            (F[ε_?NumberQ] := Re[ψOfLC[#, ε]])&[thePot];
            (* the plot *)
            pl = Plot[ψOfLC[thePot, e], {e, 0, 2}, Compiled -> False,
                    DisplayFunction -> Identity, PlotPoints -> 500];
            (* the intervals enclosing the zeros *)
            zI = zeroIntervals[Cases[pl, _Line, Infinity][[1]]];
            {α, bisectFindRoot[F, #, 10^-8]}& /@ zI, {α, 0, 3/2, 1/200}]];
```

Here is the resulting picture showing the dependence of the energy eigenvalues on α.

```
In[17]:= Show[Graphics[{{PointSize[0.003], Hue[0], Map[Point, Re[evTab1], {-2}]}}],
           PlotRange -> All, Frame -> True, AspectRatio -> Automatic]
```

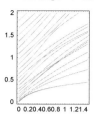

At some points, the eigenvalues seemingly cross each other [810], [809], [1818]. But in one dimension for a regular Sturm–Liouville problem, this is not possible [1243], [1857], [716]. A magnified version of the last plot near $\alpha \approx 0.35$, $\varepsilon \approx 0.85$ confirms this.

```
In[18]:= evTab2 =
    Module[{thePot, F, pl, zI},
      Table[thePot = potential[α];
         (F[ε_?NumberQ] := Re[ψOfLC[#, ε]])&[thePot];
         (* a plot for a rough sketch *)
         pl = Plot[ψOfLC[thePot, ε], {ε, 0.55, 0.75}, Compiled -> False,
                PlotPoints -> 100, DisplayFunction -> Identity];
         (* isolating intervals of the zeros *)
         zI = zeroIntervals[Cases[pl, _Line, Infinity][[1]]];
         (* refined zeros *)
         {α, bisectFindRoot[F, #, 10^-8]}& /@ zI, {α, 0.35, 0.55, 10^-3}]];

In[19]:= Show[Graphics[{PointSize[0.005], Hue[0], Map[Point, Re[evTab2], {-2}]}],
           PlotRange -> All, Frame -> True]
```

The graph above consists of single points. We could calculate a continuous function $\varepsilon_j(\alpha)$ by solving a differential equation for $\varepsilon_j(\alpha)$. Implicit differentiation of ψOfL $[L]$ as a function of $\varepsilon_j(\alpha)$ and α with respect to α gives such a differential equation. We calculate finite difference approximations for the two partial differential quotients needed.

```
In[20]:= dψdα[α_?NumberQ, ε_, δ_, prec_] :=
    (ψOfL[potential[α + δ], N[ε, prec]] - ψOfL[potential[α], N[ε, prec]])/δ

In[21]:= dψdε[α_?NumberQ, ε_, δ_, prec_] :=
    (ψOfL[potential[α], N[ε + δ, prec]] - ψOfL[potential[α], N[ε, prec]])/δ
```

To keep on the "right track", we refine the solution from time to time using FindRoot.

```
In[22]:= evIFs[j_] :=
    Module[{εStart, α, εVal, F, L = 51},
      εStart = j^2 Pi^2/L^2;
      Table[(* solve differential equation *)
        ndsol = NDSolve[{ε'[α] == -(dψdα[α, ε[α], 10^(-12), 30]/
                                     dψdε[α, ε[α], 10^(-12), 30]),
                          ε[k 1/10] == εStart}, ε,
                         {α, k 1/10, (k + 1) 1/10},
                         (* use smaller precision for plot purpose *)
                         PrecisionGoal -> 6, AccuracyGoal -> 6,
                         WorkingPrecision -> 30,
                         Method -> ExplicitRungeKutta];
        εVal = ε[(k + 1) 1/10] /. ndsol[[1]];
```

```
(* refine *)
F[ε_?NumberQ] := Re[ψOfL[potential[(k + 1) 1/10], ε]];
εStart = bisectFindRoot[F, {(1 - N[10^-2, 40]) εVal,
                            (1 + N[10^-2, 40]) εVal}, 10^-20];
ndsol, {k, 0, 14}]]
```

Here is the first eigenvalue calculated this way.

```
In[23]:= With[{nsol = evIFs[1]},
     Show[Table[Plot[Evaluate[ε[α] /. nsol[[k + 1]]],
             {α, k 1/10, (k + 1) 1/10}, DisplayFunction -> Identity],
             {k, 0, 14}], DisplayFunction -> $DisplayFunction]]
```

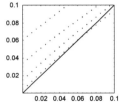

Now, let us find the value of α where the lowest energy eigenvalue crosses the maximum of the potential. The following picture shows that this happens near $\alpha \approx 0.06$. The red points are the previously calculated energy eigenvalues, and the blue line $\varepsilon = \alpha$.

```
In[24]:= Show[Graphics[{{RGBColor[0, 0, 1], Line[{{0, 0}, {3/2, 3/2}}]},
                 PointSize[0.01], Hue[0], Map[Point, Re[evTab1], {-2}]}],
         PlotRange -> {{0, 0.1}, {0, 0.1}}, Frame -> True,
         AspectRatio -> Automatic]
```

At $\varepsilon_j(\alpha) = \alpha$, the above-assumed form of the eigenfunctions is not correct and the appropriate limit form should be used. Nevertheless, we can calculate a high-precision approximation for the α-value under consideration. We choose $\varepsilon = \alpha + \epsilon$ where $\epsilon \ll \alpha$. This makes sure that the above assumed form of the eigenfunctions applies.

```
In[25]:= f[γ_, ε_] := ψOfL[potential[SetPrecision[γ, 50]], SetPrecision[γ, 50] + ε]
```

```
In[26]:= f[γ_?InexactNumberQ, ε_, prec_] := ψOfL[potential[SetPrecision[γ, prec]],
                                       SetPrecision[γ, prec] + ε]
```

Using smaller and smaller ϵ gives a result correct to about 26 digits.

```
In[27]:= Table[N[γ /. FindRoot[f[γ, 10^-(prec - 20), prec], {γ, 6/100, 7/100},
                 WorkingPrecision -> prec,
                 AccuracyGoal -> prec - 20], 30],
         {prec, 50, 150, 10}]
Out[27]= {0.063238581294718207235298039464 3,
     0.0632385812947182072352980394739, 0.0632385812947182072352980394723,
     0.0632385812947182072352980394754, 0.0632385812947182072352980394723,
     0.0632385812947182072352980394714, 0.0632385812947182072352980394723,
     0.0632385812947182072352980394798, 0.0632385812947182072352980394723,
     0.0632385812947182072352980394938, 0.0632385812947182072352980394723}
```

Now let us visualize the eigenfunctions. We normalize the eigenfunctions so that $\int_0^L |\psi(z)|^2 \, dz = 1$. ψOfzNorm returns a list of lists. Each sublist corresponds to one cell and contains the following entries: $\{x_j, x_{j+1}, k_j, c_j, s_j\}$. The slightly longish formulas for the normalization follow from straightforward integration of the absolute value squared of the wave functions.

```
In[28]:= ψOfzNorm[pot_, ε_] :=
    Module[{xList, δList, kList, cs0, cs, kj, kj1, δj, csTab, res, norm, normList},
        xList = Take[#, 2]& /@ pot;
        δList = -Apply[Subtract, xList, {1}];
        kList = Sqrt[ε - (Last /@ pot)];
        (* calculate coefficients *)
        cs0 = If[ε < pot[[1, -1]], {0, -1. I}, {0., 1.}];
        cs = cs0;
        csTab = Table[
            kj = kList[[j]]; kj1 = kList[[j + 1]]; δj = δList[[j]];
            cos = Cos[kj δj]; sin = Sin[kj δj];
            cs = {{cos, sin}, kj/kj1 {-sin, cos}}.cs,
            {j, Length[pot] - 1}];
        res = Delete[Flatten[#], 4]& /@
                Transpose[{xList, {#, 0.}& /@ kList, Prepend[csTab, cs0]}];
        (* calculate normalization coefficient *)
        normList = Apply[Function[{x1, x2, k, c, s},
            If[Im[k] == 0., (2c s + 2k(c^2 + s^2)(x2 - x1) -
            2 c s Cos[2k(x2 - x1)] + (c^2 - s^2) Sin[2k(x2 - x1)])/(4k),
                        (2c s - 2k(c^2 + s^2)(x1 - x2) -
            2 c s Cos[2k(x2 - x1)] + (c^2 - s^2) Sin[2k(x2 - x1)])/(4k)]],
                        res, {1}];
        norm = Plus @@ normList;
        (* normed eigenfunctions *)
        Apply[{#1, #2, #3, #4/Sqrt[norm], #5/Sqrt[norm]}&, res, {1}]]
```

Here is a quick check that the normalization really works.

```
In[29]:= Module[{α1, ε1}, {α1, ε1} = Re[evTab1[[111, 1]]];
    Plus @@ Apply[NIntegrate[Abs[#4 Cos[#3 (z - #1)] + #5 Sin[#3 (z - #1)]]^2,
        {z, #1, #2}]&, ψOfzNorm[potential[α1], ε1], {1}]]

Out[29]= 1.
```

The function waveFunctionLine visualizes the wave function data returned from ψOfzNorm.

```
In[30]:= waveFunctionLine[ψData_, ε0_, n_] :=
    Apply[{{Line[Table[Re[{z, ε0 + #4 Cos[#3 (z - #1)] + #5 Sin[#3 (z - #1)]}],
        {z, #1, #2, (#2 - #1)/n}]]}}&, ψData, {1}]
```

Here is an example of a wavefunction.

```
In[31]:= Module[{α1, ε1}, {α1, ε1} = Re[evTab1[[12, 1]]];
    Show[Graphics[waveFunctionLine[ψOfzNorm[potential[α1], ε1], ε1, 12]],
        PlotRange -> All, Frame -> True]]
```

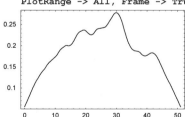

For speed reasons, we again implement a compiled version of ψOfzNorm, called ψOfzNormC. Some minor modifications on the code for ψOfzNorm allows for successful compilation.

```
In[32]:= ψOfzNormC =
    Compile[{{pot, _Real, 2}, {ε, _Real, 0}},
    Module[{xList, δList, kList, cs0, cs, cos, sin, kj, kj1, δj, csTab, res,
            normList, norm},
        xList = Take[#, 2]& /@ pot;
```

```
        δList = (#[[2]] - #[[1]])& /@ (Take[#, 2]& /@ pot);
        kList = Sqrt[ε - (Last /@ (pot + 0.0 I))]];
        (* calculate coefficients *)
        cs0 = If[ε < pot[[1, -1]], {0.0 + 0.0 I, 0.0 -1. I},
                                    {0.0 + 0.0 I, 1. + 0.0 I}];
        cs = cs0;
        csTab = Table[
            kj = kList[[j]]; kj1 = kList[[j + 1]]; δj = δList[[j]];
            cos = Cos[kj δj]; sin = Sin[kj δj];
            cs = {{cos, sin}, kj/kj1 {-sin, cos}}.cs,
                        {j, Length[pot] - 1}];
        res = Delete[Flatten[#], 4]& /@
                Transpose[{xList, {#, 0.}& /@ kList, Prepend[csTab, cs0]}];
        (* calculate normalization coefficient *)
        normList = Map[Module[{x1 = #[[1]], x2 = #[[2]],
                            k = #[[3]], c = #[[4]], s = #[[5]]},
            If[Im[k] == 0., (2c s + 2k (c^2 + s^2)(x2 - x1) -
                2 c s Cos[2k (x2 - x1)] + (c^2 - s^2) Sin[2k (x2 - x1)])/(4k),
                            (2c s - 2k (c^2 + s^2)(x1 - x2) -
                2 c s Cos[2k (x2 - x1)] + (c^2 - s^2) Sin[2k (x2 - x1)])/(4k)]&,
                            res, {1}];
        norm = Plus @@ normList;
        (* normed eigenfunctions *)
        Map[{#[[1]], #[[2]], #[[3]], #[[4]]/Sqrt[norm],
            #[[5]]/Sqrt[norm]}&, res, {1}]]];
```

The compilation was successful.

```
In[33]:= And @@ (NumberQ /@ Flatten[ψOfzNormC[[4]]])

Out[33]= True
```

Here is another example of an eigenfunction.

```
In[34]:= Module[{α1, ε1},
        {α1, ε1} = Re[evTab1[[111, 1]]];
        Show[Graphics[waveFunctionLine[ψOfzNormC[potential[α1], ε1], ε1, 12]],
                PlotRange -> All, Frame -> True]]
```

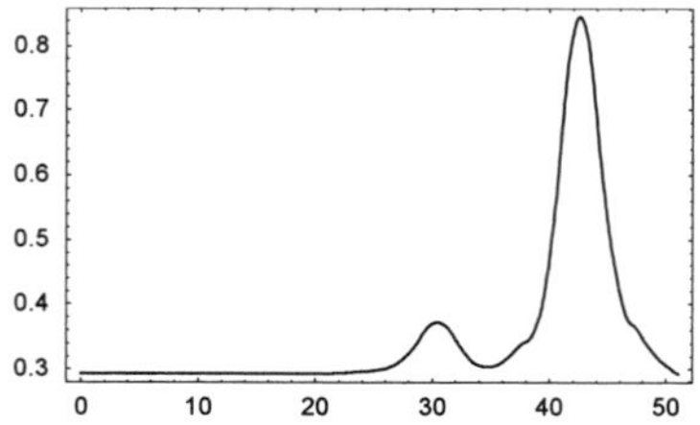

Now, we have all functions together to display a larger amount of wavefunctions quickly. potentialPolygons is an auxiliary function that renders the potential as gray background mountains in the following pictures.

```
In[35]:= potentialPolygons[pot_] :=
        With[{minP = Min[Last /@ pot], maxP = Max[Last /@ pot], ε = 10^-3},
            (* form polygons *)
            Apply[Polygon[{{#1 - ε, minP - (maxP - minP)/5}, {#1 - ε, #3},
                            {#2 + ε, #3}, {#2 + ε, minP - (maxP - minP)/5}}]&,
                pot, {1}]]
```

Here is an animation of the change of the wavefunction corresponding to the second eigenvalue as a function of α.

```
Do[{α1, ε1} = Re[N[evTab1[[k, 2]]]];
   Show[Graphics[{{GrayLevel[0.8], potentialPolygons[potential[α1]]},
               {Thickness[0.002], Hue[0],
                  waveFunctionLine[ψOfzNorm[potential[α1], ε1], ε1, 12]}}],
         Frame -> True, PlotRange -> {{0, 51}, {-1, 3}}], {k, Length[evTab1]}]
```

The following picture shows the wavefunctions shifted vertically for increasing α.

```
In[36]:= Show[Graphics[{Thickness[0.002],
         Table[{α1, ε1} = Re[N[evTab1[[k, 2]]]];
             Map[{0, k/80} + #&,
                     waveFunctionLine[ψOfzNorm[potential[α1], ε1],
                                   ε1, 12], {-2}],
             {k, 75 (* Length[evTab1] *)}]}],
         Frame -> True, PlotRange -> {{0, 51}, {-1/2, 2}}, AspectRatio -> 1]
```

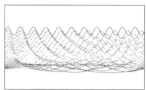

It is also instructive to view all eigenfunctions at once, as this shows some interesting envelope effects.

```
In[37]:= animationFrame[k_, opts___] :=
         Module[{data = Re[evTab1[[k]]], α0 = Re[evTab1[[k, 1, 1]]]},
             Graphics[{{GrayLevel[0.8], potentialPolygons[potential[α0]]},
                    ({α1, ε1} = #; (* eigenfunctions and background potential *)
                        {Thickness[0.002], Hue[ε1/3],
                          waveFunctionLine[ψOfzNormC[potential[α1], ε1], ε1, 6]})& /@ data},
                    opts, Frame -> True, PlotRange -> {{0, 51}, {-1, 3.1}}]]

In[38]:= Show[GraphicsArray[#]]& /@ Partition[
         animationFrame[#, FrameTicks -> None]& /@ Table[k, {k, 1, 301, 26}], 3]
```

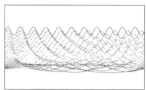
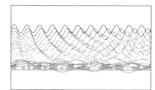
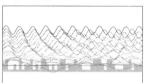
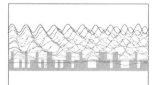
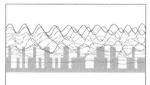
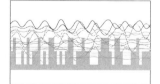

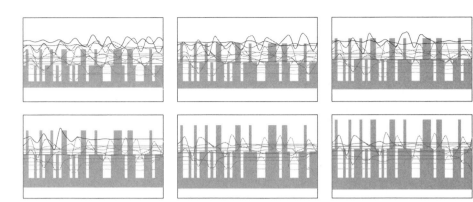

```
Do[Show @ animationFrame[k], {k, Length[evTab1]}]
```

The variation in shape of the *j*th wavefunction can be conveniently displayed as a 3D graphics object. The function
waveFunctionDevelopment shows the shape dependence of the *j*th eigenfunction on α.

```
In[39]:= Needs["Graphics`Graphics3D`"]

In[40]:= waveFunctionDevelopment[j_, opts___] :=
    Show[Graphics3D[{EdgeForm[], SurfaceColor[Hue[0.02], Hue[0.18], 2.4],
                 polys = Cases[ListSurfacePlot3D[
                     (* the data *)
                     Table[{α1, ε1} = Re[N[evTab1[[k, j]]]];
        Apply[{#1, (k - 1) 1/200, #2}&,
            Join @@ (First /@ Flatten[
                (* the single wave functions *) waveFunctionLine[
                          ψOfzNormC[potential[α1], ε1], ε1, 6]]),
            {1}], {k, Length[evTab1]}],
                 DisplayFunction -> Identity], _Polygon, Infinity],
        (* the potential in the back *)
        {SurfaceColor[RGBColor[0, 0, 1]], EdgeForm[],
        Apply[{#1, 16/10, #2}&,
        potentialPolygons[potential[1.2 Max[Abs[Last /@
                     Level[polys, {-2}]]]/2]], {-2}]}}], opts,
        BoxRatios -> {2, 1, 1}, PlotRange -> All, ViewPoint -> {0.2, -3, 1.5}]
```

Here are the first three eigenfunctions. It is interesting to note how the eigenfunctions localize in the largest, still-available
well [553].

```
In[41]:= Show[GraphicsArray[Table[waveFunctionDevelopment[k,
                          DisplayFunction -> Identity], {k, 3}]]]
```

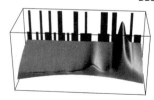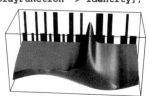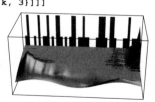

In addition to visualizing the wave function $\psi_\varepsilon(x)$, we could also visualize the corresponding Husimi function

$$\rho_\varepsilon(\xi, \kappa) \sim \left| \int_0^L \exp(((x - \xi)/(2\,\sigma))^2 - i\,\kappa\,x)\,\psi_\varepsilon(x)\,dx \right|^2$$

(here ξ and κ are the position and momentum variable; σ scales between the two) [1911], [1912], [872]. Here this is done for a selected state.

```
In[42]:= (* Husimi kernel *)
        HusimiGaussKernel[x_, ξ_, κ_, σ_] :=
        (1/(2 Pi σ^2))^(1/4) Exp[-(x - ξ)^2/(4 σ^2) - I κ x]

In[44]:= (* Husimi function within each cell xj <= x <= xjP1;
            result of exact integration contains error functions *)
        Husimiρ[{ξ_, κ_, σ_}, {xj_, xjP1_, kj_, cj_, sj_}] =
        Abs[Integrate[(cj Cos[kj (x - xj)] + sj Sin[kj (x - xj)])*
                      HusimiGaussKernel[x, ξ, κ, σ], {x, xj, xjP1}] // Simplify]^2;

In[46]:= (* Husimi function over all cells *)
        HusimiData[ψData_, σ_, κMax_, {ppξ_, ppκ_}] := Join @@@ Transpose[
        Table[Husimiρ[{ξ, κ, σ}, {#1, #2, #3, #4, #5}],
             {κ, -κMax, κMax, 2κMax/ppκ},
             {ξ, #1, #1 + (#2 - #1) (1 - 1/ppξ), (#2 - #1)/ppξ}]& @@@ ψData]

In[48]:= (* visualize Husimi function *)
        Module[{α1, ε1}, {α1, ε1} = Re[evTab1[[66, 3]]];
           ListContourPlot[Log[HusimiData[ψOfzNorm[potential[α1], ε1],
                         2, 20, {46, 90}]],
                      AspectRatio -> 1/4, FrameTicks -> False, Frame -> True,
                      Contours -> 20, ContourLines -> False,
                      PlotRange -> All, ColorFunction -> (Hue[0.8 (1 - #)]&)]]
```

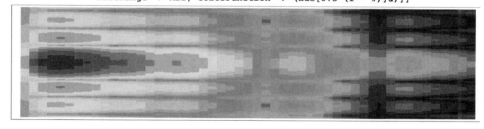

Having now implemented the normalization of the eigenfunctions opens the road for another method for the calculation of the energies as a function of α: The Hellmann–Feynman theorem [1321], [1127], [813], [877], [876], [600], [1135], [1314] (for degenerate points, see [1964], [41], and for approximate solutions, see [510]) states that for a parameter-dependent linear operator $\hat{L}_\alpha$ with the eigenvalue equation $\hat{L}_\alpha \psi_i(\alpha; z) = \varepsilon_i(\alpha) \psi_i(\alpha; z)$, the following identity holds for dependence of the eigenvalues ε_i on the parameter α for normalized eigenfunctions $\psi_i(\alpha; z)$:

$$\frac{\partial \varepsilon_i(\alpha)}{\partial \alpha} = \int_0^L \psi_i(\alpha; z) \frac{\partial \hat{L}_\alpha}{\partial \alpha} \psi_i(\alpha; z)\, dz.$$

This allows for a straightforward calculation of $\partial \varepsilon_i(\alpha)/\partial \alpha$ instead of the finite difference approximation used above. Here, the calculation of $\partial \varepsilon_i(\alpha)/\partial \alpha$ is implemented.

```
In[50]:= dεdα[α_, ε_?NumberQ] :=
        Module[{pot, xList, δList, kList, cs0, cs, kj, kj1, δj, csTab, res,
               norm, normList},
           (* data *)
           pot = potential[α];
           xList = Take[#, 2]& /@ pot;
           δList = -Apply[Subtract, xList, {1}];
           kList = Sqrt[ε - (Last /@ pot)];
           (* wave function coefficients *)
           cs0 = If[ε < pot[[1, -1]], {0, - I}, {0, 1}];
           cs = cs0;
           csTab = Table[
              kj = kList[[j]]; kj1 = kList[[j + 1]];
              δj = δList[[j]];
              cos = Cos[kj δj]; sin = Sin[kj δj];
```

```
            cs = {{cos, sin}, kj/kj1 {-sin, cos}}.cs, {j, Length[pot] - 1}];
       res = Delete[Flatten[#], 4]& /@
              Transpose[{xList, {#, 0.}& /@ kList, Prepend[csTab, cs0]}]];
       (* normalize wave functions *)
       normList = Apply[Function[{x1, x2, k, c, s},
          If[Im[k] == 0., (2c s + 2k(c^2 + s^2)(x2 - x1) -
             2 c s Cos[2k(x2 - x1)] + (c^2 - s^2) Sin[2k(x2 - x1)])/(4 k),
                 (2c s - 2k(c^2 + s^2)(x1 - x2) -
             2c s Cos[2k(x2 - x1)] +(c^2 - s^2)Sin[2k(x2 - x1)])/(4k)]], res, {1}];
       norm = Plus @@ normList;
       (* form overlap with potential *)
       1/norm Plus @@ (normList (Last /@ potential[1]))]
```

Here again, an eigenvalue, this time the sixth one as a function of α, is calculated.

```
In[51]:= (ndsol =  NDSolve[{ε'[α] ==  dεdα[α, ε[α]], ε[0] == 6^2 Pi^2/51^2},  ε,
                    {α, 0, 3/2}, PrecisionGoal -> 14, AccuracyGoal -> 14,
                    WorkingPrecision -> 25]) // Timing
Out[51]= {33.13 Second, {{ε → InterpolatingFunction[{{0, 1.5000000000000000000000000}}, <>]}}}
```

The following graphic shows a good agreement between the just-calculated eigenvalue dependence and the above one calculated using FindRoot.

```
In[52]:= Plot[(* the NDSolve solution *)
            Evaluate[ε[α] /. ndsol], {α, 0, 3/2},
            (* the data from above *)
            Prolog -> {Hue[0], PointSize[0.006], Point[#[[6]]]& /@ evTab1},
            PlotStyle -> {{Thickness[0.002], GrayLevel[0]}}]
```

We could proceed by investigating how good the energy-equipartition relation between the nodes is fulfilled [383], but we will end here. For investigations of the 1D potential based on prime numbers, see [487].

f) For the symmetric states, we can restrict ourselves to the domain $x > 0$. For a more convenient numerical evaluation of the potential $V(x)$, we rewrite it in a sum-free form.

```
In[1]:= V[x_] := Which[Abs[x] < Pi/2, -1, Abs[x] < Pi, -4,
                   True, -(Floor[Abs[x]/Pi] + 2)^2]
```

Here is a sketch of the potential at hand. In average, we have $V(x) \underset{|x|\to\infty}{\sim} -x^2$.

```
In[2]:= Plot[V[x], {x, -15, 15}, Frame -> True, Axes -> False]
```

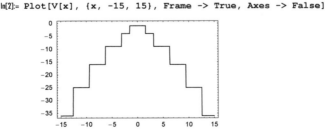

A quick way to obtain the solutions $\psi_\varepsilon(x)$ is by solving the differential equation numerically. The following graphic shows the symmetric solutions for $0 \leq x \leq 15$ and $-3 \leq \varepsilon <= 3$. (For larger negative ε-values the middle part of $\psi(x)$ is in the classically forbidden region and is exponentially damped.)

```
In[3]:= X = 15; pp = 200;
     data = Table[
     Table[Evaluate[NDSolve[{-ψ''[x] + V[x] ψ[x] == ε ψ[x],
                            ψ[0] == 1, ψ'[0] == 0},
                       ψ, {x, 0, X}, MaxSteps -> 10^5][[1, 1, 2]][x]],
               {x, 0, X, X/pp}], {ε, -3, 3, 4/pp}];
In[5]:= ListPlot3D[data, Mesh -> False, MeshRange -> {{0, X}, {-3, 3}}]
```

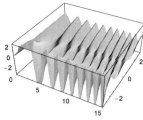

The solution shows the special role of $\varepsilon = 0$ (Levinson theorem [1343], [512], [726] or Friedel oscillations in 1D [637]). To obtain the solution $\psi(x)$ over a larger x-range, the numerical solution of the differential equation is not the most efficient way.

In the interval $j\pi \le x \le (j+1)\pi$, we can solve the differential equation exactly and have $\psi_j(x) = c_j \cos(k_j(\varepsilon) x) + s_j \sin(k_j(\varepsilon) x)$ (here $k_j(\varepsilon) = \sqrt{\varepsilon + (j+2)^2}$). At $x = (j+1)\pi$, the two functions $\psi_j(x)$ and $\psi_{j+1}(x)$ must match smoothly. Let the matrix $\mathbb{M}_j(\varepsilon)$ be defined by

$$\begin{pmatrix} c_{j+1} \\ s_{j+1} \end{pmatrix} = \mathbb{M}_j(\varepsilon) \begin{pmatrix} c_j \\ s_j \end{pmatrix}.$$

It is straightforward to calculate the following explicit form of $\mathbb{M}_j(\varepsilon)$.

```
In[6]:= M[j_, ε_] :=
     Module[{k2 = Sqrt[ε + (j + 2)^2], k3 = Sqrt[ε + (j + 3)^2],
            c2, c3, s2, s3},
            c2 = Cos[(j + 1) k2 Pi]; c3 = Cos[(j + 1) k3 Pi];
            s2 = Sin[(j + 1) k2 Pi]; s3 = Sin[(j + 1) k3 Pi];
            {{c2 c3 + k2/k3 s2 s3, c3 s2 - k2/k3 c2 s3},
             {c2 s3 - k2/k3 c3 s2, s2 s3 + k2/k3 c2 c3}}]
```

We consider only symmetric solutions, normalized to $\psi(0) = 1$. This means, we have for $0 \le x \le \pi/2$ the solution $\psi(x) = \cos(\sqrt{\varepsilon - 1}\, x)$ and for $\pi/2 \le x \le \pi$ the solution $\psi_0(x) = c_0 \cos(\sqrt{\varepsilon - 4}\, x) + s_0 \sin(\sqrt{\varepsilon - 4}\, x)$. `csStart[ε]` calculates c_0 and s_0 as a function of ε.

```
In[7]:= csStart[ε_] :=
     With[{c1 = Cos[Pi Sqrt[ε + 1]/2], c4 = Cos[Pi Sqrt[ε + 4]/2],
           s1 = Sin[Pi Sqrt[ε + 1]/2], s4 = Sin[Pi Sqrt[ε + 4]/2],
           k = Sqrt[(ε + 1)/(ε + 4)]},
           {c1 c4 + k s1 s4, c1 s4 - k c4 s1}]
```

ψPair yields the x-dependent parts of the wavefunction inside the interval $(j\pi, (j+1)\pi)$.

```
In[8]:= ψPair[j_, ε_] := If[j Pi <= x <= (j + 1) Pi, 1, 0] *
          {Cos[Sqrt[ε + (j + 2)^2] x], Sin[Sqrt[ε + (j + 2)^2] x]};
```

Iterating the multiplication with the transfer matrix $\mathbb{M}_j(\varepsilon)$ quickly yields a description of the function $\psi_\varepsilon(x)$ over a large x-range.

```
In[9]:= Φ = MapIndexed[ψPair[#2[[1]], 0].#1&,
                     Rest @ FoldList[M[#2, 0.].#1&, csStart[0.], Range[0, 10]]];

     Plot[Evaluate[Which[x <= Pi/2, Cos[x], x <= Pi, Sin[2x]/2, True, 0] +
          Φ], {x, 0, 20}, PlotRange -> All, PlotPoints -> 200]
```

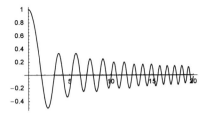

For real c_j, s_j, the normalization integral $I_j(\varepsilon) = \int_{j\pi}^{(j+1)\pi} \psi_j(x)^2 \, dx$ inside the interval $(j\pi, \ (j+1)\pi)$ is easily calculated to be $\pi(c_j^2 + s_j^2)/2$.

It is easy to see that for $\varepsilon = 0$, all c_j vanish identically and so, we have a square integrable wave function. Let us look at $I_j(\varepsilon)$ for $\varepsilon \neq 0$, say for $\varepsilon = 1$.

```
In[12]:= data = #.#& /@ FoldList[M[#2, 1.].#1&, csStart[1.], Range[0, 10^5]];
```

```
In[13]:= Show[GraphicsArray[
            Block[{$DisplayFunction = Identity},
                {ListPlot[N @ data],
                 ListPlot[N @ Log[MapIndexed[{#2[[1]], #1}&, data]]]}]]]
```

The last graphic suggests an asymptotic dependence of the form $I_j(\varepsilon) \sim 1/k$. Numerical fits support this finding. (This means that the sum $\sum_{j=1}^{\infty} I_j(\varepsilon)$ diverges, as to be expected for values of ε in the continuous spectrum). The dip of the coefficient of k^{-1} at $\varepsilon = 0$ shows the exceptional character of the zero-energy case. This single bound state is responsible for the phase shift at $\varepsilon = 0$ seen in the above 3D graphics.

```
In[14]:= h[ε_, n_] := Fit[MapIndexed[{#2[[1]] + 1, #1}&,
                #.#& /@ (* calculate coefficients *)
                FoldList[M[#2, ε].#1&, csStart[ε],
                    Range[0, n]]], {k^-1, k^-2}, k] /.
                            a_ k^-1 + b_ k^-2 :> {a, b}
```

```
In[15]:= data = Table[{ε, h[N @ ε, 200]}, {ε, -1, 1, 2/200}];
```

```
In[16]:= Show[Graphics[{PointSize[0.003],
                (* the coefficients of 1/k, 1/k^2 *)
                {Hue[0.0], Point[{#[[1]], #[[2, 1]]}]& /@ data},
                {Hue[0.8], Point[{#[[1]], #[[2, 2]]}]& /@ data}}],
                Frame -> True]
```

As expected, for a potential function $V(x) \approx -x^2$, we see a logarithmic divergence of the normalization integral [141], [1368].

For further piecewise potentials with interesting properties, see [1420] and [72]. For potentials with many bound states in the continuous spectrum, see [1662] and [999].

6. Wynn's ϵ-Algorithm, Aitken Algorithm, Numerical Regularization

a) The following first implementation is a direct translation of the formulas in the problem. Although it is possible to produce an efficient program in which the sequences are always treated as a whole, we avoid this approach to make the program more understandable.

```
In[1]:= (* procedural and treating every element by itself;
            follow definition strictly *)
        elem[n_, ord_, sequence_] := elem[n, ord, sequence] =
        elem[n + 1, ord - 2, sequence] +
        1/(elem[n + 1, ord - 1, sequence] - elem[n, ord - 1, sequence])

        elem[n_, -1, sequence_] := elem[n, -1, sequence] = 0

        elem[n_, 0, sequence_] := elem[n, 0, sequence] = sequence[[n + 1]]
```

In the following `myWynn`, all elements of the current iteration step are combined.

```
In[6]:= (* turn off messages *)
        Off[Power::infy]; Off[Infinity::indet]
In[8]:= myWynn[sequence_] := Table[elem[n, ord, sequence],
                                {ord, 0, Length[sequence] - 2},
                                {n, 0, Length[sequence] - ord - 1}]
```

Next, we give a few examples. We begin with the sequence $(1, 1/2, 1/4, 1/8, 1/16, \ldots)$.

```
In[9]:= myWynn[{1, 1/2, 1/4, 1/8}]
Out[9]= {{1, 1/2, 1/4, 1/8}, {-2, -4, -8}, {0, 0}}
```

The relevant answer is 0. If we had more terms, this simplistic implementation would not have succeeded.

```
In[10]:= myWynn[{1, 1/2, 1/4, 1/8, 1/16, 1/32}]
Out[10]= {{1, 1/2, 1/4, 1/8, 1/16, 1/32}, {-2, -4, -8, -16, -32}, {0, 0, 0, 0},
          {ComplexInfinity, ComplexInfinity, ComplexInfinity}, {Indeterminate, Indeterminate}}
```

Note that $1 - 1/3 + 1/5 - 1/7 + 1/9 + \cdots = \pi/4$.

```
In[11]:= piSeries[n_] := Table[Sum[(-1)^i/(2i + 1), {i, 0, j}], {j, 0, n}]
```

The partial sums converge very slowly to the exact result $\pi/4 = 0.78539816339744\ldots$.

`myWynn` provides a significant acceleration of the convergence.

```
In[12]:= piSeries[10] // N
Out[12]= {1., 0.666667, 0.866667, 0.72381, 0.834921,
          0.744012, 0.820935, 0.754268, 0.813091, 0.76046, 0.808079}
```

```
In[13]:= myWynn[piSeries[10]] // N
Out[13]= {{1., 0.666667, 0.866667, 0.72381, 0.834921, 0.744012, 0.820935, 0.754268, 0.813091,
           0.76046, 0.808079}, {-3., 5., -7., 9., -11., 13., -15., 17., -19., 21.},
          {0.791667, 0.783333, 0.78631, 0.784921, 0.785678, 0.78522, 0.785518, 0.785314, 0.78546},
          {-115., 329., -711., 1309., -2171., 3345., -4879., 6821.},
          {0.785586, 0.785348, 0.785416, 0.785391, 0.785402, 0.785396, 0.785399},
          {-3879.75, 14060.3, -38964.8, 90665.3, -186514., 350102.},
          {0.785404, 0.785397, 0.785399, 0.785398, 0.785398},
          {-130840., 534685., -1.72624×10^6, 4.70985×10^6},
          {0.785398, 0.785398, 0.785398}, {-4.42449×10^6, 1.94224×10^7}}
```

This answer (the ninth sublist of the last result—only the odd lists are of relevance) differs from π only in the seventh digit.

Now let us implement a functional form of the ϵ-algorithm of Wynn. The next input shows a possible form. Compared with the above implementation, now the sequences are added, subtracted, and divided at once. The `Rest` and `Drop[..., -1]` shorten the sequence at each step.

```
In[14]:= myWynnF[seq_] :=
        FixedPoint[{Rest[#[[2]]] + 1/(Rest[#[[1]]] - Drop[#[[1]], -1]),
                Drop[#[[1]], -1]}&, {seq, Table[0, {Length[seq]}]},
                SameTest -> (Length[#2[[2]]] === 1&)][[1, 1]]
```

We apply the function `myWynnF` to the above sequence `piSeries`. We consider sequence lengths up to 100. The left graphic shows the logarithm of the absolute value of the difference of the last term of `piSeries[k]` to $\pi/4$ and the right graphic shows the logarithm of the absolute value of the difference of the extrapolated value. Clearly, the ϵ-algorithm of Wynn improves the approximation of the limit by many orders of magnitude.

```
In[15]:= dataW = Table[{k, myWynnF[piSeries[k] // N[#, 100]&] - Pi/4},
                {k, 2, 100, 2}];
```

```
In[16]:= dataR = Table[{k, (Last @ piSeries[k]) - Pi/4 // N[#, 100]&},
                {k, 2, 100, 2}];
```

```
In[17]:= Show[GraphicsArray[
        ListPlot[{#1, Log[10, Abs[#2]]}]& @@@ #, AxesOrigin -> {0, 0},
                DisplayFunction -> Identity]& /@ {dataR, dataW}]]
```

For more on the ϵ-algorithm of Wynn, see [1935] and [285].

b) Here is an iterated functional implementation of the Aitken's transformation. At every iteration, we first prepare the three lists p, n, and s corresponding to the elements $\epsilon_{k-1}^{(n+1)}$, $\epsilon_{k-1}^{(n-1)}$, and $\epsilon_{k-1}^{(n)}$, so that the given transformation can be carried out in the second step with whole lists and not with the individual elements.

```
In[1]:= aitken[sequence_] := Nest[(#1 #2 - #3^2)/(#1 + #2 - 2#3)&[Drop[#, 2],
                Drop[#, -2], Drop[Drop[#, -1], 1]]&,
                sequence, Floor[(Length[sequence] - 1)/2]][[1]]
```

Another possibility would be

```
In[2]:= aitken2[sequence_] := Last /@ NestList[Apply[(#1 #3 - #2^2)/(#1 + #3 - 2#2)&,
                Partition[#, 3, 1], {1}]&, sequence,
                Floor[(Length[sequence] - 1)/2]]
```

As a test, let us again use the above sequence of partial sums of $\pi/4$.

```
In[3]:= piSeries[n_] := Rest[FoldList[Plus, 0, Array[(-1)^#/(2# + 1)&, n + 1, 0]]]
```

To better watch how the transformed sequences approach the result, we modify the above code for `aitken` in that way so that the approximation from every iteration is kept by changing `Nest` into `NestList`.

```
In[4]:= aitkenList[sequence_] :=
        Last /@ NestList[(#1 #2 - #3^2)/(#1 + #2 - 2#3)&[
                Drop[#, 2], Drop[#, -2], Drop[Drop[#, -1], 1]]&,
                sequence, Floor[(Length[sequence] - 1)/2]]
```

The numerical test shows a considerable speed up of the convergence behavior (the application of the second N @ ... prevents showing all of the digits).

```
In[5]:= N @ N[aitkenList[N[piSeries[25], 200]] - Pi/4, 100]
```

```
Out[5]= {-0.00961184, -3.99204×10^-6, -5.83883×10^-9, -1.6734×10^-11,
         -7.86489×10^-14, -5.51069×10^-16, -5.37055×10^-18, -6.81844×10^-20,
         -1.02996×10^-21, -1.19091×10^-23, 5.0397×10^-23, -1.07544×10^-22, -6.39968×10^-23}
```

For the related problem of estimating, not the limit, but the next series terms see [1841].

c) We start by calculating the first 100000 terms of the sum. We use high-precision numbers to carry out the summation.

```
In[1]:= s[k_] := (k^2 + k) Log[k]
```

```
In[2]:= v = 1 10^5;
        data = Rest[FoldList[Plus, 0, Table[N[s[k], 30], {k, v}]]];
```

Thinking on the form of the sums $\sum_{k=0}^{n} k^j$ suggests that the sum $\sum_{k=1}^{n} (k^2 + k) \ln(k)$ might contain the following types of divergent contributions.

```
In[4]:= functions[n_] = Flatten[Table[n^k Log[n]^l, {k, 0, 3}, {l, 0, 1}]]
```

$$\text{Out[4]= } \{1, \text{Log}[n], n, n \text{ Log}[n], n^2, n^2 \text{ Log}[n], n^3, n^3 \text{ Log}[n]\}$$

A numerical fit confirms this.

```
In[5]:= fit1 = Fit[data, functions[n], n]
```

$$\text{Out[5]= } 0.27878140826 + 0.08333271963213 \, n - 0.250000000043379772 \, n^2 -$$
$$0.111111111111111134997781 \, n^3 + 0.083414962105 \text{ Log}[n] + 0.666666738244388 \, n \text{ Log}[n] +$$
$$1.0000000000039423170 \, n^2 \text{ Log}[n] + 0.33333333333333333350465070 \, n^3 \text{ Log}[n]$$

Thinking again on the sums $\sum_{k=0}^{n} k^j$ lets us conjecture that the prefactors of the divergent terms are "nice" rational numbers (meaning rational numbers with small denominators). Inspecting the result fit1 more carefully confirms this conjecture for most of the divergencies.

```
In[6]:= f[n_] = Rationalize[fit1, 0.002] /. Rational[_, _?(# > 10&)] -> 0
```

$$\text{Out[6]= } -\frac{n^2}{4} - \frac{n^3}{9} + \frac{2}{3} n \text{ Log}[n] + n^2 \text{ Log}[n] + \frac{1}{3} n^3 \text{ Log}[n]$$

To find the prefactors of the remaining two divergencies n and $\ln(n)$, we subtract the known divergencies to obtain smaller numbers and to carry out a fit with fewer unknown parameters.

```
In[7]:= counterData = Table[f[N[k, 30]], {k, v}];
```

```
In[8]:= remainingFunctions = DeleteCases[functions[n],
            (Alternatives @@ f[n]) /. _Rational | _Integer f_ -> f]
```

$$\text{Out[8]= } \{1, \text{Log}[n], n\}$$

```
In[9]:= fit2 = Fit[data - counterData, remainingFunctions, n]
```

$$\text{Out[9]= } 0.279162786049 + 0.08333333321879437 \, n + 0.0833376671898 \text{ Log}[n]$$

The last fit allows now to find the prefactors of the n and $\ln(n)$ divergencies.

```
In[10]:= F[n_] = f[n] + Rationalize[fit2, 0.001] /.
                    Rational[_, _?(# > 20&)] -> 0
```

$$\text{Out[10]= } \frac{n}{12} - \frac{n^2}{4} - \frac{n^3}{9} + \frac{\text{Log}[n]}{12} + \frac{2}{3} n \text{ Log}[n] + n^2 \text{ Log}[n] + \frac{1}{3} n^3 \text{ Log}[n]$$

After subtracting all divergent terms, we are left with a convergent series. Its first terms will have the form $c_0 + c_{-1}^{(l)} \ln(n)/n + c_{-1}/n + c_{-2}^{(l)} \ln(n)/n^2 + \cdots$. Because we are interested in five correct digits and we took initially 100000 terms, interpolating the current regularized data should give about five correct digits.

```
In[11]:= counterData2 = Table[F[N[k, 30]], {k, v}];
```

```
In[12]:= SequenceLimit[Take[data - counterData2, -10000],
                    WynnDegree -> 100] // SetPrecision[#, 7]&
```

$$\text{Out[12]= } 0.2792029$$

Symbolically the above sum can be evaluated by using $\ln(k) k^\alpha = \partial k^{\alpha+\varepsilon} / \partial \varepsilon |_{\varepsilon=0}$ the above sums can be transformed into sums representing the Zeta function $\zeta(z, a)$ at negative integers [19], [20], [21], [1703], [515], [603] or by differentiating, summing, and integrating $(k^2 + k) \ln(k + \alpha)$ with respect to α [564]. Keeping ε finite and carrying out the asymptotic expansion of $\zeta(z, a)$ allows calculating the symbolic answer $\ln(\text{Glaisher}) + \zeta(3)/(4 \pi^2) = 0.27920293409\ldots$.

7. Scherk's Fifth Surface, Clebsch's Surface, Holed Dodecahedron, Smoothed Dodecahedron

a) This surface $z(x, y)$ is a minimal surface and obeys the corresponding partial differential equation:

$$\left(\left(\frac{\partial z(x, y)}{\partial y}\right)^2 + 1\right)\frac{\partial^2 z(x, y)}{\partial x^2} - 2\,\frac{\partial z(x, y)}{\partial x}\,\frac{\partial z(x, y)}{\partial y}\,\frac{\partial^2 z(x, y)}{\partial x\,\partial y} + \left(\left(\frac{\partial z(x, y)}{\partial x}\right)^2 + 1\right)\frac{\partial^2 z(x, y)}{\partial y^2} = 0.$$

```
In[1]:= Module[{x, y}, (* form derivatives *)
          With[{z = ArcSin[Sinh[x] Sinh[y]]},
               (1 + D[z, y]^2) D[z, x, x] - 2 D[z, x] D[z, y] D[z, x, y] +
               (1 + D[z, x]^2) D[z, y, y]] // Simplify]
```

Out[1]= 0

Several small problems arise in plotting this surface. We proceed as follows:

1) $z = z(x, y) = \arcsin(\sinh(x)\sinh(y))$ is real-valued only in a small part of the real x,y-plane, namely, $|\{x, y\}| - \infty \le x \le \infty$, $-\mathrm{arcsinh}(1/\sinh(x)) \le y \le \mathrm{arcsinh}(1/\sinh(x))$. We can easily plot $z(x, y)$ only over this region.

2) Let $z^* = z(x, y)$ with $\{x, y\}$ in the first quadrant. Then, $z(-x, -y) = z^*$, $z(-x, y) = -z^*$, $z(x, -y) = -z^*$. Moreover, inside the first quadrant, $z(x, y) = z(y, x)$. This allows us to calculate $z(x, y)$ only in half of the first quadrant; in the remainder of the plane, we can construct it by reflection.

3) `ArcSin[arg]` gives the value of the main branch. Because we are interested in a larger set than $-\pi/2 \le z \le \pi/2$, we have to create a continuation of this function. We do this by first reflecting the surface in the plane $z = \pi/2$ and then shifting it in the direction of the positive z-axis by a period of sine, that is, by 2π.

4) It remains to deal with the formula of the surface in half of the first quadrant. To preserve the symmetry, we choose a point on the line $y = x$, and for polar angle $5\pi/4 \le \varphi \le 3\pi/2$, we draw line segments radially from this point in the direction of the x-axis. We find the intersections of these lines with the curve $y(x) = \mathrm{arcsinh}(1/\sinh(x))$ numerically.

We now look at these steps in detail.

First, we read in the package `Graphics`Graphics3D`` in order to draw the points as polygons using `ListSurface`\`Plot3D.

```
In[2]:= Needs["Graphics`Graphics3D`"]
```

Now, we define the boundary curve.

```
In[3]:= f[x_] = ArcSinh[1/Sinh[x]];
```

Next, we define the straight lines from the point $\{x_0, y_0\}$. Here, we choose $\{x_0, y_0\} = \{3, 3\}$.

```
In[4]:= {x0, y0} = {3, 3};

        line[x_, φ_] = Tan[φ] (x - x0) + y0;
```

Now, we calculate the intersections of the straight lines with the boundary curves. `xyValues` is a list of line segments.

```
In[6]:= num = 8;

        xyValues =
        Table[If[φ =!= 3Pi/2, {{-y0/Tan[φ] + x0, 0},
                {xx = x /. FindRoot[f[x] == line[x, φ], {x, 1}], f[xx]}},
                {{x0, 0}, {x0, f[x0]}}] // N, {φ, Pi + Pi/4, 3Pi/2, Pi/4/num}];
```

Here are the corresponding line segments.

```
In[8]:= Show[Graphics[Line /@ xyValues],
            PlotRange -> All, AspectRatio -> Automatic, Axes -> True]
```

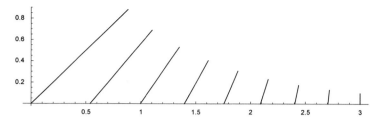

Now, we construct the surface over the above subset of the *x,y*-plane.

```
In[9]:= num2 = 8;

    fuVal[{p0_, p1_}] := Table[{x, y} = p0 + i (p1 - p0)/num2;
                               {x, y, Re[ArcSin[Sinh[x] Sinh[y]]]},
                               {i, 0, num2}]

In[11]:= ListSurfacePlot3D[fuVal /@ xyValues, PlotRange -> All]
```

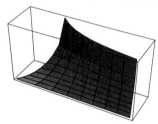

```
In[12]:= polys1 = %[[1]];
```

Next, we find the surface for the other half-quadrant.

```
In[13]:= polys2 = Map[{#[[2]], #[[1]], #[[3]]}&, polys1, {3}];
```

We now combine these polygons for the first quadrant.

```
In[14]:= Show[Graphics3D[polys3 = {polys1, polys2}]]
```

Now, we construct the polygons for the second, third, and fourth quadrants.

```
In[15]:= (* mirror on x,y-plane *)
    polys4 = Map[{-#[[1]], -#[[2]], #[[3]]}&, polys3, {-2}];
    (* mirror on x,z-plane *)
    polys5 = Map[{-#[[1]], #[[2]], -#[[3]]}&, polys3, {-2}];
    (* mirror on y,z-plane *)
    polys6 = Map[{#[[1]], -#[[2]], -#[[3]]}&, polys3, {-2}];
    polys7 = {polys3, polys4, polys5, polys6};
```

Here is the result.

```
In[22]:= Show[Graphics3D[polys7]]
```

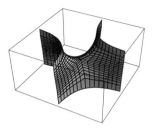

Next, we reflect the above polygons in the plane $z = \pi/2$.

```
In[23]:= polys8 = Map[{#[[1]], #[[2]], Pi/2 + Pi/2 - (#[[3]])}&, polys7, {-2}];
        Show[Graphics3D[polys9 = {polys7, polys8}]]
```

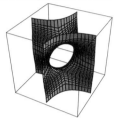

Now, we shift everything by 2π and show the result. The right graphic shows the individual pieces randomly colored.

```
In[25]:= polys10 = Map[{#[[1]], #[[2]], 2Pi + #[[3]]}&, polys9, {-2}];

        Show[GraphicsArray[
          {(* 2 periods in uniform color *)
          Graphics3D[{EdgeForm[{Thickness[0.002], Hue[0.8]}],
              SurfaceColor[Hue[0.12], Hue[0.8], 2.4],
              (* two periods *) polys11 = {polys9, polys10}},
            Boxed -> False],
          (* 2 periods; all pieces in different colors *)
          Graphics3D[Map[{EdgeForm[],
              SurfaceColor[Hue[Random[]], Hue[Random[]], 2.4], #}&,
             polys11, {-5}], Boxed -> False]}]]
```

What is unusual about Scherk's fifth surface are the "holes". There are some that join the first and third quadrants, alternating with others that join the second and fourth quadrants. We can see this with the following input.

```
In[27]:= With[{n = 4},
        Show[GraphicsArray[#]]& /@ Partition[
        Table[Graphics3D[{EdgeForm[{Thickness[0.002], Hue[φ/(Pi)]}],
            SurfaceColor[RGBColor[0, 0, 1], Hue[0.81], 2.4], polys11},
          Boxed -> False, SphericalRegion -> True,
          ViewPoint -> {2 Cos[φ], 2 Sin[φ], 0.8}],
          (* move around *) {φ, 0, Pi/2 (1 - 1/n), Pi/2/n}], 4]]
```

```
With[{n = 60},
Do[Show[Graphics3D[{EdgeForm[{Thickness[0.002], Hue[φ/Pi]}],
    SurfaceColor[Hue[φ/Pi + 1/2], Hue[φ/Pi + 1/2], 2.4], polys11}],
   Boxed -> False, SphericalRegion -> True,
   ViewPoint -> {2 Cos[φ], 2 Sin[φ], 0.8}], {φ, 0, Pi (1 - 1/n), Pi/n}]]
```

For some similar surfaces, see [885].

Due to the periodicity of the surface in *z*-direction, the Scherk surface is ideally suited to map repetitions of it along a knot. The next inputs do this.

```
In[28]:= (* map to a knot; φ1 parametrizes along the knot,
            φ2 parametrizes azimuthally in the knot's cross section,
            and r is the knot diameter *)
        toKnot[φ1_, φ2_, r_] :=
        Block[{d0, d1, d2, t, n, b, c1, c3, c4, s1, s3, s4},
             {c1, c3, c4} = Cos[{1, 3, 4} φ1]; {s1, s3, s4} = Sin[{1, 3, 4} φ1];
             (* derivatives of the space curve forming the knot *)
             d0 = { s1 + 3/2 s3,   c1 - 3/2 c3, s4};
             d1 = { c1 + 9/2 c3,  -s1 + 9/2 s3, 4 c4};
             d2 = {-s1 - 27/2 s3, -c1 + 27/2 c3, -16 s4};
             (* a local orthogonal Frenet coordinate system {t, n, b} *)
             t = d1/Sqrt[d1.d1];
             n = #/Sqrt[#.#]&[d2/d1.d1 - d1.d2/(d1.d1)^2 d1]; b = Cross[t, n];
             (* map the point *) d0 + r (Cos[φ2] n + Sin[φ2] b)]
In[30]:= (* extensions in Cartesian coordinate directions:
            [-1, 1] × [-1, 1] × [0, 1] *)
        scaledUnitPolys = With[{ρ = Max[Norm[Most[#]]& /@ Level[polys9, {-2}]]},
                          Apply[{#1/ρ, #2/ρ, (#3 + Pi/2)/(2Pi)}&, polys9, {-2}]];
In[32]:= (* polygons to be mapped along the knot *)
        tubePrePolys =
         With[{o = 120},
             Table[Apply[{#1, #2, (#3 + k)/o}&, scaledUnitPolys, {-2}],
                 {k, 0, o - 1}]];
In[34]:= (* map a polygon along the knot *)
        mapPolygonToTube[p:Polygon[l_]] :=
        Module[{mpx, mpy, mpz, arcTan = (If[{##} == {0., 0.}, 0., ArcTan[##]]&)},
             {mpx, mpy, mpz} = Plus @@ l/4;
             {(* color along knot *)
              SurfaceColor[Hue[mpz], Hue[mpz], 2.2],
              (* map polygons to knot *)
              Apply[toKnot[2Pi #3, (* add some twist *) arcTan[#1, #2] + 4Pi #3,
                     0.25 Sqrt[#1^2 + #2^2]]&, p, {-2}]}]
In[36]:= (* show the resulting knot *)
        Show[Graphics3D[{EdgeForm[], mapPolygonToTube /@ Flatten[tubePrePolys]}],
             Boxed -> False]
```

b) After changing to polar coordinates, we have the equation.

```
In[1]:= eq[r_, φ_, z_] = (32 - 216 x^2 + 648 x^2 y - 216 y^2 - 216 y^3 -
                150 z + 216 x^2 z + 216 y^2 z + 231 z^2 - 113 z^3) //.
                         {x -> r Cos[φ], y -> r Sin[φ]} // Simplify
```

Out[1]= $(-1 + z) (-32 + 216 r^2 + 118 z - 113 z^2) + 216 r^3 \sin[3\varphi]$

This shows that the surface has a threefold rotational symmetry around the z-axis. Further on, the equation obviously remains unchanged by the transformation $x \to -x$. This means that we only have to calculate one-sixth of the whole surface. Here, the calculation of one-sixth of the surface with ContourPlot3D and the construction of the other parts by reflection and rotation is implemented (the chosen limits for z avoid cutting the surface).

```
In[2]:= Needs["Graphics`ContourPlot3D`"]

In[3]:= clebschSurface =
    Show[Graphics3D[{EdgeForm[Thickness[0.002]],
    {#, Function[m, Map[m.#&, #, {-2}]][
    (* make other parts of the surface *)
        N[{{-1,  Sqrt[3], 0}, {-Sqrt[3], -1, 0}, {0, 0, 2}}/2]],
        Function[m, Map[m.#&, #, {-2}]][
        N[{{-1, -Sqrt[3], 0}, { Sqrt[3], -1, 0}, {0, 0, 2}}/2]]}&[
                Union[#, Map[{-1, 1, 1}#&, #, {-2}]]&[
        (* change to Cartesian coordinates *)
        Map[{#[[1]] Cos[#[[2]]], #[[1]] Sin[#[[2]]], #[[3]]}&,
    (* make one sixth of the surface with ContourPlot3D
        in cylindrical coordinates *)
    ContourPlot3D[eq[r, φ, z], {r, 0, 0.9}, {φ, -Pi/2., -Pi/6.},
    (* carefully selected z-boundaries to fit exactly *)
                {z, -0.975, 2.173}, MaxRecursion -> 1,
                PlotPoints -> {{8, 4}, {8, 3}, {22, 3}},
                DisplayFunction -> Identity][[1]], {-2}]]]}],
        PlotRange -> All, Boxed -> False,
        ViewPoint -> {1.85, -1.6, 0.9}, BoxRatios -> {1, 1, 2.45}]
```

Note that this surface has the remarkable property that 27 straight lines lie on it. For details, see [532], [1544], [818], [1006], [604], [1635], [1632], [29], and [840] and Exercise 27.c) of Chapter 1 of the Symbolics volume [1795].

As a graphics aside, here is a picture showing 20 Clebsch surfaces in a constellation with icosahedral symmetry.

```
Needs["Graphics`Polyhedra`"]

Module[{minx, maxx, miny, maxy, minz, maxz, cPolys, scaledPolys, f},
(* extract polygons from the 3D contour plot *)
cPolys = Cases[clebschSurface, _Polygon, Infinity];
(* size of a cube holding the polygons *)
{{minx, maxx}, {miny, maxy}, {minz, maxz}} =
  {Min[#], Max[#]}& /@ Transpose[Level[cPolys, {-2}]];
(* rescale polygons in a unit cube *)
scaledPolys = Apply[{#1, #2, 1 - #3}&,
  Map[{{0, -1, 0}, {1, 0, 0}, {0, 0, 1}}.#&,
       Apply[{(#1 - minx)/(maxx - minx) - 1/2,
              (#2 - miny)/(maxy - miny) - 1/2,
              (#3 - minz)/(maxz - minz)}&,
             cPolys, {-2}], {-2}], {-2}];
(* map scaled polygons on a face of a icosahedron *)
f[icoFace:Polygon[{p1_, p2_, p3_}]] :=
Module[{mp = (p1 + p2 + p3)/3, n = #/Sqrt[#.#]&, n1, n2, n3},
 n3 = n[mp]; n1 = n[p1 - mp]; n2 = n[# - #.n1 n1]&[p2 - mp];
 Apply[(mp + (1 + (* extend *) #3) (#1 n1 + #2 n2) + 3 #3 n3)&,
       scaledPolys, {-2}]];
(* show 20 Clebsch surfaces *)
Show[Graphics3D[{EdgeForm[], (* random color for each surface *)
 {SurfaceColor[Hue[Random[]], Hue[Random[]], 3 Random[]], #}& /@
   f /@ Take[Cases[Polyhedron[Icosahedron],
       _Polygon, Infinity], All]}], PlotRange -> All, Boxed -> False]]
```

As an alternative to ContourPlot3D, we should have the surface either in a parametric form [11] $x_1 = x_1(s, t)$, $x_2 = x_2(s, t)$, $x_3 = x_3(s, t)$ or in an explicit form $x_1 = x_1(x_2, x_3)$, (here x_1, x_2, x_3 are Cartesian coordinates in 3D space). A parametric representation is not so easy to get for this surface, but looking at eq, one recognizes that this equation is easily solved for φ to get $\varphi = \varphi(r, z)$.

```
In[4]:= Φ[{r_, z_}] := ArcSin[(-32 + 216 r^2 + 150 z - 216 r^2z - 231 z^2 +
                      113 z^3)/(216 r^3)]/3
```

(For how to parametrize implicit polynomial surfaces in rational functions, see, for instance, [1600].) Unfortunately, because of the holes in the surface, the use of Plot3D is inappropriate here and a good graphic is not directly available.

```
In[5]:= (* turn off messages because function values is not a real number *)
        Off[Plot3D::plnc]; Off[Plot3D::gval];
        Plot3D[Evaluate[Φ[{r, z}]], {r, 10^-6, 0.9}, {z, -0.975, 2.173},
               PlotPoints -> 60, AxesLabel -> {"r", "z", None}]
```

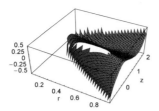

To make a picture of $\varphi = \varphi(r, z)$, we need the (r, z) domain where Φ gives a real-valued solution. With `ContourPlot`, we can get a graphic of this domain; borderlines arise from the limits of the φ-values of interest,

$$-1 \le \frac{-32 + 216\, r^2 + 150\, z - 216\, r^2\, z - 231\, z^2 + 113\, z^3}{216\, r^3} < +1,$$

that is, from `eq[r, Pi/6, z] = 0` and `eq[r, Pi/2, z] = 0`. For space reasons we will display the following r,z-plots with the r-axis vertically.

```
In[8]:= Show[ContourPlot[eq[r, #, z], {z, -0.975, 2.173}, {r, 0, 0.9},
                Contours -> {0}, AspectRatio -> Automatic,
                ContourShading -> False, PlotPoints -> 40,
                DisplayFunction -> Identity, TextStyle -> {FontSize -> 5},
                ContourStyle -> {Thickness[0.01]}]& /@ {Pi/2, Pi/6},
                DisplayFunction -> $DisplayFunction]
```

To get a numerical description of the $\{r, z\}$ domain, we first observe that the lower and the upper edge of the hole are at $\{2/3, 0\}$ and $\{2/9, 2/3\}$, respectively. The functions R1 (left border for $z < 1$), R2 (left border for $z > 1$), R3 (left side of the hole), and R4 (right side of the hole) give the r-values for the boundaries in the r,z-plane for a given z-value by numerically solving the corresponding equations.

```
In[9]:= rZeros[z_, pm_] :=
    Cases[Last /@ List @@ NRoots[32 - 216 r^2 + (-1)^pm 216r^3 -
        150z + 216r^2z + 231z^2 - 113z^3 == 0, r], _Real?Positive]
    (* the boundaries made by the zeros *)
    R1[z_] := Last @ rZeros[z, 1]; R2[z_] := Last @ rZeros[z, 2];
    R3[z_] := Min  @ rZeros[z, 2]; R4[z_] := Max  @ rZeros[z, 2];
```

We can now use R1, R2, R3, and R4 to show the r,z-domain of interest.

```
In[13]:= makeRVertical[expr_] := expr /. (pl:(Line | Polygon))[l_] :> pl[Reverse /@ l];
```

```
In[14]:= ε = 10^-10;
    pic =
    Show[makeRVertical @ Graphics[{{GrayLevel[0.8], Polygon /@
    (* the various regions glued together *)
    {Join[Table[{R1[z], z}, {z, -0.975, 0, 0.975/40}],
        Reverse @ Table[{0.9, z}, {z, -0.975, 0, 0.975/40}]],
     Join[Table[{R2[z], z}, {z, 1 + ε, 2.173, (1.173 - ε)/40}],
        Reverse @ Table[{0.9, z}, {z, 1, 2.173, 1.173/40}]],
     Join[Table[{R4[z], z}, {z, 0, 2/3, 2/3/40}],
        Reverse @ Table[{0.9, z}, {z, 0, 2/3, 2/3/40}]],
     Join[Table[{R3[z], z}, {z, 0, 2/3, 2/3/40}],
        Reverse @ Table[{R1[z], z}, {z, 0, 2/3, 2/3/40}]],
     Join[Table[{R1[z], z}, {z, 2/3, 1 - ε, (1/3 - ε)/40}],
```

```
        Reverse @ Table[{0.9, z}, {z, 2/3, 1, 1/3/40}]]},
  (* bounding lines *)
  {Thickness[0.01], GrayLevel[0], Line /@ {
      Table[{R1[z], z}, {z, -0.975, 1 - ε, (1.975 - ε)/100}],
      Table[{R2[z], z}, {z, 1 + 10^-10, 2.173, 1.173/100}],
      Table[{R4[z], z}, {z, 0, 2/3, 2/3/100}],
      Table[{R3[z], z}, {z, 0, 2/3, 2/3/100}],
      {{0, -0.975}, {0.9, -0.975}, {0.9, 2.173},
       {0, 2.173}, {0, -0.975}}}}}],
      Axes -> False, Frame -> False, AspectRatio -> Automatic]
```

Suppose we divide the gray-shaded area into smaller, simply connected pieces.

```
In[16]:= Show[{pic, makeRVertical @ Graphics[{
    Line[{{0, 1}, {0.9, 1}}], Line[{{R1[ε], 0}, {0.9, 0}}],
    Line[{{R1[2/3], 2/3}, {0.9, 2/3}}],
    (* numbers of the regions *)
    Text[StyleForm[#[[1]]], FontSize -> 5], Reverse @ #[[2]]]& /@
        {{"1", {0.7, 1.3}}, {"2", {0.7, 0.8}}, {"3", {0.7, 0.5}},
         {"4", {0.33, 0.1}}, {"5", {0.7, -0.3}}}}]}]
```

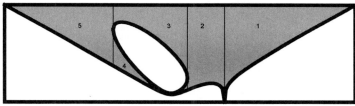

We can now easily generate a distorted rectangular grid in every one of the regions.

```
In[17]:= pr = 16;
    pz1 = 25; tab[1] = Table[Function[rMin,
            Table[{r, z}, {r, rMin + ε, 0.9, (0.9 - rMin - 2ε)/(2pr)}]][R2[z]],
            {z, 1 + ε, 2.173, (1.173 - ε)/pz1}];
    pz2 = 10; tab[2] = Table[Function[rMin,
            Table[{r, z}, {r, rMin + ε, 0.9, (0.9 - rMin - 2ε)/(2pr)}]][R1[z]],
            {z, 2/3, 1 - ε, (1/3 - ε)/pz2}];
    pz3 = 10; tab[3] = Table[Function[rMin,
            Table[{r, z}, {r, rMin, 0.9, (0.9 - rMin)/(2pr)}]][R4[z]],
            {z, 0, 2/3, 2/3/pz3}];
    pz4 = 16; tab[4] = Table[Function[{rMin, rMax},
            Table[{r, z}, {r, rMin, rMax, (rMax - rMin)/(2pr)}]][R1[z], R3[z]],
            {z, 0, 2/3, 2/3/pz3}];
    pz5 = 22; tab[5] = Table[Function[rMin,
            Table[{r, z}, {r, rMin, 0.9, (0.9 - rMin)/(2pr)}]][R1[z]],
            {z, -0.975, 0, 0.975/pz5}];
```

Here, all of these grids are shown.

```
In[23]:= Show[{pic, makeRVertical @ Graphics[{Thickness[0.002],
                    Map[Line, Array[tab, 5], {2}],
                    Map[Line, Transpose /@ Array[tab, 5], {2}]}]},
        AspectRatio -> Automatic, PlotRange -> All]
```

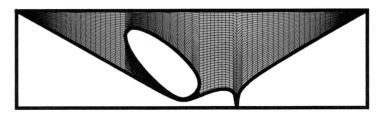

Now it is easy to rebuild one-sixth of the surface by using, for instance, `ListSurfacePlot3D`.

```
In[24]:= Needs["Graphics`Graphics3D`"]

In[25]:= (* calculate points for all five regions *)
       Do[dat[i] = Map[Flatten[{#, Re @ Φ[#]}]&, tab[i], {2}], {i, 5}]

In[27]:= (* make polygons for all five regions *)
       allPolys = Join @@ (ListSurfacePlot3D[#,
               DisplayFunction -> Identity][[1]]& /@ Array[dat, 5]);

In[29]:= (* convert to Cartesian coordinates *)
       allPolys1 = Map[{#[[1]] Cos[#[[3]]], #[[1]] Sin[#[[3]]], #[[2]]}&,
                   N[allPolys], {-2}];

In[31]:= Show[Graphics3D[{EdgeForm[{Thickness[0.002], GrayLevel[1]}],
                   allPolys1}]]
```

And here is the complete surface.

```
In[32]:= Show[Graphics3D[{EdgeForm[{}],
           (* generating other parts of the surface by reflection and rotation *)
           {#, Function[m, Map[m.#&, #, {-2}]][
               N[{{-1,  Sqrt[3], 0}, {-Sqrt[3], -1, 0}, {0, 0, 2}}/2]],
           Function[m, Map[m.#&, #, {-2}]][
               N[{{-1, -Sqrt[3], 0}, { Sqrt[3], -1, 0}, {0, 0, 2}}/2]]}&[
           Union[#, Map[{-1, 1, 1}#&, #, {-2}]]&[(* last graphic *) %[[1]]]]]},
           PlotRange -> All, Boxed -> False, ViewPoint -> {1.85, -1.6, 0.9},
           BoxRatios -> {1, 1, 2.45}]
```

For the construction of rational parametrizations of cubic surfaces see [1317], [118], and [198]; for numerical parametrizations see [796]. For an automated way to determine the topology of a surface, see [623].

c) Let us start with figuring out the shape of `thePoly[{x, y, z}] == 0`. We load the package.

```
In[1]:= Needs["Graphics`Polyhedra`"]
```

Next, we calculate.

```
In[2]:= polyWithPlatoSymmetry[plato_, {x_, y_, z_}, n_] :=
         Plus @@ (({x, y, z}.#)^n & /@  ((* midpoints of faces *)
         (Plus @@ #/Length[#]& /@ (First /@ (Polyhedron[plato][[1]]))))) - 1
```

As the name `polyWithPlatoSymmetry` suggests, this polynomial has exactly the same symmetry as the Platonic solid `plato`, in this case, as a dodecahedron. This is also easily seen from the code, which forms the sum of $(direction_i.\{x, y, z\})^{14}$, where the sum extends over all normalized directions to the midpoints of the faces of a dodecahedron. Because of the even power and the symmetry, this surface must be a closed one, including the point $\{0, 0, 0\}$, and so it represents a deformed sphere with the symmetry of a dodecahedron.

```
In[3]:= dodePoly = Chop[polyWithPlatoSymmetry[Dodecahedron, {x, y, z}, 14]]
```

$$Out[3]= -1 + (-0.760845\,x - 0.380423\,z)^{14} +$$
$$(-0.235114\,x - 0.723607\,y - 0.380423\,z)^{14} + (0.615537\,x - 0.447214\,y - 0.380423\,z)^{14} +$$
$$(0.615537\,x + 0.447214\,y - 0.380423\,z)^{14} + (-0.235114\,x + 0.723607\,y - 0.380423\,z)^{14} +$$
$$(0.760845\,x + 0.380423\,z)^{14} + (0.235114\,x - 0.723607\,y + 0.380423\,z)^{14} +$$
$$(-0.615537\,x - 0.447214\,y + 0.380423\,z)^{14} + (-0.615537\,x + 0.447214\,y + 0.380423\,z)^{14} +$$
$$(0.235114\,x + 0.723607\,y + 0.380423\,z)^{14} + 0.207754\,z^{14}$$

Remark: Symbolically `dodePoly` can be written in the following concise form *dodePoly* = 0, where

$$dodePoly =$$
$$\big(26\,(65 + 29\sqrt{5})\,(x^2 + y^2 + z^2)\,(139\,x^{12} + 484\,y^2\,x^{10} + 3135\,y^4\,x^8 + 2640\,y^6\,x^6 + 825\,y^8\,x^4 + 1100\,y^{10}\,x^2 +$$
$$154\,(x^4 - 10\,y^2\,x^2 + 5\,y^4)\,z^7\,x + 1694\,(x^2 + y^2)\,(x^4 - 10\,y^2\,x^2 + 5\,y^4)\,z^5\,x +$$
$$3850\,(x^2 + y^2)^2\,(x^4 - 10\,y^2\,x^2 + 5\,y^4)\,z^3\,x + 770\,(x^2 + y^2)^3\,(x^4 - 10\,y^2\,x^2 + 5\,y^4)\,z\,x + 125\,y^{12} +$$
$$601\,z^{12} - 594\,(x^2 + y^2)\,z^{10} + 825\,(x^2 + y^2)^2\,z^8 + 1485\,(x^2 + y^2)^3\,z^6 + 6600\,(x^2 + y^2)^4\,z^4 +$$
$$11\,(289\,x^{10} + 1095\,y^2\,x^8 + 4290\,y^4\,x^6 + 1350\,y^6\,x^4 + 1725\,y^8\,x^2 + 275\,y^{10})\,z^2) - 9765625\big) \big/ 9765625$$

We can have a quick look at this surface by changing from Cartesian to spherical coordinates and then, for given $\{\varphi, \vartheta\}$ values, numerically computing the corresponding radius with `FindRoot`.

```
In[4]:= Needs["Graphics`Graphics3D`"]
```

```
In[5]:= quickDodeGraphic[plato_, poly_, {x_, y_, z_}, {pp1_, pp2_}, opts___] :=
         Module[{l = Length[Polyhedron[plato][[1, 1, 1]]]
                 (* number of vertices of a face of plato *), f1, ρ, points, R},
         (* changing variables *)
         f1 = poly //. {x -> r Cos[φ] Sin[ϑ], y -> r Sin[φ] Sin[ϑ], z -> r Cos[ϑ]};
         ρ[φ_, ϑ_] := FindRoot[f1 == 0, {r, 1}][[1, 2]];
         (* make points of 1/5 of the surface *)
         points = Table[ρ[φ, ϑ] {Cos[φ] Sin[ϑ], Sin[φ] Sin[ϑ], Cos[ϑ]},
                    Evaluate[N[{ϑ, 0, Pi, Pi/pp1}]],
                    Evaluate[N[{φ, 0, 2Pi/l, 2Pi/l/pp2}]]];
         (* rotation matrices *)
         Do[R[i] = {{ Cos[i 2Pi/l], Sin[i 2Pi/l], 0},
                    {-Sin[i 2Pi/l], Cos[i 2Pi/l], 0},
                    {0, 0, 1}} // N, {i, 0, l - 1}];
         (* rotate in other positions and show *)
         Show[Table[ListSurfacePlot3D[Map[R[i].#&, points, {-2}],
                        DisplayFunction -> Identity],
                    {i, 0, l - 1}], opts, DisplayFunction -> $DisplayFunction]]
```

For comparison, we also display the dodecahedron. We basically get a dodecahedron with rounded edges and corners [52].

```
In[6]:= Show[GraphicsArray[
         Block[{$DisplayFunction = Identity},
         {quickDodeGraphic[Dodecahedron, dodePoly, {x, y, z}, {24, 12}, Axes -> True],
          Show[Polyhedron[Dodecahedron], Axes -> True]}]]]
```

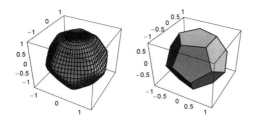

Here is the polynomial `cylinderPoly`.

```
In[7]:= cylinderAxisDirections = #/Sqrt[#.#]& /@ Chop[
           (* midpoints of the faces *) Apply[Plus,
             First /@ (* the six upper faces *)
        Take[Polyhedron[Dodecahedron][[1]], 6], {1}]/5];

        cylinderPoly = Times @@ (x^2 + y^2 + z^2 - ({x, y, z}.#)^2 -
        (* radius^2 of the cylinders *) 0.01 & /@ cylinderAxisDirections);
```

It gives the implicit representations of six narrow cylinders that go through the midpoints of opposite faces of the dodecahedron. The resulting polynomial `thePoly` consists of the product of the implicit representations of `dodePoly` and `cylinderPoly` and the added positive constant 3/100.

```
In[9]:= thePoly[{x_, y_, z_}] = dodePoly cylinderPoly + 3 10^-2;
```

Remembering our discussion from Section 3.3 of the Graphics volume [1794], this means that the cylinders drill long holes into `dodePoly`. Looking at the cross sections in the $y = 0$ and $z = 1$ planes, this reasoning is confirmed.

```
In[10]:=
        Show[GraphicsArray[
        Block[{$DisplayFunction = Identity},
        {(* x,z-plane *)
         ContourPlot[Evaluate[thePoly[{x, 0, z}]], {x, -3/2, 3/2}, {z, -3/2, 3/2},
                 Contours -> {0}, PlotPoints -> 60, PlotLabel -> "y = 0"],
          (* shifted x,y-plane *)
         ContourPlot[Evaluate[thePoly[{x, y, 1}]], {x, -1, 1}, {y, -1, 1},
                 Contours -> {0}, PlotPoints -> 60, PlotLabel -> "z = 1"]}]]]
```

Now, let us tackle the more complicated part of this exercise, making a picture of the inner, connected part of the zeros of `thePoly[{x, y, z}]` without using `ContourPlot3D`. The idea is first to use symmetry and only to calculate the 1/120 part of the whole surface to save time. Second, to be able to calculate 1/120 of the surface, we choose a modified cylindrical coordinate system along an edge and calculate the radius for a given angle and a given point on the edge. Looking at the $y = 0$ picture shows that at $x \approx 0.57$, $y \approx 0.95$, one edge is perpendicular to the plot plane.

We use the x-value and calculate the corresponding y-value.

```
In[11]:= (* geometric data from Dodecahedron for correct x/z-value
             of midpoint edge *)
         {ha, hb, hc} = Last[Apply[Plus, Partition[Append[#, First[#]]&[
           Polyhedron[Dodecahedron][[1, 1, 1]]], 2, 1], {1}]/2 // Chop];

         (* the factor 0.57 is suggested by the above picture *)
         p1 = 0.57 {1, 0, hc/ha}; (* midpoint edge *)
```

```
        p2 = {{ Cos[-2Pi/10], Sin[-2Pi/10], 0},
              {-Sin[-2Pi/10], Cos[-2Pi/10], 0}, {0, 0, 1}}.
              ({p1[[1]]/Cos[2Pi/10], 0, p1[[3]]}) // N;
      (* one vertex *) p3 = {0, 0, p1[[3]]} (* center upper face *);
```

In so doing, we can change to our polar coordinates along the edge (the edge corresponds to the *z*-axis of a cylindrical coordinate system).

```
In[17]:= dir1[s_] := #/Sqrt[#.#]&[p1 + s(p2 - p1)];
         dir2[s_] := #/Sqrt[#.#]&[p3 - (p1 + s(p2 - p1))];
```

Here, *s* ranges from 0 to 1 as the point $p1 + s(p2 - p1)$ moves from the midpoint of an edge p1 to a vertex p2; `dir1` is the direction from the origin to the point $p1 + s(p2 - p1)$, and `dir3` points from the point $p1 + s(p2 - p1)$ to the center p3 of one neighboring face. To fit exactly the symmetry of the dodecahedron, we use the two nonorthogonal directions `dir1` and `dir2` as basis vectors for a given *s* (except for $s = 0$, they are also not orthogonal to $p2 - p1$). If *s* now ranges from 0 to 1 and *p* from 0 to π, a part of a cylinder is formed that is cut at the correct angles.

```
In[19]:= op[s_, φ_, r_] := p1 + s (p2 - p1) +
                           r #/Sqrt[#.#]&[Cos[φ] dir1[s] + Sin[φ] dir2[s]]
```

```
In[20]:= ParametricPlot3D[op[s, φ, 0.1], {s, 0, 1}, {φ, 0, Pi},
                          PlotPoints -> {12, 12}, Compiled -> False,
                          ViewPoint -> {-1, 1.2, 0.5}]
```

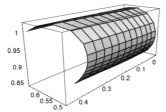

```
In[21]:= part1 = %[[1]];
```

The following calculates all other equivalent parts of the dodecahedron.

```
In[22]:= (* make second half of the edge by mirroring *)
         part2 = Join[part1, Map[{1, -1, 1}#&, part1, {-2}]];

         (* rotation matrices inside one face *)
         Do[R[i] = N[{{ Cos[i 2Pi/5], Sin[i 2Pi/5], 0},
                      {-Sin[i 2Pi/5], Cos[i 2Pi/5], 0}, {0, 0, 1}}], {i, 0, 4}];

         (* one face *)
         part3 = Join @@ Table[Map[R[i].#&, part2, {-2}], {i, 0, 4}];

         (* rotation matrices for other faces *)
         With[{amat = Table[a[j, k][i], {j, 3}, {k, 3}]},
         Do[(* the rotation matrices for rotating
               the polygons in all equivalent positions *)
            rot[i] = (amat /. Solve[Flatten[Table[Thread[
            amat.Polyhedron[Dodecahedron][[1, 1, 1, j]] ==
                Polyhedron[Dodecahedron][[1, i, 1, j]]], {j, 3}]],
                            Flatten[amat]])[[1]], {i, 12}]];

         (* rotate into other faces positions and show result *)
         Show[Graphics3D[{EdgeForm[{Thickness[0.002]}],
                          Table[Map[rot[i].#&, part3, {-2}],
                                {i, 12}]}], Axes -> True, PlotRange -> All]
```

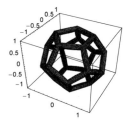

Now, we define a function `point`.

```
In[32]:= point[{r_, φ_, s_}] := p1 + s (p2 - p1) +
                    r (Cos[φ] dir1[s] + Sin[φ] dir2[s])
```

For given φ and s, we are interested in the value r, so that `thePoly[point[r, φ, s]] == 0`. Again, we use `FindRoot` for calculating this value. To speed up the calculation of `thePoly`, we use a compiled version `thePolyComp`.

```
In[33]:= SetDelayed @@ Join[Hold[thePolyComp[{x_, y_, z_}]],
            Hold[#[x, y, z]]&[Compile[{xx, yy, zz}, (* evaluate body *)
                            Evaluate[thePoly[{xx, yy, zz}]]]]]]

        aux[{r_?NumberQ, φ_?NumberQ, s_?NumberQ}] := thePolyComp[point[{r, φ, s}]]

        radius[φ_, s_] :=
        FindRoot[aux[{r, φ, s}] == 0, {r, 0.03, 0.6}, Method -> Brent][[1, 2]]
```

Now, we have everything together to calculate 1/120 of the surface. (We could use more points to get a smoother surface, but this requires more resources.)

```
In[36]:= tab = Table[Join[
            (* divide φ region for better uniformity of points *)
            Table[point[{radius[φ, s], φ, s}], {φ, 0., Pi/2., Pi/2./12}],
            Table[point[{radius[φ, s], φ, s}], {φ, Pi/2., 1. Pi, Pi/2./6}]],
                {s, 0, 1, 1/6}];
```

So, it looks like this.

```
In[37]:= ListSurfacePlot3D[tab, Axes -> True]
```

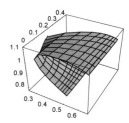

```
In[38]:= polys1 = %[[1]];
```

In analogy to the above cylinders, we get the remaining 119 parts by mirroring and rotation.

```
In[39]:= (* mirror and rotate polygons *)
        polys2 = Join[polys1, Map[{1, -1, 1}#&, polys1, {-2}]];
        polys3 = Join @@ Table[Map[ℛ[i].#&, polys2, {-2}], {i, 0, 4}];

        (* display all polygons *)
        Show[Graphics3D[{EdgeForm[{Thickness[0.002], GrayLevel[1/2]}],
                        SurfaceColor[Hue[0.12], Hue[0.4], 2.9],
                        Table[Map[rot[i].#&, polys3, {-2}], {i, 12}]}],
            Boxed -> False, PlotRange -> All]
```

d) We start with the construction of the thickened wireframe version of a dodecahedron. dodeFrame is such an object.

```
In[1]:= Needs["Graphics`Polyhedra`"]

In[2]:= (* given a polygon, "solidify" its edges *)
       solidFrame[Polygon[l_], {flo_, fli_, f2_}] :=
       Module[{mp = Plus @@ l/Length[l], l1 = Append[l, First[l]],
               pl1i, pl2i, P = Partition},
              (* points arising from contracting the polygon
                 in plane and in normal directions *)
              pl1i = (mp + fli(# - mp))& /@ l1; l2 = f2 l1;
              {pl1i, pl2i} = ((mp + fli(# - mp))& /@ #)& /@ {l1, l2};
              (* three new polygons *)
              {(Polygon[Join[#[[1]], Reverse[#[[2]]]]]& /@ Transpose[#])& /@
              {{P[l1, 2, 1], P[pl1i, 2, 1]}, {P[l2, 2, 1],
                P[pl2i, 2, 1]}, {P[pl1i, 2, 1], P[pl2i, 2, 1]}}}]

In[4]:= (* numericalized vertices and polygons *)
       nVertices = N[Vertices[Dodecahedron], 30];
       dodeFramePolys = Map[nVertices[[#]]&, Polygon /@ Faces[Dodecahedron], {-1}];

In[7]:= Show[Graphics3D[dodeFrame = N[solidFrame[#, {2/3, 2/3, 2/3}]]]& /@
            dodeFramePolys], PlotRange -> All, Boxed -> False]
```

Now, we triangulate the wireframe using the function triangulatePolygon.

```
In[8]:= (* make four triangles from a quadrilateral *)
       triangulatePolygon[Polygon[{p1_, p2_, p3_, p4_}]] :=
        With[{mp = (p1 + p2 + p3 + p4)/4},
             Polygon /@ {{p1, p2, mp}, {p2, p3, mp}, {p3, p4, mp}, {p4, p1, mp}}]

       (* make four triangles from a triangle *)
       triangulatePolygon[Polygon[{p1_, p2_, p3_}]] :=
       With[{p12 = (p1 + p2)/2, p23 = (p2 + p3)/2, p31 = (p3 + p1)/2},
            Polygon /@
            {{p1, p12, p31}, {p12, p2, p23}, {p23, p3, p31}, {p12, p23, p31}}]

       (* iterate polygon subdivision *)
       triangulatePolygon[p_Polygon, n_] :=
               Nest[Flatten[triangulatePolygon /@ #]&, {p}, n]
```

Subdividing all initial polygons three times leads to 11520 polygons.

```
In[16]:= triangles = Flatten[triangulatePolygon[#, 3]& /@ Flatten[dodeFrame]];
        Length[triangles]
Out[17]= 11520
```

In[18]:= **Show[Graphics3D[{EdgeForm[Thickness[0.002]], triangles}], Boxed -> False]**

To carry out the smoothing we need to know all neighbors of all points. Because the number and neighborhoods of the triangles do not change for each smoothing step, we have to calculate the neighbors only once. We start by constructing a list of all edges. To avoid problems with unioning machine-number lists we used high-precision numbers above.

In[19]:= **allEdges = Union[Join[#, Reverse /@ #]&[Flatten[**
          Partition[Append[#[[1]], #[[1, 1]]], 2, 1]& /@
          Cases[triangles, _Polygon, Infinity], 1]]];

In[20]:= **allPoints = Union[Level[allEdges, {2}]];**

Our polyhedral mesh has 5740 points and 34560 unoriented edges.

In[21]:= **Length /@ {allPoints, allEdges}**

Out[21]= {5740, 34560}

We numerate all points and construct symbolic versions of the list of points, edges, and triangles (they end with a P).

In[22]:= **{allPointsP, allEdgesP, trianglesP} = {allPoints, allEdges, triangles} /.**
          Dispatch[MapIndexed[(#1 -> P[#2[[1]]])&, allPoints]];

Now, we go through the list of all edges and construct all neighbors of all points.

In[23]:= **(* initialize neighbors *)**
        Do[neighbors[allPointsP[[k]]] = {}, {k, Length[allPointsP]}]
        (* take an edge and append neighbors *)
        Do[neighbors[allEdgesP[[k, 1]]] = {neighbors[allEdgesP[[k, 1]]],
                                            allEdgesP[[k, 2]]},
           {k, Length[allEdgesP]}];
        (* flatten lists of neighbors *)
        Do[neighbors[allPointsP[[k]]] = Flatten[neighbors[allPointsP[[k]]]],
           {k, Length[allPointsP]}]

This yields the following list of neighbors.

In[29]:= **smoothingRules = ((# -> (Plus @@ #/Length[#]&[neighbors[#]]))& /@ allPointsP);**

In[30]:= **Take[smoothingRules, 3]**

Out[30]= $\{$ P[1] $\rightarrow$
        $\frac{1}{12}$ (P[2] + P[3] + P[4] + P[5] + P[6] + P[7] + P[8] + P[10] + P[11] + P[14] + P[15] + P[39]),
        P[2] $\rightarrow$ $\frac{1}{6}$ (P[1] + P[3] + P[4] + P[9] + P[12] + P[13]),
        P[3] $\rightarrow$ $\frac{1}{6}$ (P[1] + P[2] + P[10] + P[12] + P[16] + P[27]) $\}$

Most points have six neighbors. Some have 4, or 8, or 12 neighbors.

In[31]:= **{First[#], Length[#]}& /@ Split[Sort[#[[2, 1]]& /@ smoothingRules]]**

Out[31]= $\{\{\frac{1}{12}, 40\}, \{\frac{1}{8}, 120\}, \{\frac{1}{6}, 5400\}, \{\frac{1}{4}, 180\}\}$

To carry out the smoothing steps quickly (we want to do it 1000 times) we construct a compiled function smoothStepCF for the smoothing step.

In[32]:= **smoothStepCF = Compile @@ (({{First[#], _Real, 1}& /@ #, Last /@ #})&[**
          smoothingRules /. (* make symbols *)
          P[k_] :> ToExpression["P" <> ToString[k]]]);

smoothStep[α, l] forms the new positions of the vertices by mixing the original positions l and the new positions smoothStepCF @@ l with the ratio $(1 - \alpha)/\alpha$.

```
In[33]:= smoothStep[α_, l_] := (1 - α) l + α (smoothStepCF @@ l)
```

For displaying the smoothed thickened dodecahedra wireframes, we use the function smoothedFrameGraphics.

```
In[34]:= smoothedFrameGraphics[data_] :=
    Graphics3D[{EdgeForm[], trianglesP /. Dispatch[Rule @@@
                            Transpose[{allPointsP, data}]]},
                PlotRange -> All, Boxed -> False, SphericalRegion -> True]
```

The following shows the thickened wireframe after the 10th smoothing step as a function of α.

```
In[35]:= Show[GraphicsArray[Function[α, smoothedFrameGraphics @
                Nest[(smoothStep[α, #])&, allPoints, 10]] /@ #]& /@
                    Partition[Table[α, {α, -1/10, 17/10, 18/10/8}], 3]]
```

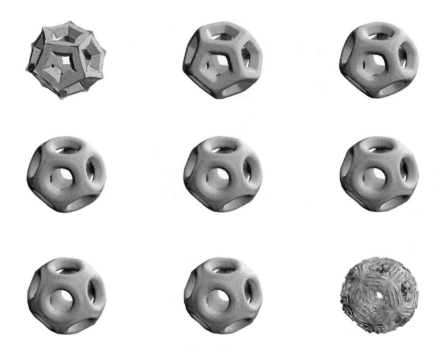

```
Do[Show[smoothedFrameGraphics @ Nest[(smoothStep[α, #])&, allPoints, 10]],
    {α, -1/10, 17/10, 1/20}]
```

Now, we apply the smoothing step 1000 times. The following graphic shows the shape of the thickened wireframe after the 10th, 100th, and 1000th steps.

```
In[36]:= Show[GraphicsArray[smoothedFrameGraphics[Nest[(smoothStep[1, #])&,
                            allPoints, #]]& /@ {10, 100, 1000}]]
```

In addition to the obvious change in shape, caused by different averaging speeds as a function of the local curvature, the wireframe as a whole shrinks [1197].

```
In[37]:= Sqrt[Max[#.#& /@ Level[Cases[#, _Polygon, Infinity], {-2}]]]& /@ %[[1]]

Out[37]= {0.944973, 0.678744, 0.0723712}
```

For different, diffusion-based smoothing procedures, see [119].

8. A Convergent Sequence for π, Contracting Interval Map

a) Here is an implementation of the recurrence. We compute all terms of the sequence with arbitrary precision *prec*.

```
In[1]:= y1[n_, prec_] := y1[n, prec] =
          N[(1 - (1 - y1[n - 1, prec]^4)^(1/4))/
            (1 + (1 - y1[n - 1, prec]^4)^(1/4)), prec]

       y1[0, prec_] := y1[0, prec] = N[Sqrt[2] - 1, prec]

       a1[n_, prec_] := a1[n, prec] =
       N[(1 + y1[n, prec])^4 a1[n - 1, prec] -
          2^(2(n - 1) + 3) y1[n, prec] (1 + y1[n, prec] + y1[n, prec]^2), prec]

       a1[0, prec_] := a1[0, prec] = N[6 - 4Sqrt[2], prec]
```

Here is a1[1, Infinity] in simplified form.

```
In[5]:= a1[1, Infinity] // Simplify
```

$$Out[5]= \frac{8\left(-7 + 4\sqrt{2} + 4\sqrt{-4 + 3\sqrt{2}}\right)}{\left(1 + \sqrt{2}\left(-4 + 3\sqrt{2}\right)^{1/4}\right)^4}$$

Next, we find the relative deviation from π as a function of the order of the iteration.

```
In[6]:= piExactness[a_, n_, prec_] := N[(Pi - 1/a[n, prec])/Pi, prec] //
                          (* display three digits *) SetPrecision[#, 3]&
```

We now numerically evaluate the first terms.

```
In[7]:= {#1, piExactness[a1, #1, #2]}& @@@
          {{1, 10}, {2, 50}, {3, 200}, {4, 800}, {5, 3000}}

Out[7]= {{1, 2.35×10^-9}, {2, 1.74×10^-41}, {3, 7.35×10^-172}, {4, 3.54×10^-695}, {5, 2.94×10^-2790}}
```

Each iteration step gives about four times as many digits of π. We have the following estimate for the accuracy: $|\alpha_n - 1/\pi| < 16 \cdot 4^n \exp((-24)^n \pi)$

```
In[8]:= Table[16 4^n Exp[-2 Pi 4^n] // N, {n, 10}]

Out[8]= {7.7834×10^-10, 5.60011×10^-42, 2.34491×10^-172, 1.126332928464978×10^-695,
         9.36800839400363×10^-2791, 7.004695457642099×10^-11173, 3.421190303778410×10^-44703,
         3.041929576118077×10^-178826, 2.970686846793133×10^-715320, 4.221898331779792×10^-2861298}
```

This means that after ten iterations, we already have more than 2.8 million (!) correct digits for π.

Here, the second formula, slightly rewritten, is implemented.

```
In[9]:= y2[n_, prec_] := y2[n, prec] =
          N[(25/((1 + (2^(1/5) (-1 + #))/((7 + Sqrt[(7 + (2 - #)^2)^2 -
            4 (-1 + #)^3] + (2 - #)^2 (-1 + #))^(1/5) +
```

```
      (2^(4/5) ((7 + Sqrt[(7 + (2 - #)^2)^2 -
      4 (-1 + #)^3] + (2 - #)^2) (-1 + #))^(1/5))/2)^2 5/#)
                                    )&[5/y2[n - 1, prec]], prec]

  y2[0, prec_] := y2[0, prec] = N[5 (Sqrt[5] - 2), prec]

  a2[n_, prec_] := a2[n, prec] =
  N[y2[n - 1, prec]^2 a2[n - 1, prec] - 5^(n - 1) *
  ((y2[n - 1, prec]^2 - 5)/2 + Sqrt[y2[n - 1, prec] (y2[n - 1, prec]^2 -
                                 2 y2[n - 1, prec] + 5)]), prec]

  a2[0, prec_] := a2[0, prec] = N[1/2, prec]
```

The first term is simplified.

```
In[13]:= a2[1, Infinity] // Simplify
```

$$\text{Out[13]= } \frac{5}{2} - 5\sqrt{-210 + 94\sqrt{5}}$$

Again, we find the relative deviation from π as a function of the order of the iteration.

```
In[14]:= {#1, piExactness[a2, #1, #2]}& @@@
             {{1, 20}, {2, 50}, {3, 200}, {4, 1000}, {5, 5000}}
```

Out[14]= $\{\{1, 0.0000177\}, \{2, 4.82\times10^{-32}\}, \{3, 8.89\times10^{-168}\}, \{4, 2.89\times10^{-849}\}, \{5, 1.66\times10^{-4259}\}\}$

Each iteration step gives about four times as many digits of π. This time, we have the following estimate for the accuracy: $|\alpha_n - 1/\pi| < 16\,5^n \exp((-5)^n \pi)$.

```
In[15]:= Table[16 5^n Exp[-Pi 5^n] // N, {n, 10}]
```

Out[15]= $\{0.0000120561, 3.10922\times10^{-32}, 5.67526\times10^{-168}, 1.839834883378971\times10^{-849},$
$1.054057368492838\times10^{-4259}, 1.040905324097306\times10^{-21313}, 1.564105773623907\times10^{-106586},$
$1.917169662865183\times10^{-532953}, 8.486970162285154\times10^{-2664791}, 2.308518158781674\times10^{-13323980}\}$

Here is an iteration such that each step contains about nine times as many correct digits of π [253].

```
In[16]:= α[0, prec_] := N[1/3, prec];

         S[1, prec_] := N[(Sqrt[3] - 1)/2, prec];

         s[1, prec_] := N[(1 - S[1, prec]^3)^(1/3), prec];

         α[n_, prec_] := α[n, prec] =
           N[m[n, prec] α[n - 1, prec] + 3 9^(n - 2) (1 - m[n, prec]), prec];

         m[n_, prec_] := m[n, prec] =
           N[27 (1 + s[n, prec] + s[n, prec]^2)/
               (t[n, prec]^2 + t[n, prec] u[n, prec] + u[n, prec]^2), prec];

         t[n_, prec_] := t[n, prec] = 1 + 2 S[n, prec];

         u[n_, prec_] := u[n, prec] =
           N[(9 S[n, prec] (1 + S[n, prec] + S[n, prec]^2))^(1/3), prec];

         s[n_, prec_] := s[n, prec] =
           N[(1 - S[n - 1, prec])^3/
               (t[n - 1, prec] + 2 u[n - 1, prec])/
               (t[n - 1, prec]^2 + t[n - 1, prec] u[n - 1, prec] +
               u[n - 1, prec]^2), prec];

         S[n_, prec_] := S[n, prec] =
               N[(1 - s[n, prec]^3)^(1/3), prec];

In[29]:= (* now calculate values *)
         SetPrecision[{1/α[2, 30], 1/α[3, 300], 1/α[4, 2000],
               1/α[5, 18000]} - Pi, 3]
```

Out[30]= $\{-2.18\times10^{-22}, -6.72\times10^{-218}, -3.55\times10^{-1985}, -2.62\times10^{-17898}\}$

b) While many arithmetic functions deal with intervals, `FractionalPart` unfortunately does not.

```
In[1]:= FractionalPart[Interval[{1/3, 5/3}]]
```

$$Out[1]= \text{FractionalPart}\left[\text{Interval}\left[\left\{\frac{1}{3}, \frac{5}{3}\right\}\right]\right]$$

But it is straightforward to define it for purely positive intervals.

```
In[2]:= fractionalPart[Interval[{a_?NonNegative, b_?NonNegative}]] :=
        Which[a <= b <= 1, Interval[{a, b}],
                 (* original interval is split into two pieces *)
                 a < 1 && b > 1, Interval[{0, FractionalPart[b]}, {a, 1}],
                 a >= 1 && b > 1, Interval[FractionalPart[{a, b}]]]

        (* an Interval-object can contain multiple intervals *)
        fractionalPart[Interval[l__]] :=
         IntervalUnion[fractionalPart[Interval[#]]& /@ {l}]
```

Using the function `fractionalPart`, it is now easy to iterate and visualize the repeated application of the map $x \to \{\alpha x + \beta\}$ to the unit interval.

```
In[6]:= iteratedMap[p_, o_] := NestList[(IntervalUnion @@
                                 Flatten[{fractionalPart[p[#]]}])&,
                 (* starting interval *) Interval[{0, 1}], o]
In[7]:= fracIntervalGraphics[p_, o_] :=
        Graphics[{Thickness[0.002],
         MapIndexed[(* iterations run upward *)
         Function[{i, p}, Line /@ Map[{#, p[[1]]}&, List @@ i, {2}]],
                 iteratedMap[p, o]]},
           PlotRange -> All, Frame -> True, FrameTicks -> None]
```

Here are some example visualizations for "random" α and β. We see that a fixed finite number of intervals emerge and that they continuously shrink.

```
In[8]:= Show[GraphicsArray[#]]& /@
           Partition[fracIntervalGraphics[#, 100]& /@
             (* random linear functions *)
             {1/12 + 65/196 #&, 174722/367323 + 24313/35392 #&,
              548/671 + 214/233 #&, 108572/423859 + 54365/63773 #&,
              184446/218429 + 389985/400717 #&,
              3620/27143 + 83848/88071 #&}, 2]
```

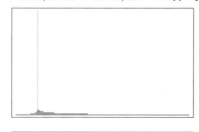

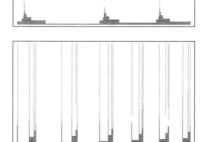

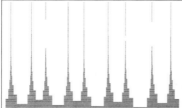

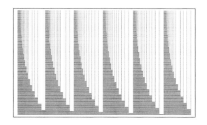 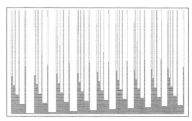

```
In[9]:= intervalLengths = N[-Apply[Subtract, List @@@
           Drop[iteratedMap[174722/367323 + 24313/35392 #&, 200], 50],
           {2}]]];
```

The next two graphics show the lengths of the intervals as a function of the iterations for the second and fourth map. All intervals have approximately the same width.

```
In[10]:= Show[GraphicsArray[Graphics[
           Line[MapIndexed[{#2[[1]] + 50, Log[10, #1]}&, #]]& /@
             Transpose[(* length of intervals *)
               N[-Apply[Subtract, List @@@ Drop[iteratedMap[#, 200],
               (* ignore early iterations *) 50], {2}]]], Frame -> True]& /@
             {174722/367323 + 24313/35392 #&, 108572/423859 + 54365/63773 #&}]]
```

9. Standard Map, Stochastic Webs, Iterated Cubics, Hénon Map, Triangle Map

a) The implementation causes no particular problems. For a better comparison of the various motions, for each K, we begin with the same $\{x_0, y_0\}$, and mark corresponding points in the path with the same color. (It is nice and interesting to animate the last sequence as a function of K.) For the mathematical and physical fundamentals and a series of similar mappings (e.g., those of Part b) of this exercise), see [1511], [1091], [1506], [1621], [391], [390], [1958], [1654], [1923], [1095], [1437], [825], and [1096].

```
In[1]:= twoPi = N[2Pi];
        start = Table[{i/11, i/11}, {i, 0, 11}];

        Show[GraphicsArray[#]]& /@
        Map[Function[K, Graphics[
          Table[(* adding some color *)
          {PointSize[0.003], Hue[i/14],
            Point /@ NestList[({p = Mod[#1 - K/twoPi Sin[twoPi #2], 1.0],
                               Mod[#2 + p, 1.0]}& @@ #)&, start[[i]], 1000]},
            {i, 12}],
                PlotRange -> All, PlotLabel -> "K = " <> ToString[K],
                Frame -> True, FrameTicks -> None]],
              (* two sets of parameters *)
          {{0.17, 0.5, 0.7, 0.95}, {1.3, 1.7, 2.3, 2.95}}, {2}]
```

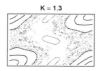
K = 1.3

K = 1.7

K = 2.3

K = 2.95

For the complex version of this map, see [574]; for a generalized version, see [1584].

b) Next, we deal with the second mapping.

```
In[1]:= map[α_, k_, x_, y_] = {y Sin[α] + Cos[α] (x + k Sin[2 Pi y]),
                              y Cos[α] - Sin[α] (x + k Sin[2 Pi y])};
```

Let $\alpha = \pi/2$, $\alpha = \pi/3$, and $\alpha = \pi/4$. To better see which points evolve from which points, we again color them.

```
In[2]:= Do[SeedRandom[999];
        Show[GraphicsArray[Table[Graphics[{PointSize[0.004],
            Table[{Hue[0.8 i/100], Point /@
              NestList[N[map[Pi/d, 0.1, ##]]& @@ #&,
               {Random[Real, {-4, 4}], Random[Real, {-4, 4}]}, 100]},
                {i, 100}]}, AspectRatio -> Automatic, PlotRange -> All], {3}]]],
          {d, 2, 4}]
```

To compute these graphics faster, we can use `Compile` for the main work carried out in `NestList`.

```
In[3]:= fastWebGraphics[k_, α_, x0_, y0_, n_, opts___] :=
        Show[Graphics[{PointSize[0.0025], Point /@
        Compile[(* zero arguments *) {},
        Module[{s, c}, s = Sin[α]; c = Cos[α];
              NestList[{s #[[2]] + c (#[[1]] + k Sin[2 Pi #[[2]]]),
                        c #[[2]] - s (#[[1]] + k Sin[2 Pi #[[2]]])}&,
```

```
                    {x0, y0}, n]]][(* zero arguments *)]}], opts,
          PlotRange -> All, AspectRatio -> Automatic]
```

The following three graphics show the result of following one point over 100000 iterations.

```
In[4]:= Show[GraphicsArray[
          Apply[fastWebGraphics[##, 10^5, DisplayFunction -> Identity]&,
          {{-0.304091,  Pi/2, -4.102058, -9.825767},
           { 0.163151,  Pi/4, -3.026477,  0.635745},
           { 0.171146, 5Pi/6, -0.862034, -5.106875}}, {1}]]]
```

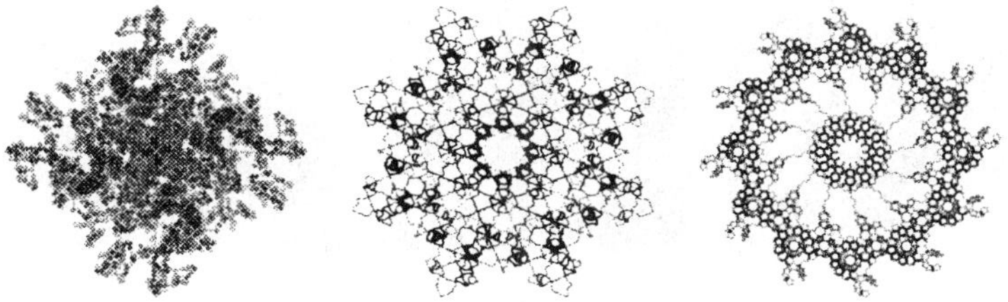

Note that other fractions of π generate interesting patterns, too.

To show the evolution of a localized, piecewise continuous density, we need a high-resolution plot. We will use an array of a few million points to follow the evolution. Because this would be too expensive for visualization, we implement a function condense that coarse graines a given pair of matrices x and y into pp bins. The matrices x and y hold the information of the x- and y-values of the finely discretized continuous distribution.

```
In[5]:= condense =
         Compile[{L, {pp, _Integer}, {x, _Real, 2}, {y, _Real, 2}},
           Module[{T = Table[0, {pp}, {pp}], d = Length[x], ξ, η, i, j},
             Do[{ξ, η} = {x[[i, j]], y[[i, j]]};
               (* if within boundaries, course grain *)
               If[Max[Abs[{ξ, η}]] < L, {i, j} = Ceiling[({ξ, η} + L)/(2L) pp];
                 T[[i, j]] = T[[i, j]] + 1], {i, d}, {j, d}];
             T]];
```

The function showMatrix displays the condensed matrix as a colored density plot.

```
In[6]:= showMatrix[m_] :=
         ListDensityPlot[m, Mesh -> False, ColorFunction -> (Hue[0.78 #]&),
                         PlotRange -> All, Frame -> False]
```

The function makeAnimation carries out the actual discretization of the initial density, the iteration of the map application, the coarse graining, and the display of the resulting densities. We choose as the initial density a square of edge length $2\,\mathcal{L}$ and follow the density in a square of edge length $2\,L$. For the calculations, we use $pp\mathcal{L}^2$ points, and for the visualization, we use ppL^2 points.

```
In[7]:= makeAnimation[d_, k_, L_, L_, ppL_, ppL_, directAnimation -> trueFalse_] :=
         Module[{x = Developer`ToPackedArray @
                       Table[N @ x, {x, -L, L, 2L/ppL}, {y, -L, L, 2L/ppL}],
                 y, m, bag = {}},
           y = Transpose[x];
           (* initial state *)
           m = condense[L, ppL, x, y];
           If[trueFalse, showMatrix[m], AppendTo[bag, m]];
           (* iterate map application *)
           Do[{x, y} = map[Pi/d, k, x, y];
             m = condense[L, ppL, x, y];
             (* display of save coarse grained matrix *)
             If[trueFalse, showMatrix[m], AppendTo[bag, m]],
             {j, 60}];
           (* potentially display the matrices as a graphics array *)
           If[Not[trueFalse],
             Show[GraphicsArray[#]]& /@
             Block[{$DisplayFunction = Identity},
                   Partition[showMatrix /@ First /@ Partition[bag, 10], 3]]]]
```

Here is the animation of a selected parameter value pair. We use about 13 million points for the discretization and about 130,000 points for the visualization.

```
In[8]:= {d0, k0} = {4, 0.3185292744651224};
        (* or, a more web-like structure:
           {d, k} = {2, 0.19197966814221262} *)

In[10]:= makeAnimation[d0, k0, 4, 4/3, 360, 3600, directAnimation -> False];
```

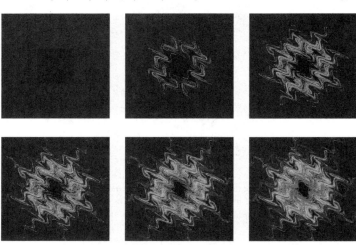

```
makeAnimation[d0, k0, 4, 4/3, 360, 3600, directAnimation -> True];
```

We end by making a more detailed graphic of a later stage in the evolution of the density. This time, we iterate the map for each point of the discretized density. This allows using many more points without a large memory usage. To speed up the calculation, we compile the procedure.

```
In[11]:= webMapDataCF =
         Compile[{L, {pp, _Integer}, {o, _Integer}, {n, _Integer}, α, k},
           Module[{T = Table[0, {pp}, {pp}], x, y, i, j,
                   s = Sin[α], c = Cos[α], ℒ = L/3},
                   (* carry out map iteration over a dense set of points *)
                   Do[{x, y} = {ξ, η};
                      Do[{x, y} = {y s + c (x + k Sin[2 Pi y]),
                                   y c - s (x + k Sin[2 Pi y])}, {n}];
                         (* monitor coarse grained result *)
                         If[Max[Abs[{x, y}]] < L,
                            {i, j} = Ceiling[({x, y} + L)/(2L) pp];
                            T[[i, j]] = T[[i, j]] + 1],
                      {ξ, -ℒ, ℒ, 2ℒ/o}, {η, -ℒ, ℒ, 2ℒ/o}];
                      (* return coarse grained result *) T]];

In[12]:= (* check for successful compilation *)
         Union[Head /@ Flatten[webMapDataCF[[4]]]]
Out[13]= {Integer, Real}
```

We will use 400 million points and iterate over 80 cycles. This calculation will only use a few megabytes of memory, but will run several hours.

```
In[14]:= (detailedData = webMapDataCF[4, 600, 20000, 80, Pi/d0, k0];) // Timing
Out[14]= {42088.1 Second, Null}

In[15]:= showMatrix[detailedData]
```

Using a more homogeneous coloring shows much more details in the resulting graphic. The red color indicates a low density.

```
In[16]:= coloring[z_] = Which @@ (* use 60 contours *)
            Flatten[MapIndexed[{z <= #1,  Hue[0.78 #2[[1]]/60]}&,
                    Last /@ Partition[Sort[Flatten[detailedData]], Round[600^2/60]]]];
In[17]:= ListDensityPlot[detailedData, Mesh -> False, ColorFunction -> coloring,
                        PlotRange -> All, Frame -> False,
                        ColorFunctionScaling -> False];
```

c) For speed reasons, we compile the map under consideration.

```
In[1]:= cf = Compile[{{t, _Real}, {l0, _Real, 1}, {K, _Real}, {n, _Integer}},
            NestList[Function[l, {#, Mod[# + 1[[2]], 2 Pi]}&[
            (1[[1]] + K Sin[#]((1 - t) + t Sign[Cos[#]])&)[1[[2]]]]], l0, n]];
```

The transition for two sets of parameters is shown next.

```
In[2]:= makeGraphics[pp_, dL_, d_, pr_] :=
        Function[ts, Show[GraphicsArray[Function[t,
        Show[Graphics[{PointSize[0.0025], (* color points consecutively *)
                    MapIndexed[{Hue[0.8 #2[[1]]/pp], #}&,
                     Point[Reverse[#]]& /@ cf[t, dL, d, pp]]}],
            Frame -> True, FrameTicks -> None, AspectRatio -> 2,
            DisplayFunction -> Identity,
            PlotRange -> {{0, 2Pi}, pr}]] /@ ts]] /@
                        Partition[Table[t, {t, 0, 1, 1/23}], 6]
In[3]:= makeGraphics[25000, {1/10, 3}, 1/100, {-0.35, 0.35}]
```

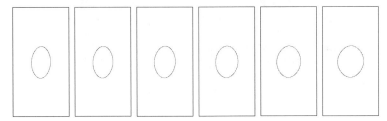

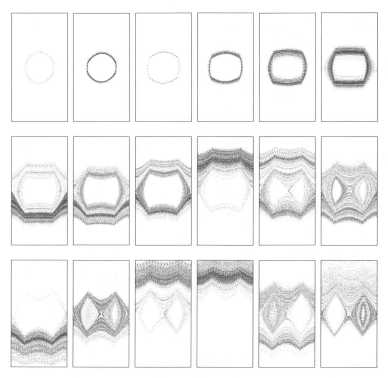

In[4]:= makeGraphics[25000, {1.9959984539060585, 0.5799489278713141},
0.01062162262514079, {1.85, 2.15}]

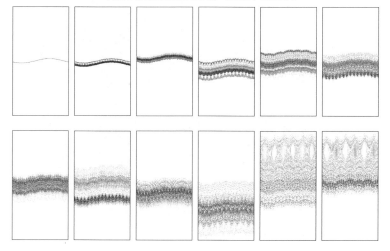

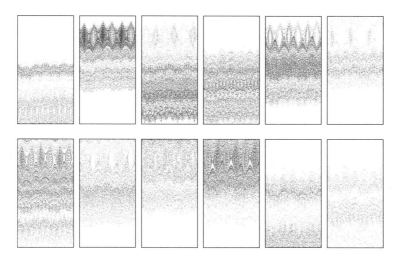

d) This is the map.

```
In[1]:= pic[{ω_, μ_}, {o_, n_}, opts___] :=
        Module[{f}, (* compile version *)
               f = Compile[{x, p},
                   { Cos[ω] x + Sin[ω](p - μ Sign[x]),
                    -Sin[ω] x + Cos[ω](p - μ Sign[x])} // N];
        (* show graphic *)
        Show[Graphics[{PointSize[0.0025],
        MapIndexed[{Hue[0.8 #2[[1]]/o], #1}&, Transpose[Table[Point /@
            NestList[f @@ #&,
            N[{Cos[2k Pi/n], Sin[2k Pi/n]}], o], {k, n}]]]]},
            opts, AspectRatio -> 1, Frame -> True, FrameTicks -> False]]
```

Using an input like the one shown below allows us to generate large amounts of possible phase space pictures in a small time. (We start the iterations from equidistant points of the unit circle in the *x,p*-plane.

```
random[scale_] := (2Random[Integer, {0, 1}] - 1) Random[]*
                  10^Random[Integer, {-1, 1} scale]
Do[{ω = random[2], μ = random[2]}; Print[{ω, μ} // InputForm];
   pic[{ω, μ}, {250, 250}], {100}]
```

Here are some of the patterns found this way.

```
In[2]:= data = (* parameter values *)
        {{ 1.69163, -0.18750}, { 0.62832,  0.33333}, {-0.57232, -0.07902},
         {-0.03407, -0.04805}, { 1.79520,  1.88372}, { 0.00000, -0.28571},
         { 1.25664,  0.36363}, { 4.15929,  0.21153}, { 4.90088,  1.21333},
         { 5.26429, -0.12698}, { 4.14561, -1.61818}, { 4.48799,  2.19512},
         { 3.00502, -1.26582}, { 2.09444,  0.98443}, { 5.71199, 0.594203},
         { 4.80479,  0.40000}, { 3.75684,  0.88709}, { 4.47697, -1.64807},
         { 4.71017, -0.82924}, { 0.61587,  0.92384}, { 3.76991,  1.00000},
         { 0.61587,  0.92384}, { 3.75684,  0.88709}, { 4.47697, -1.64807},
         { 4.71017, -0.82924}, { 0.61587,  0.92384}, { 1.59952,  9.48055},
         { 45840.5, -0.58207}, {0.0193519, -0.0182426},
         {-0.0000870841, -0.967457}};

In[3]:= Show[GraphicsArray[
        pic[#, {100, 100}, DisplayFunction -> Identity]& /@ #]]& /@ Partition[data, 5]
```

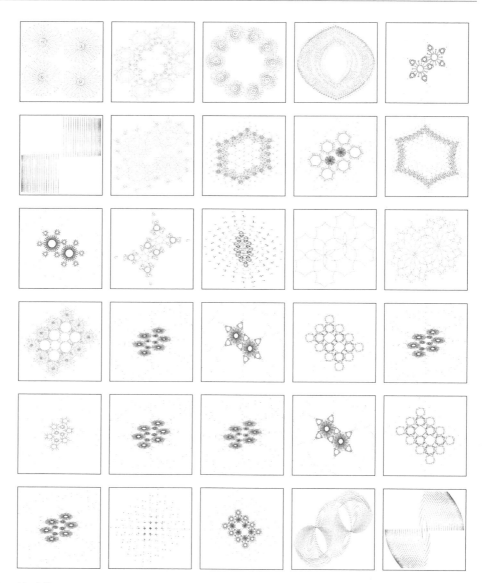

e) This is the forced logistic map.

```
In[1]:= forecedLogisticMap =
    Compile[{ω, α, ε, {n, _Integer}},
        NestList[{Mod[#[[1]] + ω, 1],
                 α #[[2]] (1 - #[[2]]) + ε Sin[2 Pi #[[1]]]}&,
                {1/Pi, 1/E}, n]];
```

Here are 18 sets of parameters.

```
In[2]:= ωαεData =
    {{0.2590734587781428, 0.127526356344035, 0.3137841105258410},
     {0.7344418996901275, 1.093831483925719, 0.089664167638380},
     {0.4170414736431796, 3.171222289673991, 0.0234672506041178},
     {0.9852007637788890, 2.998872393837978, 0.0128583093301878},
```

```
{0.0158458819044718,  3.277066099847877,   0.332149674979157},
{0.4940213195122424,  2.592025427369532,   0.2652272567598765},
{0.0016381796138206,  3.058315613083374,   0.1243607525329722},
{0.4986703873008871,  2.840976895877232,   0.153973215992413},
{0.4044093875989563,  2.947396184684467,   0.1532315252182857},
{0.2852423479966735,  3.260865537812962,   0.1812832047370404},
{0.2107516579108322,  3.234605858228152,   0.1297957317017340},
{0.2304392492070745,  3.425511003133327,   0.0407261777205752},
{0.7367304807403927,  3.395529153963405,   0.1525278961148917},
{0.6606315280209043,  2.678693424373687,   0.2567889640227677},
{0.0212082906150466,  3.433120962142957,   0.1441854444558735},
{0.1198371564588390,  3.293488607515896,   0.1025654814495575},
{0.3853210449460685,  3.436397924283918,   0.0973788191532345},
{0.1970365903310693,  3.567452915137104,   0.0830875586765750}};
```

The pattern formed in the ϑ, x-plane range from simple curves over bifurcations to strange attractors.

```
In[3]:= With[{o = 5 10^4},
        Show[GraphicsArray[#]]& /@ (Partition[#, 3]& @
        (Graphics[{PointSize[0.003],
                MapIndexed[{Hue[#2[[1]]/o], Point[#1]}&,
                        forecedLogisticMap[Sequence @@ #, o]]},
                Frame -> True, FrameTicks -> None]& /@ ωαεData))];
```

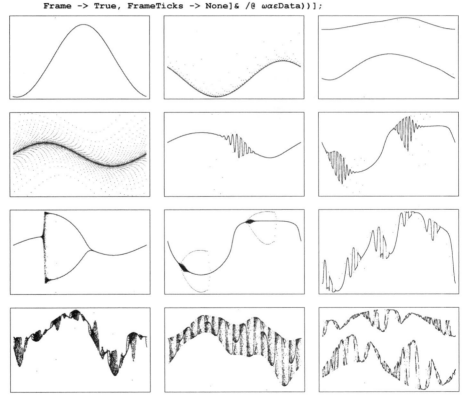

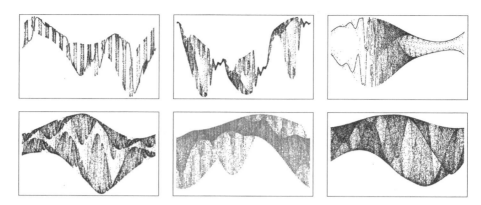

f) Here is a picture of the map under consideration.

```
In[1]:= iList[κ_, {x0_, y0_}, n_] := First /@
           NestList[{#[[2]], -#[[1]] - κ Sin[#[[2]]]}&, {x0, y0}, n]
```

```
In[2]:= ListPlot[iList[2.178805237252476,
                    {-2.9085181016961186, 0.8317263342785042}, 20000],
               Frame -> True, Axes -> False]
```

To see if the four parallel lines are real or numeric artifacts we just have to make sure that the numerics is done correctly. This can easily be done by using high-precision arithmetic with sufficiently many digits. For the first 10000 iterations, 3000 initial digits turn out to be sufficient.

```
In[3]:= nl = With[{prec = 3000},
                iList[SetPrecision[2.178805237252476, prec],
                     SetPrecision[ {-2.9085181016961186,
                                   0.8317263342785042}, prec], 10000]];
```

```
In[4]:= ListPlot[Precision /@ nl]
```

As the following graphic shows, the answer to the above question is yes and no. Yes, there really are parallel lines, but they are of different length and different positions than the ones obtained using not-so-reliable machine arithmetic.

```
In[5]:= ListPlot[nl]
```

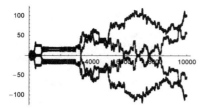

It is not difficult to find other κ, $\{x_0, y_0\}$-values so that areas of four parallel lines exist. Here are two more examples.

```
In[6]:= Show[GraphicsArray[
         ListPlot[iList[##, 20000], Frame -> True, Axes -> False,
                  DisplayFunction -> Identity]& @@@ (* parameter values *)
           {{1.6851831772164054, {-4.584884833955502, 3.786360648989101}},
            {2.1788052372524766, {-2.9085181016961186, 0.8317263342785042}}}]]
```

Carrying out more iterations shows that such parallel lines can exist over a larger number of iterations.

```
In[7]:= Show[GraphicsArray[#]]& /@ Partition[
         ListPlot[(* skip some points to avoid very large graphics *)
                  #[[Random[Integer, {1, 6}]]]& /@ Partition[
                      MapIndexed[{#2[[1]], #1}&, (* many iterations *)
                          iList[##, 200000]], 6],
                  Frame -> True, Axes -> False,
                  DisplayFunction -> Identity]& @@@ (* parameter values *)
           {{ 2.374665412407019, {-1.941779581853265, -4.439093053025886}},
            { 1.625149553648110, {-0.503881202234477,  4.715540118904786}},
            {-1.591474291419896, {-3.041198346904126, -0.162308797012832}},
            { 4.420748415245017, { 4.080793709440529, -3.785545017124877}},
            {-1.209741514831302, {-2.678100047049162, -0.017858866818805}},
            { 1.689225817520681, { 3.618891078361679,  1.322877753367511}}}, 2]
```

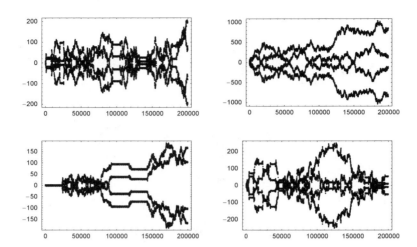

Here is another map showing parallel lines [1565].

```
In[8]:= cf = Compile[{{αβγ, _Real, 1}, {xyStart, _Real, 1}, {o, _Integer}},
        Module[{α, β, γ}, {α, β, γ} = αβγ;
            NestList[{α/(1. + #[[1]]^2) + #[[2]],
                     #[[2]] - γ #[[1]] - β}&, xyStart, o]]];

In[9]:= Show[GraphicsArray[
        (ListPlot[First /@ cf[##], PlotStyle -> {PointSize[0.0025]},
                DisplayFunction -> Identity, Frame -> True, Axes -> False,
                FrameTicks -> False])& @@@ (* parameter values *)
        {{{ 4.188695839960339,   0.00001767715505820486, 0.000046528880219864},
          { 1.949151486456914,   1.7617252938863919}, 10^5},
         {{ 3.524087855458693,   0.00004991486107317652, 0.000142554775816300},
          {-0.795876434724131,  -2.4655006746966883}, 10^5},
         {{-3.409423906138836,   0.00001308959516112935, 0.000272180351814755},
          {-0.567225700326438,   1.2067876658560852}, 10^5}}]]
```

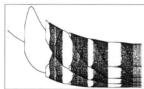 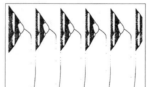 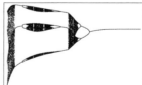

For similar lines, see [938] and [16].

g) `randomIteration` is the main function. `randomIteration[o]` generates a compiled function that implements the iteration of $\{x, y\} \longrightarrow \{p(x, y), q(x, y)\}$ where p, and q are random bivariate polynomials of total degree o. The expected argument of the returned `CompiledFunction` is the number of iterations to be carried out. To avoid that the compiled version fails because nonmachine numbers appear in the calculation, we end the iteration when x, or y becomes too small or too large.

```
In[1]:= randomIteration[o_] :=
    With[{ (* the monomials *)
          xys = Sort[Flatten[Table[x^i y^j, {i, 0, o}, {j, 0, o - i}]]]},
        Compile[{{n, _Integer}},
        Module[{x = 0., y = 0., L = Table[{0., 0.}, {n + 1}], c = 1,
            ε = 10.^-10, γ = 10.^10},
            (* iterate until too big or too small *)
            While[{x, y} = #; c <= n &&
                Abs[Max[{x, y}]] < γ && Abs[Min[{x, y}]] > ε,
```

```
                    L[[c++]] = {x, y}]; (* the coefficients *)
        Take[L, c - 1]]]&[Table[Table[Random[Real, {-1, 1}],
                                      {(o^2 + 3o + 2)/2}]].xys, {2}]]]
```

Carrying out some experiments with `randomIteration` shows that only a small fraction of all sequences shows an interesting behavior. When `randomIteration[o][n]` returns a list of length *n*, in many cases, this list will be periodic and the period will be small. We use the function `nonPeriodicCaseQ` to exclude such cases.

```
In[2]:= nonPeriodicCaseQ[points_, n_] :=
            Length[Union[SetPrecision[Take[points, -n], 6]]] == n
```

It will also happen that the resulting sequence will lie on a closed curve or all points will be occupied in a small area. We use the two functions `strangeAttractorCandidateQ` and `densePointsQ` to filter out candidates of strange attractors. `strangeAttractorCandidateQ` counts the number of points in intervals of the *x* and *y* directions and `dense↘ PointsQ` generates a bitmap version of the graphic and counts the number of black points in the bitmap.

```
In[3]:= strangeAttractorCandidateQ =
    Compile[{{points, _Real, 2}, {n, _Integer}, {fraction, _Real}},
    Module[{λ = Length[points], α = 1./n, xs, ys, xMin, xMax,
            yMin, yMax, xTab, yTab, xC, yC, c, res},
        (* x- and y-values *)
        {xs, ys} = Sort /@ Transpose[points];
        {{xMin, xMax}, {yMin, yMax}} = {First[#], Last[#]}& /@ {xs, ys};
        (* x- and y-intervals *)
        xTab = Table[x, {x, xMin, xMax, (xMax - xMin)/n}];
        yTab = Table[y, {y, yMin, yMax, (yMax - yMin)/n}];
        (* count points in x- and y-intervals *)
        xC = yC = Table[0, {n + 1}];
        c = 2;
        Do[If[xs[[k]] <= xTab[[c]], Null,
              While[c < n && xs[[k]] >= xTab[[c]], c++]];
            xC[[c]] = xC[[c]] + 1, {k, λ}];
        c = 2;
        Do[If[ys[[k]] <= yTab[[c]], Null,
              While[c < n && ys[[k]] >= yTab[[c]], c++]];
            yC[[c]] = yC[[c]] + 1, {k, λ}];
        (* form products *)
        res = Flatten[Outer[Times, Round[Rest[xC]/λ/α], Round[Rest[yC]/λ/α]]];
        (* count points *) Count[res, 0]/n^2 < fraction]];
In[4]:= (* are points dense? *)
    densePointsQ[points_, fraction_] :=
    (Count[#, 0]/Length[#] > fraction)&[
    Flatten[If[Head[#[[1, 1]]] === List, Map[First, #, {2}], #]&[
        (* make bitmap *) ImportString[ExportString[
            Graphics[{PointSize[0.0025], Point /@ thePoints}], "PGM",
                    ImageResolution -> 50], "PGM"][[1, 1]]]]]]
```

Now all needed functions are implemented and we can carry out an automated search. We will find five strange attractors for *n* = 4.

```
In[6]:= SeedRandom[123];
    allCounter = 0; goodCounter = 0; o = 4; v = 6; n1 = 10^3; n2 = 10^5;
    attractorBag = {};

    While[allCounter++;
    (* random seed *)
    r = Random[Integer, {1, 10^30}]; SeedRandom[r];
    (* calculate sequence of points *)
    cf = randomIteration[o];
    (* how many good candidates so far? *) goodCounter < v,
    thePoints = cf[n1];
    If[(* is sequence long enough? *) Length[thePoints] == n1,
        (* is sequence nonperiodic *)
        If[nonPeriodicCaseQ[thePoints, 10],
            thePoints = Drop[cf[n2], n1];
            (* does the sequence fill out the rectangle *)
            If[strangeAttractorCandidateQ[thePoints, 10, 0.7],
```

```
                    (* has resulting picture enough points *)
                    If[densePointsQ[thePoints, 0.1],
                        (* count found attractor *) goodCounter++;
                        AppendTo[attractorBag, (*  seed and graphics *)
                        {r, Graphics[{PointSize[0.0025], Point /@ thePoints},
                                Frame -> True, FrameTicks -> None]}]]]]]];
```

In[10]:= (* fraction of candidates *)
```
         goodCounter/allCounter // N
```
Out[11]= 0.00379507

In[12]:= (* the collected seeds *)
```
         First /@ attractorBag
```
Out[13]= {6262669035061989763304293879999, 6487164711641398982166012616226,
          56273180023534733961304497347, 9448127027777617153288990405,
          32838240466031080481943751687, 8191352483482590413960467912028}

In[14]:= (* the attractors *)
```
         Show[GraphicsArray[#]]& /@
                    Partition[Last /@ attractorBag, 2]
```

Here are some examples of attractors found with the code above.

In[16]:= **seededRandomIterationPicture[o_, n_, seed_, opts___] :=**
```
         ((* seed random number generator *) SeedRandom[seed];
          (* display graphic *)
          Show[Graphics[{PointSize[0.0025],
                        (* points from iteration *)
                        Point /@ randomIteration[o][n]}],
               opts, Frame -> True, FrameTicks -> None])
```
In[17]:= **Show[GraphicsArray[seededRandomIterationPicture[3, 10^5, #,**
```
                    DisplayFunction -> Identity]& /@ #]]& /@
         Partition[(* seed values *)
```

{23782946882024507695910012072S, 01895529800586611196569713919⁷,
 62079703825176181947579707349³, 675702345759412935707133176773,
 41016226634287606980545153585⁶, 49237674906448144869781779544⁵,
 314554255002996422590656895983, 48189445197157979199418747243⁹,
 479927650919489456447098137604, 612720837279176024691412231612,
 2093707983269064039831808343³⁸, 755715149992487208307145732013}, 3]

```
In[18]:= Show[GraphicsArray[seededRandomIterationPicture[4, 10^5, #,
                 DisplayFunction -> Identity]& /@ #]]& /@
         Partition[(* seed values *)
         {65466773365407942084239586574⁸, 41658977315163789941557104587¹,
          37269068503358085782268954439⁸, 35827805851118449861614025040⁵,
          360115195302390647885767512123, 044982276495944205613690884494,
          330818181198775593527384725502, 146507205385636019914454863513,
          766085634882197725430038321335⁵}, 3]
```

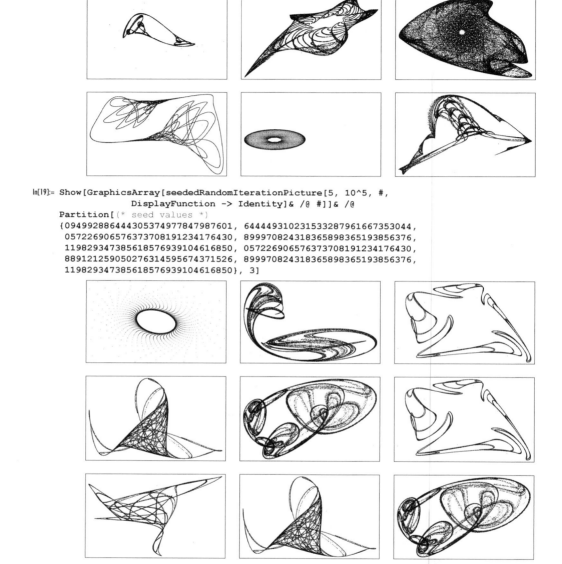

```
In[19]:= Show[GraphicsArray[seededRandomIterationPicture[5, 10^5, #,
                 DisplayFunction -> Identity]& /@ #]]& /@
      Partition[(* seed values *)
       {09499288644430537497784798601, 64444931023153328796166735304,
        05722690657637370819123417630, 89997082431836589836519385676,
        11982934738561857693910461850, 05722690657637370819123417630,
        88912125905027631459567437152, 89997082431836589836519385676,
        11982934738561857693910461850}, 3]
```

We end with some graphics that iterate the bivariate polynomial maps of this exercise until the length of the vector $\{x, y\}$ becomes larger than a given M. The function makeRandomIterationCF generates the corresponding compiled function.

```
In[20]:= makeRandomIterationCF[o_] :=
      With[{(* the monomials *)
            xys = Sort[Flatten[Table[x^i y^j, {i, 0, o}, {j, 0, o - i}]]]},
      Compile[{xMin, xMax, yMin, yMax, M, {pp, _Integer}, {d, _Integer}},
      Module[{c, x, y},
            Table[Table[c = 1; x = x0; y = y0;
            (* iterate until too big or too small *)
            While[{x, y} = #; c <= d && ({x, y}.{x, y} < M), c++]; c,
            (* x-range  and y-range *)
```

```
                    {x0, xMin, xMax, (xMax - xMin)/pp}],
                    {y0, yMin, yMax, (yMax - yMin)/pp} ]]]&[
            (* the coefficients *) XY = Table[Table[Random[Real, 3/2 {-1, 1}],
                                        {(o^2 + 3o + 2)/2}]].xys, {2}]]]
```

Here are some of the resulting graphics.

```
In[21]:= Block[{(* plot points *) pp = 600},
        Show[GraphicsArray[
        ((* seed random number generator *) SeedRandom[#1];
        ListDensityPlot[(* array of the number of iterations *)
                        makeRandomIterationCF[Random[Integer, {2, 4}]] @@ #2,
                        DisplayFunction -> Identity, Mesh -> False,
                        PlotRange -> All, FrameTicks -> False,
                        (* color from red to blue *)
                        ColorFunction -> (Hue[0.78 #]&)])& @@@ #]]]& /@
            (* 3×3 array of seeds, plotranges, and parameters *)
        {{{8029014134, {-3.8, 3.8, -3.8, 2.8, 10.^2, pp,  600}},
          {7864264117, {-1.2, 2.0, -1.6, 1.6, 10.^2, pp,  600}},
          {6738566719, {-2.0, 2.5, -2.0, 2.0, 10.^3, pp,  600}}},
         {{3112297004, {-1.0, 1.2, -1.0, 0.6, 10.^2, pp,  600}},
          {3236806203, {-2.5, 2.5, -0.8, 2.2, 10.^1, pp,  600}},
          {7884176582, {-1.6, 1.2, -1.0, 1.6, 10.^2, pp, 1200}}},
         {{ 875962147, {-4.0, 3.0, -3.0, 4.0, 10.^4, pp,  900}},
          { 279322720, {-1.5, 1.8, -1.0, 1.6, 10.^2, pp, 1200}},
          {2034679031, {-5.0, 4.0, -7.0, 7.0, 10.^3, pp, 2500}}}}]
```

h) We start by running the input.

```
In[1]:= (* parameters *)
        {P, e, x0} = {0.47629739820406214, 0.40133038775291635, 0.01011450108382423};
```

```
(* random seed *)
ℛ = 82313041463260884204;
```

In[5]:=
```
step[x_, {P_, e_}] :=
   Module[{p = If[Random[] < P, 1/2 + e, 1 - (1/2 + e)], q,
           η = If[Random[] < P, 1, 0]},
        η p/(1 - (1 - p) x) + (1 - η) (1 - p)/(1 - p x)]
```

In[6]:=
```
SeedRandom[ℛ]
nlMP = NestList[step[#, {P, e}]&, x0, 10^4];
```

This is a plot of the calculated list.

In[8]:=
```
(* option settings to be used multiple times *)
opts = Sequence[PlotRange -> All, Frame -> True, Axes -> False,
                PlotJoined -> True, PlotStyle -> {Thickness[0.002]}];
```

In[10]:=
```
ListPlot[nlMP, opts]
```

Comparison of the last plot with a high-precision version of the calculation shows that the last plot was not correct. The right graphic shows the precision of the iterated results.

In[11]:=
```
SeedRandom[ℛ];
nlHP = NestList[step[#, SetPrecision[{P, e}, 100]]&,
               SetPrecision[x0, 100], 10^4];
```

In[13]:=
```
Show[GraphicsArray[
   Block[{$DisplayFunction = Identity},
        {(* values *) ListPlot[nlHP, opts],
         (* precision *) ListPlot[Precision /@ nlHP, opts]}]]]
```

The differences between the two plots are quite visible. So why did the machine-precision calculation fail? Once the value of x is 1., it will stay 1. forever. For the machine precision calculation, this happens after 236 iterations.

In[14]:=
```
Position[nlMP, 1., {1}, 1]
```

Out[14]=
```
{{236}}
```

The high-precision calculation shows that x never takes the value 1., but comes very close to it. With \$MachineEpsilon digits, after 236 iterations 1. is reached.

In[15]:=
```
Show[GraphicsArray[
   Block[{$DisplayFunction = Identity},
        {(* difference to 1 *)
         ListPlot[Log[10, N[1 - nlHP]], opts],
         (* differences between machine and high-precision version *)
         ListPlot[Take[nlMP, 1000] - Take[nlHP, 1000], opts]}]]]
```

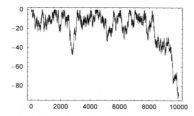 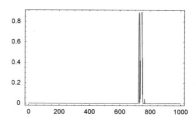

To cure the problem, we do not iterate the calculation of x, but rather $y = 1 - x$. This extends the range of applicability of machine arithmetic to a much wider range, namely until y is greater than `$MinMachineNumber`. The following compiled function `stepCF` uses x and y in parallel.

```
In[16]:= stepCF = Compile[{{xy, _Real, 1}, {Pe, _Real, 1}},
        Module[{x = xy[[1]], y = xy[[2]], P = Pe[[1]], e = Pe[[2]], p, q, η},
            p = If[Random[] < P, 1/2 + e, 1 - (1/2 + e)];
            η = If[Random[] < P, 1, 0];
            q = 1 - p;
            {If[η === 1, p/(1 - q x), q/(1 - p x)],
             If[η === 1, q y/(p + q y), (1 - η) p y/(q + p y)]}]];
```

The part of `nlCF` that uses y now agrees with the high-precision calculation.

```
In[17]:= (* seed again *) SeedRandom[ℛ];
        nlCF = NestList[stepCF[#, {P, e}]&, {x0, 1 - x0}, 10000];
In[19]:= Show[GraphicsArray[
        Block[{$DisplayFunction = Identity},
            ListPlot[#, opts]& /@ {1 - Last /@ nlCF, Log[10, Last /@ nlCF]}]]]
```

i) `cf` is a compiled function that implements *steps* iteration of the map. It returns a *steps* × *l* list of the data $u_i^{(n)}$.

```
In[1]:= cf = Compile[{{uL, _Real, 1}, {vL, _Real, 1}, r, ε, b, {steps, _Integer}},
            Module[{ul = uL, vl = vL, u', l = Length[uL], uT},
                Table[u' = Prepend[Append[ul, First[ul]], Last[ul]];
        uT = Table[(* u update *)
        If[u'[[k]] < 1, r (1/2 - Abs[u'[[k]] - 1/2]), u'[[k]]] +
        ε/2 (  If[u'[[k - 1]] < 1, r (1/2 - Abs[u'[[k - 1]] - 1/2]), u'[[k - 1]]] -
            2 If[u'[[k   ]] < 1, r (1/2 - Abs[u'[[k   ]] - 1/2]), u'[[k   ]]] +
              If[u'[[k + 1]] < 1, r (1/2 - Abs[u'[[k + 1]] - 1/2]), u'[[k + 1]]]) +
        vl[[k - 1]], {k, 2, l + 1}];
        (* v update *)vl = b (uT - ul);
        (* return u list *)ul = uT, {steps}]]];
```

To find qualitatively different-looking behavior for the $u_i^{(n)}$ we implement a function `randomGraphics` that uses the seed *seed* for the random number generator and then uses random values for the parameters r, ε, and b.

```
In[2]:= randomGraphics[seed_, {dimx_, dimt_}, opts___] :=
        ListDensityPlot[(* seed random number generator *) SeedRandom[seed];
                cf[Table[Random[Real, {0, 2}], {dimx}],
                    Table[Random[Real, {0, 2}], {dimx}],
                    Random[Real, 5{-1, 1}], Random[Real, {-1, 1}],
                    Random[Real, {-1, 1}], dimt], opts, Mesh -> False,
                FrameTicks -> None, PlotRange -> All]
```

Besides basically constant solutions, most of the nontrivial $u_i^{(n)}$ show a complicated, nonperiodic nested triangle structure. Here are some such examples.

```
In[3]:= Show[GraphicsArray[randomGraphics[#, {200, 200},
                            DisplayFunction -> Identity]& /@ #]]& /@
        (* seed values *)
        {{2278838534, 7872532848, 3354017969},
         {9940838735, 2034502289, 9655743949},
         {3525822622, 7298013434, 3829454014}}
```

j) The function `iterate` iterates the Hénon map 200 times for 200 points lying on a circle.

```
In[1]:= iterate[{{A_, B_}, {x0_, y0_}, r_}] :=
        Table[NestList[{A + B #[[2]] - #[[1]]^2, #[[1]]}&,
                        {x0, y0} + r {Cos[φ], Sin[φ]}, 200],
              {φ, 0., 2.Pi, 2.Pi/200}];
```

Frequently the iterations will lead to exponentially growing sequences. In the following, we will avoid diverging sequences.

```
In[2]:= SeedRandom[111];
        Table[iterate[{{Random[], Random[]}, {Random[], Random[]},
                       Random[]}] // Abs // Max, {10}]
        General::ovfl : Overflow occurred in computation. More…

        General::ovfl : Overflow occurred in computation. More…

        General::ovfl : Overflow occurred in computation. More…

        General::stop :
         Further output of General::ovfl will be suppressed during this calculation. More…
Out[3]= {1.70157, 1.16969, 0.605512, 1.09845, 1.0808,
         Overflow[], Overflow[], 0.980429, Overflow[], 1.46332}
```

The function showData visualizes the iterated points obtained with iterate. We color the points along the original circle and its images.

```
In[4]:= showData[data_, opts___] :=
    Show[Graphics[{PointSize[0.003], MapIndexed[{Hue[#2[[1]]/200], #1}&,
                                    Map[Point, data, {2}], {1}]}],
            opts, PlotRange -> All, AspectRatio -> 1]
```

Running a couple hundred of examples with random values for the parameters A and B and initial circles with random centers and radius shows a variety of possible shapes. In the following, we display examples from a nonexhaustive classification. Here are three examples where all images of the original circle are nonoverlapping and "spiraling in".

```
In[5]:= Υ[l1_, l2_, l3_] := Show[GraphicsArray[
        showData[iterate[#], DisplayFunction -> Identity]& /@ {l1, l2, l3}]]
```

```
In[6]:= Υ[{{-0.196873, -0.975908}, {-0.611712, -0.317721}, 0.011772},
        {{ 0.110974, -0.725533}, { 0.166260,  0.409071}, 0.074410},
        {{-0.288027, -0.847394}, {-1.008189, -0.945184}, 0.172734}]
```

In the next three examples, the images of the original circle "just touch".

```
In[7]:= Υ[{{ 0.245870, -0.666846}, { 0.111351,  0.382650}, 0.095281},
        {{-0.011114, -0.801988}, {-0.131769, -0.024045}, 0.027124},
        {{ 0.504919, -0.591557}, {-0.238026,  0.626376}, 0.211901}]
```

In many cases, the images of the circle overlap. Here are six such examples.

```
In[8]:= Υ[{{-0.113115, -0.805944}, { 0.034289, -0.082274}, 0.498386},
        {{-0.142195, -0.927490}, {-0.277595,  0.329617}, 0.216850},
        {{-0.485375, -0.886648}, {-0.211691, -0.119610}, 0.220290}]
```

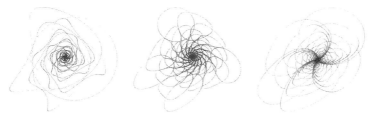

```
In[9]:= Υ[{{ 0.900252, -0.938962}, { 0.113113,  0.953987}, 0.179885},
        {{ 0.831827, -0.947450}, { 0.063354,  0.874913}, 0.175872},
        {{ 0.627669, -0.986942}, { 0.283355,  0.451676}, 0.045149}]
```

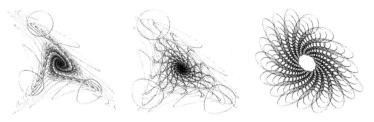

In the following three examples, the images are "highly" overlapping.

```
In[10]:= Υ[{{  0.520029, -1.002589}, {  0.081031, -0.037017}, 0.140489},
          {{  0.008762, -1.003453}, {-0.108554, -0.014908}, 0.289824},
          {{  0.036824, -0.993550}, {-0.591949, -0.076552}, 0.078160}]
```

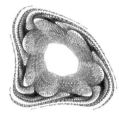

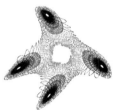

Sometimes the images separate asymptotically into two sets. Here are six such examples.

```
In[11]:= Υ[{{  0.081113,  0.918761}, {-0.099367, -0.489421}, 0.014046},
          {{  0.152268,  0.960025}, {  0.733215,  1.016126}, 0.022481},
          {{  0.142338,  0.754104}, {-0.038333, -0.647655}, 0.248361}]
```

```
In[12]:= Υ[{{  0.141785,  0.853168}, {-0.014412,  0.227796}, 0.233617},
          {{  0.106394,  0.896654}, {-0.227717,  0.192673}, 0.371535},
          {{  0.070029,  0.906877}, {  0.233056, -0.130196}, 0.3604605}]
```

We end with three limiting cases of the last type.

```
In[13]:= Υ[{{  0.064908,  0.903781}, {  0.062057,  0.531892}, 0.175994},
          {{  0.249889,  0.902063}, {-0.080825,  0.572360}, 0.140269},
          {{  0.247856,  0.978581}, {-0.285506,  0.360187}, 0.140798}]
```

Now let us search for "interesting" *A*, *B* and initial circles. The function `makeData` is an optimized version of the function `iterate`. While `iterate` always carries out all iteration steps, `makeData` ends if the generated values for one point of the original circle are too large. (This avoids the time-consuming generation of diverging sequences). In case `makeData` ends before all iterations are done it returns `$Failed`.

```
In[14]:= makeData[{{A_, B_}, {x0_, y0_}, r_}] :=
    Module[{L = Table[0, {200}], counter = 0, aux},
        While[counter < 200 && (* carry out iterations *)
                (aux = iterateR[{{A, B}, {x0, y0}, r, counter 2Pi/200}];
                Max[Abs[aux]] < 10^1), L[[counter + 1]] = aux; counter++];
        If[counter < 199, $Failed, L]]

In[15]:= (* iterate for a single point *)
    iterateR[{{A_, B_}, {x0_, y0_}, r_, φ_}] :=
    NestList[{A + B #[[2]] - #[[1]]^2, #[[1]]}&,
                {x0, y0} + r {Cos[φ], Sin[φ]}, 200]
```

`makeRandomData[]` calls `makeData` with random parameters until a "good" set of data is found.

```
In[17]:= makeRandomData[] :=
    Module[{l, data},
            While[l = {Table[Random[Real, 3/2 {-1, 1}], {2}],
                        Table[Random[Real, 3/2 {-1, 1}], {2}], Random[]/2};
                    data = makeData[l]; data === $Failed, Null];
            data]
```

Because we want to search for "interesting" graphics, we must quantify what is meant by this. A minimal requirement for being "interesting" is a sufficient amount of visible points. So, we define a function `coverage` that calculates how many rectangles of a 200×200 discretization of the graphics contain points.

```
In[18]:= coverage = Compile[{{data, _Real, 3}},
    Module[{data1 = Flatten[data, 1], data2, xMin, xMax, yMin, yMax},
        (* maximal extensions *)
        {{xMin, xMax}, {yMin, yMax}} = {Min[#], Max[#]}& /@ Transpose[data1];
        (* used rectangles *)
        data2 = Round[200 {(#[[1]] - xMin)/(xMax - xMin),
                            (#[[2]] - yMin)/(yMax - yMin)}]& /@ data1;
        (* average density *)
        Length[Union[data2]]/Length[data2]]];
```

Here are some examples of "interesting" iterated circles.

```
In[19]:= Show[GraphicsArray[
    Block[{counter, $DisplayFunction = Identity},
    (SeedRandom[#]; counter =  0;
    While[counter < 1,
            data = makeRandomData[];
            If[coverage[data] > 0.06, showData[data]; counter++]];
        showData[data])& /@ #]]& /@ (* six seed values *)
        {{918813303, 758139497, 341470813},
         {282857200, 986852526, 185943032}}]
```

For patterns resulting from initial straight line segments, see [1551].

k) To calculate points of the basins of attraction, we start with points ζ on the attractor and recursively solve the equation $\zeta = z^2 - (1 - i\lambda)\bar{z}$ with respect to z. (Compared with a straightforward forward iteration of the map, this approach yields the whole basins and a more evenly spaced set of points).

Here is a high-precision value for λ.

```
In[1]:= λhp = Solve[3 λ^2 (9 λ^4 + 3 λ^2 + 11) == 77, λ][[
            2, 1, 2]] // N[#, 60]&
Out[1]= 1.028713768218724912626343911081782827518467453693062508034234
```

P1xhp and P2xhp are the real parts of the endpoints of one line of the attractor.

```
In[2]:= P1xhp = -(3 λhp^2 + 1)/4; P2xhp = (9 λhp^4 + 6 λhp^2 + 5)/16;
```

In the following, we use reliable high-precision arithmetic and fast machine arithmetic. The high-precision numbers have the postfix hp and the machine numbers N.

```
In[3]:= {λN, P1xN, P2xN} = N[{λhp, P1xhp, P2xhp}];
```

Splitting the map $z \to z^2 - (1 - i\lambda)\bar{z}$ into real and imaginary part yields two quadratic equations with four solutions for generic λ. The solutions are quite large.

```
In[4]:= xySol = {x, y} /.
            Solve[{X, Y} == {-x + x^2 - y^2 - λ y, y + 2 x y - λ x},
              {x, y}];

          {Length[xySol], LeafCount[xySol]}
Out[6]= {4, 88497}
```

A quick glance on the solution shows that we get undefined behavior for $Y = \lambda/2$. So, we treat this case separately.

```
In[7]:= Short[xySol[[1]], 2]
```

$$Out[7]//Short= \left\{ \frac{-\dfrac{Y\,\lambda}{\sqrt{\frac{1}{12}\,(-3+4\,X-3\,\lambda^2)+\ll 2\gg+\frac{\ll 1\gg}{\ll 1\gg}}}-\frac{\lambda^2}{4\sqrt{\ll 1\gg}}+\ll 43\gg}{2\,Y-\lambda}\right.,$$

$$\left. \frac{1}{2}\sqrt{\left(\frac{1}{12}\,(-3+4\,X-3\,\lambda^2)+\frac{1}{4}\,(3-4\,X+3\,\lambda^2)+\frac{\ll 1\gg\;\ll 1\gg^2\;\ll 1\gg\;48\;\ll 1\gg\;\ll 1\gg}{6\,2^{\ll 1\gg}\;\ll 1\gg}+\frac{(\ll 1\gg)^{1/3}}{12\,2^{1/3}}\right)}-\frac{1}{2}\sqrt{\ll 1\gg}\right\}$$

```
In[8]:= xySolλ = {x, y} /.
        Solve[{X, λ/2} == {-x + x^2 - y^2 - λ y, y + 2 x y - λ x},
              {x, y}];
```

Another singular case is $9\lambda^4 + 24\,X\,\lambda^2 + 18\lambda^2 - 48\,Y\,\lambda + 16\,X^2 - 48\,Y^2 - 24\,X + 9 = 0$. In this case, the resulting equations become more complicated. Because we are only interested in a visualization here, we do not analyze this special case in detail.

To speed up the calculation of the solution of the map, we use common subexpression optimization.

```
In[9]:= optimize = (* use function from Experimental` context *)
                Experimental`OptimizeExpression[#,
                    OptimizationLevel -> 1, ExcludedForms -> {}]&;
```

The functions Fcf and Fλcf are compiled functions that calculate the preimage of the map.

```
In[10]:= {Fcf, Fλcf} = Compile[{{X, _Complex}, {Y, _Complex}},
                    Evaluate[optimize[# /. λ -> λhp]]]& /@ {xySol, xySolλ};
```

The functions Fhp and Fλhp are the extracted last elements of the compiled functions. We will use them in the following for high-precision calculations.

```
In[11]:= {Fhp, Fλhp} = Function[f, OwnValues[f][[1, 2, 5]],
                        {HoldAll, Listable}][{Fcf, Fλcf}];
```

We unite the machine precision solutions and the high-precision solution in the function preImages.

```
In[12]:= preImages[{f_, fλ_}, {X_, Y_}, λ_] :=
         Select[Chop[If[Y != λ/2, f[X, Y], fλ[X, Y]]],
             (* select real pairs *) (Im[#] == {0, 0})&]
```

For sketches of the basins of attraction, we will eliminate points that are nearby in each of the recursive solution steps. The function unitePoints selects one point from point sets having distances less than or equal ε.

```
In[13]:= unitePoints[points_, ε_] :=
             (* keep one point per group *) #[[1, 2]]& /@ Split[Sort[
              {Rationalize[#, ε], #}& /@ points], #1[[1]] == #2[[1]]&]
```

Now, we have all parts together to define the function attractorGraphics. Depending on *fs* (either {Fhp, Fλcf} or {Fhp, Fλhp}), it will calculate the real preimages *iters* times recursively, starting with *pp* + 1 points on each of the three line attractors.

```
In[14]:= attractorGraphics[fs_, λ_, {P1_, P2_}, iters_, pp_, prec___] :=
         Module[{ℛ, nl},
           Show[Table[(* rotation matrix *)
                     ℛ = N[{{ Cos[2/3(k - 1) Pi], Sin[2/3(k - 1) Pi]},
                              {-Sin[2/3(k - 1) Pi], Cos[2/3(k - 1) Pi]}}, prec];
            (nl = (* recursively calculate preimages *)
            NestList[(unitePoints[Flatten[preImages[fs, #, λN]& /@ #, 1],
                     10^-4])&, Table[ℛ.{x0, λ/2},
                         (* sample attractor *) {x0, P1, P2, (P2 - P1)/pp}], iters]);
            (* make graphics *)
            Graphics[{(* points from the three lines in red, green, and blue *)
             MapAt[1&, RGBColor[0, 0, 0], k],
             PointSize[0.0025], Map[Point[Re[#]]&, nl, {-2}]}, FrameTicks -> None,
                     PlotRange -> {{-1.7, 2.5}, {-2.3, 2.05}}, Frame -> True,
                     AspectRatio -> Automatic], {k, 1, 3}]]]
```

Using six iterated preimages and 52 points on each attractor show that the machine number calculation gives identical results to the high-precision calculations.

```
In[15]:= Off[Power::infy]; Off[Infinity::indet]; Off[Divide::infy];

         Show[GraphicsArray[
         Block[{$DisplayFunction = Identity},
         {attractorGraphics[{Fhp, Fλcf}, λN, {P1xN, P2xN}, 6, 51],
          attractorGraphics[{Fhp, Fλhp}, λhp, {P1xhp, P2xhp}, 6, 51, 60]}]]]
```

Repeating now the (faster) machine arithmetic calculation with 10 recursive preimages gives the following sketch of the basins of attraction with more than 800000 points.

```
In[17]:= attractorGraphics[{Fhp, Fλcf}, λN, {P1xN, P2xN}, 10, 51]
```

l) We start with a compiled function that generates the sequence. It takes as arguments realizations of the random variables β_0, ω_j, ω_{jk}, and the initial x_m.

```
In[1]:= cf = Compile[{s, β0, {βL, _Real, 1}, {ω0s, _Real, 1},
                      {ωM, _Real, 2}, {x0, _Real, 1},
                {o, _Integer}},
             Module[{n = Length[βL], d = Length[ωM[[1]]], X, xE, xN},
                     (* list of initial and to-be-filled-in sequence values *)
                     X = Table[If[j <= d, x0[[j]], 0.], {j, o}];
                     (* fill in sequence values *)
                     Do[xE = Reverse @ Take[X, {k - d, k - 1}];
                        xN = β0 + Sum[βL[[i]] Tanh[s ω0s[[i]] +
                                    s ωM[[i]].xE], {i, n}];
                        X[[k]] = xN, {k, d + 1, o}]; X]];
```

```
In[2]:= (* check that cf compiled successfully *)
        Union[Head /@ Flatten[cf[[4]]]]
```

```
Out[3]= {Integer, Real}
```

Next, we define a function make$\beta\omega$xData that generates concrete instances of the random variables.

```
In[4]:= makeβωxData[n_Integer, d_Integer] :=
            {Random[],
             Sqrt[n] (#/Sqrt[#.#])&[Table[Random[], {n}]],
             Table[Sqrt[2] InverseErf[0, 2 Random[] - 1], {n}],
             Table[Sqrt[2] InverseErf[0, 2 Random[] - 1], {n}, {d}],
             Table[Random[Real, {-1, 1}], {d}]}
```

In the following, we will work $n = d = 64$. So, the map under consideration has 4289 random parameters.

```
In[5]:= SeedRandom[777];
        βωxs = makeβωxData[64, 64];
```

To get an idea about the s-dependence of the map, we show the $x_m(s)$ for $m = d + 1$, ..., $d + 1 + 100$, with $m = 64$, $m = 1000$, and $m = 10000$ (we change the s-ranges from plot to plot).

```
In[7]:= Show[GraphicsArray[
        Function[{m0, sMax},
        Graphics[{PointSize[0.0025],
                Table[SeedRandom[1]; Point[{s, #}]& /@
                        Drop[cf[s, Sequence @@ βωxs, m0 + 100], m0],
```

```
                    {s, 0, sMax, sMax/1000}]},
            PlotRange -> All, Frame -> True]] @@@
 (* m- and s-values *){{64, 0.2}, {1000, 0.08}, {10000, 0.05}}]]
```

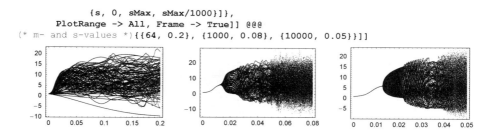

The next graphics show the sequence $x_m(s)$ for four values of s. The transition from a system with structure to a chaotic system is clearly visible.

```
In[8]:= Show[GraphicsArray[#]]& /@
        Partition[ListPlot[cf[#, Sequence @@ βωxs, 1600], PlotRange -> All,
                Frame -> True, Axes -> False, FrameTicks -> False,
                PlotLabel -> #, DisplayFunction -> Identity]& /@
                (* use four s-values *) {0.005, 0.015, 0.02, 0.03}, 2]
```

The last graphic shows that for $s \lesssim 0.024$ the resulting sequences $x_m(s)$ have some "structure". Making now the animation of the pairs $\{x_{m-1}(s), x_m(s)\}$ shows that the sequence shows a variety of pattern as a function of s. This time, we show the first 50000 $x_m(s)$ (with the first 1000 discarded).

```
In[9]:= animation2DGraphic[s_, mMax_:50000] :=
        Graphics[{PointSize[0.0025], (* color points *)
                MapIndexed[{Hue[0.78 #2[[1]]/10000], Point[#1]}&,
                        Partition[Drop[cf[s, Sequence @@ βωxs, mMax], 1000], 2]]},
                PlotRange -> All, AspectRatio -> 1, Frame -> True, FrameTicks -> None,
                PlotLabel -> s]
In[10]:= Show[GraphicsArray[animation2DGraphic /@ #]]& /@
         {{0.0124, 0.0128, 0.02}, {0.0222, 0.0264, 0.028}, {0.036, 0.0388, 0.04}}];
```

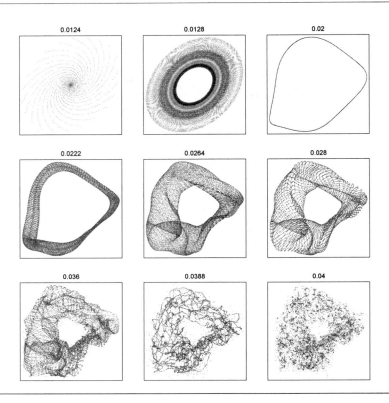

```
sMin = 0.01; sMax = 0.05;
Do[Show @ animation2DGraphic[s], {s, sMin, sMax, (sMax - sMin)/100}]
```

We end by zooming into a smaller interval of the s-range and displaying the resulting sequence as points in 3D by using triples $\{x_{m-2}(s), x_{m-1}(s), x_m(s)\}$.

```
In[11]:= animation3DGraphic[s_, mMax_:50000] :=
         Graphics3D[{PointSize[0.0025],
                 MapIndexed[{Hue[0.78 #2[[1]]/10000], Point[#1]}&,
                         Partition[Drop[cf[s, Sequence @@ βωxs, mMax], 1000], 3]]},
                 PlotRange -> All, BoxRatios -> {1, 1, 1},  PlotLabel -> s]

In[12]:= Show[GraphicsArray[animation3DGraphic /@ #]]& /@
                 {{0.03468, 0.03486, 0.03504}, {0.0351, 0.03564, 0.0366}}
```

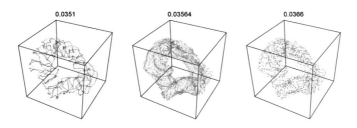

```
sMin = 0.033; sMax = 0.039;
Do[Show @ animation3DGraphic[s], {s, sMin, sMax, (sMax - sMin)/100}]
```

We end with some visualizations of high-order versions of the nonlinear univariate recurrence relation $u_{k+n+1} = (\alpha + u_k\, u_{k+n})/(u_{k+1} - 2\, u_k)$ [960].

```
In[13]:= rrsGraphic[n_, seed_, λ_] :=
         Module[{cr := Random[] Exp[ 2 Pi I Random[]], u, α},
              SeedRandom[seed];
              (* initialize *)
              α = cr; u = Table[If[k <= n, cr, 0], {k, λ}];
              (* fill in entries in list Y *)
              Do[u[[k + n + 1]] = (α + u[[k]] u[[k + n]])/(u[[k + 1]] - 2 u[[k]]),
                {k, 1, Length[u] - n - 1}];
              (* make graphics *)
              Graphics[{PointSize[0.003],
                       MapIndexed[{Hue[0.78 #2[[1]]/λ],
                                  Point[{Re[#], Im[#]}]}&, Drop[u, 4 n]]},
                     Frame -> True, FrameTicks -> False, Axes -> False,
                     AspectRatio -> Automatic]]
In[14]:= Show[GraphicsArray[rrsGraphic[##, 50000]& @@@
         (* order and seed data *) {{72, 35543}, {117, 89534}}]];
```

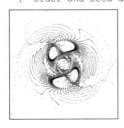 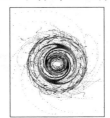

10. Movement in a 2D Periodic and Egg Crate Potential, Pearcey Integral, Subdivision of Surface, Construction of Part of Surface, Electric Field Visualizations

a) Without loss of generality, we choose $V_m = 1$.

```
In[1]:= V[x_, y_] = Cos[x] + Cos[y];
```

Newton's equations of motion for the particle read (**E** and **B** are the electric and the magnetic field strength, and we assume classical coupling. For coupling under fractional quantum Hall effect conditions, see [698].)

$$m\, \mathbf{r}''(t) = q\, \mathbf{E} + q\, \mathbf{r}'(t) \times \mathbf{B} - \mathrm{grad}V(\mathbf{r}(t)).$$

After an appropriate scaling of all quantities, we can implement the following equations (the meanings of the variables should be obvious).

```
In[2]:= NewtonEqs[e_, B_, {x0_, y0_}, {vx0_, vy0_}] =
        {x''[t] == +y'[t] B - D[V[x[t], y[t]], x[t]],
```

```
                    y''[t] == -x'[t] B - D[V[x[t], y[t]], y[t]],
                    x[0] == x0, y[0] == y0, x'[0] == vx0, y'[0] == vy0};
```

Now, we use NDSolve to compute the path, and we plot it on top of a grayscale picture of the potential to get a better idea of the movement.

```
In[3]:= trajectoryGraphics[e_, B_, {x0_, y0_}, {vx0_, vy0_}, tMax_, opts___] :=
        Module[{b1, pr},
        (* the trajectories *)
        b1 = ParametricPlot[ (* solving the equations of motions *)
                    Evaluate[{x[t], y[t]} /.
                        NDSolve[NewtonEqs[e, B, {x0, y0}, {vx0, vy0}],
                                {x[t], y[t]}, {t, 0, tMax},
                                MaxSteps -> 100000]], {t, 0, tMax},
                        PlotRange -> All, AspectRatio -> Automatic, Axes -> False,
                        DisplayFunction -> Identity, PlotPoints -> 300];
        (* underlying DensityGraphics *)
        pr = FullOptions[b1, PlotRange];
        ListDensityPlot[Table[V[x, y],
                    {y, pr[[2, 1]], pr[[2, 2]], -Subtract @@ pr[[2]]/80},
                    {x, pr[[1, 1]], pr[[1, 2]], -Subtract @@ pr[[1]]/80}],
                    opts, Mesh -> False, MeshRange -> pr,
                    Epilog -> {Hue[1], Thickness[0.01], b1[[1, 1, 1, 1]]},
                    Frame -> False, AspectRatio -> 1]]
```

We now look at a few concrete paths. In the upper two pictures, we started the particle in a potential dip, and it stays there. In the lower left picture, the particle moves along the boundaries of several valleys. Looking at the potential, the reflections are clear. The movement is disorderly, and in the lower right picture, the particle moves around a potential peak.

```
In[4]:= Show[GraphicsArray[#]]& @
        Block[{$DisplayFunction = Identity},
                (* four selected trajectories *)
                {trajectoryGraphics[0.0001, 1.4, {-2, -3}, {0.04, -0.2}, 60],
                 trajectoryGraphics[0.1, 0.4, {2, -2}, {0.4, -0.12}, 50],
                 trajectoryGraphics[0.1, 0.4, {2, -2}, {1.4, -1.12}, 70],
                 trajectoryGraphics[0.1, 1.8, {-3, 0}, {0, 1}, 60]}]
```

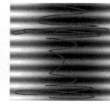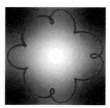

Next, we use the random number generator to generate random values for *e*, *B*, and the initial position and velocity.

```
In[5]:= Show[GraphicsArray[#]]& /@
        Block[{$DisplayFunction = Identity},
        Apply[((* seed random number generator *) SeedRandom[#1];
                trajectoryGraphics[(* random data *)
                    Random[], Random[],
                    Table[Random[Real, {-1, 1}], {2}],
                    Table[Random[Real, {-1, 1}], {2}], #2])&,
        (* seeds and integration times *)
        {{{314609943, 200}, {5416932298, 200}, {5705455867, 200}},
         {{4513316011, 400}, {4815679376, 600}, {43372625125, 400}}}, {2}]]
```

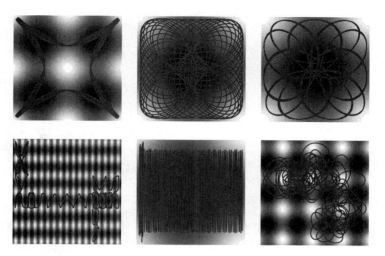

A plethora of other types of motion still exists [675].

For the nanoelectronic background for this type of calculation, see [54], [199], [1411], [1686], [211], [1046], [478], [1364], [613], [1882], [1749], [858], [1363], [1817], [183], [130], [1883], [1945], [1778], [1527], [1778], [760], [1557], and [1622]; for hard core potentials, see [1798].

For the motion of a neutral, polarizable object in crossed fields see [957], [486], [1767].

b) Below, we will use "random" values for the initial positions and velocities of the particles. A set of ranges and seeds for the random numbers are contained in the following list `initialConditions`. (The numbers were found in a random search.)

```
In[1]:= initialConditions =
        {{ 1.48623630,  0.53848783,  2.55153777,  1.20149868},
         {-1.74425110, -0.10072545,  2.94431671,  0.34481428},
         { 2.56620398,  0.46413788, -1.78449215, -1.00748510},
         {-1.57053535, -0.77327278,  2.20652869, -0.80397315},
         {-1.91300539, -1.18709673, -1.82077743,  2.37033731},
         { 2.61293202,  2.92006045,  0.73724524,  0.30891667},
         {-0.20386780, -1.19659380,  2.95867251,  1.79163109},
         { 1.02830748, -1.04954633,  0.45486299,  1.16732037},
         { 0.23031167,  0.58348899,  2.38364484, -1.11437308},
         {-0.13689325,  0.08552569,  0.01417208, -0.08969203},
         { 0.20734805,  0.28709077,  0.03910580, -0.07692804}};
```

To test the sensitivity of the trajectories from the initial conditions, we slightly perturb the initial conditions by multiplying them with $(1 - 10^{-n})$.

```
In[2]:= (* the potential *)
        V[x_, y_] := (5 + 3 (Cos[x] + Cos[y]) +  Cos[x] Cos[y])/2
In[4]:= Module[{n = 5, T = 1000, r, α},
        Do[Show[GraphicsArray[Table[
        α = 1 - 10^(-$MachinePrecision + j);
        (* solve differential equations *)
        nsol = NDSolve[{x''[t] == -D[V[x[t], y[t]], x[t]],
                        y''[t] == -D[V[x[t], y[t]], y[t]],
                        Thread[{x[0], x'[0], y[0], y'[0]} ==
                            α initialConditions[[k]]]} // Flatten,
                    {x, y}, {t, 0, T}, MaxSteps -> 10^6, MaxStepSize -> 0.1];
            (* display trajectories *)
        ParametricPlot[Evaluate[{x[t], y[t]} /. nsol], {t, 0, T},
          PlotPoints -> 2000, Frame -> True, Axes -> False, FrameTicks -> None,
          DisplayFunction -> Identity, PlotRange -> All, AspectRatio -> 1,
```

```
PlotStyle -> {{Thickness[0.002], Hue[j/n 0.8]}}], {j, 0, n}],
GraphicsSpacing -> 0.08]], {k, Length[initialConditions]}]]
```

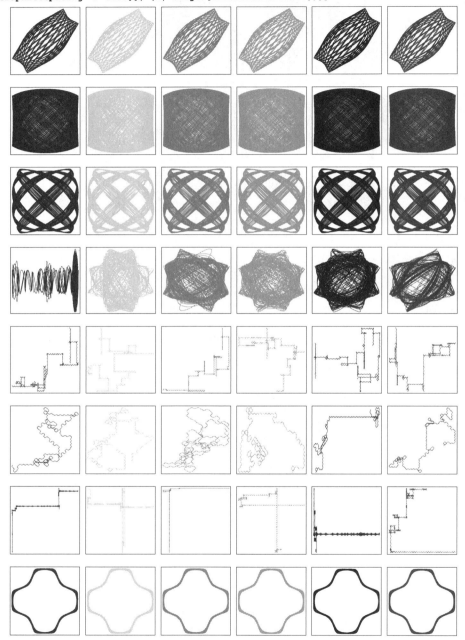

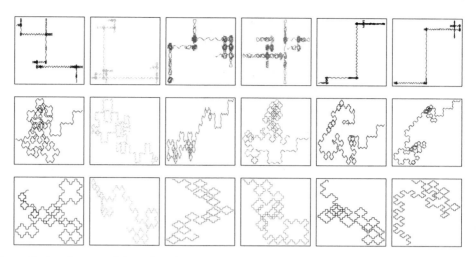

One sees that the qualitative character of the trajectories is not particularly sensitive to the initial conditions. Also, the quantitative properties of the bounded trajectories are not very sensitive to the initial conditions. But the quantitative properties of the bounded trajectories are very sensitive. For a more detailed discussion of the trajectories in this potential, see [1002]. For Levy flights in periodic potentials, see [731].

c) First, we implement the solution for the $P(x, 0) = p[x]$. Because we later also need the derivative of this function, we rewrite this second-order differential equation as two first-order ones.

```
In[1]:= {p[x_], ps[x_]} = Last[#][x]& /@
    NDSolve[(* rename variables *)
            {4 b'[x] + 2 I x b[x] + I a[x] == 0, b[x] == a'[x],
             a[0] ==  (-1)^(1/8)/2 Gamma[1/4], b[0] == I(-1)^(3/8)/2 Gamma[3/4]},
            {a, b}, {x, -15.9, 5.9}][[1]];
```

Now, we use the values of these functions as the initial values for solving for $P(x, y)$. Because we need to use the solutions of this differential equation many times later, we remember their values by using a SetDelayed[Set[...]] construction.

```
In[2]:= q[x_] := q[x] =
    NDSolve[{4 c'''[y] - 2 x c'[y] - I y c[y] == 0,
             c[0] == p[x], c'[0] == 0, c''[0] == I ps[x]},
            {c}, {y, 0, 11}][[1, 1, 2]];
```

Now, let us construct the pictures, one for the absolute value and one for the phase (appropriate values for the PlotRange are generally found after a few trials).

```
In[3]:= Do[q[x], {x, -15.9, 5.9, 0.1}];

    opts = Sequence[ContourShading -> False, DisplayFunction -> Identity,
                    ContourStyle -> {Thickness[0.002]},
                    MeshRange -> {{-15.9, 5.9}, {-10.9, 10.9}}];

    Show[GraphicsArray[{
    ListContourPlot[(* show absolute value *) Abs[tab = Transpose[
    (* the data *)
    Table[(* make one half and then mirror using symmetry *)
          Join[Reverse[#], #]&[
            Table[q[x][y], {y, 0, 10.9, 0.05}]], {x, -15.9, 5.9, 0.1}]]],
    (* set options to give a nice picture *)
                    opts, PlotRange -> {0, 2}, PlotLabel ->
                    StyleForm["Abs[x, y]]", FontFamily -> "Courier"],
                    Contours -> Table[i, {i, 0, 2, 2/10}]],
    (* show argument *)
    ListContourPlot[Map[If[# == 0, 0, Arg[#]]&, (* reuse data *) tab, {2}],
                    opts, PlotRange -> {-Pi, Pi},
```

```
ContourStyle -> {Thickness[0.002]}, PlotLabel ->
StyleForm["Arg[P[x, y]]", FontFamily -> "Courier"],
Contours -> Table[i, {i, -Pi, Pi, 2Pi/8}]]}]]
```

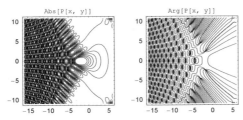

For a Borel summation-based approach for the numerical calculation of such kind of integrals, see [1310] and [1311]. For numerical integration methods of such kind of integrals, see [998].

d) Here, the parametrization of the surface is implemented.

```
In[1]:= x = (2 + Cos[s/2] Sin[t] - Sin[s/2] Sin[2t]) Cos[s];
       y = (2 + Cos[s/2] Sin[t] - Sin[s/2] Sin[2t]) Sin[s];
       z = Sin[s/2] Sin[t] + Cos[s/2] Sin[2t];
```

The surface area A of a parametrically defined surface is given as the integral over the defining s,t-region B:

$$E(s, t) = \left(\frac{\partial x(s, t)}{\partial s}\right)^2 + \left(\frac{\partial y(s, t)}{\partial s}\right)^2 + \left(\frac{\partial z(s, t)}{\partial s}\right)^2$$

$$G(s, t) = \left(\frac{\partial x(s, t)}{\partial t}\right)^2 + \left(\frac{\partial y(s, t)}{\partial t}\right)^2 + \left(\frac{\partial z(s, t)}{\partial t}\right)^2$$

$$F(s, t) = \frac{\partial x(s, t)}{\partial s} \frac{\partial x(s, t)}{\partial t} + \frac{\partial y(s, t)}{\partial s} \frac{\partial y(s, t)}{\partial t} + \frac{\partial z(s, t)}{\partial s} \frac{\partial z(s, t)}{\partial t}$$

$$A = \int_B \sqrt{E(s, t) G(s, t) - F(s, t)^2} \, ds \, dt.$$

```
In[4]:= e = D[x, s]^2 + D[y, s]^2 + D[z, s]^2;
       g = D[x, t]^2 + D[y, t]^2 + D[z, t]^2;
       f = D[x, s] D[x, t] + D[y, s] D[y, t] + D[z, s] D[z, t];

       integrand[s_, t_] = Sqrt[e g - f^2];
```

To subdivide the s,t-region, we first divide the s,t-region into stripes n_s of equal area and then subdivide each of these stripes into n_t areas. To divide these regions under interest, we rewrite the integral as a differential equation (here, the t-integration has to be carried out over the t-range for fixed s):

$$\frac{\partial A(s)}{\partial s} = \int_0^{2\pi} \sqrt{E(s, t) G(s, t) - F(s, t)^2} \, dt$$

Here, this is coded. (We turn off the `NIntegrate::precw` message, which is caused by a zero value of the integrand.)

```
In[8]:= Off[NIntegrate::precw];
       srhs[s_?NumberQ] := NIntegrate[Evaluate[N[integrand[s, t]]], {t, 0, 2Pi}]
In[10]:= ssol = NDSolve[{a'[s] == srhs[s], a[0] == 0}, a, {s, 0, 2.Pi}][[1, 1, 2]]
Out[10]= InterpolatingFunction[{{0., 6.28319}}, <>]
```

Taking `ssol[N[2Pi]]`, we get the value of the whole surface area.

```
In[11]:= A = ssol[N[2Pi]]
Out[11]= 119.72
```

Because `ssol[s]` is a monotonically increasing function, we can easily use `FindRoot` to obtain the s-values that divide the s,t-region into n_s equal parts. (Starting from here, we use specific values for n_s and n_t.)

```
In[12]:= ns = 35; nt = 18;
         sValues = Table[FindRoot[ssol[s] == i A/ns,
                                  {s, 0, 2Pi}][[1, 2]], {i, 0, ns}];
```

Now, we subdivide each of the n_s strips into n_t pieces using the same idea. Here is the right-hand side of the differential equation.

```
In[14]:= trhs[t_?NumberQ, stripNumber_] :=
         NIntegrate[Evaluate[N[integrand[s, t]]],
                    {s, sValues[[stripNumber]], sValues[[stripNumber + 1]]}]
```

Here, the differential equation is solved for every strip.

```
In[15]:= Do[tsol[i] = NDSolve[{a'[t] == trhs[t, i], a[0] == 0}, a,
                            {t, 0, N[2Pi]}][[1, 1, 2]], {i, ns}]
```

Now, the region is subdivided.

```
In[16]:= tValues = Append[#, N[2Pi]]& /@
             Table[FindRoot[tsol[i][t] == j A/(ns nt), {t, 0, 2Pi}][[1, 2]],
                   {i, ns}, {j, 0, nt - 1}];
```

Let us check that all areas are (within some precision) the same. These are the descriptions of the individual *s,t*-regions.

```
In[17]:= regions = MapThread[Function[{σ, τ}, {Prepend[σ, s], #}& /@
                                                 (Prepend[#, t]& /@ τ)],
                            {Partition[sValues, 2, 1],
                             Partition[#, 2, 1]& /@ tValues}];
```

```
In[18]:= Show[Graphics[{Thickness[0.002], Map[
         Line[{{#[[1, 2]], #[[2, 2]]}, {#[[1, 3]], #[[2, 2]]},
               {#[[1, 3]], #[[2, 3]]}, {#[[1, 2]], #[[2, 3]]},
               {#[[1, 2]], #[[2, 2]]}}]&, regions, {2}]}], Frame -> True]
```

We calculate their areas.

```
In[19]:= areas = Apply[NIntegrate[Evaluate[N[integrand[s, t]]], ##]&,
                       regions, {2}];
```

We calculate the maximal relative deviation.

```
In[20]:= Max @ Flatten[Abs[Function[mean, Map[(# - mean)/mean&, areas, {2}]][
                           (Plus @@ #/Length[#])&[Flatten[areas]]]]]
Out[20]= 6.63177 × 10^{-6}
```

e) Let us first have a look at the part of the surface to be constructed.

```
In[1]:= r[φ_, z_] = (1 + 2 Sin[z/Pi]^2 Cos[2φ]^6);

        ParametricPlot3D[Evaluate[{r[φ, z] Cos[φ], r[φ, z] Sin[φ], z,
                                   {EdgeForm[Thickness[0.002]],
                                    SurfaceColor[RGBColor[0, 1, 0],
                                                 RGBColor[1, 1, 0], 2.6]}}],
                         {φ, 0, 2Pi}, {z, 0, Pi},
                         PlotPoints -> {36, 46}, BoxRatios -> {1, 1, 2},
                         PlotRange -> {{-5/4, 5/4}, {-5/4, 5/4}, {0, Pi}},
                         Boxed -> False, Axes -> False]
```

For growing z-values, the radius grows along the axes and the use of the `PlotRange` option makes holes in the surface. We construct the visible parts of the surface by restricting the parameter φ,r-region in such a manner that exactly these holes are left out. The holes start at the following z-value.

```
In[3]:= zMin = FindRoot[r[0, z] == 5/4, {z, 0, 3}][[1, 2]]

Out[3]= 1.13527
```

In the following, we make use of the obvious fourfold rotation symmetry of the above-defined surface. For all values of z greater than zMin, the φ-range extends from a minimal value to $\pi/4$ and correspondingly rotated regions. We calculate this minimal φ value.

```
In[4]:= φMin[z_] := φMin[z] = FindRoot[r[φ, z] Cos[φ] == 5/4,
                                        {φ, (z - zMin)^(1/4)/4}][[1, 2]]
```

Using the routine `ListSurfacePlot3D` from the package `Graphics`Graphics3D``, we can so construct one-eighth of the surface.

```
In[5]:= Needs["Graphics`Graphics3D`"]

        ppz1 = 15; ppz2 = 25; ppp = 5;

        With[{ε = 10^-8},
        (* start away from the domain boundaries *)
        ListSurfacePlot3D[
        Join[Table[{r[s, z] Cos[s], r[s, z] Sin[s], z},
                    {z, 0, zMin - ε, (zMin - ε)/ppz1},
                    {s, 0, Pi/4., Pi/4./ppp}],
              Table[{r[s, z] Cos[s], r[s, z] Sin[s], z},
                    {z, zMin + ε, 1. Pi, (Pi - zMin - ε)/ppz2},
                    {s, φMin[z], Pi/4., (Pi/4 - φMin[z])/ppp}]] // N,
                          ViewPoint -> {1, 1, 2}]]
```

```
In[8]:= part = %[[1]];
```

Using the symmetry, the other parts of the surface can be generated by rotation and reflection. (To get a more interesting picture, we extend the surface in the vertical direction.)

```
In[9]:= R = N[{{Sqrt[2]/2, Sqrt[2]/2, 0}, {-Sqrt[2]/2, Sqrt[2]/2, 0}, {0, 0, -1}}];

        Show[Graphics3D[{EdgeForm[Thickness[0.002]],
                SurfaceColor[Hue[0.22], Hue[0.22], 2.6],
          {#, Map[R.#&, #, {-2}]}&[
          {#, Map[{{0, 1, 0}, {1, 0, 0}, {0, 0, 1}}.#&, #, {-2}]}&[
            {#, Map[{-1, 1, 1} #&, #, {-2}]}&[
```

```
{#, Map[{1, -1, 1} #&, #, {-2}]}&[
  {#, Map[{0, 0, 2N[Pi] - 2#[[3]]} + #&, #, {-2}]}&[part]]]]]}] /.
  (* display in horizontal direction *)
  Polygon[l_] :> Polygon[{#3, #2, #1}& @@@ l],
   BoxRatios -> {4, 1, 1}, PlotRange -> All, Boxed -> False]
```

f) This is our random trigonometric function.

```
In[1]:= (* definition for making a random trigonometric function *)
        randomTrigFunction[n_] :=
        Sum[Random[] Cos[i x + 2 Pi Random[]] Cos[j y + 2 Pi Random[]],
          {i, 0, n}, {j, 0, n}]
In[3]:= (* seed random number generator and make a random trigonometric function *)
        SeedRandom[111];
        trigF = randomTrigFunction[4];
In[6]:= pl3D = Plot3D[Evaluate[trigF], {x, -Pi, Pi}, {y, -Pi, Pi},
                 PlotPoints -> 200, Mesh -> False]
```

We find the extremal points by searching for zeros of the partial derivatives of `trigF` on a dense grid. Locally we use `FindRoot` to find the extrema.

```
In[7]:= eqs = {D[trigF, x], D[trigF, y]} // Chop;

In[8]:= pp = 100; δ = 2Pi/pp;

        (* suppress messages *)
        Internal`DeactivateMessages[
        (* use FindRoot on a grid in the x,y-plane *)
        tab = Flatten[
          Table[FindRoot[Evaluate[# == 0& /@ eqs],
                         Evaluate[{x, x0, x0 + δ, -Pi, Pi}],
                         Evaluate[{y, y0, y0 + δ, -Pi, Pi}]],
              {x0, -Pi + δ/2, Pi - δ/2, δ},
              {y0, -Pi + δ/2, Pi - δ/2, δ}], 1];
        (* select extrema *)
        extrema = {x, y} /. Select[tab, (Max[Abs[eqs /. #]] < 10^-6)&];
        (* eliminate doubles *)
        extrema = Union[extrema, SameTest -> (#.#&[#1 - #2] < 10^-6&)]];
```

We found 68 extrema.

```
In[11]:= Length[extrema]
Out[11]= 69
```

We sort the extrema in minima, maxima, and saddle points.

```
In[12]:= det[{x_, y_}] = D[trigF, x, y]^2 - D[trigF, x, x] D[trigF, y, y];
        sec[{x_, y_}] = {D[trigF, x, x], D[trigF, y, y]};

In[14]:= extremaGroups =
        Map[Last, Split[Sort[If[det[#] >= 0, {saddle, #},
             If[Min[sec[#]] >= 0, {minimum, #}, {maximum, #}]]& /@ extrema,
                          OrderedQ[{#1, #2}]&], #1[[1]] === #2[[1]]&], {2}];
```

Here is a contour plot of `trigF` and the extrema.

```
In[15]:= coloredExtrema = Table[{MapAt[1&, RGBColor[0, 0, 0], k],
                           Point /@ extremaGroups[[k]]}, {k, 3}];

In[16]:= cp = ContourPlot[Evaluate[trigF], {x, -Pi, Pi}, {y, -Pi, Pi},
               PlotPoints -> 250, ColorFunction -> (Hue[0.8 #]&),
               ContourLines -> False, Contours -> 100,
               Epilog -> {PointSize[0.016], coloredExtrema}]
```

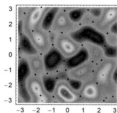

Now, we calculate the curves that follow the gradient locally. We normalize the gradient to 1. We start the integration of the differential equations on a small circle near the extrema. We start at nine points of each circle.

```
In[17]:= trigFt = trigF /. {x -> x[t], y -> y[t]};
        den = Sqrt[D[trigFt, x[t]]^2 + D[trigFt, y[t]]^2];
        odes = Thread[{x'[t], y'[t]} == {D[trigFt, x[t]], D[trigFt, y[t]]}/den];

In[20]:= makeLines[{x0_, y0_}, n_] :=
        With[{φ0 = Random[], ε = 10^-2},
        Table[(* solve differential equation *)
               nsol = NDSolve[Join[odes, {x[0] == x0 + ε Cos[φ0 + φ],
                                          y[0] == y0 + ε Sin[φ0 + φ]}],
                           {x, y}, {t, 0, 6}, MaxSteps -> 1000,
                           PrecisionGoal -> 3, AccuracyGoal -> 3][[1]];
               T = nsol[[1, 2, 1, 1, 2]];
               (* plot solution *)
               ParametricPlot[Evaluate[{x[t], y[t]} /. nsol],
                           {t, 0, T}, DisplayFunction -> Identity,
               PlotStyle -> {{GrayLevel[1/2], Thickness[0.002]}}],
               {φ, 0, (1 - 1/n)2Pi, 2Pi/n}]]

In[21]:= Off[NDSolve::mxst];
        lines = makeLines[#, 9]& /@ Take[extrema, All];
```

The next two graphics show the calculated lines on the surface `trigF`.

```
In[23]:= With[{p = Compile[{x, y}, Evaluate[trigF]]},
        Show[GraphicsArray[
        {(* 2D plots *)
        Show[{cp, lines}, PlotRange -> {{-Pi, Pi}, {-Pi, Pi}},
            DisplayFunction -> Identity],
        (* lift lines to 3D *)
        lines3D = Cases[lines, _Line, Infinity] /.
            Line[l_] :> Line[Apply[{##, p[##] + 0.01}&, l, {1}]];
        (* 3D plots *)
        Show[{pl3D, Graphics3D[lines3D]}, DisplayFunction -> Identity,
            PlotRange -> {{-Pi, Pi}, {-Pi, Pi}, All}]]]]]
```

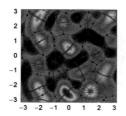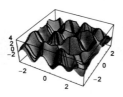

For more on the visualization of complicated 2D vector fields, see [1599].

g) Here are the line segments of a 5×5 square grid.

```
In[1]:= n = 5;
        linesPlus = Map[(# + 10^-5 Random[Real, {-1, 1}])&,
        N[Join[#, Map[Reverse, #, {2}]]&[
                Table[{{x, -1}, {x, 1}}, {x, -1, 1, 2/n}]]] //
                          Developer`ToPackedArray, {-1}];
```

The potential of a line segment was calculated in Exercise 12 of Chapter 3 of the Graphics volume [1794]. Here we use Compile to calculate the potential quickly. $\{x, y\}$ are the coordinates of the point where the potential is calculated and L is a list of the coordinates of the charged line segments.

```
In[3]:= potCF = Compile[{{x, _Real}, {y, _Real}, {L, _Real, 3}},
          Sum[With[(* coordinates of the charged lines *)
                  {x0 = L[[k, 1, 1]], y0 = L[[k, 1, 2]],
                   x1 = L[[k, 2, 1]], y1 = L[[k, 2, 2]]},
          (-Log[(x - x0)(x0 - x1) + Sqrt[(x - x0)^2 + (y - y0)^2]*
              Sqrt[(x0 - x1)^2 + (y - y1)^2] + (y - y0)(y0 - y1)] +
          Log[(x - x1)(x0 - x1) + Sqrt[(x - x1)^2 + (y - y1)^2]*
              Sqrt[(x0 - x1)^2 + (y0 - y1)^2] + (y - y1)(y0 - y1)])/
          Sqrt[(x0 - x1)^2 + (y0 - y1)^2]], {k, Length[L]}]];
```

The calculation of the potential will be proportional to the number of line segments. To avoid the calculation of the potential for many points for each rotation angle, we calculate the potential once on a dense grid and then use Interpolation to generate a high-quality interpolating function.

```
In[4]:= ppI = 501;
        (potData = Flatten[Table[{x, y, potCF[x, y, linesPlus]},
                        {y, -2., 2., 4./ppI}, {x, -2., 2., 4./ppI}], 1];)
```

We use the two mirror symmetries of the grid to calculate the potential only in the first quadrant.

```
In[6]:= ppI = 251;
        potData = Table[{x, y, potCF[x, y, linesPlus]},
                        {y, 0., 2., 2./ppI}, {x, 0., 2., 2./ppI}];

        potData = Flatten[Join[Apply[{#1, -#2, #3}&, Rest[#], {2}], #]&[
        Join[Apply[{-#1, #2, #3}&, Rest[#], {1}], #]& /@ potData], 1];

        potIpo = Interpolation[potData];
```

The potential of the rotated grid can be obtained by calculating the potential of the original grid at a rotated grid point.

```
In[10]:= potMinus[φ_, pp_] := With[{c = N[Cos[φ]], s = N[Sin[φ]]},
         Table[potIpo[c x + s y, -s x + c y],
                 {x, -Sqrt[2.], Sqrt[2.], 2.Sqrt[2.]/pp},
                 {y, -Sqrt[2.], Sqrt[2.], 2.Sqrt[2.]/pp}]];
```

For a nicer graphic, we use homogeneously distributed contour lines. We also display the two grids in black and white.

```
In[11]:= homogeneousListContourPlot[data_, φ_, cs_, colConstant_, mr_, opts___] :=
         Module[{n = Times @@ Dimensions[data]},
         ListContourPlot[data, opts, Contours ->
                 (* homogeneously spaced contour lines *)
                 (#[[Round[n/cs/2]]]& /@
                   Partition[Sort[Flatten[data]], Round[n/cs]]),
```

```
ColorFunction -> (Hue[#/colConstant]&),
ContourLines -> False, ColorFunctionScaling -> False,
(* the two grids *)
Epilog -> {Thickness[0.002], GrayLevel[1], Line /@
With[{c = Cos[φ], s = Sin[φ]},
    Map[{{c, s}, {-s, c}}.#&, linesPlus, {-2}]],
    GrayLevel[0], Line /@ linesPlus},
MeshRange -> {{-1, 1}, {-1, 1}} mr, FrameTicks -> None]]
```

Here is a view on the potential of the square grid.

```
In[12]:= ListDensityPlot[potMinus[0, 200], Mesh -> False,
                MeshRange -> {{-1, 1}, {-1, 1}} Sqrt[2]]
```

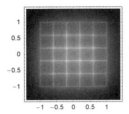

Using the function `homogeneousListContourPlot`, it is straightforward to make an animation of the equipotential lines. We show the case of opposite and equally charged grids.

```
In[13]:= makeGraphicsC[sign_] :=
    Module[{pp = 120, vP, vN},
        (* the potential *) vP = -potMinus[0, pp];
    Function[φList, Show[GraphicsArray[
    homogeneousListContourPlot[vP + (* the sign *) sign potMinus[#, pp],
                    #, 50, 4, Sqrt[2],
        DisplayFunction -> Identity]& /@ φList]]] /@
    Partition[Table[φ, {φ, Pi/30, Pi/4, (Pi/4 - Pi/30)/11}], 4]]

In[14]:= makeGraphicsC[+1]
```

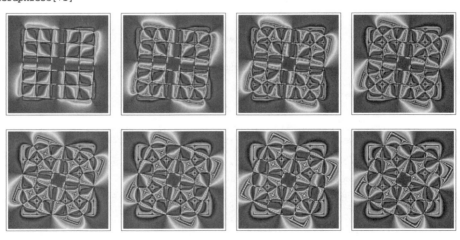

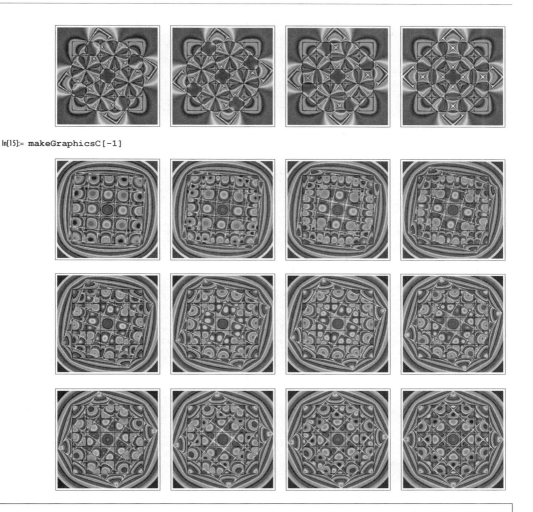

In[15]:= **makeGraphicsC[-1]**

```
Module[{pp = 120, frames = 30, vP, vN},
vP = -potMinus[0, pp];
Do[vN = potMinus[φ, pp];
   Show[GraphicsArray[Block[{$DisplayFunction = Identity},
   homogeneousListContourPlot[vP + # vN, φ, 50, 8, Sqrt[2]]& /@ {+1, -1}]]],
   {φ, Pi/frames, Pi/4, (Pi/4 - Pi/frames)/frames}]]
```

For the hexagonal grid, most of the calculations from above can be reused. We begin with the construction of the line segments that form the grid. We select a "circular" part of the grid.

```
In[16]:= r = 2.2;
        linesPlusPre =
        With[{n = Ceiling[r], (* directions *)
             a = {0, 1/2}, b = {Sqrt[3], -1}/4, c = {-Sqrt[3], -1}/4},
        (* points of a hexagonal lattice *)
        t1 = Flatten[Table[{0, 1}/2 + {i Sqrt[3], j 3}/2,
                   {i, -n, n}, {j, -n, n}], 1];
        t2 = Flatten[Table[{Sqrt[3], -1}/4 + {i Sqrt[3], j 3}/2,
                   {i, -n, n}, {j, -n, n}], 1];
```

```
        (* make lines *) Select[Union[Sort /@
          Join[{#, # + a}& /@ t1, {#, # + b}& /@ t1,
                {#, # + c}& /@ t1, {#, # + a}& /@ t2,
                {#, # + b}& /@ t2, {#, # + c}& /@ t2]],
                (Max[Sqrt[#.#]& /@ #] <= r)&]];
In[18]:= Show[Graphics[Line /@ linesPlusPre] // N,
              Frame -> True, AspectRatio -> Automatic]
```

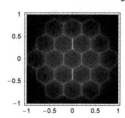

```
In[19]:= linesPlus = Map[(# + 10^-5 Random[Real, {-1, 1}])&,
                   N[linesPlusPre]/r, {-1}] // Developer`ToPackedArray ;
```

Because the part of the hexagonal grid under consideration lies inside the unit circle we can use a smaller x,y-domain than in the above square grid example.

```
In[20]:= ppI = 101;
        (potData = Table[{x, y, potCF[x, y, linesPlus]},
                {y, 0., Sqrt[2.], Sqrt[2.]/ppI}, {x, 0., Sqrt[2.], Sqrt[2.]/ppI}];)

        potData = Flatten[Join[Apply[{#1, -#2, #3}&, Rest[#], {2}], #]&[
        Join[Apply[{-#1, #2, #3}&, Rest[#], {1}], #]& /@ potData], 1];

        potIpo = Interpolation[potData];
In[25]:= potMinus[φ_, pp_] := With[{c = N[Cos[φ]], s = N[Sin[φ]]},
                          Table[potIpo[c x + s y, -s x + c y],
                                {y, -1., 1., 2./pp}, {x, -1., 1., 2./pp}]];
```

Here is a density plot of the resulting potential from one grid.

```
In[26]:= ListDensityPlot[potMinus[0, 200], Mesh -> False,
              MeshRange -> {{-1, 1}, {-1, 1}}];
```

```
In[27]:= makeGraphicsH[sign_] :=
        Module[{pp = 120, vP, vN},
          (* the potential *) vP = -potMinus[0, pp];
          Function[φList, Show[GraphicsArray[
          homogeneousListContourPlot[vP + sign potMinus[#, pp], -#, 50, 6, 1,
              DisplayFunction -> Identity]& /@ φList]]] /@
          Partition[Table[φ, {φ, Pi/30, Pi/3, (Pi/3 - Pi/30)/11}], 4]]
In[28]:= makeGraphicsH[+1]
```

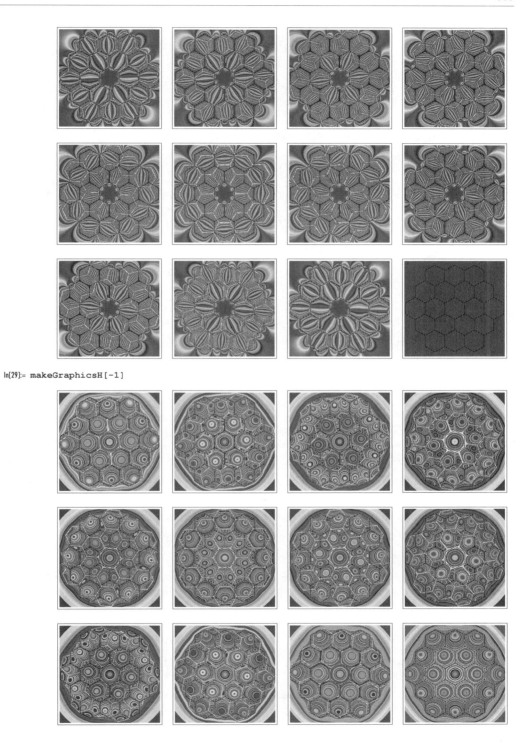

In[29]:= **makeGraphicsH[-1]**

```
Module[{pp = 120, frames = 30, vP, vN},
vP = -potMinus[0, pp];
Do[vN = potMinus[φ, pp];
   Show[GraphicsArray[Block[{$DisplayFunction = Identity},
      homogeneousListContourPlot[vP + # vN, -φ, 50, 6, 1]& /@ {+1, -1}]]],
      {φ, Pi/frames, Pi/3 - Pi/frames, (Pi/3 - 2Pi/frames)/frames}]]
```

11. A Ruler on the Fingers, Trajectories in Highly Oscillating Potential, Branched Flows

a) First, we derive Newton's equation of motion for the ruler. Here, the interesting quantities are the forces acting on the ruler. The forces are caused by static friction and kinetic friction (for details of friction, see [31], [559], [837], [1], [547], [271], [911], [1499], [1916], [272], [1432], [546], [1298], and [146]). Which of the two operates on which finger depends on the position of the center of gravity of the ruler x relative to the position of the fingers: $-s(t)$ (left finger), $s(t)$ (right finger) (here we assume the origin at the middle between the two fingers), where $s(t) = v\,t$, and v is the velocity of the finger. By statics, it follows that the normal forces F_{nl} and F_{nr} are

$$\text{left finger: } F_{nl} = m\,g\,\frac{s(t) - x(t)}{2\,s(t)}$$

$$\text{right finger: } F_{nr} = m\,g\,\frac{s(t) + x(t)}{2\,s(t)}.$$

(m is the mass of the ruler and g is the acceleration due to gravity). Let μ_k be the coefficient of kinetic friction, and let μ_s be the coefficient of static friction. Then, assuming the friction forces to be proportional to the normal forces, the ruler sticks to the left finger whenever

$$\mu_k(s(t) + x(t)) < \mu_s(s(t) - x(t))$$

and sticks to the right finger, provided

$$\mu_k(s(t) - x(t)) < \mu_s\,(s(t) + x(t)).$$

In these cases, the ruler moves with the same velocity as one of the fingers, that is, in a straight line with constant velocity. This means that the total force operating is zero. If either one of these two conditions fails, the ruler changes its motion, and it is subject to kinetic friction forces that, taking account of the two directions, add up to

$$F_{tot} = -m\,\mu_k\,\frac{x(t)}{s(t)}.$$

In the general case of an arbitrary velocity of the ruler relative to the finger, because the frictional force is independent of the direction, the force on the ruler is

$$F_{tot} = m\,\mu_k(-\text{sign}(x'(t) - v)\,F_{nl} - \text{sign}(x'(t) - v)\,F_{nr}).$$

Unfortunately, a direct computation of the movement with this force with NDSolve in the following form fails (*l0* is the initial position of the right finger). After the initial phase, we make little progress in solving the differential equation and even after one million steps, we do not get a solution evolved much further in time.

```
In[1]:= (* first straightforward try to solve the equation of motion
          for the ruler *)
       RulerMotiontTry[l0_, v_, μS_, μK_, x0_, x0p_, maxSteps_] :=
       Module[{s}, s[t_] = 10 - v t;
            NDSolve[{x''[t] == μK(-Sign[x'[t] - v] (s[t] - x[t])/(2s[t]) -
                                    Sign[x'[t] + v] (s[t] + x[t])/(2s[t])),
                        x[0] == x0, x'[0] == x0p}, x, {t, 0, 10/v},
                     MaxSteps -> maxSteps, MaxStepSize -> v/3 10]]
In[3]:= Table[{k, (* no NDSolve::mxst messages *) Internal`DeactivateMessages @
            RulerMotiontTry[20, 0.2, _, 0.35, -7, 0, 10^k]}, {k, 2, 6}]
```

```
Out[3]= {{2, {{x → InterpolatingFunction[{{0., 1.63205}}, <>]}}},
        {3, {{x → InterpolatingFunction[{{0., 1.63207}}, <>]}}},
        {4, {{x → InterpolatingFunction[{{0., 1.63213}}, <>]}}},
        {5, {{x → InterpolatingFunction[{{0., 1.63274}}, <>]}}},
        {6, {{x → InterpolatingFunction[{{0., 1.63877}}, <>]}}}}
```

This is because of the discontinuity of the force as a function of $x'(t)$ for $|x'(t)| \approx v$. We have

$$F_{tot} = \begin{cases} -m\,\mu_k & x'(t) > v \\ -m\,\mu_k\,x(t)/s(t) & v > x'(t) > -v \\ +m\,\mu_k & x'(t) < -v. \end{cases}$$

From a physical standpoint, this kind of implementation does not take into account the contact friction in the true sense. The ruler would "almost always" have a velocity different from v. From a mathematical standpoint, the force is not sufficiently smooth near the "quasi-sticking points" over a long enough time interval for the adaptive solution strategy of NDSolve to perform in a reasonable number of MaxSteps. Hence, we program the contact friction in a special way on the right-hand side of the Newton equation of motion. The choice of $\varepsilon_2 = 10^{-4}$ in the following code is largely arbitrary, and it could be reduced if we are working with a higher value of WorkingPrecision option of NDSolve. For the choice of parameters below, it does not influence the solution. (A proper mathematical treatment of this situation would require some more advanced measure differential inclusion techniques [1712], [1713], [1126], [131], [1447], [1711], [723], [724].) (Another possibility to model the discontinuous friction force would be to use a smoothened version of the sign(x) function, like tanh($L\,x$) with large L [806], [68], [454].)

```
In[4]:= RulerMotion[l0_, v_, μS_, μK_, x0_, x0p_] :=
        RulerMotion[l0, v, μS, μK, x0, x0p] =
        Module[{μ = (μS - μK)/(μS + μK), s, ε1 = 10^-8, ε2 = 10^-4},
        s[t_] = 10 - v t; (* finger motion *)
        NDSolve[{x''[t] == (* equation of motion for various situations *)
                Which[s[t] < ε1 (* ε1 and ε2 arbitrary, other
                                        "small" constant is OK too *), 0,
                    (Abs[x'[t] - v] < ε2 v) && (x[t] < μ s[t]), 0,
                    (Abs[x'[t] + v] < ε2 v) && (x[t] > -μ s[t]), 0,
                    True, (* the friction dominated terms *)
                    μK (-Sign[x'[t] - v] (s[t] - x[t])/(2s[t]) -
                        Sign[x'[t] + v] (s[t] + x[t])/(2s[t]))],
                x[0] == x0, x'[0] == x0p}, x, {t, 0, 10/v},
                MaxSteps -> 10^5, MaxStepSize -> v/3 10]]
```

We now plot the movement of the ruler. The red lines in the position-time diagram are the positions of the fingers, and the red lines in the velocity-time diagram are the velocities of the two fingers. The blue lines in the velocity-time diagram are the two stability conditions

$$\mu_k(s(t) + x(t)) < \mu_s(s(t) - x(t)) \quad \text{and} \quad \mu_k(s(t) - x(t)) < \mu_s(s(t) + x(t))$$

for sticking to one of the two fingers [1751].

```
In[5]:= MapRulerMotion[l0_, v_, μS_, μK_, x0_, x0p_, tInt_:Automatic] :=
        Module[{tm, sol, s, s1, s2, μ},
        (* ruler motion *)
        sol = RulerMotion[l0, v, μS, μK, x0, x0p];
        (* get t-interval from sol *)
        If[tInt === Automatic, tm = {t, 0, Last[Flatten[sol[[1, 1, 2, 1]]]]},
                                tm = Join[{t}, tInt]];
        Show[GraphicsArray[{
        (* the x(t) plot *)
        Plot[Evaluate[x[t] /. sol], Evaluate[tm],
            PlotRange -> All, AxesLabel -> {"t", "x(t)"},
            PlotLabel -> StyleForm["Position-time curve", FontSize -> 7],
            Prolog -> {Thickness[0.01], RGBColor[1, 0, 0],
                        {Line[{{0, 10}, {10/v, 0}}], Line[{{0, -10}, {10/v, 0}}]}},
            DisplayFunction -> Identity,
            PlotStyle -> {GrayLevel[0], Thickness[0.015]}],
        (* finger motion *)s[t_] = 10 - v t;
```

```
        μ = (μS - μK)/(μS + μK);
        s1[t_] = (x[t] /. sol)[[1]] - μ s[t];
        s2[t_] = (x[t] /. sol)[[1]] + μ s[t];
        (* the x'(t) plot *)
        Plot[Evaluate[{(x'[t] /. sol),  (* intervals of stability *)
                          If[s1[t] < 0, +v/4, 0],
                          If[s2[t] > 0, -v/4, 0]}], Evaluate[tm],
               PlotRange -> {All, {-1.2#, 1.2#}&[Max[{Abs[x0p], v}]]},
               AxesLabel -> {"t", "v(t)"}, PlotPoints -> 100,
               PlotLabel -> StyleForm["velocity-time curve", FontSize -> 7],
               (* the fingers *)
               Prolog -> {Thickness[0.01], RGBColor[1, 0, 0],
                            {Line[{{0, v}, {10/v, v}}], Line[{{0, -v}, {10/v, -v}}]}},
               DisplayFunction -> Identity,
               PlotStyle -> {{GrayLevel[0], Thickness[0.015]},
                              {RGBColor[0, 0, 1], Thickness[0.002]},
                              {RGBColor[0, 0, 1], Thickness[0.002]}}}]]]]]
```

Before presenting some plots, we find a nonadaptive solution to the equations of motions. Because we do not want local refinements here, we use the `"FixedStep"` method option setting in NDSolve. And we want the function values directly, not an interpolating function. We use the command `InterpolatingFunctionValuesOnGrid` from the package `DifferentialEquations`InterpolatingFunctionAnatomy`` to extract these values. So, we can define the function LBRK that solves the equation of motion for the ruler.

```
In[6]:= Needs["DifferentialEquations`InterpolatingFunctionAnatomy`"]
```

```
In[7]:= LBRK[10_, v_, μK_, x0_, x0p_, tr_List] :=
        Transpose[(* extract position and velocity values at fixed step times *)
                    InterpolatingFunctionValuesOnGrid /@
        {x, xv} /. NDSolve[{x'[t] == xv[t],
                      xv'[t] == μK(-Sign[xv[t] - v] (10 - v t - x[t])/(2 (10 - v t)) -
                                  Sign[xv[t] + v] (10 - v t + x[t])/(2 (10 - v t))),
                      x[0] == x0, xv[0] == x0p}, {x, xv}, Join[{t}, Most[tr]],
                      (* proper option setting *)
                      StartingStepSize -> Last[tr], Method -> {"FixedStep",
                      Method -> {"ExplicitRungeKutta", DifferenceOrder -> 4}}][[1]]]
```

Here is an example of the output of LBRK.

```
In[8]:= LBRK[20, 0.2, 0.35, -7, 0, {0, 99.9, 0.1}] // Short[#, 6]&
```

```
Out[8]//Short= {{-7., 0.}, {-6.99939, 0.0122558}, {-6.99755, 0.0245217}, {-6.99448, 0.0367956},
                {-6.99019, 0.0490753}, <<990>>, {-0.0680415, 0.193402}, {-0.0487588, 0.187236},
                {-0.0290276, 0.198706}, {-0.00945945, 0.187995}, {0.00947476, 0.186623}}
```

We package the visualization again as a function, this time called RKMapRulerMotion.

```
In[9]:= RKMapRulerMotion[10_, v_, μK_, x0_, x0p_, tr_] :=
        Module[{sol, tList, vList, xList},
          sol = LBRK[10, v, μK, x0, x0p, tr];
          (* use returned list of values *)
          tList = Range @@ tr;
          {xList, vList} = Thread[{tList,
                    #[Transpose[Take[sol, Length[tList]]]]}]& /@ {First, Last};
          (* form graphics *)
          Show[GraphicsArray[{
          ListPlot[xList, PlotRange -> All, AxesLabel -> {"t", "x(t)"},
          PlotLabel -> StyleForm["Position-time curve", FontSize -> 7],
          (* the fingers *)
          Prolog -> {Thickness[0.01], RGBColor[1, 0, 0],
                    {Line[{{0, +10}, {10/v, 0}}], Line[{{0, -10}, {10/v, 0}}]}},
          DisplayFunction -> Identity,
          PlotStyle -> {GrayLevel[0], PointSize[0.003]}],
          ListPlot[vList, PlotRange -> All, AxesLabel -> {"t", "v(t)"},
          PlotLabel -> StyleForm["Velocity-time curve", FontSize -> 7],
          (* the fingers *)
          Prolog -> {Thickness[0.01], RGBColor[1, 0, 0],
```

```
              {Line[{{0, +v}, {10/v, +v}}], Line[{{0, -v}, {10/v, -v}}]}},
     DisplayFunction -> Identity,
     PlotStyle -> {GrayLevel[0], PointSize[0.003]}]}]]]]
```

We now look at a few examples for various initial velocities and positions of the fingers. To see several changes of direction of the ruler, we choose the following parameters. Of course, others are also possible (an interpretation of the length in centimeters and the time in seconds is realistic). Here, we have only the case of kinetic friction. (The result does not depend on the step size used.)

In[10]:= **RKMapRulerMotion[20, 0.2, 0.35, -7, 0, {0, 99.9, 0.1}]**

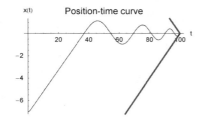
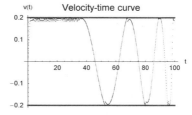

Here is the solution in case we take into account the influence of static friction. The two solutions are qualitatively the same. The solution that takes contact friction into account is quantitatively different from that where the force is

$$F_{tot} = m\,\mu_k(-\mathrm{sign}(x'(t) - v)\,F_{nl} - \mathrm{sign}(x'(t) - v)\,F_{nr})$$

in which μ_s is not involved.

In[11]:= **MapRulerMotion[20, 0.2, 0.5, 0.34, -7, 0]**

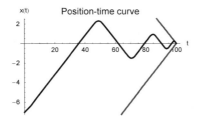
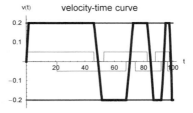

Next, we study the motion when we start with finite velocities of the ruler (once in the same direction, and once in the opposite direction of the instantaneous force on the ruler).

In[12]:= **RKMapRulerMotion[20, 0.2, 0.35, -7, 0.2, {0, 99.9, 0.1}]**

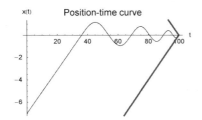
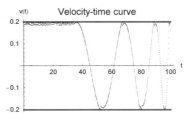

In[13]:= **MapRulerMotion[20, 0.2, 0.5, 0.35, -7, 0.2]**

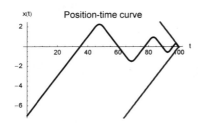 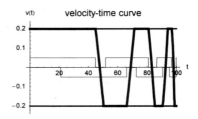

In the last example, we start with a large initial velocity.

```
In[14]:= RKMapRulerMotion[20, 0.2, 0.35, 7, -2.5, {0, 99.9, 0.1}]
```

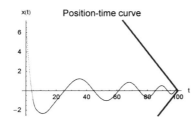 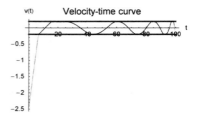

```
In[15]:= MapRulerMotion[20, 0.2, 0.5, 0.35, 7, -2.5]
```

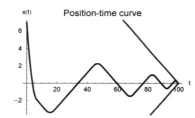 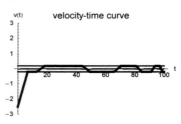

For the stick–slip motion of two coupled masses, see [1901], [1125]. For the case of the ruler slipping off the fingers, see [3].

b) Here is a visualization of the potential. It has smooth as well as wildly fluctuating regions.

```
In[1]:= V[x_, y_] := Sin[Tan[x + y] + Cot[y - x^2]]
```

```
In[2]:= cp = ContourPlot[Evaluate[V[x, y]], {x, -Pi, Pi}, {y, -Pi, Pi},
            PlotPoints -> 400, ContourLines -> False,
            ColorFunction -> GrayLevel, PlotRange -> All]
```

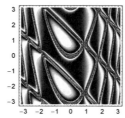

These are the equations of motions.

```
In[3]:= eqs = Thread[{x''[t], y''[t]} == ({-D[V[x, y], x], -D[V[x, y], y]} /.
                                          {x -> x[t], y -> y[t]})];
```

```
In[4]:= Off[NDSolve::mxst];
        paths = Table[
```

```
(* numerically solve equations of motion;
   start from randomly selected points *)
nsol = NDSolve[Join[eqs, {x[0] == Random[Real, {-Pi, Pi}],
                          y[0] == Random[Real, {-Pi, Pi}],
                          x'[0] == 0, y'[0] == 0}],
               {x, y}, {t, 0, 50}, MaxSteps -> 10^4];
(* visualize trajectories *)
ParametricPlot[Evaluate[{x[t], y[t]} /. nsol],
               {t, 0, nsol[[1, 1, 2, 1, 1, 2]]},
               DisplayFunction -> Identity, PlotPoints -> 500,
               PlotStyle -> {{Thickness[0.002], Hue[Random[]]}}],
               {k, 100}];
```

We display the trajectories with the contour plot of the potential.

```
In[6]:= Show[{cp, paths}, DisplayFunction -> $DisplayFunction,
        FrameTicks -> None, FrameStyle -> {Thickness[0.01]},
        PlotRange -> Pi {{-1, 1}, {-1, 1}}, Frame -> True,
        FrameStyle -> {Thickness[0.01]}]
```

For the construction of 2D vector fields with a given topology, see [1759].

c) We start with the generation of a "random" potential. The function `randomPotential` returns a pure function representing a 2D "random" potential with structures of approximate size π / *potentialHills*.

```
In[1]:= randomPotential[potentialHills_] := Function[Evaluate[
        Module[{x, y},
        (* normalize to V[0, 0] == 0 *) (# - (# /. {x -> 0, y -> 0}))&[
          Sum[Random[Real, {-1, 1}]* (* random products *)
            Cos[Random[Real, potentialHills {-1, 1}] x + 2Pi Random[]]*
            Cos[Random[Real, potentialHills {-1, 1}] y + 2Pi Random[]],
            {potentialHills}]] /. {x -> #1, y -> #2}]]];
```

Given a potential, it is straightforward to solve Newton's equations for the superposition of the "random" potential and a monotonic one. The function `downhillTrajectories` solves the equations of motions, colors them according to their local velocity, and displays the trajectories together with the full potential. We use a radial monotonic potential, imitating a single hill with potential proportional $(r^2 + 1)^{1/2}$. All particles start at the hill, with initial momentum *p0* in a dense set of all possible directions.

```
In[2]:= downhillTrajectories[V_, p0_, pathNumber_, plotPoints_,
                             α_, hillSize_, opts___] :=
        Module[{l = hillSize/2, pp = plotPoints, Vx, Vy, Vxc, Vyc,
                trajectories, nsol, speed, T, points, speedValues,
                coloredPath, min, max},
        (* avoid evaluation for symbolic x, y *)
        {Vx[x_?NumberQ, y_?NumberQ], Vy[x_?NumberQ, y_?NumberQ]} :=
                                     {Vxc[x, y], Vyc[x, y]};
        (* compiled gradient components *)
        {Vxc, Vyc} = Compile @@@ ({{x, y}, D[V[x, y], #]}& /@ {x, y});
        (* the particle paths *)
        trajectories = Table[
        (* solve equations of motion *)
        NDSolve[{x'[t] == px[t], y'[t] == py[t],
                px'[t] == -Vx[x[t], y[t]] + α x[t]/Sqrt[x[t]^2 + y[t]^2 + 1],
                py'[t] == -Vy[x[t], y[t]] + α y[t]/Sqrt[x[t]^2 + y[t]^2 + 1],
                x[0] == 0, y[0] == 0,
```

```
              (* random initial direction *)
              px[0] == p0 Cos[2Pi j/pathNumber],
              py[0] == p0 Sin[2Pi j/pathNumber]},
              {x, y, px, py}, {t, 0, 1},
              MaxSteps -> 5000, PrecisionGoal -> 4] // (nsol = #)&;
       (* speed along path *)
       speed[{ξ_, ψ_}] = {px[ξ, ψ], py[ξ, ψ]} /. nsol[[1]];
       (* maximal time *)
       T = nsol[[1, 1, 2, 1, 1, 2]];
       (* points of the path *)
       points = Table[Evaluate[{x[t], y[t]} /. nsol[[1]]], {t, 0, T, T/pp}];
       (* velocities of the path *)
       speedValues = Table[Evaluate[Sqrt[px[t]^2 + py[t]^2] /. nsol[[1]]],
                             {t, T/pp/2, T, T/pp}];
       (* color path according to local velocity *)
       coloredPath = Transpose[{GrayLevel /@ speedValues,
                                 Line /@ Partition[points, 2, 1]}],
                     {j, pathNumber}];
       (* extremal velocities *)
       {min, max} = {Min[#], Max[#]}&[First /@
                     Cases[trajectories, _GrayLevel, Infinity]];
       Show[{
        (* density plot of hill potential *)
        DensityPlot[V[x, y] - α Sqrt[x^2 + y^2 + 1],
                    {x, -1, 1}, {y, -1, 1}, PlotPoints -> 400,
                    Mesh -> False, ColorFunction -> (Hue[0.8 #]&),
                    DisplayFunction -> Identity, Frame -> True],
        (* the particle paths *)
        Graphics[{Thickness[0.002], trajectories /.
                  GrayLevel[h_] :> GrayLevel[0.8 (h - min)/(max - min)]},
            Axes -> False, Frame -> True, AspectRatio -> Automatic]},
          opts, DisplayFunction -> $DisplayFunction, FrameTicks -> False,
          PlotRange -> {{-1, 1}, {-1, 1}}]]]
```

Here are two examples of 2000 trajectories rolling down a hill, with an additional random potential.

```
In[3]:= Show[GraphicsArray[
       Block[{$DisplayFunction = Identity},
       (* two selected examples *)
       {SeedRandom[28];
        downhillTrajectories[randomPotential[36], 3, 2000, 30, 4.602, 5];
        SeedRandom[68];
        downhillTrajectories[randomPotential[36], 1, 2000, 30, 3.206, 5]}]]]
Out[3]= $Aborted
```

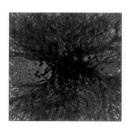 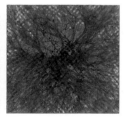

The last two graphics clearly show the branched flow known from electron flows [1773], [1772], [950], [812], [1130].

If we display very many trajectories, the last-type display is not optimal; due to the many particles trajectories the graphic gets very "crowded". In the following, we consider particles rolling down a straight slope. We discretize the configuration space and count how frequently particles pass single cells of the discretized space. This time, we let the particles start with random initial momentum and random initial positions from a small line segment.

```
In[4]:= (* add discrete points along a straight line connecting p1 and p2 *)
       addInBetweenPoints[{p1_, p2_}] :=
```

```
        If[Max[Abs[Subtract @@ {p1, p2}]] <= 1, {p1, p2},
           Table[Round[p1 + k/# (p2 - p1)], {k, 0, #}]]&[
                                Max[Abs[Subtract @@ {p1, p2}]]]
In[6]:= downhillTrajectoriesDensity[V_, pathNumber_, matrixSize_,
                              α_, slopeSize_, opts___] :=
        Module[{l = slopeSize, d = matrixSize, pp = 4 matrixSize,
                Vx, Vy, Vxc, Vyc, passageCountMatrix, nsol, T, pos,
                discretizedPoints, i, j},
          (* avoid evaluation for symbolic x, y *)
          {Vx[x_?NumberQ, y_?NumberQ],  Vy[x_?NumberQ, y_?NumberQ]} :=
                                      {Vxc[x, y], Vyc[x, y]};
          (* compiled gradient components *)
          {Vxc, Vyc} = Compile @@@ ({{x, y}, D[V[x, y], #]}& /@ {x, y});
          (* matrix for discretized paths *)
          passageCountMatrix = Table[0, {matrixSize}, {matrixSize}];
          Do[(* solve equations of motion *)
          nsol = NDSolve[{x'[t] == px[t], y'[t] == py[t],
                   px'[t] == -Vx[x[t], y[t]] + α, py'[t] == -Vy[x[t], y[t]],
                   x[0] == Random[Real, 1/10 {-1, 1}], y[0] == 0,
                       (* start into left-half plane;
                          random initial momentum and position *)
                   Sequence @@ Thread[{px[0], py[0]} ==
                                       Random[Real, {1, 4}]{Cos[#], Sin[#]}&[
                       Random[Real, Pi/2{-1, 1}]]]}, {x, y, px, py}, {t, 0, 3/2},
                   MaxSteps -> 5000, PrecisionGoal -> 4, AccuracyGoal -> 4];
          (* maximal time *)
          T = nsol[[1, 1, 2, 1, 1, 2]];
          (* points of the path *)
          points = Table[Evaluate[{x[t], y[t]} /. nsol[[1]]], {t, 0, T, T/pp}];
          (* keep only path points inside specified region *)
          pos = 1; While[pos <= pp && points[[pos, 1]] < l &&
                                      Abs[points[[pos, 2]]] < 1/2, pos++];
          (* discretized path points *)
          discretizedPoints = Select[({0, d/2} + #)& /@ Round[d/l Take[points, pos]],
                          (Min[#] > 0 && Max[#] < d)&];
          (* add intermediate points if needed *)
          discretizedPoints = Union[Flatten[addInBetweenPoints /@
                                    Partition[discretizedPoints, 2, 1], 1]];
          (* add visited path points to path point matrix *)
          Do[{i, j} = discretizedPoints[[k]],
             passageCountMatrix[[i, j]] = passageCountMatrix[[i, j]] + 1,
             {k, Length[discretizedPoints]}], {pathNumber}];
          (* display density of paths *)
          ListDensityPlot[Transpose[ArcTan[N[passageCountMatrix]]],
                          opts, FrameTicks -> False, PlotRange -> All,
                          ColorFunction -> (Hue[0.8 #]&), Mesh -> False]]
```

To avoid a too small color gradient due to the large number of visited cells near the origin, we scale the number of times a cell was visited in a nonlinear way for the final coloring.

```
In[7]:= SeedRandom[111];
        downhillTrajectoriesDensity[randomPotential[36], 2000, 500, 3, 3]
```

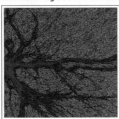

We end with some plots with the coloring determined by the time *t*.

```
In[9]:= Module[{o = 12, L = 2, T = 1, ppxy = 25, ppt = 100, α = 3},
    V = randomPotential[o];
    {Vx[x_?NumberQ, y_?NumberQ], Vy[x_?NumberQ, y_?NumberQ]} :=
                                        {Vxc[x, y], Vyc[x, y]};
    (* compiled gradient components *)
    {Vxc, Vyc} = Compile @@@ ({{x, y}, D[V[x, y], #]}& /@ {x, y});
    (* the particle paths *)
    positions = Transpose[
    Table[(* solve equations of motion *)
    (* solve equations of motions *)
    ndsol = NDSolve[{x'[t] == px[t], y'[t] == py[t],
            px'[t] == -Vx[x[t], y[t]] + α x[t]/Sqrt[x[t]^2 + y[t]^2 + 1],
            py'[t] == -Vy[x[t], y[t]] + α y[t]/Sqrt[x[t]^2 + y[t]^2 + 1],
            x[0] == ξ, y[0] == η, px[0] == 0, py[0] == 0},
            {x, y, px, py}, {t, 0, T},
            MaxSteps -> 10000, PrecisionGoal -> 4];
    (* points on the trajectories for fixed times *)
    Table[Evaluate[{x[t], y[t]} /. ndsol[[1]]], {t, 0, T, T/ppt}],
        {ξ, -L, L, 2L/ppxy}, {η, -L, L, 2L/ppxy}, {3, 2, 1}];
    (* show 2D and 3D version of the trajectories *)
    Show[GraphicsArray[
    {(* 2D plot *)
    Graphics[{PointSize[0.003],
                MapIndexed[{Hue[0.8 (#2[[1]] - 1)/ppt], #1}&,
                        Map[Point, positions, {-2}]]},
        Frame -> True, AspectRatio -> Automatic, FrameTicks -> None,
        PlotRange -> 3/2 L {{-1, 1}, {-1, 1}}],
        (* 3D plot *)
    Graphics3D[{PointSize[0.003],
                MapIndexed[{Hue[0.8 (#2[[1]] - 1)/ppt], #1}&,
                    MapIndexed[Point[Append[#, (#2[[1]] - 1)/ppt]]&,
                                    positions, {-2}]]},
        PlotRange -> {3/2 L {-1, 1}, 3/2 L {-1, 1}, {0, T}},
        BoxRatios -> {1, 1, 1}]}]]]
```

For a quantum mechanical treatment of branched flows, see [1830].

12. Maxwell's Line, Quartic Determinant

a) The equation of state of a mole of van der Waal's gas reads: $(P - a/V^2)(V - b) = RT$, where P, V, and T are the pressure, volume, and absolute temperature, and a and b are constants depending on the gas. To avoid unnecessary constants, we normalize the pressure, the volume, and the temperature to the corresponding values at the critical point where $\partial P(V)/\partial V = \partial^2 P(V)/\partial V^2 = 0$. It follows that the critical temperature, volume, and pressure are given by the following equations.

```
In[1]:= (* equations for critical point *)
    Solve[{D[R T/(V - b) - a/V^2, V] == 0,
            D[R T/(V - b) - a/V^2, {V, 2}] == 0}, {V, T}]
```

$$Out[2]= \left\{\left\{T \to \frac{8\,a}{27\,b\,R}, V \to 3\,b\right\}\right\}$$

```
In[3]:= R T/(V - b) - a/V^2 /. %
```

$$Out[3]= \left\{\frac{a}{27\,b^2}\right\}$$

Thus, $T_k = 8a/(27 bR)$, $V_k = 3b$, $p_k = a/(27 b^2)$, $v = V/V_k$, $p = P/P_k$, $t = T/T_k$. Hence, we obtain the so-called reduced van der Waal's equation $(p + 3/v^2)(3v - 1) = 8t$ with $v > 1/3$.

```
In[4]:= Factor /@ Simplify[27 b/a #& /@ (
        ((p a/(27 b^2)) + a/(v 3b)^2)((v 3b) - b) == R t 8a/(27b R))]
```

$$\text{Out[4]=} \quad \frac{(-1 + 3\,v)\,(3 + p\,v^2)}{v^2} == 8\,t$$

Without Maxwell's line, the isothermals in the p,v-diagram have the following appearance (note the nonphysical domains with $p < 0$ or $\partial p(v)/\partial v > 0$).

```
In[5]:= Plot[Evaluate[Table[8t/(3v - 1) - 3/v^2, {t, 100/200, 250/200, 6/200}]],
        {v, 0.35, 5}, PlotRange -> {{0, 5}, {-4, 5}},
        PlotStyle -> {{GrayLevel[0], Thickness[0.002]}},
        AxesOrigin -> {0, 0}, AxesLabel -> {"v", "p"}]
```

Next, we construct Maxwell's line. The real equilibrium state in the region of s-shaped curves in the p,v-diagram is described by a straight line that is distinguished by the fact that the surfaces above and below this line (and below or above $p(v)$, respectively), have the same size. Let v_1 and v_3 ($t < 1$) be the volumes in which only liquid or only gas is present, respectively. Then, Maxwell's line is determined by the solution of the following system of equations:

$$\left(p + \frac{3}{v_1^2}\right)(3v_1 - 1) = t$$

$$\left(p + \frac{3}{v_3^2}\right)(3v_3 - 1) = t$$

$$\int_{v_1}^{v_3}\left(\frac{8t}{3v - 1} - \frac{3}{v^2} - p\right)dv = -p(v_3 - v_1) + 3\left(\frac{1}{v_3} - \frac{1}{v_1}\right) + \frac{8t}{3}\ln\left(\frac{3v_3 - 1}{3v_1 - 1}\right) = 0.$$

A problem exists in trying to find the solution of this nonlinear system of equations in three variables using `FindRoot`. Even for starting values that lie very near the solution, the search quickly goes outside the search region $v_1 < 1$, $v_3 > 1$, $p < 1$. Here is the solution for $t = 199/200$.

```
In[6]:= vdWEqs[v1_, v3_, p_, t_] :=
        {(p + 3/v1^2) (3v1 - 1) == 8t, (p + 3/v3^2) (3v3 - 1) == 8t,
         -p(v3 - v1) + 3(1/v3 - 1/v1) + 8/3t Log[(3v3 - 1)/(3v1 - 1)] == 0}
In[7]:= FindRoot[Evaluate[vdWEqs[v1, v3, p, 199/200]],
                 {v1, 0.4, 1/3, 1}, {v3, 3, 1, 10^10}, {p, 0.5, 0, 1}]
Out[7]= {v1 -> 0.874706, v3 -> 1.16176, p -> 0.98012}
```

Using this result for a starting value with $t = 198/200$, we get no convergence in the following example.

```
In[8]:= FindRoot[Evaluate[vdWEqs[v1, v3, p, 198/200]],
                 {v1, 0.8747, 1/3, 1}, {v3, 1.1617, 1, 10^10},
                 {p, 0.9801, 0, 1}]
Out[8]= {v1 -> 0.830914, v3 -> 1.24295, p -> 0.960479}
```

This means that we have to put some additional effort into finding a solution of this system. (But because no parameters are involved in the scaled equation this is not too difficult.) For graphical purposes, around 150 distinct temperature values suffice. We compute the starting value of the temperature by extrapolation of the values at $t + 1/200$ and $t + 2/200$. We then get good starting values, and require relatively few iterations inside `FindRoot` (time savings). With more symbolic and numerical work, we could also find the starting values for arbitrary t. For example, for larger values of t, we could use the following code.

```
startValues[t_?(0.85 <= # < 1&)] :=
Module[{nt = N[t], v1mi, v3ma, v1s, v3s, p1s, p3s, pm},
 {v1mi, v3ma} = Drop[v /. NSolve[1 - 6v + 9v^2 - 4nt v^3 == 0, v], {1}];
 {p1mi, p3ma} = 8nt/(3v - 1) - 3/v^2 /. {{v -> v1mi}, {v -> v3ma}};
 pm = (p3ma + p1mi)/2;
 {v1s, v3s} = Drop[v /. NSolve[3 - 9v + pm v^2 + 8nt v^2 - 3pm v^3 == 0, v], {2}];
 {v1s, v3s, pm}];
```

Another possibility to avoid convergence problems would be to derive a system of coupled nonlinear ordinary differential equations for $v_1(t)$, $v_3(t)$, and $p(t)$ and to solve this system numerically. We do not bother with this here. Here is an implementation of the extrapolation.

```
In[9]:= dat[1] = {1, 1, 1};

         (* two nontrivial initial values *)
         (dat[#] =  Last /@ FindRoot[Evaluate[vdWEqs[v1, v3, p, #]],
                        {v1, 0.7, 1/3, 1}, {v3, 1.2, 1, 10^10}, {p, 0.9, 0, 1}])& /@
                        {199/200, 198/200};

         (* use already-computed data as initial conditions *)
         dat[t_] := dat[t] = Last /@ Module[{v1s, v3s, ps},
         {v1s, v3s, ps} = 2dat[t + 1/200] - dat[t + 2/200];
         FindRoot[Evaluate[vdWEqs[v1, v3, p, t]],
                        {v1, v1s, 1/3, 1}, {v3, v3s, 1, 10^10}, {p, ps, 0, 1}]]
```

We compute all data for $1/4 < t < 99/100$.

```
In[14]:= Do[dat[t], {t, 198/200, 50/200, -1/200}]
```

Now, we can define the isothermals $p_t(v)$.

```
In[15]:= pt[t_][v_] :=
         Which[t < 1, Which[v < dat[t][[1]], 8t/(3v - 1) - 3/v^2,
                            (* Maxwell's line *)
                            dat[t][[1]] <= v <= dat[t][[2]], dat[t][[3]],
                            v > dat[t][[2]], 8t/(3v - 1) - 3/v^2],
               t >= 1, 8t/(3v - 1) - 3/v^2]
```

Next, we plot them.

```
In[16]:= Plot[Evaluate[Table[pt[t][v], {t, 1/2, 5/4, 1/50}]],
             {v, 0.35, 5}, Compiled -> False, AxesLabel -> {"v", "p"},
             PlotRange -> {{0, 5}, {0, 3}}, AxesOrigin -> {0, 0},
             PlotStyle -> {{GrayLevel[0], Thickness[0.002]}},
             Prolog -> {Hue[0], Thickness[0.01],
             (* the straight line parts *)
             Line[Join[Table[{dat[t][[1]], dat[t][[3]]},
                        {t, 100/200, 200/200, 1/200}],
                    Table[{dat[t][[2]], dat[t][[3]]},
                        {t, 199/200, 1/2, -1/200}]]]}]
```

Now, we could plot the state surface of the van der Waal's gas.

```
Show[{ListPlot3D[Table[pt[t][v], {t, 1/2, 5/4, 1/50}, {v, 0.35, 5, 0.1}],
                AxesLabel -> {"v", "t", "p"}, MeshStyle -> {Thickness[0.002]},
                ClipFill -> None, DisplayFunction -> Identity,
                MeshRange -> {{0.35, 5}, {1/2, 5/4}}],
      Graphics3D[{Thickness[0.01],  (* envelope for the endpoints of Maxwell's line *)
              Line[Join[Table[{1.2 dat[t][[1]], t, 1.1 dat[t][[3]]},
                                {t, 1/2, 199/200, 1/200}],
                  Reverse @ Table[{dat[t][[2]], t, 1.05 dat[t][[3]]},
                                {t, 1/2, 199/200, 1/200}]]]]}],
      PlotRange -> {{0.4, 5}, {1/2, 5/4}, {0, 3}},
      DisplayFunction -> $DisplayFunction]
```

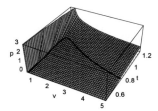

Alternatively, we could show a $p_v(t)$ diagram.

```
temperature = Interpolation[
Join[Table[{dat[t][[1]], t}, {t, 1/4, 1, 1/200}],
     Table[{dat[t][[2]], t}, {t, 1/4, 199/200, 1/200}]]];

pressure = Interpolation[Table[{t, dat[t][[3]]}, {t, 1/4, 199/200, 1/200}]];

pv[v_][t_] := Which[t >= 1, 8t/(3v - 1) - 3/v^2,
                    t > temperature[v], 8t/(3v - 1) - 3/v^2,
                    True, pressure[t]];

Plot[Evaluate[Table[pv[v][t], {v, 0.47, 4, 0.1}]], {t, 0.7, 3/2},
     PlotStyle -> {{GrayLevel[0], Thickness[0.002]}},
     AxesOrigin -> {0, 0}, PlotRange -> {{0.5, 1.5}, {0, 3}},
     AxesLabel -> {"t", "p"}, Compiled -> False]
```

For another *Mathematica* implementation of constructing the Maxwell line, see [5]; for the Bose condensation, see [1821].

b) This is the matrix under consideration.

```
In[1]:= matrix = Table[Which[i === j, 0,
                        j > i, a[i, j] x + b[i, j] y + c[i, j] z,
                        j < i, a[j, i] x + b[j, i] y + c[j, i] z],
                {i, 4}, {j, 4}];
```

This is its determinant.

```
In[2]:= det = Expand[Det[matrix]];
```

To determine the coefficients, we compare coefficients of $x^i y^j z^k$.

```
In[3]:= eqs = Union[DeleteCases[Flatten[Table[
          Factor[Plus @@ Cases[Expand[det - 2 I (x^4 + y^4 + z^4)],
              _?(FreeQ[#, x | y | z]&) x^i y^j z^k]]/(x^i y^j z^k),
              {i, 0, 4}, {j, 0, 4}, {k, 0, 4}]], 0]];
```

We got 15 equations, but have 18 variables to determine.

```
In[4]:= vars = Flatten[Cases[eqs, _a | _b | _c, Infinity]] // Union;
```

```
In[5]:= {Length[eqs], Length[vars]}
```

```
Out[5]= {15, 18}
```

To apply a numerical root finding technique (FindRoot), we substitute three values for three randomly selected variables and then numerically solve the resulting system for the 15 equations. The substitutions will not always be the right ones; so, we carry out this process until we find a set of substituted values that allows us to solve the system. We end the search when a call to FindRoot did not generate messages.

```
In[6]:= (* select randomly three out of the 18 variables *)
        takeThree := Flatten /@ Nest[Function[x,
          {{x[[1]], #}, DeleteCases[x[[2]], #]}&[x[[2]][[Random[Integer,
                {1, Length[x[[2]]]}]]]]], {{}, vars}, 3]

        (* a random value for the selected variables *)
        r := Random[Integer, {-1, 1}];
```

```
In[11]:= SeedRandom[222];
         Block[{(* avoid messages from FindRoot *) $Messages = {}},
         While[Check[(* select three variables *)
                 theThree = takeThree;
                 rep = ((# -> r)& /@ theThree[[1]]);
                 (* try to solve the resulting system *)
                 fr = FindRoot[Evaluate[eqs /. rep],
                         (* the remaining 15 variables *)
                         Evaluate[Sequence @@
                         ({#, Random[]}& /@ theThree[[2]])],
                         MaxIterations -> 50, Compiled -> False,
                         Method -> Newton, WorkingPrecision -> 30],
                         True], Null]]
```

The solution fulfills the system of equations to a high accuracy.

```
In[13]:= Chop[eqs /. rep /. fr]
```

```
Out[13]= {0, 0, 0, 0, 0, 0, 0, 0, 0, 0, 0, 0, 0, 0, 0}
```

Looking at the solution found, one naturally conjectures the real and imaginary parts of the exact coefficients to be built from $\{0, \pm 1/2, \pm 1, \pm\sqrt{2}, \pm\sqrt{2}/2\}$.

```
In[14]:= fr /. x_?InexactNumberQ :> N[x, 10] // Chop[#, 10^-20]&
```

```
Out[14]= {a[1, 2] → 1.0000000000 + 0. × 10^-11 i, a[2, 3] → 0,
          a[2, 4] → 0.5000000000 - 0.2071067812 i, a[3, 4] → -0.5000000000 - 1.2071067812 i,
          b[1, 3] → 1.0000000000 + 0. × 10^-11 i, b[1, 4] → -0.7071067812 - 0.7071067812 i,
          b[2, 3] → 1.0000000000 - 0.4142135624 i, b[2, 4] → 0.2071067812 + 0.5000000000 i,
          b[3, 4] → 1.2071067812 - 0.5000000000 i, c[1, 2] → -1.0000000000 + 0. × 10^-11 i,
          c[1, 3] → 0. × 10^-11 + 1.0000000000 i, c[1, 4] → 0, c[2, 3] → -1.0000000000 - 0.4142135624 i,
          c[2, 4] → 0.2071067812 + 0.5000000000 i, c[3, 4] → -0.5000000000 - 1.2071067812 i}
```

The function round rounds the number x to the nearest of the conjectured exact ones.

```
In[15]:= round[x_] := Module[{goodNumbers = {0, 1/2, 1, Sqrt[2], Sqrt[2]/2}},
                         Sign[x] goodNumbers[[Position[#, Min[#]]&[Abs[
                         Abs[x] - goodNumbers]][[1, 1]]]]]
```

Expressing the numerical result in the conjectured exact numbers and backsubstituting the values in the original determinant shows that the conjectured numbers are indeed a solution for the determinantal problem.

```
In[16]:= res = Join[rep,
        MapAt[(round[Re[#]] + I round[Im[#]])&, #, 2]& /@ fr]
```
$$Out[16]= \left\{a[1, 4] \to 1, a[1, 3] \to 1, b[1, 2] \to -1, a[1, 2] \to 1, a[2, 3] \to 0,\right.$$
$$a[2, 4] \to \frac{1}{2}, a[3, 4] \to -\frac{1}{2} - i, b[1, 3] \to 1, b[1, 4] \to -\frac{1+i}{\sqrt{2}},$$
$$b[2, 3] \to 1 - \frac{i}{2}, b[2, 4] \to \frac{i}{2}, b[3, 4] \to 1 - \frac{i}{2}, c[1, 2] \to -1, c[1, 3] \to i,$$
$$\left. c[1, 4] \to 0, c[2, 3] \to -1 - \frac{i}{2}, c[2, 4] \to \frac{i}{2}, c[3, 4] \to -\frac{1}{2} - i \right\}$$

Here is (one possible form of) the resulting matrix.

```
In[17]:= matrix /. res /. {(1 + I)/Sqrt[2] -> (-1)^(1/4),
                           (1 - I)/Sqrt[2] -> -(-1)^(3/4)} // Simplify
```
$$Out[17]= \left\{\left\{0, x - y - z, x + y + i z, x - \frac{(1+i) y}{\sqrt{2}}\right\}, \left\{x - y - z, 0, \left(1 - \frac{i}{2}\right) y - \left(1 + \frac{i}{2}\right) z, \frac{1}{2} (x + i (y + z))\right\},\right.$$
$$\left\{x + y + i z, \left(1 - \frac{i}{2}\right) y - \left(1 + \frac{i}{2}\right) z, 0, \left(-\frac{1}{2} - i\right) (x + i y + z)\right\},$$
$$\left. \left\{x - \frac{(1 + i) y}{\sqrt{2}}, \frac{1}{2} (x + i (y + z)), \left(-\frac{1}{2} - i\right) (x + i y + z), 0\right\}\right\}$$

Its determinant has the value that we want.

```
In[18]:= Det[%] // Expand // Simplify
```
$$Out[18]= \frac{1}{4} \left(8 i x^4 + \left((8 + 3 i) - (8 - 4 i) \sqrt{2}\right) y^4 + 2 i \left(-9 + (7 + i) \sqrt{2}\right) y^3 z + \right.$$
$$\left((-16 - 5 i) + (12 + 8 i) \sqrt{2}\right) y^2 z^2 + (2 - 2 i) \left(-4 + (2 + i) \sqrt{2}\right) y z^3 + 8 i z^4 - 4 x^3 (y + z) +$$
$$x^2 \left(((-5 + 4 i) + (2 - 6 i) \sqrt{2}) y^2 + ((6 - 8 i) - (6 - 2 i) \sqrt{2}) y z - (5 + 4 i) z^2\right) +$$
$$\left. (1 + i) x (y + z) \left(((-8 - 4 i) + (7 + 4 i) \sqrt{2}) y^2 + ((2 + 10 i) - (1 + 8 i) \sqrt{2}) y z + (2 - 2 i) z^2\right)\right)$$

13. Smoothing Functions, Secant Method Iterations, Unit Sphere Inside a Unit Cube

a) We begin with the definition of the function $g_\varepsilon(x)$. Because very small denominators quickly appear in the argument of the exponential function and create gigantic values for the exponential function, we must treat these cases specially. ($\exp(1 / 10^{-4}) \approx 8.81 \times 10^{4342}$).

We define g1 by the following two rules.

```
In[1]:= g1[ε_][x_?NumberQ] := 0 /; (Abs[x] >= ε || Abs[x^2 - ε^2] < 10^-4)

       g1[ε_][x_?NumberQ] := Exp[1/(x^2 - ε^2)] /; Abs[x] < ε
```

To avoid repeatedly computing the normalization factor $c_\varepsilon = \int_{-\varepsilon}^{\varepsilon} \exp(1/(x^2 - \varepsilon^2)) \, dx$, we compute it separately.

```
In[3]:= c[ε_] := c[ε] = NIntegrate[g1[ε][x], {x, -ε, ε}]
```

Then, $g_\varepsilon(x)$ is given by the following definitions.

```
In[4]:= g[ε_][x_?NumberQ] := 0 /; (Abs[x] >= ε || Abs[x^2 - ε^2] < 10^-4)

       g[ε_][x_?NumberQ] := 1/c[ε] Exp[1/(x^2 - ε^2)] /; Abs[x] < ε
```

All functions in the family of $g_\varepsilon(x)$ are infinitely differentiable for all x, but are not analytic in the sense of complex function theory. Here are plots for various ϵ.

```
In[6]:= (* make the single plots with Plot, and then color them;
        cannot use Plot for several functions because interval changes *)
       Show[Graphics[MapIndexed[{(* color *) Hue[#2[[1]]/12], #1}&,
           First /@ Table[Plot[g[ε][x], {x, -ε, ε},
           PlotRange -> All, DisplayFunction -> Identity],
           {ε, 0.4, 2, 0.2}]], PlotRange -> All,
           DisplayFunction -> $DisplayFunction, Axes -> True]]
```

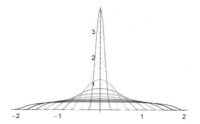

The function $f(x)$ is easy to define.

```
In[8]:= f[x_?(# <= 0 || # >= N[2Pi]&)] := 0

      f[x_?(# > 0 || # < N[2Pi]&)] := Sin[x]
```

A simplistic definition of $h_\varepsilon(x)$ leads to frequent error messages from NIntegrate, because the prescribed accuracy condition cannot be satisfied (the switch from $g_\varepsilon(y)\, f(x-y) \neq 0$ to $g_\varepsilon(y)\, f(x-y) = 0$ causes problems for NIntegrate).

```
In[10]:= h[ε_][x_] := 1/c[ε] NIntegrate[f[x - y] g[ε][y], {y, -ε, ε}]

In[11]:= h[1][0.1]
```

> NIntegrate::slwcon :
>   Numerical integration converging too slowly; suspect one of the
>   following: singularity, value of the integration being 0, oscillatory
>   integrand, or insufficient WorkingPrecision. If your integrand is
>   oscillatory try using the option Method->Oscillatory in NIntegrate. More…
>
> NIntegrate::ncvb : NIntegrate failed to converge to prescribed
>   accuracy after 7 recursive bisections in y near y = 0.1015625`. More…

```
Out[11]= 0.471846
```

These problems disappear if we restrict the domain of integration to those values for which the integrand is nonzero.

```
In[12]:= h[ε_][x_] := NIntegrate[f[x - y] g[ε][y],
                         {y, Max[-ε, x - N[2Pi]], Min[ε, x]}]

In[13]:= h[1][0.1]
Out[13]= 0.209497
```

Now, we look at this restriction on the domain for various x in $g_1(y)\, f(x-y)$. The green line is $g_\varepsilon(y)$; the red line is $f(x-y)$; the black line is the integrand. Note the different absolute lengths of the intervals.

```
In[14]:= picture[ε_, x_, opts___] :=
      Plot[{f[x - y], g[ε][y], f[x - y] g[ε][y]},
           {y, Max[-ε, x - N[2Pi]], Min[ε, x]},
           opts, AxesLabel -> {"y", None}, PlotRange -> All,
           (* color each curve differently *)
           PlotStyle -> {{Thickness[0.002], Hue[0]},
                         {Thickness[0.002], RGBColor[0, 1, 0]},
                         {Thickness[0.01], GrayLevel[0]}}];

In[15]:= Show[GraphicsArray[
           picture[1, #, DisplayFunction -> Identity]& /@ {1/2, 3, 6}]]
```

But with $h_\varepsilon(\pi)$, there is a further problem caused by the antisymmetry of the integrand, because the exact result is identically 0 and the default value of Infinity for AccuracyGoal in NIntegrate cannot be satisfied.

In[16]:= **h[1][Pi]**

> NIntegrate::ploss :
>   Numerical integration stopping due to loss of precision. Achieved
>   neither the requested PrecisionGoal nor AccuracyGoal; suspect one of
>   the following: highly oscillatory integrand or the true value of the
>   integral is 0. If your integrand is oscillatory on a (semi-)infinite
>   interval try using the option Method->Oscillatory in NIntegrate. More…

Out[16]= 1.26337×10^{-16}

Thus, we are led to define the following definition, which gives a sufficiently accurate result for a graphic.

In[17]:= **h[ε_][x_?(Abs[N[# - Pi]] < 10^-12&)] := 0**

Now, we get the exact value 0 without error messages.

In[18]:= **{h[ε][Pi], h[ε][N[Pi]]}**

Out[18]= **{0, 0}**

Now error messages are no longer generated in the computation of $h_\varepsilon(x)$ for the parameters given in the problem. We define the special case h[ε_][0.0].

Thus, we can create the desired plot with no further problems. The left graphic shows the convolution integral for various ε in the interval $(0, 2\pi)$ and the right graphic shows the convolution integral for $\varepsilon = 2$ including the tails.

In[19]:= **Show[GraphicsArray[**
     Block[{$DisplayFunction = Identity},
     {(* larger values of ε *)
     Plot[{h[4][x], h[2][x], h[1][x], h[1/2][x], Sin[x]}, {x, 0, 2Pi},
        PlotRange -> All, PlotStyle -> Table[Hue[i/10], {i, 7, 0, -7/4}]],
     (* smaller value of ε *)
     With[{ε = 2, ε = 10^-3}, Plot[h[ε][x], {x, 0 - ε + ε, 2Pi + ε - ε}]]}]]]

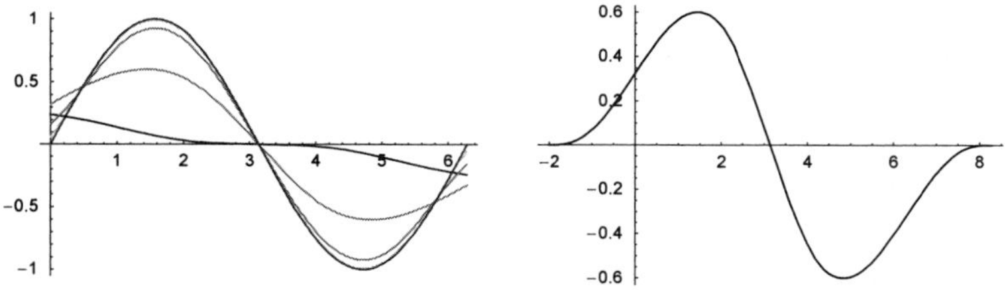

It is visible that $\lim_{\varepsilon \to 0} h_\varepsilon(x) = f(x)$.

Another frequently used smoothing function of the interval $[-1, 1]$ is

$$f(x) = \frac{1}{2}\left(1 + \text{erf}\left(\left(1 - 2\sqrt{x^2}\right)\left(1 - \left(1 - 2\sqrt{x^2}\right)^2\right)^{-1/2}\right)\right).$$

b) This is the graphic under consideration. We compile the body of FindRoot once to avoid repeated compilation in FindRoot. This does not change the graphic, but speeds up the calculation. In addition, we turn off various messages. The next graphic shows a density plot of the result of FindRoot as a function of the starting values for the iteration.

In[1]:= **(* the function *)**
     cf = Compile[x, Cos[x] - x];
     f[x_?NumberQ] := cf[x]

In[4]:= **(* carry out root-finding procedure using the secant method *)**
     f[x1_?NumberQ, x2_?NumberQ] :=
       x /. FindRoot[f[x], {x, x1, x2}, Method -> Secant,
                MaxIterations -> Round[100], Compiled -> False,
                AccuracyGoal -> {10, 10}]

     (* make density plot of the roots found;
        avoid equal starting values *)

```
dp = DensityPlot[f[x1, x2], {x1, -10, 10}, {x2, -10.001, 10.001},
                 Compiled -> False, PlotPoints -> 200,
                 Mesh -> False, ColorFunction -> (Hue[0.8 #]&),
                 DisplayFunction -> Identity]
```

Using the `All` option setting of `PlotRange` shows much less structure.

```
In[9]:= Show[GraphicsArray[
        Block[{$DisplayFunction = Identity},
              (* use Automatic and All plot range setting *)
              {dp, Show[%, PlotRange -> All]}]]]
```

Before interpreting the above graphic, let us look at some more graphics. Next, we show a magnified version of a subpart of the above graphic.

```
In[10]:= DensityPlot[x /. FindRoot[f[x], {x, x1, x2},
                                  MaxIterations -> 30, Method -> Secant],
                     {x1, 6, 8}, {x2, 2, 4},
                     Compiled -> False, PlotPoints -> 200,
                     Mesh -> False, ColorFunction -> (Hue[0.8 #]&)]
```

The number of iterations needed shows a similar pattern as the above graphic.

```
In[11]:= DensityPlot[(* number of function calls *) Length[Reap[
         FindRoot[Cos[x] - x, {x, x1, x2}, Method -> Secant,
                  MaxIterations -> 50, Compiled -> False,
                  (* monitor function calls *)
                  EvaluationMonitor :> Sow[x]]][[2, 1]]],
                  {x1, -10, 10}, {x2, -10.001, 10.001},
                  Compiled -> False, PlotPoints -> 200,
                  Mesh -> False, ColorFunction -> (Hue[3.8 #]&)]
```

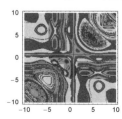

To understand the above graphic, we start by observing that (with the exception of the red line along the diagonal) all calculated function values are actual correct results from `FindRoot`. The different colors result from different distances of the FindRoot-results to the exact results. The AccuracyGoal option of FindRoot is fulfilled for all values (meaning the absolute value of the function values is approximately smaller than $10^{-(\$MachinePrecision-10)}$).

```
In[12]:= data = Cases[Flatten[dp[[1]]], _Real];
```

```
In[13]:= xExact = x /. FindRoot[Cos[x] - x, {x, 0, 1}, MaxIterations -> 30,
                WorkingPrecision -> 100, AccuracyGoal -> 80];
```

```
In[14]:= ListPlot[Sort[Log[10, (* remove zeros *) DeleteCases[#, _?(# == 0&)]&] @
                    Abs[Cases[data, _Real] - xExact]]] // N,
            PlotRange -> All, Frame -> True, Axes -> False]
```

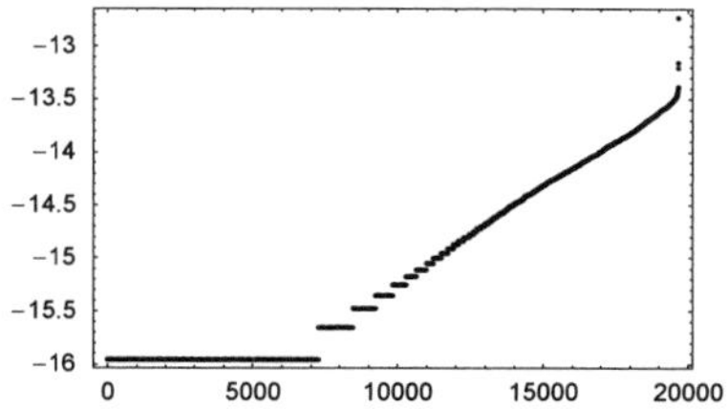

Next, we implement the secant method ourselves. The following example shows how the results are approached.

```
In[15]:= f2[x_] = Cos[x] - x;
        x[1] = -2.; x[2] = 1.;
        x[n_] := x[n] = x[n - 1] - f2[x[n - 1]](x[n - 1] - x[n - 2])/
                                (f2[x[n - 1]] - f2[x[n - 2]])
        Table[x[k], {k, 7}]
Out[18]= {-2., 1., 0.325149, 0.713324, 0.742034, 0.739068, 0.739085}
```

FindRoot calculates a similar sequence of points (it uses some selection strategy for the points to be used). This means that the pattern in the graphic results from the secant method itself and not from FindRoot's internals.

```
In[19]:= FindRoot[x - Cos[x], {x, -2, 1}, Method -> {"Secant", "StepControl" -> None},
                EvaluationMonitor :> Sow[x]] // Reap
Out[19]= {{x -> 0.739085}, {{-2., 1., 0.325149, 0.713324, 0.737776, 0.739093, 0.739085, 0.739085}}}
```

The next graphic uses high-precision arithmetic to calculate the roots. Again, we get the characteristic pattern from above. This means that the pattern in the graphic is unrelated to machine precision issues. The diagonal line stems from the failure of the secant method for identical starting points.

```
In[20]:= coloredSquare[{x1_, x2_}, δ_, frOptions___] :=
        With[{res = Check[fr = FindRoot[Cos[x] - x, {x, x1, x2}, frOptions],
                    $Failed]},
                (* analyze possible results from FindRoot *)
                {Which[Head[fr] === FindRoot, GrayLevel[1],
                        Head[res] === List, Hue[Log[10, Abs[(xExact - (x /. fr))]]/8],
                        True, GrayLevel[0]],
                    Rectangle[{x1, x2} - δ, {x1, x2} + δ]}]
In[21]:= With[{pp = 200},
        Show[Graphics[
        Table[coloredSquare[{x1, x2}, 20/pp/2, Method -> Secant,
                WorkingPrecision -> $MachinePrecision + 1, MaxIterations -> 30],
            {x1, -10, 10, 20/pp}, {x2, -10.001, 10.001, 20.002/pp}]],
            Frame -> True, AspectRatio -> 1]]
```

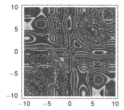

Now let us carry out *n* steps of the secant method and have a look of the resulting values in the x_1, x_2-plane.

```
In[22]:= iteratedSecantMap[f_, {x1_, x2_}, k_] :=
        Nest[{#[[2]], #[[2]] - f[#[[2]]](#[[2]] - #[[1]])/(f[#[[2]]] - f[#[[1]]])}&,
        {x1, x2}, k][[2]]
```

To speed up the highly recursive calculations, we will optimize the expression and compile the map in the following using Compile and the option setting CompileOptimizations -> True.

The following five arrays of four graphics show the *n*-times iterated application of iteratedSecantMap. The first two graphics are a density plot and a 3D plot of the last calculated value with the PlotRange -> Automatic setting. This setting nicely shows the structure of the function. The right two graphics show the same data, but this time with the PlotRange -> All setting. Whereas the left graphic nicely show how the overall pattern arise, the right graphic show how (in average) the function values decrease from iteration to iteration (although at greatly different rates as a function of x_1 and x_2). After generating the four graphics for each *k*, we print the plot ranges for the two graphics.

```
In[23]:= Do[(* compile the iterated secant map; optimize the function *)
        cf = Compile[{x1, x2}, Evaluate[iteratedSecantMap[
                        Function[x, x - Cos[x]], {x1, x2}, k]],
                        CompileOptimizations -> True];
        (* the density plot *)
        dp = DensityPlot[cf[x1, x2],
                {x1, -10 - 10^-9, 10}, {x2, -10, 10},
                Mesh -> False, ColorFunction -> (Hue[1.8 #]&),
                Compiled -> False, PlotPoints -> 201,
                DisplayFunction -> Identity, FrameTicks -> None];
        (* use default PlotRange setting *)
        Show[GraphicsArray[{dp,
        lp3d1 = ListPlot3D[dp[[1]], Mesh -> False, Axes -> False,
                        DisplayFunction -> Identity,
                        MeshRange -> {{-10, 10}, {-10, 10}}],
        (* use all values *)
        Show[dp, PlotRange -> All, DisplayFunction -> Identity],
        lp3d3 = ListPlot3D[dp[[1]], Mesh -> False, PlotRange -> All,
                        DisplayFunction -> Identity, Axes -> False,
                        MeshRange -> {{-10, 10}, {-10, 10}}]}]];
        (* print plot ranges *)
        Print[{k, Last @ FullOptions[lp3d1, PlotRange],
                Last @ FullOptions[lp3d3, PlotRange]}], {k, 1, 6}]
```

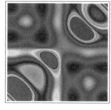

{1, {-3.94201, 4.03803}, {-63398.2, 20489.5}}

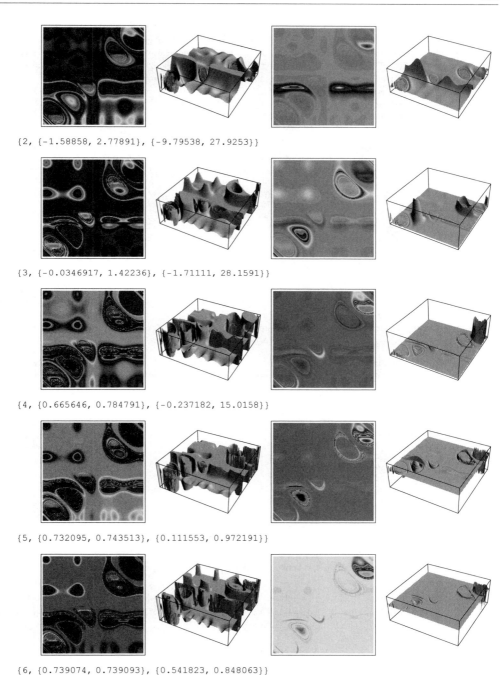

{2, {-1.58858, 2.77891}, {-9.79538, 27.9253}}

{3, {-0.0346917, 1.42236}, {-1.71111, 28.1591}}

{4, {0.665646, 0.784791}, {-0.237182, 15.0158}}

{5, {0.732095, 0.743513}, {0.111553, 0.972191}}

{6, {0.739074, 0.739093}, {0.541823, 0.848063}}

For similar pictures, based on the number of iterations, see [914] and [1836].

c) The part of the surface of a dD sphere that is inside the unit cube is obviously the whole surface for $\rho < 1/2$. At $\rho = 1/2$ parts of the sphere start to move outside at the unit cube. In general, at $k^{1/2}/2$, $k = 1, 2, \ldots, d$ a hyperface of the cube is reached and $a_d(\rho)$ will change its functional form. For the (numerical) integration to be attempted, this implies that the points $\rho = 1/2$, $\rho = 2^{1/2}/2$, ... should be treated carefully because the integrand will have a lower degree of smoothness there. In addition, the singularity of the denominator at $r = \rho$ must be dealt with to avoid that the integrand evaluates to a nonnumeric quantity. We will do this by treating all values $r < \varepsilon$ (ε being a small positive quantity) special. Given a definition of $a_{d-1}(\rho)$, the function makeaDefinition implements the defining integral for $a_d(\rho)$.

```
In[1]:= makeaDefinition[d_] := (* apply SetDelayed *)
        SetDelayed @@ (* left-hand side and right-hand side *)
        Join[Hold @@ {a[d][ρ_?NumberQ]},
        Hold[With[{ε = 10^-(2$MachinePrecision + 2)},
             (* the numerical integration *)
             NIntegrate[Evaluate[2 ρ A/If[ρ^2 - r^2 <= 0, Sqrt[1/ε],
                                    Sqrt[ρ^2 - r^2]]],
                Evaluate[Join[{r, Sqrt[Max[0, ρ^2 - 1/4]]}],
        (* intermediate points *)
        Table[Sqrt[d]/2, {d, Ceiling[4 Sqrt[Max[0, ρ^2 - 1/4]]^2],
                          Floor[4 Min[ρ, Sqrt[d]/2]^2]},
                          {Min[ρ, Sqrt[d]/2]}]],
                    PrecisionGoal -> 10, AccuracyGoal -> 10,
                    MaxRecursion -> 12, SingularityDepth -> 6]]] /.
                    (* previous area *) A -> a[d - 1][r]]
```

The 1D case is simply 2 in the unit interval and 0 elsewhere.

```
In[2]:= a[1][ρ_] := Which[0 <= ρ <= 1/2, 2, True, 0]
```

makeaDefinition makes a definition for $a_d(\rho)$. To obtain the dD area as a function of ρ, we will calculate $a_d(\rho)$ for various ρ and interpolate. To obtain automatically a uniformly good approximation, we use the function FunctionInterpolation within each of the intervals of higher smoothness $((k-1/2)^{1/2}/2, k^{1/2}/2)$, $k = 1, 2, ..., d-1$.

```
In[3]:= continuityIntervals[dim_] := Partition[Table[Sqrt[d]/2, {d, 0, dim}], 2, 1]
```

Now, we are ready for the explicit calculation of the $a_d(\rho)$. Because at $\rho = k^{1/2}$, we cannot meet all specified precision goals, we turn off the relevant messages. Calculating the first 25 $a_d(\rho)$ is now a matter of minutes.

```
In[4]:= (* turn off some warning messages *)
        Off[NIntegrate::slwcon]; Off[NIntegrate::ncvb]
        Off[FunctionInterpolation::ncvb];

        dimMax = 25;
        (* calculate a[d][ρ] for all d up to dimMax *)
        Do[(* make definition for a[dim][ρ] *)
           makeaDefinition[dim];
           (* calculate a[dim][ρ] in the dim intervals *)
           ipos = FunctionInterpolation[a[dim][ρ], Evaluate[Prepend[#, ρ]],
                    MaxRecursion -> 12, InterpolationPoints -> 25]& /@
                                            continuityIntervals[dim];
        (* make one definition for a[dim][ρ] for all ρ;
           for larger dimMax one could use a binary tree structure *)
        a[dim][ρ_] = Which @@
        Flatten[Append[Transpose[{(#[[1]] <= ρ <= #[[2]])& /@
                    continuityIntervals[dim], #[ρ]& /@ ipos]}, {True, 0}]],
        {dim, 2, dimMax}]
```

We can check the numerical quality of the solutions by calculating the integral $\int_0^{d^{1/2}/2} a_d(\rho)\, d\rho$. Using the defining equation for $a_d(\rho)$ shows immediately that this integral is 1 (geometrically speaking it measures the volume of the dD sphere inside the unit cube). The following check shows that the integral is correct to at least six places for all d.

```
In[10]:= (* return magnified error *)
         Table[Plus @@ (NIntegrate[a[d][ρ], {ρ, ##}]& @@@
                                    continuityIntervals[d]) - 1,
              {d, 2, dimMax}]/10^-7 // Round
Out[11]= {-1, -1, -1, -1, -1, -1, -1, -2, -2, -3,
         -3, -3, -4, -4, -5, -5, -5, -6, -6, -6, -7, -7, -7, -8}
```

Now, we visualize the just-calculated $a_d(\rho)$. The left graphic shows them as a function of ρ and the right graphic shows them scaled with the hypercube diameter $d^{1/2}/2$.

```
In[12]:= Show[GraphicsArray[
           Plot[##, PlotRange -> All, DisplayFunction -> Identity,
               PlotStyle -> Table[Hue[0.8 d/dimMax], {d, dimMax}]]& @@@
             (* data for ρ and scaled ρ *)
             {{Table[a[d][ρ], {d, dimMax}], {ρ, 0, Sqrt[dimMax]/2}},
              {Table[Sqrt[d]/2 a[d][Sqrt[d]/2 ρS], {d, dimMax}], {ρS, 0, 1}}}]]
```

The scaled maximum of the scaled curves seems to approach a limiting value. Here are the scaled ρ-values of the maximas.

```
In[13]:= Table[FindMinimum[Evaluate[-a[d][x (Sqrt[d]/2)]],
                           {x, 0.25, 0.75}][[2, 1, 2]], {d, 3, dimMax}]
Out[13]= {0.57735, 0.58282, 0.588548, 0.583688, 0.582677, 0.582234, 0.581532,
          0.581054, 0.580694, 0.580385, 0.58013, 0.579915, 0.57973, 0.57957, 0.57943,
          0.579307, 0.579198, 0.5791, 0.579012, 0.578932, 0.57886, 0.578794, 0.578733}
```

Using a Fourier representation of the Dirac delta function in the above definition for the $a_d(\rho)$, it is straightforward to show that for large d

$$a_d(\rho) \approx \sqrt{\frac{10}{\pi}} \ \sqrt{d} \ \frac{2\rho}{\sqrt{d}} \ \exp\left(-\frac{5}{2} d\left(\left(\frac{2\rho}{\sqrt{d}}\right)^2 - \frac{1}{3}\right)^2\right).$$

This shows that in the limit $d \to \infty$, the maximum of the areas is at $3^{-1/2} d^{1/2}/2$ and the curve itself approaches a Dirac delta function concentrated at this point.

14. Computation of Determinants, Numerical Integration, Binary Trees, Matrix Eigenvalues

a) First, we define an auxiliary function that finds the row or column, respectively, with the most zeros. We will form our expansion about such a row or column.

```
In[1]:= row[mat_] :=
         Function[trapoMat, (* look for minimum number of zeros *)
         If[# > Length[mat], {# - Length[mat], trapoMat}, {#, mat}]&[
           Position[#, Max[#]]&[ (* count all kinds of zeros in the rows *)
             Join[Count[#, 0 | 0.0 | 0.0 + 0.0 I] & /@ mat,
                  Count[#, 0 | 0.0 | 0.0 + 0.0 I] & /@ trapoMat]][[1, 1]]]][
                                                          Transpose[mat]]
```

The following function forms the adjoint of the (i, j)-element of the matrix mat.

```
In[2]:= adjoint[{i_, j_}, mat_] := MapAt[(-1)^(i + j) #& /@ #&,
                 Transpose[Drop[Transpose[Drop[mat, {i}]], {j}]], 1]
```

The function reduce reduces the computation of the determinant of a matrix of order n to the calculation of a determinant of order $n - 1$.

```
In[3]:= reduce[{pos_, mat_}, func_] :=
         (* sum of (-1)^position adjoints *)
         Plus @@ MapIndexed[#1 func[adjoint[{pos, #2[[1]]}, mat]]&, mat[[pos]]]
```

Because func will be replaced by det in reduce, and because the determinant of a 2×2 matrix can be explicitly defined, we are led to the following recursive definition.

```
In[4]:= (* initial condition *)
         det[{{a11_, a12_}, {a21_, a22_}}] := a11 a22 - a12 a21
In[6]:= (* define recursive for larger matrices *)
         det[mat_?(MatrixQ[#] && Length[#] >= 3 &)] := reduce[row[mat], det]
```

Now, we can compute the example above.

```
In[8]:= Array[Which[#2 - 2 <= #1 <= #2 + 2, #1 + #2, True, 0]&, {5, 5}] // det
Out[8]= 6090
```

The result agrees with that produced by the built-in function `Det`.

```
In[9]:= Array[Which[#2 - 2 <= #1 <= #2 + 2, #1 + #2, True, 0]&, {5, 5}] // Det
Out[9]= 6090
```

Note that the just-implemented method is (of course) not the most efficient one to compute a determinant. It just illustrates how the classical method can be implemented in a recursive manner in *Mathematica*.

b) This is the integrand.

```
In[1]:= f = Product[Sin[x/(2k + 1)]/(x/(2k + 1)), {k, 0, 7}]/Pi;
```

First, we directly integrate the function using higher than default settings for most of the options.

```
In[2]:= opts = Sequence[PrecisionGoal -> 15, WorkingPrecision -> 120,
              MaxPoints -> Infinity, Method -> GaussKronrod,
              MaxRecursion -> 10, MinRecursion -> 3];
In[3]:= 1/2 - NIntegrate[Evaluate[f], {x, 0, Infinity}, Evaluate[opts]]
Out[3]= 7.353 × 10^-12
```

As a check, we carry out a second numerical integration. This time, we subtract the value 1/2 inside the integrand in the form of the function $1/(\pi(x^2 + 1))$. Now the value of the integral is very small. This means that in addition we set the value of the option `AccuracyGoal` explicitly.

```
In[4]:= NIntegrate[Evaluate[1/(1 + x^2)/Pi - f],
              {x, 0, Infinity}, Evaluate[opts], AccuracyGoal -> 15]
Out[4]= 7.353 × 10^-12
```

Again, we confirmed that the result is not 1/2. A direct symbolical integration shows this too.

```
In[5]:= exactIntegralValue = (* faster than  Integrate[f, {x, 0, Infinity}] *)
        Module[{(* calculate indefinite integral *)
                indefInt = Expand[Integrate[#, x]& /@
                                  Take[TrigReduce[f], All]],
                S = Normal[Series[#1, {x, #2, 0}]]&, l, u},
                (* calculate definite from indefinite integral *)
                l = S[indefInt, 0];
                u = S[indefInt //.
                    {SinIntegral[a_ x] :> SinIntegral[a Infinity],
                     (Cos | Sin)[_ x] x^p_?Negative -> 0}, Infinity];
                u - l];
In[6]:= {#, N[#]}&[1/2 - exactIntegralValue]
Out[6]= { 6879714958723010531 / 935615849440640907310521750000, 7.35314 × 10^-12 }
```

c) To get an explicit numerical solution for $y(x)$ we must transform the boundary value problem to an initial value problem. This means we must find $y'(x)$ for small x. Because of the singularity of $y(x)$ for small x, we will start solving the differential equation at $x = \varepsilon$. The following shows that $\varepsilon = 10^{-10}$ is small enough to not influence the first 10 digits of the integral.

```
In[1]:= yAsymp0[x_] := -x (Log[x/4] + EulerGamma);

        int0[ε_] := NIntegrate[Evaluate[x Log[x] (1 - yAsymp0[x]^2)], {x, 0, ε}]
In[3]:= (* three sample values *)
        {int0[10^-5], int0[10^-10], int0[10^-15]}
Out[4]= {-6.00646 × 10^-10, -1.17629 × 10^-19, -1.75194 × 10^-29}
```

The function `pode3NSol` calculates a solution of the differential equation. We start at $x = \varepsilon$ and continue solving until the solution either becomes negative or starts to escape to ∞.

```
In[5]:= pode3NSol[yPrime_, {prec_, pg_, ag_}] :=
        With[{ε = 10^-10},
```

```
NDSolve[{y''[x] == 1/y[x] y'[x]^2 - 1/y[x] + y[x]^3 - 1/x y'[x],
         y[ε] == -ε (Log[ε/4] + EulerGamma),
         y'[ε] == SetPrecision[yPrime, Infinity]}, y, {x, ε, 40},
         WorkingPrecision -> prec, MaxSteps -> 10^5,
         PrecisionGoal -> pg, AccuracyGoal -> ag,
         MaxStepSize -> 1/100,
         (* no negative or growing solutions *)
         StoppingTest -> (y[x] < 0 || y[x] > 2)]]
```

The next graphic shows that, for $y'(\varepsilon) \approx 23$, the solution $y(x)$ changes its asymptotic behavior.

```
In[6]:= maxXReached[nsol_] := Max[nsol[[1, 1, 2, 1, 1]]]
```

```
In[7]:= ListPlot[Table[{yp, (y[maxXReached[#]] /. #[[1]])&[
         pode3NSol[yp, {$MachinePrecision - 1,
                        Automatic, Automatic}]]}, {yp, 0, 50}],
         Frame -> True]
```

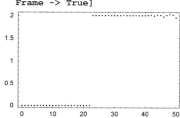

Here are some of the solutions for $y'(\varepsilon) \approx 23$.

```
In[8]:= solutionPlot[nsol_, col_:GrayLevel[0], opts___] :=
       Plot[Evaluate[y[x] /. nsol[[1]]], {x, 0, maxXReached[nsol]},
            opts, Frame -> True, Axes -> False,
            PlotStyle -> {{Thickness[0.002], col}}]
```

```
In[9]:= Show[Table[solutionPlot[pode3NSol[yp, {$MachinePrecision - 1,
                                    Automatic, Automatic}],
                         Hue[0.8 (yp - 21)/2], DisplayFunction -> Identity],
            {yp, 21, 23, 1/10}], DisplayFunction -> $DisplayFunction,
                         PlotRange -> All]
```

We could either find the actual value for the optimal $y'(\varepsilon)$ numerically using `FindRoot` or `FindMinimum`, or just by differentiating the asymptotic solution for small x.

```
In[10]:= maxInvDistSigned[yPrime_?NumberQ, prec_] :=
        Module[{ε = 10^-10, nsol, maxX, Y},
               nsol = pode3NSol[yPrime, {prec, 18, 18}];
               maxX = maxXReached[nsol];
               Y = y[maxX] /. nsol[[1]];
               (* to see search progress *)
               (* Print["y' = ", InputForm @ yPrime, " maxX = ", maxX]; *)
               1/maxX Sign[Y]];
```

```
In[11]:= FindRoot[maxInvDistSigned[yPrime, 40], {yPrime, 21, 23},
                WorkingPrecision -> 40, MaxIterations -> 100]

         FindRoot::cvmit : Failed to converge to the
             requested accuracy or precision within 100 iterations. More…
```

Out[11]= {yPrime → 22.834929626158815084407216703190183955381}

Both approaches yield the following value for $y'(\varepsilon)$.

In[12]:= yPrimeF = 22.834929626158815084407209155530956898317;

In[13]:= (yPrimeF - yAsymp0'[10^-10]) // N

Out[13]= 4.85999×10^{-16}

The next two graphics show the resulting $y(x)$ and $\ln(1 - y(x))$.

```
In[14]:= nsolF = pode3NSol[yPrimeF, {40, 18, 18}];
         δ[x_?NumberQ] := Evaluate[Log[10, 1 - y[x]] /. nsolF]

         Show[GraphicsArray[
         {solutionPlot[nsolF, GrayLevel[0], PlotRange -> All,
                      DisplayFunction -> Identity],
          Plot[Re @ δ[t], Evaluate[{t, Sequence @@ nsolF[[1, 1, 2, 1, 1]]}],
               PlotRange -> All, DisplayFunction -> Identity]}]]
```

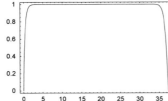 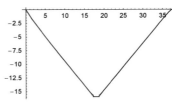

The last two graphics show that $y(x)$ is a good solution for $\varepsilon \le x \lesssim 17$. So, we obtain the following value for the integral.

```
In[17]:= nint =
         NIntegrate[Evaluate[x Log[x] (1 - y[x]^2) /. nsolF[[1]]],
                   Evaluate[{x, 10^-10, 17}], WorkingPrecision -> 40,
                   PrecisionGoal -> 15, MaxRecursion -> 8]
Out[17]= -0.09192757577472926
```

The last value is correct to about 14 digits.

In[18]:= exact = 1/4 + 7/12 Log[2] - 3 Log[Glaisher];

In[19]:= exact - nint

Out[19]= 1.124×10^{-14}

Finally, we approximate the neglected part of the integral (from $x = 17$ to infinity). Taking this contribution into account, we have agreement to nearly 16 digits.

```
In[20]:= yAsymp∞[x_] := 1 - 2/Pi Sqrt[Pi/2] Exp[-2x] Sqrt[1/(2x)];
         NIntegrate[Evaluate[x Log[x] (1 - yAsymp∞[x]^2)], {x, 17, 40},
                   WorkingPrecision -> 40]
Out[21]= $1.15767967491860517046677072533 8 \times 10^{-14}$
```

In[22]:= exact - (nint + %)

Out[22]= -3.4×10^{-16}

d) The function binaryTree generates a complete binary tree of depth n. It is built hierarchically in the form C[*parentNode*, *childNode1*, *childNode2*]. The *childNodes* at the lowest level have the head c.

```
In[1]:= binaryTree[n_] := binaryTree[n] =
        Module[{k = 1}, Nest[(# /. c[j_] :> C[j, c[k++], c[k++]])&,
                             C[1, c[2], c[3]], n]];
```

Here is the binary tree of depth 4.

In[2]:= binaryTree[4]

```
Out[2]= C[1, C[2, C[1, C[5, C[13, c[29], c[30]], C[14, c[31], c[32]]],
          C[6, C[15, c[33], c[34]], C[16, c[35], c[36]]]],
        C[2, C[7, C[17, c[37], c[38]], C[18, c[39], c[40]]],
          C[8, C[19, c[41], c[42]], C[20, c[43], c[44]]]]],
      C[3, C[3, C[9, C[21, c[45], c[46]], C[22, c[47], c[48]]],
          C[10, C[23, c[49], c[50]], C[24, c[51], c[52]]]],
        C[4, C[11, C[25, c[53], c[54]], C[26, c[55], c[56]]],
          C[12, C[27, c[57], c[58]], C[28, c[59], c[60]]]]]]]
```

Given a tree in the above form, it is straightforward to generate the corresponding adjacency matrix. The function `toAdja⁚ cencyMatrix` does this.

```
In[3]:= toAdjacencyMatrix[tree_] :=
        Module[{aux, rules, l},
         (* get all edges *)
         aux = Flatten[{{#[[1]], #[[2, 1]]}, {#[[2, 1]], #[[1]]},
             Sequence @@ If[Length[#] === 3,
             {{#[[1]], #[[3, 1]]}, {#[[3, 1]], #[[1]]}}, {}]}& /@
                  (* all C`s with nontrivial children *)
         DeleteCases[Cases[tree, _C, Infinity], C[_, ___c] | C[_]], 1];
         (* edges to position pairs *)
         rules = Apply[Rule, Transpose[{#,
                    Range[l = Length[#]]}&[Union[Flatten[aux]]]], {1}];
         (* insert 1`s in matrix of 0`s *)
         ReplacePart[Table[0, {l}, {l}], 1, Union[aux /. rules]]]
```

Here is the adjacency matrix for a binary tree of depth eight visualized.

```
In[4]:= Show[Graphics[(* show a 1 as a black square *)
                 Rectangle[# - 1/2, # + 1/2]& /@
                 Position[toAdjacencyMatrix[binaryTree[8]], 1]],
          Frame -> True, AspectRatio -> Automatic]
```

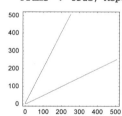

Now, we generate a random binary tree. We do this by deleting each node with a certain probability.

```
In[5]:= randomizeTree[tree_, p_] :=
        Block[{C}, C[x__] := If[Random[] < p, Sequence @@ {}, C[x]]; tree]
```

To avoid the generation of empty trees, the function `makeRandomizedTree` will call `randomizeTree` until a nontrivial random binary tree is generated.

```
In[6]:= makeRandomizedTree[tree_, p_] :=
        Module[{aux, f},
              While[aux = randomizeTree[tree, p];
                   Length[f[aux]] === 0 || Length[aux] === 1, Null];
              aux /. C -> C]
```

Here are two examples of random binary trees.

```
In[7]:= SeedRandom[111];
        makeRandomizedTree[binaryTree[4], 0.2]
Out[8]= C[1, C[3, C[3, C[9, C[22, c[47], c[48]]], C[10, C[23, c[49], c[50]], C[24, c[51], c[52]]]],
          C[4, C[11, C[25, c[53], c[54]], C[26, c[55], c[56]]],
            C[12, C[27, c[57], c[58]], C[28, c[59], c[60]]]]]]
```

```
In[9]:= SeedRandom[222];
        makeRandomizedTree[binaryTree[4], 0.2]
```

```
Out[10]= C[1, C[2, C[2, C[7, C[18, c[39], c[40]]], C[8, C[19, c[41], c[42]], C[20, c[43], c[44]]]]],
         C[3, C[3], C[4, C[12, C[27, c[57], c[58]]]]]]]
```

Now, we have all functions together to generate a random binary tree, construct its adjacency matrix, and to calculate its eigenvalues. We will start with a binary tree of depth eight. This will result in adjacency matrices of dimensions less or equal than 512×512. The horizontal axis of the graphic is the relative fraction of eigenvalues. The eigenvalues of each matrix are shown in a separate color.

```
In[11]:= SeedRandom[333];
         data = Table[Check[mat = toAdjacencyMatrix[
                         mm = makeRandomizedTree[binaryTree[8], 0.1]];
                      MapIndexed[(* scale *){#2[[1]]/Length[mat], #1}&,
                         (* sort eigenvalues *)
                         Sort[Re[Eigenvalues[N[mat]]]]], Print[mm]], {100}];

In[13]:= Show[Graphics[{PointSize[0.003], (* add color *)
                {Hue[Random[]], Point /@ #}& /@ data}],
              PlotRange -> {{0, 1}, {-3, 3}}, Frame -> True]
```

The spectrum resembles remarkably well a Cantor-like function with only small statistical deviations. For an analytical discussion of the (graph-theoretical and Laplace) spectrum of random binary trees, see [804], [147], [145], [435], [1906], [1677]. For the average shape of binary trees, see [1011]. For other graph structures with self-similar graph spectra, see [1502], [1210], [1211].

e) A[n] creates a square left triangular matrix of dimension n.

```
In[1]:= A[n_] := Table[If[1 <= i <= n - j + 1, 1, 0], {i, n}, {j, n}];
```

Here is A[10] shown.

```
In[2]:= MatrixForm[A[10]]
```

$$Out[2]//MatrixForm= \begin{pmatrix} 1 & 1 & 1 & 1 & 1 & 1 & 1 & 1 & 1 & 1 \\ 1 & 1 & 1 & 1 & 1 & 1 & 1 & 1 & 1 & 0 \\ 1 & 1 & 1 & 1 & 1 & 1 & 1 & 1 & 0 & 0 \\ 1 & 1 & 1 & 1 & 1 & 1 & 1 & 0 & 0 & 0 \\ 1 & 1 & 1 & 1 & 1 & 1 & 0 & 0 & 0 & 0 \\ 1 & 1 & 1 & 1 & 1 & 0 & 0 & 0 & 0 & 0 \\ 1 & 1 & 1 & 1 & 0 & 0 & 0 & 0 & 0 & 0 \\ 1 & 1 & 1 & 0 & 0 & 0 & 0 & 0 & 0 & 0 \\ 1 & 1 & 0 & 0 & 0 & 0 & 0 & 0 & 0 & 0 \\ 1 & 0 & 0 & 0 & 0 & 0 & 0 & 0 & 0 & 0 \end{pmatrix}$$

evs[n, k] gives the k largest in magnitude eigenvalues of A[n]. All eigenvalues of A[n] are real.

```
In[3]:= evs[n_, k_] := Take[Sort[Abs[Eigenvalues[N[A[n]]]], #1 > #2&], k]
```

Here are the 30 largest eigenvalues of A[100]. The right graphic shows the first 10 eigenvalues for various n. The linear dependence of $\lambda_k^{(n)}$ as a function of n is clearly visible.

```
In[4]:= Show[GraphicsArray[
         Block[{$DisplayFunction = Identity},
           {(* largest eigenvalues of A[100] *)
           ListPlot[evs[100, 30], PlotRange -> All, Frame -> True, Axes -> False],
           (* calculate and plot eigenvalues for various n *)
           data1 = Table[{n, #}& /@ evs[n, 10], {n, 20, 100}];
           Graphics[MapIndexed[{Hue[(#2[[2]] - 1)/13], Point[#1]}&, data1, {2}],
                Axes -> True]}]]]
```

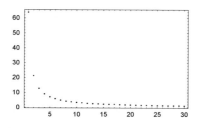 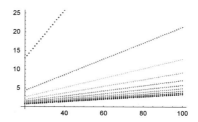

Fitting the data, data1 confirms $\lambda_k^{(n)} \sim n$ to a high precision.

```
In[5]:= Table[Fit[#[[j]]& /@ data1, {1, x, x^2, x^3}, x], {j, 3}]

Out[5]= {0.324008 + 0.636448 x + 2.11308×10⁻⁶ x² − 9.08449×10⁻⁹ x³,
         0.123229 + 0.21169 x + 6.35838×10⁻⁶ x² − 2.73435×10⁻⁸ x³,
         0.092304 + 0.126459 x + 0.0000106614 x² − 4.58737×10⁻⁸ x³}
```

Now let us look at the dependence of $\lambda_k^{(n)}$ on k.

```
In[6]:= data2 = Table[evs[n, {1, 15}], {n, 100, 200, 10}];

In[7]:= data2a = MapIndexed[{Log[10, #2[[2]]], Log[10, #1]}&, data2, {2}] // N;
```

A doubly logarithmic plot suggests $\lambda_k^{(n)} \sim k^{-\alpha}$.

```
In[8]:= Show[Graphics[MapIndexed[{Hue[(#2[[1]] − 1)/20], Point[#1]}&, data2a, {2}]],
            Axes -> True]
```

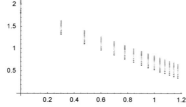

Fitting the data data2 shows $1 < \alpha < 2$.

```
In[9]:= Table[Fit[data2a[[j]], {1, x, x^2, x^3}, x], {j, 1, -1, -2}]

Out[9]= {1.80464 − 1.76435 x + 0.75213 x² − 0.261484 x³, 2.10461 − 1.76622 x + 0.758127 x² − 0.266901 x³}
```

A fit of the data suggests $\lambda_k^{(n)} \sim c_{-1} k^{-1} + c_{-2} k^{-2}$ with $c_{-2} \approx c_{-1}/2$.

```
In[10]:= Fit[Take[data2[[-1]], 15], {1, 1/x, 1/x^2}, x]

Out[10]= 1.64043 + 82.7478/x² + 43.0486/x
```

The last result suggests a series in k, and the next natural conjecture would be $\lambda_k(n) \sim n/(k - \alpha)$. Carrying out a fit for various α clearly shows $\alpha = 1/2$.

```
In[11]:= data3 = Table[MapIndexed[{n, #2[[1]]}, #1]&, evs[n, 10]],
                  {n, 200, 220}];

In[12]:= ListPlot[Table[{α, fit = Fit[Flatten[data3, 1], {n/(k - α)}, {n, k}];
                   F = Function[{n, k}, Evaluate[fit]];
                   Plus @@ ((F[##] − #3)^2& @@@ Flatten[data3, 1])},
                  {α, 0.1, 0.99, 0.01}]]
```

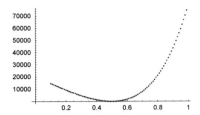

Now all that is left is to determine the constant β in $\lambda_k(n) = \beta n/(k - 1/2)$. We try to determine β from extrapolation.

```
In[13]:= Table[evs[n, 1]/n/2, {n, 10, 400, 20}] // SequenceLimit

Out[13]= {0.318335}
```

A power method calculation for the largest in magnitude eigenvalue of A[2000] gives a similar result.

```
In[14]:= cf = Compile[{{n, _Integer}, {o, _Integer}},
           Module[{matrix, vec},
                  matrix = Table[If[ 1 <= i <= n - j + 1, 1.], 0.], {i, n}, {j, n}];
                  vec = N[Table[1., {k, n}]];
                  matrix.#/#& @ Nest[((#/Max[#])&[matrix.#]&, vec, o]]];

In[15]:= cf[2000, 100][[1]]/(2000/(1 - 1/2))

Out[15]= 0.318389
```

Assuming that a nice-looking matrix has nice eigenvalues, we guess from the first four digits that $\beta = 1/\pi$. This is the correct result [129].

f) We start by implementing the differential operator obtained from the determinant.

```
In[1]:= Δ[1, z_] := D[#, z[1, 1]]&;

        Δ[n_, z_] := Δ[n, z] =
        Block[{D}, (Function @@ {Function[t, If[MemberQ[t, -1, {1}],
                          -D[#, Sequence @@ Rest[t]], D[#, Sequence @@ t]]] /@
                          Det[Table[z[i, j], {i, n}, {j, n}]]})]
```

Here are the first three of these differential operators.

```
In[3]:= Table[Δ[k, z], {k, 3}] /. z[i_, j_] :> Subscript[z, i, j] // TraditionalForm
```

$$\text{Out[3]//TraditionalForm=} \left\{ \frac{\partial \#1}{\partial z_{1,1}} \&, \ \frac{\partial^2 \#1}{\partial z_{1,1} \partial z_{2,2}} - \frac{\partial^2 \#1}{\partial z_{1,2} \partial z_{2,1}} \&, \right.$$
$$\left. \frac{\partial^3 \#1}{\partial z_{1,1} \partial z_{2,2} \partial z_{3,3}} - \frac{\partial^3 \#1}{\partial z_{1,1} \partial z_{2,3} \partial z_{3,2}} - \frac{\partial^3 \#1}{\partial z_{1,2} \partial z_{2,1} \partial z_{3,3}} + \frac{\partial^3 \#1}{\partial z_{1,2} \partial z_{2,3} \partial z_{3,1}} + \frac{\partial^3 \#1}{\partial z_{1,3} \partial z_{2,1} \partial z_{3,2}} - \frac{\partial^3 \#1}{\partial z_{1,3} \partial z_{2,2} \partial z_{3,1}} \& \right\}$$

The expectation value of the kth power of the determinant is then easily implemented as e[k, n].

```
In[4]:= e[k_, n_] := I^(k n) Nest[Δ[n, z],
                     Product[Cos[z[i, j]], {i, n}, {j, n}], k] /. _z -> 0
```

Here are the explicit values for $1 \le k, n \le 4$.

```
In[5]:= Table[e[k, n], {k, 4}, {n, 4}]

Out[5]= {{0, 0, 0, 0}, {1, 2, 6, 24}, {0, 0, 0, 0}, {1, 8, 96, 2112}}
```

For the numerical modeling, we use a compiled function. Because the most time-consuming step is the construction of the random matrix and the calculation of the determinant, we reuse these values for the various kth powers.

```
In[6]:= eModel =
        Compile[{{kMax, _Integer}, {n, _Integer}, {runs, _Integer}},
                Module[{sums = Table[0., {kMax}],
                        r = Range[kMax], mat, det},
                       Do[mat = Table[2.Random[Integer] - 1., {n}, {n}];
                          det = Det[mat];
```

```
                    sums = sums + Det[mat]^r, {runs}];
                sums/runs]];
In[7]:= SeedRandom[007];
        (* one million matrices *)
        Table[eModel[4, n, 10^6], {n, 4}] // Transpose
Out[9]= {{-0.000182, -0.00147, -0.001376, 0.002344}, {1., 1.99912, 5.98608, 23.9368},
        {-0.000182, -0.00588, -0.022016, 0.5248}, {1., 7.99646, 95.7773, 2105.66}}
```

15. Root Pictures, Weierstrass Iterations, Taylor Series Remainders

a) To recognize the number of summands, we build this sum recursively and color all points that come from the same number of summands with the same color.

```
In[1]:= RootPointPicture[{polyEqn_, var_}, m_, opts___] :=
        Function[ar, Show[Graphics[(* color roots *)
        {PointSize[0.008], Reverse @
        MapIndexed[{Hue[(#2[[1]] - 1) 0.7/If[m == 1, 1, m - 1]],
                    Point[{Re[#], Im[#]}]& /@ #1}&,
        (* form combinations of the roots *)
        FoldList[Union[Flatten[Outer[Plus, #, ar]],
                    SameTest -> (Abs[#1 - #2] < 10^-9&)]&,
                    ar, Table[ar, {m}]]]},
        opts, AspectRatio -> Automatic, PlotRange -> All]]][
                (* the roots *)  var /. NSolve[polyEqn, var]]
```

Here are two examples.

```
In[2]:= Show[GraphicsArray[
        Block[{$DisplayFunction = Identity},
            {RootPointPicture[{(x^23 - 1) == 0, x}, 2],
             RootPointPicture[{Sum[x^i, {i, 12}] == 1, x}, 3]}]]]
```

The origin of terms can be more easily observed if we draw lines (again in color) from the summands to the result.

```
In[3]:= RootLinePicture[{polyEqn_, var_}, m_, opts___] :=
        Function[ar,
        Show[Graphics[(* color roots *) {Thickness[0.002], Reverse @
        MapIndexed[{Hue[(#2[[1]] - 1) 0.7/If[m == 1, 1, m - 1]], #1}&,
        (* color according to history *)
        Apply[{Line[{{Re[#1], Im[#1]}, {Re[#3], Im[#3]}}],
              Line[{{Re[#2], Im[#2]}, {Re[#3], Im[#3]}}]}&,
        (* all combinations of integer multiples of the roots *)
        Rest[FoldList[Union[Flatten[Outer[{##, Plus[##]}&, Last /@ #, ar], 1],
          SameTest -> (Abs[#1[[3]] - #2[[3]]] < 10^-9&)]&,
                    List /@ ar, Table[ar, {m}]]], {2}]]}],
            opts, AspectRatio -> Automatic, PlotRange -> All]][
                (* the roots *) var /. NSolve[polyEqn, var]]
```

Here are a few examples.

```
In[4]:= Show[GraphicsArray[
        Block[{$DisplayFunction = Identity},
            (* root pictures of three polynomials *)
            {RootLinePicture[{Sum[(-1)^i  x^i, {i, 12}] == 1, x}, 2],
             RootLinePicture[{Sum[ -i I^i x^i, {i,  7}] == 1, x}, 2],
             RootLinePicture[{-x^4 + 5x^3 - 7x^2 + 4x - 12 == 0, x}, 3]}]]]
```

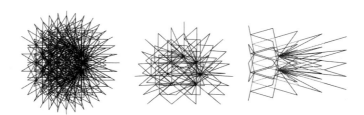

b) The implementation of the iteration of the replacement and the ongoing root finding is straightforward.

```
In[1]:= IteratedRootPicture[iter_, degree_, opts___] :=
    Module[{p},
      (* the polynomials used in the replacement *)
      Do[p[j] = Sum[z^i N[Random[] Exp[2Pi I Random[]]], {i, 0, j}], {j, degree}];
      Show[Graphics[{PointSize[0.003],
      Point[{Re[#], Im[#]}]& /@ (* root finding *)
       Cases[NRoots[# == 0, z]& /@ (* nesting the replacement *)
        NestList[Expand[# /. z^n_. :> p[n]]&,
        (* the starting polynomial *) Sum[z^i, {i, 0, degree}], iter],
        _?NumberQ, {-1}]}], opts, AspectRatio -> Automatic, Frame -> True,
          PlotRange -> Automatic, FrameTicks -> None]]
```

Here are a few pictures of the resulting roots for polynomials of various degrees.

```
In[2]:= Show[GraphicsArray[#, GraphicsSpacing -> 0.02]]& /@
    Partition[Table[IteratedRootPicture[100, i,
              DisplayFunction -> Identity, AspectRatio -> 1],
          {i, 4, 26, 2}], 4]
```

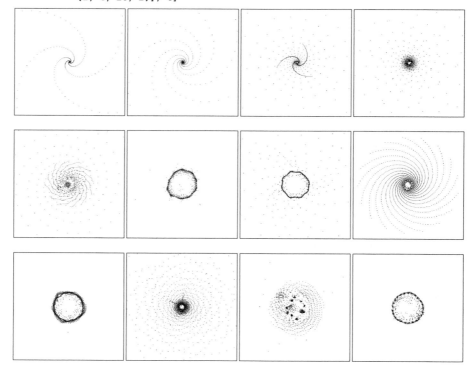

Here are some similar iterations. For the three polynomials $p[1]$, $p[2]$, and $p[3]$ in x, we use their zeros as values for the variables y.

```
In[3]:= Module[{p, roots, ppφ = 160, ρMin = 1/10, ρMax = 3, ppρ = 4},
        (* three polynomials;
           x is the polynomial variables, y and z are parameters *)
        p[1][x_, y_, z_] := -3 z^2 y + 2 z^2 x y + 2 x^2 y - 3 z^3 y^2 -
                              z x y^2 +  6 x^3 y^2 - 3 z^2 x^3 y^2 - 3 y;
        p[2][x_, y_, z_] := -2 z^2 x^2 y + 3 z^2 y^2 + 5 z^3 x y^2 +
                              z x^2 y^3 + z x^3 y^3 - 3 z^3 x^3 y^3;
        p[3][x_, y_, z_] := 3 z + z^2 + 3 z^2 x^2 - 2 z x^3 + 4 z^2 y -
                              2 z^3 y;
        (* solve polynomials with respect to x *)
        roots[j_][y_, z_] := Last /@ (List @@ NRoots[p[j][x, y, z] == 0, x]);
        Do[(* for fixed ρ, show roots of all three polynomials *)
        Show[GraphicsArray[
         Table[Graphics[{PointSize[0.003],
          Table[{Hue[φ/(2Pi)], Point[{Re[#], Im[#]}]}& /@
                (* iterate root finding for fixed z-circles;
                   use zeros as values for y *)
                Nest[Flatten[roots[j][#, ρ Exp[I φ]]& /@ #]&, {1}, 5]},
            {φ, 0, 2Pi, 2Pi/ppφ}], Frame -> True, AspectRatio -> 1,
            FrameTicks -> None, PlotRange -> All], {j, 3}]]],
        (* vary ρ *) {ρ, ρMin, ρMax, (ρMax - ρMin)/ppρ}]]
```

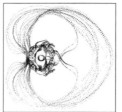
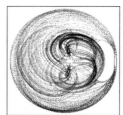

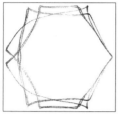
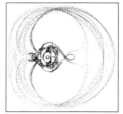
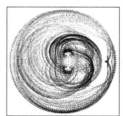

c) This is the cubic polynomial under consideration.

```
In[1]:= poly[x_, ε_] := ε x^3 + x^2 + x + 1
```

As a function of ε, the roots of this polynomial fulfill the following differential equation. (Roots are—by the inverse function theorem—(piecewise) continuous functions of the coefficients [206].)

```
In[2]:= sol = Solve[D[poly[x[ε], ε], ε] == 0, x'[ε]][[1]]
```

$$Out[2]= \left\{ x'[\varepsilon] \to -\frac{x[\varepsilon]^3}{1 + 2\,x[\varepsilon] + 3\,\varepsilon\,x[\varepsilon]^2} \right\}$$

We can find the dependence of the "third root" on ε by solving the differential equation radially outward for increasing ε. We carry out the following change of variables: $\varepsilon = R_\varepsilon \exp(i\,\varphi_\varepsilon)$, $x(\varepsilon) = \xi(R_\varepsilon)$, $\partial x(\varepsilon)/\partial\varepsilon = (1/(\partial\varepsilon/\partial R_\varepsilon))\,\partial\xi(R_\varepsilon)/\partial R_\varepsilon$.

```
In[3]:= (* parameter values *)
        RεMin = 10^-2; RεMax = 2; ppφ = 90; ppr = 60;

        ipos = Table[{N[φε],
        (* third root has largest absolute value *)
        xε = Sort[{Abs[#], #}& /@
            (x /. NSolve[poly[x, RεMin Exp[I φε]] == 0, x, $MachinePrecision + 10]),
            #1[[1]] > #2[[1]]&][[1, 2]];
        (* solve differential equation *)
        NDSolve[{ξ'[Rε]/Exp[I φε] ==
                (sol[[1, 2]] /. {x[ε] -> ξ[Rε], ε -> Rε Exp[I φε]}),
                ξ[RεMin] == xε}, ξ[Rε], {Rε, RεMin, RεMax},
                WorkingPrecision -> $MachinePrecision + 5,
                PrecisionGoal -> 15, AccuracyGoal -> 15][[1, 1, 2, 0]]},
                {φε, 0, 2Pi, 2Pi/ppφ}];
```

The resulting roots fulfill the polynomial to about six digits.

```
In[6]:= thirdRoot = Table[{Rε Cos[ipos[[k, 1]]], Rε Sin[ipos[[k, 1]]],
                        ipos[[k, 2]][Rε]}, {k, ppφ + 1},
                        {Rε, RεMin, RεMax, (RεMax - RεMin)/ppr}];

In[7]:= Apply[poly[#3, #1 + I #2]&, thirdRoot, {2}] // Abs // Max

Out[7]= 7.05514 × 10^-7
```

Here is a picture showing the real and imaginary parts of the roots.

```
In[8]:= Needs["Graphics`Graphics3D`"]

In[9]:= Show[GraphicsArray[Function[reim,
        Graphics3D[{EdgeForm[], Cases[(* extract polygons *)
```

```
ListSurfacePlot3D[Map[MapAt[reim, #, 3]&, thirdRoot, {2}],
                  DisplayFunction -> Identity], _Polygon, Infinity]},
       BoxRatios -> {1, 1, 0.8}, Axes -> True]] /@ {Re, Im}]]
```

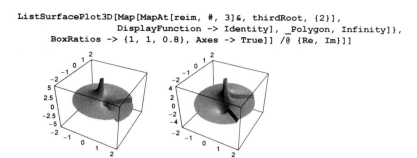

The discontinuities visible in the picture are branch cuts starting at the two branch points.

In[10]:= α /. NSolve[{poly[x, α] == 0, D[poly[x, α], x] == 0}, {x, α}]

Out[10]= {0.259259 + 0.209513 i, 0.259259 - 0.209513 i}

There is of course no unique way to continue just the third root. Depending on the integration path, one will sweep out all three sheets of the corresponding Riemann surface.

d) This is not a difficult task: Here is a possible solution.

```
In[1]:= GlobalWeierstrassIterations[poly_, x_, prec_:10^-10] :=
       (* make Rules as the result *)
       {x -> #}& /@ Sort[Chop @ Function[{degree, polyFunc},
       FixedPoint[(* the function to be iterated *)
           # - (polyFunc /@ #)/(* the denominator product *)
             Apply[Times, MapIndexed[#1[[#2[[1]]]] -
             Drop[#1, #2]&, Table[#1, {degree}]], {1}]&,
           (* some random complex starting values *)
           Array[Random[Complex, {-1 - I, 1 + I}]&, degree],
           SameTest -> ((Plus @@ Abs[#1 - #2]) < prec&)]][
               (* determine degree of poly and transform poly → Function *)
               Exponent[poly, x], Function[Evaluate[poly /. x -> #]]]] /;
           PolynomialQ[poly, x] &&
           And @@ (NumberQ /@ N[CoefficientList[poly, x]])
```

The interesting point in the above implementation is the construction of a pure function from poly via Function[
Evaluate[poly /. x -> #]] and the realization of the product with the term with equal summands dropped via
Apply[Times, MapIndexed[#1[[#2[[1]]]] - Drop[#1, #2]&, Table[#1, degree]], {1}] . Let us check
one example.

In[2]:= poly1[x_] := x^6 + 5x^5 + 4x^4 + 3x^3 + 2x^2 + x - 1

```
       {GlobalWeierstrassIterations[poly1[x], x],
        NSolve[poly1[x] == 0, x]}
```

Out[3]= {{{x → -4.19365}, {x → -0.759141 + 0.521848 i}, {x → -0.759141 - 0.521848 i},
         {x → 0.16218 + 0.83589 i}, {x → 0.16218 - 0.83589 i}, {x → 0.387568}},
         {{x → -4.19365}, {x → -0.759141 - 0.521848 i}, {x → -0.759141 + 0.521848 i},
         {x → 0.16218 - 0.83589 i}, {x → 0.16218 + 0.83589 i}, {x → 0.387568}}}

Here is a degree 12 polynomial with complex coefficients solved.

In[4]:= (x /. GlobalWeierstrassIterations[x^12 - I, x]) -
        (x /. NSolve[x^12 - I == 0, x]) // Abs // Max

Out[4]= 2.87856×10^{-15}

Now let us deal with the question if the Weierstrass iteration also produces fractals. Generating fractal graphics means generating many data points. To do this as quickly as possible, we implement a compiled version of the Weierstrass iterations. wp[*p*, *x*] generates a pure function that carries out the Weierstrass root-finding iterations for a polynomial *p* in *x*.

In[5]:= (* for each polynomial *p*, remember the compiled function *)
       wp[p_, x_] := wp[p, x] = Function[
       With[{o = 100, ε = 10.^-10},

```
(* make compiled function *)
Compile[{{z, _Complex, 1}},
       Module[{d = #2, ξ0 = z},
       FixedPointList[Function[ξ, ξ - #1[ξ]/
       Table[Product[If[k === j, 1. + 0. I, ξ[[j]] - ξ[[k]]],
                    {k, d}], {j, d}]], ξ0, o,
                SameTest -> ((Max[Abs[#1 - #2]] < ε)&)]]]][
                (* the  polynomial and the degree *)
                Function[Evaluate[p /. x -> #]], Exponent[p, x]]
```

In the following discussion, we will use a simple cubic polynomial: $x^3 - x - 1 = 0$.

In[7]:= `cf = wp[x^3 - x - 1, x]`

Out[7]= $\text{CompiledFunction}\Big[\{z\}, \text{Module}\Big[\{d\$ = 3, ξ0\$ = z\}, \text{FixedPointList}\Big[$

$\text{Function}\Big[ξ\$, ξ\$ - \dfrac{(-1 - \#1 + \#1^3 \, \&) \, [ξ\$]}{\text{Table}[\prod_{k=1}^{d\$} \text{If}[k === j, 1. + 0.\,i, ξ\$[[j]] - ξ\$[[k]]], \{j, d\$\}]}\Big],$

$ξ0\$, 100, \text{SameTest} \to (\text{Max}[\text{Abs}[\#1 - \#2]] < 1. \times 10^{-10} \, \&)\Big]\Big], \text{-CompiledCode-}\Big]$

The compiled function correctly finds the zeros of the polynomial for random starting points correctly and quickly.

In[8]:= `Table[Function[x, x^3 - x - 1] /@`
       `Last[cf[Table[Random[Complex, 3 {1 + I, -1 - I}], {3}]]],`
       `{1000}] // Abs // Max // Timing`
Out[8]= $\{0.23 \text{ Second}, 8.88178 \times 10^{-16}\}$

The following graphics show that the number of iterations needed to calculate the roots to a given precision form a fractal in the space of the starting values. In the next graphic, we choose the starting values $1 - i$, $z + i$, and $x_r + i\,x_i$. The graphics are over the x_r,x_i-plane and z varies from graphic to graphic.

In[9]:= `wpPicture[z_, pp_, opts___] :=`
       `DensityPlot[Length[cf[{1 - I, z + I, xr + I xi}]],`
       `              {xr, -3, 11}, {xi, -4, 7}, opts,`
       `              Compiled -> False, PlotPoints -> pp, Mesh -> False,`
       `              ColorFunction -> (Hue[(# - 5)/6]&), PlotRange -> All,`
       `              ColorFunctionScaling -> False, AspectRatio -> Automatic]`

In[10]:= `Show[GraphicsArray[wpPicture[#, 400, FrameTicks -> None,`
       `                    DisplayFunction -> Identity]& /@ #]]& /@`
       `              Partition[Table[z, {z, -1, 1, 2/11}], 3]`

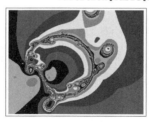 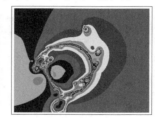 

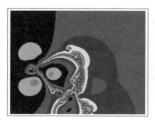 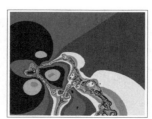 

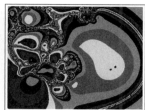

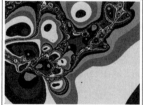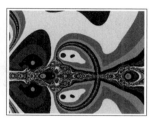

Here is an animation showing how the corresponding 2D fractal changes as a function of the real parameter z.

```
Do[wpPicture[z, 400], {z, -1, 1, 2/40}]
```

By generating a 3D table of data, we can also visualize the corresponding fractal in x_r,x_i,z-space.

```
In[11]:= (* pp×pp×pp array of data *)
        pp = 150;
        data3D = Table[Length[cf[{1 - I, z + I, xr + I xi}]],
                    {xr, -3, 11, 14/pp}, {xi, -4, 7, 11/pp}, {z, -1, 1, 2/pp}];
In[14]:= (* the following Table commands are not compilable;
            so do not even try *)
        Developer`SetSystemOptions[
                    "CompileOptions" -> {"TableCompileLength" -> Infinity}];
In[16]:= (* polygons in the x,y-, x,z-, and y,z-planes *)
        makePolysX[data_, o_] := With[{l = Length[data3D]},
        Table[If[data[[i, j, k]] === o && data[[i + 1, j, k]] =!= o,
                Polygon[{{i + 1/2, j + 1/2, k}, {i + 1/2, j, k + 1/2},
                    {i + 1/2, j - 1/2, k}, {i + 1/2, j, k - 1/2}}],
                    Sequence @@ {}], {i, l - 1}, {j, l}, {k, l}]]

        makePolysY[data_, o_] := With[{l = Length[data3D]},
        Table[If[data[[i, j, k]] === o && data[[i, j + 1, k]] =!= o,
                Polygon[{{i - 1/2, j + 1/2, k}, {i, j + 1/2, k + 1/2},
                    {i + 1/2, j + 1/2, k}, {i, j + 1/2, k - 1/2}}],
                    Sequence @@ {}], {i, l}, {j, l - 1}, {k, l}]]

        makePolysZ[data_, o_] := With[{l = Length[data3D]},
        Table[If[data[[i, j, k]] === o && data[[i, j, k + 1]] =!= o,
                Polygon[{{i - 1/2, j, k + 1/2}, {i, j + 1/2, k + 1/2},
                    {i + 1/2, j, k + 1/2}, {i, j - 1/2, k + 1/2}}],
                    Sequence @@ {}], {i, l}, {j, l}, {k, l - 1}]]
```

In the absence of volume rendering [386], we show all faces around data values 15. There are many polygons in the next graphic.

```
In[22]:= Show[Graphics3D[{EdgeForm[], #[data3D, 15]& /@ (* polygon sets *)
                    {makePolysX, makePolysY, makePolysZ}}], Axes -> False]
```

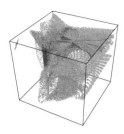

In[23]:= Count[%, _Polygon, Infinity]

Out[23]= 205682

For a multivariate generalization of the Weierstrass method, see [1558].

e) Here is a possible implementation. We form all possible combinations of the coefficients with Permutations. At every stage, the polynomials have the same degree, which allows the list of powers of the polynomials CPowers to be given only once at each stage.

```
In[1]:= iteratedCoefficientRootPicture[
            startCoefficientList_List, iter_Integer, opts___] :=
        Graphics[(* make points in R^2 from complex numbers *)
        {PointSize[0.003], Point[{Re[#], Im[#]}]}& /@
        Flatten[Rest[NestList[(* the iteration *)
          Function[{li}, (* calculate the roots *)
           With[{CPowers = Table[C^i, {i, 0, Length[li[[1]]] - 1}]},
            Cases[NRoots[CPowers.# == 0, C], _?NumberQ, {-1}]& /@
             li]][Flatten[(* all combinations *) Permutations /@ #, 1]]&,
           {startCoefficientList}, iter]]], opts, PlotRange -> Automatic,
            AspectRatio -> 1, Frame -> True] /;
            (And @@ (NumberQ /@ N[startCoefficientList])) &&
             iter <= Length[startCoefficientList] - 1
```

Here are three examples.

```
In[2]:= Show[GraphicsArray[iteratedCoefficientRootPicture @@@
            {{{1, 2, 3, 4, 5}, 2}, {{1, -2, 3, -4, 5}, 2},
             {{1, 1, 1, 1, 1, 1, 1, 1, 1}, 2}}]]
```

Here are three further examples that use random numbers from a given interval.

```
In[3]:= SeedRandom[1];

        Show[GraphicsArray[iteratedCoefficientRootPicture[
            Table[Random[Real, #], {5}], 3]& /@ {{1, 2}, {2, 3}, {8, 9}}]]
```

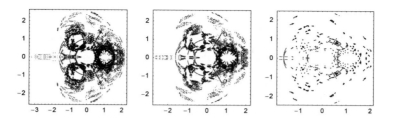

Here are three further examples that use random numbers from a given interval.

f) We start with the construction of a random polynomial with complex coefficients. For a periodic animation, we will demand $p_\tau(z) = p_{\tau+1}(z)$. This means the coefficients will be periodic functions of τ with period 1.

```
In[1]:= SeedRandom[1111];

    (* parameters for the random parametrized polynomial *)
    o = 5; d = 4; m = 2;

    f[z_] = Sum[(Sum[Random[Real, {-1, 1}] Cos[2Pi (k τ + Random[])],
                {k, 0, m}] +
                I Sum[Random[Real, {-1, 1}] Cos[(k τ + Random[])],
                {k, 0, m}]) z^k, {k, 0, d}];

    (* or simpler: f[z_] = Cos[2Pi τ] (1 - z^3) + Sin[2Pi τ] (1 - z^5) *)
```

In the graphics, we will show the number of steps needed until a fixed point is reached in the iterations. We construct a compiled function `newtonSteps` for a quick evaluation of the number of steps.

```
In[7]:= F[τ_, z_] = z - f[z]/f'[z];

    newtonSteps = Compile[{τ, {z, _Complex}},
                Module[{c = 0, ζo = 2. z + 1., ζn = z},
                    While[ζo != ζn && (* cycle case *) c < 100,
                        {ζn, ζo} = {#, ζn}; c++]; c]&[F[τ, ζn]]];
In[10]:= newtonIterationStepsPlot[τ_, pp_, opts___] :=
    DensityPlot[newtonSteps[τ, x + I y], {x, -2, 2}, {y, -2, 2},
                opts, DisplayFunction -> Identity,
                PlotPoints -> pp, Mesh -> False, Compiled -> False,
                PlotRange -> All, FrameTicks -> None,
                (* same coloring for each τ *)
                ColorFunctionScaling -> False, ColorFunction -> col]
```

Here are the resulting fractals for three different values of τ.

```
In[11]:= (* color according to number of iterations; use random colors *)
    SeedRandom[123];
    col[n_] := col[n] = Hue[Random[]]
In[14]:= Show[GraphicsArray[newtonIterationStepsPlot[#, 300]& /@ {0, 1/3, 2/3}]]
```

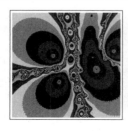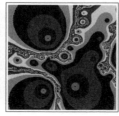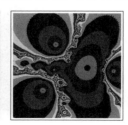

And here is the corresponding animation.

```
With[{frames = 240},
 Do[newtonIterationStepsPlot[τ, 360, DisplayFunction -> $DisplayFunction],
    {τ, 0, 2Pi (1 - 1/frames), 2Pi/frames}]]
```

g) The function `xTildeData` generates *l* random polynomials of degree *d* with real coefficients from the interval $(-cMax, cMax)$. Then it finds the corresponding $\tilde{x}$. If more than one $\tilde{x}$ is found, all are kept.

```
In[1]:= xTildeData[{x0_, X_, n_}, {d_, cMax_}, l_] :=
         Module[{poly, ser},
           Flatten[Table[(* the random polynomial *)
           poly[x_] = Sum[Random[Real, cMax {-1, 1}] x^k, {k, 0, d}];
           (* first n series terms *)
           ser[x_] = Plus @@ MapIndexed[#1/(#2[[1]] - 1)! x^(#2[[1]] - 1)&,
                            NestList[D[#, x]&, poly[x], n - 1] /. x -> x0];
           (* Lagrange remainder *)
           rem[ξ_] = D[poly[ξ], {ξ, n}]/n! (X - x0)^n;
           (* defining equation for ξ *)
           eqs = ser[X] + rem[ξ] - poly[X];
           (* solve for ξ *)
           Select[Cases[NRoots[eqs == 0, ξ], _?NumberQ, {-1}],
                 Im[#] == 0. && 0. <= # <= 1.&], {1}]]]
```

`histogramLine` generates a line corresponding to the histogram of *data* and bin size *binSize*.

```
In[2]:= histogramLine[data_, binSize_] :=
         Line[{{#[[1, 1]], #[[1, 3]]}, {#[[1, 2]], #[[1, 3]]},
              {#[[1, 2]], #[[2, 3]]}}]& /@
         Partition[({First[#] binSize, (First[#] + 1) binSize,
            Length[#]/Length[data]/binSize}& /@
                    Split[Quotient[Sort[data], binSize]]), 2, 1]
```

Here are the histograms for the $\tilde{x}$ for various *n*. We choose $x = 1$, $d = 12$, $cMax = 100$ and use 20000 random polynomials per histogram. One clearly sees how the maximum of the distribution shifts to smaller values with increasing *n* [101], [1461], [1262], [6], [7].

```
In[3]:= Show[Table[Graphics[{Hue[(o - 1)/12], Thickness[0.002],
                histogramLine[xTildeData[{0, 1, o}, {12, 100}, 2 10^4], 0.016]}],
                {o, 1, 8}], PlotRange -> All, Frame -> True]
```

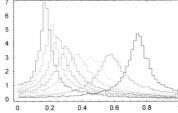

Out[3]= - Graphics -

16. Nodal Line of $\sin(24\,x)\,\sin(y) + 6/5\,\sin(x)\,\sin(24\,y) = 0$

To get an idea of the form of the curve, let us use `ContourPlot`.

```
In[1]:= f[x_, y_] = Sin[24 x] Sin[y] + 6/5 Sin[x] Sin[24 y];

       With[{ε = 10^-12},
        ContourPlot[Evaluate[f[x, y]], {x, ε, Pi}, {y, ε, Pi},
               ContourShading -> False, PlotPoints -> 50, Contours -> {0}]]
```

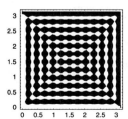

The 50 plot points used are not enough to show all details correctly. At some points, it is not possible to see if the curve does not have multiple components and if it self-intersects. That this is really never the case can be seen by zooming in at some of these points.

```
In[3]:= Show[GraphicsArray[
        Apply[ContourPlot[f[x, y], ##, ContourShading -> False,
                   FrameTicks -> None, Contours -> {0},
                   PlotPoints -> 20, DisplayFunction -> Identity]&,
            (* detailed regions *)
            {{{x, 0.570, 0.600}, {y, 0.700, 0.740}},
             {{x, 2.280, 2.305}, {y, 1.090, 1.130}},
             {{x, 0.977, 0.984}, {y, 1.630, 1.642}}}, {1}]]]
```

The picture to be constructed should then roughly look like the following graphics.

```
In[4]:= Show[%%, ContourShading -> True]
```

The idea for the determination of the curve $\sin(24\,x)\sin(y) + \frac{6}{5}\sin(x)\sin(24\,y) = 0$ is that by the implicit function theorem, it is always possible to write $y = y(x)$ or $x = x(y)$ locally. In both cases, it is easy to derive a differential equation for the curve, which can in principle be solved numerically (and will be solved in practice in a moment).

```
In[5]:= fx = D[f[x, y], x]; fy = D[f[x, y], y];

      yOfxODE[x0_, y0_] = {y'[x] == (-fx/fy /. y -> y[x]), y[x0] == y0}
```

$$\text{Out[6]}= \left\{ y'[x] == -\frac{24\,\text{Cos}[24\,x]\,\text{Sin}[y[x]] + \frac{6}{5}\,\text{Cos}[x]\,\text{Sin}[24\,y[x]]}{\frac{144}{5}\,\text{Cos}[24\,y[x]]\,\text{Sin}[x] + \text{Cos}[y[x]]\,\text{Sin}[24\,x]},\ y[x0] == y0 \right\}$$

```
In[7]:= xOfyODE[x0_, y0_] = {x'[y] == (-fy/fx /. x -> x[y]), x[y0] == x0}
```

$$\text{Out[7]}= \left\{ x'[y] == -\frac{\frac{144}{5}\,\text{Cos}[24\,y]\,\text{Sin}[x[y]] + \text{Cos}[y]\,\text{Sin}[24\,x[y]]}{24\,\text{Cos}[24\,x[y]]\,\text{Sin}[y] + \frac{6}{5}\,\text{Cos}[x[y]]\,\text{Sin}[24\,y]},\ x[y0] == x0 \right\}$$

As the above picture shows, there are many points on the curve where the tangent is parallel to the x- or y-axis. At these points, one of the two forms, either $y = y(x)$ or $x = x(y)$, breaks down and we must switch to the other one. If this switching process is properly organized, we can construct the whole curve by using a stepwise solution of a differential equation for either $y'(x)$ or for $x'(y)$. We begin in the lower right edge with a differential equation for $y'(x)$. The starting value for $x = \pi$ is determined by the solution of the equation $24\sin(y) - 6/5\sin(24\,y)$, which is obtained by series expansion of the above implicit equation. To make sure that we always stay on the true curve and do not accidentally jump to a wrong piece at the nearest of the approaching other pieces, we use *Mathematica*'s high-precision arithmetic. To keep track of whether $y'(x)$ or $x'(y)$ was solved in the ith step and if it was solved from left to right or right to left, respectively, from bottom to top or top to bottom, we use the variable horVer and direction. In this process, the points where this will happen are not known in advance.

Here, the first step is carried out (we turn off the message NDSolve::mxst because due to the infinite slope encountered at the ends of the integration interval, it is impossible to continue the solution of this differential equation).

```
In[8]:= yStart = y /. FindRoot[24 Sin[y] - 6/5 Sin[24 y], {y, 4/100},
                               WorkingPrecision -> 30, AccuracyGoal -> 25]
Out[8]= 0.0428152473428595117770433114668

In[9]:= (* turning off this NDSolve message because we only need an
           approximation for the singular point *)
        Off[NDSolve::mxst]; Off[NDSolve::ndsz];

        i = 1; horVer[i] = x; direction[i] = -1;

        sol[i] = NDSolve[yOfxODE[N[Pi, 30], yStart],
                         {y}, {x, N[Pi, 30], 0}][[1, 1, 2]]
Out[12]= InterpolatingFunction[{{0.101942, 3.14159}}, <>]
```

We will use a predictor–corrector method in the following code [40]. Now, the next step would be the solution of the above differential equation for $x'(y)$ upwards, and so on. To make sure that our starting point is always on the true curve, we refine the values we extract from the solution of the differential equation by the following routine.

```
In[13]:= refine[ξ_, η_] :=
         With[{ε = 10^-2},
          {x, y} /. (FindMinimum[Evaluate[f[x, y]^2], (* in small neighborhood *)
                                 {x, ξ, ξ - ε, ξ + ε}, {y, η, η - ε, η + ε},
                                 WorkingPrecision -> 30, AccuracyGoal -> 30,
                                 MaxIterations -> 50])[[2]]]
```

So, we can now implement one solution step in the routine step (the ε ensures that we do not use the endpoint, but a point slightly inside the solution interval).

```
In[14]:= ε = 10^-6;
         (* turning off this FindMinimum message because;
            we need a fairly accurate minimum at each step *)
         Off[FindMinimum::fmgz];

         step :=
         ((* which is the endpoint of the last step *)
          newPoint = If[direction[i] < 0, sol[i][[1, 1, 1]] + ε,
```

```
                                              sol[i][[1, 1, 2]] - ε];
        (* what is the new direction? (determined from slope of last step) *)
        dir = direction[i] * (D[sol[i][xy], xy] /. {xy -> newPoint});
        i = i + 1;
        direction[i] = Sign[dir];
        (* auxiliary function to raise precision *)
        sp = SetPrecision[#, 40]&;
    Which[(* last step was roughly parallel x-axis *)
            horVer[i - 1] === x,
            (* then next step is roughly parallel y-axis *)
            horVer[i] = y;
            (* starting point and function value for the next step
               extracted from the result of the last step *)
            ξ = newPoint; η = sol[i - 1][newPoint];
            (* refine them *)
            {ξ, η} = refine[sp @ ξ, sp @ η];
             (* solve differential equation for next step *)
            sol[i] = NDSolve[xOfyODE[ξ, η],
                    {x}, (* how far to solve in dependence of direction[i] *)
                    {y, η, If[direction[i] === 1, N[Pi - ε, 30], ε]}][[1, 1, 2]],
            (* last step was roughly parallel y-axis *)
            horVer[i - 1] === y,
            (* then next step is roughly parallel x-axis *)
            horVer[i] = x;
            (* starting point and function value for the next step
               extracted from the result of the last step *)
            η = newPoint; ξ = sol[i - 1][newPoint];
            (* refine them *)
            {ξ, η} = refine[sp @ ξ, sp @ η];
            (* solve differential equation for next step *)
            sol[i] = NDSolve[yOfxODE[ξ, η],
                    {y}, (* how far to solve in dependence of direction[i] *)
                    {x, ξ, If[direction[i] === 1, N[Pi - ε, 30], ε]}][[1, 1, 2]]])
```

Now, we iterate the alternating solution of the differential equations for $y'(x)$ and $x'(y)$ until we reach the upper left edge (which is actually detected in the next step).

```
In[18]:= While[i === 1 || Min[{ξ, η}] > 10^-3, step]
```

Using the information stored in horVer[*i*], we can assemble all pieces to get the implicitly-defined curve under consideration. By joining the various line pieces now into one polygon, we finally get the right picture to be constructed.

```
In[19]:= Show[GraphicsArray[
    Block[{$DisplayFunction = Identity},
      {(* make curves from differential equation solutions *)
       Show[ (* the single solutions *) Table[parPlot[i] =
       ParametricPlot[Evaluate[ (* parallel; x- or y-axis? *)
                    If[horVer[i] === x, {xy, sol[i][xy]}, {sol[i][xy], xy}]],
                    (* the parameter range *)
                    Evaluate[Flatten[{xy, sol[i][[1]]}]],
                    DisplayFunction -> Identity, PlotPoints -> 200],
                    {i, Length[DownValues[sol]] - 1}], Frame -> True,
                    AspectRatio -> Automatic, PlotRange -> {{0, Pi}, {0, Pi}}],
        (* form polygon *)
       Graphics[Polygon[bigPoly = Join[Join @@ (* start[i] after end[i - 1] *)
       Table[If[direction[i] < 0, Reverse[#], #]&[parPlot[i][[1, 1, 1, 1]]],
            {i, Length[DownValues[sol]] - 1}], N[{{0, Pi}, {Pi, Pi}}]]]],
            AspectRatio -> Automatic, Frame -> True]}]]]
```

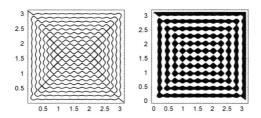

This polygon consists of more than 20000 points. The next input counts the number of points and the maximal distance between two points.

```
In[20]:= {Length[bigPoly ],
         Sort[Norm /@ Apply[Subtract, Partition[bigPoly, 2, 1], {1}]][[
              (* not counting the two sides of the square *) - 3]]}
Out[20]= {23264, 0.172923}
```

By making use of two differential equations for the coordinates $\xi = x$ and $\eta = y$ as a function of the arclength s, we can calculate a parametrization of the whole nodal line at once (for the calculation of the general intersection of two parametrized surfaces, see [655]). Here are these two differential equations.

```
In[21]:= ξxAndηyOfsODEs = Thread[{ξx'[s], ηy'[s]} == (* direction cosines *)
             (#/Sqrt[#.#]&[{D[f[ξx, ηy], ηy], -D[f[ξx, ηy], ξx]} /.
                         {ξx -> ξx[s], ηy -> ηy[s]}])];
```

We start solving these differential equations from a nondegenerate point and follow the nodal line curve in both directions.

```
In[22]:= ηyStart = ηy /. FindRoot[Evaluate[f[ξxStart = 3, ηy]], {ηy, 1/10},
                          WorkingPrecision -> 30, AccuracyGoal -> 25]
Out[22]= 0.139657702210554506037841719252
```

```
In[23]:= (* large enough S to cover the whole nodal curve *) S = 50;
         ndsol = NDSolve[Join[{ξxAndηyOfsODEs,
                        {ξx[0] == ξxStart, ηy[0] == ηyStart}],
                    {ξx, ηy}, {s, -S, S}, MaxSteps -> 10^5,
                    (* use "Projection" method for a high-precision solution *)
                    PrecisionGoal -> 10, AccuracyGoal -> 10];
```

We obtain again the nodal line from above.

```
In[25]:= ParametricPlot[Evaluate[{ξx[s], ηy[s]} /. ndsol[[1]]], {s, -S, S},
                    PlotPoints -> 1200, Frame -> True,
                    AspectRatio -> Automatic]
```

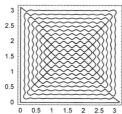

We end with an animation showing how the regions $(1 - \alpha) \sin(24\,x) \sin(y) + \alpha \sin(x) \sin(24\,y) \lessgtr 0$ depend on α.

```
In[26]:= cp[α_, {n_, pp_}] :=
         Show[{Graphics[ContourPlot[
                 (1 - α) Sin[1 x] Sin[24 y] + α Sin[24 x] Sin[1 y],
                 ##, Contours -> {0}, DisplayFunction -> Identity,
                 Frame -> False, PlotPoints -> 40]]& @@@
             (* plot domains *)
             (Table[{{x, i Pi/n, (i + 1) Pi/n}, {y, j Pi/n, (j + 1) Pi/n}},
                     {i, 0, n - 1}, {j, 0, n - 1}] // Flatten[#, 1]&),
```

```
           Graphics[{Thickness[0.01], Hue[0], (* frame *)
                   Line[{{0, 0}, {Pi, 0}, {Pi, Pi}, {0, Pi}, {0, 0}}]}]}]
In[27]:= With[{ε = 10^-3},
          Show[GraphicsArray[#]]& /@
          Partition[Table[cp[α, {12, 30}], {α, ε, 1 - ε, (1 - 2 ε)/8}], 3]]
```

```
With[{ε = 10^-3},
    Do[Show[cp[α, {12, 30}], DisplayFunction -> $DisplayFunction],
       {α, ε, 1 - ε, (1 - 2 ε)/36}]]
```

For bounds on the length of nodal lines of vibrating membranes, see [303], [1085]; for the number of domains with different sign, see [694], [1671]. For interesting nodal lines in 3D, see [485].

17. Branch Cuts of an Elliptic Curve, Strange 4D Attractors

a) These are the zeros of the polynomial under the square root.

```
In[1]:= zeros = Join[{0}, Table[1/4 Exp[2Pi I i/7]          , {i, 0, 6}],
                    Table[1/2 Exp[2Pi I i/5 + I Pi/10], {i, 0, 4}],
                    Table[3/4 Exp[2Pi I i/3 + I Pi/06], {i, 0, 2}]] // N;
```

Here is the polynomial.

```
In[2]:= poly = Times @@ ((z - #)& /@ zeros);
```

The function of interest has branch cuts when the value of `poly`, depending on the complex variable z, takes negative real values. The zeros `zeros` are the branch points of `poly`.

With `ContourPlot`, we generate a picture of the curves where the imaginary part of `poly` vanishes. We transform the so-generated `ContourGraphics` into a `BackChapterGraphics`, and the curves are then represented as `Lines`.

Because the branch cut extends only along negative real values of `poly`, we check all points of the curves with respect to this property and delete all points where `poly` has a positive real part. Here, this is implemented.

```
In[3]:= Show[Graphics[{{Thickness[0.002], GrayLevel[0],
        DeleteCases[#, (* check if real part of poly is less than 0 *)
                _?(Re[poly /. (z :> Plus @@ (# {1, I}))]) > 0 ||
                    (* display only those pieces inside a disk of radius 3/2 *)
                    #.# > 9/4&), {-2}]& /@
        (Last /@ (* transform into a Graphics-object *)Graphics[
        (* the ContourPlot of the lines with Im[poly] == 0 *)
        ContourPlot[Evaluate[Im[poly /. z -> (x + I y)]],
                {x, -3/2, 3/2}, {y, -3/2, 3/2}, Contours -> {0},
                PlotPoints -> 200, ContourShading -> False,
                DisplayFunction -> Identity]][[1]])},
        (* the branch points *)
        {PointSize[0.01], Point[{Re[#], Im[#]}]& /@ zeros}}],
        PlotRange -> All, AspectRatio -> Automatic,
        Frame -> True, Axes -> False]
```

Along the branch cuts, we have for the polynomial $p(z) = \prod_{i=0}^{n}(z(t) - z_i)$ the identity $p(z) = -t$. Here z_i are the zeros of $p(z)$ and $t \geq 0$. Differentiating this equation with respect to t gives a first-order differential equation for the branch cuts. To achieve a higher resolution near the branch points ($t \approx 0$), we change in the following, the independent variable form $t \to \exp(t)$. Writing the differential equation in the form $z'(t) = rhs$, we find by Leibniz's rule the right-hand side `rhs`.

```
In[4]:= rhs = -1/Sum[Drop[poly, {i}], {i, Length[zeros]}] Exp[t] /. {z -> z[t]};
```

Next, we determine the t-interval, in which to solve the differential equations for the branch cuts. The lower limit we determine by the condition that the first points are quite close to the branch points (in the order of at least 10^{-8}).

```
In[5]:= startExponent =
        Min[-Ceiling[Log[10, Abs[Function[deri, (deri /. z -> #)& /@ zeros][
                (* 1/(value of derivative at the branch points) *)
                1/Sum[Drop[Times @@ ((z - #)& /@ zeros), {i}],
                {i, Length[zeros]}]]]]]](* to be on the safe side *) - 8]
Out[5]= -15
```

We determined the upper limit so that the most distant point of the branch cut calculated here is, on the average, a distance 1.5 from the origin.

```
In[6]:= f[eE_?InexactNumberQ] := (* average distance *)
        (Plus @@ (Abs /@ (List @@ (Last /@ (* endPoint *)
        NRoots[poly == -10^eE, z]))))/Length[zeros])

        endExponent = endExponent /. FindRoot[f[endExponent] == 1.5,
                                        {endExponent, 1, 2},
                                        Compiled -> False]
Out[8]= 2.81735
```

Solving the equation `poly == -10^startExponent`, we get the initial conditions for the differential equations of the branch cuts.

```
In[9]:= z0List = List @@ (Last /@ NRoots[poly == -10^startExponent, z]);
```

Now, we actually solve the differential equations for the branch cuts.

```
In[10]:= Do[branchCut[i] =
          NDSolve[{z'[t] == rhs, z[N[Log[10^startExponent]]] == z0List[[i]]}, z,
                  {t, Log[10^startExponent], Log[10^endExponent]}][[1, 1, 2]],
          {i, Length[zeros]}]
```

We display the result.

```
In[11]:= bcp = Show[{Table[ (* the branch cuts *)
         ParametricPlot[Evaluate[{Re[branchCut[i][t]], Im[branchCut[i][t]]}],
                        {t, Log[10^startExponent], Log[10^endExponent]},
                        PlotRange -> All, AspectRatio -> Automatic,
                        DisplayFunction -> Identity, PlotPoints -> 250,
                        PlotStyle -> {Thickness[0.002]}], {i, Length[zeros]}],
                        (* the branch points *)
              Graphics[{PointSize[0.01], Point[{Re[#], Im[#]}]& /@ zeros}]},
         DisplayFunction -> $DisplayFunction, Frame -> True, Axes -> False,
         PlotRange -> {{-3/2, 3/2}, {-3/2, 3/2}}]
```

Using the following, we could now look at the points where two branch cuts seemingly cross in more detail.

```
In[12]:= Show[GraphicsArray[{ (* zoom into avoided crossings *)
         bcp /. {(PlotRange -> {{-3/2, 3/2}, {-3/2, 3/2}}) ->
                    (PlotRange -> {{-0.02, 0.02}, {-0.68, -0.72}}),
                 (FrameTicks -> Automatic) ->
                    (FrameTicks -> {{-0.01, 0.01}, {-0.69, -0.71}})},
         bcp /. {(PlotRange -> {{-3/2, 3/2}, {-3/2, 3/2}}) ->
                    (PlotRange -> {{-0.02, 0.02}, {0.44, 0.48}}),
                 (FrameTicks -> Automatic) ->
                    (FrameTicks -> {{-0.01, 0.01}, {0.45, 0.47}})}}]]
```

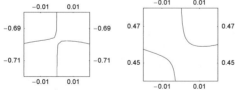

For an application of the knowledge of these branch cuts, see [848] and [1119].

b) We will carry out random searches to find such systems. The function makeODESystem generates a random system of differential equations. The parameter o indicates the total order of the polynomials on the right-hand sides and p the probability of a single monomial term to be present. For the coefficients, we use rationals with small denominators.

```
In[1]:= makeODESystem[seed_, o_, p_, {{x_, y_, z_, u_}, t_}] :=
        Module[{r, mR = 12, μR = 256},
        SeedRandom[seed]; theSeed = seed; Save["seeds", theSeed];
        (* random rational number from interval [-m, m] *)
        r[m_, μ_] := (Random[Integer, m {-#, #}]/#)&[Random[Integer, {1, μ}]];
        (* ODEs and initial conditions *)
        Join[Thread[D[{x[t], y[t], z[t], u[t]}, t] ==
            Table[Sum[If[i + j + k > o, 0, (* random products *)
```

```
          If[Random[] < p, r[mR, μR] x[t]^i y[t]^j z[t]^k u[t]^l, 0]],
            {i, 0, o}, {j, 0, o}, {k, 0, o}, {l, 0, o}], {4}]],
      Thread[{x[0], y[0], z[0], u[0]} == Table[r[mR, μR], {4}]]]]]
```

Here is an example of a system of differential equations and initial conditions.

In[2]:= `makeODESystem[1, 2, 1/8, {{x, y, z, u}, t}]`

Out[2]= $\left\{ x'[t] = -\dfrac{15\,z[t]}{7}, \right.$

$y'[t] = -\dfrac{37}{18}\,x[t]^2 + \dfrac{26}{5}\,x[t]\,y[t] + \dfrac{126}{25}\,u[t]^2\,x[t]\,y[t] + \dfrac{45}{4}\,u[t]\,z[t]^2,\ z'[t] =$

$-\dfrac{391\,u[t]}{229} + \dfrac{2511}{241}\,u[t]^2\,x[t] + \dfrac{171\,x[t]^2}{26} - \dfrac{425}{107}\,u[t]\,y[t]^2 - \dfrac{349\,z[t]}{66} + \dfrac{177}{139}\,u[t]^2\,x[t]\,z[t],$

$u'[t] = \dfrac{141}{55} - \dfrac{141}{20}\,u[t]\,x[t]\,y[t] + \dfrac{87\,z[t]}{25},\ x[0] = -\dfrac{83}{12},$

$\left. y[0] = \dfrac{165}{19},\ z[0] = -\dfrac{119}{179},\ u[0] = -\dfrac{423}{41} \right\}$

To check if a given system has a strange attractor-shaped solution, we solve the system numerically. We stop the solution in case the solution curve escapes to infinity.

In[3]:= `solveODESystem[odes_, {x_, y_, z_, u}, {t_, t1_, t2_}, tM_] :=`
```
      Module[{tc, ndsol},
        tc = TimeConstrained[
        ndsol = NDSolve[odes, {x, y, z, u}, {t, t1, t2}, MaxSteps -> 10^6,
                      (* solve with relatively low precision for efficiency *)
                      PrecisionGoal -> 4, AccuracyGoal -> 4, StoppingTest :>
                          (Max[Abs[{x[t], y[t], z[t], u[t]}]] < 10^4)], tM];
      (* return $Failed in case no solution was achieved *)
      If[tc =!= $Aborted && ndsol =!= {} && ndsol[[1, 2, 2, 1, 1, 2]] == t2,
        ndsol, $Failed]]
```

To identify strange attractor-shaped solutions, we first solve the differential equations for a time *t1* and then analyze the solution for a time *t2 − t1*. When the solution curve extends over a spatial domain of approximate size 10^0 and the solution curve fills a not-too-small fraction of the parallelepiped that bounds the solution curve, we have a candidate for a strange attractor.

In[4]:= `interestingODESolutionQ[ndsol_, {x1_, x2_, x3_}, pc_] :=`
```
      Module[{pp = 10^4, do = 100},
        (* make points along solution curve *)
        {t1, t2} = ndsol[[1, 1, 2, 1, 1]];
        points = Table[Evaluate[{x1[t], x2[t], x3[t]} /. ndsol[[1]]],
                      {t, t1, t2, (t2 - t1)/pp}];
        (* bounding intervals for solution curve *)
        {{x1l, x1u}, {x2l, x2u}, {x3l, x3u}} =
                      {Min[#], Max[#]}& /@ Transpose[points];
        {λx1, λx2, λx3} = Abs[Subtract @@@
                          {{x1l, x1u}, {x2l, x2u}, {x3l, x3u}}];
        (* solution curve is not too small and not too big and
           extends over a given fraction of the partitioned bounding cube *)
        If[Min[{λx1, λx2, λx3}] > 10^-2 && Max[{λx1, λx2, λx3}] > 1,
          fillDegree = Length[Union[Round[do (# -
              {x1l, x2l, x3l})/{λx1, λx2, λx3}]& /@ points]]/do^3 100.;
          fillDegree > pc, False]]
```

We automate the search steps in the function `findInterestingSystem`. Starting with the initial seed *seed* for the random number generator, it increases the seeds for the generation of the ODE systems until a candidate for a strange attractor solution is found. (The solutions will be qualitatively correct; for quantitatively correct solutions we should use high-precision arithmetic and larger settings for the precision and accuracy goals.)

In[5]:= `findInterestingSystem[startSeed_, o_, p_] :=`
```
      Module[{odes, ndsol, seed = startSeed},
        While[(* generate ODEs and initial conditions *)
            odes = makeODESystem[seed, o, p, {{x, y, z, u}, t}];
            (* solve ODEs and initial conditions *)
            ndsol = solveODESystem[odes, {x, y, z, u}, {t, 10, 20}, 10];
            (* did solution succeed and
               is solution probably a strange attractor? *)
```

```
Not[ndsol =!= $Failed &&  (* check filling in two subspaces *)
    interestingODESolutionQ[ndsol, {x, y, z}, 0.1] &&
    interestingODESolutionQ[ndsol, {x, y, u}, 0.1]],
  seed++]; (* remember seed *) goodSeed = seed; odes]
```

Once we have found a strange attractor, we will visualize the solution. We plot the 4D curve in the four 3D coordinate-plane projections. To solve the differential equations, we use an arc length parametrization which allows for a more controlled plotting resolution.

```
In[6]:= makeGoodGraphics[odes_, {x_, y_, z_, u_}, {t_, t1_}, Δt_, pp_] :=
  Module[{ndsolPre, r, odesScaled, ndsol, points},
    (* initial numerical solution to come to the attractor *)
    ndsolPre = NDSolve[odes, {x, y, z, u}, {t, t1, t1}, MaxSteps -> 10^6];
    (* make ODEs with arc length parametrization *)
    r = Last /@ Take[odes, 4] ;
    odesScaled = Join[Thread[D[{x[t], y[t], z[t], u[t]}, t] == r/Sqrt[r.r]],
                  Thread[{x[0], y[0], z[0], u[0]} ==
                         ({x[t1], y[t1], z[t1], u[t1]} /. ndsolPre[[1]])]];
    (* numerical solution of reparametrized ODEs *)
    ndsol = NDSolve[odesScaled, {x, y, z, u}, {t, 0, Δt}, MaxSteps -> 10^6];
    (* points along solution curve *)
    points = Table[Evaluate[{x[t], y[t], z[t], u[t]} /. ndsol[[1]]],
                   {t, 0, Δt, Δt/pp}];
    GraphicsArray[(* show all four projections in 3D *) Function[ps,
    Graphics3D[{{Thickness[0.002],   (* color along solution curve *)
               MapIndexed[{Hue[0.8 #2[[1]]/pp], #1}&,
                          Line /@ Partition[Map[#[[ps]]&,
                                                points, {-2}], 2, 1]]}},
               PlotRange -> All, Axes -> False, BoxRatios -> {1, 1, 1}]] /@
       (* coordinates for the 3D spaces *)
       {{1, 2, 3}, {1, 2, 4}, {1, 3, 4}, {2, 3, 4}}, GraphicsSpacing -> 0.01]]
```

Now, we have all ingredients together to search for a strange attractor solution curve. We start with some searches for $o = 4$, $p = 0.2$. After a few thousand trials, we obtain a strange attractor candidate. But for $pc = 0.1$, not each found solution curve is a strange attractor.

```
In[7]:= Off[NDSolve::ndsz]; Off[NDSolve::mxst]; Off[NDSolve::nderr];
  Off[NDSolve::ndcf]; Off[NDSolve::evcvmit];

  Timing[odesI = findInterestingSystem[(* seed *) # 10^6, 4, 1/5];
    Show @ makeGoodGraphics[odesI, {x, y, z, u}, {t, 10}, 10^3, 2 10^4];
    (* return seed and fill degree *) {goodSeed, fillDegree}]& /@ {2, 7, 4}
```

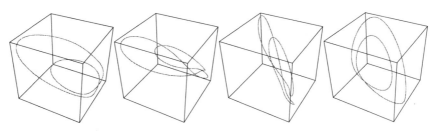

Out[9]= {{412.08 Second, {2001204, 0.3135}},
        {287.36 Second, {7000950, 0.1403}}, {575.77 Second, {4001559, 0.1168}}}

We end with some solution curves (not all being strange attractors) with the above routines.

In[10]:= (odesI = makeODESystem[#, 4, 1/5, {{x, y, z, u}, t}];
        Show @ makeGoodGraphics[odesI, {x, y, z, u}, {t, 10}, 10^3, 2 10^4])& /@
        (* the seeds for making the ODEs *)
        {(* semi-regular *)   439052578665, 196368633124, 045597599057,
         (* thin-spread *)   441841414438, 637670206592, 906002460869,
         (* medium spread *) 426422779064, 955493010259, 428100589446,
         (* dense filling *) 392994197074, 725935069823, 544122158628,
         (* "unusual" ones *) 042069052090, 846524572808, 468862754188
         (* some more seeds: 372589881752, 18139672839, 748058676295,
                              734115549094, 617985850421, 403959359454 *)}

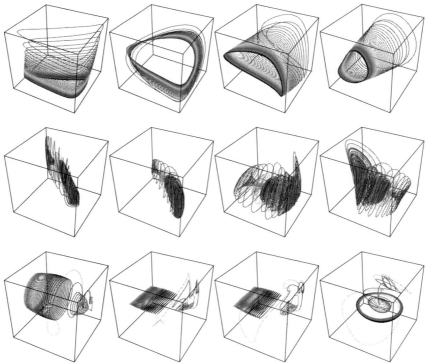

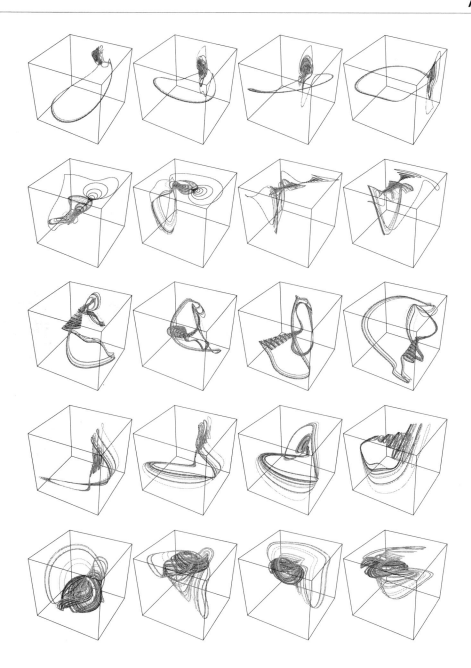

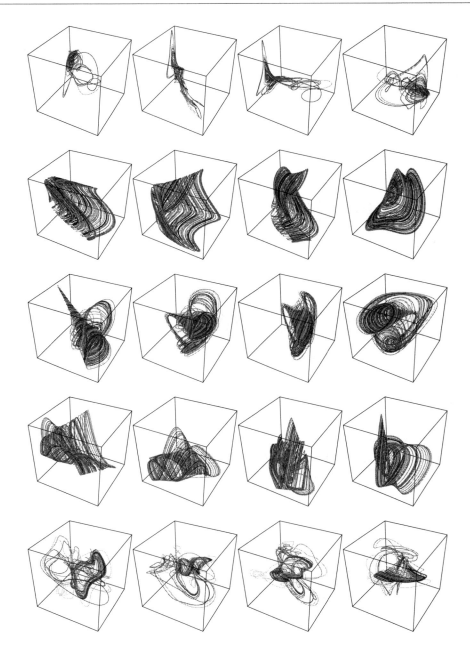

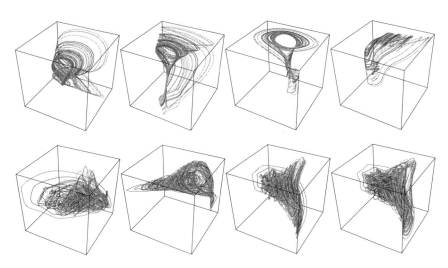

18. Differently Colored Spikes, Billiard with Gravity

a) Let us first have a look at the picture.

```
In[1]:= f[φ_, ϑ_] := Function[r, (Sign[#] Abs[#]^(5/3))& /@
                        N[r {Cos[φ] Sin[ϑ], Sin[φ] Sin[ϑ], Cos[ϑ]}]]][N @
                        Abs[(3465/16(1 - Cos[ϑ]^2)^(3/2) (221 Cos[ϑ]^6 -
                        195 Cos[ϑ]^4 + 39 Cos[ϑ]^2 - 1)) Cos[4φ]]]


     ParametricPlot3D[f[φ, ϑ], {φ, 0, 2Pi}, {ϑ, 0, Pi},
                     PlotPoints -> {41, 41}, Compiled -> False,
                     PlotRange -> All, BoxRatios -> {1, 1, 1},
                     Boxed -> False, Axes -> False]
```

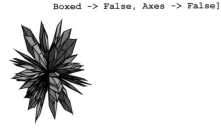

This is the ϑ-dependent radius defining function.

```
In[3]:= ϑFunc[ϑ_] = (3465/16 (1 - Cos[ϑ]^2)^(3/2)*
                     (221 Cos[ϑ]^6 - 195 Cos[ϑ]^4 + 39 Cos[ϑ]^2 - 1));
```

The spikes are caused by positive values of $rad = \text{Abs}[\vartheta\text{Func}[\vartheta]]$. The spikes are separated from each other by the zeros of rad in altitude and azimuth. Because rad factors into a function depending only on φ and a function depending only on ϑ, $rad = f(\varphi) g(\vartheta)$, the calculation of the zeros and the calculation of a tensor product mesh that separates all spikes is straightforward. The zeros for φ can be calculated analytically.

```
In[4]:= φZeros = Append[Rest[Sort[N[Table[Pi/8 + n Pi/4, {n, 0, 7}]]]],
                        N[2Pi + Pi/4]]
Out[4]= {1.1781, 1.9635, 2.74889, 3.53429, 4.31969, 5.10509, 5.89049, 7.06858}
```

For the calculation of the zeros of the ϑ dependent part, we use `FindRoot` after we have roughly located the zeros.

```
In[5]:= ϑZeros = Append[Apply[ϑ /. FindRoot[ϑFunc[ϑ] == 0, {ϑ, ##}]&,
                     (* rough search for the zeros *)
                     Map[First, Select[Partition[Table[{ϑ, ϑFunc[ϑ]},
```

```
                        {ϑ, 0, N[Pi], N[Pi]/60}], 2, 1],
                  #[[1, 2]] #[[2, 2]] < 0&], {2}], {1}], N[Pi]]
```
Out[5]= {0.683233, 1.04614, 1.39676, 1.74483, 2.09545, 2.45836, 3.14159}

Now, we choose a color for every spike randomly.

```
In[6]:= spikeColors = Table[Hue[Random[]], {7}, {8}];
```

To map the color independence of φ and ϑ, we construct a nested Which that returns one of the colors from spikeColors.

```
In[7]:= color[φ_, ϑ_] = (* list and test all cases *)
        Which @@ Flatten[Transpose[{ϑ <= #& /@ ϑZeros,
              Table[Which @@ Flatten[Transpose[{φ <= #& /@ φZeros,
                spikeColors[[i]]}]], {i, Length[spikeColors]}]}]]];
In[8]:= Short[color[φ, ϑ], 12]
```
Out[8]//Short= Which[ϑ ≤ 0.683233, Which[φ ≤ 1.1781, Hue[0.808141], φ ≤ 1.9635, Hue[0.560711], φ ≤ 2.74889,
              Hue[0.25434], φ ≤ 3.53429, Hue[0.707722], φ ≤ 4.31969, Hue[0.167227], φ ≤ 5.10509,
              Hue[0.241018], φ ≤ 5.89049, Hue[0.0577417], φ ≤ 7.06858, Hue[0.368221]], ϑ ≤ 1.04614,
              ≪9≫, ϑ ≤ 3.14159, Which[φ ≤ 1.1781, Hue[0.282595], φ ≤ 1.9635, Hue[0.0846292],
              φ ≤ 2.74889, Hue[0.853011], φ ≤ 3.53429, Hue[0.877767], φ ≤ 4.31969, Hue[0.331679],
              φ ≤ 5.10509, Hue[0.609202], φ ≤ 5.89049, Hue[0.318715], φ ≤ 7.06858, Hue[0.985084]]]]

Now, we have everything together to display the colored spikes. The right graphic uses many more plotpoints and makes individual plots for each rectangle of equal color in the φ,ϑ-parametrization plane.

```
In[9]:= fc[φ_, ϑ_] := Append[f[φ, ϑ], color[φ, ϑ]]

        (* partition into intervals *)
        parti[l_, k_] :=
         Partition[Select[N[Flatten[{0, l, k Pi}]], 0 <= # <= k Pi&], 2, 1]
In[12]:=
        Show[GraphicsArray[
        Block[{$DisplayFunction = Identity},
        {(* change color function depending on parameter values *)
        ParametricPlot3D[fc[φ, ϑ], {φ, Pi/4, 2Pi + Pi/4}, {ϑ, 0, Pi},
                        PlotPoints -> {41, 41}, Compiled -> False,
                        PlotRange -> All, Lighting -> False, Axes -> False,
                        BoxRatios -> {1, 1, 1}, Boxed -> False],
           (* plot each single-color parameter region individually *)
        Show[DeleteCases[Internal`DeactivateMessages[
        ParametricPlot3D[fc[φ, ϑ], Evaluate[Flatten[{φ, #1}]],
                        Evaluate[Flatten[{ϑ, #2}]],
                        PlotPoints -> {21, 21}, Compiled -> False,
                        DisplayFunction -> Identity]]& @@@
           Flatten[Outer[List, (* the φ,ϑ-domains of uniform color *)
                        parti[φZeros, 2], parti[ϑZeros, 1], 1], 1],
                        _ParametricPlot3D, Infinity] /. (* make surface color *)
        h_Hue :> SurfaceColor[h, h, 2.2] /. p_Polygon :> {EdgeForm[], p},
        DisplayFunction -> $DisplayFunction, PlotRange -> All,
        BoxRatios -> {1, 1, 1}, Boxed -> False, Axes -> False}]]]]
```

b) The curve that reflects the particle is built from a collection of half circles of radius $1/2$ with their centers at the x-axis at integer values. For a given x, the function $y[x]$ calculates the corresponding y-value.

```
In[1]:= 𝒴[x_] := If[EvenQ[Round[x]], -1, 1] Sqrt[1/4 - (x - Round[x])^2]
```

Here is the curve shown. At $x = 0$, we have a local minimum.

```
In[2]:= wallPlot = Plot[𝒴[x], {x, -3, 3},
           AspectRatio -> Automatic, Frame -> True, Axes -> False,
           FrameTicks -> {Table[k, {k, -4, 4}], Automatic, None, None}]
```

To model a particle bouncing off this curve, we proceed in two steps. First, we determine when and where a given particle trajectory will collide with the curve and then we carry out the ideal reflection. The function `circleParabolaIntersec tion` calculates the intersection between the particle trajectory (a parabola) and the *n*th half circle. The data for the free flight of the particle, we will use in the form `{{x0, y0}, {vx0, vy0}, t0}`. Here $\{x0, y0\}$ is the starting point, $\{vx0, vy0\}$ the starting velocity, and $t0$ the starting time. We use the optional argument *prec* to specify a working precision for solving the resulting quartic polynomial in t.

```
In[3]:= circleParabolaIntersection[n_, {{x0_, y0_}, {vx0_, vy0_}, t0_},
                                    prec___] :=
    Module[(* measure time with respect to t0 *)
    {(* real intersection times *)
     tRoots = Cases[NRoots[(* polynomial quartic in t *)
                         ((vx0 t + x0) - n)^2 +
                         (-g/2 t^2 + vy0 t + y0)^2 == 1/4, t, prec],
                  _Real?(# > 10^-12&), {-1}]},
     (* lower or upper half circle *)
     tRoots = Select[tRoots, If[EvenQ[n], # < 0, # > 0]&[
                                      -g/2 #^2 + vy0 # + y0]&];
     (* if real intersection exists, take first *)
     If[Length[tRoots] >= 1, Sort[tRoots][[1]], {}]]]
```

The function `nextCollision` updates the trajectory data from one collision with the curve to the next. Its main work is to determine at which circle the next collision happens. The two auxiliary functions `maxWallHeightIntersection` and `reflect` calculate when a given trajectory reaches the *y*-value 1/2 and carry out the actual reflection step.

```
In[4]:= (* when does the trajectory reach the highest curve point? *)
    maxWallHeightIntersection[{{x0_, y0_}, {vx0_, vy0_}, t0_},
                               prec___] :=
    If[-2 + vy0^2 + 2y0 < 0, (* highest point < 1 *) {},
       x0 + (vy0 + {+1, -1} Sqrt[vy0^2 + 2 y0 - 2]) vx0]
In[6]:= (* carry out the reflection *)
    reflect[{{x0_, y0_}, V:{vx0_, vy0_}, t0_}] :=
    With[{n = 2({x0, y0} - {Round[x0], 0})},
         {{x0, (* keep on curve *) 𝒴[x0]}, V - 2 V.n n , t0}]
In[8]:= nextCollision[d:{{x0_, y0_}, {vx0_, vy0_}, t0_}] :=
    (g = 1; prec = Precision[d];
     If[prec < $MachinePrecision, prec = Sequence[]];
     (* next collision with the corrugated wall;
        v is the well/wall number *)
     v = Round[x0]; δ = Sign[vx0];
     If[(* going down *) vy0 <= 0 || (* going up *)
        (yE1Xs = maxWallHeightIntersection[d, prec]) === {},
        (* does surely not leave the current well *)
        While[(r = circleParabolaIntersection[v, d, prec]) === {},
              v = v + δ]; {v, r},
        (* might leave the current well *)
        {x1, x2} = Sort[yE1Xs]; {v1, v2} = Round[{x1, x2}];
        (* is the wall hit when going upwards (or v1 === v2)? *)
        While[If[δ === 1, v <= v1, v >= v1] &&
              (r = circleParabolaIntersection[v, d, prec]) === {},
```

```
            v = v + δ];
    (* if not already hit, must hit when going down *)
    If[r === {},
        While[(r = circleParabolaIntersection[v2, d, prec]) === {},
            v2 = v2 + δ]]; {v, r}];
(* time and position of the next collision;
   carry out the reflection *)
reflect[{{x0, y0} + r {vx0, vy0} + {0, -g/2 r^2},
        {vx0, vy0 - r g}, t0 + r}])
```

To visualize the repeated reflections, we implement a function `ballPathGraphic` that displays the curve and the trajectories between reflections. Its first argument is a list of free flight data of the form $\{\{x0, y0\}, \{vx0, vy0\}, t0\}$.

```
In[9]:= ballPathGraphic[pathData_, opts___] :=
    Module[{pp = 100, xMin, xMax, singlePath, normalLine, wallLine},
    {xMin, xMax} = {Min[#], Max[#]}&[#[[1, 1]]& /@ pathData];
    (* trajectory between two collisions *)
    singlePath[{x0_, y0_}, {vx0_, vy0_}, δt_] :=
     Table[{x0, y0} + τ {vx0, vy0} - {0, g/2 τ^2}, {τ, 0, δt, δt/pp}];
    (* normal at the reflection point *)
    normalLine[{x0_, y0_}] := Line[
        ({Round[x0], 0} + # ({x0, y0} - {Round[x0], 0}))& /@ {1/2, 3/2}];
    (* the curve made from half circles *)
    wallLine = Line[Table[{x, y[x]},
                {x, xMin - #, xMax + #, 1/pp}]&[(xMax - xMin)/10]];
    (* show all parts; use different colors *)
    Show[Graphics[{
            {Thickness[0.008], GrayLevel[0], wallLine},
            {Thickness[0.002], RGBColor[1, 0, 0], Line[Join @@
              (singlePath[#[[1, 1]], #[[1, 2]],
                    #[[2, 3]] - #[[1, 3]]]& /@ Partition[pathData, 2, 1])]},
                (* optional curve normals *)
                If[ShowNormals /. {opts} /. {ShowNormals -> False},
                    {Thickness[0.002], RGBColor[0, 0, 1],
                    normalLine[#[[1]]]& /@ pathData}, {}]}],
            AspectRatio -> Automatic, PlotRange -> All]]
```

Here is a first example. We show 15 reflections and the normals at the collision points.

```
In[10]:= ballPathGraphic[NestList[nextCollision,
                    {{N[8/10, 200], y[8/10]}, {-1, 1/7}, 0}, 15],
                ShowNormals -> True]
```

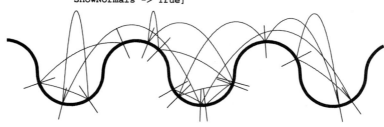

The following special initial condition makes the particle hop from well to well.

```
In[11]:= ballPathGraphic[NestList[nextCollision,
                    {{0., -1/2}, {1/2, 2}, 0}, 5]]
```

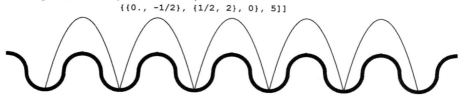

Because at each collision step, the numerical error of the flight direction effectively doubles, using machine precision arithmetic does not give quantitatively reliable trajectories for a larger number of reflections. In the next input, we compare a machine precision calculation with a 200-digit calculation for 25 reflections. The green machine precision trajectory becomes completely different after about 20 reflections.

```
In[12]:= Module[{x0 = 8/10, vx0 = -1, vy0 = 1/3, bounces = 25},
    Show[Block[{$DisplayFunction = Identity},
      {(* high-precision trajectory in red *)
      ballPathGraphic[NestList[nextCollision,
                {{N[x0, 200], 𝒴[x0]}, {vx0, vy0}, 0}, bounces]],
      (* machine-precision trajectory in green *)
      ballPathGraphic[NestList[nextCollision,
                {{N[x0], 𝒴[x0]}, {vx0, vy0}, 0}, bounces]] /.
                RGBColor[1, 0, 0] -> RGBColor[0, 1, 0]}]]]
```

In addition to the actual path, it is also interesting to look at the points $\{x0_k, vx0_k\}$, $\{y0_k, vy0_k\}$ and $\{vx0_k, vy0_k\}$. The function `ballPathRelectionGraphics` implements these graphics.

```
In[13]:= ballPathRelectionGraphics[pathData_] :=
    Show[GraphicsArray[
    Function[{i, j}, Graphics[{PointSize[0.003],
            (* {x0, vx0} and {vx0, vy0} pairs} *)
            Point[{Slot[i], Slot[j]}]& @@@ (Flatten /@ pathData)},
            PlotRange -> All, Frame -> True,
            FrameTicks -> None]] @@@ {{1, 3}, {2, 4}, {3, 4}}]]
```

We show some more possible trajectories and the corresponding $\{x0_k, vx0_k\}$, $\{y0_k, vy0_k\}$ and $\{vx0_k, vy0_k\}$ graphics. Because we are only interested in qualitative features, we use machine arithmetic in the following. The next graphic shows a trajectory that stays within one well for a long time before escaping to the next one in a creeping mode.

```
In[14]:= With[{nl = NestList[nextCollision, {{0., -1/2.},
                {0.192199, 1.423692}, 0}, 200]},
        ballPathGraphic[nl]; ballPathRelectionGraphics[nl]]
```

The next, fountain-like trajectory is a pseudoperiodic one that stays within one well.

```
In[15]:= With[{nl = NestList[nextCollision,
        {{-0.0075044852048594, -0.4999436795298},
         { 0.01110803492720627,  0.7400097011660}, 0}, 120]},
        ballPathGraphic[nl]; ballPathRelectionGraphics[nl]]
```

We end with a "generic" trajectory that randomly hops around. Here are 120 reflections shown.

```
In[16]:= With[{nl = NestList[nextCollision, {{0., -1/2.},
                            {-0.944133, 1.819119}, 0}, 120]},
            ballPathGraphic[nl]]
```

Starting with the same initial conditions, we now carry out 10000 reflections and show the $\{x0_k, vx0_k\}$, $\{vx0_k, vy0_k\}$ diagrams. The rightmost graphic shows the $x0_k$ which shows a typical random walk like behavior.

```
In[17]:= With[{nl = NestList[nextCollision, {{0., -1/2.},
                            {-0.944133, 1.819119}, 0}, 10^4]},
            Show[GraphicsArray[
            Block[{$DisplayFunction = Identity},
                {ballPathRelectionGraphics[nl],
                 (* the x-values of the reflection points *)
                 ListPlot[#[[1, 1]]& /@ nl, Frame -> True,
                                         FrameTicks -> None]}],
                 GraphicsSpacing -> -0.25]]]
```

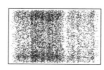

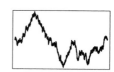

For other billiard models under the influence of gravity, see [143], [154], [184], [353], [1240], [1180], [1004], [1802], [1183], [690]. For 1D impact oscillations, see [781].

19. Schwarz-Riemann Minimal Surface, Jorge-Meeks Trinoid, Random Minimal Surfaces

a) First, we determine the region in the s,t-plane that is of interest for the parametrization of the surface. By elementary geometry, one finds that a ray originating from the origin under the angle φ intersects with one of the four circles in a distance rMax[φ].

```
In[1]:= rMax[φ_] := rMax[φ] =
        Sqrt[(4(1 + Tan[#] +   Tan[#]^2) - 2(1 + Tan[#])*
        Sqrt[ 3 + 2 Tan[#] + 3 Tan[#]^2])/(2(1 + Tan[#]^2))]&[
                                (* use symmetry *) Mod[φ, Pi/2]]
```

Here are the four circles together with their common overlap.

```
In[2]:= Show[Graphics[{ (* the overlapping region *)
            Polygon[Table[rMax[φ] {Cos[φ], Sin[φ]} // N, {φ, 0, 2Pi, 2Pi/40}]],
```

```
(* the four circles *)
Circle[#/{Sqrt[2], Sqrt[2]}, Sqrt[2]]& /@
             {{1, 1}, {-1, 1}, {1, -1}, {-1, -1}}}],
    PlotRange -> All, AspectRatio -> Automatic, Frame -> True]
```

In[3]:= `(* delete values from the graphics that are no longer necessary *)`
`DownValues[rMax] = Last[DownValues[rMax]];`

Now, we numerically integrate along rays starting from the origin of the above integrands. To avoid multiple work, we divide the ray into pieces, integrate along these pieces, and then sum the individual results. So, we get lists of the x-, y-, and z-values of the surface.

In[5]:= `ppr = 12; ppp = 4 ppp1; ppp1 = 6;`

In[6]:= `{xtab, ytab, ztab} =`
`Re[Table[FoldList[Plus, 0, Table[NIntegrate[#/Sqrt[1 - 14w^4 + w^8],`
`                (* dividing along the ray *)`
`        {w, ri/ppr rMax[φ] Exp[I φ], (ri + 1.)/ppr rMax[φ] Exp[I φ]}]],`
`            {ri, 0, ppr - 1}]], (* direction of ray *)`
`        {φ, 0, 2Pi, 2Pi/ppp}]]& /@ {1 - w^2, I(1 + w^2), 2w};`

Using the command `ListSurfacePlot3D` from the package `Graphics`Graphics3D``, we arrive at the following picture for the surface. The edges of this surface are part of a regular tetrahedron (right graphic).

In[7]:= `Needs["Graphics`Graphics3D`"]`

In[8]:= `(* edges of the enclosing tetrahedron *)`
`edges = Partition[{xtab[[#, -1]], ytab[[#, -1]], ztab[[#, -1]]}& /@`
`                {1, 1 + ppp1, 1 + 2ppp1, 1 + 3ppp1, 1 + 4ppp1}, 2, 1];`

In[10]:= `Show[GraphicsArray[`
`    Block[{$DisplayFunction = Identity, ε = 10^-5},`
`    (* surface only *)`
`    {onePiece = ListSurfacePlot3D[Transpose[{xtab, ytab, ztab}, {3, 2, 1}],`
`                    Axes -> True, PlotRange -> All],`
`    (* surface and tetrahedron edges *)`
`    Show[{onePiece, Graphics3D[{Thickness[0.01],`
`                Outer[Line[{#1, #2}]&, #, #, 1]&[`
`                (1 + ε) Union[Flatten[edges, 1]]]}]},`
`        Boxed -> False, Axes -> False, PlotRange -> All,`
`        ViewPoint -> {1.2, -2, 0.9}]}]]]`

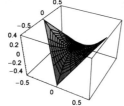

Now, let us extend the surface. We use a similar idea as in Exercise 16 of Chapter 2 of the Graphics volume [1794]: we test if a particular piece already exists by looking only at the midpoint of this piece, and in case it still does not exist, we generate the new piece by mirroring.

At the moment, we have only one piece with the midpoint {0, 0, 0}.

```
In[11]:= midPointList = {{0, 0, 0}};
```

`midPointTest` carries out the test to determine whether a particular piece already exists by looking at the distance of the new midpoint to all already existing ones. In case space still exists for a new one, the new midpoint is appended to the list `midPointList`.

```
In[12]:= midPointTest[{polys_, mp_, edges_}, i_] :=
        Module[{normal = #/Sqrt[#.#]&[Subtract @@ edges[[i]]],
                p1 = edges[[i, 1]], newMidPoint, isNewQ, ε = 10^-5},
               (* the new midpoint *)
               newMidPoint = (p1 + 2 (normal.(# - p1)) normal - (# - p1))&[mp];
               (* really a new midpoint ? *)
               isNewQ = Min[Abs[#.#& /@
                          (newMidPoint - #& /@ midPointList)]] > ε;
               If[isNewQ, AppendTo[midPointList, newMidPoint]]; isNewQ]
```

To carry out the actual reflection process, we use `lineMirror`.

```
In[13]:= lineMirror[all:{polys_, mp_, edges_}, i_] :=
        Function[{p1, normal},
                 Map[(p1 + 2 (normal.(# - p1)) normal -
                              (# - p1))&, all, {-2}]][edges[[i, 1]],
                      #/Sqrt[#.#]&[Subtract @@ edges[[i]]]]
```

The routine `mirrorAll` mirrors (in case the corresponding pieces still do not exist) along all four sides.

```
In[14]:= mirrorAll[a:{polys_, mp_, edges_}] :=
        Table[If[midPointTest[a, i], lineMirror[a, i], Sequence @@ {}], {i, 4}]
```

Now, starting with `onePiece`, we recursively carry out the mirroring process.

```
In[15]:= level[0] = {{onePiece[[1]], {0, 0, 0}, edges}};
        level[i_] := level[i] = Flatten[mirrorAll /@ level[i - 1], 1]
```

So, we can watch how the surface grows in each step. (We restrict the continuation of the surface to the first steps; it can often be continued arbitrarily.)

```
In[17]:= Show[GraphicsArray[Table[
           Graphics3D[{EdgeForm[Thickness[0.002]],
                   First /@ Join @@ Array[level, i, 0]},
               PlotRange -> All, Boxed -> False, Lighting -> True,
               ViewPoint -> {0.3, -3, 2}], {i, 2, 4}]]]
```

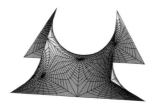  

We can identify the various pieces better if we give each piece its own surface color.

```
In[18]:= Show[GraphicsArray[
           Graphics3D[{EdgeForm[{}],
                       {SurfaceColor[Hue[Random[]], Hue[Random[]],
                                     Random[Real, {1/2, 2}]], #}& /@
                        (First /@ Join @@ Array[level, 5, 0])},
               PlotRange -> All, Boxed -> False, Lighting -> True,
               ViewPoint -> #]& /@ (* two different view points *)
             {{0.3, -3, 2}, {3, 3, 3}}]]
```

We can view a larger piece of the surface by using only four polygons in onePiece.

```
In[19]:= onePiece = Graphics3D[Polygon[Append[#, {0, 0, 0}]]]& /@ edges];
        midPointList = {{0, 0, 0}};
        (* generate neighboring pieces recursively *)
        Clear[level];
        level[0] = {{onePiece[[1]], {0, 0, 0}, edges}};
        level[i_] := level[i] = Flatten[mirrorAll /@ level[i - 1], 1]

In[25]:= (* cut a hole in a polygon *)
        makeHole[Polygon[l_], f_] :=
        Module[{mp = Plus @@ l/Length[l], ℓ}, ℓ = (mp + f(# - mp))& /@ l;
            {MapThread[Polygon[Join[#1, Reverse[#2]]]&,
                Partition[Append[#, First[#]], 2, 1]& /@ {l, ℓ}],
            Line[Append[#, First[#]]]&[ℓ]}]

In[27]:= Show[GraphicsArray[
        {(* full polygons *)
         Graphics3D[{EdgeForm[{Thickness[0.002], GrayLevel[0.8]}],
                    SurfaceColor[Hue[0.16], Hue[0.26], 2.3],
                    First /@ Join @@ Array[level, 8, 0]},
             ViewPoint -> {0.3, -3, 2}, PlotRange -> All, Boxed -> False],
         (* polygons with holes *)
         Graphics3D[{EdgeForm[{}], Thickness[0.002],
                    {SurfaceColor[Hue[Random[]], Hue[Random[]], 2.2],
                     makeHole[#, 0.66]& /@ #}& /@
                    First /@ Join @@ Array[level, 8, 0]},
             ViewPoint -> {3.3, 2.4, 2}, PlotRange -> All, Boxed -> False]}]]
```

For a general method of the construction of other minimal surfaces that can be repeatedly reflected, see [1352], [1020], [866], [1021], [954], [1049], [12], [1151], and [1152]. For many minimal surfaces calculated in *Mathematica*, see [641].

b) Let us start with discretizing the *s,t*-plane. Because the denominators of the integrands have poles at 1, $\exp(2 i \pi/3)$, and $\exp(4 i \pi/3)$, we must avoid these points and their neighborhoods. We do this by excluding circular regions around the singularities. We cover the *s,t*-plane with a tensor product grid in polar coordinates. This is implemented here in a function xyChanged that returns the coordinates of a point in the distorted mesh.

```
In[1]:= xyChanged[{x_, y_}, r_ (* radius of the disk
                                around the singularities *)] :=
       Function[R, Which[R <= 1 - r, {x, y},
                        1 - r < R < 1 + r, (* distort the mesh *)
                        R {Cos[#], Sin[#]}&[
                        Function[{φ0, φ1}, (Pi φ0 + Pi φ1 - 3 φ0 φ1)/Pi][
                            ArcTan[(1 + R^2 - r^2)/2,
                                Sqrt[2r^2 - r^4 + 2R^2 + 2r^2 R^2 - R^4 - 1]/2],
```

```
                                ArcTan[x, y]]],
                        R >= 1 + r, {x, y}]][Sqrt[x^2 + y^2]]
```

Here, the mesh with the holes around the singularities is visualized.

```
In[2]:= With[{(* rotation matrices *)
              R1 = N[{{ Cos[2Pi/3], Sin[2Pi/3]}, {-Sin[2Pi/3], Cos[2Pi/3]}}],
              R2 = N[{{ Cos[4Pi/3], Sin[4Pi/3]}, {-Sin[4Pi/3], Cos[4Pi/3]}}]},
       (* show graphic *)
       Show[Graphics[{Thickness[0.002],
         (* mirror and rotate lines *)
         {#, Map[R1.#&, #, {-2}], Map[R2.#&, #, {-2}]}&[
         {#, Map[# {1, -1}&, #, {-2}]}&[
           {Line /@ #, Line /@ Transpose[#]}&[N @ (* points *)
             Table[xyChanged[{r Cos[φ], r Sin[φ]} // N, 0.3],
                   {r, 0, 2, 1/20}, {φ, 0, N[Pi/3], N[Pi/3]/25}]]]]}],
         PlotRange -> All, AspectRatio -> Automatic, Frame -> True]]
```

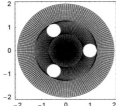

Now, we generate such a distorted mesh in the region $0 \le r < 1$, $0 \le \varphi \le \pi/3$.

```
In[3]:= xytab = Partition[#, 2, 1]& /@ Map[Complex @@ #&,
           N[Table[Table[xyChanged[{r Cos[φ], r Sin[φ]} // N, 0.4],
                   {r, 0, 1, 1/30}], {φ, 0, N[Pi/3], N[Pi/3]/45}]], {-2}];
```

We numerically carry out the integration required in the representation of the minimal surface along the mesh lines starting from $r = 0$ and going radially outward.

```
In[4]:= {xtab, ytab, ztab} = Function[f, FoldList[Plus, 0, #]& /@
           Re[Apply[(NIntegrate[f, {z, ##}])&, xytab, {-2}]]] /@
                       ({1 - z^4, I(1 + z^4), z^2}/((z^3 - 1)^2));
```

Using the command `ListSurfacePlot3D` from the package `Graphics`Graphics3D`` we look at the piece of surface generated.

```
In[5]:= Needs["Graphics`Graphics3D`"]


       lsp = ListSurfacePlot3D[Transpose[{xtab, ytab, ztab}, {3, 2, 1}],
                       Axes -> True, BoxRatios -> {1, 1, 1}]
```

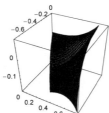

Discretizing larger regions in the *s,t*-plane or carrying out some analytical calculations of the parametric representations results in an impression of the overall shape of the surface. From the piece of the surface already generated, we can construct the whole surface by rotation around the *z*-axis by $2\pi/3$ and $4\pi/3$ and by mirroring the whole piece on a plane parallel to the *s,t*-plane. Here, this is implemented.

```
In[7]:= δz = Min[Last[Transpose[Flatten[(First /@
               Cases[lsp, _Polygon, {0, Infinity}]), 1]]]];
```

```
With[{(* rotation matrices *)
      R1 = {{-1, Sqrt[3], 0}, {-Sqrt[3], -1, 0}, {0, 0, 2}}/2.,
      R2 = {{-1, -Sqrt[3], 0}, {Sqrt[3], -1, 0}, {0, 0, 2}}/2.},
   (* display graphic *)
 Show[Graphics3D[{EdgeForm[],
     SurfaceColor[RGBColor[0.6, 0.8, 0.4], RGBColor[0.2, 0.2, 0.1], 2.1],
     (* mirror and rotate polygons *)
   {#, Map[R1.#&, #, {-2}], Map[R2.#&, #, {-2}]}&[
    {#, Map[# {1, -1, 1}&, #, {-2}]}&[
     {#, Map[({1, 1, -1}(# -{0, 0, δz}) + {0, 0, δz})&, #,
       {-2}]}&[Cases[lsp, _Polygon, {0, Infinity}]]]]},
      PlotRange -> All, BoxRatios -> {3/2, 3/2, 1}, Boxed -> False]]]
```

c) Instead of using NIntegrate, we will use NDSolve to integrate the functions under consideration. First, we will integrate radially outward from r_0 to r_{max}, and then we will integrate around circles from $\varphi = 0$ to $\varphi = \varphi_{max}$. The starting values for the integration along the circles we obtain from the radial solution. The functions which might exhibit branch cuts are the Power function with a rational second argument and the Log function. Their analytical continuation is easily done. The function minimalSurfacePicture implements this.

```
In[1]:= Needs["Graphics`Graphics3D`"]

In[2]:= minimalSurfacePicture[{f_, g_}, {r0_, r1_}, {φ0_, φ1_}, {ppr_, ppφ_}] :=
   Module[{eqs, eqsr, nsolr, eqsφ, nsolφ, data},
    (* the three terms to be integrated *)
    eqs = {(1 - g^2) f, I (1 + g^2) f, 2 f g};
    (* the radial differential equation *)
    eqsr = Join[Apply[Equal,
        Transpose[{{ψ1'[r], ψ2'[r], ψ3'[r]}/Exp[I φ],
                   eqs /. ξ -> r Exp[I φ]}], {1}] /. φ -> φ0,
        {ψ1[r0] == 0, ψ2[r0] == 0, ψ3[r0] == 0}];
    (* solving the radial differential equation *)
    nsolr = NDSolve[eqsr, {ψ1, ψ2, ψ3}, {r, r0, r1}];
    (* the azimuthal differential equation *)
    eqsφ[r_] := Join[Apply[Equal,
           Transpose[{{ψ1'[φ], ψ2'[φ], ψ3'[φ]}/(I r Exp[I φ]),
                      eqs /. ξ -> r Exp[I φ]}], {1}],
        {ψ1[φ0] == nsolr[[1, 1, 2]][r], ψ2[φ0] == nsolr[[1, 2, 2]][r],
         ψ3[φ0] == nsolr[[1, 3, 2]][r]}] /.
        (* analytical continuation of Power and Log *)
        {Exp[I φ]^n_ -> Exp[I n φ], Log[r Exp[I φ]] -> Log[r] + I φ};
    (* solving the azimuthal differential equation *)
    nsolφ[r_] := NDSolve[eqsφ[r], {ψ1, ψ2, ψ3},
                      {φ, φ0, φ1}, MaxSteps -> 10^4];
    (* the points of the surface *)
    data = Table[aux = Re[{ψ1[φ], ψ2[φ], ψ3[φ]}] /. nsolφ[r][[1]];
        Table[aux, {φ, φ0, φ1, (φ1 - φ0)/ppφ}],
                   {r, r0, r1, (r1 - r0)/ppr}];
    (* show the resulting surface *)
    Show[GraphicsArray[
     Graphics3D[{EdgeForm[], Thickness[0.001],
         (* add a random coloring *)
         SurfaceColor[Hue[Random[]], Hue[Random[]], 3 Random[]],
         (* make holes in the surface for better visibility *)
         makeHole[#, 0.72 (* more funky: Random[Real, {0.5, 0.9}] *)]& /@
```

```
        Cases[ListSurfacePlot3D[data, DisplayFunction -> Identity],
         _Polygon, Infinity]},
        Boxed -> False, PlotRange -> All, BoxRatios -> {1, 1, 1},
        ViewPoint -> #]& /@ (* two view points *)
        {{1.3, -2.4, 2.}, {4., 4., -4.}}]]]]
```

For a better view of resulting surface, we cut some holes in the individual polygons.

```
In[3]:= makeHole[Polygon[l_], factor_] :=
    Module[{mp = Mean[l], ℓ},
          ℓ = (mp + factor(# - mp))& /@ l;
         {MapThread[Polygon[Join[#1, Reverse[#2]]]&,
            Partition[Append[#, First[#]], 2, 1]& /@ {l, ℓ}],
          Line[Append[#, First[#]]]&[ℓ]}]
```

Now, we can run the given examples. We will typically use some revolutions around the origin, and we will choose the radius of the revolutions near the unit circle. The first example has no special problems.

```
In[4]:= minimalSurfacePicture[{-67/78 + 2 ξ, 44/31 + (-1 + 1/ξ)^2 + 2ξ + ξ^2},
                    {0.9, 1.1}, {0, 12 Pi}, {6, 800}]
```

In the second example, we need a continuation for $\xi^{5/3}$.

```
In[5]:= minimalSurfacePicture[{14/27 + 1/ξ^2 + ξ, 1/ξ^(5/3)},
                    {0.6, 1.1}, {0, 6 Pi}, {9, 800}]
```

In the third example, we need a continuation for $\xi^{4/3}$.

```
In[6]:= minimalSurfacePicture[{2/3 + (-1 + 1/ξ^(4/3))^2 + 2ξ + ξ^3, 112/69 + 2ξ},
                    {0.6, 1.1}, {0, 6 Pi}, {9, 800}]
```

The fifth example needs again a continuation for $\xi^{5/3}$.

```
In[7]:= minimalSurfacePicture[{-2ξ^4, -2 + ((-1 + 1/ξ^(5/3))^2 (-2 + ξ))/ξ^3},
                    {0.6, 1}, {0, 6 Pi}, {12, 800}]
```

The sixth example is easy.

```
In[8]:= minimalSurfacePicture[{-(1173/(310 ξ^3)), ξ^3},
                              {0.6, 1.3}, {0, 2 Pi}, {11, 200}]
```

The seventh example needs again a continuation for $\xi^{5/3}$.

```
In[9]:= minimalSurfacePicture[{(2 + 1/ξ^6)^2, -(28/55) ξ (1 + ξ^(5/3))^2},
                              {0.7, 0.9}, {0, 6 Pi}, {5, 600}]
```

The eighth example needs a continuation for $\xi^{4/3}$. Additionally, because of the term $\sqrt[3]{\xi + 28/9}$, the maximal radius should be smaller than $28/9$.

```
In[10]:= minimalSurfacePicture[{2 + (28/9 + ξ)^(1/3), Sinh[1/ξ^(4/3)]},
                               {0.7, 1.3}, {0, 6 Pi}, {9, 400}]
```

The next example needs a continuation for $\xi^{7/3}$.

```
In[11]:= minimalSurfacePicture[{(-(7/12) + ξ)/ξ^3, (61 ξ^(7/3))/56},
                               {0.7, 1.1}, {0, 6 Pi}, {5, 400}]
```

The tenth example contains a logarithm with a compound argument. For $r < 1$, the branch cut of the logarithm does not come into play.

```
In[12]:= minimalSurfacePicture[
            {-(Log[2]/Log[1/(1 + ξ^2)]), -(20/(17 ξ^2))},
            {0.7, 0.9}, {0, 4 Pi}, {6, 600}]
```

The eleventh example contains again a logarithm with a compound argument. Again, for $r < 1$, the branch cut of the logarithm does not come into play.

```
In[13]:= minimalSurfacePicture[
            {(I Pi)/Log[1/(1 + ξ^(5/3))^2], (4/5 + 2 ξ)^(-1 + 1/ξ^2)},
            {0.7, 0.9}, {0, 16 Pi}, {6, 600}]
```

The last example finally contains a pure logarithm and radicals, the two terms we explicitly treated in the above code.

```
In[14]:= minimalSurfacePicture[
            {1/(1 + Log[ξ]^(32/9)/(3 10^6))^2, ξ^3/(-1 + 1/(4 ξ^(4/3)))^2},
            {0.7, 0.9}, {0, 8 Pi}, {6, 800}]
```

20. Precision Modeling, `GoldenRatio` Code from the Tour, Resistor Network

a) Here, the picture under consideration is shown.

```
In[1]:= f[x_] = 2 - x - 5 x^2 + 4 x^3 + 3 x^4 - 2 x^5
```

$$Out[1]= \; 2 - x - 5\,x^2 + 4\,x^3 + 3\,x^4 - 2\,x^5$$

```
In[2]:= ListPlot[Table[{x, Precision[f[SetPrecision[x, 25]]]}, {x, -5, 5, 10/1001}],
            PlotRange -> {{-5, 5}, {20, 30}}, Frame -> True,
            PlotJoined -> True, Axes -> False]
```

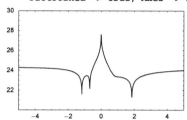

Using interval arithmetic, we see the same overall structure.

```
In[3]:= ListPlot[Table[{x, Log[10, 1/Abs[Subtract @@
            f[Interval[{x - 1/(10^25 2) x, x + 1/(10^25 2) x}]][[1]]/f[x]]]},
            {x, -5, 5, 10/1001}] // N,
            PlotRange -> {{-5, 5}, {20, 30}}, Frame -> True,
            PlotJoined -> True, Axes -> False]
```

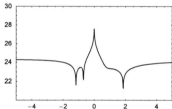

Calculating the error amplification for every term of `f` and adding these terms results in the model we are looking for.

```
In[4]:= Log[10, 1/Abs[(x Plus @@ (Abs[D[#1, x]] &) /@ List @@ f[x]/f[x])/10^25]]
```

$$Out[4]= \; \dfrac{\mathrm{Log}\Big[\dfrac{1000000000000000000000000}{\mathrm{Abs}\big[\frac{x\,(1+10\,\mathrm{Abs}[x]+12\,\mathrm{Abs}[x]^2+12\,\mathrm{Abs}[x]^3+10\,\mathrm{Abs}[x]^4)}{2-x-5\,x^2+4\,x^3+3\,x^4-2\,x^5}\big]}\Big]}{\mathrm{Log}[10]}$$

```
In[5]:= Plot[Evaluate[%], {x, -5, 5}, Frame -> True,
            Axes -> False, PlotRange -> {{-5, 5}, {20, 30}}]
```

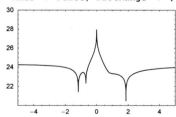

b) Here are the two definitions.

```
In[1]:= g1[k_] := FixedPoint[N[Sqrt[1 + #], k]&, 1]

        g2[k_] := 1 + FixedPoint[N[1/(1 + #), k]&, 1]
```

They really work and give the predicted precision.

In[3]:= `{g1[1000] - GoldenRatio, g2[1000] - GoldenRatio}`

Out[3]= $\{0. \times 10^{-1000}, 0. \times 10^{-1001}\}$

Let us look at what happens with an interval in such an iteration. The graphics shows the logarithm of the absolute value of the interval lengths.

In[4]:= `g2I[1_] := 1 + NestList[1/(1 + #)&, Interval[{1 - 10^-100, 1 + 10^-100}], 1]`

`ListPlot[Log[10, Abs[Apply[Subtract, First /@ g2I[120], {1}]] // N]]`

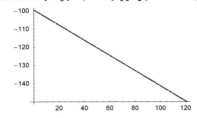

It shrinks! The reason is that the input error here does not get amplified but damped out.

In[6]:= `stab2[x_] = Abs[x D[1/(1 + x), x]/(1/(1 + x))]`

Out[6]= $\text{Abs}\left[\dfrac{x}{1 + x}\right]$

In[7]:= `Plot[Evaluate[stab2[x]], {x, 1, 2}]`

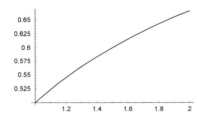

The same is true for the `g1` formulae.

In[8]:= `stab1[x_] = x D[Sqrt[1 + x], x]/Sqrt[1 + x]`

Out[8]= $\dfrac{x}{2(1 + x)}$

In[9]:= `Plot[Evaluate[stab1[x]], {x, 1, 2}]`

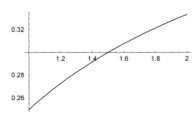

Here is the dependence of the precision as a function of the iteration number shown.

In[10]:= `Show[GraphicsArray[Function[map,`
`        ListPlot[Precision /@ NestList[Sqrt[1 + #]&, N[1, 50], 100],`
`            DisplayFunction -> Identity]] /@ {Sqrt[1 + #]&, 1/(1 + #)&}]]`

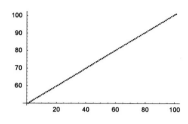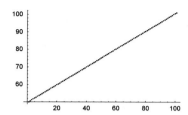

c) First, we run the code.

```
In[1]:= data = Table[If[Not[IntegerQ[n/10]],
          {n, Coefficient[Fit[Table[Plus @@ IntegerDigits[n^k], {k,  100}],
                               {1, x}, x], x]}, Sequence @@ {}], {n, 1000}];
        fit = Fit[data, {Log[10, n]}, n]
Out[2]= 1.95415 Log[n]
```

The first input calculates a list data. data consists of pairs $\{n, s(n)\}$. Here $s(n)$ is the average slope of the sum of digits of the sequence of numbers n^k, $k = 1, 2, \ldots$. An integer n has about $\log_{10}(n)$ digits in base 10. Assuming that the digits occur at random, their average value is 4.5. This means that the above fit should yield the number $4.5 / \ln(10) = 1.954\ldots$

Here is a plot of the data together with the expected line. The agreement is excellent.

```
In[3]:= Plot[1.95433 Log[10, k], {k, 1, 1000},
          Epilog -> {PointSize[0.003], Point /@ data}, PlotStyle -> {Hue[0]}]
```

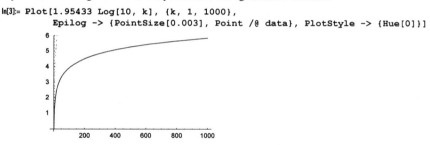

d) These are the relations of the recursion.

```
In[1]:= R[m_, m_] := R[m, m] = 4 (m - 1)/(2m - 1) R[m - 1, m - 1] -
                                 (2m - 3)/(2m - 1) R[m - 2, m - 2]

        (R[m_, n_] /; n + 1 === m) := R[m, n] = (2 R[n, n] - R[n, n - 1])

        R[m_, 0] := R[m, 0] = 4 R[m - 1, 0] - R[m - 2, 0] - 2 R[m - 1, 1]

        (R[m_, n_] /; 0 < n < m - 1) := R[m, n] =
          4 R[m - 1, n] - R[m - 2, n] - R[m - 1, n + 1] - R[m - 1, n - 1]
```

For the actual calculation of the $R_{m,n}$, we will use high-precision arithmetic. Exact arithmetic would be relatively slow due to the presence of the symbol π, and machine precision arithmetic would give wrong results for relatively small values of m, n already.

```
In[8]:= $RecursionLimit = 500;

        R[0, 0] = 0;
        (* use high-precision starting values *)
        R[1, 0] = N[1/2, 200];
        R[1, 1] = N[2/Pi, 200];
```

Here are the resistance values over the first quadrant of the m,n-plane.

```
In[13]:= data = Table[If[n1 >= n2, R[n1, n2], R[n2, n1]],
                   {n1, 0, 200}, {n2, 0, 200}];
In[14]:= Show[GraphicsArray[
           {(* 3D plot of the resistance values *)
```

```
      ListPlot3D[data, Mesh -> False, PlotRange -> All,
              DisplayFunction -> Identity],
      (* contour plot of the resistance values *)
      ListContourPlot[data, Contours -> 40,
                  DisplayFunction -> Identity]}]]
```

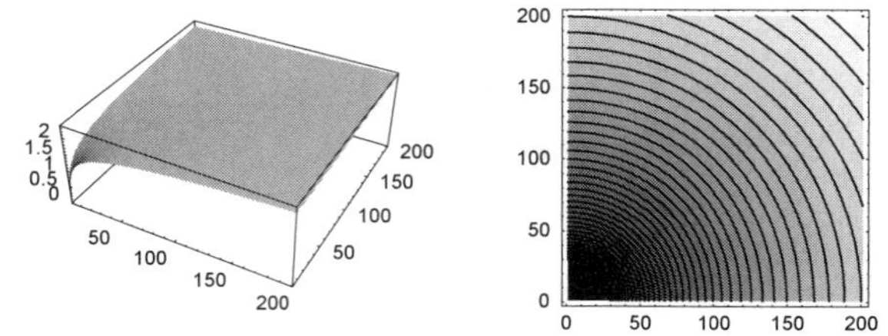

Although the last graphic suggests that $R_{m,n}$ does depend only on $(m^2 + n^2)^{1/2}$, this is not exactly true. In direction $m = n$, the resistance is slightly higher. The next graphic shows the angular dependence of the resistance.

```
In[15]:= data1 = Table[{n1, n2, If[n1 >= n2, R[n1, n2], R[n2, n1]]},
                  {n1, 0, 100}, {n2, 0, 100}];

In[16]:= ipo = Interpolation[Flatten[data1, 1]];

In[17]:= ρ = 20;
       Plot[ipo[ρ Cos[φ], ρ Sin[φ]], {φ, 0, Pi/2}, PlotRange -> All]
```

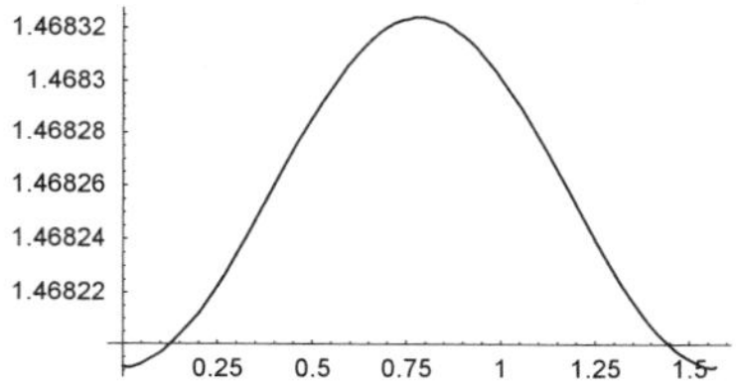

The following graphic shows the difference of the resistance in {1, 0} and {1, 1} direction.

```
In[19]:= data2 = Table[{ρ, ipo[ρ, 0] - ipo[ρ Cos[Pi/4], ρ Sin[Pi/4]]}, {ρ, 1, 100}];

In[20]:= (* show logarithm *)
       ListPlot[{#[[1]], Log[-#[[2]]]}& /@ data2]
```

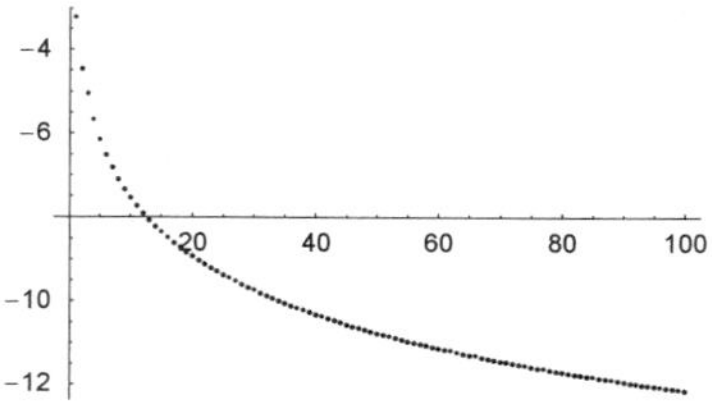

Now let us look at the asymptotic formula.

```
In[22]:= RAsymp[m_, n_] := 1/Pi (Log[Sqrt[m^2 + n^2]] + EulerGamma + Log[8]/2)

In[23]:= dataδ = Table[If[n1 == n2 == 0, 0, Log[10, Abs[
                      If[n1 >= n2, R[n1, n2], R[n2, n1]] - RAsymp[1. n1, n2]]]],
                  {n1, 0, 100}, {n2, 0, 100}];
```

The following graphic shows that near to the coordinate axes, the asymptotic value is larger than the exact value and near the diagonal, it is smaller. As a result, along two lines the asymptotic expansion is especially good.

```
In[24]:= ListPlot3D[dataδ, Mesh -> False, PlotRange -> All]
```

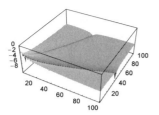

21. Auto-Compiling Functions, Card Game

a) Here is what we want to implement: We start with a set of generic definitions for *f*. First, we add a new definition (that gets matched first) to the definitions of *f*. This definition will create (when possible) a compiled version of the function *f* for actually supplied arguments. The compiled version will then be put on top of all other definitions, so that it gets matched first in further calls to *f*. In implementing this approach, we must be careful to avoid any evaluation of the definitions to be manipulated.

We start with the following function definitions.

```
In[1]:= Remove[f];

       f[x_, y_] := x^2 + y

       f[x_, y_, z_] := Sin[x] Cos[y] + z
```

We define a function `CompileAtCall`. `CompileAtCall[f]` adds a definition to *f* that will match earlier than any of the already-made definitions. `CompileAtCall[f]` is intended to be evaluated after the definitions for *f* have been made. `CompileAtCall` adds a new rule to the definitions of *f*, a rule that works for all arguments and is initially tried before any other rule is tried. The new definition `newDef` gets added to the downvalues of *f* when `makeCompiledVersion` succeeds in making a compiled version of a definition for the arguments `args`.

```
In[4]:= CompileAtCall[f_] :=
        ((* add new definition *)
        DownValues[f] = Prepend[DownValues[f],
          HoldPattern[f[args___]] :>
          With[{newDef = makeCompiledVersion[f, args]},
              (* did makeCompiledVersion generate a new definition? *)
              newDef /; Head[newDef] =!= makeCompiledVersion]];)
```

We call `CompileAtCall` with the function *f*.

```
In[5]:= CompileAtCall[f]
```

These are the definitions currently associated with *f*.

```
In[6]:= ??f

       Global`f

       f[args$___] := With[{newDef = makeCompiledVersion[f, args$]},
         newDef /; Head[newDef] =!= makeCompiledVersion]
       f[x_, y_] := x^2 + y
       f[x_, y_, z_] := Sin[x] Cos[y] + z
```

For a function definition to be compilable, its arguments must be tensors of numbers of the same type (for brevity we ignore the possibility of argument coercion here). The three functions `tensorQ`, `typedTensorQ`, and `argumentType` will determine if arguments are appropriate for a compilation.

```
In[7]:= tensorQ[l_] := TensorRank[l] === Depth[l] - 1;

       typedTensorQ[l_] := tensorQ[l] &&
       MatchQ[Union[Head /@ Flatten[{l}]], {Integer} | {Real} | {Complex}]

       argumentType[l_] := Union[Head /@ Flatten[{l}]][[1]]
```

We will also implement a three-argument version of `tensorQ` that checks both the type and the dimension of a tensor.

```
In[11]:= tensorQ[l_, type_, dim_] :=
              tensorQ[l] && TensorRank[l] === dim && argumentType[l] === type
```

Using `tensorQ`, we can implement a function `compilableArguments` that checks if a set of actual arguments can be used inside the *Mathematica* compiler.

```
In[12]:= compilableArgumentsQ[argList_] := And @@ (typedTensorQ /@ argList)
```

Here is an example. It is assumed that the exact complex arguments are coerced to machine complex numbers for compilation.

```
In[13]:= compilableArgumentsQ[{1, {2., 3.}, {{I, -I}, {-I, I}}}]
```

```
Out[13]= True
```

Symbolic arguments cannot be compiled. And no argument coercion will be tried.

```
In[14]:= compilableArgumentsQ[{Sqrt[2], X}]
```

```
Out[14]= False
```

Given a list of actual arguments `argList`, and symbol names of these arguments `varList`, the function `makeArgument Template` returns a list of argument specification as it is used by `Compile`. To avoid evaluation, we introduce now the function `pattern` that does not evaluate its arguments and does not give error messages when called with one argument.

```
In[15]:= SetAttributes[pattern, HoldAll]

         makeArgumentTemplate[argList_, varList_] :=
         If[And @@ (typedTensorQ /@ argList),
           (Hold @@ {Table[{varList[[k]], Blank[argumentType[argList[[k]]]],
                  TensorRank[argList[[k]]]}, {k, Length[argList]}]}) /.
             pattern[ξ_, ___] :> ξ, $Failed]
```

Here is an example. (We give x- and y-values to see that x and y do not get evaluated in the construction of the argument templates).

```
In[17]:= x = 1; y = 2;
         makeArgumentTemplate[{1, 2}, {pattern[x], pattern[y]}]
```

```
Out[18]= Hold[{{x, _Integer, 0}, {y, _Integer, 0}}]
```

To restrict function definitions, we will use conditions. The function `templateToTest` converts a template (generated by `makeArgumentTemplate`) to a condition. (Although `Condition` has the `HoldAll` attribute, it will evaluate its first argument, and so, we use `tensorQHeld` instead. Later, we will replace `tensorQHeld` by `tensorQ`.)

```
In[19]:= SetAttributes[tensorQHeld, HoldAll]

         templateToTest[Hold[l_List]] :=
           (And @@ (templateToTest /@ Map[Hold, Unevaluated[l], {1}]))

         templateToTest[Hold[{x_, Verbatim[Blank][type_], dim_}]] :=
                                         tensorQHeld[x, type, dim]
```

Here is again an example.

```
In[23]:= templateToTest[Hold[{{x, _Integer, 0}, {y, _Integer, 0}}]]
```

```
Out[23]= tensorQHeld[x, Integer, 0] && tensorQHeld[y, Integer, 0]
```

Now, we need one more subsidiary function. The function `buildDownvalue` makes a `RuleDelayed` as needed in the downvalues for *f*. Its arguments are the symbol *f*, the left-hand side pattern *ξηζ*, the condition on the pattern *cond*, a compiled function *cf*, and the arguments for the right-hand side *xyz*.

```
In[24]:= buildDownvalue[{f_, Verbatim[HoldPattern][_[ξηζ__]],
                cond_, cf_, pattern[xyz__]}] :=
                HoldPattern[f[ξηζ]] /; cond] :> cf[xyz]
```

Now, we come to the main part of the implementation. The function `matchingDefinition` returns the first matching definition after the `makeCompiledVersion` rule of *f* for the arguments *args*.

```
In[25]:= matchingDefinition[f_, args__] :=
         Module[{dvs, pos, genericDefinitions, lhsPatterns, matchingPattern, pos1},
                (* all current downvalues of f *)
```

```
        dvs = DownValues[f];
        (* the position of makeCompiledVersion *)
        pos = Position[dvs, makeCompiledVersion];
        (* all generic definitions of f *)
        genericDefinitions = Take[dvs, {pos[[1, 1]] + 1, Length[dvs]}];
        (* all patterns of the generic definitions *)
        lhsPatterns = (First[#]& /@ genericDefinitions) /. f -> C;
        (* the first pattern matching args *)
        matchingPattern = Select[lhsPatterns, MatchQ[C[args], #]&, 1];
        If[matchingPattern =!= {},
            pos1 = Position[lhsPatterns, matchingPattern[[1]]][[1, 1]];
            genericDefinitions[[pos1]], {}]]
```

In[26]:= **matchingDefinition[f, 1, 11]**

Out[26]= $\text{HoldPattern}[f[x\_, y\_]] :\mapsto x^2 + y$

In[27]:= **matchingDefinition[f, 1, 1., {I, -1}]**

Out[27]= $\text{HoldPattern}[f[x\_, y\_, z\_]] :\mapsto \text{Sin}[x]\,\text{Cos}[y] + z$

Now, we implement the heart of this exercise. The function makeCompiledVersion creates a compiled definition of *f* for the argument type of *args* and sticks this definition into the list of definitions for *f*.

```
In[28]:= makeCompiledVersion[f_, args___] :=
        Module[{def = matchingDefinition[f, args] /. f -> C,
                compQ = compilableArgumentsQ[{args}], matchingPattern,
                rhs, compileVars, argTemplate, cf, conds, rhsVars, newDef},
            (matchingPattern = def[[1]];
            (* the definition belonging to the matching pattern *)
            rhs = Extract[def, {2}, Hold];
            (* the variable list for Compile *)
            compileVars = List @@ First[matchingPattern /.
                                Verbatim[Pattern][ξ_, _] -> pattern[ξ]];
            (* the first argument for Compile *)
            argTemplate = makeArgumentTemplate[{args}, compileVars];
            (* the compiled function *)
            cf = Compile @@ Join[argTemplate, rhs];
            (* the conditions for the new function definition *)
            conds = templateToTest[argTemplate];
            (* the arguments for the compiled function *)
            rhsVars = Flatten[pattern @@ compileVars];
            (* the new function definition *)
            newDef = buildDownvalue[{f, matchingPattern, conds, cf, rhsVars}] /.
                                        tensorQHeld -> tensorQ;
            (* updating the list of downvalues *)
            DownValues[f] = Prepend[DownValues[f], newDef];
            (* calculate f with the new definition *)
            f[args]) /; (* were the arguments compilable, and did we find
                          a matching definition? *) def =!= {} && compQ]
```

Now, we call f with the arguments 2.0, 3.0. The existing pattern f[x_, y_] matches, and a new definition with a compiled version of x^2 + y for real scalar x and y is generated and used for the calculation.

In[29]:= **f[2., 3.]**

Out[29]= 7.

In[30]:= **??f**

```
Global`f

f[x_, y_] /; tensorQ[x, Real, 0] && tensorQ[y, Real, 0] :=
  CompiledFunction[{x, y}, x^2 + y, -CompiledCode-][x, y]

f[args$___] := With[{newDef = makeCompiledVersion[f, args$]},
  newDef /; Head[newDef] =!= makeCompiledVersion]

f[x_, y_] := x^2 + y

f[x_, y_, z_] := Sin[x] Cos[y] + z
```

If the arguments are not compilable, the newly created definitions do not fire and the original rule f[x_, y_] := x^2 + y is used.

In[31]:= **f[α, β]**

Out[31]= $\alpha^2 + \beta$

Using three vector arguments for f results in a new definition.

In[32]:= **f[{1., 2.}, {3., 4.}, {5., 6.}]**

Out[32]= {4.16695, 5.40564}

In[33]:= **First[DownValues[f]]**

Out[33]= HoldPattern[
       f[x_, y_, z_] /; tensorQ[x, Real, 1] && tensorQ[y, Real, 1] && tensorQ[z, Real, 1]] :→
       CompiledFunction[{x, y, z}, Sin[x] Cos[y] + z, -CompiledCode-][x, y, z]

Now using f with such arguments that one of the compiled definitions matches results in the use of these definitions. New definitions are only generated when new argument tensor types occur.

In[34]:= **{Length[DownValues[f]],**
       Do[f[N[i], N[+ 1]], {i, 100}]; Length[DownValues[f]]}

Out[34]= {5, 5}

In[35]:= **f @@ Table[1, {3}, {3}, {3}, {3}];**
       Length[DownValues[f]]

Out[36]= 6

By adding a further definition, *Mathematica*'s rule ordering code will still place the compiled definitions first.

In[37]:= **f[x_, y_, z_, u_, v_] := x + y + z + u + v**

In[38]:= **f[1, 2, 3, 4, 5]**

Out[38]= 15

In[39]:= **?f**

       Global`f

       f[x_, y_, z_, u_, v_] /; tensorQ[x, Integer, 0] && tensorQ[y, Integer, 0] &&
         tensorQ[z, Integer, 0] && tensorQ[u, Integer, 0] && tensorQ[v, Integer, 0] :=
       CompiledFunction[{x, y, z, u, v}, x + y + z + u + v, -CompiledCode-][x, y, z, u, v]
       f[x_, y_, z_] /;
         tensorQ[x, Integer, 3] && tensorQ[y, Integer, 3] && tensorQ[z, Integer, 3] :=
       CompiledFunction[{x, y, z}, Sin[x] Cos[y] + z, -CompiledCode-][x, y, z]
       f[x_, y_, z_] /; tensorQ[x, Real, 1] && tensorQ[y, Real, 1] && tensorQ[z, Real, 1] :=
       CompiledFunction[{x, y, z}, Sin[x] Cos[y] + z, -CompiledCode-][x, y, z]
       f[x_, y_] /; tensorQ[x, Real, 0] && tensorQ[y, Real, 0] :=
       CompiledFunction[{x, y}, x^2 + y, -CompiledCode-][x, y]
       f[args$___] := With[{newDef = makeCompiledVersion[f, args$]},
         newDef /; Head[newDef] =!= makeCompiledVersion]
       f[x_, y_] := x^2 + y
       f[x_, y_, z_] := Sin[x] Cos[y] + z
       f[x_, y_, z_, u_, v_] := x + y + z + u + v

This solution shows only the spirit of how to build compiled function definitions "on the fly" and how to stick them into the definition set of a function so that they will be used later. The above implementation can be refined in various ways. Instead of using conditions such as *lhs* /; tensorQ[*l*, Integer, 0], the pattern _Integer would be faster to test. There is no reason to restrict CompileAtCall to manipulating only downvalues. Subvalues and so on could be treated in a similar manner. For a function with many arguments, one probably does not want for all combinations of integer, real, and complex arguments, different compiled functions to be generated. We did not implement a test of the compilation succeeded. The above implementation only dealt with Blank as patterns, not BlankSequence. Using ideas from the Exercise 6-17. b), we could deal with BlankSequence and BlankNullSequence too. We leave it to the reader to refine the above implementation in all of these directions and more.

b) Here is the given implementation evaluated and tested.

```
In[1]:= cardGameSteps1 = Compile[{{A, _Integer, 1}, {B, _Integer, 1}},
        Module[{a = A, b = B, ra, rb, (* round counter *) c = 0},
              While[a != {} && b != {}, c++;
                      (* select two random cards *)
                      ra = Random[Integer, {1, Length[a]}];
                      rb = Random[Integer, {1, Length[b]}];
                      (* compare cards and add new card to one player;
                         remove second card from other player *)
                      If[a[[ra]] > b[[rb]],
                          b = Append[b, a[[ra]]]; a[[ra]] = a[[-1]]; a = Drop[a, -1],
                          a = Append[a, b[[rb]]]; b[[rb]] = b[[-1]]; b = Drop[b, -1]]];
                (* return number of rounds *) c]];

In[2]:= SeedRandom[123];
        (* test *)
        cardGameSteps1[{6, 7, 3, 1, 10}, {4, 9, 2, 8, 5}]
Out[4]= 49
```

`cardGameSteps1` did compile flawlessly.

```
In[5]:= FreeQ[cardGameSteps1[[4]], _Function, {0, Infinity}]
Out[5]= True
```

To get a reliable timing estimate, we carry out 10^3 repetitions for $n = 5$.

```
In[6]:= Table[cardGameSteps1[{6, 7, 3, 1, 10}, {4, 9, 2, 8, 5}],
          {10^3}]; // Timing
Out[6]= {0.18 Second, Null}
```

`cardGameSteps1` can be sped up by about a factor of two. In each round, the length of the two lists a and b changes. This means new lists are generated each time. This can be avoided by making both lists a and b of length $2\,n$ and adding two variables ac and bc that indicate the position of the last card owned by the player. Now no new lists have to be created and only elements of the lists have to be changed. All card values to the right of the positions ac and bc are ignored. The function `cardGameSteps2` implements this.

```
In[7]:= cardGameSteps2 = Compile[{{A, _Integer, 1}, {B, _Integer, 1}},
        Module[{a = Join[A, B], b = Join[B, A],
                ac = Length[A], bc = Length[B], ra, rb, c = 0},
              While[ac != 0 && bc != 0, c++;
                      ra = Random[Integer, {1, ac}];
                      rb = Random[Integer, {1, bc}];
                      If[a[[ra]] > b[[rb]],
                          b[[bc + 1]] = a[[ra]]; bc = bc + 1;
                          a[[ra]] = a[[ac]]; ac = ac - 1,
                          a[[ac + 1]] = b[[rb]]; ac = ac + 1;
                          b[[rb]] = b[[bc]]; bc = bc - 1]];
                c]];
```

`cardGameSteps2` produces the same results, as does `cardGameSteps1`. But `cardGameSteps2` is more than twice as fast as `cardGameSteps1`.

```
In[8]:= SeedRandom[123];
        cardGameSteps2[{6, 7, 3, 1, 10}, {4, 9, 2, 8, 5}]
Out[9]= 49
```

`cardGameSteps2` did also compile flawlessly.

```
In[10]:= FreeQ[cardGameSteps2[[4]], _Function, {0, Infinity}]
Out[10]= True

In[11]:= Table[cardGameSteps2[{6, 7, 3, 1, 10}, {4, 9, 2, 8, 5}], {10^3}]; // Timing
Out[11]= {0.07 Second, Null}
```

Next, we load the function `RandomPermutation` to generate random initial distributions of the cards.

In[12]:= `Needs["DiscreteMath`Combinatorica`"]`

Carrying out now 10^6 games with a total of 20 cards takes about 5 minutes on a 2 GHz computer.

In[13]:=
```
SeedRandom[122394];
With[{n = 20, o = 10^6},
 ListPlot[(* count cases *) {First[#], Length[#]/o}& /@
          Split[Sort[Table[rp = RandomPermutation[n];
           cardGameSteps2[Take[rp, n/2], Take[rp, -n/2]], {o}]]],
         Frame -> True, Axes -> False, PlotRange -> All]] // Timing
```

The average game length is nearly exactly $100 = (n = 10)^2$.

In[15]:= `Plus @@ (Times @@@ Map[First, First[Last[%]]], {1}]) // N`

Out[15]= `99.9375`

22. Path of Steepest Descent, Arclength of Fourier Sum, Minimum-Energy Charge Configuration

a) This was the potential function.

In[1]:= `myArg[x_, y_] := If[# >= 0, #, 2Pi + #]&[Arg[x + I y]]`

In[2]:=
```
fxy[x_, y_] :=
 Module[{φrel = myArg[x, y], r = Sqrt[x^2 + y^2], φn, φ, s},
  φn = Quotient[r - φrel, 2Pi];
  (* make spiral-like function *)
  Which[φn >= 0,
        φ = (φn + 1) 2Pi + φrel;
        s = (r - (φn 2Pi - Pi + φrel))/Pi,
        φn == -1,
        φ = φrel; s = (r - φ/2)/(φ/2)];
  φ + Tan[s Pi/2]^2]
```

Here is a graphic of this bivariate function.

In[3]:=
```
Show[GraphicsArray[
 Block[{$DisplayFunction = Identity},
 {(* 3D plot *)
  Plot3D[Log[1 + fxy[x, y]], {x, -10, 10}, {y, -10, 10},
               PlotPoints -> 100, Mesh -> False],
  (* density plot; name it for later use *)
  dpp = DensityPlot[fxy[x, y], {x, -10, 10}, {y, -10, 10},
               PlotPoints -> 100, Mesh -> False]}]]]
```

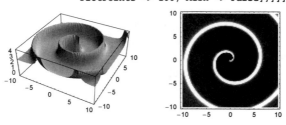

We rewrite the potential in polar coordinates and calculate the gradient vector field in Cartesian coordinates. (For a fast method to calculate the gradient, see [920].)

```
In[4]:= φrel = ArcTan[x, y];
        r = Sqrt[x^2 + y^2]; φ = (φn + 1) 2Pi + φrel;
        s = (r - ((φn 2Pi) - Pi + φrel))/Pi;
        dδ = φ + Tan[s Pi/2]^2;

In[8]:= ff1[x_, y_, φn_] = {D[dδ, x], D[dδ, y]} ;

In[9]:= r = Sqrt[x^2 + y^2]; φ = φrel[x, y];
        (* because of quadratic dependence *)
        s = (r - φ/2)/(φ/2);
        dδ = φ + Tan[s Pi/2]^2;

In[13]:= ff2[x_, y_, φn_] = {D[dδ, x], D[dδ, y]} /.
            {Derivative[i_, j_][φrel] :> Derivative[i, j][ArcTan],
             φrel[x, y] -> myArg[x, y]};

In[14]:= fxyGrad[x_, y_] :=
        Module[{φrel, r, φn, φ, s},
                (* to polar coordinates *)
                φrel = myArg[x, y]; r = Sqrt[x^2 + y^2];
                φn = Quotient[r - φrel // N, N[2Pi]];
                (* make spiral-like function *)
                Which[φn >= 0, ff1[x, y, φn], φn == -1, ff2[x, y, φn]]]

In[15]:= fxyGradx[x_?NumberQ, y_?NumberQ] := fxyGrad[x, y][[1]]
         fxyGrady[x_?NumberQ, y_?NumberQ] := fxyGrad[x, y][[2]]
```

Here, the vector field of the directions of the gradient is shown.

```
In[17]:= (* an arrow at the point {x, y} of length l *)
         arrow[{x_, y_}, l_] :=
         Module[{p1, p2, p3, normal},
                 p1 = {x, y} // N;
                 p2 = p1 + l #/Sqrt[#.#]&[fxyGrad[x, y] // N] ;
                 p3 = p1 + 2/3 (p2 - p1);
                 normal = Reverse[#/Sqrt[#.#]&[p2 - p1]]{-1, 1};
                 Polygon[{p1 + 1/20 normal, p3 + 1/20 normal,
                          p3 + 1/05 normal, p2, p3 - 1/5 normal,
                          p3 - 1/20 normal, p1 - 1/20 normal}]]

In[19]:= vpp = Graphics[{Hue[0],
           Table[arrow[{x, y}, 1/4],
                 {x, -10.012, 10, 1/3}, {y, -10.05, 10, 1/3}]},
                 AspectRatio -> Automatic, PlotRange -> All, Frame -> True];

         Show[GraphicsArray[{vpp, Show[{dpp, vpp}, DisplayFunction -> Identity]}]]
```

Now, we solve the differential equations of the steepest descent numerically until we reach the lowest point. Because the solution cannot be continued through the minimum, a message is issued.

```
In[21]:= sol = NDSolve[
           {x'[t] == -fxyGradx[x[t], y[t]], y'[t] == -fxyGrady[x[t], y[t]],
            x[0] == 8., y[0] == 8.7}, {x, y}, {t, 0, 500},
                    MaxSteps -> 50000, MaxStepSize -> 0.1];
         NDSolve::ndcf : Repeated convergence test
             failure at t == 386.1185037918339`; unable to continue. More...
```

We really reached the minimum.

```
In[22]:= tmax = Flatten[sol[[1, 1, 2, 1]]][[2]];
        {x[tmax], y[tmax]} /. sol
Out[23]= {{0.0000695508, 9.67436×10⁻⁹}}
```

This shows the path taken.

```
In[24]:= pathColor = (* better contrast for printed version *)
        If[Options[Graphics3D, ColorOutput][[1, 2]] === GrayLevel,
           GrayLevel[1/2], RGBColor[0, 0, 1]];

        Show[{dpp, vpp, (* the search path *)
              ParametricPlot[Evaluate[Flatten[{x[t], y[t]} /. sol]],
                             {t, 0, Flatten[sol[[1, 1, 2, 1]]][[2]]},
                             AspectRatio -> Automatic, PlotPoints -> 500,
                             PlotStyle -> {Thickness[0.01], pathColor},
                             DisplayFunction -> Identity]}]
```

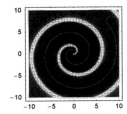

b) Here is a sketch of the Fourier series.

```
In[1]:= partialSum[n_, x_] := Sum[1/k Sin[k x], {k, n}]
```

```
In[2]:= p1 = Plot[Sum[1/k Sin[k x], {k, 100}], {x, 0, Pi},
                  PlotPoints -> 10000, Frame -> True, Axes -> False]
```

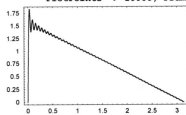

The arclength s_n is given by $s_n = \int_0^\pi (1 + f_n'(x)^2)^{1/2}\, dx$ where $f_n'(x) = \sum_{k=1}^n \sin(k\,x)/k$. After differentiation, the inner sum can be evaluated in closed form.

```
In[3]:= Sum[Cos[k x], {k, n}] // Simplify
```

$$Out[3]= \backslash[\text{Piecewise}] \quad \begin{cases} \frac{1}{2}\,(-1+n+e^{i\,n\,x}\,n + \text{Cos}[n\,x]) & \frac{x}{2\pi} \in \text{Integers} \\ \text{Cos}[\frac{1}{2}\,(1+n)\,x]\,\text{Csc}[\frac{x}{2}]\,\text{Sin}[\frac{n\,x}{2}] & \text{True} \end{cases}$$

Here is a plot of the resulting integrand.

```
In[4]:= λ[n_, x_] = Sqrt[1 + (Sin[n x/2] Csc[x/2] Cos[(n + 1) x/2])^2];
```

```
In[5]:= Plot3D[Evaluate[Log[10, Re[λ[n, x]]]], {x, -1, Pi}, {n, -3, 50},
              PlotPoints -> 200, Mesh -> False, PlotRange -> All]
```

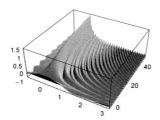

We will use `NDSolve` for the actual integration. Taking into account $\lambda[n, x] \approx n + O(x)$, we can avoid the singularity at the origin.

```
In[6]:= Off[∞::indet]; Off[NDSolve::ndnum];
        arcLength[n_, opts___] := arcLength[n, opts] =
        With[{ε = 10^-12}, (-1[#[[1, 1, 2, 1, 1, 1]]] //. #[[1]])&[
                (* solve differential equation *)
                NDSolve[{l'[x] == λ[n, x], l[Pi - ε] == 0},
                        l, {x, ε, Pi - ε}, MaxSteps -> 10^5]]]
```

The arclength of $f(x)$ is $5^{1/2}/2\,\pi$.

```
In[8]:= l0 = Sqrt[5]/2 Pi;
```

Here are the values for s_{50} and s_{150}.

```
In[9]:= {arcLength[50], arcLength[150]}
Out[9]= {6.57192, 7.27784}
```

It is straightforward to implement a binary search for the arclength. (Because we are looking for an integer result, we cannot use `FindRoot`.)

```
In[10]:= findn[l_, {n1_, n2_}, opts___] :=
         Module[{nl = Min[n1, n2], nu = Max[n1, n2], ll, lu, nm, lm},
                While[{ll, lu} = arcLength[#, opts]& /@ {nl, nu};
                        ll < l < lu && nu - nl > 1,
                        nm = Round[(nl + nu)/2]; lm = arcLength[nm, opts];
                        (* bracket correct n *)
                        If[lm > l, nu = nm, nl = nm]];
                {nl, nu}]
In[11]:= findn[2 l0, {50, 150},
              WorkingPrecision -> $MachinePrecision + 1, PrecisionGoal -> 12]
Out[11]= {101, 102}
```

This means we need 101 terms to get an arclength greater than twice the length of the curve. (A large fraction of the length comes from the steep increase at the origin, not present in $f(x)$.)

```
In[12]:= {arcLength[101], arcLength[102]} - 2 l0
Out[12]= {-0.000326151, 0.00599134}
```

c) To guarantee the symmetry group of the cube for the charge positions, we define a function `allCubePoints` that, for a given 2D point from the symmetry unit triangle $0 \le x \le 1/2$, $0 \le y \le x$ of the upper face of the unit cube, generates all 48 equivalent points. (By reflection and rotation, the whole unit cube can be constructed from the symmetry unit triangle; see Section 2.4 of the Graphics volume.)

```
In[1]:= allCubePoints[{x_, y_}] = (* reflect and rotate repeatedly *)
        Join[#, {#1, #3, #2}& @@@ #, {#3, #2, #1}& @@@ #]&[
         Join[#, {#1, #2, -#3}& @@@ #]&[Join[#, {#2, #1, #3}& @@@ #]&[
          Join[#, {#1, -#2, #3}& @@@ #]&[Join[#, {-#1, #2, #3}& @@@ #]&[
                (* map to upper face *) Append[#, +1/2]& /@ {{x, y}}]]]]];
```

The function `allOtherCubePoints` excludes the original point.

```
In[2]:= allOtherCubePoints[{x_, y_}] = Rest[allCubePoints[{x, y}]];
```

For the minimization, we would like to optimize the calculation of the $(48\,n)^2 - 48\,n$ pairs of interaction energies. Because of the symmetry of all charge positions, we can divide all charges naturally into two groups: the ones from the symmetry unit triangle and all others. The total interaction energy is then 48 times the one from all charges within the symmetry unit triangle and the one between the symmetry unit triangle and all other charges. This means we have instead only $n^2 - n + 47\,n^2$ interaction energies instead to calculate. To realize the constraint that the n charges lie within the symmetry triangle, we could use NMinimize which allows for explicitly specified constraints. But because local root-finding methods are typically faster, it would be preferable to use FindMinimum. Because FindMinimum does not support constraints currently, we could, for instance, incorporate them through a penalty function. Another possibility is mapping each point into the symmetry unit by reflection and translation. We will implement the last approach through the function rescale. (The function toUnitSphere maps a point from the symmetry triangle to the unit sphere.)

```
In[3]:= (* map a point into symmetry unit *)
        rescale[{x_ , y_}] :=
                {Mod[Abs[x], 1/2], Mod[Mod[Abs[y], 1/2], Mod[Abs[x], 1/2]]}


        (* map a point onto the unit sphere *)
        toUnitSphere[xyz_] := xyz/Sqrt[xyz.xyz]
```

We could also express the reduction in the symmetry unit triangle through analytic functions. Using the next function rescaleA gives identical results but is slightly slower because the evaluation of Mod and Abs is faster than the evaluation of Tan, ArcTan, and Sqrt.

```
In[7]:= abs[x_] := Sqrt[x^2]

        mod[x_ , y_] := y (ArcTan[Tan[Pi (x/y + 1/2)]]/Pi + 1/2)

        rescaleA[{x_ , y_}] =
                {mod[abs[x], 1/2], mod[mod[abs[y], 1/2], mod[abs[x], 1/2]]};
```

Next, we must implement the total electrostatic energy for a given set of charge positions from the symmetry triangle. Given the list of point charge positions in the symmetry triangle *li* and the 47 equivalent positions *lo*, the function, the compiled function totalEnergyC calculates the energy. This function assumes that the positions are given as explicit numerical values.

```
In[10]:= totalEnergyC =
        Compile[{{li, _Real, 2}, {lo, _Real, 2}},
                Module[{λi = Length[li], λo = Length[lo]},
                        (* within symmetry unit *)
                        48 Sum[If[j == k, 0, 1/#.#&[li[[j]] - li[[k]]]],
                                {j, λi}, {k, λi}] +
                        (* between symmetry units *)
                        48 Sum[1/#.#&[li[[j]] - lo[[k]]], {j, λi}, {k, λo}]]];
```

The function totalEnergyNumeric takes the numeric charge positions from the symmetry triangle and calculates the total electrostatic energy of the resulting charge constellation on the unit sphere.

```
In[11]:= totalEnergyNumeric[l_ ?(MatrixQ[#, NumberQ]&)] :=
        totalEnergyC[toUnitSphere /@ (Append[#, +1/2]& /@ rescale /@ l),
                toUnitSphere /@
                        Flatten[allOtherCubePoints[rescale[#]]& /@ l, 1]]
```

To use the Levenberg–Marquardt method for minimizing the total electrostatic energy, we need an explicit sum of squares. Our function to be minimized is nearly in this form. By introducing an auxiliary function *df*, we can write the total electrostatic energy as an explicit sum of squares.

```
In[12]:= (* write total energy as an explicit sum of squares *)
        totalEnergySumOfSquares[li_ , lo_] :=
                Module[{λi = Length[li], λo = Length[lo]},
                        Sum[If[j == k, 0, df[48/#.#&[li[[j]] - li[[k]]]]^2],
                                {j, λi}, {k, λi}] +
                        Sum[df[48/#.#&[li[[j]] - lo[[k]]]]^2, {j, λi}, {k, λo}]];


        (* generate equivalent charge positions and resulting energy *)
        totalEnergySumOfSquares[l_List] :=
         totalEnergySumOfSquares[toUnitSphere /@ (Append[#, +1/2]& /@ rescale /@ l),
                        toUnitSphere /@
                                Flatten[allOtherCubePoints[rescale[#]]& /@ l, 1]]
```

```
(* dummy function to compensate the explicit square *)
df[x_?InexactNumberQ] := Sqrt[x]
```

The symbolic form of the total energy finally can be easily obtained from the sum of squares form by replacing the function by the square root function. The resulting terms of the form $((\ldots)^{1/2})^2$ autoevaluate, and we have a nice form of the total energy.

```
In[19]:= totalEnergySymbolic[l_List] := totalEnergySumOfSquares[l] /. df -> Sqrt
```

These are the option settings that we will use for the minimization for the case of the objective function `totalEnergyNumeric`. In addition to the main method option name, we also specify different suboptions. This will give us a more detailed choice for selecting the best suited method. These methods cannot access the objective function in its symbolic form and so cannot form any symbolic derivatives. Any derivatives potentially needed must be calculated numerically through a finite difference approximation.

```
In[20]:= methodsNumeric = {"Gradient",
                {"ConjugateGradient",
                 StepControl -> "LineSearch", "Method" -> "PolakRibiere"},
                {"ConjugateGradient",
                 StepControl -> "LineSearch", "Method" -> "FletcherReeves"},
                {"QuasiNewton", StepControl -> "LineSearch"},
                {"QuasiNewton", StepControl -> "None"},
                {"Newton", StepControl -> "LineSearch"},
                {"Newton", StepControl -> "TrustRegion"},
                {"Newton", StepControl -> "None"},
                "PrincipalAxis"};
```

For the sum of squares version of the objective function we will use the `"LevenbergMarquardt"` option setting. (The suboption is unique and given only for consistency with the options specified above.)

```
In[21]:= methodsSumOfSquares = {{"LevenbergMarquardt", StepControl -> "TrustRegion"}};
```

And for the explicit symbolic version of the objective function, we will use the following option settings. These methods have the freedom to form symbolic derivatives.

```
In[22]:= methodsSymbolic = {"Gradient",
                {"ConjugateGradient",
                 StepControl -> "LineSearch", "Method" -> "PolakRibiere"},
                {"ConjugateGradient",
                 StepControl -> "LineSearch", "Method" -> "FletcherReeves"},
                {"Newton", StepControl -> "LineSearch"},
                {"Newton", StepControl -> "TrustRegion"},
                {"Newton", StepControl -> "None"}};
```

The function `findEnergyMinimum` minimizes the total electrostatic energy for *n* charges. The function returns the result of the minimization and the counts of the function evaluation and steps carried out.

```
In[23]:= findEnergyMinimum[tE_, n_, seed_, opts___] :=
       Module[{(* evaluation and step counters *) cE = 0, cS = 0},
         SeedRandom[seed];
         {(* potentially suppress potential messages *)
          (* Internal`DeactivateMessages *) Identity[
             FindMinimum[tE[Table[{x[k], y[k]}, {k, n}]],
                 (* starting values within the symmetry unit *)
                 Evaluate[Sequence @@ Flatten[Table[{{x[k], #1, #2},
                                      {y[k], #1 Random[], #2 Random[]}}&[
                                    Random[]/2, Random[]/2], {k, n}], 1]],
                 opts, PrecisionGoal -> 6, MaxIterations -> 250,
                 EvaluationMonitor :> (cE++), StepMonitor :> (cS++)]],
          (* return counter values *) {cE, cS}}]
```

Now, we fix a small *n*, say $n = 6$ and evaluate the minimum electrostatic energy for the various above-defined option settings. Here are the methods accessing the objective function purely numerically. (For more reliable judgments about each method, we should repeat the minimizations for different initial configurations and different values of *n*. Because we are satisfied here with a good choice, not necessary the optimal choice of the method, we will not carry out these repetitions here.) We also turn off some of the messages that would be issued repeatedly in the following because we will always supply two starting values.

```
In[24]:= (* turn off some of the messages *)
        Off[General::stop] Off[FindMinimum::fmwar]; Off[FindMinimum::lstol];

In[26]:= Print[{#, Timing[{#1[[1]], #2}& @@
            (eEnergy[N, #] = findEnergyMinimum[totalEnergyNumeric,
                              6, 1, Method -> #])}]}& /@ methodsNumeric;
```

{Gradient, {33.67 Second, {112709., {1443, 15}}}}

FindMinimum::sdprec :
  Line search unable to find a sufficient decrease in the function
    value with MachinePrecision digit precision. More…

{{ConjugateGradient, StepControl → LineSearch, Method → PolakRibiere},
 {45.23 Second, {112395., {1932, 48}}}}

{{ConjugateGradient, StepControl → LineSearch, Method → FletcherReeves},
 {28.64 Second, {112395., {1218, 34}}}}

{{QuasiNewton, StepControl → LineSearch}, {33.77 Second, {112395., {1428, 90}}}}

FindMinimum::cvmit : Failed to converge to the
  requested accuracy or precision within 250 iterations. More…

{{QuasiNewton, StepControl → None}, {82.63 Second, {112395., {3514, 250}}}}

{{Newton, StepControl → LineSearch}, {67.66 Second, {112701., {2871, 14}}}}

{{Newton, StepControl → TrustRegion}, {69.41 Second, {112395., {2928, 15}}}}

FindMinimum::cvmit : Failed to converge to the
  requested accuracy or precision within 250 iterations. More…

{{Newton, StepControl → None}, {1092.11 Second, {3.15304 × 10^8, {45764, 250}}}}

{PrincipalAxis, {50.65 Second, {112395., {2116, 17}}}}

The obtained minima all agree (with the exception of the {"Newton", StepControl -> "None"} setting that did not converge). To get an idea of the resulting charge configurations, we display them for the last results.

```
In[27]:= Needs["Graphics`Shapes`"]

In[28]:= Show[GraphicsArray[#]]& /@ Partition[
          Graphics3D[{EdgeForm[], SurfaceColor[Hue[0.22], Hue[0.12], 2.3],
            (* place a sphere at the charge positions *)
            (TranslateShape[Sphere[0.03, 12, 12], #]& /@ toUnitSphere /@
              allCubePoints[rescale[#]])& /@ (Table[{x[k], y[k]},
                {k, Length[eEnergy[N, #][[1, 2]]]/2}] /. eEnergy[N, #][[1, 2]])},
            Boxed -> False, ViewPoint -> {0, 0, 1}]& /@ methodsNumeric, 3]
```

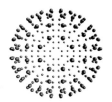 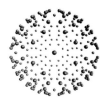 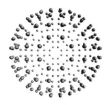

We continue with the sum of squares minimizing method. We again obtain the same minimum, but, within the 250 iterations allowed, the minimum could not be verified.

```
In[29]:= Print[{#, Timing[{#1[[1]], #2}& @@
              (eEnergy[S, #] = findEnergyMinimum[totalEnergySumOfSquares,
                                6, 1, Method -> #])}]}& /@ methodsSumOfSquares;
```

```
    FindMinimum::cvmit : Failed to converge to the
         requested accuracy or precision within 250 iterations. More…
```

```
    {{LevenbergMarquardt, StepControl → TrustRegion}, {38.23 Second, {112403., {3535, 250}}}}
```

We finish with the methods that have access to the symbolic form of the objective function.

```
In[30]:= Print[{#, Timing[{#1[[1]], #2}& @@
              (eEnergy[SS, #] = findEnergyMinimum[totalEnergySymbolic,
                                6, 1, Method -> #])}]}& /@ methodsSymbolic;
```

```
    {Gradient, {8.64 Second, {112709., {1457, 15}}}}
```

```
    {{ConjugateGradient, StepControl → LineSearch, Method → PolakRibiere},
     {11.41 Second, {112395., {1988, 48}}}}
```

```
    {{ConjugateGradient, StepControl → LineSearch, Method → FletcherReeves},
     {10.03 Second, {112395., {1498, 40}}}}
```

```
    {{Newton, StepControl → LineSearch}, {17.38 Second, {112701., {3363, 16}}}}
```

```
    {{Newton, StepControl → TrustRegion}, {15.95 Second, {112395., {2928, 15}}}}
```

```
    {{Newton, StepControl → None}, {37.07 Second, {112395., {8249, 45}}}}
```

Comparing all of the above results, the optimal method for speed and quality of the result seems to be the setting {"Conju‑gateGradient", StepControl -> "LineSearch", "Method" -> "PolakRibiere" }. To confirm our choice, we carry out four minimizations for $n = 12$ with different starting values. Three of the four results agree with each other; one is about half a percent higher.

```
In[31]:= selectedMethodSetting = {"ConjugateGradient",
                               StepControl -> "LineSearch",
                               "Method" -> "PolakRibiere"}
```

```
Out[31]= {ConjugateGradient, StepControl → LineSearch, Method → PolakRibiere}
```

```
In[32]:= Table[Timing[{#1[[1]], #2}& @@ (findEnergyMinimum[totalEnergySymbolic,
                    12, j, Method -> selectedMethodSetting])], {j, 4}]
```

```
Out[32]= {{74.56 Second, {503327., {3640, 53}}}, {112.06 Second, {503327., {4992, 77}}},
         {133.77 Second, {503327., {5720, 82}}}, {240.27 Second, {503327., {10530, 147}}}}
```

The next choice would be the suboptions to the "LineSearch" method. The following results show that the automatic value (as we see from the evaluation and step number, here this is "MoreThuente") is a good choice.

```
In[33]:= Print[{#, Timing[{#1[[1]], #2}& @@
              (findEnergyMinimum[totalEnergySymbolic,
                    6, 3, Method  -> {"ConjugateGradient",
                        StepControl -> {"LineSearch", "Method" -> #},
                        "Method" -> "PolakRibiere"}])}]}& /@
                          {Automatic, "MoreThuente", "Backtracking", "Brent"};
```

```
    {Automatic, {16.03 Second, {112701., {2310, 62}}}}
```

```
    {MoreThuente, {16.34 Second, {112701., {2310, 62}}}}
```

```
        FindMinimum::cvmit : Failed to converge to the
            requested accuracy or precision within 250 iterations. More…
    {Backtracking, {22.87 Second, {112701., {3544, 250}}}}

        FindMinimum::cvmit : Failed to converge to the
            requested accuracy or precision within 250 iterations. More…
    {Brent, {5.96 Second, {125995., {233, 3}}}}}
```

Now, we could consider changing the "MaxRelativeStepSize" suboption of the "LineSearch" method. Again, the default value seems to be a good choice.

```
In[34]:= Print[{#, Timing[{#1[[1]], #2}& @@
            (findEnergyMinimum[totalEnergySymbolic,
                6, 2, Method -> {"ConjugateGradient",
                    StepControl -> {"LineSearch", "MaxRelativeStepSize" -> #},
                        "Method" -> "PolakRibiere"}])}]}]& /@
                            {10^-1, 1, 5, 10, 50, 10^2, 10^3};
    {1/10 , {18.62 Second, {112701., {2688, 79}}}}
    {1, {15.24 Second, {112395., {2030, 47}}}}}
    {5, {13.63 Second, {112395., {1694, 34}}}}}
    {10, {12.07 Second, {112395., {1414, 35}}}}}
    {50, {14.07 Second, {112395., {1792, 39}}}}}
    {100, {11.99 Second, {112395., {1414, 34}}}}}
    {1000, {12.55 Second, {112395., {1498, 34}}}}}
```

Because this method is relatively fast, we calculate all minimal energy positions up to $n = 16$ and display the corresponding charge positions in the symmetry unit.

```
In[35]:= oMax = 16;
        (* calculate minimal energy configurations *)
        Do[Timing[{#1[[1]], #2}& @@ (eEnergyJ[j] =
            findEnergyMinimum[totalEnergySymbolic, j, 1,
                Method -> selectedMethodSetting])], {j, 1, oMax}]
In[38]:= Show[GraphicsArray[#]]& /@ Partition[
        Table[Graphics[(* symmetry triangle *)
            {{GrayLevel[0.9], Polygon[{{0, 0}, {0.5, 0}, {0.5, 0.5}}]},
            PointSize[0.03], (* the charges *) Point[rescale[#]]& /@
            (Table[{x[k], y[k]}, {k, j}] /. eEnergyJ[j][[1, 2]])},
            PlotRange -> {{-0.1, 0.6}, {-0.1, 0.6}}, Frame -> True,
            FrameTicks -> None, AspectRatio -> Automatic], {j, 1, oMax}], 4]
```

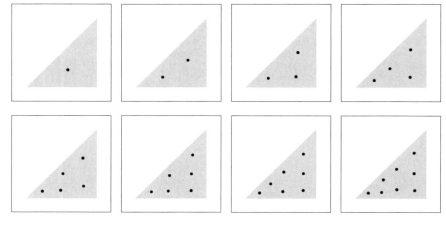

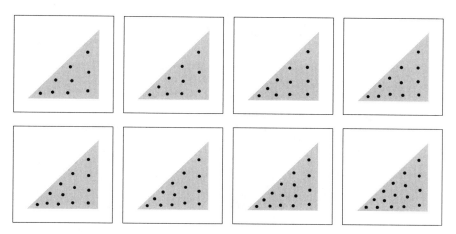

A double logarithmic plot of the minimum energy values suggests $\mathcal{E}_{\min} \sim n^{2.2}$.

```
In[39]:= logEData = Log @ MapIndexed[{#2[[1]], #1}&,
                       Table[eEnergyJ[j][[1, 1]], {j, oMax}]];
         ListPlot[logEData]
```

An interesting quantity to monitor is the ratio between the largest and the smallest minimal distance between two charges.

```
In[41]:= maxToMinDistance[eEnergyJ_] :=
         Module[{o, ℓ, ℓi, ℓo, T},
           o = Length[eEnergyJ[[1, 2]]]/2;
           ℓ = Table[{x[k], y[k]}, {k, o}] /. eEnergyJ[[1, 2]];
           ℓi = toUnitSphere /@ (Append[#, +1/2]& /@ rescale /@ ℓ);
           ℓo = toUnitSphere /@ Flatten[allOtherCubePoints[rescale[#]]& /@ ℓ, 1];
           (* all minimal distances to next neighbor *)
           T = Table[Min[{Table[If[j === k, Infinity,
                               Norm[ℓi[[j]] - ℓi[[k]]]], {k, o}],
                       Table[Norm[ℓi[[j]] - ℓo[[k]]]], {k, o 47}]}], {j, o}];
           (* ratio of largest to smallest minimal distance *)
           Max[T]/Min[T]]
In[42]:= Table[{j, maxToMinDistance[eEnergyJ[j]]}, {j, oMax}]
Out[42]= {{1, 1.}, {2, 1.03287}, {3, 1.06089}, {4, 1.15488}, {5, 1.09947}, {6, 1.10932},
         {7, 1.19619}, {8, 1.19717}, {9, 1.22494}, {10, 1.20438}, {11, 1.18517},
         {12, 1.19001}, {13, 1.23488}, {14, 1.20961}, {15, 1.18016}, {16, 1.22193}}
```

Now we will calculate the minimum energy configuration for $n = 36$. This will take about an hour.

```
In[43]:= Block[{j = 36},
         Print @ Timing[{#1[[1]], #2}& @@ (eEnergyJ[j] =
           findEnergyMinimum[totalEnergySymbolic, j, (* good seeds 12, 14 *) 8,
                   Method -> selectedMethodSetting])]]
         {3757.62 Second, {5.33192×10^6, {31524, 147}}}
```

The result is slightly lower than the result obtained by extrapolating the results for $n \le 16$.

```
In[44]:= Exp[Fit[Select[logEData, (#[[1]] > 2)&],
                 {1, lnn}, lnn] /. lnn -> Log[36]]

Out[44]= 5.40202×10^6
```

We end with three views on the resulting charge positions on the unit sphere.

```
In[45]:= Show[GraphicsArray[
           Function[vp, Graphics3D[{EdgeForm[], SeedRandom[3];
               (* color symmetry-equivalent spheres identically *)
               {SurfaceColor[Hue[Random[]], Hue[Random[]], 2.2],
                TranslateShape[Sphere[0.025, 10, 10], #]& /@ toUnitSphere /@
                allCubePoints[rescale[#]]}& /@
               (Table[{x[k], y[k]}, {k, Length[eEnergyJ[36][[1, 2]]]/2}] /.
                                                 eEnergyJ[36][[1, 2]])},
               PlotRange -> All, SphericalRegion -> False, Boxed -> True,
               ViewPoint -> vp]] /@
               (* view from above and view along face and main diagonals *)
                   {{0, 0, 2}, Sqrt[2] {1, 1, 0}, Sqrt[2/3] {1, 1, 1}}]]
```

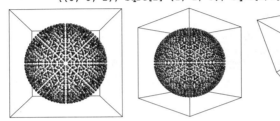

23. N[*expr*, *prec*] Questions and Compile Questions

a) Yes, N can return nonnumeric results, for instance, for indeterminate expressions or flavors of Infinity and related expressions. Here are two examples.

```
In[1]:= N[((E (E + 1) - E^2 - E))^((E (E + 1) - E^2 - E)), 111]
```

Power::indet : Indeterminate expression $0. \times 10^{-160^{0. \times 10^{-112}}}$ encountered. More…

Power::indet : Indeterminate expression $0. \times 10^{-160^{0. \times 10^{-160}}}$ encountered. More…

N::meprec : Internal precision limit $MaxExtraPrecision =
    50.` reached while evaluating $(-e - e^2 + e(1+e))^{-e-e^2+e(1+e)}$. More…

```
Out[1]= Indeterminate
```

```
In[2]:= N[Log[Sqrt[2] + Sqrt[3] - Sqrt[5 + 2 Sqrt[6]]], 222]
```

N::meprec : Internal precision limit $MaxExtraPrecision =
    50.` reached while evaluating $\text{Log}\left[\sqrt{2} + \sqrt{3} - \sqrt{5 + 2\sqrt{6}}\right]$. More…

```
Out[2]= Indeterminate
```

Here is an example where numericalization returns Underflow[], although the expression is indeterminate.

```
In[3]:= indet = ((E + 1)^2 - Expand[(E + 1)^2])/((Pi + 1)^2 - Expand[(Pi + 1)^2] +
             (Sqrt[2] + Sqrt[3] - Sqrt[5 + 2 Sqrt[6]]) 10^(E + 10)^10)
```

$$Out[3]= \frac{-1 - 2e - e^2 + (1+e)^2}{-1 + 10^{(10+e)^{10}}\left(\sqrt{2} + \sqrt{3} - \sqrt{5 + 2\sqrt{6}}\right) - 2\pi - \pi^2 + (1+\pi)^2}$$

```
In[4]:= N[indet, 22]
```

General::ovfl : Overflow occurred in computation. More…

General::ovfl : Overflow occurred in computation. More…

General::ovfl : Overflow occurred in computation. More…

General::stop :

Further output of General::ovfl will be suppressed during this calculation. More...

N::meprec : Internal precision limit $MaxExtraPrecision = 50.` reached while

evaluating $\dfrac{-1 - 2\,e - e^2 + (1+e)^2}{-1 + 10^{(10+e)^{10}}\,(\sqrt{2} + \sqrt{3} - \sqrt{\text{Plus}\,[\,\ll 2\gg\,]}\,) - 2\,\pi - \pi^2 + (1+\pi)^2}$. More...

Out[4]= Underflow[]

In[5]:= {Simplify[Numerator[indet]], FullSimplify[Denominator[indet]]}

Out[5]= {0, 0}

b) The result is 0.

In[1]:= Precision[SetAccuracy[10^+30 Pi, 50]] - Accuracy[SetPrecision[10^+30 Pi, 50]] +
    Precision[SetAccuracy[10^-30 Pi, 50]] - Accuracy[SetPrecision[10^-30 Pi, 50]]

Out[1]= -27.9065

Let us analyze in more detail the form of the summands. The first one numericalizes $10^{30} \times \pi$ to an accuracy of 50. This means the number 10^{30} is made into an approximate number with accuracy 50 and the constant π is made into an approximate number with accuracy 50 and the resulting product evaluates. The precision of 10^{30} after setting the accuracy is 80, and the precision of π after setting the accuracy is slightly above 50. This means the precision of the product is about 50.

In[2]:= Precision[SetAccuracy[10^30 Pi, 50]]

Out[2]= 50.4971

The second summand measures the accuracy of $10^{30} \times \pi$ numericalized to 50 digits. Because $10^{30} \times \pi$ has 30 digits to the left of the decimal point, the resulting accuracy is about 20.

In[3]:= Accuracy[SetPrecision[10^30 Pi, 50]]

Out[3]= 19.2018

The third summand numericalizes $10^{-30} \times \pi$ to an accuracy of 50. Because $10^{-30}\,\pi$ is roughly 3×10^{-30}, this means to achieve 50 digits to the right of the decimal point, the expression must be numericalized with precision 20.

In[4]:= Precision[SetAccuracy[10^-30 Pi, 50]]

Out[4]= 20.

The last summand measures the accuracy of $10^{-30} \times \pi$ numericalized to 50 digits. Because $10^{-30} \times \pi$ has 30 digits to the right of the decimal point, the resulting accuracy is about 80.

In[5]:= Accuracy[SetPrecision[10^-30 Pi, 50]]

Out[5]= 79.2018

Adding the four summands yields a result of about -28.

c) The result is True.

In[1]:= Log[10, Abs[N[SetPrecision[SetPrecision[Pi, 50], Infinity]/Pi - 1, 30]]] < -50

N::meprec : Internal precision limit $MaxExtraPrecision = 50.` reached while

evaluating $-1 + \dfrac{1469261698255430054843137663917859561105329 60781381}{46768052394588893382517914646921056628989841375232\,\pi}$. More...

Out[1]= True

Let us analyze the single steps involved in the above computation. We start with generating a 50 digit-precision approximation to π.

In[2]:= πApproximation50 = SetPrecision[Pi, 50]

Out[2]= 3.1415926535897932384626433832795028841971693993751

In the next step, we make an exact number from πApproximation50.

In[3]:= SetPrecision[πApproximation50, Infinity]

Out[3]= $\dfrac{1469261698255430054843137663917859561105329 60781381}{46768052394588893382517914646921056628989841375232}$

Because of the guard digits present in the internal representation, the floating-point number displayed is different from the one we see. It is actually a slightly better approximation.

In[4]:= **N[%/Pi - 1, 30]**

> N::meprec : Internal precision limit $MaxExtraPrecision = 50.` reached while
> evaluating $-1 + \dfrac{146926169825543005484313766391785956110532960781381}{46768052394588893382517914646921056628989841375232\pi}$. More…

Out[4]= $-1.3798204161495724090531778671 \times 10^{-51}$

We can see the guard digits using the function $NumberBits.

In[5]:= **(* load context of function $NumberBits *)**
       If[FreeQ[$ContextPath, #], AppendTo[$ContextPath, #]]& /@ Context /@
              Union[Join[Names["*`\*Number\*Bits\*"], Names["\*`*Bits*Number*"]]]
Out[6]= {{Global`, System`, NumericalMath`}}

In[7]:= **$NumberBits[πApproximation50]**

Out[7]= {1, {1, 1, 0, 0, 1, 0, 0, 1, 0, 0, 0, 0, 1, 1, 1, 1, 1, 1, 0, 1, 1, 0, 1, 0, 1, 0, 1,
         0, 0, 0, 1, 0, 0, 0, 1, 0, 0, 0, 0, 1, 0, 1, 1, 0, 1, 0, 0, 0, 1, 1, 0, 0, 0, 0, 1,
         0, 0, 0, 1, 1, 0, 1, 0, 0, 1, 1, 0, 0, 0, 1, 0, 0, 1, 1, 0, 0, 0, 1, 1, 0, 0, 1, 1,
         0, 0, 0, 1, 0, 1, 0, 0, 0, 1, 0, 1, 1, 1, 0, 0, 0, 0, 0, 0, 1, 1, 0, 1, 1, 1, 1, 0,
         0, 0, 0, 1, 1, 1, 1, 0, 0, 1, 1, 0, 1, 0, 0, 0, 0, 1, 0, 0, 1, 0, 1, 0, 0, 1, 0, 0, 0,
         0, 0, 0, 1, 0, 0, 1, 0, 0, 1, 1, 1, 0, 0, 0, 0, 0, 1, 0, 0, 0, 1, 0, 0, 0, 1, 0},
        {1, 0, 0, 1, 1, 0, 0, 1, 1, 1, 1, 1, 0, 0, 1, 1, 0, 0, 0, 1, 1, 1, 0, 1, 0, 0, 0, 0, 0, 0,
         0, 0, 1, 0, 0, 0, 0, 0, 1, 0, 1, 1, 1, 1, 0, 0, 0, 1, 0, 0, 0, 1, 0, 1, 1, 0, 1, 0, 1, 0}, 2}

If we use Rationalize[…, 0] instead SetPrecision[…, Infinity], we obtain a slightly lower value for the logarithm of the absolute value if the difference.

In[8]:= **Internal`DeactivateMessages @**
       {Log[10, Abs[N[SetPrecision[SetPrecision[Pi, 50], Infinity]/Pi - 1, 30]]],
        Log[10, Abs[N[Rationalize[SetPrecision[Pi, 50], 0]/Pi - 1, 30]]]}
Out[8]= {-50.8601774334179034115782393 2639, -50.335142588701425569506600 88424}

Because we have in the guard digits some extra (correct) information, the resulting effective precision of the rationalized version of πApproximation50 is greater than 50 and the above expression evaluates to True.

d) The automatic precision control model of *Mathematica* is based on the formula $|\delta f| = \left| \dfrac{f'(x)\,x}{f(x)} \dfrac{\delta x}{x} \right|$, where δx is the relative uncertainty of the input x and δf is the relative uncertainty of the output $f(x)$.

This means that the precision of the output p_{Out} is related to the precision of the input $p_{In} = -\log_{10}\left(\left|\frac{\delta x}{x}\right|\right)$ by $p_{Out} = p_{In} \pm \log_{10}\left(\left|\frac{f'(x)\,x}{f(x)}\right|\right)$. To achieve a much higher/lower precision of the output, the quantity $\frac{f'(x)\,x}{f(x)}$ must either be very large or very small. This means that $f(x)$ for a fixed x the function $f(x)$ must either be nearly a constant or must be very steep (like in the neighborhood of a singularity).

We start with a function that is very flat; for example, the function is $f(x) = 1 - e^{-(100\,x)^2}$ for $x \gtrsim 1$.

In[1]:= **f[x_] := 1 - Exp[-(100 x)^2];**

In[2]:= **Plot[f[x], {x, 0, 2}, PlotRange -> All]**

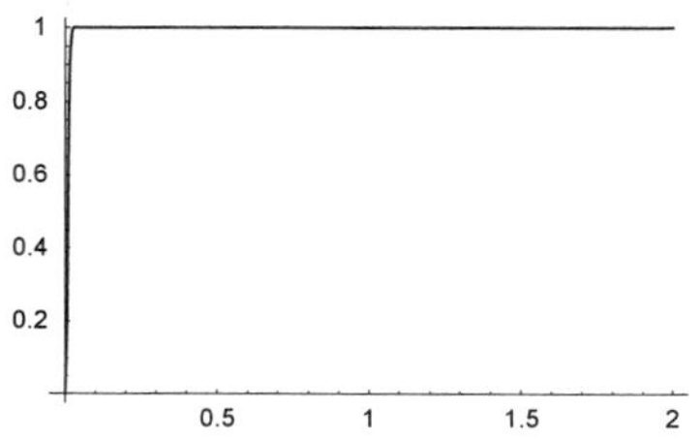

Given an input x near 1, the function value will not change much. This means the output precision will be much higher than the input precision. The quantity $\left| \dfrac{f'(x)\,x}{f(x)} \right|$ becomes quite small for large x. The next graphic uses a logarithmic scale.

In[3]:= **Plot[Evaluate[Log[10, Abs[f'[x]/f[x] x]]], {x, 0, 2}]**

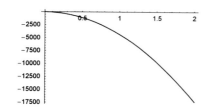

The actual precision of $f(1)$ is more than 44 times larger than the input precision.

In[4]:= **Precision[f[N[1, 100]]]**

Out[4]= 4438.64

Changing the 100th digit of the input 1 affects the 4439th digit of the output.

In[5]:= **f[N[1 + 10^-100, 5000]] - f[N[1 - 10^-100, 5000]] // N**

Out[5]= $4.541935461258944 \times 10^{-4439}$

In[6]:= **$MaxExtraPrecision = 10000;**
N[Subtract @@ f[Interval[{1 - 10^-100, 1 + 10^-100}]][[1]], 22]

Out[7]= $-4.541935461258944394164 \times 10^{-4439}$

Next, let us construct a function that lowers the precision of the input. We start with a function that is very steep around $x = 1$, such as the function $f(x) = \tan(\pi x/2)$.

In[8]:= **f[x_] := Tan[Pi /2 x];**
Plot[f[x], {x, 0, 2}, PlotRange -> All]

Given an input $x \approx 1$, the function value will change dramatically for small changes in x. This means the output precision will have a much lower value than the input precision. The quantity $f'(x) x/f(x)$ becomes now quite large for $f(x)$ near $x \approx 1$. The right picture shows a closer view (logarithmically zoomed in) near $x = 1$.

In[10]:=

```
Show[GraphicsArray[
Block[{$DisplayFunction = Identity},
 {(* overall look with linear abscissa *)
  Plot[Evaluate[Log[10, Abs[f'[x]/f[x] x]]], {x, 0, 2}],
  (* zoomed-in plot logarithmic abscissa *)
  Module[{y = 1 - 10^-x},
       Plot[Evaluate[Log[10, Abs[f'[y]/f[y] y]]], {x, 1, 15}]]}]]]
```

Changing the 90th digit of the input 1 affects the 10th digit of the output.

```
In[11]:= $MaxExtraPrecision = 1000;
        N[Log[10, Abs[f'[x]/f[x] x 10^-100/x]] /. x -> 1 - 10^-90, 50]
```

$$\text{Divide::infy : Infinite expression } \frac{1}{0. \times 10^{-50}} \text{ encountered.}$$

```
Out[12]= -10.0000000000000000000000000000000000000000000000000
```

```
In[13]:= Precision[f[N[1 - 10^-90, 100]]]
```

```
Out[13]= 10.
```

Shifting this behavior to $x = 1$, we slightly redefine $f(x)$ to $f(x) = \tan(\pi x/2 (1 - 10^{-90}))$.

```
In[14]:= Clear[f];
        f[x_] := Tan[Pi /2 x (1 - 10^-90)];
```

Now the function $f(x)$ has the desired property.

```
In[16]:= Precision[f[N[1, 100]]]
```

```
Out[16]= 10.
```

```
In[17]:= (f[N[1 + 10^-100, 200]] - f[N[1 - 10^-100, 200]])/f[N[1, 200]] // N
```

```
Out[17]= 2. × 10^-10
```

Here is a check using interval arithmetic.

```
In[18]:= $MaxExtraPrecision = 200; Off[Infinity::indet];
        N[Subtract @@ f[Interval[{1 - 10^-100, 1 + 10^-100}]][[1]],
          200]/f[N[1, 200]]
Out[19]= -2.0000000000000000002000000000000000000020000000000000000000200000000000000000019;
        99999998000000000001999999999× 10^-10
```

So the two functions $f(x) = 1 - e^{-(100 x)^2}$ and $f(x) = \tan(\pi x/2 (1 - 10^{-90}))$ are examples of functions that dramatically raise or lower the input precision.

e) To achieve the intended failure of N, we have to establish a couple of circumstances.

The automatic precision control model of *Mathematica* is based on the formula $|\delta f| = \left|\frac{f'(x)x}{f(x)}\right| \left|\frac{\delta x}{x}\right|$, where δx is the relative uncertainty of the input x, and δf is the relative uncertainty of the output $f(x)$. This means that the precision p_{Out} of the output is related to the precision of the input $p_{In} = -\log_{10}(\left|\frac{\delta x}{x}\right|)$ by $p_{Out} = p_{In} \pm \log_{10}(\left|\frac{f'(x)x}{f(x)}\right|)$. We assume in the significance arithmetic model that $f(x)$ is a continuous function. Therefore, it makes sense to form $f'(x)$. This assumption breaks down for multivalued functions along their branch cuts, e.g., for Sqrt, Power, and Log along the negative real axis. So, our expression should contain a number very near to a branch cut.

The next ingredient needed for getting a wrong result from N is nonrational arithmetic. (As long as we calculate quantities containing only rational arithmetic, everything is correct.). Therefore, we introduce in the following example a hidden zero. It is constructed from square roots and the transcendental function sin.

```
In[1]:= hiddenZero = Sin[Pi/16] - 1/2 Sqrt[2 - Sqrt[2 + Sqrt[2]]]
```

$$Out[1]= -\frac{1}{2} \sqrt{2 - \sqrt{2 + \sqrt{2}}} + \sin\left[\frac{\pi}{16}\right]$$

Using FullSimplify, we can prove that hiddenZero is identically zero.

```
In[2]:= FullSimplify[hiddenZero]
```

```
Out[2]= 0
```

hiddenZero is mathematically identically zero, but not structurally zero. Calculating N[hiddenZero, 30] indicates this.

```
In[3]:= N[hiddenZero, 30]
```

$$N::meprec : \text{Internal precision limit } \$MaxExtraPrecision = 50.\text{`}$$
$$\text{reached while evaluating } -\frac{1}{2} \sqrt{2 - \sqrt{2 + Power[\ll 2\gg]}} + \sin\left[\frac{\pi}{16}\right]. \text{ More...}$$

Out[3]= $0. \times 10^{-80}$

In[4]:= `% // InputForm`

Out[4]//InputForm= `0``79.37370823417302`

Mathematica could not calculate any correct digits of the last expression. The numerically-calculated quantity is zero to about 80 places. In addition, the sign of the quantity was negative within the arithmetic used. Inspecting the internal bit representation of the last result shows that we have no certified digit and that the sign of the uncertified digits is negative.

In[5]:= `((* load context of function $NumberBits *)`
`If[FreeQ[$ContextPath, #], AppendTo[$ContextPath, #]]& /@ Context /@`
`Union[Join[Names["*`*Number*Bits*"], Names["*`*Bits*Number*"]]])`

Out[5]= `{{Global`, System`, NumericalMath`}}`

In[6]:= `$NumberBits[%%]`

Out[6]= `{1, {}, {}, -264}`

For complex numbers, the precision is defined by its real and imaginary parts. So if we construct the expression $1 + I(\text{hiddenZero} - 10^{\wedge}-200)$ (in which the real part is dominating in magnitude), no messages are generated. However, the information about the nonvanishing imaginary part 10^{-200} has been lost within the uncertainty of order 10^{-54}, arising from the numerical determination of `hiddenZero`.

In[7]:= `N[1 + I (hiddenZero - 10^-200), 50]`

Out[7]= $1.0000000000000000000000000000000000000000000000000 + 0. \times 10^{-100} \, i$

The last number has the required precision of 50.

In[8]:= `Precision[%]`

Out[8]= `50.`

Despite the spurious nature of the imaginary part in $1 + I(\text{hiddenZero} - 10^{\wedge}-200)$, it will be used for deciding which side of the branch cut to choose when this number is an argument to `Log`. Here this effect is demonstrated for $Log[-1 \pm I$ *fuzzyZero*].

In[9]:= `Block[{$MinPrecision = - 100, fuzzyZero},`
`fuzzyZero = -1.0`-10*^-100;`
`{Log[-1 + I fuzzyZero], Log[-1 - I fuzzyZero]}]`

Out[9]= $\{0. \times 10^{-91} +$
$3.1415926535897932384626433832795028841971693993751058209749445923078164062862089934$
$8628034825 \, i, \, 0. \times 10^{-91} +$
$3.1415926535897932384626433832795028841971693993751058209749445923078164062862089934$
$8628034825 \, i\}$

When evaluating $N[1 + I(\text{hiddenZero} - 10^{\wedge}-200), 50]$, the resulting number has a very small real part and an imaginary part of the order π. If we, in a last step, apply `Exp` to "amplify" the different sign of the imaginary, we get the expression we are looking for. (To have a small real part, we used -1 in the expression.)

In[11]:= `expr = Exp[I Log[-1 + I (hiddenZero - 10^-200)]];`

In[12]:= `expr // NumericQ`

Out[12]= `True`

Calculating `N[expr, 50]` gives the following completely wrong result.

In[13]:= `N[expr, 50]`

Out[13]= $0.043213918263772249774417737171728011275728109810633 + 0. \times 10^{-52} \, i$

Calculating `N[expr, 1000]` ensures that the small imaginary part 10^{-200} is correctly treated, and so, we get the following correct result (we use ... `// N[#, 50]&` to display only the relevant digits).

In[14]:= `N[expr, 1000] // N[#, 50]&`

Out[14]= $23.140692632779269005729086367948547380266106242600 + 0. \times 10^{-49} \, i$

Here is an example where substituting a high-precision number gives a wrong result. With 30 digits, the small number 10^{-100} stays undetected. In the second substitution, we use 3000 digits and we end on the other side of the branch cut.

```
In[15]:= Log[Sin[(I (Log[1/(1 + ξ)] - Log[ξ/(1 + ξ)]) +
                 (* small quantity *) 10^-100)/2]] /.
         {{ξ -> N[-2, 30]}, {ξ -> N[-2, 3000]},
         (* exact *) {ξ -> -2}} // Simplify // N[#, 3000]& // N
Out[15]= {0.0588915 + 3.14159 i, 0.0588915 - 3.14159 i, 0.0588915 - 3.14159 i}
```

The casus irreducibilis can also be used to construct a hidden zero. When an irreducible cubic polynomial has three real roots, it is impossible to express these roots in an "I-free way" (see Chapter 1 of the Symbolics volume [1795]). Here is a cube root of the polynomial $1000\,x^3 - 4000\,x^2 + 1000\,x + 6001 = 0$.

```
In[16]:= cubeRoot = (4/3 - (65 (1 + I Sqrt[3]))/
                    (3 (1/2 (-70027 + 3 I Sqrt[431579919]))^(1/3)) -
                    1/60 (1 - I Sqrt[3]) (1/2 (-70027 + 3 I Sqrt[431579919]))^(1/3));
```

Calculating a numerical value of cubeRoot gives a term in the form $-0. \times 10^{-exp}\,i$.

```
In[17]:= N[cubeRoot, 50]
Out[17]= -1.0000833292827529859824729421561499384406839554721 5+ 0. × 10^-51 i
```

No digit of the imaginary part is certified.

```
In[18]:= $NumberBits[Im[%]]
Out[18]= {1, {}, {}, -167}
```

Using the method described above, we again arrive at an expression that has the required characteristics.

```
In[19]:= expr = Exp[I Log[cubeRoot - 10^-200 I]];
```

```
In[20]:= NumericQ[expr]
Out[20]= True
```

```
In[21]:= N[expr, 50]
Out[21]= 0.043213918113751010238387927254600090978340096359469+
         3.6008347842928322727715209848852974034700604 69× 10^-6 i
```

```
In[22]:= N[expr, 5000] // N[#, 50]&
Out[22]= 23.140692552444147082703195091104774831500236256072+
         0.0019282169799121278751542758075761156846853727 72i
```

Here is another example. This time we use the tangent and the inverse tangent function.

```
In[23]:= With[{z = 20/77 + 97/77 I},
         {N[Log[I Tan[2^8 ArcTan[z]]], 50],
          N[Log[I Tan[2^8 ArcTan[z]]], 500]} // N[#, 20]&]
Out[23]= {0. × 10^-20 + 3.1415926535897932385 i, 0. × 10^-20 - 3.1415926535897932385 i}
```

Still another possibility to force a wrong result is the following. Instead of using small quantities of the form *hiddenZero+verySmallQuantity*, one uses numerical quantities so small that they are not representable.

```
In[24]:= tooSmall = Exp[-100000000000];
         N[tooSmall, 50]
```

      General::unfl : Underflow occurred in computation. More…

      General::unfl : Underflow occurred in computation. More…

      N::mprec : Internal precision limit

         \$MaxExtraPrecision = 50.` reached while evaluating $\frac{1}{e^{100000000000}}$. More…

```
Out[25]= Underflow[]
```

```
In[26]:= expr = Exp[I Log[cubeRoot - Exp[-100000000000]]];
```

Calculating now gives no N::mprec message, but a potential General::ovfl message.

In[27]:= **N[expr, 50]**

Out[27]= 0.04321391811375101023838792725460009097834009635946\9+
3.60083478429283227277152098488529740347006046\9×10^{-6} i

This time, raising the precision at which the calculation is carried out will not help to resolve the problem.

In[28]:= **N[expr, 50]**

Out[28]= 0.04321391811375101023838792725460009097834009635946\9+
3.60083478429283227277152098488529740347006046\9×10^{-6} i

f) High-precision numbers do not necessarily have more digits than do machine numbers. On the contrary, they can have fewer. In calculating the value of a special function with a machine argument, *Mathematica* tries to calculate all of the $MachinePrecision digits correctly. This means, on most machines, about 16 digits, and this can be more expensive than calculating, say, four digits. Because elementary functions have fast hardware implementations, special functions that are completely implemented within *Mathematica* are candidates. Here, this is demonstrated for erf($(10 + \zeta)$ i).

In[1]:= (* use different function values to
            avoid the use of cached values *)
        Timing[Do[Erf[N[(10 + j/1000) I]], {j, 1000}]]

Out[2]= {4.89 Second, Null}

In[3]:= **Timing[Do[Erf[SetPrecision[(10 + j/1000) I, 5]], {j, 1000}]]**

Out[3]= {0.21 Second, Null}

The numbers agree to the precision of the result. The real part of the machine value is large in absolute magnitude, but small compared to the imaginary part.

In[4]:= **{Erf[N[10 I]], Erf[SetPrecision[10 I, 5]]}**

Out[4]= {9.33369×10^{25} + 1.52431×10^{42} i, 1.52×10^{42} i}

So, for certain special function, the calculation of low precision high-precision values can be faster than the calculation with machine numbers.

g) N::meprec messages are issued when it is not uniquely possible to decide certain properties by numeric computations. Divide::infy messages are produced when a division by a numerical zero occurs. Having a "hidden zero" zero in the expression, will make it impossible to numerically detect that zero is identical to zero. Here is a "hidden zero".

In[1]:= **zero = Sqrt[2] + Sqrt[3] - Sqrt[5 + 2 Sqrt[6]]**

Out[1]= $\sqrt{2} + \sqrt{3} - \sqrt{5 + 2\sqrt{6}}$

Now, we must find a combination of analytic functions that, also for symbolic inputs will make numeric evaluations. A good candidate for such functions are pairs *function-inverseFunction*. *function*[*inverseFunction*[z]] will typically autosimplify to z, and this is not true for *inverseFunction*[*function*[z]].

In[2]:= **{Sin[ArcSin[z]], ArcSin[Sin[z]]}**

Out[2]= {z, ArcSin[Sin[z]]}

But for numerical arguments, *Mathematica* will try to simplify *inverseFunction*[*function*[z]]. This simplification is based on testing in which part of the complex plane the numerical value of z is located. Making this location difficult by adding a "hidden zero" results in the issuing of the messages that we were looking for. Here is an example.

In[3]:= **ArcCsc[Csc[I zero]]**

    Divide::infy : Infinite expression $\dfrac{1}{0. \times 10^{-17}}$ encountered.

    Divide::infy : Infinite expression $\dfrac{1}{0. \times 10^{-43}}$ encountered.

    Divide::infy : Infinite expression $\dfrac{1}{0. \times 10^{-66}}$ encountered.

    General::stop :
      Further output of Divide::infy will be suppressed during this calculation. More…

Out[3]= i $\left(\sqrt{2} + \sqrt{3} - \sqrt{5 + 2\sqrt{6}} \right)$

Here is a similar example.

In[4]:= `ArcCot[Cot[zero]]`

> Divide::infy : Infinite expression $\dfrac{1}{0. \times 10^{-17}}$ encountered.

> Divide::infy : Infinite expression $\dfrac{1}{0. \times 10^{-43}}$ encountered.

> Divide::infy : Infinite expression $\dfrac{1}{0. \times 10^{-66}}$ encountered.

> General::stop :
> Further output of Divide::infy will be suppressed during this calculation. More…

Out[4]= $\sqrt{2} + \sqrt{3} - \sqrt{5 + 2\sqrt{6}}$

Here are two more exact inputs that give `N::meprec` messages.

In[5]:= `{Log[zero]^0, Sign[Sin[100^100]]}`

> N::meprec : Internal precision limit $MaxExtraPrecision =
> 50.` reached while evaluating $\mathrm{Log}\left[\sqrt{2} + \sqrt{3} - \sqrt{5 + 2\sqrt{6}}\right]$. More…

> N::meprec :
> Internal precision limit $MaxExtraPrecision = 50.` reached while evaluating
> Sin[1000000000000000000000000000000000000000000000000000000000000000 ≪62≫
> 00000000000000000000000000000000000000000000000000000000000000000000]. More…

Out[5]= $\Big\{ \mathrm{Log}\left[\sqrt{2} + \sqrt{3} - \sqrt{5 + 2\sqrt{6}}\right]^{0}$, Sign[Sin[
> 1000000000000000000000000000000000000000000000000000000000000000000000000000000
> 0000000000000000000000000000000000000000000000000000000000000000000000000000000
> 0000000000000000000000000000000000000000]]$\Big\}$

h) Let us carry out the two inputs.

In[1]:= `Compile[{x}, 2/Exp[x]][1000]`

> CompiledFunction::cfn : Numerical error encountered
> at instruction 4; proceeding with uncompiled evaluation. More…

Out[1]= $\dfrac{2}{e^{1000}}$

In[2]:= `Compile[{x}, Evaluate[2/Exp[x]]][1000]`

Out[2]= 0.

The first one returned a symbolic expression and the second a machine real 0. Although the two bodies of `Compile` seem to be the same, they are not. Let us compare the `FullForm` of the held and the evaluated versions of `2/Exp[x]`.

In[3]:= `FullForm[2/Exp[x]]`

Out[3]//FullForm= `Times[2, Power[E, Times[-1, x]]]`

In[4]:= `FullForm[Hold[2/Exp[x]]]`

Out[4]//FullForm= `Hold[Times[2, Power[Exp[x], -1]]]`

The last two results are different! In the evaluated case, the exponential function is called with a large negative argument. The argument gets first converted to a machine number. Then the exponential function is evaluated as `Exp[-1000.]`. Within the compiler, such a number, which is not representable as a machine real number, is automatically converted to `0.`.

In[5]:= `Compile[{}, Exp[-10000]][]`

Out[5]= 0.

Because of the `HoldAll` attribute of `Compile`, the body does not get evaluated, and in the unevaluated case, the exponential function gets called with a large positive argument. The result again is not machine-representable, and this time no equivalent of a machine zero is available.

In[6]:= `Compile[{}, Exp[10000]][]`

```
        CompiledFunction::cfn : Numerical error encountered
             at instruction 4; proceeding with uncompiled evaluation. More...
Out[6]=  e^10000
```

```
In[7]:= Compile[{}, Exp[10000.]][]
```

```
        CompiledFunction::cfn : Numerical error encountered
             at instruction 3; proceeding with uncompiled evaluation. More...
Out[7]=  8.80681822566×10^4342
```

As a result, the failed evaluation of the compiled function, the last argument of the `CompiledFunction` (the original *Mathematica* expression), goes into effect and evaluates to $\frac{2}{e^{1000}}$.

i) We start with carrying out the `Plot` command of the first input.

```
In[1]:= Plot[FractionalPart[Exp[n]], {n, 0, 200},
            Frame -> True, Axes -> False, PlotRange -> All]
```

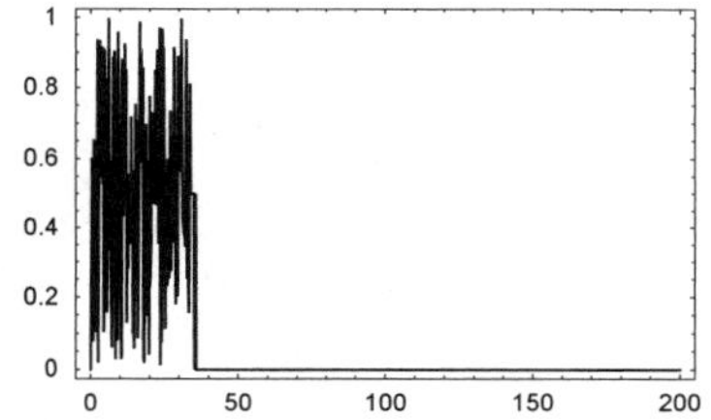

Theoretically, the picture is wrong; so let us analyze why we got a wrong picture. If `Exp[n]` is a number with significant digits to the right of the decimal point, `FractionalPart` will calculate a nontrivial result. If no nonzero digits are left to the right of the decimal point, the result will be just 0. The crossover happens at $\log(10^{\$MachinePrecision})$.

```
In[2]:= Log[10.^$MachinePrecision]
```

```
Out[2]= 36.7368
```

The second picture shows a completely different behavior.

```
In[3]:= f[n_ ?NumberQ] := FractionalPart[Exp[SetPrecision[n, Infinity]]];

        p1 = Plot[f[n], {n, 0, 200},
                Frame -> True, Axes -> False, PlotRange -> All]
```

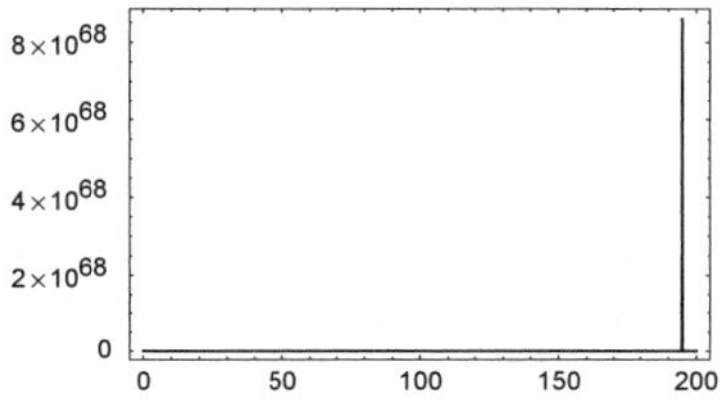

Again, theoretically, the result is wrong; so let us also here analyze why we got a wrong picture. We start our analysis by displaying the resulting curve at a logarithmic scale.

```
In[5]:= Show[% /. Line[l_] :> Line[{#[[1]], Log[10, Abs[#[[2]]]]}]& /@
                DeleteCases[l, (* remove zeros *) {_, _?(# == 0.&)}]]]
```

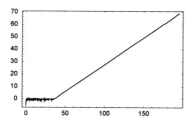

Again, for $n < \log(10^{\text{\$MachinePrecision}})$, we see the expected fluctuations. For larger n, we see this time a linear increase. The function f is not compilable (meaning that at runtime the full *Mathematica* evaluator is involved); so Plot substitutes the current value of n and sends f[n] through uncompiled evaluation after the semi-compiled try failed silently. Inside f the machine number n gets converted to an exact rational and FractionalPart returns a result in the form *largeInteger* − Exp[*aSlightlySmallerInteger*]. Here is a typical example.

```
In[6]:= (* select points where large numbers occurred *)
        badNPoints = Select[pl[[1, 1, 1, 1]],
            (#[[1]] > Log[10.^$MachinePrecision] && #[[2]] != 0.)&]
Out[7]= {{36.8665, 2.}, {37.4225, 2.}, {38.7863, 8.}, {45.013, 4096.},
         {67.1542, -1.75922×10^13}, {89.518, -1.51116×10^23}, {89.7375, 1.51116×10^23}}
```

```
In[8]:= f[badNPoints[[-1, 1]]]
Out[8]= -938598919237364683420907202163867214197 + e^{6314711774773589/70368744177664}
```

Calculating the last value within *Mathematica*'s high-precision arithmetic yields a result between 0 and 1.

```
In[9]:= N[%, 100]
Out[9]= 0.7677026611626019939969588709151786909154648908370358129739915860856465779465210215
        670831721837313713
```

Calculating the above expression using machine arithmetic yields a large number.

```
In[10]:= N[%%]
Out[10]= 1.51116×10^23
```

This is the kind of large value we see in the above picture. It remains to explain why a machine arithmetic calculation yields such a large value. We (machine-)numericalize both numbers independently.

```
In[11]:= List @@ %%%
Out[11]= {-938598919237364683420907202163867214197, e^{6314711774773589/70368744177664}}
```

```
In[12]:= N[%]
Out[12]= {-9.38599×10^38, 9.38599×10^38}
```

The two numbers are nearly the same, but looking directly at their digits, one sees small differences in the last digits. Such small differences are not unexpected. One is the result of the (easy) numericalization of an exact integer, and the other one is the result of the machine computation of the exponential function.

```
In[13]:= (* load context of function $NumberBits *)
         If[FreeQ[$ContextPath, #], AppendTo[$ContextPath, #]]& /@ Context /@
             Union[Join[Names["*`*Number*Bits*"], Names["*`*Bits*Number*"]]];
```

```
In[14]:= $NumberBits /@ %%
Out[14]= {{-1, {}, {1, 0, 1, 1, 0, 0, 0, 0, 1, 0, 0, 0, 0, 1, 1, 1, 1, 1, 1, 0, 0, 1, 0, 0, 1, 1, 1, 0,
          0, 1, 0, 0, 0, 1, 0, 0, 0, 1, 1, 0, 1, 0, 1, 1, 0, 1, 0, 1, 0, 1, 0, 0, 0}, 130},
         {1, {}, {1, 0, 1, 1, 0, 0, 0, 0, 1, 0, 0, 0, 0, 1, 1, 1, 1, 1, 1, 0, 0, 1, 0, 0, 1, 1,
          1, 0, 0, 1, 0, 0, 0, 1, 0, 0, 0, 1, 1, 0, 1, 0, 1, 1, 0, 1, 0, 1, 0, 1, 0, 0, 1}, 130}}
```

```
In[15]:= Position[Transpose[#[[3]]]& /@ %], {a_, b_} /; a != b]
Out[15]= {{53}}
```

Subtracting the two machine numbers of order 10^α yields a number of the order $10^{\alpha-\$MachinePrecision}$. As a result of this, we see the linear increase in the logarithmic graph.

Now, let us turn to the third picture. We see a single pronounced peak. Its size again shows that the picture is wrong.

In[16]:= **$MaxExtraPrecision = 1000;**

        Plot[f[n], {n, 0, 2000}, Frame -> True, Axes -> False, PlotRange -> All]

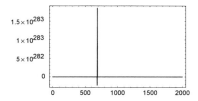

The peak is the continuation of the curve seen in the second picture. It has its maximum at the largest machine real number.

In[18]:= **Log[$MaxMachineNumber]**

Out[18]= 709.783

For all numbers n in the plot f acts in the same way as described above.

In[19]:= **f[2 %]**

Out[19]= -32317006071309472434283640110728220950641176483915449699203550419588107979531886561\
    82674935091948119949892887984738586055169674275430385087321642828380395709061482\
    67946964704606954552659636802644200644471589936675179825278602453514173159279505\
    23233105886577377485350435461095133169128169547542579535991465893514881089875276\
    56851058007091081025583591284631681051634771207172373407114140803764194713852814\
    82363966166310915775738345530172361074109842359592932915174067864039196235105568\
    96943193028908643301350497930625244552657150427895803246932691473386181652875792\
    27005954175026052503707176410749448774224864442 $6 + e^{6243314768165359/4398046511104}$

But now it is no longer possible to represent the integer as well as the exponential function in the last output as a machine number. So, *Mathematica* uses its high-precision arithmetic, and as a result, we get a "real" zero (meaning a number with no significant digits).

In[20]:= **N[%]**

Out[20]= 0.

In detail, we first get a high-precision number with zero significant digits, and then this zero gets transformed into a machine real 0.

In[21]:= **Plus @@ N[List @@ %%]**

Out[21]= $0. \times 10^{603}$

In[22]:= **N[%]**

Out[22]= 0.

Using high-precision arithmetic we can, of course, obtain a correct value.

In[23]:= **N[%%%%, 10]**

Out[23]= 0.2266098755

Using the precision-raising functionality of N, we can construct the correct graphics.

In[24]:= **SetAttributes[f, NumericFunction];**

    (* auxiliary function for numericalization *)
    ff[n_?InexactNumberQ] := N[f[n], 12]

```
In[27]:= Show[GraphicsArray[
         Block[{$DisplayFunction = Identity},
          (* the two plots from above *)
          {(* piecewise continuous plot *)
           Plot[ff[n], {n, 0, 200},
              Frame -> True, Axes -> False, PlotRange -> All],
           (* pointwise plot at integers *)
           ListPlot[Table[ff[n], {n, 2000}],
                 Frame -> True, Axes -> False, PlotRange -> All]}]]]
```

j) We are looking for cases where compilation changes the result of a calculation. We define a function `compareResults` that compares the compiled version with the uncompiled version.

```
In[1]:= SetAttributes[compareResults, HoldAll];

        compareResults[input_] := {input, Compile[{}, input][]}
```

The simplest example uses nonmachine numbers. The compiler forces the use of machine numbers. So, we get a different result in the following.

```
In[3]:= compareResults[N[Sin[1], 22]]

Out[3]= {0.8414709848078965066525, 0.841471}
```

In case we allow local variables to be used, we can get results that are quite different.

```
In[4]:= Show[GraphicsArray[
         Block[{$DisplayFunction = Identity},
            Plot[If[Hold[x[1]][[1, 1]] === 1, 1, 2], {x[1], 0, 1}, Compiled -> #]& /@
            {True, False}]]]
```

Part::partd :
  Part specification Hold[4.16667×10^{-8}][[1, 1]] is longer than depth of object. More…

Part::partd :
  Part specification Hold[4.16667×10^{-8}][[1, 1]] is longer than depth of object. More…

Part::partd :
  Part specification Hold[0.040567][[1, 1]] is longer than depth of object. More…

General::stop :
  Further output of Part::partd will be suppressed during this calculation. More…

And in the next example, we get quite different results due to the numericalization of all elements of a list in the compiled version.

```
In[5]:= compareResults[{1 + I (E^E^E)^Pi - I ((E^E^E)^Pi Sin[Pi/3]/(Sqrt[3]/2)),
                    2. I}[[1]] // Im]
```

Out[5]= {0, 0.}

In the symbolic evaluation, the imaginary parts of the expression cancel exactly. In the compiled version, the two imaginary parts are first numericalized and then subtracted.

In[6]:= **compareResults[{1 + I (E^E^E)^Pi - I ((E^E^E)^Pi Sin[Pi/3]/(Sqrt[3]/2)),**
                       2. I}]

Out[6]= {{1, 2. i}, {1. + 0. i, 0. + 2. i}}

k) Because NDSolve does not have a Hold-like attribute, let us start with the differential equation to be solved. Its right-hand side has the following form. To avoid the premature evaluation of the arguments of FindMinimum and FindRoot, we set the corresponding system option.

In[1]:= **Developer`SetSystemOptions[**
            "EvaluateNumericalFunctionArgument" -> False];

In[2]:= **rhs = (x /. FindRoot[(y /. FindMinimum[-Cos[y - x],**
                            {y, x + Pi/8}][[2]]) == w[z], {x, 0, 1}])

        FindRoot::nlnum : The function value {-5.77567×10⁻¹⁰ - 1. w[z]}
            is not a list of numbers with dimensions {1} at {x} = {0.}. More…

        ReplaceAll::reps :
          {FindRoot[(y /. FindMinimum[-Cos[≪1≫]], {y, Plus[≪2≫]}][[2]]) == w[z], {x, 0, 1}]}
            is neither a list of replacement rules nor a valid
            dispatch table, and so cannot be used for replacing. More…

        FindRoot::nlnum : The function value {-5.77567×10⁻¹⁰ - 1. w[z]}
            is not a list of numbers with dimensions {1} at {x} = {0.}. More…

        ReplaceAll::reps :
          {FindRoot[(y /. FindMinimum[-Cos[≪1≫]], {y, Plus[≪2≫]}][[2]]) == w[z], {x, 0, 1}]}
            is neither a list of replacement rules nor a valid
            dispatch table, and so cannot be used for replacing. More…

Out[2]= $x \,/.\, \text{FindRoot}\left[\left(y \,/.\, \text{FindMinimum}\left[-\text{Cos}[y-x], \left\{y, x+\frac{\pi}{8}\right\}\right][[2]]\right) == w[z], \{x, 0, 1\}\right]$

Note that because of the HoldAll attribute of Condition, the Part[..., 2] does not get evaluated and the whole expression stays unevaluated. Because w[z] was not a numeric quantity in the last input, we got some messages.

A little bit of inspection of rhs shows that this expression is equal to w[z]. A numerical experiment confirms this.

In[3]:= **Table[{w[z], rhs}, {w[z], 1, 10, 1/2}] //**
                       Internal`DeactivateMesssages

        FindMinimum::lstol :
          The line search decreased the step size to within tolerance specified
            by AccuracyGoal and PrecisionGoal but was unable to find a sufficient
            decrease in the function. You may need more than MachinePrecision
            digits of working precision to meet these tolerances. More…

        FindMinimum::lstol :
          The line search decreased the step size to within tolerance specified
            by AccuracyGoal and PrecisionGoal but was unable to find a sufficient
            decrease in the function. You may need more than MachinePrecision
            digits of working precision to meet these tolerances. More…

        FindMinimum::lstol :
          The line search decreased the step size to within tolerance specified
            by AccuracyGoal and PrecisionGoal but was unable to find a sufficient
            decrease in the function. You may need more than MachinePrecision
            digits of working precision to meet these tolerances. More…

        General::stop : Further output of
            FindMinimum::lstol will be suppressed during this calculation. More…

Out[3]= Internal`DeactivateMesssages$\left[\{\{1, 1.\}, \left\{\frac{3}{2}, 1.5\right\}, \{2, 2.\}, \left\{\frac{5}{2}, 2.5\right\}, \{3, 3.\},\right.$

$\left\{\frac{7}{2}, 3.5\right\}, \{4, 4.\}, \left\{\frac{9}{2}, 4.5\right\}, \{5, 5.\}, \left\{\frac{11}{2}, 5.5\right\}, \{6, 6.\}, \left\{\frac{13}{2}, 6.5\right\},$

$\left.\{7, 7.\}, \left\{\frac{15}{2}, 7.5\right\}, \{8, 8.\}, \left\{\frac{17}{2}, 8.5\right\}, \{9, 9.\}, \left\{\frac{19}{2}, 9.5\right\}, \{10, 10.\}\}\right]$

This means that the solution of the differential equation is $w(z) = e^z$. From this, it follows that the expected result of the input is 0. Carrying out the calculation confirms this. This time we suppress the messages.

```
In[4]:= Internal`DeactivateMessages[
         Round[E - (w[1] /. NDSolve[{w'[z] ==
            (x /. FindRoot[(y /. FindMinimum[-Cos[y - x],
                        {y, x + Pi/8}][[2]]) == w[z],
            {x, 0, 1}]), w[0] == 1}, w, {z, 0, 1}][[1]])]]
Out[4]= 0
```

l) The setting of the `SameTest` option in `Union` is by default `Automatic`. With this setting, in addition to the digits of the numbers, their precision is compared. And if their precision is different, they are treated as different numbers.

```
In[1]:= Union[{SetPrecision[1, 10], SetPrecision[1, 20]}]
Out[1]= {1.000000000, 1.0000000000000000000}
```

```
In[2]:= Union[{SetPrecision[1, 10.0], SetPrecision[1, 10.000001]}]
Out[2]= {1.000000000, 1.000000000}
```

Using this behavior of the `Automatic` setting, it is straightforward to find three real numbers a, b, and c such that three expressions $a === b$, $a === c$, and $b === c$ all give `True`, but `Union[{a, b, c}]` returns $\{a, b, c\}$. We use numbers that are the same (when tested with `SameQ`), but have different precisions. To guarantee the ordering (which takes into account guard digits), we use slightly different values for a, b, and c.

```
In[3]:= a = SetPrecision[11111/100000, 1];
        b = SetPrecision[11112/100000, 3];
        c = SetPrecision[11113/100000, 5];
```

Here is a test that the numbers fulfill the requirements of the exercise.

```
In[6]:= Union[{a, b, c}]
Out[6]= {0.1, 0.111, 0.11113}
```

```
In[7]:= {a === b, a === c, b === c}
Out[7]= {True, True, True}
```

m) This is the exact number, we start from.

```
In[1]:= x = Unevaluated[Unevaluated[10^100 + 10^-10 I]];
```

Inside the `Unevaluated`, the expression has the head `Plus`, its real part has the head `Integer`, and its imaginary part has the head `Rational`.

```
In[2]:= FullForm[x]
Out[2]//FullForm= Unevaluated[Plus[Power[10, 100], Times[Power[10, -10], \[ImaginaryI]]]]
```

Now, we effectively form `N[x, 200]`. The head `Unevaluated` is used up by the application of the pure function `N[#, 200]&`.

```
In[3]:= nx = N[#, 200]& @ x
Out[3]= 1.000000000000000000000000000000000000000000000000000000000000000000000000000000000000000000000000000000000000000000000000000000000000000000000000000000000000000000000000000000000000000000000000000000000000 × 10^100 +
        1.000000000000000000000000000000000000000000000000000000000000000000000000000000000000000000000000000000000000000000000000000000000000000000000000000000000000000000000000000000000000000000000000000000000000 × 10^-10 i
```

`N` formed one complex 200-digit-precision high-precision number from the real as well from the imaginary part.

In[4]:= `{Precision[nx], Precision[Re[nx]], Precision[Im[nx]]}`

Out[4]= `{200., 200., 200.}`

In[5]:= `{Accuracy[nx], Accuracy[Re[nx]], Accuracy[Im[nx]]}`

Out[5]= `{100., 100., 210.}`

Applying now `SetPrecision` to `nx` results again in a number with precision 200, but in a "minimalistic" version of it. Because of the large difference in the size of the magnitude of the real and imaginary parts it is possible (and necessary for uniqueness) to reduce the precision of the imaginary part. (Using `N[10^100 + 10^-10 I, 200]` would give the same number `px`.)

In[6]:= `px = SetPrecision[nx, 200];`

In[7]:= `{Precision[px], Precision[Re[px]], Precision[Im[px]]}`

Out[7]= `{200., 200.151, 90.1505}`

The last number is the result of the original input from the exercise.

In[8]:= `Precision[Im[SetPrecision[N[#, 200]& /@`
        `Unevaluated[10^100 + 10^-10 I], 200]]]`

Out[8]= `90.1505`

The accuracy of the dominating real part is the same as the one of `nx`. But the accuracy of the imaginary part dropped.

In[9]:= `{Accuracy[px], Accuracy[Re[px]], Accuracy[Im[px]]}`

Out[9]= `{100., 100.151, 100.151}`

Taking into account the definition of the accuracy of a complex number $z = x + i\,y$, accuracy$(z) = -1/2 \log_{10}(10^{-2\,\mathrm{accuracy}(x)} + 10^{-2\,\mathrm{accuracy}(y)})$ and the definition of its precision precision(z) = accuracy(z) + $\log_{10}(|z|)$, we see that reducing the accuracy of the imaginary part to the accuracy of the real part results in a precision of about 200 for the whole number and a precision of 90 for its imaginary part.

In[10]:= `Re[nx] + I SetAccuracy[Im[nx], Accuracy[Re[nx]]];`

In[11]:= `{Precision[%], Precision[Im[%]]}`

Out[11]= `{199.849, 90.}`

The following graphic shows how the precision of x depends on the precision of its imaginary part. The initial straight line is the machine arithmetic domain. Increasing the precision of the imaginary part of x increases the precision of x until we reach the point where the real part dominates. The small horizontal line in the left lower corner of the graphic is caused by the use of machine numbers.)

In[12]:= `ListPlot[Table[{k, Precision[N[10^100, 200] + N[10^-10, k] I]},`
        `{k, 250}], PlotRange -> All,`
        `Axes -> True, AxesOrigin -> {0, 0}]`

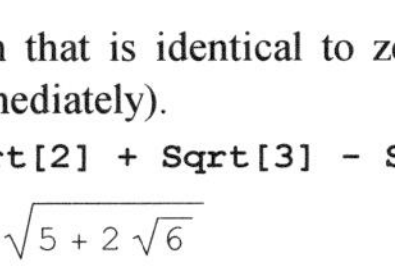

n) Here is an expression that is identical to zero, but it does not autoevaluate to 0 (squaring the first two and the last summand shows this immediately).

In[1]:= `zero = Sqrt[2] + Sqrt[3] - Sqrt[5 + 2 Sqrt[6]]`

Out[1]= $\sqrt{2} + \sqrt{3} - \sqrt{5 + 2\sqrt{6}}$

Applying `N` to zero will give a nonzero result. Typically, the result will be in *smallInteger* $10^{-\$\mathrm{MachinePrecision}}$.

In[2]:= **N[zero]**

Out[2]= 4.44089×10^{-16}

We can amplify this number by repeatedly taking a high-order root of the last expression.

In[3]:= **N[8 zero^(1/2^2^2^2)]**

Out[3]= 7.99569

o) The result will be the machine real number 2. Pi is a symbol, and as such, it represents a local variable that has nothing to do with the constant π.

In[1]:= **Compile[{{Pi, _Real}}, Pi][2]**

Out[1]= 2.

Here, the analogous behavior for Function is shown. The string "2" cannot be used by the CompiledFunction; so the last argument, the pure function, gets used and the integer 2 is returned. (Because we force a failure of the compiled code, we get a CompiledFunction::cfsa message.)

In[2]:= **Compile[{{Pi, _Real}}, Pi]["2"]**

        CompiledFunction::cfsa :
            Argument 2 at position 1 should be a machine-size real number. More…

Out[2]= 2

p) No, it will give False. (We turn off the message FindRoot::argmu, but this does not have any influence on the result.)

In[1]:= **Off[FindRoot::argmu];**

        N[FindRoot[1]] - (FindRoot[N[1]]) === 0
Out[2]= False

The first term is simply evaluated to FindRoot[1.].

In[3]:= **N[FindRoot[1]]**

Out[3]= FindRoot[1.]

The second term gives a slightly different result.

In[4]:= **FindRoot[N[1]]**

Out[4]= FindRoot[N[1]]

Because FindRoot has the attribute HoldAll, the argument N[1] will not be evaluated before FindRoot receives the argument handed over. But FindRoot with one argument does not have any rule built-in; so it too will not do any nontrivial evaluation with N[1]. As a result, FindRoot[N[1]] is returned. And FindRoot[N[1]] is different from FindRoot[1.]; so the whole input returns False.

SetPrecision will nevertheless go inside such (held) expressions and numericalize any numbers and constants to the specified precision.

In[5]:= **SetPrecision[%, MachinePrecision]**

Out[5]= FindRoot[N[1.]]

q) Here is the definition for f. The function f applied to a real number will set the precision of the argument to $Machine`Precision and return the resulting high-precision number.

In[1]:= **f[x_Real] := N[x, $MachinePrecision]**

For any "real" real number, we will not get N::meprec messages.

In[2]:= **{f[4.], f[0.], f[0``50], f[Underflow[]]}**

Out[2]= $\{4., 0., 0. \times 10^{-50}, \text{Underflow}[]\}$

There are various ways to produce a N::meprec message. Either trying to numericalize an expression that is exactly zero, or carrying out a comparison between a transcendental number and a very nearby rational approximation or .... The following input produces a N::meprec message but not when evaluating the right-hand side.

```
In[3]:= f[N[Sqrt[2] + Sqrt[3] - Sqrt[5 + 2 Sqrt[6]], 22]]
```

   N::meprec : Internal precision limit $MaxExtraPrecision =

    50.` reached while evaluating $\sqrt{2} + \sqrt{3} - \sqrt{5 + 2\sqrt{6}}$. More…

```
Out[3]=  0. × 10^-72
```

Another "cheating" way to obey the stated conditions is to supply an argument to f that changes the definition of f.

```
In[4]:= f[f[x_] := N[x, $MachinePrecision];

        f[Sqrt[2] + Sqrt[3] - Sqrt[5 + 2 Sqrt[6]]]]
```

   N::meprec : Internal precision limit $MaxExtraPrecision =

    50.` reached while evaluating $\sqrt{2} + \sqrt{3} - \sqrt{5 + 2\sqrt{6}}$. More…

```
Out[4]=  0. × 10^-72
```

To get the N::meprec message from the right-hand side "without cheating", we must slip a symbolic expression into the right-hand side. This can be done by using an argument with an "explicit" head Real.

```
In[5]:= Clear[f];
        f[x_Real] := N[x, $MachinePrecision];

        f[Real[Sqrt[2] + Sqrt[3] - Sqrt[5 + 2 Sqrt[6]]]]
```

   N::meprec : Internal precision limit $MaxExtraPrecision =

    50.` reached while evaluating $\mathrm{Real}\left[\sqrt{2} + \sqrt{3} - \sqrt{5 + 2\sqrt{6}}\right]$. More…

```
Out[7]=  Real[0. × 10^-72]
```

r) NIntegrate is a numerical routine that, after a symbolic preprocessing that checks for the presence of piecewise defined functions and for symmetries, samples its integrand in a black box like manner. To fool its algorithm we have to ensure that the integrand has some structure that will not be detected. We will start with the following integral which has the exact value π.

```
In[1]:= NIntegrate[2/(x^2 + 1), {x, 0, Infinity}]
```

```
Out[1]=  3.14159
```

We now have to add a contribution to the integrand that is about half its absolute size and not easily detected by NInte grate. For this second term of the integrand, we will use a similar, but displaced bump.

```
In[2]:= int[δ_] := 2/(x^2 + 1) - 1/((x - δ)^2 + 1)/2
```

For small displacements NIntegrate detects the bump correctly.

```
In[3]:= NIntegrate[int[10^1], {x, 0, Infinity}]
```

```
Out[3]=  1.62063
```

For displacements of the order 10^4, the function NIntegrate does not get the correct result, but the issued warning indicates that it is aware of this fact.

```
In[4]:= NIntegrate[int[10^4], {x, 0, Infinity}]
```

   NIntegrate::slwcon :
   Numerical integration converging too slowly; suspect one of the
    following: singularity, value of the integration being 0, oscillatory
    integrand, or insufficient WorkingPrecision. If your integrand is
    oscillatory try using the option Method->Oscillatory in NIntegrate. More…

   NIntegrate::ncvb :
   NIntegrate failed to converge to prescribed accuracy after 7 recursive
    bisections in x near x = 10958.922474096673`. More…

```
Out[4]=  3.13369
```

For still larger shifts, the second bump of half the absolute size of the central one goes undetected.

In[5]:= `NIntegrate[int[10^6], {x, 0, Infinity}]`

Out[5]= `3.14159`

Indicating in the second argument at which x the integrand has a nontrivial structure allows `NIntegrate` to calculate the correct result.

In[6]:= `NIntegrate[int[10^6], {x, 0, 2, 10^6 - 1, 10^6 + 1, Infinity}]`

Out[6]= `1.57079`

s) An x_k with large absolute values would produce zeros with negative accuracy.

In[1]:= `N[1, 20] + N[10^300, 20] - N[10^300, 20] // Accuracy`

Out[1]= `-280.301`

In[2]:= `N[1, 20] + N[10^300, 20] - N[10^300, 20] ==`
`N[1 + 10^300 - 10^300, 20]`

Out[2]= `True`

They would be equal to the right-hand side. So, we should look for small x_k. To shorten the following outputs we implement a function `shortenZeros` that displays a sequence of n leading and trailing zeros as 0_n (we do not compactify intermediate sequences of zeros).

In[3]:= `shortenZeros[expr_] := expr //. With[{s = Subscript},`
`{{zeros:(0 ..)} :> {s[0, Length[{zeros}]]},`
`{zeros:(0 ..), a_?(# =!= 0 &), b___} :> {s[0, Length[{zeros}]], a, b},`
`{b___, a_?(# =!= 0 &), zeros:(0 ..)} :> {b, a, s[0, Length[{zeros}]]}}]`

This is a high-precision 1.

In[4]:= `(* load context of function $NumberBits *)`
`If[FreeQ[$ContextPath, #], AppendTo[$ContextPath, #]]& /@ Context /@`
`Union[Join[Names["*`*Number*Bits*"], Names["*`*Bits*Number*"]]];`

In[6]:= `N[1, 20] // $NumberBits // shortenZeros`

Out[6]= `{1, {1, 0₆₅}, {0₆₃}, 1}`

Adding a small number to this 1 yields nontrivial guard bits.

In[7]:= `N[1, 20] + N[10^-22, 20] // $NumberBits // shortenZeros`

Out[7]= `{1, {1, 0₆₅}, {0₈, 1, 1, 1, 1, 0, 0, 0, 1, 1, 1, 0, 0, 1, 0, 0, 1, 0, 0, 0, 0, 0, 0, 0, 1, 0,`
`0, 0, 0, 0, 0, 1, 0, 1, 1, 1, 0, 1, 0, 1, 1, 1, 0, 1, 1, 1, 0, 0, 1, 0, 1, 1, 0₁}, 1}`

If we add a too small number, no nontrivial guard bit will appear.

In[8]:= `N[1, 20] + N[1 10^-39, 20] // $NumberBits // shortenZeros`

Out[8]= `{1, {1, 0₆₅}, {0₆₃}, 1}`

We search for the smallest number to be added so that the sum has nontrivial additional guard bits.

In[9]:= `δA = 1;`
`While[$NumberBits[N[1, 20]] =!=`
`$NumberBits[N[1, 20] + N[δA, 20]],`
`δA = δA/2];`
`N[2 δA]`

Out[11]= `2.93874 × 10⁻³⁹`

A number of this size was expected from counting the verified and the guard bits $\log(2^{-(2+65+63)}) \approx -39.2$. The next input shows the appearance of a new bit.

In[12]:= `N[1, 20] + N[3 10^-39, 20] // $NumberBits // shortenZeros`

Out[12]= `{1, {1, 0₆₅}, {0₆₂, 1}, 1}`

Because addition is carried out term by term, adding many (too) small numbers still does not generate a nontrivial guard bit.

```
In[13]:= (* add one million small terms that do not change the result at all *)
         ToExpression["N[1, 20] " <>
           StringJoin[Table[" + N[1 10^-39, 20]", {10^6}]]] //
                                              $NumberBits // shortenZeros
Out[14]= {1, {1, 0₆₅}, {0₆₃}, 1}
```

But by repeatedly changing guard bits, we can generate a changing verified bit.

```
In[15]:= Module[{ξ = N[2, 20]},
                Do[ξ = ξ - N[10^-22, 22], {10^6}];
                ξ]
Out[15]= 1.999999999999999000
```

If we now would add enough numbers of the form $x_k = $ N[3 10^-39, 20], we could generate False for the expression N[1, 20] + N[x_1, 20] + $\cdots$ + N[x_n, 20] == N[1 + x_1 + $\cdots$ + x_n, 20]. But we would have to add about 10^{20} terms. This would take quite a while even on a year 2005 computer.

```
In[16]:= {(* no difference yet *)
         N[1, 20] == N[1, 20] + 1 10^-18,
         (* difference visible *)
         N[1, 20] == N[1, 20] + 2 10^-18}
Out[16]= {True, False}
```

t) The function `squareRootOf3CF` basically calculates the square root of the integer 3 using Newton's method and starting from the point 1. Each arithmetic operation is slightly disturbed by multiplying it with $1 + smallRandomNumber$.

```
In[1]:= squareRootOf3CF = With[{p = $MachinePrecision - 2},
         With[{r = (# (1 + Random[Real, {-1., 1.} 10.^-p]))&},
           Compile[x, FixedPoint[
             Function[ξ, r[r[r[ξ]/r[2.]] + r[r[3.]/r[2.]/r[ξ]]]], x]]]];
```

`roots` is a list of 10000 such approximative square roots of 3. (The number of iteration steps in the fixed point iteration varies from call to call.

```
In[2]:= roots = Table[squareRootOf3CF[1.], {10^5}];
```

The last input groups the differences of the approximative square roots of 3 to the exact value and displays a histogram of the differences. Because many independent operations were carried out, we expect the histogram to resemble a Gaussian curve. To a good approximation, the resulting histogram shows a Gaussian curve.

```
In[3]:= Show[Graphics[
         Polygon[{{#1 - 1/2, 0}, {#1 + 1/2, 0}, {#1 + 1/2, #2},
                 {#1 - 1/2, #2}}]&[First[#], Length[#]]& /@
             Split[Round[(Sqrt[3] - Sort[roots])*
                         10^($MachinePrecision - 3/2)]]],
         Frame -> True]
```

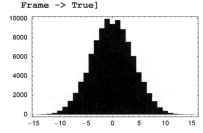

For a detailed discussion of randomized arithmetic, see [1412], [1850].

u) To optimize the function `f` we must first understand what `f` is doing. A moment of reflection about the outer `MapIn`-dexed, `FoldList`, and the inner `NestList` shows that `f` is creating a list of the following form.

$$\left\{ \mathcal{T}(x), \sqrt{\mathcal{T}(x)\,\mathcal{T}^2(x)}, \sqrt[3]{\mathcal{T}(x)\,\mathcal{T}^2(x)\,\mathcal{T}^3(x)}, \ldots, \sqrt[n]{\mathcal{T}(x)\,\mathcal{T}^2(x)\ldots\mathcal{T}^n(x)} \right\}$$

where $\mathcal{T}^n(x) = \mathcal{T}(\mathcal{T}^{n-1}(x))$ and $\mathcal{T}^0(x) = x$.

The map $\mathcal{T}$ operates on the p digits of x in base 2. `Drop[Last[`*digitList*`], Position[Last[`*digitList*`], 1, {1},` `1][[1, 1]]]` shows that $\mathcal{T}$ removes the leading zeros and the first 1 from the left. If $0.d_1 d_2 \ldots$ is the base 2 representation of a number x, and $d_j = 0$ or 1, then $\mathcal{T}(0. d_1 d_2 \ldots d_k) = \mathcal{T}(0. d_k \ldots)$ where $d_1 = d_2 = d_{k-2} = 0$ and $d_{k-1} = 1$.

```
In[1]:= f[x_?(0 < # < 1&), p_Integer, n_Integer] :=
          MapIndexed[#1^(1/#2[[1]])&, Rest[
            FoldList[Times, 1, Rest[First /@ Rest[
              NestList[{FromDigits[{Last[#], 0}, 2.],
                        Drop[Last[#], Position[Last[#], 1, {1}, 1][[1, 1]]]}&,
                        {1, RealDigits[x, 2, p][[1]]}, n]]]]]];
```

The function $\mathcal{T}$`List` represents the repeated application of the map $\mathcal{T}$.

```
In[2]:= TList[x_?(0 < # < 1&), p_Integer, n_Integer] :=
          NestList[{FromDigits[{Last[#], 0}, (* make an approximate number *) 2.],
                    Drop[Last[#], Position[Last[#], 1, {1}, 1][[1, 1]]]}&,
                    {1, RealDigits[x, 2, p][[1]]}, n]
```

Here is an example.

```
In[3]:= (* use a number with a characteristic digit pattern in base 2 *)
        TList[FromDigits[{{0, 0, 0, 1, 0, 1, 0, 0, 1, 0, 0, 0, 1, 0, 0, 0, 0, 1},
                          0}, 2], 10, 4]

Out[4]= {{1, {1, 0, 1, 0, 0, 1, 0, 0, 0, 1}}, {0.641602, {0, 1, 0, 0, 1, 0, 0, 0, 1}},
         {0.283203, {0, 0, 1, 0, 0, 0, 1}}, {0.132813, {0, 0, 0, 1}}, {0.0625, {}}}
```

Here is the result of `f[Sqrt[2]/2, 2000, 1000]`.

```
In[5]:= ListPlot[f[Sqrt[2]/2, 2000, 1000],
                 Frame -> True, Axes -> False, PlotRange -> {0.3, 0.5}]
```

A more efficient (and shorter) version of `f` can be implemented using `Compile`. To save memory and to avoid to copying large lists, we move a counter along the list of digits and we use 53 bits for `FromDigits`. Further, we form the root of the product at each step to avoid the use of nonmachine numbers. This leads to the following optimized implementation.

```
In[6]:= F[λ_] := FromDigits[{λ, 0}, 2.]

        fc = Compile[{{l, _Integer, 1}, {n, _Integer}},
                Module[{p = 1, p = 1.},
                    Table[While[l[[p]] =!= 1, p++];
                          λ = Take[l, {p + 1, p + 53}]; p++;
                          p = p^((j - 1)/j) F[λ]^(1/j),
                          {j, n}]], {{F[_], _Real}}];
```

`fc` calculates list of length 10^5 quite quickly. Here are the roots of the products for $1/\pi$, $1/\phi$, and $1/e$. Interestingly, they all converge to $1/e$ [893].

```
In[8]:= (* load context of function $NumberBits *)
        If[FreeQ[$ContextPath, #], AppendTo[$ContextPath, #]]& /@ Context /@
            Union[Join[Names["*`*Number*Bits*"], Names["*`*Bits*Number*"]]];

In[10]:= Show[Graphics[{PointSize[0.003],
                Transpose[{{Hue[0], Hue[0.2], Hue[0.8]},
                    MapIndexed[Point[{#2[[1]], #1}]&,
                        fc[$NumberBits[N[#, 10^5]][[2]], 10^5]& /@
                        (* the real numbers *) {1/Pi, 1/GoldenRatio, 1/E}}]}}],
                Frame -> True, PlotRange -> {0.3, 0.5}]
```

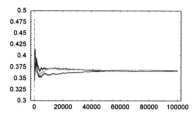

v) It is not difficult to find a discontinuous function f and numbers z_1 and z_2 such that $z_1 === z_2$ returns True, and $f(z_1) === f(z_2)$ returns False. We just position the two points z_1 and z_2 at the two sides of a branch point or a branch cut. Here is a simple example for this situation.

```
In[1]:= z1 = N[1 - 10^-10, 9.9];
        z2 = N[1 + 10^-10, 10.01]
Out[2]= 1.0000000001
```

```
In[3]:= {z1 === z2, ArcSin[z1] === ArcSin[z2]}
Out[3]= {True, False}
```

Now let us look for a continuous function. Any discontinuous step function can be approximated by a continuous step function, for instance, can be approximated using $\alpha \arctan(c\,x)$ for large c. If we then position the two points z_1 and z_2 slightly to the left and right of the step, we obtain the result False for $f(z_1) === f(z_2)$. Here is a machine-precision version of this situation.

```
In[4]:= x1 = 1 - $MachineEpsilon/2;
        x2 = 1 + $MachineEpsilon/2;
In[6]:= f1[ξ_] := ArcTan[1000 (ξ - 1)]
        {x1 === x2, f1[x1] === f1[x2], {f1[x1], f1[x2]}}
Out[7]= {True, False, {-1.11022×10^-13, 0.}}
```

And here is a high-precision version of this situation.

```
In[8]:= x1 = N[1 - 10^-20, 20];
        x2 = N[1 + 10^-20, 20];
In[10]:= (* ArcTan expressed through logarithms *)
        arcTan[ξ_] := I/2 (Log[(1 - I ξ)/(1 + I ξ)])
        f2[ξ_] = arcTan[10^10 (ξ - 1)];
        {x1 === x2, f2[x1] === f2[x2], {f2[x1], f2[x2]}}
Out[13]= {True, False, {-1.×10^-10 + 0.×10^-11 i, 1.×10^-10 + 0.×10^-11 i}}
```

Another possibility to obtain different numbers after applying a function is the situation where numbers of different types are returned. In the next input, the $f_3(x_1)$ gives a real number and $f_3(x_2)$ gives a complex number.

```
In[14]:= f3[ξ_] := Log[10^30 (ξ - 1)]
        {x1 === x2, f3[x1] === f3[x2], {f3[x1], f3[x2]}}
Out[15]= {True, False, {23. + 0. i, 23.}}
```

w) It is not difficult to find such examples. The first example uses the fact that cos is an oscillating function and to reduce $10^{100} \bmod \pi$ needs about 120 digits.

```
In[1]:= N[Cos[99^99],20]

        N::meprec :
        Internal precision limit $MaxExtraPrecision = 50.` reached while evaluating Cos[
            3697296376497267726571879056288054405956687642817411024302599724235525 ≪57≫
            967694943594516215701936440144180710606676593013849997799991592004998991].
          More…

Out[1]= 0.
```

The next input is numerically undecidable for any finite precision. (Using standard form input, the following can be shortened to $\lfloor \frac{\pi}{e} \rfloor$.)

In[2]:= `Floor[Pi/Degree]`

> Floor::meprec : Internal precision limit
> $MaxExtraPrecision = 50.` reached while evaluating Floor$\left[\frac{\pi}{\circ}\right]$. More…

Out[2]= Floor$\left[\frac{\pi}{\circ}\right]$

This input uses `Equal` instead of `Floor` above.

In[3]:= `Pi==180Degree`

> N::meprec : Internal precision limit
> $MaxExtraPrecision = 50.` reached while evaluating $-180\,° + \pi$. More…

Out[3]= $\pi == 180\,°$

$\tanh(99) \approx 1 - 2\,10^{-86}$, so the default value 50 of `$MaxExtraPrecision` is not sufficient to get 20 correct digits in the next input.

In[4]:= `N[Tanh[99]-1,20]`

> N::meprec : Internal precision limit
> $MaxExtraPrecision = 50.` reached while evaluating $-1 + Tanh[99]$. More…

Out[4]= $0. \times 10^{-70}$

Because no symbolic operations (like `Expand` or `Factor`) are carried out by `Equal`, the following identity is also numerically undecidable.

In[5]:= `E(E+1)==E^2+E`

> N::meprec : Internal precision limit $MaxExtraPrecision =
> 50.` reached while evaluating $-e - e^2 + e\,(1 + e)$. More…

Out[5]= $e\,(1 + e) == e + e^2$

Using `Less` instead of `Equal` still makes the input numerically undecidable, but saves one additional character.

In[6]:= `E(E+1)<E^2+E`

> N::meprec : Internal precision limit $MaxExtraPrecision =
> 50.` reached while evaluating $-e - e^2 + e\,(1 + e)$. More…

Out[6]= $e\,(1 + e) < e + e^2$

x) Here is the symbolic inverse of a 3×3 matrix with elements $a[i, j]$.

In[1]:= `n = 3;`
`inv = Table[a[i, j], {i, n}, {j, n}] // Inverse // Simplify;`

r calculates a random *prec* digit real number.

In[3]:= `r[prec_] := (2 Random[Integer] - 1) Random[Real, {0, 1}, prec] *`
`10^Random[Integer, {-10, 10}]`

We now construct random 3×3 matrices with complex entries and calculate their inverses using `Inverse` and by evaluating `inv`. We carry out a search until we find a matrix such that the two numbers explicitly disagree.

In[4]:= `SeedRandom[444];`
`While[tab = Table[a[i, j] = r[30] + I r[30], {i, n}, {j, n}];`
`Inverse[tab] - inv == Table[0. + 0. I, {n}, {n}], Null];`

Here is an abbreviated look at the matrix found.

In[6]:= `tab // N`

Out[6]= $\{\{-5.69776 \times 10^{-7} + 0.000467834\,i, -4.24018 \times 10^9 + 215959.\,i, 1.40213 \times 10^{-11} + 72.0271\,i\},$
$\{-5.897 \times 10^6 - 351351.\,i, -0.0190663 + 5.71224 \times 10^{-7}\,i, 7.01112 \times 10^{-7} - 5.40033 \times 10^{-9}\,i\},$
$\{-7.4084 \times 10^{-8} - 8.07978 \times 10^8\,i, 169.992 - 0.353312\,i, -0.780999 + 9.2058 \times 10^9\,i\}\}$

And here is the difference of the two matrices.

In[7]:= `Inverse[tab] - inv`

Out[7]= {{0. × 10^{-48} + 0. × 10^{-49} i, 0. × 10^{-36} + 0. × 10^{-37} i, -3. × 10^{-52} - 1.79 × 10^{-49} i},
{0. × 10^{-39} + 0. × 10^{-40} i, 0. × 10^{-46} + 0. × 10^{-45} i, 0. × 10^{-47} + 0. × 10^{-48} i},
{0. × 10^{-48} + 0. × 10^{-47} i, 0. × 10^{-37} + 0. × 10^{-38} i, 0. × 10^{-40} + 0. × 10^{-39} i}}

Here are the differing elements extracted. The differences are in the last digits.

In[8]:= {Inverse[tab][[##]]& @@@ #, inv[[##]]& @@@ #}&[
Position[%, _?(N[#] =!= 0. + 0. I&), {2}, Heads -> False]]

Out[8]= {{-8.71840553266785745480622743295 × 10^{-25} - 1.286305208673446902291500059428 × 10^{-23} i},
{-8.718405532667857454806227430 3 × 10^{-25} - 1.28630520867344690229150057 64 × 10^{-23} i}}

Raising the precision of the matrix supplied to Inverse generates a result that agrees to the precision of inv with inv.

In[9]:= Inverse[SetPrecision[tab, 2 Precision[tab]]] - inv

Out[9]= {{0. × 10^{-48} + 0. × 10^{-49} i, 0. × 10^{-36} + 0. × 10^{-37} i, 0. × 10^{-54} + 0. × 10^{-52} i},
{0. × 10^{-39} + 0. × 10^{-40} i, 0. × 10^{-46} + 0. × 10^{-45} i, 0. × 10^{-47} + 0. × 10^{-48} i},
{0. × 10^{-48} + 0. × 10^{-47} i, 0. × 10^{-37} + 0. × 10^{-39} i, 0. × 10^{-40} + 0. × 10^{-39} i}}

y) The base of the expression under consideration is approximately 10^{-100}. (This can be easily seen using the series $\arctan(x) \underset{x \to \infty}{\sim} \pi/2 - 1/x$.) When \$MaxExtraPrecision is 50, this value cannot be reliably calculated and so stays unevaluated. Raising this number to the power zero would results in an indeterminate result.

In[1]:= (ArcTan[10^100] - Pi/2)^0``100

N::meprec : Internal precision limit
\$MaxExtraPrecision = 50.` reached while evaluating $-\frac{\pi}{2}$ + ArcTan[
100000000000000000000000000000000000000000000000000000000000000000000000000000000:
00000000000000000000000]. More...

Out[1]= $\left(-\frac{\pi}{2} + \text{ArcTan}[\right.$
100000000000000000000000000000000000000000000000000000000000000000000000000000000:
$\left. 00000000000000000000000]\right)^{0. \times 10^{-101}}$

Low precision numericalization of the last expression yields an indeterminate result. High precision numericalization yields an approximate one.

In[2]:= {N[%, 20], N[%, 1000]}

Power::indet : Indeterminate expression 0. × 10^{-70} $^{0. \times 10^{-101}}$ encountered. More...

Power::indet : Indeterminate expression 0. × 10^{-70} $^{0. \times 10^{-20}}$ encountered. More...

N::meprec : Internal precision limit
\$MaxExtraPrecision = 50.` reached while evaluating $\left(-\frac{\pi}{2} + \text{ArcTan}[\right.$
10000000000000000000000000 ≪49≫ 00000000000000000000000]$\left.\right)^{0. \times 10^{-101}}$. More...

N::meprec : Internal precision limit
\$MaxExtraPrecision = 50.` reached while evaluating $\left(-\frac{\pi}{2} + \text{ArcTan}[\right.$
10000000000000000000000000 ≪49≫ 00000000000000000000000]$\left.\right)^{0. \times 10^{-101}}$. More...

Out[2]= {Indeterminate,
1.0000000000000000000000000000000000000000000000000000000000000000000000000000000:
0000000000000000 + 0. × 10^{-98} i}

z) The function f[x] is a smooth function on the real axis. Here is the function shown.

In[1]:= f[x_] := x/Sqrt[x^2] Exp[-1/Sqrt[x^2]]
f[x /; x == 0] = 0
Out[2]= 0

In[3]:= Plot[f[x/10], {x, -2, 2}, PlotRange -> All]

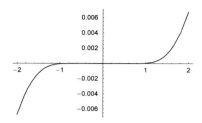

`f[x]` is real analytic everywhere and easily differentiable (with the exception of the point $x = 0$ that, when using nonsymmetric starting values, we will hit with probability 0). This means `FindRoot` should work just fine. And it does.

In[4]:= `{#, f[x/10 /. #]}& @ FindRoot[f[x/10] == 0, {x, 1}]`

> FindRoot::cvmit : Failed to converge to the
> requested accuracy or precision within 100 iterations. More...

Out[4]= `{{x → 0.0889131}, 1.42946×10⁻⁴⁹}`

To get a better result for the zero, we must now set some of the options of `FindRoot`. Because the derivatives of all orders vanish at $x = 0$ for $f(x)$, to obtain a more accurate result, we have to use the `DampingFactor` option. To accommodate for the vanishing of all orders, we use a list of exponentially increasing damping factors. This allows us to get a result with accuracy 7.

In[5]:= `FindRoot[f[x/10] == 0, {x, 1}, WorkingPrecision -> 100,`
`        DampingFactor -> (10^(Range[0, 100]))] // N`

> FindRoot::cvmit : Failed to converge to the
> requested accuracy or precision within 100 iterations. More...

Out[5]= `{x → 0.0000272998}`

Or we can use a bracketing root-finding method.

In[6]:= `FindRoot[f[x/10] == 0, {x, -E, Pi}, WorkingPrecision -> 100,`
`          Method -> Brent] // N`

> General::unfl : Underflow occurred in computation. More...

> FindRoot::nrlnum : The function value {Underflow[]} is not a
> list of real numbers with dimensions {1} at {x} = {≪111≫}. More...

Out[6]= `{x → 2.7989×10⁻¹⁰}`

α) Now, we carry out the numerical integration.

In[1]:= `Module[{c = 0.54362689559915372089486, g},`
`          g[x_] := x + c; NIntegrate[g[x] - 1/g[x], {x, 0, 1}]]`

> NIntegrate::ploss :
> Numerical integration stopping due to loss of precision. Achieved
> neither the requested PrecisionGoal nor AccuracyGoal; suspect one of
> the following: highly oscillatory integrand or the true value of the
> integral is 0. If your integrand is oscillatory on a (semi-)infinite
> interval try using the option Method->Oscillatory in NIntegrate. More...

Out[1]= `-3.81639×10⁻¹⁷`

The message suggests two possible reasons for `NIntegrate` not achieving the default precision goal: either an oscillatory integral or the value of the integral is 0. Because the integrand is the sum of two monotonic functions, the first option is improbable. But the constant c in the integrand was chosen in such a way that $\int_0^1 g(x)\,dx = \int_0^1 g^{-1}(x)\,dx$; this means within the precision of the numerical value of c, the value of the integrand was indeed 0.

In[2]:= `Integrate[(x + c) - 1/(x + c), {x, 0, 1},`
`            Assumptions -> c > 0]  // PowerExpand`

Out[2]= $\frac{1}{2} + c + \text{Log}[c] - \text{Log}[1 + c]$

In[3]:= `FindRoot[%, {c, 1/3}]`

Out[3]= $\{c \to 0.543627\}$

24. Series Expansion for Anharmonic Oscillator

a) Let us start by calculating the differential equation for $u(x)$.

```
In[1]:= y[x_] = Exp[-x^3/3] u[x];
        eq1 = Factor[-y''[x] + x^4 y[x] - λ y[x]]
```

Out[2]= $e^{-\frac{x^3}{3}} (2 x u[x] - \lambda u[x] + 2 x^2 u'[x] - u''[x])$

Substituting the power series ansatz for $u(x)$, we can derive the following recursion relation for the a_i.

```
In[3]:= eq2 = eq1[[-1]] /. u -> Function[x, a[i] x^i]
```

Out[3]= $-(-1 + i) i x^{-2+i} a[i] + 2 x^{1+i} a[i] + 2 i x^{1+i} a[i] - x^i \lambda a[i]$

```
In[4]:= eq3 = Factor[eq2 /. f_ x^k_ :> (f /. Solve[k == n, i][[1]]) x^n]
```

Out[4]= $x^n (2 n a[-1 + n] - \lambda a[n] - 2 a[2 + n] - 3 n a[2 + n] - n^2 a[2 + n])$

```
In[5]:= eq4 = Collect[eq3[[-1]] /. n -> i - 2, _a]
```

Out[5]= $2 (-2 + i) a[-3 + i] - \lambda a[-2 + i] + (-2 - 3 (-2 + i) - (-2 + i)^2) a[i]$

```
In[6]:= Solve[eq4 == 0, a[i]]
```

Out[6]= $\left\{\left\{a[i] \to \frac{-4 a[-3 + i] + 2 i a[-3 + i] - \lambda a[-2 + i]}{(-1 + i) i}\right\}\right\}$

Here is the recursion relation implemented.

```
In[7]:= a[_?Negative] = 0;
        a[0] = 1;
        a[1] = 0;
        a[i_] := a[i] = Expand[((2i - 4) a[i - 3] - λ a[i - 2])/(i^2 - i)]
```

The expression for a[100] can now easily be calculated.

```
In[11]:= Short[Timing[a[100]], 12]
```

Out[11]//Short= {0.05 Second, (20165096837591262034082619758211276332873165739285206947λ^2) /
        52299826420399074810140194321251148968950036133824755967636005790141744512002477·
        5056251438750000000 -
        (2458595369271679066339098666185803444215408512175070545294020879165221λ^5) /
        401562390147808149091141082615429521946604500738358712033021986346908613349709471·
        60240040426858766252091000000000000000 + ≪21≫ + λ^{50} /
        933262154439441526816992388562667004907159682643816214685929638952175999932299156·
        089414639761565182862536979208272237582511852109168640000000000000000000000000000}

Let us have a look at how the $\exp(-x^3/3) \sum_{i=0}^n a_i x^i$ behaves as a function of x and λ.

```
In[12]:= y1 = Exp[-x^3/3] Sum[a[i] x^i, {i, 0, 50}];
```

```
In[13]:= Plot[Evaluate[Table[y1, {λ, 1/2, 3/2, 1/10}]], {x, 0, 3},
             PlotStyle -> Table[Hue[0.7 i], {i, 0, 1, 1/10}]]
```

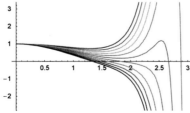

The condition that the first zero in x of $\exp(-x^3/3) \sum_{i=0}^{100} a_i x^i$ has a double zero encodes as the following in FindRoot.

```
In[14]:= y1 = Exp[-x^3/3] Sum[a[i] x^i, {i, 0, 100}];
```

```
In[15]:= y1p = D[y1, x];
```

```
In[16]:= FindRoot[Evaluate[{y1 == 0, y1p == 0}], {λ, 9/10, 11/10}, {x, 2, 3},
             WorkingPrecision -> 150, MaxIterations -> 100] //
                                    SetPrecision[#, 10]&
Out[16]= {λ → 1.060367856, x → 2.739971144}
```

The result for λ agrees with the one calculated earlier in Exercise 5) of this chapter.

b) Let us again start by calculating the differential equation for $u(x)$.

```
In[1]:= y[x_] = Exp[-x^2/2] u[x];
```

```
In[2]:= eq1 = Factor[-y''[x] + (x^2 + x^4) y[x] - λ y[x]]
```

$$Out[2]= \; e^{-\frac{x^2}{2}} \; (u[x] + x^4\, u[x] - \lambda\, u[x] + 2\, x\, u'[x] - u''[x])$$

Substituting the power series ansatz for $u(x)$, we can derive the following recursion relation for the a_i.

```
In[3]:= eq2 = eq1[[-1]] /. u -> Function[x, a[i] x^i]
```

$$Out[3]= \; -(-1 + i)\, i\, x^{-2+i}\, a[i] + x^i\, a[i] + 2\, i\, x^i\, a[i] + x^{4+i}\, a[i] - x^i\, \lambda\, a[i]$$

```
In[4]:= eq3 = Factor[eq2 /. f_ x^k_ :> (f /. Solve[k == n, i][[1]]) x^n]
```

$$Out[4]= \; x^n\, (a[-4 + n] + a[n] + 2\, n\, a[n] - \lambda\, a[n] - 2\, a[2 + n] - 3\, n\, a[2 + n] - n^2\, a[2 + n])$$

```
In[5]:= eq4 = Collect[eq3[[-1]] /. n -> i - 2, _a]
```

$$Out[5]= \; a[-6 + i] + (1 + 2\, (-2 + i) - \lambda)\, a[-2 + i] + (-2 - 3\, (-2 + i) - (-2 + i)^2)\, a[i]$$

```
In[6]:= Solve[eq4 == 0, a[i]]
```

$$Out[6]= \; \left\{\left\{a[i] \to \frac{a[-6 + i] - 3\, a[-2 + i] + 2\, i\, a[-2 + i] - \lambda\, a[-2 + i]}{(-1 + i)\, i}\right\}\right\}$$

Here, the recursion relation is implemented.

```
In[7]:= a[_?Negative] = 0;
        a[0] = 1;
        a[1] = 0;
        a[i_] := a[i] = Expand[-((a[-6 + i] +
                  (1 + 2 (-2 + i) - λ) a[-2 + i])/(i - i^2))]
```

The expression for a[100] can now easily be calculated.

```
In[11]:= Short[Timing[a[100]], 12]
```

```
Out[11]//Short= {0.11 Second,
        58601995181081281940933091909258461847659046284778762848855338455078196042566137492%
             847117799653073/
        986809955570838247322919364928773962974329356344887715783389787163669019043361846%
        3259529579952565098360404446804739589406720000000000000000000000000-
        (71056653455039116699864917308899438243330502168051726892960641784503085436654032287%
             7201892871026241 λ) /
        10401510342503430174484825738438428258378066188495915987978923511846193899165167%
        585708555031932442928663722547401309185638400000000000000000000000000+ ≪72≫ + λ^50 /
        933262154439441526816992388562667004907159682643816214685929638952175999932299156%
        089414639761565182862536979208272237582511852109168640000000000000000000000000000}
```

We numerically determine the first zero of $a_i(\lambda)$, which is around $\lambda \approx 1.3$.

```
In[12]:= Plot[Evaluate[DeleteCases[Table[a[i], {i, 5, 20}], 0]],
             {λ, 0, 2}, Frame -> True, Axes -> False]
```

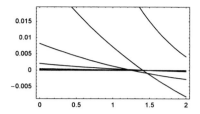

Here, a high-precision approximation of the zeros of a_i is calculated.

```
In[13]:= λValues = Table[If[a[i] === 0, Sequence @@ {},
            {i, λ /. FindRoot[Evaluate[a[i]], {λ, 13/10},
                        WorkingPrecision -> 6 i]}], {i, 30, 200}];
```

To determine the precision of a_{200}, we look at the change of the digits of the calculated λ-values.

```
In[14]:= lastλ = λValues[[-1, -1]];
```

```
In[15]:= logλValues = Apply[{#1, Log[10, Abs[#2 - lastλ]/lastλ]}&,
                        Drop[λValues, -1], {1}];
```

```
In[16]:= ListPlot[logλValues, PlotRange -> All]
```

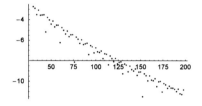

The last picture shows that (at least) the leading 11 digits are correct.

```
In[17]:= N[lastλ] // InputForm
```

```
Out[17]//InputForm= 1.3923516415221853
```

c) We find the recursion relation for the $a_i(\lambda)$ by using the differential equation: $-y''(x) + x^4\, y(x) = \lambda\, y(x)$.

```
In[1]:= ansatz = Sum[a[k] x^k, {k, 0, n}];
        s1 = Evaluate //@ ((Evaluate //@ Expand[
            (-D[ansatz, {x, 2}] + x^4 ansatz - ε ansatz)]) /.
                HoldPattern[a_ Sum[b_, c_]] :> Sum[a b, c])
```

$$Out[2]=\ \sum_{k=0}^{n} -(-1+k)\,k\,x^{-2+k}\,a[k] + \sum_{k=0}^{n} x^{4+k}\,a[k] + \sum_{k=0}^{n} -x^k\,\varepsilon\,a[k]$$

```
In[3]:= s2 = s1 /. Sum -> Σ /. (* rename exponents *)
            σ:(a_ x^e_) :> (σ /. Solve[e == λ, k][[1]]) //.
            Σ[a_, c_] + Σ[b_, c_] :> Σ[a + b, c]
```

$$Out[3]=\ \Sigma[x^\lambda\, a[-4+\lambda] - x^\lambda\, \varepsilon\, a[\lambda] - x^\lambda\, (1+\lambda)\,(2+\lambda)\, a[2+\lambda],\ \{k, 0, n\}]$$

```
In[4]:= sol = Solve[s2[[1]] == 0, a[λ + 2]] /. λ -> m - 2  // Simplify
```

$$Out[4]=\ \left\{\left\{a[m] \to \frac{a[-6+m] - \varepsilon\, a[-2+m]}{(-1+m)\,m}\right\}\right\}$$

Together with the initial conditions $a_0 = 1$, $a_1 = -\lambda/2$, $a_2 = \lambda^2/24$ (because of the symmetry for the lowest possible λ), we can implement the functions y and yPrime. To minimize memory consumption in the calculation of $y_n(x^*)$ and $y'_n(x^*)$, we form the sum after each new $a_k(\lambda)$ is calculated.

```
In[5]:= y[n_, λ_Real, x_] :=
        Module[{am6, am4, am2, aN, sum},
            (* start values *)
            {am6, am4, am2} = {1, -λ/2, λ^2/24};
```

```
        sum = am6 + am4 x^2 + am2 x^4;
        (* carry out recursion *)
        Do[aN = (am6 - λ am2)/(m(m - 1));
            {am6, am4, am2} = {am4, am2, aN};
            sum = sum + aN x^m, {m, 6, n, 2}];
        sum]

In[6]:= yPrime[n_, λ_Real, x_] :=
     Module[{am6, am4, am2, aN, sum},
        (* start values *)
        {am6, am4, am2} = {1, -λ/2, λ^2/24};
        sum = am4 2 x + am2 4 x^3;
        (* carry out recursion *)
        Do[aN = (am6 - λ am2)/(m(m - 1));
            {am6, am4, am2} = {am4, am2, aN};
            sum = sum + aN m x^(m - 1), {m, 6, n, 2}];
        sum]
```

For large n, we want the function $y_n(x)$ to vanish as $x \to \infty$. For a λ smaller than the smallest possible λ, the function $y_n(x^*)$ will not have a zero, but the function $y_n'(x^*)$ will have a zero for "suitably chosen" x^*. For a λ larger than the smallest possible λ, the function $y_n(x^*)$ will have a zero, but the function $y_n'(x^*)$ will not have a zero for "suitably chosen" x^*. The next graphic shows that for $y_{100}(x^*)$ and $y_{100}'(x^*)$, a "suitable" value of x^* is about $5/2$.

```
In[7]:= Show[GraphicsArray[
     Function[ys, Graphics[Table[{Hue[(λ - 1.06) 80],
        Line[Table[{x, ys[100, λ, x]},
                    {x, 0, 3.1, 0.03}]]}, {λ, 1.06, 1.07, 0.0002}],
        Frame -> True, Axes -> {True, False},
        PlotRange -> {-3/2, 1}]] /@ {y, yPrime}]]
```

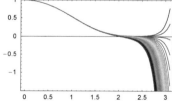 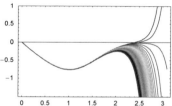

The function λBounds calculates the zeros of $y_n(x^*)$ and $y_n'(x^*)$ numerically. It returns an interval for the smallest possible λ.

```
In[8]:= λBounds[ord_, x0_, {λ0_, λ1_}, prec_, ag_] :=
     (λ /. FindRoot[#[ord, λ, x0] == 0, {λ, λ0, λ1},
                    WorkingPrecision -> prec, MaxIterations -> 100,
                    AccuracyGoal -> ag])& /@ {y, yPrime};
```

Here is an example for $n = 100$.

```
In[9]:= λBounds[100, 3, {1/2, 2}, 100, 20]

Out[9]= {1.0603621398759538800792078470554621560549454996274603764112869710616935448383344852
         196195208529795991,
         1.0603620367070596726998097240164513955316775653947019794603796280437553915117162582
         99691128678844658}
```

Analyzing at which digit the lower and the upper bounds deviate from each other allows us to form an approximative value of the smallest possible λ. The function λValue does this.

```
In[10]:= λValue[ord_, x0_, {λ0_, λ1_}, prec_, ag_] :=
     Module[{λa, λb},
            {λa, λb} = λBounds[ord, x0, {λ0, λ1}, prec, ag];
            δ = λa - λb;
            λprec = If[δ == 0, (* use accuracy *) Accuracy[δ] - 1,
                             (* use precision *)-Ceiling[Log[10, δ]]];
            SetPrecision[λa, λprec]]
```

For $n = 100$, we get about six correct digits.

```
In[11]:= λ100 = λValue[100, 3, {1/2, 2}, 100, 20]
Out[11]= 1.06036
```

We will now calculate a much more precise value for the smallest possible $λ$. We will recursively use the calculated values of $λn$ as starting values in the calculation of $λ2n$. The function `makeStartValues` forms two starting values from the $λ$-value returned by `λValue`.

```
In[12]:= makeStartValues[λ_] :=
         With[{p = Precision[λ]}, λ {1 - 10^-Round[p/2], 1 + 10^-Round[p/2]}]
```

In addition to starting values for $λ$, we need a "suitable" x^*. It turns out that the detailed choice of x^* is not critical. The numerical value of x^* determines the precision of the value of $λ$, but within broad bounds, its detailed value does not matter. The following graphic shows this for $n = 200$.

```
In[13]:= ListPlot[Table[{x0, Precision[
                 λValue[200, x0, {106/100, 107/100}, 100, 30]]},
                 {x0, 2, 6, 1/10}],
             PlotRange -> All, PlotJoined -> True, Frame -> True, Axes -> False]
```

So, we can now calculate `λValue[500, ...]`.

```
In[14]:= λ500 = λValue[500, 5, makeStartValues[λ100], 100, 50]
Out[14]= 1.0603620904841828996470460166926635
```

We can check the correctness of this value with the value $Λ$ calculated in Subsection 2.2 of the Symbolics volume [1795] and [1257].

```
In[15]:= Λ = 1.060362090484182899647046016692663545515208728528977933216245241695\
             9435630443444211268962991346717035105462443585825525581`120;
```

Now, we double the value of n repeatedly (and at the same time increase the working precision) until we reach the required 1000 digits.

```
In[15]:= λ1000 = λValue[1000, 6, makeStartValues[λ500], 1000, 200];
```

Again, all digits agree with $Λ$.

```
In[16]:= Precision[λ1000]
Out[16]= 61.
```

```
In[17]:= λ1000 - Λ
Out[17]= 0. × 10^{-61}
```

Now, we use $n = 2000$.

```
In[18]:= λ2000 = λValue[2000, 9, makeStartValues[λ1000], 2000, 300];
```

Again, all 109 digits agree with $Λ$.

```
In[19]:= Precision[λ2000]
Out[19]= 109.
```

```
In[20]:= λ2000 - Λ
Out[20]= 0. × 10^{-109}
```

The number of digits roughly doubles when doubling n. So, we double the value of n two more times to get 1000 digits of the smallest possible λ.

```
In[21]:= λ4000 = λValue[ 4000, 19/2, makeStartValues[λ2000], 3000, 300];
         Precision[λ4000]
Out[22]= 247.
```

```
In[23]:= λ8000 = λValue[ 8000, 13,   makeStartValues[λ4000], 4000, 400];
         Precision[λ8000]
Out[24]= 635.
```

```
In[25]:= λ16000 = λValue[16000, 16,   makeStartValues[λ8000], 6000, 600]
```

Out[25]= 1.060362090484182899647046016692663545515208728528977933216245241695943563044344421
2689629913467170351054624435858252558087980821029314701317683637382493578922624600
7081754469601416374884172822569059357577908880617887902636015493956902751961489009
2934873584409442694897901213971464290951923354533828347033505757615112025703988852
7202402218411030865737310913989154536584103111679405833548600092274400696311267023
8622971429699610592155832266713769355086736100008318300275179262335739139061361807
6498596961814994127928092728407079561060440722946809949136275729273872791368902798
2472226171694448895475137043806840543918778772953234245874372543178323190603810687
1604403437453014684727813918612940470431034013510716071103530089298232754276615189
6950565047160252756089526262191025688200964410287815640052705292932405076382650282
9112477362538471854714402572285438485297450458570978840249066699957047684458770917
0291243752732549071164334402302947306923981908956853745359884460160023132919330593
5869304916644281633946163324287004261461237743009952234204208597735690153565416850
0894185134879573410658547971946759646667966134676885864379526545195605682867159583
8847434670120424207149

The last result is actually more precise than we wanted.

```
In[26]:= Precision[λ16000]
Out[26]= 1184.
```

For a similar method based on the corresponding Riccati form of the differential equation, see [104] and [1269]. For the rephrased problem using differential transforms [382], see [798].

25. Gibbs Distributions, Optimal Bin Size, Rounded Sums, Odlyzko-Stanley Sequences

a) This is the noncompiled version of the modeling of the repeated collisions.

```
In[1]:= pure[n_Integer, e0_Integer, steps_Integer] :=
         Module[{list, n1, n2, e, e1, e2},
                 (* initial list of energies *)
                 list = Table[e0, {n}];
                 Do[(* the interacting pair *)
                    {n1, n2} = Table[Random[Integer, {1, n}], {2}];
                    If[n1 =!= n2,
                    (* redistribute energies *)
                    e = list[[n1]] + list[[n2]];
                    e1 = Random[Integer, {0, e}]; e2 = e - e1;
                    (* make new list of energies *)
                    list = ReplacePart[ReplacePart[list, e1, n1],
                                        e2, n2], list], {steps}]; list]
```

This is the compiled version of the modeling of the repeated collisions.

```
In[2]:= cf = Compile[{{n, _Integer}, {e0, _Integer}, {steps, _Integer}},
         Module[{list = Table[e0, {n}], n1, n2, e, e1, e2},
                 Do[{n1, n2} = Table[Random[Integer, {1, n}], {2}];
                    If[n1 =!= n2,
                    e = list[[n1]] + list[[n2]];
                    e1 = Random[Integer, {0, e}]; e2 = e - e1;
                    list = ReplacePart[ReplacePart[list, e1, n1],
                                        e2, n2], list], {steps}]; list]] ;
```

We make sure that everything was nicely compiled.

```
In[3]:= Union[Head /@ Flatten[cf[[4]]]]

Out[3]= {Integer}
```

The speed gain from compiling is obvious.

```
In[4]:= pure[1000, 500, 1000]; // Timing

Out[4]= {0.03 Second, Null}
```

```
In[5]:= cf[1000, 500, 1000]; // Timing

Out[5]= {0.01 Second, Null}
```

Here is a larger example run. The red curve is the theoretical distribution of the energies [1843], [1530]. Clearly, we see how the microcanonical Gibbs distribution is approached.

```
In[6]:= Module[{n = 5000, e0 = 50, steps = 50000, nn, data, countedData},
        nn = N[n];
        (* the simulation *)
        data = cf[n, e0, steps];
        (* analyze simulation *)
        countedData = {#, Count[data, #]}& /@ Union[data] ;
        ListPlot[countedData,
                PlotRange -> {{0, Max[data]}, {0, Max[Last /@ countedData]}},
                AxesOrigin -> {0, 0}, Frame -> True,
                PlotStyle -> {PointSize[0.005], GrayLevel[0]},
                FrameLabel -> {"energy", "number of particles"},
                (* theoretical curve *)
                Prolog -> {Hue[0], Thickness[0.002],
                  Line[Table[{i, (nn Binomial[-1 - i + nn + e0 nn, -i + e0 nn]*
                                   Gamma[e0 nn] Gamma[1 + nn])/Gamma[nn + e0 nn]},
                              {i, 0, Max[data]}]]}]]
```

For the use of such models for random number generation, see [594]. For the energy fluctuations after reaching the equilibrium, see [1514].

b) We start with implementing the repeated updating of pairs of the list of length *n*.

```
In[1]:= cf = Compile[{{ℓ, _Real, 1}, {t, _Integer}},
             Module[{l = ℓ, n = Length[ℓ], i, j, s = Sqrt[2.], r},
                    Do[(* the two numbers i and j *)
                       i = Random[Integer, {1, n}];
                       (* ensure i≠j *)
                       j = Random[Integer, {1, n - 1}];
                       j = If[j < i, j, j + 1];
                       r = 2. Random[Integer] - 1.;
                       l[[i]] = r (l[[i]] + l[[j]])/s;
                       l[[j]] = -l[[i]] + s r l[[j]], {t}];
                    l]];
```

To check cf, we generate a list of 4096 normally distributed real numbers.

```
In[2]:= With[{n = 2^12}, l = Sort[cf[Table[1., {n}], 6 n]]]; // Timing

Out[2]= {0.1 Second, Null}
```

In a graphic, the so-obtained random numbers are indistinguishable from the corresponding theoretical cumulative distribution.

```
In[3]:= (* exact distribution expressed through the inverse error function *)
        cdf[x_, n_] := InverseErf[2x/(n + 1) - 1]
```

```
In[5]:= Show[{ListPlot[l, DisplayFunction -> Identity],
        Plot[Sqrt[2] cdf[x, 2^12], {x, 1, 2^12},
            PlotStyle -> {Hue[0]}, DisplayFunction -> Identity]},
            DisplayFunction -> $DisplayFunction, Frame -> True, Axes -> False]
```

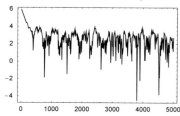

To see how the initially uniform distributed numbers approach a normal distribution, we measure the distance from the ideal distribution as a function of the number of pair updates.

```
In[6]:= dList =
        Module[{n = 2^10, L, m = 10},
            L = Table[cdf[N[k], n], {k, n}];
            l = Table[1., {n}];
            (* every m steps calculate difference *)
            Table[Do[l = cf[l, k], {k, m}];
                {j m, Plus @@ (Sort[l] - L)^2}, {j, 500}]];
```

We see that after about $3/2 \log(n) n$ (≈ 10000 for $n = 2^{10} = 1024$), we reach a stationary state.

```
In[7]:= ListPlot[{#[[1]], Log[10, #[[2]]]}& /@ dList, PlotRange -> All,
            PlotJoined -> True, Frame -> True, Axes -> False]
```

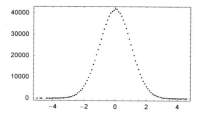

Now, let us generate one million normally distributed real numbers.

```
In[8]:= Timing[l = Module[{n = 2^20}, cf[Table[1., {n}], 6 n]];]
```

```
Out[8]= {29.12 Second, Null}
```

We bin the resulting numbers and show the distribution of the bins. δ is the size of the bins.

```
In[9]:= makeBins[l_, δ_] := {First[#] δ, Length[#]}& /@ Split[Round[Sort[l]/δ]]
```

```
In[10]:= bins = makeBins[l, 0.1];
```

```
In[11]:= ListPlot[bins, Frame -> True, Axes -> False]
```

The bins show clearly the expected $\ln(p(x)) \sim -x^2/2$ behavior.

```
In[12]:= ListPlot[Apply[{#1^2, Log[#2]}&, bins, {1}] // N,
             Frame -> True, Axes -> False]
```

Now, let us compare how long it would take to generate 10^6 normally distributed real numbers by inverting the cumulative distribution for uniformly distributed numbers.

```
In[13]:= Timing[Table[InverseErf[Random[Real, {-1, 1}]], {2^10}]][[1]] 2^10
```

```
Out[13]= 348.16 Second
```

For more detailed statistical investigation of the so obtained random numbers, see [1486] and [596].

The last result shows that it would take a few times as long. It is interesting to compare the speed of `cf` with the speed of the corresponding version for exact numbers.

```
In[14]:= f = Function[{ℓ, t},
             Module[{l = ℓ, n = Length[ℓ], i, j, s = Sqrt[2]},
                 Do[i = Random[Integer, {1, n}];
                    j = Random[Integer, {1, n - 1}];
                    j = If[j < i, j, j + 1];
                    l[[i]] = Expand[(l[[i]] + l[[j]])/s];
                    l[[j]] = Expand[-l[[i]] + s l[[j]]], {t}];
                 l]];
```

Not unexpectedly, using exact arithmetic instead of floating point arithmetic is more than two orders of magnitude slower.

```
In[15]:= (l = With[{n = 2^6}, f[Table[1, {n}], 6 n]]); // Timing
```

```
Out[15]= {0.21 Second, Null}
```

Of course, the resulting "numbers" have a much higher memory consumption.

```
In[16]:= {ByteCount[l] 2^6, Take[l, 2]}
```

$$Out[16]= \left\{1263104, \left\{-\frac{89233}{131072} - \frac{128281}{131072\sqrt{2}}, \frac{394117}{262144} + \frac{138631}{262144\sqrt{2}}\right\}\right\}$$

Note that the Box–Muller method [267], [1481], [1320], [497] can also quickly generate normally distributed random numbers.

```
In[17]:= cfBoxMuller =
         Compile[{{n, _Integer}},
                 Table[Sqrt[-2 Log[Random[]]] Cos[2 Pi Random[]], {n}]];
```

```
In[18]:= (l = cfBoxMuller[10^6];) // Timing
```

```
Out[18]= {1.46 Second, Null}
```

```
In[19]:= ListPlot[makeBins[l, 0.1], Frame -> True, Axes -> False]
```

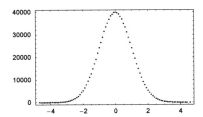

For a similar method that generates a numbers with a Gamma distribution, see [1418].

c) This is the list of data. We rescale the random sums such that we have a normalized Gaussian distribution in the limit $n \to \infty$.

```
In[1]:= SeedRandom[34952943];
       n = 10^4;
       m = 10^3;
       data = (Table[Sum[Random[], {m}], {n}] - m/2)/Sqrt[m/12];
In[5]:= ListPlot[data, PlotRange -> All, PlotStyle -> {PointSize[0.003]}]
```

The function `makeBins` partitions the data into bins of length *binSize*. `makeBins` returns a list of lists, each being of the form {*lowerBinLimit*, *upperBinLimit*, *fractionOfData*}.

```
In[6]:= makeBins[data_, binSize_] :=
       {First[#] binSize, (First[#] + 1) binSize,
        Length[#]/Length[data]/binSize}& /@ Split[Quotient[Sort[data], binSize]]
```

`makeHistogram` makes a histogram from the data returned by `makeBins`.

```
In[7]:= makeHistogram[l_] :=
       Apply[Polygon[{{#1, 0}, {#1, #3}, {#2, #3}, {#2, 0}}]&, l, {1}]
```

The following three graphics show the histograms together with the limit distribution for the bin size 0.01, 0.16, 0.5, and 1. The histogram for the bin size 0.16 fits best.

```
In[8]:= p[t_] = 1/Sqrt[2Pi] Exp[-t^2/2];
       limitDistribution = {Hue[0], Line[Table[{t, p[t]}, {t, -3.5, 3.5, 7/200}]]};
In[10]:= Show[GraphicsArray[
       Graphics[{makeHistogram[makeBins[data, #]],
                 limitDistribution}, Frame -> True]& /@ {0.01, 0.16, 0.5, 1}]]
```

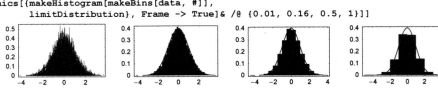

Now let us find the optimal bin size. `δSquared` measures the squared integrated difference between the histogram value and the limit distribution within a given bin.

```
In[11]:= δSquared[{t1_, t2_, c_}] = Integrate[(c - 1/Sqrt[2Pi] Exp[-t^2/2])^2,
                                             {t, t1, t2}] // Simplify
```

$$Out[11]= -c^2\, t1 + c^2\, t2 + \frac{-\mathrm{Erf}[t1] + \mathrm{Erf}[t2]}{4\sqrt{\pi}} + c\left(\mathrm{Erf}\left[\frac{t1}{\sqrt{2}}\right] - \mathrm{Erf}\left[\frac{t2}{\sqrt{2}}\right]\right)$$

We sum these squared deviations (ignoring the small contributions from the tails) and show the summed squares as a function of the bin widths (to cover a larger bin width range we use a logarithmic distribution for the bin widths).

```
In[12]:= diff[binSize_] := Plus @@ (δSquared /@ N[makeBins[data, binSize]])
```

```
In[13]:= Show[GraphicsArray[
        ListPlot[{#[[1]], Log[10, #[[2]]]}& /@ Table[{ξ, diff[10^ξ]}, {ξ, ##}],
             Frame -> True, PlotJoined -> True, Axes -> False,
             DisplayFunction -> Identity]& @@@
          (* two ξ-ranges *) {{-3, 2, 5/50}, {-1.2, -0.5, 0.7/100}}]]
```

The minimum occurs around a binsize of about $10^{-0.8} \approx 0.159$. This agrees favorably with the theoretical asymptotic value $n^{-1/3}(6/\int_{-\infty}^{\infty} p'(t)^2\, dt)^{1/3}$ [633], [1630], [555], [935], [496].

```
In[14]:= (* theoretical exponent size *)
        (6/Integrate[p'[t]^2, {t, -Infinity, Infinity}])^(1/3) n^(-1/3) //
                                                              N[Log[10, #]]&
Out[15]= -0.790405
```

For the optimal binsize of unknown distributions, see [528], [212].

d) We start by carrying out the integration for the theoretical probability numerically using NIntegrate. (We could also do the integral exactly. This would take considerably longer and yield a fraction with large numerator and denominator.)

```
In[1]:= p[n_Integer, b_:10] :=
        2/(Pi b) NIntegrate[Evaluate[(Sin[x]/x)^(n - 1) Sin[b x]^2/x^2 *
                             If[OddQ[n], 0, Cos[x]]],
                             {x, 0, Infinity}, PrecisionGoal -> 12]
```

```
In[2]:= p[100]
Out[2]= 0.767292
```

To form the random sums with a million summands, we use the following compiled function.

```
In[3]:= sumRun =
        Compile[{{n, _Integer}, L, {o, _Integer}},
        Module[{b = 10, count = 0, t},
             Do[(* random summands *)
               t = Table[Random[Real, {-L, L}], {n}];
               (* do the two sums agree within a multiple of b *)
               If[b Round[(Plus @@ t)/b] ===
                   b Round[(Plus @@ Round[t])/b], count++], {o}];
             (* frequency of coinciding sums *) count/n]];
```

For a lucky random seed, we obtain a relative error of about 4×10^{-6}.

```
In[4]:= SeedRandom[685];
        sumRun[100, 100, 10^6]
Out[5]= 7672.89
```

e) To construct the terms of the Odlyzko-Stanley sequences that are less than a given integer n, we will proceed in the following in a manner analogous to Eratosthenes sieve method for primes. We start with a list of elements -1 of length n. In this list, we mark all integers (through their position) that would form an arithmetic progression of length three. To do this, starting with the two initial elements 0 and k, for each further element we form all differences to already constructed elements and mark the larger elements that would give rise to an arithmetic progression. The next element of the sequence is then the first unmarked one to the right of the last element. The following function `OdlyzkoStanleySequence` implements this algorithm as a compiled function.

```
In[1]:= OdlyzkoStanleySequence =
       Compile[{{a1, _Integer}, {m, _Integer}},
       Module[{A = Table[0, {m}], λ = Table[-1, {m}], i, j, k, p = 0},
        A = Table[0, {m}]; λ = Table[-1, {m}];
        (* -1 is not touched yet; 0 is not usable
           (already used or would yield arithmetic sequence) *)
        λ[[1]] = 0; λ[[2]] = 0; A[[1]] = 0; A[[2]] = a1;
        λ[[A[[2]] + (A[[2]] - A[[1]])]] = 0;
        (* find next integers *) k = 3;
        While[A[[k - 1]] < m,
            j = A[[k - 1]] + 1;
            While[j <= m && Not[λ[[j]] === -1], j++];
            A[[k]] = j; λ[[k]] = 1;
            i = k; While[i > 0 && ((p = j + (j - A[[i]])) <= m),
                    λ[[p]] = 0; i--]; k++];
        If[A[[k - 1]] > m, Take[A, k - 2], Take[A, k - 1]]]];
```

With the exception of $k = 3^i$ or $k = 2 \, 3^i$, no closed form descriptions are known for the resulting sequences. Here, we check the function `OdlyzkoStanleySequence` using the closed form for $k = 3$.

```
In[2]:= (# == Table[FromDigits[IntegerDigits[k, 2], 3],
                {k, 0, Length[#] - 1}])&[OdlyzkoStanleySequence[1, 10^6]]

Out[2]= True
```

Each of the S_k with maximal element 10^6 can be calculated in a few seconds.

```
In[3]:= Do[oss[k] = OdlyzkoStanleySequence[k, 10^6], {k, 13}];
```

Here is the number of elements in the resulting sequences.

```
In[4]:= Table[{k, Length[oss[k]]}, {k, 2, 13}]

Out[4]= {{2, 8192}, {3, 8192}, {4, 4672}, {5, 4198}, {6, 8192}, {7, 4798},
         {8, 4301}, {9, 8192}, {10, 4403}, {11, 4608}, {12, 4344}, {13, 4808}}
```

Next, we display the growth rate of the sequence and their local average density. The function `fillIn` generates a list of 0's and 1's where a 1 indicates the presence of the element in the list *l*. The above-mentioned special cases for *k* show a periodic local average and the other values show a complicated fluctuating behavior.

```
In[5]:= fillIn = Compile[{{o, _Integer}, {l, _Integer, 1}},
                    Module[{L = Table[0., {o}], j = 1, p},
                        While[(p = l[[j]] + 1) <= o,
                            L[[p]] = 1.; j++]; L]];
```

```
In[6]:= Show[GraphicsArray[
        Block[{oλ = 10^3, λ = 10^5, $DisplayFunction = Identity},
            Show[Table[ListPlot[#[k],
                    PlotStyle -> {PointSize[0.003], Hue[k/16]}],
                {k, 13}], PlotRange -> All]& /@
                (* sequences and local averages *)
                {oss, ListConvolve[Table[1./oλ, {oλ}],
                            fillIn[λ, oss[#]]]&}]]]
```

26. Nesting Tan, Thompson's Lamp, Digit Jumping

a) Here, the picture under consideration is shown.

```
In[1]:= nl = NestList[Tan, N[2, 200], 2000];
        precList = Precision /@ nl;
        ListPlot[precList];
```

The steps coincide with large values of the list. The left of the next two graphics shows this. The above picture showed the successive loss of precision. Applying `Tan` to a numerical value of *x* amplifies its fuzziness in a way shown in the right picture.

```
In[4]:= Show[GraphicsArray[
           Block[{$DisplayFunction = Identity},
              {(* absolute values of the list elements *)
               ListPlot[Abs[nl]],
               (* precision-governing expression *)
               Plot[Evaluate[x Tan'[x]/Tan[x]], {x, 0, 1.4}]}]]]
```

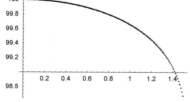

```
In[5]:= ListPlot[Table[{x, Precision[Tan[SetPrecision[x, 100]]]},
                  {x, 1/100, 15/10, 1/100}]]
```

The second fact we need to explain are the steps from the above figure is that once we have a small `x`, it will stay small for many iterations.

```
In[6]:= ListPlot[Table[{x, Plus @@ Abs[NestList[Tan, x, 20]]},
                  {x, 1/100., 15/10., 1/1000.}]]
```

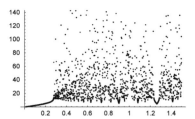

b) We start by calculating the list of tan-products.

```
In[1]:= f1  = Rest[FoldList[Times, 1., tab = Table[Tan[k], {k, 2 10^5}] // N]];
```

The products show a characteristic periodicity with a period of approximately 104000. At a finer scale, we see a periodicity with period of about 350.

```
In[2]:= Show[GraphicsArray[
           ListPlot[#, PlotRange -> All, PlotStyle -> {PointSize[0.003]},
                   DisplayFunction -> Identity]& /@ (* show two ranges *)
                   {Log @ Abs[f1], MapIndexed[{10^5 + #2[[1]], #1}&,
                   Take[Log @ Abs[f1], {100000, 104000}]]}]]]
```

The following two graphics show the values of tan(k) and the difference of k to the nearest multiple of $\pi/2$.

```
In[3]:= Show[GraphicsArray[
           Block[{$DisplayFunction = Identity},
            {(* values of Tan[k] *)
             ListPlot[Log @ Abs[tab], PlotRange -> All, Frame -> True,
                     PlotStyle -> {PointSize[0.003]}, Axes -> False],
             (* differences of Tan[k] to nearest multiple of Pi/2 *)
             ListPlot[Table[Log @ Abs[k - Pi/2. Round[k/(Pi/2.)]], {k, 2 10^5}],
                     PlotRange -> All, PlotStyle -> {PointSize[0.003]},
                     Frame -> True, Axes -> False]}]]]
```

While the values of tan(k) wildly oscillate as a function of k, after averaging over a couple thousand consecutive values (meaning about 5 to 10 small periods) we obtain a smooth curve.

```
In[4]:= tab1 = ListConvolve[Table[1., {2500}], tab];
        (* show data *)
        ListPlot[Log[Abs[tab1]], PlotRange -> All]
```

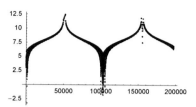

The difference of subsequences of `fl` that are p elements apart shows a characteristic behavior.

```
In[7]:= (* difference of p-apart elements *)
        diff[p_, δ_] := Plus @@ Abs[Take[fl, {p, p + δ}] - Take[fl, {1, 1 + δ}]]
        s1 = Table[{p, diff[p, 10^4]}, {p, 100000, 150000, 10}];
        (* show data *)
        ListPlot[{#[[1]], Log[#[[2]]]}& /@ s1, PlotRange -> All]
```

For the following integers k, is especially near to a multiple of $\pi/2$ [1853].

```
In[12]:= Compile[{},
         Module[{bag = {{0., 0.}}, min = 1., δ},
             Do[δ = k - Pi/2. Round[k/(Pi/2.)];
                 If[(* smaller distance than ever? *)
                     Abs[δ] < min, AppendTo[bag, {k, δ}];
                     min = Abs[δ]], {k, 10^7}];
             (* return bag *) bag]][] // InputForm
Out[12]//InputForm= {{0., 0.}, {1., -0.5707963267948966}, {2., 0.42920367320510344}, {3.,
         -0.14159265358979312},
         {11., 0.004425712435724094}, {344., -0.004395568082372847}, {355.,
         0.00003014435338855037},
         {51819., -0.00002463684359099716}, {52174., 5.507514288183302*^-6},
         {260515., -2.606800990179181*^-6}, {573204., 2.939486876130104*^-7},
         {4.846147*^6, -2.551823854446411*^-7}, {5.419351*^6, 3.818422555923462*^-8}}
```

These values can also be obtained from the continued fraction approximation of $\pi/2$.

```
In[13]:= Table[FromContinuedFraction[ContinuedFraction[Pi/2, k]],
             {k, 15}] // Numerator
Out[13]= {1, 2, 3, 11, 344, 355, 51819, 52174, 260515,
         573204, 4846147, 5419351, 37362253, 42781604, 122925461}
```

The differences of k to the nearest multiple of $\pi/2$ are uniformly distributed. The values of $\tan(k)$ and the products are strongly nonuniformly distributed.

```
In[14]:= δπs = Table[k - Pi/2. Round[k/(Pi/2.)], {k, 2 10^5}];
         Show[GraphicsArray[
         Block[{$DisplayFunction = Identity},
             (* cumulative distributions *)
             ListPlot[Sort[Abs[#]]]& /@ {δπs, tab, fl}]]]
```

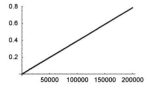 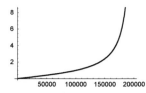 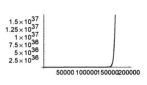

Larger periodicities can also be observed. The next graphic shows every 1000th product up to 10^8. The period 42781604 is clearly visible.

```
In[16]:= f12 = Compile[{},
            Module[{prod = 1., L = Table[{1., 1.}, {10^5}], A = 1000},
                Do[prod = prod Tan[k];
                    If[k/A - Round[k/A] == 0., L[[Round[k/A]]] = {k, prod}],
                    {k, A 10^5}]; L]][];
In[17]:= ListPlot[{{#[[1]], Log @ Abs[#[[2]]]}}& /@ f12, PlotRange -> All,
            PlotStyle -> {PointSize[0.003]}]
```

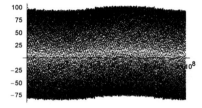

c) The `While`-loop will switch the lamp off and on as long as the inequality `t != 1``10000` holds. The inequality uses the function `Unequal` with a numeric second argument. This means the value of `Experimental`$EqualTolerance` will determine the value of `t`, such that the inequality gives `False`. With the default value, this will happen after 33213 switches.

```
In[1]:= Ceiling[-Log[2, 10^(-10000 + Experimental`$EqualTolerance)]]
Out[1]= 33213
```

Because this is an odd number and we started with the lamp on, the lamp will be out at the end.

A comparison with the actual code gives the same result.

```
In[2]:= t = 0; δt = 1/2; lampOnQ = True;
        While[t != 1``10000, t = t + δt; δt = δt/2; lampOnQ = Not[lampOnQ]]
        {lampOnQ, -1/Log[1 - t, 2]}
Out[4]= {False, 33213}
```

Now let us change the high-precision number `1``10000` to an exact 1. Obviously, now the test `t != 1` will never be fulfilled. But the code will run for a long long time. Because the only changing value is $\delta t$, which is always a negative power of 2, its storage is very efficient and the following calculation without the `TimeConstrained` would run a long time.

```
In[5]:= c = 0;
        TimeConstrained[
        t = 0; δt = 1/2; lampOnQ = True;
        While[t != 1,
            c++; t = t + δt; δt = δt/2; lampOnQ = Not[lampOnQ]], 1000];
        {c, ByteCount[δt]}
Out[7]= {376451, 47136}
```

And here is the largest power of 2 that can be currently represented.

```
In[8]:= (* get maximal power of 10 in the exponent of 2 *)
        exp = 0;
```

```
        While[IntegerQ[2^(10^exp)], exp++];
        maxExp = exp - 1
            General::ovfl : Overflow occurred in computation. More…
```

Out[11]= 9

```
In[12]:= (* get exact power of 10 *)
        n = maxExp;
        While[n >= 0,
            c[n] = 9; (* try all sums from larger to lower values *)
            While[Not[IntegerQ[2^Sum[c[k] 10^k, {k, maxExp, n, -1}]]],
                c[n] = c[n] - 1];
            n = n - 1]
            General::ovfl : Overflow occurred in computation. More…

            General::ovfl : Overflow occurred in computation. More…

            General::ovfl : Overflow occurred in computation. More…

            General::stop :
             Further output of General::ovfl will be suppressed during this calculation. More…
```

In[15]:= max2Exp = Sum[c[k] 10^k, {k, 0, maxExp}]

Out[15]= 1073741568

In[16]:= {N[2^-max2Exp], 2^-(max2Exp + 1)}

```
            General::ovfl : Overflow occurred in computation. More…
```

Out[16]= {2.758821680804834 × 10^{-323228420}, Underflow[]}

Adding all smaller powers of 2 yields the largest integer that can be represented. It is again an odd integer and so the lamp is off again: $\sum_{j=0}^{max2Exp} 2^{max2Exp-j} = 2^{max2Exp+1} - 1$.

d) The function `cf` internally does the following: First, it creates a list of random digits $\{d_1, d_2, \ldots\}$. Then, it effectively iterates the following recursion $s_k = s_{k-1} + d_{s_{k-1}}$ starting with $s_k = 1$. The process stops if a fixed point of the sequence s_k is reached (meaning $d_{s_{k-1}} = 0$). If the initial list of random digits is too short, more digits are appended as needed. The result $\{k, s_k\}$ is returned from these iterations. `cf` creates a list of length n of such pairs of integers.

```
In[1]:= cf = Compile[{{n, _Integer}},
        Table[Module[{digits, λ = 100, k = 1, sOld = 1, sNew = 1},
                    (* initial random digits sequence *)
                    digits = Join[{Random[Integer, {1, 9}]},
                            Table[Random[Integer, {0, 9}], {λ}]];
                    (* iterate *)
                    While[k++; (* more digits needed *)
                        If[sOld > λ,
                            digits = Join[digits,
                            Table[Random[Integer, {0, 9}], {sOld}]];
                            λ = Length[digits]];
                        (* next s_k *)
                        sNew = sOld + digits[[sOld]];
                        sOld =!= sNew, sOld = sNew];
                    (* return result *) {k, sNew}], {n}]];
```

`cf` compiled successfully.

In[2]:= Union[Head /@ Flatten[cf[[4]]]]

Out[2]= {Integer}

One million pairs can be calculated in minutes.

In[3]:= (data = cf[10^6]); // Timing

Out[3]= {80.87 Second, Null}

Not unexpectedly, the random jumps in the random digit sequence lead to a Poisson distribution of the k and s_k. So the following graphics show points lying approximately on lines.

```
In[4]:= Show[GraphicsArray[
        ListPlot[N[{First[#], Log @ Length[#]}]& /@ Split[Sort[# /@ data]],
            PlotRange -> All, DisplayFunction -> Identity]& /@
                                                      {First, Last}]]
```

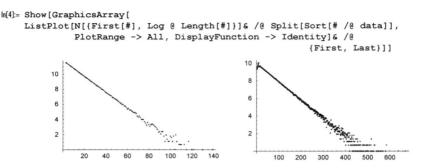

27. Parking Cars, Causal Network, Seceder Model, Run Lengths, Cycles in Random Permutations, Iterated Inner Points, Exchange Shuffling, Frog Model, Second Arcsine Law, Average Brownian Excursion Shape

a) We start with a straightforward implementation of the simulation.

The following routine `parkACar` parks a car into one of the free parking spaces `freeSpaces`. This is done by first calculating the cumulative length of all free spaces, then choosing a certain parking space, and then choosing a random position inside the chosen one. If the two so originating parking spaces before and after the newly parked car are long enough to hold another car, they are added to the list of free spaces.

```
In[1]:= parkACar[{freeSpaces_, parkedCars_}] :=
        Module[{freeParkingSpacesLengths, newCarPosition, lotNumber,
                lot, carRearPosition, newRearSpace, newFrontSpace},
            (* the length of the available parking space *)
            freeParkingSpacesLengths = FoldList[Plus, 0,
                                              -Apply[Subtract, freeSpaces, {1}]];
            (* randomly park in *)
            newCarPosition = Random[Real, {0, freeParkingSpacesLengths[[-1]]}];
            (* which lot was chosen *)
            lotNumber = Position[freeParkingSpacesLengths,
                              _?(newCarPosition > #&), {1}, 1,
                              Heads -> False][[1, 1]];
            lot = freeSpaces[[lotNumber]];
            (* randomly park inside the chosen lot *)
            carRearPosition = Random[Real, lot - {0, 1}];
            (* the new space before and after the just parked car *)
            newRearSpace = {lot[[1]], carRearPosition};
            newFrontSpace = {carRearPosition + 1, lot[[2]]};
            {Join[Delete[freeSpaces, lotNumber],
                    (* are the new spaces large enough to park another car? *)
                    If[-Subtract @@ newRearSpace  > 1, {newRearSpace }, {}],
                    If[-Subtract @@ newFrontSpace > 1, {newFrontSpace}, {}]],
              Join[parkedCars, {carRearPosition + {0, 1}}]}]]
```

Here is the first car parked in a parking lot of length 100.

```
In[2]:= SeedRandom[1999];
        parkACar[{{{0, 100}}, {}}]
Out[3]= {{{0, 61.3773}, {62.3773, 100}}, {{61.3773, 62.3773}}}
```

This parks the second car in this lot.

```
In[4]:= parkACar[%]
Out[4]= {{{62.3773, 100}, {0, 28.5319}, {29.5319, 61.3773}},
        {{61.3773, 62.3773}, {28.5319, 29.5319}}}
```

The function `fillParkingSpace` parks cars in a lot of length *l* until the lot is filled and the function returns a list of the positions (in the form of rear and front end coordinates) of the parked cars.

```
In[5]:= fillParkingSpace[length_] :=
        Module[{p = {{{0, length}}, {}}},
              While[p[[1]] =!= {}, p = parkACar[p]]; Last[p]]
```

Here is a lot of length 100 filled. 78 cars fit in (in this special run).

```
In[6]:= Length @ fillParkingSpace[100]

Out[6]= 78
```

The following graphic shows 50 slots, each of length 50 filled with cars.

```
In[7]:= Show[Graphics[{Thickness[0.002],
          MapIndexed[Function[{x, y}, (* cars as lines *)
                {Line[{{x[[1]], y[[1]]}, {x[[2]], y[[1]]}}],
                 Line[{{x[[1]], y[[1]] - 1/4}, {x[[1]], y[[1]] + 1/4}}],
                 Line[{{x[[2]], y[[1]] - 1/4}, {x[[2]], y[[1]] + 1/4}}]}],
                Table[fillParkingSpace[50], {50}],
                {2}]}], PlotRange -> All, Frame -> True]
```

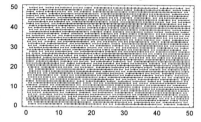

Now, let us compare with the theoretical result. We fill 100 lots, each of length 100.

```
In[8]:= tab100 = Table[Length[fillParkingSpace[100]], {100}];
```

```
In[9]:= ListPlot[{#, Count[tab100, #]}& /@ Union[tab100],
           Frame -> True, PlotRange -> All,
           PlotStyle -> {PointSize[0.01]}, Axes -> False]
```

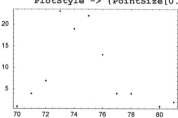

On average, we were able to fit the following number of cars.

```
In[10]:= Plus @@ tab100/100 // N

Out[10]= 74.43
```

The constant c is given by carrying out the numerical integration [1227].

```
In[11]:= c = NIntegrate[Evaluate[Exp[-2 *
              Integrate[(1 - Exp[-u])/u, {u, 0, t}]]], {t, 0, Infinity}] // Re

Out[11]= 0.747598
```

For a parking lot of length 100, this leads to the following expectation value (correct in the order 10^{-2}).

```
In[12]:= c 100 - (1 - c)

Out[12]= 74.5074
```

This is in good agreement with the above calculation.

Here are 10 parking lots of length 1000 filled.

```
In[13]:= Plus @@ Table[Length[fillParkingSpace[1000]], {10}]/10 // N

Out[13]= 745.2
```

This compares well with the theoretical prediction.

```
In[14]:= c 1000 - (1 - c)
Out[14]= 747.346
```

Now, let us implement a version of the simulation that can be compiled using `Compile`. In its current form, `parkACar` cannot be properly compiled.

```
In[15]:= (* suppress messages *)
        Internal`DeactivateMessages[
        cfTest = Function[code,
                Compile[{{freeSpaces, _Real, 2}, {parkedCars, _Real, 2}}, code],
                {HoldAll}] @@ Extract[DownValues[parkACar], {1, 2}, Hold];]
```

Here is a list of the pieces that were not successfully compiled.

```
In[17]:= DeleteCases[Union[Flatten[cfTest[[4]]]], _Integer | _Real, {1}]
Out[17]= {Apply, Block, Greater, Join, List, lotNumber, Module, newCarPosition, newFrontSpace,
         newRearSpace, Position, Random, Compile`Variable$22312, Compile`Variable$22313,
         Compile`Variable$22314, Function[{freeSpaces, parkedCars}, Real],
         Function[{freeSpaces, parkedCars}, Subtract],
         Function[{freeSpaces, parkedCars}, Subtract @@ newFrontSpace],
         Function[{freeSpaces, parkedCars}, Subtract @@ newRearSpace],
         Function[{freeSpaces, parkedCars}, Delete[freeSpaces, lotNumber]],
         Function[{freeSpaces, parkedCars},
          If[-Subtract @@ newFrontSpace > 1, {newFrontSpace}, {}]], Function[
          {freeSpaces, parkedCars}, If[-Subtract @@ newRearSpace > 1, {newRearSpace}, {}]],
         Function[{freeSpaces, parkedCars}, freeSpaces[[lotNumber]]],
         Function[{freeSpaces, parkedCars}, _ ? (newCarPosition > #1 &)],
         Function[{freeSpaces, parkedCars}, Heads → False],
         Function[{freeSpaces, parkedCars}, Compile`Variable$22312 =
           Compile`Variable$22312 + Compile`Variable$22313[[Compile`Variable$22314]]]}
```

We consider the above-generated warning messages (because of some currently uncompilable statements like `Subtract` `@@ ...`) and rewrite the corresponding code segments in such a way that they can be compiled. We also incorporate the filling process in the code to be compiled.

```
In[18]:= fillParkingLotCompiled =
        Compile[{{lotLength, _Real, 0}},
         Module[{freeParkingSpacesLengths, newCarPosition, lotNumber, lot,
                 carRearPosition, newRearSpace, newFrontSpace, freeSpaces,
                 parkedCars, newSpot, range, allLotsFilledQ = False},
           freeSpaces = {{0., lotLength}};
           (* a dummy first element in the list parkedCars for
              determining the type of parkedCars *)
           parkedCars = {{0., 0.}};
           While[Not[allLotsFilledQ],
              (* the length of the available parking space *)
           freeParkingSpacesLengths = FoldList[Plus, 0,
                   Apply[Plus, {-1, 1}#& /@ freeSpaces, {1}]];
           (* randomly park in *)
           newCarPosition = Random[Real, {0, freeParkingSpacesLengths[[-1]]}];
           (* which lot was chosen *)
           lotNumber = 1;
           While[freeParkingSpacesLengths[[lotNumber]] < newCarPosition,
                 lotNumber = lotNumber + 1];
           lotNumber = lotNumber - 1;
           lot = freeSpaces[[lotNumber]];
           (* randomly park inside the chosen lot *)
           range = lot - {0., 1.};
           carRearPosition = range[[1]] + (range[[2]] - range[[1]]) Random[];
           (* the new spaces before and after the just parked car *)
           newRearSpace = {lot[[1]], carRearPosition};
           newFrontSpace = {carRearPosition + 1, lot[[2]]};
           (* update the free spaces *)
```

```
If[newRearSpace[[2]] - newRearSpace[[1]] > 1.,
   AppendTo[freeSpaces, newRearSpace];];
If[newFrontSpace[[2]] - newFrontSpace[[1]] > 1.,
   AppendTo[freeSpaces, newFrontSpace];];
If[Length[freeSpaces] =!= 1,
   freeSpaces = Delete[freeSpaces, lotNumber],
   allLotsFilledQ = True];
(* update the parked cars *)
newSpot = carRearPosition + {0., 1.};
parkedCars = Append[parkedCars, newSpot]];
(* return the positions of the parked cars *)
Rest[parkedCars]]];
```

Looking at the compiled code, we see no Function inside; this means the compilation was complete.

```
In[19]:= Union[Head /@ DeleteCases[Flatten[fillParkingLotCompiled[[4]]],
                                   True | False]]

Out[19]= {Integer, Real}
```

Now, let us compare timing between the compiled and the uncompiled version, [1248].

```
In[20]:= Timing[Table[Length @ fillParkingSpace[100], {10}]]

Out[20]= {0.1 Second, {76, 71, 74, 72, 74, 77, 73, 71, 76, 76}}
```

```
In[21]:= Timing[Table[Length[fillParkingLotCompiled[100.]], {10}]]

Out[21]= {0.02 Second, {76, 73, 71, 75, 77, 76, 76, 76, 75, 76}}
```

The next input fills 10^4 parking lots with 100 cars and displays the resulting frequencies.

```
In[22]:= ListPlot[{First[#], Length[#]}& /@ Split[Sort[
            Table[Length[fillParkingLotCompiled[100.]], {10^4}]]],
            PlotRange -> All, Axes -> False, Frame -> True]
```

Because of this speed increase, we can now afford to fill a parking lot of length 10000.

```
In[23]:= Length[fillParkingLotCompiled[10000.]] // Timing

Out[23]= {16.95 Second, 7448}
```

For the detailed statistics of parked cars, see [521], [1509]. Concerning the application of this model for glassy behavior, see [1025], [1744], [1746], and [1745]. For a discretized version of this model, see [1001]. For the parking problem in given lots and on a 2D parking lot, see [1556], [1470], [759], and [1827]; for the selection of the optimal lot, see [1747]. For the parking of spinning cars, see [879]. For parking caravans, see [200].

b) A first attempt would be a straightforward, rule-based implementation of $\psi_n^m = 1/(\psi_{n-2}^m \, \psi_{n-1}^{m-1} \, \psi_{n-1}^m \, \psi_{n-1}^{m+1})$.

```
In[1]:= ψ[0, m_] = 1; ψ[1, m_] = 1; ψ[0, 0] = -1;
        ψ[n_, m_] := ψ[n, m] =
          1/(ψ[n - 1, m] ψ[n - 1, m + 1] ψ[n - 1, m - 1] ψ[n - 2, m])
```

Generating the ψ_n^m this way is relatively slow.

```
In[3]:= data[o_] := Table[ψ[n, k], {n, 0, o}, {k, -o, o}]
```

Visualizing the data shows that for $m \geq n + 1$ we have $\psi_n^m = 1$.

```
In[4]:= (aux = data[100]); // Timing
        ListDensityPlot[aux, Mesh -> False, AspectRatio -> Automatic]
```

Out[4]= {0.51 Second, Null}

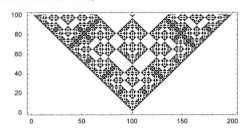

After a moment of inspection of $\psi_n^m = 1/(\psi_{n-2}^m \, \psi_{n-1}^{m-1} \, \psi_{n-1}^m \, \psi_{n-1}^{m+1})$, we translate this double recursion in a recursion with respect to n, where at each step all m are calculated at once. For a fixed n, we will store the m values in a packed array.

```
In[6]:= dataPacked[o_] :=
    Module[{lm2, lm1, new},
       {lm2, lm1} = Developer`ToPackedArray /@ Table[ψ[i, k], {i, 0, 1}, {k, -o, o}];
       Join[{lm2}, {lm1},
          Table[new = lm1 RotateRight[lm1] RotateLeft[lm1] lm2;
             lm2 = lm1; lm1 = new; new, {o}]]]
```

Now the calculation up to $n = 1000$ is quite fast.

```
In[7]:= dataPacked[1000]; // Timing
```

Out[7]= {0.09 Second, Null}

The next graphic shows the resulting pattern after 500 steps.

```
In[8]:= Show[Graphics[Rectangle[# - 1/2, # + 1/2]& /@
                   Reverse /@ Position[dataPacked[500], -1]],
          AspectRatio -> Automatic]
```

c) The implementation of an update step is straightforward. We use `Compile` to run the resulting model as fast as possible.

```
In[1]:= cf = Compile[{{l, _Real, 1}},
    Module[{list = l, λ = Length[l], i1, i2, i3,
            s1, s2, s3, Fs, maxF, parent, offspring},
    Do[(* select three random elements from the list *)
       {i1, i2, i3} = Table[Random[Integer, {1, λ}], {3}];
       {s1, s2, s3} = l[[{i1, i2, i3}]];
       (* difference to average *)
       Fs = Abs[{s1, s2, s3} - (s1 + s2 + s3)/3.];
       (* select element most distant to average *)
       maxF = Max[Fs];
       parent = Which[maxF == Fs[[1]], s1, maxF == Fs[[2]], s2, True, s3];
       (* add random number *)
       offspring = parent + Random[Real, {-1, 1}];
       (* replace randomly chosen list element *)
       list[[Random[Integer, {1, λ}]]] = offspring,
       (* repeat λ times *) {λ}];
    list]];
```

The compilation was successful.

```
In[2]:= Union[Head /@ Flatten[cf[[4]]]]
Out[2]= {Integer, Real}
```

Here is a quick test of `cf`. We start with a list of length 250 and carry out 500 update steps.

```
In[3]:= nl = With[{steps = 500, members = 250},
            NestList[cf, Array[0.&, {members}], steps]];
```

Here are the resulting data visualized.

```
In[4]:= Show[Graphics[{PointSize[0.003],
            MapIndexed[Point[{#2[[1]], #1}]&, nl, {2}]}],
        Frame -> True, PlotRange -> All]
```

Running the model with a starting list of length 1000 and for 1000 update steps results in one million data points. To avoid the generation of a graphics with one million points, we identify very nearby points. This reduces the number of points by about 90%.

```
In[5]:= steps = 1000; members = 1000;
        {l, L} = {Table[{C}, {steps}], Array[0.&, {members}]};
        Do[L = cf[L]; l[[k]] = Union[Round[L]], {k, steps}] // Timing
Out[7]= {5.32 Second, Null}
```

```
In[8]:= Length[Flatten[l]]
Out[8]= 102739
```

```
In[9]:= Show[Graphics[{PointSize[0.003],
                MapIndexed[Point[{#2[[1]], #1}]&, l, {2}]}],
        Frame -> True, PlotRange -> All]
```

A list length of 10000 and 10000 update steps would produce 10^8 data points. 10% of these points are still too many for a visualization. So, we must further reduce the data to be visualized. Instead of using points, we will use vertical lines for clustered points. This is possible because the points tend be present in a small number of clusters. Estimating that $\max|\pm l| \approx 10^3$ for this simulation and aiming for a vertical resolution of about 700 points we split all vertical points that have a larger distance than 3. The time needed to carry out this $10^4 \times 10^4$ modeling will be about 100 larger than for the $10^3 \times 10^3$ modeling (ten times as many update steps and each update step is ten times larger; the list length itself does not matter for taking three elements). Because the `Split[Sort[L], (Abs[#1 - #2] < 3)&]` has to unpack the list and carry out 10^4 comparisons, the modeling and animation will take about 200 times more time than the above $10^3 \times 10^3$ modeling.

```
steps = 10000; members = 10000;
SeedRandom[123454321];

L = Array[0.&, {members}]; l = Table[{C}, {steps}];
Do[L = cf[L]; new = Line[{{k, First[#]}, {k, Last[#]}}]& /@
                          Split[Sort[L], (Abs[#1 - #2] < 3)&];
   l[[k]] = new, {k, steps}] // Timing

Show[Graphics[l]]
```

d) To do the calculation quickly, we will use Compile for the relevant functions. The following routine implements the generation of a random permutation of the numbers 1, 2, 3, ..., $n - 1$, n.

```
In[1]:= randomPermutation :=
    Compile[{{n, _Integer}},
       Module[{l = Range[n], tmp1, tmp2, j},
          Do[(* randomly swap elements *)
             tmp1 = l[[i]];
             j = Random[Integer, {i, n}];
             tmp2 = l[[j]];
             l[[i]] = tmp2; l[[j]] = tmp1, {i, Length[l]}]; l]]
```

The function runDistribution analyzes a random permutation and counts the number of runs it contains.

```
In[2]:= runDistribution :=
    Compile[{{l, _Integer, 1}},
    Module[{del, i0 = 1, i, left, right, maxI = 1, count, n = Length[l]},
             (* counter for runs of length l *) count = Table[0, {n}];
             While[i0 < n,
                 i = i0;
                 (* still a run? *)
                 While[i < n && (left = l[[i]]; right = l[[i + 1]];
                                 left < right), i++];
                 (* length of the run *) del = i - i0 + 1;
                 count[[del]] = count[[del]] + 1; i++; i0 = i];
         Do[If[count[[i]] =!= 0, maxI = i;], {i, n}];
             (* return result in form {runLength, numberOfRuns} *)
         Rest[MapIndexed[{#2[[1]], #1}&, Take[count, maxI]]]]]
```

Here is a simple test for the function runDistribution.

```
In[3]:= runDistribution[
    {1, 2, 3, 4, 5, 6, 1, 2, 3, 4, 5, 1, 2, 3, 4, 1, 2, 3, 1, 2, 3}]
Out[3]= {{2, 0}, {3, 2}, {4, 1}, {5, 1}, {6, 1}}
```

The above two compiled functions randomPermutation and runDistribution enable a fast generation and analysis of a random permutation of length 10000.

```
In[4]:= runDistribution[randomPermutation[10000]] // Timing
Out[4]= {0.01 Second, {{2, 2063}, {3, 929}, {4, 270}, {5, 60}, {6, 5}, {7, 1}}}

In[5]:= SeedRandom[1111]
```

Here are 1000 random permutations analyzed.

```
In[6]:= With[{n = 1000.},
        {#[[1, 1]], 1/n Plus @@ Last /@ #}& /@
         Split[Sort[Flatten[Table[runDistribution[randomPermutation[1000]], {n}],
           1]], #1[[1]] == #2[[1]]&]]
Out[6]= {{2, 207.194}, {3, 92.22}, {4, 26.217},
         {5, 5.821}, {6, 0.966}, {7, 0.156}, {8, 0.02}, {9, 0.003}}
```

For a detailed analysis of the distribution of runs of length *l* in a random distribution of length *n*, see [1633]. For the (much) more complicated issue of the distribution of increasing subsequences, see [108], [1782], [109], and [880].

e) To carry out this calculation in a reasonable amount of time we must care about efficiency: First, for a good complexity, we want the generation and the analysis of the permutations to be $O(n)$. Second, for a quick calculation, we will use Compile. For the fast creation of a random permutation, we use the function RandomPermutation from the package DiscreteMath`Combinatorica`.

```
In[1]:= Needs["DiscreteMath`Combinatorica`"]
```

```
In[2]:= RandomPermutation[10^5]; // Timing
```

```
Out[2]= {0.08 Second, Null}
```

DiscreteMath`Permutations` provides the function ToCycles to decompose a given permutation into cycles. It is quite fast.

```
In[3]:= Table[l = RandomPermutation[2^k];
         Chop @ Timing[ToCycles[l];][[1]]/Second, {k, 16}]
Out[3]= {0, 0, 0, 0, 0, 0, 0, 0.01, 0, 0.02, 0.02, 0.05, 0.1, 0.21, 0.43, 0.87}
```

Now, we have all of the functions to carry out the required calculation in a few seconds on a 2 GHz computer.

```
In[4]:= SeedRandom[123];
        (data1 = Table[Length @ ToCycles[RandomPermutation[1000]], {1000}];) // Timing
Out[5]= {14.04 Second, Null}
```

Next, we implement the generation of the random permutation through a compiled function randomPermutation.

```
In[6]:= randomPermutation :=
        Compile[{{n, _Integer}},
        Module[{lTemp = Range[n], l = n, tmp1, tmp2, j},
             Do[tmp1 = lTemp[[i]];
                j = Random[Integer, {i, l}]; tmp2 = lTemp[[j]];
                lTemp[[i]] = tmp2; lTemp[[j]] = tmp1, {i, l}];
             lTemp]]
```

We also implement the counting of the number of cycles through a compiled function.

```
In[7]:= numberOfCycles =
        Compile[{{l, _Integer, 1}},
                Module[{(* working list *)
                        lTemp = Table[0, {Length[l]}],
                        pos = 0, counter = 0, start, n, n1,
                        L = Length[l]},
                        While[(* go through the list *)
                              pos++; pos <= L,
                              (* new cycle? *)
                              If[lTemp[[pos]] === 0,
                                 (* follow through the cycle *)
                                 lTemp[[pos]] = 1;
                                 start = l[[pos]];
                                 n = start;
                                 While[n1 = l[[n]]; n1 =!= start,
                                       lTemp[[n]] = 1; n = n1];
                                 counter++]];
                        (* return number of cycles *) counter]];
```

The function numberOfCycles calculates results that agree with ToCycles.

```
In[8]:= SeedRandom[123];
       data1 === Table[numberOfCycles[RandomPermutation[1000]], {1000}]
Out[9]= True
```

Calculating now the number of cycles can be done about ten times as fast as above.

```
In[10]:= SeedRandom[123];
        (data2 = Table[numberOfCycles[randomPermutation[1000]], {1000}];) // Timing
Out[11]= {1.47 Second, Null}
```

We now generate 100000 random permutations and count the number of cycles.

```
In[12]:= SeedRandom[123];
        (data3 = Table[numberOfCycles[randomPermutation[1000]], {100000}];) // Timing
Out[13]= {147.35 Second, Null}
```

The following plot shows the distribution of the number of cycles.

```
In[14]:= ListPlot[{First[#], Length[#]}& /@ Split[Sort[data3]], PlotJoined -> True]
```

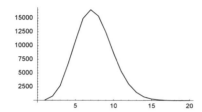

The average number of cycles is about 7.5 (the expected number of cycles is log(*n*) + *O*(1) [1902]).

```
In[15]:= Mean[data3] // N
Out[15]= 7.48879
```

f) We start by implementing a function `patternFreeQ` that generates a compiled function. This compiled function tests if a given permutation is free of the pattern {α, β, γ, δ}. It does this by checking all 4-tuples of a permutation. For efficiency, we use `Or` to stop after a matching 4-tuple was found.

```
In[1]:= patternFreeQ[{α_, β_, γ_, δ_}, n_] := patternFreeQ[{α, β, γ, δ}, n] =
       Module[{a, b, c, d},
        (* order of the pattern integers *)
        {a, b, c, d} = Last /@ Sort[Transpose[{{α, β, γ, δ}, Range[4]}]];
        (* make compiled function *)
        Compile[{{σ, _Integer, 1}},
         (* check all sub-4-tuples of the permutation *)
         Evaluate[(* relative order of the pattern integers *)
         Or @@ ((#[[a]] < #[[b]] < #[[c]] < #[[d]])& /@
         Flatten[Table[{σ[[i]], σ[[j]], σ[[k]], σ[[l]]},
                  {i, n}, {j, i + 1, n}, {k, j + 1, n}, {l, k + 1, n}], 3])]]];
```

Here are two examples.

```
In[2]:= (* turn off Part message coming from the symbolic
           evaluation of the Compile body *) Off[Part::partd];
       {patternFreeQ[{1, 2, 3, 4}, 10][Range[10]],
        patternFreeQ[{1, 2, 3, 4}, 10][Reverse @ Range[10]]}
Out[3]= {True, False}
```

The function `patternFreeQ[{1, 2, 3, 4}, 10]` compiled successfully.

```
In[4]:= DeleteCases[Flatten[patternFreeQ[{1, 2, 3, 4}, 10][[4]]],
                    True] // (Head /@ #)& // Union
Out[4]= {Integer}
```

Now, it is straightforward to check all permutations for a given n. For a faster calculation, we pack the permutations returned by `Permutations`.

```
In[5]:= Table[
          perms = Developer`ToPackedArray @ Permutations[Range[n]];
          {#[[1, 1]], Last /@ #}& /@
          Split[Sort[{Count[patternFreeQ[#, n] /@ perms, False], #}& /@
              Permutations[Range[4]]], #1[[1]] === #2[[1]]&], {n, 7, 8}]
Out[5]= {{{2740, {{1, 3, 4, 2}, {1, 4, 2, 3}, {2, 3, 1, 4}, {2, 4, 1, 3}, {2, 4, 3, 1},
            {3, 1, 2, 4}, {3, 1, 4, 2}, {3, 2, 4, 1}, {4, 1, 3, 2}, {4, 2, 1, 3}}}},
          {2761, {{1, 2, 3, 4}, {1, 2, 4, 3}, {1, 4, 3, 2}, {2, 1, 3, 4}, {2, 1, 4, 3}, {2, 3, 4, 1},
            {3, 2, 1, 4}, {3, 4, 1, 2}, {3, 4, 2, 1}, {4, 1, 2, 3}, {4, 3, 1, 2}, {4, 3, 2, 1}}}},
          {2762, {{1, 3, 2, 4}, {4, 2, 3, 1}}}}},
          {{15485, {{1, 3, 4, 2}, {1, 4, 2, 3}, {2, 3, 1, 4}, {2, 4, 1, 3}, {2, 4, 3, 1},
            {3, 1, 2, 4}, {3, 1, 4, 2}, {3, 2, 4, 1}, {4, 1, 3, 2}, {4, 2, 1, 3}}}},
          {15767, {{1, 2, 3, 4}, {1, 2, 4, 3}, {1, 4, 3, 2}, {2, 1, 3, 4}, {2, 1, 4, 3}, {2, 3, 4, 1},
            {3, 2, 1, 4}, {3, 4, 1, 2}, {3, 4, 2, 1}, {4, 1, 2, 3}, {4, 3, 1, 2}, {4, 3, 2, 1}}}},
          {15793, {{1, 3, 2, 4}, {4, 2, 3, 1}}}}}}
```

For the $n = 10$ case, we proceed slightly different. While generating all 10! permutations with `Permutations[Range[10]]` works, it uses about 400 MB of memory. We can reduce the amount of memory needed by packing the permutations (the result of `Permutations` is not packed by default) and generating the permutations not all at once, but, for instance, by generating the 9! permutations of $\{1, 2, ..., 9\}$ and adding the 10 to subsets of these permutations in all 10 possible positions. The function `insert10s` does this.

```
In[6]:= prePermutations = Developer`ToPackedArray @
                            Permutations[Range[9]]; 1;
In[7]:= insert10s = Compile[{{l, _Integer, 2}},
                    Flatten[Table[Insert[l[[k]], 10, j],
                    {k, Length[l]}, {j, 10}], 1]];
```

The function `countPatternFreePermutations10` can now be used to calculate the permutations of $\{1, 2, ..., 9\}$ that are free of the pattern $\{1, 2, 3, 4\}$ in about 8 minutes on a 2 GHz computer using a meager 50 MB of memory.

```
In[8]:= countPatternFreePermutations10[pattern_] :=
        Module[{Λ, sum = 0},
          Do[Λ = insert10s[Take[prePermutations, {k 8! + 1, (k + 1) 8!}]];
            (* to watch progress: Print @ *) Timing[
            Do[sum = sum + If[patternFreeQ[pattern, 10][Λ[[j]]] == True, 0, 1],
              {j, 10 8!}]], {k, 0, 8}];
          sum]
In[9]:= countPatternFreePermutations10[{1, 2, 3, 4}]
Out[9]= 586590
In[10]:= MaxMemoryUsed[]/10.^6 "MB"
Out[10]= 51.2825 MB
```

A moment of reflection shows that of the 24 possible patterns, only three are really different (as was already visible in the above $n = 7$ and $n = 8$ examples). Here are the counts for the remaining two for $n = 10$.

```
In[11]:= countPatternFreePermutations10 /@ {{1, 3, 4, 2}, {1, 3, 2, 4}}
Out[11]= {555662, 591950}
```

For more on pattern-avoiding permutations, see [236], [1213], [1214], [1903], [235], [87], [545], [645], [321].

g) Given a permutation l, the function `cutSequenceSpecifications` will calculate a list of pairs, each pair being the starting and ending position of a sublist of the cut sequence. This process can be compiled.

```
In[1]:= cutSequenceSpecifications =
        Compile[{{l, _Integer, 1}},
        Module[{λ = Length[l], c = 1, min = Length[l], nMin, cutl, r},
        cutl = Table[0, {λ}];
        (* go through the list; start at the end *)
        Table[nMin = Min[min, l[[k]]];
            If[nMin =!= min, min = nMin; cutl[[c++]] = k, 0], {k, λ, 1, -1}];
```

```
(* fix beginning *)
r = Drop[Take[cutl, c], -1]; If[Last[l] =!= λ, r = Rest[r]];
(* sequence specifications *)
{#[[1]] + 1, #[[2]]}& /@
                    Partition[Join[{0}, Reverse[r], {Length[l]}], 2, 1]]];
```

The compilation was successful.

In[2]:= `Union[Head /@ Flatten[cutSequenceSpecifications[[4]]]]`

Out[2]= `{Integer}`

Here is an example.

In[3]:= `cutSequenceSpecifications[{5, 4, 1, 3, 2, 6, 7}]`

Out[3]= `{{1, 3}, {4, 5}, {6, 6}, {7, 7}}`

Given the sequence specification of the sublists of the cut sequence, it is straightforward to calculate the cut sequence itself.

In[4]:= `CutSequence[l_] := Take[l, #]& /@ cutSequenceSpecifications[l]`

Here are three examples of cut sequences.

In[5]:= `CutSequence[{5, 4, 1, 3, 2, 6, 7}]`

Out[5]= `{{5, 4, 1}, {3, 2}, {6}, {7}}`

In[6]:= `CutSequence[{1, 2, 6, 7, 4, 5, 3}]`

Out[6]= `{{1}, {2}, {6, 7, 4, 5, 3}}`

In[7]:= `CutSequence[{1, 2, 3, 4, 5, 6, 7}]`

Out[7]= `{{1}, {2}, {3}, {4}, {5}, {6}, {7}}`

We calculate the length of the subsequences for 10000 cut sequences of random permutations of 1000 integers.

```
In[8]:= randomPermutation :=
    Compile[{{n, _Integer}},
      Module[{l = Range[n], tmp1, tmp2, j},
        Do[tmp1 = l[[i]]; j = Random[Integer, {i, n}]; tmp2 = l[[j]];
           l[[i]] = tmp2; l[[j]] = tmp1, {i, Length[l]}]; l]]
In[9]:= data = Table[l - Apply[Subtract, cutSequenceSpecifications[
                               randomPermutation[1000]], {1}],
                {10000}];
```

In average, we have $7\frac{1}{2}$ subsequences.

In[10]:= `Plus @@ #/Length[#]&[Length /@ data] // N`

Out[10]= `7.504`

Here is a plot of the frequency distribution of the number of subsequences.

```
In[11]:= ListPlot[data1 = {First[#], Length[#]}& /@ Split[Sort[Length /@ data]],
            PlotRange -> All, PlotJoined -> True, AxesOrigin -> {0, 0},
            Epilog -> {PointSize[0.02], Point /@ data1}]
```

h) For an efficient manipulation of the 0-matrix, we implement a compiled version that checks the neighbors of all elements and performs the 0 → 1 flip. We will use periodic boundary conditions here.

```
In[1]:= cf = Compile[{{t, _Real, 2}},
            Module[{n = Length[t], kp, kn, lp, ln},
                Table[Which[k == 1, kp = k + 1; kn = n,
                            k == n, kp = 1; kn = k - 1,
                            True,   kp = k + 1; kn = k - 1];
                      Which[l == 1, lp = l + 1; ln = n,
                            l == n, lp = 1; ln = l - 1,
                            True,   lp = l + 1; ln = l - 1];
                      If[t[[kp, l]] == 0 && t[[kn, l]] == 0 &&
                         t[[k, lp]] == 0 && t[[k, ln]] == 0, 0, 1],
                      {k, n}, {l, n}]]];
```

To monitor how the 0's evolve into 1's we will use an additional matrix **T** whose elements count after how many iterations the corresponding element of the matrix t became 1.

```
In[2]:= switch = Compile[{{t, _Real, 2}, {T, _Integer, 2}, {c, _Integer}},
                    Module[{n = Length[t], T1 = T},
                        Do[If[t[[k, l]] == 0, T1[[k, l]] = c],
                           {k, n}, {l, n}]; T1]];
```

The function `transitionMatrix` generates the just-described **T** matrix for an initial $n \times n$ matrix that has (on average) $p\%$ 0's.

```
In[3]:= transitionMatrix[n_, p_] :=
    Module[{(* initial random 0-1 matrix *)
            t = Table[If[Random[] < p, 0, 1], {n}, {n}],
            T = Table[0, {n}, {n}], c = 0},
        While[t = cf[t]; c++; T = switch[t, T, c];
        MemberQ[t, 0, {2}], Null]; T]
```

Now, we calculate the iterated inner points for $p = 90$, $p = 99$, and $p = 99.9$. For each percentage, we do two simulations and return the number of iterations needed.

```
In[4]:= pData = Function[p, Table[transitionMatrix[256, p], {3}]] /@
                    {0.9, 0.99, 0.999};
```

```
In[5]:= Show[GraphicsArray[
    ListDensityPlot[#, Mesh -> False, FrameTicks -> None,
                    ColorFunction -> GrayLevel,
                    DisplayFunction -> Identity]& /@ #]]& /@ pData
```

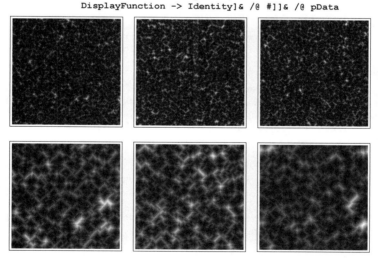

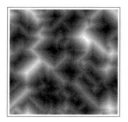 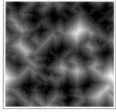 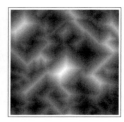

```
In[6]:= Map[Max, pData, {2}]
Out[6]= {{9, 8, 8}, {26, 25, 32}, {72, 77, 86}}
```

For some applications of such pictures in physics see [447], [396].

i) shuffleCF will carry out *o* complete exchange shuffles of list of length *n*. We use Compile for shuffling quickly.

```
In[1]:= shuffleCF = Compile[{{n, _Integer}, {o, _Integer}},
        Module[{A = Range[n], a = Range[n], l},
            Table[(* starting list of integers *) a = A;
                (* do the shuffling *)
                Do[l = Random[Integer, {1, n}];
                    {a[[k]], a[[l]]} = {a[[l]], a[[k]]}, {k, n}]; a, {o}]]];
```

shuffleFrequencies and shuffleData will analyze the frequencies of the permutations found by shuffleCF. It returns lists showing how the frequencies developed as more runs were carried out.

```
In[2]:= shuffleData[n_, o_, p_] :=
        Table[{First[#], Length[#]}& /@ Split[Sort[shuffleCF[n, o]]], {p}];
```

```
In[3]:= shuffleFrequencies[data_] :=
        Module[{allOccurringPermutations, c, data1, n, frequencies},
        (* all occurring permutations *)
        allOccurringPermutations = Union[Level[data, {-2}]];
        n = Plus @@ (Last /@ data[[1]]);
        (* supplement with zeros *)
        data1 = (c = Complement[allOccurringPermutations, First /@ #];
                Sort[If[c =!= {}, Join[#, {#, 0}& /@ c], #]])& /@ data;
        (* the frequencies *)
        frequencies = MapIndexed[{#2[[2]], #1/(n #2[[2]])}&, Transpose[
                    Rest[FoldList[Plus, 0, Map[Last, data1, {2}]]]], {2}];
        (* return permutations and frequencies *)
        Transpose[{Sort[Permutations[data[[1, 1, 1]]]], frequencies}]]
```

Given the result of shuffleFrequencies, the two functions mostFrequentPermutations and shuffleGraph ⸞ ics return the four most frequent permutations and show how the probabilities evolved as more runs were carried out.

```
In[4]:= mostFrequentPermutations[res_] := (First /@ Take[#, 4])& @
                Sort[res, #1[[-1, -1, -1]] > #2[[-1, -1, -1]]&]
```

```
In[5]:= shuffleGraphics[res_] :=
            Graphics[{Hue[Random[]], Line[Last[#]]}& /@ res,
                Frame -> True, PlotRange -> All];
```

Now, we carry out 10^6 shuffles for $n = 4$, $n = 5$, and $n = 6$. We see that the frequencies of the resulting permutations are quite nonuniform. We display the frequencies and return the four most common permutations.

```
In[6]:= {res1, res2, res3} =
            Table[shuffleFrequencies[shuffleData[k, 10^3, 10^3]], {k, 4, 6}];
```

```
In[7]:= Show[GraphicsArray[shuffleGraphics /@ {res1, res2, res3}]]
```

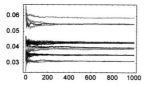 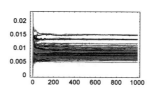 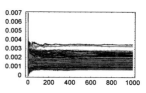

Now let us calculate the exact frequencies. There are n^n possible shuffles for a list of length n. We have to take all of them into account. Given a list of lists *ls*, the function `shuffleAll` will exchange the nth elements of all sublists of *ls* with the first n other elements. In the following, the sublists have $n + 1$ elements, the first n being the numbers to be shuffled and the last being the frequency. Having the frequency instead of the lists avoids carrying out duplicated shuffling. To do this process quickly and economically, we will use `Compile` again.

```
In[8]:= shuffleAll =
    Compile[{{ls, _Integer, 2}, {n, _Integer}},
        Module[{newDecks, λ = Length[ls[[1]]] - 1, w, c, wc, parts, L},
            (* all new configurations *)
            newDecks = Sort[Flatten[Function[l,
                    Table[L = l; (* do the shuffling *)
                        {L[[n]], L[[k]]} = {L[[k]], L[[n]]}; L,
                    {k, λ}]] /@ ls, 1]];
            (* emulate Split *)
            w = Table[0, {Length[newDecks]}];
            w[[1]] = 1; c = 1; wc = 2;
            Do[If[Drop[newDecks[[c]], -1] =!= Drop[newDecks[[c + 1]], -1],
                    w[[wc]] = c + 1; c++; wc++, c++],
                {Length[newDecks] - 1}];
            parts = ({0, -1} + #)& /@ Partition[Join[Take[w, wc - 1],
                    {Length[newDecks] + 1}], 2, 1];
            (* condense and update counts *)
            Join[Take[#[[1]], λ], {Plus @@ (Last /@ #)}]&[
                Take[newDecks, #]]& /@ parts]];
```

Here is an example.

```
In[9]:= shuffleAll[{{1, 2, 3, 4, 5, 6}}, 3]

Out[9]= {{1, 2, 3, 4, 5, 6}, {1, 2, 4, 3, 5, 6},
    {1, 2, 5, 4, 3, 6}, {1, 3, 2, 4, 5, 6}, {3, 2, 1, 4, 5, 6}}
```

Recursively using `shuffleAll` gives the theoretical frequencies of the permutations.

```
In[10]:= exactShuffleFrequencies[n_] :=
    {Drop[#, -1], Last[#]/n^n}& /@
        Sort[Fold[shuffleAll, {Append[Range[n], 1]}, Range[n]],
            #1[[-1]] > #2[[-1]]&]
```

Here are the four most frequently found permutations for $n = 4$, $n = 5$, and $n = 6$.

```
In[11]:= First /@ Take[exactShuffleFrequencies[4], 4]

Out[11]= {{2, 1, 4, 3}, {2, 3, 4, 1}, {2, 3, 1, 4}, {1, 3, 4, 2}}
```

```
In[12]:= First /@ Take[exactShuffleFrequencies[5], 4]

Out[12]= {{2, 3, 1, 5, 4}, {2, 1, 4, 5, 3}, {2, 3, 4, 5, 1}, {2, 3, 4, 1, 5}}
```

```
In[13]:= First /@ Take[exactShuffleFrequencies[6], 4]

Out[13]= {{2, 3, 1, 5, 6, 4}, {2, 3, 4, 1, 6, 5}, {2, 1, 4, 5, 6, 3}, {2, 3, 4, 5, 6, 1}}
```

For theoretical considerations for larger n, see [700], [710], [298].

We could now continue shuffling. For instance, starting with a randomly ordered list, how many times do we have to randomly exchange two elements until we reach the sorted state? For a given initial list, the function `shuffleCountUntilOrdered` counts the number of needed transpositions.

```
In[14]:= shuffleCountUntilOrdered = Compile[{{l, _Integer, 1}},
       Module[{n = Length[l], ℓ = l, counter = 0,
             L = Range[Length[l]], α, β},
             While[(* ordered form reached? *) ℓ =!= L,
                   (* do one shuffle *)
                   {α, β} = Table[Random[Integer, {1, n}], {2}];
                   {ℓ[[α]], ℓ[[β]]} = {ℓ[[β]], ℓ[[α]]}; counter++];
             counter]];
```

On average, one needs for large *n* about *n*! random transpositions [612]. But for single runs, one could need much more or much less. Here are two quite different runs for *n* = 10. The first is a minimal run and the second needs more than five million times as many transpositions.

```
In[15]:= SeedRandom[884516];
       shuffleCountUntilOrdered[Reverse[Range[10]]]

Out[16]= 5

In[17]:= SeedRandom[1826];
       shuffleCountUntilOrdered[Reverse[Range[10]]]

Out[18]= 28525731
```

For slower shuffle methods, see [912].

j) We will use periodic boundary conditions in the following (meaning that all frogs are sitting on a torus).

```
In[1]:= jumpFrogsJump =
       Compile[{{mat, _Integer, 2}},
         Module[{n = Length[mat], m, k, i, j},
               m = Table[0, {n}, {n}];
               (* jumping frogs *)
               Do[If[mat[[i, j]] =!= -1, k = mat[[i, j]];
                     Do[i = If[Random[] < 0.5, i - 1, i + 1];
                        j = If[Random[] < 0.5, j - 1, j + 1];
                        Which[i == 0, i = n, i == n + 1, i = 1];
                        Which[j == 0, j = n, j == n + 1, j = 1];
                        m[[i, j]] = m[[i, j]] + 1, {k}]],
                     {i, n}, {j, n}];
               (* add activated and still sleeping frogs *)
               Do[Which[mat[[i, j]] === -1 && m[[i, j]] === 0,
                     m[[i, j]] = -1,
                     mat[[i, j]] === -1 && m[[i, j]] =!= 0,
                     m[[i, j]] = m[[i, j]] + 1], {i, n}, {j, n}];
               (* return new frog distribution *)m]];
```

In the following, we will use a 129 × 129 lattice. The animation shows that after about 150 jumps the frogs are uniformly distributed.

```
In[2]:= makeMat[n_] := Module[{mat0 = Table[-1, {n}, {n}]},
                          mat0[[(n - 1)/2, (n - 1)/2]] = 1; mat0];

In[3]:= mat = makeMat[128 + 1];

In[4]:= (* gray level proportional number of frogs *)
       lPlot[m_, opts___] :=
       ListDensityPlot[m, opts, Mesh -> False, PlotRange -> All,
                    FrameTicks -> None, ColorFunctionScaling -> False,
                    ColorFunction -> (GrayLevel[1 - (# + 1)/12]&)]

In[6]:= Show[GraphicsArray[#]]& /@ Partition[
       Table[mat = jumpFrogsJump[mat];
             If[k == 1 || IntegerQ[k/10],
                lPlot[mat, DisplayFunction -> Identity], Sequence @@ {}],
             {k, 150}], 3]
```

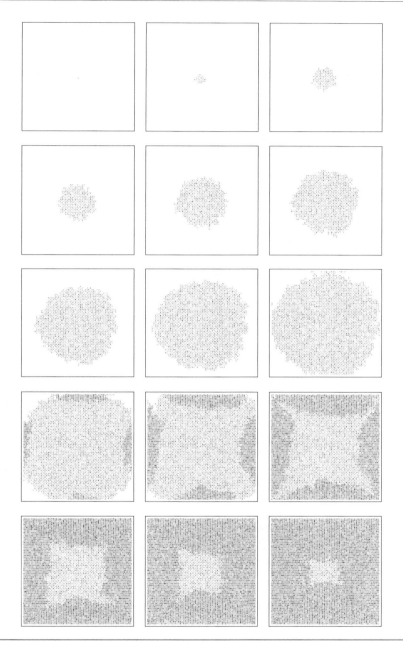

```
SeedRandom[12345];
Do[mat = jumpFrogsJump[mat]; lPlot[mat], {k, 150}]
```

Next, we will carry out 1000 further jump time steps. After each step, we analyze the distribution of frogs per site. The function frogsPerSite generates a list of occurrences of *k* frogs on a site.

```
In[7]:= frogsPerSite[m_] := With[{λ = Times @@ Dimensions[m]},
         {First[#], 1. Length[#]/λ}& /@
            Split[Sort[DeleteCases[Flatten[m], -1]]]]
```

```
In[8]:= SeedRandom[123];
        frogsPerSiteData = Table[frogsPerSite[mat = jumpFrogsJump[mat]], {1000}];
```

The next graphic shows how the distribution of frogs per site varies with time. The cumulative data are, of course, much smoother and are shown in the right graphic.

```
In[10]:= (* calculate cumulative data *)
         cumulativeData = DeleteCases[
         MapIndexed[{#2[[1]], Log[10, #1/#2[[1]]]}&, (* sum up *)
                        Rest[FoldList[Plus, 0, #]]]& /@
              Transpose[(Flatten[Table[Last /@ Cases[#, {k, _}],
                {k, 0, 10}] /. {} -> {0.}])& /@ frogsPerSiteData],
                                       {_, Indeterminate}, {2}];
```

```
In[12]:= Show[GraphicsArray[
           {(* raw data *)
            Graphics[MapIndexed[{Hue[#1[[1]]/14],
                             Point[{#2[[1]], Log[10, #1[[2]]]}]}]&,
                        frogsPerSiteData, {2}], Frame -> True],
           (* cumulative data *)
            Graphics[MapIndexed[{Hue[#2[[1]]/14], Line[#]}&,
                            cumulativeData], Frame -> True]}]]
```

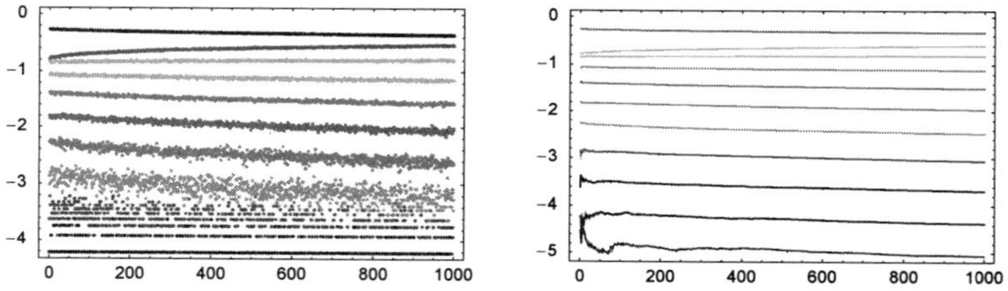

These are the so-obtained probabilities for the number of frogs per site.

```
In[13]:= MapIndexed[{#2[[1]], #1}&,
               10^Map[Last, cumulativeData, {1, 2}] // SetPrecision[#, 3]&]
```

```
Out[13]= {{1, 0.472}, {2, 0.250}, {3, 0.154}, {4, 0.0771}, {5, 0.0319}, {6, 0.0111},
        {7, 0.00333}, {8, 0.000867}, {9, 0.000204}, {10, 0.0000433}, {11, 8.59×10⁻⁶}}
```

k) The function `randomWalk1DZeroCrossingNormalizedTimesCF` is a compiled function that will carry out r random walks, each having *maxSteps* steps. After each step, the rounded value of the variable τ/T is recorded. The last argument d governs the discretization coarseness for τ/T.

```
In[1]:= randomWalk1DZeroCrossingNormalizedTimesCF =
        Compile[{{maxSteps, _Integer}, {r, _Integer}, {d, _Integer}},
        Module[{x, steps, τ0, tab = Table[0, {d}], t, sum = 0., data},
                (* carry out r random walks *)
                Do[x = 0; steps = 0; τ0 = 0;
                   (* carry out maxSteps steps *)
                   Do[steps++; x = x + 2 Random[Integer] - 1;
                   (* potentially update time origin was visited *)
                   If[x == 0, τ0 = steps];
                   (* add count to discretized counts of τ/T *)
                   t = 1 + Round[1. τ0/steps (d - 1)];
                   tab[[t]] = tab[[t]] + 1, {maxSteps}], {r}];
                (* normalized probability density *)
                sum = 1. Plus @@ tab;
                Table[{j/d + 1/2/d, d tab[[j]]/sum}, {j, d}]]];
```

The last input compiled successfully.

```
In[2]:= Union[Head /@ Flatten[randomWalk1DZeroCrossingNormalizedTimesCF[[4]]]]
```

```
Out[2]= {Integer, Real}
```

Here is the theoretical probability density.

```
In[3]:= theoreticalProbabilityDensity =
    Plot[(* density from differentiation of distribution *)
        Evaluate[D[2/Pi ArcSin[Sqrt[x]], x]], {x, 0, 1},
        PlotStyle -> {{Hue[0], Thickness[0.008]}},
        DisplayFunction -> Identity];
```

The function shows the modeling results together with the theoretical density.

```
In[4]:= arcSineLawModel[maxSteps_, r_, d_] :=
    Show[{(* theoretical curve *)
          theoreticalProbabilityDensity,
          (* data shown as points *)
          ListPlot[randomWalk1DZeroCrossingNormalizedTimesCF[
                    maxSteps, r, d],
              PlotStyle -> {PointSize[0.01], GrayLevel[0]},
              DisplayFunction -> Identity]},
        PlotRange -> {0, 3}, Axes -> False, Frame -> True]
```

Now, we carry out two simulations. In the first, we use 10 walks, each having a million steps. In the second, we use 3000 walks, each having 3000 steps. Because the maximal fluctuations are smaller in the shorter walks, the second data set fits considerably better the theoretical curve

```
In[5]:= Show[GraphicsArray[{arcSineLawModel[1000000, 10, 200],
                            arcSineLawModel[3000, 3000, 200]}]]
```

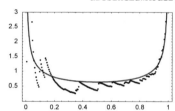 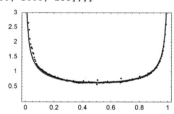

l) Because excursions will come in greatly different lengths and because of the use of a constant step size, we must interpolate the path of the Brownian particle. The function stretchedScaledInterpolatedExcursion carries out a 1D Brownian motion with constant step size until an excursion of length greater or equal to *minLength* has been found. Then it rescales and interpolates (linearly) this excursion and returns a list of *pp* positions from the excursion.

```
In[1]:= stretchedScaledInterpolatedExcursion[pp_, minLength_:1] :=
    Module[{excursion, λ, scaledStretchedExcursion, interpolatedExcursionIF},
        (* carry out Brownian motion until excursion of length minLength is found *)
        While[excursion = Abs[Last /@
            NestWhileList[{#[[1]] + 1, (#[[2]] + 2 Random[Integer] - 1)}&,
                          {0, 0}, (#[[1]] == 0 || #[[2]] =!= 0)&]];
            (λ = Length[excursion]) < minLength];
        (* rescale excursion *)
        scaledStretchedExcursion =
            MapIndexed[{(#2[[1]] - 1)/(λ - 1), #1}&, N[excursion]/Sqrt[λ]];
        (* interpolate between steps *)
        interpolatedExcursionIF = Interpolation[scaledStretchedExcursion,
                                                InterpolationOrder -> 1];
        (* rescaled and interpolated excursion *)
        Table[interpolatedExcursionIF[x], {x, 0, 1, 1/(pp - 1)}]]
```

Here are some excursions shown.

```
In[2]:= SeedRandom[1];
    Show[Graphics[
        Table[{Hue[Random[]], Line[MapIndexed[{#2[[1]], #1}&,
                stretchedScaledInterpolatedExcursion[120]]]}, {120}]],
        Frame -> True, AspectRatio -> 1/3, FrameTicks -> False,
        PlotRange -> All]
```

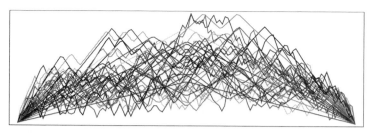

Many excursions have a very small length (three or five steps). To better model a scale-invariant Brownian motion, we next display 120 random excursions, each having at least 24 steps. The average shape becomes less pointy at $\tau = 1/2$.

```
In[4]:= SeedRandom[1];
     Show[Graphics[
         Table[{Hue[Random[]], Line[MapIndexed[{#2[[1]], #1}&,
                     stretchedScaledInterpolatedExcursion[120, 24]]]}, {120}]],
         Frame -> True, AspectRatio -> 1/3, FrameTicks -> False,
         PlotRange -> All]
```

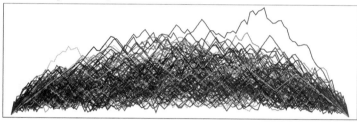

Averaging 240 such excursions shows a good agreement with the theoretical average excursion shape.

```
In[6]:= SeedRandom[1];
     Module[{pp = 120, minLength = 24, o = 240, data},
             (* average o excursions *)
             data = Sum[stretchedScaledInterpolatedExcursion[pp, minLength], {o}]/o;
             Show[Graphics[
               {(* theoretical average excursion shape in red *)
                {Hue[0], Line[Table[{τ, Sqrt[8/Pi τ (1 - τ)]}, {τ, 0, 1, 1/100}]]},
                (* modeled average excursion shape as black points *)
                {GrayLevel[0], PointSize[0.006],
                 Point /@ MapIndexed[{(#2[[1]] - 1)/(pp - 1), #1}&, data]}}] // N,
             Frame -> True, AspectRatio -> 1/3, FrameTicks -> True,
             PlotRange -> All]]
```

For the area distribution of Brownian excursions, see [1312], [966].

28. Poincaré Sections, Random Stirring, ABC-System, Vortices on a Sphere, Oscillations of a Triangular Spring Network, Lorenz System

a) Here, the differential equation under consideration is implemented.

```
In[1]:= odeSol[{{x0_, y0_, z0_}, {t0_, t1_}}, opts___] :=
     NDSolve[{x'[t] == y[t], y'[t] == -x[t] + y[t] z[t], z'[t] == 1 - y[t]^2,
```

```
                    x[t0] == x0, y[t0] == y0, z[t0] == z0},
                   {x, y, z}, {t, t0, t1}, opts,
                   MaxSteps -> 100000, MaxStepSize -> 0.01,
                   Method -> "ExplicitRungeKutta"];
```

Let us look at the solution in 3D for some randomly chosen initial conditions.

In[2]:= `nsol = odeSol[{{-1, -0.01, 0.2}, {0, 100}}]`

Out[2]= `{{x → InterpolatingFunction[{{0., 100.}}, <>],`
`        y → InterpolatingFunction[{{0., 100.}}, <>],`
`        z → InterpolatingFunction[{{0., 100.}}, <>]}}`

In[3]:= `pp3D = ParametricPlot3D[Evaluate[{x[t], y[t], z[t]} /. nsol],`
`                          {t, 0, 100}, PlotPoints -> 5000, PlotRange -> All]`

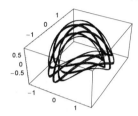

Here, $z(t)$ is shown. Every time $z(t)$ equals 0, we have to pick the corresponding x- and y-values.

In[4]:= `Plot[Evaluate[z[t] /. nsol], {t, 0, 50}, PlotRange -> All, PlotPoints -> 200]`

We start with a programming exercise and find the crossings with the x,y-plane from the interpolating functions returned from NDSolve. The following routine poincareSectionPoints calculates the points in the x,y-plane for a given solution of the differential equation.

In[5]:= `poincareSectionPoints[if_] :=`
`    Module[{ifz = z /. if[[1]], zRange, zR, tInterval, tValues},`
`        (* the t range *)`
`        zRange = ifz[[1, 1]];`
`        zR = zRange[[2]] - zRange[[1]];`
`        (* scan values at t-distance 0.1 *)`
`        scanValues = Table[{t, ifz[t]}, {t, zRange[[1]], zRange[[2]],`
`                        zR/Ceiling[zR/0.1]}];`
`        (* t-start values for root search *)`
`        tIntervals = Map[First, Select[Partition[`
`            scanValues, 2, 1], #[[1, 2]] #[[2, 2]] < 0&], {2}];`
`        (* zeros of z *)`
`        tValues = Apply[(t /. FindRoot[Evaluate[ifz[t] == 0],`
`                        {t, ##}])&, tIntervals, {1}];`
`        (* {x, y} coordinates *)`
`        Evaluate[{if[[1, 1, 2]][#], if[[1, 2, 2]][#]}]& /@ tValues]`

For the above solution, we get the following points.

In[6]:= `Short[poincareSectionPoints[nsol], 8]`

```
Out[6]//Short= {{0.462597, 1.59855}, {1.46459, 0.0251216}, {0.44656, -1.62295}, {-0.978186, 0.239918},
                {0.491053, 1.27488}, {1.17828, 0.608295}, {0.101201, -1.87552}, <<58>>,
                {0.981989, -0.351005}, {-0.377795, -1.24632}, {-1.08849, -0.6371},
                {0.0063887, 1.88722}, {1.09746, -0.637234}, {0.394986, -1.24806}}

In[7]:= Show[{pp3D, Graphics3D[{Hue[0], PointSize[0.01], Point[Append[#, 0]]& /@ %}]},
            ViewPoint -> {4, 0, 0}]
```

Now, let us collect a larger number of points.

To avoid storing large amounts of data contained in the InterpolatingFunction-objects returned by NDSolve, we solve the differential equations for time intervals of length 10 and use the results for constructing the initial conditions for the next time interval. We follow every orbit for a time length of at most 1000.

```
In[8]:= PoincareSection[numberOfPointsGoal_] :=
         Module[{points = {}, counter = 0, data, nsol, T, T = 1000, ΔT = 10},
           While[(* until enough points have been collected;
                     start a new orbit *)
                 counter < numberOfPointsGoal,
                 (* random initial conditions *)
                 data = {Table[Random[Real, {-1, 1}], {3}], {0, ΔT}};
                 nsol = 0;  T = 0;
             While[(* follow every orbit up to time T *)
                   T < T && nsol =!= $Failed &&
                   counter < numberOfPointsGoal,
                   (* check of numerical solution of ode worked fine *)
                   nsol = odeSol[data];
                   nsol = If[Max[nsol[[1, 1, 2, 1]]] - T == ΔT, nsol, $Failed];
                   If[nsol =!= $Failed,
                      (* calculate {x, y} points *)
                      newPoints = poincareSectionPoints[nsol];
                      points = {points, newPoints};
                      counter = counter + Length[newPoints];
                      (* construct new initial conditions *)
                      T = T + ΔT;
                      data = {{x[T], y[T], z[T]} /. nsol[[1]], {T, T + ΔT}},
                      Null]]];
             Partition[Flatten[points], 2]]
```

Here, the result for 10000 points is shown (this corresponds to around 35 orbits followed).

```
In[9]:= (graphic1 = Graphics[{{PointSize[0.002], Point /@ PoincareSection[10000]},
                              PlotRange -> All, Frame -> True,
                              AspectRatio -> Automatic]); // Timing
Out[9]= {119.81 Second, Null}
```

Now, we repeat a version of the last calculation and use the event detection mechanism.

```
In[10]:= poincaréData[{{x0_, y0_, z0_}, {t0_, t1_}}, opts___] :=
          (* extract crossings *) Reap[
            NDSolve[{x'[t] == y[t], y'[t] == -x[t] + y[t] z[t],
                     z'[t] == 1 - y[t]^2,
                     x[t0] == x0, y[t0] == y0, z[t0] == z0},
                     (* do not return any interpolating functions *) {},
                    {t, t0, t1}, opts, MaxSteps -> Infinity,
                    PrecisionGoal -> 12, AccuracyGoal -> 12,
```

```
(* detect event z[t] == 0 and interpolate t-value *)
Method -> {EventLocator, "Event" -> z[t],
  "EventAction" :> Sow[{x[t], y[t]}],
  (* interpolate t-value *)
  "EventLocationMethod" -> "LinearInterpolation",
  "Method" -> "BDF"}]][[2, 1]]
```

Here are the crossings with the *x,y*-plane shown. The next graphic contains approximately 37300 points. The calculations are clearly faster than the above ones.

```
In[11]:= (* suppress messages arising from no functions specified *)
        Off[NDSolve::noout];

        (graphic2 = Graphics[{PointSize[0.004],
            Table[{Hue[Random[]], (* repeat for 100 starting values *)
                Point /@ poincaréData[{Table[Random[Real, {-1, 1}], {3}],
                                       {0, 1000}}]}, {100}]},
                    PlotRange -> All, AspectRatio -> Automatic,
                    Frame -> True]); // Timing
Out[13]= {579.24 Second, Null}
```

We display the two Poincaré sections calculated.

```
In[14]:= Show[GraphicsArray[{graphic1, graphic2}]]
```

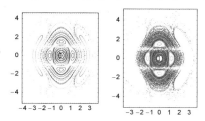

b) The compiled function *rotate* rotates the points *points* around the point *center* by an angle $\varphi \exp(-r)$ where r is the distance of a point from *center*. This means points near to center are rotated by a larger amount than points further away.

```
In[1]:= rotate = Compile[{{center, _Real, 1}, φ, {points, _Real, 2}},
            Module[{r, p, φ, ℛ}, Function[point,
                (* rotate (as a function of distance) *)
                p = point - center; r = Sqrt[p.p]; φ = Exp[-r] φ;
                ℛ = {{Cos[φ], Sin[φ]}, {-Sin[φ], Cos[φ]}};
                ℛ.p + center] /@ points]];
```

The second set of inputs creates a list $\mathcal{L}$ of *pp* points along a circle centered at the origin of radius 3/4 and three points centers at the vertices of a regular triangle (the vertices have distance 1 from the origin).

```
In[2]:= pp = 10000; R = 3/4; n = 3; m = 3;
        centers = Table[N[{Cos[j/n 2Pi], Sin[j/n 2Pi]}], {j, 0, n - 1}];
        ℒ0 = Table[R {Cos[φ], Sin[φ]}, {φ, 0, 2Pi, 2Pi/pp}] // N;
```

The third part of the input finally generates an animation. First, a vertex from the triangle is chosen at random. Then a positive rotation angle φ is chosen at random. Then all points from the set $\mathcal{L}$ are rotated in *m* steps around the chosen point by φ (the points at intermediate steps are denoted by $\mathbb{L}$). The points are connected in their original order and displayed as a polygon. Because the individual rotations rotate all points at a fixed distance in a rigid way, no intersections arise for smooth curves. Due to the discretization of the original circle, we obtain intersections of nonadjacent line segments after a finite number of rotations. Due to the randomly chosen rotations, the polygons will be stretched and folded and it is natural to assume that ultimately the iterated map of the original circle will cover the whole plane densely in thin filaments.

The following graphics show the resulting polygon after $9\,k$ rotations.

```
In[5]:= SeedRandom[333];
        (* a list to collect every ninth graphic *)
        bag = Table[{}, {12}]; ℒ = ℒ0;
        (* carry out all rotations *)
```

```
Do[If[IntegerQ[j/9], (* keep every 9th step *) bag[[j/9 + 1]] =
    Graphics[Polygon[L],
        PlotRange -> Max[Abs[L]]{{-1, 1}, {-1, 1}},
        Frame -> True, FrameTicks -> None, AspectRatio -> Automatic]];
    (* next rotation *)
    center = centers[[Random[Integer, {1, n}]]];
    φ = Random[Real, Pi {0, 1}];
    L = rotate[center, φ, L], {j, 0, 100}];
In[10]:= Show[GraphicsArray[#]]& /@ Partition[bag, 4]
```

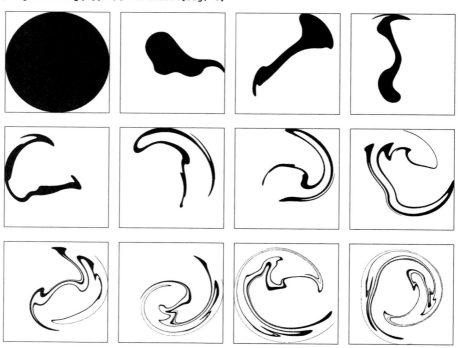

After carrying out 900 more random rotations, we display the resulting positions of the initial points. This time we do not display them as a polygon because they are already too mixed together.

```
In[11]:= Do[L = rotate[centers[[Random[Integer, {1, n}]]],
        Random[Real, Pi {0, 1}], L], {900}];
In[12]:= Show[Graphics[{PointSize[0.0025],
            MapIndexed[{Hue[#2[[1]]/pp], #}&, Point /@ L]},
        Frame -> True, AspectRatio -> Automatic]]
```

Here is the corresponding animation.

```
SeedRandom[333];
ℒ = ℒ0; m = 3;
Do[center = centers[[Random[Integer, {1, n}]]];
    φ = Random[Real, Pi {0, 1}];
    Do[L = rotate[center, k/m φ, ℒ];
        Show[Graphics[Polygon[L],
            PlotRange -> Max[Abs[L]]{{-1, 1}, {-1, 1}}, Frame -> True,
            FrameTicks -> None, AspectRatio -> Automatic]], {k, 0, m - 1}];
    ℒ = rotate[center, φ, ℒ], {j, 100}]
```

For a fluid dynamics-based example of a similar, more physical model, see [63], [982], [276], [772], [689], [1322], [1358], [64], [526], and [1188]; for the symplectic restrictions, see [466], [467]. For the shape changes of an elastic body in a stirred fluid, see [635].

Now let us deal with the generalization to 3D. All steps are quite straightforward to generalize to this case (actually to the *n*D case). Instead of three rotation centers at the vertices of a regular triangle, we now choose four rotation centers at the vertices of a regular tetrahedron. And, instead of an initial disk, we choose an initial ball. The random 2D rotations we generalize to random rotations in 3D.

Here are the centers of rotation and the initial ball.

In[13]:= **ppφ = 240; ppθ = 120; R = 3/4;**

```
(* rotation centers at vertices of tetrahedron *)
centers = {{0, 0, 3}, {0, 2Sqrt[2], -1}, {-Sqrt[6], -Sqrt[2], -1},
            {Sqrt[6], -Sqrt[2], -1}}/3 // N;

(* initial ball *)
ℒ0 = Table[R {Cos[φ] Sin[θ], Sin[φ] Sin[θ], Cos[θ]} // N,
            {φ, 0., 2.Pi, 2Pi/ppφ}, {θ, 0., 1. Pi, Pi/ppθ}];
```

The function *rotate*3D rotates the points *points* around the point *center* using rotation angles φ, ϑ, ω for the rotations around the *x*-, *y*-, and *z*-axes.

In[19]:= **rotate3D =**
```
    Compile[{{center, _Real, 1}, φ, θ, ω, {points, _Real, 3}},
        Module[{r, p, φ1, θ1, ω1, ℛ},
            Map[Function[point,
                p = point - center; r = Sqrt[p.p];
                {φ1, θ1, ω1} = Exp[-r] {φ, θ, ω};
                ℛ = {{Cos[φ1], Sin[φ1], 0.}, {-Sin[φ1], Cos[φ1], 0.}, {0., 0., 1.}}.
                    {{Cos[θ1], 0., Sin[θ1]}, {0., 1., 0.}, {-Sin[θ1], 0., Cos[θ1]}}.
                    {{1., 0., 0.}, {0., Cos[ω1], Sin[ω1]}, {0., -Sin[ω1], Cos[ω1]}};
                ℛ.p + center], points, {2}]]];
```

For the display of the deformed sphere, we implement a function showSurface. To better see the deformed sphere, we remove every second polygon in a checkerboard-like fashion.

In[20]:= **showSurface[L_, opts___] := Show[Graphics3D[{EdgeForm[],**
```
        (* remove every second polygon *)
        MapIndexed[If[(-1)^(Plus @@ #2) === 1,
                {SurfaceColor[#, #, 2]&[Hue[Sin[#2[[1]]/ppφ]]],
                #1}, {}]&, #, {2}]]], opts, Boxed -> False,
            BoxRatios -> {1, 1, 1}, PlotRange -> All]&[
        (* the polygons *)
        Table[Polygon[{L[[i, j]], L[[i + 1, j]],
                    L[[i + 1, j + 1]], L[[i, j + 1]]}],
            {i, Length[L] - 1}, {j, Length[L[[1]]] - 1}]]
```

Here are the resulting surfaces after 5, 10, ..., 45 random rotations.

In[21]:= **SeedRandom[12321];**
```
    bag = {}; ℒ = ℒ0;
    Do[center = centers[[Random[Integer, {1, Length[centers]}]]];
        {φ, θ, ω} = Table[Random[Real, Pi {0, 1}], {3}];
```

```
         𝓛 = rotate3D[center, φ, 𝜗, ω, 𝓛];
         If[IntegerQ[j/5], AppendTo[bag, 𝓛]], {j, 0, 45}]
In[24]:= Show[GraphicsArray[showSurface[#, DisplayFunction -> Identity]& /@
                              #]]& /@ Partition[Rest[bag], 3]
```

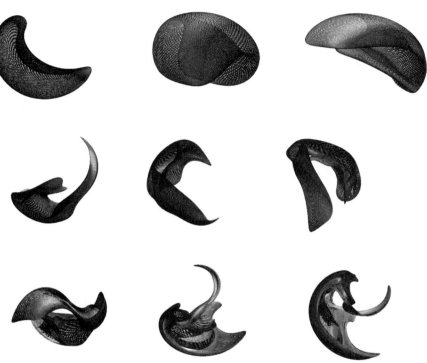

By simulating the individual random rotations with *m* steps, we get a smooth animation.

```
SeedRandom[12321];
m = 4; 𝓛 = 𝓛0;
Do[center = centers[[Random[Integer, {1, Length[centers]}]]];
   {φ, 𝜗, ω} = Table[Random[Real, Pi {0, 1}], {3}];
   Do[L = rotate3D[center, k/m φ, k/m 𝜗, k/m ω, 𝓛];
      showSurface[L, SphericalRegion -> True], {k, 0, m - 1}];
   𝓛 = rotate3D[center, φ, 𝜗, ω, 𝓛], {j, 0, 45}]
```

c) nsol solves the system of differential equations for given initial conditions $x(0) = x0$, $y(0) = y0$, and $z(0) = z0$.

```
In[1]:= nsol[{α_, β_, γ_}, {x0_, y0_, z0_}, T_] :=
        NDSolve[{x'[t] == α Sin[z[t]] + γ Cos[y[t]],
                 y'[t] == β Sin[x[t]] + α Cos[x[t]],
                 z'[t] == γ Sin[y[t]] + β Cos[x[t]],
                 x[0] == x0, y[0] == y0, z[0] == z0},
                {x, y, z}, {t, 0, T}, MaxSteps -> 10^4];
```

To avoid too complicated graphics and to better see the symmetry in the solutions, we do not display $\{x(t), y(t), z(t)\}$ as a space curve, but rather show the three projections on the coordinate planes. The function solutionGraphics implements the generation of the three projections for a given γ. As starting points, we use $o^3 - o$ points on a tensor product grid from $(-\pi, \pi) \times (-\pi, \pi) \times (-\pi, \pi)$. We choose the parameters $\alpha = \beta = 1$.

```
In[2]:= solutionGraphics[γ_, T_, o_, opts___] :=
    Module[{}, tab =
    Table[(* solve differential equations *)
            ns = nsol[{1, 1, γ}, {a, b, c}, 100][[1]];
            (* plot solution *)
            ParametricPlot[Evaluate[{x[t], y[t], z[t]}[[#]] /. ns],
                    {t, 0, T}, PlotPoints -> 40, Axes -> False,
                    PlotRange -> 2Pi {{-1, 1}, {-1, 1}},
                    PlotStyle -> {Thickness[0.002],
                        (* color according to starting positions *)
                     RGBColor[(a + Pi)/(2Pi), (b + Pi)/(2Pi),
                                (c + Pi)/(2Pi)]}, Frame -> True,
                    AspectRatio -> Automatic,
                    FrameTicks -> False, DisplayFunction -> Identity]& /@
                    {{1, 2}, {1, 3}, {2, 3}},
            {a, -Pi, Pi, 2Pi/o}, {b, -Pi, Pi, 2Pi/(o + 1)},
            {c, -Pi, Pi, 2Pi/(o + 2)}];
        (* the three projections on the coordinate planes *)
    Show[GraphicsArray[
            Show[Table[# /. Line[l_] :> Line[(2. Pi {i, j} + #)& /@ l],
            {i, -1, 1}, {j, -1, 1}]]& /@ Transpose[Flatten[tab, 2]]],
        opts]]
```

Here are the projections for some values of γ and the corresponding animation.

```
In[3]:= Do[solutionGraphics[γ, 10, 2, DisplayFunction -> $DisplayFunction],
        {γ, 0, 1, 1/3}]
```

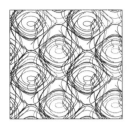 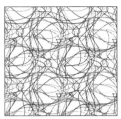 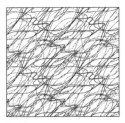

```
Do[solutionGraphics[γ, 10, 2, DisplayFunction -> $DisplayFunction],
    {γ, 0, 1, 1/60}]
```

Here is a solution for a larger value of γ.

In[4]:= **solutionGraphics[10, 10, 2]**

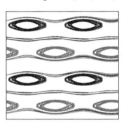 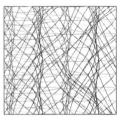 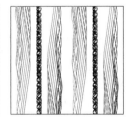

d) For brevity, let $V[x, t]$ stand for $\{x[1][t], x[2][t], x[3][t]\}$.

In[1]:= **V[x_, t_] := {x[1][t], x[2][t], x[3][t]}**

The function equationsOfMotion generates the equations of motion.

```
In[2]:= equationsOfMotion[n_] := Flatten[Thread /@
    Table[D[V[X[j], t], t] ==
        Sum[If[j === k, 0,
            Γ[k] Cross[V[X[k], t], V[X[j], t]]/
                    (1 - V[X[k], t].V[X[j], t])], {k, n}],
        {j, n}]]
```

initialConditions generates random initial conditions.

```
In[3]:= initialConditions[X_] := Thread[V[X, 0] ==
                (* random points on a sphere *)
                {Cos[#1] Sin[#2], Sin[#1] Sin[#2], Cos[#2]}&[
                                    Pi Random[], 2Pi Random[]]]
In[4]:= sphereVorticesGraphics[n_, T_, seed_, opts___] :=
    Module[{eqs, nsol, tMax},
    SeedRandom[seed];
    eqs = Flatten[{(* equations of motion *) equationsOfMotion[n],
                (* initial conditions *)
                Table[initialConditions[X[k]], {k, n}]}] /.
            Table[Γ[k] -> Random[Real, {-1, 1}], {k, n}];
    (* solve equations numerically *)
    nsol = NDSolve[eqs, Flatten[Table[X[j][k], {j, n}, {k, 3}]],
                {t, 0, T}, MaxSteps -> 25000,
                (* use larger values for PrecisionGoal and
                    AccuracyGoal settings *)
                PrecisionGoal -> 10, AccuracyGoal -> 10,
                (* use appropriate method to keep solutions
                    on the sphere *)
                Method -> {"Projection", "Invariants" ->
```

```
                        Table[Sum[X[j][k][t]^2, {k, 3}], {j, n}]}];
        (* in case integration did not succeed until T *)
        tMax = nsol[[1, 1, 2, 1, 1, 2]];
        (* make graphics *)
        Graphics3D[Table[(* random coloring *)
                        {Hue[0.8 (j - 1)/(n - 1)], Thickness[0.002],
        Line[Table[Evaluate[Table[X[j][k][t], {k, 3}] /. nsol[[1]]],
                        {t, 0, tMax, tMax/10^4}]]}, {j, n}], opts,
                        PlotRange -> 1.1 {{-1, 1}, {-1, 1}, {-1, 1}}]]
```

Here are some example trajectories for three and four vortices.

```
In[5]:= Show[GraphicsArray[sphereVorticesGraphics[3, 100, #]& /@
                {4326151752, 8659830492, 9832138988}]]
```

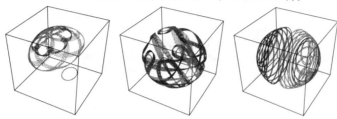

```
In[6]:= Show[GraphicsArray[sphereVorticesGraphics[4, ##]& @@@
                {{100, 6184868154}, {10, 4617323998}, {100, 4691930450}}]]
```

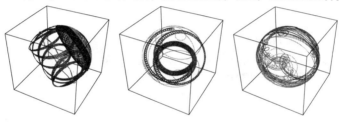

For vortices on a rotating sphere, see [1108].

e) Without loss of generality, we assume the outermost point masses to be located at $\{-1, 0\}$, $\{1, 0\}$, and $\{0, 3^{1/2}\}$. We subdivide the triangle formed by the point masses three times recursively.

```
In[1]:= {Π[1], Π[2], Π[3]} = {{-1, 0}, {1, 0}, {0, Sqrt[3]}};

       subDivisionOrder = 3;

       triangles = Nest[Flatten[# /. {p1_, p2_, p3_} :>
           Module[{p12 = (p1 + p2)/2, p23 = (p2 + p3)/2, p31 = (p3 + p1)/2},
                   (* four new triangles *)
                   {{p1, p12, p31}, {p12, p2, p23},
                    {p23, p3, p31}, {p12, p23, p31}}], 1]&,
                       {{Π[1], Π[2], Π[3]}}, subDivisionOrder];
```

The edges of the resulting 64 triangles are the 108 springs.

```
In[4]:= equilibriumSprings = Union[Sort /@
           Flatten[Partition[Append[#, First[#]], 2, 1]& /@ triangles, 1]];

       Length[equilibriumSprings]
Out[6]= 108
```

We form two lists of rules `coordRules` and `vertexRules` connecting the numbered point masses with their positions.

```
In[7]:= coordRules = MapIndexed[(#1 -> P[#2[[1]]])&,
                               Union[Flatten[equilibriumSprings, 1]]];
```

```
                vertexRules = Reverse /@ coordRules;
                Take[vertexRules, 4]
```

Out[9]= $\{P[1] \to \{-1, 0\}, P[2] \to \{-\frac{7}{8}, \frac{\sqrt{3}}{8}\}, P[3] \to \{-\frac{3}{4}, 0\}, P[4] \to \{-\frac{3}{4}, \frac{\sqrt{3}}{4}\}\}$

`fixedVertices` are the three outer, fixed point masses. `movingVertices` are all other point masses.

```
        In[10]:= fixedVertices = Cases[vertexRules, HoldPattern[Rule][_, Π[1] | Π[2] | Π[3]]]
```

Out[10]= $\{P[1] \to \{-1, 0\}, P[25] \to \{0, \sqrt{3}\}, P[45] \to \{1, 0\}\}$

```
        In[11]:= movingVertices = DeleteCases[vertexRules, Alternatives @@ fixedVertices];
```

`springGroups` is the list of lists of springs. The kth sublist of `springGroups` contains the springs attached to the kth point mass P_k.

```
        In[12]:= springGroups = Split[Sort[Join[#, Reverse /@ #]&[equilibriumSprings /. coordRules]],
                            #1[[1]] == #2[[1]]&];
```

Many lists of `springGroups` have length six; these are the inner point masses. The length four sublists are the point masses at the boundaries. The three length two sublists are the three point masses at the vertices.

```
        In[13]:= {First[#], Length[#]}& /@ Split[Sort[Length /@ springGroups]]
```

Out[13]= $\{\{2, 3\}, \{4, 21\}, \{6, 21\}\}$

Here are the springs attached to P_{22}.

```
        In[14]:= springGroups[[22]]
```

Out[14]= $\{\{P[22], P[14]\}, \{P[22], P[17]\}, \{P[22], P[18]\},$
          $\{P[22], P[26]\}, \{P[22], P[27]\}, \{P[22], P[31]\}\}$

`singleSprings` is a list of the 108 springs of the system.

```
        In[15]:= singleSprings = Union[Sort /@ Flatten[springGroups, 1]];
```

For visualizing this system, we implement some functions to draw the springs. We color the springs according to their deformation. Equilibrium springs are red, compressed or elongated springs become successively yellow, green, and blue with increasing deformation.

```
        In[16]:= l0 = 2/(2^subDivisionOrder);

                (* a schematic drawing of a spring *)
                springLine[{p1_, p2_}, w_, r_] :=
                Module[{q1 = p1 + 1/6 (p2 - p1), q2 = p2 + 1/6 (p1 - p2), dq, l, n},
                        (* directions and points *)
                        dq = q2 - q1; l = Sqrt[dq.dq]; n = Reverse[dq/l]{-1, 1};
                        (* the spring as a line *)
                        Line[{p1, q1, ##, q2, p2}]& @@ Table[q1 +
                            (k/w + 1/2/w) (q2 - q1) + (-1)^k n r l, {k, 0, w - 1}]]
        In[19]:= (* color the spring *)
                springColor[{p1_, p2_}] :=
                    Hue[0.8 ArcTan[Abs[(Sqrt[#.#&[p2 - p1]] - l0)/l0]]/(Pi/2)]
        In[21]:= (* enclosing plot range *)
                pr = With[{δ = l0/2}, {{-1 - δ, 1 + δ}, {-δ, Sqrt[3] + δ}}];

                (* graphic of the set of springs;
                    PRules are the point mass coordinates *)
                springGraphic[PRules_, opts___] :=
                Module[{s = Fold[ReplaceAll, singleSprings, PRules]},
                    Graphics[{(* the point masses *)
                            {GrayLevel[0], PointSize[0.02], Map[Point, s, {2}]},
                            (* the colored springs *)
                            {springColor[#], springLine[#, 12, 0.2]}& /@ s},
                            AspectRatio -> Automatic, PlotRange -> pr]]
```

The following two graphics show the equilibrium state and a state with random small elongations of the springs.

```
        In[25]:= Block[{r := l0/2 Random[Real, {-1, 1}]},
                Show[GraphicsArray[
```

```
{Graphics[springGraphic[vertexRules, PlotRange -> All]] // N,
  Graphics[springGraphic[Join[(* three fixed vertices *) fixedVertices,
                  MapAt[(# + r)&, #, 2]& /@ movingVertices],
            PlotRange -> All]]}]]]
```

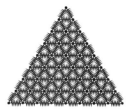 

To model the oscillations of the spring system for given initial conditions, we need the equations of motions. Assuming that the force of a spring is proportional to its length deviation from the equilibrium length and the force acts along the direction of the spring, we can implement the following function makeEquationOfMotion that, for a given spring *l* with spring constant α, generates equations of motions according to Newton's second law [1761]. The time-dependent Cartesian coordinates of the point P[k] are {x[k][t], y[k][t]}. (We set the mass of the point masses to 1.)

```
In[26]:= makeEquationOfMotion[l_, α_, {x_, y_}] :=
    Module[{(* symbolic coords *) L = l /. P[k_] :> {x[k][t], y[k][t]},
      (* equilibrium coords *) ℒ = l /. vertexRules, n = Sqrt[#.#]&},
      D[L[[1, 1]], {t, 2}] == (* sum over all attached springs *)
      α Sum[(* force direction *) #/n[#]&[L[[k, 2]] - L[[1, 1]]]*
            (* force magnitude *) (n[L[[k, 2]] - L[[1, 1]]] -
              n[ℒ[[k, 2]] - ℒ[[1, 1]]]), {k, Length[l]}] // Thread]
```

eoms is a list of the 84 equations of motions.

```
In[27]:= eoms = Flatten @ Delete[makeEquationOfMotion[#, α, {x, y}]& /@ springGroups,
                  (* remove three fixed point masses *)
                  {#[[1, 1]]}& /@ fixedVertices] //.
        Flatten[Thread /@ (fixedVertices /. P[k_] :> {x[k][t], y[k][t]})];
```

Here is one of the equations of motions. Because the magnitude of the spring force is proportional to the length change, the equations of motions are nonlinear.

```
In[28]:= eoms[[1]]
```

$$
Out[28]= \; x[2]''[t] == \alpha \left(\frac{(-1 - x[2][t])\left(-\frac{1}{4} + \sqrt{(-1 - x[2][t])^2 + y[2][t]^2}\right)}{\sqrt{(-1 - x[2][t])^2 + y[2][t]^2}} + \right.
$$

$$
\frac{(-x[2][t] + x[3][t])\left(-\frac{1}{4} + \sqrt{(-x[2][t] + x[3][t])^2 + (-y[2][t] + y[3][t])^2}\right)}{\sqrt{(-x[2][t] + x[3][t])^2 + (-y[2][t] + y[3][t])^2}} +
$$

$$
\frac{(-x[2][t] + x[4][t])\left(-\frac{1}{4} + \sqrt{(-x[2][t] + x[4][t])^2 + (-y[2][t] + y[4][t])^2}\right)}{\sqrt{(-x[2][t] + x[4][t])^2 + (-y[2][t] + y[4][t])^2}} +
$$

$$
\left. \frac{(-x[2][t] + x[5][t])\left(-\frac{1}{4} + \sqrt{(-x[2][t] + x[5][t])^2 + (-y[2][t] + y[5][t])^2}\right)}{\sqrt{(-x[2][t] + x[5][t])^2 + (-y[2][t] + y[5][t])^2}} \right)
$$

xys are the 84 dependent coordinates x[k], y[k] of the movable point masses appearing in eoms.

```
In[29]:= xys = Map[#[[0]]&, Cases[eoms, x[_][t] | y[_][t], Infinity] // Union, {1}];
```

To actually solve the equations of motions, we have to supplement initial conditions. We will assume all movable point masses rotated by $\ell_0/4$ around the center of the system and that they have vanishing initial velocities.

```
In[30]:= rotate[p_, δ_] := p + δ Reverse[#/Sqrt[#.#]&[
                  p - (Π[1] + Π[2] + Π[3])/3]]{-1, 1}
In[31]:= xyICs = Flatten[Join[(* initial point mass positions *)
        Thread /@ Equal @@@
              Map[MapAt[rotate[#, ℓ0/4]&, #, 2]&,
                Delete[vertexRules, {#[[1, 1]]}& /@ fixedVertices] /.
                      P[k_] :> {x[k][0], y[k][0]}],
```

```
                          (* initial point mass velocities *)
            Thread /@ Equal @@@ Map[MapAt[0.&, #, 2]&,
                  Delete[vertexRules, {#[[1, 1]]}& /@ fixedVertices] /.
                              P[k_] :> {x[k]'[0], y[k]'[0]}]]];
```

Now, it is straightforward to solve the resulting system of coupled nonlinear differential equations numerically. We set the spring constant to 1.

```
In[32]:= α = 1; T = 50;
        nsol = NDSolve[Join[eoms, xyICs], xys, {t, 0, T}, MaxSteps -> 10^4];
```

We choose unique random colors for the point mass trajectories and show the positions of the point masses in the left graphic. The right graphic shows the velocities of the point masses.

```
In[34]:= (* color associated to vertices *)
        colors = Table[{Thickness[0.002], Hue[Random[]]},
                      {k, Length[movingVertices]}];
        (* common option settings *)
        commonOpts = Sequence[FrameTicks -> None, PlotRange -> All,
                              PlotStyle -> colors, PlotRange -> All,
                              AspectRatio -> Automatic, Axes -> False,
                              DisplayFunction -> Identity, Frame -> True];

        Show[GraphicsArray[{
         (* trajectories *)
         ParametricPlot[Evaluate[({x[#][t], y[#][t]}& /@ (#[[1, 1]]& /@
                              movingVertices)) /. nsol[[1]]],
                      {t, 0, T}, Evaluate[commonOpts]],
         (* velocities *)
         Plot[Evaluate[(#.#&[{x[#]'[t], y[#]'[t]}]& /@ (#[[1, 1]]& /@
                              movingVertices)) /. nsol[[1]]],
                      {t, 0, T}, AspectRatio -> 1,
                      Evaluate[commonOpts]]}]]
```

 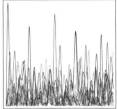

We quickly check the above calculations by verifying energy conservation. The following graphic shows the potential energy of the springs in red, the kinetic energy of the point masses in blue, and the sum of both in black. As expected, energy conservation holds accurately.

```
In[39]:= VPot[t_] = (* sum of squares of neighbor distances *)
                α/2 Plus @@ ((Sqrt[#.#]&[#[[1]] - #[[2]]] - ℓ0)^2& /@
                    ((singleSprings /. coordRules /. fixedVertices /.
                        P[k_] :> {x[k][t], y[k][t]}) /. nsol[[1]]));

In[40]:= Tkin[t_] = (* sum of squares of velocities *)
                (Plus @@ (#.#&/2& /@ (Delete[First /@ vertexRules ,
                    {#[[1, 1]]}& /@ fixedVertices] /.
                        P[k_] :> {x[k]'[t], y[k]'[t]}))) /. nsol[[1]];

In[41]:= Plot[{VPot[t], Tkin[t], Tkin[t] + VPot[t]}, {t, 0, T},
            PlotStyle -> {Hue[0], Hue[0.8], GrayLevel[0]},
            Compiled -> False, PlotRange -> All, AspectRatio -> 0.3]
```

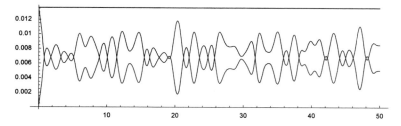

The following graphics show some of the resulting configurations of our spring system at various times.

```
In[42]:= Show[GraphicsArray[#]]& /@
         Partition[Table[springGraphic[{{fixedVertices,
                        P[k_] :> ({x[k][t], y[k][t]} /. nsol[[1]])}}],
                  {t, 0, T, T/7}], 4]
```

Here is the corresponding animation.

```
Do[Show @ springGraphic[{{fixedVertices,
                P[k_] :> ({x[k][t], y[k][t]} /. nsol[[1]])}}],
   {t, 0, T, T/12}]
```

Now let us look at the linearized version of our spring system. We assume the elongation of the point masses from their equilibrium positions to be small. This means we can expand the nonlinear right-hand sides in a Taylor series around the equilibrium positions and only keep constant and linear terms in the elongations. The function `makeLinearRHS` does this.

```
In[43]:= makeLinearRHS[l_] :=
         Module[{(* symbolic coordinates *) L = l /. P[k_] :>
                        (* elongations relative to equilibrium positions *)
                              (P[k] /. vertexRules) + λ {ξ[k][t], η[k][t]},
                  (* equilibrium coords *) ℒ = l /. vertexRules,
                  ε, n = Sqrt[#.#]&},
         Sum[ε = #/n[#]&[L[[k, 2]] - L[[1, 1]]]*
                  (n[L[[k, 2]] - L[[1, 1]]] - n[ℒ[[k, 2]] - ℒ[[1, 1]]]);
             (* Taylor expansion *) Expand[D[ε, λ] /. λ -> 0], {k, Length[l]}]]
```

Here is the linearized version of the above nonlinear equation of motion.

```
In[44]:= makeLinearRHS[springGroups[[1]]]
```

$$Out[44]= \left\{ -\frac{1}{4}\sqrt{3}\,\eta[1][t] + \frac{1}{4}\sqrt{3}\,\eta[2][t] - \frac{5}{4}\,\xi[1][t] + \frac{1}{4}\,\xi[2][t] + \xi[3][t], \right.$$
$$\left. -\frac{3}{4}\,\eta[1][t] + \frac{3}{4}\,\eta[2][t] - \frac{1}{4}\sqrt{3}\,\xi[1][t] + \frac{1}{4}\sqrt{3}\,\xi[2][t] \right\}$$

`linearRhs` is a list of all 90 linear right-hand sides.

```
In[45]:= linearRhs = Flatten[makeLinearRHS /@ springGroups];
```

```
In[46]:= Length[linearRhs]
```

```
Out[46]= 90
```

To obtain the eigenoscillations, we assume a harmonic time-dependence of the ξ_k and η_k in the form $\{\xi_k(t), \eta_k\}(t) \sim \cos(\omega_j t)\{\xi_k(0), \eta_k(0)\}$. This means the eigenoscillations follow from the eigenvalues of the coefficient matrix of the equations `linearRhs`. The function `makeCoefficientMatrix` constructs this coefficient matrix. We order the coefficients as $\xi_1, \eta_1, \xi_2, ..., \eta_{90}$.

```
In[47]:= makeCoefficientMatrix[rhss_] :=
    Module[{dim = Length[rhss], mat, terms},
        (* initialize matrix *) mat = Table[0, {dim}, {dim}];
        Do[terms = List @@ rhss[[i]];
            Do[(* extract coefficient and matrix position *)
                f = terms[[m]] /. (ξ | η)[_][t] -> 1;
                ξη = Cases[terms[[m]], (ξ | η)[_][t], {0, Infinity}][[1, 0]];
                j = 2 (ξη[[1]] - 1) + If[ξη[[0]] === ξ, 1, 2];
                (* add coefficient *)
                mat[[i, j]] = mat[[i, j]] + f, {m, Length[terms]}],
            {i, dim}]; mat]
```

We delete the six rows and columns corresponding to the x- and y-coordinates of three fixed point masses and obtain the matrix $\mathcal{M}$.

```
In[48]:= (* initial coefficient matrix *)
    M = makeCoefficientMatrix[linearRhs];

    (* row and column numbers of ξ and η coordinates
       of the three fixed vertices *)
    fixedRowColumns =
    Flatten[{2# - 1, 2#}& /@ (#[[1, 1]]& /@ fixedVertices)];

    (* reduced coefficient matrix *)
    M = Delete[#, List /@ fixedRowColumns]& /@
                        Delete[M, List /@ fixedRowColumns];
```

The following two graphics visualize the matrix elements and the eigenvalues. Due to the reflection and rotation symmetry of our triangular spring system, the eigenvalues appear in degenerate groups.

```
In[54]:= evals = Eigenvalues[N[M]];
```

```
In[55]:= (* form and count groups of eigenfrequencies *)
    {First[#], Length[#]}& /@
     Split[Sort[Length /@ Split[evals, Abs[#1 - #2] < 10^-6&]]]
```

```
Out[56]= {{1, 28}, {2, 28}}
```

```
In[57]:= Show[GraphicsArray[
    Block[{$DisplayFunction = Identity},
            {ListDensityPlot[M, Mesh -> False, PlotRange -> All],
             ListPlot[-evals, PlotRange -> All, Frame -> True, Axes -> False]}]]]
```

We end by visualizing some of the eigenoscillations. We calculate the eigenvalues and implement a function `vertexRules` `Linear` that generates the rules of the point mass positions for a given $\beta = \omega_j t$.

```
In[58]:= eVectors = Eigenvectors[N[M]];

In[59]:= vertexRulesLinear[ev_, β_] :=
         Rule @@@ Transpose[{First /@ vertexRules, N[Last /@ vertexRules] +
             Cos[β] /@ Partition[Fold[Insert[#1, 0., #2]&,
                                 ev, List /@ fixedRowColumns], 2]}]

In[60]:= eigenOscillationsGraphicsArray[ev_] :=
         With[{evβ = vertexRulesLinear[ev, β]},
         GraphicsArray[Table[
         (* the springs and point masses *)
         Graphics[{springColor[#], springLine[#, 12, 0.2]}& /@
                                 (singleSprings /. evβ),
             Frame -> True, FrameTicks -> None,
             AspectRatio -> Automatic, PlotRange -> pr],
             {β, 0, 4/5 2Pi, 2Pi/5}]]]

In[61]:= eigenOscillationsGraphicsArray[ev_] :=
         With[{evβ = vertexRulesLinear[ev, β]},
             GraphicsArray[Table[springGraphic[evβ], {β, 0, 4/5 2Pi, 2Pi/5}]]]
```

The next three graphics arrays show snapshots of three eigenoscillations. The first one is the most energetic one and the inner springs contract and expand opposite. The lowest energy mode moves the spring array basically as a whole.

```
In[62]:= Show @ eigenOscillationsGraphicsArray[eVectors[[ 1]]]
```

```
In[63]:= Show @ eigenOscillationsGraphicsArray[eVectors[[ 28]]]
```

```
In[64]:= Show @ eigenOscillationsGraphicsArray[eVectors[[- 1]]]
```

Here are again the corresponding animations.

```
With[{frames = 24},
 evβA = vertexRulesLinear[eVectors[[1]], β];
 evβB = vertexRulesLinear[eVectors[[28]], β];
 evβC = vertexRulesLinear[#/Sqrt[#.#]&[eVectors[[-1]] + eVectors[[-2]]], β];
 (* make graphics for animation *)
 Do[Show[GraphicsArray[springGraphic /@ {evβA, evβB, evβC}]],
    {β, 0, 2Pi (1 - 1/frames), 2Pi/frames}]]
```

We could now go on and compare the above numerical solution of the nonlinear one with a superposition solution of the linearized one, or calculate the time-dependence of the linearized system through exponentiating the matrix M, but we will end here. For symbolic results on triangular spring structures, see [1389] and [1390]; for hexagonal structures, see [446].

For normal mode-like calculations for nonlinear systems, see [376], [377], [378], [379], and [1577]. For the Cauchy–Born hypothesis of spring networks, see [639]. For the limit of small length springs, see [1552]. For random spring networks, see [799]. The interested reader could also deform such spring systems as 3D objects [943], [944], [717].

f) We start with a sketch of the classic Lorenz attractor. We display a set of solution curves for initial conditions on the unit sphere centered at the origin. (For the initial conditions of selected periodic solutions, see [1855].)

```
In[1]:= Show[Graphics3D[{Thickness[0.002],
          Table[{(* color curves randomly *) Hue[Random[]], Line @
            Table[Evaluate[With[{σ = 10, ρ = 28, β = 8/3, r = 1},
                {x[t], y[t], z[t]} /. (* solve odes *)
                NDSolve[{x'[t] == σ (y[t] - x[t]),
                        y'[t] == ρ x[t] - y[t] - x[t] z[t],
                        z'[t] == x[t] y[t] - β z[t],
                        (* start on unit sphere *)
                        x[0] == r Cos[φ] Sin[ϑ], y[0] == r Sin[φ] Sin[ϑ],
                        z[0] == r Cos[ϑ]}, {x, y, z}, {t, 0, 20},
                        MaxSteps -> 10^5][[1]]]], {t, 0, 20, 20/2500}]},
              {φ, 0, 2Pi, 2Pi/4}, {ϑ, 0, Pi, Pi/4}]}],
          PlotRange -> All, Axes -> True]
```

The last graphic suggests considering the visualizations in a domain of approximate size $-10^2 \leq x, y, z \leq 10^2$. In a first try to visualize the surface $x(t; x_0, y_0, z_0)$, we just solve the Lorenz system for a set of points on the y,z-plane as a function of t backwards and display the resulting surface. Here this is done for three different times t.

```
In[2]:= Σ[{y0_, z0_}, T_] :=
        With[{σ = 10, ρ = 28, β = 8/3}, {x[T], y[T], z[T]} /.
            NDSolve[{x'[t] == σ (y[t] - x[t]),
                    y'[t] == ρ x[t] - y[t] - x[t] z[t],
                    z'[t] == x[t] y[t] - β z[t],
                    x[0] == 0, y[0] == y0, z[0] == z0},
                    {x, y, z}, {t, 0, T}][[1]]]

In[3]:= L = 120;
        Show[GraphicsArray[Function[T,
        ParametricPlot3D[Σ[{y0, z0}, T], {y0, -120, 120}, {z0, -120, 120},
                PlotPoints -> 30, Compiled -> False, BoxRatios -> {1, 1, 1},
                PlotRange -> L {{-1, 1}, {-1, 1}, {-1, 1}},
                Axes -> False, Boxed -> True,
                DisplayFunction -> Identity]] /@ {-0, -0.08, -0.12}]]
```

The last graphic shows two problems: First, the surface stretches very inhomogeneously and we would need very many points to resolve it properly for larger t. Second, the surface evolves with a nearly unpredictable speed as a function of t. To circumvent the latter problem, we will rescale the independent variable t, such that each solution curve has unit speed, meaning $\mathbf{r}'(t) = \{x'(t), y'(t), z'(t)\} \longrightarrow \mathbf{r}'(\tau(t)) = \mathbf{r}'(t)/|\mathbf{r}'(t)|$ (to avoid radicals and multivalued solutions, we will express $|\mathbf{r}'(t)|$ in $x'(t)$, $y'(t)$, and $z'(t)$). We solve the former problem by evolving the Lorenz system forward in time on a grid of points in the x_0, y_0, z_0-space and then display the equi-surface $x(\tau; x_0, y_0, z_0) = 0$. The function $\chi[\{x0, y0, z0\}]$ calculates a list of values $x(\tau; x_0, y_0, z_0)$ for $pp\,T\,\tau$-values between 0 and T.

```
In[5]:= X[{x0_, y0_, z0_}, T_, ppT_] :=
      With[{σ = 10, ρ = 28, β = 8/3},
        Module[{norm, ndsol}, (* to ensure unit speed *)
            norm = 1/Sqrt[(σ (y[τ] - x[τ]))^2 +
                          (ρ x[τ] - y[τ] - x[τ] z[τ])^2 +
                          (x[τ] y[τ] - β z[τ])^2];
            ndsol = NDSolve[(* solve rescaled odes *)
                {x'[τ] == norm (σ (y[τ] - x[τ])),
                 y'[τ] == norm (ρ x[τ] - y[τ] - x[τ] z[τ]),
                 z'[τ] == norm (x[τ] y[τ] - β z[τ]),
                 x[0] == x0, y[0] == y0, z[0] == z0},
                {x, y, z}, {τ, 0, T}, MaxSteps -> 10^5,
                (* good enough for visualization purposes *)
                PrecisionGoal -> 4, AccuracyGoal -> 4][[1, 1, 2]];
            ndsol /@ Table[τ, {τ, 0, T, T/ppT}]]]]
```

$x(\tau; x_0, y_0, z_0)$ can be calculated in a few minutes for more than 100000 points in the space of initial conditions. We record $x(\tau; x_0, y_0, z_0)$ for 10 τ-values.

```
In[6]:= pp = 47; T = 500; ppT = 10;
      (* x[t] on points in the space of the initial conditions *)
      (data = Table[X[{x0, y0, z0}, T, ppT], {z0, -L, L, 2L/pp},
                    {y0, -L, L, 2L/pp}, {x0, -L, L, 2L/pp}];) // Timing
Out[8]= {523.4 Second, Null}
```

The following graphic shows slices of the resulting surface in the x,y-plane, x,z-plane, and y,z-plane for the τ-values from data.

```
In[9]:= With[{data = Map[#[[Round[ppT/2]]]&, data, {-2}]},
      Show[GraphicsArray[Show[Table[(* contour lines for all t *)
      ListContourPlot[#[[k]], Contours -> {0}, PlotRange -> All,
            ContourShading -> False, DisplayFunction -> Identity,
            ContourStyle -> {{Thickness[0.002], Hue[0.8 k/pp]}}],
                        {k, pp}], FrameTicks -> False]& /@
          (* transpose data *)
          (Transpose[data, #]& /@ {{1, 2, 3}, {2, 1, 3}, {3, 2, 1}})]]]]
```

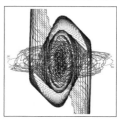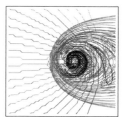

Because the surface to be sketched is quite complicated for larger t, we color the surface so that it is easy to follow through its windings. The function `colorPolys` takes a set of polygons *polys*, selects the polygons nearest to the origin, and recursively forms layers of polygons that share one (or more) vertex with already selected polygons.

```
In[10]:= colorPolys[polys_] :=
       Module[{ε = 10^-10,
               rationalPolys, vertices, polygons, polygonAvailable, λ,
               layerLayerList, nextLayer, layerList, aux, colorLayers},
           (* polygons with exact vertices *)
           rationalPolys = Select[Rationalize[polys, ε], (* degenerate cases *)
                               Length[Union[#[[1]]]] > 2&];
           (* all vertices *)
           vertices = Union[Level[rationalPolys, {-2}]];
           (* polygon around the vertices *)
```

```
          (polygons[#] = {})& /@ vertices;
          Do[(polygons[#] = Append[polygons[#], rationalPolys[[k]]])& /@
                                        rationalPolys[[k, 1]],
             {k, Length[rationalPolys]}];
          (* polygon still available? *)
          (polygonAvailable[#] = True)& /@ rationalPolys;
          (* loop over all connected components *)
          layerLayerList = {};
          While[Length[DownValues[polygonAvailable]] > 0,
          (* the next connected set of polygons *)
          nextLayer = Last /@ (* polygons nearest to origin *)
          Split[Sort[{Sqrt[#.#]&[Plus @@ N[#[[1]]]/Length[#[[1]]]], #}& /@
                       (#[[1, 1, 1]]& /@ DownValues[polygonAvailable])],
                     (#[[1]] == #2[[1]])&][[1]];
          (* next layer of polygons *)
          (polygonAvailable[#] =.)& /@ nextLayer; layerList = {nextLayer};
          While[aux = Select[Union[Flatten[polygons /@
                            Union[Flatten[First /@ nextLayer, 1]]]],
                polygonAvailable]; aux =!= {},
                (* update layer-related data *)
                nextLayer = aux; layerList = {layerList, nextLayer};
                (polygonAvailable[#] =.)& /@ nextLayer];
          AppendTo[layerLayerList, layerList]];
          (* color polygons of each component by layer distance *)
          colorLayers[layerList_] :=
          (λ = Length[theLayers = Level[layerList, {-5}]];
           MapIndexed[{SurfaceColor[#, #, 2.3&[Hue[0.78 #2[[1]]/λ]],
                        #1}&, theLayers]);
          (* return colored polygons *)
          colorLayers /@ layerLayerList]
```

To see remote parts of the surface better, we cut holes into the individual polygons.

```
In[11]:= makeHoledPolygon[Polygon[l_], factor_] :=
         Module[{mp = Plus @@ l/Length[l], λ, L, Λ},
                λ = (mp + factor(# - mp))& /@ l;
                {L, Λ} = Partition[Append[#, First[#]]&[#], 2, 1]& /@ {l, λ};
                MapThread[Polygon[Join[#1, Reverse[#2]]]&, {L, Λ}]]
```

And to see into the center of the resulting surface, we remove the polygons from the octant $x > 0$, $y < 0$, $z > 0$.

```
In[12]:= removeOctantPolygon[p:Polygon[l_]] :=
         With[{d = Sqrt[#.#]&[Plus @@ l/Length[l]]},
              If[(* cut corners *) d > 1.2 L, {},
                 If[Max[#[[1]]] > 0 && Min[#[[2]]] < 0 && Max[#[[3]]] > 0,
                    {}, p]&[Transpose[l]]]]
```

Now, we have all functions together to visualize the surfaces $x(t; x_0, y_0, z_0)$. The function xOftSurface generates a Graphics3D-object of the kth set of values of data.

```
In[13]:= Needs["Graphics`ContourPlot3D`"]
```

```
In[14]:= xOftSurface[k_] := Module[{
         (* make contour plot to get polygons *)
         lcp3D = ListContourPlot3D[Map[#[[k]]&, data, {-2}],
                        MeshRange -> {{-L, L}, {-L, L}, {-L, L}},
                        DisplayFunction -> Identity]},
         (* make 3D Graphics *)
         Graphics3D[{EdgeForm[], (* color polygons and cut holes *)
         (colorPolys @ Cases[lcp3D, _Polygon, Infinity]) /.
         p_Polygon :> (removeOctantPolygon /@ makeHoledPolygon[p, 0.75])},
         Boxed -> True, BoxRatios -> {1, 1, 1}, Axes -> False,
         PlotRange -> {{-L, L}, {-L, L}, {-L, L}}]]
```

Here finally are sketches of the surface $x(\tau) = 0$.

```
In[15]:= xOftSurface[1]
```

```
In[16]:= Show[GraphicsArray[xOftSurface /@ #]]& /@ {{1, 2, 3}, {4, 5, 6}}
```

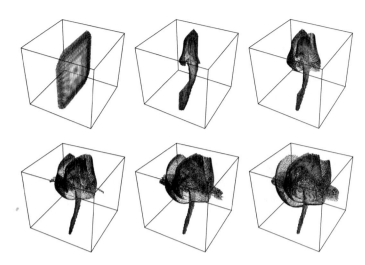

For $\tau = 500$, we have a quite complicated surface that curls around the y-axis and around the z-axis.

```
In[17]:= Show[xOfτSurface[ppT], Boxed -> False]
```

For other complicated surfaces related to the Lorenz system, see [747], [1394], [1057], [1395], and [1056]. For similar attractors, see [384], [1181], [1182], [1941], [380]; for the periodically forced Lorenz system, see [1590].

29. Polynomial Coefficient, Fourier Differentiation

a) The coefficient of z^n in the polynomial $p(z)$ is given as the residue of $p(z) z^{-n-1}$ at $z = 0$ [1371], [1143].

```
In[1]:= NIntegrate[Evaluate[(((x + 1)^2000 (x^2 + x + 1)^1000 *
        (x^4 + x^3 + x^2 + x + 1)^500)/x^3001 /. x -> E^(I φ)) I E^(I φ)],
                {φ, 0, 2 Pi}]
Out[1]= -7.457752023015746×10^1413 + 2.496901565467728×10^1427 i
```

```
In[2]:= Im[%/(2 Pi)]
Out[2]= 3.973942265580806×10^1426
```

We use high-precision arithmetic to ensure that we get 13 correct digits. We also use the trapezoidal rule because the integrand is periodic.

```
In[3]:= exactResult = N[Im[NIntegrate[Evaluate[
        (((x + 1)^2000 (x^2 + x + 1)^1000*
        (x^4 + x^3 + x^2 + x + 1)^500)/x^3001 /.
        x -> E^(I φ)) I E^(I φ)], {φ, 0, 2 Pi},
        (* use efficient Method option for periodic integrand *)
        Method -> Trapezoidal, WorkingPrecision -> 400,
        PrecisionGoal -> 16, MaxRecursion -> 10]]/(2 Pi), 15]
Out[3]= 3.97394226558004×10^1426
```

Now, let us calculate the integral in saddle point approximation. For doing this, we rewrite the integral in the following manner:

$$\oint \frac{p(x)}{x^{3001}}\,dx = \oint \exp\!\left(\ln\!\left(\frac{p(x)}{x^{3001}}\right)\right)dx.$$

Then, we look for zeros of the first derivative of $\ln(p(x)/x^{3001})$ and approximate the integral locally by using the Taylor expansion of $\ln(p(x)/x^{3001})$. This leads to Gaussian integrals that can be integrated exactly (and as usual in saddle point approximations, one neglects the rest of the integral).

```
In[4]:= poly = ((1 + x)^2000 (1 + x + x^2)^1000*
              (1 + x + x^2 + x^3 + x^4)^500)/x^3001;
```

```
In[5]:= logPoly[x_] = PowerExpand[Log[poly]]
```

$$Out[5]= -3001\,\text{Log}[x] + 2000\,\text{Log}[1+x] + 1000\,\text{Log}[1+x+x^2] + 500\,\text{Log}[1+x+x^2+x^3+x^4]$$

```
In[6]:= num = Numerator[Together[D[logPoly[x], x]]] // Factor
```

$$Out[6]= -3001 - 5503\,x - 5005\,x^2 - 1506\,x^3 + 1494\,x^4 + 4995\,x^5 + 5497\,x^6 + 2999\,x^7$$

These are the potential saddle points.

```
In[7]:= (saddles = NSolve[num == 0, x, 80]) // N
```

$$Out[7]= \{\{x \to -0.890414 - 0.455215\,i\}, \{x \to -0.890414 + 0.455215\,i\},$$
$$\{x \to -0.666684 - 0.745375\,i\}, \{x \to -0.666684 + 0.745375\,i\},$$
$$\{x \to 0.140395 - 0.990144\,i\}, \{x \to 0.140395 + 0.990144\,i\}, \{x \to 1.00046\}\}$$

The phases of the zeros of the derivative of a polynomial are visible as endpoints of "branch cuts" in a contour plot. The right graphic shows a density plot of $\sin(\arg(p(x)))$. The special status of the rightmost saddle point (largest distance from the origin) is clearly visible.

```
In[8]:= Show[GraphicsArray[
         Block[{$DisplayFunction = Identity},
          {(* contour plot of argument numerator of derivative *)
           ContourPlot[Evaluate[Arg[num] /. x -> x + I y],
                       {x, -1.2, 1.2}, {y, -1.2, 1.2},
                       PlotPoints -> 100, ContourShading -> False,
                       ContourStyle -> {{GrayLevel[0], Thickness[0.002]}},
                       Contours -> 50, Epilog -> ({Hue[0], PointSize[0.025],
                                     Point[{Re[#], Im[#]}]}& /@ (x /. saddles))],
           (* density plot of sine of argument of original polynomial *)
           DensityPlot[Evaluate[Sin[Arg[poly /. x -> x + I y]]],
                       {x, -1.2, 1.2}, {y, -1.2, 1.2}, (* use many points *)
                       PlotPoints -> 600, Mesh -> False,
                       Epilog -> ({Hue[0], PointSize[0.025],
                                   Point[{Re[#], Im[#]}]}& /@ (x /. saddles))]}]]] //
                                               Internal`DeactivateMessages
```

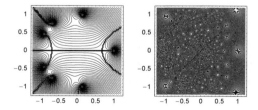

Now, we make a series expansion around a saddle point. (For a detailed discussion about the use of `Series`, see Chapter 1 of the Symbolics volume [1795].)

```
In[9]:= fser1 = Series[f[x], {x, x0, 6}] /. f'[_] -> 0
```

In[9]:= f[x0] + $\frac{1}{2}$ f''[x0] (x - x0)^2 + $\frac{1}{6}$ f$^{(3)}$[x0] (x - x0)^3 +

$\frac{1}{24}$ f$^{(4)}$[x0] (x - x0)^4 + $\frac{1}{120}$ f$^{(5)}$[x0] (x - x0)^5 + $\frac{1}{720}$ f$^{(6)}$[x0] (x - x0)^6 + O[x - x0]^7

In[10]:= **fser2 = Normal[Series[Normal[Exp[fser1 - (f[x0] + 1/2 f''[x0](x - x0)^2)]],**
{x, x0, 8}]] /. (x - x0)^(_?OddQ) -> 0

Out[10]= 1 + $\frac{1}{24}$ (x - x0)^4 f$^{(4)}$[x0] + (x - x0)^6 $\left(\frac{1}{72}$ f$^{(3)}$[x0]^2 + $\frac{1}{720}$ f$^{(6)}$[x0]$\right)$

So the main contribution to the integral at the saddle at x0 is given by the following expression. We also calculate the first correction term by integrating the next term in the series (see [169], [902], [1666], [1575], and [904]).

In[11]:= **{mainTerm, corrTerm} =**
Integrate[Exp[1/2 f2[x0] x^2] #, {x, -Infinity, Infinity},
Assumptions -> f2[x0] < 0]& /@ {1, x^4}

Out[11]= $\left\{ \frac{\sqrt{2\pi}}{\sqrt{-f2[x0]}} , \frac{3\sqrt{2\pi}}{(-f2[x0])^{5/2}} \right\}$

Calculating numerical values for these two terms, we arrive at the following approximations for the coefficient under consideration.

In[12]:= (* main part *)
fu1 = Function[x0, Evaluate[Exp[logPoly[x0]]*
Sqrt[2 Pi]/Sqrt[-logPoly''[x0]]]];

In[14]:= **(t1 = Plus @@ (fu1 /@ (x /. saddles))) // N**

Out[14]= -4.357121181057191$\times 10^{363}$ - 2.497711987531419$\times 10^{1427}$ i

In[15]:= (* correction part *)
fu2 = Function[x0, Evaluate[Exp[logPoly[x0]] logPoly''''[x0]*
3/24 Sqrt[2 Pi]/(-logPoly''[x0])^(5/2)]]

Out[16]= Function$\left[$x0, $\left(\sqrt{\frac{\pi}{2}}\right.$ (1 + x0)^{2000} (1 + x0 + x0^2)^{1000} (1 + x0 + x0^2 + x0^3 + x0^4)^{500} $\left(\frac{18006}{x0^4} - \frac{12000}{(1 + x0)^4} - \right.$

$\frac{6000 (1 + 2 x0)^4}{(1 + x0 + x0^2)^4} + \frac{24000 (1 + 2 x0)^2}{(1 + x0 + x0^2)^3} - \frac{12000}{(1 + x0 + x0^2)^2} - \frac{3000 (1 + 2 x0 + 3 x0^2 + 4 x0^3)^4}{(1 + x0 + x0^2 + x0^3 + x0^4)^4} +$

$\frac{6000 (2 + 6 x0 + 12 x0^2) (1 + 2 x0 + 3 x0^2 + 4 x0^3)^2}{(1 + x0 + x0^2 + x0^3 + x0^4)^3} - \frac{1500 (2 + 6 x0 + 12 x0^2)^2}{(1 + x0 + x0^2 + x0^3 + x0^4)^2} -$

$\left. \frac{2000 (6 + 24 x0) (1 + 2 x0 + 3 x0^2 + 4 x0^3)}{(1 + x0 + x0^2 + x0^3 + x0^4)^2} + \frac{12000}{1 + x0 + x0^2 + x0^3 + x0^4} \right) \right) /$

$\left(4 x0^{3001} \left(-\frac{3001}{x0^2} + \frac{2000}{(1 + x0)^2} + \frac{1000 (1 + 2 x0)^2}{(1 + x0 + x0^2)^2} - \frac{2000}{1 + x0 + x0^2} + \right.\right.$

$\left.\left. \frac{500 (1 + 2 x0 + 3 x0^2 + 4 x0^3)^2}{(1 + x0 + x0^2 + x0^3 + x0^4)^2} - \frac{500 (2 + 6 x0 + 12 x0^2)}{1 + x0 + x0^2 + x0^3 + x0^4} \right)^{5/2} \right) \right]$

In[17]:= **t2 = Plus @@ (fu2 /@ (x /. saddles))**

Out[17]= -6.53713142774643895660980922856316827943493584725975342677634647747518133$\times 10^{363}$ - 1.35185118166632749410663029243018237141606977080155755977448564219218820820 69$\times 10^{1424}$ i

In[18]:= **-{Im[t1/(2 Pi)],** (* sum of main part and correction part *)
Im[(t1 + t2)/(2 Pi)]} // N

Out[18]= {3.975232092355078$\times 10^{1426}$, 3.977383630333946$\times 10^{1426}$}

The fourth digit is different, which is roughly what we would expect taking into account that we are effectively dealing in the saddle point approximation with a power series in 1/3001.

In[19]:= **{(-exactResult + Im[-t1/(2 Pi)])/exactResult, 1/3001.} //**
SetPrecision[#, 4]&

Out[19]= {0.0003246, 0.0003332}

b) To express the νth component of the discrete Fourier transform through the continuous Fourier transform, we approximate the sum with an integral for small L/n. Using the definition of the discrete Fourier transform, we obtain:

$$\hat{f}_v = \frac{1}{\sqrt{n}} \sum_{\mu=1}^{n} f_\mu \exp(2\pi i (\mu - 1)(v - 1)/n) \approx \frac{\sqrt{n}}{L} \int_0^L f(x) \exp\left(i x \frac{2\pi(v-1)}{L} \right) dx$$

$$= \frac{\sqrt{2\pi n}}{L} \exp\left(2\pi i \frac{1-v}{n} \right) \hat{f}\left(\frac{2\pi}{L}(v-1) \right).$$

Using now $\exp(2\pi i (\mu - 1)(v - 1)/n) = \exp(2\pi i (\mu - 1)(-(n - v) - 1)/n)$ we obtain, for $v > (n + 1)/2$ the similar formula

$$\hat{f}_v = \frac{\sqrt{2\pi n}}{L} \exp\left(2\pi i \frac{n-v+1}{n} \right) \hat{f}\left(\frac{2\pi}{L}(v-n-1) \right).$$

We implement the last formula together with the direct sum definition of an element of the discrete Fourier transform. (For a given index v, this will be more efficient than a call to `Fourier`.)

```
In[1]:= FourierComponent[v_, n_, L_] := 1/Sqrt[n] *
          Sum[Evaluate[f[µ/n L] Exp[2 Pi I (µ - 1)(v - 1)/n]], {µ, n}]


      FourierComponentApprox[v_, n_, L_] := Sqrt[2 Pi n]/L*
              Exp[2 Pi I/n If[v <= (n + 1)/2, 1 - v, n - v + 1]]*
              fF[2 Pi/L If[v <= (n + 1)/2, v - 1, v - n - 1]]
```

Now let us consider an example. We choose $f(x) = \exp(-2(x - L/2)^2) \sin(36\, 2\pi/L\,(x - L/2))$. This function is strictly periodic with period L.

```
In[3]:= f[x_] := Exp[-2(x - L/2)^2] Sin[36 2Pi/L (x - L/2)]


      fF[k_] = 1/Sqrt[2 Pi] Integrate[f[x] Exp[I k x], {x, 0, L},
                               GenerateConditions -> False];
```

For $n = 512$, $L = 40$ the agreement between the direct discrete Fourier transform values and the ones calculated using the continuous transform seems to be quite good.

```
In[5]:= Block[{L = 40, n = 512, reIm = Re},
         Show[Graphics[{
           {Hue[0], PointSize[0.01], (* exact values *)
            MapIndexed[Point[{#2[[1]], reIm @ #1}]&,
                      Fourier[Table[f[k/n L], {k, n}]]]] // N},
           {GrayLevel[0], PointSize[0.005], (* approximate values *)
            Table[Point[{v, reIm @ FourierComponentApprox[v, n, L]}],
             {v, 1, n}]}} // N, PlotRange -> All, Frame -> True]]]
```

Here is the case $v = 40$, $n = 128 \times 2^j$ studied quantitatively. The convergence is quite fast.

```
In[6]:= Block[{L = N[40, 500], v = 40},
             {#, Abs[FourierComponent[v, #, L] -
                   FourierComponentApprox[v, #, L]]}& /@
             {128, 256, 512, 1024}] // N
Out[6]= {{128., 0.0000306192}, {256., 3.28551×10^-45},
         {512., 5.6457×10^-257}, {1024., 1.586950168774445×10^-352}}
```

For a continuous Fourier transform of a function that is not periodic, the difference would decrease and then stay constant due to nonmatching periodic continuations at $z = k\,L$, $k \in \mathbb{Z}$.

For the approximation of the sum through the integral to hold, L/n must be small and we must be sure to capture all oscillations of the functions. For varying L, the following graphic shows that for the function $f(x)$ used here, all oscillations are captured for $L \leq 200$.

```
In[7]:= ListPlot[Abs[
         Block[{n = 512, v0 = 40},
             Table[{L, FourierComponent[v0 L/40, n, N[L, 22]] -
                     FourierComponentApprox[v0 L/40, n, N[L, 22]]},
                 {L, 10, 1200, 10}]]], PlotRange -> All]
```

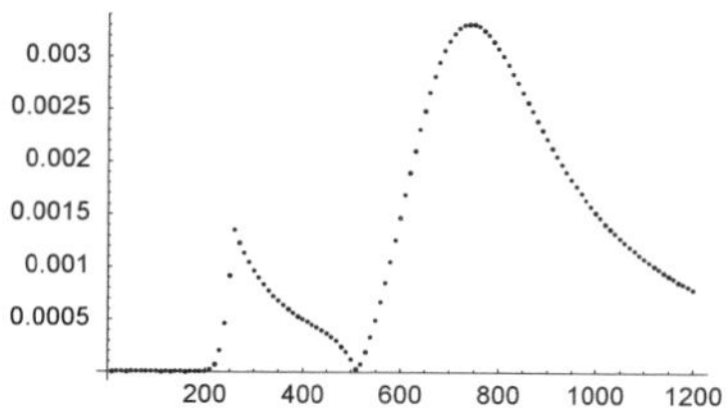

We now consider the discrete version of differentiation. From the above formulas that express the vth component of the discrete Fourier transform through the continuous Fourier transform we immediately can write down the equivalent to multiplication by $(-i\,k)$.

```
In[8]:= FourierD[l_] := InverseFourier[
           I MapIndexed[(If[#2[[1]] <= (Length[l] + 1)/2,
                           1 - #2[[1]],
                           Length[l] - #2[[1]] + 1] #1)&, l]]
```

We use a list of odd length to avoid imaginary parts in the derivatives. To demonstrate the differentiation, we start with a periodic function.

```
In[9]:= o = 256;
         (* the x points *)
         xPoints = Table[x, {x, 0, 2Pi (1 - 1/o), 2Pi/o}] // N;
         (* the 2π-periodic function *)
         f[x_] := Cos[x] + 1/2 Cos[2x] - 1/3 Cos[3x] + 1/4 Cos[4x];
         (* function values *)
         fs = f /@ xPoints;
         (* discrete Fourier transform *)
         fF = Fourier[fs];
```

fDExactList is a list of the exact derivatives of f and fDFList is a list of the derivatives calculated through multiplication of the discrete Fourier transform followed by an inverse Fourier transform.

```
In[18]:= (* exact derivative at the x points *)
          fDExactList = Table[df = Derivative[k][f]; df /@ xPoints, {k, 0, 6}];

          (* approximate derivatives at the x points *)
          fDFList = NestList[FourierD[Fourier[#]]&, fs, 5];
```

The function showDerivatives[] produces a pair of graphics. The left graphic shows the functions and its first two derivatives. The right graphic shows the relative error in the first four derivatives calculated using the discrete Fourier transform.

```
In[23]:= showDerivatives[] := Show[GraphicsArray[{
          Graphics[(* approximate derivatives *)
          Table[{Hue[(k - 1)/6], Line[Transpose[{xPoints, Re @ fDFList[[k]]}]]},
              {k, 3, 1, -1}], PlotRange -> {-10, 10}, Frame -> True],
          Graphics[(* relative error of the approximate derivatives *)
          Table[{Hue[(k - 1)/6], Point /@ DeleteCases[
                  Transpose[{xPoints,
                   Log[10, Abs[1 - Abs[fDExactList[[k]]/fDFList[[k]]]]]}],
                  (* zero error cases *) {_, Indeterminate}]},
              {k, 2, 5}], Frame -> True, PlotRange -> All]}]]
```

For the periodic function, the error in the derivatives is overall quite small for the first derivatives.

In[24]:= `showDerivatives[]`

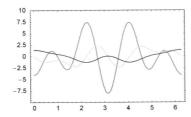 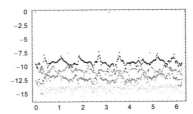

Because of the harmonic form of f and its strict periodicity, derivatives of all orders are actually exact. The apparent decrease in accuracy visible in the last results from the use of machine arithmetic. Using high-precision arithmetic, the higher order derivatives agree exactly. (Here we take care to use enough initial precision.)

```
In[25]:= Function[prec,
           (* avoid messages from division by low-precision zeros *)
           Internal`DeactivateMessages[
           (* high precision x-values *)
           xPointsHP = N[Table[x, {x, 0, 2Pi (1 - 1/o), 2Pi/o}], prec];
           (* high precision Fourier derivatives *)
           fsHP = f /@ xPointsHP; fFHP = Fourier[fsHP];
           df1HP = Nest[FourierD[Fourier[#]]&, fsHP, 24];
           (* exact derivatives *)
           dfHP = Derivative[24][f] /@ xPointsHP;
           (* absolute error *)
           Max[Abs[dfHP - df1HP]]]] /@ (* precisions *) {200, 500}
Out[25]= {Indeterminate, 0. × 10^-120}
```

Now, we repeat the last calculations with a function that is not even continuous, but is periodically continued. The derivatives contain large high-frequency oscillations and the derivatives are in average quite poorly approximated.

```
In[26]:= f[x_] := Sin[x/3] + 1/2 Sin[x/2] - 1/3 Sin[x] + 1/4 Sin[4x];
         fs = f /@ xPoints; fF = Fourier[fs];
         fDFList = NestList[FourierD[Fourier[#]]&, fs, 5];
         fDExactList = Table[df = Derivative[k][f]; df /@ xPoints,
                             {k, 0, 6}];
         showDerivatives[]
```

For explicit error estimations, see [1741]. For another *Mathematica* implementation of differentiation through Fourier transformation, see [1872], for the optimal extension of nonperiodic functions to periodic ones, see [274]. For the corresponding Fourier differentiation matrices, see [702].

30. Abel Differentials, Implicit Parameter Dependence

a) This is the polynomial under consideration.

In[1]:= `poly = y^6 - 4 x^4 + y^2 x^3 - 2 x + y - 3;`

We cannot simply numerically solve for *y* for a given *x* because we would not know how to connect the roots properly. To get $y_i(x)$ as a *smooth* function of *x*, we derive a differential equation for the roots.

In[2]:= `Solve[D[poly /. y -> y[x], x] == 0, y'[x]]`

In[2]:= $\{\{y'[x] \rightarrow \dfrac{2 + 16\,x^3 - 3\,x^2\,y[x]^2}{1 + 2\,x^3\,y[x] + 6\,y[x]^5}\}\}$

These are the starting values for the differential equation.

In[3]:= `yi0s = y /. NSolve[(y^6 - 4 x^4 + y^2 x^3 - 2 x + y - 3 /. x -> 0) == 0, y]`

Out[3]= `{-1.27391, -0.567906 - 1.10869 i, -0.567906 + 1.10869 i,`
`0.648981 - 0.971557 i, 0.648981 + 0.971557 i, 1.11176}`

We now solve the differential equations corresponding to the roots.

In[4]:= `nSols1 = NDSolve[{y'[x] == -((-2 - 16 x^3 + 3 x^2 y[x]^2)/`
`(1 + 2 x^3 y[x] + 6 y[x]^5)),`
`y[0] == #}, y[x], {x, 0, 2}]& /@ yi0s;`

The resulting roots are smooth functions of *x*.

In[5]:= `Show[GraphicsArray[`
`Plot[Evaluate[#[y[x]] /. nSols1], {x, 0, 2},`
`PlotRange -> All, Frame -> True, Axes -> False,`
`DisplayFunction -> Identity]& /@ {Re, Im}]]`

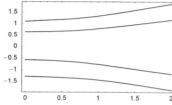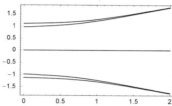

Instead of numerically integrating $\sum_{i=1}^{6}(y_i(x)-1)/(y_i(x)+1)$ for various *x*-values of the upper limit, we solve the corresponding differential equation.

In[6]:= `diffs = ((#[[1, 1, 2]] - 1)/(#[[1, 1, 2]] + 1)& /@ nSols1) /. x -> Re[x];`

In[7]:= `nSols2 = (S[x] /. NDSolve[{S'[x] == #, S[0] == 0},`
`S[x], {x, 0, 2}][[1]])& /@ diffs;`

We finally arrive at the integral under consideration.

In[8]:= `int[x_] = Total[nSols2];`

In[9]:= `Show[GraphicsArray[`
`Plot[Evaluate[#[int[x]]], {x, 0, 2},`
`PlotLabel -> #, DisplayFunction -> Identity]& /@ {Re, Im}]]`

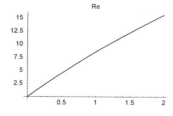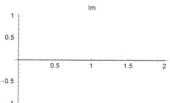

For a theoretical treatment of such kinds of integrals, see [714].

b) One possibility to solve this problem would be the calculation of $w_\zeta(1)$ for $\zeta \approx 1$ and numerical differentiation of the result. To get 20 correct digits in the second derivative would require a very high-precision solution of the differential equation and would be very slow. So, we try to obtain a differential equation for $\partial^2 w_\zeta(z)/\partial\zeta^2$ instead. This can be achieved by first making the ζ-dependence explicit and introducing two new functions $w\zeta = \partial w_\zeta(z)/\partial\zeta$ and $w\zeta\zeta = \partial^2 w_\zeta(z)/\partial\zeta^2$, both being functions of *z*.

In[1]:= `eq = wζ'[z] == Cos[wζ[z]] + Sin[ζ] /.`
`{wζ[z] :> w[ζ, z], wζ'[z] :> D[w[ζ, z], z]}`

Out[1]:= $w^{(0,1)}[\zeta, z] == \text{Cos}[w[\zeta, z]] + \text{Sin}[\zeta]$

In[2]:= {eq, D[eq, ζ], D[eq, ζ, ζ]}

Out[2]= $\{w^{(0,1)}[\zeta, z] == \text{Cos}[w[\zeta, z]] + \text{Sin}[\zeta], w^{(1,1)}[\zeta, z] == \text{Cos}[\zeta] - \text{Sin}[w[\zeta, z]] w^{(1,0)}[\zeta, z],$
$w^{(2,1)}[\zeta, z] == -\text{Sin}[\zeta] - \text{Cos}[w[\zeta, z]] w^{(1,0)}[\zeta, z]^2 - \text{Sin}[w[\zeta, z]] w^{(2,0)}[\zeta, z]\}$

In[3]:= % /. {w[ζ, z] :> w[z], Derivative[0, 1][w][ζ, z] -> w'[z],
Derivative[1, kz_][w][ζ, z] :> Derivative[kz][wζ][z],
Derivative[2, kz_][w][ζ, z] :> Derivative[kz][w$\zeta\zeta$][z]}

Out[3]= $\{w'[z] == \text{Cos}[w[z]] + \text{Sin}[\zeta], w\zeta'[z] == \text{Cos}[\zeta] - \text{Sin}[w[z]] w\zeta[z],$
$w\zeta\zeta'[z] == -\text{Sin}[\zeta] - \text{Cos}[w[z]] w\zeta[z]^2 - \text{Sin}[w[z]] w\zeta\zeta[z]\}$

We supplement the last equations with the corresponding initial conditions obtained from differentiating the initial condition $w_\zeta(0) = \zeta^2$.

In[4]:= eqsζ = Join[%, {w[0] == ζ^2, wζ[0] == 2ζ, w$\zeta\zeta$[0] == 2}]

Out[4]= $\{w'[z] == \text{Cos}[w[z]] + \text{Sin}[\zeta], w\zeta'[z] == \text{Cos}[\zeta] - \text{Sin}[w[z]] w\zeta[z],$
$w\zeta\zeta'[z] == -\text{Sin}[\zeta] - \text{Cos}[w[z]] w\zeta[z]^2 - \text{Sin}[w[z]] w\zeta\zeta[z], w[0] == \zeta^2, w\zeta[0] == 2\zeta, w\zeta\zeta[0] == 2\}$

Now, we can solve this system of differential equations numerically and get an approximate value for $\partial^2 w_\zeta(1)/\partial\zeta^2 |_{\zeta=1}$.

In[5]:= nsol = NDSolve[eqsζ /. ζ -> 1, {w, wζ, w$\zeta\zeta$}, {z, 0, 1},
WorkingPrecision -> 30,
PrecisionGoal -> 25, AccuracyGoal -> 25]

Out[5]= $\{\{w \to \text{InterpolatingFunction}[\{\{0, 1.00000000000000000000000000000\}\}, <>],$
$w\zeta \to \text{InterpolatingFunction}[\{\{0, 1.00000000000000000000000000000\}\}, <>],$
$w\zeta\zeta \to \text{InterpolatingFunction}[\{\{0, 1.00000000000000000000000000000\}\}, <>]\}\}$

In[6]:= der$\zeta\zeta$N = w$\zeta\zeta$[1] /. nsol[[1]]

Out[6]= 0.1338635544772364561304283105070

The problem can be solved in closed form and we obtain the following solution for $w_\zeta(z)$.

In[7]:= wExact =
With[{α = ArcCosh[(1 + Cos[ζ^2] Sin[ζ])/(Cos[ζ^2] + Sin[ζ])] + z Cos[ζ]},
-2 ArcCos[-(Sqrt[(-Cosh[α/2]^2) (-1 + Sin[ζ]) (Cosh[α] + Sinh[α])]/
Sqrt[(Cosh[α] - Sin[ζ]) (Cosh[α] + Sinh[α])])] + 2Pi];

The numerical value agrees to 25 digits with the exact value.

In[8]:= der$\zeta\zeta$N - (D[wExact /. z -> 1, ζ, ζ] /. ζ -> 1)

Out[8]= 7.4755×10^{-26}

31. Modular Function Fourier Series, Differential Equation for $J(z)$, Singular Moduli

a) The terms of the series $\sum_{k=1}^{\infty} a_k e^{k i \pi z/12}$ vanish rapidly when the imaginary part of z increases. We will use this fact to calculate the series terms. Starting from a high-precision numerical value η of $\eta(z)$, we will subtract the largest integer power of $e^{i\pi z/12}$ contained in η. From the remaining difference, we will again subtract the largest integer power of $e^{i\pi z/12}$ and so on. Here, this is implemented. With Rationalize[..., 100], we check the integer value of k. We collect the a_k and the k w in the list 1. We determine the sign of the a_k by testing both possibilities ± 1 and selecting the one with the smaller resulting difference.

In[1]:= z = SetPrecision[2 I, 6000];
η = DedekindEta[z];

In[3]:= k = 0; d[0] = η; exp[0] = 1; l = {};

```
While[k = k + 1; a[k] = Exp[Pi I z exp[k - 1]/12];
      (* positive and negative difference *)
      auxM = d[k - 1] - a[k]; auxP = d[k - 1] + a[k];
      (* decide if term must be added or subtracted *)
      If[Abs[auxM] < Abs[auxP], d[k] = auxM, d[k] = auxP];
      (* collect data *)
      AppendTo[l, If[Abs[auxM] < Abs[auxP],
                     {+1, exp[k - 1]}, {-1, exp[k - 1]}]];
      (* proceed as long as the difference has enough precision *)
```

```
           Precision[d[k]] > 100 && k <= 51,
           (* get the next exponent *)
           exp[k] = Rationalize[N[-12 I Log[Abs[Re[d[k]]]]/(Pi z), 100], 10^-1]]
```

Here, the resulting series for $\eta(x)$ is defined as dedekindEta[x].

```
In[5]:= dedekindEta[x_] = Plus @@ Apply[#1 Exp[#2 I Pi x/12]&, 1, {1}];
```

```
          (* abbreviated version *) Short[dedekindEta[x], 2]
```
Out[6]//Short= $e^{\frac{i\pi x}{12}} - e^{\frac{25 i\pi x}{12}} - e^{\frac{49 i\pi x}{12}} + e^{\frac{121 i\pi x}{12}} + \ll 71\gg + e^{\frac{24025 i\pi x}{12}}$

A numerical check of the result for a "random" value of z.

```
In[7]:= With[{z = 1 + I/2, prec = 1000},
          N[DedekindEta[z], prec] - N[dedekindEta[z], prec]]
```
Out[7]= $0. \times 10^{-1000} + 0. \times 10^{-1000} i$

Now, let us deal with converting the product representation to a series. The direct multiplication of products of the form $(1 - x^k)$ is slow and gives terms with very high powers. Because we are only interested in terms with a maximal power of about $\lceil \frac{21025}{2 \times 12} \rceil = 877$, we use a power series to truncate the polynomials at every step. (Here we use the functions Series and Normal; we will discuss them in detail in Chapter 1 of the Symbolics volume [1795].)

```
In[8]:= dedekindEtaP[x_] = Expand[Exp[Pi I x/12]*
          Normal[Product[1 - X^k, {k, 1, 900}] + O[X]^900] /. X -> Exp[2 Pi I x]];
```

```
          (* abbreviated version *) Short[dedekindEtaP[x], 2]
```
Out[10]//Short= $e^{\frac{i\pi x}{12}} - e^{\frac{25 i\pi x}{12}} - e^{\frac{49 i\pi x}{12}} + e^{\frac{121 i\pi x}{12}} + \ll 65\gg + e^{\frac{20449 i\pi x}{12}} + e^{\frac{21025 i\pi x}{12}}$

The resulting series is the same as above for the first 50 terms.

```
In[11]:= dedekindEta[x] - dedekindEtaP[x]
```
Out[11]= $-e^{\frac{22201 i\pi x}{12}} - e^{\frac{22801 i\pi x}{12}} + e^{\frac{24025 i\pi x}{12}}$

A closed form of the series is the following one [314].

```
In[12]:= dedekindEtaSeries[z_, n_] :=
          Sum[With[{m = Mod[k, 12]},
              Which[m == 1 || m == 11, 1, m == 5 || m == 7, -1,
                  True, 0] Exp[I Pi z k^2/12]], {k, 1, n}]
In[13]:= dedekindEta[x] - dedekindEtaSeries[x, 150]
```
Out[13]= $-e^{\frac{22801 i\pi x}{12}} + e^{\frac{24025 i\pi x}{12}}$

Another way to write the formula from the last input is to use the JacobiSymbol (to be discussed in the next chapter) $\eta(z) = \sum_{k=0}^{\infty} \left(\frac{3}{2k+1}\right) \exp(\pi i z (2k+1)^2 / 12)$.

b) The first method we will implement is the same as above: We calculate a high-precision value for $J(z)$ and recursively subtract the exponential parts. The coefficients a_k are obtained by forming integers from the corresponding approximative high-precision numbers (using Rationalize).

```
In[1]:= zHP = SetPrecision[25 I, 10000];

       l = {};
       d[0] = 12^3 KleinInvariantJ[zHP] - (Exp[-2 Pi I zHP] + 744);

       k = 0;
       While[k = k + 1;
           (* get the integer coefficient *)
           f[k] = Rationalize[N[Exp[-2 Pi I zHP k] d[k - 1], 200], 10^-1];
           (* add obtained coefficient to the list *)
           AppendTo[l, {k, f[k]}];
           (* the new difference *)
           d[k] = d[k - 1] - f[k] Exp[2 Pi I zHP k];
           (* proceed as long as the difference has enough precision *)
           Precision[d[k]] > 100 && k < 100, Null]
```

Here are the obtained Fourier coefficients.

In[6]:= **Short[1, 12]**

Out[6]//Short= {{1, 196884}, {2, 21493760}, {3, 864299970}, {4, 20245856256}, {5, 333202640600},
{6, 4252023300096}, {7, 44656994071935}, {8, 401490886656000}, {9, 3176440229784420},
{10, 22567393309593600}, {11, 146211911499519294}, {12, 874313719685775360},
≪77≫, {90, 1435071724672834538855152223427829911923532076032000},
{91, 2755010426167891537490806178938367969511333929783496},
{92, 527036058053281764188089220041629201191975505756160},
{93, 10047304534409390428438989653654129816903071458278400},
{94, 19088640983213103024886047390986184059389384773797958400},
{95, 361443217930446268187967680912046468497513083620525000},
{96, 6821306832689380776546629825653465084003418476904448},
{97, 1283156845093056623704915719101710486121743363428996000},
{98, 24060143444937604997591586090380473418086401696839680},
{99, 44972195698011806740150818275177754986409472910549646},
{100, 837988311107074769127519503847574527038091833907200}}

The coefficients calculated fulfill known congruences [56], [751].

In[7]:= **Map[Mod[(#[[1]] + 1) #[[2]], 24] == 0 &&**
        Which[IntegerQ[#[[1]]/ 2], Mod[#[[2]], 2048],
                IntegerQ[#[[1]]/ 3], Mod[#[[2]], 243],
                IntegerQ[#[[1]]/ 5], Mod[#[[2]], 25],
                IntegerQ[#[[1]]/ 7], Mod[#[[2]], 7],
                IntegerQ[#[[1]]/11], Mod[#[[2]], 11],
                True, 0] == 0&, 1] // Union

Out[7]= {True}

The second method will be based on the numerical calculation of the Fourier coefficients of $1728 J(z) - e^{-2i\pi z} - 744 = \sum_{k=1}^{\infty} a_k e^{2ki\pi z}$ using NIntegrate. We will carry out the integration along a line segment parallel to the real axis from $0 + i\beta(k)$ to $1 + i\beta(k)$. For speed, we choose the imaginary part of the contour to be a function of k. The oscillatory behavior of $1728 J(z) - e^{-2i\pi z} - 744$ becomes more pronounced by approaching the real axis. Because the integrand is a periodic function, we use the option setting Method -> Trapezoidal.

In[8]:= **Show[GraphicsArray[**
        Block[{z, $DisplayFunction = Identity},
        Plot[Evaluate[Re[(12^3 KleinInvariantJ[z] - (Exp[-2 Pi I z] + 744)) /.
                                                        z -> t + 1/2 I]],
            {t, 0, 1}, PlotRange -> {-1, 1}]& /@ {1/2, 1/12}]]]

This means that for calculating the higher Fourier coefficients, we should use a smaller imaginary part for the integration segment. We choose the following explicit form of $\beta(k)$.

In[9]:= **β[k_] := I (1/2 - 5/12 k/100)**

So, we can implement the calculation of the Fourier coefficients in the following way. With increasing k, we increase the working precision and the precision goal too (because the resulting coefficients a_k have more and more digits).

In[10]:= **a[k_] :=**
        NIntegrate[(12^3 KleinInvariantJ[z] -
                (Exp[-2 Pi I z] + 744)) Exp[-2 k Pi I z], {z, β[k], 1 + β[k]},
                PrecisionGoal -> 20 + 2k/3, WorkingPrecision -> 40 + 2k/3,
                Method -> Trapezoidal, MaxRecursion -> 8] // Rationalize

In[11]:= **lNIntegrate = Table[{k, a[k]}, {k, 100}];**

Comparing with the above methods shows that the two results agree.

```
In[12]:= lNIntegrate == l
Out[12]= True
```

Now, we implement the calculation of a_{100} using the Bessel series. We start with $l(j, h)$.

```
In[13]:= H = Compile[{{j, _Integer}, {h, _Integer}},
              Module[{η = 0}, While[Mod[h η, j] =!= j - 1, η = η + 1]; η]];
```

The following input implements the $A_j(k)$ and the series terms of the a_k. We calculate a_{100} using this formula.

```
In[14]:= A[j_, k_] := Sum[If[GCD[h, j] === 1,
                      N[Exp[-2 Pi I(h k + H[j, h])/j], 200], 0],
                      {h, 0, j - 1}]
```

```
In[15]:= b[k_, j_] := 2 Pi/Sqrt[k] 1/j A[j, k] BesselI[1, 4 Pi Sqrt[k]/j]
```

Here, the convergence of the series to a_{100} is shown.

```
In[16]:= δ = FoldList[Plus, 0, Table[b[100, j], {j, 1, 120}]] - l[[100, 2]];
```

```
In[17]:= ListPlot[δ, PlotRange -> {-2, 2}, Frame -> True, Axes -> False]
```

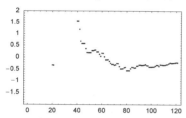

Here is a symbolic method to calculate the series terms under consideration [1614].

```
In[18]:= G[k_, n_, q_] := Sum[DivisorSigma[k, j] q^j, {j, n}]
```

```
In[19]:= symbolicCoeffs =
          With[{n = 120},
          CoefficientList[-1/q +
              1728 Normal[(1 + 240 G[3, n, q])^3/
              ((1 + 240 G[3, n, q])^3 - (1 - 504 G[5, n, q])^2) +
                O[q]^(n + 1)], q] // Take[#, {2, 101}]&];
```

```
In[20]:= symbolicCoeffs === Last /@ l
Out[20]= True
```

The next input calculates the series coefficients using the function Series (to be discussed in Chapter 1 of the Symbolics volume [1795]).

```
In[21]:= (DeleteCases[#[[3]], 0]& @
          With[{ϑ0 = EllipticTheta[#1, 0, #2]&},
          1728 (Series[(4 (ϑ0[2, q]^8 - ϑ0[2, q]^4 ϑ0[3, q]^4 + ϑ0[3, q]^8)^3)/
                  (27 ϑ0[2, q]^8 ϑ0[3, q]^8 (ϑ0[2, q]^4 - ϑ0[3, q]^4)^2),
                  {q, 0, 200}] - 1/(1728 q^2) - 31/72)]) == Last /@ l
Out[21]= True
```

We end with a recursive definition for the coefficients [1193], [990], [939].

```
In[22]:= a[-1] = 1;
         a[ 0] = 744;
         a[ 1] = 196884;
         a[ 2] = 21493760;
         a[ 3] = 864299970;
         a[ 4] = 20245856256;
         a[ 5] = 333202640600;

         a[v_] /; Mod[v, 4] == 0 := a[v] =
```

```
       With[{n = v/4}, a[2n + 1] + (a[n]^2 - a[n])/2 +
           Sum[a[k] a[2n - k], {k, n - 1}]]

       a[v_] /; Mod[v, 4] == 1 := a[v] =
       With[{n = (v - 1)/4}, a[2n + 3] - a[2] a[2n] +
           (a[n + 1]^2 - a[n + 1])/2 + (a[2n]^2 + a[2n])/2 +
           Sum[a[k] a[2n - k + 2], {k, n}] -
           Sum[(-1)^(k - 1) a[k] a[4n - k], {k, 2n - 1}] +
           Sum[a[k] a[4n - 4k], {k, n - 1}]]

       a[v_] /; Mod[v, 4] == 2 := a[v] =
       With[{n = (v - 2)/4}, a[2n + 2] +
           Sum[a[k] a[2n - k + 1], {k, n}]]

       a[v_] /; Mod[v, 4] == 3 := a[v] =
       With[{n = (v - 3)/4}, a[2n + 4] - a[2] a[2n + 1] -
           (a[2n + 1]^2 - a[2n + 1])/2 +
           Sum[a[k] a[2n - k + 3], {k, n + 1}] -
           Sum[(-1)^(k - 1) a[k] a[4n - k + 2], {k, 2n}] +
           Sum[a[k] a[4n - 4k + 2], {k, n}]]
```

In[36]:= `Table[a[k], {k, 100}] == Last /@ l`

Out[36]= `True`

Asymptotically we have $a_k \sim 2^{-1/2}\, n^{-3/4} \exp(4\pi n^{1/2})$ [1496]. Here is the ratio between the calculated values and the asymptotic value shown.

In[37]:= `j[n_] := 1/Sqrt[2] n^(-3/4) Exp[4 Pi Sqrt[n]]`
`ListPlot[Table[l[[k, 2]]/j[k], {k, 100}] // N,`
`        PlotRange -> All, Frame -> True, Axes -> False]`

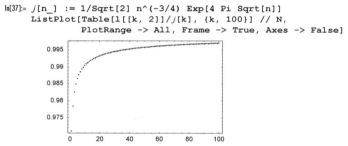

We could now continue and use the calculated coefficients a_k to investigate $J(z)$ on the real axis. At certain z of the form $z = -\exp(-n^{1/2}\pi)$ the function $1728\, J(z)$ takes on integer values (Heegner numbers) [807].

In[39]:= `J[q_, o_] := 1/q + 744 + Sum[a[k] q^k, {k, o}]`
`probablyInteger[x_, ε_] := Abs[x - Round[x]] < ε`

In[41]:= `Table[If[probablyInteger[J[-Exp[-Sqrt[N[n, 100]] Pi], 200], 10^-20],`
`        n, Sequence @@ {}], {n, 200}]`

Out[41]= `{1, 3, 7, 11, 19, 27, 43, 67, 163}`

In[42]:= `Res = {#, Rationalize[J[-Exp[-Sqrt[N[#, 200]] Pi], 200], 0]}& /@ %`

Out[42]= `{{1, 1728}, {3, 0}, {7, -3375}, {11, -32768}, {19, -884736}, {27, -12288000},`
`        {43, -884736000}, {67, -147197952000}, {163, -262537412640768000}}`

c) We will determine the c_k in a purely numerical manner. We will use the given formula and substitute high-precision values for the $w^{(n)}(z)$ for random z. This yields a linear system for the c_k. We start by implementing a numerical derivative. d is a very straightforward implementation of a numerical first derivative. Then for the higher derivatives, we just iterate d.

In[1]:= `d[f_, x0_, ε_] = (f[x0 + ε/2] - f[x0 - ε/2])/ε;`

In[2]:= `j[0][x_, ε_] = KleinInvariantJ[x];`

`Do[j[i][x_, ε_] = Simplify[d[j[i - 1][#, ε]&, x, ε]], {i, 1, 5}]`

This is the form of the differential equation for the `KleinInvariantJ` function.

```
In[4]:= odeJAnsatz =
          c[1] w'[z]^4 + c[2] w[z] w'[z]^4 + c[3] w[z]^2 w'[z]^4 +
          c[4] w[z]^2 w''[z]^2 + c[5] w[z]^3 w''[z]^2 +
          c[6] w[z]^4 w''[z]^2 + c[7] w[z]^2 w'[z] w'''[z] +
          c[8] w[z]^3 w'[z] w'''[z] + c[9] w[z]^4 w'[z] w'''[z];
```

These are the nine constants to be determined.

```
In[5]:= cs = Cases[odeJAnsatz, _c, Infinity]

Out[5]= {c[1], c[2], c[3], c[4], c[5], c[6], c[7], c[8], c[9]}
```

The function `randomSubstitution` generates a list of substitution rules for the derivatives $w^{(n)}(z)$ for randomly chosen z. We use a high precision in the calculation of the derivatives.

```
In[6]:= randomSubstitution :=
          With[{z0 = SetPrecision[
                  Random[] + I Random[Real, {1, 2}], 1000]},
                Table[Derivative[k][w][z] -> j[k][z0, 10^-200], {k, 0, 3}]]
```

Without loss of generality, we fix the coefficient c_1 to be 1.

```
In[7]:= odeJAnsatz1 = odeJAnsatz /. c[1] -> 1;
```

Using eight random substitutions `randomSubstitution`, we obtain a linear system of eight equations in the eight unknowns $c_2 \ldots c_9$.

```
In[8]:= SeedRandom[1111];
          eqs = Table[odeJAnsatz1 /. randomSubstitution, {k, 8}];
```

The solutions of the linear system indicate that all of the c_k are rational.

```
In[10]:= (sol = Solve[# == 0& /@ eqs, Rest[cs]]) /. (c_ -> n_) :> c -> N[n, 60]

Out[10]= {{c[2] → -1.28125000000000000000000000000000000000000000000000000000000 + 0. × 10^-61 i,
          c[3] → 1.12500000000000000000000000000000000000000000000000000000000 + 0. × 10^-61 i,
          c[4] → -3.37500000000000000000000000000000000000000000000000000000000 + 0. × 10^-60 i,
          c[5] → 6.75000000000000000000000000000000000000000000000000000000000 + 0. × 10^-60 i,
          c[6] → -3.37500000000000000000000000000000000000000000000000000000000 + 0. × 10^-60 i,
          c[7] → 2.25000000000000000000000000000000000000000000000000000000000 + 0. × 10^-60 i,
          c[8] → -4.50000000000000000000000000000000000000000000000000000000000 + 0. × 10^-60 i,
          c[9] → 2.25000000000000000000000000000000000000000000000000000000000 + 0. × 10^-60 i}}
```

We rationalize the solutions and plug them into the ansatz for the differential equation.

```
In[11]:= rsol = Rationalize[sol[[1]]]
```

$$Out[11]= \left\{c[2] \to -\frac{41}{32}, \; c[3] \to \frac{9}{8}, \; c[4] \to -\frac{27}{8}, \right.$$
$$\left. c[5] \to \frac{27}{4}, \; c[6] \to -\frac{27}{8}, \; c[7] \to \frac{9}{4}, \; c[8] \to -\frac{9}{2}, \; c[9] \to \frac{9}{4}\right\}$$

```
In[12]:= Factor[32 (odeJAnsatz /. rsol /. c[1] -> 1)]
```

$$Out[12]= 32\,w'[z]^4 - 41\,w[z]\,w'[z]^4 + 36\,w[z]^2\,w'[z]^4 - 108\,w[z]^2\,w''[z]^2 + 216\,w[z]^3\,w''[z]^2 -$$
$$108\,w[z]^4\,w''[z]^2 + 72\,w[z]^2\,w'[z]\,w^{(3)}[z] - 144\,w[z]^3\,w'[z]\,w^{(3)}[z] + 72\,w[z]^4\,w'[z]\,w^{(3)}[z]$$

The so obtained differential equation is fulfilled for any z in the upper half-plane.

```
In[13]:= Table[% /. randomSubstitution, {10}]

Out[13]= {0. × 10^-392 + 0. × 10^-391 i, 0. × 10^-382 + 0. × 10^-382 i,
          0. × 10^-382 + 0. × 10^-382 i, 0. × 10^-382 + 0. × 10^-382 i,
          0. × 10^-394 + 0. × 10^-394 i, 0. × 10^-385 + 0. × 10^-385 i, 0. × 10^-385 + 0. × 10^-385 i,
          0. × 10^-390 + 0. × 10^-390 i, 0. × 10^-393 + 0. × 10^-393 i, 0. × 10^-391 + 0. × 10^-391 i}
```

d) To find numerical approximations of the $q_{n,J}^*$ we use `FindRoot` with random starting values.

```
In[1]:= δ[n_, q_?NumberQ, prec_] :=
          With[{v = SetPrecision[n, prec], q = SetPrecision[q, prec]},
               KleinInvariantJ[q] - KleinInvariantJ[n q]]

In[2]:= findSingularModul[n_, prec_] :=
          q /. FindRoot[δ[n, q, prec], {q, Random[Complex, {0, 2 + 1I}]},
```

```
                                    Random[Complex, {0, 2 + 1I}]},
            WorkingPrecision -> prec, MaxIterations -> prec]
```

To check if we have found a singular value (`FindRoot` might not converge), we use the function `isSingularModulQ`.

```
In[3]:= isSingularModulQ[n_, q_, prec_] := δ[n, q, prec] == 0
```

To convert the approximative values for the singular moduli into exact values, we use the function `Recognize` from the package `NumberTheory`Recognize``.

```
In[4]:= Needs["NumberTheory`Recognize`"]
```

Using the just-implemented functions, it is straightforward to implement a search for singular moduli. We call the function `findSingularModul` until we find a singular moduli.

```
In[5]:= findSingularModul[n_Integer?(# > 1&)] :=
    Module[{q, poly, t, q, qExact},
        While[q = findSingularModul[n, 50];
            poly = Recognize[q, 2, q];
            If[PolynomialQ[poly, t] &&
                (* want small coefficients *)
                Max[Abs[Cases[poly, _Integer, Infinity]]] < 10^3,
                (* exact value of the potential singular modul *)
                qExact = Select[q /. Solve[poly == 0, q], (Im[#] > 0)&][[1]];
                (* did we find a singular value *)
                Not[isSingularModulQ[n, qExact, 200]], True], Null];
    (* return singular value *)
    qExact]
```

Because the Klein modular function is invariant under the modular group, the singular moduli found by `findSingular`
`Modul` are not unique. Here are six different values for the case $n = 2$.

```
In[6]:= SeedRandom[111];

       (* suppress messages *)
       Internal`DeactivateMessages[Table[findSingularModul[2], {10}] // Union]
```
$$Out[8]= \left\{ \frac{1}{2} + \frac{i}{2}, \frac{47}{34} + \frac{i}{34}, \frac{3}{2} + \frac{i}{2}, \frac{1}{4}\left(3 + i\sqrt{7}\right), \right.$$
$$\left. \frac{1}{4}\left(5 + i\sqrt{7}\right), \frac{1}{4}\left(7 + i\sqrt{7}\right), \frac{1}{44}\left(31 + i\sqrt{7}\right), \frac{1}{28}\left(49 + i\sqrt{7}\right)\right\}$$

We end with calculating some singular moduli for $n = 2, ..., 20$.

```
In[9]:= SeedRandom[222];
       Internal`DeactivateMessages[
         smi = Table[{k, Simplify[Mod[findSingularModul[k], 1]]}, {k, 2, 20}]]
```
$$Out[10]= \left\{\left\{2, \frac{i}{\sqrt{2}}\right\}, \left\{3, \frac{1}{6}\left(5 + i\sqrt{11}\right)\right\}, \left\{4, \frac{1}{16}\left(7 + i\sqrt{15}\right)\right\}, \left\{5, \frac{4}{5} + \frac{2i}{5}\right\}, \left\{6, \frac{1}{12}\left(3 + i\sqrt{15}\right)\right\},\right.$$
$$\left\{7, \frac{1}{14}\left(13 + 3i\sqrt{3}\right)\right\}, \left\{8, \frac{1}{64}\left(45 + i\sqrt{23}\right)\right\}, \left\{9, \frac{1}{18}\left(1 + i\sqrt{35}\right)\right\}, \left\{10, \frac{1}{40}\left(23 + i\sqrt{31}\right)\right\},$$
$$\left\{11, \frac{1}{22}\left(17 + i\sqrt{19}\right)\right\}, \left\{12, \frac{1}{60}\left(23 + i\sqrt{11}\right)\right\}, \left\{13, \frac{1}{26}\left(23 + i\sqrt{43}\right)\right\},$$
$$\left\{14, \frac{1}{14}\left(1 + i\sqrt{13}\right)\right\}, \left\{15, \frac{1}{15}\left(2 + i\sqrt{11}\right)\right\}, \left\{16, \frac{1}{48}\left(15 + i\sqrt{15}\right)\right\}, \left\{17, \frac{1}{17}\left(3 + 2i\sqrt{2}\right)\right\},$$
$$\left.\left\{18, \frac{1}{72}\left(41 + i\sqrt{47}\right)\right\}, \left\{19, \frac{1}{76}\left(55 + i\sqrt{15}\right)\right\}, \left\{20, \frac{1}{140}\left(81 + i\sqrt{19}\right)\right\}\right\}$$

```
In[11]:= (δ[#1, #2, Infinity]& @@@ N[smi, 1000]) // Abs // Max
```
$$Out[11]= 0. \times 10^{-998}$$

32. Curve Thickening, Textures, Seed Growth

a) This is the spiral to be "thickened".

```
In[1]:= f[t_] = t Exp[I t];
```

The function `conjugate` carries out a "symbolic conjugation", by explicitly conjugating all numbers with head `Complex`.

```
In[2]:= conjugate[f_] := f /. c_Complex -> Conjugate[c]
```

The inner integral is given as the following expression.

```
In[3]:= int1[t_] = Integrate[conjugate[f[τ]] D[f[τ], τ] -
                          f[τ] D[conjugate[f[τ]], τ], {τ, 0, t}]
```

$$Out[3]= \frac{2\,i\,t^3}{3}$$

The integrand of the outer integral is then given as the following expression.

```
In[4]:= int2 = With[{ρ = 1}, Exp[-f[t] conjugate[f[t]]/ρ^2 +
                       2 z conjugate[f[t]]/ρ^2 + 1/ρ^2 int1[t]] D[f[t], t]]
```

$$Out[4]= e^{-t^2 + \frac{2\,i\,t^3}{3} + 2\,e^{-i\,t}\,t\,z}\,(e^{i\,t} + i\,e^{i\,t}\,t)$$

For a symbolic z and T, *Mathematica* cannot integrate this function in closed form.

```
In[5]:= Integrate[int2, {t, 0, T}]
```

$$Out[5]= \int_0^T e^{-t^2 + \frac{2\,i\,t^3}{3} + 2\,e^{-i\,t}\,t\,z}\,(e^{i\,t} + i\,e^{i\,t}\,t)\,dt$$

So, we will resort to a numerical calculation of this integral with `NIntegrate`. For every numeric value of z, we have to call `NIntegrate`. Because the integrand for each z is of the same form, we do not want `NIntegrate` to compile the integrand each time. So, we compile the integrand before the (many) calls to `NIntegrate`.

```
In[6]:= compiledIntegrand = Compile[{{t, _Real}, {z, _Complex}}, Evaluate[int2]]
```

$$Out[6]= CompiledFunction\Big[\{t, z\}, Block\Big[\{\$\$449, \$\$450\}, \$\$449 = i\,t;$$
$$\$\$450 = e^{\$\$449};\ e^{-t^2 + \frac{2\,i\,t^3}{3} + 2\,e^{-i\,t}\,t\,z}\,(\$\$450 + i\,\$\$450\,t)\Big], -CompiledCode-\Big]$$

Because we are interested in a visualization, it will be enough to have three valid digits. This leads to the following option settings in `NIntegrate`.

```
In[7]:= (* avoid evaluation of the compiled body *)
        Developer`SetSystemOptions["EvaluateNumericalFunctionArgument" -> False];
In[9]:= F[z_] := Exp[-z Conjugate[z]] *
        NIntegrate[compiledIntegrand[t, z], {t, 0, N[2 Pi]},
                   PrecisionGoal -> 3, AccuracyGoal -> 3,
                   Compiled -> False, MaxRecursion -> 10]
```

Now, let us generate a picture. To do this, we calculate a dense set (say, 200×200 points) of function values within the square $-7 \le \mathrm{Re}(z)$, $\mathrm{Im}(z) \le 7$. Here is an estimation of how long this will take.

```
In[10]:= Round[200^2/200 Timing[Do[F[Random[Complex, 7 {-1 - I, 1 + I}]],
                            {200}]][[1]]/(60 Second)] Minutes
Out[10]= 2 Minutes

In[11]:= spiralData = Table[Table[F[x + I y], {x, -7, 7, 7/100}],
                       {y, -7, 7, 7/100}];
```

Here are the real and imaginary parts.

```
In[12]:= opts =  Sequence[Mesh -> False, ColorFunction -> Hue,
                     MeshRange -> {{-7, 7}, {-7, 7}}];
In[13]:= Show[GraphicsArray[
        Block[{$DisplayFunction = Identity},
            ListDensityPlot[#[spiralData], opts]& /@ {Re, Im}]]]
```

The absolute value represents a thickened version of the original spiral and the argument shows interesting behavior.

```
In[14]:= Show[GraphicsArray[
           Block[{$DisplayFunction = Identity},
                 ListDensityPlot[#[spiralData], opts]& /@ {Abs, Arg, Arg[#]^2&}]]]
```

Here are contour plots with 25 contour lines of the absolute value and the argument.

```
In[15]:= contours = 25; pp = 200;
         n1 = Round[pp^2/contours]; n2 = Round[pp^2/contours/2];
         (* start values *)
         opts := Sequence[Contours -> contourValues, ContourStyle -> False,
                          ContourLines -> False, ColorFunction -> Hue,
                          MeshRange -> {{-7, 7}, {-7, 7}}];

In[19]:= Show[GraphicsArray[
           Block[{$DisplayFunction = Identity},
             ((* homogeneous contour values *)
             contourValues = #[[n2]]& /@ Partition[Sort[Flatten[#[spiralData]]], n1];
             ListContourPlot[#[spiralData], opts])& /@
             (* show absolute vale and argument *) {Abs, Arg}]]]
```

Now let us deal with the circle. Proceeding as before, we get the expression $i \exp(3\,i\,t + 2\,z\,e^{-i\,t} - 1)$ for the integrand.

```
In[20]:= compiledIntegrand = Compile[{{t, _Real}, {z, _Complex}},
                                     I Exp[-1 + 3 I t + 2 z Exp[-I t]]];
```

We again calculate numerical values using NIntegrate.

```
In[21]:= circleData =
         Table[Table[F[N[x + I y]], {x, -3, 3, 6/200}], {y, -3, 3, 6/200}];
```

The following pictures show the behavior of $f_1[e^{i\,t}]\,(z)$. The absolute value gives the thickened circle line.

```
In[22]:= Show[GraphicsArray[#]]& /@ Partition[
         ListDensityPlot[#[circleData], DisplayFunction -> Identity,
                         Mesh -> False, ColorFunction -> Hue,
                         MeshRange -> {{-3, 3}, {-3, 3}}]& /@ {Re, Im, Abs, Arg}, 4]
```

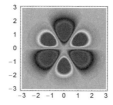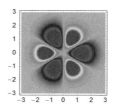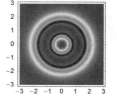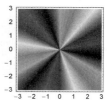

b) Before giving some examples, we define the function `arg`, which basically is identical to `Arg`, but evaluates to `0.` for a vanishing argument.

```
In[1]:= arg = Compile[{{𝒯, _Complex, 2}},
              Map[If[# == 0. + 0. I, 0., Arg[#]]&, 𝒯, {2}]];
```

Further on, we set some options for `ListDensityPlot`.

```
In[2]:= SetOptions[ListDensityPlot, Mesh -> False,
              DisplayFunction -> Identity, FrameTicks -> False];
```

Experimenting with various kernels for the convolution, operations on and between the matrices and their Fourier transforms as well as using combinations of `Re`, `Im`, `Abs`, and `Arg` to visualize the resulting matrices it is not difficult to find some "nice" textures. Here are some examples.

```
In[3]:= SeedRandom[333];
        𝒯 = Table[N[Tan[i + j] + i/j], {i, 512}, {j, 512}];
        𝒦 = Table[Abs[Cos[i + j]], {i, 23}, {j, 23}];
        Show[GraphicsArray[ListDensityPlot[Arg @ #]& /@ #]]& @
        Table[ℱ = Fourier[𝒯]; 𝒯 = ListConvolve[𝒦, 𝒯 ℱ, {2, 2}];
              𝒯 = 𝒯/Max[Abs[𝒯]], {5}][[{3, 4, 5}]]
```

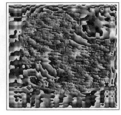

```
In[7]:= SeedRandom[555];
        𝒯 = Table[Random[], {i, 256}, {j, 256}];
        𝒦 = Table[Random[Real], {23}, {23}];
        Show[GraphicsArray[ListDensityPlot[Arg @ #]& /@ #]]& @
        Table[ℱ = Fourier[𝒯]; 𝒯 = ListConvolve[𝒦, 1/𝒯 + ℱ, {2, 2}],
              {10}][[{7, 8, 10}]]
```

```
In[11]:= SeedRandom[999];
         𝒯 = Table[Random[], {k, 256}, {l, 256}];
         𝒦 = Table[Random[Integer], {11}, {11}];
         Show[GraphicsArray[ListDensityPlot[Arg @ #]& /@ #]]& @
         Table[ℱ = Fourier[𝒯]; 𝒯 = ListConvolve[𝒦, 𝒯 + ℱ^2, {1, 1}],
               {6}][[{3, 4, 6}]]
```

 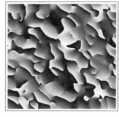 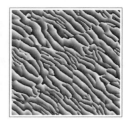

```
In[15]:= SeedRandom[333];
        𝒯 = Table[Random[], {k, 256}, {l, 256}];
        𝒦 = Table[Random[Real], {21}, {21}];
        Show[GraphicsArray[ListDensityPlot[Arg @ #]& /@ #]]& @
        Table[ℱ = Fourier[𝒯]; 𝒯 = ListConvolve[𝒦, 1/𝒯 + ℱ, {20, 20}],
            {21}][[{5, 9, 21}]]
```

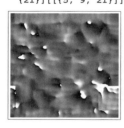 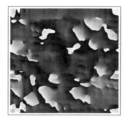 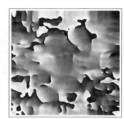

```
In[19]:= SeedRandom[111];
        𝒯 = Table[Random[Complex, {-1 - I, 1 + I}], {k, 256}, {l, 256}];
        𝒦 = Table[Random[Integer], {15}, {15}];
        Show[GraphicsArray[ListDensityPlot[arg @ #]& /@ #]]& @
        Table[ℱ = Fourier[𝒯]; 𝒯 = ListConvolve[𝒦, Chop[ℱ, 0.5], {2, 2}],
            {14}][[{5, 10, 14}]]
```

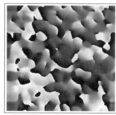 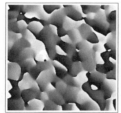 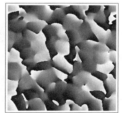

```
In[23]:= SeedRandom[345];
        𝒯 = Table[If[Random[] <= 1/100, 1./(k + 1), 0.],
                {k, 256}, {l, 256}];
        𝒦 = Table[Random[Real, {0, 1}], {i, 5}, {j, 5}];
        Show[GraphicsArray[ListDensityPlot[arg @ #]& /@ #]]& @
        Table[ℱ = Fourier[𝒯]; 𝒯 = ListConvolve[𝒦, ℱ, {2, 2}],
            {16}][[{12, 13, 14}]]
```

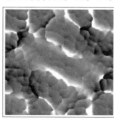 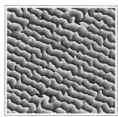 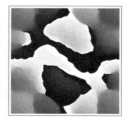

The next three graphics show variants of the Fourier and inverse Fourier transform of a 2D Sierpinski sponge.

```
In[27]:= Module[{m = {{1, 1, 1}, {1, 0, 1}, {1, 1, 1}},
          z = {{0, 0, 0}, {0, 0, 0}, {0, 0, 0}}, s},
          (* iterate substitution to generate Sierpinski fractal *)
          s = Nest[Flatten[MapThread[Join, {##}]& @@@
                    (# /. {1 -> m, 0 -> z}), 1]&, m, 4];
          (* DFT *) sFT = Fourier[s];
          Show[GraphicsArray[
          ListDensityPlot[#, Mesh -> False, PlotRange -> All,
                    FrameTicks -> False, DisplayFunction -> Identity]& /@
                    (* three functions of the Fourier transform *)
                    {Log[Abs[sFT]], Abs[InverseFourier[Tan[sFT]]],
                     Abs[InverseFourier[Sign[sFT]]]}]]]
```

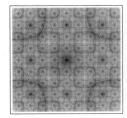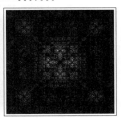

The following function implements a generalization of the game of life [567], [568], [1191].

```
In[28]:= LtLStepC =
          Compile[{{m0, _Integer, 2}, {ρ, _Integer},
                   {β1, _Integer}, {β2, _Integer}, {δ1, _Integer}, {δ2, _Integer}},
          Module[{𝒦 = Table[1, {2ρ + 1}, {2ρ + 1}], m1, d = Length[m0]},
                   (* count living neighbors *)
                   m1 = ListConvolve[𝒦, m0, ρ + 1];
                   (* keep alive or make alive *)
                   Table[If[m0[[i, j]] == 1, If[δ1 <= m1[[i, j]] <= δ2, 1, 0],
                                            If[β1 <= m1[[i, j]] <= β2, 1, 0]],
                   {i, d}, {j, d}]]];
```

Starting with random initial conditions in the long run a complicated maze of horizontal and vertical stripes of living cells forms for the parameter values $\rho = 5$, $\beta_1 = 24$, $\beta_2 = 41$, $\delta_1 = 34$, and $\delta_2 = 58$.

```
In[29]:= m0 = With[{d = 240}, Table[Random[Integer], {d}, {d}]];

         bag = {};
         Do[If[MemberQ[{20, 50, 100, 200, 500, 1000}, k],
                AppendTo[bag, m0]];
            m0 = LtLStepC[m0, 5, 34, 41, 34, 58], {k, 1000}];
In[32]:= Show[GraphicsArray[#]]& /@
         Partition[ListDensityPlot[#, Mesh -> False, FrameTicks -> None,
                                DisplayFunction -> Identity]& /@ bag, 3]
```

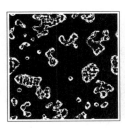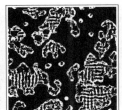

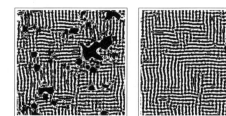 

The last "Larger than Life" steps allow a wide variety of patterns to be generated. The following code generates random instances of "Larger than Life" and shows the resulting nontrivial state after wdh steps.

```
Module[{pictureCounter = 0, wdh = 100,
        counter, m0, ρ, X, β1, β2, δ1, δ2},
While[pictureCounter < 200,
  (* random parameters *)
  ρ = Random[Integer, {2, 12}];
  X = (2ρ + 1)^2;
  {{β1,  β2}, {δ1,  δ2}} =
    Table[Rest[NestList[Random[Integer, {#, X}]&, 0, 2]], {2}];
  (* random start *)
  m0 = With[{d = 200}, Table[Random[Integer], {d}, {d}]];
  (* iterate LtLStepC application *)
  counter = 0;
  While[counter < wdh && MemberQ[m0, 1, {2}],
      m0 = LtLStepC[m0, ρ, β1, β2, δ1, δ2]; counter++];
  (* display result if there are survivors *)
  If[counter === wdh, Print[{ρ, β1, β2, δ1, δ2}]; pictureCounter++;
    ListDensityPlot[m0, Mesh -> False, FrameTicks -> None]]]]
```

Here are some of the patterns found with the above code.

```
In[33]:= (* parameter values and number of steps *)
         LtLData =
             {{{7, 130, 218, 25, 160}, {50}},  {{7, 3, 127, 60, 223}, {5}},
              {{3, 7, 33, 15, 48}, {12, 100}},  {{6, 52, 148, 89, 134}, {10}},
              {{9, 48, 134, 170, 342}, {10}},  {{7, 54, 145, 103, 224}, {10}},
              {{6, 56, 92, 36, 150}, {25}},  {{3, 8, 49, 9, 34}, {31}}};

In[35]:= Show[GraphicsArray[#]]& /@ Partition[Flatten[
         Table[{ρ, β1, β2, δ1, δ2} = LtLData[[j, 1]];
                   (* start with identical initial conditions *) SeedRandom[123];
                   m0 = With[{d = 256}, Table[Random[Integer], {d}, {d}]];
                   (* apply LtLStepC *)
                   Do[m0 = LtLStepC[m0, ρ, β1, β2, δ1, δ2], {LtLData[[j, 2, h]]}];
                   (* make graphic *)
                   ListDensityPlot[m0, Mesh -> False, FrameTicks -> None,
                                   DisplayFunction -> Identity],
               {j, Length[LtLData]}, {h, Length[LtLData[[j, 2]]]}]], 3]
```

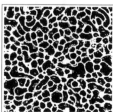 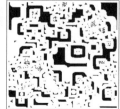 

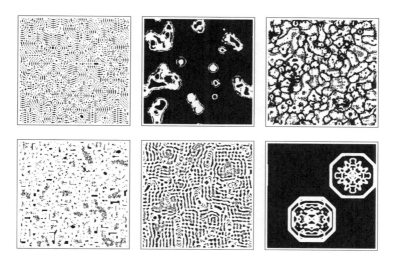

The last graphic shows an interesting symmetric pattern. The following animation shows the first 90 steps in the life of such patterns.

```
In[36]:= Show[GraphicsArray[#]]& /@ Partition[
         SeedRandom[16];
         m0 = With[{d = 256}, Table[Random[Integer], {d}, {d}]];
         Table[m0 = LtLStepC[m0, 3, 8, 49, 9, 34];
               If[(* selected frames *)
                  MemberQ[{7, 11, 19, 29, 43, 47}, j],
                  ListDensityPlot[m0, Mesh -> False, FrameTicks -> None,
                                     DisplayFunction -> Identity],
                  Sequence @@ {}], {j, 50}], 3]
```

```
SeedRandom[16];
m0 = With[{d = 256}, Table[Random[Integer], {d}, {d}]];
Do[m0 = LtLStepC[m0, 3, 8, 49, 9, 34];
   If[j >= 2,
      ListDensityPlot[m0, Mesh -> False, FrameTicks -> None,
                      PlotLabel -> j]], {j, 75}]
```

The next graphics show examples of coupled sine-circle maps [1922].

```
In[37]:= coupledSineCircleMapGraphics[d_, {Ω_, k_, ε_}, iters_] :=
      Module[{f, g, k, M, step, ms, dp = Developer`ToPackedArray},
         (* direct step and coupling function *)
         f[m_] := m + Ω + k/(2 Pi) Sin[2. Pi m]; g[m_] := -Sin[2. Pi m];
         (* interacting neighbors *)
         k = {{1., 1., 1.}, {1., 0., 1.}, {1., 1., 1.}};
         (* number of nontrivial neighbors *)
         M = dp @ ListConvolve[k, Table[1., {d}, {d}], {2, 2}, 0.];
         (* evolution step *)
         step[m_] := Mod[f[m] + ε (ListConvolve[k,  g[m], {2, 2}, 0.] - M m), 1];
         (* apply step repeatedly, keep copies after iters steps *)
         ms = FoldList[Nest[step[#]&, #1, #2]&, dp @ Table[Random[], {d}, {d}],
                       iters] // Rest;
         (* show resulting matrices *)
         Show[GraphicsArray[ListDensityPlot[#, Mesh -> False, FrameTicks -> False,
                            DisplayFunction -> Identity, PlotRange -> All]& /@ ms]]]

In[38]:= SeedRandom[111];
      coupledSineCircleMapGraphics[256, #, {500, 500, 1000}]& /@
         (* four parameter sets *)
      {{0.198533510264941, 0.921092880667925, 0.00677045483929650},
       {0.839844198217592, 0.549554151913025, 0.00276897531626948},
       {0.381613090551368, 0.966377514218281, 0.00517068726947331},
       {0.725121932951714, 0.874520148656763, 0.05218079560737359}
       (* some more parameter sets that yield structured results *)
      (*{-1.122400003109135,  0.52822153941290, 0.0160059174285709}
       { 0.673024056984886,  1.17875795757175, 0.0306293071275626}
       { 1.078861755166273,  1.95482550400827, 0.0050523900596807}
       { 0.641241740868246,  1.03142774418203, 0.0200855768805339}
       { 1.164995645400203, -1.24241188782262, 0.0200731073004374}
       *)}
```

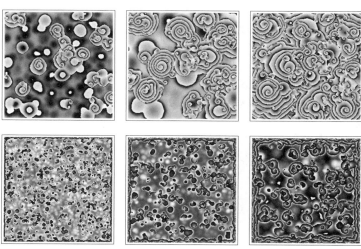

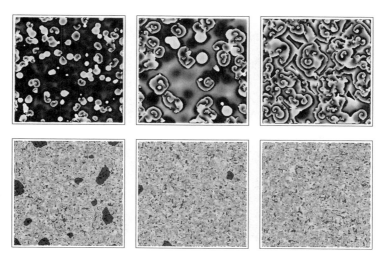

Here are some instances of coupled logistic maps. The convolution kernels and the parameters are created through (seeded) calls to Random.

```
In[40]:= Module[{d = 200, f, g, step, r := Random[Real, {-1, 1}],
          mS, r, ε, mStart, tp = Developer`ToPackedArray},
      (* logistic map and cubic nonlinearity *)
      f[m_, r_] := r m (1 - m); g[m_] := m^3;
      (* one step for the matrix m *)
      step[m_, m_, k_, {r_, ε_}, kd_] :=
          Mod[f[m, r] + ε (ListConvolve[k, g[m], kd + 1] - m m), 1];
      (* list of coupled elements *)
      mS = ((* seed random number generator *)
          SeedRandom[#1]; kd = Random[Integer, {1, 5}];
          (* random parameter and coupling *)
          {r, ε} = {Random[Real, {-6, 6}],
                (2 Random[Integer] - 1) 10^Random[Real, {-9/5, -4}]};
          (* random kernel; central element is zero *)
          k = Table[r, {2 kd + 1}, {2 kd + 1}]; k[[kd + 1, kd + 1]] = 0.;
          m = tp @ ListConvolve[k, Table[1., {d}, {d}], kd + 1];
          (* initial matrix *)
          mStart = tp @ Table[Random[Real, {-3, 3}], {d}, {d}];
          (* iterate the map *)
          Nest[step[#, m, k, {r, ε}, kd]&, mStart, #2])& @@@
          (* seeds and iteration numbers *)
          {{20655051550348424502,  50}, {49891185317788053985, 200},
          {94211589627949494617,  60}, {48194176649231493479, 120},
          {76726471197372414773, 100}, {13969063396213191371,  30},
          {37161662882663241419, 140}, {64885051256101825987, 100},
          {84150698566826738338, 100}};
      (* visualize resulting matrices *)
      Show[GraphicsArray[#]]& /@ Partition[
          ListDensityPlot[#, Mesh -> False, FrameTicks -> False,
            PlotRange -> All, DisplayFunction -> Identity]& /@ mS, 3]]
```

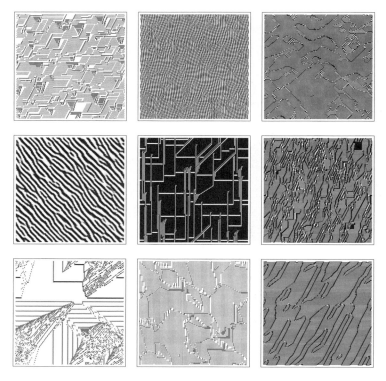

Instead of iterating just one matrix, we could go on and iterate a pair of matrices. The function `step` implements a matrix version of a coupled logistic map [1685], [410].

```
In[41]:= step[{A_, B_}, c1_, c2_, μ_, β_] :=
    With[{𝒦 = {{0., 1., 0.}, {1., -4., 1.}, {0., 1., 0.}}},
        Abs[{μ A (1. - A) Exp[-β B] + c1 ListConvolve[𝒦, A, {2, 2}],
            A (1. - Exp[-β B])    + c2 ListConvolve[𝒦, B, {2, 2}]}]]
```

For various parameters $c1$, $c2$, μ, and β, interesting patterns arise already after a few iterations. Here are two sets of examples.

```
In[42]:= SeedRandom[111];
    (* initial matrices *)
    AB = Table[Table[Random[], {256}, {256}], {2}];
    (* bag to collect the matrices to be displayed *)
    bag = {};
    (* list of iterations after which to display matrix A *)
    L = {90, 170, 230, 530, 1050, 1780};
    Do[AB = step[AB, 0.001, 0.2, 4, 5];
        If[MemberQ[L, k], AppendTo[bag, First[AB]]], {k, 1780}];

In[50]:= (* display the collected matrices *)
    Show[GraphicsArray[ListDensityPlot[#,
        Mesh -> False, FrameTicks -> None,
        DisplayFunction -> Identity]& /@ #]]& /@ Partition[bag, 3]
```

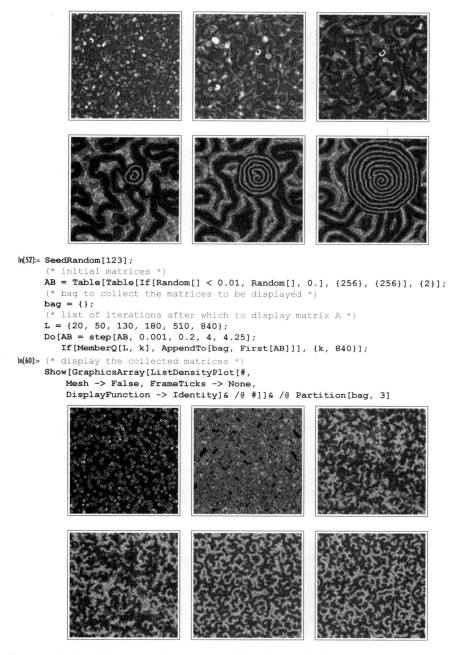

```
In[52]:= SeedRandom[123];
        (* initial matrices *)
        AB = Table[Table[If[Random[] < 0.01, Random[], 0.], {256}, {256}], {2}];
        (* bag to collect the matrices to be displayed *)
        bag = {};
        (* list of iterations after which to display matrix A *)
        L = {20, 50, 130, 180, 510, 840};
        Do[AB = step[AB, 0.001, 0.2, 4, 4.25];
           If[MemberQ[L, k], AppendTo[bag, First[AB]]], {k, 840}];
In[60]:= (* display the collected matrices *)
        Show[GraphicsArray[ListDensityPlot[#,
           Mesh -> False, FrameTicks -> None,
           DisplayFunction -> Identity]& /@ #]]& /@ Partition[bag, 3]
```

Here is a generalization of the last map for larger kernels. We randomly select kernels and parameters.

```
iteratedExpMap[d_, {seed_, o_, j_}] :=
Module[{step, cf, kd1, kd2, K1, K2, paras, AB},
(* one step *)
step[{A_, B_}, {K1_, K2_}, {d1_, d2_}, {α_, c1_, c2_, μ_, β_}] :=
      cf[A, B, K1, K2, d1, d2, α, c1, c2, μ, β];
(* compiled version of a single step *)
Developer`SetSystemOptions["CatchMachineUnderflow" -> False];
cf = Compile[{{A, _Real, 2}, {B, _Real, 2}, {K1, _Real, 2}, {K2, _Real, 2},
              {d1, _Integer}, {d2, _Integer}, α, c1, c2, μ, β},
With[{expB = Exp[-β B]}, (* keep bounded *) α Sin[Abs[
  {μ A (1. - A) expB + c1 ListConvolve[K1, A, {d1 + 1, d1 + 1}],
   A (1. - expB) +    c2 ListConvolve[K2, B, {d2 + 1, d2 + 1}]}]/α]]];
(* seed random number generator *)
SeedRandom[seed];
(* random kernels and parameters *)
kd1 = Random[Integer, {1, 5}]; kd2 = Random[Integer, {1, 5}];
paras = {10^Random[Real, {-2, 2}],
         10^Random[Real, {0, -6}], 10^Random[Real, {0, -6}],
         Random[Real, {0, 6}], Random[Real, {0, 6}]};
(#1 = Table[Random[Real, {0, 1}], {2 #2 + 1}, {2 #2 + 1}];
 (* sum of kernel elements vanishes *)
 #1[[#2 + 1, #2 + 1]] = #1[[#2 + 1, #2 + 1]] - Plus @@ Flatten[#1];
 #1 = Developer`ToPackedArray @ #1)& @@@ {{K1, kd1}, {K2, kd2}};
(* random initial matrices *)
AB = Table[Table[Random[Real, {-3, 3}], {d}, {d}], {2}];
(* iterate the map *)
Do[AB = step[AB, {K1, K2}, {kd1, kd2}, paras], {o}];
(* make density plot of matrix *)
ListDensityPlot[AB[[j]], PlotRange -> All, FrameTicks -> False,
                         Mesh -> False]]

Show[GraphicsArray[
Block[{$DisplayFunction = Identity},
     (iteratedExpMap[256, #])& /@ #]]]& /@
Partition[
{{73246262326297097051,  100, 1}, {54933538759263413590, 1200, 2},
 {25204249815386110422,  600, 2}, {42150212803414984098,  400, 1},
 {17894704366925879557, 1000, 1}, {70240617244524700237,  600, 2},
 {94864641349600396625, 1400, 2}, {18109549243474921419,  600, 1},
 {73801373522241679655,  200, 1}, {67581968782809763424,  500, 2},
 {87350874984692361099,  200, 2}, {88458315009275327744,  600, 1},
 {14302159700639933524,  200, 2}, {34582973883770424131,  700, 2},
 {33838170911407170384,  200, 1}, {26970896108452198075,  400, 2},
 {22097797493184066040,  100, 2}, {00072863763923609326, 3000, 2},
 {25513319394572996769,  700, 2}, {73444095762667569397, 1000, 1},
 {62780463356523461139,  200, 1}}, 3]
```

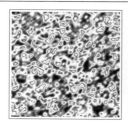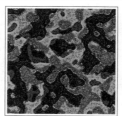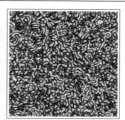

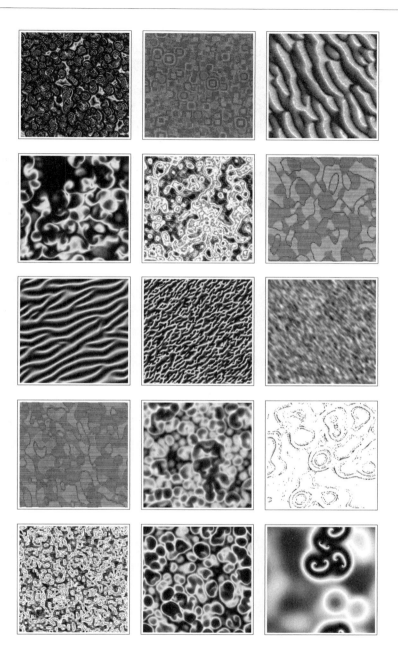

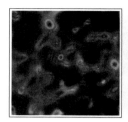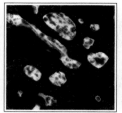

Here is a still more general map. This time, we add the Fourier transform of the second matrix to a nonlinear function of the first matrix that was and convolved with a random kernel.

```
iteratedGeneralizedExpFourierMap[d_, {seed_, o_, j_}] :=
Module[{step, gC, hC, r, kd1, kd2, K1, K2, paras1, paras2, AB},
(* one step *)
step[{A_, B_}, {𝒦1_, 𝒦2_}, {d1_, d2_},
     {r1a_, r1b_, ε1_}, {r2a_, r2b_, ε2_}] :=
{hC[ListConvolve[𝒦1, gC[A, r1a, r1b], d1 + 1] + ε1 Fourier[B]],
 hC[ListConvolve[𝒦2, gC[B, r2a, r2b], d2 + 1] + ε2 Fourier[A]]};
(* compiled version of a single step *)
Developer`SetSystemOptions["CatchMachineUnderflow" -> False];
(* apply discontinuous map *)
gC = Compile[{{m, _Complex, 2}, {ra, _Complex}, {rb, _Complex}},
             Map[If[Arg[#] >= 0, Exp[ra #], rb/(2 + #)]&, m, {2}]];
(* normalize *)
hC = Compile[{{m, _Complex, 2}}, m/Sqrt[Plus @@ (Abs[Flatten[m]]^2)]];
(* seed random number generator and define random complex number *)
SeedRandom[seed]; r := Random[] Exp[2 Pi I Random[]];
(* random kernels and parameters *)
kd1 = Random[Integer, {1, 5}]; kd2 = Random[Integer, {1, 5}];
{paras1, paras2} = Table[{3 r, 3 r, 10^Random[Real, {1, -4}] r}, {2}];
(#1 = Table[r, {2 #2 + 1}, {2 #2 + 1}]; (* sum of kernel elements vanishes *)
 #1[[#2 + 1, #2 + 1]] = -Plus @@ Flatten[#1]; #1)& @@@ {{K1, kd1}, {K2, kd2}};
(* random initial matrices *)
AB = Table[Table[r, {d}, {d}], {2}];
(* iterate the map *)
Do[AB = step[AB, {K1, K2}, {kd1, kd2}, paras1, paras2], {o}];
(* make density plot of matrix *)
ListDensityPlot[ArcTan[Abs[AB[[j]]]],
                PlotRange -> All, FrameTicks -> False, Mesh -> False]]

Show[GraphicsArray[
Block[{$DisplayFunction = Identity},
      iteratedGeneralizedExpFourierMap[256, #]& /@ #]]& /@
Partition[
{{41579617359245739672, 20, 1}, {59346381404383797803, 40, 1},
 {19623361188578420788, 80, 2}, {16657354678322857612, 90, 2},
 {71309179983063865646, 70, 1}, {17957367256622297849, 60, 2},
 {41201656448708510745, 40, 1}, {04316187455520639200, 20, 1},
 {83157801450080512057, 60, 2}}, 3]
```

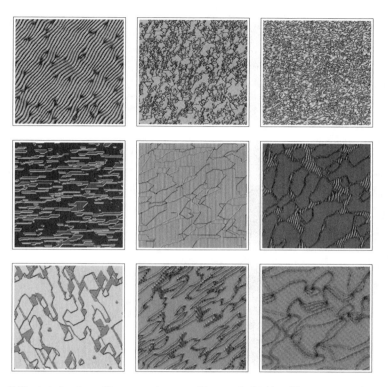

Still another possibility is to iterate multicomponent vectors. Here we deal with a 1D quantum random walk [1037], [1034], [1035], [25], [1335], [1785], [1039], [1036], [971], [347], [1012], [1786], [1038], [1048] with cyclic boundary conditions. The object to be iterated is a list ℓ of two-component elements. Here we must use `ListConvolve` with all its seven potential arguments.

```
In[62]:= (* left-stepping, right-stepping and zero-stepping *)
        {P, Q, O} = {{{1, 1}, {0, 0}}, {{0, 0}, {1, -1}},
                      {{0, 0}, {0, 0}}}/Sqrt[2];

        quantumRandomWalkStep[ℓ_] := ListConvolve[{P, O, Q}, ℓ, 2, ℓ, Dot, Plus, 1]
```

The next input shows the first few steps of an initially localized wave function.

```
In[66]:= NestList[Expand[quantumRandomWalkStep[#]]&,
         {{0, 0}, {0, 0}, {0, 0}, {α, β}, {0, 0}, {0, 0}, {0, 0}}, 4]
```

$$
\begin{aligned}
\text{Out[66]= } &\Big\{\{\{0, 0\}, \{0, 0\}, \{0, 0\}, \{\alpha, \beta\}, \{0, 0\}, \{0, 0\}, \{0, 0\}\}, \\
&\{\{0, 0\}, \{0, 0\}, \{\tfrac{\alpha}{\sqrt{2}} + \tfrac{\beta}{\sqrt{2}}, 0\}, \{0, 0\}, \{0, \tfrac{\alpha}{\sqrt{2}} - \tfrac{\beta}{\sqrt{2}}\}, \{0, 0\}, \{0, 0\}\}, \\
&\{\{0, 0\}, \{\tfrac{\alpha}{2} + \tfrac{\beta}{2}, 0\}, \{0, 0\}, \{\tfrac{\alpha}{2} - \tfrac{\beta}{2}, \tfrac{\alpha}{2} + \tfrac{\beta}{2}\}, \{0, 0\}, \{0, -\tfrac{\alpha}{2} + \tfrac{\beta}{2}\}, \{0, 0\}\}, \\
&\{\{\tfrac{\alpha}{2\sqrt{2}} + \tfrac{\beta}{2\sqrt{2}}, 0\}, \{0, 0\}, \{\tfrac{\alpha}{\sqrt{2}}, \tfrac{\alpha}{2\sqrt{2}} + \tfrac{\beta}{2\sqrt{2}}\}, \{0, 0\}, \{-\tfrac{\alpha}{2\sqrt{2}} + \tfrac{\beta}{2\sqrt{2}}, -\tfrac{\beta}{\sqrt{2}}\}, \\
&\quad \{0, 0\}, \{0, \tfrac{\alpha}{2\sqrt{2}} - \tfrac{\beta}{2\sqrt{2}}\}\}, \{\{0, -\tfrac{\alpha}{4} + \tfrac{\beta}{4}\}, \{\tfrac{3\alpha}{4} + \tfrac{\beta}{4}, \tfrac{\alpha}{4} + \tfrac{\beta}{4}\}, \\
&\quad \{0, 0\}, \{-\tfrac{\alpha}{4} - \tfrac{\beta}{4}, \tfrac{\alpha}{4} - \tfrac{\beta}{4}\}, \{0, 0\}, \{\tfrac{\alpha}{4} - \tfrac{\beta}{4}, -\tfrac{\alpha}{4} + \tfrac{3\beta}{4}\}, \{\tfrac{\alpha}{4} + \tfrac{\beta}{4}, 0\}\}\Big\}
\end{aligned}
$$

To carry out more steps quickly, we use numerical versions of the three matrices P, Q, and O.

```
In[67]:= {P, Q, O} = N[{P, Q, O}];
        quantumRandomWalkStep[ℓ0_, n_] := NestList[quantumRandomWalkStep, ℓ0, n]
```

The following two graphics show the patterns resulting from a random and a pseudo-random initial wave function.

```
In[69]:= (* normalize a list of two-component elements *)
         normalize[l_] := If[# == {0, 0}, {0, 0}, #/#.Conjugate[#]]& /@ l;
In[71]:= With[{o = 240, r = 400},
        Show[GraphicsArray[
        ListDensityPlot[Map[(* show norm *) #.Conjugate[#]&,
                            quantumRandomWalkStep[normalize[#], r], {-2}],
                        Mesh -> False, ColorFunction -> (Hue[0.8 #]&),
                        FrameTicks -> None, DisplayFunction -> Identity]& /@
            {(* random initial wave function *)
             SeedRandom[111];
             Table[Random[Integer, 2 {-1, 1}], {j, o}, {2}] // N,
             (* pseudo-random initial wave function *)
             Table[{Sin[j], Tan[j]}, {j, o}] // N}]]]
```

We end with a model for dendritic growth [975], [1458], [1161], [1517]. The function dendritGrowthCF carries out *o* dendritic growth steps of a $d \times d$ grid. In each step, the diffusion equation is approximately solved, and then particles are aggregated, beginning at $(2m + 1)^2$ particles at the center. β and $\beta 0$ are the boundary and initial concentration [1571], [1572], [1574]

```
In[72]:= dendritGrowthCF = ReleaseHold[
        Hold[Compile[{{d, _Integer}, {o, _Integer}, {m, _Integer}, {aMax, _Integer},
                      uMin, α, Δt, β, β0},
        Module[{u, x, a, aS, aC, g, i = 1, j = 1, mnnl,
                borderMatrix, innerMatrix, uTemp, counter = 1, counterA = 0.,
                nearestNeighborList = {{0, 0}}, nextNearestNeighborList = {{0, 0}}},
        (* initialize matrices *)
        (* count aggregated particles *)
        a = Table[If[Abs[i - (d + 1)/2] <= m &&
                     Abs[j - (d + 1)/2] <= m, N[counter], 0.],
                 {i, d}, {j, d}]; aS = Sign[a]; aC = 1. - aS;
        counterA = Plus @@ Flatten[aS];
        (* auxiliary matrix for interface aggregation *)
        x = Table[If[a[[i, j]] == 1, 0., 0.], {i, d}, {j, d}];
        (* diffusion field matrix *)
        u = Table[Which[i == 1 || i == d || j == 1 || j == d, β,
                        aS[[i, j]] == 1., 0., True, β0], {i, d}, {j, d}];
        (* minimal values for aggregation *)
        g = Table[Random[Real, {uMin, 1.}], {i, d}, {j, d}];
        (* matrices for the inner and edge part *)
        borderMatrix = Table[If[i == 1 || i == d || j == 1 || j == d, 1., 0.],
                             {i, d}, {j, d}];
        innerMatrix = Table[1., {i, d}, {j, d}] - borderMatrix;
        (* current nearest and next nearest neighbors *)
        nearestNeighborList = Union[Flatten[
          Table[If[a[[i, j]] != 1 && nearestNeighborTest,
                {i, j}, {0, 0}], {i, 2, d - 1}, {j, 2, d - 1}], 1]];
        nextNearestNeighborList = Union[Flatten[
          Table[If[a[[i, j]] != 1 && nextNearestNeighborTest,
                {i, j}, {0, 0}], {i, 2, d - 1}, {j, 2, d - 1}], 1]];
        nextNearestNeighborList = Prepend[Complement[nextNearestNeighborList,
                                          nearestNeighborList], {0, 0}];
        (* do o "growth" steps *)
```

```
While[nearestNeighborList != {{0, 0}} &&
      nextNearestNeighborList != {{0, 0}} &&
      counterA < aMax && counter < o,  counter++;
 (* diffusion step; keep gradient at edges constant *)
 u = u + Δt ListConvolve[{{0., 1., 0.}, {1., -4., 1.}, {0., 1., 0.}},
                         u, {2, 2}];
 u = (u innerMatrix + β borderMatrix) aC;
 (* surfaceStep *)
 Do[(* nearest neighbors to an aggregated site *)
    {i, j} = nearestNeighborList[[k]];
    uTemp = u[[i, j]]; u[[i, j]] == 0.; x[[i, j]] = x[[i, j]] + uTemp,
    {k, 2, Length[nearestNeighborList]}];
  Do[(* next nearest neighbors to an aggregated site *)
     {i, j} = nextNearestNeighborList[[k]];
     uTemp = u[[i, j]]; u[[i, j]] = (1 - α) uTemp;
     x[[i, j]] = x[[i, j]] + α uTemp,
     {k, 2, Length[nextNearestNeighborList]}];
  (* aggregate step *)
  If[Max[x - g] > 0,
     Do[If[a[[i, j]] == 0. && x[[i, j]] > g[[i, j]],
           a[[i, j]] = N[counter]; u[[i, j]] = 0; x[[i, j]] = 0],
        {i, 2, d - 1}, {j, 2, d - 1}]; aS = Sign[a]; aC = 1 - aS;
     (* update neighbor list *)
     nearestNeighborList = Union[Flatten[
       Table[If[aS[[i, j]] != 1 && nearestNeighborTest,
         {i, j}, {0, 0}], {i, 2, d - 1}, {j, 2, d - 1}], 1]];
     nextNearestNeighborList = Union[Flatten[
       Table[If[aS[[i, j]] != 1 && nextNearestNeighborTest &&
                                   Not[nearestNeighborTest],
              {i, j}, {0, 0}], {i, 2, d - 1}, {j, 2, d - 1}], 1]];
     counterA = Plus @@ Flatten[aS]]];
     (* for intermediate progress information:
        If[counter < o, Print["exited at step ", counter,"."] *)
  (* return aggregated particle matrix *) a]]] /.
 {(* the repeatedly used test for nearest and next nearest neighbors *)
  nearestNeighborTest :> aS[[i + 1, j]] == 1 || aS[[i - 1, j]] == 1 ||
                         aS[[i, j + 1]] == 1 || aS[[i, j - 1]] == 1,
  nextNearestNeighborTest :> aS[[i + 1, j + 1]] == 1 ||
                             aS[[i + 1, j - 1]] == 1 ||
                             aS[[i - 1, j + 1]] == 1 ||
                             aS[[i - 1, j - 1]] == 1}];
```

The last function was a compiled function to speed up the calculations. All steps, with the exception of the `ListConv`
volve call to solve the diffusion equation approximately compiled fine.

```
In[73]:= DeleteCases[Flatten[dendritGrowthCF[[4]]], _Integer | _Real | True | False]

Out[73]= {ListConvolve}
```

Here are two typical, basically 1D dendritic pattern.

```
In[74]:= Module[{steps = 20000},
    Show[GraphicsArray[
     (SeedRandom[#1];
      ListDensityPlot[1 - Sign[dendritGrowthCF @@ #2],
                      DisplayFunction -> Identity, PlotRange -> All,
                      Mesh -> False, FrameTicks -> False])& @@@
     (* two sets of parameters and seeds *)
     {{4504201162,
       {128, steps, 2, 1200, 0.8761372169858113, 0.75370192429737671,
        0.01138171709483295, 0.0005052576511337, 0.04172021713631082}},
      {9808714780,
       {128, steps, 2, 1200, 0.9154428206208689, 0.34118610239038644,
        0.03795465482207016, 0.0948027031923239, 0.01677197879405640}}}]]]]
```

The next input produces a denser, branched pattern.

```
In[75]:= Module[{a, aMax, n = 85000},
         SeedRandom[4203675447];
         (* carry out the growth process *)
         a = dendritGrowthCF @@
                 {600, 5000, 2, n, 0.287259948632231, 1.0663413850451138,
                  0.12936842836973722, 0.000855863698000357, 0.2859847122296571};
         (* (* another set of parameters giving a nice, slightly less dense figure *)
             SeedRandom[4203675447];
             a = dendritGrowthCF @@
                 {256, 20000, 2, 8000, 0.5189513615515259, 0.625565041762247,
                  0.1293084654783588, 0.0009500316643086716, 0.2662514824713898};
         *)
         (* color corresponds to aggregation time *)
         aMax = Max[a];
         Show[Graphics[
         MapIndexed[If[#1 == 0., {}, {Hue[0.76 + 0.40 #1/aMax],
                     Rectangle[#2 - 1/2, #2 + 1/2]}]&,
                 a, {2}]], AspectRatio -> Automatic]]
```

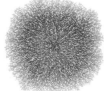

```
In[76]:= DeleteCases[Flatten[dendritGrowthCF[[4]]], _Integer | _Real | True | False]

Out[76]= {ListConvolve}
```

For further examples, see, for instance, [871], [18] and [1899]. For the modeling of hallucination patterns as seen after ingestion of LSD, see [284].

c) Here is a possible implementation of the function `growSeeds`. We step repeatedly through all elements $a_{i,j}$ of the matrix and replace zeros by the value of a randomly selected nonzero neighbor elements until all elements are nonzero. To guarantee overall randomness in the growth process around the seeds with value k_m, we traverse the elements $a_{i,j}$ in a random order, a different order for each matrix traversal. ($\mathcal{T}$ is a list of matrix element indices in random order.)

```
In[1]:= growSeeds = Compile[{{m, _Integer, 2}},
        Module[{c = Plus @@ Sign[Flatten[m]], m = m, n = Length[m],
                i, j, t, τ, 𝒯, ij, L},
          (* matrix of matrix indices *)
          𝒯 = Flatten[Table[{i, j}, {i, n}, {j, n}], 1];
          While[ (* add elements until no 0's are left *) c < n^2,
          (* recursive form random permutation of all n^2 matrix elements *)
          Do[t = 𝒯[[i]]; τ = 𝒯[[j = Random[Integer, {i, n^2}]]];
            {𝒯[[i]], 𝒯[[j]]} = {τ, t}, {i, n^2}];
          Do[(* go through all matrix elements *)
            If[{i, j} = 𝒯[[k]]; m[[i, j]] == 0,
```

```
(* neighbors of the element m[[i, j]] *)
ij = Select[{{i - 1, j}, {i + 1, j}, {i, j - 1}, {i, j + 1}},
            (Min[#] > 0 && Max[#] < n + 1)&];
(* are there nontrivial neighbors? *)
L = Table[{i, m[[ij[[i, 1]], ij[[i, 2]]]]}, {i, Length[ij]}];
If[Max[Last /@ L] > 0, (* select a nontrivial neighbor randomly *)
   L = Select[L, (Last[#] =!= 0)&];
   m[[i, j]] = L[[Random[Integer, {1, Length[L]}], 2]];
   (* count nontrivial elements *) c++]], {k, n^2}]]; m]];
```

The function `growSeeds` compiled successfully.

```
In[2]:= Head /@ DeleteCases[Flatten[growSeeds[[4]]], True | False] // Union

Out[2]= {Integer}
```

Next, we generate random start matrices with seeds. The function `randomSeedMatrix` generates a $n \times n$ matrix with approximately n^2 *seedDensity* nonzero elements that are numbered consecutively.

```
In[3]:= randomSeedMatrix[n_, seedDensity_] :=
    Module[{c = 0, m},
           (* until at least one seed is present *)
           While[m = Table[If[Random[] < seedDensity, c++, 0],
                           {i, n}, {j, n}]; Max[m] == 0, Null]; m]
```

Here are three start matrices with random initial seed distribution of various sizes.

```
In[4]:= SeedRandom[777];
    startMatrix[1] = randomSeedMatrix[100, 1/010];
    startMatrix[2] = randomSeedMatrix[200, 1/400];
    startMatrix[3] = randomSeedMatrix[400, 1/400];
```

For a better visual discrimination of the regions with the same k_m, we define a function `randomize` that replaces the k_m with random real values. (If we would not do this, then, due to the construction of the random seed matrices, neighboring regions would have very similar colors.)

```
In[8]:= randomize[m_] := m //. (* fast replacements *) Dispatch[Rule @@@
        Transpose[{#, Table[Random[], {Length[#]}]}]&[Union[Flatten[m]]]]
```

Here are the three matrices after all seeds have (randomly) full grown.

```
In[9]:= Show[GraphicsArray[
    Table[ListDensityPlot[
        randomize[finalMatrix[k] = growSeeds[startMatrix[k]]],
                      Mesh -> False, FrameTicks -> False,
                      DisplayFunction -> Identity,
                      ColorFunction -> (Hue[0.8 #]&)], {k, 3}]]]
```

  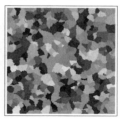

By using the initial start matrices, we can color all elements of the fully-grown matrices according to their distance from their original seed.

```
In[10]:= scaledDistanceColoring[startMatrix_, finalMatrix_, f_] :=
    Module[{seedPosition, n = Length[startMatrix]},
           (* position of seed for a given k_m *)
           (seedPosition[#1] = #2)& @@@
             DeleteCases[Flatten[MapIndexed[{#1, #2}&,
                         startMatrix, {2}], 1], {0, _}];
           (* calculate distance to seed for all elements *)
```

```
            MapIndexed[(f Sqrt[#.#]&[#2 - seedPosition[#1]]/n)&,
                    finalMatrix, {2}]]
```

Now let use construct three matrices with a regular seed distribution. The first matrix hexLatticeMatrix is a 201×201 matrix that has seed near the lattice points of a hexagonal lattice.

```
In[11]:= hexLatticeMatrix =
        Module[{n = 100, as, m, c = 1},
         (* lattice points *)
         as = Select[{Re[#], Im[#]}& /@
                      Flatten[Table[Round[16 (i + j Exp[I Pi/3])],
                                   {i, -20, 20}, {j, -20, 20}], 1],
                 (Min[#] > 0 && Max[#] <= (2n + 1))&];
         m = Table[0, {i, -n, n}, {j, -n, n}];
         (* insert seeds into the matrix m *)
         Do[(m[[as[[j, 1]], as[[j, 2]]]]) = c++, {j, Length[as]}]; m];
```

The second matrix spiralMatrix is a 201×201 matrix that has seeds along a spiral curve.

```
In[12]:= spiralMatrix =
        Module[{n = 100, spiral, as, m, c = 1},
         (* points along a spiral *)
         spiral = Union[Table[Round[2 φ {Cos[φ], Sin[φ]}] + (n + 1),
                          {φ, 0, 15 2Pi, 15 2Pi/295}]];
         as = Select[spiral, Min[#] > 0 && Max[#] <= (2n + 1)&];
         m = Table[0, {i, -n, n}, {j, -n, n}];
         (* insert seeds into the matrix m *)
         Do[(m[[as[[j, 1]], as[[j, 2]]]]) = c++, {j, Length[as]}]; m];
```

The third matrix dragonMatrix is a 301×301 matrix whose seeds are at rationalized positions of a Goffinet dragon.

```
In[13]:= dragonMatrix =
        Module[{n = 150, o = 8, zs, α, as, m, c = 1},
         (* points of the Goffinet dragon *)
         zs = {Re[#], Im[#]}& /@ Flatten[
           Table[Sum[α[k] (0.65 - 0.3 I)^k, {k, 0, o}],
             Evaluate[Sequence @@ Table[{α[k], 0, 1}, {k, 0, o}]]]];
         (* scaled dragon points *)
         {{xMin, xMax}, {yMin, yMax}} = {Min[#], Max[#]}& /@ Transpose[zs];
         as = Round[2n {(#1 - xMin)/(xMax - xMin),
                       (#2 - yMin)/(yMax - yMin)}& @@@ zs] + 1;
         m = Table[0, {i, -n, n}, {j, -n, n}];
         (* insert seeds into the matrix m *)
         Do[(m[[as[[j, 1]], as[[j, 2]]]]) = c++, {j, Length[as]}]; m];
```

Here are the seeds of the three matrices hexLatticeMatrix, spiralMatrix, and dragonMatrix.

```
In[14]:= Show[GraphicsArray[
          ListDensityPlot[1 - Sign[#], Mesh -> False, FrameTicks -> False,
                          DisplayFunction -> Identity]& /@
              {hexLatticeMatrix, spiralMatrix, dragonMatrix}]]
```

 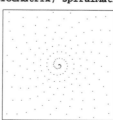 

We emphasize the boundaries between two regions with matrix elements k_m and k_n by drawing lines between them. The function varietyBoundaries calculates these lines for a given matrix m.

```
In[15]:= varietyBoundaries[m_] := Module[{lines},
          (* all indexed line segments between matrix elements *)
          lines = Flatten[{Line[#1, Sort[{#2, #2 + {1, 0}}]],
```

```
                            Line[#1, Sort[{#2, #2 + {0, 1}}]],
                            Line[#1, Sort[{#2 + {1, 0}, #2 + {1, 1}}]],
                            Line[#1, Sort[{#2 + {0, 1}, #2 + {1, 1}}]]]& @@@
                    Flatten[MapIndexed[{#1, #2 - 1}&, Transpose[m], {2}], 1]];
            (* keep only once appearing line segments *)
        Union[Rest[First[#]]& /@ Select[Split[Sort[lines]], Length[#] == 1&]]]
```

Here are the full-grown matrices arising from `hexLatticeMatrix`, `spiralMatrix`, and `dragonMatrix` together with the boundaries between seed varieties shown. Again, we use the function `randomize` for a random coloring of the resulting regions.

```
In[16]:= Show[GraphicsArray[
        ListDensityPlot[randomize[#], DisplayFunction -> Identity,
                Mesh -> False, FrameTicks -> False, ColorFunction -> Hue,
                Epilog -> {Thickness[0.002], varietyBoundaries[#]}]&[
        growSeeds[#]]& /@ (* the three matrices *)
        {hexLatticeMatrix, spiralMatrix, dragonMatrix}]]
```

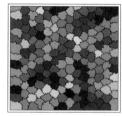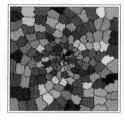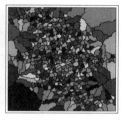

d) The function `squareDistanceMatrix` calculates the matrix of 1s, 2s, ... Starting with the matrix *mat*, it finds the 1s and then iteratively replaces all 0s in existing neighborhoods with k ($k = 2, 3, ...$). The second argument of `squareDistanceMatrix` is a list *neighbors* of the relative coordinates of the points forming a neighborhood. We use `Compile` for speed reasons.

```
In[1]:= squareDistanceMatrix =
        Compile[{{mat, _Integer, 2}, {neighbors, _Integer, 2}},
         Module[{dimy = Length[mat], dimx = Length[mat[[1]]],
                M = mat, nN = Length[neighbors], oldC = 0,
                oldNeighborList, newNeighborList,
                level, newC, iO, jO, iN, jN},
            (* initialize lists of current outermost squares *)
            oldNeighborList = newNeighborList = Table[{0, 0}, {dimx dimy}];
            (* find initially marked cells *)
            Do[If[M[[i, j]] == 1,
                oldC++; oldNeighborList[[oldC]] = {i, j}],
              {i, dimy}, {j, dimx}];
            (* form neighbors as long as there are unused squares *)
            level = 2;
            While[oldC =!= 0,
            newC = 0;
            (* loop over all outermost squares of the last round *)
            Do[{iO, jO} = oldNeighborList[[c]];
              Do[{iN, jN} = {iO, jO} + neighbors[[h]];
                If[Min[{iN, jN}] >= 1 && iN <= dimy && jN <= dimx,
                  If[M[[iN, jN]] == 0,
                    M[[iN, jN]] = level;
                    newC++; newNeighborList[[newC]] = {iN, jN}]],
                  {h, nN}],
              {c, oldC}];
            (* make last neighbor list the old one *)
            Do[oldNeighborList[[k]] = newNeighborList[[k]],
              {k, newC}]; oldC = newC;
            level++]; M]];
```

`squareDistanceMatrix` compiled successfully.

```
In[2]:= (Head /@ DeleteCases[Flatten[squareDistanceMatrix[[4]]],
                                                True | False]) // Union
Out[2]= {Integer}
```

In the following, we use three neighborhoods: the four-element von Neumann neighborhood, the eight-element Moore neighborhood, and a randomly chosen eight-element "neighborhood".

```
In[3]:= (* von Neumann neighborhood *)
        vonNeumannNeighborhood = {{-1, 0}, {1, 0}, {0, -1}, {0, 1}};
        (* Moore neighborhood *)
        mooreNeighborhood = {{-1,  0}, { 1,  0}, { 0, -1}, { 0,  1},
                             { 1,  1}, { 1, -1}, {-1,  1}, {-1, -1}};
        (* a random neighborhood (with "holes") *)
        randomNeighborhood = {{-4, -3}, {-4, -1}, {-3,  0}, {-2, -4},
                              { 0, -4}, { 0,  3}, { 2,  0}, { 4,  1}};

        neighborhoods = {vonNeumannNeighborhood, mooreNeighborhood,
                         randomNeighborhood};
```

We start with visualizing the resulting matrices for the case of a random 300×300 matrix, that is sparsely filled with 1s.

```
In[10]:= Function[M, Show[GraphicsArray[
         ListDensityPlot[squareDistanceMatrix[M, #], Mesh -> False,
                      FrameTicks -> None, DisplayFunction -> Identity]& /@
             (* the three neighborhoods *) neighborhoods]]][
             (* sparse random initial matrix *) SeedRandom[123];
             Table[If[Random[] < 0.005, 1, 0], {i, 300}, {j, 300}]]
```

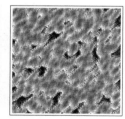

Here are the last resulting matrices colored in an alternative way.

```
In[11]:= Show[% /. DensityGraphics[m_, rest___] :>
                   DensityGraphics[Mod[m, 7], rest]]
```

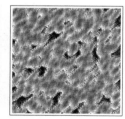

We continue with a more structured matrix. The points of the following 301×301 matrix lie along a spiral.

```
In[12]:= spiralM =
         Module[{n = 150, spiral, a, m, c = 1},
           (* points along a spiral *)
           a = Select[Union[Table[Round[φ {Cos[φ], Sin[φ]}] + (n + 1),
                            {φ, 0, 40 2Pi, 40 2Pi/384}]],
                    (Min[#] > 0 && Max[#] <= 2n + 1)&];
           m = Table[0, {i, -n, n}, {j, -n, n}];
           (* insert seeds into the matrix m *)
           (m[[##]] = 1)& @@@ a; m];
```

This time, we color the resulting matrices in a rainbow- like fashion.

```
In[13]:= Show[GraphicsArray[
        ListDensityPlot[squareDistanceMatrix[spiralM, #],
                    Mesh -> False, ColorFunction -> (Hue[3.4 #]&),
                    DisplayFunction -> Identity, FrameTicks -> None]& /@
                    neighborhoods]]
```

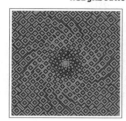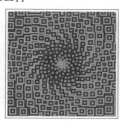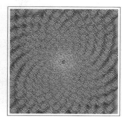

We end with a still more structured matrix, namely one containing a pixelated version of the *Mathematica* letters.

```
In[14]:= mathematica =
       {{{ 26,   29}, { 27,   90}}, {{ 27,   90}, { 37,   57}},
        {{ 37,   57}, { 51,   89}}, {{ 51,   89}, { 49,   32}},
        {{ 55,   30}, { 68,   87}}, {{ 68,   87}, { 78,   34}},
        {{ 61,   57}, { 74,   56}}, {{ 92,   35}, { 93,   86}},
        {{ 78,   89}, {106,   88}}, {{114,   89}, {112,   37}},
        {{137,   91}, {132,   37}}, {{112,   63}, {134,   62}},
        {{166,   86}, {150,   88}}, {{150,   88}, {147,   35}},
        {{147,   35}, {166,   35}}, {{148,   63}, {163,   63}},
        {{174,   36}, {175,   86}}, {{175,   86}, {187,   60}},
        {{187,   60}, {197,   87}}, {{197,   87}, {199,   36}},
        {{205,   36}, {216,   86}}, {{216,   86}, {227,   37}},
        {{209,   57}, {222,   57}}, {{241,   88}, {239,   39}},
        {{226,   90}, {251,   89}}, {{260,   89}, {256,   38}},
        {{287,   82}, {283,   88}}, {{283,   88}, {278,   90}},
        {{278,   90}, {272,   85}}, {{272,   85}, {270,   75}},
        {{270,   75}, {269,   65}}, {{269,   65}, {269,   54}},
        {{269,   54}, {270,   45}}, {{270,   45}, {276,   42}},
        {{276,   42}, {282,   43}}, {{282,   43}, {287,   47}},
        {{287,   47}, {288,   53}}, {{295,   40}, {306,   89}},
        {{306,   89}, {316,   39}}, {{300,   63}, {311,   63}}};
In[15]:= mathematicaM = Table[0, {120}, {350}];
       Do[(* points along line segments *)
         {i, j} = Round[#1 + (#2 - #1) t]; mathematicaM[[j, i]] = 1,
         {t, 0, 1, 1/50}]& @@@ mathematica;
In[17]:= ListDensityPlot[1 - mathematicaM, FrameTicks -> None,
                    Mesh -> False, AspectRatio -> Automatic]
```

After calculating the resulting matrices for the von Neumann and Moore neighborhoods, we display a little animation of the resulting matrices.

```
In[18]:= {mathematicaMvN, mathematicaMM} =
            squareDistanceMatrix[mathematicaM, #]& /@
               {vonNeumannNeighborhood, mooreNeighborhood};
```

```
In[19]:= Do[Show[GraphicsArray[
            ListDensityPlot[Mod[#, k], Mesh -> False, Frame -> False,
                            AspectRatio -> Automatic, PlotRange -> All,
                            ColorFunctionScaling -> True,
                            DisplayFunction -> Identity,
                            ColorFunction -> GrayLevel]& /@
            (* von Neumann and Moore neighborhoods *)
            {mathematicaMvN, mathematicaMM}]], {k, 2, 12}]
```

33. First Digit Frequencies in Mandelbrot Set Calculation

The function cf[c] calculates the Mandelbrot set and the frequency of the digits. To do this as fast as possible, we generate an appropriate CompiledFunction-object. The *i*th element of the list digitCounter represents the number of occurrences of the first digit *i*. It would be natural for cf to return something like {*listofDigitOccurences*, *numberOfIterations*}, but the *Mathematica* compiler does not allow such a return type in the moment. So, we will add the integer *numberOfIterations* as the tenth element into the list of integers digitCounter. To get the first digit of the real number *x*, we will not use RealDigits. The reason being, again, that the returned expression of RealDigits is not a tensor. Instead, we use $\lfloor 10^{-\lfloor \log_{10}(x) \rfloor} x \rfloor$, which can be compiled and quickly evaluated.

```
In[1]:= cf = Compile[{{c, _Complex}},
    Module[{iterCounter = 1, digitCounter = Table[0, {10}],
        z = 0. + 0.I, firstDigit = 1},
      While[(* iterate maximal 100 times or as long as we are certain
              that after squaring we still have a machine number *)
        iterCounter < 100 && Abs[z] < 10.^100,
        iterCounter = iterCounter + 1;
        (* the iteration *)
        z = z^2 - c;
        (* first digit of the real part *)
        If[(* no first digit *)Re[z] != 0.0,
          (* get the first digit *)
          firstDigit = Floor[#/10.^Floor[Log[10, #]]]&[Abs[Re[z]]];
          If[(* a safety check for the improbable,
                but possible, situation z = 0.9999999999999 *)
            firstDigit != 0,
            (* increase counter for the corresponding first digit *)
            digitCounter[[firstDigit]] = digitCounter[[firstDigit]] + 1]];
        (* first digit of the imaginary part *)
        If[(* no first digit *)Im[z] != 0.0,
          (* get the first digit *)
          firstDigit = Floor[#/10.^Floor[Log[10, #]]]&[Abs[Im[z]]];
          If[firstDigit != 0,
            (* increase counter for the corresponding first digit *)
```

```
                    digitCounter[[firstDigit]] = digitCounter[[firstDigit]] + 1]]];
            (* the number of iterations carried out *)
            digitCounter[[10]] = If[Abs[z] < 10.^100,
                                    0, iterCounter];
            digitCounter]];
```

Now, we calculate a 200×200 array of digit occurrences.

```
In[2]:= pp = 200;
       (data = Table[cf[N[cr + I ci]],
                   {ci, -3/2, 3/2, 3/pp}, {cr, -1, 2, 3/pp}]); // Timing
Out[3]= {5.53 Second, Null}
```

Here is the resulting Mandelbrot set.

```
In[4]:= ListDensityPlot[Map[Last, data, {2}],
                     Mesh -> False, MeshRange -> {{-1, 2}, {-3/2, 3/2}}]
```

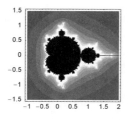

This plot shows the distribution of 1s and 2s in the calculation of the zs.

```
In[5]:= Show[GraphicsArray[
       ListDensityPlot[Map[#, data, {2}], Mesh -> False,
                     MeshRange -> {{-1, 2}, {-3/2, 3/2}},
                     DisplayFunction -> Identity]& /@ {#[[1]]&, #[[2]]&}]]
```

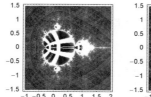
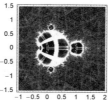

Here are the relative probabilities for the first digits.

```
In[6]:= #/Plus @@ #&[Plus @@ Flatten[Map[Drop[#, -1]&, data, {2}], 1]] // N
Out[6]= {0.293567, 0.19377, 0.159627, 0.106836,
         0.0612205, 0.0525333, 0.0485696, 0.0431112, 0.0407659}
```

Because we are only interested in the cumulative number of first 1's, 2's, ... and to save memory (by about a factor of 10), we slightly change the function above to take as its input a list similar to `digitCounter` as well.

```
In[7]:= cf2 = Compile[{{c, _Complex}, {dC, _Integer, 1}},
       Module[{iterCounter = 1, digitCounter = Table[0, {10}]},
              z = 0. + 0.I, firstDigit = 1},
          While[(* iterate maximal 100 times or as long as we are
                  certain that after squaring we still have a machine number *)
                iterCounter < 100 && Abs[z] < 10.^100,
                iterCounter = iterCounter + 1;
                (* the iteration *)
                z = z^2 - c;
                (* first digit of the real part *)
                If[(* no first digit *)Re[z] != 0.0,
                   (* get the first digit *)
```

```
                    firstDigit = Floor[#/10.^Floor[Log[10, #]]]&[Abs[Re[z]]];
                    If[(* a safety check for the improbable,
                        but possible situation z = 0.9999999999999 *)
                       firstDigit != 0,
                       (* increase counter for the corresponding first digit *)
                       digitCounter[[firstDigit]] = digitCounter[[firstDigit]] + 1]];
                 (* first digit of the imaginary part *)
                 If[(* no first digit *)Im[z] != 0.0,
                    (* get the first digit *)
                    firstDigit = Floor[#/10.^Floor[Log[10, #]]]&[Abs[Im[z]]];
                    If[firstDigit != 0,
                       (* increase counter for the corresponding first digit *)
                       digitCounter[[firstDigit]] = digitCounter[[firstDigit]] + 1]]];
              (* the number of iterations carried out *)
              digitCounter[[10]] = If[Abs[z] < 10.^100, 0, iterCounter];
              ReplacePart[dC, 0, 10] + digitCounter]];
```

Now let us deal with an example containing about five times as many points. This will take less than a minute on a 2 GHz computer.

```
In[8]:= pp = 450; dC = Table[0, {10}];
        (data = Table[dC = cf2[N[cr + I ci], dC]; Last[dC],
                      {ci, -3/2, 3/2, 3/pp}, {cr, -1, 2, 3/pp}]); // Timing
Out[9]= {27.92 Second, Null}
```

```
In[10]:= ListDensityPlot[data, Mesh -> False, MeshRange -> {{-1, 2}, {-3/2, 3/2}}]
```

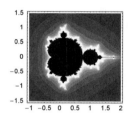

Here is the number of occurrences of the digits 1 to 9.

```
In[11]:= Drop[dC, -1]
Out[11]= {3291191, 2177306, 1792170, 1203751, 694195, 596524, 547679, 487833, 451797}
```

Together, we have examined about 11 million numbers for their first digit.

```
In[12]:= Plus @@ %
Out[12]= 11242446
```

Here are the resulting frequencies of occurrences.

```
In[13]:= N[%%/%]
Out[13]= {0.292747, 0.193668, 0.159411, 0.107072,
          0.0617477, 0.05306, 0.0487153, 0.0433921, 0.0401867}
```

The last numbers are in good agreement with the $\log_{10}(1 + 1/k)$ predicted by Benford's law.

```
In[14]:= N[Table[Log[10, 1 + 1/k], {k, 1, 9}]]
Out[14]= {0.30103, 0.176091, 0.124939, 0.09691,
          0.0791812, 0.0669468, 0.0579919, 0.0511525, 0.0457575}
```

In a similar way, we could now investigate the mantissa distribution of the numbers occurring in a larger calculation.

```
In[15]:= cf3 = Compile[{{c, _Complex}},
         Module[{iterCounter = 1, mantissaBag = Table[0., {200}],
                 z = 0. + 0.I, mantissaBagCounter = 1},
                While[(* iterate maximal 100 times or as long as we are certain
                          that after squaring we still have a machine number *)
```

```
                    iterCounter < 100 && Abs[z] < 10.^100,
                    iterCounter = iterCounter + 1;
                    (* the iteration *)
                    z = z^2 - c;
                    (* mantissa of the real part *)
                    mantissaBag[[mantissaBagCounter]] = If[Re[z] != 0.,
                         Abs[Re[z]]/10.^Ceiling[Log[10, Abs[Re[z]]]], 0.];
                    mantissaBagCounter ++;
                    mantissaBag[[mantissaBagCounter]] = If[Im[z] != 0.,
                         Abs[Im[z]]/10.^Ceiling[Log[10, Abs[Im[z]]]], 0.];
                    mantissaBagCounter ++;
                    Take[mantissaBag, mantissaBagCounter - 1]]];
```

In[16]:= `pp = 120;`
```
        (data = Table[cf3[N[cr + I ci + Random[] 10^-12 + I Random[] 10^-12]],
                    {ci, -3/2, 3/2, 3/pp}, {cr, -1, 2, 3/pp}]) ; // Timing
```
Out[17]= `{2.18 Second, Null}`

The distribution of the mantissa is conjectured to be logarithmic [1596].

In[18]:= `With[{ll = Length[Flatten[data]]},`
```
        ListPlot[(* prepare data *)
                    MapIndexed[{#1, #2[[1]]}&, Sort[Flatten[data]]],
                    PlotStyle -> {PointSize[0.003], GrayLevel[0]},
                    PlotRange -> All, Frame -> True, Axes -> False,
                    Prolog -> {Hue[0], Thickness[0.01],
                    (* Benford's law *)
                    Line[Table[{x, ll (1 - 10 x + Log[10] + 10 x Log[x])/
                                    (-9 + Log[10])}, {x, 0.1, 1, 0.1}]]}]]
```

34. Interesting Jerk Functions

Because in the following, we will deal in a "rough" way with the function NDSolve, for instance, we will input functions that possess or develop singularities, we turn off some messages associated with NDSolve. Messages would be generated because its default number of number of steps would have been used or because the convergence of the methods could not be guaranteed anymore.

In[1]:= `Function[m, Off[MessageName[NDSolve, m]], {HoldAll, Listable}][`
```
                    (* message tags *) {"ndsz", "nderr", "mxst", "ndcf"}];
```

To get an idea of possible shapes of the solution, we will look at the phase portrait of some very simple jerk functions, namely, $x'''(t) = -x(t)^j$, $x'''(t) = -x'(t)^j$, $x'''(t) = -x''(t)^j$, where $j = -3, -2, -1, 0, 1, 2, 3$. We choose the initial conditions $x(0) = x'(0) = x''(0) = 1$ (of course, other ones are possible, too). We use $y(t) = x'(t)$, $z(t) = x''(t)$ to solve a system of differential equations instead of solving a third-order differential equation to avoid the numerical differentiation of the InterpolatingFunction returned by NDSolve in the generation of the phase portrait. This strategy also gives a higher precision and accuracy for the result.

In[2]:= `Module[{nsol, rhs},`
```
        Show[GraphicsArray[#]]& /@
        Table[(* solve the differential equation *)
        nsol = NDSolve[{x'[t] == y[t], y'[t] == z[t],
                    z'[t] == (rhs = -Which[i == 1, x[t]^j,
                                            i == 2, y[t]^j,
                                            i == 3, z[t]^j]),
                    x[0] == 1, y[0] == 1, z[0] == 1},
                    {x, y, z}, {t, 0, 100}, MaxSteps -> 2500];
```

```
(* graph of the phase portrait *)
ParametricPlot[Evaluate[{x[t], y[t]} /. nsol],
            {t, 0, nsol[[1, 1, 2, 1, 1, 2]]},
            AspectRatio -> 1, DisplayFunction -> Identity,
            Frame -> True, Axes -> False, FrameTicks -> None,
            PlotLabel -> StyleForm[
               x'''[t] == (rhs /. {y -> x', z -> x''}), FontSize -> 3]],
               {i, 1, 3}, {j, -3, 3}]]
```

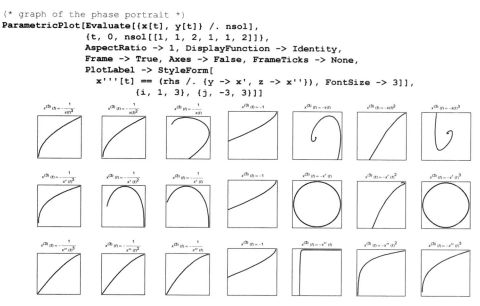

Two kinds of solutions are recognizable in the last graphic array: Periodic ones and monotonously increasing or decreasing ones, including those that tend to a straight line.

Let us look at some more examples, this time of the form $x'''(t) = -x(t)^i\, x'(t)^j\, x''(t)^k$, i, j, $k = -1, 0, 1, 2, 3$. Here, we rescale the phase portraits and position them in a 3D i, j, k-grid.

```
In[3]:= Module[{nsol, line, minx, maxx, miny, maxy},
    Show[Table[
    (* solve the differential equation *)
    nsol = NDSolve[{x'[t] == y[t], y'[t] == z[t],
                z'[t] == -x[t]^i y[t]^j z[t]^k,
                x[0] == 1, y[0] == 1, z[0] == 1},
                {x, y, z}, {t, 0, 50}, MaxSteps -> 2500];
    (* graph of the phase portrait *)
    line = Cases[ParametricPlot[Evaluate[{x[t], y[t]} /. nsol],
                        {t, 0, nsol[[1, 1, 2, 1, 1, 2]]},
                        AspectRatio -> 1, PlotPoints -> 200,
                        DisplayFunction -> Identity], _Line, Infinity];
    (* rescale the graph *)
    {{minx, maxx}, {miny, maxy}} =
            {Min[#], Max[#]}& /@ Transpose[line[[1, 1]]];
    Graphics3D[{RGBColor[(i + 1)/4, (j + 1)/4, (k + 1)/4],
                Thickness[0.002],
                (* lift graph into 3D *)
                Apply[{0.1 + 0.8(#1 - minx)/(maxx - minx) + i,
                    0.1 + 0.8(#2 - miny)/(maxy - miny) + j,
                    k}&, line, {-2}]}],
        {i, -1, 3}, {j, -1, 3}, {k, -1, 3}], ViewPoint -> {0, -3, 1}]]
```

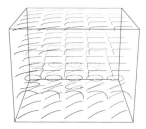

In the last graphic, we basically recognize the same solutions as above. Now let us try to find some more interesting, nonmonotonous solutions. To avoid trying many jerk functions by hand, we will automate the search. This automated search is implemented in the function `interestingJerk`.

`interestingJerk[`*ord*, *terms*, *rhsVars*, *tMax*`]` does the following:

It starts by generating a random jerk function `rhs` with up to *terms* summands, containing the variables from *rhsVars*, each raised to a randomly chosen integer power between 0 and *ord*. Then, the differential equation system corresponding to `x'''[t] ==` *jerk* is solved numerically. The resulting `InterpolatingFunction`-object is used to generate a graph of the phase portrait. The `Line` forming the phase portrait is extracted, and the $x'(t)$ values are extracted. Then, the local minima and maxima are extracted from the list of $x'(t)$ values. If there are at least local minima or maxima and they are "reasonably" pronounced (this avoids counting spurious extrema originating from the numerical method for solutions of jerk functions that approach a constant), the jerk function and its phase portrait are returned.

```
In[4]:= interestingJerk[ord_, terms_, rhsVars_: {t, x[t], y[t], z[t]}, tMax_:10] :=
        Module[{r, rhs, nsol, plot, yValues, localExtremaTriples, pronouncedExtrema},
          (* define random exponents *)
          r := Random[Integer, {0, ord}];
          (* generate the random jerk function--
             the right-hand side of the ode x'''[t] == jerk *)
          (While[rhs = Sum[Random[Integer, {-10, 10}] *
                         Product[rhsVars[[i]]^r, {i, Length[rhsVars]}],
                         {Random[Integer, {1, terms}]}];
          (* solve the differential equation *)
          nsol = NDSolve[{x'[t] == y[t], y'[t] == z[t], z'[t] == rhs,
                         x[0] == 1, y[0] == 1, z[0] == 1},
                        {x, y, z}, {t, 0, tMax}, MaxSteps -> 5000,
                        PrecisionGoal -> 8, AccuracyGoal -> 8];
          (* graph the phase portrait *)
          plot = ParametricPlot[Evaluate[{x[t], y[t]} /. nsol],
                              {t, 0, nsol[[1, 1, 2, 1, 1, 2]]},
                              AspectRatio -> 1, PlotRange -> All,
                              Frame -> True, Axes -> False,
                              DisplayFunction -> Identity];
          (* extract x'[t] values of the phase portrait *)
          yValues = Last /@ plot[[1, 1, 1, 1, 1]];
          If[(* is solution bounded? *)Max[Abs[yValues]] < 100,
            (* does the solution have a few pronounced oscillations? *)
          localExtremaTriples = Select[Abs[Partition[yValues, 3, 1]],
                              (#[[1]] < #[[2]] > #[[3]]) ||
                              (#[[1]] > #[[2]] < #[[3]])&];
          (* the pronounced extrema *)
          pronouncedExtrema = DeleteCases[Chop[Abs[Apply[Subtract,
                Partition[#[[2]]& /@ localExtremaTriples, 2, 1], {1}]], 10^-2], 0];
          Length[pronouncedExtrema] < 5, True], Null];
          (* return right-hand side and phase portrait *)
          {rhs, plot})]
```

Here, `interestingJerk` is used to find some "interesting" jerks. We allow the right-hand side to contain the independent and dependent variables. For space-efficiency, we display two solutions at once.

```
In[5]:= SeedRandom[987654321];

       Do[Function[{jerks},
```

```
Print[x'''[t] == (#[[1]] /. {y -> x', z -> x''})]& /@ jerks;
Show[GraphicsArray[#[[2]]& /@ jerks]]][
{interestingJerk[3, 5], interestingJerk[3, 5]}], {6}]
```

$x^{(3)}[t] = -5 t^2 x[t]^3 x'[t]^3 - 7 t x'[t]^3 x''[t]^3$

$x^{(3)}[t] = -2 t^2 x[t] + 2 t^3 x'[t] - 8 t^3 x[t]^3 x'[t] + 6 x[t]^2 x''[t] - 7 x[t]^3 x'[t]^2 x''[t]$

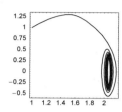 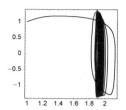

$x^{(3)}[t] = 3 t^3 x[t]^2 x'[t]^3 - 8 t^2 x[t]^3 x'[t]^3 + x[t] x''[t]^2$

$x^{(3)}[t] = -9 t^3 x[t]^3 x'[t] + t x[t]^3 x''[t]^2$

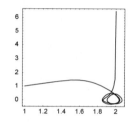 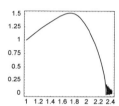

$x^{(3)}[t] = 3 t x[t] - 10 x[t] x'[t] - 3 t^3 x'[t] x''[t]^2 - 5 t x[t]^2 x'[t]^3 x''[t]^2$

$x^{(3)}[t] = -9 t^2 x'[t]^3$

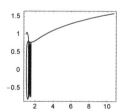 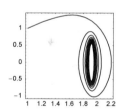

$x^{(3)}[t] = -6 t x[t] x'[t]^3 - 5 t^2 x[t]^2 x'[t]^3 x''[t]^2$

$x^{(3)}[t] = -9 t^2 x[t]^3 x'[t]^3$

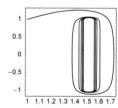 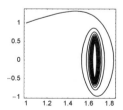

$x^{(3)}[t] = -5 x[t]^3 x'[t]^3 - t x[t] x'[t]^3 x''[t] - 4 t x[t]^3 x'[t]^2 x''[t]^2 - 8 x[t]^2 x'[t]^3 x''[t]^2$

$x^{(3)}[t] = 3 t x[t]^3 - 5 t x[t] x'[t]^3 + 7 x[t] x''[t] - 5 t x[t] x''[t]$

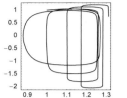 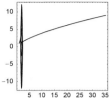

$x^{(3)}[t] =$
$\quad -8\,t^3\,x[t]^2\,x'[t]^2 - 2\,x[t]^2\,x'[t]^3 - 6\,t^3\,x[t]^2\,x'[t]^3 + 9\,t\,x[t]\,x'[t]^2\,x''[t] - 5\,x[t]\,x''[t]^3$
$x^{(3)}[t] = 9\,t^2\,x[t]\,x'[t] - 3\,t\,x[t]\,x'[t]^2$

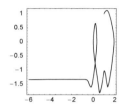 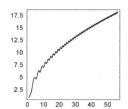

In the next input, we do not allow the right-hand side to contain the independent variable.

```
In[7]:= SeedRandom[2222222222];

Do[Function[{jerks},
        Print[x'''[t] == (#[[1]] /. {y -> x', z -> x''})]& /@ jerks;
        Show[GraphicsArray[#[[2]]& /@ jerks]]][
    Table[interestingJerk[3, 5, {x[t], y[t], z[t]}], {2}]], {3}]
```

$x^{(3)}[t] = -4\,x[t]\,x'[t]$

$x^{(3)}[t] = -5\,x[t]^2\,x'[t] - 8\,x[t]\,x'[t]^3 + 7\,x[t]\,x'[t]^3\,x''[t]^2 - 8\,x[t]^2\,x''[t]^3$

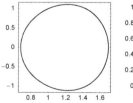 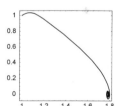

$x^{(3)}[t] = -4\,x[t]\,x'[t]^2 - 10\,x'[t]^3 - 7\,x[t]\,x'[t]\,x''[t]$
$x^{(3)}[t] = -10\,x'[t]^3$

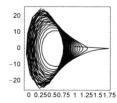 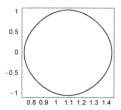

$x^{(3)}[t] = -5\,x[t]^3\,x'[t]$

$$x^{(3)}[t] == -2\,x[t]^2\,x'[t]^3 - 7\,x'[t]^3\,x''[t]^2$$

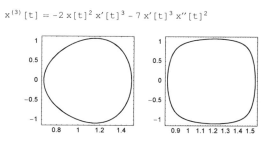

Now, let us look at the extended phase portraits. Although we will proceed as above, now we test for oscillations with respect to $x(t)$, $x'(t)$, and $x''(t)$. The function interestingJerk3D[*ord*, *terms*, *rhsVars*, *tMax*] returns a random jerk function with up to *terms* summands, containing the variables from *rhsVars*, each raised to a randomly chosen power less than *ord*.

```
In[9]:= interestingJerk3D[ord_, terms_, rhsVars_: {t, x[t], y[t], z[t]}, tMax_:10] :=
    Module[{(* random exponents *)
            r := Random[Integer, {0, ord}], c := Random[Integer, {-10, 10}],
            odeSystem, rhs, nsol, T, ε = 10^-2, xValues, yValues, zValues,
            interestingStructureQ, localExtremaTriples, pronouncedExtrema},
      (* search until *)
      While[rhs = Sum[c Product[rhsVars[[i]]^r, {i, Length[rhsVars]}],
                {Random[Integer, {1, terms}]}]];
      (* solve the differential equation *)
      nsol = NDSolve[odeSystem =
                    {x'[t] == y[t], y'[t] == z[t], z'[t] == rhs,
                     x[0] == c, y[0] == c, z[0] == c},
                    {x, y, z}, {t, 0, tMax}, MaxSteps -> 5000,
                    PrecisionGoal -> 8, AccuracyGoal -> 8];
      (* interesting structure present? *)
      interestingStructureQ[xyzValues_] :=
      (localExtremaTriples = Select[Abs[Partition[xyzValues, 3, 1]],
                         (#[[1]] < #[[2]] > #[[3]]) ||
                         (#[[1]] > #[[2]] < #[[3]])&];
       pronouncedExtrema = DeleteCases[Chop[Abs[Apply[Subtract,
          Partition[#[[2]]& /@ localExtremaTriples, 2, 1], {1}]], ε], 0];
       Length[pronouncedExtrema] < 5);
       (* x[t]-, x'[t]-, and x''[t]-values *)
      {xValues, yValues, zValues} =
      (T = nsol[[1, 1, 2, 1, 1, 2]];
       Transpose[Table[Evaluate[{x[t], y[t], z[t]} /. nsol[[1]]],
                    {t, 0, T, T/500}]]);
      If[(* is solution bounded? *)
         Max[Abs[{xValues, yValues, zValues}]] < 1000,
         (* does the solution have a few pronounced oscillations ? *)
      interestingStructureQ[xValues] || interestingStructureQ[yValues] ||
      interestingStructureQ[zValues], True], Null];
      (* return right-hand side and phase portrait *)
      {odeSystem, nsol}]
```

Here are three functions, all with "interesting" extended phase portraits.

```
In[10]:= SeedRandom[13579];

    jerk3DList = Table[interestingJerk3D[4, 4, {t, x[t], y[t], z[t]}, 60], {3}];

    First /@ jerk3DList
Out[13]= {{x'[t] == y[t], y'[t] == z[t], z'[t] == -8 y[t]^3 + 7 y[t] z[t], x[0] == 3, y[0] == 2, z[0] == 5},
    {x'[t] == y[t], y'[t] == z[t], z'[t] == -5 t x[t] - x[t]^2 z[t]^3, x[0] == -8, y[0] == 5, z[0] == 0},
    {x'[t] == y[t], y'[t] == z[t], z'[t] == -7 t^2 x[t]^3 y[t]^3, x[0] == 7, y[0] == 10, z[0] == 5}}

In[14]:= Show[GraphicsArray[Function[jerk,
    ParametricPlot3D[Evaluate[{x[t], y[t], z[t],
```

```
              {Hue[0], Thickness[0.002]}} /. jerk[[2]]],
           {t, 0, jerk[[2, 1, 1, 2, 1, 1, 2]]},
           PlotRange -> All, PlotPoints -> 2000,
           DisplayFunction -> Identity, Axes -> False,
           BoxRatios -> {1, 1, 1}]] /@ jerk3DList]]
```

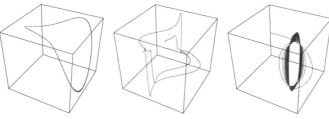

For simple nonpolynomial jerk functions, see [1156].

Related types of differential equations are the following continuous cellular automata-type ODEs [401].

```
In[15]:= odeCA[x_, i_, {b1_, b2_, b3_, c1_, c2_, z0_, z1_, z2_}] :=
        x[i]'[t] == (-x[i][t] + Abs[x[i][t] + 1] - Abs[x[i][t] - 1]) +
             (z2 + c2 Abs[z1 + c1 Abs[z0 + b1 x[i - 1][t] +
                                      b2 x[i][t] + b3 x[i + 1][t]]])
```

The function `odeCAGraphics` creates the *n* ODEs and solves them using the parameters *bczParamaters* and the initial conditions *ics*.

```
In[16]:= odeCAGraphics[n_Integer?Positive, T_?Positive,
                       bczParamaters_List, ics_List] :=
        Module[{Tmax, odes, ndsol},
           (* the n differential equations *)
           odes = Table[odeCA[x, i, bczParamaters], {i, 0, n - 1}] /.
                                     x[k_] :> x[Mod[k, n]];
           (* solve the n differential equations *)
           ndsol = NDSolve[Join[odes,
                           Thread[Table[x[k][0], {k, 0, n - 1}] == ics]],
                       Table[x[k], {k, 0, n - 1}], {t, 0, T},
                       MaxSteps -> 10^5];
           (* maximal t achieved *)
           Tmax = ndsol[[1, 1, 2, 1, 1, 2]];
           (* show colored solution curves *)
           Plot[Evaluate[Table[x[k][t], {k, 0, n - 1}] /. ndsol[[1]]],
              {t, 0, Tmax}, PlotRange -> All, Frame -> True,
              Axes -> False, PlotPoints -> 200,
              PlotStyle -> Table[{Thickness[0.003], Hue[0.8 k/(n - 1)]},
                                 {k, 0, n - 1}]]]
```

For most parameter values and initial conditions, the resulting solutions either diverge quickly or approach a constant value. Here are nine random examples.

```
In[17]:= (* create parameter values and initial conditions using a seed *)
        randomOdeCAGraphics[n_, T_, seed_] :=
        (SeedRandom[seed];
         odeCAGraphics[n, 100, Table[Random[Integer, {-2, 2}], {8}],
                       Table[Random[Real, {-2, 2}], {k, 0, n - 1}]])
In[19]:= Show[GraphicsArray[#]]& /@
           Block[{$DisplayFunction = Identity},
                 (* display 3×3 examples *)
                 Partition[#, 3]& @ Table[randomOdeCAGraphics[12, 100, j^2],
                                          {j, 9}]]
```

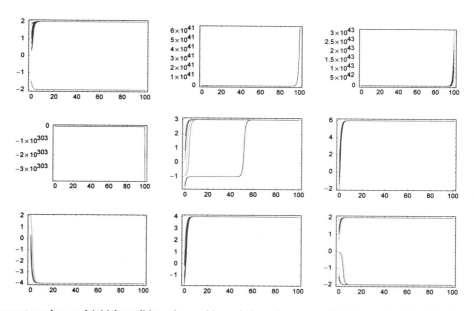

But for some parameter values and initial conditions the resulting solutions show a complicated quasiperiodic behavior. Here are 15 such solutions shown.

```
In[20]:= Show[GraphicsArray[# /. (FrameTicks -> _) -> (FrameTicks -> False)]]& /@
          Block[{$DisplayFunction = Identity},
          Partition[#, 3]& @ (randomOdeCAGraphics[12, 10, #]& /@
          (* seed values from a random search *)
          {7205456156767762967, 24117479060546962385, 36865963467501954110,
          79902080186667294937, 50554249030447327737, 70929784843952762134,
          38822477109576963078, 79938893467770579021, 55093541639391835481,
          36405997048803330286, 81835354711511251179, 88321322355360999277,
          64926519924445834190, 53684681849582355410, 71761111538887630322})]
```

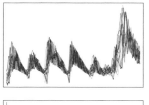 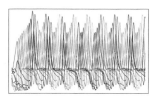 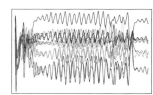

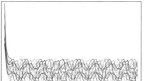 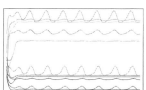 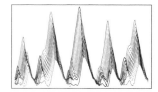

35. Initial Value Problems for the Schrödinger Equation

The separation of variables for the partial differential equation [833]

$$i \frac{\partial \psi(\xi, \tau)}{\partial \tau} = -\frac{1}{4\pi} \frac{\partial^2 \psi(\xi, \tau)}{\partial \xi^2}$$

gives the expansion $\psi(\xi, \tau) = \sum_{k=1}^{\infty} c_k \psi_k(\xi, \tau)$, where the complete system of eigenfunctions are given by [1692], [590], [1805]

$$\psi_k(\xi, \tau) = \frac{1}{\sqrt{2}} \sin(k \pi \xi) e^{-\frac{1}{4} \pi i k^2 \tau}.$$

The Fourier coefficients c_k are determined by the initial condition $\psi(\xi, 0)$ as $c_k = \int_0^1 \psi_k(\xi, \tau) \psi(\xi, 0) \, d\xi$.

Here is a short test of the eigenfunctions and their normalization.

```
In[1]:= ψEF[k_, ξ_, τ_] := Sqrt[2] Sin[k Pi ξ] Exp[-Pi/4 I k^2 τ];

{(* is ψEF an eigenfunction? *)
Factor[I D[ψEF[k, ξ, τ], τ] - (-1/(4 Pi)) D[ψEF[k, ξ, τ], ξ, ξ]],
(* is ψEF normalized to 1? *)
Integrate[ψEF[k, ξ, 0]^2, {ξ, 0, 1}] /. Sin[2 k π] -> 0}
Out[2]= {0, 1}
```

Here are the first six eigenfunctions.

```
In[3]:= Plot[Evaluate[Table[ψEF[k, ξ, 0], {k, 6}]], {ξ, 0, 1},
        PlotStyle -> Table[{Hue[i/10]}, {i, 10}]]
```

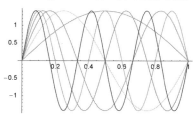

Here is a picture of the initial wave form for the first problem.

```
In[4]:= Plot[Evaluate[1/Sqrt[2] Sum[ψEF[k, ξ, 0], {k, 5}]], {ξ, 0, 1}]
```

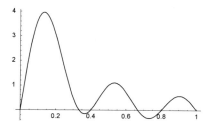

Because the terms of the sum that form the initial conditions are eigenfunctions (up to the overall normalization $\sqrt{2}$), the time development is simple [703], [948]. We have $\psi(\xi, \tau) = 2^{-1/2} \sum_{k=1}^{\infty} \psi_n(\xi, \tau)$. For a fast summation, we compile the sum (for a still faster way, we could use numerical Fourier and inverse Fourier transform,).

```
In[5]:= ψcf = Compile[{{ξ, _Real, 0}, {τ, _Real, 0}, {n, _Integer}},
            Sum[Sin[k Pi ξ] Exp[-Pi/4 I k^2 τ], {k, n}]];
```

We calculate a 200×200 array of function values and visualize their absolute values.

```
In[6]:= ppξ = 200; ppτ = 200;
        (dataSum = Table[ψcf[ξ, τ, 5], {τ, 0, 1, 1/ppτ}, {ξ, 0, 1, 1/ppξ}];) //
                                                                              Timing
Out[7]= {0.56 Second, Null}
```

```
In[8]:= ListDensityPlot[Abs[dataSum], Mesh -> False, ColorFunction -> (Hue[0.8 #]&),
            MeshRange -> {{0, 1}, {0, 1}}]
```

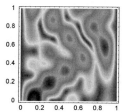

The numerical solution nsol using NDSolve is straightforward for the first initial condition. The initial condition is smooth and slowly varying; so use 300 points for the spatial discretization. We strive for about 4 correct digits in the result.

```
In[9]:= (nsol =
    NDSolve[{I D[Φ[ξ, τ], τ] + 1/(4 Pi) D[Φ[ξ, τ], {ξ, 2}] == 0,
            Φ[ξ, 0] == 1/Sqrt[2] Sum[ψEF[k, ξ, 0], {k, 5}],
            Φ[0, τ] == 0, Φ[1, τ] == 0},
            Φ[ξ, τ], {ξ, 0, 1}, {τ, 0, 1},
            AccuracyGoal -> 5, PrecisionGoal -> 4, MaxSteps -> 10^5,
            Method -> {"MethodOfLines", "SpatialDiscretization" ->
                        {"TensorProductGrid", "DifferenceOrder" -> 4,
                        "MaxPoints" -> 400, "MinPoints" -> 400}}]) // Timing
    NDSolve::eerr :
    Warning: Scaled local spatial error estimate of 753.0608060050768` at τ = 1.` in
        the direction of independent variable ξ is much greater than prescribed error
        tolerance. Grid spacing with 400 points may be too large to achieve the desired
        accuracy or precision.  A singularity may have formed or you may want to
        specify a smaller grid spacing using the MaxStepSize or MinPoints options. More…
Out[9]= {3.1 Second, {{Φ[ξ, τ] → InterpolatingFunction[{{0., 1.}, {0., 1.}}, <>][ξ, τ]}}}
```

The timing for the solution with NDSolve is similar to the timing for the summation of the series from the separation of variables method. No visible differences are in the two solutions for the first initial condition.

```
In[10]:= ppξ = 200; ppτ = 200;
        (dataNDSolve = Table[Evaluate[Φ[ξ, τ] /. nsol[[1]]],
                            {τ, 0, 1, 1/ppτ}, {ξ, 0, 1, 1/ppξ}]);
```

In[12]:= `ListDensityPlot[Abs[dataNDSolve], Mesh -> False, ColorFunction -> (Hue[0.8 #]&),`
                `MeshRange -> {{0, 1}, {0, 1}}]`

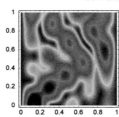

The next picture shows the relative difference between the two solutions.

In[13]:= `ListPlot3D[(Abs[dataSum - dataNDSolve]/`
          `Map[If[# == 0., 1., #]&, Abs[dataSum], {2}]),`
          `Mesh -> False, MeshRange -> {{0, 1}, {0, 1}},`
          `ViewPoint -> {-3, -4, 1}]`

Let us check how well the `NDSolve` solution conserves the normalization integral. It is conserved with about 0.01%.

In[14]:= `{(* exact value *)`
      `Integrate[Abs[1/Sqrt[2] Sum[`ψ`EF[k, `ξ`, 0], {k, 5}]]^2, {`ξ`, 0, 1}],`
      `(* difference to exact value *)`
      `5/2 - NIntegrate[Evaluate[Abs[`Φ`[`ξ`, `τ`]]^2 /. nsol[[1]] /. `τ` -> 1], {`ξ`, 0, 1},`
          `MaxRecursion -> 10, PrecisionGoal -> 2]}`

Out[14]= $\left\{ \dfrac{5}{2}, -0.0359099 \right\}$

For comparison, let us also implement the Fourier transform method [1263], [1268]. We carry out a Fourier transform with respect to the space variable. After this, the time development becomes a simple multiplication. The time-dependent wave function is then easily obtained by carrying out the inverse Fourier transform.

In[15]:= `fourierMethod[m_, n_] :=`
      `Module[{fourier, `ψ`, data, dataSum1},`
      `(* Fourier transform of the eigenfunctions *)`
      `Do[fourier[k] = Fourier[Table[`ψ`EF[k, (r - 1)/(n - 1), 0], {r, 1, n}]],`
       `{k, m}];`
      `(* time-dependent wave function *)`
      ψ`[`τ`_] := Sum[InverseFourier[Exp[-Pi/4 I k^2 `τ`] fourier[k]], {k, m}];`
      `dataSum1 = Table[`ψ`[`τ`], {`τ`, 0, 1, 1/n}];`
      `(* display result *)`
      `ListDensityPlot[Abs[dataSum1], Mesh -> False, ColorFunction -> (Hue[#]&),`
          `MeshRange -> {{0, 1}, {0, 1}}]];`

The fast available algorithm to perform a Fourier transform of lists of length 2^n makes this method quite fast.

In[16]:= `fourierMethod[5, 256] // Timing`

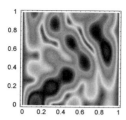

The initial condition in the second case is much more centered on the left boundary (this behavior is expected from the relation $\sum_{k=1}^{n} \sin(k\,\pi\,\xi) = \csc(\pi\,\xi/2)\sin(n\,\pi\,\xi/2)\sin((n+1)\,\pi\,\xi/2)$).

```
In[17]:= ψEF[k_, ξ_, τ_] := Sqrt[2] Sin[k Pi ξ] Exp[-Pi/4 I k^2 τ];
```

```
In[18]:= ψcf = Compile[{{ξ, _Real, 0}, {τ, _Real, 0}, {n, _Integer}},
              Sum[Sin[k Pi ξ] Exp[-Pi/4 I k^2 τ], {k, n}]];
```

```
In[19]:= Plot[Evaluate[Sum[ψEF[k, ξ, 0], {k, 20}]], {ξ, 0, 1}, PlotRange -> All]
```

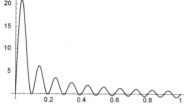

Again, the initial condition is a superposition of eigenfunctions and we can proceed as in case 1.

```
In[20]:= ppξ = 200; ppτ = 200;
        (dataSum = Table[ψcf[ξ, τ, 20], {τ, 0, 1, 1/ppτ}, {ξ, 0, 1, 1/ppξ}];) //
                                                                    Timing
```

```
Out[21]= {1.53 Second, Null}
```

The picture looks more interesting this time. A couple of "lines" are shooting out from the left and right ξ endpoints [1542].

```
In[22]:= ListDensityPlot[Abs[dataSum], Mesh -> False, ColorFunction -> (Hue[#]&),
              MeshRange -> {{0, 1}, {0, 1}}]
```

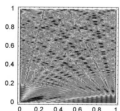

For the NDSolve solution nsol, we use now a lower accuracy and precision goal.

```
In[23]:= (nsol =
        NDSolve[{I D[Φ[ξ, τ], τ] + 1/(4 Pi) D[Φ[ξ, τ], {ξ, 2}] == 0,
              Φ[ξ, 0] == 1/Sqrt[2] Sum[ψEF[k, ξ, 0], {k, 20}],
              Φ[0, τ] == 0, Φ[1, τ] == 0},
              Φ[ξ, τ], {ξ, 0, 1}, {τ, 0, 1},
              AccuracyGoal -> 2.5, PrecisionGoal -> 2.5, MaxSteps -> 10^5,
              Method -> {"MethodOfLines", "SpatialDiscretization" ->
                          {"TensorProductGrid", "DifferenceOrder" -> 4,
                          "MaxPoints" -> 251, "MinPoints" -> 251}}]) // Timing
```

```
NDSolve::eerr :
  Warning: Scaled local spatial error estimate of 21.133389553802044` at τ = 1.` in
    the direction of independent variable ξ is much greater than prescribed error
    tolerance. Grid spacing with 251 points may be too large to achieve the desired
    accuracy or precision.  A singularity may have formed or you may want to
    specify a smaller grid spacing using the MaxStepSize or MinPoints options. More…
```

Out[23]= {44.09 Second, {{Φ[ξ, τ] → InterpolatingFunction[{{0., 1.}, {0., 1.}}, <>][ξ, τ]}}}

As to be expected in this case, the calculation of nsol was slower than was the separation of variable method because the initial condition has many structures that had to be accurately resolved by NDSolve, but those structures emerge from the separation of variable method "automatically". NDSolve warns that the resulting error is much greater than were required by the AccuracyGoal and PrecisionGoal options. By using an even smaller StartingStepSize we could get a better solution, but the calculation would take much longer and use much more memory. Here is the memory used for carrying out the last calculation.

In[24]:= {MemoryInUse[], MaxMemoryUsed[]} /10.^6 "MB"

Out[24]= {794.408 MB, 1257.24 MB}

Here is the resulting picture from the nsol solution. It is remarkably identically-looking to the above graphic.

In[25]:= ppξ = 200; ppτ = 200;
        (dataNDSolve = Table[Evaluate[Φ[ξ, τ] /. nsol[[1]]],
                       {τ, 0, 1, 1/ppτ}, {ξ, 0, 1, 1/ppξ}]);

In[27]:= ListDensityPlot[Abs[dataNDSolve],
                        Mesh -> False, ColorFunction -> (Hue[#]&),
                        MeshRange -> {{0, 1}, {0, 1}}]

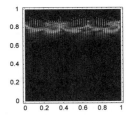

Although the last picture looks qualitatively similar to the one obtained by the separation of variables method, the error of the NDSolve solution is much larger this time.

In[28]:= ListPlot3D[Abs[dataSum - dataNDSolve]/
                   Map[If[# == 0., 1., #]&, Abs[dataSum], {2}],
                   Mesh -> False, MeshRange -> {{0, 1}, {0, 1}},
                   ViewPoint -> {-3, -3, 2}, PlotRange -> {0, 2}]

For a detailed discussion of the "lines" in the last pictures, see [741], [947], and [1842].

Again, the Fourier method is the fastest.

In[29]:= fourierMethod[20, 256]; // Timing

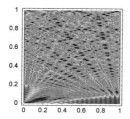

Out[29]= {0.93 Second, Null}

Using the "Pseudospectral" options setting for the "DifferenceOrder" suboption yields a better result, but we need many more time steps.

```
In[30]:= (nsol =
        NDSolve[{I D[Φ[ξ, τ], τ] + 1/(4 Pi) D[Φ[ξ, τ], {ξ, 2}] == 0,
                Φ[ξ, 0] == 1/Sqrt[2] Sum[ψEF[k, ξ, 0], {k, 20}],
                Φ[0, τ] == 0, Φ[1, τ] == 0},
                Φ[ξ, τ], {ξ, 0, 1}, {τ, 0, 1},
                AccuracyGoal -> 2, PrecisionGoal -> 2, MaxSteps -> 10^5,
                Method -> {"MethodOfLines", "SpatialDiscretization" ->
                          {"TensorProductGrid", "DifferenceOrder" -> "Pseudospectral",
                          "MaxPoints" -> 101, "MinPoints" -> 101}}]) // Timing
        NDSolve::mxst : Maximum number of 100000
            steps reached at the point τ == 0.22703255444574885`. More…
Out[30]= {50.66 Second, {{Φ[ξ, τ] → InterpolatingFunction[{{0., 1.}, {0., 0.227033}}, <>][ξ, τ]}}}
```

Here is the envelope of initial conditions for the third problem shown. It is a smooth function being identically zero in the right half of the ξ-domain. The function is infinitely many times differentiable at $\xi = 1/2$.

```
In[31]:= Plot[If[ξ < 1/2, Exp[1/((ξ - 1/4)^2 - 1/16)], 0], {ξ, 0, 1},
            PlotRange -> All, PlotStyle -> {{Thickness[0.01]}}]
```

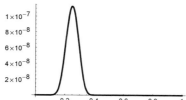

Including the modulation from the $e^{5i\xi}$ term, we get the following form for the real and imaginary parts of the initial conditions.

```
In[32]:= ψ0[ξ_] := If[ξ < 1/2, Exp[5I ξ] Exp[1/((ξ - 1/4)^2 - 1/16)], 0]
```

```
In[33]:= Show[GraphicsArray[
        Plot[#[ψ0[ξ]], {ξ, 0, 1}, PlotRange -> All,
            PlotStyle -> {{Thickness[0.01]}},
            DisplayFunction -> Identity]& /@ {Re, Im}]]
```

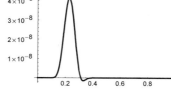 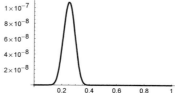

This time, it is not possible to get closed-form analytic expressions for the coefficients c_k; the following symbolic integration fails.

```
In[34]:= TimeConstrained[Integrate[Exp[1/((ξ - 1/4)^2 - 1/16)] Sin[k ξ], ξ], 100]
```

$$Out[34]= \int e^{-\frac{1}{16}+\left(-\frac{1}{4}+ξ\right)^2} \; Sin[k\,ξ]\,dξ$$

So, we will calculate the coefficients numerically, using NIntegrate. Because of the exponential vanishing of the initial condition, we will carry out the numerical integration between 0.01 and 0.49 instead of between 0 and 0.5. At 0.01, the initial condition is of order 10^{-89}, which is more than 80 orders of magnitude smaller than its maximum value. The restriction to the interval $(0.01, 0.49)$ allows us to use the *Mathematica* compiler for the calculation of the function values needed in the numerical integration.

```
In[35]:= cf = Compile[{{k, _Real}, {ξ, _Real}},
            Which[ξ <= 0.01, 0., 10^-2 <= ξ <= 0.49,
                Exp[5I ξ] Exp[1/((ξ - 1/4)^2 - 1/16)] Sin[k Pi ξ], True, 0.]];
```

Because of the highly oscillatory nature of $\psi_n(ξ, 0)$ for larger n, we split the integration intervals into subintervals at each of the zeros of $\psi_n(ξ, 0)$.

```
In[36]:= integrationIntervals[k_] :=
            Join[{0.01}, Table[i/k, {i, Ceiling[0.01 k], Floor[0.49 k]}], {0.49}]
```

This calculates a numerical approximation for the k-th Fourier coefficient c_k.

```
In[37]:= (* avoid the evaluation of the compiled function within NIntegrate *)
            Developer`SetSystemOptions["EvaluateNumericalFunctionArgument" -> False];

            c[k_] := Sqrt[2](Plus @@ Apply[NIntegrate[cf[k, x], {x, ##},
                                            Compiled -> False]&,
                            Partition[integrationIntervals[k], 2, 1], {1}])
```

Here is a list of the first 100 c_k's.

```
In[40]:= (cs = Table[c[k], {k, 100}]); // Timing
```
```
Out[40]= {1.41 Second, Null}
```

A logarithmic plot of the absolute values of the amplitudes shows that the coefficients vanish quite rapidly and that taking for the time development the first 100 into account is enough to achieve at least ten digits of accuracy of the solution.

```
In[41]:= ListPlot[Log[10, Abs[cs]], Frame -> True, Axes -> False]
```

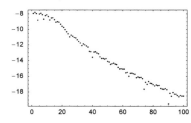

```
In[42]:= ψcf = Compile[{{ξ, _Real, 0}, {τ, _Real, 0}, {l, _Complex, 1}},
                Sqrt[2] Sum[l[[k]] Sin[k Pi ξ] Exp[-I Pi/4 k^2 τ],
                        {k, Length[l]}]];
```

The difference between the given initial condition and the approximating sum is overall in the order of 10^{-18}, when the first 100 Fourier coefficients are taken into account.

```
In[43]:= Plot[Abs[ψ0[ξ] - ψcf[ξ, 0, cs]], {ξ, 0, 1}, PlotRange -> All]
```

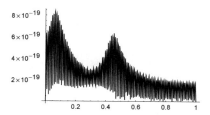

Here is the numerical solution of the third initial value problem, generated by series summation.

```
In[44]:= ppξ = 200; ppτ = 200;
     (dataSum = Table[ψcf[ξ, τ, cs], {τ, 0, 1, 1/ppτ}, {ξ, 0, 1, 1/ppξ}];) //
                                                              Timing
Out[45]= {7.16 Second, Null}

In[46]:= ListDensityPlot[Abs[dataSum], Mesh -> False, ColorFunction -> (Hue[#]&),
              MeshRange -> {{0, 1}, {0, 1}}]
```

To find the numerical solution using NDSolve, we rescale the maximal value $\psi(1/4, 0)$ of $\psi(\xi, 0)$ to 1. This and the fact that too small-scale structures are present in the initial conditions makes sure that our typical setting AccuracyGoal -> 3 and PrecisionGoal -> 3 are appropriate settings.

```
In[47]:= Φ0[ξ_] := Which[ξ <= 0.01, 0.,
                  10^-2 <= ξ <= 0.49,
                  Exp[16] Exp[5I ξ] Exp[1/((ξ - 1/4)^2 - 1/16)],
                  ξ >= 0.49, 0]

In[48]:= (nsol =
        NDSolve[{I D[Φ[ξ, τ], τ] + 1/(4 Pi) D[Φ[ξ, τ], {ξ, 2}] == 0,
            Φ[ξ, 0] == Φ0[ξ],
            Φ[0, τ] == 0, Φ[1, τ] == 0},
            Φ[ξ, τ], {ξ, 0, 1}, {τ, 0, 1},
            AccuracyGoal -> 3, PrecisionGoal -> 3, MaxSteps -> 10^5,
            Method -> {"MethodOfLines", "SpatialDiscretization" ->
                        {"TensorProductGrid", "DifferenceOrder" -> 4,
                         "MaxPoints" -> 101, "MinPoints" -> 101}}]) // Timing
        NDSolve::eerri :
         Warning: Estimated initial error on the specified spatial grid in the direction
            of independent variable ξ exceeds prescribed error tolerance. More…

        NDSolve::eerr :
         Warning: Scaled local spatial error estimate of 91.8907972429531` at τ = 1.` in
            the direction of independent variable ξ is much greater than prescribed error
            tolerance. Grid spacing with 101 points may be too large to achieve the desired
            accuracy or precision. A singularity may have formed or you may want to
            specify a smaller grid spacing using the MaxStepSize or MinPoints options. More…
Out[48]= {1.92 Second, {{Φ[ξ, τ] → InterpolatingFunction[{{0., 1.}, {0., 1.}}, <>][ξ, τ]}}}
```

This time, the resulting picture looks for larger times different from the one obtained by the separation of variable method. For $\tau \gtrsim 1/2$, the error of nsol becomes too large. For $\tau \lesssim 1/2$, the solution looks similar to the solution obtained by the separation of variable method.

```
In[49]:= ppξ = 200; ppτ = 200;
        (dataNDSolve1 = Table[Evaluate[Ψ[ξ, τ] /. nsol[[1]]],
                              {τ, 0, 1, 1/ppτ}, {ξ, 0, 1, 1/ppξ}]);
In[51]:= ppξ = 200; ppτ = 100;
        (dataNDSolve2 = Table[Evaluate[Ψ[ξ, τ] /. nsol[[1]]],
                              {τ, 0, 1/2, 1/2/ppτ}, {ξ, 0, 1, 1/ppξ}]);
In[53]:= Show[GraphicsArray[
        Block[{$DisplayFunction = Identity},
          {(* full time domain *)
          ListDensityPlot[Abs[dataNDSolve1], ColorFunction -> (Hue[#]&),
                          Mesh -> False, MeshRange -> {{0, 1}, {0, 1}}],
            (* restricted time domain *)
          ListDensityPlot[Abs[dataNDSolve2],
                          Mesh -> False, ColorFunction -> (Hue[#]&),
                          MeshRange -> {{0, 1}, {0, 1/2}},
                          AspectRatio -> Automatic]}]]]
```

Although the last picture looks qualitatively similar to the one obtained by the separation of variable method, the error of the NDSolve solution is much larger this time. Again, we could reduce the starting step size to obtain a better result.

Graphing the difference between the two solutions shows clearly that the error in the NDSolve solution grows dramatically for $\tau \gtrsim 1/4$.

```
In[54]:= Map[If[# < 0, 1., #]&, Abs[Exp[16] dataSum], {2}];

In[55]:= ListPlot3D[Abs[Exp[16] dataSum - dataNDSolve1]/
                    (* ignore function values below accuracy goal *)
                    Map[If[# <= 10^-3, 1., #]&, Abs[Exp[16] dataSum], {2}],
                    Mesh -> False, MeshRange -> {{0, 1}, {0, 1/2}},
                    ViewPoint -> {-1, -3, 1}, PlotRange -> {0, 10}]
```

Because the use of the Fourier transform method to this problem is straightforward, we do not carry it out here.

For the fourth problem, it is possible to calculate the Fourier coefficients c_k in closed form. They take the following simple form.

```
In[56]:= c[k_] = Integrate[Sqrt[2] Sin[Pi ξ k] 1, {ξ, 0, 1}] /.
                 Cos[k π] -> (-1)^k // Simplify
```

$$Out[56]= -\frac{\sqrt{2}\ (-1 + (-1)^k)}{k\,\pi}$$

The last output means the even Fourier coefficients vanish and the odd ones are of size $2\sqrt{2}/(k\pi)$. The $1/k$ behavior shows that resulting series will converge slowly. To get reliable values for the sums $\psi(\xi, \tau) = \sum_{k=1}^{\infty} c_k\,\psi_n(\xi, \tau)$, the following function calculates a list of partial sums of length m, starting from the nth term. To avoid unnecessary calculations, we drop the overall factor $2\sqrt{2}/\pi$.

```
In[57]:= ψcf = Compile[{{ξ, _Real}, {τ, _Real}, {n, _Integer}, {m, _Integer}},
         Module[{ψ0},
             ψ0 = Sum[1/(2k + 1) Sin[(2k + 1) Pi ξ] Exp[-I Pi/4 (2k + 1)^2 τ],
                 {k, 0, n}];
             ψ0 + FoldList[Plus, 0, Table[1/(2k + 1)*
                         Sin[(2k + 1) Pi ξ] Exp[-I Pi/4 (2k + 1)^2 τ],
                         {k, n + 1, n + m}]]]];
```

The following graphic shows the convergence for some selected points in the ξ,τ-plane.

```
In[58]:= Show[Graphics[{Thickness[0.002],
             Table[{Hue[Random[]],
                 Line[MapIndexed[{#2[[1]], #1}&, Abs[ψcf[ξ, τ, 0, 1000]]]]},
                 {ξ, 0.1, 0.5, 0.1}, {τ, 0.1, 0.9, 0.1}]}],
         PlotRange -> All, Frame -> True]
```

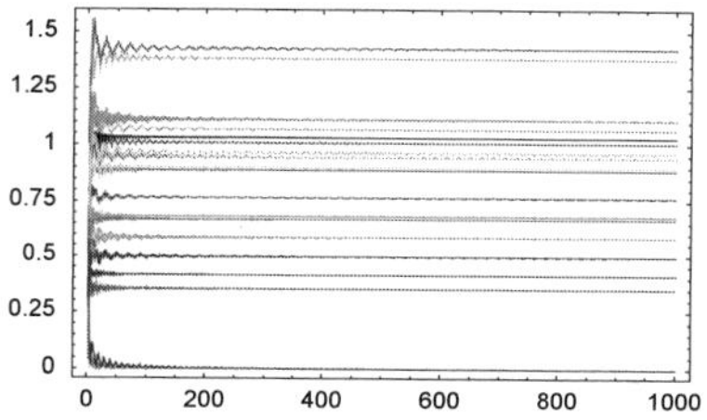

The last picture shows that by using brute force summation of about 10^3 terms, we will get a result correct to two digits. To further enhance the convergence, we will use SequenceLimit on 50 partial sums after summing the first 1000 terms. The following plots show the partial sums, together with the sequence limit for some example points.

```
In[59]:= convergencePlot[ξ_, τ_, n_, m_, ns_, ms_] :=
         ListPlot[Abs[ψcf[ξ, τ, n, m]],
                 PlotStyle -> {PointSize[0.002]},
                 Prolog -> {{Thickness[0.002], Hue[0], Line[{{1, #}, {m, #}}]}&[
                 Abs[SequenceLimit[ψcf[ξ, τ, ns, ms]]]]}}]
In[60]:= Show[GraphicsArray[
         Block[{$DisplayFunction = Identity},
             {convergencePlot[0.3, 0.6, 0, 1000, 1000, 50],
              convergencePlot[0.5, 1, 0, 1000, 1000, 50]}]]]
```

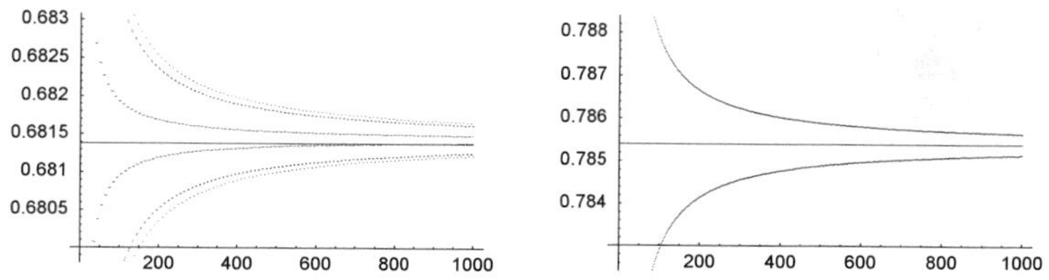

Because of the symmetry $\psi(1/2 + \xi, 0) = \psi(1/2 - \xi, 0)$ and the invariance of the governing differential equation with respect to mirroring on the $\xi = 1/2$ line, we will restrict the calculation of the numerical values to the domain $0 \le \xi \le 1/2$ in the following input.

```
In[61]:= Off[SequenceLimit::seqlim];
         ppξ = 100; ppτ = 200;
         (data = Table[Table[SequenceLimit[ψcf[ξ, τ, 1000, 50]],
                     {ξ, 0, 1/2, 1/2/ppξ}], {τ, 0, 1, 1/ppτ}];) // Timing
Out[63]= {93.89 Second, Null}

In[64]:= ListDensityPlot[Abs[Join[#, Reverse[#]]& /@ data],
                     Mesh -> False, ColorFunction -> Hue]
```

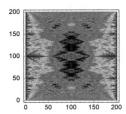

Because the initial conditions and the boundary conditions are not compatible, the resulting $\psi(\xi, \tau)$ is not a smooth, differentiable function, but a fractal [197], [1235], [1167], [1914], [771], [1166], [1913], [1618], [1367], [1716], [1398], [240], [46], [1587]. (For the relation of this fractal to the boundary of analyticity of the function $\vartheta_3(z, q)$, see [1619], [644], [1396], [946], [656]; for measurement induced fractal wave functions, see [171].) So a solution with NDSolve will not give a useful result. Already for small times τ, the global structure is not reflected in the solution.

```
In[65]:= (nsol =
    NDSolve[{I D[Φ[ξ, τ], τ] + 1/(4 Pi) D[Φ[ξ, τ], {ξ, 2}] == 0,
            Φ[ξ, 0] == 1,
            Φ[0, τ] == 0, Φ[1, τ] == 0},
            Φ[ξ, τ], {ξ, 0, 1}, {τ, 0, 1},
            AccuracyGoal -> 3, PrecisionGoal -> 3,
            Method -> {"MethodOfLines", "SpatialDiscretization" ->
                        {"TensorProductGrid", "DifferenceOrder" -> 4,
                         "MaxPoints" -> 200, "MinPoints" -> 200}}]) // Timing

    NDSolve::ibcinc : Warning: Boundary and initial conditions are inconsistent. More…

Out[65]= {0.02 Second, {{Φ[ξ, τ] → InterpolatingFunction[{{0., 1.}, {0., 1.}}, <>][ξ, τ]}}}

In[66]:= ppξ = 200; ppτ = 200;
    (dataNDSolve = Table[Evaluate[Φ[ξ, τ] /. nsol[[1]]],
                    {τ, 0, 1, 1/ppτ}, {ξ, 0, 1, 1/ppξ}]);

In[68]:= ListDensityPlot[Abs[dataNDSolve], Mesh -> False, ColorFunction -> Hue,
                    MeshRange -> {{0, 1}, {0, 1}}, AspectRatio -> Automatic,
                    PlotRange -> Automatic]
```

Because numerical methods do not operate well with inconsistent boundary and initial conditions, we change the initial conditions slightly at the boundaries of the domain to be compatible with the boundary conditions.

```
In[69]:= (nsol =
    NDSolve[{I D[Φ[ξ, τ], τ] + 1/(4 Pi) D[Φ[ξ, τ], {ξ, 2}] == 0,
            Φ[ξ, 0] == If[ξ < 10^-8 || ξ > 1 - 10^-8, 0, 1],
            Φ[0, τ] == 0, Φ[1, τ] == 0},
            Φ[ξ, τ], {ξ, 0, 1}, {τ, 0, 1},
            AccuracyGoal -> 2, PrecisionGoal -> 2, MaxSteps -> 10^5,
            Method -> {"MethodOfLines", "SpatialDiscretization" ->
                        {"TensorProductGrid", "DifferenceOrder" -> 4,
                         "MaxPoints" -> 201, "MinPoints" -> 201}}]) // Timing

Out[69]= {11.64 Second, {{Φ[ξ, τ] → InterpolatingFunction[{{0., 1.}, {0., 1.}}, <>][ξ, τ]}}}
```

For not too large times, the solution is now remarkably similar to the above one.

```
In[70]:= ppξ = 200; ppτ = 200;
         (dataNDSolve = Table[Evaluate[Φ[ξ, τ] /. nsol[[1]]],
                       {τ, 0, 1, 1/ppτ}, {ξ, 0, 1, 1/ppξ}]);
In[72]:= ListDensityPlot[Abs[dataNDSolve], Mesh -> False, ColorFunction -> Hue,
                       MeshRange -> {{0, 1}, {0, 1}}, AspectRatio -> Automatic]
```

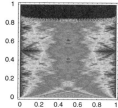

Now let us deal with the fifth subproblem. We chose $\bar{k} = 1000$, $\delta k = 150$, and $\sigma = 10$. $\psi 0$cf is a compiled function to quickly calculate the initial conditions and the time-dependent superposition.

```
In[73]:= ψ0cf = Compile[{{ξ, _Real, 0}, {τ, _Real, 0}},
                   Module[{k0 = 10^3, δk = 150, σ = 10},
                       Sum[Exp[-(k - k0)^2/(4 σ^2)]*
                               Sin[k Pi ξ] Exp[-Pi/4 I k^2 τ],
                           {k, k0 - δk, k0 + δk}]]];
```

The absolute value of the initial conditions is a smooth function with a localized bump.

```
In[74]:= Plot[Abs[ψ0cf[ξ, 1/10^3]], {ξ, 0, 1}, PlotRange -> All]
```

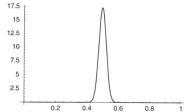

Using $\psi 0$cf, it is straightforward to calculate and visualize the time dependence of the bump. The initial bump is a coherent state and behaves largely as a classical particle that bounces back and forth between the left and right wall [626], [627]. At later times, the spreading of the initial bump becomes visible.

```
In[75]:= ppξ = 400; ppτ = 400;
         dataSum = Table[ψ0cf[ξ, τ 10^-3], {τ, 1, 40, 39/ppτ}, {ξ, 0, 1, 1/ppξ}];
In[77]:= Show[GraphicsArray[
         ListDensityPlot[#[dataSum], ColorFunction -> (Hue[#]&),
                       MeshRange -> {{0, 1}, {1, 40}}, Mesh -> False,
                       DisplayFunction -> Identity, AspectRatio -> 2]& /@ {Abs, Re}]]
```

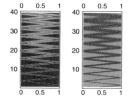

Now, we will try to use NDSolve on the time-dependent problem. The function solveODEAndShowSolution implements the numerical solution and its visualization for a given value of the initial spatial discretization δx.

```
In[78]:= ΦInit[ξ_?NumberQ] := If[ξ == 1, 0, ψ0cf[ξ, 10^-3]]
```

```
In[79]:= solveODEAndShowSolution[δx_, opts___] :=
       Module[{nsol, pp = Round[1/δx], ppξ = 200, ppτ = 200, T = 40 10^-3, tMax},
           (* solve differential equation *)
           nsol = NDSolve[{I D[Φ[ξ, τ], τ] + 1/(4 Pi) D[Φ[ξ, τ], {ξ, 2}] == 0,
                         Φ[ξ, 10^-3] == ΦInit[ξ],
                         Φ[0, τ] == 0, Φ[1, τ] == 0},
                         Φ[ξ, τ], {ξ, 0, 1}, {τ, 1 10^-3, T},
                         AccuracyGoal -> 2, PrecisionGoal -> 2,
                           Method -> {"MethodOfLines", "SpatialDiscretization" ->
                         {"TensorProductGrid", "DifferenceOrder" -> 4,
                         "MaxPoints" -> pp, "MinPoints" -> pp}}];
           (* reached max time (in case MaxSteps was too small to reach T *)
           tMax = nsol[[1, 1, 2, 0, 1, 2, 2]];
           (* list of function values *)
           dataNDSolve = Table[Evaluate[Φ[ξ, τ] /. nsol[[1]]],
                         {τ, 1 10^-3, tMax, (tMax - 10^-3)/ppτ},
                         {ξ, 0, 1, 1/ppξ}];
           (* make density plot *)
           ListDensityPlot[Abs[dataNDSolve], opts, PlotRange -> All,
                         Mesh -> False, ColorFunction -> (Hue[#]&),
                         MeshRange -> {{0, 1}, {10^-3, tMax}}]]
```

The following four graphics show the resulting $\Phi[\xi, \tau]$ for $\delta x = 0.01/1$, $0.01/4$, $0.01/5$, and $0.01/6$. The second and fourth pictures show a localized bump moving with a wrong period. The first and third graphics are basically structureless. For the first two calls to NDSolve, we also get a warning message that the result is probably wrong. The timings and memory usages suggest not to decrease δx further.

```
In[80]:= res = {Timing[(* suppress surely emitted
                       NDSolve::eerr and NDSolve::eerri messages *)
                Internal`DeactivateMessages @
                solveODEAndShowSolution[0.01/#,
                       DisplayFunction -> Identity]],
                MaxMemoryUsed[]/10.^6 "MB"}& /@ {1, 4, 5, 6}
In[81]:= Show[GraphicsArray[#[[1, 2]]& /@ res]]
```

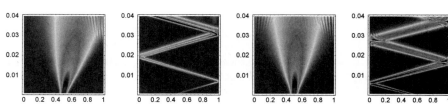

The reason for the failure of NDSolve becomes obvious if we look at the real and imaginary parts of the initial conditions. While the absolute value of the initial conditions is a smooth function, its real and imaginary parts are highly oscillating functions. The period of the highest oscillations is in the order $2/\overline{k} \approx 0.002$, which is much smaller than the δx-values used above. So, the fine structure of the initial conditions was not resolved and the subtle interference between all fundamental oscillations leading to the localized structure was not observed. For some of the δx a few, but no all of the oscillations were felt and as a result, a localized structure moving with another period arose.

```
In[82]:= Show[GraphicsArray[
         Plot[#[ψ0cf[ξ, 1/10^3]], {ξ, 0.4, 0.6},
              PlotRange -> All, PlotPoints -> 500, Frame -> True,
              Axes -> False, DisplayFunction -> Identity]& /@ {Re, Im}]]
```

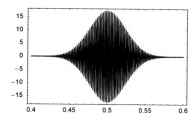 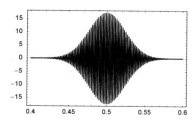

For further methods for solving the time-dependent Schrödinger equation, see [1043] and [857].

36. Initial Value Problems for the Wave Equation

We start with the 1D case. This is the initial condition.

```
In[1]:= u0[r_] := If[Abs[r] < 1, Cos[r Pi/2]^2, 0]
```

For the numerical solution of the partial differential equation, we choose such a spatial domain that we have known conditions at the boundary. The wave equation is a hyperbolic differential equation, and the velocity of disturbances is 1. This means that to calculate the solution in the space-time domain $0 \le t \le 3$, $0 \le |x| \le 3$, we use the spatial domain $-4 \le |x| \le 4$. (For artificial boundary conditions for $t > 3$, see [33].)

```
In[2]:= X = 4; T = X - 1;
```

The arguments for NDSolve are obvious. We start by using the "Pseudospectral" method option setting for the spatial discretization.

```
In[3]:= nsol =
        NDSolve[{D[u[t, r], {t, 2}] == D[u[t, r], {r, 2}],
                u[0, r] == u0[r], Derivative[1, 0][u][0, r] == 0,
                u[t, -X] == 0, u[t, X] == 0}, u, {t, 0, T}, {r, -X, X},
                PrecisionGoal -> 3, AccuracyGoal -> 4,
                Method -> {"MethodOfLines",
                    "SpatialDiscretization" -> {"TensorProductGrid",
                    "DifferenceOrder"-> "Pseudospectral",
                    "MinPoints" -> 121, "MaxPoints" -> 121}}];

In[4]:= ByteCount[nsol]/10.^6 "MB"

Out[4]= 14.883 MB
```

The function generates a GraphicsArray of three graphics showing a 3D plot and a contour plot of the solution $u(t, r)$ over the t,r-plane and a plot of the solution for $t = 2$.

```
In[5]:= solutionPlots[] :=
        With[{opts = Sequence[PlotRange -> All, DisplayFunction -> Identity]},
        Show[GraphicsArray[{
        (* a 3D plot *)
        Plot3D[Evaluate[u[t, r] /. nsol[[1]]], {r, 0, X - 1}, {t, 0, T},
                opts, PlotPoints -> 200, Mesh -> False,
                AxesLabel -> {"r", "t", None}, ClipFill -> None],
        (* a contour plot *)
        ContourPlot[Evaluate[u[t, r] /. nsol[[1]]], {r, 0, X - 1}, {t, 0, T},
                PlotPoints -> 200, Contours -> 30,
                ColorFunction -> (Hue[0.8 #]&), opts],
        (* a plot for t==2 *)
        Plot[Evaluate[u[2, r] /. nsol[[1]]], {r, 0, X - 1}, opts]}]]]
```

Here are these three plots for the 1D case.

```
In[6]:= solutionPlots[]
```

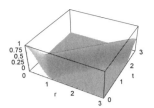 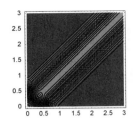 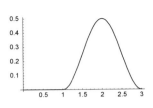

Here is a comparison of the NDSolve solution with the exact solution. The overall error is quite small. The largest error occurs (not unexpectedly) at the transition from a nonzero to an identical-zero solution. The absolute size of the error is in the order 10^{-3}, as expected by the above used setting for the PrecisionGoal option.

```
In[7]:= uExact1D[t_, r_] := (u0[r + t] + u0[r - t])/2
```

```
In[8]:= Plot3D[Evaluate[uExact1D[t, r] - (u[t, r] /. nsol[[1]])],
          {r, 0, X - 1}, {t, 0, T},
          PlotPoints -> 200, PlotRange -> All, Mesh -> False]
```

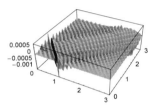

We redo the numerical solution of the PDE, this time using a finite difference spatial discretization. This time, we get an error message indicating that the requested solution quality might not be realized in the returned solution. Again, the error is in the prescribed order of magnitude, but is much more localized at the light cone boundaries. (We do not display the calculated solution again.)

```
In[9]:= nsol =
        NDSolve[{D[u[t, r], {t, 2}] == D[u[t, r], {r, 2}],
            u[0, r] == u0[r], Derivative[1, 0][u][0, r] == 0,
            u[t, -X] == 0, u[t, X] == 0}, u, {t, 0, T}, {r, -X, X},
            PrecisionGoal -> 3, AccuracyGoal -> 4,
            Method -> {"MethodOfLines",
                "SpatialDiscretization" -> {"TensorProductGrid",
                "DifferenceOrder"-> 4,
                "MinPoints" -> 1200, "MaxPoints" -> 1200}}];
```

```
NDSolve::eerri :
  Warning: Estimated initial error on the specified spatial grid in the direction
     of independent variable r exceeds prescribed error tolerance. More…

NDSolve::eerr :
  Warning: Scaled local spatial error estimate of 73.54933297521723` at
     t = 3.` in the direction of independent variable r is much greater
     than prescribed error tolerance. Grid spacing with 1200 points
     may be too large to achieve the desired accuracy or precision.  A
     singularity may have formed or you may want to specify a smaller
     grid spacing using the MaxStepSize or MinPoints options. More…
```

```
In[10]:= ByteCount[nsol]/10.^6 "MB"
```

```
Out[10]= 72.9023 MB
```

```
In[11]:= Plot3D[Evaluate[uExact1D[t, r] - (u[t, r] /. nsol[[1]])],
          {r, 0, X - 1}, {t, 0, T},
          PlotPoints -> 200, PlotRange -> All, Mesh -> False]
```

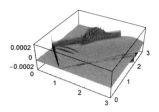

Now let us consider the 2D case. For the numerical solution, we use again the fact that the velocity of disturbances is 1. Because of the symmetry of the initial conditions, we use a polar coordinate system and calculate only the r-dependence of the solution. We continue the solution symmetrically to negative r. For the Laplace operator, we use the radial part in cylindrical coordinates. To avoid division by zero caused by the term $\partial u(t, r)/\partial r/r$, we discretize the spatial domain in such a way that we avoid the $r = 0$ singularity. This leads to the following input for NDSolve. We use again a finite difference discretization.

```
In[12]:= nsol =
    NDSolve[{D[u[t, r], {t, 2}] == D[u[t, r], {r, 2}] + 1/r D[u[t, r], r],
        u[0, r] == u0[r], Derivative[1, 0][u][0, r] == 0,
        u[t, -X] == 0, u[t, X] == 0}, u, {t, 0, T}, {r, -X, X},
        PrecisionGoal -> 3, AccuracyGoal -> 4,
        Method -> {"MethodOfLines",
            "SpatialDiscretization" -> {"TensorProductGrid",
            "DifferenceOrder"-> 4,
            "MinPoints" -> 1200, "MaxPoints" -> 1200}}];
```
NDSolve::eerri :
   Warning: Estimated initial error on the specified spatial grid in the direction
      of independent variable r exceeds prescribed error tolerance. More…

NDSolve::eerr :
   Warning: Scaled local spatial error estimate of 68.28316445124355` at
      t = 3.` in the direction of independent variable r is much greater
      than prescribed error tolerance. Grid spacing with 1200 points
      may be too large to achieve the desired accuracy or precision. A
      singularity may have formed or you may want to specify a smaller
      grid spacing using the MaxStepSize or MinPoints options. More…

```
In[13]:= ByteCount[nsol]/10.^6 "MB"
```
Out[13]= 74.8467 MB

This time, the solution is more interesting: In the middle region, u does not reduce to 0, but rather to a slowly decaying function of t [498], [1707], [963], [1305], [1306], [406], [1752], [1753], [469]. (The nonlinear shallow water wave equations can be reduced to a similar PDE [942], [346] where the $1/r$ singularity does not arise from the polar coordinate system.)

```
In[14]:= solutionPlots[]
```

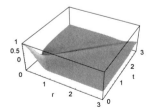 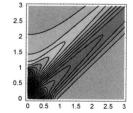 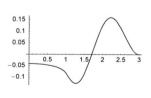

Because of the $1/r$ factor in the differential equation, a pseudospectral discretization does not perform well in the 2D case.

```
In[15]:= Block[{nsol =
    NDSolve[{D[u[t, r], {t, 2}] == D[u[t, r], {r, 2}] + 1/r D[u[t, r], r],
```

```
            u[0, r] == u0[r], Derivative[1, 0][u][0, r] == 0,
            u[t, -X] == 0, u[t, X] == 0}, u, {t, 0, T}, {r, -X, X},
            PrecisionGoal -> 3, AccuracyGoal -> 4,
            Method -> {"MethodOfLines",
                "SpatialDiscretization" -> {"TensorProductGrid",
                "DifferenceOrder"-> "Pseudospectral",
                "MinPoints" -> 61, "MaxPoints" -> 61}}]},
       Plot3D[Evaluate[u[t, r] /. nsol[[1]]], {r, 0, X - 1}, {t, 0, 1},
            PlotPoints -> 200, Mesh -> False, AxesLabel -> {"r", "t", None},
            ClipFill -> None]]
```

For the 2D case, the calculation of the solution of the wave equation based on the integral is more complicated. We have to calculate a time derivative of a 2D integral. We will carry out the integral numerically. To avoid a numerical differentiation or the numerical calculation of more than one integral over nonrectangular domains, we will rewrite the double integral in such a manner that the integration limits have no t-dependence and we will differentiate the integrand before carrying out the numerical integration:

$$\frac{\partial}{\partial t} \int_{z_1(t)}^{z_2(t)} f(z, t)\, dz = \int_0^1 \frac{\partial f((z_2(t) - z_1(t))\, \zeta + z_1(t), t)\, (z_2(t) - z_1(t))}{\partial t}\, d\zeta.$$

Here is an implementation of this change of variables for a double integral. We will use H and Ξ as the new integration variables.

In[16]:= `newIntegrand[{f_, {z_, z1_, z2_}}, ζ_] :=`
            `(f /. z -> (z2 - z1) ζ + z1) (z2 - z1)`

In[17]:= `newDoubleIntegrateIntegrand[`
            `{f_, {ξ_, ξ1_, ξ2_}, {η_, η1_, η2_}}, {Ξ_, H_}] :=`
            `newIntegrand[{newIntegrand[{f, {η, η1, η2}}, H], {ξ, ξ1, ξ2}}, Ξ]`

For the numerical integration, we will differentiate four cases.

Case 1: $u_0(\xi)$ and $B_t(x)$ do not overlap. This is the case for $r - t > R$. Then, $u = 0$. This is the far away situation.

Case 2: $u_0(\xi)$ contains $B_t(x)$ completely. This is the case for $r + t < R$. This is the small distance, small time situation.

In[18]:= `u0NonZero[r_] := Cos[r Pi/2]^2`

Mathematica cannot carry out the integral in closed form.

In[19]:= `Integrate[u0NonZero[Sqrt[ξ^2 + η^2]]/Sqrt[t^2 - (r - ξ)^2 - η^2],`
            `{ξ, r - t, r + t},`
            `{η, -Sqrt[t^2 - (r - ξ)^2], Sqrt[t^2 - (r - ξ)^2]}]`

Out[19]= $\displaystyle \int_{r-t}^{r+t} \int_{-\sqrt{t^2-(r-\xi)^2}}^{\sqrt{t^2-(r-\xi)^2}} \frac{\text{Cos}\left[\frac{1}{2}\,\pi\,\sqrt{\eta^2 + \xi^2}\right]^2}{\sqrt{t^2 - \eta^2 - (r-\xi)^2}}\, d\eta\, d\xi$

This is the form of the integrand before differentiation.

In[20]:= `ndi2 = newDoubleIntegrateIntegrand[`
            `{u0NonZero[Sqrt[ξ^2 + η^2]]/Sqrt[t^2 - (r - ξ)^2 - η^2],`
            `{ξ, r - t, r + t},`
            `{η, -Sqrt[t^2 - (r - ξ)^2], Sqrt[t^2 - (r - ξ)^2]}},`
            `{Ξ, H}] // Simplify`

Out[20]= $\displaystyle \frac{2\,t\,\sqrt{-t^2\,(-1 + \Xi)\,\Xi}\,\text{Cos}\left[\frac{1}{2}\,\pi\,\sqrt{r^2 + 2\,r\,t\,(-1 + 2\,\Xi) + t^2\,(1 + 16\,H\,(-1 + \Xi)\,\Xi - 16\,H^2\,(-1 + \Xi)\,\Xi)}\right]^2}{\sqrt{t^2\,(-1 + H)\,H\,(-1 + \Xi)\,\Xi}}$

And this is a simplified form of the final integrand.

```
In[21]:= integrand2[t_, r_, Ξ_, H_] = D[ndi2, t] // Simplify
```

$$Out[21]= \left(2\sqrt{-t^2(-1+\Xi)\,\Xi}\,\cos\left[\tfrac{1}{2}\pi\sqrt{r^2+2\,r\,t\,(-1+2\,\Xi)+t^2\,(1+16\,H\,(-1+\Xi)\,\Xi-16\,H^2\,(-1+\Xi)\,\Xi)}\,\right]\right.$$
$$\left(\sqrt{r^2+2\,r\,t\,(-1+2\,\Xi)+t^2\,(1+16\,H\,(-1+\Xi)\,\Xi-16\,H^2\,(-1+\Xi)\,\Xi)}\right.$$
$$\cos\left[\tfrac{1}{2}\pi\sqrt{r^2+2\,r\,t\,(-1+2\,\Xi)+t^2\,(1+16\,H\,(-1+\Xi)\,\Xi-16\,H^2\,(-1+\Xi)\,\Xi)}\,\right]+$$
$$\pi\,t\,(r-2\,r\,\Xi+t\,(-1-16\,H\,(-1+\Xi)\,\Xi+16\,H^2\,(-1+\Xi)\,\Xi))$$
$$\left.\sin\left[\tfrac{1}{2}\pi\sqrt{r^2+2\,r\,t\,(-1+2\,\Xi)+t^2\,(1+16\,H\,(-1+\Xi)\,\Xi-16\,H^2\,(-1+\Xi)\,\Xi)}\,\right]\right)\right)\Big/$$
$$\left(\sqrt{t^2\,(-1+H)\,H\,(-1+\Xi)\,\Xi}\,\sqrt{r^2+2\,r\,t\,(-1+2\,\Xi)+t^2\,(1+16\,H\,(-1+\Xi)\,\Xi-16\,H^2\,(-1+\Xi)\,\Xi)}\right)$$

Case 3: $B_t(x)$ contains $u_0(\xi)$ completely. This is the case for $R < r + t \wedge r - t < -R$. This is the large time approximation. Again, the integral cannot be done in closed form.

```
In[22]:= R = 1;
         Integrate[u0NonZero[Sqrt[ξ^2 + η^2]]/Sqrt[t^2 - (r - ξ)^2 - η^2],
                   {ξ, -R, R}, {η, -Sqrt[R^2 - ξ^2], Sqrt[R^2 - ξ^2]}]
```

$$Out[23]= \int_{-1}^{1}\int_{-\sqrt{1-\xi^2}}^{\sqrt{1-\xi^2}}\frac{\cos\left[\tfrac{1}{2}\pi\sqrt{\eta^2+\xi^2}\,\right]^2}{\sqrt{t^2-\eta^2-(r-\xi)^2}}\,d\eta\,d\xi$$

Proceeding similarly to case 2, we derive the form of the integrand suitable for numerical integration.

```
In[24]:= ndi3 = newDoubleIntegrateIntegrand[
             {u0NonZero[Sqrt[ξ^2 + η^2]]/Sqrt[t^2 - (r - ξ)^2 - η^2],
              {ξ, -R, R}, {η, -Sqrt[R^2 - ξ^2], Sqrt[R^2 - ξ^2]}},
             {Ξ, H}] // Simplify
```

$$Out[24]= \frac{8\sqrt{-(-1+\Xi)\,\Xi}\,\cos\left[\tfrac{1}{2}\pi\sqrt{1+16\,H\,(-1+\Xi)\,\Xi-16\,H^2\,(-1+\Xi)\,\Xi}\,\right]^2}{\sqrt{-1-r^2+t^2+16\,H\,\Xi-16\,H^2\,\Xi-16\,H\,\Xi^2+16\,H^2\,\Xi^2+r\,(-2+4\,\Xi)}}$$

```
In[25]:= integrand3[t_, r_, Ξ_, H_] = D[ndi3, t]
```

$$Out[25]= -\frac{8\,t\,\sqrt{-(-1+\Xi)\,\Xi}\,\cos\left[\tfrac{1}{2}\pi\sqrt{1+16\,H\,(-1+\Xi)\,\Xi-16\,H^2\,(-1+\Xi)\,\Xi}\,\right]^2}{(-1-r^2+t^2+16\,H\,\Xi-16\,H^2\,\Xi-16\,H\,\Xi^2+16\,H^2\,\Xi^2+r\,(-2+4\,\Xi))^{3/2}}$$

Case 4: $B_t(x)$ overlaps with $u_0(\xi)$ partially. This is the case for $-R < x - t < R$. The overlapping domain is bounded on the left by an arc of the boundary of $B_t(x)$ and on the right by an arc of the boundary of $u_0(\xi)$. The two intersection points of the two boundaries are as follows.

```
In[26]:= Solve[{ξ^2 + η^2 == R^2, (r - ξ)^2 + (0 - η)^2 == t^2}, {ξ, η}] // Simplify
```

$$Out[26]= \left\{\left\{\eta\to-\sqrt{1-\frac{(1+r^2-t^2)^2}{4\,r^2}}\,,\ \xi\to\frac{1+r^2-t^2}{2\,r}\right\},\ \left\{\eta\to\sqrt{1-\frac{(1+r^2-t^2)^2}{4\,r^2}}\,,\ \xi\to\frac{1+r^2-t^2}{2\,r}\right\}\right\}$$

```
In[27]:= ξ0[t_, r_] = (1 - t^2 + r^2)/(2 r);
         η0[t_, r_] = Sqrt[1 - ξ0[t, r]^2];
```

We split the integration region into its left and right parts and calculate the corresponding integrands.

```
In[29]:= ndi4a = newDoubleIntegrateIntegrand[
             {u0NonZero[Sqrt[ξ^2 + η^2]]/Sqrt[t^2 - (r - ξ)^2 - η^2],
              {ξ, r - t, ξ0[t, r]},
              {η, -Sqrt[t^2 - (r - ξ)^2], Sqrt[t^2 - (r - ξ)^2]}},
             {Ξ, H}] // Simplify
```

Out[29]= $-\left(\left(-1 + r^2 - 2\,r\,t + t^2\right) \sqrt{t^2 - \frac{\left(-2\,r\,t\,\left(-1 + \Xi\right) + r^2\,\Xi + \left(-1 + t^2\right)\,\Xi\right)^2}{4\,r^2}}\right.$

$\qquad \mathrm{Cos}\left[\frac{1}{2}\,\pi\,\sqrt{\left(\left(-r + t + \frac{\left(-1 + r^2 - 2\,r\,t + t^2\right)\,\Xi}{2\,r}\right)^2 + }\right.}$

$\qquad\qquad \left.\left.\left.\left(1 - 2\,H\right)^2\,\left(t^2 - \frac{\left(-2\,r\,t\,\left(-1 + \Xi\right) + r^2\,\Xi + \left(-1 + t^2\right)\,\Xi\right)^2}{4\,r^2}\right)\right)\right]\right]^2\right) \bigg/$

$\qquad \left(r \sqrt{\frac{\left(-1 + r^2 - 2\,r\,t + t^2\right)\,\left(-1 + H\right)\,H\,\Xi\,\left(-2\,r\,t\,\left(-2 + \Xi\right) + r^2\,\Xi + \left(-1 + t^2\right)\,\Xi\right)}{r^2}}\right)$

In[30]:= `integrand4a[t_, r_, Ξ_, H_] = D[ndi4a, t];`

In[31]:= `ndi4b = newDoubleIntegrateIntegrand[`
`    {u0NonZero[Sqrt[ξ^2 + η^2]]/Sqrt[t^2 - (r - ξ)^2 - η^2],`
`    {ξ, ξ0[t, r], R}, {η, -Sqrt[R^2 - ξ^2], Sqrt[R^2 - ξ^2]}},`
`    {Ξ, H}] // Simplify`

Out[31]= $-\left(\left(1 - 2\,r + r^2 - t^2\right) \sqrt{1 - \frac{\left(-r^2\,\left(-1 + \Xi\right) + \left(-1 + t^2\right)\,\left(-1 + \Xi\right) + 2\,r\,\Xi\right)^2}{4\,r^2}}\right.$

$\qquad \mathrm{Cos}\left[\frac{1}{2}\,\pi\,\sqrt{\left(\frac{\left(-r^2\,\left(-1 + \Xi\right) + \left(-1 + t^2\right)\,\left(-1 + \Xi\right) + 2\,r\,\Xi\right)^2}{4\,r^2} + \right.}\right.}$

$\qquad\qquad \left.\left.\left.\left(1 - 2\,H\right)^2\,\left(1 - \frac{\left(-r^2\,\left(-1 + \Xi\right) + \left(-1 + t^2\right)\,\left(-1 + \Xi\right) + 2\,r\,\Xi\right)^2}{4\,r^2}\right)\right)\right]\right]^2\right) \bigg/$

$\qquad \left(r \sqrt{\left(t^2 - \frac{\left(\left(-1 + t^2\right)\,\left(-1 + \Xi\right) + 2\,r\,\Xi - r^2\,\left(1 + \Xi\right)\right)^2}{4\,r^2} - \left(1 - 2\,H\right)^2\right.}\right.$

$\qquad\qquad \left.\left.\left(1 - \frac{\left(-r^2\,\left(-1 + \Xi\right) + \left(-1 + t^2\right)\,\left(-1 + \Xi\right) + 2\,r\,\Xi\right)^2}{4\,r^2}\right)\right)\right)$

In[32]:= `integrand4b[t_, r_, Ξ_, H_] = D[ndi4b, t];`

Because `integrand4a` and `integrand4b` are to be integrated over the unit square, we can unite the two integrals into one.

In[33]:= `integrand4[t_, r_, Ξ_, H_] =`
`          integrand4a[t, r, Ξ, H] + integrand4b[t, r, Ξ, H];`

To speed up the further calculations, we optimize and compile the integrands.

In[34]:= `cf2D = With[{R = R},`
`    (* compile the integrand *) Compile @@`
`    compile[{t, r, Ξ, H}, (* optimize the integrand *)`
`    Experimental`OptimizeExpression @`
`    Which[(* nonoverlapping *) r - t > R, 0,`
`          (* Bt in u0 *) r + t < R, #1,`
`          (* u0 in Bt *) R < r + t && r - t < -R, #2,`
`          (* nontrivial overlap *) True, #3]]&[`
`          integrand2[t, r, Ξ, H], integrand3[t, r, Ξ, H],`
`          integrand4[t, r, Ξ, H]]];`

In[35]:= `(* evaluate only for numerical values`
`       of the integration variables *)`
`cf2DH[t_, r_, Ξ_?NumberQ, H_?NumberQ] := cf2D[t, r, Ξ, H]`

Now, we have integrands for all possible cases and we can implement the numerical integration.

In[37]:= `(* wrapper function *)`
`uExact2D[t_?NumberQ, r_?NumberQ] :=`
`With[{ε = 10^-8},`
`    (* exclude endpoints to avoid singularities *)`
`    1/(2Pi) NIntegrate[cf2DH[t, r, Ξ, H], {Ξ, ε, 1 - ε}, {H, ε, 1 - 2ε},`
`                       PrecisionGoal -> 4, AccuracyGoal -> 4,`
`                       Method -> MultiDimensional, Compiled -> False]]`

We check the quality of the NDSolve-solution on a 50×50 grid in the *r,t*-plane. We avoid the domain boundaries (they could easily be included by using special cases of the above-derived integrands). The precision of the NDSolve solution is quite good. One has to keep in mind that the above calculation of nsol issued messages about an estimated error that is too large.

```
In[39]:= ε = 10^-6;
       Plot3D[uExact2D[t, r] - (u[t, r] /. nsol[[1]]),
              {r, ε, X - 1 - ε}, {t, ε, T - ε},
              PlotPoints -> 50, PlotRange -> All, Mesh -> False]
```

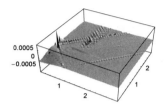

Now, let us consider the 3D case. For the numerical solution, we use again the fact that the velocity of disturbances is 1. Because of the symmetry of the initial conditions, we use a spherical coordinate system and calculate only the *r*-dependence of the solution. We continue the solution symmetrically to negative *r*. For the Laplace operator, we use the radial part in spherical coordinates. And we use again a finite difference approximation for the spatial derivatives. And again, we obtain a warning message that indicates that the requested quality of the solution might have not been achieved.

```
In[41]:= nsol =
       NDSolve[{D[u[t, r], {t, 2}] ==
               D[u[t, r], {r, 2}] + 2/r D[u[t, r], r],
               u[0, r] == u0[r], Derivative[1, 0][u][0, r] == 0,
               u[t, -X] == 0, u[t, X] == 0}, u, {t, 0, T}, {r, -X, X},
               PrecisionGoal -> 3, AccuracyGoal -> 4,
               Method -> {"MethodOfLines",
                  "SpatialDiscretization" -> {"TensorProductGrid",
                  "DifferenceOrder"-> 4,
                  "MinPoints" -> 1200, "MaxPoints" -> 1200}}];

       NDSolve::eerri :
        Warning: Estimated initial error on the specified spatial grid in the direction
           of independent variable r exceeds prescribed error tolerance. More…

       NDSolve::eerr :
        Warning: Scaled local spatial error estimate of 175.57144503968615` at
           t = 3.` in the direction of independent variable r is much greater
           than prescribed error tolerance. Grid spacing with 1200 points
           may be too large to achieve the desired accuracy or precision.  A
           singularity may have formed or you may want to specify a smaller
           grid spacing using the MaxStepSize or MinPoints options. More…
```

```
In[42]:= ByteCount[nsol]/10.^6 "MB"
Out[42]= 78.1737 MB
```

This time, the solution unites features from the 1D and the 2D case: We have areas of negative *u*, and we have again a $u = 0$ region in the middle.

```
In[43]:= solutionPlots[]
```

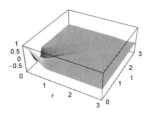 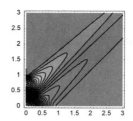 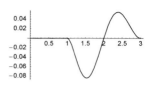

To implement the solution based on the integral formula, we differentiate between the same four cases as in the 2D case.

Case 1: $u_0(x + t\xi)$ and $\partial B_1(x)$ do not overlap. This is the case for $r - t > R$. Then, $u = 0$.

Case 2: $u_0(x + t\xi)$ contains $\partial B_1(x)$ completely. This is the case for $r + t < R$. This is the small distance, small time situation. We use a spherical coordinate system to carry out the integration over σ_ξ. The z-axis of this coordinate system points from the point x toward the origin.

```
In[44]:= integrand2[t_, r_, θ_] =
            D[t u0NonZero[Sqrt[t^2 + r^2 - 2 r t Cos[Pi θ]]] Sin[Pi θ] Pi, t];
```

Case 3: $\partial B_1(x)$ contains $u_0(x + t\xi)$ completely. This is the case for $R < r + t \wedge r - t < -R$. But in distinction with the 2D case, now we have to integrate only over the surface; as the result, we have $u = 0$ again. This is the characteristic behavior of the far away situation for odd-dimensional spaces.

Case 4: $\partial B_1(x)$ overlaps with $u_0(x + t\xi)$ partially. This is the case for $-R < x - t < R$. Now, the integration has to be carried out over a spherical cap with opening angle ΘMax$[t, r]$.

```
In[45]:= θMax[t_, r_] = ArcTan[(r - ξ0[t, r]), η0[t, r]];

In[46]:= integrand4[t_, r_, θ_] =
            D[t u0NonZero[Sqrt[t^2 + r^2 - 2 r t Cos[θ θMax[t, r]]]]*
                            Sin[θ θMax[t, r]] θMax[t, r], t];

In[47]:= cf3D = With[{R = R},
            (* compile the integrand *) Compile @@
            compile[{t, r, θ}, (* optimize the integrand *)
            Experimental`OptimizeExpression @
            Which[(* nonoverlapping *) r - t > R, 0,
                  (* Bt in u0 *) r + t < R, #1,
                  (* u0 in Bt *) R < r + t && r - t < -R, 0,
                  (* nontrivial overlap *) True, #2]]&[
                  integrand2[t, r, θ], integrand4[t, r, θ]]];

In[48]:= (* evaluate only for numerical values
             of the integration variables *)
         cf3DH[t_, r_, θ_?NumberQ] := cf3D[t, r, θ]

In[50]:= (* wrapper function *)
         uExact3D[t_?NumberQ, r_?NumberQ] :=
           1/2 NIntegrate[cf3DH[t, r, θ], {θ, 0, 1}, Compiled -> False,
                     PrecisionGoal -> 4, AccuracyGoal -> 4]
```

The overall quality of the NDSolve solution is reasonable, although not quite as good as in the 2D case. The largest errors occur at the transition to the identical 0 solution.

```
In[52]:= ε = 10^-8;
         Plot3D[uExact3D[t, r] - (u[t, r] /. nsol[[1]]),
                 {r, ε, X - 1 - ε}, {t, ε, T - ε},
                 PlotPoints -> 50, PlotRange -> All, Mesh -> False]
```

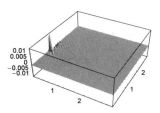

We end with a calculation of the spread of our initial bump in seven dimensions. Again, we have a sharp light cone with no perturbations inside. But now the shape of the propagating wave changes substantially over time [139].

```
In[54]:= dim = 7;

      nsol =
      NDSolve[{D[u[t, r], {t, 2}]} ==
              1/r^(dim - 1) D[r^(dim - 1) D[u[t, r], r], r],
              u[0, r] == u0[r], Derivative[1, 0][u][0, r] == 0,
              u[t, -X] == 0, u[t, X] == 0}, u, {t, 0, T}, {r, -X, X},
              PrecisionGoal -> 3, AccuracyGoal -> 4,
              Method -> {"MethodOfLines",
                  "SpatialDiscretization" -> {"TensorProductGrid",
                  "DifferenceOrder"-> 12,
                  "MinPoints" -> 1200, "MaxPoints" -> 1200}}];
```

```
      NDSolve::eerri :
       Warning: Estimated initial error on the specified spatial grid in the direction
          of independent variable r exceeds prescribed error tolerance. More…
```

```
      NDSolve::eerr :
       Warning: Scaled local spatial error estimate of 26453.93509229168` at
          t = 3.` in the direction of independent variable r is much greater
          than prescribed error tolerance. Grid spacing with 1200 points
          may be too large to achieve the desired accuracy or precision.  A
          singularity may have formed or you may want to specify a smaller
          grid spacing using the MaxStepSize or MinPoints options. More…
```

```
In[56]:= Plot3D[Evaluate[u[t, r] /. nsol[[1]]], {r, X/100, X - 1}, {t, 0, T},
              PlotPoints -> 200,  PlotRange -> {-1, 3/2}, Mesh -> False,
              AxesLabel -> {"r", "t", None}]
```

For interesting initial conditions for the 3D wave equation, see [1879]. For interesting results on the flux through hypersurfaces, see [1508].

37. Kleinian Group, Continued Fractions, Lüroth Expansions, Lehner Fractions, Brjuno Function, Bolyai Expansion

a) These are the two matrices.

```
In[1]:= γ = Exp[2 Pi I/7];
      {a, b, c} = {γ^5 - γ^2, γ^3 - γ^4, γ^6 - γ}/Sqrt[-7];

      p[1] = {{γ, 0, 0}, {0, γ^4, 0}, {0, 0, γ^2}};
      p[2] = {{a, b, c}, {b, c, a}, {c, a, b}};
```

We will use *Mathematica*'s significance arithmetic to generate numerical approximations of the other group elements. At the same time, to the calculation of the group elements, we will calculate the group table. To do this, we will store all group elements in the list `bag`, and, for each newly added group element, we will calculate where its product with every other element appears in `bag`. The resulting number will be stored with the variable `pDone`.

```
In[5]:= bag = N[{p[1], p[2]}, 30];
```

```
In[6]:= oldBagLength = 0;
      While[goOnQ = oldBagLength < Length[bag];
            oldBagLength = Length[bag]; goOnQ,
            n = Length[bag];
            Do[If[(* has bag[[j]].bag[[i]] already been carried out? *)
                  Head[pDone[i, j]] =!= Integer,
                  (* do group multiplication *)
                  newProd = bag[[j]].bag[[i]];
                  pos = Position[bag, _?(# == newProd&), {1}, 1];
                  If[pos === {},
                      (* a new group element *) AppendTo[bag, newProd],
                      pDone[i, j] = pos[[1, 1]]]], {i, n}, {j, n}]]
```

The group has 168 elements. As a result, the group table has $168^2 = 18224$ elements.

```
In[8]:= {Length[bag], Length[DownValues[pDone]]}
```

```
Out[8]= {168, 28224}
```

Every group element appears exactly once in the group table.

```
In[9]:= Length /@ Split[Sort[Flatten[Table[pDone[i, j],
                                    {i, 168}, {j, 168}]]]]] // Union
```

```
Out[9]= {168}
```

Here is a density plot visualizing the group table.

```
In[10]:= ListDensityPlot[Table[pDone[i, j], {i, 168}, {j, 168}], Mesh -> False]
```

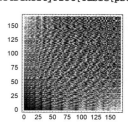

b) The solution involves a simple iterative process whereby at each step the largest divisor of the form $1/x$ is removed. (The solution could be found in [1918].)

```
In[1]:= continuedFraction[x_Real?Positive, ord_Integer?Positive] :=
            Floor[NestList[(1/(# - Floor[#]))&, x, ord - 1]]
```

The result then has the form $\{n_1, n_2, ..., n_{ord}\}$, and the associated continued fraction is $n_1 + (1/(n_2 + 1/\cdots))$. Let us look at an example.

```
In[2]:= continuedFraction[N[Pi, 50], 20]
```

```
Out[2]= {3, 7, 15, 1, 292, 1, 1, 1, 2, 1, 3, 1, 14, 2, 1, 1, 2, 2, 2, 2}
```

The result agrees with the built-in function.

```
In[3]:= ContinuedFraction[N[Pi, 50], 20]
```

```
Out[3]= {3, 7, 15, 1, 292, 1, 1, 1, 2, 1, 3, 1, 14, 2, 1, 1, 2, 2, 2, 2}
```

Here are the corresponding approximations of π by fractions.

```
In[4]:= Fold[(#2 + 1/#1)&, First[#], Rest[#]]& /@
                (Reverse /@ Table[Take[%, i], {i, Length[%]}])
```

Out[4]= $\{3, \dfrac{22}{7}, \dfrac{333}{106}, \dfrac{355}{113}, \dfrac{103993}{33102}, \dfrac{104348}{33215}, \dfrac{208341}{66317}, \dfrac{312689}{99532}, \dfrac{833719}{265381},$
$\dfrac{1146408}{364913}, \dfrac{4272943}{1360120}, \dfrac{5419351}{1725033}, \dfrac{80143857}{25510582}, \dfrac{165707065}{52746197}, \dfrac{245850922}{78256779},$
$\dfrac{411557987}{131002976}, \dfrac{1068966896}{340262731}, \dfrac{2549491779}{811528438}, \dfrac{6167950454}{1963319607}, \dfrac{14885392687}{4738167652}\}$

This is the relative deviation from π.

```
In[5]:= (N[%, 50] - N[Pi, 50])/N[Pi, 50] // N[#, 5]&
```

Out[5]= $\{-0.045070, 0.00040250, -0.000026490, 8.4914 \times 10^{-8}, -1.8395 \times 10^{-10},$
$1.0556 \times 10^{-10}, -3.8947 \times 10^{-11}, 9.2766 \times 10^{-12}, -2.7742 \times 10^{-12}, 5.1271 \times 10^{-13},$
$-1.2862 \times 10^{-13}, 7.0489 \times 10^{-15}, -1.8433 \times 10^{-16}, 5.2229 \times 10^{-17}, -2.4885 \times 10^{-17},$
$6.1637 \times 10^{-18}, -9.7724 \times 10^{-19}, 1.7551 \times 10^{-19}, -2.4276 \times 10^{-20}, 9.9414 \times 10^{-21}\}$

The continued fraction expansion of the `GoldenRatio` is particularly interesting.

```
In[6]:= continuedFraction[N[GoldenRatio, 50], 20]
```

Out[6]= $\{1, 1, 1, 1, 1, 1, 1, 1, 1, 1, 1, 1, 1, 1, 1, 1, 1, 1, 1, 1\}$

Again, the result agrees with the built-in function.

```
In[7]:= ContinuedFraction[N[GoldenRatio, 50], 20]
```

Out[7]= $\{1, 1, 1, 1, 1, 1, 1, 1, 1, 1, 1, 1, 1, 1, 1, 1, 1, 1, 1, 1\}$

These are the continued fraction representations for 10000 random numbers.

```
In[8]:= data = Flatten[Table[Rest @ continuedFraction[Random[], 15], {100}]];
```

By counting the appearances of the first 20 integers in these data, we get the following probabilities.

```
In[9]:= result = {#, Count[data, #]}& /@ Range[20];
        numberOfDigitsCounted = Length[data];
        Take[{#[[1]], N[#[[2]]/numberOfDigitsCounted]}& /@ result, 20]
```

Out[11]= $\{\{1, 0.412143\}, \{2, 0.192143\}, \{3, 0.0892857\}, \{4, 0.0607143\}, \{5, 0.0371429\},$
$\{6, 0.0178571\}, \{7, 0.0257143\}, \{8, 0.0135714\}, \{9, 0.0135714\}, \{10, 0.0114286\},$
$\{11, 0.0121429\}, \{12, 0.00571429\}, \{13, 0.00714286\}, \{14, 0.00857143\}, \{15, 0.00428571\},$
$\{16, 0.005\}, \{17, 0.005\}, \{18, 0.005\}, \{19, 0.00357143\}, \{20, 0.00285714\}\}$

The exact asymptotic probabilities for a randomly chosen real number are given by the following expression (see [234], [450] and [983]).

```
In[12]:= Table[-Log[2, 1 - 1/(k + 1)^2], {k, 1, 20}] // N
```

Out[12]= $\{0.415037, 0.169925, 0.0931094, 0.0588937, 0.040642, 0.0297473, 0.0227201,$
$0.0179219, 0.0144996, 0.0119726, 0.0100537, 0.00856201, 0.00737953, 0.00642627,$
$0.00564656, 0.00500068, 0.00445965, 0.00400193, 0.00361125, 0.00327513\}$

In a very similar way, we can define the list of remainders.

```
In[13]:= remainders[x_] := DeleteCases[#, ComplexInfinity | Indeterminate]& @
                 Rest[FixedPointList[(1/(# - Floor[#]))&, x]]
```

The sequence $\log(1 + 1/remainder)$ is uniformly distributed. The next graphic shows the first 20 moments divided by the value for a uniform distribution for five "random" numbers.

```
In[14]:= Show[With[{p = 1000, o = 20},
        (* avoid messages from division of low accuracy zeros *)
        Internal`DeactivateMessages[
        Module[{data = 1/remainders[N[#, p]], scaledData, l},
              (* scale data *)
              l = Length[scaledData = Log[2, 1 + data]];
            (* display moments *)
            ListPlot[Table[(k + 1) Plus @@ (scaledData^k), {k, o}]/l,
              PlotJoined -> True, PlotRange -> All, Axes -> False,
              DisplayFunction -> Identity, Frame -> True,
              PlotStyle -> {Hue[Random[]]}]]]& /@
              (* "random" numbers *)
```

```
{Pi, Log[3], E^Pi, 1/Pi, Random[Real, {0, 1}, p]}],
DisplayFunction -> $DisplayFunction]
```

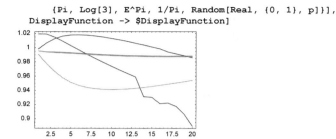

For more on continued fractions in *Mathematica*, see the package `Number Theory`ContinuedFractions`` and [915] and [1173]. For some especially neat continued fractions, see [1825].

We could now refine the definition of `continuedFraction` for the terminating case (meaning termination after less than *ord* digits).

In[15]:= `continuedFraction[N[1/3, 100], 4]`

Power::infy : Infinite expression $\frac{1}{0. \times 10^{-100}}$ encountered. More...

∞::indet :
    Indeterminate expression ComplexInfinity + ComplexInfinity encountered. More...

Out[15]= {0, 3, ComplexInfinity, Indeterminate}

We will not do this here, but in the following function `continuedInverseRoot`.

It is straightforward to change the above code to roots instead of division. Now, we also prepare for the situation that the expansion might terminate before *ord* terms are generated. We do this by using `NestWhileList` instead of `NestList`. We allow for a general root α, not only of the form $1 / positiveInteger$.

In[16]:= `continuedInverseRoot[x_?Positive, α_, ord_?Positive] :=`
      `Floor[Last /@ Drop[NestWhileList[`
        `{#[[1]] + 1, If[# != 0, #^(-1/α), C]&[#[[2]] - Floor[#[[2]]]]}&,`
        `{(* count digits *) 0, x}, (#[[1]] < (ord + 1) && #[[2]] =!= C)&], -1]]`

The function `fromContinuedInverseRoot` takes the list generated by `continuedInverseRoot` and forms the corresponding nested root expression.

In[17]:= `fromContinuedInverseRoot[l_List, α_?Positive] :=`
      `Fold[(#2 + #1^(-α))&, First[Reverse[l]], Rest[Reverse[l]]]`

Here are two symbolic examples using `fromContinuedInverseRoot`.

In[18]:= `{fromContinuedInverseRoot[Table[Subscript[c, k], {k, 0, 6}], 1/2],`
      `fromContinuedInverseRoot[Table[Subscript[c, k], {k, 0, 3}], 1/3]}`

Out[18]= $\left\{c_0 + \sqrt{\cfrac{1}{c_1 + \sqrt{\cfrac{1}{c_2 + \sqrt{\cfrac{1}{c_3 + \cfrac{1}{c_4 + \cfrac{1}{\sqrt{c_5 + \cfrac{1}{\sqrt{c_6}}}}}}}}}}}, \; c_0 + \cfrac{1}{\left(c_1 + \cfrac{1}{\left(c_2 + \cfrac{1}{c_3^{1/3}}\right)^{1/3}}\right)^{1/3}}\right\}$

Here are the first 20 expansion coefficients of 77/37 using the power 7/11.

In[19]:= `continuedInverseRoot[77/37, 7/11, 20]`

Out[19]= {2, 51, 1, 5, 8, 1, 10, 1, 11, 13, 1, 4, 1, 1, 21, 2, 1, 2, 5, 3, 5}

In[20]:= `fromContinuedInverseRoot[%, 7/11] //`
                `(* for a space-efficient representation *) InputForm`

Out[20]//InputForm= 2 +
          (51 +
          (1 +
          (5 +
            (8 + (1 + (10 + (1 + (11 + (13 + (1 + (4 + (1 + (1 + (21 + (2 + (1 + (2 + (5
            + (3 +
```
```

```
5^(-7/11))^(-7/11))^(-7/11))^(-7/11))^(-7/11))^(-7/11))^(-7/11))^

(-7/11))^(-7/11))^(-7/11))^(-7/11))^(-7/11))^(-7/11))^(-7/11))^(-7/11))^
        (-7/11))^(-7/11))^(-7/11))^(-7/11))^(-7/11))^(-7/11)
```

In[21]:= **N[% - 77/37, 22]**

Out[21]= $-5.999238771607294570093 \times 10^{-25}$

Next, we expand π in eth roots. 20 terms in the expansion yield only about 7 correct digits in π.

In[22]:= **continuedInverseRoot[Pi, E, 20]**

Out[22]= {3, 2, 2, 1, 4, 1, 1, 1, 1, 1, 1, 1, 1, 1, 1, 1, 2, 1, 1, 1, 1}

In[23]:= **fromContinuedInverseRoot[%, E] - Pi // N[#, 22]&**

Out[23]= $-2.470073180365115861384 \times 10^{-7}$

Here are some terminating cases.

In[24]:= **continuedInverseRoot[2/3, 1/2, 20]**

Out[24]= {0, 2, 16}

In[25]:= **continuedInverseRoot[2/9, 1/3, 20]**

Out[25]= {0, 91, 512}

In[26]:= **(* quick check *)**
     {fromContinuedInverseRoot[{0, 2, 16}, 1/2],
     fromContinuedInverseRoot[{0, 91, 512}, 1/3]}

Out[27]= $\left\{ \frac{2}{3}, \frac{2}{9} \right\}$

The following list density plots indicate the fractions i/j in the i,j-plane that have terminating expansions for $\alpha = 1/2$ and $\alpha = 1/3$.

In[28]:= **ListDensityPlot[**
        Table[Length[continuedInverseRoot[N[i/j, 200], 5/7, 50]],
            {i, 50}, {j, 50}], Mesh -> False, PlotRange -> All]

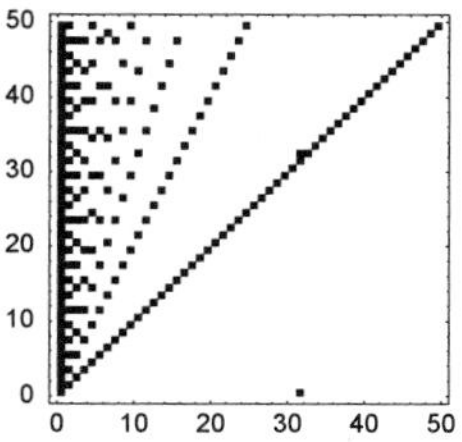

Similarly to the continued fraction expansion, we could now investigate the digit distributions of the root expansions. Here are the results of expanding 1000 fractions into 100 terms for $\alpha = 1/2$, $\alpha = 1/3$, and $\alpha = 1/99/100$.

In[29]:= **Show[GraphicsArray[**
     Block[{$DisplayFunction = Identity},
      ListPlot[(* make log-log-plot *)
            N[{Log @ First[#], Log @ Length[#]}]& /@
      (* use 1000 fractions, each 100 terms *)
      Split[DeleteCases[Sort[Flatten[
       Table[continuedInverseRoot[(* high-precision random number *)
                      Random[Real, {0, 1}, 300], #, 100],
            {1000}]]], 0]], PlotRange -> All, Frame -> True,
        Axes -> False]& /@ (* the α-values *) **{ 1/2, 1/3, 99/100}]]]**

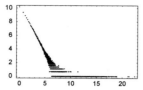 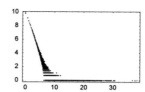 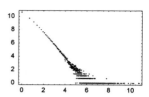

c) We will use random algebraic numbers of the form $i^j - \lfloor i^j \rfloor$ as the random numbers from the interval $(0, 1)$. (To avoid error messages from `ContinuedFraction` we exclude integers.)

```
In[1]:= randomRoot[] :=
    Module[{r}, While[IntegerQ[r =
      ((# - IntegerPart[#])&[Random[Integer, {10^2, 10^6}]^
        (Random[Integer, {1, 100}]/Random[Integer, {1, 10^4}])])]], Null]; r]
```

For 1000 random numbers `randomRoots` we calculate 100 terms of their continued fraction expansion (hoping that 100 is large enough for a numerical confirmation of the $n \to \infty$ statement).

```
In[2]:= l = 100;
    SeedRandom[173303579];
    cfs = Table[ContinuedFraction[randomRoot[], 100], {l}];
```

Given a list l, the function `cumulativeMax` will construct a list of the cumulative maxima of l.

```
In[5]:= cumulativeMax[list_] := Rest[FoldList[Max, 0, list]]
```

So, we have the following data from our numerical experiment.

```
In[6]:= data = Transpose[cumulativeMax /@ cfs];
```

Displaying a graphic of the theoretical curve and the values obtained from our numerical curve shows excellent agreement.

```
In[7]:= theoreticalCurve = Graphics[{{GrayLevel[0.6], Thickness[0.02],
                            Line[Table[{X, Exp[-1/X]}, {X, 100}]]}}];
In[8]:= probabilityCurve[cfTerms_, n_] := Point /@
    With[{l = Length[cfTerms]},
    Rest[FoldList[{#2[[1]], #1[[2]] + #2[[2]]}&, {0, 0},
        {(First[#] - 1)/n Log[2], Length[#]/l}& /@
            Split[Sort[cfTerms[[n]]]]]]]];
In[9]:= Show[{theoreticalCurve,
              (* use small colored points for the numerical values *)
              Table[Graphics[{PointSize[0.003], Hue[0.8 n/100],
                probabilityCurve[data, n]}], {n, 1, 100}]} // N,
              PlotRange -> {{1, 100}, {0, 1.2}}, Frame -> True]
```

We repeat the experiment for random high-precision real numbers. Again, we obtain good agreement with the theoretical curve.

```
In[10]:= SeedRandom[282];
    cfs = Table[ContinuedFraction[Random[Real, {0, 1}, 200], 100], {l}];
    data = Transpose[cumulativeMax /@ cfs];
    Show[{theoreticalCurve,
              (* use small colored points for the numerical values *)
              Table[Graphics[{PointSize[0.003], Hue[0.8 n/100],
                probabilityCurve[data, n]}], {n, 1, 100}]} // N,
              PlotRange -> {{1, 100}, {0, 1.2}}, Frame -> True]
```

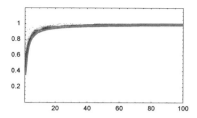

d) We start by implementing the two functions α and $\mathcal{T}$.

```
In[1]:= T[x_] := Floor[1/x](Floor[1/x] + 1) x - Floor[1/x]
        α[x_] := Floor[1/x] + 1
```

To avoid the repeated calculation of powers of $\mathcal{T}$, we introduce the sequence t. This leads to the following implementation for the Lüroth terms a_n.

```
In[3]:= a[x_, 0] = Infinity;
        t[x_, 0] = x;
        t[x_, n_] := t[x, n] = Expand[T[t[x, n - 1]]]
        a[x_, n_] := a[x, n] = α[t[x, n - 1]]
```

We set $MaxExtraPrecision to a value higher than the default value to allow for the calculation of the required floor functions.

```
In[7]:= $MaxExtraPrecision = 2000;
```

Now the first 1200 terms of the Lüroth terms of $1/\pi$ can be calculated quite quickly.

```
In[8]:= (πList = Table[a[1/Pi, k], {k, 1200}];) // Timing
Out[8]= {0.33 Second, Null}
```

```
In[9]:= Take[πList, 12]
Out[9]= {4, 2, 2, 4, 3, 13, 4, 2, 2, 4, 2, 4}
```

Similarly to continued fractions, the distribution of the Lüroth terms is strongly centered at small integers.

```
In[10]:= ListPlot[{First[#], Length[#]}& /@ Split[Sort[πList]],
            PlotJoined -> True]
```

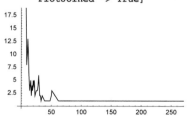

Given the Lüroth terms, the function `FromLürothList` returns a list of the successive approximations.

```
In[11]:= FromLürothList[l_List] :=
        Rest[FoldList[Plus, 0, Prepend[1/Rest[l]*
            Drop[1/Rest[FoldList[Times, 1, # (# - 1)& /@ l]], -1], 1/First[l]]]]
```

A plot shows that we need about 1100 Lüroth terms to approximate $1/\pi$ to 1000 digits.

```
In[12]:= δList = N[FromLürothList[πList] - 1/Pi, 22];
```

```
In[13]:= ListPlot[N[Log[10, -δList], 22]]
```

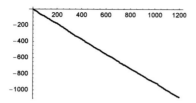

It turns out that we need exactly 1100 Lüroth terms to approximate $1/\pi$ to 1000 digits.

```
In[14]:= Position[-δList, _?(# < 10^-1000&), {1}, 1]
Out[14]= {{1100}}
```

Now let us investigate the geometric mean of the a_k. We take 20 random high-precision real numbers and calculate the geometric mean for the first 2000 a_k. The picture suggests that the mean approaches the number 3.49... [893], [451].

```
In[15]:= With[{prec = 3000, o = 2000},
        Show[Graphics[Table[
        (* a random fraction *)
        ζ = SetPrecision[Random[Real, {0, 1}, prec], Infinity];
        (* remove old special definitions *)
        DownValues[a] = Take[DownValues[a], -2];
        (* new data *)
        ll = Table[a[ζ, k], {k, o}];
        (* make graphics primitives *)
        {Hue[Random[]], PointSize[0.003],
         MapIndexed[Point[{#2[[1]], #^(1/#2[[1]])}]&,
                    Rest[FoldList[Times, 1, N[ll]]]]}, {20}]],
        Frame -> True, PlotRange -> {3, 4}]]
```

A related expansion is the Cantor expansion [518], [331]. For $0 < x < 1$ and positive integers b_1, b_2, ... we have $x = \prod_{k=1}^{\infty} (\alpha_k / \prod_{j=1}^{k} b_k)$.

```
In[16]:= CantorExpansion[α_, bList_] := Last /@
        FoldList[Function[{a, b}, {#, Floor[#]}&[Expand[b Subtract @@ a]]],
                 {#, Floor[#]}&[α bList[[1]]], Rest[bList]]
```

Here are the Cantor digits for $x = 1/\pi$ and $b_k = 2^k$.

```
In[17]:= $MaxExtraPrecision = 10000;
        CantorExpansion[1/Pi, 2^Range[12]]
Out[18]= {0, 2, 4, 5, 30, 24, 27, 114, 228, 261, 595, 3972}

In[19]:= (* difference of the approximation to 1/Pi *)
        N[Fold[Plus, 0, %/Rest[FoldList[Times, 1, 2^Range[12]]]] - 1/Pi, 22]
Out[20]= -0.3183098861837906715378 + 0.6416325606551538662938 Null
```

e) It is straightforward to implement the map $\mathcal{L}$ and the two functions `LehnerContinuedFraction` and `fromLehner ContinuedFraction`. To avoid rounding problems, we use rational arithmetic.

```
In[1]:= L[x_] := If[1 <= x < 3/2, 1/(2 - x), 1/(x - 1)]

       β[r_] := If[1 <= r < 3/2, {2, -1}, {1, 1}]
```

```
        𝓛List[x_] := FixedPointList[𝓛, x]

        LehnerContinuedFraction[x_] := Append[β /@ 𝓛List[x], {1, 1}]
In[5]:= fromLehnerContinuedFraction[l_] :=
        Fold[#2[[1]] + #2[[2]]/#1&, #[[1, 2]], Drop[#, 1]]&[Reverse[l]]
```

Here is the Lehner continued fraction of the number $173/100$.

```
In[6]:= LehnerContinuedFraction[173/100]

Out[6]= {{1, 1}, {2, -1}, {1, 1}, {1, 1}, {2, -1}, {1, 1},
         {2, -1}, {1, 1}, {1, 1}, {1, 1}, {1, 1}, {2, -1}, {2, -1}, {1, 1}}
```

`fromLehnerContinuedFraction` recovers the fraction $173/100$.

```
In[7]:= fromLehnerContinuedFraction[%]

Out[7]= 173/100
```

The results of repeatedly applying $\mathcal{L}$ to 1000-digit approximations of the numbers $\sqrt{3}$, $e/2$, $\tan(1)$, $\ln(7)$, $3^{1/3}$, $3^{1/4}$, $\pi/2$, and a random number are shown below.

```
In[8]:= With[{prec = 1000},
        Show[GraphicsArray[ListPlot[𝓛List[Rationalize[N[#, prec], 0]],
                            DisplayFunction -> Identity,
                            PlotRange -> {1, 2}]& /@ #]]& /@
        {{Sqrt[3], E/2}, {Tan[1], Log[7]},
         {3^(1/3), 3^(1/4)}, {Pi/2, Random[Real, {1, 2}]}}]]
```

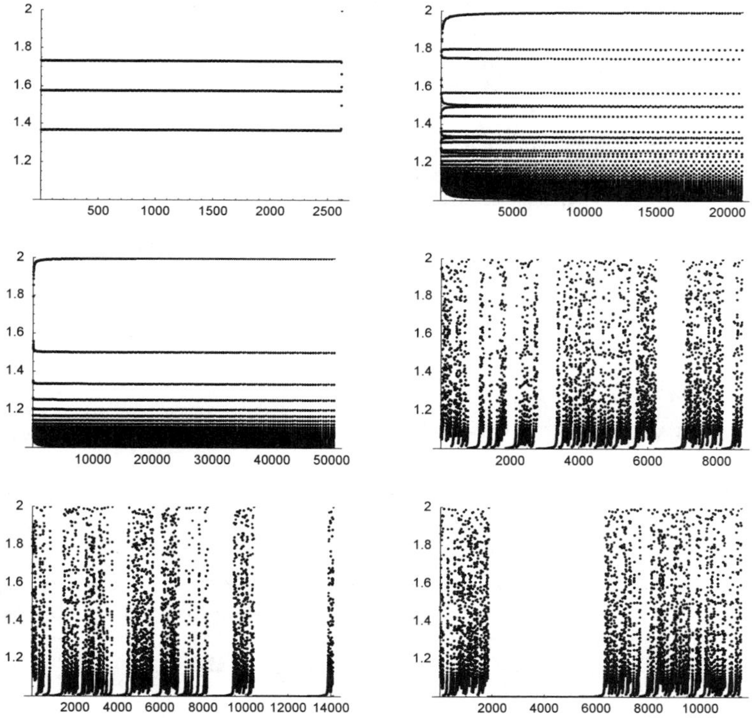

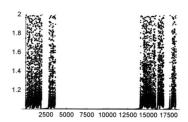 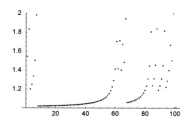

Not unexpectedly for continued fractions, square roots and rational functions of *e* exhibit regular patterns.

We end by calculating the cumulative probability of $\{b_k, c_{k+1}\} = \{1, 1\}$. To carry out this investigation, we slightly change the definition of `LehnerContinuedFraction`. The above definition first repeatedly applies $\mathcal{L}$ and then β. For very high-precision numbers the first calculated list contains large fractions and needs a lot of memory to be stored. The function `LehnerContinuedFraction1` avoids the intermediate storage of the list of fractions.

```
In[9]:= LehnerContinuedFraction1[x_] := Append[Last /@
            FixedPointList[{#, β[#]}&[L[#[[1]]]]&, {x, β[x]},
                SameTest -> (#1[[1]] === #2[[1]]&)], {1, 1}]
```

`LehnerContinuedFraction` and `LehnerContinuedFraction1` generate identical results.

```
In[10]:= LehnerContinuedFraction [Rationalize[N[Pi/E, 22], 0]] ===
             LehnerContinuedFraction1[Rationalize[N[Pi/E, 22], 0]]
Out[10]= True
```

```
In[11]:= plot[l_, opts___] :=
          ListPlot[(* cumulative probability *)MapIndexed[#1/#2[[1]]&,
                  FoldList[If[#2 === {1, 1}, #1 + 1, #1]&, 0, l]],
              opts, PlotStyle -> {PointSize[0.003]},
              PlotRange -> {0, 0.25}]
```

Next, we calculate high-order Lehner continued fractions for $\pi/2$ and $e/2$.

```
In[12]:= {πlcf, elcf} = LehnerContinuedFraction1[
                        Rationalize[N[#/2, 5000], 0]]& /@ {Pi, E};

        Show[GraphicsArray[{plot[πlcf, DisplayFunction -> Identity],
                        plot[elcf, DisplayFunction -> Identity]}]]
```

The last graphics suggest that the relative occurrence of $\{b_k, c_{k+1}\} = \{1, 1\}$ tends to zero [452].

f) To avoid the calculation of the convergent for each truncated continued fraction we start by implementing a function `ConvergentsNumDen`. It will return a list of all numerators and denominators of the convergents of the continued fraction *cf* at once. It is based on recursion relations [1186].

```
In[1]:= ConvergentsNumDen[cf_] := (Last /@
          (* given the continued fraction [c0, c1, c2, …]
             the numerators and denominators of the convergents
             obey the following recursion relations:
             p[n] = c[n] p[n - 1] + p[n - 2] && p[-2] == 0 && p[-1] == 1
             q[n] = c[n] q[n - 1] + q[n - 2] && q[-2] == 1 && q[-1] == 0 *)
          FoldList[{#1[[2]], #1[[3]], #2 #1[[3]] + #1[[2]]}&,
                  #, Rest[cf]])& /@ {{0, 1, cf[[1]]}, {1, 0, 1}}
```

The function $\mathrm{Brjuno}\mathcal{B}[x, \delta]$ calculates an approximation to $\mathcal{B}(x)$. We sum the terms of the continued fraction expansion of x until the ratio of the terms to the sum become smaller than δ.

```
In[2]:= BrjunoB[x_Real, δ_] :=
    Module[{cf = ContinuedFraction[x], n, p, q, term, sum, k, t},
        n = Length[cf];
        (* the numerators and denominators of the convergents *)
        {p, q} = ConvergentsNumDen[cf];
        (* the terms of the sum *)
        summand[j_] := Abs[p[[j]] - q[[j]] x] Log[(p[[j]] - q[[j]] x)/
                        (q[[j + 1]] x - p[[j + 1]])];
        sum = -Log[x - Floor[x]]; k = 1;
        (* sum until ratio is small enough *)
        While[k < n && (s = summand[k]; (k < 5 || Abs[s/sum] > δ)),
                sum = sum + s; k++]; sum]
```

To sketch the Brjuno function $\mathcal{B}(x)$ we will use two ways. First, we will generate random high-precision numbers and second we will use numerical approximations to rational numbers with small denominators. Both pictures agree in their global appearance. (Strictly speaking, the Brjuno function is not defined for rational numbers.)

```
In[3]:= With[{n = 10^4, denMax = 120},
    Show[GraphicsArray[
    ListPlot[#, DisplayFunction -> Identity, PlotRange -> {0, 4},
            Frame -> True, PlotStyle -> {PointSize[0.003]},
            Axes -> False]& /@
    {(* random numbers *) Sort[Table[{#, BrjunoB[#, 20]}&[
        Random[Real, {0, 1}, 50]], {n}],
        #1[[1]] < #2[[1]]&],
    (* small fractions *) {#, BrjunoB[#, 0]}& /@
        N[Union[Flatten[Table[i/j, {j, denMax}, {i, j - 1}]]]]}]]]]
```

We end by checking the functional equation for some irrational x to 100 digits.

```
In[4]:= With[{prec = 200},
    Function[x, BrjunoB[N[x, prec], 0] -
            (-Log[x] + x BrjunoB[N[1/x, prec], 0])] /@
        {1/Pi, 1/E, 1/Sqrt[2], 3^(-1/3), Log[2]}]
Out[4]= {0. × 10^-101, 0. × 10^-102, 0. × 10^-101, 0. × 10^-101, 0. × 10^-101}
```

g) We use the efficient implementation for the convergents from the last subexercise.

```
In[1]:= Convergents[x_?NumberQ] :=
    With[{cf = ContinuedFraction[x]},
    Divide @@@ Transpose[(Last /@
        FoldList[{#1[[2]], #1[[3]], #2 #1[[3]] + #1[[2]]}&, #,
            Rest[cf]])& /@ {{0, 1, cf[[1]]}, {1, 0, 1}}]]
```

This defines the function $\mathcal{P}(x)$.

```
In[2]:= P[x_?NumericQ] := Plus @@ (x - Convergents[x])
```

Obviously, $\mathcal{P}(x)$ is periodic with period 1. So, we display $\mathcal{P}(x)$ in the interval $(0, 1)$. Here are two pictures for $\mathcal{P}(x)$. The left graphic uses all fractions with denominators less than 10000 and the right graphic uses random numbers.

```
In[3]:= With[{n = 10^4},
    Show[GraphicsArray[
    ListPlot[Table[{k/n, P[k/n]}, {k, 0, n}], Frame -> True, Axes -> False,
```

```
        DisplayFunction -> Identity,
        PlotStyle -> {PointSize[0.003]}]& /@
{Table[{k/n, P[k/n]}, {k, 0, n}], (* random values *)
   SeedRandom[110]; Table[{#, P[#]}&[Random[]], {n}]}]]]
```

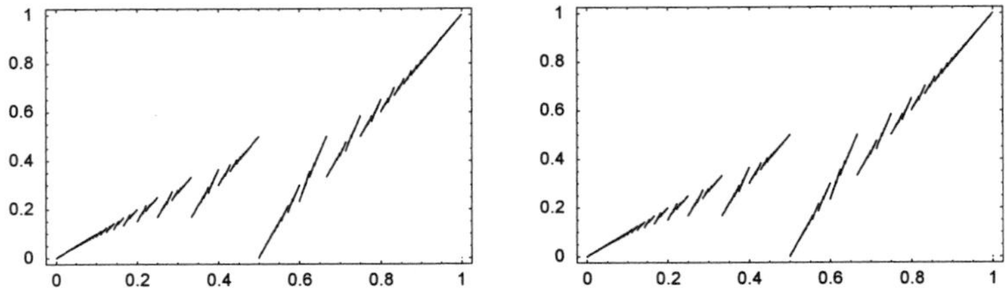

The function $P(x)$ turns out to be continuous at all irrational x. This gives some hope for numerically integrating $\mathcal{P}(x)$.

In[4]:= **Off[NIntegrate::slwcon];**

```
NIntegrate[P[x], {x, 0, 1},
        MaxRecursion -> 8, PrecisionGoal -> 10] // InputForm
```
NIntegrate::ncvb : NIntegrate failed to converge to prescribed accuracy
        after 9 recursive bisections in x near x = 0.3330078125`. More…

Out[5]//InputForm= 0.3750000000000001

The last result clearly suggests that $\int_0^\infty \mathcal{P}(x)\,dx = 3/8$ [1529].

Now let us use the function `Convergents` to establish that the average scaled error of the continued fraction approximations is $1/\ln(16)$. We calculate about 100 continued fraction terms of the "random" numbers $|\sin(k)|$, $k = 1, \ldots, 10^4$, the corresponding scaled error, and the cumulative averages for each k. For each k, we start with a 150-digit approximation of $|\sin(k)|$.

In[6]:= **o = 10000;**
```
        tab = Table[x = N[Abs[Sin[k]], 150];
        (* scaled errors *)
        θs = (Denominator[#]^2 Abs[x - #])& /@ Rest[Convergents[x]];
        (* cumulative averages *)
        MapIndexed[#1/#2[[1]]&, Rest[FoldList[Plus, 0, N[θs]]]],
                {k, 1, o}];
```

The resulting cumulative averages have greatly different lengths, ranging from 110 to 179. The following graphics show the cumulative averages for $k = 1, \ldots, 20$ and the distribution of the length of the continued fraction.

In[8]:= **Show[GraphicsArray[**
```
        {(* first 20 cumulative averages *)
        Graphics[{Hue[Random[]], Thickness[0.003],
                Line[MapIndexed[{#2[[1]], #1}&, #]]}& /@ Take[tab, 20],
                Frame -> True, PlotRange -> {0.3, 0.4}],
        (* length distribution of the continued fractions *)
        ListPlot[{First[#], Length[#]}& /@ Split[Sort[Length /@ tab]],
                DisplayFunction -> Identity]}]]
```

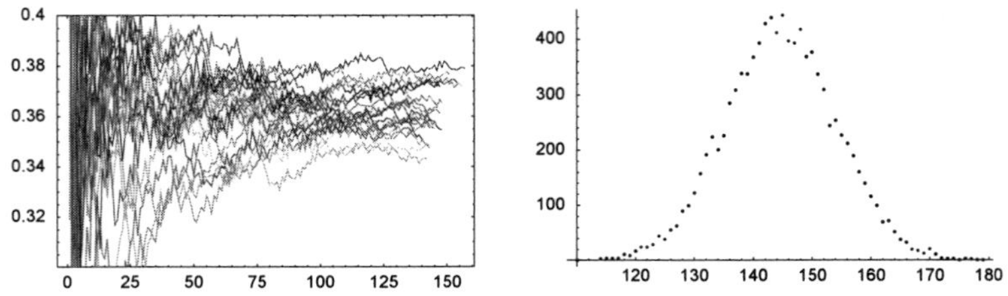

We now form the ensemble average over all 10^4 sample sequences.

```
In[9]:= lMin = Min[Length /@ tab];
        θsAv = Plus @@ (Take[#, lMin]& /@ tab)/o;
In[11]:= ℒ = 1/Log[16];
```

The following graphic shows the result. We display $\log_{10}(|\overline{\theta}_n(x) - 1/\ln(16)|)$. The average agrees to more than three digits with the theoretical value $1/\ln(16)$.

```
In[12]:= ListPlot[Log[10, Abs[θsAv - ℒ]],
           Frame -> True, PlotRange -> All, Axes -> False]
```

h) The function g calculates n steps of the recursion. ξ is the starting value and p the precision to be used.

```
In[1]:= g[ξ_, n_, p_] := g[ξ, n, p] =
        Block[{$MaxPrecision = p},
         Module[{x = SetPrecision[ξ, p]}, Table[x = (1 + 1/x)^k, {k, 1, n}]]]
```

Here are the 60th terms for the starting values 1 and 2.

```
In[2]:= Take[g[1, 60, 22], -2]
Out[2]= {5.764607523034166440×10^17, 1.00000000000000010408}
```

```
In[3]:= Take[g[2, 60, 22], -2]
Out[3]= {1.00000000000000020470, 1.152921504606839896×10^18}
```

A moment of reflection shows that for large n, the value of x_n will be either very large or nearly 1. The following density plot shows the values of the x_n as a function of n and ξ.

```
In[4]:= ListDensityPlot[Log[Table[g[x0, 100, 22], {x0, 1/2, 2, 3/2/60}]],
           Mesh -> False, MeshRange -> {{1, 60}, {1/2, 2}}]
```

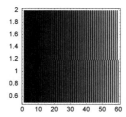

The graphic clearly shows that near $\xi \approx 1.2$ large values, and values near to 1 get inverted. At $\tilde{\xi}$, the sequence will be nonoscillating. It is straightforward to implement a bisection call to bracket $\tilde{\xi}$. The function step does this.

```
In[5]:= step[{x1_, x2_}, n_, p_] :=
        Module[{x12 = (x1 + x2)/2, s1, s2, s3},
           {s1, s2, s12} = g[#, n, p]& /@ {x1, x2, x12};
           If[Last[s12] > 2, {x1, x12}, {x12, x2}]]
```

We iterate the bisection and obtain $\tilde{\xi}$ to 100 digits.

```
In[6]:= Nest[step[#, 300, 300]&, {1, 13/10}, 400] // N[#, 100]&
Out[6]= {1.18745235112650105459548015839651935121569268158586035301010412619878041872352540 7·
          387024657606086579,
         1.18745235112650105459548015839651935121569268158586035301010412619878041872352540 7·
          387024657606086579}
```

The following graphic shows that for $\tilde{\xi}$ to 100 digits the first 220 terms of the x_n are monotonic increasing.

```
In[7]:= ListPlot[Log[g[%[[1]], 300, 300]]]
```

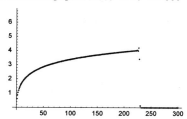

i) The implementation of the functions $\mathcal{T}$ and α is straightforward.

```
In[1]:= α[x_] := Which[0 <= x < Sqrt[2] - 1, 0,
                Sqrt[2] - 1 <= x < Sqrt[3] - 1, 1,
                Sqrt[3] - 1 <= x <= 1, 2]

      𝒯[x_] := Together[(x + 1)^2 - 1 - α[x]]

      𝒯List[x_, n_] := NestList[𝒯, x, n]

      BolyaiDigits[x_, n_] := α /@ 𝒯List[x, n]
```

Obviously, the numbers $\phi - 1$ and 1 have simple Bolyai expansions.

```
In[6]:= bdG = BolyaiDigits[N[GoldenRatio - 1, 30], 30]

Out[6]= {1, 1, 1, 1, 1, 1, 1, 1, 1, 1, 1, 1, 1, 1, 1, 1, 1, 1, 1, 1, 1, 1, 1, 1, 1, 1, 1, 1, 1, 1, 1}

In[7]:= bd2 = BolyaiDigits[N[1, 100], 100]

Out[7]= {2, 2, 2, 2, 2, 2, 2, 2, 2, 2, 2, 2, 2, 2, 2, 2, 2, 2, 2, 2, 2, 2, 2,
        2, 2, 2, 2, 2, 2, 2, 2, 2, 2, 2, 2, 2, 2, 2, 2, 2, 2, 2, 2, 2, 2, 2, 2,
        2, 2, 2, 2, 2, 2, 2, 2, 2, 2, 2, 2, 2, 2, 2, 2, 2, 2, 2, 2, 2, 2, 2, 2, 2,
        2, 2, 2, 2, 2, 2, 2, 2, 2, 2, 2, 2, 2, 2, 2, 2, 2, 2, 2, 2, 2, 2, 2, 2, 2}
```

Here we calculate 2500 Bolyai digits of $1/\pi$.

```
In[8]:= bdπ = BolyaiDigits[N[1/Pi, 2500], 2500];
```

The function `nestedFraction` forms a nested square root from the Bolyai digits *bolyaiDigits*.

```
In[9]:= nestedFraction[bolyaiDigits_] :=
            -1 + Fold[Sqrt[#1 + #2]&, 0, Reverse[bolyaiDigits]]
```

The direct numericalization of `nestedFraction[Take[bdπ, 1000]]` makes problems. In addition to the expected `N::meprec` messages, we get `$RecursionLimit::reclim` messages. These messages arise from the deeply nested structure of the square root and the way adaptive precision control works. To avoid these problems we numericalize all integers and fractions and let the resulting expression collapse to an approximate number. Because *Mathematica*'s high-precision arithmetic keeps track of correct digits automatically, we get the correct result.

```
In[10]:= Map[(* use 1000 digits for numericalization *) N[#, 1000]&,
             1/Pi - nestedFraction[Take[bdπ, 1000]], {-1}] // N
Out[10]= 2.774992250213482×10^-462
```

So, we implement another function `nestedFractionN` that numericalizes "automatically". We insert an approximative zero with accuracy *prec* in the innermost square root and let the whole expression then collapse to an approximative number. The result will be correct, but in distinction to using `N`, we do not know in advance the number of correct digits.

```
In[11]:= nestedFractionN[bolyaiDigits_, prec_] :=
             -1 + Fold[Sqrt[#1 + #2]&, SetAccuracy[0, prec], Reverse[bolyaiDigits]]
```

Using `nestedFractionN`, we see that 2500 correct Bolyai digits correspond to about 1146 decimal digits.

```
In[12]:= (πRecovered = nestedFractionN[bdπ, 30]) - 1/Pi

Out[12]= -3.344321×10^-1146
```

Fortunately, the intermediate condition numbers are such that we do not lose precision per step, but rather gain about half a digit per square root step.

```
In[13]:= Precision[πRecovered]
```
```
Out[13]= 1152.39
```

And 2185 Bolyai digits are needed to approximate the first 1000 decimal digits of $1/\pi$.

```
In[14]:= nestedFractionN[Take[bdπ, 2185], 30] - 1/Pi
```
```
Out[14]= -5.2920051622756279213007340689 3×10⁻¹⁰⁰¹
```

Similarly, we need 2193 Bolyai to approximate the first 1000 decimal digits of $1/e$.

```
In[15]:= bde = BolyaiDigits[N[1/E, 2500], 2500];
         nestedFractionN[Take[bde, 2193], 30] - 1/E
```
```
Out[16]= -1.4251222313887075068938819299 3×10⁻¹⁰⁰¹
```

Zeros are the most common Bolyai digits.

```
In[17]:= {First[#], Length[#]}& /@ Split[Sort[bdπ]]
```
```
Out[17]= {{0, 1155}, {1, 782}, {2, 564}}
```

```
In[18]:= {First[#], Length[#]}& /@ Split[Sort[bde]]
```
```
Out[18]= {{0, 1173}, {1, 766}, {2, 562}}
```

About 46% of all Bolyai digits of random numbers are zeros.

```
In[19]:= data = Table[{First[#], Length[#]}& /@ Split[Sort[
                 BolyaiDigits[Random[Real, {0, 1}, 1000], 1000]]], {100}];
```
```
In[20]:= {#[[1, 1]], Plus @@ (Last /@ #)}& /@
           Split[Sort[Flatten[data, 1]], #1[[1]] === #2[[1]]&]
```
```
Out[20]= {{0, 46018}, {1, 30531}, {2, 23551}}
```

```
In[21]:= With[{sum = Plus @@ (Last /@ %)}, Last /@ %/sum] // N
```
```
Out[21]= {0.45972, 0.305005, 0.235275}
```

Now let us calculate closed form expressions for the numbers with repeating sequences of Bolyai digits: {1, 2}, {2, 1}, and {0, 1, 2}. Taking into account that the sequence of Bolyai digits continues forever, we can write down the following self-consistent equation for the radical part of a number x with repeating Bolyai digits $\{a_1, \ldots, a_n\}$.

$$\sqrt{a_1 + \sqrt{a_2 + \sqrt{a_3 + \cdots + \sqrt{a_n + x + 1}}}} = x + 1$$

The number with repeating Bolyai digits {1,2} is the root of a cubic polynomial.

```
In[22]:= bd12R = -1 + x /. Solve[Sqrt[1 + Sqrt[2 + x]] == x, x][[1]]
```

$$Out[22]= -1 + \frac{1}{2} \sqrt{\frac{1}{3} \left(4 - 8 \left(\frac{2}{-133 + 3\sqrt{2193}}\right)^{1/3} + \left(\frac{1}{2}\left(-133 + 3\sqrt{2193}\right)\right)^{1/3}\right)} +$$

$$\frac{1}{2} \sqrt{\left(\frac{8}{3} + \frac{8}{3}\left(\frac{2}{-133 + 3\sqrt{2193}}\right)^{1/3} - \frac{1}{3}\left(\frac{1}{2}\left(-133 + 3\sqrt{2193}\right)\right)^{1/3} + \right.}$$

$$\frac{2}{\sqrt{\frac{1}{3}\left(4 - 8\left(\frac{2}{-133 + 3\sqrt{2193}}\right)^{1/3} + \left(\frac{1}{2}\left(-133 + 3\sqrt{2193}\right)\right)^{1/3}\right)}}$$

This is the cubic polynomial.

```
In[23]:= (bd12R // RootReduce)[[1]][x]
```
$$Out[23]= \ -3 - x + 4\,x^2 + 4\,x^3 + x^4$$

Here is a quick check for the first 1000 Bolyai digits.

```
In[24]:= BolyaiDigits[N[bd12R, 1000], 1000] // Partition[#, 2]& // Union
```
$$Out[24]= \ \{\{1, 2\}\}$$

The number with repeating Bolyai digits {2,1} is also the root of a cubic polynomial.

```
In[25]:= bd21R = -1 + x /. Solve[Sqrt[2 + Sqrt[1 + x]] == x, x][[1]]
```

$$Out[25]= \ -1 + \frac{1}{2} \sqrt{\frac{8}{3} + \frac{1}{3}\left(\frac{763}{2} - \frac{3\sqrt{2193}}{2}\right)^{1/3} + \frac{1}{3}\left(\frac{1}{2}\left(763 + 3\sqrt{2193}\right)\right)^{1/3}} +$$

$$\frac{1}{2}\sqrt{\left(\left(\frac{1}{3}\left(16 - \left(\frac{763}{2} - \frac{3\sqrt{2193}}{2}\right)^{1/3} - \left(\frac{1}{2}\left(763 + 3\sqrt{2193}\right)\right)^{1/3} + \frac{6}{\sqrt{\frac{8}{3} + \frac{1}{3}\left(\frac{763}{2} - \frac{3\sqrt{2193}}{2}\right)^{1/3} + \frac{1}{3}\left(\frac{1}{2}\left(763 + 3\sqrt{2193}\right)\right)^{1/3}}}\right)\right)\right)}$$

Again, here is a quick check for the first 1000 Bolyai digits.

```
In[26]:= BolyaiDigits[N[bd21R, 1000], 1000] // Partition[#, 2]& // Union
```
$$Out[26]= \ \{\{2, 1\}\}$$

The number with repeating Bolyai digits {0,1,2} is a root of an irreducible polynomial of degree eight. (In Chapter 1 of the Symbolics volume [1795], we will discuss the function Root.)

```
In[27]:= bd012R = -1 + x /. Solve[Sqrt[0 + Sqrt[1 + Sqrt[2 + x]]] == x, x][[1]]
```
$$Out[27]= \ -1 + Root[-1 - \#1 - 2\,\#1^4 + \#1^8 \ \&, \ 2]$$

```
In[28]:= N[%]
```
$$Out[28]= \ 0.295333$$

Here are the two other expressions expressed as a root of a quartic polynomial.

```
In[29]:= {RootReduce[bd12R], RootReduce[bd21R]}
```
$$Out[29]= \ \{Root[-3 - \#1 + 4\,\#1^2 + 4\,\#1^3 + \#1^4 \ \&, \ 2], \ Root[-1 - 5\,\#1 + 2\,\#1^2 + 4\,\#1^3 + \#1^4 \ \&, \ 2]\}$$

Again, here is a quick check for the first 999 Bolyai digits of bd012R.

```
In[30]:= BolyaiDigits[N[bd012R, 1000], 999] // Partition[#, 3]& // Union
```
$$Out[30]= \ \{\{0, 1, 2\}\}$$

j) By lightly adapting the above function continuedFraction, it is straightforward to implement the function symmet￢ ricContinuedFraction. (We iterate either as long as needed or as long as possible.)

```
In[1]:= symmetricContinuedFraction[x_?(# > 1&), ord_Integer?Positive] :=
            DeleteCases[Floor[NestWhileList[(Floor[#]/(# - Floor[#]))&, x,
                (Not[IntegerQ[#]] && NumericQ[#])&, 1, ord - 1]],
                            0 | _?(Not[NumericQ[#]]&)]
```

Here are the first 10 and the last three integers of the symmetric continued fraction expansion of π.

```
In[2]:= $MaxExtraPrecision = 10^4;

        scfπ = symmetricContinuedFraction[Pi, 100];
        {Take[scfπ, 10], Take[scfπ, -3]}
```

```
Out[4]= {{3, 21, 111, 113, 158, 160, 211, 216, 525, 1634},
         {5541000097844351429447917235101 0487252,
          1951903430141377492189313228485 87364459, 888232035871304639615488967054813983622}}
```

To calculate the precision of the just-calculated symmetric continued fraction expansion, we implement a function `fromSymmetricContinuedFraction` that converts a list of integers representing the symmetric continued fraction expansion to an explicit fraction.

```
In[5]:= fromSymmetricContinuedFraction[scf_] :=
            Fold[(#2 + #2/#1)&, Last[scf], Rest[Reverse[scf]]]
```

The following two graphics show the logarithm of the $d_k(\pi)$ and the logarithm of the difference of the approximations to π.

```
In[6]:= Show[GraphicsArray[
          Block[{$DisplayFunction = Identity},
                {(* the terms of the symmetric continued fraction expansion *)
                 ListPlot[Log[10, scfπ] // N, Frame -> True],
                 (* the difference of the symmetric continued fraction expansion *)
                 ListPlot[Table[{k, Log[10, Abs[N[fromSymmetricContinuedFraction[
                           Take[scfπ, k]] - Pi, 22]]]}, {k, 100}] // N,
                   Frame -> True]}]]]
```

Taking into account the first 100 digits of π yields about 1933 correct base 10 digits.

```
In[7]:= N[fromSymmetricContinuedFraction[scfπ] - Pi, 22]

Out[7]= 7.530218834761384602980 × 10^{-2011}
```

The symmetric continued fraction expansions of $\sqrt{2}$ and e have a regular structure. But the quality of their approximation is much smaller.

```
In[8]:= symmetricContinuedFraction[Sqrt[2], 100]

Out[8]= {1, 2, 4, 4, 4, 4, 4, 4, 4, 4, 4, 4, 4, 4, 4, 4, 4, 4, 4, 4, 4, 4, 4,
         4, 4, 4, 4, 4, 4, 4, 4, 4, 4, 4, 4, 4, 4, 4, 4, 4, 4, 4, 4, 4, 4, 4, 4,
         4, 4, 4, 4, 4, 4, 4, 4, 4, 4, 4, 4, 4, 4, 4, 4, 4, 4, 4, 4, 4, 4, 4, 4,
         4, 4, 4, 4, 4, 4, 4, 4, 4, 4, 4, 4, 4, 4, 4, 4, 4, 4, 4, 4, 4, 4, 4, 4}
```

```
In[9]:= N[fromSymmetricContinuedFraction[%] - Sqrt[2], 22]

Out[9]= 7.877853966269712754948 × 10^{-77}
```

```
In[10]:= symmetricContinuedFraction[E, 100]

Out[10]= {2, 2, 2, 3, 4, 5, 6, 7, 8, 9, 10, 11, 12, 13, 14, 15, 16, 17, 18, 19, 20,
          21, 22, 23, 24, 25, 26, 27, 28, 29, 30, 31, 32, 33, 34, 35, 36, 37, 38, 39,
          40, 41, 42, 43, 44, 45, 46, 47, 48, 49, 50, 51, 52, 53, 54, 55, 56, 57, 58, 59,
          60, 61, 62, 63, 64, 65, 66, 67, 68, 69, 70, 71, 72, 73, 74, 75, 76, 77, 78, 79,
          80, 81, 82, 83, 84, 85, 86, 87, 88, 89, 90, 91, 92, 93, 94, 95, 96, 97, 98, 99}
```

```
In[11]:= N[fromSymmetricContinuedFraction[%] - E, 22]

Out[11]= 7.611447614771579559039 × 10^{-162}
```

Contrary to the normal continued fraction expansion, not all square roots have simple symmetric continued fraction expansions.

```
In[12]:= symmetricContinuedFraction[Sqrt[3], 20]

Out[12]= {1, 1, 2, 2, 2, 2, 2, 2, 2, 2, 2, 2, 2, 2, 2, 2, 2, 2, 2, 2}
```

In[13]:= **symmetricContinuedFraction[Sqrt[5], 20]**

Out[13]= {2, 8, 16, 16, 16, 16, 16, 16, 16, 16, 16, 16, 16, 16, 16, 16, 16, 16, 16, 16}

The terms of the symmetric continued fraction expansion of $\sqrt{7}$ grow exponentially.

In[14]:= **symmetricContinuedFraction[Sqrt[7], 20]**

Out[14]= {2, 3, 30, 34, 111, 235, 3775, 5052, 7352, 9091, 34991, 35530,
     53424, 57290, 66023, 1409179, 1519111, 1725990, 1812396, 4370835}

In[15]:= **N[fromSymmetricContinuedFraction[
                    symmetricContinuedFraction[Sqrt[7], 100]] - Sqrt[7], 22]**

Out[15]= $1.672350762354551058270 \times 10^{-1952}$

Here is the symmetric continued fraction expansion of a rational number.

In[16]:= **symmetricContinuedFraction[6563440628747948887/4714400135799462226, 30]**

Out[16]= {1, 2, 3, 4, 5, 6, 7, 8, 9, 10, 11, 12, 13, 14, 15, 16, 17, 18, 19, 20}

In[17]:= **fromSymmetricContinuedFraction[%]**

Out[17]= $\dfrac{6563440628747948887}{4714400135799462226}$

The finite symmetric continued fraction expansion with $d_k = k$ for $k < k_{max}$ approaches the following limiting value as $k_{max} \to \infty$.

In[18]:= **{N[%, 22], N[fromSymmetricContinuedFraction[Range[100]], 20]}**

Out[18]= {1.392211191177332814377, 1.3922111911773328144}

References

1   M. Abadie in B. Brogliato (ed.). *Impacts in Mechanical Systems*, Springer-Verlag, Berlin, 2000.

2   E. Abadoglu, M. Bruschi, F. Calogero. *Theor. Math. Phys.* 128, 835 (2001).

3   V. A. Abalmassow, D. A. Maljutin. *arXiv:physics*/0409154 (2004).

4   M. M. Abbott in K. C. Chao, R. L. Robinson, Jr. *Equations of State in Engineering and Research*, American Chemical Society, Washington, 1972.

5   P. Abbott. *The Mathematica Journal* 8, 27 (2001).

6   U. Abel. *Am. Math. Monthly* 110, 627 (2003).

7   U. Abel, M. Ivan, T. Riedel. *J. Math. Anal. Appl.* 295, 1 (2004).

8   O. Aberth. *Proc. Am. Math. Soc.* 30, 151 (1971).

9   S. S. Abhyankar, T. Moh. *J. reine angew. Math.* 260, 47 (1973).

10   S. S. Abhyankar, T. Moh. *J. reine angew. Math.* 261, 29 (1973).

11   S. S. Abhyankar, C. Bajaj. *Comput. Aided Design* 19, 449 (1987).

12   F. F. Abi-Khuzam. *Proc. Am. Math. Soc.* 123, 3837 (1995).

13   M. J. Ablowitz, B. M. Herbst, C. M. Scober. *J. Phys.* A 34, 10671 (2001).

14   E. Abramochkin, V. Volostnikov. *Opt. Commun.* 125, 302 (1996).

15   E. G. Abramochkin, V. G. Volostnikov. *Usp. Fiz. Nauk* 174, 1273 (2004).

16   G. Abramson. *arXiv:nlin.AO*/0010049 (2000).

17   D. J. Acheson. *Eur. J. Phys.* 21, 269 (2000).

18   A. Adamatzky. *Computing in Nonlinear Media and Automata Collectives*, Institute of Physics, Bristol, 2001.

19   V. Adamchik. *J. Comput. Appl. Math.* 100, 191 (1998).

20   V. Adamchik in B. Mourrain (ed.). *ISSAC 2001*, ACM, Baltimore, 2001.

21   V. Adamchik. *arXiv:math.CA*/0308074 (2003).

22   V. V. Afanasiev, G. M. Zaslavsky in G. Györgyi, I. Kondor, L. Sasvári, T. Tél (eds.). *From Phase Transitions to Chaos*, World Scientific, Singapore, 1992.

23   R. P. Agnew. *Am. Math. Monthly* 54, 273 (1947).

24   J. L. Agudin, A. M. Platzek. *Phys. Rev.* D 26, 1923 (1982).

25   D. Aharonov, A. Ambainis, J. Kempe, U. Vazirani. *arXiv:quant-ph*/0012090 (2000).

26   R. Airapetyan, A. G. Ramm, A. Smirnova. *Math. Models Meth. Appl. Sci.* 9, 463 (1999).

27   R. Airapetyan. *Appl. Anal.* 73, 464 (1999).

28   L. V. Ahlfors, L. Sario. *Riemann Surfaces*, Princeton University Press, Princeton, 1960.

29   R. W. Ahrens, G. Szekeres. *Austral. Math. J.* 10, 485 (1969).

30   A. C. Aitken. *Proc. R. Soc. Edinb.* 46, 289 (1926).

31   N. Akerman. *The Necessity of Friction*, Physica, Heidelberg, 1993.

32   N. Akhmediev, J. M. Soto–Crespo, A. Ankiewicz in F. K. Abdullaev, V. V. Konotop (eds.). *Nonlinear Waves: Classical and Quantum Aspects*, Kluwer, Dordrecht, 2004.

33   U. E. Aladl, A. S. Deakin, H. Rasmussen. *J. Comp. Appl. Math.* 138, 309 (2002).

34   D. J. Albers, J. C. Sprott. *arXiv:nlin.CD*/0408017 (2004).

35   E. L. Albuquerque, M. G. Cottam. *Phys. Rep.* 376, 225 (2003).

36   G. Alefeld, J. Herzberger. *Introduction to Interval Computations*, Academic Press, New York, 1983.

37   G. E. Alefeld, F. A. Porta, X. Shi. *ACM Trans. Math. Softw.* 21, 327 (1995).

38   J. C. Alexander, B. R. Hunt, I. Kan, J. A. Yorke. *Ergod. Th. Dynam. Sys.* 16, 651 (1996).

39   T. Alieva. *J. Opt. Soc. Am.* A 13, 1189 (1996).

40   E. L. Allgower, K. Georg in P. G. Ciarlet, J. L. Lions (eds.). *Techniques of Scientific Computing* v.V, Elsevier, Amsterdam, 1997.

41   O. E. Alon, L. S. Cederbaum. *Phys. Rev.* B 68, 033105 (2003).

42   J. M. Alonso, R. Ortega. *J. Diff. Eq.* 143, 201 (1998).

43   V. Allori, N. Zhanghí. *arXiv:quant-ph*/0112008 (2001).

44   J. S. Alper, M. Bridger, J. Earman, J. D. Norton. *Synthese* 114, 355 (1998).

45   O. S. M. Alves, F. P. Machado, S. Y. Popov. *arXiv:math.PR*/0102182 (2001).

46   E. J. Amanatidis, D. E. Katsanos, S. N. Evangelou. *Phys. Rev.* B 69, 195107 (2004).

47   V. Amar, M. Pauri, A. Scotti. *J. Math. Phys.* 34, 3343 (1993).

48   A. Ambrosetti, G. Prodi. *A Primer of Nonlinear Analysis*, Cambridge University Press, Cambridge, 1993.

49   M. Amit, Y. Schmerler, E. Eisenberg, M. Abraham, N. Shnerb. *Fractals* 2, 7 (1994).

50   P. Amore. *arXiv:math-ph*/0409036 (2004).

51   J. G. Analytis, S. J. Blundell, A. Ardavan. *Am. J. Phys.* 72, 613 (2004).

52   S. Andersson, M. Jacob, S. Lidin. *Z. Kristallogr.* 210, 3 (1995).

53   M. Ando, S. Ito, S. Katsumoto, J. Iye. *J. Phys. Soc. Jpn.* 68, 3462 (1999).

54   T. Ando, Y. Arakawa, K. Furuya, S. Komiyama, H. Nakashima (eds.). *Mesoscopic Physics and Electronics*, Springer-Verlag, Berlin, 1998.

55   M. Andrews. *Am. J. Phys.* 66, 252 (1998).

56   T. M. Apostol. *Modular Functions and Dirichlet Series in Number Theory*, Springer-Verlag, New York, 1997.

57   T. Apostol, M. A. Mnatsakanian. *Math. Mag.* 75, 307 (2002).

58   N. Aquino. *J. Math. Chem.* 18, 349 (1995).

59   I. S. Aranson, L. Kramer. *arXiv:cond-mat*/0106115 (2001).

60   K. Arbter. Ph.D. thesis, Hamburg-Harburg, 1990.

61   H. Aref, N. Pomphrey. *Phys. Lett.* A 78, 297 (1980).

62   H. Aref. *Ann. Rev. Fluid Mech.* 15, 345 (1983).

63   H. Aref. *J. Fluid Mech.* 143, 1 (1984).

64   H. Aref. *Phys. Fluids* 14, 1315 (2002).

65   M. Argentina, P. Coullet. *Phys. Rev.* E 56, R2359 (1997).

66   M. Argentina, P. Coullet in A. Carbone, M. Gromov, P. Prusinkiewicz (eds.). *Pattern Formation in Biology, Vision and Dynamics*, World Scientific, Singapore, 2000.

67   J. Argyris, H.-P. Mlejneh. *Die Methode der Finiten Elemente*, v. III, Vieweg, Braunschweig, 1988.

68   B. Armstrong-Hélouvry, P. Dupont, C. Canudas de Wit. *Automatica* 30, 1083 (1994).

69   J. Arndt, C. Haenel. π *Unleashed*, Springer-Verlag, Berlin, 2001.

70   V. I. Arnold. *Singularities of Caustics and Wave Fronts*, Kluwer, Dordrecht, 1978.

71   V. I. Arnold, B. A. Khesin. *Topological Methods in Hydrodynamics*, Springer-Verlag, New York, 1998.

72   N. Aronszajn. *Am. J. Math.* 79, 597 (1957).

73   G. Arreaga–García, H. Villegas–Brena, J. Saucedo–Morales. *J. Phys.* A 37, 9419 (2004).

74   J. Arriaga, V. R. Velasco. *J. Phys.* C 9, 8031 (1997).

75   F. M. Arscott. *Periodic Differential Equations*, MacMillan, New York, 1964.

76   S. J. Arseth, J. P. Anosova, V. V. Orlov. *Celest. Mech. Dynam. Astron.* 58, 1 (1994).

77   D. A. Aruliah, R. M. Corless in J. Gutierrez (ed.). *ISSAC 04*, ACM Press, New York, 2004.

78   N. S. Asaithambi. *Numerical Analysis*, Saunders College Publishing, Fort Worth, 1995.

79   F. Asakura. *Methods Appl. Anal.* 6, 477 (1999).

80   Y. Ashkenazy, L. P. Horwitz. *arXiv:chao-dyn*/9901015 (1999).

81   A. Ashtekar, A. Corichi. *ESI Preprint* 363 (1997). ftp://ftp.esi.ac.at:/pub/Preprints/esi363.ps

82   N. M. Atakishiyev, L. E. Vincent, K. B. Wolf. *J. Comput. Appl. Math.* 107, 73 (1999).

83   L. Atanassova. *J. Comput. Appl. Math.* 50, 99 (1994).

84   L. Y. Atanassova, J. P. Herzberger in V. Lakshmikantham. *World Congress of Nonlinear Analysis '92*, de Gruyter, Berlin, 1996.

85   F. M. Atay, J. Jost. *arXiv:nlin.AO*/0312026 (2003).

86   D. Atkinson, F. J. van Steenwijk. *Am. J. Phys.* 67, 486 (1999).

87   M. D. Atkinson, T. Stitt. *Discr. Math.* 259, 19 (2002).

88   A. Atoyan, J. Patera. *J. Math. Phys.* 45, 2468 (2004).

89   D. Aubin, A. D. Dalmedico. *Hist. Math.* 29, 273 (2002).

90   H. Auda. *IEEE Trans. Microwave Th. Techn.* MIT-31, 515 (1983).

91   M. Audin. *Spinning Tops*, Cambridge University Press, Cambridge, 1996.

92   D. Auerbach. *Am. J. Phys.* 62, 522 (1994).

93   B. W. Augenstein. *Chaos, Solitons, Fractals* 10, 1087 (1999).

94   G. Aumann, K. Spitzmüller. *Computerorientierte Geometrie*, BI, Mannheim, 1993.

95   L. Auslander, R. Tolimieri. *Bull. Am. Math. Soc.* 6, 847 (1979).

96   M. O. Avdeeva. *Funct. Anal. Appl.* 38, 79 (2004).

97   J. E. Avron, D. Osadchy. *arXiv:math-ph*/0110026 (2001).

98   J. E. Avron. *arXiv:math-ph*/0308030 (2003).

99   M. Y. Azbel. *Europhys. Lett.* 24, 623 (1993).

100   A. Azizi. *J. Math. Phys.* 43, 299 (2002).

101   A. G. Azpeitia. *Am. Math. Monthly* 89, 311 (1982).

102   M. Baake, U. Grimm, D. H. Warrington. *J. Phys.* A 27, 2669 (1994).

103   M. Baake, R. V. Moody, P. A. Pleasants. *arXiv:math.MG*/9906132 (1999).

104   B. Bacus, Y. Meurice, A. Soemadi. *J. Phys.* A 28, L381, (1995).

105   A. Bäcker. *arXiv:nlin.CD*/0204061 (2002).

106   E. Bänsch, P. Morin, R. H. Nochetto. *J. Comput. Phys.* 203, 321 (2005).

107   M. Bär, R. Hegger, H. Kantz. *Phys. Rev.* E 59, 337 (1999).

108   J. Baik, P. Deift, K. Johansson. *arXiv:math.CO*/9810105 (1998).

109   J. Baik, E. M. Rains. *arXiv:math.CO*/9910019 (1999).

110   H. Bai-lin. *Elementary Symbolic Dynamics*, World Scientific, Singapore, 1989.

111   D. H. Bailey, P. N. Swarztrauber. *SIAM Rev.* 33, 389 (1991).

112   D. H. Bailey, P. N. Swarztrauber. *SIAM J. Sci. Comput.* 15, 1105 (1994).

113   D. H. Bailey, J. M. Borwein and R. E. Crandall. *CECM Preprint* 36/95 (1995).
       http://www.cecm.sfu.ca/ftp/pub/CECM/Preprints/Postscript/95:036-Bailey-Borwein-Crandall.ps.gz

114   D. H. Bailey, S. Plouffe. *Can. Math. Soc.* 20, 73 (1997).

115   D. H. Bailey. *ACM SIGNUM* 33, 17 (1998).

116   D. H. Bailey, D. J. Broadhurst.*arXiv:math.NA*/9905048 (1999).

117   D. H. Bailey. *Comput.Sci. Eng.* 2, n1, 24 (2000).

118   C. L. Bajaj, R. J. Holt, A. N. Netravali. *ACM Trans. Graphics* 17, 1 (1998).

119   C. L. Bajaj, G. Xu in M. Sarfaz (ed.). *Geometric Modeling: Techniques, Applications, Systems, and Tools*, Kluwer, Dordrecht, 2004.

120   I. N. Baker, P. J. Rippon. *Ann. Acad. Sci. Fenn.* A Math. 9, 49 (1984).

121   A. Baldassari, F. Colaiori, C. Castellano. *Phys. Rev. Lett.* 90, 060601 (2003).

122   M. L. Balinski, S. T. Rachev. *Num. Funct. Anal. Optim.* 14, 475 (1993).

123   M. L. Balinski, S. T. Rachev. *Math. Sci.* 22, 1 (1997).

124   N. J. Balmforth, C. Pasquero, A. Provenzale. *Physica* D 138, 1 (2000).

125   W. Balser. *J. Math. Sc.* 124, 5085 (2004).

126   S. Bandyopadhyay, D. Home. *arXiv:quant-ph*/9907029 (1999).

127   S. Bandyopadhyay, A. S. Majumdar, D. Home. *arXiv:quant-ph*/0103001 (2001).

128   S. Banerjee, J. A. Yorke, C. Grebogi. *arXiv:chao-dyn*/9803001 (1998).

129   M. Bank. *arXiv:nlin.CD*/0009020 (2000).

130   L. Banyai, S. W. Koch. *Semiconductor Quantum Dots*, World Scientific, Singapore, 1993.

131   D. Baraff. *Algorithmica* 10, 292 (1993).

132   J. L. M. Barbosa, A. G. Colares. *Minimal Surfaces in* $\mathbb{R}^3$, Springer-Verlag, Berlin, 1986.

133   A. Barelli, J. Bellissard, R. Rammal. *J. Phys. France* 51, 2167 (1990).

134   A. Barelli, R. Fleckinger. *Int. J. Appl. Sc. Comput.* 2, 362 (1995).

135   A. Barelli, R. Fleckinger. *Phys. Rev.* B 46, 11559 (1992).

136   E. Barouch, B. M. McCoy, T. T. Wu. *Phys. Rev. Lett.* 31, 1409 (1973).

137   J. Barré, F. Bouchet, T. Dauxois, S. Ruffo. *arXiv:cond-mat*/0204407 (2002).

138   D. F. Barrow. *Am. Math. Monthly* 43, 150 (1936).

139   J. D. Barrow. *Proc. R. Soc.* A 310, 337 (1983).

140   W. Barth. *J. Alg. Geom.* 5, 173 (1996).

141   G. Barton. *Ann. Phys.* 166, 322 (1986).

142   J. C. Barton, C. J. Eliezer. *J. Austral. Math. Soc.* B 41, 358 (2000).

143   L. Basnarkov, V. Urumov. *Progr. Theor. Phys.* S 150, 321 (2003).

144   M. Batista, M. Lakner, J. Peternelj. *Eur. J. Phys.* 25, 145 (2004).

145   M. Bauer, O. Golinelli. *J. Stat. Phys.* 103, 301 (2001).

146   T. Baumberger, C. Caroli. *Commun. Cond. Matt.* 17, 307 (1995).

147   M. Bauer, O. Golinelli. *arXiv:cond-mat*/0007127 (2000).

148   A. D. Baute, I. L. Egusquiza, J. G. Muga. *arXiv:quant-ph*/0007079 (2000).

149   B. F. Bayman. *Am. J. Phys.* 44, 671 (1976).

150 J. Beck. *Discr. Math.* 229, 29 (2001).

151 L. Beeckmans. *J. Number Th.* 43, 173 (1993).

152 H. Behnke, F. Sommer. *Theorie der analytischen Funktionen einer komplexen Veränderlichen*, Springer-Verlag, Berlin, 1962.

153 J. W. Belcher, S. Olbert. *Am. J. Phys.* 71, 220 (2003).

154 V. V. Beletsky, G. V. Kasatkin E. L. Starostin. *Chaos, Solitons, Fractals* 7, 1145 (1996).

155 F. J. Belinfante. *A Survey of Hidden-Variable Theories*, Pergamon Press, Oxford, 1973.

156 J. Bell. *The Speakable and the Unspeakable in Quantum Mechanics*, Cambridge University Press, Cambrige, 1987.

157 J. Bellissard in S. Albeverio, J. E. Fenstad, H. Holden, T. Lindstrøm (eds.). *Ideas and Methods in Quantum and Statistical Physics*, Cambridge University Press, Cambridge, 1992.

158 J. Bellissard in M. Waldschmidt, P. Moussa, J.-M. Luck, C. Itzykson (eds.). *From Number Theory to Physics*, Springer-Verlag, Berlin, 1992.

159 J. Bellissard in S. Graffi (ed.). *Transitions to Chaos in Classical and Quantum Physics*, Springer-Verlag, Berlin, 1994.

160 R. Bellman. *Quart. J. Appl. Math.* 13, 322 (1955).

161 R. Bellman, B. Kotkin. *Math. Comput.* 16, 473 (1962).

162 E. D. Belokolos, A. I. Bobenko, V. Z. Enol'skii, A. R. Its, V. B. Matveev. *Algebro-Geometric Approach to Nonlinear Integrable Equations*, Springer-Verlag, Berlin, 1994.

163 D. W. Belousek. *Found. Sci.* 8, 109 (2003).

164 E. G. Beltrametti, G. Casinelli. *The Logic of Quantum Mechanics*, Addison-Wesley, Reading, 1976.

165 J. V. Beltran, J. Monterde in A. Laganá, M. L. Gavrilova, V. Kumar, Y. Mun, C. J. K. Tan, O. Gervasi (eds.). *Computational Science and Its Applications* v1, Springer-Verlag, Berlin, 2004.

166 I. Bena, C. van den Broeck. *Europhys. Lett.* 48, 498 (1999).

167 G. P. Benacerraf. *J. Phil.* 59, 765 (1962).

168 D. ben-Avraham, E. Bollt, C. Tamon. *arXiv:cond-mat/0409514* (2004).

169 C. M. Bender, S. A. Orszag. *Advanced Mathematical Methods for Scientists and Engineers*, McGraw-Hill, New York, 1978.

170 C. M. Bender, K. A. Milton, S. S. Pinsky, L. M. Simmons, Jr. *J. Math. Phys.* 30, 1447 (1989).

171 C. M. Bender, D. C. Brody, B. K. Meister. *arXiv:quant-ph/0309119* (2003).

172 L. Bendersky. *Acta Math.* 61, 263 (1933).

173 E. Ben–Jacob, I. Cohen, O. Shochet, A. Tenenbaum, A. Czirók, T. Vicsek. *Phys. Rev. Lett.* 75, 2899 (1995).

174 E. Ben-Jacob, I. Cohen, H. Levine. *Adv. Phys.* 49, 395 (2000).

175 E. Ben–Naim, P. L. Krapivsky. *arXiv:nlin.AO/0108047* (2001).

176 E. Ben–Naim, P. L. Krapivsky. *arXiv:cond-mat/0403157* (2004).

177 V. L. Berdichevsky. *Thermodynamics of Chaos and Order*, Longman, Essex, 1997.

178 I. A. Beresnev, V. N. Nikolaevskii. *Physica* D 66, 1 (1993).

179 N. Berestycki. *arXiv:math.PR/0411011* (2004).

180 A. A. Berezin. *Nature* 315, 104 (1985).

181 A. Berger, L. A. Bunimovich, T. P. Hill. *Trans. Am. Math. Soc.* 357, 197 (2004).

182 L. Berggren, J. Borwein, P. Borwein. *Pi: A Source Book*, Springer, New York, 1997.

183 N. Berglund, A. Hansen, E. H. Hauge, J. Piasecki. *Phys. Rev. Lett.* 77, 2149 (1996).

184 N. Berglund, H. Kunz. *J. Stat. Phys.* 83, 81 (1996).

185   M. K. Berkenbusch, I. Claus, C. Dunn, L. P. Kadanoff, M. Nicewicz, S. C. Venkataramani. *arXiv:cond-mat*/0310257 (2003).

186   K. Berndl, M. Daumer, D. Dürr, S. Goldstein, N. Zanghi. *arXiv:quant-ph*/9504010 (1995).

187   A. Bernhart. *Scripta Math.* 24, 189 (1959).

188   D. N. Bernshtein. *Funct. Anal. Appl.* 9, 183 (1975).

189   D. N. Bernstein, A. G. Kouchnirenko, A. G. Khovanskii. *Funct. Anal. Appl.* 31, 201 (1976).

190   G. C. Berresford. *Analysis* 41, 1 (1981).

191   A. Berretti, G. Gentile. *mp_arc* 01-406 (2001). http://rene.ma.utexas.edu/mp_arc/c/01/01-406.ps.gz

192   M. Berry, C. Upstill in E. Wolf (ed.). *Progress in Optics XVIII*, North Holland, Amsterdam, 1980.

193   M. V. Berry. *J. Phys.* A 18, 15 (1985).

194   M. Berry in H. Blok, H. A. Ferwerda, H. K. Kuiken (eds.). *Huygen's Principle 1690-1990: Theory and Applications*, Elsevier, Amsterdam, 1992.

195   M. V. Berry. *J. Phys.* A 27, L391 (1994).

196   M. V. Berry, M. A. Morgan. *Nonlinearity* 9, 787 (1996).

197   M. V. Berry. *J. Phys.* A 29, 6617 (1996).

198   T. G. Berry, R. R. Patterson. *Comput. Aided Geom. Design* 18, 723 (2001).

199   G. Berthold, J. Smoliner, V. Rosskopf, E. Gornik, G. Bohn, G. Weimann. *Phys. Rev.* B 45, 11350 (1992).

200   J. Bertoin, G. Miermont. *arXiv:math.PR*/0502220 (2005).

201   D. Bertrand, W. Zudilin. *arXiv:math.NT*/0006176 (2000).

202   E. H. Berwick. *Proc. Lond. Math. Soc.* 28, 53 (1928).

203   D. Bessis, J. D. Fournie in G. Chaohao, L. Yishen , T. Guizhang (eds.). *Visualization in Teaching and Learning Mathematics*, Springer, Berlin, 1990.

204   M. Bestehorn, E. V. Grigorieva, S. A. Kaschenko. *Phys. Rev.* E 70, 026202 (2004).

205   R. A. Beth. *J. Appl. Phys.* 37, 2568 (1966).

206   R. Bhatia. *Matrix Analysis*, Springer-Verlag, New York, 1997.

207   S. Bhomwick, R. Bhattacharya, D. Roy. *Comput. Phys. Commun.* 54, 31 (1989).

208   V. N. Biktashev, J. Brindley, A. V. Holden, M. A. Tsyganov. *arXiv:nlin.PS*/0406012 (2004).

209   D. Bini, G. Fiorention in V. Keränen, P. Mitic (eds.). *Mathematics with Vision*, Computational Mechanics Publications, Southampton, 1995.

210   D. A. Bini, L. Gemignani, V. Y. Pan. *Comput. Math. Appl.* 47, 447 (2004).

211   J. P. Bird, R. Akis, D. K. Ferry. *Adv. Imaging Electron Phys.* 107, 1 (1999).

212   L. Birgé, Y. Rozenholz. *Prépublication* PMA-721 (2002).http://www.proba.jussieu.fr/mathdoc/preprints/lb.Fri_Apr_12_10_23_45_CEST_2002.html

213   E. R. Bittner. *arXiv:quant-ph*/0001119 (2000).

214   P. Bizón. *arXiv:math-ph*/0411041 (2004).

215   Å. Björck. *Numerical Methods for Least Square Problems*, SIAM, Philadelphia, 1996.

216   J. D. Bjorken, S. D. Drell. *Relativistic Quantum Mechanics*, McGraw, New York, 1964.

217   J. A. Blackburn, H. J. T. Smith. *Chaos, Solitons, Fractals* 15, 783 (2002).

218   H.-Blatt, M. Götz. *J. Approx. Th.* 106, 276 (2000).

219   S. Bleher, C. Grebogi, E. Ott. *Physica* D 46, 87 (1990).

220   M. N. Bleicher. *J. Number Th.* 4, 342 (1972).

221 M. N. Bleicher. *J. Number Th.* 8, 168 (1976).

222 R. P. Boas, Jr. *Am. Math. Monthly* 84, 237 (1977).

223 S. Boccaletti, J. Kurths, G. Osipov, D. L. Valladares, C. S. Zhou. *Phys. Rep.* 366, 1 (2002).

224 E. Bogomolny, O. Bohigas, P. Leboeuf. *J. Stat. Phys.* 85, 639 (1996).

225 D. Bohm. *Phys. Rev.* 85, 166 (1952).

226 D. Bohm. *Phys. Rev.* 85, 180 (1952).

227 D. Bohm, B. J. Hiley, P. N. Kaloyerou. *Phys. Rep.* 144, 321 (1987).

228 D. J. Bohm, B. J. Hiley. *The Undivided Universe: An Ontological Interpretation of Quantum Mechanics*, Routledge, London, 1993.

229 T. Bohr, M. H. Jensen, G. Paladin, A. Vulpiani. *Dynamical Systems Approach to Turbulence*, Cambridge University Press, Cambridge, 1998.

230 T. Bohr, M. van Hecke, R. Mikkelsen, M. Ipsen. *Phys. Rev. Lett.* 86, 5482 (2001).

231 B. D. Bojanov, H. A. Hakopian, A. A. Sahakian. *Spline Functions and Multivariate Interpolations*, Kluwer, Dordrecht, 1993.

232 E. M. Bollt, A. Klebanoff. *Int. J. Bifurc. Chaos* 12, 1843 (2002).

233 B. M. Bolotovskiĭ, A. V. Serov. *Usp. Fiz. Nauk* 173, 667 (2003).

234 E. Bombieri, A. J. van der Poorten in W. Bosma, A. van der Poorten (eds.). *Computational Algebra and Number Theory*, Kluwer, Amsterdam, 1995.

235 M. Bóna. *Discr. Math.* 175, 55 (1997).

236 M. Bóna. *Combinatorics of Permutations*, Chapman & Hall, Boca Raton, 2004.

237 D. Bonatsos, C. Daskaloyannis. *arXiv:nucl-th*/9999003 (1999).

238 J. D. Bondurat, S. A. Fulling. *arXiv:math-ph*/0408054 (2004).

239 P. Bonfert in H. Kriete (ed.). *Progress in Holomorphic Dynamics*, Longman, Harlow, 1998.

240 R. Bonifacio, I. Marzoli, W. P. Schleich. *J. Mod. Opt.* 47, 2891 (2000).

241 E. P. Borges. *J. Phys.* A 31, 5281 (1998).

242 Y. Le Borgne, D. Rossin. *Discr. Math.* 256, 775 (2002).

243 F. Borgonovi. *arXiv:chao-dyn*/9801032 (1998).

244 F. Bornemann. *arXiv:math.NA*/0211049 (2002).

245 J. M. Borwein, P. B. Borwein. *Pi and the AGM*, Wiley, New York, 1987.

246 J. M. Borwein, P. B. Borwein in G. E. Andrews, R. A. Askey, B. C. Berndt, K. G. Ramanathan, R. A. Rankin (eds.). *Ramanujan Revisited*, Academic Press, New York, 1988.

247 J. M. Borwein, P. Borwein, D. H. Bayley. *Am. Math. Monthly* 96, 201 (1989).

248 J. M. Borwein, P. Borwein. *Notices Am. Math. Soc.* 39, 825 (1992).

249 J. M. Borwein, D. M. Bradley. *Am. Math. Monthly* 99, 622 (1992).

250 J. M. Borwein, P. B. Borwein. *J. Comput. Appl. Math.* 46, 281 (1993).

251 D. Borwein, J. M. Borwein. *CECM Preprint* 142/99 (1999).
http://www.cecm.sfu.ca/ftp/pub/CECM/Preprints/Postscript/99:142-Borwein-Borwein.ps.gz

252 D. Borwein, J. M. Borwein, B. A. Mares, Jr. *Ramanujan J.* 6, 189 (2002).

253 J. M. Borwein, F. G. Garvan. *CECM Preprint* 64/96 (1999).
http://www.cecm.sfu.ca/ftp/pub/CECM/Preprints/Postscript/96:064-Borwein-Garvan.ps.gz

254 J. M. Borwein, P. Lisonek. *CECM Preprint* 104/97 (1997).
http://www.cecm.sfu.ca/ftp/pub/CECM/Preprints/Postscript/97:104-Borwein-Lisonek.ps.gz

255 D. Borwein, J. M. Borwein. *Ramanujan J.* 5, 73 (2001).

256   J. H. Borwein, D. H. Bailey. *Experimentation in Mathematics: Computational Paths to Discovery*, A K Peters, Dordrecht, Wellesley, 2003.

257   J. Borwein, D. Bailey. *Mathematics by Experiment*, A K Peters, Nautick, 2004.

258   W. Bosma, H. Jager, F. Wiedijk. *Indag. Math.* 45, 281 (1983).

259   W. Bosma. *Math. Comput.* 54, 421 (1990).

260   W. Bosma, C. Kraaikamp. *J. Austral. Math. Soc.* A 50, 481 (1991).

261   W. Bosma, K. Dajani, C. Kraaikamp. *Technical Report* 9925/99 University of Nijmegen (1999).
       http://www-math.sci.kun.nl/math/onderzoek/reports/rep1999/rep9925.ps.gz

262   J. B. Bost in M. Waldschmidt, P. Moussa, J.-M. Luck, C. Itzykson (eds.). *From Number Theory to Physics*, Springer-Verlag, Berlin, 1992.

263   K. Boström. *arXiv:quant-ph*/0005024 (2000).

264   S. Bottani, B. Delamotte. *Physica* D 103, 430 (1997).

265   U. Bottazzini. *arXiv:math.HO*/0305022 (2003).

266   O. Bottema, B. Roth. *Theoretical Kinematics*, North-Holland, Amsterdam, 1979.

267   G. E. P. Box, M. E. Muller. *Annals Math. Stat.* 29, 610 (1958).

268   N. Bouleau. *Error Calculus for Finance and Physics: The Language of Dirichlet Forms*, Academic Press, Boston, 1989.

269   C. L. Bouton, M. Bocher. *Ann. Math.* 12, 1 (1898).

270   A. Bovier, J.-M. Ghez. *Commun. Math. Phys.* 158, 45 (1993).

271   F. P. Bowden, D. Tabor. *The Friction and Lubrication of Solids* v.1, Clarendon Press, Oxford, 1950.

272   F. P. Bowden, D. Tabor. *The Friction and Lubrication of Solids* v.4, Clarendon Press, Oxford, 1954.

273   W. E. Boyce, R. C. DiPrima. *Elementary Differential Equations and Boundary Value Problems*, Wiley, New York, 1992.

274   J. P. Boyd. *J. Comput. Phys.* 178, 118 (2002).

275   T. H. Boyer. *arXiv:physics*/0405090 (2004).

276   P. L. Boyland, H. Aref, M. S. Stremler. *J. Fluid Mech.* 403, 277 (2000).

277   A. J. Bracken, G. F. Melloy. *J. Phys.* A 27, 2197 (1994).

278   D. H. Bradley, D. J. Broadhurst. *arXiv:math.CA*/9906134 (1999).

279   D. M. Bradley. *Analysis* 22, 219 (2002).

280   Y. Braiman, J. F. Lindner, W. L. Ditto. *Nature* 378, 465 (1995).

281   J. W. Braun, Q. Su, R. Grobe. *Phys. Rev.* A 59, 604 (1999).

282   M. Braverman. *arXiv:cs.CC*/0502066 (2005).

283   R. P. Brent. *J. ACM* 23, 242 (1976).

284   P. C. Bressloff, J. C. Dowan in J. Hogan, A. Champneys, B. Krauskopf, M. di Bernando, E. Wilson, H. Osinga, M. Homer. *Nonlinear Dynamics and Chaos: Where do we go from here?*, Institute of Physics, Bristol, 2003.

285   C. Brezinski. *Padé-Type Approximation and General Orthogonal Polynomials*, Birkhäuser, Basel, 1980.

286   C. Brezinski, M. Redivo Zaglia. *Extrapolation Methods: Theory and Practice*, North-Holland, Amsterdam, 1991.

287   C. Brezinski. *History of Continued Fractions and Padé Approximations*, Springer-Verlag, Berlin, 1991.

288   C. Brezinski. *Appl. Num. Math.* 20, 299 (1996).

289   C. Brezinski. *J. Comput. Appl. Math.* 122, 1 (2000).

290   C. Brezinski in C. Brezinski, L. Wuytack (eds.). *Numerical Analysis: Historical Developments in the 20th Century*, Elsevier, Amsterdam, 2001.

291   M. Bridger, J. S. Alper. *Synthese* 114, 355 (1998).

292   M. Bridger, J. S. Alper. *Synthese* 119, 325 (1999).

293   W. L. Briggs, V. E. Henson. *The DFT*, SIAM, Philadelphia, 1995.

294   W. L. Briggs, V. E. Henson. *Appl. Num. Math.* 21, 375 (1996).

295   A. Brill. *Vorlesungen über ebene algebraische Kurven und algebraische Funktionen*, Vieweg, Braunschweig, 1925.

296   D. C. Brody, B. K. Meister in O. Hirota, A. S. Holevo, C. M. Caves (eds.). *Quantum Communication, Computing, and Measurement*, Plenum Press, New York, 1997.

297   D. C. Brody, B. K. Meister. *J. Phys.* A 32, 4921 (1999).

298   S. Brofferio, W. Woes. *Proc. Am. Math. Soc.* 129, 1513 (2000).

299   I. N. Bronshtein, K. A. Semendyayev. *Handbook of Mathematics*, van Nostrand, New York, 1985.

300   J. C. Bronski, J. N. Kutz. *Phys. Lett.* A 254, 325 (1999).

301   R. Brout. *Phase Transitions*, North Holland, Amsterdam, 1965.

302   R. W. Brown, S. M. Shvartsman. *Phys. Rev. Lett.* 83, 1946 (1999).

303   J. Brünning, D. Gromes. *Math. Z.* 124, 79 (1972).

304   A. D. Bruno. *The Restricted 3-Body Problem*, de Gruyter, Berlin, 1994.

305   A. Bruno–Alfonso, M. deDios–Leyva, L. E. Oliveira. *Phys. Rev.* B 57, 6573 (1998).

306   M. Bruschi, R. G. Campos, E. Pace. *Nuov. Cim.* B 105, 131 (1990).

307   M. Bruschi, F. Calogero. *arXiv:quant-ph*/0403011 (2004).

308   J. Bub. *Interpreting the Quantum World*, Cambridge University Press, Cambridge, 1997.

309   H. Buchholz. *Elektrische und magnetische Potentialfelder*, Springer-Verlag, Berlin, 1957.

310   V. M. Buchstaber, D. V. Leykin, M. V. Pavlov. *Funct. Anal. Appl.* 37, 251 (2003).

311   A. S. Budhiraja, A. J. Sommese. *Preprint MSRI* 2000-008 (2000). ftp://ftp.msri.org/pub/publications/preprints/2000/2000-008/2000-008.ps.gz

312   X. Buff, C. Henriksen. *Nonlinearity* 16, 989 (2003).

313   Y. Bugeaud, J.-P. Conze in M. Planat (ed.). *Noise, Oscillators and Algebraic Randomness*, Springer-Verlag, Berlin, 2000.

314   D. Bump. *Automorphic Forms and Representations*, Cambridge University Press, Cambridge, 1997.

315   E. B. Burger, R. Tubbs. *Making Transcendence Transparent*, Springer-Verlag, New York, 2004.

316   J. Burm, P. Fishback. *Math. Mag.* 74, 47 (2001).

317   A. M. Burns. *Math. Mag.* 67, 258 (1994).

318   W. Bürger. *Physik in unserer Zeit* 14, 140 (1983).

319   R. Burioni, D. Casi, A. Vezzani in V. A. Kaimanovich (ed.). *Random Walks and Geometry*, de Gruyter, Berlin, 2004.

320   R. Burioni, D. Cassi. *J. Phys.* A 38, R45 (2005).

321   A. Burstein, T. Mansour. *Ann. Combinat.* 7, 1 (2003).

322   E. I. Butikov. *Eur. J. Phys.* 26, 157 (2005).

323   J. L. Cabrera, J. G. Milton. *Phys. Rev. Lett.* 89, 158702 (2002).

324   D. Cai, D. W. McLaughlin, K. T. R. McLaughlin in T. Fiedler (ed.). *Handbook of Dynamical Systems*, v. 2, Elsevier, Amsterdam, 2002.

325   M. C. Calkin. *Lagrangian and Hamiltonian Mechanics*, World Scientific, Singapore, 1996.

326   F. Calogero, E. Franco. *Nuov. Cim.* B 89, 161 (1985).

327   F. Calogero. *J. Phys.* A 35, 4249 (2002).

328 F. Calogero, V. I. Inozemtsev. *J. Phys.* A 35, 10365 (2002).

329 F. Calogero, J.-P. Francoise. *Theor. Math. Phys.* 137, 1663 (2003).

330 F. Calogero. *J. Nonlin. Math. Phys.* 11, 208 (2004).

331 C. S. Calude, L. Staiger, K. Svozil *arXiv:nlin. CD*/0304019 (2003).

332 S. L. Campbell, C. D. Meyrer, Jr. *Generalized Inverses of Linear Transformations*, Pitman. London, 1979.

333 R. G. Campos. *Bol. Soc. Mat. Mex.* 3, 279 (1997).

334 R. G. Campos, L. O. Pimentel. *arXiv:quant-ph*/0008120 (2000).

335 R. G. Campos, C. Meneses. *arXiv:math.NA*/0407020 (2004).

336 E. J. Candès, J. Romberg. *arXiv:math.CA*/0411273 (2004).

337 J. R. Cannon, P. DuChateau. *SIAM J. Num. Anal.* 14, 473 (1977).

338 J. Cano in D. L. Tê, K. Saito, B. Teissier (eds.). *Singularity Theory*, World Scientific, Singapore, 1995.

339 J. Cantarella, D. De Turck, H. Gluck. *J. Math. Phys.* 42, 876 (2001).

340 J. Cantarella, D. DeTurck, H. Gluck. *Am. Math. Monthly* 109, 409 (2002).

341 C. Canuto, M. Y. Hussaini, A. Quarteroni, T. A. Zang. *Spectral Methods in Fluid Dynamics*, Springer-Verlag, New York, 1988.

342 F. Cao, L. Moisan. *SIAM J. Num. Anal.* 39, 624, (2001).

343 F. Cao. *Geometric Curve Evolution and Image Processing*, Springer-Verlag, Berlin, 2003.

344 J. P. Carini, J. T. Londergan, D. P. Murdock. *Phys. Rev.* B 55, 9852 (1997).

345 T. Carletti. *mp_arc* 03-248 (2001). http://rene.ma.utexas.edu/mp_arc/c/03/03-248.ps.gz

346 G. F. Carrier, T. T. Wu, H. Yeh. *J. Fluid Mech.* 475, 79 (2003).

347 H. A. Carteret, M. E. H. Ismail, B. Richmond. *arXiv:quant-ph*/0303105 (2003).

348 J. H. E. Cartwright. *arXiv:nlin.PS*/0211001 (2002).

349 R. Carvalho, R. V. Mendes, J. Seixas. *Physica* D 126, 27 (1999).

350 A. R. R. Carvalho, R. L. de Matos Filho, L. Davidovich. *Phys. Rev.* E 70, 026211 (2004).

351 J. R. Cary, R. G. Littlejohn. *Ann. Phys.* 151, 1 (1983).

352 E. Casas-Alvero. *Singularities of Plane Curves*, Cambridge University Press, Cambridge, 2000.

353 G. Castaldi, V. Fiumara, V. Galdi, V. Pierro, I. M. Pinto. *arXiv:nlin.CD*/0208017 (2002).

354 D. Castellanos. *Math. Mag.* 61, 67 (1988).

355 D. Castellanos. *Math. Mag.* 61, 148 (1988).

356 D. P. L. Castrigiano, S. A. Hayes. *Catastrophe Theory*, Westview Press, Boulder, 2004.

357 M. Castrillón, A. Valdés. *Int. J. Comput. Vision* 42, 191 (2001).

358 M. Ceberio, L. Granvilliers. *Computing* 69, 51 (2002).

359 H. D. Ceniceros, F.-R. Tian. *Phys. Lett.* A 306, 25 (2002).

360 J. M. Cerveró, A. Rodríguez. *arXiv:cond-mat*/0206486 (2002).

361 J.-L. Chabert (ed.). *A History of Algorithms*, Springer-Verlag, Berlin, 1999.

362 R. Chacón. *arXiv:nlin.CD*/0406066 (2004).

363 G. D. Chagelishvili, A. G. Tevzadze, G. T. Gogoberidze, J. G. Lominadze. *Phys. Rev. Lett.* 84, 1619 (2000).

364 A. Chakrabarti, S. N. Karmakar, R. K. Moitra. *arXiv:cond-mat*/9407038 (1994).

365 A. Chakraborti, B. K. Chakrabarti. *arXiv:cond-mat*/0004256 (2000).

366 J.-M.-F. Chamayou. *Math. Comput.* 27, 197 (1973).

367 H. H. Chan, W.-C. Liaw. *Pac. J. Math.* 192, 219 (2000).

368 S. Chandrasekhar. *Rev. Mod. Phys.* 15, 1 (1943).

369 T. Chandrasekaran, T. D. Wilkerson. *Phys. Rev.* 181, 329 (1969).

370 S. Chapman. *Am. J. Phys.* 28, 705 (1960).

371 H. Chaté, P. Manneville. *Phys. Rev. Lett.* 58, 112 (1987).

372 H. Chaté, P. Manneville. *Physica* D 32, 409 (1988).

373 H. Chaté. *Nonlinearity* 7, 185 (1994).

374 H. Chaté, A. Pikovsky, O. Rudzick. *Physica* D 131, 17 (1999).

375 A. Chatterjee. *Phys. Rev.* E 59, 5912 (1999).

376 G. M. Chechin, V. P. Sakhnenko. *Physica* D 117, 43 (1998).

377 G. M. Chechin, V. P. Sakhnenko, H. T. Stokes, A. D. Smith, D. M. Hatch. *Int. J. Nonl. Mech.* 35, 497 (2000).

378 G. M. Chechin, N. V. Novikova, A. A. Abramenko. *Physica* D 166, 208 (2002).

379 G. M. Chechin, K. G. Zhukov, D. S. Ryabov. *arXiv:nlin.PS*/0403040 (2004).

380 G. M. Chechin, D. S. Ryabov. *Phys. Rev.* E 69, 036202 (2004).

381 A. Y. Cheer, D. A. Goldston. *Math. Comput.* 55, 129 (1990).

382 C.-K. Chen, S.-H. Ho. *Appl. Math. Comput.* 79, 173 (1996).

383 G. Chen, A. Fulling, J. Zhou. *J. Math. Phys.* 38, 5350 (1997).

384 G. Chen, T. Ueta. *Int. J. Bifurc. Chaos* 9, 1465 (1999).

385 K.-C. Chen. *Ergod. Th. Dynam. Sys.* 23, 1691 (2003).

386 M. Chen, A. E. Kaufman, R. Yagel (eds.). *Volume Graphics*, Springer-Verlag, London, 2000.

387 X. Chen, W. V. Li. *Ann. Prob.* 31, 1052 (2003).

388 A. Chenciner, J. Gerver, R. Montgomery, C. Simó in P. Newton, P. Holmes, A. Weinstein (eds.). *Geometry, Mechanics, and Dynamics*, Springer-Verlag, New York, 2002.

389 H. Cheng, L. Reemgard, V. Rokhlin. *J. Comput. Phys.* 155, 468 (1999).

390 A. A. Chernikov, A. V. Rogalsky. *Chaos* 4, 35 (1994).

391 A. A. Chernikov, R. Z. Sagdeev, D. A. Usikov, G. M. Zaslavsky. *Comput. Math. Appl.* 17, 17 (1989).

392 C. Chicone, B. Mashhoon, D. G. Retzloff. *Class. Quantum Grav.* 16, 507 (1999).

393 C. Chicone. *arXiv:gr-qc*/0110001 (2001).

394 A. M. Childs, E. Farhi, S. Gutmann. *arXiv:quant-ph*/0103020 (2001).

395 A. M. Childs, R. Cleve, E. Deotto, E. Farhi, S. Gutmann, D. A. Spielman. *arXiv:quant-ph*/0209131 (2002).

396 B. Chopard, M. Droz. *Cellular Automata Modeling of Physical Systems*, Cambridge University Press, Cambridge, 1998.

397 D. L. Chopp, J. A. Sethian. *Interfaces Free Boundaries* 1, 1 (1999).

398 P. Chossat, M. Golubitsky. *Physica* D 32, 423 (1988).

399 O. Christensen, K. L. Christensen. *Approximation Theory*, Birkhauser, Boston, 2004.

400 L. O. Chua in J. L. Huertas, W.-K. Chen, R. N. Madan (eds.). *Visions of Nonlinear Science in the 21st Century*, World Scientific, Singapore, 1999.

401 L. O. Chua, V. I. Sbitnev, S. Yoon. *Int. J. Bifurc. Chaos* 13, 2377 (2003).

402 D. V. Chudnovsky, G. V. Chudnovsky. *J. Complexity* 3, 1 (1987).

403 D. V. Chudnovsky, G. V. Chudnovsky. *Proc. Natl. Acad. Sci. USA* 95, 2744 (1998).

404 R. F. Churchhouse, S. T. E. Muir. *J. Inst. Math. Appl.* 5, 318 (1969).

405   B. Cipra. *Science* 265, 473 (1994).

406   M. A. Cirone, J. P. Dahl, M. Fedorov, D. Greenberger, W. P. Schleich. *J. Phys.* B 35, 191 (2002).

407   P. A. Clarkson. *J. Diff. Eq.* 124, 225 (1996).

408   I. Claus, P. Gaspard. *arXiv:cond-mat*/0204264 (2002).

409   J. C. Claussen, J. Nagler, H. G. Schuster. *Phys. Rev.* E 70, 032101 (2004).

410   D. Coca, S. A. Billings. *Phys. Lett.* A 287, 65 (2001).

411   P. Cohen. *Math. Proc. Camb. Phil. Soc.* 95, 389 (1984).

412   E. R. Cohen in L. Crovini, T. J. Quinn (eds.). *Metrology at the Frontiers of Physics and Technology*, North Holland, Amsterdam, 1992.

413   A. M. Cohen. *Num. Math.* 68, 225 (1994).

414   D. Cohen, N. Lepore, E. J. Heller. *arXiv:nlin.CD*/0108014 (2001).

415   D. Cohen, N. Lepore, E. J. Heller. *J. Phys.* A 37, 2139 (2004).

416   E. Cohen, R. F. Riesenfeld, G. Elber. *Geometric Modeling with Splines: An Introduction*, A K Peters, Natick, 2001.

417   P. Cohen in M. Planat (ed.). *Noise, Oscillators and Algebraic Randomness*, Springer-Verlag, Berlin, 2000.

418   H. Cohn. *Math. Comput.* 51, 787 (1988).

419   H. Cohn. *Math. Comput.* 61, 155 (1993).

420   H. Cohn. *Acta Arithm.* 75, 297 (1996).

421   F. Colaiori, A. Baldasarri, C. Castellano. *arXiv:cond-mat*/0402285 (2004).

422   C. Colijn, E. R. Vrscay. *Phys. Lett.* A 300, 334 (2002).

423   C. Colijn, E. R. Vrscay. *J. Phys.* A 36, 4689 (2003).

424   A. Comtet, J. Desbois. *arXiv:cond-mat*/0302269 (2003).

425   J. N. L. Connor, P. R. Curtis. *J. Phys.* A 15, 1179 (1982).

426   J. N. L. Connor, P. R. Curtis, D. Farrelly. *Molec. Phys.* 48, 1305 (1983).

427   J. N. L. Connor, D. Farrelly. *J. Chem. Phys.* 75, 2831 (1981).

428   J. N. L. Connor, C. A. Hobbs. *arXiv:physics*/0411015 (2004).

429   F. Cooper, A. Khare, U. Sukhatme. *Phys. Rep.* 251, 267 (1995).

430   B. J. Copeland. *Theor. Comput. Sc.* 317, 251 (2004).

431   R. Coquereaux, A. Grossmann, B. E. Lautrup. *IMA J. Numer. Anal.* 10, 119 (1990).

432   R. Cori, D. Rossin, B. Salvy. *Preprint INRIA* n 3946 (2000). http://www.inria.fr/RRRT/RR-3946.html

433   R. M. Corless, D. J. Jeffrey. *SIGSAM Bull.* 32, 11 (1998).

434   P. P. Corso, E. Fiordilino, F. Persico. *Phys. Rev.* A 67, 063402 (2003).

435   M. G. Cosenza, K. Tucci. *Phys. Rev.* E 64, 026208 (2001).

436   B. Costa, W. S. Don. *Appl. Num. Math.* 33, 151 (2000).

437   R. Courant, D. Hilbert. *Methods of Mathematical Physics*, v. II, Wiley, New York, 1953.

438   D. A. Cox. *Primes of the Form $x^2 + n\,y^2$*, Wiley, New York, 1989.

439   D. Cox, J. Little, D. O'Shea. *Using Algebraic Geometry*, Springer, New York, 1998.

440   M. Creutz. *Comput. Phys.* 5, 198 (1992).

441   M. Creutz. *arXiv:cond-mat*/0401302 (2004).

442   H. T. Croft, K. J. Falconer, R. K. Guy. *Unsolved Problems in Geometry*, Springer-Verlag, New York, 1991.

443   J. Cserti. *Am. J. Phys.* 68, 896 (2000).

444   J. Cserti, G. Dávid, A. Piróth. *arXiv:cond-mat*/0107362 (2001).

445   J. Cserti, G. Dávid, A. Piróth. *Am. J. Phys.* 70, 153 (2002).

446   J. Cserti, G. Tichy. *arXiv:cond-mat*/0307616 (2003).

447   E. Cuansing, H. Nakanishi. *arXiv:cond-mat*/0008294 (2000).

448   J. Daboul, M. Hickman. *Phys. Lett.* A 267, 232 (2000).

449   H. Daido. *Progr. Theor. Phys.* 102, 197 (1999).

450   K. Dajani, C. Kraaikamp. *Trans. Am. Math. Soc.* 351, 2055 (1999).

451   K. Dajani, C. Kraaikamp. *J. Th. Nombres Bordeaux* 8, 331 (1996).

452   K. Dajani, C. Kraaikamp. *Colloq. Math.* 84/85, 109 (2000).

453   R. H. Dalling, M. E. Goggin. *Am. J. Phys.* 62, 563 (1994).

454   H. Dankowicz. *ZAMM* 79, 399 (1999).

455   S. Datta, R. Ramaswamy, A. Prasad. *Phys. Rev.* E 70, 046203 (2004).

456   T. Dauxois, P. Holdsworth, S. Ruffo. *Eur. J. Phys.* B 16, 659 (2000).

457   E. B. Davies. *Brit. J. Phil. Sci.* 52, 671 (2001).

458   P. J. Davis. *Am. Math. Monthly* 79, 252 (1972).

459   P. J. Davis, P. Rabinowitz. *Methods of Numerical Integration*, Academic Press, Orlando, 1984.

460   M. Davison, A. Doeschl. *SIAM Rev.* 46, 115 (2004).

461   O. F. de Alcantara Bonfim, J. Florencio, F. C. Sá Barreto. *Phys. Lett.* A 277, 129 (2000).

462   L. Debnath. *Integral Transforms and their Applications*, CRC Press, Boca Raton, 1995.

463   C. A. A. de Carvalho, H. M. Nussenzweig. *Phys. Rep.* 364, 83 (2002).

464   W. Decker, T. de Jong in B. Buchberger, F. Winkler (eds.). *Gröbner Bases and Applications,* Cambridge University Press, Cambridge, 1998.

465   L. H. de Figueiredo, J. Stolfi. *Num. Alg.* 37, 147 (2004).

466   M. de Gosson. *J. Phys.* A 34, 10085 (2001).

467   M. de Gosson. *arXiv:math-ph*/0207034 (2002).

468   P. de Groen. *Nieuw Archief Wiskunde* 14, 237 (1996).

469   G. Gzyl. *Diffusions and Waves*, Kluwer, Dordrecht, 2002.

470   J. P. Delahaye, B. Germanin-Bonne. *SIAM J. Num. Anal.* 19, 840 (1982).

471   J.-P. Delahaye. *Sequence Transformations*, Springer-Verlag, Berlin, 1988.

472   R. de la Llave in K. Meyer, D. Schmidt (eds.). *Computer Aided Proofs in Analysis*, Springer-Verlag, New York, 1991.

473   A. C. de la Torre, D. Goyeneche. *arXiv:quant-ph*/0205159 (2002).

474   M. Dellnitz, A. Hohmann in H. W. Broer, S. A. van Gils, I. Hoveijn, F. Takens (eds.). *Nonlinear Dynamical Systems and Chaos*, Birkhäuser, Basel, 1996.

475   M. Dellnitz, O. Junge in B. Fiedler, G. Iooss and N. Kopell (eds.). *Handbook of Dynamical Systems III: Towards Applications*, World Scientific, Singapore, 2001.

476   M. Dellnitz, G. Froyland, O. Junge in B. Fiedler (ed.). *Ergodic Theory, Analysis, and Efficient Simulation of Dynamical Systems*, Springer-Verlag, Berlin, 2001.

477   M. Dellnitz, O. Junge in T. Fiedler (ed.). *Handbook of Dynamical Systems*, v. 2, Elsevier, Amsterdam, 2002.

478   M. del Mar Espinosa, M. A. Martín-Delgado, A. Niella, D. Páramo, J. Rodríguez-Laguna. *arXiv:chao-dyn*/9808012 (1998).

479   R. A. De Millo, R. J. Lipton, A. J. Perlis. *Commun. ACM* 22, 271 (1979).

480   C. R. de Oliveira, M. V. Lima. *Rep. Math. Phys.* 45, 431 (2000).

481   J. W. Demmel. *Applied Numerical Linear Algebra*, SIAM, Philadelphia, 1997.

482   J. Demmel, Y. Hida. *SIAM J. Sci. Comput.* 25, 1214 (2003).

483   B. Denardo, J. Earwood, V. Sazonova. *Am. J. Phys.* 67, 981 (1999).

484   Z.-T. Deng, H. J. Caulfield, M. Schamschula. *Opt. Lett.* 21, 1430 (1996).

485   M. R. Dennis. *New J. Phys.* 5, a134 (2003).

486   J. Denschlag, D. Cassettari, J. Schmiedmayer. *Phys. Rev. Lett.* 82, 2014 (1999).

487   C. R. de Oliveira, G. Q. Pellegrino. *arXiv:cond-mat*/0103313 (2001).

488   J. A. de Sales, J. Florencio. *Phys. Rev.* E 67, 016216 (2003).

489   J.-A. Désidéri. *Preprint INRIA* n 4561 (2002). http://www.inria.fr/RRRT/RR-4561.html

490   P. Deuflhard, F. Bornemann. *Scientific Computing with Ordinary Differential Equations*, Springer-Verlag, Berlin, 2002.

491   J. M. Deutch, F. E. Low. *Ann. Phys.* 228, 184 (1993).

492   B. d'Espagnat. *Veiled Reality*, Addison-Wesley, Reading, 1985.

493   P. Deuflhard, A. Hohmann. *Numerische Mathematik I*, de Gruyter, Berlin, 1993.

494   J. I. Deutsch. *Comput. Math. Appl.* 42, 1293 (2001).

495   R. L. Devaney in B. Aulbach, S. Elaydi, G. Ladas (eds.). *Proceedings of the Sixth integrnational Conference on Difference Equations*, CRC Press, Boca Raton, 2004.

496   L. Devroye, L. Györfi. *Nonparametric Density Estimation*, Wiley, New York, 1985.

497   L. Devroye. *Non-Uniform Random Variate Generation*, Springer-Verlag, New York, 1986

498   A. K. Dewdney. *The Planiverse*, Poseidon Press, New York, 1984.

499   F. De Zela. *arXiv:physics*/0406037 (2004).

500   A. De Wit, G. Dewel, P. Borckmans. *Phys. Rev.* E 48, R4191 (1993).

501   A. Dhar. *Am. J. Phys.* 61, 58 (1993).

502   P. Diaconis. *Bull. Am. Math. Soc.* 40, 155 (2003).

503   P. Diamond, P. Kloeden, A. Pokrovski, A. Vladimirov. *Physica* D 86, 559 (1995).

504   B. W. Dickinson. *IEEE Trans. Acoust., Speech, Signal Proc.* 30, 25 (1982).

505   G. Diener. *Phys. Lett.* A 223, 327 (1996).

506   I. Diener in R. Horst, P. M. Pardalos (eds.). *Handbook of Global Optimization*, Kluwer, Amsterdam, 1995.

507   T. Digernes, E. Husstad, V. S. Varadarajan in S. C. Lim, R. Abd–Shukor, H. K. Kwek (eds.). *Frontiers in Quantum Physics*, World Scientific, Singapore, 1998.

508   T. Digernes, V. S. Varadarajan, R. S. Varadhan. *Rev. Math. Phys.* 6, 621 (1994).

509   P. Dittrich, F. Liljeros, A. Soulier, W. Banzhaf. *Phys. Rev. Lett.* 84, 3205 (2000).

510   M. Di Ventra, S. T. Pantelides. *Phys. Rev.* B 61, 16207 (2000).

511   M. A. Doncheski, R. W. Robinett. *Eur. J. Phys.* 20, 29 (1999).

512   S.-H. Dong, Z.-Q. Ma, M. Klaus. *arXiv:quant-ph*/9903016 (1999).

513   D. L. Donoho, P. B. Stark. *SIAM J. Appl. Math.* 49, 906 (1989).

514   J. S. Dowker, K. Kirsten. *arXiv:hep-th*/0205029 (2002).

515   J. S. Dowker, K. Kirsten. *arXiv:hep-th*/0301143 (2003).

516   A. Drăgulescu, V. M. Yakovenko. *arXiv:cond-mat*/0001432 (2000).

517   P. G. Drazin. *Nonlinear Physics*, Cambridge University Press, Cambridge, 1992.

518   S. Drobot *Real Numbers*, Prentice-Hall, Englewood Cliffs, 1964.

519   B. Drossel. *arXiv:cond-mat*/0205658 (2002).

520   M. Drton, U. Schwingenschlögl. *MPS: Applied mathematics*/0306018 (2003).
       http://www.mathpreprints.com/math/Preprint/Drton/20020520/1/mdus.preprint.pdf

521   F. Dunlop, T. Huillet. *arXiv:math.PR*/0211046 (2002).

522   D. Dürr, S. Goldstein, N. Zanghì. *arXiv:quant-ph*/9511016 (1995).

523   D. Dürr. *Bohmsche Mechanik als Grundlage der Quantenmechanik*, Springer-Verlag, Berlin, 2001.

524   A. Dutt, V. Rokhlin. *SIAM J. Sci. Comput.* 14, 1368 (1993).

525   D. Duval. *Compositio Math.* 70, 119 (1989).

526   W. M. B. Duval. *Chaos* 14, 716 (2004).

527   W. Dyck (ed.). *Katalog mathematischer und mathematisch physikalischer Modelle, Apparate und Instrumente*, Georg Olms Verlag, Heidelberg, 1994.

528   W. T. Eade, D. Drijard, F. E. James, M. Roos, B. Sadoulet. *Statistical Methods in Experimental Physics*, North Holland, Amsterdam, 1971.

529   J. Earman, J. D. Norton in A. Morton, S. P. Stich (eds.). *Benacerraf and his Critics*, Blackwell, Oxford, 1996.

530   W. Ebeling, T. Pöschel. *Europhys. Lett.* 26, 241 (1994).

531   D. S. Ebert, F. K. Musgrave, D. Peachey, K. Perlin, S. Worley. *Texturing and Modeling*, Academic Press, San Diego, 1998.

532   F. E. Eckhardt. *Math. Annalen* 10, 227 (1876).

533   W. L. Edge. *Proc. Cambr. Phil. Soc.* 34, 6 (1938).

534   A. Eftekhari. *arXiv:cs.CL*/0408041 (2004).

535   R. B. Eggleton. *Electr. J. Combinat.* 8, n2, R6 (2001).

536   S. Ehrich. *SIAM J. Num. Anal.* 35, 78 (1998).

537   R. Eichhorn, S. J. Linz, P. Hänggi. *Phys. Rev. E* 58, 7151 (1998).

538   R. Eichhorn, S. J. Linz, P. Hänggi. *Chaos, Solitons, Fractals* 13, 1 (2001).

539   S. N. Elaydi. *Discrete Chaos*, Chapman & Hall, Boca Raton, 2000.

540   G. Elekes. *Acta Arithm.* 81, 365 (1997).

541   G. Elekes, I. Z. Ruzsa. *Studia Scient. Math. Hungarica* 40, 301 (2003).

542   V. M. Eleonskiĭ, V. G. Korolev, N. E. Kulagin. *JETP Lett.* 76, 728 (2002).

543   V. M. Eleonskii, V. G. Korolev, N. E. Kulagin. *Theor. Math. Phys.* 136, 1131 (2003).

544   D. Elliott. *Int. J. Math. Edu. Sci. Technol.* 10, 1 (1979).

545   S. Elizalde. *arXiv:math.CO*/0212221 (2002).

546   F.-J. Elmer. *J. Phys. A* 30, 6057 (1997).

547   F.-J. Elmer. *arXiv:chao-dyn*/9711024 (1997).

548   M. El–Mikkawy. *Appl. Math. Comput.* 138, 403 (2003).

549   M. El-Mikkawy. *Appl. Math. Comput.* 139, 503 (2003).

550   M. E. A. El–Mikkawy. *J. Comput. Appl. Math.* 166, 581 (2004).

551   H. F. El-Nashar, Y. Zhang, H. A. Cerdeira, I. A. Fuwape. *Chaos* 13, 1216 (2003).

552   C. Elsner, J. W. Sander, J. Steuding. *Math. Slov.* 51, 281 (2001).

553   U. Elsner, V. Mehrmann, F. Milde, R. A. Römer, M. Schreiber. *arXiv: physics*/9802009 (1998).

554   J. Elsrodt. *Math. Semesterber.* 41, 167 (1994).

555 J. D. Emerson, D. C. Hoaglin in D. C. Hoaglin, F. Mosteller, J. W. Tukey (eds.). *Understanding Robust and Exploratory Data Analysis*, Wiley, New York, 1983.

556 M. Encinosa, F. Sales–Mayor. *arXiv:quant-ph*/0204047 (2003).

557 H. Engels. *Numerical Quadrature and Cubature*, Academic Press, New York, 1980.

558 D. Eppstein. *Mathematica Edu. Res.* 4, n2, 5 (1995).

559 M. Erdmann. *Int. J. Robotics* 13, 240 (1994).

560 S. Erkoc, H. Oymak. *Phys. Lett.* A 290, 28 (2001).

561 T. Ernst. *J. Nonlin. Math. Phys.* 10, 487 (2003).

562 O. Ersoy. *Fourier-Related Transforms, Fast Algorithms and Applications*, Prentice-Hall, Upper Saddle River, 1997.

563 T. O. Espfield. *SIAM Rev.* 37, 603 (1995).

564 O. Espinosa, J. Gamboa, S. Lepe, F. Méndez. *Phys. Lett.* B 520, 421 (2001).

565 H. Essén. *arXiv:physics*/0308007(2003).

566 G. Evans. *Practical Numerical Integration*, Wiley, Chichester, 1993.

567 K. M. Evans. *Discr. Math. Theor. Comput. Sci.* Proc. AA DM–CCG 177 (2001).
http://www.emis.ams.org/journals/DMTCS/proceedings/html/dmAA0113.abs.html

568 K. M. Evans. *Physica* D 183, 45 (2003).

569 L. C. Evans. *Partial Differential Equations*, American Mathematical Society, Providence, 1998.

570 N. W. Evans, M. A. Berger in H. K. Moffatt, G. M. Zaslavsky, P. Comte, M. Tabor (eds.) *Topological Aspects of the Dynamics of Fluids and Plasmas*, Kluwer, Dordrecht, 1987.

571 C. J. Everett, E. D. Cashwell in G.-C. Rota (ed.). *Science and Computers*, Academic Press, Orlando, 1986.

572 S. P. Eveson, C. J. Fewster, R. Verch. *arXiv:math-ph*/0312046 (2003).

573 J. Ewing, C. Foias in C. S. Calude, G. Păun (eds.). *Finite Versus Infinite*, Springer-Verlag, London, 2000.

574 N. Fagella. *J. Math. Anal. Appl.* 229, 1 (1999).

575 N. Fagella, A. Garijo. *arXiv:math.DS*/0207103 (2002).

576 C. Faivre. *Acta Arithm.* 82, 119 (1997).

577 C. Faivre. *Arch. Math.* 70, 455 (1998).

578 P. E. Falloon, J. B. Wang. *Comput. Phys. Commun.* 134, 167 (2001).

579 P. Falsaperla, G. Fonte. *Phys. Lett.* A 316, 382 (2003).

580 H. Fangohr, A. R. Price, S. J. Cox, P. A. J. de Groot, G. J. Daniell. *arXiv:physics*/0004013 (2000).

581 J. Farago, C. van den Broeck. *Europhys. Lett.* 54, 411 (2001).

582 H. M. Farkas, I. Kra. *Riemann Surface*, Springer-Verlag, New York, 1992.

583 F. Farrelly, A. Petri. *Int. J. Mod. Phys.* C 9, 927 (1998).

584 J.-M. Farto. *J. Pure Appl. Alg.* 108, 203 (1996).

585 A. Fdil. *Appl. Num. Math.* 25, 21 (1997).

586 G. J. Fee, M. B. Monagan. *SIGSAM Bull.* 31, n1, 22 (1997).

587 B. F. Feeny, R. Kappagantu. *J. Sound Vibr.* 211, 607 (1998).

588 C. Fefferman. *Commun. Pure Appl. Math.* 39, S67 (1986).

589 C. L. Fefferman, L. A. Seco in R. B. Kearfott, V. Kreinovich (eds.). *Applications of Interval Computations*, Kluwer, Amsterdam, 1996.

590 R. Feinerman, D. J. Newman. *Mich. Math. J.* 15, 305 (1968).

591 D. Felbacq, A. Moreau. *J. Opt.* A 5, L9 (1991).

592   J. F. Fernández, J. Rivero. *Nuov. Cim.* 109 B, 1135 (1994).

593   F. M. Fernández, R. Guardiola, J. Ros. *Comput. Phys. Commun.* 115, 170 (1998).

594   J. F. Fernández, C. Criado. *arXiv:cond-mat*/9901202 (1999).

595   J. F. Fernández, C. Criado. *Phys. Rev.* E 60, 3361 (1999).

596   J. F. Fernández, C. Criado. *Phys. Rev.* E 63, 058702 (2001).

597   D. L. Ferrario. *arXiv:math.DS*/0208188 (2002).

598   D. L. Ferrario, S. Terracini. *arXiv:math-ph*/0302022 (2003).

599   T. Fevens, H. Jiang. *SIAM J. Sci. Comput.* 21, 255 (2000).

600   R. P. Feynman. *Phys. Rev.* 56, 340 (1939).

601   A. Fiasconaro, D. Valenti, B. Spagnolo. *arXiv:cond-mat*/0403104 (2004).

602   M. Field in B. Fiedler, K. Gröger, J. Sprekels (eds.). *International Conference on Differential Equations* v.1, World Scientific, Singapore, 2000.

603   J. A. Fill, P. Flajolet, N. Kapur. *arXiv:math.CO*/0306225 (2003).

604   G. Fischer (ed.). *Mathematical Models: Commentary*, Vieweg, Braunschweig, 1986.

605   G. Fischer. *Ebene algebraische Kurven*, Vieweg, Wiesbaden, 1994.

606   P. Flajolet, R. Schott. *Eur. J. Combinat.* 11, 421 (1990).

607   P. Flajolet, R. Sedgewick. *Preprint INRIA* n 4103 (2001). http://www.inria.fr/RRRT/RR-4103.html

608   P. Flajolet, R. Sedgewick. *Analytic Combinatorics—Symbolic Combinatorics* (2002). http://pauillac.inria.fr/algo/flajolet/Publications/FlSe02.ps

609   P. Flajolet, W. Szpankowski, B. Vallée. *Preprint INRIA* (2002). http://pauillac.inria.fr/algo/flajolet/Publications/FlSzVa02.ps.gz

610   H. Flanders. *IEEE Trans. Circuit Th.* 18, 326 (1971).

611   H. Flanders. *J. Math. Anal. Appl.* 40, 30 (1972).

612   L. Flatto, A. M. Odlyzko, D. B. Wales. *Ann. Prob.* 13, 154 (1885).

613   R. Fleischmann, T. Geisel, R. Ketmerick. *Phys. Rev. Lett.* 68, 1367 (1992).

614   N. H. Fletcher. *Am. J. Phys.* 70, 1205 (2002).

615   S. Flügge, H. Marshall. *Practical Quantum Mechanics*, Springer-Verlag, Berlin, 1974.

616   A. Fodgen. *Coll. de Phys.* C7, n23, 149 (1990).

617   A. Fogdon, S. T. Hyde. *Acta Cryst.* A 48, 575 (1992).

618   G. B. Folland. *Introduction to Partial Differential Equations*, Princeton University Press, Princeton, 1995.

619   S. K. Foong. *Phys. Rev.* A 46, R707 (1992).

620   G. W. Forbes, M. A. Alonso. *Am. J. Phys.* 69, 1091 (2001).

621   G. W. Forbes, M. A. Alonso, A. E. Siegman. *J. Phys.* A 36, 7027 (2003).

622   B. Fornberg. *A Practical Guide to Pseudospectral Methods*, Cambridge University Press, Cambridge, 1996.

623   E. Fortuna, P. Gianni, P. Parenti, C. Traverso. *J. Symb. Comput.* 36, 343 (2003).

624   S. Foteinopoulou, E. N. Economou, C. M. Soukoulis. *Phys. Rev. Lett.* 90, 107402 (2003).

625   R. Fox, C. G. Kuper, S. G. Lipson. *Proc. R. Soc. Lond.* A 316, 515 (1970).

626   R. F. Fox, M. H. Choi. *Phys. Rev.* A 61, 032107 (2000).

627   R. F. Fox, M. H. Choi. *Phys. Rev.* A 64, 042104 (2001).

628   M. France. *Coll. Math. J. Bolyai.* 13, 183 (1974).

629   M. M. France in P. Turán (ed.). *Topics in Number Theory*, North Holland, Amsterdam, 1976.

630   H. W. Franke, H. S. Helbig. *Leonardo* 25, 291 (1992).

631   H. W. Franke in A. Dress, G. Jäger (eds.). *Visualisierung in der Mathematik, Technik und Kunst*, Vieweg, Braunschweig, 1999.

632   M. Frasca. *arXiv:quant-ph*/0202067 (2002).

633   D. Freedman, P. Diaconis. *Z. Wahrscheinlichkeitsth. verw. Geb.* 57, 453 (1981).

634   T. E. Freeman. *Am. J. Phys.* 63, 273 (1995).

635   G. Frenkel, M. Schwartz. *arXiv:cond-mat*/0112195 (2001).

636   E. Frey, K. Kroy. *Ann. Physik* 14, 20 (2005).

637   J. Friedel. *Phil. Mag.* 43, 153 (1952).

638   J. H. Friedman, J. L. Bentley, R. A. Finkel. *ACM Trans. Math. Softw.* 3, 209 (1977).

639   G. Friesecke, F. Theil. *J. Nonlin. Sci.* 12, 445 (2002).

640   U. Frisch, J. Bec. *arXiv:nlin.CD*/0012033 (2000).

641   U. Fuchs. *Periodische Minimalflächen über regulären Riemannschen Flächen*, Logos, Berlin, 1997.

642   T. Fujiwara in Z. M. Stadnik (ed.). *Physical Properties of Quasicrystals*, Springer-Verlag, Berlin, 1999.

643   S. Fukao, T. Tsuda. *J. Plasma Phys.* 9, 409 (1973).

644   S. A. Fulling, K. S. Güntürk. *Am. J. Phys.* 71, 55 (2002).

645   M. Fulmek. *Adv. Appl. Math.* 30, 607 (2003).

646   Y. Funato, J. Makino, P. Hut, E. Kokubo, D. Kinoshita. *Nature* 427, 518 (2004).

647   N. Furuichi. *Progr. Theor. Phys.* 105, 525 (2001).

648   S. Furuta. *Phys. Rev.* A 64, 042110 (2001).

649   J. Galambos. *Quart. J. Math. Oxford* 23, 147 (1972).

650   J. Galambos. *Representations of Real Numbers by Infinite Series*, Springer-Verlag, Berlin, 1976.

651   A. Galántai. *J. Comput. Appl. Math.* 124, 25 (2000).

652   J. A. C. Gallas. *Phys. Rev.* E 48, R4156 (1993).

653   J. A. C. Gallas. *Europhys. Lett.* 47, 649 (1999).

654   G. Galperin. *Reg. Chaotic Dynam.* 8, 375 (2003).

655   A. Gálvez, J. Puig–Pey, A. Igelesias in A. Laganá, M. L. Gavrilova, V. Kumar, Y. Mun, C. J. K. Tan, O. Gervasi (eds.). *Computational Science and Its Applications* v2, Springer-Verlag, Berlin, 2004.

656   A. Gammal, A. M. Kamchatnov. *arXiv:cond-mat*/0402554 (2004).

657   C. Ganatsiou. *Int. J. Math. Math. Sci.* 28, 367 (2001).

658   W. Gander, G. H. Golub, R. Strebel. *BIT* 34, 556 (1994).

659   W. Gander, D. Gruntz. *ETH Technical Report 305* (1998).
       ftp://ftp.inf.ethz.ch/pub/publications/tech-reports/3xx/305.abstract

660   F. Garciá–Moliner, V. R. Velasco. *Theory of Single and Multiple Interfaces*, World Scientific, Singapore, 1992.

661   A. Garriga, J. Kurchan, F. Ritort. *J. Stat. Phys.* 106, 109 (2002).

662   T. Garrity. *arXiv:math.NT*/9906016 (1999).

663   C. Gasquet, P. Witomski. *Fourier Analysis with Applications*, Springer, New York, 1999.

664   O. Gat, J. E. Avron. *arXiv:cond-mat*/0212647 (2002).

665   W. Gautschi in G. V. Milovanović (ed.). *Numerical Methods and Approximation Theory*, Niš, 1987.

666   N. Gavrilov, A. Shilnikov. *Am. Math. Soc. Transl.* 200, 99 (2000).

667   P. Gawronski, K. Kulakowski. *Int. J. Mod. Phys.* C 11, 247 (2000).

668   H. Geiger, G. Obermaier, C. Helm. *arXiv:quant-ph*/9905068 (1999).

669   H. Geiger, G. Obermair, C. Helm. *arXiv:quant-ph*/9906082 (1999).

670   T. Geisel in M. F. Shlesinger, G. M. Zaslavsky, U. Frisch (eds.). *Lévy Flights and related Topics in Physics*, Springer-Verlag, Berlin, 1995.

671   I. M. Gel'fand, M. M. Kapranov, A. V. Zelevinsky. *Resultants and Multi-Dimensional Determinants*, Birkhäuser, Basel, 1994.

672   J. E. Gentle. *Random Number Generation and Monte Carlo Methods*, Springer-Verlag, New York, 1998.

673   G. J. L. Gerhardt, M. Frichembruder, R. B. Rizzato, S. R. Lopes. *Chaos, Solitons, Fractals* 13, 1269 (2002).

674   R. R. Gerhardts in H. D. Doebner, W. Scherer, F. Schroeck, Jr. (eds.). *Classical and Quantum Systems*, World Scientific, Singapore, 1993.

675   R. R. Gerhardts, S. D. M. Zwerschke. *arXiv:cond-mat*/0105603 (2001).

676   A. Gersten. *Ann. Phys.* 262, 47 (1998).

677   J. L. Gerver. *Math. Comput.* 40, 667 (1983).

678   J. L. Gerver. *Exper. Math.* 12, 187 (2003).

679   V. A. Geyler, I. Y. Popov, A. V. Popov, A. A. Ovechkina. *Chaos, Solitons, Fractals* 11, 281 (2000).

680   G. Giacomelli, R. Hegger, A. Politi, M. Vassalli. *Phys. Rev. Lett.* 85, 3616 (2000).

681   D. Gieseker, H. Knörrer, E. Trubowitz. *The Geometry of Algebraic Fermi Curves*, Academic Press, Boston, 1992.

682   R. P. Gilbert. *Appl. Anal.* 47, 103 (1992).

683   T. Gilbert, J. R. Dorfman. *arXiv:nlin.CD*/0003012 (2000).

684   T. Gilbert, J. R. Dorfman, P. Gaspard. *arXiv:nlin.CD*/0007008 (2000).

685   P. E. Gill, W. Murray, M. H. Wright. *Practical Optimization*, Academic Press, London, 1981.

686   M. Gilli, T. Roska, L. O. Chua, P. P. Civalleri. *Int. J. Bifurc. Chaos* 12, 2051 (2002).

687   R. Gilmore, M. Lefranc. *The Topology of Chaos*, Wiley, New York, 2002.

688   C. A. Ginsburg, E. W. Montroll. *J. Math. Phys.* 19, 336 (1978).

689   M. Giona, A. Adrover, F. J. Muzzio, S. Cerbelli, M. M. Alvarez. *Physica* D 132, 298 (1999).

690   S. Giusepponi, F. Marchesoni. *Europhys. Lett.* 64, 36 (2003).

691   P. Ghose, A. S. Majumdar, S. Guha, J. Sau. *arXiv:quant-ph*/0102071 (2001).

692   P. Ghose, A. S. Majumdar, S. Guha, J. Sau. *Phys. Lett.* A 290, 205 (2001).

693   S. Ghosh. *arXiv:hep-th*/9708045 (1997).

694   G. M. L. Gladwell, H. Zhu. *ZAMM* 83, 275 (2003).

695   C. Godrèche, J. M. Luck, A. Janner, T. Janssen. *J. Physique* I 3, 1921 (1993).

696   J. S. Golan. *Foundations of Linear Algebra*, Kluwer, Dordrecht, 1995.

697   D. Goldberg. *ACM Comput. Surv.* 23, 1, March 1991.

698   A. S. Goldhaber. *arXiv:cond-mat*/9806008 (1998).

699   R. Goldman in R. Goldman, R. Krasauskas (eds.). *Topics in Algebraic Geometry and Geometric Modeling*, American Mathematical Society, Providence, 2003.

700   D. Goldstein, D. Moews. *Aequ. Math.* 65, 3 (2003).

701   F. J. Gomez, J. Sesma. *arXiv:quant-ph*/0503087 (2005).

702   J. Goodman, T. Hou, E. Tadmor. *Num. Math.* 67, 93 (1994).

703   F. Gori, D. Ambrosini, R. Borghi, V. Mussi, M. Santarsiero. *Eur. J. Phys.* 22, 53 (2001).

704   A. Goryachev, R. Kapral. *arXiv:chao-dyn*/9901024 (1999).

705   R. W. Gosper in F. Garvan, M. Ismail (eds.). *Symbolic Computation, Number Theory, Special Functions, Physics and Combinatorics*, Kluwer, Dordrecht, 2001.

706   P. K. Ghosh, P. K. Jain. *IEEE Comput. Graphics Appl.* n9, 50 (1993).

707   R. Ghrist. *Lett. Math. Phys.* 55, 193 (2001).

708   M. Glück, A. Kolovsky, H. J. Korsch. *Phys. Rev.* E 58, 6835 (1998).

709   M. Goldberg. *Scripta Math.* 21, 253 (1955).

710   D. Goldstein, D. Moews. *arXiv:math.CO*/0010066 (2000).

711   S. Goldstein. *arXiv:quant-ph*/9901005 (1999).

712   S. Golin. *J. Phys.* A 22, 4573 (1989).

713   A. A. Golovin. *Phys. Rev.* E 67, 056202 (2003).

714   X. Gómez-Mont in S. Sertöz (ed.). *Algebraic Geometry*, Marcel Dekker, New York, 1997.

715   L. Gonzáles-Vega, G. Trujillo. *Revista Math.* 10 Suppl., 119 (1997).

716   E. A. Gonzáles-Velasco. *Fourier Analysis and Boundary Value Problems*, Academic Press, San Diego, 1995.

717   A. Gopinathan, T. A. Witten, S. C. Venkataramani. *Phys. Rev.* E 65, 036613 (2002).

718   F. Gori in J. C. Dainty (ed.). *Current Trends in Optics*, Academic Press, London, 1994.

719   H. P. W. Gottlieb. *Am. J. Phys.* 66, 903 (1998).

720   H. P. W. Gottlieb. *J. Sound Vibr.* 271, 671 (2004).

721   J. L. Goupy. *Methods for Experimental Design*, Elsevier, Amsterdam, 1993.

722   C. Goutis, G. Casella. *Am. Statist.* 53, 216 (1999).

723   S. Goyal, A. Ruina, J. Papadopoulos. *Wear* 143, 307 (1991).

724   S. Goyal, A. Ruina, J. Papadopoulos. *Wear* 143, 331 (1991).

725   I. Grabec, S. Mandelj. *Progr. Theor. Phys.* S 150, 81 (2003).

726   N. Graham, R. L. Jaffe, M. Quandt, H. Weigel. *Ann. Phys.* 293, 240 (2001).

727   C. Grama, N. Grama, I. Zamfirescu. *Phys. Rev.* A 61, 032716 (2000).

728   P. Grasberger. *Phys. Rev.* E 49, 2436 (1994).

729   M. A. Grayson. *J. Diff. Geom.* 26, 285 (1987).

730   M. W. Green, A. J. Korsak, M. C. Pease. *SIAM Rev.* 18, 501 (1976).

731   A. A. Greenenko, A. V. Chechkin, N. F. Shul'ga. *Phys. Lett.* A 324, 82 (2004).

732   D. Greenspan. *n-Body Problems and Models*, World Scientific, Singapore, 2004.

733   R. T. Gregory, D. L. Karney. *A Collection of Matrices for testing Computational Algorithms*, Wiley, Chichester, 1969.

734   D. J. Griffith, T. A. Abbott. *Am. J. Phys.* 60, 951 (1992).

735   M. Grinfeld, P. A. Knight, H. Lamba. *J. Phys.* A 29, 8035 (1996).

736   M. I. Grivich, D. P. Jackson. *Am. J. Phys.* 68, 469 (2000).

737   D. Gronau. *Aequ. Math.* 65, 306 (2003).

738   D. Gronau. *Am. Math. Monthly* 111, 230 (2004).

739   D. Gronau. *Aequ. Math.* 68, 230 (2004).

740   B. H. Gross, D. B. Zagier. *J. reine angew. Math.* 355, 191 (1985).

741   F. Großmann, J.-M. Rost, W. P. Schleich. *J. Phys.* A 30, L277 (1997).

742   M. J. Grote. *Int. J. Numer. Model.* 13, 397 (2000).

743 J. Grotendorst in C. Brezinski, U. Kulisch (eds.). *Computational and Applied Mathematics* v.1, North Holland, Amsterdam, 1992.

744 A. Grünbaum in W. C. Salmon (ed.). *Zeno's Paradoxes*, Bobbs–Merrill, Indianapolis, 1970.

745 F. A. Grünbaum. *Appl. Comput. Harmon. Anal.* 15, 163 (2003).

746 H. Grussbach, M. Schreiber. *Phys. Rev.* B 48, 6650 (1993).

747 J. Guckenheimer, A. Vladimirsky. *SIAM J. Appl. Dynam. Syst.* 3, 232 (2004).

748 J. P. Guillement, B. Helffer, P. Treton. *J. Phys. France* 50, 2019 (1989).

749 C. Guilpin, J. Gacougnolle, Y. Simon. *Appl. Num. Math.* 48, 27 (2004).

750 G. Gumbs, P. Kekete. *Phys. Rev.* B 56, 3787 (1994).

751 S. D. Gupta, X. She. *J. Indian Math. Soc.* 65, 197 (2000).

752 S. D. Gupta, X. She. *Adv. Stud. Contemp. Math.* 4, 43 (2001).

753 T. Gustafson, T. Abe. *Math. Intell.* 20, n1, 63 (1998).

754 B. Gutkin. *arXiv:nlin.CD*/0301031 (2003).

755 M. W. Gutowski. *arXiv:math.SC*/0108163 (2001).

756 M. W. Gutowski. *arXiv:physics*/0302034 (2003).

757 M. C. Gutzwiller. *Rev. Mod. Phys.* 70, 589 (1998).

758 C. J. Guu. *Discr. Math.* 213, 163 (2000).

759 M. Ha, M. den Nijs. *Phys. Rev.* E 66, 036118 (2002).

760 G. Hackenbroich, F. von Oppen. *Europhys. Lett.* 29, 151 (1995).

761 B. Hafskjøld, S. Ratke in J. S. Shiner (ed.). *Entropy and Entropy Generation*, Kluwer, Dordrecht, 1996..

762 K. Hagenbuch. *Am. J. Phys.* 45, 693 (1977).

763 Y. Hagihara. *Celestial Mechanics*, v.1, MIT Press, Cambridge, 1970.

764 Y. Hagihara. *Celestial Mechanics*, v.IV/1, Kojimachi, Tokyo, 1975.

765 F. A. Haight, M. A. Breuer. *Biometrika* 47, 143 (1960).

766 J. M. Haile. *Molecular Dynamics Simulations*, Wiley, New York, 1992.

767 T. Hakioglu. *arXiv:quant-ph*/9809074 (1998).

768 O. H. Hald, J. R. McLaughlin. *Inverse Nodal Problems: Finding the Potential from Nodal Lines*, American Mathematical Society, Providence, 1995.

769 M. Halibard, I. Kanter. *Physica* A 249, 525 (1998).

770 M. Hall. *Ann. Math.* 48, 966 (1947).

771 M. J. W. Hall, M. S. Reineker, W. P. Schleich. *J. Phys.* A 32, 8275 (1999).

772 G. Haller, G. Yuan. *Physica* D 147, 352 (2000).

773 F. R. Halpern. *J. Math. Phys.* 14, 219 (1973).

774 J. Hamilton. *Aharonov-Bohm and other Cyclic Phenomena*, Springer-Verlag, Berlin, 1997.

775 J. D. Hamkins. *arXiv:math.LO*/0212049 (2002).

776 S. M. Hammel, C. K. R. T. Jones, J. V. Moloney. *J. Opt. Soc. Am.* B , 552 (1985).

777 R. W. Hamming. *Numerical Methods for Scientists and Engineers*, McGraw-Hill, New York, 1973.

778 D. Han, X. Wang. *Chin. Sci. Bull.* 42, 1849 (1997).

779 J. H. Han, D. J. Thouless, H. Hiramoto, M. Kohmoto. *Phys. Rev.* B 50, 11365 (1994).

780 X. W. D. Han in Z.-C. Shi, T. Ushijima (eds.). *Proceedings of the First China-Japan Seminar on Numerical Mathematics*, World Scientific, Singapore, 1993.

781   R. P. S. Han, A. C. J. Luo. *J. Sound Vibr.* 181, 231 (1995).

782   S. Handa, H. Kashiwagi, T. Takada. *J. Visual. Comput. Anim.* 12, 167 (2001).

783   J. H. Hannay. *J. Phys.* A 18, 221 (1985).

784   G. Hanrot, J. Rivat, G. Tenenbaum, P. Zimmermann. *Theor. Comput. Sc.* 291, 135 (2003).

785   E. Hansen in K. Nickel (ed.). *Interval Mathematics*, Springer-Verlag, New York, 1975.

786   E. Hansen, M. Patrick. *Num. Math.* 27, 257 (1977).

787   P. Hanusse, M. Gomez-Gesteira. *Physica Scripta* T 67, 117 (1996).

788   B.-L. Hao, W.-M. Zheng. *Applied Symbolic Dynamics and Chaos*, World Scientific, Singapore, 1998.

789   D. N. Hào, T. D. Van, R. Gorenflo. *Banach Center Publ.* 27, 111 (1992).

790   G. Harburn, C. A. Taylor, T. R. Welberry. *Atlas of Optical Transforms*, Cornell University Press, Ithaca, 1975.

791   U. Harder, M. Paczuski. *arXiv:cs.PF*/0412027 (2004).

792   G. H. Hardy. *Orders of Infinity*, Cambridge University Press, Cambridge, 1910.

793   R. Harish, K. P. N. Murthy. *Physica* A 287, 161 (2000).

794   J. Harnad. *arXiv:solv-int*/9902013 (1999).

795   P. G. Harper. *J. Phys. Cond. Matter* 3, 3047 (1991).

796   E. Hartmann. *Comput. Aided Geom. Design* 17, 251 (2000).

797   J. Hass, R. Schalfly. *arXiv:math.DG*/0003157 (2000).

798   I. H. A.-H. Hassan. *Appl. Math. Comput.* 127, 1 (2002).

799   M. B. Hastings. *arXiv:cond-mat*/0212303 (2002).

800   M. Hata. *Acta Arithm.* 63, 335 (1993).

801   Y. Hatsugai. *Phys. Rev.* B 48, 11851 (1993).

802   H. Hayakawa, H. Kuninaka. *arXiv:cond-mat*/0312005 (2003).

803   Y. Hayase, T. Ohta. *Phys. Rev.* E 62, 5998 (2000).

804   L. He, X. Liu, G. Strang. *Stud. Appl. Math.* 110, 123 (2003).

805   X. T. He. *Phys. Rev.* E 66, 037201 (2002).

806   D. A. Heassig, Jr., B. Friedland. *J. Dynamic Syst., Measur. Contr.* 113, 354 (1991).

807   K. Heegner. *Math. Z.* 56, 227 (1952).

808   S. Heinrich. *arXiv:quant-ph*/0112152 (2001).

809   W. D. Heiss, W.-H. Steeb. *J. Math. Phys.* 32, 3003 (1991).

810   W. D. Heiss. *arXiv:quant-ph*/9901023 (1999).

811   E. J. Heller. *Phys. Scr.* 40, 354 (1989).

812   E. Heller, S. Shaw. *Int. J. Mod. Phys.* B 17, 3977 (2003).

813   H. Hellmann. *Einführung in die Quantenchemie*, Deuticke, Vienna, 1937.

814   G. Helmberg. *J. Approx. Th.* 81, 389 (1995).

815   G. Helmberg. *J. Approx. Th.* 89, 308 (1997).

816   P. C. Hemmer, M. Kac, G. E. Uhlenbeck. *J. Math. Phys.* 5, 60 (1964).

817   P. C. Hemmer, G. E. Uhlenbeck, M. Kac. *J. Math. Phys.* 5, 75 (1964).

818   A. Henderson. *The Twenty Seven Lines Upon the Cubic Surface*, Cambridge University Press, Cambridge, 1911.

819   M. Hendrey, K. Nam, P. Guzdar, E. Ott. *Phys. Rev.* E 62, 7627 (2000).

820   M. Hénon. *Astron. Astrophys.* 28, 415 (1973).

821   M. Hénon. *Commun. Math. Phys.* 50, 69 (1976).

822   M. Henón. *Generating Families in the Restricted Three-Body Problem*, Springer-Verlag, Berlin, 1997.

823   P. Henrici. *SIAM Rev.* 21, 481 (1979).

824   M. S. Henry. *Am. Math. Monthly* 91, 497 (1984).

825   D. Herceg, S. Tričković, M. Petković. *Nonlinear Anal. Methods Appl.* 30, 83 (1997).

826   O. Herrmann. *J. reine angew. Math.* 275, 187 (1975).

827   F. Herrmann, P. Schmälzle. *Am. J. Phys.* 49, 761 (1981).

828   F. Herrmann, M. Seitz. *Am. J. Phys.* 50, 977 (1982).

829   F. Herrmann, H. Hauptmann, M. Suleder. *Am. J. Phys.* 68, 171 (2000).

830   J. Herzberger in J. Herzberger (ed.). *Topics in Validated Computing*, Elsevier, Amsterdam, 1994.

831   K. J. Heuvers, D. S. Moak, B. Boursaw in T. M. Rassias (ed.). *Functional Equations and Inequalities*, Kluwer, Dordrecht, 2000.

832   N. J. Higham. *SIAM J. Sci. Comput.* 14, 783 (1993).

833   F. B. Hildebrand. *Advanced Calculus for Applications*, Prentice-Hall, Englewood Cliffs, 1976.

834   B. J. Hiley. *arXiv:quant-ph*/0010020 (2000).

835   R. N. Hill. *J. Math. Phys.* 8, 201 (1967).

836   E. Hille. *Analytic Function Theory* v.II, Chelsea, New York, 1973.

837   N. Hinrichs, M. Oestreich, K. Popp. *J. Sound Vibr.* 216, 435 (1998).

838   H. Hinrichsen, O. Stenull, H.–K. Janssen. *arXiv:cond-mat*/0110343 (2001).

839   S. Hiramoto, M. Kohmoto. *Int. J. Mod. Phys.* B 6, 281 (1992).

840   J. W. P. Hirschfeld. *Rendiconti Mat.* 26, 115 (1967).

841   T. Hisakado, K. Okumura. *Phys. Lett.* A 292, 263 (2002).

842   E. Hlawka in H. Hlawka, R. F. Tichy (eds.). *Number-Theoretic Analysis*, Springer-Verlag, Berlin, 1990.

843   J. Hoefkens, M. Berz, K. Makino. *Adv. Comput. Math.* 19, 231 (2003).

844   J. Hofbauer, S. J. Schreiber. *Nonlinearity* 17, 1393 (2004).

845   T. Höfer, J. S. Sherratt, P. K. Maini. *Proc. R. Soc. Lond.* B 259, 249 (1995).

846   T. Höfer, P. K. Maini. *Phys. Rev.* E 56, 2074 (1997).

847   W. A. Hofer. *arXiv:quant-ph*/9909009 (1999).

848   F. Hofmann. *Methodik der stetigen Deformation von zweiblättrigen Riemannschen Flächen*, Louis Neubert, Halle, 1888.

849   G. Hofmeister, P. Stoll. *J. reine angew. Math.* 362, 141 (1985).

850   D. Hofstadter. *Phys. Rev.* B 14, 2239 (1976).

851   P. R. Holland. *The Quantum Theory of Motion*, Cambridge University Press, Cambridge, 1993.

852   D. Home. *Conceptual Foundations of Quantum Theory*, Plenum, New York, 1997.

853   J. Honerkamp. *Statistical Physics*, Springer-Verlag, Berlin, 1998.

854   M. Hopkins, B. Marshall, G. Schmidt, S. Zlobec. *ZAMM* 74, 295 (1994).

855   S. Hornus, A. Angelidis, M.-P. Cani. *Visual Comput.* 19, 94 (2003).

856   X. Hu, Y. Shen, X. Liu, R. Fu, J. Zi. *Phys. Rev.* E 69, 030201 (2004).

857   X.-G. Hu. *Phys. Rev.* E 39, 2471 (1999).

858   D. Huang, G. Gumbs, A. H. MacDonald. *Phys. Rev.* B 48, 2843 (1993).

859   B. Huber, B. Sturmfels. *Discr. Comput. Geom.* 17, 137 (1997).

860   T. E. Hull, R. Mathon. *ACM Trans. Math. Softw.* 22, 261 (1996).

861   Y. Hurmuzlu, V. Ceanga in B. Brogliato (ed.). *Impacts in Mechanical Systems*, Springer-Verlag, Berlin, 2000.

862   M. Hutchings, F. Morgan, M. Ritoré; An. Ros. *Electron. Res. Announc. Amer. Math. Soc.* 6, 45 (2000). http://www.ams.org/journal-getitem?pii=S1079-6762-00-00079-2

863   S. Hutzler, G. Delaney, D. Weaire, F. MacLeod. *Am. J. Phys.* 72, 1508 (2004).

864   C.-L. Hwang. *Tamkang J. Math.* 35, 305 (2004).

865   J. S. Hwang. *Ann. Prob.* 15, 1203 (1987).

866   S. T. Hyde, S. Andersson. *Z. Kristallogr.* 174, 237 (1986).

867   R. A. Hyman, B. Doporcyk, J. Tetzlaff. *arXiv:physics*/0307119 (2003).

868   F. Iachello, A. Del Sol Mesa. *J. Math. Chem.* 25, 345 (1999).

869   M. Igarashi in T. M. Rassias, H. M. Srivastava, A. Yanushauskas (eds.). *Topics in Polynomials of One and Several Variables*, World Scientific, Singapore, 1993.

870   K. Ikeda. *Opt. Commun.* 30, 257 (1979).

871   A. Ilachinski. *Cellular Automata*, World Scientific, Singapore, 2001.

872   G.-L. Ingold, A. Wobst, C. Aulbach, P. Hänggi. *arXiv:cond-mat*/0212035 (2002).

873   M. Iosifescu, C. Kraaikamp. *Metrical Theory of Continued Fractions*, Kluwer, Dordrecht, 2002.

874   M. Ipsen, M. van Hecke. *arXiv:nlin.PS*/0104067 (2001).

875   A. Iserles, S. P. Nørsett, E. B. Saff. *Rocky Mountain J. Math.* 21, 331 (1991).

876   M. E. H. Ismail, R. Zhang in N. K. Thakare, K. C. Sharma, T. T. Raghunatahn (eds.). *Ramanujan International Symposium on Analysis*, MacMillan India, Delhi, 1987.

877   M. E. H. Ismail, R. Zhang. *Adv. Appl. Math.* 9, 439 (1988).

878   H. Ito. *Proc. Jpn. Acad.* 71A, 48 (1995).

879   Y. Itoh, L. Shepp. *J. Stat. Phys.* 97, 209 (1999).

880   A. R. Its, C. A. Tracy, H. Widom. *arXiv:nlin.SI*/0004018 (2000).

881   T. Iwai, N. Katayama. *J. Math. Phys.* 35, 2914 (1994).

882   T. Iwai, N. Katayama. *J. Math. Phys.* 36, 1790 (1995).

883   K. Iwasaki, Y. Dobashi, T. Nishita. *Comput. Graphics Forum* 22, 601 (2003).

884   A. D. Jackson. *arXiv:physics*/0009055 (2000).

885   M. Jacob. *J. Physique* II 7, 1035 (1997).

886   C. Jacoboni, P. Bordone. *Rep. Progr. Phys.* 67, 1033 (2004).

887   D. Jacobson. *The Mathematica Journal* 2, n3, 42 (1992).

888   M. A. Jafarizadeh. *arXiv:cond-mat*/0005227 (2000).

889   M. A. Jafarizadeh, M. Mirzaee, H. Aghlara. *arXiv:cond-mat*/0005338 (2000).

890   M. A. Jafarizadeh. *Physica* A 287, 1 (2000).

891   M. Jafarpour, D. Afshar. *J. Phys.* A 35, 87 (2002).

892   S. Jaffard. *J. Fourier Anal. Appl.* 3, 1 (1997).

893   H. Jager, C. de Vroedt. *Indag. Math.* 31, 31 (1968).

894   J. F. James. *A Student's Guide to Fourier Transforms*, Cambridge University Press, Cambridge, 1995.

895   M. Jammer. *The Philosophy of Quantum Mechanics*, Wiley, New York, 1974.

896 M. S. Janaki, G. Gash. *J. Phys.* A 20, 3679 (1987).

897 W. Janke, H. Kleinert. *Phys. Lett.* A 199, 287 (1995).

898 T. Janssen in F. Axel, D. Gratias (eds.). *Beyond Quasicrystals*, Springer, Berlin, 1995.

899 G. Jaroszkiewicz. *arXiv:gr-qc*/0004026 (2000).

900 L. Jaulin, M. Kieffer, O. Didrit, É. Walter. *Applied Interval Analysis*, Springer-Verlag, London, 2003.

901 E. Jeandel. *Preprint INRIA* 0242/2000 (2000). http://www.inria.fr/RRRT/RT-0242.html

902 H. Jeffreys. *Asymptotic Approximations*, Clarendon Press, Cambridge, 1962.

903 M. Jeng. *Am. J. Phys.* 68, 37 (2000).

904 J. L. Jensen. *Saddlepoint Approximations*, Clarendon Press, Oxford, 1995.

905 U. D. Jentschura, P. J. Mohr, G. Soff, E. J. Weniger. *arXiv:Math.NA* 9809111 (1998).

906 S.-O. Jeong, T.-W. Ko, H.-T. Moon. *Phys. Rev. Lett.* 89, 154104 (2002).

907 A. J. Jerri. *Linear Difference Equations with Discrete Transform Methods*, Kluwer, Dordrecht 1996.

908 F. Jézéquel. *Appl. Num. Math.* 50, 147 (2004).

909 F. Jézéquel, J.-M. Chesneaux. *Num. Alg.* 36, 265 (2004).

910 S. Jiang, L. Greengard. *Comput. Math. Appl.* 47, 955 (2004).

911 A. Johansen, D. Sornette. *arXiv:cond-mat*/9901350 (1999).

912 J. Jonasson. *arXiv:math.PR*/0501401 (2005).

913 G. A. Jones, D. Singerman. *Complex Functions—An Algebraic and Geometric ViewPoint*, Cambridge University Press, Cambridge, 1987.

914 J. D. Jones. *Comput. Graphics* 15, 451 (1991).

915 W. B. Jones, W. J. Thron. *Continued Fractions—Analytic Theory and Applications*, Addison-Wesley, Reading, 1980.

916 W. B. Jones, W. J. Thron. *Appl. Num. Math.* 4, 143 (1988).

917 H. T. Jongen, P. Jonker, F. Twilt in R. Martini (ed.). *Geometrical Approaches to Differential Equations*, Springer-Verlag, Berlin, 1980.

918 H. T. Jongen, P. Jonker, F. Twilt. *Acta Appl. Math.* 13, 81 (1988).

919 A. Jorba, J. C. Tatjer. *J. Approx. Th.* 120, 85 (2003).

920 S. P. Jordan. *arXiv:quant-ph*/0501051 (2005).

921 J. V. José, E. J. Saletan. *Classical Dynamics: A Contemporary Approach*, Cambridge University Press, Cambridge, 1998.

922 N. Joshi, M. D. Kruskal in P. A. Clarkson (ed.). *Applications of Analytic and Geometric Methods to Nonlinear Differential Equations*, Plenum Press, New York, 1993.

923 C. Jung, H.-J. Scholz. *J. Phys.* A 21, 2301 (1988).

924 H. W. Jung. *Einführung in die Theorie der algebraischen Funktionen einer Veränderlichen*, de Gruyter, Berlin, 1923.

925 S.-M. Jung, P. K. Sahoo. *Appl. Math. Lett.* 15, 435 (2002).

926 W. Jung. *Preprint* mp_arc 96-691 (1996). http://rene.ma.utexas.edu/mp_arc-bin/mpa?yn=96-691

927 W. Just, H. Kantz. *J. Phys.* A 33, 163 (2000).

928 M. Kac, G. E. Uhlenbeck, P. C. Hemmer. *J. Math. Phys.* 4, 216 (1963).

929 H.-H. Kairies. *Aequ. Math.* 53, 207 (1997).

930 H.-H. Kairies. *Aequ. Math.* 58, 183 (1999).

931 G. Kälbermann. *arXiv:quant-ph*/9912042 (1999).

932   D. I. Kamenev, G. P. Berman *Quantum Chaos: A Harmonic Oscillator in Monochromatic Wave*, Rinton Press, Princeton, 2001

933   D. Kaminski, R. B. Paris. *J. Comput. Appl. Math.* 107, 31 (1999).

934   D. D. Kana, D. J. Fox. *Chaos* 5, 298 (1995).

935   Y. Kanazawa. *Stat. Prob. Lett.* 17, 293 (1993).

936   T. R. Kane, D. A. Levinson. *J. Guidance Contr.* 3, 99 (1980).

937   K. Kaneko in K. Kawasaki, M. Suzuki, A. Onuki (eds.). *Formation, Dynamics and Statistics of Patterns* v.1, World Scientific, Singapore, 1990.

938   K. Kaneko, I. Tsuda. *Complex Systems: Chaos and Beyond*, Springer-Verlag, Berlin, 2001.

939   S.-J. Kang, M.-H. Kim. *J. Algebra* 183, 560 (1996).

940   Y. Kannai. *arXiv:math.AP*/9909139 (1999).

941   S. Kanno, N. V. Kjurkchiev, T. Yamamoto. *Jpn. J. Industr. Appl. Math.* 13, 267 (1996).

942   U. Kânoglu. *J. Fluid Mech.* 513, 363 (2004).

943   Y. Kantor, M. Kardar, D. R. Nelson. *Phys. Rev. Lett.* 57, 791 (1986).

944   Y. Kantor, M. Kardar, D. R. Nelson. *Phys. Rev.* A 35, 3056 (1987).

945   M.-Y. Kao, J. Wang. *arXiv:cs.DS*/9907015 (1999).

946   L. Kapitanski, I. Rodnianski. *arXiv:quant-ph*/9711062 (1997).

947   A. E. Kaplan, P. Stifter, K. A. H. van Leeuwen, W. E. Lamb, Jr., W. P. Schleich. *Physica Scripta* T 76, 93 (1998).

948   A. E. Kaplan, I. Marzoli, W. E. Lamb, Jr., W. P. Schleich. *Phys. Rev.* A 61, 032101 (2000).

949   D. Kaplan, L. Glass. *Understanding Nonlinear Dynamics*, Springer-Verlag, New York, 1995.

950   L. Kaplan. *arXiv:nlin.CD*/0206040 (2002).

951   G. M. Kapoulitsas. *ZAMP* 46, 709 (1995).

952   E. A. Karatsuba in N. Papamichael, S. Ruscheweyh, E. B. Saff (eds.). *Computational Methods and Function Theory*, World Scientific, Singapore, 1999.

953   I. Karatzas, S. E. Shreve. *Brownian Motion Stochastic Calculus* Springer-Verlag, New York, 1991.

954   H. Karcher. *Manuscripta Math.* 64, 291 (1989).

955   Z. P. Karkuszewski. *arXiv:quant-ph*/0412073 (2004).

956   E. Karpov, G. Ordonez, T. Petrosky, I. Prigogine, G. Pronko. *arXiv:quant-ph*/0003075 (2000).

957   A. P. Kasantsev, A. M. Dykhne, V. L. Pokrovsky. *JETP Lett.* 63, 694 (1996).

958   P. Kastanek, J. Kosek, D. Snita, I. Schreiber, M. Marek. *Physica* D 84, 79 (1995).

959   E. Katz, R. H. Marty, S. D. Silliman. *Int. J. Math. Edu. Sci. Technol.* 23, 723 (1992).

960   L. H. Kauffman. *New J. Phys.* 6, 173 (2004).

961   S. Kaufmann. *Mathematica as a Tool*, Birkhäuser, Basel, 1994.

962   R. K. Kaul. *arXiv:hep-th*/9907119 (1999).

963   P. O. Kazinski, S. L. Lyakhovich, A. A. Sharapov. *Phys. Rev.* D 66, 025017 (2002).

964   R. B. Kearfott (ed.). *Applications of Interval Computations*, Kluwer, Amsterdam, 1996.

965   B. R. Kearfott. *Euromath Bull.* 2, 95 (1996).

966   M. J. Kearney, S. N. Majumdar. *arXiv:cond-mat*/0501445 (2005).

967   J. Keiper. *MathSource* 0204-028 (1992).

968   J. Keiper. *MathSource* 0204-646 (1992).

969   J. B. Keiper. *Interval Comput.* n3, 76 (1993).

970   J. B. Keller. *Proc. Natl. Acad. Sci.* 101, 1120 (2004).

971   J. Kempe. *quant-ph*/0303081 (2003).

972   V. Kendon, B. C. Sanders in S. M. Barnett, E. Andersson, P. Öhberg, O. Hirota (eds.). *Quantum Communication, Measurement and Computing*, American Institute of Physics, New York, 2004.

973   V. Kendon, B. C. Sanders. *Phys. Rev.* A 71, 022307 (2005).

974   I. O. Kerner. *Num. Math.* 8, 290 (1966).

975   D. A. Kessler, H. Levine, W. N. Reynolds. *Phys. Rev.* A 42, 6125 (1997).

976   R. Ketzmerick. *Chaos, fraktale Spektren und Quantendynamik in Halbleiter-Mikrostrukturen*, Verlag Harry Deutsch, Frankfurt am Main, 1992.

977   R. Ketzmerick, K. Kruse, F. Steinbach, T. Geisel. *arXiv:cond-mat*/9806070 (1998).

978   J. Kevorkian. *Partial Differential Equations*, Springer-Verlag, New York, 2000.

979   G. Keying. *Ann. Diff. Eq.* 14, 120 (1998).

980   A. Z. Khachatrian. *Int. J. Mod. Phys.* C 15, 1105 (2004).

981   H. G. Khajah, E. L. Ortiz. *Appl. Num. Math.* 21, 431 (1996).

982   D. V. Khakar, H. Rising, J. M. Ottino. *J. Fluid Mech.* 172, 419 (1986).

983   A. Y. Khinchine. *Continued Fractions*, University of Chicago Press, Chicago, 1964.

984   D. Khosenevisan, Z. Shi. *Trans. Am. Math. Soc.* 350, 4253 (1998).

985   A. N. Khovanskii. *The Application of Continued Fractions and their Generalizations to Problems in Approximation Theory*, Noordhoff, Groningen, 1963.

986   A. Khovanskii in E. Bierstone, B. Khesin, A. Khovanskii, J. E. Marsden (eds.). *The Arnoldfest*, American Mathematical Society, Providence, 1999.

987   R. Kidambi, P. K. Newton. *Physica* D 116, 143 (1998).

988   T. Kilgore in G. A. Anastassiou (ed.). *Applied Mathematics Reviews*, v.1 World Scientific, Singapore, 2000.

989   J. P. Killingbeck, N. A. Gordon, M. R. M. Witwit. *Phys. Lett.* A 206, 279 (1995).

990   C. H. Kim, J. K. Koo. *J. Pure Appl. Math.* 160, 53 (2001).

991   D. Kim, J. K. Koo. *Acta Arithm.* 100, 105 (2001).

992   P.-J. Kim, T.-W. Ko, H. Jeong, H.-T. Moon. *arXiv:nlin.PS*/0406008 (2004).

993   S.-Y. Kim, B. Hu. *Phys. Rev.* A 41, 5431 (1990).

994   G. Kirchhoff. *Ann. Phys. Chem.* 72, 497 (1847).

995   S. Kirkpatrick. *Rev. Mod. Phys.* 45, 574 (1973).

996   O. Kirillova. *arXiv:cond.mat*/0010231 (2000).

997   N. Kiriushcheva, S. Kuzmin. *Am. J. Phys.* 66, 867 (1998).

998   N. P. Kirk, J. N. L. Connor, C. A. Hobbs. *Comput. Phys. Commun.* 132, 142 (2000).

999   A. Kiselev. *mp_arc* 01-425 (2001). http://rene.ma.utexas.edu/mp_arc/c/01/01-425.ps.gz

1000   N. Kjurkchiev. *J. Comput. Appl. Math.* 58, 233 (1995).

1001   C. A. J. Klaassen, J. T. Runnenburg. *arXiv:math.PR*/0112056 (2001).

1002   J. Klafter, G. Zumofen. *Phys. Rev.* E 49, 4873 (1994).

1003   J. Klafter, G. Zumofen, M. F. Shlesinger in M. F. Shlesinger, G. M. Zaslavsky, U. Frisch (eds.). *Lévy Flights and related Topics in Physics*, Springer-Verlag, Berlin, 1995.

1004   R. Klages, I. F. Barna, L. Mátyás. *Phys. Lett.* A 333, 79 (2004).

1005   M. S. Klamkin (ed.). *Mathematical Modelling: Classroom Notes in Applied Mathematics*, SIAM, Philadelphia, 1995.

1006   F. Klein. *Math. Annalen* 14, 551 (1873).

1007   P. E. Kloeden, E. Platen. *Numerical Solution of Stochastic Differential Equations*, Springer-Verlag, Berlin, 1999.

1008   R. Knapp, S. Wagon. *Mathematica Edu. Res.* 7, n4, 76 (1998).

1009   R. Knapp. *The Mathematica Journal* 4, 542 (2000).

1010   K. Kneisl. *Chaos* 11, 359 (2001).

1011   C. Knessl, W. Szpankowski. *Theor. Comput. Sc.* 289, 649 (2002).

1012   P. L. Knight, E. Roldán, J. E. Sipe. *arXiv:quant-ph*/0304201 (2003).

1013   H. E. Knoepfel. *Magnetic Fields*, Wiley, New York, 2000.

1014   A. Knopfmacher, J. Knopfmacher. *Int. J. Math. Math. Sci.* 12, 603 (1989).

1015   G. D. Knott. *Interpolating Cubic Splines*, Birkhäuser, Boston, 2000.

1016   D. E. Knuth. *The Art of Computer Programming*, v. 1, Addison-Wesley, Reading, 1969.

1017   L. Kocarev, Z. Tasev, T. Stojanovski, U. Parlitz. *Chaos* 7, 635 (1997).

1018   L. Kocarev, Z. Tasev, U. Parlitz. *Phys. Rev. Lett.* 79, 51 (1997).

1019   A. J. Koch, M. Meinhardt. *Rev. Mod. Phys.* 66, 1481 (1994).

1020   E. Koch, W. Fischer. *Acta Cryst.* A 46, 33 (1990).

1021   E. Koch, W. Fischer in H. T. Davis, J. C. C. Nitsche (eds.). *Statistical Thermodynamics and Differential Geometry of Microstructured Materials*, Springer-Verlag, New York, 1993.

1022   G. Köhler, J. Spilker. *Elem. Math.* 59, 89 (2004).

1023   M. Kohmoto, Y. Oono. *Phys. Lett.* A 102, 145 (1984).

1024   M. Kohmoto, B. Sutherland. *Phys. Rev.* B 35, 1020 (1987).

1025   A. J. Kolan, E. R. Nowak, A. V. Tkachenko. *arXiv:cond-mat*/9809434 (1998).

1026   A. Y. Kolesov, N. K. Rozov, V. A. Sadovnichiy. *J. Math. Sc.* 120, 1372 (2004).

1027   P. Kolinko, D. R. Smith. *Opt. Express* 11, 640 (2003).

1028   M. Kollar. *Int. J. Mod. Phys.* B 16, 3491 (2002).

1029   V. N. Komarov. *Tech. Phys.* 46, 117 (2001).

1030   N. L. Komarova, I. Rivin. *arXiv:math.PR*/0105236 (2001).

1031   T. Komatsu. *Proc. Cambr. Phil. Soc.* 134, 1 (2003).

1032   T. Komatsu. *J. Number Th.* 109, 27 (2004).

1033   A. I. Komech, N. J. Mauser, A. P. Vinnichenko. *Russ. J. Math. Phys.* 11, 289 (2004).

1034   N. Konno. *arXiv:quant-ph*/0206053 (2002).

1035   N. Konno. *arXiv:quant-ph*/0206103 (2002).

1036   N. Konno. *arXiv:quant-ph*/0210011 (2002).

1037   N. Konno, T. Namiki, T. Soshi. *arXiv:quant-ph*/0205065 (2002).

1038   N. Konno, K. Mitsuda, T. Soshi, H. J. Yoo. *arXiv:quant-ph*/0403107 (2004).

1039   N. Konno, T. Namiki, T. Soshi, A. Sudbury. *arXiv:quant-ph*/0208122 (2002).

1040   T. H. Koornwinder, R. F. Swarttouw. *Trans. Am. Math. Soc.* 333, 445 (1992).

1041   H. J. Korsch, H.-J. Jodl. *Chaos*, Springer-Verlag, Berlin, 1999.

1042   K. Koschdon, M. Schäfer in K. Hutter, N. Kichner (eds.). *Dynamic Response of Granular and Porous Materials under Large and Catastrophic Deformations.*, Springer-Verlag, Berlin, 2003.

1043   R. Kosloff in J. Broeckhove, L. Lathouwers (eds.). *Time-Dependent Quantum Molecular Dynamics*, Plenum Press, New York, 1992.

1044 R. Kotlyar, C. A. Stafford, S. Das Sarma. *Phys. Rev.* B 58, 3989 (1998).

1045 O. Kotovych, J. C. Bowman. *J. Phys.* A 35, 7849 (2002).

1046 J. P. Kotthaus in S. Namba, C. Hamaguchi, T. Ando (eds.). *Science and Technology of Mesoscopic Structures*, Springer-Verlag, Berlin, 1992.

1047 D. Kozen, K. Stefansson. *J. Symb. Comput.* 24, 125 (1997).

1048 J. Košík, V. Buzek. *arXiv:quant-ph*/0410154 (2004).

1049 V. V. Kozlov. *Am. Math. Soc. Transl.* (2) 168 (1995).

1050 C. Kraaikamp. *Acta Arithm.* 57, 1 (1991).

1051 P. Kramer, T. Kramer. *arXiv:math-phys*/9903009 (1999).

1052 I. Krasikov, G. J. Rodgers, C. E. Tripp. *J. Phys.* A 37, 2365 (2004).

1053 I. V. Krasovsky. *Phys. Rev. Lett.* 85, 4920 (2000).

1054 W. Kratz. *Lin. Alg. Appl.* 337, 1 (2001).

1055 K. Kraus. *Physik in unserer Zeit* 14, 81 (1983).

1056 B. Krauskopf, H. M. Osinga. *Chaos* 9, 768 (1999).

1057 B. Krauskopf, H. M. Osinga. *Nonlinearity* 17, C1 (2004).

1058 P. Kravanja, M. Van Barel. *Computing the Zeros of Analytic Functions*, Springer-Verlag, Berlin, 2000.

1059 Y. A. Kravtsov. *Sov. Phys. Usp.* 35, 606 (1992).

1060 Y. A. Kravtsov, Y. I. Orlov. *Caustics, Catastrophes and Wave Fields*, Springer-Verlag, Berlin, 1993.

1061 C. Kreft, R. Seiler. *J. Math. Phys.* 37, 5207 (1996).

1062 S. Kreidl, G. Grübl, H. G. Embacher. *arXiv:quant-ph*/0305163 (2003).

1063 S. Kreidl. *arXiv:quant-ph*/0406045 (2004).

1064 K. Kreith. *Georgian Math. J.* 1, 173 (1994).

1065 P. Krekora, Q. Su, R. Grobe. *Phys. Rev.* A 70, 054101 (2004).

1066 H. M. Krenzlin, J. Budczies, K. W. Kehr. *Ann. Physik* 7, 732 (1998).

1067 E. Kreuzer. *Numerische Untersuchung nichtlinearer dynamischer Systeme*, Springer-Verlag, Berlin, 1987.

1068 L. Krlín, M. Zápotocký. *Czech. J. Phys.* 54, 759 (2004).

1069 H. Kröger. *Phys. Rev.* A 65, 052118 (2002).

1070 A. R. Krommer, C. W. Ueberhuber. *Computational Integration* SIAM, Philadelphia, 1998.

1071 A. S. Kronrod. *Nodes and Weights for Quadrature Formulae. Sixteen-Place Tables*, Consultants Bureau, New York, 1965.

1072 M. D. Kruskal, N. Joshi, R. Halburd. *arXiv:solv-int*/9710023 (1997).

1073 R. Kubo. *Thermodynamics*, North-Holland, Amsterdam, 1968.

1074 A. Kudrolli, M. C. Abraham, J. P. Gollub. *Phys. Rev.* E 63, 026208 (2001).

1075 U. Kuhl, H.-J. Stöckmann. *Phys. Rev. Lett.* 80, 3232 (1998).

1076 O. Kühn, V. Fessatidis, H. L. Cui, P. E. Selbmann, N. J. M. Horing. *Phys. Rev.* B 47, 13019 (1993).

1077 K. Kumar. *Eur. J. Phys.* 20, 501 (1999).

1078 A. Kunold, M. Torres. *arXiv:cond-mat*/9912378 (1999).

1079 A. Kunold, M. Torres. *arXiv:cond-mat*/0409579 (2004).

1080 A. Kunold, M. Torres. *Ann. Phys.* 315, 532 (2005).

1081 L. G. Kurakin, V. I. Yudovich. *Chaos* 12, 574 (2002).

1082   M. Kurata, L. Books. *Numerical Analysis for Semiconductor Devices*, Heath, Toronto, 1982.

1083   J.-P. Kuska. *Phys. Rev.* B 46, 5000 (1992).

1084   M. A. Kutay, H. M. Ozaktas. *Nonl. Dynam.* 29, 157 (2002).

1085   J. R. Kuttler, V. G. Sigillito. *SIAM Rev.* 26, 163 (1984).

1086   A. P. Kuznetsov, S. P. Kuznetsov, E. Mosekilde, L. V. Turukina. *Physica* A 300, 367 (2001).

1087   N. V. Kyurkchiev. *Initial Approximations and Root Finding Methods*, Wiley–VCH, Berlin, 1998.

1088   O. Labs. *arXiv:math.AG/*0409348 (2004).

1089   Y. Lacroix, A. Thomas. *J. Number Th.* 49, 308 (1994).

1090   M. Lakshmanan, P. Muruganandam. *arXiv:nlin.PS/*0011006 (2000).

1091   M. Lakshmanan, S. Rajasekar. *Nonlinear Dynamics*, Springer-Verlag, Berlin, 2003.

1092   H. Länger. *Elem. Math.* 51, 75 (1996).

1093   C. S. Lam in C. C. Bernido, K. Nakamura, M. V. Carpio-Bernido, K. Watanabe (eds.). *Mathematical Methods of Quantum Physics*, Gordon and Breach, Amsterdam, 1999.

1094   C. S. Lam. *J. Math. Phys.* 41, 4497 (1999).

1095   J. S. W. Lamb. *J. Phys.* A 26, 2921 (1993).

1096   J. S. W. Lamb in J. Seimenis (ed.). *Hamiltonian Mechanics*, Plenum Press, New York, 1994.

1097   P. Lancaster. *Theory of Matrices*, Academic Press, New York, 1969.

1098   L. D. Landau, E. M. Lifschitz. *Course of Theoretical Physics, v. 8: Electrodynamics of Continuous Media*, Pergamon Press, Oxford, 1984.

1099   L. D. Landau, E. M. Lifschitz. *Course of Theoretical Physics, v. 7: Theory of Elasticity*, Pergamon Press, Oxford, 1986.

1100   S. Lang. *Elliptic Functions*, Addison-Wesley, Reading, 1973.

1101   P. K. Lange. *Numerical Analysis for Statisticians*, Springer-Verlag, New York, 1999.

1102   J. P. Laraudogoitia. *Mind* 105, 81 (1996).

1103   J. P. Laraudogoitia. *Synthese* 135, 339 (2003).

1104   J. Mc Laughlin, D. Bowman. *arXiv:math.NT/*0107043 (2001).

1105   J. Mc Laughlin. *On the Convergence and Divergence of q-Continued Fractions on and off the Unit Circle*, PhD thesis, Urbana, 2002.

1106   J. Mc Laughlin, N. J. Wyshinski. *arXiv:math.NT/*0402462 (2004).

1107   F. Laurent–Polz. *arXiv:math.DS/*0109202 (2001).

1108   F. Laurent–Polz. *arXiv:math.DS/*0301360 (2003).

1109   D. P. Laurie. *Math. Comput.* 66, 1133 (1997).

1110   P. Laurence, E. Stredulinsky. *J. Math. Phys.* 41, 3170 (2000).

1111   D. P. Laurie in W. Gautschi, G. H. Golub, G. Opfer (eds.). *Applications and Computation of Orthogonal Polynomials*, Birkhäuser, Basel, 1999.

1112   D. P. Laurie in C. Brezinski, L. Wuytack (eds.). *Numerical Analysis: Historical Developments in the 20th Century*, Elsevier, Amsterdam, 2001.

1113   H. A. Lauwerier in A. V. Holden (ed.). *Chaos*, Princeton University Press, Princeton, 1986.

1114   L. Lavatelli. *Am. J. Phys.* 40, 1246 (1972).

1115   B. Lavenda. *arXiv:cond-mat/*0401024 (2004).

1116   C. K. Law. *Phys. Rev.* A 70, 062311 (2004).

1117   N. N. Lebedev. *Special Functions and Their Applications*, Prentice-Hall, Englewood Cliffs, 1965.

1118 M. Le Berre, E. Ressayre, A. Tallet, Y. Pomeau, L. Di Menza. *Phys. Rev.* E 66, 026203 (2002).

1119 J.-E. Lee. *Theor. Math. Phys.* 101, 1281 (1994).

1120 J. Lega, T. Passot. *Chaos* 14, 562 (2004).

1121 M. Lehman. *Opt. Commun.* 195, 11 (2001).

1122 D. H. Lehmer. *Am. Math. Monthly* 45, 656 (1938).

1123 D. H. Lehmer. *Duke Math. J.* 4, 323 (1938).

1124 J. Lehner in W. Abikoff, J. S. Birman, K. Kuiken (eds.). *The Mathematical Legacy of Wilhelm Magnus*, American Mathematical Society, Providence, 1994.

1125 R. I. Leine, D. H. van Campen. *Math. Comput. Model.* 36, 259 (2002).

1126 R. Leine, B. Brogliato. *Preprint INRIA* n 4322 (2001). http://www.inria.fr/rrrt/rr-4322.html

1127 J. C. Lemm, J. Uhlig, A. Weiguny. *arXiv:quant-ph*/0312191 (2003).

1128 A. Lenard. *Proc. Am. Math. Soc.* 108, 951 (1990).

1129 T. Leppänen, M. Karttunen, K. Kaski. *arXiv:cond-mat*/02306121 (2003).

1130 B. J. LeRoy. *J. Phys.* Cond. Mat. 15, R1835 (2003).

1131 C. Letellier, S. Elaydi, L. A. Aguirre, A. Alaoudi. *Physica* D 195, 29 (2004).

1132 D. Levi, J. Negro, M. A. del Olmo. *J. Phys.* A 37, 3459 (2004).

1133 Y. Levin, J. J. Arenzon. *Europhys. Lett.* 63, 415 (2003).

1134 H. Levine, E. J. Moniz, D. H. Sharp. *Am. J. Phys.* 45, 75 (1977).

1135 R. D. Levine. *Proc. R. Soc. Lond.* 294, 467 (1966).

1136 R. D. Levine. *Physica* E 9, 591 (2001).

1137 B. M. Levitan, I. S. Sargsjan. *Operatory Sturm-Lioville i Diraka*, Moscow, Nauka, 1988.

1138 F. Leyvraz, M.-C. Firpo, S. Ruffo. *arXiv:cond-mat*/0204255 (2002).

1139 C.-F. Li, Q. Wang. *Phys. Lett.* A 275, 287 (2000).

1140 D. Li, J. Xu. *J. Sound Vibr.* 275, 1 (2004).

1141 J.-R. Li, L. Greengard. *J. Comput. Phys.* 198, 295 (2004).

1142 M. Li, Y. Liu, Z.-Q. Zhang. *Phys. Rev.* B 61, 16193 (2000).

1143 T.-Y. Li. *SIAM J. Num. Anal.* 20, 865 (1983).

1144 T. Y. Li. *Math. Intell.* 9, n3, 33 (1987).

1145 T. Y. Li. *Acta Numerica* 6, 399 (1997).

1146 T.-Y. Li. *Taiwanese J. Math.* 3, 251 (1999).

1147 T. Y. Li in F. Cucker (ed.). *Handbook of Numerical Analysis. Special Volume: Foundations of Computational Mathematics*, Elsevier, Amsterdam, 2003.

1148 W.-R. Li, S. S. Cheng, T. T. Lu. *Appl. Math. E-Notes* 1, 1 (2001). http://math2.math.nthu.edu.tw/~amen/papers/2001.html

1149 X. Li in Z. U. Ahmed, N. K. Govil, P. K. Jain (eds.). *Fourier Analysis, Approximation*, New Age International, New Dehli, 1997.

1150 S. S. Liaw, C. C. Yang, R. T. Liu, J. T. Hong. *Phys. Rev.* E 64, 041909 (2001).

1151 S. Lidin, S. T. Hyde. *J. de Phys.* 48, 1585 (1987).

1152 S. Lidin, S. T. Hyde. *J. Phys. France* 51, 801 (1990).

1153 M. N. Limber, J. H. Curry. *Commun. Appl. Nonlinear Anal.* 1, 1 (1994).

1154 S. J. Linz. *Am. J. Phys.* 65, 523 (1997).

1155   S. J. Linz. *Am. J. Phys.* 66, 1109 (1998).

1156   S. J. Linz, J. C. Sprott. *Phys. Lett.* A 259, 240 (1999).

1157   O. Lipan. *arXiv:cond-mat*/9910225 (1999).

1158   H. Lipson (ed.). *Optical Transforms*, Academic Press, London, 1972.

1159   S. Lise, M. Paczuski. *Phys. Rev.* E 63, 036111 (2001).

1160   A. G. Litvak, V. A. Mironov, E. M. Sher. *JETP* 91, 1268 (2000).

1161   F. Liu, N. Goldenfeld. *Physica* D 47, 124 (1991).

1162   Y. Liu, R. Riklund. *Phys. Rev.* B 35, 6034 (1987).

1163   Y. Liu, Z. Hou, P. M. Hui, W. Sritrakool. *Phys. Rev.* B 60, 13444 (1999).

1164   U. Locatelli, C. Froeschlé, E. Lega, A. Morbidelli. *Physica* D 139, 48 (2000).

1165   G. Lochs. *Abh. Math. Seminar Univ. Hamburg* 27, 142 (1964).

1166   W. Loinaz, T. J. Newman. *arXiv:quant-ph*/9902039 (1999).

1167   W. Loinaz, T. J. Newman. *J. Phys.* A 32, 8889 (1999).

1168   A. W. Lohmann, D. Mendlovic, Z. Zalevsky in E. Wolf (ed.). *Progress in Optics* v. 38, Elsevier, Amsterdam, 1998.

1169   M. Y. Loktev, V. G. Volostnikov. *Proc. SPIE* 3487, SPIE, Bellingham, 1997.

1170   M. J. Longo. *Am. J. Phys.* 72, 1312 (2004).

1171   F. J. López, F. Martín. *Publ. Mat.* 43, 341 (1999).

1172   A. Lorek, A. Ruffing, J. Wess. *Z. Phys.* C 74, 369 (1997).

1173   L. Lorentzen, H. Waadeland. *Continued Fractions with Applications*, North Holland, Amsterdam, 1992.

1174   E. N. Lorenz. *J. Atmospheric Sc.* 20, 130 (1963).

1175   W. S. Loud. *Proc. Am. Math. Soc.* 2, 440 (1951).

1176   J. Lowenstein, S. Hatjispyros, F. Vivaldi. *Chaos* 7, 49 (1997).

1177   F. Luca. *Can. Math. Bull.* 45, 115 (2002).

1178   F. Luccio. *Theor. Comput. Sc.* 282, 223 (2002).

1179   J. M. Luck, D. Petritis. *J. Stat. Phys.* 42, 289 (1986).

1180   J. M. Luck. *Phys. Rev.* E 48, 3988 (1993).

1181   J. Lü, G. Chen. *Int. J. Bifurc. Chaos* 12, 659 (2002).

1182   J. Lü, G. Zhang, S. Celikovský. *Int. J. Bifurc. Chaos* 12, 2917 (2002).

1183   G. A. Luna–Acosta. *Phys. Rev.* A 42, 7155 (1990).

1184   D. A. MacDonald. *BIT* 36, 766 (1996).

1185   M. H. Macfarlane. *Ann. Phys.* 271, 159 (1999).

1186   E. P. Machikina, B. Y. Ryabko. *Discr. Math. Appl.* 9, 497 (1999).

1187   T. D. Mackay, S. D. Bartlett, L. T. Stephenson, B. C. Sanders. *arXiv:quant-ph*/0108004 (2001).

1188   R. S. MacKay. *Proc. Trans. R. Soc. Lond.* A 359, 1479 (2001).

1189   D. Madden. *Pac. J. Math.* 198, 123 (2001).

1190   J. J. V. Maestri, R. H. Landau, M. J. Páez. *Am. J. Phys.* 68, 1113 (2000).

1191   M. Magnier, C. Lattaud, J.-C. Heudin. *Complex Systems* 11, 419 (1997).

1192   M. Mahajan, V. Vinay. *SIAM J. Discr. Math.* 12, 474 (1999).

1193   K. Mahler. *J. Austral. Math. Soc.* 22, 65 (1976).

1194   F. M. Mahomed, F. Vawda in N. H. Ibragimov, F. M. Mahomed, D. P. Mason, D. Sherwell (eds.). *Differential Equations and Chaos*, New Age International, New Delhi, 1998.

1195   F. M. Mahomed, F. Vawda. *Nonl. Dynam.* 21, 307 (2000).

1196   J. J. Mahony, J. J. Shepard. *J. Austral. Math. Soc.* B 23, 17, 136, 310 (1981).

1197   J. Maillot, J. Stam. *Comput. Graphics Forum* 20, C-471 (2001).

1198   A. J. Makowski, S. Konkel. *Phys. Rev.* A 58, 4975 (1998).

1199   A. J. Makowski, M. Fracckowiak. *arXiv:quant-ph*/0111155 (2001).

1200   A. J. Makowski, K. Z. Górska. *Phys. Rev.* A 66, 062103 (2002).

1201   J.-M. Malasoma. *Phys. Lett.* A 264, 383 (2000).

1202   J.-M. Malasoma. *Chaos, Solitons, Fractals* 13, 1835 (2002).

1203   J.-M. Malasoma. *Phys. Lett.* A 305, 52 (2002).

1204   M. Malavasi, S. Marmi. *J. Phys.* A 22, L563 (1989).

1205   A. Malevanets, R. Kapral in A. Stasiak, V. Katritch, L. H. Kauffman (eds.). *Ideal Knots*, World Scientific, Singapore, 1998.

1206   G. B. Malikin, S. A. Karlomov. *Usp. Fiz. Nauk* 173, 985 (2003).

1207   R. K. Mallik. *Lin. Alg. Appl.* 325, 109 (2001).

1208   B. A. Malomed. *Phys. Rev.* A 45, 1009 (1992).

1209   B. A. Malomed, J. Yang. *arXiv:nlin.PS*/0209017 (2002).

1210   L. Malozemov, A. Teplyaev. *Math. Phys. Anal. Geom.* 6, 201 (2003).

1211   L. Malozemov, A. Teplyaev. *J. Funct. Anal.* 129, 390 (1994).

1212   P. Mannesville. *Dissipative Structures and Weak Turbulence*, Academic Press, New York, 1990.

1213   T. Mansour. *arXiv:math.CO*/0401218 (2004).

1214   T. Mansour. *arXiv:math.CO*/0401217 (2004).

1215   R.-M. Mantel, D. Barkley. *Warwick Preprints* 28/1996 (1996).

1216   C. Marchal. *The Three-Body Problem*, Elsevier, Amsterdam, 1990.

1217   V. A. Marchenko. *Sturm-Liouville Operators and Applications*, Birkhäuser, Basel, 1986.

1218   P. A. Markovich. *The Stationary Semiconductor Device Equations*, Springer-Verlag, Wien, 1986.

1219   P. A. Markovich, C. A. Ringhofer, C. Schmeiser. *Semiconductor Equations*, Springer-Verlag, Wien, 1990.

1220   L. Markus, K. R. Meyer. *Am. J. Math.* 102, 25 (1980).

1221   A. I. Markushevich. *The Theory of Functions of a Complex Variable*, Prentice-Hall, Englewood Cliffs, 1965.

1222   S. M. Marlov in G. Alefeld, A. Frommer, B. Lang (eds.). *Scientific Computing and Validated Numerics*, Akademie Verlag, Berlin, 1996.

1223   A. R. Marlow. *Am. J. Phys.* 59, 951 (1991).

1224   S. Marmi, P. Moussa, J.-C. Yoccoz. *Mathematical Physics Preprint* 467/99 (1999). http://rene.ma.utexas.edu/mp_arc-bin/mpa?yn=99-467

1225   S. Marmi. *J. Phys.* A 23, 3447 (1990).

1226   S. Marmi. *arXiv:math.DS*/0009232 (2000).

1227   G. Marsaglia, A. Zaman, J. Marsaglia. *Math. Comput.* 53, 191 (1989).

1228   J. E. Marsden, J. Scheurle, J. M. Wendtlandt in K. Kirchgässner, O. Mahrenholtz, R. Mennicken (eds.). *ICIAM 95*, Akademie Verlag, Berlin, 1996.

1229   J. E. Marsden, T. S. Ratiu. *Introduction to Mechanics and Symmetry*, Springer-Verlag, New York, 1998.

1230   A. Marshakov. *Seiberg-Witten Theory and Integrable Systems*, World Scientific, Singapore, 1999.

1231   G. Martin. *arXiv:math.NT*/0206166 (2002).

1232   G. Martin. *J. Austral. Math. Soc.* 77, 305 (2004).

1233   E. Martinenko, B. K. Shivamoggi. *Phys. Rev.* A 69, 052504 (2004).

1234   A. Martínez, H. Míguez, A. Griol, J. Martí. *Phys. Rev.* B 69, 165119 (2004).

1235   I. Marzoli, F. Saif, I. Bialynicki-Birula, O. M. Friesch, A. E. Kaplan, W. P. Schleich. *Acta Physica Slovaca* 48, 323 (1998).

1236   G. Mastroianni, D. Occorsio. *J. Comput. Appl. Math.* 134, 325 (2001).

1237   J. H. Mathews, R. W. Howell. *Complex Analysis for Mathematics and Engineering*, Jones and Bartlett, Boston, 1997.

1238   J. Matoušek. *Geometric Discrepancy*, Springer-Verlag, Berlin, 1999.

1239   V. B. Matveev. *Inverse Problems* 17, 633 (2001).

1240   L. Mátyás, R. Klages. *arXiv:nlin.CD*/0211023 (2002).

1241   K. Maurin. *The Riemann Legacy* Kluwer, Dordrecht, 1997.

1242   J. D. Maynard. *Rev. Mod. Phys.* 73, 401 (2001).

1243   W. L. McCubbin, F. A. Teemull. *Phys. Rev.* A 6, 2478 (1972).

1244   J. McDonald. *J. Pure Appl. Math.* 104, 213 (1995).

1245   J. McDonald. *Disc. Comput. Geom.* 27, 501 (2002).

1246   K. T. McDonald. *arXiv:physics*/0006075 (2000).

1247   J. McGarva, G. Mullineux. *Appl. Math. Model.* 17, 213 (1993).

1248   C. C. McGeoch, B. M. E. Moret. *SIGACT News* 30, 85 (1999).

1249   H. McKean, V. Moll. *Elliptic Curves*, Cambridge University Press, Cambridge, 1997.

1250   J. R. McLaughlin, O. H. Hald. *Bull. Am. Math. Soc.* 32, 241 (1995).

1251   A. McLennan. *arXiv:math.PR*/9904120 (1999).

1252   J. M. McNamee. *SIGSAM Bull.* 38, 1 (2004).

1253   A. B. Medvinski, C. V. Petrovski, I. A. Tichonov, D. A. Tichonov, B.-L. Li, E. Venturino, C. Malcho, G. R. Ivanitzki. *Usp. Fiz. Nauk* 172, 31 (2002).

1254   R. Meester, F. Redig, D. Znamenski. *arXiv:cond-mat*/0301481 (2003).

1255   E. Meijering. *Proc. IEEE* 90, 319 (2002).

1256   M. Meinhardt. *Rep. Progr. Phys.* 55, 797 (1992).

1257   H. Meißner, E. O. Steinborn. *Phys. Rev.* A 56, 1189 (1997).

1258   J. B. M. Melissen. *Ph.D. thesis*, Utrecht 1997.

1259   R. Melko, R. B. Mann. *arXiv:gr-qc*/0011004 (2000).

1260   S. S. Melnyk, O. U. Usatenko, V. A. Yampol'skii, V. A. Golick. *arXiv:physics*/0402042 (2004).

1261   M. Menzel. *Ph.D. thesis*, Kaiserslautern, 1997.

1262   R. Mera. *Am. Math. Monthly* 99, 56 (1992).

1263   B. Mercier. *An Introduction to the Numerical Analysis of Spectral Methods*, Springer-Verlag, Berlin, 1989.

1264   R. Merks, A. Hoekstra, J. Kaandorp, P. Sloot. *J. Theor. Biol.* 224, 153 (2003).

1265   P. W. Messer, P. F. Arndt, M. Lässig. *arXiv:bio.GN*/0501010 (2005).

1266   N. Metropolis, R. L. Ashenhurst. *J. ACM* 6, 415 (1959).

1267   N. Metropolis, S. M. Tanny. *Comput. Math. Appl.* 3, 77 (1977).

1268   G. Meurant. *Computer Solution of Large Linear Systems*, Elsevier, Amsterdam, 1999.

1269   Y. Meurice. *arXiv:quant-ph*/0202047 (2002).

1270   R. E. Mickens. *Application of Nonstandard Finite Difference Schemes*, World Scientific, Singapore, 2000.

1271   G. Micula, M. Micula. *Rev. d'Anal. Numer. Th. Approx.* 26, 117 (1997).

1272   G. Micula, S. Micula. *Handbook of Splines*, Kluwer, Dordrecht, 1999.

1273   A. A. Middleton, C. Tang. *Phys. Rev. Lett.* 74, 742 (1995).

1274   M. Mignotte, D. Stefănescu. *Polynomials*, Springer-Verlag, Singapore, 1999.

1275   P. Mikulík in F. Axel, D. Gratias (eds.). *Beyond Quasicrystals*, Springer, Berlin, 1995.

1276   P. D. Miller, S. Kamvissis. *Phys. Lett.* A 247, 75 (1998).

1277   G. N. Milstein, M. V. Tretyakov. *Stochastic Numerics for Mathematical Physics*, Springer, Berlin, 2004.

1278   A. Minguzzi, P. Vignolo, M. P. Tosi. *Phys. Rev.* A 63, 063604 (2001).

1279   K. Mischaikow, M. Mrozek. *Math. Comput.* 67, 1023 (1998).

1280   A. M. Mitichkina, A. A. Ryabenko. *Program.* 27, 10 (2001).

1281   M. Mitzenmacher. *Internet Math.* 1, 226 (2004). http://www.internetmathematics.org/volumes/1/3/Mitzenmacher.pdf

1282   T. Miyamoto. *Nonlin. Anal.* 54, 755 (2003).

1283   G. R. Mocken, C. H. Keitel. *J. Comput. Phys.* 199, 558 (2004).

1284   J. R. Moffitt, P. Macdonald, J. F. Lindner. *Phys. Rev.* E 70, 016203 (2004).

1285   R. A. Mollin. *Fundamental Number Theory with Applications*, CRC Press, Boca Raton, 1998.

1286   R. A. Mollin. *Nieuw Archief Wiskunde* 17, 383 (1999).

1287   A. F. Monna. *Dirichlet's Principle*, Ososthoek, Scheltema & Holkema, Utrecht, 1975.

1288   R. Montagne, E. Hernández–García. *arXiv:nlin.CD*/0002053 (2000).

1289   M. A. Montemurro, D. H. Zanette. *arXiv:cond-mat*/0109218 (2001).

1290   M. A. Montemurro, P. A. Pury. *Fractals* 10, 451 (2002).

1291   R. Montgomery. *Notices Am. Math. Soc.* 48, 471 (2001).

1292   T. K. Moon, W. C. Stirling. *Mathematical Methods and Algorithms for Signal Processing*, Prentice Hall, Upper Saddle River, 2000.

1293   R. E. Moore. *Interval Analysis*, Prentice-Hall, Englewood Cliffs, 1966.

1294   R. E. Moore. *Methods and Applications of Interval Analysis*, Prentice-Hall, SIAM, Philadelphia, 1979.

1295   P. Moretti, A. Agresti. *Nuov. Cim.* 110 B, 905 (1995).

1296   A. Morgan. *Solving Polynomial Systems using Continuation for Engineering and Scientific Problems*, Prentice-Hall, Englewood Cliffs, 1987.

1297   K. Morita. *Applied Fourier Transform*, IOS Press, Ohmsha, 1995.

1298   S. Morita, S. Fujisawa, Y. Sugawara. *Surf. Sc. Rep.* 23, 1 (1996).

1299   R. E. Moritz. *Ann. Math.* 4, 179 (1903).

1300   W. J. Morokoff, R. E. Caflisch. *SIAM J. Sci. Comput.* 15, 1251 (1994).

1301   W. J. Morokoff, R. E. Caflisch. *J. Comput. Phys.* 122, 218 (1995).

1302   P. Morton, F. Vivaldi. *Nonlinearity* 8, 571 (1995).

1303   P. Mosseri, J. F. Sadoc. *Coll. de Phys.* C7, n23, 257 (1990).

1304   R. Mosseri. *J. Phys.* A 25, L 25 (1992).

1305   W. A. Moura–Melo, J. A. Helayël–Neto. *arXiv:hep-th*/0111116 (2001).

1306   W. A. Moura–Melo, J. A. Helayël–Neto. *Phys. Rev.* D 63, 065013 (2001).

1307   P. Moussa, J.-C. Yoccoz. *Commun. Math. Phys.* 186, 265 (1997).

1308   P. Moussa, S. Marmi. *Mathematical Physics Preprint* 466/99 (1999).
       http://rene.ma.utexas.edu/mp_arc-bin/mpa?yn=99-466

1309   P. Moussa, A. Cassa, S. Marmi. *J. Comput. Appl. Math.* 105, 403 (1999).

1310   A. I. Mudrov, K. B. Varnashev. *arXiv:cond-mat*/9805081 (1998).

1311   A. I. Mudrov, K. B. Varnashev. *arXiv:hep-th*/9811125 (1998).

1312   S. N. Mujamdar, A. Comtet. *arXiv:cond-mat*/0409566 (2004).

1313   J. G. Muga, J. P. Palao, B. Navarro, I. L. Egusquiza. *Phys. Rep.* 395, 357 (2004).

1314   S. Mukhopadhyay, K. Bhattacharyya. *Int. J. Quant. Chem.* 101, 27 (2004).

1315   O. Mülken, A. Blumen. *arXiv:quant-ph*/0410243 (2004).

1316   O. Mülken, A. Blumen. *arXiv:quant-ph*/0502004 (2005).

1317   R. Müller. *Comput. Aided Geom. Design* 19, 479 (2002).

1318   J. G. Muga, I. L. Egusquiza, J. A. Damborenea, F. Delgado. *arXiv:quant-ph*/0206181 (2002).

1319   G. B. Muravskii. *J. Sound Vibr.* 274, 653 (2004).

1320   K. P. N. Murthy. *arXiv:cond-mat*/0104215 (2001).

1321   J. I. Musher. *Am. J. Phys.* 34, 267 (1966).

1322   F. J. Muzzio, P. D. Swanson. *Phys. Fluids* A 3, 822 (1991).

1323   S. Nagahiro, Y. Hayakawa. *arXiv:cond-mat*/0210374 (2002).

1324   J. Nagler. *Phys. Rev.* E 71, 026227 (2005).

1325   H. Nakada. *Monatsh. Math.* 105, 131 (1988).

1326   H. Nakada. *Acta Arithm.* 56, 279 (1990).

1327   M. Nakagawa in M. Tokuyama, H. E. Stanley (eds.). *Statistical Physics*, American Institute of Physics, Melville, 2000.

1328   T. Nakaki. *Nonlin. Anal.* 47, 3849 (2001).

1329   V. Namias. *J. Inst. Math. Appl.* 25, 241 (1980).

1330   W. Narkiewicz. *Coll. Math.* 58, 151 (1989).

1331   C. Nash in I. M. James (ed.). *History of Topology*, Elsevier, Amsterdam, 1999.

1332   I. P. Natanson. *Constructive Function Theory*, v 3, Fredrick Ungar, New York, 1964.

1333   M. B. Nathanson. *Proc. Am. Math. Soc.* 125, 9 (1997).

1334   M. Nauenberg. *arXiv:nlin.CD*/0112003 (2001).

1335   A. Nayak, A. Vishwanath. *arXiv:quant-ph*/0010117 (2000).

1336   N. S. Nedialkov, V. Kreinovich, S. A. Starks. *Num. Alg.* 36, 325 (2004).

1337   G. H. Nedzhibov, M. G. Petkov. *Appl. Math. Comput.* 162, 427 (2005).

1338   S. S. Negi, R. Ramaswamy. *arXiv:nlin.CD*/0104054 (2001).

1339   D. Neshev, A. Nepomnyashchy, Y. S. Kishvar. *arXiv:nlin.PS*/0107005 (2001).

1340   J. Neuberger, J. Gruenebaum. *Eur. J. Phys.* 3, 22 (1982).

1341   N. D. Newby, Jr. *Am. J. Phys.* 47, 161 (1960).

1342   P. K. Newton. *The N-Vortex Problem*, Springer-Verlag, New York, 2001.

1343   R. G. Newton. *Scattering Theory of Waves and Particles*, Springer-Verlag, New York, 1982.

1344   P. Niamsup. *J. Math. Anal. Appl.* 250, 598 (2000).

1345 P. Nicodéme, B. Salvy, P. Flajolet. *INRIA Preprint* 3606/99 (1999). http://www.inria.fr/RRRT/RR-3606.html

1346 A. M. Niell. *Comput. Math. Appl.* 41, 1 (2001).

1347 I. N. Nikitin, J. de Luca. *Int. J. Mod. Phys.* C 12, 739 (2001).

1348 H. Nikolić. *arXiv:gr-qc*/9904078 (1999).

1349 J. Nievergelt, P. Schorn. *Informatik Spektrum* 16, 39 (1993).

1350 Y. Nievergelt. *SIAM Rev.* 36, 258 (1994).

1351 J. C. C. Nitsche. *Lectures on Minimal Surface* vI. Cambridge University Press, Cambridge, 1989.

1352 J. C. C. Nitsche in H. T. Davis, J. C. C. Nitsche (eds.). *Statistical Thermodynamics and Differential Geometry of Microstructured Materials*, Springer-Verlag, New York, 1993.

1353 H. Nitta, T. Kuda, H. Minowa. *Am. J. Phys.* 67, 966 (1999).

1354 J. Nocedal, S. J. Wright. *Numerical Optimization*, Springer-Verlag, New York, 1999.

1355 Y. Nogami, F. M. Toyama, W. van Dijk. *arXiv:quant-ph*/0005109 (2000).

1356 E. Novak. *arXiv:quant-ph*/0008124 (2000).

1357 E. A. Novikov. *JETP* 41, 937 (1976).

1358 C. R. Nugent, W. M. Quarles, T. H. Solomon. *Phys. Rev. Lett.* 93, 218301 (2004).

1359 H. N. Nunez-Yepez, A. L. Salas-Brito, C. A. Vargas, L. Vicente. *Phys. Lett.* A 145, 101 (1990).

1360 K. J. Nurmeal. *J. Phys.* A 31, 1035 (1998).

1361 M. A. Nyblom. *Am. Math. Monthly* 109, 559 (2002).

1362 J. F. Nye. *J. Opt.* A 5, 495 (2003).

1363 R. B. S. Oakeshott, A. MacKinnon. *J. Phys.* CM 5, 6991 (1993).

1364 R. B. S. Oakeshott, A. MacKinnon. *J. Phys.* CM 5, 9355 (1993).

1365 K. O'Bryant, B. Reznick, M. Serbinowska. *arXiv:math.NT*/0308028 (2003).

1366 J. R. Ockendon. *arXiv:math-ph*/0308024 (2003).

1367 R. O'Connell. *arXiv:quant-ph*/0212092 (2002).

1368 D. H. J. O'Dell. *arXiv:quant-ph*/0011105 (2000).

1369 A. Ogawa. *The Mathematica Journal* 2, n1, 59 (1992).

1370 T. Odagaki in B. Gruber, T. Otsuka (ed.). *Symmetries in Science VII*, Plenum, New York, 1994.

1371 A. Odlyzko in R. Graham, M. Grötschel, L. Lovász (eds.). *Handbook of Combinatorics*, Elsevier, Amsterdam, 1995.

1372 G.-Y. Oh. *arXiv:cond-mat*/9901156 (1999).

1373 G.-Y. Oh. *Phys. Rev.* B 63, 087301 (2001).

1374 Z. Olami, H. J. S. Feder, K. Christensen. *Phys. Rev. Lett.* 68, 1244 (1992).

1375 M. H. Oliveira, J. A. Miranda. *Eur. J. Phys.* 22, 31 (2001).

1376 V. S. Olkhovsky, E. Recami. *Phys. Rep.* 214, 339 (1992).

1377 V. S. Olkhovsky, E. Recami, F. Raciti, A. K. Zaichenko. *J. Physique* I 5, 1351 (1995).

1378 V. S. Olkhovsky, E. Recami, J. Jakiel. *Phys. Rep.* 398, 133 (2004).

1379 F. Ollendorff. *Potentialfelder in der Elektrotechnik*, Springer-Verlag, Berlin, 1932.

1380 F. Ollendorff. *Berechnung magnetischer Felder*, Springer-Verlag, Wien, 1952.

1381 M. A. Ol'shanetskiĭ, V.-B. K. Rogov. *Sbornik Math.* 190, 717 (1999).

1382 F. W. J. Olver. *SIAM Rev.* 12, 228 (1970).

1383 R. E. O'Malley, Jr. *Introduction to Singualr Perturbations*, Academic Press, New York, 1974.

1384 R. E. O'Malley. *J. Austral. Math. Soc.* B 40, 469 (1999).

1385 R. E. O'Malley, Jr. *J. Math. Anal. Appl.* 251, 433 (2000).

1386 R. Omnés. *The Interpretation of Quantum Mechanics*, Princeton University Press, Princeton, 1994.

1387 E. T. Ong, K. M. Lim, K. H. Lee, H. P. Lee. *J. Comput. Phys.* 192, 244 (2003).

1388 E. Onofri, M. Pauri. *J. Math. Phys.* 19, 1850 (1995).

1389 A. H. Opie, J. Gridlay. *Phys. Rev.* E 51, 724 (1995).

1390 A. H. Opie, J. Gridlay. *Physica* A 303, 119 (1995).

1391 D. Osadchy, J. E. Avron. *arXiv:math-ph*/0101019 (2001).

1392 W. F. Osgood. *Lehrbuch der Funktionentheorie*, Teubner, Leipzig, 1923.

1393 S. Osher, R. Fedkiw. *Level Set Methods and Dynamic Implicit Surfaces*, Springer-Verlag, New York, 2003.

1394 H. M. Osinga, B. Krauskopf. *Comput. Graphics* 26, 815 (2002).

1395 H. M. Osinga, B. Krauskopf. *Math. Intell.*. 26, n4, 25 (2004).

1396 K. I. Oskolkov. *ESI Preprint* 1356 (2003). ftp://ftp.esi.ac.at:/pub/Preprints/esi1356.ps

1397 N. B. Ouchi, K. Kaneko. *Chaos* 10, 359 (2000).

1398 Y. B. Ovchinnikov. *Opt. Commun.* 182, 35 (2000).

1399 H. Oymak, S. Erkoc. *Int. J. Mod. Phys.* C 11, 891 (2000).

1400 H. M. Ozaktas, M. A. Kutay, D. Mendlovic in P. Hawkes (ed.). *Advances in Imaging and Electron Physics* 106, Academic Press, San Diego, 1999.

1401 H. M. Ozaktas, Z. Zalevsky, M. Alper Kutay. *The Fractional Fourier Transform*, Wiley, Chichester, 2001.

1402 C. Pabst, R. Naulin. *Rev. Columb. Mater.* 30, 25 (1996).

1403 J. J. Palacios, C. Tejedor. *Phys. Rev.* B 48, 5386 (1993).

1404 V. Y. Pan. *SIAM Rev.* 39, 187 (1997).

1405 A. Papageorgiou, J. F. Traub. *arXiv:physics*/0011053 (2000).

1406 A. Papageorgiou, H. Wozniakowski. *arXiv:quant-ph*/0502054 (2005).

1407 A. Papoulis. *Systems and Transforms with Applications in Optics*, McGraw-Hill, New York, 1968.

1408 J. Paradís, P. Viader. *J. Math. Anal. Appl.* 253, 107 (2001).

1409 M. Pardy. *arXiv:math-ph*/0503003 (2005).

1410 R. P. Paris. *Proc. R. Soc. Lond.* A 432, 391 (1991).

1411 B.-Y. Park, G. M. Shim, G. Ihm, H. Lee. *J. Korean Phys. Soc.* 36, 135 (2000).

1412 D. S. Parker, B. Pierce, P. R. Eggert. *Comput. Sci. Eng.* July/August 58 (2000).

1413 M. C. Parker, S. D. Walker. *Opt. Commun.* 229, 23 (2004).

1414 R. L. Parker. *Proc. R. Soc. Lond.* 291, 60 (1966).

1415 B. Partoens, P. S. Deo. *arXiv:cond-mat*/0401207 (2004).

1416 S. Paszkowski. *Num. Alg.* 32, 193 (2003).

1417 R. K. Pathria. *Statistical Mechanics*, Pergamon Press, Oxford, 1972.

1418 M. Patriarca, K. Kaski, A. Chakraborti. *arXiv:cond-mat*/0402200 (2004).

1419 P. Patrício. *arXiv:physics*/04020361 (2004).

1420 D. B. Pearson. *Commun. Math. Phys.* 60, 13 (1978).

1421 J. E. Pearson. *Science* 261,189 (1993).

1422 T. Pearcey. *Lond. Edinb. Dubl. Phil. Mag.* 37, 311 (1946).

1423 S. Pekarsky, J. E. Marsden. *J. Math. Phys.* 39, 5894 (1998).

1424 S. Peleš. *Progr. Theor. Phys.* Suppl. 139, 496 (2000).

1425 P. Pellat-Finet. *Opt. Lett.* 19, 1388 (1994).

1426 M. D. Penrose. *Commun. Math. Phys.* 218, 153 (2001).

1427 J. M. Pérez–Jordá, W. Yang. *Chem. Phys. Lett.* 247, 484 (1995).

1428 O. Perron. *Die Lehre von den Kettenbrüchen*, v.1, v.2 Teubner, Stuttgart, 1954.

1429 O. Perron. *Irrationalzahlen*, de Gruyter, Berlin, 1960.

1430 P. Pérez–Álvarez, C. Trallero–Herrero, F. García–Moliner. *Eur. J. Phys.* 22, 275 (2001).

1431 P. Petek, M. Rugelj. *J. Math. Anal. Appl.* 222, 38 (1998).

1432 R. D. Peters, T. Pritchett. *Am. J. Phys.* 65, 1067 (1997).

1433 M. S. Petković. *Iterative Methods for Simultaneous Inclusion for Polynomial Zeros*, Springer-Verlag, Berlin, 1989.

1434 M. S. Petković, C. Carstensen, M. Trajkovíc. *Num. Math.* 69, 353 (1995).

1435 M. S. Petković, N. Kjurkchiev. *J. Comput. Appl. Math.* 80, 163 (1997).

1436 M. S. Petković, S. B. Tričković. *Comput. Math. Appl.* 31, 85 (1996).

1437 M. Petković, S. Ilić, S. Tričković. *Comput. Math. Appl.*. 34, n10, 49 (1997).

1438 M. S. Petkovič, L. D. Petkovič. *Complex Interval Arithmetic and Its Applications*, Wiley, Berlin 1998.

1439 M. S. Petković, S. Tričković, D. Herceg. *Jpn. J. Industr. Appl. Math.* 15, 295 (1998).

1440 M. S. Petković, D. Herceg. *J. Comp. Appl. Math.* 136, 283 (2001).

1441 V. Petrillo, L. Refaldi. *Opt. Commun.* 186, 35 (2000).

1442 E. Petrisor. *Physica* D 112, 319 (1998).

1443 P. G. Petropoulos. *Int. J. Numer. Model.* 13, 483 (2000).

1444 S. S. Petrova, A. D. Solov'ev. *Hist. Math.* 24, 361 (1997).

1445 T. J. Pickett. *Comm. Anal. Th. Contin. Fractions.* 11, 81 (2003).

1446 D. Pfannkuche, R. R. Gerhardts. *Phys. Rev.* B 46, 12606 (1992).

1447 F. Pfeiffer, C. Glocker. *Multibody Dynamics with Unilateral Contacts*, Wiley, New York, 1996.

1448 A. Pfluger. *Theorie der Riemannschen Fläche*, Springer-Verlag, Berlin, 1957.

1449 P. Pieranski, J. Baranska, A. Skjeltorp. *Eur. J. Phys.* 25, 613 (2004).

1450 T. A. Pierce. *Am. Math. Monthly* 36, 523 (1929).

1451 E. Piña, T. Ortiz. *J. Phys.* A 21, 1293 (1988).

1452 D. Pingel, P. Schmelcher, F. K. Diakonos. *Phys. Rep.* 400, 67 (2004).

1453 R. J. Pinnington. *J. Sound Vibr.* 268, 343 (2003).

1454 A. Pinkus. *J. Approx. Th.* 107, 1 (2000).

1455 G. Piranian, C. J. Titus, G. S. Young. *Mich. Math. J.* 1, 69 (1952).

1456 I. Pitowsky. *Jerusalem Phil. Quart.* 39, 81 (1990).

1457 I. Pitowsky. *Stud. Hist. Phil. Mod. Phys.* 27, 161 (1996).

1458 M. Plapp, J.-F. Gouyet. *Phys. Rev.* E 55, 45 (1997).

1459 E. Platen. *Acta Numerica* 8, 191 (1999).

1460 R. Plato. *Concise Numerical Mathematics*, American Mathematical Society, Providence, 2003.

1461 E. I. Poffald. *Am. Math. Monthly* 97, 205 (1990).

1462 W. F. Pohl. *J. Math. Mech.* 17, 975 (1968).

1463 H. Pollard. *Mathematical Introduction to Celestial Mechanics*, Prentice-Hall, Englewood Cliffs, 1966.

1464 G. Pólya, S, Szegö. *Aufgaben und Lehrsaetze aus der Analysis.* Springer-Verlag, Berlin, 1964.

1465 A. D. Polyanin. *Handbook of Linear Partial Differential Equations for Engineers and Scientists*, Chapman & Hall, Boca Raton, 2002.

1466 Y. Pomeau. *J. Phys.* A 13, L 193 (1981).

1467 S. Y. Popov. *J. Stat. Phys.* 102, 191, (2001).

1468 E. D. Popova, C. P. Ullrich in O. Gloor (ed.). *ISSAC 1998*, ACM Press, New York, 1998.

1469 I. R. Porteous. *Geometric Differentiation*, Cambridge University Press, Cambridge, 1994.

1470 M. Porto, H. E. Roman. *Phys. Rev.* E 62, 100 (2000).

1471 H. A. Posch, W. Thirring. *ESI Preprint* 843 (2000). http://ftp.esi.ac.at:/Abstracts/abs843.html

1472 H. A. Posch, W. Thirring. *J. Math. Phys.* 41, 3430 (2000).

1473 T. Pöschl, N. V. Brilliantov. *arXiv:cond-mat*/9906138 (1999).

1474 T. Poston, I. Stewart. *Catastrophe Theory and its Applications*, Pitman, London, 1978.

1475 L. Potin. *Formules et tables numériques relatives aux fonctions circulaires, hyperboliques, elliptiques*, Gauthier-Villars, Paris, 1925.

1476 A. D. Poularikas (ed.). *The Transforms and Application Handbook*, CRC Press, Boca Raton, 1995.

1477 M. B. Pour–El, I. Richards. *Ann. Math. Logic* 17, 61 (1979).

1478 M. B. Pour-El in E. R. Griffor (ed.). *Handbook of Computability Theory*, Elsevier, Amsterdam, 1999.

1479 V. V. Prasalov. *Polynomials*, Springer, Berlin, 2004.

1480 V. Pratt. *Comput. Graphics* 21, 145 (1987).

1481 W. H. Press, B. P. Flannery, S. A. Teukolsky, W. T. Vetterling. *Numerical Recipes, The Art of Scientific Computing*, Cambridge University Press, Cambridge, 1986.

1482 J. Prestin, E. Quak. *Rev. D'Anal. Numer. Theor. L'Approx.* 30, 219 (2001).

1483 L. Prigozhin, B. Zaltzman. *arXiv:cond-mat*/0005239 (2000).

1484 A. Pringsheim. *Math. Ann.* 22, 109 (1883).

1485 A. Pringsheim. *Math. Ann.* 44, 41 (1892).

1486 M. I. J. Probert. *Phys. Rev.* E 63, 058701 (2001).

1487 J. D. Pryce. *Numerical Solution of Sturm-Liouville Problems*, Oxford University Press, Oxford, 1993.

1488 S. P. Pudasaini, W. Eckart, K. Hutter. *Math. Models Meth. Appl. Sci.* 13, 1019 (2003).

1489 F. Qi, Z. Hou, H. Xin. *Phys. Rev. Lett.* 91, 064102 (2003).

1490 Y. Qi. *Stat. Prob. Lett.* 62, 93 (2003).

1491 Y.-Y. Qi, X.-S. Liu, P.-Z. Ding. *Int. J. Quant. Chem.* 101, 21 (2004).

1492 G. Qi, S. Du, G. Chen, Z. Chen, Z. Yuan. *Chaos, Solitons, Fractals* 23, 1671 (2004).

1493 M. Queffélec. *Substitution Dynamical Systems. Spectral Analysis*, Springer-Verlag, Berlin, 1987.

1494 M. I. Rabinovich, A. B. Ezersky, P. D. Weidman *The Dynamics of Patterns*, World Scientific, Singapore, 2000.

1495 S. Rabinovich. *Measurement Errors*, American Physics Institute, New York, 1993.

1496 H. Rademacher. *Am. J. Math.* 61, 237 (1939).

1497 S. Rahmann, E. Rivals. *Combinat., Prob. Comput.* 12, 73 (2003).

1498 S. Rajasekar, V. Chinnathambi. *Physica* A 282, 137 (2000).

1499 S. Ramanathan, A. E. Lobkovsky. *arXiv:cond-mat*/9901030 (1999).

1500 S. Ramanujan in G. H. Hardy, P. V. Seshu Aiyar, B. M. Wilson (eds.). *Collected Papers of Srinivasa Ramanujan*, Chelsea, New York, 1962.

1501 G. Ramharter. *Proc. Am. Math. Soc.* 99, 596 (1987).

1502 R. Rammal. *J. Physique* 45, 191 (1984).

1503 R. Rammal, J. Bellissard. *J. Phys. France* 51, 1803 (1990).

1504 A. F. Rañada, J. L. Trueba. *Phys. Lett.* A 232, 25 (1997).

1505 D. C. Rapaport. *The Art of Molecular Dynamic Simulations*, Cambridge University Press, Cambridge, 1997.

1506 S. N. Rasband. *Chaotic Dynamics of Nonlinear Systems*, Wiley, Chichester, 1990.

1507 H. E. Rauch, A. Lebowitz. *Elliptic Functions, Theta Functions and Riemann Surfaces*, Williams & Wilkins, Baltimore, 1973.

1508 J. Rauch. *Arch. Rat. Mech. Anal.* 174, 83 (2004).

1509 S. Rawal, G. J. Rodgers. *Physica* A 346, 621 (2005).

1510 E. Recami. *Found. Phys.* 31, 1119 (2001).

1511 L. E. Reichl. *The Transition to Chaos*, Springer-Verlag, New York, 1992.

1512 C. Reick, E. Mosekilde. *Phys. Rev.* E 52, 1418 (1995).

1513 C. Reid, S. Whineray. *Phys. Lett.* A 199, 49 (1995).

1514 R. Reigada, A. H. Romero, A. Sarmiento, K. Lindenberg. *arXiv:cond-mat*/9905003 (1999).

1515 M. Reinsch. *Am. J. Phys.* 62, 271 (1994).

1516 M. W. Reinsch, J. J. Morehead. *arXiv:math-ph*/9906007 (1999).

1517 C. A. Reiter. *Chaos, Solitons, Fractals* 23, 1111 (2005).

1518 A. Rényi. *Acta Math.* 8, 477 (1957).

1519 M. Revers. *Monatsh. Math.* 131, 215 (2000).

1520 E. Reyssat. *Quelques Aspects des Surfaces de Riemann*, Birkhäuser, Basel, 1989.

1521 P. Ribenboim. *Catalan's Conjecture: Are 8 and 9 the only Consecutive Powers?*, Academic Press, Boston, 1994.

1522 P. Ribenboim. *The New Book of Prime Number Records*, Springer-Verlag, New York, 1996.

1523 P. E. Ricci. *Operator Theory: Advances and Applications* 110, 257 (1999).

1524 G. Richardson, J. R. King. *J. Phys.* A 35, 9857 (2002).

1525 T. J. Richardson. *Ann. Math. Artif. Intell.* 31, 127 (2001).

1526 T. J. Richardson. *Ann. Math. Artif. Intell.* 31, 147 (2001).

1527 K. Richter. *Europhys. Lett.* 29, 151 (1995).

1528 N. Richert. *Am. Math. Monthly* 99, 101 (1992).

1529 J. N. Ridley, G. Petruska. *Indag. Math.* 11, 273 (2000).

1530 M. Ripoll, P. Español, M. H. Ernst. *arXiv:cond-mat*/9902121 (1999).

1531 P. J. Rippon. *Math. Gaz.* 67, 189 (1983).

1532 H. Risken. *The Fokker–Planck Equation*, Springer-Verlag, Berlin, 1984

1533 J. J. Risler. *Mathematical Methods for CAD*, Cambridge University Press, Cambridge, 1992.

1534 J. F. Ritt. *Trans. Am. Math. Soc.* 24, 21 (1922).

1535 K. Ritter. *Average-Case Analysis of Numerical Problems*, Springer-Verlag, Berlin, 2000.

1536 R. Rivera, D. Villarroel. *J. Math. Phys.* 43, 5026 (2002).

1537 F. B. Rizzato, G. I. de Oliveira, R. Erichsen. *Phys. Rev.* E 57, 2776 (1998).

1538 D. P. Robbins, E. D. Bolker. *Aequ. Math.* 22, 268 (1981).

1539 A. Robert. *Am. Math. Monthly* 101, 420 (1994).

1540 T. G. Robertazzi, S. C. Schwartz. *ACM Trans. Math. Softw.* 14, 101 (1988).

1541 L. Robinett. *Phys. Rev.* D 18, 3610 (1978).

1542 R. W. Robinett. *arXiv:quant-ph*/0401031 (2004).

1543 A. M. Rocket, P. Szüsz. *Continued Fractions*, World Scientific, Singapore, 1992.

1544 C. Rodenberg. *Math. Annalen* 14, 46 (1879).

1545 D. F. Rogers, J. A. Adams. *Mathematical Elements for Computer Graphics*, McGraw-Hill, New York, 1990.

1546 J. M. Rojas, X. Wang. *J. Complexity* 12, 116 (1996).

1547 J. M. Rojas. *arXiv:math.AG*/0212309 (2002).

1548 F. L. Román. *Am. J. Phys.* 70, 847 (2002).

1549 J. S. Roman, L. Roso, L. Plaja. *J. Phys.* B 37, 435 (2004).

1550 J. M. Romero, J. A. Santiago, J. D. Vergara. *arXiv:hep-th*/0211165 (2002).

1551 M. Romera, V. Bañuls, G. Pastor, G. Álvarez, F. Montoya. *Comput. Graphics* 25, 529 (2001).

1552 P. Rosenau. *Phys. Rev.* B 36, 5868 (1987).

1553 P. L. Rosin. *Pattern Recogn. Lett.* 14, 799 (1993).

1554 C. Ross, J. Sorensen. *Coll. Math. J.* 31, 1 (2000).

1555 W. Rossman. *J. Math. Soc. Jap.* 52, 25, (2000).

1556 H. Rost. *Math. Semesterber.* 46, 97 (1999).

1557 P. Rotter, U. Rössler, H. Silberbauer, M. Suhrke. *Physica* B 212, 231 (1995).

1558 O. Ruatta in B. Mourrain (ed.). *ISSAC 2001*, ACM, Baltimore, 2001.

1559 R. T. Rubin, R. Zwanzig. *J. Math. Phys.* 2, 861 (1961).

1560 O. Rudzick. *arXiv:nlin.PS*/0407017 (2004).

1561 K. Ruedenberg, J.-Q. Sun. *J. Chem. Phys.* 100, 5836 (1994).

1562 A. Ruffing, M. Simon. *Progr. Theor. Phys.* S 150, 401 (2003).

1563 A. Ruffing, M. Simon in B. Aulbach, S. Elaydi, G. Ladas (eds.). *Proceedings of the Sixth integrnational Conference on Difference Equations*, CRC Press, Boca Raton, 2004.

1564 S. M. Ruiz, J. Sondow. *arXiv:math.NT*/0210312 (2002).

1565 N. F. Rulkov. *arXiv:nlin.CD*/0011028 (2000).

1566 B. W. Rust. *Comput. Sc. Eng.* 4, n4, 72 (2002).

1567 M. Ruzzi. *J. Phys.* A 35, 1763 (2002).

1568 D. G. Saari, Z. J. Xia. *Notices Am. Math. Soc.* 42, 538 (1995).

1569 P. G. Saffman. *Vortex Dynamics*, Cambridge University Press, Cambridge, 1992.

1570 M. Saint Jean, C. Even, C. Guthman. *arXiv:cond-mat*/0101285 (2001).

1571 H. Sakaguchi. *J. Phys. Soc. Jpn.* 67, 96 (1998).

1572 H. Sakaguchi, M. Ohtaki. *Physica* A 272, 300 (1999).

1573 H. Sakaguchi. *Progr. Theor. Phys.* Suppl. 103, 703 (2000).

1574 H. Sakaguchi, M. Ohtaki. *arXiv:nlin.PS*/0405045 (2004).

1575 T. Sakaguchi. *arXiv:math-ph*/0412020 (2004).

1576 T. Sakajo. *Physica* D 196, 243 (2004).

1577 V. P. Sakhnenko, G. M. Chechin. *Phys. Dokl.* 39, 42 (1994).

1578 L. Salasnich. *J. Math. Phys.* 41, 8016 (2000).

1579 A. Salat. *Z. Naturf.* 40 a, 959 (1985).

1580 W. Salejda, P. Szyszuk. *Physica* A 252, 547 (1998).

1581 A. Salam. *J. Comput. Appl. Math.* 46, 455 (1993).

1582 R. Salem, A. Zygmund. *Duke Math. J.* 12, 569 (1945).

1583 P. Sancho. *Open Sys. Inform. Dyn.* 7, 157 (2000).

1584 R. Sankaranarayanan, A. Lakshminarayn, V. B. Sheorey. *arXiv:nlin.CD*/0005035 (2000).

1585 D. E. Santos, Jr., R. Viera Martins, O. C. Winter, R. R. Cordeiro. *Astron. Astrophys.* 331, 1108 (1998).

1586 A. S. Sanz, F. Borondo. *Phys. Rev.* B 61, 7743 (2000).

1587 A. S. Sanz. *arXiv:quant-ph*/0412050 (2004).

1588 J. M. Sanz-Serma in P. L. Garrido, J. Marro (eds.). *Third Granada Lectures in Computational Physics*, Springer-Verlag, Heidelberg, 1995.

1589 G. Sapiro, V. Caselles. *J. Diff. Eq.* 135, 238 (1997).

1590 R. Saravanan, O. Narayan, K. Banerjee, J. K. Bhattacharjee. *Phys. Rev.* A 31, 520 (1984).

1591 P. Saucez, W. E. Schiesser, A. V. Wouver. *Math. Comput. Simul.* 56, 171 (2001).

1592 T. Sauter. *Comput. Phys. Commun.* 125, 119 (2000).

1593 A. Schädle. *Wave Motion* 35, 181 (2002).

1594 A. C. Schaeffer. *Duke Math. J.* 21, 383 (1954).

1595 R. W. Scharstein. *IEEE Trans. Antennas Prop.* 52, 452 (2004).

1596 P. Schatte. *ZAMM* 53, 553 (1973).

1597 A. Schenkel, J. Zhang, Y.-C. Zhang. *Fractals* 1, 47 (1993).

1598 A. Schenkel, J. Wehr, P. Wittwer. *Math. Phys. Electr. J.* 6 (2000). http://www.ma.utexas.edu/mpej/MPEJ.html

1599 G. Scheuermann, H. Krüger, M. Menzel, A. P. Rockwood. *IEEE Trans. Visual. Comput. Graphics* 4, 109 (1998).

1600 J. Schicho in O. Gloor (ed.). *ISSAC 1998*, ACM Press, New York, 1998.

1601 W. E. Schiesser. *The Numerical Method of Lines of Partial Differential Equations*, Academic Press, San Diego, 1991.

1602 K. Schittkowski. *J. Comput. Appl. Math.* 163, 29 (2004).

1603 D. Schleicher. *Ann. Acad. Sci. Fenn.* Math. 28, 3 (2003).

1604 T. Schlösser, K. Ensslin, J. P. Kotthaus, M. Holland. *Europhys. Lett.* 33, 683 (1996).

1605 A. L. Schmidt. *Acta Math.* 134, 1 (1975).

1606 A. L. Schmidt. *Monatsh. Math.* 93, 39 (1982).

1607 A. L. Schmidt in A. D. Pollington, W. Moran (eds.). *Number Theory with an Emphasis on the Markoff Spectrum*, Marcel Dekker, New York, 1993.

1608 F. Schmidt, R. Simion. *Aequ. Math.* 44, 11 (1992).

1609 J. H. Schmidt. *Stud. Hist. Phil. Mod. Phys.* 28, 433 (1997).

1610 J. H. Schmidt. *Stud. Hist. Phil. Mod. Phys.* 29, 81 (1998).

1611 M. Schmidt, F. Selleri. *Found. Phys. Lett.* 4, 1 (1991).

1612 M. Schmidt in F. Selleri (ed.). *Wave–Particle Duality*, Plenum Press, New York, 1992.

1613 B. Schmuland. *Am. Math. Monthly* 110, 407 (2003).

1614 B. Schoeneberg. *Elliptic Modular Functions*, Springer-Verlag, Berlin, 1974.

1615   A. Schönhage. *Jber. Dt. Math.-Ver.* 92, 1 (1990).

1616   S. H. Schot. *Am. J. Phys.* 46, 1090 (1978).

1617   M. R. Schroeder. *Number Theory in Science and Technology*, Springer-Verlag, Berlin, 1984.

1618   L. S. Schulman. *arXiv:quant-ph*/0109149 (2001).

1619   L. S. Schulman. *Chaos, Solitons, Fractals* 14, 823 (2002).

1620   R. L. Schult, D. G. Ravenhall, H. W. Wyld. *Phys. Rev.* B 39, 5476 (1989).

1621   H. G. Schuster. *Deterministic Chaos*, Physik-Verlag, Weinheim, 1984.

1622   R. Schuster, K. Ensslin in R. Helbig (ed.). *Festkörperprobleme 34*, Vieweg, Braunschweig, 1995.

1623   G. M. Schütz, S. Trimper. *arXiv:cond-mat*/0406593 (2004).

1624   D. Schwalbe, S. Wagon. *VisualDSolve*, TELOS/ Springer, Santa Clara, 1997.

1625   C. Schwartz. *J. Math. Phys.* 26, 411 (1985).

1626   H. A. Schwarz. *Gesammelte Mathematische Abhandlungen*, Chelsea, New York, 1972.

1627   F. Schweiger. *Ergodic Theory of Fibred Systems and Metric Number Theory*, Clarendon Press, Oxford, 1995.

1628   J. Schwenk in L. Castellani, J. Wess (eds.). *Quantum Groups and Their Application in Physics*, IOS Press, Amsterdam, 1996.

1629   A. J. Scott, C. A. Holmes, G. J. Milburn. *arXiv:nlin.CD*/0001037 (2000).

1630   D. W. Scott. *Biometrika* 66, 605 (1979).

1631   R. E. Sears. *Am. J. Phys.* 63, 854 (1995).

1632   T. W. Sederberg, J. P. Snively in R. R. Martin (ed.). *The Mathematics of Surfaces II*, Clarendon Press, Oxford, 1987.

1633   R. Sedgewick, P. Flajolet. *An Introduction to the Analysis of Algorithms*, Addison-Wesley, Reading, 1996.

1634   D. M. Sedrakian, A. Z. Khachatrian. *arXiv:cond-mat*/9909275 (1999).

1635   B. Segre. *The Nonsingular Cubic Surfaces*, Oxford University Press, Oxford, 1942.

1636   K. Semerdzhiev. *Mathematica Balkanica* 8, 311 (1994).

1637   S. Sen, M. Manciu. *Phys. Rev.* E 64, 056605 (2001).

1638   B. Sendov, A. Andreev, N. Kjurkchiev in P. G. Ciarlet, J. L. Lions (eds.). *Handbook of Numerical Analysis* III, North Holland, Amsterdam, 1994.

1639   M. Senechal. *Quasicrystals and Geometry*, Cambridge University Press, Cambridge, 1995.

1640   H. Serizawa, K. Hongo. *Wave Motion* 36, 103 (2002).

1641   R. Séroul. *Programming for Mathematicians*, Springer-Verlag, Berlin, 2000.

1642   J. A. Sethian. *Level Set Methods and Fast Marching Methods*, Cambridge University Press, Cambridge, 1999.

1643   A. P. Seyranian. *Dokl. Phys.* 394, 338 (2004).

1644   O. Shagrir. *Theor. Comput. Sc.* 317, 105 (2004).

1645   J. Shallit. *J. Res. NBS* B 80, 285 (1976).

1646   D. Shanks. *J. Math. Phys.* 34, 1 (1953).

1647   J. V. Shebalin. *Physica* D 66, 381 (1993).

1648   T. Shibata, K. Kaneko. *arXiv:nlin.AO*/0204024 (2002).

1649   L. Shifren, R. Akis, D. K. Ferry. *Phys. Lett.* A 274, 75 (2000).

1650   K. Shihara, T. Sasaki. *Jpn. J. Industr. Appl. Math.* 13, 107 (1996).

1651   N. Shimoyama, K. Sugawara, T. Mizuguchi, Y. Hayakawa, M. Sano. *Phys. Rev. Lett.* 76, 3870 (1996).

1652   K. C. Shin. *arXiv:math-ph*/0404015 (2004).

1653 M. Shishikura in T. Lei (ed.). *The Mandelbrot Set, Theme and Variations*, Cambridge University Press, Cambridge, 2000.

1654 M. F. Shlesinger, G. M. Zaslavsky, J. Klafter. *Nature* 363, 31 (1993).

1655 N. Shnerb, E. Eisenberg. *Phys. Rev. E* 49, R 1005 (1994).

1656 Y. I. Shokin in G. Alefeld, A. Frommer, B. Lang (eds.). *Scientific Computing and Validated Numerics*, Akademie Verlag, Berlin, 1996.

1657 J.-G. Si, X.-P. Wang. *Comput. Math. Appl.* 43, 81 (2002).

1658 K. Siddiqi, S. Bouix, A. Tannenbaum, S. W. Zucker. *Int. J. Comput. Vision* 48, 215 (2002).

1659 W. Sillitto. *J. Mod. Opt.* 48, 459 (2001).

1660 C. Simó in C. Casacuberta, R. M. Miró–Roig, J. Verdera, S. Xambó–Descamps (eds.). *European Congress of Mathematics*, Birkhäuser, Basel, 2001.

1661 B. Simon. *Ann. Phys.* 146, 209 (1983).

1662 B. Simon. *Proc. Am. Math. Soc.* 125, 203 (1997).

1663 T. E. Simos. *Comput. Chem.* 23, 513 (1999).

1664 S. Sinha, W. L. Ditto. *Phys. Rev. Lett.* 81, 2156 (1998).

1665 C. Sire, P. L. Krapivsky. *arXiv:cond-mat*/0106457 (2001).

1666 L. Sirovich. *Techniques of Asymptotic Analysis*, Springer, New York, 1971.

1667 R. D. Skeel, J. Keiper. *Elementary Numerical Computing with Mathematica*, McGraw-Hill, Englewood Cliffs, 1993.

1668 J. Slepian. *Am. J. Phys.* 19, 87 (1951).

1669 I. H. Sloan. *ANZIAM* 42, 3 (2000).

1670 N. J. A. Sloane in C. Ding, T. Helleseth, N. Niederreiter (eds.). *Sequences and their Applications*, Springer-Verlag, London, 1999.

1671 U. Smilansky, R. Sankaranarayanan. *arXiv:nlin.SI*/0503002 (2005).

1672 H. V. Smith. *Numerical Methods of Integration*, Chartwell-Bratt, Lund, 1993.

1673 S. J. Smith. *Expos. Math.* 18, 389 (2000).

1674 W. R. Smythe. *Static and Dynamic Electricity*, McGraw-Hill, New York, 1968.

1675 P. M. Soardi. *Potential Theory of Infinite Networks*, Springer-Verlag, Berlin, 1994.

1676 S. L. Sobolev, V. L. Vaskevich. *The Theory of Cubature Formulas*, Kluwer, Dordrecht, 1997.

1677 A. V. Sobolev, M. Solomyak. *Rev. Math. Phys.* 14, 421 (2002).

1678 S. L. Sobolevskii. *Diff. Equ.* 40, 807 (2004).

1679 M. Sodin, B. Tsirelson. *arXiv:math.CV*/0210090 (2003).

1680 M. Sodin, B. Tsirelson. *arXiv:math.CV*/0312258 (2003).

1681 M. Sodin. *arXiv:math.CV*/0410343 (2004).

1682 A. D. Sokal. *arXiv:math.NA*/0212035 (2002).

1683 D. Sokolovski in J. G. Muga, R. Sala Mayato, I. L. Egusquiza (eds.). *Time in Quantum Mechanics*, Springer-Verlag, Berlin, 2002.

1684 H. G. Solari, R. Gilmore. *Phys. Rev. A* 37, 3096 (1988).

1685 R. V. Solé, J. Valls, J. Bascompte. *Phys. Lett. A* 166, 123 (1992).

1686 L. Solimany. *arXiv:cond-mat*/0003061 (2000).

1687 F. J. Solis, L. Jódar, B. Chen. *Appl. Math. Lett.* 18, 85 (2005).

1688 A. M. Solunin, M. A. Solunin, S. A. Solunin. *Russ. Phys. J.* 46, 1010 (2003).

1689   J. C. Sommerer, E. Ott. *Nature* 365, 138 (1993).

1690   J. C. Sommerer, E. Ott. *Phys. Lett.* A 214, 243 (1996).

1691   I. Soprounov. *arXiv:math.AG*/0310168 (2003).

1692   Y. Sosov, C. E. Theodosiou. *J. Math. Phys.* 43, 2831 (2002).

1693   A. Soulier, T. Halpin–Healy. *arXiv:cond-mat*/0209451 (2002).

1694   D. T. Spasic, T. M. Atanackovic. *Arch. Appl. Mech.* 71, 327 (2001).

1695   H. Späth. *One Dimensional Spline Interpolation Algorithms*, A. K. Peters, Wellesley, 1995.

1696   H. Späth. *Computing* 57, 179 (1996).

1697   J. C. Sprott. *Strange Attractors*, M&T Books, New York, 1993.

1698   J. C. Sprott. *Comput. Graphics* 17, 325 (1993).

1699   J. C. Sprott. *Phys. Rev.* E 50, R 647 (1994).

1700   J. C. Sprott. *Am. J. Phys.* 65, 537 (1997).

1701   J. C. Sprott. *Am. J. Phys.* 68, 758 (2000).

1702   J. C. Sprott, S. J. Linz. *Int. J. Chaos Theory and Appl.* 5, 3 (2000).

1703   H. M. Srivastava, J. Choi. *Series Associated with the Zeta and Related Functions*, Kluwer, Dordrecht, 2001.

1704   J. J. Stamnes, B. Spjelkavik. *Optica Acta* 30, 1331 (1983).

1705   H. M. Stark in A. O. L. Atkin, B. J. Birch (eds.). *Computers in Number Theory*, Academic Press, New York, 1971.

1706   A. J. Starobin, S. W. Teitsworth. *arXiv:nlin.CD*/0010028 (2000).

1707   I. P. Stavroulakis, S. A. Tersian. *Partial Differential Equations*, World Scientific, Singapore, 1999.

1708   K. Stefanski. *Open Sys. Inform. Dyn.* 8, 89 (2001).

1709   H. J. Stetter. *Numerical Polynomial Algebra*, SIAM, Philadelphia, 2004.

1710   B. M. Stewart. *Theory of Numbers*, MacMillan, New York, 1964.

1711   D. E. Stewart. *SIAM Rev.* 42, 3 (2000).

1712   D. E. Stewart in A. Guran (ed.). *Impact and Friction of Solids, Structures, and Machines*, Birkhäuser, Boston, 2000.

1713   D. E. Stewart, J. C. Trinkle in M. C. Ferris, J.-S. Pang (eds.). *Complementary and Variational Problems*, SIAM, Philadelphia, 1997.

1714   I. Stewart. *Sci. Am.* n6, 122 (1992).

1715   E. U. Stickel. *SIAM J. Num. Anal.* 14, 971 (1987).

1716   P. Stifter, C. Leichtle, W. P. Schleich, J. Marklof. *Z. Naturf.* 52a, 377 (1997).

1717   D. L. Stock. *Math. Mag.* 73, 339 (2000).

1718   R. G. Stomphorst. *arXiv:quant-ph*/0111045 (2001).

1719   W. Strampp, V. Ganzha. *Differentialgleichungen mit Mathematica*, Vieweg, Braunschweig, 1995.

1720   G. Strang. *Calculus*, Wellesley, Wellesley, 1991.

1721   G. Strang. *SIAM Rev.* 41, 135 (1999).

1722   Y. M. Strelniker, R. Berkovits, A. Frydman, S. Havlin. *Phys. Rev.* E 69, 065105(R) (2004).

1723   R. S. Strichartz. *J. Fourier Anal. Appl.* 6, 533 (2000).

1724   S. H. Strogatz. *Nature* 378, 444 (1995).

1725   W. J. Stronge. *Impact Mechanics*, Cambridge University Press, Cambridge, 2000.

1726   B. Sturmfels. *Am. Math. Monthly* 105, 907 (1998).

1727   S. P. Suetin. *Russ. Math. Surv.* 57, 43 (2002).

1728 M. Sugihara in Z.-C. Shi, M. Mori (eds.). *Proceedings of Third China–Japan Seminar on Numerical Analysis*, Science Press, Bejing, 1998.

1729 M. Sugihara. *Num. Math.* 75, 379 (2001).

1730 S. J. Sullivan, B. G. Zorn. *ACM Trans. Math. Softw.* 21, 266 (1995).

1731 F. Sun, X. Zhang. *Appl. Math. Comput.* 137, 15 (2003).

1732 D. Sundarajan. *The Discrete Fourier Transform*, World Scientific, Singapore, 2001.

1733 S. K. Suslov. *J. Approx. Th.* 115, 289 (2002).

1734 S. K. Suslov. *An Introduction to Basic Fourier Series*, Kluwer, Dordrecht, 2003.

1735 P. M. Sutcliffe, A. T. Winfree. *Phys. Rev.* E 68, 016218 (2003).

1736 A. Sütö in F. Axel, D. Gratias (eds.). *Beyond Quasicrystals*, Springer, Berlin, 1995.

1737 K. Svozil. *arXiv:quant-ph*/9710052 (1997).

1738 N. Szpak. *arXiv:qgr-qc*/0411050 (2004).

1739 G. G. Szpiro. *Phys. Rev.* E 47, 4560 (1993).

1740 M. Tabor, J. Weiss. *Phys. Rev.* A 24, 2157 (1981).

1741 E. Tadmor. *SIAM J. Num. Anal.* 23, 1 (1986).

1742 H. Takahasi, M. Mori. *Publ. RIMS Kyoto Univ.* 9, 721 (1974).

1743 L. Takács. *Methol. Comput. Appl. Prob.* 1, 7 (1999).

1744 J. Talbot, G. Tarjus, P. Viot. *J. Phys.* A 32, 2997 (1999).

1745 J. Talbot, G. Tarjus, P. R. Van Tassel, P. Viot. *arXiv:cond-mat*/9906428 (1999).

1746 J. Talbot, G. Tarjus, P. Viot. *arXiv:cond-mat*/0008183 (2000).

1747 M. Tamaki. *J. Appl. Prob.* 19, 803 (1982).

1748 J. Tan, P. Jiang. *J. Comput. Appl. Math.* 163, 219 (2004).

1749 R. W. Tank, R. B. Stinchcombe. *J. Phys.* CM 5, 5623 (1993).

1750 S. Tasaki, P. Gaspard. *J. Stat. Phys.* 81, 935 (1995).

1751 K. Taubert. *Num. Math.* 26, 379 (1976).

1752 M. Tegmark. *arXiv:gr-qc*/9704009 (1997).

1753 M. Tegmark. *Class. Quantum Grav.* 14, L69 (1997).

1754 B. Teissier in D. L. Tê, K. Saito, B. Teissier (eds.). *Singularity Theory*, World Scientific, Singapore, 1995.

1755 A. F. F. Teixera. *arXiv:physics*/0310081 (2003).

1756 V. V. Ternovskiǐ, A. M. Khapaev. *JETP Lett.* 78, 272 (2003).

1757 P. K. Thakur, P. Biswas. *arXiv:cond-mat*/9811350 (1998).

1758 B. Thaller. *arXiv:quant-ph*/0409079 (2004).

1759 H. Theisel. *Comput. Graphics Forum* 21, 595 (2002).

1760 W. Thirring. *A Course in Mathematical Physics*, v.1 *Mechanics*, Springer-Verlag, New York, 1983.

1761 J. Thomchick, J. P. McKelvey. *Am. J. Phys.* 46, 40 (1978).

1762 J. Thomson. *Analysis* 15, 1 (1954/55).

1763 A. S. Thorndike, C. R. Cooley, J. F. Nye. *J. Phys.* A 11, 1455 (1978).

1764 D. J. Thouless. *Commun. Math. Phys.* 127, 187 (1990).

1765 D. J. Thouless, Y. Tan. *J. Phys.* A 24, 4055 (1991).

1766 W. J. Thron. *Am. Math. Monthly* 68, 734 (1961).

1767 J. H. Thywissen, M. Olshanii, G. Zabow, M. Drndić, K. S. Johnson, R. M. Westervelt, M. Prentiss. *Eur. Phys. J.* D 7, 361 (1999).

1768 D. A. Tikhonov, J. Enderlein, H. Malchow, A. B. Medvinsky. *Chaos, Solitons, Fractals* 12, 277 (2001).

1769 P. Tilli. *Calcolo* 35, 3 (1998).

1770 J. Todd. *Am. Math. Monthly* 66, 517 (1949).

1771 K. Tomizawa. *Numerical Simulation of Submicron Semiconductor Devices*, Artech House, Boston, 1993.

1772 M. A. Topinka, B. J. LeRoy, R. M. Westervelt, S. E. J. Shaw, R. Fleischmann, E. J. Heller, K. D. Maranowski, A. C. Gossard. *Nature* 410, 183 (2001).

1773 M. A. Topinka, R. M. Westervelt, E. J. Heller. *Physics Today* 56, n12, 47 (2003).

1774 R. Toral, C. R. Mirasso, E. Hernández-García, O. Piro. *arXiv:nlin.CD*/0002054 (2000).

1775 R. Toral, C. R. Mirasso, E. Hernández–García, O. Piro. *arXiv:nlin.CD*/0104044 (2001).

1776 R. Toral, C. R. Mirasso, E. Hernández-García, O. Piro in D. Abbott, L. B. Kish (eds.). *Unsolved Problems of Noise and Fluctuations*, American Institute of Physics, Melville, 2000.

1777 S. Torii, T. Sakurai, H. Sugiura in Z.-C. Shi, T. Ushijima (eds.). *Proceedings of the First China-Japan Seminar on Numerical Mathematics*, World Scientific, Singapore, 1993.

1778 M. Tornow, D. Weiss, K. von Klitzing, K. Eberl. *Phys. Rev. Lett.* 77, 147 (1996).

1779 I. M. Torrens. *Interatomic Potentials*, Pergamon Press, New York, 1972.

1780 G. F. Torres del Castillo. *J. Math. Phys.* 36, 3413 (1995).

1781 R. Town (ed.). *Lab. Laser Ener.* 69, 24 (1996).

1782 C. A. Tracy, H. Widom. *arXiv:math-ph*/9909001 (1999).

1783 J. F. Traub, A. G. Werschulz. *Complexity and Information*, Cambridge University Press, Cambridge, 1998.

1784 J. F. Traub. *Physics Today* 52, n5, 39 (1999).

1785 B. C. Travaglione, G. J. Milburn. *arXiv:quant-ph*/0109076 (2001).

1786 B. Tregenna, W. Flanagan, R. Maile, V. Kendon. *arXiv:quant-ph*/0304204 (2003).

1787 M. I. Tribelsky, K. Tsuboi. *Phys. Rev. Lett.* 76, 1631 (1996).

1788 M. I. Tribelsky, M. G. Velarde. *Phys. Rev.* E 54, 4973 (1996).

1789 H. Triebel. *Höhere Analysis*, Verlag der Wissenschaften, Berlin, 1972.

1790 M. Trott. *Mathematica Edu. Res.* 6, n4, 15 (1997).

1791 M. Trott. *The Mathematica Journal* 4, 465 (2000).

1792 M. Trott. *The Mathematica Journal* 8, 409 (2002).

1793 M. Trott. *The Mathematica GuideBook for Programming*, Springer-Verlag, New York, 2004.

1794 M. Trott. *The Mathematica GuideBook for Graphics*, Springer-Verlag, New York, 2004.

1795 M. Trott. *The Mathematica GuideBook for Symbolics*, Springer-Verlag, New York, 2005.

1796 M. Trott. *arXiv:quant-ph*/0012147 (2000).

1797 M. Trott. *The Mathematica Journal* 8, 50 (2001).

1798 S. A. Trugman. *Physica* D 83, 271 (1995).

1799 M. A. Trump, W. C. Schieve. *Classical Relativistic Many-Body Dynamics*, Kluwer, Dordrecht, 1999.

1800 W. Tucker. *Found. Comput. Math.* 2, 53 (2002).

1801 N. B. Tufillaro, H. G. Solari, R. Gilmore. *Phys. Rev.* A 41, 5717 (1990).

1802 N. B. Tufillaro. *Phys. Rev.* E 50, 4509 (1994).

1803 R. Tumulka. *arXiv:quant-ph*/0408113 (2004).

1804 J. Tupper. *M.Sc. Thesis* Department of Computer Science, University of Toronto (1996). http://www.dgp.toronto.edu/~mooncake/msc.html

1805 L. Turko. *arXiv:quant-ph*/0310128 (2003).

1806 J. E. Turner, K. Fox. *J. Phys.* A 1, 118 (1968).

1807 V. N. Tutubalin. *Physics-Uspekhi* 36, 628 (1993).

1808 F. Twilt in A. van der Burgh, J. Simonis (eds.). *Topics in Engineering Mathematics*, Kluwer, Amsterdam, 1992.

1809 C. J. Tymczak, G. S. Japaridze, C. R. Hardy, X.-Q. Wang. *Phys. Rev. Lett.* 80, 3673 (1998).

1810 C. J. Tymczak, G. S. Japaridze, C. R. Hardy, X.-Q. Wang. *Phys. Rev.* A 58, 2708 (1998).

1811 C. Tzara. *Ann. Phys.* 276, 12 (1999).

1812 C. W. Ueberhuber. *Numerical Computations I, II*, Springer, Berlin, 1997.

1813 T. Ueta in S. Namba, C. Hamaguchi, T. Ando (eds.). *Science and Technology of Mesoscopic Structures*, Springer-Verlag, Berlin, 1992.

1814 G. E. Uhlenbeck, P. C. Hemmer, M. Kac. *J. Math. Phys.* 4, 229 (1963).

1815 S. Ulam in D. Mauldin (ed.). *The Scottish Book*, Birkhäuser, Basel, 1981.

1816 C. Upstill. *Proc. R. Soc. Lond.* A 365, 95 (1979).

1817 S. Uryu, T. Ando. *Physica* B 227, 138 (1996).

1818 A. G. Ushveridze. *J. Phys.* A 21, 955 (1988).

1819 R. A. Usmani. *Indian J. Math.* 22, 23 (1980).

1820 A. V. Ustinov. *Math. Notes* 73, 97 (2003).

1821 J. J. Valencia, M. de Llano. *arXiv:cond-mat*/0304286 (2003).

1822 A. Valentini, H. Westman. *arXiv:quant-ph*/0403034 (2004).

1823 P. M. van den Berg, J. T. Fokkema. *IEEE Antennas Prop.* 27, 577 (1979).

1824 A. Vande Wouwer, P. Saucez, W. E. Schiesser in A. Vande Wouwer, P. Saucez, W. E. Schiesser (eds.). *Adapative Methods of Lines*, Chapman & Hall, Boca Raton, 2001.

1825 A. J. van der Poorten. *Nieuw Archief Wiskunde* 14, 221 (1996).

1826 B. L. van der Waerden. *Quellen und Studien zur Geschichte der Mathematik, Astronomie und Physik* 4, 359 (1938).

1827 N. Vanderwalle, S. Galam, M. Kramer. *arXiv:cond-mat*/0004271 (2000).

1828 M. van Hecke. *arXiv:cond-mat*/0110068 (2001).

1829 S. Van Huffel, P. Lemmerling (eds.). *Total Least Squares and Error-in-Variables Modeling*, Kluwer, Dordrecht, 2002.

1830 J. Vaníček, E. J. Heller. *arXiv:nlin.CD*/0209001 (2002).

1831 D. C. van Leijenhorst. *Nieuw Archief Wiskunde* 14, 255 (1996).

1832 W. van Saarloos. *arXiv:cond-mat*/0308540 (2003).

1833 J. M. Varah. *BIT* 36, 842 (1996).

1834 A. Vardi, J. R. Anglin. *arXiv:physics*/0007054 (2000).

1835 C. Varea, J. L. Aragón, R. A. Barrio. *Phys. Rev.* E 60, 4588 (1994).

1836 J. L. Varona. *Math. Intell.* 24, n1, 37 (2002).

1837 H. Varvoglis, C. Vozikis, K. Wodnar. *Celest. Mech. Dynam. Astron.* 89, 343 (2004).

1838 V. M. Vasyliunas. *J. Geophys. Res.* 77, 6271 (1972).

1839 J. L. Vega, T. Uzer, J. Ford. *Phys. Rev.* E 52, 1490 (1995).

1840 J. Veit. *ZAMM* 83, 51 (2003).

1841 D. Vekemans. *J. Comput. Appl. Math.* 85, 181 (1997).

1842 A. Venugopalan, G. S. Agarwal. *arXiv:quant-ph*/9811012 (1998).

1843 S. Velasco, J. A. White, J. Gümez. *Eur. J. Phys.* 14, 166 (1993).

1844 G. Venezian. *Am. J. Phys.* 62, 1000 (1994).

1845 G. Venezian. *BIT* 36, 302 (1996).

1846 J. Verschelde, R. Cools. *AAECC* 4, 169 (1993).

1847 J. Verschelde. *arXiv:math.NA.*/9907060 (1999).

1848 J. Vidal, P. Butaud, B. Doucot, R. Mosseri. *arXiv:cond-mat*/0103611 (2001).

1849 J. Vidal, R. Mosseri. *arXiv:cond-mat*/0209037 (2002).

1850 J. Vignes. *Math. Comput. Simul.* 35, 233 (1993).

1851 P. Vignolo, A. Minguzzi, M. P. Tosi. *Phys. Rev.* A 64, 023421 (2001).

1852 J. Villavicencio. *J. Phys.* A 33, 6061 (2000).

1853 J. Vinson. *Exper. Math.* 10, 337 (2001).

1854 D. Viswanath. *Math. Comput.* 69, 1131 (2000).

1855 D. Viswanath. *Nonlinearity* 16, 1035 (2003).

1856 D. Viswanath. *Physica* D 190, 115 (2004).

1857 V. S. Vladimirov. *Equations of Mathematical Physics*, Mir, Moscow, 1984.

1858 H. Voderberg. *Jahresber. DMV* 46, 229, (1936).

1859 H. Voderberg. *Jahresber. DMV* 47, 159 (1937).

1860 A. Voros. *J. Phys.* A 27, 4653 (1994).

1861 A. Voros. *arXiv:math-ph*/9902016 (1999).

1862 A. Vourdas. *J. Comput. Appl. Math.* 133, 657 (2001).

1863 A. Vourdas. *Rep. Progr. Phys.* 67, 267 (2004).

1864 J. E. Vuillemin. *IEEE Trans. Comput.* 39, 1087 (1990).

1865 S. Wagon. *Mathematica in Action*, W. H. Freeman, New York, 1990.

1866 S. Wagon. *Am. Math. Monthly* 102, 195 (1995).

1867 R. Walde, P. Russo. *Am. Math. Monthly* 101, 318 (1994).

1868 J. S. Walker. *Fast Fourier Transforms*, CRC Press, Boca Raton, 1999.

1869 R. J. Walker. *Algebraic Curves*, Princeton University Press, Princeton, 1950.

1870 H. S. Wall. *Analytic Theory of Continued Fractions*, van Nostrand, Princeton, 1948.

1871 J. Walter in E. A. Dold, B. Eckmann (eds.). *Ordinary Differential Equations and Operators*, Springer-Verlag, Berlin, 1983.

1872 J. Wang. *The Mathematica Journal* 8, 383 (2002).

1873 Y. Wang, K. Hutter, S. P. Pudasaini. *ZAMM* 84, 507 (2004).

1874 T. Watanabe, Y, Choyal, K. Minami, V. L. Granatstein. *Phys. Rev.* E 69, 056606 (2004).

1875 G. I. Watson. *J. Phys.* A 24, 4999 (1991).

1876 E. Weber. *Electromagnetic Fields* v.1, Wiley, London, 1950.

1877 G. W. Wei. *J. Phys.* B 33, 343 (2000).

1878 K. Weierstraß. *Mathematische Werke* v. III, p. 251, Mayer & Müller, Berlin, 1903.

1879 K. Weihrauch, N. Zhong. *Proc. Lond. Math. Soc.* 85, 312 (2002).

1880 S. Weinberg. *The Quantum Theory of Fields* v.1, Cambridge University Press, Cambridge, 1995.

1881 S. Weinzierl. *arXiv:hep-ph*/0006269 (2000).

1882 D. Weiss, K. Richter, A. Menschig, R. Bergmann, H. Schweizer, K. von Klitzing, G. Weimann. *Phys. Rev. Lett.* 70, 4118 (1993).

1883 D. Weiss, K. Richter, E. Vasiliadou, G. Gütjering. *Surf. Sci.* 305, 408 (1994).

1884 G. H. Weiss. *Physica* A 311, 381 (2002).

1885 A. Weisshaar, S. M. Goodnick, V. K. Tripathi. *IEEE Trans. Microwave Th. Techn.* 40, 2200 (1992).

1886 B. D. Welfert. *SIAM J. Num. Anal.* 34, 1640 (1997).

1887 E. J. Weniger. *Comput. Phys. Rep.* 10, 189 (1989).

1888 E. J. Weniger. *arXiv:math.CA*/0002111 (2000).

1889 E. J. Weniger. *arXiv:math.NA*/0003227 (2000).

1890 D. H. Werner. *IEEE Trans. Antennas Prop.* 44, 157 (1996).

1891 D. H. Werner. *IEEE Trans. Antennas Prop.* 47, 1351 (1999).

1892 D. H. Werner in D. H. Werner, R. Mittra (eds.). *Frontiers in Electromagnetics*, IEEE Press, New York, 2000.

1893 A. G. Werschulz. *J. Complexity* 16, 377 (2000).

1894 J. Wess. *arXiv:math-ph*/9910013 (1999).

1895 J. Wess in H. Gausterer, H. Grosse, L. Pittner (eds.). *Geometry and Quantum Physics*, Springer-Verlag, Berlin, 2000.

1896 H. Weyl. *Die Idee der Riemannschen Fläche*, Teubner, Stuttgart, 1955.

1897 B. Widom, J. S. Rowlinson *J. Chem. Phys.* 52, 1670 (1970).

1898 P. B. Wiegmann. *arXiv:hep-th*/9909083 (1999).

1899 A. L. Wiens, B. J. Ross. *Comput. Graphics* 26, 75 (2002).

1900 J. Wiersig, G. G. Carlo. *arXiv:nlin.CD*/0212011 (2002).

1901 B. Wikiel, J. M. Hill. *Int. J. Nonl. Mech.* 35, 953 (2001).

1902 H. S. Wilf. *Generatingfunctionology*, Academic Press, New York, 1994.

1903 H. S. Wilf. *Discr. Math.* 257, 575 (2002).

1904 J. H. Wilkinson in G. H. Golub (ed.). *Studies in Mathematics* 24, American Mathematical Society, Providence, 1984.

1905 L. T. Wille, J. Vennik. *J. Phys.* A 18, L1113 (1985).

1906 F. W. Williams, W. P. Howson, A. Watson. *Proc. R. Soc. Lond.* A 460, 1243 (2004).

1907 H. C. Williams. *Utilitas Math.* 28, 201 (1985).

1908 J. C. Wilson. *Acta Arithm.* 66, 1 (1994).

1909 J. Wimp. *Sequence Transformations and Their Applications*, Academic Press, New York, 1981.

1910 D. A. Wisniacki, E. R. Pujals. *arXiv:quant-ph*/0502108 (2005).

1911 A. Wobst, G.-L. Ingold, P. Hänggi, D. Weinmann. *arXiv:cond-mat*/0110028 (2001).

1912 A. Wobst, G.-L. Ingold, P. Hänggi, D. Weinmann. *Phys. Rev.* B 68, 085103 (2003).

1913 D. Wójcik, I. Bialnicki-Birula, K. Zyczkowski. *arXiv:quant-ph*/0005060 (2000).

1914 D. Wójcik, K. Zyczkowski. *arXiv:math-ph*/0107030 (2001).

1915 A. Wolf, S. J. van Hook, E. R. Weeks. *Am. J. Phys.* 64, 714 (1996).

1916 D. E. Wolf, P. Grassberger (eds.). *Friction, Arching, Contact Dynamics*, World Scientific, Singapore, 1997.

1917 K. B. Wolf. *Integral Transforms in Science and Engineering*, Plenum Press, New York, 1979.

1918 S. Wolfram. *Mathematica: A System for Doing Mathematics by Computer*, Addison-Wesley, Redwood City, 1992.

1919 S. Wolfram. *The Mathematica Book*, Cambridge University Press and Wolfram Media, Cambridge, 1999.

1920   E. Wong. *arXiv:math.PR*/0412291 (2004).

1921   C. Wongtawatnugool, S. Y. Wu, C. C. Shih. *J. Math. Phys.* 22, 633 (1981).

1922   S.-J. Woo, J. Lee, K. J. Lee. *Phys. Rev.* E 68, 016208 (2003).

1923   R. Worg. *Deterministisches Chaos*, BI, Mannheim, 1993.

1924   J. W. Wrench, D. Shanks. *Math. Comput.* 20, 444 (1966).

1925   C. W. Wu. *Phys. Lett.* A 296, 105 (2002).

1926   F. Y. Wu. *J. Phys.* A 37, 6653 (2004).

1927   H. Wu, D. W. L. Sprung. *Phys. Lett.* A 261, 150 (1999).

1928   H. Wu, D. W. L. Sprung, J. Martorell, S. Klarsfeld. *Phys. Rev.* B 44, 6351 (1991).

1929   J. Wu. *Bull. London Math. Soc.* 34, 16, 2002.

1930   J. Wu. *Monatsh. Math.* 134, 337 (2002).

1931   T. T. Wu, B. M. McCoy, C. A. Tracy, E. Barouch. *Phys. Rev.* B 13, 316 (1976).

1932   X. Wu. *Appl. Math. Comput.* 162, 539 (2005).

1933   Z. Wu, E. López, S. V. Buldyrev, L. A. Braunstein, S. Havlin, H. E. Stanley. *arXiv:cond-mat*/0411062 (2004).

1934   Z.-B. Wu, J.-Y. Zeng. *arXiv:quant-ph*/9812077 (1998).

1935   P. Wynn. *Math. Tables Aids Comput.* 10, 91 (1956).

1936   R. Wyss. *Elem. Math.* 45, 37 (1990).

1937   H. Xi, J. D. Gunton, M. I. Tribelsky. *arXiv:cond-mat*/0002118 (2000).

1938   H. Xi, J. D. Gunton, J. Vinals. *Phys. Rev. Lett.* 71, 2030 (1993).

1939   Z. Xia. *Ann. Math.* 135, 411 (1992).

1940   W. Xinghua, Z. Shiming. *Kexue Tongbao* 33, 1235 (1988).

1941   W. G. Xu, Q. S. Li. *Chaos, Solitons, Fractals* 15, 663 (2003).

1942   X. Xu, B. You, X. Lin. *Int. J. Infrared Millim. Waves* 20, 1825 (1999).

1943   T. Yalcinkaya, Y.-C. Lai. *Comput. Phys.* 9, 511 (1995).

1944   T. Yamamoto, K. Kaneko. *Progr. Th. Phys.* 100, 1089 (1998).

1945   T. Yamauchi. *Phys. Lett.* A 191, 317 (1994).

1946   S. Y. Yang, Y. C. Zhou, G. W. Wei. *Comput. Phys. Commun.* 143, 113 (2002).

1947   W. Yao, C. Essex, P. Yu, M. Davison. *Phys. Rev.* E 69, 066121 (2004).

1948   S. Yarmukhamedov. *Sib. Math. J.* 45, 580 (2004).

1949   T. Yogi. *J. Phys. Soc. Jpn.* 65, 765 (1996).

1950   M. Yor. *Some Aspcts of Brownian Motion*, Birkhäuser, Basel, 1992.

1951   D. York, N. M. Evensen, M. L. Martínez, J. D. B. Delgado. *Am. J. Phys.* 72, 367 (2004).

1952   L. Y. Yu, R. H. Parmenter. *Chaos* 2, 581 (1992).

1953   G. Yuan, J. A. Yorke. *Physica* D 136, 18 (2000).

1954   G.-C. Yuan, K. Nam., T. M. Antonsen, E. Ott, P. N. Guzdar. *Chaos* 10, 39 (2000).

1955   N. Yui. *J. reine angew. Math.* 300, 185 (1978).

1956   C. G. Zagouras, E. Perdios, O. Ragos. *Astrophys. Space Sci.* 240, 273 (1996).

1957   G. M. Zaslavsky. *Chaos* 1, 1 (1991).

1958   G. M. Zaslavsky, R. Z. Sagdeev, D. A. Usikov, A. A. Chernikov. *Weak Chaos and Quasi Regular Patterns*, Cambridge University Press, Cambridge, 1991.

1959 G. M. Zaslavsky, M. Edelman, B. A. Niyazov. *Chaos* 7, 159 (1997).

1960 G. M. Zaslavsky. *Phys. Rep.* 371, 461 (2002).

1961 H. Zemanian. *Proc. IEEE* 64, 6 (1976).

1962 A. H. Zemanian. *IEEE Trans. Circuits Syst.* 35, 1346 (1988).

1963 A. H. Zemanian. *Transfiniteness for Graphs, Electrical Networks, and Random Walks*, Birkhäuser, Boston, 1996.

1964 G. P. Zhang, T. F. George. *Phys. Rev.* B 69, 167102 (2004).

1965 A. Zhedanov. *arXiv:math.CA*/0401086 (2004).

1966 Z. Zhou, C. Whiteman. *Nonl. Anal.* 28, 1177 (1996).

1967 C. Zhu. *Phys. Lett.* A 172, 169 (1992).

1968 M. Znojil. *J. Phys.* A 31, 3349 (1998).

1969 M. Zöckler, D. Stalling, H.-C. Hege. *Proc. Visualization '96*, IEEE, New York, 1996.

1970 P. Zupanović, D. Juretić, S. Botrić. *Phys. Rev.* E 70, 056108 (2004).

1971 A. Zvonkin in D. Krob, A. A. Mikhalev, A. V. Mikhalev (eds.). *Formal Power Series and Algebraic Combinatorics*, Springer-Verlag, Berlin, 2000.

1972 D. Zwillinger. *Handbook of Integration*, Jones and Bartlett, Boston, 1992.

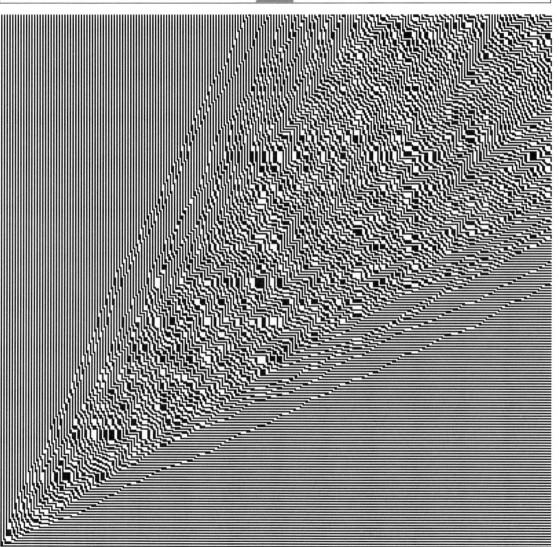

Computations with Exact Numbers

2.0 Remarks

Although computing with integers or rational numbers is numerical calculation, it differs from working with "ordinary" (approximative) real numbers because roundoff errors never occur. In this chapter, we focus on numerical computations and functions which typically have integer arguments, and that produce integer (or rational) values for integer arguments. Functions (the so-called special functions of mathematical physics) of real and complex numbers are discussed in the Chapters 2 and 3 of the Symbolics volume [556]. Of course, all of the functions discussed here can also be regarded as functions of purely symbolic variables; in which case, *Mathematica* carries out purely symbolic calculations. Moreover, many of the functions in this chapter have analytic continuations to the complex plane and, thus, in a certain sense also belong in Chapter 3 of the Symbolics volume. Indeed, there is no unique way to decide which functions belong here, and our choices between this chapter and Chapter 3 of the Symbolics volume are based mostly on historical grounds and on the most frequent use of the functions. The functions of this chapter are typically evaluating to integers for integer arguments. And they are less frequently used for symbolic operations than for numeric evaluations.

Because the amount of material presented here is not too large, we are able to go into some detail on several examples. So, we will occasionally take the opportunity to compare various implementations and programming styles for a given problem. In addition, we will give various identities for number theoretic and combinatorial functions and check them for small arguments.

Besides addition and multiplication, the most important operation in rational number calculations is the reduction of the rational number to canonical form (canceling common factors). This is done by calculating the greatest common divisor between the numerator and the denominator, a function discussed in this chapter. Calculations with integer and rational numbers can be carried out in an exact manner. In the last chapter, we discussed the functions Floor, Ceiling, Round, Equal, Unequal, Sign, Re, Im, ... We recall that these functions work for approximative, as well as for exact input and use internally numerical techniques for exact nonrational arguments. But because of the significance arithmetic used inside *Mathematica*, their results are guaranteed to be correct.

```
In[1]:= {Floor[Sqrt[2]], Ceiling[E^GoldenRatio], Mod[Log[Pi], 1/2]}
Out[1]= {1, 6, -1 + Log[π]}
```

But such calculations use internally numerical techniques to establish the result and so may give rise to N::meprec messages, and sometimes they may return unevaluated.

```
In[2]:= Sign[Pi - FromContinuedFraction[ContinuedFraction[Pi, 20]]]
Out[2]= -1
```

```
In[3]:= Sign[Pi - FromContinuedFraction[ContinuedFraction[Pi, 100]]]
```

$$N::meprec : \text{Internal precision limit \$MaxExtraPrecision = 50.` reached while}$$
$$\text{evaluating } -\frac{4170888101980193551139105407396069754167439670144501}{1327634917026642108692848192776111345311909093498260} + \pi. \text{ More...}$$

Out[3]= $\mathrm{Sign}\left[-\dfrac{4170888101980193551139105407396069754167439670144501}{1327634917026642108692848192776111345311909093498260}+\pi\right]$

Certain operations with roots of univariate polynomials with rational coefficients can also be carried out exactly; we will discuss these operations in the first chapter of the Symbolics volume [556] when dealing with `Root`-objects.

Even without defining any special integer functions, *Mathematica*'s ability to work with arbitrarily large integers and rational numbers allows us to investigate interesting problems.

Let us consider the map m defined by $\mathbb{Z}^2 \to \mathbb{Z}^2 : \{m,\, n\} \to \{\lfloor \alpha\, m \rfloor - n,\, m\}$ (here, α is a fixed rational number) [75].

```
In[4]:= m[α_][{m_, n_}] := {Floor[α m] - n, m}
```

The map m is periodic. For a given initial pair $\{m,\, n\}$, the function `cycle` returns the cycle of pairs.

```
In[5]:= cycle[α_, {m_, n_}] :=
    Module[{ξ = {m, n}, bag = {{m, n}}},
        (* apply m[α] until we find {m, n} again *)
        While[(ξ = m[α][ξ]) =!= {m, n},
            bag = {bag, ξ}]; Level[bag, {-2}]]
```

Here is an example.

```
In[6]:= cycle[2/5, {1, 3}]
```

Out[6]= {{1, 3}, {-3, 1}, {-3, -3}, {1, -3}, {3, 1}, {0, 3}, {-3, 0},
{-2, -3}, {2, -2}, {2, 2}, {-2, 2}, {-3, -2}, {0, -3}, {3, 0}}

We define a function `allCycles` that calculates all cycles with starting pairs $\{m,\, n\}$ such that $0 \le m,\, n \le r$.

```
In[7]:= allCycles[α_Rational, r_Integer] :=
    Module[{c, cycleBag = C[], newCycle},
        Do[If[c[k, l] =!= "done",
                (* a new cycle *) newCycle = cycle[α, {k, l}];
                cycleBag = C[cycleBag, newCycle];
                (* hash used pairs *) Apply[(c[##] = "done")&, newCycle, {1}]],
            {k, 0, r}, {l, 0, r}];
        Level[cycleBag, {-3}]];
```

Here is a graphic of the resulting cycle length and number of cycle elements as a function of the fraction α. For α, we use all proper fractions with denominators less than 120 and we use $r = 10$.

```
In[8]:= Show[GraphicsArray[
    Block[{$DisplayFunction = Identity},
    Function[f, (* show number of cycles and number of elements *)
        ListPlot[{#, Length[f[allCycles[#, 10]]]}& /@
                Union[Flatten[Table[i/j, {j, 2, 120}, {i, j - 1}]]],
            PlotRange -> All, Frame -> True, Axes -> False]] /@
        {Identity, Flatten}]]]
```

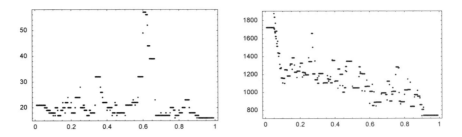

We end with a small animation showing the cycles in the plane. The function `cycleGraphic` calculates the cycles and colors them according to their length. This gives an interesting pattern in the *m,n*-plane.

```
In[9]:= cycleGraphic[α_, r_: 25] :=
    Module[{Λ = allCycles[α, r], λ, ζ},
    (* count cycles *)
    {λ1, λ2} = {Min[#], Max[#]}&[Length /@ Λ];
    (* largest occurring {m, n} *)
    ζ = Max[Abs[Transpose[Flatten[Λ, 1]]]];
    (* make graphic of cycles; color  cycles *)
    Graphics[Map[{Hue[0.78 (Length[#1] - λ1)/(λ2 - λ1)],
            Rectangle[# - 1/2, # + 1/2]& /@ #}&, Λ],
        PlotRange -> (ζ + 1){{-1, 1}, {-1, 1}}, Frame -> True,
        AspectRatio -> Automatic, FrameTicks -> None]]
```

Here are the resulting cycle patterns for all proper fractions with denominators less than 50.

```
In[10]:= manyFractions = With[{o = 49},
        Union[Flatten[Table[i/j, {j, 2, o}, {i, j - 1}]]]];
```

```
In[11]:= Show[GraphicsArray[#]]& /@
    Partition[cycleGraphic[#, 50]& /@
        (* show cycle graphic for selected fractions *)
        manyFractions[[{1, 39, 182, 252, 453, 466, 628, 735}]], 4]
```

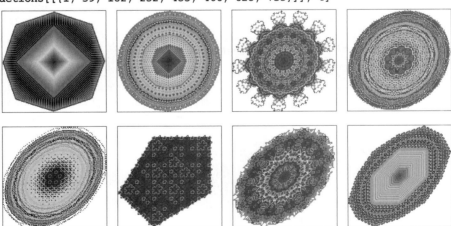

```
(Show[cycleGraphic[#, 25]]; (* do not return graphics *) Null) & /@
                                                              manyFractions
```

Here is another example. Let us consider the following continued fraction.

Here is another example. Let us consider the following continued fraction. A "level" n approximation is implemented by the function `neatContinuedFraction`. The optional last argument allows specifying the innermost denominator to "freeze" the expression to better see the structure of the expression.

```
In[12]:= neatContinuedFraction[n_, nLast_:Automatic] :=
         1/(1 + Fold[#2/(1 + #1)&,
                     If[nLast === Automatic, n, nLast], Range[n - 1, 1, -1]])

         neatContinuedFraction[10, …]
```

Out[14]=
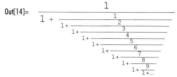

Calculating the value of the level 10000 approximation as a single rational number takes less than a second on a 2 GHz computer.

```
In[15]:= (nCF = neatContinuedFraction[10^4]); // Timing
Out[15]= {0.28 Second, Null}
```

The resulting rational number has quite large integers as numerator and denominator.

```
In[16]:= N[{Numerator[nCF], Denominator[nCF]}]
Out[16]= {2.936997629074093×10^17112, 4.479318690102064×10^17112}
```

In the limit $n \to \infty$, the value of this continued fraction is $\sqrt{\pi e / 2}\ \text{erfc}(1/\sqrt{2})$ [306]. (We will discuss the function erfc in Chapter 3 of the Symbolics volume [556].)

```
In[17]:= Block[{$MaxExtraPrecision = 100},
              N[Sqrt[Pi E/2] Erfc[1/Sqrt[2]] - nCF, 22]]
Out[17]= -5.629433376385423612627×10^-87
```

Similar to *Mathematica*'s capability to operate with floating point numbers with arbitrary many digits, the capability to carry out computations with integers and rational numbers of arbitrary size gives the possibility to experimentally investigate a variety of mathematical identities and to come up with conjectures about new identities and theorems.

But integer operations sometimes have a misleading suggestive power. Consider the following recursion for fixed positive integer d:

$$x_0^{(d)} = 1$$

$$x_n^{(d)} = \frac{1}{n}\left(1 + \sum_{i=0}^{n-1} \left(x_i^{(d)}\right)^d\right)$$

```
In[18]:= x[0, d_] = 1;
```

```
x[n_Integer?Positive, d_Integer?Positive] :=
    x[n, d] = (1 + Sum[x[i, d]^d, {i, 0, n - 1}])/n
```

Are the computed $x_n^{(2)}$ always integers? The first few terms certainly are integers.

In[20]:= `Table[x[i, 2], {i, 0, 12}]`

Out[20]= {1, 2, 3, 5, 10, 28, 154, 3520, 1551880, 267593772160,
71606426901226633501504, 4661345794146064133843098964919305264116096,
18106787177169334423257416302750040844148654208985912235226820224474389280191\
72629856}

$x_{26}^{(2)}$ is an integer with more than one million digits.

In[21]:= `{IntegerQ[x[26, 2]], N[x[26, 2]]}`

Out[21]= {True, $1.425676019571794 \times 10^{1361736}$}

The computed values are not always integers. But one has to go to `x[43, 2]` to see this. We compare this to the numerical calculations of the last chapter. There we could check that two numbers were identical to many thousand of digits. In the example under consideration, we have just 26 data points that are integers. In fact, also the first terms of this sequence for the analogous sums of cubes, fourth powers, etc. are integers. For instance, this is what happens for $d = 12$:

In[22]:= `Table[x[i, 12], {i, 0, 4}]`

Out[22]= {1, 2, 2049, 1825501718375633359283402425726847864833,
3423946434621324753923611936331042300923372469845999657985856570576101261730\.
741652349604662025458159130613079297874579034344194825462758558470975491583\.
235397468163559010400874607131379225205694998779306522969903844431158567006\.
621932210731936523546467971462264836099558441144401108838024436138715216195\.
0952915963224685080571320847678794304365074623707515065275068020199151065389\.
4019844516155228939332718984565961094952392278340408572380980013326583673840\.
01565713009665}

$x_8^{(12)}$ is an integer with nearly ten million digits. Its calculation takes a few seconds on a 2 GHz computer.

In[23]:= `N[x[8, 12]] // Timing`

Out[23]= {3.4 Second, $1.078613965644093 \times 10^{9755673}$}

For an explanation of why many members of this sequence are integers, see [233] and [587].

We also want to remind the reader that the function `Compile` (discussed in the last chapter) can deal with integer arguments. `Compile` allows doing number-theory calculations that deal with relatively small numbers (meaning machine integers) quite efficiently. The following function `countPrimes[P]` generates a compiled function that calculates the number of prime less than or equal to P taken by the polynomial $x^2 + a x + b$. (We will discuss the two functions `Prime` and `PrimePi` shortly.)

In[24]:= `countPrimes[P_] := countPrimes[P] =`
```
Compile[{{a, _Integer}, {b, _Integer}},
        Module[{x1, x2, values}, (* the roots *) {x1, x2} =
              (({-1, 1} Sqrt[4 (P - b) + 0.I + a^2] - a)/2;
              If[Im[x1] =!= 0. && Im[x2] =!= 0., 0,
              (* the values *)
              values = Table[x^2 + a x + b,
                              {x, Ceiling[Re[x1]], Floor[Re[x2]]}];
              (* count primes *)
              Length[Intersection[values, #]]]]]&[
              (* primes to be used *) Prime[Range[PrimePi[P]]]];
```

A graphic of `countPrimes[P][a, b]` shows that most primes occur around the curve $b = (a^2 - 163)/4$ [179], [188], [493], [292].

```
In[25]:= ListDensityPlot[(* the countPrimes data *)
               Table[countPrimes[100][a, b],
                     {b, -250, 1000}, {a, -100, 100}],
               Mesh -> False, ColorFunction -> (Hue[0.8 #]&),
               MeshRange -> {{-100, 100}, {-250, 1000}},
               PlotRange -> All, Epilog ->
          (* the curve b == (a^2 -163)/4 *)
          Plot[(a^2 -163)/4, {a, -Sqrt[4163], Sqrt[4163]},
              DisplayFunction -> Identity,
              PlotStyle -> {GrayLevel[1/2], Thickness[0.01],
                          Dashing[{0.01, 0.03}]}][[1]]]
```

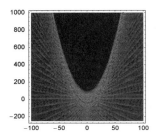

2.1 Divisors and Multiples

In many applications, particularly for certain simplifications, we need to factor integers ([30], [466], [390], and [596]) (i.e., find their prime divisors) and to find greatest common divisors or least common multiples. For the first of these purposes, the following *Mathematica* command is useful. (For some details on how to factor an integer, see [408], [605], [351], and [158].)

`FactorInteger[`*integer*`]`

computes a list of the prime factors of *integer* in the form $\{\{optionalSignOrIFactor_1, optionalOne\},$ $\{primeFactor_2, power_2\}, \ldots \{primeFactor_n, power_n\}\}$, where

$$n = (optionalSignOrIFactor_1)^{optionalOne} \prod_{k=2}^{n} (primeFactor_k)^{power_k}.$$

Positive integers are factored "as expected"; the factorizations of negative integers contain the factor -1.

```
In[1]:= {FactorInteger[+2376459222], FactorInteger[-2376459222]}
Out[1]= {{{2, 1}, {3, 1}, {23, 1}, {17220719, 1}},
         {{-1, 1}, {2, 1}, {3, 1}, {23, 1}, {17220719, 1}}}
```

Here is a check of the last results.

```
In[2]:= Times @@@ Apply[Power, %, {2}]
Out[2]= {2376459222, -2376459222}
```

`FactorInteger` has two options.

```
In[3]:= Options[FactorInteger]
Out[3]= {FactorComplete → True, GaussianIntegers → False}
```

In addition to `GaussianIntegers`, which we have already discussed in Chapter 5 of the Programming volume [554], we have `FactorComplete`.

`FactorComplete`

is an option for `FactorInteger` that determines whether a complete prime factorization should be computed.
Default:
> True

Admissible:
> False (only some factors will be determined)

Using `FactorInteger`, it is possible to factor integers with up to, say, about 20 digits in a few seconds. Larger numbers are (much) more difficult to factorize. The practical problems to factoring integers (on a classical computer) are the basis of many modern cryptosystems.

We now give several examples involving numbers up to 40 digits. As the timing shows, some of them take much longer than others. (Factorization of integers is probably exponential in the number of digits.)

```
In[4]:= Chop[Timing[FactorInteger[#]]]& /@ (* a few selected big integers *)
            {10119, 53502, 437950, 1417730066516, 5464080459997,
             24983298079949049, 693947612191892844546726253436430,
             29809257105416205324446280575524190557391}
Out[4]= {{0, {{3, 1}, {3373, 1}}}, {0, {{2, 1}, {3, 1}, {37, 1}, {241, 1}}},
        {0, {{2, 1}, {5, 2}, {19, 1}, {461, 1}}}, {0, {{2, 2}, {81119, 1}, {4369291, 1}}},
        {0, {{79, 1}, {232153, 1}, {297931, 1}}},
        {0, {{3, 1}, {31, 1}, {2131, 1}, {126061761503, 1}}},
        {0.15 Second, {{2, 1}, {5, 1}, {916827101, 1}, {756901286442112758343, 1}}},
        {0.75 Second, {{43024211467, 1}, {692848423922428033663938400973, 1}}}}
```

We see an overall increase of the time needed with increasing number of digits, but the time needed to factor an integer strongly depends on the actual number, not only on the number of digits.

Note the result of evaluating `FactorInteger` of 0 or 1.

```
In[5]:= {FactorInteger[0], FactorInteger[1]}
Out[5]= {{{0, 1}}, {}}
```

`FactorInteger` can be used to define two important number theory functions, $\omega(n)$ and $\Omega(n)$ [322]. The function $\omega(n)$ is the number of distinct prime factors of n, whereas $\Omega(n)$ is the total number of prime factors of n, counting each prime factor according to its multiplicity.

```
In[6]:= (* do not count any sign *)
        discardSign[{{-1 | I | - I, 1}, rest___}] := {rest}
        discardSign[l_] := l

        ω[n_Integer | n:Complex[_Integer, _Integer], opts___] :=
                     Length[discardSign[FactorInteger[n, opts]]];
```

```
Ω[n_Integer | n:Complex[_Integer, _Integer], opts___] :=
    Plus @@ Last[Transpose[discardSign[FactorInteger[n, opts]]]]
```

Here is a plot of both functions. We show the averages over *k* points for *k* = 1, 3, ..., 51 as points colored from red to blue.

```
In[11]:= Module[{data, averages, nMax = 4096, d = 25},
    Show[GraphicsArray[
      ((* the divisor counts *)
      data = Table[#[n], {n, 2, nMax}];
      (* form local averages *)
      averages = Table[ListConvolve[Array[1&, 2k + 1]/(2k + 1),
                        N @ data, k + 1, 0], {k, 0, d}];
      (* show averaged values from red to blue *)
      Graphics[{PointSize[0.005],
        MapIndexed[{Hue[0.78 #2[[1]]/d], Point[{#2[[2]], #1}]}&,
                   averages, {2}]}, Frame -> True,
                   PlotRange -> {{0, nMax - d}, All},
                   PlotLabel -> StyleForm[#, TraditionalForm],
                   AxesLabel -> {StyleForm["n", TraditionalForm],
                                 None}])& /@ {ω, Ω}]]]
```

We can also look at these functions in the complex plane, that is, with numbers of the form *n* + *i m* as arguments, where *n* and *m* are real integers [528]. (To save time, we look only at the first quadrant; by symmetry, the plot looks the same in the other quadrants.) Here we use the option `GaussianIntegers` discussed in Chapter 5 of the Programming volume [554].

```
In[12]:= With[{o = 100},
    Show[GraphicsArray[
      ListDensityPlot[Table[#1, {nr, o}, {ni, o}],
              PlotRange -> All, MeshRange -> {{1, o}, {1, o}},
              Mesh -> False, DisplayFunction -> Identity, PlotLabel -> #2,
              AxesLabel -> {StyleForm["n", TraditionalForm], None}]& @@@
        (* ω[n] and Ω[n] *)
        {{ω[nr + I ni, GaussianIntegers -> True], "ω(n)"},
         {Ω[nr + I ni, GaussianIntegers -> True], "Ω(n)"}}]]]
```

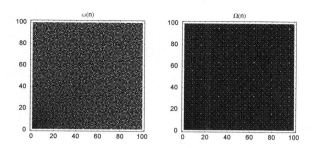

For some interesting relations between $\omega(n)$ and $\Omega(n)$, see [387]. Note that although `ListPlot` is usually the right choice for functions that are given only at a discrete set of points (such as ω and Ω above), it is possible to use `Plot` by writing `Plot[integerFunction[Round[n]], {n, n_0, n_1}]`.

If we are interested in all divisors rather than just those that are prime, we can use the command `Divisors`.

`Divisors[integer]`

    produces a list of all divisors of *integer*.

Here is an example.

```
In[13]:= Divisors[315216000] // Short[#, 4]&
```
Out[13]//Short=

```
{1, 2, 3, 4, 5, 6, 8, 9, 10, 11, 12, 15, 16, 18, 20, 22, 24, 25, 30, 32, 33,
  36, 40, 44, 45, 48, <<332>>, 6567000, 7004800, 7164000, 7880400, 8756000,
  9552000, 9850500, 10507200, 12608640, 13134000, 14328000, 15760800,
  17512000, 19701000, 21014400, 26268000, 28656000, 31521600, 35024000,
  39402000, 52536000, 63043200, 78804000, 105072000, 157608000, 315216000}
```

The Fourier transform of the number of divisors of a number shows a self-similar structure. The right graphic shows the 2D Fourier transform for Gaussian integers.

```
In[14]:= Show[GraphicsArray[
       Block[{$DisplayFunction = Identity, n1D = 2^14, n2D = 2^9},
         {(* divisor counts of positive integers *)
          ListPlot[Abs[Fourier[Table[Length[Divisors[n]], {n, n1D}]]],
                 PlotRange -> {0, 600}, PlotJoined -> True],
          (* divisor counts of Gaussian integers *)
          ListDensityPlot[Abs[Fourier[Table[Length[Divisors[nr + I ni,
                                           GaussianIntegers -> True]],
                   {nr, n2D}, {ni, n2D}]]], Mesh -> False,
                   (* scale values *)
                   ColorFunction -> (GrayLevel[ArcTan[#]/(Pi/2)]&),
                   ColorFunctionScaling -> False, PlotRange -> All]}]]]
```

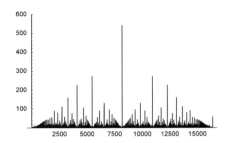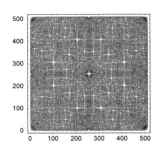

A famous relationship exists between the number of divisors of an integer *n* and the number of ways to represent *n* as a sum of four squares [144], [510].

```
In[15]:= r4[n_] := 8 Plus @@ Select[Divisors[n], Mod[#, 4] =!= 0&]
```

For the number 100 we get the result 744.

```
In[16]:= r4[100]
Out[16]= 744
```

Most of these 744 possibilities are permutations. The following one-liner generates the sum of squares explicitly. (We could also use the function SumOfSquaresRepresentations from the package NumberTheory` : NumberTheoryFunctions`.)

```
In[17]:= squaresThatSum[d_, n_, repetitionsQ_:True] :=
    Union[DeleteCases[Level[#, {-2}], {}]]& @
    Which[(* stop (too large sum or too many terms or no terms left) *)
        Length[#[[1]]] > d || #[[2]] > n ||
        (#[[2]] =!= n && #[[3]] === {}) ||
        (#[[2]] === n && Length[#[[1]]] =!= d), {},
        (* found representation *)
        Length[#[[1]]] === d && #[[2]] === n, #[[1]],
        (* recursive calls *)
        True, Function[x, Function[r,
        #0[(* {integersToBeSquared, sum, remainingIntegers} *)
          {Append[#[[1]], First[x]], #[[2]] + First[x]^2, r}]] /@
        (* use square again or not *)
        If[repetitionsQ, {x, Rest[x]}, {Rest[x]}]] /@
      NestList[Rest, #[[3]], Length[#[[3]]] - 1]]&[
        (* initial list *) {{}, 0, Range[0, Sqrt[n]]}]
```

```
In[18]:= squaresThatSum[4, 100]
Out[18]= {{0, 0, 0, 10}, {0, 0, 6, 8}, {1, 1, 7, 7},
    {1, 3, 3, 9}, {1, 5, 5, 7}, {2, 4, 4, 8}, {5, 5, 5, 5}}
```

```
In[19]:= addSignsAndPermute[ls_] := Flatten[Function[l, Flatten[Permutations /@
    Union[Sort /@ Flatten[Outer[List, Sequence @@ Transpose[{l, -l}]],
                                            Length[l] - 1]], 1]] /@ ls, 1]
```

```
In[20]:= addSignsAndPermute[%%] // Length
Out[20]= 744
```

The divisors can also be used to determine the number of different possibilities to express *n* as a sum of two, three, or four squares [259] or more [396]. The next input calculates the number of possibilities to represent 5525 as a sum of two different squares; the order of the two summands does not matter.

```
In[21]:= δ[n_] := If[IntegerQ[Sqrt[n]], 1, 0]
        δ[r_, m_, n_] := Count[Divisors[n], _?(Mod[#, m] == r&)]
        ε[r_, s_, m_, n_] := δ[r, m, n] - δ[s, m, n]
        p2[n_] := (ε[1, 3, 4, n] + δ[n] + δ[n/2])/2
In[25]:= p2[5525]
Out[25]= 6
```

Here are these six possibilities.

```
In[26]:= squaresThatSum[2, 5525, False]
Out[26]= {{7, 74}, {14, 73}, {22, 71}, {25, 70}, {41, 62}, {50, 55}}
```

The number of divisors of an integer varies considerably along the real axis. Cumulatively, we have [238], [169]:

$$\sum_{i \leq n} numberOfDivisors(i) = n \ln(n) + n(2\gamma - 1) + O(\sqrt{n}).$$

The following two pictures show the number of divisors and their cumulative sum for integers ≤ 150.

```
In[27]:= Module[{t, b0, b1, b2, b3},
        (* the number of divisors *)
        t = Table[DivisorSigma[0, i], {i, 150}];
        b0 = ListPlot[t, DisplayFunction -> Identity, PlotRange -> All];
        (* the cumulative sum *)
        b1 = ListPlot[Rest[FoldList[Plus, 0, t]], DisplayFunction -> Identity,
                      PlotStyle -> {PointSize[0.003]}];
        (* the asymptotic value *)
        b2 = Plot[x Log[x] + (2 EulerGamma - 1) x, {x, 1, 150},
                  DisplayFunction -> Identity, PlotStyle -> {GrayLevel[1/2]}];
        b3 = Show[{b1, b2}];
        (* display all graphics *)
        Show[GraphicsArray[{b0, b3}]]]
```

Here is a little programming example. Using `FactorInteger`, it is not difficult to create our own program for `Divisors`. We have to find all possible products of the form $factor_1^{j_1} \; factor_2^{j_2} \cdots factor_n^{j_n}$. In this connection, it is interesting to employ the automatic iterator generation discussed in Chapter 6 of the Programming volume [554].

```
In[28]:= myDivisors1[n_Integer?Positive] :=
    Module[{(* factors of n *) factors = FactorInteger[n], num},
           num = Length[factors];
           Sort[Flatten[
             Table[(* the factors of the divisors *)
                   Evaluate[Times @@ Table[factors[[i, 1]]^j[i], {i, num}]],
                   (* generation of the iterator *)
```

```
                    Evaluate[Sequence @@ Table[{j[i], 0, factors[[i, 2]]},
                    {i, num}]]]]]]
```

Here is a test.

```
In[29]:= myDivisors1[8000]

Out[29]= {1, 2, 4, 5, 8, 10, 16, 20, 25, 32, 40, 50, 64, 80, 100, 125,
          160, 200, 250, 320, 400, 500, 800, 1000, 1600, 2000, 4000, 8000}

In[30]:= myDivisors1[8000] === Divisors[8000]

Out[30]= True
```

Introducing the auxiliary variable `factors` makes the program easier to follow, and the factorization has to be computed only once. The program can still be shortened somewhat.

```
In[31]:= myDivisors2[n_Integer?Positive] :=
            Sort[Flatten[
                Table[(* divisor construction *)
                        Evaluate[Times @@ Table[#[[i, 1]]^j[i], {i, Length[#]}]],
                        (* iterator construction *)
                        Evaluate[Sequence @@ Table[{j[i], 0, #[[i, 2]]},
                        {i, Length[#]}]]]]]&[FactorInteger[n]]

In[32]:= myDivisors2[8000]

Out[32]= {1, 2, 4, 5, 8, 10, 16, 20, 25, 32, 40, 50, 64, 80, 100, 125,
          160, 200, 250, 320, 400, 500, 800, 1000, 1600, 2000, 4000, 8000}
```

Now, using a purely functional style, it can be written still much shorter (and will run faster).

```
In[33]:= myDivisors3[1] = {1};

         myDivisors3[n_] := Sort[Flatten[Apply[Outer[Times, ##]&,
                        Apply[#1^Range[0, #2]&, FactorInteger[n], {1}]]]]

In[35]:= myDivisors3[8000]

Out[35]= {1, 2, 4, 5, 8, 10, 16, 20, 25, 32, 40, 50, 64, 80, 100, 125,
          160, 200, 250, 320, 400, 500, 800, 1000, 1600, 2000, 4000, 8000}
```

(To avoid the repeated calculation of powers, instead of `#1^Range[0, #2]&`, we could also use `Function[{base, maxPower}, FoldList[Times[#, base]&, 1, Range[maxPower]]`   or `Function[{base, maxPower}, NestList[base #&, 1, maxPower]`   in   the   last   definition   of `myDivisors3` for a further speedup.)

Using `Divisors`, we can easily find two numbers such that the sum of the divisors of the first number is equal to the second number, whereas the sum of the divisors of the second number is equal to the original number. (See, e.g., [247] and [488].) Of course, the numbers themselves should not be counted in the lists of their divisors. We now find all such number pairs less than 10000.

```
In[36]:= d[x_] := Drop[Divisors[x], -1];
         p[x_] := Plus @@ d[x];

         (* format identity *)
         makeId[i_, l_] := HoldForm[i == l] /. List -> Plus

         pBag = {};
         Do[(* the search *)
          If[p[p[i]] == i && p[i] =!= i,
```

```
If[FreeQ[pBag, i], (* avoid doubles *) pBag = Join[pBag, {i, p[i]}];
   Print[makeId[p[i], d[i]]]; Print[makeId[i, d[p[i]]]]]], {i, 2, 10000}]
```

$$284 == 1 + 2 + 4 + 5 + 10 + 11 + 20 + 22 + 44 + 55 + 110$$

$$220 == 1 + 2 + 4 + 71 + 142$$

$$1210 == 1 + 2 + 4 + 8 + 16 + 32 + 37 + 74 + 148 + 296 + 592$$

$$1184 == 1 + 2 + 5 + 10 + 11 + 22 + 55 + 110 + 121 + 242 + 605$$

$$2924 == 1 + 2 + 4 + 5 + 10 + 20 + 131 + 262 + 524 + 655 + 1310$$

$$2620 == 1 + 2 + 4 + 17 + 34 + 43 + 68 + 86 + 172 + 731 + 1462$$

$$5564 == 1 + 2 + 4 + 5 + 10 + 20 + 251 + 502 + 1004 + 1255 + 2510$$

$$5020 == 1 + 2 + 4 + 13 + 26 + 52 + 107 + 214 + 428 + 1391 + 2782$$

$$6368 == 1 + 2 + 4 + 8 + 19 + 38 + 41 + 76 + 82 + 152 + 164 + 328 + 779 + 1558 + 3116$$

$$6232 == 1 + 2 + 4 + 8 + 16 + 32 + 199 + 398 + 796 + 1592 + 3184$$

Next, we will use `Divisors` to interpret the question: How "quadratic" are the natural numbers? Here, we will measure the "quadraticity" $q(x)$ of natural numbers in an average sense using [473]:

$$q(x) = \frac{1}{x} \sum_{k \leq x} \sum_{\substack{1 \leq n \leq m \\ n\,m = k}} \frac{n}{m}.$$

```
In[42]:= With[{x = 10^3},
     ListPlot[MapIndexed[{#2[[1]], #1/#2[[1]]}&, Rest[FoldList[Plus, 0,
        Table[Plus @@ Select[Divisors[n]^2/n, # <= 1&], {n, x}]]]],
        Frame -> True, Axes -> False, PlotStyle -> PointSize[0.003]]]
```

One clearly conjectures the limit $\lim_{x \to \infty} q(x) = 1/2$.

Here is another small application of the divisors of a number. The derivative of a positive integer n, denoted by n', can be defined as fulfilling the Leibniz product rule $n' = (j\,k)' = j'\,k + j\,k'$ for all pairs of integers $\{j, k\}$, such that $j\,k = n$ [37]. In addition, we postulate that the derivative of prime numbers vanishes. Using now `Divi`·`sors` to write down the 7069 equations arising from differentiating the first 1000 integers, we seem to obtain a unique solution. (In the input, we denote the derivative by d.)

```
In[43]:= With[{o = 1000},
     Solve[(* for equation by taking all pairs of factors *)
           Flatten[Table[(d[n] == #1 d[#2] + d[#1] #2)& @@@ ({#, n/#}& /@
                       Divisors[n]), {n, o}]] /. d[k_?PrimeQ] -> 1,
```

```
(* solve for derivatives of nonprimes *)
DeleteCases[Table[d[k], {k, o}], d[k_?PrimeQ]]]] // Take[#[[1]], 32]&
```

Out[43]= {d[1] → 0, d[4] → 4, d[6] → 5, d[8] → 12, d[9] → 6, d[10] → 7, d[12] → 16,
          d[14] → 9, d[15] → 8, d[16] → 32, d[18] → 21, d[20] → 24, d[21] → 10, d[22] → 13,
          d[24] → 44, d[25] → 10, d[26] → 15, d[27] → 27, d[28] → 32, d[30] → 31,
          d[32] → 80, d[33] → 14, d[34] → 19, d[35] → 12, d[36] → 60, d[38] → 21,
          d[39] → 16, d[40] → 68, d[42] → 41, d[44] → 48, d[45] → 39, d[46] → 25}

Using the unique factorization of a positive integer $n = \prod_{k=1}^{o} p_k^{\alpha_k}$, we can write the derivative as $n' = n \prod_{k=1}^{o} \alpha_k / p_k$ [558]. This allows for the following more efficient definition for calculating the derivative of an integer.

In[44]:= (* save factorizations *)
```
Σ[n_] := Σ[n] = Plus @@ (#2/#1& @@@ FactorInteger[n])

d[n_Integer] := Which[n > 0, n Σ[n], n < 0, -n Σ[-n], n == 0, 0]
```

The next graphic shows the nonvanishing values of the first 50 derivatives of the first 1000 integers. The colors change from red (the original integer), to violet (the 50th derivative).

In[47]:= Module[{o = 50, n},
```
       (* calculate multiple derivatives *)
       nl = NestList[(d /@ #)&, Range[0, 1000], o];
       (* visualize with color according to derivative order *)
       Show[Graphics[{PointSize[0.002], Table[{Hue[0.78 k/o],
                     MapIndexed[If[#1 =!= 0,
                                Point @ N[{#2[[1]], Log[#1]}], {}]&, nl[[k]]]},
                   {k, o}]}], PlotRange -> All, Frame -> True]]
```

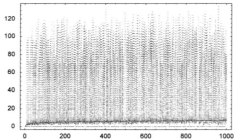

One can even continue the differentiation operation to rational numbers. Here are the numerators and denominators modulo 2 of derivatives of rationals i / j shown.

In[48]:= d[r_Rational] := r (Σ[Numerator[#]] - Σ[Denominator[#]])&[Abs[r]]

In[49]:= Module[{dQs = Table[d[i/j], {i, 120}, {j, 120}]},
```
       Show[GraphicsArray[
           ListDensityPlot[Mod[#[dQs], 2], Mesh -> False,
                       FrameTicks -> False, DisplayFunction -> Identity]& /@
                                             {Numerator, Denominator}]]]
```

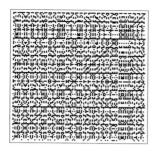

In number theory calculations, we often need the sum $\sigma_k(n)$ of the kth powers of the divisors of a number n. To this end, we have the function DivisorSigma.

DivisorSigma[*power*, *integer*]

  produces the sum of the *power*th powers of the divisors of *integer*.

Thus, DivisorSigma[0, *integer*] produces the number of divisors of integer.

```
In[50]:= {Divisors[384] // Length, DivisorSigma[0, 384]}
Out[50]= {16, 16}
```

Here is an example for various powers.

```
In[51]:= Table[DivisorSigma[i, 315216000], {i, 0, 8}] // TableForm
Out[51]//TableForm=
    384
    1241136000
    156321269256304880
    37497264289490267026128000
    10680240753673712045433740325202736
    3226726258099063367235096637686792920400000
    997960002140330605296449524559274425109755100968240
    31179434274780941320937389990051238647409543113130026 5168000
    9786629030756208657159565848520465915569663443567429865 1067473123376
```

We are already familiar with the option of DivisorSigma.

```
In[52]:= Options[DivisorSigma]
Out[52]= {GaussianIntegers → False}
```

We now look at the absolute value of the sum of powers of divisors of numbers of the form *integer* + *i integer'* in the complex plane. We display the real part, the absolute value, and the argument of the Fourier transforms. Not unexpectedly, we obtain similar plots as above.

```
In[53]:= Show[GraphicsArray[
    Table[ListDensityPlot[(* form Fourier transform *) # @ Fourier @
    Table[If[x + I y == 0, 0,
            Abs[DivisorSigma[j, x + I y, GaussianIntegers -> True]]],
        {x, 0, 256}, {y, 0, 256}] /. (* special case *)
        DivisorSigma[0, 0, GaussianIntegers -> True] -> 0,
                ColorFunction -> (Hue[0.8 #]&), Mesh -> False,
                FrameTicks -> None, DisplayFunction -> Identity],
```

```
{j, 4}]]]& /@
(* use real part, absolute value, and argument *) {Re, Abs, Arg}
```

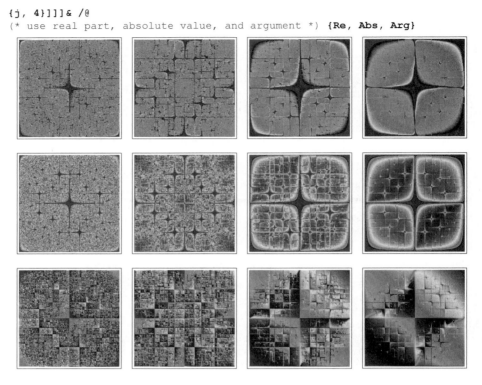

As just mentioned, cumulative we have $\sum_{j=1}^{n} \sigma_0(j) \propto \ln(n)\, n + (2\,\gamma - 1)\, n$. This means, on average an integer n has $\ln(n) + 2\,\gamma$ divisors. The following graphics show this theoretical average with the local average over 50 and 500 integers for the first 100000 integers. Using DivisorSigma instead of counting the divisors explicitly allows easily analyzing much more integers in the same time.

```
In[54]:= Module[{o = 10^5},
    Show[GraphicsArray[
    Block[{$DisplayFunction = Identity},
    Show[{(* the calculated averages *)
        ListPlot[(* form averages at once *)
                ListConvolve[Table[1, {#}],
                    Table[DivisorSigma[0, n], {n, o}]]/#],
        (* theoretical average *)
        Plot[Log[x] + 2EulerGamma, {x, 1, o},
            PlotStyle -> {{Thickness[0.01], Hue[0]}},
            PlotRange -> All]}]& /@
            (* average over *) {50, 500} (* integers *)]]]]
```

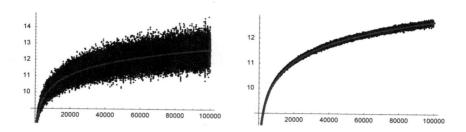

The cumulative differences between the sums and their asymptotics has a similar asymptotic as the cumulative sums [501].

```
In[55]:= Module[{o = 10^3, sumsOfδs,
            (* function to form cumulative sums *)
            rfp = Function[l, Rest[FoldList[Plus, 0., l]]],
            (* cumulative sum *)
            f = Function[k, k Log[k] + k (2 EulerGamma - 1)],
            (* asymptotic of cumulative sum of differences *)
            g = Function[k, k Log[k]/2 + k (EulerGamma - 1/4)]},
       (* cumulative sum of differences *) sumsOfδs =
       MapIndexed[(#1 - f[#2[[1]]])&, rfp[Table[DivisorSigma[0, n], {n, o}]]];
   Show[GraphicsArray[
     {(* show numerical and asymptotic cumulative sum of differences *)
     Graphics[{{Hue[0], Line[N @ MapIndexed[{#2[[1]], #1}&, rfp[sumsOfδs]]]},
        (* difference between the two curves *)
        {Hue[0.8], Line[N @ Table[{k, g[k]}, {k, o}]]}}, Frame -> True],
     ListPlot[N @ MapIndexed[(#1 - g[#2[[1]]])&, rfp[sumsOfδs]],
              DisplayFunction -> Identity, PlotRange -> All]}]]]
```

For $n \to \infty$, the quantity $\sigma_n(n+1)/\sigma_n(n+1)$ approaches e [545]. Here is the difference to e for $n = 10^6$.

```
In[56]:= With[{n = 10^6}, DivisorSigma[N[n], n + 1]/DivisorSigma[N[n], n]] - E
Out[56]= -1.4 × 10^{-6}
```

The next function to be discussed is of great importance in many fields. We have encountered this command earlier, but we now officially introduce it. The function Mod plays a major role in connection with decomposition of numbers.

Mod[a, b]

   produces a mod b, that is, the remainder after dividing a by b.

The maximum number of times that b appears as a factor of a is computed with Quotient.

Quotient[a, b]

    produces the maximal number occurrences of b's in a.

Clearly, a = Quotient[a, b] b + Mod[a, b].

Here is an example of the use of Quotient and Mod.

```
In[57]:= Module[{a = 8},
    Table[{a, b, Mod[a, b], Quotient[a, b], b Quotient[a, b] + Mod[a, b]},
        {b, 10}] // TableForm[#, TableHeadings -> {None,
            StyleForm[#, FontWeight -> "Bold"]& /@
                {"a", "b", "Mod[a, b]", "Quotient[a, b]", "8"}}]&]
```

Out[57]//TableForm=

a	b	Mod[a, b]	Quotient[a, b]	8
8	1	0	8	8
8	2	0	4	8
8	3	2	2	8
8	4	0	2	8
8	5	3	1	8
8	6	2	1	8
8	7	1	1	8
8	8	0	1	8
8	9	8	0	8
8	10	8	0	8

We get the following results for a negative first argument.

```
In[58]:= (Cases[DownValues[In],    (* reuse last input *)
            HoldPattern[Verbatim[HoldPattern][In[#]] :> _]&[$Line - 1]] /.
        {8 :> -8, "8" :> "-8"})[[1, 2]]
```

Out[58]//TableForm=

a	b	Mod[a, b]	Quotient[a, b]	-8
-8	1	0	-8	-8
-8	2	0	-4	-8
-8	3	1	-3	-8
-8	4	0	-2	-8
-8	5	2	-2	-8
-8	6	4	-2	-8
-8	7	6	-2	-8
-8	8	0	-1	-8
-8	9	1	-1	-8
-8	10	2	-1	-8

Here is a plot of Mod. For each point with integer coordinates $\{x, y\}$ in the x,y-plane, we build a tower of height x mod y.

```
In[59]:= modTower[{a_, b_, mod_, n_}] :=
    Module[{p1 = {a - 1/3, b - 1/3, mod}, p2 = {a + 1/3, b - 1/3, mod},
            p3 = {a + 1/3, b + 1/3, mod}, p4 = {a - 1/3, b + 1/3, mod},
            p5 = {a + 1/3, b - 1/3, 0 }, p6 = {a - 1/3, b - 1/3, 0 },
            p7 = {a + 1/3, b + 1/3, 0 }, p8 = {a - 1/3, b + 1/3, 0 }},
        {Hue[mod/n 78/100], EdgeForm[{Thickness[0.001]}], (* the five faces *)
        Polygon /@ {{p1, p2, p3, p4}, {p1, p2, p5, p6},
                {p5, p2, p3, p7}, {p6, p8, p4, p1}, {p4, p8, p7, p4}}}]
```

We display this plot together with a density plot over larger argument ranges.

```
In[60]:= Show[GraphicsArray[
    Block[{$DisplayFunction = Identity},
    {With[{n = 25},
      (* 3D graphics of the towers *)
      Show[Graphics3D[Map[modTower, Table[{a, b, Mod[a, b]}, n},
                {a, 0, n}, {b, (* modulo 0 does not exist *) n}], {2}]],
          Lighting -> False, PlotRange -> All, ViewPoint -> {-1, -3/2, 1},
          AxesLabel -> (StyleForm[#, TraditionalForm]& /@
                  {"a", "b", " a mod b"}), Axes -> True]],
      (* 2D density plot *)
    ListDensityPlot[Table[Mod[i, j], {i, 0, 100}, {j, 100}],
              ColorFunction -> (Hue[# 0.8]&),
              MeshRange -> {{0, 100}, {1, 100}},
              MeshStyle -> {Thickness[0.002]}, Mesh -> False]}]]]
```

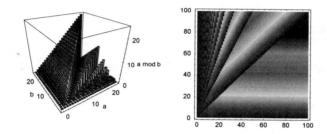

Here is a slightly larger example. We form $d \times d$ matrices with random rational elements. Then we invert the matrices and show a density plot of the numerators and denominators modulo 3. Characteristic stripes are visible.

```
In[61]:= Module[{d = 60, r, m, inv},
      (* a random rational number *)
    r[n_] := (2 Random[Integer] - 1)*
            Random[Integer, {1, n}]/Random[Integer, {1, n}];
      (* a random rational number *)
    Show[GraphicsArray[
    Function[{seed, mod, numden},
        (* seed random number generator *) SeedRandom[seed];
        (* form matrix and invert it *)
      m = Table[r[8], {d}, {d}]; inv = Inverse[m];
      ListDensityPlot[Mod[numden[inv], mod], PlotRange -> All, Mesh -> False,
                DisplayFunction -> Identity, FrameTicks -> None]] @@@
        (* data for the four plots *)
      {{13, 3, Numerator},    {28, 3, Numerator},
        { 1, 3, Denominator}, { 8, 3, Denominator}}]]]
```

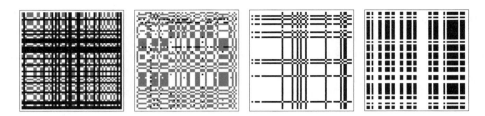

Via $c = a\,b \bmod n$, two numbers a, $b < n$ can be combined to form a new number less than n. The mathematical objects $\mathbb{Z}_n$ created by this operation play an important role for algorithms (especially in computer algebra, see, e.g., [64]). We now look at this operation graphically).

```
In[62]:= With[{n1 = 100, n2 = 65, n3 = 73, o = 128},
     Show[GraphicsArray[#]]& @
     Block[{$DisplayFunction = Identity},
     {(* Mod[i j, n] colored according to absolute value *)
       DensityPlot[Mod[i j, n1], {i, 1, n1 - 1}, {j, 1, n1 - 1},
                Frame -> False, Mesh -> False, PlotPoints -> n1 - 1],
       (* Mod[i j, n] colored according to parity *)
       ListDensityPlot[Table[Mod[i j, n2], {i, 1, n2 - 1}, {j, 1, n2 - 1}],
                Frame -> False, Mesh -> False,
                ColorFunction -> (GrayLevel[If[EvenQ[Round[#]], 1, 0]]&),
                PlotRange -> All, ColorFunctionScaling -> False],
       (* Mod[i j, n] colored according to primeness *)
       ListDensityPlot[Table[Mod[i j, n3], {i, 1, n3 - 1}, {j, 1, n3 - 1}],
                Frame -> False, Mesh -> False,
                ColorFunction -> (GrayLevel[If[PrimeQ[Round[#]], 0, 1]]&),
                PlotRange -> All, ColorFunctionScaling -> False],
       (* scaled modular inverses *)
     Graphics[Table[{Hue[0.8 n/o],
           Table[If[Mod[i j, n] === 1, Point[{i, j}/o], {}],
               {i, 1, n - 1}, {j, 1, n - 1}]}, {n, 2, o}],
           AspectRatio -> Automatic, PlotRange -> All,
           Frame -> True, FrameTicks -> False]}]]
```

The next three graphics visualize averaged integer-valued functions calculated modulo a $m = 479$.

```
In[63]:= Show[GraphicsArray[
     Block[{$DisplayFunction = Identity, i, j, o = 480, d = 13, m = 479},
           ListDensityPlot[ListConvolve[(* average *) Table[1, {d}, {d}],
                               Table[Mod[#, m], {i, o}, {j, o}]],
                     Mesh -> False, PlotRange -> All,
                     FrameTicks -> None]& /@
           (* three two-argument functions *)
           {j^i + i^j, i^i + j^j, (i j)^(i + j)}]]]
```

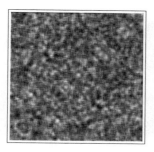 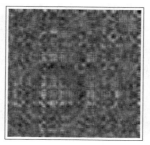 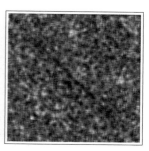

The arguments of Mod can be arbitrary numerical expression, including exact symbolic expressions. (Meaning expressions that, after applying N, evaluate to a number, or, saying it in another way, NumericQ[*expr*] would return True.) Here is a simple example.

```
In[64]:= Mod[Range[10] Pi, 1]
```

$$Out[64]= \{-3 + \pi, -6 + 2\pi, -9 + 3\pi, -12 + 4\pi, -15 + 5\pi,$$
$$-18 + 6\pi, -21 + 7\pi, -25 + 8\pi, -28 + 9\pi, -31 + 10\pi\}$$

The next input is concerned with the three gap theorem: For any irrational α the sequence $\{s_n(\alpha)\}_{n=1,\ldots,\infty}$ where

$$s_n(\alpha) = (((n + 1)\alpha) \bmod 1) - ((n\alpha) \bmod 1)$$

takes at least two different and at most three different values [567], [356], [563]. We use six different values of α and check the first 10^5 n.

```
In[65]:= Length[Union[Subtract @@@ Partition[
                Table[Mod[n #, 1], {n, 10^5}], 2, 1]]]& /@
                    {Pi, E, Sqrt[2], 3^(1/3), Log[2], Cos[1], Pi^E - E^Pi}
Out[65]= {2, 2, 2, 2, 2, 2, 3}
```

The functions Quotient and Mod also work for exact and inexact complex arguments. They are *defined* in such a manner to keep the identity $a =$ Quotient[*a, b*] $b +$ Mod[*a, b*] alive. Here is a quick numerical check for this identity for 100000 random numbers.

```
In[66]:= Table[a = Random[Complex, {-10 - 10 I, 10 + 10 I}];
            b = Random[Complex, {-10 - 10 I, 10 + 10 I}];
            a - (b Quotient[a, b] + Mod[a, b]), {10^5}] // Chop // Union
Out[66]= {0}
```

While the function mod seems to be a quite simple function, because of its discontinuity it can, together with other functions, lead to quite complicated behaviors. In the following graphics, we take an initial set of regularly arranged points from the square $(0, 1) \times (0, 1)$, iteratively rotate it, and fold it back in the unit square using mod. As a function of the rotation angle, complicated patterns arise for the multiple rotated and folded points [205], [206], [3], [87], [312].

```
In[67]:= multiRotatedMod1Graphics[φ_, pp_:30, iters_:30, opts___] :=
    Module[{initialPoints, (* rotation matrix *)
            R = {{Cos[φ], Sin[φ]}, {-Sin[φ], Cos[φ]}} // N},
            initialPoints = (* starting points *)
            Flatten[Table[N[{x, y}], {x, 0, 1, 1/pp}, {y, 0, 1, 1/pp}], 1];
            Graphics[{PointSize[0.002],
            MapIndexed[{RGBColor[(#2[[2]] - 1)/(pp + 1), 0,
                                (#2[[3]] - 1)/(pp + 1)], Point[#1]}&,
                    Partition[#, pp + 1]& /@ (* iterate rotating and mod-ing *)
```

```
        NestList[(Mod[#, 1]& @ (R.#& /@ #))&,
                initialPoints, iters], {3}]]},
     AspectRatio -> Automatic, PlotRange -> All]];
```

In[68]:= `Show[GraphicsArray[multiRotatedMod1Graphics /@ #]]& /@`
         `{{0, 1, 15}, {30, 31, 45}, {60, 72, 73}}`

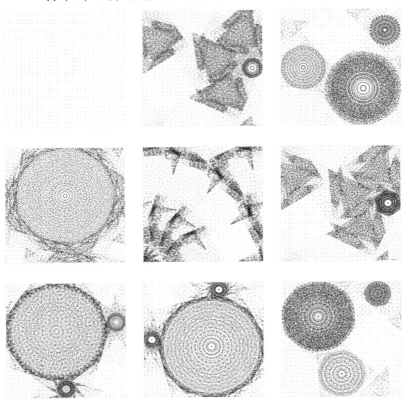

Here is the corresponding animation.

```
    Do[Show @ multiRotatedMod1Graphics[φ], {φ, 0, Pi/2, Pi/2/90}]
```

In the following, we will carry out 10^4 rotations. Instead of displaying all resulting points, we discretize the unit square and show the number of points per cell. For speed reasons, we carry out these operations in a compiled function.

In[69]:= `cf = Compile[{φ, {n, _Integer}, {p, _Integer}, {iters, _Integer}},`
         `   Module[{points = Flatten[Table[{x, y}, {x, 0, n}, {y, 0, n}]/n, 1],`
         `         tab = Table[0, {p + 1}, {p + 1}],`
         `         (* rotation matrix *)`
         `         R = {{Cos[φ], Sin[φ]}, {-Sin[φ], Cos[φ]}}, i, j},`
         `     Do[points = R.#& /@ points;`
         `        (* bin and count *)`
         `        Do[{i, j} = Floor[p Mod[points[[k]], 1]] + 1;`
```
```

```
       tab[[i, j]] = tab[[i, j]] + 1, {k, (n + 1)^2}],
    {iters}]; Drop[#, -1]& /@ Drop[tab, -1]]];
```

Here are two examples.

```
In[70]:= Show[GraphicsArray[
    ListDensityPlot[Log[10, # + 1], ColorFunction -> (Hue[0.8 #]&),
                Mesh -> False, PlotRange -> All, FrameTicks -> None,
                DisplayFunction -> Identity]& /@
            {cf[30 Degree, 25, 100, 10^4], cf[59 Degree, 25, 400, 10^4]}]]
```

Next, we present a simple programming example: Using the two commands Mod and Quotient, we can easily reprogram the command IntegerDigits. Here is the two-argument form.

```
In[71]:= myIntegerDigits1[n_Integer?Positive, base_Integer?Positive] :=
    myBaseForm[base, Flatten[First /@ Rest[
    FoldList[(* pair of quotient and remainder *)
            {Quotient[Last[#1], #2], Mod[Last[#1], #2]}&,
            {0, n}, (* needed base^i *)
            base^#& /@ Reverse[Range[0, Floor[N[Log[base, n]]]]]]]]]]
```

Another, slightly shorter, and more efficient way to program myIntegerDigits would be the following.

```
In[72]:= myIntegerDigits2[n_Integer?Positive, base_Integer?Positive] :=
    myBaseForm[base, Reverse[Mod[Drop[FixedPointList[Quotient[#, base]&, n],
                    -2], base]]]
```

The result of myIntegerDigits*k* is an object with head myBaseForm. Its first argument is the base, and the second contains the digits in the base.

```
In[73]:= myIntegerDigits1[1000, 10]
Out[73]= myBaseForm[10, {1, 0, 0, 0}]
```

```
In[74]:= myIntegerDigits1[1024, 2]
Out[74]= myBaseForm[2, {1, 0, 0, 0, 0, 0, 0, 0, 0, 0, 0}]
```

```
In[75]:= myIntegerDigits1[547342935348523, 2254]
Out[75]= myBaseForm[2254, {21, 462, 1545, 631, 2085}]
```

myIntegerDigits2 gives the same result.

```
In[76]:= myIntegerDigits2[547342935348523, 2254]
Out[76]= myBaseForm[2254, {21, 462, 1545, 631, 2085}]
```

For comparison, consider the following results from the built-in function IntegerDigits.

```
In[77]:= IntegerDigits[1000, 10]
```

In[77]:= `{1, 0, 0, 0}`

In[78]:= `IntegerDigits[1024, 2]`

Out[78]= `{1, 0, 0, 0, 0, 0, 0, 0, 0, 0, 0}`

In[79]:= `IntegerDigits[547342935348523, 2254]`

Out[79]= `{21, 462, 1545, 631, 2085}`

We can convert back to a decimal number as follows.

In[80]:= `toDecimalNumber[myBaseForm[base_, l_]] :=`
`            Reverse[base^#& /@ Range[0, Length[l] - 1]].l`

In[81]:= `toDecimalNumber[%%%%]`

Out[81]= `547342935348523`

If one has to calculate a^n mod b, there is a *much* faster way than calling $\text{Mod}[a^n, b]$, namely, the command PowerMod.

`PowerMod[a, b, n]`

    produces a^n mod b.

The function `PowerMod` can be used to calculate the nth digit of a proper fraction without calculating all previous digits [201]. The following function `NthDigitOfFraction` calculates the nth digit of a "decimal" fraction in base b of a proper fraction p/q.

In[82]:= `NthDigitOfProperFraction[`(* check arguments *)
`        r:Rational[p_Integer?Positive, q_Integer?Positive]?(0 < # < 1&),`
`            n_Integer?Positive, base_Integer?Positive] :=`
`    Floor[base Mod[p PowerMod[base, n - 1, q], q]/q]`

Here are two small tests for the fraction $17/123$ in base 10 and for the fraction $171/222$ in base 25. We compare the digits with the output of `RealDigits`.

In[83]:= `Array[NthDigitOfProperFraction[17/123, #, 10]&, 25]`

Out[83]= `{1, 3, 8, 2, 1, 1, 3, 8, 2, 1, 1, 3, 8, 2, 1, 1, 3, 8, 2, 1, 1, 3, 8, 2, 1}`

In[84]:= `RealDigits[N[17/123, 25]][[1]]`

Out[84]= `{1, 3, 8, 2, 1, 1, 3, 8, 2, 1, 1, 3, 8, 2, 1, 1, 3, 8, 2, 1, 1, 3, 8, 2, 1}`

In[85]:= `Array[NthDigitOfProperFraction[171/222, #, 25]&, 13]`

Out[85]= `{19, 6, 10, 11, 20, 15, 5, 1, 17, 5, 18, 14, 13}`

In[86]:= `RealDigits[171/222, 25, 13][[1]]`

Out[86]= `{19, 6, 10, 11, 20, 15, 5, 1, 17, 5, 18, 14, 13}`

Now, the 10^{100}th digit of $1234/5678$ is calculated. The calculation happens basically instantaneously.

In[87]:= `NthDigitOfProperFraction[1234/5678, 10^100, 10] // Timing`

Out[87]= `{0. Second, 8}`

Using the built-in function `RealDigits`, it is straightforward to implement the following function `NthDigit` OfProperFraction1 that returns nth digit of a proper fraction.

In[88]:= `NthDigitOfProperFraction1[`
`        r:Rational[p_Integer?Positive, q_Integer?Positive]?(0 < # < 1&),`
`            n_Integer?Positive, base_Integer?Positive] :=`
`    Module[{rd = RealDigits[r], a, p, α, λ},`

```
(* initial and periodic part *)
a = Drop[rd[[1]], -1]; p = rd[[1, -1]];
{α, λ} = Length /@ {a, p};
(* extract digit *)
If[n <= α, a[[n]], p[[Mod[n - α, λ, 1]]]]]]
```

In[89]:= **NthDigitOfProperFraction1[1234/5678, 10^100, 10]**

Out[89]= 8

Using the function Mod and the function Floor (which was introduced in the previous chapter, and which also works for rational arguments), we now look at Schönberg's version of a Peano curve.

Mathematical Remark: Schönberg's Peano curve

This version of a Peano curve is popular because the proof that it is area-filling is simple. (However, in contrast to the Peano curves shown in Chapter 1 of the Graphics volume [555], the nth-order approximation of this curve is self-intersecting.) For more on Schönberg's Peano curve, see [479]. At the mth step, the 3^m points $\{x_i^{(m)}, y_i^{(m)}\}$ on this curve are given by

$$x_i^{(m)} = \sum_{k=1}^{\lfloor(m+1)/2\rfloor} 2^{-k}\, \varphi\!\left(\frac{3^{2k-2}\, i}{3^m}\right) + 2^{-\lfloor(m+1)/2\rfloor}\left(i - 2\left\lfloor\frac{i}{2}\right\rfloor\right)$$

$$y_i^{(m)} = \sum_{k=1}^{\lfloor(m+1)/2\rfloor} 2^{-k}\, \varphi\!\left(\frac{3^{2k-1}\, i}{3^m}\right) + 2^{-\lfloor(m+1)/2\rfloor}\left(i - 2\left\lfloor\frac{i}{2}\right\rfloor\right)$$

where $m = 1, 2, 3, \ldots$ and $i = 1, 2, \ldots 3^m$. Here, $\lfloor z \rfloor$ is the integer part of z, and $\varphi(t)$ is the 2-periodic even function:

$$\varphi(t) = \begin{cases} 0 & \text{for } 0 \le t < \dfrac{1}{3} \\[2mm] 3\,t - 1 & \text{for } \dfrac{1}{3} \le t < \dfrac{2}{3} \\[2mm] 1 & \text{for } \dfrac{2}{3} \le t \le 1. \end{cases}$$

First, we implement the function φ.

In[90]:= **φ[t_] := Which[0 <= # < 1/3, 0,**
        1/3 <= # < 2/3, 3# - 1,
        2/3 <= # <= 1, 1]&[If[# > 1, 1 - (# - 1), #]&[
    (* use pure function to avoid multiple evaluation of Mod *) **Mod[t, 2]]]**

Here is a rewritten form of the function $\varphi(t)$.

In[91]:= **φ[t] // PiecewiseExpand**

Out[91]= $\begin{cases} 1 & \frac{2}{3} \le \mathrm{Mod}[t, 2] \le \frac{4}{3} \\[1mm] 5 - 3\,\mathrm{Mod}[t, 2] & \frac{4}{3} < \mathrm{Mod}[t, 2] \le \frac{5}{3} \\[1mm] -1 + 3\,\mathrm{Mod}[t, 2] & \frac{1}{3} \le \mathrm{Mod}[t, 2] < \frac{2}{3} \end{cases}$

And here is a plot of $\varphi(t)$.

In[92]:= `Plot[φ[t], {t, -4, 4}]`

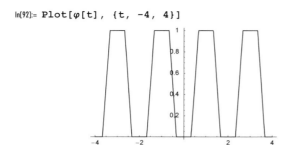

We now compute the points $\{x_i^{(m)}, y_i^{(m)}\}$.

In[93]:=
```
x[m_, i_] := Sum[φ[(3^(2k - 2) i)/3^m]/2^k, {k, 1, Floor[(m + 1)/2]}] +
                (i - 2 Floor[i/2])/2^Floor[(m + 1)/2]

y[m_, i_] := Sum[φ[(3^(2k - 1) i)/3^m]/2^k, {k, 1, Floor[(m + 1)/2]}] +
                (i - 2 Floor[i/2])/2^Floor[(m + 1)/2]
```

To best understand a Peano curve (if it is at all possible to "understand" such a curve), we need a plot. Because we will use the result twice, we use a `SetDelayed[Set[...]]` construction.

In[95]:=
```
SchoenbergPeanoCurve[ord_Integer?Positive] := (* remember values *)
    SchoenbergPeanoCurve[ord] = Table[{x[ord, i], y[ord, i]}, {i, 3^ord}]
```

To better follow the order of the evaluations inside the unit square, we color the curve.

In[96]:=
```
ColoredSchoenbergPeanoCurvePicture[ord_, opts___] :=
    Show[Graphics[MapIndexed[(* add color *)
                {Hue[#2[[1]]/3^ord 0.7], Line[#1]}&,
                Partition[SchoenbergPeanoCurve[ord], 2, 1]]],
        opts, AspectRatio -> Automatic]
```

Here are the first six levels.

In[97]:=
```
Show[GraphicsArray[#]]& @
    Block[{opts = Sequence[DisplayFunction -> Identity,
                            PlotRange -> {{0, 1}, {0, 1}}]},
        Table[ColoredSchoenbergPeanoCurvePicture[k, opts], {k, 6}]]
```

We implement a different visualization technique for Schönberg's curves of higher order. We discretize all lines and count how often a line passes over a subsquare of size $1/pp \times 1/pp$.

In[98]:=
```
lineSegmentsDensityMatrixCF =
    Compile[{{segments, _Real, 3}, {pp, _Integer}},
        Module[{T = Table[0, {pp + 1}, {pp + 1}], p1, p2, λ, i = 1, j = 1},
                (* loop over all line segments *)
                Do[{p1, p2} = segments[[k]];
                    (* discretize line segments according to its length *)
                    λ = Round[Sqrt[(p2 - p1).(p2 - p1)] pp];
                    Do[{i, j} = Round[pp (p1 + k/λ (p2 - p1))] + 1;
```

```
                               (* count passes *) T[[i, j]] = T[[i, j]] + 1, {k, λ}],
               {k, Length[segments]}]; (* return counts *) T]];
```

Here are the resulting frequencies shown for order 13. We use different colorings.

```
In[99]:= Show[GraphicsArray[
         Module[{(* prepare data *)
                 densityData = lineSegmentsDensityMatrixCF[
                              Partition[SchoenbergPeanoCurve[13], 2, 1], 401]},
                 (* show data using two coloring schemes *)
                 Function[f, ListDensityPlot[(* scale logarithmically *)
                              Log[densityData + 1],
                     Mesh -> False, PlotRange -> All,
                     ColorFunction -> (Hue[f (1 - #)]&), FrameTicks -> None,
                     DisplayFunction -> Identity]] /@
                 (* blue to red and other coloring *) {0.78, 3.8}]]]
```

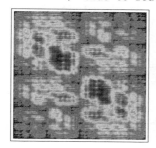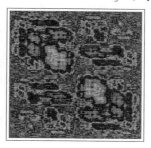

For the description of a Schönberg curve filling an *n*-dimensional unit cube, see [294].

The last two commands to be discussed in this section find the least common multiple and the greatest common divisor.

$\text{LCM}[n_1, n_2, \ldots, n_m]$

    calculates the least common multiple (lcm) of the numbers n_i.

$\text{GCD}[n_1, n_2, \ldots, n_m]$

    calculates the greatest common divisor (gcd) of the numbers n_i.

Here are two examples with quite large numbers.

```
In[100]:= LCM[73474597256439, 90267659905, 836549984, 2835649]
Out[100]= 15733080535976895764268934230254508834720
```

```
In[101]:= GCD[0876, 6787675, 568564647, 974692, 86585958]
Out[101]= 1
```

The function LCM modulo an integer shows some symmetric patterns.

```
In[102]:= myLCMPicture[mod_Integer?(# >= 3&), pp_Integer?(# >= 4&)] :=
         ListDensityPlot[DeleteCases[(* cut out zeros *)
         Table[If[Mod[a, mod] === 0 || Mod[b, mod] === 0,
                 Sequence @@ {}, Mod[LCM[a, b], mod]],
                     {a, -pp, pp}, {b, -pp, pp}], {}],
                     Mesh -> False, FrameTicks -> None] /; pp > mod
```

```
In[103]:= Show[GraphicsArray[
    Block[{$DisplayFunction = Identity},
        (* use four mod values *)
        Table[myLCMPicture[mod, 64], {mod, 3, 6}]]]]
```

The least common multiple of the first n integers growth exponentially with n. The next (left) graphic shows the difference. The right graphic shows the difference scaled by $2\,x^{1/2} \ln^2(x)$, which asymptotically should be less than 1 if the Riemann hypothesis holds [176]).

```
In[104]:= lcmLogData[o_] :=
    Module[{T = Table[0, {o}], lcm = 1},
        Do[lcm = LCM[lcm, k]; T[[k]] = N[Log[lcm] - k], {k, 2, o}]; T]

In[105]:= Show[GraphicsArray[
    Block[{$DisplayFunction = Identity},
        (* differences and scaled differences *)
        {ListPlot[#], ListPlot[MapIndexed[If[#2 === {1}, 0,
            #1/(2 Sqrt[#2[[1]]] Log[#2[[1]]]^2)]&, #]]}&[lcmLogData[10^5]]]]]
```

Using the prime factorization of the integers up to n and keeping only the exponents of the prime factors allows reducing the computational work to calculate $\ln(\mathrm{lcm}(1, 2, \ldots, n))$ much faster.

```
In[106]:= NLogLCM[n_, prec_:MachinePrecision] :=
    Module[{L = {}, k = 1, e},
        (* prime exponents other than 1 *)
        While[(e = Floor[Log[Prime[k], n]]) > 1, L = {L, e}; k++];
        (* exponents of prime factors *)
        L = Join[Flatten[L], Table[1, {PrimePi[n] - k + 1}]];
        (* calculate Log[LCM[1, 2, …, n]] *)
        Plus @@ MapIndexed[N[#1 Log[Prime[#2[[1]]]], MachinePrecision]&, L]]
```

Now, we can evaluate the deviation from the exponential growth for $n = 10^6$ in about a second.

```
In[107]:= With[{n = 10^6}, n - NLogLCM[n]] // Timing

Out[107]= {0.71 Second, 413.403}
```

We now discuss an application of these commands: For the Euclidean algorithm (which was mentioned in Chapter 2 of the Programming volume [554]), see, e.g., [572] and [561]; for Gaussian integers, see [45].

Although it is beyond the scope of this book to treat all of the algorithms of computer algebra, we make an exception for this relatively old, simple but very important algorithm. The Euclidean algorithm permits a very fast computation of the greatest common divisor of two numbers.

```
In[108]:= SeedRandom[333];
        largeNumbers = Table[Random[Integer, {1, 10^100}], {20}];
```

The probability of getting a number with fewer than 98 digits is quite small.

```
In[110]:= Length[IntegerDigits[#]]& /@ largeNumbers
Out[110]= {98, 100, 99, 100, 100, 100, 100, 100, 100,
          99, 100, 100, 100, 100, 100, 100, 100, 100, 100, 99}
```

We now find their greatest common divisor. This can be done very fast.

```
In[111]:= {GCD @@ largeNumbers, Timing[Do[GCD @@ largeNumbers, {10^4}]]}
Out[111]= {1, {0.34 Second, Null}}
```

The probability that n randomly chosen positive integers are relatively prime is $1/\zeta(n)$ [117], [121], [550], where $\zeta(z)$ is Riemann's zeta function (see Chapter 3 of the Symbolics volume [556]). (For Gaussian primes, see [118].) So the above result is not unexpected.

```
In[112]:= N[1/Zeta[20], 10]
Out[112]= 0.9999990460
```

We could now investigate the greatest common divisors of all number pairs by using GCD. The relative fraction of relatively prime numbers is $1/\zeta(2) \approx 0.607....$ Here, we choose 1000 random numbers and find the greatest common divisor of all pairs.

```
In[113]:= SeedRandom[111];
        (* list of 1000 random integers *)
        largeNumbers = Table[Random[Integer, {1, 10^21}], {1000}];

        (* frequency of pairs that have a nontrivial common divisor *)
        Count[#, 1]/Length[#]&[ (* gcd of all pairs *)
         Flatten[Table[GCD[largeNumbers[[i]], largeNumbers[[j]]],
                      {i, Length[largeNumbers]},
                      {j, i + 1, Length[largeNumbers]}]]] // N
Out[117]= 0.600989
```

And in the next input, we generate lists of ten integers until they have a nontrivial common divisor. In average, we have to generate about 1006 lists until this happens.

```
In[118]:= {(* theoretical value *)
        1/(1 - N[1/Zeta[10], 10]),
        (* average value for 10 trials *)
        Module[{counter}, SeedRandom[117];
              Table[counter = 0;
                    While[(* make lists of ten integers until they
                              have a nontrivial common divisor *)
                          (GCD @@ Table[Random[Integer, {1, 10^100}],
                                       {10}]) === 1,
                      (* return number of trials *)
                      counter++]; counter, {10}] // Mean // N]}
Out[118]= {1006.454, 1006.6}
```

Next, we form tuples $\mathbf{t} = \{t_1, t_2, \ldots, t_n\}$ of random integers t_i and calculate their greatest common divisor. Then, for two random vectors $\mathbf{x}$ and $\mathbf{y}$, we form the greatest common divisor of $\mathbf{x.t}$ and $\mathbf{y.t}$. In average the two greatest common divisors agree in $6/\pi^2 \approx 0.60793$ cases [578].

```
In[119]:= Module[{o = 10^6, t, x, y},
    Sum[SeedRandom[k];
        t = Table[Random[Integer, {1, 10^10}], {6}];
        (* use smaller integers for elements of x and y *)
        {x, y} = Table[Random[Integer, {1, 10}], {2}, {6}];
        If[GCD[x.t, y.t] === (GCD @@ t), 1., 0], {k, o}]/o]
Out[119]= 0.607578
```

Here is a small number theoretical example where extrapolating from thousands of consecutive data leads to a wrong conclusion. We consider the sequence $g_k = \gcd(2^k - 3, 3^k - 2)$ [546].

```
In[120]:= gcd23[k_] := GCD[2^k - 3, 3^k - 2]
```

Inspecting the first 20 values of g_k leads naturally to the conjecture $g_k = 5\,\delta_{3, k \bmod 4} + (1 - \delta_{3, k \bmod 4})$. (We discuss the function Mod in a moment.)

```
In[121]:= Table[{k, gcd23[k]}, {k, 20}]
Out[121]= {{1, 1}, {2, 1}, {3, 5}, {4, 1}, {5, 1}, {6, 1},
    {7, 5}, {8, 1}, {9, 1}, {10, 1}, {11, 5}, {12, 1}, {13, 1},
    {14, 1}, {15, 5}, {16, 1}, {17, 1}, {18, 1}, {19, 5}, {20, 1}}

In[122]:= gcd23Conjectured[k_] := 4 KroneckerDelta[3, Mod[k, 4]] + 1
```

And the first 2500 integers seem to confirm the conjecture.

```
In[123]:= And @@ Table[gcd23[k] == gcd23Conjectured[k], {k, 2500}]
Out[123]= True
```

But the conjecture breaks the first time at $k = 3783$ (and more generally, at $k = 3783 + j\,5332$, $j \in \mathbb{N}$).

```
In[124]:= Table[gcd23[k], {k, 3783, 3783 + 10 5332, 5332}]
Out[124]= {26665, 26665, 26665, 26665, 26665, 26665, 26665, 26665, 26665, 26665, 26665}
```

Mathematical Remark: The Euclidean Algorithm

The Euclidean algorithm divides a fraction p/q by the largest common factor contained in p and q. The algorithm proceeds as follows: Let k and g be the smaller and the larger of the two numbers p and q, respectively. Let $r = g \bmod k$. This remainder r is smaller than k, and after identifying $k' = r$, $g' = k$, this procedure can be repeated until it stops. The final remainder r ($\neq 0$) is the greatest (common) divisor of p and q.

We now implement the Euclidean algorithm. By direct application of the above formulas (including the avoidable auxiliary variables g_{new} and k_{new}), and using Quotient (to improve its comprehensibility), we get the following code. We collect the g's and k's in a list 1. At the end, we include the shortened form of the result from *Mathematica*.

```
In[125]:= Euclid1[p_Integer?Positive, q_Integer?Positive] :=
    Module[{l, gnew, knew},
    If[IntegerQ[Max[p, q]/Min[p, q]],
        ToString[Max[p, q]] <> "/" <> ToString[Min[p, q]] <>
                " = " <> ToString[Max[p, q]/Min[p, q]],
        For[(* initialization *) g = Max[p, q]; k = Min[p, q]; l = {},
```

```
        Mod[g, k] != 0      (* until *), Null,
        knew = Mod[g, k];  (* one step *) gnew = k;
        (* prepare for nice output form *)
        AppendTo[l, (* save intermediate results *)
        {ToString[g] <> " = " <> ToString[Quotient[g, k]] <> " * " <>
         ToString[k] <> " + " <> ToString[knew]}];
        g = gnew; k = knew];
        AppendTo[l, (* compare with built-in result *)
           {ToString[p/knew] <> "/" <> ToString[q/knew], p/q}];
        (* output in TableForm *)TableForm[l]]]
```

To better understand how the Euclidean algorithm works, we write out the decompositions. Here are a few examples.

In[126]:= **Euclid1[124, 42]**

Out[126]//TableForm=

$$124 = 2 * 42 + 40$$
$$42 = 1 * 40 + 2$$
$$62/21 \qquad\qquad \frac{62}{21}$$

In[127]:= **Euclid1[64685639, 96409]**

Out[127]//TableForm=

$$64685639 = 670 * 96409 + 91609$$
$$96409 = 1 * 91609 + 4800$$
$$91609 = 19 * 4800 + 409$$
$$4800 = 11 * 409 + 301$$
$$409 = 1 * 301 + 108$$
$$301 = 2 * 108 + 85$$
$$108 = 1 * 85 + 23$$
$$85 = 3 * 23 + 16$$
$$23 = 1 * 16 + 7$$
$$16 = 2 * 7 + 2$$
$$7 = 3 * 2 + 1$$
$$64685639/96409 \qquad\qquad \frac{64685639}{96409}$$

In[128]:= **Euclid1[3595197375, 3048192]**

Out[128]//TableForm=

$$3595197375 = 1179 * 3048192 + 1379007$$
$$3048192 = 2 * 1379007 + 290178$$
$$1379007 = 4 * 290178 + 218295$$
$$290178 = 1 * 218295 + 71883$$
$$218295 = 3 * 71883 + 2646$$
$$71883 = 27 * 2646 + 441$$
$$8152375/6912 \qquad\qquad \frac{8152375}{6912}$$

In[129]:= **Euclid1[51, 6756]**

Out[129]//TableForm=

$$6756 = 132 * 51 + 24$$
$$51 = 2 * 24 + 3$$
$$17/2252 \qquad\qquad \frac{17}{2252}$$

The above implementation of Euclid was such that the algorithm was straightforward. A faster implementation, also returning a list of how the numbers are composed, is given by the following.

In[130]:= **Euclid2[p_Integer?Positive, q_Integer?Positive] :=**
    Module[{l = If[Abs[p] > Abs[q], {p, q}, {q, p}], t},

```
While[(t = Mod @@ Take[1, -2]) != 0, AppendTo[1, t]];
Append[MapThread[{#1 <> " = " <> #2 <> " * " <> #3 <> " + " <> #4}&,
  {Drop[#, -2], ToString /@ ((Quotient @@ #&) /@ Partition[Drop[1, -1], 2, 1]),
          Drop[Drop[#, 1], -1], Drop[#, 2]}&[ToString /@ 1]],
  {#1 <> "/" <> #2, p/q}& @@ (ToString /@ ({p, q}/Last[1]))] // TableForm]
```

Euclid2 gives the same result as Euclid1.

In[131]:= `Euclid2[51, 6756]`

Out[131]//TableForm=

```
6756 = 132 * 51 + 24
51 = 2 * 24 + 3
17/2252
```
$$\frac{17}{2252}$$

Although no built-in function returns the steps of the Euclidean algorithm, the terms of the continued fraction of p/q ($p > q$) turn out to be just the steps. Here, we calculate the continued fraction of the above three pairs of integers.

In[132]:= `ShortEuclid3[a_, b_] := Drop[ContinuedFraction[a/b], -1]`

In[133]:= `ShortEuclid3[64685639, 96409]`

Out[133]= `{670, 1, 19, 11, 1, 2, 1, 3, 1, 2, 3}`

In[134]:= `ShortEuclid3[3595197375, 3048192]`

Out[134]= `{1179, 2, 4, 1, 3, 27}`

In[135]:= `ShortEuclid3[6756, 51]`

Out[135]= `{132, 2}`

Because of the recursive nature, the most natural way to implement Euclid is probably the following (this just computes the result; we do not create a list of the intermediate values).

In[136]:= `Euclid4GCD[p_Integer?Positive, q_Integer?Positive] :=`
          `Euclid4GCD[#2, Mod[##]]& @@ Reverse[Sort[{p, q}]]`

          `Euclid4GCD[p_Integer?Positive, 0] = p;`

Again, we use Euclid4GCD to calculate the greatest common divisor for the three examples from above.

In[138]:= `{Euclid4GCD[64685639, 96409], Euclid4GCD[3595197375, 3048192],`
          `Euclid4GCD[6756, 51]}`

Out[138]= `{1, 441, 3}`

For a discussion of its speed (complexity), see [385], [566], and [251]. The classical Euclid algorithm is not the fastest possible way to calculate the greatest common divisor of two numbers; for a discussion of faster methods, see [42] and [525]. If one replaces Mod in the definition of Euclid4GCD by Subtract, one also obtains the greatest common divisor of two numbers, but in more steps [320], [106].

The following graphic visualizes the dynamics of the Euclidean algorithm and its subtractive version by connecting the point $\{p, q\}$ with the points obtained after one step. The second pair of graphics uses the function lineSegmentsDensityMatrixCF from above.

In[139]:= `Show[GraphicsArray[Function[ms,`
          `Graphics[{Thickness[0.002],`
          `    Table[Line[{{p, q}, {#2, ms[##]}& @@ Reverse[Sort[{p, q}]]}],`
          `        {p, 50}, {q, 50}]}, AspectRatio -> Automatic]] /@`
          `                            {Mod, Subtract}]]`

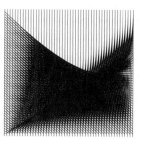

```
In[140]:= (* show density of lines *)
          Show[GraphicsArray[#]]& @
          Block[{o = 200, $DisplayFunction = Identity},
          Function[ms,
            ListDensityPlot[(* discretize connecting lines and count
                              hits over rasterized square *)
                        Transpose[lineSegmentsDensityMatrixCF[Flatten[
                          Table[({{p, q}, {#2, ms[##]}}& @@
                                          Reverse[Sort[{p, q}]]})/o,
                             {p, o}, {q, o}], 1], o]],
                      Mesh -> False, PlotRange -> All,
                      FrameTicks -> None]] /@ {Mod, Subtract}]
```

By adding a side effect in the recursive `Euclid4GCD` calls, we could now go on and monitor and analyze all intermediate *p,q*-pairs. In the following, we choose four pairs of two large integers and display the cumulative distribution of their ratio. The four curves nearly coincide.

```
In[142]:= Euclid4GCD[p_Integer?Positive, q_Integer?Positive] :=
          ((* keep {p, q} pair by adding to bag *) bag = {bag, {p, q}};
          (Euclid4GCD[#2, Mod[##]]& @@ Reverse[Sort[{p, q}]]))

In[143]:= $IterationLimit = Infinity; $RecursionLimit = Infinity;
          SeedRandom[111];
          Do[bag = {}; (* two large integers *)
            {pLarge, qLarge} = Table[Random[Integer, {1, 10^(10^4)}], {2}];
            Euclid4GCD @@ {pLarge, qLarge}; data[k] = bag, {k, 4}];

In[146]:= Show[Table[(* make list plot of  size distribution *)
                ListPlot[Sort @ Log[10, Divide @@@
                    (Sort /@ N[Partition[Flatten[data[k]], 2]])],
                    PlotStyle -> {PointSize[0.003], Hue[(k - 1)/5]},
                    DisplayFunction -> Identity],
```

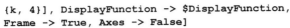

```
              {k, 4}], DisplayFunction -> $DisplayFunction,
              Frame -> True, Axes -> False]
```

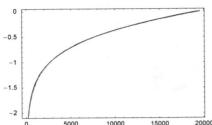

A purely recursive definition (using only elementary arithmetic operations, but no Mod) of the greatest common divisor function is Stein's algorithm [326].

```
In[147]:= With[{(* abbreviate right-hand side *)
           rhs = SteinGCD[p_Integer?Positive, q_Integer?Positive]},
           (* recursive definition *)
           rhs := 2 SteinGCD[p/2, q/2] /; EvenQ[p] && EvenQ[q];
           rhs := SteinGCD[p/2, q] /; EvenQ[p];
           rhs := SteinGCD[p - q, q] /; p > q;
           rhs := SteinGCD[p, q - p] /; q > p;
           rhs := p /; q == p]
```

```
In[148]:= {SteinGCD[64685639, 96409], SteinGCD[3595197375, 3048192],
           SteinGCD[6756, 51]}
```

```
Out[148]= {1, 441, 3}
```

Other methods for calculating greatest common divisors exist. The next two definitions implement the binary and the left-shift binary greatest common divisor algorithm [534], [524], [83], [505]. (The step taking the logarithm could be carried out more efficiently).

```
In[149]:= bGCD[p_, 0] := p

         bGCD[p_, q_] := bGCD[q, p] /; p < q

         bGCD[p_, q_] := 2 bGCD[p/2, q/2] /; EvenQ[p] && EvenQ[q]

         bGCD[p_, q_] := bGCD[p/2, q] /; EvenQ[p] && OddQ[q]

         bGCD[p_, q_] := bGCD[p, q/2] /; OddQ[p] && EvenQ[q]

         bGCD[p_, q_] := bGCD[q, (p - q)/2]
```

```
In[155]:= {bGCD[64685639, 96409], bGCD[3595197375, 3048192],
           bGCD[6756, 51]}
```

```
Out[155]= {1, 441, 3}
```

```
In[156]:= lsbGCD[p_, 0] := p

         lsbGCD[p_, q_] := lsbGCD[q, p] /; p < q

         lsbGCD[p_, q_] := lsbGCD @@ Reverse[Sort[{q, Min[p - 2^# q, 2^(# + 1) q - p]&[
                           Floor[Log[2, p/q]]]}]] /; p >= q
```

```
In[159]:= {lsbGCD[64685639, 96409], lsbGCD[3595197375, 3048192],
           lsbGCD[6756, 51]}
```

Out[159]= {1, 441, 3}

We now present two graphics gems, the roses of P. M. Maurer and the friezes of N. G. deBruijn. Both graphics make extensive use of Mod, GCD, and LCM.

Mathematical Remark: Maurer Roses

The basic idea of these images is simple (see [382], and [581]). Suppose we are given a closed curve in the plane with n-fold rotational symmetry (later on, we will also apply the idea to curves without this symmetry), that is, a rose with n leaves (petals): $r(\varphi) = |\cos(n\varphi/2)|$. (This definition of a rose [354], [236], deviates somewhat from the original paper, but the difference does not affect the resulting images.) We now divide the circle into m parts (with m an integer), and starting with 0 (of course, we can also choose a different starting value), we move through d subangles ($d < m$) along straight lines with endpoints on the above boundary curve, until we get back to the starting point. This requires exactly lcm(m, d) steps. The same procedure is repeated starting at 1, 2, ..., gcd(m, d) − 1 instead of 0. Finally, we plot all of these closed curves on a single image.

We begin with the implementation of the initial curve. (It is not a rose in the classical mathematical sense [13]; roses satisfy $r(\varphi) = \sin(m/n\varphi)$, m, n integer.)

```
In[160]:= rose[n_, φ_] = Abs[Cos[n φ/2]];
```

Here are some plots of these "roses".

```
In[161]:= Show[GraphicsArray[Table[
        ParametricPlot[rose[n, φ] {Cos[φ], Sin[φ]}, {φ, 0, 2Pi},
                AspectRatio -> 1, Axes -> False,
                DisplayFunction -> Identity], {n, 2, 6}]]]
```

The above construction leads immediately to the following program.

```
In[162]:= MaurerRose[n_, d_, m_] :=
    Module[{α = N[2Pi/m] (* for speed *), φn},
        Table[ (* all equivalent lines *)
                Line[ (* one line *)
                Table[φn = φ α; rose[n, φn] {Cos[φn], Sin[φn]},
                        {φ, φ0, LCM[m, d] + d, d}]],
                {φ0, 0, GCD[m, d] - 1}]]
```

Here is a large bouquet of Maurer roses.

```
In[163]:= With[{MR = MaurerRose},
    Show[Graphics[{Thickness[0.002],
     {(* center rose *)
     Map[# + {0, 0}&, MR[4, 30, 360], {3}],
     (* first layer *)
     Map[# + 2.1{Cos[ Pi/6], Sin[ Pi/6]}&, MR[166, 221, 235], {3}],
     Map[# + 2.1{Cos[ Pi/2], Sin[ Pi/2]}&, MR[ 74, 181, 229], {3}],
     Map[# + 2.1{Cos[ 5Pi/6], Sin[ 5Pi/6]}&, MR[ 78,  10, 229], {3}],
     Map[# + 2.1{Cos[ 7Pi/6], Sin[ 7Pi/6]}&, MR[234,  56, 234], {3}],
     Map[# + 2.1{Cos[ 3Pi/2], Sin[ 3Pi/2]}&, MR[140, 101, 336], {3}],
     Map[# + 2.1{Cos[11Pi/6], Sin[11Pi/6]}&, MR[145,  42, 299], {3}],
     (* second layer *)
     Map[# + 3.6{1,             0}&, MR[178, 145, 360], {3}],
     Map[# + 3.6{Cos[ Pi/3], Sin[ Pi/3]}&, MR[181,  90, 359], {3}],
     Map[# + 3.6{Cos[ 2Pi/3], Sin[ 2Pi/3]}&, MR[194,  10, 197], {3}],
     Map[# + 3.6{Cos[ Pi ], Sin[ Pi ]}&, MR[195,  35, 326], {3}],
     Map[# + 3.6{Cos[ 4Pi/3], Sin[ 4Pi/3]}&, MR[259,  10, 195], {3}],
     Map[# + 3.6{Cos[ 5Pi/3], Sin[ 5Pi/3]}&, MR[298,  40, 293], {3}]}}],
       AspectRatio -> Automatic, PlotRange -> All]]
```

With this principle of periodic evaluation of points on a given curve, we can generate a variety of similar images. Here is an implementation of the key computational step, which we introduce as a function GeneralizedMaurerRose.

```
In[164]:= GeneralizedMaurerRose[func_, d_, m_] :=
    (* func is now arbitrary *)
     Table[Line[Table[func[φ #], {φ, φ0, LCM[m, d] + d, d}]],
       {φ0, 0, GCD[m, d] - 1}]&[N[2Pi/m]]
```

Here are two Lissajous figures created in this way.

```
In[165]:= Show[GraphicsArray[{Graphics[
    {Thickness[0.002], (* a Lissajou curve *)
     GeneralizedMaurerRose[{Sin[3#], Cos[5#]}&, 100, 400]},
        PlotRange -> All, AspectRatio -> Automatic],
     Graphics[{Thickness[0.002],
     (* now a linear combination of two trigs *)
     GeneralizedMaurerRose[{Sin[7#] + Cos[3#], Cos[9#] + Sin[2#]}&, 20, 600]},
        PlotRange -> All, AspectRatio -> Automatic]}]]
```

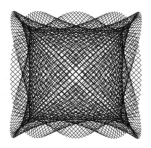 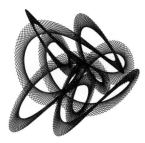

Here are two three-dimensional (3D) Maurer rose images, both arising from a threefold path around a torus.

```
In[166]:= Show[GraphicsArray[{Graphics3D[
    {Thickness[0.002],
     GeneralizedMaurerRose[(* now func returns a list with 3 elements *)
      {Cos[3#] + 1/3 Cos[3#] Cos[#], Sin[3#] + 1/3 Sin[3#] Cos[#],
       1/3 Sin[#]}&, 226, 351]},
     PlotRange -> All, AspectRatio -> Automatic, Boxed -> False],
    Graphics3D[{Thickness[0.002],
     (* same curve, different number of lines *)
     GeneralizedMaurerRose[(* now func returns a list with 3 elements *)
      {Cos[3#] + 1/3 Cos[3#] Cos[#], Sin[3#] + 1/3 Sin[3#] Cos[#],
       1/3 Sin[#]}&, 55, 330]},
     PlotRange -> All, AspectRatio -> Automatic, Boxed -> False]}]]
```

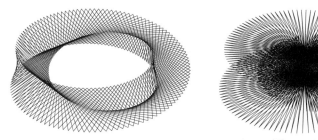

As with several examples in Chapters 1 and 2 of the Graphics volume [555], with some imagination, there is no limit to the number of interesting images that can be generated based on the idea of the original Maurer roses. For example, we could replace the lines (in the two-dimensional (2D) case) with colored polygons. As another alternative, we could replace all of the individual line segments by circles whose diameters are the lengths of the line segments.

```
Show[GraphicsArray[
{(* first rose *)
 Graphics[Function[x, {(* make polygons *)
 MapIndexed[{Hue[(#2[[1]] - 1)/Length[x]], Polygon @@ #1}&, x],
        {GrayLevel[0], Polygon @@ Last[x]}}][GeneralizedMaurerRose[
            {Cos[2 #] + Sin[7 #]/2, Sin[2 #] + Cos[7 #]/2}&, 256, 6720]],
  PlotRange -> All, AspectRatio -> Automatic, Background -> GrayLevel[0]],
 (* second rose *)
 Graphics[Function[x,
 MapIndexed[{Hue[(#2[[1]] - 1)/Length[x]], Polygon[#1]}&, x]][
 Function[pair, Module[{perp}, (* perpendicular direction *)
              perp = {#[[2]], -#[[1]]}&[(Subtract @@ pair)/50];
              {pair[[1]] + perp, pair[[2]] + perp,
               pair[[2]] - perp, pair[[1]] - perp}]] /@
 (* the points grouped in pairs *)
 Flatten[Partition[#, 2, 1]& /@ (First /@
 GeneralizedMaurerRose[{Sin[9#]/3 + Sin[12#], Sin[3#]}&, 358, 329]), 1]],
       AspectRatio -> Automatic, PlotRange -> All],
 (* third rose *)
 Graphics[{Thickness[0.002], (* making circles *)
 Circle[Plus @@ #/2, Sqrt[#.#]&[Subtract @@ #]/2]& /@
 (* the points *) Flatten[Partition[#[[1]], 2, 1]& /@
 GeneralizedMaurerRose[{Sin[#] + Sin[2 #]/2 + Sin[3 #]/3,
                    Cos[#] + Cos[2 #]/2 + Cos[3 #]/3}&, 64, 129], 1]},
       AspectRatio -> Automatic, PlotRange -> All] ,
 (* generalized 3D rose *)
 Graphics3D[{Thickness[0.002],
  MapIndexed[{Hue[#2[[1]]/31 0.75], #1}&,
   Line /@ Partition[#[[1]], 2, 1]& /@
   GeneralizedMaurerRose[(* now func returns a list with three elements *)
                    {Cos[4 #], Sin[5 #], Cos[7 #]}&, 31, 31 21]},
       PlotRange -> All, AspectRatio -> Automatic, Boxed -> False]}]]
```

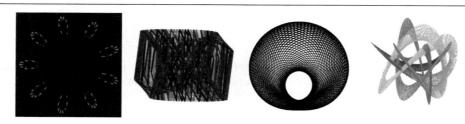

The second graphics set of applications that make heavy use of Mod are de Bruijn medallions and friezes.

Mathematical Remark: de Bruijn Medallions and Friezes

The construction rests on the following formula. Suppose we are given a fraction of the form p/q with $0 < p < 2q$, and suppose that $T_n = n(n+1)/2$ and $K_n = p\,T_n \bmod 2\,q$.

Let $\epsilon_n(K_n)$ be 0 for $0 \le K_n < q$ and 1 otherwise. Starting with these 0's and 1's, we generate $\delta_n = 2\epsilon_n - 1$ and set $x_k = x_{k-1} + (\delta_{2k-1} + \delta_{2k})/2$, $y_k = y_{k-1} + (\delta_{2k-1} - \delta_{2k})/2$ with $\{x_0, y_0\} = \{0, 0\}$ (without loss of generality). If p is odd, the resulting curves close after $4q$ steps. If p is even, we get translation invariant structures after $4q$ steps. For a proof that the structures are closed or periodic, see [134].

```mathematica
In[167]:= deBruijneSeries[pq_Rational?(Numerator[#] < 2 Denominator[#]&)] :=
    Module[{p = Numerator[pq], q = Denominator[pq], K},
        K[n_, p_, q_] := Mod[p 1/2 n (n + 1), 2q];
        (2 If[# < q, 0, 1] - 1)& /@ Table[K[i, p, q], {i, 4q}]]
```

Here is an example of the resulting numbers (all are +1 or −1) of the δ_n sequence.

```mathematica
In[168]:= deBruijneSeries[13/29]
```

```mathematica
Out[168]= {-1, 1, -1, -1, -1, 1, -1, -1, -1, -1, 1, -1, -1, 1, 1, -1, -1, -1, 1, -1, 1, 1, 1, -1,
    1, 1, 1, -1, 1, -1, -1, -1, 1, -1, -1, -1, 1, -1, 1, 1, 1, -1, -1, 1, 1, -1, 1,
    1, 1, 1, -1, 1, 1, 1, -1, 1, 1, 1, 1, -1, 1, 1, 1, 1, -1, 1, 1, -1,
    -1, 1, 1, 1, -1, 1, -1, -1, -1, 1, -1, -1, -1, 1, -1, 1, 1, 1, -1, 1, 1, 1, -1,
    1, -1, -1, -1, 1, 1, -1, -1, 1, -1, -1, -1, -1, 1, -1, -1, -1, 1, -1, -1, -1}
```

In a second step, we generate the sequence of points $\{x_k, y_k\}$.

```mathematica
In[169]:= deBruijnPicture[ser_] :=
    With[{par = Partition[ser, 2]},
    Graphics[{Thickness[0.002],
        Line[ (* make {x, y} pairs *)
            Transpose[{FoldList[Plus, (* add up values for x *)
                        0, (Apply[Plus, #]/2)& /@ par],
                    FoldList[Plus, (* add up values for y *)
                        0, (Apply[Subtract, #]/2)& /@ par]}]]}]];
```

We now look at the medallions and friezes we have constructed. Here are three examples.

```mathematica
In[170]:= Show[GraphicsArray[
        deBruijnPicture[deBruijneSeries[#]]& /@ {501/15001, 1002/7001, 340/12567}]]
```

We can use the line segments from the last picture and arrange them in a "thickened" way along the surface of a torus.

```mathematica
In[171]:= Module[{lines, linesSegments, scaledLinesSegments, torus,
            normaltorus, spindle},
        (* the line from the last picture *)
        lines = Cases[%, _Line, Infinity];
        (* line segments of the last picture *)
        linesSegments = Union[Sort /@ Partition[lines[[1, 1]], 2, 1]];
        (* maximal extensions *)
        {{xMin, xMax}, {yMin, yMax}} =
            {Min[#], Max[#]}& /@ Transpose[Flatten[linesSegments, 1]];
        (* scale segments into (0, 2Pi)x(0, 2Pi) area *)
```

```
scaledLinesSegments =
Apply[2. Pi {(#1 - xMin)/(xMax - xMin),
             (#2 - yMin)/(yMax - yMin)}&, linesSegments, {-2}];
(* a parametrization of a torus *)
torus[{φ1_, φ2_}, r1_, r2_] = {r1 Cos[φ1] + r2 Cos[φ1] Cos[φ2],
                               r1 Sin[φ1] + r2 Sin[φ1] Cos[φ2],
                                    r2           Sin[φ2]};
(* normal on a torus *)
normaltorus[{φ1_, φ2_}] = {Cos[φ1] Cos[φ2], Cos[φ2] Sin[φ1], Sin[φ2]};

(* a spindle *)
spindle[{{φ11_, φ12_}, {φ21_, φ22_}}, {r1_, r2_, δr_}] :=
Module[{P1, P2, mP, normal, dir, nPs},
    P1 = torus[{φ11, φ12}, r1, r2]; P2 = torus[{φ21, φ22}, r1, r2];
    mP = (P1 + P2)/2;
    (* normalized direction vectors *)
    normal = normaltorus[{φ11 + φ21, φ21 + φ22}/2];
    dir = #/Sqrt[#.#]&[Cross[mP - P1, normal]];
    nPs = Table[mP + δr (Cos[φ] normal + Sin[φ] dir),
              {φ, 0., 2. Pi, Pi/2.}];
    {Polygon[Append[#, P1]]& /@ Partition[nPs, 2, 1],
     Polygon[Append[#, P2]]& /@ Partition[nPs, 2, 1]}];
    (* show spindles along the line segments mapped onto a torus *)
    Show[Graphics3D[{EdgeForm[],
        {SurfaceColor[Hue[Random[]], Hue[Random[]], 3Random[]],
         spindle[#, {3, 1, 0.02}]}] & /@ scaledLinesSegments}],
        Boxed -> False]]
```

The following inputs will produce further examples of medallions and friezes. (Using constructions like `Prime[Random[Integer, 5, 2000]]` for the generation of numbers for the denominators and related constructions for the numerators can produce such pictures more automatically and randomly.)

```
(* medallions *)
Show[GraphicsArray[
 deBruijnPicture[deBruijn∈Series[#]]& /@
             {23/8819, 2673/13337, 7753/7759, 4751/7911}]]

(* friezes *)
Show[GraphicsArray[
 deBruijnPicture[deBruijn∈Series[#]]& /@
             {1000/5007, 6/50003, 3346/13597}]]
```

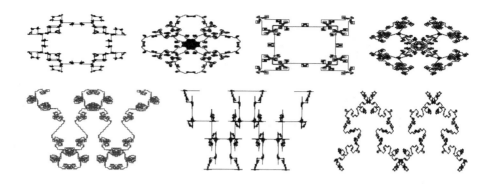

2.2 Number Theory Functions

Number theory has the reputation of being the least "practical" of all mathematical disciplines. Although in the past there has been some truth to this, the subject has gained a lot of practical importance in connection with cryptosystems [72], [568], [127], [328] and coding theory. For a discussion of them and other applications, see [491], [186], [327], [116], [492], [291], [333], [537], [58], [57], [361], [379], [383], [39], [386], and [511]. A *Mathematica*-based course on number theory is [84].

Another application of number theory that has only recently emerged involves calculations using *p*-adic numbers in physical situations (for a survey, see [80] and [313]). These applications suggest that a short discussion of the functions of number theory is worthwhile. For an excellent compilation of many number theoretic identities, see [402].

Let us start with the primes. Prime numbers play an especially important role in number theory [234], [467], [415], [541] (for a nice review of "prime records", see [462]). We discussed PrimeQ in Chapter 5 of the Programming volume [554]. One can test quite large numbers for primality. Here is one of the longest known arithmetic progressions [462] that gives primes.

```
In[1]:= Table[PrimeQ[11410337850553 + j 4609098694200], {j, 0, 22}]

Out[1]= {True, True, True, True, True, True, True, True, True, True, True,
         True, True, True, True, True, True, True, True, True, True, True, False}
```

When arranged on an integer spiral the prime numbers often fall on straight diagonal lines [491].

```
In[2]:= makeIntegerSpiral[n_] :=
    Module[{d = {0, 1}, k = 1, j = 1, xy = {0, 0}, bag = {{0, 0}}},
    (* go outwards in steps 1, 1, 2, 2, 3, 3, ...;
       turn left after k steps *)
    Do[Do[j++; xy = xy + d; bag = {bag, xy}, {Floor[k/2]}];
       d = {{0, -1}, {1, 0}}.d, {k, 2, Ceiling[Sqrt[1 + 4 n]]}];
    (* take n points visited *) Take[Level[bag, {-2}], n]]

In[3]:= With[{(* a 200×200 point graphic *) spiral = makeIntegerSpiral[40000]},
    Show[Graphics[MapIndexed[If[PrimeQ[#2[[1]]] + 40],
                         Rectangle[#1 - 1/2, #1 + 1/2], {}]&, spiral]],
       AspectRatio -> Automatic, Frame -> True, FrameTicks -> None]]
```

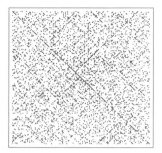

In addition to `PrimeQ`, here are two other functions of interest: `Prime` and `PrimePi`.

`Prime[positiveInteger]`

    produces the *positiveInteger*th prime number.

`PrimePi[positiveInteger]`

    gives the number of prime numbers that are smaller than or equal to *positiveInteger*, that is, $\pi(\text{positiveInteger})$.

Here is a list of the first 20 prime numbers.

```
In[4]:= Table[Prime[i], {i, 20}]
Out[4]= {2, 3, 5, 7, 11, 13, 17, 19, 23, 29, 31, 37, 41, 43, 47, 53, 59, 61, 67, 71}
```

The kth prime number p_k ($k \geq 6$) can be bounded in the following way [153], [102], [476], [577], [69]:

$$k\,(\log(k) + \log(\log(k))) \leq p_k \leq k\,(\log(k) + \log(\log(k)) - 1)$$

Here are these bounds for the first 10000 primes visualized.

```
In[5]:= With[{n = 1000},
    Show[Graphics[{Thickness[0.002],
      (* lower bound *)
      {Hue[0.0], Line[Table[{k, k (Log[k] + Log[Log[k]])}, {k, 2, n}]]},
      (* upper bound *)
      {Hue[0.3], Line[Table[{k, k (Log[k] + Log[Log[k]] - 1)}, {k, 2, n}]]},
      (* the primes *)
      {GrayLevel[0], Line[Table[{k, Prime[k]}, {k, 2, n}]]}}],
        PlotRange -> All, Frame -> True] // N]
```

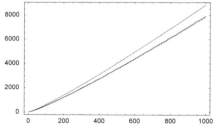

More precisely, we can approximate p_k in the following way. The right picture shows the relative error.

```
In[6]:= pApprox[k_] = k (Log[k] + Log[Log[k]] - 1 + (Log[Log[k]] - 2)/Log[k] -
            (Log[Log[k]]^2 - 6 Log[Log[k]] + 11)/Log[k]^2);
```

```
With[{n = 10^4},
Show[GraphicsArray[
Block[{$DisplayFunction = Identity},
   {Plot[pApprox[k], {k, 2, n}, PlotStyle -> {Thickness[0.02]},
        (* show primes as points *)
      Epilog -> Table[{Hue[0], Point[{k, Prime[k]}]},
                    {k, 2, n}], PlotRange -> All],
     ListPlot[Table[N[{k, (pApprox[k] - Prime[k])/Prime[k]}],
                {k, 2, 10000}]]}]]]]]
```

We have the following asymptotic series for large n [399], [173]:

$$p_n \propto n\left(\ln(n) + \ln(\ln(n)) - 1 + \sum_{j=1}^{o}(-1)^{j-1} P_j(\log(\log(n))) \log^{-j}(n)\right) + o(n \ln(\ln(n)) \log^{-o}(n)).$$

The $P_j(x)$ are polynomials of degree j. The following inputs calculate the first 100 $P_j(x)$.

```
In[9]:= (* first polynomial *)
      CipollaP[1][x_] := x - 2;

      (* implicit definition of P_j[0] *)
      CipollaPZeroValue[k_] := With[{p = CipollaP},
              -(k - 1) (Sum[Binomial[k - 2, j] p[j][0]*
                          (p[k - 1 - j][0] + p[k - 1 - j]'[0]), {j, k - 2}] +
                  (p[k - 1][0] + p[k - 1]'[0])) -  k p[k - 1][0] - p[k]'[0]]

In[13]:= (* ODE for the polynomials *)
      CipollaPODE[k_] := CipollaP[k]'[x] ==
                      k (k - 1) CipollaP[k - 1][x] - k CipollaP[k - 1]'[x];

In[15]:= Table[If[FreeQ[#[[1, 1, 0, 1]]]& /@ SubValues[CipollaP], m],
         CipollaP[m][x_] = CipollaP[m][x] /. (* solve ODE *)
         DSolve[{CipollaPODE[m], (* initial conditions *)
                 CipollaP[m][0] == (CipollaPZeroValue[m] /. CipollaP[m]'[0] ->
                                        (CipollaPODE[m][[2]] /. x -> 0))},
               CipollaP[m][x], x][[1]] // ExpandAll, CipollaP[m][x]],
            {m, 1, 100}] // (* display first eight *) Take[#, 8]&
Out[15]= {-2 + x, 11 - 6 x + x², -131 + 84 x - 21 x² + 2 x³, 2666 - 1908 x + 588 x² - 92 x³ + 6 x⁴,
      -81534 + 62860 x - 22020 x² + 4380 x³ - 490 x⁴ + 24 x⁵,
      3478014 - 2823180 x + 1075020 x² - 246480 x³ + 35790 x⁴ - 3084 x⁵ + 120 x⁶,
      -196993194 + 165838848 x - 66811920 x² + 16775640 x³ - 2838570 x⁴ +
         322224 x⁵ - 22428 x⁶ + 720 x⁷, 14297456816 - 12358329648 x + 5177983104 x² -
         1381360960 x³ + 257567520 x⁴ - 34369776 x⁵ + 3186848 x⁶ - 185184 x⁷ + 5040 x⁸}
```

The next graphic visualizes the logarithm of the coefficients $c_k^{(j)}$ of $P_j(x)$ (we display $\ln\!\left(c_k^{(j)}\right)$ as a function of k/j).

```
In[16]:= Show[Graphics[{PointSize[0.005], Table[{Hue[k/120], Point /@
            MapIndexed[(* scale coordinate and tale logarithm *)
              {(#2[[1]] - 1)/k, Log[10, #1/(k - 1)!]}&,
              Abs[CoefficientList[CipollaP[k][x], x]]]} // N,
            {k, 100}]}], Frame -> True, PlotRange -> All]
```

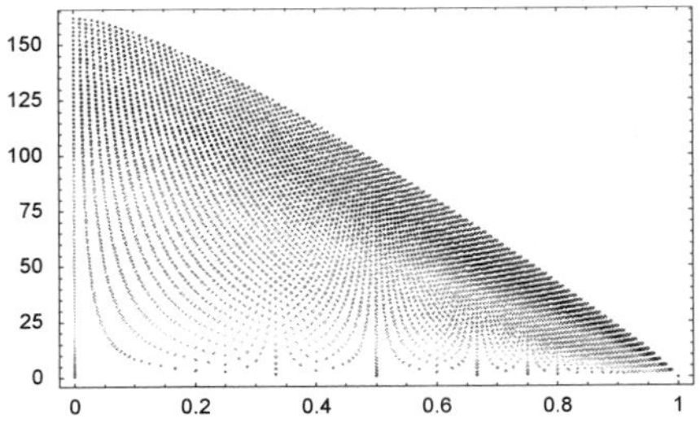

For the billionth prime, the error can be reduced from about 0.2% to less than 0.01% by taking into account the two first correction terms.

```
In[17]:= Module[{n = 10^9, o = 10, p},
           p = Prime[n];
           Table[{k, n (Log[n] + Log[Log[n]] - 1 +
              Sum[(-1)^(j - 1) CipollaP[j][Log[Log[n]]]/Log[n]^j,
                {j, k}])/p - 1.},
             {k, 0, o}]]
Out[17]= {{0, -0.0020718}, {1, 0.000110633}, {2, -0.0000937092},
         {3, -0.000160865}, {4, -0.000215398}, {5, -0.000290594}, {6, -0.000433701},
         {7, -0.000803644}, {8, -0.00204427}, {9, -0.0072793}, {10, -0.0344114}}
```

Now, we continue with counting the primes. Here is a list of the number of prime numbers for numbers less than 21.

```
In[18]:= Table[PrimePi[i], {i, 20}]
Out[18]= {0, 1, 2, 2, 3, 3, 4, 4, 4, 4, 5, 5, 6, 6, 6, 6, 7, 7, 8, 8}
```

Determining a good approximation of $\pi(n)$ is one of the most interesting problems in mathematics. Here are some estimates:

$$\text{Hadamard } \pi(n) \quad \approx \frac{n}{\ln(n)}$$

$$\text{Gauss } \pi(n) \quad \approx \int_2^n \frac{1}{\ln(x)}\,dx$$

$$\text{Guiasu } \pi(n) \quad \approx \int_2^n \lambda_n(u)\,du$$

$$\lambda_x(u) \quad = \begin{cases} \frac{1}{\lfloor \log_2 x \rfloor}\ln u & \text{if } 2 \le u \le x^{1/\lfloor \log_2 x \rfloor} \\ \frac{1}{k\ln(u)}, & \text{if } x^{1/(k+1)} < u \le x^{1/k} \quad k = 1, \ldots, \lfloor \log_2 x \rfloor - 1. \end{cases}$$

There are many more formulas approximating $\pi(x)$, for instance Gram's [51] which involves the Bernoulli number B_k (to be discussed below)

$$\pi(x) \approx \int_0^\infty \frac{\ln^t(x)}{t\,t!\,\zeta(t+1)}\,dt = -\frac{4}{\pi}\sum_{k=1}^\infty \frac{(-1)^k\,k}{(2\,k-1)\,B_{2\,k}}\left(\frac{\ln(x)}{2\,\pi}\right)^{2\,k-1}.$$

For further formulas, see [228], [229], [468], [209], [541], [608], [156], [480], [302], [581], [428], [23], [40], [315], and [282].)

We now look at a plot.

```
In[19]:= exactπ[n_] := (* modeling step curve *)
    Line[(Sequence @@ {# - {0, 1}, #})& /@
            Thread[{Table[Prime[i], {i, n}], Table[i, {i, n}]}]];
    (* Hadamard's approximation *)
    Hadamardπ[n_] := Line[Table[N[{x, x/Log[x]}],
                            {x, 2, Prime[n], Prime[n]/120}]];
    (* Gauss's approximation *)
    Gaussπ[n_] := Line[Table[N[{x, LogIntegral[x] - LogIntegral[2]}],
                            {x, 2, Prime[n], Prime[n]/120}]];
    (* Guiasu's approximation *)
    Guiasuπ[n_] := Line[Table[N[{x,
    Function[h, 1/h (LogIntegral[x^(1./h)] - LogIntegral[2.]) +
            Plus @@ ((LogIntegral[x^(1/#)] - LogIntegral[
            x^(1/(# + 1))])/#& /@ Reverse[N @ Range[h - 1]])][
                                    Floor[Log[2, x] // N]]},
                {x, 2, Prime[n], Prime[n]/120}]];
```

The exact curve exhibits a characteristic step shape; the other three are good approximations for large numbers.

```
In[26]:= PrimeNumberSet[n_, opts___] :=
    Show[Graphics[(* the four curves *)
                {exactπ[n],
                {AbsoluteDashing[{7, 7}], Hadamardπ[n]},
                {AbsoluteDashing[{5, 5}], Gaussπ[n]},
                {AbsoluteDashing[{3, 3}], Guiasuπ[n]}}], opts, Axes -> True]

In[27]:= Needs["Graphics`Legend`"]

In[28]:= styledText[text_] := StyleForm[text, FontSize -> 6];

    ShowLegend[PrimeNumberSet[50, DisplayFunction -> Identity],
            (* this is all for the legend *)
        {{{Graphics[Line[{{0, 1}, {1, 1}}]], styledText["exact"]},
        {Graphics[{AbsoluteDashing[{5, 5}],
                Line[{{0, 0}, {1, 0}}]}], styledText["Hadamard"]},
        {Graphics[{AbsoluteDashing[{3, 3}],
                Line[{{0, -1}, {1, -1}}]}], styledText["Gauss"]},
        {Graphics[{AbsoluteDashing[{2, 2}],
                Line[{{0, -2}, {1, -2}}]}], styledText["Guiasu"]}},
            LegendPosition -> {0.2, -0.4}, LegendSize -> {0.65, 0.4}}]
```

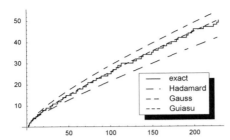

In connection with prime numbers, we should mention Euclid's famous proof that no largest prime number exists. Consider the following sequence: $2 + 1$, $2 \cdot 3 + 1$, $2 \cdot 3 \cdot 5 + 1$, ..., $1 + \prod_{k=1}^{n} p_k$. Not all numbers in this sequence are prime numbers, but the decomposition of the ith number into prime factors always involves a prime number that is larger than any prime number involved in the $(i - 1)$st step [128], [9]. The following graphic shows the prime factors of $1 + \prod_{k=1}^{n} p_k$. (For comparison, the right graphic shows the prime factors of $1 + \sum_{k=1}^{n} p_k$.)

```
In[31]:= Show[GraphicsArray[
        Graphics[{{GrayLevel[0.5], Thickness[0.008],
          Line[Table[{x, Log[x]}, {x, 1, #2, 0.1}]]]},
          (* kth prime appearing in the decomposition
             of the jth product/sum *)
          Table[Point[{k, Log @ PrimePi[First[#]]}]& /@
                FactorInteger[#1[Prime[j], {j, k}] + 1], {k, #2}]},
            PlotRange -> All, Frame -> True]& @@@
          (* do for products and sums *) {{Product, 12}, {Sum, 1200}}]]
```

For Gaussian primes, a similar remark holds.

```
In[32]:= Module[{o = 12, gaussianPrimes, δs},
        (* Gaussian primes *)
        gaussianPrimes = DeleteCases[Union @
                    Flatten[Table[If[PrimeQ[i + j I], i + j I, 0],
                            {i, -o, o}, {j, -o, o}]], 0];
        (* distances of the Gaussian primes from the origin *)
        δs = Union[Abs[gaussianPrimes]];
        (* prime factors of the products plus 1 + I *)
        Table[First /@ FactorInteger[1 + I + Times @@
                        Select[gaussianPrimes, Abs[#] <= δs[[k]]&],
                            GaussianIntegers -> True], {k, 10}]]
Out[32]= {{i, 1 + i, 8 + 7 i}, {-1, 1 + i, 35 + 68 i, 2285 + 7132 i},
        {i, 1 + i, 24152613220125000 + 24152613220124999 i},
        {-i, 1 + i, 161 + 80 i, 1619656263237838446357383595965491480+
```

544365798017696739232149673332800041i}, {1 + i, 4 + 21 i, 67 + 28 i, 286 + 159 i,
36452845313452080275477728874337521925491121146445467260063209413580821920655
168830409959210279907747 +
2569572178592377746084198911001906579061161468431122025282862561290838001499
02123672436769666111118i}, {-i, 1 + i, 2 + 3 i}, {1 + i, 10 + 17 i, 90 + 347 i},
{i, 1 + i, 10 + 9 i, 11 + 14 i, 2379 + 2974 i, 3167 + 17650 i},
{i, 1 + i, 2 + 17 i, 5 + 4 i, 74 + 115 i, 8222 + 1295 i, 8762553 + 6569462 i},
{1 + i, 363 + 62 i, 362199914 + 421018441 i, 28550916440 + 114619759849 i}}

Here are the prime numbers in the Gaussian plane shown. The right picture shows a convolution of the primes with a small neighborhood. An interesting symmetric pattern arises.

```
In[33]:= primeData = With[{n = 100}, Table[If[PrimeQ[i + I j], 0, 1],
                        {i, -n, n}, {j, -n, n}]];
      Show[GraphicsArray[
      Block[{$DisplayFunction = Identity},
        {ListDensityPlot[primeData, Mesh -> False, FrameTicks -> False],
         ListDensityPlot[ListConvolve[Table[1, {3}, {3}], primeData],
                Mesh -> False, FrameTicks -> False,
                ColorFunction -> (GrayLevel[Round[#]]&)]}]]]
```

Using the functions `Prime` and `PrimePi`, we can implement a quick visualization of the following interesting limit: $e = \lim_{n \to \infty} \left(\prod_{p_i < n} p_i \right)^{1/n}$ [349].

```
In[35]:= FoldList[(#2[[2]] (#1^#2[[1]]))^(1/#2[[2]])&, 1,
               N @ Partition[Prime[Range[PrimePi[10^6]]], 2, 1]] //
                  ListPlot[#, Frame -> True, Axes -> False]&
```

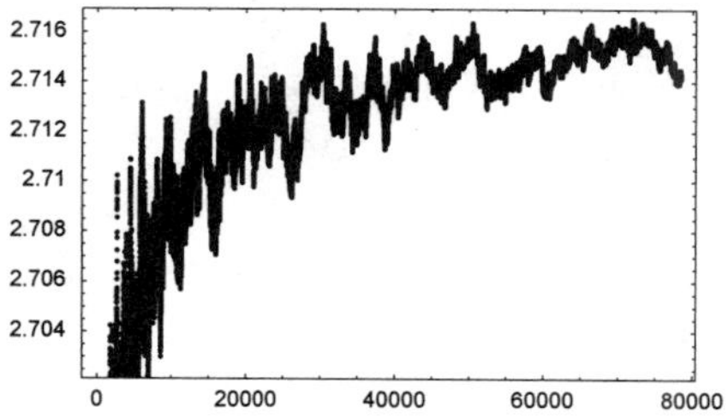

Another important number theory function is Euler's divisor function $\phi(n)$, which gives the number of integers that are smaller than n that do not divide n.

`EulerPhi[`*positiveInteger*`]`

produces the value of the Euler function $\phi(\text{positiveInteger})$.

(An interesting survey of the applications of the Euler function ϕ can be found in [516], [518], and [481].)

Here are the first few values of ϕ.

```
In[36]:= Table[{i, EulerPhi[i]}, {i, 20}]
Out[36]= {{1, 1}, {2, 1}, {3, 2}, {4, 2}, {5, 4}, {6, 2},
         {7, 6}, {8, 4}, {9, 6}, {10, 4}, {11, 10}, {12, 4}, {13, 12},
         {14, 6}, {15, 8}, {16, 8}, {17, 16}, {18, 6}, {19, 18}, {20, 8}}
```

In the following figure (on the left), we show a few more. The function itself varies greatly. For large n, its cumulative sum is approximately $3\,n^2/\pi^2$ (the application of `FoldList` avoids repeated calculations).

```
In[37]:= Show[GraphicsArray[{
         (* EulerPhi itself *)
         ListPlot[Table[EulerPhi[i], {i, 200}],
                 DisplayFunction -> Identity, PlotLabel -> "EulerPhi[n]"],
         (* the cumulative sum *)
         Graphics[{{PointSize[0.015], Point /@
         MapThread[List, (* generate the sum *)
         {#, Rest[FoldList[Plus, 0, EulerPhi /@ #]]}&[Table[n, {n, 50}]]]},
           (* the asymptotic of the cumulative sum *)
           {GrayLevel[1/2], Line[Table[N[{n, 3/Pi^2 n^2}], {n, 50}]]}},
             PlotRange -> All, Axes -> True,
             PlotLabel -> "EulerPhi[n] cumulative"}]]]
```

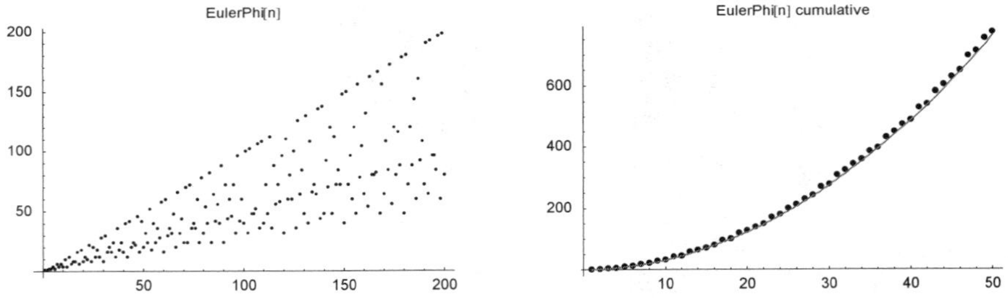

(For the derivation of this and other similar cumulative asymptotic relations, see [131], [20], and [144].)

Using the function `GCD` introduced earlier, we can create a (not particularly efficient, but working) program for $\phi(n)$.

```
In[38]:= ourEulerPhi[n_Integer?Positive] :=
         Length[Select[Table[i, {i, n}], GCD[#, n] == 1&]]
```

Here is a "check".

```
In[39]:= Table[EulerPhi[n] - ourEulerPhi[n], {n, 25}] // Union
Out[39]= {0}
```

The following important relationship holds for $\phi(n)$: $\sum_{k|n}^{n} \phi(k) = n$ where the sum is taken over all divisors k of n. Here is a "check" of this formula.

```
In[40]:= sumtest[n_] := Plus @@ (EulerPhi /@ Divisors[n])

         Table[sumtest[k] - k, {k, 1000}] // Union
Out[41]= {0}
```

$\phi(n)$ has an interesting representation: $\phi(n) = |\mathbf{A}|$ where the elements A_{ij} of the $n \times n$ matrix $\mathbf{A}$ are given by

$$A_{ij} = \begin{cases} i & \text{if } j = n \\ 1 & \text{if } j \text{ is a divisor of } n \text{ .} \\ 0 & \text{otherwise.} \end{cases}$$

We now examine this identity for $1 \le n \le 12$.

```
In[42]:= Table[Table[Which[j == n, i, IntegerQ[i/j], 1, True, 0],
                {i, n}, {j, n}] // Det, {n, 12}]
Out[42]= {1, 1, 2, 2, 4, 2, 6, 4, 6, 4, 10, 4}
```

```
In[43]:= Table[EulerPhi[n], {n, 12}]
Out[43]= {1, 1, 2, 2, 4, 2, 6, 4, 6, 4, 10, 4}
```

The fast integer arithmetic of *Mathematica* allows us to check the identity $|\gcd(i, j)|_{i,j=n} = \prod_{k=1}^{n} \phi(k)$ [430], [522] for $n = 100$ quickly.

```
In[44]:= With[{n = 100},
            Det[Array[GCD, {n, n}]] - Product[EulerPhi[k], {k, n}]] // Timing
Out[44]= {0.11 Second, 0}
```

Using the Euler function $\phi(n)$, we can say more about the number of fractions p/q with $0 \le p/q < 1$ and $q < n$ as a function of n. This is of interest for Hofstadter's butterfly (discussed in the last chapter). This sequence of numbers is called the Farey sequence. In Chapter 1, we discussed the following direct construction [330].

```
In[45]:= FareySequenceLength[n_] :=
            Length[Union[Flatten[Table[p/q, {q, n}, {p, 0, q - 1}]]]]
```

```
In[46]:= Table[FareySequenceLength[n], {n, 20}]
Out[46]= {1, 2, 4, 6, 10, 12, 18, 22, 28, 32, 42, 46, 58, 64, 72, 80, 96, 102, 120, 128}
```

Concerning the importance of the function $\phi(n)$, it is easy to see that `FastFareySequenceLength[n]` = $\sum_{k=1}^{n} \phi(k)$ produces the same result.

```
In[47]:= FastFareySequenceLength[n_] := Sum[EulerPhi[k], {k, n}]
```

```
In[48]:= Table[FastFareySequenceLength[n], {n, 20}]
Out[48]= {1, 2, 4, 6, 10, 12, 18, 22, 28, 32, 42, 46, 58, 64, 72, 80, 96, 102, 120, 128}
```

For more on Farey sequences, see [237] and [420].

Although there is no obvious pattern in $\phi(n)$ itself, its iterated differences of $\phi(n)^p$ show a pattern known from cellular automata.

```
In[49]:= With[{n = 100},
         Module[{coloredDiamond},
         (* a colored, diamond-shaped polygon at the point p *)
         coloredDiamond[c_, p_] := {Hue[Mod[c, 3]/3],
                 Polygon[p + #& /@ {{0, 2}, {-1, 0}, {0, -2}, {1, 0}}/2]};
         (* the picture *)
         Show[GraphicsArray[(* use first three powers *)
         Table[Graphics[
          Map[(* make colored  diamonds *)
             coloredDiamond[#[[1]], #[[2]]]&,
           MapIndexed[(* center list horizontally *)
                   {#1, {#2[[2]] + #2[[1]]/2, #2[[1]]}}&,
```

```
(* list of iterated differences *)
NestList[Abs[Apply[Subtract, Partition[#, 2, 1], {1}]]&,
          EulerPhi[Range[n]]^p, n], {2}],
  {2}], Frame -> True, AspectRatio -> 1], {p, 3}]]]]]
```

Iterating the `EulerPhi` functions leads after a few iterations to 1. Here this is visualized. The right graphic shows a randomized version that uses the map $n \to \phi(a\,n + b) + c$ with small random integers a, b, and c.

```
In[50]:= Show[GraphicsArray[
      {(* n → EulerPhi[n] *)
       Graphics[{Thickness[0.002],
       Table[Line[MapIndexed[{#1, #2[[1]]}&,
                 (* iterate *)
                 FixedPointList[EulerPhi, n]]], {n, 200}]},
         PlotRange -> All, Frame -> True, AspectRatio -> 0.8],
       (* n → EulerPhi[a n + b] +c *)
       SeedRandom[1];
       Graphics[{Thickness[0.002],
         Table[Line[MapIndexed[{#1, #2[[1]]}&,
                 (* iterate *) FixedPointList[(
                 EulerPhi[Random[Integer, {1, 3}] # +
                         Random[Integer, {1, 3}]] +
                         Random[Integer, {1, 3}])&,
                             n, 10]]], {n, 200}]},
           PlotRange -> {{0, 1000}, All}, Frame -> True,
           AspectRatio -> 0.8]}]]
```

A theoretically very challenging problem is the distribution of the values of $\phi(n)$ and the number of solutions of $\phi(m) = n$ with respect to m for a given n. The following function `InverseEulerPhi` solves $\phi(m) = n$ and returns all solutions [432], [120] (for a more efficient algorithm, see M. Rytin's *MathSource* package http://library.wolfram.com/infocenter/MathSource/696).

```
In[52]:= InverseEulerPhi[n_] :=
    Function[𝛿, Sort[Select[Times @@@ Flatten[
          FoldList[(* form relatively prime tuples *)
          Function[{o, n}, Union[Sort /@ (
          Function[{l, n}, If[Union[GCD[#, n[[1]]]& /@ l] === {1},
                     Append[l, n[[1]]], Sequence @@ {}]] @@@
             Flatten[Outer[List, o, n, 1], 1])]],
             List /@ 𝛿, Table[List /@ 𝛿, {Length[𝛿]}]], 1],
             (* final selection *) EulerPhi[#] === n&]]][
       (* all potential factors *)
    Function[d, Flatten[DeleteCases[Flatten /@
    Table[(* EulerPhi[primePower] == divisor *)
    Function[p, Select[p^(1/#), PrimeQ[#]&]^#& /@
          Range[Floor[Log[2, 2 d[[j]]]]]]][First /@
    Cases[{#, EulerPhi[#]}& /@ Range[2 d[[j]]], {_, d[[j]]}]],
          {j, Length[d]}], {}]]][
       (* select divisors *) Select[Divisors[n],
          # == 1 || # == n || (EvenQ[#] && EvenQ[n/#])&]]]
```

If a solution exists, it has at least two elements. For an n with many small primes, many solutions exist.

```
In[53]:= {InverseEulerPhi[2 5 829], InverseEulerPhi[2^2 3 5 7]}

Out[53]= {{8291, 16582}, {421, 473, 497, 539, 633, 639, 842, 844, 946, 994, 1078, 1266, 1278}}
```

By calculating all $\phi(n)$ for all n less than 10^6, we obtain all solutions for $m \le 10^5$.

```
In[54]:= phiSolutions = {#[[1, 1]], Last /@ #}& /@
    Split[Select[Sort[Table[{EulerPhi[k], k}, {k, 10^6}]],
             #[[1]] < 10^5&], #1[[1]] === #2[[1]]&];
```

For $n = 420$, we have again the 13 from above.

```
In[55]:= Cases[phiSolutions, {2^2 3 5 7, _}]

Out[55]= {{420, {421, 473, 497, 539, 633, 639, 842, 844, 946, 994, 1078, 1266, 1278}}}
```

The first two graphics show the number of solutions as a function of m, and its average of 5000 consecutive m. The second two graphics show the 525 solutions of $\phi(n) = 2^8\, 3^2\, 5\, 7 = 80640$ and the consecutive differences between the sorted values of n.

```
In[56]:= Show[GraphicsArray[#]]& /@
    Block[{$DisplayFunction = Identity, o = PlotRange -> All,
             (* count solution multiplicities *)
             L = {#1, Length[#2]}& @@@ phiSolutions,
             (* averaging kernel *)
             kernel = Table[1, {500}],
             (* solutions of EulerPhi[n] == 80640 *)
             l = Cases[phiSolutions, {80640, _}][[1, 2]]},
      {{ListPlot[L, o],
        ListPlot[Transpose[{ListConvolve[kernel, First /@ L],
                       ListConvolve[kernel, Last /@ L]}/500]]},
       {ListPlot[l, o],
        ListPlot[Sort[Subtract @@@ Partition[l, 2, 1]], o]}}]
```

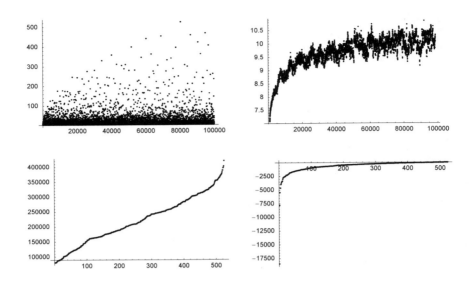

Having the list `phiSolutions` allows easily to show the number of different values taken by ϕ less than or equal to x. The left graphic shows these numbers and the right graphic shows the ratio of the actual number to the first terms of the asymptotic number [175].

```
In[57]:= Show[GraphicsArray[
         Block[{$DisplayFunction = Identity, f},
                (* relevant terms of asymptotics *)
                f[x_] := x/Log[x] Exp[0.8178 (Log[Log[Log[x]]] -
                                              Log[Log[Log[Log[x]]]])^2];
           {(* number of different values taken *)
            ListPlot[data = N @ MapIndexed[{#1, #2[[1]]}&,
                                    First /@ phiSolutions]],
            (* ratio of actual to asymptotic number *)
            ListPlot[Apply[If[#1 > 1000, {#1, #2/f[#1]},
                           Sequence @@ {}]&, data, {1}],
                PlotRange -> All]}]]]
```

The following interesting numerical quadrature formulas of the function f are based on the `EulerPhi` function [442], [454]. The integration range is the interval (0,1). (For related integral formulas, see [100]).

```
In[58]:= intApp[f_, n_, prec_] := 1/EulerPhi[n] Plus @@ (f /@
                      N[Select[Range[n], GCD[#, n] == 1&]/n, prec])
```

Let us take the cos function as a simple example.

```
In[59]:= Integrate[Cos[x], {x, 0, 1}] // N
Out[59]= 0.841471
```

Here, the difference of intApp[Cos, i, 15] to the exact result is shown.

```
In[60]:= ListPlot[N[Table[intApp[Cos, i, 15], {i, 1000}] - Sin[1]]]
```

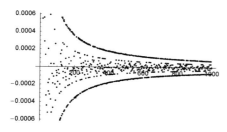

The Euler function has the property $\phi(m\,n) = \phi(m)\,\phi(n)$ for relatively prime m and n. An additive version of the Euler function with the property $\Phi_{add}(m\,n) = \Phi_{add}(m) + \Phi_{add}(n)$ for relatively prime m and n plays an important role in determining the possible rotational symmetries of a dD crystals [257], [594], [29].

```
In[61]:= ΦAdditive[n_Integer?Positive] :=
    Plus @@ (If[{##} == {2, 1}, 0, EulerPhi[Power[##]]]& @@@ FactorInteger[n])

    ΦAdditive[2] = 1;
```

Here are the values of the additive Euler function for the first 10^5 integers.

```
In[64]:= ΦAdditiveData = Table[{k, ΦAdditive[k]}, {k, 10^5}];
```

```
In[65]:= Show[GraphicsArray[
    ListPlot[ΦAdditiveData, PlotStyle -> {PointSize[0.003]}, PlotRange -> #,
                DisplayFunction -> Identity]& /@
        (* show all and lower part *) {All, {0, 100}}]]
```

The possible degrees of rotation axes in a dD crystal are all m, such that $\Phi_{add}(m) \le d$ [257]. The following input shows the possible values for $d \le 10$ (the possible degrees in odd-dimensional spaces of dimension $2\,n + 1$ are identical to the ones in dimension $2\,n$).

```
In[66]:= {#[[1, 1]], Flatten[Rest /@ #]}& /@
    Take[Split[Sort[Reverse /@ ΦAdditiveData], #1[[1]] === #2[[1]]&], 11]
Out[66]= {{0, {1}}, {1, {2}}, {2, {3, 4, 6}}, {4, {5, 8, 10, 12}},
    {6, {7, 9, 14, 15, 18, 20, 24, 30}}, {8, {16, 21, 28, 36, 40, 42, 60}},
    {10, {11, 22, 35, 45, 48, 56, 70, 72, 84, 90, 120}},
    {12, {13, 26, 33, 44, 63, 66, 80, 105, 126, 140, 168, 180, 210}},
    {14, {39, 52, 55, 78, 88, 110, 112, 132, 144, 240, 252, 280, 360, 420}},
```

```
    {16, {17, 32, 34, 65, 77, 99, 104, 130, 154,
        156, 165, 198, 220, 264, 315, 330, 336, 504, 630, 840}},
    {18, {19, 27, 38, 51, 54, 68, 91, 96, 102, 117, 176, 182, 195, 231,
        234, 260, 308, 312, 390, 396, 440, 462, 560, 660, 720, 1260}}}}
```

The Euler ϕ function can be used to construct an absolutely abnormal number (meaning its digits are nonuniformly distributed in any integer base [241]) σ that is transcendental [377], [378]: $\sigma = \prod_{j=2}^{\infty}(1 - 1/d_j)$ where $d_j = j^{\phi(d_{j-1})}$ and $d_2 = 2$.

```
In[67]:= d[2] = 2;
         d[j_] := d[j] = j^EulerPhi[d[j - 1]]

In[69]:= α[k_] := Product[1 - 1/d[j], {j, 2, k}]
```

Because σ contains long sequences of identical digits, we implement a function `shortenedDecimalNum⌐ber` that displayed such sequences in subscript notation.

```
In[70]:= shortenedDecimalNumber[z_] :=
         Module[{chars, l},
         (* the digits *)
         chars = ToExpression /@ Characters[
         StringTake[#, {StringPosition[#, "."][[1, 1]] + 1,
                   StringPosition[#, "`"][[1, 1]] - 1}]]&[
                                ToString[InputForm[z]]];
         (* compactify strings of identical digits *)
         l = If[Length[#] === 1, First[#],
                 Subscript[First[#], Length[#]]]& /@ Split[chars];
         (* display number *)
         StyleBox[DeleteCases[MakeBoxes[#, StandardForm]&[l],
                       "{" | "}" | ",", Infinity] /.
                   RowBox[{b__}] :> RowBox[{"0.", b}],
                   ZeroWidthTimes -> True] // DisplayForm]
```

Here are the first 243180 digits of σ.

```
In[71]:= shortenedDecimalNumber[N[α[6], 243180]]
```
Out[71]//DisplayForm=

$0.31249_219_{243165}830525632692130437189$

Next, we consider the Möbius function $\mu(n)$.

MoebiusMu[*positiveInteger*]

    produces the value of the Möbius function $\mu(positiveInteger)$.

It is defined as follows:

$$\mu(n) = \begin{cases} 0 & \text{if } a \text{ prime factor appears more than once in the prime number factorization of } n \\ (-1)^k & \text{if the prime number decomposition of } n \text{ consists of } k \text{ distinct prime numbers,} \\ & \text{each of which appears only once} \\ 1 & \text{if } n = 1. \end{cases}$$

We now examine $\mu(n)$ and the cumulative sum graphically. (There is no closed-form asymptotic for the cumulative sum here, but it is known that $\sum_{i=1}^{n} \mu(i) > \sqrt{n}$ for some (very big) n's [424], [91], [519], [497], and [303].)

```
In[72]:= Show[GraphicsArray[
    Block[{$DisplayFunction = Identity},
     {(* MoebiusMu itself *)
      ListPlot[Table[MoebiusMu[i], {i, 100}],
          Axes -> {False, True}, PlotLabel -> "μ"],
      (* the cumulative sum *)
      ListPlot[FoldList[Plus, 0, Table[MoebiusMu[i], {i, 1000}]],
          Axes -> True, AxesOrigin -> {0, -13},
          PlotRange -> All, PlotLabel -> "Σμ"]}]]]
```

Although the last picture does not show much regularity in $\mu(n)$, the sums $\sum_{k=N}^{N+\delta} \mu(k)$ are conjectured to obey a normal distribution with variance $6\sigma/\pi^2$ [214]. We will calculate the first one million values for $\mu(n)$ (what takes about 25 s on a 2 GHz computer) and show the distribution of the sums as a function of δ.

```
In[73]:= (moebiusMuList = MoebiusMu[Range[10^6]];) // Timing

Out[73]= {1.91 Second, Null}
```

```
In[74]:= normalizedDistribution[l_, n_] :=
    Module[{(* frequency of sum values *) d = {#[[1]], Length[#]}& /@
             Split[Sort[Apply[Plus, Partition[l, n], {1}]]], sum},
        (* normalize distribution *)
        sum = Plus @@ (Last /@ d); Apply[{#1, #2/sum}&, d, {1}]]
```

```
In[75]:= Show[Graphics[Reverse[
    Table[{Hue[0.8 n/1000], Line[normalizedDistribution[moebiusMuList, n]]},
        {n, 100, 1000, 100}]]], Frame -> True, PlotRange -> All]
```

The sum $\sum_{i=1}^{n} \mu(i)$ is also the value of the Redheffer matrix [459], whose entries $a_{i,j}$ are 1 of j divides i and for $i = 1$.

```
In[76]:= Table[Det[Table[If[i === 1 || IntegerQ[i/j], 1, 0], {j, n}, {i, n}]] ===
        Sum[MoebiusMu[k], {k, n}], {n, 100}] // Union

Out[76]= {True}
```

The Möbius μ function is of special importance for the inversion of number theory sums (see [240], [482], [562], [279], for some special cases of the following relation). Defining $F(n)$ by $F(n) = \sum_{d|n} f(d)$, we have $f(n) = \sum_{d|n} F(d)\,\mu(n/d)$.

Here, $F(n)$ takes complex values and is defined over the integers. More generally, if we write a complex-valued function defined for real arguments $F(x)$ in the form $F(x) = \sum_{n=1}^{\infty} f(n^{\alpha} x)$ for $\alpha \in \mathbb{R}$, $\alpha \neq 0$ $x \in \mathbb{R}$ then $f(x) = \sum_{n=1}^{\infty} F(n^{\alpha} x)\,\mu(n)$. (For more on this and related transformations, see [461], [276], and [93].)

We now implement the computation of $f(n)$ for a given function $F(n)$.

```
In[77]:= MoebiusInverse[n_, F_] := Plus @@ ((F[#] MoebiusMu[n/#])& /@ Divisors[n])
```

For $F(n) = n^2$, $f(n)$ is not very smooth.

```
In[78]:= ListPlot[save = (* to be used later *)
            MoebiusInverse[#, Function[x, x^2]]& /@ Table[i, {i, 100}],
            AxesLabel -> {"n", "f(n)"}, PlotJoined -> True]
```

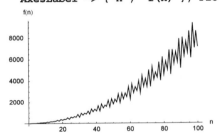

Next, we reconstruct $F(n) = n^2$ from the above $f(n)$ (= save).

```
In[79]:= Table[Plus @@ (save[[#]]& /@ Divisors[i]), {i, 100}]
```

```
Out[79]= {1, 4, 9, 16, 25, 36, 49, 64, 81, 100, 121, 144, 169, 196, 225, 256, 289, 324, 361,
         400, 441, 484, 529, 576, 625, 676, 729, 784, 841, 900, 961, 1024, 1089, 1156, 1225,
         1296, 1369, 1444, 1521, 1600, 1681, 1764, 1849, 1936, 2025, 2116, 2209, 2304,
         2401, 2500, 2601, 2704, 2809, 2916, 3025, 3136, 3249, 3364, 3481, 3600, 3721,
         3844, 3969, 4096, 4225, 4356, 4489, 4624, 4761, 4900, 5041, 5184, 5329, 5476,
         5625, 5776, 5929, 6084, 6241, 6400, 6561, 6724, 6889, 7056, 7225, 7396, 7569,
         7744, 7921, 8100, 8281, 8464, 8649, 8836, 9025, 9216, 9409, 9604, 9801, 10000}
```

For the use of Möbius transformations to solve some interesting physical problems (which remained unsolved for many years), along with generalizations, see [108], [110], [112], [400], [585], [548], [421], [373], [357], [603], [111], [272], [109], and [526] for some supersymmetric applications.

A plethora of fascinating formulas involving the Möbius μ function exist. The Möbius μ function can be used to express e as the following infinite product $e = \prod_{k=1}^{\infty} k^{-\mu(k)/k}$ [329]. We now look at the convergence of this product.

```
In[80]:= ListPlot[(* form cumulative product *)
            E - FoldList[Times, 1., Table[k^(-MoebiusMu[k]/k), {k, 10^5}]],
            Frame -> True, Axes -> False, PlotRange -> {-1/2, 1/2},
            PlotStyle -> {PointSize[0.003]}]
```

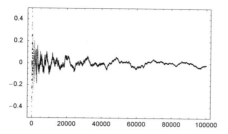

More generally, we have $\exp(z) = \prod_{k=1}^{\infty} (1 - z^k)^{-\mu(k)/k}$ for $|z| < 1$. The following plots show the truncated product over the complex z-plane. The dense set of singularities at the unit circle is clearly visible.

```
In[81]:= exp[z_ , n_] := Product[(1 - z^k)^(-MoebiusMu[k]/k), {k, n}];
```

```
In[82]:= Module[{ppr = 121, ppφ = 241, δ = 1/12, data},
        (* list of points *)
        data = Table[N[{r Cos[φ], r Sin[φ], exp[N[r Exp[I φ]], 50]}],
                   {r, 0, (1 + δ), (1 + δ)/ppr}, {φ, 0, 2Pi, 2Pi/ppφ}];
        (* show two 3D graphics *)
        Show[GraphicsArray[
        Graphics3D[{EdgeForm[], (* the polygons *)
        Table[Polygon[{data[[i, j]], data[[i + 1, j]],
                     data[[i + 1, j + 1]], data[[i, j + 1]]}],
             {i, Length[data] - 1}, {j, Length[data[[1]]] - 1}] /. #},
             ViewPoint -> {-3, -2, 1}, BoxRatios -> {1, 1, 0.6}]& /@
          (* shows absolute value colored according to argument and
             real part colored according to scaled imaginary part *)
          {Polygon[l_] :> {SurfaceColor[#, #, 2.2]&[
                          Hue[Plus @@ (((Arg /@ (Last /@ l)) + Pi)/(2Pi))/4]],
                          Polygon[Apply[{#1, #2, Abs[#3]}&, l, {-2}]]}},
           Polygon[l_] :> {SurfaceColor[#, #, 2.2]&[
                          Hue[Plus @@ (ArcTan[Im /@ (Last /@ l)]/(Pi/2))/4]],
                          Polygon[Apply[{#1, #2, Re[#3]}&, l, {-2}]]}}]]]
```

Here is a representation of $\sin(z)$ through a limit containing the Möbius μ function [119].

$$\sin(z) = -\pi \lim_{n \to \infty} \sum_{k=1}^{n} \mu(k) \frac{\ln(n) - \ln(k)}{k \ln(n)} \left(\frac{k z}{2\pi} - \left\lfloor \frac{k z}{2\pi} \right\rfloor \right).$$

Compiling parts of the summation and precalculating the needed values for $\mu(k)$, we can quickly generate a graphic of 5000 points, each being the sum of 10000 values.

```
In[83]:= With[{n = 10^4, pp = 5 10^3},
         (* precalculate values of MoebiusMu *)
         μTab = N[Table[MoebiusMu[k] Log[n/k]/Log[n]/k, {k, n}]];
         (* compile summation *)
         sin = Compile[{{v, _Integer}, z, {μList, _Real, 1}},
                      -Pi Table[FractionalPart[k z/(2Pi)], {k, v}].μList];
         (* values of sin *)
         sinPoints = Table[Point[{α, sin[n, α, μTab]}], {α, 0, 2Pi, 2Pi/pp}];
         Show[Graphics[{{GrayLevel[0.8], Thickness[0.02],
               (* Sin curve *)
               Line[Table[{x, Sin[x]} // N, {x, 0, 2Pi, 2Pi/pp}]]},
               {GrayLevel[0], PointSize[0.003], sinPoints // N}}], Frame -> True]]
```

An important numerical analysis application of the Möbius inversion and so the Möbius function is the calculation of Fourier coefficients of a function $f(x)$ [597], [271], [114]. For the case of a band-limited function

$$f(x) = \sum_{k=0}^{o} c_k^{(\cos)} \cos\left(k x \frac{2\pi}{L}\right) + \sum_{k=0}^{o} c_k^{(\sin)} \sin\left(k x \frac{2\pi}{L}\right)$$

with period L, the Fourier coefficients $c_k^{(\cos)}$ and $c_k^{(\sin)}$ can be calculated with the following sums. (c[Cos] stands for $c_k^{(\cos)}$ and c[Sin] stands for $c_k^{(\sin)}$; f is assumed to be a (pure) function.)

```
In[84]:= c[Cos][f_, L_, n_, o_] :=
              Sum[MoebiusMu[k] BrunsB[f, L, 2 k n, 0], {k, 1, Floor[o/n], 2}]

         c[Sin][f_, L_, n_, o_] :=
              Sum[MoebiusMu[k] (-1)^((k - 1)/2) BrunsB[f, L, 2 k n, 1/(4 k n)],
                  {k, 1, Floor[o/n], 2}]

         BrunsB[f_, L_, n_, α_] := 1/n Sum[(-1)^k f[L (k/n + α)], {k, 0, n - 1}]
```

Here is a small example showing the functions c[Cos] and c[Sin] at work.

```
In[89]:= f[x_] = Sum[1/k Cos[k x] + k Sin[k x], {k, 6}]
```

$$Out[89]= \cos[x] + \frac{1}{2} \cos[2x] + \frac{1}{3} \cos[3x] + \frac{1}{4} \cos[4x] + \frac{1}{5} \cos[5x] + \frac{1}{6} \cos[6x] +$$
$$\sin[x] + 2\sin[2x] + 3\sin[3x] + 4\sin[4x] + 5\sin[5x] + 6\sin[6x]$$

```
In[90]:= Table[c[Cos][f, N[2 Pi, 30], n, 10], {n, 8}] // Chop // N
```

$$Out[90]= \{1., 0.5, 0.333333, 0.25, 0.2, 0.166667, 0., 0.\}$$

```
In[91]:= Table[c[Sin][f, N[2 Pi, 30], n, 10], {n, 8}] // Chop // N
```

$$Out[91]= \{1., 2., 3., 4., 5., 6., 0., 0.\}$$

We turn now to the last function in this section: the Jacobi symbol $\left(\frac{a}{P}\right)$. It is of relatively little importance in analysis, but it plays an important role in the quadratic Gaussian reciprocity law, which C. F. Gauss proved in eight different ways [196].

```
JacobiSymbol[a, P]
```
    produces the Jacobi symbol $\left(\frac{a}{P}\right)$.

The Jacobi symbol $\left(\frac{a}{P}\right)$ is defined as follows: If P is an even number, $\left(\frac{a}{P}\right)$ is not defined. Let P be an odd number. If P is a divisor of a, $\left(\frac{a}{P}\right) = 0$. If this is not the case, let $p_0\, p_1\, p_2 \ldots p_n$ be the prime factorization of P. Then,

$$\left(\frac{a}{P}\right) = \left(\frac{a}{p_1}\right)\left(\frac{a}{p_2}\right)\cdots\left(\frac{a}{p_n}\right).$$

On the other hand, for odd primes p,

$$\left(\frac{a}{p}\right) = \begin{cases} +1 & \text{if } \exists\, x \text{, such that } x^2 = a \bmod p \\ -1 & \text{otherwise.} \end{cases}$$

Here are four examples. One stays unevaluated and issues a message.

In[92]:= **Apply[JacobiSymbol, {{7, 5}, {5, 4}, {9, 3}, {7, 3}}, {1}]**

        JacobiSymbol::jcpo :
            Second argument 4 in JacobiSymbol[5, 4] should be an odd, positive integer. More…

Out[92]= {-1, JacobiSymbol[5, 4], 0, 1}

This slightly nonconstructive definition of the Legendre symbol can be made constructive using the following (for practical purposes very slow) definition ($p > 1$, p prime):

$$\left(\frac{a}{p}\right) = \begin{cases} +1 & \text{if } p \text{ divides } a^{\frac{p-1}{2}} - 1 \\ -1 & \text{if } p \text{ divides } a^{\frac{p+1}{2}} + 1 \\ 0 & \text{otherwise.} \end{cases}$$

Here this definition is implemented.

In[93]:= **bfJacobiSymbol[a_, p_?(OddQ[#] && Positive[#] && # =!= 2&)] :=**
    Switch[{ (# - 1)/p, (# + 1)/p}&[a^((p - 1)/2)],
        {_Integer , _Rational}, +1, {_Rational, _Integer }, -1,
        {_Rational, _Rational}, 0]

    bfJacobiSymbol[a_, 1] := 1

The values calculated with `bfJacobiSymbol` are identical to the values given by `JacobiSymbol`.

In[95]:= **Union[Flatten[**
    Table[bfJacobiSymbol[a, Prime[i]] === JacobiSymbol[a, Prime[i]],
        {a, 1, 50}, {i, 2, 50}]]]

Out[95]= {True}

We now have a look at the function $\left(\frac{a}{P}\right)$ in the "*a,P*-plane". Here, red represents $\left(\frac{a}{P}\right) = 1$, green represents $\left(\frac{a}{P}\right) = 0$, blue represents $\left(\frac{a}{P}\right) = -1$, and white means that $\left(\frac{a}{P}\right)$ is not defined. In the right graphic, we omit the "even" P and display the Fourier transform of the absolute value of $\left(\frac{a}{P}\right)$.

```
In[96]:= Module[{n1 = 50, n2 = 400},
    Show[GraphicsArray[
    {(* Value graphic *)
    Graphics[
    Apply[ (* color according to the various
              possible values: 0, +1, -1, or undefined *)
        {Which[#3 == "not defined", GrayLevel[1],
               #3 ==  1, RGBColor[1, 0, 0],
               #3 ==  0, RGBColor[0, 1, 0],
               #3 == -1, RGBColor[0, 0, 1]],
          Rectangle[{#1, #2} - 1/2, {#1, #2} + 1/2]}&,
      (* the table of the values of the Jacobi symbol *)
    Table[{a, P, If[OddQ[P] && P > 2, JacobiSymbol[a, P], "not defined"]},
          {a, n1}, {P, n1}], {2}],
        AspectRatio -> Automatic, Frame -> True, FrameLabel -> {"a", "P"}],
    (* Fourier transform graphic *)
    ListDensityPlot[Abs[Fourier[Abs[
                   Table[JacobiSymbol[a, P], {P, 1, 2n2, 2}, {a, n2}]]]],
              FrameTicks -> None, DisplayFunction -> Identity,
              Mesh -> False, ColorFunction -> (Hue[1 - 0.5#]&)]}]]]
```

Despite the somewhat difficult-looking definition of the Jacobi symbol, it is not too difficult to derive a set of identities that allow (for arguments that are not too large) an easy implementation for positive arguments [24], [333], [126], [504].

```
In[97]:= js[1, b_?OddQ] = 1;

    js[0, b_?OddQ] = 0;

    js[a_, b_?(OddQ[#] && # > 1&)] := js[Mod[a, b], b] /; a >= b

    js[a_?OddQ, b_?OddQ] := js[b, a] /; Mod[a, 4] =!= 3 || Mod[b, 4] =!= 3

    js[a_?OddQ, b_?OddQ] := -js[b, a] /; Mod[a, 4] === 3 && Mod[b, 4] === 3

    js[a_?EvenQ, b_?OddQ] := js[a/4, b] /; IntegerQ[a/4]

    js[a_, b_?OddQ] :=  js[a/2, b] /; !IntegerQ[a/4] && Mod[b, 8] === 1

    js[a_?EvenQ, b_?OddQ] := js[a/2, b] /;
                    !IntegerQ[a/4] && (Mod[b, 8] === 1 || Mod[b, 8] === 7)

    js[a_?EvenQ, b_?OddQ] := -js[a/2, b] /;
                    !IntegerQ[a/4] && (Mod[b, 8] === 3 || Mod[b, 8] === 5)
```

We test whether our so-defined Jacobi symbol `myJacobiSymbol` agrees with the built-in one. Here is a quick check for the first 900 argument pairs.

```
In[106]:= Table[js[i, k] === js[i, k], {i, 1, 32, 2}, {k, 1, 32, 2}] //
                                                     Flatten // Union
Out[106]= {True}
```

For odd a, P, and $a < P$, the value of the Jacobi symbol can be inferred from the continued fraction expansion of a/P [452], [172], [407]. Here this is implemented and checked for all $P \le 100$.

```
In[107]:= jsCF[a_Integer?Positive, P_Integer?Positive] :=
     If[GCD[a, P] =!= 1, 0,
     Module[{cf = If[OddQ[Length[#]] && Last[#] =!= 1,
                 (* make continued fraction of even length *)
                 Join[Drop[#, -1], {Last[#] - 1, 1}], #]&[
                                        ContinuedFraction[a/P]],
             (* modular inverse *) m = PowerMod[a, -1, P]},
         (-1)^((3 - m - a +
         P Plus @@ MapIndexed[(-1)^#2[[1]] #1&, Rest[cf]])/4)]] /;
                     (* for off a and P *) OddQ[a] && OddQ[P] && a < P
In[108]:= Table[jsCF[a, P] === JacobiSymbol[a, P],
             {P, 1, 100, 2}, {a, 1, P - 2, 2}] // Flatten // Union
Out[108]= {True}
```

Now, we mention the Reciprocity law [350]. It states:

$$\left(\frac{P}{Q}\right)\left(\frac{Q}{P}\right) = (-1)^{\frac{P-1}{2}\frac{Q-1}{2}}.$$

Here P and Q are distinct, positive, odd prime numbers (actually, the weaker condition that P and Q are odd and have no common divisor is sufficient). Here is the *Mathematica* implementation for P, Q pairs.

```
In[109]:= ReciprocityLaw[{Q_, P_}] :=
     JacobiSymbol[P, Q] JacobiSymbol [Q, P] == (-1)^((P - 1)(Q - 1)/4)
```

Here is a list of lists of such prime number pairs.

```
In[110]:= oddPrimeNumbers = Table[Prime[i], {i, 2, 20}];

     pairs = Outer[List, oddPrimeNumbers, oddPrimeNumbers] //
                 (* get rid of the diagonal *) DeleteCases[#, {x_, x_}, {2}]&;
```

The Reciprocity law holds.

```
In[112]:= Union[Flatten[Map[ReciprocityLaw, pairs, {2}]]]
Out[112]= {True}
```

For a much more detailed discussion of the number theoretical functions discussed here, see the mentioned literature and [517].

2.3 Combinatorial Functions

Combinatorial functions arise in connection with problems involving the sorting of n objects into m bins and in related problems [419], [105]. The simplest problem of this type is the following: How many ways are there to arrange a set of n objects? The answer is given by the factorial function.

Factorial[*number*]

   or

number!

   produces *number*!. Here, *number* need not necessarily be a positive integer. Note that
   number != Γ(*number* + 1) (Γ(z) is the Gamma function; see Chapter 3 of the Symbolics volume [556]). For
   positive integers, *number*! = $\prod_{j=1}^{number} j$.

A closely related function is Factorial2.

Factorial2[*number*]

   or

number!!

   produces *number*!!. Here, *number* need not necessarily be a positive integer. If *number* is an even positive
   integer, *number*!! = $\prod_{j=1}^{number/2} (2\,j)$. If it is an odd positive integer, *number*!! = $\prod_{j=0}^{(number-1)/2} (2\,j+1)$.

Here are the first eight values of Factorial[i].

```
In[1]:= Do[Print[If[ i < 10, " ", ""] <> ToString[i] <> " != ", i!], {i, 8}]
```

```
        1 != 1

        2 != 2

        3 != 6

        4 != 24

        5 != 120

        6 != 720

        7 != 5040

        8 != 40320
```

But *Mathematica*, of course, can calculate much larger factorials. Here, we test if $((p-1)!+1)/p$ is an integer. (This is true if and only if p is a prime, which is Wilson's theorem; see [507].)

```
In[2]:= With[{o = 4000},
          Flatten[Position[Table[IntegerQ[((p - 1)! + 1)/p],
                             {p, 2, o}], True]] + 1 ==
          Table[Prime[p], {p, PrimePi[o]}]] // Timing
Out[2]= {3.48 Second, True}
```

The factorial numbers $n!$, $n \in \mathbb{N}$ have the interesting property that any integer m appears as the first digits for some n [381]. The following code calculates the factorials up to $n = 10^4$ and analyzes the frequency of the one, two, and three digits numbers in the first and (nonzero) last digits of $n!$.

```
In[3]:= Module[{o = 10, m = 10^4, fac, ids, lLeft, lRight},
       (* recursion start *) fac = o!;
       (* list bags *) lLeft = Table[0, {m}]; lRight = lLeft;
       (* form recursively all n! (store only last)
           and extract first and last three nonzero digits *)
      Do[fac = fac n; ids = IntegerDigits[fac, 10];
         lLeft[[n]] = Take[ids, 3];
         lRight[[n]] = Take[DeleteCases[ids, 0], -3], {n, m}];
     (* show digits distributions *)
     Function[{l, s}, Show[GraphicsArray[Table[
     ListPlot[{FromDigits[#[[1]]], Length[#]}& /@
              (* count occurrences of the various digit patterns *)
              Split[Take[#, s j]& /@ Sort[Take[l, {o + 1, -1}]]],
         PlotRange -> All, Frame -> True, Axes -> False,
         PlotStyle -> {PointSize[0.008]},
         DisplayFunction -> Identity], {j, 3}]]]] @@@
         (* first and last digits *) {{lLeft, 1}, {lRight, -1}}]
```

Whenever possible, functions in *Mathematica* allow for arbitrary complex arguments as inputs. Although `Prime` does not do this, the function `Factorial2` does. Here, this is shown.

```
In[4]:= Show[GraphicsArray[
     Block[{$DisplayFunction = Identity},
     {Plot[Re[Factorial2[x]], {x, -5, 5}, PlotLabel -> "Re[Factorial2[x]]"],
      Plot[Im[Factorial2[x + I]], {x, -5, 5},
                           PlotLabel -> "Im[Factorial2[x + I]]"}]]]]
```

Although the process of continuation into the complex plane is not unique for a function defined only at a set without a finite accumulation point, it is often possible to keep many "nice" properties (like functional relations) intact for certain continuations. Frequently, it is necessary to compute $n!$ for very large values of n (that is in statistical mechanics and in probabilistic calculations). Stirling's formula for the approximate computation of large values of the factorial functions is $n! \approx (n/e)^n \sqrt{2\pi n}$. Although Stirling's formula is quite accurate, better approximations are known, for example, $n! \approx \sqrt{2\pi/(n+1)}\,((n+1)/e)^{n+1}\,\exp(1/(12(n+1)))$. For a detailed discussion of these and similar formulas, see [353], [431], [530], [414], [135], [388], [376], [529], [56], [449], [78], and [494].

We now examine the relative accuracy of Stirling's formula in comparison to the above estimate (note the values of the ordinates) and the approximation which results from the first two terms of the asymptotic expansion of $n!$.

```
In[5]:= opts = Sequence[DisplayFunction -> Identity,
                 PlotRange -> All, PlotJoined -> True];
     Show[GraphicsArray[{
     (* classical Stirling *)
     ListPlot[Table[{n, N[n!]/N[(n/E)^n Sqrt[2 Pi n]]} - 1},
              {n, 500, 1000, 10}], opts,
           PlotLabel -> "good old Stirling"],
     (* the first two terms of the series expansion *)
     ListPlot[Table[{n, N[n!]/N[Sqrt[2Pi/(n + 1)]((n + 1)/E)^(n + 1)*
                                (1 + 1/(12(n + 1)))]} - 1},
              {n, 500, 1000, 10}], opts,
           PlotLabel -> "two series terms"],
     (* better approximation *)
     ListPlot[Table[{n, N[n!]/N[Sqrt[2 Pi/(n + 1)] ((n + 1)/E)^(n + 1)*
                                Exp[1/(12 (n + 1))]]} - 1},
              {n, 500, 1000, 10}], opts,
           PlotLabel -> "better"]}]]
```

This comparison is rather difficult using traditional programming languages, because intermediate results are very large numbers.

The second classical problem of combinatorics is the following: How many ways are there to choose m elements from a set of n objects ($n \geq m$)? The answer is given by the binomial coefficient [469].

Binomial[n, m]

  gives the binomial coefficients $\dbinom{n}{m}$.

Here is a table of the first several binomial coefficients.

```
In[7]:= Table[Binomial[n, m], {n, 0, 6}, {m, 0, n}] // TableForm
```

Out[7]//TableForm=
```
   1
   1    1
   1    2    1
```

1	3	3	1			
1	4	6	4	1		
1	5	10	10	5	1	
1	6	15	20	15	6	1

This is the Pascal triangle because we have:

$$(a + b)^n = \sum_{m=1}^{n} \binom{n}{m} a^m\, b^{n-m}.$$

A close relationship exists between the Pascal triangle and the Sierpinski triangle constructed in Chapter 1 of the Graphics volume [555]. If we mark all odd numbers in the Pascal's triangle with a point (for better fitting of the forms, we use a hexagon for each point), we get a Sierpinski triangle. For a discussion of the origin of this relationship, see [535], [97], [440], [1], [599], [591], [514], [515], [588], [88], [536], [338], and [71]; for counts of the results, see [539] and [113].

```
In[8]:= (* hexagons fit nicely together in a large triangle *)
       hexagon[x_, y_] =
       Polygon[{x, y} + #& /@ (1/Sqrt[3] Table[{Cos[φ], Sin[φ]},
                         {φ, Pi/2, 2Pi + Pi/2, 2Pi/6}])] // N;

In[10]:= Show[GraphicsArray[#]]& /@
       Partition[
       Graphics[Table[{GrayLevel[Mod[# @ Binomial[y, x], 2]],
                   hexagon[x - y/2, -y Sqrt[3]/2] // N},
               {y, 0, 90}, {x, 0, y}],
           AspectRatio -> Automatic, PlotRange -> All]& /@
       (* use binomial itself and rounded logarithm and digit sum;
          product of digits, and primeness of neighbors,
          and rounded tangent function *)
               {Identity, Round[Log[#]]&, DigitCount[#, 2, 1]&,
                Mod[Times @@ IntegerDigits[#, PrimePi[10^11]], 2]&,
                Boole[PrimeQ[# - 1] || PrimeQ[# + 1]]&, Round[Tan[#]]&}, 3]
```

Computing modulo larger numbers also leads to interesting patterns in the Pascal triangle.

```
In[11]:= Show[GraphicsArray[
        Graphics[Table[{Hue[0.8 Mod[Binomial[y, x], #]/#],
            hexagon[x - y/2, -y Sqrt[3]/2]}, {y, 0, 60}, {x, 0, y}],
        AspectRatio -> Automatic, PlotRange -> All]& /@
                            (* three values for mod *) {3, 5, 8}]]
```

Note that related nested triangle patterns also exist in other situations [74], for instance, in the iterated differences for prime numbers [513], [343], [26] or in the expansion of integer powers [280] (we encountered a similar pattern already earlier in this chapter).

```
Show[GraphicsArray[
  Function[f, Graphics[MapIndexed[{GrayLevel[1 - #1],
  (* the hexagon for the picture *)
  hexagon[#2[[1]] - #2[[2]]/2, #2[[2]] Sqrt[3]/2]}&,
    #/Max[Flatten[#]]&[Reverse @ Rest[f @
  (* list of iterated differences of the primes *)
  NestList[Abs[(Subtract @@ #)& /@ Partition[#, 2, 1]]&,
    (* the primes *) Array[Prime, 60], 59]]], {2}],
      AspectRatio -> Automatic, PlotRange -> All]] /@
      (* function to be applied to the iterated differences *)
      {Identity, Mod[#, 3]&, Mod[#, 4]&}]]
```

 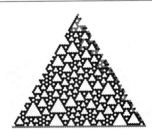 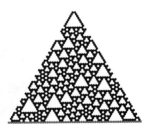

We can define the binomial coefficients by multiplying out $(a + b)^n$.

```
In[12]:= myBinomial[n_, m_] := (* works, but is extremely ineffective *)
        First[Cases[Expand[(a + b)^n], _.a^m b^(n - m)]/(a^m b^(n - m))]
```

The results agree with those of the built-in binomial function, but myBinomial is, of course, much slower for large values of n.

```
In[13]:= Timing[myBinomial[1200, 30]]
```

```
Out[13]= {0.01 Second, 6209253645350212024703382205911041114544827035775203925568890
```

```
In[14]:= Timing[Binomial[1200, 30]]
```

Out[14]= {0. Second, 62092536453502120247033822059110411145448270357752039255688Q

For some special symbolic cases, *Mathematica* simplifies the binomial coefficients.

In[15]:= **Clear[n]**
        {Binomial[n, 1], Binomial[n, 2], Binomial[n - 1, 3]}
Out[16]= $\left\{ n,\ \frac{1}{2}\ (-1+n)\ n,\ \frac{1}{6}\ (-3+n)\ (-2+n)\ (-1+n) \right\}$

Statistically for large integers n and k, the prime factors of the binomial coefficients $\binom{n}{k}$, $0 \le k \le n$ contain any prime [31]. The following graphic shows the counts of the primes for all binomial coefficients with $0 \le k \le 1000$.

In[17]:= **Module[{o = 1000, data},**
        data = { (* the nth prime *) PrimePi[First[#]],
                (* number of occurrences *) Length[#]}& /@
          Split[Sort[Flatten[(* get prime factors *)
          (First /@ FactorInteger[#]) & /@
            DeleteCases[Union[Flatten[(* the binomial coefficients *)
              Table[Binomial[n, k], {n, 0, o}, {k, 0, n}]]], 0 | 1]]]];
        ListPlot[data, PlotRange -> All]]

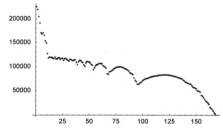

The zeros of the polynomials $p_{n,k}(z) = \sum_{j=0}^{k} \binom{n}{j} z^j$ form an interesting pattern in the complex z-plane [168].

In[18]:= **Module[{μ = 36, z},**
      Show[Graphics[
        Table[{RGBColor[n/μ, 0, k/μ], PointSize[0.006],
          Point[{Re[#], Im[#]}]}& /@
          N[z /. Solve[Sum[Binomial[n, j] z^j, {j, 0, k}] == 0, z], 22]},
          {n, μ}, {k, μ}]],
        Frame -> True, PlotRange -> {{-0.51, 0.5}, {-3, 3}}]]

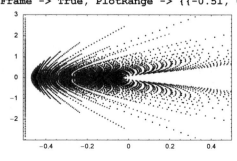

Here is a slightly unusual application of the binomial coefficients. The well-known quadratically converging Newton iterations $x_{n+1} = (x_n - f(x_n)/f'(x_n))$ for calculating a root of the equation $f(x) = c$ can be generalized to a faster converging method through [252]

$$x_{n+1} = x_n - \left(\sum_{k=0}^{o-2} \binom{1/2}{k} (-1)^k \, 2^{k+1} \left(\frac{f(x_n)\, f''(x_n)}{f'(x_n)^2} \right)^{k+1} \right) \frac{f(x_n)}{f'(x_n)}, \quad o = 2, 3, \dots .$$

The following function `higherOrderNewton` implements this method for the simple quadratic equation $x^2 = c$ with starting value x_0.

```
In[19]:= higherOrderNewton[c_, o_, x0_] :=
           FixedPointList[Function[x, If[(* fixed point reached? *) c != x^2,
                   (* sum can be expressed as Gauss 2F1 function *)
                   x Sum[Binomial[1/2, k] (-1)^k (1 - c/x^2)^k,
                       {k, 0, o - 1}], x]], x0, 100]
```

For the parameter values $c = x_0 = 3$, we use the method for different values of o and show the number of iterations needed until a fixed point has been reached as well as the precision loss.

```
In[20]:= Show[GraphicsArray[
         Block[{(* carry out iterations;
                 store number of steps and precision of result *)
             data = Table[{k, {Length[#], 1000 - Precision[#]}& @
                     higherOrderNewton[3, k, N[3, 1000]]}, {k, 2, 120}],
             $DisplayFunction = Identity},
                 (* show number of iterations and precision loss *)
                 Function[p, ListPlot[{#1, #2[[p]]}& @@@ data,
                     PlotRange -> All, AxesOrigin -> {0, 0}]] /@ {1, 2}]]]
```

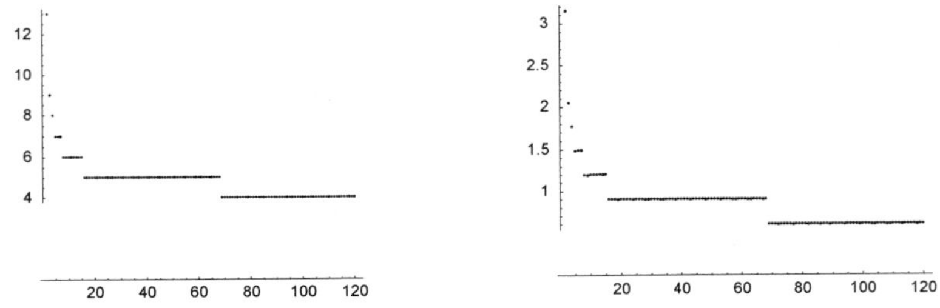

The following graphics show which of the two roots $\pm 3^{1/2}$ has been reached when using with complex starting values for the iterations.

```
In[21]:= Show[GraphicsArray[
         Block[{$DisplayFunction = Identity, L = 2},
             DensityPlot[Re[Round[Last @ higherOrderNewton[4, #, N[x + I y]] - 2]],
                 {x, -L, L}, {y, -L, L}, FrameTicks -> None,
                 PlotPoints -> 600, Mesh -> False]& /@
                 (* orders of the methods to use *) {2, 4, 8}]]] // Timing
```

An immediate generalization of the binomial coefficients is the multinomial coefficient $(n; n_1, n_2, ..., n_m)$. It counts the number of divisions of n objects into m subsets of n_i objects ($i = 1, ..., m$) with $n_1 + n_2 + \cdots + n_m = n$.

Multinomial$[n_1, n_2, ..., n_m]$

    produces the multinomial coefficient $(n; n_1, n_2, ..., n_m) = (n_1 + n_2 + \cdots + n_m)! / (n_1! \, n_2! \cdots n_m!)$.

The multinomial coefficient satisfies a formula analogous to the one for binomial coefficients:

$$(a_1 + a_2 + \cdots + a_m)^n = \sum_{\substack{n_1, n_2, ..., n_m \\ n_1 + n_2 + \cdots + n_m = n}} (n; n_1, n_2, ..., n_m) \, a_1^{n_1} \, a_2^{n_2} \cdots a_m^{n_m}.$$

Sierpinski-type figures also arise from multinomial coefficients modulo 2 (or other numbers). Here is an example involving trinomial coefficients modulo 2.

```
In[22]:= Show[GraphicsArray[
    Block[{$DisplayFunction = Identity, n = 64},
    {ListDensityPlot[(* Mod[Multinomial[64, i, j], 2] *)
    Table[Mod[Multinomial[n, i, j], 2],
         {i, 0, n}, {j, 0, n}], FrameTicks -> None,
         PlotLabel -> StyleForm["Mod[Multinomial[64, i, j], 2]",
                FontFamily -> "Courier", FontSize -> 6]],
    (* Mod[Multinomial[n, i, j], 2] now with three arguments *)
    Show[Graphics3D[{EdgeForm[Thickness[0.002]],
    Table[Cuboid[#, # + {1, 1, 1}]& /@ Rest /@ Cases[Flatten[
    Table[{Mod[Multinomial[n, i, j], 2], i, j, -n},
        {i, 0, n}, {j, 0, n - i}], 1], (* taking only the 1 values *)
                        {1, _, _, _}], {n, 32}]}],
        ViewPoint -> {1.3, 1.5, 1.2}, Boxed -> False]}]]]
```

Mod[Multinomial[64, i, j], 2]

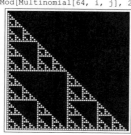 

Many other typical combinatorial problems can be solved in terms of simple combinations of `Factorial[n]`, `Binomial[n, m]`, and `Multinomial[n_1, n_2, ... , n_m]`. In addition to these classical combinatorial functions, *Mathematica* also includes four others: `StirlingS1`, `StirlingS2`, `PartitionsP`, and `PartitionsQ`. The importance of Stirling numbers in combinatorics is discussed in [602], [46], [81], and [443] (for complex arguments, see [171] and [310]). For a collection of identities for Stirling numbers, see [295], [104] and for some interesting continuations, see [2] and [94].

`StirlingS1[n, m]`
> produces the Stirling number $S_n^{(m)}$ of the first kind.

$(-1)^{n-m} S_n^{(m)}$ is the number of permutations of n elements with exactly m cycles. The generating function can be written as:

$$x(x-1)(x-2) \cdots (x-m+1) = \sum_{n=1}^{m} S_n^{(m)} x^n.$$

Here is a look at this formula.

```
In[23]:= With[{n = 12}, Expand[Product[(x - i), {i, 0, n - 1}]]]
```
$$Out[23]= \ -39916800\,x + 120543840\,x^2 - 150917976\,x^3 + 105258076\,x^4 - 45995730\,x^5 +$$
$$13339535\,x^6 - 2637558\,x^7 + 357423\,x^8 - 32670\,x^9 + 1925\,x^{10} - 66\,x^{11} + x^{12}$$

Here is a comparison with `StirlingS1`.

```
In[24]:= Table[StirlingS1[12, i], {i, 0, 12}]
```
```
Out[24]= {0, -39916800, 120543840, -150917976, 105258076,
          -45995730, 13339535, -2637558, 357423, -32670, 1925, -66, 1}
```

We now look at the dependence of the function `StirlingS1` on its two arguments.

```
In[25]:= opts = Sequence[PlotRange -> All, PlotJoined -> True,
                    DisplayFunction -> Identity];
    Show[GraphicsArray[{
    (* StirlingS1 picture *)
    ListPlot[N @ Table[Log[Abs[StirlingS1[10, m]]], {m, 10}], opts,
            PlotLabel -> StyleForm["Log[Abs[StirlingS1[10, m]]]",
                            FontFamily -> "Courier"],
            AxesLabel -> {StyleForm["m", TraditionalForm], None}],
    (* StirlingS2 picture *)
    ListPlot[N @ Table[Log[Abs[StirlingS1[n, 5]]], {n, 5, 15}], opts,
            PlotLabel -> StyleForm["Log[Abs[StirlingS2[n, 5]]]",
                            FontFamily -> "Courier"],
            AxesLabel -> {StyleForm["m", TraditionalForm], None}]}]]
```

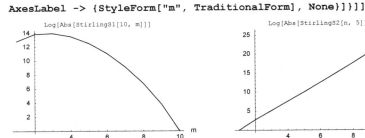
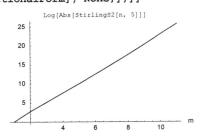

For large numbers asymptotically, we have

$$S_n^{(m)} \approx \frac{(n-1)! \, (\gamma + \ln n)^{m-1}}{(m-1)!}.$$

In addition to the Stirling numbers of the first kind, there are also Stirling numbers of the second kind $S_n^{(m)}$ [391]. They describe the number of ways to divide a set of n elements into m nonempty sets.

`StirlingS2[n, m]`

   produces the Stirling number $S_n^{(m)}$ of the second kind.

We again look at the first few values of $S_n^{(m)}$.

```
In[27]:= Table[StirlingS2[n, m], {n, 6}, {m, 6}] // TableForm
```
Out[27]//TableForm=

1	0	0	0	0	0
1	1	0	0	0	0
1	3	1	0	0	0
1	7	6	1	0	0
1	15	25	10	1	0
1	31	90	65	15	1

Here is an explicit formula for the Stirling numbers of the second kind:

$$S_n^{(m)} = \frac{1}{m!} \sum_{k=0}^{m} \frac{(-1)^{m-k} \, m!}{k! \, (m-k)!} \, k^n.$$

Once again, we check them.

```
In[28]:= myStirlingS2[n_, m_] :=
            1/m! Sum[(-1)^(m - k) Binomial[m, k] k^n, {k, 0, m}]
```

```
In[29]:= Table[StirlingS2[n, m], {n, 10}, {m, 10}] ==
    Table[myStirlingS2[n, m], {n, 10}, {m, 10}]
```
Out[29]= True

Like `Binomial` and `Multinomial`, the Stirling numbers show nested triangle patterns modulo an integer [593].

```
In[30]:= Show[GraphicsArray[
    ListDensityPlot[Table[Mod[#[a, b], 2], {a, 0, 50}, {b, 3, 50}],
                    Mesh -> False, DisplayFunction -> Identity]& /@
                {StirlingS1, StirlingS2}]]
```

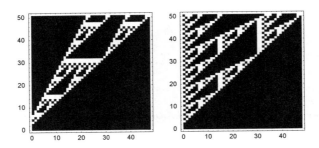

Using the Stirling numbers of the second kind, it is possible to sum powers of integers [495]:

$$\sum_{k=1}^{m} k^n = \sum_{k=1}^{n+1} (k-1)! \binom{m}{k} S_{n+1}^{(k)}.$$

Here is a short check of this formula for n, $m < 100$.

```
In[31]:= With[{o = 100},
    Table[Sum[k^n, {k, 1, m}] ==
        Sum[(k - 1)! Binomial[m, k] StirlingS2[n + 1, k], {k, 1, n + 1}],
            {m, 0, o}, {n, 0, o}]] // Flatten // Union

Out[31]= {True}
```

For a fixed first argument, the Stirling numbers $S_n^{(m)}$ are oscillating functions of the second argument.

```
In[32]:= With[{n = 50},
    ListPlot[Sign[#]&[Table[N[StirlingS1[n, k]], {k, 0, n}]],
        PlotJoined -> True, Frame -> True, Axes -> False,
        PlotRange -> All]]
```

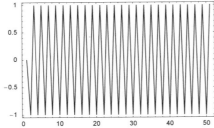

Here are density plots of $S_n^{(m)}$ and the Fourier transform over the n,m-plane.

```
In[33]:= Module[{m = 200},
    data = Table[#/Max[Abs[#]]&[Table[N[StirlingS1[n, k]],
                                {k, 0, m}]], {n, m}];
    (* set too small numbers to 0 *)
    data = data /. _?(Abs[#] < $MinMachineNumber&) -> 0.;
    Show[GraphicsArray[(* data and Fourier transform of data *)
        ListDensityPlot[#, DisplayFunction -> Identity,
            Mesh -> False, ColorFunction -> (Hue[6.2 #]&)]& /@
                                {data, Im[Fourier[data]]}]]]
```

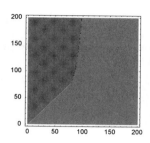
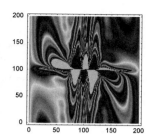

The Stirling numbers have many applications in combinatorics and other fields. Here we want to briefly discuss their use in connecting differentials with finite differences. If we define the kth order finite difference of a function $f(x)$ at x recursively through $\Delta_{x,h}^k f(x) = \Delta_{x,h}^{k-1}(\Delta_{x,h} f(x))$ where $\Delta_{x,h} f(x) = f(x+1) - f(x)$, then for a smooth function $f(x)$, the kth order finite difference and the kth derivative can be expressed as [296], [180], [311], [298], [255], [244]

$$\Delta_{x,h}^k f(x) = \sum_{j=0}^{\infty} \frac{h^{k+j}}{(k+1)_j} \, \mathcal{S}_{k+j}^{(k)} \, \frac{\partial^{k+j} f(x)}{\partial x^{k+j}}$$

$$\frac{\partial^k f(x)}{\partial x^k} = \sum_{j=0}^{\infty} \frac{h^{-k}}{(k+1)_j} \, \mathcal{S}_{k+j}^{(k)} \, \Delta_{x,h}^{k+j} f(x).$$

We implement truncated versions of these infinite sums.

```
In[34]:= (* the kth finite difference *)
         Δ[k_][x_, h_][f_, x_] := Nest[((# /. x -> x + h) - #)&, f, k]
In[36]:= (* avoid evaluation of finite differences
            and derivatives for explicitly given functions *)
         SetAttributes[differenceToDifferential, HoldFirst];
         SetAttributes[derivativesToDifference, HoldFirst];

         (* finite difference to derivatives *)
         differenceToDifferential[HoldPattern[Δ[k_][x_, h_][f_, x_]], o_] :=
         Sum[h^(k + j) StirlingS2[k + j, k] D[f, {x, k + j}] k!/(k + j)!,
            {j, 0, o}]

         (* derivative to finite differences *)
         derivativesToDifference[HoldPattern[D[f_, {x_, k_}]], h_, o_] :=
         h^-k Sum[StirlingS1[k + j, k] Δ[k + j][x, h][f, x] k!/(k + j)!,
            {j, 0, o}]
```

For polynomials $f(x)$, the infinite sums are naturally truncated. The next two inputs use $f(x) = x^7$.

```
In[44]:= Function[Δ, Δ - differenceToDifferential[Δ, 7], {HoldAll}][
                    Δ[5][x, h][f[x], x]] /. f -> Function[x, x^7] // Expand
Out[44]= 0

In[45]:= Function[d, d - derivativesToDifference[d, h, 7], {HoldAll}][
                    D[f[x], {x, 5}]] /. f -> Function[x, x^7] // Expand
Out[45]= 0
```

The last two functions of this section deal with the decomposition of a prescribed positive integer into a sum of positive integers.

`PartitionsP[n]`

> produces $p(n)$, which is the number +1 of the ways to express n as a sum of positive integers that counts different orderings of the integers only once.

`PartitionsQ[n]`

> produces $q(n)$, the number +1 of the ways to express n as a sum of distinct positive integers.

We now examine the sizes of these quantities.

```
In[46]:= Show[GraphicsArray[
    ListPlot[Table[#[n], {n, 40}], PlotRange -> All,
            DisplayFunction -> Identity, PlotLabel ->
                StyleForm[ToString[#] <> "[n]", FontFamily -> "Courier"],
            AxesLabel -> {"n", None}]& /@ {PartitionsP, PartitionsQ}]]
```

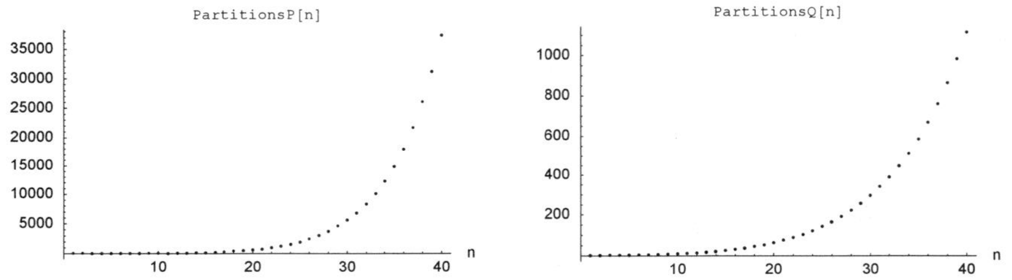

Asymptotically, we have [324], [465]:

$$p(n) \approx \frac{1}{4\,n\,\sqrt{3}} \, \exp\!\left(\pi \sqrt{n} \, \sqrt{\frac{2}{3}}\right), \quad q(n) \approx \frac{1}{4\,n^{3/4}\,3^{1/4}} \, \exp\!\left(\pi \sqrt{n} \, \sqrt{\frac{1}{3}}\right).$$

Here is a plot of the relative error of these asymptotic approximations.

```
In[47]:= Show[GraphicsArray[
    ListPlot[Table[{n, N[#1[n]/#2] - 1},
                {n, 200, 250}], Frame -> True, Axes -> False, PlotLabel ->
                StyleForm[ToString[#1] <> "[n]", FontFamily -> "Courier"],
            FrameLabel -> {"n", None},
            RotateLabel -> True, DisplayFunction -> Identity]& @@@
                (* the two partitions functions *)
                {{PartitionsP, (1/(4 n Sqrt[3]) Exp[Pi Sqrt[n 2/3]])},
                 {PartitionsQ, (1/(4 n^(3/4) 3^(1/4))  Exp[Pi Sqrt[n 1/3]])}}]]
```

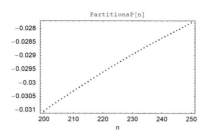 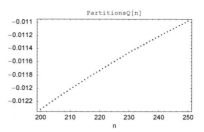

So, what if one needs the partitions itself and not only how many of them are there? There is no function in the System` context giving the partitions. So, let us have a look in other contexts.

In[48]:= **Names["*`*Partitions*"]**

Out[48]= {Internal`DirectPartitionsP,
    Internal`DirectPartitionsQ, Experimental`IntegerPartitions,
    Internal`IntegerPartitions, PartitionsP, PartitionsQ}

The last search revealed the function IntegerPartitions, but it is from the context Experimental`. We add this context to the default context path.

In[49]:= **AppendTo[$ContextPath, "Experimental`"];**

Although one might find many neat functions in this context, its use is generically discouraged. There is no guarantee that these functions will be available in later versions, or that they behave properly for any input; actually, they typically do not have any documentation at all.

Anyway, the name is so suggestive, let us try it.

In[50]:= **IntegerPartitions[6]**

Out[50]= {{6}, {1, 5}, {2, 4}, {3, 3}, {1, 1, 4}, {1, 2, 3}, {2, 2, 2},
    {1, 1, 1, 3}, {1, 1, 2, 2}, {1, 1, 1, 1, 2}, {1, 1, 1, 1, 1, 1}}

In[51]:= **{Length[%], PartitionsP[6]}**

Out[51]= {11, 11}

The form of the output of Experimental`IntegerPartitions (it does not resembles the output from FactorInteger) is a list of lists of the explicit parts of the partitions. Let us check it.

In[52]:= **Apply[Plus, %%, {1}]**

Out[52]= {6, 6, 6, 6, 6, 6, 6, 6, 6, 6, 6}

We define a function myIntegerPartitions that has the structure of the result of FactorInteger. It counts how often an integer occurs in a partition.

In[53]:= **myIntegerPartitions[n_Integer?Positive] := Reverse /@**
    (({First[#], Length[#]}& /@ Split[#])& /@ IntegerPartitions[n])

Here is an example.

In[54]:= **myIntegerPartitions[6]**

Out[54]= {{{6, 1}}, {{5, 1}, {1, 1}}, {{4, 1}, {2, 1}},
    {{3, 2}}, {{4, 1}, {1, 2}}, {{3, 1}, {2, 1}, {1, 1}}, {{2, 3}},
    {{3, 1}, {1, 3}}, {{2, 2}, {1, 2}}, {{2, 1}, {1, 4}}, {{1, 6}}}

Looking at the summands of larger *n* is not very enlightening. So, we graphically represent some partitions. The left graphic shows how often an integer occurs in the partition of the integers 1 to 24 and the middle and right right graphics show the scaled total counts of the occurring integers.

```
In[55]:= With[{o = 36},
    Show[GraphicsArray[
     {(* 3D plot of the number of occurrences of each integer *)
      Graphics3D[Table[{Hue[0.78 n/o],
                      Map[Line[{{{#[[1]], n, 0}, {#[[1]], n, #[[2]]}}}]&,
                        myIntegerPartitions[n], {2}]}, {n, o}],
               Axes -> True, PlotRange -> All, BoxRatios -> {1, 1, 0.6}],
      (* scaled number of occurrences of each integer without multiplicity *)
      Graphics[Table[{Hue[0.78 n/o],
       Function[l, Point[{(#1 - 1)/(Length[l] - 1), #2/l[[1, 2]]}]]& @@@ l] @
        ({{#[[1, 1]], Length[#]}& /&
        Split[Sort[Level[myIntegerPartitions[n], {2}]], #1[[1]] == #2[[1]]&])},
                    {n, 2, o}], Frame -> True, PlotRange -> All],
      (* logarithmically scaled number of occurrences of each integer *)
      Graphics[Table[{Hue[0.78 n/o],
       Function[l, Point[{(#1 - 1)/(Length[l] - 1), Log[#2]}]]& @@@ l] @
        ({{#[[1, 1]], Total[Last /& #]}& /&
        Split[Sort[Level[myIntegerPartitions[n], {2}]], #1[[1]] == #2[[1]]&])},
                    {n, 2, o}], Frame -> True, PlotRange -> All]}]]]
```

For explicit algorithms to compute the partitions itself, see [533].

The function `partitionsLengthCount` counts the length of all possible partitions of the integer *n*.

```
In[56]:= partitionsLengthCount[n_] :=
    {First[#], Length[#]}& /&
          Split[Sort[Length /& Experimental`IntegerPartitions[n]]]
```

Here is the number of partitions as a function of their length for *n* = 20 and *n* = 50. The average length of a partition is $n^{1/2} \log(n)/(\pi (2/3)^{1/2})$ [586].

```
In[57]:= Show[GraphicsArray[
          ListPlot[partitionsLengthCount[#], PlotRange -> All,
               DisplayFunction -> Identity,
               Frame -> True, Axes -> False]& /& {20, 50}]]
```

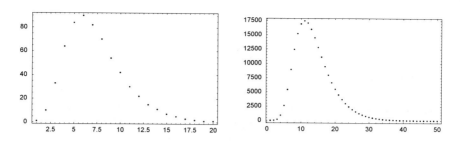

The partitions fulfill many identities involving functions discussed in this chapter [170], [273], [441], [412]. Let us denote all partitions (not taking the order into account) of n by $P(n)$. A single partition is of the form $\{\{1,\ k_1\},\ \{2,\ k_2\},\ \ldots,\ \{n,\ k_n\}\}$ where some of the k_j might be zero and the corresponding $\{j,\ k_j\}$ is suppressed by *Mathematica*. Using this notation, here are seven identities involving the list of partitions and the factorial, multinomial, Möbius μ, Euler ϕ function, and the Fibonacci numbers (to be discussed in the next section).

$$\sum_{P(n)} \left(\prod_{j=1}^{n} k_j! \, j^{k_j} \right)^{-1} = 1$$

$$\sum_{P(n)} n! \left(\prod_{j=1}^{n} \frac{1}{k_j!} \left(\frac{x}{k} \right)^{k_j} \right) = \prod_{k=0}^{n-1} (x+k)$$

$$\sum_{P(n)} \left(\sum_{j=1}^{n} k_j;\ k_1,\ \ldots,\ k_n \right) = 2^{n-1}$$

$$\sum_{P(n)} \left(\sum_{j=1}^{n} \mu(j)\, k_j \right) = p(n-1)$$

$$\sum_{P(n)} \delta_{k_1,0} \, \frac{\left(\sum_{j=1}^{n} k_j \right)!}{\prod_{j=1}^{n} k_j!} = F_{n-1}$$

$$\sum_{P(n)} \left(\sum_{j=1}^{n} \phi(j)\, k_j \right) = \sum_{j=0}^{n} (n-j)\, p(j)$$

$$\prod_{P(n)} \left(\prod_{j=1}^{n} j^{k_j} \right) = \prod_{P(n)} \left(\prod_{j=1}^{n} k_j! \right)$$

Here is a quick check of these identities.

```
In[58]:= factorialId[n_] := Plus @@
              (1/(Function[l, Times @@ Flatten[{Apply[Power, l, {1}],
              (Last /@ l)!}]] /@ myIntegerPartitions[n])) === 1
```

```
In[59]:= factorialId2[n_] :=
         Expand[n! Plus @@ (Function[1, Times @@ ((x/#[[1]])^#[[2]]/#[[2]]!& /@ 1)]
         /@
                      myIntegerPartitions[n])] === Expand[Product[x + k, {k, 0, n - 1}]]

In[60]:= multinomialId[n_] := Plus @@ Multinomial @@@ Map[Last,
                         myIntegerPartitions[n], {2}] === 2^(n - 1)

In[61]:= moebiusMuId[n_] := (Plus @@ Flatten[Apply[MoebiusMu[#1] #2&,
                     myIntegerPartitions[n], {2}]]) === PartitionsP[n - 1]

In[62]:= FibonacciId[n_] := Plus @@ Map[(Plus @@ #)!/(Times @@ (#!))&,
             Map[Last, Select[myIntegerPartitions[n + 1],
                     FreeQ[#, {1, _}]&], {2}]] == Fibonacci[n]

In[63]:= EulerPhiId[n_] := (Plus @@ Flatten[Apply[EulerPhi[#1] #2&,
                     myIntegerPartitions[n], {2}]]) ===
                         Sum[(n - j) PartitionsP[j], {j, 0, n}]

In[64]:= factorialProductId[n_] := Equal @@ Times @@@ Transpose[
         {Times @@ ((Last /@ #)!), Times @@ Flatten[Table[#1, {#2}]& @@@ #]}& /@
                                         myIntegerPartitions[n]]

In[65]:= (* check the last six identities for 1 ≤ n ≤ 25 *)
         Table[{factorialId[n], factorialId2[n], multinomialId[n],
                moebiusMuId[n], FibonacciId[n], EulerPhiId[n],
                factorialProductId[n]}, {n, 25}] // Flatten // Union

Out[66]= {True}
```

2.4 Euler, Bernoulli, and Fibonacci Numbers

Euler and Bernoulli numbers [495], [463] belong in this chapter because they involve calculations with exact numbers. They are both of great importance for the summation of series: the Euler numbers for exact symbolic summation, and the Bernoulli numbers for numerical summation using the Euler-MacLaurin formula.

Euler numbers are the coefficients of the powers of x in the Euler polynomials.

EulerE[n]

   gives the Euler number E_n.

Here are the first few.

```
In[1]:= Table[EulerE[n], {n, 30}]
Out[1]= {0, -1, 0, 5, 0, -61, 0, 1385, 0, -50521, 0, 2702765, 0,
         -199360981, 0, 19391512145, 0, -2404879675441, 0, 370371188237525, 0,
         -69348874393137901, 0, 15514534163557086905, 0, -4087072509293123892361,
         0, 1252259641403629865468285, 0, -441543893249023104553682821}
```

The Bernoulli numbers are the only functions discussed in this chapter that return fractions.

BernoulliB[n]

   gives the Bernoulli number B_n.

Here are the first few.

```
In[2]:= Table[BernoulliB[n], {n, 30}]
```

$$\text{Out[2]= } \left\{-\frac{1}{2}, \frac{1}{6}, 0, -\frac{1}{30}, 0, \frac{1}{42}, 0, -\frac{1}{30}, 0, \frac{5}{66}, 0, -\frac{691}{2730}, \right.$$

$$0, \frac{7}{6}, 0, -\frac{3617}{510}, 0, \frac{43867}{798}, 0, -\frac{174611}{330}, 0, \frac{854513}{138}, 0,$$

$$\left.-\frac{236364091}{2730}, 0, \frac{8553103}{6}, 0, -\frac{23749461029}{870}, 0, \frac{8615841276005}{14322}\right\}$$

The denominators of the Bernoulli numbers of even order can be expressed as a product of modified divisors of the order [463].

```
In[3]:= Table[Denominator[BernoulliB[2 k]] - Times @@
            (Select[Divisors[2 k], PrimeQ[# + 1]&] + 1), {k, 500}] // Union
Out[3]= {0}
```

Bernoulli numbers fulfill many interesting identities. Here is an example.

```
In[4]:= N[FromContinuedFraction[BernoulliB[Range[100]]], 30]
Out[4]= -0.500000000000000000000000000000
```

Interestingly, Bernoulli numbers can be calculated similarly to binomial coefficients using a pyramidal scheme [299].

```
In[5]:= AkiyamaTanigawaAlgorithm[n_] := MapAt[-#&,
        First /@ (NestList[MapIndexed[First[#2] Subtract @@ #1&,
                        Partition[#, 2, 1]]&, 1/Range[n], n - 1]), 2]
In[6]:= AkiyamaTanigawaAlgorithm[100] -
        Table[BernoulliB[k], {k, 0, 99}] // Union
Out[6]= {0}
```

Using the Bernoulli numbers, we can program the Euler–Maclaurin formula.

Mathematical Remark: Euler-Maclaurin Formula

The Euler–Maclaurin formula can be regarded as a method either for the computation of definite integrals or for summation. The appropriate formula for integration is:

$$\int_a^{a+m\,h} f(x)\, dx \approx$$

$$h\left(\frac{f(a)}{2} + \sum_{i=1}^{m-1} f(a + ih) + \frac{f(a + mh)}{2}\right) + \sum_{k=1}^{n} \frac{(-1)^k |B_{2k}| h^{2k}}{(2k)!} \left(f^{(2k-1)}(a + mh) - f^{(2k-1)}(a)\right).$$

Thus, the integral is approximated by an asymptotic sum of the function values at the data points along with the values of the derivatives at the endpoints of the interval of integration. For sufficiently smooth functions, this integration formula can achieve surprising accuracy. For a derivation of the Euler-Maclaurin formula, see, for example, [323], [344], [538], [218], [22], [425], [156], [250], [395], [210], [427], [457], [345], and [162]; for infinite intervals, see [235]. (In the limit $n \to \infty$, the Euler–Maclaurin formula is an asymptotic series [178], [79], [256], [544], [540].)

Let $\{a, b\}$ be the interval of integration, m be the number of data points, n be the integration order, and f be the function to be integrated. Then the above integration formula can be directly programmed as follows.

```
In[7]:= EulerMaclaurin[{a_ , b_}, m_Integer?Positive,
                   n_Integer?Positive, f_Function | f_Symbol] :=
     Module[{h = (b - a)/m},
       h (f[a]/2 + Sum[f[a + i h], {i, 1, m - 1}] + f[b]/2) +
        Sum[h^(2k) (-1)^k Abs[BernoulliB[2k]]/(2k)!*
             Subtract @@ (Derivative[2k - 1][f] /@ {b, a}), {k, 1, n}]]
```

The Euler-Maclaurin formula is a generalization of the trapezoidal rule for numerical integration. The extension involves the higher derivatives at the endpoints of the interval.

We now look at a symbolic example.

```
In[8]:= EulerMaclaurin[{a, b}, 5, 4, f]
```

$$\text{Out[8]}= \frac{1}{5} (-a + b) \left(\frac{f[a]}{2} + \frac{f[b]}{2} + f\left[a + \frac{1}{5} (-a + b)\right] + \right.$$
$$\left. f\left[a + \frac{2}{5} (-a + b)\right] + f\left[a + \frac{3}{5} (-a + b)\right] + f\left[a + \frac{4}{5} (-a + b)\right] \right) -$$
$$\frac{1}{300} (-a + b)^2 (-f'[a] + f'[b]) + \frac{(-a + b)^4 (-f^{(3)}[a] + f^{(3)}[b])}{450000} -$$
$$\frac{(-a + b)^6 (-f^{(5)}[a] + f^{(5)}[b])}{472500000} + \frac{(-a + b)^8 (-f^{(7)}[a] + f^{(7)}[b])}{472500000000}$$

The integral $\int_0^1 x^{19}\, dx$ is computed exactly by the ninth-order `EulerMaclaurin` formula.

```
In[9]:= Table[{k, EulerMaclaurin[{0, 1}, 3, k, #^19&]}, {k, 12}]
```

$$\text{Out[9]}= \left\{\left\{1, -\frac{42347669}{4649045868}\right\}, \left\{2, \frac{4211220271}{46490458680}\right\}, \left\{3, \frac{8879888737}{325433210760}\right\}, \left\{4, \frac{19293652279}{325433210760}\right\},\right.$$
$$\left\{5, \frac{15436702819}{325433210760}\right\}, \left\{6, \frac{16412949001}{325433210760}\right\}, \left\{7, \frac{16259110561}{325433210760}\right\},$$
$$\left.\left\{8, \frac{2034012401}{40679151345}\right\}, \left\{9, \frac{1}{20}\right\}, \left\{10, \frac{1}{20}\right\}, \left\{11, \frac{1}{20}\right\}, \left\{12, \frac{1}{20}\right\}\right\}$$

```
In[10]:= N[% - 1/20]
```

$$\text{Out[10]}= \{\{0.95, -0.0591089\}, \{1.95, 0.0405825\}, \{2.95, -0.0227136\}, \{3.95, 0.00928606\},$$
$$\{4.95, -0.00256568\}, \{5.95, 0.000434155\}, \{6.95, -0.0000385639\},$$
$$\{7.95, 1.34796 \times 10^{-6}\}, \{8.95, 0.\}, \{9.95, 0.\}, \{10.95, 0.\}, \{11.95, 0.\}\}$$

In the presence of singularities in the function to be integrated, the Euler–MacLaurin formula must be adapted; see [417], [337], [163], [512], and [607] for details.

Here is a version of the Euler–MacLaurin formula using a truncated version of the symbolic differentiation operator $(\partial. / \partial x / 2)/(\tanh(\partial. / \partial x / 2))$ [305], [4].

```
In[11]:= (* truncated expanded differential operator *)
     𝓛[d_, o_Integer] := Normal[Series[d/2/Tanh[d/2], {d, 0, o}]]
     𝓛𝓓[h_, o_Integer] := Function @@ ((Hold @@ {# + 𝓛[d, o] - 1}) /.
                                 d^e_. :> D[#, {h, e}])
```

```
(* Euler-MacLaurin formula in the form
    sum == integral + (derivative terms of integration endpoints) *)
EML[f_, {x_, a_, b_}, o_] :=
Module[{α, β},
  (f/2 /. x -> a) + Sum[f, {x, a + 1, b - 1}] + (f/2 /. x -> b) ==
    𝓛𝓓[α, o] @ 𝓛𝓓[β, o] @ Integrate[f, {x, a - α, b + β}] /.
                                    {α -> 0, β -> 0}]
```

In[16]:= **EML[f[x], {x, a, b}, 12]**

Out[16]= $\dfrac{f[a]}{2} + \dfrac{f[b]}{2} + \displaystyle\sum_{x=1+a}^{-1+b} f[x] ==$

$\displaystyle\int_{a}^{b} f[x]\, dx - \dfrac{f'[a]}{12} + \dfrac{f'[b]}{12} + \dfrac{1}{720}\, f^{(3)}[a] - \dfrac{1}{720}\, f^{(3)}[b] - \dfrac{f^{(5)}[a]}{30240} + \dfrac{f^{(5)}[b]}{30240} +$

$\dfrac{f^{(7)}[a]}{1209600} - \dfrac{f^{(7)}[b]}{1209600} - \dfrac{f^{(9)}[a]}{47900160} + \dfrac{f^{(9)}[b]}{47900160} + \dfrac{691\, f^{(11)}[a]}{1307674368000} - \dfrac{691\, f^{(11)}[b]}{1307674368000}$

A little known application of Bernoulli polynomials is the expansion of a function in Bernoulli polynomials. A sufficiently smooth, nonperiodic function $f(x)$ defined on the interval $[a,\ b]$ can be expanded in the form [122], [123], [136], [137], [559], [124]

$$f(x) = f(a) + \sum_{k=1}^{\infty} \mathcal{B}_k\!\left(\frac{x-a}{b-a}\right) \frac{(b-a)^{k-1}}{k!}\, (f^{(k-1)}(b) - f^{(k-1)}(a)).$$

Here the $B_k(x)$ are shifted Bernoulli polynomials $\mathcal{B}_k(x) = B_k(x) - B_k$.

In[17]:= **𝓑[k_][x_] := BernoulliB[k, x] - BernoulliB[k]**

In[18]:= **TruncatedBernoulliExpansion[f_, o_, {a_, b_}] :=**
f[a] + Sum[𝓑[k][(x - a)/(b - a)] (b - a)^(k - 1)/k! *
(Derivative[k - 1][f][b] - Derivative[k - 1][f][a]),
{k, 1, o}]

For polynomials f and o greater than their degrees, `TruncatedBernoulliExpansion` is an exact expansion.

In[19]:= **TruncatedBernoulliExpansion[#^6&, 8, {-2, 2}]**

Out[19]= $64 + 768\left(\dfrac{1}{4}\,(-2-x) + \dfrac{1}{16}\,(2+x)^2\right) + 5120\left(\dfrac{1}{16}\,(2+x)^2 - \dfrac{1}{32}\,(2+x)^3 + \dfrac{1}{256}\,(2+x)^4\right) +$

$4096\left(-\dfrac{1}{32}\,(2+x)^2 + \dfrac{5}{512}\,(2+x)^4 - \dfrac{3\,(2+x)^5}{1024} + \dfrac{(2+x)^6}{4096}\right)$

In[20]:= **Expand[%]**

Out[20]= x^6

The following graphics show the truncated Bernoulli expansions and the logarithm of the absolute error for the function cos in the interval $[0,\ 3/2\,\pi]$ up to order 30.

In[21]:= **Show[GraphicsArray[Function[f,**
Plot[Evaluate[Table[f[TruncatedBernoulliExpansion[Cos[#]&,
k, {0, 3/2Pi}]], {k, 0, 30}]],
{x, 0, 3/2Pi}, PlotStyle -> Table[{Hue[k/38]}, {k, 0, 30}],
Frame -> True, Axes -> False, DisplayFunction -> Identity]] /@
{(* approximation itself *) Identity,
(* logarithm of error *) Log[10, Abs[# - Cos[x]]]&}]]

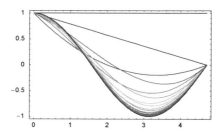 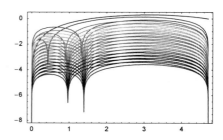

In the limit $b \to a$, the expansion coefficients become the Taylor coefficients at a. (The next input uses the function Limit, discussed in Chapter 1 of the Symbolics volume [556].)

In[22]:= **TruncatedBernoulliExpansion[Cos, 12, {0, ε}] // Limit[#, ε -> 0]&**

Out[22]= $1 - \dfrac{x^2}{2} + \dfrac{x^4}{24} - \dfrac{x^6}{720} + \dfrac{x^8}{40320} - \dfrac{x^{10}}{3628800} + \dfrac{x^{12}}{479001600}$

Bernoulli polynomials are also of importance for two-point interpolations. For a sufficiently smooth function $f(x)$ in $[0, 1]$ with $f^{(n)}(0) = O(p_1^n)$ and $f^{(n)}(1) = O(p_2^n)$, $p_1, p_2 < \pi$ [125] for large n, the series [6], [5], [7]

$$f_{L,o} = \sum_{k=1}^{o-1} (f^{(2k)}(0)\, \Lambda_k(1-x) + f^{(2k)}(1)\, \Lambda_k(x))$$

converges uniformly to $f(x)$. Here the Lidstone polynomial $\Lambda_n(x)$ can be expressed through Bernoulli polynomials as $\Lambda_n(x) = 2^{2n+1} B_{2n+1}((x+1)/2)/(2n+1)!$.

The next input defines the Lidstone polynomials.

In[23]:= **LidstoneΛ[n_, t_] := 2^(2n + 1)/(2n + 1)! BernoulliB[2n + 1, (1 + t)/2]**

They fulfill the differential equation $\Lambda_n''(x) = \Lambda_{n-1}(x)$.

In[24]:= **(D[LidstoneΛ[n, t], t, t] - LidstoneΛ[n - 1, t]) // FullSimplify**

Out[24]= 0

The next graphic shows the first few scaled Lidstone polynomials on the left and the scaled differences of the higher Lidstone polynomials on the right.

In[25]:= **Show[GraphicsArray[**
    Block[{$DisplayFunction = Identity, o = 12},
      {(* scaled Lidstone polynomials *)
      Plot[Evaluate[Table[LidstoneΛ[k, t]/LidstoneΛ[k, 1/2],
                     {k, 1, o}]], {t, 0, 1},
          PlotStyle -> Table[Hue[0.78 k/o], {k, o}]],
      (* difference of scaled Lidstone polynomial to first one *)
      Plot[Evaluate[Table[LidstoneΛ[k, t]/LidstoneΛ[k, 1/2] -
                     8/3 t (1 - t) (1 + t),
                     {k, 1, o}]], {t, 0, 1},
          PlotStyle -> Table[Hue[0.78 k/o], {k, o}]]}]]]

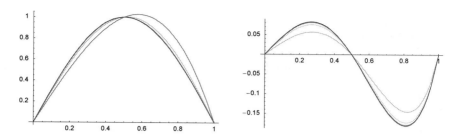

Here is an example. We model the function $\cos(3x)$ using the first 48 Lidstone approximations and display the logarithm of the pointwise approximation error.

```
In[26]:= Module[{f, fΛA, o = 48, pp = 101},
    (* the function to be approximated *)
    f[x_] := Cos[3 x];
    (* form the approximations *)
    Do[fΛA[m][t_] = Sum[Derivative[2 k][f][0] LidstoneΛ[k, 1 - t] +
                        Derivative[2 k][f][1] LidstoneΛ[k, t], {k, 0, m - 1}] //
                                                    Expand, {m, o}];
    (* display logarithm of absolute approximation error *)
    Show[Graphics[{PointSize[0.004],
        Table[{Hue[0.78 m/o], (* use high-precision arithmetic *)
            Table[Point[{t, Log[10, Abs[N[fΛA[m][t] - f[t], 20]]]}],
                {t, 1/pp, 1 - 1/pp, 1/pp}]}, {m, o}]}], Frame -> True]]
```

Using the Euler numbers, we can program the Boole summation formula.

Mathematical Remark: Boole Summation Formula

The Boole summation formula can be regarded either as a summation formula or as a method for extrapolation. The appropriate formula for summation is:

$$\sum_{k=0}^{n} (-1)^k f(x+k+\delta) \approx \frac{1}{2} \sum_{k=0}^{m-1} \frac{E_k(\delta)}{k!} \left((-1)^n f^{(k)}(n+x+1) + f^{(k)}(x)\right).$$

Here, $0 \le \delta \le 1$. Thus, the sum over n terms is approximated by an asymptotic sum of the derivatives of the function values near the beginning and end of the summation. For a derivation of the Boole formula, see for example, [547], [422], and [243].

Let n be the number of terms in the sum on the left-hand side, m be the summation order on the right-hand side, and f be the function to be summed. Then, the above summation formula can be directly programmed as follows.

```
In[27]:= BooleSum[{x_, δ_}, m_Integer?Positive, n_Integer?Positive,
                f_Function | f_Symbol] :=
         1/2 Sum[EulerE[k, δ]/k! ((-1)^n Derivative[k][f][x + n + 1] +
                                   Derivative[k][f][x]), {k, 0, m - 1}]
```

The following input shows the result of the Boole summation formula for $f(x) = x^{19}$. Taking only 7 nontrivial terms in the Boole summation formula results in more than 20 correct digits.

```
In[28]:= Module[{f = #^19&, x = 1/7, δ = 1/3, n = 10^4, exact},
                exact = Sum[(-1)^k f[x + δ + k], {k, 0, n}];
                Table[BooleSum[{x, δ}, m, n, f], {m, 7}]/exact - 1] // N
Out[28]= {0.000317111, 3.79917×10^-7, -1.16741×10^-10,
          -1.0523×10^-13, 2.89654×10^-17, 2.23917×10^-20, -5.35223×10^-24}
```

It is interesting to write the Euler–Maclaurin summation formula and the Boole summation formula in a uniform way that resembles the Taylor expansion of a function $f(x) = \sum_{k=0}^{n} f^{(k)}(a)/k!\,(x-a)^k + R_{n+1}$. This can be done by using the integral average and the arithmetic average of $f^{(k)}(a)$ and the Euler and Bernoulli polynomials of $(x-a)$ instead of powers [542].

```
In[29]:= arithmeticAverage[f_, x_] := (f[x + 1] + f[x])/2
         integralAverage[f_, x_] := Integrate[f[x + t], {t, 0, 1}]
         identityAverage[f_, x_] := f[x]

In[32]:= TaylorSeries[f_, {x_, a_, n_}] :=
           Sum[identityAverage[Derivative[k][f], a]/k! Power[x - a, k],
               {k, 0, n}]

         EulerMacLaurinSeries[f_, {x_, a_, n_}] :=
           Sum[arithmeticAverage[Derivative[k][f], a]/k! EulerE[k, x - a],
               {k, 0, n}]

         BooleSeries[f_, {x_, a_, n_}] :=
           Sum[integralAverage[Derivative[k][f], a]/k! BernoulliB[k, x - a],
               {k, 0, n}]
```

The following graphics show the three series for $f(x) = \cos(x)$, $a = 0$, and $n = 28$. Not unexpectedly, because derivative values at $x = 1$ (instead of only at $x = 0$) are taken into account, the Euler–Maclaurin and the Boole series are more precise for larger x. The right graphic displays the logarithm of the relative error and shows that, near the origin (the expansion point), the Taylor series is clearly the most precise approximation. Last but not least, we see that, overall, the Boole series (which averages derivative information of the interval [0, 1]) is most precise.

```
In[36]:= Module[{f = Cos, o = 28, pp = 100, ts, ems, bs, dataS, dataf},
              (* the three expansions *)
              ts = TaylorSeries[f, {x, 0, o}];
              ems = EulerMacLaurinSeries[f, {x, 0, o}];
              bs = BooleSeries[f, {x, 0, o}];
           (* values of the three expansions;
              use high-precision arithmetic to calculate values *)
```

```
dataS = Table[Evaluate[{{x, ts}, {x, ems}, {x, bs}}],
              {x, -4Pi, 4Pi, 8Pi/pp}] // N[#, 80]& // Transpose;
(* values of Cos[x] *)
dataf = Table[f[x], {x, -4Pi, 4Pi, 8Pi/pp}];
Show[GraphicsArray[Function[{δ},
Show[ListPlot[DeleteCases[δ[#1], {_, Indeterminate}],
          PlotRange -> All, PlotStyle -> #2,
          PlotJoined -> True, PlotRange -> All,
          DisplayFunction -> Identity]& @@@
    (* Taylor series in red, Euler-Maclaurin in green,
       and Boole in red *)
    Table[{dataS[[k]], MapAt[1&, RGBColor[0, 0, 0], k]},
          {k, 3}]]] /@ (* value and logarithm of error *)
    {Identity, Transpose[{First /@ #,
          Log[10, Abs[(Last /@ #) - dataf]]}]&}]]]
```

Mathematica also implements a two-argument version of `EulerE` and `BernoulliB`—the Euler and Bernoulli polynomials. They have many interesting properties. Let us mention one here: The Hankel determinant of the matrix with elements $E_{i+j}(x)$ or $B_{i+j}(x)$ is independent from x. Here we check this statement for $1 \le n \le 6$.

In[37]:= `HankelDet[f_, n_] := Det[Table[f[i + j, x], {i, 0, n}, {j, 0, n}]]`

In[38]:= `Table[Expand[HankelDet[BernoulliB, n]], {n, 6}]`

Out[38]= $\left\{ -\dfrac{1}{12}, -\dfrac{1}{540}, \dfrac{1}{42000}, \dfrac{1}{3215625}, -\dfrac{4}{623959875}, -\dfrac{64}{213746467935} \right\}$

Closed form expressions for the determinants involve factorials [15].

In[39]:= `Table[(-1)^(n(n + 1)/2) Product[i!, {i, n}]^6/`
        `Product[i!, {i, 2n + 1}], {n, 6}]`

Out[39]= $\left\{ -\dfrac{1}{12}, -\dfrac{1}{540}, \dfrac{1}{42000}, \dfrac{1}{3215625}, -\dfrac{4}{623959875}, -\dfrac{64}{213746467935} \right\}$

The zeros of the scaled Euler and Bernoulli polynomials $E_n(nz)$ and $B_n(nz)$ approach characteristic curves in the complex z-plane as $n \to \infty$ [207]. The following two graphics show the zeros for $n = 1, \ldots, 80$.

In[40]:= `Show[GraphicsArray[`
        `Graphics[Table[{Hue[k/100], PointSize[0.01],`
                `Point[{Re[z], Im[z]}] /.`
                    `N[Solve[#[k, k z] == 0, z]]}, {k, 80}],`
            `PlotRange -> All, Frame -> True, PlotLabel -> #]& /@`
                                    `{EulerE, BernoulliB}]]`

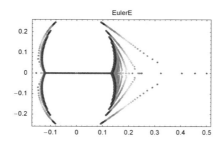
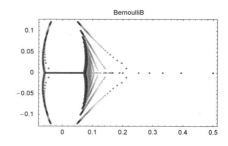

Another set of numbers that one frequently encounters in combinatorial problems, in the analysis of the complexity of algorithms, and in other areas are the famous Fibonacci numbers. (For many properties and uses of Fibonacci numbers, see [579], [580], [564], [264], [435], [436], [437], [49], [438], [152], and [265].)

Fibonacci[n]

    gives the Fibonacci number F_n.

Here are the first 20 Fibonacci numbers. (For their first and last digits, see [527].)

```
In[41]:= Table[Fibonacci[n], {n, 0, 20}]
```
```
Out[41]= {0, 1, 1, 2, 3, 5, 8, 13, 21, 34, 55, 89,
           144, 233, 377, 610, 987, 1597, 2584, 4181, 6765}
```

Fibonacci numbers obey the following recursion relation $F_n = F_{n+1} + F_{n+2}$ with the initial conditions $F_0 = 0$ and $F_0 = 1$.

```
In[42]:= fibonacci[0] = 0;
         fibonacci[1] = 1;
         fibonacci[n_Integer?Positive] := fibonacci[n] =
                 fibonacci[n - 1] + fibonacci[n - 2];
```
```
In[45]:= Table[fibonacci[n], {n, 0, 20}]
```
```
Out[45]= {0, 1, 1, 2, 3, 5, 8, 13, 21, 34, 55, 89,
           144, 233, 377, 610, 987, 1597, 2584, 4181, 6765}
```

While the linear defining relation for Fibonacci numbers $F_n = F_{n-1} + F_{n-2}$ is nice and simple, it is slow for calculating F_n for larger n. Using the properties of the Fibonacci numbers, one can derive the following relations that express F_n through F_m with $m \approx n/2$ [98].

```
In[46]:= fib[n_?EvenQ] := fib[n] = fib[n/2] (2 fib[n/2 + 1] - fib[n/2])
         fib[n_?OddQ] := fib[n] = fib[(n - 1)/2]^2 + fib[(n - 1)/2 + 1]^2
         fib[0] = 0; fib[1] = 1; fib[2] = 1;
```

Using fib we achieve about the speed of the built-in function Fibonacci. We calculate F_{10^7}, an integer with more than two million digits.

```
In[49]:= {fib[10^7]; // Timing, Fibonacci[10^7]; // Timing}
```
```
Out[49]= {{1.28 Second, Null}, {1.13 Second, Null}}
```

```
In[50]:= (* check if the two Fibonacci numbers are identical *)
         {fib[10^7] === Fibonacci[10^7], N @ fib[10^7]}
```
```
Out[51]= {True, 1.129834378225400×10^2089876}
```

The recursion relation (difference equation) can be used in the following closed form for the nth Fibonacci number

$$F_\nu = \frac{\phi^\nu - \cos(\nu\pi)\,\phi^{-\nu}}{\sqrt{5}}.$$

In[52]:= `fibonacciContinued[v_] = (GoldenRatio^v - Cos[v Pi]/GoldenRatio^v)/Sqrt[5];`

For positive integers ν, this formula again reduces to the above numbers.

In[53]:= `Table[Cancel[ExpandNumerator[ExpandDenominator[`
`                Together[fibonacciContinued[n]] /.`
`                    GoldenRatio -> (1 + Sqrt[5])/2]]], {n, 0, 20}]`

Out[53]= `{0, 1, 1, 2, 3, 5, 8, 13, 21, 34, 55, 89,`
`       144, 233, 377, 610, 987, 1597, 2584, 4181, 6765}`

The above formula is used in *Mathematica* to continue the Fibonacci numbers into the complex plane.

In[54]:= `Plot3D[Re[Fibonacci[x + I y]], {x, -4, 4}, {y, -2, 2},`
`            PlotPoints -> 60, PlotRange -> All, Mesh -> False]`

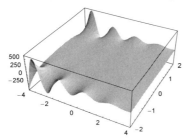

Fibonacci polynomials modulo a prime show interesting periodicity properties.

In[55]:= `Show[GraphicsArray[`
`      ListDensityPlot[Table[Mod[Fibonacci[i, j], #], {i, 128}, {j, 128}],`
`                Mesh -> False, ColorFunction -> (Hue[0.8 #]&),`
`                DisplayFunction -> Identity]& /@`
`                {13, 23, 37}]]`

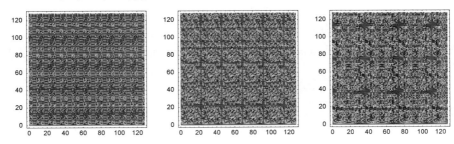

Iterating the function `fibonacciContinued` yields a fractal structure over the complex ν-plane [460]. The right graphic shows the iteration of the functions $\nu \to \ln(F_\nu)$.

In[56]:= `Show[GraphicsArray[`
`      Block[{$DisplayFunction = Identity},`
`        {(* iteration of Fibonacci until large value is reached *)`
`        DensityPlot[(* escape steps *)`
`                Module[{z = x + I y, c = 0, φ = GoldenRatio},`

```
                    (* iterate fibonacciContinued; use explicit form,
                        not Fibonacci, for compilation *)
                    While[z = (φ^z - Cos[z Pi] φ^-z)/Sqrt[5];
                          Abs[z] < 200. && c < 400, c++]; c],
            {x, -8, 8}, {y, -2, 2},
            ColorFunction -> (Hue[0.78 #]&), FrameTicks -> False,
            PlotPoints -> 240, Mesh -> False],
      (* iteration of Log[Fibonacci[v]]five times *)
      DensityPlot[Abs @ Nest[Log[Fibonacci[#]]&, x + I y, 5],
            {x, -12, 12}, {y, -6, 6},
            FrameTicks -> False, PlotPoints -> 600,
            Mesh -> False, ColorFunction -> (Hue[0.78 #]&)]}]]]
```

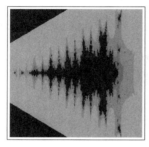 

Fibonacci numbers have many beautiful properties. One of them is the nice pattern in the continued fraction expansion of the ratio of successive Fibonacci numbers.

In[57]:= **ContinuedFraction /@**
          Apply[#2/#1&, Partition[Fibonacci[Range[20]], 2, 1], {1}]

Out[57]= {{1}, {2}, {1, 2}, {1, 1, 2}, {1, 1, 1, 2}, {1, 1, 1, 1, 2},
          {1, 1, 1, 1, 1, 2}, {1, 1, 1, 1, 1, 1, 2}, {1, 1, 1, 1, 1, 1, 1, 2},
          {1, 1, 1, 1, 1, 1, 1, 1, 2}, {1, 1, 1, 1, 1, 1, 1, 1, 1, 2},
          {1, 1, 1, 1, 1, 1, 1, 1, 1, 1, 2}, {1, 1, 1, 1, 1, 1, 1, 1, 1, 1, 1, 2},
          {1, 1, 1, 1, 1, 1, 1, 1, 1, 1, 1, 1, 2}, {1, 1, 1, 1, 1, 1, 1, 1, 1, 1, 1, 1, 1, 2},
          {1, 1, 1, 1, 1, 1, 1, 1, 1, 1, 1, 1, 1, 1, 2},
          {1, 1, 1, 1, 1, 1, 1, 1, 1, 1, 1, 1, 1, 1, 1, 2},
          {1, 1, 1, 1, 1, 1, 1, 1, 1, 1, 1, 1, 1, 1, 1, 1, 2},
          {1, 1, 1, 1, 1, 1, 1, 1, 1, 1, 1, 1, 1, 1, 1, 1, 1, 2}}

Fibonacci numbers also appear in the continued fraction expansion of the sum $\sum_{k=0}^{\infty} 2^{-\lfloor k/\phi \rfloor}$ [18].

$$\sum_{k=0}^{\infty} 2^{-\left\lfloor \frac{k}{\phi} \right\rfloor} = 2 + \cfrac{1}{1 + \cfrac{2^{F_0}}{1 + \cfrac{2^{F_1}}{1 + \cfrac{2^{F_2}}{1 + \cfrac{2^{F_3}}{1 + \cdots}}}}}$$

Calculating the value of the sum to 10000 digits allows us to see the first 19 Fibonacci numbers.

In[58]:= **sum = Sum[N[1/2^Floor[n/GoldenRatio], 10000], {n, 1, 25000}];**

In[59]:= **Log[2, Take[ContinuedFraction[sum], 20]]**

Out[59]= {1, 0, 1, 1, 2, 3, 5, 8, 13, 21, 34, 55, 89, 144, 233, 377, 610, 987, 1597, 2584}

```
In[60]:= Drop[%, 2] == Table[Fibonacci[k], {k, 18}]
Out[60]= True
```

Similarly to the binomial coefficients, one can define so-called Fibonacci coefficients $\begin{bmatrix} n \\ k \end{bmatrix}$ [340]:

$$\begin{bmatrix} n \\ k \end{bmatrix} = \frac{\prod_{j=n-k+1}^{n} F_j}{\prod_{j=1}^{k} F_j}.$$

```
In[61]:= FibonacciCoefficient[n_, k_] :=
            Product[Fibonacci[j], {j, n - k + 1, n}]/
                Product[Fibonacci[j], {j, k}]
```

Not unexpectedly, modulo small integers, the Fibonacci numbers show a nested triangle structure as did the binomial coefficients.

```
In[62]:= Show[GraphicsArray[
        Table[Graphics[
            Table[{GrayLevel[Mod[FibonacciCoefficient[n, k], mod]/(mod - 1)],
                Point[{k - n/2, -n}]}, {n, 0, 120}, {k, 0, n}]],
            (* use three moduli *) {mod, 2, 4}]]]
```

And, after introducing the Fibonacci equivalent $\mathcal{PF}$ of powers, we can state even a Fibonacci-Binomial theorem [191], [368].

```
In[63]:= PF[k_, x_] := Product[-x (((1 - Sqrt[1 + 4])/2)^(j - 1)) +
                            (((1 + Sqrt[1 + 4])/2)^(j - 1)), {j, k}]
```

```
In[64]:= FibonomialTheorem[n_, x_, y_] :=
        PF[n, x y] == Sum[FibonacciCoefficient[n, k] y^k*
                PF[k, x] PF[n - k, y], {k, 0, n}]
```

Here is a quick check of the Fibonacci-Binomial theorem for small n.

```
In[65]:= Table[FibonomialTheorem[n, α, β], {n, 0, 10}] // Simplify // Union
Out[65]= {True}
```

As a small application of the Fibonacci numbers, let us look at the discretized version of the Arnold cat map [70]. The continuous version is a map from the unit square to the unit square given by

$$x_{i+1} = (x_i + y_i) \bmod 1$$
$$y_{i+1} = (x_i + 2 y_i) \bmod 1.$$

Dividing the x- and y-intervals into n parts and numbering the parts consecutively, the natural discretized version is [289], [194], [195], [190]

$$x_{i+1} = (x_i + y_i) \bmod n$$
$$y_{i+1} = (x_i + 2 y_i) \bmod n.$$

Taking into account the kth power of the transformation matrix can be calculated in closed form using Fibonacci numbers [293]

$$\begin{pmatrix} 1 & 1 \\ 1 & 2 \end{pmatrix}^k = \begin{pmatrix} F_{2k-1} & F_{2k} \\ F_{2k} & F_{2k+1} \end{pmatrix}.$$

It is straightforward to implement a function `period` that calculates after how many steps we return to the initial configuration.

```
In[66]:= period[k_] :=
    Module[{j = 1},
        While[Mod[{{Fibonacci[2 j - 1], Fibonacci[2 j]},
                {Fibonacci[2 j], Fibonacci[2 j + 1]}}, k] =!=
            IdentityMatrix[2], j++]; j]
```

The following graphic shows that the period increases in average linearly with k, but has large fluctuations [289], [155], [8].

```
In[67]:= ListPlot[Table[{k, period[k]}, {k, 2, 1000}], PlotRange -> All]
```

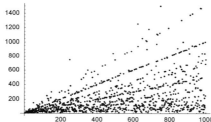

Let us also have a look at the iterations itself. The compiled function `catMap` carries out one step of the cat map.

```
In[68]:= catMap = Compile[{{m, _Integer, 2}},
        Module[{n = Dimensions[m][[1]], m},
            m = Table[0, {n}, {n}];
            Do[m[[Mod[i + j, n, 1], Mod[i + 2 j, n, 1]]] = m[[i, j]],
                {i, n}, {j, n}]; m]];
```

We will apply `catMap` 75 times to a matrix that represents a 202×202 discretization to a graphic of concentric circles. We display nine of the iterations. We recognize the cut-and-reassembled first few and last few pictures. The middle pictures seem basically random. Interestingly, the picture from step 25 shows a partial recurrence.

```
In[69]:= dp = DensityPlot[If[Sin[x^2 + y^2] <= 0, 0, 1], {x, -5, 5}, {y, -5, 5},
                Mesh -> False, PlotPoints -> 202,
                DisplayFunction -> Identity]
```

```
In[70]:= m = Round[dp[[1]]]; mBag = {};
    (* list of integers for the iterations to be shown *)
    iterationsToShow = {0, 1, 2, 3, 6, 25, 34, 74, 75};
    Do[If[MemberQ[iterationsToShow, k], AppendTo[mBag, m]];
        m = catMap[m], {k, 0, 75}]
```

Here is a simple animation showing the 75 steps. (We leave it to the reader to build a more refined animation that first shears the picture, then cuts it into four pieces and moves each piece into its new position or display the resulting matrices over a torus, which is, due to the mod operation, the natural phase space of the cat map.)

```
In[74]:= Show[GraphicsArray[
    ListDensityPlot[#, Mesh -> False, FrameTicks -> None,
                    DisplayFunction -> Identity]& /@ #]]& /@
                                   Partition[mBag, 3]
```

```
m = Round[dp[[1]]];
Do[ListDensityPlot[m, Mesh -> False, FrameTicks -> None, PlotLabel -> k];
    m = catMap[m], {k, 0, 75}]
```

Exercises

1.[L1] `DivisorSigma`, **Primes, Maximum Formula,** `StirlingS2`, `LegendreSymbol`, GCD**-Iterations, Isenkrahe Algorithm, Multinomial Terms, Prime Divisors**

a) The divisor function $\sigma_k(n)$ can be written in the form: If n has the prime factorization (the p_i are prime numbers) $n = \prod_{i=1}^{l} p_i^{\alpha_i}$, then

$$\sigma_k(n) = \prod_{i=1}^{l} \frac{p_i^{(\alpha_i+1)k} - 1}{p_i^k - 1}.$$

Implement this method of calculating $\sigma_k(n)$ in a one-liner.

b) Calculate the first six terms of the following sequence [608]:

$$p_{n+1} = \left\lfloor 1 - \log_2\left(\frac{1}{2} + \sum_{r=1}^{n} \sum_{1 \le i_1 < \cdots < i_r \le n} \frac{(-1)^r}{2^{p_{i_1} \cdots p_{i_r}} - 1}\right)\right\rfloor$$

$$p_1 = 2.$$

Here, $\lfloor x \rfloor$ denotes the integer part of x.

c) The maximum of n integers (real numbers) x_1, x_2, ..., x_n can be expressed through the minimum of these numbers in the following (inefficient) way [355]:

$$\max(x_1, x_2, \ldots, x_n) = \sum_{1 \le k_1 \le n} x_{k_1} - \sum_{1 \le k_1 < k_2 \le n} \min(x_{k_1}, x_{k_2}) + \sum_{1 \le k_1 < k_2 < k_3 \le n} \min(x_{k_1}, x_{k_2}, x_{k_3}) - $$

$$\ldots + (-1)^{j+1} \sum_{1 \le k_1 < k_2 < \ldots k_j \le n} \min(x_{k_1}, x_{k_2}, \ldots, x_{k_j}) + $$

$$\ldots + (-1)^{n+1} \sum_{1 \le k_1 < k_2 < \ldots k_{n-1} \le n} \min(x_{k_1}, x_{k_2}, \ldots, x_{k_n}).$$

Implement the above formula. How many calls to the function min are needed to calculate the maximum of the list $\{1, 2, 3, \ldots 13, , 14, 15\}$?

d) The Stirling numbers of the second kind $S_n^{(m)}$ have the following neat representation [219], [602]:

$$S_n^{(m)} = \sum_{1 \le k_1 \le k_2 \le \cdots \le k_{n-m} \le m} k_1\, k_2 \cdots k_{n-m}.$$

Here, $n \ge m$ is assumed. Implement this sum representation as a "one-liner".

e) In Section 2.2, we gave the following definition for the Legendre symbol $\left(\frac{a}{p}\right)$: Let p be an odd prime. If p divides a, $\left(\frac{a}{p}\right) = 0$. If this is not the case:

$$\left(\frac{a}{p}\right) = \begin{cases} +1 & \text{if } \exists x, \text{ such that } x^2 = a \bmod p \\ -1 & \text{otherwise.} \end{cases}$$

Find an "explicit formula" for a function `LegendreSymbol[a, p]` that agrees with the built-in function `JacobiSymbol[a, p]` when p is an odd prime and a is a positive integer, and implement the formula as an "one-liner". Do not use boolean functions such as `PrimeQ`, procedural constructs such as `If`, quantifiers such as `Exists`, etc.

f) Analyze the number of steps need in carrying out Euclid's algorithm for determining $\gcd(p, q)$ for $1 \le p, q \le 256$.

g) The least common multiple of r positive integers numbers n_1, n_2, ..., n_k can be expressed through the minimum of these numbers in the following (inefficient) way [347], [500]:

$$\mathrm{lcm}(n_1, n_2, \ldots, n_r) = \frac{\displaystyle\prod_{i=1}^{r} \gcd(n_i) \prod_{i<j<k}^{r} \gcd(n_i, n_j, n_k) \cdots}{\displaystyle\prod_{i<j1}^{r} \gcd(n_i, n_j) \prod_{i<j<k<l}^{r} \gcd(n_i, n_j, n_k, n_l) \cdots}.$$

Implement the above formula. How many calls to the function GCD are needed to calculate the least common multiple of the list $\{1, 2, 3, 4, 5, 6, 7, 8, 9, 10, 11, 12\}$?

h) Guess the limit of the following sequence [470], [14]:

$$\frac{1}{2}, \quad \frac{\frac{1}{2}}{\frac{3}{4}}, \quad \frac{\frac{\frac{1}{2}}{\frac{3}{4}}}{\frac{\frac{5}{6}}{\frac{7}{8}}}, \quad \ldots$$

i) Let $f_n(\xi)$ be the relative number of the sorted divisors $\{d_1, d_2, \ldots, d_{k(n)}\}$ of n that are smaller or equal to n^ξ.

$$f_n(\xi) = \frac{l(n)}{k(n)}, \quad d_{l(n)} \le n^\xi < d_{l(n)+1}$$

Then, the averaged version of $f_n(\xi)$ (denoted by $F(\xi)$)

$$F_m(\xi) = \frac{1}{m} \sum_{n=2}^{m} f_n(\xi)$$

has asymptotically the form [238], [141]

$$F(\xi) = \lim_{m \to \infty} F_m(\xi) = \frac{2}{\pi} \arcsin\left(\sqrt{\xi}\right).$$

Use the first one million positive integers to see how well $F(\xi)$ is approximated.

j) Let $a^\dagger$, a two noncommuting variables that obey the commutation relation $a\, a^\dagger - a^\dagger a = 1$. Consider the expansion

$$(a^\dagger a)^n = \sum_{j=0}^{n} c_{n,j} (a^\dagger)^j a^j.$$

Carry out this expansion for small n, and guess a closed form for the $c_{n,j}$.

k) Let $\lambda(p/q)$ be the length of the continued fraction expansion of the fraction p/q. For fixed denominator n, the cumulative average

$$l_q = \sum_{\substack{p=0 \\ \gcd(p,q)=1}}^{q} \lambda\left(\frac{p}{q}\right)$$

has asymptotically the form [444], [245], [565]

$$l_q \underset{q\to\infty}{\approx} \varphi(q)\left(\frac{12}{\pi^2}\ln(2)\ln(q) + c\right).$$

Calculate l_q for $q \le 1000$ and l_{10^6} and estimate the value of c.

l) The Isenkrahe algorithm [415] for the calculation for next prime p after the prime q is given as the fixed point of the map

$$n \longrightarrow \frac{n!}{\psi(q,\,n)} + \frac{\psi(q,\,n)}{(n-1)!} - \left\lfloor\frac{(n-1)!}{\psi(q,\,n)}\right\rfloor.$$

The function $\psi(m,\,n)$ (for fixed n) is defined as the following product over all primes p less than or equal to m

$$\psi(m,\,n) = \prod_{p\le m} p^{\nu_p(n)}$$

$$\nu_p(n) = \sum_{k=1}^{\infty}\left\lfloor\frac{n}{p^k}\right\rfloor.$$

Implement the Isenkrahe algorithm as a one-liner.

m) The multinomial theorem describes the expansion of $(\sum_{j=1}^{m} a_j)^n$. It can be written in the following form [369]:

$$\left(\sum_{j=1}^{m} a_j\right)^n = \sum_{k=1}^{\max(m,n)} \Gamma_k^{(n)}(a_1, a_2, \ldots, a_m)$$

Here $\Gamma_k^{(n)}(a_1, a_2, \ldots, a_m)$ is the sum of all terms containing products of k different a_j.

$$\Gamma_k^{(n)}(a_1, a_2, \ldots, a_m) = \sum_{\mu_1=1}^{m-(k-1)}\sum_{\mu_2=\mu_1+1}^{m-(k-2)}\cdots\sum_{\mu_k=\mu_{k-1}+1}^{m}$$

$$\sum_{\nu_1=1}^{n-(k-1)}\sum_{\nu_2=1}^{n-\nu_1-(k-2)}\cdots\sum_{\nu_{k-1}=1}^{n-\nu_1-\nu_2-\ldots-\nu_{k-2}-1}\sum_{\nu_k=n-\nu_1-\nu_2-\cdots-\nu_{k-1}}^{n-\nu_1-\nu_2-\cdots-\nu_{k-1}}(n;\,\nu_1, \nu_2, \ldots, \nu_m)\,a_{\mu_1}^{\nu_1}\,a_{\mu_2}^{\nu_2}\cdots a_{\mu_k}^{\nu_k}.$$

Implement the calculation of the $\Gamma_k^{(n)}(a_1, a_2, \ldots, a_m)$ and the expansion of $(\sum_{j=1}^{m} a_j)^n$ based on the $\Gamma_k^{(n)}$.

n) Visualize that on average, an integer n has about $\omega(n) \propto \ln(\ln(n)) + B_1 + \gamma/\ln(n)$ different prime factors [62], [63], [557]. Here $B_1 = \sum_{k=2}^{\infty} k^{-1}\mu(k)\log(\zeta(k))$ is the Mertens constant.

o) The Fibonacci numbers F_n with $n \geq 3$ can be characterized through the following property: an integer k is a Fibonacci number if and only if the interval $[k\varphi - 1/\varphi, k\varphi + 1/\varphi]$ contains an integer [331]. Implement a recursive definition for the calculation of the Fibonacci numbers based on this property.

p) Consider the Kimberling sequence $a_k = (2^{-v_2(k)} + 1)/2$ [316], where $v_2(k)$ is the largest power of 2 that occurs as a factor in k (in *Mathematica* `IntegerExponent[k, 2]`). The Kimberling sequence has the remarkable property that deleting the first occurrence of each positive integer in it results in the original sequence. Check this property for the first million terms of the Kimberling sequence.

q) Let $C(x)$ be the classical Cantor function. Is $\int_0^\infty C(x)^{C(x)} \, dx > 3/4$?

2.^{L1} Cattle Problem of Archimedes

Solve the following cattle problem of Archimedes (see [249], [352]): The sun god Helios had a herd of cattle (steer and cows) that were either white, or black, or spotted, or brown. The number of white steer was equal to $1/2 + 1/3$ the number of black steer more than the number of brown steer. The number of black steer was $1/4 + 1/5$ the number of spotted steer more than the number of brown steer. The number of spotted steer was $1/6 + 1/7$ the number of white steer more than the number of brown steer. The number of white cows was $1/3 + 1/4$ the number of black cattle. The number of black cows was $1/4 + 1/5$ the number of spotted cattle. The number of spotted cows was $1/5 + 1/6$ the number of brown cattle, and the number of brown cows was $1/6 + 1/7$ of the number of white cattle. How many of each type were in the herd of Helios?

3.^{L1} Complicated Identity, π-Formula

a) Interactively check the following identity:

$$\sqrt{16 - 2\sqrt{29} + 2\sqrt{55 - 10\sqrt{29}}} = \sqrt{22 + 2\sqrt{5}} - \sqrt{11 + 2\sqrt{29}} + \sqrt{5}$$

(from [132]; for a listing of related identities, see [132], [52], [53], and [346]).

b) Can a finite n truncation of the following limit [380] be used for calculating π to 10 digits? How long would it approximately take?

$$\pi = \lim_{n \to \infty} \sqrt{6 \frac{\ln(F_1 F_2 \cdots F_n)}{\ln(\mathrm{lcm}(F_1, F_2, \ldots, F_n))}}$$

4.^{L1} Mirror Charges

Consider the following electrostatic problem: Given two metal plates that meet at an angle of θ, a point charge is located at an angle of φ from the first metal plate. Here is a sketch.

Which of the following problems can be solved with the reflection charge method (see [290], [549], and [487]), and which cannot? Look at the solutions graphically, along with the failures of the construction:

$\theta = 90°$, $\varphi = 45°$

$\theta = 60°$, $\varphi = 30°$

$\theta = 90°$, $\varphi = 22.5°$

$\theta = 11.25°$, $\varphi = 5.625°$

$\theta = 20°$, $\varphi = 5°$

$$\theta = 50°, \varphi = 10°$$
$$\theta = 120°, \varphi = 40°$$
$$\theta = 30°, \varphi = 15°$$

5.^{L1} Periodic Decimal Numbers, Digit Sequences, Numbered Permutations, Binomial Values, Smith's Sturmian Word Theorem

a) Built-in *Mathematica* functions can handle (exact) rational numbers and (approximate) real numbers, but not "exact real numbers with a decimal point". Write a routine `ExactDecimalNumber[`*fraction*`]` along the classical, school taught line of how to divide two integers writes the fraction in an exact way (possibly with a periodically repeating part). (The function `RealDigits` provides a similar functionality.)

b) Given a real number x with decimal expansion $d_1 d_2 \ldots d_k.d_{k+1} \ldots d_1 \neq 0$, form a sequence of integers $\{a_1, a_2, \ldots\}$ such that a_k is the smallest integer formed by concatenating the consecutive d_k that have not already occurred. For example, for $x = \pi = 3.1415926535897932384\ldots$, the corresponding sequence of integers is $\{3, 1, 4, 15, 9, 2, 6, 5, 35, 8, 97, 93, 23, 84, \ldots\}$. Implement a function `smallestIntegerSequence` that calculates the sequence $\{a_1, a_2, \ldots\}$ for a given real number x. What is the approximate growth rate of the a_k for a random x?

b) The $n!$ permutations of a list of n different elements e_k, $k = 1, \ldots, n$ can be conveniently numbered in the following way [159], [499]: The kth permutation is obtained by first writing k in the form $k = \sum_{k=1}^{n} c_k (n-k)!$, where each c_k is chosen nonnegative and as large as possible (factorial representation). Then, starting with the list $\ell = \{e_1, e_2, \ldots, e_n\}$, successively take the $(c_k + 1)$th element of the list, store it in a new list, and delete it from the original list. The list of the so-extracted list is the kth permutation. Implement a function `kthPermuta‑tion` which calculates the kth permutation of a given list.

d) Let $\mathcal{B}(o)$ be the integer counting the number of times the integer m occurs as a binomial coefficient $\binom{n}{k}$ for nonnegative n and k [514], [1]. Implement a function that calculates the $\mathcal{B}(o)$ pairs $\{n_j, k_j\}$ such that $\binom{n_j}{k_j} = o$ for a given o. Find an o, such that $\mathcal{B}(o) = 8$.

e) Write a one-liner, that, for a given irrational number α ($0 < \alpha < 1$) tests the first n digits in Smith's Sturmian word theorem [520], [86], [365]. Smith's Sturmian word theorem says that the sequence of 0's and 1's defined through $\lfloor (k+1)\alpha \rfloor - \lfloor k\alpha \rfloor$, $k = 1, 2, \ldots$ is identical to the sequence $g = \lim_{k \to \infty} g_k$, where $g_0 = 0$, $g_1 = 0^{c_1-1}1$, $g_k = g_{k-1}^{c_k} g_{k-2}$, where g_j^m means m repetitions of g_j, c_k is the kth digits in the ordinary continued fraction expansion of α (not counting the leading 0), and concatenation of digits in the g_k is implicitly understood.

6.^{L2} Galton Board, Ehrenfest Urn Model, Ring Shift, Longest Common Subsequence, Riffle Shuffles

a) Model a Galton Board [187] and compare the resulting distribution with the expected theoretical distribution.

b) Let two dogs A and B share n numbered fleas. Initially, m fleas are on dog A and $n - m$ fleas are on dog B. In each step, a random number r between 1 and n is chosen. As a result, the flea number r switches its dog. Model how the number of fleas approaches their equilibrium distribution when initially 10000 fleas are on dog A. When initially 100 fleas are on dog A, carry out 10^7 steps and analyze how often k fleas are on dog A. Compare with the theoretical distribution for the equilibrium [16], [157], [389], [203], [204], [446].

c) Given a list of n nonnegative integers, possible "moves" are

$$\{k_1, k_2, \ldots, k_j, k_{j+1}, \ldots, k_n\} \longrightarrow \{k_1, k_2, \ldots, k_j - 1, k_{j+1} + 1, \ldots, k_n\}$$

if $k_j \geq k_{j+1}$ and

$$\{k_1, k_2, \ldots, k_n\} \longrightarrow \{k_1 + 1, k_2, \ldots, k_n - 1\}$$

if $k_n \geq k_1$ [274], [211], [212].

For the starting list $\{m, 0, \ldots, 0\}$ of length n, for which $2 \leq m, n \leq 10$ does repeatedly carrying out moves leads to configurations so that no further moves are possible?

d) Consider the following sandpile model: Given a list of positive integers at each step, all elements that are greater than a given value c are randomly split into two integers, and added to its neighbors [277]. Assuming periodic boundary conditions, implement this process and visualize the position of changed elements for various integer lists and values of c.

e) Given two list of random integers $\{s_1, s_2, \ldots, s_n\}$ and $\{t_1, t_2, \ldots, t_n\}$ $(1 \leq s_k, t_k \leq c)$, two (possibly empty) subsequences $\{s_{i_1}, s_{i_2}, \ldots, s_{i_l}\}$ and $\{t_{j_1}, t_{j_2}, \ldots, t_{j_l}\}$ with $s_{i_k} = t_{j_k}$ of identical elements can be found. The length of the longest possible common subsequence $l_{n,n}$ can be found by the following recursive definition [258], [92]:

$$l_{i,j} = \max(l_{i-1,j-1} + \delta_{s_i,t_j}, l_{i-1,j}\, l_{i,j-1})$$
$$l_{i,0} = l_{0,j} = 0.$$

For two random strings of length $n = 10^4$ and $c = 2$, calculate and visualize $l_{k,k}/k$ $(k = 1, \ldots, n)$.

f) The popular riffle shuffle of a deck of cards can be modeled in the following way [10], [41], [551], [185], [531]: Split a deck of card into two piles. The probability that the two piles have k and $n - k$ cards is $\binom{n}{k} 2^{-n}$.

Then riffle the two piles together in such a way that the probability of taking a card from a given pile is proportional to the size of this pile.

A simple measure for the mixing-degree of a deck of cards is the number of rising subsequences contained in the deck [41], [531]. Without loss of generality, we assume the initial deck in the form $\{1, 2, \ldots, n\}$. A rising subsequence of a list of integers is the maximal length subsequence of successive values. (For example, the sequence $\{8, 11, 1, 4, 3, 5, 10, 6, 7, 2, 9, 12\}$ has the six rising subsequences $\{8, 9\}$, $\{11, 12\}$, $\{1, 2\}$, $\{4, 5, 6, 7\}$, $\{3\}$, and $\{10\}$.)

Carry out 1000 riffle shuffles for decks of length 52, 100, and 1000. Visualize the increase of the number of rising sequences of the repeatedly shuffled decks.

7.^{L1} Friday the 13th and Easter

a) What is the probability that the 13th day of a month is a Friday? The days of the week (Sunday = 1, Monday = 2, Tuesday = 3, Wednesday = 4, Thursday = 5, Friday = 6, Saturday = 7) are given in our (Gregorian) calendar by a date of the form month/day/year, which can be computed by ([309], [48], [32], [246], [181], [359])

$$weekday = \left(\left\lfloor \frac{23\,month}{9} \right\rfloor + day + year + 4 + \left\lfloor \frac{z}{4} \right\rfloor - \left\lfloor \frac{z}{100} \right\rfloor + \left\lfloor \frac{z}{400} \right\rfloor\right) \bmod 7 - \delta$$

with

$$\delta = \begin{cases} 2 & month \geq 3 \\ 0 & otherwise \end{cases}$$

$$z = \begin{cases} year - 1 & month < 3 \\ year + 0 & otherwise. \end{cases}$$

Here, $\lfloor x \rfloor$ is the largest integer contained in x.

b) What are the earliest and latest days (and the year in which they occur) for Easter? According to Gauss, Easter Sunday is the $i + j + 1$th day after the 21st of March, where

$$a = year \bmod 19$$
$$b = year \bmod 4$$
$$c = year \bmod 7$$
$$d = \left\lfloor \frac{(8[year/100] + 13)}{25} \right\rfloor - 2$$
$$e = \left\lfloor \frac{year}{100} \right\rfloor - \left\lfloor \frac{year}{400} \right\rfloor - 2$$
$$f = (15 + e - d) \bmod 30$$
$$g = (6 + e) \bmod 7$$
$$h = (f + 19\,a) \bmod 30$$
$$i = \begin{cases} 28 & h = 29 \\ 27 & h = 28 \text{ and } a \geq 11 \\ h & \text{otherwise} \end{cases}$$
$$j = (2\,b + 4\,c + 6\,i + g) \bmod 7.$$

(See [543], [61], [372], [216], [358], [50], [560], and [248] and on-line at http://www.uni-bamberg.de/~ba1lw1/fkal.html). Again, $\lfloor x \rfloor$ is the largest integer contained in x.

For more on calendars, see also the package `Miscellaneous`Calendar`` of I. Vardi, [569] and [140].

8.$^{\text{L1}}$ Number of Lattice Points, Binomial Digits, Decreasing Partitions, Partition Moments

a) Determine the number n and the distance (from the origin) of all grid points of a d-dimensional simple cubic lattice [498] inside of a sphere of given radius r for moderate-sized radii, and plot the result $n(r)$ for various d.

b) Write a one-liner that, for a given positive integer R, counts the number $p(R)$ of relatively prime integers m and n such that $m^2 + n^2 \leq R^2$ [611]. Visualize $p(R)$ for $R \leq 1000$. Asymptotically we have $p(R) = 6/\pi R + O(R^\alpha)$ for $R \to \infty$ [67], [336]. Estimate a value of α from the calculated data.

c) Analyze the mean and the fluctuations of the digits sum (in base 10) of $\binom{n}{\lfloor n/2 \rfloor}$ for $1 \leq n \leq 10000$. From this analysis, does one expect the fluctuations asymptotically to have a normal distribution?

d) Let $\lambda = \{n_1, n_2, \ldots, n_m\}$ be a strictly decreasing partition of the integer μ, meaning $n_k > n_{k+1}$ and $\sum_{k=1}^{m} n_k = \mu$. For such a partition λ, let $\rho(\lambda, x) = \sum_{k=1}^{\infty} \theta(n_k - x)$. For $\mu = 100$, form the average of the $\rho(\lambda, x)$ over all partitions λ of μ and compare this average with the function $g_\mu(x) = \ln(1 + \exp(-\sigma x))/\sigma$ where $\sigma = \pi/(2\,(3\,\mu)^{1/2})$ [604], [362], [573], [574].

e) Let $p_n(m)$ be the number of partitions of the positive integer n into m summands (ignoring the order of the summands). $p_n(m)$ obeys the recursion $p_n(m) = \sum_{j=1}^{\min(n-m,m)} p_{n-m}(j)$ [19].

For the moments $\mu_n^{(k)} = \sum_{j=1}^{m} m^k\, p_n(m)$, we have the following asymptotics for large n [465], [266]:

$$\mu_n^{(k)} \underset{n\to\infty}{\sim} \frac{1}{4\sqrt{3}\,n} \exp\left(\pi \sqrt{\frac{2\,n}{3}}\right) \left(\frac{\sqrt{6\,n}}{\pi}\right)^k \Sigma_k\left(\ln\left(\frac{\sqrt{6\,n}}{\pi}\right) + \gamma, \, 1!\,\zeta(2), \, \ldots, \, (k-1)!\,\zeta(k)\right)$$

where $\zeta(z)$ is the Riemann Zeta function (in *Mathematica* `Zeta[z]`) and $\Sigma_n(x_1, x_2, \ldots, x_n)$ is

$$\Sigma_n(x_1, x_2, \ldots, x_n) = \sum_{j_1, j_2, \ldots, j_n} \frac{n!}{j_1! \, j_2! \ldots j_n!} \left(\frac{x_1}{1!}\right)^{j_1} \left(\frac{x_2}{2!}\right)^{j_2} \cdots \left(\frac{x_n}{n!}\right)^{j_n}$$

and the sum extends over all n-tuples $\{j_1, j_2, \ldots, j_n\}$, such that $\sum_{i=1}^{n} i\, j_i = n$. Compare the exact values of $\mu_n^{(k)}$ with the asymptotic values for $n = 1000$ and $k = 0, \ldots, 10$.

9.[L1] 15 and 6174

What do the following pieces of code compute?

a)

```
Fifteen[n_Integer?(#>1&)] :=
 FixedPoint[Plus @@ Flatten[IntegerDigits /@ Divisors[#]]&, n]
```

b)

```
TwoOrThreeOrFourOrFiveOrSeven[n_Integer?(#>1&)] :=
  FixedPoint[Plus @@ Flatten[IntegerDigits /@
        Flatten[Apply[Table[#1,{#2}]&, FactorInteger[#], {1}]]]&, n]
```

c)

```
f6174[n_Integer?(0<=#<=9999&)] :=
 Drop[FixedPointList[(#1.#2 - #1.Reverse[#2]& @@
  {10^Range[0, Length[#] - 1], #}&[
      Sort[IntegerDigits[#, 10, 4]]])&, n], -1]
```

Generalize the process carried out by `f6174` an arbitrary base b.

10.[L3] Selberg Identity, Kluyver Identity, Goodwyn Property, Guiasu Formula, Prime Sums, Farey–Brocot Sequence, Divisor Sum Identities

a) Write a one-liner that proves the Selberg identity [20]

$$\Lambda(n) \ln n + \sum_{d|n} \Lambda(n) \Lambda\left(\frac{n}{d}\right) = \sum_{d|n} \mu(n) \ln^2\left(\frac{n}{d}\right)$$

for a given $n \geq 0$ (the sum goes over all divisors d of n), where $\Lambda(n)$ is the Mangoldt function defined via

$$\Lambda(n) = \begin{cases} \ln n & n = p^m \text{ for some prime } p \\ 0 & \text{otherwise.} \end{cases}$$

for a given n by direct calculation of both sides.

b) Show numerically for some positive integers n that the so-called Kluyver identities [318] hold:

$$\sum_{\substack{v \\ \gcd(v,n)=1}} \cos(2\pi v/n) = \mu(n)$$

$$\sum_{\substack{v \\ \gcd(v,n)=1}} 2\sin(2\pi v/n) = e^{\Lambda(n)}$$

where $\mu(n)$, $\Lambda(n)$ are the Möbius function and the Mangoldt function (from part a).

c) Verify for some maximal denominator n the following property: Every member (excluding the first and the last term) of the ordered Farey sequence of fractions less than 1 is equal to (see [144], [341], and [267]):

$$\frac{\text{numerator}(leftNeighbor) + \text{numerator}(rightNeighbor)}{\text{denominator}(leftNeighbor) + \text{denominator}(rightNeighbor)}$$

Make a picture of the Ford circles [177] and [47] of a Farey sequence. A Ford circle of the fraction $\frac{h}{k}$ is a circle with radius $1/(2\,k^2)$ and center $(h/k,\, 1/(2\,k^2))$.

d) The elements $\left\{\mathcal{F}_k^{(n)}\right\}_{k=0,\ldots,2^n}$ of the Farey–Brocot sequence $\mathcal{F}^{(n)}$ [364], [576], [200] of order n are recursively defined as

$$\mathcal{F}_{2k+1}^{(n)} = \frac{\text{numerator}\left(\mathcal{F}_k^{(n)}\right) + \text{numerator}\left(\mathcal{F}_k^{(n)}\right)}{\text{denominator}\left(\mathcal{F}_k^{(n)}\right) + \text{denominator}\left(\mathcal{F}_k^{(n)}\right)}, \quad k = 0, 2, \ldots, 2^{n-1}$$

$$\mathcal{F}_{2k}^{(n)} = \mathcal{F}_k^{(n-1)}, \quad k = 0, 1, 2, \ldots, 2^{n-1}$$

$$\mathcal{F}_0^{(0)} = 0$$

$$\mathcal{F}_1^{(0)} = 1$$

This means that the sequence $\mathcal{F}^{(n+1)}$ is constructed from the sequence $\mathcal{F}^{(n)}$ by inserting mediants between all consecutive pairs and numbering (lower index) the resulting points consecutively. Given an interval $[a, b] \subset [0, 1]$, express this interval as the smallest union of disjoint intervals $\left[\mathcal{F}_{k_j}^{(j)}, \mathcal{F}_{k_j+1}^{(j)}\right]$. Calculate the number of Farey–Brocot intervals $\left[\mathcal{F}_{k_j}^{(j)}, \mathcal{F}_{k_j+1}^{(j)}\right]$ needed to cover the uniform subdivision of $[0, 1]$ in 2^{12} intervals of length 2^{-12}.

e) Implement the following relation for the calculation of the number of primes [228], [229] less than x in an efficient manner:

$$\pi(x) = -\sum_{k=1}^{\lfloor \log_2 x \rfloor} \mu(k) \sum_{n=2}^{\left\lfloor \sqrt[k]{x} \right\rfloor} \mu(n)\, \Omega(n) \left\lfloor \frac{\sqrt[k]{x}}{n} \right\rfloor.$$

Here $\lfloor x \rfloor$ is the integer part of x, $\mu(n)$ is the Möbius μ function, and $\Omega(n)$ is the number of prime factor of n. Is there a big advantage of an efficient implementation in comparison to the direct one?

f) The nth prime p_n (n even) can be represented as the addition and subtraction of all earlier primes in the form $p_n = \pm 1 \pm p_1 \pm \ldots \pm p_{n-1}$. Find such formulas for all primes p_2, $p_4 \ldots$ less than 100 [486], [144].

g) Define the functions [90]

$$Q_d(z) = \sum_{k=1}^{\infty} \sigma_d(k)\, z^k$$

where $\sigma_d(k)$ are the divisor sums. Visualize some $Q_d(z)$ as a function of the complex variable z.

For some small integers α_k, β_k, and γ_k, relations of the following form exists:

$$\sum_{k=0}^{3} a_{\alpha_k}\, Q_{\alpha_k}(z) = \sum_{k=0}^{3} b_{\beta_k,\gamma_l}\, Q_{\beta_k}(z)\, Q_{\gamma_l}(z).$$

Use an "experimental mathematics" approach to find such relations.

11.^{12} Choquet Approximation, Magnus Expansion

a) Given two fractions a_1/b_1 and a_2/b_2, then, for the fraction $(a_1 + a_2)/(b_1 + b_2)$, we have $a_1/b_1 < (a_1 + a_2)/(b_1 + b_2) < a_2/b_2$ [174]. Use this relation to construct a rational approximation of N[Pi, 50], N[E, 50], N[5^(1/3), 50], and N[Sqrt[2], 50], starting with the nearest integers. Visualize how the upper and lower bounds move toward the given irrational number.

b) The solution of a system of linear, homogeneous differential equations

$$\mathbf{x}(t) = \mathbf{A}(t)\, \mathbf{x}(t)$$
$$\mathbf{x}(0) = \mathbf{x}_0$$

where $\mathbf{x}(t) = \{x_1(t), \ldots, x_n(t)\}$ and $\mathbf{A}(t)$ is a $n \times n$ matrix with elements $a_{i\,j}(t)$ can be expressed as

$$\mathbf{x}(t) = e^{\Omega(t)}.\mathbf{x}_0.$$

The matrix $\Omega(t)$ can be calculated in the following way (Magnus expansion [284], [374], [263], [584], [589], [65], [445], [404], [283], [166], [401], [161], [285], [165], [286]):

$$\Omega(t) = \sum_{k=1}^{\infty} \Omega_k(t)$$

$$\Omega_k(t) = \sum_{j=0}^{n-1} \frac{B_j}{j!} \int_0^t S_n^{(j)}(\tau)\, d\tau$$

$$\mathbf{S}_n^{(0)} = \mathbf{A}(t)$$

$$\mathbf{S}_n^{(j)} = \begin{cases} 0 & n > 1 \\ \sum_{m=1}^{n-j} \left(\Omega_k(t).\mathbf{S}_{n-m}^{(j-1)} - \mathbf{S}_{n-m}^{(j-1)}.\Omega_k(t) \right). \end{cases}$$

How many terms $\Omega_k(t)$ are needed to obtain 20 correct digits for $\{x_1(1), x_2(1)\}$ for the solution of the system

$$x_1'(t) = t\, x_1(t) + x_2(t)$$
$$x_2'(t) = x_1(t) + t^2\, x_2(t)$$

with initial conditions $x_1(0) = x_2(0) = 1$?

12.^{L1} Rademacher Identity, Goldbach Problem, Optical Factoring

a) The so-called Rademacher identity ([450], [451], [447], [19]) for the partition function is (this is a generalization of the asymptotic expansion given in Section 2.3):

$$p(n) = \frac{1}{\sqrt{2}\,\pi} \sum_{k=1}^{\infty} A_k(n)\,\sqrt{k}\,\frac{d}{dn}\left(\frac{\exp\left(\pi\sqrt{\frac{2}{3}}\,\sqrt{n-\frac{1}{24}}\right)}{\sqrt{n-\frac{1}{24}}}\right)$$

$$A_k(n) = \sum_{\substack{h=1 \\ \gcd(h,k)=1}}^{n} \omega_{h,k}\,\exp\left(-\frac{2\pi i h n}{k}\right)$$

$$\omega_{h,k} = \exp\left(\pi i \sum_{m=1}^{k-1} \frac{m}{k}\left(\frac{h m}{k} - \left\lfloor\frac{h m}{k}\right\rfloor - \frac{1}{2}\right)\right).$$

Here, $\lfloor x \rfloor$ denotes the greatest integer less than or equal to x.

How many terms of the k sum must be included to get, after rounding to the nearest integer, the correct result for $p(1000)$?

b) Write a one-liner that, for a given positive integer n, returns a list of all pairs of prime number indices $\{k_1,\ k_2\}$ such that $p_k + q_k = n$, where p_{k_1}, and p_{k_2} are prime numbers and $(p_{k_1} \le p_{k_2})$, $k = 1, \ldots, m(n))$. Calculate all such pairs for $n = 10^6$. Can one calculate the number of prime pairs for $n = 10^9$? Compare the result with the asymptotic conjecture [239], [220], [582], [142], [128], [33], [167], [288]:

$$m(n) \underset{n\to\infty}{=} \left(\prod_{k=2}^{\infty} \frac{p_k(p_k-2)}{(p_k-1)^2}\right)\left(\prod_{k=2,\,p_k|n}^{\infty} \frac{p_k-1}{p_k-2}\right)\int_2^n \frac{1}{\ln(x)^2}\,dx.$$

c) Consider the sums $s_{n,m}(x) = \sum_{k=(1-n)/2}^{(n-1)/2} \exp(i\pi(k-x)^2/m)$ for odd n and m. Conjecture a relation between the shape of the function $s_{n,m}(x)$ as a function of the real variable x and the property that m is a divisor of n.

13.^{L1} Bernoulli and Fibonacci Numbers, Zeckendorf Representation

a) A number of identities involving Bernoulli numbers B_n can be conveniently written in a symbolic way, where after expanding powers they are replaced by suffices $B^n \to B_n$ (see, for instance, [477], [455], [197], [77], [231], [458], [99], and [474] for details on this kind of symbolic rewriting). One of these identities is ($n > 1$) (we use $\doteq$ to emphasize that this identity is a not a literal one): $B_n \doteq (B+1)^n$. Use this identity to implement the calculation of Bernoulli numbers.

In a similar way, the Bernoulli polynomial $B_n(x)$ can be represented in the following manner: $B_n(x) \doteq (B+x)^n$. Use this identity to implement the calculation of the first Bernoulli polynomials.

b) A number of identities involving Fibonacci numbers F_n can be conveniently written in a symbolic way, where after expanding powers they are replaced by suffices $F^n \to F_{n+1}$, (see, for instance, [144]). One of these identities is ($n, m > 0$) (we again use $\doteq$ to emphasize that this identity is a symbolic one):

$$F^{n+m} \doteq F^{n-m}(F+1)^m.$$

Check this identity for $0 \le n, m \le 12$.

c) Any nonnegative integer n can be uniquely written as a sum of Fibonacci numbers (Zeckendorf representation) [609], [503]:

$$n = \sum_{k=1}^{s_n} F_{l_k^{(n)}}.$$

Here $l_1 \geq 2$ and $l_{k+1}^{(n)} \geq l_k^{(n)} + 2$. Implement a function `ZeckenDorfDigits` that for a given n calculates the list of the $l_k^{(n)}$. Calculate `ZeckenDorfDigits[`10^{1000}`]`.

The function

$$\Sigma_n = \frac{1}{n} \sum_{k=1}^{n} \sum_{l=1}^{k} (-1)^{s_l}$$

has the asymptotic form

$$\Sigma_n = G(\log_\phi(n)) + O\!\left(\frac{\sqrt{\ln(n)}}{n}\right)$$

where G is a periodic function (period 1) [146]. Use the Zeckendorf representation of the first 100000 integers to visualize the function G.

Every real number z, $0 < z < 1$ can be uniquely written as (the Sylvester–Fibonacci expansion) [182]

$$z = \sum_{k=3}^{\infty} \frac{c_k}{F_k}$$

where $c_k \in \{0, 1\}$. Implement a function `SylvesterFibonacciDigits[`z, n`]` that calculates a list of the first n nonvanishing "digits" c_k. Calculate `SylvesterFibonacciDigits[1/Pi, 100]`.

14.^{12} lcm–gcd Iteration, Generalized Multinomial Theorem, Ramanujan τ Function

a) Starting with a list of integers $\{l_1, l_2, \ldots, l_n\}$, one randomly selects a pair of integers $\{l_i, l_j\}$ from this list and replaces l_i by $\text{lcm}(l_i, l_j)$ and l_j by $\gcd(l_i, l_j)$. Repeating this process with all pairs until a fixed point is reached leads uniquely to a set of integers, such that one number of every pair of integers from the resulting list divides the other. Implement the construction of this set of integers.

b) For a positive integer n, integers p_k, and $1 \leq k \leq m$, $m \geq 2$, Hurwitz gave the following generalization of the multinomial theorem [278], [469], [44], [610]:

$$A_n(x_1, x_2, \ldots, x_m, p_1, p_2, \ldots, p_m) = \sum_{k=0}^{n} \binom{n}{k} k! \, (k+x) \, A_{n-k}(x_1 + k, x_2, \ldots, x_m, p_1 - 1, p_2, \ldots, p_m).$$

The functions $A_n(x_1, x_2, \ldots, x_m, p_1, p_2, \ldots, p_m)$ are defined via the formula

$$A_n(x_1, x_2, \ldots, x_m, p_1, p_2, \ldots, p_m) = \sum_{k_1, k_2, \ldots k_m = 0}^{n} (n; k_1, k_2, \ldots, k_m) \prod_{j=1}^{m} (x_j + k_j)^{k_j + p_j}.$$

Implement the calculation of the $A_n(x_1, x_2, \ldots, x_m, p_1, p_2, \ldots, p_m)$. Check the validity of the above identity for some values of n and $p_1, p_2, \ldots, p_m$.

c) Let the Ramanujan τ function $\tau(n)$ be implicitly defined by

$$\sum_{k=0}^{\infty} \tau(n)\, z^n = z \prod_{k=1}^{\infty} (1 - z^k)^{24}.$$

$\tau(n)$ fulfills the following identity [21], [232]:

$$\tau(m)\, \tau(n) = \sum_{d | \gcd(m,n)} d^{11}\, \tau\!\left(\frac{m\, n}{d^2}\right).$$

Verify this identity for all m, n, and $m\, n \le 100$.

d) In [154], the following nice formula for the Ramanujan τ function was given:

$$\tau(n) = \sum_{\substack{a,b,c,d,e=-\infty \\ C}}^{\infty} \frac{(a-b)(a-c)(a-d)(a-e)(b-c)(b-d)(b-e)(c-d)(c-e)(d-e)}{1!\,2!\,3!\,4!}.$$

The summation variables a, b, c, d, $e \in \mathbb{Z}$ have to fulfill the additional conditions C:

$$a + b + c + d + e = 0$$
$$a^2 + b^2 + c^2 + d^2 + e^2 = 10\, n$$
$$\{a,\ b,\ c,\ d,\ e\} \bmod 5 = \{1, 2, 3, 4, 0\}.$$

Use this formula to calculate $\tau(1000)$. How many 5-tuples $\{a,\ b,\ c,\ d,\ e\}$ contribute to the sum? (The Ramanujan τ function is implemented in the package `NumberTheory`Ramanujan`` as `RamanujanTau`.)

The Ramanujan τ function can be expressed through divisor sums in the following form [396], [342]:

$$\tau(n) = \sum_{k} \beta_{c_k}\, \sigma_{c_k}(n) + \sum_{k,l} \beta_{c_k,c_l} \sum_{i=1}^{n-1} \sigma_{c_k}(i)\, \sigma_{c_l}(n-i).$$

Here the β_{c_k}, β_{c_k,c_l} are integers and the c_k are small integers. Find such a representation.

Still another (recursive) representation of the Ramanujan τ function is $\tau(n) = F_{n-1}(12, \tau(2), \ldots, \tau(n-1))$ [426]. Here the multivariate polynomials F_n are defined by

$$F_n(x_1, x_2, \ldots, x_n) = -\frac{2}{n}\, x_1\, \sigma_1(n) + \sum_{\substack{m_1,m_2,\ldots,m_{n-1} \ge 0 \\ m_1+2m_2+\ldots+(n-1)m_{n-1}=n}} (-1)^{\sum_{j=1}^{n-1} m_j} \frac{(-1 + \sum_{j=1}^{n-1} m_j)!}{\prod_{j=1}^{n-1}(m_j!)} \prod_{j=1}^{n-1} x_{j+1}^{m_j}.$$

Use this formula to calculate $\tau(2)$, $\ldots$, $\tau(25)$.

15.¹³ Cross-Number Puzzle

The *Berlin Intelligencer* (special issue of the Springer-Verlag publication *Mathematical Intelligencer* for the International Congress of Mathematicians, 1998; published in cooperation with the DMV) gave this cross-number puzzle:

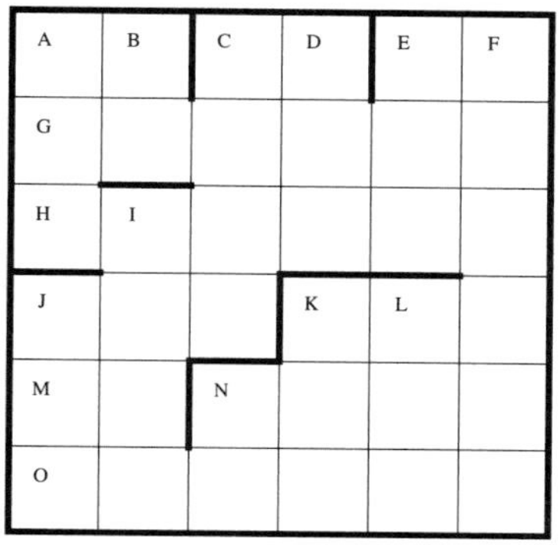

A	B	C	D	E	F
G					
H	I				
J			K	L	
M		N			
O					

HORIZONTAL

A) Prime number
B) The backnumber is a square
E) Prime number
G) The digits of this number are distinct
H) G)-horizontal - O)-horizontal
J) Square root of O)-horizontal
K) The cross-product is a square
M) Prime number
N) Multiple of E-vertical
O) The cross-sum is a square

VERTICAL

A) The backnumber divides G)-horizontal
B) A prime number to the power 4
C) The cross-sum equals C)-horizontal
D) M)-horizontal plus the backnumber of L)-vertical
E) The cross-product is prime
F) K)-horizontal times C)-vertical
I) Multiple of the backnumber of D)-vertical
J) The backnumber is a multiple of the backnumber of A)-vertical
K) A prime number to the power 5
L) Multiple of M)-horizontal
N) Prime number

The backnumber of a number is the number read backwards. Example: 5492 is the backnumber of 2945. The cross-sum is the sum of the digits and the cross-product is the product of the digits. There are no zeros.
Solve the cross-number puzzle using *Mathematica*.

16.¹² Cyclotomic Polynomials, Generalized Bell Polynomials

a) The cyclotomic polynomials $C_n(z)$ ($n \in \mathbb{N}$) satisfy the following relations for prime p [491]:

$$C_p(z) = \sum_{k=0}^{p-1} z^k$$

$$C_{n\,p^k}(z) = C_{n\,p}\left(z^{p^{k-1}}\right)$$

$$C_{n\,p}(z) = \frac{C_n(z^p)}{C_n(z)}$$

Together with $C_0(z) = 1$ and $C_1(z) = z - 1$, these relations completely define the cyclotomic polynomials. Based on the formulas given, implement the calculation of the first 100 cyclotomic polynomials.

b) The nth generalized Bell polynomial of order m, $Y_n^{(m)}(f_{1,1}, f_{2,1}, \ldots, f_{m+1,1}; f_{1,2}, f_{2,2}, \ldots, f_{m+1,2}; \ldots; f_{1,n}, f_{2,n}, \ldots, f_{m+1,n})$ with $n = 1, 2, \ldots, m = 2, 3, \ldots$ is recursively defined through [416]

$$Y_n^{(m)}(f_{1,1}, f_{2,1}, \ldots, f_{m+1,1}; f_{1,2}, f_{2,2}, \ldots, f_{m+1,2}; \ldots; f_{1,n}, f_{2,n}, \ldots, f_{m+1,n}) =$$

$$\sum_{P(n)} \frac{n!}{r_1!\,r_2!\,\ldots\,r_n!}\, f_{1,r} \left(\frac{Y_1^{(m-1)}(f_{2,1}, \ldots, f_{m+1,1})}{1!} \right)^{r_1}$$

$$\left(\frac{Y_2^{(m-1)}(f_{2,1}, \ldots, f_{m+1,1}; f_{2,2}, \ldots, f_{m+1,2}; \ldots; f_{1,n})}{2!} \right)^{r_2} \times \ldots \times$$

$$\left(\frac{Y_n^{(m-1)}(f_{2,1}, \ldots, f_{m+1,1}; f_{2,2}, \ldots, f_{m+1,2}; \ldots; f_{2,n}, \ldots, f_{m+1,n})}{n!} \right)^{r_n}$$

with the initial term

$$Y_n^{(1)}(f_{1,1}, f_{2,1}; f_{1,2}, f_{2,2}; \ldots; f_{1,n}, f_{2,n}) = \sum_{P(n)} \frac{n!}{r_1!\,r_2!\,\ldots\,r_n!}\, f_{1,r} \left(\frac{f_{2,1}}{1!} \right)^{r_1} \left(\frac{f_{2,2}}{2!} \right)^{r_2} \times \ldots \times \left(\frac{f_{2,n}}{n!} \right)^{r_n}.$$

Here the sums extend over the set of all partitions $P(n) = \{\{r_1, r_2, \ldots, r_n\} \mid 1\,r_1 + 2\,r_2 + \ldots + n\,r_n = n\}$ of the integer n and $r = r_1 + r_2 + \ldots + r_n$. Implement the calculation of the Bell polynomials using the given recursion relation.

The generalized Bell polynomials allow to express derivatives of nested functions via

$$\frac{d\,f_1(f_2(\ldots f_m(x)))}{d\,x^n} = Y_n^{(m-1)}\left(\tilde{f}_1', \tilde{f}_2', \ldots, \tilde{f}_m'; \tilde{f}_1'', \tilde{f}_2'', \ldots, \tilde{f}_m''; \ldots; \tilde{f}_1^{(n)}, \tilde{f}_2^{(n)}, \ldots, \tilde{f}_m^{(n)}\right)$$

where we use the abbreviation $\tilde{f}_j^{(k)} = f_j^{(k)}(f_{j+1}(f_{j+2}(\ldots f_m(x))))$. Check this relation for various n and m.

17.¹² Factorization of $n!$ into n Factors, Bin Packing, Composition Multiplicities

a) Write a function that can construct all possible factorizations of $n!$ in exactly n factors. This means a factorization of the form $n! = \prod_{j=1}^{n} p_j^{m_j}$, where all p_j are prime numbers. For example, for $10! = 2^8\,3^4\,5^2\,7$, we have

$10! = 2^1\,2^1\,2^1\,2^1\,2^1\,2^1\,2^2\,3^4\,5^2\,7^1 = 2^1\,2^1\,2^1\,2^1\,2^1\,2^1\,2^3\,3^1\,3^3\,5^2\,7^1 = 2^1\,2^1\,2^1\,2^1\,2^1\,2^1\,2^3\,3^2\,3^2\,5^2\,7^1 = \ldots$ Try to avoid using the built-in function `FactorInteger`. How many possible factorizations of this kind exist for 25!?

b) Rods of length $1/3$, $2/3$, and 1 are randomly generated (each with probability $1/3$) and stuffed into boxes of length 1 (the so-called on-line bin packing problem [348]). The stuffing happens in such a way that, whenever possible, a new rod is put in a box that has still enough space and, if possible, a box is filled. After the generation of n rods, the average cumulative rod length λ_n in not completely filled boxes is [448], [319]

$$\lambda_n = \frac{1}{3^{n+1}} \left((2n+1) \binom{n, 3}{n} + (2n+1) \binom{n-1, 3}{n-1} + \binom{n-1, 3}{n-2} - 3^n \right).$$

Here $\binom{n, 3}{k}$ is the trinomial coefficient

$$\binom{n, 3}{k} = [x^k] \left((1 + x + x^2)^n \right).$$

Model the stuffing process for $n = 1000$ by carrying out 10^4 random realizations and compare the average cumulative rod length with the theoretical value. Model the generation of 10 trials for $n = 10^6$ rods and visualize how the average evolves as a function of n.

c) A composition of the positive integer n into p parts is a list of positive integers $\{k_1, k_2, \ldots, k_p\}$ that sum to n [260], [261], [262], [253] (meaning a partition where the order of the summands matters). Write a program that, for a given integer n, forms all possible compositions. The multiplicity count of a composition $\lambda_j(\{k_1, k_2, \ldots, k_p\})$ is the number of k_l having the value j. Calculate the cumulative multiplicity count for all compositions of $n = 50$ in a memory-efficient way.

18.^{L2} Level Spacing, Subset Sums

a) The following function `levelSpacings` calculates the distribution of the differences of the sorted list of sums of two squares $k^2 + l^2 \le n$.

```
levelSpacings[n_] := {First[#], Length[#]}& /@ Split[
  Sort[-Apply[Subtract, Partition[Sort[Flatten[
    Table[k^2 + l^2, {k, n},
      {l, 1, Floor[Sqrt[n^2 - k^2]]}]]], 2, 1], {1}]]];
```

Rewrite this function in a form that uses less memory. Calculate `levelSpacings[1000]`, and visualize the result.

b) Choose n random integers $k_1, k_2, \ldots, k_n$ from the interval $(1, N)$ and form all possible $2^n - 1$ sums of these integers containing each integer at most once (meaning $k_1, k_2, \ldots, k_n, k_1 + k_2, k_1 + k_3, \ldots, k_{n-1} + k_n, \ldots, k_1 + k_2 + \ldots + k_n$). Analyze the frequency of the values of the sums and compare with the theoretical distribution for large n (s is the value of the sum) [483], [484]

$$w(s) = \frac{1}{2^n - 1} \frac{\exp\left(\sum_{j=1}^n \ln(1 + e^{\beta k_j}) + \beta \sum_{j=1}^n k_j(1 + e^{\beta k_j})^{-1} \right)}{\sqrt{2\pi \sum_{j=1}^n k_j^2 (1 + e^{\beta k_j})^{-1} (1 + e^{-\beta k_j})^{-1}}}$$

where β is implicitly defined through $s = \sum_{j=1}^n k_j(1 + e^{\beta k_j})^{-1}$. Use $n \ge 20$ and $N \ge 100$.

Solutions

1. DivisorSigma, **Primes, Maximum Formula,** StirlingS2, LegendreSymbol, GCD**-Iterations, Isenkrahe Algorithm,**
Multinomial Terms, Prime Divisors

a) Here is a possible implementation. We factorize the given integer n with FactorInteger and then calculate the numerators and denominators of the factors. The product itself is formed at the end.

```
In[1]:= myDivisorSigma[k_, n_] :=
         Times @@ ((Apply[Power[#1, (#2 + 1) k]&, #, {1}] - 1)/
             ((First /@ #)^k - 1)&[FactorInteger[n]])
```

Here is a quick check for myDivisorSigma.

```
In[2]:= Table[myDivisorSigma[i, j] == DivisorSigma[i, j],
         {i, 100}, {j, 100}] // Flatten // Union
Out[2]= {True}
```

b) The interesting part of this exercise is the inner sum that extends over all n-tuple of indices with the restriction that they are ordered. We generate these n-tuples by first calculating iterators with Join[{{j[1], 1, n}}, Table[{j[i], j[i - 1] + 1, n}, {i, 2, r}]] and then use these iterators with the body Product[p[j[i]], {i, r}]. After rewriting the given definition in the form $p_n = p_n(p_1, \ldots, p_{n-1})$, we can implement the following code. (Because we use j explicitly and Evaluate inside Table, the variable r cannot be scoped and so, we enclose the whole routine in a Module.)

```
In[1]:= p[1] = 2;

         p[n_Integer?Positive] := p[n] =
         Module[{j, r},
           (* calculate all previous values *) Array[p, n - 1];
           Floor[N[1 - Log[2, 1/2 + Plus @@ Flatten[Table[(-1)^r/
             (2^Flatten[(* the inner sum *)
              Table[Evaluate[Product[p[j[i]], {i, r}]],
                (* the various iterators *)
                Evaluate[Sequence @@ Join[{{j[1], 1, n - 1}},
                  Table[{j[i], j[i - 1] + 1, n - 1}, {i, 2, r}]]]]],
                  r - 1] - 1), {r, n - 1}]]]]]]]
```

And here are the first terms of the sequence.

```
In[3]:= {p[1], p[2], p[3], p[4], p[5], p[6], p[7]}
Out[3]= {2, 3, 5, 7, 11, 13, 17}
```

This sequence defines the prime numbers.

c) We start by implementing the function min[*list*, *parts*], which gives the minimum of the parts *parts* of the list *list*. As a side effect, we increase a counter *counter*.

```
In[1]:= min[list_, parts_] := (counter = counter + 1; Min[list[[parts]]])
```

The function sum[*list*, n] carries out the summation for each of the nested sums of depth n.

```
In[2]:= sum[list_, n_] :=
         Module[{C},
           With[{parts = Table[C[i], {i, n}]},
             Sum[min[list, parts], (* the iterator for the increasing k_i *)
               Evaluate[Sequence @@ Table[{C[i], If[i == 1, 1, C[i - 1] + 1],
                 Length[list]}, {i, n}]]]]]
```

Here, the maximum of the list {1, 2, 3, 4, 5, 6, 7, 8, 9, 10, 11, 12, 13, 14, 15} is calculated.

```
In[3]:= testList = Range[15]; counter = 0;
         Sum[(-1)^(k + 1) sum[testList, k], {k, Length[testList]}]
Out[4]= 15
```

In the last calculation, $2^{15} - 1 = 32767$ calls to min were carried out.

```
In[5]:= {counter, 2^15 - 1}
Out[5]= {32767, 32767}
```

d) The implementation of the formula is straightforward. We generate a body and an iterator for `Sum` with $n - m$ terms and then evaluate the sum.

```
In[1]:= stirlingS2[n_, m_] :=
        Sum[Evaluate[Product[i[k], {k, n - m}]],
            Evaluate[Sequence @@ Table[{i[k], If[k == 1, 1, i[k - 1]]},
                {k, n - m}]]] /; n >= m
```

The so-defined function `stirlingS2` gives the same results as `StirlingS2`. Here is a quick check for small n.

```
In[2]:= Union[Flatten[Table[stirlingS2[n, m] - StirlingS2[n, m],
                    {n, 1, 12}, {m, 1, n}]]]
Out[2]= {0}
```

e) We start by expressing the property of being prime in a formula. p is a prime if and only if $(\exp(2\pi i (p - 1)!/p) - 1) - (\exp(2\pi i/p) - 1) = 1$ [464]. Here is a quick check of this formula.

```
In[1]:= Table[If[PrimeQ[p], # === 1, # =!= 1]&[
            (Exp[2Pi I (p - 1)!/p] - 1)/(Exp[-2Pi I /p] - 1)],
         {p, 2, 1000}] // Union
Out[1]= {True}
```

Now, we have to express the property that p divides a. This means $\frac{a}{p}$ is an integer and from this follows that the difference between $\frac{a}{p}$ and $\lfloor \frac{a}{p} \rfloor$ is zero if and only if p divides a. So, we are led to the expression $1 - \delta_{\frac{a}{p} - \lfloor \frac{a}{p} \rfloor}$. Here is a quick check.

```
In[2]:= Table[If[IntegerQ[a/p], 1, 0] == KroneckerDelta[a/p - Floor[a/p]],
         {a, 10}, {p, 10}] // Flatten // Union

Out[2]= {True}
```

Now, the last condition—if an n exists, such that $k^2 = a \bmod p$—has to be dealt with. We have the periodicity $(k + p)^2 \bmod p = k^2 \bmod p$. This means we have to test the numbers $k = 1, \ldots, p$. We select a match via $\delta_{k^2 \bmod p}$. Because more than one match might happen, we sum them and take the sign. The result of the sum is either 0 or 1. We map this result to ± 1 using the function $x \to 2x - 1$. Putting all of the above together, we have the following code for our function `LegendreSymbol`.

```
In[4]:= LegendreSymbol[a_Integer, p_Integer] :=
        (1 - KroneckerDelta[a/p - Floor[a/p]])*
        (2 Sign[Sum[KroneckerDelta[Mod[k^2, p], Mod[a, p]],
                    {k, p}]] - 1) /; (* restrict argument *)
                    p > 2 && a > 0 && (* p is prime *)
                    (Exp[2Pi I (p - 1)!/p] - 1)/(Exp[-2Pi I/p] - 1) == 1
```

Here is a check that the function works as advertised.

```
In[5]:= Table[LegendreSymbol[n, Prime[k]] - JacobiSymbol[n, Prime[k]],
            {k, 2, 100}, {n, 100}] // Flatten // Union
Out[5]= {0}
```

Using some of the "strange" analytic functions from Exercise 7 of Chapter 2 of the Programming volume [554], we could now continue to construct `LegendreSymbol` as an "analytic" function containing mainly differential algebraic constants.

f) We use the recursive code given in Section 2.1 and increase the value of a local counter at each recursive call. The function `GCDSteps` returns a list of two elements, the first being the greatest common divisor and the second being the number of steps carried out in Euclid's algorithm. (As mentioned above, we could use `ContinuedFraction` here.)

```
In[1]:= GCDSteps[p_Integer?Positive, q_Integer?Positive] :=
        Module[{counter = 0},
                (* Euclid's algorithm *)
                gcd[r_Integer?Positive, s_Integer?Positive] :=
                ((* count calls to gcd *) counter = counter + 1;
                  gcd[#2, Mod[##]]& @@ Reverse[Sort[{r, s}]]);
                gcd[r_Integer?Positive, 0] = r;
```

```
(* return gcd and number of calls *)
{gcd[p, q], counter}]
```

Next, we calculate the array of the number of steps for calculating gcd(p, q) for $1 \le p, q \le 2^8$.

```
In[2]:= data = Table[GCDSteps[a, b][[2]], {a, 2^8}, {b, 2^8}];
```

Here is a density plot of the resulting values.

```
In[3]:= ListDensityPlot[data, Mesh -> False, ColorFunction -> (Hue[0.7 #]&)]
```

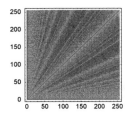

A list plot shows the absolute number of steps better.

```
In[4]:= ListPlot[Sort[Flatten[data]], PlotRange -> All, Frame -> True, Axes -> False]
```

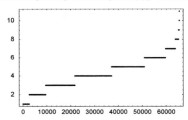

At most, 11 steps were needed.

```
In[5]:= Max[data]
Out[5]= 11
```

```
In[6]:= Position[data, %]
Out[6]= {{144, 233}, {233, 144}}
```

The 11 steps were needed for gcd(144, 233) = gcd(233, 144). For this pair of numbers, we have the following chain of reduction steps in Euclid's algorithm:

$233 = 1 \times 144 + 89 \Longrightarrow 144 = 1 \times 89 + 55 \Longrightarrow 89 = 1 \times 55 + 34 \Longrightarrow 55 = 1 \times 34 + 21 \Longrightarrow$
$34 = 1 \times 21 + 13 \Longrightarrow 21 = 1 \times 13 + 8 \Longrightarrow 13 = 1 \times 8 + 5 \Longrightarrow 8 = 1 \times 5 + 3 \Longrightarrow 5 = 1 \times 3 + 2 \Longrightarrow 3 = 1 \times 2 + 1$

The two integers 144 and 233 are Fibonacci numbers. The Euclidean algorithm takes most steps for pairs of Fibonacci numbers [31], [321].

```
In[7]:= Fibonacci[{12, 13}]
Out[7]= {144, 233}
```

For an in-depth analysis of the complexity of Euclid's algorithm, see [326].

g) The function gcdProduct calculates a single product of the form $\prod_i^r \mathrm{gcd}(n_i)$, and $\prod_{i,j}^r \mathrm{gcd}(n_i, n_j)$, ... Here m is the number of terms inside the gcd. To later count the function calls, we give the function *gcd* as an argument.

```
In[1]:= gcdProduct[gcd_, l_List, m_] :=
    Module[{i, n = Length[l], v},
    Product[gcd @@ l[[Table[v[i], {i, m}]]],
        (* make nested iterators *)
        Evaluate[Sequence @@ Table[{v[i], If[i === 1, 1, v[i - 1] + 1],
                                    n}, {i, m}]]]]
```

Here is the list of all three-element subsequences with nondecreasing indices.

```
In[2]:= gcdProduct[f, Array[a, 5], 3]
Out[2]= f[a[1], a[2], a[3]] f[a[1], a[2], a[4]] f[a[1], a[2], a[5]]
        f[a[1], a[3], a[4]] f[a[1], a[3], a[5]] f[a[1], a[4], a[5]]
        f[a[2], a[3], a[4]] f[a[2], a[3], a[5]] f[a[2], a[4], a[5]] f[a[3], a[4], a[5]]
```

Using `gcdProduct`, the implementation of the formula under consideration is straightforward.

```
In[3]:= lcm[gcd_, l_List] := Product[gcdProduct[gcd, l, k], {k, 1, Length[l], 2}]/
                             Product[gcdProduct[gcd, l, k], {k, 2, Length[l], 2}]
```

The values calculated by `lcm` agree with the ones from the built-in function LCM.

```
In[4]:= Table[(lcm[GCD, #] === LCM @@ #)&[
              Table[Random[Integer, {1, 100}], {Random[Integer, {1, 10}]}]],
              {100}] // Union
Out[4]= {True}
```

Now let us count the number of calls to the built-in function GCD. We count the application as a side effect by supplying the pure function `(counter ++; GCD[##])&` as the first argument for `lcm`.

```
In[5]:= Table[counter = 0; lcm[(counter ++; GCD[##])&, Range[k]];
              counter, {k, 12}]
Out[5]= {1, 3, 7, 15, 31, 63, 127, 255, 511, 1023, 2047, 4095}
```

The number of function calls is $2^r - 1$.

```
In[6]:= Table[2^k - 1, {k, 12}]
Out[6]= {1, 3, 7, 15, 31, 63, 127, 255, 511, 1023, 2047, 4095}
```

h) Calculating a few terms of the series leads to the conjecture $\sqrt{2}/2$. (For a more "complicated" real number, we could use http://www.lacim.uqam.ca/pi/ or http://www.cecm.sfu.ca/projects/ISC/ISCmain.html to find a closed form.)

```
In[1]:= fraction[k_] := Nest[Apply[Divide, Partition[#, 2], {1}]&,
                            Range[k], Log[2, k]];
In[2]:= Table[fraction[2^j], {j, 6}]
```

$$Out[2]= \left\{\left\{\frac{1}{2}\right\}, \left\{\frac{2}{3}\right\}, \left\{\frac{7}{10}\right\}, \left\{\frac{286}{405}\right\}, \left\{\frac{144305}{204102}\right\}, \left\{\frac{276620298878}{391202754597}\right\}\right\}$$

```
In[3]:= N[%]
Out[3]= {{0.5}, {0.666667}, {0.7}, {0.706173}, {0.707024}, {0.707102}}
```

To "verify" the conjecture, we calculate 16 terms of the sequence.

```
In[4]:= fractionBoxes[k_] := Nest[Apply[FractionBox, Partition[#, 2], {1}]&,
                                 ToString /@ Range[k], Log[2, k]];
```

Here is the resulting fraction displayed.

```
fractionBoxes[k_] :=
 Nest[Apply[FractionBox, Partition[#, 2], {1}]&, Range[k], Log[2, k]]

NotebookPut[Notebook[{Cell[BoxData[fractionBoxes[2^16][[1]]], "Output",
    ShowCellBracket -> False]},           (* to see all levels *)
    WindowElements -> {"VerticalScrollBar"},
    WindowSize -> {200, 600}, WindowFrameElements -> {"CloseBox"}]];
```

```
In[5]:= f = fraction[2^16];
In[6]:= {Numerator[f], Denominator[f]} // N
Out[6]= {{2.51689814943806×10^15420}, {3.55943149804705×10^15420}}
```

The small differences indicate that $\sqrt{2}/2$ is the limit. For an analytic proof, see [470], and [506].

```
In[7]:= f - Sqrt[2]/2 // N[#, 22]&

Out[7]= {-8.419387419944729869994×10⁻³²}
```

i) Here is a list of the divisors of the first million integers. They can be calculated in about 1 minute on a 2 GHz computer. To keep the memory usage low, we pack each list of divisors. We deal with nearly 14 million divisors.

```
In[1]:= o = 10^6;
        L = Table[{0}, {o}];
        Do[L[[n]] = Developer`ToPackedArray[Divisors[n]], {n, 2, o}] // Timing
Out[3]= {22.27 Second, Null}

In[4]:= {ByteCount[L], Plus @@ (Length /@ L)}
Out[4]= {115880148, 13970034}
```

Now, we must implement the calculation of $f_n(\xi)$. We could use
`Length[Select[L[[n]], # <= N[n^ξ]&]]/DivisorSigma[0, n]`
but this would be quite inefficient. So instead, we code a compiled version that avoids the repeated evaluation of n^ξ and ends the selection process as soon as possible.

```
In[5]:= f = Compile[{{ξ, _Real}, {l, _Integer, 1}},
                 Module[{ν = Floor[Last[l]^ξ], k = 1, λ = Length[l]},
                     If[ξ == 1., 1.,
                        (* step through the list of divisors
                           as long as divisor < n^ξ *)
                        While[l[[k]] <= ν && k <= λ, k++];
                        (k - 1.)/λ]]];
```

`F[ξ]` can now be defined straightforwardly.

```
In[6]:= F[ξ_] := 1/(o - 1) Sum[f[ξ, L[[k]]], {k, 2, o}]
```

The following graphic shows the resulting values for $F(\xi)$ as points on the background of the asymptotic curve. The approximation is fairly good.

```
In[7]:= Show[Block[{$DisplayFunction = Identity},
                  {(* limit distribution *)
                   Plot[2/Pi ArcSin[Sqrt[u]], {u, 0, 1},
                       PlotStyle -> {{Thickness[0.015], GrayLevel[0.6]}}],
                   (* data *)
                   ListPlot[Table[{ξ, F[ξ]}, {ξ, 0, 1, 1/20}],
                       PlotStyle -> {PointSize[0.015]}]}]]
```

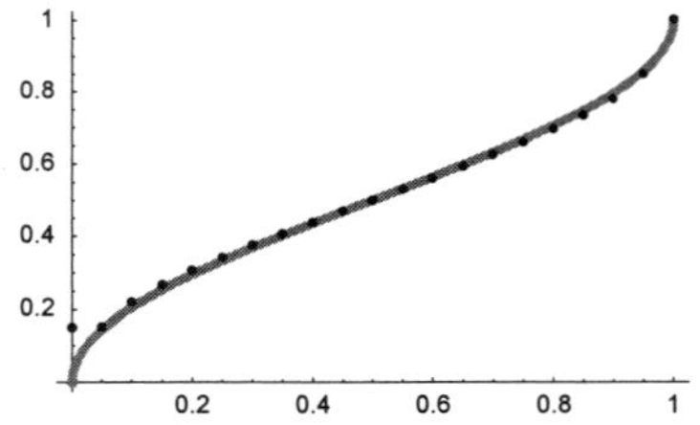

This implementation for F is very memory efficient, but the calculation for each ξ will take about a minute. We could have traded memory for time by flattening all divisors and generating a cumulative list of discontinuities of the approximating curve for $F(\xi)$

j) For the purpose of this exercise, the following crude implementation of a noncommutative multiplication t for the canonical commutation relation is sufficient. (A recursive approach would be faster, but because here we are interested in some small *n*, this effort is not needed here.)

```
In[1]:= (* bring ap to the left *)
        t[l___, a, ap, r___] := t[l, a, ap, r] = t[l, ap, a, r] + t[l, r]
        t[] := 1
```

The function `rewriteProduct` writes strings of consecutive $a^\dagger$'s and a's as powers.

```
In[4]:= rewriteProduct[n_] :=
            t[Sequence @@ Flatten[Table[{ap, a}, {n}]]]  /.
              t[aps:ap.., as:a..] :> t[ap^Length[{aps}], a^Length[{as}]]
```

Here are the first five products $(a^\dagger a)^n$ rewritten as sums of terms $(a^\dagger)^j a^j$.

```
In[5]:= Table[rewriteProduct[n], {n, 5}]
```
```
Out[5]= {t[ap, a], t[ap, a] + t[ap^2, a^2], t[ap, a] + 3 t[ap^2, a^2] + t[ap^3, a^3],
          t[ap, a] + 7 t[ap^2, a^2] + 6 t[ap^3, a^3] + t[ap^4, a^4],
          t[ap, a] + 15 t[ap^2, a^2] + 25 t[ap^3, a^3] + 10 t[ap^4, a^4] + t[ap^5, a^5]}
```

The coefficients are well known. They are the Stirling numbers of the second kind $S_n^{(j)}$ [307], [308], [208], [433], [184].

```
In[6]:= coefficients[expr_] := CoefficientList[expr /. t[ap^e_., a^e_.] :> C^e, C]
```

```
In[7]:= Table[coefficients[rewriteProduct[n]], {n, 10}]
```
```
Out[7]= {{0, 1}, {0, 1, 1}, {0, 1, 3, 1}, {0, 1, 7, 6, 1}, {0, 1, 15, 25, 10, 1},
          {0, 1, 31, 90, 65, 15, 1}, {0, 1, 63, 301, 350, 140, 21, 1},
          {0, 1, 127, 966, 1701, 1050, 266, 28, 1}, {0, 1, 255, 3025, 7770, 6951, 2646, 462, 36, 1},
          {0, 1, 511, 9330, 34105, 42525, 22827, 5880, 750, 45, 1}}
```

```
In[8]:= % == Table[StirlingS2[k, j], {k, 10}, {j, 0, k}]
```
```
Out[8]= True
```

We check one more case, namely $n = 50$.

```
In[9]:= $RecursionLimit = Infinity;
        With[{n = 50}, coefficients[rewriteProduct[n]] ===
                       Table[StirlingS2[n, j], {j, 0, n}]]
```
```
Out[10]= True
```

The following data shows that the above implementation for t demands many resources.

```
In[11]:= {Length[DownValues[t]], ByteCount[DownValues[t]]}
```
```
Out[11]= {20827, 104574968}
```

The operators $a^\dagger$ and a are typically interpreted as creation and annihilation operators in quantum mechanics (although such an interpretation should not be taken too literal [403]). For a similar expansion, see [268], [598], [394], [523], [149], [601], [217], [66], [489], [571], [575], and [508].

k) The list equalDenominatorLists contains lists of fractions with equal denominators.

```
In[1]:= qMax = 1000;
        equalDenominatorLists = Prepend[#, 0]& /@ Map[Last,
        Split[Sort[{Denominator[#], #}& /@
                       Union[Flatten[Table[Range[0, j]/j, {j, qMax}]]]],
             #1[[1]] == #2[[1]]&], {2}]; // Timing
```
```
Out[2]= {5.72 Second, Null}
```

There are 305193 fractions in equalDenominatorLists.

```
In[3]:= Length[Flatten[equalDenominatorLists]]
```
```
Out[3]= 305193
```

allCFLength gives the sum of the length of the continued fraction expansions of the list l.

```
In[4]:= allCFLength[l_] := Plus @@ (Length[ContinuedFraction[#]]& /@ l)
```

cS calculates the value of $c(q)$ from the exact l_q.

```
In[5]:= cS[q_] := allCFLength[equalDenominatorLists[[q]]]/EulerPhi[q] -
                12. Log[2]/Pi^2 Log[q]
```

cData finally is a list of the values of c for the denominators less than or equal to 1000.

```
In[6]:= cData = cS /@ Range[qMax];
```

The *c*-values are quite fluctuating. We smooth the data by forming the cumulative average. The resulting data points are shown in the right graphic.

```
In[7]:= (* averaging function *)
        cumulativeAverage[l_] :=
        Rest[FoldList[1/#2[[1]]((#2[[1]] - 1) #1 + #2[[2]])&, 0,
            MapIndexed[{#2[[1]], #1}&, l]]]
```

```
In[9]:= Show[GraphicsArray[
        Block[{$DisplayFunction = Identity},
            {(* original data *)
            ListPlot[cData],
            (* averaged data *)
            ListPlot[cumulativeAverage[cData], PlotRange -> {1, 2}]}]]]
```

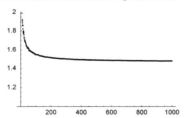

We smooth the data by forming the cumulative average.

```
In[10]:= ListPlot[cumulativeAverage[cData], PlotRange -> {1, 2}]
```

The last plot suggests $c \approx 1.47$. The theoretical value for *c* is given by the following expression [325], [423].

```
In[11]:= 6 Log[2]/Pi^2(3 Log[2] + 4 EulerGamma - 24/Pi^2 Zeta'[2] - 2) - 1/2 // N
Out[11]= 1.46708
```

Using larger denominators, we obtain a more accurate estimation of *c*.

```
In[12]:= Sum[With[{q = 10^4 + k}, allCFLength[Select[Range[0, q - 1]/q,
                    Denominator[#] === q&]]/EulerPhi[q] -
                        12. Log[2]/Pi^2 Log[q]], {k, 10}]/10
Out[12]= 1.46704
```

The calculation of l_{10^6} needs a few seconds and gives again a value for *c* correct to three digits.

```
In[13]:= With[{q = 10^6}, allCFLength[Select[Range[0, q - 1]/q,
                    Denominator[#] === q&]]/EulerPhi[q] -
                    12. Log[2]/Pi^2 Log[q]]
Out[13]= 1.46712
```

l) A rule-based implementation of the Isenkrahe algorithm is straightforward. (We truncate the infinite sum appropriately.)

```
In[1]:= v[p_, n_] := Sum[Floor[n/p^k], {k, Ceiling[Log[p, n!]]}]

        ψ[m_, n_] := Times @@ (#^v[#, n]& /@ Prime[Range[PrimePi[m]]])

        d[n_, p_] := (n!/# + #/(n - 1)! - Floor[(n - 1)!/#])&[ψ[p, n]]
```

```
nextPrime[p_?PrimeQ] := FixedPoint[d[#, p]&, p]
```

Here is a quick check that `nextPrime` really calculates the next prime.

```
In[5]:= (nextPrime /@ Prime[Range[20]]) == Prime[Range[2, 21]]

Out[5]= True
```

Rewriting now all definition as pure functions and renaming some of the variables gives the following one-liner.

```
In[6]:= nextPrimeOneliner[p_?PrimeQ] :=
        FixedPoint[Function[l, Function[ψ, (l!/ψ + ψ/(l - 1)! -
            Floor[(l - 1)!/ψ])][Function[{m, n},
              Times @@ (#^Function[{q, n},
                Sum[Floor[n/q^k], {k, Ceiling[Log[q, n!]]}]]]][#, n]& /@
                                Prime[Range[PrimePi[m]]])][p, l]]], p]
```

`nextPrimeOneliner` gives the same result as `nextPrime`.

```
In[7]:= (nextPrimeOneliner /@ Prime[Range[20]]) == Prime[Range[2, 21]]

Out[7]= True
```

m) Here is a straightforward implementation of the $\Gamma_k^{(n)}(a_1, a_2, ..., a_m)$. The $2k$ iterators involving $\mu_1,...,\mu_k, \nu_1,...,\nu_k$ are first constructed as a list and then spliced into `Sum`.

```
In[1]:= Γ[k_, n_, A_List] :=
        Module[{m = Length[A]},
        Sum[Multinomial[Sequence @@ Table[v[j], {j, k}]]]*
                     Product[A[[μ[j]]]^v[j], {j, k}],
        Evaluate[Sequence @@
        (* the μ[j] and ν[j] iterators *)
        Join[Table[{μ[j], 1 + If[j === 1, 0, μ[j - 1]], m - (k - j)}, {j, 1, k}],
            Table[{v[j], 1, n - Sum[v[l], {l, j - 1}] - (k - j)}, {j, 1, k - 1}],
            {{v[k], n - Sum[v[l], {l, k - 1}], n - Sum[v[l], {l, k - 1}]}}]]]]
```

Here are some simple examples: $\Gamma_k^{(2)}(a_1, a_2)$, $\Gamma_k^{(3)}(a_1, a_2)$, and $\Gamma_k^{(i)}(a_1, a_2, a_3)$.

```
In[2]:= Table[Γ[i, 2, {a, b}], {i, 5}]

Out[2]= {a² + b², 2 a b, 0, 0, 0}
```

```
In[3]:= Table[Γ[i, 3, {a, b}], {i, 5}]

Out[3]= {a³ + b³, 3 a² b + 3 a b², 0, 0, 0}
```

```
In[4]:= Table[Γ[i, 3, {a, b, c}], {i, 5}]

Out[4]= {a³ + b³ + c³, 3 a² b + 3 a b² + 3 a² c + 3 b² c + 3 a c² + 3 b c², 6 a b c, 0, 0}
```

Using Γ, we now implement the expansion of $(\sum_{j=1}^m a_j)^n$.

```
In[5]:= expandedPower[HoldPattern[Plus[summands__]^n_.]] :=
            Sum[Γ[k, n, List[summands]], {k, Max[n, Length[{summands}]]}] /;
            Length[{summands}] >= 2 && IntegerQ[n] && n > 0
```

Here is an example.

```
In[6]:= expandedPower[(a + b + c)^3]

Out[6]= a³ + 3 a² b + 3 a b² + b³ + 3 a² c + 6 a b c + 3 b² c + 3 a c² + 3 b c² + c³
```

We check some larger examples of `expandedPower` and compare against the result of `Expand`.

```
In[7]:= Table[(expandedPower[#] - Expand[#])&[Sum[α[j], {j, i}]^k],
            {i, 2, 8}, {k, 8}] // Flatten // Union

Out[7]= {0}
```

n) We start by calculating a numerical value for B_1.

```
In[1]:= B1 = EulerGamma + NSum[MoebiusMu[k] Log[Zeta[k]]/k, {k, 2, Infinity}]

Out[1]= 0.261497
```

For visualizing the number of prime divisors, we calculate the number of distinct prime factors for the first 200000 integers.

```
In[2]:= data = Table[Length @ FactorInteger[n], {n,  2 10^5}];
```

After averaging over a couple of thousand integers, we see an excellent agreement with the theoretical curve.

```
In[3]:= Show[GraphicsArray[
          Function[{h, pr}, Show[{(* the calculated averages *)
              ListPlot[(* form averages at once *)
                      MapIndexed[{#2[[1]] + h/2, #1}&,
                      ListConvolve[Table[1, {h}], data]/h],
                      DisplayFunction -> Identity],
              (* theoretical average *)
              Plot[Log[Log[x]] + B1 + EulerGamma/Log[x], {x, E, 2 10^5},
                  PlotStyle -> {{Thickness[0.01], Hue[0]}},
                  DisplayFunction -> Identity]},
                  FrameTicks -> {{10^5, 2 10^5}, Automatic, None, None},
                  Frame -> True, Axes -> False,
                  PlotRange -> {{3 10^4 , 2 10^5}, pr}]] @@@
                  (* average over how many integers *)
                  {{1, All}, {100, {2, 3}}, {2000, {2.6, 2.9}}}]]]
```

For the fluctuations round the asymptotic value $\ln(\ln(n))$, a Gaussian error law holds [557]: For a given n, the number of integers k with $\omega(k) \leq \ln(\ln(n))/\ln(\ln(n))^{1/2}$ is equal to $(\mathrm{erf}(2^{-1/2} x) + 1)/2$. Taking $n = 10^6$ shows a glimpse of this theorem.

```
In[4]:= With[{o = 10^6},
          Show[{(* data *)
              Graphics[Line[{#1[[1]], #1[[-1]]}]& /@ Split[
              MapIndexed[{#1, (#2[[1]] + 1.)/o}&, Sort[
                Table[(Length[FactorInteger[n]] - Log[Log[N[o]]])/
                      Sqrt[Log[Log[N[o]]]], {n, 2, o}]]],
                              #1[[1]] == #2[[1]]&]],
              (* Gauss curve *)
              Plot[(1 + Erf[x/Sqrt[2]])/2, {x, -2, 2},
                  DisplayFunction -> Identity, PlotStyle -> {Hue[0]}]},
              DisplayFunction -> $DisplayFunction, Frame -> True]]
```

o) It is straightforward to implement a recursive function definition based on the property $\lceil k\,\varphi - 1/\varphi \rceil = \lfloor k\,\varphi + 1/\varphi \rfloor$. We use `NestWhileList` and test all numbers starting from F_{n-1} for this property.

```
In[1]:= F[1] = 1; F[2] = 1;
        F[n_] := F[n] = F[n - 1] + Length[NestWhileList[(# + 1)&, F[n - 1] + 1,
                                    Ceiling[GoldenRatio # - 1/#] =!=
                                    Floor[GoldenRatio # + 1/#]&]]
```

Here is a quick check showing that F indeed calculates the Fibonacci numbers.

```
In[3]:= {#, # == Table[Fibonacci[k], {k, 12}]}& @ Table[F[k], {k, 12}]
```

Out[3]= {{1, 1, 2, 3, 5, 8, 13, 21, 34, 55, 89, 144}, True}

p) The function `KimberlingSequence` calculates the first *n* terms of the Kimberling sequence. To avoid recomputing and storing too many values, we use this slightly complicated-looking definition.

```
In[1]:= KimberlingSequence[n_] :=
    If[# > n, (* take from existing sequence *) Take[KimberlingSequence[#], n],
        (* calculate more terms *) KimberlingSequence[n] =
    Join[If[# > 0, KimberlingSequence[#], (* first time *) {}],
            Table[(k 2^-IntegerExponent[k, 2] + 1)/2, {k, Max[0, #] + 1, n}]]]&[
    Max[DeleteCases[#[[1, 1]]& /@ (First /@ DownValues[KimberlingSequence]),
                    HoldPattern[Pattern][_, Blank[]]]]]
```

Here are the first 20 terms of the Kimberling sequence.

```
In[2]:= KimberlingSequence[20]
```

Out[2]= {1, 1, 2, 1, 3, 2, 4, 1, 5, 3, 6, 2, 7, 4, 8, 1, 9, 5, 10, 3}

The following graphic shows the sequence and the real part of its Fourier transform.

```
In[3]:= Show[GraphicsArray[
    ListPlot[N[# @ KimberlingSequence[2^14]],
            PlotStyle -> {PointSize[0.003]}, PlotRange -> All,
            DisplayFunction -> Identity, Frame -> True, Axes -> False]& /@
        (* scaled value and scaled real part of the Fourier transform *)
            {Log[#]&, Log[Abs[Re[Fourier[#]]]]&}]]
```

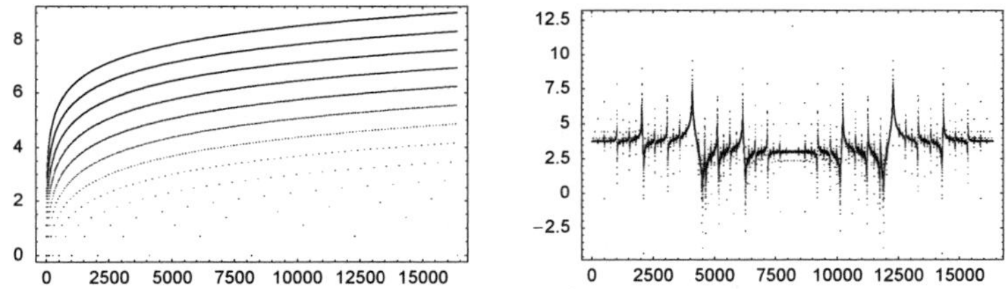

To show the property that the sequence with the first occurrence of each integer deleted reproduces the original sequence for the first million elements, we must take care about efficiency. We first number the elements, order them with respect to the sequence elements, delete the first element, and then restore the original order of the remaining elements.

```
In[4]:= reducedKimberlingSequence[n_] :=
    Last /@ Sort[Reverse /@ Flatten[Rest /@ Split[Sort[
                    Transpose[{KimberlingSequence[n], Range[n]}]],
                        #1[[1]] == #2[[1]]&], 1]]
```

The actual check can now be carried out quite quickly.

```
In[5]:= KimberlingSequence[10^6];
    (KimberlingSequence[Length[#]] === #)&[
    reducedKimberlingSequence[10^6]] // Timing
Out[6]= {6.68 Second, True}
```

Similar to the Kolakoski sequence from Chapter 6 of the Programming volume [554], can we find a number γ, such that a power series with the Kimberling sequence as coefficients equals the continued fraction.

```
In[7]:= Module[{γ = 0.56562077432015242215685655334782097782567120171716322960422 1909},
        X = X = FromContinuedFraction[KimberlingSequence[500]] // N[#, 500]&;
        X - KimberlingSequence[500].(γ^Range[500])]
Out[7]= 0. × 10^-63
```

q) We start by defining the Cantor function for an explicit numerical argument.

```
In[1]:= CantorC[x_?NumericQ] :=
    Module[{l, r, Λ, L, R, p1, p2},
    Which[x == 0, 0, x == 1, 1, True,
        {l, r} = {0, 1}; Λ = 1; {L, R} = {0, 1};
        flag = True; (* subdivide interval *)
```

```
            While[flag && l != x && r != x, print[{l, r}];
                p1 = l + Λ/3; p2 = l + 2Λ/3;
                Which[l <= x <= p1,
                        r = p1; Λ = Λ/3; R = (L + R)/2,
                      p1 <= x <= p2, flag = False,
                      p2 <= x <= r,
                        l = p2; Λ = Λ/3; L = (L + R)/2]];
        (L + R)/2]]
```

Here is a plot of the integral to be calculated.

In[2]:= **Plot[CantorC[x]^CantorC[x], {x, 0, 1}]**

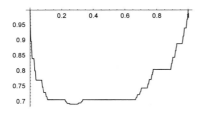

A numerical integration, using two different methods, strongly suggests that the integral is greater than $3/4$. But because of the nonsmooth nature of the derivative of the function and the messages from NIntegrate, we cannot really rely on the last results.

In[3]:= **NIntegrate[CantorC[x]^CantorC[x], {x, 0, 1}, Method -> #,**
         PrecisionGoal -> 6, MaxRecursion -> 16]& /@
                        {GaussKronrod, Trapezoidal}

    NIntegrate::slwcon :
    Numerical integration converging too slowly; suspect one of the
        following: singularity, value of the integration being 0, oscillatory
        integrand, or insufficient WorkingPrecision. If your integrand is
        oscillatory try using the option Method->Oscillatory in NIntegrate. More…

    NIntegrate::slwcon :
    Numerical integration converging too slowly; suspect one of the
        following: singularity, value of the integration being 0, oscillatory
        integrand, or insufficient WorkingPrecision. If your integrand is
        oscillatory try using the option Method->Oscillatory in NIntegrate. More…

Out[3]= {0.750373, 0.750373}

Because all the contributions from all intervals of the constant function value are positive, it is possible to estimate the integral from below. We calculate the 524287 largest intervals. The function CantorIntervalValues returns the intervals and the function values after o recursive subdivisions of the interval $[0, 1]$.

In[4]:=
```
(* make constant middle interval and
   to-be-subdivided boundary intervals *)
subDivideCantorInterval[cantorInterval[{l_, r_}, {L_, R_}]] :=
With[{p1 = l + (r - l)/3, p2 = l + 2(r - l)/3, LR = (L + R)/2},
    {cantorInterval[{l, p1}, {L, LR}], interval[{p1, p2}, LR],
     cantorInterval[{p2, r}, {LR, R}]}]

(* recursively carry out subdivision *)
subDivideStep[l_] :=
Module[{λ = Flatten[l /. cI_cantorInterval :>
                        subDivideCantorInterval[cI]]},
    {Cases[λ, _cantorInterval], Cases[λ, _interval]}]

CantorIntervalValues[o_Integer] :=
Last /@ Rest[NestList[subDivideStep[First[#]]&,
              {cantorInterval[{0, 1}, {0, 1}], {}}, o]]
```
In[11]:= **ints = CantorIntervalValues[19];**
         Length[Flatten @ ints]

In[12]:= 524287

Estimating now the integral from below shows conclusively that $\int_0^\infty C(x)^{C(x)}\,dx > 3/4$.

```
In[13]:= powerIntegral[l_List] := powerIntegral /@ l
        powerIntegral[interval[{a_, b_}, v_]] := N[v, 22]^v (b - a)
In[15]:= sum = 0;
        Do[sum = sum + Plus @@ powerIntegral[ints[[k]]];
           (* to see progress:  Print[{k, sum}] *),
          {k, Length[ints]}];

        (* return sum *)
        sum
Out[19]= 0.7500186365938695315156
```

For closed form expressions of the integrals $\int_0^1 C(x)^\lambda\,dx$, see [215].

2. Cattle Problem of Archimedes

We denote the steer with capital letters and the cows with lowercase first letters, in which variables are chosen to be the first letters of the associated color. The solution is given by the following linear system of equations.

```
In[1]:= preSol = Solve[
            {W   ==  Br  + (1/2 + 1/3) Bl, Bl ==  Br  + (1/4 + 1/5) S,
             S   ==  Br  + (1/6 + 1/7) W,  w  == (1/3 + 1/4) (Bl + bl),
             bl  == (1/4 + 1/5) (S + s),   s  == (1/5 + 1/6) (Br + br),
             br  == (1/6 + 1/7) (W + w)},
            {W   (* White steer   *),  Bl (* Black steer  *),
             S   (* Spotted steer *),  Br (* Brown steer   *),
             w   (* white cows    *),  bl (* black cows     *),
             s   (* spotted cows  *),  br (* brown cows     *)}]
        Solve::svars : Equations may not give solutions for all "solve" variables. More…
```

$$Out[1]= \left\{\left\{W \to \frac{3455494\,br}{1813071},\ Bl \to \frac{828946\,br}{604357},\ S \to \frac{7358060\,br}{5439213},\right.\right.$$
$$\left.\left.Br \to \frac{461043\,br}{604357},\ w \to \frac{2402120\,br}{1813071},\ bl \to \frac{543694\,br}{604357},\ s \to \frac{1171940\,br}{1813071}\right\}\right\}$$

Because the number of cows of each type must be a whole number, we need to compute the least common multiple of the denominators.

In[2]:= NC = LCM @@ (Denominator[#[[2]]]& /@ preSol[[1]])

Out[2]= 5439213

We select the undetermined variable from the solution and assign it to the symbol undetermined. (We do this automatically instead of picking it out "by eye" because the variable that is undetermined may change from one version of *Mathematica* to the next.)

In[3]:= undetermined = Union[Cases[#, _Symbol][[1]]& /@ Last /@ preSol[[1]]][[1]]

Out[3]= br

We get the following result.

```
In[4]:= sol = Append[preSol[[1]] /. (undetermined -> NC PositiveInteger),
                     undetermined -> NC PositiveInteger]
Out[4]= {W → 10366482 PositiveInteger, Bl → 7460514 PositiveInteger,
         S → 7358060 PositiveInteger, Br → 4149387 PositiveInteger, w → 7206360 PositiveInteger,
         bl → 4893246 PositiveInteger, s → 3515820 PositiveInteger, br → 5439213 PositiveInteger}
```

The exercise we gave is not the complete story. In the library of Wolfenbüttel, G. E. Lessing found two more conditions for the cattle problem [352], [496]: The sum of the white steer and the brown steer should be a square number. The sum of the spotted steer and the brown steer should be a triangular number. The first condition is easy to satisfy.

In[5]:= FactorInteger[((W + Bl)/PositiveInteger) /. sol]

Out[5]= {{2, 2}, {3, 1}, {11, 1}, {29, 1}, {4657, 1}}

The result has to be the product of the primes and a square number.

```
In[6]:= rule1 = PositiveInteger -> (3 11 29 4657) PositiveInteger1^2

Out[6]= PositiveInteger → 4456749 PositiveInteger1²
```

The second condition is much harder to satisfy. A triangular number is a number of the form $n(n+1)/2$ [314]. (Whether Archimedes really had this condition in mind is unclear; see [144], [145], [490], [583], and [590] for a discussion of this subject.) Then we would have the following condition.

```
In[7]:= Factor[8# + 1]& /@ ((S + Br /. sol /. rule1) == n (n + 1)/2)

Out[7]= 1 + 410286423278424 PositiveInteger1² == (1 + 2 n)²
```

This is an equation of the form $p^2 - d\,q^2 = 1$, with $d = 410286423278424$. It is called a Pell equation (see [148], [193], [85], [405], [406], [35], [456], [226], [570], [107], and [297] for a quick way to solve Pell equations numerically). It always has solutions in the integers, and thanks to Lagrange, we can also calculate p and q constructively. We do not discuss how this goes in detail in this solution; here is a straightforward *Mathematica* implementation for the smallest p and q, which solve the Pell equation for a given d. This function makes heavy use of the built-in functions `ContinuedFraction` and `FromContinuedFraction`.

```
In[8]:= PellSolve[d_Integer] := {Numerator[#], Denominator[#]}&[
          FromContinuedFraction[Drop[Flatten[ContinuedFraction[Sqrt[d]]], -1]]];
```

Let us try one example.

```
In[9]:= PellSolve[d = 2 3 5 7 11 13 17 19]

Out[9]= {6915878018249487671919, 22205900901368228}
```

```
In[10]:= %[[1]]^2 - d %[[2]]^2

Out[10]= 1
```

Letting now `PellSolve` run yields (after more than 200000 iterations inside `PellSolve`) the pair $\{p, q\}$ we are looking for.

```
In[11]:= PositiveInteger1 = (ps = PellSolve[410286423278424])[[2]];
```

This results in two integers with 103273 and 103266 digits, the first one containing 10305 zeros, 10408 ones, and so on.

```
In[12]:= N[ps]

Out[12]= {3.765344502347206×10^103272, 1.858921901386913×10^103265}
```

Using the function `Reduce` (to be discussed in the Chapter 1 of the Symbolics volume [556]), we can even get a complete solution for the Pell equation at hand.

```
In[13]:= redSol = Reduce[p^2 - 410286423278424 q^2 == 1 &&
                         p > 0 && q > 0, {p, q}, Integers];

           (* for brevity, display N of solution;
              use enough precision for numericalization *)
           $MaxExtraPrecision = 300000;
           N[redSol, 4] // N[#, 4]&

Out[17]= C[1] ∈ Integers && C[1] ≥ 1.000 && p == 0.5000 (1.328×10^{-103273 C[1]} + 7.531×10^{103272 C[1]}) &&
         q == -2.468×10^{-8} (1.328×10^{-103273 C[1]} - 1.000 7.531×10^{103272 C[1]})
```

Here is the smallest of the possible solutions.

```
In[18]:= redSol /. C[1] -> 1 // Short[#, 4]&

Out[18]//Short= p ==
              3765344502347205884018786193550576892856610111171560836237851390311234167566480643⸲
              694312349186281391904414227178867936750151393527439345809909970799 <<102979>>
              673643240595499598030609181360646829210094787991040818520976114141057706314714705⸲
              096647940993314916269513956239940831835922232584777023371728320049&& <<1>>
```

Then, we get the results for the smallest solution of the complete cattle problem of Archimedes. For obvious (space) reasons, we give the result for `N[sol /. rule1, 50]` and not for `sol /. rule1`.

```
In[19]:= N[sol /. rule1, 50]
```

In[19]:= {W → 1.5965108046711445314355261943712480861390650863455× 10²⁰⁶⁵⁴⁴,
Bl → 1.14897138772828999971235982182513002425641611335 38× 10²⁰⁶⁵⁴⁴,
S → 1.1331927544386380771195558792033117592541432328957× 10²⁰⁶⁵⁴⁴,
Br → 6.3903464823090286500855967618363973259205165855070× 10²⁰⁶⁵⁴³,
w → 1.1098298923733190397239602158253096236533390088978× 10²⁰⁶⁵⁴⁴,
bl → 7.5359414205454263981442911958968647343502218761388× 10²⁰⁶⁵⁴³,
s → 5.4146089457145667802361994210710262615701718402398× 10²⁰⁶⁵⁴³,
br → 8.3767688241852443869222198410845833874526793424198× 10²⁰⁶⁵⁴³}

This means that the minimal size of the Helios' herd was roughly 7.7602710^{206544} cattle.

In[20]:= Plus @@ (Last /@ %)

Out[20]= 7.7602714064868182695302328332138866642323224059234× 10²⁰⁶⁵⁴⁴

Remarkably, the first and last few dozen digits of this result were calculated by the end of the last century by [17] from Gotha (Thuringia) and a few years later by [43] from Illinois. The first computer solution was obtained in 1965 [595] using a computer that had 32 kB memory. For a recent detailed study of the Archimedes cattle problem, see [225].

3. Complicated Identity, π-Formula

a) Here is the equation to be established. Because *Mathematica* uses numerical techniques to establish the (in)equality, we get the $MaxExtraPrecision messages.

In[1]:= equation = Sqrt[16 - 2 Sqrt[29] + 2Sqrt[55 - 10 Sqrt[29]]] ==
Sqrt[22 + 2 Sqrt[5]] - Sqrt[11 + 2 Sqrt[29]] + Sqrt[5]

N::meprec :
Internal precision limit $MaxExtraPrecision = 50.` reached while evaluating
$-\sqrt{5} - \sqrt{22 + 2\sqrt{5}} + \sqrt{11 + 2\sqrt{29}} + \sqrt{16 - 2\sqrt{29} + 2\sqrt{55 - 10\,\text{Power}[\ll 2\gg]}}$. More…

Out[1]= $\sqrt{16 - 2\sqrt{29} + 2\sqrt{55 - 10\sqrt{29}}} = \sqrt{5} + \sqrt{22 + 2\sqrt{5}} - \sqrt{11 + 2\sqrt{29}}$

The two sides are numerically identical, at least up to the numerical accuracy of our machine.

In[2]:= N @ equation

Out[2]= True

Of course, this is not a proof. Analytically, we can proceed as follows. Using Cases, we find parts containing large roots, move them to one side of the equation, and square the equation. We continue until no more roots remain in the equation. This could proceed as follows.

In[3]:= Off[N::meprec];
g1 = Expand[Factor //@ (Subtract @@ (Expand[#^2]& /@ equation))]

Out[4]= $-22 - 2\sqrt{5} - 4\sqrt{29} - 2\sqrt{10\left(11 + \sqrt{5}\right)} + 2\sqrt{5\left(11 - 2\sqrt{29}\right)} +$
$2\sqrt{5\left(11 + 2\sqrt{29}\right)} + 2\sqrt{2\left(121 + 11\sqrt{5} + 22\sqrt{29} + 2\sqrt{145}\right)}$

In[5]:= var1 = 2 Sqrt[5 (11 + 2 Sqrt[29])] +
2 Sqrt[2 (11 + Sqrt[5]) (11 + 2 Sqrt[29])]

Out[5]= $2\sqrt{5\left(11 + 2\sqrt{29}\right)} + 2\sqrt{2\left(11 + \sqrt{5}\right)\left(11 + 2\sqrt{29}\right)}$

In[6]:= g2 = Expand[Factor //@ Expand[(g1 - var1)^2 - var1^2]]

Out[6]= $440 + 40\sqrt{5} - 80\sqrt{29} + 40\sqrt{2\left(11 + \sqrt{5}\right)} - 40\sqrt{11 - 2\sqrt{29}} -$
$88\sqrt{5\left(11 - 2\sqrt{29}\right)} - 16\sqrt{145\left(11 - 2\sqrt{29}\right)} - 40\sqrt{2\left(121 + 11\sqrt{5} - 22\sqrt{29} - 2\sqrt{145}\right)}$

In[7]:= var2 = -40 Sqrt[11 - 2 Sqrt[29]] - 88 Sqrt[5 (11 - 2 Sqrt[29])] -
16 Sqrt[145 (11 - 2 Sqrt[29])] -
40 Sqrt[2 (11 + Sqrt[5]) (11 - 2 Sqrt[29])];

In[8]:= g3 = Expand[Factor //@ Expand[(g2 - var2)^2 - var2^2]]

Out[8]= 0

Because we have squared an equation a few times, we should additionally make sure (for instance, by looking at the numerical values of the two sides of an equation within a sufficient precision [485], [392], [393], and [434]) that we did not prove something like $-1 = 1$. For the identity under consideration, it is easy to check that all is fine.

Note that we could write a routine to do this, but we do not check such equations very frequently.

Also, note that it is often interesting to find out which polynomial equations with integer coefficients have such algebraic numbers as solutions. The function `Recognize` in the package `NumberTheory`Recognize`` answers this question for the numerical value of the algebraic number. We now look at a few such algebraic equations of various orders that have approximate solutions `N[a]`.

```
In[9]:= Needs["NumberTheory`Recognize`"]
```

```
In[10]:= ?Recognize
```

> Recognize[x, n, t] finds a polynomial of degree at most n in the variable t, which
>   is approximately satisfied by the number x. Recognize[x, n, t, k] also finds a
>   polynomial of degree at most n in the variable t, but with a penalty of weight
>   k against higher degree polynomials. k must be a nonnegative integer. More...

We define a short function `recogAlg` that, for a given accuracy `prec` of the above algebraic number, generates polynomials of order at most `ord`.

```
In[11]:= recogAlg[ord_, prec_] :=
       Module[{a, i, poly, δ},
       (* the number to recognize *)
       a = N[Sqrt[16 - 2 Sqrt[29] + 2 Sqrt[55 - 10 Sqrt[29]]], prec];
       Do[(* find best polynomial *)
          poly = Recognize[a, i, x];
          δ = N[Min[Abs[(x /. NSolve[poly == 0, x, prec]) - a]], 5];
          (* form and print formatted cell *)
          CellPrint[Cell[TextData[
            {"○ Maximal order of polynomial: ", ToString[i],
             ". Best polynomial: ",
             Cell[BoxData[FormBox[MakeBoxes[#, TraditionalForm]& @ poly,
                      TraditionalForm]]],
             " with deviation ",
             Cell[BoxData[FormBox[MakeBoxes[#, TraditionalForm]& @ δ,
                      TraditionalForm]]], "."}],
             "PrintText"]], {i, ord}]]
```

Here are a few examples.

```
In[12]:= recogAlg[5, 12]
```

> ○ Maximal order of polynomial: 1. Best polynomial: $1170941 - 431237\,x$ with deviation $0. \times 10^{-12}$.

> ○ Maximal order of polynomial: 2. Best polynomial: $1135\,x^2 - 5823\,x + 7443$ with deviation $0. \times 10^{-10}$.

> ○ Maximal order of polynomial: 3. Best polynomial: $11\,x^3 - 452\,x^2 + 884\,x + 712$ with deviation $0. \times 10^{-11}$.

> ○ Maximal order of polynomial: 4. Best polynomial: $-42\,x^4 + 137\,x^3 - 61\,x^2 + 10\,x - 37$ with deviation $0. \times 10^{-11}$.

> ○ Maximal order of polynomial: 5. Best polynomial: $-7\,x^5 + 4\,x^4 + 52\,x^3 - 2\,x^2 - 65\,x - 34$ with deviation $1. \times 10^{-11}$.

```
In[13]:= recogAlg[3, 100]
```

> ○ Maximal order of polynomial: 1. Best polynomial:
> $16958279270456088009094693417025115162349453923539x - 46046940515859924752395900578263487780823656626872$
> with deviation $0. \times 10^{-100}$.

> ○ Maximal order of polynomial: 2. Best polynomial:
> $11945607667012148245992677274041 2x^2 -$ with deviation $0. \times 10^{-99}$.
> $54814668509740028485148405406187 5x + 6076497303805722049658987582988 34$

> ○ Maximal order of polynomial: 3. Best polynomial: $-13814732701432024042427 18x^3 + 21442298609723737708007 69x^2 +$
> $6183153500026020799535736x - 49416979881440760220925 33$
> with deviation $0. \times 10^{-100}$.

Note that the exact polynomials for `Sqrt[16 - 2Sqrt[29] + 2Sqrt[55 - 10Sqrt[29]]]` and `Sqrt[22 + 2Sqrt[5]] - Sqrt[11 + 2Sqrt[29]] + Sqrt[5]` are given by the following input. (This can be checked by increasing the order in and/or by solving this biquartic equation exactly.)

```
In[14]:= Recognize[N[Sqrt[16 - 2Sqrt[29] + 2Sqrt[55 - 10Sqrt[29]]], 40], 10, x]
```
$$Out[14]= 6400 - 4160\, x^2 + 864\, x^4 - 64\, x^6 + x^8$$

```
In[15]:= Recognize[N[Sqrt[22 + 2Sqrt[5]] - Sqrt[11 + 2Sqrt[29]] + Sqrt[5], 40], 10, x]
```
$$Out[15]= 6400 - 4160\, x^2 + 864\, x^4 - 64\, x^6 + x^8$$

Also, note that the function `RootReduce` (to be discussed in Chapter 1 of the Symbolics volume [556]) rewrites expressions containing radicals in a canonical form using Root-objects. Their first argument is just the corresponding irreducible polynomial.

```
In[16]:= RootReduce[Sqrt[16 - 2 Sqrt[29] + 2Sqrt[55 - 10 Sqrt[29]]]]
```
$$Out[16]= Root[6400 - 4160\, \#1^2 + 864\, \#1^4 - 64\, \#1^6 + \#1^8 \&, 7]$$

```
In[17]:= RootReduce[Sqrt[22 + 2 Sqrt[5]] - Sqrt[11 + 2 Sqrt[29]] + Sqrt[5]]
```
$$Out[17]= Root[6400 - 4160\, \#1^2 + 864\, \#1^4 - 64\, \#1^6 + \#1^8 \&, 7]$$

Of course, one can use `RootReduce` directly to establish the truth of the equality under consideration.

```
In[18]:= RootReduce[Sqrt[16 - 2 Sqrt[29] + 2 Sqrt[55 - 10 Sqrt[29]]]] ==
          RootReduce[Sqrt[22 + 2 Sqrt[5]] - Sqrt[11 + 2 Sqrt[29]] + Sqrt[5]]
```
```
Out[18]= True
```

The function `FullSimplify` (to be discussed in Chapter 3 of the Symbolics volume [556]) internally calls the above-used function `RootReduce`.

```
In[19]:= FullSimplify[Sqrt[16 - 2 Sqrt[29] + 2 Sqrt[55 - 10 Sqrt[29]]] ==
                      Sqrt[22 + 2 Sqrt[5]] - Sqrt[11 + 2 Sqrt[29]] + Sqrt[5]]
```
```
Out[19]= True
```

b) In the following, we will truncate at $n = 2500$. For analyzing the growth rate as a function of n, we will form cumulative lists of products and least common multiples. The products can be calculated relatively quickly.

```
In[1]:= nMax = 2500;
```

```
In[2]:= {(* calculate Fibonacci numbers (fast) *)
         Timing[fibs = Table[Fibonacci[k], {k, nMax}];],
         (* calculate product *)
         Timing[productw = Rest[FoldList[Times, 1, fibs]];]}
```
```
Out[2]= {{0.25 Second, Null}, {9.96 Second, Null}}
```

The calculation of the least common multiples is by far the most time-consuming operation. (One call to `LCM` with all n arguments would be slightly faster.)

```
In[3]:= Timing[lcmsAndTimes =
            Rest[FoldList[Timing[LCM[#1[[2]], #2]]&, {0, 1}, fibs]];]
```
```
Out[3]= {20.35 Second, Null}
```

The resulting numbers are of size $\exp(2^{-1} \ln(\phi)\, n^2)$ and $\exp(3\, \pi^{-2} \ln(\phi)\, n^2)$.

```
In[4]:= (* absolute sizes *)
        N[{Last[productw], Last[lcmsAndTimes][[2]]}]
```
$$Out[5]= \{9.69570323679209 \times 10^{652473}, 4.679924404814568 \times 10^{397050}\}$$

```
In[6]:= With[{n = N @ nMax},
         {(Log[GoldenRatio]/2 n^2)/Log[Last[productw]],
          (Log[GoldenRatio]3/Pi^2 n^2)/Log[Last[lcmsAndTimes][[2]]]}]
```
```
Out[6]= {1.00094, 0.999945}
```

Using the recorded timings for the calculations of the least common multiples, the time needed to calculate $\ln(\operatorname{lcm}(F_1, F_2, ..., F_n))$ can be estimated to have the complexity n^4.

```
In[7]:= (* timings for least common multiples *)
        timings = First /@ lcmsAndTimes/Second;
        (* log-log timings for least common multiples *)
        logTimings = Select[Log[10, MapIndexed[{#2[[1]], #1}&, timings]],
                            (#[[2]] > -3/2)&];
```

```
(* cumulative log-log timings for least common multiples *)
logCumulTimings =
Select[Log[10, MapIndexed[{#2[[1]], #1}&,
                    Rest[FoldList[Plus, 0, timings]]]],
       (#[[2]] > -1)&];

Show[GraphicsArray[
Block[{$DisplayFunction = Identity},
     ListPlot[#, PlotRange -> All]& /@
          (* the three timing data *)
          {timings, logTimings, logCumulTimings}]]]
```

```
In[15]:= (* fit individual and cumulative lcm timings *)
      {Fit[logTimings, {1, n}, n], TLog[n_] = Fit[logCumulTimings, {1, n}, n]}
Out[16]= {-1.39794 - 9.78751×10^-14 n, -11.4926 + 3.76353 n}
```

With $n = 1000$, we have nearly three correct digits of π.

```
In[17]:= lcms = Last /@ lcmsAndTimes;
      f[p_, l_] := Sqrt[6 Log[p]/Log[l]];
      f[N[Last[productw]], N[Last[lcms]]]
Out[19]= 3.14003333927178953468
```

Analyzing the error of the π-approximation suggests *error* $\propto n^{-1}$.

```
In[20]:= Off[Power::infy]; Off[Infinity::indet]; Off[Graphics::gptn];
      (* cumulative π-approximations *)
      δList = Abs[f[N[productw], N[lcms]] - Pi];
      (* log-log cumulative π-approximations *)
      δLogList = Cases[Log[10, MapIndexed[{#2[[1]] - 1, #1}&, δList]],
                    {_?NumericQ, _?NumericQ}];

Show[GraphicsArray[
  Block[{$DisplayFunction = Identity},
        {ListPlot[δList // N], ListPlot[δLogList // N]}]]]
```

```
In[26]:= (* fit log-log cumulative π-approximations *)
      δπLog[ε_] = Fit[δLogList, {1, ε}, ε]
Out[27]= 0.632873 - 0.973296 ε
```

This means calculating 10 correct digits of π would take the astronomically large time of about 10^{32} seconds.

```
In[28]:= nTenDigits = Exp[δπLog[Log[10^-10]]];
      Exp[TLog[Log[nTenDigits]]] Seconds
Out[29]= 4.71588×10^32 Seconds
```

4. Mirror Charges

Because we are working only with exact input parameters, we construct an exact solution. We use a "simplistic" approach; that is, independent of the given parameters, we continue the process of reflection until either we get a solution or we run into a definite problem. Making use of some relatively easily obtained analytical results (e.g., the only problems that are solvable with the reflection method are those for which $q = 2\pi/k$ with k even) [73], it is possible to make this implementation more efficient.

The function `mirror` reflects the charge; the value of n determines which of the two plates is involved.

```
In[1]:= mirror[φ_, θ_, n_Integer] :=
            Mod[Which[OddQ[n], -φ, EvenQ[n], θ + (θ - φ)], 360]
```

The problem can only be solved exactly using the mirror method if after a finite (or infinite) number of steps; we get back to the original charge with the correct polarity. This could fail as follows.

- After finitely many steps, we do not get back to the initial charge. We restrict ourselves to a maximum of `max` reflections.

- During the reflection, we end up on one of the two metal plates.

- During the reflections, the reflected charge ends up in the sector $(0, \theta)$.

- We end up with the opposite polarity.

```
In[2]:= MirrorCharges[start_, θ_, max_] :=
        Module[{li, counter, φ},
        For[li = {start}; counter = 1; φ = start,
            (* it still makes sense to go on with mirroring *)
            counter < max &&
              φ != 0 && φ != 180 && φ != θ && φ != θ + 180 &&
              (φ != start || counter == 1) && (!(0 < φ < θ) || counter == 1),
            counter = counter + 1,
            φ = mirror[φ, θ, counter];
            AppendTo[li, φ]];
            (* until we come to the start again or something bad happens *)
            Which[counter == max, {"Does not stop", li},
              φ == 0 || φ == 180 || φ == θ || φ == θ + 180,
                {ToString[counter] <> "th charge lies on plate", li},
              0 < φ < θ && φ != start,
                {ToString[counter] <> "th reflected charge inside", li},
              φ == start && EvenQ[counter],
                {"at the starting point with wrong polarity", li},
              φ == start && OddQ[counter], {" ", li}]]
```

Here is a possibility for a plot.

```
In[3]:= mirrorChargePlot[start_, θ_, max_, opts___] :=
        Module[{data, charges, thickHorizontal, thinHorizontal,
                thickSlanted, thinSlanted, i, points, connectingLines,
                θD = Degree θ, θD180 = Degree (θ + 180)},
        (* the various lines for a nice picture *)
        data = MirrorCharges[start, θ, max];
        charges = N[Degree data[[2]]];
        (* the metal plates *)
        thickHorizontal = {RGBColor[0, 1, 0], Thickness[0.015],
                          Line[{{0, 0}, {1.3, 0}}]};
        thinHorizontal  = {RGBColor[0, 1, 0], Thickness[0.005],
                          Line[{{0, 0}, {-1.3, 0}}]};
        thickSlanted    = {RGBColor[0, 1, 0], Thickness[0.015],
                          Line[{{0, 0}, 1.3 {Cos[θD], Sin[θD]}}]};
        thinSlanted     = {RGBColor[0, 1, 0], Thickness[0.005],
                          Line[{{0, 0}, 1.3 {Cos[θD180], Sin[θD180]}}]};
        i = 0;
        (* all mirrored charges *)
        points = {i = i + 1; If[EvenQ[i], RGBColor[0, 0, 1],
                RGBColor[1, 0, 0]], PointSize[0.02],
```

```
                    Point[{Cos[#], Sin[#]}]}]& /@ charges;
        (* the order of mirroring *)
        connectingLines = {GrayLevel[0], Line[#]}& /@
                        Partition[{Cos[#], Sin[#]}& /@ charges, 2, 1];
        (* display charges, plates and mirror order *)
        Show[Graphics[{thickHorizontal, thinHorizontal,
                thickSlanted, thinSlanted, points, connectingLines,
                Text[StyleForm[data[[1]], FontFamily -> "Helvetica",
                    FontSize -> 6], {0, 0}]}], opts,
            AspectRatio -> Automatic, PlotRange -> All]] /; 0 < start < θ
```

We now look at the cases requested in the problem. (The plot labels are both the starting angle and the angle between the metal plates.)

```
In[4]:= Show[GraphicsArray[#]]& /@ Partition[
        mirrorChargePlot[##, DisplayFunction -> Identity]& @@@
            (* the example data *)
            {{ 45, 90, 30, PlotLabel -> {"45°", "90°"}},
             { 30, 60, 30, PlotLabel -> {"30°", "60°"}},
             { 45/2, 45, 30, PlotLabel -> {"45/2°", "90°"}},
             { 360/64, 360/32, 60, PlotLabel -> {"360/64°", "360/32°"}},
             { 5, 20, 30, PlotLabel -> {"5°", "20°"}},
             { 10, 50, 30, PlotLabel -> {"10°", "50°"}},
             { 40, 120, 30, PlotLabel -> {"40°", "120°"}},
             { 15, 30, 30, PlotLabel -> {"15°", "30°"}}}, 2]
```

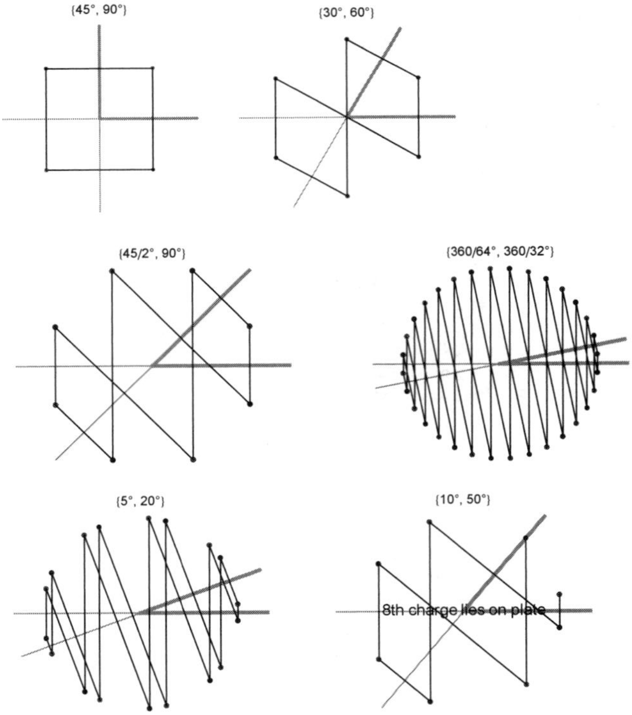

(45°, 90°)

(30°, 60°)

(45/2°, 90°)

(360/64°, 360/32°)

(5°, 20°)

(10°, 50°)

8th charge lies on plate

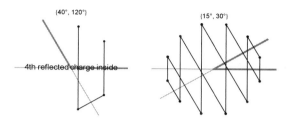

For a general solution, see [133]; for a detailed discussion of the periodic paths, see [230].

5. Periodic Decimal Numbers, Digit Sequences, Numbered Permutations, Binomial Values, Smith's Sturmian Word Theorem

a) We implement the usual algorithm (taught in elementary school) for dividing one integer by another. In the list `remain-der`, we collect the remainders after each step. We stop the division when a remainder appears for the second time.

```
In[1]:= ExactDecimalNumber[fraction_Rational] :=
    Module[{denominator, integerPart, remainder, l, remainders,
            period, digit, pos, previous},
        (* initializing all variables *)
        denominator = Denominator[fraction];
        integerPart = Floor[fraction];
        remainder = (fraction - integerPart) denominator;
        l = {}; (* list of remainders to recognize periodicity *)
        remainders = {remainder};
        While[(* carry out the repeated division until end
                 or until we meet same situation we already had *)
            digit = Floor[10 remainder/denominator];
            AppendTo[l, digit];
            remainder = (10 remainder/denominator - digit) denominator;
            (* this situation we already had--end here *)
            !MemberQ[remainders, remainder],
            AppendTo[remainders, remainder]];
        (* format the result nicely *)
        pos = Position[remainders, remainder][[1, 1]];
        period = StringJoin[ToString[OverBar[#], StandardForm]& /@
                        Take[l, {pos, Length[l]}]];
        previous = StringJoin[ToString /@ Take[l, {1, pos - 1}]];
        (* the periodic part *)
        ToString[integerPart] <> "." <> previous <>
                        If[period != "0", period, ""]]
```

Here are a few examples.

```
In[2]:= {ExactDecimalNumber[1/3], ExactDecimalNumber[345/564]}
```

```
Out[2]= {0.3̄, 0.6117021276595744680851063829787234042553191489363̄6̄}
```

The last result agrees with the result of the built-in function `RealDigits`.

```
In[3]:= RealDigits[345/564]
```

```
Out[3]= {{6, 1, {1, 7, 0, 2, 1, 2, 7, 6, 5, 9, 5, 7, 4, 4, 6, 8, 0, 8, 5, 1, 0, 6,
        3, 8, 2, 9, 7, 8, 7, 2, 3, 4, 0, 4, 2, 5, 5, 3, 1, 9, 1, 4, 8, 9, 3, 6}}, 0}
```

If a remainder 0 appears at the end, we, of course, do not write it. In the implementation above, we could easily take it out.

```
In[4]:= {ExactDecimalNumber[1/16], ExactDecimalNumber[2/5]}
```

```
Out[4]= {0.062506̄, 0.40̄}
```

Next, we present an especially interesting number: The numbers 01, 02, 03, …, 95, 96, 97, (not 98) and 99 appear sequentially in the periodic part. (For similar examples, see [202], [96]; for fractions containing the Fibonacci numbers in their periodic parts, see [521], [360].)

```
In[5]:= ExactDecimalNumber[1/99^2]
```

In[5]:= 0.000102030405060708091011121314151617181920212223242526272829303132333435363738394041424344454647484950515253545556575859606162636465666768697071727374757677787980818283848586878889909192939495969799

To make sure our calculations were correct, let us compare the last result with the function `RealDigits`, which returns an exact result for fractions. (Notice the different positions of the leading/trailing periodic zeros.)

In[6]:= **RealDigits[1/99^2]**

Out[6]= {{{1, 0, 2, 0, 3, 0, 4, 0, 5, 0, 6, 0, 7, 0, 8, 0, 9, 1, 0, 1, 1, 1, 2, 1, 3, 1, 4, 1, 5, 1, 6, 1,
        7, 1, 8, 1, 9, 2, 0, 2, 1, 2, 2, 2, 3, 2, 4, 2, 5, 2, 6, 2, 7, 2, 8, 2, 9, 3, 0, 3,
        1, 3, 2, 3, 3, 3, 4, 3, 5, 3, 6, 3, 7, 3, 8, 3, 9, 4, 0, 4, 1, 4, 2, 4, 3, 4, 4, 4,
        5, 4, 6, 4, 7, 4, 8, 4, 9, 5, 0, 5, 1, 5, 2, 5, 3, 5, 4, 5, 5, 5, 6, 5, 7, 5, 8, 5,
        9, 6, 0, 6, 1, 6, 2, 6, 3, 6, 4, 6, 5, 6, 6, 6, 7, 6, 8, 6, 9, 7, 0, 7, 1, 7, 2, 7,
        3, 7, 4, 7, 5, 7, 6, 7, 7, 7, 8, 7, 9, 8, 0, 8, 1, 8, 2, 8, 3, 8, 4, 8, 5, 8, 6, 8,
        7, 8, 8, 8, 9, 9, 0, 9, 1, 9, 2, 9, 3, 9, 4, 9, 5, 9, 6, 9, 7, 9, 9, 0, 0, 0}}, -3}

The following number has the decimal expansion $0.1\,2\,3\ldots997\,998\,999\ldots$ [96].

In[7]:= **10/81 - 334/3267 10^-8 - 1099022/120758121 10^-187 -**
       ToExpression["0." <> StringJoin[Table[ToString[k], {k, 999}]]]

Out[7]= $1. \times 10^{-2889}$

Here is an interesting property of the periodic part for a fraction of the form m/p with even period length (p a prime and m coprime to p) [199].

In[8]:= **Plus @@ (Partition[#, Length[#]/2]& @**
                RealDigits[2^5/Prime[50]][[1, -1]])

Out[8]= {9, 9, 9, 9, 9, 9, 9, 9, 9, 9, 9, 9, 9, 9, 9, 9, 9, 9, 9, 9, 9, 9, 9, 9, 9, 9, 9, 9, 9, 9, 9,
        9, 9, 9, 9, 9, 9, 9, 9, 9, 9, 9, 9, 9, 9, 9, 9, 9, 9, 9, 9, 9, 9, 9, 9, 9, 9, 9, 9, 9,
        9, 9, 9, 9, 9, 9, 9, 9, 9, 9, 9, 9, 9, 9, 9, 9, 9, 9, 9, 9, 9, 9, 9, 9, 9, 9, 9, 9, 9,
        9, 9, 9, 9, 9, 9, 9, 9, 9, 9, 9, 9, 9, 9, 9, 9, 9, 9, 9, 9, 9, 9, 9, 9, 9, 9, 9, 9, 9}

We could make `ExactDecimalNumber` much more elegant. It could be written to only format the numbers in this way, but leave them internally unchanged for further use, for instance. A function similar to `ExactDecimalNumber` is the function `PeriodicForm` from the package `NumberTheory`ContinuedFractions``.

In[9]:= **Needs["NumberTheory`ContinuedFractions`"]**

In[10]:= **PeriodicForm[%%]**

Out[10]= PeriodicForm[{9, 9, 9, 9, 9, 9, 9, 9, 9, 9, 9, 9, 9, 9, 9, 9, 9, 9, 9, 9, 9, 9, 9, 9, 9, 9, 9, 9, 9,
        9, 9, 9, 9, 9, 9, 9, 9, 9, 9, 9, 9, 9, 9, 9, 9, 9, 9, 9, 9, 9, 9, 9, 9, 9, 9, 9, 9, 9, 9,
        9, 9, 9, 9, 9, 9, 9, 9, 9, 9, 9, 9, 9, 9, 9, 9, 9, 9, 9, 9, 9, 9, 9, 9, 9, 9, 9, 9, 9, 9,
        9, 9, 9, 9, 9, 9, 9, 9, 9, 9, 9, 9, 9, 9, 9, 9, 9, 9, 9, 9, 9, 9, 9, 9, 9, 9, 9, 9, 9, 9}]

If we only want to determine the number of nonperiodic and periodic digits after the decimal point, we do not have to calculate all digits explicitly. The following function `numberOfDigitsAfterDot` returns a list of the number of nonperiodic and periodic digits after the decimal point for the input fraction b [201].

In[11]:= **numberOfDigitsAfterDot[b_Rational?(# < 1&)] :=**
       Function[q, If[Complement[First /@ q, {2, 5}] === {},
             {Max[Last /@ q], 0}, {If[# == -Infinity, 0, #]&[
             Max[Last /@ Cases[q, {2 | 5, _}]]],
       Module[{n = 1}, While[PowerMod[10, n, #] =!= 1,
             n = n + 1]; n]&[Times @@ Apply[Power,
                Cases[q, {_?(# =!= 2 && # =!= 5&), _}], 1]]}]][
                                    FactorInteger[Denominator[b]]]

       numberOfDigitsAfterDot[b_Rational?(# >= 1&)] :=
                numberOfDigitsAfterDot[b - Floor[b]]

       numberOfDigitsAfterDot[i_Integer] := {0, 0}

For our examples from above, we get the following number of nonperiodic and periodic digits after the decimal point.

In[14]:= **numberOfDigitsAfterDot /@ {1/3, 1/16, 2/5, 1/99^2}**

Out[14]= {{0, 1}, {4, 0}, {1, 0}, {0, 198}}

Next, we visualize the period for 10000 fractions i/j.

```
In[15]:= ListDensityPlot[Table[numberOfDigitsAfterDot[i/j][[2]],
                    {i, 100}, {j, 100}], Mesh -> False]
```

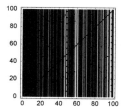

Again, we could have used `RealDigits` here.

```
In[16]:= numberOfDigitsAfterDot1[b_Rational?(# < 1&)] :=
                    {Length[#] - 1, Length[Last[#]]}& @ RealDigits[b][[1]]
In[17]:= numberOfDigitsAfterDot1 /@ {1/3, 1/16, 2/5, 1/99^2}
Out[17]= {{0, 1}, {2, 0}, {0, 0}, {0, 198}}
```

This returns a list with the number of nonrecurring digits and the number of recurring digits. For more details on periodic decimal fractions, see [240], [144], [25], [450], [453], [267], [150], [475], [189], and [201].

b) We start by implementing a function `smallestIntegerSequence` that, for a given list of digits, returns the above-described sequence of integers.

```
In[1]:= smallestIntegerSequence[digits_, base_:10] :=
        Module[{pos = 1, counter = 1, δ, occurred},
        (* continue as long as enough digits are left *)
        While[δ = 0;
              (* extract next digit sequence *)
              While[pos + δ < Length[digits] && (* new? *)
                    NumberQ[occurred[seq = Take[digits, {pos, pos + δ}]]], δ++];
              If[pos + δ < Length[digits],
                 (* new digit sequence *) occurred[seq] = counter; counter++];
              pos = pos + δ + 1; pos < Length[digits], Null];
        (* all occurring digit sequences *)
        Last /@ Sort[{#2, FromDigits[#1[[1, 1]], base]}& @@@
                                        DownValues[occurred]]]
```

Here are three examples: the first integers of the sequences of π, φ, and $1/7$. The first two sequences show a moderate growth rate and the last shows a faster one (due to the periodicity of the decimal expansion).

```
In[2]:= smallestIntegerSequence @ RealDigits[N[Pi, 50]][[1]]
Out[2]= {3, 1, 4, 15, 9, 2, 6, 5, 35, 8, 97, 93, 23,
         84, 62, 64, 33, 83, 27, 95, 0, 28, 841, 971, 69, 39, 937}
In[3]:= smallestIntegerSequence @ RealDigits[N[GoldenRatio, 50]][[1]]
Out[3]= {1, 6, 18, 0, 3, 39, 8, 87, 4, 9, 89, 48, 482,
         4, 5, 86, 83, 43, 65, 63, 81, 17, 7, 2, 3, 9, 179, 80}
In[4]:= smallestIntegerSequence @ RealDigits[N[1/7, 50]][[1]]
Out[4]= {1, 4, 2, 8, 5, 7, 14, 28, 57, 142, 85, 71, 42, 857, 1428, 571, 428, 5714, 285, 714, 2857}
```

For larger sequences, we do not want to see the sequences themselves, but only some of their properties. The function `realAnalyser` plots and fits $f(data)$.

```
In[5]:= realAnalyser[l_List, f_:Identity, base_:10] :=
        Module[{data, λ, data1},
               (* the sequences *)
               data = smallestIntegerSequence[RealDigits[#, base][[1]],
                                              base]& /@ l;
               λ = Min[Length /@ data];
```

```
(* list of k-th elements *)
data1 = (Plus @@@ Transpose[Take[#, λ]& /@ data])/Length[l];
(* plot and fit *)
{ListPlot[N[f @ data1], DisplayFunction -> Identity],
 Fit[f @ data1, {x}, x]}]
```

For the two "random" sequences $p_j^{1/2}$ and $\cos(j)$ and base 10, we obtain $a_k \approx 4.5\,k$.

```
In[6]:= {ra1 = realAnalyser[Table[N[Sqrt[Prime[j]], 1000], {j, 1000}]],
         ra2 = realAnalyser[Table[N[Cos[j], 1000], {j, 1000}]]}
        Show[GraphicsArray[{ra1[[1]], ra2[[1]]}]]
```

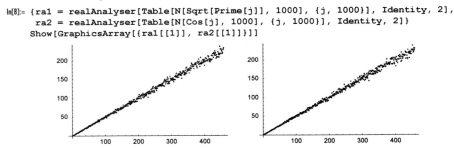

Here is the last calculation repeated in base 2, we get $a_k \approx 0.5\,k$.

```
In[8]:= {ra1 = realAnalyser[Table[N[Sqrt[Prime[j]], 1000], {j, 1000}], Identity, 2],
         ra2 = realAnalyser[Table[N[Cos[j], 1000], {j, 1000}], Identity, 2]}
        Show[GraphicsArray[{ra1[[1]], ra2[[1]]}]]
```

Repeating the last calculation in base 3, we get $a_k \approx k$. This leads to the general conjecture $a_k = (base - 1)/2\,k$.

```
In[10]:= {ra1 = realAnalyser[Table[N[Sqrt[Prime[j]], 1000], {j, 1000}], Identity, 3],
          ra2 = realAnalyser[Table[N[Cos[j], 1000], {j, 1000}], Identity, 3]}
         Show[GraphicsArray[{ra1[[1]], ra2[[1]]}]]
```

For fractions, we get an exponential increase, $a_k \sim \ln(k)$. Instead of averaging over various digits sequences, we could, of course, also average over long runs of one sequence. The resulting graphic is shown at the right side.

```
In[12]:= ra3 = realAnalyser[Table[N[1/Prime[j], 1000], {j, 1000}], Log]
```

```
In[13]:= data = smallestIntegerSequence[RealDigits[N[Pi, 10^5]][[1]]];
```

```
In[14]:= ra4 = With[{k = 1000},
                 {(* for a smoother graphics, average *)
                  ListPlot[ListConvolve[Table[1./k, {k}], data],
                           PlotStyle -> {PointSize[0.003]},
                           DisplayFunction -> Identity],
                  Fit[data, {x}, x]}]
```

In[15]:= (* display the two last graphics *)
        Show[GraphicsArray[{ra3[[1]] // N, ra4[[1]] // N}]]

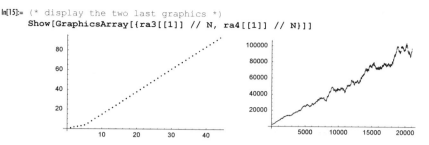

c) We start by implementing a function `FactorialBaseForm` that generates the digits c_k of an integer n in the factorial representation with base b.

```
In[1]:= FactorialBaseForm[n_, b_] := First /@ Rest[
        FoldList[Function[{r, f}, (* remainder and digit *)
                 {Floor[#], f (# - Floor[#])}&[r[[2]]/f]],
                 {0, n}, Range[b - 1, 0, -1]!]] /; n < b!
```

The inverse function, `FromFactorialBaseForm` is straightforward to implement.

```
In[2]:= FromFactorialBaseForm[digits_, b_] := digits.Range[b - 1, 0, -1]! /;
                                               Length[digits] == b
```

Here is an example: we write 1000 in the factorial representation with base 10.

```
In[3]:= FactorialBaseForm[100, 10]

Out[3]= {0, 0, 0, 0, 0, 4, 0, 2, 0, 0}
```

```
In[4]:= FromFactorialBaseForm[%, 10]

Out[4]= 100
```

The large number 10^{1000} can be represented in base 1000, in a fraction of a second.

```
In[5]:= FromFactorialBaseForm[FactorialBaseForm[10^1000, 1000],
                              1000]^(1/1000) // Timing

Out[5]= {0.21 Second, 10}
```

Next, we define the function `factorialDigitsToPermutations` that, given a list of factorial representation digits of length n, forms the corresponding permutation of the numbers $\{1, 2, ..., n\}$.

```
In[6]:= factorialDigitsToPermutations[digits_List] :=
        Fold[{Delete[#1[[1]], #2], Append[#1[[2]], #1[[1, #2]]]}&,
             {Range[Length[digits]], {}}, digits + 1][[2]]
```

Here are the first and the last permutation of the sequence $\{1, 2, 3, 4, 5\}$.

```
In[7]:= {factorialDigitsToPermutations[{0, 0, 0, 0, 0}],
         factorialDigitsToPermutations[{4, 3, 2, 1, 0}]}

Out[7]= {{1, 2, 3, 4, 5}, {5, 4, 3, 2, 1}}
```

Now, we have all functions to implement `nthPermutation`.

```
In[8]:= nthPermutation[l_List, n_Integer?Positive] :=
        l[[factorialDigitsToPermutations[
           FactorialBaseForm[n - 1, Length[l]]]]] /; n <= Length[l]!
```

The following input shows that `nthPermutation` generates the 720 permutations of six symbols in the same order as the built-in function `Permutations`.

```
In[9]:= Table[nthPermutation[{a, b, c, d, e, f}, n], {n, 6!}] ===
        Permutations[{a, b, c, d, e, f}]

Out[9]= True
```

But with `nthPermutation`, we can generate single permutations of many more symbols than the built-in function `Permutations` with generates all permutations.

```
In[10]:= {nthPermutation[Range[100], 10^100],
         MemoryConstrained[Permutations[Range[100]][[10^100]], 10^8]}
Out[10]= {{1, 2, 3, 4, 5, 6, 7, 8, 9, 10, 11, 12, 13, 14, 15, 16, 17, 18, 19, 20, 21, 22, 23,
         24, 25, 26, 27, 28, 29, 30, 89, 61, 44, 74, 63, 54, 81, 31, 59, 39, 47, 42, 73,
         91, 99, 100, 83, 51, 95, 66, 98, 60, 71, 49, 96, 64, 80, 37, 56, 36, 62, 40, 90,
         50, 85, 72, 34, 55, 97, 94, 32, 87, 69, 57, 82, 78, 68, 45, 84, 75, 88, 48, 53,
         67, 33, 52, 43, 46, 76, 70, 86, 41, 92, 93, 79, 38, 65, 58, 77, 35}, $Aborted}
```

We end by defining the function `permutationToPermutationNumber` that for a given permutation returns its number.

```
In[11]:= permutationToPermutationNumber[l_] :=
         FromFactorialBaseForm[
         Nest[Function[ℓ, {Rest[ℓ[[1]]], Append[ℓ[[2]],
                          Count[Rest[ℓ[[1]]], _?(# < ℓ[[1, 1]]&)]]}],
              {l, {}}, Length[l]][[2]], Length[l] + 1

In[12]:= permutationToPermutationNumber[%%[[1]]]

Out[12]= 1000000000000000000000000000000000000000000000000000000000000000000000000000000000000\
         00000000000000000000
```

d) Because $\binom{n}{0} = 1$, we will ignore the trivial case $\mathcal{B}(1) = \infty$ in the following. It is straightforward to implement a function `BinomialValueArguments` that returns the $\mathcal{B}(o)$ pairs $\{n_j, k_j\}$. For a given o, we must have $n_j \leq o$. For each n, we increase k, as long as $\binom{n}{k} \leq o$. If we encounter the situation $\binom{n}{k} = o$, we add the pairs $\{n, k\}$ to the bag `knValues` (and, depending on the parity of n, also $\{n - k, k\}$).

```
In[1]:= BinomialValueArguments[o_] :=
        Module[{knValues = {}, n = 1, k = 1, b},
               While[n <= o,
                   While[k <= n/2 && (b = Binomial[n, k]) <= o,
                         If[b == o, AppendTo[knValues, {n, k}]];
                                    (* unique middle element? *)
                                    If[Not[EvenQ[n]] && k == n/2,
                                       AppendTo[knValues, {n, n - k}]]];
                                    k++]; n++; k = 1];
               knValues]
```

Calculating the six pairs $\{n_j, k_j\}$ for $o = 120$ happens basically instantaneously.

```
In[2]:= {#, Binomial @@@ #}&[BinomialValueArguments[120]] // Timing
Out[2]= {3.88958×10⁻¹⁸ Second, {{{10, 3}, {10, 7}, {16, 2}, {16, 14}, {120, 1}, {120, 119}},
        {120, 120, 120, 120, 120, 120}}}
```

Calculating the two pairs $\{n_j, k_j\}$ for $o = 10^5$ takes a couple of seconds.

```
In[3]:= {#, Binomial @@@ #}&[BinomialValueArguments[100000]] // Timing
Out[3]= {1.82 Second, {{{100000, 1}, {100000, 99999}}, {100000, 100000}}}
```

To find pairs $\{n_j, k_j\}$ such that $\mathcal{B}(o) = 8$, we slightly modify our function `BinomialValueArguments`. Instead of looping over o (which would be quadratically slow), we count at once the values of all binomial coefficients less than or equal to *max*.

```
In[4]:= BinomialValueCount[max_] :=
        Module[{count, n = 1, k = 1},
               (* list of counters for each o <= max *)
               countList = Table[0, {o, 2, max}];
               While[n <= max,
                     While[k <= n/2 && (b = Binomial[n, k]) <= max,
                           countList[[b - 1]] = countList[[b - 1]] +
                                                If[EvenQ[n] && k == n/2, 1, 2];
                           k++]; n++; k = 1];
               (* return counts for each o *)
               Transpose[{Range[2, max] + 1, countList}]]
```

Calculating $\mathcal{B}(o)$ for all $o \le 10^6$ takes about a minute. The result shows that, in the vast majority of cases, $\mathcal{B}(o) = 2$. Only ten cases of $\mathcal{B}(o) = 3$ and even only a single case of $\mathcal{B}(o) = 8$ are found.

```
In[5]:= (res = BinomialValueCount[10^6]); // Timing
Out[5]= {30.74 Second, Null}

In[6]:= {First[#], Length[#]}& /@ Split[Sort[Last /@ res]]
Out[6]= {{1, 1}, {2, 998266}, {3, 10}, {4, 1715}, {6, 6}, {8, 1}}
```

Here are the eight pairs $\{n_j, k_j\}$ yielding $\binom{n_j}{k_j} = 3003$.

```
In[7]:= Position[res, {_, 8}][[1, 1]] + 1
Out[7]= 3003

In[8]:= {#, Binomial @@@ #}&[BinomialValueArguments[3003]]
Out[8]= {{{14, 6}, {14, 8}, {15, 5}, {15, 10}, {78, 2}, {78, 76}, {3003, 1}, {3003, 3002}},
        {3003, 3003, 3003, 3003, 3003, 3003, 3003, 3003}}
```

e) The implementation of a function SmithTheoremTest is straightforward. The left-hand side is trivial to implement. The right-hand side uses NestWhile to concatenate 0's and 1's as long as we have fewer than n digits.

```
In[1]:= SmithTheoremTest[α_?((0 < # < 1 && Head[#] =!= Rational)&),
                         n_Integer?Positive] :=
        Array[(Floor[(# + 1) α] - Floor[# α])&, {n}] ===
        Take[NestWhile[Append[#, Flatten[{Table[#[[-1]],
            {ContinuedFraction[α, Length[#] + 1][[-1]]}], #[[-2]]}]]&,
            {{0}, Flatten[{Table[0, {ContinuedFraction[α, 2][[-1]] - 1}], 1}]},
                        (Length[Last[#]] < n)&][[-1]], n]
```

Here we check Smith's Sturmian word theorem for seven irrational numbers to 10^5 digits.

```
In[2]:= SmithTheoremTest[#, 10^5]& /@
             {1/Sqrt[2], 3^(-1/3), Log[2], 1/Pi, 1/E^2, -Cos[E Pi], Cot[1]}
Out[2]= {True, True, True, True, True, True, True}
```

6. Galton Board, Ehrenfest Urn Model, Ring Shift, Longest Common Subsequence, Riffle Shuffles

a) Here is the result (including all intermediate results) of a model of the Galton board with rows rows of nails into which balls balls are dropped.

```
In[1]:= GaltonBoard[rows_, balls_] :=
        MapIndexed[Function[{x, y}, (* analyze result *)
        {#, Count[x, #]}& /@ Union[x]], Transpose[Table[(* balls trials *)
        NestList[(* generate value at next row *)
                (# + (* + or - 1 *) 2Random[Integer, {0, 1}] - 1)&, 0,
                                   rows], {balls}]]];
```

Here is an example.

```
In[2]:= GaltonBoard[12, 5]
Out[2]= {{{0, 5}}, {{-1, 4}, {1, 1}}, {{-2, 2}, {0, 2}, {2, 1}}, {{-3, 1}, {-1, 3}, {3, 1}},
        {{-4, 1}, {-2, 1}, {0, 2}, {2, 1}}, {{-5, 1}, {-1, 3}, {3, 1}},
        {{-4, 1}, {-2, 1}, {0, 2}, {4, 1}}, {{-3, 1}, {-1, 2}, {1, 1}, {3, 1}},
        {{-4, 1}, {-2, 2}, {2, 2}}, {{-5, 1}, {-3, 1}, {-1, 1}, {1, 2}},
        {{-4, 1}, {-2, 1}, {0, 3}}, {{-5, 1}, {-1, 2}, {1, 2}}, {{-6, 1}, {-2, 1}, {0, 1}, {2, 2}}}

In[3]:= {Length[%] - 1, Plus @@ #& /@ Map[Last, %, {-2}]}
Out[3]= {12, {5, 5, 5, 5, 5, 5, 5, 5, 5, 5, 5, 5, 5}}
```

To create a plot, we use Line primitives. (For a large number of nails and balls, the plotting of polygons would take a long time.) We plot the distribution in all rows, starting at the beginning.

```
In[4]:= galtonGraphics[data_] :=
        Graphics3D[{Thickness[0.002],
        MapIndexed[Function[{x, y}, {Hue[y[[1]]/50], (* add color *)
                    Line @ (Insert[#, -y[[1]] + 1, 2]& /@
```

```
                    (* make shifted 3D lines *)
                    Flatten[{{{#[[1]] - 1/2, 0}, # - {1/2, 0},
                        # + {1/2, 0}, {#[[1]] + 1/2, 0}}& /@ x, 1])}], data]},
              BoxRatios -> {2, 1, 1}, PlotRange -> All,
              ViewPoint -> {0.1, -2, 0.6}, Axes -> True];
```

For comparison, we now show the theoretically expected result—the binomial distribution. The very improbable tails in the distribution are clearly visible in the theoretical distribution.

```
In[5]:= Show[GraphicsArray[
            {(* experimental distribution *)
             galtonGraphics[GaltonBoard[120, 50]],
             (* theoretical distribution *)
             galtonGraphics[Table[Thread[{Table[i, {i, -j, j, 2}],
                     50 #/(Plus @@ #)&[(* the large number result *)
                 Table[Binomial[j, i], {i, 0, j}]]}], {j, 0, 120}]]}]]
```

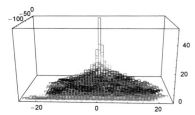 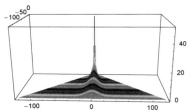

It is much "broader" than is the experimental result, but the probability of the appearance of a ball in the edge bins—in a 50-ball experiment—is extremely small.

```
In[6]:= 50 Binomial[50, 0]/Sum[Binomial[50, i], {i, 0, 50}] // N

Out[6]= 4.44089×10^-14
```

For a more detailed simulation of a Galton board, see [409], [269], [339], [115], [367], [89], [332], and [270].

b) To carry out the simulation efficiently, we should use `Compile`. The implementation of the simulation is straightforward.

```
In[1]:= cf = Compile[{{t, _Integer}, {n, _Integer}, {m, _Integer}},
        Module[{fleaCount, fleaPosition, fleasA, r},
            (* list of number of fleas at dog A *)
            fleaCount = Table[0, {t}];
            (* a 1 means dog A and a -1 means dog B *)
            fleaPosition = Join[Table[1, {m}], Table[-1, {n - m}]];
            (* number of fleas on dog A *)
            fleasA = m;
            Do[(* the chosen flea *)
               r = Random[Integer, {1, n}];
               (* update flea count for dog A *)
               If[fleaPosition[[r]] == 1, fleasA--, fleasA++];
               (* switch flea to other dog *)
               fleaPosition[[r]] = -fleaPosition[[r]];
               (* store flea count in list fleaCount *)
               fleaCount[[k]] = fleasA, {k, t}];
            (* return fleaCount *)
            fleaCount]];
```

The function compiled successfully.

```
In[2]:= Union[Head /@ Flatten[cf[[4]]]]

Out[2]= {Integer}
```

Here, we model the distribution of initially 10000 fleas on dog A for 100000 steps. After about 20000 steps, the equilibrium distribution is reached.

```
In[3]:= ListPlot[cf[10^5, 10^4, 0], PlotRange -> {0, 10^4},
            PlotStyle -> {Thickness[0.002]},
            PlotJoined -> True, Frame -> True, Axes -> False]
```

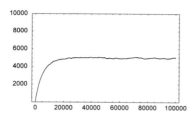

To model the distribution of 1000 fleas over 10 million time steps, we will change the above function `cf` slightly. It is much more memory efficient not to store the 10 million different flea counts (one for each time step) but instead count how often exactly k fleas were on dog A. This means we only have to store a list of length 1000 instead of a list of length 10000000.

```
In[4]:= cf = Compile[{{t, _Integer}, {n, _Integer}, {m, _Integer}},
          Module[{fleaCount, fleaPosition, fleasA, r},
            (* list of number of the n + 1 possible values
               for the number of fleas on dog A *)
            fleaCount = Table[0, {n + 1}];
            (* initially, m fleas are on dog A *)
            fleaCount[[m + 1]] = m;
            fleasA = m;
            (* a +1 means dog A and a -1 means dog B *)
            fleaPosition = Join[Table[1, {m}], Table[-1, {n - m}]];
            Do[(* the chosen flea *)
               r = Random[Integer, {1, n}];
               (* update flea count for dog A *)
               If[fleaPosition[[r]] == 1, fleasA--, fleasA++];
               (* switch flea to other dog *)
               fleaPosition[[r]] = -fleaPosition[[r]];
               (* update fleaCount *)
               fleaCount[[fleasA]] = fleaCount[[fleasA]] + 1, {k, t}];
            (* return fleaCount *)
            fleaCount]];
```

Again, the function compiled successfully.

```
In[5]:= Union[Head /@ Flatten[cf[[4]]]]
```

```
Out[5]= {Integer}
```

Carrying out the simulation for 1000 fleas and 10 million steps takes about 20 s on a 2 GHz computer.

```
In[6]:= Timing[data = cf[10^7, 1000, 0];]
```

```
Out[6]= {9.4 Second, Null}
```

The resulting distribution of fleas is strongly peaked near 500 fleas per dog. The theoretical equilibrium distribution is a binomial distribution. The red points in the following graphic represent the theoretical distribution.

```
In[7]:= ListPlot[MapIndexed[{#2[[1]] - 1, #1}&, data/10^7],
          PlotRange -> All, Frame -> True, Axes -> False,
          PlotJoined -> True, PlotStyle -> {Thickness[0.002], GrayLevel[0]},
          Prolog -> {PointSize[0.008], Hue[0],
            Table[Point[{k, Binomial[1000, k] 2^-1000}], {k, 0, 1000}]}]]
```

For an analysis of the fluctuations from the average, see [242]; for multidog generalizations, see [300].

c) We start by implementing the moves. Given a list $\{k_1, k_2, ..., k_n\}$, the function `nextLists` will generate a list of all lists that can be obtained by all allowed moves.

```
In[1]:= nextLists[l_] :=
           With[{λ = Length[l]},
               MapAt[# + 1&, MapAt[# - 1&, 1, #], If[# < λ, # + 1, 1]]& /@
                   Flatten[First /@ Select[MapIndexed[{#2, #1}&,
                                   Partition[Append[l, First[l]], 2, 1]],
                                       Greater @@ #[[2]]&]]]
```

Here is an example.

```
In[2]:= nextLists[{6, 1, 5, 2, 0, 4, 1}]
Out[2]= {{5, 2, 5, 2, 0, 4, 1}, {6, 1, 4, 3, 0, 4, 1}, {6, 1, 5, 1, 1, 4, 1}, {6, 1, 5, 2, 0, 3, 2}}
```

Starting list $\{m, 0, ..., 0\}$ we will recursively carry out all possible moves. Let $\Lambda[k]$ be the list of all lists after k moves. The function `step` carries out all moves on a given list from $\Lambda[k]$, eliminates doubles, and divides the resulting lists into ones already encountered in earlier moves and new ones.

```
In[3]:= step[l_, Λ_] :=
           Module[{newLists, oldLists, newOld, newNew},
               (* all already encountered lists *)
               oldLists = #[[2, 1]]& /@ DownValues[Λ];
               (* where were newlists already encountered? *)
               newLists = {Position[oldLists, #], #}& /@ nextLists[l];
               (* already encountered and newly encountered lists *)
               newOld = Last /@ Cases[newLists, {_?(# =!= {}&), _}];
               newNew = Complement[Last /@ newLists, newOld];
               {newNew, newOld}]
```

`stepΛ` carries out `step` on all lists of $\Lambda[k]$ and eliminates doubles.

```
In[4]:= stepΛ[{l_, lOld_}, Λ_] :=
           Union[Flatten[#, 1]]& /@ Transpose[step[#, Λ]& /@ l]
```

The following example shows how the repeated application of `stepΛ` to the initial list `{3, 0, 0}` naturally ends after three steps.

```
In[5]:= Λ1[0] = {{{3, 0, 0}}, {}};

In[6]:= Λ1[1] = stepΛ[Λ1[0], Λ1]
Out[6]= {{{2, 1, 0}}, {}}

In[7]:= Λ1[2] = stepΛ[Λ1[1], Λ1]
Out[7]= {{{1, 2, 0}, {2, 0, 1}}, {}}

In[8]:= Λ1[3] = stepΛ[Λ1[2], Λ1]
Out[8]= {{{1, 1, 1}}, {}}

In[9]:= Λ1[4] = stepΛ[Λ1[3], Λ1]
Out[9]= {{}, {}}
```

Starting with the list `{3, 0}` leads to the list `{2, 1}`, which was already earlier encountered.

```
In[10]:= Λ2[0] = {{{3, 0}}, {}};

In[11]:= Λ2[1] = stepΛ[Λ2[0], Λ2]
Out[11]= {{{2, 1}}, {}}

In[12]:= Λ2[2] = stepΛ[Λ2[1], Λ2]
Out[12]= {{{1, 2}}, {}}

In[13]:= Λ2[3] = stepΛ[Λ2[2], Λ2]
Out[13]= {{}, {{2, 1}}}
```

Now, we will carry out the repeated application of stepΛ automatically. We apply stepΛ until all possible configurations have been reached. The function doAsLongAsPossible does this for the initial list *l* and returns the number of applications of stepΛ, the length of the second list of the last Λ[*k*] and an optional value based on the downvalues of Λ.

```
In[14]:= doAsLongAsPossible[l_, f_: ((Sequence @@ {}) &)] :=
         Module[{Λ, c = 1},
                Λ[0] = {{1}, {}};
                While[(* carry out recursion *)
                      Λ[c] = stepΛ[Λ[c - 1], Λ];
                      Λ[c][[1]] =!= {}, c = c + 1];
                      (* return counter, last step, and optional result *)
                      {c, Length[Last[Λ[c]]], f[DownValues[Λ]]}]]
```

doAsLongAsPossible reproduces the two results from above.

```
In[15]:= {doAsLongAsPossible[{3, 0, 0}], doAsLongAsPossible[{3, 0}]}

Out[15]= {{4, 0}, {3, 1}}
```

```
In[16]:= do[m_, n_] := doAsLongAsPossible[Flatten[{m, Table[0, {n - 1}]}]]
```

```
In[17]:= Table[do[m, n], {n, 10}, {m, 10}]

Out[17]= {{{1, 0}, {1, 0}, {1, 0}, {1, 0}, {1, 0}, {1, 0}, {1, 0}, {1, 0}, {1, 0}, {1, 0}},
          {{2, 1}, {2, 0}, {3, 1}, {3, 0}, {4, 1}, {4, 0}, {5, 1}, {5, 0}, {6, 1}, {6, 0}},
          {{3, 1}, {4, 1}, {4, 0}, {6, 1}, {7, 1}, {7, 0}, {9, 1}, {10, 1}, {10, 0}, {12, 1}},
          {{4, 1}, {6, 1}, {7, 1}, {7, 0}, {10, 1}, {12, 1}, {13, 1}, {13, 0}, {16, 1}, {18, 1}},
          {{5, 1}, {8, 1}, {10, 1}, {11, 1}, {11, 0}, {15, 1}, {18, 1}, {20, 1}, {21, 1}, {21, 0}},
          {{6, 1}, {10, 1}, {13, 1}, {15, 1}, {16, 1}, {16, 0}, {21, 1}, {25, 1}, {28, 1}, {30, 1}},
          {{7, 1}, {12, 1}, {16, 1}, {19, 1}, {21, 1}, {22, 1}, {22, 0}, {28, 1}, {33, 1}, {37, 1}},
          {{8, 1}, {14, 1}, {19, 1}, {23, 1}, {26, 1}, {28, 1}, {29, 1}, {29, 0}, {36, 1}, {42, 1}},
          {{9, 1}, {16, 1}, {22, 1}, {27, 1}, {31, 1}, {34, 1}, {36, 1}, {37, 1}, {37, 0}, {45, 1}},
          {{10, 1}, {18, 1}, {25, 1}, {31, 1}, {36, 1}, {40, 1}, {43, 1}, {45, 1}, {46, 1}, {46, 0}}}}
```

From these results, we conjecture that the moves naturally terminate if $n = k\,m$ for some integer *k*. This is indeed the case [274], [211].

Using the optional third argument on doAsLongAsPossible, we can display the number of different sublists of each step. Here this is shown for $m = n = 10$.

```
In[18]:= ListPlot[doAsLongAsPossible[{10, 0, 0, 0, 0, 0, 0, 0, 0, 0},
                   (Length[#[[2, 1]]] & /@ #) &][[-1]]]
```

d) The function step implements the updating step for the list of integers *l*.

```
In[1]:= step[l_, c_] :=
        Module[{λ = Table[0, {Length[l]}], n = Length[l], r},
               Do[If[(* no change *)
                     l[[k]] < c, λ[[k]] = λ[[k]] + l[[k]],
                     (* split value randomly *)
                     r = Random[Integer, {1, l[[k]]}];
                     (* distribute split value *)
                     λ[[Mod[k - 1, n, 1]]] = λ[[Mod[k - 1, n, 1]]] + r;
                     λ[[Mod[k + 1, n, 1]]] =
                        λ[[Mod[k + 1, n, 1]]] + (l[[k]] - r)], {k, n}]; λ]
```

The function changeList applies step as long as possible and then compares pairs of lists to find the changed elements.

```
In[2]:= changeList[l0_, c_, maxSteps_: Infinity] :=
       1 - Abs[Sign[Subtract @@@ Partition[
                     FixedPointList[step[#, c]&, l0, maxSteps], 2, 1]]]
```

Here are three examples for lists of length 100 and $c = 100$.

```
In[3]:= Show[GraphicsArray[
       Block[{$DisplayFunction = Identity},
        Table[(* reproducible seed *) SeedRandom[k];
              ListDensityPlot[changeList[Table[Random[Integer, {1, 110}],
                                               {400}], 100], Mesh -> False],
              {k, 3}]]]]
```

For very long evolvements, it is inefficient to store the zeros indicating no change. The following routine `changeList·`
`Points` records only the position of changes.

```
In[4]:= changeListPoints[l0_, c_, maxSteps_: Infinity] :=
       Module[{lOld = l0, bag = {}, k = 1, lNew, changedPoints},
        While[k <= maxSteps && (lNew = step[lOld, c]];
         (* points where change occurred *)
         changedPoints = Rectangle[{# - 1/2, k - 1/2}, {# + 1/2, k + 1/2}]& /@
                         Flatten[Position[Abs[Sign[lNew - lOld]], 1]];
         changedPoints =!= {}),
         (* update variables *) lOld = lNew; bag = {bag, changedPoints}; k++];
         (* return bag *) bag]
```

In the following example, 2915 applications of `step` are needed to reach a state where all elements are smaller than *c*.

```
In[5]:= SeedRandom[100];
       clp = changeListPoints[Table[Random[Integer, {1, 120}], {200}], 100];
```

For clarity, we will display the result with interchanged axes.

```
In[7]:= Show[Graphics[clp] /. Rectangle[l1_, l2_] :>
                              Rectangle[Reverse[l1], Reverse[l2]],
              AspectRatio -> 1/4, PlotRange -> All, Frame -> True]
```

e) Here is a straightforward, rule-based definition of $l_{i,j}$ using caching.

```
In[1]:= longestCommonSubsequenceLength[s1_, s2_, l_] :=
       (Clear[l];
        l[i_, 0] = 0; l[0, j_] = 0;
        (* recursive definition for l *)
        l[i_, j_] := l[i, j] =
           Max[l[i - 1, j - 1] + If[s1[[i]] === s2[[j]], 1, 0],
                                   l[i - 1, j], l[i, j - 1]];
```

```
                       (* length of longest possible subsequence *)
                     l[Length[s1], Length[s1]])
```

Here is the result for a simple example.

```
In[2]:= longestCommonSubsequenceLength[{1,  0, 2,  0, 3,  0, 4,  0, 5},
                                        {1, -1, 2, -1, 3, -1, 4, -1, 5}, ℓ]

Out[2]= 5
```

Because of the term $\delta_{i,j}$ for a sequence of length n, we have to carry out n^2 comparisons. This means that doubling the length of the sequences will roughly quadruple the timings.

```
In[3]:= $RecursionLimit = Infinity;

       n = 300; SeedRandom[111];

       s1 = Table[Random[Integer, {1, 2}], {n}];
       s2 = Table[Random[Integer, {1, 2}], {n}];

       longestCommonSubsequenceLength[s1, s2, l] // Timing
Out[7]= {1.56 Second, 236}

In[8]:= $RecursionLimit = Infinity;

       n = 600; SeedRandom[111];

       s1 = Table[Random[Integer, {1, 2}], {n}];
       s2 = Table[Random[Integer, {1, 2}], {n}];

       longestCommonSubsequenceLength[s1, s2, l] // Timing
Out[12]= {7.36 Second, 492}
```

Here are visualizations of all the calculated $l_{i,j}$. ($l_{i,j}$ is the length of the longest possible subsequence from the two strings $\{s_1, s_2, \ldots, s_i\}$ and $\{t_1, t_2, \ldots, t_j\}$.)

```
In[13]:= Show[GraphicsArray[
        Block[{$DisplayFunction = Identity},
        {(* 3D plot of values *)
         ListPlot3D[Table[l[i, j], {i, 0, n}, {j, 0, n}], Mesh -> False],
         (* mod 2 density plot *)
         ListDensityPlot[Table[Mod[l[i, j], 2], {i, 0, n}, {j, 0, n}], Mesh -> False]}]]]
```

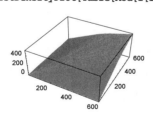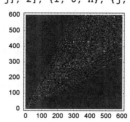

To deal with $n = 10^4$, we will use a compiled function. While this does not change the complexity of the algorithm, it speeds up the calculations by about a factor of 20. (Because positions of list elements start with 1, we have to renumber the indices in the above definition where we started with 0.)

```
In[14]:= longestCommonSubsequenceListCF =
        Compile[{{l1, _Integer, 1}, {l2, _Integer, 1}},
        Module[{n = Length[l1], pre, ℒ, L, bag, x},
              pre = Table[0, {n + 1}]; ℒ = pre; bag = pre;
              Do[L = pre;
                 Do[x = Max[ℒ[[j - 1]] + If[l1[[i]] === l2[[j - 1]], 1, 0],
                          ℒ[[j]], L[[j - 1]]]; L[[j]] = x,
                 {j, 2, n + 1}]; ℒ = L; bag[[i + 1]] = ℒ[[i + 1]], {i, n}];
              bag]];
```

```
In[15]:= Union[Head /@ Flatten[longestCommonSubsequenceListCF[[4]]]]
```

```
Out[15]= {Integer}
```

```
In[16]:= longestCommonSubsequenceListCF[s1, s2][[-1]] // Timing
```

```
Out[16]= {0.56 Second, 492}
```

Now, we can calculate $l_{k,k}$ for $n = 10^4$. We carry out the calculation three times.

```
In[17]:= c = 2; n = 10000;
     Do[lcs[i] = longestCommonSubsequenceListCF[
                 Table[Random[Integer, {1, c}], {n}],
                 Table[Random[Integer, {1, c}], {n}]], {i, 3}];
```

Here is a plot of the resulting $l_{k,k} / k$. As expected, $l_{k,k} / k$ approaches a limiting value, the so-called Chvátal–Sankoff constant [11], [317]. Due to finite size-effects, the sequences $l_{k,k} / k$ approach the limit from below.

```
In[19]:= Show[ListPlot[MapIndexed[#1/#2[[1]]&, lcs[#]],
               DisplayFunction -> Identity,
               PlotStyle -> {Hue[(# - 1)/4], PointSize[0.003]}]& /@ {1, 2, 3},
          DisplayFunction -> $DisplayFunction, Frame -> True, Axes -> False]
```

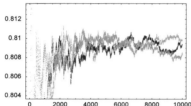

For a more advanced algorithm to calculate the longest common subsequence, see [221]; for an algorithm for the number of longest common subsequences, see [222].

f) For an efficient implementation of the riffle shuffle, we will use compiled functions whenever appropriate. The first step is the splitting of the initial desk into two piles according to the binomial distribution. The function `pileSplittingFunc tion` generates a nested `If`-construct that, for a given random number between 0 and 1, returns the size of one of the two piles. Nesting the `If` functions in a binary tree-like way needs only $\approx \log_2 n$ comparisons versus n for a consecutive comparison. At the end, the `If`-construct is compiled.

```
In[1]:= pileSplittingFunction[n_] := pileSplittingFunction[n] =
     Module[{p, if, half},
            (* cumulative splitting probabilities *)
            MapIndexed[(p[#2[[1]] - 1] = #1)&,
             FoldList[Plus, 0., Table[Binomial[n, k], {k, 0, n}]]/2^n];
            (* approximately half a list *)
            half[{x_?NumberQ}] := x;
            half[{x_?NumberQ, y_?NumberQ}] := if[ξ < p[y], x, y];
            half[l_List] := With[{v = Ceiling[Length[l]/2]},
                      (* recursive calls to half *)
                      if[ξ < p[l[[v + 1]]], half @ Take[l, v],
                        half @ Take[l, {v + 1, Length[l]}]]] /.
                      (* make nested If *)p -> p /. if -> If;
            (* compile nested If construct *)
            Compile @@ {{ξ}, half[Range[0, n]]}]
```

Here is the resulting function for $n = 8$.

```
In[2]:= pileSplittingFunction[8]
```

```
Out[2]= CompiledFunction[{ξ}, If[ξ < 0.636719,
         If[ξ < 0.144531, If[ξ < 0.0351563, If[ξ < 0.00390625, 0, 1], 2], If[ξ < 0.363281, 3, 4]],
         If[ξ < 0.964844, If[ξ < 0.855469, 5, 6], If[ξ < 0.996094, 7, 8]]], -CompiledCode-]
```

The following graphic shows the sizes of the piles for one million splittings of a desk with 128 cards. The underlying curve is the binomial distribution.

```
In[3]:= With[{n = 128, o = 10^6},
    ListPlot[{First[#], Length[#]/o}& /@ Split[Sort[
      Table[pileSplittingFunction[n][Random[]], {o}]]],
          PlotStyle -> {GrayLevel[0], PointSize[0.01]},
          PlotRange -> All, Frame -> True, Axes -> False,
          (* theoretical distribution *)
          Prolog -> {Hue[0], Thickness[0.002], Line[
                    Table[{k, Binomial[n, k]/2^n}, {k, 0, n}]]}]]]
```

Now, we must unite two piles of cards. The function `unitePilesC` does this.

```
In[4]:= unitePilesC = Compile[{{l1, _Integer, 1}, {l2, _Integer, 1}},
    Module[{n1 = Length[l1], n2 = Length[l2], j1 = 0, j2 = 0, n, l},
        n = n1 + n2; (* empty united pile *) l = Table[0, {n}];
        (* add new card from pile 1 or 2 *)
        Do[If[Random[] <= (* probability *) (n1 - j1)/(n - j + 1),
            j1++; l[[j]] = l1[[j1]],
            j2++; l[[j]] = l2[[j2]]], {j, n}];
        (* return united pile *) l]];
```

Using the two functions `pileSplittingFunction` and `unitePilesC`, we can now define the function `riffleShuffle` that carries out one riffle shuffle of the desk *l*.

```
In[5]:= riffleShuffle[l_] :=
    unitePilesC[Take[l, #], Drop[l, #]]&[
                pileSplittingFunction[Length[l]][Random[]]]
```

The next output shows the result of one riffle shuffle of an initially ordered set of 52 cards.

```
In[6]:= sortedCards[n_] := sortedCards[n] = Range[n];
```

```
    riffleShuffle[sortedCards[52]]
Out[7]= {1, 2, 26, 27, 28, 29, 30, 3, 31, 32, 33, 34, 4, 5, 35, 36, 6,
    7, 8, 9, 10, 37, 11, 38, 39, 12, 40, 13, 41, 42, 14, 43, 44, 15, 16,
    17, 45, 46, 18, 19, 20, 47, 21, 22, 48, 49, 23, 24, 25, 50, 51, 52}
```

It remains to implement a function that counts the number of rising sequences of a list of integers. The function `risingSequenceCountC` steps through the list *l* and checks for each element of its predecessor already occurred earlier.

```
In[8]:= risingSequenceCountC :=
    Compile[{{l, _Integer, 1}},
    Module[{L = Table[0, {Length[l] + 1}], c = 1, x = 0},
        (* "zeroth" element *) L[[1]] = c;
        (* did predecessor already occur? *)
        Do[x = l[[j]]; If[L[[x]] =!= 0, L[[x + 1]] = L[[x]],
                          c++; L[[x + 1]] = c],
          {j, Length[l]}];
        (* return number of increasing sequences *) c]]
```

A sequence ordered increasingly has 1 rising subsequence and a sequence ordered decreasingly has *n* rising subsequences.

```
In[9]:= {risingSequenceCountC[Range[52]],
    risingSequenceCountC[Reverse @ Range[52]]}
Out[9]= {1, 52}
```

A random permutation of integers has typically $n/2$ rising subsequences.

```
In[10]:= Needs["DiscreteMath`Combinatorica`"]

       Table[risingSequenceCountC[RandomPermutation[52]], {12}]
Out[11]= {27, 27, 26, 26, 27, 27, 30, 26, 26, 26, 24, 23}
```

Now, we will analyze how the number of rising subsequences evolves with repeated riffle shuffles. `risingSequenceAv` `erage` carries out *o* repetitions of *m* shuffles of a desk with *n* cards.

```
In[12]:= risingSequenceAverage[n_, m_, o_] :=
       Sum[risingSequenceCountC /@
         NestList[riffleShuffle, sortedCards[n], m], {o}]/o
```

Carrying out 2000 shuffles for a desk of 52 cards takes only a few seconds.

```
In[13]:= Timing[(sum52 = risingSequenceAverage[52, 20, 1000]);]

Out[13]= {2.4 Second, Null}
```

In the beginning, the number of increasing subsequences doubles with each shuffle. After about $3/2 \log_2 n$ shuffles the number of rising subsequences is about $n/2$.

```
In[14]:= {Take[sum52, 5], Take[sum52, -5]} // N

Out[14]= {{1., 2., 4., 7.948, 14.088}, {26.369, 26.357, 26.305, 26.38, 26.451}}
```

The following graphics show how the number of rising subsequences evolves with more shuffles.

```
In[15]:= Show[GraphicsArray[
       ListPlot[risingSequenceAverage[#, 20, 1000],
         PlotLabel -> "n = " <> ToString[#],
         DisplayFunction -> Identity]& /@ {52, 100, 1000}]]
```

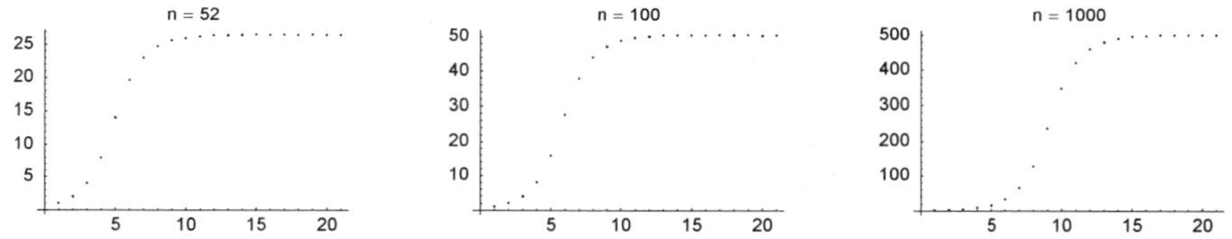

7. Friday the 13th and Easter

a) Here is an implementation of the formula.

```
In[1]:= weekday[day_, month_, year_] :=
     {Sunday, Monday, Tuesday, Wednesday, Thursday, Friday, Saturday}[[
     (* which day *) Mod[Floor[23 month/9] + day + 4 + year +
                 (Floor[#/4] - Floor[#/100] +
                 Floor[#/400])&[If[month < 3, year - 1, year]] -
                 If[month >= 3, 2, 0], 7] + 1]]
```

Here is a test.

```
In[2]:= (* Programming volume of the GuideBooks was released *)
     weekday[12, 10, 2004]
Out[3]= Tuesday
```

Next, we check the frequency of the various days of the week over 400 years. (Our calendar has this period.)

```
In[4]:= (* all 13th of a month over 400 years *)
     data = Flatten[Table[weekday[13, m, j], {m, 12}, {j, 1900, 2299}]];
     {#, Count[data, #]}& /@ {Sunday, Monday, Tuesday, Wednesday,
                 Thursday, Friday, Saturday}
Out[6]= {{Sunday, 687}, {Monday, 685}, {Tuesday, 685},
     {Wednesday, 687}, {Thursday, 684}, {Friday, 688}, {Saturday, 684}}
```

This shows that the 13th day of the month falls on Friday more often than on any other day of the week.

b) Here is a direct procedural implementation of the Gauss formula.

```
In[1]:= EasterData[year_Integer?Positive] :=
    Module[{a, b, c, d, e, f, g, h, i, j},
        a = Mod[year, 19];
        b = Mod[year, 4];
        c = Mod[year, 7];
        d = Floor[(8 Floor[year/100] + 13)/25] - 2;
        e = Floor[year/100] - Floor[year/400] - 2;
        f = Mod[15 + e - d, 30];
        g = Mod[6 + e, 7];
        h = Mod[f + 19 a, 30];
        i = Which[h == 29, 28, h == 28 && a >= 11, 27, True, h];
        j = Mod[2b + 4c + 6i + g, 7];
        i + j + 1]
```

The following function computes the date.

```
In[2]:= EasterSunday[year_Integer?Positive] :=
    (* write Easter date in a date form *)
    Which[# <= 10, ToString[21 + #] <> ". March " <> ToString[year],
          # > 10, ToString[# - 10] <> ". April " <> ToString[year]]&[
          EasterData[year]]
```

The dates for Easter for a few years are shown.

```
In[3]:= EasterSunday /@ {2001, 2002, 2003, 2004, 2005, 2006}

Out[3]= {15. April 2001, 31. March 2002, 20. April 2003,
         11. April 2004, 27. March 2005, 16. April 2006}
```

The calendar package that comes with *Mathematica* gives the same dates for Easter. Because this package also defined a function EasterSunday, we refer to the package function by including the context.

```
In[4]:= Needs["Miscellaneous`Calendar`"]

    EasterSunday::shdw : Symbol EasterSunday appears in multiple
        contexts {Miscellaneous`Calendar`, Global`}; definitions in context
        Miscellaneous`Calendar` may shadow or be shadowed by other definitions. More…
```

```
In[5]:= Miscellaneous`Calendar`EasterSunday /@ {2001, 2002, 2003, 2004, 2005, 2006}

Out[5]= {{2001, 4, 15}, {2002, 3, 31}, {2003, 4, 20}, {2004, 4, 11}, {2005, 3, 27}, {2006, 4, 16}}
```

Now, we find the earliest and latest occurrences of Easter in this century.

```
In[6]:= Flatten[{Position[#, Min[#]], Position[#, Max[#]]}&[
                 Array[EasterData, 100, 1900]]] + 1899

Out[6]= {1913, 1943}
```

```
In[7]:= EasterSunday /@ %

Out[7]= {23. March 1913, 25. April 1943}
```

A plot of the dates for Easter exhibits some interesting structure. (It is a quasicrystal; see the references cited in the problem.)

```
In[8]:= EasterPlot[from_, until_, opts___] :=
    With[{textOptions = Sequence[FontWeight -> "Bold", FontSize -> 6]},
    ListPlot[Table[{y, EasterData[y]}, {y, from, until}],
             opts, PlotRange -> All, Axes -> None,
             (* make nice date ticks *)
             FrameTicks -> {Automatic, Join[
             Table[{i, ToString[21 + i] <> ". 3."}, {i, 0, 10}],
             Table[{10 + i, ToString[i] <> ". 4."}, {i, 30}]]},
             TextStyle -> {FontSize -> 4},
             PlotLabel -> StyleForm["EASTER", textOptions],
             Frame -> True, FrameLabel -> {StyleForm["year", textOptions],
                                StyleForm["Easter Sunday", textOptions]},
             RotateLabel -> True]]
```

```
In[9]:= Show[GraphicsArray[
    Block[{$DisplayFunction = Identity},
```

```
{EasterPlot[1900, 2000],
 EasterPlot[1500, 2500, PlotStyle -> {PointSize[0.003]}]}]]]
```

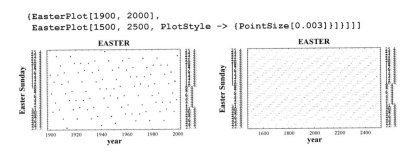

8. Number of Lattice Points, Binomial Digits, Decreasing Partitions, Partition Moments

a) Here is one possible compact implementation.

```
In[1]:= numberOfLatticePoints[dim_Integer?Positive, rMax_?(NumberQ[#]&)] :=
    Thread[{Sqrt[First[#2]], Plus @@ (#1 Last[#2])}]&[
     Array[(* how many parts due to symmetry *)
          Binomial[dim, #] 2^#&, dim], {Last[#],
      (* analyze one part *)
     Function[p, Count[p, #]& /@ Last[#]] /@ First[#]}&[
       {#, Select[Union[Flatten[#]],
         (* select all lattice points inside
            sphere of radius rMax *)# <= N[rMax^2]&]}&[
         (* the distances *)
        Table[Flatten[Outer[Plus @@ Flatten[List[##]^2]&, ##]& @@
        Table[#, {j, i}]], {i, dim}]&[
           (* all points in the hypercube *) Range[rMax]]]]]];
```

For the *d*-dimensional grid, there are $2^i d!/((d-i)!\, i!)\,(i = 1, \ldots, d)$ equal(-valued), *i*-dimensional parts (hyperplanes). For computational reasons, we only investigate the part consisting of points whose coordinates are all ≥ 0, and then multiply by the above degeneration factor. We generate the possible lattice points themselves with Outer. Then, we use Select to find those with distance smaller than rMax. Here is an example for the 2D lattice.

```
In[2]:= numberOfLatticePoints[2, 4.03]
```

$$\text{Out[2]= } \{\{1, 4\}, \{\sqrt{2}, 4\}, \{2, 4\}, \{\sqrt{5}, 8\}, \{2\sqrt{2}, 4\}, \{3, 4\}, \{\sqrt{10}, 8\}, \{\sqrt{13}, 8\}, \{4, 4\}\}$$

For distances smaller than 1, no points exists.

```
In[3]:= numberOfLatticePoints[3, 0.03]
```

Out[3]= {}

We now look at more data graphically.

```
In[4]:= Show[GraphicsArray[#]]& /@ Map[
    Graphics[Line[{{{#[[1]], 0}, #} // N]& /@ (* the data *)
          numberOfLatticePoints[Sequence @@ #],
        PlotRange -> All, Axes -> True, AxesOrigin -> {0, 0},
        TextStyle -> {FontSize -> 5}, DisplayFunction -> Identity,
        PlotLabel -> StyleForm[ToString[#[[1]]] <> "D",
                        FontWeight -> "Bold", FontSize -> 6]]&,
                    {{{1, 100}, {2, 20}}, {{3, 12}, {4, 9}}}, {2}]
```

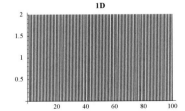
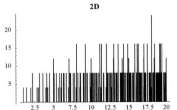

3D
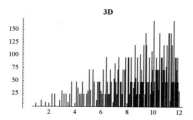

4D
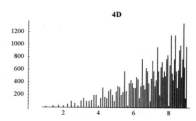

For more on lattice point counting problems, see [334], [103], [67], [38], and [335]. For nonrectangular n-dimensional lattices, see [257].

b) It is straightforward to write down a corresponding one-liner. Instead of dealing with all pairs of integer, we use the eightfold symmetry of the problem and calculate the ordered ones from the first quadrant with the pair's second element less than its first. Because $\gcd(0, y) = 0$ and $\gcd(\xi, \xi) = \xi$, we do not have to take into account the coordinate axis points and the diagonal points (with the exception of the pair $\{1, 1\}$). The function coPrimeLatticePointsInCircle returns a list of the form $\{\{\rho_1, c_1\}, ..., \{\rho_n, c_n\}\}$. For $\rho_k^2 \le R^2 < \rho_{k+1}^2$, the integer c_k denotes the number $p(R)$ of relatively prime integer pairs. The function coPrimeLatticePointsInCircle returns a list of data to avoid repeated calls to the function when analyzing the R-dependence.

```
In[1]:= coPrimeLatticePointsInCircle[R_] :=
    FoldList[(* account for symmetric parts *)
        {#2[[1]], #1[[2]] + 8 Length[#2]}&, {2, 4},
        (* all pairs in lower half of first quadrant *)
        Split[Sort[First /@ Select[
        Flatten[Table[{i^2 + j^2, {i, j}}, {i, R},
                    {j, Min[i - 1, Sqrt[R^2 - i^2]]}], 1],
                        GCD @@ #[[2]] == 1&]]]]
```

Here is the result of coPrimeLatticePointsInCircle for $R = 6$.

```
In[2]:= cpc6 = coPrimeLatticePointsInCircle[6]
```
```
Out[2]= {{2, 4}, {5, 12}, {10, 20}, {13, 28}, {17, 36}, {25, 44}, {26, 52}, {29, 60}, {34, 68}}
```

The nested list structure can easily be converted to an explicit function. The function toFunction does this.

```
In[3]:= toFunction[cpc_] := Function[Evaluate[Which @@
    Flatten[Reverse[Function[{x, y}, {# >= x, y}] @@@
        Join[{{-Infinity, 0}}, Drop[cpc, -1],
            {{cpc[[-1, 1]], NaN}}]]]]]
```

The following graphic shows $p(R)$ for $R < 10$.

```
In[4]:= Plot[toFunction[coPrimeLatticePointsInCircle[11]][x],
        {x, 0, 100}, (* use many points to get steep parts *)
        PlotPoints -> 500]
```

Now, we will deal with the $R = 1000$ situation. We have $p(1000) = 1909772$ and this value is reached after 132966 steps.

```
In[5]:= cplpc = coPrimeLatticePointsInCircle[1000];
    {Length[cplpc], cplpc[[-1]]}
```
```
Out[6]= {132966, {999997, 1909772}}
```

Analyzing the step size shows that (modulo the factor 8 due to symmetry) about as many single pairs as double pairs contribute. About one quarter consists of four pairs and eight pairs contribute only 699 times at once.

```
In[7]:= {First[#], Length[#]}& /@ Split[Sort[Subtract @@@
                Partition[Reverse[Last /@ cplpc], 2, 1]]]
Out[7]= {{8, 60083}, {16, 57843}, {32, 14340}, {64, 699}}
```

Subtracting now the average value $6/\pi R$ gives a quite oscillating remainder. The following two graphics show $p(R)$ and $p(R) - 6/\pi R$.

```
In[8]:= cplpcR = Apply[{#1, #2 - 6./Pi #1}&, cplpc, {1}];
```

```
In[9]:= Show[GraphicsArray[
            ListPlot[#, PlotRange -> All,
                DisplayFunction -> Identity]& /@ {cplpc, cplpcR}]]
```

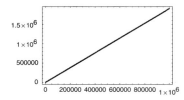
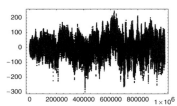

The slope of the calculated points agrees excellently with $6/\pi$.

```
In[10]:= {Fit[cplpc, {1, x}, x], 6./Pi}
Out[10]= {5.26756 + 1.90986 x, 1.90986}
```

To estimate α in the $O(R^\alpha)$ term we extract the consecutive largest fluctuations.

```
In[11]:= bag = {}; max = -Infinity;
         Do[(* current element of cplpcR *)
            ξ = Abs[cplpcR[[k]]];
            (* compare with current maximum *)
            If[ξ[[2]] > max, bag = {bag, ξ}; max = ξ[[2]]],
            {k, Length[cplpcR]}]
```

Assuming the functional form $O(R^\alpha)$, a double logarithmic plot should show a straight line.

```
In[13]:= ListPlot[N[cplpcR2 = Log[Rest[Level[bag, {-2}]]]]]
```

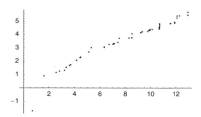

Ignoring the fluctuations for small R, we obtain $\alpha \approx 0.36$, in excellent agreement with theoretical estimations of $\alpha = 221/608 = 0.363648\ldots$ [600].

```
In[14]:= Fit[Select[cplpcR2, #[[1]] > 6&], {1, x}, x]
Out[14]= 0.726771 + 0.366187 x
```

For the number of different circles that enclose n integer lattice points in the plane, see [612].

c) We start by calculating the 10000 values of the digits sums of the binomials. This will take less than a minute.

```
In[1]:= n = 10000;
        Do[b[k] = Binomial[k, Floor[k/2]];
           ds[k] = Plus @@ IntegerDigits[b[k]], {k, n}] // Timing
```

Out[2]= {30.31 Second, Null}

Using the Stirling formula, we see immediately $\left(\genfrac{}{}{0pt}{}{n}{\lfloor n/2\rfloor}\right) \underset{n\to\infty}{\sim} n$. From this and the "belief" that there is no structure in this digit sequence, we expect $s_n \underset{n\to\infty}{\sim} n$. The following two graphics confirm these two statements.

```
In[3]:= Show[GraphicsArray[
         Block[{$DisplayFunction = Identity},
              {ListPlot[Table[Log[2, N @ b[k]], {k, n}]],
               ListPlot[tab1 = Table[ds[k], {k, n}]]}]]]
```

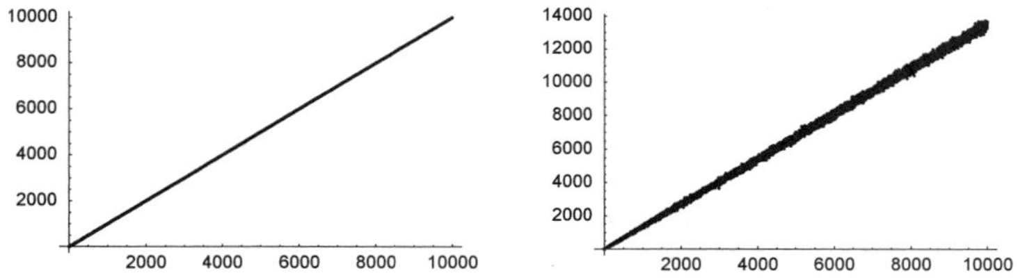

We subtract the main term and are left with the fluctuations.

```
In[4]:= fit[x_] = Fit[tab1, {1, x}, x]
```

Out[4]= -11.6246 + 1.35327 x

```
In[5]:= tab2 = Table[(fit[k] - ds[k]), {k, n}];
```

The fluctuations are a growing function of k. The right graphics shows the fluctuations scaled with $k^{-1/2}$.

```
In[6]:= Show[GraphicsArray[
         Block[{$DisplayFunction = Identity},
           {(* remaining fluctuations *)
            ListPlot[N[tab2 = Table[(fit[k] - ds[k]), {k, n}]]],
            (* fit to k^(-1/2) *)
            ListPlot[N[tab3 = Table[{Log[k], Log[Abs[fit[k] - ds[k]]]}, {k, n}]],
                 PlotRange -> All]}]]]
```

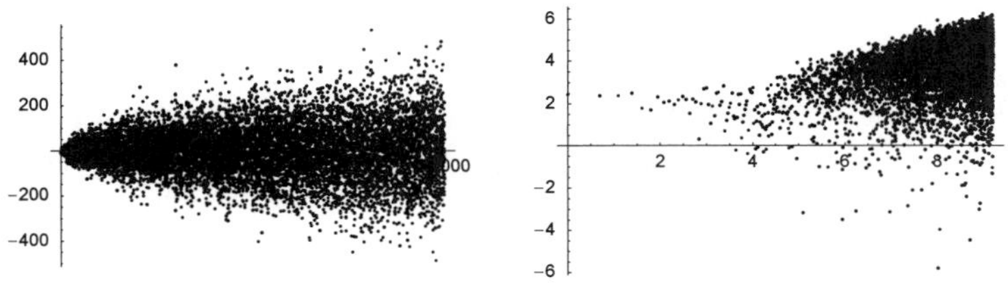

A numerical fit shows that a rescaling $\sim k^{-1/2}$ is appropriate.

```
In[7]:= ListPlot[N[tab3 = Table[{Log[k], Log[Abs[fit[k] - ds[k]]]}, {k, n}]],
             PlotRange -> All]
```

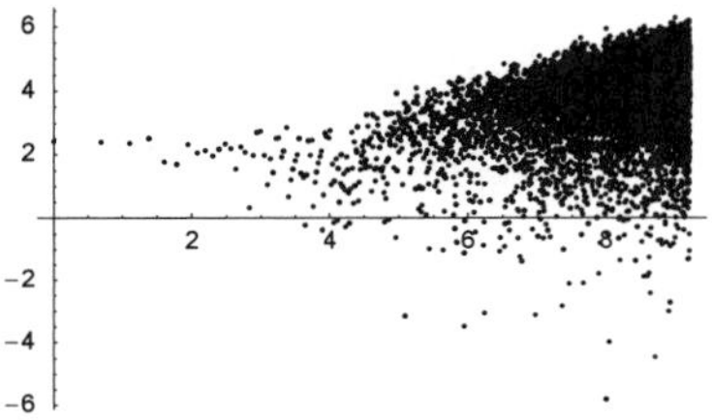

```
In[8]:= Fit[tab3, {1, x}, x]
```

Out[8]= -0.0658465 + 0.488031 x

After this rescaling, the magnitude of the fluctuations seems independent of k.

```
In[9]:= ListPlot[N[tab4 = Table[(fit[k] - ds[k])/Sqrt[k], {k, n}]]]
```

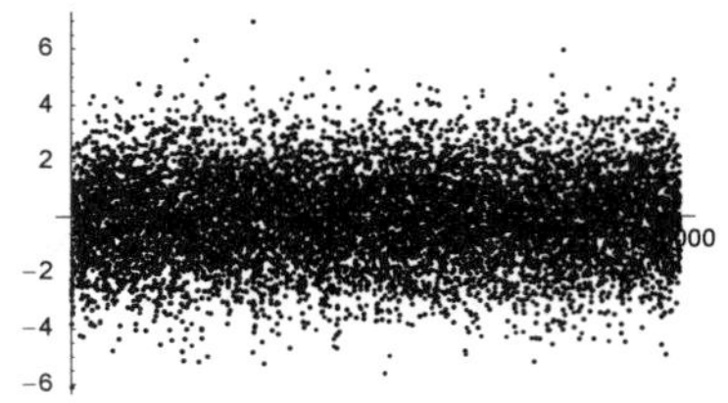

Now, we bin the fluctuations `tab4`. Not unexpectedly, they show a Gaussian-like distribution.

```
In[10]:= makeBinRectangles[data_, binSize_] :=
         Rectangle[{#1, 0}, {#2, #3}]& @@@
          ({First[#] binSize, (First[#] + 1) binSize,
            Length[#]/Length[data]/binSize}& /@
                      Split[Quotient[Sort[data], binSize]])

In[11]:= Show[Graphics[makeBinRectangles[tab4, 0.2]],
             Frame -> True, PlotRange -> All]
```

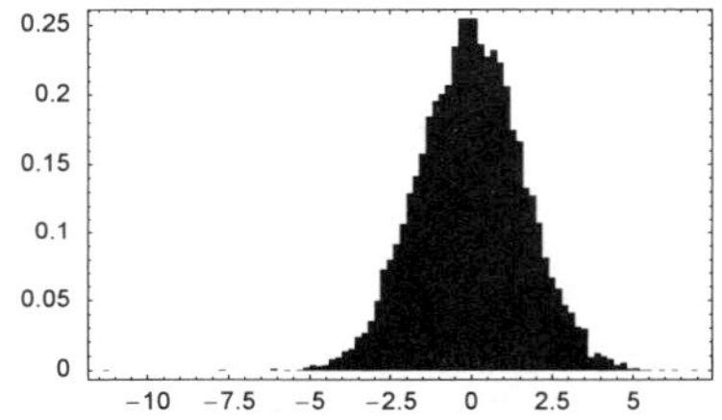

For the theoretical distributions of digitsums of linear recurrences, see [147], [151].

d) We start by implementing a function `decreasingPartitions` that generates all strictly decreasing partitions of the integer μ.

```
In[1]:= decreasingPartitions[μ_Integer?(# > 1&)] :=
        Module[{f},
          (* conditions for ending the recursion *)
          f[{a___, _}] := Sequence[] /; Length[{a}] > Length[Union[{a}]] ||
                                        Not[OrderedQ[Reverse[{a}]]];
          (* split last integer *)
          f[{a___, b_}] := Table[f[{a, b - k, k}], {k, 0, b - 2}] /; b > 1;
          (* carry out the recursion and delete 0's *)
          DeleteCases[Sequence @@@ Flatten[f[{μ}]], 0, Infinity]]
```

Here are the 15 partitions of the integer 12 shown.

```
In[2]:= decreasingPartitions[12]

Out[2]= {{12}, {11, 1}, {10, 2}, {9, 3}, {9, 2, 1}, {8, 4}, {8, 3, 1}, {7, 5},
         {7, 4, 1}, {7, 3, 2}, {6, 5, 1}, {6, 4, 2}, {6, 3, 2, 1}, {5, 4, 3}, {5, 4, 2, 1}}

In[3]:= {Length[%], PartitionsQ[12]}

Out[3]= {15, 15}
```

The function ρPF forms the pure function corresponding to $\rho(\lambda, x)$.

```
In[4]:= ρPF[partition_] := Function @@ {Which @@
           Flatten[MapIndexed[Function[{k, p}, {# >= k, p[[1]] - 1}],
                      Append[partition, 0]]]}
```

The following graphic shows the 64 functions $\rho(\lambda, x)$ corresponding to $\mu = 20$.

```
In[5]:= Plot[Evaluate[ρPF[#][x]& /@ decreasingPartitions[20]],
        {x, 1, 20}, PlotPoints -> 200, PlotRange -> All,
        Axes -> False, Frame -> True]
```

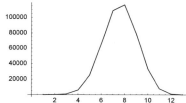

Now, we form the 444793 strictly decreasing partitions of $\mu = 100$.

```
In[6]:= μ = 100;
        {Length[partisμ = decreasingPartitions[μ]], PartitionsQ[μ]}
Out[7]= {444793, 444793}
```

The next graphic shows the length distribution of the partitions. Their average length is 7.61....

```
In[8]:= ListPlot[{First[#], Length[#]}& /@ Split[Sort[Length /@ partisμ]],
        PlotJoined -> True]
```

We form the average of $\rho(\lambda, x)$ over all partitions. To save memory, we add the results incrementally instead of using Sum.

```
In[9]:= (* form average; collapse after each partition *)
        tempSum = Table[0, {x, 0, μ}];
        Do[f = ρPF[partisμ[[k]]];
           tempSum = tempSum + Table[f[x], {x, 0, μ}], {k, Length[partisμ]}];
        average = tempSum/Length[partisμ];
        (* add x-coordinates *)
        points = MapIndexed[{(#2[[1]] - 1), #1}&, average];
```

The following graphic shows the averages (black points) and the curve $g_\mu(x)$ (in red). The agreement of the two is quite good.

```
In[15]:= (* scaling constant *) σ = (Pi/(2 Sqrt[3 μ]));
         ListPlot[points, PlotRange -> All,
                 PlotStyle -> {GrayLevel[0], PointSize[0.01]},
                 Prolog -> {Thickness[0.002], Hue[0], Line @
                            Table[N[{t, Log[1 + Exp[-σ t]]/σ}], {t, 0, μ, 1/2}]}]
```

For related results for general partitions, see [139].

e) We start by implementing the exact calculation of the $p_n(m)$. A straightforward recursive definition is the following.

```
In[1]:= $RecursionLimit = Infinity;
        p[n_, n_] = 1;
        p[n_, m_] := p[n, m] = Sum[p[n - m, k], {k, Min[n - m, m]}]
```

We can speed up the calculation of the $p_n(m)$ by storing the values in a matrix. Retrieving the already calculated elements from a matrix will be faster than from the downvalues of p.

```
In[4]:= PnmTable[d_] :=
        Module[{M = Table[0, {d}, {d}]},
               Do[M[[n, 1]] = 1; M[[n, n]] = 1, {n, d}];
               Do[M[[n, m]] = Plus @@ Take[M[[n - m]], Min[n - m, m]],
                  {n, 2, d}, {m, 2, n - 1}]; M]
```

The following timings show that PnmTable is clearly faster.

```
In[5]:= With[{d = 100},
             {Timing[Sum[p[d, m], {m, d}]], Timing[Plus @@ PnmTable[d][[d]]],
              (* check *) PartitionsP[d]}]
Out[5]= {{0.1 Second, 190569292}, {0.08 Second, 190569292}, 190569292}
```

Here are the values of $p_{1000}(m)$ calculated.

```
In[6]:= Pnm1000 = PnmTable[1000];
```

Next, we implement the asymptotical formulas for the moments.

```
In[7]:= (* asymptotics of the partitions *)
        PartitionsMomentsAsymp[0][n_] = 1/(4 Sqrt[3] n) Exp[Pi Sqrt[2n/3]];

        PartitionsMomentsAsymp[k_Integer?Positive][n_] :=
        PartitionsMomentsAsymp[0][n]*
        (Sqrt[6n]/Pi)^k partitionSum[Join[{Log[Sqrt[6n]/Pi] + EulerGamma},
                        Table[j Zeta[j + 1], {j, k - 1}]]]
```

We calculate the sum of the partitions using the explicit partitions returned by Experimental`IntegerPartitions.

```
In[10]:= (* define integer partitions in the style of result of FactorInteger *)
         myIntegerPartitions[n_Integer?Positive] := Reverse /@
           (({First[#], Length[#]}& /@ Split[#])& /@
                        Experimental`IntegerPartitions[n])
```

```
In[12]:= partitionSum[l_] :=
         With[{partis = myIntegerPartitions[Length[l]]},
              Length[l]! Plus @@ (Function[p, 1/(Times @@ (Last /@ p)!) Times @@
                    ((l[[First /@ p]]/(First /@ p!))^(Last /@ p))] /@ partis)]
```

Here are the first few partition sums.

```
In[13]:= Table[partitionSum[Table[Subscript[ρ, j], {j, o}]], {o, 5}] // Expand
Out[13]= {ρ_1, ρ_1^2 + ρ_2, ρ_1^3 + 3 ρ_1 ρ_2 + ρ_3, ρ_1^4 + 6 ρ_1^2 ρ_2 + 3 ρ_2^2 + 4 ρ_1 ρ_3 + ρ_4,
          ρ_1^5 + 10 ρ_1^3 ρ_2 + 15 ρ_1 ρ_2^2 + 10 ρ_1^2 ρ_3 + 10 ρ_2 ρ_3 + 5 ρ_1 ρ_4 + ρ_5}
```

Now, we can compare the exact moments with the asymptotical expressions. For $k \leq 2$, the error is less than 1%, but increases quickly with k.

```
In[14]:= Table[1 - PartitionsMomentsAsymp[k][1000.]/
                   Pnm1000[[1000]].(Range[1000.]^k), {k, 10}]
Out[14]= {0.00259773, -0.00186806, -0.0247115, -0.055932,
          -0.0823989, 0.0194806, 0.0922358, 0.204917, 0.327032, 0.473348}
```

9. 15 and 6174

a) This computes the result of the iterated application of the function [384]: $\sum_{divisors}$ digitsum(*divisor*).

```
In[1]:= fifteen[n_Integer?(# > 1&)] :=
        FixedPoint[Plus @@ Flatten[IntegerDigits /@ Divisors[#]]&, n]
```

The fixed point of this mapping is always the number 15 (for $n > 1$) (see [34]).

```
In[2]:= Array[fifteen, 12, 2]
Out[2]= {15, 15, 15, 15, 15, 15, 15, 15, 15, 15, 15, 15}
```

The convergence to the number 15 can be seen in the following graphic.

```
In[3]:= Show[Graphics[{Thickness[0.002], Line /@ Union @
        Flatten[Table[(* add second dimension and make lines *)
        Partition[MapIndexed[{#1, -#2[[1]]}&, Reverse[
        (* the intermediate values *)
        FixedPointList[Plus @@ Flatten[IntegerDigits /@
                      Divisors[#]]&, n]]], 2, 1], {n, 2, 300}], 1]}],
        (* set options for plot *)
        AspectRatio -> 1, PlotRange -> All, Axes -> True,
        AxesOrigin -> {0, -22}, Ticks -> {Automatic,
        {{-5, "5"}, {-10, "10"}, {-15, "15"}, {-20, "20"}}},
        AxesLabel -> {"n", "Iteration"}, GridLines -> {{15}, None}]
```

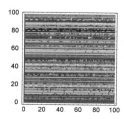

For a base *b* different from 10, the process becomes periodic.

```
In[4]:= digitSum[n_, b_] := Plus @@ IntegerDigits[n, b]

        step[n_, b_] := Plus @@ (digitSum[#, b]& /@ Divisors[n])
```

In the following example, the period is 2.

```
In[6]:= NestList[step[#, 5]&, 424, 20]

Out[6]= {424, 42, 28, 20, 14, 12, 16, 15, 8, 11, 4, 7, 4, 7, 4, 7, 4, 7, 4, 7, 4}
```

In the following example, the period is 6.

```
In[7]:= NestList[step[#, 56]&, 244, 12]

Out[7]= {244, 49, 57, 25, 31, 32, 63, 49, 57, 25, 31, 32, 63}
```

The function period calculates the length of the period.

```
In[8]:= periodicity[l_] := Subtract @@ Reverse[Flatten[Position[l, Last[l]]]]

        period[n0_, b_Integer] :=
                  periodicity[NestWhileList[step[#, b]&, n0, Unequal, All]]
```

Here is a density plot for various numbers and bases.

```
In[10]:= data = Table[period[n0, b], {b, 2, 100}, {n0, 1, 100}];

In[11]:= ListDensityPlot[data, Mesh -> False, ColorFunction -> (Hue[0.8 #]&)]
```

For Gaussian integers, we can treat the real and imaginary part separately in the calculation of the digitsum.

```
In[12]:= step[n_, b_] := (Function[divisors,
              Plus @@ (digitSum[#, b]& /@ Re[divisors]) +
            I Plus @@ (digitSum[#, b]& /@ Im[divisors])])[Divisors[n]]
```

```
In[13]:= NestList[step[#, 10]&, (2 + 2I) (3 + 5I), 16]
```

Out[13]= {-4 + 16 i, 38 + 16 i, 67 + 76 i, 22 + 22 i, 18 + 9 i, 30 + 13 i, 4 + 4 i, 14 + 7 i,
         15 + 8 i, 11 + 9 i, 5 + 11 i, 15 + 6 i, 15 + 8 i, 11 + 9 i, 5 + 11 i, 15 + 6 i, 15 + 8 i}

```
In[14]:= period[(2 + 2I) (3 + 5I), 10]
```

Out[14]= 4

The next graphics visualize the periods for the bases 2 and 10.

```
In[15]:= Show[GraphicsArray[
           Block[{$DisplayFunction = Identity},
             ListDensityPlot[Table[period[nr + I ni, #], {nr, 1, 30}, {ni, 1, 30}],
                           Mesh -> False]& /@ {2, 10}]]]
```

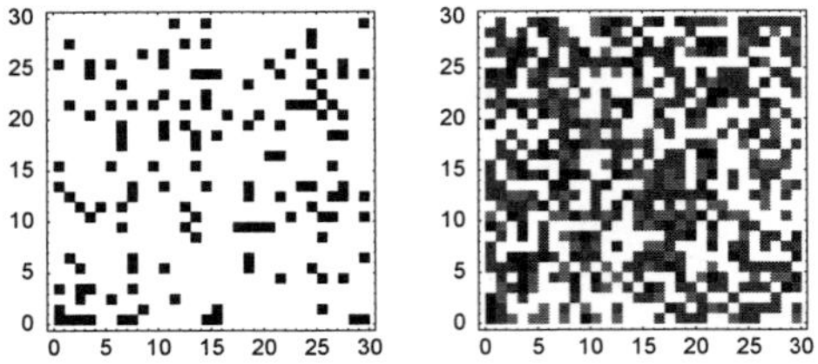

b) This computes the result of the iterated application of the function [606]: $\sum_{primeFactors(n)}$ digitsum(*primeFactor*).

```
In[1]:= TwoOrThreeOrFourOrFiveOrSeven[n_Integer?(# > 1&)] :=
          FixedPoint[Plus @@ Flatten[IntegerDigits /@
          Flatten[Apply[Table[#1, {#2}]&, FactorInteger[#], {1}]]]&, n]
```

The fixed point of this mapping is always the number 2 , 3 , 4, 5, or 7 (for $n > 1$).

```
In[2]:= Array[TwoOrThreeOrFourOrFiveOrSeven, 12, 2]
```

Out[2]= {2, 3, 4, 5, 5, 7, 5, 5, 7, 2, 7, 4}

```
In[3]:= Array[TwoOrThreeOrFourOrFiveOrSeven, 1000, 2] // Union
```

Out[3]= {2, 3, 4, 5, 7}

The convergence to the numbers can be seen in the following graphics. We use the bases 2, 3, 7, 10, and 13.

```
In[4]:= digitSum[n_, b_] := Plus @@ IntegerDigits[n, b]
```

```
In[5]:= primeFactors[n_] := Flatten[Apply[Table[#1, {#2}]&, FactorInteger[n], {1}]]
```

```
In[6]:= step[n_, b_] := Plus @@ (digitSum[#, b]& /@ primeFactors[n])
```

```
In[7]:= convGraphics[data_] := Graphics[{Thickness[0.002],
          Line /@ MapIndexed[{#2[[2]] - 1, #1}&, data, {2}]},
          Frame -> True, PlotRange -> All, AspectRatio -> 2];
```

```
In[8]:= Show[GraphicsArray[
          Function[b, convGraphics[Table[FixedPointList[step[#, b]&, i],
                        {i, 100}]]] /@ {2, 3, 7, 10, 13}]]
```

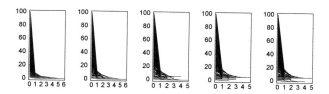

c) This is the function under consideration.

```
In[1]:= f6174[n_Integer?(0 <= # <= 9999&)] :=
          Drop[FixedPointList[(#1.#2 - #1.Reverse[#2]& @@
          {10^Range[0, Length[#] - 1], #}&[
                            Sort[IntegerDigits[#, 10, 4]]])&, n], -1]
```

f6174 computes the result of the iterated application of the function
largestIntegerThatCanBeFormedFromAllDigitsOfn − smallestIntegerThatCanBeFormedFromAllDigitsOfn
to a four-digit integer. (The construction
1.#2 − #1.Reverse[#2])& @@ {10^Range[0, Length[#] − 1], #}&[*sortedListOfIntegers*]
constructs the difference.) The fixed point of this mapping is for most numbers in the specified range 6174 ([301], [267], and [552]).

```
In[2]:= f6174[0081]
```
Out[2]= {81, 8082, 8532, 6174}

```
In[3]:= f6174[3547]
```
Out[3]= {3547, 4086, 8172, 7443, 3996, 6264, 4176, 6174}

```
In[4]:= f6174[4536]
```
Out[4]= {4536, 3087, 8352, 6174}

The following *n* are exceptional; they give the result 0.

```
In[5]:= exceptionList = {};

        tab = Table[(If[Last[#] =!= 6174,
                     AppendTo[exceptionList, {i, Last[#]}]]; #)&[f6174[i]],
                {i, 0, 9999}];

        exceptionList
```
Out[7]= {{0, 0}, {1111, 0}, {2222, 0}, {3333, 0}, {4444, 0},
        {5555, 0}, {6666, 0}, {7777, 0}, {8888, 0}, {9999, 0}}

For this example, let us also have a view on the convergence of the map. After at most seven steps, we obtain the number 6174.

```
In[8]:= Show[Graphics[{Thickness[0.002],
                     Line[MapIndexed[{#2[[1]] - 1, #1}&, #]]& /@ tab}],
          PlotRange -> All, Frame -> True]
```

The next two graphics show the argument of the Fourier transform of the number of steps and 2D density plot of the partitioned number of steps [143].

```
In[9]:= Show[GraphicsArray[
        Block[{$DisplayFunction = Identity},
        {(* argument of the Fourier transform of the lengths *)
        ListPlot[Arg[Fourier[Length /@ tab]]],
        (* density plot of the partitioned lengths *)
        ListDensityPlot[Partition[Length /@ tab, 100],
                        PlotRange -> All, ColorFunction -> (Hue[0.8 #]&),
                        Mesh -> False, FrameTicks -> False, PlotRange -> All]}]]]
```

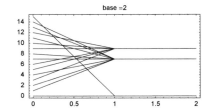

Now let us implement the same process in a base *b*. As numerical experiments show, typically one does not have a fixed point, but rather the process becomes periodical. So, we use `NestWhileList` in the following implementation.

```
In[10]:= fInOtherBase[n_, d_:Automatic, b_] :=
         With[{d = If[d === Automatic, Length @ IntegerDigits[n, b], d]},
             NestWhileList[(FromDigits[Reverse[#], b] - FromDigits[#, b])&[
                           Sort[IntegerDigits[#, b, d]]]&, n, Unequal, All]]
In[11]:= fInOtherBase[n_, b_] :=
             NestWhileList[(FromDigits[Reverse[#], b] - FromDigits[#, b])&[
                           Sort[IntegerDigits[#, b]]]&, n, Unequal, All]
```

Here, the sequences for the starting number 6174 in various bases are shown.

```
In[12]:= Table[{b, fInOtherBase[6174, b]}, {b, 8, 12}]

Out[12]= {{8, {6174, 26082, 22491, 17388, 25074, 30681, 26586, 21987, 21483, 25578, 26586}},
          {9, {6174, 5968, 5456, 3712, 5168, 5456}}, {10, {6174, 6174}},
          {11, {6174, 9420, 12300, 9860, 4100, 12520, 9420}}, {12, {6174, 12089,
              11022, 3454, 17402, 18326, 14344, 9031, 10626, 13816, 12617, 7172, 14344}}}
```

The following pictures show visualizations of the sequences for starting numbers between 0 and $b^4 - 1$ and bases b between 2 and 11.

```
In[13]:= Show[GraphicsArray[#]]& /@ Partition[
         Table[Graphics[{Thickness[0.002], Line /@ MapIndexed[{#2[[2]] - 1, #1}&,
                         Table[fInOtherBase[n, 4, b], {n, 0, b^4 - 1}], {2}]},
                         Frame -> True, PlotRange -> All, AspectRatio -> 1/2,
                         PlotLabel -> ("base =" <> ToString[b])],
             {b, 2, 11, 3}], 2]
```

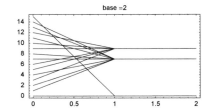
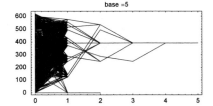

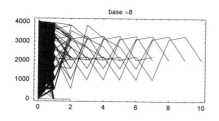

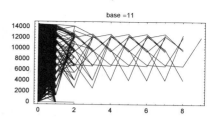

And here are the corresponding density plots of the number of steps until a fixed point (or a cycle) is reached for the bases 9, 11, and 15.

```
In[14]:= Show[GraphicsArray[Function[b,
         ListDensityPlot[Partition[Length /@ Table[fInOtherBase[n, 4, b],
                {n, 0, b^4 - 1}], b^2],
              PlotRange -> All, ColorFunction -> (Hue[0.8 #]&),
              DisplayFunction -> Identity,
              Mesh -> False, FrameTicks -> False,
              PlotRange -> All]] /@ (* three bases *) {9, 11, 15}]]
```

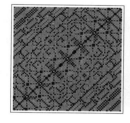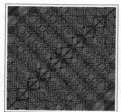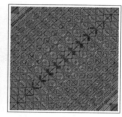

10. Selberg Identity, Kluyver Identity, Goodwyn Property, Guiasu Formula, Prime Sums, Farey–Brocot Sequence, Divisor Sum Identities

a) Here, the definition of the Mangoldt function is implemented.

```
In[1]:= MangoldtΛ[n_] := If[Length[#] === 1, Log[#[[1, 1]]], 0]&[FactorInteger[n]]
```

Also, the implementation of the Selberg identity is straightforward; we calculate the divisors of n only once. After the direct evaluation of the Selberg identity, we simplify the logarithms of nonprimes.

```
In[2]:= SelbergIdentity[n_Integer?(# >= 1&)] := Expand @ ((
       MangoldtΛ[n] Log[n] + Plus @@ (MangoldtΛ[#] MangoldtΛ[n/#]& /@ #) -
       Plus @@ (MoebiusMu[#] Log[n/#]^2 & /@ #))&[Divisors[n]] //.
       (* simplification of logarithms *)
          Log[c_] :> Plus @@ (#[[2]] Log[#[[1]]]& /@ FactorInteger[c]))
```

Here, the Selberg identity is verified for some n (the Expand simplifies SelbergIdentity).

```
In[3]:= Union @ Array[SelbergIdentity, 120]

Out[3]= {0}
```

b) The implementation is straightforward in both cases.

```
In[1]:= KluyverCosIdentity[n_Integer?Positive] :=
            Plus @@ Cos[2Pi Select[Range[n], GCD[#, n] == 1&]/n]

In[2]:= KluyverCosIdentity[17]
```

$$Out[2]= \cos\left[\frac{2\pi}{17}\right] + \cos\left[\frac{4\pi}{17}\right] + \cos\left[\frac{6\pi}{17}\right] + \cos\left[\frac{8\pi}{17}\right] + \cos\left[\frac{10\pi}{17}\right] +$$
$$\cos\left[\frac{12\pi}{17}\right] + \cos\left[\frac{14\pi}{17}\right] + \cos\left[\frac{16\pi}{17}\right] + \cos\left[\frac{18\pi}{17}\right] + \cos\left[\frac{20\pi}{17}\right] +$$
$$\cos\left[\frac{22\pi}{17}\right] + \cos\left[\frac{24\pi}{17}\right] + \cos\left[\frac{26\pi}{17}\right] + \cos\left[\frac{28\pi}{17}\right] + \cos\left[\frac{30\pi}{17}\right] + \cos\left[\frac{32\pi}{17}\right]$$

```
In[3]:= Array[{Chop[N[KluyverCosIdentity[#]]], MoebiusMu[#]}&, 32]
```

```
Out[3]= {{1., 1}, {-1., -1}, {-1., -1}, {0, 0}, {-1., -1}, {1., 1}, {-1., -1},
        {0, 0}, {0, 0}, {1., 1}, {-1., -1}, {0, 0}, {-1., -1}, {1., 1}, {1., 1}, {0, 0},
        {-1., -1}, {0, 0}, {-1., -1}, {0, 0}, {1., 1}, {1., 1}, {-1., -1}, {0, 0},
        {0, 0}, {1., 1}, {0, 0}, {0, 0}, {-1., -1}, {-1., -1}, {-1., -1}, {0, 0}}

In[4]:= KluyverSinIdentity[n_Integer?(# > 2&)] :=
        Times @@ (2 Sin[Pi Select[Range[n], GCD[#, n] == 1&]/n])

In[5]:= MangoldtΛ[n_] := If[Length[#] === 1,
                          Log[#[[1, 1]]], 0]&[FactorInteger[n]]

In[6]:= Array[{Chop[N[KluyverSinIdentity[#]]], Exp[MangoldtΛ[#]]}&, 32, 3]

Out[6]= {{3., 3}, {2., 2}, {5., 5}, {1., 1}, {7., 7}, {2., 2}, {3., 3}, {1., 1},
        {11., 11}, {1., 1}, {13., 13}, {1., 1}, {1., 1}, {2., 2}, {17., 17}, {1., 1},
        {19., 19}, {1., 1}, {1., 1}, {1., 1}, {23., 23}, {1., 1}, {5., 5}, {1., 1},
        {3., 3}, {1., 1}, {29., 29}, {1., 1}, {31., 31}, {2., 2}, {1., 1}, {1., 1}}
```

In Chapter 1 of the Symbolics volume [556], we will discuss symbolic techniques for proving such-type identities.

c) This exercise is also straightforward.

```
In[1]:= showGoodwinPropertyOfFareySequence[maxDenominator_] :=
        {(Numerator[ #[[1]]] + Numerator[ #[[3]]])/
         (Denominator[#[[1]]] + Denominator[#[[3]]]), #[[2]]}& /@
            Partition[Union[Flatten[Table[j/i, {i, maxDenominator},
                                             {j, i - 1}]]], 3, 1]
```

Here is an example.

```
In[2]:= showGoodwinPropertyOfFareySequence[5]
```

$$Out[2]= \left\{\left\{\frac{1}{4}, \frac{1}{4}\right\}, \left\{\frac{1}{3}, \frac{1}{3}\right\}, \left\{\frac{2}{5}, \frac{2}{5}\right\}, \left\{\frac{1}{2}, \frac{1}{2}\right\}, \left\{\frac{3}{5}, \frac{3}{5}\right\}, \left\{\frac{2}{3}, \frac{2}{3}\right\}, \left\{\frac{3}{4}, \frac{3}{4}\right\}\right\}$$

Here we test the first 100 Farey sequences.

```
In[3]:= GoodwinPropertyOfFareySequenceQ[n_] :=
        And @@ Apply[SameQ, showGoodwinPropertyOfFareySequence[n], {1}]

In[4]:= Table[GoodwinPropertyOfFareySequenceQ[n], {n, 100}] // Union

Out[4]= {True}
```

The Ford circles are also easy to implement.

```
In[5]:= Show[Graphics[{Thickness[0.002],
        Circle[{#, 1/2/Denominator[#]^2}, 1/2/Denominator[#]^2]& /@
        Union[Flatten[Table[i/j, {j, 20}, {i, j - 1}]]]}],
            AspectRatio -> Automatic, PlotRange -> All, Axes -> False]
```

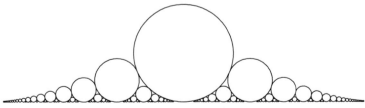

The interesting property of these circles is that they belong to consecutive fractions of a Farey sequence touching at exactly one point.

d) mediants implements the calculation of the mediants of two fractions.

```
In[1]:= mediant[{pq1_, pq2_}] := (Numerator[pq1] + Numerator[pq2])/
                                (Denominator[pq1] + Denominator[pq2])
```

To express an interval $[a, b] \in [0, 1]$ as the smallest union of disjoint intervals $\left[\mathcal{F}_{k_j}^{(j)}, \mathcal{F}_{k_j+1}^{(j)}\right]$, we will proceed in the following way: First, we will find the smallest interval $\left[\mathcal{F}_{k_j}^{(l)}, \mathcal{F}_{k_j+1}^{(l)}\right]$ that contains the given interval $[a, b]$. This means $[a, b] \subset \left[\mathcal{F}_{k_j}^{(l)}, \mathcal{F}_{k_j+1}^{(l)}\right]$. If $[a, b] = \left[\mathcal{F}_{k_j}^{(l)}, \mathcal{F}_{k_j+1}^{(l)}\right]$, we are done. Else, we carry out one more subdivision step. The point $\mathcal{F}_{2k_j+1}^{(l+1)}$ lies then in the interval $[a, b]$: $\mathcal{F}_{2k_j+1}^{(l+1)} \in [a, b]$. At this point, we split the interval $[a, b]$ into two subintervals, the left

$[a, \mathcal{F}_{2k_l+1}^{(l+1)}]$ and the right $[\mathcal{F}_{2k_l+1}^{(l+1)}, b]$. Inside each of these subintervals, we recursively carry out further subdivision steps until we reach $\mathcal{F}_{k_m}^{(m)} = a$ and $\mathcal{F}_{k_n}^{(n)} = b$. Whenever in this subdivision process we encounter the situation $[\mathcal{F}_{k_j}^{(j)}, \mathcal{F}_{k_j+1}^{(j)}] \subset [a, b]$, we collect this subinterval and ignore it in further subdivision steps.

The function `fareyEncapsulateStep` carries out one step in encapsulating $[a, b]$, and the function `fareyEncapsu`-`late` carries out the first step described above. `fareyEncapsulate` returns `C[`l`, {`$\mathcal{F}_{k_l}^{(l)}$`, `$\mathcal{F}_{k_l+1}^{(l)}$`}]`. (We use the head `C` here instead of `List` to more conveniently apply `Flatten` in the nested lists `intBagL` and `intBagR` below.)

```
In[2]:= fareyEncapsulateStep[initialFareyInterval:Interval[{pq1_, pq2_}], int_] :=
          Module[{newPoint = mediant @@ initialFareyInterval},
            Which[(* left interval *)
                  IntervalMemberQ[Interval[{pq1, newPoint}], int],
                  Interval[{pq1, newPoint}],
                  (* right interval *)
                  IntervalMemberQ[Interval[{newPoint, pq2}], int],
                  Interval[{newPoint, pq2}],
                  (* mediant is inside [a, b] *)
                  True, $Failed]]


        fareyEncapsulateStep[$Failed, _] := $Failed
In[5]:= fareyEncapsulate[int_] :=
          Module[{c = 0, fareyInterval = Interval[{0, 1}], aux},
            While[c++; aux = fareyEncapsulateStep[fareyInterval, int];
                  aux =!= $Failed, fareyInterval = aux];
          (* return encapsulating interval *)
          C[c - 1, fareyInterval]]
```

The encapsulating Farey–Brocot interval for the interval $[3/13, 4/13]$ is $[0, 1/3]$.

```
In[6]:= fareyEncapsulate[Interval[{3/13, 4/13}]]
```

$$\text{Out[6]= } C\left[2, \text{Interval}\left[\left\{0, \tfrac{1}{3}\right\}\right]\right]$$

The next mediant would fall into the interval $[3/13, 4/13]$.

```
In[7]:= 3/13 <= mediant[{0, 1/3}] <= 4/13
```

Out[7]= True

The two functions `zoomInLeft` and `zoomInRight` implement the second of the above-described steps, the actual covering step. The intervals $[\mathcal{F}_{k_j}^{(j)}, \mathcal{F}_{k_j+1}^{(j)}]$ are collected in the form `C[`j`, {`$\mathcal{F}_{k_j}^{(j)}$`, `$\mathcal{F}_{k_j+1}^{(j)}$`}]` in the container `intBagLOrR`.

```
In[8]:= (* xl < a < xu *)
        zoomInLeft[{xlStart_, xuStart_}, k0_, a_] :=
        Module[{xl = xlStart, xu = xuStart, k = k0,
                intBagL = {}, notDone = True},
          While[(* until Farey-Brocot point is endpoint *) notDone,
            k++; xNew = mediant[{xl, xu}];
            Which[a < xNew, (* add interval *)
                        fInt = {xNew, xu};
                        intBagL = {intBagL, C[k, Interval[fInt]]};
                        xu = xNew,
                  a > xNew,
                        xl = xNew,
                  a == xNew,
                        fInt = {xNew, xu};
                        intBagL = {intBagL, C[k, Interval[fInt]]};
                        notDone = False]];
          Flatten[intBagL]]
```

Here is the example from above continued.

```
In[10]:= zoomInLeft[{0, 1/3}, 2, 3/13]
```

$$\text{Out[10]= } \left\{C\left[3, \text{Interval}\left[\left\{\tfrac{1}{4}, \tfrac{1}{3}\right\}\right]\right], C\left[6, \text{Interval}\left[\left\{\tfrac{3}{13}, \tfrac{1}{4}\right\}\right]\right]\right\}$$

```
In[11]:= (* xl < b < xu *)
     zoomInRight[{xlStart_, xuStart_}, k0_, b_] :=
     Module[{xl = xlStart, xu = xuStart, k = k0,
            intBagR = {}, notDone = True},
       While[(* until Farey-Brocot point is endpoint *)notDone,
         k++; xNew = mediant[{xl, xu}];
         Which[b < xNew,
                  xu = xNew,
               b > xNew, (* add interval *)
                  fInt = {xl, xNew};
                  intBagR = {intBagR, C[k, Interval[fInt]]};
                  xl = xNew,
               b == xNew,
                  fInt = {xl, xNew};
                  intBagR = {intBagR, C[k, Interval[fInt]]};
                  notDone = False]];
       Flatten[intBagR]]
```

We need four further subdivision steps until we reach the right endpoint $4/13$ for the example interval $[3/13, 4/13]$.

```
In[13]:= zoomInRight[{0, 1/3}, 2, 4/13]
```

$$\text{Out[13]}= \left\{C\left[3, \text{Interval}\left[\left\{0, \tfrac{1}{4}\right\}\right]\right], C\left[4, \text{Interval}\left[\left\{\tfrac{1}{4}, \tfrac{2}{7}\right\}\right]\right],\right.$$
$$\left. C\left[5, \text{Interval}\left[\left\{\tfrac{2}{7}, \tfrac{3}{10}\right\}\right]\right], C\left[6, \text{Interval}\left[\left\{\tfrac{3}{10}, \tfrac{4}{13}\right\}\right]\right]\right\}$$

Putting now all of the functions together, we can implement `FareyBrocotUnion`. `FareyBrocotUnion[` *int* `]` returns the minimal covering of the interval *int* in the form of a list of $C[j, \{\mathcal{F}_{k_j}^{(j)}, \mathcal{F}_{k_j+1}^{(j)}\}]$.

```
In[14]:= FareyBrocotUnion[int_] :=
     Module[{a, b, fe, α, β, k0, xN, fl, fr, xlStart, xuStart},
       {a, b} = int[[1]];
       fe = fareyEncapsulate[int];
       {α, β} = fe[[2, 1]];
       k0 = fe[[1]];
       Which[(* both endpoints match *)
              {a, b} === {α, β},
               {fe},
              (* left endpoint matches *)
              a === α,
               xN = mediant[{α, β}];
               fl = C[k0, Interval[{a, xN}]];
               xlStart = mediant[{α, β}]; xuStart = β;
               fr = zoomInRight[{xlStart, xuStart}, k0 + 1, b];
               Flatten[{fl, fr}],
              (* right endpoint matches *)
              b === β,
               xlStart = α; xuStart = mediant[{α, β}];
               fl = zoomInLeft[{xlStart, xuStart}, k0 + 1, a];
               xN = mediant[{α, β}];
               fr = C[k0, Interval[{xN, b}]];
               Flatten[{fl, fr}],
              (* int is inside *)
              α < a < b < β,
               xlStart = α;  xuStart = mediant[{α, β}];
               fl = zoomInLeft[{xlStart, xuStart}, k0 + 1, a];
               xlStart = mediant[{α, β}]; xuStart = β;
               fr = zoomInRight[{xlStart, xuStart}, k0 + 1, b];
               Flatten[{fl, fr}]]]
```

```
In[15]:= FareyBrocotUnion[Interval[{3/13, 4/13}]]
```

$$\text{Out[15]}= \left\{C\left[6, \text{Interval}\left[\left\{\tfrac{3}{13}, \tfrac{1}{4}\right\}\right]\right], C\left[4, \text{Interval}\left[\left\{\tfrac{1}{4}, \tfrac{2}{7}\right\}\right]\right],\right.$$
$$\left. C\left[5, \text{Interval}\left[\left\{\tfrac{2}{7}, \tfrac{3}{10}\right\}\right]\right], C\left[6, \text{Interval}\left[\left\{\tfrac{3}{10}, \tfrac{4}{13}\right\}\right]\right]\right\}$$

Using the functions `IntervalUnion` and `IntervalIntersection`, we can quickly check that the intervals returned by `FareyBrocotUnion` are disjoint and cover *int*.

```
In[16]:= checkFareyBrocotUnion[l_] :=
         {(* the union of the intervals *)
          IntervalUnion @@ (Last /@ l),
          (* the intersections of the intervals *)
          If[Length[l] === 1, Interval[], IntervalIntersection @@ (Last /@ l)]}
In[17]:= checkFareyBrocotUnion[%%]
Out[17]= {Interval[{ 3/13 , 4/13 }], Interval[]}
```

Next, we cover the larger interval (37/1357, 7345/9879).

```
In[18]:= fbu = FareyBrocotUnion[Interval[{37/1357, 7349/9879}]];
In[19]:= {Length[fbu], checkFareyBrocotUnion[fbu]}
Out[19]= {167, {Interval[{ 37/1357 , 7349/9879 }], Interval[]}}
```

The following graphic shows the original intervals and the 167 covering Farey–Brocot intervals.

```
In[20]:= Show[Graphics[{
           {Hue[0], Thickness[0.01],
            Line[{{37/1357, 0}, {7345/9879, 0}}]},
           {GrayLevel[0], Thickness[0.002],
            Line[{{#2[[1, 1]], #1}, {#2[[1, 2]], #1}}]& @@@ fbu}}],
          PlotRange -> All, Frame -> True]
```

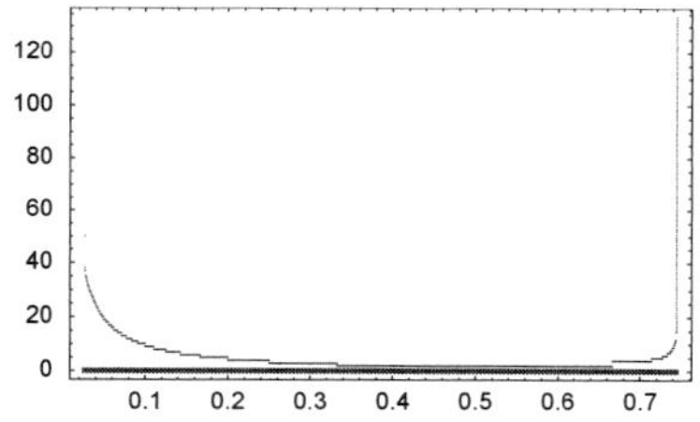

Here, we cover an interval of length about 10^{-6} with Farey–Brocot intervals.

```
In[21]:= FareyBrocotUnion[Interval[{111/2^20, 112/2^20}]]; // Timing
Out[21]= {0.42 Second, Null}
```

The Farey–Brocot coverings for the uniform subdivision of [0, 1] in 4096 subintervals can now be quickly calculated.

```
In[22]:= tab = Table[{#, FareyBrocotUnion[#]}&[Interval[{v, v + 1}/2^12]],
                     {v, 0, 2^12 - 1}]; // Timing
Out[22]= {10.46 Second, Null}
```

88764 Farey–Brocot intervals are needed to cover the 4096 intervals.

```
In[23]:= Flatten[Last /@ tab] // Length
Out[23]= 88764
```

The next graphic shows the distribution of the number of Farey–Brocot intervals needed.

```
In[24]:= Show[GraphicsArray[
           Function[f, ListPlot[N[f[Length[#[[2]]]]]& /@ tab,
                   PlotRange -> All, DisplayFunction -> Identity,
                   PlotStyle -> {PointSize[0.003]}]] /@
               {Identity, Log[10, #]&}]]
```

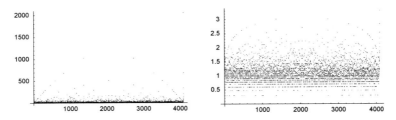

Here are the Farey–Brocot intervals themselves.

```
In[25]:= CToLine[C[h_, Interval[{x1_, x2_}]]] :=
            {Hue[0.8 h/1000], Line[{{x1, h}, {x2, h}}]}
```

```
In[26]:= Show[Graphics[{Thickness[0.002],
                CToLine /@ Flatten[Last /@ tab]}],
            Frame -> True, PlotRange -> {0, 1000}]
```

Here, the distribution of the j in $C[j, \{\mathcal{F}_{k_j}^{(j)}, \mathcal{F}_{k_{j+1}}^{(j)}\}]$ for all intervals.

```
In[27]:= ListPlot[N[{Log[10, First[#]], Length[#]}]& /@
            Split[Sort[First /@ Flatten[Last /@ tab]]], PlotRange -> All]
```

```
In[28]:= FareyBrocotMeasure[C[j_, _]] := 2^-j
        FareyBrocotMeasure[l_List] := Plus @@ (FareyBrocotMeasure /@ l)
```

A further question of interest is the sum $\sum_j 2^{-j}$ *numberOfIntervalsAtLevel n* [439], [363], [304].

```
In[30]:= Off[Graphics::gptn];
        ListPlot[FareyBrocotMeasure /@ Last /@ tab,
            PlotRange -> All, PlotStyle -> {PointSize[0.003]}]
```

For correlations of the Farey sequence, see [68].

e) Here is the straightforward implementation.

```
In[1]:= GuiasuPrimePi[x_] := -Sum[MoebiusMu[k]*
            Sum[MoebiusMu[n] (Plus @@ (Last /@ FactorInteger[n]))*
                                    Floor[x^(1/k)/n // N],
               {n, 2, Floor[x^(1/k)] // N}], {k, 1, Floor[Log[2, x] // N]}]
```

Here is a test.

```
In[2]:= Array[GuiasuPrimePi, 25] == Array[PrimePi, 25]
Out[2]= True
```

This is a timing for a specific example.

```
In[3]:= GuiasuPrimePi[10^5] // Timing
Out[3]= {2.75 Second, 9592}
```

Because of the double sum, some of the needed values of $\mu(n)$, $\Omega(n)$ are calculated more than once. The following implementation avoids this inefficiency.

```
In[4]:= GuiasuPrimePiFaster[x_] := Function[r, Function[{t, a},
            -Apply[Plus, Rest /@ (Floor[(1/(Range[#])& /@  t) (x^(1/r))]
                        (Take[a, #]& /@ t)), {1}].(MoebiusMu /@ Round[r])][
            Floor[x^(1/r)], Array[MoebiusMu[#] (Plus @@
            (Last /@ FactorInteger[#]))&, x]]][N[Range[Floor[Log[2, N[x]]]]]]
```

Nevertheless, it is not much faster because the relative number of values calculated more than one time is small.

```
In[5]:= GuiasuPrimePiFaster[10^5] // Timing
Out[5]= {2.26 Second, 9592}
```

f) Here is a quick look at the possible sums one can form by adding and subtracting the first six primes.

```
In[1]:= FoldList[Union[Flatten[{#1 + #2, #1 - #2}]]&, {1}, Prime[Range[6]]]
Out[1]= {{1}, {-1, 3}, {-4, 0, 2, 6}, {-9, -5, -3, 1, 5, 7, 11},
         {-16, -12, -10, -6, -2, 0, 2, 4, 8, 12, 14, 18}, {-27, -23, -21, -17,
          -13, -11, -9, -7, -5, -3, -1, 1, 3, 5, 7, 9, 11, 13, 15, 19, 23, 25, 29},
         {-40, -36, -34, -30, -26, -24, -22, -20, -18, -16, -14, -12, -10, -8, -6,
          -4, -2, 0, 2, 4, 6, 8, 10, 12, 14, 16, 18, 20, 22, 24, 26, 28, 32, 36, 38, 42}}
```

We implement two functions `step` and `union` that add and subtract the next prime and that keep only one representative of every sum.

```
In[2]:= step[C[sum_, list_], new_] :=
            {C[sum + new, {list, new}], C[sum - new, {list, -new}]}
In[3]:= union[l_] := First /@ Split[Sort[l, #1[[1]] < #2[[1]]&],
                        #1[[1]] == #2[[1]]&]
```

Now, we form the sums and differences for the first 25 primes.

```
In[4]:= FoldList[Function[{x, y}, union[Flatten[step[#, y]& /@ x]]],
               {C[1, {1}]}, Prime[Range[25]]] // Rest;
```

Finally, we extract the primes under consideration and have the following representations.

```
In[5]:= (Plus @@ # == (HoldForm[#] /. List -> Plus)) & /@
        Flatten /@ Last /@ Flatten[MapIndexed[If[OddQ[#2[[1]]],
               Cases[#, C[Prime[#2[[1]] + 1], _]], {}]&, %]]
```

Out[5]= $\{3 == 1 + 2, 7 == 1 - 2 + 3 + 5, 13 == 1 + 2 - 3 - 5 + 7 + 11, 19 == 1 + 2 - 3 + 5 + 7 + 11 + 13 - 17,$
$29 == 1 + 2 + 3 + 5 + 7 + 11 - 13 + 17 + 19 - 23, 37 == 1 - 2 + 3 + 5 + 7 + 11 + 13 + 17 + 19 + 23 - 29 - 31,$
$43 == 1 + 2 - 3 + 5 + 7 + 11 + 13 - 17 + 19 + 23 + 29 + 31 - 37 - 41,$
$53 == 1 + 2 + 3 + 5 - 7 + 11 + 13 + 17 + 19 + 23 + 29 + 31 + 37 - 41 - 43 - 47,$
$61 == 1 + 2 + 3 + 5 + 7 + 11 + 13 + 17 + 19 + 23 + 29 - 31 + 37 + 41 + 43 - 47 - 53 - 59,$
$71 == 1 - 2 + 3 + 5 - 7 + 11 + 13 + 17 + 19 + 23 + 29 + 31 + 37 + 41 + 43 + 47 - 53 - 59 - 61 - 67,$
$79 == 1 - 2 + 3 + 5 + 7 + 11 + 13 + 17 + 19 + 23 + 29 + 31 + 37 + 41 - 43 + 47 + 53 + 59 - 61 - 67 - 71 - 73,$
$89 == 1 + 2 - 3 + 5 + 7 + 11 + 13 - 17 + 19 + 23 + 29 + 31 + 37 + 41 + 43 + 47 +$
$\quad 53 + 59 + 61 - 67 - 71 - 73 - 79 - 83, 101 == 1 + 2 + 3 + 5 + 7 + 11 + 13 + 17 + 19 +$
$\quad 23 + 29 + 31 + 37 + 41 + 43 + 47 + 53 - 59 + 61 + 67 + 71 - 73 - 79 - 83 - 89 - 97\}$

In[6]:= `ReleaseHold //@ %`

Out[6]= {True, True, True, True, True, True, True, True, True, True, True, True, True}

In[7]:= `Prime[Range[2, 25, 2]]`

Out[7]= {3, 7, 13, 19, 29, 37, 43, 53, 61, 71, 79, 89}

g) We start with the visualizations. For the calculation of numerical values of $Q_d(z)$, we implement the following compiled function. We sum until the fifth digit does not change anymore. For a visualization, this surely is good enough. (Using the function `InverseEllipticNomeQ`, it would be possible to express $Q_d(z)$ as a finite sum [397].)

```
In[1]:= cf[d_] := cf[d] = Compile[{{z, _Complex}},
        Module[{sum = z, Z = z, old = 0. + 0. I, k = 1, δ = 10.^-5, l = #},
                (* sum until result is good enough *)
                While[k < 10000 && (k < 10 || Abs[sum/old - 1] > δ),
                      k = k + 1; Z = Z z;
                      (* in case we do not get enough digits *)
                      If[k > 9998, Print[{d, Z, "no divisor sums left"}]];
                      old = sum;
                      sum = old + l[[k]] Z];
                  sum]]&[(* numericalize list of divisors *)
                  Table[N[DivisorSigma[d, k]], {k, 10000}]];
```

The functions $Q_d(z)$ are defined inside the unit disk. On the unit disk, they have a dense set of singularities. We will calculate and display $Q_1(z)$, $Q_2(z)$, and $Q_5(z)$ for $|z| < 0.98$. For $Q_1(z)$, we choose a density plot-like style.

```
In[2]:= ε1 = 10^-6; ε2 = 10.^-2;

        {ppr, ppφ} = {120, 240};
        δr = (1. - ε1 - ε2)/ppr; δφ = 2Pi/ppφ;
        polys = Table[{(* color according to scaled real part *)
                        Hue[(ArcTan[Re[cf[1][r Exp[I φ]]]] + Pi/2)/Pi],
                      Polygon[{r {Cos[φ], Sin[φ]}, (r + δr){Cos[φ], Sin[φ]},
                          (r + δr){Cos[φ + δφ], Sin[φ + δφ]},
                          r{Cos[φ + δφ], Sin[φ + δφ]}}]},
                  {r, ε1, 1 - ε2, (1 - ε1 - ε2)/ppr}, {φ, 0, 2Pi, 2Pi/ppφ}];

In[6]:= Show[Graphics[polys], AspectRatio -> Automatic]
```

$Q_2(z)$ we will display as a 3D graphic. The height represents the (scaled) real part and the color represents the imaginary part.

```
In[7]:= ParametricPlot3D[{r Cos[φ], r Sin[φ], ArcTan[Re[#]],
                          {EdgeForm[], SurfaceColor[#, #, 3]&[
                (* color according to scaled imaginary part *)
                Hue[(ArcTan[Im[#]] + Pi/2)/Pi]]}}&[cf[3][r Exp[I φ]]],
```

```
            {r, ε1, 1 - ε2}, {φ, 0, 2Pi},
    PlotPoints -> {120, 240}, Compiled -> False,
    BoxRatios -> {1, 1, 1/3}, Boxed -> False, Axes -> False]
```

$Q_3(z)$ finally, we will visualize using a contour plot. We first calculate a contour plot in rectangular coordinates and the map into the unit disk. For a better picture, we do not use the default contour scaling.

```
In[8]:= Show[GraphicsArray[
    Function[{reIm},
    Module[{cp, cls, lcp, pp = 180, cl = 180, ε1 = 10^-8, ε2 = 10.^-2},
    (* original contour plot *)
    cp = ContourPlot[reIm[cf[5][r Exp[I φ]]],
                {r, ε1, 1 - 2 ε2}, {φ, ε1, 2Pi - ε1},
                PlotPoints -> {pp, 2pp}, Compiled -> False,
                DisplayFunction -> Identity];
    (* homogeneous contour spacing *)
    cls = #[[Round[pp^2/cl/2]]]& /@ Partition[Sort[Flatten[cp[[1]]]],
                                        Round[pp^2/cl]];
    (* new contour plot *)
    lcp = ListContourPlot[cp[[1]], Contours -> cls,
            MeshRange -> {{ε1, 1 - 2 ε2}, {ε1, 2Pi - ε1}},
            DisplayFunction -> Identity, ContourShading -> False,
            ContourStyle -> Table[{Hue[0.8 k/(2/3 pp)]}, {k, 20}]];
    (* Cartesian -> polar coordinates *)
    Show[{Graphics[{GrayLevel[0], Disk[{0, 0}, 1]}],
            Graphics[lcp] /. Line[l_] :>
            Line[#[[1]] {Cos[#[[2]]], Sin[#[[2]]]}& /@ l],
            Graphics[{GrayLevel[0.8], Thickness[0.02], Circle[{0, 0}, 1]}]},
            AspectRatio -> Automatic, Frame -> False,
            DisplayFunction -> Identity]]] /@
                            (* use real and imaginary part *) {Re, Im}]]
```

Now, let us search for some quadratic relations of the described form. We start by implementing a power series for the functions $Q_d(z)$. (For convenience, we use here the `SeriesData` functions from Chapter 1 of the Symbolics volume [556]. This could be avoided by an implementation similar to the one used in Solution 14c.)

```
In[9]:= Qs[d_, z_, n_] := Sum[DivisorSigma[d, k] z^k, {k, n}] + O[z]^n
```

Here are some examples.

```
In[10]:= Qs[1, z, 10]
```

$$Out[10]= z + 3 z^2 + 4 z^3 + 7 z^4 + 6 z^5 + 12 z^6 + 8 z^7 + 15 z^8 + 13 z^9 + O[z]^{10}$$

```
In[11]:= Qs[3, z, 10]
```

In[11]:= $z + 9 z^2 + 28 z^3 + 73 z^4 + 126 z^5 + 252 z^6 + 344 z^7 + 585 z^8 + 757 z^9 + O[z]^{10}$

In[12]:= `Qs[5, z, 10]`

Out[12]= $z + 33 z^2 + 244 z^3 + 1057 z^4 + 3126 z^5 + 8052 z^6 + 16808 z^7 + 33825 z^8 + 59293 z^9 + O[z]^{10}$

Now, we make a general quadratic ansatz with unknown coefficients A.

In[13]:= `eq[l_, n_] := (* generic ansatz *)`
          `Sum[A[l[[i]]] Qs[l[[i]], z, n], {i, Length[l]}] +`
          `Sum[A[l[[i]], l[[j]]] Qs[l[[i]], z, n] Qs[l[[j]], z, n],`
          `{i, Length[l]}, {j, i, Length[l]}];`

Using now the truncated $Q_d(z)$ in the ansatz gives a linear system for the unknown coefficients A. If we use enough terms in the series, we will get a highly overdetermined system. Getting a solution for such a system is a candidate relation. We will systematically try all α_k, β_k, and γ_k less than 15.

In[14]:= `Off[Solve::svars];`
          `tab = Table[ (* equations *)`
          `cl = CoefficientList[Expand[eq[{i, j, k}, 20]], z];`
          `(* variables *) vars = Cases[cl, _A, Infinity];`
          `sol = Solve[# == 0& /@ cl, vars];`
          `If[(* nontrivial solution? *) Union[Last /@ sol[[1]]] =!= {0},`
          `{{i, j, k}, sol}, {}], {i, 15}, {j, i}, {k, j}];`

Eliminating doubles, we end with the following two candidate solutions.

In[16]:= `eqRes[l_, n_] :=`
          `Sum[A[l[[i]]] Subscript[Q, l[[i]]], {i, Length[l]}] +`
          `Sum[A[l[[i]], l[[j]]] Subscript[Q, l[[i]]]*`
          `Subscript[Q, l[[j]]],`
          `{i, Length[l]}, {j, i, Length[l]}];`

In[17]:= `Union[Numerator[Together[#]]& /@`
          `Flatten[eqRes[#[[1]], Length[#[[1]]]]] /. #[[2]]& /@`
          `DeleteCases[Flatten[tab, 2], {}, Infinity]] /. _A -> 1]`

Out[17]= $\{Q_3 + 120 Q_3^2 - Q_7, -10 Q_3 + 21 Q_5 + 5040 Q_3 Q_5 - 11 Q_9, 21 Q_5 - 20 Q_7 + 10080 Q_5 Q_7 - Q_{13},$
          $-10 Q_3 + 11 Q_9 + 2640 Q_3 Q_9 - Q_{13}, -21 Q_5 + 22 Q_9 - 2904 Q_9^2 - Q_{13} + 504 Q_5 Q_{13}\}$

Using now a power series with many more terms still shows complete cancellation. This strongly suggests that the found relations hold to all orders.

In[18]:= `% /. Subscript[Q, d_] :> Qs[d, z, 200]`

Out[18]= $\{O[z]^{200}, O[z]^{200}, O[z]^{200}, O[z]^{200}, O[z]^{200}\}$

Comparing coefficients in z in the found relations transforms products of Q_d functions into convolution sums of σ_d, and we end with the following relations for the divisor sums [21], [232], [275].

$$\sigma_3(n) - \sigma_7(n) + 120 \sum_{k=1}^{n-1} \sigma_3(n-k)\,\sigma_3(k) = 0$$

$$-10\,\sigma_3(n) + 21\,\sigma_5(n) - 11\,\sigma_9(n) + 5040 \sum_{k=1}^{n-1} \sigma_3(n-k)\,\sigma_5(k) = 0$$

$$21\,\sigma_5(n) - 20\,\sigma_7(n) - \sigma_{13}(n) + 10080 \sum_{k=1}^{n-1} \sigma_5(n-k)\,\sigma_7(k) = 0$$

$$-10\,\sigma_3(n) + 11\,\sigma_9(n) - \sigma_{13}(n) + 2640 \sum_{k=1}^{n-1} \sigma_3(n-k)\,\sigma_9(k) = 0$$

$$-21\,\sigma_5(n) + 22\,\sigma_9(n) - \sigma_{13}(n) - 2904 \sum_{k=1}^{n-1} \sigma_9(n-k)\,\sigma_9(k) + 504 \sum_{k=1}^{n-1} \sigma_5(n-k)\,\sigma_{13}(k) = 0$$

11. Choquet Approximation, Magnus Expansion

a) The routine `ChoquetApproximation` calculates new intervals for the given number z until the interval boundaries are, within the precision of z, the same as z.

```
In[1]:= ChoquetApproximation[z_?Positive] :=
    Module[{li, z1, z2},
    For[(* first bracketing of z *)
        {z1, z2} = {Floor[z], Ceiling[z]}; li = {{z1, z2}},
        !(z1 == z || z2 == z) (* test *),
    (* new values *)
    {z1, z2} = If[# > z, {z1, #}, {#, z2}]&[
                (Numerator[ z1] + Numerator[ z2])/
                (Denominator[z1] + Denominator[z2])];
    (* collect new values *)
        AppendTo[li, {z1, z2}], Null]; li]
```

Here is an example.

```
In[2]:= ChoquetApproximation[13/7]
```

$$Out[2]= \left\{\{1, 2\}, \left\{\tfrac{3}{2}, 2\right\}, \left\{\tfrac{5}{3}, 2\right\}, \left\{\tfrac{7}{4}, 2\right\}, \left\{\tfrac{9}{5}, 2\right\}, \left\{\tfrac{11}{6}, 2\right\}, \left\{\tfrac{13}{7}, 2\right\}\right\}$$

For the graphical representation, we use the logarithm of base 10 distance of the two fractions representing the approximation to the exact value.

```
In[3]:= picture[iL_, z_, opts___] :=
    Graphics[{Thickness[0.002],
        (* the smaller fraction *) Line[MapIndexed[{#2[[1]], #1}&,
        Log[10, -Drop[First[N[Transpose[iL], Precision[z]] - z], -1]]]],
        (* the larger fraction *) Line[MapIndexed[{#2[[1]], #1}&,
        Log[10, Drop[Last[N[Transpose[iL], Precision[z]] - z], -1]]]]},
            opts, Axes -> True, PlotRange -> All, AxesOrigin -> {0, 0}]
```

Here are the three larger examples.

```
In[4]:= lPi = ChoquetApproximation[pi      = N[Pi,       50]];
    1E  = ChoquetApproximation[e       = N[E,        50]];
    15  = ChoquetApproximation[cubeRoot = N[5^(1/3), 50]];
    12  = ChoquetApproximation[root     = N[Sqrt[2], 50]];
```

We do not look at the results, because they are fairly large due to large integers in the fractions.

```
In[8]:= ByteCount /@ {lPi, 1E, 15, 12}
```

$$Out[8]= \{119520, 66736, 639424, 25728\}$$

The last approximations are as follows.

```
In[9]:= Last[lPi]
```

$$Out[9]= \left\{ \frac{19838350385690370130755600}{6314743053343261487206311}, \frac{43668160973609007601087l}{139000073493649328482341} \right\}$$

```
In[10]:= Last[1E]
```

$$Out[10]= \left\{ \frac{13633221032043148263477779}{50153817346347875356067l}, \frac{6713353765303144209805l3}{246970483156591884266112} \right\}$$

```
In[11]:= Last[15]
```

$$Out[11]= \left\{ \frac{15754627587487757546258l4}{9213362104950395102235l9}, \frac{1331624026750446891684517}{778738454969750155604233} \right\}$$

```
In[12]:= Last[12]
```

$$Out[12]= \left\{ \frac{65138564066681764252300l}{4605992036830504954l5105}, \frac{2698127666992833483072l3}{190786436983767147107902} \right\}$$

So the intervals shrink.

In[13]:= `Show[GraphicsArray[#]]& /@ {{picture[lPi, pi],    picture[lE, e]},`
`                            {picture[15, cubeRoot], picture[12, root]}}`

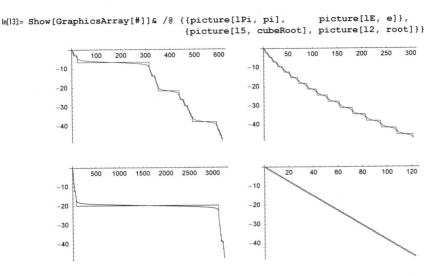

One interesting fact about this approximation is that for a rational number, it always terminates after finite steps (as in the example above).

Note that the difference in the approximation property of the three numbers π, e, and $2^{1/2}$ is ultimately connected to the continued fraction representation of these numbers; see [85], [281], [532]. The four continued fractions are shown here.

In[14]:= `ContinuedFraction[N[Pi, 100], 30]`

Out[14]= `{3, 7, 15, 1, 292, 1, 1, 1, 2, 1, 3, 1, 14, 2, 1, 1, 2, 2, 2, 2, 1, 84, 2, 1, 1, 15, 3, 13, 1, 4}`

In[15]:= `ContinuedFraction[N[E, 100], 30]`

Out[15]= `{2, 1, 2, 1, 1, 4, 1, 1, 6, 1, 1, 8, 1, 1, 10, 1, 1, 12, 1, 1, 14, 1, 1, 16, 1, 1, 18, 1, 1, 20}`

In[16]:= `ContinuedFraction[N[5^(1/3), 100], 30]`

Out[16]= `{1, 1, 2, 2, 4, 3, 3, 1, 5, 1, 1, 4, 10, 17, 1, 14, 1, 1, 3052, 1, 1, 1, 1, 1, 1, 2, 2, 1, 3, 2}`

In[17]:= `ContinuedFraction[N[Sqrt[2], 100], 30]`

Out[17]= `{1, 2, 2, 2, 2, 2, 2, 2, 2, 2, 2, 2, 2, 2, 2, 2, 2, 2, 2, 2, 2, 2, 2, 2, 2, 2, 2, 2, 2, 2}`

They can be observed in the simpler approximation of these numbers by rationals of the form p/q with fixed q. The following picture shows the absolute value of the difference between the nearest p/q and the above numbers (with q scaled) [76], [410], [411].

In[18]:= `Show[GraphicsArray[#]]& /@`
`      Partition[Function[s, ListPlot[`
`      Table[q * (* the scaled difference *)`
`            Min[Abs[{#/q - s, (# + 1)/q - s}]&[`
`               (* the optimal numerator or + 1 of this number *)`
`               Floor[s/(1/q)]]], {q, 5000}],`
`               PlotStyle -> {PointSize[0.003]}, Frame -> True,`
`               PlotRange -> All, DisplayFunction -> Identity]] /@`
`         (* the three numbers to investigated *)N[{Pi, E, 5^(1/3), Sqrt[2]}], 2]`

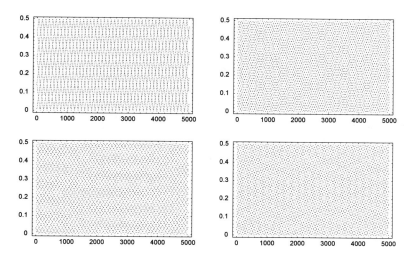

Also, in the following sunflower-like picture [28], [418], we can see the different continued fraction expansions of the three numbers.

```
In[19]:= With[{tpi = N[2Pi]},
        Show[GraphicsArray[#]]& /@
        Partition[Graphics[{PointSize[0.003],
        (* the points of the spiral *)
        Table[Point[N[Sqrt[n + 1/2] {Cos[tpi n #], Sin[tpi n #]}]],
            {n, 2500}]}, PlotRange -> All, FrameTicks -> None,
            AspectRatio -> Automatic, DisplayFunction -> Identity,
            PlotLabel -> ToString[InputForm[#]], Frame -> True]& /@
                (* four numbers *) {Pi, E, 5^(1/3), Sqrt[2]}, 4]]
```

b) To compare, we start with a high-precision numerical calculation of the system.

```
In[1]:= solExact = {x1[1], x2[1]} /.
        NDSolve[{x1'[t] == t x1[t] + x2[t], x2'[t] == x1[t] + t^2 x2[t],
            x1[0] == 1, x2[0] == 1}, {x1, x2}, {t, 0, 1},
            AccuracyGoal -> 20, PrecisionGoal -> 20,
            WorkingPrecision -> 40][[1]]
Out[1]= {4.2712893080044541487434723243526011541, 3.9927987384304049365050775316591183552}
```

The implementation of the Magnus method is straightforward. To reduce the size of intermediate expression we use Expand.

```
In[2]:= A = {{t, 1}, {1, t^2}};

        S[1, 0][t_] = A;

        S[n_, 0][t_] = {{0, 0}, {0, 0}};

        (* use pattern on the right-hand side of a
            SetDelayed[..., Set[...]]] construction *)
```

```
Off[RuleDelayed::rhs];
S[n_, j_][t_] := S[n, j][t_] = Expand[
  Sum[Ω[m][t].S[n - m, j - 1][t] - S[n - m, j - 1][t].Ω[m][t], {m, 1, n - j}]]

Ω[n_][t_] := Ω[n][t] = Expand[
  Sum[BernoulliB[j]/j! Integrate[S[n, j][τ], {τ, 0, t}], {j, 0, n - 1}]]
```

$\Omega S[o][t]$ is the sum of the first o of the $\Omega_k(t)$.

```
In[9]:= Do[ΩS[o][t_] = Sum[Ω[n][t], {n, o}], {o, 22}]
```

Here are the first few terms of one element of the resulting $\Omega(t)$.

```
In[10]:= ΩS[8][t][[1, 1]]
```

$$\text{Out[10]=}\ \frac{t^2}{2} - \frac{t^5}{180} + \frac{t^7}{1890} - \frac{t^8}{30240} + \frac{19\,t^9}{302400} - \frac{11\,t^{10}}{100800} + \frac{t^{11}}{83160} + \frac{83\,t^{12}}{3801600} -$$
$$\frac{71399\,t^{13}}{9340531200} + \frac{311\,t^{14}}{106142400} - \frac{433\,t^{15}}{179625600} + \frac{11\,t^{16}}{11430720} - \frac{2269\,t^{17}}{15157134720}$$

The following data shows that for 17 terms, we obtain 20 correct digits.

```
In[11]:= ListPlot[Table[{o, Log[10, Max[Abs[MatrixExp[N[ΩS[o][1], 22]].{1, 1} -
              solExact]]]}, {o, 22}], Frame -> True, Axes -> False]
```

12. Rademacher Identity, Goldbach Problem, Optical Factoring

a) Here, the approximation is implemented. (To avoid unnecessary work, we have carried out the n differentiation and remember the values of $A[k, n]$.)

```
In[1]:= RademacherPartitionPApproximation[n_, max_] :=
  With[{v = Pi Sqrt[2/3] Sqrt[n - 1/24]/k},
    (1/(Pi Sqrt[2]) Sum[A[k, n] Sqrt[k] *
          (Pi Cosh[v]/(Sqrt[6] k (n - 1/24)) -
          Sinh[v]/(2(n - 1/24)^(3/2))), {k, max}])]

A[k_, n_] := A[k, n] = Plus @@ (ω[#, k] Exp[-2Pi I # n/k]& /@
    (* a reduced residue system *) Select[Range[k], GCD[#, k] == 1&])

ω[h_, k_] := Exp[Pi I Sum[m/k (h m/k - Floor[h m/k] - 1/2), {m, k - 1}]]
```

Exactly eight terms must be included to reproduce all 32 digits of $p(1000)$.

```
In[4]:= (Table[Re @ N[RademacherPartitionPApproximation[1000, i], 100],
            {i, 8}] - PartitionsP[1000]) // N
Out[4]= {-4.08974×10^13, 3.02545×10^7, -83598.7, -2338.49, 94.9297, 15.0296, -2.69326, 0.0455524}
```

A more exact treatment would require us to derive some bounds on the terms of the series, which are not always monotonically decreasing; we do not do this here.

b) It is straightforward to implement a function `primePairIndices` that generates all index pairs: We step through all index combinations and select the relevant ones.

```
In[1]:= primePairIndices[n_] :=
  Module[{λ = {}},
          Do[If[PrimeQ[n - Prime[k]], λ = {λ, C[k, PrimePi[n - Prime[k]]]}],
              {k, PrimePi[Floor[n/2]]}]; List @@@ Flatten[λ]]
In[2]:= primePairIndices[100]
Out[2]= {{2, 25}, {5, 24}, {7, 23}, {10, 20}, {13, 17}, {15, 16}}
```

We rewrite `primePairIndices` as a one-liner by eliminating the local variable λ.

```
In[3]:= primePairIndices[n_] := Partition[Flatten[Last[Nest[{# + 1,
            If[PrimeQ[n - Prime[#]],
               {#2, {#, PrimePi[n - Prime[#]]}}, #2]}& @@ #&,
            {1, {}}, PrimePi[Floor[n/2]]]]], 2]
```

Calculating all prime pairs for $n = 10^6$ takes less than a second. There are 5402 pairs.

```
In[4]:= (ppi1000000 = primePairIndices[10^6]); // Timing

Out[4]= {0.61 Second, Null}
```

```
In[5]:= Length[ppi1000000]

Out[5]= 5402
```

Here is a quick check that the pairs are correct.

```
In[6]:= Union[Plus @@@ Prime[ppi1000000]]

Out[6]= {1000000}
```

Here is a graphic of the sum of the indices as a function of their position.

```
In[7]:= ListPlot[Plus @@@ ppi1000000]
```

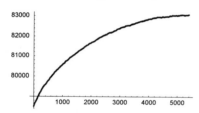

To calculate the number of prime pairs for $n = 10^9$, we can simplify the code from above slightly. The explicit index of the second prime is now no longer needed.

```
In[8]:= primePairCount[n_] :=
        Module[{c = 0},
               Do[If[PrimeQ[n - Prime[k]], c++], {k, PrimePi[Floor[n/2]]}]; c]
```

```
In[9]:= primePairCount[10^6]

Out[9]= 5402
```

Now, we can calculate `primePairCount[10^9]`. On a fast year-2005 computer, this input will take less than an hour and return the number 2274205.

Let us compare with the asymptotic conjecture. The first product (the Hardy–Littlewood constant) we calculate to about 10 digits by explicit multiplication.

```
In[10]:= hwc = Module[{pn = 1., k = 2},
                      While[po = pn;
                            pn = pn Prime[k] (Prime[k] - 2)/(Prime[k] - 1)^2;
                            Abs[pn - po]/pn > 10^-12, k++]; pn]

Out[10]= 0.660162
```

For the second product, only the prime divisor 5 matters.

```
In[11]:= Select[Divisors[10^9], PrimeQ]

Out[11]= {2, 5}
```

This gives an asymptotic estimation that is off by about 0.4%.

```
In[12]:= hwc (5 - 1)/(5 - 2) NIntegrate[1/Log[x]^2, {x, 2, 10^9}]

Out[12]= 2.28354 × 10^6
```

c) For a fast calculation of the sum $s_{n,m}(x)$, we implement a direct version s and a compiled version sC.

```
In[1]:= s[x_, n_, m_] = Sum[Exp[I Pi (k - x)^2/m], {k, (1 - n)/2, (n - 1)/2}];
```

```
In[2]:= sC = Compile[{x, n, m}, Sum[Exp[I Pi (k - x)^2/m],
                        {k, (1 - n)/2., (n - 1)/2.}]];
```

We fix an odd integer $n = \nu = 1155$ and display $|s_{\nu,77}(x)|$, $|s_{\nu,105}(x)|$, and $|s_{\nu,199}(x)|$.

```
In[3]:= ν = 3 5 7 11;
```

```
Show[GraphicsArray[
     Plot[Abs @ sC[x, ν, #], {x, 0, 6},
          PlotRange -> All, PlotPoints -> 1000,
          DisplayFunction -> Identity]& /@ {77, 105, 199}]]
```

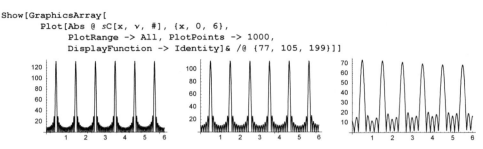

The last graphics show the following properties of the sums $s_{n,m}(x)$: They are highly oscillating functions with $|s_{n,m}(x)|$ having pronounced maxima at $x = 1/2 + j$, $j \in \mathbb{Z}$. They have the obvious periodicity $s_{n,m}(x + m) = s_{n,m}(x)$. And the overall shapes of the three graphics suggest the following conjecture: If m divides n, then the large peaks are of equal height, meaning $s_{n,m}(1/2) = s_{n,m}(j + 1/2)$, $j = 1, \ldots, m$. A quick exact check for the small divisors and other small odd integers seems to confirm this conjecture.

```
In[5]:= (Equal @@ Table[s[x, ν, #], {x, 1/2, # + 1/2}])& /@ {3, 5, 7, 11, 15, 21}
```

```
Out[5]= {True, True, True, True, True, True}
```

```
In[6]:= (Equal @@ Table[s[x, ν, #], {x, 1/2, # + 1/2}])& /@ {9, 13, 17, 19, 23, 25}
```

```
Out[6]= {False, False, False, False, False, False}
```

Here is a numerical check of this conjecture for the above ν. For m divides n, the large maxima of $|s_{n,m}(x)|$ are all of the same height.

```
In[7]:= (* the large maxima *)
        periodValues[n_, m_?OddQ] := Table[sC[1/2 + k, n, m], {k, 0, m}]
```

```
In[9]:= (* difference of the large maxima *)
        δm[l_] := (Max[#] - Min[#])/(Plus @@ #/Length[#])&[Abs[l]]
```

```
In[11]:= ListPlot[Table[{m, δm @ periodValues[ν, m]}, {m, 3, ν, 2}],
              PlotJoined -> True, GridLines -> {Divisors[ν], None}]
```

The property of equal size maxima could be used to factor integers using optical diffraction gratings [116], [27], [370]; for optical searching, see [60]; for a related wave packet version, see [371].

13. Bernoulli and Fibonacci Numbers, Zeckendorf Representation

a) Here is a possible implementation. We start by expanding the expression $(C + 1)^{\wedge}n$ (we use C instead of myBernoulliB to avoid recursion). Then, we carry out the substitution $C^{\wedge}n \rightarrow C[n]$ and substitute all earlier values of myBernoulliB[i], $1 \leq i < n$. (The earlier values we remember via the SetDelayed[myBernoulliB[n_], Set[myBernoulliB[n], ...]] construction.) Finally, we solve the resulting linear equation for myBernoulliB[n].

```
In[1]:= myBernoulliB[0] = 1;

      myBernoulliB[n_] := myBernoulliB[n] =
      (Solve[(* generate equations *)
          ((Expand[(C + 1)^(n + 1) - C^(n + 1)] /.
          (* power -> index *) C^exp_. -> C[exp]) /.
          C[i_?(# < n&)] :> myBernoulliB[i]) == 0, C[n]])[[1, 1, 2]];
```

The numbers calculated this way are exactly the Bernoulli numbers.

```
In[3]:= Table[myBernoulliB[i], {i, 0, 30}]
```

$$Out[3]= \left\{1, -\frac{1}{2}, \frac{1}{6}, 0, -\frac{1}{30}, 0, \frac{1}{42}, 0, -\frac{1}{30}, 0, \frac{5}{66}, 0, -\frac{691}{2730}, 0, \frac{7}{6}, 0, -\frac{3617}{510}, 0, \frac{43867}{798}, 0,\right.$$
$$\left.-\frac{174611}{330}, 0, \frac{854513}{138}, 0, -\frac{236364091}{2730}, 0, \frac{8553103}{6}, 0, -\frac{23749461029}{870}, 0, \frac{8615841276005}{14322}\right\}$$

```
In[4]:= Table[BernoulliB[i], {i, 0, 30}]
```

$$Out[4]= \left\{1, -\frac{1}{2}, \frac{1}{6}, 0, -\frac{1}{30}, 0, \frac{1}{42}, 0, -\frac{1}{30}, 0, \frac{5}{66}, 0, -\frac{691}{2730}, 0, \frac{7}{6}, 0, -\frac{3617}{510}, 0, \frac{43867}{798}, 0,\right.$$
$$\left.-\frac{174611}{330}, 0, \frac{854513}{138}, 0, -\frac{236364091}{2730}, 0, \frac{8553103}{6}, 0, -\frac{23749461029}{870}, 0, \frac{8615841276005}{14322}\right\}$$

```
In[5]:= % === %%
```

```
Out[5]= True
```

In a similar way, we can implement the calculation of the Bernoulli polynomials.

```
In[6]:= myBernoulliB[0, x] = 1;

      myBernoulliB[n_, x_] := myBernoulliB[n, x] =
          Expand[(C + x)^n] /. C^i_. :> BernoulliB[i]
In[8]:= Table[myBernoulliB[i, x] - BernoulliB[i, x], {i, 0, 20}]
```

```
Out[8]= {0, 0, 0, 0, 0, 0, 0, 0, 0, 0, 0, 0, 0, 0, 0, 0, 0, 0, 0, 0, 0}
```

b) The following implementation parallels the one from Part a) of this exercise.

```
In[1]:= fibonacciIdentity[n_, m_, F_] :=
          Expand[F^(n + m) - F^(n - m) (F + 1)^m] /. F^(l_.) :> F[l + 1]
```

Here is one example identity.

```
In[2]:= fibonacciIdentity[3, 4, F]
```

```
Out[2]= -4 - F[0] - 6 F[2] - 4 F[3] - F[4] + F[8]
```

After replacing the F's by `Fibonacci`, we see that the identity holds.

```
In[3]:= % /. F -> Fibonacci
```

```
Out[3]= 0
```

The identity holds for all *n* and *m*.

```
In[4]:= Table[fibonacciIdentity[n, m, F] /. F -> Fibonacci,
          {n, 0, 100}, {m, 0, 100}] // Flatten // Union
```

```
Out[4]= {0}
```

c) To calculate the digits of Zeckendorf representation of a number n, we recursively subtract the largest Fibonacci number F_k smaller or equal to in n.

```
In[1]:= $RecursionLimit = Infinity;

      ZeckendorfDigits[0] = {};

      ZeckendorfDigits[n_] := ZeckendorfDigits[n] =
          Flatten[{δ = f[n], ZeckendorfDigits[n - Fibonacci[δ]]}]
```

For small n, the following function `greatestFibonacci` implements the greatest Fibonacci number smaller or equal to n.

```
In[4]:= greatestFibonacci[n_] :=
        Module[{k = 1}, While[Fibonacci[k] <= n, k++]; k - 1];
            Do[f[k] = greatestFibonacci[k], {k, 1, 100}]
```

For larger n, we can use the representation $F_k = 5^{-1/2} \left(((1 + 5^{1/2})/2)^k - (-2/(1 + 5^{1/2}))^k \right)$ (or $F_k = \lfloor \phi^k \, 5^{-1/2} \rceil$) to calculate the greatest Fibonacci number smaller or equal to n.

```
In[6]:= f[n_] := With[{k = Floor[Log[Sqrt[5] n]/Log[(1 + Sqrt[5])/2]]},
                    If[Fibonacci[k + 1] <= n, k + 1, k]]
```

The Zeckendorf representation of 10^{1000} can be quickly calculated. It has 1316 terms.

```
In[7]:= (zdd = ZeckendorfDigits[10^1000]); // Timing
```

```
Out[7]= {0.7 Second, Null}
```

```
In[8]:= {Length[zdd], Max[zdd]}
```

```
Out[8]= {1316, 4786}
```

Here is a check for the correctness of the calculated representation.

```
In[9]:= {(* the number itself *)
        10^1000 - Plus @@ Fibonacci[zdd],
        (* difference between digits *)
        Min[Subtract @@@ Partition[zdd, 2, 1]]}
```

```
Out[9]= {0, 2}
```

To calculate the Zeckendorf representation of the first 100000 integers, we implement a compiled function for calculating the greatest Fibonacci number smaller or equal to n. It is based on a recursive search.

```
In[10]:= which[{{c_, v_}}] := v;
        which[l_] := Module[{aux = Partition[l, Length[l]/2]},
                            which[aux[[1, -1, 1]]], which @ aux[[1]],
                            aux[[2, -1, 1]]], which @ aux[[2]]]]
```

```
In[12]:= Clear[f];
        f = Compile[{{z, _Integer}}, Evaluate[
        (which[Table[{z < Fibonacci[k + 1], k}, {k, 32}]] /.
            which -> Which) //. (* must have a True for compilation *)
            HoldPattern[Which[c1_, r1_, c2_, r2_?(# =!= 0&)]] :>
                            Which[c1, r1, c2, r2, True, 0]]];
```

```
In[14]:= ZeckendorfDigits[100]
```

```
Out[14]= {11, 6, 4}
```

The Zeckendorf representation of the first 10000 integers can be quickly computed and verified.

```
In[15]:= (data = Table[ZeckendorfDigits[k], {k, 100000}]); // Timing
```

```
Out[15]= {2.61 Second, Null}
```

```
In[16]:= Table[k - Plus @@ Fibonacci[ZeckendorfDigits[k]], {k, 100000}] //
                                                Union // Timing
```

```
Out[16]= {2.88 Second, {0}}
```

The following graphic shows the $\sum_{k=1}^{s_n} l_k$ as a function of n.

```
In[17]:= ListPlot[Plus @@@ data, PlotStyle -> {PointSize[0.003]}]
```

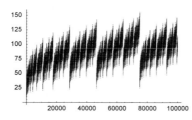

For a nontrivial application of the Zeckendorf digits, see [192]. We display about three periods of the function ϕ. ϕ is a continuous, but nowhere differentiable, function.

```
In[18]:= ListPlot[MapIndexed[{Log[GoldenRatio, #2[[1]]]/6, #1}&,
            MapIndexed[#1/#2[[1]]&, FoldList[Plus, 0,
                FoldList[Plus, 0, (-1)^(Length /@ data)]]]]] // N,
            Frame -> True, Axes -> False, PlotRange -> All]
```

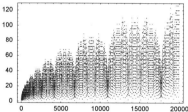

The Zeckendorf digits can be used to calculate the number of representations of an integer n as a sum of distinct Fibonacci numbers (order does not matter) efficiently [160]. (For a digit-minimizing expansion in Fibonacci numbers, see [254].))

```
In[19]:= FibonacciSumRepresentations[n_] :=
    Module[{s = Reverse @ ZeckendorfDigits[n], t},
            t[0] := 1; t[1] = Floor[s[[1]]/2];
            t[j_] := t[j] = If[OddQ[s[[j - 1]] + s[[j]]],
                ⌊(s[[j]] - s[[j - 1]] + 2)/2⌋ t[j - 1],
                ⌊(s[[j]] - s[[j - 1]] + 2)/2⌋ t[j - 1] - t[j - 2]];
            t[Length[s]]]
```

Here is the number of representations for the first 20000 integers visualized.

```
In[20]:= ListPlot[Table[FibonacciSumRepresentations[k], {k, 20000}],
            PlotRange -> All, Frame -> True,
            PlotStyle -> {PointSize[0.003]}]
```

Now let us deal with the Sylvester–Fibonacci expansion. The expansion in terms $1/F_k$ is similar to the terms in the normal continued fraction expansion of a number. This suggests recursively to extract the largest fraction of the form $1/F_k$ from a given number z. The smallest F_k that is larger or equal to n is easily calculated by using the above implemented function greatestFibonacci.

```
In[21]:= smallestFibonacci[n_] :=
    Module[{m = greatestFibonacci[n]},
            Which[m <= 2, 3, Fibonacci[m] === n, m, True, m + 1]]
```

Making use of smallestFibonacci the implementation of SylvesterFibonacciDigits is straightforward.

```
In[22]:= SylvesterFibonacciDigits[z_, n_] := First /@ Rest[
    NestWhileList[Function[{d, r}, {#, r - 1/Fibonacci[#]}&[
                smallestFibonacci[Ceiling[1/r]]]] @@ #&,
            {0, z}, #[[2]] != 0&, 1, n]]
```

Here is an example expansion of a rational number (this special expansion is finite; the Sylvester–Fibonacci expansion for most rational numbers is infinite).

```
In[23]:= SylvesterFibonacciDigits[23/24, 10]

Out[23]= {3, 4, 6}
```

```
In[24]:= Plus @@ (1/Fibonacci[{3, 4, 6}])
```

Out[24]= $\dfrac{23}{24}$

To calculate the first hundred nonvanishing Sylvester–Fibonacci digits of $1/\pi$, we have to set $MaxExtraPrecision to a higher value.

```
In[25]:= $MaxExtraPrecision = 1000;
         SylvesterFibonacciDigits[1/Pi, 100]
```

Out[26]= {5, 7, 9, 11, 17, 21, 25, 27, 29, 31, 33, 35, 37, 39, 45, 49, 54, 58, 73, 75,
         77, 79, 81, 84, 86, 89, 91, 93, 95, 100, 102, 105, 109, 112, 114, 120, 122,
         128, 131, 136, 142, 149, 153, 158, 161, 163, 167, 170, 172, 174, 181, 184,
         186, 188, 191, 195, 197, 200, 205, 207, 213, 217, 226, 233, 237, 240, 242, 244,
         248, 250, 254, 256, 258, 260, 262, 264, 267, 270, 272, 276, 278, 281, 286, 288,
         290, 293, 296, 298, 302, 305, 308, 313, 320, 322, 324, 326, 334, 336, 338, 340}

The difference between the corresponding sum and $1/\pi$ is about 10^{-71}.

```
In[27]:= N[Plus @@ (1/Fibonacci[%]) - 1/Pi, 22]
```

Out[27]= $-9.529167043687048994211 \times 10^{-72}$

14. lcm–gcd Iteration, Generalized Multinomial Theorem, Ramanujan τ Function

a) The function LCMGCDStep[*l*] replaces the first found pair of numbers from the list *l* by their least common divisor and their greatest common denominator. We sort the list after every step to keep the numbers in the list increasing from the beginning to the end.

```
In[1]:= LCMGCDStep[l_List] :=
        Module[{i = 1, j = 2, n = Length[l], makeNoChange = True,
                o1 = 1, o2 = 1, n1 = 1, n2 = 1, lAux = Sort[l]},
            While[i < n && j <= n &&
                    ((* old pair *)
                    {o1, o2} = lAux[[{i, j}]];
                    (* new pair *)
                    {n1, n2} = {LCM[o1, o2], GCD[o1, o2]};
                    (* did something change? *)
                    makeNoChange = {n1, n2} === {o2, o1}),
                    (* advance i and j for stepping through the list *)
                    If[(* increase j *) j < n, j = j + 1,
                       (* increase i, reset j *) i = i + 1; j = i + 1]];
            (* build new list *)
            Sort[If[makeNoChange, lAux,
                    ReplacePart[lAux, {n1, n2}, {{i}, {j}}, {{1}, {2}}]]]]
```

In the following example, the first two elements are exchanged from {2, 3} to {lcm(2, 3), gcd(2, 3)} = {6, 1}. Then the resulting list {6, 1, 5, 30} gets sorted.

```
In[2]:= LCMGCDStep[{2, 3, 5, 3}]
```

Out[2]= {1, 3, 5, 6}

Operating on the last output with LCMGCDStep exchanges the second and the third elements and replaces them by 1 and 30.

```
In[3]:= LCMGCDStep[{1, 5, 6, 30}]
```

Out[3]= {1, 1, 30, 30}

Because of the sorting used, the result of the repeated application of LCMGCDStep does not depend on the order of the numbers. In the following example, we randomly permute the list after each application of LCMGCDStep.

```
In[4]:= randomPermutation[l_List] :=
        Module[{λ = l, L = Length[l], σ, τ},
                Do[σ = λ[[i]]; j = Random[Integer, {i, L}]; τ = λ[[j]];
                   λ[[i]] = τ; λ[[j]] = σ, {i, Length[l]}]; λ]
In[5]:= SeedRandom[112];
        Table[FixedPoint[LCMGCDStep[randomPermutation[#]]&,
                                        {2, 11, 23, 8, 4, 9}], {6}]
```

Out[6]= {{1, 1, 1, 2, 4, 18216}, {1, 1, 1, 2, 4, 18216}, {1, 1, 1, 2, 4, 18216},
        {1, 1, 1, 2, 4, 18216}, {1, 1, 1, 2, 4, 18216}, {1, 1, 1, 2, 4, 18216}}

Let us take a longer list, say, one with 100 integers.

In[7]:= SeedRandom[111];
        l = Table[Random[Integer, {1, 1000}], {100}]

Out[8]= {143, 748, 250, 140, 235, 764, 954, 755, 809, 76, 681, 562, 522, 808, 307, 166, 378,
         82, 89, 795, 596, 113, 131, 29, 283, 94, 574, 909, 878, 768, 178, 987, 817, 730,
         166, 958, 382, 997, 594, 471, 799, 302, 823, 701, 956, 909, 690, 868, 957, 768,
         465, 356, 189, 164, 280, 321, 792, 946, 300, 969, 820, 725, 184, 406, 549, 421,
         743, 252, 395, 425, 780, 561, 550, 956, 400, 280, 660, 406, 6, 786, 1, 39, 411,
         386, 764, 6, 960, 898, 736, 568, 749, 679, 835, 60, 209, 735, 713, 102, 584, 277}

It takes 236 applications of LCMGCDStep until the final list is reached.

In[9]:= fpl = FixedPointList[LCMGCDStep, l];

In[10]:= Length[fpl]

Out[10]= 237

The resulting list is as follows.

In[11]:= Last[fpl]

Out[11]= {1, 1, 1, 1, 1, 1, 1, 1, 1, 1, 1, 1, 1, 1, 1, 1, 1, 1, 1, 1, 1, 1, 1, 1, 1, 1, 1, 1, 1, 1,
         1, 1, 1, 1, 1, 1, 1, 1, 1, 1, 1, 1, 2, 2, 2, 2, 2, 2, 2, 2, 2, 2, 2, 2, 2, 2, 2, 2,
         2, 2, 2, 2, 2, 2, 6, 6, 6, 6, 6, 12, 12, 12, 12, 12, 12, 60, 60, 60, 60, 60, 60,
         60, 60, 60, 420, 420, 840, 840, 27720, 27720, 27720, 27720, 68329800, 136659600,
         230161825000800, 9555080201661933194256000, 26697434353436601763869742085902380883200,
         763657932005112174778387828553887736528984489951968150240183387289107824348952240711·
         26590943155296000}

Each pair in the resulting list has one number that divides the other.

In[12]:= Union[Flatten[
          Table[Head[Last[fpl][[j]]/Last[fpl][[i]]],
            {i, Length[fpl[[1]]]}, {j, i + 1, Length[fpl[[1]]]}]]]

Out[12]= {Integer}

The following picture shows how the final list forms.

In[13]:= ListDensityPlot[Transpose[#/Max[#]& /@ fpl], AspectRatio -> Automatic,
              Mesh -> False, ColorFunction -> (Hue[0.8 #]&)]

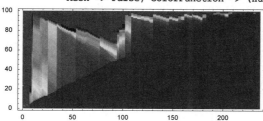

The next inputs repeat the calculation for the list {1, 2, ..., 100}.

In[14]:= fpl = FixedPointList[LCMGCDStep, Range[100]];

In[15]:= Last[fpl]

Out[15]= {1, 1, 1, 1, 1, 1, 1, 1, 1, 1, 1, 1, 1, 1, 1, 1, 1, 1, 1, 1, 1, 1, 1, 1, 1, 1, 1, 1, 1, 1,
         1, 1, 1, 1, 1, 1, 1, 1, 1, 1, 1, 1, 1, 1, 1, 1, 1, 1, 1, 2, 2, 2, 2, 2, 2, 2, 2, 2, 2, 2,
         2, 2, 2, 2, 2, 2, 6, 6, 6, 6, 6, 6, 6, 6, 12, 12, 12, 12, 12, 60, 60, 60, 60, 60, 60,
         420, 420, 840, 2520, 2520, 27720, 27720, 360360, 720720, 232792560, 26771144400,
         144403552893600, 3099044504245996706400, 697203752297124771645338089535312303556800}

```
In[16]:= Union[Flatten[
         Table[Head[Last[fpl][[j]]/Last[fpl][[i]]],
              {i, Length[fpl[[1]]]}, {j, i + 1, Length[fpl[[1]]]}]]]

Out[16]= {Integer}

In[17]:= ListDensityPlot[Transpose[#/Max[#]& /@ fpl], AspectRatio -> Automatic,
                Mesh -> False, ColorFunction -> (Hue[0.8 #]&)]
```

b) Here is the calculation of the $A_n(x_1, x_2, ..., x_m, p_1, p_2, ..., p_m)$ implemented. To avoid carrying out more summation steps than needed, we carry out the summation only over $m - 1$ iterators by choosing k_m such that $k_1, k_2, ..., k_m = n$.

```
In[1]:= (* special case *)
      A[0, xList_, pList_] :=
        Product[(xList[[j]])^(pList[[j]]), {j, 1, Length[pList]}];


      A[n_, xList_, pList_] :=
      Module[{m = Length[xList], k, iterVars, iter},
            (* all iterator variables *)
            iterVars = Table[k[i], {i, m}];
            (* the actual iterators for the summation *)
            iter = MapIndexed[{k[#2[[1]]], 0, #1}&,
                   n - FoldList[Plus, 0, Drop[iterVars, -2]]];
            (* giving k[m] a value *)
            Evaluate[Last[iterVars]] = n - Plus @@ Drop[iterVars, -1];
            (* doing the summation *)
            Sum[Evaluate[(Multinomial @@ iterVars)*
                  Product[(xList[[j]] + iterVars[[j]])^
                          (iterVars[[j]] + pList[[j]]), {j, 1, m}]],
                Evaluate[Sequence @@ iter]]]
```

For special values of the p_k, the function $A_n(x_1, x_2, ..., x_m, p_1, p_2, ..., p_m)$ assumes simple values. Here are three examples.

```
In[4]:= Table[A[n, {x, y}, {-1, 0}], {n, 1, 5}] // Together // Factor
```

$$Out[4]= \left\{ \frac{1 + x + y}{x}, \frac{(2 + x + y)^2}{x}, \frac{(3 + x + y)^3}{x}, \frac{(4 + x + y)^4}{x}, \frac{(5 + x + y)^5}{x} \right\}$$

```
In[5]:= Table[A[n, {x, y, z}, {-1, -1, 0}], {n, 1, 5}] // Together // Factor
```

$$Out[5]= \left\{ \frac{1 + x + y + z}{x\,y}, \frac{(2 + x + y + z)^2}{x\,y}, \frac{(3 + x + y + z)^3}{x\,y}, \frac{(4 + x + y + z)^4}{x\,y}, \frac{(5 + x + y + z)^5}{x\,y} \right\}$$

```
In[6]:= Table[A[n, {x, y, z}, {-1, -1, -1}], {n, 1, 5}] // Together // Factor
```

$$Out[6]= \left\{ \frac{x + y + z}{x\,y\,z}, \frac{(x + y + z)\,(2 + x + y + z)}{x\,y\,z}, \frac{(x + y + z)\,(3 + x + y + z)^2}{x\,y\,z}, \right.$$
$$\left. \frac{(x + y + z)\,(4 + x + y + z)^3}{x\,y\,z}, \frac{(x + y + z)\,(5 + x + y + z)^4}{x\,y\,z} \right\}$$

Now, let us test the Hurwitz identity. HurwitzIdentityQ returns True if the identity holds for given x_k and p_k.

```
In[7]:= HurwitzIdentityQ[n_, xList_, pList_] :=
      Together[A[n, xList, pList] -
        Sum[Binomial[n, k] k! (First[xList] + k) *
          A[n - k, MapAt[(# + k)&, xList, 1],
                   MapAt[(# - 1)&, pList, 1]], {k, 0, n}]] === 0
```

Here is an example.

```
In[8]:= HurwitzIdentityQ[3, {x, y}, {-1, 0}]
```

Out[8]= True

We carry out five randomly chosen examples for symbolic x_k.

```
In[9]:= SeedRandom[1111];

     Table[With[{m = Random[Integer, {2, 6}]},
             HurwitzIdentityQ[Random[Integer, {1, 6}], Table[x[i], {i, m}],
                     Table[Random[Integer, {-6, 6}], {i, m}]]],
         {5}]
Out[10]= {True, True, True, True, True}
```

For numerical x_k, we can test larger values of n and p_k.

```
In[11]:= SeedRandom[1111];

     Table[With[{m = Random[Integer, {2, 10}]},
             HurwitzIdentityQ[Random[Integer, {1, 10}],
                     Table[Random[Integer, {1, 10}], {i, m}],
                     Table[Random[Integer, {-10, 10}], {i, m}]]],
         {5}]
Out[12]= {True, True, True, True, True}
```

c) We start by calculating the first 100 terms of the series $\sum_{k=0}^{\infty} \tau(n)\, z^n$ by expanding the product. (Making use of the function `Series` from Chapter 1 of the Symbolics volume [556], this could be implemented slightly more efficiently.)

```
In[1]:= poly = With[{o = 200},
             Fold[(Expand[#1 #2] /. x^_?(# > o&) -> 0)&, x,
                     Table[Expand[(1 - x^k)^24] /. x^_?(# > o&) -> 0, {k, o}]]];
```

Taking the coefficients of this polynomial results in the first 100 values for $\tau(n)$.

```
In[2]:= Apply[Set[τ[#1], #2]&, (# /. c_. x^e_. -> {e, c})& /@
                     (List @@ Take[poly, 100]), {1}];
```

$\tau\Pi$ implements the multiplication rule for $\tau(n)$.

```
In[3]:= τΠ[i_, j_] := Plus @@ ((#^11 τ[i j/#^2])& /@ Divisors[GCD[i, j]])
```

The multiplication rule is fulfilled for all calculated values of $\tau(n)$.

```
In[4]:= Table[τ[i] τ[j] - τΠ[i, j], {i, 100}, {j, Floor[100/i]}] // Flatten // Union
Out[4]= {0}
```

d) To avoid carrying out a five-dimensional sum, we use the two relations $a + b + c + d + e = 0$ and $a^2 + b^2 + c^2 + d^2 + e^2 = 10\,n$ to determine the values of d and e for given values of a, b, and c. This leaves only a 3D sum to be carried out. The limits for a, b, and c we determine from $a^2 + b^2 + c^2 + d^2 + e^2 = 10\,n$. The five conditions $\{a, b, c, d, e\} \bmod 5 = \{1, 2, 3, 4, 0\}$ we take into account for the starting values of a, b, and c and their increments of 5.

```
In[1]:= (* the next integer y after x, such that Mod[y, 5] == m *)
     nearestMod[x_Integer, m_] :=
     Block[{y = x}, While[Mod[y, 5] =!= m, y = y + 1]; y];

     (* an auxiliary quantity arising from solving
         the two equations for d and e *)
     root[a_, b_, c_, n_] = Sqrt[20n - 3 (a^2 + b^2 + c^2) - 2 (b c + a (b + c))];

     (* fulfill d and e, the subsidiary conditions *)
     gooddeQ[d_, e_] := IntegerQ[d] && IntegerQ[e] &&
                     Mod[d, 5] == 4 && Mod[e, 5] == 0;

     (* the individual summand *)
     term[a_, b_, c_, d_, e_] = (a - b)(a - c)(a - d)(a - e)(b - c)*
                     (b - d)(b - e)(c - d)(c - e)(d - e);
```

It is straightforward to implement the sum under consideration using the just-defined functions.

```
In[10]:= τ[n_] := Module[{d, e, r},
     Sum[r = root[a, b, c, n];
```

```
If[(* is r an integer *) IntegerQ[r],
   (* one solution pair for {d, e} *)
   {d, e} = (-(a + b + c) + r {-1, 1})/2;
   (* fulfill {d, e} the subsidiary conditions *)
   If[gooddeQ[d, e], term[a, b, c, d, e], 0] +
   (* exchange d and e *)
   If[gooddeQ[e, d], term[a, b, c, e, d], 0], 0],
(* a, b, c iterators *)
{a, nearestMod[-Floor[Sqrt[10n]], 1], Floor[Sqrt[10n]], 5},
{b, nearestMod[-Floor[Sqrt[10n - a^2]], 2],
   Floor[Sqrt[10n - a^2]], 5},
{c, nearestMod[-Floor[Sqrt[10n - a^2 - b^2]], 3],
   Floor[Sqrt[10n - a^2 - b^2]], 5}]/288]
```

Now, let us calculate $\tau(1000)$.

```
In[11]:= τ[1000] // Timing
```

```
Out[11]= {1.4 Second, -30328412970240000}
```

The value obtained for $\tau(1000)$ agrees with the one from the package. And the timing is reasonably good.

```
In[12]:= Needs["NumberTheory`Ramanujan`"]
```

```
In[13]:= RamanujanTau[1000] // Timing
```

```
Out[13]= {0.46 Second, -30328412970240000}
```

Now, let us determine the number of nontrivial 5-tuples that contribute to the result. To do this, we reuse the above code for $\tau[n]$ and add counters c1 and c2, which we increment. c1 counts all tuples, and c2 counts all tuples that contribute to the result.

```
In[14]:= τAndCount[n_] :=
   Module[{d, e, r, (* counters *) c1 = 0, c2 = 0},
      {Sum[r = root[a, b, c, n]; c1 = c1 + 1;
         If[IntegerQ[r], {d, e} = (-(a + b + c) + r {-1, 1})/2;
            If[gooddeQ[d, e], c2 = c2 + 1; term[a, b, c, d, e], 0] +
            If[gooddeQ[e, d], c2 = c2 + 1; term[a, b, c, e, d], 0], 0],
         {a, nearestMod[-Floor[Sqrt[10n]], 1], Floor[Sqrt[10n]], 5},
         {b, nearestMod[-Floor[Sqrt[10n - a^2]], 2],
            Floor[Sqrt[10n - a^2]], 5},
         {c, nearestMod[-Floor[Sqrt[10n - a^2 - b^2]], 3],
            Floor[Sqrt[10n - a^2 - b^2]], 5}]/288, c1, c2}]
```

Redoing the calculation for $\tau[1000]$ shows that only 625 terms are needed.

```
In[15]:= τAndCount[1000]
```

```
Out[15]= {-30328412970240000, 33536, 625}
```

Now, we will search for the formula that expresses the Ramanujan τ function through divisor sums. To find the c_k and the corresponding β_{c_k} and $\beta_{c_k c_l}$ we make a generic ansatz of the form described with unknown coefficients.

```
In[16]:= σ[k_, n_] := σ[k, n] = DivisorSigma[k, n];

   sum[o_][n_] := Sum[β[k] σ[k, n], {k, o}] +
      Sum[β[k, 1] Sum[σ[k, i] σ[1, n - i], {i, n - 1}], {k, o}, {1, k, o}]
```

Terminating the sums at, say, 15 gives 135 unknowns β_{c_k} and $\beta_{c_k c_l}$.

```
In[18]:= o = 15;
   λ = Length[vars = Cases[sum[o][n], _β, Infinity]]
```

```
Out[19]= 135
```

Now, we generate more equations than unknowns in the hope to uniquely identify the β_{c_k} and $\beta_{c_k c_l}$.

```
In[20]:= theSums = Table[sum[o][j] - RamanujanTau[j], {j, λ + 10}];
```

```
In[21]:= sol = Solve[# == 0& /@ theSums, vars];
   Take[sol[[1]], 12]
```

```
   Solve::svars : Equations may not give solutions for all "solve" variables. More…
```

Out[22]= $\{\beta[1] \to 0, \beta[2] \to 0, \beta[3] \to \frac{691}{1800} + \frac{1}{120}\beta[3, 3] - \frac{1}{504}\beta[3, 5] - \frac{1}{264}\beta[3, 9] +$

$\frac{1}{600}\beta[5, 5] + \frac{691\,\beta[5, 9]}{163800} - \frac{691\,\beta[7, 7]}{229320} + \frac{1}{528}\beta[7, 9] - \frac{3617\,\beta[7, 11]}{480480},$

$\beta[4] \to 0, \beta[5] \to \frac{1}{240}\beta[3, 5] - \frac{1}{252}\beta[5, 5] + \frac{1}{480}\beta[5, 7] - \frac{1}{264}\beta[5, 9] -$

$\frac{5\,\beta[7, 11]}{1056} + \frac{7}{968}\beta[9, 9] - \frac{3617\,\beta[9, 11]}{274560}, \beta[6] \to 0,$

$\beta[7] \to \frac{691}{900} - \frac{1}{120}\beta[3, 3] + \frac{1}{300}\beta[5, 5] - \frac{1}{504}\beta[5, 7] + \frac{1}{240}\beta[7, 7] -$

$\frac{1}{264}\beta[7, 9] + \frac{691\,\beta[7, 11]}{65520} + \frac{7\,\beta[9, 11]}{1056}, \beta[8] \to 0,$

$\beta[9] \to -\frac{11\,\beta[3, 5]}{5040} + \frac{1}{240}\beta[3, 9] - \frac{1}{504}\beta[5, 9] + \frac{1}{480}\beta[7, 9] - \frac{1}{132}\beta[9, 9] + \frac{691\,\beta[9, 11]}{65520},$

$\beta[10] \to 0, \beta[11] \to$

$-\frac{91}{600} - \frac{13\,\beta[5, 5]}{12600} + \frac{1}{600}\beta[5, 9] - \frac{1}{840}\beta[7, 7] + \frac{1}{480}\beta[7, 11] - \frac{1}{264}\beta[9, 11], \beta[12] \to 0\}$

Some of the $\beta_{c_k c_l}$ remained.

In[23]:= **res = Expand[sum[o][n] /. HoldPattern[**
        β[k_, l_] Sum[σ[k_, i] σ[ℓ_, n - i], {i, n - 1}]] :>
        β[k, l] Sum[σ[k, i] σ[l, n - i], {i, n - 1}] /. sol[[1]]];

In[24]:= **βs = Cases[res, _β, ∞] // Union**

Out[24]= $\{\beta[3, 3], \beta[3, 5], \beta[3, 9], \beta[5, 5], \beta[5, 7],$
        $\beta[5, 9], \beta[7, 7], \beta[7, 9], \beta[7, 11], \beta[9, 9], \beta[9, 11]\}$

All of the multipliers of the undetermined $\beta_{c_k c_l}$ vanish.

In[25]:= **coefficient[expr_, c_] := Plus @@ Cases[Expand[expr], c _]**

In[26]:= **Table[{βs[[k]], Union[Table[Evaluate[coefficient[res, βs[[k]]]],**
                                    {n, 3}]]}, {k, Length[βs]}]

Out[26]= $\{\{\beta[3, 3], \{0\}\}, \{\beta[3, 5], \{0\}\}, \{\beta[3, 9], \{0\}\},$
        $\{\beta[5, 5], \{0\}\}, \{\beta[5, 7], \{0\}\}, \{\beta[5, 9], \{0\}\}, \{\beta[7, 7], \{0\}\},$
        $\{\beta[7, 9], \{0\}\}, \{\beta[7, 11], \{0\}\}, \{\beta[9, 9], \{0\}\}, \{\beta[9, 11], \{0\}\}\}$

So, we are lead to the following result.

In[27]:= **MakeBoxes[RamanujanTau[n], TraditionalForm] =**
                    RowBox[{"τ", "(", "n", ")"}];

        (id[n_] = RamanujanTau[n] == (res /. ((# -> 0)& /@ βs)) /.
                    σ -> DivisorSigma) // TraditionalForm

Out[29]//TraditionalForm=

$$\tau(n) = \frac{691\,\sigma_3(n)}{1800} + \frac{691\,\sigma_7(n)}{900} - \frac{91\,\sigma_{11}(n)}{600} + \frac{2764}{15}\sum_{i=1}^{n-1}\sigma_3(i)\,\sigma_7(n-i)$$

Here is a quick check for the result for larger n.

In[30]:= **Table[id[n], {n, 500}] // Union**

Out[30]= $\{True\}$

The above displayed result is not unique; the coefficient of β_{55} contains the same divisor sum convolution and appropriate linear combinations could be formed.

In[31]:= **Select[Coefficient[res, #]& /@ βs,**
            MemberQ[#, HoldPattern[Sum[DivisorSigma[3, _]*
                                    DivisorSigma[7, _], _]] |
                    HoldPattern[Sum[σ[3, _] σ[7, _], _]],
                    Infinity]&] /. σ -> DivisorSigma // TraditionalForm

Out[31]//TraditionalForm=

$$\left\{\frac{\sigma_3(n)}{600} - \frac{\sigma_5(n)}{252} + \frac{\sigma_7(n)}{300} - \frac{13\,\sigma_{11}(n)}{12600} + \sum_{i=1}^{n-1}\sigma_5(i)\,\sigma_5(n-i) + \frac{4}{5}\sum_{i=1}^{n-1}\sigma_3(i)\,\sigma_7(n-i)\right\}$$

In a similar way, we might search for identities like $(n-1)\,\tau(n) = -24\sum_{k=1}^{n-1}\sigma_1(n)\,\tau(n-k)$ [471].

Now, we deal with the representation of the Ramanujan τ function through the polynomials F_n. Here is a straightforward implementation for the F_n. (We carry out a $(n-2)$-dimensional sum and calculate the resulting value of m_{n-1}.)

```
In[32]:= F[{x_}] := -2x

        F[varList_?(Length[#] > 1&)] :=
        Module[{n = Length[varList], X = varList, m},
               Expand[-2 X[[1]] DivisorSigma[1, n]/n +
               Sum[(* calculate m[n - 1] *)
                   m[n - 1] = (n - Sum[k m[k], {k, n - 2}])/(n - 1);
                   If[IntegerQ[m[n - 1]] && m[n - 1] >= 0,
                      (* form summands *)
                      M = Table[m[k], {k, n - 1}];
                      (-1)^(Plus @@ M) ((Plus @@ M) - 1)!/
                      (Times @@ (M!)) (Times @@ (Rest[X]^M)), 0],
                   (* make iterators for the m[1], m[2], ..., m[n -2] *)
                   Evaluate[Sequence @@ Table[{m[j], 0, (n - Sum[k m[k],
                            {k, 1, j - 1}])/Max[(j - 1), 1]}, {j, n - 2}]]]]]
```

Here are the first few of the F_n.

```
In[34]:= Table[F[Table[Subscript[x, k], {k, n}]], {n, 5}]
```

$$\text{Out[34]}= \left\{ -2\,x_1, \; -3\,x_1 + \frac{x_2^2}{2}, \; -\frac{8\,x_1}{3} - \frac{x_2^3}{3} + x_2\,x_3, \right.$$

$$\left. -\frac{7\,x_1}{2} + \frac{x_2^4}{4} - x_2^2\,x_3 + \frac{x_3^2}{2} + x_2\,x_4, \; -\frac{12\,x_1}{5} - \frac{x_2^5}{5} + x_2^3\,x_3 - x_2\,x_3^2 - x_2^2\,x_4 + x_3\,x_4 + x_2\,x_5 \right\}$$

The values of $\tau(n)$ calculated using the F_n agree with the built-in values. But the calculation is much slower.

```
In[35]:= (Rest[Nest[Append[#, F[#]]&, {12}, 24]] ==
         Table[RamanujanTau[n], {n, 2, 25}]) // Timing
Out[35]= {16.27 Second, True}
```

We end with a representation of the Ramanujan coefficient based on the partitions of an integer [82].

```
In[36]:= (* define integer partitions in the style of result of FactorInteger *)
         myIntegerPartitions[n_Integer?Positive] := Reverse /@
         (({First[#], Length[#]}& /@ Split[#])& /@
                          Experimental`IntegerPartitions[n])
In[38]:= RamanujanTauPartitionsBased[n_Integer?(# > 1&)] :=
         Module[{lλ, zλ, ζλ},
                (* functions defined for a partition of n *)
                lλ[parti_] := Plus @@ (Last /@ parti);
                zλ[parti_] := Times @@ ((#1^#2 #2!)& @@@ parti);
                ζλ[parti_] := Times @@ ((DivisorSigma[1, #1]^#2)& @@@ parti);
                Plus @@ (((-24)^lλ[#] ζλ[#]/zλ[#])& /@ myIntegerPartitions[n - 1])]
```

While this representation is interesting, it cannot compete for speed with the above representation.

```
In[39]:= Table[RamanujanTauPartitionsBased[n] - RamanujanTau[n], {n, 2, 20}]
Out[39]= {0, 0, 0, 0, 0, 0, 0, 0, 0, 0, 0, 0, 0, 0, 0, 0, 0, 0, 0}
```

15. Cross-Number Puzzle

We will tackle the problem of solving the cross-number puzzle in the following way: In the first step, we calculate possible fillings for the various horizontal and vertical fields (as much as this is possible). Then, we start at a letter with a relatively small amount of possible realizations (for this cross-number puzzle, it turns out that this is the letter K). We generate all possible cross-number puzzle realizations at each step. A realization of a cross-number puzzle will be a 6×6 array of letters with head CNP. Entries in the 6×6 that are not filled in will have the value 0. Then, we try to fill in possible digit sequences for other horizontal and vertical letters, possibly those that have a nonzero overlap with the already filled in fields. At each stage of this process, we keep a list of all possible realizations.

The number of possible realizations will initially grow, but after reaching a maximum, will decrease. At the end, we will have a list of all possible solutions of the cross-number puzzle. The actual order in which the letters are filled in below is not the optimal one. Many other ones are possible too. As long as one has enough memory to hold all combinations, the actual order is not very important, although the resulting timings may vary considerably. The order does not influence the result, but determines how many possible realizations are generated in intermediate steps. (This strategy for solving the cross-number puzzle is quite general and can be applied to many related puzzles [553].)

Let us start by implementing a few transformations of numbers and tests that will be needed later on. backNumber[*n*] generates the backnumber of *n*.

```
In[1]:= backNumber[n_] := FromDigits[Reverse[IntegerDigits[n]]]
```

crossSum[*n*] generates the cross-sum of *n*.

```
In[2]:= crossSum[n_] := Plus @@ IntegerDigits[n]
```

crossProduct[*n*] generates the cross-product of *n*.

```
In[3]:= crossProduct[n_] := Times @@ IntegerDigits[n]
```

For a more straightforward generation of numbers, we implement the function nDigitNumbers[*n*], which gives a list of all zero-free, *n*-digit numbers.

```
In[4]:= nDigitNumbers[n_] := Select[Range[10^(n - 1), 10^n - 1], zeroFreeQ]
```

The function zeroFreeQ[*n*], used on the right-hand side of the last definition, tests whether any of the digits of *n* is 0.

```
In[5]:= zeroFreeQ[n_] := FreeQ[IntegerDigits[n], 0]
```

nDigitNumbersQ[*m*, *n*] tests if the number *m* has *n* digits.

```
In[6]:= nDigitNumberQ[m_, n_] := Length[IntegerDigits[m]] == n
```

nDigitQ[*m*, *n*, *p*] tests if the *n*th digit of *m* is *p*.

```
In[7]:= nthDigitQ[m_, n_, p_] := IntegerDigits[m][[n]] == p
```

We consider the cross-number puzzle to be a matrix and will count rows from the top and columns from the left. The function fields[{*horizontalOrVertical*, "*letter*"}] gives the position of the word belonging to *horizontalOrVertical letter*.

```
In[8]:= fields[{horizontal, "A"}] = {{1, 1}, {1, 2}};
        fields[{horizontal, "C"}] = {{1, 3}, {1, 4}};
        fields[{horizontal, "E"}] = {{1, 5}, {1, 6}};
        fields[{horizontal, "G"}] = {{2, 1}, {2, 2}, {2, 3}, {2, 4}, {2, 5}, {2, 6}};

        fields[{horizontal, "H"}] = {{3, 1}, {3, 2}, {3, 3}, {3, 4}, {3, 5}, {3, 6}};
        fields[{horizontal, "J"}] = {{4, 1}, {4, 2}, {4, 3}};
        fields[{horizontal, "K"}] = {{4, 4}, {4, 5}, {4, 6}};
        fields[{horizontal, "M"}] = {{5, 1}, {5, 2}};
        fields[{horizontal, "N"}] = {{5, 3}, {5, 4}, {5, 5}, {5, 6}};
        fields[{horizontal, "O"}] = {{6, 1}, {6, 2}, {6, 3}, {6, 4}, {6, 5}, {6, 6}};

In[18]:= fields[{vertical, "A"}] = {{1, 1}, {2, 1}, {3, 1}};
        fields[{vertical, "B"}] = {{1, 2}, {2, 2}};
        fields[{vertical, "C"}] = {{1, 3}, {2, 3}, {3, 3}, {4, 3}};
        fields[{vertical, "D"}] = {{1, 4}, {2, 4}, {3, 4}};
        fields[{vertical, "E"}] = {{1, 5}, {2, 5}, {3, 5}};
        fields[{vertical, "F"}] = {{1, 6}, {2, 6}, {3, 6}, {4, 6}, {5, 6}, {6, 6}};

        fields[{vertical, "I"}] = {{3, 2}, {4, 2}, {5, 2}, {6, 2}};
        fields[{vertical, "J"}] = {{4, 1}, {5, 1}, {6, 1}};
        fields[{vertical, "K"}] = {{4, 4}, {5, 4}, {6, 4}};
        fields[{vertical, "L"}] = {{4, 5}, {5, 5}, {6, 5}};
        fields[{vertical, "N"}] = {{5, 3}, {6, 3}};
```

We implement the function show[*crossNumberPuzzle*], which displays the current form of the array in a nice way. This step will better show how the digits fill in.

```
In[29]:= show[CNP[m_], opts___] :=
        Module[{grid, dividers, outerFrame, letters},
          (* the outline *)
          grid = {Table[Line[{{0, y}, {6, y}}], {y, 0, -6, -1}],
                Table[Line[{{x, 0}, {x, -6}}], {x, 0, 6}]};
          dividers = {Line[{{2, 0}, {2, -1}}], Line[{{4, 0}, {4, -1}}],
```

```
            Line[{{1, -2}, {2, -2}}], Line[{{0, -3}, {1, -3}}],
            Line[{{2, -5}, {2, -4}, {3, -4}, {3, -3}, {5, -3}}]}];
      outerFrame = Line[{{0, 0}, {6, 0}, {6, -6}, {0, -6}, {0, 0}}];
      (* the letters indicating the start of a number *)
      letters = Apply[Text[StyleForm[#1, FontSize -> 5, FontFamily -> "Times"],
                      #2 + {0.25, -0.25}]&,
          {{"A", {0, 0}}, {"B", {1, 0}}, {"C", {2, 0}},
           {"D", {3, 0}}, {"E", {4, 0}}, {"F", {5, 0}},
           {"G", {0, -1}}, {"H", {0, -2}}, {"I", {1, -2}},
           {"J", {0, -3}}, {"K", {3, -3}}, {"L", {4, -3}},
           {"M", {0, -4}}, {"N", {2, -4}}, {"O", {0, -5}}}, {1}];
      (* display outline, letters, and numbers *)
      Show[Graphics[
       {{GrayLevel[0], Thickness[0.002], grid},
        {GrayLevel[0], Thickness[0.012], dividers, outerFrame},
        letters, (* the actual filling *)
        MapIndexed[{Hue[0], Text[StyleForm[#1, FontSize -> 8,
                   FontFamily -> "Times", FontWeight -> "Plain"],
                   {#2[[2]] - 1, - #2[[1]] + 1} + {0.5, -0.5}]}&,
               m /. 0 -> " ", {2}] }],
        opts, AspectRatio -> Automatic, PlotRange -> All]]
```

Here is an example.

```
In[30]:= show[CNP[{{9, 9, 9, 0, 0, 0},
                   {4, 0, 0, 0, 0, 0},
                   {4, 0, 0, 0, 0, 0},
                   {4, 0, 0, 2, 1, 2},
                   {0, 0, 0, 4, 0, 0},
                   {0, 0, 0, 3, 0, 0}}]]
```

^{A}9	^{B}9	^{C}9	D	E	F
^{G}4					
^{H}4	I				
^{J}4		^{K}2	^{L}1	2	
M		N	4		
O			3		

We will compute the possible fillings as much as they can be computed in a moment and use them later. (This means the fillings that do not relate in an overly complicated way to other fillings.) Here are the possible fillings for horizontal A.

```
In[31]:= horizontal["A"] = Select[nDigitNumbers[2], PrimeQ]
```

Out[31]= {11, 13, 17, 19, 23, 29, 31, 37, 41, 43, 47, 53, 59, 61, 67, 71, 73, 79, 83, 89, 97}

The following fillings come from a straightforward translation of the given conditions.

```
In[32]:= horizontal["C"] = Select[nDigitNumbers[2],
             (IntegerQ[Sqrt[backNumber[#]]] && zeroFreeQ[#])&]
```

Out[32]= {18, 46, 52, 61, 63, 94}

```
In[33]:= horizontal["E"] = Select[nDigitNumbers[2], PrimeQ]
```

Out[33]= {11, 13, 17, 19, 23, 29, 31, 37, 41, 43, 47, 53, 59, 61, 67, 71, 73, 79, 83, 89, 97}

```
In[34]:= horizontal["K"] = Select[nDigitNumbers[3],
             (IntegerQ[Sqrt[crossProduct[#]]] && zeroFreeQ[#])&]
```

```
Out[34]= {111, 114, 119, 122, 128, 133, 141, 144, 149, 155, 166, 177, 182, 188, 191,
          194, 199, 212, 218, 221, 224, 229, 236, 242, 248, 263, 281, 284, 289, 292, 298,
          313, 326, 331, 334, 339, 343, 362, 368, 386, 393, 411, 414, 419, 422, 428, 433,
          441, 444, 449, 455, 466, 477, 482, 488, 491, 494, 499, 515, 545, 551, 554, 559,
          595, 616, 623, 632, 638, 646, 661, 664, 669, 683, 696, 717, 747, 771, 774, 779,
          797, 812, 818, 821, 824, 829, 836, 842, 848, 863, 881, 884, 889, 892, 898, 911,
          914, 919, 922, 928, 933, 941, 944, 949, 955, 966, 977, 982, 988, 991, 994, 999}

In[35]:= horizontal["M"] = Select[nDigitNumbers[2], PrimeQ]

Out[35]= {11, 13, 17, 19, 23, 29, 31, 37, 41, 43, 47, 53, 59, 61, 67, 71, 73, 79, 83, 89, 97}
```

Taking into account horizontal J, we have these possible values of horizontal O.

```
In[36]:= horizontal["O"] = Select[Range[Floor[Sqrt[111111]], Ceiling[Sqrt[999999]]]^2,
          (IntegerQ[Sqrt[crossSum[#]]] && zeroFreeQ[Sqrt[#]] &&
          zeroFreeQ[#])&]

Out[36]= {113569, 114244, 119716, 126736, 132496, 133225, 139129, 145924, 149769,
          152881, 153664, 173889, 174724, 175561, 178929, 182329, 183184, 214369,
          215296, 222784, 223729, 231361, 232324, 268324, 278784, 286225, 288369,
          294849, 315844, 327184, 337561, 346921, 389376, 413449, 414736, 436921, 438244,
          439569, 459684, 467856, 471969, 473344, 474721, 511225, 512656, 524176, 525625,
          579121, 589824, 591361, 617796, 633616, 664225, 665856, 675684, 685584, 692224,
          695556, 746496, 751689, 753424, 755161, 767376, 777924, 788544, 793881, 799236,
          853776, 859329, 887364, 915849, 927369, 938961, 944784, 956484, 968256, 974169}

In[37]:= horizontal["J"] = Sqrt[horizontal["O"]]

Out[37]= {337, 338, 346, 356, 364, 365, 373, 382, 387, 391, 392, 417, 418, 419, 423,
          427, 428, 463, 464, 472, 473, 481, 482, 518, 528, 535, 537, 543, 562, 572,
          581, 589, 624, 643, 644, 661, 662, 663, 678, 684, 687, 688, 689, 715, 716,
          724, 725, 761, 768, 769, 786, 796, 815, 816, 822, 828, 832, 834, 864, 867, 868,
          869, 876, 882, 888, 891, 894, 924, 927, 942, 957, 963, 969, 972, 978, 984, 987}

In[38]:= vertical["C"] = Select[nDigitNumbers[4],
          (MemberQ[horizontal["C"], crossSum[#]] && zeroFreeQ[#])&];
          Short[vertical["C"], 6]

Out[39]//Short= {1179, 1188, 1197, 1269, 1278, 1287, 1296, 1359, 1368, 1377, 1386, 1395, 1449,
          1458, 1467, 1476, 1485, 1494, 1539, 1548, 1557, 1566, 1575, 1584, 1593, 1629,
          1638, 1647, 1656, 1665, 1674, 1683, 1692, 1719, 1728, 1737, 1746, 1755, 1764,
          1773, 1782, 1791, 1818, 1827, 1836, «366», 8352, 8361, 8415, 8424, 8433, 8442,
          8451, 8514, 8523, 8532, 8541, 8613, 8622, 8631, 8712, 8721, 8811, 9117, 9126,
          9135, 9144, 9153, 9162, 9171, 9216, 9225, 9234, 9243, 9252, 9261, 9315, 9324,
          9333, 9342, 9351, 9414, 9423, 9432, 9441, 9513, 9522, 9531, 9612, 9621, 9711}
```

Vertical B has only two possible fillings.

```
In[40]:= vertical["B"] = Select[nDigitNumbers[1],
          (PrimeQ[#] && nDigitNumberQ[#^4, 2]) &]^4

Out[40]= {16, 81}

In[41]:= vertical["E"] = Select[nDigitNumbers[3], PrimeQ[crossProduct[#]]&]

Out[41]= {112, 113, 115, 117, 121, 131, 151, 171, 211, 311, 511, 711}
```

Vertical K is unique. There is only one possible way to fill it.

```
In[42]:= vertical["K"] = Select[Prime[nDigitNumbers[1]]^5, nDigitNumberQ[#, 3]&]

Out[42]= {243}
```

The last result narrows down the possible choices for horizontal K and horizontal O.

```
In[43]:= horizontal["K"] = Select[horizontal["K"], nthDigitQ[#, 1, 2]&]

Out[43]= {212, 218, 221, 224, 229, 236, 242, 248, 263, 281, 284, 289, 292, 298}

In[44]:= horizontal["O"] = Select[horizontal["O"], nthDigitQ[#, 4, 3]&]
```

```
Out[44]= {182329, 214369, 231361, 232324, 268324, 288369,
         389376, 473344, 591361, 767376, 859329, 887364, 927369}
```

Taking into account the possible numbers for horizontal M, we can obtain the possible fillings for vertical L.

```
In[45]:= vertical["L"] =
         Select[Flatten[Table[i horizontal["M"], {i, 100}]],
               (nDigitNumberQ[#, 3] && zeroFreeQ[#])&];
         Short[vertical["L"], 6]
```

```
Out[46]//Short= {118, 122, 134, 142, 146, 158, 166, 178, 194, 111, 123, 129, 141, 159, 177, 183, 213,
                219, 237, 249, 267, 291, 116, 124, 148, 164, 172, 188, 212, 236, 244, 268, 284, 292,
                316, 332, 356, 388, 115, 145, 155, 185, 215, 235, 265, 295, 335, 355, 365, 395, 415,
                445, 485, 114, 138, 174, 186, «345», 884, 988, 583, 689, 594, 918, 715, 935, 616,
                728, 952, 627, 741, 969, 638, 754, 986, 649, 767, 671, 793, 682, 693, 819, 832,
                715, 845, 726, 858, 737, 871, 748, 884, 759, 897, 781, 923, 792, 936, 949, 814,
                962, 825, 975, 836, 988, 847, 858, 869, 891, 913, 924, 935, 946, 957, 968, 979}
```

```
In[47]:= horizontal["N"] =
         Select[Flatten[Table[vertical["E"] i, {i, 100}]],
               (nDigitNumberQ[#, 4] && nthDigitQ[#, 2, 4] && zeroFreeQ[#])&]
```

```
Out[47]= {1422, 1477, 2488, 1441, 3421, 1452, 1456, 1469, 1495, 2416, 2489, 2415,
         2457, 4431, 2464, 2486, 3473, 7464, 4446, 5486, 3472, 5472, 4454, 5436,
         4477, 4446, 6498, 4485, 6493, 5445, 9495, 5499, 5424, 5488, 6419, 6413, 6435,
         8456, 6441, 7467, 6496, 7488, 7475, 7458, 8424, 9432, 8475, 9438, 9477, 9492}
```

We will use CNP[*sixTimesSixMatrix*] as the data structure for the cross-number puzzle. The function fill
CNP[{*horizontalOrVertical*, *letter*}, *realizations*] generates a list of CNPs where the letters corresponding to *horizontalOr-Vertical letter* are replaced with the digits corresponding to the integers of *realizations*.

```
In[48]:= fillCNP[{hv_, abc_}, realizations_] :=
         CNP /@ (ReplacePart[Table[0, {6}, {6}], #, fields[{hv, abc}],
                 Table[{i}, {i, Length[IntegerDigits[realizations[[1]]]]}]]& /@
                 Map[IntegerDigits, realizations])
```

Here is an example.

```
In[49]:= fillCNP[{horizontal, "C"}, horizontal["C"]]
```

```
Out[49]= {CNP[{{0, 0, 1, 8, 0, 0}, {0, 0, 0, 0, 0, 0}, {0, 0, 0, 0, 0, 0},
          {0, 0, 0, 0, 0, 0}, {0, 0, 0, 0, 0, 0}, {0, 0, 0, 0, 0, 0}}],
        CNP[{{0, 0, 4, 6, 0, 0}, {0, 0, 0, 0, 0, 0}, {0, 0, 0, 0, 0, 0},
          {0, 0, 0, 0, 0, 0}, {0, 0, 0, 0, 0, 0}, {0, 0, 0, 0, 0, 0}}],
        CNP[{{0, 0, 5, 2, 0, 0}, {0, 0, 0, 0, 0, 0}, {0, 0, 0, 0, 0, 0},
          {0, 0, 0, 0, 0, 0}, {0, 0, 0, 0, 0, 0}, {0, 0, 0, 0, 0, 0}}],
        CNP[{{0, 0, 6, 1, 0, 0}, {0, 0, 0, 0, 0, 0}, {0, 0, 0, 0, 0, 0},
          {0, 0, 0, 0, 0, 0}, {0, 0, 0, 0, 0, 0}, {0, 0, 0, 0, 0, 0}}],
        CNP[{{0, 0, 6, 3, 0, 0}, {0, 0, 0, 0, 0, 0}, {0, 0, 0, 0, 0, 0},
          {0, 0, 0, 0, 0, 0}, {0, 0, 0, 0, 0, 0}, {0, 0, 0, 0, 0, 0}}],
        CNP[{{0, 0, 9, 4, 0, 0}, {0, 0, 0, 0, 0, 0}, {0, 0, 0, 0, 0, 0},
          {0, 0, 0, 0, 0, 0}, {0, 0, 0, 0, 0, 0}, {0, 0, 0, 0, 0, 0}}]}
```

In the part of the calculation shown below, we will often have to "unite" two given CNPs. The function uniteCNPs does this. When two CNPs are compatible (this means all already filled in numbers present in both CNPs are identical), the resulting CNP with the same common entries and all noncommon entries will be returned. If the two CNPs are incompatible, the result will be Null.

```
In[50]:= uniteCNPs[cnp1_CNP, cnp2_CNP] :=
         Block[{m1 = cnp1[[1]], m2 = cnp2[[1]], posis1, posis2, common},
               (* the position of the nonzero entries *)
               {posis1, posis2} = Position[#, _?(# != 0&)]& /@ {m1, m2};
               (* the position of the common nonzero entries *)
               common = Intersection[posis1, posis2];
               If[common === {}, CNP[m1 + m2],
                  (* test if the common nonzero entries are identical *)
                  If[And @@ Apply[m1[[##]] == m2[[##]]&, common, {1}],
                     (* fill in new CNP *)CNP[-Sqrt[m1 m2] + m1 + m2]]]]
```

If there is more than one possible CNP, we will collect all CNPs in lists. The function uniteCNPs will carry out the process described above for each possible pair of CNPs.

```
In[51]:= uniteCNPs[l1_List, l2_List] :=
          DeleteCases[Flatten[Outer[uniteCNPs, l1, l2]], Null]
```

Now, let us actually start. There is one entry in the CNP with only one possibility—horizontal K. We will form all possible CNPs of vertical K and horizontal K.

```
In[52]:= s1 = uniteCNPs[fillCNP[{horizontal, "K"}, horizontal["K"]],
                        fillCNP[{vertical, "K"}, vertical["K"]]];
         Length[s1]
Out[53]= 14
```

```
In[54]:= Length[s1]
Out[54]= 14
```

```
In[55]:= s1[[1]]
Out[55]= CNP[{{0, 0, 0, 0, 0, 0}, {0, 0, 0, 0, 0, 0}, {0, 0, 0, 0, 0, 0},
             {0, 0, 0, 2, 1, 2}, {0, 0, 0, 4, 0, 0}, {0, 0, 0, 3, 0, 0}}]
```

Here is one of the possibilities.

```
In[56]:= show[s1[[1]]]
```

A	B	C	D	E	F
G					
H	I				
J			K 2	L 1	2
M		N	4		
O			3		

Let us unite the possible CNPs from s1 with the ones from horizontal N. The construction of the resulting 700 possible realizations takes a few seconds.

```
In[57]:= s2 = uniteCNPs[s1, fillCNP[{horizontal, "N"}, horizontal["N"]]];
         Length[s2]
Out[58]= 700
```

Here is one of the possibilities.

```
In[59]:= show[s2[[1]]]
```

A	B	C	D	E	F
G					
H	I				
J			K 2	L 1	2
M		N 1	4	2	2
O			3		

Let us stop filling in the lower right corner and go to the upper left corner. (As mentioned above, the actual order of filling in the numbers does not matter.) At C, we obtain 306 possibilities.

```
In[60]:= s3 = uniteCNPs[fillCNP[{horizontal, "C"}, horizontal["C"]],
                        fillCNP[{vertical,"C"}, vertical["C"]]];
         Length[s3]
Out[61]= 306
```

This is the first one.

In[62]:= **show[s3[[-1]]]**

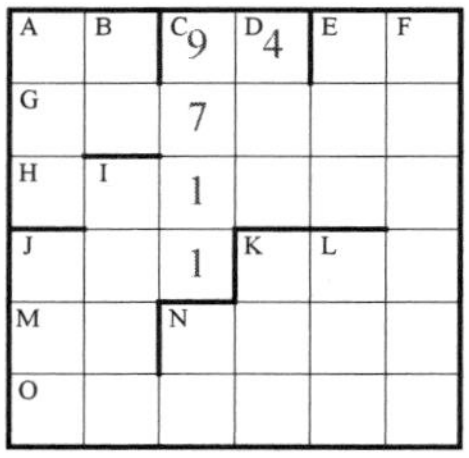

A	B	C$_9$	D$_4$	E	F
G		7			
H	I	1			
J		1	K	L	
M		N			
O					

Horizontal C and vertical C depend on each other. Taking this dependence into account, we get 57 possibilities.

In[63]:= **s4 = Union[Select[s3,**
          (FromDigits[{#[[1, 1, 3]], #[[1, 1, 4]]}] ==
                    #[[1, 1, 3]] + #[[1, 2, 3]] +
                    #[[1, 3, 3]] + #[[1, 4, 3]])&]];

Length[s4]

Out[64]= 57

Here is one of the possibilities.

In[65]:= **show[s4[[1]]]**

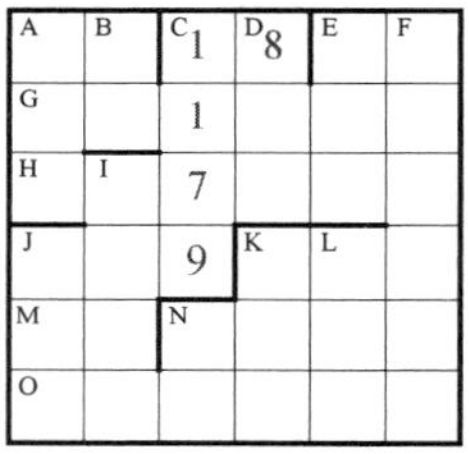

A	B	C$_1$	D$_8$	E	F
G		1			
H	I	7			
J		9	K	L	
M		N			
O					

Vertical F can be constructed from horizontal K and vertical C. Both of them are already filled. So, we can fill vertical F.

In[66]:= **fillVerticalF[cnp1_, cnp2_] :=**
    Block[{verticalC, horizontalK, cnp3},
    (* get the actual realizations of vertical C and horizontal K *)
    verticalC = Apply[cnp2[[1, ##]]&, fields[{vertical,"C"}], {1}];
    horizontalK = Apply[cnp1[[1, ##]]&, fields[{horizontal,"K"}], {1}];
    (* construct vertical F *)
    verticalF = IntegerDigits[
            FromDigits[verticalC] FromDigits[horizontalK]];
    If[(* is the calculated vertical F compatible with the existing cnp1 *)
        verticalF[[4]] === cnp1[[1, 4, 6]] &&
        verticalF[[5]] === cnp1[[1, 5, 6]] &&
        Length[verticalF] == 6 && FreeQ[verticalF, 0],
        (* form new CNP *)
        cnp3 = uniteCNPs[cnp1, cnp2];
        (* fill vertical F *)
        Do[cnp3[[1, k, 6]] = verticalF[[k]], {k, 6}]; cnp3]]

In[67]:= **s5 = DeleteCases[Flatten[Outer[fillVerticalF, s2, s4]], Null];**
    Length[s5]

Out[68]= 293

Our cross-number puzzle really starts to fill in. Here, two possible realizations are shown.

In[69]:= **Show[GraphicsArray[**
            show[s5[[#]], DisplayFunction -> Identity]& /@ {1, -1}]]

A	B	C 1	D 8	E	F 3
G		5			2
H	I	3			6
J		9	K 2	L 1	2
M		N 1	4	5	6
O			3		8

A	B	C 1	D 8	E	F 5
G		8			5
H	I	7			7
J		2	K 2	L 9	8
M		N 8	4	7	5
O			3		6

Now, let us fill in horizontal E.

```
In[70]:= s6 = uniteCNPs[s5, fillCNP[{horizontal, "E"}, horizontal["E"]]];
         Length[s6]
Out[71]= 690
```

Adding the possible vertical E's, we obtain 1265 possibilities.

```
In[72]:= s7 = uniteCNPs[s6, fillCNP[{vertical, "E"}, vertical["E"]]];
         Length[s7]
Out[73]= 1265
```

```
In[74]:= show[s7[[1]]]
```

A	B	C 1	D 8	E 1	F 3
G		5		1	2
H	I	3		2	6
J		9	K 2	L 1	2
M		N 1	4	5	6
O			3		8

Let us stop here and go to the empty squares around A.

```
In[75]:= s8 = uniteCNPs[fillCNP[{horizontal, "A"}, horizontal["A"]],
                  fillCNP[{vertical, "B"}, vertical["B"]]];
         Length[s8]
Out[76]= 5
```

```
In[77]:= s9 = uniteCNPs[s7, s8];
         Length[s9]
Out[78]= 6325
```

At this stage of the calculation, we have 6325 possible CNPs.

```
In[79]:= show[s9[[1]]]
```

A 1	B 1	C 1	D 8	E 1	F 3
G	6	5		1	2
H	I	3		2	6
J		9	K 2	L 1	2
M		N 1	4	5	6
O			3		8

Now, let us deal with horizontal G. First, we know that all digits of horizontal G are distinct. We can sort out some of the existing realizations.

```
In[80]:= s10 = Select[s9, Union[DeleteCases[#[[1, 2]], 0]] ==
                        Sort[DeleteCases[#[[1, 2]], 0]]&];
        Length[s10]
Out[81]= 2570
```

This reduced the number of possibilities by more than a factor of 2. Now, we take into account that all digits of G are different and that the backnumber of vertical A must divide horizontal G. The function `horizontalG` fills in both, horizontal G and vertical A.

```
In[82]:= horizontalG[cnp_] :=
        Block[{cnpAux = cnp, unUsedGDigits, gPairs, candidateGDigits, G,
               verticalADigits},
         (* possible digits for G horizontal *)
         unUsedGDigits = Complement[Range[0, 9], cnp[[1, 2]]];
         (* possible pairs for G horizontal *)
         gPairs = Flatten[Table[If[i != j, {unUsedGDigits[[i]], unUsedGDigits[[j]]},
                          Sequence @@ {}], {i, 5}, {j, 5}], 1];
         Table[pair = gPairs[[i]];
               (* possible G horizontal *)
               candidateGDigits = cnp[[1, 2]];
               candidateGDigits[[1]] = pair[[1]];
               candidateGDigits[[4]] = pair[[2]];
               G = FromDigits[candidateGDigits];
               Table[(* vertical A realizations *)
                     verticalADigits = {cnp[[1, 1, 1]], candidateGDigits[[1]], j};
                     A = FromDigits[verticalADigits];
                     If[(* divides backnumber vertical A horizontal G? *)
                        IntegerQ[G/backNumber[A]],
                        cnpAux[[1, 2]] = candidateGDigits;
                        cnpAux[[1, 3, 1]] = j;
                        (* return new CNP *) cnpAux], {j, 9}],
               {i, Length[gPairs]}]]]
```

The following calculation takes about a quarter of a minute on a 2 GHz computer. For each of the 2570 CNPs contained in s10, we have to try 25 possible G's × 9 possible vertical A's, totaling to 578250 possible combinations. All of them have to be tested for compatibility.

```
In[83]:= s11 = DeleteCases[Flatten[horizontalG /@ s10], Null];
        Length[s11]
Out[84]= 1476
```

Here, two of the resulting 1476 realizations are displayed.

```
In[85]:= Show[GraphicsArray[
             show[s11[[#]], DisplayFunction -> Identity]& /@ {1, -1}]]
```

^{A}4	^{B}1	^{C}1	^{D}8	^{E}1	^{F}3
^{G}4	6	5	3	1	2
^{H}4	^{I}	3		2	6
^{J}		9	^{K}2	^{L}1	2
^{M}		^{N}1	4	5	6
^{O}			3		8

^{A}7	^{B}1	^{C}1	^{D}8	^{E}1	^{F}3
^{G}8	6	1	5	7	4
^{H}2	^{I}	7		1	4
^{J}		9	^{K}2	^{L}9	2
^{M}		^{N}6	4	9	6
^{O}			3		8

Now, let us fill in horizontal O.

```
In[86]:= s12 = uniteCNPs[s11, fillCNP[{horizontal, "O"}, horizontal["O"]]];
        Length[s12]
Out[87]= 488

In[88]:= show[s12[[1]]]
```

A7	B1	C1	D8	E1	F3
G8	6	4	7	3	1
H2	I	5		1	7
J		8	K2	L1	8
M		N2	4	6	4
O2	3	2	3	2	4

Now, that horizontal O is filled, we can also determine horizontal J.

```
In[89]:= horizontalJ[cnp_] :=
             uniteCNPs[{cnp}, fillCNP[{horizontal, "J"},
                  {Sqrt[FromDigits[cnp[[1, 6]]]]}]]
In[90]:= s13 = DeleteCases[Flatten[horizontalJ /@ s12], Null];
In[91]:= Length[s13]
Out[91]= 244

In[92]:= show[s13[[1]]]
```

A7	B1	C1	D8	E1	F3
G8	6	4	7	3	1
H2	I	5		1	7
J5	1	8	K2	L1	8
M		N2	4	6	4
O2	6	8	3	2	4

To find vertical J, we need to determine one more digit. The function `verticalJ` does this.

```
In[93]:= verticalJ[cnp_] :=
         Block[{Hs = Table[FromDigits[{cnp[[1, 4, 1]]], k, cnp[[1, 6, 1]]}], {k, 9}],
                  backA = FromDigits[{cnp[[1, 3, 1]], cnp[[1, 2, 1]], cnp[[1, 1, 1]]}]},
           Hs = Select[Hs, IntegerQ[backNumber[#]/backA]&];
           If[Hs =!= {}, uniteCNPs[{cnp}, fillCNP[{vertical, "J"}, Hs]]]]
In[94]:= s14 = Union[DeleteCases[Flatten[verticalJ /@ s13], Null]];
         Length[s14]
Out[95]= 60
```

Our cross-number puzzle is almost completely filled.

```
In[96]:= show[s14[[1]]]
```

A3	B1	C1	D8	E1	F3
G3	6	4	9	5	2
H1	I	7		1	6
J8	7	6	K2	L2	1
M9		N1	4	6	9
O7	6	7	3	7	6

Meanwhile, we have horizontal G and horizontal O. So, we can easily find horizontal H.

```
In[97]:= horizontalH[cnp_] :=
             uniteCNPs[{cnp}, fillCNP[{horizontal, "H"}, {FromDigits[cnp[[1, 2]]] -
                                         FromDigits[cnp[[1, 6]]]}]]
```

```
In[98]:= s15 = Union[DeleteCases[Flatten[horizontalH /@ s14], Null]];
         Length[s15]
Out[99]= 4
```

Only four possible cross-number puzzles are left now.

```
In[100]:= show[s15[[1]]]
```

$^A7$	$^B1$	$^C1$	$^D8$	$^E1$	$^F3$
$^G9$	6	4	5	1	2
$^H1$	$^I9$	7	1	3	6
$^J8$	7	6	$^K2$	$^L2$	1
$^M8$		$^N1$	4	6	9
$^O7$	6	7	3	7	6

Filling in now horizontal M will complete the puzzle.

```
In[101]:= s16 = uniteCNPs[s15, fillCNP[{horizontal, "M"}, horizontal["M"]]];
          Length[s16]
Out[102]= 8
```

We are left with eight realizations. But we have not actually used (and fulfilled) all conditions; so let us test the remaining eight candidates with respect to all conditions. To do this, we must extract the numbers from our constructed CNPs. The function getNumber[cnp, {horizontalOrVertical, letter}] will extract the number *horizontalOrVertical letter* from the cross-number puzzle *cnp*.

```
In[103]:= getNumber[cnp_CNP, {hv_, abc_}] :=
              FromDigits[Apply[cnp[[1, ##]]&, fields[{hv, abc}], {1}]]
In[104]:= completeTest[cnp_CNP] :=
          (* horizontal *)
          PrimeQ[getNumber[cnp, {horizontal, "A"}]] &&
          IntegerQ[backNumber[getNumber[cnp, {horizontal, "C"}]]] &&
          PrimeQ[getNumber[cnp, {horizontal, "E"}]] &&
          Length[Union[IntegerDigits[getNumber[cnp, {horizontal,"G"}]]]] == 6 &&
          getNumber[cnp, {horizontal,"H"}] ==
                  getNumber[cnp, {horizontal, "G"}] -
                  getNumber[cnp, {horizontal, "O"}] &&
          getNumber[cnp, {horizontal, "J"}] ==
                  Sqrt[getNumber[cnp, {horizontal, "O"}]] &&
          IntegerQ[Sqrt[crossProduct[getNumber[cnp, {horizontal, "K"}]]]] &&
          PrimeQ[getNumber[cnp, {horizontal, "M"}]] &&
          IntegerQ[getNumber[cnp, {horizontal, "N"}]/
                  getNumber[cnp, {vertical, "E"}]] &&
          IntegerQ[Sqrt[crossSum[getNumber[cnp, {horizontal, "O"}]]]] &&
          (* vertical *)
          IntegerQ[getNumber[cnp, {horizontal, "G"}]/
                  backNumber[getNumber[cnp, {vertical, "A"}]]] &&
          PrimeQ[getNumber[cnp, {vertical, "B"}]^(1/4)] &&
          crossSum[getNumber[cnp, {vertical, "C"}]] ==
                  getNumber[cnp, {horizontal, "C"}] &&
          getNumber[cnp, {vertical, "D"}] ==
           getNumber[cnp, {horizontal, "M"}] +
           backNumber[getNumber[cnp, {vertical, "L"}]] &&
          PrimeQ[crossProduct[getNumber[cnp, {vertical, "E"}]]] &&
          getNumber[cnp, {vertical, "F"}] ==
           getNumber[cnp, {horizontal, "K"}] getNumber[cnp, {vertical, "C"}] &&
          IntegerQ[getNumber[cnp, {vertical,"I"}]/
                  backNumber[getNumber[cnp, {vertical,"D"}]]] &&
          IntegerQ[backNumber[getNumber[cnp, {vertical, "J"}]]/
                  backNumber[getNumber[cnp, {vertical, "A"}]]] &&
          PrimeQ[getNumber[cnp, {vertical, "K"}]^(1/5)] &&
```

```
                IntegerQ[getNumber[cnp, {vertical, "L"}]]/
                     getNumber[cnp, {horizontal, "M"}]] &&
                PrimeQ[getNumber[cnp, {vertical, "N"}]]]
```

Applying the final test `completeTest`, we end with one possible solution.

```
In[105]:= s17 = Select[s16, completeTest];
         Length[s17]
Out[106]= 1
```

```
In[107]:= show[s17[[1]]]
```

$^A7$	$^B1$	$^C1$	$^D8$	$^E1$	$^F3$
$^G9$	6	4	5	1	2
$^H1$	$^I9$	7	1	3	6
$^J8$	7	6	$^K2$	$^L2$	1
$^M8$	9	$^N1$	4	6	9
$^O7$	6	7	3	7	6

This shows how to use *Mathematica* for solving puzzles. For a collection of other puzzles solved along the lines described above, see [553].

16. Cyclotomic Polynomials, Generalized Bell Polynomials

a) The following inputs implement the recursive calculation of the cyclotomic polynomials $C_n(z) = $ `c[n][z]`. First, we use the rule $C_p(z) = \sum_{k=0}^{p-1} z^k$, then the rule $C_{n\,p^k}(z) = C_{n\,p}(z^{p^{k-1}})$, and lastly the rule $C_{n\,p}(z) = C_n(z^p)/C_n(z)$.

```
In[1]:= c[0][z_] = 1;

       c[1][z_] = z - 1 ;

       c[p_?PrimeQ][z_] := Sum[z^k, {k, 0, p - 1}];

       c[n_][z_] :=
       Module[{(* the factors of n *)
               factors = FactorInteger[n], p, k, m},
               ((* the first multiple factor *)
               {p, k} = Cases[factors, {_, _?(# > 1&)}, {1}, 1][[1]];
               (* the remaining part *)
               m = Times @@ Apply[Power, DeleteCases[factors, {p, k}], {1}];
               c[m p][z^p^(k - 1)]) /;
               (* is a multiple prime factor present? *)
               Max[Last /@ factors] > 1];


       c[n_][z_] :=
       Module[{(* the factors of n *)
               factors = FactorInteger[n], p, k, m},
               (* the first factor *)
               p = factors[[1, 1]];
               (* the remaining part *)
               m = Times @@ Apply[Power, Rest[factors], {1}];
               (* make a polynomial *)
               Factor[c[m][z^p]]/Factor[c[m][z]]]
```

To reuse already-calculated cyclotomic polynomials, we cache the results for *z* when it is a symbol.

```
In[8]:= cyclotomic[n_, z_] :=
       Module[{res = c[n][z], x},
       If[Head[z] === Symbol, (* for symbols z cache the cyclotomics *)
           cyclotomic[n, x_] = res /. z -> x; res, res]]
```

Here is the resulting list of cyclotomic polynomials. For brevity, we do not show them all.

```
In[9]:= res = Table[C[k, z] == cyclotomic[k, z], {k, 0, 100}];
      TableForm[Take[res, 12]]
```

Out[10]//TableForm=

$C[0, z] = 1$

$C[1, z] = -1 + z$

$C[2, z] = 1 + z$

$C[3, z] = 1 + z + z^2$

$C[4, z] = 1 + z^2$

$C[5, z] = 1 + z + z^2 + z^3 + z^4$

$C[6, z] = 1 - z + z^2$

$C[7, z] = 1 + z + z^2 + z^3 + z^4 + z^5 + z^6$

$C[8, z] = 1 + z^4$

$C[9, z] = 1 + z^3 + z^6$

$C[10, z] = 1 - z + z^2 - z^3 + z^4$

$C[11, z] = 1 + z + z^2 + z^3 + z^4 + z^5 + z^6 + z^7 + z^8 + z^9 + z^{10}$

```
In[11]:= Take[res, -4]
```

Out[11]= $\{C[97, z] = 1 + z + z^2 + z^3 + z^4 + z^5 + z^6 + z^7 + z^8 + z^9 + z^{10} + z^{11} + z^{12} + z^{13} + z^{14} + z^{15} + z^{16} +$
$z^{17} + z^{18} + z^{19} + z^{20} + z^{21} + z^{22} + z^{23} + z^{24} + z^{25} + z^{26} + z^{27} + z^{28} + z^{29} + z^{30} + z^{31} + z^{32} +$
$z^{33} + z^{34} + z^{35} + z^{36} + z^{37} + z^{38} + z^{39} + z^{40} + z^{41} + z^{42} + z^{43} + z^{44} + z^{45} + z^{46} + z^{47} + z^{48} +$
$z^{49} + z^{50} + z^{51} + z^{52} + z^{53} + z^{54} + z^{55} + z^{56} + z^{57} + z^{58} + z^{59} + z^{60} + z^{61} + z^{62} + z^{63} + z^{64} +$
$z^{65} + z^{66} + z^{67} + z^{68} + z^{69} + z^{70} + z^{71} + z^{72} + z^{73} + z^{74} + z^{75} + z^{76} + z^{77} + z^{78} + z^{79} + z^{80} +$
$z^{81} + z^{82} + z^{83} + z^{84} + z^{85} + z^{86} + z^{87} + z^{88} + z^{89} + z^{90} + z^{91} + z^{92} + z^{93} + z^{94} + z^{95} + z^{96},$
$C[98, z] = 1 - z^7 + z^{14} - z^{21} + z^{28} - z^{35} + z^{42},$
$C[99, z] = 1 - z^3 + z^9 - z^{12} + z^{18} - z^{21} + z^{27} - z^{30} + z^{33} - z^{39} + z^{42} - z^{48} + z^{51} - z^{57} + z^{60},$
$C[100, z] = 1 - z^{10} + z^{20} - z^{30} + z^{40}\}$

The polynomials cyclotomic[k, z] agree with the built-in polynomials Cyclotomic[k, z].

```
In[12]:= Table[Cyclotomic[k, z] == cyclotomic[k, z], {k, 0, 10}] // Union
```

Out[12]= {True}

For square-free *n*, a closed form of the coefficients can be given as [227]:

$$C_n(z) = \sum_{k=0}^{\phi(n)} c_{n,k} \, z^{\phi(n)-k}$$

$$c_{n,k} = -\frac{\mu(n)}{k} \sum_{j=0}^{k-1} c_{n,j} \, \mu(\gcd(n, k-j)) \, \phi(\gcd(n, k-j))$$

$$c_{n,0} = 1.$$

b) We begin by making the function IntegerPartitions visible. This will allow to form the partitions of an integer easily.

```
In[1]:= (* define integer partitions in the style of result of FactorInteger *)
      myIntegerPartitions[n_Integer?Positive] := Reverse /@
        (({First[#], Length[#]}& /@ Split[#])& /@
                    Experimental`IntegerPartitions[n])
```

The implementation of the lowest order Bell polynomials is straightforward.

```
In[3]:= BellY[1][ℓ_] :=
      Plus @@ (Times[Length[ℓ]!/(Times @@ ((Last /@ #)!)),
                  ℓ[[Plus @@ (Last /@ #), 1]],
                  (Times @@ ((ℓ[[#1, 2]]/#1!)^#2& @@@ #))]& /@
                      myIntegerPartitions[Length[ℓ]]) /;
                  MatrixQ[ℓ] && Length[Transpose[ℓ]] === 2
```

Here are the first three polynomials in explicit form.

```
In[4]:= {BellY[1][{{f1, g1}}], BellY[1][{{f1, g1}, {f2, g2}}],
      BellY[1][{{f1, g1}, {f2, g2}, {f3, g3}}]}
```

```
Out[4]= {f1 g1, f2 g1² + f1 g2, f3 g1³ + 3 f2 g1 g2 + f1 g3}
```

```
In[5]:= BellY[n_][/_] := Expand[
          Plus @@ (Times[Length[/]!/(Times @@ ((Last /@ #)!)),
                         /[[Plus @@ (Last /@ #), 1]],
                   (Times @@ ((BellY[n - 1][Rest /@ Take[/, #1]]/#1!)^#2& @@@ #))]& /@
                             myIntegerPartitions[Length[/]])] /;
                        MatrixQ[/] && Length[Transpose[/]] === n + 1
```

```
In[6]:= {BellY[2][{{f1, g1, h1}}], BellY[2][{{f1, g1, h1}, {f2, g2, h2}}],
         BellY[2][{{f1, g1, h1}, {f2, g2, h2}, {f3, g3, h3}}]}
```

```
Out[6]= {f1 g1 h1, f2 g1² h1² + f1 g2 h1² + f1 g1 h2,
         f3 g1³ h1³ + 3 f2 g1 g2 h1³ + f1 g3 h1³ + 3 f2 g1² h1 h2 + 3 f1 g2 h1 h2 + f1 g1 h3}
```

```
In[7]:= {BellY[3][{{f1, g1, h1, i1}}],
         BellY[3][{{f1, g1, h1, i1}, {f2, g2, h2, i2}}],
         BellY[3][{{f1, g1, h1, i1}, {f2, g2, h2, i2}, {f3, g3, h3, i3}}]}
```

```
Out[7]= {f1 g1 h1 i1, f2 g1² h1² i1² + f1 g2 h1² i1² + f1 g1 h2 i1² + f1 g1 h1 i2,
         f3 g1³ h1³ i1³ + 3 f2 g1 g2 h1³ i1³ + f1 g3 h1³ i1³ + 3 f2 g1² h1 h2 i1³ + 3 f1 g2 h1 h2 i1³ +
           f1 g1 h3 i1³ + 3 f2 g1² h1² i1 i2 + 3 f1 g2 h1² i1 i2 + 3 f1 g1 h2 i1 i2 + f1 g1 h1 i3}
```

```
In[8]:= BellY[4][{{f1, g1, h1, i1, j1}}]
```

```
Out[8]= f1 g1 h1 i1 j1
```

We use the abbreviation $f_{i,j}$ for $f_i^{(j)}(f_{i+1}(f_{i+2}(\ldots f_m(x))))$.

```
In[9]:= D[Fold[#2[#1]&, x, Reverse @ Table[f[k], {k, 4}]], {x, 4}] /.
              Derivative[k_][f[j_]][_] :> Subscript[f, j, k]
```

```
Out[9]= f₁,₄ f²₂,₁ f⁴₃,₁ f⁴₄,₁ + 6 f₁,₃ f²₂,₁ f₂,₂ f⁴₃,₁ f⁴₄,₁ + 3 f₁,₂ f²₂,₁ f²₂,₂ f⁴₃,₁ f⁴₄,₁ + 4 f₁,₂ f²₂,₁ f₂,₃ f⁴₃,₁ f⁴₄,₁ +
         f₁,₁ f₂,₄ f⁴₃,₁ f⁴₄,₁ + 6 f₁,₃ f³₂,₁ f²₃,₁ f₃,₂ f⁴₄,₁ + 18 f₁,₂ f₂,₁ f₂,₂ f²₃,₁ f₃,₂ f⁴₄,₁ + 6 f₁,₁ f₂,₃ f³₃,₁ f₃,₂ f⁴₄,₁ +
         3 f₁,₂ f²₂,₁ f²₃,₂ f⁴₄,₁ + 3 f₁,₁ f₂,₂ f²₃,₂ f⁴₄,₁ + 4 f₁,₂ f²₂,₁ f₃,₃ f₃,₃ f⁴₄,₁ + 4 f₁,₁ f₂,₁ f₃,₃ f₃,₃ f⁴₄,₁ +
         f₁,₁ f₂,₁ f₃,₄ f⁴₄,₁ + 6 f₁,₃ f³₂,₁ f³₃,₁ f²₄,₁ f₄,₂ + 18 f₁,₂ f₂,₁ f₂,₂ f³₃,₁ f²₄,₁ f₄,₂ +
         6 f₁,₁ f₂,₃ f³₃,₁ f²₄,₁ f₄,₂ + 18 f₁,₂ f²₂,₁ f₃,₁ f₃,₂ f²₄,₁ f₄,₂ + 18 f₁,₁ f₂,₂ f₃,₁ f₃,₂ f²₄,₁ f₄,₂ +
         6 f₁,₁ f₂,₁ f₃,₃ f²₄,₁ f₄,₂ + 3 f₁,₂ f²₂,₁ f²₃,₁ f²₄,₂ + 3 f₁,₁ f₂,₂ f²₃,₁ f²₄,₂ + 3 f₁,₁ f₂,₁ f₃,₂ f²₄,₂ +
         4 f₁,₂ f²₂,₁ f²₃,₁ f₄,₁ f₄,₃ + 4 f₁,₁ f₂,₂ f²₃,₁ f₄,₁ f₄,₃ + 4 f₁,₁ f₂,₁ f₃,₂ f₄,₁ f₄,₃ + f₁,₁ f₂,₁ f₃,₁ f₄,₄
```

```
In[10]:= % - BellY[3][Table[Subscript[f, j, k], {k, 4}, {j, 4}]]
```

```
Out[10]= 0
```

We end with a check of the generalized Faà di Bruno formula for *m* and *n* up to eight.

```
In[11]:= Table[(D[Fold[#2[#1]&, x, Reverse @ Table[f[k], {k, μ}]], {x, ν}] /.
                    Derivative[k_][f[j_]][_] :> Subscript[f, j, k]) -
                       BellY[μ - 1][Table[Subscript[f, j, k], {k, ν}, {j, μ}]],
                {ν, 8}, {μ, 2, 8}] // Flatten // Union
```

```
Out[11]= {0}
```

17. Factorization of *n*! into *n* Factors, Bin Packing, Composition Multiplicities

a) A factorial number is relatively easy to factor. We know all possible primes present in the factorization, `Prime[Range[ PrimePi[n]]]`. We just count the number of times these primes occur in the number k, $1 \le k \le n$ [183]. The function `FactorialPrimeDecomposition[n]` gives the same result as `FactorInteger[n!]` would give, only much faster.

```
In[1]:= FactorialPrimeDecomposition =
          Compile[{{n, _Integer}},
                  Function[p, {p, Plus @@ FixedPointList[Floor[#/p]&,
                      Floor[n/p]]}] /@ Prime[Range[PrimePi[n]]],
                         {{Prime[Range[PrimePi[n]]], _Integer, 1}}];

          (* or:
           Function[n, {#, (n - Plus @@ IntegerDigits[n, #])/(# - 1)}& /@
                                         Prime[Range[PrimePi[n]]]]
          *)
```

```
In[3]:= FactorialPrimeDecomposition[10]
```

```
In[3]:= {{2, 8}, {3, 4}, {5, 2}, {7, 1}}

In[4]:= FactorInteger[10!]

Out[4]= {{2, 8}, {3, 4}, {5, 2}, {7, 1}}
```

Here is a "closed-form" for the exponents of the primes.

```
In[5]:= FactorialPrimeDecompositionR[n_] :=
         {#, (n - Plus @@ IntegerDigits[n, #])/(# - 1)}& /@ Prime[Range[PrimePi[n]]]

In[6]:= FactorialPrimeDecompositionR[10]

Out[6]= {{2, 8}, {3, 4}, {5, 2}, {7, 1}}
```

The result of `FactorInteger` is a list of lists. The second elements of the sublist are the power degrees. To write $n!$ as a product of exactly n factors, we have to use some primes more than once. Given a list of numbers $\{g_1, g_2, \ldots, g_k\}$, we first construct all lists of n numbers $\{h_1, h_2, \ldots, h_n\}$ such that $\sum_{j=1}^{k} g_j = \sum_{j=1}^{l} h_j$. The function `listPartition` is doing this for a list *gList* with nondecreasing elements.

```
In[7]:= listPartition[gList_, n_] :=
         Module[{l = Length[gList], gs, iter, tab},
             (* the body of the Table *)
             gs = Table[If[i === l, n - Sum[g[j], {j, 1 - 1}]], g[i]], {i, 1}];
             (* the iterator *)
             iter = Table[If[i == 1,
                         (* let g[i] vary in strict bounds *)
                         {g[i], 1, Min[gList[[i]], n - (1 - 1)]},
                         {g[i], 1, Min[gList[[i]],
                                    n - Sum[g[j], {j, i - 1}]]}],
                         {i, 1 - 1}];
             tab = Flatten[Table[Evaluate[gs], Evaluate[Sequence @@ iter]], 1 - 2];
             (* delete lists where last element is impossible to realize *)
             Select[tab, (Last[#] =!= 0 && Last[#] <= Last[gList])&]]
```

The complete factorization of 10! results in 15 factors.

```
In[8]:= FactorInteger[10!]

Out[8]= {{2, 8}, {3, 4}, {5, 2}, {7, 1}}
```

We have the following eight possibilities to distribute the number of factors.

```
In[9]:= listPartition[Last /@ Reverse[%], 10]

Out[9]= {{1, 1, 1, 7}, {1, 1, 2, 6}, {1, 1, 3, 5}, {1, 1, 4, 4},
         {1, 2, 1, 6}, {1, 2, 2, 5}, {1, 2, 3, 4}, {1, 2, 4, 3}}
```

The last sublist of this output means that for the powers of 2 we have to make three factors out from 2^8. There are various possibilities to do this, such as $2^8 = 2^1 \, 2^1 \, 2^6 = 2^1 \, 2^2 \, 2^5 = 2^1 \, 2^3 \, 2^4 \ldots$. The function `integerPartition[l, n]` partitions the integer l into n integers h_j such that $\sum_{j=1}^{n} h_j = l$. Because $2^1 \, 2^1 \, 2^6 = 2^1 \, 2^6 \, 2^1$, we can restrict ourselves to the construction of all nondecreasing sequences of exponents.

```
In[10]:= integerPartition[l_, 1] = {{l}};

          integerPartition[l_, n_] :=
          Module[{gs, iter, tab},
              (* the body of the Table *)
              gs = Table[If[i === n, 1 - Sum[g[j], {j, n - 1}]], g[i]], {i, n}];
              (* the iterator *)
              iter = Table[{g[i], If[i == 1, 1, g[i - 1]],
                          If[i == 1, 1 - (n - 1), 1 - Sum[g[j],
                          {j, i - 1}] - (n - i)]}, {i, n - 1}];
              tab = Flatten[Table[Evaluate[gs], Evaluate[Sequence @@ iter]], n - 2];
              (* delete lists where last element is too small *)
              Select[tab, #[[-1]] >= #[[-2]]&]]
```

We split 8 into three factors.

```
In[12]:= integerPartition[8, 3]
```

In[12]:= `{{1, 1, 6}, {1, 2, 5}, {1, 3, 4}, {2, 2, 4}, {2, 3, 3}}`

Now, we must build the actual factorization. The function `singlePrimeFactorization[`*prime*`, `*l*`, `*n*`]` will return a list of all possible factorizations of *prime*^{l} into *n* factors. Instead of using the built-in functions `Times` and `Power`, we will use the functions `times` and `power` to avoid evaluation.

In[13]:= `SetAttributes[times, Flat]`

In[14]:= `singlePrimeFactorization[prime_, l_, n_] :=`
`    Apply[times, Map[power[prime, #]&, integerPartition[l, n], {2}], {1}]`

Here is the above example 2^8 again.

In[15]:= `singlePrimeFactorization[2, 8, 5]`

Out[15]= `{times[power[2, 1], power[2, 1], power[2, 1], power[2, 1], power[2, 4]],`
`    times[power[2, 1], power[2, 1], power[2, 1], power[2, 2], power[2, 3]],`
`    times[power[2, 1], power[2, 1], power[2, 2], power[2, 2], power[2, 2]]}`

Now, we have all ingredients together to implement the function `factorialFactorization[`*n*`]`, which will return a list of all possible factorizations of *n*! into exactly *n* prime factors.

In[16]:= `factorialFactorization[n_] :=`
`    Module[{primeFactors, partitions},`
`        (* factor n! *)`
`        primeFactors = Reverse[FactorialPrimeDecomposition[n]];`
`        (* partition list of factors in n lists *)`
`        partitions = listPartition[Last /@ primeFactors, n];`
`        (* for each partition form all partitions for all sublists`
`           and form outer product of all possibilities *)`
`        Flatten[Flatten[Outer[times, Sequence @@`
`                Reverse[Apply[singlePrimeFactorization,`
`                    Transpose[{First /@ primeFactors,`
`                        Last /@ primeFactors, #}],`
`                    {1}]], 1]]& /@ partitions]]`

Here are all possible factorizations of 10! into 10 prime factors.

In[17]:= `(factorialFactorization[10] /.`
`    (* for a shorter output *) {power -> p, times -> t}) // Take[#, 4]&`

Out[17]= `{t[p[2, 1], p[2, 1], p[2, 1], p[2, 1], p[2, 1], p[2, 1], p[2, 2], p[3, 4], p[5, 2], p[7, 1]],`
`    t[p[2, 1], p[2, 1], p[2, 1], p[2, 1], p[2, 1], p[2, 3], p[3, 1], p[3, 3], p[5, 2], p[7, 1]],`
`    t[p[2, 1], p[2, 1], p[2, 1], p[2, 1], p[2, 1], p[2, 3], p[3, 2], p[3, 2], p[5, 2], p[7, 1]],`
`    t[p[2, 1], p[2, 1], p[2, 1], p[2, 1], p[2, 2], p[2, 2], p[3, 1], p[3, 3], p[5, 2], p[7, 1]]}`

There are 31 possible factorizations.

In[18]:= `Length[%]`

Out[18]= `4`

To make the list of factorizations more readable, we will implement a function `factorizationForm` that will make use of unevaluated powers and products.

In[19]:= `factorizationForm[prod_] := HoldForm[prod] //. {t -> Times, p -> Power}`

Here are the above factorizations, displayed in a nicer form.

In[20]:= `(factorizationForm /@ %%%) // TableForm`

Out[20]//TableForm=
$2^1 \, 2^1 \, 2^1 \, 2^1 \, 2^1 \, 2^1 \, 2^2 \, 3^4 \, 5^2 \, 7^1$
$2^1 \, 2^1 \, 2^1 \, 2^1 \, 2^1 \, 2^3 \, 3^1 \, 3^3 \, 5^2 \, 7^1$
$2^1 \, 2^1 \, 2^1 \, 2^1 \, 2^1 \, 2^3 \, 3^2 \, 3^2 \, 5^2 \, 7^1$
$2^1 \, 2^1 \, 2^1 \, 2^1 \, 2^2 \, 2^2 \, 3^1 \, 3^3 \, 5^2 \, 7^1$

All factorizations really are equal to 10!.

In[21]:= `ReleaseHold[%] // Union`

Out[21]= `{3628800}`

Now, let us calculate the possible factorizations of 25!.

```
In[22]:= fF = factorialFactorization[25];
```

There are exactly 211426 of them.

```
In[23]:= Length[fF]
Out[23]= 211426
```

All found factorizations are truly different.

```
In[24]:= Length[Union @ fF]
Out[24]= 211426
```

Here, the first and last five factorizations are shown.

```
In[25]:= factorizationForm /@ (Take[fF, 5] /. {power -> p, times -> t})
```

$$Out[25]= \{2^1\,2^1\,2^1\,2^1\,2^1\,2^1\,2^1\,2^1\,2^1\,2^1\,2^1\,2^1\,2^1\,2^1\,2^1\,2^6\,3^{10}\,5^6\,7^3\,11^2\,13^1\,17^1\,19^1\,23^1,$$
$$2^1\,2^1\,2^1\,2^1\,2^1\,2^1\,2^1\,2^1\,2^1\,2^1\,2^1\,2^1\,2^1\,2^1\,2^2\,2^5\,3^{10}\,5^6\,7^3\,11^2\,13^1\,17^1\,19^1\,23^1,$$
$$2^1\,2^1\,2^1\,2^1\,2^1\,2^1\,2^1\,2^1\,2^1\,2^1\,2^1\,2^1\,2^1\,2^3\,2^4\,3^{10}\,5^6\,7^3\,11^2\,13^1\,17^1\,19^1\,23^1,$$
$$2^1\,2^1\,2^1\,2^1\,2^1\,2^1\,2^1\,2^1\,2^1\,2^1\,2^1\,2^1\,2^1\,2^2\,2^2\,2^4\,3^{10}\,5^6\,7^3\,11^2\,13^1\,17^1\,19^1\,23^1,$$
$$2^1\,2^1\,2^1\,2^1\,2^1\,2^1\,2^1\,2^1\,2^1\,2^1\,2^1\,2^1\,2^1\,2^2\,2^3\,2^3\,3^{10}\,5^6\,7^3\,11^2\,13^1\,17^1\,19^1\,23^1\}$$

```
In[26]:= factorizationForm /@ (Take[fF, -5] /. {power -> p, times -> t})
```

$$Out[26]= \{2^{10}\,2^{12}\,3^1\,3^1\,3^1\,3^1\,3^1\,3^1\,3^1\,3^3\,5^1\,5^1\,5^1\,5^1\,5^1\,5^1\,7^1\,7^1\,7^1\,7^1\,11^1\,11^1\,13^1\,17^1\,19^1\,23^1,$$
$$2^{10}\,2^{12}\,3^1\,3^1\,3^1\,3^1\,3^1\,3^1\,3^2\,3^2\,5^1\,5^1\,5^1\,5^1\,5^1\,5^1\,7^1\,7^1\,7^1\,7^1\,11^1\,11^1\,13^1\,17^1\,19^1\,23^1,$$
$$2^{11}\,2^{11}\,3^1\,3^1\,3^1\,3^1\,3^1\,3^1\,3^1\,3^3\,5^1\,5^1\,5^1\,5^1\,5^1\,5^1\,7^1\,7^1\,7^1\,11^1\,11^1\,13^1\,17^1\,19^1\,23^1,$$
$$2^{11}\,2^{11}\,3^1\,3^1\,3^1\,3^1\,3^1\,3^1\,3^2\,3^2\,5^1\,5^1\,5^1\,5^1\,5^1\,5^1\,7^1\,7^1\,7^1\,11^1\,11^1\,13^1\,17^1\,19^1\,23^1,$$
$$2^{22}\,3^1\,3^1\,3^1\,3^1\,3^1\,3^1\,3^1\,3^1\,3^2\,5^1\,5^1\,5^1\,5^1\,5^1\,5^1\,7^1\,7^1\,7^1\,11^1\,11^1\,13^1\,17^1\,19^1\,23^1\}$$

Here is a quick check that these factorizations are correct.

```
In[27]:= {Union[fF /. {times -> Times, power -> Power}], 25!}
Out[27]= {{15511210043330985984000000}, 15511210043330985984000000}
```

For some interesting considerations about the large *n* limit of the factorization discussed here, see [12], [366]. For the general Dirichlet problem of writing an integer as a product of exactly *k* integers, see [413], [95].

b) We start by implementing the trinomial coefficient. From the definition of the multinomial coefficient $(n; n_1, n_2, \dots n_m)$ with $n_1 + n_2 + n_3 = n$ and $n_2 + 2 n_3 = k$, the following sum representation follows straightforwardly.

```
In[1]:= trinomial[n_, 3, k_] := (* == (-1)^k GegenbauerC[k, -n, 1/2] *)
                Sum[Multinomial[n + j - k, k - 2 j, j], {j, 0, n}]
```

Here is a quick check for `trinomial`.

```
In[2]:= Table[Expand[Sum[trinomial[n, 3, k] x^k, {k, 0, 2 n}] -
                (1 + x + x^2)^n], {n, 10}]
Out[2]= {0, 0, 0, 0, 0, 0, 0, 0, 0, 0}
```

$\lambda[n]$ implements the average cumulative rod length λ_n in unfilled boxes.

```
In[3]:= λ[n_] := 3^(-n - 1) ((2n + 1) trinomial[n, 3, n] +
                (2n + 1) trinomial[n - 1, 3, n - 1] +
                trinomial[n - 1, 3, n - 2] - 3^n)
```

For an efficient simulation, we should use `Compile`. We will use exact arithmetic and, to avoid fractions, we will multiply the rod lengths by 3. The function `fillBoxes` will fill length 3 boxes by using *n* rods.

```
In[4]:= fillBoxes = Compile[{{n, _Integer}},
            Module[{completelyFilledBoxNumbers = 0, oneThirdFilledBoxNumbers = 0,
                    twoThirdFilledBoxNumbers = 0, newRodLength},
                Do[(* random generation of a new rod *)
                    newRodLength = Random[Integer, {1, 3}];
                    Which[(* new rod just fits in one box *)
                        newRodLength === 3,
                        completelyFilledBoxNumbers = completelyFilledBoxNumbers + 1,
```

```
                        (* new rod is 2/3 of the box length *)
                        newRodLength === 2,
                        If[(* is there a 1/3 filled box where it would fit? *)
                           oneThirdFilledBoxNumbers =!= 0,
                           oneThirdFilledBoxNumbers = oneThirdFilledBoxNumbers - 1,
                           (* start filling a new box *)
                           twoThirdFilledBoxNumbers ++],
                        (* new rod is 1/3 of the box length *)
                        newRodLength === 1,
                        Which[(* is there a 2/3 filled box where it would fit? *)
                              twoThirdFilledBoxNumbers =!= 0,
                              twoThirdFilledBoxNumbers = twoThirdFilledBoxNumbers - 1,
                              (* is there a 1/3 filled box
                                 where it would fit? *)
                              oneThirdFilledBoxNumbers =!= 0,
                              oneThirdFilledBoxNumbers = oneThirdFilledBoxNumbers - 1;
                              twoThirdFilledBoxNumbers = twoThirdFilledBoxNumbers + 1,
                              (* no empty container left *)
                              True,
                              oneThirdFilledBoxNumbers = 1]],
                  {n}]; (* current number of 1/3 and 2/3 filled boxes *)
                     {oneThirdFilledBoxNumbers, twoThirdFilledBoxNumbers}]];
```

Now, we generate 1000 times 1000 rods.

```
In[5]:= SeedRandom[999];
        data = Table[{1/3, 2/3.}.fillBoxes[10^3], {10^4}];
```

On average, the cumulative length of the rods in unfilled boxes is given by the following value.

```
In[7]:= Plus @@ data/Length[data]
Out[7]= 13.55
```

This value agrees favorable with the theoretical value.

```
In[8]:= λ[1000] // N
Out[8]= 13.4086
```

Here is a plot of the distribution of the average length.

```
In[9]:= ListPlot[{First[#], Length[#]/10^4}& /@ Split[Sort[data]]]
```

To monitor how the cumulative rod length in the unfilled boxes evolves as a function of generated rods, we slightly modify the above function fillBoxes. We monitor the current length every δ rod generations. The local variable average is updated after each rod generation.

```
In[10]:= fillBoxesHistory = Compile[{{n, _Integer}, {δ, _Integer}},
         Module[{completelyFilledBoxNumbers = 0, oneThirdFilledBoxNumbers = 0,
                 twoThirdFilledBoxNumbers = 0, newRodLength, newData = 0,
                 c = 0, average = 0.},
         Table[(* monitor average every δ rod generations *)
         Do[c++; newRodLength = Random[Integer, {1, 3}];
            Which[newRodLength === 3,
                  completelyFilledBoxNumbers =
                  completelyFilledBoxNumbers + 1,
                  newRodLength === 2,
```

```
If[oneThirdFilledBoxNumbers =!= 0,
   oneThirdFilledBoxNumbers = oneThirdFilledBoxNumbers - 1,
   twoThirdFilledBoxNumbers ++],
newRodLength === 1,
Which[twoThirdFilledBoxNumbers =!= 0,
      twoThirdFilledBoxNumbers = twoThirdFilledBoxNumbers - 1,
      oneThirdFilledBoxNumbers =!= 0,
      oneThirdFilledBoxNumbers = oneThirdFilledBoxNumbers - 1;
      twoThirdFilledBoxNumbers = twoThirdFilledBoxNumbers + 1,
       True,
        oneThirdFilledBoxNumbers = 1]];
(* current cumulative length of rods in not completely filled boxes *)
newData =  oneThirdFilledBoxNumbers + 2 twoThirdFilledBoxNumbers;
(* current average *)
average = 1/c ((c - 1) average) + newData/c, {δ}];
(* returned results *)
{c, average/3}, {n}]]];
```

Here are 10 trials with each containing 10^6 rod generations. We monitor the cumulative rod length in unfilled boxes every 100 rod generations.

```
In[11]:= SeedRandom[111];
        Show[Graphics[{Thickness[0.002],
          Table[{Hue[Random[]], Line[fillBoxesHistory[10^4, 10^2]]}, {10}]}],
            PlotRange -> All, Frame -> True, Frame -> True]
```

From the above exact formula for λ, we get the following asymptotics. For $n = 10^6$, we get an average length of about 434.

```
In[13]:= λA[n_] := 4/(3 Sqrt[3]) Sqrt[n/Pi] - 1/3

        λA[10.^6]
Out[14]= 433.98
```

For complexity considerations of on-line box packings, see [129], [130].

c) Using the function `Experimental`IntegerPartitions` mentioned in Section 2.4, we can immediately write a one-liner that returns the compositions of a positive integer *n*.

```
In[1]:= (* define integer partitions in the style of result of FactorInteger *)
       myIntegerPartitions[n_Integer?Positive] := Reverse /@
         (({First[#], Length[#]}& /@ Split[#])& /@
                       Experimental`IntegerPartitions[n])
In[3]:= compositions[n_Integer?Positive] :=
       Flatten[Permutations /@ (Join @@@ (Apply[Table[#1, {#2}]&,
                                 myIntegerPartitions[n], {2}])), 1]
```

Here is an example.

```
In[4]:= compositions[5]
Out[4]= {{5}, {4, 1}, {1, 4}, {3, 2}, {2, 3}, {3, 1, 1}, {1, 3, 1}, {1, 1, 3}, {2, 2, 1}, {2, 1, 2},
        {1, 2, 2}, {2, 1, 1, 1}, {1, 2, 1, 1}, {1, 1, 2, 1}, {1, 1, 1, 2}, {1, 1, 1, 1, 1}}
```

For a given *n*, there are 2^{n-1} different compositions.

```
In[5]:= Table[{k, Length @ compositions[k]}, {k, 1, 16}]
```

Out[5]= {{1, 1}, {2, 2}, {3, 4}, {4, 8}, {5, 16}, {6, 32}, {7, 64}, {8, 128}, {9, 256},
{10, 512}, {11, 1024}, {12, 2048}, {13, 4096}, {14, 8192}, {15, 16384}, {16, 32768}}

Now let us deal with the multiplicity count. To count the multiplicities in a memory-efficient way, we surely should not generate all 204226 partitions of 50 and surely not the $2^{49} \approx 5.63 \cdot 10^{14}$ different compositions.

In[6]= {v = PartitionsP[n = 50], 2^49}

Out[6]= {204226, 562949953421312}

In the hope of finding another useful function, we have a look at all built-in functions dealing with integer partitions.

In[7]= Names["*`*Partition*"]

Out[7]= {Internal`DirectPartitionsP, Internal`DirectPartitionsQ,
Internal`IntegerPartitionQ, Experimental`IntegerPartitions,
Internal`IntegerPartitions, myIntegerPartitions, Internal`NextIntegerPartition,
Partition, Developer`PartitionMap, PartitionsP, PartitionsQ,
Internal`PreviousIntegerPartition, NDSolve`SymplecticPartitionedRungeKutta,
NDSolve`SymplecticPartitionedRungeKuttaCoefficients,
Internal`TransposeIntegerPartition}

Given a partition $\{\{j_1, \lambda_{j_1}\}, \ldots \{j_l, \lambda_{j_l}\}\}$, the function System`Private`NextIntegerPartition generates the "next" partition ("next" meaning the j_i are weakly nondecreasing). Here are two examples.

In[8]= (* add the context of NextIntegerPartition to context path *)
AppendTo[$ContextPath, #]& /@
Union[Context /@ Names["*`*IntegerPartition*"]];

In[10]= NextIntegerPartition[{{4, 1}}]

Out[10]= {{3, 1}, {1, 1}}

In[11]= NextIntegerPartition[{{3, 1}, {1, 1}}]

Out[11]= {{2, 2}}

Once the last partition $\{\{1, n\}\}$ is reached, the results become periodic.

In[12]= NextIntegerPartition[{{1, 4}}]

Out[12]= {{1, 4}}

For a given partition $\{\{j_1, \lambda_{j_1}\}, \ldots \{j_l, \lambda_{j_l}\}\}$, the number of compositions arising from permutations is $\left(\sum_{\mu=1}^{l} \lambda_{j_\mu}\right)! / \prod_{\mu=1}^{l} \lambda_{j_\mu}!$.

In[13]= numberOfPermutations[partition_] :=
(Plus @@ (Last /@ partition))!/Times @@ ((Last /@ partition)!);

Here is an example.

In[14]= {numberOfPermutations[{{3, 3}, {2, 2}, {1, 1}}],
(* for comparison *)
Length[Permutations[{3, 3, 3, 2, 2, 1}]]}

Out[14]= {60, 60}

Using numberOfPermutations, we can implement the function multiplicityCounts that for a given permutation returns the result {{j, multiplicityOfj}, numberOfThisMultiplicityCount}.

In[15]= multiplicityCounts[partition_] :=
With[{n = numberOfPermutations[partition]},
{{First[#], Last[#]}, n}& /@ partition]

Here is again an example.

In[16]= multiplicityCounts[{{1, 3}, {2, 2}}]

Out[16]= {{{1, 3}, 10}, {{2, 2}, 10}}

The last result can be read from the following explicit form of the corresponding compositions.

In[17]= Permutations[{1, 1, 1, 2, 2}]

```
Out[17]= {{1, 1, 1, 2, 2}, {1, 1, 2, 1, 2}, {1, 1, 2, 2, 1}, {1, 2, 1, 1, 2}, {1, 2, 1, 2, 1},
         {1, 2, 2, 1, 1}, {2, 1, 1, 1, 2}, {2, 1, 1, 2, 1}, {2, 1, 2, 1, 1}, {2, 2, 1, 1, 1}}
```

Generating now all partitions sequentially and analyzing the multiplicity counts of their permutations is a very memory-efficient procedure. The next input carries this out.

```
In[18]:= Do[counter[{int, intCount}] = 0, {int, n}, {intCount, n}];
         partition = {{n, 1}}; k = 1;
         While[(* loop through all partitions *) k <= v,
               (* analyze current partition *)
               data = multiplicityCounts[partition];
               (* increase multiplicity counters *)
               (counter[First[#]] = counter[First[#]] + Last[#]) & /@ data;
               (* form next partition *)
               partition = NextIntegerPartition[partition];
               k++;]
```

Here are the most and least common multiplicity counts.

```
In[21]:= resultingCount =
         Sort[#, #1[[2]] > #2[[2]]&]& @ DeleteCases[
         Table[{{int, intCount}, counter[{int, intCount}]},
               {int, n}, {intCount, n}] // Flatten[#, 1]&, {_, 0}];
```

```
In[22]:= Take[resultingCount, 6]
```

```
Out[22]= {{{5, 1}, 214095114418258}, {{4, 1}, 193293250152002}, {{4, 2}, 157509916326265},
          {{6, 1}, 151695994687575}, {{3, 3}, 138777210492508}, {{3, 2}, 122581923329223}}
```

```
In[23]:= Take[resultingCount, -6]
```

```
Out[23]= {{{50, 1}, 1}, {{25, 2}, 1}, {{10, 5}, 1}, {{5, 10}, 1}, {{2, 25}, 1}, {{1, 50}, 1}}
```

Here is a logarithmic plot of the resulting multiplicity counts.

```
In[24]:= ListPlot3D[Table[Log[10, counter[{i, j}] + 1], {i, n}, {j, n}]]
```

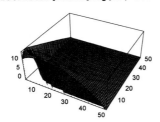

The memory usage for calculating the multiplicities for $n = 50$ was quite small.

```
In[25]:= MaxMemoryUsed[]
```

```
Out[25]= 4390432
```

18. Level Spacing, Subset Sums

a) Let us start by analyzing what `levelSpacings[n]` does. `Table[k^2 + l^2, {k, n}, {l, 1, Floor[Sqrt[n^2 - k^2]]}]` creates a nested list of all sums of squares $\leq n$. Then, this list gets flattened and sorted. Pairs of adjacent elements are formed, and their (negative) difference is calculated by `-Apply[Subtract, sortedListOfSumOfSquares, {1}]`. `Split` generates sublists of equal differences, and `{First[#], Length[#]} & /@ splittedListOfDifferences` determines how often each difference occurs.

Because all numbers appearing in the calculation are machine integers, the first step for a more efficient implementation is to use `CompiledFunction`. This saves about a factor of three of memory.

```
In[1]:= ByteCount /@ {Table[k, {k, 99}], Compile[{}, Table[k, {k, 99}]][]}
```

```
Out[1]= {2000, 452}
```

Because `CompiledFunction` can only deal with tensors and the list `Table[k^2 + l^2, {k, n}, {l, 1, Floor[ ⸴ Sqrt[n^2 - k^2]]]` is not a tensor, we will instead build one long list inside `Compile`. Another way to save memory is to avoid using `Partition[list, 2, 1]`. This operation generates a list roughly twice the size as *list*. Instead, calculating a list of differences uses just as much memory as *list*. A final possibility is not to use `Split`. Because, generically, the result of `Split` will not be a tensor (it is not in the case under consideration), we will go through the list of sorted differences and watch out when a new difference shows up. Taking all of these said remarks into account, we get the following implementation. We repeatedly assign to `l` to avoid multiple large lists.

```
In[2]:= levelSpacingsC =
    Compile[{{n, _Integer}},
        Module[{l = Table[1 + l^2, {l, 1, Floor[Sqrt[n^2 - 1]]}]},
                u, c, posList, j, old, new, counts},
                (* the list of all sums of two squares *)
                Do[l = Join[l, Table[k^2 + l^2,
                        {l, 1, Floor[Sqrt[n^2 - k^2]]}]], {k, 2, n - 1}];
                (* sorted list of differences of all sums of two squares *)
                l = Sort[l];
                l = Sort[Table[l[[j + 1]] - l[[j]], {j, Length[l] - 1}]];
                (* all occurring differences *)
                u = Union[l];
                (* a list with positions where a new difference starts *)
                (* initialization *)
                posList = Table[0, {Length[u] + 1}];
                j = 1; old = -1;
                (* go through the list diffs *)
                Do[new = l[[k]];
                    If[old =!= new, posList[[j]] = k; j++];
                    old = new, {k, Length[l]}];
                posList[[Length[u] + 1]] = Length[l] + 1;
                (* how often does each difference occur? *)
                counts = Map[(Last[#] - First[#])&, Partition[posList, 2, 1]];
                (* differences and counts of differences *)
                Transpose[{u, counts}]]];
```

The function compiled `levelSpacingsC` successfully.

```
In[3]:= Union[Head /@ Flatten[levelSpacingsC[[4]]]]
Out[3]= {Integer}
```

Calculating the level spacing distribution for *n* = 500 uses 6 MB of memory.

```
In[4]:= levelSpacingsC[500]
Out[4]= {{0, 138625}, {1, 11322}, {2, 5237}, {3, 11941}, {4, 7657}, {5, 5415}, {6, 2651},
    {7, 3714}, {8, 3840}, {9, 1942}, {10, 470}, {11, 833}, {12, 906}, {13, 339},
    {14, 161}, {15, 295}, {16, 218}, {17, 76}, {18, 34}, {19, 35}, {20, 56},
    {21, 32}, {22, 4}, {23, 15}, {24, 7}, {25, 2}, {27, 4}, {28, 2}, {29, 2}, {31, 1}}
```

```
In[5]:= MaxMemoryUsed[]
Out[5]= 6064856
```

The original function `levelSpacings` needs more than three times as much memory.

```
In[6]:= levelSpacings[n_] := {First[#], Length[#]}& /@ Split[
    Sort[-Apply[Subtract, Partition[Sort[Flatten[
        Table[k^2 + l^2, {k, n},
            {l, 1, Floor[Sqrt[n^2 - k^2]]}]]], 2, 1], {1}]]];
```

```
In[7]:= levelSpacings[500]
Out[7]= {{0, 138625}, {1, 11322}, {2, 5237}, {3, 11941}, {4, 7657}, {5, 5415}, {6, 2651},
    {7, 3714}, {8, 3840}, {9, 1942}, {10, 470}, {11, 833}, {12, 906}, {13, 339},
    {14, 161}, {15, 295}, {16, 218}, {17, 76}, {18, 34}, {19, 35}, {20, 56},
    {21, 32}, {22, 4}, {23, 15}, {24, 7}, {25, 2}, {27, 4}, {28, 2}, {29, 2}, {31, 1}}
```

```
In[8]:= MaxMemoryUsed[]
```

Out[8]= 25375136

Using this amount of memory, `levelSpacingsC` can cope with $n = 1000$. Here, we use it to generate a graphic of the distribution of differences. The numbers obtained above for the differences suggests using a logarithmic scale.

```
In[9]:= ListPlot[N[{#[[1]], Log[10, #[[2]]]}]& /@ levelSpacingsC[1000]],
            PlotRange -> All, PlotJoined -> True,
            Frame -> True, Axes -> False]
```

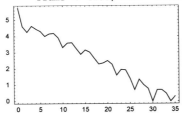

```
In[10]:= MaxMemoryUsed[]
```
Out[10]= 25375136

The last picture shows a nearly exponential dependence of the number of occurrences of a given spacing as a function of the spacing [509]. When interpreting $k^2 + l^2$ as the eigenvalues of the Laplace operator on a square (using Dirichlet boundary conditions), this is the result to be expected for an integrable system [54], [198], [164], [478], [472], [398], [502] [592], as much as this is asymptotically possible here [55]. For many more details on this simple, but quite interesting system, see [287], [101], and [213]; for random walks based on the eigenvalues, see [375]. For approximations of the number of $\sum_j k_k^2 \le K$, see [138], [59]. For the spacings of an nD harmonic oscillator, see [429], [223], [224].

b) Let $\{s\}_m$ be the set of all sums of m different elements k_j. Using T to denote a symbolic table (similar as $\sum$ stands for sum), we have to form the following set:

$$\{s\}_m = \mathop{\mathrm{T}}_{i_1=1}^{n} \mathop{\mathrm{T}}_{i_2=i_1+1}^{n} \cdots \mathop{\mathrm{T}}_{i_m=i_{m-1}+1}^{n} (k_{i_1} + k_{i_2} + \ldots + k_{i_m}).$$

A straightforward generation of the n sets $\{s\}_m$, $k = 1, \ldots, n$ is used in the function `allSums1`. K is the list $\{k_1, k_2, \ldots, k_n\}$.

```
In[1]:= allSums1[K_] :=
        Module[{body, iterators, table, flatten, part, n = Length[K]},
          Table[(* the body of the iterated table *)
                body = Sum[part[K, i[j]], {j, m}];
                (* the iterators *)
                iterators = Table[{i[j], If[j === 1, 1, i[j - 1] + 1], n}, {j, m}];
                Developer`ToPackedArray @ (* the nested table *)
                Flatten[ReleaseHold[Hold[Table[b, i]] /.
                        i -> Sequence @@ iterators /.
                        b -> body /. part -> Part]], {m, 1, n}]];
```

Here is a simple example showing `allSums1` at work.

```
In[2]:= allSums1[{1, 1, 1, 1, 1}]
```
Out[2]= {{1, 1, 1, 1, 1}, {2, 2, 2, 2, 2, 2, 2, 2, 2, 2},
         {3, 3, 3, 3, 3, 3, 3, 3, 3, 3}, {4, 4, 4, 4, 4}, {5}}

The following graphic shows the distribution of the sums for four random realizations of K and $n = 12, 13, 14, 15, 16$. For increasing n, the number of possible sums forms increasingly pronounced curves. The explicit shape and position of the curve depends on the list of integers K.

```
In[3]:= sumGraphics[Ks_, opts___] :=
        Graphics[{PointSize[0.01],
            Table[{(* color each realization differently *)
                    Hue[(j - 1)/Length[Ks]],
                    Point[{First[#], Length[#]}]& /@
```

```
                    Split[Sort[Flatten[allSums1[Ks[[j]]]]]]},
                {j, Length[Ks]}]}, opts, Frame -> True];
```

```
In[4]:= SeedRandom[111];
      Table[Show[GraphicsArray[
          Table[sumGraphics[Table[Developer`ToPackedArray @
                  Table[Random[Integer, {1, 100}], {n + δ}],
                      {4}], PlotLabel -> n + δ], {δ, 0, 1}]]],
        {n, 11, 15, 2}]
```

To deal with larger lists K, we make some changes in allSums1 to achieve a better memory efficiency. The nested table used in allSums1 generates many empty lists accounting for up to 50% of the memory use. The function allSums2 does not use one Table command with multiple iterators, but m Table commands, each with a single iterator. To avoid the generation of empty lists, we flatten all intermediate results. Compared with allSums1 the function allSums2 is slower, but much more memory efficient.

```
In[6]:= allSums2[K_] :=
      Module[{body, iterators, table, flatten, part, n = Length[K]},
      Table[(* the body of the iterated tables *)
            body = Sum[part[K, i[j]], {j, m}];
            (* the iterators for the m tables *)
            iterators = Table[{i[j],
            If[j === 1, 1, i[j - 1] + 1], n}, {j, m}];
                Developer`ToPackedArray @ ReleaseHold[
                  (Hold @@ {Fold[flatten[table[#1, #2]]&,
                              body, Reverse @ iterators]}) /.
            table -> Table /. (* flatten out *) flatten -> Flatten /.
            part -> Part], {m, 1, n}]];
```

Now, we use allSums2 to deal with a list of 21 integers. This results in about two million sums.

```
In[7]:= SeedRandom[111];
      n = 21;
```

```
𝒦 = Developer`ToPackedArray @ Table[Random[Integer, {1, 100}], {n}];
sums = allSums2[𝒦];
```

Here is the frequency distribution of the resulting sums.

In[11]:= `lp = ListPlot[sumData = {First[#], Length[#]/(2^n - 1)}& /@`
                                    `Split[Sort[Flatten[sums]]]]`

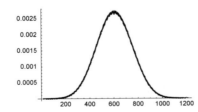

To compare with the asymptotic distribution $w(s)$ we implement the function w. The value of β is found using a numerical root-finding procedure.

In[12]:= `w[s_?NumberQ, K_] := Module[{fr, β},`
          `(* determine β *)`
          `fr = FindRoot[Evaluate[Plus @@ (#/(1 + Exp[β #])& /@ K) == s],`
                  `{β, 0}, MaxIterations -> 100];`
          `(* determine w(s) *)`
          `Exp[(Plus @@ (Log[1 + Exp[-β #]]& /@ K)) +`
             `β (Plus @@ (#/(1 + Exp[β #])& /@ K))]/Sqrt[2 Pi (Plus @@`
                `(#^2/((1 + Exp[β #])(1 + Exp[-β #]))& /@ K))]/(2^n - 1) /. fr]`

The following graphic shows that $w(s)$ excellently approximates the actual frequencies overall. The right graphic shows the relative error. At the small and large sum tails, the fluctuations dominate, but in the center, the average error is quite small (in the order of a few percent).

In[13]:= `Show[GraphicsArray[`
          `Block[{$DisplayFunction = Identity},`
          `(* frequencies and approximation *)`
          `{Show[{Plot[w[s, 𝒦], {s, 0, 1200},`
                  `PlotStyle -> {{Thickness[0.002], Hue[0]}}], lp}],`
          `(* relative error *)`
          `ListPlot[Transpose[{ (First /@ sumData), (* use occurring sums *)`
                  `(w[#, 𝒦]& /@ (First /@ sumData))/(Last /@ sumData) - 1}],`
                  `Frame -> True, Axes -> False, PlotRange -> {-1, 1}]}]]]`

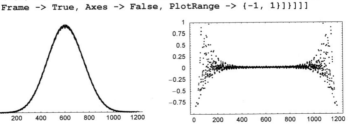

References

1 H. L. Abbott, P. Erdös, D. Hanson. *Am. Math. Monthly* 81, 256 (1974).

2 V. Adamchik. *J. Comput. Appl. Math.* 79, 119 (1997).

3 R. Adler, B. Kitchens, C. Tresser in J.-M. Gambaudo, P. Hubert, P. Tisseur, S. Vaienti (eds.). *Dynamical Systems*, World Scientific, Singapore, 2000.

4 J. Agapita, J. Weitsman. *arXiv:math.CO*/0411457 (2004).

5 R. P. Agarwal, P. J. Y. Wong. *Comput. Math. Appl.* 17, 1397 (1989).

6 R. P. Agarwal, P. J. Y. Wong. *Error Inequalities in Polynomial Interpolation and Their Applications*, Kluwer, Dordrecht, 1993.

7 R. P. Agarwal, P. J. Y. Wong in N. K. Govil, R. N. Mohapatra, Z. Nashed, A. Sharma, J. Szabados (eds.). *Approximation Theory*, Dekker, New York, 1998.

8 P. Akritas, I. E. Antoniou, G. P. Pronko. *Chaos, Solitons, Fractals* 12, 2805 (2001).

9 J. M. Aldaz, A. Bravo. *Am. Math. Monthly* 110, 63 (2003).

10 D. Aldous, P. Diaconis. *Am. Math. Monthly* 93, 333 (1986).

11 K. S. Alexander. *Ann. Appl. Prob.* 4, 1074 (1994).

12 K. Alladi, C. Grinstead. *J. Number Th.* 9, 452 (1977).

13 S. C. Althoen, M. F. Wyneken. *Am. Math. Monthly* 97, 907 (1990).

14 J.-P. Allouche, J. Shallit in C. Ding, T. Helleseth, N. Niederreiter (eds.). *Sequences and their Applications*, Springer-Verlag, London, 1999.

15 W. A. Al–Salam, L. Carlitz. *Portug. Math.* 18, 91 (1959).

16 V. Ambegaokar, A. Clerk. *arXiv:cond-mat*/9907423 (1999).

17 A. Amthor. *Z. Math. Phys.* 25, 153 (1880).

18 P. G. Anderson, T. C. Brown, P. J.-S. Shiue. *Proc. Am. Math. Soc.* 123, 2005 (1995).

19 G. Andrews. *The Theory of Partitions*, Addison-Wesley, Reading, 1976.

20 T. M. Apostel. *Introduction to Analytic Number Theory*, Springer-Verlag, New York, 1976.

21 T. M. Apostol. *Modular Functions and Dirichlet Series in Number Theory*, Springer-Verlag, New York, 1997.

22 T. M. Apostol. *Am. Math. Monthly* 106, 409 (1999).

23 T. Apostol in R. P. Bambah, V. C. Dumir, R. J. Hans–Gill. *Number Theory*, Birkhäuser, Basel, 2000.

24 R. G. Archibald. *An Introduction to the Theory of Numbers*, Charles E. Merrill, Columbus, 1970.

25 H. Ardahan. *Tr. J. Math.* 20, 335 (1996).

26 S. Ares, M. Castro. *arXiv:cond-mat*/0310148 (2003).

27 V. Arizón, J. Ojeda–Castañeda. *Appl. Optics* 33, 5925 (1994).

28 K. Azukawa, T. Yuzawa. *Math. J. Toyama Univ.* 13, 165 (1990).

29 M. Baake, U. Grimm. *Z. Kristallogr.* 219, 72 (2004).

30 E. Bach. *Annu. Rev. Comput. Sci.* 4, 119 (1990).

31 E. Bach, J. Shallit. *Algorithmic Number Theory*, v.1, MIT Press, Cambridge, 1996.

32 H. Bachmann. *Kalenderarithmetik*, Juris, Zurich, 1984.

33 M. Baica. *Ital. J. Pure Appl. Math.* 9, 187 (2001).

34 V. Balakrishnan, B. Sury. *Nieuw Archief Wiskunde* 11, 97 (1993).

35   C. Baltus. *Commun. Anal. Th. Cont. Fractions* III, 4 (1994).

36   G. Barat, P. J. Grabner. *J. Lond. Math. Soc.* 64, 523 (2001).

37   E. J. Barbeau. *Can. Math. Bull.* 4, 117 (1961).

38   A. Barvinok, J. E. Pommersheim in L. J. Billera, A. Björner, C. Greene, R. E. Simion, R. P. Stanley (eds.). *New Perspectives in Algebraic Combinatorics*, Cambridge University Press, Cambridge, 1999.

39   F. L. Bauer. *Kryptologie*, Springer-Verlag, Berlin, 1993.

40   F. L. Bauer. *Math. Intell.* 25, n3, 7 (2003).

41   D. Bayer, P. Diaconis. *Ann. Appl. Prob.* 2, 294 (1992).

42   A. Belenkye, R. Vidunas. *Int. J. Alg. Comput.* 8, 617 (1998).

43   A. H. Bell. *Am. Math. Monthly* 2, 140 (1895).

44   J. Bennies, J. Pitman. *Comb., Prob. Comput.* 10, 203 (2001).

45   S. Benson. *Coll. Math. J.* 25, 118 (1995).

46   C. Berge. *Principles of Combinatorics*, Academic Press, New York, 1971.

47   M. Berger, P. Pansu, J.-P. Berry, X. Saint-Raymond. *Problems in Geometry*, Springer-Verlag, New York, 1984.

48   W. Bergmann. *J. Gen. Phil. Sci.* 22, 15 (1991).

49   G. E. Bergum, A. N. Philippou, A. F. Horodam (eds.). *Applications of Fibonacci Numbers*, Kluwer, Dordrecht, 1991.

50   F. R. Berlekamp, J. H. Conway, R. K. Guy. *Winning Ways*, v.2, Academic Press, New York, 1982.

51   B. C. Berndt. *Ramanujan's Notebooks* IV, Springer-Verlag, New York, 1994.

52   B. C. Berndt, H. H. Chan, L.-C. Zhang. *Am. Math. Monthly* 104, 905 (1997).

53   B. C. Berndt, Y.-S. Choi, S.-Y. Kang in B. C. Berndt, F. Gesztesy (eds.). *Continued Fractions: From Analytic Number Theory to Constructive Approximation*, American Mathematical Society, Providence, 1999.

54   M. V. Berry, M. Tabor. *Proc. R. Soc. Lond.* A 356, 375 (1977).

55   M. V. Berry. *Ann. Phys.* 131, 163 (1981).

56   M. V. Berry. *Proc. R. Soc. Lond.* A 434, 465 (1991).

57   Beth, M. Frisch, G. J. Simmons. *Public-Key Cryptography: State of the Art and Future Directions*, Springer-Verlag, Berlin, 1991.

58   A. Beutelspacher. *Cryptology: An Introduction to the Art and Science of Enciphering, Encrypting, Concealing, Hiding and Safeguarding, described Without any Arcane Skullduggery but not Without Cunning Waggery for the Delectation and Instruction of the General Public*, American Mathematical Society, Washington, 1994.

59   V. B. Bezerra, A. N. Chaba. *J. Phys.* A 18, 3381 (1985).

60   N. Bhattacharya, H. B. van Linden van den Heuvell, R. J. C. Spreeuw. *Phys. Rev. Lett.* 88, 137901 (2002).

61   R. Bien. *Arch. Hist. Exact Sci.* 58, 439 (2004).

62   P. Billingsley. *Am. Math. Monthly* 76, 132 (1969).

63   P. Billingsley. *Am. Math. Monthly* 80, 1099 (1973).

64   I. F. Blahe, X. Gao, R. C. Mullin, S. A. Vanstone, T. Yaghoobian. *Applications of Finite Fields*, Kluwer, Dordrecht, 1993.

65   S. Blanes, F. Casas, J. A. Oteo, J. Ros. *J. Phys.* A 31, 259 (1998).

66   P. Blasiak, K. A. Penson, A. I. Solomon. *quant-ph*/0303030 (2003).

67   P. Bleher in D. A. Hejhal, J. Friedman, M. C. Gutzwiller, A. M. Odlyzko (eds.). *Emerging Applications of Number Theory*, Springer-Verlag, New York, 1999.

68   F. P. Boca, A. Zaharescu. *arXiv:math.NT*/0404114 (2004).

69   A. E. Bojarintzev. *Mat. Sapiski* 6, t U1, 21 (1967).

70   A. Bokulich. *Phil. Sci.* 70, 609 (2003).

71   B. A. Bondarenko. *Generalized Pascal Triangles and Pyramids: Their Fractals, Graphs, and Applications*, Fibonacci Association, Santa Clara, 1995.

72   D. Boneh. *Notices Am. Math. Soc.* 46, 203 (1999).

73   G. F. Borgonovi. *Eur. J. Phys.* 17, 216 (1996).

74   P. Borwein, L. Jörgenson. *Am. Math. Monthly* 108, 897 (2001).

75   D. Bosio, F. Vivaldi. *Preprint mp_arc* 99-479 (1999). http://rene.ma.utexas.edu/mp_arc-bin/mpa?yn=99-479

76   W. Bosma. *Math. Comput.* 54, 421 (1990).

77   A. Bottreau, A. Di Bucchianico, D. E. Loeb. *Discr. Math.* 180, 65 (1998).

78   W. G. C. Boyd. *Proc. R. Soc. Lond.* 447, 609 (1994).

79   J. P. Boyd. *Acta Appl. Math.* 56, 1 (1999).

80   I. L. Brekke, G. O. Freund. *Phys. Rep.* 233, 1 (1993).

81   D. Branson. *Math. Scientist* 25, 1 (2000).

82   B. Brent. *arXiv:math.NT/0405083* (2004).

83   R. P. Brent in J. F. Traub (ed.). *Algorithms and Complexity*, Academic Press, New York, 1976.

84   D. M. Bressoud, S. Wagon. *A Course in Computational Number Theory*, Springer-Verlag, New York, 2000.

85   C. Brezinski. *History of Continued Fractions and Padé Approximants*, Springer-Verlag, Berlin, 1991.

86   T. C. Brown. *Theor. Comput. Sc.* 273, 5 (2002).

87   H. Bruin, A. Lambert, G. Poggiaspalla, S. Vaienti. *Chaos* 13, 558 (2003).

88   W. A. Brumhead. *Math. Gaz.* 66, 267 (1972).

89   L. Bruno, A. Calvo, I. Ippolito. *Eur. J. Phys.* E 11, 131 (2003).

90   D. Bump. *Automorphic Forms and Representations*, Cambridge University Press, Cambridge, 1997.

91   P. Bundschuh. *Einführung in die Zahlentheorie*, Springer-Verlag, Berlin, 1988.

92   R. Bundschuh. *arXiv:cond-mat/0106326* (2001).

93   E. O. Burton. *Stud. Univ. Babes–Bolyai Math.* 46, 47 (2001).

94   P. L. Butzer, M. Hauss in S. Baron, D. Leviatan (eds.). *Israel Mathematical Conference Proceedings*, 1991.

95   C. Calderón, J. Zárate. *Acta Math. Hungar.* 85, 287 (1999).

96   F. Calogero. *Math. Intell.* 25, n4, 73 (2003).

97   I. J. Calvo, J. M. Masquè. *Acta Appl. Math.* 42, 139 (1996).

98   R. M. Capocelli, G. Gerbone, P. Cull, J. L. Holloway in R. M. Capocelli (ed.). *Sequences*, Springer-Verlag, New York, 1990.

99   C. Carathéodory. *Theory of Functions of a Complex Variable*, Birkhäuser, Basel, 1950.

100 R. E. Carr. *Proc. Am. Math. Soc.* 2, 925 (1951).

101 G. Casati, I. Guarneri, F. Valz–Gris. *Phys. Rev.* A 30, 1586 (1984).

102 M. E. Cesaro. *Comput. Rend.* 119, 848 (1895).

103 F. Chamizo, H. Iwaniec. *Rev. Mat. Iberoamericana* 11, 417 (1995).

104 C. A. Charalambides, J. Singh. *Commun. Stat. Th. Meth.* 17, 1533 (1988).

105 C. A. Charalambides. *Enumerative Combinatorics*, Chapman & Hall, Boca Raton, 2002

106 L. Chastkofsky. *Fibon. Quart.* 39, 320 (2001).

107 C. C. Chen. *Bull. ICA* 13, 45 (1995).

108   N. X. Chen. *Phys. Rev. Lett.* 64, 1193 (1990).

109   N. Chen. *J. Math. Phys.* 35, 3099 (1994).

110   N. X. Chen, G. B. Ren. *Phys. Lett.* A 160, 319 (1991).

111   N. Chen, Z. Chen, Y. Shen, S. Liu, M. Li. *Phys. Lett.* A 184, 347 (1994).

112   B. Chen. *Discr. Math.* 169, 211 (1997).

113   Y. G. Chen, C. G. Ji. *Acta Math. Sinica* 18, 647 (2002).

114   Z. Chen, Y. Shen, J. Ding. *Appl. Math. Comput.* 117, 161 (2001).

115   A. D. Chepelianskii, D. L. Shepelyansky. *Phys. Rev. Lett.* 87, 034101 (2001).

116   J. F. Clauser, J. P. Dowling. *Phys. Rev.* A 53, 4587 (1996).

117   G. L. Cohen. *Int. J. Math. Edu. Sci. Technol.* 24, 417 (1993).

118   G. E. Collins, J. R. Johnson in P. M. Gianni (ed.). *ISSAC 1988*, ACM, Baltimore, 1988.

119   J. B. Conrey, G. Myerson. *Preprint*, Am. Inst. Math. Preprint 3/1 (2000). http://www.aimath.org/preprints_00.html

120   S. Contini, E. Croot, I. E. Shparlinski. *arXiv:math.NT/*0404116 (2004).

121   G. Cooperman, S. Feisel, J. von zur Gathen, G. Havas in T. Asano, H. Imai, D. T. Lee, S. Nakano, T. Tokuyama (eds.). *Computing and Combinatorics*, Springer-Verlag, Berlin, 1999.

122   F. A Costabile, F. Dell'Accio. *BIT* 41, 451 (2001).

123   F. Costabile, F. Dell'Accio. *Num. Alg.* 28, 63 (2001).

124   F. A. Costabile, F. Dell'Accio, R. Luceri. *J. Comput. Appl. Math.* 176, 77 (2005).

125   F. A. Costabile, F. Dell'Accio. *Appl. Num. Math.* 52, 339 (2005).

126   D. A. Cox. *Am. Math. Monthly* 95, 442 (1988).

127   S. C. Coutinho. *The Mathematics of Ciphers: Number Theory and RSA Cryptography*, A K Peters, Natick, 1999.

128   R. Crandall, C. Pomerance. *Prime Numbers*, Springer-Verlag, New York, 2001.

129   J. Csirik, G. J. Woeginger in A. Fiat, G. J. Woeginger (eds.). *Online Algorithms*, Springer-Verlag, Berlin, 1998.

130   J. Csirik, D. S. Johnson. *Algorithmica* 31, 115 (2001).

131   N. Daili. *Approximations asymptotiques. Calcul de densities asymptotiques de certaintes fonctions arithmetiques*, Publication de l'Institut de Recherche Mathematique Avance, Strasbourg, 1990.

132   J. H. Davenport, Y. Siret, E. Tournier. *Computer Algebra*, Academic Press, London, 1993.

133   L. C. Davis, J. R. Reitz. *J. Math. Phys.* 16, 1219 (1975).

134   N. G. de Bruijn. *Nieuw Archief Wiskunde* 9, 361 (1991).

135   M. A. B. Deakin. *Austral. Math. Soc. Gaz.* 19, 41 (1992).

136   L. Dedić, M. Matić, J. Pečarić. *Bull. Belg. Math. Soc.* 8, 479 (2001).

137   L. J. Dedić, M. Matić, J. Pečarić, A. Vukelić. *J. Inequal. Appl.* 7, 787 (2002).

138   U. de Freitas, A. N. Chaba. *J. Phys.* A 16, 2205 (1983).

139   A. Dembo, A. Vershik, O. Zeitouni. *Markoc Processes Rel. Fields* 6, 147 (2000).

140   N. Dershowitz, E. M. Reingold. *Calendrical Calculations*, Cambridge University Press, Cambridge, 1997.

141   J. M. Deshouillers, F. Dress, G. Tenenbaum. *Acta Arith.* 34, 273 (1979).

142   J.-M. Deshouillers, H. J. J. te Riele, Y. Saouter in J. P. Buhler (ed.). *Algorithmic Number Theory*, Springer-Verlag, Berlin, 1998.

143   D. Deutsch, B. Goldman. *Math. Teacher* 98, 234 (2004).

144   L. E. Dickson. *History of the Theory of Numbers*, Chelsea, New York, 1952.

145   H. Dörrie. *100 Great Problems of Elementary Mathematics*, Dover, New York, 1965.

146 M. Drmota, M. Skalba. *Manuscripta Math.* 101, 361 (2000).

147 M. Drmota, J. Gajdosik. *J. Th. Nombres Bordeaux* 10, 17 (1998).

148 A. Dubickas, J. Steuding. *Elem. Math.* 59, 133 (2004).

149 G. Duchamp, K. A. Penson, A. I. Solomon, A. Horzela, P. Blasiak. *arXiv:quant-ph*/0401126 (2004).

150 U. Dudley. *Elementary Number Theory*, W. H. Freeman, San Francisco, 1969.

151 J. M. Dumont, A. Thomas. *J. Number Th.* 62, 19 (1997).

152 R. A. Dunlap. *The Golden Ratio and Fibonacci Numbers*, World Scientific, Singapore, 1998.

153 P. Dusart. *Math. Comput.* 48, 411 (1999).

154 F. J. Dyson. *Bull. Am. Math. Soc.* 78, 635 (1972).

155 F. J. Dyson, H. Falk. *Am. Math. Monthly* 99, 603 (1992).

156 H. M. Edwards. *Riemann's Zeta Function*, Academic Press, Boston, 1974.

157 P. Ehrenfest, T. Ehrenfest. *Phys. Zeit.* 8, 311 (1907).

158 R.-M. Elkenbracht–Huizing. *Nieuw Archief Wiskunde* 14, 375 (1996).

159 E. Elston, P. J. Kok. *Am. Math. Monthly* 70, 174 (1963).

160 D. A. Englund. *Fibon. Quart.* 39, 250 (2001).

161 I. J. Epstein. *Proc. Am. Math. Soc.* 14, 266 (1963).

162 R. Estrada, R. P. Kanwal. *Asymptotic Analysis: A Distributional Approach*, Birkhäuser, Basel, 1994.

163 R. Estrada. *Bol. Soc. Mat. Mex.* 3, 117 (1997).

164 A. J. Fendrik, M. J. Sánchez. *J. Phys.* A 33, 2345 (2000).

165 F. M. Fernández. *Phys. Rev.* A 67, 022104 (2003).

166 F. M. Fernández. *Eur. J. Phys.* 26, 151 (2005).

167 J. C. Ferreira. *arXiv:math.GM*/0209232 (2002).

168 M. Filaseta, D. V. Pasechnik. *arXiv:math.NT*/0409523 (2004).

169 S. Finch. *Preprint* (2003). http://people.bu.edu/srfinch/asym.pdf

170 N. J. Fine. *Basic Hypergeometric Series and Applications*, American Mathematical Society, Providence, 1988.

171 P. Flajolet, H. Prodinger. *SIAM J. Discr. Math.* 12, 155 (1999).

172 P. Flajolet, B. Vallee. *INRIA Technical Report* RR-4072 (2000). http://www.inria.fr/RRRT/RR-4072.html

173 F. Flajolet, S. Gerhold, B. Salvy. *arXiv:math.CO*/0501379 (2005).

174 C. Flegg, C. Hay, B. Moss (eds.). *N. Chuquet, Renaissance Mathematician*, Reidel, Dordrecht, 1985.

175 K. Ford. *arXiv:math.NT*/9907204 (1999).

176 K. Ford, D. Konyagin. *Duke Math. J.* 113, 313 (2002).

177 L. R. Ford. *Am. Math. Monthly* 45, 586 (1938).

178 W. B. Ford. *Studies on Divergent Series and Summability and the Asymptotic Developments of Functions Defined by MacLaurin Series*, Chelsea, New York, 1960.

179 R. Forman. *Am. Math. Monthly* 99, 548 (1992).

180 T. Fort. *Finite Differences*, Clarendon Press, Oxford, 1948.

181 A. Fraenkel. *J. reine angew. Math.* 138, 133 (1910).

182 H. T. Freitag, G. M. Phillips in F. T. Howard (eds.). *Applications of Fibonacci Numbers* v8, Kluwer, Dordrecht, 1999.

183 R. J. Friedlander. *Coll. Math. J.* 12, 12 (1981).

184 K. Fujii, T. Suzuki. *arXiv:quant-ph*/0304094 (2003).

185   J. Fulman. *arXiv:math.CO*/0102105 (2001).

186   J. A. C. Gallas. *Physica* A 222, 125 (1995).

187   F. Galton. *Natural Inheritance*, Macmillan, New York, 1894.

188   B. Garrison. *Am. Math. Monthly* 97, 316 (1990).

189   G. Garza, J. Young. *Math. Mag.* 77, 314 (2004).

190   G. Gaspari. *Physica* D 73, 352 (1994).

191   G. Gasper, M. Rahman. *Basic Hypergeometric Series*, Cambridge University Press, Cambridge, 1990.

192   J.-P. Gazeau, B. Champagne in Y. Sant–Aubin, L. Vinet (eds.). *Algebraic Methods in Physics*, Springer-Verlag, New York, 2001.

193   A.O. Gelfond. *The Solution of Equations in Integers*, Noordhoff, Groningen, 1960.

194   B. Georgeot, D. L. Shepelyansky. *arXiv:quant-ph*/0105149 (2001).

195   B. Georgeot, D. L. Shepelyansky. *Phys. Rev. Lett.* 86, 5393 (2001).

196   M. Gerstenhaber. *Am. Math. Monthly* 70, 397 (1963).

197   I. M. Gessel. *arXiv:math.CO*/0108121 (2001).

198   M.-J. Giannoni, A. Voros, J. Zinn-Justin (eds.). *Chaos and Quantum Physics*, North-Holland, Amsterdam, 1991.

199   B. D. Ginsberg. *Coll. Math. J.* 35, 26 (2004).

200   R. Girgensohn. *J. Math. Anal. Appl.* 203, 127 (1996).

201   K. Girstmair in A. Beutelspacher, S. D. Chatterji, U. Kulisch, R. Liedl (eds.) *Jahrbuch Überblicke Mathematik 1995*, Vieweg, Braunschweig, 1995.

202   J. W. L. Glaisher. *Messenger Math.* 2, 41 (1873).

203   C. Godrèche, J. M. Luck. *arXiv:cond-mat*/0106272 (2001).

204   G. Godréche, J. M. Luck. *arXiv:cond-mat*/0109213 (2001).

205   A. Goetz. *Ill. J. Math.* 44, 465 (2000).

206   A. Goetz in J.-M. Gambaudo, P. Hubert, P. Tisseur (eds.). *From Crystal to Chaos*, World Scientific, Singapore, 2000.

207   W. M. Y. Goh, R. Boyer. *arXiv:math.CO*/0409062 (2004).

208   H. J. Goldstein. *Am. Math. Monthly* 41, 565 (1934).

209   L. J. Goldstein. *Am. Math. Monthly* 80, 599 (1973).

210   H. Goldstine. *A History of Numerical Analysis from the 16th through the 19th Century*, Springer-Verlag, New York, 1977.

211   E. Goles, M. Morvan, H. D. Phan. *arXiv:cs.DM*/0010036 (2000).

212   E. Goles, M. Morvan, H. D. Phan. *Disc. Appl. Math.* 117, 51 (2002).

213   J. M. G. Gómez, R. A. Molina, A. Relaño, J. Retamosa. *Phys. Rev.* E 66, 036209 (2002).

214   I. J. Good, R. F. Churchhouse. *Math. Comput.* 22, 857 (1968).

215   E. A. Gorin, B. N. Kukushkin. *St. Petersburg Math. J.* 15, 449 (2003).

216   M. Gossler. *Astron. Nachr.* 302, 161 (1981).

217   J. Gough. *arXiv:quant-ph*/0311161 (2003).

218   H. W. Gould, W. Squire. *Am. Math. Monthly* 70, 44 (1963).

219   H. W. Gould. *Jahresber. DMV* 73, 149 (1971).

220   A. Granville, J. van de Lune, H. J. J. te Riele in R. A. Mollin (ed.). *Number Theory and Applications*, Plenum, New York, 1989.

221   R. I. Greenberg. *arXiv:cs.DS*/0211001 (2002).

222   R. I. Greenberg. *arXiv:cs.DS*/0301034 (2003).

223   C. Greenman. *J. Phys.* A 29, 4065 (1996).

224   C. Greenman. *J. Phys.* A 30, 915 (1997).

225   C. C. Grosjean, H. E. D. Meyer in T. M. Rassias (ed.). *Constantin Carathéodory: An International Tribute*, World Scientific, Singapore, 1991.

226   C. C. Grosjean in T. M. Rassias, H. M. Srivastava, A. Yanushauskas (eds.). *Topics in Polynomials of One and Several Variables and Their Applications*, World Scientific, Singapore 1993.

227   A. Grytczuk, B. Tropak. in A. Pethö, M. E. Pohst, H. C. Williams, H. G. Zimmer (eds.). *Computational Number Theory*, de Gruyter, Berlin, 1991.

228   S. Guiasu in *Short Commun. ICM 1994*, Zürich, 1994.

229   S. Guiasu. *Math. Mag.* 68, 110 (1995).

230   B. Guilfoyle, W. Klingenberg, S. Sen. *arXiv:math.DG*/0406399 (2004).

231   A. Guinand. *Am. Math. Monthly* 86, 187 (1979).

232   R. C. Gunning. *Lectures on Modular Forms*, Princeton University Press, Princeton, 1962.

233   R. K. Guy. *Am. Math. Monthly* 95, 697 (1988).

234   R. K. Guy. *Unsolved Problems in Number Theory*, Springer-Verlag, New York, 1994.

235   S. Haber. *SIAM J. Num. Anal.* 14, 668 (1977).

236   L. M. Hall. *Coll. Math. J.* 23, 35 (1992).

237   R. R. Hall. *J. Lond. Math. Soc.* 2, 139 (1970).

238   R. R. Hall, G. Tenenbaum. *Divisors*, Cambridge University Press, Cambridge, 1988.

239   G. H. Hardy, J. E. Littlewood. *Acta Math.* 44, 1 (1923).

240   G. H. Hardy, E. M. Wright. *An Introduction to the Theory of Numbers*, Oxford University Press, Oxford, 1979.

241   G. Harman in M. A. Bennett, B. C. Berndt, N. Boston, H. G. Diamond, A. J. Hildebrand, W. Philipp. *Number Theory for the Millennium*, A K Peters, Natick, 2002.

242   C. Hauert, J. Nagler, H. G. Schuster. *J. Stat. Phys.* 116, 1453 (2004).

243   M. Hauss in P. L. Butzer, E. T. Jongen, W. Oberschelp (eds.). *Charlemagne and his Heritage—1200 Years of Civilization and Science in Europe*, Brepols, Turnhout, 1998.

244   T. X. He, L. C. Hsu, P. J.-S. Shiue, D. C. Torney. *J. Comput. Appl. Math.* 177, 17 (2005).

245   H. Heilbronn in P. Turán (eds.). *Number Theory and Analysis*, Plenum Press, New York, 1969.

246   H. Hemme. *Bild der Wissenschaft* n5, 155 (1987).

247   H. Hemme. *Bild der Wissenschaft* n6, 130 (1990).

248   H. Hemme. *Bild der Wissenschaft* n3, 132 (1991).

249   H. Hemme. *Bild der Wissenschaft* n6, 111 (1992).

250   P. Henrici. *Applied and Computational Complex Analysis*, v2, Wiley, New York, 1977.

251   D. Hensley. *J. Number Th.* 49, 142 (1994).

252   M. A. Hernández, N. Romero. *J. Comput. Appl. Math.* 177, 225 (2005).

253   S. Heubach, T. Mansour. *arXiv:math.CO*/0310197 (2003).

254   C. Heuberger. *Period. Math. Hung.* 49, 65 (2004).

255   F. B. Hildebrand. *Finite-Difference Equations and Simulations*, Prentice-Hall, Englewood Cliffs, 1968.

256   F. H. Hildebrand. *Introduction to Numerical Analysis*, Dover, New York, 1974.

257   H. Hiller. *Acta Cryst.* A 41, 541 (1985).

258 D. S. Hirschberg. _J. ACM_ 24, 664 (1977).

259 M. D. Hirschhorn. _Discr. Math._ 211, 225 (2000).

260 P. Hitczenko, C. Rousseau, C. D. Savage. _arXiv:math.CO_/0110181 (2001).

261 P. Hitczenko, G. Stengle. _arXiv:math.CO_/0110189 (2001).

262 P. Hitczenko, C. Rosseau, C. D. Savage. _J. Comput. Appl. Math._ 142, 107 (2002).

263 M. Hochbruck, C. Lubich. _SIAM J. Num. Anal._ 41, 945 (2003).

264 V. E. Hoggatt, Jr. _Fibonacci and Lucas Numbers_, Houghton Mifflin, New York, 1969.

265 V. E. Hoggatt, Jr, M. Bricknell-Johnson. (eds.). _A Collection of Manuscripts Related to the Fibonacci Sequence_, Fibonacci Association, Santa Clara 1980.

266 M. Holthaus, E. Kalinowski. _Ann. Phys._ 270, 198 (1998).

267 R. Honsberger. _Ingenuity in Mathematics_, Random House, New York, 1970.

268 A. Horzela, P. Blasiak, G. E. H. Duchamp, K. A. Penson, A. I. Solomon. _arXiv:quant-ph_/0409152 (2004).

269 W. G. Hoover, B. Moran. _Phys. Rev._ A 40, 5319 (1989).

270 W. G. Hoover in M. Mareschal, B. L. Holian (eds.). _Microscopic Simulations of Complex Hydrodynamic Phenomena_, Plenum Press, New York, 1992.

271 C.-C. Hsu, I. S. Reed, T. K. Truong. _IEEE Trans. Signal Proc._ 42, 2823 (1994).

272 L. C. Hsu, J. Wang. _Tamkang J. Math._ 29, 89 (1998).

273 L. C. Hsu, P. J.-S. Shiue. _Ann. Combinat._ 5, 179 (2001).

274 S.-T. Huang. _ACM Trans. Progr. Lang._ 15, 563 (1993).

275 J. G. Huard, Z. M. Ou, B. K. Spearman, K. S. Williams in M. A. Bennett, B. C. Berndt, N. Boston, H. G. Diamond, A. J. Hildebrand, W. Philipp. _Number Theory for the Millennium_, A K Peters, Natick, 2002.

276 B. D. Hughes, N. E. Frankel, B. W. Ninham. _Phys. Rev._ A 42, 3643 (1990).

277 D. Hughes, M. Paczuski. _arXiv:cond-mat_/0105408 (2001).

278 A. Hurwitz. _Z. Math. Phys._ 35, 56 (1890).

279 A. Hurwitz. _Übungen zur Zahlentheorie_, Schriftenreihe der ETH-Bibliothek 32, ETH-Bibliothek, Zürich, 1993.

280 Y. Imai, Y. Seto, S. Tanaka, H. Yutani. _Fibon. Quart._ 26, 33 (1988).

281 M. C. Irwin. _Am. Math. Monthly_ 96, 696 (1989).

282 F. Ischebek. _Math. Semesterber._ 40, 212 (1993).

283 A. Iserles, S. P. Nørsett, A. F. Rasmussen. _Appl. Num. Math._ 39, 379 (2001).

284 A. Iserles. _Notices Am. Math. Soc._ 49, 430 (2002).

285 A. Iserles. _Appl. Num. Math._ 43, 145 (2002).

286 A. Iserles. _BIT_ 44, 473 (2004).

287 C. Itzykson, J. M. Luck. _J. Phys._ A 19, 211 (1986).

288 H. Iwaniec, E. Kowalski. _Analytic Number Theory_, American Mathematical Society, Providence, 2004.

289 E. A. Jackson. _Perspectives on Nonlinear Dynamics_, Cambridge University Press, Cambridge, 1991.

290 J. D. Jackson. _Classical Electrodynamics_, Wiley, New York, 1975.

291 T. Jackson. _From Number Theory to Secret Codes_, IOP Publishing, Bristol, 1987.

292 M. J. Jacobson, Jr., H. C. Williams. _Math. Comput._ 72, 499 (2003).

293 E. Jacobsthal. _Sitzungsber. Berl. Math. Ges._ 17, 43 (1920).

294 R.-S. Jih. _Bull. Inst. Math. Acad. Sinica_ 5, 333 (1977).

295 C. Jordan. _Tôhoku Math. J._ 37, 254 (1933).

296 C. Jordan. *Calculus of Finite Differences*, Chelsea, New York, 1965.

297 R. Jozsa. *Ann. Phys.* 306, 241 (2003).

298 V. G. Kadyshevskiĭ, R. M. Mir–Kasimov, N. B. Skachkov. *Sov. J. Nucl. Phys.* 9, 125 (1969).

299 M. Kaneko. *J. Integer Sequences* 3, 00.2.9 (2000).
http://www.emis.ams.org/journals/JIS/VOL3/KANEKO/AT-kaneko.html

300 Y.-M. Kao, P.-G. Luan. *arXiv:cond-mat*/0210338 (2002).

301 D. R. Kaprekar. *Scripta Math.* 15, 244 (1949).

302 A. A. Karatsuba, S. M. Voronin. *The Riemann Zeta-Function*, de Gruyter, Berlin, 1992.

303 A. A. Karatsuba. *Complex Analysis in Number Theory*, CRC Press, Boca Raton, 1995.

304 P. Kargaev, A. Zhigljavsky. *J. Number Th.* 65, 130 (1997).

305 Y. Karshon, S. Sternberg, J. Weitsman. *arXiv:math.CO*/0307127 (2003).

306 V. S. Katchmar, B. P. Rusin, V. I. Shmoilov. *J. Vis. Mat. Mat. Fis.* 38, 1436 (1998).

307 J. Katriel. *Lett. Nuov. Cim.* 10, 565 (1975).

308 J. Katriel. *J. Opt.* B 4, S200 (2002).

309 M. Keith, T. Craver. *J. Recreat. Math.* 22, 280 (1990).

310 G. Kemkes, C. F. Lee, D. Merlini, B. Richmond. *SIAM J. Discr. Math.* 16, 179 (2003).

311 I. R. Khan, R. Ohba, N. Hozumi. *J. Comput. Appl. Math.* 150, 303 (2003).

312 B. Khang. *Adv. Math.* 185, 178 (2004).

313 A. Y. Khrennikov. *J. Math. Sc.* 73, 243 (1995).

314 H. K. Kim. *Proc. Am. Math. Soc.* 131, 65 (2002).

315 Y.-C. Kim. *arXiv:math.CA*/0502062 (2005).

316 C. Kimberling. *Acta Arith.* 73, 103 (1995).

317 M. Kiwi, M. Loebl, J. Matoušek. *arXiv:math.CO*/0308234 (2003).

318 J. C. Kluyver. *Verslag. Wiss. Ak. Wetenschapen* 15, 423 (1906).

319 W. Knödel. *Elektron. Informationsverarb. Kybernet.* 19, 427 (1983).

320 A. Knopfmacher. *Fibon. Quart.* 30, 80 (1992).

321 A. Knopfmacher, M. E. Mays. *Fibon. Quart.* 33, 153 (1995).

322 J. Knopfmacher, W.-B. Zhang. *Number Theory Arising From Finite Fields*, Marcel Dekker, New York, 2001.

323 K. Knopp. *Infinite Sequences and Series*, Dover, New York, 1956.

324 M. I. Knopp. *Modular Functions in Analytic Number Theory*, Chelsea, New York, 1993.

325 D. E. Knuth. *Comput. Math. Appl.* 2, 137 (1976).

326 D. E. Knuth. *The Art of Computer Programming*, v.2, Addison-Wesley, Reading, 1998.

327 N. Koblitz. *A Course in Number Theory and Cryptography*, Springer-Verlag, New York, 1987.

328 N. Koblitz. *Algebraic Aspects of Cryptography*, Springer-Verlag, Berlin, 1998.

329 N. Koblitz. *p-adic Numbers, p-adic Analysis and Zeta Functions*, Springer-Verlag, New York, 1998.

330 L. Kocić, L. Stefanovska in Z. Li, L. Vulkov, J. Wasniewski (ed.). *Numerical Analysis and Its Applications*, Springer-Verlag, Berlin, 2005.

331 T. Komatsu. *Fibon. Quart.* 41, 3 (2003).

332 V. V. Kozlov, M. Y. Mitrofanova. *Reg. Chaotic Dynam.* 8, 431 (2003).

333 E. Kranakis. *Primality and Cryptography*, Teubner, Stuttgart, 1986.

334   E. Krätzel. *Lattice Points*, Verlag der Wissenschaften, Berlin, 1988.

335   E. Krätzel. *Math. Nachr.* 163, 257 (1993).

336   E. Krätzel. *Analytische Funktionen in der Zahlentheorie*, Teubner, Stuttgart, 2000.

337   B. A. Kryzien. *Liet. Matem. Rink* 36, 7 (1996).

338   R. P. Kubelka. *Fibon. Quart.* 42, 70 (2004).

339   K. Kumičák in D. Abbott, L. B. Kish (eds.). *Unsolved Problems of Noise and Fluctuations*, American Institute of Physics, Melville, 2000.

340   A. K. Kwasniewski. *arXiv:math.CO*/0402344 (2004).

341   J. C. Lagarias, C. P. Tresser. *IBM J. Res. Dev.* 39, 283 (1995).

342   D. B. Lahiri. *Bull. Austral. Math. Soc.* 1, 307 (1969).

343   A. Lakhtakia. *Comput. Graphics* 13, 57 (1989).

344   V. Lampret. *Math. Mag.* 74, 109 (2001).

345   V. Lampret. *SIAM Rev.* 46, 311 (2004).

346   S. Landau. *Math. Intell.* 16, n2, 49 (1994).

347   V.-A. Le Besgue. *Introduction á la théorie des nombres*, Mallet-Bachelier, Paris, 1862.

348   C. C. Lee, D. T. Lee. *J. ACM* 32, 562 (1985).

349   F. Le Lionnais. *Les nombres remarquables*, Hermann, Paris, 1983.

350   F. Lemmermeyer. *Reciprocity Laws*, Springer-Verlag, Berlin, 2000.

351   A. K. Lenstra. *Designs, Codes and Cryptography* 19, 101 (2000).

352   H. W. Lenstra, Jr. *Notices Am. Math. Soc.* 49, 182 (2002).

353   C. Leubner. *Eur. J. Phys.* 6, 299 (1985).

354   J. Leutenegger. Ph. D. thesis, Basel, 1904.

355   W. J. LeVeque. *Fundamentals of Number Theory*, Addison-Wesley, Reading, 1977.

356   F. M. Liang. *Discr. Math.* 28, 325 (1979).

357   M. Li, S. J. Liu, N. X. Chen. *Phys. Lett.* A 177, 134 (1993).

358   H. Lichtenberg. *Hist. Math.* 24, 441 (1997).

359   H. Lichtenberg. *Math. Semesterber.* 50, 45 (2003).

360   P.-Y. Lin. *Fibon. Quart.* 22, 229 (1984).

361   M. E. Lines. *Think of a Number*, Adam Hilger, Bristol, 1990.

362   B. F. Logan, L. A. Shepp. *Adv. Math.* 26, 206 (1977).

363   M. P. Losada, E. Cesaratto. *Revista Mat. Argentina* 41, 51 (1998).

364   M. P. Losada, S. Grynberg. *Int. J. Bifurc. Chaos* 8, 1095 (1998).

365   M. Lothaire. *Algebraic Combinatorics on Words*, Cambridge University Press, Cambridge, 2002.

366   F. Luca, P. Stănică. *arXiv:math.NT*/0304272 (2003).

367   A. Lue, H. Brenner. *Phys. Rev.* E 47, 3128 (1993).

368   A. Lupas. *Octogon* 7, 2 (1999).

369   N. Ma. *Appl. Math. Comput.* 124, 365 (2001).

370   H. Mack, M. Bienert, F. Haug, F. S. Straub, M. Freyberger, W. P. Schleich. *arXiv:quant-ph*/0204040 (2002).

371   H. Mack, M. Bienert, F. Haug, M. Freyberger, W. P. Schleich. *arXiv:quant-ph*/0208021 (2002).

372   A. L. Mackay. *Mod. Phys. Lett.* B 4, 989 (1990).

373   J. Maddox. *Nature* 344, 377 (1990).

374   W. Magnus. *Commun. Pure Appl. Math.* 7, 649 (1954).

375   J. Marklof. *Commun. Math. Phys.* 199, 169 (1998).

376   G. Marsaglia, J. C. W. Marsaglia. *Am. Math. Monthly* 97, 826 (1990).

377   G. Martin. *arXiv:math.NT*/0006089 (2000).

378   G. Martin. *Am. Math. Monthly* 108, 746 (2001).

379   J. Mathews. *Am. Math. Monthly* 97, 15 (1990).

380   Y. V. Matiyasevich. *Am. Math. Monthly* 93, 631 (1986).

381   J. E. Maxfield. *Math. Mag.* 43, 64 (1970).

382   P. Maurer. *Am. Math. Monthly* 94, 631 (1987).

383   J. H. McClellan, C. M. Radar. *Number Theory in Digital Signal Processing*, Prentice-Hall, Englewood Cliffs, 1979.

384   W. L. McDaniel. *Nieuw Archief Wiskunde* 9, 103 (1991).

385   A. R. Meijer. *Int. J. Math. Edu. Sci. Technol.* 23, 324 (1992).

386   A. J. Menezes, I. F. Blake, X. Gao, C. Mullin, S. A. Vanstone, T. Yaghoobian. *Applications of Finite Fields*, Kluwer, Dordrecht, 1993.

387   A. Mercier. *Am. Math. Monthly* 97, 503 (1990).

388   N. D. Mermin. *Am. J. Phys.* 52, 362 (1984).

389   R. Metzler, W. Kinzel. *arXiv:cond-mat*/0007382 (2000).

390   C. Meyer. *Public. Inst. Rech. Math. Avancée* 017/1994 (1994).

391   R. Michel. *Math. Semesterber.* 43, 81 (1996).

392   M. Mignotte in B. Buchberger, G. E. Collins, R. Loos, R. Albrecht (eds.). *Computer Algebra—Symbolic and Algebraic Computations*, Springer-Verlag, Wien, 1983.

393   M. Mignotte. *AAECC* 6, 327 (1995).

394   V. V. Mikhailov. *J. Phys.* A 18, 231 (1985).

395   S. Mills. *Arch. Hist. Exact Sci.* 33, 1 (1985).

396   S. C. Milne. *arXiv:math.NT*/0008068 (2000).

397   S. C. Milne. *arXiv:math.NT*/0009130 (2000).

398   N. Minami. *mp_arc* 00-163 (2000).   http://rene.ma.utexas.edu/mp_arc-bin/mpa?yn=00-163

399   G. Mincu. *J. Ineq. Pure Appl. Math.* 4, n2, A30 (2003).   http://jipam.vu.edu.au/v4n2/153_02.html

400   D. Ming, T. Wen, J. Dai, X. Dai, W. E. Evenson. *Phys. Rev.* E 62, R3019 (2000).

401   H. N. Minh, G. Jacob. *Discr. Math.* 210, 87 (2000).

402   D. S. Mitrinovic, J. Sándor, B. Crstici. *Handbook of Number Theory*, Kluwer, Dordrecht, 1996.

403   S. S. Mizrahi, V. V. Dodonov. *J. Phys.* A 35, 8847 (2002).

404   P. C. Moan, J. A. Oteo. *J. Math. Phys.* 42, 501 (2001).

405   R. A. Mollin. *Quadratics*, CRC Press, Boca Raton, 1996.

406   R. A. Mollin. *Fundamental Number Theory with Applications*, CRC Press, Boca Raton, 1997.

407   R. A. Mollin. *Far East J. Math. Sci.* 6, 355 (1998).

408   P. L. Montgomery. *CWI Quart.* 7, 337 (1994).

409   B. Moran, W. G. Hoover. *J. Stat. Phys.* 48, 709 (1987).

410   P. Moussa, J.-C. Yoccoz. *Commun. Math. Phys.* 186, 265 (1997).

411 P. Moussa, A. Cassa, S. Marmi. *J. Comput. Appl. Math.* 105, 403 (1999).

412 J. Müller. *J. Combinat. Th.* A 101, 271 (2003).

413 H. Nakaya. *Nagoya Math. J.* 122, 149 (1991).

414 V. Namias. *Am. Math. Monthly* 93, 25 (1986).

415 W. Narkiewicz. *The Development of Prime Number Theory*, Springer-Verlag, Berlin, 2000.

416 P. Natalini, P. E. Ricci. *Comput. Math. Appl.* 47, 719 (2004).

417 I. Navot. *SIAM J. Num. Anal.* 2, 259 (1965).

418 M. Naylor. *Math. Mag.* 75, 163 (2002).

419 E. Netto. *Lehrbuch der Combinatorik*, Teubner, Leipzig, 1927.

420 E. H. Neville. *Proc. Lond. Math. Soc.* 51, 132 (1949).

421 B. W. Ninham, B. D. Hughes, N. E. Frenkel, M. L. Glasser. *Physica* A 186, 441 (1992).

422 N. E. Nörlund. *Vorlesungen über Differenzenrechnungen*, Springer, Berlin, 1924.

423 G. H. Norton. *J. Symb. Comput.* 10, 53 (1990).

424 A. M. Odlyzko, H. J. J. te Riele. *J. reine angew. Math.* 357, 138 (1985).

425 F. W. J. Olver. *Asymptotics and Special Functions*, Academic Press, San Diego, 1974.

426 K. Ono. *CRM Proc.* 36, 229 (2004).

427 A. M. Ostrowski. *J. reine angew. Math.* 239/240, 268 (1970).

428 L. Panaitopol. *Nieuw Archief Wiskunde* 5, 55 (2000).

429 A. Pandey, R. Ramaswamy. *Phys. Rev.* A 43, 4237 (1991).

430 E. Pascal. *Die Determinanten*, Teubner, Leipzig, 1900.

431 S. V. Pavlovic. *Mitt. Math. Ges. Hamburg* 10, n5, 349 (1977).

432 L. L. Pennisi. *Am. Math. Monthly* 64, 497 (1957).

433 K. A. Penson, A. I. Solomon. *arXiv:quant-ph*/0211028 (2002).

434 M. Petkovic, M. Mignotte, M. Trajkovic. *ZAMM* 75, 551 (1995).

435 A. N. Philippou, A. F. Horodam, G. E. Bergum (eds.). *Applications of Fibonacci Numbers*, Kluwer, Dordrecht, 1986.

436 A. N. Philippou, A. F. Horodam, G. E. Bergum (eds.). *Applications of Fibonacci Numbers*, Kluwer, Dordrecht, 1988.

437 A. N. Philippou, A. F. Horodam, G. E. Bergum (eds.). *Applications of Fibonacci Numbers*, Kluwer, Dordrecht, 1990.

438 A. N. Philippou, A. F. Horodam, G. E. Bergum (eds.). *Applications of Fibonacci Numbers*, Kluwer, Dordrecht, 1991.

439 M. Piacquadio, E. Cesaratto. *arXiv:math-ph*/0002046 (2000).

440 C. A. Pickover in M. Emmer (ed.). *The Visual Mind: Art and Mathematics*, MIT, Cambridge, 1993.

441 S. Ping. *Fibon. Quart.* 40, 287 (2002).

442 G. Pólya, G. Szegö. *Problems and Theorems in Analysis*, Springer, Heidelberg, 1972.

443 G. Pólya, R. E. Tarjan, D. R. Woods. *Notes on Introductory Combinatorics*, Birkhäuser, Basel, 1983.

444 J. W. Porter. *Mathematika* 22, 20 (1975).

445 D. Prato, P. W. Lamberti. *J. Chem. Phys.* 106, 4640 (1997).

446 S. Prestipino. *Eur. J. Phys.* 26, 137 (2005).

447 W. D. A. Pribitkin. *Ramanujan J.* 4, 455 (2001).

448 H. Prodinger. *Elektron. Informationsverarb. Kybernet.* 21, 3 (1985).

449 J. R. Quine, R. R. Song. *Proc. Am. Math. Soc.* 119, 373 (1993).

450   H. Rademacher. *Proc. Lond. Math. Soc.* 43, 241 (1937).

451   H. Rademacher, E. Grosswald. *Dedekind Sums,* American Mathematical Society, Washington, 1972.

452   H. Rademacher. *Topics in Analytic Number Theory,* Springer-Verlag, New York, 1973.

453   H. Rademacher, O. Toeplitz. *The Enjoyment of Mathematics,* Princeton University Press, Princeton, 1957.

454   C. Radoux. *Ann. Soc. Sci. Bruxelles* I 91, 13, 8283 (1977).

455   E. D. Rainville. *Special Functions,* Macmillan, New York, 1960.

456   A. M. S. Ramasamy. *Indian J. Pure Appl. Math.* 25, 577 (1994).

457   V. V. Rane. *Proc. Indian Acad. Sci. (Math. Sci.)* 113, 213 (2003).

458   M. Razpet. *J. Math. Anal. Appl.* 149, 1 (1990).

459   R. M. Redheffer. *Internationale Schriftenreihe für Numerische Mathematik* 36, Springer-Verlag, Berlin, 1977.

460   C. A. Reiter. *Comput. Graphics* 28, 297 (2004).

461   S. Y. Ren, J. D. Dow. *Phys. Lett.* A 154, 215 (1991).

462   P. Ribenboim. *Nieuw Archief Wiskunde* 12, 53 (1994).

463   P. Ribenboim. *Classical Theory of Algebraic Numbers,* Springer-Verlag, New York, 2001.

464   P. Ribenboim. *The New Book of Prime Number Records,* Springer-Verlag, New York, 1996.

465   L. B. Richmond. *Acta Arithm.* 26, 411 (1975).

466   H. Riesel. *Mitt. Math. Ges. Hamburg* 12, 253 (1991).

467   H. Riesel. *Prime Numbers and Computer Methods of Factorization,* Birkhäuser, Basel, 1994.

468   H. Riesel, G. Göhl. *Math. Comput.* 24, 969 (1970).

469   J. Riordan. *Combinatorial Identities,* Wiley, New York, 1968.

470   D. Robbins. *Am. Math. Monthly* 86, 394 (1979).

471   N. Robbins. *Ars Combinat.* 60, 219 (2001).

472   M. Robnik, G. Veble. *arXiv:nlin.CD/*0003049 (2000).

473   F. Roesler. *Arch. Math.* 73, 193 (1999).

474   S. M. Roman, G.-C. Rota. *Adv. Math.* 27, 95 (1978).

475   K. H. Rosen. *Elementary Number Theory and Its Application,* Addison-Wesley, Reading, 1993.

476   J. B. Rosser, L. Schoenfeld. *Ill. J. Math.* 6, 64 (1962).

477   G.-C. Rota, B. D. Taylor. *SIAM J. Math. Anal.* 25, 694 (1994).

478   Z. Rudnick, P. Sarnak. *Commun. Math. Phys.* 194, 61 (1998).

479   H. Sagan. *Am. Math. Monthly* 93, 361 (1986).

480   J. W. Sander. *Math. Semesterber.* 39, 185 (1992).

481   G. Sandor, R. Sivaramakrishnan. *Nieuw Archief Wiskunde* 11, 195 (1993).

482   J. Sándor, A. Bege. *Adv. Stud. Contemp. Math.* 6, 77 (2003).

483   T. Sasamoto, T. Toyoizumi, H. Nishimori. *arXiv:cond-mat/*0106125 (2001).

484   T. Sasamoto. *Physica* A 321, 369 (2003).

485   E. R. Scheinerman. *Am. Math. Monthly* 107, 489 (2000).

486   H. F. Scherk. *J. Math.* 10, 201 (1833).

487   R. Schirzinger, P. A. A. Laura. *Conformal Mapping: Methods and Applications,* Elsevier, Amsterdam, 1991.

488   E. A. Schmidt. *J. Undergrad. Math.* 24, 317 (1992).

489   M. Schork. *J. Phys.* A 36, 4651 (2003).

490   P. Schreiber. *Hist. Math.* 20, 304 (1993).

491   M. R. Schroeder. *Number Theory in Science and Technology*, Springer-Verlag, Berlin, 1984.

492   M. R. Schröder. *Math. Intell.* 7, n4, 18 (1985).

493   A. Schultz. *Elem. Math.* 54, 64 (1999).

494   W. Schuster. *Arch. Math.* 77, 170 (2001).

495   I. J. Schwatt. *An Introduction to the Operations with Series*, Chelsea, New York, 1961.

496   F. Schwarz. *MathPAD* 10, 42 (2001).

497   W. Schwarz in J.-P. Pier (ed.). *Development of Mathematics 1900-1950*, Birkhäuser, Basel, 1994.

498   R. L. E. Schwarzenberger. *N-dimensional Crystallography*, Pitman, San Francisco, 1980.

499   R. Sedgewick. *Comput. Surveys* 9, 137 (1977).

500   N. Sedrakian, J. Steinig. *L'Enseigm. Math.* 44, 3 (1998).

501   S. L. Segal. *Duke Math. J.* 32, 278 (1965).

502   R. A. Serota, J. M. A. S. P. Wickramasinghe. *Int. J. Mod. Phys.* B 16, 4649 (2002).

503   R. Séroul. *Programming for Mathematicians*, Springer-Verlag, Berlin, 2000.

504   J. Shallit. *J. Symb. Comput.* 10, 59 (1990).

505   J. Shallit, J. Sorenson in L. M. Adleman, M.-D. Huang (eds.). *Algorithmic Number Theory*, Springer-Verlag, Berlin, 1994.

506   J. O. Shallit, J.-P. Allouche in C. Ding. T. Helleseth, and H. Niederreiter (eds.). *Sequences and Their Applications*, Springer-Verlag, New York, 1999.

507   D. Shanks. *Solved and Unsolved Problems in Number Theory*, Spartan Books, New York, 1962.

508   R. A. Sharipov. *arXiv:math.RA/*0009194 (2000).

509   G. B. Shaw. *J. Phys.* A 7, 1537 (1974).

510   G. Shimura. *Am. J. Math.* 124, 1059 (2002).

511   I. E. Shparlinski. *Computational and Algorithmic Problems in Finite Fields*, Kluwer, Dordrecht, 1993.

512   A. Sidi. *Num. Math.* 98, 371 (2004).

513   W. Sierpinski. *A Selection of Problems in the Theory of Numbers*, Pergamon, New York, 1964.

514   D. Singmaster. *Am. Math. Monthly* 78, 385 (1971).

515   D. Singmaster. *Fibon. Quart.* 13, 295 (1975).

516   R. Sivaramakrishnan. *Nieuw Archief Wiskunde* 4, 175 (1986).

517   R. Sivaramakrishnan. *Classical Theory of Arithmetic Functions*, Marcel Dekker, New York, 1989.

518   R. Sivaramakrishnan. *Nieuw Archief Wiskunde* 8, 169 (1990).

519   M. Skalba. *Coll. Math.* 69, 143 (1995).

520   H. J. S. Smith. *Messenger Math.* 6, 1 (1876).

521   J. Smoak. *Coll. Math. J.* 34, 58 (2003).

522   S. Solak, R. Türkmen, D. Bozkurt. *Appl. Math. Comput.* 146, 595 (2003).

523   A. I. Solomon. *arXiv:quant-ph/*0310174 (2003).

524   J. P. Sorenson. *J. Algor.* 16, 110 (1994).

525   J. Sorenson in A. H. M. Levelt (ed.). *Proc. ISSAC '95* ACM Press, New York, 1995.

526   D. Spector. *Commun. Math. Phys.* 127, 239 (1990).

527 J. Spilker. *Elem. Math.* 58, 26 (2003).

528 R. Spira. *Am. Math. Monthly* 68, 120 (1961).

529 J. L. Spouge. *SIAM J. Num. Anal.* 31, 931 (1994).

530 J. Stalker. *Complex Analysis*, Birkhäuser, Boston, 1998.

531 D. Stark, A. Ganesh, N. O'Connel. *Combinat., Prob. Comput.* 11, 79 (2002).

532 H. M. Stark. *An Introduction to Number Theory*, Markham, Chicago, 1970.

533 A. G. Starling. *Congr. Numerantium* 131, 49 (1998).

534 D. Stehlé, P. Zimmermann in D. A. Buell (eds.). *Algorithmic Number Theory, 6th International Symposium*, Springer-Verlag, Berlin, 2004.

535 I. Stewart. *Game, Set and Math: Enigmas and Conundrums*, Penguin, Harmondsworth, 1991.

536 I. Stewart. *Math. Intell.* 17, n1, 52 (1995).

537 G. R. Stibitz. *Am. Math. Monthly* 45, 22 (1938).

538 J. Stoer, R. Bulirsch. *Introduction to Numerical Analysis*, Springer-Verlag, Berlin, 1993.

539 K. B. Stolarsky. *SIAM J. Appl. Math.* 32, 717 (1977).

540 R. Stone. *Pac. J. Math.* 217, 331 (2004).

541 J. Stopple. *A Primer of Analytic Number Theory*, Cambridge University Press, Cambridge, 2003.

542 W. Strodt. *Am. Math. Monthly* 67, 452 (1960).

543 F. Suagher, J. P. Parisot. *Astronomie* 103, 107 (1989).

544 M. Sugihara in S. Saitoh, N. Hayashi, M. Yamamoto (eds.). *Analytic Extension Formulas and their Applications*, Kluwer, Dordrecht, 2001.

545 M. Sugunamma. *Ann. Pol. Math.* 8, 173 (1960).

546 K. Szymiczek. *Funct. Approx.* 28, 221 (2000).

547 N. M. Temme. *Special Functions*, John Wiley, New York, 1996.

548 G. Tenenbaum. *Introduction to Analytic and Probabilistic Number Theory*, Cambridge University Press, Cambridge, 1995.

549 R. Terras. *Am. J. Phys.* 48, 526 (1980).

550 L. Tóth. *Fibon. Quart.* 40, 13 (2002).

551 L. N. Trefethen, L. M. Trefethen. *Proc. R. Soc. Lond.* A 456, 2561 (2000).

552 C. W. Trigg. *Math. Mag.* 121 (1972).

553 M. Trott. *The Mathematica Journal* 7, 291 (1999).

554 M. Trott. *The Mathematica GuideBook for Programming*, Springer-Verlag, New York, 2004.

555 M. Trott. *The Mathematica GuideBook for Graphics*, Springer-Verlag, New York, 2004.

556 M. Trott. *The Mathematica GuideBook for Symbolics*, Springer-Verlag, New York, 2005.

557 P. Turán. *J. Lond. Math. Soc.* 9, 274 (1934).

558 V. Ufnarovski, B. Åhlander. *J. Integer Seq.* 6, A 03.3.4 (2003).
http://www.math.uwaterloo.ca/JIS/VOL6/Ufnarovski/ufnarovski.html

559 A. V. Ustinov. *Math. Notes* 71, 851 (2002).

560 O. Vaduvescu. *arXiv:math.GM*/0404109 (2004).

561 B. L. van der Waerden. *Algebra* II, Springer-Verlag, Berlin, 1971.

562 J. H. van Lint, R. M. Wilson. *A Course in Combinatorics*, Cambridge University Press, Cambridge, 1992.

563 T. van Ravenstein. *Fibon. Quart.* 27, 18 (1989).

564   S. Vajda. *Fibonacci & Lucas Numbers, and the Golden Section*, Ellis Horwood, New York, 1989.

565   B. Vallée. *J. Th. Nombres Bordeaux* 12, 531 (2000).

566   B. Vallée. *Theor. Comput. Sc.* 297, 447 (2003).

567   T. van Ravenstein. *J. Austral. Math. Soc.* A 45, 360 (1988).

568   H. C. A. van Tilborg. *Fundamentals of Cryptology*, Kluwer, Boston, 2000.

569   I. Vardi. *Mathematica Recreations*, Addison-Wesley, Reading, 1990.

570   I. Vardi. *Am. Math. Monthly* 105, 305 (1998).

571   A. Varvak. *arXiv:math.CO*/0402376 (2004).

572   J.-L. Verley in J. Dieudonné. *Geschichte der Mathematik*, Deutscher Verlag der Wissenschaften, Berlin, 1985.

573   A. M. Vershik, S. V. Kerov. *Sov. Math. Dokl.* 18, 527 (1977).

574   A. Vershik, Y. Yakubovich. *Moscow Math. J.* 1, 457 (2001).

575   O. V. Viskov. *Dokl. Math.* 70, 931 (2004).

576   D. Viswanath. *Math. Comput.* 69, 1131 (2000).

577   P. Vlamos. *Int. J. Math. Math. Sci.* 31, 31 (2002).

578   J. von zur Gathen, I. E. Shparlinski in R. Fleischer, G. Trippen (eds.). *Algorithms and Computation*, Springer, Berlin, 2004.

579   N. N. Vorob'ev. *Fibonacci Numbers*, Blaisdell, New York, 1961.

580   N. N. Vorobiev. *Fibonacci Numbers*, Birkhäuser, Boston, 2002.

581   S. Wagon. *Mathematica in Action*, W. H. Freeman, New York, 1991.

582   Y. Wang. *The Goldbach Conjecture*, World Scientific, Singapore, 2002.

583   W. C. Waterhouse. *Hist. Math.* 22, 186 (1995).

584   J. Wei. *J. Math. Phys.* 4, 1337 (1963).

585   Y. Wei, G. Yan, Q. Zhang. *Comput. Math. Appl.* 41, 641 (2001).

586   C. Weiss, M. Holthaus. *Europhys. Lett.* 59, 486 (2002).

587   D. Wells. *The Penguin Dictionary of Curious and Interesting Numbers*, Penguin Books, Harmondsworth, 1986.

588   D. L. Wells. *Fibon. Quart.* 32, 111 (1994).

589   J. Wensch, M. Däne, W. Hergert, A. Ernst. *Comput. Phys. Commun.* 160, 129 (2004).

590   E. E. Whitford. *Pell Equation*, Columbia University Press, New York, 1912.

591   W. M. A. Whitworth. *Choice and Chance*, Hafner, New York, 1965.

592   J. M. A. S. P. Wickramasinghe, R. A. Serota. *arXiv:cond-mat*/02306110 (2003).

593   H. S. Wilf. *Generatingfuntionology*, Academic Press, Boston, 1990.

594   D. E. Williams. *Acta Cryst.* A 29, 408 (1973).

595   H. C. Williams, R. A. German, C. R. Zarnke. *Math. Comput.* 19, 671 (1965).

596   H. C. Williams, J. O. Shallit in W. Gautschi (ed.). *Mathematics of Computation 1943-1993: A Half-Century of Computational Mathematics*, American Mathematical Society, Providence, 1994.

597   A. Wintner. *An Arithmetical Approach to Ordinary Fourier Series*, Waverly Press, Baltimore, 1945.

598   W. Witschel. *J. Phys.* A 8, 143 (1975).

599   S. Wolfram. *Am. Math. Monthly* 91, 566 (1984).

600   J. Wu. *Monatsh. Math.* 135, 69 (2002).

601   A. Wurm, M. Berg. *arXiv:physics*/0212061 (2002).

602   R. Wyss. *Elem. Math.* 51, 102 (1996).

603   Q. Xie, M. Huang. *Phys. Lett.* A 184, 119 (1993).

604   Y. Yakubovich. *J. Math. Sc.* 107, 4296 (2001).

605   S. Y. Yan. *Amicable and Sociable Numbers*, World Scientific, Singapore, 1996.

606   S. Yates in R. A. Mollin (ed.). *Number Theory*, de Gruyter, Berlin, 1990.

607   H. M. Ymeri. *Punjab Univ. J. Math.* 28, 69 (1995).

608   D. Zagier. *Beihefte Z. Elem. Math.* n15, 3 (1977).

609   E. Zeckendorf. *Bull. Soc. Roy. Sci. Liège* 41, 179 (1972).

610   J. Zeng. *Discr. Math.* 160, 219 (1996).

611   W. G. Zhai, X. D. Zhao. *Acta Arithm.* 90, 1 (1999).

612   J. Zunić. *J. Math. Imag. Vision* 21, 199 (2004).

Index

The alphabetization is character-by-character, including spaces. Numbers and symbols come first, with the exception of $. All fonts are treated equally.

The index entries refer to the sections or subsections and are hyperlinked. The index entry for a subject from within the exercises and solutions are hyperlinked mostly to the exercises and not to the corresponding solutions.

"*subject* in action" refers to examples or solutions of exercises making very heavy use of *subject*, or could be considered archetypical use of *subject*.

Index entries are grouped at most one level deep. Index entries containing compound names, such as Riemann–Siegel, are mentioned on their own and not as a subentry under the first name.

Built-in functions are referenced to the section in which they are first discussed. Built-in functions and functions defined in the standard packages appear in the font Courier bold (example: **Plot**); functions defined in *The Mathematica GuideBooks* appear in the font Courier plain (example: DistributionOfBends).